D1572901

THE UNIVERSITY OF ARIZONA SPACE SCIENCE SERIES

RICHARD P. BINZEL, GENERAL EDITOR

Comparative Climatology of Terrestrial Planets
Stephen J. Mackwell, Amy A. Simon-Miller, Jerald W. Harder,
and Mark A. Bullock, editors, 2013, 610 pages

Europa
Robert T. Pappalardo, William B. McKinnon,
and Krishan K. Khurana, editors, 2009, 727 pages

The Solar System Beyond Neptune
M. Antonietta Barucci, Hermann Boehnhardt, Dale P. Cruikshank,
and Alessandro Morbidelli, editors, 2008, 592 pages

Protostars and Planets V
Bo Reipurth, David Jewitt, and Klaus Keil, editors, 2007, 951 pages

Meteorites and the Early Solar System II
D. S. Lauretta and H. Y. McSween, editors, 2006, 943 pages

Comets II
M. C. Festou, H. U. Keller,
and H. A. Weaver, editors, 2004, 745 pages

Asteroids III
William F. Bottke Jr., Alberto Cellino, Paolo Paolicchi,
and Richard P. Binzel, editors, 2002, 785 pages

TOM GEHRELS, GENERAL EDITOR

Origin of the Earth and Moon
R. M. Canup and K. Righter, editors, 2000, 555 pages

Protostars and Planets IV
Vincent Mannings, Alan P. Boss,
and Sara S. Russell, editors, 2000, 1422 pages

Pluto and Charon
S. Alan Stern and David J. Tholen, editors, 1997, 728 pages

**Venus II—Geology, Geophysics, Atmosphere,
and Solar Wind Environment**
S. W. Bougher, D. M. Hunten,
and R. J. Phillips, editors, 1997, 1376 pages

Cosmic Winds and the Heliosphere
J. R. Jokipii, C. P. Sonett,
and M. S. Giampapa, editors, 1997, 1013 pages

Neptune and Triton
Dale P. Cruikshank, editor, 1995, 1249 pages

Hazards Due to Comets and Asteroids
Tom Gehrels, editor, 1994, 1300 pages

Resources of Near-Earth Space
John S. Lewis, Mildred S. Matthews,
and Mary L. Guerrieri, editors, 1993, 977 pages

Protostars and Planets III
Eugene H. Levy and Jonathan I. Lunine, editors, 1993, 1596 pages

Mars
Hugh H. Kieffer, Bruce M. Jakosky, Conway W. Snyder,
and Mildred S. Matthews, editors, 1992, 1498 pages

Solar Interior and Atmosphere
A. N. Cox, W. C. Livingston,
and M. S. Matthews, editors, 1991, 1416 pages

The Sun in Time
C. P. Sonett, M. S. Giampapa,
and M. S. Matthews, editors, 1991, 990 pages

Uranus
Jay T. Bergstralh, Ellis D. Miner,
and Mildred S. Matthews, editors, 1991, 1076 pages

Asteroids II
Richard P. Binzel, Tom Gehrels,
and Mildred S. Matthews, editors, 1989, 1258 pages

Origin and Evolution of Planetary and Satellite Atmospheres
S. K. Atreya, J. B. Pollack,
and Mildred S. Matthews, editors, 1989, 1269 pages

Mercury
Faith Vilas, Clark R. Chapman,
and Mildred S. Matthews, editors, 1988, 794 pages

Meteorites and the Early Solar System
John F. Kerridge and Mildred S. Matthews, editors, 1988, 1269 pages

The Galaxy and the Solar System
Roman Smoluchowski, John N. Bahcall,
and Mildred S. Matthews, editors, 1986, 483 pages

Satellites
Joseph A. Burns and Mildred S. Matthews, editors, 1986, 1021 pages

Protostars and Planets II
David C. Black and Mildred S. Matthews, editors, 1985, 1293 pages

Planetary Rings
Richard Greenberg and André Brahic, editors, 1984, 784 pages

Saturn
Tom Gehrels and Mildred S. Matthews, editors, 1984, 968 pages

Venus
D. M. Hunten, L. Colin, T. M. Donahue,
and V. I. Moroz, editors, 1983, 1143 pages

Satellites of Jupiter
David Morrison, editor, 1982, 972 pages

Comets
Laurel L. Wilkening, editor, 1982, 766 pages

Asteroids
Tom Gehrels, editor, 1979, 1181 pages

Protostars and Planets
Tom Gehrels, editor, 1978, 756 pages

Planetary Satellites
Joseph A. Burns, editor, 1977, 598 pages

Jupiter
Tom Gehrels, editor, 1976, 1254 pages

Planets, Stars and Nebulae, Studied with Photopolarimetry
Tom Gehrels, editor, 1974, 1133 pages

Comparative Climatology of Terrestrial Planets

Comparative Climatology of Terrestrial Planets

Edited by

Stephen J. Mackwell, Amy A. Simon-Miller,
Jerald W. Harder, and Mark A. Bullock

With the assistance of

Renée Dotson

With 69 collaborating authors

THE UNIVERSITY OF ARIZONA PRESS
Tucson

in collaboration with

LUNAR AND PLANETARY INSTITUTE
Houston

About the front cover:

Obscuring the martian sun. A murky landscape on Mars during the early stages of one of the martian global dust storms (such as the one that obscured nearly all surface features during the arrival of the Mariner 9 orbiter in 1971). Distant features are mostly obscured by airborne dust, raised by dust devils. Patches of dusty snow hint that the view may represent a different climatic era, perhaps at somewhat higher obliquity. Painting by William K. Hartmann, Planetary Science Institute, Tucson, Arizona.

About the back cover:

Polar vortices on several solar system bodies. *Upper left:* The cyclonic vortex at Saturn's north pole, shown in a false-color image from NASA's Cassini spacecraft; the red colors indicate deeper clouds. Credit: NASA/JPL-Caltech/SSI. *Upper right:* True color view of Titan's south polar vortex as seen by the Cassini spacecraft; this vortex is seasonal and its formation heralds the arrival of south pole winter. Credit: NASA/JPL-Caltech/Space Science Institute. *Lower left:* Polar vortex on Earth, shown in a simulated map of ozone abundance and air mixing. Credit: Eugene Cordero, San Jose State University. *Lower right:* A thermal infrared view of Venus' south polar vortex; white regions show cooler, high-altitude clouds. Credit: ESA/VIRTIS-VenusX/INAF-IASF/LESIA-Obs. de Paris (G. Piccioni, INAF-IASF).

The University of Arizona Press
in collaboration with the Lunar and Planetary Institute

∞ This book is printed on acid-free, archival-quality paper.
Manufactured in the United States of America

18 17 16 15 14 13 6 5 4 3 2 1

Library of Congress Cataloging-in-Publication Data

Comparative climatology of terrestrial planets / edited by Stephen J. Mackwell, Amy A. Simon-Miller, Jerald W. Harder, and Mark A. Bullock ; with the assistance of Renée Dotson ; with 69 collaborating authors ; foreword by James E. Hansen.
pages cm — (The University of Arizona space science series)
Includes bibliographical references and index.
ISBN 978-0-8165-3059-5 (cloth : alk. paper)
1. Planetary meteorology. 2. Planets—Atmospheres. 3. Climatology. I. Mackwell, Stephen J., editor of compilation. II. Simon-Miller, Amy A., 1971–, editor of compilation. III. Harder, Jerald W., editor of compilation. IV. Bullock, Mark A., editor of compilation. V. Dotson, Renée. VI. Lunar and Planetary Institute. VII. Series: University of Arizona space science series.
QB603.A85C65 2013
551.60999—dc23 2013022120

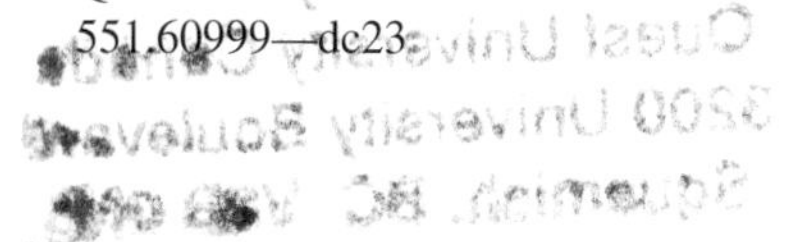

Contents

PART I: FOUNDATIONS

PART II: GREENHOUSE EFFECT AND ATMOSPHERIC DYNAMICS

PART III: CLOUDS AND HAZES

PART IV: SURFACE AND INTERIOR

PART V: SOLAR INFLUENCES

List of Contributing Authors

Acknowledgment of Reviewers

The editors gratefully acknowledge the following individuals, along with several anonymous reviewers, for their time and effort in reviewing chapters for this volume:

Nathan Bridges
Gary Chapman
Todd Clancy
Tom Cravens
Roger Davies
Anthony Del Genio
Timothy E. Dowling
Peter Gierasch
Stephen M. Griffies
Virginia Gulick
Christiane Helling
Sarah M. Horst
Alan Howard
David L. Huestis
Brian Jackson
Ralph A. Kahn
James Kasting
Walter S. Kiefer
Jeffrey Kiehl
Ravi Kopparapu
Jae N. Lee
Ralph D. Lorenz
Janet Luhmann
Kevin McGouldrick
Jay Melosh
Jonathan L. Mitchell
Franck Montmessin
Jim Murphy
Hans Nilsson
Francis Nimmo
Tyler D. Robinson
Gavin A. Schmidt
Tapio Schneider
Sue Smrekar
Aymeric Spiga
Fred Taylor
Gary E. Thomas
Nicholas Tosca
David M. Tratt
Channon Visscher
Thomas Widemann

Foreword

Comparative climatology of the terrestrial planets, including their evolution, provides a broad perspective that helps us understand consequences of human activity and assess actions needed to preserve the life and life support systems of our remarkable planet. Indeed, analysis of the range of climates that can occur within the habitable zone around any star helps us understand how difficult it is to achieve and maintain habitability, as illustrated by an example.

Several decades ago planetary scientists puzzled over an enigma. How had Earth avoided the "snowball" fate when the solar system was young and the Sun was much dimmer than today? Energy-balance climate models showed that when a planet becomes cold enough for ice to reach the subtropics, the resulting increase of planetary albedo (reflectivity) is so large as to cause a runaway amplifying feedback, with ice and snow thus descending to the equator.

Such a snowball state should be stable. So how could life have developed or survived? *Kirschvink* (1992) proposed the answer, after first showing from geologic evidence that Earth had indeed plummeted into snowball conditions several times with ice descending to sea level at the equator. Kirschvink's solution to the riddle was that the weathering process would be nearly absent when Earth was in the snowball state. Chemical reactions associated with the weathering of terrestrial rock normally take CO_2 from the air and deposit it as carbonates on the ocean floor. The amount of CO_2 in the air, on long timescales and in the absence of human contributions, is thus largely determined by the balance between volcanic injection of CO_2 into the air and weathering removal of CO_2. Plate tectonics (continental drift) and associated volcanic activity would continue on snowball Earth, thus causing CO_2 to accumulate in the atmosphere until there was enough to cause a large greenhouse effect. When ice began to melt at low latitudes the albedo feedback would work in the reverse sense, causing rapid planetary deglaciation. Weathering would then resume and begin to draw down the excessive atmospheric CO_2, depositing it as carbonates on the ocean floor.

Hoffman et al. (1998) and *Hoffman and Schrag* (2000) confirmed the general picture painted by Kirschvink. They showed, for example, that thick carbonate layers were deposited on the ocean floor following snowball events, the expected deposition of accumulated atmospheric CO_2. The most recent snowball Earth occurred about 600 million years ago, by which time the Sun's brightness was only about 6% lower than today. The simple life forms that existed on Earth prior to the final snowball, such as protozoa and algae, were surely stressed and winnowed by the harsh snowball conditions, but some survived. The final snowball event was soon followed by the Cambrian explosion of multicellular life from which all of today's plants and animals have descended.

Snowball conditions are not expected to occur again on Earth, as our Sun, a rather ordinary "main-sequence" star, will continue to brighten over the next several billion years. What about the other climate extreme? Can a hothouse Earth with Venus-like conditions occur in our future? The conditions on Venus today provide a useful perspective.

Venus has a surface temperature of several hundred degrees and a surface pressure of about 90 bars as a result of its huge atmosphere, which is mainly CO_2. Earth has a comparable amount of carbon, but most of Earth's carbon is in the crust, as weathering continually removes CO_2 from the air. Venus and Earth must have formed from the same interstellar gas and dust, and thus the initial composition of Venus included enough water to form an ocean.

However, the ocean on Venus was lost over time. Ultraviolet radiation is continually breaking up any water vapor (into its component hydrogen and oxygen atoms) in the outer reaches of a planetary atmosphere.

If the atmosphere is sufficiently hot, the thermal motion of some of the hydrogen atoms can be fast enough for hydrogen to overcome the planet's gravitational field and escape to space. Hydrogen escape from Venus is confirmed by measurement of a large overabundance of heavy hydrogen (deuterium) relative to normal hydrogen, as heavy hydrogen escapes less readily. The early atmosphere of Venus was much warmer than Earth, because the intensity of solar radiation at Venus is twice as large as it is at Earth's distance from the Sun. Nevertheless, thermal escape of hydrogen is a slow process, so it required at least hundreds of millions of years for Venus to lose its ocean, a necessary condition before the carbon in the crust could be baked into the atmosphere and result in 90-bar surface pressure and a surface hot enough to melt lead.

Could such a fate befall Earth? Yes. On a timescale of several billion years, as the Sun brightens and eventually expands into a red giant, it is expected that escape of hydrogen will accelerate. The ocean would then be gradually depleted and crustal carbon could accumulate in the atmosphere as CO_2, producing a Venus-like state. However, such a timescale is of no practical concern. Indeed, by several billion years from now, Earthlings will probably have either exterminated themselves or developed technology to try to save themselves. For example, they may try to slow Earth in its orbit, so that it would move to a greater distance from the Sun, thereby prolonging their existence.

How hot would Earth become if humanity burned all the fossil fuels? No matter how hot, the fossil fuel CO_2 would be removed from the air over tens of thousands of years and deposited on the ocean floor. The fraction of the ocean lost in that period would be negligible, so surface pressure would remain close to 1 bar and Venus-like surface conditions would not occur. However, large warming would be expected, because the sum of all fossil fuels including methane hydrates could produce as much as 5–10 times the preindustrial CO_2 amount.

Simulations with a global climate model (*Russell et al.,* 2013; *Hansen et al.,* 2013) indicate that Earth's climate becomes more responsive, i.e., climate sensitivity increases, when the forcing reaches levels as great as 5–10 times the preindustrial amount of CO_2, but extreme "runaway" warming (*Ingersoll,* 1969) does not occur. The model yields an increasing climate sensitivity as the tropopause rises and becomes less pronounced — thus increasing the amount of stratospheric water vapor — and tropospheric cloud cover decreases. Paleoclimate data, including the rapid 5°–6°C global warming that occurred about 55 million years ago during the Paleocene-Eocene Thermal Maximum (PETM) (*Dunkley Jones et al.,* 2010), support the existence of higher climate sensitivity in a warmer world (*Hansen et al.,* 2013).

Portions of planet Earth would become practically uninhabitable by humans if CO_2 rose to 5–10 times its preindustrial amount (280 ppm). *Hansen et al.* (2013) find that summer wet bulb temperatures in substantial areas would be near 35°C, which is so hot that the human body in the open air could not dispel its internal heat (*McMichael and Dear,* 2010). And long before this ultimate warming occurred, life on Earth would be whipsawed through tumultuous climatic conditions as the ice sheets disintegrated.

Terrestrial life survived the PETM relatively unscathed except for widespread dwarfism (*Sherwood and Huber,* 2010), which was presumably an evolutionary response to the higher temperature and perhaps to a reduced food supply. However, if exponentially increasing fossil fuel use continued, we could in principle burn through the fossil fuels in only 1–2 centuries. That compares with the several millennia over which PETM carbon injection occurred. It is likely that many species would not survive such unprecedented rapid climate change and shifting of climate zones, especially in view of the other stresses that humankind are placing on planetary life. Humans surely would find such an overheated planet to be more desolate and less livable.

These extreme potential consequences should help wake humanity up to the folly of the continued rush to extract and burn every fossil fuel. Earth, with its spectacular varied life, is the Garden of Eden of the planets. It is still just conceivable that fossil fuel emissions could be phased out fast enough to keep Earth's climate close to that of the Holocene (*Hansen et al.,* 2013) — the past 10,000 years, the period in which civilization developed. Indeed, that is necessary if we are to preserve our coastlines and all the history contained in thousands of coastal cities. With the accompanying elimination of air and water pollution from mining and burning of fossil fuels, we have a vision of a planetary utopia that is surely worth fighting for.

REFERENCES

Dunkley Jones T., Ridgwell A., Lunt D. J., Maslin M. A., Schmidt D. N., and Valdes P. J. (2010) A Palaeogene perspective on climate sensitivity and methane hydrate instability. *Philos. Trans. R. Soc. A, 368,* 2395–2415.

Hansen J., Sato M., Russell G., and Kharecha P. (2013) Climate sensitivity, sea level, and atmospheric CO_2. *Philos. Trans. R. Soc. A*, in press.

Hoffman P. F., Kaufman A. J., Halverson G. P., and Schrag D. P. (1998) A Neoproterozoic snowball Earth. *Science, 281(5381),* 1342–1346.

Hoffman P. F. and Schrag D.P. (2000) Snowball Earth. *Sci. Am., 282,* 68–75.

Ingersoll A. P. (1969) Runaway greenhouse — a history of water on Venus. *J. Atmos. Sci., 26,* 1191–1198.

Kirschvink J. L. (1992) Late Proterozoic low-latitude global glaciation: The snowball Earth. In *The Proterozoic Biosphere: A Multidisciplinary Study* (J. W. Schopf and C. Klein, eds.), pp. 51–52. Cambridge Univ., Cambridge.

McMichael A. J. and Dear K. B. (2010) Climate change: Heat, health, and longer horizons. *Proc. Natl. Acad. Sci. USA, 107,* 9483–9484.

Russell G. L., Rind D. H., Colose C., Lacis A. A., and Opstbaum R. F. (2013) Fast atmosphere-ocean model run with large changes in CO_2. *Geophys. Res. Lett.,* in press.

Sherwood S. C. and Huber M. (2010) An adaptability limit to climate change due to heat stress. *Proc. Natl. Acad. Sci. USA, 107,* 9552–9555.

James E. Hansen
New York, New York
May 2013

Preface

Public awareness of climate change on Earth is currently very high, promoting significant interest in atmospheric processes. We are fortunate to live in an era where it is possible to study the climates of many planets, including our own, using spacecraft and groundbased observations as well as advanced computational power that allows detailed modeling. Planetary atmospheric dynamics and structure are all governed by the same basic physics. Thus differences in the input variables (such as composition, internal structure, and solar radiation) among the known planets provide a broad suite of natural laboratory settings for gaining new understanding of these physical processes and their outcomes. Diverse planetary settings provide insightful comparisons to atmospheric processes and feedbacks on Earth, allowing a greater understanding of the driving forces and external influences on our own planetary climate. They also inform us in our search for habitable environments on planets orbiting distant stars, a topic that was a focus of Exoplanets, the preceding book in the University of Arizona Press Space Sciences Series.

Quite naturally, and perhaps inevitably, our fascination with climate is largely driven toward investigating the interplay between the early development of life and the presence of a suitable planetary climate. Our understanding of how habitable planets come to be begins with the worlds closest to home. Venus, Earth, and Mars differ only modestly in their mass and distance from the Sun, yet their current climates could scarcely be more divergent. Our purpose for this book is to set forth the foundations for this emerging science and to bring to the forefront our current understanding of atmospheric formation and climate evolution. Although there is significant comparison to be made to atmospheric processes on nonterrestrial planets in our solar system — the gas and ice giants — here we focus on the terrestrial planets, leaving even broader comparisons to a future volume.

Our authors have taken on the task to look at climate on the terrestrial planets in the broadest sense possible — by comparing the atmospheric processes at work on the four terrestrial bodies, Earth, Venus, Mars, and Titan (Titan is included because it hosts many of the common processes), and on terrestrial planets around other stars. These processes include the interactions of shortwave and thermal radiation with the atmosphere, condensation and vaporization of volatiles, atmospheric dynamics, chemistry and aerosol formation, and the role of the surface and interior in the long-term evolution of climate. Chapters herein compare the scientific questions, analysis methods, numerical models, and spacecraft remote sensing experiments of Earth and the other terrestrial planets, emphasizing the underlying commonality of physical processes. We look to the future by identifying objectives for ongoing research and new missions. Through these pages we challenge practicing planetary scientists, and most importantly new students of any age, to find pathways and synergies for advancing the field.

In Part I, Foundations, we introduce the fundamental physics of climate on terrestrial planets. Starting with the best studied planet by far, Earth, the first chapters discuss what is known and what is not known about the atmospheres and climates of the terrestrial planets of the solar system and beyond. In Part II, Greenhouse Effect and Atmospheric Dynamics, we focus on the processes that govern atmospheric motion and the role that general circulation models play in our current understanding. In Part III, Clouds and Hazes, we provide an in-depth look at the many effects of clouds and aerosols on planetary climate. Although this is a vigorous area of research in the Earth sciences, and very strongly influences climate modeling, the important role that aerosols and clouds play in the climate of all planets is not yet well constrained. This section is intended to stimulate further research on this critical subject.

The study of climate involves much more than understanding atmospheric processes. This subtlety is particularly appreciated for Earth, where chemical cycles, geology, ocean influences, and biology are considered in most climate models. In Part IV, Surface and Interior, we look at the role that geochemical cycles, volcanism, and interior mantle processes play in the stability and evolution of terrestrial planetary climates.

There is one vital commonality between the climates of all the planets of the solar system: Regardless of the different processes that dominate each of the climates of Earth, Mars, Venus, and Titan, they are all ultimately forced by radiation from the same star, albeit at variable distances. In Part V, Solar Influences, we discuss how the Sun's early evolution affected the climates of the terrestrial planets, and how it continues to control the temperatures and compositions of planetary atmospheres. This will be of particular interest as models of exoplanets, and the influences of much different stellar types and distances, are advanced by further observations.

Comparisons of atmospheric and climate processes between the planets in our solar system has been a focus of numerous conferences over the past decade, including the Exoclimes conference series. In particular, this book project was closely tied to a conference on Comparative Climatology of Terrestrial Planets that was held in Boulder, Colorado, on June 25–28, 2012. This book benefited from the opportunity for the author teams to interact and obtain feedback from the broader community, but the chapters do not in general tie directly to presentations at the conference. The conference, which was organized by a diverse group of atmospheric and climate scientists led by Mark Bullock and Lori Glaze, sought to build connections between the various communities, focusing on synergies and complementary capabilities. Discussion panels at the end of most sessions served to build connections between planetary, solar, astrophysics and Earth climate scientists. These presentations and discussions allowed broadening of the author teams and tuning of the material in each chapter.

Comparative Climatology of Terrestrial Planets is the 38th book in the University of Arizona Press Space Sciences Series. The support and guidance from General Editor Richard Binzel has been critical in timely production of a quality volume. Renée Dotson of the Lunar and Planetary Institute, with support from Elizabeth Cunningham and Katy Buckaloo, provided outstanding help in the management of the book project and especially in the preparation of the chapters for publication. Her quiet reminders and attention to detail are critical in making the Space Science Series such an asset for the planetary science community. As for so many other books in this series, William Hartmann used his artistic skills to masterfully capture the book's theme. Much gratitude is owed to Adriana Ocampo of NASA Headquarters for her support of both the conference and book projects and her shepherding of the NASA contributions from the diverse groups within the Science Mission Directorate. Equally, James Green and Jonathan Rall of NASA Headquarters provided the financial resources and corporate oversight that helped make this book project such a success.

Mark A. Bullock
Amy A. Simon-Miller
Jerald W. Harder
Stephen J. Mackwell
July 2013

Part I: Foundations

Del Genio A. D. (2013) Physical processes controlling Earth's climate. In *Comparative Climatology of Terrestrial Planets* (S. J. Mackwell et al., eds.), pp. 3–18. Univ. of Arizona, Tucson, DOI: 10.2458/azu_uapress_9780816530595-ch001.

Physical Processes Controlling Earth's Climate

Anthony D. Del Genio
NASA Goddard Institute for Space Studies

As background for consideration of the climates of the other terrestrial planets in our solar system and the potential habitability of rocky exoplanets, we discuss the basic physics that controls the Earth's present climate, with particular emphasis on the energy and water cycles. We define several dimensionless parameters relevant to characterizing a planet's general circulation, climate, and hydrological cycle. We also consider issues associated with the use of past climate variations as indicators of future anthropogenically forced climate change, and recent advances in understanding projections of future climate that might have implications for Earth-like exoplanets.

1. INTRODUCTION

The amazing diversity of atmospheric behavior seen in our own solar system represents a challenge to our fundamental understanding of atmospheric physics. The weather and climate of Earth, explored extensively during the past century, provide an invaluable basis for understanding this diversity. Yet Earth's atmosphere occupies only one part of the parameter space of factors that can affect weather and climate. The extent to which our terrestrial assumptions carry over to other planets must continually be evaluated. A planetary perspective on Earth can help us decide which features that we take for granted are unique to our planet and which it has in common with subsets of other planets. The rapidly growing list of discovered exoplanets, and the need to assess which of these might lie in the habitable zone (see Domagal-Goldman and Segura, this volume), stretches our assumed understanding of Earth. Furthermore, increasing evidence of conditions during Earth's own past, as well as the urgency to be able to predict important features of future anthropogenically forced climate change, has led to a resurgence in fundamental studies of Earth's present climate in recent years. The science of comparative planetary climatology is indeed coupled to questions of our own planet's past and future habitability (see Foreword to this volume).

Although there is still much we do not understand about our own climate system, orders of magnitude more theoretical, observational, and modeling effort has been devoted to our planet than any other. Thus, for solar system or exoplanet researchers reading this book, we hope that starting with a discussion of Earth will provide a useful, even necessary, foundation for thinking about conditions on other planets. For terrestrial scientists, we hope that the planetary perspective of Earth presented in this chapter will stimulate thinking about the features that make Earth unique in our solar system. The time has never been riper for such thinking, since the advent of general circulation models (GCMs) (see Dowling, this volume) run on high-performance computing platforms now makes it possible to ask almost any imaginable "What if . . .?" question about Earth and the physics that control its climate. This chapter assumes a knowledge of basic atmospheric science; derivations and further discussion of many of the principles presented here can be found in textbooks such as *Houghton* (2002) and *Wallace and Hobbs* (2006).

2. THE ENERGY AND WATER CYCLES

New estimates of the solar irradiance of Earth (*Kopp and Lean,* 2011) indicate a value S_o = 1360.8 W/m^2 at solar minimum, considerably lower than previous estimates (see Harder and Woods, this volume). This so-called solar constant actually fluctuates by ~0.1 W/m^2 over the 11-year solar cycle, an inconsequential variation for Earth's climate. Averaged over the Earth, the absorbed shortwave (SW) flux equals the emitted longwave (LW) flux in equilibrium (see Covey et al., this volume):

$$\frac{S_o(1-A)}{4d^2} = \sigma T_e^4 \tag{1}$$

where A = 0.293 is the planetary albedo (*Loeb et al.,* 2009), d the planet-Sun distance in astronomical units (=1 for Earth), σ the Stefan-Boltzmann constant, and T_e the effective temperature of a blackbody that would produce the observed LW flux. Actually, Earth is currently out of equilibrium by ~0.6 W/m^2 due to anthropogenic emissions of greenhouse gases and other climate forcings (*Lyman et al.,* 2010).

Figure 1 shows the contributions to the global energy balance at the top-of-atmosphere (TOA), within the atmospheric column, and at the surface (*Stephens et al.,* 2012). Clouds account for almost half of Earth's planetary albedo; the other half is made up of almost equal contributions from

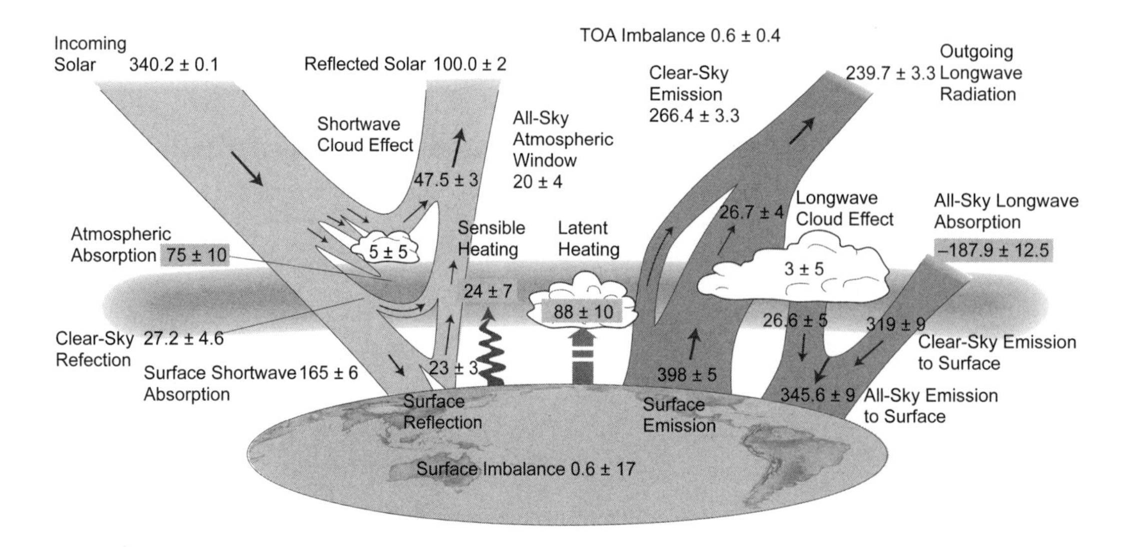

Fig. 1. See Plate 1 for color version. Estimate of the components of the current annual mean energy balance of Earth at TOA (upper row), within the atmosphere (middle row), and at the surface (bottom row). SW fluxes are in yellow, LW fluxes in magenta, and surface turbulent fluxes in red and violet. From *Stephens et al.* (2012). ©Copyright Nature Publishing Group; reprinted with permission.

the clear atmosphere (aerosol scattering and gaseous Rayleigh scattering) and the surface. The outgoing LW flux to space comes almost entirely from the atmosphere, with only an 8% contribution from the surface that escapes through the atmospheric thermal infrared window. Even in clear skies, the surface only contributes ~25% of the emission to space because of the LW opacity of greenhouse gases (*Costa and Shine,* 2012).

Earth's atmosphere is fairly transparent to our Sun's radiation, which peaks in the visible; ~22% is absorbed within the atmosphere, primarily due to O_2 and O_3 in the ultraviolet, H_2O in the near-infrared, and absorbing aerosols in the visible. The SW flux that reaches the surface is primarily (~88%) absorbed. The surface energy budget is balanced largely by upward surface turbulent latent and heat fluxes and partly by the net (up–down) LW flux (see Showman et al., this volume). The LW contribution is secondary because atmospheric absorbers induce a downward LW flux that offsets much of the upward emission from the surface.

T_e in equation (1) = 255 K, 33 K colder than Earth's surface temperature T_s = 288 K. The difference is a measure of Earth's greenhouse effect (see Covey et al., this volume) due to LW absorbers. Water vapor accounts for ~50% of the greenhouse effect, followed by clouds (~25%), CO_2 (~19%), and other absorbing gases and particulates (~7%) (*Schmidt et al.,* 2010). These relative contributions obscure the fact that CO_2 is actually the primary driver of Earth's climate, since water vapor is controlled (largely via the Clausius-Clapeyron equation of thermodynamics) by the heating due to CO_2 and other noncondensable greenhouse gases (*Lacis et al.,* 2010). Figure 2, for example, shows that when CO_2 and other noncondensable greenhouse gases are removed from an Earth GCM, the planet cools by ~35°C and the water vapor content is reduced by 90%. On Titan (see Griffith et al., this volume), the noncondensable absorber H_2 exerts a similar control on the climate by regulating the concentration of the condensable gas CH_4 (*McKay et al.,* 1991).

Figure 3 shows the geographic distribution of the TOA absorbed SW and outgoing LW flux in clear skies,

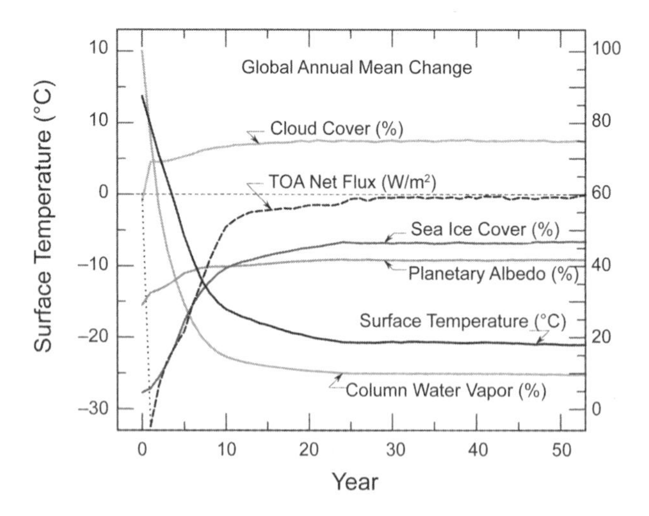

Fig. 2. Temporal evolution of surface temperature, column water vapor, TOA net radiative flux, cloud cover, sea ice cover, and planetary albedo in a GCM coupled to a 250-m Q-flux ocean after the concentrations of CO_2 and other noncondensing greenhouse gases were set to zero. From *Lacis et al.* (2010). ©Copyright American Association for the Advancement of Science; reprinted with permission.

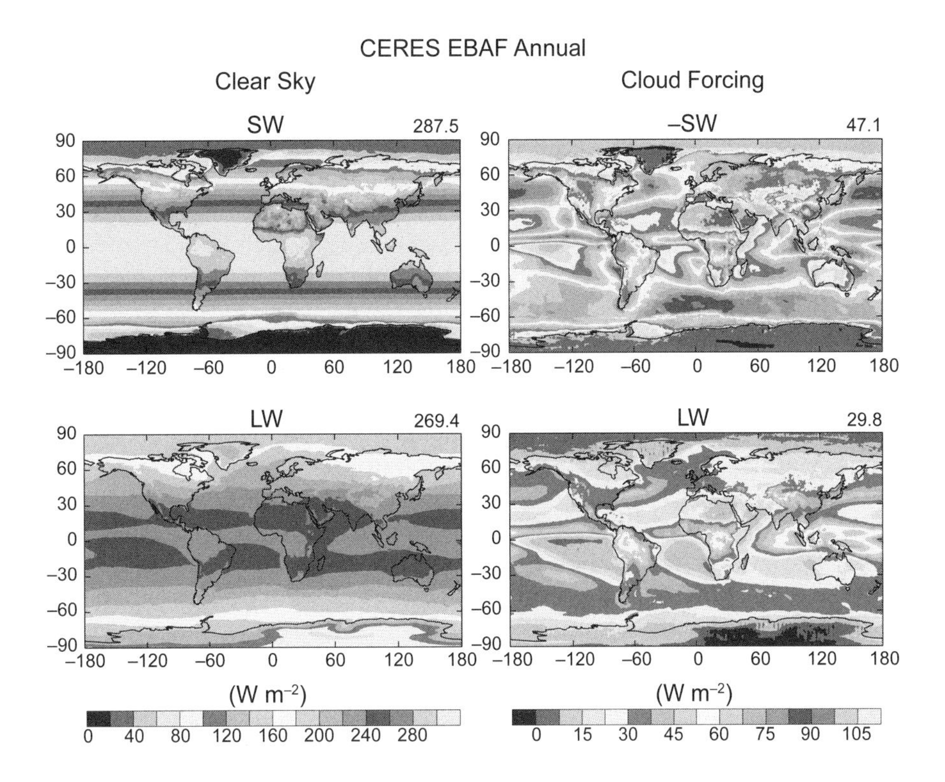

Fig. 3. See Plate 2 for color version. TOA Earth SW (upper panels) and LW (lower panels) annual mean clear-sky fluxes (left panels) and cloud forcing (right panels) derived from the NASA Clouds and the Earth's Radiant Energy System Energy Balanced and Filled (CERES EBAF) data product. The global mean values are given in the upper right corner of each panel. The SW cloud forcing has been multiplied by –1 for display purposes. Figure courtesy of J. Jonas.

and the difference between that and the total fluxes in all conditions, known as the "cloud forcing" or "cloud radiative effect" (*Loeb et al.,* 2009). The clear sky SW flux is dominated by the solar zenith angle variation with latitude, but departures from the zonal mean at each latitude reveal radiatively important details of Earth's surface: minima over bright ice-covered Greenland and Antarctica and to a lesser extent over the Arctic and Antarctic sea ice packs, maxima over the dark oceans, and intermediate values over other land surfaces.

The other fluxes are regulated primarily by the atmosphere and are thus indicators of the general circulation. Earth's "tropical" and "extratropical" dynamical regimes are determined by its radius, rotation period, and thermodynamic structure (see Showman et al., this volume). Low-level moisture convergence and deep convection in the rising branch of the Hadley cell maximize the greenhouse effect near the equator and produce a minimum in clear sky LW flux to space there relative to the flux maxima in the surrounding subtropical dry subsidence regions. Otherwise the clear sky flux reflects primarily the decrease in temperature and water vapor with latitude. Longwave cloud forcing (which is positive because clouds reduce LW emission to space and thus warm the planet) peaks coincide with the tropical locations of deep convection; secondary maxima exist in the extratropical storm tracks. Shortwave cloud forcing (which is negative because clouds are brighter than most of Earth's surface and thus cool the planet) also is large in magnitude in these regions. Other SW cloud forcing maxima exist off the west coasts of the continents but with no corresponding LW signal. These are locations of low-level marine stratocumulus decks whose tops are below the clear sky emission to space level. Globally averaged, SW cooling by clouds exceeds LW warming by ~17 W/m^2, i.e., the net effect of clouds is to cool the current climate.

Aerosols (Fig. 4) have three qualitatively different effects on the climate: (1) *Direct effects* due to their interaction with (mostly SW) radiation (*Myhre,* 2009). Since most aerosol types (sulfates, nitrates, sea salt, and organic carbon) are bright, they cool Earth. However, black carbon and mineral dust absorb SW radiation; this cools the surface by preventing sunlight from reaching it but warms the atmosphere and the planet as a whole. This is analogous to the "anti-greenhouse" effect of absorbing hydrocarbon hazes on Titan (*McKay et al.,* 1991; Griffith et al., this volume), but with a smaller magnitude. (2) *Indirect effects* due to the interaction of aerosols with clouds (*McComiskey and Feingold,* 2012). Aerosols act as cloud condensation nuclei and affect the number concentration and size of cloud droplets. With an increase in aerosols, the available water is shared among more but smaller droplets, increasing their area/volume ratio and thus making the cloud more reflective. Smaller cloud droplets may also affect the microphysics and dynamics of clouds, but the magnitude of this effect is quite uncertain. (3) *Semi-direct effects* due to the influence of warming by absorbing aerosols on the atmospheric thermodynamic structure (*Koch and Del Genio,* 2010). For example, marine stratocumulus clouds that occur under strong inversions may become more widespread if an absorbing aerosol advected from an adjacent continent increases the inversion strength.

The geographic distribution of aerosols (*Bauer et al.,* 2008) is determined by natural and anthropogenic sources, tropospheric chemical reactions, wet and dry deposition, and advection by the general circulation. Sulfate and nitrate are largely of anthropogenic origin and are concentrated near and downwind of industrial, transportation, and agricultural sources. Organic and black carbon are most prevalent in regions of biomass burning, but also have important industrial and transportation sources. Mineral dust is mostly natural in origin, the largest source region being the Sahara and Arabian deserts. Sea salt is produced mainly in oceanic regions of strong surface winds such as the extratropical storm tracks.

Although the TOA-absorbed SW and outgoing LW radiative fluxes must balance globally for the climate to be in equilibrium, this is not true locally. In fact, there is a latitudinally varying imbalance of net radiative heating on Earth (Fig. 5): The tropics receives more sunlight than the heat it emits to space, while higher latitudes emit more than they absorb. In equilibrium this must be offset by the transport of heat from low to high latitudes. This is the fundamental driver of the circulations of the atmosphere and ocean (*Trenberth and Caron,* 2001). The atmospheric component of heat transport is accomplished primarily by the Hadley circulation in the tropics and by synoptic-scale eddies created by baroclinic instability in the extratropics (see Showman et al., this volume) and dominates at high latitudes. The ocean circulation contribution to transport dominates at low latitudes (Fig. 6).

The ocean circulation has two fundamental components: a surface circulation in the top ~100 m driven by the frictional interaction with atmospheric surface winds and the Coriolis force, and a thermohaline circulation driven by density gradients due to temperature and salinity differences. The surface winds created by Earth's three-cell mean meridional circulation produce tropical easterly and extratropical westerly ocean currents. At longitudinal continental boundaries, the easterlies are deflected poleward and the westerlies equatorward, forming roughly circular ocean gyres in each ocean basin and hemisphere. The gyres transport warm water poleward off the east coasts of the continents and cold water equatorward off the west coasts, effecting a net poleward heat transport (*Klinger and Marotzke,* 2000). The wind-driven transport may be an important factor in limiting sea ice expansion to the tropics in "snowball Earth" scenarios (*Poulsen and Jacob,* 2004; *Rose and Marshall,* 2009). At high latitudes, dense (cold and/or salty) water sinks to the ocean bottom and slowly circulates to lower latitudes. High-salinity water is produced by net evaporation (i.e., in excess of precipitation) and the formation of sea ice, which cannot easily accommodate salt and thus produces salty ocean water where it occurs. The

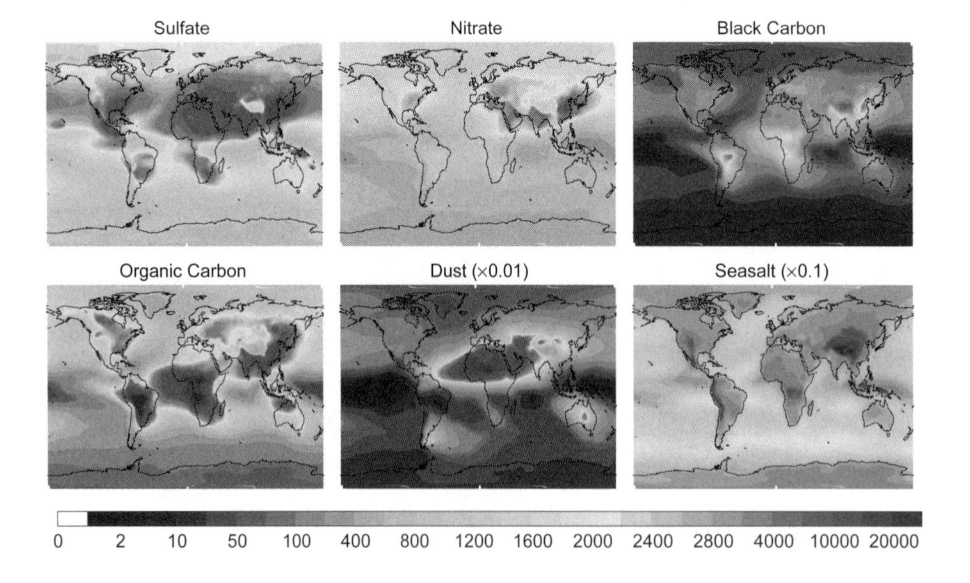

Fig. 4. See Plate 3 for color version. Annual mean column mass concentrations (μg/m^2) of the six major Earth aerosol types, calculated from the MATRIX model in the Goddard Institute for Space Studies GCM (*Bauer et al.,* 2008). The dust and sea salt concentrations have been multiplied by factors of 0.01 and 0.1, respectively. Figure courtesy of S. Bauer.

"deep water" that forms at high latitudes must be replaced by warm surface water transported from lower latitudes, adding to the poleward heat transport (*Broecker,* 1991). The removal of dense surface water to the deep ocean takes it out of thermal contact with the atmosphere until it returns to the surface at low latitudes much later. This implies that the effects of changes in greenhouse gas concentrations will not fully be reflected in surface temperatures for centuries, perhaps even a millennium or more (*Hansen et al.,* 1985).

Stone (1978a) argued that the total meridional heat transport is controlled by external parameters (solar constant, planet size, and obliquity) and the planetary albedo, and thus that the details of partitioning between atmosphere and ocean are unimportant. This would be good news for exoplanet science, since nothing is known about possible oceans on exoplanets. However, *Enderton and Marshall* (2009) show that in an idealized "aquaplanet" Earth GCM, Stone's conclusion is valid only for warm, ice-free climates. *Ferreira et al.* (2011) find that multiple equilibria, from ice-free to ice-covered ("snowball") states are possible in such models. Figure 7 compares two versions of an Earth atmospheric GCM coupled to a mixed-layer ocean with specified ocean heat transports. One has transports calculated to produce observed sea surface temperatures — a "Q-flux" model (*Hansen et al.,* 1984) — and another has ocean heat transports turned off. The atmosphere has clearly not compensated for the absence of ocean heat transport, leading to much more extensive sea ice coverage and a clearly different climate for the same external forcing.

Vertical heat transport is just as important in determining the Earth's climate. The simplest assumption for the vertical thermodynamic structure is to assume that it is in radiative equilibrium, i.e., the temperature profile is such that the divergence of the radiative flux is zero at all levels. Simple pedagogical models that assume an atmosphere transparent to SW radiation but with a nonzero gray (i.e., wavelength-independent) LW optical thickness τ^* predict a surface temperature given by

$$T_s^4 = \left(1 + c\tau^*\right) T_e^4 \quad (2a)$$

where c is a constant that depends on the details of the model. This equation shows that a planet with an atmosphere containing greenhouse gases has a surface temperature that exceeds its effective temperature, to a greater degree as the atmosphere becomes optically thicker. (An exception to this rule is a planet such as Mars with large day-night T_s contrasts; see Rafkin et al. and Covey et al., this volume). Equation (2a) can be modified to incorporate atmospheric SW absorption due, e.g., to an ozone layer or an absorbing haze (*McKay et al.,* 1999), giving

$$\sigma T_s^4 = (1-\gamma) F_s \left(1 + c\tau^*\right) + \frac{\gamma F_s}{2} \quad (2b)$$

where $F_s = S_0(1-A)/4d^2$ is the SW flux absorbed by the planet and γ is the fraction of this flux that is absorbed within the atmosphere (see Covey et al., this volume). Equations (2a) and (2b) are not quantitatively accurate but capture the basic physics of how the atmosphere radiatively modulates surface temperature.

Models that lead to equation (2a) inherently produce a surface-atmosphere temperature discontinuity because the surface absorbs all the incident SW radiation. Furthermore, the T_s they predict considerably exceeds the observed T_s for realistic τ^*. More sophisticated one-dimensional radiative equilibrium models also indicate that the lower troposphere lapse rate of Earth in radiative equilibrium is superadiabatic (*Manabe and Strickler,* 1964). Thus an adjustment, either

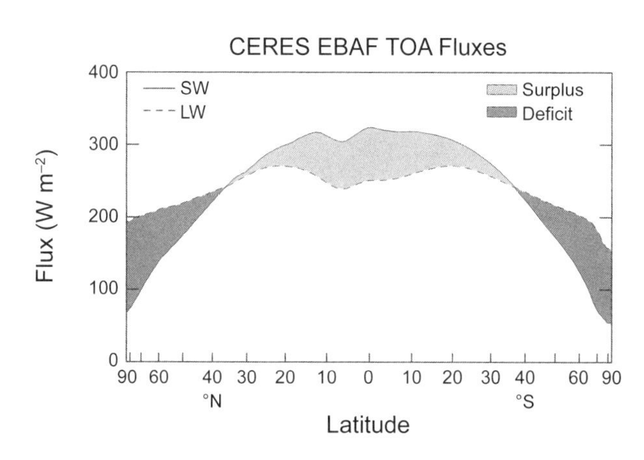

Fig. 5. Zonal and annual mean Earth TOA absorbed SW (solid) and outgoing LW (dashed) radiative fluxes from CERES EBAF data. The light shaded area shows latitudes at which the net radiative energy input to the Earth system is positive, the darker area where it is negative. Figure courtesy of J. Jonas.

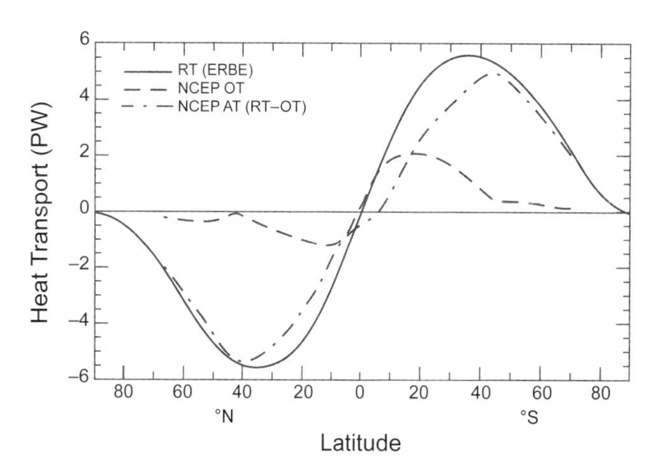

Fig. 6. Latitudinal profiles of the annual mean Earth total energy transport (solid) required by the observed latitudinal profiles of TOA radiative fluxes, and its estimated partitioning into ocean (OT, dashed) and atmosphere (AT, dash-dot) components. From *Trenberth and Caron* (2001). ©Copyright 2001 American Meteorological Society; reprinted with permission.

to a specified threshold lapse rate (dry or moist adiabatic) or determined by a more detailed model of convection, is applied to limit lapse rates and produces a realistic T_s and temperature profile. The implied upward transport of heat by convection reduces the sensitivity of T_s to changes in atmospheric composition or insolation.

Observations indicate that Earth's tropical lapse rate in the free troposphere is close to moist adiabatic up to ~400 hPa (*Zelinka and Hartmann,* 2011), indicating that moist convection is the controlling process up to that level (Fig. 8). This is possible because over the warm tropical oceans, the boundary layer is sufficiently humid for moist convection to occur frequently enough to adjust conditionally unstable lapse rates to near-neutral stability. This is not the case over the cooler tropical ocean regions, but the weak tropical Coriolis force allows advection to adjust the free troposphere lapse rate there to a value close to that in the warmer ocean regions (see Showman et al., this volume). Over tropical land, where moisture is not always available and the surface temperature responds more quickly to solar heating, more unstable lapse rates can sometimes occur; consequently, continental convective updraft speeds are stronger than those over ocean (*Zipser and Lutz,* 1994). Above the 400-hPa level, convective influence is more sporadic because entrainment of dry air often limits convection penetration depth (*Del Genio,* 2012).

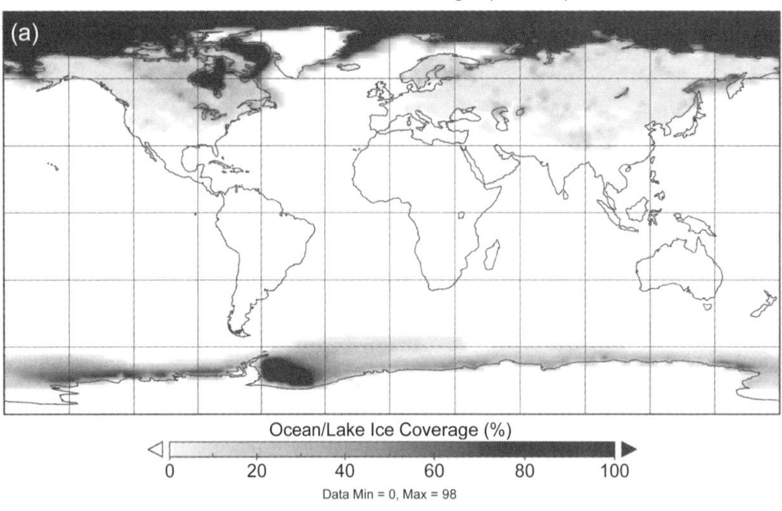

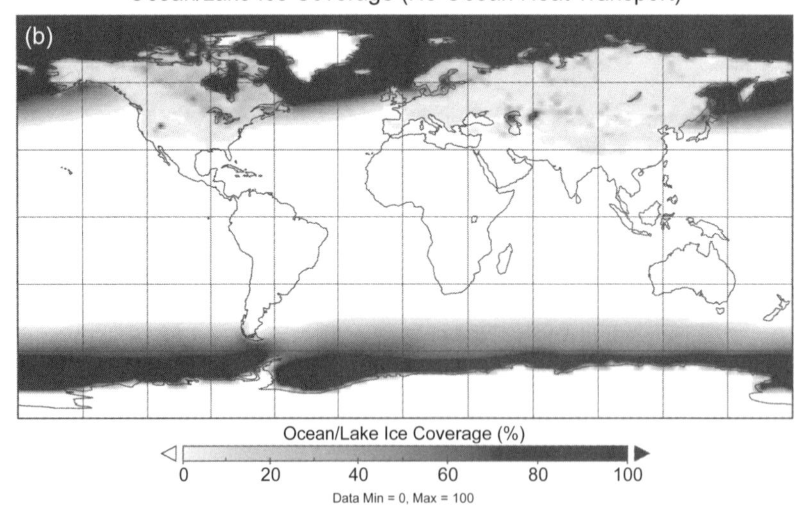

Fig. 7. December-January-February sea ice and lake ice coverage simulated by the Goddard Institute for Space Studies atmospheric GCM coupled to a mixed layer ocean with specified ocean heat transports. **(a)** Ocean heat transports are specified to reproduce observed SSTs ("Q-flux" model). **(b)** Ocean heat transports are turned off, so SSTs are determined solely by the surface flux the atmosphere provides to the ocean. Figures courtesy of M. Way.

Poleward of ~30° latitude, the tropospheric lapse rate is more stable than predicted by the moist adiabatic lapse rate. Surprisingly, extratropical static stability (proportional to the difference between the dry adiabatic and actual lapse rate) is not understood as well as tropical static stability despite our firmer understanding of extratropical quasi-geostrophic dynamics. *Stone* (1978b) suggested that the lapse rate is set by the requirement that dry baroclinic eddies associated with latitudinal temperature gradients adjust the atmosphere to a baroclinically neutral state. More recent work indicates that the extratropical lapse rate is affected by synoptic transports by both baroclinic eddies and moist convection (*Frierson,* 2008), but there is not universal agreement on the details. An effective static stability that is based on the moist adiabat but also accounts for dry downwelling circulations and meridional temperature gradients appears to hold promise (*O'Gorman,* 2011).

Equilibrium of the water cycle requires that on climate timescales, global mean precipitation P must balance global mean evaporation E. Just as for the energy cycle, though, the local water cycle is not balanced between sources and sinks (*Trenberth et al.,* 2007). Instead, the surface water balance E-P reflects the general circulation, with P > E in the moist equatorial rising branch of the Hadley cell and the extratropical storm tracks, and E > P in the dry subtropical descending branches of the Hadley cell (Fig. 9). The general circulation offsets these imbalances, with net moisture convergence into the equatorial region (hence the name Intertropical Convergence Zone) and net moisture divergence from the subtropics. Thus a significant amount of tropical precipitation does not originate from locally evaporated water.

The existence of tropical and extratropical dynamical regimes (see Showman et al., this volume) on Earth, plus the presence of preferred upwelling and downwelling regions in each, creates four basic climate or habitability zones: (1) *equatorial*, characterized by weak stratification, high humidity, and regular though intermittent strong precipitation; (2) *subtropical*, with weak stratification (except for low-level inversions) but low humidity, and therefore weak precipitation; (3) *midlatitude*, with stronger stratification and high humidity, and therefore regular slowly varying moderate precipitation; and (4) *polar*, characterized by strong stratification and low humidity, and thus weak precipitation. Life exists in each of these zones, but the more extreme conditions and life forms of the subtropical and polar desert regimes are less likely to be detected from space. The more abundant life that thrives in the more heavily precipitating equatorial and midlatitude regimes is more likely to provide an exoplanet biosignature.

A similar water cycle imbalance applies to different surface types. The oceans evaporate ~10% more water than they precipitate, and vice versa for land (Fig. 10). Thus there is a net atmospheric transport of water from ocean to land, implying that in some regions, continental precipitation is provided more by remote ocean sources than by local evapotranspiration. Water returns to the ocean via surface runoff and subsurface groundwater flow.

At TOA, the energy cycle is purely radiative, as given by equation (1). However, at the surface and within the atmosphere, the energy cycle (Fig. 1) and water cycle (Fig. 10) are directly coupled. At the surface, the turbulent fluxes that balance most of the absorbed sunlight (Fig. 1) are

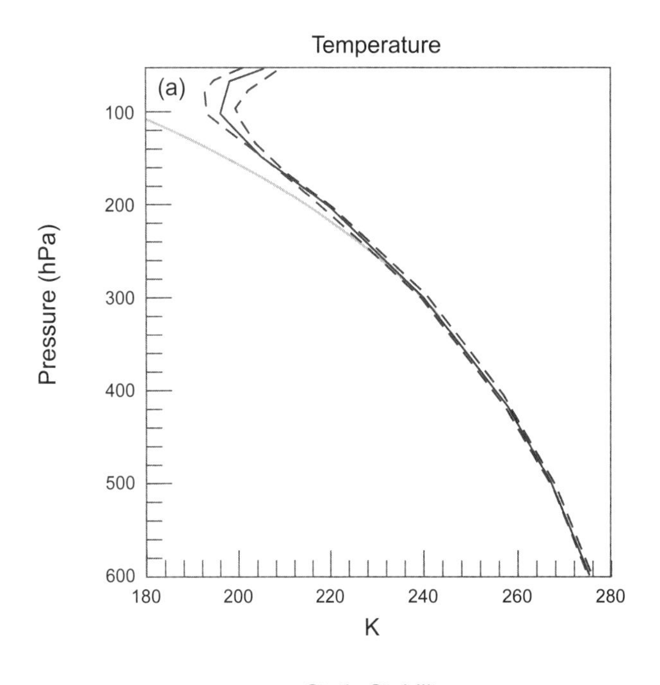

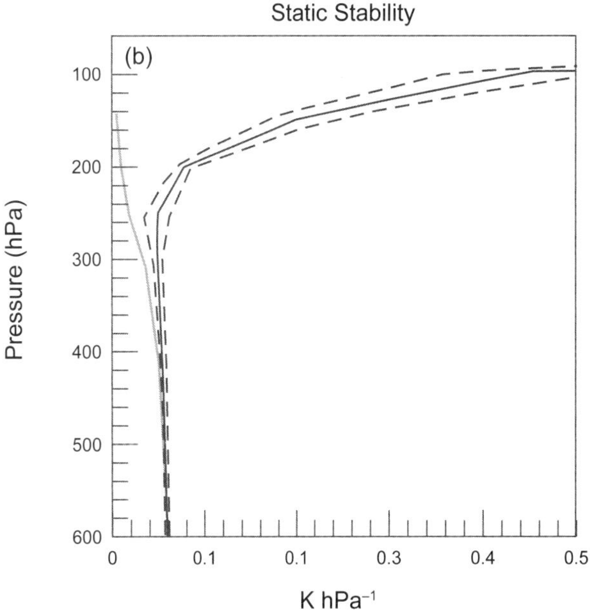

Fig. 8. Observed tropical mean vertical profiles of **(a)** temperature and **(b)** static stability for Earth's middle and upper troposphere (solid black curves). The dashed black curves represent the 2σ range of variability of the tropical means. The solid gray curves are the temperature and static stability profiles of a moist adiabat with an 850-hPa temperature equal to that observed. Adapted from *Zelinka and Hartmann* (2011). ©Copyright American Geophysical Union; reprinted with permission.

dominated by the latent heat flux (evaporation) over ocean; i.e., the Bowen ratio of sensible to latent heat flux is small, while over land the ratio varies considerably according to the soil moisture. This, in addition to the greater thermal inertia of the ocean, causes surface temperatures to respond much more strongly to perturbations over land than over ocean. Within the atmosphere, the net radiative cooling of the column (outgoing LW to space + downward surface LW–upward surface LW–SW absorption) is balanced by turbulent fluxes from the surface, mostly by the latent heat flux when water vapor condenses and precipitates (see Showman et al., this volume). Thus, precipitation in GCMs depends not only directly on the physical processes that produce precipitation (e.g., moist convection), but indirectly on phenomena whose direct influences are only radiative (e.g., the areal coverage of nonprecipitating stratocumulus clouds).

3. DIMENSIONLESS NUMBERS AS CLIMATE INDICATORS

To translate this knowledge to other planets and anticipate similarities and differences, we need to understand why particular processes control particular aspects of the climate. In general, when multiple processes occur, the fastest process dominates, and so it is instructive to define characteristic time or length scales for different processes and form appropriate dimensionless ratios to assess their importance.

This approach is well known in atmospheric dynamics, where quantities such as the Rossby number (see Showman et al., this volume), the ratio of the advective to Coriolis forces, differentiate the tropical from the extratropical dynamical regime. Another useful dimensionless number is the Richardson number $Ri = [N/(du/dz)]^2$, where N, the Brunt-Väisälä frequency, describes the vertical stratification of the atmosphere, and the vertical wind shear dU/dz for large-scale gradient wind-balanced flows is related to the meridional temperature gradient via the thermal wind equation (see Showman et al., this volume). *Allison et al.* (1995) showed that the ratio of the potential temperature (the temperature of a parcel moved adiabatically to a given reference pressure) contrasts $\Delta_V\theta$ and $\Delta_H\theta$ over the vertical and horizontal scales of the flow (the slope of the isentropes) is

$$\frac{\Delta_V\theta}{\Delta_H\theta} \sim \frac{Ri}{D^*}\left(1+\frac{1}{Ro}\right)^{-1} \tag{3}$$

where D^* is the depth of the flow in scale heights, $Ro = U/\Omega a$ is a global Rossby number based on the planet angular rotation frequency Ω and radius a, and Ri is expressed in log pressure coordinates. Figure 11 shows that Ri and Ro separate the planets in our solar system into three dynamical regimes. The slowly rotating planets Venus (Bullock and Grinspoon, this volume) and Titan (Griffith et al., this volume) have $\Delta\theta_v \gg \Delta\theta_h$, due to stabilizing upper level aerosol

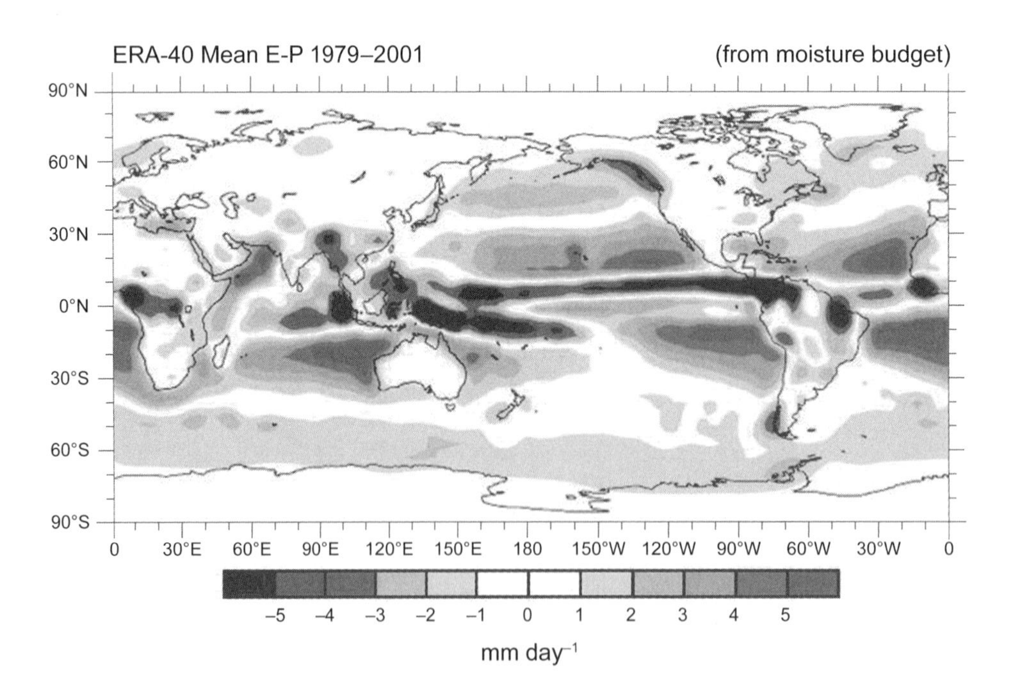

Fig. 9. See Plate 4 for color version. The annual mean Earth surface water budget E-P calculated from monthly means of the vertically integrated moisture budget of the atmosphere in the European Centre for Medium Range Weather Forecasts 40-year reanalysis (ERA-40). From *Trenberth et al.* (2007). ©Copyright 2007 American Meteorological Society; reprinted with permission.

haze decks and the broad Hadley cells that accompany slow rotation and efficiently transport heat poleward (see also Showman et al., this volume). On the jovian planets, $\Delta\theta_v \ll \Delta\theta_h$ because of their internal heat sources and convective interiors that effectively transport heat upward (*Gierasch,* 1976). Only Earth and Mars (Rafkin et al., this volume) share the intermediate regime $\Delta\theta_v \sim \Delta\theta_h$, for which both meridional and vertical transport matter. No known terrestrial planets occupy the high Ro, low Ri corner of Fig. 11, but GCMs suggest that a slowly rotating Earth with few clouds or hazes, significant surface SW absorption, and thus a convective troposphere might reside in this part of parameter space (*Del Genio et al.,* 1993).

To organize our thinking about the importance of moist processes and the combined temperature and water conditions that are most conducive to supporting life, it is useful to define several energy and water-related timescales that might be relevant to assessments of habitability:

1. The radiative time constant $t_{rad} = pc_pT/gF$ is the ratio of atmospheric heat content to the rate of radiative cooling to space F of a layer of depth one scale height. An analogous time constant with suitable values of the heat capacity for different thicknesses of fluid can be used to explain the timescales on which the ocean mixed layer, thermocline, and deep ocean respond to radiative perturbations at the surface.

2. The residence time of water t_{res} = PW/P, similarly defined as the ratio of the column water vapor content PW to the precipitation rate P, measures the vigor of the hydrologic cycle.

3. The dynamical timescales for synoptic scale weather variability (t_{ds} = L/U), where L is a typical length scale for synoptic flow; for planetary scale heat transport (t_{dp} = a/V), where V is a typical mean meridional wind speed; and for convective adjustment of the lapse rate (t_{con} = H/W), where H is the scale height and W a typical convective updraft speed, assess the importance of transports by different dynamical mechanisms.

4. External astronomical timescales such as the diurnal (t_{day}) and seasonal (t_{year}) timescales measure the importance of variations in solar/stellar forcing.

From these time constants a variety of potentially useful dimensionless numbers can be constructed. For example, numerical weather prediction models of Earth need not acknowledge the rising and setting of the Sun to predict the arrival of a low pressure center and rain in 24 hr because $t_{ds}/t_{rad} \ll 1$, but the general circulation and climate of Earth vary nonnegligibly with the seasons because t_{dp}/t_{year} and $t_{rad}/t_{year} \sim 1$. Seasonally varying temperatures lag seasonally varying insolation by a phase shift of $\tan^{-1}(2\pi t_{rad}/t_{year})$, and the amplitude of the seasonal cycle of temperature is t_{year}/t_{rad} when this ratio is small (*Conrath and Pirraglia,* 1983). Thus the impact of large orbital eccentricity on exoplanet habitability will be more muted in a thicker atmosphere.

A useful measure of the role of moist processes in a planet's climate might be given by the ratio t_{res}/t_{rad}, with a small value being diagnostic of a vigorous hydrologic cycle (presumably of interest for habitability) and its importance to the circulation and surface energy balance. For Earth

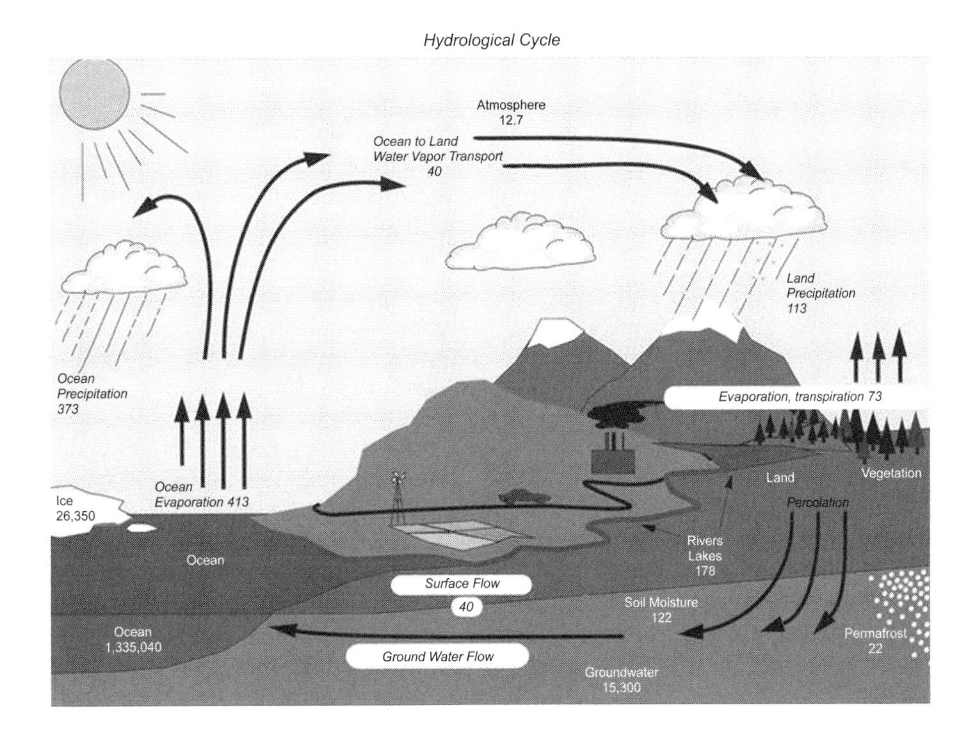

Fig. 10. See Plate 5 for color version. Global annual mean components of Earth's hydrological cycle. Storage amounts in the reservoirs are given in roman font, and flows/exchanges between reservoirs in italics (units are 1000 km^3). From *Trenberth et al.* (2007). ©Copyright 2007 American Meteorological Society; reprinted with permission.

(to date, a habitable planet) t_{res} ~ 8 d and t_{rad} ~ 2 months, giving t_{res}/t_{rad} ~ 0.1. For Titan, t_{rad} ~ 300 yr in the lower troposphere (*Strobel et al.,* 2009), while for methane, the condensable gas there, t_{res} ~ 75 yr (*Mitchell,* 2012). This gives t_{res}/t_{rad} ~ 0.25, consistent with its occasional observed storms and apparent control of its general circulation by moist processes (*Turtle et al.,* 2011). The long timescales for Titan are a consequence of its cold temperature, which would make it uninhabitable even if the condensable gas there was not methane. But their ratio nonetheless suggests a strong hydrologic cycle. The most habitable exoplanets might thus be defined not only by their global mean surface temperature, but also by small t_{res}/t_{rad}, and for those in highly eccentric orbits, modest spatial-temporal variability in temperature ($t_{rad}/t_{year} > 1$, $t_{dp}/t_{year} < 1$).

Dimensionless numbers can also correct misconceptions even about Earth's atmosphere. For example, one-dimensional models assume radiative-convective equilibrium, i.e., superadiabatic lapse rates are adjusted to a neutrally stable lapse rate. This is only valid if $t_{con}/t_{rad} \ll 1$, i.e., convection stabilizes the atmosphere much faster than radiation destabilizes it. This may not be the case if the condensable gas is present at too low a relative humidity (another dimensionless ratio) to trigger convection or to allow it to penetrate sufficiently deep to stabilize the thermal structure. "Quasi-equilibrium" theories that underlie many GCM cumulus parameterizations assume that $t_{con}/t_{ds} \ll 1$, i.e., moist convection is a "slave" to the large-scale dynamical forcing and quickly equilibrates to neutralize the large-scale state. Yet diurnal departures from quasi-equilibrium occur because t_{con}/t_{day} is not sufficiently small (*Jones and Randall,* 2011). Sometimes convection organizes into a mesoscale cluster with "memory," i.e., the current state depends on its prior evolution, lengthening t_{con} and invalidating $t_{con}/t_{ds} \ll 1$ (*Zimmer et al.,* 2011).

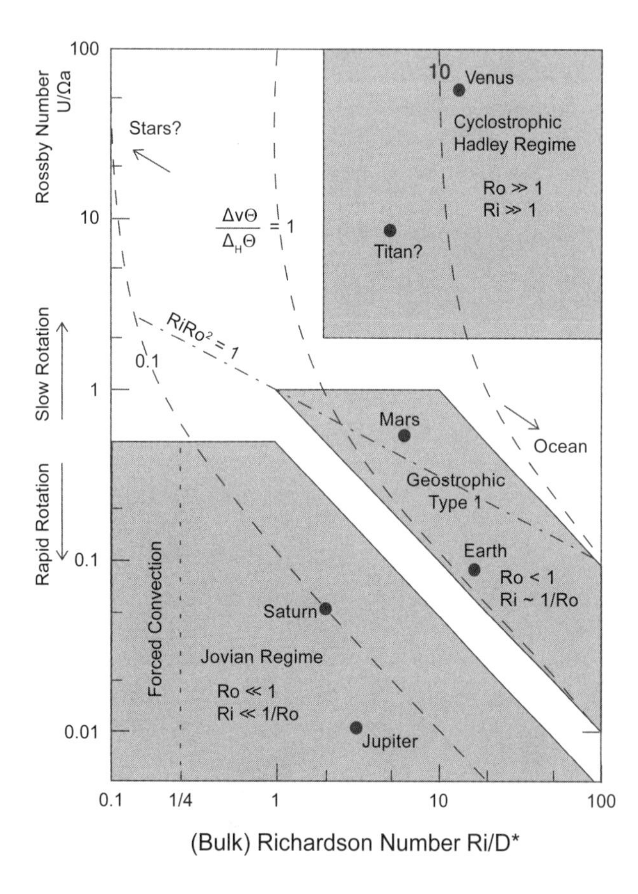

Fig. 11. Richardson-Rossby number diagram for defining atmospheric dynamical regimes of solar system planets. The three primary regimes (jovian, geostrophic, cyclostrophic/Hadley) are shaded. Dashed lines represent factor of 10 intervals in isentropic slope. The dash-dot line represents a value of the Burger number B = $RiRo^2$ = 1, which for the geostrophic regime indicates a characteristic length scale equal to the Rossby radius of deformation (see Showman et al., this volume). From *Allison et al.* (1995). ©Copyright American Geophysical Union; reprinted with permission.

4. LESSONS FROM TERRESTRIAL CLIMATE CHANGE

Most readers of this book will think of other planets when they hear the term "comparative climatology." However, Earth's own history, and its projected near-term future, are also exercises in comparative climatology. These have considerable relevance for issues such as the limits of the habitable zone for exoplanets. For in-depth analysis of specific climate change issues we refer the reader to the chapters in this volume by Covey et al. (greenhouse effect and feedbacks), Zent (orbital variability effects), Harder and Woods (solar variability effects), and Tian et al. (climate evolution). In this section we instead consider what we have and have not yet learned about Earth in particular, and climate change processes in general, from studying past and future climates.

The most relevant climate change topic for a planetary audience is the sensitivity of Earth's climate to external forcings. The only way for the *global* climate to change is via an imbalance between absorbed SW radiation (Q) and outgoing LW radiation (F). From equation (1), this can only be initiated by a change in solar (or stellar) luminosity, changing S_o; a forced change in planetary albedo A (due to a volcanic eruption, aerosol emissions, deforestation, etc.); or a change in greenhouse gas concentrations (which changes the LW opacity and thus moves the emission-to-space level to a different altitude, temporarily changing T_e). For the purpose of evaluating habitable zone limits for exoplanets, the same thing is accomplished by changing d in equation (1) (e.g., *Abe et al.,* 2011). Once a climate change is externally forced, internal feedbacks (see Covey et al., this volume) further alter A and T_e as the system tries to equilibrate. For a gradual climate perturbation such as anthropogenic emission of greenhouse gases, the most rapid response is a cooling of the stratosphere, which is approximately in radiative equilibrium, followed shortly by warming of the troposphere and land surface (*Hansen et al.,* 2005). Tropopause height and emission level rise, but since the troposphere has warmed, T_e at the new higher emission level is not very different from that before the onset of the climate change. Thus, as mentioned in section 2, today's climate is only out of balance by ~0.6 W/m^2. The surface temperature warms as the deeper troposphere convectively adjusts on the longer timescales associated with the thermal inertia of the ocean and climate feedbacks (*Hansen et al.,* 1985).

Historically an idealized "equilibrium climate sensitivity" has been estimated in GCMs by instantaneously doubling the concentration of CO_2 ("$2\times CO_2$") in an atmospheric GCM coupled to a Q-flux ocean. This decreases F because the more opaque atmosphere radiates to space from a higher colder altitude, causing an imbalance Q > F that causes the planet to warm. When the climate equilibrates and Q = F once again, the resulting surface temperature change ΔT_s is referred to as the equilibrium climate sensitivity. In the current generation of GCMs being run for the fifth Coupled Model Intercomparison Project (CMIP5), a different way of estimating climate sensitivity is being employed that uses a regression based on the transient climate change of a fully coupled atmosphere-ocean model, which sometimes gives a different result.

4.1. Is the Past Prologue?

Unlike numerical weather prediction models, whose short-term weather forecasts are evaluated every day, we have no simple way to define observational metrics that are useful as indicators of the fidelity of climate change projections. Such metrics have been proposed (e.g., *Reichler and Kim,* 2008) but tend to be based on available datasets rather than a consideration of the factors that matter most to climate change. Even TOA radiative fluxes, which are ultimately the source of climate change, are useless as climate change metrics when only their current climate mean state is utilized (*Pincus et al.,* 2008; *Collins et al.,* 2011; *Klocke et al.,* 2011). Most previous studies have used *static metrics* such as monthly geographical distributions of climate parameters, as opposed to *process-based metrics,* which measure how some important aspect of the climate changes with time or as a function of some independent controlling parameter.

One possible way to constrain climate sensitivity is to use the changes that have occurred in Earth's past. Earth has experienced multiple cold (glacial) and warm (interglacial) periods in its history, which in principle can provide useful information. Some of the more dramatic possible climate changes occurred in the distant past, e.g., the hypothesized (and debated) snowball Earth episodes of the Neoproterozoic Era (1000–542 Ma) (*Sohl and Chandler,* 2007; *Abbot et al.,* 2012), which are relevant to the issue of the outer edge of the habitable zone. The paleoclimatic evidence for such distant epochs, however, is sparse. More recent periods such as the mid-Holocene (6 ka) are better documented but involve small (<1°C) climate changes that make them less useful for insights into the future (*Hargreaves and Annan,* 2009).

The epoch with the best combination of extensive documentation and a large magnitude of climate change is the Last Glacial Maximum (LGM) (21 ka). It has often been assumed that the LGM, and earlier glacials and interglacials associated with the Milankovitch cycles of orbital variations (see Zent, this volume), can be used to infer the sensitivity of a future warmed climate. One problem with this assumption is that the LGM climate is less well known than is needed to provide a simple constraint. Early attempts to simulate the LGM with GCMs used prescribed sea surface temperature (SST) anomalies derived from ocean proxies by the CLIMAP Project (see *Rind,* 2008). The CLIMAP SSTs had some unusual features, such as a mid-Pacific *warming* signal and only a small tropical ocean cooling signal of ~1.5°C, despite 10°C cooling at high latitudes and significant cooling based on different proxies in tropical land areas. These results implied a fairly low climate sensitivity, and multiple GCMs could not reproduce them. More recent LGM reconstructions (*Waelbroeck et al.,* 2009) have used multiple proxies instead, finding larger tropical cooling that implies a higher climate sensitivity. Today's GCMs do not use the proxy data directly but rather run coupled atmosphere-ocean models and use the reconstructions for evaluation. Nonetheless, some glaring inconsistencies between the models and the data remain, calling the interpretation of the proxies into question (*Brady et al.,* 2013).

The second problem with the LGM as a proxy for future climate sensitivity is that GCMs provide conflicting results. *Crucifix* (2006) found that four GCMs that spanned almost the full range of ΔT_s uncertainty for a doubling of CO_2 concentration also simulated almost identical climate sensitivities to each other in response to LGM climate forcings (Fig. 12). The discrepancies were mainly attributed to cloud feedbacks that changed sign between the colder and warmer climates in some of the models. This should not be a surprise given the nonlinearity of the climate system. *Ye et al.* (1998), for example, showed that convective available potential energy, which affects moist convection, varies differently in response to +2 and –2°C SST changes, because in the warmer climate, the current balance between convective heating and radiative cooling just intensifies, while in the cooler climate, convection decreases dramatically, yielding to the large-scale dynamics as the process that balances radiative cooling to first order. *Hargreaves et al.* (2012) have suggested that LGM tropical cooling is a good indicator of equilibrium climate sensitivity based on a larger ensemble of GCMs, but this test has not yet been applied to future climate change predictions, and inferences drawn from it are still subject to uncertainties in the interpretation of the observational proxies.

The message is not that past climates are useless as constraints on future climate, but rather that continued thought must be given to how we can best use what we know about them to differentiate responses that are robust to the type of climate change examined from those that are sensitive to different types of climate change or to uncertainties in the observations themselves. *Schmidt et al.* (2013) propose several useful strategies for doing this.

The most commonly used "metric" of the past for demonstrating the realism of climate models is the global warming that has occurred over the twentieth century. Older generations of climate GCMs generally matched the twentieth century warming quite well. However, *Kiehl* (2007) showed that they did this despite a wide variety of climate forcings and climate sensitivities (within the observational uncertainties) because the forcings and sensitivities were negatively correlated. For example, a model with more aerosol cooling might also have a higher sensitivity to CO_2. The current CMIP5 generation of models does not show this behavior (*Forster et al.,* 2013) (Fig. 13), although a subset of the models that agrees best with the observed temperature trend does, perhaps because most current models now try to include the very uncertain aerosol indirect effect on clouds. This causes some models to depart from the observed trend to an extent not seen in older models. Observations are not yet good enough to tell us the correct combination of forcing and feedback, hence we cannot use the twentieth century temperature record to "validate" climate models. On the other hand, no GCM with only natural climate forcings has ever been able to reproduce the observed warming; this is perhaps the best evidence we have that the warming of recent decades is due to anthropogenic influences (*Knutti,* 2008). In this sense today's climate models are "wrong but useful" (*Schmidt,* 2009).

A caveat to the statements above is the chaotic nature of the climate system. Weather forecasts are sensitive to errors in the initial conditions and the parameterization of unresolved processes such as moist convection and thus must be considered probabilistic (*Slingo and Palmer,* 2011). One can then ask to what extent long-term climate change is predictable. *Lorenz* (1968) considered a possible "almost intransitive" climate system whose statistics were insensitive to initial conditions on infinite timescales but sensitive to initial conditions over long but finite time intervals. *Hasselmann* (1976) showed how short-term weather acts as a stochastic forcing whose effect is integrated by slow components of the climate system (e.g., the upper ocean) to produce longer timescale variations in sea surface temperatures. The twentieth century climate appears to have unforced decadal variations, perhaps due to episodes of deep ocean heat storage (and little surface warming) alternating with episodes of greater near-surface warming. To date, however, ensemble climate model integrations over many centuries do not exhibit the longer-term, larger-amplitude natural variability required for the twentieth century to be explained as the behavior of an almost intransitive system. Uncertainty in future climate projections is dominated instead by uncertainty in model physics (*Slingo and Palmer,* 2011), as we discuss next.

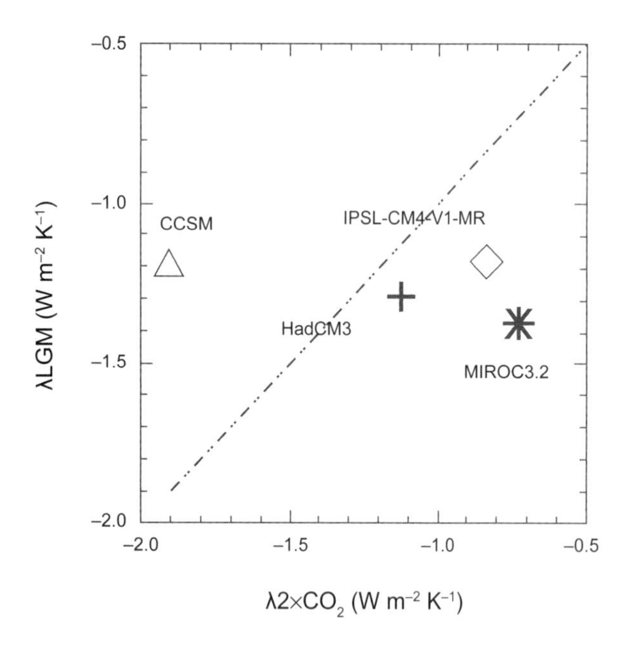

Fig. 12. The climate sensitivity parameter λ simulated by four GCMs for the LGM vs. the same sensitivity parameter of each model for $2{\times}CO_2$. In equilibrium the sensitivity parameter $\lambda = \Delta T_s/G$, where G is the external radiative forcing applied and ΔT_s is the equilibrium surface temperature response to the forcing. If the sensitivity parameter were identical to LGM and $2{\times}CO_2$ forcing for each GCM, the points would lie along the 45° (dash-double dot) line. From *Crucifix* (2006). ©Copyright American Geophysical Union; reprinted with permission.

4.2. How Well Can We Project Future Climate Warming?

Climate GCMs have been used to project future climate change for four decades, but the uncertainty in the $2CO_2$ ΔT_s has not changed much. A doubling of CO_2 concentration by itself warms the surface temperature by ~1.2°C, yet for much of the history of climate modeling, the $2{\times}CO_2$ ΔT_s of different climate models has ranged from ~1.5° to 4.5°C. The reason for the difference is feedbacks that either amplify (positive feedback) or mitigate (negative feedback) the original CO_2-forced climate change (see Covey et al., this volume, for a detailed discussion). Briefly, the primary feedbacks and their sign are the Planck feedback due to increased emission to space from a warmer atmosphere (negative); the water vapor feedback due to the greenhouse effect of the water vapor added by evaporation from a warmer surface (positive); the lapse rate feedback due primarily to the temperature dependence of the moist adiabatic lapse rate, which increases warming at altitude at the expense of that at the surface (negative); the snow/ice-albedo feedback from the decrease in planetary albedo caused by melting snow and sea ice as climate warms (positive); and the cloud feedback due to changes in cloud cover, height, and optical thickness.

Quantitative differences exist among models in the Planck, water vapor, lapse rate, and snow/ice-albedo feedbacks (Fig. 14), but their sign is not in doubt. Most of the spread in ΔT_s comes from uncertainty in the sign and magnitude of the cloud feedback. It is often claimed that as climate warms, more water evaporates from the oceans, making more cloud, which reflects more sunlight, a negative feedback. But clouds do not depend on the mixing ratio of water vapor (which changes dramatically with temperature, given the Clausius-Clapeyron equation) but rather the relative humidity (the mixing ratio relative to its saturation value), static stability, and other environmental factors, which do not obviously vary with warming in a direction that would either increase or decrease cloud. Furthermore, clouds at different altitudes have radiative effects of opposite sign. Low stratocumulus clouds, which reflect considerable sunlight but have little effect on outgoing LW radiation (Fig. 3), cool the planet, so increasing/decreasing them would be a negative/positive feedback. Cirrus clouds, on the other hand, are thinner and reflect less sunlight, but lie above the clear sky emission to space level and thus reduce LW emission to space, so increasing/decreasing them would be a positive/negative feedback. Changes in cloud height and optical thickness contribute to the cloud feedback as well, more so for some cloud types in some places than others. This makes representing cloud effects in one-dimensional models of exoplanet habitable zones (*Kitzman et al.,* 2010; *Zsom et al.,* 2012) a tremendous challenge.

For the CMIP3 generation of climate models, cloud feedback as evaluated by the climate change in cloud forcing varied from negative to near-neutral to positive, according to *Soden et al.* (2008) (Fig. 14), consistent with the canonical wisdom that we do not know the sign of cloud feedback. *Soden and Held* (2006), however, note that the cloud forcing method of calculating feedback is biased because the cloud effect on radiation is differenced against the clear sky radiative fluxes, which also change as the climate changes. By calculating "radiative kernels" that represent the partial derivatives of TOA fluxes with respect to different feedback parameters, they derive an adjusted cloud forcing change that is a more accurate depiction of the cloud feedback. Determined in this way, all the CMIP3 models have either a near-neutral or positive cloud feedback (Fig. 14), the opposite of the canonical wisdom. Furthermore, the range of CMIP3 climate sensitivities is 2.1°–4.4°C, i.e., the lowest sensitivities are no longer being simulated by any model.

Early indications from the current CMIP5 GCMs are that the situation has not changed dramatically, but now the sources of the net positive cloud feedback are being understood (*Zelinka et al.,* 2013): a positive LW cloud feedback due to an upward shift in the altitude of high clouds, and a positive SW feedback due to a decrease in total cloudiness outside the polar regions. The former is a robust consequence of the deepening of convection as the surface warms. The latter is a combination of two effects: a poleward shift of the storm tracks as the Hadley cell expands, also a robust result, and a decrease in low-latitude stratocumulus and shallow cumulus cloud cover, which may be understood (*Brient and Bony,* 2012) but is not thought to be reliably simulated by GCMs with coarse vertical resolution. At high latitudes, cloud optical thickness tends to increase with warming, a negative feedback. The response of climate model clouds to warming can be summarized as an "upward and outward" shift of clouds from the tropical lower troposphere to higher

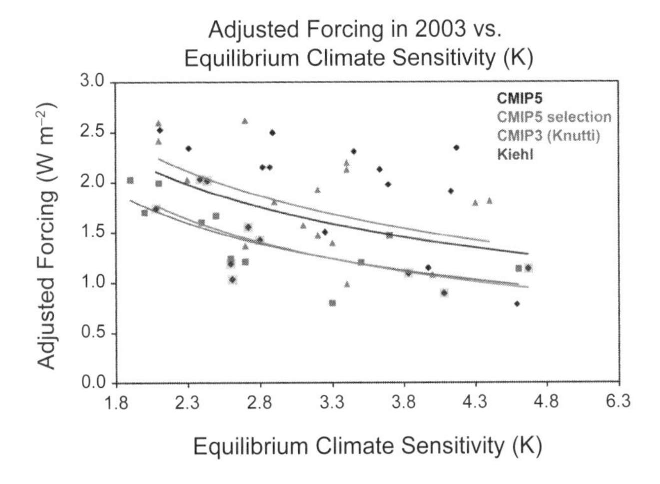

Fig. 13. See Plate 6 for color version. The relationship between 2003 climate forcing and equilibrium climate sensitivity for the turn-of-century GCMs analyzed by *Kiehl* (2007) (red), the CMIP3 GCMs analyzed by *Knutti* (2008) (blue), the CMIP5 GCMs (black), and a subset of CMIP5 GCMs that are within the 90% uncertainty range of the observed 100-year linear temperature trend (green). From *Forster et al.* (2013). ©Copyright American Geophysical Union; reprinted with permission.

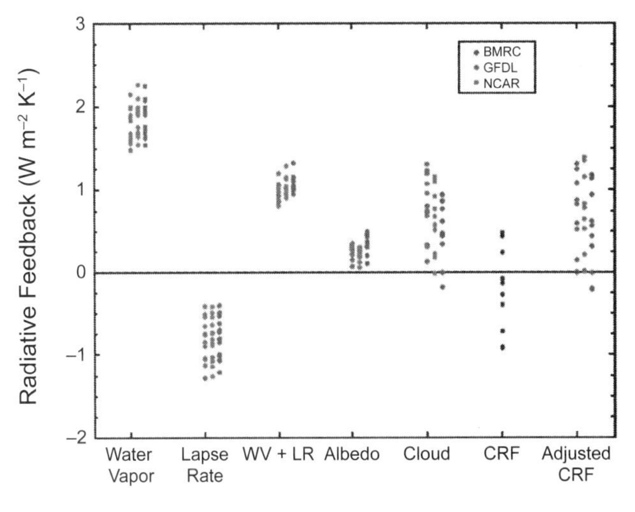

Fig. 14. Radiative feedbacks in response to 2×CO_2 for a large selection of GCMs. The symbols in a given column portray the range of feedbacks over all the models. The three different columns represent the same feedbacks calculated using radiative kernels obtained from three different GCMs as the baseline. The first five columns show the water vapor, lapse rate, combined water vapor + lapse rate, surface albedo (primarily snow/ice), and cloud feedbacks. The sixth column shows cloud feedback calculated using the traditional method of the change in cloud forcing, and the seventh column shows the results for an adjusted cloud forcing method that takes account of clear sky changes. From *Soden et al.* (2008). ©Copyright 2008 American Meteorological Society; reprinted with permission.

altitudes and latitudes (*Zelinka et al.,* 2013).

The CMIP5 models are an "ensemble of opportunity," i.e., a sample of the models that happen to exist today with "structural" differences (resolution, parameterization approaches) but that do not represent the full range of feedback uncertainty. A different approach is the "perturbed parameter ensemble," a set of thousands of simulations using a single climate model but with free parameters altered from one simulation to the next over the range of their uncertainty. This produces a probability density function of climate sensitivity whose 5–95% range of probability is not very different from that of the ensemble of opportunity except for some probability of climate sensitivities much greater than any ever simulated by an operational model (*Stainforth et al.,* 2005). With the perturbed parameter approach one can identify which parameters have the greatest effect on the model's sensitivity. Analysis of this ensemble suggests that ΔT_s (in the context of one particular GCM) is most sensitive to assumptions about the rate of entrainment of dry air into convective updrafts, the fall speed of ice crystals, and parameters that affect the formation of clouds and precipitation (*Sanderson et al.,* 2010).

The climate sensitivities simulated by today's GCMs incorporate only "fast" feedbacks due to atmospheric or near-surface processes. On longer timescales, "slow" feedbacks due to continental ice sheet growth and retreat, vegetation changes, and carbon fluxes into and out of the atmosphere, soil, biosphere, and ocean can produce an "Earth system sensitivity" quite different from the climate sensitivity due to fast feedbacks. Estimates of this sensitivity suggest that it could be 50–100% larger than the ~3°C median fast feedback sensitivity (*Lunt et al.,* 2010; *Park and Royer,* 2011). Furthermore, even for the fast feedbacks, climate sensitivity is expected to depend on atmospheric state. Thus it may vary from that portrayed by the CMIP5 models when larger climate excursions are considered and thresholds for qualitatively different behaviors of climate system components are crossed (*Hansen et al.,* 2013).

The apparent preponderance of positive feedbacks in the Earth system thus makes continued future warming highly probable. Furthermore, with the caveat that exoplanet science considers variations much larger and nonlinear than those discussed here, what we know about Earth also may have implications for assessments of the habitable zone of Earth-like planets:

1. Ocean heat transport, which retards the sea ice growth (Fig. 7) and tropical/polar heating imbalance that helps determine habitable zone limits, but which is missing from most exoplanet models (e.g., *Abe et al.* 2011), may broaden the habitable zone.

2. The negative net cloud forcing of Earth's climate (Fig. 3) may shift the habitable zone inward relative to estimates that use a cloud-free atmosphere (*Kopparapu et al.,* 2013).

3. The likelihood that many fast and slow feedbacks on Earth are positive offsets some part of the ocean- and cloud-forcing impacts, and would shrink the habitable zone.

5. CONCLUSIONS

Much remains to be better understood about Earth's climate: how convection interacts with the general circulation in different dynamic and thermodynamic environments; how aerosols, radiation, and clouds interact with turbulence to determine the presence or absence of marine stratocumulus clouds; how stratospheric circulations affect the troposphere; how stable continental ice sheets are to perturbations; and so on. Even when we understand the basic physics, we are not yet able to always portray it properly in GCMs that do not resolve the processes, and this limits our predictive ability.

Nonetheless, our accumulated knowledge about Earth can serve as a valuable resource for solar system and exoplanet atmospheric scientists. For exoplanets, inferences are limited by our inability to resolve spatial details of observed planets. Thus, one might observe Earth remotely as if it were an exoplanet (*Karalidi et al.,* 2012), not to learn anything new about Earth, but to use our detailed knowledge of its spatially varying radiative properties to understand how best to interpret a disk-integrated exoplanet spectrum. Likewise, a chronic sampling problem plagues solar system science because missions visit specific planets only occasionally and often do not monitor them globally and continuously. Is the composition and thermodynamic structure observed by a single planetary probe typical of the global mean? Do differences between snapshots of a planet taken by a past flyby and a current orbiter indicate a climate change or random samples of natural variability? Our sparse observations of other planets might be placed into an appropriate context by sampling Earth satellite datasets or reanalyses in a similar way and asking how representative those samples are of conditions elsewhere or at other times.

More importantly, our understanding of the mechanisms that conspire to produce Earth's weather and climate should probably be used as the starting point for trying to interpret any new observation of another planet. Often Earth is a valuable guide to the physics operating on other terrestrial planets: baroclinic waves on Mars, deep moist convection on Titan, photochemical sulfuric acid aerosols on Venus. When this is not the case, asking why not brings deeper insights than considering either planet in isolation. The same applies in the opposite direction: The terrestrial atmospheric community would benefit greatly from thinking more often about other planets and testing the depth of their understanding by bringing it to a different part of parameter space. Such a planetary perspective on our own planet will help us identify the fundamental unanswered questions that will advance the science the furthest.

REFERENCES

Abbot D. S., Voigt A., Branson M., Pierrehumbert R. T., Pollard D., Le Hir G., and Koll D. D. B. (2012) Clouds and snowball Earth deglaciation. *Geophys. Res. Lett., 39,* L20711, DOI: 10.1029/2012GL052861.

Abe Y., Abe-Ouchi A., Sleep N. H., and Zahnle K. J. (2011) Habitable zone limits for dry planets. *Astrobiology, 11,* 443–460.

Allison M., Del Genio A. D., and Zhou W. (1995) Richardson number constraints for the Jupiter and outer planet wind regime. *Geophys. Res. Lett., 22,* 2957–2960.

Bauer S. E., Wright D., Koch D., Lewis E. R., McGraw R., Chang L.-S., Schwartz S. E., and Ruedy R. (2008) MATRIX (Multi-configuration Aerosol TRacker of mIXing state): An aerosol microphysical module for global atmospheric models. *Atmos. Chem. Phys.*, 8, 6003–6035.

Brady E. C., Otto-Bliesner B. L., Kay J. E., and Rosenbloom N. (2013) Sensitivity to glacial forcing in the CCSM4. *J. Climate, 26,* 1901–1925.

Brient F. and Bony S. (2012) Interpretation of the positive low-cloud feedback predicted by a climate model under global warming. *Climate Dynam., 40,* 2415–2431, DOI: 10.1007/s00382-011-1279-7.

Broecker W. S. (1991) The great ocean conveyor. *Oceanography, 4,* 79–89.

Collins M., Booth B. B. B., Bhaskaran B., Harris G. R., Murphy J. M., Sexton D. M. H., and Webb M. J. (2011) Climate model errors, feedbacks and forcings: A comparison of perturbed-physics and multi-model ensembles. *Climate Dynam., 36,* 1737–1766.

Conrath B. J. and Pirraglia J. A. (1983) Thermal structure of Saturn from Voyager infrared measurements: Implications for atmospheric dynamics. *Icarus, 53,* 286–292.

Costa S. M. S. and Shine K. P. (2012) Outgoing longwave radiation due to directly transmitted surface emission. *J. Atmos. Sci., 69,* 1865–1870.

Crucifix M. (2006) Does the last glacial maximum constrain climate sensitivity? *Geophys. Res. Lett., 33,* L18701, DOI: 10/1029/2006GL027137.

Del Genio A. D. (2012) Representing the sensitivity of convective cloud systems to tropospheric humidity in general circulation models. *Surv. Geophys., 33,* 637–656.

Del Genio A. D., Zhou W., and Eichler T. P. (1993) Equatorial superrotation in a slowly rotating GCM: Implications for Titan and Venus. *Icarus, 101,* 1–17.

Enderton D. and Marshall J. (2009) Explorations of atmosphere-ocean-ice climates on an aquaplanet and their meridional energy transports. *J. Atmos. Sci., 66,* 1593–1611.

Ferreira D., Marshall J., and Rose B. (2011) Climate determinism revisited: Multiple equilibria in a complex climate model. *J. Climate, 24,* 992–1012.

Forster P. M., Andrews T., Good P., Gregory J. M., Jackson L. S., and Zelinka M. (2013) Evaluating adjusted forcing and model spread for historical and future scenarios in the CMIP5 generation of climate models. *J. Geophys. Res.–Atmos., 118,* 1–12.

Frierson D. M. W. (2008) Midlatitude static stability in simple and comprehensive general circulation models. *J. Atmos. Sci., 65,* 1049–1062.

Gierasch P. J. (1976) Jovian meteorology: Large-scale moist convection. *Icarus, 29,* 445–454.

Hansen J., Lacis A., Rind D., Russell G., Stone P., Fung I., Ruedy R., and Lerner J. (1984) Climate sensitivity: Analysis of feedback mechanisms. In *Climate Processes and Climate Sensitivity* (J. E. Hansen and T. Takahashi, eds.), pp. 130–163. AGU Geophys. Monogr. 29, Maurice Ewing Vol. 5, American Geophysical Union, Washington, DC.

Hansen J., Russell G., Lacis A., Fung I., Rind D., and Stone P. (1985) Climate response times: Dependence on climate sensitivity and ocean mixing. *Science, 229,* 857–859.

Hansen J., Sato M., Ruedy R., et al. (2005) Efficacy of climate forcings. *J. Geophys. Res., 110,* D18104, DOI: 10.1029/2005JD005776.

Hansen J., Sato M., Russell G., and Kharecha P. (2013) Climate sensitivity, sea level, and atmospheric CO_2. *Philos. Trans. R. Soc. A,* in press, arXiv: 1211.4846v2.

Hargreaves J. C. and Annan J. D. (2009) On the importance of paleoclimate modeling for improving predictions of future climate change. *Climate Past, 5,* 803–814.

Hargreaves J. C., Annan J. D., Yoshimori M., and Abe-Ouchi A. (2012) Can the Last Glacial Maximum constrain climate sensitivity? *Geophys. Res. Lett., 39,* L24702, DOI: 10.1029/2012GL053872.

Hasselmann K. (1976) Stochastic climate models. *Tellus, 28,* 473–485.

Houghton J. T. (2002) *The Physics of Atmospheres, 3rd edition.* Cambridge Univ., Cambridge. 340 pp.

Jones T. R. and Randall D. A. (2011) Quantifying the limits of convective parameterizations. *J. Geophys. Res., 116,* D08210, DOI: 10.1029/2010JD014913.

Karalidi T., Stam D. M., Snik F., Bagnulo S., Sparks W. B., and Keller C. U. (2012) Observing the Earth as an exoplanet with LOUPE, the lunar observatory for unresolved polarimetry of Earth. *Planet. Space Sci., 74,* 202–207.

Kiehl J.T. (2007) 20th century climate model response and climate sensitivity. *Geophys. Res. Lett., 34,* L22710.

Kitzman D., Patzer A. B. C., von Paris P., Godolt M., Stracke B., Gebauer S., Grenfell J. L., and Rauer H. (2010) Clouds in the atmospheres of extrasolar planets. 1. Climatic effects of multi-layer clouds for Earth-like planets and implications for habitable zones. *Astron. Astrophys., 511,* A66.

Klinger B. A. and Marotzke J. (2000) Meridional heat transport by the subtropical cell. *J. Phys. Ocean., 30,* 696–705.

Klocke D., Pincus R., and Quaas J. (2011) On constraining estimates of climate sensitivity with present-day observations through model weighting. *J. Climate, 24,* 6092–6099.

Knutti R. (2008) Why are global climate models reproducing the observed global surface warming so well? *Geophys. Res. Lett.*, 35, L18704, DOI: 10.1029/2008GL034932.

Koch D. and Del Genio A. D. (2010) Black carbon semi-direct effects on cloud cover: Review and synthesis. *Atmos. Chem. Phys., 10,* 7685–7696.

Kopp G. and Lean J. L. (2011) A new, lower value of total solar irradiance: Evidence and climate significance. *Geophys. Res. Lett., 38,* L01706, DOI: 10.1029/2010GL045777.

Kopparapu R. K., Ramirez R., Kasting J. F., Eymet V., Robinson T. D., Mahadevan S., Terrien R., Domagal-Goldman S., Meadows V., and Deshpande R. (2013) Habitable zones around main-sequence stars: New estimates. *Astrophys. J., 765,* A131.

Lacis A. A., Schmidt G. A., Rind D., and Ruedy R. A. (2010) Atmospheric CO_2: Principal control knob governing Earth's temperature. *Science, 330,* 356–359.

Loeb N. G., Wielicki B. A., Doelling D. R., Smith G. L., Keyes D. F., Kato S., Manalo-Smith N., and Wong T. (2009) Toward optimal closure of the Earth's top-of-atmosphere radiation budget. *J. Climate, 22,* 748–766.

Lorenz E. L. (1968) Climatic determinism. *Meteorol. Monogr. Am. Meteorol. Soc., 25,* 1–3.

Lunt D. J., Haywood A. M., Schmidt G. A., Salzmann U., Valdes P. J., and Dowsett H. J. (2010) Earth system sensitivity inferred from Pliocene modeling and data. *Nature Geosci., 3,* 60–64.

Lyman J. M., Good S. A., Gouretski V. V., Ishii M., Johnson G. C.,

Palmer M. D., Smith D. M., and Willis J. K. (2010) Robust warming of the global upper ocean. *Nature, 465,* 334–337.

Manabe S. and Strickler R. F. (1964) Thermal equilibrium of the atmosphere with a convective adjustment. *J. Atmos. Sci., 21,* 361–385.

McComiskey A. and Feingold G. (2012) The scale problem in quantifying aerosol indirect effects. *Atmos. Chem. Phys., 12,* 1031–1049.

McKay C. P., Pollack J. B., and Courtin R. (1991) The greenhouse and antigreenhouse effects on Titan. *Science, 253,* 1118–1121.

McKay C. P., Lorenz R. D., and Lunine J. I. (1999) Analytic solutions for the antigreenhouse effect: Titan and the early Earth. *Icarus, 137,* 56–61.

Mitchell J. L. (2012) Titan's transport-driven methane cycle. *Astrophys. J. Lett., 756,* L26.

Myhre G. (2009) Consistency between satellite-derived and modeled estimates of the direct aerosol effect. *Science, 325,* 187–190.

O'Gorman P. A. (2011) The effective static stability experienced by eddies in a moist atmosphere. *J. Atmos. Sci.*, 68, 75–90.

Park J. and Royer D. L. (2011) Geologic constraints on the glacial amplification of Phanerozoic climate sensitivity. *Am. J. Sci., 311,* 1–26.

Pincus R., Batstone C. P., Hofmann R. P. J., Taylor K. E., and Gleckler P. J. (2008) Evaluating the present-day simulation of clouds, precipitation, and radiation in climate models. *J. Geophys. Res., 113,* D14209, DOI: 10.1029/2007JD009334.

Poulsen C. and Jacob R. (2004) Factors that inhibit snowball Earth simulation. *Paleoceanography, 19,* PA4021, DOI: 10.1029/2004PA001056.

Reichler T. and Kim J. (2008) How well do coupled models simulate today's climate? *Bull. Am. Meteorol. Soc., 89,* 303–311.

Rind D. (2008) The consequences of not knowing low- and high-latitude climate sensitivity. *Bull. Am. Meteorol. Soc., 89,* 855–864.

Rose B. E. J. and Marshall J. (2009) Ocean sea ice, heat transport, and multiple climate states: Insights from energy balance models. *J. Atmos. Sci., 66,* 2828–2843.

Sanderson B. M., Shell K. M., and Ingram W. (2010) Climate feedbacks determined using radiative kernels in a multi-thousand member ensemble of AOGCMs. *Climate Dynam., 30,* 175–190.

Schmidt G. (2009) Wrong but useful. *Phys. World, 13,* 33–35.

Schmidt G. A., Ruedy R., Miller R. L., and Lacis A. A. (2010) The attribution of the present-day total greenhouse effect. *J. Geophys. Res., 115,* D20106, DOI: 10.1029/2010JD014287.

Schmidt G. A., Annan J. D., Bartlein P. J., Cook B. I., Guilyardi E., Hargreaves J. C., Harrison S. P., Kageyama M., LeGrande A. N., Konecky B., Lovejoy S., Mann M. E., Masson-Delmotte V., Risi C., Thompson D., Timmermann A., Tremblay L.-B., and Yiou P. (2013) Using paleo-climate comparisons to constrain future projections in CMIP5. *Climate Past,* in press, DOI: 10.5194/cpd-9-775-2013.

Slingo J. and Palmer T. (2011) Uncertainty in weather and climate prediction. *Philos. Trans. R. Soc. A, 369,* 4751–4767.

Soden B. J. and Held I. M. (2006) An assessment of climate feedbacks in coupled ocean-atmosphere models. *J. Climate, 19,* 3354–3360.

Soden B. J., Held I. M., Colman R., Shell K. M., Kiehl J. T., and Shields C. A. (2008) Quantifying climate feedbacks using radiative kernels. *J. Climate, 21,* 3504–3520.

Sohl L. E. and Chandler M. A. (2007) Reconstructing Neoproterozoic palaeoclimates using a combined data/modelling approach. In *Deep-Time Perspectives on Climate Change: Marrying the Signal from Computer Models and Biological Proxies* (M. Williams et al., eds.), pp. 61–80. Micropalaeontological Society Spec. Publ. #2.

Stainforth D. A., Aina T., Christensen C., Collins M., Faull N., Frame D. J., Kettleborough J. A., Knight S., Martin A., Murphy J. M., Piani C., Sexton D., Smith L. A., Spicer R. A., Thorpe A. J., and Allen M. R. (2005) Uncertainty in predictions of the climate response to rising levels of greenhouse gases. *Nature, 433,* 403–406.

Stephens G. L., Li J., Wild M., Clayson C. A., Loeb N., Kato S., L'Ecuyer T., Stackhouse P. W. Jr., Lebsock M., and Andrews T. (2012) An update on Earth's energy balance in light of the latest global observations. *Nature Geosci., 5,* 691–696.

Stone P. H. (1978a) Constraints on dynamical transports of energy on a spherical planet. *Dynam. Atmos. Oceans, 2,* 123–139.

Stone P. H. (1978b) Baroclinic adjustment. *J. Atmos. Sci., 35,* 561–571.

Strobel D. F., Atreya S. K., Bézard B., Ferri F., Flasar F. M., Fulchignoni M., Lellouch E., and Muller-Wodarg I. (2009) Atmospheric structure and composition. In *Titan from Cassini-Huygens* (R. H. Brown et al., eds.), Chapter 10, pp. 235–257, DOI: 10.1007/978-1-4020-9215-2_10. Springer Dordrecht, Heidelberg-London-New York.

Trenberth K. E. and Caron J. M. (2001) Estimates of meridional atmosphere and ocean heat transports. *J. Climate, 14,* 3433–3443.

Trenberth K. E., Smith L., Qian T., Dai A., and Fasullo J. (2007) Estimates of the global water budget and its annual cycle using observational and model data. *J. Hydrometeor., 8,* 758–769.

Turtle E. P., Del Genio A. D., Barbara J. M., Perry J. E., Schaller E. L., McEwen A. S., West R. A., and Ray T. L. (2011) Seasonal changes in Titan's meteorology. *Geophys. Res. Lett., 38,* L03203, DOI: 10.1029/2010GL046266.

Waelbroeck C. and MARGO Project Members (2009) Constraints on the magnitude and patterns of ocean cooling at the Last Glacial Maximum. *Nature Geosci., 2,* 127–132.

Wallace J. M. and Hobbs P. V. (2006) *Atmospheric Science: An Introductory Survey, 2nd edition.* Academic, New York. 504 pp.

Ye B., Del Genio A. D., and Lo K. K.-W. (1998) CAPE variations in the current climate and in a climate change. *J. Climate, 11,* 1997–2015.

Zelinka M. D. and Hartmann D. L. (2011) The observed sensitivity of high clouds to mean surface temperature anomalies in the tropics. *J. Geophys. Res., 116,* D23103, DOI: 10.1029/2011JD016459.

Zelinka M. D., Klein S. A., Taylor K. E., Andrews T., Webb M. J., Gregory J. M., and Forster P. M. (2013) Contributions of different cloud types to feedbacks and rapid adjustments in CMIP5. *J. Climate,* in press, DOI: 10.1175/JCLI-D-12-00555.1.

Zimmer M., Craig G. C., Keil C., and Wernli H. (2011) Classification of precipitation events with a convective response timescale and their forcing characteristics. *Geophys. Res. Lett., 38,* L05802, DOI: 10.1029/2010GL046199.

Zipser E. J. and Lutz K. R. (1994) The vertical profile of radar reflectivity of convective cells: A strong indicator of storm intensity and lightning probability? *Mon. Weather Rev., 122,* 1751–1759.

Zsom A., Kaltenegger L., and Goldblatt C. (2012) A 1D microphysical cloud model for Earth, and Earth-like exoplanets: Liquid water and water ice clouds in the convective troposphere. *Icarus, 221,* 603–616.

Bullock M. A. and Grinspoon D. H. (2013) The atmosphere and climate of Venus. In *Comparative Climatology of Terrestrial Planets* (S. J. Mackwell et al., eds.), pp. 19–54. Univ. of Arizona, Tucson, DOI: 10.2458/azu_uapress_9780816530595-ch002.

The Atmosphere and Climate of Venus

Mark A. Bullock
Southwest Research Institute

David H. Grinspoon
NASA Kluge Center, Library of Congress

Venus lies just sunward of the inner edge of the Sun's habitable zone. Liquid water is not stable. Like Earth and Mars, Venus probably accreted at least an ocean's worth of water, although there are alternative scenarios. The loss of this water led to the massive, dry CO_2 atmosphere, extensive H_2SO_4 clouds (at least some of the time), and an intense CO_2 greenhouse effect. This chapter describes the current understanding of Venus' atmosphere, established from the data of dozens of spacecraft and atmospheric probe missions since 1962, and by telescopic observations since the nineteenth century. Theoretical work to model the temperature, chemistry, and circulation of Venus' atmosphere is largely based on analogous models developed in the Earth sciences. We discuss the data and modeling used to understand the temperature structure of the atmosphere, as well as its composition, cloud structure, and general circulation. We address what is known and theorized about the origin and early evolution of Venus' atmosphere. It is widely understood that Venus' dense CO_2 atmosphere is the ultimate result of the loss of an ocean to space, but the timing of major transitions in Venus' climate is very poorly constrained by the available data. At present, the bright clouds allow only 20% of the sunlight to drive the energy balance and therefore determine conditions at Venus' surface. Like Earth and Mars, differential heating between the equator and poles drives the atmospheric circulation. Condensable species in the atmosphere create clouds and hazes that drive feedbacks that alter radiative forcing. Also in common with Earth and Mars, the loss of light, volatile elements to space produces long-term changes in composition and chemistry. As on Earth, geologic processes are most likely modifying the atmosphere and clouds by injecting gases from volcanos as well as directly through chemical reactions with the surface. The sensitivity of Venus' atmospheric energy balance is quantified in this chapter in terms of the initial forcing due to a perturbation, radiative response, and indirect responses, which are feedbacks — either positive or negative. When applied to one Venus climate model, we found that the albedo-radiative feedback is more important than greenhouse forcing for small changes in atmospheric H_2O and SO_2. An increase in these gases cools the planet by making the clouds brighter. On geologic timescales the reaction of some atmospheric species (SO_2, CO, OCS, S, H_2O, H_2S, HCl, HF) with surface minerals could cause significant changes in atmospheric composition. Laboratory data and thermochemical modeling have been important for showing that atmospheric SO_2 would be depleted in ~10 m.y. if carbonates are available at the surface. Without replenishment, the clouds would disappear. Alternatively, the oxidation of pyrite could add SO_2 to the atmosphere while producing stable Fe oxides at the surface. The correlation of near-infrared high emissivity (dark) surface features with three young, large volcanos on Venus is strong evidence for recent volcanic activity at these sites, certainly over the timescale necessary to support the clouds. We address the nature of heterogeneous reactions with the surface and the implications for climate change on Venus. Chemical and mineralogical signatures of past climates must exist at the surface and below, so *in situ* experiments on the composition of surface layers are vital for reconstructing Venus' past climate. Many of the most Earth-like planets found around other stars will probably resemble Venus or a younger version of Venus. We finish the chapter with discussing what Venus can tell us about life in the universe, since it is an example of a planetary climate rendered uninhabitable. It also resembles our world's likely future. As with the climate history of Venus, however, the timing of predictable climate transitions on the Earth is poorly constrained by the data.

1. INTRODUCTION

Venus is the same size as Earth and a little closer to the Sun. Its orbital semimajor axis is close to $1/\sqrt{2}$ AU, so it receives about twice as much solar energy as Earth. However, the bright H_2SO_4 clouds make Venus' albedo more than twice Earth's. Consequently there is less solar forcing of Venus' climate than of Earth's in spite of its greater proximity to the Sun.

Venus has an extremely dense CO_2/N_2 atmosphere, with an average surface pressure of 92 bar (*Avduevskii et al.,* 1976; *Marov et al.,* 1973). The atmosphere is 96.5% CO_2 and 3.5% N_2 by volume, which is the same as the molar ratio (*Oyama et al.,* 1979, 1980). It has trace amounts of water vapor, S gases, and halogens that play important roles in the formation of the global cloud layers and in the chemistry and energy balance of the atmosphere. The atmosphere has three main cloud layers from 48 to 68 km above the surface (366 K and 1.4 bar to 235 K and 50 mbar) consisting of liquid H_2SO_4/water aerosols (Fig. 1) (*Esposito et al.,* 1983; *Knollenberg and Hunten,* 1980). There are variable hazes above and below the main cloud decks. Large spatial and temporal variabilities in the middle and lower clouds, including aerosol populations, are evident from Galileo's Near Infrared Mapping Spectrometer (NIMS) (*Carlson et al.,* 1993; *Grinspoon et al.,* 1993) and from observations by the European Space Agency's Venus Express (VEX) spacecraft (*Barstow et al.,* 2012; *Wilson et al.,* 2008). The atmosphere near the surface is about one-tenth as dense as water, and the CO_2 is supercritical (P_{crit},T_{crit} for CO_2 is 72.9 bar, 304.25 K). So while Venus' atmosphere, like Earth's, is a fluid, it is neither a gas nor a liquid. As a mixture of gas and supercritical fluid, it has some properties of both. The fluid dynamics and heat transport operate in a medium with a density somewhere between water and air, without significant Coriolis forces. Relative to Earth and Mars, Venus rotates slowly. Strong zonal winds at the cloud tops are assumed to reflect cyclostrophic balance, in which the equatorward component of centrifugal force is balanced by meridional pressure gradient (*Gierasch,* 1975). High winds circle the planet from east to west, at speeds 60 times faster than rotation of the solid planet (*Schubert,* 1983). Peaking at 120 m s^{-1} at the equatorial cloud tops, this superrotating zonal flow decreases almost linearly with depth down to the surface. With its speed and mass, the atmosphere has 0.1% of the angular momentum of the entire planet, peaking at 21 km above the surface. Large-scale cells that arise from equatorial heating of the clouds transport heat and momentum to high latitudes. A deeper return flow, perhaps in the form of eddies beneath the clouds, must exist but has never been observed (*Gierasch et al.,* 1997). Zonal winds decrease with increasing latitude until they transition to the giant polar vortices (*Piccioni et al.,* 2007; *Limaye et al.,* 2009; *Suomi and Limaye,* 1978; *Luz et al.,* 2011) at about the same latitude as the descending branch of the hemispheric meridional circulation (*Imamura and Hashimoto,* 1998). Above the clouds, the east-west zonal flow gives way to the subsolar-antisolar winds generated by

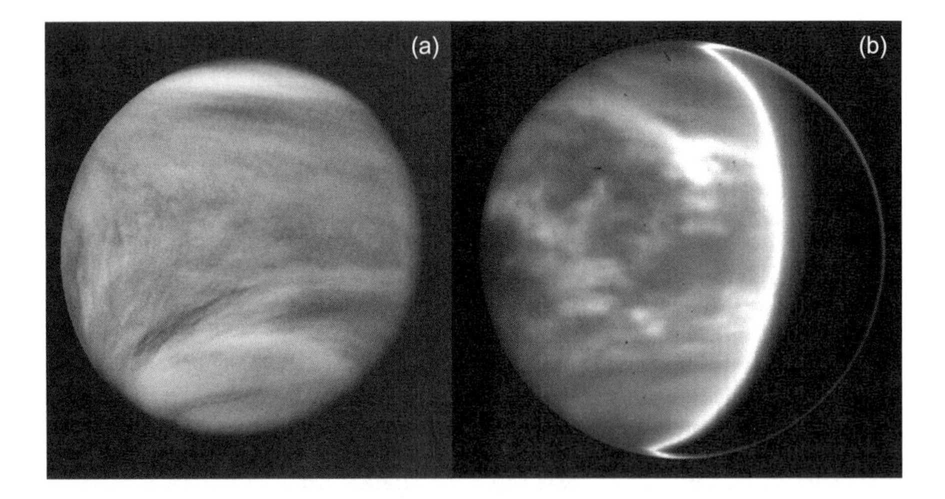

Fig. 1. See Plate 7 for color version. **(a)** Venus in reflected UV sunlight. Acquired on February 26, 1979, by the Pioneer Venus Orbiter. The distinctive tilted bands at the cloudtops (68 km) are produced by low wavenumber planetary waves that circle the planet in about four days. **(b)** The nightside of Venus glowing at 2.3 μm (*Young et al.,* 2007, 2010). This image was taken with the SpeX imager/spectrometer at NASA's 3-m Infrared Telescope Facility on Mauna Kea, Hawaii (*Rayner et al.,* 2003) on May 6, 2004. Images and spectra are taken simultaneously; the dark vertical line is the spectrometer slit. The plate scale is 0.115″/pixel. Seeing that morning was about 0.5″, with about 2 pr. μm of H_2O vapor above the observatory — superb conditions for infrared observations. Cloud features as small as 80 km were resolved. The lower clouds are silhouetted by upwelling heat radiation from below. Bright regions are holes in the lower cloud, while dark regions are where the lower cloud is thickest. The haze that makes up the upper cloud and the atmosphere are transparent to radiation at this wavelength. Venus' clouds are highly variable from night to night at this level of the atmosphere (about 52 km).

solar ultraviolet heating of the Venus mesosphere (*Gierasch et al.,* 1997; *Sanchez-Lavega et al.,* 2008). General circulation models (GCMs) of Venus, informed by VEX data, can replicate the thermal tides and some aspects of the strong east-west circulation (*Lebonnois et al.,* 2010). A complete understanding of the superrotation of Venus' atmosphere remains an enduring mystery, but there is evidence for transfer of angular momentum between the surface and atmosphere (*Margot,* 2011; *Bengtsson et al.,* 2013).

In spite of the dramatic difference in atmospheric composition between Earth and Venus, there is actually quite a bit of symmetry between the atmospheres of the three inner planets with atmospheres. Most of the C in Earth's surface environment is in the form of massive carbonates on the continental margins. If these carbonates were decomposed to CO_2, Earth's atmosphere would be 100 bar of CO_2 with 0.7 bar of N_2 (*Sleep and Zahnle,* 2001; *Zahnle et al.,* 2007) — really quite close to Venus' atmospheric composition and mass. Furthermore, the atmosphere of Mars, although it is 10,000 times less massive, has a CO_2/N_2 atmosphere with a ratio quite similar to Venus' (Table 1).

TABLE 1. Major atmospheric gases in the atmospheres of Mars, Earth, and Venus.

	CO_2 (%)	N_2 (%)	Ar (%)	Surface P (bars)
Mars*	95.3	2.7	1.6	0.007
Earth†	100	0.7	0.9	1.01
Venus‡	96.5	3.5	0.007	92

*Data from *Owen* (1992).
†Assuming that all carbonate is converted to CO_2 (*Sleep and Zahnle,* 2001).
‡Data from *Oyama et al.* (1980).

The main agents of Venus' powerful greenhouse effect are CO_2, H_2O, and the clouds, just as on Earth (*Pollack,* 1969a; *Sagan,* 1960). However, it isn't just the intrinsic radiative properties of CO_2 and H_2O that make the Venus greenhouse effect so powerful. Pressure broadening of both CO_2 and H_2O infrared absorption lines results in almost complete blockage of thermal radiation to space, except for a small and extremely important window between 2.3 and 2.7 μm (*Pollack,* 1969a; *Pollack et al.,* 1980). About 3% of outgoing thermal radiation escapes to space through this window. The result is that the global mean surface temperature is three times higher than it would be without an atmosphere. For comparison, Earth's surface is only 13% warmer, Mars' is only 1.5% warmer, and Titan's is only 17% warmer than they would be without atmospheres.

The global mean temperature of Venus is 735 K at an average surface height of 6051 km from the center of the planet (*Marov et al.,* 1973; *Seiff,* 1983). Surface heights are referenced to the mean radius defined by *Seidelmann et al.* (2007). Venus has a unimodal surface hypsometry with only 1% of the surface area 5 km or more above the mean (Fig. 2) (*Ford and Pettengill,* 1992). At the highest point on Venus, in the Maxwell Montes at 12 km, the surface temperature is 650 K (*Fegley et al.,* 1997; *Tanaka et al.,* 1997). This is due to the dry adiabatic lapse rate of the CO_2/N_2 atmospheric mixture (*Crisp and Titov,* 1997; *Seiff,* 1983). Lowlands represent a large fraction of Venus' surface, with 20% more than 1 km below the datum (*Tanaka et al.,* 1997). In the lowest feature on Venus, 2.5 km below the datum in Diana Chasma, the surface temperature is 765 K. At this temperature, the surface could be seen glowing at night with the unaided eye.

The climates of Earth and Mars are driven by seasonable variability due to the tilt of their spin axes relative to the axes of their orbits about the Sun (their obliquity) (e.g., *Peixoto and Oort,* 1992). Life on Earth thrives on the gradients and material fluxes forced by these cyclical changes. The landscape and atmosphere of Mars also change in response to the seasons (*Lowell,* 1909), with an intensity that bears some resemblance to seasonal changes on Earth. Earth's obliquity of 23.5° and Mars' obliquity of 25.2° result in similar patterns of solar energy input over a year. Venus, however, has an obliquity of 1.7°, and hence little or no seasonal variability in solar input. In addition, the atmosphere is so massive that there is little variability in temperature of the surface, either with time or latitude. Adjusting for altitude, the polar and equatorial surfaces are within 4 K of each other, and do not vary more than that over a year (*Seiff et al.,* 1980).

Venus is the only planet in the solar system that has a retrograde diurnal motion. The original spin state of the terrestrial planets was the sum of the angular momentum imparted by all the impactors that formed each of them, but was dominated by the last one or two large impactors (*Dobrovolskis,* 1980). However, core-mantle friction or the chaotic evolution of its spin state may also be responsible for Venus' backward spin (*Correia and Laskar,* 2001). Earth and Mars coincidentally have almost the same length of day, with Earth at 24 hours and Mars at 24 hours 40 minutes. However, Earth's last large impact created the Moon and left Earth spinning with a four-hour day (*Canup and Asphaug,* 2001). The length of day has gradually lengthened as the Moon has receded. It is possible that Mars' last giant impact stripped it of its atmosphere, including the CO_2 and noble gases (*Brain and Jakosky,* 1998; *Owen,* 1992). This story will become more clear as the Mars Atmosphere and Volatile Evolution (MAVEN) spacecraft explores Mars' upper atmosphere beginning in 2015 (*Jakosky,* 2011). The ancient impact history of Venus did not form a moon, but instead left it in a slow, retrograde rotating state completing one rotation every 243 days (τ_d). Since Venus' year (τ_y) is less than this, at 224.7 days, the solar day, from sunrise to sunrise (τ_s), is less than either of them

$$\frac{1}{\tau_s} = \frac{1}{\tau_y} + \frac{1}{\tau_d}$$

which is 116 days, 18 hours. The Sun rises in the west and moves eastward across the sky, perhaps sometimes visible through holes in the lower clouds, to set in the east 58 days

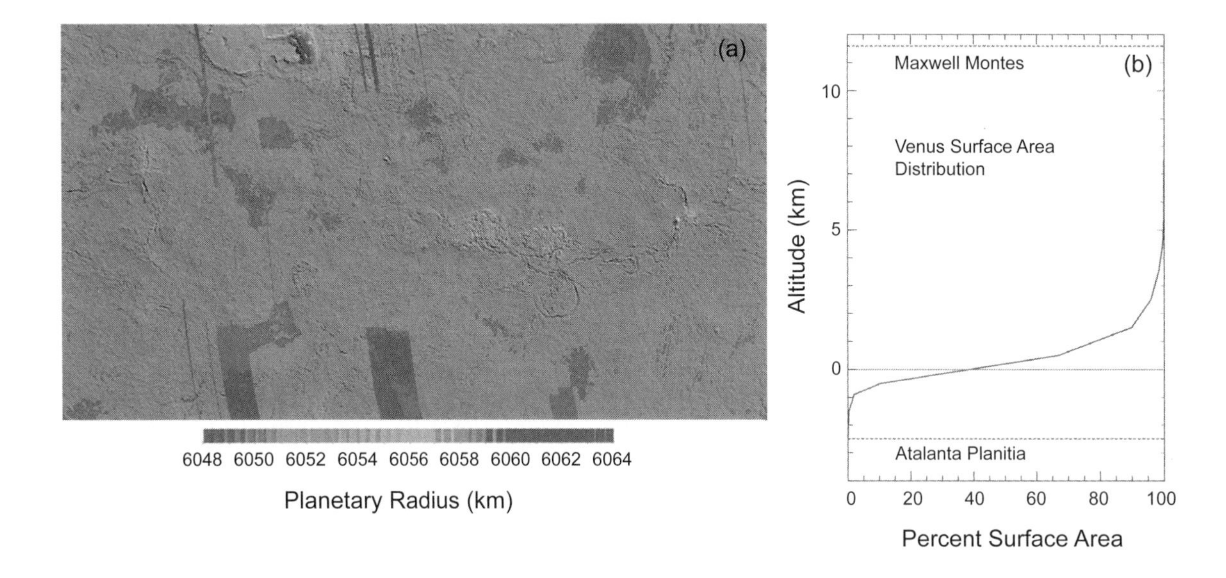

Fig. 2. See Plate 8 for color version. **(a)** Altimetry of the surface of Venus obtained by the Magellan radar altimeter (*Ford and Pettengill,* 1992). "Sea level" datum (shown in green) is 6051 km from the center of the planet. Blue areas are lowland volcanic plains, and yellow regions are the two major continent-like highlands Aphrodite (center, equator) and Ishtar Terra (upper left). The highest region on Venus is Maxwell Montes on top of Ishtar, shown in orange. Large volcanic shields that form at the intersections of major rift systems are also yellow. **(b)** Cumulative hypsometric curve for Venus' topography. It shows the percentage of the surface area above a given altitude. Venus' topography is unimodal, with deep chasmata 2 km below the datum and Maxwell Montes 12 km above the datum. This is unlike Earth, which has a bimodal topography consisting of distinct oceanic and continental crust. Data from *Ford and Pettengill* (1992).

later. Possible variations in Venus' rotation rate between Magellan and VEX have been reported (*Mueller et al.,* 2012), due to angular momentum exchanged with either the core or atmosphere (*Cottereau et al.,* 2011).

The VEX mission, still orbiting the planet as of this writing, is probing the atmosphere and surface of the planet in greater detail than ever before (*Svedhem et al.,* 2007). Some of VEX's discoveries and their significance for Venus' evolution are discussed in the chapter by Baines et al. in this volume.

Section 2 of this chapter describes the current state of knowledge of the Venus atmosphere. Beginning with the composition and structure of the atmosphere and clouds, this section goes on to discuss what is known and unknown about Venus' atmospheric circulation, and then describes the greenhouse effect as it is manifested on Venus. Section 3 discusses the early evolution of Venus and the loss of its water due to a runaway greenhouse effect. Section 4 describes the present understanding of the processes that affect Venus' climate, quantitatively casting climate sensitivity in terms of radiative forcing and feedbacks. To this end, the role of surface sources and sinks of climatologically important gases is described, and their implications for the state of the clouds and climate of Venus. The search for habitable planets in our galaxy is underway, and Venus provides us with the one other example of the evolution of an Earth-sized, rocky planet. Section 5 addresses the importance of Venus for understanding planets around other stars and the search for another habitable planet.

The extreme conditions of Earth's sister planet invites the question of how Earth and Venus evolved along such divergent paths. In spite of apparently similar initial conditions, one has been continuously habitable and the other is extremely inhospitable to life as we know it. Their atmospheres today are vastly different, yet the radiative, thermodynamic, and geologic processes that shape their climates are similar. Like Earth, Venus' atmospheric temperatures are influenced by cloud-radiative feedbacks and by long-term geologic and geochemical processes that slowly or episodically alter the atmosphere. Large-scale volcanic events on both planets have the potential to disrupt the climate and to push it to a new stable state (*Bullock and Grinspoon,* 1996; *Hashimoto and Abe,* 2000).

2. VENUS' ATMOSPHERE

2.1. Structure and Composition

2.1.1. Atmospheric fluids. The proportion of the major gases CO_2 and N_2 in the atmospheres of Venus, Mars, and Earth (if Earth's carbonates were returned to the atmosphere) are roughly the same (Table 1). However, the proportion of minor gases that carry H, S, and Cl are very different in each of these atmospheres (*von Zahn et al.,* 1983), due largely to differences in temperature and history of loss to space. Atmospheric S cycles that form the clouds on Venus are linked with atmospheric C cycles on short timescales (*Krasnopolsky,* 2007) and coupled to surface minerals on long timescales (*Esposito et al.,* 1997; *Fegley and Treiman,* 1992; *Prinn,* 1985).

The quantity of water vapor in Venus' atmosphere is about the same as in Earth's atmosphere, but there is no large surface reservoir as there is on Earth. Earth's surface environment has about 100,000 times more water than Venus'. While the oceans are the primary reservoir of H on Earth, the primary reservoir of H on Venus is the massive H_2SO_4 cloud system that globally occupies the atmosphere from 49 to 65 km altitude. The clouds contain about 10 times more H than the atmospheric water vapor does, although about 1% of the atmospheric water condenses with the hygroscopic H_2SO_4 as cloud aerosols are formed. As discussed by *Kasting* (1988) and *Kasting et al.* (1984), the extreme desiccation of the surface of Venus is because at some time in the past, Venus lost the majority of its initial endowment of H to rapid hydrodynamic escape following a runaway water greenhouse.

Pinning down the water abundance of Venus' atmosphere has been the subject of numerous investigations in observational planetary astronomy. It is now accepted that the abundance of water vapor in the lowest scale height of the Venus atmosphere is about 30 ppmv. *In situ* measurements from the Venera 4 mission indicated a water abundance of 0.1–0.5%, and greenhouse models were able to explain the surface temperature with this amount of water (*Pollack et al.,* 1980). However, the Pioneer Venus (PV) Gas Chromatograph/Mass Spectrometer (GCMS) measured a value of 200 ppmv beneath the clouds. Although drier than expected, improved spectral databases and methods were able to successfully reproduce Venus surface temperatures because they accounted for more spectral lines at these temperatures and pressures (*Pollack et al.,* 1980).

Meanwhile, groundbased observers began exploiting near-infrared windows in Venus' spectrum that were sensitive to water abundance. These were first discovered by *Allen and Crawford* (1984), who realized that the lower cloud structure on the nightside could be seen silhouetted by hot upwelling radiation from below, and that the windows could be used for remote sounding of trace gases (*Taylor et al.,* 1997). *Pollack et al.* (1993) analyzed nightside near-infrared spectra of Venus taken by the Near Infrared Mapping Spectrometer (NIMS) on Galileo during a Venus flyby. By forward modeling using synthetic spectra calculated from the HITRAN spectral database developed by the U.S. Air Force (*Rothman et al.,* 1992), they retrieved a water abundance of 30 ± 10 ppmv between 10 and 40 km. Using a very-high-resolution Fourier Transform spectrometer at the Canada France Hawaii Telescope, *de Bergh et al.* (1995) retrieved H_2O abundances at altitudes of 30–40 km, 15–25 km, and 0–15 km of 30 ± 15 ppmv. The most recent data on water abundance in the Venus lower atmosphere come from the high-resolution channel of the Visible and Infrared Thermal Imaging Spectrometer (VIRTIS) on VEX (*Drossart et al.,* 2007). The VIRTIS team determined that the average atmospheric water abundance on Venus from 30 to 40 km is 31 ± 2 ppmv (*Marcq et al.,* 2008). Complementary observations of the 1.18-μm window with the medium-resolution VIRTIS-M yielded a slightly higher H_2O mixing ratio in the bottom 15 km of the atmosphere, 44 ± 9 ppmv (*Bezard et al.,* 2009). Recent groundbased infrared spectra of Venus' nightside from the Anglo-Austrialian Telescope (AAT), also at 1.18 μm, were also used to retrieve atmospheric water vapor abundance from 0 to 15 km (*Chamberlain et al.,* 2013). The spectral resolution of this instrument, R ~ 2400, was higher than VIRTIS-M's. They obtained an H_2O mixing ratio of 31 ± 9/6 ppmv, which agrees with the VIRTIS retrievals.

Dayside near-infrared spectra were also acquired with VIRTIS-H to retrieve the abundance of atmospheric water vapor above the clouds. Using CO_2 and H_2O absorption lines between 2.48 and 2.6 μm, *Cottini et al.* (2012) determined that cloud top water abundance was 3 ± 1 ppmv between ±40° latitude, increasing to an average of 5 ppmv at high latitudes. Beyond 40° latitude, cloud top water vapor was highly variable, between 1 and 15 ppmv. *Federova et al.* (2008), for example, determined the mixing ratio of water vapor above the northern polar clouds using VEX's Solar Occultations in the Infrared (SOIR) spectrometer, a part of VEX's Spectroscopy for Investigation of Characteristics of the Atmosphere of Venus (SPICAV) instrument. The 2.61-μm absorption line of H_2O during solar occultations yielded a value of 1.16 ± 0.24 ppmv. A summary of deep atmosphere water vapor measurements in the Venus atmosphere is shown in Fig. 3.

The ratio of D/H in a planetary atmosphere is often indicative of the fractionating loss of H from some initial reservoir. The Earth's Standard Mean Ocean Water (SMOW) D/H is 155.8 ± 0.1 ppmv, about the same as the D/H of chondritic water. Water in martian meteorites has measured D/H that ranges between 1 and 5 times SMOW, indicative of the mixing of fractionated and less-fractionated sources (*Leshin et al.,* 1996; *Watson et al.,* 1994). The D/H in Venus' H_2SO_4 aerosols was measured *in situ* by the PV Large Probe to be 0.025 ± 0.005, or 157 ± 30 SMOW. This has been interpreted as either being wholly due to the primordial loss of water from a reservoir that initially had a D/H value equal to SMOW (*Donahue et al.,* 1982) or as reflecting subsequent steady-state evolution of the atmospheric water reservoir where fractionating loss is balanced by an exogenous (cometary) or endogenous (volcanic) source (*Grinspoon,* 1993; *Donahue and Hodges,* 1992). High-resolution spectra of Venus' nightside from the Canada France Hawaii Telescope yielded a D/H in H_2O vapor beneath the clouds of 0.0187 ± 0.0062, or 120 ± 40 times SMOW (*Bezard et al.,* 1990; *de Bergh et al.,* 1991). In the VEX era, *Federova et al.* (2008) reported an average D/H between 70 and 95 km using SOIR's measurement of the 3.58-μm HDO line. This value was 0.038 ± 0.004, or 240 ± 25 times SMOW. SOIR found the D/H in Venus' atmosphere to increase with altitude from 0.025 in the deep atmosphere to 0.06 at 90 km (*Bertaux et al.,* 2007). This is not well understood but may be due to some combination of fractionating condensation and photochemical and

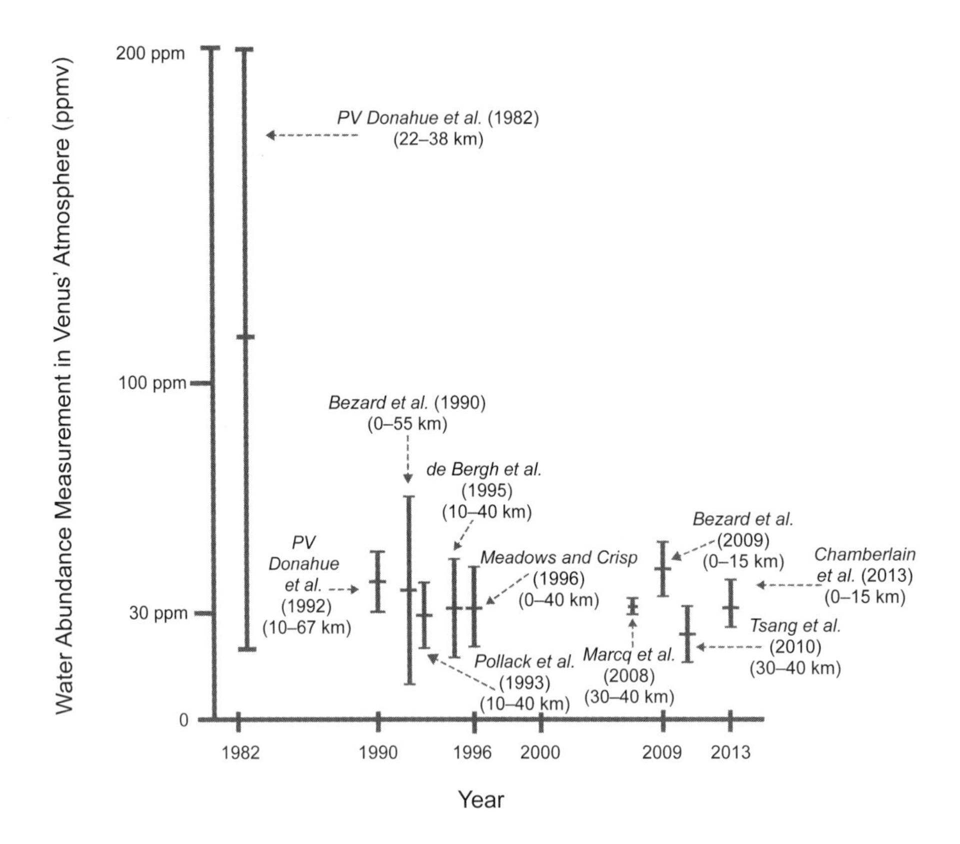

Fig. 3. Water vapor is the most potent variable greenhouse gas in Venus' atmosphere, as it is in Earth's. Observations from telescopes and then later from Venus-orbiting spacecraft led to evolving estimates of this quantity, with important implications for the fidelity of greenhouse models. After the discovery of windows between pressure-broadened CO_2 bands in the near-IR (*Allen and Crawford,* 1984), groundbased observers employed near-IR spectroscopy within the windows to sound for atmospheric water vapor at different altitudes. VIRTIS-H on Venus Express subsequently retrieved water vapor abundances at several altitudes with much higher signal to noise. This chart shows the key measurements that have led to the currently accepted bulk molar mixing ratio of 31 ± 1 ppmv H_2O in Venus' atmosphere (*Marcq et al.,* 2008), as well as 44 ± 9 ppmv (*Bezard et al.,* 2009) and 31 ± 9/6 ppmv at 15 km from the 1.18-μm window (*Chamberlain et al.,* 2013). Figure by J. DesMarines, Denver Museum of Nature & Science.

dynamical processes that vary with altitude and longitude (*Liang and Yung,* 2009). Groundbased echelle spectroscopy in the 2.3-μm spectral region on the dayside of Venus yielded a D/H of 140 ± 20 SMOW (*Matsui et al.,* 2012). Sensing HDO at 62–67 km, these IRTF observations are consistent with the lower values seen below the clouds by VEX nightside spectroscopy and the higher values seen by SOIR at 80–100 km.

Sulfur gases play a prominent role in the gas phase chemistry of the atmosphere, and perhaps in heterogeneous reactions with likely surface minerals (*Mills et al.,* 2007). The clouds are the main reservoir of S, but it also exists as SO_2, COS, H_2S, and S_x throughout the atmosphere. This remarkable range of oxidation states is due to the participation of S in redox and photochemical cycles involving C. Venera 11 and 12 scanning spectrophotometer measurements showed the total S_x vapor density increasing from 3×10^{-11} g cm^{-3} at 50 km to 2×10^{-9} g cm^{-3} at the surface (*San'ko,* 1980). Below 20 km elemental S exists as S_2. In the clouds, S_8 is the stable form with smaller allotropes beneath the clouds. At the low pressure of the cloud tops, smaller S molecules again become stable with altitude (see Fig. 17).

The abundance of SO_2 in the deep atmosphere, and therefore Venus' total atmospheric inventory, is still somewhat of a mystery. The PV Large Probe GCMS *in situ* measurements indicated a concentration of 185 ± 35 ppmv at 22 km, but the instrument failed below this altitude (*Oyama et al.,* 1980). However, reanalysis of the Vega descent probe ultraviolet spectrometer data, acquired *in situ* by ingesting the atmosphere and measuring the absorption in a cell, yielded quantitative data all the way to the surface (*Bertaux et al.,* 1996). The data from both Vega probes showed a steep decrease in SO_2 abundance toward the surface, with concentrations at 12 km of 20–27 ppmv. This profile is difficult to understand unless there is an active mineral sink at the surface. There were several difficulties with interpreting the Vega ultraviolet spectrometer results because of anomalies with the spacecraft, but laboratory experiments showed that reactions of SO_2 with the surface are not implausible (*Fegley and Prinn,* 1989). The discrepancy between PV and Vega results amount to fivefold uncertainty in atmospheric SO_2 column density, with profound implications for surface/atmosphere interactions, the mineralogy of weathered material at the surface, and the longevity of the clouds.

Solar occultations with VEX SOIR measured the vertical structure of gases and aerosols above the clouds to high precision (*Belyaev et al.,* 2008). Sulfur dioxide abundance from 85 to 105 km was retrieved in the 190–230-nm band by SPICAV simultaneously with SO in the same region (*Belyaev et al.,* 2012). At 65–80 km, SO_2 was measured by SOIR at 4 μm. Sulfur dioxide was found to exist in two layers above the clouds: at 0.02–0.5 ppmv just above the clouds, and at higher abundance in the warmer (higher) mesosphere. As the atmospheric temperature increases with altitude above 85 km, so does the SO_2 abundance. Sulfur dioxide increased from 0.1 ppmv at 165–170 K to 0.5–1 ppmv at 190–192 K. At low latitudes, SO_2 above the clouds is more abundant than at polar latitudes, with scale heights of 3 ± 1 km and 1 ± 0.4 km, respectively (*Belyaev et al.,* 2008). These observations support the idea that SO_2 is produced above the clouds by the photodissociation of H_2SO_4 vapor. This has implications for the widely reported decline of SO_2 above the clouds seen by the PV ultraviolet spectrometer over the first five years of the mission (*Esposito,* 1984). When PV first arrived in 1978, above-cloud SO_2 abundance was 10 times higher than previously measured, but similar in magnitude to what had been observed in the 1950s. Esposito surmised that episodic volcanism could have been responsible for these large changes in above-cloud SO_2. However, long-term changes in vertical mixing could not be ruled out, and a variable source of S at 85–100 km is implied by the extreme variability of SO_2 and SO seen in groundbased submillimeter measurements by the James Clerk Maxwell Telescope (JCMT) (*Sandor et al.,* 2010). Sulfuric acid vapor may not be the whole story, however, since submillimeter JCMT observations from 2004 to 2008 put an upper limit of this gas at 3 to 44 ppbv above the clouds (*Sandor et al.,* 2012). These workers concluded that H_2SO_4 is an insufficient source of S to explain the SO_2 and SO variations in the mesosphere, and so the likely culprit is elemental S_n. Using photochemical calculations, *Zhang et al.* (2012) was able to duplicate the vertical profile of S gases measured by SPICAV and SOIR at these altitudes. They made the intriguing suggestion that 1.7-μm emission from SO_3 and SO ($a^1\Delta \rightarrow X^3\Sigma^-$) could be used to determine whether the mesospheric source of S is H_2SO_4 or S_n, respectively.

The distribution of CO above the nighttime clouds was determined by fitting emission models of CO in the 4.7-μm band to VIRTIS spectra (*Irwin et al.,* 2008). Little variability was observed around the retrieved 40 ± 10 ppmv between 65 and 70 km. The polar above-cloud atmosphere appears to be enhanced in CO, however. The high resolving power of VIRTIS-H ($\lambda/\Delta\lambda \sim 2000$) was put to use to remotely retrieve the below-cloud abundances of CO, OCS, H_2O, and SO_2. Using the 2.3-μm window, *Marcq et al.* (2008) determined that CO at 36 km increases with latitude, while OCS decreases with latitude. Measured CO abundances were 24 ± 3 ppmv to 31 ± 2 ppmv from equator to 60°. These values are very consistent with the 23 ± 2 ppmv at the equator and maximum 32 ± 2 ppmv at 60° obtained by retrievals using VIRTIS-M, also in the 2.3 μm window (*Tsang et al.,* 2008). Carbonyl sulfide was found to be 2.5 ± 1 ppmv to 4 ± 1 ppmv at 33 km. The complementarity of these two gases is probably reflective of a globally averaged vertical circulation. Water and SO_2 at these altitudes in the Venus atmosphere were found by VIRTIS-M to be 31 ± 2 ppmv and 130 ± 3 ppmv, respectively.

Table 2 lists the molar mixing ratio of the nine most abundant gases in Venus atmosphere at 47 levels, and Fig. 4 plots their mixing ratios from the Venus International Reference Atmosphere (VIRA) (*Kliore et al.,* 1986). The alternate SO_2 abundance profile from the Vega ultraviolet spectrometers (*Bertaux et al.,* 1996) is shown with a dashed line.

2.1.2. Clouds and hazes. The clouds of Venus are really two separate systems, although they both primarily consist of liquid H_2SO_4/H_2O aerosols. The upper cloud, which ranges from 57–63 km above the surface, is a photochemical haze produced by the reaction of H_2O with photochemically oxidized SO_2 and SO_3 (*Esposito et al.,* 1983). The resulting

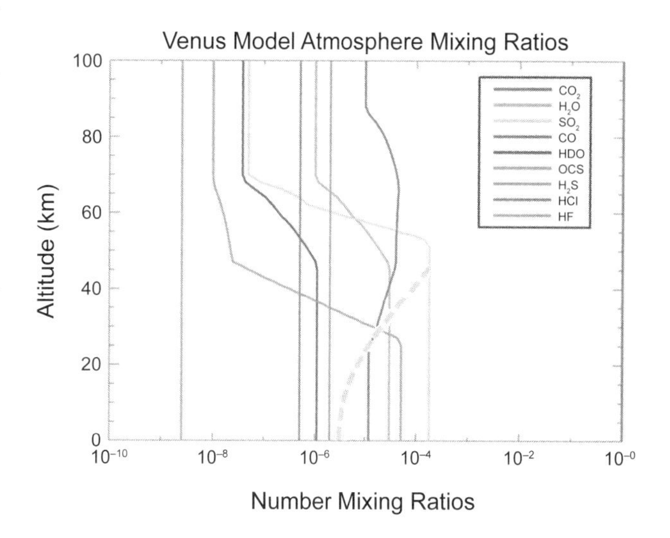

Fig. 4. See Plate 9 for color version. Molar mixing ratios of nine radiatively active gases in Venus' atmosphere. Data from Table 2, with N_2 removed and deuterated water (HDO) added. The plotted mixing ratios are from the Pioneer Venus Gas Chromatograph (*Oyama et al.,* 1980) and several groundbased spectroscopic datasets. The data are from *Kliore et al.* (1986), updated to include the best available retrievals since then. Clouds occupy the region from 48 to 68 km. Alternate Vega SO_2 abundance (*Bertaux et al.,* 1996) is shown by the dashed line, as in *Taylor and Grinspoon* (2009). Note that H_2O and SO_2 are both depleted in the clouds as they are removed by condensation to form H_2SO_4/H_2O cloud aerosols. Note also that the CO and OCS mixing ratios are complementary, with CO increasing with altitude from well below the clouds to well above them, and OCS decreasing over the same range.

submicrometer particles are close to a 1:1 stoichiometric ratio of H_2O to H_2SO_4, which is an H_2SO_4 mass ratio of 86.4% and a pH of 0. These particles can grow to diameters of up to 2 μm in the upper clouds, and have lifetimes of about a month. The upper clouds therefore store a month's worth of photochemical production of H_2SO_4.

The lower and middle cloud layers from 49 to 52 km and 52 to 55 km, respectively, are condensation clouds created by atmospheric upwelling of H_2SO_4 vapor-rich atmosphere below the clouds (*James et al.,* 1997). The convective supperrotating atmosphere, along with radiative-dynamic feedbacks, creates a highly variable cloud deck that has been likened to marine stratocumulus clouds on Earth (*McGouldrick and Toon,* 2008).

One way to make the comparison between the clouds of Earth and Venus is to examine their respective cloud

TABLE 2. Atmospheric molar mixing ratios of gases in the Venus atmosphere (*Kliore et al.,* 1986).

P (mbar)	CO_2	N_2	H_2O	SO_2	CO	OCS	H_2S	HCl	HF
0.17	0.965	0.035	1.00E-06	5.00E-08	1.00E-05	1.00E-08	2.00E-06	5.00E-07	2.50E-09
0.30	0.965	0.035	1.00E-06	5.00E-08	1.00E-05	1.00E-08	2.00E-06	5.00E-07	2.50E-09
0.50	0.965	0.035	1.00E-06	5.00E-08	1.00E-05	1.00E-08	2.00E-06	5.00E-07	2.50E-09
0.84	0.965	0.035	1.00E-06	5.00E-08	1.00E-05	1.00E-08	2.00E-06	5.00E-07	2.50E-09
1.39	0.965	0.035	1.00E-06	5.00E-08	1.41E-05	1.00E-08	2.00E-06	5.00E-07	2.50E-09
2.27	0.965	0.035	1.00E-06	5.00E-08	1.81E-05	1.00E-08	2.00E-06	5.00E-07	2.50E-09
3.64	0.965	0.035	1.00E-06	5.00E-08	2.20E-05	1.00E-08	2.00E-06	5.00E-07	2.50E-09
5.74	0.965	0.035	1.00E-06	5.00E-08	2.58E-05	1.00E-08	2.00E-06	5.00E-07	2.50E-09
8.91	0.965	0.035	1.00E-06	5.00E-08	2.94E-05	1.00E-08	2.00E-06	5.00E-07	2.50E-09
13.63	0.965	0.035	1.00E-06	5.00E-08	3.29E-05	1.00E-08	2.00E-06	5.00E-07	2.50E-09
20.61	0.965	0.035	1.00E-06	5.00E-08	3.64E-05	1.00E-08	2.00E-06	5.00E-07	2.50E-09
30.83	0.965	0.035	1.00E-06	5.00E-08	3.97E-05	1.00E-08	2.00E-06	5.00E-07	2.50E-09
45.69	0.965	0.035	1.00E-06	5.00E-08	4.29E-05	1.00E-08	2.00E-06	5.00E-07	2.50E-09
67.09	0.965	0.035	1.37E-06	1.37E-07	4.47E-05	1.06E-08	2.00E-06	5.00E-07	2.50E-09
97.65	0.965	0.035	2.40E-06	3.82E-07	4.40E-05	1.21E-08	2.00E-06	5.00E-07	2.50E-09
140.75	0.965	0.035	3.40E-06	6.20E-07	4.32E-05	1.36E-08	2.00E-06	5.00E-07	2.50E-09
200.80	0.965	0.035	4.37E-06	8.51E-07	4.25E-05	1.51E-08	2.00E-06	5.00E-07	2.50E-09
279.60	0.965	0.035	5.76E-06	4.05E-06	4.18E-05	1.65E-08	2.00E-06	5.00E-07	2.50E-09
389.10	0.965	0.035	8.22E-06	1.39E-05	4.14E-05	1.79E-08	2.00E-06	5.00E-07	2.50E-09
531.40	0.965	0.035	1.07E-05	4.48E-05	4.09E-05	1.93E-08	2.00E-06	5.00E-07	2.50E-09
710.90	0.965	0.035	1.38E-05	0.000141	4.02E-05	2.05E-08	2.00E-06	5.00E-07	2.50E-09
934.70	0.965	0.035	1.77E-05	0.00018	4.00E-05	2.18E-08	2.00E-06	5.00E-07	2.50E-09
1213.00	0.965	0.035	2.22E-05	0.00018	4.00E-05	2.31E-08	2.00E-06	5.00E-07	2.50E-09
1556.00	0.965	0.035	2.80E-05	0.00018	4.00E-05	2.52E-08	2.00E-06	5.00E-07	2.50E-09
1979.00	0.965	0.035	3.00E-05	0.00018	3.74E-05	5.02E-08	2.00E-06	5.00E-07	2.50E-09
2499.00	0.965	0.035	3.00E-05	0.00018	3.34E-05	1.00E-07	2.00E-06	5.00E-07	2.50E-09
3135.00	0.965	0.035	3.00E-05	0.00018	3.01E-05	2.08E-07	2.00E-06	5.00E-07	2.50E-09
3903.00	0.965	0.035	3.00E-05	0.00018	2.71E-05	4.45E-07	2.00E-06	5.00E-07	2.50E-09
4822.00	0.965	0.035	3.00E-05	0.00018	2.41E-05	9.50E-07	2.00E-06	5.00E-07	2.50E-09
5917.00	0.965	0.035	3.00E-05	0.00018	2.20E-05	1.98E-06	2.00E-06	5.00E-07	2.50E-09
7211.00	0.965	0.035	3.00E-05	0.00018	2.00E-05	4.35E-06	2.00E-06	5.00E-07	2.50E-09
8729.00	0.965	0.035	3.00E-05	0.00018	1.80E-05	9.21E-06	2.00E-06	5.00E-07	2.50E-09
10500.00	0.965	0.035	3.00E-05	0.00018	1.60E-05	1.98E-05	2.00E-06	5.00E-07	2.50E-09
12560.00	0.965	0.035	3.00E-05	0.00018	1.44E-05	4.41E-05	2.00E-06	5.00E-07	2.50E-09
14930.00	0.965	0.035	3.00E-05	0.00018	1.30E-05	5.10E-05	2.00E-06	5.00E-07	2.50E-09
17660.00	0.965	0.035	3.00E-05	0.00018	1.20E-05	5.10E-05	2.00E-06	5.00E-07	2.50E-09
20790.00	0.965	0.035	3.00E-05	0.00018	1.20E-05	5.10E-05	2.00E-06	5.00E-07	2.50E-09
24360.00	0.965	0.035	3.00E-05	0.00018	1.20E-05	5.10E-05	2.00E-06	5.00E-07	2.50E-09
28430.00	0.965	0.035	3.00E-05	0.00018	1.20E-05	5.10E-05	2.00E-06	5.00E-07	2.50E-09
33040.00	0.965	0.035	3.00E-05	0.00018	1.20E-05	5.10E-05	2.00E-06	5.00E-07	2.50E-09
38260.00	0.965	0.035	3.00E-05	0.00018	1.20E-05	5.10E-05	2.00E-06	5.00E-07	2.50E-09
44160.00	0.965	0.035	3.00E-05	0.00018	1.20E-05	5.10E-05	2.00E-06	5.00E-07	2.50E-09
50810.00	0.965	0.035	3.00E-05	0.00018	1.20E-05	5.10E-05	2.00E-06	5.00E-07	2.50E-09
58280.00	0.965	0.035	3.00E-05	0.00018	1.20E-05	5.10E-05	2.00E-06	5.00E-07	2.50E-09
66650.00	0.965	0.035	3.00E-05	0.00018	1.20E-05	5.10E-05	2.00E-06	5.00E-07	2.50E-09
76010.00	0.965	0.035	3.00E-05	0.00018	1.20E-05	5.10E-05	2.00E-06	5.00E-07	2.50E-09
86450.00	0.965	0.035	3.00E-05	0.00018	1.20E-05	5.10E-05	2.00E-06	5.00E-07	2.50E-09

extinction depths. On Earth, it is not unusual to fly through clouds so thick that you cannot see the tip of an airplane wing 10 m away. Compare this with a typical extinction depth of about 1 km in Venus' lower clouds. By Earth standards, the clouds of Venus are quite diffuse but very extensive, since they are persistently 20 km thick. The result is a typical total cloud extinction comparable to a thickly overcast sky on Earth, but with a high degree of variability on timescales of hours to days.

The structure of Venus' clouds has been explored intensively by nine Soviet descent probes and the four descent probes of the PV mission. Each obtained a snapshot profile of the atmospheric temperature, pressure, clouds, and winds at a single latitude and local time of day. Figure 5 maps the entry location of each of the probes. The energetics of atmospheric insertion from an Earth-originating trajectory favors one side of the planet. As a result, no probes have entered on the evening side, from noon to midnight local time. So this half of the atmosphere has never been explored *in situ*. All the probes acquired temperature and pressure on the way down, and radio tracking provided horizontal wind speeds. Instruments to measure the size and number density of cloud particles were on Veneras 9, 10, and 11 (*Marov et al.,* 1980) and on the PV probes (*Ragent and Blamont,* 1980). Nephelometers onboard the Veneras and all four instrumented PV probes measured the scattering of light off the particles, which allows for reconstruction of the index of refraction and scattering phase function, as well as model-dependent determination of number density and size. More detailed particle size measurements were made by the PV Large Probe Particle size Spectrometer (LCPS) as it descended through the atmosphere at 5°N latitude at about 7:30 a.m. local time. Scattering data were acquired every 50 m in the main cloud decks. The lower size limit for these measurements was 0.6 μm diameter, but the LCPS used a shadow graph imaging technique to measure particles as large as 500 μm. Extensive hazes were seen both above and below the clouds, up to 70 km and down to 30 km.

In a spectacular technological feat, two Soviet Vega balloons were deployed in the Venus middle cloud layer at

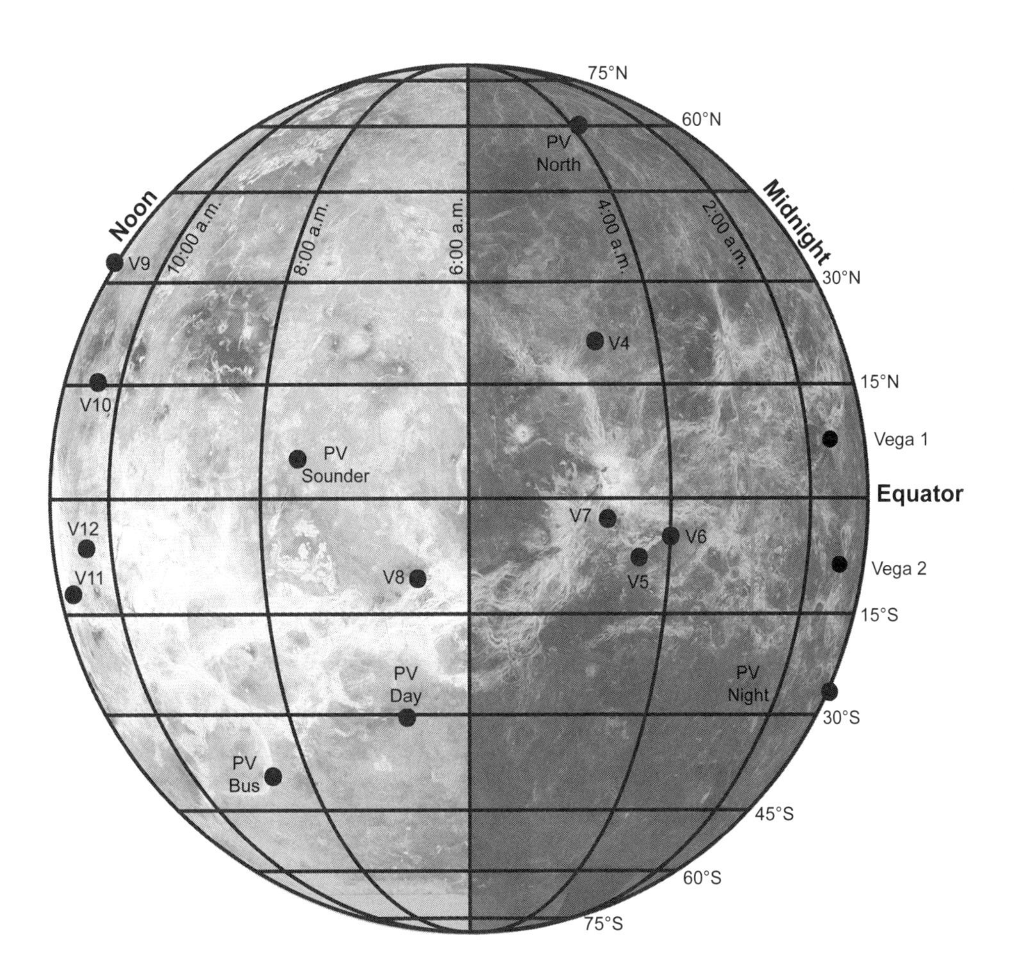

Fig. 5. Entry locations of all probes that have descended through the atmosphere of Venus. Map is centered on the morning terminator; no probes have explored the afternoon side of the planet from noon to 11:00 p.m. local time. Veneras are designated "Vn," Vegas are "Ven," and Pioneer Venus Probes are "PV." The PV bus was not an entry probe and burned up at about 110 km. The Vega probes entered on the backside of the image, near the right edge at 11:00 p.m. Adapted from *Schubert* (1983) and produced by J. DesMarines.

55 km and flew for up to 48 hours (*Sagdeev et al.,* 1986a). These first planetary balloons returned extraordinary information on vertical winds, turbulence, temperature, pressure, and wind velocity within the clouds until their batteries ran out (*Linkin et al.,* 1986; *Sagdeev et al.,* 1986b).

Comparisons of particle sizes and number densities measured on descent reveal that while the upper cloud is uniform and constant, the lower clouds vary considerably in thickness and sometimes disappear completely, as do the hazes above and below the clouds (*Knollenberg and Hunten,* 1980). VEX VIRTIS observations enabled a more detailed study of Venus' cloud structure. *Wilson et al.* (2008) reported from VIRTIS-M data that cloud aerosols near the poles are larger than at low latitudes. This conclusion was confirmed by *Barstow et al.* (2012), who were also able to retrieve aerosol H_2SO_4 concentration and cloud base altitude. They observed that H_2SO_4 concentration was generally higher where the optical depth of clouds was larger. *Cottini et al.* (2012), using VIRTIS-H, retrieved the H_2SO_4 concentration of 1-µm aerosols and saw a latitudinal trend. Polar aerosols were 75–83% H_2SO_4, while equatorial 1-µm particles were more acidic, at 80–83% H_2SO_4. Cloud base altitude varies with latitude from 47 to 50 km, reaching the maximum height at a latitude of 50° (*Barstow et al.,* 2012). Variations in water vapor abundance below the clouds retrieved from Earth-based near-infrared spectra of the nightside were difficult to see (*Bell et al.,* 1991). These observations saw the emission of water vapor from 35 to 45 km. Higher signal-to-noise data from VIRTIS-M enabled *Tsang et al.* (2010) to report on the correlation of cloud thickness and water vapor abundance from 30 to 45 km, a result confirmed by the analysis of additional VEX data by *Barstow et al.* (2012).

Using VIRTIS-H spectra at 2.48 and 2.6 µm, *Cottini et al.* (2012) found that cloud tops at the equator were at 69.5 ± 2 km, and declined steadily to 64 km at the edge of the south polar vortex at 80°. Cloud top altitudes and aerosol scale heights were also derived by combining VIRTIS spectrometer data with VEX radio science (VeRa) atmospheric temperature sounding (*Lee et al.,* 2012). VIRTIS 4–5-µm spectra and VeRa temperature profiles showed that cloud tops were at 67.2 ± 1.9 km at low southern latitudes and 62.8 ± 4.1 km near the poles. Aerosol scale heights at the cloud tops were 3.8 ± 1.6 km and 1.7 ± 2.4 km, respectively. Cloud top altitudes were also determined from the depth of the 1.6-µm CO_2 absorption band in VIRTIS spectra, reaching a minimum in the eye of the southern vortex at the pole (*Ignatiev et al.,* 2009). The cloud tops were seen to change in altitude by as much as 1 km in less than 10 hours, with larger-amplitude variations at the highest latitudes. A minimum in the atmospheric temperature inversion above the clouds was observed over the high-latitude cold collar (*Taylor et al.,* 1980), consistent with radiative cooling at the cloud tops. The upper hazes were explored with 200 SOIR solar occultations at 3 µm from September 2006 to September 2010 (*Wilquet et al.,* 2012). Vertical profiles of these hazes showed variability in extinction on timescales of a few days, with an additional periodicity of about 80 days.

Erard et al. (2009) reported the discovery with VIRTIS of several new faint infrared windows through which thermal emission of underlying atmospheric layers can be observed. They are at 1.51, 1.55, 1.78, and 1.82 µm. The authors use a multivariate method called independent component analysis — similar in some respects to principal component analysis — to optimize retrievals through infrared windows. In particular, they found that the polar vortex clouds had alternating layers of cloud particle sizes.

In the upper cloud, H_2SO_4 is produced photochemically. First, SO_2 is photochemically oxidized. A parallel photochemical branch simultaneously produces elemental S

$$CO_2 + SO_2 + h\nu \rightarrow CO + SO_3$$

$$3SO_2 + 2h\nu \rightarrow 2SO_3 + \Sigma$$

H_2SO_4 is produced by

$$SO_3 + H_2 \rightarrow H_2SO_4$$

that condenses with H_2O onto activated cloud condensation nuclei (*Pruppacher and Klett,* 1997). A downward flux of liquid H_2SO_4/H_2O aerosols evaporates to form a reservoir of H_2SO_4 vapor (*James et al.,* 1997; *Krasnopolsky and Pollack,* 1994). Upward vertical motion, caused by convection or atmospheric waves, cools the rising H_2SO_4 gas-rich atmosphere where it condenses with H_2O to reform the lower cloud aerosols.

The Venus Monitoring Camera (VMC) on VEX is making a comprehensive study of Venus' clouds at four wavelengths: 365, 513, 965, and 1010 nm (*Titov et al.,* 2012). These wavelengths image the clouds at different depths, with a resolution from VEX orbit apoapsis of 50 km/pixel to several hundred meters per pixel near periapsis. SO_2 contrast features at the cloud tops are seen on the dayside at 365 nm. Close in, VMC has observed cumulus columns and small-scale convection near the equator, giving way to streaks of linear, more featureless clouds above 50° latitude. Convection and wave activity are strongest at the subsolar point and wave trains with wavelengths of ~200 km are common. Above 60°S is the bright polar hood, featureless except for narrow (300 km) circular or spiral dark lanes. Several times in the past four years, VMC has observed an expansion of the polar hood to lower latitudes, resulting in a days-long temporary increase in global brightness. The southern polar hood has extended up to 35° in latitude in under 24 hours (*Markiewicz et al.,* 2007).

In the lower and middle clouds, the condensation of H_2SO_4 and H_2O vapor together on nucleating particles creates cloud aerosols and droplets that grow up to 7 µm in diameter as they fall through the cloud. The lower and middle clouds are mixtures of several particle populations, from the submicrometer cloud condensation nuclei (CCNs), to the 2-µm particles from the upper clouds, to the larger 7-µm droplets that are concentrated largely near the bottom of the lower cloud. The smallest population of particles, seen throughout the clouds, has a modal radius of 0.3 μm and is designated mode 1. Two distinct populations with modal radii of 1.0 and 1.4 μm are called modes 2 and 2′. The largest particles, concentrated near the bottom of the lower cloud, have a modal radius of 3.65 μm and are called mode 3 particles. Their distributions are modeled to fit the PV data with log-normal distributions and standard deviations given in Table 3.

A schematic of the Venus cloud structure and chemistry is shown in Fig. 6. Properties of the various cloud regions are tabulated in Table 4.

The H_2SO_4 vapor reservoir is confined to altitudes above 38 km. Below this the gas thermally decomposes

$$H_2SO_4 \rightarrow SO_3 + H_2O$$

SO_3 oxidizes whatever it can find in the lower atmosphere

$$SO_3 + CO \rightarrow SO_2 + CO_2$$

$$SO_3 + 2OCS \rightarrow CO_2 + SO_2 + S_2$$

Near-infrared observations of the nightside of Venus at the AAT confirmed the conclusion of *Tsang et al.* (2009) that CO concentration below the clouds is enhanced at 60° latitude (*Cotton et al.*, 2012). These recent observations indicate that the enhancement of CO exists near 45 km, but not at 35 km. This is consistent with the downwelling of CO at midlatitudes and its removal via the above reaction with SO_3.

OCS is an important intermediary in atmospheric S cycles, which is reduced by

$$OCS \rightarrow CO + S$$

$$OCS + S \rightarrow CO + S_2$$

$$OCS + S_2 \rightarrow CO + S_3$$

$$OCS + H \rightarrow CO + SH$$

where

$$SH + SH \rightarrow H_2S + S$$

S interconverts between different allotropes of S_x and all the other S species in the Venus atmosphere via numerous other reactions in the lower atmosphere. S cycles are also coupled with N and Cl reactions that circulate these elements between different reservoirs (i.e., Cl_2, $ClSO_2$, HCl, NO_x) (*Krasnopolsky,* 2007; *Mills et al.,* 2007).

Cloud aerosols composed of liquid H_2SO_4 in solution with water contain about 100 times more S than all the S gases in the atmosphere combined. In addition to trace S species, the halogen gases HCl and HF were detected in the spectra of *Pollack et al.* (1993). Along with the S species, these have a volcanic source on Venus as they do on Earth.

TABLE 3. Venus cloud aerosol size populations (from *Knollenberg and Hunten,* 1980).

	Mode 1	Mode 2	Mode 2′	Mode 3
r_m (μm)	0.30	1.00	1.40	3.65
σ (μm)	0.52	0.43	0.41	0.32
$\%H_2SO_4$	84.5	84.5	95.6	95.6
r_{min} (μm)	0.01	0.01	0.01	0.01
r_{max} (μm)	15.0	15.0	15.0	15.0

Absorption of sunlight by the upper clouds can be seen in images of Venus taken in reflected ultraviolet (Fig. 1). SO_2, S_8, and other S gases are responsible for absorbing sunlight, but over half the radiation is absorbed by an unknown constituent (*Esposito,* 1980). Dark patches in the near-ultraviolet and visible are correlated with SO_2 distribution, but have a lifetime of hours, and therefore the "unknown absorber" is believed to have a similar lifetime. *Toon et al.* (1982) identified allotropes of S as being the likely agent responsible for the mysterious absorption, pointing out that S_4 is metastable with respect to S_3 on a timescale of several hours. Sulfur allotropes absorb in the near-ultraviolet and S_4 absorbs in a continuum from ultraviolet to blue wavelengths, with no discrete spectral features. Toon et al. used a cloud chemical/microphysical model to calculate the amount of S_x that could be incorporated into aerosols. They found that enough metastable S_4 could condense in the lower part of the upper cloud to explain the remaining ultraviolet-blue absorption before decaying to S_3.

The condensation clouds are thickest near the equator and thinner and banded at mid- and high latitudes. These patterns may be tracers of a globally organized pattern of upward convection near the equator and subsidence at high latitudes. Analysis of VEX VIRTIS-M nightside images at 1.74 μm over 407 Earth days showed that Venus' middle and lower clouds have steadily became thinner (*McGouldrick et al.,* 2012). Both the variability in cloudiness and the magnitude of the change are greater in the polar regions. This work also demonstrated that individual cloud features at these levels, from 48 to 55 km, persist up to 30 hours. However, advection alone cannot explain how the features change. Vorticity and divergence show changes that are due to vertical motion driven by radiative-dynamical feedback. The 1.71-μm edge of the 1.74-μm window was used by *Satoh et al.* (2009) to map the opacities of the lower clouds (optical depth 20–50) and the opacity of hazes below the main cloud deck (optical depth 1–4).

Most of the solar energy incident on Venus is absorbed in the clouds, with only about 17% making it to the surface (*Tomasko et al.,* 1979). H_2SO_4/H_2O aerosols are highly reflective and have a single-scattering albedo very close to 1 at wavelengths below 2.7 μm (*Toon et al.,* 1982). Thus the clouds mostly scatter conservatively at visible wavelengths. Since the clouds are optically thick, solar photons undergo many scattering events, effectively increasing their path

TABLE 4. Properties of the Venus cloud layers (*Esposito et al.*, 1983).

Layer	Altitude (km)	Temperature (K)	Pressure (bars)	Optical Depth τ at 0.63 μm	Mode (see Table 3)
Upper Haze	70–90	225–190	0.0267–0.0028	0.2–1.0	m1
Upper Cloud	56.5–70	286–225	0.406–0.0267	6–8	m1, m2
Middle Cloud	50.5–56.5	345–286	0.981–0.406	8–10	m1,m2′,m3
Lower Cloud	47.5–50.5	367–345	1.391–0.981	6–12	m1,m2,m3
Lower Haze	31–47.5	482–367	8.14–1.39	0.1–0.2	m1

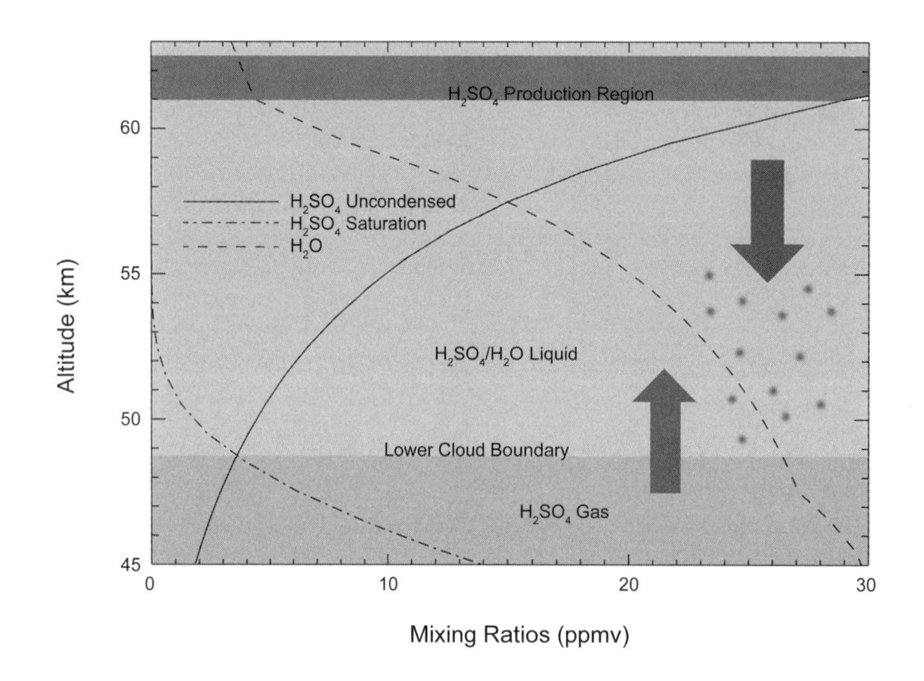

Fig. 6. See Plate 10 for color version. Schematic of the physical and chemical processes that create the Venus clouds. H_2SO_4 is produced photochemically between 62 and 64 km (red band). The saturation vapor pressure curves for H_2SO_4 (dot dashed line, lower left) and H_2O (dashed line, center) determine how cloud aerosols grow as they fall through the three cloud layers. The lower cloud boundary is formed where the vapor pressure of H_2SO_4 equals its saturation vapor pressure, at about 48 km (boundary between green and yellow regions). There is an ~200-m break in the clouds at about 52 km (not shown) between the lower and middle cloud layers. The flux of liquid H_2SO_4 is balanced by an upward flux of H_2SO_4 gas, resupplying vapor from which cloud particles can grow. The rate of vapor mixing and hence the resupply of H_2SO_4 is highly dependent on the eddy diffusion coefficient in this one-dimensional model. In reality, three-dimensional advection by all processes controls the distribution of vapor constituents.

lengths and the probability that they will be absorbed by the gaseous constituents of the cloud-level atmosphere. Liquid H_2SO_4 aerosols absorb radiation at thermal wavelengths, which is reradiated both upward and downward. The clouds therefore act as a greenhouse agent.

2.1.3. Lightning. Direct detection of flashes interpreted to be lightning in Venus' atmosphere have been reported but are controversial (*Grebowsky et al.*, 1997; *Hansell et al.*, 1995). In Earth's atmosphere, lightning usually requires ice aerosols, and evidence for solid particles in Venus' clouds is slim. Concentrated H_2SO_4 aerosols freeze at extremely low temperatures, but it is possible that some mixture of frozen and liquid aerosols exist in the upper clouds (*McGouldrick et al.*, 2011). Whistler mode waves detected by magnetometers on PV strongly suggested lightning, and data from the plasma wave instrument on the Galileo spacecraft during its flyby of Venus was interpreted as a detection of lightning in the Venus atmosphere (*Grebowsky et al.*, 1997). The magnetometer on VEX has now provided definitive evidence for lightning from whistler mode waves, indicating lightning discharges with about the same energy as seen in Earth's atmosphere but about one-fifth as frequent.

2.1.4. Thermal profile. There is no substitute for direct *in situ* measurements, so the profiles obtained by entry probes are invaluable. However, radio occultation experiments with orbiting spacecraft can produce very accurate temperature, pressure, and H_2SO_4 vapor abundance retrievals. The PV Orbiter Radio Occultation experiments produced temperature profiles of Venus' atmosphere over a wide range of latitudes and local times of day (*Kliore and Patel,* 1980). Magellan occultations with 3.6 and 13 cm wavelength were performed at several opportunities during the mission, obtaining data down to 35 km (*Jenkins et al.,* 1994). One hundred and eighteen radio occultations between July 2006 and June 2007 by VEX VeRa were inverted to obtain temperature, pressure, and static stability from 40 to 90 km (*Tellmann et al.,* 2009). These data showed that above 65 km, temperatures were within 5 K of those established in VIRA (*Kliore et al.,* 1986). A 30-K contrast between the equator and poles was seen at the 1-bar level. A future orbital mission optimized for radio occultations could obtain temperature, pressure, turbulence, and H_2SO_4 vapor profiles down to 35 km at several hundred locations on the planet, extending the timescale over which Venus atmospheric soundings have been performed.

The temperature-dependent opacity bands of CO_2 from 3.8 to 5.0 μm were used to retrieve temperatures from VIRTIS-M spectra from 64 to 95 km (*Grassi et al.,* 2008, 2010). Temperatures were consistent with VeRa radio occultation retrievals and had an accuracy of 1 K between 66 and 77 km.

VIRTIS-H temperature retrievals were possible from 4 to 100 mbar, or 65–80 km. The cold collar stands out at the 100-mbar level at 65°–70°S, exhibiting temperatures lower than at surrounding latitudes and with an asymmetry between evening and morning temperatures (*Migliorini et al.,* 2012).

Groundbased spectroscopy has also been used to determine atmospheric temperatures. High-resolution heterodyne spectroscopy at 10.4 μm was used to retrieve the temperature at 110 km in Venus' atmosphere at NASA's Infrared Telescope Facility (IRTF) with an accuracy of 10 K (*Sonnabend et al.,* 2012). Similar observations 19 years later at Kitt Peak Observatory indicated mesospheric temperatures 10–20 K lower. These lower temperatures agree with SOIR-derived temperatures at the same altitude.

From the surface to the clouds, the temperature profile of the Venus atmosphere follows the dry adiabat of the CO_2/N_2 mixture. The atmosphere is statically stable almost everywhere except for a region around 25 km and within the lower and middle clouds (Fig. 7). As expected, there is a sharp decrease in static stability just beneath the convective clouds. The clouds scatter and absorb thermal radiation from below, and reemit that radiation upward and downward. The downward-propagating thermal radiation gives rise to the slight bump in the temperature profile just beneath the clouds (Fig. 8). As discussed above, deep atmospheric temperatures vary by about 30 K. This is less than 5% of the surface temperature, so the atmosphere is sometimes treated mathematically as one-dimensional. The latent heat released by the condensation of H_2O vapor and H_2SO_4 vapor onto cloud particles has a much smaller impact on the atmospheric temperature than does the condensation of water in Earth's atmosphere. This is mostly due to the fact that the density of Venus' clouds is much lower than the density of Earth water clouds. While terrestrial cumulus clouds may have thousands of 20-μm particles per cm^3, Venus' clouds typically consist of 2-μm particles with densities of several hundred per cm^3 (Table 4). Above the clouds, the atmosphere is largely in radiative equilibrium; i.e., Venus' stratosphere is above the clouds. Large temperature gradients don't arise until the mesosphere, where temperature inversions can develop at high latitudes. Often, the polar mesosphere is warmer than the equator. In the summer hemisphere, VeRa found that the atmospheric temperature inversion just above the clouds extended to higher altitudes at the poles than at midlatitudes (*Patzold et al.,* 2007).

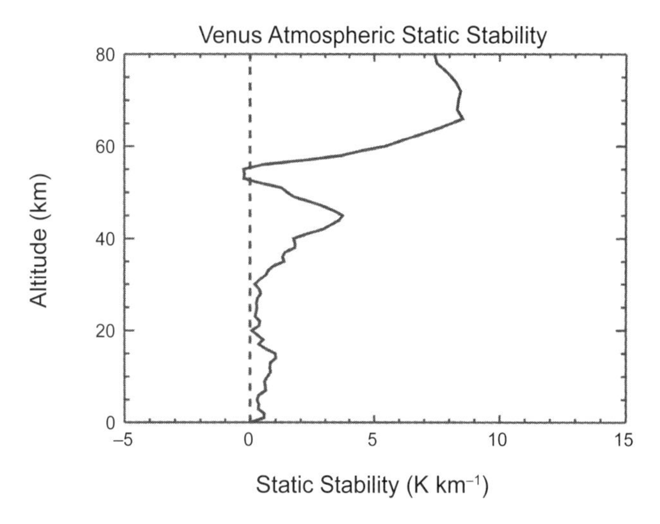

Fig. 7. Static stability of the Venus atmosphere, observed by PV descent through the atmosphere. If the measured lapse rate is $\Gamma = \partial T/\partial z$, and the adiabatic lapse rate is $\Gamma_a = g/c_p$, then the static stability is $\Gamma_s = g/c_p - \Gamma$. For $\Gamma_s > 0$ the atmosphere is stable with respect to vertical perturbations and gravity waves can propagate. If $\Gamma_s < 0$, the atmosphere is statically unstable and convection ensues. The Venus atmosphere has low static stability from 20 to 30 km and again in the middle cloud at about 52 km. It is in this latter region of instability that convection aids in forming the clouds. The Vega balloons experienced vertical wind speeds of up to 5 m s^{-1} in this region, consistent with turbulent convection (*Linkin et al.,* 1986).

A detailed radiative transfer model that uses high-temperature spectral data and line shape models that incorporate the physics of the far wings of CO_2 absorption lines and line mixing is necessary to replicate the Venus temperature profile (*Bullock and Grinspoon,* 2001; *Lee et al.,* 2005). Multiple scattering of short- and longwave radiation in the clouds must also be treated in order to get the proper coupling of the cloud and radiation field. Temperature profiles obtained from probes that descended through Venus' atmosphere, by radio occultation, and models with a radiative transfer code are shown in Fig. 8. The VIRA (*Kliore et al.,* 1986), tabulating temperature, pressure, specific heat, and nonideality coefficient for the equatorial region (±30° latitude) at 24 levels in the atmosphere, is shown in Table 5.

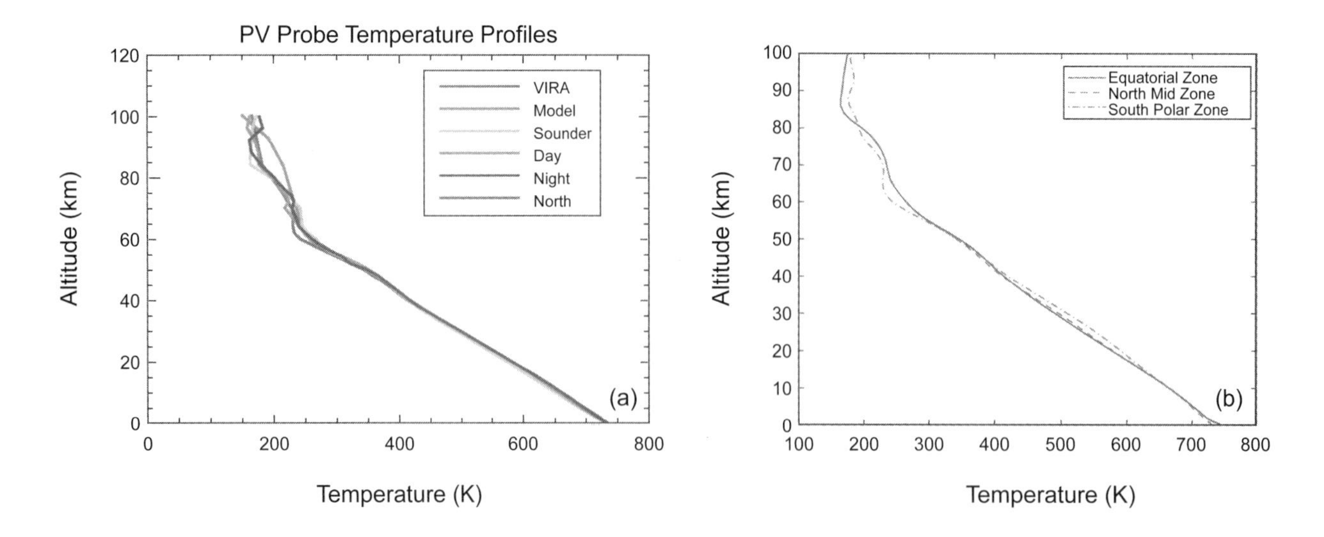

Fig. 8. See Plate 11 for color version. **(a)** The globally averaged equatorial atmospheric temperature profile of Venus (red line) (*Kliore et al.,* 1986) compared with the radiative transfer calculations of *Bullock and Grinspoon* (2001) (orange line). *In situ* temperatures measured by the four Pioneer Venus atmospheric entry probes are shown in other colors. Yellow: sounder probe, green: day probe, purple: night probe, and magenta: north probe. **(b)** Temperature retrievals from Magellan occultations at three latitude regions (*Jenkins et al.,* 2002).

TABLE 5. The Venus International Reference Atmosphere for ±30° latitude (*Kliore et al.,* 1986).

Altitude (km)	Temperature (K)	Pressure (bars)	Density (kg/m^3)	Adiabat (K/km)	Gravity (m s^{-2})	ζ (PV = ζnk_BT)	Sp. Head (J/kg-K)
100	167.2	2.66E-05	7.89E-05	11.25	8.583	1.0000	738.199
92	175.4	2.19E-04	6.84E-04	11.25	8.605	1.0000	731.094
84	183.8	1.73E-03	4.93E-03	11.25	8.628	1.0000	745.477
76	212.1	1.08E-02	2.66E-02	11.25	8.650	0.9998	771.679
68	235.4	5.45E-02	0.1210	11.11	8.673	0.9994	794.430
64	245.4	0.1156	0.2443	10.98	8.684	0.9988	807.720
60	262.8	0.2357	0.4694	10.75	8.696	0.9982	821.000
56	291.8	0.4559	0.8183	10.38	8.708	0.9976	848.899
52	333.3	0.8167	1.2840	9.91	8.719	0.9972	887.942
48	366.4	1.3750	1.9670	9.62	8.730	0.9966	918.579
46	379.7	1.7560	2.4260	9.52	8.736	0.9961	929.810
44	391.2	2.2260	2.9850	9.44	8.742	0.9957	941.044
42	403.5	2.8020	3.6460	9.35	8.748	0.9951	952.520
40	417.6	3.5010	4.4040	9.26	8.753	0.9946	964.000
38	432.5	4.3420	5.2760	9.17	8.759	0.9941	976.830
36	448.0	5.3460	6.2740	9.08	8.765	0.9937	989.668
32	479.9	7.9400	8.7040	8.91	8.776	0.9931	1016.040
28	513.8	11.4900	11.7700	8.77	8.788	0.9928	1042.780
24	547.5	16.2500	15.6200	8.64	8.800	0.9929	1067.800
20	580.7	22.5200	20.3900	8.52	8.811	0.9935	1091.000
16	613.3	30.6600	26.2700	8.41	8.823	0.9944	1111.320
12	643.2	41.1200	33.5400	8.32	8.834	0.9958	1129.180
8	673.6	54.4400	42.2600	8.24	8.846	0.9990	1146.380
4	704.6	71.2000	52.6200	8.14	8.858	1.0034	1163.460
0	735.3	92.1000	64.7900	8.06	8.869	1.0100	1181.000

2.2. The General Circulation

The motion of Venus' atmosphere is driven by the disparate energy input between the equator and the polar regions, just as on Earth, Mars, and Titan (Schubert and Mitchell, this volume). However, Venus rotates only once every 243.023 Earth days, so the flow that is established to redistribute the heat is subject to much smaller Coriolis forces than on Earth or Mars. Large-scale atmospheric motion in Earth's atmosphere is determined by the balance of Coriolis and pressure forces, a condition known as geostrophic balance. It is what gives rise to the circulation around pressure highs and lows that characterize midlatitude winds, and to trade winds and hurricanes in the tropics. Without a force to bend meridional winds, the zonal flow regime of Venus' atmosphere is quite different from Earth's. Pressure forces are instead balanced by centrifugal forces, resulting in a balance known as cyclostrophic flow. Small-scale rotating flows on Earth, such as dust devils and tornados, are cyclostrophically balanced. The manifestation of cyclostrophic balance on Venus is a strong, steady zonal circulation at mid- and low latitudes (*Gierasch et al.,* 1997).

The tracking of 13 probes as they descended through the Venus atmosphere at various latitudes and local times of day (Fig. 5) gives us some limited snapshots of the altitude profile of the zonal winds. Zonal wind speeds peak at the top of the upper clouds at about 120 m s^{-1}, a few kilometers above the depth of maximum solar heating. They decrease with depth almost linearly until they are zero at the surface (Fig. 9a). Actual wind speeds at the surface were measured by the successful Veneras and were shown to be light and variable at about 1 m s^{-1} (*Marov et al.,* 1973). The dominant motion of Venus' atmosphere from cloud tops to the surface is a superrotating flow from east to west. At the cloud tops, the atmosphere is rotating 60 times faster than the solid body. This zonal wind decreases as latitude increases, until it gives way to the polar vortices at each pole. Above the clouds, the zonal winds decrease with altitude, until they begin to mix with the subsolar-antisolar flow that is dominant at 120 km and above.

An important difference between the solar forcing of Earth's atmospheric dynamics and Venus' is the location of energy deposition. On Earth, most of the solar energy is deposited at the ground, so that the deep convection that develops in the tropics involves the entire troposphere. On Venus, most of the solar energy is deposited in the upper clouds, which reradiates it both upward and downward. It is reasonable to expect that the latitudinal dependence of solar energy deposition gives rise to an organized hemispheric flow, rising along the equator and descending at high latitudes. Direct evidence for hemispheric Hadley cells on Venus is lacking, largely because the return flow from polar regions to low latitudes is below the clouds and has never been observed. However, cloud top meridional motion measured by VEX clearly shows equator to pole motion that could be the top of hemispheric Hadley cells (*Sanchez-Lavega et al.,* 2008). Interpreting cloud base meridional motion is more difficult, partly because the speeds are slower and therefore measured uncertainties are higher. In addition, cloud base winds sometimes appear to move from equator to pole, and occasionally the other direction, as if these altitudes occasionally dip down into the return branch of the Hadley cell, but more often participate with the direct flow poleward (*Hueso et al.,* 2012). *Schubert* (1983) hypothesized that there may be several stacked alternating direct-indirect Hadley cells in each hemisphere.

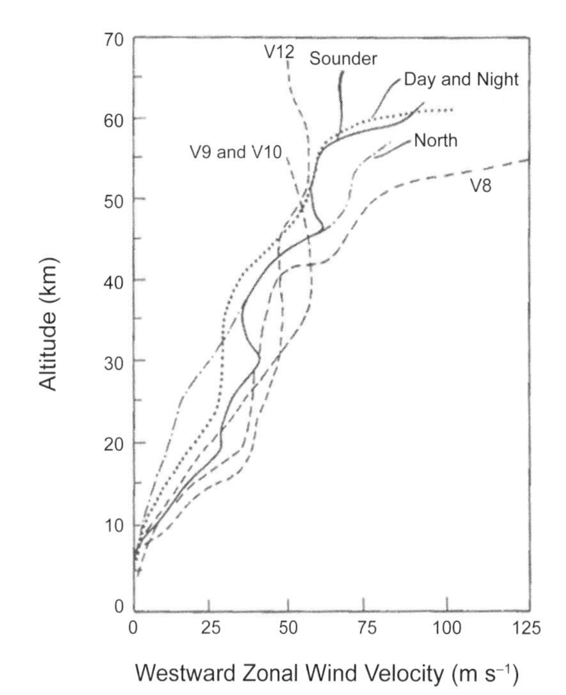

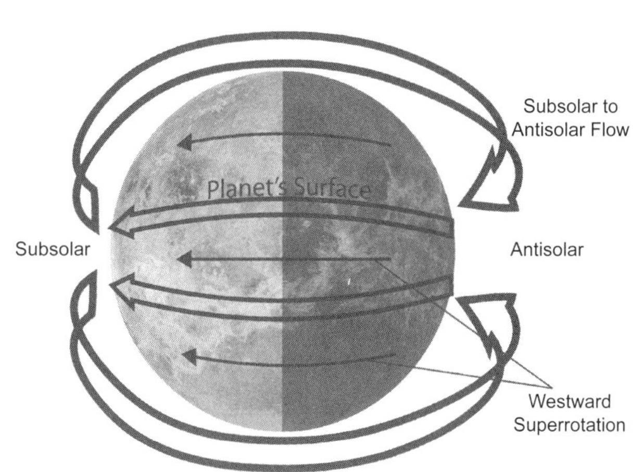

Fig. 9. **(a)** All entry probes experienced steadily decreasing zonal winds as they descended, from over 100 m s^{-1} at the cloud tops to zero at the surface. Variations along each descent path were probably due to transient waves. From *Schubert* (1983). **(b)** Schematic of the general circulation of Venus' atmosphere as it is currently understood; figure produced by J. DesMarines. Zonal winds are depicted as arrows from right to left. On top of these winds is the subsolar to antisolar flow that dominates above about 110 km.

Cloud tracking of upper cloud features on VMC images shows a zonal flow of 90–100 m s^{-1} between ±40°. Zonal velocities decrease toward the poles, becoming 50 m s^{-1} at 75° (*Markiewicz et al.,* 2007).

VIRTIS-M is an imaging spectrograph and therefore adds images of atmospheric layers not seen by the VMC. VIRTIS dayside images at 380 nm see the cloud tops at 66–72 km in reflected sunlight, and 1.74-μm images of the nightside sees thermal radiation from 44 to 48 km. Tracking cloud features on 860 Earth days of VIRTIS images yields cloud top zonal velocities consistent with those derived from VMC, and cloud base average zonal velocities of approximately 60 m s^{-1} at 0°–50°S (*Hueso et al.,* 2012). Individual tracked winds were estimated to have a 5 m s^{-1} error. Cloud base zonal winds equatorward of 50°S were stable with a standard deviation in each 10° latitude bin of 9 m s^{-1}. However, high-latitude wind speed variability seemed to be related to activity of the south polar vortex.

Meridional winds at the cloud tops were toward the south pole at 10 m s^{-1}, as would be expected for Hadley circulation. However, cloud top meridional winds vary with time of day, indicating a coupling with the bottom of subsolar-antisolar flow in the mesosphere. Cloud base meridional motion was seen to be highly variable between 0 and ±15 m s^{-1}, with no net polar or equatorial motion (*Hueso et al.,* 2012).

The Doppler-shifted fully resolved spectroscopy of the non-long-term-evolution (LTE) 10.423-μm line of CO_2 from Kitt Peak Observatory enables the determination of winds at 110 ± 10 km (*Sornig et al.,* 2012). This is an important region because it is expected to be a transition from the dominant zonal flow to the dominant subsolar-antisolar flow above. These observations showed a zonal velocity of 0 m s^{-1} at the equator increasing to 43 ± 13 m s^{-1} at 33°N.

Doppler wind measurements have also been made using submillimeter lines of CO, observed from 2001 to 2009 by JCMT (*Clancy et al.,* 2012b). These observations probe even lower down than the thermal infrared spectroscopy, from 95 to 115 km. Spanning ±60° in latitude and from 8:00 p.m. to 4:00 a.m. local time, the purpose of these observations was to determine the relative contribution of zonal and subsolar-antisolar flow at these altitudes. At the 1 μbar level (108–110 km), *Clancy et al.* (2012a) found that zonal winds were stronger than the tidal winds, 85 m s^{-1} to 65 m s^{-1}, respectively. However, the magnitude of these components have a lot of variability at this altitude, between day and night and on daily and weekly timescales. Wave activity may be partly responsible for the variability of winds in the mesosphere as it transports zonal momentum into the mesosphere (*Alexander,* 1992).

The glow of excited O_2 at 1.27 μm has long been recognized as a tracer of atmospheric motion and convergence on the nightside of Venus (*Crisp et al.,* 1996). Atomic O produced by the photolysis of CO_2 on the dayside crosses the terminator and recombines to O_2 ($a^1\Delta_g$) near midnight via three-body collisions. Maps of 1.27-μm O_2 ($a^1\Delta_g$) emission, using VIRTIS, show that the O_2 emission is highly variable and centered on the midnight meridian (*Hueso et al.,* 2008). It also appears over the south pole. The emission is from 95 to 107 km. Bright features at these altitudes were tracked, showing that typical zonal velocities at these altitudes were 60 m s^{-1} prograde to –50 m s^{-1} retrograde. Meridional velocities were –20 m s^{-1} to +100 m s^{-1} equatorward. More recently, VIRTIS was used to create three-dimensional maps of atomic O on the nightside, providing detailed information on the circulation from 70 to 120 km (*Brecht et al.,* 2012). Observed density contours agree with Venus thermospheric circulation models (*Bougher et al.,* 2006) (VTGCM) within 30° of the antisolar point. Discrepancies at the lower O densities suggest that wave drag effects used in the VTGCM to simulate the variation of thermospheric winds are insufficient for a faithful reproduction of Venus thermospheric circulation. Atomic O is also a precursor of ozone, O_3, so excited OH $\Delta v = 1$ emission at 3 μm should also be produced in Venus' atmosphere via

$$H + O_3 \rightarrow OH^* + O_2$$

Indeed, *Gerard et al.* (2012) compared the morphologies of the O_2 ($a^1\Delta_g$) 1.27-μm and OH 3-μm Meinel nightglow observed by VIRTIS and found them well correlated.

NO airglow, originating from a mean altitude of 113 km, was observed by acquiring SPICAV limb measurements in the ultraviolet (*Gerard et al.,* 2008). O and N are photochemically produced on the dayside, and transported to the nightside by solar to antisolar flow where they radiatively recombine. The mean altitude of emission drops with increasing latitude, from 95 to 132 km. Daytime limb observations by VIRTIS have mapped CO_2 fluorescence at 4.3 and 2.7 μm, between 130 and 160 km, and CO fluorescence at 4.7 μm, from 120 km (*Gilli et al.,* 2008). A schematic of the general circulation of Venus as it is currently understood is drawn in Fig. 9b.

Large-scale subsidence, which may be the downward branch of a direct Hadley cell, is manifested as a cold collar in each hemisphere at a latitude of about 60°. From space, Venus looks like a sphere of uniform temperature (about 235 K), with 10°-wide bands in each hemisphere at 60° that are about 10° cooler. At higher latitudes, giant, complex vortices cover each pole, and are fed by the poleward flow within the clouds (*Suomi and Limaye,* 1978; *Luz et al.,* 2011). The cold, shallow river of atmosphere at 60° encircles each vortex. The south pole of Venus has a double-eyed vortex just as in the north (*Piccioni et al.,* 2007). This dipole is the site of strong downwelling, with clouds dipping down to 50 km toward the center. The vortices are usually bipolar, and sometimes break into three or more centers of rotation. Wind speeds seem to vary in both vortices, which draw down the atmosphere to create sloping pressure surfaces that define the vortices (*Limaye et al.,* 2009). The southern dipole is rotating slightly faster than the northern dipole did as measured by PV 35 years ago. The south polar vortex is shown imaged by VIRTIS on VEX at two wavelengths in Fig. 10.

On a slowly rotating planet like Venus or Titan, strong zonal winds at the cloud top can be assumed to be in cyclostrophic balance, in which equatorward component of centrifugal force is balanced by meridional pressure gradient. If Venus' low-latitude winds are in cyclostrophic balance, a jet should form at midlatitudes. The jet is the result of barotropic instability poleward of the jet, where the vortex forms. It is possible that eddies form at the

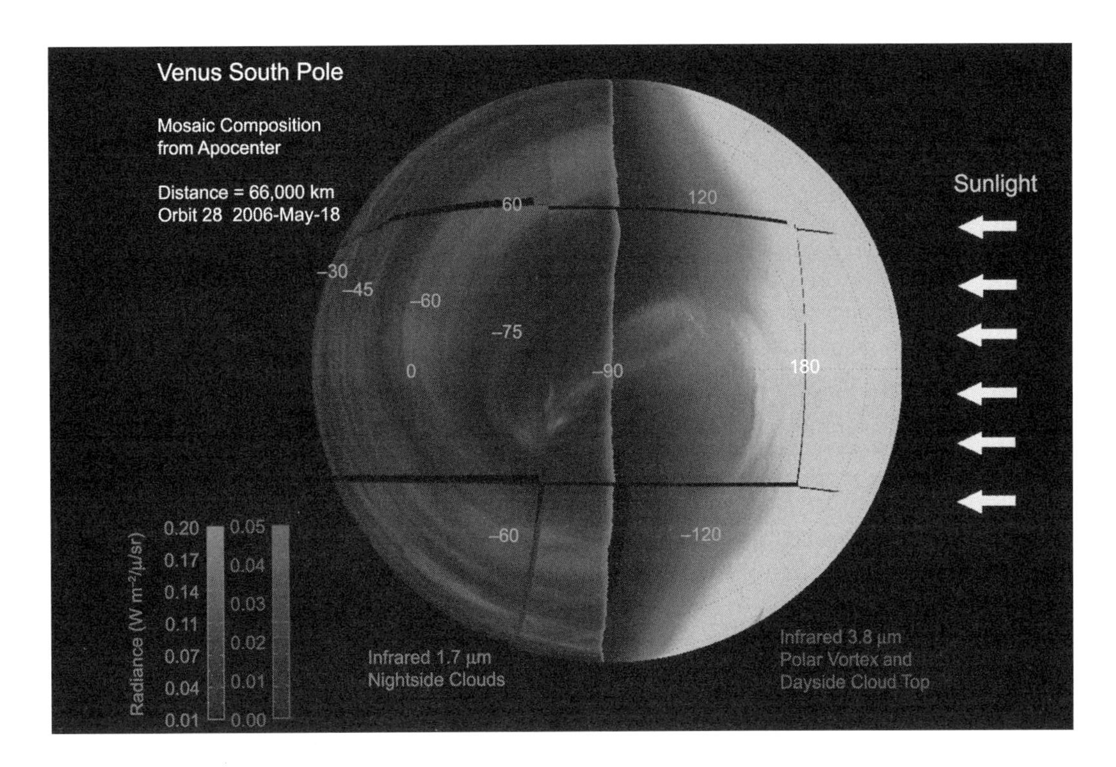

Fig. 10. See Plate 12 for color version. The south pole of Venus as imaged by VIRTIS at two different wavelengths. The vortex on May 18, 2006, was bipolar, rotating about once every four days. The sunlit (blue) image of the cloudtops was acquired at 3.8 μm, and the nightside (red) image at 1.7 μm is of the lower clouds silhouetted by infrared radiation from below. The nightside measurement probes the same level of the atmosphere as the groundbased infrared image in Fig. 1b. The red image shows the clouds at a level that is 10–15 km below that shown in the blue image. Press release image ESA/VIRTIS-VenusX IASF-INAF, Observatoire de Paris (A. Cardesín Moinelo, IASF-INAF).

equatorward edge of the jets and could be involved in the return flow to the equator (*Gierasch,* 1975). No midlatitude jets were initially seen by VMC in the southern hemisphere (*Markiewicz et al.,* 2007). However, groundbased images of the lower clouds of Venus at the transparent wavelength of 2.3 μm showed that midlatitude jets are transient and may occur in either or both hemispheres (*Young et al.,* 2008). These observations were made by viewing the lower clouds silhouetted by upwelling thermal radiation from below (Fig. 1). Since 2007, tracking of ultraviolet features with VMC has yielded interesting variabilities in the dominant zonal flow (*Moissl et al.,* 2008). Zonal winds can vary on timescale of days, and midlatitude jets come and go, as they do in groundbased nightside infrared images. Using the constraint of cyclostrophically balanced winds, *Piccialli et al.* (2008) derived winds at 60–95 km from temperature retrievals produced by VIRTIS data. Midlatitude jets accompany the cold collar, and zonal winds sharply decrease with latitude until they disappear at the edge of the vortex at 70°. Following up on this work, atmospheric temperature and pressure profiles at several different latitudes from VeRa radio occultations were used to calculate cyclostrophically balanced winds near the cloud tops (*Piccialli et al.,* 2012). They found that a jet from 20°–50° latitude forms with wind speeds of 140 ± 15 m s^{-1}. The atmosphere is adiabatic between 45 and 60 km, and winds in the jet, which decrease above 70 km, are stable.

In order to understand the transition from solar-tide-dominated flow to zonal flow and cyclostrophic balance, GCMs of Venus' atmosphere are becoming more and more useful. Having been greatly aided by the atmospheric data delivered by VEX, these models are now able to describe at least some of the momentum transport required to keep the atmosphere superrotating (*Lebonnois et al.,* 2010). However, like earlier GCMs, unrealistically high thermal fluxes are required deep in the atmosphere to sustain the superrotation profile that has been observed (*Hollingsworth et al.,* 2007). A Venus mesosphere/thermosphere GCM modeling the 80–180-km range was used to study winds in the transition region (*Hoshino et al.,* 2012). These workers imposed a thermal tide and Rossby and Kelvin waves at the lower boundary. They found that subsolar-antisolar winds dominated over zonal winds above about 90 km, and that a weak antisolar-subsolar return flow develops under 90 km. Diurnal and semidiurnal tides were damped below 90 km, but Rossby and Kelvin waves propagated up to 130 km. Rossby waves were highly attenuated at this altitude, with winds less than 1 m s^{-1}. Kelvin waves, however, exhibited wind speeds of ~6 m s^{-1}, accelerating to over 10 m s^{-1} near the terminator. Comparisons of GCM results with the infrared heterodyne Doppler wind measurements, however, showed that wind speed variability at 110 km is far too large and random to be due to four-day period planetary waves (*Nakagawa et al.,* 2013). This new modeling and analysis showed that breaking gravity waves are the most likely source of the high variability of winds at these altitudes.

As with Earth and Mars, atmospheric wave activity from a few kilometers to planetary scale are a dominant feature of the atmospheric motion. Topographically generated waves, propagating both upward and horizontally, are readily observed on both Mars and Earth. On Venus, waves are likely to be launched from the level of the clouds, propagating downward, upward, and horizontally. Small-scale gravity waves are common at the cloud tops, as observed by VMC on VEX (*Markiewicz et al.,* 2007). Monitoring upper cloud wave activity, VMC saw an enhancement over Ishtar Terra, suggesting an orographic standing wave cloud (*Titov et al.,* 2012). Gravity waves at two confined levels of the Venus atmosphere were studied by *Peralta et al.* (2008). Trains of waves at 66 km were seen by VIRTIS-M in reflected light, with wavelengths of 60–150 km. The waves are nearly parallel with latitude circles, and propagate westward with low phase velocities relative to the zonal flow. Alternating bands of clouds at 47 km on the nightside were seen as silhouettes at 1.74 μm with VIRTIS, with the same wavelengths and phase velocities. In both cases the characteristics of the waves correspond to gravity waves propagating in confined stable layers of the atmosphere. Vertical wavelengths are 2–5 km in the lower cloud and 5–15 km in the upper cloud.

The coherent sideways "Y" structure seen in the upper clouds in reflected solar ultraviolet (Fig. 1) appears to be the juxtaposition of two planetary-scale waves, one propagating faster than the mean zonal wind, and one propagating backward (*Belton et al.,* 1976). These Rossby and mixed Rossby-Kelvin waves, respectively, are persistent features of the circulation at cloud top altitudes. The midlatitude jets are likely cyclostrophic instabilities in the flow, from which waves in the form of eddies may be launched. Although these have not yet been directly observed, *Yamamoto and Takahashi* (2012), using a middle-atmosphere GCM, tested the idea that a strong planetary-scale wave could sustain the atmosphere's superrotation. They found that a 5.5-day period wave could force zonal wind speeds of 100 m s^{-1}, and also produced a 4-day planetary wave at the cloud tops along with the 5.5-day planetary wave at the cloud base. Together these waves produced a Y-shaped pattern maintained by amplitude modulation in the presence of strong thermal tides. The GCM simulations also indicated that the polar vortices are unstable and change to monopole or tripole configurations when divergent eddies with high wavenumbers occur at the poles. The cold collars are enhanced by a slowly propagating wavenumber 1 planetary wave.

2.3. The Greenhouse Effect

Carbon dioxide absorbs throughout the thermal infrared, although its major bands are at 4.4 and 15 μm. The 15-μm band figures prominently in the greenhouse warming of Earth's troposphere, and the cooling of its mesosphere. Due to Venus' higher temperatures, the 4.4-μm band of CO_2 absorbs a large percentage of the outgoing thermal radiation. More importantly, collisions between the densely packed and rapidly moving molecules redistribute energy emitted from excited states, broadening the absorption lines. Water absorbs throughout the Venus thermal infrared and its broadened lines close some of the gaps through which radiation can escape to space. Adjacent absorption lines due to deuterated water serve to enhance the infrared absorption due to water. Carbon dioxide and H_2O together leave only a few narrow windows at 0.76, 0.85, 0.95, 1.01,1.08, 1.10, 1.18,1.32, 1.74, and 2.3 to 2.7 μm. Although H_2O absorbs in many of these windows, there are several other radiatively active gases in Venus' atmosphere that absorb primarily in the 2.3- to 2.7-μm region. They are SO_2, CO, OCS, HCl, and HF (*Crisp and Titov,* 1997).

Greenhouse models must solve the radiative transfer equation over the entire emitting spectrum of the planet. Triatomic molecules have extremely complex vibrational-rotational spectra, so the transmission of radiation varies rapidly as a function of wavelength. Spectral databases such as HITRAN (*Rothman et al.,* 2003) and CDSD (*Tashkun et al.,* 2003) typically contain line position, line strength, temperature and pressure dependence, and partition coefficients for millions of H_2O absorption lines in the infrared. Similarly, hundreds of thousands of CO_2 absorption lines must be included to properly account for the transmission of radiation through CO_2. There are a large number of hot bands, and very weak absorption lines that have gradually been included as high-temperature databases for CO_2 and H_2O have improved. Comparisons of radiative balance calculations for Venus' atmosphere using different databases have shown that HITEMP (*Rothman et al.,* 2010) is an improvement over HITRAN 2004 (*Rothman et al.,* 2005), but CDSD includes more lines and produces opacities that can better explain the current radiative balance of Venus' atmosphere.

Pollack (1969a,b) developed nongray models of Venus' atmosphere from laboratory spectral data for CO_2 and H_2O that was available at the time. They were motivated by radio telescope data that indicated a very high surface temperature (*Mayer et al.,* 1958), and then by centimeter wavelength emission measured by the very first interplanetary probe, Mariner 2, that indicated the same thing (*Barath et al.,* 1963). The models of *Pollack* (1969a,b) were one-dimensional radiative-convective models that included pressure broadening of CO_2 lines, and used the adiabatic temperature profile beneath a layer of optically thick clouds. *Pollack* (1969a) was able to explain the Venera 4 atmospheric temperature data by assuming that the Venus atmosphere was 0.5% H_2O (*Nature Editorial Board,* 1967). However, *in situ* atmospheric H_2O measurements by the Soviet Venera 11 and 12 spectrophotometers in 1977 (*Moroz et al.,* 1978), and then by the PV GCMs in 1980 (*Oyama et al.,* 1979, 1980), put the abundance of water at about 100 ppmv (Fig. 3).

The PV mission characterized the composition, solar flux deposition, temperature, and structure of the Venus atmosphere and clouds from 1978 to 1990. The first results allowed *Pollack et al.* (1980) to develop a far more accurate model of the energy transport within the Venus atmosphere. They incorporated infrared opacity due to CO_2, SO_2, H_2O, OCS, H_2S, HCl, and HF, extinction due to cloud particles, and an improved spectral database. Solar flux in the Venus atmosphere was measured *in situ* by the solar flux radiometer on PV's sounder probe (*Tomasko et al.,* 1979, 1980a). Using the additional constraint of the observed solar flux deposition profile, *Pollack et al.* (1980) calculated the resulting temperature profile with their greenhouse model to be in excellent agreement with temperatures measured on descent. The modeled atmospheric static stability profile was also in good agreement with PV measurements. A schematic of the processes included in greenhouse models of Earth and Venus is shown below in Fig. 11.

As described above, groundbased infrared telescopic observations of Venus' nightside eventually converged on a globally averaged H_2O abundance of 30 ppmv. In order to explain the high temperature of Venus' surface with so little water in the atmosphere, a more detailed radiative transfer model than that of *Pollack et al.* (1980) was required. *Bullock and Grinspoon* (2001) developed a one-dimensional two-stream radiative-convective model of the Venus atmosphere that used high-temperature databases and treated multiple scattering of thermal radiation in the clouds. It included the optical properties of nine gases, pressure broadening, water vapor and CO_2 continuum, and Rayleigh scattering. Reordering the gas absorption coefficients within 68 spectral intervals over the thermal infrared produced correlated-k coefficients that were used in the solution to the radiative transfer equations (*Lacis and Oinas,* 1991). Solar energy input was determined using the solar absorption data from the PV radiometer (*Tomasko et al.,* 1980a). The hemispheric mean method of solving the equations with multiple scattering was used to calculate the radiative equilibrium state (*Toon et al.,* 1989). The optical properties of H_2SO_4/H_2O solutions (*Palmer and Williams,* 1975) were used with the aerosol size and number densities from PV to calculate the scattering properties of the cloud using Mie theory (*Hansen and Travis,* 1974). This model was sufficient to reproduce the details of Venus' measured temperature profile, as shown in Fig. 8. To assess the importance of each gas to Venus' greenhouse effect, they were deleted from the model one at a time, while calculating the resulting cooler temperature. The results of these experiments are shown in Table 6.

3. ORIGIN AND EARLY EVOLUTION: RUNAWAY GREENHOUSE

One of the most compelling scientific questions of our time is, "Are we alone in the universe?" The search for habitability, both within and outside our solar system, has

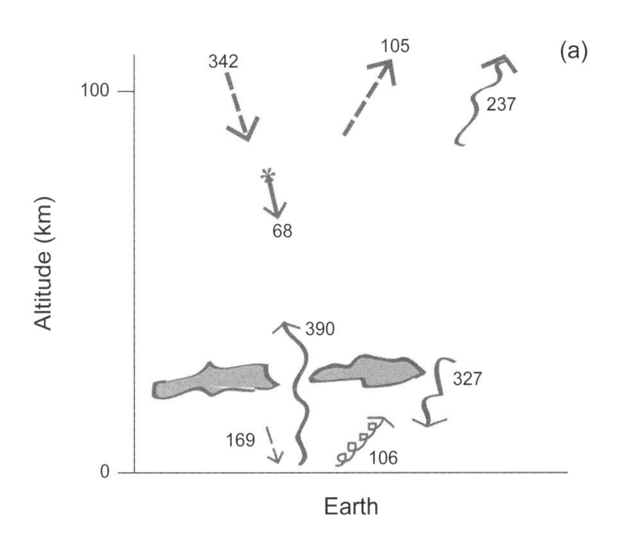

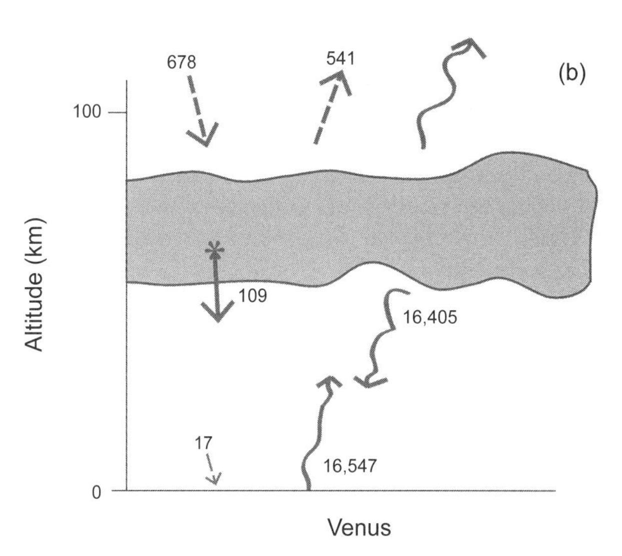

Fig. 11. The balance of energy in the atmospheres of Earth and Venus. **(a)** Earth receives 342 W m^{-2} of solar energy, of which 105 W m^{-2} is reflected to space. 68 W m^{-2} is absorbed in the atmosphere, and 169 W m^{-2} is absorbed at the surface. The surface of Earth radiates 390 W m^{-2} of thermal energy and a further 106 W m^{-2} is transported upward due to latent and sensible heat. 327 W m^{-2} of thermal energy is absorbed and reradiated downward. The net thermal flux radiated to space by Earth is 237 W m^{-2}, balancing the incoming solar flux. **(b)** Venus receives about twice the solar energy that Earth does, at 678 W m^{-2}, but reflects 541 W m^{-2} back to space. This leaves 137 W m^{-2} to power its greenhouse effect, which is ultimately radiated to space as thermal energy. Most of the solar energy (109 W m^{-2}) is absorbed in the clouds and atmosphere. Only 17 W m^{-2} is absorbed at the surface. The surface of Venus radiates 16,547 W m^{-2} of thermal energy, which is largely balanced by the 16,405 W m^{-2} that is absorbed and reradiated toward the surface. Figure produced by J. DesMarines.

TABLE 6. Venus greenhouse model sensitivity calculations (*Bullock and Grinspoon,* 2001).

Source Deleted	Decrease in Surface T
CO_2	422.7 K
H_2O	68.8 K
SO_2	2.5 K
OCS	12.0 K
CO	3.3 K
HCl	1.5 K
Clouds	142.8 K

been the driving force behind much of the solar system robotic exploration undertaken by the U.S., Europe, Russia, and Japan in the last decade. A vigorous program of *in situ* experiments is seeking habitable niches on the surface of Mars and evidence for past habitable conditions. The discovery of planets around nearby stars vastly accelerated when Kepler observations began. As of this writing, enough Kepler data have accumulated that the discovery of Earth-sized planets in stellar habitable zones has begun. Science is on the brink of finding another planet that is habitable in the way that Earth is habitable. Only the crudest measurements of these planets, such as size, mass, effective emitting temperature, global albedo, and some restricted, opportunistic abundance determinations are possible with currently available ground- and spacebased telescopes. A new generation of space telescopes will be necessary to obtain the spectra of potentially habitable planets around other stars at a range of phase angles. These data will be used to characterize their atmospheres and surfaces, and to look for direct and indirect evidence for life. In looking for habitable planets around other stars, it would be extremely helpful to know what to expect. In order to anticipate the kinds of atmospheres, clouds, and surfaces that may be found on these planets, it is important to understand how terrestrial planets evolve. We need to know by what mechanisms a planet in the habitable zone remains habitable, and what processes might operate to veer the planet toward an extreme, inhospitable state for life.

A cursory comparison of the fates of Earth and Venus should give us pause. What caused the evolution of the Earth and Venus to diverge so dramatically, and when did it happen? In spite of their proximity and similarity in size, Earth has always had liquid water at its surface, and Venus has evolved to a state so hot and dry that any remnants of C-based life (other than isotopic) on the surface would have been destroyed. It is possible that Venus formed in a drier part of the solar nebula and never received an endowment of water comparable to Earth's (*Lewis,* 1974). However, the high D/H in Venus' atmosphere is a clear indicator that Venus once had much more water than it does today. Given the uncertainties in the escape mechanisms of the past, and hence the efficiency of H fractionation, the present D/H can yield a lower bound for Venus' initial endowment of water. *Donahue and Hodges* (1992) calculated that Venus had, at the least, an amount of water equal to a global layer 10 m in depth. However, *Grinspoon* (1993) pointed out that the present D/H could also be greatly altered by fractionation during a phase of steady-state evolution of water with exospheric loss balanced by a supply from comets and/or juvenile water from the interior.

N-body dynamical simulations of the formation and early evolution of the solar system have been successful in simulating the formation of the Kuiper belt and Oort cloud, as well as the late heavy bombardment in the inner solar system (*Morbidelli,* 2002). Scattering of icy planetesimals formed between Uranus and Neptune during the migration of the giant planets gave rise to the modern comet reservoirs in the outer solar system. The inner planets were formed by the collisional aggregation of chondritic bodies scattered through the inner solar system, resulting in the broadly similar bulk compositions of Mars, Earth, and Venus. Isotopic evidence favors a chondritic source for most of Earth's H_2O (*Dreibus and Wanke,* 1989). Icy planetesimals were also scattered inward, probably bringing volatile ices such as H_2O, CO_2, N_2, and Ar to the inner solar system (*Morbidelli et al.,* 2000; *Owen and Bar-Nun,* 1995). Given the approximate similarities of the bulk C and N_2 abundance between Earth and Venus, it is likely that their initial endowments of H_2O were also similar. However, some accretional models calculate that Venus received between 5 and 30 terrestrial oceans of water (*Chassefiere et al.,* 2012).

Ingersoll (1969) examined the thermodynamics of a planetary atmosphere with a radiatively active, condensable species in it. The infrared opacity of the gas warms the surface due to the greenhouse effect. He found that such an atmosphere can be forced by solar flux up to a critical point, after which thermal equilibrium is impossible. Any increase in solar flux results in increased greenhouse warming, which leads to more radiatively active vapor, which leads to more warming . . . a positive climate feedback that rapidly puts all the greenhouse gas into the atmosphere. Ingersoll approximated the infrared opacity of H_2O with a gray model (a single absorption coefficient), and used a moist adiabat to describe the temperature profile of the early Venus atmosphere. Using the lower solar flux of the early Sun, he concluded that Venus' atmosphere probably experienced a runaway greenhouse early in its history. Photodissociation of H_2O by the higher solar ultraviolet fluxes of the early Sun and the loss of H to space could have been responsible for the loss of an Earth's ocean's worth of water.

Kasting (1988) concluded that Venus would have held onto its oceans for at least 600 m.y. after its origin. But his models did not include the effects of clouds, and this lifetime is a lower bound. The upper bound for the lifetime of Venus' oceans was more than the age of the solar system. Furthermore, Kasting discovered that the runaway greenhouse calculated by *Ingersoll* (1969) did not adequately explain the high D/H ratio in Venus' massive CO_2 atmosphere. Fractionating loss of an entire ocean would have resulted in a much higher ratio than observed, but it is not possible for H_2O to be dissociated and H lost in a dense CO_2 background atmosphere. *Kasting* (1988) suggested instead that Venus underwent a "moist runaway greenhouse" under a warm, evaporating ocean. With abundant surface water, most of the CO_2 presently in Venus' atmosphere would have been locked up as carbonates, as it is on Earth. The dense steam atmosphere would have had a saturated upper atmosphere while the warm lower atmosphere followed a dry adiabat. With a wet stratosphere, water would be lost due to relatively nonfractionating hydrodynamic escape to space. As the oceans evaporated and water vapor pressure fell below about 4 bar, surface carbonates would begin to decompose and along with ongoing volcanism, the atmosphere would begin filling up with CO_2. The more massive atmosphere would inhibit the transport of H to the exobase, and fractionating thermal and nonthermal escape mechanisms would take over. In this scenario, Venus' D/H is the signature of the loss of the final 4 bar of H_2O, and says nothing about what the initial inventory was. But it does explain how water was lost early in Venus' history in spite of the fact Venus today has a thick CO_2 atmosphere that would otherwise have made the loss of an ocean's worth of water impossible. A schematic of the runaway and moist runaway processes adapted from *Kasting et al.* (1988) is shown in Fig. 12.

In an attempt to refine the upper bound calculations for the lifetime of an ocean on Venus, *Grinspoon and Bullock* (2007) used a gray one-dimension two-stream greenhouse model coupled to a microphysical water cloud model. Using a nonideal equation of state for water vapor, they modeled the onset of a moist runaway, with clouds, as the early Sun increased in luminosity. Cloud coverage in a planetary atmosphere is highly dependent on details that are difficult to model, such as the circulation, and lifting due to convection or land mass distribution. This is partly due to the fact that clouds are involved in multiple feedbacks with water vapor distribution, the radiation field, surface moisture, and surface albedo. At present, Earth has an average cloudiness of about 50%, but it is not known if the average cloud cover would increase or decrease with an increase in surface temperature. Since the climate effects of clouds are highly dependent on their altitude, it is also not known if a change in cloudiness would have a net positive or net negative effect on forcing. So understanding the cloudiness of the Venus atmosphere as it approached a runaway state would seem to be hopeless. In order to get around this problem, *Grinspoon and Bullock* (2007) calculated radiative-convective equilibrium temperature profiles for a dense H_2O/N_2 atmosphere and a range of cloudiness from 0 to 100%. These results, assuming a young Sun 25% dimmer than today, are shown in Fig. 13. *Grinspoon and Bullock* (2007) found that if early Venus had no clouds, it would have entered a runaway state by about 500 m.y. after formation, in agreement with *Kasting* (1988). They calculated that if early Venus was completely covered in clouds, it would have never reached a moist runaway state. With 50% clouds, it would have been about 2 G.y. after formation before the onset of a runaway greenhouse. The evolution of surface temperature for this case is shown in Fig. 14.

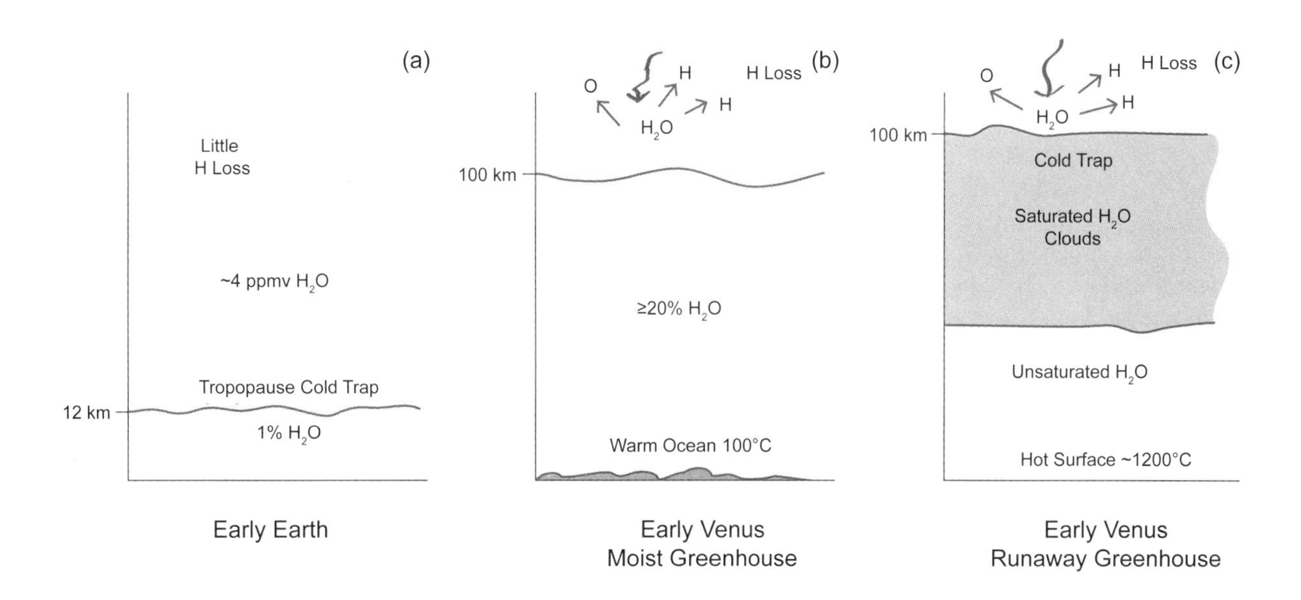

Fig. 12. Conditions for the early Earth and two possible scenarios for early Venus (<600 m.y. old) according to *Kasting* (1988). **(a)** With a cold trap at the tropopause, Earth's stratosphere is dry and little water vapor is available for photodissociation and hence loss of H to space. **(b)** The moist greenhouse effect, which best explains the current D/H of the Venus atmosphere, suggests a warm ocean overlain with a very wet atmosphere of >20% water vapor. In this case, the cold trap is so high that water can be dissociated by solar UV and H and O lost to space. **(c)** The more extreme runaway greenhouse requires a very hot, dry surface overlain by a hot unsaturated atmosphere. Above this layer is a saturated, cloudy region with a cold trap above 100 km. H and O are lost to space via photodissociation of water vapor. Adapted from *Kasting* (1988) by J. DesMarines.

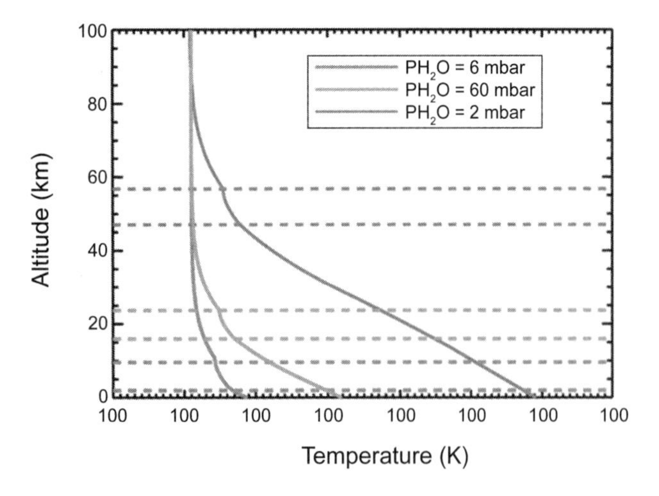

Fig. 13. See Plate 13 for color version. Radiative-convective equilibrium calculations for a 1-bar N_2 atmosphere with 350 ppmv of CO_2, available surface water, and cloud formation. The model is a one-dimensional two-stream radiative calculation with a two-component gray absorption. Calculations were performed for 6 (green), 60 (blue), and 2000 mbar (orange) of H_2O vapor in the atmosphere under present-day solar luminosity at 1 AU. Clouds form where the vapor pressure of water equals its saturation vapor pressure, and grow until they are no more than one scale height thick (horizontal dashed lines). A wet adiabatic lapse rate prevails in the cloud and the Eddington approximation is used to calculate the reflection and absorption of sunlight in the clouds. The 6-mbar case predicts a surface temperature of 285 K with cloud formation in the bottom 10 km of the atmosphere, much like Earth's present-day atmosphere. 2 bar of H_2O in the atmosphere results in a Venus-like surface temperature and clouds from 45 to 55 km.

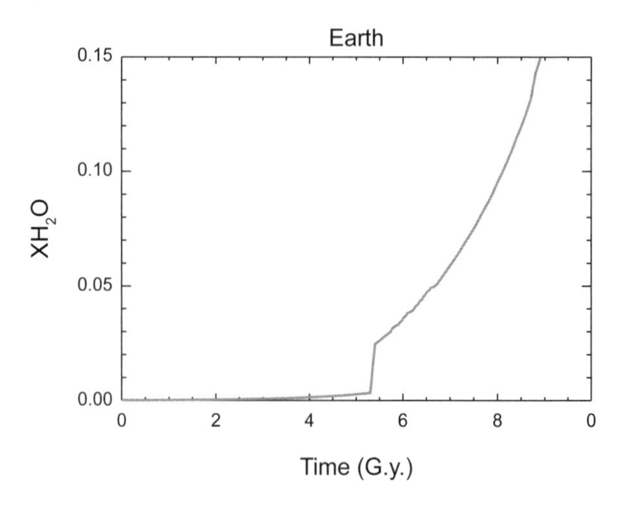

Fig. 14. The evolution of atmospheric water abundance for the planetary climate model described in Fig. 15. For this case, the luminosity of the Sun was assumed to be low right after the solar system formed and evolved according to *Gough* (1981). The 1-bar N_2, 350-ppmv CO_2 atmosphere at 1 AU is assumed to have a surface relative humidity of 77% and achieves an average cloudiness of 50%. About 4.7 G.y. after formation, atmospheric H_2O reaches a critical level and a runaway greenhouse ensues. The one-dimensional model uses a primitive radiative-convective scheme but incorporates radiative-cloud albedo feedback to investigate the effect of clouds on the runaway greenhouse. Without the help of burning fossil fuels, it is unlikely that Earth will undergo a runaway greenhouse in 160 m.y.

4. CLIMATE OF VENUS

Due to the obscuring global cloud layers, the surface of Venus was first explored by radar returns from the 85-ft Goldstone radio dish during the 1961, 1962, and 1964 inferior conjunctions (*Carpenter,* 1964, 1966; *Muhleman,* 1961). Venus' solid-body rotation rate, radius, and the dry nature of the surface were determined from these studies. While PV made a thorough investigation of Venus' atmosphere, the surface was revealed in only limited regions at kilometer resolution by Venera 15 and 16's radar imagers. Complementing these spacecraft radar images were Aricebo and Goldstone radio reflection observations, which were used to characterize the potential volcanic surface on Venus through comparisons with similar observations of Earth and Mars (*Mouginis-Mark et al.,* 1985). So although an understanding of the atmosphere was being developed, almost nothing was known about the potential sources and sinks of atmospheric gases at the surface. The Magellan mission, launched in 1989, changed all that by obtaining synthetic aperture radar images of almost the entire planet at a resolution of 200 m. What Magellan found was a planetary surface dominated by basaltic volcanism, but with no evidence for plate tectonics. Without oceans and plate tectonics, the transport of elements between atmospheric, surface, and interior reservoirs of Venus must be quite different from Earth. About 80% of Venus is volcanic plains, studded with fields of shield volcanos, and rent by a global system of rifts. Perched on top of the plains, often at the intersection of huge rift valleys, are giant shield volcanos. The scale of Venus' geologic features and their relative timing suggest that large injections of H_2O and SO_2 into the atmosphere may have occurred in the past. The large volcanos are stratigraphically the youngest features on the planet, with ages of 100 to 200 m.y. estimated from the cratering record (*McKinnon et al.,* 1997). Therefore the rates and timing of perturbations to Venus' atmosphere due to volcanism and outgassing are not well constrained.

Nevertheless, the wealth of information on possible sources of radiatively active gases to Venus' atmosphere made it possible for the first time to construct a numerical model of Venus' climate. The surface of Venus has about 1000 impact craters on it — about the same number as Earth. But unlike Earth, they are randomly distributed and almost all are unaffected by volcanism or tectonism. This implies that the surface was paved over by 300–700 Ma (*McKinnon et al.,* 1997). The geologic record from the first 80% of Venus' existence is gone. While still controversial, the simplest explanation is large-scale magmatic activity from the catastrophic foundering of the lithosphere because there are no plate tectonics to cool the interior (*Turcotte,* 1995). Using a one-dimensional two-stream radiative-convective model coupled to a cloud microphysical model, *Bullock and Grinspoon* (2001) calculated the effects that volcanism, the loss of water to space, and surface-atmosphere reactions have on the Venus climate system. They found that intense magmatism, such as that associated with the global foundering of the lithosphere, would have been sufficient to cause a dramatic greenhouse-forced change in surface temperature of up to 100 K. The pulse of temperature change would propagate into the surface over 100 m.y., adding to existing lithospheric stresses and simultaneously placing a fabric of wrinkle ridges all across the planet (*Solomon et al.,* 1999). This is in fact observed in Venus' surface geology, as are other enigmatic features that point to large changes in climate in Venus' past.

Earth's climate is affected by the many chemical cycles that transport elements between the atmosphere, oceans, crust, and mantle. Variations in Earth's orbit changes the solar forcing quasi-periodically, and its climate reacts through many feedbacks that involve the thermodynamics and microphysics of condensable species, radiative transfer, atmospheric chemistry, and the chemical cycles between reservoirs. Venus is the only other planet in the solar system that has a climate with feedbacks like these, although it has evolved along a completely different path. Terrestrial planets discovered around other stars will no doubt have climates equally complex, so it is fundamental to the search for habitability that we understand how climate works on Venus, as well as on Earth.

The condensation of minor species in the atmosphere produces climate feedbacks on Venus, just as on Earth and Mars (*Hashimoto and Abe,* 2001). In addition, perturbations to the climate due to volcanic activity are possible, as are climate feedbacks due to chemical reactions between the surface and atmosphere (*Bullock and Grinspoon,* 2001; *Hashimoto and Abe,* 2000). One way to compare the climates of Earth and Venus is to quantitatively assess the nature of direct climate forcing, responses, and feedbacks for each physical process that affects the state of the atmosphere. In terrestrial climate models, for example, a common test is to calculate the global mean increase in temperature due to doubling of atmospheric CO_2 (*Houghton et al.,* 1990).

On average, Earth's tropospheric temperature profile is established by the adiabatic lapse rate, which is determined by the atmospheric composition. The stratosphere is in radiative equilibrium, so it is the temperature at the bottom of this layer, the tropopause, that establishes the global mean surface temperature. The effective radiating temperature of Earth, T_e, is approximately the average tropopause temperature, so the balance of energy in Earth's atmosphere is

$$\frac{S_0}{4}(1-a) = \sigma T_e^4$$

where S_0 is the flux of solar energy at 1 AU, about 1368 W m^{-2}; a is Earth's Bond albedo, about 0.3; and σ is the Stefan-Boltzmann constant. The Bond albedo is the total amount of solar energy reflected divided by the total energy incident, and is therefore integrated over all possible phase angles. In practice, the Bond albedo is difficult to determine precisely because it is derived from many geometric albedo measurements and/or assumptions and measurements of the reflection phase function. The factor of 4 in the denominator is the ratio of the emitting surface area to the area that receives sunlight. This yields an effective temperature of T_e = 255 K. Earth's global mean surface temperature is 288 K; the greenhouse effect warms the surface of the Earth by an average of 33 K. At T_s = 288 K, Earth's surface emits a flux of

$$\sigma T_s^4 = 390 \ \mathrm{W/m^2}$$

The amount of flux that is received from the Sun is only 238 W m^{-2}, so the amount radiated to space must be the same. That leaves 153 W m^{-2} trapped in the atmosphere: the greenhouse effect.

For Venus, S_{V0} = 2600 W m^{-2}, but the Bond albedo has been determined to be a = 0.8 ± 0.1 (*Moroz et al.,* 1985; *Tomasko et al.,* 1980b). This means that the total incoming flux is 164 W m^{-2} — less than for Earth! Therefore, Venus' effective radiating temperature, at 232 K, is lower than Earth's. This is approximately the temperature of Venus' cloud tops.

The energy balance of Earth is affected by how a and T_e are altered by changes in temperature, water vapor, clouds, surface albedo due to water, land, and ice, as well as by changes to the surface due to ground cover or ground use that changes the surface albedo or moisture. Changes in radiatively active noncondensing gases, such as CO_2 and CH_4, alter the amount of thermal energy trapped in the atmosphere. Therefore geologic processes and geochemical cycles involving C in the land and oceans also have long term effects on climate.

Since these physical and chemical processes are related through feedbacks that alter the forcing of Earth's climate, it is helpful to quantitatively look at the effects of these processes. We wish to compare the magnitudes of the direct forcing, response, and resulting feedback strength due to small changes in each process. Together, these linearized sensitivity terms can be used to test how different climate models respond to changes in forcing. In addition, the ultimate net response in surface temperature due to realistic changes in atmospheric parameters can be compared to the historical (direct and proxy) temperature record.

The change in radiative forcing is designated ΔF, and has units of W m^{-2}. The change due to altering atmospheric CO_2 concentration, $[CO_2]$, is thus

$$\Delta F = \frac{\partial F}{\partial [CO_2]} \Delta [CO_2]$$

(*Ramanathan,* 1987). For Earth, doubling CO_2 results in trapping an additional 4 W m^{-2} (from 153 W m^{-2}) in the atmosphere. This is the direct forcing. ΔF must adjust to 0, however, so the physical response is to increase the temperature at the tropopause by ΔF. Thus the response is

$$\frac{\partial F}{\partial T}\Delta T$$

and the total change in flux due to doubling CO_2 is

$$\Delta F = \frac{\partial F}{\partial [CO_2]}\Delta [CO_2] + \frac{\partial F}{\partial T}\Delta T$$

Earth's atmosphere adjusts to the increased trapping of thermal energy on its radiative timescale of a few days, so that $\Delta F \rightarrow 0$ and

$$\Delta T = -\frac{\frac{\partial F}{\partial [CO_2]}}{\frac{\partial F}{\partial T}}\Delta [CO_2]$$

We know that the numerator is 4 W m^{-2}. Since the atmosphere and surface heat up in order to radiate more, $\partial F/\partial T$, the response, is a negative (stabilizing) feedback. $\partial F/\partial T$ is easy to calculate

$$\frac{\partial F}{\partial T} = \frac{\partial(\sigma T^4)}{\partial T} = 4\sigma T^3$$

For T_e = 255 K, $\partial F/\partial T$ = 3.3 W m^{-2}-K, and therefore

$$\Delta T = \frac{4 \text{ W m}^{-2}}{3.3 \text{ W m}^{-2}\text{-K}} = 1.2 \text{ K}$$

This includes the direct forcing due to doubling CO_2, and the resulting thermal response, which is a negative feedback.

The Intergovernmental Panel on Climate Change (IPCC) (*Albritton et al.,* 2001) assessed about a dozen major Earth climate models and concluded that doubling Earth's atmospheric CO_2 would increase the global mean temperature by 2.0 to 4.5 K, depending upon the model. In all these models, the temperature increase is greater than that calculated by considering just the direct forcing and response. The reason for this is that there are important indirect affects that the above analysis ignores. An obvious one is the role of atmospheric water vapor, a far more potent greenhouse gas than CO_2. The indirect response due to water vapor is

$$\frac{\partial F}{\partial [H_2O]}\frac{\partial [H_2O]}{\partial T}\Delta T$$

The concentration of H_2O in Earth's atmosphere is set largely by the surface temperature. H_2O concentration is an exponential function of surface temperature via the Clausius-Clapeyron equation. $-\partial F/\partial[H_2O]$ is larger than $-\partial F/\partial[CO_2]$ due to stronger absorption of thermal radiation by water vapor than by CO_2. Water vapor produces a positive feedback to the climate. With water vapor feedback included, the next change in radiative flux is thus

$$\Delta F = \underbrace{\frac{\partial F}{\partial [CO_2]}\Delta [CO_2]}_{\text{initial forcing}} + \underbrace{\frac{\partial F}{\partial T}\Delta T}_{\text{direct response}} + \underbrace{\frac{\partial F}{\partial [H_2O]}\frac{\partial [H_2O]}{\partial T}\Delta T}_{\text{indirect response}}$$

Since $\Delta F \rightarrow 0$, the net change in temperature is

$$\Delta T = \frac{-\frac{\partial F}{\partial [CO_2]}\Delta [CO_2]}{\frac{\partial F}{\partial T} + \frac{\partial F}{\partial [H_2O]}\frac{\partial [H_2O]}{\partial T}}$$

or

$$\Delta T = \frac{-4 \text{ W m}^{-2}}{3.3 \text{ W m}^{-2}\text{-K} + (-0.9 \text{ to } -2.0) \text{ W m}^{-2}\text{-K}}$$

So that

$$\Delta T = 1.7 \text{ to } 3.0 \text{ K}$$

This still underestimates the change predicted by complex, fully three-dimensional climate models due to doubling of atmospheric CO_2.

In general, the change in surface temperature can be written

$$\Delta T = \frac{\Delta Q}{\lambda}$$

where ΔQ is the forcing and λ is the total net feedback. $\partial F/\partial T$ in the denominator is positive and greater than 1, and is therefore a negative feedback. $\partial[H_2O]/\partial T$ is positive and $\partial F/\partial[H_2O]$, so the water vapor contribution is negative, and therefore a positive feedback.

Altering the atmosphere can result in changes in cloud cover, and hence albedo. A change in surface temperature can change the albedo of the surface dramatically, for example, by increasing or decreasing polar ice cover. A colder surface will result in more ice cover and hence a higher surface albedo, so ice albedo is a positive feedback. Any positive feedback has the potential to "run away" as it takes over the dynamics of the climate system. Indeed, an ice albedo runaway is something that Earth has experienced several times, most recently during the late Precambrian, when the planet entered a "snowball Earth" phase (*Hoffman et al.,* 1998).

The effects of cloud albedo are less clear than those of surface albedo. Low clouds on Earth radiate at almost the same temperature as the ground, but increase the albedo locally. They therefore decrease radiative forcing. High clouds are cold, however, and reradiate upwelling thermal energy both upward and downward. While they increase the albedo, they also act as a greenhouse agent and can have a net increase in radiative forcing. Due to the complex horizontal and vertical distribution of clouds at any instant in Earth's atmosphere, the net forcing due to clouds is poorly constrained. The albedo and greenhouse effects of clouds partially cancel each other, and the clouds in total can result in either a slight net increase or net decrease in radiative forcing.

The temperature dependence of albedo's effect on the energy balance of the atmosphere is

$$\Delta F = \frac{S_0}{4}\left(\frac{\partial a}{\partial T}\right)\Delta T$$

Therefore the total change in outgoing flux due to the doubling of CO_2 includes the initial forcing, the direct response, the indirect water vapor response, and the indirect albedo response. It is

$$\Delta F = \frac{\partial F}{\partial [CO_2]}\Delta[CO_2] + \frac{\partial F}{\partial T}\Delta T + \frac{\partial F}{\partial [H_2O]}\frac{\partial [H_2O]}{\partial T}\Delta T + \frac{S_0}{4}\left(\frac{\partial a}{\partial T}\right)$$

Since $\Delta F \rightarrow 0$ in a few days, the net change in global mean surface temperature is

$$\Delta T = \frac{-\dfrac{\partial F}{\partial [CO_2]}\Delta[CO_2]}{\dfrac{\partial F}{\partial T} + \dfrac{\partial F}{\partial [H_2O]}\dfrac{\partial [H_2O]}{\partial T} + \dfrac{S_0}{4}\left(\dfrac{\partial a}{\partial T}\right)}$$

Putting in the numbers for the range of these parameters calculated by the climate models considered by the IPCC

$$\Delta T = \frac{4\ \mathrm{W\ m^{-2}}}{3.3\ \mathrm{W\ m^{-2}\text{-}K} + (-1.3\ \text{to}\ -2.4)\ \mathrm{W\ m^{-2}\text{-}K} + (-0.3\ \text{to}\ -0.5)\ \mathrm{W\ m^{-2}\text{-}K}}$$

or

$$\Delta T = 2 - 4.5\ \mathrm{K}$$

By linearizing the rates of change described by the direct response to CO_2 doubling, the indirect thermal response, and the water vapor and cloud albedo responses, we have captured the main short-term sensitivities of Earth's climate. The effects of these three simple feedbacks adequately explain the results from the climate models considered by the IPCC.

Considering Venus' atmosphere in a similar way, we can gain considerable insight into how its climate works by investigating the nature and magnitude of feedbacks that may arise. Water and SO_2 are both greenhouse gases and are particularly effective on Venus because they have absorption lines in the 2.3- to 2.7-μm region of CO_2. However, they are both constituents of the clouds, and so must also have an effect on the albedo. Variations in both of these atmospheric gases, both on short and long timescales, are almost certain. Carbon dioxide, H_2O, and SO_2 are the most common compounds outgassed by basaltic volcanism (*Kaula et al.,* 1981), and there is widespread evidence for a long history of this activity across the planet (*Head et al.,* 1992). At least three of the nine potential volcanic "hotspots" on Venus (*Stofan et al.,* 2009) appear to have a higher near-infrared surface emissivity than most of the planet (*Smrekar et al.,* 2010). These measurements were made by the VIRTIS infrared imaging spectrometer, on the nightside of the planet through the 1.01-μm window, which allows us to see all the way to the surface. The clouds scatter this emitted radiation so that the smallest patch that can be observed is about 70 km. After accounting for the altitude dependence of the emitted flux using Magellan altimetry, *Mueller et al.* (2008) produced maps of surface emissivity that were clearly correlated with geologic structures from Magellan radar images. *Smrekar et al.* (2010) argued that these correlations indicated recent volcanism that emplaced fresh lavas that have yet to weather to the composition of older terrain.

About half the SO_2 that enters Earth's atmosphere comes from volcanism; the other half is anthropogenic. Large volcanic eruptions spew SO_2 high into the atmosphere, where it is photochemically converted in the presence of water vapor, into H_2SO_4 aerosols. This Junge layer is Earth's temporary version of a Venus cloud deck. In 1991, Mount Pinatubo erupted, creating a thin stratospheric H_2SO_4 cloud that increased Earth's albedo by 0.1 for almost two years (*Stowe et al.,* 1992). Interestingly, the resulting cooler temperatures offset the trend of increasing temperature attributed to anthropogenic activity, obfuscating the debate about human-induced climate change. With the passage of time (and more volcanos), the cooling blip caused by Mount Pinatubo and other volcanos can be seen against the background of rising temperatures. On a more dramatic scale, the rapid emplacement of igneous provinces in Earth's past implies times of rapid and sustained injection of greenhouse gases into the atmosphere (*Coffin and Eldholm,* 1994). It is likely that the climate responded through changes in cloud cover and greenhouse forcing, resulting in perturbations in global mean temperature. Venus gives us an example where some of the same processes are operating in a very different context.

Using a one-dimensional climate model that calculates the radiative-convective equilibrium temperature profile from a two-stream correlated-k coupled radiative/cloud microphysics model, we calculated the equilibrium temperature profile for Venus' atmosphere as H_2O was varied from 0.3 to 3000 ppmv, and as SO_2 was varied from 1.8 to 18,000 ppmv (one-one-hundredth to 100 times the present abundances). These models use present-day photochemical production rates of H_2SO_4, scaled to cloud top H_2O and SO_2 abundance to grow H_2SO_4/H_2O condensate that descends through the cloud at the Stokes' velocity. Multiple scattering of incoming sunlight and outgoing thermal radiation by the resulting cloud aerosols couples the cloud to the radiative transfer calculation. In this way, the role of H_2O and SO_2 in forming the clouds and influencing albedo, as well as contributing to the greenhouse effect, was explored numerically. The global mean surface temperatures from these model calculations are shown in Fig. 15.

The effectiveness of water vapor as a greenhouse gas in Venus' atmosphere can be seen by the magnitude of the initial forcing, given by

$$\frac{\partial F}{\partial [H_2O]} \Delta [H_2O]$$

Recall that for CO_2 in Earth's atmosphere, this term was 4 W m^{-2}. The Venus climate model calculates a value of 24.1 W m^{-2} for a doubling of H_2O content. Without considering the cloud albedo feedback, the rise in surface temperature due to the initial forcing and thermal response is

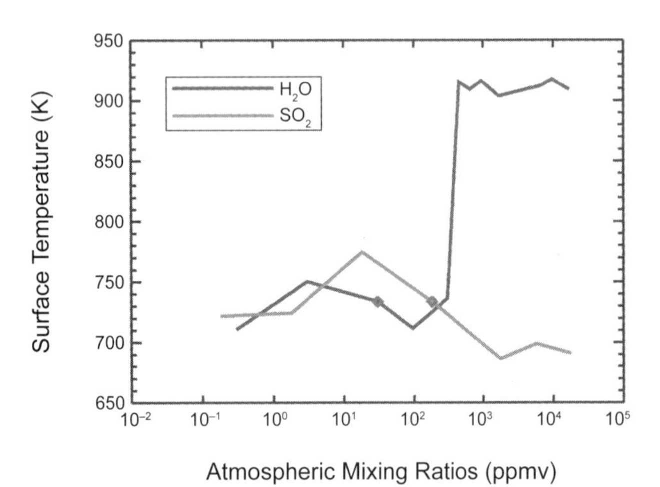

Fig. 15. See Plate 14 for color version. A suite of several hundred runs of the Venus climate model of *Bullock and Grinspoon* (2001). Atmospheric H_2O is varied from 1/100th the present abundance to 1000 times the present abundance (blue curve), and atmospheric SO_2 is varied from 1/1000th the present abundance to 100 times the present abundance (orange curve). The model calculates that the clouds will disappear if atmospheric water drops below about 2 ppmv, or if SO_2 drops below about 5 ppmv (not shown). The red dots indicate current conditions. These data are used in section 4 to assess the sensitivity of the climate model to perturbations in H_2O and SO_2. Around present-day conditions the slope is negative for both species, indicating that albedo feedback is more important than the greenhouse effect to Venus' energy balance. However, if atmospheric water abundance is increased by more than about 10 times its present value, the enhanced greenhouse initiates a cloud collapse runaway. Warmer temperatures raise the cloud base, thinning the clouds, which lowers the albedo and increases radiative forcing. This in turn warms the atmosphere, further thinning the clouds. The clouds rapidly erode from below until all that is left is a thin, high water cloud with a stable surface temperature that is 200 K hotter than Venus today.

$$\Delta T = \frac{-\frac{\partial F}{\partial [H_2O]} \Delta [H_2O]}{\frac{\partial F}{\partial T}}$$

As for Earth, $\partial F/\partial T$ can be calculated and is 2.8 W m^{-2}-K. So the increase in Venus' surface temperature due to a doubling of atmospheric H_2O, for the greenhouse effect only, is

$$\Delta T = 8.6 \text{ K}$$

A change in atmospheric H_2O will cause a change in the cloud albedo, which acts as a negative feedback. The change in forcing due to the sensitivity of cloud albedo is

$$\Delta F = \frac{S_0}{4} \left(\frac{\partial a}{\partial T} \right)_{H_2O}$$

The Venus climate model calculates that forcing due to albedo with respect to H_2O is –11.1 W m^{-2} K. The change in surface temperature as a result of doubling atmospheric H_2O, the thermal response, and the cloud albedo feedback is

$$\Delta T = \frac{-\frac{\partial F}{\partial [H_2O]} \Delta [H_2O]}{\frac{\partial F}{\partial T} + \frac{S_0}{4} \left(\frac{\partial a}{\partial T} \right)_{H_2O}}$$

which yields a net change in surface temperature due to varying atmospheric water abundance of

$$\Delta T = -2.9 \text{ K}$$

For the case of water in Venus' atmosphere, cloud albedo dominates over the greenhouse effect so that an increase produces a cooling, and a decrease produces warming.

A similar analysis may be done for SO_2, where doubling the SO_2 abundance yields a net change in surface temperature

$$\Delta T = \frac{-\frac{\partial F}{\partial [SO_2]} \Delta [SO_2]}{\frac{\partial F}{\partial T} + \frac{S_0}{4} \left(\frac{\partial a}{\partial T} \right)_{SO_2}}$$

Sulfur dioxide is a far weaker greenhouse gas than H_2O, which is reflected in the change in forcing resulting from a doubling of atmospheric SO_2, which is 1.1 W m^{-2}. The thermal response is still 2.8 W m^{-2} K, so the change in temperature due to the doubling of SO_2, for the greenhouse effect only, is

$$\Delta T = 0.4 \text{ K}$$

However, when the albedo feedback is included, with a

value of –2.9 W m^{-2} K, a negative feedback, the net change in temperature is

$$\Delta T = -8.0 \text{ K}$$

As with water, an increase in SO_2 warms the atmosphere due to the slightly enhanced greenhouse effect. This is overwhelmed by the increase in albedo, which reduces solar forcing.

Table 7 lists the values of the forcing and feedback terms for water in Earth's atmosphere, and water and SO_2 in Venus' atmosphere. Here we see direct comparisons between the changes in climate forcing due to the doubling of CO_2 on Earth and the doubling of H_2O or SO_2 on Venus. On Earth, the water vapor feedback enhances the effect of CO_2 doubling (a positive feedback) but cloud albedo feedback also limits the increase (negative feedback). On Venus, with the present climate model, albedo feedback is stronger than greenhouse forcing.

TABLE 7. Climate sensitivity of Earth and Venus (direct forcing, response, and feedbacks due to doubling).

	Earth CO_2	Venus H_2O	Venus SO_2
Forcing (W m^{-2})	–4.0	–24.1	–1.2
Response (W m^{-2}-K)	3.5	2.8	2.8
ΔT no feedback (K)	+1.2 K	+8.6 K	+0.4 K
Feedback (W m^{-2}-K)	–2.1	–11.1	–2.9
ΔT (K)	+2.9 K	–2.9 K	–8.0 K

4.1. Atmospheric Loss and Long-Term Climate

The loss of up to a terrestrial ocean's worth of water may have had the most dramatic impact on Venus' evolution. It is not known if a planetary magnetic field ever existed and therefore what role it may have played in the loss of volatiles from Venus' atmosphere. Today, however, Venus does not have a magnetic field, and the escape rates of H and O are significant. *Grinspoon* (1993) pointed out that pre-VEX H escape rates implied a H residence time in the atmosphere of Venus between 74 and 680 m.y. Assuming an average H residence time of 160 m.y., Venus' atmosphere would have had 10 times more H_2O at 1 Ga. Climate simulations that model both the clouds and greenhouse effect show that a dramatically different climate would have existed then, even if no H_2O had been outgassed from the interior (*Bullock and Grinspoon,* 2001).

The Analyzer for Space Plasmas and Energetic Atoms (ASPERA-4) instrument on VEX showed that the dominant escaping ions are O^+, He^+, and H^+. The ratio of H^+/O^+ escape rates is 1.9, so water is escaping stoichiometrically (*Barabash et al.,* 2007a) . H^+ escapes by collisions with hot atoms, which are heated by photochemistry. ASPERA-4 measured the He^+/O^+ escape rate ratio to be 0.07. The escaping ions leave Venus through the central portion of the plasma wake (the plasma sheet), and in the boundary layer of the induced magnetosphere. ASPERA-4 was also used to characterize the tailward flow of energetic neutral H observed on the nightside by providing global images of energetic neutral atom intensity (*Galli et al.,* 2008). The images showed a highly concentrated tailward flow of H tangential to the Venus limb around the Sun's direction. No hot O above the instrument threshold was detected, although O^+ and O must be lost to ion pickup and sputtering in the plasma wake during coronal mass ejections (*Luhmann et al.,* 2008). Distortion of the bow shock was observed during a coronal mass ejection event (*Zhang et al.,* 2008). In the plasma wake and beneath the induced magnetosphere, O^+ beams are highly variable and exhibit a ray-like outflow pattern around the center of the wake (*Szego et al.,* 2009). The H in the tailward flow originates from shocked solar wind protons that charge exchange with the neutral H exosphere.

Proton cyclotron waves were detected with VEX's two magnetometers upstream of the solar wind, indicating pickup of planetary H from Venus' exosphere and loss to space (*Delva et al.,* 2008). These waves were found up to 9 Venus radii out, and over a wide range of angles between the solar wind and magnetic field. Therefore, pickup loss of neutral H from the exosphere is efficient over a large volume of space upstream of the planet.

Solar wind at solar minimum does not enter the atmosphere. VEX magnetic field measurements (MAG) in the plasma environment surrounding Venus show that the induced magnetosphere almost completely shields the atmosphere from the solar wind (*Zhang et al.,* 2007). There is a well-defined outer boundary of the induced magnetosphere, and therefore an induced magnetopause multiscale turbulence is observed at the magnetosheath boundary layer and near the bow shock (*Voros et al.,* 2008).

Venus' H Ly-α corona was imaged by SPICAV at altitudes from 1000 to 6000 km on the dayside (*Chaufray et al.,* 2012). A cold population of H exists below 2000 km, while a hot population produced by neutral-ion interactions exists above 4000 km. There is a larger H exobase density on the dawn side, driving a strong dawn-dusk asymmetry. These results are qualitatively similar to those seen by the PV ultraviolet spectrometer; the dawn-dusk contrast measured by SPICAV is lower than that seen by PV 35 years ago.

4.2. Role of Geology and Geochemistry in Sustaining Clouds

The abundance of S gases in Venus' atmosphere and their wide range of oxidation states has important implications for Venus' past climate. Sulfur dioxide, H_2S, OCS, and S_x are abundant and involved in several chemical cycles in the atmosphere. Laboratory work by *Fegley and Prinn* (1989) showed that at Venus surface temperature and pressure, SO_2 reacts rapidly with calcite to form anhydrite

$$CaCO_3 + SO_2 \leftrightarrow CaSO_4 + CO$$

Carbonates are likely to exist on Venus, as different species are in equilibrium with the CO_2 atmosphere at different altitudes (*Fegley and Treiman,* 1992; *Hashimoto et al.,* 1997). The somewhat weak experimental evidence is the interpretation of Venera X-Ray Fluorescence spectrometer data as indicating the amount of Ca as 7% CaO, which may in fact be calcite.

Pioneer Venus' GCMS measured 180 ppmv SO_2 at an altitude of 22 km. The other carriers of S are in lower abundance and roughly constant in mixing ratio beneath the clouds, so this value was taken as the bulk mixing ratio of SO_2 in the atmosphere. However, this is two orders of magnitude greater than the amount of SO_2 that should be in equilibrium with calcite. So either there is no available calcite at the surface for SO_2 to react with, or SO_2 has been injected into the atmosphere faster than it has reacted out. To complicate the situation, a reanalysis of Vega *in situ* ultraviolet spectrometer data indicated that SO_2 beneath the clouds rapidly decreases to the surface, where it achieves a value of 2 ppmv. This is just the surface concentration expected if atmospheric SO_2 is in equilibrium with surface minerals.

Scaling their laboratory kinetic data to Venus, *Fegley and Prinn* (1989) calculated that all the SO_2 could be removed from Venus' atmosphere by reaction with calcite in 2 m.y. They took the surface reaction rate determined in the laboratory and applied this to the Venus surface. However, diffusion of atmospheric SO_2 to reach new calcite reaction sites in the subsurface would certainly slow the uptake of atmospheric SO_2. *Bullock and Grinspoon* (2001) employed a temperature-dependent reaction-diffusion model to calculate a lifetime of about 20 m.y. for SO_2 in Venus' atmosphere, assuming available calcite in the top few hundred meters of Venus crust. Since the clouds cannot be maintained if atmospheric SO_2 drops below about 10 ppmv (*Bullock and Grinspoon,* 2001), this means that active volcanism in the past 20 m.y. must be sustaining them. *Fegley and Prinn* (1989) determined from their reaction rates and reasonable assumptions about the S content of Venus magmas that on average, between 0.4 and 11 km^3 of lava must be erupted on Venus each year to sustain the clouds. This is rather remarkable in that two entirely different methods, one from a detailed study of the geomorphology (*Head et al.,* 1992), and one from an analysis of the cratering record (*Bullock et al.,* 1993), yield very similar values. These rates are about the same as all the volcanism on Earth not associated with plate boundaries, i.e., all the intraplate volcanism such as Hawaii.

The high radar reflectivity of Venus mountain tops led *Klose et al.* (1992) to suggest that pyrite was abundant and that atmospheric SO_2 is controlled by its oxidation to magnetite

$$3FeS_2 + 16CO_2 \leftrightarrow Fe_3O_4 + 6SO_2 + 16CO$$

Zolotov (1991) pointed out that the mixing ratio of SO_2 from PV is in approximate equilibrium with this reaction.

5. WHAT VENUS TELLS US ABOUT LIFE IN THE UNIVERSE

5.1. The Inner Edge of the Habitable Zone

We know that the surface of Venus is far too hot for carbon-based life. It is even too hot for any remnants of carbon-based life to have survived, other than isotopic signatures. This does not mean that Venus has nothing to teach us about the origin of life, however. On the contrary, an Earth-sized planet with a CO_2/N_2 atmosphere perched on the inner edge of the habitable zone is something that probably exists around many stars (Fig. 16). We are scouring the Earth and its history for an understanding of the origin of life, but when we have the capability to look for habitable planets, it is likely that we will find something similar to Venus. It is therefore crucial that we understand how the evolution of the Earth and Venus diverged so dramatically, and the implications of this history for the habitability of planets around other stars.

After the Viking missions to Mars, and the growing evidence that Mars may have once been warm and wet, quantitative arguments about the origin of life became possible. The earliest fossils on Earth have been dated to 3.6 Ga,

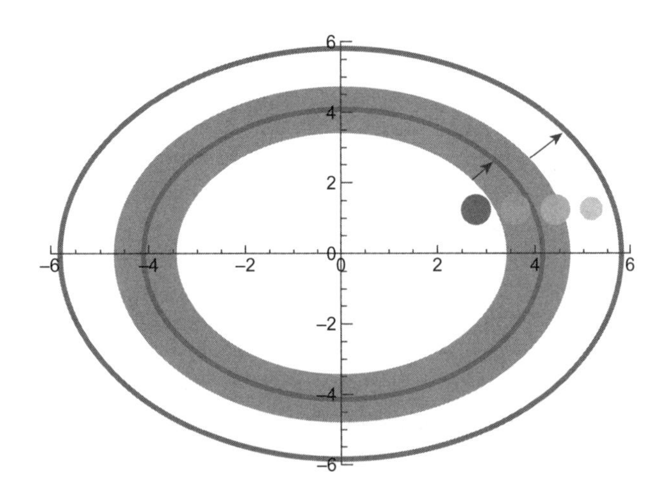

Fig. 16. See Plate 15 for color version. Evolution of the habitable zone around an F star. The orange ring represents the region of the stellar nebula where liquid water was stable on a forming planet. As the star brightens, the habitable zone moves outward (black arrows) to the zone outlined by the red ellipses. The intersection of these two regions is the continuously habitable zone. The blue circle represents an Earth-like planet that remains in the continuously habitable zone for billions of years. The dark brown circle depicts a planet that would undergo a runaway greenhouse early in its history, while the lighter brown circle is more representative of a Venus-like planet that undergoes a moist greenhouse. The gray circle is a planet that is frozen for much of its existence, like Mars, but where liquid water may one day become stable at the surface as the luminosity of the star increases over time. Units on the axes are in AU.

just 850 m.y. after the formation of Earth. Isotopic carbon fractionations in Greenland shales have been interpreted to mean that carbon-fractionating life was extant on the Earth at 3.8 Ga, or 650 m.y. after its formation. These remarkable data tell us that life can arise on a suitable planet within 650 m.y. of its formation. About half that time was punctuated by planet-sterilizing impacts, including the one that caused the formation of the Moon. We may conservatively estimate that life's origin on Earth took somewhere around 300 Ma. Does this mean that life generally arises on planets that have stable liquid water at their surfaces for at least this duration? The geologic record shows unambiguously that liquid water flowed on the surface of Mars for at least this long. Furthermore, studies in Antarctica's Dry Valleys show that water remains liquid beneath ice-covered lakes even if globally annual temperatures are many degrees below freezing, due to a very small trickle of liquid from nearby locally warm regions. Sedimentary rocks show evidence of large-scale precipitation of salts from standing and subsurface water on Mars, indicating that briny fluids with suppressed freezing points flowed on Mars well beyond the time at which the globally averaged temperature fell below the freezing point of water.

From first principles, *De Duve* (1995) estimated that the onset of replicating molecules that could participate in natural selection may have taken as little as 10,000 years, once the environment became conducive and stable for the origin of life. Even if this is hopelessly optimistic, the above consideration suggests that Mars and Earth may have had roughly equal ability to host the origin of life. What about Venus? If Venus had an ocean that lasted 500 m.y. or 2 G.y., it may well have been an ideal place for life to begin. The current state implies that life or even life's remnants can't exist at the surface. But what if life did arise on Venus, and migrated into the only remaining habitable niche when things got really bad? Aside from the chemistry, clement conditions for life exist in Venus' clouds. The clouds themselves are vast, heterogeneous reservoirs that could host life if it solved the problems of low pH, low nutrient availability, and a harsh ultraviolet flux. *Schulze-Makuch et al.* (2004) suggested that organisms with a sulfur-based survival strategy might both harness energy and protect themselves from ionizing radiation in the perennial aerosols that are suspended above Venus' surface. A graphic of elemental S available for such microbes is shown in Fig. 17.

As discussed above, the D/H ratio in Venus' atmosphere strongly implies that it has lost a lot of H through fractionating escape, and an unknown amount to less-fractionating escape. *Kasting*'s (1988) models showed that the amount of energy deposited into Venus' atmosphere just after the planet formed was greater than the critical amount that would lead to a runaway water greenhouse (*Ingersoll,* 1969). Thus, water would reside in the atmosphere. If the Earth's oceans began evaporating (as they will someday), the increased water vapor would increase the greenhouse effect, which would make the oceans evaporate faster. This is the runaway greenhouse effect. Run to completion, the atmosphere would consist of 240 bar of water vapor, in addition to the 0.7 bar of N_2 and 0.3 bar of O_2. There is controversy over what the initial water endowments of Mars, Earth, and Venus were, but we can be fairly sure that Venus had at least a few tens of bars of water, and perhaps as much or more than Earth. With this much water in the atmosphere, there is no longer a functioning "cold trap" at the tropopause, and water flows into the stratosphere. There it is vulnerable to photodisassociation into H and O. Presently, Earth's stratosphere is very dry, limiting the amount of H that can be produced and make it to the exobase to be thermally lost to space.

So with a very wet atmosphere, Venus would have lost H via hydrodynamic escape, which doesn't fractionate D from H very efficiently. As mentioned earlier, *Kasting* (1988) calculated that when the atmospheric water abundance fell to about 4 bar, fractionating escape took over, resulting in the D/H signature that we see today. In this scenario, all we can say is that Venus had at least 4 bar of water, but it may have had much more. One of the problems with this picture is what happened to all the O that was left from the

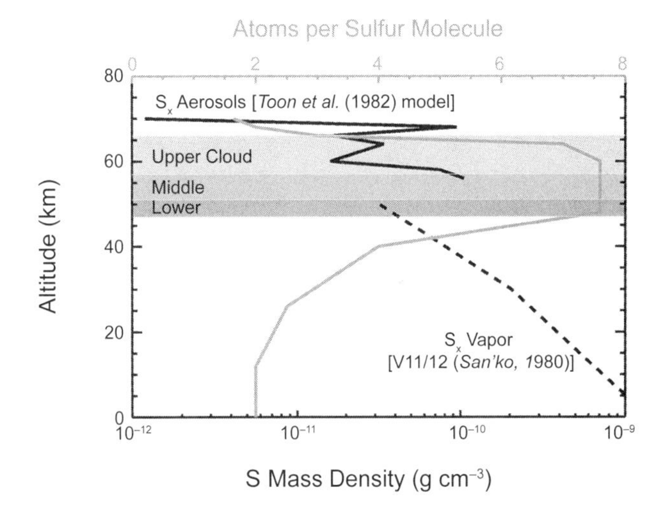

Fig. 17. See Plate 16 for color version. Elemental S in the Venus atmosphere. The only measurements of elemental S on Venus were made with the scanning spectrophotometers on Veneras 11 and 12 (*San'ko,* 1980) (black dashed line). These detected only the total amount of S_x vapor. The thermodynamics of S allotropes require that S is mostly in the form of S_2 near the surface and above the clouds (orange line). Within the clouds, however, S_8 is stable and other allotropes also appear. *Toon et al.* (1982) used a cloud chemical/microphysical model to estimate the abundance of S allotropes in the upper Venus cloud. Their data showed that conversion between S_4 and S_3 in the upper clouds is plausible (where the near-UV absorber must reside), and could explain the absorption of almost half the sunlight that enters the clouds. *Schulze-Makuch et al.* (2004) hypothesized that the mild and persistent conditions in the Venus clouds make them a possible abode for life that uses the abundant S cleverly. From *Schulze-Makuch et al.* (2004).

loss of H. It has been widely suggested that this O went to oxidizing the crust. However, it is known from the geologic record in the Magellan radar images that the present crust is no older than about 1 G.y. So any era of wholesale oxidation of a primary crust is lost from view. The present crust appears not to be quite as oxidized as Mars', based on the interpretation of color images from Venera 13 (*Pieters et al.,* 1986). One surprising result from the VEX mission is that O is lost, via nonthermal mechanisms, at about half the rate of H. In other words, stoichiometrically, water is lost from Venus to space at the present time. It appears that the present crust has not had to absorb much excess O, consistent with the effective wholesale loss of H_2O to space.

5.2. The Fate of Earth

Main-sequence stars slowly brighten over time as the products of H fusion collect in their cores. Much of the research on the climate evolution of the terrestrial planets has centered on how both Mars and Earth could have had stable liquid at their surfaces when the Sun was 70% as bright as it is today. As the Sun increases in luminosity, increased solar forcing will inevitably lead to a water runaway greenhouse on Earth. This may be precipitated by the decarbonation of the continental margins, dramatically raising the pressure of the atmosphere and enhancing the greenhouse effect through pressure broadening of atmospheric absorption lines. Ultimately, the Earth will go the way of Venus, as the oceans evaporate and 240 bar of H_2O enter the atmosphere. Overwhelming the stratospheric cold trap, water will be dissociated by solar ultraviolet, and the atmosphere will enter a phase of hydrodynamic escape. Along with H and O, N_2 and CO_2 will be lost, but what will remain is a hot, dense CO_2 atmosphere with no surface water, most likely H_2SO_4 and hydrochloric acid clouds.

Acknowledgments. This chapter benefited from many discussions with K. Baines, G. Hashimoto, C. Tsang, F. Taylor, C. Wilson, P. Drossart, H. Svedhem, and many members of the VEX science team. The authors gratefully acknowledge the support of the National Science Foundation through a Planetary Astronomy research grant to M.A.B. Support for the climate sensitivity work was provided by a NASA Planetary Atmospheres grant to M.A.B. Support for D.H.G. was provided by a Library of Congress Fellowship.

REFERENCES

Albritton D. L., et al. (2001) *Climate Change: The Scientific Basis, Summary for Policymakers, A Report of Working Group I of the Intergovernmental Panel on Climate Change.* Cambridge Univ., Cambridge. 20 pp.

Alexander M. J. (1992) A mechanism for the Venus thermospheric superrotation. *Geophys. Res. Lett., 19,* 2207–2210.

Allen D. A. and Crawford J. W. (1984) Cloud structure on the dark side of Venus. *Nature, 307,* 222–224.

Avduevskii V. S., Borodin N. F., Burtsev V. P., Malkov I. V., Marov M. I., Morozov S. F., Rozhdestvenskii M. K., Romanov R. S., Sokolov S. S., and Fokin V. G. (1976) Automatic probes Venera 9 and Venera 10 — Functioning of descent vehicles and measurement of atmospheric parameters. *Kosmicheskie Issledovaniia, 14,* 655–666.

Barabash S., Sauvaud J. A., Gunell H., Andersson H., Grigoriev A., Brinkfeldt K., Holmström M., Lundin R., Yamauchi M., Asamura K., Baumjohann W., Zhang T. L., Coates A. J., Linder D. R., Kataria D. O., Curtis C. C., Hsieh K. C., Sandel B. R., Fedorov A., Mazelle C., Thocaven J. J., Grande M., Koskinen H. E. J., Kallio E., Säles T., Riihela P., Kozyra J., Krupp N., Woch J., Luhmann J., McKenna-Lawlor S., Orsini S., Cerulli-Irelli R., Mura M., Milillo M., Maggi M., Roelof E., Brandt P., Russell C. T., Szego K., Winningham J. D., Frahm R. A., Scherrer J., Sharber J. R., Wurz P., and Bochsler P. (2007) The Analyser of Space Plasmas and Energetic Atoms (ASPERA-4) for the Venus Express mission. *Planet. Space Sci., 55,* 1772–1792.

Barath F. T., Barrett A. H., Copeland J., Jones D. C., and Lilley A. E. (1963) Microwave radiometers, part of Mariner II: Preliminary reports on measurements of Venus. *Science, 139,* 908–909.

Barstow J. K., Tsang C. C. C., Wilson C. F., Irwin P. G. J., Taylor F. W, McGouldrick K., Drossart P., Piccioni G., and Tellmann S. (2012) Models of the global cloud structure on Venus derived from Venus Express observations. *Icarus, 217,* 542–560.

Bell J. F., Crisp D., Lucey P. G., Ozorosky T. A., Sinton W. A., Willis S. C., and Campbell B. A. (1991) Spectroscopic observations of bright and dark emission features on the night side of Venus. *Science, 252,* 1293–1296.

Belton M. J. S., Smith G., Schubert G., and Del Genio A. D. (1976) Cloud patterns, waves and convection in the Venus atmosphere. *J. Atmos. Sci., 33,* 1394–1417.

Belyaev D., Korablev O., Fedorov A., Bertaux J. L., Vandaele A. C., Montmessin F., Mahieux A., Wilquet V., and Drummond R. (2008) First observations of SO_2 above Venus' clouds by means of solar occultation in the infrared. *J. Geophys. Res., 113(E5),* DOI: 10.1029/2008JE003143.

Belyaev D. A., Montmessin F., Bertaux J.-L., Mahieux A., Fedorova A. A., Korablev O. I., Marcq E., Yung Y. L., and Zhang X. (2012) Vertical profiling of SO_2 and SO above Venus' clouds by SPICAV/SOIR solar occultations. *Icarus, 217,* 740–751.

Bengtsson L., Bonnet R. M., Grinspoon D. H., Koumoutsaris S., Lebonnois S., and Titov D. (2013) *Towards Understanding the Climate of Venus.* Springer, Berlin. 185 pp.

Bertaux J. L., Widemann T., Hauchecorne A., Moroz V. I., and Ekonomov A. P. (1996) Vega-1 and Vega-2 entry probes: An investigation of UV absorption (220–400 nm) in the atmosphere of Venus. *J. Geophys. Res., 101,* 12709–12745.

Bertaux J.-L., Vandaele A.-C., Korablev O., Villard E., Fedorova A., Fussen D., Quemerais E., Belyaev D., Mahieux A., Montmessin F., Muller C., Neefs E., Nevejans D., Wilquet V., Dubois J. P., Hauchecorne A., Stepanov A., Vinogradov I., and Rodin A. (2007) A warm layer in Venus' cryosphere and high-altitude measurements of HF, HCl, H_2O and HDO. *Nature, 450,* 646–649.

Bezard B., de Bergh C., Crisp D., and Maillard J.-P. (1990) The deep atmosphere of Venus revealed by high-resolution nightside spectra. *Nature, 345,* 508–511.

Bezard B., Tsang C. C. C., Carlson R. W., Piccioni G., Marcq E., and Drossart P. (2009) Water vapor abundance near the surface of Venus from Venus Express/VIRTIS observations. *J. Geophys. Res.–Planets, 114,* E00B39.

Bougher S. W., Rafkin S., and Drossart P. (2006) Dynamics of the Venus upper atmosphere: Outstanding problems and new constraints expected from Venus Express. *Planet. Space Sci., 54,* 1371–1380.

Brain D. A. and Jakosky B. M. (1998) Atmospheric loss since the onset of the martian geologic record: Combined role of impact erosion and sputtering. *J. Geophys. Res., 103,* 22689–22694.

Brecht A. S., Bougher S. W., Gerard J. C., and Soret L. (2012) Atomic oxygen distributions in the Venus thermosphere: Comparisons between Venus Express observations and global model simulations. *Icarus, 217,* 759–766.

Bullock M. A. and Grinspoon D. H. (1996) The stability of climate on Venus. *J. Geophys. Res., 101,* 7521–7529.

Bullock M. A. and Grinspoon D. H. (2001) The recent evolution of climate on Venus. *Icarus, 150,* 19–37.

Bullock M. A., Grinspoon D. H., and Head J. W. (1993) Venus resurfacing rates: Constraints provided by 3-D Monte Carlo simulations. *Geophys. Res. Lett., 20,* 2147–2150.

Canup R. M. and Asphaug E. (2001) Origin of the Moon in a giant impact near the end of the Earth's formation. *Nature, 412,* 708–712.

Carlson R. W., Kamp L. W., Baines K. H., Pollack J. B., Grinspoon D. H., Encrenaz T., Drossart P., Lellouch E., and Bezard B. (1993) Variations in Venus cloud particle properties: A new view of Venus's cloud morphology as observed by the Galileo near-infrared mapping spectrometer. *Planet. Space Sci., 41,* 477–486.

Carpenter I. R. L. (1966) Study of Venus by CW radar — 1964 results. *Astron. J., 71,* 142.

Carpenter R. L. (1964) Symposium on radar and radiometric observations of Venus during the 1962 conjunction: Study of Venus by CW radar. *Astron. J., 69,* 2.

Chamberlain S., Bailey J., Crisp D., and Meadows V. (2013) Ground-based near-infrared observations of water vapor in the Venus troposphere. *Icarus, 222,* 364–378.

Chassefiere E., Wieler R., Marty B., and Leblanc F. (2012) The evolution of Venus: Present state of knowledge and future exploration. *Planet. Space Sci., 63,* 15–23.

Chaufray J. Y., Bertaux J. L., Quemerais E., Villard E., and Leblanc F. (2012) Hydrogen density in the dayside venusian exosphere derived from Lyman-α observations by SPICAV on Venus Express. *Icarus, 217,* 767–778.

Clancy R. T., Sandor B. J., and Moriarty-Schieven G. (2012a) Circulation of the Venus upper mesosphere/lower thermosphere: Doppler wind measurements from 2001–2009 inferior conjunction, sub-millimeter CO absorption line observations. *Icarus, 217,* 794–812.

Clancy R. T., Sandor B. J., and Moriarty-Schieven G. (2012b) Thermal structure and CO distribution for the Venus mesosphere/lower thermosphere: 2001–2009 inferior conjunction sub-millimeter CO absorption line observations. *Icarus, 217,* 779–793.

Coffin M. F. and Eldholm O. (1994) Large igneous provinces: Crustal structure, dimension, and external consequences. *Rev. Geophys., 32,* 1–36.

Correia A. C. M. and Laskar J. (2001) The four final rotation states of Venus. *Nature, 411,* 767–770.

Cottereau L., Rambaux N., Lebonnois S., and Souchay J. (2011) The various contributions in Venus rotation rate and LOD. *Astron. Astrophys., 531,* 45.

Cottini V., Ignatiev N. I., Piccioni G., Drossart P., Grassi D., and Markiewicz W. J. (2012) Water vapor near the cloud tops of Venus from Venus Express/VIRTIS dayside data. *Icarus, 217,* 561–569.

Cotton D. V., Bailey J., Crisp D., and Meadows V. S. (2012) The distribution of carbon monoxide in the lower atmosphere of Venus. *Icarus, 217,* 570–584.

Crisp D. and Titov D. (1997) The thermal balance of the Venus atmosphere. In *Venus II* (S. W. Bougher et al., eds.), pp. 353–384. Univ. of Arizona, Tucson.

Crisp D., Meadows V. S., Bezard B., de Bergh C., Maillard J. P., and Mills F. P. (1996) Ground-based near-infrared observations of the Venus nightside: 1.27-mm $O_2(a\Delta g)$ airglow from the upper atmosphere. *J. Geophys. Res., 101,* 4577–4593.

de Bergh C., Bezard B., Owen T., Crisp D., Maillard J. P., and Lutz B. L. (1991) Deuterium on Venus: Observations from Earth. *Science, 251,* 547–549.

de Bergh C., Bezard B., Crisp D., Maillard J. P., Owen T., Pollack J. B., and Grinspoon D. H. (1995) Water in the deep atmosphere of Venus from high-resolution spectra of the night side. *Adv. Space Res., 15, 4,* 79–88.

De Duve C. (1995) *Vital Dust: Life as a Cosmic Imperative.* Basic, New York. 362 pp.

Delva M., Zhang T. L., Volwerk M., Voros Z., and Pope S. A. (2008) Proton cyclotron waves in the solar wind at Venus. *J. Geophys. Res.–Planets, 113(E9),* DOI: 10.1029/2008JE003148.

Dobrovolskis A. R. (1980) Atmospheric tides and the rotation of Venus II. Spin evolution. *Icarus, 41,* 18–35.

Donahue T. M. and Hodges R. R. (1992) Past and present water budget of Venus. *J. Geophys. Res., 97,* 6083–6091.

Donahue T. M., Hoffman J. H., Hodges R. R., and Watson A. J. (1982) Venus was wet: A measurement of the ratio of D to H. *Science, 216,* 630–633.

Dreibus G. and Wanke H. (1989) Supply and loss of volatile constituents during the accretion of terrestrial planets. In *Origin and Evolution of Planetary and Satellite Atmospheres* (S. K. Atreya et al., eds.), pp. 487–512. Univ. of Arizona, Tucson.

Drossart P., Piccioni G., Adriani A., Angrilli F., Arnold G., Baines K. H., Bellucci G., Benkhoff J., Bézard B., Bibring J. P., Blanco A., Blecka M. I., Carlson R. W., Coradini A., Di Lellis A., Encrenaz T., Erard S., Fonti S., Formisano V., Fouchet T., Garcia R., Haus R., Helbert J., Ignatiev N. I., Irwin P. G. J., Langevin Y., Lebonnois S., Lopez-Valverde M. A., Luz D., Marinangeli L., Orofino V., Rodin A. V., Roos-Serote M. C., Saggin B., Sanchez-Lavega A., Stam D. M., Taylor F. W., Titov D., Visconti G., Zambelli M., Hueso R., Tsang C. C. C., Wilson C. F., and Afanasenko T. Z. (2007) Scientific goals for the observation of Venus by VIRTIS on ESA/Venus Express mission. *Planet. Space Sci., 55,* 1653–1672.

Erard S., Drossart P., and Piccioni G. (2009) Multivariate analysis of Visible and Infrared Thermal Imaging Spectrometer (VIRTIS) Venus Express nightside and limb observations. *J. Geophys. Res., 114,* E00B27.

Esposito L. W. (1980) Ultraviolet contrasts and the absorbers near the Venus cloud tops. *J. Geophys. Res., 85,* 8151–8157.

Esposito L. W. (1984) Sulfur dioxide: Episodic injection shows evidence for active Venus volcanism. *Science, 223,* 1072–1074.

Esposito L. W., Knollenberg R. G., Marov M. Y., Toon O. B., and Turco R. P. (1983) The clouds and hazes of Venus. In *Venus* (D. M. Hunten et al., eds.), pp. 484–564. Univ. of Arizona, Tucson.

Esposito L. W., Bertaux J. L., Krasnopolsky V. A., Moroz V. I., and Zasova L. V. (1997) Chemistry of lower atmosphere and clouds. In *Venus II* (S. W. Bougher et al., eds.), pp. 415–458. Univ. of Arizona, Tucson.

Federova A., Korablev O., Vandaele A. C., Bertaux J. L., Belyaev D., Mahieux A., Neefs E., Wilquet V., Drummond R., Montmessin F., and Villard E. (2008) HDO and H_2O vertical distributions and isotopic ratio in Venus mesopshere by SOIR spectrometer on board Venus Express. *J. Geophys. Res., 113,* DOI: 10.1029/2008JE003146.

Fegley B. and Prinn R. G. (1989) Estimation of the rate of volcanism on Venus from reaction rate measurements. *Nature, 337,* 55–58.

Fegley B. and Treiman A. H. (1992) Chemistry of atmosphere-surface interactions on Venus and Mars. In *Venus and Mars: Atmospheres, Ionospheres and Solar Wind Interactions* (J. G. Luhmann et al., eds.), pp. 7–71. AGU Geophys. Monogr. 66, American Geophysical Union, Washington, DC.

Fegley B., Klingelhofer G., Lodders K., and Widemann T. (1997) Geochemistry of surface-atmosphere interactions on Venus. In *Venus II* (S. W. Bougher et al., eds.), pp. 591–636. Univ. of Arizona, Tucson.

Ford P. G. and Pettengill G. H. (1992) Venus topography and kilometer-scale slopes. *J. Geophys. Res., 97,* 13103–13114.

Galli A., Fok M. C., Wurz P., Barabash S., Grigoriev A., Futaana Y., Holmstrom M., Ekenback A., Kallio E., and Gunell H. (2008) Tailward flow of energetic neutral atoms observed at Venus. *J. Geophys. Res.–Planets, 113(E9),* DOI: 10.1029/2008JE003096.

Gerard J. C., Cox C., Saglam A., Bertaux J. L., Villard E., and Nehme C. (2008) Limb observations of the ultraviolet nitric oxide nightglow with SPICAV on board Venus Express. *J. Geophys. Res.—Planets, 113,* E00B03.

Gerard J. C., Soret L., Piccioni G., and Drossart P. (2012) Spatial correlation of OH Meinel and O_2 infrared atmospheric nightglow emissions observed with VIRTIS-M on board Venus Express. *Icarus, 217,* 813–817.

Gierasch P. J. (1975) Meridional circulation and the maintenance of the Venus atmospheric circulation. *J. Atmos. Sci., 32,* 1038–1044.

Gierasch P. J., Goody R. M., Young R. E., Crisp D., Edwards C., Kahn R., Rider D., Del Genio A., Greeley R., Hou A., Leovy C. B., McCleese D., and Newman M. (1997) The general circulation of the Venus atmosphere: An assessment. In *Venus II* (S. W. Bougher et al., eds.), pp. 459–500. Univ. of Arizona, Tucson.

Gilli G., Lopez-Valverde M. A., Drossart P., Piccioni G., Erard S., and Moinelo C. (2008) Limb observations of CO_2 and CO non-LTE emissions in the Venus atmosphere by VIRTIS/Venus Express. *J. Geophys. Res., 113,* DOI: 10.1029/2008JE003112.

Gough D. O. (1981) Solar interior structure and luminosity variations. *Solar Phys., 74,* 21–34.

Grassi D., Drossart P., Piccioni G., Ignatiev N. I., Zasova L. V., Adriani A., Moriconi M. L., Irwin P. G. J., Negrao A., and Migliorini A. (2008) Retrieval of air temperature profiles in the venusian mesosphere from VIRTIS-M data: Description and validation of algorithms. *J. Geophys. Res.–Planets., 113(E9),* DOI: 10.1029/2008JE003075.

Grassi D., Migliorini A., Montabone L., Lebonnois S., Cardesìn-Moinelo A., Piccioni G., Drossart P., and Zasova L. V. (2010) Thermal structure of Venusian nighttime mesosphere as observed by VIRTIS-Venus Express. *J. Geophys. Res.–Planets, 115,* E09007.

Grebowsky J. M., Strangeway R. J., and Hunten D. M. (1997) Evidence for Venus lightning. In *Venus II* (S. W. Bougher et al., eds.), pp. 125–157. Univ. of Arizona, Tucson.

Grinspoon D. H. (1993) Implications of the high D/H ratio for the sources of water in Venus' atmosphere. *Nature, 363,* 428–431.

Grinspoon D. H. and Bullock M. A. (2007) Searching for evidence of past oceans on Venus. *Bull. Am. Astron. Soc., 39,* 540.

Grinspoon D. H., Pollack J. B., Sitton B. R., Carlson R. W., Kamp L. W., Baines K. H., Encrenaz T., and Taylor F. W. (1993) Probing Venus' cloud structure with Galileo NIMS. *Planet. Space Sci., 41,* 515–542.

Hansell S., Wells S. A., and Hunten D. M. (1995) Optical detection of lightning on Venus. *Icarus, 117,* 345–351.

Hansen J. E. and Travis L. D. (1974) Light scattering in planetary atmospheres. *Space Sci. Rev., 16,* 527–610.

Hashimoto G. L. and Abe Y. (2000) Stabilization of Venus' climate by a chemical-albedo feedback. *Earth Planets Space, 52,* 197–202.

Hashimoto G. L. and Abe Y. (2001) Predictions of a simple cloud model for water vapor cloud albedo feedback on Venus. *J. Geophys. Res., 106,* 14675–14690.

Hashimoto G. L., Abe Y., and Sasaki S. (1997) CO_2 amount on Venus constrained by a criterion of topographic greenhouse instability. *Geophys. Res. Lett., 24,* 289–292.

Head J. W., Crumpler L. S., Aubele J. C., Guest J. E., and Saunders R. S. (1992) Venus volcanism: Classification of volcanic features and structures, associations, and global distribution from Magellan data. *J. Geophys. Res., 97,* 13153–13198.

Hoffman P. F., Kaufman A. J., Halverson G. P., and Schrag D. P. (1998) A neoproterozoic snowball Earth. *Science, 281,* 1342–1346.

Hollingsworth J. L., Young R. E., Schubert G., Covey C., and Grossman A. S. (2007) A simple-physics global circulation model for Venus: Sensitivity assessments of atmospheric superrotation. *Geophys. Res. Lett., 34,* DOI: 10.1029/2006GL028567.

Hoshino N., Fujiwara H., Takagi M., Takahashi Y., and Kasaba Y. (2012) Characteristics of planetary-scale waves simulated by a new venusian mesosphere and thermosphere general circulation model. *Icarus, 217,* 818–830.

Houghton J. T., Jenkins G. J., and Ephraums J. J. (1990) *Climate Change: The IPCC Scientific Assessment.* Cambridge Univ., Cambridge. 365 pp.

Hueso R., Sánchez-Lavega A., Piccioni G., Drossart P., Gerard J. C., Khatuntsev I., Zasova L., and Migliorini A. (2008) Morphology and dynamics of Venus oxygen airglow from Venus Express/Visible and Infrared Thermal Imaging Spectrometer observations. *J. Geophys. Res.–Planets, 113(E5),* DOI: 10.1029/2008JE003081.

Hueso R., Peralta J., and Sánchez-Lavega A. (2012) Assessing the long-term variability of Venus winds at cloud level from VIRTIS-Venus Express. *Icarus, 217,* 585–598.

Ignatiev N. I., Titov D. V., Piccioni G., Drossart P., Markiewicz W. J., Cottini V., Roatsch T., Almeida M., and Manoel N. (2009) Altimetry of the Venus cloud tops from the Venus Express observations. *J. Geophys. Res., 114,* E00B43.

Imamura T. and Hashimoto G. L. (1998) Venus cloud formation in the meridional circulation. *J. Geophys. Res., 103,* 31349–31366.

Ingersoll A. P. (1969) The runaway greenhouse: A history of water on Venus. *J. Atmos. Sci., 26,* 1191–1198.

Irwin P. G. J., Teanby N. A., de Kok R., Fletcher L. N., Howett C. J. A., Tsang C. C. C., Wilson C. F., Calcutt S. B., Nixon C. A., and Parrish P. D. (2008) The NEMESIS planetary atmosphere radiative transfer and retrieval tool. *J. Quant. Spectrosc. Radiat. Transfer, 109,* 1136–1150.

Jakosky B. M. (2011) The 2013 Mars Atmosphere and Volatile Evolution (MAVEN) mission to Mars. Abstract P23E-01 presented at 2011 Fall Meeting, AGU, San Francisco, California.

James E. P., Toon O. B., and Schubert G. (1997) A numerical microphysical model of the condensational Venus cloud. *Icarus, 129,* 147–171.

Jenkins J. M., Steffes P. G., Hinson D. P., Twicken J. D., and Tyler G. L. (1994) Radio occultation studies of the Venus atmosphere with the Magellan spacecraft 2. Results from the October-1991 experiments. *Icarus, 110,* 79–94.

Jenkins J. M., Kolodner M. A., Butler B. J., Suleiman S. H., and Steffes P. G. (2002) Microwave remote sensing of the temperature and distribution of sulfur compounds in the lower atmosphere of Venus. *Icarus, 158,* 312–328.

Kasting J. F. (1988) Runaway and moist greenhouse atmospheres and the evolution of Earth and Venus. *Icarus, 74,* 472–494.

Kasting J. F., Pollack J. B., and Ackerman T. P. (1984) Response of Earth's atmosphere to increases in solar flux and implications for loss of water from Venus. *Icarus, 57,* 335–355.

Kasting J. F., Toon O. B., and Pollack J. B. (1988) How climate evolved on the terrestrial planets. *Sci. Am., 258,* 90–97.

Kaula W. M., Head J. W., Merrill R. B., Pepin R. O., Solomon S. C., Walker D., and Wood C. A. (1981) *Basaltic Volcanism on the Terrestrial Planets.* Pergamon, New York. 1286 pp.

Kliore A. J. and Patel I. R. (1980) Vertical structure of the atmosphere of Venus from Pioneer Venus orbiter radio occultations. *J. Geophys. Res., 85,* 7957–7962.

Kliore A., Moroz V. I., and Keating G. M. (1986) *The Venus International Reference Atmosphere.* Pergamon, Oxford.

Klose K. B., Wood J. A., and Hashimoto A. (1992) Mineral equilibria and the high radar reflectivity of Venus mountaintops. *J. Geophys. Res., 97,* 16353–16369.

Knollenberg R. G. and Hunten D. M. (1980) The microphysics of the clouds of Venus: Results of the Pioneer Venus particle size spectrometer experiment. *J. Geophys. Res., 85,* 8039–8058.

Krasnopolsky V. A. (2007) Chemical kinetic model for the lower atmosphere of Venus. *Icarus, 191,* 25–37.

Krasnopolsky V. A. and Pollack J. B. (1994) H_2O-H_2SO_4 system in Venus' clouds and OCS, CO and H_2SO_4 profiles in Venus' troposphere. *Icarus, 109,* 58–78.

Lacis A. A. and Oinas V. (1991) A description of the correlated k method for modeling nongray gaseous absorption, thermal emission and multiple scattering in vertically inhomogeneous atmospheres. *J. Geophys. Res., 96,* 9027–9063.

Lebonnois S., Hourdin F., Eymet V., Crespin A., Fournier R., and Forget F. (2010) Superrotation of Venus' atmosphere analyzed with a full general circulation model. *J. Geophys. Res.–Planets, 115,* 6006.

Lee C., Lewis S. R., and Read P. L. (2005) A numerical model of the atmosphere of Venus. *Adv. Space Res., 36,* 2142–2145.

Lee Y. J., Titov D. V., Tellmann S., Piccialli A., Ignatiev N., Pätzold M., Häusler B., Piccioni G., and Drossart P. (2012) Vertical structure of the Venus cloud top from the VeRa and VIRTIS observations onboard Venus Express. *Icarus, 217,* 599–609.

Leshin L. A., Epstein S., and Stolper E. M. (1996) Hydrogen isotope geochemistry of SNC meteorites. *Geochem. Cosmochim. Acta, 60,* 2635–2650.

Lewis J. S. (1974) Volatile element influx on Venus from cometary impacts. *Earth Planet. Sci. Lett., 22,* 239–244.

Liang M. C. and Yung Y. L. (2009) Modeling the distribution of H_2O and HDO in the upper atmosphere of Venus. *J. Geophys. Res., 113,* E00B28.

Limaye S. S., Kossin J. P., Rozoff C., Piccioni G., Titov D. V., and Markiewicz W. J. (2009) Vortex circulation on Venus: Dynamical similarities with terrestrial hurricanes. *Geophys. Res. Lett., 36,* 04204.

Linkin V. M., Kerzhanovich V. V., Lipatov A. N., Pichkadze K. M., Shurupov A. A., Terterashvili A. V., Ingersoll A. P., Crisp D., Grossman A. W., Young R. E., Seiff A., Ragent B., Blamont J. E., Elson L. S., and Preston R. A. (1986) Vega balloon dynamics and vertical winds in the Venus middle cloud region. *Science, 231,* 1417–1419.

Lowell P. (1909) Mars 1909. *Lowell Obs. Bull., 1,* 219–219.

Luhmann J. G., Fedorov A., Barabash S., Carlsson E., Futaana Y., Zhang T. L., Russell C. T., Lyon J. G., Ledvina A., and Brain D. A. (2008) Venus Express observations of atmospheric oxygen escape during the passage of several coronal mass ejections. *J. Geophys. Res.–Planets, 113(E9),* E00B04, DOI: 10.1029/2008JE003092.

Luz D., Berry D. L., Piccioni G., Drossart P., Politi R., Wilson C. F., Erard S., and Nuccilli F. (2011) Venus's southern polar vortex reveals precessing circulation. *Science, 332,* 577–580.

Marcq E., Bezard B., Drossart P., Piccioni G., Reess J. M., and Henry F. (2008) A latitudinal survey of CO, OCS, H_2O, and SO_2 in the lower atmosphere of Venus: Spectroscopic studies using VIRTIS-H. *J. Geophys. Res.–Planets, 113,* DOI: 10.1029/2008JE003074.

Margot J.-L. (2011) Probing planetary interior structure and processes with high-precision spin measurements. *Bull. Am. Astron. Soc., 43.*

Markiewicz W. J., Titov D. V., Limaye S. S., Keller H. U., Ignatiev N., Jaumann R., Thomas N., Michalik H., Moissl R., and Russo P. (2007) Morphology and dynamics of the upper cloud layer of Venus. *Nature, 450,* 633–636.

Marov M. Y., Avdnevsky V. S., Kerzhanovich V. V., Rozhdestvensky M. K., Borodin N. F., and Ryabov O. L. (1973) Venera 8: Measurements of temperature, pressure, and wind velocity on the illuminated side of Venus. *J. Atmos. Sci., 30,* 1210–1214.

Marov M. Y., Lystsev V. E., Lebedev V. N., Lukashevich N. L., and Shari V. P. (1980) The structure and microphysical properties of the Venus clouds: Venera 9, 10, and 11 data. *Icarus, 44,* 608–639.

Matsui H., Iwagami N., Hosouchi M., Ohtsuki S., and Hashimoto G. L. (2012) Latitudinal distribution of HDO abundance above Venus' clouds by ground-based 2.3 μm spectroscopy. *Icarus, 217,* 610–614.

Mayer C. H., McCullough T. P., and Sloanaker R. M. (1958) Observations of Venus at 3.15 cm wavelength. *Astrophys. J., 127,* 1–10.

McGouldrick K. and Toon O. B. (2008) Observable effects of convection and gravity waves on the Venus condensational cloud. *Planet. Space Sci., 56,* 1112–1131.

McGouldrick K., Toon O. B., and Grinspoon D. H. (2011) Sulfuric acid aerosols in the atmospheres of the terrestrial planets. *Planet. Space Sci., 59,* 934–941.

McGouldrick K., Momary T. W., Baines K. H., and Grinspoon D. H. (2012) Quantification of middle and lower cloud variability and mesoscale dynamics from Venus Express/VIRTIS observations at 1.74 μm. *Icarus, 217,* 615–628.

McKinnon W. B., Zahnle K. J., Ivanov B. A., and Melosh H. J. (1997) Cratering on Venus: Models and observations. In *Venus II* (S. W. Bougher et al., eds.), pp. 969–1014. Univ. of Arizona, Tucson.

Meadows V. S. and Crisp D. (1996) Ground-based near-infrared observations of the Venus night side: The thermal structure and water abundance near the surface. *J. Geophys. Res., 101,* 4595–4622.

Migliorini A., Grassi D., Montabone L., Lebonnois S., Drossart P., and Piccioni G. (2012) Investigation of air temperature on the nightside of Venus derived from VIRTIS-H on board Venus-Express. *Icarus, 217,* 640–647.

Mills F. P., Esposito L. W., and Yung Y. L. (2007) Atmospheric composition, chemistry, and clouds. In *Exploring Venus as a Terrestrial Planet* (L. W. Esposito et al., eds.), pp. 73–100. AGU Geophys. Monogr. 176, American Geophysical Union, Washington, DC.

Moissl R., Khatuntsev I., Limaye S. S., Titov D., Markiewicz W. J., Ignatiev N., Roatsch T., Matz K. D., Jaumann R., Almeida M., Portyankina G., Behnke T., and Hviid S. (2008) Venus cloud top winds from tracking UV features in VMC images. *J. Geophys. Res., 113,* E00B01, DOI: 10.1029/2008JE003087.

Morbidelli A. (2002) Modern integrations of solar system dynamics. *Annu. Rev. Earth Planet. Sci., 30,* 89–112.

Morbidelli A., Chambers J., Lunine J. I., Petit J. M., Robert F., Valsecchi G. B., and Cyr R. E. (2000) Source regions and timescales for the delivery of water to Earth. *Meteoritics & Planet. Sci., 35,* 1309–1320.

Moroz V. I., Moshkin B. E., Ekonomov A. P., San'ko N. F., Parfent'ev N. A., and Golovin Y. M. (1978) *Spectrophotometric Experiment On Board the Venera-11, -12 Descenders: Some Results of the Analysis of the Venus Day-Sky Spectrum.* Publ. Space Res. Institute Academy Sci., Leningrad.

Moroz V. I., Ekonomov A. P., Moshkin B. E., Revercomb H. E., Sromovsky L. A., and Schofield J. T. (1985) Solar and thermal radiation in the Venus atmosphere. *Adv. Space Res., 5,* 197–232.

Mouginis-Mark P. J., Gaddis L. R., Blake P. L., Fryer P., and Ferrall C. (1985) Planning for VRM: Radar and sonar studies of volcanic terrains on Earth, Venus and Mars (abstract). In *Terrestrial Planets: Comparative Planetology,* p. 18. LPI Contrib. No. 569, Lunar and Planetary Institute, Houston.

Mueller N., Helbert J., Hashimoto G. L., Tsang C. C. C., Erard S., Piccioni G., and Drossart P. (2008) Venus surface thermal emission at one micrometer in VIRTIS imaging observation — Evidence for variation of crust and mantle differentiation conditions. *J. Geophys. Res., 1117,* DOI: 10.1117/1.JRS.6.063580.

Mueller N. T., Helbert J., Erard S., Piccioni G., and Drossart P. (2012) Rotation period of Venus estimated from Venus Express VIRTIS images and Magellan altimetry. *Icarus, 217,* 474–483.

Muhleman D. O. (1961) Early results of the 1961 JPL Venus Radar Experiment. *Astronom. J., 66,* 292.

Nakagawa H., Hoshino N., Sornig M., Kasaba Y., Sonnabend G., Stupar D., Aoki S., and Murata I. (2013) Comparison of general circulation model atmospheric wave simulations with wind observations of Venusian mesosphere. *Icarus, 225,* 840–849.

Nature Editorial Board (1967) News and Views: More news from Venus. *Nature, 216,* 427–428.

Owen T. (1992) The composition and early history of the atmosphere of Mars. In *Mars* (H. H. Kieffer et al., eds.), pp. 818–834. Univ. of Arizona, Tucson.

Owen T. and Bar-Nun A. (1995) Comets, impacts and atmospheres. *Icarus, 116,* 215–226.

Oyama V. I., Carle G. C., Woeller F., and Pollack J. B. (1979) Venus lower atmosphere composition: Analysis by gas chromatography. *Science, 203,* 802–804.

Oyama V. I., Carle G. C., Woeller F., Pollack J. B., Reynolds R. T., and Craig R. A. (1980) Pioneer Venus gas chromatography of the lower atmosphere of Venus. *J. Geophys. Res., 85,* 7891–7902.

Palmer K. F. and Williams D. (1975) Optical constants of sulfuric acid: Application to the clouds of Venus? *Appl. Opt., 14,* 208–219.

Patzold M., Hausler B., Bird M. K., Tellmann S., Mattei R., Asmar S. W., Dehant V., Eidel W., Imamura T., Simpson R. A., and Tyler G. L. (2007) The structure of Venus' middle atmosphere and ionosphere. *Nature, 450,* 657–660.

Peixoto J. P. and Oort A. H. (1992) *Physics of Climate.* American Institute of Physics, New York. 520 pp.

Peralta J., Hueso R., Sanchez-Lavega A., Piccioni G., Lanciano O., and Drossart P. (2008) Characterization of mesoscale gravity waves in the upper and lower clouds of Venus from VEX-VIRTIS images. *J. Geophys. Res., 113,* E00B18, DOI: 10.1029/2008JE003185.

Piccialli A., Titov D. V., Grassi D., Khatuntsev I., Drossart P., Piccioni G., and Migliorini A. (2008) Cyclostrophic winds from the Visible and Infrared Thermal Imaging Spectrometer temperature sounding: A preliminary analysis. *J. Geophys. Res.–Planets, 113,* DOI: 10.1029/2008JE003127.

Piccialli A., Tellmann S., Titov D. V., Limaye S. S., Khatuntsev I. V., Patzold M., and Hausler B. (2012) Dynamical properties of the Venus mesosphere from the radio-occultation experiment VeRa onboard Venus Express. *Icarus, 217,* 669–681.

Piccioni G., Drossart P., Sanchez-Lavega A., Hueso R., Taylor F. W., Wilson C. F., Grassi D., Zasova L., Moriconi M., Adriani A., Lebonnois S., Coradini A., Bezard B., Angrilli F., Arnold G., Baines K. H., Bellucci G., Benkhoff J., Bibring J. P., Blanco A., Blecka M. I., Carlson R. W., Di Lellis A., Encrenaz T., Erard S., Fonti S., Formisano V., Fouchet T., Garcia R., Haus R., Helbert J., Ignatiev N. I., Irwin P. G. J., Langevin Y., Lopez-Valverde M. A., Luz D., Marinangeli L., Orofino V., Rodin A. V., Roos-Serote M. C., Saggin B., Stam D. M., Titov D., Visconti G., and Zambelli M. (2007) South-polar features on Venus similar to those near the north pole. *Nature, 450,* 637–640.

Pieters C. M., Head J. W., Patterson W., Pratt S., Garvin J., Barsukov V. L., Basilevsky A. T., Khodakovsky I. L., Selivanov A. S., Panfilov A. S., Gektin Y. M., and Narayeva Y. M. (1986) The color of the surface of Venus. *Science, 234,* 1379–1383.

Pollack J. B. (1969a) A nongray CO_2-H_2O greenhouse model of Venus. *Icarus, 10,* 314–341.

Pollack J. B. (1969b) Temperature structure of nongray planetary atmospheres. *Icarus, 10,* 301–313.

Pollack J. B., Dalton J. B., Grinspoon D. H., Wattson R. B., Freedman R., Crisp D., Allen D. A., Bezard B., de Bergh C., Giver L. P., Ma Q., and Tipping R. H. (1993) Near infrared light from Venus' nightside: A spectroscopic analysis. *Icarus, 103,* 1–42.

Pollack J. B., Toon O. B., and Boese R. (1980) Greenhouse models of Venus' high surface temperature, as constrained by Pioneer Venus measurements. *J. Geophys. Res., 85,* 8223–8231.

Prinn R. G. (1985) The photochemistry of the atmosphere of Venus. In *The Photochemistry of Atmospheres* (J. S. Levine, ed.), pp. 281–336. Academic, New York.

Pruppacher H. R. and Klett J. D. (1997) *Microphysics of Clouds and Precipitation.* Kluwer, Dordrecht. 954 pp.

Ragent B. and Blamont J. E. (1980) The structure of the clouds of Venus: Results of the Pioneer Venus nephelometric experiment. *J. Geophys. Res., 85,* 8089–8105.

Ramanathan V. (1987) The role of Earth radiation budget studies in climate and general circulation research. *J. Geophys. Res.,*

92, 4075–4095.

Rayner J. T., Toomey D. W., Onaka P. M., Denault A. J., Stahlberger W. E., Vacca W. D., Cushing M. C., and Wang S. (2003) SpeX: A medium-resolution 0.8–5.5 micron spectrograph and imager for the NASA Infrared Telescope Facility. *Publ. Astron. Soc. Pac., 115,* 362–382.

Rothman L. S., Gamache R. R., Tipping R. H., Rinsland C. P., Smith M. A. H., Benner D. C., Devi V. M., Flaud J. M., Camy-Peyret C., Perrin A., Goldman A., Massie S. T., Brown L. R., and Toth R. A. (1992) The HITRAN molecular database: Editions of 1991 and 1992. *J. Quant. Spectrosc. Radiat. Transfer, 48,* 469–507.

Rothman L. S., Gordon I. E., Barber R. J., Dothe H., Gamache R. R., Goldman A., Perevalov V. I., Tashkun S. A., and Tennyson J. (2010) HITEMP, the high-temperature molecular spectroscopic database. *J. Quant. Spectrosc. Radiat. Transfer, 111,* 2139–2150.

Rothman L. S., Barbe A., Benner D. C., Brown L. R., Camy-Peyret C., Carleer M. R., Chance K. V., Clerbaux C., Dana V., Devi V. M., Fayt A., Flaud J. M., Gamache R. R., Goldman A., Jacquemart D., Jucks K. W., Lafferty W. J., Mandin J. Y., Massie S. T., Nemtchinov V., Newnham D. A., Perrin A., Rinsland C. P., Schroeder J., Smith K. M., Smith M. A. H., Tang K., Toth R. A., Auwera J. V., Varanasi P., and Yoshino K. (2003) The HITRAN molecular spectroscopic database: Edition of 2000 including updates through 2001. *J. Quant. Spectrosc. Radiat. Transfer, 82,* 5–44.

Rothman L. S., Jacquemart D., Barbe A., Benner D. C., Birk M., Brown L. R., Carleer M. R., Chackerian C., Chance K. V., Coudert L. H., Dana V., Devi V. M., Flaud J. M., Gamache R. R., Goldman A., Hartmann J.-M., Jucks K. W., Maki A. G., Mandin J. Y., Massie S. T., Orphal J., Perrin A., Rinsland C. P., Smith M. A. H., Tennyson J., Tolchenov R. N., Toth R. A., Auwera J. V., Varanasi P., and Wagner G. (2005) The HITRAN 2004 molecular spectroscopic database. *J. Quant. Spectrosc. Radiat. Transfer, 96,* 139–204.

Sagan C. (1960) *The Radiation Balance of Venus.* JPL Tech. Rept. 32-34, Pasadena, California.

Sagdeev R. V., Linkin V. M., Blamont J. E., and Preston R. A. (1986a) The VEGA Venus balloon experiment. *Science, 231,* 1407–1408.

Sagdeev R. Z., Linkin V. M., Kerzhanovich V. V., Lipatov A. N., Shurupov A. A., Blamont J. E., Crisp D., Ingersoll A. P., Elson L. S., Preston R. A., Hildebrand C. E., Ragent B., Seiff A., Young R. E., Petit G., Boloh L., Alexandrov Y. N., Armand N. A., Bakitko R. V., and Selivanov A. S. (1986b) Overview of VEGA Venus balloon in situ meteorological measurements. *Science, 231,* 1411–1414.

San'ko N. F. (1980) Gaseous sulphur in the venusian atmosphere. *Cosmic Res., 18,* 437–443.

Sanchez-Lavega A., Hueso R., Piccioni G., Drossart P., Peralta J., Perez-Hoyos S., Wilson C. F., Taylor F. W., Baines K. H., Luz D., and Lebonnois S. (2008) Variable winds on Venus mapped in three dimensions. *Geophys. Res. Lett., 35,* L13204, DOI: 10.1029/2008GL033817.

Sandor B. J., Clancy R. T., Moriarty-Schieven G., and Mills F. P. (2010) Sulfur chemistry in the Venus mesosphere from SO_2 and SO microwave spectra. *Icarus, 208,* 49–60.

Sandor B. J., Clancy R. T., and Moriarty-Schieven G. (2012) Upper limits for H_2SO_4 in the mesosphere of Venus. *Icarus, 217,* 839–844.

Satoh T., Imamura T., Hashimoto G. L., Iwagami N., Mitsuyama K., Sorahana S., Drossart P., and Piccioni G. (2009) Cloud structure in Venus middle-to-lower atmosphere as inferred from VEX/VIRTIS 1.74 μm data. *J. Geophys. Res., 114(E9),* DOI: 10.1029/2008JE003184.

Schubert G. (1983) General circulation and the dynamical state of the Venus atmosphere. In *Venus* (D. M. Hunten et al., eds.), pp. 681–765. Univ. of Arizona, Tucson.

Schulze-Makuch D., Grinspoon D. H., Abbas O., Irwin L. N., and Bullock M. A. (2004) A sulphur-based survival strategy for putative phototrophic life in the venusian atmosphere. *Astrobiology, 4,* 11–18.

Seidelmann P. K., Archinal B. A., A'Hearn M. F., Conrad A., Consolmagno G. J., Hestroffer D., Hilton J. L., Krasinsky G. A., Neumann G., Oberst J., Stooke P., Tedesco E. F., Tholen D. J., Thomas P. C., and Williams I. P. (2007) Report of the IAU/IAG Working Group on cartographic coordinates and rotational elements: 2006. *Cel. Mech. Dyn. Astron., 98,* 155–180.

Seiff A. (1983) The thermal structure of the atmosphere. In *Venus* (D. M. Hunten et al., eds.), pp. 215–279. Univ. of Arizona, Tucson.

Seiff A., Kirk D. B., Young R. E., Blanchard R. C., Findlay J. T., Kelly G. M., and Sommer S. C. (1980) Measurements of thermal structure and thermal contrasts in the atmosphere of Venus and related dynamical observations: Results from the four Pioneer Venus probes. *J. Geophys. Res., 85,* 7903–7933.

Sleep N. H. and Zahnle K. (2001) Carbon dioxide cycling and implications for climate on ancient Earth. *J. Geophys. Res., 106,* 1373–1399.

Smrekar S. E., Stofan E. R., Mueller N., Treiman A., Elkins-Tanton L., Helbert J., Piccioni G., and Drossart P. (2010) Recent hotspot volcanism on Venus from VIRTIS emissivity data. *Science, 328,* 605–8.

Solomon S. C., Bullock M. A., and Grinspoon D. H. (1999) Climate change as a regulator of tectonics on Venus. *Science, 286,* 87–89.

Sonnabend G., Krötz P., Schmülling F., Kostiuk T., Goldstein J., Sornig M., Stupar D. A., Livengood T., Hewagama T., Fast K., and Mahieux A. (2012) Thermospheric/mesospheric temperatures on Venus: Results from ground-based high-resolution spectroscopy of CO_2 in 1990/1991 and comparison to results from 2009 and between other techniques. *Icarus, 217,* 856–862.

Sornig M., Livengood T. A., Sonnabend G., Stupar D., and Kroetz P. (2012) Direct wind measurements from November 2007 in Venus' upper atmosphere using ground-based heterodyne spectroscopy of CO_2 at 10 μm wavelength. *Icarus, 217,* 863–874.

Stofan E. R., Smrekar S. E., Helbert J., Martin P., and Mueller N. (2009) Coronae and large volcanoes on Venus with unusual emissivity signatures in VIRTIS-Venus Express Data (abstract). In *Lunar Planet. Sci. XL,* Abstract #1033, Lunar and Planetary Institute, Houston.

Stowe L. L., Carey R. M., and Pellegrino P. P. (1992) Monitoring the Mt. Pinatubo aerosol layer with NOAA/11 AVHRR data. *Geophys. Res. Lett., 19,* 159–162.

Suomi V. E. and Limaye S. S. (1978) Venus — Further evidence of vortex circulation. *Science, 201,* 1009–1011.

Svedhem H., Titov D. V., McCoy D., Lebreton J. P., Barabash S., Bertaux J. L., Drossart P., Formisano V., Häusler B., Korablev O., Markiewicz W. J., Nevejans D., Pätzold M., Piccioni G., Zhang T. L., Taylor F. W., Lellouch E., Koschny D., Witasse O., Eggel H., Warhaut M., Accomazzo A., Rodriguez-Canabal J., Fabrega J., Schirmann T., Clochet A., Coradini M. (2007) Venus Express — The first European mission to Venus. *Planet.*

Space Sci., 55, 1636–1652.

Szego K., Bebesi Z., Dobe Z., Fränz M., Fedorov A., Barabash S., Coates A., and Zhang T. L. (2009) The O^+ ion flow below the magnetic barrier at Venus post terminator. *J. Geophys. Res.–Planets, 114(E9),* DOI: 10.1029/2008JE003170.

Tanaka K. L., Senske D. A., Price M., and Kirk R. L. (1997) Physiography, geomorphic/geologic mapping, and stratigraphy of Venus. In *Venus II* (S. W. Bougher et al., eds.), pp. 667–694. Univ. of Arizona, Tucson.

Tashkun S. A., Perevalov V. I., Teffo J. L., Bykov A. D., and Lavrentieva N. N. (2003) CDSD-1000, the high-temperature carbon dioxide spectroscopic databank. *J. Quant. Spectrosc. Radiat. Transfer, 82,* 165–196.

Taylor F. and Grinspoon D. (2009) Climate evolution of Venus. *J. Geophys. Res.–Planets, 114(E9),* DOI: 10.1029/2008JE003316.

Taylor F. W., Beer R., Chahine M. T., Diner D. J., Elson L. S., Haskins R. D., McCleese D. J., Martonchik J. V., Reichley P. E., Bradley S. P., Delderfield J., Schofield J. T., Farmer C. B., Froidevaux L., Leung J., Coffey M. T., and Gille J. C. (1980) Structure and meteorology of the middle atmosphere of Venus: Infrared remote sensing from the pioneer orbiter. *J. Geophys. Res., 85,* 7963–8006, DOI: 10.1029/JA085iA13p07963.

Taylor F. W., Crisp D., and Bezard B. (1997) Near-infrared sounding of the lower atmosphere of Venus. In *Venus II* (S. W. Bougher et al., eds.), pp. 325–351. Univ. of Arizona, Tucson.

Tellmann S., Pätzold M., Häusler B., Bird M. K., and Tyler G. L. (2009) Structure of the Venus neutral atmosphere as observed by the Radio Science experiment VeRa on Venus Express. *J. Geophys. Res.–(Planets), 114,* DOI: 10.1029/2008JE003204.

Titov D. V., Markiewicz W. J., Ignatiev N. I., Song L., Limaye S. S., Sanchez-Lavega A., Hesemann J., Almeida M., Roatsch T., Matz K.-D., Scholten F., Crisp D., Esposito L. W., Hviid S. F., Jaumann R., Keller H. U., and Moissl R. (2012) Morphology of the cloud tops as observed by the Venus Express Monitoring Camera. *Icarus, 217,* 682–701.

Tomasko M. G., Doose L. R., Smith P. H., and Odell A. P. (1979) Absorption of sunlight in the atmosphere of Venus. *Science, 205,* 80–82.

Tomasko M. G., Doose L. R., Smith P. H., and Odell A. P. (1980a) Measurements of the flux of sunlight in the atmosphere of Venus. *J. Geophys. Res., 85,* 8167–8186.

Tomasko M. G., Smith P. H., Suomi V. E., Sromovsky L. A., Revercomb H. E., Taylor F. W., Martonchik D. J., Seiff A., Boese R., Pollack J. B., Ingersoll A. P., Schubert G., and Covey C. C. (1980b) The thermal balance of Venus in light of the pioneer Venus mission. *J. Geophys. Res., 85,* 8223–8231, DOI: 10.1029/JA085iA13p08223.

Toon O. B., Turco R. P., and Pollack J. B. (1982) The ultraviolet absorber on Venus: Amorphous sulfur. *Icarus, 51,* 358–373.

Toon O. B., McKay C. P., and Ackerman T. P. (1989) Rapid calculation of radiative heating rates and photodissociation rates in inhomogeneous multiple scattering atmospheres. *J. Geophys. Res., 94,* 16287–16301.

Tsang C. C. C., Irwin P. G. J., Wilson C. F., Taylor F. W., Lee C., de Kok R., Drossart P., Piccioni G., Bezard B., and Calcutt S. B. (2008) Tropospheric carbon monoxide concentrations and variability on Venus from Venus Express/VIRTIS-M observations. *J. Geophys. Res., 113,* DOI: 10.1029/2008JE003089.

Tsang C. C. C., Taylor F. W., Wilson C. F., Liddell S. J., Irwin P. G. J., Piccioni G., Drossart P., and Calcutt S. B. (2009) Variability of CO concentrations in the Venus troposphere from Venus Express/VIRTIS using a band ratio technique. *Icarus, 201,* 432–443.

Tsang C. C. C., Wilson C. F., Barstow J. K., Irwin P. G. J., Taylor F. W., McGouldrick K., Piccioni G., Drossart P., and Svedhem H. K. (2010) Correlations between cloud thickness and sub-cloud water abundance on Venus. *Geophys. Res. Lett., 37,* 02202.

Turcotte D. L. (1995) How did Venus lose heat? *J. Geophys. Res., 100,* 16931–16940.

von Zahn U., Kumar S., Niemann H., and Prinn R. G. (1983) Composition of the Venus atmosphere. In *Venus* (D. M. Hunten et al., eds.), pp. 299–430. Univ. of Arizona, Tucson.

Voros Z., Zhang T. L., Leubner M. P., Volwerk M., Delva M., and Baumjohann W. (2008) Intermittent turbulence, noisy fluctuations and wavy structures in the venusian magnetosheath and wake. *J. Geophys. Res.–Planets, 113(E12),* DOI: 10.1029/2008JE003159.

Watson L. L., Hutcheon I. D., Epstein S., and Stolper E. M. (1994) Water on Mars: Clues from deuterium/hydrogen and water contents of hydrous phases in SNC meteorites. *Science, 265,* 86–90.

Wilquet V. R., Drummond R., Mahieux A., Robert S. V., Vandaele A. C., and Bertaux J.-L. (2012) Optical extinction due to aerosols in the upper haze of Venus: Four years of SOIR/VEX observations from 2006 to 2010. *Icarus, 217,* 875–881.

Wilson C. F., Guerlet S., Irwin P. G. J., Tsang C. C. C., Taylor F. W., Carlson R. W., Drossart P., and Piccioni G. (2008) Evidence for anomalous cloud particles at the poles of Venus. *J. Geophys. Res.–Planets, 113(E9),* DOI: 10.1029/2008JE003108.

Yamamoto M. and Takahashi M. (2012) Venusian middle-atmospheric dynamics in the presence of a strong planetary-scale 5.5-day wave. *Icarus, 217,* 702–713.

Young E. F., Bullock M. A., Tavenner T., Coyote S., and Murphy J. (2008) Temporal variability and latitudinal jets in Venus' zonal wind profiles. *Bull. Am. Astron. Soc., 40,* 513.

Young E., Bullock M., McGouldrick K., and Tsang C. (2010) Venus' lower atmospheric winds from the south pole to 60°N. *Geophys. Res. Abstr., 12,* EGU2010-13414.

Zahnle K., Arndt N., Cockell C., Halliday A., Nisbet E., Selsis F., and Sleep N. H. (2007) Emergence of a habitable planet. *Space Sci. Rev., 129,* 35–78.

Zhang T. L., Delva M., Baumjohann W., Auster H. U., Carr C., Russell C. T., Barabash S., Balikhin M., Kudela K., Berghofer G., Biernat H. K., Lammer H., Lichtenegger H., Magnes W., Nakamura R., Schwingenschuh K., Volwerk M., Voros Z., Zambelli W., Fornacon K. H., Glassmeier K. H., Richter I., Balogh A., Schwarzl H., Pope S. A., Shi J. K., Wang C., Motschmann U., and Lebreton J. P. (2007) Little or no solar wind enters Venus' atmosphere at solar minimum. *Nature, 450,* 654–656.

Zhang T. L., Pope S., Balikhin M., Russell C. T., Jian L. K., Volwerk M., Delva M., Baumjohann W., Wang C., Cao J. B., Gedalin M., Glassmeier K. H., and Kudela K. (2008) Venus Express observations of an atypically distant bow shock during the passage of an interplanetary coronal mass ejection. *J. Geophys. Res.–Planets, 113(E9),* DOI: 10.1029/2008JE003128.

Zhang X., Liang M. C., Mills F. P., Belyaev D. A., and Yung Y. L. (2012) Sulfur chemistry in the middle atmosphere of Venus. *Icarus, 217,* 714–739.

Zolotov M. Y. (1991) Pyrite stability on the surface of Venus (abstract). In *Lunar Planet. Sci. XXII,* p. 1569, Abstract #1779. Lunar and Planetary Institute, Houston.

Rafkin S. C. R., Hollingsworth J. L., Mischna M. A., Newman C. E., and Richardson M. I. (2013) Mars: Atmosphere and climate overview. In *Comparative Climatology of Terrestrial Planets* (S. J. Mackwell et al., eds.), pp. 55–89. Univ. of Arizona, Tucson, DOI: 10.2458/azu_uapress_9780816530595-ch003.

Mars: Atmosphere and Climate Overview

S. C. R. Rafkin
Southwest Research Institute

J. L. Hollingsworth
NASA Ames Research Center

M. A. Mischna
California Institute of Technology/Jet Propulsion Laboratory

C. E. Newman and M. I. Richardson
Ashima Research Corporation

Mars has the most observed planetary atmosphere beyond Earth. Although the understanding of Mars' weather and climate is still in its infancy when compared to Earth, the understanding gained from over four decades of spacecraft observations is unprecedented when compared to the terrestrial planetary atmospheres of Venus and Titan. An overview of the weather and climate of Mars is presented in this chapter. It is not possible to be exhaustive in content, but does cover the basic composition and structure of the atmosphere; the circulations and atmospheric waves of all scales that are known to exist; the cycles of water, CO_2, and dust; and the evolution of the atmosphere to its present state. Where possible, similarities to and differences from the terrestrial planetary atmospheres of Earth, Venus, and Titan are noted.

1. INTRODUCTION

Mars' atmosphere shares a great number of similar physical properties, characteristics, and circulations with the atmospheres of Earth and the other terrestrial planets in the solar system. At the same time, differences in atmospheric gas, aerosol composition, mass, planetary size, and orbital characteristics, together with radiative processes, yield different climates. This chapter provides an overview of the current structure and climate of the atmosphere of Mars, including present-day structure and composition, circulation of the atmosphere from the micro- to large scale, dust and water cycles, and climate evolution. The goal of this chapter is to present an overview of the Mars climate system as a foundation for other chapters that provide a more detailed look at the comparative climatology of Mars.

Mars has predominantly a CO_2 atmosphere with the notable property that temperatures and pressures are sufficient to force a significant condensation cycle of this primary atmospheric constituent (e.g., *Leighton and Murray,* 1966). The martian CO_2 condensation cycle produces one of the most dominant annual climate signals ever directly detected through surface pressure measurements (*Tillman,* 1988). This process is unique to Mars compared to other terrestrial planetary atmospheres: Earth is far too warm for the dominantly nitrogen and oxygen atmosphere to condense; Venus is much too warm for CO_2 condensation; and Titan has clouds composed of volatile organics, but these are minor gas components of the dominant N_2 atmosphere. The structure and composition of the atmosphere, including the CO_2 cycle, is reviewed in section 2.

Superimposed on the seasonal CO_2 pressure cycle are large-scale mean and eddy circulations. Most of these circulations are also present on Earth and perhaps Titan and Venus. The globally asymmetric seasonal heating of Mars' atmosphere produces a mean overturning direct thermal circulation (the Hadley cell) like Earth, but the mean meridional circulation of Mars is nearly global in extent. Also, like Earth, Mars has a diurnal atmospheric thermal tide, but its magnitude is much larger compared to the terrestrial tide. The dynamics of these phenomena and other large-scale circulations, including baroclinic eddies, Kelvin waves, Rossby waves, and barotropic waves are covered in section 3.1.

Section 3.2 describes mesoscale and microscale circulations that contribute in important ways to the dynamics, structure, and transport in the atmosphere. Like the large-scale circulations, all the known Mars mesoscale circulations are found on Earth, but the strength, distribution, and frequency of occurrence are quite different. On Mars, near the surface, thermally driven mesoscale circulations can dominate over the large-scale flow, particularly in regions of complex topography or large-amplitude topographic relief. Mesoscale circulations may also play an important role in maintaining the background atmospheric dust loading and contribute to the perturbations of this background during dust storms. Additional mesoscale phenomena, such as gravity waves, cloud circulations, bore waves, and certain aspects of baroclinic fronts, are also elements of the mesoscale zoo. Microscale circulations are associated with small-scale and short-lived turbulent flows, including the well-known Mars dust devils. These circulations are a key component of the dust cycle and provide a major mechanism by which lifted dust is mixed and transported vertically in approximately the lowest-scale height of the atmosphere. Microscale turbulence also contributes to the overall transport of energy.

After CO_2, dust is the most significant radiatively active component of the present-day atmosphere. It absorbs and scatters solar radiation and contributes to the greenhouse effect through infrared absorption and emission (*Pollack et al.,* 1979). Dust thus affects atmospheric temperatures, temperature gradients, and, hence, winds. Dust is lifted into the atmosphere by a variety of phenomena ranging from dust devils and wind gusts to larger, organized wind systems associated with fronts. Local and large-scale positive feedbacks that are not fully understood can drive the expansion of local storms toward regional and, on occasion, global events. The decay mechanisms and interannual variability of global storms are also not well understood.

The water cycle is closely coupled to the dust cycle. Dust may serve as condensation nuclei for ice clouds, so that the distribution of dust has an influence on the location and microphysical properties of the clouds. At the same time, cloud condensation scavenges dust, and the dust distribution may be altered by subsequent sedimentation and sublimation of cloud particles. Water vapor is in very low abundance on Mars compared to Earth due to the much colder temperatures and the exponential dependence of the saturation vapor pressure curve on temperature. The low water vapor abundance produces only trivial radiative or diabatic effects. Once that water vapor condenses, however, the radiative properties of the clouds are important, although the latent heat release is still essentially negligible. Due to the often-strong intercoupling of the dust and water cycles, both are covered consecutively in sections 3.3 and 3.4.

Finally, in section 4, the evolution of the atmosphere to its present state is addressed. Following the conclusion of the late heavy bombardment (~3.8 Ga), there has been a gradual cooling of the martian surface environment, from a combination of decreased internal heat flux and a steady erosion of the atmosphere such that, by ~3.5 Ga, the character of the surface changed substantially. In contrast to the cold and dry contemporary climate, Noachian-aged surfaces appear to have required abundant water (and hence presumably warmer temperatures) to form. There are substantially fewer terrains of younger, Hesperian age that show the clear signature of enduring water. The presence of these younger fluvial features on the surface indicates at least the temporary presence of liquid water at the surface on either a regional or global scale through the Hesperian. Episodes of large-scale volcanism and impacts through Mars' history may have contributed to punctuated warm environments that may have produced the observed fluvial features. Alternatively, such features may have been controlled by processes independent of the surface environment, and the martian climate may simply have undergone a secular cooling over several billion years.

2. COMPOSITION AND STRUCTURE

2.1. Gases and Aerosols

Direct *in situ* measurement of the composition of the Mars atmosphere was first carried out by the Viking landers (see *Owen,* 1992). The inventory and concentration of major and minor species (Table 1) has remained essentially unaltered since the Viking experiments. However, recent reports from the Sample Analysis at Mars (SAM) experiment on the Mars Science Laboratory (MSL) suggest that, in contrast to that measured by Viking, argon may be more abundant than nitrogen (*Atreya et al.,* 2013). One possible explanation for the difference is a decadal-scale evolution of composition. This seems unlikely. The more plausible explanation is that either Viking or SAM (or both) measurements were in error. Of the major gases, only CO_2 is volatile at Mars temperatures and pressures, and the condensation and sublimation cycle drives a seasonal variation in total atmospheric mass of approximately 25% (Fig. 1). The molar fraction of other gases depends on the total atmospheric mass, and because this changes so dramatically due to the CO_2 phase change, the fractional abundance of other gases also exhibit a seasonal cycle. For example, the noncondensable gases argon and nitrogen can be locally enhanced by as much as a factor of six (*Sprague et al.,* 2004, 2007). Unlike on Earth, where water is a minor constituent of up to several percent, water vapor is just a trace constituent (typically <10 ppm) on Mars. Water vapor abundance is throttled by the cold martian temperatures that provide an upper bound at the saturation vapor pressure.

TABLE 1. The gaseous composition of the atmosphere of Mars (*Kieffer et al.,* 1992).

Species	Molar Fraction (%)*
CO_2	95.32
N_2	2.7
Ar	1.6
O_2	0.13
CO	0.07
NO	0.013
H_2O	2.5 ppm

*Because CO_2 abundance varies by ~25% seasonally, values represent typical, annual mean conditions.

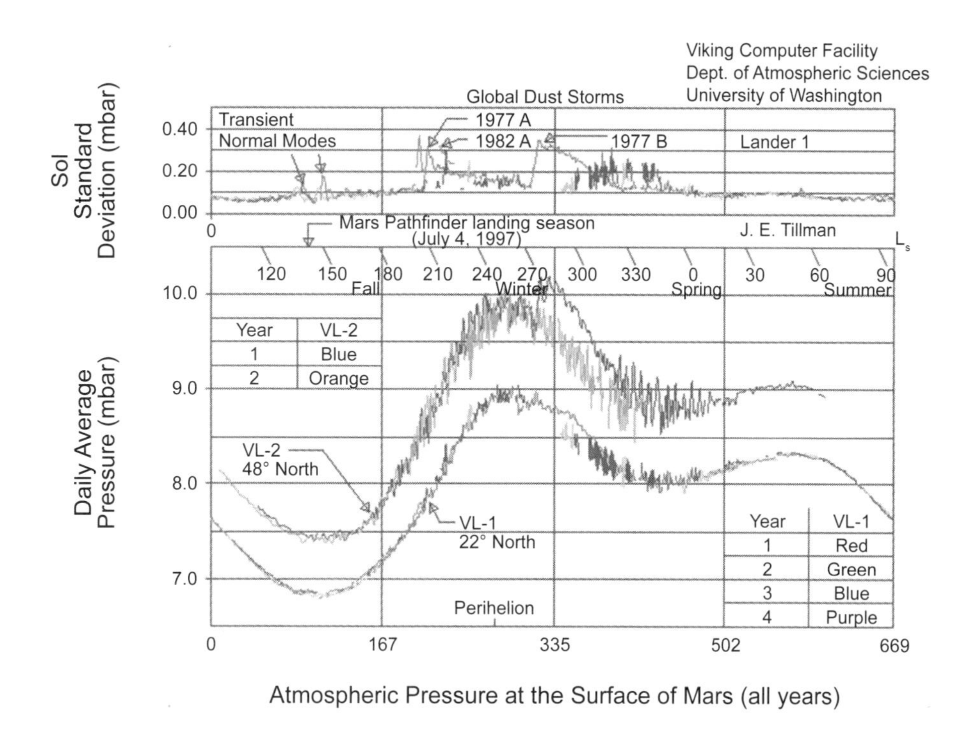

Fig. 1. Due to condensation and sublimation of the main atmospheric constituent, CO_2, the total atmospheric mass, as measured by the surface pressure, varies by up to 25% over a Mars year. Available in color at *www-k12.atmos.washington.edu/k12/resources/mars_data-information/overlay.gif.*

Almost all other known trace gases on Mars are the result of the photodissociation and photochemistry of CO_2, H_2O, and N_2 (*Krasnopolsky,* 2011). Of notable exception, the relatively recent purported detections of methane at concentrations of 1–100 ppm (*Formisano et al.,* 2004; *Krasnopolsky et al.,* 2004; *Mumma et al.,* 2009) are surprising for at least two reasons. First, photochemical and dynamical modeling cannot explain such high concentrations without a modest surface source. Second, the methane distribution exhibits spatial variations that indicate a localized source and/or localized destruction mechanisms; the photochemical lifetime of methane is much longer than the mixing timescale, so that if methane were present it should be well mixed. The importance of methane is that, if present, it signals either unexpected geophysical activity or biological processes (*Atreya et al.,* 2007). The possibility of a biological origin strongly influenced the development of the former joint NASA/European Space Agency (ESA) Mars Trace Gas Orbiter (*Zurek et al.,* 2011), now led by ESA alone, and the selection of instrumentation on the MSL rover capable of measuring methane gas. Initial results from MSL (*Webster et al.,* 2013) have found no methane at concentrations above the detection limit of ~1 ppb. These latest nondetection results are consistent with photochemical and dynamical modeling (*Lefèvre and Forget,* 2009; *Mischna et al.,* 2011) that make any significant amount of methane unlikely.

Besides gases, Mars is well known for atmospheric dust loading that can occasionally obscure the entire surface of the planet. Water vapor can also condense on dust nuclei to produce clouds of radiative importance. In this way, the dust and water cycles are coupled, with water ice clouds scavenging dust through condensation and sedimentation, but owing their existence to the presence of the dust. The most comprehensive climatology of column dust, water vapor, and water ice cloud opacity was obtained for several Mars years by the Thermal Emission Spectrometer on the Mars Global Surveyor (*Smith,* 2004) (Fig. 2). Water vapor is tied strongly to the polar source region, particularly the north. Over time, this water makes its way to the tropics where the aphelion season cloud belt is produced. High-altitude (primarily mesospheric) CO_2 clouds are also known to exist (*Clancy et al.,* 2007; *Montmessin et al.,* 2007; *Määttänen et al.,* 2010).

From Fig. 2 it is clear that there are times and locations that favor periods of high dust loading. Globally, dust opacity increases dramatically around $L_s = 180°$, where L_s is the aerocentric longitude defined so that $L_s = 0°$ is the northern hemisphere equinox and $L_s = 180°$ is equinox in the southern hemisphere. In some years, it is short-lived, but in other years the dust remains globally extensive for nearly an entire season. Also revealed is the presence of small dust storms at the receding edge of the seasonal CO_2 ice caps. These events are thought to result from strong katabatic winds and perhaps baroclinic disturbances (*Cantor et al.,* 2001).

While the total column opacity of dust and water ice is relatively well mapped, the vertical distribution of these aerosols is only now coming to light. The Mars Climate Sounder (MCS) on the Mars Reconnaissance Orbiter has

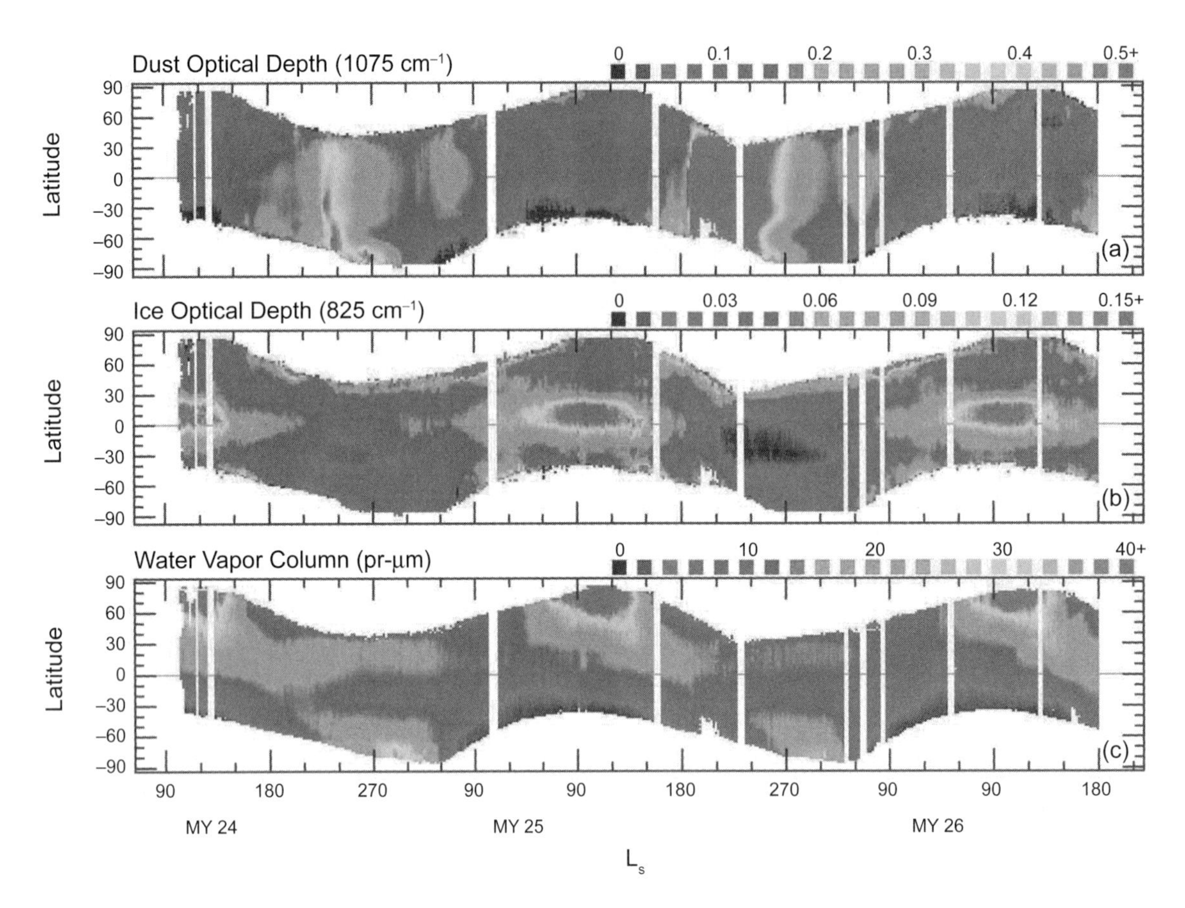

Fig. 2. See Plate 17 for color version. Zonally averaged dust optical depth at 1075 cm^{-1} scaled to an equivalent 6.1-mbar pressure surface (top), water ice optical depth at 825 cm^{-1} (middle), and water vapor column abundance in precipitable micrometers (bottom), as a function of season (L_s) and latitude.

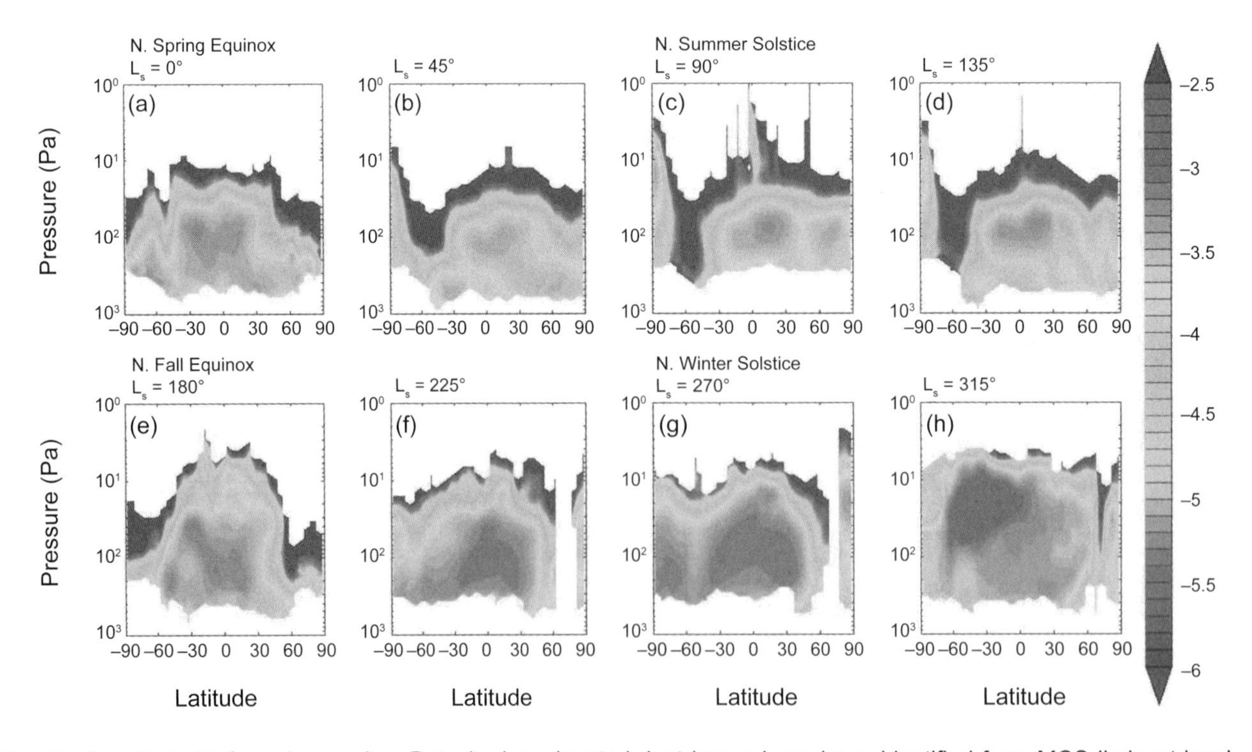

Fig. 3. See Plate 18 for color version. Detached or elevated dust layers have been identified from MCS limb retrievals. These layers persist throughout the year and are in contrast to the monotonically decreasing profile described by the Conrath-υ profile. Shaded values are $\log_{10}$ of the density scale opacity. From *McCleese et al.* (2010).

made definitive detections of elevated or detached dust layers (Fig. 3) through limb scan retrievals (*Heavens et al.,* 2011; *McCleese et al.,* 2010). The Compact Reconnaissance Imaging Spectrometer for Mars (CRISM) has recently done the same (*Smith et al.,* 2013). The amount of dust in these layers varies with location and season, but it is generally omnipresent. The ubiquitous and persistent nature of these layers is in contrast to previously assumed profiles of dust that have a well-mixed layer capped by a rapidly decreasing dust mixing ratio as a function of height (*Conrath,* 1975). The importance and possible mechanisms producing these layers are discussed more fully in section 3.2.

2.2. Thermal Profile

Previous orbital missions produced a climatology of thermal profiles, the most recent of which was obtained by MCS with approximately one-half scale height (~5 km) resolution (Fig. 4). For most of the year, the warmest temperatures are found in the summer high latitudes where solar insolation is maximum. Coldest temperatures are found over the winter poles and also at high altitudes in the tropics. There is also notable polar warming at altitude poleward of 60°–70° latitude, and this results in a reversed temperature gradient aloft. A strong baroclinic zone is found in the winter middle latitudes. During the brief equinoctial period, the warmest temperatures are found at the equator, with cold temperatures over the poles. The martian temperature patterns that exhibit large latitudinal excursions as a function of season may be contrasted with Earth, where the warmest tropospheric temperatures migrate with latitude over the seasons, but always remain in the tropics; the enormous thermal heat capacity of Earth's oceans and the lack of oceans on Mars underlie this difference. The large-scale meridional temperature gradients on Mars are thought to produce zonal winds roughly in dynamical balance with the thermal structure (i.e., geostrophic balance, as is generally the case on Earth), although such global wind systems have never been directly observed or measured.

Near the surface, measurements of temperature from the Viking landers and especially the short-lived Mars Pathfinder mission indicate that during the afternoon, a very strong superadiabatic layer is present (*Schofield et al.,* 1997). Vertical variations of 10 K or more were recorded over less than a meter by the Mars Pathfinder Atmospheric Structure Investigation/Meteorology (ASI/MET) experiment (Fig. 5). The variations were indicative of a highly turbulent, convective atmosphere. The superadiabatic layer likely transitions to a more neutral profile that can extend to 10 km or more above ground level (*Hinson et al.,* 2008; *Smith et al.,* 2006) during daytime. At night, strong radiative cooling produces a very strong and shallow nocturnal inversion. While an unstable daytime convective layer and stable nocturnal inversion are also typical of Earth, Mars swings through far more extreme end members (*Sutton et al.,* 1978; *Tillman et al.,* 1994). The Mini-Thermal Emission Spectrometer (Mini-TES) provided low-level vertical scans of the atmospheric thermal structure consistent with the previously measured superadabiatic lapse rate. The broad weighting functions of Mini-TES that increase dramatically with height preclude a detailed measurement of the highly structured convective layer near the surface or the structure

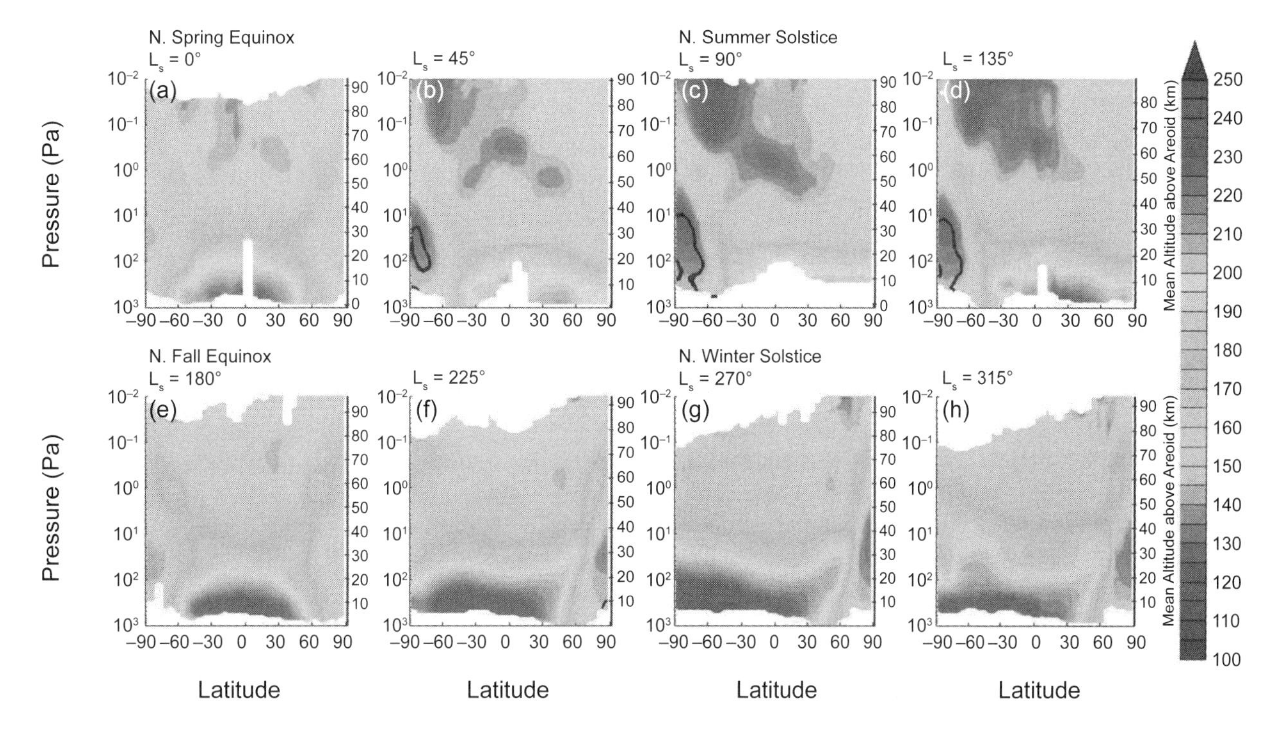

Fig. 4. See Plate 19 for color version. Zonally averaged temperature profiles obtained from MCS limb retrievals.

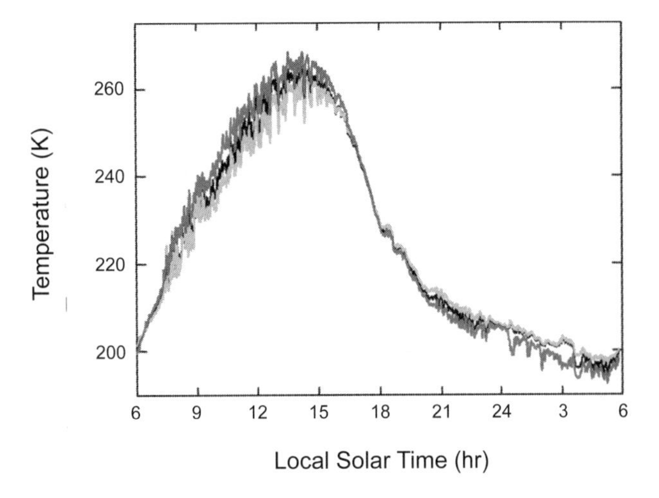

Fig. 5. The diurnal temperature cycle measured during the ASI/MET experiment on the Mars Pathfinder mission shows a very strong superadiabatic lapse rate during the day and a strong inversion at night. The magnitude of the superadiabatic layer and the nocturnal inversion are much greater than typically found on Earth. Temperatures are taken at a height of 1 m (light gray), 50 cm (black), and 25 cm (dark gray). From *Schofield et al.* (1997).

much above 2 km. The Rover Environmental Monitoring Stations (REMS) on MSL may provide additional information on the surface meteorology (*Gómez-Elvira et al.*, 2012), but a damaged wind sensor and mechanical and thermal contamination from the rover may limit the value of this data. Also, unlike Pathfinder, REMS lacks the vertical configuration of sensors needed to directly measure vertical gradients.

2.3. Carbon Dioxide Cycle

As previously mentioned, the large seasonal variations are due to the CO_2 condensation cycle. Like all planetary atmospheres, the large-scale circulation of Mars acts to redistribute heat from regions of excess energy input (i.e., the summer hemisphere) to the regions of energy deficit (i.e., the winter hemisphere). In the case of Mars, this transport is unable to keep the winter polar regions above the CO_2 condensation temperature. In the winter hemisphere, the surface of the planet cools radiatively to the CO_2 saturation temperature and any further cooling results in CO_2 ice deposition on the surface. The amount of ice deposited is an amount sufficient to liberate latent heat of condensation that exactly balances the infrared cooling. This basic process was described by *Leighton and Murray* (1966) long before Viking pressure measurements were obtained. Numerous studies (e.g., *Kieffer et al.*, 1992; *Pollack et al.*, 1990; *James and North*, 1982; *Forget*, 1998) all point to the same basic physical processes explained by *Leighton and Murray* (1966).

Earth, of course, has polar water ice caps that form under similar circumstances. However, on Earth water is a minor gas constituent with a very large source (i.e., the oceans). As Earth's water ice caps are produced, the water vapor can be relatively easily replaced via evaporation from the oceans. Furthermore, because water is a minor constituent, the loss of this gas from the atmosphere has a very small effect on the total atmospheric pressure.

A unique aspect of Mars' condensation cycle is that CO_2 is the primary constituent that lacks a large, semi-infinite source like water on Earth. The loss of atmospheric CO_2 ice to the surface does have a significant impact on the total pressure. When a polar cap reaches its maximum volume the global atmospheric pressure reaches a local minimum. The relatively large eccentricity of Mars' orbit compared to Earth accentuates seasonal asymmetries; the southern Mars winter is noticeably colder than in the north and the southern CO_2 ice cap is larger. The result of the seasonal asymmetry is an absolute minimum of pressure at around $L_s = 150°$. As the Sun returns to the southern pole, polar temperatures rise as the CO_2 sublimates and mass is returned to the atmosphere. The global pressure reaches an absolute maximum just before the onset of northern winter ($L_s = 270°$) when the northern cap begins to grow and drag the global pressure down again. The southern CO_2 is massive enough that a residual cap remains throughout the southern summer. In contrast, in the north, the CO_2 ice cap completely sublimates in the northern summer where it exposes an underlying water ice cap.

Although condensation and sublimation occurs locally at the pole, there is global communication of the corresponding regional pressure changes. The component of the atmospheric circulation that accomplishes the CO_2 transport that produces this communication is called the condensation flow. The condensation is a global, nearly pole-to-pole circulation that transports CO_2 from a sublimating region to a condensing region. Consequently, the condensation flow generally opposes the mean, global-scale direct thermal circulation (i.e., the so-called Hadley cell). The direct thermal circulation is characterized by a mean near-surface flow from the cold hemisphere toward the warm.

3. CONTEMPORARY METEOROLOGY AND CLIMATE

3.1. Global and Large-Scale Circulations and Dynamics

There are only limited *in situ* observations of combined winds, temperature, water vapor and ice particulates, and atmospheric dust loading in the martian atmosphere. The Viking landers (VL1 and VL2) of the mid-1970s/1980s are by far the most substantive and quantitative long-term surface meteorological measurements to date (*Leovy,* 1979). With sophisticated analyses (e.g., spectral techniques), the impressions of various large-scale components of Mars' atmospheric circulation have been deduced; furthermore, the broad impacts of these circulations on the dust and water cycle have been qualitatively assessed (*Zurek et al.,* 1992). The VL1 and VL2 surface weather stations were able to detect various elements of the planet's global circulation. Retrievals of the thermal structure from Mars orbiters have also provided important information about the balanced circulations consistent with these structures. Additional analyses of the temperature structures have further characterized the multitude of waves that are present. Because of the lack of sufficient global measurements of winds on the planet, much of what we understand of the global atmospheric circulation of Mars comes from atmospheric models (e.g., *Haberle et al.,* 1982; *Wilson and Hamilton,* 1996; *Forget et al.,* 1999; *Richardson and Wilson,* 2002a) of various sophistication that have been constrained by the temperatures retrieved through remotely sensed radiance observations (e.g., thermal structure, atmospheric aerosols, etc.).

3.1.1. Mean meridional circulation. Fundamental to theories of the global circulation of rapidly rotating, shallow, terrestrial atmospheres (e.g., Earth and Mars) are the longitudinally averaged (i.e., *zonally symmetric*) mean meridional overturning circulations (MMCs): the so-called Hadley cell in the tropics/subtropics, the Ferrel cell in middle and high latitudes, and the polar cell in high latitudes. Such MMCs, and the statistical construction of various circulation cells, are relevant when discussing decomposition of the global circulation, in addition to formulating diagnostics of energy and angular momentum budgets (*Peixoto and Oort,* 1992). They are also of relevance when depicting circulation component roles and how global quantities (e.g., angular momentum) are transported and distributed in latitude and height in a mean zonal sense.

A mean mass stream function is fundamentally a two-dimensional construct (*Peixoto and Oort,* 1992). When a shallow, stratified, rotating atmosphere is subject to external forcings, it may respond by a change in the structure and strength of its MMC and/or its zonally averaged temperature and wind structure (*Garcia,* 1987). In particular, north-south mean thermal gradients, together with constraints of hydrostatic and dynamical balance, ultimately determine the structure and strength of a terrestrial planet's wintertime polar vortex. In a nearly inviscid, nearly absolute angular momentum-conserving regime, the zonal wind structure is primarily controlled by gradient wind balance with the thermal field, and the strength of the MMC by the intensity and meridional gradient of the net diabatic heating rate (*Held and Hou,* 1980; *Peixoto and Oort,* 1992).

For both Earth and Mars, solar differential heating drives seasonal MMCs, as theory predicts. The Hadley circulation is associated with warm air rising in low latitudes, poleward motion aloft, cool air sinking at higher latitudes, and a low-level return flow toward the rising branch. At best, such a schematic of this thermally direct circulation is a simplified modeling construct; it truly only makes such a "coherent" circulation cell in a mean zonal sense. Because global atmospheric circulation patterns are fundamentally and richly three-dimensional, there are both spatial and temporal deviations from this two-dimensional construct. During solstice periods, atmospheric global circulation models (GCMs) indicate that Mars' Hadley cells are hemispherically quite asymmetric, both in a north-south and in an east-west sense (Fig. 6). The bulk of the hemispheric asymmetries arise due to the large-scale and large-amplitude variations in the planet's surface topography (*Holingsworth and Barnes,* 1996; *Barnes et al.,* 1996; *Joshi et al.,* 1994). Moreover, during Mars' northern winter, an intense and deep, cross-equatorial single Hadley circulation cell dominates the subtropics and encroaches into the extratropical atmosphere, with a rising motion in the summer hemisphere and a sinking motion in the winter hemisphere (Fig. 6). During global or near-hemispheric dusty episodes that often occur around boreal (northern hemisphere) winter on Mars, the zonally symmetric overturning circulation is highly vigorous with enhanced sinking motion in the middle and high latitudes aloft and an associated adiabatic compressional heating of the atmosphere. A rather distinct kinematic warming through a deep region of the polar and high-latitude atmosphere results (*Wilson,* 1997), and such events have been termed martian "polar warmings."

Shown in Fig. 7 (cf. Fig. 4) are mean zonal structures of key climatological fields during the four cardinal seasons on Mars (i.e., boreal spring, summer, autumn, winter) from a Mars GCM (*Hollingsworth et al.,* 2011). The winter hemisphere exhibits a rather sharp north-south temperature contrast (i.e., mean baroclinicity) between low and high latitudes that extends deep into the atmosphere from near the surface. The depth of this strong meridional thermal gradient is several scale heights. Because of Mars' thin CO_2 atmosphere, the maximum temperatures occur near the summer pole, which is different from that of Earth (on an annual-mean basis the terrestrial polar regions are cooler than the tropics). During solstice, when the bulk of the atmosphere is dominated by zonal mean zonal motion, there are significant deep, westerly jets in the winter hemisphere with maximum wind speeds well over 100 m s^{-1}). In contrast, the meridional winds are weaker, ~10 m s^{-1}, interhemispheric, and also rather deep; i.e., the bulk of the upper-level flow is from the warm, summer hemisphere toward the cold winter hemisphere, with a weak and shallow low-level return flow from the cold winter hemisphere toward the warm summer hemisphere. It can be seen that the Hadley circulation, in the zonally symmetric sense, shows a rising in the southern (northern) subtropics and a sinking in the northern (southern) extratropics during boreal (austral) winter. Furthermore, the MMC appears stronger during boreal winter compared to austral winter, and this feature has been attributed to the mean-zonal slope (bulk negative) of the planet's topography (*Richardson and Wilson,* 2002b), despite the increased incoming solar flux during Mars' aphelion season, and regardless of a more vigorous overturning circulation associated with enhanced atmospheric dust loading during austral summer.

During equinox periods, the mean zonal temperature is mostly symmetric about the equator, with the coldest temperatures at high latitudes. This thermal structure

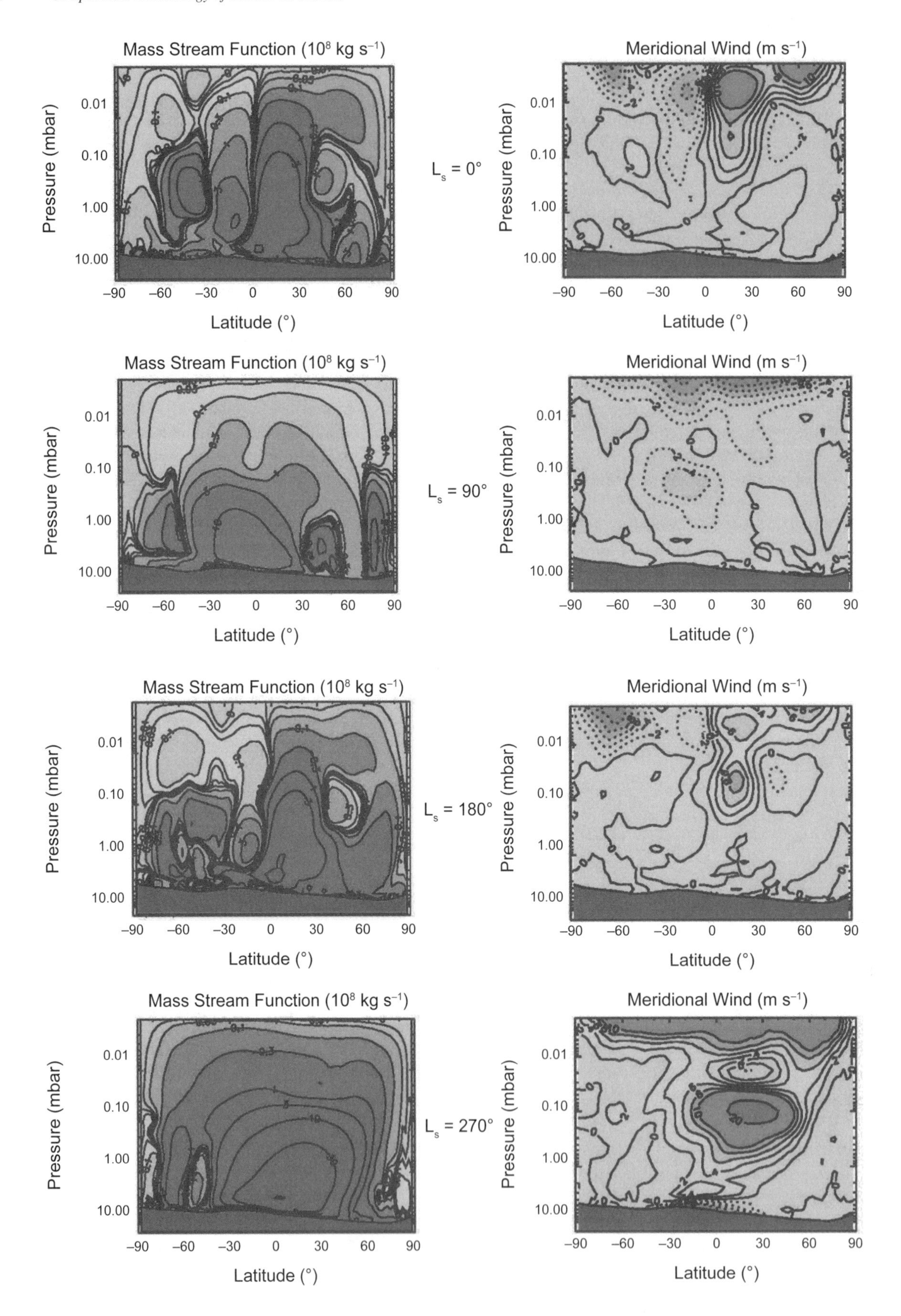

Fig. 6. See Plate 20 for color version. The mean meridional stream function and mean meridional wind at the four cardinal seasons as determined from the NASA Ames GCM (*Hollingsworth et al.*, 2011).

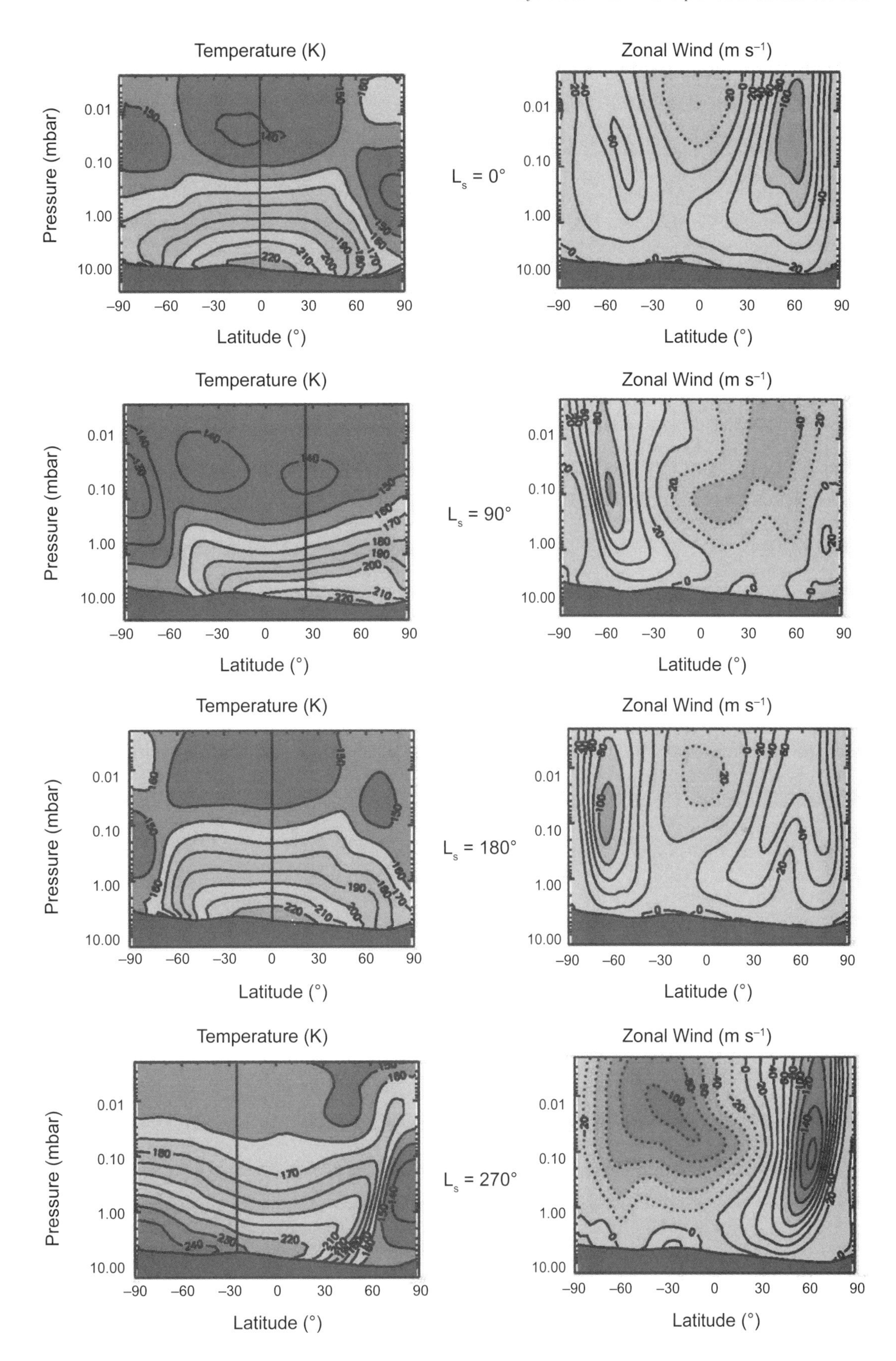

Fig. 7. See Plate 21 for color version. The zonally averaged temperature and zonal wind at the four cardinal seasons as determined from the NASA Ames GCM (*Hollingsworth et al.,* 2011).

supports deep and less-intense westerly jets in the middle and high latitudes of both hemispheres. In the tropics, there are indications of equatorial superrotation in the lowest few scale heights from the surface that has been shown to be a result of equatorward momentum fluxes associated with diurnally varying, global thermal tidal modes (*Lewis and Read,* 2003). It can also be seen that the Hadley circulation during equinox is weaker and more symmetric about the equator than at solstice.

Mars' atmospheric MMC is, in contrast to Earth's atmosphere, more intense and interhemispheric. This two-dimensional perspective is in many ways too simplistic, as can be assessed by looking more carefully at the whole three-dimensional mean seasonal circulation. Shown in Fig. 8, for example, are longitude-pressure slices during northern autumn (L_s ~ 215°) from a Mars GCM simulation (*Hollingsworth and Kahre,* 2010; *Hollingsworth et al.,* 2011) of different seasonal mean fields at 40°N: mean potential temperature, zonal wind, meridional wind, and vertical wind. In this longitude-pressure cross section there is a significant change in the static stability (i.e., buoyancy frequency, N); namely, the western hemisphere exhibits a different concavity of isentropic surfaces in an east-west sense (i.e., it is more statically unstable) than that of the eastern hemisphere. There are also significant changes in the potential temperature's vertical structure. The ramifications of these subtle but important longitudinal changes in thermal gradients and bulk static stability can give rise to vastly different corridors of momentum, and thus variations in atmospheric transport of tracer quantities.

It can be seen in Fig. 8 that the bulk of the northward motion of the northern hemisphere thermally direct MMC (Hadley cell) occurs aloft in the western hemisphere. There is a weaker southward (return) flow in the eastern hemisphere. At low levels of the atmosphere, the signatures of boundary currents (*Joshi et al.,* 1994) on the eastward-facing edge of the steep relief can be seen. This low-level flow "ducts" the bulk of the return branch of the thermally direct MMC. Furthermore, the mean vertical wind shows even finer longitudinal structures (Fig. 8d), where deep and vigorous vertical motion occurs in the western hemisphere over the Tharsis highlands. This hemisphere exhibits oscillating "plume-like" structures of upward, downward motion, but it is important to note that there are not yet observations of sufficient fidelity to determine whether such modeled structures are representative of the real atmosphere. The bulk of the atmosphere indicates predominant descending motion at middle and upper levels, consistent with the downward branch of the zonally symmetric view of the Hadley circulation.

Associated with Mars' seasonal CO_2 polar ice caps that wax and wane in both hemispheres on an annual basis, its seasonal-mean atmosphere exhibits pronounced north-south temperature contrasts (*Zurek et al.,* 1992). These strong thermal contrasts exhibit deep vertical structures in the middle and high latitudes. In the presence of such temperature gradients and from large-scale balance constraints (*Peixoto and Oort,* 1992), intense and deep westerly zonal jets exceeding 100 m s^{-1} form in its extratropical atmosphere (*Conrath et al.,* 2000; *Smith et al.,* 2001; *Smith,* 2004). Most GCMs of Mars can simulate the basic character of such polar jets, an example of which is shown in Fig. 7.

3.1.2. Large-scale, topographically forced waves. Mars also exhibits continental-scale orographic complexes (*D. E. Smith et al.,* 2001). In Mars' northern middle latitudes, Tharsis in the western hemisphere, and Arabia Terra and Elysium in the eastern hemisphere, are the primary large-scale topographic features. In the southern middle latitudes, the Hellas region, a ~3000-km-diameter and ~10-km-deep circular impact basin, is the predominant topographic feature. It is roughly 180° east of the massive Tharsis highlands, which rise >10 km above the zero topographic datum (defined to correspond to the global and annual mean surface pressure of 6.11 mbar). With regard to the planet's orography, Mars' northern extratropics are more or less rather unimpressive (i.e., of small relative amplitude perturbations), whereas its southern extratropics exhibit extremely large (~10–15 km) topographic perturbations. In Fig. 9, these large-scale topographic features are clearly indicated in terms of a seasonal mean surface pressure longitude-latitude map during northern mid-autumn (L_s ~ 215°) (*Hollingsworth et al.,* 2011).

Past investigations have considered the forced, planetary atmospheric wave response to large-scale surface orography. Such studies have been based on quasi-geostrophic theory (*Mass and Sagan,* 1976), linear primitive equation models (*Webster,* 1977; *Hollingsworth and Barnes,* 1996); shallow-water models in spherical geometry (*Keppenne and Ingersoll,* 1995), and fully nonlinear Mars GCMs (*Pollack et al.,* 1981; *Barnes et al.,* 1996). Both linear and nonlinear primitive equation models (e.g., *Hollingsworth and Barnes,* 1996; *Barnes et al.,* 1996) have predicted the occurrence of very large amplitude forced stationary planetary (i.e., Rossby) modes in Mars' southern middle latitudes between late autumn and early spring, with predominant zonal scales of wavenumbers 1–2. These planetary waves are excited by large (~10 km) east-west excursions in surface elevation embedded within zones of enhanced potential vorticity gradients accompanying the westerly polar vortex. The signature of such waves has been detected by various atmospheric sounding instruments onboard the Mars Global Surveyor (MGS) spacecraft (*Smith et al.,* 2001; *Banfield et al.,* 2003; *Hinson et al.,* 2003). Furthermore, forced Rossby modes have been thought to provide influence on the southern seasonal polar ice cap east-west asymmetry (*Colaprete et al.,* 2005). Although the extratropical east-west excursions in the planet's northern hemisphere relief are not as pronounced as in the southern hemisphere, they are still significant enough to force similar stationary Rossby modes with predominant zonal scales of wavenumbers 1–3 (*Hollingsworth and Barnes,* 1996; *Barnes et al.,* 1996; *Hinson et al.,* 2001).

All in all, the direction and intensity of Mars' lower and middle atmosphere's (westerly) jet stream can be significantly altered by large-scale topography. The occur-

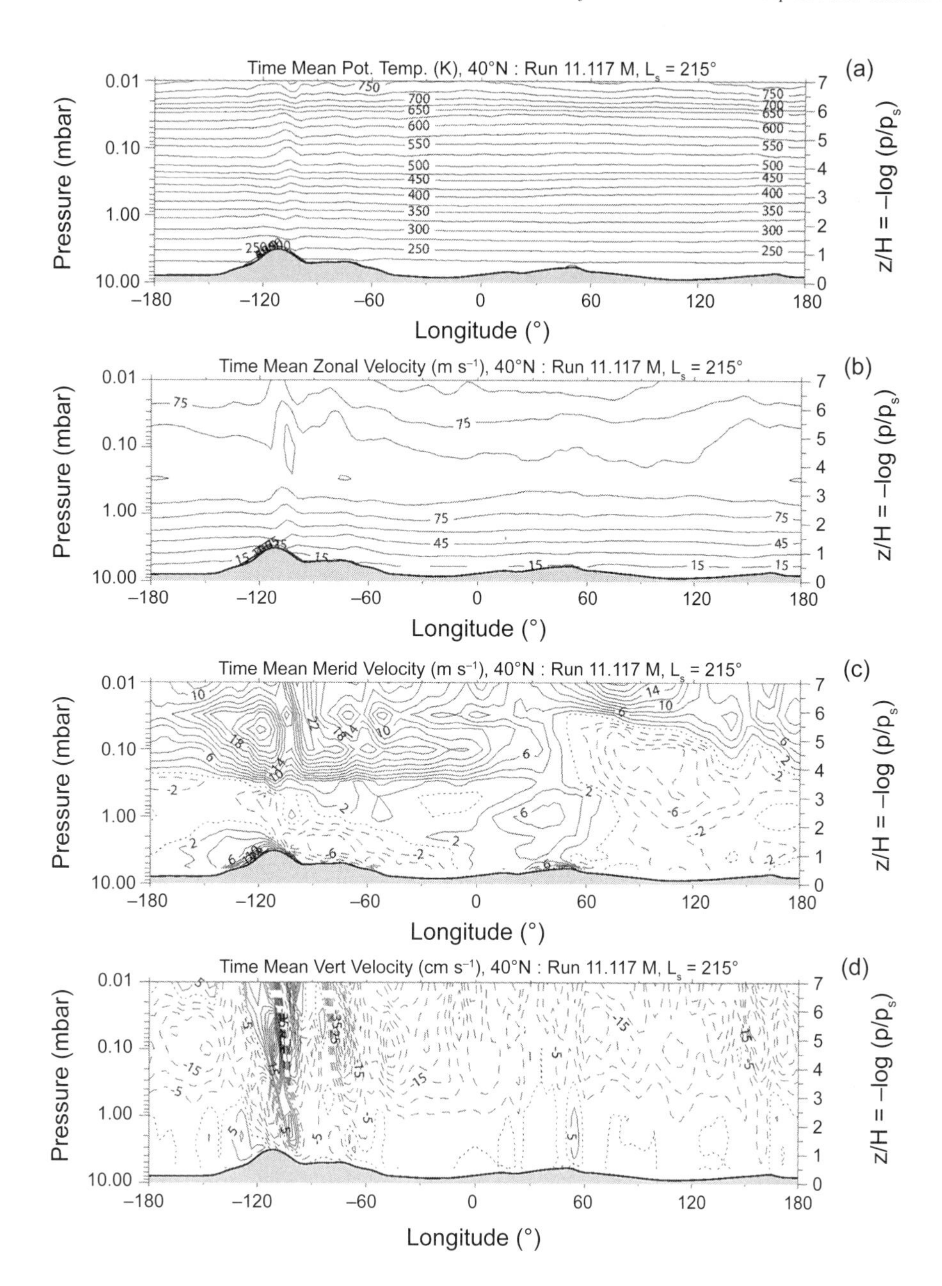

Fig. 8. For the northern autumn season (L_s ~ 215°), a longitude-pressure slice at 40°N of **(a)** time mean potential temperature (K); **(b)** the time mean zonal wind (m s^{-1}); **(c)** time mean meridional wind (m s^{-1}); and **(d)** time mean vertical wind (cm s^{-1}). In **(b)** through **(d)** solid (dashed) contours correspond to positive (negative) values: **(b)** westerly (easterly) zonal wind with a contour interval of 15 m s^{-1}; **(c)** northerly (southerly) meridional wind with a contour interval of 2 m s^{-1}; **(d)** upward (downward) vertical wind with a contour interval of 5 cm s^{-1}.

rence of forced Rossby modes and their influences on the seasonal mean circulation can be seen in Fig. 10, which is a latitude-pressure "slice" from a Mars GCM simulation (*Hollingsworth and Kahre,* 2010; *Hollingsworth et al.,* 2011) at –45° E for a variety of meteorological fields from a model. At this particular longitude, the stationary geopotential height field (Fig. 10a) shows significant cyclones (i.e., deep, low-pressure anomalies) in the extratropics of both hemispheres, collocated within a significant mean westerly jet (Fig. 10b). The vertical and horizontal shears associated with this mean zonal jet support a planetary waveguide (*Hollingsworth and Barnes,* 1996) that is favorable to forced Rossby wave propagation for the largest zonal wavenumbers (s = 1–3). It can also be seen that the mean meridional wind is highly varying at this longitude (Fig. 10c), with poleward motion at upper levels in the northern hemisphere, and equatorward motion in middle and high latitudes. At low levels, the meridional motion is somewhat counterintuitive, with poleward motion in the southern extratropics (i.e., consistent with the return branch

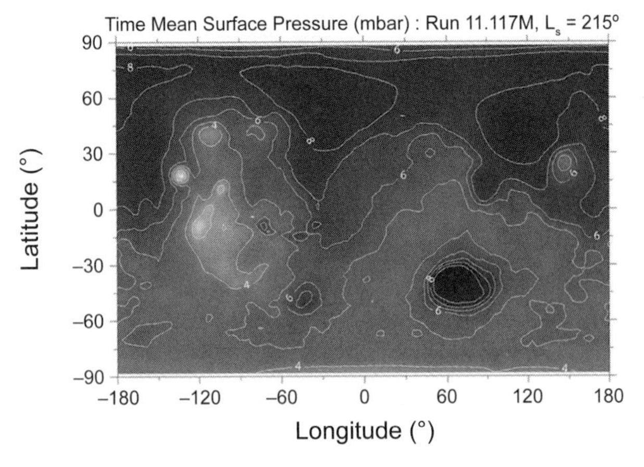

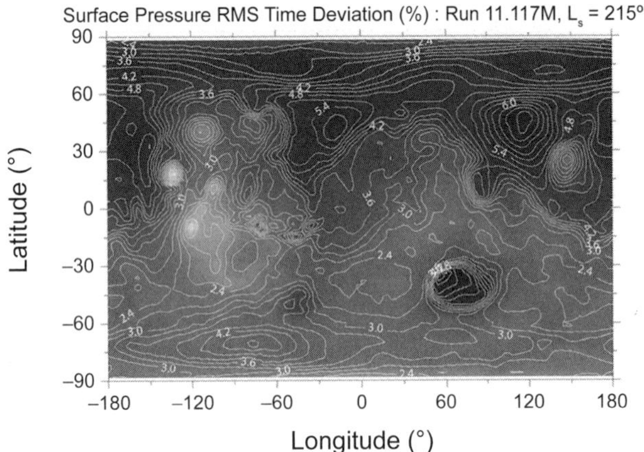

Fig. 9. Same as in Fig. 8 with longitude-latitude cross section of the **(a)** time mean surface pressure (mbar) with a contour interval of 2 mbar and **(b)** root-mean squared (RMS) surface pressure time deviation, normalized by a globally averaged value (%). The gray shading corresponds to the planetary orography (dark is low elevation; light is high elevation).

of the thermally direct MMC), but equatorward motion in the northern extratropics. The latter is not consistent with the return branch of the zonally symmetric Hadley cell. Thus, it is evident that a simple two-dimensional view of a mean meridional overturning circulation has limitations; the three-dimensional circulation is far more complicated. Finally, the mean vertical wind field shows significant, narrow upward-motion "plumes" associated with the equatorial high-topographic peaks.

In addition, the meridional scale and amplitude of northern and southern hemisphere traveling weather systems, and their preferred geographic regions (see the next two subsections below), are also highly influenced by Mars' large-scale topography. As depicted in Fig. 10, the longitudinal variations in the time mean wind and (potential) temperature fields in northern middle latitudes arise predominantly from nontrivial forced, stationary Rossby wave modes; in terms of geopotential height anomalies, a wavenumber 1–2 disturbance is embedded in the extratropical atmospheric circulation (not shown).

3.1.3. Traveling synoptic-period weather systems. The thermally indirect (i.e., eddy-driven) circulation cells are also robust in both middle and high latitudes of Mars and Earth. Missions to Mars such as Viking and the Mars Reconnaissance Orbiter (MRO) have unquestionably demonstrated that Mars' winter atmosphere — as with Earth's — exhibits traveling extratropical weather systems and frontal waves on day-to-day timescales (i.e., synoptic periods) associated with north-south mean temperature contrasts (i.e., sheared westerly flow within such a temperature gradient, i.e., baroclinic instability). Having more seemingly "regular" lifecycles compared to those on Earth (*Collins et al.,* 1996), Mars' traveling disturbances and their accompanying poleward transports of heat, momentum, and tracer species (e.g., water vapor and aerosols such as dust and water ice) strongly influence the global atmospheric energy budget. Mars' extratropical cyclogenesis and frontal wave disturbances have a much larger planetary-scale expression for the size of the planet (i.e, the Rossby deformation radius, ~2000 km, is similar to Earth, such that a fewer number of weather disturbances encompass the globe at a given instance of time). One essential aspect of transient baroclinic eddies is that they tend to weaken the baroclinicity of the mean circulation via their associated low-level poleward heat and upper-level momentum fluxes. Within such weather systems, atmospheric transport processes are complex and associated with *secondary* circulations embedded within such disturbances (*Holton,* 1992): This has implications for dust transport on Mars, and for water vapor and clouds.

During late autumn through early spring, extratropical regions on Mars exhibit profound equator-to-pole thermal contrasts. The imposition of this strong meridional temperature variation supports intense eastward-traveling weather systems (i.e., transient synoptic-period waves) associated with the dynamical process of baroclinic instability (*Barnes,* 1980, 1981; *Banfield et al.,* 2004; *Hinson and Wilson,* 2002; *Hinson,* 2006). Such disturbances grow, mature, and decay within the east-west varying seasonal mean midlatitude jet stream (i.e., the polar vortex). Near the surface, the weather disturbances indicate large-scale spiraling "comma"-shaped dust cloud structures and scimitar-shaped dust fronts in the northern extratropical and subtropical environment (*James and Cantor,* 2001; *Wang et al.,* 2003, 2005; *Hinson and Wang,* 2010). Their occurrences in the southern hemisphere are distinctly more limited in spatial scale, especially meridionally compared to their northern hemisphere counterparts.

3.1.4. Extratropical storm zones. The large-scale topographic features of Tharsis in the western hemisphere and Arabia Terra and Elysium in the eastern hemisphere set up preferred geographic regions of synoptic-period weather variability (referred to as "storm tracks" or "storm zones") in northern midlatitudes (*Hollingsworth et al.,* 1996, 1997). During late southern winter, a weaker and more longitudinally confined storm zone occurs in the midlatitudes between Tharsis and Argyre (*Hinson and Wilson,* 2002). Furthermore, the low-topographic-relief regions in the northern hemisphere correspond to preferred regions of

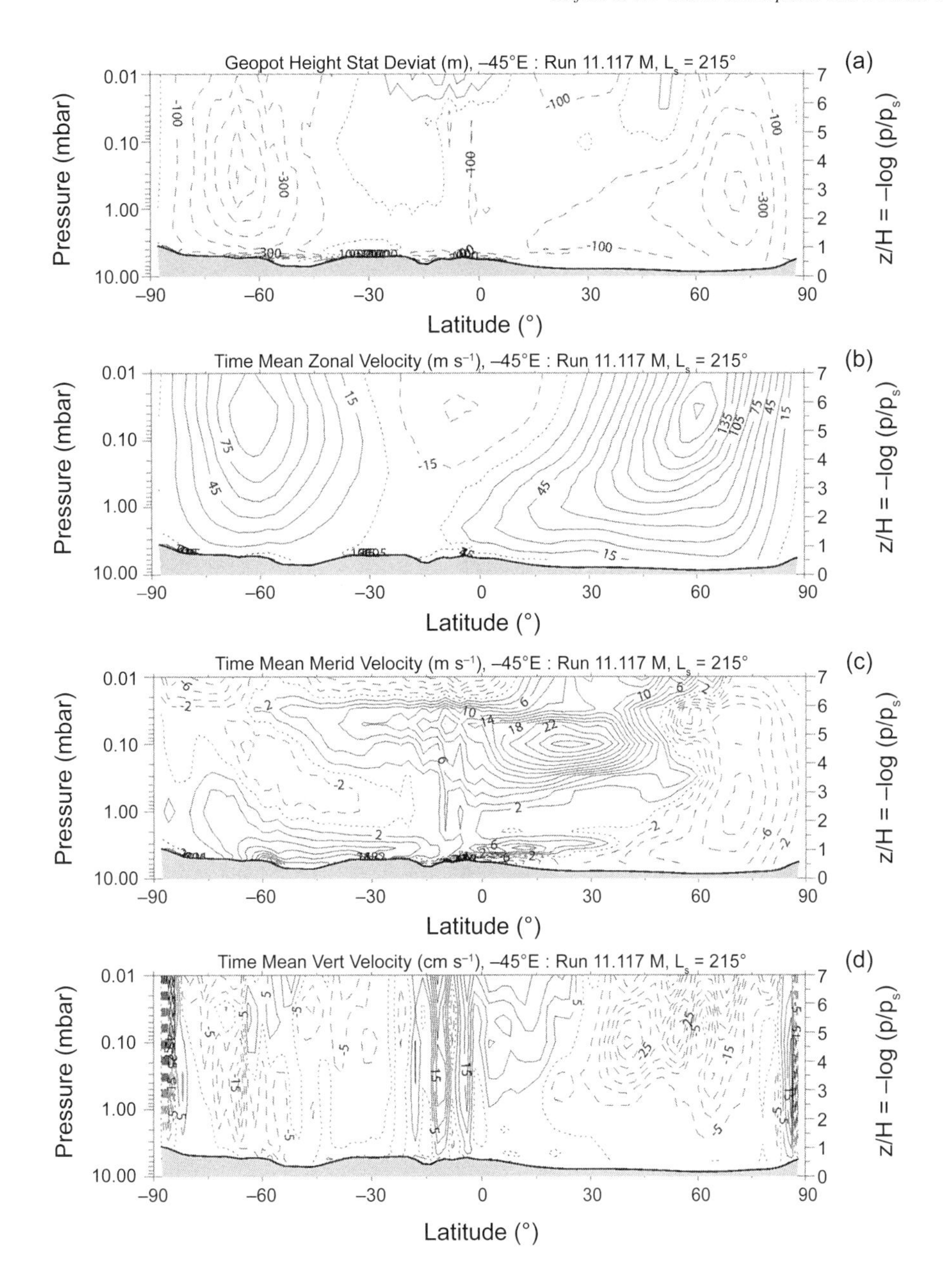

Fig. 10. Same as in Fig. 8 but with a latitude-pressure slice at –45°E of **(a)** geopotential height (m) stationary deviations; **(b)** time mean zonal wind (m s^{-1}); **(c)** time mean meridional wind (m s^{-1}); and **(d)** time mean vertical wind (cm s^{-1}). In **(a)** through **(d)** the solid (dashed) contours correspond to positive (negative) values: **(a)** positive (negative) geopotential height stationary anomaly with a contour interval of 100 m; **(b)** westerly (easterly) zonal wind with a contour interval of 15 m s^{-1}; **(c)** northerly (southerly) meridional wind with a contour interval of 2 m s^{-1}; and **(d)** upward (downward) vertical wind with a contour interval of 5 cm s^{-1}.

synoptic-period, weather system variability as can be seen in Fig. 9b. In terms of root-mean-squared surface pressure anomalies [i.e., band-pass filtered time deviations in surface pressure (*Hollingsworth et al.,* 1996, 1997)], there is an enhancement within the low-topographic-relief regions of the northern extratropics corresponding to the Acidalia, Arcadia, and Utopia regions. In the southern extratropics, there is also at this season an enhancement of synoptic variability in the western hemisphere south of Tharsis and Argyre. Such geographic regions are the baroclinic "storm zones" on Mars, and these storm zones show significant intraseasonal variations (*Hollingsworth et al.,* 1997).

Furthermore, baroclinic waves in the southern hemisphere are much less intense during southern late winter/early spring than their northern hemisphere counterparts (*Barnes et al.,* 1993). Exactly why this is the case remains unresolved, but it may likely have to do with the very large east-west (i.e., zonal) asymmetries in topographic relief of this hemisphere (*Hollingsworth,* 2003; *Hollingsworth et al.,* 2008). Quasi-stationary, forced Rossby waves can cause significant latitudinal excursions of the seasonally averaged atmospheric flow (*Hollingsworth and Barnes,* 1996; *Barnes et al.,* 1996). As on Earth, Mars' baroclinic wave activity and storm track intensity in the extratropical baroclinic zone may be highly modulated by large-scale surface asymmetries of each hemisphere (*Hoskins and Ambrizzi,* 1993; *Son et al.,* 2009) in addition to structural differences of the background (i.e., seasonal mean) zonal flow.

3.1.5. Thermal tidal planetary wave modes. Mars, as Earth, presents externally forced global wave modes of variability associated with diurnal heating of its rapidly rotating, shallow atmosphere from incoming solar insolation. Such atmospheric forcing is periodic on diurnal timescales. As a result of this periodic heating/cooling (i.e., dayside to nightside), the atmosphere of Mars exhibits global thermal tidal modes. [The wave modes described here are not to be confused with *gravitational tides* associated with the Sun or the planet's moons Phobos and Deimos, which induce far weaker perturbations upon the geophysical fluid envelope than those at Earth (i.e., primarily its oceans) due to Earth's large Moon.] The global atmospheric modes that are excited by the diurnally varying heating (i.e., referenced to a fixed position on the planetary surface from east to west) are commonly termed "migrating" (westward-propagating) thermal tides. Without global-scale dust storms, a sizeable amount of incoming solar radiation falls upon the surface of Mars, where it heats up and exchanges with the atmosphere associated with radiative-convective processes. The day-night diurnal cycle on Mars is particular strong, ~100 K, due to weak thermal inertia of the planet's atmosphere and the strong solar heating per unit mass during the daytime (*Leovy,* 1979).

From classical tidal theory — i.e., for an isothermal, motionless, adiabatic, and inviscid atmosphere — the linearized primitive equations of meteorology are separable with respect to the vertical and horizontal (i.e., meridional) directions. Upon separation, the linearized equations yield Laplace's tidal and vertical structure equations (both homogeneous) (*Andrews et al.,* 1987; *Volland,* 1988). As such, they form the prototypical equations governing *a range* of dynamical phenomena: atmospheric thermal tides, Rossby waves, equatorial waves, and inertia-gravity waves. Diabatic effects (e.g., realistic differential solar heating and infrared cooling) render the vertical structure equation nonhomogeneous. A single parameter known as Lamb's parameter, ε (i.e., a rotational Froude number), couples the equations that are inversely proportional to the so-called "equivalent depth," h. For purely vertically propagating modes (i.e., those that are not trapped in height), the vertical wavelength is proportional to both the square root of the Lamb parameter and the static stability, N. Under the above assumptions, and for a given wavenumber and frequency, solutions to Laplace's tidal equations can be expressed in terms of a series expansion of special functions (i.e., Hough functions that are composed of associated Legendre polynomials) that satisfy an eigenvalue problem for each wave mode. Dispersion diagrams of the Lamb's parameter (ε) for a given wave frequency (σ), zonal wavenumber (s), and meridional index (n) represent the eigenfrequencies of Laplace's tidal equation and occur upon two separate manifolds: type I waves (or gravity modes) with $\varepsilon > 0$ for all σ, and type II waves (or Rossby modes) with $\varepsilon > 0$ for $\sigma < 0$ (westward propagation) and $\varepsilon < 0$ for $\sigma > 0$ (eastward propagation) (*Andrews et al.,* 1987).

Application and extension of classical tidal theory to Mars' atmosphere for the diurnal tide was carried out during the Viking era by *Zurek* (1976) for a variety of relatively clear and dust-laden atmospheres, assuming representative global mean (i.e., basic state) temperature profiles and Newtonian cooling rates, and including the effects of spatial variations in the planet's large-scale topography. A generalized extension to both short-period (subdiurnal and up to several days) forced and free planetary waves under realistic background thermal and diabatic forcing conditions was investigated by *Zurek* (1988). These early, seminal studies offered the first quantitative estimates of meridional and vertical wind magnitudes and their variations with altitude. Furthermore, estimates of near-surface, diurnally varying horizontal winds, together with the potentials and conditions for short-period, planetary-wave resonances, were provided.

As mentioned above, classical theory of thermal tides assumes linearized disturbances upon a basic (thermal) state without spatial (longitude, latitude, height) variations. Under such assumptions, migrating tides of diurnal and semidiurnal periods can be sought from the Laplace tidal equations for a given equivalent depth, h. The predominant migrating tides within Earth's and Mars' atmosphere are quite similar, in that the wavenumber s = 1 diurnal component can primarily vertically propagate in the tropics with a vertical wavelength of ~30 km, and become vertically trapped at low altitudes within middle and high latitudes (*Andrews et al.,* 1987; *Wilson and Hamilton,* 1996). The wavenumber s = 2 semidiurnal component has a much larger vertical wavelength of ~100 km with very little phase variation with altitude.

Shown in Fig. 11 are examples of the diurnal and semidiurnal migrating tides in terms of surface pressure amplitude globally, and phase at the VL1 lander location from a Mars GCM simulation (*Wilson and Hamilton,* 1996) during northern summer (L_s ~ 90°).

Atmospheric temperature measurements retrieved from the MGS TES and MRO MCS observations for Mars' lower and middle atmosphere have demonstrated the existence of thermal tidal modes, particularly the diurnal mode (*Banfield et al.,* 2003; *Lee et al.,* 2009). In particular, MGS TES measurements have provided an opportunity to assess a variety of forced planetary and thermal tidal waves for the entire annual cycle (*Banfield et al.,* 2003). Shown in Fig. 12 are seasonal "maps" of latitude-vertical structures of the temperature anomaly (i.e., 1400–0200 LT time difference) that are dominated by the migrating diurnal thermal tide. It can be seen that the NH winter extratropical hemisphere has maximum diurnal temperature excursions of O(4–6 K) over several scale heights from the near surface.

In addition to the Sun-synchronous migrating thermal tidal modes, there are differential heating contributions from regional insolation upon east-west (zonal) asymmetries in surface topography, thermal inertia, and albedo. Such global modes excited by east-west asymmetries in surface heating/cooling excite "nonmigrating" (eastward-propagating) thermal tides. On a perfectly smooth terrestrial-like, shallow atmosphere that is rapidly rotating, without any sur-

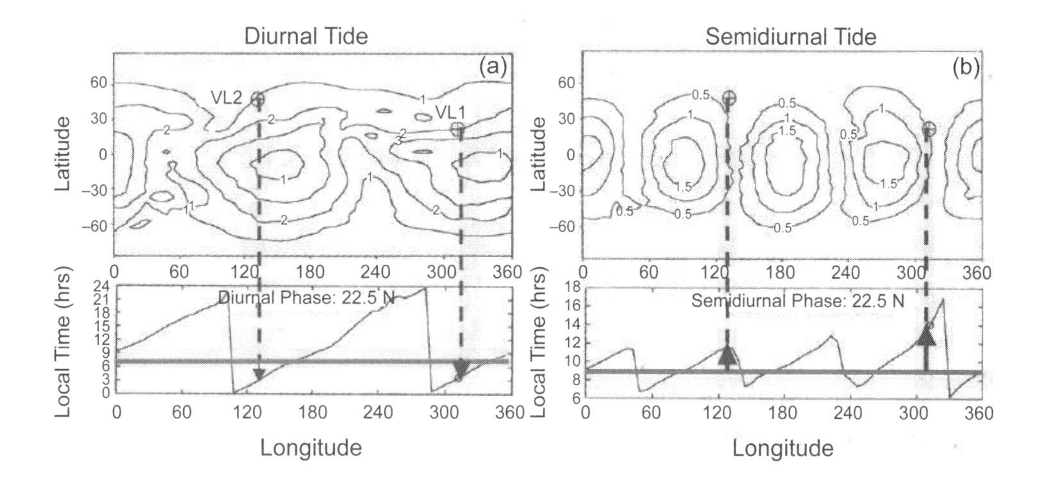

Fig. 11. The spatial variation of **(a)** diurnal and **(b)** semidiurnal surface pressure amplitude from a Mars GCM simulation for northern hemisphere summer (L_s = 90°). Amplitudes are normalized to show magnitude as a percentage of the local diurnal-mean surface pressure. Locations of VL1 and VL2 landers are indicated. The **(a)** diurnal and **(b)** semidiurnal thermal tidal phases at VL1 latitude (22.5°N) are indicated in the bottom row. Adapted from *Wilson and Hamilton* (1996) and a modified figure.

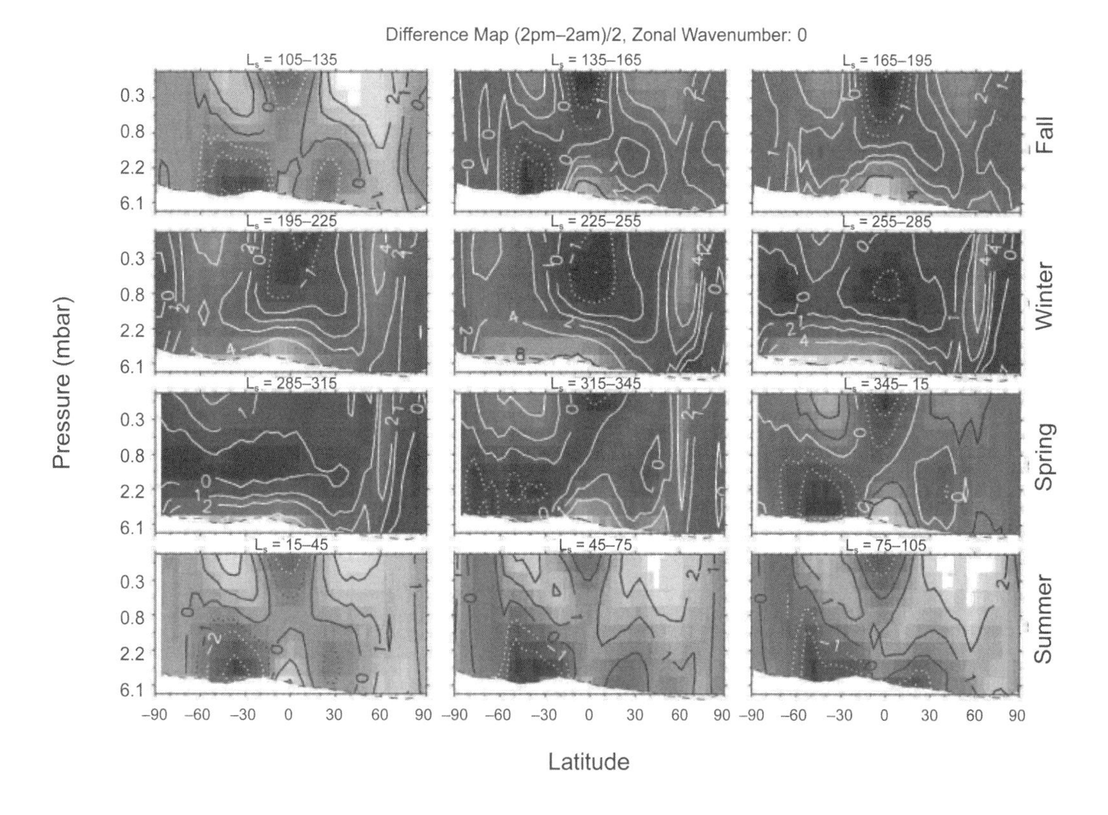

Fig. 12. Temperature difference maps (1400–0200)/2 for m = 0 (zonal mean). Contours are labeled in K, with dotted contours less than zero. Brightness indicates high positive values. The fields are dominated by the Sun-synchronous diurnal tide. Adapted from *Banfield et al.* (2003).

face zonal asymmetries in planetary orography or thermal properties [e.g., surface albedo and thermal inertia (or land surface or ocean contrasts for Earth)], the latter would be identically zero (i.e., vanish). Earth and Mars are far from "billiard ball" in surface properties, and for Mars, the global thermal tidal modes (both migrating and nonmigrating) are extremely important in the bulk atmospheric circulation.

Under linear theory, such nonmigrating thermal tidal modes were predicted by *Zurek* (1976) for simplified atmospheric thermal basic states and dissipation rates. Full Mars GCM simulations extended the predictions and assessments of such modes within spatially/temporally varying mean background zonal mean zonal flows and temperatures (*Wilson and Hamilton,* 1996).

From MGS TES measurements, evidence for nonmigrating thermal tidal modes within Mars' atmosphere has been verified, and in particular, the wavenumber 1 diurnal Kelvin wave (DK1) has been assessed (*Banfield et al.,* 2003). It has been found that the seasonal variations of the amplitude of the diurnal Kelvin wave is far weaker than the Sun-synchronous migrating diurnal tide, at least during the first full mapping year of MGS (*Banfield et al.,* 2003). Furthermore, from MGS aerobraking observations of atmospheric density and temperature perturbations, the implications of nonmigrating thermal tides on upper-atmospheric variability have been assessed and modeled by *Forbes et al.* (2002).

Finally, during episodes of enhanced atmospheric dust loading accompanying hemispheric and/or global dust storms on Mars, the thermal tidal modes are much enhanced, particularly the semidiurnal component (*Zurek,* 1976, 1988; *Wilson and Hamilton,* 1996). As such, thermal tides and other free and forced planetary wave modes are important wave components of the atmosphere of Mars and have significant dynamical implications for its present climate.

3.2. Mesoscale and Microscale Meteorology

Mesoscale circulations generally fall into the temporal-spatial domain ranging from ~1 hour to ~1 day and ~1 km to ~100 km. Thus, valley-slope circulations, gravity waves, and many dust storms are dominantly mesoscale in nature. At these scales, the Coriolis force can be important, but it is not usually dominant, as it is in the case of longer-lived and larger planetary circulations. As a result, the great majority of mesoscale circulations have a significant dynamically unbalanced component, which is to say the flows cannot be fully described by steady-state dynamics. Unbalanced circulations exhibit significant ageostrophic flows, or winds, that flow parallel to the pressure gradient rather than parallel to the isobars. Gravity waves are an ideal example of the unbalanced nature of mesoscale circulations. Slope flows are another example where winds flow (nearly) directly from high to low pressure.

At even smaller scales, Mars exhibits a highly turbulent daytime convective boundary layer that spawns dust devils, some of which can be an order of magnitude taller than those on Earth. The ferocity of the turbulence is a direct result of the low density of the atmosphere and the direct radiative heating of the very lowest layers of the atmosphere by the surface. Although dynamically vigorous, the actual turbulent fluxes of energy produce only a modest perturbation to what would be expected through pure radiative equilibrium. Nonetheless, the turbulent circulations are directly responsible for the lifting of dust and for the transport and mixing of heat, momentum, and other species within the boundary layer.

One of the first in-depth mesoscale studies of the atmosphere of Mars compared the strength of slope flows to those of Earth (*Ye et al.,* 1990). Due to the radiative and eddy flux heating peculiarities of Mars' atmosphere (described below), isotherms tend to follow topography on Mars, whereas on Earth they tend to be more independent of topography. This difference results in the generation of very strong mesoscale circulations on Mars driven by horizontal thermal gradients much stronger than those typically found on Earth. For a slope of only ~0.5°, the Mars circulations can be as much as 2.5× stronger, while the depth of the circulation can be up to 4× greater than on Earth. Mars mesoscale circulations can more easily dominate over large-scale circulations at the local and regional scale when compared to their equivalents on Earth, particularly in regions with complex topography or large topographic relief. It is important to note that none of the theorized or modeled strong slope winds on Mars have ever been directly measured, although the MSL rover has arrived at Gale Crater, which has significant topographic variations that may provide additional information (Fig. 13). Surface wind measurements have only been obtained in the relatively benign topographic areas of VL1 and VL2, Pathfinder, and the Phoenix lander (*Hess et al.,* 1976, 1977; *Holstein-Rathlou et al.,* 2009; *Murphy et al.,* 1990; *Schofield et al.,* 1997). All these data showed some indication of gentle slope flows superimposed on a larger-scale circulation. Several proposed landing sites for the Mars Exploration Rovers, Spirit and Opportunity, were rejected due to mesoscale model predictions of strong winds that exceeded landing safety tolerances (*Golombek et al.,* 2003; *Kass et al.,* 2003). The high science priority site within Valles Marineris is the most notable example. Within the canyon systems, cross-canyon winds accelerated toward the canyon walls and then turned skyward, jetting several kilometers above the surrounding plateau (*Rafkin and Michaels,* 2003; *Toigo and Richardson,* 2003). A strong along-canyon flow was also present, possibly strengthened by the phasing of the thermal tide that accelerated the winds through the dominantly east-west-facing valley. Interestingly, while the previous mesoscale model studies all predicted strong canyon winds, the phasing in time of the peak winds was different, which indicates that there is still a great deal to be understood about these types of mesoscale wind systems. Likewise, given the limited available data, there is still much to be understood about the performance and validity of Mars mesoscale models.

Another mesoscale feature that is likely common on Mars is the density current. These cold, shallow air masses

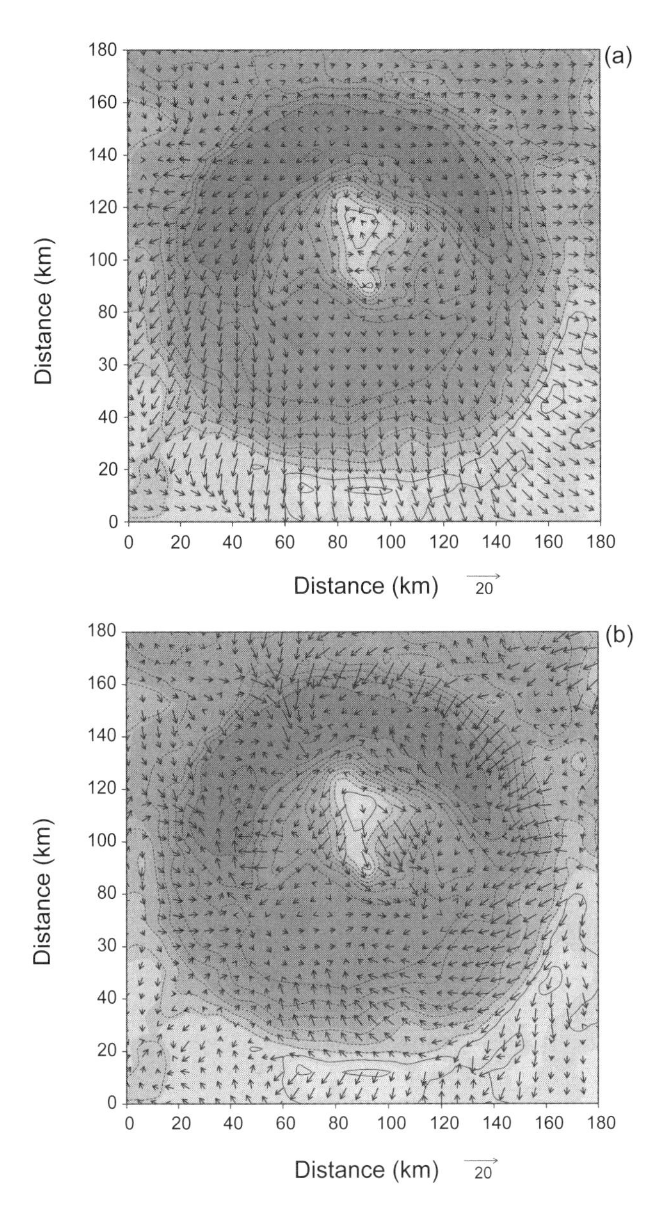

Fig. 13. The mesoscale circulation at Gale Crater during the **(a)** day and **(b)** night, as simulated by the Mars Regional Atmospheric Modeling System (*Rafkin et al.,* 2001). Winds are generally upslope during the day as warm air rises along the slopes. The larger-scale wind system adds to the mesoscale flow. During the day, this results in stronger upslope winds on the south rim of the crater and weaker winds on the north side. At night, the flows reverse as cold, dense air flows downhill. The flows can be complicated due to the production of local slope flows and the collision of these flows in and around the crater. The arrow at the bottom of each panel is a wind vector scale in m s^{-1}. Contours indicate topography.

may be generated at night through radiative cooling and through air mass modification over the ice-covered poles. As this cold, dense air moves downhill, it accelerates. The leading edge of the cold air produces a katabatic front with strong winds that can lift dust, especially as the circulation emerges from the polar cap onto the ice-free regolith (*Parish,* 2002; *Spiga,* 2011; *Tyler and Barnes,* 2005). In some cases, the front will impinge upon existing inversion layers to produce bore waves (*Hunt et al.,* 1981; *Sta. Maria et al.,* 2006) visible only through water ice clouds generated in the rising frontal air and in the wave oscillations that trail behind.

Gravity waves — buoyancy oscillations in the stably stratified atmosphere — are yet another common mesoscale feature on Mars. In the lower atmosphere, the stable nocturnal inversion provides ideal conditions for wave generation (*Holton,* 1992). Waves are easily triggered as the cold, dense night air is perturbed as it flows over topography. Under the right conditions, gravity waves can become trapped and reflected to produce long trains of wave disturbances downwind of the triggering topographic feature, such as a crater (*Pickersgill and Hunt,* 1979, 1981; *Kahn and Gierasch,* 1982; *Pirraglia,* 1976). During the daytime, the unstable convective boundary layer cannot support vertically propagating gravity waves, but the waves can be triggered by the penetration of vigorous convective plumes into the stable air above (*Rafkin et al.,* 2004; *Spiga et al.,* 2008). Waves that are not trapped, but which propagate vertically, must grow in amplitude in the absence of dissipating forces in order to conserve energy under the background of decreasing atmospheric density with height. The energy density of long vertical wavelength gravity waves has been computed using MGS radio science profiles (*Creasey et al.,* 2006). This analysis, however, does not account for the full spectrum of waves, although shorter waves are generally expected to be filtered with height (*Holton,* 1992). The longest waves — those on the planetary scale — can penetrate deep into the thermosphere where the density variations may be observed by (and affect) spacecraft during aerobraking maneuvers (*Fritts et al.,* 2006; *Keating et al.,* 1998). Gravity waves growing in amplitude with height can produce thermal perturbations sufficient to saturate the air not only with respect to water, but also CO_2, and thus produce high-altitude CO_2 clouds (*Spiga et al.,* 2012). Breaking gravity waves can deposit momentum that acts to provide a drag on the mean flow. Gravity wave breaking is an important forcing process in the atmosphere of Earth, and this may be the case for Mars as well (*Barnes,* 1990; *Collins et al.,* 1997; *Joshi et al.,* 1995; *Medvedev et al.,* 2011), but its relative contribution is still debated.

The very strong mesoscale circulations, particularly those associated with the larger topographic features, may also influence the overall dust and water cycles (*Michaels et al.,* 2006; *Rafkin et al.,* 2002). Water ice clouds are often observed around the highest topographic features (*Benson et al.,* 2006; *Wang and Ingersoll,* 2002; *Wilson et al.,* 2008). The relative importance of mesoscale circulations to the dust and water cycle is still uncertain, however. Mesoscale simulations of the Tharsis Montes suggest that the strong thermal circulations can pump water and dust high into the atmosphere in amounts that rival the assumed transport by the large-scale circulation (Fig. 14). These strong circula-

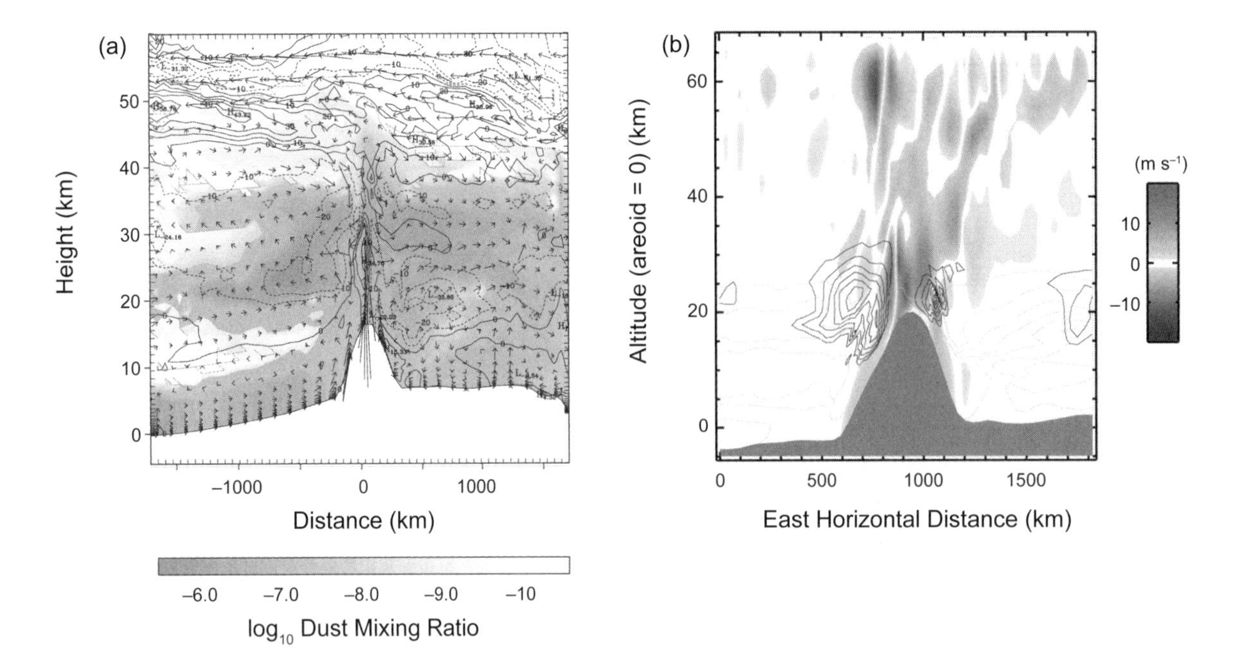

Fig. 14. See Plate 22 for color version. **(a)** Dust and **(b)** water transport at Olympus Mons, Mars, as predicted from a mesoscale model (*Rafkin et al.,* 2002; *Michaels et al.,* 2006). Dust (brown shading) is lifted into the boundary layer and advected vertically along the slopes of the mountain where it is ejected in upper level outflow. Water vapor (green contours) follows a similar trajectory, except that the rising moist air (red shades are positive vertical velocity, blue is negative) can also produce clouds (black contours). The thermal circulation from the mountain produces elevated or detached layers of dust and water vapor.

tions, as well as those associated with the numerous other hills, ridges, and crater valleys, also show evidence of contributing to the vertical transport of dust and water (*Rafkin,* 2012). This mesoscale transport process is not unlike the venting of tropospheric material through deep convection, or the topographic venting of pollutants by deep thermal circulations on Earth (*Cotton et al.,* 1995; *Wakimoto and McElroy,* 1986). More theoretical, observational, and modeling work is required to refine the importance of mesoscale phenomena in the martian climate system. New observations and data analysis are particularly important to provide validation of the models. Direct observation of many modeled features, such as dusty updraft plumes, may be difficult to observe directly (*Määttänen et al.,* 2009), but the indirect effects of the mesoscale circulations can provide valuable clues (*Rafkin,* 2012). Elevated detached dust layers are one such example of observations that may be consistent with transport by mesoscale systems (*Heavens,* 2010; *Heavens et al.,* 2011; *McCleese et al.,* 2010). *Spiga et al.* (2013) modeled a process where dusty plumes can produce detached dust layers by gaining buoyancy through radiative heating.

Many dust storms begin at the local and regional scale as a result of lifting within mesoscale disturbances. Some storms have clear origins: lifting along fronts (*Wang,* 2007) or lifting from strong katabatic winds emanating from the polar caps (*Toigo et al.,* 2002). Other storms have no obvious genesis mechanism, and these storms often appear very convective in nature. Positive feedbacks and instabilities in mesoscale circulations may play a role in determining the growth of these unexplained storms. The basic feedback mechanism for this was envisioned decades ago in the form of a dusty hurricane (*Gierasch and Goody,* 1973), and has recently been refined with the aid of mesoscale numerical modeling (*Rafkin,* 2009). The basic process involves the radiative heating of an existing small dust disturbance. This heating hydrostatically lowers the surface pressure, which leads to an acceleration of surface winds that can lift more dust into the system. The increase in atmospheric dust leads to further heating an overall amplification of the system. Additional observations that can measure pressure, temperature, winds, and dust within these systems are needed to test the feedback hypothesis.

The strong superadiabatic layer near the surface in the afternoon (Fig. 5) results in vigorous turbulent motion. Large eddy simulations and parameterizations indicate that eddy covariances may be an order of magnitude or larger than those on Earth (*Michaels and Rafkin,* 2004; *Sorbjan,* 2007; *Spiga et al.,* 2010; *Toigo et al.,* 2003; *Tyler et al.,* 2008). Even so, the magnitudes of the turbulent fluxes are much smaller than on Earth due to the much lower density; the Mars atmosphere works harder but accomplishes less. The work that the turbulence does do, however, is opposite to that of Earth. On Earth, turbulent eddies act to heat the atmosphere through heat flux convergence. It is primarily through this mechanism that the heat obtained through conduction by the air in immediate contact with the warming ground is transported through the boundary layer. In contrast, Mars' atmosphere is heated dominantly by radiation (*Savijärvi and Siili,* 1993; *Savijärvi and Kauhanen,* 2008) and the turbulent eddies act to cool the atmosphere (*Haberle et al.,* 1993; *Michaels and Rafkin,* 2004). The different behavior between the two planets is explained by the substantial differences in density and the much greater importance of direct radiative heating of the atmosphere on Mars compared to Earth. As the Sun rises and heats the surface of Mars, the atmosphere absorbs an increasing amount of upward-directed infrared radiation. The flux convergence of this radiation in the thin martian atmosphere results in massive direct heating of the near-surface layer. Turbulent eddies respond to this heating and attempt to transport the heat upward (i.e., the eddies act to cool the radiatively heated layers), but the low mass of the martian atmosphere makes convective transport inefficient. In contrast, the atmospheric surface layer of Earth is not heated as strongly by radiation. The higher atmospheric mass implies a smaller increase in temperature for a given energy input, and Earth's atmosphere is not nearly as opaque to infrared radiation as Mars. Furthermore, the higher density means that turbulent eddies can effectively heat the atmosphere directly. On Earth, eddy flux convergence in the atmospheric surface layer is of the opposite sign of that on Mars, and eddies heat the atmosphere. Above the surface layer on Mars (typically a few hundred meters in a fully developed convective layer), the radiative heating falls to a value where eddies can compete with radiative forcing and the sign of the eddy flux convergence reverses; eddies begin to heat the atmosphere, as they do on Earth.

Given the fundamental differences in the behavior of turbulence between Earth and Mars, it is somewhat surprising that the atmospheric surface layer turbulence between the two planets seems to follow similar energy and enstrophy cascade relationships (*Larsen et al.,* 2002; *Tillman et al.,* 1994). Caution must still be exercised, however. Eddy fluxes have yet to be properly measured on Mars, and because the turbulent fluxes on Mars represent only a modest correction to the radiative equilibrium solution, the formulations and parameterizations used to infer the turbulence on Mars could still be significantly in error. In particular, the Monin-Obukhov similarity theory parameterization of surface fluxes developed for terrestrial models is based on the application of Buckingham π-theory (*Stull,* 1988), and there is no *a priori* reason to expect that the dimensional parameters selected for Earth should also apply to Mars. For example, the radiative heating rate (or similar) may be an extremely important parameter for Mars, but it is not considered a relevant parameter in terrestrial atmospheric surface layer flux theory.

3.3. Dust Cycle

Dust raised from the surface has a major impact on the radiative balance of the thin martian atmosphere, affecting absorption and scattering of visible and solar radiation (*Kahn et al.,* 1992), and therefore also atmospheric temperatures and, hence, winds. For example, significant amounts of lofted dust can greatly strengthen the Hadley cell, also broadening it such that it extends from pole to pole at its upper levels (*Haberle et al.,* 1982; *Wilson,* 1997).

Instruments on Mars orbiters, including Mariner 9, Vikings 1 and 2, Phobos, MGS, Mars Observer (MO), Mars Express (MEx), and MRO, have been used to study the seasonally evolving distribution of temperature and column dust opacity (e.g., *Smith,* 2004), and in many cases also the variation of dust with height (e.g., *Heavens et al.,* 2011) and/or the distribution and evolution of particle sizes and optical properties (e.g., *Wolff et al.,* 2009). Such datasets provide vital constraints on our description of both the evolving dust distribution and its impact on atmospheric heating rates. *In situ* measurements (from the Viking landers to the Mars Exploration Rovers) have provided insight into dust particle properties (e.g., *Tomasko et al.,* 1999), dust devils, and (if meteorological instruments are included in the payload) atmospheric conditions related to dust phenomena, as well as surface monitoring of the evolution of dust opacity over time (*Lemmon et al.,* 2004).

Throughout a martian year, the background 9-μm dust opacity typically ranges from <0.05 to ~ 0.2 (*Martin and Richardson,* 1993; *Smith,* 2004), and is maintained by small-scale dust lifting processes such as dust devils and wind gusts (e.g., *Basu et al.,* 2004; *Ferri et al.,* 2003). Stronger or more prolonged high wind stresses may lift enough dust to form local, regional, or global dust storms. As defined in *Zurek and Martin* (1993), local dust storms are those under ~2000 km in length, regional storms are those over ~2000 km in length but not covering all longitudes, and global storms are regional storms that spread to cover all longitudes (and sometimes all latitudes as well). The largest regional storms typically consist of several dust-lifting centers, and occasionally the dust cloud(s) produced will trigger more widespread lifting and expand to become global in extent, blanketing most or all latitudes in dust with opacities $\gg 1$, although the dust lifting itself does not occur everywhere on the surface.

The record of spacecraft observations of global storms began with Mariner 9 in 1971, and telescopic observations of events now interpreted as global storm events go back to ~1800, with the 1956 (MY1) global storm being the first to be observed in any detail (for an excellent review, see, e.g., *McKim,* 1999). More recent "mapping" observations from orbit (from instruments such as the MGS Mars Orbiter Camera) have resulted in hugely improved monitoring and resolution of dust storm onset, expansion, and decay phases (e.g., *Cantor et al.,* 2001; *Strausberg et al.,* 2005), and of the resultant change in surface albedo due to dust rearrangement by storms (*Szwast et al.,* 2006), as well as providing additional statistics on dust devils, including the surface tracks they leave behind if dust removal produces detectable albedo changes (e.g., *Fisher et al.,* 2005). The most recent global storms occurred in 2001 (MY25, onset at $L_s \sim 183°$) and 2007 (MY28, onset at $L_s \sim 265°$) and were observed in unprecedented detail by the MGS, MEx, and MRO spacecraft (e.g., *Cantor,* 2007; *Smith,* 2009; *Smith et al.,* 2002).

Global storms have been detected in roughly one out of every three Mars years (*Zurek and Martin,* 1993; *McKim,* 1999; *Liu et al.,* 2003), and all have occurred during the southern spring and summer "dust storm season" (L_s ~ 180°–330°). This is a time of peak circulation strength due to a combination of (1) the hemispheric dichotomy in topography; (2) perihelion (the time of peak solar insolation) occurring at L_s ~ 251°, shortly before southern summer solstice; and (3) higher background dust opacities and air temperatures that result from the climate feedback between insolation, dust lifting, and opacity (*Richardson and Wilson,* 2002b). This all results in stronger near-surface wind stresses and dust lifting, which are further enhanced by predominantly positive feedbacks on additional lifting. On the local scale, for example, strong winds are driven by large temperature gradients at the edge of dust clouds, while on larger scales, increased atmospheric dustiness strengthens the global circulation (*Newman et al.,* 2002a,b). Such positive feedbacks lead to the expansion of local storms to become regional, and on occasion global, events.

Dust injection, dust storm decay mechanisms, and why global storms develop in some years but not others are probably the least-well-understood aspects of the martian dust cycle. The dust injection processes controlling even the "background" dust distribution, which varies significantly — although rather predictably — over much of the year, are unclear. There is little information on what fraction of this dust is raised by small convective systems vs. larger-scale systems (e.g., lifting associated with strong surface winds at the cap edge), or how it is then mixed through the boundary layer and into the lower levels of the "free" atmosphere. As described above and shown in Fig. 3, recent MCS observations have shown that this "background" dust is not uniformly mixed up to some altitude, as previously assumed (*Heavens et al.,* 2011). This is particularly true in northern summer, when a significant dust mass mixing ratio maximum occurs above ~10–15 km over the tropics, suggesting a deep emplacement mechanism not associated with dust storm events.

Properly representing dust injection, including the massive amount that occurs during dust storms, requires both understanding of dust lifting/mixing processes *and* knowledge of surface dust source type and availability (*Pankine and Ingersoll,* 2002). Unlike Earth, for which dust-lifting thresholds and fluxes have been measured *in situ*, and for which surface dust abundances and size distributions are reasonably well known, on Mars such data are either very limited or entirely absent, although changes in surface albedo and thermal inertia may be used to at least qualitatively infer the changing surface distribution of dust (e.g., *Szwast et al.,* 2006). *Ruff and Christensen* (2002) also used a short wavelength MGS TES absorption feature associated with small (<<100 μm) surface silicate particles to define a dust cover index, low values of which indicate low emissivity over this wavelength band and hence more surface dust.

Studies examining changes in Mars surface albedo, etc., over multiple years indicate that surface dust is redistributed by storm events, and it seems improbable that regions with the highest wind stresses are able to retain enough surface dust to act as source regions over tens to thousands of Mars years. Indeed, the MGS TES albedo data indicate that individual large dust storms can deplete major dust source sites, and also demonstrate that fallout of dust following large dust storms can be subsequently removed from large areas of the surface prior to the next storm season (*Szwast et al.,* 2006). Thus source dust availability and its evolution over a multiyear period likely plays a major role in the dust cycle. Dust storm decay may be related to the depletion of source dust, with observations suggesting that full depletion of key source regions may be the cause of at least some observed storms "shutting off" (*Szwast et al.,* 2006). Another possibility is that high atmospheric opacities "decouple" the near-surface circulation from that lofting, reducing the surface wind stress and shutting off further dust lifting.

While it is unclear what fraction of the observed interannual variability in major storms is due to surface dust availability, observations indicate that intrinsic atmospheric variability also plays a role. For example, northern autumn and winter storms that propagate from northern high latitudes toward the equator appear to require that tides and planetary waves be "in phase" (*Wang et al.,* 2003). In either case, the prediction of martian dust storms seems to require detailed knowledge of the system (the atmospheric state and/or surface dust distribution), and modeling of storms on Mars also suggests that the onset and growth of major storms is highly sensitive to initial conditions (e.g., *Wang et al.,* 2003; *Newman et al.,* 2005). To date, however, most modeling of Mars dust storms and the dust cycle have assumed an infinite reservoir of surface dust, and studies that attempt to consider a limited supply of surface dust are still in their infancy.

On Earth, restricted surface dust availability limits the magnitude of terrestrial dust storms (although roughly a third of the land surface is dust-producing), and the presence of widespread oceans and liquid water not only contributes to the lack of surface dust but also means that much of the atmospheric dust is "scavenged" by serving as cloud condensation nuclei. In addition, airborne dust has far less ability to significantly impact temperatures in Earth's thick, moist atmosphere. For such reasons, Earth does not experience global dust storms, although each year as much as an estimated 2 billion metric tons (2×10^{12} kg) of dust are lofted into the atmosphere (e.g., *Tegen et al.,* 1996; *Ginoux et al.,* 2001). However, many desert regions (e.g., the Sahara and the deserts of Mongolia and northern China) produce large regional storms and dust clouds that can travel around the globe (e.g., *Swap et al.,* 1992; *Sun et al.,* 2001). Such storms generally depend very sensitively on atmospheric and surface conditions at high resolution, from meteorology to surface roughness to surface sources and their particle size distributions. Predicting dust storms on Earth thus remains challenging, even with Earth's abundant space- and groundbased datasets (e.g., *Schulz et al.,* 1998; *Nickovic et al.,* 2001; *Perez et al.,* 2006; *Ridley et al.,* 2012). This further highlights the difficulty of predicting such events on Mars. As the dust cycle is crucial to the atmospheric circulation, and also to the CO_2 and water cycles via radiative feedbacks, scavenging, etc., the challenges in predicting dust lifting and distributions have a major impact on the prediction of the martian climate in general.

3.4. Water Cycle

Water influences the martian climate through the radiative impact of water ice — in the form of surface ices that significantly impact the surface albedo and atmospheric cloud ice, which can significantly impact the heating of both the surface and atmosphere in both the visible and thermal infrared (*Zurek et al.,* 1992; *Wilson et al.,* 2008; *Madeleine et al.,* 2012). Atmospheric cloud ice may be of secondary importance as a means of sequestering of dust as cloud condensation nuclei, consequently providing changes in radiative heating (*Rodin et al.,* 1999). As a result, in the martian climate system, water plays a subset of the roles it plays on Earth. This is a consequence of both the lower temperatures and pressures on Mars (which mostly preclude the presence of liquid water at the surface) and the much lower relative abundance of water.

As with the dust cycle, the water cycle is strongly influenced by orbital eccentricity and the timing of perihelion (*Jakosky,* 1983a,b; *Richardson and Wilson,* 2002a). The exposure of extensive surface water ice from beneath seasonal CO_2 only during northern summer leads to a significant increase in atmospheric vapor during this season (*Jakosky and Farmer,* 1982; *Smith,* 2002). Thus, this is also the season of greatest water ice (cloud) opacity, visible in both spacecraft and groundbased images of Mars (*Wang and Ingersoll,* 2002; *Clancy et al.,* 1996). The cloud distribution shows a strong diurnal variation, associated with the propagation of the thermal tide (*Lee et al.,* 2009; *Kleinböhl et al.,* 2013), but shows a generally lower and thicker tropical belt in northern summer (*McCleese et al.,* 2010). Although the variables necessary to adequately constrain the water cycle have not all been measured — and those that have been obtained were not measured simultaneously — observations relevant to the water cycle were made by the Viking orbiters, and more recently have been obtained nearly continuously by MGS, MEx, and MRO. These observations are sufficient to draw some general conclusions about the water cycle, but only with the aid of numerical models.

Simple atmospheric diffusion models and GCMs show that it is the longer and cooler northern summer that makes the northern pole the preferred location for the residual water ice cap. The existence of only one residual cap and the biasing of its location to the aphelion summer pole is fundamentally related to the nonlinear dependence of vapor pressure upon temperature. As such, a reservoir experiencing a longer period at lower temperature (the aphelion summer pole) is stable relative to one that experiences a shorter period at higher temperatures (the perihelion summer pole) after integrating over the whole year (*Jakosky,* 1983a,b; *Richardson and Wilson,* 2002a). This biasing may have an analog in Titan's climate system with surface liquid methane deposits.

While models show that the bulk of the water cycle can be explained to first order as the consequence of annual thermal forcing of the northern residual water ice cap and the consequent involvement of water in the seasonal ice caps (*Jakosky,* 1983a,b; *Richardson and Wilson,* 2002a), there are potentially significant roles that might be played by clouds and/or regolith exchange (*Richardson et al.,* 2002; *Montmessin et al.,* 2004; *Böttger et al.,* 2005). Clouds may provide an additional mechanism for containing water within the aphelion summer atmosphere through the depression of the hydropause, i.e., the restriction of atmospheric transport of vapor by the formation and sedimentation of cloud ice particles primarily within the Inter Tropical Convergence Zone (ITCZ, or the "tropical cloud belt") (*Clancy et al.,* 1996; *Richardson et al.,* 2002; *Montmessin et al.,* 2004). Similarly, the regolith may impede equatorward transport of water vapor from the subliming polar cap through adsorption (*Jakosky,* 1983b; *Böttger et al.,* 2005). A major argument for the plausibility of a significant role for regolith exchange on global vapor abundance is the significant drop in vapor during the 1977b global dust storm. Global circulation model simulations suggest that the cooling of daytime surface temperatures due to elevated atmospheric visible optical depths during the storm could have yielded increased water uptake by regolith adsorption (*Böttger et al.,* 2004). Indeed, both clouds and the regolith are likely essential for explaining the specific amounts and distributions of water vapor observed in the atmosphere, in addition to the primary ice cap source/sinks. For example, the Phoenix lander observations of near-surface vapor and precipitating atmospheric ice suggest a very dynamic interchange of atmospheric and surface/subsurface water, at least at Phoenix's northern high latitude landing site (*Whiteway et al.,* 2009; *Zent et al.,* 2010).

Major challenges remain for extracting further understanding of the water cycle. One major problem is how best to use and combine the disparate observations to provide decisive constraints on the system. Most models have been only rather loosely and qualitatively tested against column vapor observations. It seems at least plausible that more constraints can be extracted from the data through application of more rigorous testing against multiple simultaneous datasets, potentially within the framework of data assimilation that yields information on the patterns of model biases relative to observations (*Lee et al.,* 2011; *Greybush et al.,* 2012). The water cycle on Mars is also a highly transport-dependent system, and observations of other gases in the martian atmosphere with much simpler source/sink terms [namely argon as observed from the Mars Odyssey orbiter (*Sprague et al.,* 2007)] suggest that current Mars GCMs do a rather poor job with atmospheric transport, especially in locations of large gradients such as may occur across the polar front, across the hygropause, and in other critical locations for the water cycle (*Lian et al.,* 2012). Finally, most modeling to date has not coupled the water and dust cycles. While it is debatable whether either system is adequately (i.e., the basic mechanisms are modeled) understood to accomplish this, it now seems very likely that the radiative effects of each system affects the other, and that dust and ice microphysics are intimately coupled in that ice appears only to nucleate on dust grains.

While the well-known temperature differences between Mars and Earth produce a significant effect in terms of the (effective, if not complete) elimination of the liquid phase from the climate system (*Zurek et al.,* 1992), the huge role played by the enormous differences in the climatically available inventory of water is often overlooked. Even if raised to terrestrial temperatures, the martian climate system would remain remarkably arid, and both clouds and precipitation (especially rain) would be extremely restricted. In this regard, the martian water cycle is profoundly different from that of Earth. While thermal and dynamical constraints determine the hydrological cycle on Earth, with effectively infinite water available at the surface at all latitudes (i.e., oceans do not become exhausted), on Mars the restricted presence of water is the first-order constraint over both thermal and dynamical considerations.

4. PAST CLIMATE AND CLIMATE EVOLUTION

The long-term evolution of the martian climate is a remarkable example of the interplay of slowly evolving exogenic and endogenic forces as they modify a planet's environment on a global scale over billions of years. The visual and geochemical evidence is compelling that Mars had liquid water at its surface for significant periods of time early in its history. As liquid water is unstable in Mars' present environment, this points to a climate both warmer and wetter than the present day.

The widespread presence of fluvial surface features such as valley networks and bedrock-incising outflow channels (*Carr,* 1996) point to periodic, if not long-term, episodes where liquid water freely flowed across the surface. Geomorphic evidence also suggests the possibility of widespread surface water in the form of a northern ocean or large sea (*Clifford and Parker,* 2001). Chemical evidence of liquid water comes in the form of the presence of clay minerals during the earliest period of martian history, and the presence of both sulfates and aluminum-rich clay minerals in the stratigraphy (*Ehlmann et al.,* 2009). Such clays require significant amounts of liquid water to develop to maturity. This evidence is consistent with a hydrologic cycle of some form earlier in martian history — a clear departure from today's cold and dry climate.

It is thought that the early atmospheres of Earth, Mars, and Venus began in a largely similar state, from the same building blocks, and that through time, the atmospheric compositions diverged, influenced by the planets' respective geology, chemistry, orbital location, and (in the case of Earth) biology. We can clearly see this divergence at play when we examine the isotopic ratios of atmospheric constituents of the planets. Assuming the terrestrial planets all formed from the same "primordial" building blocks with a similar abundance of volatile isotopes, the ratio of "heavy" to "light" isotopes in an atmospheric sample (on any body) suggests the strength of the processes that have modified it over time. Simply stated, within an atmosphere, a variety of intrinsic and extrinsic processes, such as thermal and hydrodynamic escape of individual molecules and sputtering by the solar wind, preferentially remove (fractionate) the lighter isotope from the system, resulting in enrichment of the heavier isotope. Both thermal and nonthermal processes are at play at Mars, and have resulted in a present-day atmosphere quite removed from its primitive state. In the absence of a global magnetic field, thought to have been lost around 3.5 to 3.7 Ga (*Acuña et al.,* 1999), the solar wind will erode the uppermost layers of the martian atmosphere, already preferentially populated by lighter species and isotopes. Over the history of the planet, this "distillation effect" leads to an enrichment of the heavier isotopes of all elements, but most noticeably that of hydrogen, which can be reflected in the ratio of deuterium to hydrogen (the D/H ratio). If we assume that the primordial composition of both Earth and Mars was similar, and that the present-day composition of the terrestrial oceans remains close to that of ancient mixed water in the solar system (*Fisher,* 2007; *Yung et al.,* 1988), then the observed D/H ratio on Mars, which is about 5.5× that observed in terrestrial standard mean ocean water (SMOW), reflects significant loss of water from the atmosphere and a concomitant loss of atmospheric mass. Models of atmospheric loss indicate that such a 5.5× enrichment factor corresponds to a loss of 60–75% of martian water to space. Comparatively, on Venus, the D/H enrichment is about 100× that of SMOW, indicating near-total loss of atmospheric water. Both Mars and Venus lack a global magnetic field, which would act to offset the impinging solar wind. Additionally, for Mars, its small size makes it easier for an atmospheric molecule to achieve escape velocity and depart the system.

Similarly, enrichment of other atmospheric constituents is consistent with significant atmospheric loss to varying degrees (*Jakosky and Phillips,* 2001). As seen in Table 2,

TABLE 2. Isotopic ratios of several elements on Mars and their enrichment factor.

Ratio	Earth	Mars	Enrichment Factor
D/H	1.56×10^{-4}	9×10^{-4}	~5.5
$^{12}C/^{13}C$	89	90	1.05
$^{16}O/^{18}O$	489	490	1.025
$^{14}N/^{15}N$	272	170	0.6
$^{36}Ar/^{38}Ar$	5.3	4.2	0.75
$^{129}Xe/^{132}Xe$	0.97	2.5	2.5

Data from *Jakosky and Phillips* (2001), *Chassefière and Leblanc* (2004), and *Owen* (1992).

enrichment of heavy carbon and oxygen isotopes is small, which is consistent with the idea that most volatile carbon and oxygen are bound in atmospheric CO_2, a much heavier molecule than other atmospheric gases, leading to less fractionation. Enrichment of nitrogen and argon support the argument that the primitive atmosphere must have been thicker. Most recently, results from the SAM instrument on MSL have refined the enrichment factor of argon to a value of ~4.2 (S. Atreya, personal communication), substantially below the solar (and also terrestrial) ratio of ~5.3–5.5. Argon is an excellent indicator of atmospheric loss since, as a noble gas, it has few chemical reactions and does not condense or otherwise sequester at or below the surface once in the atmosphere.

The implication of this widespread enrichment is that the early martian atmosphere must have been substantially thicker than at present and composed of many of these simple (and infrared-absorbing) gases. This bolsters the argument that early Mars was a warmer environment than today. The Mars Atmosphere and Volatile Evolution (MAVEN) mission, scheduled to launch in 2013, has been designed to address these issues and understand the evolution of the martian atmosphere and what role the loss of volatiles has played through time.

Combined with the geologic, geochemical, and geomorphological evidence mentioned above, isotopic signatures provide compelling evidence that the earliest period of martian history, the Noachian (>3.7 Ga), had a more clement environment than today. However, the mechanism by which this early climate was not only obtained, but *sustained*, remains unclear. As noted above, a number of hypotheses have been developed positing various mechanisms by which early Mars could have supported a warm climate; however, none appear to be wholly satisfactory. To begin with, the faint young Sun paradox immediately puts many of these theories at a disadvantage, with reduced insolation at the martian surface making it all the more difficult to obtain liquid water (the melting temperature of pure water, 273 K, can be considered a nominal "warm" temperature required for liquid water to exist on Mars; however, the presence of salty brine solutions can decrease this temperature by many tens of Kelvin, partially offsetting the challenge of this paradox).

Given the present composition of Mars' atmosphere, a thicker (multibar) CO_2 atmosphere was originally suggested as a logical means by which warm temperatures could be obtained, due to the strong greenhouse warming generated by CO_2 (e.g., *Pollack,* 1979; *Cess et al.,* 1980; *Hoffert et al.,* 1981, *Postawko and Kuhn,* 1986; *Pollack et al.,* 1987). However, *Kasting* (1991) showed that CO_2 alone would be incapable of providing the necessary warming as, at the multibar pressures required, CO_2 condensation would occur in the middle atmosphere, limiting the greenhouse effect, while Rayleigh scattering by CO_2 molecules would increase significantly, dominating over any greenhouse absorption and resulting in net *cooling* of the atmosphere at pressures above ~5 bar. Precluding this argument are recent analyses of the martian surface morphology (*Craddock and Greeley,* 2009) and crust formation modeling efforts (*Grott et al.,* 2011) that suggest a likely outgassed CO_2 inventory of only between several hundred millibars to 1 bar, well below the capacity of CO_2 alone to provide adequate warming.

Contributions from a variety of other atmospheric gases and aerosols have been proposed as alternative secondary greenhouse contributors, but, like CO_2, all suffer from fatal flaws that have thus far precluded them from providing the needed warming. Gases such as ammonia and methane (*Kuhn and Atreya,* 1979; *Kasting,* 1982) have short photochemical lifetimes in the martian atmosphere. Organic hazes (*Sagan and Chyba,* 1997) are unlikely to be produced in the oxidizing martian environment, while sulfur dioxide (*Postawko and Kuhn,* 1986; *Yung et al.,* 1997, *Halevy et al.,* 2007; *Tian et al.,* 2010) is highly soluble and may precipitate quickly or form sulfate aerosols. Infrared scattering by CO_2 ice clouds has been suggested (*Forget and Pierrehumbert,* 1997; *Mischna et al.,* 2000), but would require a large fractional cloud cover. Recent GCM studies (*Forget et al.,* 2013) show that while CO_2 clouds may be widespread across the planet, they are nonetheless unable to produce suitable levels of warming.

That liquid water did indeed flow across the early martian surface is hardly in doubt, yet the inability of climate models to produce the requisite temperatures remains a vexing problem with no present solution. Considering the wealth of observational evidence for liquid water, it is likely that future paleoclimate modeling efforts will be increasingly focused on providing explanatory answers to the fate of water over time. There still remain large holes in our understanding of early martian climate, and many interconnected processes of unknown importance are absent from even the most sophisticated of climate models. Atmospheric microphysics and photochemistry are two important areas of active climate model development (e.g., *Madeleine et al.,* 2012; *Daerden et al.,* 2010; *Lefèvre et al.,* 2004, 2008, *Lefèvre and Forget,* 2009). Surface and near-surface ice deposits, both of water ice and CO_2, play an uncertain role in buffering the climate (*Phillips et al.,* 2011; *Kahre and Haberle,* 2010; *Haberle et al.,* 2008), and the condensation, flow, and sublimation processes of these deposits may likewise be important. The role of local slopes on preserving such ice deposits will certainly be an important factor as well (*Conway and Mangold,* 2013; *Kreslavsky and Head,* 2003).

During the Noachian, the volcanic flux was many orders of magnitude greater than the present day, and was likely the primary regulator of the environment. The rise of the Tharsis province during this period attests to the sheer magnitude of volcanic activity that was extant. Following the most significant eruptive events, volcanic gases in large quantities were likely regularly and abundantly released into the atmosphere (*Phillips et al.,* 2001; *Jakosky and Phillips,* 2001) — among which the most prominent would have been CO_2, H_2O, SO_2, H_2S, and CH_4 — all greenhouse gases of varying strength. *Wilson and Head* (2002) suggest, for example, that a single large dike intruded radial to one of the Tharsis Montes could inject as much as 60,000 km^3 of magma in a few days' time. This could result in the release of ~10^{14} kg of SO_2 in a single large intrusive event (*Johnson et al.,* 2008). The detection of sulfate salts on the martian surface makes SO_2 and/or H_2S attractive choices of greenhouse gases for continued examination (*Halevy et al.,* 2007; *Johnson et al.,* 2008, 2009; *Tian et al.,* 2010). Recent modeling studies (*Mischna et al.,* 2013) have found that melting temperatures can, in fact, be achieved on early Mars under optimal conditions with assorted trace gases (Fig. 15), although it is not clear how likely it is to obtain such conditions.

The ability to obtain these warm temperatures appears to be tightly linked to the local surface properties, which, for early Mars, are unknown. Present-day surface properties such as albedo and thermal inertia can serve as a proxy, but are likely inadequate to capture the true state of the early Mars environment. The widespread presence of dark, unweathered basalt or liquid water, both of which would be more likely early in martian history, are examples of surfaces with thermophysical properties that deviate from present-day values, and which, incidentally, are better at maintaining longer-term warm conditions. The presence of liquid water on the surface by atmospheric effects alone remains a contentious point to this day (*Forget et al.,* 2013; *Wordsworth et al.,* 2013), but cannot yet be excluded. Its existence may prove to be somewhat self-sustaining given its dark appearance and high thermal inertia, depending on its depth and distribution, and the extent of any fractional ice cover, which itself depends in part on the orientation of the planet.

Perhaps it is the case, however, that the early morphological evidence for liquid water did not require warm temperatures, but rather could have formed during brief, warm interludes within an otherwise cold climate. It has been postulated that hydrothermal convection may have been responsible for much of the fluvial erosion observed on early Mars, enabled by a stronger geothermal heat source (from a warmer, early martian interior) and increased impact flux (*Squyres and Kasting,* 1994; *Gulick,* 2001). The role of impacts on producing periodic (or sustained) episodes of surface warming would be increasingly important as one travels further back in time. During the early Noachian, especially through the late heavy bombardment (~4.1–3.8 Ga), the impact flux was substantially greater than today. Regular impacts of moderate- to large-sized bodies (tens to hundreds of kilometers) would have provided enough heat to produce up to hundreds of meters of hot ejecta, which would have kept the surface temperatures warm for potentially significant periods. The largest impacts (>100 km) may have caused melting of extant polar ice and produced both liquid water reservoirs on the surface and an enhanced abundance of water vapor globally in the atmosphere (*Segura et al.,* 2002). The resulting warm climate would have subsisted from tens to thousands of years and may have experienced periods of heavy precipitation, potentially contributing to the formation of Noachian-aged valley networks. Smaller impactors (~30–100 km) would have resulted in more localized warming conditions that may have existed for decades or centuries (*Segura et al.,* 2008). Through the late Noachian and into the Hesperian and beyond, the impact flux decreased and impacts, especially from larger bodies, became a much less important process for martian climate.

Whether Noachian Mars supported stable liquid water at its surface for extended periods or merely intermittent episodes, it is clear that liquid was once present, whereas today it is not. This climatic dichotomy suggests a period of transition was necessary through the middle (Hesperian) period of martian history, with decreasing frequency of "warm" conditions, and a slow but steady migration into the cold/dry state of Mars today. The presence of liquid water on the martian surface requires greater atmospheric pressure than today (~6.1 mbar), if only to ensure that surface conditions remain above the triple point of water (*Haberle et al.,* 2001). Estimates of the early Mars surface pressure vary significantly, from tens of millibars to several bars, but, as noted above, these estimates are slowly converging on an answer between several hundred to a couple thousand millibars of CO_2 (*Craddock and Greeley,* 2009; *Grott et*

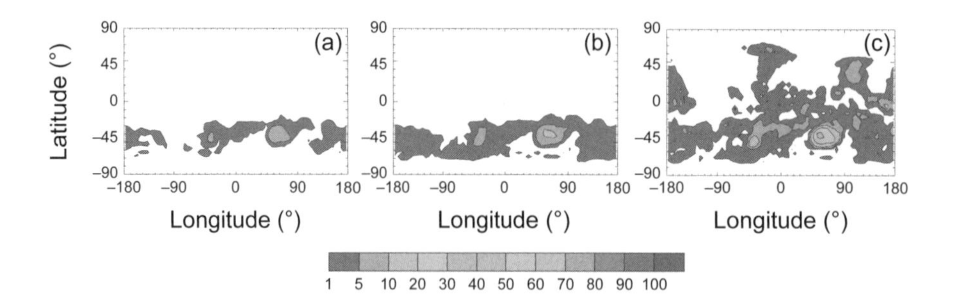

Fig. 15. See Plate 23 for color version. Map view of fraction of the year for which surface temperatures exceed 273 K on Mars for various atmospheric compositions. **(a)** 500-mbar CO_2 atmosphere; **(b)** 500-mb CO_2 atmosphere with water vapor from nominal martian water cycle; **(c)** 500-mbar CO_2 atmosphere with water vapor from nominal martian water cycle plus 245 ppm SO_2. All runs use present-day orbital configuration. Results from *Mischna et al.* (2013).

al., 2011). This quantity of CO_2 must have been lost (or buried) to achieve Mars' present condition. Estimates of loss to space of CO_2 (*Kass and Yung,* 1995) suggest as much as 3 bar of CO_2 could reasonably be removed via sputtering in 3.5 G.y., since the time the planetary dynamo ceased and the global magnetic field was lost (*Acuña et al.,* 1999). The upcoming MAVEN mission will refine this value by measuring the present-day escape rates of various atmospheric gases and determining the net integrated loss to space through time.

During the Hesperian, there is less compelling evidence of *sustained* liquid water, but the presence of the Hesperian-aged outflow channels and valley networks does suggest that liquid water may have flowed, albeit intermittently, across the surface, and therefore temperatures were at least periodically warm (*Mangold et al.,* 2004, 2008). Such catastrophic outflow events may be linked to episodic volcanism and a resulting release of liquid water from the subsurface. Hesperian-aged hydrated silicates such as opal in regions adjacent to Valles Marineris (*Milliken et al.,* 2008) require the presence of water to form, but the low levels of quartz on the surface along with the enhanced levels of opal-A and opal-CT indicate that the conversion to quartz was incomplete, and that the duration of water interaction at the surface was limited after opal emplacement. This is consistent with the idea that the Hesperian achieved only periodic warming episodes that would have supported liquid water at the surface. *Mangold et al.* (2004) have identified valley networks that cut through Hesperian-aged terrain north of Valles Marineris, which suggests precipitation (and warmer conditions) at least periodically through this period. The presence of these younger fluvial features on the surface indicates at least the temporary presence of liquid water at the surface, although the extent and duration of such events appears likely to be much less than during the Noachian. Valley networks dated to the late Hesperian–early Amazonian boundary are thought to be due to more regional warming events (e.g., impacts, local hydrothermal activity) rather than global warming activity (*Mangold et al.,* 2004; *Parsons et al.,* 2013), suggesting a decreased role for global volcanism later in martian history, and a transition to a more obliquity-controlled climate like at present.

General circulation model studies support the idea that global transient warming is possible during this period, following, for example, a large volcanic event (*Mischna et al.,* 2013). Naturally, the duration of such warm events depends on the magnitude and timing of the volcanic activity, but also on the orbital state of the planet and other atmospheric considerations. It becomes more difficult, however, for the Hesperian climate, with its thinner atmosphere, to remain warm.

Although present-day Mars is a cold, dry desert, it is adorned with thick, bright ice caps covering each pole. In the north, the permanent (residual) water ice cap and underlying polar layered deposits are several kilometers thick and composed of extremely pure water ice interspersed with layered material of greater compositional heterogeneity (*Phillips et al.,* 2008). In the south, a smaller permanent water ice cap may be found, coated year-round with a thin (~10 m) veneer of CO_2 ice. The means of maintaining this CO_2 veneer are unknown. It is possible that the permanent CO_2 cap in the south represents the last remnants of a once thicker CO_2 ice layer, or perhaps it is merely a transient occurrence that is nearly fully eroded. Orbital observations of the cap show it to be decaying over time, producing the so-called "swiss cheese terrain" (*Thomas et al.,* 2000; *Byrne and Ingersoll,* 2003). The presence of the CO_2 cap in the south sets south polar surface temperatures at the CO_2 frost point — about 148 K — making it, in effect, a cold trap for atmospheric water vapor. Growth and recession of the polar caps and movement of water vapor into and out of the caps during the year has been discussed previously in section 3.4.

Under early martian conditions, when the atmosphere was assumedly thicker than today, warmer temperatures would have likely precluded the existence of this permanent CO_2 veneer in the south. This would allow for a somewhat greater symmetry in the water cycle (the spatial extent of the two polar caps is unknown, although the polar layered deposits in both hemispheres show a broader ice coverage that was more equal). Warmer temperatures overall would have led to a more humid environment where the role of clouds and precipitation may have been more extensive than we find today, where clouds are thin and precipitation essentially nonexistent.

Throughout the entire early period of Mars' history, changes in Mars' orbit and axial tilt provided a secondary influence on the climate, largely masked by the significance of the ongoing volcanic and impactor activity. Only late in its history did orbital and spin-axis variability become primary drivers of martian climate. Chief among these are planetary obliquity (axial tilt), eccentricity, and the argument of perihelion. While presently at a moderate 25°, the obliquity of Mars oscillates with a period of ~124,000 years and, within the past 20 m.y., has ranged between 15° and 45° (*Laskar et al.,* 2004). In the recent martian history of the Amazonian, changes in obliquity have had the most significant impact on the martian climate. When obliquity is low (<15°), the polar regions are in perpetual twilight and remain cold year-round, serving as the planetary cold traps at the CO_2 frost point temperature. Atmospheric CO_2 will condense at the poles, forming permanent ice caps and resulting in the ultimate collapse of the atmosphere to very low pressures (the noncondensable fraction of the current martian atmosphere is ~4.7%, yielding a nominal minimum pressure of ~0.25 mbar). Atmospheric loss at low obliquity would be limited by the latitudinal growth of a CO_2 cap and the allowable basal pressure and temperature that such a cap could support. *Mellon* (1996) found that the polar caps at their current size could not support more than ~250 mbar of CO_2, corresponding to a depth of ~1 km — sufficient to support the present atmospheric inventory but perhaps not the case earlier in Mars' history. Under a putative thicker, early atmosphere (>250 mbar), total collapse would therefore not take place, as basal melting of the polar caps and equatorward flow at the distal cap edges would result in a quasi-steady-state with CO_2 condensing at the pole and resublimating at the cap edge.

At the other extreme, high obliquity values (>45°) push ice deposits to the tropics where they are more thermodynamically stable. At high obliquities, larger fractions of the polar and mid-latitudes experience alternating seasons of perpetual sunlight and perpetual darkness — a more expansive version of the "midnight Sun" experienced in the polar regions here on Earth. During the height of summer, polar temperatures will, in fact, be greater than those in the tropics; conversely, during winter, temperatures in the polar night are maintained at the CO_2 frost point temperature, well below temperatures in the lower latitudes. However, the cause of the migration of ice to the tropics at high obliquity is not strictly due to these extremes in temperature as much as it is due to the differential water vapor holding capacity of the atmosphere across this range of temperatures (tropical temperatures, annually averaged, remain higher than their polar counterparts, which would otherwise seem to suggest the poles would be colder and should remain as the "cold trap" for water vapor). Water vapor holding capacity is an exponential, not linear, function of atmospheric temperature, where small changes in temperature can lead to substantial changes in the amount of water vapor that can be held in the atmosphere, thus the annually averaged amount of water vapor that can be held in the atmosphere at any location is most heavily influenced by the peak summertime conditions, when the holding capacity is greatest by an exponential factor (*Jakosky and Carr,* 1985; *Mischna et al.,* 2003). So, although annually averaged temperatures in the tropics are greater than at the poles, it is the holding capacity of the polar atmosphere that is, on average, greater, and ice is more likely to form where the average holding capacity is smaller. At high obliquities, this occurs in the tropics. In fact, this represents an application to very different orbital states of the same mechanism found to determine the location of the residual water ice cap in the current climate (*Jakosky,* 1983a,b; *Richardson and Wilson,* 2002a).

Evidence for the migration of ice in the tropics when Mars is at high obliquity is widespread, and locations of glacial deposit remnants are consistent with where numerical models would suggest ice to be deposited. Using a global climate model (*Forget et al.,* 1999), *Forget et al.* (2006) have demonstrated that glacier-like landforms observed on the western flanks of the Tharsis Montes may be the result of adiabatic cooling of ascending air over these large topographic features, followed by atmospheric condensation and precipitation. Accumulation of precipitated ice over thousands of years may build glacial deposits several hundreds of meters thick. The modeled locations of these deposits are in excellent agreement with observed, fan-shaped deposits seen on the mountain flanks (e.g., *Head and Marchant,* 2003).

The eccentricity of Mars' orbit has a secondary influence on recent martian climate (*Laskar et al.,* 2002; *Mischna et al.,* 2003). Unlike variations in obliquity, for which the distance to the Sun and net insolation do not vary, oscillations in eccentricity do affect the amount of incident radiation on the surface. Martian eccentricity cycles every 96,000 years and ranges from near circular (~0.0) to elliptic (~0.13). Presently it is at 0.093. Changes in eccentricity have the effect of changing the Mars-Sun distance at aphelion, with an increase in eccentricity leading to colder (and longer) aphelion climate. For the present-day condition, where aphelion occurs at $L_s = 71°$ (northern summer), an increase in eccentricity will result in longer northern summers/southern winters, and shorter northern winters/southern summers. It also makes the southern summer warmer and northern summer cooler (although the winters are no cooler or warmer due to the limitations of the CO_2 condensation temperature). Due to the polar processes controlling the water cycle described above (*Jakosky,* 1983a,b; *Richardson and Wilson,* 2002a), this asymmetry has the effect of driving volatiles toward the northern hemisphere (which is the cooler of the two hemispheres under this configuration). For the present day, this may also indicate a loss of the permanent CO_2 ice veneer in the south (although the processes controlling the stability of the residual CO_2 ice cap are very poorly understood) and a reduction in the seasonal sublimation of the permanent water ice cap in the north. Conversely, by reducing the eccentricity, northern summers become warmer while southern summers become cooler. The net migration of volatiles trends southward with the northern residual water ice cap becoming less strongly favored. Note, however, that in an ideal system, a change in the hemisphere of the residual water ice cap will not be forced by a change in the eccentricity alone (the eccentricity merely determines how strongly a given hemisphere is favored). In the real system, and for very low eccentricity, the effect of zonal mean topography may become dominant in determining which cap retains the permanent water ice cap. Residual water ice caps at both poles would require substantially more water than is present in the current active climate system.

The strong control of polar temperatures on the stability of caps (*Jakosky,* 1983a,b; *Richardson and Wilson,* 2002a) immediately suggests that the hemisphere with the residual cap will switch with transition in the argument of perihelion. The timing of perihelion dictates the relative length and intensity of the hemispheric summer seasons and has a cycle of about 51,000 years. Under the present configuration, polar cap sublimation in the north is at a near minimum, as the wintertime-at-perihelion configuration ensures cooler temperatures in the north. Under a reverse configuration, with perihelion occurring closer to northern summer solstice, we should expect the northern polar cap to more rapidly be sublimating and water vapor migrating to the colder southern hemisphere. Indeed, GCM simulations by *Montmessin et al.* (2007) show that synchronization of the precession cycle with the northern (southern) summer would drive atmospheric water vapor to the southern (northern) pole. The present configuration of Mars (with perihelion near southern summer) favors ice deposition at the northern pole. When the precession cycle was reversed (about 25,000 years ago), the southern pole would be the preferred water ice trap.

It is the combination of these three components that ultimately drives the climate system through cycles in the distribution of insolation across the planet and, most importantly for the historical climate record, the polar regions. As demonstrated in *Laskar et al.* (2002), over the past few million years, insolation levels in the northern polar region have varied by as much as 30%, oscillating in accordance with the convolved state of the three cycles (Fig. 16). The widely banded polar layer deposits (Fig. 17) may be indicative of periodicity in the martian climate over multiple timescales, and there is evidence that they are linked to this variation in insolation (*Milkovich and Head,* 2005; *Perron and Huybers,* 2009). It has been suggested that alternating bright and dark bands are linked to net cooler and warmer conditions, respectively. Brighter regions, consisting of a larger fraction of water ice, occur during periods of net deposition. Darker regions come about following the sublimation of previously deposited water ice during warmer polar conditions, which leaves behind a dusty lag deposit and an increased fraction of dust in the remaining material (*Mischna and Richardson,* 2005). This admixture of dust and ice appears visually darker. The periodicity in these layered deposits, as well as indications of recently emplaced volatile-rich deposits in the lower and mid latitudes dated to recent times (e.g., *Mustard et al.,* 2001; *Christensen,* 2003; *Head et al.,* 2003, 2005, 2006a,b) provide compelling evidence of the role of the martian orbital cycles on modifying the near-surface environment on Mars in recent time.

5. CONCLUSIONS

Most of the atmospheric phenomena found on Mars also occur on the other terrestrial planetary bodies: mean, global-scale overturning circulations, large-scale storm systems, Rossby waves, mesoscale slope flows, microscale turbulence, clouds, etc. The manifestation of these phenomena in terms of strength, distribution, and frequency can be quite different due to differences in radiative forcing and planetary rotation rate. Unlike the other terrestrial planets, Mars is unique in its strong, seasonal CO_2 condensation cycle. Only Mars has an atmosphere where its dominant gas is volatile. Earth, Titan, and Venus all have clouds, and

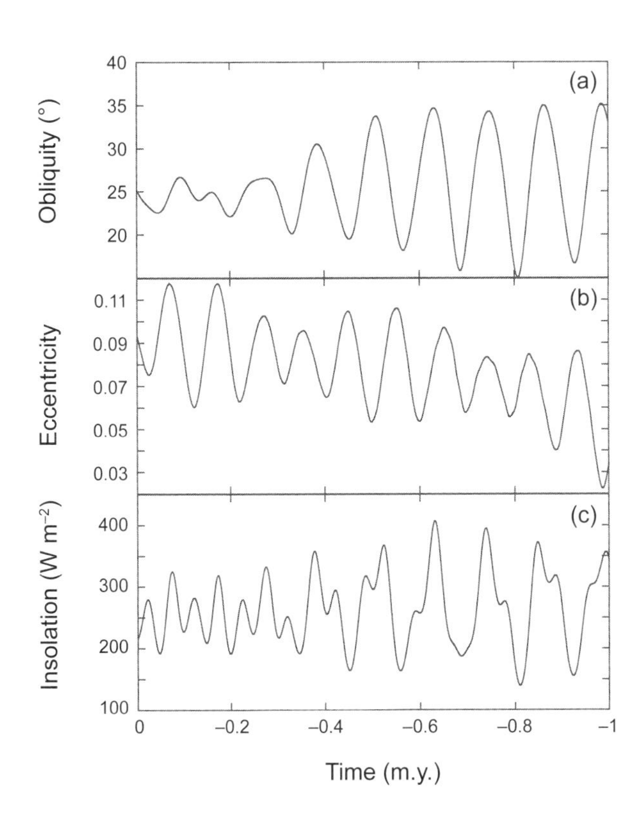

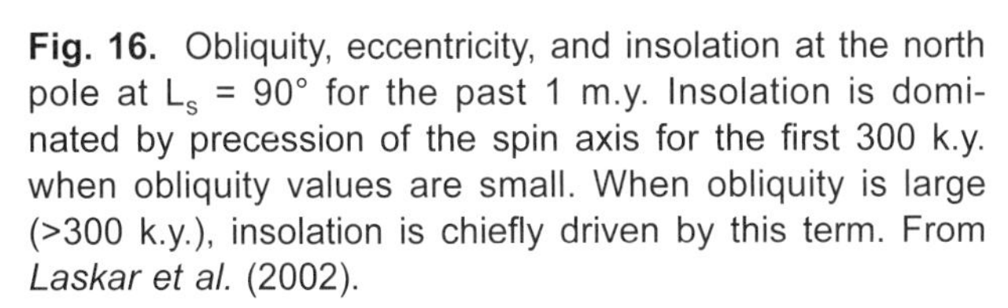

Fig. 16. Obliquity, eccentricity, and insolation at the north pole at L_s = 90° for the past 1 m.y. Insolation is dominated by precession of the spin axis for the first 300 k.y. when obliquity values are small. When obliquity is large (>300 k.y.), insolation is chiefly driven by this term. From *Laskar et al.* (2002).

Fig. 17. Mars Global Surveyor Mars Observer Camera-Narrow Angle (MOC-NA) image M00-02100 showing polar layered deposits exposed on a trough wall. Individual bands of alternating bright and dark coloring suggest periodicity in the climate that created these layers. From *Milkovich and Head* (2005).

in the case of Earth, polar caps, but these all result from the condensation of minor gas constituents. Mars is also distinguished from the other terrestrial bodies in that its total atmospheric mass is substantially lower. The comparatively thin atmosphere and isotopic composition are indicators that Mars has undergone an evolutionary process that differs significantly from its terrestrial counterparts. It appears Mars had an atmosphere that, at one time, could support large, stable bodies of water. That is no longer the case due to the cold temperatures and large swaths of surface topography that are at pressures lower than the triple point.

Comparing and contrasting Mars' weather systems, climate, and climate evolution to other terrestrial atmospheres provides an opportunity to test hypotheses about how atmospheres work and evolve over geologic timescales. Several outstanding questions remain to be addressed in future studies of the martian climate and its meteorology. Apart from the aforementioned differences between the martian atmosphere and those of other terrestrial bodies, fundamental questions remain about basic components of Mars' atmospheric composition and circulation. The atmospheric dust cycle remains a perplexing behavior that warrants additional attention. Mechanisms for the initiation, growth, and decay of global dust storms remain unknown, as does the semi-periodic nature of this behavior (which occurs during some years but not others, with no clear "trigger"). The presence of a residual CO_2 veneer in the southern hemisphere remains perplexing as well, as climate modeling suggests it should not be a stable feature at the surface. While it is possible that we are observing the last remnants of a once greater surface feature, we know from albedo observations that there are basic aspects of the CO_2 cycle of which we are ignorant. In a grander sense, the mechanism(s) by which the planet has evolved from one with a thicker (and warmer) atmosphere to the present day remain unclear. Whether or not Mars was able to sustain liquid water for long periods or only for intermittent epochs remains an open question. Suggestions that early Mars was not "warm and wet" but rather influenced by cold-based phenomena (surface glaciers and ice) reflect a planet yet again different from the prevailing (although inconclusive) paradigm.

Further advances in understanding the climate and weather of Mars will depend on continued observations from both orbit and the surface. Observations of winds are particularly lacking and acquisition of these data would be beneficial to the community. A new class of innovative meteorological sensors — properly competed and accommodated on future spacecraft — that provide information on not just the state of the atmosphere, but the forcing mechanisms as well, are also needed. For example, by measuring turbulent fluxes of heat and momentum and radiative fluxes, the energy inputs that collectively drive the atmosphere can be simultaneously measured for the first time. More of the same, previously acquired standard measurements are helpful, but are not likely to substantially advance the science. Given new information about winds and forcing mechanisms, it becomes possible to validate models, and to provide greater insight into the processes driving the atmosphere of Mars.

Acknowledgments. Part of the research was carried out at the Jet Propulsion Laboratory, California Institute of Technology, under a contract with the National Aeronautics and Space Administration.

REFERENCES

Acuña M. H., et al. (1999) Global distribution of crustal magnetization discovered by the Mars Global Surveyor MAG/ER experiment. *Science, 284,* 790–793.

Andrews D. G., Holton J. R., and Leovy C. B. (1987) *Middle Atmosphere Dynamics.* Academic, San Diego. 489 pp.

Atreya S. K., Mahaffy P. R., and Wong A.-S. (2007) Methane and related trace species on Mars: Origin, loss, implications for life, and habitability. *Planet. Space Sci., 55,* 358–369.

Atreya S., Squyres S., Mahaffy P., Leshin L., Franz H., Trainer M., Wong M., McKay C., Navarro-Gonzalez R., and the Mars Science Laboratory ScienceTeam (2013) MSL/SAM measurements of non- condensable volatiles, comparison with Viking lander, and implications for seasonal cycle. *EGU General Assembly 2013, 15,* EGU2013-1649.

Baker V. R., Strom R. G., Gulick V. C., Kargel J. S., Komatsu G., and Kale V. S. (1991) Ancient oceans, ice sheets and hydrological cycle on Mars. *Nature, 352,* 589–594.

Banfield D., Conrath B. J., Smith M. D., Christensen P. R., and Wilson R. J. (2003) Forced waves in the martian atmosphere from MGS TES nadir data. *Icarus, 161,* 319–345.

Banfield D., Conrath B. J., Gierasch P. J., Wilson R. J., and Smith M. D. (2004) Traveling waves in the martian atmosphere from MGS TES nadir data. *Icarus, 170,* 365–403.

Barnes J. R. (1980) Time spectral analysis of midlatitude disturbances in the martian atmosphere. *J. Atmos. Sci., 37,* 2002–2015.

Barnes J. R. (1981) Midlatitude disturbances in the martian atmosphere: A second Mars year. *J. Atmos. Sci., 38,* 225–234.

Barnes J. R. (1990) Possible effects of breaking gravity waves on the circulation of the middle atmosphere of Mars. *J. Geophys Res., 95,* 1401–1421.

Barnes J. R., Pollack J. B., Haberle R. M., Zurek R. W., Leovy C. B., Lee H., and Schaeffer J. (1993) Mars atmospheric dynamics as simulated by the NASA-Ames general circulation model II. Transient baroclinic eddies. *J. Geophys. Res., 98,* 3125–3148.

Barnes J. R., Haberle R. M., Pollack J. B., Lee H., and Schaeffer J. (1996) Mars atmospheric dynamics as simulated by the NASA Ames general circulation model, 3. Winter quasi-stationary eddies. *J. Geophys. Res., 101,* 12753–12776.

Basu S., Richardson M. I., and Wilson R. J. (2004) Simulation of the martian dust cycle with the GFDL Mars GCM. *J. Geophys. Res., 109,* E11006, DOI: 10.1029/2004JE002243.

Benson J. L., James P. B., Cantor B. A., and Remigio R. (2006) Interannual variability of water ice clouds over major martian volcanoes observed by MOC. *Icarus, 184,* 365–371.

Böttger H. M., Lewis S. R., Read P. L., and Forget F. (2004) The effect of a global dust storm on simulations of the martian water cycle. *Geophys. Res. Lett., 31,* L22702, DOI: 10.1029/2004GL021137.

Böttger H. M., Lewis S. R., Read P. L., and Forget F. (2005) The

effects of the martian regolith on GCM water cycle simulations. *Icarus, 177,* 174–189.
Byrne S. and Ingersoll A. P. (2003) A sublimation model for martian south polar ice features. *Science, 299,* 1051–1053.
Cantor B. (2007) MOC observations of the 2001 Mars planet-encircling dust storm. *Icarus, 186,* 60–96.
Cantor B. A., James P. B., Caplinger M., and Wolff M. J. (2001) Martian dust storms: 1999 Mars Orbiter Camera observations. *J. Geophys. Res., 106,* 23653–23687.
Carr M. H. (1996) *Water on Mars.* Oxford Univ., New York. 229 pp.
Cess R. D., Ramanathan V. and Owen T. (1980) The martian paleoclimate and enhanced atmospheric carbon dioxide, *Icarus, 41,* 159–165.
Chassefière E. and Leblanc F. (2004) Mars atmospheric escape and evolution; interaction with the solar wind. *Planet. Space Sci., 52,* 1039–1058.
Christensen P. R. (2003) Formation of recent martian gullies through melting of extensive water-rich snow deposits. *Nature, 422,* 45–48.
Clancy R. T., Grossman A. W., Wolff M. J., James P. B., Billawala Y. N., Sandor B. J., Lee S. W., and Rudy D. J. (1996) Water vapor saturation at low altitudes around Mars aphelion: A key to Mars climate? *Icarus, 122,* 36–62.
Clancy R. T., Wolff M. J., Whitney B. A., Cantor B. A., and Smith M. D. (2007) Mars equatorial mesospheric clouds: Global occurrence and physical properties from Mars Global Surveyor Thermal Emission Spectrometer and Mars Orbiter Camera limb observations. *J. Geophy. Res.–Planets, 112(E4),* E04004.
Clifford S. M. and Parker T. J. (2001) The evolution of the martian hydrosphere: Implications for the fate of a primordial ocean and the current state of the northern plains. *Icarus, 154,* 40–79.
Colaprete A., Barnes J. R., Haberle R. M., Hollingsworth J. L., Kieffer H. H., and Titus T. N. (2005) Albedo of the south pole on Mars determined by topographic forcing of atmosphere dynamics. *Nature, 435(7039),* 184–188.
Collins M., Lewis S. R., Read P. L., and Hourdin F. (1996) Baroclinic wave transitions in the martian atmosphere. *Icarus, 120,* 344–357.
Collins M., Lewis S. R., and Read P. L. (1997) Gravity wave drag in a global circulation model of the martian atmosphere: Parameterisation and validation. *Adv. Space Res., 19,* 1245–1254.
Conrath B. J. (1975) Thermal structure of the martian atmosphere during the dissipation of the dust storm of 1971. *Icarus, 24,* 36–46.
Conrath B. J., Pearl J. C., Smith M. D., Maguire W. C., Christensen P. R., Dason S., and Kaelberer M. S. (2000) Mars Global Surveyor Thermal Emission Spectrometer (TES) observations: Atmospheric temperatures during aerobraking and science phasing. *J. Geophys. Res., 105,* 9509–9519.
Conway S. J. and Mangold N. (2013) Evidence for Amazonian mid-latitude glaciation on Mars from impact crater asymmetry. *Icarus, 225,* 413–423.
Cotton W. R., Alexander G. D., Hertenstein R., Walko R. L., McAnelly R. L., and Nicholls M. (1995) Cloud venting — A review and some new global annual estimates. *Earth-Sci. Rev., 39,* 169–206.
Craddock R. A. and Greeley R. (2009) Minimum estimates of the amount and timing of gases released into the martian atmosphere from volcanic eruptions. *Icarus, 204,* 512–526.
Creasey J. E., Forbes J. M., and Hinson D. P. (2006) Global and seasonal distribution of gravity wave activity in Mars' lower atmosphere derived from MGS radio occultation data. *Geophys. Res. Lett., 33,* L01803, DOI: 10.1029/2005GL024037.
Daerden F., Whiteway J. A., Davy R., Verhoeven C., Komguem L., Dickinson C., Taylor P. A., and Larsen N. (2010) Simulating observed boundary layer clouds on Mars. *Geophys. Res. Lett., 37,* L04203, DOI: 10.1029/2009GL041523.
Ehlmann B. L., et al. (2009) Identification of hydrated silicate minerals on Mars using MRO-CRISM: Geologic context near Nili Fossae and implications for aqueous alteration. *J. Geophys. Res., 114,* E00D08, DOI: 10.1029/2009JE003339.
Ferri F., Smith P. H., Lemmon M., and Rennó N. O. (2003) Dust devils as observed by Mars Pathfinder. *J. Geophys. Res., 108,* 5133, DOI: 10.1029/2000JE001421.
Fisher D. (2007) Mars' water isotope (D/H) history in the strata of the north polar cap: Inferences about the water cycle. *Icarus, 187,* 430–441.
Fisher J. A., Richardson M. I., Newman C. E., Szwast M. A., Graf C., Basu S., Ewald S. P., Toigo A. D., and Wilson R. J. (2005) A survey of martian dust devil activity using Mars Global Surveyor Mars Orbiter Camera images. *J. Geophys. Res., 110,* E03004, DOI: 10.1029/2003JE002165.
Forbes J. M., Bridger A. F. C., Bougher S. W., Hagan M. E., Hollingsworth J. L., Keating G. M., and Murphy J. (2002) Nonmigrating tides in the thermosphere of Mars. *J. Geophys. Res., 107,* 5113, DOI: 10.1029/2001JE001582.
Forget F. (1998) Mars CO_2 ice polar caps. In *Solar System Ices* (B. Schmitt et al., eds.), pp. 477–507. Astrophysics and Space Science Library, Vol. 227, DOI: 10.1007/978-94-011-5252-5_20.
Forget F. and Pierrehumbert R. T. (1997) Warming early Mars with carbon dioxide clouds that scatter infrared radiation. *Science, 278,* 1273–1276.
Forget F., Hourdin F., Fournier R., Hourdin C., Talagrand O., Collins M., Lewis S. R., and Read P. L. (1999) Improved general circulation models of the martian atmosphere from the surface to above 80 km. *J. Geophys. Res., 104,* 24155–24176.
Forget F., Haberle R. M., Montmessin F., Levrard B., and Head J. W. (2006) Formation of glaciers on Mars by atmospheric precipitation at high obliquity. *Science, 311,* 368–371.
Forget F., Wordsworth R., Millour E., Madeleine J.-B., Kerber L., and Leconte J. (2013) 3D modeling of the early martian climate under a denser CO_2 atmosphere: Temperatures and CO_2 ice clouds. *Icarus, 222,* 81–99.
Formisano V., Atreya S., Encrenaz Th., Ignatiev N., and Giuranna M. (2004) Detection of methane in the atmosphere of Mars. *Science, 306,* 1758–1761.
Fritts D. C., Wang L., and Tolson R. H. (2006) Mean and gravity wave structures and variability in the Mars upper atmosphere inferred from Mars Global Surveyor and Mars Odyssey aerobraking densities. *J. Geophys. Res., 111,* A12304, DOI: 10.1029/2006JA011897.
Garcia R. R. (1987) On the mean meridional circulation of the middle atmosphere. *J. Atmos. Sci., 44,* 3559–3609.
Gierasch P. J. and Goody R. M. (1973) A model of a martian great dust storm. *J. Atmos. Sci., 30,* 169–179.
Ginoux P., Chin M., Tegen I., Prospero J., Holben B., Dubovik O., and Lin S.-J. (2001) Sources and global distributions of dust aerosols simulated with the GOCART model. *J. Geophys. Res., 106,* 20255–20273.
Golombek M. P., et al. (2003) Selection of the Mars Exploration Rover landing sites. *J. Geophys. Res., 108,* 8072, DOI: 10.1029/2003JE002074.
Gómez-Elvira J., et al. (2012) REMS: The environmental sensor

suite for the mars science laboratory rover. *Space Sci. Rev., 170(1-4),* 583–640.

Greybush S. J., Wilson R. J., Hoffman R. N., Hoffman M. J., Miyoshi T., Ide K., McConnochie T., and Kalnay E. (2012) Ensemble Kalman filter data assimilation of thermal emission spectrometer temperature retrievals into a Mars GCM. *J. Geophys. Res.–Planets, 117(E11),* E11008.

Grott M., Morschhauser A., Breuer D., and Hauber E. (2011) Volcanic outgassing of CO_2 and H_2O on Mars. *Earth Planet. Sci. Lett., 308,* 391–400.

Gulick V. (2001) Origin of the valley networks on Mars: A hydrological perspective. *Geomorphology, 37,* 241–268.

Haberle R. M., Leovy C. B., and Pollack J. B. (1982) Some effects of global dust storms on the atmospheric circulation of Mars. *Icarus, 50,* 322–367.

Haberle R. M., Houben H. C., Hertenstein R., and Herdtle T. (1993) A boundary layer model for Mars: Comparison with Viking entry and Lander data. *J. Atmos. Sci., 50,* 1544–1559.

Haberle R. M., McKay C. P., Schaeffer J., Cabrol N. A., Grin E. A., Zent A. P., and Quinn R. (2001) On the possibility of liquid water on present-day Mars. *J. Geophys. Res., 106,* 23317–23326.

Haberle R. M., Forget F., Colaprete A., Schaeffer J., Boynton W. V., Kelly N. J. and Chamberlain M. A. (2008) The effect of ground ice on the martian seasonal CO_2 cycle. *Planet. Space Sci., 56,* 251–255.

Halevy I., Zuber M. T., and Schrag D. P. (2007) A sulfur dioxide climate feedback on early Mars. *Science, 318,* 1903–1907.

Head J. W. and Marchant D. R. (2003) Cold-based mountain glaciers on Mars: Western Arsia Mons. *Geology, 31,* 641–644.

Head J. W., Mustard J. F., Kreslavsky M. A., Milliken R. E., and Marchant D. R. (2003) Recent ice ages on Mars. *Nature, 426,* 797–802.

Head J. W., et al. (2005) Tropical to mid-latitude snow and ice accumulation, flow and glaciation on Mars. *Nature, 434,* 346–351.

Head J. W., Marchant D. R., Agnew M. C., Fassett C. I., and Kreslavsky M. A. (2006a) Extensive valley glacier deposits in the northern mid-latitudes of Mars: Evidence for Late Amazonian obliquity-driven climate change. *Earth Planet. Sci. Lett., 241,* 663–671.

Head J. W., Nahm A. L., Marchant D. R., and Neukum G. (2006b) Modification of the dichotomy boundary on Mars by Amazonian mid-latitude regional glaciation. *Geophys. Res. Lett., 33,* L08S03, DOI: 10.1029/2005GL024360.

Heavens N.G. (2010) The impact of mesoscale processes on the atmospheric circulation of Mars. Ph.D. thesis, California Institute of Technology, Pasadena.

Heavens N. G., McCleese D. J., Richardson M. I., Kass D. M., Kleinböhl A., and Schofield J. T. (2011) Structure and dynamics of the martian lower and middle atmosphere as observed by the Mars Climate Sounder: 2. Implications of the thermal structure and aerosol distributions for the mean meridional circulation. *J. Geophys. Res., 116,* E01010.

Held I. M. and Hou A. Y. (1980) Nonlinear axially symmetric circulations in a nearly inviscid atmosphere. *J. Atmos. Sci., 37,* 515–533.

Hess S. L., Henry R. M., Greene G. C., Leovy C. B., Tillman J. E., Ryan J. A., Chamberlain T. E., Cole H. L., Dutton R. G., and Simon W. E. (1976) Preliminary meteorological results on Mars from the Viking 1 Lander. *Science, 193,* 788–791.

Hess S. L., Henry R. M., Leovy C. B., Ryan J. A., and Tillman J. E. (1977) Meteorological results from the surface of Mars: Viking 1 and 2. *J. Geophys. Res., 82,* 4559–4574.

Hinson D. P. (2006) Radio occultation measurements of transient eddies in the northern hemisphere of Mars. *J. Geophys. Res., 111,* E05002, DOI: 10.1029/2005JE002612.

Hinson D. P. and Wang H. (2010) Further observations of regional dust storms and baroclinic eddies in the northern hemisphere of Mars. *Icarus, 206,* 290–305.

Hinson D. P. and Wilson R. J. (2002) Transient eddies in the southern hemisphere of Mars. *Geophys. Res. Lett., 29,* 1154, DOI: 10.1029/2001GL014103.

Hinson D. P., Tyler G. L., Hollingsworth J. L., and Wilson R. J. (2001) Radio occultation measurements of forced atmospheric waves on Mars. *J. Geophys. Res., 106,* 1463–1480.

Hinson D. P., Wilson R. J., Smith M. D., and Conrath B. J. (2003) Stationary planetary waves in the atmosphere of Mars during southern winter. *J. Geophys. Res., 108,* 5004, DOI: 10.1029/2002JE001949.

Hinson D. P., Patzold M., Tellmann S., Hausler B., and Tyler G. L. (2008) The depth of the convective boundary layer on Mars. *Icarus, 198,* 57–66.

Hoffert M. I., Callegari A. J., Hsieh T., and Ziegler W. (1981) Liquid water on Mars — an energy balance climate model for CO_2/H_2O atmospheres. *Icarus, 47,* 112–129.

Hollingsworth J. L. (2003) Cyclogenesis and frontal waves on Mars. In *Workshop on Mars Atmosphere Modelling and Observations,* Granada, Spain. Available online at *www-mars.lmd.jussieu.fr/granada2003/abstract/hollingsworth.pdf.*

Hollingsworth J. L. and Barnes J. R. (1996) Forced, stationary planetary waves in Mars' winter atmosphere. *J. Atmos. Sci., 53,* 428–448.

Hollingsworth J. L. and Kahre M. A. (2010) Extratropical cyclones, frontal waves, and Mars dust: Modeling and considerations. *Geophys. Res. Lett., 37,* L22202, DOI: 10.1029/2010GL044262.

Hollingsworth J. L., Haberle R. M., Barnes J. R., Bridger A. F. C., Pollack J. B., Lee H., and Schaeffer J. (1996) Orographic control of storm zones on Mars. *Nature, 380,* 413–416.

Hollingsworth J. L., Haberle R. M., and Schaeffer J. (1997) Seasonal variations of storm zones on Mars. *Adv. Space Res., 19,* 1237–1240.

Hollingsworth J. L., Kahre M. A., and Haberle R. M. (2008) Mars' southern hemisphere: Influences of the great impact basins on extratropical weather and the water cycle (abstract). In *Third International Workshop on the Mars Atmosphere: Modeling and Observations,* Abstract #9117. Lunar and Planetary Institute, Houston.

Hollingsworth J. L., Kahre M. A., Haberle R. M., and Montmessin F. (2011) Radiatively-active aerosols within Mars' atmosphere: Implications on the weather and climate as simulated by the NASA ARC Mars GCM. In *Fourth International Workshop on the Mars Atmosphere: Modelling and Observations*, Paris, France. Available online at *www-mars.lmd.jussieu.fr/paris2011/abstracts/hollingsworth2_paris2011.pdf.*

Holstein-Rathlou C., et al. (2009) Winds at the Phoenix landing site. *J. Geophys. Res., 115,* E00E18, DOI: 10.1029/2009JE003411.

Holton J. R. (1992) *An Introduction to Dynamic Meteorology, 3rd edition.* Academic, San Diego. 511 pp.

Hoskins B. J. and Ambrizzi T. (1993) Rossby wave propagation on a realistic longitudinally varying flow. *J. Atmos. Sci., 50,* 1661–1671.

Hunt G. E., Pickersgill A. O., James P. B., and Evans N. (1981)

Daily and seasonal Viking observations of martian bore wave systems. *Nature, 293,* 630–633.
Jakosky B. M. (1983a) The role of seasonal reservoirs in the Mars water cycle: I. Seasonal exchange of water with the regolith. *Icarus, 55,* 1–18.
Jakosky B. M. (1983b) The role of seasonal reservoirs in the Mars water cycle: II. Coupled models of the regolith, the polar caps, and atmospheric transport. *Icarus, 55,* 19–39.
Jakosky B. M. and Carr M. H. (1985) Possible precipitation of ice at low latitudes of Mars during periods of high obliquity. *Nature, 315,* 559–561.
Jakosky B. M. and Farmer C. B. (1982) The seasonal and global behavior of water vapor in the Mars atmosphere — Complete global results of the Viking atmospheric water detector experiment. *J. Geophys. Res., 87,* 2999–3019.
Jakosky B. M. and Philips R. J. (2001) Mars' volatile and climate history. *Nature, 412,* 237–244.
James P. B. and Cantor B. A. (2001) Martian north polar cap recession: 2000 Mars Orbiter Camera observations. *Icarus, 154,* 131–144.
James P. B. and North G. R. (1982) The seasonal CO_2 cycle on Mars: An application of an energy balance climate model. *J. Geophys. Res., 87,* 10271–10283.
Johnson S. S., Mischna M. A., Grove T. L., and Zuber M. T. (2008) Sulfur-induced greenhouse warming on early Mars. *J. Geophys. Res., 113,* E08005, DOI: 10.1029/2007JE002962.
Johnson S. S., Pavlov A. A., and Mischna M. A. (2009) Fate of SO_2 in the ancient martian atmosphere: Implications for transient greenhouse warming. *J. Geophys. Res., 114,* E11011, DOI: 10.1029/2008JE003313.
Joshi M. M., Lewis S. R., Read P. L., and Catling D. C. (1994) Western boundary currents in the atmosphere of Mars. *Nature, 367,* 548–552.
Joshi M. M., Lawrence B. N., and Lewis S. R. (1995) Gravity wave drag in three-dimensional atmospheric models of Mars. *J. Geophys. Res., 100,* 21235–21245.
Kahn R. and Gierasch P. (1982) Long cloud observations on Mars and implications for boundary layer characteristics over slopes. *J. Geophys. Res., 87,* 867–880.
Kahn R. A., Martin T. Z., Zurek R. W., and Lee S. W. (1992) The martian dust cycle. In *Mars* (H. H. Kieffer et al., eds.), pp. 1017–1053. Univ. of Arizona, Tucson.
Kahre M. A. and Haberle R. M. (2010) Mars CO_2 cycle: Effects of airborne dust and polar cap ice emissivity. *Icarus, 207,* 648–653.
Kass D. M. and Yung Y.L. (1995) Loss of atmosphere from Mars due to solar wind-induced sputtering. *Science, 268,* 697–699.
Kass D. M., Schofield J. T., Michaels T. I., Rafkin S. C. R., Richardson M. I., and Toigo A. D. (2003) Analysis of atmospheric mesoscale models for entry, descent, and landing. *J. Geophys. Res., 108,* 8090.
Kasting J. F. (1982) Stability of ammonia in the primitive terrestrial atmosphere. *J. Geophys. Res., 87,* 3091–3098.
Kasting J. F. (1991) CO_2 condensation and the climate of early Mars. *Icarus, 94,* 1–13.
Keating G. M., et al. (1998) The structure of the upper atmosphere of Mars: In situ accelerometer measurements from Mars Global Surveyor. *Science, 279,* 1672–1676.
Keppenne C. L. and Ingersoll A. P. (1995) High-frequency orographically forced variability in a single-layer model of the martian atmosphere. *J. Atmos. Sci., 52,* 1949–1958.
Kieffer H. H., Jakosky B. M., and Snyder C. W. (1992) The planet Mars — From antiquity to the present. In *Mars* (H. H. Kieffer et al., eds.), pp. 1–33. Univ. of Arizona, Tucson.
Kleinböhl A., Wilson R. J., Kass D., Schofield J. T., and McCleese D. J. (2013) The semidiurnal tide in the middle atmosphere of Mars. *Geophys. Res. Lett., 40(10),* 1952–1959.
Krasnopolsky V. A. (2011) Atmospheric chemistry on Venus, Earth, and Mars: Main features and comparison. *Planet. Space Sci., 59,* 952–964.
Krasnopolsky V. A., Maillard J. P., and Owen T. C. (2004) Detection of methane in the martian atmosphere: Evidence for life? *Icarus, 172,* 537–547.
Kreslavsky M. A. and Head J. W. (2003) North-south topographic slope asymmetry on Mars: Evidence for insolation-related erosion at high obliquity. *Geophys. Res. Lett., 30,* 1815, DOI: 10.1029/2003GL017795.
Kuhn W. R. and Atreya S. K. (1979) Ammonia photolysis and the greenhouse effect in the primordial atmosphere of the Earth. *Icarus, 37,* 207–213.
Larsen S. E., Jørgensen H. E., Landberg L., and Tillman J. E. (2002) Aspects of the atmospheric surface layers on Mars and Earth. *Boundary-Layer Meteorol., 105,* 451–470.
Laskar J., Levrard B., and Mustard J. F. (2002) Orbital forcing of the martian polar layered deposits. *Nature, 419,* 375–377.
Laskar J., Gastineau M., Joutel F., Robutel P., Levrard B., and Correia A. (2004) Long term evolution and chaotic diffusion of the insolation quantities of Mars. *Icarus, 170,* 343–364.
Lee C., Lawson W. G., Richardson M. I., Heavens N. G., Kleinboehl A., Banfield D., McCleese D. J., Zurek R., Kass D., Schofield J. T., Leovy C. B., Taylor F. W., and Toigo A. D. (2009) Thermal tides in the martian middle atmosphere as seen by the Mars Climate Sounder. *J. Geophys. Res., 114,* E03005, DOI: 10.1029/2008JE003285.
Lee C., Lawson W. G., Richardson M. I., Anderson J. L., Collins N., Hoar T., and Mischna M. (2011) Demonstration of ensemble data assimilation for Mars using DART, MarsWRF, and radiance observations from MGS TES. *J. Geophys. Res.–Planets, 116(E11),* E11011.
Lefèvre F. and Forget F. (2009) Observed variations of methane on Mars unexplained by known atmospheric chemistry and physics. *Nature, 460,* 720–723.
Lefèvre F., Lebonnois S., Montmessin F., and Forget F. (2004) Three-dimensional modeling of ozone on Mars. *J. Geophys. Res., 109,* E07004, DOI: 10.1029/2004JE002268.
Lefèvre F., Bertaux J.-L., Clancy R.T., Encrenaz Th., Fast K., Forget F., Lebonnois S., Montmessin F., and Perrier S. (2008) Heterogeneous chemistry in the atmosphere of Mars. *Nature, 454,* 971–975.
Leighton R. B. and Murray B.C. (1966) Behavior of carbon dioxide and other volatiles on Mars. *Science, 153,* 136–144.
Lemmon M. T. and 14 co-authors (2004) Atmospheric imaging results from the Mars exploration rovers: Spirit and Opportunity. *Science, 306,* 1753–1756, DOI: 10.1126/science.1104474.
Leovy C. B. (1979) Martian meteorology. *Annu. Rev. Astron. Astron., 17,* 387–413.
Lewis S. R. and Read P. L. (2003) Equatorial jets in the dusty martian atmosphere. *J. Geophys. Res., 108,* 5034.
Lian Y., Richardson M. I., Newman C. E., Lee C., Toigo A. D., Mischna M. A., and Campin J. (2012) The Ashima/MIT Mars GCM and argon in the martian atmosphere. *Icarus, 218,* 1043–1070, DOI: 10.1016/j.icarus.2012.02.012.
Liu J., Richardson M. I., and Wilson R. J. (2003) An assessment of the global, seasonal, and interannual spacecraft record of

martian climate in the thermal infrared. *J. Geophys. Res., 108,* 5089, DOI: 10.1029/ 2002JE001921.

Määttänen A., et al. (2009) A study of the properties of a local dust storm with Mars Express OMEGA and PFS data. *Icarus, 201,* 504–516.

Määttänen A., et al. (2010) Mapping the mesospheric CO_2 clouds on Mars: MEx/OMEGA and MEx/HRSC observations and challenges for atmospheric models. *Icarus, 209,* 452–469.

Madeleine J.-B., Forget F., Millour E., Navarro T., and Spiga A. (2012) The influence of radiatively active water ice clouds on the martian climate. *Geophys. Res. Lett., 39,* L23202, DOI: 10.1029/2012GL053564.

Mangold N., Quantin C., Ansan V., Delacourt C., and Allemand P. (2004) Evidence for precipitation on Mars from dendritic valleys in the Valles Marineris area. *Science, 305,* 78–81.

Mangold N., Ansan V., Masson Ph., Quantin C., and Neukum G. (2008) Geomorphic study of fluvial landforms on the northern Valles Marineris plateau, Mars. *J. Geophys. Res., 113,* E08009, DOI: 10.1029/2007JE002985.

Martin T. Z. and Richardson M. I. (1993) New dust opacity mapping from Viking Infrared Thermal Mapper data. *J. Geophys. Res., 98,* 10941–10949.

Mass C. and Sagan C. (1976) A numerical circulation model with topography for the martian southern hemisphere. *J. Atmos. Sci., 33,* 1418–1430.

McCleese D. J., et al. (2010) Structure and dynamics of the martian lower and middle atmosphere as observed by the Mars Climate Sounder: Seasonal variations in zonal mean temperature, dust and water ice aerosols, *J. Geophys. Res., 115,* E12016, DOI: 10.1029/2010JE003677.

McKim R. J. (1999) Telescopic martian dust storms: A narrative and catalogue. *Mem. British Astron. Soc., 44,* 13–124.

Medvedev A. S., Yi it E., and Hartogh P. (2011) Estimates of gravity wave drag on Mars: Indication of a possible lower thermospheric wind reversal. *Icarus, 211,* 909–912.

Mellon M. T. (1996) Limits on the CO_2 content of the martian polar deposits. *Icarus, 124,* 268–279.

Michaels T. I. and Rafkin S. C. R. (2004) Large-eddy simulation of atmospheric convection on Mars. *Q. J. R. Meteor. Soc., 130,* 1251–1274.

Michaels T. I., Colaprete A., and Rafkin S. C. R. (2006) Significant vertical water transport by mountain-induced circulations on Mars. *Geophys. Res. Lett., 33,* L16201, DOI: 10.1029/2006GL026562.

Milkovich S. M. and Head J. W. III (2005) North polar cap of Mars: Polar layered deposit characterization and identification of a fundamental climate signal. *J. Geophys. Res., 110,* E01005, DOI: 10.1029/2004JE002349.

Milliken R. E., et al. (2008) Opaline silica in young deposits on Mars. *Geology, 36,* 847–850.

Mischna M. A. and Richardson M. I. (2005) A reanalysis of water abundances in the martian atmosphere at high obliquity. *Geophys. Res. Lett., 32,* L03201, DOI: 10.1029/2004GL021865.

Mischna M. A., Kasting J. F., Pavlov A., and Freedman R. (2000) Influence of carbon dioxide clouds on early martian climate. *Icarus, 145,* 546–554.

Mischna M. A., Richardson M. I., Wilson R. J., and McCleese D. J. (2003) On the orbital forcing of martian water and CO_2 cycles: A general circulation model study with simplified volatile schemes. *J. Geophys. Res., 108,* 5062, DOI: 10.1029/ 2003JE002051.

Mischna M. A., Allen M., Richardson M. I., Newman C. E., and Toigo A. D. (2011) Atmospheric modeling of Mars methane surface releases. *Planet. Space Sci., 59,* 227–237.

Mischna M. A., Baker V., Milliken R., Richardson M., and Lee C. (2013) Effects of obliquity and water vapor/trace gas greenhouses in the early martian climate. *J. Geophys. Res., 118,* 560–576.

Montmessin F., Forget F., Rannou P., Cabane M., and Haberle R. M. (2004) Origin and role of water ice clouds in the martian water cycle as inferred from a general circulation model. *J. Geophys. Res., 109,* E10004, DOI: 10.1029/2004JE002284.

Montmessin F., Haberle R. M., Forget F., Langevin Y., Clancy R. T., and Bibring J.-P. (2007) On the origin of perennial water ice at the South Pole of Mars: A precession-controlled mechanism? *J. Geophys. Res., 112,* E08S17, DOI: 10.1029/2007JE002902.

Mumma M. J., Villanueva G. L., Novak R. E., Hewagama T., Bonev B. P., DiSanti M. A., Mandell A. M., and Smith M. D. (2009) Strong release of methane on Mars in northern summer 2003. *Science, 323,* 1041–1045.

Murphy J. R., Leovy C. B., and Tillman J. E. (1990) Observations of martian surface winds at the Viking Lander 1 site. *J. Geophys. Res., 95,* 14555–14576.

Mustard J. F., Cooper C. D., and Rifkin M. K. (2001) Evidence for recent climate change on Mars from the identification of youthful near-surface ground ice. *Nature, 412,* 411–414.

Newman C. E., Lewis S. R., Read P. L., and Forget F. (2002a) Modeling the martian dust cycle 1. Representations of dust transport processes. *J. Geophys. Res., 107,* 5123, DOI: 10.1029/2002JE001910.

Newman C. E., Lewis S. R., Read P. L., and Forget F. (2002b) Modeling the martian dust cycle 2. Multiannual radiatively active dust transport simulations. *J. Geophys. Res., 107,* 5124, DOI: 10.1029/2002JE001920.

Newman C. E., Lewis S. R., and Read P. L. (2005) The atmospheric circulation and dust activity in different orbital epochs on Mars. *Icarus, 174,* 135–160.

Nickovic S., Kallos G., Papadopoulos A., and Kakaliagou O. (2001) A model for prediction of desert dust cycle in the atmosphere. *J. Geophys. Res., 106,* 18113–18129.

Owen T. (1992) The composition and early history of the atmosphere of Mars. In *Mars* (H. H. Kieffer et al., eds.), pp. 818–834. Univ. of Arizona, Tucson.

Pankine A. A. and Ingersoll A. P. (2002) Interannual variability of martian global dust storms: Simulations with a low-order model of the general circulation. *Icarus, 155(2),* 299–323.

Parish T. R. (2002) Katabatic sinds. In *Encyclopedia of Atmospheric Sciences* (J. R. Holton et al., eds.), pp. 1057–1061. Academic, San Diego.

Parsons R. A., Moore J. M., and Howard A. D. (2013) Evidence for a short period of hydrologic activity in Newton crater, Mars, near the Hesperian-Amazonian transition. *J. Geophys. Res., 118,* 1082–1093.

Peixoto J. and Oort A. H. (1992) Physics of climate. *Rev. Mod. Phys., 56,* 365.

Pérez C., Nickovic S., Pejanovic G., Baldasano J. M., and Özsoy E. (2006) Interactive dust-radiation modeling: A step to improve weather forecasts. *J. Geophys. Res., 111,* D16206, DOI: 10.1029/2005JD006717.

Perron J. T. and Huybers P. (2009) Is there an orbital signal in the polar layered deposits on Mars? *Geology, 37,* 155–158.

Phillips R. J. et al. (2001) Ancient geodynamics and global-scale hydrology on Mars. *Science, 291,* 2587–2591.

Phillips R. J., et al. (2008) Mars north polar deposits: Stratigraphy, age and geodynamical response. *Science, 320,* 1182–1185.

Phillips R. J., et al. (2011) Massive CO_2 ice deposits sequestered in the south polar layered deposits of Mars. *Science, 332,* 838–841.

Pickersgill A. O. and Hunt G. E. (1979) The formation of martian lee waves generated by a crater. *J. Geophys Res., 84,* 8317–8331.

Pickersgill A. O. and Hunt G. E. (1981) An examination of the formation of linear lee waves generated by giant martian volcanoes. *J. Atmos. Sci., 38,* 40–51.

Pirraglia J. A. (1976) Martian atmospheric lee waves. *Icarus, 27,* 517–530.

Pollack J. B. (1979) Climatic change on the terrestrial planets. *Icarus, 37,* 479–553.

Pollack J. B., Colburn D. S., Flasar F. M., Kahn R., Carlston C. E., and Pidek D. (1979) Properties and effects of dust particles suspended in the martian atmosphere. *J. Geophys. Res., 84,* 2929–2945.

Pollack J. B., Leovy C. B., Greiman P. W., and Mintz Y. (1981) A martian general circulation experiment with large topography. *J. Atmos. Sci., 38,* 3–29.

Pollack J. B., Kasting J. F., Richardson S. M., and Poliakoff K. (1987) The case for a wet, warm climate on early Mars. *Icarus, 71,* 203–224.

Pollack J. B., Haberle R. M., Schaeffer J., and Lee H. (1990) Simulations of the general circulation of the martian atmosphere: 1. Polar processes. *J. Geophys. Res., 95,* 2156–2202, DOI: 10.1029/JB095iB02p01447.

Postawko S. E. and Kuhn W. R. (1986) Effect of the greenhouse gases (CO_2, H_2O, SO_2) on martian paleoclimate. *Proc. Lunar Planet. Sci. Conf. 16th,* in *J. Geophys. Res., 91,* D431–D438.

Rafkin S. C. R. (2009) A positive radiative-dynamic feedback mechanism for the maintenance and growth of martian dust storms. *J. Geophys. Res., 114,* E01009, DOI: 10.1029/2008JE003217.

Rafkin S. C. R. (2012) The potential importance of non-local, deep transport on the energetics, momentum, chemistry, and aerosol distributions in the atmospheres of Earth, Mars, and Titan. *Planet. Space Sci., 60,* 147–154.

Rafkin S. C. R. and Michaels T. I. (2003) Meteorological predictions for 2003 Mars Exploration Rover high-priority landing sites. *J. Geophys. Res., 108,* 8091, DOI: 10.1029/2002JE002027.

Rafkin S. C. R., Haberle R. M., and Michaels T. I. (2001) The Mars Regional Atmospheric Modeling System (MRAMS): Model description and selected simulations. *Icarus, 151,* 228–256.

Rafkin S. C. R., Sta. Maria M. R. V., and Michaels T. I. (2002) Simulation of the atmospheric thermal circulation of a martian volcano using a mesoscale numerical model. *Nature, 419,* 697–699.

Rafkin S. C. R., Michaels T. I., and Haberle R. M. (2004) Meteorological predictions for the Beagle 2 mission to Mars. *Geophys. Res. Lett., 31,* L01703, DOI: 10.1029/2003GL018966.

Richardson M. I. and Wilson R. J. (2002a) Investigation of the nature and stability of the martian seasonal water cycle with a general circulation model. *J. Geophys. Res., 107,* 5031, DOI: 10.1029/2001JE001536.

Richardson M. I. and Wilson R. J. (2002b) A topographically forced asymmetry in the martian circulation and climate. *Nature, 416,* 298–301.

Richardson M. I., Wilson R. J., and Rodin A. V. (2002) Water ice clouds in the martian atmosphere: General circulation model experiments with a simple cloud scheme. *J. Geophys. Res., 107,* 5064, DOI: 10.1029/ 2001JE001804.

Ridley D. A., Heald C. L., and Ford B. (2012) North African dust export and deposition: A satellite and model perspective. *J. Geophys. Res., 117,* D02202, DOI: 10.1029/2011JD016794.

Rodin A. V., Clancy R. T., and Wilson R. J. (1999) Dynamical properties of Mars water ice clouds and their interactions with atmospheric dust and radiation. *Adv. Space Res., 23(9),* 1577–1585.

Ruff S. W. and Christensen P. R. (2002) Bright and dark regions on Mars: Particle size and mineralogical characteristics based on Thermal Emission Spectrometer data. *J. Geophys. Res.–Planets, 107(E12),* 5119.

Sagan C. and Chyba C. (1997) The early faint Sun paradox: Organic shielding of ultraviolet-labile greenhouse gases. *Science, 276,* 1217–1221.

Savijärvi H. and Kauhanen J. (2008) Surface and boundary-layer modelling for the Mars Exploration Rover sites. *Q. J. R. Meteor. Soc., 134(632),* 635–641.

Savijärvi H. and Siili T. (1993) The martian slope winds and the nocturnal PBL jet. *J. Atmos. Sci., 50(1),* 77–88.

Schofield J. T., Barnes J. R., Crisp D., Haberle R. M., Larsen S., Magalhães J. A., Murphy J. R., Seiff A., and Wilson G. (1997) The Mars Pathfinder Atmospheric Structure Investigation/Meteorology (ASI/MET) experiment. *Science, 278,* 1752–1758.

Schulz M., Balkanski Y. J., Guelle W., and Dulac F. (1998) Role of aerosol size distribution and source location in a three-dimensional simulation of a Saharan dust episode tested against satellite-derived optical thickness. *J. Geophys. Res., 103,* 10579–10592, DOI: 10.1029/97JD02779.

Segura T. L., Toon O. B., Colaprete A., and Zahnle K. (2002) Environmental effects of large impacts on Mars. *Science, 298,* 1977–1980.

Segura T. L., Toon O. B., and Colaprete A. (2008) Modeling the environmental effects of moderate-sized impacts on Mars. *J. Geophys. Res., 113,* E1107, DOI: 10.1029/2008JE003147.

Smith D. E., et al. (2001) Mars Orbiter Laser Altimeter (MOLA): Experiment summary after the first year of global mapping of Mars. *J. Geophys. Res., 106,* 23689–23722.

Smith M. D. (2002) The annual cycle of water vapor on Mars as observed by the Thermal Emission Spectrometer. *J. Geophys. Res., 107,* 5115, DOI: 10.1029/2001JE001522.

Smith M. D. (2004) Interannual variability in TES atmospheric observations of Mars during 1999–2003. *Icarus, 167,* 148–165.

Smith M. D. (2009) THEMIS observations of Mars aerosol optical depth from 2002–2008. *Icarus, 202,* 444–452.

Smith M. D., Pearl J. C., Conrath B. J., and Christensen P. R. (2001) Thermal Emission Spectrometer results: Mars atmos-, pheric thermal structure and aerosol distribution. *J. Geophys. Res., 106,* 23929–23945.

Smith M. D., Conrath B. J., Pearl J. C., and Christensen P. R. (2002) Thermal Emission Spectrometer observations of martian planet-encircling dust storm 2001A. *Icarus, 157,* 259–263.

Smith M. D., Wolff M. J., Spanovich N., Ghosh A., Banfield D., Christensen P. R., Landis G. A., and Squyres S. W. (2006) One martian year of atmospheric observations using MER Mini-TES. *J. Geophys. Res.–Planets, 111(E12),* E12S13.

Smith M. D., Wolff M. J., Clancy R. T., Kleinböhl A., and Murchie S. L. (2013) Vertical distribution of dust and water ice aerosols from CRISM limb-geometry observations. *J. Geophys. Res.,*

118(2), 321–334, DOI: 10.1002/jgre.20047.

Son S.-W., Ting M., and Polvani L. M. (2009) The effect of topography on storm-track intensity in a relatively simple general circulation model. *J. Atmos. Sci., 66,* 393–411.

Sorbjan Z. (2007) Statistics of shallow convection on Mars based on large-eddy simulations. Part 1: Shearless conditions. *Boundary-Layer Meteorol., 123,* 121–142.

Spiga A. (2011) Elements of comparison between martian and terrestrial mesoscale meteorological phenomena: Katabatic winds and boundary layer convection. *Planet. Space Sci., 59,* 915–922.

Spiga A., Teitelbaum H., and Zeitlin V. (2008) Identification of the sources of inertia-gravity waves in the Andes Cordillera region. *Ann. Geophys.-Italy, 26,* 2551–2568.

Spiga A., Forget F., Lewis S. R., and Hinson D. P. (2010) Structure and dynamics of the convective boundary layer on Mars as inferred from large-eddy simulations and remote-sensing measurements. *Q. J. R. Meteor. Soc., 136(647),* 28.

Spiga A., González-Galindo F., López-Valverde M. Á., and Forget F. (2012) Gravity waves, cold pockets and CO_2 clouds in the martian mesosphere. *Geophys. Res. Lett., 39,* L02201, DOI: 10.1029/2011GL050343.

Spiga A., Faure J., Madeleine J.-B., Määttänen A., and Forget F. (2013) Rocket dust storms and detached dust layers in the martian atmosphere. *J. Geophys. Res., 118(4),* 746–767.

Sprague A. L., Boynton W. V., Kerry K. E., Janes D. M., Hunten D. M., Kim K. J., Reedy R. C., and Metzger A. E. (2004) Mars' south polar Ar enhancement: A tracer for south polar seasonal meridional mixing. *Science, 306,* 1364–1367.

Sprague A. L., Boynton W. V., Kerry K. E., Janes D. M., Kelly N. J., Crombie M. K., Nelli S. M., Murphy J. R., Reedy R. C., and Metzger A. E. (2007) Mars' atmospheric argon: Tracer for understanding martian atmospheric circulation and dynamics. *J. Geophys. Res., 112,* E03S02.

Sta. Maria M. R. V., Rafkin S. C. R., and Michaels T. I. (2006) Numerical simulation of atmospheric bore waves on Mars. *Icarus, 185,* 383–394.

Strausberg M. J., Wang H., Richardson M. I., Ewald S. P., and Toigo A. D. (2005) Observations of the initiation and evolution of the 2001 Mars global dust storm. *J. Geophys. Res., 110,* E02006.

Stull R. B. (1988) *An Introduction to Boundary Layer Meteorology.* Kluwer, Dordrecht. 684 pp.

Squyres S. W. and Kasting J. F. (1994) Early Mars: How warm and how wet? *Science, 265,* 744–749.

Sun J., Zhang M., and Liu T. (2001) Spatial and temporal characteristics of dust storms in China and its surrounding regions, 1960–1999: Relations to source area and climate. *J. Geophys. Res., 106,* 10325–10333.

Sutton J. L., Leovy C. B., and Tillman J. E. (1978) Diurnal variations of the martian surface layer meteorological parameters during the first 45 Sols at two Viking Lander Sites. *J. Atmos. Sci., 35(12),* 2346–2355.

Swap R., Garstang M., Greco S., Talbot R., and Kallberg P. (1992) Saharan dust in the Amazon Basin. *Tellus, 44B,* 133–149.

Szwast M. A., Richardson M. I., and Vasavada A. R. (2006) Surface dust redistribution on Mars as observed by the Mars Global Surveyor and Viking orbiters. *J. Geophys. Res., 111,* E11008, DOI: 10.1029/ 2005JE002485.

Tegen I., Lacis A. A., and Fung I. (1996) The influence on climate forcing of mineral aerosol from disturbed soils. *Nature, 380,* 419–422.

Thomas P. C., Malin M. C., Edgett K. S., Carr M. H., Hartmann W. K., Ingersoll A. P., James P. B., Soderblom L. A., Veverka J., and Sullivan R. (2000) North-south geological differences between the residual polar caps on Mars. *Nature, 404,* 161–164.

Tian F., Claire M. W., Haqq-Misra J. D., Smith M., Crisp D. C., Catling D., Zahnle K., and Kasting J. F. (2010) Photochemical and climate consequences of sulfur outgassing on early Mars. *Earth Planet. Sci. Lett., 295,* 412–418.

Tillman J. E. (1988) Mars global atmospheric oscillations: Annually synchronized, transient normal-mode oscillations and the triggering of global dust storms. *J. Geophys. Res., 93,* 9433–9451.

Tillman J. E., Landberg L., and Larsen S. E. (1994) The boundary layer of Mars: Fluxes, stability, turbulent spectra, and growth of the mixed layer. *J. Atmos. Sci., 51,* 1709–1727.

Toigo A. D. and Richardson M. I. (2003) Meteorology of proposed Mars Exploration Rover landing sites. *J. Geophys. Res., 108,* 8092, DOI: 10.1029/2003JE002064.

Toigo A. D., Richardson M. I., Wilson R. J., Wang H., and Ingersoll A. P. (2002) A first look at dust lifting and dust storms near the South Pole of Mars with a mesoscale model. *J. Geophys. Res., 107,* 5050, DOI: 10.1029/2001JE001592.

Toigo A. D., Richardson M. I., Ewald S. P., and Gierasch P. J. (2003) Numerical simulation of martian dust devils. *J. Geophys. Res., 108,* 5047, DOI: 10.1029/2002JE002002.

Tomasko M. G., Doose L. R., Lemmon M., Smith P. H., and Wegryn E. (1999) Properties of dust in the martian atmosphere from the Imager on Mars Pathfinder. *J. Geophys. Res., 104(E4),* 8987–9007, DOI: 10.1029/1998JE900016.

Tyler D. and Barnes J. R. (2005) A mesoscale model study of summertime atmospheric circulations in the north polar region of Mars. *J. Geophys. Res., 110,* E06007, DOI: 10.1029/2004JE002356.

Tyler D. Jr., Barnes J. R., and Skyllingstad E. D. (2008) Mesoscale and large-eddy simulation model studies of the martian atmosphere in support of Phoenix. *J. Geophys. Res., 113(E3),* E00A12.

Volland H. (1988) *Atmospheric Tidal and Planetary Waves.* Kluwer, Dordrecht. 348 pp.

Wakimoto R. M. and McElroy J. L. (1986) Lidar observation of elevated pollution layers over Los Angeles. *J. Climate Appl. Meteorol., 25,* 1583–1599.

Wang H. (2007) Dust storms originating in the northern hemisphere during the third mapping year of Mars Global Surveyor. *Icarus, 189,* 325–343.

Wang H. and Ingersoll A. P. (2002) Martian clouds observed by Mars Global Surveyor Mars Orbiter Camera. *J. Geophys. Res., 107,* 5078, DOI: 10.1029/2001JE001815.

Wang H., Richardson M. I., Wilson R. J., Ingersoll A. P., Toigo A. D., and Zurek R. W. (2003) Cyclones, tides, and the origin of a cross-equatorial dust storm on Mars. *Geophys. Res. Lett., 30,* 1488, DOI: 10.1029/2002GL016828.

Wang H., Zurek R. W., and Richardson M. I. (2005) Relationship between frontal dust storms and transient eddy activity in the northern hemisphere of Mars as observed by Mars Global Surveyor. *J. Geophys. Res., 110,* E07005, DOI: 10.1029/2005JE002423.

Webster C. R., Mahaffy P. R., Atreya S. K., Flesch G. J., Christensen L. E., Farley K. A., and the MSL Science Team (2013) Measurements of Mars methane at Gale Crater by the SAM Tunable Laser Spectrometer on the Curiosity Rover (abstract). In *Lunar Planet. Sci. Conf. XLIV,* Abstract #1366. Lunar and

Planetary Institute, Houston.
Webster P. J. (1977) The low-latitude circulation of Mars. *Icarus, 30,* 626–649.
Whiteway J. A., et al. (2009) Mars water-ice clouds and precipitation. *Science, 325,* 68–70.
Wilson L. and Head J. W. (2002) Tharsis-radial graben systems as the surface manifestation of plume-related dike intrusion complexes: Models and implications. *J. Geophys. Res., 107,* 5057, DOI: 10.1029/2001JE001593.
Wilson R. J. (1997) A general circulation model simulation of the martian polar warming. *Geophys. Res. Lett., 24,* 123–126.
Wilson R. J. and Hamilton K. (1996) Comprehensive simulation of thermal tides in the martian atmosphere. *J. Atmos. Sci., 53,* 1290–1326.
Wilson R. J., Lewis S. R., Montabone L., and Smith M. D. (2008) Influence of water ice clouds on martian tropical atmospheric temperatures. *Geophys. Res. Lett., 35,* L07202, DOI: 10.1029/2007GL032405.
Wolff M. J., Smith M. D., Clancy R. T., Arvidson R., Kahre M., Seelos F. IV, Murchie S., and Savijärvi H. (2009) Wavelength dependence of dust aerosol single scattering albedo as observed by the Compact Reconnaissance Imaging Spectrometer. *J. Geophys. Res., 114,* E00D04, DOI: 10.1029/2009JE003350.
Wordsworth R., Forget F., Millour E., Head J. W., Madeleine J.-B., and Charnay B. (2013) Global modelling of the early martian climate under a denser CO_2 atmosphere: Water cycle and ice evolution. *Icarus, 222,* 1–19.
Ye Z. J., Segal M., and Pielke R. A. (1990) A comparative study of daytime thermally induced upslope flow on Mars and Earth. *J. Atmos. Sci., 47,* 612–628.
Yung Y. L., Nair H., and Gerstell M. F. (1997) CO_2 greenhouse in the early martian atmosphere: SO_2 inhibits condensation. *Icarus, 130,* 222–224.
Yung Y. L, Wen J.-S., Pinto J. P., Pierce K. K., and Allen M. (1988) HDO in the martian atmosphere — Implications for the abundance of crustal water. *Icarus, 76,* 146–159.
Zent A. P., Hecht M. H., Cobos D. R., Wood S. E., Hudson T. L., Milkovich S. M., DeFlores L. P., and Mellon M. T. (2010) Initial results from the Thermal and Electrical Conductivity Probe (TECP) on Phoenix. *J. Geophys. Res., 115,* E00E14, DOI: 10.1029/2009JE003420.
Zurek R. W. (1976) Diurnal tide in the martian atmosphere. *J. Atmos. Sci., 33,* 321–337.
Zurek R. W. (1988) Free and forced modes in the martian atmosphere. *J. Geophys. Res., 93,* 9452–9462.
Zurek R. W. and Martin L. J. (1993) Interannual variability of planet-encircling dust activity on Mars. *J. Geophys. Res., 98,* 3247–3259.
Zurek R. W., Barnes J. R., Haberle R. M., Pollack J. B., Tillman J. E., and Leovy C. B. (1992) Dynamics of the atmosphere of Mars. In *Mars* (H. H. Kieffer et al., eds.), pp. 835–933. Univ. of Arizona, Tucson.
Zurek R. W., Chicarro A., Allen M. A., Bertaux J.-L., Clancy R. T., Daerden F., Formisano V., Garvin J. B., Neukum G., and Smith M. D. (2011) Assessment of a 2016 mission concept: The search for trace gases in the atmosphere of Mars. *Planet. Space Sci., 59,* 284–291.

Griffith C. A., Mitchell J. L., Lavvas P., and Tobie G. (2013) Titan's evolving climate. In *Comparative Climatology of Terrestrial Planets* (S. J. Mackwell et al., eds.), pp. 91–119. Univ. of Arizona, Tucson, DOI: 10.2458/azu_uapress_9780816530595-ch004.

Titan's Evolving Climate

Caitlin A. Griffith
University of Arizona

Jonathan L. Mitchell
University of California, Los Angeles

Panayotis Lavvas
Centre National de la Recherche Scientifique/Université Reims Champagne-Ardenne

Gabriel Tobie
Centre National de la Recherche Scientifique/Université de Nantes

---

Titan uniquely resembles Earth, with a volatile cycle involving clouds, rain and seas. Yet instead of water, it is methane (CH_4), the second most abundant atmospheric constituent, that exists as a gas, liquid, and solid. Methane clouds the size of Earth's hurricanes (~2000 km) occasionally engulf Titan's small globe (radius 2575 km). Usually, however, the skies are clear and the expected forecast is ~1% cloud coverage. Branching drainages and dark surface patches that appear after storms indicate intermittent rainfall. The surface is largely devoid of liquids except for the polar regions, where lakes can reach the size of the Great Lakes that border Canada and the United States. However, this methane-rich world is evolving, as solar ultraviolet (UV) photolysis continually depletes methane at a rate that would render Titan dry in 20 m.y. if not resupplied.

1. INTRODUCTION

Climatic conditions on Titan conjure an odd mixture of traits borrowed from Venus, Earth, and Mars. Titan has a N_2-based atmosphere with a surface pressure of 1.45 bar, similar to Earth and distinct from other satellites. The total column density of Titan's atmosphere (U) exceeds Earth's and its surface temperature (T_{surf}) varies little with season and region. The difference between Titan's pole to equator surface temperature (ΔT_{surf}) is more like that of Venus (Table 1). Titan's observable volatile inventory (atmosphere and seas) is equivalent to a global precipitable methane layer of ~7 m. It is more similar to the water abundance on Mars, ~20 m, rather than the 2.7 km of water found on Earth. Yet unlike Mars and Earth, most of Titan's volatile inventory (5 m of methane) is found in the atmosphere rather than on the surface. Surface features, while largely formed by erosion (redolent of Earth), appear more highly organized, as do the clouds. Unlike the current volatile inventory on any of the terrestrial planets, Titan's climate is not stable. Its methane abundance is continually dwindling and its climate evolving, unless a continual surface supply of CH_4 precisely matches the methane destruction rate by UV photolysis.

Starting with the basic observations that establish our level of understanding of Titan's atmosphere, one can ask whether it is possible to explain Titan's climate and weather as resulting from the different fundamental properties, such as Saturn's orbital eccentricity (e), Titan's smaller radius (R), longer spin period (P), larger distance from the Sun (a), higher atmospheric column abundance (U), effective temperature (T_{eff}), and disparate chemistry (Tables 1 and 2). This chapter explores the processes by which these fundamental parameters establish the contrast between Titan's climate and those on other planetary bodies.

2. TITAN'S COMPOSITION

Most of the composition of Titan's atmosphere is established by photodissociation and ionization of the two main atmospheric components, N_2 and CH_4 (*Strobel,* 1982; *Yung et al.,* 1984). This chemistry produces mainly H_2, ethane (C_2H_6), acetylene (C_2H_2), hydrogen cyanide (HCN), propane (C_3H_8), and ethylene (C_2H_4) (Table 2, Fig. 1). Notable exceptions are CO, CO_2, and H_2O, which require a source of oxygen, possibly due to ablation of micrometeorites and/or the deposition of energetic O^+ (*Hörst et al.,* 2008). As understood through UV to infrared (IR) spectroscopy of Titan's atmosphere, plasma measurements of the ionosphere, and *in situ* mass spectrometer measurements extending from the surface to the thermosphere, multiple energy sources contribute to this critical first step. Energetic photons and photoelectrons deposit their energy in the upper

TABLE 1. Titan's characteristics.

Quantity	Venus	Earth	Mars	Titan
a (AU)	0.72	1.00	1.52	9.54
R (km)	6052	6378	3397	2575
e	0.0068	0.0167	0.0933	0.0565
P (days)	–243	1.0	1.026	15.95
g (m s^{-1})	8.87	9.82	3.73	1.354
H (km)	15.9	8.42	11.07	20
A_{bond}	0.90	0.30	0.25	0.30
U (km am)	535	8	0.088	92
T_{eff} (K)	232	255	210	82
T_{surf} (K)	735	288	214	94
ΔT_{surf} (K)	0	150	440	4

TABLE 2. Most abundant constituents.

Species	Abundance	Origin/Evolution
N_2	0.98	Interior/Stable
CH_4	0.014	Interior/Photodissociates
H_2	0.004	CH_4 photolysis/Escapes
C_2H_6	2×10^{-5}	CH_4 photolysis/Sediments
C_2H_2	3×10^{-6}	CH_4 photolysis/Sediments
HCN	7×10^{-7}	CH_4 photolysis/Sediments
C_3H_8	5×10^{-7}	CH_4 photolysis/Sediments
C_3H_4	1×10^{-8}	CH_4 photolysis/Sediments
C_2H_4	7×10^{-7}	CH_4 photolysis/Sediments
HC_3N	4×10^{-8}	CH_4 photolysis/Sediments
CO	4.7×10^{-5}	CH_4 photolysis/Meteorites
CO_2	1.5×10^{-8}	CO/Photodissociates
H_2O	8×10^{-9}	Meteorites/Photodissociates

Average values for the lower stratosphere, taken from *Vinatier et al.* (2010b) (C_2H_6, C_2H_2, HCN, C_3H_8, C_3H_4, C_2H_4, C_2H_4, HC_3N), *Coustenis et al.* (1998) (H_2O), and *deKok et al.* (2007) (CO).

atmosphere (above 700 km) and are the dominant energy source in the daytime (*Lavvas et al.,* 2011c). Energetic particles and electrons from Saturn's magnetosphere supply a smaller and temporally variable energy source to the upper atmosphere (*Galand et al.,* 2013). Galactic cosmic rays, which penetrate deeper in the atmosphere, close to 70 km, are the only source of N_2 dissociation in the stratosphere (*Molina-Cuberos et al.,* 1999). Meteoroid ablation occurs between 500 and 800 km, depending on the properties of the incoming meteoroids, and mainly acts as a local source of ionization without impacting the background chemistry (*Molina-Cuberos et al.,* 2001). Among these contributions, it is the photons (with their corresponding photoelectrons) and the cosmic rays that primarily control the photochemistry and thus the production and loss of the main species affecting the atmospheric thermal balance.

The photolysis of N_2 and CH_4 provides radicals and ions, which rapidly react to produce more complex hydrocarbons and nitrogen-containing molecules (Appendix A, Tables 3 and 4). Although nitrogen's abundance exceeds that of methane, the photolysis rate of the latter is significantly larger than that of the former (Fig. 2). This effect occurs because methane has a larger absorption cross section at longer wavelengths than does N_2. Methane photolysis ex-

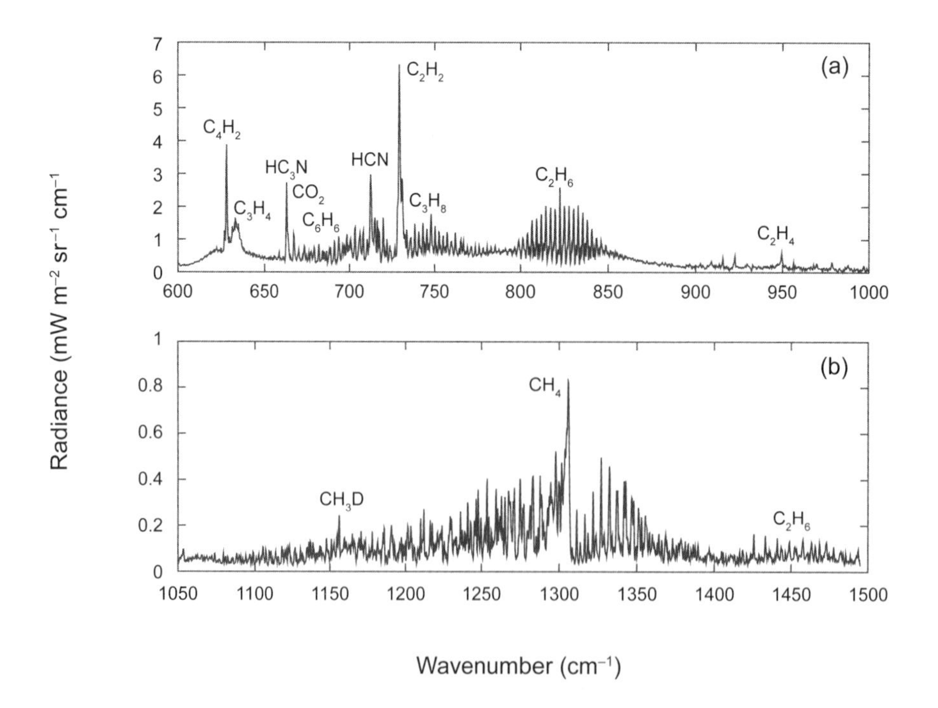

Fig. 1. Cassini limb spectra of Titan recorded between January 2005 and July 2007 at altitudes between 100 and 200 km and latitudes in the range 55°–90°N display emission bands of methane and its photochemical byproducts. From *Bézard* (2009).

TABLE 3. Destruction rates of methane and nitrogen.

Process	N_2 (cm^{-2} s^{-1})	CH_4 (cm^{-2} s^{-1})	Mass Flux (g cm^{-2} s^{-1})
Photons	2.8×10^8	3.1×10^9	9.4×10^{-14}
Photo e⁻	1.6×10^8	2.6×10^7	8.1×10^{-15}
Cosmic rays	6.0×10^7	1.2×10^5	2.8×10^{-15}
Meteoroids			$\sim 10^{-17}$
Chemistry		7.1×10^9	
Total	5.0×10^8	1.0×10^{10}	1.0×10^{-13}

tends to ~1500 Å, and involves primarily radiation from the strongest solar line, Lyman-α (Ly-α), at 1215 Å. Nitrogen photolysis involves wavelengths only below 1000 Å, as well as extreme ultraviolet (EUV) photons that have a much smaller flux than Ly-α photons. Consequently, the products of methane photochemistry are more abundant than the pure nitrogen photochemical products, while the chemical reactivity of carbon enables the evolution of a structurally complex chemistry.

The mixing ratios of the photochemical species are largest in the upper atmosphere, where most of the solar energy that induces the chemical processes through the dissociation of N_2 and CH_4 is deposited. Diffusion and mixing transport these species to lower altitudes where their mixing ratios decrease, away from the production region. Secondary production processes can affect the rate of decrease for each species, and for some cases the cosmic rays in the lower atmosphere can locally increase the mixing ratios. Eventually the cooler temperatures in the lower stratosphere and troposphere force most of the photochemical products to condense, thereby rapidly decreasing their mixing ratios to small values. Most of the byproducts end up on Titan's surface as liquid (e.g., ethane and propane) and solid sediments, e.g., C_2H_2, HCN, and the refractory haze (*Lunine et al.,* 1983). If methane has been present in Titan's atmosphere over most of the course of its history, the total accumulation of organic sediments would amount to half a kilometer, although not all necessarily exposed at the surface (*Lunine and Stevenson,* 1985a; *Kossacki and Lorenz,* 1996; *Mousis and Schmitt,* 2008; *Choukroun and Sotin,* 2012).

TABLE 4. Chemical production and loss rates (cm^{-2} s^{-1}).

Molecule	Production	Loss	Condensation Flux
C_2H_6	2.8×10^9	7.3×10^8	2.1×10^9
C_2H_2	8.5×10^9	8.4×10^9	1.3×10^8
HCN	1.0×10^9	8.9×10^8	1.3×10^8

Globally averaged, C_2H_6 accumulates on Titan's surface at a rate of 5.7×10^6 cm yr^{-1}.

One important component of Titan's atmosphere is its vast photochemically produced haze. These complex organic molecules form an optically thick aerosol layer, which veils the planet at optical wavelengths (Fig. 3) and provides a strong radiative forcing of Titan's climate. The mass flux of aerosols implied by *in situ* Huygens Descent Imager-Spectral Radiometer (DISR) observations (3.0×10^{-14} g cm^{-2} s^{-1}), indicate that 29% of the photolyzed N_2 and CH_4 mass ends up as aerosols. *In situ* UV to IR observations (*Koskinen et al.,* 2011; *Porco et al.,* 2005; *Vinatier et al.,* 2010a; *Anderson et al.,* 2011; *Tomasko et al.,* 2008) indicate that the aerosols are aggregates of ~3000 smaller quasi-spherical primary particles of 40 nm radius that can be traced back to the upper atmosphere as high up as the ionosphere (*Vuitton et al.,* 2007; *Waite et al.,* 2007; *Coates et al.,* 2007). The mass flux of aerosols in the thermosphere at ~1000 km, 3.2×10^{-15} g cm^{-2} s^{-1} (*Wahlund et al.,* 2009), is a significant fraction of that inferred for the main haze layer in the stratosphere at 200 km altitude (3.2×10^{-14} g cm^{-2} s^{-1}) (*Tomasko et al.,* 2008; *Lavvas et al.,* 2011b), which is comparable to that at 500 km (*Lavvas et al.,* 2009). Thus, a large fraction of the aerosols are formed in the upper atmosphere, yet the production of aerosols below 500 km cannot be ruled out.

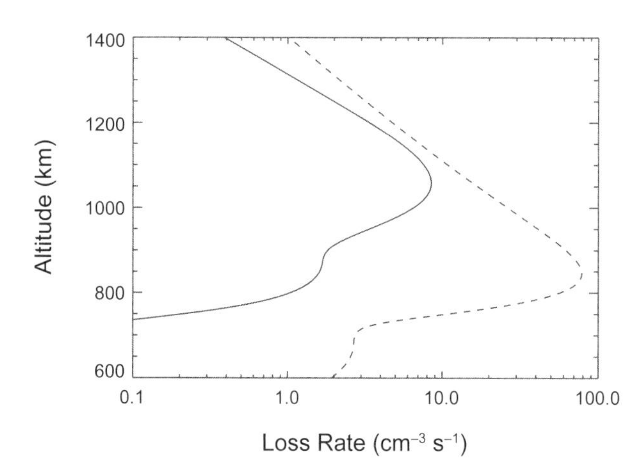

Fig. 2. Loss rates for N_2 (solid) and CH_4 (dashed) by photons and photoelectrons in the upper atmosphere.

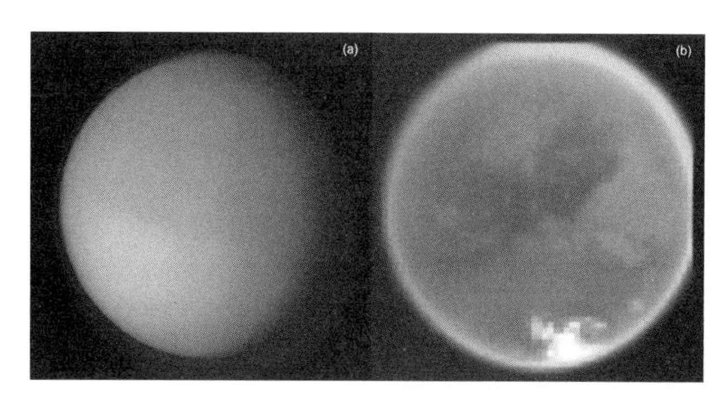

Fig. 3. See Plate 24 for color version. **(a)** At optical wavelengths, Titan's atmosphere below 100 km is shrouded by a complex organic haze. **(b)** At near-IR wavelengths between strong methane bands, Titan's surface and south polar tropospheric clouds (shown in white) can be seen through the haze. Credit: Voyager, Cassini, University of Arizona.

In situ pyrolysis of Titan's aerosol indicates the presence of nitrogen and carbon (*Israel et al.,* 2005). However, the detailed composition of the aerosol is unknown. The reactivity of the different molecules observed in Titan's atmosphere as well as laboratory experiments suggest pathways that include polymers of aliphatic compounds such as polyynes, or copolymers with nitrile species, as well as aromatic structures (*Lebonnois et al.,* 2002; *Wilson et al.,* 2003; *Lavvas et al.,* 2008, and references therein). The recent detection of abundant benzene in Titan's thermosphere (*Vuitton et al.,* 2008), considered together with aerosol growth models (*Lavvas et al.,* 2011a), suggest that polycyclic aromatic compounds (PAHs) are important, a scenario further supported by the structural elements implied by the far-IR optical properties of the aerosols (*Anderson et al.,* 2011). Indeed, all these pathways are potentially valid; aerosols do not necessarily result from a single chemical mechanism, but more likely from multiple chemical processes working in parallel.

The production of organic material through CH_4 photolysis and the continual depletion of CH_4 are possible only because CH_4 leaks into the stratosphere where UV photons break it apart. The culprit is the thermal profile (next section), which ironically is established by methane through its radiative forcing and that of its photochemical byproducts. In the lower atmosphere, both Earth's and Titan's temperatures reach minimum values at ~0.1 bar. Here, at the tropopause, the volume mixing ratio of each condensable is constrained to lie below its saturation value, which is ~3×10^{-6} for water on Earth and ~2×10^{-2} for methane on Titan. These mixing ratios establish the stratospheric values. At the tropopause, Earth's water abundance is small enough that water is not critically eroded by its small leakage into the upper atmosphere, where it is exposed to UV radiation. However, on Titan, significant methane reaches the upper stratosphere, where it is dissociated and constantly depleted. It is not clear what processes supply the current methane inventory in Titan's atmosphere and on its surface. The atmospheric methane is probably related to a subsurface reservoir, but the nature and depth of this reservoir as well as the involved exchange processes remain unconstrained (see section 8.3). Better understood are the depletion rate of methane and the climatic implications of methane fluctuations in Titan's atmosphere (sections 9–10).

3. TEMPERATURE PROFILE

Three different platforms have measured the temperature profiles of Titan's lower atmosphere. In 1980, Voyager 1 transmitted radio signals through Titan's atmosphere toward Earth, as the spacecraft passed behind Titan (egress) and reemerged (ingress). These occultations yielded the thermal structure of the atmosphere above 8.5°N latitude, 256°W longitude, and 6.2°N latitude, 77°W longitude, at dawn and dusk respectively, from the refraction of the radio signal by Titan's atmosphere (*Lindal et al.,* 1983). Cassini similarly measured a dozen temperature profiles at latitudes extending from tropics to poles (*Schinder et al.,* 2011, 2012). Huygens determined the temperature in greater detail during its descent through Titan's atmosphere at 10°S latitude, 192°W longitude (*Fulchignoni et al.,* 2005). Surface temperature maps are obtained from IR brightness temperatures, measured by Voyager and Cassini (*Flasar,* 1983; *Jennings et al.,* 2009, 2011).

Perhaps the most notable characteristic of Titan's thermal profile is its lack of variability (Fig. 4). Voyager egress and ingress measurements of the dawn and dust atmospheres indicate the same temperature pressure profiles within 0.7 K. Occultations by the ongoing Cassini mission, during 2004–2008, measured six profiles within ~35°N and S latitude circles; all these profiles match the Voyager profiles to within 1 K. The Huygens probe also measured the same profile, on January 14, 2005, during its descent. However, outward from the equator, temperatures decrease by ~1 K at 50°N and S latitudes, and by 3–4 K poleward of 70° latitude circles. A subtle ~1-K surface contrast exists between the summer and winter pole (Fig. 4).

Titan's overall thermal structure resembles that of Earth: There is a well-formed troposphere, stratosphere (*Lindal et al.,* 1983), and mesosphere (*Griffith et al.,* 2005b). Within ~2.5 km of the surface, Titan's temperature profile follows a dry lapse rate, i.e., the change of temperature with pressure (or height) of adiabatically rising dry parcels (*Tokano et al.,* 2006; *Schinder et al.,* 2011). This region defines the well-mixed "boundary layer," caused by the heating of the surface, which gives rise to dry convection. At 3–15 km, Titan's atmosphere is conditionally unstable; it is less steep than the dry lapse rate, but steeper than the moist lapse rate. Here parcels are unstable only if saturated, and therefore the stability of the atmosphere depends on the humidity profile. Convective systems must originate below 15 km (*Griffith et al.,* 2000). Cloud formation can initiate in the upper troposphere (15–45 km) as well, yet such stratiform clouds, unlike convective systems, would exhibit little vertical evolution. Without tracking the vertical evolutions of clouds, it is difficult to distinguish stratiform clouds from convective systems that form below 15 km and appear at higher altitudes because of convective upwelling.

4. METHANE PROFILE

The methane profile of Titan's atmosphere has been measured only once, in 2005 by the Huygens probe during its descent to the surface at 10°S latitude and 192°W longitude (*Niemann et al.,* 2005, 2010). This profile is markedly simple. Cassini's Gas Chromatograph Mass Spectrometer (GCMS) determined a CH_4 mixing ratio of 0.057 (50% humidity) above a damp surface (*Niemann et al.,* 2010). The mixing ratio is essentially constant below ~7 km altitude, above which the atmosphere is cold enough to be saturated. Between 9 km and 16 km altitude, the atmosphere is saturated (or nearly so) with respect to the ambient condensate, a binary solution of liquid CH_4 and ~20% dissolved N_2. At 25–30 km altitude, it is saturated with respect to pure CH_4

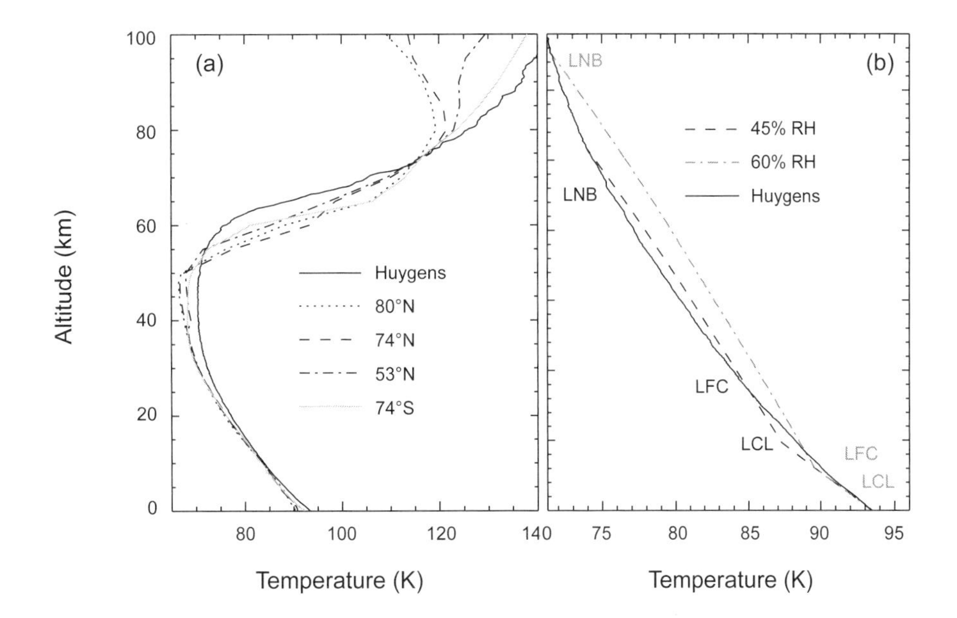

Fig. 4. (a) Titan's temperature profile at several latitudes indicates a 3 K cooler surface temperature and a more stable atmosphere at the poles (*Schinder et al.,* 2012). **(b)** The path of convectively rising parcels at Titan's landing site, assuming surface relative humidities of 45% and 60% are compared to the measured temperature profile.

ice, and an intermediate abundance exists between these two regions (*Tokano et al.,* 2006; *Lavvas et al.,* 2011b). Such a profile clearly indicates that methane originates from the surface and mixes with the air up to 7 km, above which condensation of the ambient condensate controls its abundance. The simplicity of the profile further suggests that it may be representative of average conditions, and therefore actually useful for climatic studies. High-resolution Keck observations of weak nonsaturated CH_3D lines imply that the methane abundance below 10 km does not change by more than 20% between 32°S and 18°N latitudes (*Penteado and Griffith,* 2010). The potential uniformity of Titan's tropical humidity profile is consistent with the paucity of tropical surface liquids and the roughly constant surface temperature.

The combined methane and temperature profiles measured at the landing site indicate the stability of the atmosphere with respect to convection. Consider the path of surface parcels, which have the most amount of methane and therefore latent heat to fuel convection. Assume that the parcel updrafts rapidly enough to be adiabatic. Then, since the atmosphere is undersaturated at the surface, the temperature of an updrafted parcel follows the dry lapse rate until it reaches the altitude, called the lifting condensation level (LCL), where the methane within the parcel condenses. Above this point the parcel follows the wet adiabatic lapse rate and cools less quickly during its assent because of the release of latent heat. This transition can be seen as a kink in the parcels paths in Fig. 4b. The parcel becomes buoyant when its temperature exceeds the atmospheric temperature and thus its density is smaller than the ambient values. Above this level, called the level of free convection (LFC), the parcel buoys upward until its temperature is smaller than ambient. As a result of the parcel's momentum, it will rise a few kilometers above this level of neutral buoyancy (LNB). Parcel paths calculated for the Huygens landing site conditions indicate that the parcel's temperature does not differ significantly from the ambient values (*Barth and Rafkin,* 2007; *Griffith et al.,* 2008), as shown in Fig. 4. Thus the convective available potential energy (CAPE) is quite low (~120 J kg^{-1}). This CAPE is much smaller than, for example, that of a typical terrestrial thunderstorm, where the CAPE is ~1000 J kg^{-1}, which would be achieved if the humidity were raised to 60% at the landing site. Such stable conditions with low and largely unchanging CAPE are seen over most of Earth. Here a large CAPE is readily created through humidification and large-scale forcing and readily destroyed, or stabilized, by moist convection due mainly to the convective downdraught flux of low-entropy air (*Emanuel et al.,* 1994). However, in Titan's tropical atmosphere, the methane humidity and thus the CAPE do not obviously change as does the humidity on Earth, because Titan has a high atmospheric methane content (5 m) and a lack of methane over the tropical surface. It is still not clear how the CAPE changes, particularly in the tropics, to produce the one observed tropical rainstorm (*Turtle et al.,* 2011a) and the surface erosional features that point to rainfall (*Soderblom et al.,* 2007).

Convective processes and large-scale dynamics are connected, and must be in general studied together (e.g., *Arakawa and Schubert,* 1973). For Titan, however, the long radiative time constant, which is tens of years just considering the lower 5 km of troposphere (*Mitchell et al.,* 2009), enables studies of convective systems that occur on timescales of a few hours or days without considering the radiative processes (*Hueso and Sánchez-Lavega,* 2006; *Barth and Rafkin,* 2007, 2010). Local cloud models take advantage of this characteristic to probe nonadiabatic processes and the microphysics of methane condensation, such as coagulation, sedimentation, and evaporation, as well as the effects of latent heat transfer, all of which are needed to investigate rainfall. Research by *Barth and Rafkin* (2007, 2010), which does not include large-scale forcing that can act to increase the CAPE, indicates that for a 60% humidity atmosphere, up to 1 m of precipitation may be produced in only a few hours. However, rainfall is a rare event; only two events have been witnessed during Cassini's now nine-year mission (*Turtle et al.,* 2009, 2011a). For the conditions measured at the probe landing site, most raindrops evaporate before reaching the surface (*Toon et al.,* 1988; *Lorenz,* 1993; *Barth and Toon,* 2004; *Graves et al.,* 2008, *Lavvas et al.,* 2011b).

5. ATMOSPHERIC ENERGY PARTITIONING

Titan's main constituent, N_2, contributes to the moon's greenhouse effect and chemistry and provides a background gas that broadens the methane lines. Yet it is the second most abundant constituent, methane, and its photochemical byproducts that establish the atmosphere's structure and composition.

Titan's haze and methane heat its stratosphere and limit the radiation reaching the surface (Fig. 5). These roles are witnessed from optical images of Titan, which reveal a hazy featureless moon (Fig. 3) and optical spectra that resemble a methane transmission spectrum. Although the haze particle concentration is not large by comparison to that on Earth, the great size of Titan's atmosphere, with a high scale height ~3 times larger than Earth's, renders the overlying integrated column of haze significantly higher. Titan's haze is optically thick throughout optical wavelengths and optically thin in the infrared (Fig. 3), precisely the opposite of a greenhouse gas (*McKay et al.,* 1991). This antigreenhouse component absorbs ~40% of the incident radiation, thereby cooling the surface by 9 K and heating the stratosphere ~9 K. The resulting radiative forcing is complicated because the haze drifts from one hemisphere to the other with the seasonal winds, causing temporal changes in the haze density with altitude and latitude. This haze migration in turn affects the circulation (*Rannou et al.,* 2006, 2012).

Still, analogies exist between the atmospheric radiative balance of Titan and Earth. Earth's atmosphere also experiences an antigreenhouse effect due to O_3, smoke, and dust, albeit one of much smaller magnitude. Titan's atmosphere also experiences a greenhouse effect. Pressure-induced absorption of N_2-N_2, N_2-CH_4, CH_4-CH_4, and N_2-H_2 absorbs the 30–800 cm^{-1} radiation emitted by Titan (Fig. 6), which warms the surface by 21 K on average (*McKay et al.,* 1991), comparable to the 33 K greenhouse effect on Earth (Table 1). Considering the cooling effect of the haze, the net difference between Titan's effective temperature, T_{eff}, and its surface temperature, T_{surf}, is 12 K (Table 1). Titan's greenhouse effect is highly sensitive to the abundance of the minor constituent H_2, which absorbs in the strong IR window at 400–600 cm^{-1} (Fig. 6). Its role parallels that of terrestrial CO_2; the vibrational bending mode band of CO_2 caps the 15-μm window of Earth's IR spectrum (*McKay et al.,* 1991). Like H_2O on Earth, CH_4 absorbs broadly in the IR and, as a condensable, its abundance is subject to climatic conditions. It is interesting to note that without methane, Titan's atmosphere would lack a stratosphere, strong greenhouse warming, and antigreenhouse cooling. An optically thin nitrogen atmosphere would remain.

Most radiative transfer derivations of the tropical temperature profile assume the global average incident solar radiation (3.7 W m^{-2}) rather than the local insolation (*Samuelson et al.,* 1983; *McKay et al.,* 1991) and yield temperature profiles that match those measured in Titan's tropical latitudes. These studies found that only 10% of the incident flux reaches the surface, and 1% goes into nonradiative fluxes (such as sensible heat and evaporation). However, the average insolation at equatorial latitudes is $4/\pi$ greater than the global average. Equatorial temperatures are therefore cooler than the radiative equilibrium solution that assumes the local insolation. A more recent study (*Williams et al.,* 2012) evaluates the surface energy balance at the Huygens landing site (10°S, 191°W) at the time of the landing (summer with a solar longitude of L_s = 300°). At this equatorial summer spot, the diurnally averaged insolation at the landing site (5.4 W m^{-2}) is significantly higher than the global mean (3.7 W m^{-2}), which establishes the atmospheric thermal profile. Since Titan's surface is not highly conductive, the extra energy at the surface goes into the atmosphere such that at 10°S latitude in the summer, 6.5% of the incident top-of-atmosphere (TOA) insolation goes into nonradiative fluxes (*Williams et al.,* 2012; *Griffith*

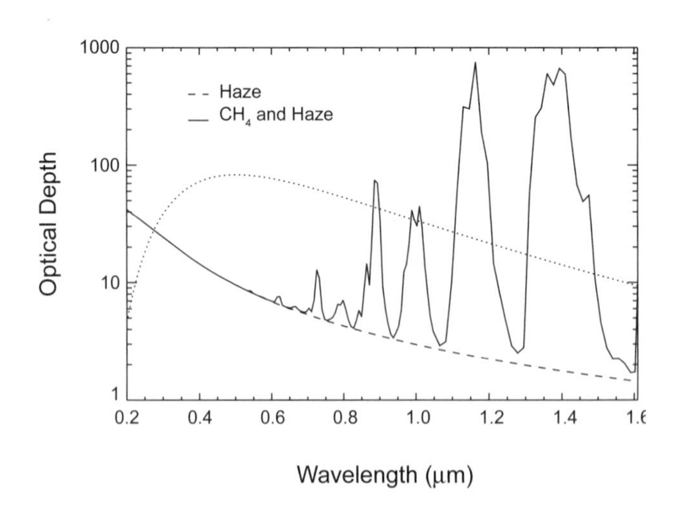

Fig. 5. Conservative scattering by Titan's haze (single scattering albedo of ~0.95) and methane absorption dominate Titan's optical opacity. Shown as dots is the 5778 K Planck function in units of 10^{-6} W m^{-2} $μm^{-1}$.

et al., 2013). In contrast, during winter, the radiative fluxes are fairly well balanced at the landing site surface. The lack of radiative equilibrium in Titan's atmosphere on an average annual scale indicates that the circulation redistributes Titan's heat, which, as further discussed in section 7, gives rise to nonradiative surface fluxes particularly in the polar summers (*Mitchell et al.,* 2012). Global heat transport plays a larger role in establishing the local partitioning of energy on Titan than it does on Earth, where conditions are established more by the local insolation (section 7). Titan's energy partitioning contrasts that on Earth, where on average 29% of the incident radiation is channeled to nonradiative processes, with ~7% going into sensible heat and ~22% into evaporation.

6. METHANE CYCLE

Observations of Titan's clouds and surface characteristics reveal the workings of the moon's weather and climate. These measurements point to distinct patterns.

6.1. Methane Clouds

Detected from backscattered optical and near-IR incident sunlight (*Griffith et al.,* 1998), Titan's methane clouds are fastidiously organized by latitude and season. Observations of Titan's clouds now span a period (1999–2012) of almost half a Titan year. During Titan's southern summer (1999–2004), clouds appeared exclusively at latitudes poleward of ~60°S (*Brown et al.,* 2002; *Roe et al.,* 2002; *Bouchez and Brown,* 2005), as shown in Fig. 3. In 2004, bands of clouds emerged surrounding 40°S latitude (*Roe et al.,* 2005). Roughly a year later the population of southern clouds diminished (*Schaller et al.,* 2006b), and disappeared altogether in 2008. The approach of vernal equinox in August 2009 caused equatorial clouds to become slightly more common, but still rare. From late 2006 to 2008, five small clouds and one massive cloud, covering 5–7% of the disk, appeared (*Griffith et al.,* 2008; *Schaller et al.,* 2009). By 2007, northern polar clouds were detected through ground-based and Cassini data from the Visual and Infrared Mapping Spectrometer (VIMS) (*Brown et al.,* 2009); however, it is not clear from observations alone whether northern clouds existed previously when the northern polar region was shrouded in darkness. The general seasonal trends of Titan's clouds, summarized in several papers (*Rodriguez et al.,* 2009, 2011; *Brown et al.,* 2010; *Turtle et al.,* 2011a), indicate the seasonal migration of clouds from the southern to northern polar regions as Titan moves from south summer solstice, in October 2002, to north summer solstice, in May 2017.

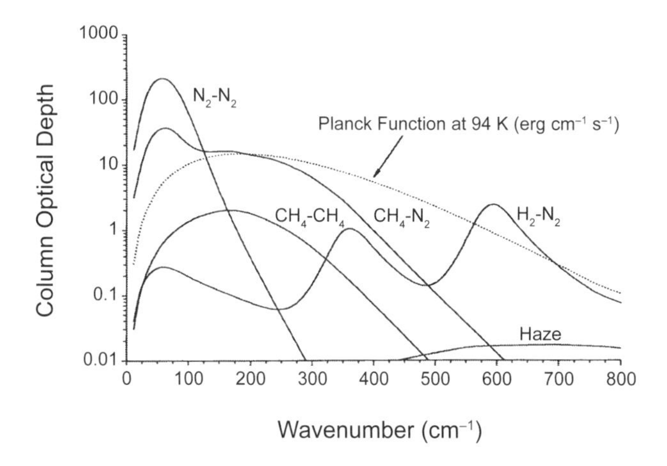

Fig. 6. The total column optical depth in the thermal infrared for Titan's atmosphere is due primarily to collision induced absorption. The dotted line indicates the black body flux emitted from a surface of 94 K. From *McKay et al.* (1991).

6.2. Cloud Evolution

Convective processes were initially indicated in Titan's atmosphere by the hourly variations of clouds and their common residence at the estimated LNB level (*Griffith et al.,* 2000). More recently, Cassini VIMS spectral images trace the vertical evolution of a handful of Titan's clouds. Radiative transfer analyses of four clouds at 41°S–61°S show that they rose from ~20 to ~40 km altitude at a rate of 2–10 m s^{-1}, consistent with convective updraft rates, and dissipated at a rate consistent with a rain falling at 5.5 m s^{-1} (*Griffith et al.,* 2005a), close to that expected of millimeter-sized drops (*Toon et al.,* 1988, 1992; *Lorenz, 1993*). Three larger clouds, also at high southern latitudes, show a similar evolution (*Ádámkovics et al.,* 2010). These observations can be explained as resulting from deep convection of surface air with a humidity of ~65%, assuming the tropical temperature profile (*Barth and Rafkin,* 2010); they suggest a more methane-rich atmosphere than that measured at the Huygens landing site.

Convection is also indicated by images of the cumuli structure and evolution of small polar cloud systems (*Porco et al.,* 2005). Studies of small tropical clouds display a similar cumuli structure, but, in contrast, appear lower in the atmosphere; none have been detected above 26 km altitude (*Griffith et al.,* 2009). These clouds are consistent with the Huygens temperature and methane profiles, for which convective clouds lie below ~26 km altitude (*Griffith et al.,* 2008).

Usually less than 1% of Titan's disk is cloudy. This sparse coverage of clouds agrees with that expected, if one regards the atmosphere as a heat engine fueled by solar insolation that drives convective motion that is opposed by dissipation (*Lorenz et al.,* 2005). Here the dissipation is assumed to depend on the convective flux scaled by a value that is based on shallow atmosphere models of homogeneous and isotropic turbulence on Earth (*Renno and Ingersoll,* 1996). The areal coverage of deep convection, which carries air aloft, is limited by the atmosphere's radiative cooling rate, which controls the downward flux of air, since on average the upward flux of air must equal the downward flux. Yet occasionally cloud systems emerge that are large enough to brighten Titan's disk by 9–66% (*Griffith et al.,* 1998; *Schaller et al.,* 2006a, 2009; *Turtle et al.,* 2011b). No analysis of the vertical evolution exists for such large systems, thus it is unclear whether these systems are convective. One series of large clouds, monitored through consecutive Keck images, displayed a coupled evolution of large equatorial and polar cloud systems, suggesting that both systems were triggered by the same planetary wave phenomena (*Schaller et al.,* 2009). Another large tropical storm (*Turtle et al.,* 2011a) showed a morphology characteristic of Kelvin wave clouds (*Mitchell et al.,* 2011). Convectively coupled waves are predicted by general circulation models (GCM), which have reproduced the observed characteristics of the large tropical systems (*Mitchell et al.,* 2011; *Schneider et al.,* 2012), yet the processes that drive cloud formation and the prevalence of deep convection are supported by only the handful of systems analyzed in detail so far.

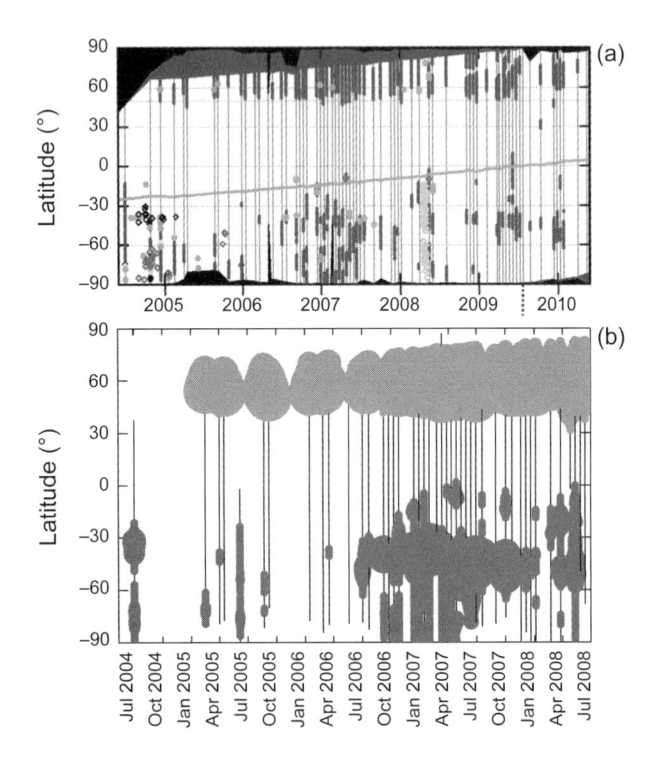

Fig. 7. See Plate 25 for color version. Time evolution of cloud latitudes from July 2004 to April 2010. Vertical thin lines mark the time and latitude coverage of VIMS observations. **(a)** Thicker blue lines show the latitude extension of detected clouds, summed over all longitudes. Dots and diamonds are Cassini and Earth-based detections (*Porco et al.,* 2005; *Baines et al.,* 2005; *Griffith et al.,* 2005a, 2006, 2009; *LeMouelic et al.,* 2012; *Turtle et al.,* 2009; *Roe et al.,* 2005; *Schaller et al.,* 2006a, 2009; *de Pater et al.,* 2006; *Hirtzig et al.,* 2006). The green line indicates the latitude of the maximum of insolation, with north spring equinox in August 2009. Gray areas are night; black areas no observations. From *Rodriguez et al.* (2011). **(b)** Circles show latitudes of tropospheric methane clouds (red) and the northern tropopause ethane clouds (green) recorded by VIMS; the size is proportional to areal coverage. From *Brown et al.* (2010).

6.3. Ethane Clouds

Ethane combined with other byproducts of methane photolysis are mixed down to Titan's troposphere, preferentially steered to the winter pole by Titan's circulation. Most of these species condense below 90 km altitude, where temperatures cool in the lower stratosphere. Photochemical byproducts therefore arrive to Titan's surface predominantly at the poles during the winter, where they accumulate as solids and liquids on the surface (*Griffith et al.,* 2006; *Rannou et al.,* 2006, 2012; *Le Mouélic et al.,* 2012).

Cassini images of Titan's northern (winter) hemisphere in 2005 display a uniform cloud poleward of 50°N (*Griffith et al.,* 2006). This cloud cap (Fig. 7b) reflects sunlight preferentially at wavelengths less than 3 μm, indicating that it consists of particles 2–3 μm in size; its spectral signature indicates an altitude of ~52 km and a column mass of ~60,000 cm^{-2} (*Griffith et al.,* 2006). The inferred particle size matches that expected of a minor species like ethane; its altitude matches that where ethane condenses, and its winter residence matches GCM predictions of ethane and the other less-abundant photochemical byproducts (*Rannou et al.,* 2006; *Griffith et al.,* 2006). More recent observations show that this cloud has progressively faded since 2005, consistent with the dissipation of the downwelling circulation branch upon the arrival of northern spring (*Le Mouélic et al.,* 2012; *Rannou et al.* 2012).

While IR spectra indicate abundant stratospheric ethane (*Vinatier et al.,* 2010b), cloud observations indicate ethane precipitation (*Griffith et al.,* 2006), and lake spectra suggest an ethane presence (*Brown et al.,* 2008), its abundance in the troposphere and surface are unconstrained. Although several orders of magnitude less abundant than methane in the atmosphere, ethane, if mixed with the methane liquid at the surface-atmosphere interface of a pond or lake, can limit the exposure of methane and therefore its saturation vapor pressure. Ultimately it can control the stability of the atmosphere with respect to convection (*Griffith,* 2009). In vapor pressure equilibrium, a liquid containing 39% methane, 54% ethane, and 7% nitrogen is in vapor pressure equilibrium with atmospheric mixing ratios of ~1.4×10^{-5} and 0.049 for ethane and methane, respectively (*Thompson et al.,* 1992; *Griffith et al.,* 2008). In Titan's polar regions, ethane, if comprising a large fraction of the polar lakes, could limit the methane humidity. Yet measurements of seasonal changes in the southern polar lake levels suggest that surface liquid ethane is not abundant enough to throttle the evaporation (*Cassini RADAR Team,* 2011; *Turtle et al.,* 2011c). In Titan's tropics, the Huygens probe upon landing detected a puff of methane and ethane, suggesting an ethane and methane damp soil (*Niemann et al.,* 2005). Yet the characteristics of the particulate population above the landing site point to an ethane abundance close to the surface that is significantly undersaturated (*Lavvas et al.,* 2011b). Thus one of the largest uncertainties in our understanding of Titan's methane cycle hinges on the abundance of this minor species, which can affect evaporation processes (*Thompson et al.,* 1992), control the CH_4 humidity (*Griffith,* 2009), and even control the circulation (*Tokano,* 2009).

6.4. Rainfall

Rainstorms on Titan potentially involve higher precipitation rates than terrestrial storms, because Titan's atmosphere contains ~200 times more methane than Earth's atmosphere has water. Cassini imaged the effects of rainfall on two occasions (*Turtle et al.,* 2009, 2011a). Observations taken a year apart (in 2004 and 2005) reveal that a ~34,000-km² area of Titan's surface near the south pole (80°S and 120°W) darkened (at 0.938 μm wavelength) after the passage of a large cloud system (*Turtle et al.,* 2009). More recently, in 2010, a large cloud system appeared at tropical latitudes, followed by an ~10% darkening of ~500,000 km² of the surface, most of which brightened within two weeks (*Turtle et al.,* 2011a). In both cases the depth of the surface moistening, whether a wetting or more substantial ponding, is unknown, as is the process by which the surface brightened. The liquid could have seeped into the subsurface or evaporated.

6.5. Surface Erosion

Titan's weather extensively erodes the surface, rendering the surface largely devoid of craters (*Neish and Lorenz,* 2012) and reminiscent of Earth. Distinct terrains appear at the poles and the equator and suggest different climates. Yet distributed fairly evenly with latitude across Titan's globe are long-branching flow features carved by running liquid, the source of which is not understood (*Elachi et al.,* 2005; *Lopes et al.,* 2010; *Langhans et al.,* 2012).

6.5.1. Polar surface. Only poleward of the 55° latitude lines have lakes been detected by Cassini RADAR measurements (*Stofan et al.,* 2007). Lakes are identified by a low reflectivity surface at 2.2 cm, which is "black" within the noise (Fig. 8), that requires at least 8 m of liquid (*Hayes et al.,* 2010). In addition, features morphologically similar to the lakes, but not black, dot the polar surface, indicating the presence of shallow and dry lakes (Fig. 9). Covering a region of roughly 500,000 km² in the north polar region and 15,000 km² in the south polar region, lakes contain a total surface liquid volume that is equivalent to a 1–2-m globally uniform layer of methane (*Lorenz et al.,* 2008). As further discussed in section 7, lakes are predicted to reside at the poles, because Titan's circulation readily transports methane to high latitudes (*Rannou et al.,* 2006; *Mitchell,* 2008; *Schneider et al.,* 2012).

Repeated observations of the lake boundaries demonstrate surface exchange of volatiles on seasonal timescales. Radar and optical measurements of the south polar lake Ontario Lacus were recorded from 2005 to 2009, during which time southern polar clouds diminished until they disappeared entirely in 2008 (*Schaller et al.,* 2006b; *Rodriguez et al.,* 2009). During these progressively drier years in the southern hemisphere, the level of Ontario Lacus may have dropped by ~5 m based on estimates of shoreline retreat (*Cassini RADAR Team et al.,* 2011; *Turtle et al.,* 2011c). However, measurements of the location of the shoreline are currently debated (*Cornet et al.,* 2012; *Ventura et al.,* 2012), and it is not clear that shoreline variations are actually occurring. Radar measurements detect no similar lake level drops in northern lakes (*Hayes et al.,* 2010). Further evidence of evaporation comes from the bright 5-μm albedo in the dry lake beds, which suggests the exposure of evaporites too recent to be filled by haze deposits (*Barnes et al.,* 2011). Taken together, these observations imply a seasonal drying of Titan's south polar region by evaporation and/or surface infiltration.

These measurements bear on the seasonal variations of the average humidity in Titan's tropics. If a 1-m yr^{-1} drop in the lake depth is typical of the seasonal drying in the south polar region, the evaporation rate, assuming a lake coverage roughly twice that of Ontario Lacus, 30,000 km², is roughly 3×10^{10} m^3 yr^{-1}. This value is consistent with the upper limit (2×10^{11} m^3 yr^{-1}) estimated from the seasonal difference (integrated summer minus winter) in the surface insolation, integrated poleward of 60° latitude (*Griffith et al.,* 2008). Both evaporation rates are too small to significantly enhance humidity in the tropical latitudes with changes in the seasons.

The distribution of Titan's lakes displays a hemispherical asymmetry (Fig. 9). Lakes predominate in the north; here the lake areal coverage is ~25 times that in the south (*Aharonson et al.,* 2009). This difference cannot be explained by seasonal

Fig. 8. Cassini RADAR images of Titan's surface indicate north polar lakes (Ligeia Mare shown here) and a highly eroded surface. Credit: Cassini RADAR Team.

methane migration because of the limited temperature variations (*Schinder et al.,* 2011). A seasonal solution is also not substantiated by the hemispherically asymmetric distribution of dry lakebeds (Fig. 9), the surface area of which is three times greater in the north compared to the south (*Aharonson et al.,* 2009). The north-south lake asymmetry is postulated to be caused instead by the differing insolation in the north and south polar regions that result from the eccentricity of Titan's orbit about the Sun (*Aharonson et al.,* 2009; *Schneider et al.,* 2012). Titan's inclination (0.33°) and obliquity (0.6°) to Saturn's equatorial plane are much smaller than Saturn's obliquity (26.7°). Insolation patterns are therefore determined mainly by Saturn's obliquity along with the orbital eccentricity (e = 0.054). The perihelion passage lies presently at $L_{s,p}$ = 277.7° near northern winter solstice; thus the southern summers are shorter and the insolation more intense (24% higher at peak values) than the northern summers (*Aharonson et al.,* 2009). The resulting asymmetry in the seasonal cycles of haze transport and methane evaporation/precipitation is hypothesized to cause a liquid accumulation in the northern terrains (*Aharonson et al.,* 2009). A recent three-dimensional GCM study predicts an accumulation of liquid methane on the polar surface with a latitudinal distribution that agrees with the observations (*Schneider et al.,* 2012). This work predicts the preponderance of surface liquids near the north pole, rather than near the south pole: Methane is cold-trapped in the north pole, where summers, albeit longer, are cooler than the summers in the south (*Schneider et al.,* 2012). The insolation function varies on longer timescales due to precession of the perihelion passage (~45 k.y.), variations of eccentricity, and the position of the spin axis (~270 k.y.). The seasonal effects are therefore hypothesized to be modulated by dynamical variations in the orbit that vary on timescales of tens of thousands of years, potentially causing climatic changes similar in scale to Earth's climatic cycles (*Aharonson et al.,* 2009).

6.5.2. Equatorial surface. On January 14, 2005, the Huygens probe parachuted to the surface at 10°S and 192°W, a region that shows no standing liquid, but that is extensively carved by running liquid. Only here has Titan's surface been investigated in detail. Huygens landed in a flood plain scoured by eastward-flowing liquid (*Tomasko et al.,* 2005), strewn with cobbles, 5–15 cm in size over a finer-grained terrain. At touchdown, Huygens measured increased levels of methane and other carbon compounds, indicating surface evaporation and the presence of damp terrain (*Niemann et al.,* 2010). Roughly 5 km north of the landing site reside hills ~180 m high with two kinds of drainage systems (*Tomasko et al.,* 2005). Highly branched networks (Fig. 10) resemble terrestrial washes formed by rainfall (*Soderblom et al.,* 2007). Short and stubby channels mirror terrestrial valleys at the heads of closed canyons, which are caused by spring sapping (*Soderblom et al.,* 2007). This heavily-liquid-eroded terrain lies in the middle of a vast dune field. In Titan's tropics (30°S–30°N), dunes are common and wrap around the globe, extending thousands of kilometers (*Lorenz et al.,* 2006; *Barnes et al.,* 2007; *Radebaugh et al.,* 2008).

Despite the wealth of data, Titan's tropical climate remains elusive. Rainfall is indicated by the landing site's drainage systems and by a surface darkening following a tropical storm (*Soderblom et al.,* 2007; *Turtle et al.,* 2011a). Current GCMs (*Mitchell,* 2008; *Schneider et al.,* 2012) predict, consistent with one-dimensional models (*Hueso and Sánchez-Lavega,* 2006; *Barth and Rafkin,* 2007, 2010), the presence of tropical rainfall for atmospheres that are more humid everywhere than that measured. However, it is not yet clear how the tropical atmosphere humidifies and where the methane that flooded the landing site came from. It is particularly difficult to understand how the surface could be damp, since surface liquid is unstable at the tropics; it readily evaporates and is transported to Titan's poles

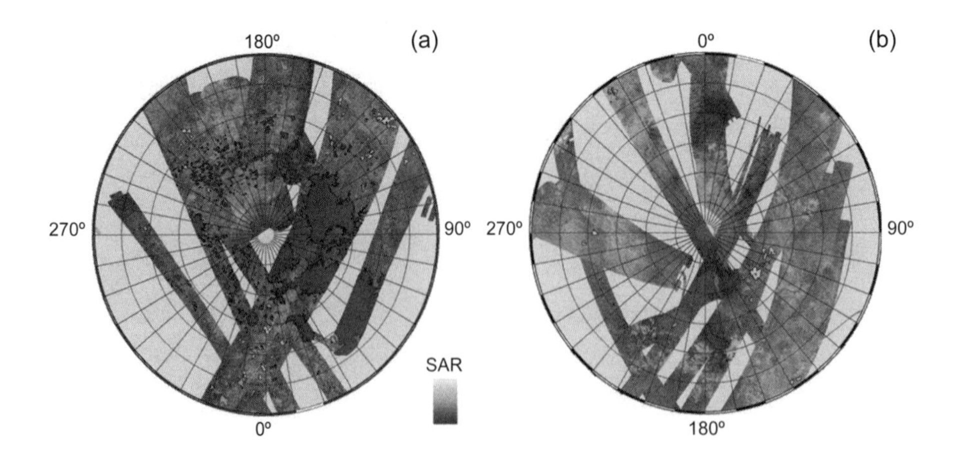

Fig. 9. See Plate 26 for color version. Distribution of Titan's lakes for the **(a)** northern and **(b)** southern hemispheres. The colors indicate whether the lake is filled (blue), intermediate (cyan), or empty (pale blue). The backscatter brightness of Cassini's RADAR is shown in shades of brown. The diameter of the image is ~2400 km. From *Aharonson et al.* (2009).

(*Rannou et al.,* 2006; *Mitchell,* 2008). One can postulate that Huygens landed right after a rainstorm. Yet rainfall has been evidenced only once, and during the tropical wet season; it has never been detected during the dry season when Huygens landed. A source of methane is still needed to dampen, flood, and carve the landing site.

Radar images have not detected a black surface indicative of liquids between 50°N and 50°S latitudes. Yet recent near-IR Cassini VIMS spectral images reveal several black surfaces indicative of surface liquid (*Griffith et al.,* 2012). Near-IR images have the advantage of detecting shallow surface liquids, because methane absorbs stronger at near-IR compared to radar wavelengths (*Griffith et al.,* 2012). The largest potential tropical lake, with an area of 2400 km^2, appears in VIMS data dating back to 2004, before the rainy season. The existence of a tropical lake points to a subsurface source. Subsurface methane is also potentially responsible for the landing site's damp surface and flood plain, and perhaps explains how such a highly eroded terrain might exist in a region surrounded by dune fields. Additional observations are needed to explore these possibilities. The presence of subterranean methane is particularly interesting because such sources could provide the long-sought-out supply of methane that furnishes the atmospheric methane, which is depleted by photolysis at a rate of ~2 m per 10^7 yr.

7. TITAN'S CIRCULATION

Titan's circulation is responsible for the highly uniform temperature field in the troposphere, the global distribution

Fig. 10. The Huygens landing site (as imaged by the Huygens DISR instrument) shows the highly branched drainage systems. From *Tomasko et al.* (2005).

of methane, and the dynamics that trigger cloud formation. This section discusses the general physical mechanisms underlying Titan's atmospheric circulation, with the goal of highlighting fundamental processes and focusing on order-of-magnitude calculations rather than detailed derivations. The aim is to synthesize related phenomena into an overarching theoretical framework, without repeating the thorough review of previous work on Titan's circulation given in *Griffith et al.* (2013). The discussion is broken into three broad categories: atmospheric dynamics, timescales and seasons, and their coupling to the methane cycle.

7.1. Atmospheric Dynamics

The context of Titan's atmosphere, i.e., its vertical extent and the body on which it resides, already distinguishes it from Earth and the approximations possible toward an understanding of its physics. The terrestrial atmosphere represents a thin layer on a rotating sphere in the presence of a gravitational field. Its aspect ratio, a measure of this "thinness" given by the ratio, H/a, with scale height H and planetary radius a, is ~10^{-3}. Because this ratio is small, only the local vertical component of planetary rotation, the Coriolis parameter ($f = 2\Omega \sin \theta$, where Ω is the rotation rate and θ the latitude) influences fluid motion. The rotation axis does not project onto the equator (the Coriolis parameter vanishes), therefore the equatorial atmosphere behaves in a fundamentally different manner than the middle and high latitudes, as discussed further below. In addition, large-scale atmospheric motion with small aspect ratio is very nearly hydrostatic. Much of our general intuition of Earth's atmosphere derives from these approximations.

Titan's lower atmosphere (including the stratosphere and mesosphere) extend hundreds of kilometers; its aspect ratio, ~10^{-1}, is quite high. However, Titan's climate and weather are both primarily governed by tropospheric processes. The troposphere extends up to ~40 km, where the aspect ratio, ~10^{-2}, is still sufficiently small for thin approximations to apply.

Hydrostatic balance provides a fundamental constraint on the vertical structure of an atmosphere. When combined with the ideal gas equation of state, hydrostatic balance tells us that at a given altitude, a warm column of air has higher pressure than a cold column. This difference gives rise to a pressure gradient force on a global scale from the warm equator toward the cold poles. Steady-state winds are maintained by balancing this pressure gradient force against the coriolis force, which we now describe.

7.1.1. Influence of rotation. A standard measure of the influence of planetary rotation is the Rossby number, Ro = U/(fL), with characteristic horizontal wind speed U, Coriolis parameter f, and characteristic horizontal length L. Physically, the Rossby number is the time it takes an air parcel to undergo an inertial oscillation relative to the time for the air parcel to travel a distance L. On Earth in the middle and high latitudes, Ro is on the order of 10^{-1}, and therefore parcels are strongly influenced by rotation through the Coriolis effect. On large scales, there is a balance between Coriolis and pressure gradient forces, or geostrophic balance. At the global scale, geostrophic and hydrostatic balance links the equator-to-pole temperature gradient to the vertical wind shear through the thermal wind equation.

The momentum transported poleward by the Hadley circulation of air cannot proceed all the way to the poles, otherwise upper-level tropospheric winds would become unrealistically large. Instead, air sinks and cools at intermediate latitudes, reenters the frictional boundary layer where surface friction resets the momentum, and follows a "return flow" near the surface. The Coriolis force acting on the equatorially convergent winds gives rise to the trade winds. Near the equator, air is lofted into the upper troposphere by convection, thereby completing a mass circuit. This overturning circulation is called the Hadley circulation, and its extent roughly corresponds to the region of horizontally uniform temperatures on Earth referred to as "the tropics." The poleward edges of the Hadley circulation can be established by consideration of fluid instabilities (*Held,* 2000) and/or energy and angular momentum conservation (*Held and Hou,* 1980). Both approaches give reasonable estimates of the size of Earth's Hadley circulation. Hadley circulation width is proportional to a power of the Rossby number, $Ro = U/(2\Omega a)$. The Rossby number at global scales for planetary bodies can be estimated by setting $U = U_{th}$, the upper-level wind speed required for thermal wind balance with the mean equator-to-pole temperature gradient. These theories therefore predict widening tropics for smaller and slower-rotating planetary bodies. On Titan, the Hadley circulation is nearly global in extent — Titan is "all tropics."

7.1.2. Important length scales. The length scale at which an atmosphere adjusts to geostrophic balance is the deformation radius $L_d = NH/f$. This is the scale over which gravity waves must travel to feel the influence of rotation. We immediately see the deformation radius is poorly defined at the equator. The equatorial deformation radius, $L_\beta = (c_g/\beta)^{1/2}$, instead quantifies the characteristic scale of geostrophic adjustment in low latitudes. Here $\beta = (2\Omega/a)\cos(\phi)$ is the meridional gradient of the coriolis parameter. β is greatest at the equator, where the Coriolis parameter vanishes, and their combined effect creates an equatorial waveguide that can trap waves at the equator (*Matsuno,* 1966) or prevent them from propagating into middle and high latitudes (see Showman et al., this volume).

7.1.3. Weak temperature gradients in the tropics. In the tropics, the Rossby number is on the order of 1, geostrophic balance does not hold, and we are forced to look for other approximations. If we consider the heat pulse again, now near the equator, we see that the bulge will spread out evenly over the entire region until the bulge no longer remains. By this process, gravity waves are extremely efficient at removing temperature gradients in the tropics. This has led to the weak temperature gradient (WTG) approximation for idealized studies of tropical weather and climate (*Sobel and Bretherton,* 2000).

In the WTG approximation, it is assumed that gravity waves effectively remove all horizontal temperature gradients, $\nabla_h T = 0$. The utility of this approximation is quite different than the mid-latitude geostrophic approximation. However, the qualitative aspects of WTG having horizontally uniform temperatures will suffice for our purposes. The question of WTG being a valid approximation over how large a region in latitude remains. Although an exact answer to this is somewhat subtle, the qualitative trend is obvious: Slower-rotating and/or smaller planets will have larger regions where the WTG approximation is valid. Because Titan is both smaller and rotates slower than Earth, we expect the WTG approximation to hold over a very wide range of latitudes.

The WTG extends through much of the troposphere, as indicated by the "flat" outgoing longwave radiation (OLR) at the top of Titan's atmosphere (Fig. 11). Such a featureless climate cannot be explained by the large heat capacity of Titan's atmosphere alone. Instead, the latitude-independence of Titan's climate is indicative of atmospheric heat transport (*Mitchell,* 2012), which fluxes heat away from the equator and toward the poles much as it does on Earth (*Trenberth and Stepaniak,* 2003).

Titan's atmospheric heat transport has profound consequences for the climate, and in particular the strength of the methane cycle as measured by precipitation. Latitudinal heat transport by the atmosphere allows the local, instantaneous climate to be comparatively far from radiative equilibrium. The TOA radiative imbalance, the difference between insolation and OLR, indicates how radiatively out of balance the local atmosphere is. Here insolation refers to the diurnal-average sunlight with the reflected portion (one albedo) removed. Enhanced atmospheric absorption at slant angles (*Lora et al.,* 2011) and shielding by a polar hood are not accounted for. Observations indicate a ~0.5 W m^{-2} excess of insolation at the equator in the annual mean (Fig. 11) and more than 1 W m^{-2} instantaneous local imbalance in the summer hemisphere (*Li et al.,* 2011). Climate simulations indicate that most of the TOA imbalance is transferred to a surface radiative imbalance, which is then available to drive evaporation and sensible heat (*Mitchell,* 2012). At the time of the Huygens landing, for instance, the estimated surface radiative imbalance is 0.35 W m^{-2} (*Williams et al.,* 2012; *Griffith et al.,* 2013), which is significantly greater than the value (0.04 W m^{-2}) derived from the assumption of disk-averaged radiative-convective equilibrium (*McKay et al.,* 1989).

Thus in comparison to Earth we expect Titan's atmosphere to (1) have a large Rossby number; (2) have a large tropical, WTG region; and (3) not be in geostrophic balance (*Flasar,* 1998). The first and second points are confirmed in observations; Titan's winds are strong compared to planetary rotation (e.g., *Bird et al.,* 2005; *Achterberg et al.,* 2008) and Titan's surface temperature and OLR are relatively "flat" (independent of latitude) (*Jennings et al.,* 2009; *Li et al.,* 2011). The third point is somewhat problematic, since we argued large-scale winds on Earth are in thermal wind balance, and Titan's winds must be in force

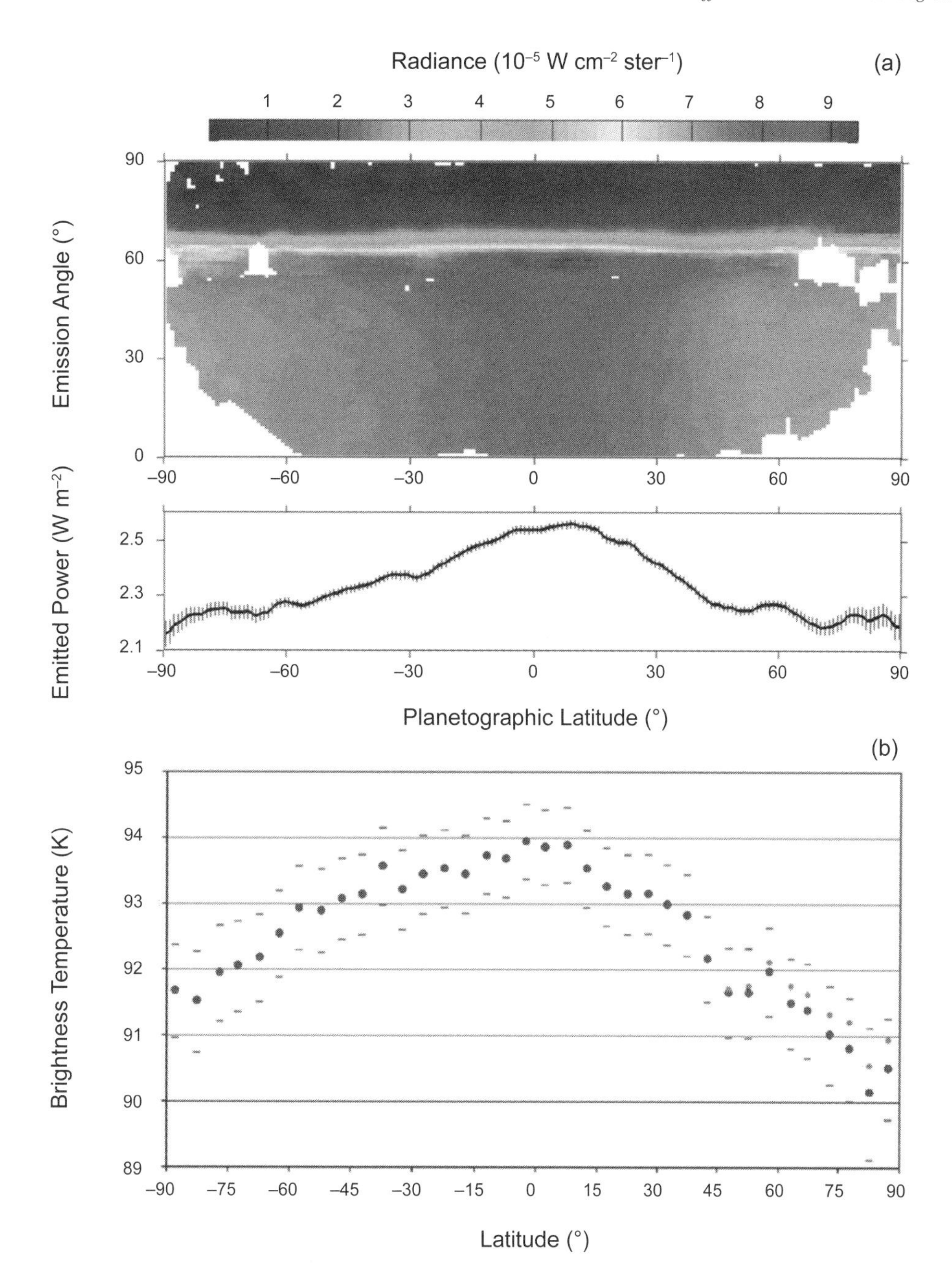

Fig. 11. See Plate 27 for color version. **(a)** Titan's outgoing longwave radiation as measured by Cassini CIRS (*Li et al.,* 2011). **(b)** The brightness temperature of Titan's surface (*Jennings et al.,* 2009).

balance if they are to be in steady-state. Observations of Titan show that superrotating winds are in cyclostrophic balance; roughly speaking, the extension of geostrophic balance when the relative rotation of the atmosphere itself supplies the "Coriolis effect."

Superrotation is itself somewhat of a persistent enigma. Defined as an atmosphere that rotates faster than the equatorial surface, superrotation is actually common in the solar system, with both Titan and Venus as terrestrial examples. Superrotation requires there to be an angular momentum maximum somewhere in the interior, a situation that is not expected if angular momentum is diffused down its mean gradient (*Hide,* 1969). Some process involving the transport of angular momentum up the mean gradient is therefore necessary (*Gierasch,* 1975). Many studies have aimed at identifying the exact mechanism of this transport, but a consensus remains to be achieved.

7.1.4. Waves and instabilities. One implication of a "globally tropical" Titan is that tropical waves play a major role in the weather and/or climate. Tropical wave theory (*Matsuno,* 1966) has been successfully applied to observations of Earth's tropics (*Wheeler and Kiladis,* 1999) to identify several prominent wave modes that organize small-scale convective storms into mesoscale storm complexes. Recent storm activity near Titan's equator (*Schaller et al.,* 2009; *Turtle et al.,* 2011a) indicates the same sort of "convective coupling" between tropical waves and convective storms (*Mitchell et al.,* 2011). Both Kelvin and Rossby waves have been identified in Cassini and groundbased cloud observations. However, it is not clear if other waves and instabilities are present and how they might be detected. In fact, it is uncertain what energy source or instability drives large-scale waves and the role that they play in maintaining the global circulation and superrotation of the atmosphere.

7.2. Seasons

Having established the climatic consequences of Titan's small radius and slow rotation rate, we now turn our attention to another feature that distinguishes Titan from Earth: a strong seasonal cycle. The processes that influence the strength of a seasonal cycle are worth careful consideration. This is especially true for Titan, where the atmosphere provides the dominant source of heat capacity to the climate system. The situation is reversed on Earth, where the surface (oceans) provides heat capacity to the climate system. This reversal of roles leads to unexpected seasonal behavior.

Observations of clouds indicate a dominant role for the convergence zone associated with Titan's global meridional overturning circulation. Like in Earth's intertropical convergence zone (ITCZ), surface trade winds sweep over large swaths of the surface acting as a drying agent on the flanks of the convergence-driven updraft. Convective clouds are generated within the updraft, and methane is returned to the surface as precipitation.

Unlike Earth's ITCZ, however, Titan's convergence zone migrates away from the equator in latitude and deep into the summer hemisphere. This "seasonal oscillation" in Titan's large-scale circulation has profound consequences for the surface climate. During long summers, methane evaporates into surface-level air parcels, which are carried long distances to the convergence zone and precipitated in the summer hemisphere. As a result, the equatorial region experiences net drying on average while accumulation of methane occurs at higher latitudes (Fig. 12).

Most GCMs predict a strong seasonal cycle in the latitudinal position of Titan's cloud-forming region. We can understand the strong seasonality of Titan's lower atmosphere by considering the timescales involved in the driving and response of the climate system by analogy with a forced-damped oscillator, with the forcing provided by insolation and the damping provided by radiative cooling. Titan's seasons are set by Saturn's 29.5-year orbit around the Sun, so that each season takes just over 7 years. The response of Titan's climate is quantified by thermal inertia, which we express as a timescale given the cooling flux and heat capacity of the layers experiencing the seasons. If the cooling time is short compared to seasons, then the system responds in lock-step with the forcing, but if the cooling time is long compared to a full year, the seasonal cycle is muted and lags by ~7 years.

Titan's seasonal cycle is forced by surface warming by sunlight, is communicated to near-surface air, and drives convection through an atmospheric layer that is dependent on the buoyancy of the air. This process is fast, happening essentially instantaneously, and thus the heat capacity of Titan's climate system is the sum of the surface and convecting atmospheric layer. Buoyancy of an air parcel is dependent on both heat and methane content, however. It is by now well established that Titan's dry-convective boundary layer is ~2–3 km deep, extending through the lowest $\Delta p \sim 200$ hPa (*Fulchignoni et al.,* 2005; *Griffith et al.,* 2008; *Schinder et al.,* 2011, 2012). The heat capacity per unit area, $C_A = C_p\Delta p/g$, combined with the cooling flux sets the response timescale, $t_r = (\tau C_A)/(4\ \sigma T^3)$, where τ is the infrared optical depth of the convecting layer, which for the dry boundary layer is ~1. Titan's solid surface has a relatively small amount of heat capacity and therefore the heat capacity of the boundary layer of the atmosphere dominates. In the first 2–3 km of the atmosphere, the radiative timescale, t_r ~2.5 years, establishes the longest time for heat transfer because, as indicated by the temperature profiles, heat is also transferred by the quicker process of dry convection. Thus a strong seasonal cycle is possible in the dry boundary layer and surface.

The radiative timescale of the entire troposphere is much longer for two reasons. First, there is more mass, with $\Delta p \sim 1000$ hPa. Second, the infrared optical depth is $\tau \sim 10$. These effects combine to increase the timescale by a factor of 50 relative to the dry boundary layer to ~100 years, an estimate consistent with that measured (*Strobel,* 2009). Such a large timescale compared to the length of Titan's year (29.5 Earth years) led to speculation that Titan's atmosphere would exhibit a very muted seasonal cycle (*Flasar et al.,* 1981).

A long radiative damping time in the troposphere also affects the atmospheric dynamics by limiting the rate of overturning in the Hadley circulation. The overturning rate directly influences the seasonal cycle and mean state of the zonal mean winds through the advection of momentum (section 7.1.1). Its timescale is estimated by considering thermal balance of air parcels in the descending branch, where radiative cooling and adiabatic heating are in balance, $w\ (\Gamma_d\text{-}\Gamma) = J$, with vertical velocity w, dry adiabatic lapse rate Γ_d, environmental lapse rate Γ, and radiative cooling rate J in K s^{-1}. For the troposphere, we estimate $w = H/t_d$, with H = 40 km, and the overturning time t_d. The radiative cooling rate is approximately $J \sim F_{IR}/C_A \sim 5 \times 10^{-9}$ K s^{-1}, with C_A as defined previously and $F_{IR} \sim 5$ W m^{-2} is a characteristic infrared radiative flux. We estimate the difference in lapse rates to be $(\Gamma_d\text{-}\Gamma) \sim 0.5$ K km^{-1} (*Fulchignoni et al.,* 2005), and thus an overturning time of $t_d > 100$ years. This number is confirmed by atmospheric calculations with a deep overturning circulation (*Mitchell et al.,* 2009). Because of the sluggish overturning in the troposphere, mid-tropospheric winds should not significantly change with seasons.

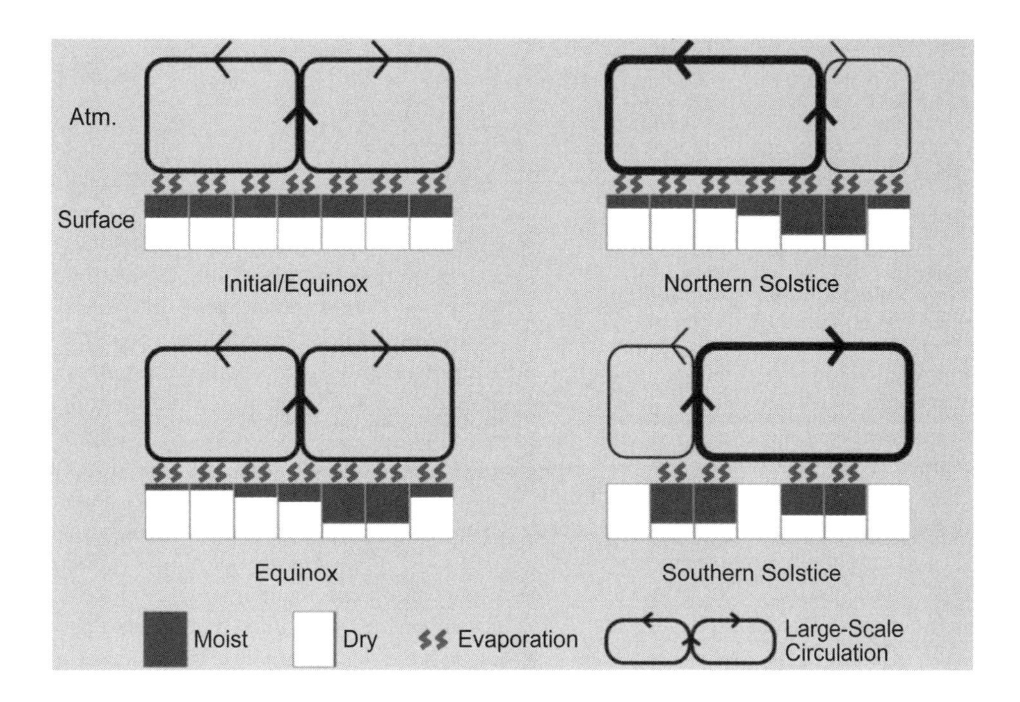

Fig. 12. See Plate 28 for color version. Titan's global overturning circulation transports methane to the convergence zone, where lifting deposits it locally as precipitation. A strong seasonal cycle leads climatologically dry conditions at the equator.

7.3. Coupling to the Methane Cycle

Since the radiative time (thermal inertia) of the entire troposphere is $t_r > 100$ years, we conclude that a key parameter in the seasonal response of Titan's atmosphere is the depth of the layer through which convection is penetrates, Δp (*Mitchell et al.,* 2006, 2009; *Mitchell,* 2008). The presence of methane vapor creates the possibility of increasing Δp by destabilizing the atmosphere toward moist convection. Under the most favorable conditions with abundant evaporating surface liquids, moist convection on Titan can penetrate throughout the depth of the troposphere. Therefore moist convection driven by surface evaporation plays an important role in regulating the strength of Titan's seasonal cycle.

General circulation models indicate that nonradiative fluxes are sufficient to fuel surface evaporation convection and precipitation at the poles during summer and the equator near equinox, provided that there is a sufficient surface supply of liquid methane (*Mitchell,* 2012). The persistent drying of Titan's equatorial region likely supports the widespread occurrence of dunes there. General circulation models predict ephemeral storms over Titan's tropical deserts, and if humid enough, brief equinoctial downpours (*Mitchell,* 2008; *Schneider et al.,* 2012); one such storm has been observed (*Turtle et al.,* 2011a). Yet some throttling of the methane cycle must be occurring due to limited evaporation from surface reservoirs.

Cassini RADAR measurements indicate the equivalent of roughly a 1–2-m layer globally distributed (*Lorenz et al.,* 2008); however, the amount of subsurface methane and its proximity to the surface is unknown. To explore the interaction of a limited surface supply of methane with the atmospheric methane cycle, climate simulations were performed with a variable but fixed total methane inventory (*Mitchell,* 2008; *Schneider et al.,* 2012). Since Titan's atmospheric methane likely outgasses episodically with extended periods of drawdown of the global reservoir (section 9), the question arises as to how the climate changes with the combined atmosphere and surface methane inventory.

Mitchell (2008) and *Mitchell et al.* (2009) investigated the effects of Titan's global methane abundance on its climate with a series of GCMs that all start with a uniform surface reservoir of pure liquid methane, but of different depths ranging from 7 to 30 m. To simplify the effects investigated, the microphysics of precipitation are parameterized rather than detailed (*Graves et al.,* 2008; *Lavvas et al.,* 2011b); the potential effects of ethane (*Tokano,* 2009) and the radiative coupling to the haze (*Rannou et al.,* 2006) are not included. Each simulation is initialized with a methane-free isothermal atmosphere, zero winds, and a uniform surface reservoir of liquid methane, the depth of which establishes the total methane abundance. The climate is found to depend on the total methane abundance (Fig. 13). All the models produce climatologically dry conditions at the equator, between ~30°N and 30°S latitudes. The seasonal migration of Titan's ITCZ is responsible for the dry equatorial surface. However, although climatologically dry conditions persist at the equator, seasonal precipitation is predicted to occur. Four distinct bands of surface accumulation are present at middle and high latitudes if the reservoir depth is >10 m, but only two, the polar accumulation zones, are present for smaller reservoirs. For the 7-m model, that of Titan's present atmosphere, most of the methane is contained in the atmosphere, and precipitation is sporadic and bursty and largely confined to the summer pole. This model also finds that the magnitude of surface evaporation can locally exceed 1 Wm^{-2}:

Timescale	Time (years)
Seasonal	7
Boundary layer	2.5
Troposphere	100
Overturning	100

The transition between a "moist" climate with four surface accumulation zones and a "dry" climate with only two happens quite abruptly as the total reservoir approaches 7 m, as hinted at by earlier infinite reservoir models [*Mitchell et al.* (2006, 2009); although simulations by *Schneider et al.* (2012) have monotonically increasing surface methane toward the poles]. The response of the circulation reveals the importance of methane in surface-atmosphere coupling and the resulting thermal damping timescale "experienced" by the climate system. In the dry limit, only the dry boundary layer participates in the seasonal cycle, and large seasonal excursions of Titan's ITCZ were produced in-phase with the pattern of insolation. The bottom row in Fig. 13 roughly corresponds to this dry case. The moist case, on the other hand, experiences persistent, deep convection, and the ITCZ migrates only to summer middle latitudes and responds out-of-phase with the insolation. An "intermediate" case bridges the two extremes. The 30–10-m cases (Fig. 13) share common characteristics with both the moist and intermediate cases. As described in the previous section, Titan's climate timescale as measured by atmospheric radiative damping is strongly influenced by the presence or absence of moist convection, and has a substantial impact on the resulting seasonal cycle.

In the context of Titan's evolving climate, we might then expect that all these possible climatic states were sampled during the drawdown of methane by photolysis, perhaps even several times. The supply of methane may be episodic, with extended periods of drawdown by photolysis (section 9).

Despite considerable progress in our understanding of Titan's atmospheric circulation and methane cycle, it is still not clear whether underground reservoirs participate in the methane cycle, or how the high polar cloud opacity (*Rannou et al.,* 2012) affects the resultant circulation. In addition, it is uncertain whether erosional surface features are a result of today's climatic state or the remnant of a "wetter" Titan. The interaction of the methane cycle with Titan's superrotation is also an open question.

8. TITAN'S ORIGIN AND EVOLUTION

Titan's atmosphere is not stable, because it is constantly losing methane. This unique condition raises the broader questions of how common large methane atmospheres may be in the outer planetary systems and how such an atmosphere can persist despite its instability. To address these questions we briefly explore the chemical and dynamical processes that granted Titan an atmosphere and now sustain it.

Titan, like the other icy satellites of Jupiter and Saturn, accreted from a complex assemblage of various ices, hydrates, silicate minerals, and organics. The composition of the present-day atmosphere does not directly reflect the composition of the primordial building blocks. Rather it is the result of complex evolutionary processes involving internal chemistry and outgassing, impact cratering, photochemistry, escape, crustal storage and recycling, and other processes. This history is partially constrained by the isotopic ratios of major atmospheric species, as well as by the noble gas abundances. A comparison of Titan's composition and internal structure with those of the other icy moons of Saturn and Jupiter also permits us to better understand what makes Titan so unique.

8.1. Comparison with Other Moons

Although Titan is similar in terms of mass and size to Jupiter's moons, Ganymede and Callisto, it is the only one harboring a massive atmosphere. Moreover, unlike the jovian system populated with four large moons, Titan is the only large moon around Saturn. The other saturnian moons are much smaller and have an average density at least 25% less that Titan's uncompressed density and much below the density expected for solar composition (*Johnson and Lunine,* 2005), although with a large variation from satellite to satellite. In addition, unlike the jovian moons, Saturn's system does not exhibit a strong gradient in ice-to-rock ratio with distance from the planet. Even though both Jupiter's and Saturn's moon systems are thought to have formed in a disk around the growing giant planet, these differences in architecture probably reflect different disk characteristics and evolution (e.g., *Estrada et al.,* 2009; *Sasaki et al.,* 2010), and in the case of Saturn, possibly the catastrophic loss of one or more Titan-sized moons during the accretion phase (*Canup,* 2010). Moreover, the presence of a massive atmosphere on Titan as well as the emission of gases from Enceladus' active south polar region (*Waite et al.,* 2009) suggest that the primordial building blocks that comprise the saturnian system were more volatile-rich than Jupiter's.

Titan's internal structure also provides pertinent clues on the moon's origin and evolution. The moment of inertia factor estimated from Titan's gravity field, $C/MR^2 \sim$ 033–0.34 (*Iess et al.,* 2010) is intermediate between that of Ganymede, 0.31 (*Anderson et al.,* 1996), and Callisto, 0.355 (*Anderson et al.,* 2001). This indicates that the internal temperature during Titan's accretion and the subsequent evolution was probably elevated enough to separate most of the primordial ice and rock to form a rock-rich core, but it was not warm enough to lead the formation of an iron core similar to Ganymede. Titan's moment of inertia

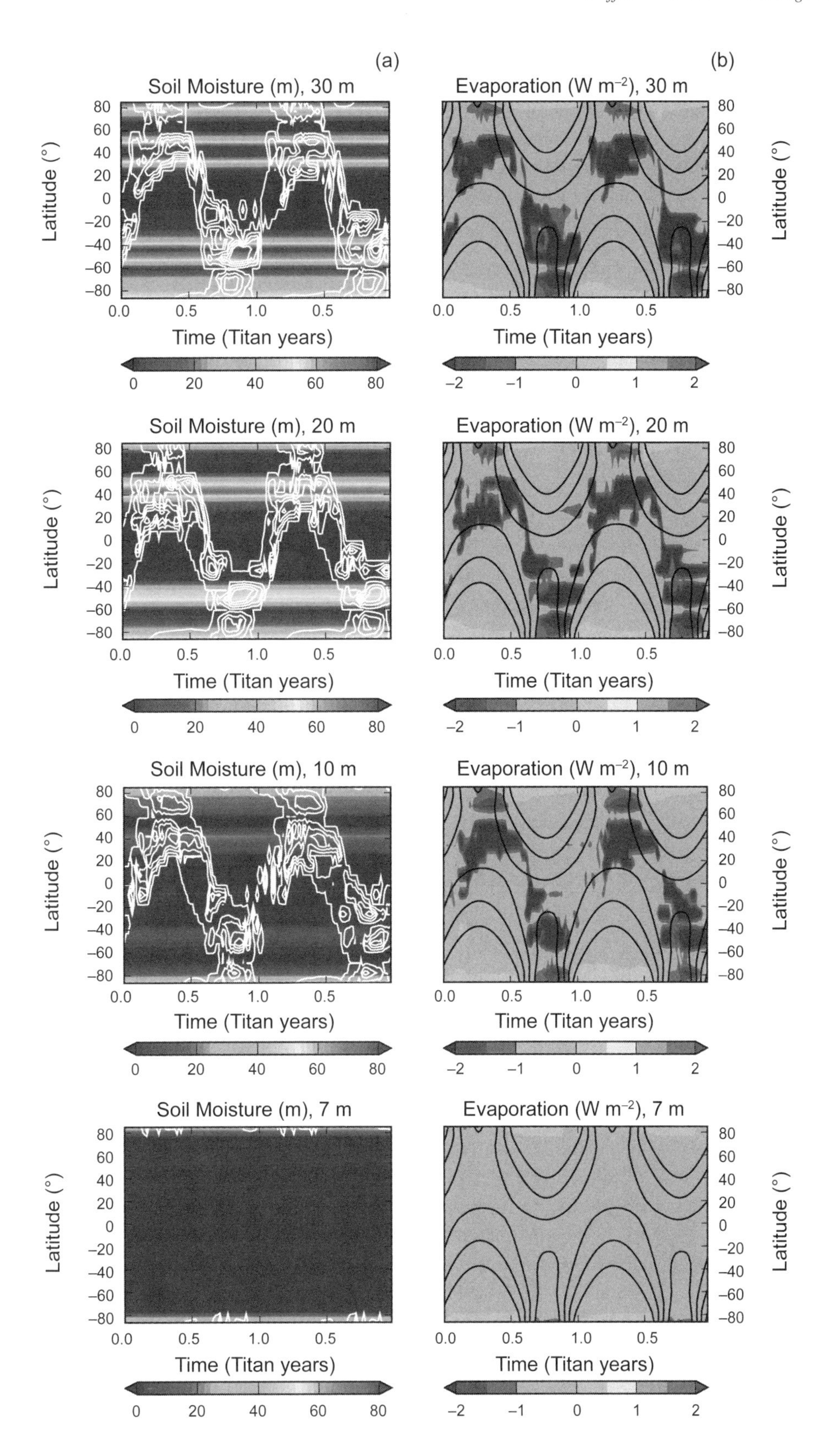

Fig. 13. See Plate 29 for color version. Climate simulations show the influence of a limited methane reservoir, with inventories ranging between 30 and 7 m global liquid equivalent (*Mitchell*, 2008). **(a)** Seasonal pattern of precipitation (white) and surface reservoir depth (colorbar; m) over the final two years of a 45-Titan-year simulation. **(b)** Seasonal patterns of insolation (black) and surface evaporation (colorbar; W m^{-2}) over the same two Titan years.

can be explained if a layer of rock-ice mixture still exists between a rocky core and a high-pressure ice layer, or if the rocky core consists mainly of highly hydrated silicate minerals (*Castillo-Rogez and Lunine,* 2010; *Fortes,* 2012). In both cases, this is indicative of prolonged interactions between silicate minerals and H_2O liquid or ice in Titan's interior (*Tobie et al.,* 2013). Such water-rock interactions probably have affected the thermochemical evolution of the interior and the composition of the atmosphere through outgassing. Differences in primordial composition, accretion processes, and water-ice-rock interactions may explain why Titan's evolutionary path differs significantly from that of Ganymede and Callisto.

8.2. Generation of a Massive Nitrogen Atmosphere

The low $^{36}Ar/N_2$ ratio measured by the GCMS on Huygens (*Niemann et al.,* 2005, 2010) suggests that nitrogen was not brought to Titan in the form of N_2, but rather in the form of easily condensable compounds, e.g., NH_3 (*Owen,* 1982) or N-bearing organic matter (*Tobie et al.,* 2012). The argument goes as follows: Argon and N_2 have similar volatility and affinity with water ice. Thus, if the primary carrier of Titan's nitrogen were N_2, the Ar/N_2 ratio on Titan should be within an order of magnitude of the solar composition ratio of about 0.1 (*Owen,* 1982; *Lunine and Stevenson,* 1985b), which is about 500,000 times larger than the observed ratio (*Niemann et al.,* 2010). This implies that nitrogen was converted *in situ* by either photochemistry (*Atreya et al.,* 1978, 2010), impact-driven chemistry (*Jones and Lewis,* 1987; *McKay et al.,* 1988; *Sekine et al.,* 2011; *Ishimaru et al.,* 2011), or endogenic processes (*Glein et al.,* 2009; *Tobie et al.,* 2012). Each conversion process requires restrictive conditions to occur, and therefore specific evolutionary scenarios for Titan.

Photochemical conversion requires the presence of a significant amount of ammonia in the atmosphere, perhaps possible during the accretion phase. Toward the end of accretion, the surface temperature of the growing Titan likely exceeded the melting point of water ice (*Lunine and Stevenson,* 1987; *Kuramoto and Matsui,* 1994). The presence of ammonia aids melting by depressing the melting point. Under such conditions, a primitive atmosphere in direct contact with a global water ocean doped with ammonia may form (*Lunine and Stevenson,* 1987; *Lunine et al.,* 1989). Photochemistry during this phase potentially produced more than 10 bar of nitrogen (*Atreya et al.,* 2010). However, this depends on how fast the primordial atmosphere-ocean system cooled down and the surface solidified, isolating the ammonia-enriched reservoir from the atmosphere. Preliminary estimates suggest that this warm phase did not last more than 10 to 100 m.y. (*Lunine et al.,* 1989, 2010).

An impact-driven conversion may have operated on longer timescales, from the accretion phase to the late heavy bombardment (LHB), about 600–800 m.y. after accretion (*Jones and Lewis,* 1987; *McKay et al.,* 1988; *Zahnle et al.,* 1992; *Sekine et al.,* 2011). Impact heating of a primitive atmosphere with high ammonia and low water content can rapidly produce several bars of nitrogen. However, such an ammonia-rich atmosphere is difficult to reconcile with recent models of Saturn's system formation, which predict a bulk ammonia concentration of only a few percent relative to water (e.g., *Alibert and Mousis,* 2007; *Hersant et al.,* 2008). Moreover, large high-velocity impacts would also lead to significant atmospheric erosion (*Shuvalov,* 2009; *Sekine et al.,* 2011), so that the survival of a primitive massive atmosphere remains questionable.

Alternatively, if the crust contains several percent of nitrogen-bearing solids (ammonia ice, ammonia hydrate, or ammonium sulfate), the production of N_2 may have been active over a long period of time, with a maximum production rate during the LHB (*Fukuzaki et al.,* 2010; *Sekine et al.,* 2011). An atmosphere of about 1.5 bar of nitrogen may have formed from a crust containing about 2% ammonia, if impactors were large enough ($\geq$20 km) and had a total mass of at least 1.5–2 $\times$ 10^{20} kg (*Sekine et al.,* 2011). In such a scenario, the formation of a massive early atmosphere is not needed; the atmosphere may have been relatively tenuous before the LHB.

Another mechanism is the endogenous production of N_2 from the oxydation of NH_3 during water-rock interactions (*Glein et al.,* 2009) or from the decomposition of N-bearing organic matter brought by the chondritic phase (*Tobie et al.,* 2012). These two endogenic processes are controlled by the thermal evolution of the rocky interior, and may have started early in Titan's evolution in the case of warm accretion or relatively late in the case of relatively cold accretion. N_2 produced in the deep warm interior would have been progressively stored in the H_2O mantle and then transported to the surface, where it could be released through outgassing events. The release of endogenic nitrogen to the atmosphere may still be occurring, and may be associated with methane outgassing (*Tobie et al.,* 2012).

Although several different mechanisms may have converted ammonia to nitrogen on Titan, there is still no clear evidence for ammonia as a main constituent on Titan. A small fraction of ammonia has been detected in Enceladus' plumes (*Waite et al.,* 2009), but the observed fraction barely exceeds 1%. An ammonia-rich primitive atmosphere thus seems quite unlikely. The conversion mechanism must therefore be efficient for low ammonia content. Another challenge is to explain the high $^{15}N/^{14}N$ observed in Titan's nitrogen, which is 1.5 times terrestrial (*Niemann et al.,* 2010), while the other isotopic ratios in H, C, and O are very close to the terrestrial values. This enrichment has been previously attributed to atmospheric escape (e.g., *Lunine et al.,* 1999), implying that the initial nitrogen mass was 2 to 10 times the present-day mass (*Niemann et al.,* 2005). However, more recent estimates indicate that the fractionation by atmospheric escape is probably not efficient enough to have significantly changed the initial $^{15}N/^{14}N$ (*Mandt et al.,* 2009). This suggests that Titan may have accreted ammonia or other N-bearing molecules that were initially enriched in ^{15}N. If confirmed, this would imply that a massive early atmosphere is no longer needed and any process capable of producing the present-day mass of nitrogen would be sufficient. Future measurements of $^{15}N/^{14}N$ in ammonia and other N-bearing molecules in comets, which are possible analogs of Titan's building blocks, may provide useful constraints on the primordial value. In the absence of additional constraints, the origin of this massive atmosphere remains undetermined. Precise measurements of noble gas abundances in the future will be probably the best way to understand why Titan, besides the Earth, is the only planetary object with a massive N_2-rich atmosphere.

8.3. The Long-Term Carbon Cycle

At present, Titan has an active methane cycle. It is characterized by the photochemical production of complex hydrocarbon species, cloud formation, rainfall, and the accumulation of hydrocarbon liquids in near-polar lakes and seas. However, it is still unclear if this cycle has existed during most of Titan's evolution or if it was triggered only relatively recently. Indeed, contrary to $^{15}N/^{14}N$ in nitrogen, the $^{13}C/^{12}C$ ratio measured in methane by the GCMS on Huygens (*Niemann et al.,* 2010), and derived from the INMS and CIRS on Cassini (*Nixon et al.,* 2012; *Mandt et al.* 2012), is close to the ratio measured on Earth and several other solar system objects, indicating that the atmospheric methane inventory has not been affected by atmospheric escape (*Mandt et al.,* 2009, 2012; *Nixon et al.,* 2012). This nearly unmodified ratio suggests that today's methane was injected in the atmosphere not more than 0.5–1 G.y. ago (*Mandt et al.,* 2012; *Nixon et al.,* 2012) and that any methane released during the early stage of Titan's evolution should have been totally lost or rapidly retrapped in the subsurface without being affected by atmospheric fractionation processes.

The building blocks that formed Titan probably contained a significant amount of methane, together with other carbon compounds, such as CO_2, CH_3OH, and more complex organic material (e.g., *Hersant et al.,* 2008; *Mousis et al.,* 2009; *Tobie et al.,* 2012). The low abundance of CO in the present-day atmosphere suggests that the building blocks were depleted in CO (and N_2), which is consistent with the low $^{36}Ar/N_2$ ratio, or that CO was efficiently converted during the accretion phase (*Trigo-Rodriguez and Martin-Torres,* 2012). The lack of extensive liquid reservoirs of ethane or methane at the surface suggests either that relatively little methane has been released from the interior through time, or that the ethane produced by methane-driven photochemistry, a liquid on Titan's surface, has been efficiently retrapped in the crust as liquids and/or clathrates (*Osegovic and Max,* 2005; *Mousis and Schmitt,* 2008; *Choukorun and Sotin,* 2012).

The detection of ^{40}Ar, the decay product of ^{40}K mostly contained in silicate minerals, is the most convincing evidence for internal outgassing (*Niemann et al.,* 2010; *Tobie et al.,* 2012). Its present-day atmospheric abundance suggests that at least 7–9% of ^{40}Ar potentially contained in the rocky core (assuming a chondritic composition) has been outgassed through time (*Tobie et al.,* 2012). The possible detection of ^{22}Ne by GCMS (*Niemann et al.,* 2010) also suggests efficient internal outgassing (*Tobie et al.,* 2012). Assuming that Titan initially contained about 0.1–0.3 wt% of methane, an outgassing efficiency similar to ^{40}Ar would result in the injection of a total mass of CH_4 equivalent to about 50–180 times the present-day atmospheric inventory (*Tobie et al.,* 2012). The crust may potentially accommodate most of this released methane, either in the form of methane-ethane clathrate reservoir (*Osegovic and Max,* 2005; *Tobie et al.,* 2006; *Mousis and Schmitt,* 2008; *Choukroun and Sotin,* 2012) or in the form of liquid hydrocarbons in crustal pores (*Kossacki and Lorenz,* 1996; *Hayes et al.,* 2008; *Griffith et al.,* 2012). The exchange efficiency with the crustal reservoir, which is unknown at present, controls the abundance of atmospheric methane and Titan's climate through time.

Therefore while the nitrogen in Titan's atmosphere was likely present early on, at least after the LHB era, the methane content in the current atmosphere, based on the mostly unmodified $^{13}C/^{12}C$ ratio, is likely recent. Methane may have been injected massively, but not more than 1 G.y. ago, or may be currently renewed by internal outgassing. This suggests that the methane abundance may have significantly changed during Titan's history, alternating between methane-poor (dry) and methane-rich (wet) periods.

9. PAST AND FUTURE CLIMATES

Titan's current state may be just a brief snapshot in a range of past sultry and arid climates, whose existences depend on the supply of methane from the interior. Fluctuations in the methane content of Titan's atmosphere affect its radiative forcing, general circulation, temperature structure, and stability, as well as the mass of the atmosphere. The evolution of the Sun's EUV to optical flux also influenced Titan's climate by affecting the photochemistry and insolation. Titan's atmospheric conditions thus depend on the methane supply and the Sun's radiation, which change on timescales of ~10 m.y. and ~1 b.y. respectively.

9.1. Methane Destruction Rate

Considering only the loss of CH_4 due to the energy deposition (dominated by Ly-α photons), methane is destroyed at a rate of 3.1×10^9 cm^{-2} s^{-1} (Table 3), and the current inventory has a lifetime of 85 m.y. (*Lavvas et al.,* 2008, 2011a; *Yelle et al.,* 2010; *Vuitton et al.,* 2012). The lifetime is reduced to 26 m.y. by the inclusion of the chemical loss of methane, and further reduced by the methane escape, which, at a rate of $2.5–3 \times 10^9$ cm^{-2} s^{-1}, is required to interpret the methane density profiles in the thermosphere (*Yelle et al.,* 2008). Taking all these processes into account, the methane lifetime is currently 21 m.y., short by comparison to Titan's age of 4600 m.y.

On timescales of ~20 m.y., Titan's climate evolves with the destruction of methane. On longer timescales of billions of years, Titan's climate was affected by the changes in the Sun's luminosity and Ly-α flux.

9.2. Evolution with Varying Methane Abundance

While the history of Titan's methane inventory cannot be established (section 8), two end scenarios are possible: an atmosphere devoid of methane, and one nearly saturated but perhaps not entirely, because methane condenses out of the atmosphere at the cooler poles.

The total extinction of methane in Titan's atmosphere may cool the atmosphere sufficiently to condense nitrogen, leading to surface accumulation of liquid nitrogen. If the surface albedo is high enough, nitrogen will freeze, causing Titan's atmosphere to collapse. A moon more similar to Triton would remain *(Lorenz et al.,* 1997). The metamorphosis proceeds first with the depletion of surface methane reservoirs. The tropospheric methane then declines, within the following ~16 m.y., until the entire troposphere attains the constant methane mixing ratio (~0.0148) set by the tropopause temperature, which at this point has not changed significantly. During this period the stratospheric haze and gas abundances would not significantly change. Nor would the H_2 abundance decrease, which regulates (along with CH_4 and N_2) the greenhouse effect (*McKay et al.,* 1989; *Lunine and Rizk,* 1989). The troposphere would become more stable due to the low humidity and less-prevalent rainfall. The tropospheric haze density would likely increase from a lack of clearing by rain. Assuming no change in the surface albedo, the reduction of the methane content to 0.0148 lowers the tropical surface temperature by roughly $\Delta T_s \sim 3$ K (Fig. 14a). To illustrate thermal profiles in mostly past investigations, changes in Titan's thermal profile were calculated with the one-dimensional radiative convective modeled after *McKay et al.* (1989) but updated with the Cassini-derived haze and infrared opacity. The incident insolation is set to the disk-averaged value, which accurately reproduces the temperature profile of the large tropical region on Titan, extending from ~40°S to ~40°N. This model is perturbed to estimate tropical thermal profiles for different atmospheric compositions. Considering the maximum entropy production principal, the tropical to polar temperature contrast would be similar to that measured for the current atmosphere, i.e., 3–4 K (*Lorenz et al.,* 2001), and the polar temperature would be well below the freezing point of pure CH_4 (90.4 K).

Within the next ~4 m.y., stratospheric methane is depleted from the entire atmosphere. Although both the stratospheric composition and temperature profiles change during this period, the rate of photolysis remains constant until the methane abundance is depleted by 4 orders of magnitude. This constant methane photolysis rate arises for two reasons. Methane is optically thick at Ly-α wavelengths, therefore all the photons are absorbed, i.e., methane photolysis is photon-limited. Also, at Ly-α wavelengths, the methane absorption dominates, with a cross-section of 6 orders of magnitude greater than that of Rayleigh scattering. Therefore methane absorption is shielded only when methane is depleted by 4 orders of magnitude, since N_2 is 100 times more abundant.

Since at this point the methane abundance is diminishing at all altitudes, the pressure levels at which UV radiation occurs increases. Our sensitivity studies indicate that if the stratospheric CH_4 is halved (or doubled), the C_2H_2 profile is not strongly affected below 600 km altitude, yet the abundances of C_2H_6 and HCN are at most halved (or doubled). Therefore, Titan's haze composition presumably changes from the decrease in the stratospheric methane in such a way that depends on the as-yet-unknown chemical pathways that lead to its production. As the methane dwindles, the haze and H_2 productions decline. The absence of CH_4, haze, and H_2 gives rise to a tropical surface temperature

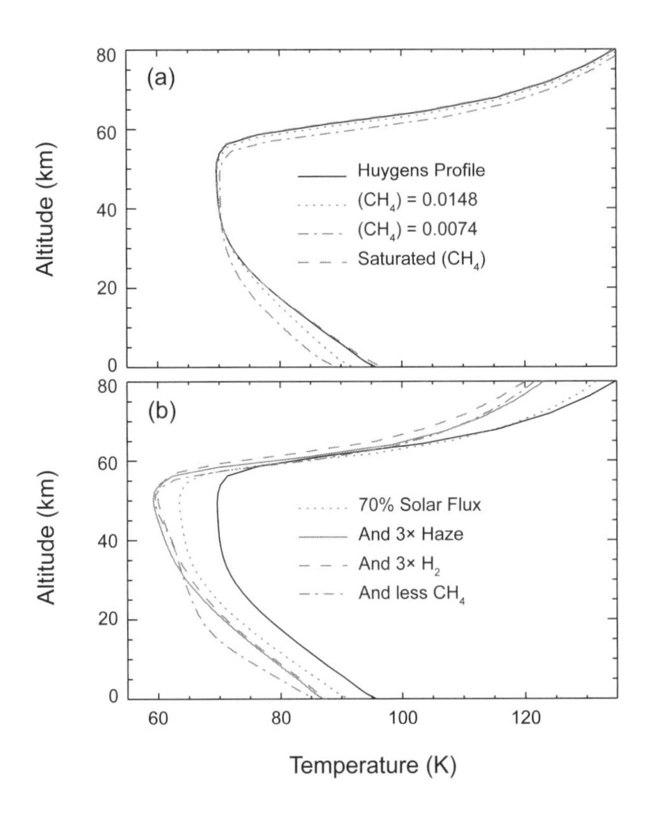

Fig. 14. Titan's tropical thermal profile as measured by Huygens (black solid line) (*Fulchignoni et al.,* 2005) is compared to those for different **(a)** methane abundances and **(b)** insolations and photochemical byproduct compositions. **(a)** A drop in the CH_4 abundance to that set by the tropopause temperature (dotted line) causes the surface temperature to cool by 3 K (*McKay et al.,* 1989). Additional cooling occurs with further depletion of CH_4. **(b)** A lower luminosity of 70% of the present value, as expected for the 1-G.y.-old Sun, causes a cooler troposphere [dotted line (*Lunine et al.,* 1989)]. An increased Lyman-α flux of the young Sun causes a cooler atmosphere through the increased in the haze abundance (solid gray line), which is slightly offset by the warming caused by an increase in H_2 (dashed line). The resultant lower CH_4 vapor pressure (dot-dashed line) also cools the lower troposphere.

of 86 K, slightly warmer than Titan's T_{eff}, because of the remaining small greenhouse effect due to N_2 *(Lorenz et al.,* 1997). The state of the now largely pure N_2 atmosphere depends on the planet's circulation, which might taper with the atmospheric mass sufficiently to create a cold trap in the poles; it also depends on the surface albedo, which is difficult to constrain *(Lorenz et al.,* 1997). Yet it is interesting to note that a scaling of Triton's temperature to that of Titan's insolation indicates a surface temperature of 63 K, at which the N_2 vapor pressure is 0.1 bar *(Lorenz et al.,* 1997). Conceivably a large outgassing, impact, or heating event would be needed to restore such a frozen Titan to the thick N_2- and CH_4-rich atmosphere present today.

Titan potentially experienced an episodic delivery of methane, causing a waxing and waning of its atmosphere (*Lunine et al.,* 1998). The presence of one or more periods of a tenuous atmosphere could in theory be revealed by small craters in Titan's crater record, since the current atmosphere would disrupt small bolides, causing a cutoff in the crater size of ~6 km (*Lunine et al.,* 1998). However, Cassini has identified only ~10 features as certain craters, all with diameters larger than 10 km; since smaller craters may have been eroded away, it is too soon to infer anything about Titan's past atmospheric density (*Neish et al.,* 2012). The implied age of the surface by the crater density is confounded by the lack of constraints on the crater production rate. Current estimates, based on two separate production rates, place the age at 200 Ma and 2 Ga (*Korycansky and Zahnle,* 2005; *Artiemieva and Lunine,* 2005). These ages are consistent with the predicted thickening of Titan's crust (*Tobie et al.,* 2006) and the age (0.5–1.0 Ga) of the current CH_4 inventory (*Mandt et al.,* 2012).

If Titan's atmosphere were fully saturated, rather than 50% humidity as measured at the surface by Huygens, the increased greenhouse effect would increase the surface temperature by only ~1 K (Fig. 14a) from the slightly higher atmospheric opacity at 400–600 cm^{-1}, the region in which Titan's surface radiates most strongly to space (Fig. 6). A higher humidity would render the atmosphere less stable to convection and potentially capable of more vigorous convective cloud systems (*Hueso and Sánchez-Lavega,* 2006; *Barth and Rafkin,* 2007; *Griffith et al.,* 2008). In addition, a larger surface coverage of liquid methane would enable a more vigorous methane cycle (*Mitchell et al.,* 2012). However, Titan is arguably as warm and humid as possible. The measured humidity of 50% at the landing site (*Niemann et al.,* 2010) is an interesting value, because it corresponds to a mixing ratio (0.057) near that of saturation at the polar temperatures. Either this is a coincidence or the poles control Titan's atmospheric methane, which is therefore roughly in vapor pressure equilibrium with the polar lakes (*Griffith et al.,* 2013). If the methane abundance is established, on average, by polar surface temperatures, then as long as the surface liquid is confined mainly to the poles, the atmospheric methane abundance would be expected to remain constant and fairly stable. However, there is a caveat: The production of ethane and the accompanied changes in the lake composition affect the partial pressure of CH_4 and, potentially, the atmospheric mass of N_2. These considerations were studied in detail before Titan's lakes were detected by Cassini.

Well before the arrival of Cassini it was realized that a methane-rich atmosphere like that of Titan's could exist in a relatively stable state if sufficient methane outgassed early on in Titan's history to form a large methane ocean (*Lunine et al.,* 1983; *Lunine and Rizk,* 1989). It was realized that Titan's photochemistry depletes atmospheric CH_4, and thereby depletes any surface liquids of methane that supply the atmosphere. The methane surface liquid is slowly replaced with the main photochemical byproduct, ethane, which is also a liquid on Titan's surface (*Lunine et al.,* 1983). However, as long as methane exists in the atmosphere, so does the haze and H_2, and thus the atmosphere is relatively stable (*Lunine and Rizk,* 1989). The major changes in the atmosphere result from the evolving composition of the large ocean as methane is destroyed and ethane produced. The methane abundance, if in vapor pressure equilibrium, depends on the methane mole fraction in the ocean. The N_2 abundance and thus surface pressure depends on the ocean abundance, because the solubility of N_2 in CH_4 is about 10 times higher than it is in C_2H_6 (*Lunine and Rizk,* 1989). Depending on the temperature, pressure, and ethane fraction, a large fraction of the nitrogen could have been dissolved in the ocean (*Lunine and Rizk,* 1989). However, additional studies point out that its evolution is unstable to small increases in the current insolation, which vaporize CH_4, thereby releasing another greenhouse gas, N_2, with both gases further heating the surface (*Lorenz et al.,* 1999). This feedback eventually leads to a runaway greenhouse effect where the atmosphere's temperature irreversibly increases to a warmer stable state, which depends on the supply of methane (*Lorenz et al.,* 1999).

This coupled atmosphere-ocean system was not revealed to us by Cassini. Moreover, the surface ethane, which Titan most certainly produces, is still unconstrained. Nonetheless, a massive ethane-methane ocean could have existed in the past. If so, the methane must have been largely destroyed, because the $^{13}C/^{12}C$ ratio of the current atmospheric methane suggests a recent source (section 8.3). In addition, the ethane must have been modified chemically or trapped or buried in the crust, because no large ethane oceans exist today (*Osegovic and Max,* 2005; *Mousis and Schmitt,* 2008; *Choukorun and Sotin,* 2012). An inventory of organic material on Titan's surface would test this possibility.

9.3. Evolution in an Astronomical Context

Both Titan's accretion and the Sun's evolution affected the chemistry and structure of Titan's early atmosphere. Preliminary studies of the accretion period indicate that surface heat flows of 0.5 W m^{-2} (500 ergs cm^{-2} s^{-1}) can raise surface temperatures by 20 K and more than double the pressure if there is sufficient methane on the surface (*McKay et al.,* 1993). During the late stages of accretion the energy released is expected to be much higher, potentially enough to form a relatively brief NH_3 and H_2O atmosphere *(Lunine and Stevenson,* 1985b). However, without measurements of that epoch, Titan's early atmosphere is highly unconstrained.

The Sun's evolution is better established. The solar radiation budget changes with time at rates that differ with wavelength, depending on its source. Best understood is the Sun's photosphere. Observations and stellar evolution models indicate that the solar luminosity has been increasing with time over the past 4.5 G.y. such that 4 G.y. ago it was at 70% of its current value (Fig. 15). Such a low value would cool the atmosphere by ~7 K, comparable to the expected drop in the effective temperature of the moon (*Lunine et al.,* 1989; *McKay et al.,* 1993). Assuming the current surface albedo, composition, and aerosol abundance, we find that the tropopause temperature is lowered to 62 K (Fig. 14b), which has two consequences: Nitrogen condenses in the upper troposphere, and the saturation pressure of methane decreases. Both of these effects lower the greenhouse warming of the surface. Under these conditions methane exists as an ice on Titan's surface. Also lowered is the mixing ratio of methane in the stratosphere. While this depletion of methane might affect the chemistry, the overall methane photolysis rate in Titan's atmosphere would not strongly change, because essentially all the photons act toward breaking apart methane (section 9.2).

The solar flux and its evolution at UV wavelengths, although not fully understood because it emanates from the chromospheric and coronal regions, has been constrained through observations of the Sun as well as other G stars that sample different stages of stellar evolution. These measurements indicate that the Sun's Ly-α emission has changed with time according to the power law

$$F_{Ly\alpha} = 19.2\left[\tau\left(G_{yr}\right)\right]^{-0.72} \text{ ergs}^{-1}\text{cm}^{-2}$$

(*Ribas et al.,* 2005). The Sun's Ly-α emission was three times its current value when 1 G.y. old (Fig. 15). An increased Ly-α emission causes two competing effects on Titan's climate: (1) It increases the abundances of the haze, which further shields and cools the surface; and (2) it increases the H_2 gas, which raises the 400–600 cm^{-1} opacity, thereby enhancing the greenhouse effect and the surface temperature. To explore these competing effects, we estimate the temperature profile when Titan was 1 G.y. old by assuming that it had a significant stratospheric mixing ratio of methane and a saturated atmosphere, and the Sun emitted three times its current Ly-α emission and 70% its current luminosity (Fig. 14b). Tests of the effects of three-fold variations in the C_2H_6 and C_2H_2 abundances cause stratospheric temperatures to change by a few Kelvin, while leaving tropospheric temperatures unchanged. The most important effects are the cooling of the lower atmosphere due to the decrease in the solar flux (Fig. 14b, dashed line) and the increase in the haze (Fig. 14b, solid line). The increase in the H_2 (Fig. 14b, dot-dashed line) only slightly warms the lower atmosphere, while the lower CH_4 abundance cools the troposphere considerably (Fig. 14b, dotted line). The temperature at the poles are expected to be 3–4 K, similar to the current temperatures based on the principal of maximum entropy production, and therefore the atmosphere will freeze out at the poles (*Lorenz et al.,* 2001). Outside of periods of intense bombardment and geological resurfacing, Titan's early surface was significantly cooler than the freezing point of pure methane.

In about 7 b.y., the Sun will become a red giant star. Its radius will increase to perhaps 214 times its present value, thereby potentially engulfing Earth. The Sun's luminosity will increase steadily to 2.73 $L_\odot$ and then precipitously to ~17 $L_\odot$ when it is a little over 12 G.y. old. During this time the Sun will redden from an effective temperature of 5780 K to 4900 K. The climatic state of Titan during this warmer period is unclear, although it is speculated that there might be a short period of several hundred million years when liquid water and ammonia can form oceans on the surface (*Lorenz et al.,* 1997).

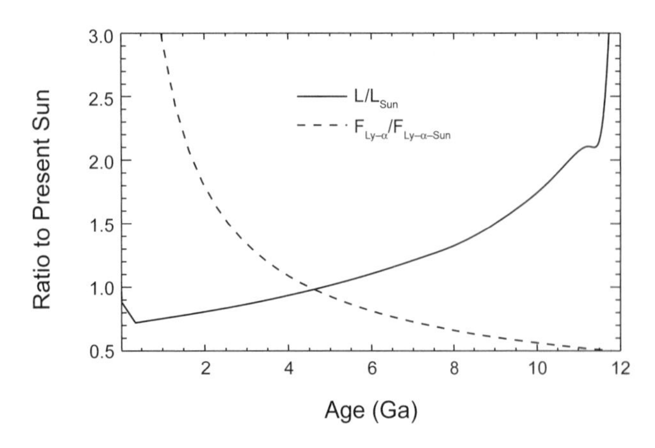

Fig. 15. Evolution of the Sun's luminosity (L) and Lyman-α flux ($F_{Ly\text{-}\alpha}$). From *Ribas et al.* (2010).

10. WHITHER TITAN?

Currently, conditions are warm enough to support liquid methane on the surface and cirrus plus rain drops in the atmosphere. Clouds follow — slightly lagging — the latitude of high seasonal insolation. The occurrence of rainfall, although highly unconstrained, appears to be infrequent. The presence of convective clouds and wave-induced clouds, as well as their seasonal migrations, are well explained to first order by GCMs. However, the details of cloud formation, including the great heights of some convective clouds, the presence of tropical rainstorms, and the nature of the spring polar clouds (now in the north), are not well understood. In addition, it is not clear what role Titan's surface plays in the methane cycle — whether it supplies the atmosphere with methane from subsurface reservoirs. Nor has the transfer of methane surface liquids to and from the atmosphere been quantified. This coupling depends on the unknown ethane abundance in Titan's atmosphere and surface.

It is also not clear where the methane, which controls Titan's climate, derives. Yet it has been established that photochemistry depletes Titan's methane at a rate that will destroy the atmospheric component in 20 m.y. Part of the problem in these investigations is that there is only one measurement of Titan's methane mixing ratio profile. Arguably, it is typical of the tropopause, and potentially indicates that the polar lakes primarily supply the atmospheric methane, because the measured tropical surface mixing ratio is similar to the near-saturation value at the poles. However, we also know that the situation is more complicated than this. Analyses of the 40°–60°S latitude clouds indicate convective systems that rise up to the troposphere (~45 km altitude), which would only be possible if the surface humidity, according to convective models, were 65% or more. Thus potentially the ambient humidity is often higher at 40° latitude than in the tropics, or the atmosphere potentially acquires 0.5–1.0 m more methane locally from an as-yet-unknown source.

In fact, if indeed we are witnessing Titan in its humid state, it is a bit perplexing that there is no obvious detection of a great reservoir of methane. Only about 1–2 m of methane exists on the surface, compared to the 5 m of methane in the atmosphere. The detection of a tropical lake potentially hints at a vast subsurface reservoir of methane. In that case, we have not coincidentally arrived at a wet Titan at just the moment when the surface surplus of methane is almost gone.

Presently, the Cassini mission and groundbased measurements are tracking the seasonal variations in Titan's clouds, occasional rain, temperature profiles, microphysics, chemistry, and — through the hemispherical migration of haze and gases — circulation. Groundbased high-resolution spectroscopy may succeed in determining the broad latitudinal variations in the humidity of the lower atmosphere. The presence of liquids, organic solids, and exposed ice bedrock will be extensively mapped, as well the seasonal variations of the lakes. These and prior studies indicate the broad workings of Titan's climate

Yet significant questions remain. We are still left with no clear path toward the formation of Titan's haze and the expected amino acids within. We cannot detail the processes that establish the local weather, in part because there exist no measurements of the local temperatures, pressures, winds, and surface conditions under which clouds are triggered and evolve. Titan's surface sediments — a fossil layer of past atmospheres, surface geology, and chemistry — have not been probed. Ultimately we cannot explain how Titan acquired and retains a thick atmosphere. While the Cassini mission has revolutionized our understanding of Titan, this moon is still an enigmatic exception in the solar system.

APPENDIX A: CHEMICAL PATHWAYS

This is a brief discussion of the chemical pathways involved in the production and loss of the major gases (C_2H_6, C_2H_2, HCN, H_2, CH_4, N_2) that affect Titan's climate. The summary is cast in terms of the photolysis products of methane and nitrogen. Cosmic rays and energetic electrons also dissociate and ionize nitrogen and methane, as do photons, and thus result in similar chemical pathways. For a more detailed description of the photochemistry in Titan's atmosphere, the reader is referred to the latest review of Titan's chemistry by *Vuitton et al.* (2013).

Neutral photodissociation of CH_4 by solar Ly-α photons produces primarily the radicals

$$\begin{aligned} CH_4 + h\nu &\rightarrow {}^1CH_2 + H_2 && 48\% \\ &\rightarrow {}^3CH_2 + 2\,H && 3\% \\ &\rightarrow CH_3 + H && 42\% \\ &\rightarrow CH + H_2 + H && 7\% \end{aligned}$$

where the numbers represent the yield of each channel (*Gans et al.,* 2011). Reactions among these radicals and the background main species produce more complex hydrocarbons. Ethane, the main byproduct, is produced by the three-body recombination of methyl radicals:

$$2\,CH_3 + M \rightarrow C_2H_6 + M$$

Ethylene is produced from

$$\begin{aligned} CH + CH_4 &\rightarrow C_2H_4 + H \\ {}^3CH_2 + CH_3 &\rightarrow C_2H_4 + H \end{aligned}$$

while acetylene results from

$$2\,{}^{1,3}CH_2 \rightarrow C_2H_2 + 2\,H$$

and the photolysis of ethylene

$$C_2H_4 + h\nu \rightarrow C_2H_2 + 2H/H_2$$

Similarly, N_2 photolysis forms ions and atomic nitrogen in the ground (^4S) or excited state (^2D)

$$\begin{aligned} N_2 + h\nu &\rightarrow N(^2D) + N(^4S) \\ &\rightarrow N_2^+ + e^- \\ &\rightarrow N^+ + N + e^- \\ &\rightarrow N^+ + N(^2D) + e^- \end{aligned}$$

The yield of each channel depends on the photon energy (*Lavvas et al.,* 2011c). The nitrogen radicals generate more complex species. For example, reaction of atomic nitrogen with a methyl radical leads to the production of HCN through the H_2CN intermediate

$$\begin{aligned} N(^4S) + CH_3 &\rightarrow H_2CN + H \\ H_2CN + H &\rightarrow HCN + H_2 \end{aligned}$$

These species are chemically deleted in reactions with radicals and by photolysis, forming even more complex molecules. Photolysis of HCN forms cyanogen radicals that rapidly react with other unsaturated hydrocarbons to form nitrile species

$$HCN + h\nu \rightarrow H + CN$$
$$CN + C_2H_2 \rightarrow HC_3N + H$$
$$CN + C_2H_4 \rightarrow C_2H_3CN + H$$

while in the lower atmosphere, a major loss of HCN comes directly from

$$HCN + C_2H_3 \rightarrow C_2H_3CN + H$$

Photolysis of acetylene produces ethynyl radicals that readily react with acetylene to produce diacetylene

$$C_2H_4 + h\nu \rightarrow C_2H + H$$
$$C_2H + C_2H_2 \rightarrow C_4H_2 + H$$

and the reaction of ethylene with methylidyne forms allene

$$CH + C_2H_4 \rightarrow CH_2CCH_2 + H$$

Ethane, as a saturated hydrocarbon, is not very reactive with other species. In addition, since ethane's photo-absorption cross section is similar to methane's, the photolysis rate of the former is limited through the efficient shielding of the latter. The main chemical loss of C_2H_6 is also through reaction with ethynyl radicals

$$C_2H + C_2H_6 \rightarrow C_2H_2 + C_2H_5$$

followed by

$$H + C_2H_5 \rightarrow 2\ CH_3$$

which recycles ethane, and also by

$$CH_3 + C_2H_5 + M \rightarrow C_3H_8 + M$$

which is the major pathway for the production of propane. Ethane is also lost through reaction with excited nitrogen atoms

$$N(^2D) + C_2H_6 \rightarrow CH_2NH + CH_3$$

The same mechanism proceeds with methane

$$N(^2D) + CH_4 \rightarrow CH_2NH + H$$

These pathways differentiate the amine chemistry, which is generated by the former reactions, from the nitrile pathways introduced above and provide two different roots for the incorporation of nitrogen into the aerosols; the amine compounds are particularly interesting because they may have astrobiological implications.

Note that methane can also be destroyed through reactions with the radicals produced during the photolysis of the new molecules produced. The most characteristic example is the reaction with ethynyl

$$C_2H + CH_4 \rightarrow CH_3 + C_2H_2$$

which is the major mechanism for the catalytic destruction of methane.

These schemes are only the first steps in a far more complex and highly nonlinear system that generates the observed chemical composition of Titan's atmosphere. The chemistry is complicated by the presence of both gas and complex organic solids that affect the chemical composition of the gaseous species by absorbing UV radiation and shadowing the photolysis of other species, and by their reactions with atomic hydrogen, for example (*Sekine et al.,* 2008a,b; *Lavvas et al.,* 2008). A full simulation of Titan's chemistry involving ~100 species currently includes thousands of reactions. Although we still do not have a complete set of parameters for Titan's conditions, the impact of the uncertainty for each parameter in the overall evolution of the chemistry has been quantified with the aid of stochastic models (*Hebrard et al.,* 2009). Still, current models benefit from a far more accurate database compared to older simulations, and consequently represent Cassini observations fairly well (*Vuitton et al.,* 2013).

Acknowledgments. The work by C.A.G. is supported under NASA's Cassini Data Analysis Program. Part of the research leading to these results has received funding from the European Research Council under the European Community's Seventh Framework Programme (FP7/2007-2013 Grant Agreement No. 259285). We thank A. Rodrigues for her invaluable help with the manuscript typesetting.

REFERENCES

Achterberg R. K., Conrath B. J., Gierasch P. J., et al. (2008) Titan's middle-atmospheric temperatures and dynamics observed by the Cassini Composite Infrared Spectrometer. *Icarus, 194,* 263–277.

Ádámkovics M., Barnes J. W., Hartung M., and de Pater I. (2010) Observations of a stationary mid-latitude cloud system on Titan. *Icarus, 208,* 868–877.

Aharonson O., Hayes A. G., Lunine J. I., et al. (2009) An asymmetric distribution of lakes on Titan as a possible consequence of orbital forcing. *Nature Geosci., 2,* 851–854.

Alibert Y. and Mousis O. (2007) Formation of Titan in Saturn's subnebula: Constraints from Huygens probe measurements. *Astron. Astrophys., 465,* 1051–1060.

Anderson C. M. and Samuelson R. E. (2011) Titan's aerosol and stratospheric ice opacities between 18 and 500 μm: Vertical and spectral characteristics from Cassini CIRS. *Icarus, 212,* 762–778.

Anderson J. D., Lau E. L., Sjogren W. L., et al. (1996) Gravitational constraints on the internal structure of Ganymede. *Nature, 384,* 541–543.

Anderson J. D., Jacobson R. A., McElrath T. P., et al. (2001) Shape, mean radius, gravity field, and interior structure of Callisto. *Icarus, 153,* 157–161.

Arakawa A. and Schubert W. H. (1973) Interaction of a cumulus cloud ensemble with the large-scale environment, Part 1. *J. Atmos. Sci., 31,* 674–701.

Artemieva N. and Lunine J. I. (2005) Impact cratering on Titan II. Global melt, escaping ejecta, and aqueous alteration of surface organics. *Icarus 175,* 522–533.

Atreya S. K., Donahue T. M., and Kuhn W. R. (1978) Evolution of a nitrogen atmosphere on Titan. *Science, 201,* 611–613.

Atreya S., Lorenz R., and Waite H. (2010) Volatile origin and cycles: Nitrogen and methane. In *Titan from Cassini-Huygens* (R. H. Brown et al., eds.), pp. 177–199. Springer-Verlag, Berlin.

Baines K. H., Drossart P., Momary T. W., et al. (2005) The atmospheres of Saturn and Titan in the near-infrared first results of Cassini/VIMS. *Earth Moon Planets, 96,* 119–147.

Barnes J. W., Brown R. H., Soderblom L., et al. (2007) Global-scale surface spectral variations on Titan seen from Cassini/VIMS. *Icarus, 186,* 242–258.

Barnes J. W., Bow J., Schwartz J., et al. (2011) Organic sedimentary deposits in Titan's dry lakebeds: Probable evaporite. *Icarus, 216,* 136–140.

Barth E. L. and Rafkin S. C. R. (2007) TRAMS: A new dynamic cloud model for Titan's methane clouds. *Geophys. Res. Lett., 34,* 3203.

Barth E. L. and Rafkin S. C. R. (2010) Convective cloud heights as a diagnostic for methane environment on Titan. *Icarus, 206,* 467–484.

Barth E. L. and Toon O. B. (2004) Properties of methane clouds on Titan: Results from microphysical modeling. *Geophys. Res. Lett., 31,* 17.

Bézard B. (2009) Composition and chemistry of Titan's stratosphere. *Philos. Trans. R. Soc. London Ser. A, 367,* 683–695.

Bird M. K., Allison M., Asmar S. W., et al. (2005) The vertical profile of winds on Titan. *Nature,* 438, 800–802.

Bouchez A. H. and Brown M. E. (2005) Statistics of Titan's south polar tropospheric clouds. *Astrophys. J. Lett., 618,* L53–L56.

Brown M. E., Bouchez A. H., and Griffith C. A. (2002) Direct detection of variable tropospheric clouds near Titan's south pole. *Nature, 420,* 795–797.

Brown M. E., Schaller E. L., Roe H. G., et al. (2009) Discovery of lake-effect clouds on Titan. *Geophys. Res. Lett., 36,* 1103.

Brown M. E., Roberts J. E., and Schaller E. L. (2010) Clouds on Titan during the Cassini prime mission: A complete analysis of the VIMS data. *Icarus, 205,* 571–580.

Brown R. H., Soderblom L. A., Soderblom J. M., et al. (2008) The identification of liquid ethane in Titan's Ontario Lacus. *Nature, 454,* 607–610.

Canup R. M. (2010) Origin of Saturn's rings and inner moons by mass removal from a lost Titan-sized satellite. *Nature, 468,* 943–926.

Cassini RADAR Team, Hayes A. G., Aharonson O., et al. (2011) Transient surface liquid in Titan's polar regions from Cassini. *Icarus, 211,* 655–671.

Castillo-Rogez J. C. and Lunine J. I. (2010) Evolution of Titan's rocky core constrained by Cassini observations. *Geophys. Res. Lett., 37,* L20205.

Choukroun M. and Sotin C. (2012) Is Titan's shape caused by its meteorology and carbon cycle? *Geophys. Res. Lett., 39,* 4201.

Coates A. J., Crary F. J., and Lewis G. R. (2007) Discovery of heavy negative ions in Titan's ionosphere, *Geophys. Res. Lett., 34,* CiteID L 22103.

Cornet T., Bourgeois O., Le Mouélic S., et al. (2012) Edge detection applied to Cassini images reveals no measurable displacement of Ontario Lacus' margin between 2005 and 2010. *J. Geophys. Res., 117(E7),* E07005.

Coustenis A., Salama A., Lellouch E., et al. (1998) Evidence for water vapor in Titan's atmosphere from ISO/SWS data. *Astron. Astrophys., 336,* L85–L89.

deKok R., Irwin P.G., Teanby N. A., et al. (2007) Oxygen compounds in Titan's stratosphere as observed by Cassini CIRS. *Icarus, 186,* 354–363.

de Pater I., et al. (2006) Titan imagery with Keck adaptive optics during and after probe entry. *J. Geophys. Res., 111,* E07S05, DOI: 10.1029/2005JE002620.

Elachi C., et al. (2005) Cassini RADAR views the surface of Titan. *Science, 308,* 970–974.

Emanuel K. A., Neelin D., and Bretherton C. S. (1994) On large scale circulations in convecting atmospheres. *Q. J. R. Meteor. Soc., 120,* 1111–1143.

Estrada P. R., Mosqueira I., Lissauer J. J., et al. (2009) Formation of Jupiter and conditions for accretion of the Galilean satellites. In *Europa* (R. T. Pappalardo et al., eds.), pp. 27–58. Univ. of Arizona, Tucson.

Flasar F. M. (1983) Oceans on Titan? *Science, 221,* 55–57.

Flasar F. M. (1998) The dynamical meteorology of Titan. *Planet. Space Sci., 46,* 1125–1147.

Flasar F. M., Samuelson R. E., and Conrath B. J. (1981) Titan's atmosphere: Temperature and dynamics. *Nature, 290,* 693–698.

Fortes A. D. (2012) Titan's internal structure and the evolutionary consequences. *Planet. Space Sci., 60,* 10–17.

Fukuzaki S., Sekine Y., Genda H., et al. (2010) Impact-induced N_2 production from ammonium sulfate: Implications for the origin and evolution of N_2 in Titan's atmosphere. *Icarus, 209,* 715–722.

Fulchignoni M., et al. (2005) In situ measurements of the physical characteristics of Titan's environment. *Nature, 438,* 785–791.

Galand M., Coates A., Cravens T., et al. (2013) Titan's ionosphere. In *Titan: Interior, Surface, Atmosphere, and Space Environment* (I. Müller-Wodarg et al., eds.), in press. Cambridge Univ., New York.

Gans B., Boye-Peronne S., Broquier M., et al. (2011) Photolysis of methane revisited at 121.6 nm and at 118.2 nm: Quantum yields of the primary products, measured by mass spectrometry. *Phys. Chem. Chem. Phys., 13,* 8140–8152.

Gierasch P. J. (1975) Meridional circulation and the maintenance of the Venus atmospheric rotation. *J. Atmos. Sci., 32,* 1038–1044.

Glein C. R., Desch S. J., and Shock E. L. (2009) The absence of endogenic methane on Titan and its implications for the origin of atmospheric nitrogen. *Icarus, 204,* 637–644.

Graves S. D. B., McKay C. P., Griffith C. A., et al. (2008) Rain and hail can reach the surface of Titan. *Planet. Space Sci,. 56,* 346–357.

Griffith C. A. (2009) Storms, polar deposits and the methane cycle in Titan's atmosphere. *Philos. Trans. R. Soc. London Ser. A, 367,* 713–728.

Griffith C. A., Owen T., Miller G., et al. (1998) Transient clouds in Titan's lower atmosphere. *Nature, 395,* 575–578.

Griffith C., Hall J. L., and Geballe T. R. (2000) Detection of daily clouds on Titan. *Science, 290,* 509–513.

Griffith C. A., Penteado P., Baines K., et al. (2005a) The evolution of Titan's mid-latitude clouds. *Science, 310,* 474–477.

Griffith C. A., Penteado P., Greathouse T., et al. (2005b) Observations of Titan's mesosphere. *Astrophys. J. Lett.*, 629, L57–L60.

Griffith C. A., Penteado P., Rannou P., et al. (2006) Evidence for a polar ethane cloud on Titan. *Science, 313,* 1620–1622.

Griffith C. A., McKay C. P., and Ferri F. (2008) Titan's tropical storms in an evolving atmosphere. *Astrophys. J., 687,* L41–L44.

Griffith C. A., Penteado P., Rodriguez S., et al. (2009) Characterization of clouds in Titan's tropical atmosphere. *Astrophys. J. Lett., 702,* L105–L109.

Griffith C. A., et al. (2012) Possible tropical lakes on Titan from observations of dark terrain. *Nature, 486,* 237–239.

Griffith C. A., Rafkin, S., Rannou, P. et al. (2013) Storms, clouds and weather. In *Titan: Interior, Surface, Atmosphere and Space Environment* (I. Müller-Wodarg et al., eds.), pp. 144–171. Cambridge Univ., New York.

Hayes A., Aharonson O., Callahan P., et al. (2008) Hydrocarbon lakes on Titan: Distribution and interaction with a porous regolith. *Geophys. Res. Lett., 35,* L09204.

Hayes A. G., Wolf A. S., Aharonson O., et al. (2010) Bathymetry and absorptivity of Titan's Ontario Lacus. *J. Geophys. Res.–Planets, 115,* 9009.

Hébrard E., Dobrijevic M., Pernot P., et al. (2009) How measurements of rate coefficients at low temperature increase the predictivity of photochemical models of titan's atmosphere. *J. Phys. Chem. A, 113,* 11227–11237.

Held I. M. (2000) *The General Circulation of the Atmosphere.* Woods Hole Program in Geophysical Fluid Dynamics. Available online at *www.whoi.edu/fileserver.do?id=21464&pt=10&p=17332.*

Held I. M. and Hou A. Y. (1980) Nonlinear axially symmetric circulations in a nearly inviscid atmosphere. *J. Atmos. Sci., 37,* 515–533.

Hersant F., Gautier D., Tobie G., et al. (2008) Interpretation of the carbon abundance in Saturn measured by Cassini. *Planet. Space Sci., 56,* 1103–1111.

Hide R. (1969) Dynamics of the atmospheres of major planets with an appendix on the viscous boundary layer at the rigid boundary surface of an electrically conducting rotating fluid in the presence of a magnetic field. *J. Atmos. Sci, 26,* 841–853.

Hirtzig M., Coustenis A., Gendron E., et al. (2006) Monitoring atmospheric phenomena on Titan. *Astron. Astrophys., 456,* 761–774.

Hörst S. M., Vuitton V., and Yelle R. V. (2008) Origin of oxygen species in Titan's atmosphere. *J. Geophys. Res., 113(E10),* E10006.

Hueso R. and Sánchez-Lavega A. (2006) Methane storms on Saturn's moon Titan. *Nature, 442,* 428–431.

Iess L., Rappaport N. J., Jacobson R. A., et al. (2010) Gravity field, shape, and moment of inertia of titan. *Science, 327,* 1367–1369.

Ishimaru R., Sekine Y., Matsui T., et al. (2011) Oxidizing protoatmosphere on Titan: Constraint from N_2 formation by impact shock. *Astrophys. J. Lett., 741,* L10.

Israël G., Szopa C., Raulin F., et al. (2005) Complex organic matter in Titan's atmospheric aerosols from in situ pyrolysis and analysis. *Nature, 438,* 796–799.

Jennings D. E., Flasar F. M., Kunde V. G., et al. (2009) Titan's surface brightness temperatures. *Astrophys. J. Lett., 691,* L103.

Jennings D. E., Cottini V., Nixon C. A., et al. (2011) Seasonal changes in Titan's surface temperatures. *Astrophys. J. Lett., 737,* L15–L17.

Johnson T. V. and Lunine J. I. (2005) Saturn's moon Phoebe as a captured body from the outer solar system. *Nature, 435,* 69–71.

Jones T. D. and Lewis J. S. (1987) Estimated impact shock production of N_2 and organic compounds on early Titan. *Icarus, 72,* 381–393.

Korycansky D. G. and Zahnle K. J. (2005) Modeling crater populations on Venus and Titan. *Planet. Space Sci., 53,* 695–710.

Koskinen T. T., Yelle R. V., Snowden D. S., et al. (2011) The mesosphere and lower thermosphere of Titan revealed by Cassini/UVIS stellar occultations. *Icarus, 216,* 507–534.

Kossacki K. J. and Lorenz R. D. (1996) Hiding Titan's ocean: Densification and hydrocarbon storage in an icy regolith. *Planet. Space Sci., 44,* 1029–1037.

Kuramoto K. and Matsui T. (1994) Formation of a hot protoatmosphere on the accreting giant icy satellite: Implications for the origin and evolution of Titan, Ganymede, and Callisto. *J. Geophys. Res., 99,* 21183–21200.

Langhans M. H., Jaumann R., Stephan K., et al. (2012) Titan's fluvial valleys: Morphology, distribution, and spectral properties. *Planet. Space Sci., 60,* 34–51.

Lavvas P. P., Coustenis A., and Vardavas I. M. (2008) Coupling photochemistry with haze formation in Titan's atmosphere, Part 2: Results and validation with Cassini/Huygens data. *Planet. Space Sci., 56(1),* 67–99.

Lavvas P., Yelle R. V., and Vuitton V. (2009) The detached haze layer in Titan's mesosphere. *Icarus, 201(2),* 626–633.

Lavvas P., Galand M., Yelle R. V., et al. (2011a) Energy deposition and primary chemical products in Titan's upper atmosphere. *Icarus, 213,* 233–251.

Lavvas P., Griffith C. A., and Yelle R. V. (2011b) Condensation in Titan's atmosphere at the Huygens landing site. *Icarus, 215,* 732–750.

Lavvas P., Sander M., Kraft M., et al. (2011c) Surface chemistry and particle shape: Processes for the evolution of aerosols in Titan's atmosphere. *Astrophys. J., 728,* 80.

Lebonnois S., Bakes E. L. O., and McKay C. P. (2002) Transition from gaseous compounds to aerosols in Titan's atmosphere. *Icarus, 159,* 505–517.

Le Mouélic S., Rannou P., Rodriguez S., et al. (2012) Dissipation of Titan's north polar cloud at northern spring equinox. *Planet. Space Sci. 60,* 86–92.

Li L., Nixon C. A., Achterberg R. K., et al. (2011) The global energy balance of Titan. *Geophys. Res. Lett., 38,* L23201.

Lindal G. F., Wood G. E., Hotz H. B., et al. (1983) The atmosphere of Titan — An analysis of the Voyager 1 radio occultation measurements. *Icarus, 53,* 348–363.

Lopes R. M. C., Stofan E. R., Peckyno R., et al. (2010) Distribution and interplay of geologic processes on Titan from Cassini RADAR data. *Icarus, 205,* 540–558.

Lora J. M., Goodman P. J., Russell J. L., and Lunine J. I. (2011) Insolation in Titan's troposphere. *Icarus, 216,* 116–119.

Lorenz R. D. (1993) The life, death and afterlife of a raindrop on Titan. *Planet. Space Sci., 41,* 647–655.

Lorenz R. D., Lunine J. I., and McKay C. P. (1997) Titan under a red giant Sun: A new kind of 'habitable' moon. *Geophys. Res. Lett., 24,* 2905.

Lorenz R. D., McKay C. P., and Lunine J. I. (1999) Analytic investigation of climate stability on Titan: Sensitivity to volatile inventory. *Planet. Space Sci., 47,* 1503–1515.

Lorenz R. D., Lunine J. I., Withers, P. G., and McKay C. P. (2001) Titan, Mars and Earth: Entropy production by latitudinal heat transport. *Geophys. Res. Lett., 28,* 415–418.

Lorenz R. D., Griffith C. A., Lunine J. I., et al. (2005) Convective plumes and the scarcity of Titan's clouds. *Geophys. Res. Lett., 32,* 1201.

Lorenz R. D., et al. (2006) The sand seas of Titan: Cassini

RADAR observations of longitudinal dunes. *Science, 312,* 724–727.

Lorenz R. D., Mitchell K. L., Kirk R. L., and the Cassini RADAR Team (2008) Titan's inventory of organic surface materials. *Geophys. Res. Lett., 35,* L02206.

Lunine J. I. and Rizk B. (1989) Thermal evolution of Titan's atmosphere. *Icarus, 80,* 370–389.

Lunine J. I. and Stevenson D. J. (1985a) Thermodynamics of clathrate hydrate at low and high pressures with application to the outer solar system. *Astrophys. J. Suppl. Ser., 58,* 493–531.

Lunine J. I. and Stevenson D. J. (1985b) Evolution of Titan's coupled ocean-atmosphere system and interaction of ocean with bedrock. In *Ices in the Solar System* (J. Klinger et al., eds.), pp. 741–757. Reidel, Dordrecht.

Lunine J. I. and Stevenson D. J. (1987) Clathrate and ammonia hydrates at high pressure — Application to the origin of methane on Titan. *Icarus, 70,* 61–77.

Lunine J. I., Stevenson D. J., and Yung Y. L. (1983) Ethane ocean on Titan. *Science, 222,* 1229.

Lunine J. I., Atreya S. K., and Pollack J. B. (1989) Present state and chemical evolution of the atmospheres of Titan, Triton, and Pluto. In *Origin and Evolution of Planetary and Satellite Atmospheres* (S. K. Atreya et al., eds.), pp. 605–665. Univ. of Arizona, Tucson.

Lunine J. I., Lorenz R. D., and Hartmann W. K. (1998) Some speculations on Titan's past, present, and future. *Planet. Space Sci., 46,* 1099–1107.

Lunine J. I., Yung Y. L., and Lorenz R. D. (1999) On the volatile inventory of Titan from isotopic abundances in nitrogen and methane. *Planet. Space Sci., 47,* 1291–1303.

Lunine J. I., Choukroun M., Stevenson D. J., et al. (2010) The origin and evolution of Titan. In *Titan from Cassini-Huygens* (R. H. Brown et al., eds.), pp. 35–59. Springer-Verlag, Berlin.

Mandt K. E., Waite J. H., Lewis W., et al. (2009) Isotopic evolution of the major constituents of Titan's atmosphere based on Cassini data. *Planet. Space Sci., 57,* 1917–1930.

Mandt K. E., Waite J. H., Teolis B., et al. (2012) The $^{12}C/^{13}C$ ratio on Titan from Cassini INMS measurements and implications for the evolution of methane. *Astrophys. J., 749,* 160.

Matsuno, T. (1966) Quasi-geostrophic motions in the equatorial area. *J. Meteor. Soc. Japan, 44,* 25-43.

McKay C. P., Scattergood T. W., Pollack J. B., et al. (1988) High-temperature shock formation of N_2 and organics on primordial Titan. *Nature, 332,* 520–522.

McKay C. P., Pollack J. B., and Courtin R. (1989) The thermal structure of Titan's atmosphere. *Icarus, 80,* 23–53.

McKay C. P., Pollack J. B., and Courtin R. (1991) The greenhouse and antigreenhouse effects on Titan. *Science, 253,* 1118–1121.

McKay C. P., Pollack J. B., Lunine J. I., et al. (1993) Coupled atmosphere-ocean models of Titan's past. *Icarus, 102,* 88–98.

Mitchell J. L. (2008) The drying of Titan's dunes: Titan's methane hydrology and its impact on atmospheric circulation. *J. Geophys. Res.–Planets, 113,* 8015.

Mitchell J. L. (2012) Titan's transport-driven methane cycle. *Astrophys. J. Lett., 756(2),* L26.

Mitchell J. L., Pierrehumbert R. T., Frierson D. M. W., and Caballero R. (2006) The dynamics behind Titan's methane clouds. *Proc. Natl. Acad. Sci., 103,* 18421–18426.

Mitchell J. L., Pierrehumbert R. T., Frierson D. M. W., and Caballero R. (2009) The impact of methane thermodynamics on seasonal convection and circulation in a model Titan atmosphere. *Icarus. 203,* 250–264.

Mitchell J. L., Ádámkovics M., Caballero R., et al. (2011) Locally enhanced precipitation organized by planetary-scale waves on Titan. *Nature Geosci., 4,* 589–592.

Molina-Cuberos G. J., López-Moreno J. J., Rodrigo R., et al. (1999) Ionization by cosmic rays of the atmosphere of Titan. *Planet. Space Sci., 47,* 1347–1354.

Molina-Cuberos G. J., Lammer H., Stumptner W., et al. (2001) Ionospheric layer induced by meteoric ionization in Titan's atmosphere. *Planet. Space Sci., 49,* 143–153.

Mousis O. and Schmitt B. (2008) Sequestration of ethane in the cryovolcanic subsurface of Titan. *Astrophys. J. Lett., 677,* L67–L70.

Mousis O., Lunine J. I., Thomas C., et al. (2009) Clathration of volatiles in the solar nebula and implications for the origin of Titan's atmosphere. *Astrophys. J., 691,* 1780–1786.

Neish C. D. and Lorenz R. D. (2012) Titan's global crater population: A new assessment. *Planet. Space Sci., 60,* 26–33.

Niemann H. B., Atreya S. K., Bauer S. J., et al. (2005) The abundances of constituents of Titan's atmosphere from the GCMS instrument on the Huygens probe. *Nature, 438,* 779–784.

Niemann H. B., Atreya S. K., Demick J. E., et al. (2010) Composition of Titan's lower atmosphere and simple surface volatiles as measured by the Cassini-Huygens probe gas chromatograph mass spectrometer experiment. *J. Geophys. Res.–Planets, 115(E14),* E12006.

Nixon C. A., Temelso B., Vinatier S., et al. (2012) Isotopic ratios in Titan's methane: Measurements and modeling. *Astrophys. J., 749,* 159–174.

Osegovic J. P. and Max M. D. (2005) Compound clathrate hydrate on Titan's surface. *J. Geophys. Res., 110(E9),* 8004.

Owen T. (1982) The composition and origin of Titan's atmosphere. *Planet. Space Sci., 30,* 833–838.

Penteado P. F. and Griffith C. A. (2010) Groundbased measurements of the methane distribution on Titan. *Icarus, 206,* 345–351.

Porco C. C., et al. (2005) Imaging of Titan from the Cassini spacecraft. *Nature, 434,* 159–168.

Radebaugh J., Lorenz R. D., Lunine J. I., et al. (2008) Dunes on Titan observed by Cassini RADAR. *Icarus, 194,* 690–703.

Rannou P., Montmessin F., Hourdin F., et al. (2006) The latitudinal distribution of clouds on Titan. *Science, 311,* 201–205.

Rannou P., Le Mouélic S., Sotin C., et al. (2012) Cloud and haze in the winter polar region of Titan observed with visual and infrared mapping spectrometer on board Cassini. *Astrophys. J., 748,* 4.

Rennò N. O. and Ingersoll A. P. (1996) Natural convection as a heat engine: A theory for CAPE. *J. Atmos. Sci.,* 53, 572–585.

Ribas I., Guinan E. F., Güdel M., and Audard M. (2005) Evolution of the solar activity over time and effects on planetary atmospheres: I. High-energy irradiances (1–1700 Å). *Astrophys. J., 622,* 680–694.

Ribas I., Porto de Mello G. F., Ferreira L. D., et al. (2010) Evolution of the solar activity over time and effects on planetary atmospheres. II. κ^{-1} Ceti, an analog of the Sun when life arose on Earth. *Astrophys. J., 714,* 384–395.

Rodriguez S., Le Mouélic S., Rannou P., et al. (2009) Global circulation as the main source of cloud activity on Titan. *Nature, 459,* 678–682.

Rodriguez S., Le Mouélic S., Rannou P., et al. (2011) Titan's cloud seasonal activity from winter to spring with Cassini/VIMS. *Icarus, 216,* 89–110.

Roe H. G., de Pater I., Macintosh B. A., et al. (2002) Titan's clouds from Gemini and Keck adaptive optics imaging. *Astrophys. J., 581,* 1399–1406.

Roe H. G., Bouchez A. H., Trujillo C. A., et al. (2005) Discovery of temperate latitude clouds on Titan. *Astrophys. J. Lett., 618,* L49–L52.

Samuelson R. E. (1983) Radiative equilibrium model of Titan's atmosphere. *Icarus, 53,* 364–387.

Sasaki T., Stewart G. R., and Ida S. (2010) Origin of the different architectures of the jovian and saturnian satellite systems. *Astrophys. J., 714,* 1052–1064.

Shuvalov V. (2009) Atmospheric erosion induced by oblique impacts. *Meteoritics & Planet. Sci., 44,* 1095–1105.

Schaller E. L., Brown M. E., Roe H. G., et al. (2006a) A large cloud outburst at Titan's south pole. *Icarus, 182,* 224–229.

Schaller E. L., Brown M. E., Roe H. G., et al. (2006b) Dissipation of Titan's south polar clouds. *Icarus, 184,* 517–523.

Schaller E. L., Roe H. G., Schneider T., et al. (2009) Storms in the tropics of Titan. *Nature, 460,* 873–875.

Schinder P. J., Flasar F. M., Marouf E. A., et al. (2011) The structure of Titan's atmosphere from Cassini radio occultations. *Icarus, 215,* 460–474.

Schinder P. J., Flasar F. M., Marouf E. A., et al. (2012) The structure of Titan's atmosphere from Cassini radio occultations: Occultations from the prime and equinox missions. *Icarus, 221,* 1020–1031.

Schneider T., Graves S. D. B., Schaller E. L., et al. (2012) Polar methane accumulation and rainstorms on Titan from simulations of the methane cycle. *Nature, 481,* 58–61.

Sekine Y., Imanaka H., Matsui T., et al. (2008a) The role of organic haze in Titan's atmospheric chemistry. I. Laboratory investigation on heterogeneous reaction of atomic hydrogen with Titan tholin. *Icarus, 194,* 186–200.

Sekine Y., Lebonnois S., Imanaka H., et al. (2008b) The role of organic haze in Titan's atmospheric chemistry. II. Effect of heterogeneous reaction to the hydrogen budget and chemical composition of the atmosphere. *Icarus, 194,* 201–211.

Sekine Y., Genda H., Sugita S., et al. (2011) Replacement and late formation of atmospheric N_2 on undifferentiated Titan by impacts. *Nature Geosci., 4,* 359–362.

Shuvalov V. (2009) Atmospheric erosion induced by oblique impacts. *Meteoritics & Planet. Sci., 44,* 1095–1105.

Sobel A. H. and Bretherton C. S. (2000) Modeling tropical precipitation in a single column. *J. Climate, 13,* 4378-4392.

Soderblom L. A., Tomasko M. G., Archinal B. A., et al. (2007) Topography and geomorphology of the Huygens landing site on Titan. *Planet. Space Sci., 55,* 2015–2024.

Stofan E. R., et al. (2007) The lakes of Titan. *Nature, 445,* 61–64.

Strobel D. F. (1982) Chemistry and evolution of Titan's atmosphere. *Planet. Space Sci., 30,* 839–848.

Strobel D. F. (2009) Titan's hydrodynamically escaping atmosphere: Escape rates and the structure of the exobase region. *Icarus, 202,* 632–641.

Thompson W. R., Zollweg J. A., and Gabis D. H. (1992) Vapor-liquid equilibrium thermodynamics of $N_2 + CH_4$ — Model and Titan applications. *Icarus, 97,* 187–199.

Tobie G., Lunine J. I., and Sotin C. (2006) Episodic outgassing as the origin of atmospheric methane on Titan. *Nature, 440,* 61–64.

Tobie G., Gautier D., and Hersant F. (2012) Titan's bulk composition constrained by Cassini- Huygens: Implications for internal outgassing. *Astrophys. J., 752,* 125.

Tobie G., Lunine J. I., Monteux J., et al. (2013) The origin and evolution of Titan. In *Titan: Surface, Atmosphere and Magnetosphere* (I. Müller-Wodarg et al., eds.), in press. Cambridge Univ., New York.

Tokano T. (2009) Impact of seas/lakes on polar meteorology of Titan: Simulation by a coupled GCM-Sea model. *Icarus, 204,* 619–636.

Tokano T., McKay C. P., Neubauer F. M., et al. (2006) Methane drizzle on Titan. *Nature, 442,* 432–435.

Toon O. B., McKay C. P., Courtin R., and Ackerman T. P. (1988) Methane rain on Titan. *Icarus, 75,* 255–284.

Toon O. B., McKay C. P., Griffith C. A., and Turco R. P. (1992) A physical model of Titan's aerosols. *Icarus, 95,* 24–53.

Tomasko M. G., et al. (2005) Rain, winds and haze during the Huygens probe's descent to Titan's surface. *Nature, 438,* 765–778.

Tomasko M. G., Doose L., Engel S., et al. (2008) A model of Titan's aerosols based on measurements made inside the atmosphere. *Planet. Space Sci., 56,* 669–707.

Trenberth K. E. and Stepaniak D. P. (2003) Covariability of components of poleward atmospheric energy transports on seasonal and interannual timescales. *J. Climate, 16,* 3691–3705.

Trigo-Rodriguez J. M. and Martín-Torres J. (2012) Clues on the importance of comets in the origin and evolution of the atmospheres of Titan and Earth. *Planet. Space Sci., 60,* 3–9.

Turtle E. P., Perry J. E., McEwen A. S., et al. (2009) Cassini imaging of Titan's high-latitude lakes, clouds, and south-polar surface changes. *Geophys. Res. Lett., 36,* 2204.

Turtle E. P., Perry J. E., Hayes A. G., et al. (2011a) Rapid and extensive surface changes near Titan's equator: Evidence of April showers. *Science, 331,* 1414–1417.

Turtle E. P., Del Genio A. D., Barbara J. M., et al. (2011b) Seasonal changes in Titan's meteorology. *Geophys. Res. Lett., 38,* 3203.

Turtle E. P., Perry J. E., Hayes A. G., et al. (2011c) Shoreline retreat at Titan's Ontario Lacus and Arrakis Planitia from Cassini Imaging Science Subsystem observations. *Icarus, 212,* 957–959.

Ventura B., Notarnicola C., Casarano D., et al. (2012) Electromagnetic models and inversion techniques for Titan's Ontario Lacus depth estimation from Cassini. *Icarus, 221,* 960–969.

Vinatier S., Bézard B., deKok R., et al. (2010a) Analysis of Cassini/CIRS limb spectra of Titan acquired during the nominal mission II: Aerosol extinction profiles in the 600–1420 cm^{-1} spectral range. *Icarus, 210,* 852–866.

Vinatier S., Bézard B., Nixon, C. N., et al. (2010b) Analysis of Cassini/CIRS limb spectra of Titan acquired during the nominal mission I: Hydrocarbons, nitriles, CO_2 vertical mixing ratio profiles. *Icarus, 205,* 559–570.

Vuitton V., Yelle R. V., and McEwan M. J. (2007) Ion chemistry and N-containing molecules in Titan's upper atmosphere. *Icarus, 191,* 722–742.

Vuitton V., Yelle R. V., and Cui J. (2008) Formation and distribution of benzene on Titan. *J. Geophys. Res., 113,* E05007, 1–18.

Vuitton V., Yelle R. V., Lavvas P., et al. (2012) Rapid association reactions at low pressure: Impact on the formation of hydrocarbons on Titan. *Astrophys. J., 744,* 11.

Vuitton V., Dutuit O., Smith M. A., and Balucani N. (2013) Chemistry of Titan's atmosphere. In *Titan: Interior, Surface, Atmosphere and Space Environment,* (I. Müller-Wodarg et al., eds.), pp. 144–171. Cambridge Univ., New York.

Wahlund J. E., Galand M., Müller-Wodarg I., et al. (2009) On the amount of heavy molecular ions in Titan's ionosphere. *Planet.*

Space Sci., 57(14–15), 1857–1865.

Waite J. H., Young D. T., Cravens T. E., et al. (2007) The process of tholin formation in Titan's upper atmosphere. *Science, 316(5826),* 870.

Waite J. H. Jr., Lewis W. S., Magee B. A., et al. (2009) Liquid water on Enceladus from observations of ammonia and ^{40}Ar in the plume. *Nature, 460,* 487–490.

Wheeler M. and Kiladis G. N. (1999) Convectively coupled equatorial waves: Analysis of clouds and temperature in the wavenumber-frequency domain. *J. Atmos. Sci., 56,* 374–399.

Williams K. E., McKay C. P., and Persson F. (2012) The surface energy balance at the Huygens landing site and the moist surface conditions on Titan. *Planet. Space Sci., 60,* 376–385.

Wilson E. H. and Atreya S. K. (2003) Chemical sources of haze formation in Titan's atmosphere. *Planet. Space Sci., 51,* 1017–1033.

Yelle R. V., Cui J., and Müller-Wodarg I. C. F. (2008) Methane escape from Titan's atmosphere. *J. Geophys. Res.–Planets, 113,* 10003.

Yelle R. V., Vuitton V., Lavvas P., et al. (2010) Formation of NH_3 and CH_2NH in Titan's upper atmosphere. *Faraday Discussions, 147,* 31.

Yung Y. L., Allen M., and Pinto J. P. (1984) Photochemistry of the atmosphere of Titan: Comparison between model and observations. *Astrophys. J. Suppl. Ser., 55,* 465–506.

Zahnle K., Pollack J. B., Grinspoon D., et al. (1992) Impact-generated atmospheres over Titan, Ganymede, and Callisto. *Icarus, 95,* 1–23.

Domagal-Goldman S. D. and Segura A. (2013) Exoplanet climates. In *Comparative Climatology of Terrestrial Planets* (S. J. Mackwell et al., eds.), pp. 121–135. Univ. of Arizona, Tucson, DOI: 10.2458/azu_uapress_9780816530595-ch005.

Exoplanet Climates

Shawn D. Domagal-Goldman
NASA Goddard Space Flight Center

Antígona Segura
Universidad Nacional Autónoma de México

This chapter discusses how potential habitability can be assessed for approximately terrestrial-sized exoplanets (extrasolar planets). We review the concept of the habitable zone and various definitions of this term. These discussions include limits for "water-worlds," planets with haze- or cloud-rich atmospheres, "dune-like" planets with limited water reservoirs, and planets with "exotic" greenhouse gases composed of H_2 and He. Limits on the mass and radius of the planet are also discussed as they relate to the ability of a planetary atmosphere to retain liquid surface water. Finally, this chapter reviews relatively recent research on specific exoplanets and the implications known exoplanets have for planetary climate.

1. INTRODUCTION

The great scientific and public interest in the possibilities for life beyond Earth has caused research on the climates of terrestrial-sized exoplanets to focus on the determination of surface habitability. More specifically, the majority of research in this area has attempted to answer a specific question: Could liquid water be stable at the surface of a given exoplanet? The pursuit of answers to this question began long before potentially habitable exoplanets had been observed. This early work (e.g., *Kasting et al.,* 1993) defined "the habitable zone," the region around a star in which a planet could maintain surface oceans of water. It continues today in the examinations of individual planets (e.g., *Selsis et al.,* 2007), expansions of the habitable zone for certain cases or exotic atmospheres dissimilar from that currently found on Earth (e.g., *Abe et al.,* 2011), or even from any observed in our solar system. But nearly all the work on this subject has historically had the same focus: water stability at the surface of a planet.

The only example of life and a habitable planet we have is Earth. Independently of our definition of life, we found two properties that are the case for all the organisms that live on this planet: They are made of carbon and require liquid water. Water and carbon compounds are present in the molecular clouds where planetary systems form (e.g., *Ehrenfreund and Charnley,* 2000; *Herbst and van Dishoeck,* 2009). Simulations of planet-system formation indicate that water-rich planets may be common (*Raymond et al.,* 2007), and therefore our generalization about life on Earth may be applied to exoplanets [for more detailed discussion about water properties and planetary habitability, see, e.g., *Segura and Kaltenegger* (2010) and *Lineweaver and Chopra* (2012)]. The reason for the focus on surface water is because the measurement techniques are only suitable to detect planets with high rates of biological activity, enough to change the expected composition of the planetary atmosphere. This focus is often criticized as being overly "Earth-centric." However, this strategy is not driven by an inherent assumption that all extraterrestrial life will be "Earth-like," but instead by practical concerns. The search for life on exoplanets is being planned around spectroscopy, using telescopes that could detect absorption features from biogenic gases such as oxygen (O_2) and methane (CH_4). This will be an extremely difficult measurement to make, as the parent star has a brightness that is orders of magnitude greater than the planet [see *Levine et al.* (2006) for an overview of many of the technical challenges faced by such a mission]. Biospheres based on subsurface water, tiny surface water reservoirs, or alternative solvents are not thought to be robust enough to produce a signature that is observable across interstellar distances against the background of the star. In other words, the focus of exoplanet habitability research on stable surface water is driven by biosignature detection technologies. This also explains the misleading nature of the term "habitable zone." Moons well outside the habitable zone in our solar system — including Europa, Titan, and Enceladus — may harbor subsurface life in organic solvents. However, if a biosphere exists on any of these objects, it is not robust enough to have been detected remotely from across interplanetary distances, much less interstellar distances. Thus, the way the term "habitable zone" has been historically applied is really the "remotely detectable biosphere zone," or to be explicit about all the assumptions involved (but less poetically), the "potential for surface water ocean zone."

2. EXOPLANET OBSERVATION TECHNIQUES

Our knowledge of individual exoplanets is much more limited than it is for any of the planets in our solar system. Prior to NASA's Kepler mission, most planets had been detected via the radial velocity technique, wherein the planet's gravitational pull on the star is detected via the Doppler effect as the star moves toward us and away from us (see *exoplanets.org* to view information about all the known exoplanets). With this technique, we can determine the size of the planet's orbit, as well as a minimum mass for the planet. (The minimum mass is the mass of the planet if the plane of its orbit is in our line of sight; if the orbit is inclined, the planet must have a greater mass to cause the same back-and-forth motion in the star.) The Kepler mission has used a transit search to detect thousands of planet candidates (*Batalha et al.,* 2013), dozens of which have been confirmed with independent observations. This technique detects planets by measuring the total light coming from a system, and observing a dip when the planet passes in front of the star and blocks some of the star's light. This technique delivers orbital information, as well as the radius of the planet. If multiple planets exist in the same system, their masses can be measured by transit timing variations (TTVs): observing the effects their mutual gravitational pull had on the timing of their transits. A handful of planets have been detected with other methods such as direct imaging and microlensing. Direct imaging is of particular interest because of its power to learn significantly more information about planets. To date, this method has only been applied to giant planets far from their host stars, with temperatures that are hot enough for the planet to radiate significant amounts of infrared energy (e.g., *Kalas et al.,* 2008; *Currie et al.,* 2011). Future flagship-scale missions to characterize exoplanets are designed around this principle (*Levine et al.,* 2006). In the meantime, information on a rocky planet's atmosphere will be limited to transit spectroscopy, which astronomers can use to observe the light passing through a planet's atmosphere. This has been done for a handful of super-Earth objects, with more expected from the James Webb Space Telescope (*Deming et al.,* 2009) and future groundbased observatories. Although these observations are mostly limited to delivering information on the uppermost parts of an atmosphere, that is still enough for broad characterization of a planet's bulk properties, the scale height of its atmosphere, and to detect the presence of high-altitude clouds and particles (*Bean et al.,* 2011).

3. THE HABITABLE ZONE

Definitions and applications of the habitable zone (*Kasting et al.,* 1993; *Selsis et al.,* 2007; *Kaltenegger and Sasselov,* 2011; *Kopparapu et al.,* 2013) have leveraged our knowledge of the climate history of planets in our solar system. The resulting definitions are usually given in terms of planet's orbiting distance (semimajor axis) and the parent star's luminosity (or the star's mass or size). The habitable zone is bound on the inner edge by the orbital distances for which most of the water (H_2O) goes into the stratosphere due to increased stellar flux, and is lost to space rapidly via photodissociation and hydrogen (H) escape. On the outer edge, the habitable zone is bound by the orbital distance for which "snowball Earth" conditions are unavoidable. Both of these limits are inherently tied to surface temperature of the planet.

Exoplanet observers have often defined habitable zones based on estimates of the equilibrium temperature for a planet (*Batalha et al.,* 2013), using that as a proxy for surface water stability. However, the calculation of equilibrium temperature requires knowledge of the planet's albedo, which is rarely known. Furthermore, translating a planet's equilibrium temperature into a surface temperature requires estimations of greenhouse and antigreenhouse effects that are also unknown. Based on our knowledge of planets in our own solar system, we know that albedo effects and greenhouse/antigreenhouse effects can vary greatly (Fig. 1). Venus presents a particularly stunning example of both of these problems: It has a much higher Bond albedo (0.75) than the value commonly assumed for Earth-like exoplanets (0.3), and its surface temperature (735 K) is hundreds of degrees higher than its equilibrium temperature (230 K) due to a significant greenhouse effect. The uncertainty in equilibrium temperature is so large that error bars representing the uncertainty imparted by the unknown albedo can run the entire width of the habitable zone (see Fig. 2 in *Kaltenegger and Sasselov,* 2011). This demonstrates that without knowledge of the albedo of a planet or the magnitude of the greenhouse effect, equilibrium temperature is

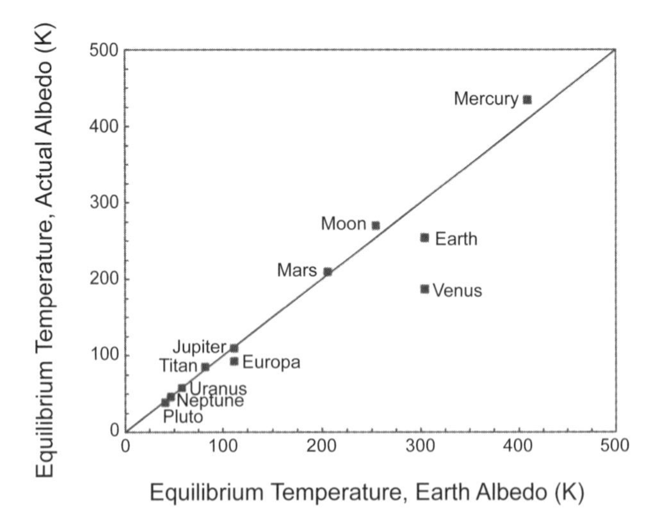

Fig. 1. Solar system planets plotted as a function of equilibrium temperature based on an assumed Earth-like albedo of 0.3, and based on the actual albedo of the planet. The diagonal line represents a one-to-one mapping from one value to the other. Deviations from this line represent errors in calculating a planet's equilibrium temperature resulting from the assumption of an Earth-like albedo.

an unknown quantity that provides unreliable estimates of surface temperature.

For these reasons, atmospheric modelers have incorporated the effects of albedo and of greenhouse effects into past definitions of the habitable zone (*Kasting et al.,* 1993; *Selsis et al.,* 2007; *Kaltenegger and Sasselov,* 2011; *Kopparapu et al.,* 2013). Climate simulations and detailed knowledge of objects in our solar system have helped determine the placement of these boundaries, and have included planets with a variety of atmospheric compositions, greenhouse effects, and albedos. Similar to *Kaltenegger and Sasselov* (2011) and *Kopparapu et al.* (2013), we apply this "classical" definition of the habitable zone in Fig. 2, which plots the known exoplanets [from *exoplanets.org* and *Wright et al.* (2011)] and exoplanet candidates (*Batalha et al.,* 2013) in terms of their orbiting distance and their host star's luminosity. Planets that fall within the dark gray regions of Fig. 2 could maintain liquid water on their surfaces if they have cloud coverage roughly similar to that found on modern-day Earth. Planets that fall into the lighter shade of gray are not habitable in these model results, but could be habitable if Venus was cooler and habitable in its recent past (inner edge) or if early Mars was warm and wet (outer edge). This lighter gray area of space around a star is called the "optimistic" habitable zone in the most recent habitable zone calculations (*Kopparapu et al.,* 2013), but we refer to it here as the "empirical habitable zone" because we consider that more descriptive in the context of even more optimistic boundaries we shall discuss later in the chapter.

3.1. Inner Edge of the Classic Habitable Zone

The inner edge of the habitable zone is the distance within which one of two processes prevents the global stability of liquid water at the surface: (1) absorption of more energy than can be emitted by a planet with an atmosphere containing water vapor, or (2) increased advection of water into the stratosphere, followed by photolysis of H_2O in the upper atmosphere and irreversible escape of H to space. Once either of these conditions is triggered, the other is likely to follow as a result of positive feedback loops. Venus is thought to have gone through both of these processes. This led to the planet ending up in its current state, with a massive CO_2-rich atmosphere, tremendously large greenhouse effect, highly reflective clouds and particles, and a very hot surface despite only absorbing a small amount of incoming sunlight. Venus is thought to be the "end state" of planets that are too hot to be habitable. Because of the runaway nature of the feedbacks involved, it is expected that planets too close to their host stars may resemble Venus is many ways.

On planets with liquid water at the surface, the consequent presence of water vapor limits the amount of mid-infrared radiation that can be emitted by the lower atmosphere to ~291 W m^{-2} (*Kopparapu et al.,* 2013). At this limit, water vapor absorption features are broad enough to close all the "windows" in the infrared region of the spectrum through which radiation otherwise escapes to space. Beyond this limit, atmospheres are unstable: Increases in the temperature of the surface will not be stabilized by increased emission of energy in the mid-infrared. This causes an unstable, runaway condition that leads to surface temperatures above 1400 K, at which point the planet emits enough energy in long-infrared and visible wavelengths to restore and maintain energy balance (*Kasting et al.,* 1993). In the process, water vapor may freely enter the upper atmosphere where it is photolyzed, allowing irreversible loss of H to space. This is what is widely thought to have caused the dry yet uninhabitable greenhouse currently present on Venus. Furthermore, for most conditions this presents a "tighter" constraint on habitability than the runaway greenhouse limit, and is therefore the limiting factor at the inner edge of the habitable zone (*Kopparapu,* 2013; *Kopparapu et al.,* 2013). For planets with an Earth-like albedo (~0.3), the inner edge is 0.99 AU from the Sun.

3.2. Outer Edge of the Classic Habitable Zone

At least three times, Earth's surface has been covered by ice from pole to pole (*Hoffman et al.,* 1998). This phenomenon, known as the snowball Earth, lasted only a few million years because it was counteracted by the accumulation in the atmosphere of CO_2 from volcanos, but water-rich terrestrial planets in the outer boundary of the habitable zone may be frozen for longer periods (*Tajika,* 2008). On these planets the geothermal heat flow may be

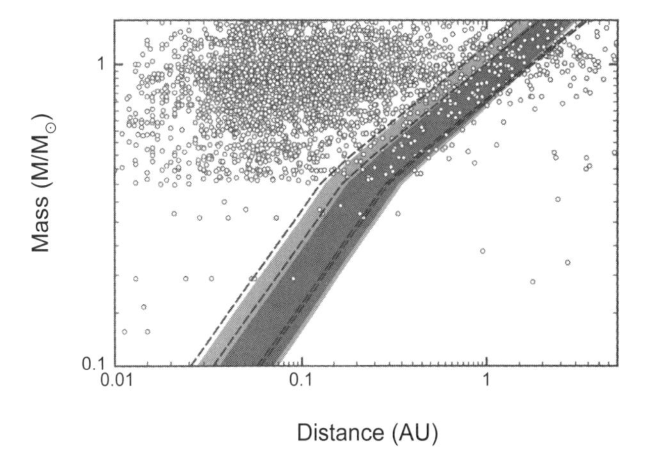

Fig. 2. Known planets, plotted as a function of the host star's mass and the semimajor axis of the planet's orbit. The dark and light gray areas respectively represent the habitable zones for the theoretical (conservative) and empirical (optimistic) habitable zones. The dashed lines are the same habitable zone boundaries, but are calculated on instellation alone and ignore the effects of stellar energy distribution on the planet's albedo. In other words, the dashed lines assume the total amount of energy hitting the top of the planet's surface is the sole driver of habitability, and the wavelength distribution of that energy is irrelevant.

enough to keep a liquid ocean preserved underneath the ice cover. For example, the internal ocean of the snowball Earth would survive as far as ~ 4 AU from the Sun (*Tajika,* 2008). These planets may be recognized by their high albedos and further characterized by time-resolved, multiband optical photometry (*Cowan et al.,* 2011).

On Earth, the snowball period started because of the ice-albedo positive feedback. If about half the Earth's surface area were to become ice covered, then the high albedo of the ice (nearly 1 in the visible) would enhance the cooling of the surface, promoting the creation of more ice (*Hoffman and Schrag,* 2002). If the planet is not geologically active or does not produce enough CO_2 — or other greenhouse gases like CH_4 — then the ice-albedo positive feedback may drive the planetary climate, provided that there is no other heat source for the planet. Recently, *Joshi and Haberle* (2012) proposed that the water-ice and snow-albedo feedback may not be efficient to cool planets around M dwarfs. The reflectivity of ice and snow is high in the visible (>0.7) but very low for infrared radiation (<0.2). Because M dwarfs emit most of their radiation in the near-infrared wavelength region, Joshi and Haberle pointed out that snow and ice might not have the cooling effect they have for planets around Sun-like stars that have their maximum emission in the visible. More detailed calculations were carried out by *Shields et al.* (2013) using a one-dimensional energy-balance climate model and a general circulation model (GCM) for planets around F, G, K, and M main-sequence stars. Their results indicate that the spectral dependence of water ice and snow albedo does play a role in affecting climate. At a given instellation, planets around M dwarfs exhibit higher global mean surface temperatures than planets orbiting stars with more visible and near-ultraviolet radiation output.

The outer edge of the habitable zone is set by the ability of a planet to avoid such planet-wide glaciations (*Kasting et al.,* 1993). Greenhouse gases can prevent this condition, but condensation pressures limit the abundance of such gases. Once these gases begin to condensate, they form clouds that can reflect incoming stellar radiation. This has the potential to increase the albedo of the planet, and thereby decrease surface temperatures. (Although it should be noted that clouds can also warm the surface, depending on their composition and height in the atmosphere.) At 1.4 AU from the Sun, the CO_2 concentrations required to keep the surface from freezing over are high enough for CO_2 clouds to condense. While CO_2 clouds may end up warming the surface, the treatment of the competing warming/cooling effects has not been fully considered in most prior work, due to the limitations by the one-dimensional models historically used (but see below for an exception to this). Such models have tremendous difficulty incorporating the effects of clouds. Future work on this would be best done with three-dimensional global climate models similar to those used to study anthropogenic climate change. Such models have been applied to ancient Mars (*Kahre and Haberle,* 2010; *Johnson et al.,* 2008; *Hoffman et al.,* 2010; *Forget et al.,* 2013) and to specific exoplanets (*Wordsworth et al.,* 2011), but not to the habitable zone in a general sense. Because of these limitations, the most conservative distance for the outer edge of the habitable zone is the distance at which CO_2 begins to condense. If the radiative effects of CO_2 clouds are totally ignored, the habitable zone can go out to ~1.67 AU. Increases to CO_2 past the levels required at this distance lead to net cooling, as each additional CO_2 molecule imparts more of a cooling effect from Rayleigh scattering than a warming one from greenhouse effects.

Similar to the inner edge of the habitable zone, the history of known planets can be used as a guide to defining the habitable zone's outer edge. There is extensive evidence in the Earth's rock record for surface liquid water and a biosphere back to at least 3.5 b.y. ago (*Holland,* 1984). Based on stellar evolution models (*Gough,* 1981), the amount of energy received at this time corresponds to a planet orbiting the modern Sun with a semimajor axis of ~1.14 AU. If controversial evidence for extensive surface water during the martian Noachian Eon (which lasted until ~3.5 b.y. ago) were accepted, the empirical outer limit would be 1.77 AU (see Table 1 in *Kopparapu et al.,* 2013). The most complex modeling of planets at the outer edge of the habitable zone is that of *Wordsworth et al.* (2011), which showed that Gliese 581d is potentially habitable. The orbit of Gliese 581d gives it an incoming energy level that corresponds to an orbiting distance around the Sun of ~1.93 AU (*Kaltenegger et al.,* 2011). These orbiting distances all fall within the habitable zone boundaries defined above. Taken together, the above habitable zone constraints place the outer edge somewhere between ~1.67 and ~2.0 AU.

3.3. Instellation as a Habitable Zone Metric

In almost every discussion of the habitable zone, the limits of stability of water at the surface (and thus the habitable zone) are set by the amount of energy absorbed at or near the planetary surface. As discussed above, these boundaries are usually set in terms of the luminosity (or mass) of the host star, and the orbiting distance (semimajor axis) of the planet. In many discussions by the observational community, the (unknown) albedo is combined with these two variables to produce the effective temperature of the planet. However, the two known properties can be combined into another single term — instellation, the amount of stellar energy reaching the top of the planetary atmosphere — without introducing much inaccuracy or uncertainty in the calculation. This metric has previously been used by R. Pierrehumbert in the introduction to his textbook on planetary atmospheres (*Pierrehumbert,* 2010), as well as in recent analyses of the habitability of known and candidate planets (*Gaidos,* 2013; *Dressing and Charbonneau,* 2013; *Kopparapu et al.,* 2013; *Kopparapu,* 2013).

The instellation (S) is given in terms of the star's luminosity (L) with respect to the Sun's ($L_\odot$), the semimajor axis of the planet (a) with respect to that of Earth ($a_\oplus$), and the instellation currently received at the top of Earth's atmosphere ($S_{\odot}$, normally referred to as "Earth insolation")

$$S = S_{\odot} \times L/L_{\odot} \times (a_{\oplus}/a)^2$$

where all these properties are either known ($S_{\odot}$, $L_{\odot}$, and $a_{\oplus}$) or are capable of being ascertained by current planet detection surveys (a, L). If we use units scaled to the Earth-Sun system (S in terms of $S_{\odot}$, L in terms of $L_{\odot}$, and a in terms of $a_{\oplus}$), this reduces to a very simple relationship

$$S = L/a^2$$

These reduced units will be employed in the rest of this chapter, except where otherwise noted.

The "classic" habitable zone, as defined by Kasting (*Kasting et al.,* 1993, 1997; *Kasting,* 1997) and applied to exoplanets by *Selsis et al.* (2007), *Kaltenegger et al.* (2011), and *Kopparapu et al.* (2013) includes secondary effects on habitability beyond the instellation. Specifically, it includes the effect of the star's temperature on the planetary albedo: Cooler stars are red-shifted, and preferentially emit longer wavelength photons that are less likely to be scattered by the planetary atmosphere and more likely to be absorbed. Thus, if we take a planet around a Sun-type star and move it into an orbit around a cooler star, the planetary albedo will decrease (Fig. 6 in *Kopparapu et al.,* 2013). Definitions of the habitable zone that include these effects are given by the luminosity (L) and the difference (T^*) between the temperature of the star (T_{eff}) and the Sun (5780 K). The total energy received by the planet, S_{eff}, is calculated from $S_{eff} = S_{eff_{\odot}} + aT^* + bT^{*2} + cT^{*3} + dT^{*4}$, where $S_{eff_{\odot}}$ is 1.0146 for the inner limit of the habitable zone (for a moist greenhouse) and 0.3507 for the outer limit of the habitable zone (for a maximum greenhouse) and the values for the constants a, b, c, and d represent the given in Table 3 of *Kopparapu et al.* (2013). The corresponding distances can be calculated using the relation

$$a = \left(\frac{L/L_{\odot}}{S_{eff}}\right)^{0.5} \text{AU}$$

This habitable zone is defined for main-sequence stars between 2600 and 7200 K (this corresponds to stellar masses ~0.1 to ~1.4 times the mass of the Sun).

The T-dependent terms in these equations represent the corrections to the habitable zone required by this albedo effect. Ignoring them leads to relative errors up to ~25% at the outer edge of the habitable zone and <10% at the inner edge of the habitable zone. This means that errors can be constrained to these levels if we rely on instellation alone, and ignore secondary factors such as luminosity/energy distribution effects on albedo. A graphical demonstration of this is shown in Fig. 2, where a luminosity-only definition of the habitable zone is compared to traditional ones that also include secondary effects.

The relative accuracy of this metric compared to equilibrium temperature (which is unconstrained due to the lack of albedo measurements) allows us to accurately collapse the two axes of "traditional habitable zone" plots into a single axis: instellation. This quantity can be scaled to the amount of energy the modern-day Earth receives from the Sun (1367 W m^{-2}). Doing this provides us with an intuitive, convenient metric that is capable of being calculated from quantities that are measured for most known exoplanets and exoplanet candidates. Furthermore, it allows direct comparison with past habitable zone calculations, most of which use the Earth system as a starting point for calculating the bounds of the habitable zone.

3.4. Expansions of the Habitable Zone

The above discussion follows the majority of research on the topic, and focuses on finding and modeling planets relatively similar to Earth. But other possibilities exist, and have been explored with the same theoretical models used to plot the "classic habitable zone." Some prominent examples, considered here, are cloudy/hazy planets, dry planets without significant water vapor in their atmospheres, and planets with atmospheres dominated by hydrogen. These expansions are plotted in Fig. 3, which also shows the effects of planet size on habitability (see section 3.6).

3.4.1. Cloudy/hazy planets. A planet with significantly higher cloud coverage than Earth could exist closer to the star. According to *Selsis et al.* (2007), a planet with an additional cloud layer that covered 50% of the planet could have a planetary albedo ~0.6, and a planet with 100% coverage by that additional cloud layer could have a planetary albedo of ~0.8. For these planets, the inner edge of the habitable zone around the Sun is 0.68 AU and 0.46 AU, respectively. These "cloudy habitable zones" are respectively represented by medium-gray and light-gray regions in Fig. 2.

Is it possible to have an additional cloud layer that has essentially global coverage? Observations of other objects in our solar system suggest this is possible, at least for planets that are thought to be uninhabited. Venus has a high-altitude sulfate haze that reflects the vast majority of incoming energy back to space. An organic-rich haze similar to the one that currently exists on Titan has been proposed as an explanation for geochemical data from the middle of Earth's Archean Eon (*Zerkle et al.,* 2012; *Domagal-Goldman et al.,* 2008; *Pavlov et al.,* 2001; *Haqq-Misra et al.,* 2008; *Trainer et al.,* 2004, 2006). This is a period when Earth was inhabited, so a "Titan-like" haze presents one way for an inhabited planet to have a secondary global cloud/haze layer. While such a planet has not been modeled at the inner edge of the habitable zone, we can use Titan as a rough guide for comparisons to the 50% and 100% cloud layer models by *Selsis et al.* (2007). Titan does not have a very high albedo, but it does have a significant antigreenhouse effect resulting from absorption of incoming stellar radiation high in its atmosphere (*McKay et al.,* 1991). As a result, only 20% of Titan's incoming stellar energy is absorbed at or near the surface; the rest is reflected or absorbed high in the atmosphere as part of an "antigreenhouse effect."

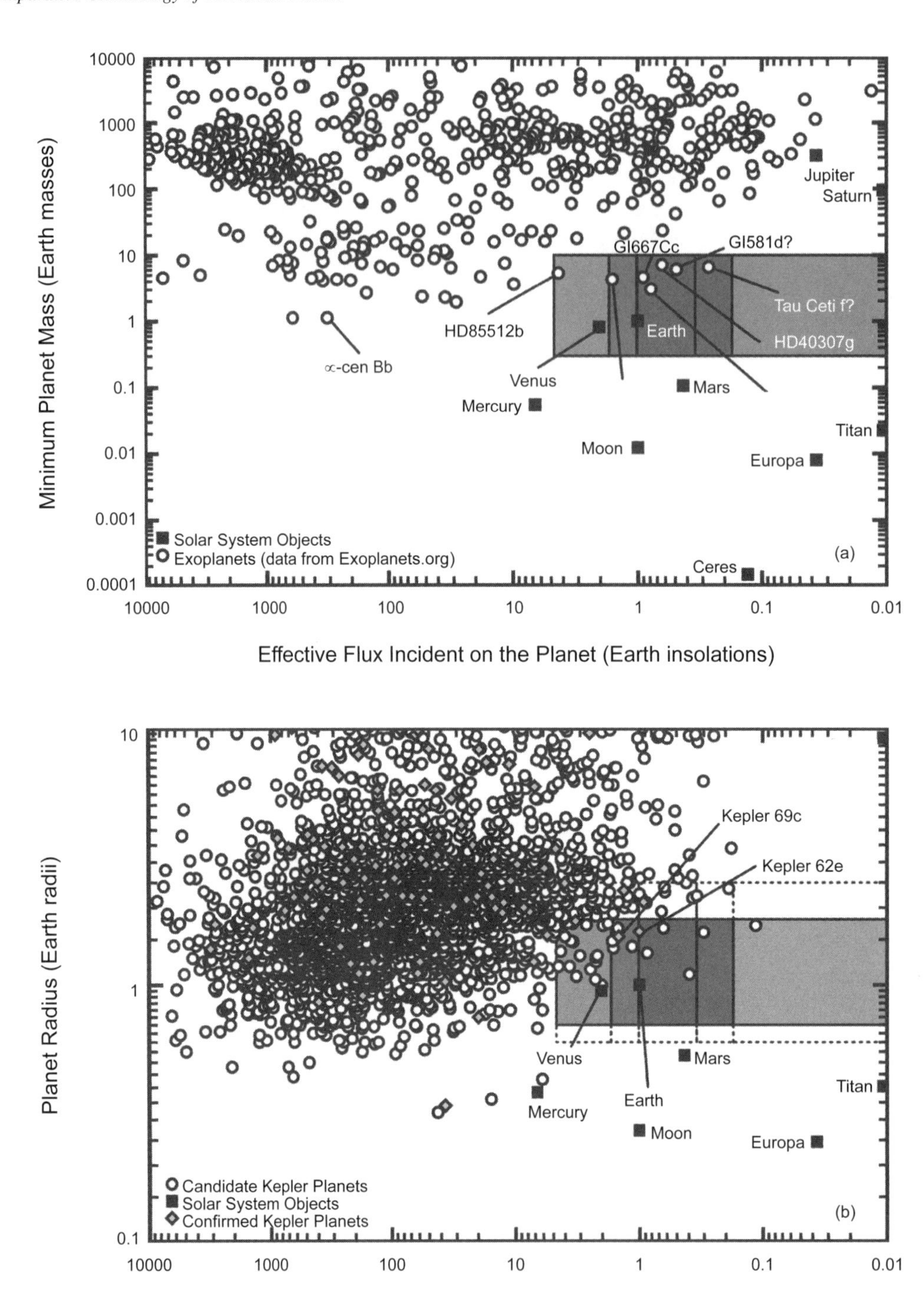

Fig. 3. See Plate 30 for color version. Solar system objects (squares), exoplanets and candidate Kepler planets (circles), and confirmed Kepler planets (diamonds) plotted as a function of energy incident upon the top of the planetary atmosphere and either **(a)** planet mass or **(b)** planet radius. All units are scaled to Earth. In both panels, the colored boxes are representative of different habitable zones defined in the literature. Blue areas represent the "conservative habitable zone," according to updated calculations (*Kopparapu et al.,* 2013). Brown areas show an extension of the inner edge of the habitable zone for dry planets (*Abe et al.,* 2011). Orange areas represent further extension inward based on the cloud effects, and represent the "100% cloud inner limit" (*Selsis et al.,* 2007). Similarly, red boxes represent an extension of the outer part of the habitable zone by clouds according to the "100% cloud outer limit" (*Selsis et al.,* 2007). Gray boxes show an extension of the outer edge of the habitable zone for H_2-He greenhouse planets (*Pierrehumbert and Gaidos,* 2011). In **(b),** vertical extensions of these boxes are also included with dashed lines, showing planetary radii for which objects could be potentially habitable, but only if planets in those areas have compositions that are "just right" (see text).

Thus, the antigreenhouse and reflective effects combine to give an effective albedo of ~0.8, the same value as for Selsis' "100% water cloud cover" case. This is the lowest-proposed proportion of energy absorbed by the surface of any inhabited planet, past or present, so it is used here as the albedo for defining the inner edge of the habitable zone (which depends on absorbing as little energy as possible). This is the most optimistic inner edge for the habitable zone shown in Figs. 2 and 3. However, these processes have not been modeled in great detail. Until such detailed models are published, it is best to focus on planets that would be habitable without (nearly) global cloud coverage; this is the approach taken by *Kopparapu et al.* (2013).

Because clouds can induce a net warming influence, they can also expand the habitable zone further outward. Carbon dioxide clouds can impart their own greenhouse effect, and do not efficiently scatter in the visible, and thus their presence might allow planets to be habitable at greater orbiting distances. Previous work (*Forget and Pierrehumbert,* 1997) shows that the outer edge of the habitable zone around the Sun for a planet with 100% CO_2 clouds could be as far out as 2.4 AU. We can use this limit for the "100% cloud cover" case on the outer edge, and a similarly derived limit of 1.95 AU for a "50% cloud cover" outer edge (Figs. 2 and 3). Similar to the inner edge, planets that demand high cloud coverage should be distinguished from planets that could be habitable without such effects being invoked. Furthermore, these cloud treatments should be viewed as optimistic, as three-dimensional GCM simulations suggest the warming effects from clouds on ancient Mars peak at CO_2 pressures of ~1 bar, and their net effects are fairly small at CO_2 pressures of 5 bar (*Colaprete and Toon,* 2003).

3.4.2. Dry planets. Both the inner and outer edges of the habitable zone are determined by conditions that cause runaway feedbacks that involve water. At the inner edge, water absorbs outgoing infrared radiation, raising the temperature and the amount of water in the atmosphere (warmer air can carry more water), which in turn absorbs more outgoing infrared radiation, and so on. At the outer edge, a drop in surface pressure causes water to freeze, which increases the reflectivity of the surface and decreases the amount of energy being absorbed by the surface, thereby further cooling the surface. In either case, water plays a major role in feedbacks that worsen conditions for life. On a planet with water reservoirs that are present, but limited, these feedbacks will be weakened and the habitable zone thereby expanded. These effects were elucidated, discussed in detail, and quantified by *Abe et al.* (2011). Their effects on the inner edge of the habitable zone are shown in Fig. 3.

3.4.3. Water worlds. Water-rich worlds ("water worlds") may be common in the universe, and have unique constraints on habitability. Planets a few times the size of Earth ("super-Earths") are the most commonly found planets in surveys (e.g., *Howard et al.,* 2010); planets in this size range are suggested to have high volatile contents by dynamical simulations (*Raymond et al.,* 2007). The first estimates of the bulk densities of super-Earths (*exoplanets.org*) (*Fortney et al.,* 2008; *Seager et al.,* 2008; *Seager,* 2011), as well as the atmospheric composition of super-Earths (*Berta and Charbonneau,* 2012), corroborate this suggestion. Because water is one of the most abundant volatile species in the universe, planets with large water reservoirs may be common.

Are these potentially common worlds habitable? Surely water worlds would provide plenty of water molecules for the formation of liquid oceans. However, they may have too much of a good thing: If the oceans are too deep, then the pressures at the ocean floor may be high enough to form an layer of crystalline ice phase (ice VII) that is denser than water. Such an ice layer could reduce the efficiency of the recycling of nutrients that are essential to life, or it could reduce the efficiency of the carbonate-silicate feedback cycle that has served as a themostat in Earth's history and significantly expands the habitable zone. This latter issue — slow tectonic recycling of CO_2 — could also arise on water worlds without an ice VII layer, but with global ocean coverage that breaks this cycle by eliminating the weathering of continents (*Abbot et al.,* 2013).

3.4.4. Hydrogen super-greenhouse planets. As explained above, the fundamental limit at the outer edge of the "classical habitable zone" is the distance at which global glaciation is unavoidable. Past habitable zone work has focused on CO_2-H_2O greenhouses. But both these gases are condensable, and therefore prone to forming clouds at high partial pressures. A more idealized outer limit would come from a greenhouse gas that does not condense, such as H_2 (*Pierrehumbert and Gaidos,* 2011). Because H_2 is a noncondensable greenhouse gas, its pressures — and corresponding warming effects — can increase virtually without the limits imposed by cloud formation. This could significantly extend the habitable zone further outward, to a distance of 10 AU. This is past the orbit of Saturn! The implication is that a terrestrial-sized planet (or moon) with an H_2-rich atmosphere could maintain liquid water at tremendous distances from the parent star. This is a relatively new idea, so there are many details that have yet to be worked out (*Wordsworth and Pierrehumbert,* 2013). The biological details for a biosphere under an H_2-dominated atmosphere of a planet have not been modeled. Future work could explore potential metabolisms, biosignatures, and molecular chemistries for such a biosphere, and could study self-consistent atmospheres composed of CO_2, H_2O, H_2, and CH_4. Additional work could explore the climatic implications of a CO_2-H_2O-H_2-CH_4 greenhouse for early Mars (*Ramirez and Kasting,* 2011) and early Earth (*Wordsworth and Pierrehumbert,* 2013). This can be thought of more broadly as an extension of the outer edge of the habitable zone for such "exotic" greenhouse gases; this expansion is demonstrated in Fig. 3.

4. EFFECTS OF PLANET SIZE ON HABITABILITY AND CLIMATE

Liquid water stability requires the presence of water (H_2O) molecules, and a temperature and pressure such that liquid water is the stable phase. Most of the above discussion has focused on surface temperature; however, the availability of water molecules and the atmospheric surface pressure must also be considered. These two factors are the main determinants of habitability constraints based on the size or mass of the planet. Some planets are too small to retain enough atmospheric pressure, others so large as to have surface pressures too high for liquid water to be stable at the surface, and yet others will have a density such that water molecules are likely to be too rare in the planet's bulk composition for liquid oceans to form.

Current exoplanet surveys can usually determine either a minimum planetary mass or measure the planet's radius; in rare cases we can obtain both mass and radius, and thereby ascertain the planet's density. While the limits on habitability placed by these physical parameters are not as well defined as the astronomical/climatic considerations, this is an active area of theoretical research (*Barnes et al.,* 2009, 2013; *Jackson et al.,* 2008; *Valencia et al.,* 2006, 2008; *Van Heck and Tackley,* 2011), spurred on by the discovery of a series of "super-Earth" planets with masses between ~2 $M_\oplus$ and ~10 $M_\oplus$ (*Batalha et al.,* 2013; *Howard et al.,* 2010). Thus the geophysical boundaries presented below are expected to shift as this research (and further observations) furthers our understanding of objects of varying sizes.

Of these three measureable physical properties, the most straightforward habitability constraints come from knowledge of planetary mass. Planets that are too small cannot hold onto a thick atmosphere, without which liquid water will not be stable at the surface of the planet. Within the solar system, Earth and Venus (with masses of 1.0 and 0.8 $M_\oplus$) have thick atmospheres, Mars (with a mass of 0.10 $M_\oplus$) has a thin atmosphere and evidence of liquid water in its past, and Mercury (with a mass of 0.05 $M_\oplus$) has a tenuous atmosphere, and is not thought to have ever harbored liquid water at its surface. However, the presence or absence of an atmosphere cannot be a function of planetary mass alone, as Titan has a lower mass than Mercury but also has a relatively thick atmosphere. Based on a more thorough analysis of known planets (*Catling and Zahnle,* 2009; *Zahnle and Catling,* 2013), K. Zahnle and D. Catling have suggested that the presence of an atmosphere is controlled by both the planet's size as well as the amount of energy the planet receives from its parent star. Larger planets will have higher escape velocities and perhaps greater inventories for outgassing; planets that receive more energy will have higher escape rates. For the amount of energy that reaches planets in the habitable zone, and the mass-radius relationship for planets with Earth-like compositions (*Seager et al.,* 2008; *Fortney et al.,* 2008), planets with at least ~0.3 $M_\oplus$ should be able to retain their atmospheres.

Uranus provides us with an empirical upper limit of 15 $M_\oplus$ for "Earth-like" planets. Above this mass, planets may be more "Uranus-like" and thus absent of solid surfaces in contact with liquid water oceans (considered a habitability requirement because of the need to recycle nutrients). Models suggest an upper limit of 10 $M_\oplus$, as planets more massive than this would undergo rapid gas accumulation during accretion, and would instead become a gas (nonterrestrial) planet (*Ida and Lin,* 2004; *Ikoma and Genda,* 2006). The upper limit to habitability is 10 $M_\oplus$ in Fig. 3a, which plots exoplanets with known masses (from *exoplanets.org*) (*Wright et al.,* 2011) as a function of both instellation and planetary mass for various definitions of the habitable zone.

By utilizing these mass limits, bounds to the radius of Earth-like, habitable planets can also be set. Based on multiple theoretical models (*Fortney et al.,* 2008; *Seager et al.,* 2008), planets with masses <10 $M_\oplus$ must have radii <1.8 $R_\oplus$ if they have an Earth-like composition. Planets larger than 1.8 $R_\oplus$ would either have a mass larger than 10 $M_\oplus$ or would have a composition that is less rocky and more water-rich Earth. Planets with radii between 1.8 $R_\oplus$ and 2.5 $R_\oplus$ would be water worlds. Such planets would have their own potential pitfalls for habitability. For example, the extremely high pressures at the bottom of the ocean on such a planet would lead to the presence of ice VII layer that would separate the planet's silicate and liquid water reservoirs (*Valencia et al.,* 2008). Thus, they may have liquid water at their surfaces, yet still be uninhabitable. Another potential concern is dilution of nutrients: If Earth-like nutrient fluxes enter a water world, they will be spread out in a larger volume of water; furthermore, the lack of land masses means those nutrients will not be concentrated by upwelling at or near continental shelves. Due to these concerns, such water world planets should be classified differently than ones with rocky, Earth-like compositions.

This approach can also be used to set lower radius limits on habitability. Planets with Earth-like compositions and masses greater than 0.3 $M_\oplus$ should have radii larger than 0.7 $R_\oplus$. Planets with radii between 0.6 and 0.7 $R_\oplus$ could be above the 0.3-$M_\oplus$ limit, but may be too water-poor to support oceans. A firm lower limit on radius can be placed at ~0.6 $R_\oplus$, below which planets are either below the lower mass limit of 0.3 $M_\oplus$ or composed of pure iron.

The Kepler data field (*Batalha et al.,* 2013) is plotted in Fig. 3b as a function of instellation and planetary radius, with the above habitability constraints shown. Planets with radii between 0.7 $R_\oplus$ and 1.8 $R_\oplus$ could have masses between 0.3 $M_\oplus$ and 10 $M_\oplus$ and a composition similar to that of Earth. Thus they are plotted in dark colors, corresponding to "potentially Earth-like." Planets with radii between 1.8 $R_\oplus$ and 2.5 $R_\oplus$, and planets with radii between 0.6 $R_\oplus$ and 0.7 $R_\oplus$, could be of the correct mass to be habitable, but would have a composition that might preclude habitability in an "Earth-like" sense. They are therefore shown as dotted boundaries to mark them as "water worlds" (high radius) or "Dune worlds" (lower radius).

5. LESSONS FROM KNOWN EXOPLANETS

5.1. Modeling the First Potentially Habitable Planet: Gliese 581d

Gliese 581 (Gl 581) is a main-sequence M star at a distance of 6.3 pc. (M stars are cooler than G stars such as the Sun. M stars can also be referred to as "red dwarfs," "M dwarfs," or "late-type stars.") The first planet detected around this star was a Neptune-mass planet with an orbital semimajor axis of 0.041 AU (*Bonfils et al.,* 2005). In 2007, two more planets were found, Gl 581c, a ~5-$M_\oplus$ planet at 0.073 AU and Gl 581d with ~8 $M_\oplus$ at 0.25 AU from its star (*Udry et al.,* 2007). Udry et al. made a rough estimate of the equilibrium temperature for Gl 581c, and concluded that the planet was in the habitable zone of its star. This back-of-the-envelope calculation used the albedo of Earth and Venus to illustrate the range of equilibrium temperatures and neglected the greenhouse warming effect of an atmosphere. That same year, two groups (*Von Bloh et al.,* 2007; *Selsis et al.,* 2007) analyzed the conditions needed for the planets in the Gl 581 system for being potentially habitable. Although the approaches used by the groups were different, they reached the same conclusion: Gl 581c was not habitable, but Gl 581d was close to the outer limit of the habitable zone. *Von Bloh et al.* (2007) applied the photosynthesis-sustaining habitable limits, where the presence of CO_2 is not only considered for being a greenhouse gas that keeps the surface warm, but is required to maintain biological productivity. *Selsis et al.* (2007) used the classical limits of the habitable zone calculated by *Kasting et al.* (1993), and found that Gl 581d needs to be 100% covered by CO_2 clouds to be habitable.

Another planet was detected around Gl 581 in 2009 (*Mayor et al.,* 2009): a 1.9-$M_\oplus$ planet with a semimajor axis of 0.03 AU. The new observations allowed a more accurate calculation of the orbital parameters of Gl 581d. Its updated semimajor axis was 0.22 AU. This meant that the planet was in the habitable zone according to the results of *Von Bloh et al.* (2007) and *Selsis et al.* (2007). Several papers published between 2010 and 2011 calculated the atmosphere needed to keep Gl 581d surface above the freezing point of water. The assumptions and results of papers that studied Gl 581d habitability using one-dimensional models are summarized in Table 1.

Wordsworth et al. (2011) used a three-dimensional model for assessing the potential habitability of Gl 581d. They considered rocky and ocean (50% water) planets with orbit-rotation resonances of 1:1, 1:2, and 1:10, and atmospheres composed of CO_2, H_2O, and N_2 atmospheres. While some one-dimensional models found that 5-bar, CO_2 rich atmospheres were enough to keep the average surface temperature above the water freezing point, three-dimensional simulations showed that these atmospheres will collapse for rocky and ocean planets regardless of the orbit-rotation resonance considered, i.e., the atmosphere will freeze permanently in the nonilluminated hemisphere. Ocean planets required atmospheric pressures of at least 20 bar to not collapse, and rocky planets required at least 10 bar of atmospheric pressure, except for the planet with 1:1 resonance.

All the models assumed that the planet was either a rocky or a water planet, but this may not be the case. No transit has been reported for this planet, and therefore its radius is not constrained. According to models of planetary interiors, a 7-$M_\oplus$ planet may be composed of a mixture of an iron core, a rocky mantle, and an envelope of water or H/He, or it may have a H_2 atmosphere (*Baraffe et al.,* 2010). Even for cases where the mass and radius has been measured, it is not possible to constrain the planet bulk composition (*Rogers and Seager,* 2010). In order to confirm whether Gl 581d is habitable will require spectroscopic observations.

5.2. Carbon Planets

Solid planetary bodies are usually considered to be composed of a mixture of iron and silicates, as is the case for solid bodies in the solar system. Another interesting possibility for solid bodies is a composition dominated by carbon. These planets may be formed in environments with a C/O ratio twice the solar value (~0.5) or larger (*Cameron et al.,* 1988; *Gaidos,* 2000; *Kuchner and Seager,* 2005). Carbon-rich stars are candidates for having carbon planets (*Madhusudhan et al.,* 2012), and locally high C/O or low H_2O zones of otherwise low C/O protoplanetary disks may also result in carbon-rich planets. The mass-radius relationships for these planets are very similar to those composed of iron-silicate mixtures (*Seager et al.,* 2008).

Unfortunately, the C/O ratio of a planet is not easily observable. The composition of the parent star C/O can be used as an initial estimate for a planet's composition. However, mixing within a system can lead to a wide variety of planet compositions, even for a given stellar composition (*Bond et al.,* 2010). Observations of exoplanetary atmospheres may help address this uncertainty because carbon-planet atmospheres are expected to differ from those of silicate planets, although there are no simulations available for atmospheres of solid-carbon planets. For the atmospheres of hot Jupiters, photochemical models show that H_2O, CH_4, CO_2, and CO abundances are highly dependent on the C/O ratio (*Kopparapu et al.,* 2012; *Moses et al.,* 2013) Carbon-rich planets should also have hydrogen cyanide (HCN) and acetylene (C_2H_2) in relatively high abundances, with mole fractions from ~10^{-6} to 10^{-4} (*Moses et al.,* 2013). Spitzer photometry for secondary eclipses for WASP-12b, XO-1b, and CoRoT-2b is consistent with the atmospheric disequilibrium chemistry for a planet with C/O ~1 (*Moses et al.,* 2013).

5.3. Tidally Heated Planets

The internal heat of a planet can be as important for climate as the incoming energy from its star. On Earth, mantle heating by radiogenic elements drives plate tectonics, which play a key role on recycling our atmosphere and maintaining the long-term stability of the planet's climate. Along with radiogenic elements, internal heating can be driven by tidal forces. Jupiter's moon, Io, is an extreme example of a planet heated by tides. For M dwarfs, the habitable zone is close enough to the star for the effect of tides to become an important heat source. A planetary orbit that starts with a high eccentricity is eventually damped to near-zero eccentricity after billions of years. Because the

TABLE 1. Assumptions and results of studies of the habitability potential of Gliese 581d.

Authors	Atmospheric Composition	Stellar Properties	Planet Parameters	Model Characteristics	Atmosphere Needed for Habitability
Von Bloh et al. (2007)	CO_2	$L = 0.013\ L_\odot$	$M = 5$ and $8\ M_\oplus$ Rocky planets Variable distance to star	Photosynthesis-sustaining habitable zone	4–10 bars CO_2
Selsis et al. (2007)	CO_2, H_2O, N_2	$T_{eff} = 3200$ K, $L = 0.013\ L_\odot$	$M = 8.3\ M_\oplus$ Rocky planet at 0.253 AU	*Kasting et al.* (1993) limits revisited	With 100% CO_2 cloud coverage
Wordsworth et al. (2011)	CO_2, H_2O, N_2	AD Leo (T_{eff} = 3400 K), $L = 0.0135\ L_\odot$	Surface gravity 10–30 m s^{-1}	One-dimensional radiative model, k-coefficients	5 bars CO_2
Von Paris et al. (2010)	CO_2, H_2O, N_2	Measured IUE + NextGen model ($T_{eff} = 3200$ K)	$M = 7.09\ M_\oplus$ Rocky planet at 0.22 AU	One-dimensional cloud-free radiative model, k-coefficients	5 bars surface pressure, 95% CO_2
Hu and Ding (2011)	CO_2, H_2O	Blackbody at 3200 K	$M = 8.3\ M_\oplus$ $R = 2.0\ R_\oplus$ at 0.21 AU	One-dimensional VPL radiative model	6.7 bars of CO_2 for albedo = 0.13; 11.5 bars of CO_2 for albedo = 0.3
Kaltenegger et al. (2011)	CO_2, H_2O, N_2, CH_4 (radiative model)	Model M dwarf $T_{eff} = 3600$ K	$M = 7\ M_\oplus$ $R = 1.7\ R_\oplus$ rocky planet at 0.20 AU	One-dimensional coupled photochemical, radiative models	7.6 bars surface pressure 90% CO_2

timescale for this heating mechanism is long, it is relevant for the planet's long-term climatological evolution (*Jackson et al.,* 2008). Tidal heating may affect climates by (1) driving plate tectonics, (2) outgassing volatiles counteracting atmospheric loss; (3) making the planet uninhabitable due to extreme volcanism (*Jackson et al.,* 2008), or (4) making the planet uninhabitable by triggering a runaway greenhouse effect (*Barnes et al.,* 2013).

Carbon dioxide contributed to the weathering of continental rocks, and a byproduct of this process is the burial of calcium carbonate ($CaCO_3$, the primary mineral in limestone) on ocean floors. This effectively stores CO_2 below the planetary surface, only to be released on long timescales by volcanism after the crust with the sequestered carbon is subducted and heated in the mantle. The weathering process is temperature-dependent: Higher surface temperatures lead to higher rates of weathering and CO_2 burial. This creates a strong (but very long-term) stabilizing feedback loop: Increases in temperature will increase weathering rates, draw down CO_2, and offset the initial temperature increase.

This process is called the carbonate-silicate cycle, and is the one that controls Earth's atmospheric CO_2 abundance over long timescales, and in turn the climate of the planet (*Kasting and Catling,* 2003). For example, this cycle is what allowed the planet to escape "snowball Earth" episodes during which glaciers advanced to the equator. The cold temperatures associated with these periods of Earth history — and the slowdown of the planet's hydrological cycle — decreased the rate at which CO_2 was removed from the atmosphere. This allowed CO_2 to accumulate to high enough concentrations to warm the planet and melt the glaciers. There is geological evidence for this sequence of events: Immediately after the snowball Earth episodes, the planet's rock record contained massive calcium carbonate deposits consistent with a CO_2-rich atmosphere. Had Earth not been tectonically active, this recycling of CO_2 would not have occurred, and the planet may not have escaped from an ice-covered state.

Other planets may have additional linkages between heating rates and climate. *Jackson et al.* (2008) suggest that planets with extreme tidal heating rates may lead to global volcanism. Such planets may have difficulty holding onto certain volatiles if this massive volcanism leads to either direct escape, or to photolysis followed by escape of individual elements such as hydrogen.

If high volcanism rates do not present a barrier to habitability, extreme tidal heating rates could directly cause a climate calamity. Simulations suggest some planets may have tidal heating rates greater than the rate at which a water-rich atmosphere can efficiently radiate energy to space. Such planets will undergo a runaway surface heating, and their surface temperatures will not stabilize until the planet loses its water to hydrogen escape or becomes hot enough to efficiently radiate at shorter wavelengths, or if the surface melts and lowers tidal heating rates. In any case, the planet would not be habitable under any classical definition of the term.

Habitable zone limits have been revised by including the effect of tides, resulting in an insolation-tidal habitable zone (ITHZ) (*Barnes et al.,* 2009). The ITHZ is a function of planetary mass, orbital eccentricity, the planet's semimajor axis (*Barnes et al.,* 2009), and the orbits and masses of other planets and stars in the system. For many configurations around cool, low-mass stars, the effects of tides will place tighter habitability constraints than insolation. This means that a planet that may be considered habitable using only the insolation limit could be uninhabitable due to a runaway greenhouse triggered by tidal heating (*Barnes et al.,* 2009).

5.4. Hot Earths and Venus-Like Planets

In addition to habitability metrics, there have also been models of uninhabitable planets. Because there is an observational bias toward close-in planets, the first of these papers (*Schaefer and Fegley,* 2009, 2011; *Leger et al.,* 2009, 2011; *Barnes et al.,* 2010; *Miguel et al.,* 2011; *Treiman and Bullock,* 2012) has focused on planets too close to their stars and far too hot to be habitable. Silicate planets with surface temperatures larger than 1500 K may lose their volatile compounds like H, C, and N, and their atmospheres would be composed of Na gas, O_2, O, and SiO gas (*Schaefer and Fegley,* 2009). Extended clouds of Na and K are expected on these planets as a result of the interaction of the interaction of the stellar wind with the planetary atmosphere (*Schaefer and Fegley,* 2009). Sodium clouds have been observed around Mercury (*Potter and Morgan,* 1987) and Io (*Spencer and Schneider,* 1996). These observations guide models of hot super-Earths. *Miguel et al.* (2011) classify hot rocky planet atmospheres in five groups, assuming a komatiite magma composition (Table 2).

The first detected hot rocky planet was Corot-7b, a ~1.7-$R_\oplus$ planet at 0.017 AU from a Sun-like star (0.93 $M_\odot$). The calculated planet surface temperature ranges from 1800 to 2600 K at the substellar point on the dayside (*Léger et al.,* 2009). Dynamical simulations (*Jackson et al.,* 2010; *Barnes et al.,* 2010) suggest that Corot-7b may be the remnant core of a gas giant that lost its volatiles during migration toward its parent star. However, thermal mass loss calculations by *Leitzinger et al.* (2011) indicate that it is not possible for this planet to be the remnant of a hydrogen-rich gas giant or a Uranus-type ice giant. Regardless of its origin, models of planet interiors confirm the rocky nature of Corot-7b (*Wagner et al.,* 2012; *Swift et al.,* 2012), and it is expected to have an atmosphere mainly composed by silicates (*Léger et al.,* 2009, 2011), as well as a tail composed of Na atoms. The tail is a result of the enormous amount of radiation pressure impinging upon the planet (*Mura et al.,* 2011). According to models by *Barnes et al.* (2010), a slight eccentricity in the orbit of CoRoT-7b can result in enough tidal heating for it to be a super-Io planet with very high volcanism rates. If the orbital eccentricity of Corot-7b was initially larger than zero, this planet may have been a super Io until the energy dissipated by tides circularized its orbit.

Hot, Venus-like planets are also expected to occur in other planetary systems. These planets are characterized by CO_2-rich atmospheres with minor compounds like water, N_2, SO_2, HCl, HF, and CO (*Schaefer and Fegley,* 2011; *Treiman and Bullock,* 2012). Two different approaches have been used for these atmospheres. The first is to assume that CO_2 is controlled by buffering reactions with certain minerals at the planet's surface. *Schaefer and Fegley* (2011) calculate surface temperatures and CO_2 partial pressures for three possible mineral compositions: hot mafic, cold felsic, and cold mafic. Only in the hot mafic case do predicted temperatures and CO_2 pressures exceed those on Venus, for which the surface temperature is 740 K and the CO_2 pressure is ~92 bar. For all other cases, this buffering will cause "Venus-like planets" to be cooler than modern-day Venus.

The other approach to this problem contends that mineral buffers do not control the abundance of CO_2 and H_2O on Venus or Venus-like planets. For these planets, perturbations to T or CO_2 pressure (for example, a volcanic eruption that adds a significant mass of CO_2) can produce catastrophic expansion or collapse of the atmosphere (*Treiman and Bullock,* 2012, and references therein). Under this approach, the CO_2 abundance of a Venus-like exoplanet would not be useful to constrain the composition of its surface. In this model, species with lower concentrations, such as SO_2, could still be controlled by mineral buffers, and therefore

TABLE 2. Atmospheric composition of hot rocky planets (*Miguel et al.,* 2011).

Atmosphere Type	Planet Equilibrium Temperature (T_p)	Composition
I	$T_p < 2033$ K	Monatomic Na, O_2, monatomic O, and monatomic Fe in order of abundance.
II	2033 K $< T_p <$ 2588 K	SiO becomes more abundant than monatomic Fe. Atmosphere characterized by monatomic Na, O_2, monatomic O, SiO, monatomic Fe, and monatomic Mg from the most to the least abundant.
III	2588 K $< T_p <$ 2890 K	Monatomic Mg becomes more abundant than monatomic Fe. The gases with the highest column densities are monatomic Na, O_2, monatomic O, SiO, monatomic Mg, and monatomic Fe. Column densities of O_2, monatomic O, and SiO are almost equal.
IV	2890 K $< T_p <$ 3168 K	SiO becomes more abundant thanO_2 and the atmosphere becomes silicate and monatomic Na dominated.
V	$T_p > 3168$ K	Dominated by silicate oxide, followed closely by monatomic O, O_2, and monatomic Na.

the general approach linking atmospheric composition to the chemical characteristics of planetary surfaces (*Treiman and Bullock,* 2012) is still valid.

The other line of work on extrasolar "Venus-like" planets involved the planetary cloud deck. On Venus, ultraviolet radiation drives the formation of H_2SO_4 clouds. But cooler-type stars have significantly more ultraviolet radiation. *Schaefer and Fegley* (2011) point out that planets around such stars (specifically, F stars) may develop thicker clouds. This would have important climatic implications, as well as consequences for the observed spectrum and broadband albedo of the planet.

6. CONCLUDING REMARKS

One of the great joys of working on terrestrial exoplanet climates is that this area of research exemplifies the spirit of this book. Because we know very few details about these planets — usually just their bulk physical and orbital properties — we are forced to make estimates based on extrapolations from other known objects. In practice, this means we have to apply the knowledge gained from the detailed observations and models of Earth and the other terrestrial-sized atmosphere-bearing worlds inside the solar system.

However, this era of extrapolation will soon be at an end. We are already obtaining information about the climates of Jupiter-sized exoplanets (*Madhusudhan and Seager,* 2009) and a small handful of super-Earth-sized objects (*Berta and Charbonneau,* 2012). The launch of the James Webb Space Telescope will expand our knowledge of this latter group of planets even further (*Deming et al.,* 2009; *Soummer et al.,* 2009). And when we fly a telescope that can probe the atmospheres of terrestrial-sized planets (*Levine et al.,* 2006; *Des Marais et al.,* 2002), we will begin to learn about these objects as planets unto themselves.

Our experiences with exoplanets thus far have taught us to expect the unexpected. We have found planets with orbits previously thought to be implausible, and discovered a multitude of planets in a size range previously thought to be rare and with densities previously thought to be impossible. For a review of these discoveries and what we have learned about the possibilities for planetary climate, there are a number of excellent textbooks that have been recently published (*Mason,* 2008; *Barnes,* 2010; *Seager,* 2011). Similarly, the book on planetary climate by R. Pierrehumbert is a must for anyone that wishes to research this subject (*Pierrehumbert,* 2010).

Based on the surprises outlined above, we should expect that future discoveries about terrestrial-sized exoplanets will contradict many of the assumptions and assertions made in this chapter and the work it reviews. And by adapting our theories on planetary atmospheres and climate to account for the surprises we find, our theories will more completely explain the terrestrial planets of our solar system. When that happens, information will begin to flow in the opposite direction, from exoplanet observations to our understanding of planets closer to home.

REFERENCES

Abe Y., Ayako A.-O., Sleep N. H., and Zahnle K. J. (2011) Habitable zone limits for dry planets. *Astrobiology, 11(5),* 443–460.

Abbot D. S., Cowan N. B., and Ciesla F. J. (2012) Indication of insensitivity of planetary weathering behavior and habitable zone to surface land fraction. *Astrophys. J., 756(2),* 178.

Baraffe I., Chabrier G., and Barman T. (2010) The physical properties of extra-solar planets. *Rept. Prog. Phys., 73(1),* DOI: 10.1088/0034-4885/73/1/016901.

Barnes R., ed. (2010) *Formation and Evolution of Exoplanets.* Wiley, New York.

Barnes R., Jackson B., Greenberg R., and Raymond S. N. (2009) Tidal limits to planetary habitability. *Astrophys. J. Lett., 700,* L30–L33, DOI: 10.1088/0004-637X/700/1/L30.

Barnes R., Raymond S. N., Greenberg R., Jackson B., and Kaib N. A. (2010) CoRoT-7b: Super-Earth or super-Io? *Astrophys. J. Lett., 709,* L95–L98, DOI: 10.1088/2041-8205/709/2/L95.

Barnes R., Mullins K., Goldblatt C., Meadows V. S., Kasting J. F., and Heller R. (2013) Tidal Venuses: Triggering a climate catastrophe via tidal heating. *Astrobiology, 13(3),* 225–250.

Batalha N. M., Rowe J. F., Bryson S. T., et al. (2013) Planetary candidates observed by Kepler III : Analysis of the first 16 months of data. *Astrophys. J. Suppl. Ser., 204,* 24.

Bean J. L., Désert J. M., and Kabath P. (2011) The optical and near-infrared transmission spectrum of the super-Earth GJ 1214b: Further evidence for a metal-rich atmosphere. *Astrophys. J.,(1),* 1–13.

Berta Z. K. and Charbonneau D. (2012) The flat transmission spectrum of the super-Earth GJ1214b from Wide Field Camera 3 on the Hubble Space Telescope. *Astrophys. J., 747(1),* 35.

Bond J. C., O'Brien D. C., and Lauretta D. S. (2010). The compositional diversity of extrasolar terrestrial planets. I. In situ simulations. *Astrophys. J., 715(2),* 1050.

Bonfils X., Forveille T., Delfosse X., Udry S., Mayor M., Perrier C., Bouchy F., Pepe F., Queloz D., and Bertaux J.-L. (2005) The HARPS search for southern extra-solar planets. VI. A Neptune-mass planet around the nearby M dwarf Gl 581. *Astron. Astrophys., 443,* L15–L18, DOI: 10.1051/0004-6361:200500193.

Cameron A. G. W., Benz W., Fegley B. Jr., and Slattery W. L. (1988) The strange density of Mercury — Theoretical considerations. In *Mercury* (F Vilas et al., eds.), pp. 692–708. Univ. of Arizona, Tucson.

Catling D. C. and Zahnle K. J. (2009) The planetary air leak. *Sci. Am., 300(5),* 36–43.

Colaprete A. and Toon O. B. (2003) Carbon dioxide clouds in an early dense martian atmosphere. *J. Geophys. Res., 108(E4),* 5025.

Cowan N. B., Robinson T., Livengood T. A., et al. (2011) Rotational variability of Earth's polar regions: Implications for detecting snowball planets. *Astrophys. J., 731,* 76, DOI: 10.1088/0004-637X/731/1/76.

Currie T., Thalmann C., Matsumura S., Madhusudhan N., Burrows A., and Kuchner M. (2011) A 5 μm image of β Pictoris b at a sub-Jupiter projected separation: Evidence for a misalignment between the planet and the inner, warped disk. *Astrophys. J. Lett., 736,* L33, DOI: 10.1088/2041-8205/736/2/L33.

Deming D., Seager S., Winn J., et al. (2009) Discovery and characterization of transiting super Earths using an all-sky transit survey and follow-up by the James Webb Space Telescope. *Publ. Astron. Soc. Pac., 121(883),* 952–967.

Des Marais D. J., Harwit M. O., Jucks K. W., Kasting J. F., Lin D. N. C., Lunine J. I., Schneider J., Seager S., Traub W. A., and Woolf N. J. (2002) Remote sensing of planetary properties and biosignatures on extrasolar terrestrial planets. *Astrobiology, 2(2),* 153–181.

Domagal-Goldman S. D., Kasting J. F., Johnston D. T., and Farquhar J. (2008) Organic haze, glaciations and multiple sulfur isotopes in the Mid-Archean era. *Earth Planet. Sci. Lett., 269(1–2),* 29–40, DOI: 10.1016/j.epsl.2008.01.040.

Dressing C. D. and Charbonneau D. (2013) The occurrence rate of small planets around small stars. *Astrobiology, 13(3),* 225–250, DOI: 10.1089/ast.2012.0851.

Ehrenfreund P. and Charnley S. B. (2000) Organic molecules in the interstellar medium, comets, and meteorites: A voyage from dark clouds to the early Earth. *Annu. Rev. Astron. Astrophys., 38(1),* 427–483, DOI: 10.1146/annurev.astro.38.1.427.

Forget F. and Pierrehumbert R. T. (1997) Warming early Mars with carbon dioxide clouds that scatter infrared radiation. *Science, 278(5341),* 1273–1276.

Forget F., Wordsworth R., Millour E., Madeleine J. B., Kerber L., Leconte J., Marcq E., and Haberle R M. (2013) 3D modelling of the early martian climate under a denser CO_2 atmosphere: Temperatures and CO_2 ice clouds. *Icarus, 222(1),* 81–99, DOI: 10.1016/j.icarus.2012.10.019.

Fortney J. J., Marley M. S., and Barnes J. W. (2008) Planetary radii across five orders of magnitude in mass and stellar insolation: Application to transits. *Astrophys. J., 659(2),* 1661.

Gaidos E. J. (2000) Note: A cosmochemical determinism in the formation of Earth-like planets. *Icarus, 145,* 637–640, DOI: 10.1006/icar.2000.6407.

Gaidos E. (2013) Candidate planets in the habitable zones of Kepler stars. *Astrophys. J.,* in press, arXiv:1301.2384.

Gough D. O. (1981) Solar interior structure and luminosity variations. *Solar Phys., 74(1),* 21–34.

Haqq-Misra J. D., Domagal-Goldman S. D., Kasting P. J., and Kasting J. F. (2008) A revised, hazy methane greenhouse for the archean Earth. *Astrobiology, 8(6),* 1127–1137, DOI: 10.1089/ast.2007.0197.

Herbst E. and van Dishoeck E. F. (2009) Complex organic interstellar molecules. *Annu. Rev. Astron. Astrophys., 47(1),* 427–480, DOI: 10.1146/annurev-astro-082708-101654.

Hoffman M. J., Greybush S. J., Wilson R. J., Gyarmati G., HoffmanR. N., Kalnay E., Ide K., Kostelich E. J., Miyoshi T., and Szunyogh I. (2010) An ensemble Kalman filter data assimilation system for the martian atmosphere: Implementation and simulation experiments. *Icarus, 209(2),* 470–481.

Hoffman P. F. and Schrag D. P. (2002) The snowball Earth hypothesis: Testing the limits of global change. *Terra Nova, 14,(3),* 129–155, DOI: 10.1046/j.1365-3121.2002.00408.x.

Hoffman P. F., Kaufman A. J., Halverson G. P., and Schrag D. P. (1998) A Neoproterozoic snowball Earth. *Science, 281,* 1342, DOI: 10.1126/science.281.5381.1342.

Holland H. D. (1984) *The Chemical Evolution of the Atmosphere and Oceans.* Princeton Univ., Princeton.

Howard A. W., Marcy G. W., Johnson J. A., Fischer D. A., Wright J. T., Isaacson H., Valenti J. A., Anderson J., Lin D. N. C., and Ida S. (2010) The occurrence and mass distribution of close-in super-Earths, Neptunes, and Jupiters. *Science, 330(6004),* 653–655.

Hu Y. and Ding F. (2011) Radiative constraints on the habitability of exoplanets Gliese 581c and Gliese 581d. *Astron. Astrophys., 526,* A135.

Ida S. and Lin D. N. C. (2004) Toward a deterministic model of planetary formation. I. A desert in the mass and semimajor axis distributions of extrasolar planets. *Astrophys. J.,* 604, 388–413, DOI: 10.1086/381724.

Ikoma M. and Genda H. (2006) Constraints on the mass of a habitable planet with water of nebular origin. *Astrophys. J., 648,* 696–706, DOI: 10.1086/505780.

Jackson B., Greenberg R., and Barnes R. (2008) Tidal evolution of close-in extrasolar planets. *Astrophys. J., 678,* 1396–1406, DOI: 10.1086/529187.

Jackson B., Miller N., Barnes R., Raymond S. N., Fortney J. J., and Greenberg R. (2010) The roles of tidal evolution and evaporative mass loss in the origin of CoRoT-7 b. *Mon. Not. R. Astron. Soc., 407,* 910–922, DOI: 10.1111/j.1365-2966.2010.17012.x.

Johnson S. S., Mischna M. A., Grove T. L., and Zuber M. T. (2008) Sulfur-induced greenhouse warming on early Mars. *J.Geophys. Res., 113(E8),* E08005.

Joshi M. M. and Haberle R. M. (2012) Suppression of the water ice and snow albedo feedback on planets orbiting red dwarf stars and the subsequent widening of the habitable zone. *Astrobiology, 12,* 3–8, DOI: 10.1089/ast.2011.0668.

Kahre M. A. and Haberle R. M. (2010) Mars CO_2 cycle: Effects of airborne dust and polar cap ice emissivity. *Icarus, 207(2),* 648–653, DOI: 10.1016/j.icarus.2009.12.016.

Kalas P., Graham J. R., Chiang E., Fitzgerald M. P., Clampin M., Kite E. S., Stapelfeldt K., Marois C., and Krist J. (2008) Optical images of an exosolar planet 25 light-years from Earth. *Science, 322,* 1345, DOI: 10.1126/science.1166609.

Kaltenegger L. and Sasselov D. (2011) Exploring the habitable zone for Kepler planetary candidates. *Astrophys. J. Lett., 736(2),* L25.

Kaltenegger L., Segura A., and Mohanty S. (2011) Model spectra of the first potentially habitable super-Earth — GL581d. *Astrophys. J., 733(2),* 35.

Kasting J. F. (1997) Habitable zones around low mass stars and the search for extraterrestrial life. *Origins Life Evol. Biosph., 27(1–3),* 291–307.

Kasting J. F. and Catling D. (2003). Evolution of a habitable planet. *Annu. Rev. Astron. Astrophys., 41,* 429–463.

Kasting J. F., Whitmire D. P., and Reynolds R. T. (1993) Habitable zones around main-sequence stars. *Icarus, 101(1),* 108–128.

Kasting J. F., Whittet D. C. B., and Sheldon W. R. (1997) Ultraviolet radiation from F and K stars and implications for planetary habitability. *Origins Life Evol. Biosph., 27(4),* 413–420.

Kopparapu R. K. (2013) A revised estimate of the occurrence rate of terrestrial planets in the habitable zones around Kepler M-dwarfs. *Astrophys. J. Lett., 767(1),* L8.

Kopparapu R. K., Kasting J. F., and Zahnle K. J. (2012) A photochemical model for the carbon-rich planet WASP-12b. *Astrophys. J., 745,* 77, DOI: 10.1088/0004-637X/745/1/77.

Kopparapu R. K., Ramirez R., Kasting J. F., Eymet V., Robinson T. D., Mahadevan S., Terrien R. C., Domagal-Goldman S., Meadows V., and Deshpande R. (2013) Habitable zones around main-sequence stars: New estimates. *Astrophys. J., 765(2),* 131.

Kuchner M. J. and Seager S. (2005) Extrasolar carbon planets. *ArXiv preprints,* arXiv:astro-ph/0504214.

Léger A., Rouan D., Schneider J., et al. (2009) Transiting exoplanets from the CoRoT space mission. VIII. CoRoT-7b: The first super-Earth with measured radius. *Astron. Astrophys., 506,* 287–302, DOI: 10.1051/0004-6361/200911933.

Léger A., Grasset O., Fegley B., et al. (2011) The extreme physical properties of the CoRoT-7b super-Earth. *Icarus, 213,* 1–11, DOI: 10.1016/j.icarus.2011.02.004.

Leitzinger M., Odert P., Kulikov Y. N., et al. (2011) Could CoRoT-7b and Kepler-10b be remnants of evaporated gas or ice giants? *Planet. Space Sci., 59,* 1472–1481, DOI: 10.1016/j.pss.2011.06.003.

Levine H., Shaklan S., and Kasting J. (2006) *Terrestrial Planet Finder Coronagraph Science and Technology Definition Team(STDT) Report.* JPL Document D-34923, Pasadena, California.

Lineweaver C. H. and Chopra A. (2012) The habitability of our Earth and other Earths: Astrophysical, geochemical, geophysical, and biological limits on planet habitability. *Annu. Rev. Earth Planet. Sci., 40,* 597–623, DOI: 10.1146/annurev-earth-042711-105531.

Madhusudhan N. and Seager S. (2009) A temperature and abundance retrieval method for exoplanet atmospheres. *Astrophys. J., 707,* 24–39, DOI: 10.1088/0004-637X/707/1/24.

Madhusudhan N., Lee K. K. M., and Mousis O. (2012) A possible carbon-rich interior in super-Earth 55 Cancri e. *Astrophys. J. Lett., 759,* L40, DOI: 10.1088/2041-8205/759/2/L40.

Mason J., ed. (2008) *Exoplanets: Detection, Formation, Properties, Habitability.* Springer, Berlin (e-Book).

Mayor M., Bonfils X., Forveille T., Delfosse X., Udry S., Bertaux J.-L., Beust H., Bouchy F., Lovis C., and Pepe F. (2009) The HARPS search for southern extra-solar planets. *Astron. Astrophys., 507(1),* 487–494.

McKay C. P., Pollack J. B., and Courtin R. (1991) The greenhouse and antigreenhouse effects on Titan. *Science, 253(5024),* 1118–1121.

Miguel Y., Kaltenegger L., Fegley B., and Schaefer L. (2011) Compositions of hot super-Earth atmospheres: Exploring Kepler candidates. *Astrophys. J. Lett., 742,* L19, DOI: 10.1088/2041-8205/742/2/L19.

Moses J. I., Madhusudhan N., Visscher C., and Freedman R. S. (2013) Chemical consequences of the C/O ratio on hot Jupiters: Examples from WASP-12b, CoRoT-2b, XO-1b, and HD 189733b. *Astrophys. J., 763,* 25, DOI: 10.1088/0004-637X/763/1/25.

Mura A., Wurz P., Schneider J., Lammer H., et al. (2011) Comet-like tail-formation of exospheres of hot rocky exoplanets: Possible implications for CoRoT-7b. *Icarus, 211,* 1–9, DOI: 10.1016/j.icarus.2010.08.015.

Pavlov A. A., Brown L. L., and Kasting J. F. (2001) UV shielding of NH_3 and O_2 by organic hazes in the archean atmosphere. *J. Geophys. Res.–Planets, 106(E10),* 23267–23287.

Pierrehumbert R. T. (2010) *Principles of Planetary Climate.* Cambridge Univ., Cambridge.

Pierrehumbert R. and Gaidos E. (2011) Hydrogen greenhouse planets beyond the habitable zone. *Astrophys. J. Lett., 734(1),* L13.

Potter A. E. and Morgan T. H. (1987) Variation of sodium on Mercury with solar radiation pressure. *Icarus, 71,* 472–477, DOI: 10.1016/0019-1035(87)90041-8.

Ramirez R. M. and Kasting J. F. (2011) Greenhouse warming on early Mars and extrasolar planets plus a critique of the martian valley impact hypothesis. In *AGU Fall Meeting Abstracts,* Abstract #P24B-04.

Raymond S. N., Quinn T., and Lunine J. I. (2007) High-resolution simulations of the final assembly of Earth-like planets. 2. Water delivery and planetary habitability. *Astrobiology, 7(1),* 66–84.

Rogers L. A. and Seager S. (2010) Three possible origins for the gas layer on GJ 1214b. *Astrophys. J., 716,* 1208–1216, DOI: 10.1088/0004-637X/716/2/1208.

Schaefer L. and Fegley B. (2009) Chemistry of silicate atmospheres of evaporating super-Earths. *Astrophys. J. Lett., 703,* L113–L117, DOI: 10.1088/0004-637X/703/2/L113.

Schaefer L. and Fegley B. Jr. (2011) Atmospheric chemistry of Venus-like exoplanets. *Astrophys. J., 729,* 6, DOI: 10.1088/0004-637X/729/1/6.

Seager S., ed. (2011) *Exoplanets.* Univ. of Arizona, Tucson.

Seager S., Kuchner M., Hier-Majumder C. A., and Militzer B. (2008) Mass-radius relationships for solid exoplanets. *Astrophys. J., 669(2),* 1279.

Segura A. and Kaltenegger L. (2010) Search for habitable planets. In *Astrobiology: Emergence, Search and Detection of Life* (V. Basiuk, ed.), p. 500. American Scientific Publishers, Valencia, California.

Selsis F., Kasting J. F., and Levrard B. (2007) Habitable planets around the star Gliese 581? *Astron. Astrophys., 476(3),,* 1373–1387, DOI: 10.1051/0004-6361.

Shields A., Meadows V., Robinson T. D., Bitz C. M., Pierrehumbert R. T., and Joshi M. M. (2013) The effect of host star spectral energy distribution and ice-albedo feedback on the climate of extrasolar planets. In *221st American Astronomical Society Meeting Abstracts,* Abstract #333.07.

Soummer R., Cash W., Brown R. A., Jordan I., Roberge A., Glassman T., Lo A., Seager S., and Pueyo L. (2009) A starshade for JWST: Science goals and optimization. In *Techniques and Instrumentation for Detection of Exoplanets IV* (S. B. Shaklan, ed.), Proc. SPIE Vol. 7440, International Society for Optical Engineering.

Spencer J. R. and Schneider N. M. (1996) Io on the eve of the galileo mission. *Annu. Rev. Earth Planet. Sci., 24,* 125–190, DOI: 10.1146/annurev.earth.24.1.125.

Swift D. C., Eggert J. H., Hicks D. G., Hamel S., Caspersen K., Schwegler E., Collins G. W., Nettelmann N., and Ackland G. J. (2012) Mass-radius relationships for exoplanets. *Astrophys. J., 744,* 59, DOI: 10.1088/0004-637X/744/1/59.

Tajika E. (2008) Snowball planets as a possible type of water-rich terrestrial planet in extrasolar planetary systems. *Astrophys. J. Lett., 680,* L53–L56, DOI: 10.1086/589831.

Trainer M. G., Pavlov A. A., Curtis D. B., McKay C. P., Worsnop D. R., Delia A. E, Toohey D. W., Toon O. B., and Tolbert M. A. (2004) Haze aerosols in the atmosphere of early Earth: Manna from heaven. *Astrobiology, 4(4),* 409–19, DOI: 10.1089/ast.2004.4.409.

Trainer M. G., Pavlov A. A., DeWitt H. L., Jimenez J. L., McKay C. P., Toon O. B., and Tolbert M. A. (2006) Organic haze on Titan and the early Earth. *Proc. Natl. Acad. Sci., 103(48),* 18035–18042, DOI: 10.1073/pnas.0608561103.

Treiman A. H. and Bullock M. A. (2012) Mineral reaction buffering of Venus atmosphere: A thermochemical constraint and implications for Venus-like planets. *Icarus, 217,* 534–541, DOI: 10.1016/j.icarus.2011.08.019.

Udry S., Bonfils X., Delfosse X., et al. (2007) The HARPS search for southern extra-solar planets. XI. Super-Earths (5 and 8 M_{earth}) in a 3-planet system. *Astron.Astrophys., 469,* L43–L47, DOI: 10.1051/0004-6361:20077612.

Valencia D., O'Connell R. J., and Sasselov D. (2006) Internal structure of massive terrestrial planets. *Icarus, 181(2),* 545–554.

Valencia D., Sasselov D. D., and O'Connell R. J. (2008) Detailed models of super-Earths: How well can we infer bulk properties? *Astrophys. J., 665(2),* 1413.

Van Heck H. J. and Tackley P. J. (2011) Plate tectonics on super-Earths: Equally or more likely than on Earth. *Earth Planet. Sci. Lett., 310(3),* 252–261.

Von Bloh W., Bounama C., Cuntz M., and Franck S. (2007) The habitability of super-Earths in Gliese 581. *Astron. Astrophys., 476,* 1365–1371, DOI: 10.1051/0004-6361:20077939.

Von Paris P. (2010) *The Atmospheres of Super-Earths.* Technischen Universität, Berlin.

Wagner F. W., Tosi N., Sohl F., Rauer H., and Spohn T. (2012) Rocky super-Earth interiors: Structure and internal dynamics of CoRoT-7b and Kepler-10b. *Astron. Astrophys., 541,* A103, DOI: 10.1051/0004-6361/201118441.

Wordsworth R. and Pierrehumbert R. (2013) Hydrogen-nitrogen greenhouse warming in Earth's early atmosphere. *Science, 339(6115),* 64–67, DOI: 10.1126/science.1225759.

Wordsworth R. D., Forget F., Selsis F., Millour E., Charnay B., and Madeleine J. B. (2011) Gliese 581d is the first discovered terrestrial-mass exoplanet in the habitable zone. *Astrophys. J. Lett., 733,* L48, DOI: 10.1088/2041-8205/733/2/L48.

Wright J. T., Fakhouri O., Marcy G. W., Han E., Feng Y., Johnson J. A., Howard A. W., Fischer D. A., Valenti J. A., and Anderson J. (2011) The exoplanet orbit database. *Publ. Astron. Soc. Pac., 123(902),* 412–422.

Zahnle K. J. and Catling D. C. (2013) The cosmic shoreline (abstract). In *44th Lunar Planet. Sci. Conf.,* Abstract #2787. LPI Contrib. No. 1719, Lunar and Planetary Institute, Houston.

Zerkle A. L., Claire M. W., Domagal-Goldman S. D., Farquhar J., and Poulton S. W. (2012) A bistable organic-rich atmosphere on the Neoarchaean Earth. *Nature Geosci., 5(5),* 359–363. DOI: 10.1038/ngeo1425.

Baines K. H., Atreya S. K., Bullock M. A., Grinspoon D. H., Mahaffy P., Russell C. T., Schubert G., and Zahnle K. (2013) The atmospheres of the terrestrial planets: Clues to the origins and early evolution of Venus, Earth, and Mars. In *Comparative Climatology of Terrestrial Planets* (S. J. Mackwell et al., eds.), pp. 137–160. Univ. of Arizona, Tucson, DOI: 10.2458/azu_uapress_9780816530595-ch006.

The Atmospheres of the Terrestrial Planets: Clues to the Origins and Early Evolution of Venus, Earth, and Mars

Kevin H. Baines
University of Wisconsin-Madison

Sushil K. Atreya
The University of Michigan

Mark A. Bullock
Southwest Research Institute

David H. Grinspoon
Denver Museum of Nature and Science

Paul Mahaffy
NASA Goddard Space Flight Center

Christopher T. Russell and Gerald Schubert
University of California, Los Angeles

Kevin Zahnle
NASA Ames Research Center

We review the current state of knowledge of the origin and early evolution of the three largest terrestrial planets — Venus, Earth, and Mars — setting the stage for the chapters on comparative climatological processes to follow. We summarize current models of planetary formation, as revealed by studies of solid materials from Earth and meteorites from Mars. For Venus, we emphasize the known differences and similarities in planetary bulk properties and composition with Earth and Mars, focusing on key properties indicative of planetary formation and early evolution, particularly of the atmospheres of all three planets. Finally, we review the need for future *in situ* measurements for improving our understanding of the origin and evolution of the atmospheres of our planetary neighbors and Earth, and suggest the accuracies required of such new *in situ* data.

I. INTRODUCTION

1.1. Venus Spacecraft Data

In the last decade, the exploration of Venus and Mars has experienced a renaissance as space agencies in Europe, Japan, and America have developed and executed a number of missions to Earth's neighbors. Since April 11, 2006, the European Space Agency's (ESA) Venus Express mission has been in orbit, scrutinizing Venus from the ground up. As such, it has accurately determined the thermal structure at hundreds of radio occultation sites from pole-to-pole (e.g., *Tellman et al.,* 2008); followed the movements, spatial morphologies, and vertical structures of clouds (e.g., *Sánchez-Lavega et al.,* 2008; *Hueso et al.,* 2012; *Markiewicz et al.,* 2007; *Titov et al.,* 2012; *McGouldrick et al.,* 2008, 2012; *Barstow et al.,* 2012); studied the composition of reactive and dynamically-diagnostic gases in the lower and upper atmosphere and their spatial and temporal variability (e.g., *Belyaev et al.,* 2008; *Bézard et al.,* 2008; *Marcq et al.,* 2008; *Irwin et al.,* 2008; *Piccioni et al.,* 2008; *Tsang et al.,* 2008, 2009; *Vandaele et al.,* 2008; *Cottini et al.,* 2012); and obtained constraints on the composition of its surface and its spatial variability, particularly associated with major geologic features (e.g., *Mueller et al.,* 2008; *Smrekar et al.,* 2010). Venus Express has also obtained solid evidence for lightning (*Russell et al.,* 2007, 2008) and has sensed and characterized the solar-induced leakage of planetary materials — in particular, H and O — into space (*Barabash et al.,* 2007; *Luhmann et al.,* 2008; *Fedorov et al.,* 2011). As this mission winds down in 2015, current plans are to perform a second attempt to place the Japanese Akatsuki spacecraft [also known as Venus Climate Orbiter (VCO)] in orbit, to continue studies of Venus' circulation and dynamics, particularly the planet's global superrotating wind structure.

1.2. Mars Spacecraft Data

The twenty-first century has brought an abundance of both orbiting and surface explorers to Mars, with several additional missions planned for launch by 2020. Starting in early 2004, three heavily instrumented rovers have explored the Red Planet, conducting detailed observations of the planet's geology, surface properties, atmospheric composition, and climate, in particular sampling surface materials and atmospheric trace gases for signs of ancient water and other biologically related materials. Two Mars Exploration Rovers landed on January 3 and 25, 2004, and quickly discovered sulfur-bearing minerals formed within ancient standing bodies of water on the planet's surface (e.g., *Squyres and Knoll,* 2005). On August 6, 2012, the Curiosity rover began exploring the strata of the 5.5-km-tall layered Mount Sharp to elucidate the geologic history of the planet. The Sample Analysis at Mars (SAM) instrument (*Mahaffy et al.,* 2012) onboard Curiosity is of particular interest, as it acquires *in situ* samples of a host of atmospheric gases, including noble gases and their isotopes, light isotopes of C, O, and N, and methane [previously reported by *Mumma et al.* (2009), *Krasnapolsky et al.* (2004), and *Formisano et al.* (2004), but still a matter of much debate], all to assess Mars' history — its origin, evolution, and biological record (*Grotzinger et al.,* 2012). Since their arrival in December 2003, October 2001, and March 2006, respectively, three orbital missions — ESA's Mars Express and NASA's Mars Odyssey and Mars Reconnaissance Orbiter missions — have been continuously surveying the Red Planet for, in particular, evidence of past and current surface and/or subsurface water (*Squyres et al.,* 2004; *Knoll et al.,* 2008; *Arvidson et al.,* 2010; *Cull et al.,* 2010; *Picardi et al.,* 2005; *Feldman et al.,* 2002; *Boynton et al.,* 2002; *Mitrafanov et al.,* 2002).

The next decade is expected to witness the arrival at Mars of a number of missions launched by a variety of the world's space agencies. In September 2014, NASA's Mars Atmosphere and Volatile Evolution (MAVEN) mission is scheduled to enter orbit, with a principal objective of exploring Mars' climate history through measurements of ionospheric properties and by sampling trace materials in the upper reaches of the atmosphere to determine the rate of leakage into space, over the eons, of CO_2, N_2, Ar, and H_2O (*Jakosky,* 2011). Other missions from Europe and Asia to search for signs of life are the Indian Mangalyaan mission planned for launch in November 2013, and the joint ESA/Russian Federal Space Agency dual-launch mission currently planned for liftoff in 2016 and 2018. The 2016 launch will (1) deliver the Trace Gas Orbiter to globally map important minor species, in particular, gaseous methane, in the atmosphere, and (2) land the Entry, Descent and Landing Demonstrator Module (EDM) that will characterize the local dust environment. The 2018 launch will deliver a rover via a Russian-developed landing system. Meanwhile, the U.S. is set to launch the InSight surface lander mission in 2016 to investigate the planet's interior, principally via seismometry and heat flow measurements. NASA is also developing plans to launch a follow-on nuclear-powered Curiosity-style rover mission in 2020 that will continue the long-term exploration of the martian surface.

1.3. Future Exploration of the Origin and Early Evolution of the Terrestrial Planets

A major impetus for the continued revival of Venus exploration and a major expansion of Mars exploration — including proposed sample return missions — has come from the U.S. planetary science community. In 2003 and 2011, under the auspices of the Space Studies Board of the National Research Council, the community's Solar System Exploration Survey (SSES) produced "Decadal Surveys" summarizing and prioritizing scientific objectives and missions to the planets for the ensuing decade. A major finding of both reports (*National Research Council,* 2003, 2011) is the need for direct, *in situ* sampling of both Mars and Venus, both their atmospheres and surfaces. For Mars, many of the *in situ* rover, lander, and orbiter experiments currently being conducted are direct and valuable responses by NASA and other space agencies across the world to such community recommendations.

Extraordinary *in situ* experiments at the surface of Venus by the Venera and Vega landers in the 1970s and 1980s have provided the basis of what we know about the elementary composition of the surface (*Surkov et al.,* 1984; *Barsukov et al.,* 1986; *Barsukov,* 1992). In the same era, Pioneer Venus (*Oyama et al.,* 1980) and the Vega probes (*Krasnopolsky,* 1989) directly sampled atmospheric gases and clouds successfully above 22 km altitude. In the 1990s, fundamental new measurements were provided during the fleeting (less than a few hours) Galileo and Cassini flybys, including the first close-up near-infrared images and spectra of Venus' lower atmosphere and surface(e.g., *Carlson et al.,* 1991, 1993a,b; *Collard et al.,* 1993; *Baines et al.,* 2000; *Hashimoto et al.,* 2008). The ongoing Venus Express, in orbit since April 2006, has been sending back data nearly continuously, with over 4 terrabits of science measurements returned as of mid 2013, much of it composed of near-infrared imagery and spectra pioneered by Galileo and Cassini. The Venus Express mission is currently expected to end in 2015 upon the exhaustion of the spacecraft's fuel, needed to maintain orbit. Near that time, Akatsuki, Japan's first spacecraft to Venus, is currently planned to make a second attempt to enter orbit, having failed on its first attempt in December 2010. Taking the reins from Venus Express, Akatsuki will continue to monitor Venus' sulfur-based chemistry and circulation, particularly in the middle cloud layer.

Despite the successes and expectations of the flybys of the 1990s and the orbiters of the early twenty-first century, we have yet to fly experiments that provide the necessary accuracy in *in situ* sampling and other measurements to answer fundamental questions dealing with the origin and evolution of Venus, including the role of the local solar nebula in creating planets inside Earth's orbit. Required measurements include those currently being conducted on Mars by SAM: the abundances and isotopic ratios of inert noble gases — in particular the heavier constituents Xe and Kr — and isotopic ratios of the light elements C, N, and O. As is the case for Mars, such measurements bear the telltale fingerprints of the planet's origin and evolution, as is well illustrated in the remainder of this chapter.

1.4. Overview

In section 2, we present a general review of the state of knowledge of the origin and evolution of the trio of rocky, atmosphere-enshrouded planets Earth, Mars, and Venus. Here, we focus on what is known based on solid samples collected on Earth, including information from asteroidal and martian meteorites. Additional information comes from lunar samples.

As noted above, fundamental additional evidence comes from atmospheric samples, in particular from noble gases, their isotopes, and the isotopes of light gases. In section 3 we use Venus as a case study, examining what is known from spacecraft data of its isotopic record. We note that compared to Earth and Mars, Venus is rather poorly covered; only Ar and bulk Ne have thus far been adequately measured. No solid samples of Venus have been returned to Earth or are known in our meteorite collections, although γ-ray spectrometers on the Vega landers and Veneras 8, 9, and 10 measured the abundances of the radioactive isotopes of K, Th, and U (*Surkov et al.,* 1987). The available data lead to few constraints, allowing several possible hypotheses on the origin and early evolution of the venusian atmosphere. Since the origin of Earth's atmosphere is itself poorly understood, discovering isotopic evidence that supports a unique theory for the origin and evolution of Venus' atmosphere will likely have very large implications for understanding the origin of our own atmosphere.

Section 4 describes how noble gases in planetary atmospheres systematically constrain their origins and the processes that alter relative isotope abundances. These include atmospheric loss to space, interaction with the solar wind, interactions with the surface and interior, and impacts from space. Each process alters the abundance of noble gases in the atmosphere differently, and each process imprints a unique fractionation pattern of noble-gas isotopes. Radiogenic noble-gas isotopes reflect processes that occurred over one or two half-lives of the parent species, and therefore a suite of isotopes sensitive to different timescales can be used to probe the evolution of planetary atmospheres.

In section 5 we discuss what the totality of spacecraft and groundbased data and theoretical models say about the long-term evolution of terrestrial planetary climates. By acquiring the right kinds of data from Venus and Mars in the future, it will be possible to reconstruct the climate evolution of both these planets in comparison with Earth's. When we begin to characterize the atmospheres of planets around other stars, it will be crucial to have this general understanding of planetary climate evolution in order to understand what these data are telling us about planetary habitability.

2. TOWARD UNDERSTANDING PLANETARY ORIGIN AND EVOLUTION

2.1. Initial Formation

As byproducts of the formation of the Sun, the formation and early evolution of the inner planetary trio Venus, Earth, and Mars are thought to largely follow the development of our central star. According to the standard model of planetary formation (*Wetherill,* 1990), the Sun formed out of a dense interstellar molecular cloud comprised predominantly of H, He, H_2O, and refractory materials (e.g., Fe,Mg-silicates and metallic Fe) that crystallized in the form of small-particle (<0.1 µm) dust. As the cloud collapsed due to gravitational instabilities, conservation of angular momentum dictated that some 2–10% of the gas and dust maintain the bulk of the system's angular momentum by orbiting the developing Sun in a flattened disk. Within this "solar nebula," over the next ~10^5 years collisions between dust particles created a large population (~10^{12}) (*Greenberg et al.,* 1978) of kilometer-sized rocky planetesimals. Accretion then proceeded mainly due to mutual gravitational perturbations among these bodies, creating, in another ~10^5 years, approximately 10^3 moon-sized (10^{25}–10^{26} g) "planetary embryos" in the inner 3 AU or so (e.g., models of *Greenberg et al.,* 1978). Due to further mutual gravitational interactions, over the next 10^7–10^8 years several hundred of the planetesimals then coalesced to form the bulk of the 1.18×10^{28} g composing the masses of the three terrestrial planets we see today (e.g., *Wetherill,* 1980). In this scenario, the gaseous component of the nebula was swept out of the inner solar system by the powerful stellar winds of the pre-main-sequence T-Tauri phase of the young Sun. This occurred during the first ~3 m.y., approximately 2 m.y. prior to the establishment of the dominant embryos of the two largest terrestrial planets, Earth and Venus, in their near-circular, low-inclination orbits (e.g., *Haisch et al.,* 2001).

Stellar evolution models predict that the T-Tauri phase operates until the star's convective envelope is depleted, at which point the star joins the main sequence, about 12 m.y. after stellar nuclear ignition (*Zahnle and Walker,* 1982). As discussed in more detail in section 4.2, if the planets were formed before the end of the Sun's T-Tauri phase, their atmospheres would have been substantially lost due to the extreme T-Tauri winds at that time. The resulting atmospheres seen today would therefore consist predominantly of outgassed volatiles from the mantle subsequent to the end of the T-Tauri phase. Since ^{129}I, the parent isotope of ^{129}Xe, has a half-life of 15.7 m.y., i.e., close to the duration of the T-Tauri phase, the atmosphere that accumulated after the end of the T-Tauri phase should be depleted in ^{129}I-produced ^{129}Xe relative to the unaltered mantle. Indeed, this is what is observed for Earth (*Allegre and Schneider,* 1994). This in turn implies that the primary atmospheres of the terrestrial planets were subjected to ultraviolet fluxes 10^4 times greater than today (*Zahnle and Walker,* 1982), and thus, along with dramatically enhanced escape rates, also experienced significantly increased photochemical reactions that would have dramatically altered the chemistry and compositional character of the original atmospheres.

A significant variant in the standard model is the Kyoto model (*Hayashi et al.,* 1985), wherein the loss of nebular gas in the inner solar system occurs after, and not before, the final stages of terrestrial planet formation. In this gas-rich scenario, the atmosphere of Earth is predicted to be 10^5 more massive than the present atmosphere (*Hayashi et al.,* 1979). An alternative theory to the standard model is one invoking "gaseous giant protoplanets" that form in parallel to the formation of the Sun, as a result of massive gas-dust instabilities in the solar nebula (e.g., *von Weizsäcker,* 1944; *Kuiper,* 1951; *Cameron,* 1978). In both of these scenarios, the massive primordial atmospheres of the terrestrial planets are mostly H_2 with smaller (but still substantial) amounts of H_2O and CO_2.

2.2. Clues from ^{182}Hf-^{182}W Isotopes

Observational evidence of the timescales for Earth's formation and early differentiation are provided by isotopic variations produced by the decay of extinct nuclides, in particular the decay of the short-lived (9-m.y. half-life) hafnium isotope ^{182}Hf to the tungsten isotope ^{182}W (e.g., *Harper and Jacobsen,* 1996; *Jacobsen,* 2005; *Kleine et al.,* 2009). Given the affinity of Hf for O, Hf remained enriched in Earth's silicate mantle during differentiation, in contrast to the denser, more iron-loving W compounds that sank to the core. Thus, a relatively high value of ^{182}W abundance in the crust/mantle compared to the abundance of nonradiogenic W, as observed [with more than 90% of terrestrial W going into Earth's core during formation (*Halliday,* 2000b)], indicates that significant ^{182}W was formed after core differentiation was well along, but before ^{182}Hf became extinct. The short ^{182}Hf decay half-life thus implies that Earth experienced early and rapid accretion and core formation that lasted just ~30 m.y., with most of the accumulation occurring in ~10 m.y. (*Jacobsen,* 2005).

For Mars, W-isotopic measurements from martian meteorites indicate a significantly shorter accretion and core-formation period, perhaps as short as 15 m.y. (*Lee and Halliday,* 1997; *Halliday,* 2001), about one-half the period noted above and one-fourth as long as their own analyses for Earth. Since Mars has about one-eighth the mass of Earth, this result implies that the two planets had similar accretion rates during the Mars formation period. Noting a correlation between Th/Hf and $^{176}Hf/^{177}Hf$ in chondrites, however, *Dauphas and Pourmand* (2011) estimated the mobility of Th in martian meteorites. This permitted a much more accurate accretion rate from the Hf/W ratio, indicating that Mars accreted to half its final size in 1.8(+9,–1) m.y., again consistent with the notion that Mars and Earth had similar accretion rates during the relatively short Mars formation period. Thus, rather than the rate of accretion, a major difference in formation between the two planets is the longer period of accretion for Earth.

Meteorites from Mars exhibit marked heterogeneities in W-isotopic abundances, which then places an upper limit on the date of the last global-scale impact on Mars. These heterogeneities are significantly larger than found on Earth, where mantle convection over the eons has smoothed Hf-W variability (*Halliday et al.,* 2000b). Indeed, the earliest isotopic heterogeneities on Earth are less than 2 b.y. old, while the martian mantle heterogeneities must have occurred during the first 30 m.y. of the solar system. Thus, unlike Earth, Mars suffered no major, moon-forming impacts sufficiently energetic to effectively homogenize the isotopic W composition later than 4.53 Ga (*Lee and Halliday,* 1997). These heterogeneities also indicate that large-scale convective mantle-mixing processes, such as those that drive Earth's plate tectonics, are absent on Mars. While the large shield constructs such as Tharsis and Elysium are indicative of some convective upwelling, there is little observational evidence for large-scale convective overturn (*Breuer et al.,* 1997), and there are few if any plate-tectonic-like features (*Sleep,* 1994).

2.3. Evidence and Impact of Large Collisions Near the End of Planetary Formation

Many of the unique characteristics of the planets we see today are thought to result from unpredictable, stochastic events near the end of their formation processes — such as the "giant impacts" of the last protoplanetary collisions likely experienced by each planet (*Kaula,* 1990). For example, the prevailing theory of lunar genesis holds that Earth's Moon was formed from the collision of a Mars-sized object with the forming Earth some 50–70 m.y. after the start of planetary formation (e.g., *Boss,* 1986; *Stevenson,* 1987; *Newsom and Taylor,* 1989; *Benz et al.,* 1986, 1987, 1989; *Cameron and Benz,* 1991; *Halliday,* 2000a,b) after 50–95% of Earth was accreted (*Halliday,* 2000a; *Canup,* 2004). A somewhat earlier date of ~30 m.y. after the start of planetary formation is reported by *Jacobsen* (2005) from analysis of Hf-W data from Apollo lunar samples.

A number of the Moon's gross features, such as its low bulk Fe abundance, the large depletion of volatile elements, and the similarity of its overall silicate chemistry to that of Earth's mantle, are natural consequences of the giant collision model (*Hartmann,* 1986; *Wänke and Dreibus,* 1986). This giant collision also likely created a global terrestrial magma ocean (*Melosh,* 1990; *Benz and Cameron,* 1990; *Stevenson,* 1987) from the melting of 30–55% of Earth (*Tonks and Melosh,* 1993) to a depth of ~2000 km (*Canup,* 2008) and, perhaps along with other major collisions, helped tear away Earth's original atmosphere (*Cameron,* 1983).

A persistent problem with this picture has been the similarity of O isotopes in the mantle silicates of Earth and the Moon. Impact simulations show that most of the Moon's mantle should consist of the mantle of the impactor (*Canup,* 2004). However, isotopic ratios of O vary smoothly with distance from the Sun, reflecting the temperature regime during silicate condensation. A large, Mars-sized impactor would have to have come from outside Earth's accretional feeding zone, and therefore should have had a different isotopic O pattern than that found in Earth's mantle (*Canup,* 2004). Recently, impact simulations incorporating smooth particle hydrodynamics have shown that an oblique impact by a larger impactor would have led to more mantle mixing and a thorough homogenization of mantle material, including O isotopes (*Canup,* 2012). This oblique impact likely gave Earth a rapid rotation rate (*Cameron and Ward,* 1976), subsequently altered by both tidal interactions between the Moon and Earth and an orbital resonance between the Sun and the Moon (*Cuk and Stewart,* 2012), that slowed Earth's spin from its initial ~4.1-hour period (e.g., *Dones and Tremaine,* 1993) to the 24-hour period we see today.

On Venus, the last impact by a large planetesimal is thought to be responsible for the planet's unusually slow — and retrograde — spin rate, resulting in a solar day that is equivalent to 117 Earth days between sunrises and — due to its backward spin — a sidereal day 18 days longer than its orbital period of 224.7 days. This exceedingly slow rate of rotation may be the reason why Venus does not have an intrinsic magnetic field, although *Stevenson* (2003) concludes that slow rotation may be more favorable for a magnetic dynamo than fast rotation. Whether due to the lack of plate tectonics (discussed below) or its spin state, Venus' lack of a magnetic field exposes the planet to the ravages of the solar wind, which has proceeded to drain it of most of its H and H_2O. This turned the planet into a dry world of sulfur-based clouds and meteorology and starved it of the lubricating and chemical effects of mantle water needed to operate major Earth-like geologic processes, such as plate tectonics, resulting in a geologic history radically different from Earth's.

2.4. From Oceans of Magma to Seas of Water

For all the terrestrial planets, the immense amount of gravitational energy released by accretion during planetary formation resulted in temperatures sufficiently hot to cause global melting, the differentiation of materials, and the formation of a hot, dense core. Core formation involves a positive feedback whereby the mantle is heated by the release of gravitational potential energy from the fall of Fe and other dense materials to the center. The warmer, more fluid mantle speeds the collection of these dense materials at the core, releasing yet more potential energy (*Turcotte and Schubert,* 1982). Several additional mechanisms may have contributed to the trapping of heat to prolong near-surface melting to create a variety of possible magma oceans (*Abe,* 1997), including (1) the blanketing effect of a protoatmosphere (e.g., *Abe and Matsui,* 1985, 1986; *Hayashi et al.,* 1979; *Matsui and Abe,* 1986; *Zahnle et al.,* 1988), resulting in a surface ocean of magma in chemical equilibrium with the atmosphere, and (2) the deposition of underground heat by planetesimal impacts, producing a subsurface magma ocean (e.g., *Safronov,* 1969, 1978; *Kaula,* 1979; *Davies,* 1985; *Coradini et al.,* 1983). As noted for the Moon-forming collision, any giant impact would likely generate a completely molten deep magma ocean (e.g., *Melosh,* 1990), with temperatures exceeding several thousand Kelvin over the entire Earth. Indeed, this may have happened several times to Earth during its latter stages, forming a magma ocean each time (*Wetherill,* 1988). A recent interpretation of Earth's Ne- and Xe-isotopic record, however, implies that these impact events never succeeded in rehomogenizing the mantle (*Mukhopadhyay,* 2012). Regardless, without an atmospheric blanketing effect, the deep magma ocean would be transient, cooling via radiation into space and solidifying the lower mantle within a few thousand years (*Abe,* 1997).

With an atmospheric blanketing effect — due to either the gravitationally captured solar-type, H_2-dominated protoatmosphere or the impact-produced steam atmosphere — a magma ocean can be sustained for ~5 m.y., as indicated by the analytical modeling of *Elkins-Tanton* (2008) for both Earth and Mars. These models include the effects of (1) the partitioning of H_2O and CO_2 between mineral assemblages that accumulate in the solidifying mantle, (2) evolving liquid compositions, and (3) a growing atmosphere. Salient results are that mantle solidification is 98% complete in less than 5 m.y. for all magma oceans investigated for both planets, and is less than 100,000 years for low-volatile magma oceans. For all cases, Elkins-Tanton found that subsequent cooling to clement surface conditions occurs in 5–10 m.y.

The modeling results of Elkins-Tanton further indicate that, for both Earth and Mars, atmospheres significantly thicker than those that exist today could have been created through solidification of a magma ocean with low initial volatile content. Mars is relatively dry today because as much as 99% of Mars' original volatile content is thought to have been lost by 3.8 Ga through hydrodynamic escape, impacts, and sputtering by solar wind.

Direct measurements of terrestrial materials largely confirm these theoretical results, which depict a rapid evolution to clement surface conditions able to support a hydrosphere. While no rocks exist from the "dark ages" of the Hadean era 4.0–4.5 Ga (*Harrison et al.,* 2008; *Harrison,* 2009) — due presumably to the late heavy bombardment ca. 3.9 Ga (*Gomes et al.,* 2005) — detrital zircons from this period have been found in metamorphosed sediments at Jack Hills in western Australia. These hydrologically formed minerals were created by the remelting and resolidification of ocean sediments to granite and zircon, dating variously to 4.404 ± 8 Ga (*Wilde et al.,* 2001; *Peck et al.,* 2001) and ~4.36 Ga (*Harrison et al.,* 2008). Additional evidence for their hydrologic heritage comes from two measures of temperature derived from the zircons: (1) the crystallization temperature of Ti found within the zircons, which are found clustered over a narrow range of temperatures, 680 ± 25°C (*Watson and Harrison,* 2005, 2006), and (2) the high ^{18}O content of zircons indicative of relatively cool temperatures (*Valley et al.,* 2002). Both results substantiate the conclusion that the zircons encountered wet, minimum-melting conditions during their formation. Together, these results reveal that Earth cooled rapidly enough to form both continental crust and liquid water oceans by 4.36–4.40 Ga, just ~160–210 m.y. after the condensation of the first solid particles in the solar system at 4.56 Ga (*Valley et al.,* 2002) and less than 140 m.y. after the Moon-forming impact (*Touboul et al.,* 2007). The crust itself may have started forming much earlier: prior to 4.5 Ga, as inferred by the crustal-type Lu/Hf environment of zircons determined by *Harrison et al.* (2008), just ~60 m.y. after the solid-particle formation in the solar nebula. By ~4.36 Ga, the crust had evolved to the point of taking on continental characteristics (*Harrison et al.,* 2008), including the current pattern of crust formation, erosion, and sediment recycling behavior of plate tectonics (*Watson and Harrison,* 2005).

Two mechanisms have been noted to explain this early appearance of a water ocean on terrestrial planets (*Elkins-Tanton,* 2011). First, a solidifying magma ocean may be water-enriched to the point that excess water extrudes onto the planet's surface at the end of the period of silicate mineral solidification. Second, a water ocean may form when a thick supercritical fluid and steam atmosphere collapses into a water ocean upon cooling past the critical point of water (*Abe,* 1997; *Elkins-Tanton,* 2008).

According to the analysis of *Elkins-Tanton* (2011), the mechanism of water extrusion from solidifying magma may occur for planets with water comprising 1–3% of its bulk mass, while planets with an initial water content of 0.01% by mass or even somewhat less can form oceans hundreds of meters deep from a collapsing steam atmosphere. The Elkins-Tanton models include conditions for Earth, where the total mass of frozen and liquid water on the planet's surface today is about 1.4×10^{21} kg, while the mass of Earth's mantle is 6.0×10^{24} kg. Thus, in bulk, water in the silicate Earth is at least 0.02% by mass, or 200 ppm (*Elkins-Tanton,* 2008). Although there are good constraints on the water abundance of the upper mantle (*Saal et al.,* 2002), the total concentration of water in Earth's interior is poorly contrained but undoubtedly somewhat larger than 200 ppm. These models also should apply to both Venus and Mars, since they likely formed out of similar water-rich chondritic material that largely formed Earth, containing up to 20% water by mass (*Wood,* 2005).

These models then indicate that water oceans may have formed on Earth, Venus, and Mars early in each of their histories. Also, as noted by *Elkins-Tanton* (2011), water oceans may be commonplace on rocky planets throughout the universe, produced by the collapse of their steam atmospheres within tens to hundreds of millions of years of their last major accretionary impact.

As discussed in the chapter by Bullock and Grinspoon (this volume), the lifetime of such oceans is uncertain, depending on atmospheric escape mechanisms, including atmospheric erosion by stellar fluxes of ultraviolet light and charged particles, and blowoff by large residual impacts. On Earth, the late bombardment between 3.8 and 4.0 Ga — during which time the impact rate was 10^3 greater than today (*Valley et al.,* 2002) — may have temporarily vaporized a significant fraction of Earth's oceans. However, the constancy of the $^{18}O/^{16}O$-isotopic ratio found throughout the Archeaen (4.4–2.6 Ga) (*Valley et al.,* 2002) suggests that conditions conducive to oceans may not have been entirely eliminated from the globe during the bombardment period.

2.5. The Dawn of Life

The existence of liquid water around 4.4 Ga has important implications for the evolution of life (*Sleep et al.,* 2011). Ancient sediments that have undergone some metamorphism and carbonaceous materials — including carbonaceous inclusions within grains of apatite in West Greenland (*Moorbath et al.,* 1986) — 3.8 G.y. old have been found to hold distinctively biogenic C-isotope ratios (*Hayes et al.,* 1983; *Schidlowski,* 1988; *Mojzsis et al.,* 1996), suggesting that life arose near the conclusion of the late heavy bombardment (>3.8 Ga). Microfossils as old as 3.5 Ga (*Schopf,* 1993) show structural complexity, indicating that photosynthetic life able to leave durable fossils may have evolved over ~0.3 G.y. or perhaps longer (Fig. 1). Studies of ancient metasediments and their fossils are notoriously difficult, however, and both the C-isotope and microfossil records have been challenged (e.g., *Brasier et al.,* 2002). As noted by *Wilde et al.* (2001), primitive life may have arisen in ancient oceans as early as ~4.4 Ga, but may have been globally extinguished and subsequently arose more than once by the high rate of impacts during the late heavy bombardment (*Sleep and Zahnle,* 1998). Alternatively, as suggested by the ^{18}O analysis of *Valley et al.* (2002) for oceans, life may have survived the bombardment cataclysm. This view is consistent with the time of molecular divergence among archaebacteria determined by *Battistuzzi et al.* (2004), which is consistent with 4.1 Ga. Thus life may have risen within 0.3 Ga after standing bodies of water appeared on Earth.

Figure 2 summarizes the current understanding of the possible eras of oceans based on the bombardment history of Earth. During periods of low bombardment, i.e., within approximately two orders of magnitude of today's rate of about one major impact every 50 m.y., conditions would be sufficiently clement to sustain oceans. Such relatively quiescent, cool, wet periods appear likely not only after the late heavy bombardment that occurred from 3.8 to 4.0 Ga,

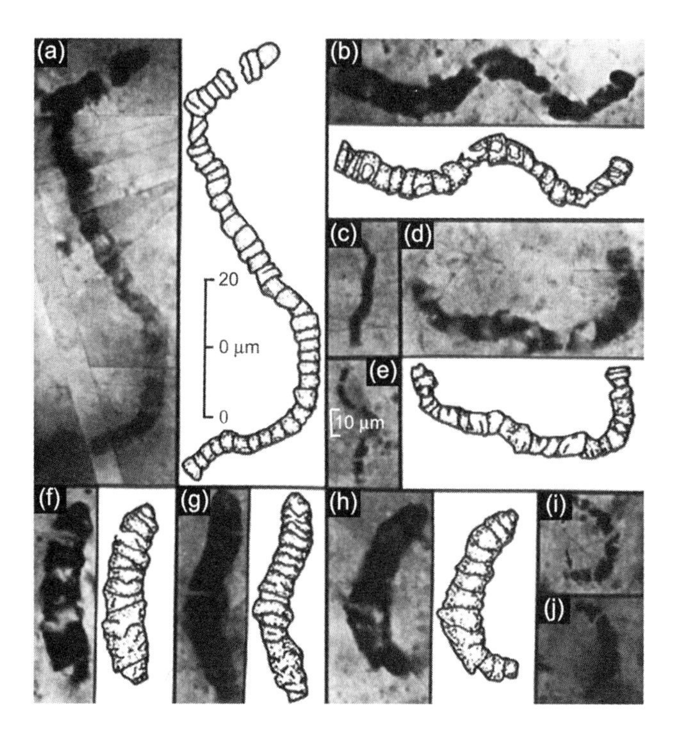

Fig. 1. Microfossils and interpretive drawings of the earliest life found on Earth. These carbonaceous fossils are from the Apex chert in North Pole, Australia, and are 3.465 G.y. old. These multicellular organisms were cyanobacteria and thus harnessed solar energy through photosynthesis. Fossils seen in **(a)**, **(b)**, **(c)**, **(d)**, and **(e)** are archea, Primaevifilum amoenum; those seen in **(f)**, **(g)**, **(h)**, **(i)**, and **(j)** are Primaevifilum conicoterminatum. Adapted from *Schopf* (1992).

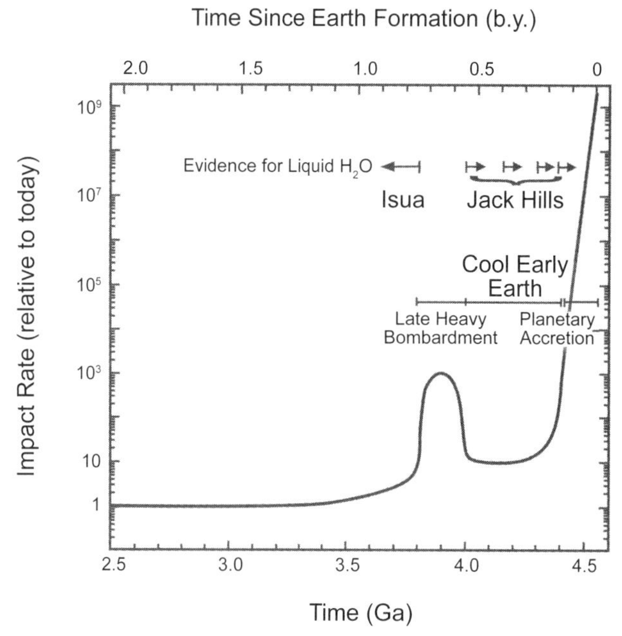

Fig. 2. The estimated meteorite-impact rate during the first 2 b.y. of Earth's history, as proposed by *Valley et al.* (2002). The cool early Earth hypothesis of Valley et al. indicates that the meteoritic flux dropped precipitously over the ~0.15-G.y. period of accretion, so that by 4.3 Ga, Earth was cool enough to support liquid water. Evidence for clement conditions during this era before the late heavy bombardment at ~3.9 Ga comes from the analysis of ^{18}O in zircon samples. Post late heavy bombardment, "Isua" denotes the oldest known water-lain chemical sediments, found in the Isua supracrustal belt in western Greenland, dating to 3.6–3.85 Ga (*Nutmann et al.,* 1997; *Whitehouse et al.,* 1999). Adapted from *Valley et al.* (2002).

but also during the ~0.4-G.y. period from 4.4 to 4.0 Ga. As noted by *Valley et al.* (2002), it may then be a misnomer to include this 0.4-G.y. era in the Hadean ("hell-like") period from 4.0–4.5 Ga. Indeed, given the real possibility and importance of the birth of life on Earth during this period, "The Arcadian" may be a more appropriate title.

2.6. Planetary Samples

Our picture of the early history of Earth and Mars is beginning to be completed in some detail, due largely to the existence of solid materials such as martian meteorites embedded with fruitful clues to their conditions of formation. The mild gravity and low atmospheric pressure of Mars favor the expulsion of surface materials into space by kilometer-sized impactors (*Melosh,* 1988). Even more fortuitously, gravitational perturbations due to Jupiter tend to steer such ejected material into Earth-crossing trajectories (*Gladman,* 1997). The process is so efficient that meteorites from Mars are in transit to Earth for an average of only 15 m.y., based on cosmic-ray-exposure ages. The isotopic ratios of gases captured within glassy inclusions of these meteorites match the atmosphere of Mars precisely, as measured by the Viking landers (*Clark et al.,* 1976). These samples from our neighboring planet have taught us much about the petrology and mineralogy of martian rocks, as well as about processes that have altered them since they were formed.

Venus, for which we have no known solid samples in hand, is more poorly constrained. The extreme conditions of its surface — with temperatures exceeding 470°C and pressures greater than 94 bar — together with its significant gravity (surface gravity ~90% of Earth) prevent the ready return of rock samples via surface lander missions. In addition, we know of no Venus rocks in our meteorite collections. Thus, to paint the picture of Venus' early history, we are left with the analysis of its bulk properties — including its lack of both a moon and a magnetic field, and the unusually slow speed and retrograde direction of its rotation — and with the interpretation of atmospheric samples acquired by planetary probes, all of which, of course, aid in our understanding of Earth and Mars as well.

3. THE EARLY EVOLUTION OF VENUS: EVIDENCE FROM BULK PROPERTIES AND RADAR IMAGERY

Venus and Earth differ by less than 20% in their size, mass, and bulk density and are at comparable distances from the Sun, suggesting that these two planets had similar early histories of accretion and internal differentiation. Nevertheless, Venus and Earth proceeded along different evolutionary paths, as is evident from a comparison of the planets' satellites (*Alemi and Stevenson,* 2006), magnetic fields (*Nimmo and Stevenson,* 2000), tectonics (*Solomon et al.,* 1992), rotational states (*Del Genio and Suozzo,* 1987), and atmospheres (*Kasting,* 1988).

3.1. Natural Satellites

For reasons that are poorly understood, Venus lacks a moon. This could simply be by chance, of course, but as noted earlier, Venus was likely subject to major impacts in its early evolution comparable to Earth's Moon-forming impact. The lack of a venusian moon has been attributed to the inward spiraling of a previously existing satellite that eventually collided with the planet. Satellite escape is another possible explanation (*Stevenson,* 2006). However, neither hypothesis is particularly plausible. The inward-spiraling scenario is unlikely to produce Venus' retrograde motion, since such a satellite would likely have been in a prograde orbit. The escape scenario is unlikely for reasonable Q values and satellite masses (*Stevenson,* 2006).

3.2. Planetary Magnetic Field

Today, Venus lacks an intrinsic magnetic field (*Russell,* 1980; *Phillips and Russell,* 1987; *Donahue and Russell,* 1997; *Nimmo,* 2002), as may reasonably be explained through an analysis of the planet's thermal evolution. Like Earth, Venus was probably initially hot and differentiated, and thus also probably had a liquid metallic core that, to some extent, likely still exists today. An indication that at least some portion of Venus' core is currently still liquid is provided by the tidal Love number, 0.295 ± 0.066, determined from Doppler radio tracking of Magellan and Pioneer Venus (*Konopliv and Yoder,* 1996; *Sjogren et al.,* 1997; *Yoder,* 1997). Hence it is quite possible that the planet had an early dynamo-generated magnetic field that later disappeared, its lifetime dependent upon the cooling history of the core. In particular, for a magnetic field to exist, the rate of core cooling, controlled by the mantle's ability to extract heat from the core, must have been strong enough to support convection and dynamo action in the core. One explanation then for the lack of a magnetic field today is that Venus underwent a transition in the efficiency of the core to transfer heat, as has been proposed as the explanation for the lack of a martian magnetic field (*Nimmo and Stevenson,* 2000). Venus potentially evolved from efficient core cooling by a plate-tectonic-like style of mantle convection to an inefficient style of stagnant or sluggish lid mantle convection that cooled the core slower than required to support dynamo activity. One potential proof of the existence of an ancient dynamo would be the detection of remanent magnetization from this epoch. However, since Venus' high surface temperature is close to the Curie point of most natural ferromagnetic minerals, it seems unlikely that surface materials could have retained such evidence over the eons.

A second explanation for the lack of a magnetic field is also related to inefficient core cooling by the mantle and the lower pressure in Venus' core compared with the pressure in Earth's core, due to the smaller mass of Venus. It is believed that dynamo action in Earth's core is substantially facilitated by compositional convection driven by inner core solidification and growth. On Venus, it is possible that the core has not yet cooled sufficiently to initiate such inner core growth, but has cooled enough to prevent the operation of a purely thermally driven dynamo (*Stevenson et al.,* 1983).

3.3. Radar Imaging of Venus' Surface

Radar global imagery by Magellan strongly indicates that Venus is a one-plate planet, devoid of plate tectonics, as suggested by the almost total absence of a global system of ridges and trenches as found in Earth's oceans (*Kaula and Phillips,* 1981). Mantle convection in Venus must therefore be of the sluggish or stagnant lid mode, which, as noted earlier, is a relatively inefficient way to cool the core (*Schubert et al.,* 1997). The absence of plate tectonics and its style of mantle convection could be directly related to the lack of water on Venus (*Sleep,* 2000). It is widely understood that the asthenosphere, the ductile layer of the upper mantle just below the lithosphere, serves to lubricate plate motion on Earth and facilitate plate tectonics. The physical properties of this layer are due to its water content (*Turcotte and Schubert,* 1988). It is therefore possible that atmospheric evolution, by way of an intense greenhouse effect and runaway loss of water (Bullock and Grinspoon, this volume), had a controlling influence on the evolution of the entire solid planet, shutting down plate tectonics, core cooling, dynamo action and a magnetic field. Thus a relatively minor part of Venus by mass — the atmosphere — could have controlled the thermal and rotational history of the entire planet.

Venus rotates in a retrograde sense with a planetary spin period of 243 Earth days, more than 236 times that of Mars and 335 times longer than any of the gas/ice giant planets. As discussed earlier, how this state of slow retrograde spin came about is unknown, but it might be maintained by a balance between solid body tides raised by the Sun's gravity and a frictional torque on the solid planet exerted by the superrotating atmosphere (*Gold and Soter,* 1969, 1979; *Kundt,* 1977; *Ingersoll and Dobrovolskis,* 1978; *Dobrovolskis and Ingersoll,* 1980). Such a balance invokes the solar torque on the atmospheric thermal tide as the source of atmospheric angular momentum. Alternatively, the angular momentum of the superrotating atmosphere, some 1.6×10^{-3} of the solid-body angular momentum (*Schubert,* 1983), could be derived from the solid planet, resulting in a secular variation in the planet's spin. Recent groundbased high-precision long-baseline radar measurements have detected length-of-day variations of 50 ppm over ~8 years (*Margot et al.,* 2012), thus confirming that such an angular momentum exchange likely plays a meaningful role in powering the superrotation of Venus' atmosphere.

3.4. Atmospheric Evolution

Today, the composition and structure of the atmosphere of Earth bears little resemblance to that of Venus or Mars (see Table 1 for a comparison of the atmospheric compositions of the terrestrial planets). For example, the venusian atmosphere is almost 100 times as massive as Earth's, which in turn is nearly 100 times more massive than Mars. Also, while N is the dominant gas in Earth's atmosphere, composing some 78% of the atmosphere by volume, it composes less than 4% of the content of both Venus and Mars. Indeed, N was recently found to be less abundant on Mars than the noble element Ar (volume mixing ratio of 0.0193 for ^{40}Ar vs. 0.0189 for N_2) (*Mahaffy et al.,* 2013).

As this last result indicates, since arriving at Mars in August 2012, the SAM instrument on the Mars Science Laboratory (MSL) Curiosity rover has been fundamentally refining our understanding of the composition of the martian atmosphere. Besides finding the surprisingly large ^{40}Ar abundance, SAM has found a ^{40}Ar/N_2 ratio that is nearly a factor of two greater than that measured by Viking (*Mahaffy et al.,* 2013). Considering the noncondensable and inert nature of these volatiles, their ratio is not expected to change with time or seasons, thus implying an unrecognized instrumental effect or unknown atmospheric process (*Atreya et al.,* 2013). Very similar values of the ^{13}C isotope are obtained by the quadrupole mass spectrometer (QMS) and the tunable laser spectrometer (TLS) subsystems on SAM, namely, δ^{13}C of 45‰ (*Mahaffy et al.,* 2013), whereas for the O-isotope ^{18}O, the TLS measures δ^{19}O of 48‰ in CO_2 (*Webster et al.,* 2013a). These C and O fractionations showing enhancement in the heavy isotopes imply loss of a substantial fraction of the original atmosphere from Mars due to escape processes. Similarly, loss of a substantial amount of water from Mars is implied by the TLS measurement of an enhanced D component, with a δD of 5000‰ (*Webster et al.,* 2013a). The isotopic fractionation carries the geologic and climate history of all terrestrial planets, as discussed in sections 4 and 5.

Another significant result of the initial SAM measurements is the lack of methane in the atmosphere of Mars. As methane is a potential biomarker — 90–95% of the methane in Earth's atmosphere is biologically derived — and since previous Mars orbital and groundbased observations indicated 10–70 ppbv of the gas in the atmosphere (*Formisano et al.,* 2004; *Mumma et al.,* 2009), the *in situ* SAM measurements were eagerly anticipated. Somewhat surprisingly, the highly precise TLS measurements obtained at the surface of Mars found no evidence of methane, with an upper limit of 3 ppbv (*Webster et al.,* 2013b). If methane were present at such low levels, it could result from any combination of processes, including (1) near-surface-rock reactions such as serpentinization, followed by metal-catalyzed Fischer-Tropsch reactions; (2) ultraviolet/charged particle degradation of surface organics; or (3) methanogenesis (*Atreya et al.,* 2007, 2011).

In contrast to the thin, cold atmosphere of Mars (average surface pressure and temperature of 6 mbar and 218 K), and Earth's habitable ground-level conditions (1 bar and 288 K surface pressure and temperature, respectively), the surface environment of Venus is extreme: about 94 bar and 740 K (*Seiff,* 1983). Venus' relatively thick atmosphere is predominantly CO_2, roughly equivalent to the amount of CO_2 tied up in carbonate rocks on Earth. Also, the total amount of water in Venus' atmosphere is within about an order of magnitude of the total water content of Earth's atmosphere, but is spread over a column number density of gas that is nearly two orders of magnitude greater. Thus the atmosphere of Venus is extremely dry, with a H_2O volume mixing ratio of about 30 ppm, corresponding to ~10^{-3} of the molar fraction of water found in Earth's atmosphere — and several orders of magnitude less if Earth's oceans are included. Consequently, Venus' ubiquitous cloud system is composed not of water, but of sulfuric acid, indicative of a complex cycle of sulfur-based corrosive chemistry throughout the atmosphere. How did the atmosphere evolve to its present state if it was initially similar to Earth's atmosphere with a substantial complement of water? As noted in Bullock and Grinspoon (this volume), a strong possibility is that an intense greenhouse effect led to the vaporization of all water on early Venus, followed by photodissociation of the water vapor at high altitudes and the escape of the released H into space. Without liquid water on its surface, CO_2 could then not be sequestered in the surface rocks. This change in Venus' atmosphere, if it occurred, would have triggered changes in the planet reaching all the way to its core. As noted earlier, the loss of water would have stopped plate tectonics. This would have changed the style of mantle convection, slowed the rate of core cooling, and turned off the magnetic field.

How can these evolutionary story lines be checked? As discussed in detail below, one way is the measurement of noble gases, their isotopic compositions, and the isotopic compositions of light elements measured in planetary atmospheres.

4. NOBLE GASES: THE KEY TO THE PAST

4.1. The Significance of Atmospheric Noble Gases

The earliest record of the atmospheres of the terrestrial planets is contained in the noble gases and their isotopes. This is because these elements do not react with the surface or with other gases, and isotopic fractionation records cataclysmic events (such as collisions with planetesimals or

TABLE 1. Atmospheric compositions of the terrestrial planets.

Constituent	Primary Diagnostics	Venus	Earth	Mars
Noble-Gas Abundance (VMR)				
^{132}Xe	Planetary origin: Atmospheric blow-offs Cometary/planetesimal impacts	Not measured; expected: ~1.9×10^{-9}[a,b]	8.7×10^{-8}	8.0×10^{-8}[c]
^{84}Kr	Planetary origin: Role of cold comets	$(7.0 \pm 3.5) \times 10^{-7}$ (Venera)[d] or $(5.0 \pm 2.5) \times 10^{-8}$ (PV)[d]	1.14×10^{-6}	3×10^{-7}[c]
$^{36+38}Ar$	Planetary origin: Roles of comets and planetesimals	~$(7.5 \pm 3.5) \times 10^{-5}$[f]	3.7×10^{-5} [g]	5.3×10^{-6}[c]
^{40}Ar	Evolution: Interior outgassing	$(7 \pm 2.5) \times 10^{-5}$[d]	0.0093	0.0193 ± 0.0002[sam]
^{20}Ne	Planetary origin: Common origin of Venus, Earth, Mars?	$(7 \pm 3) \times 10^{-6}$[d]	1.82×10^{-5}	2.5×10^{-6}[c]
^{4}He	Evolution: Interior outgassing	$[1.2 (+2.4,-0.8)] \times 10^{-5}$[d]	5.24×10^{-6}	$(1.1 \pm 0.4) \times 10^{-6}$[h]
Noble-Gas Isotopic Ratio				
$^{129}Xe/^{132}Xe$	Early evolution: Large atmospheric blow-off	Not measured; expected: ~3	0.983 ± 0.001[j]	$2.5(+2,-1)$[k]
$^{136}Xe/^{132}Xe$	Origin/evolution: U–Xe hypothesis	Not measured; expected: ~1	0.3294 ± 0.0004[j]	Not measured
$^{40}Ar/^{36}Ar$	Early history: Interior outgassing	1.03 ± 0.04[l] or 1.19 ± 0.07[m]	298.56 ± 0.31[g]	$1.9 (\pm 0.3) \times 10^{3}$[sam]
$^{36}Ar/^{38}Ar$	Late formation: Large impact	5.6 ± 0.6[l] or 5.08 ± 0.05[n]	5.304[g]	5.5 ± 1.5[k]
$^{21}Ne/^{22}Ne$	Origins: Common planet origin hypothesis	Not measured; expected: <0.067	0.029	Not measured
$^{20}Ne/^{22}Ne$	Early evolution: Hydrodynamic escape	11.8 ± 0.7[a]	9.78	Not measured
$^{3}He/^{4}He$	Early evolution: Impact of solar wind	Not measured; expected: $<3 \times 10^{-4}$[d]	1.37×10^{-6}	Not measured
Light-Element Isotopic Ratio				
D/H	Evolution: Loss of H/water	0.016 ± 0.002[p] 0.064–0.08 above 70 km alt.[q]	1.56×10^{-4}	$(9.3 \pm 2.2) \times 10^{-4}$[sam]
$^{14}N/^{15}N$	Evolution: Atmospheric loss since planetary formation	273 (+70,–46)[r]	272	170 ± 15[k]
$^{16}O/^{17}O$	Origin: Common kinship of terrestrial planets	Not measured	2520	2577 ± 12[sam]
$^{16}O/^{18}O$	Origin: Common kinship of terrestrial planets	500 ± 25[l]	489	462 ± 2.5[sam]
$^{33}S/^{32}S$	Past/current volcanic activity, magmatic composition	Not measured; expected: ~8×10^{-3}	8.01×10^{-3}	Not measured
$^{34}S/^{32}S$	Past/current volcanic activity, magmatic composition	Not measured; expected: ~0.04	0.045	Not measured
$^{12}C/^{13}C$	Biological marker	88 (+2,–1)[n]	89 (inorganic)	85.1 ± 0.3[sam]
CO_2	Dominant original atmospheric constituent	0.965 ± 0.08[d]	3.9×10^{-4}	0.9597[sam]
N_2	Second most prevalent original atmospheric constituent	0.035 ± 0.08[d]	0.78	0.0189 ± 0.0004[sam]
O_2	Disequilibrium species; indicator of biology	$<1 \times 10^{-6}$ @60 km alt. 4×10^{-5} @50 km 1.6×10^{-5} near surface[d,s]	0.21	0.00146 ± 0.00001[sam]
Bulk Abundances (VMR)				
CO	Venus: Hot thermochemistry Tracer of meridional circulation	2.5×10^{-5} @35 km alt. $(3–20) \times 10^{-5}$ @60–150 km alt.[d,s]	1.9×10^{-7}	$5.57 (\pm 0.01) \times 10^{-4}$[sam]
H_2O	Geologic styles; Earth, Venus: Condensable Earth: Weather cycle	$(4.4 \pm 0.9) \times 10^{-5}$ @0–40 km 2×10^{-4} @50 km 1×10^{-6} @60 km[d,s]	4×10^{-3} avg. 0.01–0.04 near surface	2×10^{-4} (seas. var.)[t]
H_2	Photochemistry	$(2.5 \pm 1.0) \times 10^{-6}$ at 50–60 km[s]	5.5×10^{-7}	$(1.5 \pm 0.5) \times 10^{-5}$[v]
H_2O_2	Photochemistry	None detected	$(0.1–6) \times 10^{-8}$	4×10^{-8} max (seas. var.)[w]
O_3	Photochemistry	None detected	$(0–7) \times 10^{-8}$	$(1–80) \times 10^{-8}$ (seas. var.)[x,sam]
CH_4	Disequilibrium species; biological marker	None detected	1.81×10^{-6}	$<3 \times 10^{-9}$[sam]

Data sources: [a] *Donahue* (1986); [b] *Pepin* (1991); [c] *Owen et al.* (1977); [d] *von Zahn et al.* (1983), recommended values; [f] derived from *von Zahn et al.* (1983) values; [g] *Lee et al.* (2006); [h] *Krasnopolsky et al.* (1994); [j] *Ozima et al.* (1983); [k] *Owen et al.* (1977), *Owen* (1992); [l] *Hoffman et al.* (1980); [m] *Istomin et al.* (1980a); [n] *Istomin et al.* (1980b); [p] *Donahue et al.* (1982); [q] *Bertaux et al.* (2007); [r] *Hoffman et al.* (1979); [s] *Svedhem et al.* (2010); [t] *Smith* (2002); [v] *Krasnopolsky and Feldman* (2001); [w] *Encrenaz et al.* (2004); [x] *Barth* (1974); [sam] *Mahaffy et al.* (2013), *Webster et al.* (2013a,b), *Atreya et al.* (2013).

solar-induced blow-off of the atmosphere) and geologic upheavals that degas the interior. Therefore, noble-gas isotopes in the atmosphere can point to past and present-day geologic activity (*Pepin,* 1991). The noble-gas elements Xe, Kr, Ar, and Ne and their isotopes provide an accessible historical record of ancient events, pertinent to planetary formation and early evolutionary processes. Radiogenic isotopes of He, Ar, and Xe also provide dating constraints on geologic processes that over the eons may have delivered materials from the deep interior to the surface (*Zahnle,* 1993).

The remarkable property that renders the noble gases and their isotopes so valuable in determining ancient events is their stability against chemical alterations. Unfortunately, this stability against chemical reactions manifests itself as well in very weak coupling to electromagnetic radiation. Thus no strong spectral features exist. Consequently, the abundances of these elements cannot be readily assessed by remote sensing techniques, such as those used by Venus Express and the Mars Reconnaissance Orbiter. Only *in situ* sampling can do this. How the terrestrial planets originated and initially formed, the nature of cataclysmic events in their early histories, and insights into major geologic events throughout their evolution are examples of fundamental insights potentially provided through accurate measurements of noble gases and their isotopes. The record for Earth and Mars, particularly given the new SAM measurements, is quite complete. What is missing to complete the broad history of planetary atmospheres throughout the inner solar system are accurate measurements for Venus.

4.2. Xenon and Krypton: Coded Messages of Ancient Cataclysms on Nascent Planets

Xenon, which has never been measured on Venus, is of special interest because of its role in understanding the formation and evolution of early atmospheres on all three inner planets. On both Earth and Mars, its abundance and isotopic distribution bear little resemblance to any known source material in the solar system, which points to major globe-changing events in the early histories of both planets. As shown in Fig. 3, one difference is that on Earth and Mars, the eight nonradiogenic (primordial) isotopes of Xe are strongly mass fractionated (with a gradient of ~4%/amu) compared to the solar wind or any plausible solar system source such as chondrites (*Pepin,* 1991; *Zahnle,* 1993). Another difference is that the bulk abundance of Xe on Earth and Mars is depleted with respect to Kr by a factor of ~20 when compared to typical chondritic meteorites (*Pepin,* 1991), as depicted in Fig. 4 (blue and red curves vs. brown curve). As discussed below, Kr may have been preferentially released from the interiors of Earth and Mars over the eons, or it may not have been severely affected by solar-wind-induced erosional processes. In either case, the strong mass fractionation of Xe observed on Earth and Mars implies that nonradiogenic Xe isotopes are remnants of atmospheric escape that occurred during the formation of Earth and Mars.

A similar story is evident for the radiogenic isotopes of Xe, which are also markedly depleted compared to the measured abundances of their radioactive parents in primitive solar system materials. Thus, all three Xe abundance characteristics — the fractionation of primordial Xe, and the low abundances of both bulk and radiogenic Xe — indicate that one or more significant cataclysmic events on both Earth and Mars occurred late in the planetary formation process (*Pepin,* 2000).

The decay of short-lived isotopes of I (^{129}I, half-life 15.7 m.y.) and Pu (^{244}Pu, half-life 82 m.y.) are significant sources of radiogenic Xe. Iodine produces just a single isotope, ^{129}Xe, thus enabling straightforward measurement of I-decay-produced Xe. For both Earth and Venus, the atmospheric ^{129}Xe abundance is much less than expected if escape had not occurred. More specifically, Earth's atmosphere has retained only ~0.8% of the complement of ^{129}Xe produced by the decay of I, while today Mars has just 0.1% (*Porcelli and Pepin,* 2000; *Ozima and Podosek,* 2002). In other words, Earth and Mars have lost more than 99% of the^{129}Xe produced from the radioactive decay of I. Given the relatively short 15.7-m.y. half-life of ^{129}I, much of this escape must have occurred within planetesimals during accretion. But its half-life is also long enough that much of it must have also been lost after Earth and Mars were planets, and, in the case of Earth, some of it must have been lost after the Moon-forming impact. Thus multiple impacts

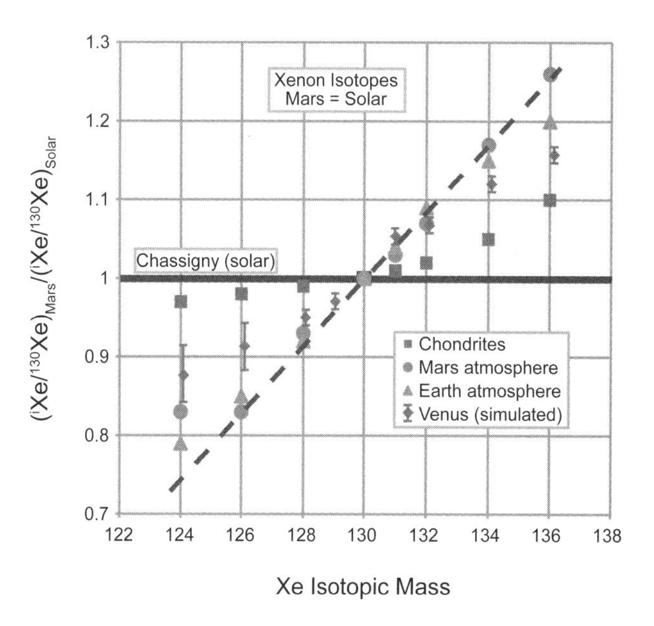

Fig. 3. See Plate 31 for color version. Fractionation of Xe isotopes. Possible Venus Xe fractionation pattern observed by a future *in situ* mission is depicted (green) compared to the patterns of Earth, Mars, chondrites, and the Sun (after *Bogard et al.,* 2001). A common Venus/Earth/Mars pattern would bolster the hypothesis that a common source of comets or large planetesimals delivered volatiles throughout the inner solar system. A different pattern for Venus would strengthen the solar EUV blowoff theory.

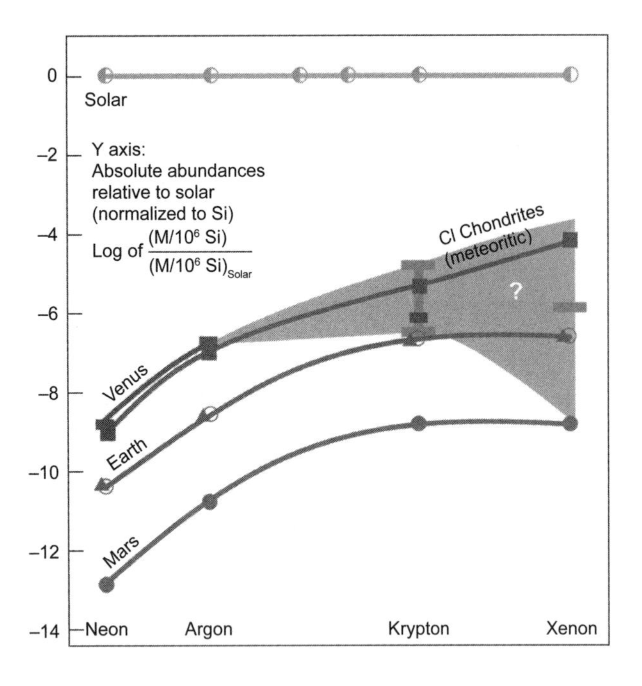

Fig. 4. See Plate 32 for color version. Noble gas abundances for Earth, Mars, Venus, chondrites, and the Sun (after *Pepin,* 1991). Missing Xe and poorly constrained Kr data for Venus are critical for understanding the history of its early atmosphere. For Ne and Ar, and likely Kr, Venus is more enhanced in noble gases vs. Earth and Mars, likely indicating (1) a large blowoff of the original atmospheres of Mars and Earth, and/or (2) enhanced delivery of noble elements to Venus from comets and/or solar-wind-implanted planetesimals. For Venus, the unknown/poorly constrained Xe/Kr and Kr/Ar ratios — denoted by the range of slopes of the aqua area — are consistent with all solar system objects shown, including (1) a solar-type (Jupiter and cometary) composition (orange line), (2) a chondritic composition (brown line), and (3) the composition of Earth and Mars (blue and red lines).

or other escape events — such as energized by the T-Tauri phase of the Sun — are indicated for Earth, spread over perhaps ~100 m.y. As discussed earlier in section 2.1, if the planets formed before the end of the T-Tauri phase, then the dominant escape process was likely atmospheric blow-off due to the strong T-Tauri solar wind. This would explain the enrichment of ^{129}Xe in Earth's mantle compared with the atmosphere (*Allegre and Schneider,* 1994). For Venus, ^{129}Xe measurements would indicate whether it also lost much of its original atmosphere during its first ~100 m.y.

The spontaneous fission of ^{244}Pu provides somewhat similar insights. Plutonium is especially noteworthy because its 82-m.y. half-life probes the early evolution of the planets for ~200 m.y. after accretion. Fission Xe has been detected both in Earth's mantle and likely in its atmosphere (*Pepin,* 2000). The precise amount is somewhat uncertain, but the upper limit is just 20% of what should have been generated by Earth's primordial Pu. The uncertain process that removed fissiogenic Xe from the atmosphere (or perhaps kept it sequestered in the interior, preventing its venting into the atmosphere) occurred some 200 m.y. after the formation of Earth.

Four general scenarios can be invoked to explain the loss and fractionation of Xe, as illustrated schematically in Fig. 5. In one scenario, illustrated in Fig. 5a, the terrestrial planets experienced different degrees of blowoff of atmospheric H, driven by solar extreme ultraviolet (EUV) radiation more than 100 times stronger than today (*Zahnle and Walker,* 1982). As H escaped, it dragged other gases with it, the lighter gases preferentially. What stayed behind was isotopically heavy, as observed today (*Zahnle and Kasting,* 1986; *Hunten et al.,* 1987; *Sasaki and Nakazawa,* 1988; *Zahnle et al.,* 1990a; *Pepin,* 1991). A variant attributes escape to the Moon-forming impact (*Pepin,* 1997). In both of *Pepin*'s (1991, 1997) models, escape was followed by degassing of the lighter noble gases from the interior — in particular Kr but not Xe, due to its stronger affinity for mantle melts than the lighter noble gases. Thus Earth's atmosphere was replenished with the amounts of lighter noble gases seen today, resulting in a relative depletion of atmospheric Xe. In both models there is relatively little escape from Venus with its much thicker atmosphere, so that Venus holds a larger and less altered portion of its original complement of noble gases.

A related model invokes impacts during accretion as the cause of the loss of atmospheric Xe (*Zahnle,* 1993) (Fig. 5b). Impact erosion is expected to expel all gases equally, regardless of mass. Thus, due to the sequestration of condensibles in the lower reaches of planetary atmospheres, any fractionation would occur between gases and condensed materials rather than among gases of different molecular mass. Consequently, one expects impact erosion to expel well-mixed noble gases much more efficiently than water sequestered in oceans or as vapor in the lower atmosphere. This, then, may have been the means by which planets lost their radiogenic Xe without losing an ocean of water. But impact erosion cannot account for the mass fractionation of the Xe isotopes. Thus, impact erosion is not the whole story, although this process may have contributed to the loss of bulk Xe.

A third scenario posits that Earth's current Xe-isotopic pattern is largely that of an external source such as large planetesimals or large comets formed in the outer solar system (Fig. 5c). Large planetesimals can fractionate isotopes by gravitational segregation (*Ozima and Podosek,* 2002; *Zahnle et al.,* 1990b). Extremely porous cold bodies, such as large comets, immersed in the cold nebular gas cloud in the outer reaches of the solar system could perhaps incorporate Xe-isotopic signatures that differ substantially from those in the inner nebula. Such bodies could then deliver these signatures to Earth (and likely other nearby bodies) to produce the puzzling distribution of Xe isotopes observed on our world today. To account for the depletions of bulk and radiogenic Xe, this hypothesis also requires that much of Earth's primordial Xe — including radiogenic products from Pu and I decay — remain sequestered deep inside the planet.

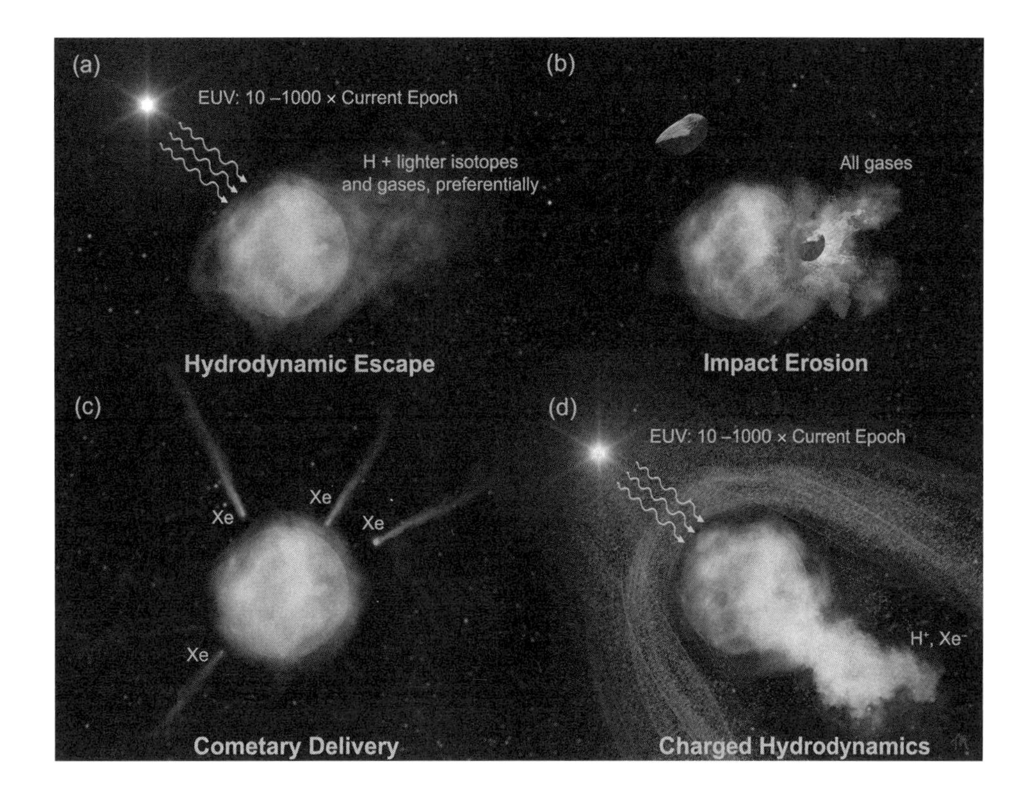

Fig. 5. See Plate 33 for color version. Schematics showing major features of the four main hypotheses of early planetary evolution that led to atmospheric Xe-isotopic abundances on the terrestrial planets. **(a)** Hydrodynamic escape. Early solar EUV was 100–1000 times more intense than today, heating the atmosphere to the point where H began to flow out of the exosphere. It dragged lighter molecules with it, fractionating the atmosphere so that it was isotopically heavier. **(b)** Impact erosion. Small impacts both added material and eroded the atmosphere. Large impacts blew off most of the atmosphere above the plane of impact, indiscriminate of mass, so no fractionation occurred. **(c)** Cometary delivery. Cold comets from the outer solar system with trapped Xe incorporating the isotopic ratios of the cold outer solar system impacted the surface. **(d)** Charged hydrodynamics. Hydrogen and Xe are both easily ionized, unlike other molecules. Thus ionization by the UV or solar wind preferentially allowed H and Xe to escape, while also allowing somewhat smaller portions of other ionized atoms and molecules to go along for the ride. The process is fractionating, dependent upon both mass and ionization energy.

A fourth hypothesis invokes a unique property of Xe. Unlike the other noble gases, Xe ionizes more easily than H. Consequently, when embedded in a partially ionized H wind that is flowing upward to space, Xe tends to ionize while the other noble gases remain neutral. The Coulomb force between Xe ions and protons is large and thus Xe ions are relatively easily dragged away by the escaping wind, while the lighter noble gases, being neutral, remain behind (Fig. 5d). This describes a variant on atmospheric blowoff that leaves the bulk of the atmosphere intact while allowing Xe to preferentially escape (*Zahnle,* 1993).

Sampling Xe and its isotopes on Venus can help resolve which mechanism or combination of mechanisms are responsible for the similar Xe-isotopic patterns found on Earth and Mars (cf. Fig. 3), thus providing key insights into the early histories of all three planets. A common fractionation signature on all three planets would indicate a common source of Xe isotopes, strengthening the hypothesis that large comets or planetesimals were a prime source of volatiles throughout the inner solar system (*Zahnle,* 1993). Alternatively, due to the difference in the power of H blowoff between Venus and the outermost inner planets, the solar EUV blowoff theory would be bolstered if the Xe-isotopic fractionation pattern on Venus were found to be different than that found for Earth and Mars. The Earth-blowoff theory would be further strengthened if relatively large amounts of radiogenic ^{129}Xe were found on Venus, rather than the small amounts found on Earth. This would indicate that (1) blowoff almost completely stripped Earth's atmosphere early in its history, and (2) impact erosion was relatively unimportant on Venus.

The bulk ratios of the heavy noble gases provide additional clues about the source of the materials that formed the inner planets. The Galileo probe's discovery of a uniform enhancement of Xe, Kr, and Ar in Jupiter (*Niemann et al.,* 1998; *Atreya et al.,* 2003) implies that "solar composition" planetesimals ("cold comets") were abundant in the solar system and comprise a significant fraction of the giant planets. Krypton/xenon and Ar/Kr on Venus thus provide tests on whether these solar-like planestimals also reached Venus in significant numbers. However, as depicted in Fig. 4, existing Kr measurements for Venus differ by more than an order of magnitude (*Von Zahn et al.,* 1983). If the lower estimate is correct, then Ar/Kr on Venus more closely resembles Jupiter's atmosphere and the solar wind, which strengthens the cold comet hypothesis (lower boundary between Ar and Kr in Fig. 4). If the higher estimate is correct, then, as depicted by the high boundary of the aqua area between Ar and Kr in Fig. 4, the Venus Ar/Kr ratio instead resembles other objects, including meteorites, Earth, and Mars (i.e, the "planetary pattern") as well as laboratory measurements of gases trapped in cold ice (*Notesco et al.,* 2003). The measured Kr/Xe ratio can then be invoked to help discriminate between the different sources and events that formed the inner planets. For Earth's mantle, *Holland et al.* (2006) note that processes unique to Earth determine the abundance of Ar, Kr, and Xe. Seawater subduction controls the composition of heavy noble gases in the mantle, so direct bulk comparisons of these gases from the interiors of the planets is likely to be difficult. For planetary atmospheres, however, a significantly improved determination of the bulk Kr abundance in Venus at the 5% level would provide fundamental insight into whether or not common sources supplied the atmospheres of all of the inner planets.

The isotopic distribution of Xe at Venus can also test the controversial "U-Xe" hypothesis invoked for Earth (*Pepin,* 1991). This theory maintains that the original source of Earth's Xe, known as "U-Xe," had an isotopic composition that was distinctively depleted in the heavy isotopes, with 8% less $^{136}Xe/^{130}Xe$ than in the solar wind and meteorites. The Xe-isotopic distribution in Earth's atmosphere can be fit by the sum of mass-fractionated U-Xe with a small addition of heavy isotopes from ^{244}Pu decay. Without U-Xe, Earth's Xe-isotopic distribution is difficult to explain. If U-Xe is in fact an accurate description of Earth's primordial Xe, then it must have come from a source that was isotopically distinct from the solar nebula as a whole. Venus also then likely accreted its Xe from this same source. Indeed, according to the U-Xe hypothesis, Venus' atmosphere is the most likely place in the solar system to find U-Xe. If such a discovery were to be made there, the implications for Earth and Xe-isotopic heterogeneity in the inner solar system would be major.

To resolve these various scenarios of the history of the inner solar system, relatively precise Xe-isotopic ratio measurements of at least 5% accuracy are needed, sufficient to resolve well the ~20% fractionation displayed by terrestrial and martian Xe. Somewhat higher precision — ~3% — is required to measure (1) the difference in $^{136}Xe/^{130}Xe$ between Xe derived from hypothetical U-Xe and solar wind Xe (8%), or (2) the abundance of radiogenic ^{129}Xe (7% on Earth). Fissogenic Xe is perhaps more difficult, as its reported detection in Earth's atmosphere, at the 4% level, is model-dependent (*Pepin,* 2000). However, up to 10% of the ^{136}Xe on Venus could be fissiogenic, so that a ~3% measurement accuracy in the $^{136}Xe/^{130}Xe$ ratio would likely be sufficient to determine the relative contributions of Xe sources.

4.3. Neon and Oxygen Isotopic Ratios: A Common Heritage for the Terrestrial Planets?

Neon in Earth's atmosphere has an isotopic composition that is mass-fractionated with respect to that in the mantle, with a larger proportion of heavier Ne isotopes borne in the atmosphere. This suggests that isotopically-light terrestrial Ne as well as bulk Ne has escaped into space. The ~100-times-smaller bulk Ne/Ar ratio seen on Earth and Mars today (*Zahnle,* 1993), compared with parent nebular values (e.g., compare the Ne/Ar values for Earth and Mars against the solar values in Fig. 4), provides further evidence of Ne escape. Models of both fractionated and bulk Ne developed by *Sasaki and Nakazawa* (1988) demonstrate this behavior, showing that in principle, Earth could generate the observed fractionation while also depleting the bulk Ne from an originally solar nebula composition. However, *Ballentine et al.* (2005) showed that Ne-isotope abundances in magmatic CO_2 well gases strongly point to the subduction of volatiles implanted in late-accreting material as the original source of Earth's Ne and He.

Neon has three stable isotopes, providing a means by which to test whether (1) Ne on the inner planets is related by mass fractionation of a common ancient Ne reservoir located in the same region of the parent nebula, or (2) Venus, Earth, and Mars accreted Ne from different sources. Specifically, if the ($^{22}Ne/^{20}Ne$, $^{21}Ne/^{20}Ne$) ratios for two planets fall on the same mass fractionation line, it would imply that they shared the same source of Ne (and perhaps other noble gases and volatiles), and then evolved to their present-day ratios via escape processes. If, instead, the observed ratios do not both fall on this line, then the two planets likely began with distinct isotopic compositions originating from different source reservoirs, thus indicating that the inner planets were built from different realms of the parent nebula.

The three isotopes of O can be used in a similar manner to distinguish differences in original compositions among solar system objects (*Clayton,* 1993; *Yurimoto et al.,* 2006). For example, Earth and the Moon share the same distinctive O-isotope composition, which is clearly inconsistent with the make-up of nearly all meteorites, including Vesta and martian meteorites (e.g., *Stevenson,* 2005). Specifically, for Vesta meteorites, the $^{17}O/^{16}O$ ratio lies about 0.02% below the Earth-Moon mass-dependent fractionation line, while the martian value lies 0.04% above it. That Earth and the Moon share nearly the same O-isotope ratio as well as the isotopic ratios of many refractory elements (*Cuk and Stewart,* 2012) has been somewhat puzzling, since detailed simulations suggest that the Moon formed almost wholly from the mantle of the Earth-striking Mars-sized object (*Canup,* 2004). As discussed in section 2.3, however, more recent simulations (*Canup,* 2012) seem to indicate that the oblique impact of an Earth-sized planetesimal would have homogenized the protolunar accretion disk and Earth's mantle. We would thus expect that terrestrial mantle and lunar O, Cr, and Ti isotopes would be similar. Further isotopic comparisons will no doubt refine or refute this interesting twist on the Moon's formation.

For Venus, if it is found that its O isotopes are distinct from Earth and the Moon, then Venus accreted from a different pool of planetesimals than Earth and the Moon. If the O-isotopic ratios for Venus are similar to those found on the Earth/Moon system, it would not only show that Venus accreted from the same reservoir of materials as Earth and the Moon, but it would also suggest that the equivalence of Earth and the Moon is not so mysterious. For example, this result could suggest that the Moon may have formed from nebular materials located inside Earth's orbit.

Currently, measurements are needed for all three isotopes of Ne and O on Venus (e.g., *Pepin,* 2006; *Zahnle,* 1993). For Ne, isotopic ratios to within 5% are needed to discriminate between early evolution models. For O, an accuracy of 0.02% for the $^{17}O/^{16}O$ ratio is required.

4.4. Argon and Neon: Potential Evidence of Large Impacts

In contrast to the dearth of Xe measurements and order-of-magnitude variation in Kr results reported between the Pioneer Venus and Venera probes (*Donahue and Russell,* 1997) (cf. Fig. 4), reasonably precise abundances have been determined for the values of primordial (nonradiogenic) Ne and Ar. As depicted in Fig. 4, the Pioneer Venus probes showed unexpectedly high abundances of Ne and primordial Ar (^{36}Ar and ^{38}Ar, some 80 times that found on Earth), indicative of an unusual source of Ne and Ar not seen in the other inner planets (*Kaula,* 1999).

There are three leading theories for the high abundances of Ne and Ar on Venus, two of which involve an anomalous large impact event unique to the planet. As illustrated in Fig. 6a, the first hypothesis holds that the solar wind implanted noble gases into meter-sized particles in Venus' neighborhood that later assembled into a large, several-hundred-kilometer-scale planetesimal that eventually struck the planet (*McElroy and Prather,* 1981; *Wetherill,* 1981). Thus the Sun is the source of enhanced Ne and Ar, delivered via a very large, "solar-contaminated" impactor. *Sasaki* (1991) showed that this model can work in the more realistic context of an optically thick, vertically extended dust disk generated by collisions between planetesimals.

A second theory maintains that a large comet from the outer solar system delivered Ar and Ne trapped in cold cometary ices (*Owen et al.,* 1992). The impactor would need to have been ≥200 km in diameter to supply the large Ar content found in Venus, assuming the bolide had the same Ar-rich make-up as that of the hypothetical polluters of Jupiter's atmosphere (*Zahnle,* 1993). More recently, *Gomes et al.* (2005) estimate that ~10^{23} g of cometary material was delivered to Venus and Earth during the late heavy bombardment, corresponding to the mass of a dozen 200-km-diameter comets (cf. Fig. 6b). Thus, if such cold comets existed — as suggested by the enhanced noble-gas abundances on Jupiter measured by the Galileo probe (*Mahaffy et al.,* 2000) — the *Gomes et al.* (2005) result implies that there is little difficulty in such objects delivering enough Ar from the outer solar system.

A third theory holds that the terrestrial planets gravitationally captured Ar and Ne directly from the local solar nebula (Fig. 6c). Due to its thicker, more protective atmosphere, Venus was significantly better than Earth and Mars at preserving its atmosphere against giant impacts (*Genda and Abe,* 2005).

These theories can be distinguished via precise (1–3%) *in situ* measurements of $^{36}Ar/^{38}Ar$ in Venus' atmosphere as well as the telltale Xe-isotopic ratios noted earlier. In particular, such sampling can reveal whether the solar wind ratio, currently under analysis by the Genesis mission (e.g., *Grimberg et al.,* 2007), is found on Venus. Prior Pioneer Venus probes measured the $^{36}Ar/^{38}Ar$ ratio on Venus to an accuracy of only 10% (*Oyama et al.,* 1980), too crude to determine if the planet's signature is that of the solar wind, based on early Genesis results.

4.5. Radiogenic Argon and Helium: A Record of Planetary Degassing

Radiogenic Ar and radiogenic and nonradiogenic He are powerful probes of mantle degassing. Created underground by the decay of potassium, ^{40}K (half-life of 1.3 G.y.), radiogenic ^{40}Ar diffuses upward through the mantle and is emitted into the atmosphere. Compared to Earth's atmosphere, Venus has about 25% as much atmospheric radiogenic ^{40}Ar. This strongly implies that, compared to Earth, either (1) Venus has markedly less K, or (2) the planet has degassed less (*Turcotte and Schubert,* 1988). A plausible scenario for (2) would be a large decrease in Venus' rate of degassing approximately 1 b.y. after formation. The low ^{40}Ar abundance coupled with the very high ^{36}Ar content results in a $^{40}Ar/^{36}Ar$ ratio near unity, significantly less than the range of ~150–2000 measured for Earth, Mars, and Titan (*Owen,* 1992; *Atreya et al.,* 2006, 2007) indicative of active geologies. However, recent laboratory studies on the compatibility of Ar in silicate melt indicates that ^{40}Ar may not be indicative of interior degassing efficiency (*Watson et al.,* 2007). Instead, the abundance of ^{40}Ar in Earth's atmosphere may be due largely to the hydrological weathering of the crust. If this is the case, the amount of ^{40}Ar seen in Venus' atmosphere may indicate that it has experienced about one-fourth of the aqueous weathering that Earth has over the age of the solar system. Further laboratory work on the compatibility and diffusion of Ar in the mantle is needed to understand exactly what processes are responsible

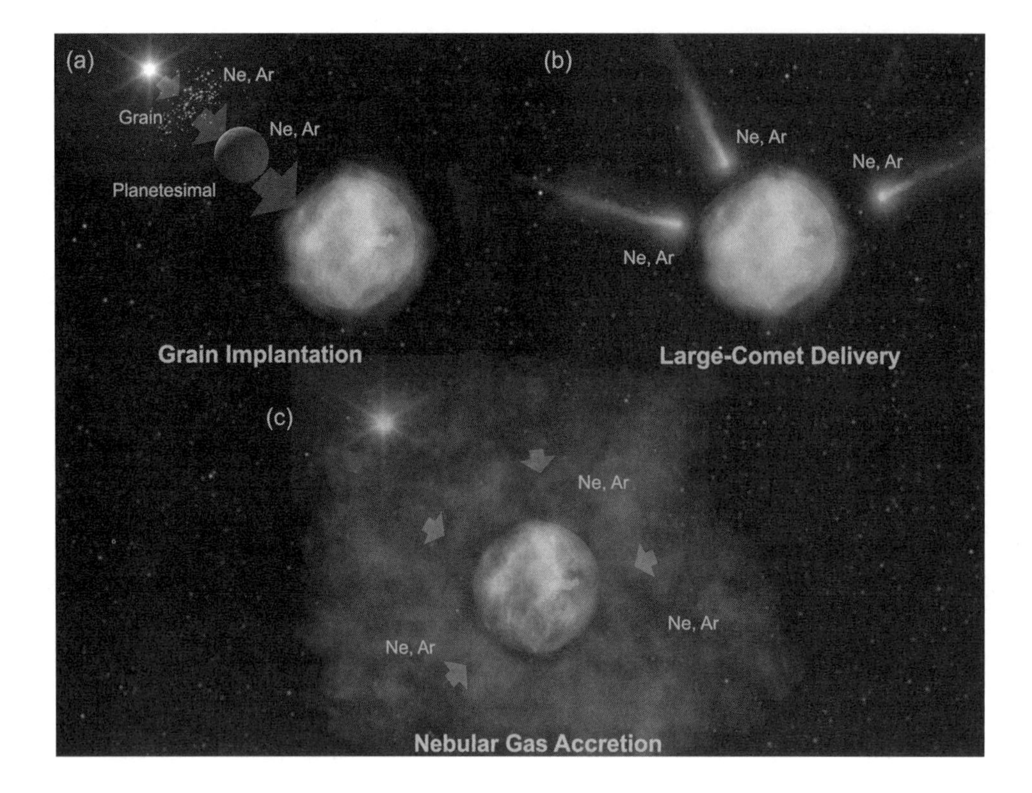

Fig. 6. See Plate 34 for color version. Schematics of the three main hypotheses for the origin of large Ne and Ar abundances in Venus' atmosphere. **(a)** Grain implantation. Early solar wind fluxes were many orders of magnitude higher than today. Solar-wind Ar and Ne implanted into grains around the Sun, which coelesced into Ar,Ne-rich planetesimals that impacted Venus. **(b)** Large-comet delivery. Argon and Ne frozen and trapped in cometary ice latices was delivered by one or more large comets (≥200 km diameter) from the cold reaches of the outer solar system, leaving Venus with an Ar,Ne-rich atmosphere. **(c)** Nebular gas. Venus' present atmosphere came directly from the nebula in which it was formed. The atmosphere survived relatively intact against subsequent impactors that blew off the original atmospheres of Earth and Mars, resulting in a relatively Ar,Ne-rich atmosphere on Venus.

for the ^{40}Ar abundance in terrestrial planetary atmospheres.

Helium provides another probe of planetary degassing through time. There are two stable He isotopes: the abundant ^{4}He and the much rarer ^{3}He. On Earth, more than 90% of the ^{4}He is radiogenic, primarily created by the decay of U and Th (with relevant half-lives of 0.7, 4.5, and 12 b.y.), while most of the terrestrial ^{3}He is primordial. Earth's continents are granitic and are strongly enriched in Th and U, and thus they are a major source of ^{4}He. The mantle is a source of both isotopes, with a ratio of 90,000:1 (*Jambon,* 1994). Degassing of primordal ^{3}He indicates that Earth's interior still retains a noble-gas imprint that dates to its formation. For Mars, EUV observations of the 584-Å He line indicate that the He mixing ratio in the lower atmosphere is 4 ± 2 ppm (*Krasnopolsky and Gladstone,* 1996). Coupled with models of the production and outgassing of ^{4}He and its escape to space, these authors conclude that He outgassing rate on Mars is 2–4 times smaller than its escape. Similarly, measurements of ^{3}He and ^{4}He on Venus should provide insights into the degree of mantle degassing and associated active geology on Earth's sister world (*Namiki and Solomon,* 1998).

While He escapes quickly from Earth's atmosphere, loss rates are significantly slower from Venus (*Donahue and Russell,* 1997). If there really has been very little atmospheric escape over the eons on Venus, then about twice as much radiogenic ^{4}He would be expected as radiogenic ^{40}Ar (assuming an Earth-like K/U ratio). Yet less ^{4}He is observed. If ^{4}He escaped, then the corresponding ^{3}He escape rate should be greater, thereby reducing the ^{3}He/^{4}He ratio. Alternatively, a high ^{3}He/^{4}He ratio in the current atmosphere implies either (1) a relatively small He escape rate, (2) a relatively recent and large mantle degassing event that released primordial ^{3}He, or (3) a dearth of near-surface ^{4}He-producing granitic material (*Turcotte and Schubert,* 1988). New insights into the geology, outgassing history, and evolution of Venus would be provided by better information on both the isotopic ratios and escape rates of He, the former to a measurement accuracy of approximately 20%. The He escape rates are currently under investigation by the Venus Express Analyser of Space Plasma and Energetic Atoms (ASPERA) experiment (e.g., *Galli et al.,* 2008). What remains, then, is more accurate atmospheric measurements of the He-isotopic abundances.

5. ISOTOPES OF LIGHT GASES: ATMOSPHERIC LOSS THROUGH TIME

5.1. Isotopes of Hydrogen and Nitrogen: Possible Loss of an Ancient Ocean

Primarily due to (1) the lack of a protective internally-generated magnetic field and (2) a strong greenhouse effect, the atmosphere of Venus today is vastly different from that at the end of its formation period some 3.8 G.y. ago. Perhaps the single most significant change on Venus was the loss of the bulk of its water over the eons, as evidenced by the large ratio of deuterated (heavy) water (HDO), relative to H_2O, observed today that indicates that the D/H ratio is some 150 times that found on Venus' sister planet, Earth (*Donahue,* 1999). Potential mechanisms responsible for this high D/H ratio range from the rapid loss of a primordial ocean into space to steady-state mechanisms promoting the idea that atmospheric water is supplied by volcanic outgassing and cometary infall (*Grinspoon,* 1993). The D/H of Earth's water is similar to the chondritic value, suggesting that chondritic bodies were the main source of Earth's volatiles (*Alexander et al.,* 2012). However, if Venus' primordial water came from comets, the D/H of the source may have been higher than Earth's present value. Most comets exhibit a D/H of about twice that found on Earth (*Mousis et al.,* 2000), although *Hartogh et al.* (2011) reported a Jupiter-family comet with an Earth-like chondritic D/H. When water in Venus' atmosphere was photodissociated, the O was lost via the oxidation of Fe-bearing crustal minerals and H escape out of the atmosphere. The escape of H is facilitated by the absence of a magnetic field that then allows the solar wind to energize and drag away H atoms from the top of the atmosphere.

A fundamental question is the abundance of water at the end of the planet's formation period, estimated to be equivalent to a global ocean between 5 and 500 m in depth (*Donahue et al.,* 1997). The large uncertainty arises from several factors, including imprecise estimates of the global mean D/H ratio, which the Venus Express mission found to vary significantly with altitude (*Federova et al.,* 2008), possibly due to photochemical fractionation effects (*Liang and Yung,* 2009). Another source of uncertainty is the fractionation factor for the escape of H and D and how this factor has varied with time, as differing loss processes have dominated. Over the full range of measurement uncertainties, the D/H ratio ranges from 0.013 to 0.038, based on data from the Pioneer Venus probe mass spectrometer and infrared spectrometer (*Donahue,* 1999) and the Ultraviolet and Infrared Atmospheric Spectrometer (SPICAV/SOIR) onboard Venus Express (*Bertaux et al.,* 2007; *Fedorova et al.,* 2008). If escape fluxes are in the upper end of the range suggested by pre-Venus Express data, $>3 \times 10^7$ cm^{-2} s^{-1}, then the greatly enhanced (over terrestrial) D/H ratio must reflect loss over the last 0.5 G.y., masking a primordial signature (*Grinspoon,* 1993; *Donahue,* 1999). The escape flux and ocean volume is presumably even greater for the most recent Venus Express D/H ratio of 240 ± 25 times the terrestrial ocean value reported by *Federova et al.* (2008) for Venus' upper atmosphere. Also, if a large amount of H and D loss occurred during an early phase of fractionating hydrodynamic escape, driven by an intense early solar EUV flux (*Chassefière,* 1996), then the original water inventory could have been many times larger than the values indicated by the D/H ratio seen today.

Additional information on Venus' atmospheric loss comes from the isotopic ratio of N, $^{15}N/^{14}N$. Currently the venusian ratio ($3.8 \pm 0.8 \times 10^{-3}$) is known to ±20%, comparable to the terrestrial value (3.7×10^{-3}) and broadly similar to N in meteorites (*Donahue and Pollack,* 1983), but at variance with the atmospheres of Mars ($5.9 \pm 0.5 \times 10^{-3}$) and Jupiter ($2.2 \pm 0.3 \times 10^{-3}$) (*Owen et al.,* 1977; *Owen,* 1992; *Abbas et al.,* 2004). The consensus scenario is that Earth and Mars accreted their N from a common meteoritic or chondritic source. However, on low-gravity Mars, the light N preferentially escaped, resulting in a relatively high $^{15}N/^{14}N$ ratio. In contrast, the low $^{15}N/^{14}N$ ratio on Jupiter pertaining to a relatively high light N component means that Jupiter's N was supplied by a cometary or nebular source (*Owen and Bar-Nun,* 1995). Since Venus and Earth have approximately the same $^{15}N/^{14}N$ ratio, the expectation is that Venus also accreted its N from the same common source as Earth and Mars. Comparison of the N isotopes between Venus and Earth should help determine whether N escape was significant on at least one of the two planets. In summary, *in situ* measurements of the isotopic ratios of $^{15}N/^{14}N$ and HDO/H_2O to 5% — the latter over a range of altitudes to assess photochemical and escape fractionation effects — would provide insights into understanding (1) present and past escape rates, (2) the nature of Venus' likely water-rich ancient climate, and (3) the role of comets in supplying volatiles to the inner planets.

5.2. Sulfur Isotopes: Probes of Current Degassing

The isotopic ratios of S, $^{33}S/^{32}S$ and $^{34}S/^{32}S$, are established in volcanic and interior processes. Measurements of these ratios in the venusian atmosphere therefore potentially provide information on interior degassing during recent geologic times. However, S-isotopic ratios are susceptible to modifications via (1) photochemical fractionation processes involved in the atmospheric sulfur cycle and (2) surface chemistry involving Fe sulfides. Indeed, anomalous fractionations — deviations from simple mass-dependent fractionation — as large as 7% have been measured in laboratory experiments at 193 nm (*Farquhar et al.,* 2004). Yet in practice, natural samples that have not been biologically fractionated show anomalous fractionations of a few tenths of a percent at most and conventional fractionations less than 1%. For Mars, the largest anomalous fractionation measured in an escaped meteorite is 0.1% (*Farquhar et al.,* 2007). Precise S-isotopic measurements at the surface of Mars at the 0.1% level will be obtained shortly by the Mars Science Laboratory (*Mahaffy et al.,* 2012).

For Venus, useful measurements of the $^{34}S/^{32}S$- and $^{33}S/^{32}S$-isotopic ratios begin at the 1% level, sufficient to provide quantitative constraints on anomalous isotopic effects in photochemical hot spots. To discrimate reaction pathways, 0.1% precision is needed. Due to the atmospheric photochemical fractionation process, accurate characterization of S isotopes requires multiple samples obtained in a variety of thermochemical environments, during both day and night and over a variety of altitudes.

Water and its isotopologue HDO are also potential tracers of volcanic activity. Measurements of the abundances of magmatic H_2O and HDO released in a venusian volcanic eruption would provide unique information distinguishing recent volcanic sources from primordial or exogenic sources. This would provide insights into (1) the current rate and history of volcanic activity, (2) the efficacy of present theories of global tectonics, (3) constraints on the oxidation rate of the crust, and (4) the overall evolution of the current H_2O-poor atmosphere.

6. CONCLUSIONS

An alien visitor to our young solar system 4.1 b.y. ago would have likely found a very intriguing trio of planets situated between 0.7 and 1.7 AU from the Sun. Only ~450 m.y. old at that time, each would probably have been enshrouded by a substantial greenhouse atmosphere overlaying a surface adorned with seas of liquid water. After 300 m.y. of relative tranquility following the end of planetary accretion, our visitor would likely find primitive one-celled life flourishing on at least the middle planet, Earth. But conditions on the others would seem conducive to life as well, with atmospheres composed — as on Earth — mostly of CO_2, nitrogen, and water. The alien explorer would note that Earth has two major physical differences that set it apart from the others: First, it is enshrouded by a significant magnetic field that protects it from the Sun's onslaught of charged particles, and second, it is attended by a large satellite orbiting just 25,000 km away (*Ida et al.,* 1997) that keeps the planet's spin axis remarkably stable. Given these differences, the wondering visitor may have contemplated what would evolve on these planets during the ensuing eons. Our investigator may have wondered whether, given other potentially important conditions — such as the solar flux of both light and charged-particle radiation, the rotation rate and inclination of planetary spin, and the possibility of occasional bombardment by the remaining detritus of asteroids and comets —their atmospheres, hydrospheres, and geologies would all significantly change over the eons. And if they did, would they evolve in a similar or disparate manner? That intriguing story is the subject of the remainder of this book.

Acknowledgments. Funding for K.H.B., S.K.A., D.G., and C.T.R. was largely provided by NASA for their participation in support of ESA's Venus Express Mission. M.A.B. was supported by a Planetary Astronomy grant from the NSF. The authors would like to thank T. W. Momary of the Jet Propulsion Laboratory for his valuable contribution in creating the artwork for many figures in this work.

REFERENCES

Abbas M. M., LeClair A., Owen T., Conrath B. J., Flasar F. M., Kunde V. G., Nixon C. A., Achterburg R. K., Bjoraker G., Jennings D. J., Orton G., and Roman P. N. (2004) The nitrogen isotopic ratio in Jupiter's atmosphere from observations by the Composite Infrared Spectrometer on the Cassini spacecraft. *Astrophys. J., 602,* 1063–1074.

Abe Y. (1997) Thermal and chemical evolution of the terrestrial magma ocean. *Phys. Earth. Planet. Inter., 100,* 27–39.

Abe Y. and Matsui T. (1985) The formation of an impact-generated H_2O atmosphere and its implications for the thermal history of the Earth. *Proc. Lunar Planet. Sci. Conf. 15th,* in *J. Geophys. Res., 90,* C545–C559.

Abe Y. and Matsui T. (1986) Early evolution of the Earth: Accretion, atmosphere formation, and thermal history. *Proc. Lunar Planet. Sci. Conf. 17th,* in *J. Geophys. Res., 91,* E291–E302.

Alemi A. and Stevenson D. (2006) Why Venus has no moon. *Bull. Am. Astron. Soc., 38,* 491.

Alexander C. M. O'D., Bowden R., Fogel M. L., Howard K. T., and Herd C. D. K. (2012) The origin of water in chondrites and volatiles in the terrestrial planet region (abstract). In *Lunar Planet. Sci. Conf. 43rd,* Abstract #1929. Lunar and Planetary Institute, Houston.

Allegre C. J. and Schneider S. H. (1994.) The evolution of the Earth. *Sci. Am., 271,* 66–75.

Arvidson R. E., Bell J. F. III, Bellutta P., et al. (2010) Spirit Mars Rover mission: Overview and selected results from the northern Home Plate winter haven to the side of Scamander Crater. *J. Geophys. Res., 115,* E7, DOI: 10.1029/2010JE003633.

Atreya S. K., Mahaffy P. R., Niemann H. B., Wong M. H., and Owen T. C. (2003) Composition and origin of the atmosphere of Jupiter: An update, and implications for extrasolar giant planets. *Planet. Space Sci., 51,* 105–112.

Atreya S. K., Adams E. Y., Niemann H. B., Demick-Montelara J. E., Owen T. C., Fulchignoni M., Ferri F., and Wilson E. (2006) Titan's methane cycle. *Planet. Space Sci., 54,* 1177–1187.

Atreya S. K., Mahaffy P. R., and Wong A.-S. (2007) Methane and related trace species on Mars: Origin, loss, implications for life, and habitability. *Planet. Space Sci., 55,* 358–369.

Atreya S. K., Witasse O., Chevrier V. F., Forget F., Mahaffy P. R., Price P. B., Webster C. R., and Zurek R. W. (2011) Methane on Mars: Current observations, interpretation, and future plans. *Planet. Space Sci., 59,* 133–136.

Atreya S. K., Squyres S. W., Mahaffy P. R., et al. (2013) MSL/SAM measurements of non-condensable volatiles in the atmosphere of Mars — Possibility of seasonal variations (abstract). In *Lunar Planet. Sci. Conf. 44th,* Abstract #2130, Lunar and Planetary Institute, Houston.

Baines K. H., Bellucci G., Bibring J.-P., et al. (2000) Detection of sub-micron radiation from the surface of Venus by Cassini/VIMS. *Icarus, 148,* 307–311.

Ballentine C. J., Marty B., Sherwood Lollar B., and Cassidy M.

(2005) Neon isotopes constrain convection and volatile origin in the Earth's mantle. *Nature, 433,* 33–38.

Barabash S., Fedorov A., Sauvaud J. J., et al. (2007) The loss of ions from Venus through the plasma wake. *Nature, 450,* 650–653.

Barstow J. K., Tsang C. C. C., Wilson C. F., Irwin P. G. J., Taylor F. W., McGouldrick K., Drossart P., Piccioni G., and Tellmann S. (2012) Models of the global cloud structure on Venus derived from Venus Express observations. *Icarus, 217,* 542–560.

Barsukov V. L. (1992) Venusian igneous rocks. In *Venus Geology, Geochemistry, and Geophysics — Research Results from the USSR* (V. L. Barsukov et al., eds.), pp. 165–176. Univ. of Arizona, Tucson.

Barsukov V. L., Surkov Y. A., Dimitriyev L. V., and Khodakovsky I. L. (1986) Geochemical studies on Venus with the landers from the Vega 1 and Vega 2 probes. *Geochem. Intl., 23,* 53–65.

Barth C. A. (1974) The atmosphere of Mars. *Annu. Rev. Earth Planet. Sci., 2,* 333–367.

Battistuzzi F. U., Feijao A., and Hedges S. B. (2004) A genomic timescale of prokaryote evolution: Insights into the origin of methanogenesis, phototrphy, and the colinazation of land. *BMC Evol. Biol., 4,* 44.

Belyaev D., Korablev O., Fedorova A., Bertaux J.-L., Vandaele A.-C., Montmessin F., Mahieux A., Wilquet V., and Drummond R. (2008) First observations of SO_2 above Venus' clouds by means of solar occultation in the infrared. *J. Geophys. Res., 113,* E00B25.

Benz W. and Cameron A. G. W. (1990) Terrestrial effects of the giant impact. In *Origin of the Earth* (J. H. Jones and H. Newsom, eds.), pp. 61–68. Oxford Univ., New York.

Benz W., Slattery W. L., and Cameron A. G. W. (1986) The origin of the Moon and the single impact hypothesis I. *Icarus, 66,* 515–535.

Benz W., Slattery W. L., and Cameron A. G. W. (1987) The origin of the Moon and the single impact hypothesis II. *Icarus, 71,* 30–45.

Benz W., Cameron A. W. G., and Melosh H. J. (1989) The origin of the moon and the single impact hypothesis III. *Icarus, 81,* 113–131.

Bertaux J.-L., Vandaele A.-C., Korablev O., Villard E., Fedorova A., Fussen D., Quémerais E., Belyaev D., Mahieux A., Montmessin F., Muller C., Neefs E., Nevejans D., Wilquet V., Dubois J. P., Hauchecorne A., Stepanov A., Vinogradov I., Rodin A., and the SPICAV/SOIR Team (2007) A warm layer in Venus' cryosphere and high-altitude measurements of HF, HC_l, H_2O and HDO. *Nature, 450,* 646–649.

Bézard B., Tsang C. C. C., Carlson R. W., Piccioni G., and Marcq E. (2008) Water vapor abundance near the surface of Venus from Venus Express/VIRTIS observations. *J. Geophys. Res., 113,* E00B39.

Bogard D. D., Clayton R. N., Marti K., Owen T., and Turner G. (2001) Martian volatiles. *Space Sci. Rev., 96,* 425–458.

Boss A. P. (1986) The origin of the Moon. *Science, 231,* 341–345.

Boynton W. V., Feldman W. C., Squyres S. W., Prettyman T. H., Brückner J., Evans L. G., Reedy R. C., Starr R., Arnold J. R., Drake D. M., Englert P. A. J., Metzger A. E., Mitrofanov I., Trombka, J. I., d'Uston C., Wänke H., Gasnault O., Hamara D. K., Janes D. M., Marcialis R. L., Maurice S., Mikheeva I., Taylor G. J., Tokar R., and Shinohara C. (2002) Distribution of hydrogen in the near-surface of Mars: Evidence for subsurface ice deposits. *Science, 297,* 81-85.

Brasier M. D., Green O. R., Jephcoat A. P., Kleppe A. K., Van Kranendonk M., Lindsay J. F., Steele A., and Grassineau N. V. (2002) Questioning the evidence for Earth's oldest fossils. *Nature, 416,* 76–81.

Breuer D., Yuen D. A., and Spohn T. (1997) Phase transitions in the martian mantle: Implications for partially layered convection. *Earth Planet. Sci. Lett., 148,* 457–469.

Cameron A. G. W. (1978) Physics of the primitive solar accretion disk. *Earth Moon Planets, 18,* 5–40.

Cameron A. G. W. (1983) Origin of the atmospheres of the terrestrial planets. *Icarus, 56,* 195–201.

Cameron A. G. W. and Benz W. (1991) The origin of the Moon and the single impact hypothesis IV. *Icarus, 92,* 204–216.

Cameron A. G. W. and Ward W. R. (1976) The origin of the Moon (abstract). In *Lunar Sci. Conf. 7th,* Abstract #1041. Lunar Science Institute, Houston.

Canup R.M. (2004) Simulations of a late lunar-forming impact. *Icarus, 168,* 433–456.

Canup R. (2008) Accretion of the Earth. *Philos. Trans. R. Soc. London, Ser. A, 3366,* 4061.

Canup R. M. (2012) Forming a moon with an Earth-like composition via a giant impact. *Science, 338,* 1052–1055.

Carlson R. W., Baines K. H., Encrenaz Th., et al. (1991) Galileo infrared imaging spectroscopy measurements at Venus. *Science, 253,* 1541–1548.

Carlson R. W., Baines K. H., Girard M. A., Kamp L. W., Drossart P.,Encrenaz Th., and Taylor F. W. (1993a) Galileo/NIMS near-infrared thermal imagery of the surface of Venus (abstract). In *Lunar Planet. Sci. Conf.,* Abstract #1127. Lunar and Planetary Institute, Houston.

Carlson R. W., Kamp L. W., Baines K. H., Pollack J. B., Grinspoon D. H., Encrenaz Th., Drossart P., and Taylor F. W. (1993b) Variations in Venus cloud particle properties: A new view of Venus's cloud morphology as observed by the Galileo Near-Infrared Mapping Spectrometer. *Planet. Space Sci., 41,* 477–485.

Chassefière E. (1996) Hydrodynamic escape of hydrogen from a hot water-rich atmosphere: The case of Venus. *J. Geophys Res., 101,* 26039–26056.

Clark B. C., Castro A. J., Rowe C. D., Baird A. K., Evans P. H., Rose H. J. Jr., Toulmin P. III, Keil K., and Kelliher W. C. (1976) Inorganic analyses of martian surface samples at the Viking landing sites. *Science, 194,* 1283–1288.

Clayton R. N. (1993) Oxygen isotopes in meteorites. *Annu. Rev. Earth Planet. Sci., 21,* 115–149.

Collard A. D., Taylor F. W., Calcutt S. B., Carlson R. W., Kamp L. W., Baines K. H., Encrenaz Th., Drossart P., Lellouch E., and Bézard B. (1993) Latitudinal distribution of carbon monoxide in the deep atmosphere of Venus. *Planet. Space Sci., 41,* 487–494.

Coradini A., Federico C., and Lanciano P. (1983) Earth and Mars: Early thermal profiles. *Phys. Earth Planet. Inter., 31,* 145–160.

Cottini V., Ignatiev N. I., Piccioni G., Drossart P., Grassi D., and Markiewicz W. J. (2012) Water vapor near the cloud tops of Venus from Venus Express/VIRTIS dayside data. *Icarus, 217,* 561–589.

Cuk M. and Stewart S. T. (2012) Making the Moon from a fast-spinning Earth: A giant impact followed by resonant despinning. *Science, 338,* 1047–1052.

Cull S. C., Arvidson R. E., Catalano J. G., et al. (2010) Concentrated perchlorate at the Mars Phoenix landing site: Evidence for thin film liquid water on Mars. *Geophys. Res. Lett., 37,* DOI: 10.1029/2010GL045269.

Dauphas N. and Pourmand A. (2011) Hf-W-Th evidence for rapid growth of Mars and its status as a planetary embryo. *Nature, 473,* 489–492.

Davies G. F. (1985) Heat deposition and retention in a solid planet growing by impacts. *Icarus, 63,* 45–68.

Del Genio A. D. and Suozzo R. J. (1987) A comparative study of rapidly and slowly rotating dynamical regimes in a terrestrial general circulation model. *J. Atmos. Sci., 44,* 973–986.

Dobrovolskis A. R. and Ingersoll A. P. (1980) Atmospheric tides and the rotation of Venus. I. Tidal theory and the balance of torques. *Icarus, 41,* 1–17.

Donahue T. M. (1986) Fractionation of noble gases by thermal escape from accreting planetesimals. *Icarus, 66,* 195–210.

Donahue T. M. (1999) New analysis of hydrogen and deuterium escape from Venus. *Icarus, 141,* 226–235.

Donahue T. M. and Pollack J. B. (1983) Origin and evolution of the atmosphere of Venus. In *Venus* (D. M. Hunten et al., eds.), pp. 1003–1036. Univ. of Arizona, Tucson.

Donahue T. M. and Russell C. T. (1997) The Venus atmosphere and ionosphere and their interaction with the solar wind: An overview. In *Venus II* (S. W. Bougher et al., eds.), pp. 3–31. Univ. of Arizona, Tucson.

Donahue T. M., Grinspoon D. H., Hartle R. E., and Hodges R. R. (1997) Ion/neutral escape of hydrogen and deuterium: Evolution of water. In *Venus II* (S. W. Bougher et al., eds.), pp. 385–414. Univ. of Arizona, Tucson.

Donahue T. M., Hoffman J. H., Hodges R. R. Jr., and Watson A. J. (1982) Venus was wet: A measurement of the ratio of D/H. *Science, 216,* 630–633.

Dones L. and Tremaine S. (1993) On the origin of planetary spin. *Icarus, 103,* 67–92.

Elkins-Tanton L. T. (2008) Linked magma ocean solidification and atmospheric growth for Earth and Mars. *Earth Planet. Sci. Lett., 271,* 181–191.

Elkins-Tanton L. T. (2011) Formation of early water oceans on rocky planets. *Astrophys. Space Sci., 332,* 359–364.

Encrenaz Th., Bézard B., Greathouse T., Richter M., Lacy J., Atreya S. K., Wong A. S., Lebonnois S., Lefevre F., and Forget F. (2004) Hydrogen peroxide on Mars: Spatial distribution and seasonal variations. *Icarus, 170,* 424–429.

Farquhar J., Johnston D. T., Calvin C., and Condie K. (2004) Implications of sulfur isotopes for the evolution of atmospheric oxygen (abstract). In *Lunar Planet. Sci. Conf. 35th,* Abstract #1920. Lunar and Planetary Institute, Houston.

Farquhar J., Kim S.-T., and Masterson A. (2007) Sulfur isotope analysis of the Nakhla meteorite: Implications for the origin of sulfate and the processing of sulfur in the meteorite parent (abstract). In *Lunar Planet. Sci. Conf. 38th,* Abstract #1438. Lunar and Planetary Institute, Houston.

Fedorova A., Korablev O., Vandaele A.-C., Bertaux J.-L., Belyaev D., Mahieux A., Neefs E., Wilquet W. V., Drummond R., Montmessin F., and Villard E. (2008) HDO and H_2O vertical distributions and isotopic ratio in the Venus mesosphere by Solar Occultation at Infrared spectrometer on board Venus Express. *J. Geophys. Res.–Planets, 113,* EO0B22, DOI: 10.1029/2008JE003146.

Fedorova A., Barabash S., Sauvaud J.-A., Futaana Y., Zhang T. L., Lundin R., and Ferrier C. (2011) Measurements of the ion escape rates from Venus for solar minimum. *J. Geophys. Res.–Planets, 116,* A07220.

Feldman W. C., Boynton W. V., Tokar R. L., Prettyman T. H., Gasnault O., Squyres S. W., Elphic R. C., Lawrence D. J., Lawson S. L., Maurice S., McKinney G. W., Moore K. R., and Reedy R. C. (2002) Global distribution of neutrons from Mars: Results from Mars Odyssey. *Science, 297,* 75–78.

Formisano V., Atreya S. K., Encrenaz T., Ignatiev N., and Giuranna M. (2004) Detection of methane in the atmosphere of Mars. *Science, 306,* 1758–1761.

Galli A., Fok M. C., Wurz P., Barabash S., Grigoriev A., Futaana Y., Holmstrom M., Ekenback A., Kallio E., and Gunell H. (2008) Tailward flow of energetic neutral atoms observed at Venus. *J. Geophys. Res.–Planets, 113,* E00B15, DOI: 10.1029/2008JE003096.

Genda H. and Abe Y. (2005) Enhanced atmospheric loss on protoplanets at the giant impact phase in the presence of oceans. *Nature, 433,* 842–844.

Gladman B. J. (1997) Destination: Earth. Martian meteorite delivery. *Icarus, 130,* 228–246.

Gold T. and Soter S. (1969) Atmospheric tides and the resonant rotation of Venus. *Icarus, 11,* 356–366.

Gold T. and Soter S. (1979) Theory of Earth-synchronus rotation of Venus. *Nature, 277,* 280–281.

Gomes R., Levison H., Tsiganis K., and Morbidelli A. (2005) Origin of the cataclysmic late heavy bombardment period of the terrestrial planets. *Nature, 435,* 466–469.

Greenberg R., Wacker J. F., Hartmann W. K., and Chapman C. R. (1978) Planetesimals to planets: Numerical simulation of collisional evolution. *Icarus, 35,* 1–26.

Grimberg A., Baur H., Burnett D. S., Bochsler P., and Wieler R. (2007) The depth distribution of neon and argon in the bulk metallic glass flown on Genesis (abstract). In *Lunar Planet. Sci. Conf. 38th,* Abstract #1270. Lunar and Planetary Institute, Houston.

Grinspoon D. H. (1993) Implications of the high D/H ratio for the sources of water in Venus' atmosphere. *Nature, 363,* 428–431.

Grotzinger J. P., et al. (2012) Mars Science Laboratory mission and science investigation. *Space Sci. Rev., 170,* 5–56.

Haisch K. E. Jr., Lada E. A., and Lada C. J. (2001) Disk frequencies and lifetimes in young clusters. *Astrophys J. Lett., 553,* L153–L156.

Halliday A. N. (2000a) HF-W chronometry and inner solar system accretion rates. *Space Sci. Rev., 92,* 355–370.

Halliday A. N. (2000b) Terrestrial accretion rates and the origin of the Moon. *Earth Planet. Sci. Lett., 176,* 17–30.

Halliday A. N., Wänke H., Birck J. L., and Clayton R. N. (2001) The accretion, composition, and early differentiation of Mars. *Space Sci. Rev., 96,* 197–230.

Harper C. L. and Jacobsen S. B. (1996) Evidence for ^{182}Hf in the early solar system and constraints on the timescale of terrestrial core formation. *Geochim. Cosmochim. Acta, 60,* 1131–1153.

Harrison T. M. (2009) The Hadean crust: Evidence from >4 Ga zircons. *Annu. Rev. Earth Planet. Sci., 37,* 479–505.

Harrison T. M., Schmitt A. S., McCulloch M. T., and Lovera O. M. (2008) Early (≥4.5 Ga) formation of terrestrial crust Lu-Hf, δ^{18}O, and Ti thermometry results for Hadrean zircons. *Earth Planet. Sci. Lett., 268,* 476–486.

Hartmann W. K. (1986) Moon origin: The impact-trigger hypothesis. In *Lunar Planet. Sci. Conf. 16th,* Abstract #1164. Lunar and Planetary Institute, Houston.

Hartogh P., Lis D. C., Bockelee-Morvan D., de Val-Borro M., Biver N., Kuppers M., Emprechtinger M., Bergin E. A., Crovisier J., Rengel M., Moreno R., Szutowicz S., and Blake G. A. (2011) Ocean-like water in the Jupiter-family Comet 103P/Hartley 2. *Nature, 478,* 218–220.

Hashimoto G. L., Roose-Serote M., Sugita S., Gilmore M. S., Kemp L. W., Carlson R. W., and Baines K. H. (2008) Feslic highland crust on Venus suggested by Galileo near-infrared mapping spectrometer data. *J. Geophys. Res.–Planets, 113,* E00B24.

Hayashi C., Nakazawa K., and Mizuno H. (1979) Earth's melting due to the blanketing effect of the primordial dense atmosphere. *Earth Planet. Sci. Lett., 43,* 22–28.

Hayashi C., Nakazawa K., and Nakagawa Y. (1985) Formation of the solar system. In *Protostars and Planets II* (D. C. Black et al., eds), pp. 1100–1153. Univ. of Arizona, Tucson.

Hayes J. M., Kaplan I. R., and Wedeking K. W. (1983) In *Earth's Earliest Biosphere: Its Origin and Evolution* (J. W. Schopf, ed.), pp. 93–134. Princeton Univ., New Jersey.

Hueso R., Peralta J., and Sánchez-Lavega A. (2012) Assessing the long-term variability of Venus winds at cloud level from VIRTIS-Venus Express. *Icarus, 217,* 585–598.

Hoffman J. H., Hodges R. R. Jr., McElroy M. B., Donahue T. M., and Kolpin M. (1979) Composition and structure of the Venus atmosphere: Results from Pioneer Venus. *Science, 205,* 49–52.

Hoffman J. H., Hodges R. R. Jr., Donahue T. M., and McElroy M. B. (1980) Composition of the Venus lower atmosphere from the Pioneer Venus spectrometer. *J. Geophys. Res., 85,* 7882–7890.

Holland G. and Ballentine C. J. (2006) Seawater subduction controls the heavy noble gas composition of the mantle. *Nature, 441,* 186–191.

Hunten D. M., Pepin R. O., and Walker J. C. G. (1987) Mass fractionation in hydrodynamic escape. *Icarus, 69,* 532–549.

Ida S., Canup R. M., and Stewart G. R. (1997) Lunar accretion from an impact-generated disk. *Nature, 389,* 353–357.

Ingersoll A. P. and Dobrovolskis A. R. (1978) Venus rotation and atmospheric tides. *Nature, 275,* 37–38.

Irwin P. G. J., de Kok R., Negrão A., Tsang C. C. C., Wilson C. F., Drossart P., Piccioni G., Grassi D., and Taylor F. W. (2008) Spatial variability of carbon monoxide in Venus' mesosphere from Venus Express/Visible and Infrared Thermal Imaging Spectrometer measurements. *J. Geophys. Res.–Planets, 113,* E00B01.

Istomin V. G., Grechnev K. V., and Kochnev C. A. (1980a) Mass spectrometer measurements of the lower atmosphere of Venus. *Space Res., 20,* 215–218.

Istomin V. G., Grechnev K. V., and Kochnev C. A. (1980b) Mass spectrometry of the lower atmosphere of Venus: Krypton isotopes and other recent results of the Venera-11 and -12 data processing. In *23rd COSPAR Meeting,* Budapest, Hungary.

Jacobsen S. B. (2005) The Hf-W isotopic system and the origin of the Earth and Moon. *Annu. Rev. Earth Planet. Sci., 33,* 531–570.

Jakosky B. M. (2011) The 2013 Mars Atmosphere and Volatile Evolution (MAVEN) mission to Mars (abstract). In *American Geophysical Union, Fall Meeting,* Abstract #P23E01. American Geophysical Union, Washington, DC.

Jambon A. (1994) Earth degassing and large-scale geochemical cycling of volatile elements. In *Volatiles in Magmas* (M. R. Carroll et al., eds.), pp. 479–517. Mineralogical Society of America, Washington, DC.

Kasting J. F. (1988) Runaway and moist greenhouse atmospheres and the evolution of Earth and Venus. *Icarus, 74,* 472–494.

Kaula W. M. (1979) Thermal evolution of Earth and Moon growing by planetesimal impacts. *J. Geophys. Res., 84,* 999–1008.

Kaula W. M. (1990) Difference between the Earth and Venus arising from origin by large planetesimal infall. In *Origin of the Earth* (J. H. Jones and H. Newsom, eds.), pp. 45–57. Oxford Univ., New York.

Kaula W. M. (1999) Constraints on Venus evolution from radiogenic argon. *Icarus, 139,* 32–39.

Kaula W. M. and Phillips R. J. (1981) Quantitative tests for plate-tectonics on Venus. *Geophys. Res. Lett., 8,* 1187–1190.

Kleine T., Touboul M., Bourdon B., Nimmo F., Mezger K., Palme H., Jacobsen S. B., Yin Q. Z., and Halliday A. N. (2009) Hf-W chronology of the accretion and early evolution of asteroids and terrestrial planets. *Geochim. Cosmochim. Acta, 73,* 5150–5188.

Knoll A. H., Jolliff B. L., Farrand W. H., et al. (2008) Veneers, rinds, and fracture fills: Relatively late alteration of sedimentary rocks at Meridiani Planum, Mars. *J. Geophys. Res., 113,* E06S16, DOI: 10.1029/2007JE002949.

Konopliv A. S. and Yoder C. F. (1996) Venusian k(2) tidal love number from Magellan and PVO tracking data. *Geophys. Res. Lett., 23,* 1857–1860.

Krasnopolsky V. A. (1989) Vega mission results and chemical composition of venusian clouds. *Icarus, 80,* 202–210.

Krasnopolsky V.A. and Feldman P.D. (2001) Detection of molecular hydrogen in the atmosphere of Mars. *Science, 294,* 1914–1917.

Krasnopolsky V. A. and Gladstone G. R. (1996) Helium on Mars: EUVE and PHOBOS data and implications for Mars' evolution. *J. Geophys. Res., 101,* 15765–15772.

Krasnopolsky V. A., Bowyer S., Chakrabarti S., Gladstone G., and McDonald J. S. (1994) First measurement of helium on Mars: Implications for the problem of radiogenic gases on the terrestrial planets. *Icarus, 109,* 337–351.

Krasnopolsky V. A., Maillard J. P., and Owen T. C. (2004) Detection of methane in the martian atmosphere: Evidence for life? *Icarus, 172,* 537–547.

Kuiper G. P. (1951) On the origin of the solar system. In *Astrophysics* (J. A. Hynek, ed.), Chapter 8. McGraw-Hill, New York.

Kundt W. (1977) Spin and atmospheric tides of Venus. *Astron. Astrophys., 60,* 85–91.

Lee D.-C. and Halliday A. N. (1997) Core formation of Mars and differentiated asteroids. *Nature, 388,* 854–857.

Lee J.-Y., Kurt M., Severenghaus J. P., Kawamura K., Yoo H.-S, Lee J. B., and Kim J. S. (2006) A redetermination of the isotopic abundances of atmospheric Ar. *Geochim. Cosmochim. Acta, 70,* 4507–4512.

Liang M.-C. and Yung Y. L. (2009) Modeling the distribution of H_2O and HDO in the upper atmosphere of Venus. *J. Geophy. Res., 114,* E00B28, DOI: 10.1029/2008JE003095.

Luhmann J. G., Fedorov A., Barabash S., Carlsson E., Futaana T., Zhang T. L., Russell C. T., Lyon J. G., Ledvina S. A., and Brain D. A. (2008) Venus Express observations of atmospheric oxygen escape during the passage of several corinal mass ejections. *J. Geophys. Res.–Planets, 113,* E00B04.

Mahaffy P. R., Niemann H. B., Alpert A., Atreya S. K., Demick J., Donahue T. M., Harpold D. N., and Owen T. C. (2000) Noble gas abundance and isotope ratios in the atmosphere of Jupiter from the Galileo Probe Mass Spectrometer. *J. Geophys. Res., 105,* 15061–15072.

Mahaffy P. R., et al. (2012) The Sample Analysis at Mars investigation and instrument suite. *Space Sci. Rev., 70,* 401–478.

Mahaffy P. R., Cabane M., Webster C. R., et al. (2013) Sample Analysis at Mars (SAM) investigation: Overview of results from the first 120 sols on Mars (abstract). In *Lunar Planet. Sci. Conf. 44th,* Abstract #1395. Lunar and Planetary Institute, Houston.

Marcq E., Bézard B., Drossart P., Piccioni G., Reess J. M., and Henry F. (2008) A latiudinal survey of CO, OCS, H_2O, and SO_2 in the lower atmosphere of Venus: Spectroscopic studies by VIRTIS-H. *J. Geophys. Res.–Planets, 113,* E00B07.

Margot J.-L., Campbell D. B., Peale S. J., and Ghigo F. D. (2012) Venus length-of-day variations (abstract). In *American Astronomical Society/Division for Planetary Sciences 44,* Abstract #507.02.

Markiewicz W. J., Titov D. V., Limaye S. S., Keller H. U., Ignatiev N., Jaumann R., Thomas N., Michalik H., Moissl R., and Russo P. (2007) Morphology and dynamics of the upper cloud layers of Venus. *Nature, 450,* 633–636.

Matsui T. and Abe Y. (1986) Evolution of an impact-induced atmosphere and magma ocean on the accreting Earth. *Nature, 319,* 303–305.

McElroy M. B. and Prather M. J. (1981) Noble gases in the terrestrial planets. *Nature, 293,* 535–539.

McGouldrick K., Momary T. W., Baines K. H., and Grinspoon D. H. (2008) Venus Express/VIRTIS observations of middle and lower cloud variability and implications for dynamics. *J. Geophys. Res.–Planets, 113,* E00B14.

McGouldrick K., Baines K. H., Momary T. W., and Grinspoon D. H. (2012) Quantification of middle and lower cloud variability and mesoscale dynamics from Venus Express/VIRTIS observations at 1.74 μm. *Icarus, 217,* 615–6238.

Melosh H. J. (1988) The rocky road to panspermia. *Nature, 332,* 687–688.

Melosh H. J. (1990) Giant impacts and the thermal state of the early Earth. In *Origin of the Earth* (J. H. Jones and H. Newsom, eds.), pp. 69–84. Oxford Univ., New York.

Mitrofanov I., Anfimov D., Kozyrev A., Litvak M., Sanin A., Tret'yakov V., Krylov A., Shvetsov V., Boynton W., Shinohara C., Hamara D., and Saunders R. S. (2002) Maps of subsurface hydrogen from the high energy neutron detector, Mars Odyssey. *Science, 297,* 78–81.

Mojzsis S. J., Arrhenius G., McKeegan K. D., Harrison T. M., Nutman A. P., and Friend R. L. (1996) Evidence for life on Earth before 3,800 million years ago. *Nature, 384,* 55–59.

Moorbath S., Taylor P. N., and Jones N. W. (1986) Dating the oldest terrestrial rocks — Facts and fiction. *Chem. Geol., 57,* 63–80.

Mousis O., Gautier D., Bockelée-Morvan D., Robert F., Dubrulle B., and Drouart A. (2000) Constraints on the formation of comets from D/H ratios measured in H_2O and HCN. *Icarus, 148,* 513–525.

Mueller N., Helbert J., Hashimoto G. L., Tsang C. C. C., Erard S., Piccioni G., and Drossart P. (2008) Venus surface thermal emissions at 1 μm in VIRTIS imaging observations: Evidence for variations of crust amd mantle differentiaton conditions. *J. Geophys. Res.–Planets, 113,* E00B17.

Mukhopadhyay S. (2012) Early differentiation and volatile accretion recorded in deep-mantle neon and xenon. *Nature, 486,* 101–104.

Mumma M. J., Villanueva G. L., Novak R. E., Hewagama T., Bonev B. P., DiSanti M. A., Mandell A. M., and Smith M. D. (2009) Strong release of methane on Mars in northern summer 2003. *Science, 323,* 1041–1045.

Namiki N. and Solomon S. C. (1998) Volcanic degassing of argon and helium and the history of crustal production on Venus. *J. Geophys. Res., 103,* 3655–3678.

National Research Council (NRC) of the National Academies, Solar System Survey Space Studies Board (2003) *New Frontiers in the Solar System: An Integrated Exploration Strategy.* National Academies, Washington, DC.

National Research Council (NRC) of the National Academies, Committee on the Planetary Science Decadal Survey, Space Studies Board (2011) *Vision and Voyages for Planetary Science in the Decade 2013–2022.* National Academies, Washington, DC.

Newsom H. E. and Taylor S. R. (1989) Geochemical implications for the formation of the Moon by a single giant impact. *Nature, 338,* 29–34.

Niemann H. B., Atreya S. K., Carignan G. R., Donahue T. M., Haberman J. A., Harpold D. N., Hartle R. E., Hunten D. M., Kasprzak W. T., Mahaffy P. R., Owen T. C., and Way S. H. (1998) The composition of the jovian atmosphere as determined by the Galileo probe mass spectrometer. *J. Geophys. Res., 103,* 22831–22846.

Nimmo F. (2002) Why does Venus lack a magnetic field? *Geology, 30,* 987–990.

Nimmo F. and Stevenson D. J. (2000) Influence of early plate tectonics on the thermal evolution magnetic field of Mars. *J. Geophys. Res.–Planets, 105,* 11969–11979.

Notesco G., Bar-Nun A., and Owen T. (2003) Gas trapping in water ice at very low deposition rates and implications for comets. *Icarus, 162,* 183–189.

Nutman A. P., Mojzsis S. J., and Friend C. R. L. (1997) Recognition of ≥3850 Ma water-lain sediments in West Greenland and their significance for the early archaen Earth. *Geochim. Cosmochim. Acta, 61(12),* 2475–2474.

Owen T. C. (1992) The composition and early history of the atmosphere of Mars. In *Mars* (H. Kieffer et al., eds.), pp. 818–834. Univ. of Arizona, Tucson.

Owen T. and Bar-Nun A. (1995) Comets, impacts and atmospheres II. Isotopes and noble gases. In *Volatiles in the Earth and Solar System* (K. A. Farley, ed.), pp. 123–138. AIP Conf. Proc. 341, American Institute of Physics, New York.

Owen T. C., Biemann K., Rushneck D. R., Biller J. E, Howath D. W., and Lafleur A. L. (1977) The composition of the atmosphere at the surface of Mars. *J. Geophys. Res., 82,* 4635–4639.

Owen T., Bar-Nun A., and Kleinfeld I. (1992) Possible cometary origin of heavy noble gases in the atmospheres of Venus, Earth and Mars. *Nature, 358,* 43–46.

Oyama V. I., Carle G. C., Woeller F., Pollack J. B., Reynolds R. T., and Craig R. A. (1980) Pioneer Venus gas chromatography of the lower atmosphere of Venus. *J. Geophys. Res., 85,* 7891–7902.

Ozima M. and Podosek F. A. (1983) *Noble Gas Geochemistry.* Cambridge Univ., New York. 367 pp.

Ozima M. and Podosek F. A. (2002) *Noble Gas Geochemistry.* Cambridge Univ., New York.

Peck W. H., Valley J. W., Wilde S. A., and Graham C. M. (2001) Oxygen isotope ratios and rare earth elements in 3.3 to 4.4 Ga zircons: Ion microprobe evidence for high $\delta^{18}O$ continental crust in the early Archean. *Geochim. Cosmochim. Acta, 65,* 4215–4229.

Pepin R. O. (1991) On the origin and early evolution of terrestrial planet atmospheres and meteoritic volatiles. *Icarus, 92,* 2–79.

Pepin R. O. (1997) Evolution of the Earth's noble gases: Consequences of assuming hydrodynamic loss driven by giant impact. *Icarus, 126,* 148–156.

Pepin R. O. (2000) On the isotopic composition of primordial xenon in terrestrial planet atmospheres. *Space Sci. Rev., 92,* 371–395.

Pepin R. O. (2006) Atmospheres on the terrestrial planets: Clues to origin and evolution. *Earth Planet. Sci. Lett., 252,* 1–14.

Phillips J. L. and Russell C. T. (1987) Upper limits on the intrinsic magnetic field of Venus. *J. Geophys. Res.–Space Phys., 92,* 2253–2263.

Picardi G., Plaut J. J., Biccari D., et al. (2005) Radar soundings of the subsurface of Mars. *Science, 310,* 1925–1928.

Piccioni G., Zasova L. Migliorini A., Drosssrt P., Sjhakun A., Garcia-Muñoz A., Mills F. P., and Cardeson-Moinelo A. (2008) Near-IR oxygen nightglow observed by VIRTIS in the Venus upper atmosphere. *J. Geophys. Res.–Planets, 113,* E00B38.

Porcelli D. and Pepin R. O. (2000) Rare gas constraints on early Earth history. In *Origin of the Earth and Moon* (R. M. Canup and K. Righter, eds.), pp. 435–458. Univ. of Arizona, Tucson.

Russell C. T. (1980) Planetary magnetism. *Rev. Geophys. Space Phys., 18,* 77–106.

Russell C. T., Zhang T. I., Delva M., Magnes W., Strangeway R. J., and Wei H. Y. (2007) Lightning on Venus inferred from whistler-mode waves in the ionopshere. *Nature, 450,* 661–662.

Russell C. T., Zhang T. I., and Wei H. Y. (2008) Whistler mode waves from lightning on Venus: Magnetic control of ionospheric access. *J. Geophys. Res.–Planets, 113(E5),* DOI: 10.1029/2008JE003137.

Saal A. E., Hauri E. H., Langmuir C. H., and Perfit M. R. (2002) Vapour undersaturation in primitive mid-oceanridge basalt and the volatile content of Earth's upper mantle. *Nature, 419,* 451–455.

Safronov V. S. (1969) *Evolution of the Protoplanetary Cloud and Formation of the Earth and Planets* (in Russian). Nauka, Moscow.

Safronov V. S. (1978) The heating of the Earth during its formation. *Icarus, 33,* 3–12.

Sánchez-Lavega A., Hueso R., Piccioni G., Drossart P., Peralta J., Perez-Hoyos S., Wilson C. F., Taylor F. W., Baines K. H., Luz D., Erard S., and Lebonnois S. (2008) Variable winds on Venus mapped in three dimensions. *Geophys. Res. Lett., 35,* DOI: 10.1029.2008GL033817.L13204.

Sasaki S. (1991) Off-disk penetration of ancient solar wind. *Icarus, 91,* 29–38.

Sasaki S. and Nakazawa K. (1988) Origin of isotopic fractionation of terrestrial Xe: Hydrodynamic fractionation during escape of the primordial H_2-He atmosphere. *Earth Planet. Sci. Lett., 89,* 323–334.

Schidlowski M. (1988) A 3,800-million-year isotopic record of life from carbon in sedimentary rocks. *Nature, 333,* 313–318.

Schopf J. W. (1992) The oldest fossils and what they mean. In *Major Events in the History of Life* (J. W. Schopf, ed.), pp. 29–63. Jones and Bartlett, Boston.

Schopf J. W. (1993) Microfossils in the early Archean Apex chert: New evidence for the antiquity of life. *Science, 260,* 640–646.

Schubert G. (1983) General circulation and the dynamical state of the Venus atmosphere. In *Venus* (D. M. Hunten et al., eds.), pp. 681–765. Univ. of Arizona, Tucson.

Schubert G., Solomatov V. S., Tackley P. J., and Turcotte D. L. (1997) Mantle convection and the thermal evolution of Venus. In *Venus II* (S. W. Bougher et al., eds.), pp. 1245–1287. Univ. of Arizona, Tucson.

Seiff A. (1983) Thermal structure of the atmosphere of Venus. In *Venus* (D. M. Hunten et al., eds.), pp. 154–158. Univ. of Arizona, Tucson.

Sjogren W. L., Banerdt W. B., Chodas P. W., Konopliv A. S., Balmino G., Barriot J. P., Arkani-Hamed J., Colvin T. R., and Davies M. E. (1997) The Venus gravity field and other geodetic parameters. In *Venus II* (S. W. Bougher et al., eds.), pp. 1125–1161. Univ. of Arizona, Tucson.

Sleep N. H. (1994) Martian plate tectonics. *J. Geophys. Res., 99,* 5639–5655.

Sleep N. H. (2000) Evolution of the mode of convection within terrestrial planets. *J. Geophys. Res., 105,* 17563–17578.

Sleep N. H. and Zahnle K. (1998) Refugia from asteroid impacts on early Mars and the early Earth. *J. Geophys. Res., 103,* 28529.

Sleep N. H., Bird D. K., and Pope E. C. (2011) Serpentinite and the dawn of life. *Philos. Trans. R. Soc. London, Ser. B, 366,* 2857–2869.

Smrekar S. E., Stofan E. R., Nueller N., Treiman A., Elkins-Tanton L., Helbert J., Piccioni G., and Drossart P. (2010) Recent hotspot volcanism on Venus from VIRTIS emissivity data. *Science, 328,* 605–608.

Smith M. D. (2002) The annual cycle of water vapor on Mars as observed by the Thermal Emission Spectrometer. *J. Geophys. Res., 107(E11),* 5115, DOI: 10.1029/2001JE001522.

Solomon S. C., Smrekar S. E., Bindschadler D. L., Grimm R. E., Kaula W. M., McGill G. E., Phillips R. J., Saunders R. S., Schubert G., Squyres S. W., and Stofan E. R. (1992) Venus tectonics: An overview of Magellan observations. *J. Geophys. Res., 97,* 13199–13256.

Squyres S. W. and Knoll A. H. (2005) Sedimentary rocks at Meridiani Planum: Origin, diagetiesis, and implications for life on Mars. *Earth Planet. Sci. Lett., 240,* 1–10.

Squyres S.W., Grotzinger J. P., Arvidson R. E., et al. (2004) In-situ evidence for an ancient aqueous environment at Meridiani Planum, Mars. *Science, 306,* 1709–1714.

Stevenson D. J. (1987) Origin of the Moon — The collision hypothesis. *Annu. Rev. Earth Planet. Sci., 15,* 271–315.

Stevenson D. J. (2003) Planetary magnetic fields. *Earth Planet. Sci. Lett., 208,* 1–11.

Stevenson D. J. (2005) The oxygen isotope similarity between the Earth and Moon — Source region or formation process? (abstract). In *Lunar Planet. Sci. Conf. 36th*, Abstract #2382. Lunar and Planetary Institute, Houston.

Stevenson D. J. (2006) Evolution of Venus: Initial conditions, internal dynamics and rotational state (abstract). In *AGU Chapman Conference on Exploring Venus as a Terrestrial Planet,* American Geophysical Union, Washington, DC.

Stevenson D. J., Spohn T., and Schubert G. (1983) Magnetism and thermal evolution of the terrestrial planets. *Icarus, 54,* 466–489.

Surkov Y. A., Barsukov V. L., Moskalyeva L. P., Kharyukova V. P., and Kemurdzhian A. L. (1984) New data on the composition, structure and properties of Venus rock obtained by Venera 13 and Venera 14. *Proc. Lunar Planet. Sci. Conf. 14th,* in *J. Geophys. Res., 89,* pp. B393–B402.

Surkov I. A., Kirnozov F. F., Glazov V. N., Dunchenko A. G., and Tatsy L. P. (1987) Uranium, thorium, and potassium in the Venusian rocks at the landing sites of Vega 1 and 2. *J. Geophys. Res., 92,* 537.

Svedhem H, Titov D. V., Taylor F. W., and Witasse O. (2010) Venus Express mission. *J Geophys.Res., 114,* E00B33, DOI: 1029/2008JE003290.

Tellmann S., Pätzold M., Häusler B., Bird M. K., and Tyler G. L. (2008) Structure of the Venus neutral atmosphere as observed by the radio science experiment VeRa on Venus Express. *J. Geophys Res., 114,* E00B36.

Titov D. V., Markiewicz W. J., Ignatiev N. I., Song L., Limaye

S. S., Sanchez-Lavega A., Hesemann J., Almeida M., Roatsch T., Matz K.-D., Scholten F., Crisp D., Esposito L. W., Hviid S. F., Jaumann R., Keller H. U., and Moissl R. (2012) Morphology of the cloud tops as observed by the Venus Express Monitoring Camera. *Icarus, 217,* 682–701.

Tonks W. B. and Melosh H. J. (1993) Magma ocean formation due to giant impacts. *J. Geophys. Res., 98,* 5319–5333.

Touboul M., Kleine T., Bourdon B., Palme H., and Wieler R. (2007) Late formation and prolonged differentiation of the Moon inferred from W isotopes in lunar metals. *Nature, 450,* 1206.

Tsang C. C. C., Irwin P. G. J., Wilson C. F., Taylor F. W., Lee C., de Kok R., Drossart P., Piccioni G., Bezard B., and Calcutt S. (2008) Tropospheric carbon monoxide concentrations and variability on Venus from Venus Express/VIRTIS-M observations. *J. Geophys. Res.–Planets, 113,* E00B08.

Tsang C. C. C., Taylor F. W., Wilson C. F., Liddell S. J., Irwin P. G. J., Piccioni G., Drossart P., and Calcutt S. B. (2009) Variability of CO concentrations in the Venus troposphere from Venus Express/VIRTIS using a band ratio tchnique. *Icarus, 201,* 432–443.

Turcotte D. L. and Schubert G. (1982) *Geodynamics.* Wiley, New York. 450 pp.

Turcotte D. L. and Schubert G. (1988) Tectonic implications of radiogenic noble gases in planetary atmospheres. *Icarus, 74,* 36–46.

Valley J. W., Peck W. H., King E. M., and Wilde S. A. (2002) A cool early Earth. *Geology, 30,* 351–354.

Vandaele A. C., De Mazière M., Drummond R., Mahieux A., Neefs E., Wilquet V., Korablev O., Fedorova A., Belyaev D., Montmesssin F., and Bertaux J.-L. (2008) Composition of the Venus mesosphere measured by Solar Occultation at infrared on board Venus Express. *J. Geophys. Res.–Planets, 113,* E00B23.

Von Weizsäcker C. F. (1944) Über die Entstehung des Planetensystems. Mit 2 Abbildungen. *Z. Astrophys., 22,* 319.

Von Zahn U., Komer S., Wieman H., and Prinn R. (1983) Composition of the Venus atmosphere. In *Venus* (D. M. Hunten et al., eds.), pp. 297–430. Univ. of Arizona, Tucson.

Wänke H. and Dreibus G. (1986) Geochemical evidence for the formation of the Moon by impact-induced fission of the proto-Earth. In *Origin of the Moon* (W. K. Hartmann et al., eds.), pp. 649–672. Lunar and Planetary Institute, Houston.

Watson E. B. and Harrison T. M. (2005) Zircon thermometer reveals minimum melting conditions on earliest Earth. *Science, 308,* 841–844.

Watson E. B. and Harrison T. M. (2006) Response to comments on "Zircon thermometer reveals minimum melting conditions on earliest Earth." *Science, 311,* 779.

Watson E. B., Thomas J. B., and Cherniak D. J. (2007) ^{40}Ar retention in the terrestrial planets. *Nature, 449,* 299–304.

Webster C. R., Mahaffy P. R., Leshin L. A., et al. (2013a) Mars atmospheric escape recorded by H, C and O isotope ratios in carbon dioxide and water measured by the SAM Tunable Laser Spectrometer on the Curiosity Rover (abstract). In *Lunar Planet. Sci. Conf. 44th,* Abstract #1365. Lunar and Planetary Institute, Houston.

Webster C. R., Mahaffy P.R., Atreya S. K., Flesch G. J., and Farley K. A. (2013b) Measurements of Mars methane at Gale Crater by the SAM Tunable Laser Spectrometer on the Curiosity Rover (abstract). In *Lunar Planet. Sci. Conf. 44th,* Abstract #1366. Lunar and Planetary Institute, Houston.

Wetherill G. W. (1980) Formation of the terrestrial planets. *Annu. Rev. Earth Planet. Sci., 13,* 201–240.

Wetherill G. W. (1981) Solar wind origin of ^{36}Ar on Venus. *Icarus, 46,* 70–80.

Wetherill G. W. (1988) Formation of the Earth. In *Origin and Extinctions* (D. E. Osterbrock et al., eds.), pp. 43–81. Oxford Univ., New York.

Wetherill G. W. (1990) Formation of the Earth. *Annu. Rev. Earth Planet. Sci., 118,* 205–256.

Whitehouse M. J., Kamber B. S., and Moorbath S. (1999) Age significance of U-Th-Pb zircon data from early Archean rocks of west Greenland: A reassessment based on combined ion-microprobe and imaging studies. *Chem. Geol., 160,* 201–224.

Wilde S. A., Valley J. W., Peck W. H., and Graham C. M. (2001) Evidence from detrital zircons for the existence of continental crust and oceans on Earth 4.4 Gyr ago. *Nature, 409,* 175–178.

Wood J. A. (2005) The chondrite types and their origins. In *Chondrites and the Protoplanetary Disk* (A. N. Krot et al., eds.), pp. 953–971. ASP Conf. Ser. 341, Astronomical Society of the Pacific, San Francisco.

Yoder C. F. (1997) Venusian spin dynamics. In *Venus II* (S. W. Bougher et al., eds.), pp. 1087–1124. Univ. of Arizona, Tucson.

Yurimoto H., Kuramoto K., Krot A. N., Scott E. R. D., Cuzzi J. N., Thiemans M. H., and Lyons J. R. (2006) Origin and evolution of oxygen isotopic compositions of the solar system. In *Protostars and Planets V* (B. Reipurth et al., eds.), pp. 849–862. Univ. of Arizona, Tucson.

Zahnle K. J. (1993) Planetary noble gases. In *Protostars and Planets III* (E. H. Levy et al., eds.), pp. 1305–1338. Univ. of Arizona, Tucson.

Zahnle K. and Kasting J. F. (1986) Mass fractionation during transonic escape and implications for loss of water from Mars and Venus. *Icarus, 68,* 462–480.

Zahnle K. J. and Walker J. C. G. (1982) Evolution of solar ultraviolet luminosity. *Rev. Geophys. Space Phys., 20,* 280–292.

Zahnle K., Kasting J. F., and Pollack J. B. (1988) Evolution of a steam atmosphere during Earth's accretion. *Icarus, 74,* 62–97.

Zahnle K., Kasting J., and Pollack J. (1990a) Mass fractionation of noble gases in diffusion-limited hydrodynamic hydrogen escape. *Icarus, 84,* 502–527.

Zahnle K., Pollack J. B., and Kasting J. F. (1990b) Xenon fractionation in porous planetesimals. *Geochim. Cosmochim. Acta, 54,* 2577–2586.

Part II:
Greenhouse Effect and Atmospheric Dynamics

Covey C., Haberle R. M., McKay C. P., and Titov D. V. (2013) The greenhouse effect and climate feedbacks. In *Comparative Climatology of Terrestrial Planets* (S. J. Mackwell et al., eds.), pp. 163–179. Univ. of Arizona, Tucson, DOI: 10.2458/azu_uapress_9780816530595-ch007.

The Greenhouse Effect and Climate Feedbacks

Curt Covey
Lawrence Livermore National Laboratory

Robert M. Haberle and Christopher P. McKay
NASA Ames Research Center

Dmitri V. Titov
European Space Agency

This chapter reviews the theory of the greenhouse effect and climate feedback. It also compares the theory with observations, using examples taken from all four known terrestrial worlds with substantial atmospheres: Venus, Earth, Mars, and Titan. The greenhouse effect traps infrared radiation in the atmosphere, thereby increasing surface temperature. It is one of many factors that affect a world's climate. (Others include solar luminosity and the atmospheric scattering and absorption of solar radiation.) A change in these factors — defined as climate forcing — may change the climate in a way that brings other processes — defined as feedbacks — into play. For example, when Earth's atmospheric carbon dioxide increases, warming the surface, the water vapor content of the atmosphere increases. This is a positive feedback on global warming because water vapor is itself a potent greenhouse gas. Many positive and negative feedback processes are significant in determining Earth's climate, and probably the climates of our terrestrial neighbors.

1. INTRODUCTION

The greenhouse effect occurs when short-wavelength solar radiation penetrates an atmosphere more readily than the long-wavelength infrared radiation (IR) emanating from the surface and from the atmosphere itself. *Fourier* (1827) pointed out that this effect would warm Earth's surface by trapping heat supplied by the Sun. He also noted that freely rising warm air — natural convection — would counteract the effect to some extent (so the greenhouse analogy is somewhat misleading). Subsequent nineteenth- and twentieth-century work beginning with laboratory measurements by *Tyndall* (1873) quantified IR absorption by water vapor, carbon dioxide (CO_2), and other atmospheric constituents. *Arrhenius* (1896) and *Calendar* (1938) then computed surface temperature increases due to atmospheric CO_2 produced by fossil fuel burning. These were only hypothetical increases because the ocean could potentially absorb nearly all the CO_2. But when *Revelle and Suess* (1957) demonstrated that a substantial fraction of human-produced CO_2 remains in the atmosphere, longstanding theory then implied that global warming would follow.

Although human-induced global warming (a change of temperature) is logically distinct from the natural greenhouse effect and its influence on equilibrium steady-state temperature, the two processes are related. Indeed, as observations of human-produced atmosphere CO_2 accumulated (*Keeling et al.,* 1976), spacecraft observations of Venus found both a massive CO_2 atmosphere and an extremely hot surface (*Hunten and Goody,* 1969; *Pollack et al.,* 1980). Modern observations have confirmed natural greenhouse effects on Earth and other worlds (*Houghton,* 2002) as well as "very likely" anthropogenic global warming on Earth during recent decades (*Solomon et al.,* 2007).

The greenhouse effect is only one of many factors that determine climate. Solar luminosity is the ultimate energy source of a world unless its internal heat sources are significant. Atmospheric gases, cloud particles, aerosols (other suspended particles), and the surface absorb and scatter solar radiation as well as IR; at the temperatures considered in this chapter they also emit significant amounts of IR. Therefore any change of solar behavior, atmospheric constituents, or surface properties can potentially change weather and climate. Traditionally such changes are divided into "forcing factors" external to the system under consideration (e.g., solar luminosity variations, asteroid and comet impacts, volcanic eruptions) and "feedback processes" that result from the initial climate change due to the forcing factors. Feedback can either be positive or negative, i.e., either reinforcing or counteracting the initial climate change.

This chapter applies the above concepts to the four known terrestrial worlds possessing substantial atmospheres: Venus, Earth, Mars, and Titan. In this chapter a "terrestrial" world is defined as mainly rocky in composition, thus excluding the large gaseous and watery planets, and "substantial atmosphere" means (rather arbitrarily) a surface pressure ≥1 mbar, thus excluding Triton. It so happens that none of the four selected worlds possesses internal heat sources large enough to directly affect its global weather and climate.

2. RADIATIVE BALANCE AND GLOBAL TEMPERATURE

2.1. The Greenhouse Effect and Global Energy Balance

2.1.1. Wavelength dependence of infrared emission. Spacecraft observing Earth and other terrestrial worlds detect atmospheric IR absorption at the wavelengths and intensities predicted by molecular physics. Figures 1–3 follow *Pierrehumbert* (2011) in illustrating this principle for Earth, Venus, and Mars respectively. Figure 1 shows an early weather satellite's observation of IR looking down at the Mediterranean Sea in the absence of clouds (*Hanel et al.,* 1971). Sufficiently thick clouds (e.g., Earth's liquid water clouds) absorb and emit IR as blackbodies, interacting with IR to the maximum possible extent. In contrast, the main gaseous components of an atmosphere often do not interact with IR significantly. Interaction is normally weak for diatomic molecules like oxygen (O_2) and nitrogen (N_2) that have relatively few rotational and vibrational degrees of freedom. More complex molecules like CO_2, methane (CH_4), ozone (O_3), nitrous oxide (N_2O), and water vapor (H_2O) can dominate the greenhouse effect under clear-sky conditions, even if they are present only in trace concentrations. This process is evident in Fig. 1.

In addition to observed outgoing IR intensity (energy flux per unit area per unit solid angle), Fig. 1 plots the Planck blackbody function $2hc^2\bar{\nu}^3/\exp(hc\bar{\nu}/kT)-1$ as a function of inverse wavelength ($\bar{\nu}$) for several values of temperature (T). Infrared intensities observed in "window" regions of the spectrum, at wavelengths for which trace gas absorption is negligible, are consistent with blackbody emission directly from a surface with T > 280 K. In contrast, intensities where absorption from various trace gases is significant — as marked in the figure — are lower and correspond to colder temperatures. The most prominent minimum in effective radiating temperature is T = 220 K in the CO_2 absorption band centered at $\bar{\nu}$ = 670 cm^{-1} (around 15 μm wavelength). "That dip represents energy that would have escaped to space were it not for the [IR] opacity of CO_2" (*Pierrehumbert,* 2011). T = 220 K is a temperature characteristic of the lower and middle stratosphere. Figure 1 thus shows that at wavelengths near 15 μm, IR emission originates mainly from the stratosphere, because emission from lower levels of the atmosphere is prevented from escaping to space. This is direct evidence of the CO_2 greenhouse effect.

Figure 2 shows analogous spacecraft data from Venus. (Note the scale expansion for near-IR frequencies $\bar{\nu}$ > 1500 cm^{-1}, corresponding to wavelengths <7 μm. This part of the spectrum is an informative diagnostic of lower atmospheric and surface conditions, although its energy fluxes are small compared with the longer wavelength thermal-IR region.) *In situ* observations by a long series of American and Soviet space probes have found that T exceeds 700 K near the surface under an atmosphere nearly

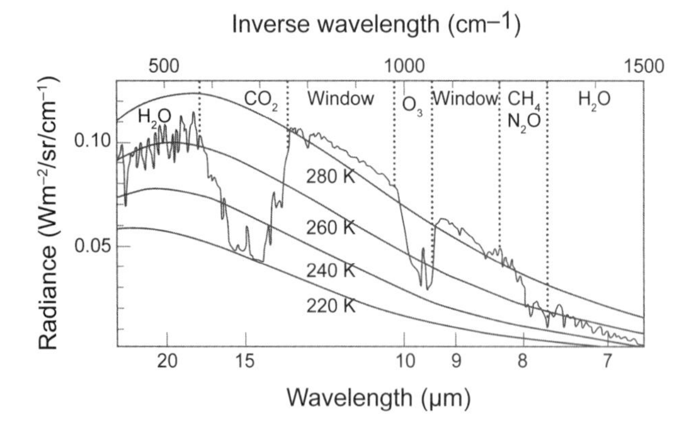

Fig. 1. Intensity of outgoing IR as a function of wavelength, as observed by the Nimbus 4 satellite over the Mediterranean Sea. Clouds were not present in this area at the time of observation. Smooth lines give Planck blackbody IR emission intensities at selected temperatures. Principal molecular IR-absorption bands are indicated. In the "window" regions outside these bands, outgoing IR intensity comes directly from the sea surface with temperature slightly greater than 280 K. Redrawn from *Houghton* (2009).

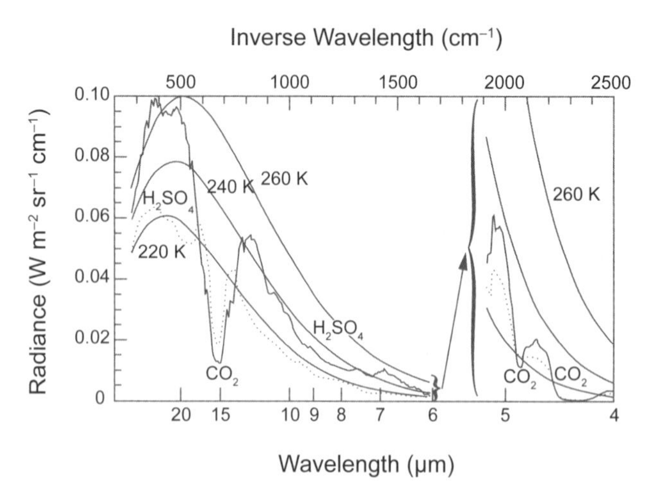

Fig. 2. Same as Fig. 1 for Venus, observed by the Venera 15 and Galileo spacecraft over the equator (solid lines) and higher latitudes (dotted lines). Outgoing intensities from near-IR wavelengths (right side) are multiplied by a factor of 100. Redrawn from *Titov et al.* (2007).

100 times more massive than Earth's (*Schubert,* 1983). The observed effective radiating temperatures, however, lie in the range 220–260 K, characteristic of the top levels of Venus' planet-wide sulfuric acid (H_2SO_4) clouds. Nearly all IR from the hot surface and lower atmosphere is absorbed before it reaches the cloud tops. The clouds themselves are responsible for a good deal of this absorption, but the sharp minimum of IR intensity around 670 cm^{-1} in Fig. 2 reveals that CO_2 plays a significant role as well — as expected since CO_2 is the major component in Venus' atmosphere (96 vol%).

Indeed, radiative-convective computations (section 2.1.3) suggest that CO_2 is the most important factor in the Venus greenhouse. Removing factors one at a time, *Titov et al.* (2007) find that by far the greatest greenhouse warming in their model is due to CO_2 (see their Table 5.1). Applying same procedure to Earth, *Schmidt et al.* (2010) identify H_2O as the most important greenhouse gas, with CO_2 and clouds tied for second place (see their Table 1). But Schmidt et al. also note that one-at-a-time removal of factors can be misleading in the presence of spectral overlaps. After taking this complication into account, they conclude "that water vapor is responsible of just over half, clouds around a quarter, and CO_2 about a fifth of the present-day total greenhouse effect" on Earth.

Satellite data for Mars analogous to Figs. 1 and 2 show the same sharp minimum due to CO_2 absorption. For example, a summer afternoon spectrum finds blackbody emission directly from the surface with T ~ 270 K in most wavelengths, but the effective emission T drops below 180 K at the center of the 670-cm^{-1} CO_2 absorption band (Fig. 3). This result is expected since the martian atmosphere is 95 vol% CO_2.

Analogous data for Titan — obtained by the same instrument that provided the Earth data in Fig. 1 — reveal only a

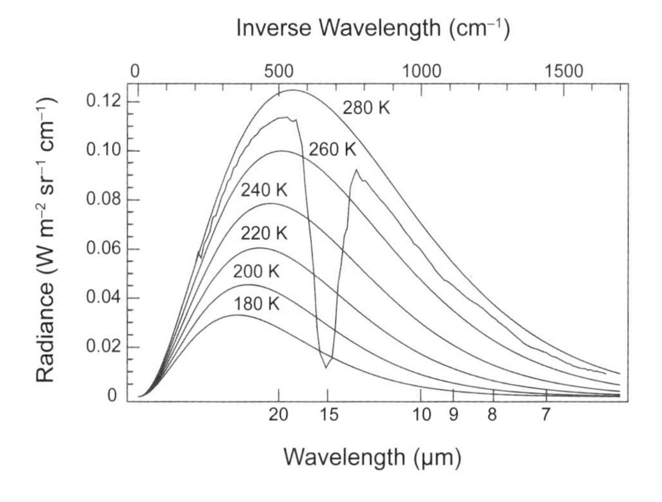

Fig. 3. Same as Figs. 1 and 2 for Mars, observed by the Mars Global Surveyor spacecraft over a cloud-free area in summer (*Christensen et al.,* 1998).

trace of CO_2 but do find strong emission bands from CH_4 and other hydrocarbons, as well as pressure-induced absorption bands of molecular hydrogen (*Hanel et al.,* 1981, 1982; *Samuelson et al.,* 1983). The hydrocarbons and molecular hydrogen (H_2) make up small but important percentages of Titan's atmosphere, which consists mainly of N_2. The pressure-induced absorption effect occurs because frequent collisions of H_2 with N_2 molecules lead to an enhancement of their IR absorption beyond the weak amounts normally expected of diatomic molecules (*Samuelson et al.,* 1981).

2.1.2. Global mean energy balance. Integrating the IR intensity escaping an atmosphere over all wavelengths and over all outward directions produces the energy flux of outgoing longwave radiation (OLR). Absent significant internal heat sources, global area- and time-averaged OLR must balance absorbed solar energy flux in a steady-state equilibrium climate. Spacecraft observations of terrestrial worlds confirm this balance. Even during the recent period of global warming on Earth, the imbalance is only a fraction of a percent (*Trenberth and Fasullo,* 2010).

The assumption of global mean energy balance leads to a simple assessment of the greenhouse effect. Given an incoming solar energy flux S and a direction- and wavelength-weighted planetary albedo α, absorbed solar energy flux is $(S/4)(1-\alpha)$ per unit area. (The factor 4 is the ratio of total surface area, $4\pi r^2$, to the surface area of a disk intercepting S, πr^2.) Setting OLR = σT_e^4 then defines an effective global mean radiating temperature $T_e = \sqrt[4]{(S/4\sigma)(1-\alpha)}$. One would expect the global mean surface temperature T_{sfc} to equal T_e in the absence of an atmosphere (given the same value of α). Table 1 compares T_e with observed T_{sfc} for Venus, Earth, Mars, and Titan. In all cases except Mars, T_{sfc} exceeds T_e, indicating greenhouse warming. Judging the magnitude of the greenhouse effect by $T_{sfc}-T_e$, Venus exhibits by far the largest, but Earth's natural greenhouse effect is also important. Without it, our world's surface would be a frigid 255 K and probably devoid of life. Titan's surface pressure (proportional to mass per unit area) is greater than Earth's and its greenhouse effect is evidently appreciable, as expected for an atmosphere including CH_4 and other greenhouse gases. Surface pressure on Mars is just 1% the value on Earth, so only a small greenhouse warming might be expected even though the martian atmosphere is nearly all CO_2.

Of course there is more than one way to define the magnitude of the greenhouse effect. Rather than $T_{sfc}-T_e$ one might consider the amount of energy flux reaching the surface that is due to IR from the atmosphere. This amount is about one-third the total energy flux reaching the surface for Mars, a significant effect. For Earth, Titan, and Venus the fractions are about 2/3, 9/10, and 999/1000 respectively (*Courtin et al.,* 1992). By this measure Titan's greenhouse is second only to that of Venus.

The Mars numbers in Table 1 actually show T_{sfc} slightly less than T_e, but this is an artifact of averaging nonlinear functions. The average of the second power of a quantity $(\overline{x^2})$ always exceeds the average value raised to the second

TABLE 1. Earthlike worlds with substantial atmospheres.

Planet/Moon	Solar Flux (S)	Planetary Albedo (α)	Effective T (T_e)	Surface T (T_{sfc})
Venus	2599 W m^{-2}	0.76	230 K	730 K
Earth	1361 W m^{-2}	0.30	255 K	288 K
Mars*	586 W m^{-2}	0.24	210 K	204 K
Titan	15 W m^{-2}	0.3	82 K	94 K

*Mars α and T_{sfc} values are from the NASA ARC Mars GCM; $T_{sfc} < T_e$ due to strong diurnal cycle and nonlinear Planck function (*Haberle,* 2013).

power ($\overline{x}^2$). The average of the fourth power ($\overline{x^4}$) exceeds the average value raised to the fourth power ($\overline{x}^4$) as long as the asymmetry of variations is modest, which is the case for planetary surface temperature. Thus in general $\overline{\sigma T^4} > \sigma \overline{T}^4$, and the space-time averaging that produced the global mean numbers in Table 1 underestimates IR emission from the surface. Defining an effective surface temperature $\sqrt[4]{\overline{T_{sfc}^4}}$ overcomes the problem. This correction is evidently not needed for the more massive atmospheres of Venus, Earth, and Titan, which moderate surface temperature changes, but for Mars it reverses the sign of $T_{sfc}-T_e$ (*Haberle,* 2013). The seasonal, day-to-night and equator-to-pole temperature contrasts on Mars are so large that direct observation of global means is difficult, and the martian numbers in Table 1 come from a general circulation model (section 3) whose output is consistent with available observations (*Haberle et al.,* 2010).

The varying number of digits in the other entries of Table 1 reflects varying accuracy of the observations. Values of S are obtained by scaling precise measurements of Earth satellites (*Loeb et al.,* 2009) by the known distances to other worlds. Values of α are less accurate due to uncertainties in the angular distribution of outward reflected solar energy flux and its space-time dependence. For Venus, spacecraft measurements give a range of α from 0.75 to 0.82, with the lower values obtained more recently (*Titov et al.,* 2013). For Earth, extensive measurements suggest a 95% confidence interval of α = 0.28–0.32 (*Covey and Klein,* 2010; but see *Kim and Ramanathan,* 2012, for a more precise claim). Other energy flux error bars are more difficult to quantify. With the exception of Mars, the temperature values in Table 1 come from *in situ* measurements: once for Titan (*Fulchignoni et al.,* 2005) several times for Venus (*Schubert,* 1983), and hundreds of thousands of times for Earth (*Jones et al.,* 1999). The observations are sufficiently accurate to unequivocally establish natural greenhouse warming on Venus, Earth, and Titan, and to suggest a small greenhouse warming in the thin martian atmosphere.

2.1.3. Radiative-convective equilibrium. The simpler climate models work with space-time averaged data. One can start with the latitude-, longitude-, and time-averaged solar energy flux incident on the atmosphere (the single number S/4; see section 2.1.2). This input and the approximations described below allow computation of temperature T as a function of altitude z only: a one-dimensional climate model (*Ramanathan and Coakley,* 1978). Given well-known interactions of greenhouse gas molecules with IR (section 2.1.1) together with some reasonable assumptions about surface and cloud-particle interactions with IR and with solar radiation, upward and downward radiative energy fluxes are computed as a function of z. Natural convection is included as an additional upward energy flux; often this is done by simply adjusting the lapse rate $|\partial T/\partial z|$ to avoid exceeding an instability criterion. The critical lapse rate is generally reached in the lower part of a sufficiently thick atmosphere because the blackbody emission function σT^4 increases linearly downward with optical depth (see, e.g., equation (2.15) in *Houghton,* 2002) and most optical depth is concentrated near the surface. The z-derivative of the total net energy flux gives a heating or cooling rate and thus the local rate of temperature change $\partial T/\partial t$. Temperature is then updated, the fluxes are recomputed, and the process continued until convergence to a radiative-convective equilibrium T(z) profile is obtained.

One might question the usefulness of a model that ignores the three-dimensional structure and circulation of an atmosphere, and accounts for turbulent vertical heat transport by the simplest possible convective adjustment. But one-dimensional models often produce results in agreement with space-time averaged observations (perhaps because vertical gradients are typically much larger than horizontal gradients in atmospheres). For example, *Manabe and Weatherald* (1967) created a one-dimensional model of Earth's atmosphere assuming current CO_2, H_2O, O_3, and cloudiness. They obtained an equilibrium temperature T(z) decreasing linearly from 288 K at the surface to about 210 K at z ~ 15 km, then increasing slowly through z > 40 km. The model's surface temperature agrees with averaged surface observations (*Jones et al.,* 1999) (see Table 1). Surface temperature would be much higher if convection did not carry heat upward (*Lindzen and Emanuel,* 2002), so the convective adjustment procedure evidently represents this heat transport appropriately. The altitude and the value of the model's T(z) minimum agree with modern observations to within a few kilometers and a few Kelvins respectively (see Fig. 3.1a in *Erying et al.,* 2010).

Analogous models for Mars (*Marinova et al.,* 2005) and Venus reproduce the much smaller and much larger greenhouse effects, respectively, on these two worlds compared with Earth. The case of Titan includes an "anti-greenhouse" effect and will be discussed in the following section. Venus provides a severe challenge to climate models due to its atmosphere's extreme temperatures, its mass (nearly two orders of magnitude greater than Earth's atmosphere), and its ubiquitous cloud cover, which absorbs nearly all the incident solar radiation. The challenge is simplified, however, by the massive atmosphere's tendency to equalize temperatures in space and time. One-dimensional models have obtained excellent agreement with T(z) observed by numerous space probes between the surface and about 70 km altitude (the level of the cloud tops), although unresolved issues persist involving undetected or poorly measured aerosols and trace gases (*Pollack et al.,* 1980; *Crisp and Titov,* 1997; *Bullock and Grinspoon,* 2001). The consistent message of the models is that Venus' high surface temperature (Table 1) arises primarily from the large amount of CO_2 in its atmosphere. Indeed, Table 1 shows that the effective radiating temperature T_e of Venus is less than that of Earth. By definition, this means that Venus absorbs *less* solar energy than Earth despite its orbit closer to the Sun, due to the high planetary albedo α arising from its clouds. The greenhouse potential of atmospheric CO_2 is thus incontestable.

2.2. Greenhouse Versus Anti-Greenhouse Effects

What if the upper atmosphere absorbed all the solar radiation that it did not reflect back to space, but at the same time had very little opacity in the IR? In that case the upper atmosphere would radiate only a small amount of outgoing longwave radiation upward to space and an equal amount downward. Energy balance at the surface would then imply that σT_{sfc}^4 = OLR, and essentially all the surface IR emission would pass through the upper atmosphere and escape to space. Energy balance at the top of the atmosphere would then imply that $(S/4)(1-\alpha) \equiv \sigma T_e^4 = \text{OLR} + \sigma T_{sfc}^4$. Elimination of OLR between the two energy balance equations gives $\sigma T_e^4 = 2\ \sigma T_{sfc}^4$: Surface temperature is reduced from the effective radiating temperature by a factor $1/2^{1/4}$ = 0.84. This reasoning is idealized not only by the assumed complete dominance of solar over IR opacity, but also by the implicit separation of the atmosphere into two distinct layers. Nevertheless, the conclusion that surface temperature is reduced 16% below T_e indicates the possibility of a significant cooling effect. *McKay et al.* (1991) called this scenario an "anti-greenhouse effect" because it reverses the greenhouse scenario, in which atmospheric optical depth is greater in the IR than in solar energy wavelengths.

Figure 4 compares the anti-greenhouse effect with the greenhouse effect in schematic examples. In these examples we assume $\alpha = 0$ for simplicity. In the no-atmosphere case, all three units of incoming solar radiation are absorbed by the surface and reradiated as OLR, and $T_{sfc} = T_e$. If an idealized purely greenhouse atmosphere is introduced, blocking IR with no effect on α, then an equilibrium is eventually obtained in which three units of OLR are still emitted to space and T_e is unchanged. But now it is the atmosphere that emits the OLR, together with an equal amount of IR downward to the surface. Thus the surface receives a total of six units of energy flux from the combination of solar and IR — twice the amount received in the no-atmosphere case — and in equilibrium $T_{sfc} = 2^{1/4}\ T_e = 1.19\ T_e$. As shown above, an idealized purely anti-greenhouse scenario gives $T_e = 0.84\ T_e$. The schematic example in the bottom picture in Fig. 4 depicts an anti-greenhouse combined with a greenhouse effect. Again we assume that $\alpha = 0$, so at equilibrium the OLR and T_e are unchanged. But now we assume an upper atmosphere absorbs two of the three units of incoming solar radiation while a lower atmosphere blocks IR, reradiating two units upward to space and an equal amount downward to the surface. As a result, the surface absorbs a total of three units of energy flux and $T_{sfc} = T_e$: In this example the greenhouse and anti-greenhouse effects cancel exactly.

McKay et al. (1991) found that on Titan, the greenhouse effect dominates, but the anti-greenhouse effect is significant. They obtained realistic numbers for Titan from Voyager spacecraft data together with a one-dimensional climate model; the numbers are consistent with subsequent Cassini-Huygens spacecraft data (*Fulchignoni et al.,* 2005; *Griffith,* 2009). On Titan the greenhouse effect is primarily due to pressure-induced IR absorption by N_2, CH_4, and H_2, while the anti-greenhouse effect is due to solar absorption by a high-altitude hydrocarbon haze layer. Titan's effective radiating temperature is T_e = 82 K. The idealized extreme anti-greenhouse would reduce this by $(1-0.84) \times 82$ K = 13 K at the surface, but the actual anti-greenhouse on Titan reduces it by only 9 K. At the same time the greenhouse effect raises surface temperature by 21 K. The net result is that T_{sfc} = 82–9 + 21 = 94 K.

2.3. Radiative Forcing, Feedback, and Climate Response

Changing the factors that affect global energy balance (Table 1) will in general change surface temperature and other aspects of a planet's climate. In studying this process, it is useful to define radiative forcing (ΔF) as the change in net energy flux at the top of the atmosphere that *would* occur *if* a given factor changed more or less instantaneously with all other climate conditions (temperature, winds, etc.) remaining constant. For example, ~0.1% peak-to-peak variations of solar luminosity have been observed over the last few sunspot cycles, implying $\Delta F \sim (0.001)(S/4)(1-\alpha) \sim 0.2$ W m^{-2} averaged over Earth's surface. This number may be compared with $\Delta F \equiv \Delta F_{2\times CO_2} \sim 4$ W m^{-2} for a hypothetical sudden doubling of atmospheric CO_2 (*Forster et al.,* 2007).

Some cautionary and clarifying points are in order. First, for most factors ΔF is a hypothetical number not subject to direct observation (although it is computable from laboratory experiments and the standard theory of electromagnetic radiation). Second, as a single number, ΔF necessarily omits variations in space-time and other details. For example, the above-noted 0.1% variations of S include much larger relative variations in the ultraviolet range of the spectrum; these can affect O_3 and temperature in the stratosphere, which in turn may affect the climate near the surface (*Gerber et al.,* 2012). Finally, the phrase "more or less instantaneous" in the definition of ΔF is deliberately chosen. It allows ΔF to incorporate rapid processes that could be reasonably considered a part of climate forcing, such as interactions of aerosols and cloud particles (*Forster et al.,* 2007) or stratospheric temperature adjustment (*Hansen et al.,* 1981). Global climate change may then be separated conceptually into short-term forcing ΔF and long-term response, with the latter produced by a combination of ΔF and a variety of feedback processes. In the following subsections we consider examples of both forcing and feedback.

2.3.1. Human perturbation of Earth's energy balance. Observations demonstrate that CO_2 and other greenhouse gases in Earth's atmosphere are increasing at unprecedented rates. Figure 5 shows time series of CO_2, CH_4, and N_2O over the last 10,000 years. Each time series combines *in situ* atmospheric measurements for the recent past with measurements of air bubbles trapped in continental ice sheets for the more distant past. Unlike the components of urban smog, CO_2, CH_4, and N_2O are chemically stable enough to be well mixed globally, so the difference between local

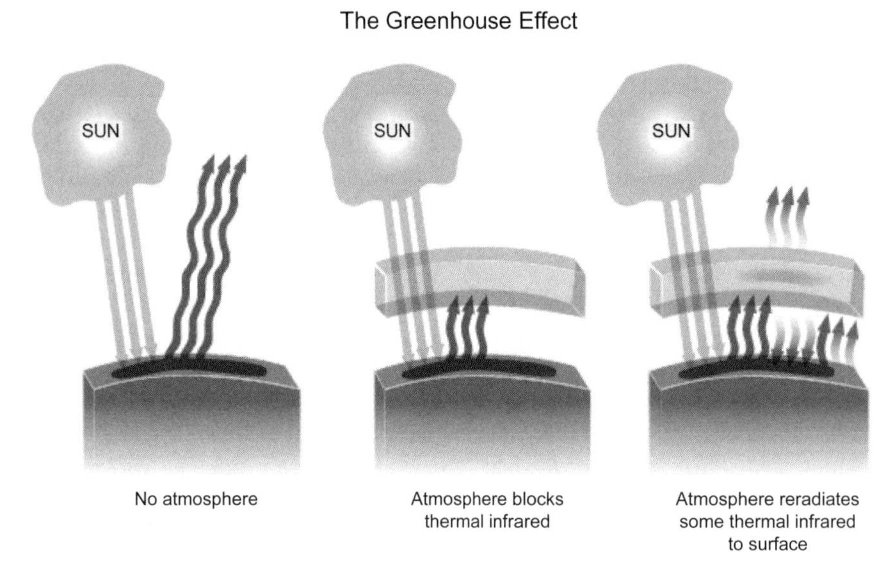

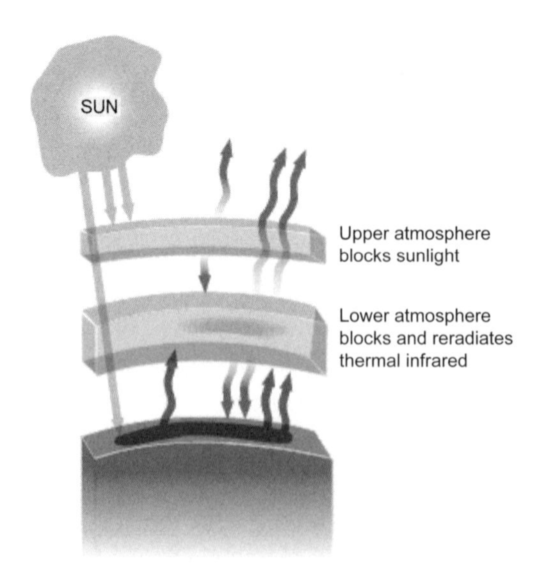

Fig. 4. Schematic comparison of the greenhouse effect (top pictures) with the anti-greenhouse effect (bottom picture). Gray arrows denote solar radiation, black arrows denote infrared radiation, and the number of arrows indicate the relative sizes of energy fluxes.

measurements and the global average is not an issue. The sharp concentration upturns of all three gases around the time of the Industrial Revolution is prima facie evidence of a human origin. The data are consistent with fossil fuel burning and other human activities such as forest clearing and agriculture. For CO_2, the record of fossil fuel use provides more than enough carbon emission to account for the atmospheric increase. The total amount of carbon produced by fossil fuel burning over the last two centuries corresponds to roughly twice the atmospheric CO_2 increase during that time. This implies that about half the emitted CO_2 has been absorbed by the ocean and the biosphere (or more than half, if forest clearing and agriculture contributed an appreciable fraction of CO_2 emissions).

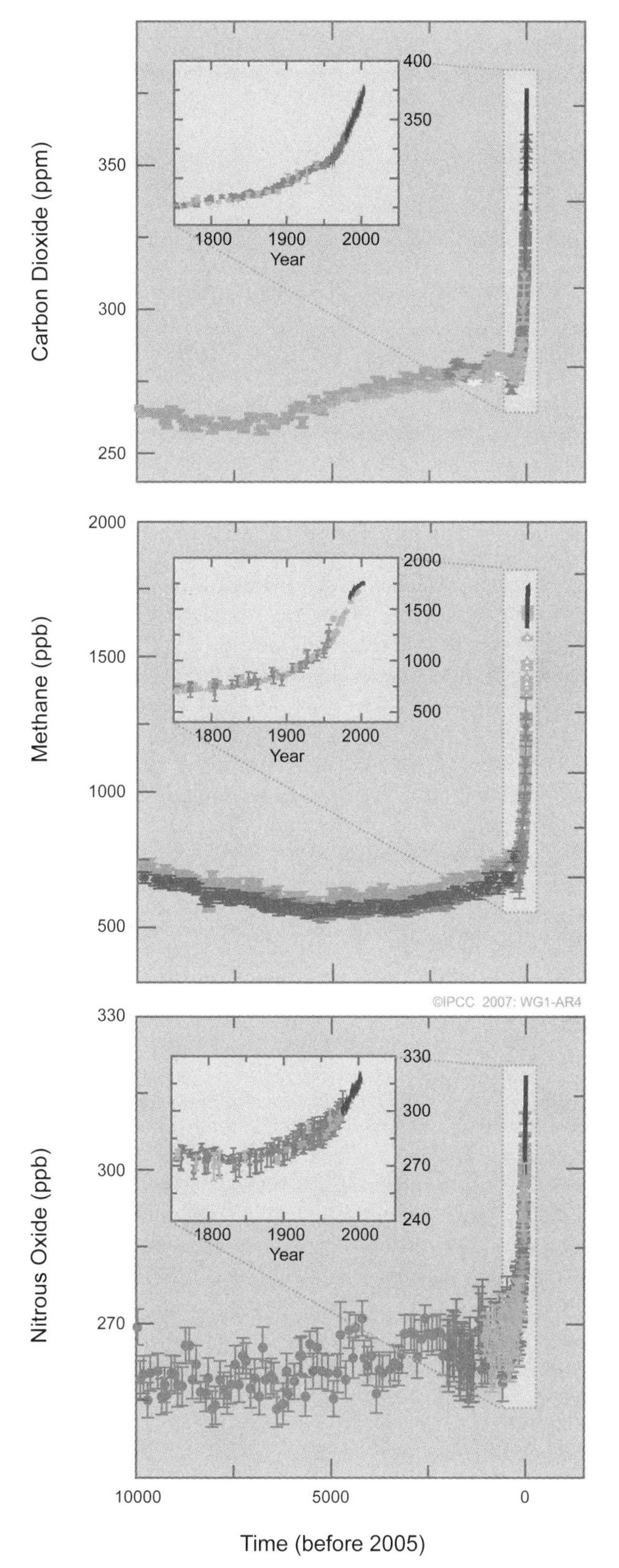

The increasing greenhouse gases shown in Fig. 5 imply a radiative forcing ΔF associated with each gas. By this measure, CO_2 is the most important greenhouse gas emitted by human activity. Taking the beginning of the Industrial Revolution (ca. 1750) as a starting point, ΔF from CO_2 exceeds ΔF from CH_4 by more than a factor of 3, which in turn exceeds ΔF from N_2O by about a factor of 3 (see Table 2). Although CO_2 absorption is saturated near the center of the 15-μm band, adding CO_2 to the atmosphere increases heat trapping by broadening the absorption band to include a wider range of wavelengths (see Fig. 1). This effect leads to an approximate logarithmic dependence of ΔF on atmospheric CO_2 increase, in contrast to a stronger (square-root or linear) dependence for CH_4, N_2O, O_3, and chlorofluorocarbons (CCl_2F_2, etc.) whose IR absorption is less saturated at band centers (*Dickinson and Cicerone,* 1986).

Table 2 shows that human greenhouse gas emissions have put Earth well on the way to an effective doubling of atmospheric CO_2 ($\Delta F_{2\times CO_2}$ ~ 4 W m^{-2}). The combined ΔF due to CO_2, CH_4, and N_2O is 2.30 ± 0.23 W m^{-2} (90% confidence interval) (*IPCC,* 2007). Also — as shown in Fig. 1 — O_3 is a greenhouse gas, and human activities affect it. Air pollution has increased O_3 in the lower atmosphere (the troposphere) while human-produced chlorofluorocarbons have decreased O_3 in the stratosphere. ΔF from O_3 is less precisely quantified than ΔF due to CO_2, CH_4, and N_2O because O_3 is not well mixed in the atmosphere, but its human-induced changes together with the chlorofluorocarbons (themselves potent greenhouse gases) may contribute up to an additional 1 W m^{-2}. Finally, carbon-rich sooty pollutants probably exert a small warming effect when deposited onto snow and ice surfaces (ΔF ~ 0.1 W m^{-2}).

Some other human activities exert a cooling effect on climate. Over the past few centuries an appreciable fraction

Fig. 5. Atmospheric concentrations of the primary well-mixed greenhouse gases in Earth's atmosphere over the last 10,000 years (large panels) and since 1750 (inset panels). Points with error bars are measurements from ice cores, with different gray shading indicating different studies. Black lines beginning ca. 1950 are direct atmospheric measurements. Redrawn from Fig. 6.4 of *Jansen et al.* (2007); the original color figure extends over the last 20,000 years, identifies the different studies, and also shows the corresponding radiative forcing (see text).

TABLE 2. Human radiative forcing of Earth's climate (*Forster et al.,* 2007).

Forcing Agent	Best Estimate	90% Confidence Interval
CO_2	1.66 W m^{-2}	1.49 to 1.83 W m^{-2}
CH_4	0.48 W m^{-2}	0.43 to 0.53 W m^{-2}
N_2O	0.16 W m^{-2}	0.14 to 0.18 W m^{-2}
Chlorofluorocarbons	0.34 W m^{-2}	0.31 to 0.37 W m^{-2}
Tropospheric O_3	0.35 W m^{-2}	0.25 to 0.65 W m^{-2}
Stratospheric O_3	–0.04 W m^{-2}	–0.15 to +0.05 W m^{-2}
Land use change	–0.20 W m^{-2}	–0.40 to 0.00 W m^{-2}
Direct H_2SO_4 aerosol	–0.40 W m^{-2}	–0.60 to –0.20 W m^{-2}
Aerosol/cloud albedo	–0.70 W m^{-2}	–1.10 to +0.40 W m^{-2}

Considering increases from the beginning of the Industrial Revolution (ca. 1750) to 2005.

of Earth's land surface area has been converted from forests to croplands and pastures (see Fig. 2.15 in *Forster et al.,* 2007). Consequent surface albedo increases may subtract up to 0.4 W m^{-2} from ΔF. Aerosol droplets containing H_2SO_4, originating from sulfur-containing pollutants, also exert cooling effects — both by direct backscattering of solar radiation to space, and by indirect effects that brighten and otherwise alter clouds. Aerosol/cloud interactions remain extremely uncertain because they involve both natural processes and measuring techniques that span many orders of magnitude in spatial scale (*Penner et al.,* 2011; *McComiskey and Feingold,* 2012; Stevens *and Boucher,* 2012). The combined direct and indirect effects of aerosols could subtract between ~0.5 and 2 W m^{-2} from ΔF.

The grand total of human-induced ΔF is positive with a 90% confidence interval of 0.5–2.5 W m^{-2} and a most likely value of 1.6 W m^{-2} (*Forster et al.,* 2007). The most likely value is just 0.7% of absorbed solar energy flux (S/4) $(1-\alpha)$, but this value is substantially larger than ΔF due to variations in solar luminosity. These variations are periodic, following the 11-year sunspot cycle, with positive and negative phases largely canceling. (The lack of a readily apparent 11-year periodicity in observed surface temperatures rules out the possibility of substantial effects on global climate that are indirectly catalyzed by phenomena connected with the solar cycle, such as magnetic and cosmic-ray fluxes.) Long-term solar variability probably contributes a ΔF of only ~0.1 W m^{-2}. Large volcanic eruptions, however, can temporarily create enough H_2SO_4 droplets in the stratosphere to reverse the human-induced upward trend in ΔF for a year or two and induce a small but measureable cooling effect for decades thereafter (*Gleckler et al.,* 2006).

In short, human perturbation of Earth's energy balance is almost certainly exerting a net warming effect ($\Delta F > 0$) and, over multidecadal timescales, is larger than natural perturbations from solar variability and volcanic eruptions.

2.3.2. Water vapor feedback and related processes. A subtle assumption in the one-dimensional model of *Manabe and Weatherald* (1967) and many related Earth climate models was to fix not atmospheric H_2O concentration *per se*, but relative humidity: the ratio of H_2O concentration to its saturation value. Correspondingly, the threshold value of lapse rate that initiates convective adjustment was reduced in magnitude to a value appropriate for moist convection. In moist convection, condensation of water vapor or another volatile substance releases latent heat above the surface and thereby adds to upward heat transport (see, e.g., section 3.2 in *Houghton,* 2002). This reduction in lapse rate would by itself tend to reduce the surface warming due to an enhanced greenhouse effect. But constant relative humidity implies that the average water vapor content of the atmosphere increases with increasing T. Since H_2O is a potent greenhouse gas, a positive feedback effect would then occur. The combined lapse rate and water vapor feedback in climate models is invariably positive. *Manabe and Weatherald* (1967) found that with constant relative humidity the global mean equilibrium warming due to doubled CO_2 in Earth's atmosphere ($\Delta T_{2\times CO_2}$) would be about 2 K. Under the assumption that to first approximation a climate response depends on the value of ΔF but not on its physical origin, $\Delta T_{2\times CO_2}$ has become a standard measure of global climate sensitivity to any external forcing. The value of $\Delta T_{2\times CO_2}$ obtained by Manabe and Weatherald lies within the likely range inferred by a variety of modern techniques, including more realistic three-dimensional general circulation modeling (see section 3 below).

In the early history of Venus, water vapor feedback may have been strong enough to preclude any equilibrium climate short of boiling away oceans and losing the hydrogen in H_2O to space: a "runaway greenhouse." This concept was first quantified by *Komabayashi* (1967) and *Ingersoll* (1969) using one-dimensional climate models and later elaborated by many investigators (e.g., *Shaviv et al.,* 2011; *Kurokawa and Nakamoto,* 2012). Although it is necessarily speculative, the possibility of a runaway greenhouse follows naturally from inspection of a phase diagram for water (*Ingersoll,* 1969). The same phase diagram illustrates why open bodies of liquid water can be common on Earth but must be essentially nonexistent on (present-day) Mars. On Earth, atmospheric pressures and temperatures lie well within the domain of liquid water's existence. On Mars, temperatures rarely exceed the freezing point and atmospheric pressures lie near the lower limit for liquid water's existence. Since the pressure is not due to H_2O vapor itself (but rather comes from CO_2), any liquid water exposed to atmosphere must quickly evaporate (e.g., *Hartmann,* 2005, Chapter 13).

Water vapor feedback is one example of a general process in which significant IR opacity comes from a gas whose atmospheric concentration depends on cycles of evaporation and condensation. This means that the gas concentration is controlled by coexistence with solid or liquid phases through the strong exponential dependence of vapor pressure on temperature. Both water vapor and carbon dioxide are involved in this process on Mars (see the following subsection). A similar positive feedback arises in Titan's atmosphere from the trace gas CH_4, which plays a role analogous to H_2O in Earth's atmosphere (*McKay et al.,* 1991; *Griffith,* 2009). Both liquid and vapor phases of CH_4 exist on Titan with ~45% relative humidity in the equatorial atmosphere (*Fulchignoni et al.,* 2005) while water vapor is frozen out. *McKay et al.* (1993) found that CH_4 feedback on Titan is about twice as strong as water vapor feedback on Earth due to (1) a steeper dependence of vapor pressure on temperature and (2) stronger pressure-induced IR absorption.

2.3.3. Orbital effects and feedbacks. A planet's orbit about the Sun and the orientation of its spin axis (obliquity) can vary due to gravitational perturbations of other planets. Orbital variations give rise to variations both of global- and annual-mean S (from eccentricity changes) and the distribution of incoming solar radiation as a function of latitude and season (from obliquity and precession changes). These energy flux variations have long been suspected of triggering Earth's ice ages, although it is not obvious how such very small forcing amplifies to produce dramatic climate change (see FAQ 6.1 in *Jansen et al.,* 2007). Mars may provide a more clear-cut example of orbital effects on climate (*Toon et al.,* 1980).

Three research developments encourage the study of ice age analogs on Mars (*Haberle et al.,* 2013). First, spacecraft have increasingly scrutinized the planet's geologic record, revealing a variety of ice-related deposits that cannot be produced in today's martian climate. Second, the increasing realism of atmospheric general circulation models (section 3) applied to Mars permits a more sophisticated interpretation of the geologic record. Third, variations in martian orbital properties are predictable and much larger than for other solar system bodies (due to the lack of a large stabilizing moon, and the proximity of massive Jupiter). Calculations show that martian obliquity has varied between 15° and 45° and the eccentricity between 0 and 0.12 during the past 20 m.y. [*Laskar et al.* (2004); by definition obliquity = 0° when the planet's spin axis is perpendicular to its orbit plane]. Beyond that point, the calculations become chaotic but obliquity excursions as high as 80° are possible. These variations lead to large changes in the latitudinal and seasonal distribution of solar radiation that can mobilize and redistribute volatile reservoirs.

The two main exchangeable volatiles are CO_2 and H_2O. CO_2 may be adsorbed at the surface (*Keiffer and Zent,* 1992) or buried as ice below the surface near the south pole (*Phillips et al.,* 2011). Taken together, these two sources can store the equivalent of between ~5 and 30 mbar of CO_2 (i.e., ~1–4 times the current atmospheric mass). Water ice is known to exist at the surface near the north pole, and in the subsurface at middle and high latitudes (*Boynton et al.,* 2002). These two sources represent a combined equivalent of at least a ~1-m-thick layer of globally distributed H_2O ice. Mobilization of CO_2 and H_2O as the martian climate warms would be roughly analogous to the H_2O and CH_4 vapor feedback processes discussed above for Earth and Titan respectively.

Obliquity variations produce the largest variations in received solar radiation and have received most of the attention in studies of recent martian climate change. At times of high obliquity (>35°) the polar regions warm, volatile reservoirs are destabilized, and surface pressures and humidities rise. The magnitude of the climate change is model dependent, but a doubling of surface pressure and an order of magnitude increase in the H_2O content of the atmosphere are plausible outcomes. Under these circumstances the greenhouse effect is strengthened and mean annual surface temperatures increase. However, the increase including only these vapor feedbacks is limited to a few Kelvins since the thin atmosphere does not significantly pressure-broaden the IR absorption lines, and atmospheric H_2O content is limited by the relatively low temperatures.

Cloud feedback could further amplify warming at high obliquity. Martian clouds tend to form at high cold altitudes, and climate models predict that they would increase in the high-obliquity atmosphere. This could lead to an increasing cloud greenhouse effect. Modeling studies suggest that the clouds can potentially warm surface temperatures by tens of Kelvins (*Haberle et al.,* 2012; *Madeleine et al.,* 2013). But as with Earth (section 3.2), cloud feedbacks are a major source of uncertainty in martian climate models. Another complication is dust storms, which should increase in frequency at high obliquity. How the dust cycle couples to the water and CO_2 cycles at times of high obliquity is unclear.

The opposite situation prevails at low obliquity. The atmosphere condenses at the poles and permanent polar CO_2 ice caps form. The surface pressure is then dictated by the polar energy balance (*Leighton and Murray,* 1966) and can fall to less than 1 mbar at 15° obliquity (*Wood et al.,* 2012). In this regime, atmospheric water vapor declines and dust storms cease, so the atmosphere should be relatively aerosol-free. The climate at low obliquity is therefore dictated mostly by the radiative balance of an airless body.

2.3.4. Time variations. The term "equilibrium" has occurred frequently in the discussion so far. It means that time variations are slow enough that departures from global mean energy balance (section 2.1.2) may be neglected. If this assumption is not valid, then one must consider the heat capacity of the climate system and (at least conceptually) the energy balance equation $(S/4)(1-\alpha) = OLR$ must be replaced by a temperature evolution equation. Such cases arise particularly on Earth, where the thermal inertia of the ocean is huge. One example is provided by volcanic aerosols (q.v.), which are removed from the atmosphere faster than the climate system can adjust to them. Two more dramatic examples are discussed below.

Pollack et al. (1983) computed the temporary cooling caused by a debris cloud following impact of a large asteroid or comet with Earth, which would block solar energy from the surface. This scenario is a massively enhanced version of the volcanic cooling mentioned above. Pollack et al. interpreted their one-dimensional model output to "represent a region near the impact site" rather than a global horizontal average. They considered two very different types of lower boundary condition: land surface that adjusts instantaneously to near-surface air temperature, and a well-mixed upper ocean layer of depth $\Delta z = 75$ m. In the ocean case the surface temperature response is buffered by the heat capacity of water $\rho c = 4 \times 10^6$ J m^{-3} K^{-1}, and the timescale for initial cooling is $(\rho c \Delta z)(288 \text{ K})/(S/4)(1-\alpha) \sim 10$ yr. Numerical results of Pollack et al. featured 40 K cooling in the land case, peaking two to five months after the impact, but only a few Kelvins cooling in the ocean case, extending well beyond two years. These results are plausible for mid-continental and open ocean areas respectively. Quantitative consideration of important issues such as land-sea heat transport and the effect of seasons, however, require three-dimensional modeling (*Toon et al.,* 1997).

Similar one-dimensional computations led to the concept of "nuclear winter." *Turco et al.* (1983) argued that worldwide smoke would arise (literally) from cities burning in the wake of a major nuclear war. The smoke could intercept over 90% of solar energy flux well above the surface, so that land areas would cool sharply within a few weeks. The amount of land surface cooling predicted by the *Pollack et al.* (1983) one-dimensional climate model varied from 5 K to 40 K depending on the war scenario and assumptions about smoke production and smoke optical properties. The same inherent limitations of one-dimensional modeling noted above for impact scenarios also apply to nuclear winter. Three-dimensional general circulation simulations have not changed the basic conclusion that mid-continental areas would suffer rapid cooling; indeed they imply that even a "minor" nuclear war could lead to drastic climate change (*Robock,* 2011). Fortunately these predictions remain untested by direct real-world observations.

Temporary cooling in the asteroid impact and nuclear winter scenarios is related to the permanent cooling of the anti-greenhouse effect (section 2.2). *Pollack et al.* (1983) noted that their land climate "cools because it receives virtually no sunlight . . . despite the greatly enhanced infrared opacity of the atmosphere." The net result is that "surface temperature tends asymptotically toward the effective temperature at which the planet radiates to space" $T_e \sim 290 \text{ K} - 40 \text{ K} \sim 250$ K (see Table 1). This is not, however, as much cooling as in the "pure" anti-greenhouse scenario, for which infrared opacity is negligible and surface temperature falls below T_e.

2.4. Nonlinear Combination of Feedback Terms

Any realistic examination of a world's climate must deal with several different feedback processes at once. In addition to water vapor and analogous volatiles (section 2.3.2), climate models of Earth and other terrestrial worlds have incorporated a long list of both positive and negative feedback ingredients. Rather than attempting to enumerate them, the following discussion emphasizes how they are distinguished from forcing and how they interact with each other.

By definition, feedback processes arise from the climate change induced by a separate initial forcing ΔF, amplifying the initial change (positive feedback) or diminishing it (negative feedback). The distinction between forcing and feedback is somewhat arbitrary. For example, one may assume an increase of atmospheric CO_2 as input to a climate model, as in section 2.3.1 and *Pollack et al.* (1987) for Earth and Mars respectively. In this case atmospheric CO_2 provides the forcing. But it is a feedback in the Mars climate models of section 2.3.3, and an "Earth system model" that includes the biosphere and ocean chemistry may take human CO_2 emissions as input, producing atmospheric CO_2 concentration together with climate as output. In this case the carbon cycle becomes a feedback process [likely positive; see *Denman et al.* (2007)]. On very long timescales, natural CO_2 emission from volcanoes and the removal of atmospheric CO_2 by rock weathering can together create a negative feedback loop (*Walker et al.,* 1981; *Goudie and Viles,* 2012).

Albedo provides one example of how feedback may be quantified. Figure 6 schematically shows how global mean absorbed solar radiation $(S/4)(1-\alpha)$ and OLR (section 2.1.2) may vary as a function of global mean surface temperature. One would normally expect that OLR increases with increasing T_{sfc} and also curves upward, i.e., both $\partial(\text{OLR})/\partial T_{sfc}$ and $\partial^2(\text{OLR})/\partial T_{sfc}^2$ are positive (unless a runaway greenhouse happens). The figure shows this behavior for both present-day and twice present-day concentrations of atmospheric CO_2, with the latter curve lowered by $\Delta F = 4$ W m^{-2} (section 2.3.1). Considering worlds like Earth and Mars on which bright surface ice melts or evaporates with increasing T_{sfc}, one might guess that absorbed solar energy flux $(S/4)(1-\alpha)$ also increases with increasing T_{sfc}. The figure shows this behavior under the additional assumption that $(S/4)(1-\alpha)$ approaches a maximum value when T_{sfc} becomes very warm and all ice disappears. With these assumptions, $\partial[(S/4)(1-\alpha)]/\partial T_{sfc}$ is positive but $\partial^2[(S/4)(1-\alpha)]/\partial T_{sfc}^2$ is negative. Cloud feedback (q.v.) also affects albedo and thus $(S/4)(1-\alpha)$, although the resulting signs of $\partial[(S/4)(1-\alpha)]/\partial T_{sfc}$ and $\partial^2[(S/4)(1-\alpha)]/\partial T_{sfc}^2$ are not obvious.

Figure 6 thus illustrates how climate forcing due to doubled CO_2 in Earth's atmosphere leads to global mean warming $\Delta T_{2\times CO_2}$ in the presence of positive surface-albedo feedback. Energy balance occurs whenever the OLR and $(S/4)(1-\alpha)$ curves intersect. Two intersections occur for each OLR curve, giving four equilibrium climate states. The pair

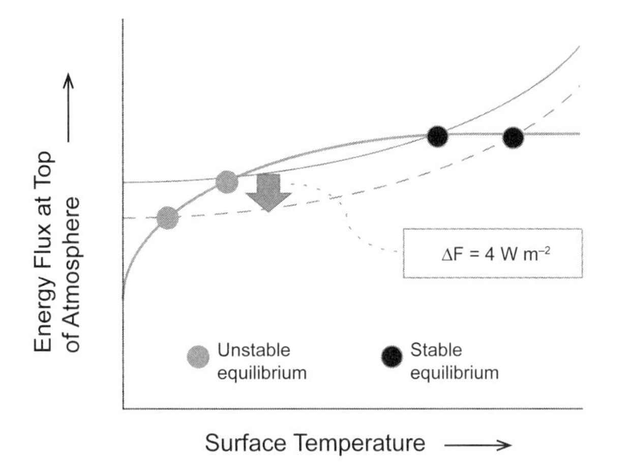

Fig. 6. Schematic of outgoing IR energy fluxes (upward-curving lines) and absorbed solar energy fluxes (downward-curving line) as a function of global mean surface temperature. Outgoing IR curves upward with increasing temperature unless a runaway greenhouse exists. Absorbed solar energy increases with temperature but asymptotes at high temperatures if snow- and ice-albedo feedback dominates. A hypothetical sudden doubling of atmospheric carbon dioxide lowers outgoing IR by ~4 W m^{-2}, as shown in the change from solid to dashed line for outgoing IR.

of intersections at lower values of T_{sfc}, however, represents unstable climate equilibrium states. In these states, slight decreases of T_{sfc} lead to OLR exceeding $(S/4)(1-\alpha)$ and slight increases of T_{sfc} lead to $(S/4)(1-\alpha)$ exceeding OLR, so that small perturbations are amplified. Since they are not physically attainable, these states can be ignored. The opposite situation occurs for the pair of intersections at higher T_{sfc}, which represents stable climate equilibrium states. For these states the figure shows that lowering the OLR curve (enhancing the greenhouse effect) leads to increasing T_{sfc}. After the climate system is perturbed by ΔF, equilibrium is restored by compensating changes in OLR and $(S/4)(1-\alpha)$. If changes in T_{sfc} are small compared with the thermodynamic absolute value of T_{sfc} then one may linearize the global energy balance condition $(S/4)(1-\alpha)$ = OLR and obtain

$$\Delta F = \frac{\partial(\text{OLR})}{\partial T_{sfc}} \Delta T_{sfc} + \frac{S}{4}\frac{\partial\alpha}{\partial T_{sfc}} \Delta T_{sfc} \Rightarrow$$
$$\Delta T_{sfc} = \frac{\Delta F}{\dfrac{\partial(\text{OLR})}{\partial T_{sfc}} + \dfrac{S}{4}\dfrac{\partial\alpha}{\partial T_{sfc}}} \qquad (1)$$

Equation (1) implies that equilibrium ΔT_{sfc} is proportional to ΔF. The proportionality constant enters as a sum of feedback terms in the denominator. In the absence of an atmosphere, only the Planck term $\partial(\text{OLR})/\partial T_{sfc} = \partial(\sigma T_{sfc}^4)/\partial T_{sfc} = 4\sigma T_{sfc}^3 = 4\sigma T_e^3$ would appear in the denominator. Substituting T_{sfc} = 255 K for Earth (Table 1) gives a Plank term 3.8 W m^{-2} K^{-1}. If this were the only feedback process operating, then taking |ΔF| ~ 4 W m^{-2} would give $\Delta T_{2\times CO_2}$ ~ 1 K. Including albedo feedback adds the term $(S/4)\ \partial\alpha/\partial T_{sfc}$ to the denominator. This term is negative in the case of surface albedo feedback discussed above, decreasing the magnitude of the denominator and increasing ΔT_{sfc}. Separately, the water vapor feedback discussed above would reduce $\partial(\text{OLR})/\partial T_{sfc}$ and further increase ΔT_{sfc}. Evidently a combination of processes in the *Manabe and Weatherald* (1967) model produced a net positive feedback that doubled ΔT_{sfc} ($=\Delta T_{2\times CO_2}$) from about 1 K to 2 K. Cloud feedbacks (not considered by Manabe and Weatherald) will in general affect both the $\partial(\text{OLR})/\partial T_{sfc}$ and $(S/4)\ \partial\alpha/\partial T_{sfc}$ terms in equation (1).

An important feature of equation (1) is that separate feedback processes combine *nonlinearly*. For example, the effect of albedo feedback is more pronounced if water vapor feedback is strongly positive, i.e., if $\partial(\text{OLR})/\partial T_{sfc}$ is relatively small. This makes identifying and quantifying different feedback processes a difficult endeavor (*Roe and Baker,* 2007). Nowadays the endeavor is pursued using detailed observations, often acquired by spacecraft, together with climate models that are considerably more complex and hopefully more sophisticated than those discussed above.

3. MODERN CLIMATE MODELS

3.1. The Need for Complexity

There are limits to the ability of one-dimensional climate models to simulate a real three-dimensional world. Some limits are obvious. A model with no geographical variation over Earth's surface cannot say much about different regional effects of global warming, for example. But other limitations are more subtle. Although a long-run global mean balance between absorbed solar and emitted OLR energy fluxes exists (section 2.1.2), the two fluxes are often far out of balance locally. On both Earth and Venus, satellite observations clearly show that absorbed solar radiation exceeds OLR at low latitudes and that the reverse is true at high latitudes (*Schofield and Taylor,* 1982). These observations imply that the atmosphere (and in Earth's case the ocean) transports heat northward and southward away from the tropics and toward the poles. Much of Earth's poleward heat transport is accomplished by longitude-varying baroclinic waves that apparently have analogs on Mars (*Read and Lewis,* 2004) and perhaps on Titan as well (*Mitchell,* 2012). In Venus' atmosphere, a strong equatorial superrotation is the most prominent observed feature of the circulation; it cannot be explained without invoking longitude-varying waves (*Gierasch et al.,* 1997). On Earth, land-sea heat transport plays a significant role in the seasonal cycle (*Fasullo and Trenberth,* 2008) and other transient climate changes (section 2.3.4). Also, the distributions of clouds and other atmospheric constituents such as local pollutants vary strongly. Such considerations have led to application of three-dimensional atmospheric general circulation models (GCMs) to all four terrestrial worlds discussed in this chapter, including details of clouds, surface terrain, etc., despite the consequent strain on computer resources.

General circulation models are closely related to numerical weather prediction models of Earth's atmosphere, but they are used for a different purpose (*Washington and Parkinson,* 2005). Predicting the weather on Earth involves carefully initializing a model with the best observations of current weather and then simulating the atmosphere's evolution over a period of at most a week or two. The object is a detailed forecast of the weather at each location on Earth. A GCM, in contrast, runs for much longer periods of (simulated) time, and the weather systems it predicts are considered valid only in a statistical sense. Initial conditions are not as important as in weather prediction. The sort of question addressed by Earth GCM climate simulations is "Will my home state suffer drought more often later in this century?" rather than "Will it rain in my hometown later this week?" General circulation models are also used to explore fundamental questions concerning the maintenance of atmospheric thermal structure and circulation on Earth and other planets.

3.2. Conservation Laws and Subgrid-Scale Parameterizations

General circulation models are computer programs that solve the partial differential equations expressing physical conservation laws for mass, momentum, and energy. These include vertical heat transport: Essentially a GCM creates a one-dimensional climate model (section 2.1.3) at each of several thousand grid-points around the world and computes radiative and convective contributions to $\partial T/\partial t$. At the same time, the GCM determines other contributions to $\partial T/\partial t$ arising from horizontal and vertical temperature advection. This leads to a self-consistent computation of winds and temperatures. A GCM simulation comes closer to "first principles" than simpler models. For example, instead of assuming any particular value of relative humidity, Earth GCMs compute the concentration of atmospheric water vapor using thermodynamic laws and the conservation of water mass.

It must not be thought, however, that GCM simulations are based only on precisely known laws of classical physics. The equations that express these laws are nonlinear owing to the advection terms in which wind velocity **v** multiplies other dynamical variables, e.g. temperature advection proportional to $\mathbf{v} \bullet \nabla T$. Solution of the equations therefore entails numerical approximation in which discrete points on a three-dimensional lattice replace spatially continuous fields, and discrete time steps replace continuous evolution in time. Roughly 100-km horizontal spacing, vertical spacing of a few kilometers, and time steps a few minutes long are used in today's GCMs. This numerical discretization in turn requires something to be assumed about phenomena at finer space-time scales.

Clouds in Earth's atmosphere provide the most studied and perhaps the most challenging example of subgrid-scale parameterization. They often vary over horizontal scales on the order of 1 km or less, so they cannot be explicitly simulated with horizontal grid points separated by ~10–100 km. Therefore Earth GCMs must use empirical parameterizations to derive the large-scale effects of clouds from large-scale explicitly simulated fields like **v**, T, and water vapor concentration. One important feedback effect of clouds involves planetary albedo. Since clouds are typically brighter than the underlying surface, they can produce either positive or negative climate feedback via the term $(S/4)\ \partial\alpha/\partial T_{sfc}$ in equation (1) depending on whether bright clouds decrease or increase, respectively, in a warmer climate. Cloud feedback can also operate by affecting OLR (both on Earth and — as discussed in section 2.3.3 — on Mars). The *Manabe and Weatherald* (1967) one-dimensional simulation of equilibrium climate sensitivity assumed fixed cloudiness and obtained $\Delta T_{2\times CO_2} = 2$ K (section 2.3.2). Different assumptions embedded in different cloud parameterizations are the primary reason that $\Delta T_{2\times CO_2}$ varies roughly between 2 and 5 K among different climate models (see Box 10.2 in *Meehl et al.,* 2007).

3.3. Successes and Limitations

Like their one-dimensional cousins, GCMs are subject to the following general warning about modeling complex systems (*Box and Draper,* 1987, p. 424): "Essentially, all models are wrong, but some are useful. However, the approximate nature of the model must always be borne in mind." General circulation models attempt to simulate details of climate and climate change on spatial scales on the order of a few grid-point spacings, or roughly a few hundred kilometers in horizontal dimensions. Subgrid-scale parameterizations, which all GCMs incorporate, aim to represent processes with space-time scales as small as micrometers and seconds. All this is done over the full global extent of a world, and typically continued for at least several Earth-years of simulated time. Obviously models this complicated will make mistakes. Of course, the mistakes are not just simple ones that are readily corrected; they include the consequences (not all of them known) of inevitable approximations.

Judging whether a GCM is useful despite its mistakes depends, of course, on the problem to which the model is applied. As the name "general circulation model" indicates, one goal is fundamental understanding of the mechanisms creating and maintaining an atmosphere's general large-scale circulation. A useful start toward this goal was achieved by the first very Earth GCM (*Phillips,* 1956) and has now been achieved by GCMs for the very un-Earthlike environments of Titan and Venus (e.g., see *Friedson et al.,* 2009; *Lebonnois et al.,* 2010; and references therein). The case of Mars exemplifies a more practical goal. A credible model bridging the gap between local meteorology and global climatology aids both the choice of spacecraft landing sites on Mars, and the interpretation of on-the-ground findings (e.g., *Haberle et al.,* 1999). Considering Earth, the most important challenge that GCMs face is predicting future climate over the next few decades and beyond. Put more precisely, the challenge is to gauge the strengths and weaknesses of GCMs, providing scientific background to the economic and political decisions that will be made about global warming. The models have already made and will continue to make predictions. Should we take these predictions seriously? Many considerations enter attempts to answer this question.

A necessary condition for believing predictions of future climate is a model's ability to simulate the present-day climate. Modern Earth GCMs produce temperature output including a seasonal cycle and large-scale geographical variations that match observations to the level of 95% correlation; precipitation is more problematic, typically a 50–60% correlation with the observations, but simulations are improving and their overall global patterns are recognizably similar to observed patterns (see *Bader et al.,* 2008, especially their Fig. 5.1). This provides some confidence in the largest-scale GCM predictions of future precipitation trends (Fig. SPM.7 in *IPCC,* 2007). In addition, GCMs produce a multiyear El Nino/Southern Oscillation (ENSO) with increasing accuracy (*AchutaRao and Sperber,* 2006). El Nino/Southern Oscillation arises from natural ocean-atmosphere interactions in the tropics. Its spontaneous appearance in climate model output is important because ENSO is the leading mode of Earth's internally generated climate variability. Alongside naturally forced variations like those caused by volcanic eruptions (q.v.), internally generated variations like ENSO form the background "noise" against which human impacts on global climate must be assessed.

A more directly pertinent criterion for judging Earth GCMs is their ability to simulate global climate *changes* during the past. These include both natural phenomena such as the ice ages (*Rohling et al.,* 2012) and more recent changes, in particular the global warming observed over recent decades and thought "very likely due to the observed increase in anthropogenic greenhouse gas concentrations" (*IPCC,* 2007). Figure 7 shows both observations and GCM simulations of zonally averaged temperature trends since 1979. The observations include three different groups' processing of microwave emissions recorded by a series of satellites. Different observing groups obtain somewhat divergent results from the same satellite data due to different assumptions regarding instrument calibrations, orbit decay effects, etc. Three broad levels of the atmosphere are observed: the lower troposphere, mid- to upper-troposphere, and lower stratosphere [z ~ 0–8 km, 0–18 km, and 14–29 km respectively; see Table 2 in *Karl et al.* (2006)]. The model output has been processed to give the temperatures that satellites would observe over exactly the same altitude ranges, thus enabling an "apples to apples" comparison with the observations.

Figure 7 shows all models and all observations exhibiting a warming trend ($dT/dt > 0$) in the lower troposphere and a cooling trend ($dT/dt < 0$) in the lower stratosphere over more than 30 years at virtually all latitudes. Stratospheric cooling along with near-surface warming is the expected signature of an enhanced greenhouse effect, since increased greenhouse gases in the upper atmosphere lead to increased cooling to space. The lower tropospheric trends in Fig. 7 represent better consistency than earlier comparisons in which surface observations and GCM simulations indicated significant warming, while satellite observations of the lower troposphere indicated no significant temperature change (*Christy and McNider,* 1994) — leading to suggestions that one of the two classes of observation was systematically erroneous. The qualitative consistency of lower tropospheric temperature trends implied by Fig. 7 is confirmed by analyses of *in situ* surface observations (most recently by *Wickham et al.,* 2013).

Despite this qualitative agreement between models and observations, Fig. 7 shows that throughout the troposphere within ~40° latitude of the equator, nearly all models simulate a faster rate of warming than indicated by the observations. Time series of globally averaged near-surface temperature also show this quantitative discrepancy: Both models and observations exhibit fairly steady warming over the past 35 years, but the overall rate of warming is greater in the models (J. R. Christy, personal communication, 2012). Many different explanations for the discrepancy are possible. We believe the following possibilities warrant special attention:

1. The models generally overestimate positive feedbacks and/or underestimate negative feedbacks, thereby overestimating Earth's equilibrium climate sensitivity $\Delta T_{2\times CO_2}$. This would be a serious indictment of the models and would make global warming a less worrisome problem that conventionally thought.

2. Climate models estimate $\Delta T_{2\times CO_2}$ correctly but overestimate the rate of warming as Earth adjusts to a new equilibrium state. This would also be a serious indictment of the models, possibly connected with an underestimate of natural internal oscillations of the climate system, e.g., those arising in the ocean with its enormous heat capacity. This model deficiency could make global warming less worrisome, although warming of the magnitude that models predict could still be "in the pipeline."

3. Forcing factors are missing or inaccurately represented in the models, causing the net climate forcing ΔF to be overestimated. Some models do not include volcanic effects on climate, which leads to a noticeable overestimate of ocean warming (see Fig. 1c in *Gleckler et al.,* 2012). Inaccurate accounting of O_3 depletion in the lower stratosphere may also affect model-simulated temperature trends (*Santer et al.,* 2013). This type of error would be a much less serious indictment of the models because it could be corrected by more accurate input. It might or might not make global warming less worrisome. For example, if model input needs to be corrected by increasing the magnitude of negative forcing due to sulfate pollution (section 2.3.1),

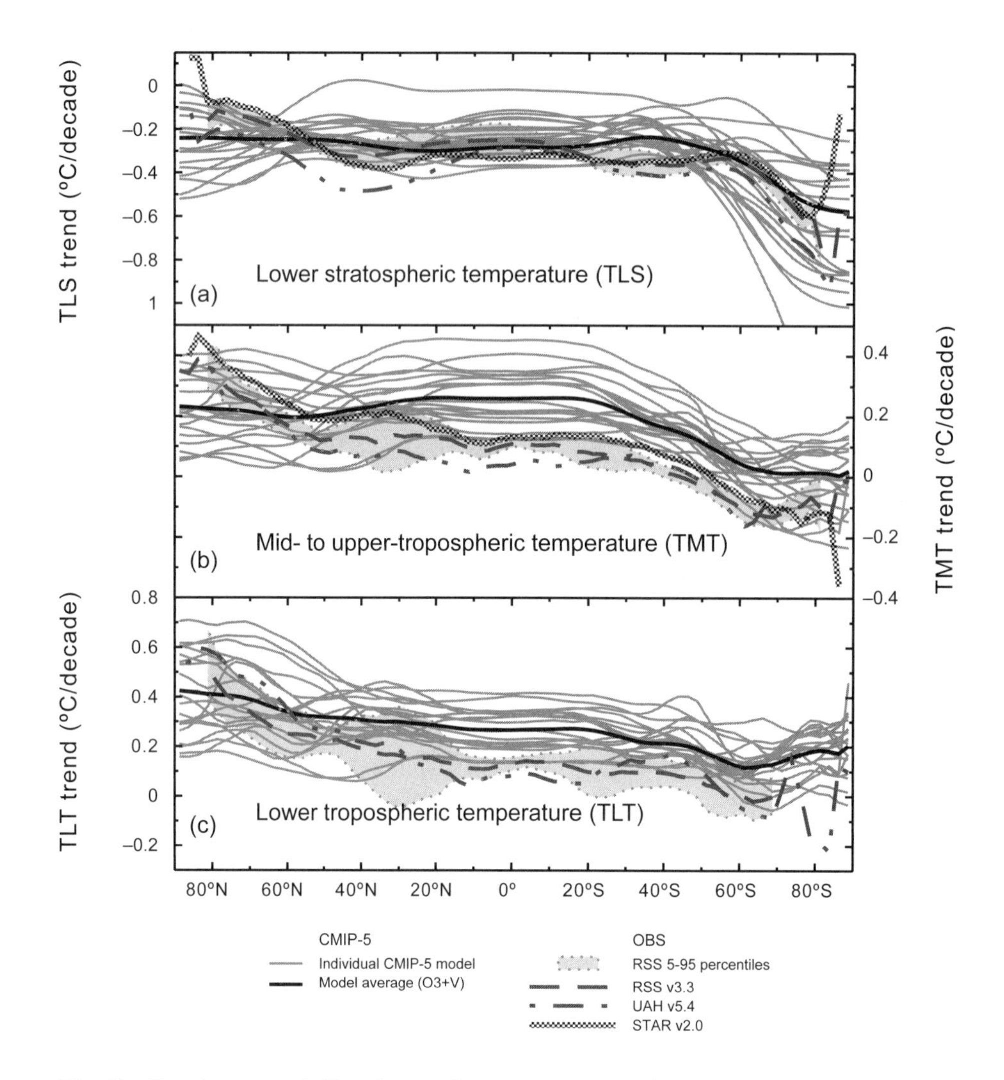

Fig. 7. Zonal mean satellite-observed temperature trends, 1979–2011 (OBS), shown as a function of latitude for three atmospheric levels together with corresponding model-simulated trends. Redrawn from Fig. 3 of *Santer et al.* (2013); the original color figure identifies the different model simulations.

then simulated current rates of warming would become less, but future rates of warming would become greater if the pollution is reduced.

4. The observational errors are underestimated. Figure 7 — representing the most recent work at the time of writing this chapter — includes error estimates for only one of the three versions of the observations (5% to 95% for RSS). This suggests that identification and quantification of observational uncertainties is still in its early days. Future work may well expand the observational "error bars." This would lessen the degree of discrepancy between models and observations.

4. CONCLUSIONS

The subject of the greenhouse effect and climate feedbacks, put in the context of comparative climatology of terrestrial planets, is broad enough for several books like this volume. In this chapter we have focused on basic principles. Observations of Earth and other worlds in the solar system leave no doubt that the greenhouse effect can significantly increase surface temperature. The theory of human-induced global warming on Earth follows naturally from the observations.

The same theory, however, concludes that poorly known processes such as those involving clouds translate into a wide margin of uncertainty about Earth's future climate. Although the power of computers and with it the complexity of climate models has increased enormously over the past few decades, uncertainty in the fundamental sensitivity parameter $\Delta T_{2\times CO_2}$ has not decreased — despite gratifying improvements in simulating the present-day climate,short-term climate variations, recent (~100-yr) global warming, and paleoclimates. Current best estimates for the possible range of $\Delta T_{2\times CO_2}$ (see Box 10.2 in *Meehl et al.,* 2007) are not radically different from the 1.5 to 4.5-K range estimated by the first official report on human-induced global warming (*Charney,* 1979). The lower ends of this range might be tolerable for humanity; the higher estimate would put Earth into a state not seen for millions of years.

Reducing these uncertainties may require a different approach than incremental refinement of detailed Earth-centric models. To some extent this process is already underway. Global warming simulations are increasingly informed by paleoclimate simulations, which test the ability of models to explain a "different Earth" (*Jansen et al.,* 2007). Many extraterrestrial weather and climate models have evolved from Earth GCMs (section 3). The extraterrestrial simulations certainly benefit from decades of Earth climate modeling. Now may be the time to expand the connections by building a "coherent" climate model that *simultaneously* represents several worlds at once (*Schmidt,* 2012).

Acknowledgments. We thank A. Henke for graphic arts, J. Christy for discussion of model evaluation, P. Christensen for supplying the data for Fig. 3, and B. Santer for both discussions and preparing a special grayscale version of Fig. 7. This work was supported in part by NASA and conducted under the auspices of the Office of Science, U.S. Department of Energy, by Lawrence Livermore National Laboratory under Contract DE-AC52-07NA27344.

REFERENCES

AchutaRao K. M. and K. R. Sperber (2006) ENSO simulation in coupled ocean-atmosphere models: Are the current models better? *Climate Dynamics, 27,* 1–15.

Arrhenius S. (1896) On the influence of carbonic acid in the air upon the temperature of the ground. *Philos. Mag., 41,* 237–276.

Bader D. C., Covey C., Gutowski W. J., Held I. M., Kunkel K. E., Miller R. L, Tokmakian R. T., and Zhang M. H. (2008) *Climate Models: An Assessment of Strengths and Limitations.* U.S. Department of Energy, Washington, DC. Available online at *www.climatescience.gov/Library/sap/sap3-1/final-report/sap3-1-final-all.pdf.*

Boynton W. V., Feldman W. C., Squyres S. W., et al. (2002) Distribution of hydrogen in the near surface of Mars: Evidence for subsurface ice deposits. *Science, 297,* 81–85.

Box G. E. P. and Draper N. R. (1987) *Empirical Model-Building and Response Surfaces.* Wiley, New York. 688 pp.

Bullock M. A. and Grinspoon D. H. (2001) The recent evolution of climate on Venus. *Icarus, 150,* 19–37.

Calendar G. S. (1938) The artificial production of carbon dioxide and its influence on temperature. *Q. J. R. Meteorol. Soc., 64,* 223–237.

Charney J. G. (1979) *Carbon Dioxide and Climate: A Scientific Assessment.* National Academy of Sciences, Washington, DC. 22 pp.

Christensen P. R., Anderson D. L., Chase S. C., Clancy R. T., Clark R. N., Kieffer H. H., Kuzmin R. O., Malin M. C., Pearl J. C., Roush T. L., and Smith M. D. (1998) Results from the Mars Global Surveyor thermal emission spectrometer. *Science, 279,* 1692–1698.

Christy J. R. and McNider R. T. (1994) Satellite greenhouse signal. *Nature, 367,* 325–325.

Courtin R., McKay C. P., and Pollack J. (1992) L'effet de serre dans le systeme solaire. *La Recherche, 23,* 542–549.

Covey C. and Klein S. (2010) *Plausible Ranges of Observed TOA Fluxes.* Lawrence Livermore National Laboratory Technical Report LLNL-TR-435351. Available online at *https://e-reports-ext.llnl.gov/pdf/402766.pdf.*

Crisp D. and Titov D. (1997) The thermal balance of the Venus atmosphere. In *Venus II: Geology, Geophysics, Atmosphere, and Solar Wind Environment* (S. W. Bougher et al., eds.), pp. 353–384. Univ. of Arizona, Tuscon.

Denman K. L., Brasseur G., Chidthaisong A., et al. (2007) Couplings between changes in the climate system and biogeochemistry. In *Climate Change 2007: The Physical Science Basis. Contribution of Working Group I to the Fourth Assessment Report of the Intergovernmental Panel on Climate Change* (S. Solomon et al., eds.). Cambridge Univ., Cambridge.

Dickinson R. E. and Cicerone R. J. (1986) Future global warming from atmospheric trace gases. *Nature, 319,* 109–115.

Eyring V., Shepherd T. G., and Waugh D. W., eds. (2010) *Chemistry-Climate Model Validation.* SPARC Rept. 5, WCRP-30, WMO/TD-40. 194 pp. Available online at *www.sparc-climate.org/publications/sparc-reports/sparc-report-no5.*

Fasullo J. T. and Trenberth K. E. (2008) The annual cycle of the energy budget. Part II: Meridional structures and poleward transports. *J. Climate, 21,* 2313–2325.

Forster P., Ramaswamy V., Artaxo P., et al. (2007) Changes in atmospheric constituents and in radiative forcing. In *Climate Change 2007: The Physical Science Basis. Contribution of Working Group I to the Fourth Assessment Report of the Intergovernmental Panel on Climate Change* (S. Solomon et al., eds.). Cambridge Univ., Cambridge.

Fourier J. (1827) Memoire sur la temperature du globe terrestre et des espaces planetaires. *Memoirs of the Royal Academy of Sciences of the Institut de France,* pp. 569–604. English translation available at *www.wmconnolley.org.uk/sci/fourier_1827/fourier_1827.html.*

Friedson A. J., West R. A., Wilson E. H., Oyafuso F., and Orton G. S. (2009) A global climate model of Titan's atmosphere and surface. *Planet. Space Sci., 57,* 1931–1949.

Fulchignoni M., Ferri F., Angrilli F., et al. (2005) In situ measurements of the physical characteristics of Titan's environment. *Nature, 438,* 785–791.

Gerber E. P., Butler A., Calvo N., et al. (2012) Assessing and understanding the impact of stratospheric dynamics and variability on the Earth system. *Bull. Am. Meterol. Soc., 93,* 845–859.

Gierasch P. J., Goody R. M., Young R. E., Crisp D., Edwards C., Kahn R., McCleese D., Rider D., Del Genio A., Greeley R., Hou A., Leovy C. B., and Newman M. (1997) The general circulation of the Venus atmosphere: An assessment. In *Venus II: Geology, Geophysics, Atmosphere, and Solar Wind Environment* (S. W. Bougher et al., eds.), pp. 459–500. Univ. of Arizona, Tucson.

Gleckler P. J., Wigley T. M. L., Santer B. D., Gregory J. M., AchutaRao K., and Taylor K. E. (2006) Volcanoes and climate: Krakatoa's signature persists in the ocean. *Nature, 439,* 675.

Gleckler P. J., Santer B. D., Domingues C. M., Pierce D. W., Barnett T. P., Church J. A., Taylor K. E., AchutaRao K. M., Boyer T. P., Ishii M., and Caldwell P. M. (2012) Human-induced global ocean warming on multidecadal timescales. *Nature Climate Change, 2,* 524–529.

Goudie A. S. and Viles H. A. (2012) Weathering and the global carbon cycle: Geomorphological perspectives. *Earth-Sci. Rev., 113,* 59–71.

Griffith C. A. (2009) Storms, polar deposits and the methane cycle in Titan's atmosphere. *Philos. Trans. R. Soc., A367,* 713–728.

Haberle R. M. (2013) Estimating the power of Mars' greenhouse effect. *Icarus,* in press.

Haberle R. M., Joshi M. M., Murphy J. R., Barnes J. R., Schofield J. T., Wilson G. R., Lopez-Valverde M., Hollingsworth J. L., Bridger A. F. C., and Schaeffer J. (1999). GCM simulations of the Mars Pathfinder ASI/MET data. *J. Geophys. Res., 104,* 8957–8974.

Haberle R. M. and the NASA/Ames Mars General Circulation Modeling Group (2010) Modeling the seasonal water cycle on Mars: Implications for sources and sinks. AGU Fall Meeting Abstract #P51E-01. Available online at *www.agu.org/meetings/fm10/fm10-sessions/fm10_P51E.html.*

Haberle R. M., Kahre M. A., Hollingsworth J. L., Schaeffer J., Montmessin F., and Phillips R. J. (2012) A cloud greenhouse effect on Mars: Significant climate change in the recent past? In *Lunar and Planetary Science XLIII,* Abstract #1665. Lunar and Planetary Institute, Houston.

Haberle R. M., Forget F., Head J., Kahre M. A., and Kreslavsky M. (2013) Summary of the Mars recent climate change workshop, NASA Ames Research Center, May 15–17, 2012. *Icarus, 222,* 415–418.

Hanel R. A., Schlachman B., Rogers D., and Vanous D. (1971) Nimbus 4 Michelson interferometer. *Appl. Optics, 10,* 1376–1882.

Hanel R., Conrath B., Flasar F. M., Kunde V., Maguire W., Pearl J., Pirraglia J., Samuelson R., Herath L., Allison M., Cruikshank D., Gautier D., Gierasch P., Horn L., Koppany R., and Ponnamperuma C. (1981) Infrared observations of the saturnian system from Voyager 1. *Science, 212,* 192–200.

Hanel R., Conrath B., Flasar F. M., Kunde V., Maguire W., Pearl J., Pirraglia J., Samuelson R., Cruikshank D., Gautier D., Gierasch P., Horn L., and Ponnamperuma C. (1982) Infrared observations of the saturnian system from Voyager 2. *Science, 215,* 544–548.

Hansen J., Johnson D., Lacis A., Lebedeff S., Lee P., Rind D., and Russell G. (1981) Climate impact of increasing atmospheric carbon dioxide. *Science, 213,* 957–966.

Hartmann W. K. (2005) *Moons and Planets, 5th edition.* Books/Cole, Belmont, California. 428 pp.

Houghton J. (2002) *The Physics of Atmospheres, 3rd edition.* Cambridge Univ., Cambridge. 340 pp.

Houghton J. (2009) *Global Warming: The Complete Briefing, 4th edition.* Cambridge Univ., Cambridge. 456 pp.

Hunten D. M. and Goody R. M. (1969) Venus: The next phase of planetary exploration. *Science, 165,* 1317–1323.

Ingersoll A. P. (1969) The runaway greenhouse: A history of water vapor on Venus. *J. Atmos. Sci., 26,* 1191–1198.

IPCC (2007) Summary for policymakers. In *Climate Change 2007: The Physical Science Basis. Contribution of Working Group I to the Fourth Assessment Report of the Intergovernmental Panel on Climate Change* (S. Solomon et al., eds.). Cambridge Univ., Cambridge.

Jansen E., Overpeck J., Briffa K. R., et al. (2007) Palaeoclimate. In *Climate Change 2007: The Physical Science Basis. Contribution of Working Group I to the Fourth Assessment Report of the Intergovernmental Panel on Climate Change* (S. Solomon et al., eds.). Cambridge Univ., Cambridge.

Jones P. D., New M., Parker D. E., Martin S., and Rigor I. G. (1999) Surface air temperature and its variations over the last 150 years. *Rev. Geophys., 37,* 173–199.

Karl T. R., Hassol S. J., Miller C. D., and Murray W. L., eds. (2006) *Temperature Trends in the Lower Atmosphere: Steps for Understanding and Reconciling Differences.* National Oceanic and Atmospheric Administration, Washington, DC. Available online at *www.climatescience.gov/Library/sap/sap1-1/finalreport/sap1-1-final-frontmatter.pdf.*

Keeling C. D., Bacastow R. B., Bainbridge A. E., et al. (1976) Atmospheric carbon dioxide variations at Mauna Loa Observatory, Hawaii. *Tellus, 28,* 538–551.

Kieffer H. and Zent A.P. (1992) Quasi-periodic climate change on Mars. In *Mars* (H. H. Kieffer et al., eds.), pp. 1180–1218. Univ. of Arizona, Tucson.

Kim M. and Ramanathan V. (2012) Improved estimates and understanding of global albedo and atmospheric solar absorption. *Geophys. Res. Lett., 39,* L24704.

Komabayashi M. (1967) Discrete equilibrium temperatures of a hypothetical planet with the atmosphere and the hydrosphere of one component-two phase system under constant solar radiation. *J. Meteor. Soc. Japan, 45,* 137–139.

Kurokawa H. and Nakamoto T. (2012) Effects of atmospheric absorption of incoming radiation on the radiation limit of the troposphere. *J. Atmos. Sci., 69,* 403–413.

Laskar J., Correia A.C.M., Gastineau M., Joutel F., Levrard B., and Robutel P. (2004) Long term evolution and chaotic diffusion of the insolation quantities of Mars. *Icarus, 170,* 343–364.

Lebonnois S., Hourdin F., Eymet V., Crespin A., Fournier R., and Forget F. (2010) Superrotation of Venus' atmosphere analyzed with a full general circulation model. *J. Geophys. Res.–Planets, 115,* E06006.

Leighton R. B. and Murray B.C. (1966) Behavior of carbon dioxide and other volatiles on Mars. *Science, 153,* 136–144.

Lindzen R. S. and Emanuel K. (2002) The greenhouse effect. In the *Encyclopedia of Global Change* (J. R. Holton, ed.), pp. 562–566. Oxford Univ., New York.

Loeb N. G., Wielicki B. A., Doelling D. R., Smith G. L., Keyes D. F., Kato S., Manalo-Smith N., and Wong T. (2009) Toward optimal closure of the Earth's top-of-atmosphere radiation budget. *J. Climate, 22,* 748–766.

Madeleine J.-B., Head J. W., Forget F., Navarro T., Millour E., Spiga A., Colaitis A., Montmession F., and Määttänen A. (2013) What defines a martian glacial state? Analysis of the Mars climate system under past conditions using the new LMD global climate model. In *Lunar Planet. Sci. XLIV,* Abstract #1895. Lunar and Planetary Institute, Houston.

Manabe S. and Weatherald R. T. (1967) Thermal equilibrium of the atmosphere with a given distribution of relative humidity. *J. Atmos. Sci., 24,* 241–259.

Marinova M. M., McKay C. P., and Hashimoto H. (2005) Radiative-convective model of warming Mars with artificial greenhouse gases. *J. Geophys. Res.–Planets, 110,* E03002, DOI: 10.1029/2004JE002306.

McKay C. P., Pollack J. B., and Courtin R. (1991) The greenhouse and antigreenhouse effects on Titan. *Science, 253,* 1118–1121.

McKay C. P., Pollack J. B., Lunine J. I., and Courtin R. (1993) Coupled atmosphere-ocean models of Titan's past. *Icarus, 102,* 88–98.

McComiskey A. and Feingold G. (2012) The scale problem in quantifying aerosol indirect effects. *Atmos. Chem. Phys., 12,* 1031–1049.

Meehl G.A., Stocker T. F., Collins W. D., et al. (2007) Global climate projections. In *Climate Change 2007: The Physical Science Basis. Contribution of Working Group I to the Fourth*

Assessment Report of the Intergovernmental Panel on Climate Change (S. Solomon et al., eds.). Cambridge Univ., Cambridge.
Mitchell J. L. (2012) Titan's transport-driven methane cycle. *Astrophys. J. Lett., 756,* L26.
Penner J. E., Xu L., and Wang M. H. (2011) Satellite methods underestimate indirect climate forcing by aerosols. *Proc. Natl. Acad. Sci. USA, 108,* 13404–13408.
Phillips N. A. (1956) The general circulation of the atmosphere: A numerical experiment. *Q. J. R. Meteor. Soc., 82,* 123–164.
Phillips R., Davis B. J., Tanaka K. L., et al. (2011) Massive CO_2 ice deposits sequestered in the south polar layered deposits of Mars. *Science, 332,* 838–841.
Pollack J. B., Toon O. B., and Boese R. (1980) Greenhouse models of Venus' high surface temperature, as constrained by Pioneer Venus measurements. *J. Geophys. Res., 85,* 8223–8231.
Pollack J. B., Toon O. B., Ackerman T. P., and McKay C. P. (1983) Environmental effects of an impact-generated dust cloud: Implications for the Cretaceous-Tertiary extinctions. *Science, 219,* 287–289.
Pollack J. B., Kasting J. F., Richardson S. M., and Poliakoff K. (1987) The case for a wet, warm climate on early Mars. *Icarus, 71,* 203–224.
Pierrehumbert R. T. (2011) Infrared radiation and planetary temperature. *Phys. Today, 64,* 33–38.
Ramanathan V. and Coakley J. A. (1978) Climate modeling through radiative-convective models. *Rev. Geophys. Space Phys., 16,* 465–489.
Read P. L. and Lewis S. R. (2004) *The Martian Climate Revisited: Atmosphere and Environment of a Desert Planet.* Springer, Berlin. 352 pp.
Revelle R. and Suess H. (1957) Carbon dioxide exchange between atmosphere and ocean and the question of an increase of atmospheric CO_2 during the past decades. *Tellus, 9,* 18–27.
Robock A. (2011) Nuclear winter is a real and present danger. *Nature, 473,* 275–276.
Roe G. H. and Baker M. B. (2007) Why is climate sensitivity so unpredictable? *Science, 318,* 629–632.
Rohling E. J., Sluijs A., Dijkstra H. A., et al. (2012) Making sense of paleoclimate sensitivity. *Nature, 491,* 683–691.
Samuelson R. E., Hanel R. A., Kunde V. G., and Maguire W. C. (1981) Mean molecular weight and hydrogen abundance of Titan's atmosphere. *Nature, 292,* 688–693.
Samuelson R. E., Maguire W. C., Hanel R. A., Kunde V. G., Jennings D. E., Yung Y. L., and Aikin A. C. (1983) CO_2 on Titan. *J. Geophys. Res., 88,* 8709–8715.
Santer B. D., Painter J. F., Mears C. A., et al. (2013) Identifying human influences on atmospheric temperature. *Proc. Natl. Acad. Sci. USA, 110,* 26–33.
Schmidt G. A. (2012) Issues in building a coherent climate model for terrestrial planets. In *Comparative Climatology of Terrestrial Planets,* Abstract #8088. Available online at *www.lpi.usra.edu/meetings/climatology2012/pdf/8088.pdf.*
Schmidt G. A., Ruedy R. A., Miller R. M., and Lacis A. A. (2010) Attribution of the present-day total greenhouse effect. *Geophys. Res. Lett., 115,* D20106.
Schofield J. T. and Taylor F. W. (1982) Net global thermal emission for the venusian upper atmosphere. *Icarus, 52,* 245–262.
Schubert G. (1983) General circulation and the dynamical state of the Venus atmosphere. In *Venus* (D. M. Hunten et al., eds.), pp. 681–765. Univ. of Arizona, Tucson.
Shaviv N. J., Shaviv G., and Wehrse R. (2011) The maximal runaway temperature of Earth-like planets. *Icarus, 216,* 403–415.
Solomon S., Qin D., Manning M., et al. (2007) Technical summary. In *Climate Change 2007: The Physical Science Basis. Contribution of Working Group I to the Fourth Assessment Report of the Intergovernmental Panel on Climate Change* (S. Solomon et al., eds.). Cambridge Univ., Cambridge.
Stevens B. and Boucher O. (2012) The aerosol effect. *Nature, 490,* 40–41.
Titov D. V., Bullock M. A., Crisp D., Renno N. O., Taylor F. W., and Zasova L. V. (2007) Radiation in the atmosphere of Venus. In *Exploring Venus as a Terrestrial Planet* (L. W. Esposito et al., eds.), pp. 121–138. AGU Geophys. Monogr. 176, American Geophysical Union, Washington DC.
Titov D. V., Drossart P., Piccioni G., and Markiewicz W. J. (2013) Radiative energy balance in the Venus atmosphere. In *Towards Understanding the Climate of Venus: Applications of Terrestrial Models to Our Sister Planet* (L. Bengtsson et al., eds.), pp. 23–49. ISSI Scientific Report Ser., Vol. 11. Springer, Berlin.
Toon O. B., Pollack J. B., Ward W., Burns J. A., and Bilski K. (1980) The astronomical theory of climate change on Mars. *Icarus, 44,* 552–607.
Toon O. B., Zahnle K., Morrison D., Turco R., and Covey C. (1997) Environmental perturbations caused by the impacts of asteroids and comets. *Rev. Geophys., 35,* 41–78.
Trenberth K. E. and Fasullo J. T. (2010) Tracking Earth's energy. *Science, 328,* 316–317.
Turco R., Toon O. B., Ackerman T. P., Pollack J. B., and Sagan C. (1983) Nuclear winter: Global consequences of multiple nuclear explosions. *Science, 222,* 1283–1292.
Tyndall J. (1873) *Contributions to Molecular Physics in the Domain of Radiant Heat.* Appleton and Co., New York. 462 pp. Available online at *archive.org/details/contributionsto02tyndgoog.*
Walker J. C. G., Hays P. B., and Kasting J. F. (1981) A negative feedback mechanism for the long-term stabilization of Earth's surface temperature. *J. Geophys. Res., 86,* 9776–9782.
Washington W. M. and Parkinson C. L. (2005) *An Introduction to Three-Dimensional Climate Modeling, 2nd edition.* Univ. Science Books, Sausalito, California. 353 pp.
Wickham C., Rohde R., Muller R., et al. (2013) Influence of urban heating on the global temperature land average using rural sites identified from MODIS classifications. *Geoinfor. Geostat., 1,* 1–6. Available online at *www.scitechnol.com/2327-4581/2327-4581-1-104.pdf.*
Wood S. F., Griffiths S. D., and Bapst J. N. (2012) Mars at low obliquity: Perennial CO_2 caps, atmospheric collapse, and subsurface warming. In *Mars Recent Climate Change Workshop,* pp. 44–47. NASA Conf. Publ. 20120216054. Available online at *spacescience.arc.nasa.gov/mars-climate-workshop-2012/documents/extendedabstracts/Wood_SE_ExAbst.pdf.*

Schubert G. and Mitchell J. L. (2013) Planetary atmospheres as heat engines. In *Comparative Climatology of Terrestrial Planets* (S. J. Mackwell et al., eds.), pp. 181–191. Univ. of Arizona, Tucson, DOI: 10.2458/azu_uapress_9780816530595-ch008.

Planetary Atmospheres as Heat Engines

G. Schubert
University of California, Los Angeles

J. L. Mitchell
University of California, Los Angeles

We review the workings of Earth's atmospheric heat engine and describe the energy and entropy exchanges that occur to support the atmospheric circulation. The heat absorbed by the atmosphere increases its internal and gravitational potential energies. A very small percentage of potential energy is converted into kinetic energy to maintain the circulation against dissipation, which irreversibly converts it to internal energy. The thermodynamic efficiency of the atmospheric heat engine can be defined as the fraction of the radiative imbalance at the surface converted to the kinetic energy of the motions. This is equivalent to the ratio of the frictional energy dissipation to the convective heat flux. Estimates of this ratio for Earth are several percent, much less than the Carnot efficiency of about 13%. We apply these concepts to the atmospheres of Venus, Mars, and Titan. The rate of dissipation of atmospheric kinetic energy is one of the main quantities entering the energy budgets of these planetary atmospheres. For Earth, frictional dissipation in rainfall is comparable to the turbulent dissipation of kinetic energy. Rainfall might also be a significant source of dissipation on Titan but it is not likely to be important for Mars or Venus. Frictional dissipation in dust storms might be a uniquely martian phenomenon. The breaking of upward propagating internal gravity waves generated by convection and flow over surface topography is another source of dissipation and is possibly dominant on Venus. The fluxes of radiative entropy are estimated for Earth, Venus, Mars, and Titan. The net radiative entropy flux of a planet is negative because atmospheres absorb solar radiation at a higher temperature than the temperature at which they reemit an equal amount of longwave radiation to space. If in a state of statistical equilibrium, the entropy of an atmosphere is constant; the radiative entropy loss must be balanced by the entropy production associated with thermally direct heat transports, frictional dissipation, and other processes. If estimates of entropy production through thermally direct heat transports and other processes can be obtained and combined with estimates of the temperature at which frictional or turbulent dissipation occurs, then the rate at which frictional dissipation generates heat can be constrained. Using this approach, it is estimated that entropy production due to frictional dissipation in the atmospheres of Venus, Earth, Mars, and Titan occurs at a rate, respectively, of less than about 23, 29, 2, and 0.1 mW m^{-2} K^{-1}. If frictional dissipation in Earth's atmosphere occurs between the temperatures of 250 K and 288 K, the rate must be less than 7.3–8.4 W m^{-2}. This upper bound is much larger than observationally based estimates of the rate of frictional dissipation because other sources of entropy production are difficult to evaluate and are not taken into account. Carnot efficiencies of the atmospheres of Venus, Earth, Mars, and Titan are, respectively, less than about 27.5%, 13.2%, 4.4%, and 4.1%.

1. INTRODUCTION, BASIC CONCEPTS, AND EARTH'S ATMOSPHERIC HEAT ENGINE

Earth's atmosphere has long been described as a heat engine (*Peixoto and Oort,* 1992). It absorbs energy at higher temperatures than it reemits back to space and thus it can do work. The work generates motions that transport energy from hot regions at low latitudes and near the surface to cold regions at high latitudes and altitudes. In steady state, the time rate of generation of the kinetic energy of the motions is balanced by the rate of frictional or turbulent dissipation. The atmosphere is heated by direct absorption of sunlight, by absorption of infrared radiation from the land and the ocean, by release of latent heat through the condensation of water, and by sensible heat transport from land and ocean. The heat increases the internal and gravitational potential energy of the atmosphere. A very small percentage of this is converted into kinetic energy to maintain the circulation against friction. All the heat absorbed by the atmosphere is reradiated to space at long wavelengths.

The analogy of the atmospheric engine with a heat engine is somewhat subtle. In thermodynamics, a heat engine is supplied with heat Q_{in}, rejects heat Q_{out} to its surroundings, and does work $W = Q_{in} - Q_{out}$ on its surroundings. The efficiency of the engine η can be defined as $\eta = W/Q_{in}$. In the classic Carnot cycle, Q_{in} derives from a hot reservoir at temperature T_H and Q_{out} is transferred to a cold reservoir at temperature T_C during isothermal stages. Adiabatic compression and expansion stages connect the isotherms. The cycle is assumed to be ideal, such that the change in entropy over the cycle, $\Delta S_{cycle} = 0$, whence, the Carnot efficiency is $\eta_C = 1-(T_C/T_H)$. The atmospheric heat engine is therefore a somewhat curious one in the sense that it does not absorb any net heat. The atmosphere emits to space as much energy as it absorbs from the Sun, but the emission and absorption occur at different temperatures. The Carnot efficiency applied to radiant energy fluxes sets an upper limit to the efficiency of Earth's atmospheric heat engine. If we assume that the absorption of solar radiation in Earth's atmosphere occurs at about 288 K (the average surface temperature) and emission to space occurs at about 250 K, then η_C for Earth's atmosphere is about 13%. Since absorption and emission of radiation in Earth's atmosphere take place over a range of temperatures, the Carnot efficiency of the atmosphere is necessarily an approximate quantity.

Imbalances in net radiant energy fluxes are responsible for driving kinetic energy fluxes through instabilities. An order-of-magnitude derivation of convective energy fluxes in a dry boundary layer illustrates this process. The imbalance of net (solar and infrared) radiative energy fluxes at the surface R_s is offset by turbulent exchange of sensible energy between the surface and atmosphere. The exchange results in buoyant instability of surface-level air and convection develops through a depth Z_{bl}, which we assume resets temperatures to the dry adiabat with lapse rate $\Gamma = g/C_p$. The enthalpy flux from convection $F_{conv} \sim w\rho C_p dT$ carries the energy supplied by the radiative imbalance, $F_{conv} = R_s$, with convective velocity w, density ρ, specific heat C_p, and temperature anomaly dT. Surface parcels with temperature anomaly dT have buoyancy, $b = g\ dT/T$, with surface gravity g and ambient temperature T. Convection releases the potential energy as the parcel rises through the boundary layer depth, Z_{bl}, converting it into kinetic energy, such that $b \sim w^2/Z_{bl}$. Combining this with our expression for the convective energy flux, we find $F_{conv} \sim \rho w^3 (T/\Delta T_{bl})$, with the dry-adiabatic temperature drop across the boundary layer $\Delta T_{bl} = Z_{bl}\ (g/C_p)$. Identifying the kinetic energy flux $F_{KE} \sim \rho w^3$, we find that $F_{KE} \sim R_s\ (T/\Delta T_{bl}) = Q_{in}\eta_{bl}$. In words, the kinetic energy flux from boundary layer convection is equivalent to the surface radiative flux imbalance times the Carnot efficiency of the boundary layer, assuming heat $Q_{in} = R_s$ is absorbed at the surface, carried through the boundary layer by convection, and rejected as radiation at the top of the boundary layer.

To pursue the notion of an Earth atmospheric heat engine we need to look more closely at the exchanges of energy in Earth's atmosphere that are depicted in Fig. 1 (*Kiehl and Trenberth,* 1997). The energy fluxes in Fig. 1 are in W m^{-2} and the numbers in parentheses are percentages of the incoming solar radiation. The solar energy incident on Earth amounts to 342 W m^{-2} and 107 W m^{-2} (31%) is reflected directly to space from clouds and Earth's surface. Of the 235 W m^{-2} absorbed by Earth, 67 W m^{-2} (20%) is absorbed by the atmosphere and 168 W m^{-2} (49%) is absorbed by the surface. Earth reradiates 235 W m^{-2} (69%) to space, of which 195 W m^{-2} (57%) comes from the atmosphere and 40 W m^{-2} (12%) from the surface through an atmospheric window. The energy balance at the surface involves a number of contributions. We have already noted that the surface absorbs sunlight at the rate 168 W m^{-2} (49%). It also receives 324 W m^{-2} (95%) in the form of longwave radiation from the atmosphere, for a total of 492 W m^{-2}. The total energy absorbed by the surface is balanced by 390 W m^{-2} (114%) reradiated to space and to the atmosphere and 24 W m^{-2} (7%) transferred to the atmosphere as sensible heat together with 78 W m^{-2} (23%) lost to the atmosphere in the form of latent heat.

The energies absorbed in and emitted from the atmosphere are identified in the gray band in Fig. 1. They include 67 W m^{-2} (20%) of absorbed sunlight, 350 W m^{-2} (102%) of absorbed longwave radiation from the surface, 24 W m^{-2} (7%) of sensible heat absorbed from the surface, and 78 W m^{-2} (23%) of latent heat absorbed from the surface. The total of these absorbed energies is 519 W m^{-2} (152%). The atmosphere emits 195 W m^{-2} (57%) of longwave radiation to space and 324 W m^{-2} (95%) of longwave radiation to the surface, for a total of 519 W m^{-2} (152%) to complete the balance.

1.1. Earth's Lorenz Energy Cycle

The energies involved in the atmospheric heat engine are identified in the black box within the gray band of Fig. 1 and are only small quantities compared to the radiative energy flows we have been discussing. Only 4–7 W m^{-2} (1.2–2%) of potential energy in the atmosphere is converted into kinetic energy of motion. Generation of kinetic energy is balanced by frictional or turbulent dissipation that returns the energy in the form of heat to the internal energy reservoir of the atmosphere. Frictional dissipation adds heat to the atmosphere, which is radiated to space.

A more detailed look at the global energy cycle of Earth's atmosphere as pioneered by *Lorenz* (1967) is shown in Fig. 2. The figure divides the available potential energy of the global atmosphere into mean (P_M) and eddy (P_E) parts, and the global atmospheric kinetic energy is treated similarly (K_M, K_E). The energy contents within each category are indicated by the numbers within the boxes (units are 10^5 J m^{-2}). Energy is transferred between the forms, e.g., mean potential energy P_M is converted to eddy potential energy P_E and the latter is converted to eddy kinetic energy K_E. These conversions are indicated by the arrows and the conversion rates are listed (units W m^{-2}). The arrows pointing into the P_M and P_E boxes represent the generation rates

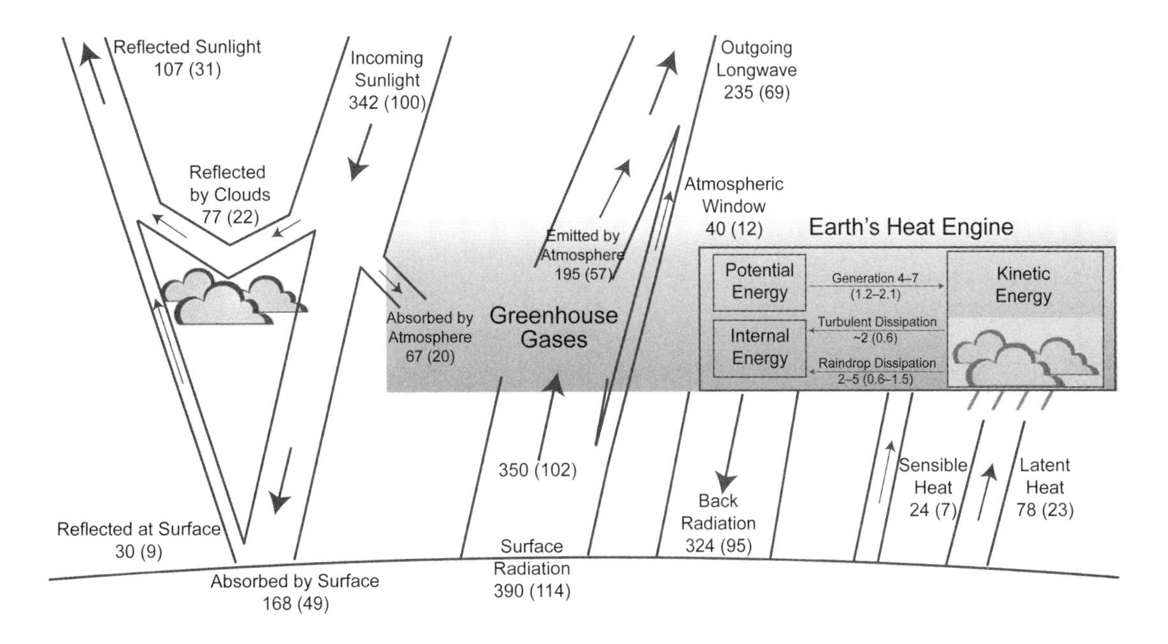

Fig. 1. Flow of energy in Earth's atmosphere. Fluxes are in units of W m^{-2}. Numbers in parentheses are percentages of the absorbed solar radiation. Numerical values are based on *Peixoto and Oort* (1984) and *Kiehl and Trenberth* (1997).

of these forms of energy (values are given in W m^{-2}), while the arrows pointing out of the K_M and K_E boxes represent the frictional or turbulent rates of dissipation of kinetic energy (values are given in W m^{-2}). Numbers in italics are based on an observationally limited early dataset (*Oort and Peixoto,* 1983) modified by *Peixoto and Oort* (1992). Numbers in bold are based on two reanalysis satellite datasets presented in *Li et al.* (2007). The reanalysis datasets are of better quality and are more complete.

The estimate of frictional or turbulent dissipation from *Peixoto and Oort* (1992) is 1.9 W m^{-2} ((1.7 + 0.18) W m^{-2}). The estimates from the two reanalysis datasets are 2.6 W m^{-2} ((2.04 + 0.51) W m^{-2}) and 2.1 W m^{-2} ((1.81 + 0.25) W m^{-2}). General circulation model-based estimates are also ~2 W m^{-2} (see *Becker,* 2003). The reanalysis data show that the early estimate of conversion of K_M to P_M is probably incorrect. In fact, the conversion occurs in the opposite direction — mean available potential energy is transferred to mean kinetic energy. Other notable differences between the early and reanalysis datasets include the reduction in size of the mean available potential energy reservoir by a factor of 2 and the near doubling in size of the mean kinetic energy reservoir.

Frictional dissipation not only adds heat to the atmosphere, it also irreversibly adds entropy. Other (reversible) entropy sources for the atmosphere shown in Fig. 1 are the thermally direct circulations such as the sensible and latent heat fluxes from the surface to the atmosphere. Yet additional entropy sources within the atmosphere, not shown in Fig. 1, are associated with horizontal heat transport from low to high latitudes (reversible), diffusion of heat and water vapor (irreversible) (*Pauluis,* 2011), and irreversible phase changes that occur when liquid water evaporates in unsaturated air or when water vapor condenses in supersaturated air (*Pauluis and Held,* 2002a). We constrain the total rate of entropy production from thermally direct circulations, frictional dissipation, and other processes based on the top-of-atmosphere radiative entropy loss, assuming the system is in statistical equilibrium such that the entropy is steady state. This requires the total rate of entropy generation to exactly offset the loss by radiative processes.

Earth's entropy budget is diagrammed in Fig. 3. The incident and reflected solar energies, 342 W m^{-2} and 107 W m^{-2}, respectively, are indicated in the figure. The entropy increase associated with atmospheric absorption of 67 W m^{-2} of sunlight at 250 K is 268 mW m^{-2} K^{-1}. The entropy increase associated with surface absorption of 168 W m^{-2} of sunlight at 288 K is 583 mW m^{-2} K^{-1}. The transfer of 24 W m^{-2} and 78 W m^{-2} of sensible and latent heat, respectively, from the surface to the atmosphere at a temperature of 288 K represents losses of entropy from the surface in the amounts 83 and 271 mW m^{-2} K^{-1}. The direct loss of 40 W m^{-2} of longwave radiation from the surface to space results in a loss of entropy from the surface of 139 mW m^{-2} K^{-1}. The surface also loses 90 mW m^{-2} K^{-1} in longwave radiative exchange with the atmosphere. The entropy budget of the surface is thus seen to be in balance. The atmosphere emits longwave radiation to space in the amount 195 W m^{-2} corresponding to an entropy loss from the atmosphere of 195/250 W m^{-2} K^{-1} = 780 mW m^{-2} K^{-1}. The total longwave radiative entropy loss from Earth is thus (780 + 139) mW m^{-2} K^{-1} = 919 mW m^{-2} K^{-1}. Incident

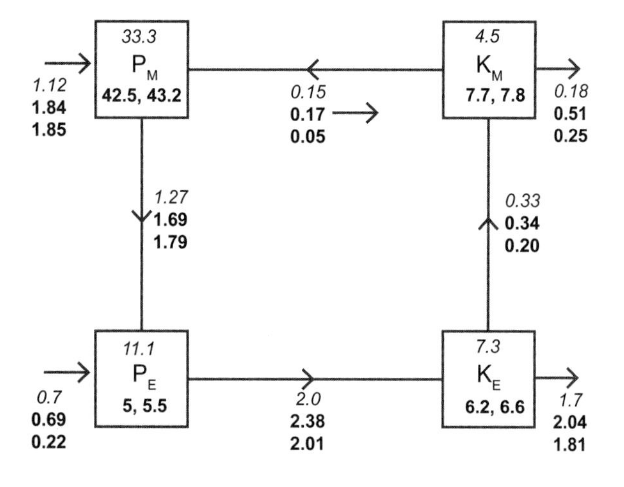

Fig. 2. The energy cycle in the global atmosphere for annual mean conditions. P and K refer to available potential energy and kinetic energy reservoirs, respectively. Subscripts M and E refer to mean and eddy parts. The units of P and K are 10^5 J m^{-2}. The arrows denote conversion of energy among the reservoirs. Units for energy conversions are W m^{-2}. Numbers in italics are from *Peixoto and Oort* (1992). Numbers in bold are from *Li et al.* (2007).

shortwave solar radiation carries a negligible entropy flux since the radiation is emitted from the Sun at a high temperature. The radiative entropy balance for Earth represents a loss of 919 mW m^{-2} K^{-1}, in approximate agreement with other estimates in the literature (e.g., *Wu and Liu,* 2010). If Earth's entropy is constant in the long term, then this entropy loss must be offset by an equal amount of entropy production by the Earth system. Thermalization and absorption of the solar radiation leads to an entropy gain of 268 mW m^{-2} K^{-1} by the atmosphere and 583 mW m^{-2} K^{-1} by the surface. Exchange of longwave radiation between the atmosphere and surface results in a gain of (104–90) mW m^{-2} K^{-1} = 14 mW m^{-2} K^{-1} by the Earth system. Adding all the radiative entropy sources and sinks for Earth as a whole yields (268 + 583 + 14–919) mW m^{-2} K^{-1} = –54 mW m^{-2} K^{-1}. Thus, other nonradiative processes must supply 54 mW m^{-2} K^{-1} to the Earth system to achieve entropy balance. As shown in Fig. 3, latent heat and sensible heat exchanges between the atmosphere and surface produce ((293–271) + (86–83)) mW m^{-2} K^{-1} = 25 mW m^{-2} K^{-1}. This leaves 29 mW m^{-2} K^{-1} that must be accounted for by processes not explicitly included in Fig. 3. Since our estimates of entropy fluxes involve entropy balance for the surface, 29 mW m^{-2} K^{-1} must be produced in the atmosphere. This requirement for additional entropy production in the atmosphere can be obtained by analyzing the entropy sources and sinks separately for the atmosphere. Entropy sources for the atmosphere include the absorption of the latent and sensible heat fluxes and the net absorption of longwave radiation in the exchange with the surface. We assume that these energy absorptions occur at the temperatures 266 K, 280 K, and 250 K, respectively, resulting in entropy production rates of 293, 86, and 104 mW m^{-2} K^{-1}. Summation of all entropy sources and sinks for the atmosphere shown in Fig. 3 yields (268 + 293 + 86 + 104–780) = –29 mW m^{-2} K^{-1}.

The 29 mW m^{-2} K^{-1} that must be produced in the atmosphere can arise through several mechanisms. One is frictional or turbulent dissipation of atmospheric kinetic energy. Another is the horizontal sensible and latent heat transport between equatorial and polar regions. Other entropy-generating processes already mentioned are frictional dissipation in the boundary layers of falling raindrops and diffusion of water vapor. While we are mainly interested in constraining turbulent dissipation, it is difficult to quantify the magnitudes of all the processes that contribute to making up the 29 mW m^{-2} K^{-1}. If we accept the estimates of *Peixoto and Oort* (1992) and *Li et al.* (2007) that frictional or turbulent dissipation is about 2 W m^{-2} and if this dissipation occurs between temperatures of 250 K and 288 K, then this source of entropy production contributes between about 7 and 8 mW m^{-2} K^{-1} or about 26% of the total requirement. This is reasonable given the probable importance of equator-to-pole heat transport and rainfall to entropy production.

1.2. Available Potential Energy (APE) Generation and Thermodynamic Efficiency of the Atmosphere

It is well understood that only a small fraction of the potential energy reservoir can be converted to kinetic energy. This has led to the distinction of available potential energy (APE), which is the fraction that can be converted to kinetic energy by the agency of work. Therefore in diagnosing the efficiency of an atmosphere's heat engine, it is important to identify the rate at which APE is generated by radiation. For the example of Earth's heat engine, only a few tens of percent of the incoming solar radiation is converted to APE, and we identify the latter as the "energy absorbed" by the atmospheric heat engine when estimating efficiencies.

Kinetic energy of the atmosphere is produced by the work done by pressure forces and destroyed by the work done against friction. In steady state, and for the atmosphere as a whole, these effects balance each other. Internal energy is increased by frictional heating and decreased by the work done by pressure forces. In steady state and for the atmosphere as a whole these terms are also in balance. The work done by pressure forces to increase kinetic energy is equal to the pressure work term that decreases internal energy (*Gierasch,* 1971).

The atmosphere is heated nonuniformly and its circulation redistributes this heat. The energy for the motions is borrowed from the internal energy and the frictional heating generated by the motions returns this heat to the internal energy. These internal exchanges of energy maintain the motions in the atmosphere without any net atmospheric heating.

Given the complexity of all the energy exchanges in Earth's atmosphere, how are we then to define the efficiency of the atmospheric heat engine? We can use the formula given earlier $\eta = W/Q_{in}$ as a basis by identifying W with the rate of generation of kinetic energy, which from Fig. 1 and the above discussion is the frictional or turbulent dissipation rate. For Q_{in} we use the nonradiative energy flux, i.e., the surface radiative imbalance, consisting of the sum of the sensible heat flux and latent heat flux (see Fig. 1), essentially the convective energy flux. The efficiency of Earth's atmospheric heat engine is then (see Figs. 1 and 2) about (2 W m^{-2})/((24 + 78) W m^{-2}) = 2%, based on the *Peixoto and Oort* (1992) and *Li et al.* (2007) estimates of frictional dissipation. Previous estimates of the efficiency of Earth's atmospheric heat engine are in agreement (*Lorenz,* 1967; *Peixoto and Oort,* 1992). Our estimate of the efficiency is about 15% of the Carnot efficiency of 13.2% derived above. Assuming the Carnot efficiency and a convective heat flux of 102 W m^{-2}, the maximum frictional dissipation rate in the atmosphere would be 11.7 W m^{-2}, almost a factor of 6 larger than the *Peixoto and Oort* (1992) and *Li et al.* (2007) estimate. This is also the maximum rate at which the atmosphere could do work to generate motion. *Lorenz* (1967) and *Peixoto and Oort* (1992) define the efficiency of the atmospheric heat engine as the ratio of the dissipation rate to the mean incoming solar radiation, which, if adopted, would give efficiencies about a factor of 2 smaller than those estimated above.

The rate of dissipation in the atmosphere is the more uncertain of the two quantities whose ratio determines the efficiency of the atmospheric heat engine. Only recently has it been shown from satellite observations that frictional dissipation in the microphysical shear zones surrounding falling raindrops is comparable to the turbulent dissipation of kinetic energy (*Pauluis and Dias,* 2012). The latter has been estimated from observations and modeling (*Lorenz,* 1967; *Peixoto et al.,* 1991; *Peixoto and Oort,* 1992; *Becker,* 2003; *Li et al.,* 2007), with values ranging between about 1 and 2 W m^{-2}. Dissipation from falling raindrops offsets the work done by the atmosphere in lifting water vapor from the surface to its condensation level. This process is likely to be sufficiently slow to maintain hydrostatic balance, so that the work does not contribute to the generation of kinetic energy. It does, however, represent an avenue by which the atmosphere does work against its environment and perhaps should be included when considering the efficiency of the atmospheric

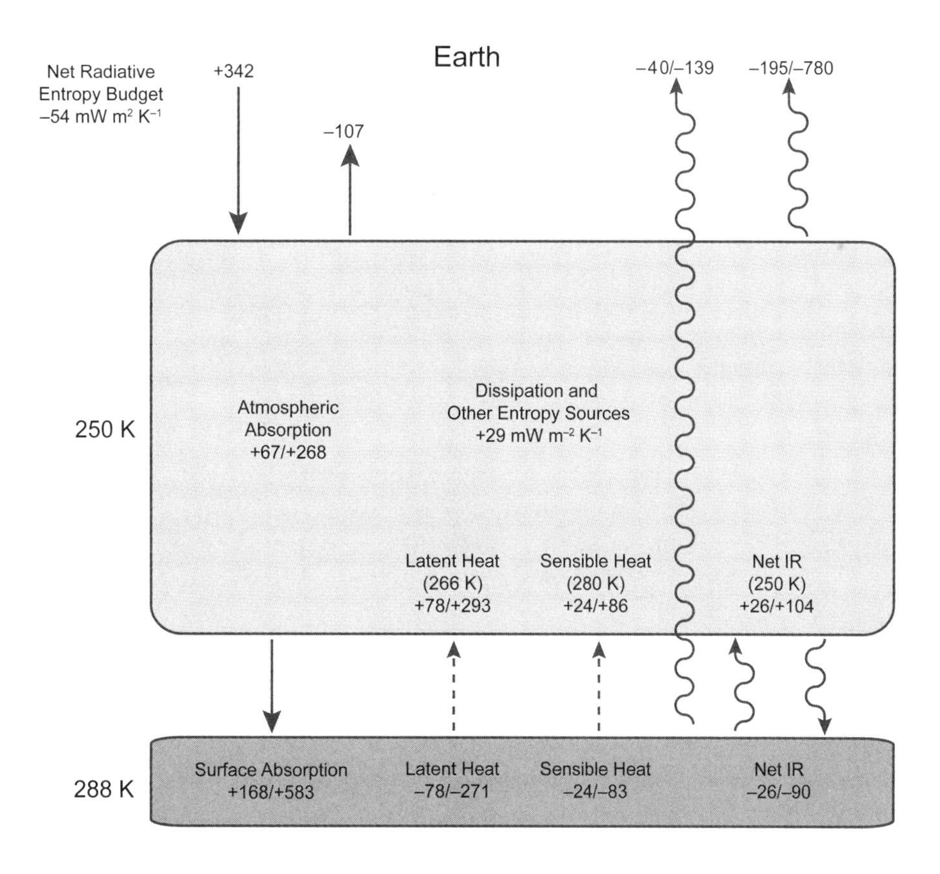

Fig. 3. Earth entropy budget. Solid vertical lines with arrows refer to the solar flux in units of W m^{-2}. Curved vertical lines with arrows refer to the longwave radiation. Numbers separated by "/" are values of energy flux and entropy flux. Energy flux is the first number in units of watts per square meter. Units of entropy flux are mW m^{-2} K^{-1}. Vertical dashed lines with arrows indicate the exchanges of sensible and latent heat between the surface and atmosphere. Some temperatures associated with heat absorption in the atmosphere are listed in parentheses.

heat engine. *Pauluis and Held* (2002a,b) treat the water cycle of the atmosphere as a separate "dehumidification" process.

The definition of efficiency we have adopted assumes that the mechanical work done by the atmosphere and dissipation occur mainly near the surface and involve the nonradiative transfer of energy from the surface to the atmosphere through the fluxes of sensible heat and latent heat. Entropy production associated with the latitudinal transport of heat by atmospheric motions is assumed to be relatively less important. However, in Earth's atmosphere and in other planetary atmospheres, horizontal heat transport might be as important or even more important than vertical heat transfer in doing work.

In the following sections we apply the concept of a heat engine to the atmospheres of Venus, Mars, and Titan.

2. VENUS

The energy and entropy budgets of Venus, illustrated in Fig. 4, have been discussed by *Titov et al.* (2007). Because of its high albedo (about 76%), Venus absorbs only 150 W m^{-2} of the incident solar flux of 625 W m^{-2}. Thus, Venus absorbs only about 65% of the sunlight absorbed by Earth even though it is closer to the Sun than is Earth. About 100 W m^{-2} of solar absorption takes place in the upper cloud region above 57 km altitude, about 30 W m^{-2} of solar radiation is absorbed in the lower atmosphere between about 20 and 40 km altitude, and about 20 W m^{-2} of solar energy is absorbed at the surface. Figure 4 shows 150 W m^{-2} of longwave radiation being emitted to space from the upper cloud deck to balance the incoming absorbed solar radiation. Venus actually emits a small amount of infrared radiation directly to space from its surface and lower atmosphere. Unlike the atmospheres of Earth and Mars, Venus' atmosphere is heated mainly from above.

Figure 4 indicates the approximate temperatures at which solar energy absorption and longwave radiative cooling take place. The entropy fluxes listed on the figure have been obtained by simply dividing the energy fluxes by the listed temperatures. Based on the values in Fig. 4, the globally averaged temperature for solar energy absorption is about 345 K. With the energy to space being radiated at about 250 K, the Carnot efficiency is about 27.5%, roughly a factor of 2 larger than the Carnot efficiency estimated for Earth.

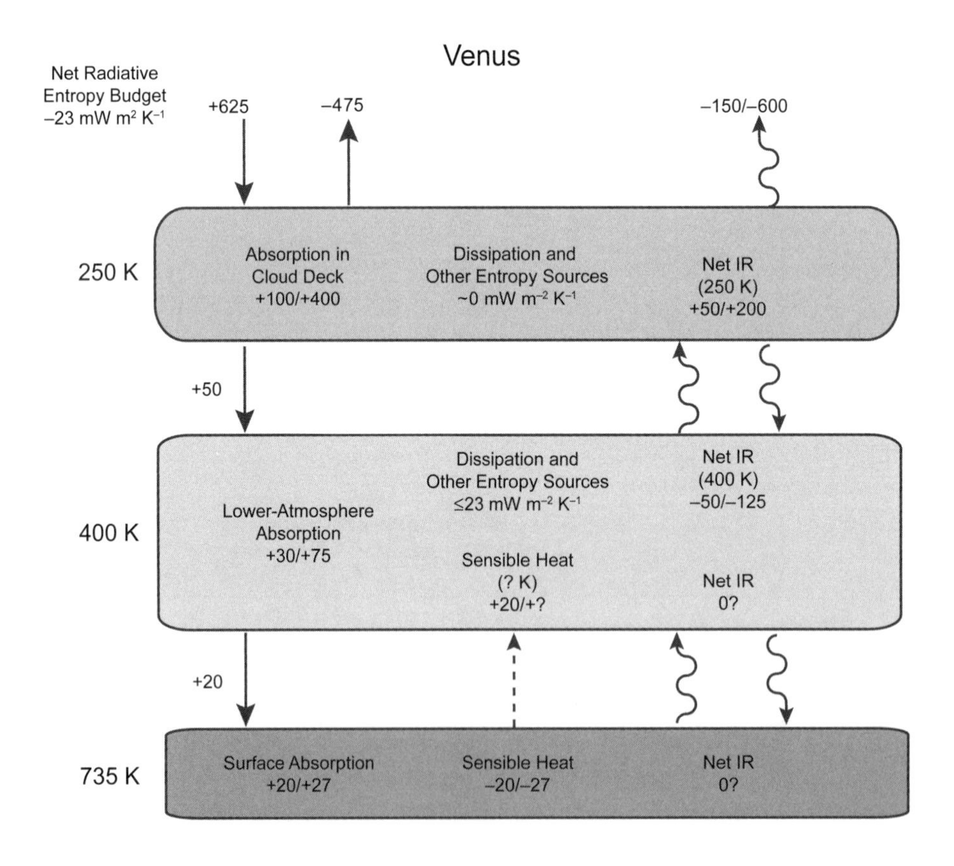

Fig. 4. Venus energy and entropy budgets. Solid vertical lines with arrows refer to the solar flux in units of W m^{-2}. Curved vertical lines with arrows refer to the longwave radiation. Numbers separated by "/" are values of energy flux and entropy flux. Energy flux is the first number in units of W m^{-2}. Units of entropy flux are mW m^{-2} K^{-1}. The vertical dashed line with an arrow indicates the exchange of sensible heat between the surface and atmosphere. Some temperatures associated with heat absorption in the atmosphere are listed in parentheses.

The entropy fluxes of absorbed solar radiation in Fig. 4 add to 502 mW m^{-2} K^{-1}, while the entropy flux of emitted longwave radiation is –600 mW m^{-2} K^{-1}. Thus there is a negative entropy flux of 98 mW m^{-2} K^{-1} associated with radiative processes in the Venus atmosphere. On Earth, we have seen that the net flux of radiative entropy is –68 mW m^{-2} K^{-1}. The radiative entropy loss from Venus' atmosphere must be balanced by the production of entropy through viscous and turbulent dissipation and thermally direct circulations. For Earth's atmosphere, the radiative entropy loss is balanced by entropy production from not only viscous and turbulent dissipation of atmospheric motions, but also by water phase transitions and dissipation associated with rainfall and of course the thermally direct motions. On Venus we do not expect significant entropy production associated with precipitation in the thin sulfuric acid clouds (*Lorenz and Renno,* 2002).

In estimating the efficiency of Earth's atmospheric heat engine, we focused on the radiative imbalance at the surface and the upward mechanical transport of heat. Venus' atmospheric heat engine is primarily located in the cloud region where it is known from Vega balloon measurements that the upward convective heat flux in the low stability region from 50 to 55 km altitude is about 40 W m^{-2} (*Crisp and Titov,* 1997). Below this convective layer, Venus' atmosphere is close to being stable against convection, although the nature of the surface atmospheric boundary layer is unknown. The upward convective heat flux in the low stability layer is comparable to the upward radiative flux from the deeper atmosphere (about 50 W m^{-2}), so convection is responsible for the bulk of the upward heat transport through the 50–55-km cloud region. According to *Titov et al.* (2007), observations of turbulence in the clouds of Venus put an upper bound of 1 mW m^{-2} K^{-1} on the entropy production from this process. We can get a dissipation rate from this value if we know the temperature at which the bulk of the dissipation occurs. Taking this temperature to be 250 K, we get a dissipation rate of 0.25 W m^{-2} and division by the convective heat flux gives an efficiency of 0.625%. The cloud level heat engine appears to be relatively inefficient. The estimated efficiency is much less than the Carnot efficiency of the cloud layer heat engine, which is about 25% based on a temperature difference of 90 K across the cloud layer and an ambient temperature of about 360 K (*Lorenz and Renno,* 2002). It can be seen in Fig. 4 that the cloud level atmosphere is essentially in entropy balance without turbulent dissipation.

The upper limit on the cloud level entropy production leaves open the question of what processes, acting where in the atmosphere, are responsible for the required entropy production rate estimated from the radiative energy fluxes and temperatures in Venus' atmosphere. Turbulent and viscous dissipation somewhere in Venus' massive lower atmosphere together with the net entropy production from surface-atmosphere sensible heat transfer must produce an entropy flux of about (125–27–75) = 23 mW m^{-2} K^{-1} (Fig. 4) to balance the radiative entropy loss of the atmosphere [part of the 98 mW m^{-2} K^{-1} net radiative entropy budget (Fig. 4) is balanced by the net entropy production associated with longwave radiative exchange between the lower atmosphere and the cloud region (200–125 = 75 mW m^{-2} K^{-1}) (Fig. 4)]. Since the temperature difference over which the sensible heat exchange between the surface and lower atmosphere occurs is unknown, only an upper bound of 23 mW m^{-2} K^{-1} can be placed on the entropy production due to turbulent dissipation. A likely candidate is the breaking of upward-propagating internal gravity waves generated by convection and flow over the surface topography, which is possibly dominant on Venus (*Izakov,* 2010).

3. MARS

Global energy and entropy budgets for Mars similar to those for Earth and Venus shown above do not appear to be available in the literature. A schematic of the processes involved in the global energy balance of Mars is shown in Fig. 5. The solar flux at Mars' orbit is 152 W m^{-2}. Mars absorbs about 117 W m^{-2} of solar energy and reemits it as longwave radiation (*Goody,* 2007). This is only about 50% of Earth's solar energy absorption. The solar energy absorbed by Mars is absorbed by the CO_2 constituting the bulk of the atmosphere, by dust suspended in the atmosphere, and by the surface. Longwave radiation emitted by the surface is also absorbed by atmospheric dust. A uniquely martian feature of the energy budget is the latent heat exchange between the surface and atmosphere due to sublimation and condensation of CO_2 at the poles. The atmosphere and surface also exchange energy via sensible heat. The observed emission temperature for Mars is 220 K and the mean surface temperature is about 230 K. If we use the latter as the temperature at which solar energy is absorbed, then the Carnot efficiency is 4.35%. The Carnot efficiency of Mars is only about one-third that of Earth.

If Mars absorbs 112 W m^{-2} at 230 K and 5 W m^{-2} at 220 K and reemits this energy at 220 K, then the entropy flux of this radiative exchange is –22 mW m^{-2} K^{-1} (Fig. 5). Dissipation and thermally direct sources of entropy production in the atmosphere of Mars must provide an entropy production of 22 mW m^{-2} K^{-1} to balance the radiative loss of entropy. With the sensible and latent heat flux estimates given in Fig. 4 (values are from A. Zent, personal communication, 2012), these thermally direct sources of entropy production are negligible. The net exchange of longwave radiation between the surface and atmosphere provides an entropy source of (495–474) = 21 mW m^{-2} K^{-1}, leaving only 1 mW m^{-2} K^{-1} to be provided by turbulent dissipation. There are a number of mechanisms that could contribute to dissipation in the martian atmosphere, including turbulence, breaking of gravity waves, and losses encountered in the lifting and settling of dust. If the dissipation occurred at the temperature of 225 K, then the dissipative flux of energy would be about 0.23 W m^{-2}.

Lorenz and McKay (2003) have estimated vertical convective heat fluxes of about 5 W m^{-2} based on model calculations (about twice as large as the sensible heat flux in Fig. 5). If this heat flux could be converted into mechanical energy with an efficiency of 10% it would yield 0.5 W m^{-2}, about a factor of 2 larger than the above estimate, and large enough to drive dust devils (perhaps 0.23 W m^{-2} is also sufficient). An even more generous estimate of the rate of generation of mechanical energy from *Lorenz and Renno* (2002) is based on the inference from Viking lander measurements that the upward convective heat transport near the surface is about 20 W m^{-2} during the afternoon. With a possible conversion efficiency of 10%, this gives an available work of less than 1 W m^{-2}, averaged over the day. One implication of these admittedly uncertain estimates is that on Mars horizontal heat transport might perform more work than does vertical heat transfer. The possible dominance of horizontal heat transport over vertical heat transport on Mars is also suggested by the thin atmosphere and the weakness of the greenhouse effect. This is substantiated by widespread aeolian activity and the movement of sand dunes (*Bridges et al.,* 2012), which are evidence of the ability of the martian heat engine to do work.

4. TITAN

Like Venus, Titan possesses an optically thick layer in its upper atmosphere, the primary difference being it is a photochemical haze. The energetics of haze production results in statically stable conditions there, so it is unlikely that mechanical dissipation plays a large role in this region, although photochemistry might. Figure 6 illustrates energy and entropy fluxes as understood from observations (*Li et al.,* 2011) and modeling (*McKay et al.,* 1989; *Mitchell,* 2012). Titan absorbs a total of 2.6 W m^{-2} of sunlight in the stratosphere, in the troposphere, and at the surface at an average temperature of 85.3 K. It emits the same flux of longwave radiation at an average temperature of 81.8 K. The Carnot efficiency based on these layer temperatures is 4.1%.

The solar flux and outgoing longwave radiation are the only observationally verified, global-mean quantities presented here. The remainder of the atmospheric fluxes in Fig. 6 are updated from a model by *McKay et al.* (1989). The stratosphere absorbs 1.49 W m^{-2} at roughly 85 K, which increases entropy at a rate of +17.5 mW m^{-2} K^{-1}.

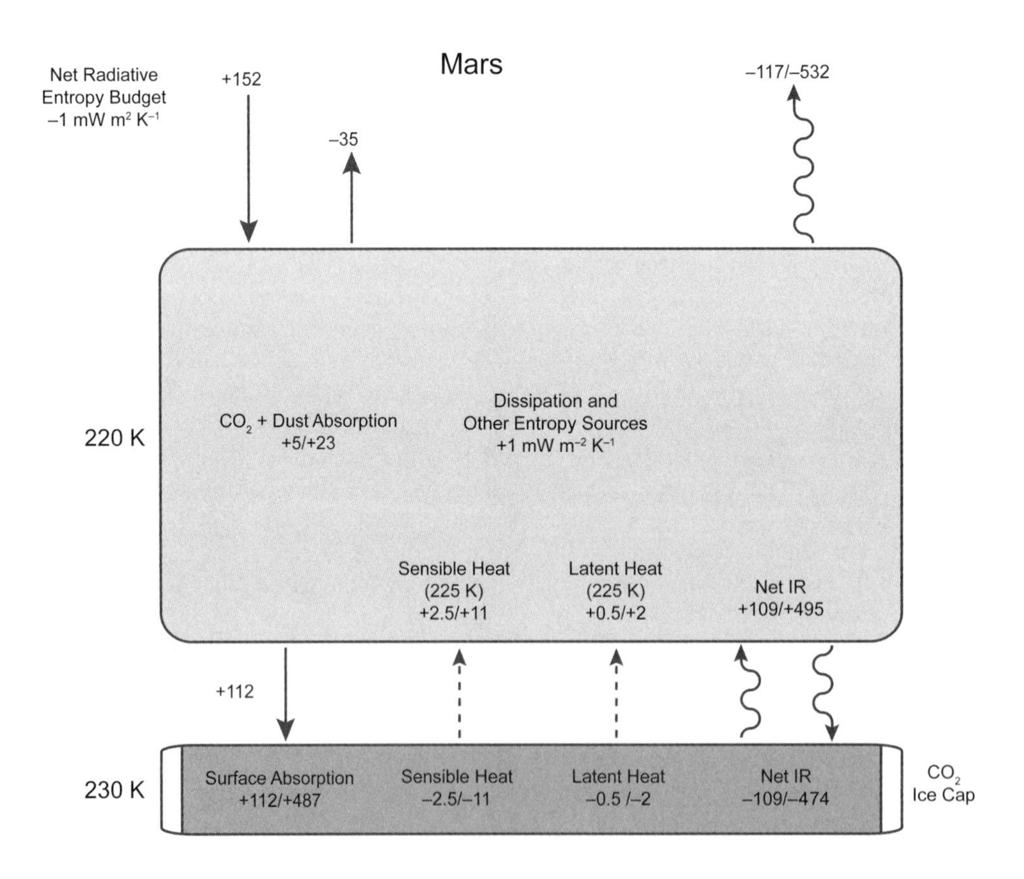

Fig. 5. Entropy and energy budgets for Mars. Solid vertical lines with arrows refer to the solar flux in units of W m^{-2}. Curved vertical lines with arrows refer to the longwave radiation. Numbers separated by "/" are values of energy flux and entropy flux. Energy flux is the first number in units of W m^{-2}. Units of entropy flux are mW m^{-2} K^{-1}. Vertical dashed lines with arrows indicate the exchanges of sensible and latent heat between the surface and atmosphere. Some temperatures associated with heat absorption in the atmosphere are listed in parentheses.

It returns 0.93 W m^{-2} directly back to space as outgoing longwave radiation, and if we assume this radiation emerges from the 85 K layer, we infer a reduction in entropy at a rate of –10.9 mW m^{-2} K^{-1}. The stratosphere also radiates downward longwave radiation to the troposphere at a rate of 0.56 W m^{-2}, and at 85 K this reduces the entropy by –6.6 mW m^{-2} K^{-1}. The troposphere receives 1.48 W m^{-2} solar flux, of which it reflects 0.37 W m^{-2} back to the stratosphere. It also receives 0.56 W m^{-2} of infrared from the stratosphere, which is absorbed at a temperature of 80 K thereby increasing the tropospheric entropy by 7 mW m^{-2} K^{-1}.

Up to this point, our discussion is roughly consistent with model results from *McKay et al.* (1989). The surface energy balance we present, however, is quite different than has been previously assumed. *Mitchell* (2012) argues based on idealized GCM calculations that horizontal heat transport dramatically alters Titan's surface energetics, and the numbers we present here reflect this new understanding. First, the troposphere on average transmits 0.47 W m^{-2} at solar wavelengths to the surface. Direct absorption of solar wavelengths in the troposphere is therefore 0.64 W m^{-2}. The absorption occurs at roughly 80 K ambient temperatures, and leads to an increase in entropy at a rate of +8 mW m^{-2} K^{-1}. The troposphere returns 1.67 W m^{-2} to space as outgoing longwave radiation that reduces the entropy at a rate of –20.8 mW m^{-2} K^{-1}. In the model of *Mitchell* (2012), all 0.47 W m^{-2} at solar wavelengths are absorbed by the surface. If we assume a 93.5 K surface temperature, this leads to an increase in entropy at a rate of +5 mW m^{-2} K^{-1}. The surface exchanges 0.31 W m^{-2} directly with the atmosphere through sensible and latent heat fluxes, reducing the surface entropy by –3.3 mW m^{-2} K^{-1}. The net difference between upward and downward infrared fluxes between the surface and troposphere is 0.16 W m^{-2}, and this balances the surface entropy budget by reducing it at a rate of –1.7 mW m^{-2} K^{-1}.

The entropy of the troposphere decreases at a rate of –3.8 mW m^{-2} K^{-1} due to the net radiative energy fluxes (both shortwave and longwave). In addition, the entropy increases at a rate of 3.4 mW m^{-2} K^{-1} if sensible and latent heat is absorbed near the surface at an average temperature

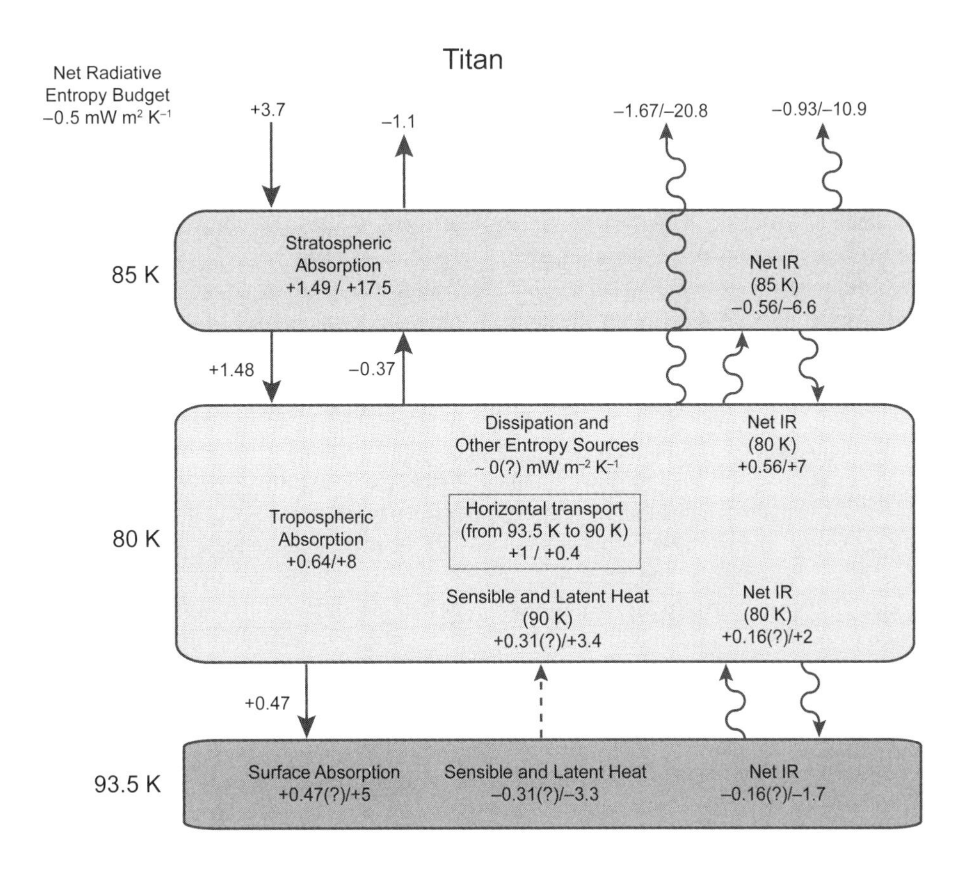

Fig. 6. Entropy and energy budgets for Titan. Solid vertical lines with arrows refer to the solar flux in units of W m^{-2}. Curved vertical lines with arrows refer to the longwave radiation. Numbers separated by "/" are values of energy flux and entropy flux. Energy flux is the first number in units of W m^{-2}. Units of entropy flux are mW m^{-2} K^{-1}. The vertical dashed line with an arrow indicates the exchange of sensible and latent heat between the surface and atmosphere. Some temperatures associated with heat absorption in the atmosphere are listed in parentheses. The box around "Horizonal transport" in the middle of the figure indicates fluxes associated with thermally direct horizontal heat transport by the atmosphere.

of 90 K. For the troposphere to be in statistical equilibrium, we require an additional entropy source of 0.4 mW m^{-2} K^{-1}. If due to dissipation primarily in the planetary boundary layer at a temperature of 90 K, this implies a dissipation rate of 0.036 W m^{-2} and an efficiency of 12%. This is three times higher than the estimate of the Carnot efficiency (4.1%) based on radiative energy fluxes. Clearly there is a missing entropy source.

In addition to vertical fluxes of energy, Titan's atmosphere has a seasonally reversing, thermally direct heat transport (*Mitchell,* 2012) that significantly contributes to entropy production in the atmosphere. For the purpose of estimation, we assume ~1 W m^{-2} is fluxed in latitude from the summer hemisphere at ~93.5 K to the winter hemisphere at ~90 K, and contributes ~0.4 mW m^{-2} K^{-1} to the entropy. Therefore dissipation might not even be necessary for Titan to be in statistical equilibrium. Given the small residual involved, this number is quite uncertain. Irreversible sources of entropy generation may include kinetic energy dissipation, raindrop dissipation, photochemistry, and breaking of large-scale waves.

5. DISCUSSION AND CONCLUSIONS

The rate at which mechanical or kinetic energy is dissipated in a planetary atmosphere is the key parameter in characterizing the efficiency with which the atmosphere acts as a heat engine. Unfortunately, the dissipation rate, or the rate of generation of mechanical energy, is poorly constrained observationally, perhaps even for Earth. Many phenomena can contribute to entropy production, including turbulence and viscous friction, gravity wave breaking, resistance encountered by falling raindrops in Earth's atmosphere and possibly on Titan, friction associated with dust storms on Mars, and photochemical processes in Titan's atmosphere.

If, in the long term, the entropy of an atmosphere is constant (statistical equilibrium), then the entropy flux associated with the absorption of solar radiation and the reradiation of longwave energy to space provides an important constraint on entropy production and dissipation. The radiative entropy flux of an atmosphere is negative essentially because an atmosphere absorbs energy at a higher temperature than it reradiates it to space. Dissipation within the atmosphere and entropy production associated with thermally direct heat transports and other processes provide positive contributions to the entropy balance that offset the radiative entropy loss. If all the entropy sources can be identified and quantified, then the entropy production from dissipation can be determined. If in addition the temperature at which dissipation occurs can be determined or estimated, then the magnitude of the dissipative entropy source multiplied by this temperature gives the dissipative energy flux. Because of uncertainties in estimating rates of entropy production from the many sources of entropy active in Earth's atmosphere, we have accepted the value of dissipative energy flux of about 2 W m^{-2} inferred from observations. For Venus we are only able to put an upper bound of 23 mW m^{-2} K^{-1} on the entropy production in the lower atmosphere due to turbulent dissipation. For Mars we have estimated a dissipation rate of less than 1 W m^{-2}, but this value is uncertain. For Titan the dissipation rate is also uncertain but perhaps ~0.01 W m^{-2}, and this may imply that the atmospheric heat engine on Titan is operating close to the Carnot efficiency. It is emphasized that the constraints on dissipation rate are approximate because required entropy production occurs by a variety of processes and over a range of temperatures.

The dissipation rate estimated for convective activity in the cloud deck of Venus is small compared with the entropy sources required to offset radiative entropy loss. Other processes in the lower atmosphere must supply the required entropy production in Venus' atmosphere. Gravity wave breaking is a likely dissipative phenomenon of importance.

The literature on Mars' atmosphere is largely devoid of discussion regarding its energy and entropy budgets. We suggest that the entropy production due to turbulent dissipation and other processes in the martian atmosphere must be about 1 mW m^{-2} K^{-1}.

Titan's surface energy budget has only recently been recognized to require significant sensible and/or latent heat transfer (*Mitchell,* 2012). In addition, horizontal, thermally direct heat transport by the atmosphere contributes an important source of entropy. Our estimates of the energy and entropy budgets are necessarily based on model calculations. Although our estimate of dissipation might appear small, so is the generation of APE, and we infer a rather efficient atmospheric heat engine.

While we have adopted a definition of atmospheric efficiency equal to the ratio of turbulent dissipation to convective energy flux, it is by no means certain that this definition should be the preferred one. For example, in the case of Earth, it could be argued that rainfall dissipation should be included since the atmosphere does work in lifting the moisture. This might also apply to Titan, and in the case of Mars, the dissipation in dust storms might also be included, since the atmosphere does work in lifting dust thereby altering the opacity and the absorption of solar and infrared radiation.

Our main conclusions are:

1. The radiative entropy loss from planetary atmospheres determines the required entropy production and constrains the amount of turbulent dissipation in them. This conclusion rests on the assumption that the atmosphere is in a statistically steady state. Turbulent dissipation cannot be determined directly since a variety of processes contribute to entropy production and there is uncertainty in their quantitative estimation.

2. Turbulent dissipation in Earth's atmosphere is observed at about 2 W m^{-2}. This amounts to about ¼ of the entropy production (in excess of that due to latent heat and sensible heat transfer between the atmosphere and surface) required by Earth's radiative entropy imbalance.

3. Carnot efficiencies of the atmospheres of Venus, Earth, Mars and Titan are, respectively, about 27.5, 13.2, 4.4 and 4.1%.

4. Entropy production rates (in excess of that associated with latent and sensible heat transfer between the atmosphere and surface) in the atmospheres of Venus, Earth, Mars and Titan are, respectively, about 23, 29, 1, and 0.1 mW m^{-2} K^{-1}. Turbulent dissipation of atmospheric kinetic energy is only one of the processes that contribute to this required entropy production. On Earth and perhaps on Titan rainfall dissipation is important. Other sources of entropy, perhaps important on all the planets with atmospheres, include horizontal heat transport, dissipation of gravity waves in the upper atmosphere, photochemical processes, diffusion of water vapor or other minor constituents, and, in the case of Mars, generation and dissipation of dust storms.

REFERENCES

Becker E. (2003) Frictional heating in global climate models. *Mon. Weath. Rev., 131,* 508–520.

Bridges N. T., Ayoub F., Avouac J.-P., Leprince S., Lucas A., and Mattson S. (2012) Earth-like sand fluxes on Mars. *Nature, 485,* 339–342, DOI: 10.1038/nature11022.

Crisp D. and Titov D. (1997) The thermal balance of the Venus atmosphere. In *Venus II: Geology, Geophysics, Atmosphere, and Solar Wind Environment* (S. W. Bougher et al., eds.), pp. 353–384. Univ. of Arizona, Tucson.

Gierasch P. J. (1971) Dissipation in atmospheres: The thermal structure of the martian lower atmosphere with and without viscous dissipation. *J. Atmos. Sci., 28,* 315–324.

Goody R. (2007) Maximum entropy production in climate theory. *J. Atmos. Sci., 64,* 2735–2739.

Izakov M. N. (2010) Dissipation of buoyancy waves and turbulence in the atmosphere of Venus. *Solar System Res., 44,* 475–486.

Kiehl J. T. and Trenberth K. E. (1997) Earth's annual global mean energy budget. *Bull. Am. Meteorol. Soc., 78,* 197–208.

Li L., Ingersoll A. P., Jiang X., Feldman D., and Yung Y. L. (2007) Lorenz energy cycle of the global atmosphere based on reanalysis datasets. *Geophys. Res. Lett., 34,* L16813, DOI: 10.1029/2007GL029985.

Li L., Nixon C. A., Achterberg R. K., Smith M. A., Gorius N. J. P., Jiang X., Conrath B. J., Gierasch P. J., Simon-Miller A. A., Flasar F. M., Baines K. H., Ingersoll A. P., West R. A., Vasavada A. R., and Ewald S. P. (2011) The global energy balance of Titan. *Geophys. Res. Lett., 38,* L23201, DOI: 10.1029/2011GL050053.

Lorenz E. N. (1967) *The Nature and Theory of the General Circulation of the Atmosphere.* World Meteorological Organization, Geneva. 161 pp.

Lorenz R. D. and McKay C. P. (2003) A simple expression for vertical convective fluxes in planetary atmospheres. *Icarus, 165,* 407–413.

Lorenz R. D. and Renno N. O. (2002) Work output of planetary atmospheric engines: Dissipation in clouds and rain. *Geophys. Res. Lett., 29,* DOI: 10.1029/2001GL013771.

McKay C. P., Pollack J. B., and Courtin R. (1989) The thermal structure of Titan's atmosphere. *Icarus, 80,* 23–53.

Mitchell J. L. (2012) Titan's transport-driven methane cycle. *Astrophys. J. Lett., 756,* L26, DOI: 10.1088/2041-8205/756/2/L26.

Oort A. H. and Peixoto J. P. (1983) Global angular momentum and energy balance requirements from observations. *Adv. Geophys., 25,* 355–490.

Pauluis O. (2011) Water vapor and mechanical work: A comparison of Carnot and steam cycles. *J. Atmos. Sci., 68,* 91–102, DOI: 10.1175/2010JAS3530.1.

Pauluis O. and Dias J. (2012) Satellite estimates of precipitation-induced dissipation in the atmosphere. *Science, 335,* 953–956, DOI: 10.1126/science.1215869.

Pauluis O. and Held I. M. (2002a) Entropy budget of an atmosphere in radiative-convective equilibrium. Part II: Latent heat transport and moist pressures. *J. Atmos. Sci., 59,* 140–149.

Pauluis O. and Held I. M. (2002b) Entropy budget of an atmosphere in radiative-convective equilibrium. Part I: Maximum work and frictional dissipation. *J. Atmos. Sci., 59,* 125–139.

Peixoto J. P. and Oort A. H. (1984) Physics of climate. *Rev. Mod. Physics, 56,* 365–429.

Peixoto J. P. and Oort A. H. (1992) *Physics of Climate.* American Institute of Physics, New York.

Peixoto J. P., Oort A. H., Dealmeida M., and Tome A. (1991) Entropy budget of the atmosphere. *J. Geophys. Res., 96,* 10981–10988.

Titov D. V., Bullock M. A., Crisp D., Renno N. O., Taylor F. W., and Zasova L. V. (2007) Radiation in the atmosphere of Venus. In *Exploring Venus as a Terrestrial Planet* (L. W. Esposito et al., eds.), pp. 121–138. AGU Geophys. Monogr. 176, American Geophysical Union, Washington, DC.

Wu W. and Liu Y. (2010) Radiation entropy flux and entropy production of the Earth system. *Rev. Geophys., 48,* RG2003, DOI: 1029.2008RG000275.

Dowling T. E. (2013) Earth general circulation models. In *Comparative Climatology of Terrestrial Planets* (S. J. Mackwell et al., eds.), pp. 193–211. Univ. of Arizona, Tucson, DOI: 10.2458/azu_uapress_9780816530595-ch009.

Earth General Circulation Models

Timothy E. Dowling
University of Louisville

The development of Earth general circulation models (GCMs) is rapidly evolving on all fronts, with today's nonhydrostatic and global cloud resolving models (GCRMs) making an impact on par with the original introduction of GCMs in the 1960s. Here we take a look at the organizational structure of these models, including the dynamical core and physics layers, the latest horizontal and vertical grids, standardized frameworks for software components, data and metadata, and the manner in which the international scientific community systematically compares climate models. Data assimilation and the quantification of forecast skill are two well-developed Earth-GCM concepts that are beginning to see utility in planetary science. Also discussed are the philosophical questions that arise when an analysis, an optimal blend of data and model, is used in place of a pure dataset for scientific work. Fully unstructured, dynamically adapting grids optimized for stratified fluids are predicted to be the preferred framework for GCMs in 25 years.

Don't hardwire anything.
— Andrew P. Ingersoll

1. INTRODUCTION

Atmospheres and oceans present marvelous systems for study — protective, tempestuous, composed of thousands of interlocking fluid parts — these are the spheres of a planet we take most personally. Fidelity to a planet's outer fluid system is the overarching goal of general circulation models (GCMs). As far removed from day-to-day planetary-science research as it may seem, there is a group of GCM modelers who study an atmosphere that has people living *inside* it — lots of them, in fact. All of them, in fact, except for the three or so on the International Space Station. This is clearly a liability, but it is also an asset.

Here, we peer over the shoulders of these geocentric GCMers to see how their modeling capabilities are coordinated and applied to the home planet, and how the components that make up Earth GCMs are organized. A great deal of Earth GCM technology has been successfully adapted to planetary research, and some planetary GCM developments have returned the favor. As Earth modelers begin to develop surface-to-thermosphere global models that include space weather [e.g., the Whole Atmosphere Community Climate Model (WACCM)], one can expect to see an increase in interdisciplinary collaborations.

This chapter is aimed at the planetary-science reader who, on the one hand, is more likely to be studying an atmosphere with 960,000 ppmv CO_2, rather than Earth's exiguous 390 ppmv CO_2, and on the other hand, has a standard for a respectable data volume set by planetary flagship missions like Cassini, which averages 0.8 Gbits/day, rather than (to pick one component of one Earth observing system) the GOES-R Earth-weather satellite system and its High Rate Information Transmission (HRIT) network for emergency managers, which runs at 34.6 Gbits/day, or more than 40× faster. Consequently, the focus here is away from the latest Earth climate results, their causes and consequences, and toward the advanced strategies developed by Earth GCM developers for model discretization, GCM component interoperability, data assimilation, quantification of forecast skill, and systematic model intercomparison. The following chapter by François Forget and Sebastien Lebonnoi deals specifically with planetary GCMs and the particular questions motivating planetary scientists; the redundancy between these two chapters is intentional, and we hope, beneficial.

1.1. Does GCM stand for General Circulation Model, Global Circulation Model, or Global Climate Model?

The definition of the acronym "GCM" has evolved over time, and even split. It originally signified a realistic model of a planet's complete outer fluid system with no artificial lateral boundary conditions. Early appearances of the phrase "general circulation" refer to Earth's planetary-scale winds (e.g., *Jeffreys,* 1922): "The winds of the globe, before their respective positions in the above dynamical classification are assigned, may be roughly classified according to their horizontal extent. a. World-wide phenomena, including the general circulation and its seasonal extent." Examples of models that are extremely valuable but are not global include quasi-geostrophic models, which are not valid in the equatorial region, regional atmospheric models like the original mesoscale Weather Research & Forecasting model [WRF, pronounced "wharf," which now includes a global option and a planetary version, planetWRF (*Richardson et al.,* 2007), single-basin ocean models, or any idealized model on a doubly periodic domain.

In the 1980s, "GCM" was brandished with bravado, not unlike the modifier "nonhydrostatic" is used today. All throughout its young history, the acronym has also signaled to many highly respected detractors a model that leaves in its wake a lack of understanding, a void that must be spanned by process studies supported by a hierarchy of models (*Held,* 2005; *Randall et al.,* 2007).

Today, a complete Earth GCM is presumed to model the planet's atmosphere, oceans, land, and ice as a coupled system, and hence we have the notion of a coupled atmosphere-ocean GCM (AOGCM, see Table 1), but the phrases atmospheric GCM (AGCM) and oceanic GCM (OGCM) are not ambiguous and have utility. For the planets, in the middle of a talk by the author in the mid 1990s about the then-new, global EPIC atmospheric model applied to Jupiter, at Walsh Cottage in Woods Hole, Massachusetts, the following sequence generated a big laugh: Melvin Stern, scratching his beard, asked, "Does EPIC simulate Jupiter's interior?" "No." "Then, it isn't a GCM."

Regarding the hierarchy of global models, on the less-complex end of the spectrum are simple climate models (SCMs), e.g., energy-balance box models, and Earth system models of intermediate complexity (EMICs), which use reduced grid resolution to enable simulations to run for centuries. On the high end are global cloud-resolving models (GCRMs), which are ushering in a major advance in GCM capability, on par with the invention of the discipline itself (*Randall,* 2011).

For regional models, the term regional climate model (RCM) has been coined to indicate, so to speak, a local GCM. These typically use a GCM's output for their boundary conditions, and they add value by downscaling to higher spatial and temporal resolution, often employing additional, or more realistic, physics. (The term "downscaling" implies the addition of capabilities and is an optimistic, local-hero term; it is not to be confused with the similar sounding "downscoping," which is the shedding of capabilities and is a pessimistic term.)

TABLE 1. Weather and climate model types.

Acronym	Definition
AGCM	Atmospheric GCM
AOGCM	Coupled Atmosphere-Ocean GCM
EMIC	Earth System Model of Intermediate Complexity
ESM	Earth System Model
GCM	General Circulation Model, Global Circulation Model
GCM	Global Climate Model
GCRM	Global Cloud Resolving Model
NWP	Numerical Weather Prediction Model
OGCM	Oceanic GCM
RCM	Regional Climate Model
SCM	Simple Climate Model

Traditionally, Earth weather and climate models have been designed differently, reflecting their different purposes. Generally speaking, weather models are operational tools used to make forecasts, and climate models are research tools used to study global, long-term processes, such as the benchmark CO_2-doubling experiment. Interestingly, since the 1960s the computational cost of running numerical weather prediction (NWP) models has risen by orders of magnitude more than the cost of running climate models (*Randall,* 2011).

One of the distinguishing features of a good climate model is that its globally conserved quantities, such as total energy, do not drift with time when the model is run for decades and centuries. A researcher who expects this of a GCM is using the acronym in its newer sense to mean global climate model. On the other hand, short-range weather forecasts are done with models that do not need to be strictly globally conservative, since they are periodically adjusted by observations via data assimilation (see below). There is a trend for NWP models to expand their domains to Earth's full extent, to be GCMs in the original sense of the word, and hence the need for a phrase that heralds "general circulation" and/or "global circulation" is fading in Earth research, and is giving way to a need for a phrase that implies "not weather forecasting." Thus, in the context of Earth applications, the acronyms NWP and GCM are now typically used to indicate models that are employed in operational weather forecasting and global climate research, respectively.

The Meteorological Office (Met Office) of the United Kingdom is leading a new trend to use the same nonhydrostatic, full bells-and-whistles global model for both weather prediction and climate research (*Davies et al.,* 2005). The conservation requirements of climate models must still be satisfied, but this approach underscores the exciting possibilities that are opening up as a consequence of the steady rise in parallel-computer execution speed and data storage capacity.

In planetary research, which trots alongside geocentric research like a younger sibling, operational-meteorology products are rarely the point, except for adjusting to upper-atmospheric solar-weather conditions during the occasional probe entry or aerobraking maneuver, and increasingly in support of *in situ* exploration of Mars (*Read et al.,* 2006). Hence there is not much need in planetary science for an antonym to NWP. The global conservation properties of climate models are just as important as in Earth research, e.g., for Venus superrotation spin-up experiments, but the gas giants are so large (Jupiter's surface area is 120 times that of Earth's) and zonally symmetric (banded) that regional or channel configurations of circulation models are often referred to as GCMs, without causing much confusion. Based on the way the acronym is used in the planetary science literature and at scientific meetings, the truest translation of "GCM" in today's planetary lexicon is "good circulation model"; in other words, a realistic model, one with all the rights, privileges, and detractions thereunto pertaining. But, younger siblings eventually grow up, and one can expect to see the more descriptive acronyms in Table 1 being integrated into planetary science in due course.

1.2. Unpredictable Earth

It is wise in any research endeavor to identify and start with the easiest problems first, to the extent possible, and only after these are well understood, to move onward and upward to harder problems. In 2000, neurologist E. Kandel shared the Nobel Prize in Physiology or Medicine in part because he chose not to study the most complex nervous systems available, people and higher mammals, but rather the simplest, the California sea hare. The first 50 years of planetary exploration has taught us that of all the atmospheres in the solar system, Earth's is the most complex, the most unpredictable. This theme echoes back and forth across the pages of this book. In this light, it is important that students of Earth's atmosphere appreciate that when they chose to focus on Earth, this was the moment when they decided to tackle, in most theoretical aspects, the *hardest* atmosphere available for study. Accordingly, this decision needs to be justified.

Of course, it is straightforward to do so. At the time of this writing, Earth is the only known habitable planet; more to the point, it is the only accessible, habitable planet. We need an army of Earth researchers studying it, and we need a second army collaborating with the first to write, and update, a user's manual for Earth.

Just how unpredictable is Earth's weather and climate compared to the planets? Baroclinic instability on Venus is largely confined to its middle cloud layer (*Young et al.,* 1984), and seasonal forcing is naught because Venus has a negligible axial tilt (obliquity). Mars is currently tilted like Earth and has analogous baroclinic instability and seasons, but its weather has a rhythm that is predictable and coherent (*Read et al.,* 2006). Titan has monsoons and cloudy processes, and hence is Earth-like in its options, but nevertheless its mountains are just hills compared to Earth (*Radebaugh et al.,* 2007), its oceans just lakes. Expanding out to the gas giants, the locations of long-lived storms on Jupiter, Saturn, Uranus, and Neptune can be accurately predicted months in advance using no more than a ruler and a stopwatch. In fact, the Voyager 1 and 2 missions would have been utter failures had this not been the case, since for logistical reasons each narrow-angle camera sequence had to be uploaded more than a week before it was executed.

Imagine programming an Earth weather satellite to take a sequence of close-up images of a hurricane's eye, and uploading the detailed pointing instructions a week or two before the first picture is snapped. On August 25, 1989, Voyager 2 did just this, unerringly snapping one narrow-angle, close-up bull's eye after another of Neptune's fast-paced storms, which are embedded in the strongest wind shears ever discovered. Less than a month after Voyager 2 achieved this feat, back on Earth, 1-to-3-day (not 7-to-10-day) forecasts of the landfall of Hurricane Hugo proved to be insufficient to distinguish between Savannah, Georgia, and Charleston, South Carolina, and so both cities were evacuated. The latter got the hurricane, on September 22, 1989, and the former got a faint drizzle, which is what Voyager 2 would have photographed, had it been assigned to image Hugo, or Bob, or any other swerving cyclone on Earth.

What is it about the home planet that makes its weather and climate so unpredictable? This is a fundamental question in Earth GCM research. A climate model can reveal statistical responses to external forcing more faithfully than it can generate an accurate forecast from initial conditions, so the way climate questions are posed is pivotal to their tractability. *Palmer and Hagedorn* (2006) have edited a book that provides a comprehensive survey of the mathematics and physics of the predictability of Earth's atmosphere and oceans. Going forward, there are many exciting opportunities available to test these theories on the planets.

From a comparative planetology standpoint, Earth is positioned only 108 Sun diameters from the Sun, and consequently its atmosphere receives more watts per square meter than any other in the solar system (except for Mercury's tenuous atmosphere), including the atmosphere of Venus, whose clouds are more reflective. Add to this copious energy density the fact that Earth has tall mountain ranges oriented disruptively in the north-south direction, and it has water, water, everywhere, with oceans confined to nearly closed basins. Even Earth's size is a complicating factor: A good analogy is to think of a gas giant's elegantly spinning cyclones and anticyclones as being like a line of ballerinas pirouetting across a grand stage ("bolshoi" means "large"), whereas Earth's cyclones and anticyclones are like the same ballerinas, but trying to perform *Swan Lake* inside a large elevator. In short, Earth GCM modelers have far and away the hardest weather and climate system to simulate. But as we shall see, they are starting to meet the challenge.

1.3. Overview

This and the other chapters in this book pertaining to Earth's climate, if treated to their fullest extent, would expand to be several times the size of this book. In fact they already have been so treated, in the form of the *Fourth Assessment Report: Climate Change 2007* (*AR4*) by the Intergovernmental Panel on Climate Change (IPCC). The reader may recall that the IPCC was awarded the Nobel Peace Prize in 2007. The full *AR4* spans four volumes, available free online as html or downloadable .pdf files (*www.ipcc.ch/publications_and_data/publications_and_data_reports.shtml*), or the books can be purchased. The volume by Working Group I is the pertinent one here, a 996-page work entitled *Climate Change 2007: The Physical Science Basis* (hereafter *AR4-Physical*). Six hundred nineteen (619) authors contributed to *AR4-Physical*; a few are authors on this book as well. We draw on the *AR4* work below, and add some highlights from the five years between then and the time of this writing (2012), which is about two years before the release of IPCC's fifth assessment report (AR5), due out in 2014.

We start with a look at the organization of the modern Earth GCM, its dynamical core and its physics layers, paying particular attention to the standardized data and software frameworks being developed to enhance interoperability. Next, we examine the optimal blending of observations and theory, called data assimilation, and its resulting products, called analyses and reanalyses, followed by the quantification of forecast skill. Data assimilation and forecast skill are emerging concepts in planetary science. At this point in the chapter, we get some comic relief in the form of analogies between the game of golf and GCM modeling. We then review the international efforts to systematically compare Earth climate models, and end with a handful of short-range, medium-range, and long-range forecasts on the future of Earth GCMs, topping off with a satirical jab at the Earth GCM modeler's still favorite, yet archaic programming language (whose moniker begins with an "F").

2. DYNAMICAL CORE

In GCM jargon, the dynamical core (nicknamed dycore or simply the core) is the software component that handles Newton's second law (conservation of momentum), written in the fluid-dynamicist's preferred per-mass form, $\mathbf{a} = \mathbf{f}/m$, plus all relevant continuity equations (conservation of mass for total atmospheric mass, water in all its phases, precipitation and any trace gases or aerosols), plus the first law of thermodynamics (conservation of energy). The etymology of "core" traces to the Latin "cor" and French "coeur," both meaning heart; the GCM's dynamical core is its beating heart.

All other components of a GCM, in particular radiation and subgrid-scale processes such as turbulence and cloud microphysics that require parameterization, but also any chemistry, are unabashedly referred to as the model's physics layers (or physics). *Donner et al.* (2011) have edited an engaging and comprehensive review of the history of Earth GCM components.

Although the physics layers give a GCM its outward personality, a great deal is going on inside. Students of geophysical fluid dynamics (GFD) quickly learn that the acceleration terms that sum to give the net acceleration, $\mathbf{a}$, are surprisingly numerous; they include the velocity tendencies and advection terms that together comprise the relative accelerations, relative to the planet's rotating frame of reference, and the Coriolis, or rotating-frame (aka planetary) accelerations. The advection terms of fluid dynamics are challenging because they describe how the winds (or ocean currents) blow themselves around, and hence are intrinsically nonlinear. In the two horizontal directions, the primary force in a GCM is the pressure-gradient force, and both it and gravity are the dominant forces in the vertical direction. The job of the dynamical core is to handle these conservation laws with precision, in both space and time.

2.1. Horizontal Discretization

To do this, the calculus of the conservation equations is first converted into an approximate algebraic system, because to date, and at least for the near future, computers are adept at solving planet-sized algebraic systems in the floating-point sense, but are miserable at solving planet-sized calculus problems in the analytical sense. Approximating calculus with algebra is called discretizing the system, and it is the choice of how this is done that most distinguishes one GCM core from another. A history of the discretization of GCMs is given in the chapter by *Randall* (2011), in the Donner et al. book cited above, which we draw on here.

The chairman, co-founder, and CEO of Apple Inc., Steve Jobs, liked to point out that the 1960s happened in the early 1970s — the author was nine years old in 1971, living in California, so he can corroborate — but what in fact did happen in the 1960s germane to the discussion is the first true GCMs were written. The architects of these inaugural models were educated in the U.S. and Japan, but the early-model birthplaces were all in the U.S., specifically at the Geophysical Fluid Dynamics Laboratory (GFDL) in Princeton, New Jersey; the University of California at Los Angeles (UCLA); and the National Center for Atmospheric Research (NCAR) in Boulder, Colorado. Soon afterward came the British invasion, and in fact the European Centre for Medium Range Weather Forecasting (ECMWF) has led the world in GCM forecast accuracy since the 1980s. Today, Earth GCM research and development is an international endeavor, with a large organizational role played by the United Nations (UN).

Turning to the nuts and bolts, the first GCMs discretized the governing equations using the finite-difference approach on a longitude-latitude grid. A major research theme today is to find the optimal horizontal grid to use for a sphere (*Staniforth and Thuburn,* 2012), or to be precise, an oblate spheroid. The main problem with the longitude-latitude grid is that the longitude spacing, or east-west or x-direction spacing, shrinks to zero at the poles, such that the grid cells become too skinny and too misshapen to provide a properly supportive trellis. For a number of reasons, this has driven GCM developers to search for and find better grids, the most serious problem being a severe restriction on the size of the numerically stable time step.

An elegant solution to the pole problem is to not use finite differencing at all, but rather to employ spherical harmonics, where the discretization comes from the need to truncate the infinite spherical-harmonic series that represent each prognostic variable. Several GCMs went this route in the 1970s and 1980s, following the lead of the GFDL (Princeton) and ECMWF (European) teams. However, working exclusively in spectral or wavenumber space is difficult when it comes to nonlinear processes like advection, which is ubiquitous, and so a hybrid approach was developed called the transform method. In this scheme, the GCM maintains both physical and wavenumber versions of key variables, and attempts a best-of-both-worlds strategy.

The spectral method has an Achilles heel, however. Sharp gradients, including rugged mountains and cloud boundaries, to name an old and a new computational challenge, are difficult to model with spherical harmonics without ringing (Gibb's ripples), and advection tends to produce spurious negative mass. A way to kill two birds with one stone was subsequently developed, the semi-Lagrangian method for advection (*Ritchie et al.,* 1995), which both does not ring and does not require small time steps for numerical stability. This scheme was first implemented in Canada (March 1991) and at the ECMWF (September 1991), after which most operational groups followed suit.

But today, the spectral approach has run its course in GCM research laboratories. Its ultimate downfall has been the relentless, heady improvement of computer hardware. Spectral transformations fall into the class of algorithms that require global operations on their variable fields. In contrast, grid-point models are local by nature, and hence run efficiently on massively parallel computers, which for wiring-topology reasons hold only a portion of the entire model in the memory of each node, i.e., use distributed-memory architecture. The new push is to find horizontal grids that provide quasi-uniform resolution everywhere on the sphere, such that they have no misshapen grid cells. We revisit this subject in section 7.1 below.

2.2. Vertical Coordinate

Tabling modern horizontal grids for the moment, we now concentrate on the vertical coordinate, which has had an equally intriguing, and largely independent, history. Scaling analyses showed early that some terms in the governing equations related to accelerations and motions in the vertical dimension could be neglected at the synoptic (continental) and planetary scales, because Earth's atmosphere is shallow, leading to a set of equations called the primitive equations (*Holton,* 2004); this is the set discretized by most GCMs to date.

That is, until recently. Lately, the steady increase in grid resolution has altered these scaling arguments — Earth's atmosphere is no longer computationally shallow, because mesoscale (city-sized) processes are now being resolved (comparative note: Titan's atmosphere has never qualified as shallow, not even on the planetary scale; see the next chapter) — and so the modern trend is to retain all the vertical terms, most significantly the nonhydrostatic vertical accelerations [which expose a GCM to numerical instabilities associated with sound waves, and therefore must be treated judiciously (see *www.ecmwf.int/publications/library/do/references/list/201010*)], but also the "nontraditional" Coriolis terms that couple to vertical velocity (e.g., *Hayashi and Itoh,* 2012).

What should one use for the vertical coordinate in a GCM? If you ask a planetary fluid-interior/dynamo modeler what he uses for his vertical coordinate, the question will strike him as fatuous; he will pause, then politely oblige with "radius," delivered with the level look of a chess champion who is comfortable with all the legal moves available to his opponent. Geopotential height is similar to radius, but folds in the rotating planet's centrifugal force. Subtract a reference height (a reference ellipsoid that sufficiently approximates the geoid) and you have altitude, z, which the earliest NCAR model used for its vertical coordinate, i.e., the early NCAR developers made the same self-evident choice. Oceanographers regularly use z as well, particularly when density fluctuations can be neglected except for the buoyancy force (the Boussinesq approximation). So far, so good; use height for the height coordinate in GCMs. Except no one in meteorology actually does this, not even close, not since the first NCAR model.

In fact, if you ask a modeling group that has developed a GCM that simulates both Earth's atmosphere and oceans with the same dynamical core (e.g., the MITgcm group) this vertical-coordinate question, you will get only partially the same answer: *Marshall et al.* (2004) point out that z as used in ocean models is isomorphic to p, pressure, as used in atmospheric models when vertical accelerations can be neglected (the hydrostatic approximation). Why do atmospheric modelers often use p for the vertical coordinate when the hydrostatic approximation is valid? Meteorologists have long recognized that there are several advantages to isobaric coordinates (*Holton,* 2004), mostly related to the fact that the fluid between two pressure surfaces maintains constant weight per area even as it moves up and down and side to side, so long as its vertical accelerations are negligible, which squelches many of the symbolic-manipulation complications of compressibility in an atmosphere, particularly the exponential fall-off of density with height. The key point is that one has options and trade-offs when it comes to choosing the vertical coordinate in GCMs, and if some new choice, A, routinely yields a better forecast than good old z, it is hard to argue with those who switch to A (the pros and cons of various vertical coordinates are discussed in the online MetED education modules, *https://www.meted.ucar.edu/*).

A common misconception, which is important to clear up at this juncture, is the notion that a GCM's horizontal directions are locally orthogonal to its vertical coordinate when a nonheight choice has been made for the latter. This is not the case; in fact, the zonal and meridional winds, u and v, respectively, point perpendicular to the gravity — the gradient of the geopotential — no matter what vertical coordinate is employed. Think of the GCM's variables as being arrayed horizontally on ergonomically tilted racks; that is all that is happening. In this sense, meteorologists use nonorthogonal coordinates whenever they elect to use a nonheight variable for the vertical coordinate (strictly, a nongeopotential-height variable).

Many vertical-coordinate choices have been made during

the history of GCM design, including the above mentioned isobaric coordinates, but also isentropic coordinates (entropy or potential temperature), terrain-following (or sigma) coordinates, and hybrid coordinates that are terrain following near the surface but transition into either isentropic or isobaric coordinates aloft. The choice of vertical coordinate is a major distinguishing feature between GCM dynamical cores. The current upgrade to nonhydrostatic models may well be the knight's move (using the chess analogy) that lands GCMs back to using height as the vertical coordinate (*Randall,* 2011). Peering farther into the misty future, this author can see the outlines of GCMs with unstructured, dynamically adapting grids ultimately winning the match.

When it comes to discretization, the vertical and horizontal dimensions in a GCM are almost, but not entirely, independent. For example, the most important horizontal force, the pressure-gradient force, is difficult to discretize accurately for advanced vertical coordinates, particularly in the presence of rugged terrain, if one starts with the partial differential equation (PDE) form. Mathematicians call the PDE form the strong formulation, because theorems that can be proved in PDE form are strong. But fortunately for modeling, the pressure-gradient force can also be discretized, once and for all, in three-dimensional finite-volume form (e.g., *Bradley and Dowling,* 2012), even if the rest of the GCM is in finite-difference form — the weak formulation renders the pressure-gradient term rugged.

A cottage industry spanning both Earth and planetary GCM research has been the design of quasi-Lagrangian vertical coordinates, ones that move up and down at least approximately with the fluid (*www.esrl.noaa.gov/outreach/events/hybridmodeling08*). The idea is to reduce vertical truncation errors by suppressing the need to track rising-and-falling motions of the fluid relative to the grid, because such errors have proved to be a major source of inaccuracy in dynamical cores. The isentropic coordinate is a popular choice because most atmospheric processes are adiabatic (moist processes in thunderstorms are diabatic, but even then the water-vapor fuel is delivered thousands of kilometers to the storm site almost adiabatically). Isentropic coordinates are usually implemented in a hybrid form so that they become terrain following near the surface. The first operational (weather forecasting) model for Earth's atmosphere to employ a hybrid-isentropic coordinate was the NOAA Rapid Update Cycle (RUC) (*Benjamin et al.,* 2004).

But the Earth-atmosphere forecasting world is a pressure cooker, and RUC has already come and gone. It was replaced by the Rapid Refresh (RAP) model, which is based on the above-mentioned WRF model and employs a terrain-following (sigma) coordinate instead of a hybrid-isentropic coordinate. The reasons why WRF/RAP trumped RUC are interesting, and will provide the segue into the next section.

Before moving from, in the vernacular of GCMs, the dynamics to the physics, consider the features of another new GCM, one for which all the dynamical-core themes in this section converge: the Flow-following finite-volume Icosahedral Model (FIM) (*fim.noaa.gov*). In the horizontal, FIM employs an icosahedral grid, meaning it covers the globe Buckminster Fuller-style, with all hexagons except for 12 pentagons, such that it has no misshapen grid cells anywhere. Meanwhile in the vertical, it uses a hybrid isentropic-terrain following coordinate. It also features a finite-volume advection algorithm. FIM is slated to be part of the multimodel ensemble used by the U.S. National Centers for Environmental Prediction (NCEP), and one hopes to see planetary applications shortly.

So, why did WRF trump RUC? The primary technical advantage of the WRF model is its nonhydrostatic option. RAP/WRF (and RUC before it) is the weather-forecasting model used in the U.S. by pilots, severe weather forecasters, and anyone needing forecasts updated every hour. Consequently, it is a regional NWP model that uses tight horizontal, vertical, and temporal resolution. Computers are fast enough today that such models can retain all the vertical terms, and hence can be nonhydrostatic, and WRF has one of the most advanced nonhydrostatic cores available. But WRF also has a meta-advantage: Like with many models, a single group developed RUC, but in contrast, the WRF project is a community-wide development effort [as is the Whole Atmosphere Community Climate Model (WACCM) mentioned in the introduction]. In other words, how WRF was designed and assembled is just as pioneering, and as significant to the future of GCM development, as its shiny-new parts.

3. PHYSICS LAYERS

We now turn to a brief introduction of the physics components that give a GCM its character; the next chapter goes into the details. For a stimulating introduction to the physical and chemical processes that affect planetary climate, consult *Pierrehumbert* (2010), and in addition to the following chapter, see also the sections in this book that discuss clouds, hazes, aerosols, precipitation, hydrology, and land-air interactions.

3.1. GCM Spec-Sheet

The most efficient way to convey the look-and-feel of the latest Earth GCM physics components is to show a snapshot from which one can discern at a glance what was uppermost in the minds of Earth modelers in 2012. Accordingly, the following is the spec-sheet for Version 1 of the NCEP (U.S. National Centers for Environmental Prediction) operational RAP model (*rapidrefresh.noaa.gov*) just mentioned:

Core:	WRFv3.2.1+
Convection:	Grell-G3
Microphysics:	Thompson/NCAR
Longwave Radiation:	RRTM
Shortwave Radiation:	Goddard
Turbulence:	MYJ mixing
Land Model:	RUC-Smirnova
Data Assimilation:	Gridpoint Statistical Interpolation (GSI 2010)

GCMs that are used for long-range forecasts make tradeoffs between numerical accuracy and efficiency, hence they use different submodels than above for each physics layer, and they typically include a model for Earth's cryosphere (sea ice and ice sheets). Chapter 8 of *AR4-Physical*, entitled "Climate Models and Their Evaluation" (*Randall et al.*, 2007), includes a comprehensive review of the physics-layer capabilities of Earth GCMs ca. 2006, covering processes in Earth's atmosphere, oceans, land, and cryosphere; Table 8.1 of that work lists modeling details and references for 23 specific AOGCMs (coupled atmosphere-ocean GCMs). A few Earth GCMs that have appeared since *AR4* are also mentioned in this chapter.

3.2. Modeling Clouds and Aerosols

Arguably the most difficult physical process to parameterize in a GCM is cumulus convection. This umbrella process spans a large dynamic range; has rapidly evolving, heterogeneous structures with sharp boundaries (clouds); deals with all manner of vertical motions, including convection and precipitation, and the challenges they present to precise mass continuity; and accommodates the prodigious, pulsating energy transfers associated with water phase changes. *Randall* (2011) and references therein provide a critical review of the cumulus parameterization problem. This is an area that is rapidly evolving as GCM grids get denser. In fact, wholesale replacement of long-established schemes is occurring at several research institutions and centers, and this will be a major research and development area for decades.

In the IPCC Third Assessment Report of 2001 (TAR), new attention was drawn to the fact, long appreciated in the climate community, that clouds have an enormous impact on climate via their interaction with radiation, both shortwave (solar) and longwave (terrestrial). Earth's global albedo value is currently 29% (*Wielicki et al.*, 2005), fully two-thirds of which is the result of its water clouds. A drop to 28%, i.e., a drop of just one point, would have approximately the same direct radiative effect (DRE) as a doubling of atmospheric CO_2. Several observational campaigns have been mounted to study cloud-radiation effects in detail, including a comprehensive Atmospheric Radiation Measurement (ARM) program, a study of convective systems called the Coupled Ocean-Atmosphere Response Experiment (COARE), and a stratocumulus study called the Atlantic Stratocumulus Transition Experiment (ASTEX). For a broad historical overview of cloud field campaigns, see chapter 1 in *AR4-Physical*.

How GCMs model the radiative properties of clouds has been isolated as a major source of disagreement between models. *Senior and Mitchell* (1993) found that changing just the cloud physics-layer properties from one reasonable radiation-interaction parameterization to another resulted in a surface-temperature increase from 1.9°C to 5.4°C. This is a large model sensitivity; it has since been confirmed by other groups and is addressed in the AR5 design of experiments. For further details on the cloud-radiation problem as it stood around 2006, see section 1.5.2 of *AR4-Physical*, "Model Clouds and Climate Sensitivity." Excitingly, model resolution has improved to the point where cloud systems are beginning to be resolved, as shown in Fig. 1, and global models (GCRMs) that resolve individual cumulus cells are beginning to appear.

Another GCM-physics area of particular interest is the modeling of aerosols, which is addressed in chapter 2.4 of *AR4-Physical*. The Aerosol Model Intercomparison (AeroCom) project compares standardized experiments between many independent groups (*Kinne et al.*, 2006). They have found that simulated aerosol life cycles depend strongly on modeling details, including advection, chemistry, and aerosol microphysics, and less so on the distribution of sources in space and time (*Textor et al.*, 2007).

3.3. Interoperability

Many GCM software development teams around the world today operate as proprietary institutions, i.e., they maintain complete, closed-software models that they use for their own purposes, and they lease their executable code to interested operational and research centers. But the modern trend is away from this mode of operation, which functions like a collection of city states, and toward an open, global software framework.

Engineers are the professionals when it comes to standardization; they do it early and often. Although such a culture may appear to run the risk of stifling innovation, in practice just the opposite occurs, chiefly because the process is actively managed (the author developed an appreciation for this way of doing business by watching his father chair an IEEE standards working group for years). Think of how easy it is today to swap and upgrade one or more hardware components in a personal computer, not to mention how easy it is to assemble a blazingly fast, complete PC from off-the-shelf components. In the software realm this is called object-oriented programming, a decades-old concept (a majority of the languages on the TIOBE Programming Community Index are object-oriented) that is nevertheless new to the culture of GCM design, where getting rival institutions to standardize their component interfaces is like herding cats.

The payoffs of increasing interoperability in GCM design are starting to roll in, and the WRF and WACCM models are good early examples. In 2009, the U.S. National Oceanic and Atmospheric Administration (NOAA) organized a Global Interoperability Program (GIP) (*gip.noaa.gov*) to promote and support this effort, which is rapidly evolving. Three areas of particular interest are (1) data formats, (2) software frameworks, and (3) science workflow gateways.

The quickest and easiest way for a group to make its model more interoperable with the global research community is to replace any proprietary data formats (especially unformatted Fortran files) with a standard, self-describing,

Fig. 1. See Plate 35 for color version. Cloud-system resolving Earth GCM. Shown is a snapshot (simulated time: September 8, 2008) of the Geophysical Fluid Dynamics Laboratory (GFDL) High Resolution Atmosphere Model (HiRAM), run in 2011 at 12.5-km (0.1°) resolution to simulate the peak of the 2008 hurricane-typhoon season (the full movie is available at www.gfdl.noaa.gov/visualizations-hurricanes). Both large-scale and vortex-scale features are nudged toward the analysis (see section 4) as the simulation proceeds. These results indicate that the Atlantic hurricane season has significant predictability, as long as ocean temperatures can be forecast precisely (*Chen and Lin,* 2011). Image supplied by Shian-Jiann Lin, GFDL.

portable data format. "Portable" means, e.g., no big-endian vs. little-endian (byte ordering) or similar issues between computer platforms. A self-describing output file can be read, plotted, and analyzed by software that is unaware of the model that produced the output. In atmospheric and oceanic research, the leading such data format is netCDF (Network Common Data Form, .nc files), which is developed in the U.S. by Unidata (*www.unidata.ucar.edu/software/netcdf*). It is being used successfully in many disciplines (Earth and planetary GCMs, fusion energy data, neural imagery, molecular dynamics, etc.) and is intended as a general-purpose data format. At last count, there were more than 100 plotting and analysis packages (*www.unidata.ucar.edu/software/netcdf/software.html*) that work with .nc files; a good starting plotter for atmospheric work is Panoply, which is free from the Goddard Institute for Space Studies (GISS).

In meteorology and oceanography, in addition to standardized data formatting, there is also standardized metadata formatting, e.g., the Climate and Forecast (CF) conventions, which specify standard variable (or field) names, units, and attributes. These are similar to the Data Dictionary in NASA's Planetary Data System (PDS). For netCDF, the LibCF software library is currently the primary metadata tool. The GIP Global Interoperability Program (GIP) is working to expand the metadata concept to Earth system models in general, with a set of conventions called GRIDSPEC. GCM modelers participating in the current, fifth phase of the international Coupled-Model Intercomparison Project (CMIP5, see below) are required to use netCDF and CF for their data and metadata formats, respectively, for depositing their results in the official Program for Climate Model Diagnosis and Intercomparison (PCMDI) archives. To take this a step farther, e.g., to establish interoperability between large archiving systems like PCMDI for climate science and PDS for planetary science, will require that all parties follow the rules set by the International Organization for Standardization (ISO), not the least of which is guaranteed backward compatibility.

In the Earth climate community, the metadata concept is being broadened to include not just standards for data variables, but also for the descriptions of all aspects of GCM modeling, including model grid types and the design of experiments. This is called CIM (Common Information Model; the concept's descriptive name has been adopted as its implementation name). CIM is being developed in Europe by the Common Metadata for Climate Modelling Digital Repositories (METAFOR) project, as commissioned by the World Climate Research Programme (WCRP) for CMIP5.

In terms of making GCM cores and physics layers swappable like PC parts, high-frequency, standardized couplings are required. To meet this goal, the Earth System Modeling Framework (ESMF) (*Sandgathe et al.,* 2009) and the Partnership for Research Infrastructures in Earth System Modeling (PRISM) have formed in the U.S. and Europe, respectively. These frameworks couple components slightly differently, and the WCRP and NASA have both sponsored collaborations between the ESMF and PRISM groups to work toward an industry standard. Much of the synergy achieved since 2005 is organized into the Earth System Curator project and the Earth System Grid. It is hard to think of reasons why broadening these into a Planetary System Curator project and a Planetary System Grid would not be beneficial to everyone.

4. DATA ASSIMILATION

We now turn to the interface between theory and observations. In meteorology and oceanography, an analysis is an optimal blending of observations and GCM. Today, in the GCM world the analysis has replaced the dataset as the guide to what is true. The first question that may come to the reader's mind is, how is this not cheating? We will consider this core philosophical question first, and then highlight the leading data assimilation techniques.

Regarding the role of an analysis vs. a dataset, it comes down to the subtle differences between getting as close as possible to the truth now, vs. finding ways to get closer to the truth. To say a model is scientific means it produces a prediction that one can test against observations, against nature. If a model prediction cannot be compared against observations, for whatever reason, then that model is strictly not scientific. It may be mathematical, it may be true, but it is not scientific. What if a prediction cannot be tested now, but probably will be testable later? Strictly, at the moment it is not a scientific prediction. Let us call it a potentially scientific prediction.

Now, what if the main purpose of a model is to predict the future? Until we reach the future, such a model is only potentially scientific. Its immediate purpose is in fact not closed-loop science, but the equally noble endeavor of forecasting. We are using the model to extrapolate the atmospheric or oceanic state into the future. For Earth (and Mars), usually we are trying to save lives and property. The point is that forecasting is different than closed-loop science, even though the ultimate goal of both is to get at the truth.

4.1. Extrapolation vs. Interpolation

Now we come to the crux of the matter. We have two related processes, extrapolation, which is what yields a forecast, and interpolation, which is what yields an analysis. Their similarity lies in the fact that each involves the use of a model, usually the same model. Extrapolation is the crazy cousin that extends out past observations, and consequently it is only a matter of time before it gets into trouble. Interpolation is the well-behaved cousin, like a child holding hands with both mom and dad as they walk together. What meteorologists are doing when they make an analysis is using a model, a GCM, to interpolate across data points, spatially and temporally, but also across dynamical and thermodynamical constraints. The model provides self-consistency that today's best global orchestration of observations cannot provide. When done well (and there is a specialized industry called data assimilation that provides optimal procedures specifically for this purpose), the results come as near to the truth as possible — they have smaller errors than either the data or the model. Thus, interpolating with a GCM is like forecasting, but much safer. But that brings us to the main point in terms of what constitutes closed-loop science: Even though it involves interpolation rather than extrapolation, an analysis is never more than potentially scientific.

Because a grid of observations in both space and time pins down an analysis, one can argue that using it in place of observations will not lead to a widening drift away from nature. This is correct. However, the danger is that we must eventually end up orbiting around the truth and stop spiraling in closer to it, unless the quality of the data gets better with time. Because we are currently in a period when GCM enhancements are outpacing observational improvements to drive down errors, a period that has now lasted for several decades, generations of graduate students have been raised in a culture that compares models to models fruitfully, and for which an analysis is always more accurate than a dataset. But we will not always be in this modeler's market, and the timing for when the observer's market reemerges in the various subdisciplines is dicey to predict, leastwise by its practitioners. Arguably, one can already see mistakes being made.

The way to not to get stuck in this trap is to realize that the observational grid must continuously be refined, spatially, temporally, and variable-wise, and the data errors must continuously be reduced. Absent this, and the asymptote for GCM error will not be the zero line. The good news is that *if* we are steadily refining the observational grid and reducing the errors in the data, then data assimilation's ability to get us as close as possible to the truth, given all the information in hand, is an honest tool in the scientific process. But absent that *if*, our collective efforts to continue spiraling inward toward zero error will stagnate into circles. Keeping an eye, then, on the path of our error trajectory, which of late is cleaning up respectably, we now take a look at how data assimilation is practiced by the professionals.

4.2. Data Assimilation Resources

Data assimilation is big business, because nonlinear dynamical systems used to realistically simulate nature are big business, and there are several helpful resources available. The techniques work with both streamlined climate models and complex GCMs. An example of the latter is the first GCM with interactive chemistry to be coupled to a data assimilation system, the Canadian Middle Atmosphere Model (*Polavarapu et al.,* 2005). For groups just getting started, the Data Assimilation Research Testbed (DART) provides algorithms, software-engineering guidance, and tutorials (*Anderson et al.,* 2009). *Kalnay* (2003) has written a highly regarded textbook on the subject, *Swinbank et al.* (2003) provide an inclusive set of NATO-sponsored lectures by several authors, *Lewis et al.* (2006) discuss the broad hierarchy of mathematical approaches across a host of disciplines, and *Lahoz et al.* (2010) update *Swinbank et al.* (2003) to include novel applications, including a full chapter on data assimilation applied to planets.

4.3. Model Adjoint

There are two main data assimilation schemes in current practice in Earth meteorology: the variational approach, called 4DVAR, which stands for four-dimensional variational, and the ensemble approach, which employs hybrid Kalman filters to optimally reduce uncertainties from all sources. When observations are infrequent (which characterizes most planetary data to date) the former tends to do the better job (*Kalnay et al.,* 2007), but the latter has recently moved into the top position in Earth operational meteorology. Since a goal of this chapter is to identify techniques most useful for planetary research, we focus on the main requirements of a GCM for 4DVAR [see also *Lorenc and Rawlins* (2005) for a comparison of 4DVAR to 3DVAR; the latter is more relevant for many applications].

Because data assimilation is computationally expensive, to make efficient use of 4DVAR a GCM must have, and maintain, a companion model called an adjoint. In conversations between GCM developers, one often hears compliments such as "that is a good model, *and* it has an adjoint." *Errico* (1997) provides a clear explanation of what this companion model is, and what it is used for.

First, the shortcomings of the 4DVAR approach reveal themselves at the outset. It is a shortsighted concept, literally; i.e., a linear model that is accurate only for small perturbations. However, if most of what a user knows about the behavior of a GCM comes from running it manually, the rapid, automatic insights enabled by the adjoint model will be an eye opener.

In terms of the mathematics, the sensitivity of a simple function, J(p), to its input variable, p, is given by its first derivative, dJ/dp, since for a small variation, Δp, the leading terms of the Taylor series yield $\Delta J = (dJ/dp)_0 \Delta p + O(\Delta p)^2 \approx (dJ/dp)_0 \Delta p$. Thus, a large or small first derivative implies a large or small sensitivity, respectively, and the sign of the derivative, evaluated at the reference point, indicates a correlation or anticorrelation. So too, the sensitivity of a GCM to its inputs is dictated by a first derivative, but one where the bookkeeping for the millions-of-variables aspect of a GCM is handled by linear algebra.

There are five actors on the stage, the first four of which are (1) the GCM inputs, organized as a state vector, p_i ("p" for "present"); (2) the GCM's tangent linear model, organized as a matrix, M_{ji} ("M" for "model"); (3) the tangent linear model's outputs, a state vector f_j ("f" for "future"); we have $f_j = M_{ji} p_i$ (a repeated index implies summation); and (4) an inner product (dot product) that reduces the output state vector to a diagnostic scalar, $J = f_j\,(\partial J/\partial f_j)_0$.

This diagnostic scalar, J (standard notation), may be a root-mean-square error, a variance, a peak value of temperature (global or surface) or wind speed, or any other reduction of the copious GCM output into a single value. In the forum of variational data assimilation, J speaks for the GCM's entire output. The vector $(\partial J/\partial f_j)_0$ represents the weights used to reduce the model output to J, evaluated at the reference state. The beauty of this approach is that the researcher may (and usually does) focus on a single aspect of the full output state vector, f_j, by using these weights to zero out the rest. For example, for Earth the scientific question might revolve around the model's temperatures, but for Venus the focus might be the zonal (east-west) winds. Notice that $(\partial J/\partial f_j)_0$ is straightforward to calculate since it is defined entirely in terms of output quantities and the reference state (for convenience, we henceforth drop the zero subscript denoting evaluation at the reference state).

The sensitivity of J to the model's input variables is given by its first variation, but now with a million or so inputs, which are handled with the chain rule of calculus

$$\Delta J \approx \frac{\partial J}{\partial p_1}\Delta p_1 + \frac{\partial J}{\partial p_2}\Delta p_2 + \ldots = \frac{\partial J}{\partial p_j}\Delta p_j \tag{1}$$

As with the simple function above, equation (1) is an approximate expression for ΔJ since it only retains the first-order (linear) terms of the full Taylor series. The computational key is to find an efficient way to calculate $\partial J/\partial p_j$, the sensitivity vector of the output scalar J to each and every input of the GCM (not to be confused with the above, easier-to-compute weights on the output, $\partial J/\partial f_j$). The sensitivity information contained in $\partial J/\partial p_j$ provides the answers to many — some would argue to most — of the scientific questions that can be fruitfully asked of a GCM, and the power of the adjoint method is that we can calculate this vector once and for all, given the GCM's tangent linear model about the reference state, and a choice of J. The researcher may then calculate the inner product equation (1) with all manner of different initial-condition Δp_j vectors, which is efficient because such calculations do not require running the GCM over and over.

To execute this strategy, we need the GCM's adjoint model, the fifth actor on the stage

$$\frac{\partial J}{\partial p_j} = \frac{\partial f_i}{\partial p_j}\frac{\partial J}{\partial f_i} \tag{2}$$

The term adjoint stems from the fact that the matrix $\partial f_i/\partial p_j$, called the Jacobian matrix, is being summed in equation (2) over the index in its numerator, the future values, rather than its denominator, the present values. There will be a different adjoint model for each choice of diagnostic scalar, J, and for each reference state, but these are straightforward to construct once we have the Jacobian matrix. The Jacobian matrix is the computationally demanding component of the variational approach to data assimilation. There exist programs that can read an appropriately written GCM line-by-line and build the corresponding adjoint model automatically. For full details on this and other powerful analysis techniques that make use of a GCM's adjoint, consult *Errico* (1997), *Anderson et al.* (2009), and the references therein.

4.4. Reanalyses

With a data assimilation scheme, one can combine observations and models in a manner that optimally minimizes errors. This is routinely done while operational GCMs are running to increase the accuracy of the forecast (operational models essentially never stop running). It is also done after the fact, to get as close as possible to the past state of an atmosphere or ocean. When old data are assimilated with a new GCM, the result is called a reanalysis. This is the product that is used to characterize the state of an atmosphere, instead of pure observations (advisedly, see above). For example, a 50-year reanalysis by the U.S. National Centers for Environmental Prediction (NCEP) in collaboration with NCAR is available in CD-ROM format (*Kistler et al.,* 2001), and regular updates are maintained online at the NCEP-NCAR reanalysis website (*www.esrl.noaa.gov/psd/data/reanalysis/reanalysis.shtml*).

5. FORECAST SKILL

With an optimal means of getting as close as possible to the true state of the atmosphere, the next step is to keep score on how the various Earth GCM models are doing. The metrics for forecast skill provide critical, quantitative feedback for the Earth GCM-development process, and are emerging in planetary modeling (*Lahoz et al.,* 2010). And inevitably, keeping score adds a gaming element to any endeavor. In fact, there are many similarities between the realms of weather-and-climate modeling and, of all sports, professional golf, starting with a mutual interest in approaching thunderstorms.

5.1. Analogies with Professional Golf

Consider that GCM research and golf are both led by Europe and North America (Canada and the U.S.), notably the United Kingdom for the former, but many European countries sport top players. Furthermore, there are always keen competitors from Japan and Australia, with several other nationalities investing in both pursuits (this analogy would seem to suggest that South Africa needs to catch up on its climate modeling, and Russia on its golf). Both climate modeling and golf are precision endeavors that obey the physical laws of forward trajectories; are played outdoors (more horizontally than vertically); have water and sand in play, with ridges and troughs abounding; and have the goal of minimizing cumulative error over multiple days. What is "climate" to one is "par" to the other, "model crash" to one is "unplayable lie" to the other, and while there is the rare faultless five-day forecast or double eagle, no one harbors any illusions that the mathematically perfect score will ever be attained for an entire tournament, let alone an entire season. Both are played on the clock, require many forward steps, use overhead cameras and satellite link-ups for observations, and endlessly reanalyze past performances, and both bottom-line budgets are about the same, albeit with different philosophies for payout. About the only major difference (for the men anyway) is that one group shops in the khaki pants aisle, and the other does not. But above all, perfection in the field is the mutual goal, and there is always the next forecast or the next hole, which is the collective charm. "Let the dead past bury its dead," and as the poet Longfellow also wrote, "but to act, that each tomorrow find us farther than today." Then again, Bobby Jones once observed, "some emotions cannot be endured with a golf club in your hands." GCM developers, read "coffee mug."

5.2. Anomaly Correlation Coefficient

Several statistical measures of forecast skill have proven to be useful for weather and climate forecasting. These include the root-mean-square (rms) error of geopotential-height on standard pressure levels (e.g., 850, 500, 200 hPa), errors for vector wind, temperature and precipitation, the duration of skill, and correctly predicted departures from climatology. The height of the 500-hPa pressure level has physical significance for Earth because it marks the center of mass of an atmospheric column. Hence, when a single metric is needed to distinguish the skill levels of different Earth models (see Fig. 2), it is often taken to be the monthly-averaged, 500-hPa height anomaly correlation (not including the tropics, where the Coriolis acceleration is negligible and hence the dynamics and statistics are different than in the extratropics).

An anomaly correlation coefficient (ACC) is a pattern correlation in which seasonal variation has been removed by subtracting the climate average. This practice avoids generating misleadingly good values, in essence by making sure that there is no credit given for predicting that the weather will act like northern summer in July or like northern winter in February. The World Meteorological Organization (WMO) subtracts mean error as well as the climate average, and defines (*www.ecmwf.int/products/forecasts/guide/Measure_of_skill_the_anomaly_correlation_coefficient.html*)

$$\mathrm{ACC} = \frac{\overline{\left[(\mathrm{f}-\mathrm{c})-\overline{(\mathrm{f}-\mathrm{c})}\right]\left[(\mathrm{a}-\mathrm{c})-\overline{(\mathrm{a}-\mathrm{c})}\right]}}{\left\{\overline{\left[(\mathrm{f}-\mathrm{c})-\overline{(\mathrm{f}-\mathrm{c})}\right]^2}\;\overline{\left[(\mathrm{a}-\mathrm{c})-\overline{(\mathrm{a}-\mathrm{c})}\right]^2}\right\}^{1/2}} \tag{3}$$

where the overbars indicate spatial averages, and f, c, and a are the pointwise values of the forecast, climate average (30-year averages are standard for Earth data), and analysis, respectively. The bigger the ACC, the better the skill. The value 60% has been empirically established as the minimum threshold for usefulness, and 50% is the base score one gets by taking the limit $\mathrm{f} \rightarrow \mathrm{c}$, in other words using the climate as the forecast. Further details on forecast verification may be

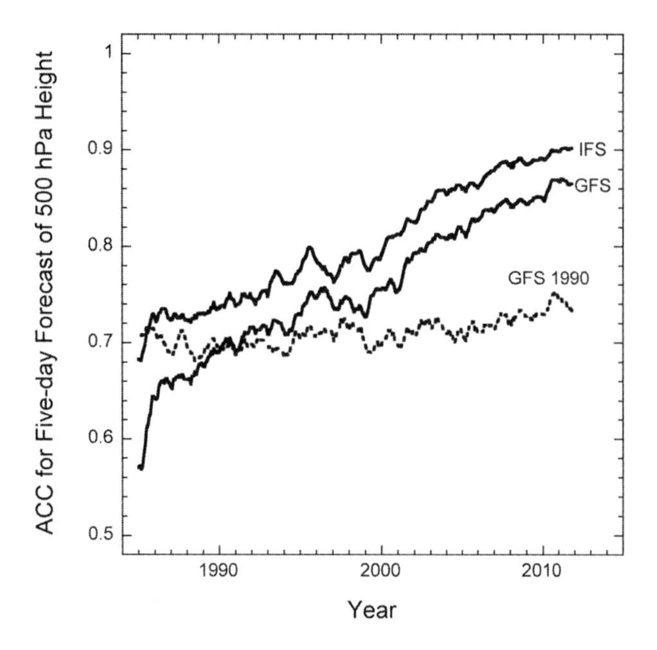

Fig. 2. Earth GCM forecast skill vs. time. Shown is monthly mean anomaly correlation coefficient (ACC) for 5-day forecasts of the height of the 500-hPa pressure surface (northern hemisphere, 12-month running average). Seasonal variations are removed by subtracting the 30-year climate time series. An ACC of 100% indicates a perfect forecast, higher than 60% implies a useful forecast, and 50% is the value obtained by using the climate as the forecast. The European and U.S. profiles are labeled IFS and GFS, respectively. The increase in skill due to improved observations alone may be discerned by comparing with the dotted-line profile, which employs only the 1990 GFS model for the full range. The GCM skill scores are improving steadily with time, predominantly due to model improvements. In this plot, the Europeans consistently lead the U.S. by 2 to 5 years development time. Data supplied by Fanglin Yang, NCEP.

found in *Jolliffe and Stephenson* (2012), *Wilks* (2005), and the special issue of *Meteorological Applications, Vol. 15,* based on the *3rd International Verification Workshop,* 2007 (Reading, U.K).

5.3. Selected Standout Institutions

The current leader board in numerical weather prediction (NWP) includes the Integrated Forecast System (IFS) (Europe, ECMWF), the Canadian Meteorological Centre (CMC), the Global Forecast System (GFS) (U.S., NCEP), and the Japan Meteorological Agency (JMA). In Japan, a host of institutions are collaborating on improving climate prediction, including the JMA, the Meteorological Research Institute (MRI) in the Tsukuba Science City, the University of Tokyo's Center for Climate System Research (CCSR), and the Japan Agency for Marine-Earth Science and Technology (JAMSTEC), the latter of which is housed in six campuses across the country. In 2002, a state-of-the-art, 60-billion-yen supercomputer, the Earth Simulator, was opened to support the work of JAMSTEC, as well as the Japanese aerospace and atomic energy research efforts. The Earth Simulator 2 (ES2) superseded this in 2009; the mesoscale WRF model developed by the U.S. has been optimized to run on the ES2.

If you ask an observational astronomer what he or she thinks of GCMs, the likely answer is that, when it comes to predicting night-sky observing conditions, the CMC's forecasts are the most useful. They use an enhanced-resolution, regional version of their global forecasting system to produce night-sky charts (*www.cleardarksky.com/csk* and *www.weatheroffice.gc.ca/astro/index_e.html*) that forecast ground-level wind, temperature, humidity, hour-by-hour cloud cover, sky transparency (from moisture), and seeing (from turbulence).

France is another major player, with two primary Earth GCM development efforts, both participants in Intergovernmental Panel on Climate Change (IPCC) research. The Institut Pierre-Simon Laplace (IPSL) of the Laboratoire de Meteorologie Dynamique (LMD) (the home institute of the authors of the following chapter) develops the LMDZ model, which has advanced zooming and physics capabilities (e.g., *Hourdin et al.,* 2012). In addition, the national meteorology agency, Météo-France, in collaboration with the European Centre for Medium-Range Weather Forecasting (ECMWF) and IPSL, develops the ARPEGE-Climate model, which can zoom in on the Europe-Mediterranean-North Africa region with 50-km resolution.

In weather and climate forecasting, the current player with honors is the ECMWF, in particular their Integrated Forecast System (IFS in Fig. 2). All major-center skill scores have risen significantly over the past several decades, as the designs of the GCM dynamical cores and physics layers, and the data-inversion techniques used in satellite remote sensing, have steadily improved, and as computers have steadily sped up.

Nevertheless, one cannot help but wonder why there have not been any lead changes in decades, with so many big players on the course. One hears the view that the culture of technology transfer between researchers and operational forecasters in Europe has been more synergistic than elsewhere. *Randall* (2011) advocates for two-track modeling centers (operational and research) as being probably the best way to organize GCM development — the U.S.'s new National Center for Weather and Climate Prediction in Riverdale, Maryland, is just such a facility. Others opine that the Europeans have been able to maintain more sophisticated satellite-data inversion algorithms. For their part, the Europeans are decorous and demure, but will point out that starting new collaborations with U.S. research laboratories and centers has been noticeably harder since the events of 9/11. In any event, so long as forecast skill keeps trending upward, this international competition is beneficial to the world at large. (And of course, after not losing the Ryder Cup for all of the 1960s and 1970s, the Yanks have it coming to them.)

6. MODEL INTERCOMPARISON

We turn now to the systematic comparison of Earth climate models, one of the bright spots in all of international scientific collaboration. The leading enterprise is the Climate Model Intercomparison Project (CMIP).

6.1. CMIP World

CMIP establishes the accepted protocol by which Earth GCMs are methodically compared to each other and to analyses. Reporting to the UN via the UN Educational, Scientific and Cultural Organization (UNESCO) and the WMO, the cognizant organizer for CMIP is the WCRP (*wcrp@wmo.int*). WCRP is steered by its decadal strategy for 2005–2015 entitled *Coordinated Observation and Prediction of the Earth System* (COPES) (WMO/TD-No. 1291). As is appreciated in the planetary and astronomical communities, such a decadal survey is an effective means of identifying and overcoming deficiencies in coordinated research on an international scale. Several researchers in the planetary community are also active in the "CMIP world."

Perhaps the biggest customer of CMIP results is the IPCC assessment report process itself, but the entire Earth-climate community makes heavy use of the simulations; they are the accepted baseline for the discipline. Starting with its third phase, the multimodel output from the CMIP project is now released as soon as possible, which has resulted in a surge of peer-reviewed publications.

One of the main results to come from CMIP is the identification of four top sources of error in AOGCMs: (1) cloud-radiation processes, (2) the cryosphere, (3) the deep ocean, and (4) ocean-atmosphere interactions. An in-depth discussion is given in section 1.5 of *AR4-Physical*, "Examples of Progress in Modelling the Climate." One of the major milestones achieved in *AR4* is that most participating AOGCMs no longer require artificial flux corrections to maintain a stable climate. Predictions of sea surface temperature and pressure both show significant improvement in *AR4* compared to past phases of CMIP. Tropical processes still present some of the greatest challenges, including tropical precipitation and the Madden-Julian Oscillation (MJO).

6.2. CMIP5

Taylor et al. (2012) provide an overview of the fifth phase of CMIP (CMIP5), which is underway at the time of this writing. Currently, over 50 GCMs are employed in CMIP5, with experiments being carried out by more than 20 groups. CMIP5 is designed to address three major gaps in knowledge identified after release of *AR4* in 2007, namely (1) determining which GCM details cause models to differ the most in their handling of difficult physics, in particular feedbacks related to Earth's carbon cycle and clouds; (2) what level of predictability is possible on decadal timescales; and (3) when different Earth GCMs are given similar initial conditions and forcing, which modeling details are most responsible for inconsistencies in the results.

Just as in planetary science, CMIP5 is a flagship whose design of experiments is necessarily limited and prioritized, but whose reach is expanded by well-organized auxiliary projects. One such project is the Coordinated Regional Climate Downscaling Experiment (CORDEX), which uses CMIP5 simulations as boundary conditions for local, increased-resolution runs that add value by bringing climate research down to the personal level, including new three hourly fields that permit investigation of temporal anomalies in precipitation (*Evans,* 2011). Complex topography has been identified as a primary cause of discrepancy between large-scale predictions of precipitation vs. local observations. Another auxiliary project is the Geoengineering Model Intercomparison Project (GeoMIP), which specifies forcing scenarios to systematically study how proposals for planet-scale engineering of Earth's environment, e.g., manipulating sulfate aerosols in the stratosphere, behave in CMIP5 models (*Kravitz et al.,* 2011).

Two distinct categories of experiments are being executed for CMIP5, long-term and near-term runs, with simulation times spanning several centuries and several decades, respectively. The near-term experiments are new in CMIP5, and are attempting to characterize the actual trajectory of Earth's climate over the next few decades. The CMIP5 protocol consists of a core set of runs that each participating group is expected to execute, plus tier 1 and tier 2 experiments that may be chosen from respective lists, depending on each group's interests and resources.

Coupled atmosphere-ocean-land-ice GCMs (AOGCMs) are the standard platform, but Earth Models of Intermediate Complexity (EMICS) are also being employed in CMIP5 for long-term experiments, in the manner of the modeling hierarchy advocated by *Held* (2005). For a listing of eight specific EMIC models, with details on how they handle Earth's atmosphere, oceans, land surface, cryosphere, and biosphere, see Table 8.3 in *AR4-Physical.*

On the high end of complexity, some of the long-term runs in CMIP5 are for the first time including biogeochemical components designed to close the carbon cycle; these are called Earth System Models (ESMs). The CMIP5 protocol allows groups the option of trading between full coupling and better atmospheric chemistry or higher resolution, called the time-slice option; this is enabling more weather forecast centers to contribute.

The models being used in CMIP5 have horizontal grid spacing in the range 0.5° to 4°, which for Earth is on the order of 100 km. The protocol for CMIP5 core experiments is given by *Taylor et al.* (2009) (*cmip-pcmdi.llnl.gov/cmip5/docs/Taylor_CMIP5_design.pdf*), and specifies the amount of atmospheric CO_2 to include each year, except for the ESM experiments, which simulate this component (their surface fluxes of CO_2 are being saved in each type of run for later comparison). It also specifies what to do with the sea-surface temperature (SST) and related air-ocean couplings. Four different Representative Concentration Pathways (RCPs) are laid out in CMIP5, labeled RCP2.6, RCP4.5, RPC6, and RCP8.5, where the numbers indicate the approximate radiative forcing in Wm^{-2} in 2100 compared to the preindustrial baseline. The cheeriest one is RCP2.6, which is called a peak-and-decay profile because it assumes a maximum around 2050 before dropping to 2.6 Wm^{-2} by 2100.

For the near-term experiments, forecast skill is being assessed via 10-year hindcasts, which are simulations generated by forecast models using only past data (thereby encompassing a closed scientific loop). These are initialized with analyses in five-year increments, from 1960 to 2005. In addition, there are two 30-year hindcasts in the protocol that span 1960–1990 and 1980–2010, and one 30-year forecast that spans 2005–2035. Specified parameters include the nature of volcanic eruptions, both real and hypothetical, and how to incorporate the improvements to observations of ocean temperature and salinity resulting from the introduction of Argo floats (the deployment of the first 3000 probes was completed relatively recently in November 2007). The Argo experiments are needed because there is not yet a consensus on how best to assimilate ocean data; in fact, a distinguishing feature of CMIP5 is its expanded list of oceanic output.

The manager for CMIP5 archiving is the Program for Climate Model Diagnosis and Intercomparison (PCMDI), which is located at the U.S. Lawrence Livermore National Laboratory (LLNL), near San Francisco, California. The volume of data is peta/exascale, and hence is too large for a single repository; consequently it is being spread out globally across linked nodes located at or near the sites of the various simulations. The archiving system is the Earth System Grid (ESG), an open-source project managed by the Earth System Grid Federation (ESGF) (*esgf.org*). The goal is to have the multimodel, multinational output, the CMIP5 ensemble, be nearly as easy to work with as simulations from a single model.

7. FUTURE TRENDS

To this point, we have reviewed the history, make-up, and global organization of Earth GCMs from a planetary-science perspective. Now, we turn to a consideration of future trends. It has been said that the best way to see the future is to visit research laboratories and look around. We start with this anchor, and then drift farther and farther out to sea with some medium-range and long-range forecasts.

7.1. Short-Range: Quasi-Uniform Grids

One interesting aspect of just about every Earth GCM today is that the home planet's geopotential is approximated as spherical. Earth's true geopotential is made oblate by its rotation, an effect that becomes severe with the gas giants, most notably Saturn, whose gravity increases by 30% from equator to pole, to be compared with Earth's 0.5% (to appreciate dramatic flattening, turn an image of Saturn by 90°). Consequently, gas-giant models cannot leave the starting gate without addressing this problem. For Earth, *White et al.* (2008) have found that using nested, similar oblate spheroids with geographic (planetographic) latitude [the scheme used in EPIC (*Dowling et al.,* 1998)] yields a viable oblate-spheroidal coordinate system for Earth GCMs. Our first short-range forecast is that this will be adapted into operational models in the near future, by 2017, because it easily removes a small but systematic error that accumulates over time. One hopes that in making such upgrades, all map factors in Earth GCMs will be replaced by ones that are not hardwired.

Staniforth and Thuburn (2012) have written a thorough review of modern horizontal grids for GCMs. One of the banes of discretized systems is the existence of extra solutions, called computational modes, which do not correspond to physical solutions of the original conservation laws. Another is grid imprinting, wherein the pattern of a grid unnaturally affects, and is detectable, in the output (this a longstanding problem for terrain-following vertical coordinates). Enormous GCM development time is spent taming these gremlins.

Fortunately, progress is being made at research laboratories and centers on two types of quasi-uniform grids: the cubed sphere, which was inspired by the paneling on a football (a soccer ball), and the icosahedral grid, which is the one made popular by B. Fuller. As detailed by *Randall* (2011), GCMs based on the cubed sphere are currently being employed in the U.S. at GFDL, GISS, and the Massachusetts Institute of Technology (MIT), and icosahedral grids are being employed in Germany at the Deutsche Wetterdienst (German Weather Service) and Max Planck Institute for Meteorology, in Japan at the Frontier Research Center for Global Change (FRCGC), and in the U.S. at Colorado State University, the Earth System Research Laboratory (ESRL), and NCAR.

At the time of this writing, only the German Weather Service has such a grid deployed for operational weather forecasts, their icosahedral GME model, but the FIM model mentioned above is about to come online. The Japanese Nonhydrostatic Icosahedral Atmosphere Model (NICAM) is a GCRM that has been run on the Earth Simulator for aquaplanet conditions with 3.5-km resolution, which is high enough to reduce the need for cumulus parameterization (*Miura et al.,* 2005). Given the many benefits of quasi-uniform grids, our second short-range prediction is that most operational GCMs will adopt them in the next decade, by 2022.

7.2. Medium-Range: Unstructured Grids

The structured grids used in weather and climate research, from the first hand-calculated NWP effort by Richardson's group (*Lynch,* 2006) to today's GCMs, have all experienced problems in both the horizontal and vertical directions (*Staniforth and Thuburn,* 2012). In the horizontal, much effort has been spent overcoming the topological issues associated with gridding the sphere, with the promising cubed-sphere and icosahedral grids mentioned above currently winding their way into operational centers.

Stepping back, the impression one gets is that the ever-increasing software development, runtime bookkeeping and computational resources committed to making three-dimensional structured-grid GCMs work better, might well be better spent on developing and running three-dimensional unstructured-grid GCMs. Engineers made this switch years ago. Both engineers and physicists are making rapid progress implementing such models efficiently on massively parallel computers (e.g., *Shephard et al.,* 2007).

Unstructured-grid generators produce meshes composed of tetrahedra in three dimensions (triangles in two dimensions) by employing Delaunay and related algorithms that avoid skinny finite volumes and concentrate resolution where it is needed. These grids are readily adapted on the fly as local fluid structures evolve and decay. In other words, they optimize both the spatial and temporal resolution of a model. This is the opposite of grid imprinting, where the grid fights the physics; this is physics imprinting, where the grid keeps up with the physics.

In the Earth GCM realm, oceanographers have been the most enthusiastic proponents of unstructured grids to date, because they have the same complicated-boundary challenges as engineers. As with any discretization scheme, current unstructured-grid algorithms have their own gremlins, e.g., dispersion issues (*Slingo et al.,* 2009), and such grids will need to be optimized for stratification. Nevertheless, the potential advantages are enormous, as is the leverage from the efforts of the larger engineering and physics communities, which suggests that technical problems are likely to be significantly reduced in the next decade.

One of the first major atmospheric models to employ an unstructured grid with dynamically adapting resolution for its horizontal mesh, but still a regular grid in the vertical (because of stratification), is the Operational Multiscale Environment Model with Grid Adaptivity (OMEGA) (*Bacon et al.,* 2000, 2007). An example of a hurricane simulation by OMEGA is shown in Fig. 3. Several other atmospheric groups are working on unstructured-grid GCMs [see the references in section 1.3 of *Staniforth and Thuburn* (2012)].

Although Earth's atmosphere does not have a closed basin or an abrupt surface like its oceans, it does have mountains, and heterogeneous, intermittent mesoscale structures in the form of cumulus towers, fronts, clouds, and precipitation. Once three-dimensional unstructured, dynamically adapting grids are in place and working well in GCMs, there are not likely to be any reasons to return to structured grids, because to do so would be to forfeit computer resources. Therefore, our medium-range prediction is that within 25 years, by 2037, most operational GCMs will employ fully adaptive, three-dimensional unstructured grids.

7.3. Long-Range: Direct Numerical Simulation

Engineers and meteorologists have traded their best fluid-dynamical ideas back and forth every decade or so for over a century. A big idea in turbulence modeling that started with GCM work (*Smagorinsky,* 1963) and has since become popular in the engineering realm is the large eddy simulation (LES) approach. For the reader familiar with Reynolds stresses such as $\rho\overline{(u_i'u_j')}$, where the overbar indicates a time average, the LES approach produces similar eddy correlations via local spatial averaging. In a nutshell, the nonlinear fluid dynamics is fully resolved for structures larger than the grid spacing; these are the large eddies, but subgrid scales are parameterized by closure schemes that relate the eddy correlations to the resolved variables. Large eddy simulation models, which include GCMs, have enjoyed success over a wide range of turbulent fluid applications because of their computational efficiency.

However, as computers become faster, LES models are slowly but methodically being replaced by direct numerical simulation (DNS) models, which, as the name implies, resolve all spatial and temporal scales of turbulence, and hence require no turbulence closure scheme. More and more frequently, one hears the conjecture that the only acceptable turbulence model will ultimately prove to be the Navier-Stokes equations themselves, i.e., no turbulence model. How soon might DNS be viable for GCMs?

The key nondimensional variable is the Reynold's number, Re, the ratio of inertia force to viscous force, which is also the ratio of the scale of the domain to the viscous scale; the larger the Re, the more turbulent the flow. The biggest DNS simulations to date have been performed with the Earth Simulator 2 (ES2) in Japan. *Kaneda et al.* (2010) describe reaching Re = 2560 in a DNS run using a 2048 × 1536 × 2048 grid to simulate incompressible turbulent channel flow, running at 6.1 Tflops on 64 nodes, which is 11.7% of the ES2 peak performance.

To estimate the order of magnitude of Re in Earth's atmosphere, consider an atmospheric jet speed $U = 60\ ms^{-1}$ (Earth has slow jets), at an altitude H = 8 km, which we use to get a challengingly realistic scale of interest, L = H, and combine with a representative kinematic viscosity, $\nu = 2 \times 10^{-5}\ m^2s^{-1}$, to obtain $Re = UL/\nu \sim 2 \times 10^{10}$. This is a factor of about 2^{23} (about 8×10^6) larger than the Reynolds number in the ES2 simulation just cited. According to CDF-Online [*www.cfd-online.com/Wiki/Direct_numerical_simulation_(DNS)*], the number of floating-point operations required for DNS scales as Re^3, so we need a platform that can run about 2^{69} (or 6×10^{20}) times faster than Japan's Earth Simulator 2 ran in 2010, i.e., we need about 4×10^9 Yflops (Y = yotta = 10^{24}).

Could Moore's Law possibly stretch that far? It is impossible to say for certain, but if we assume the reasons Moore's Law overestimates computer speed a century from now (e.g., the limitations imposed by the finite speed of light) will be neatly undermined by the reasons it underestimates computer speed a century from now (e.g., the reinvention of the analog-digital hybrid computer in quantum terms; "plasmonics" is the current buzzword), then GCMs will be running as DNS models in about 69 × 2 = 138 yr. Accordingly, the long-range prediction is that by 2150, operational GCMs will resolve all dynamical scales of motion directly.

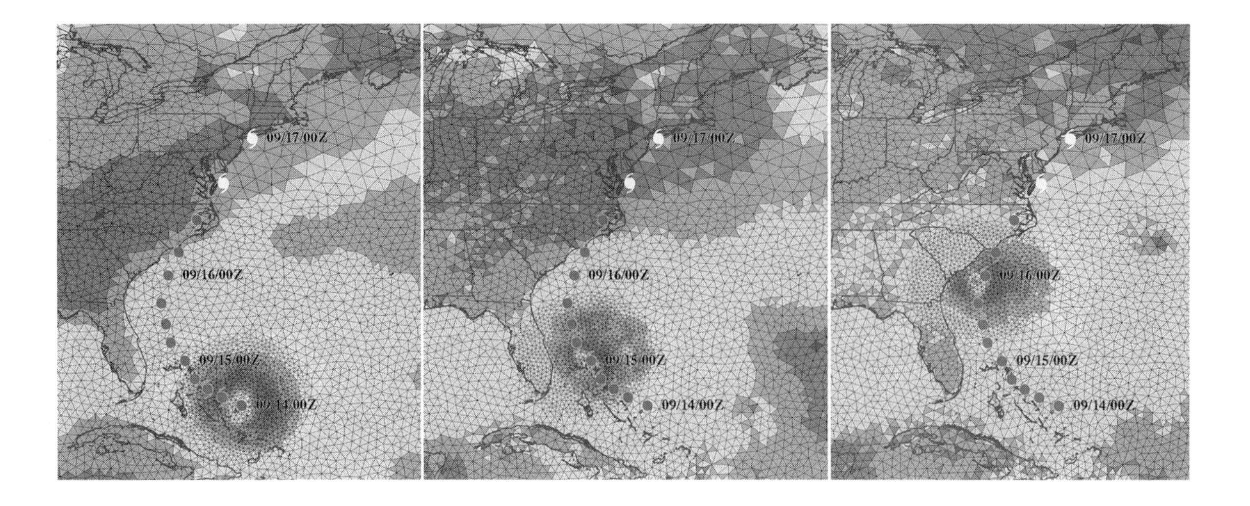

Fig. 3. See Plate 36 for color version. The OMEGA atmospheric model's unstructured, triangular horizontal mesh with dynamic adaptation optimizes grid efficiency based on local conditions in space and time. Shown is an OMEGA simulation of Category 4 Hurricane Floyd from 1999, with wind speed indicated by color. The cyclone symbols mark the actual storm track (time is indicated in MM/DD/HH format at zero hour UTC).

This still leaves microphysical scales, which may take a bit longer. But it is clear that models will certainly be doing stunningly better than they are doing today, even within the next few decades (*Randall,* 2011). This is already having an impact on GCM development, particularly with regard to parameterization schemes that must be rewritten to handle nonhydrostatics and resolved cumulus cells. *Wyngaard* (2004) studies an example of how this squeeze on model parameterizations, which is at once welcome and demanding, is affecting turbulence modeling.

Drifting now into the sunset, one can sense the challenges in Earth GCM modeling at some point shifting away from the limitations set by computer hardware and software and toward finding ways to improve the characterization of external forcing of Earth's atmosphere and oceans, which in the long run include changes in human, solar, and geophysical activity. In today's jargon, to say a GCM is coupled means it actively models Earth's atmosphere, oceans, land, and ice, with the present IPCC fifth phase experiments starting to include interactions with the biosphere. But in the future, to say a GCM is coupled will mean it actively models Earth's atmosphere, oceans, land, ice, biosphere, human enterprise, economics, the Sun's dynamics, geophysics, and celestial mechanics.

At least, we do not have to contend with the additional extreme factors experienced by many extrasolar planets. But, in about 1 b.y., we do have to deal with a Sun-Earth connection that will become critical, well before our star's widely reported 5-b.y.-from-now red-giant stage, which is coincidentally also the start of the head-on collision between the Milky Way and Andromeda Galaxies. In about 1 b.y., the Sun's rise in luminosity as it naturally creeps along the stellar main sequence is forecast to exceed the habitability level of 10% (*Schröder and Smith,* 2008). At this point, one way or another, there will be something new under the Sun.

7.4. A Modest Defense of Fortran (and Fahrenheit)

Finally, paddling all the way back to the present and its benevolent Sun, we note that today most Earth GCMs and their components are written in either a high or low variant of Fortran, hence a few practical, closing words are in order in defense of the Fortran programing language (and while we are at it, the Fahrenheit temperature scale, °F). In 1982, the author cut his undergraduate programming teeth on Fortran (.f), and remembers how fast, fun to learn, and mechanically impressive modern card readers were (in the blink of an eye they vacuumed multiple, colorful card decks through their scanners; it was sad to see them carted off the premises the following year). While it is true that Fortran retains the images of cards, to borrow a line from Mark Twain, the reports of its death are greatly exaggerated. The reason is that .f works, and works well (so too, °F works well — ranging from ~0 on Earth's surface in cold winters, to ~100 in its hot summers — and it is used not only in the U.S., but also in Belize and on the Cayman Islands). In fact, Fortran ranks in the top 30 in terms of popularity of programming languages, as witnessed by its twenty-ninth place at the time of this writing on the TIOBE Programming Community Index (one below COBOL, one above R). While it is true that for all his GCM development work, the straightforward technique of using shift macros for multidimensional arrays allowed the author to switch cold turkey to C in 1991 (C is currently first on the TIOBE Index), it is unimpeachable that Fortran programs compile and run well, and this is the most important trait of any

GCM component. (Granted, it is a different question to ask what are the most important traits of a best-practices, large software-development environment.)

Consider that many GCM physics layers in use today have descended in one form or another from dusty-deck Fortran 77 programs written in the 1980s or before. Even if one realizes that the modern CPU core is starved for data in a manner not envisioned then, and in particular is sensitive to cache misses caused by storing arrays separately and in a padded manner, rather than in a striped, cache-aware manner, the fact that arrays of structures in C, as well as dynamic memory allocation, naturally produce the latter while Fortran 77 encourages the former falls short of providing a compelling reason to migrate, especially with the advent of high-performance variants like Fortran 90, not to mention Fortran 2003 and its advanced structure types and interoperability with C. For anyone considering such a switch, the open-source, 18,000-line manual rewrite of the DISORT radiative-transfer package from Fortran 77 to C (*Buras et al.,* 2011, Appendix A) may be used as a Rosetta Stone; however, one must be cautioned that that particular rewrite yielded considerably less than 2 orders of magnitude reduction in execution time, in fact only a factor of 12, although it did result in increased numerical stability.

What of powerful modern software-development tools that do not work with Fortran, e.g., the clang static analyzer? What of graphics processing units (GPUs) and their noncard-heritage software like CUDA and OpenCL? What of the distribution of massive problems on screen savers and modern video games, which are not written in Fortran, to a willing public? Can such gimmicks be relevant to the future of climate modeling? No, not likely. After all, GCMs are not as complex or realistic as modern video games. As we peer into the future, there does not appear to be, based on rigorous comparisons of GCMs written in Fortran against one another for the better part of a century, any reason to convert now, or likely ever, from .f to .c (or for that matter, from °F to °C), or to any other non-Fortran software development environment. *Semper fortis.* Instead, we will do well to keep our eye on the goal of planning for the tomorrow of GCM modeling, as we have invested in it today.

Acknowledgments. The author thanks the anonymous reviewers and the book's editors for their insightful comments, which helped improve the chapter. Valuable input was provided by colleagues D. Bacon, R. Beebe, S. Benjamin, S. Boll, M. Bradley, J. Du-Caines, K. Emanuel, F. Forget, H. Houben, L. Huber, A. Ingersoll, D. Johnson, Y. Kaspi, J. Kielkopf, S. Lebonnois, S.-J. Lin, B. MacCall, J. Marshall, B. Mayer, K. Mountain, S. Osprey, D. Randall, P. Read, R. Rew, T. Robinson, T. Schaack, A. Showman, L. Spilker, N. Sugimoto, M. Sussman, J. Thuburn, L. Uccellini, J. Wilson, and F. Yang. The opinions expressed are solely those of the author.

REFERENCES

Anderson J., Hoar T., Raeder K., Liu H., Collins N., Torn R., and Avellano A. (2009) The Data Assimilation Research Testbed: A community facility. *Bull. Am. Meteor. Soc., 90,* 1283–1296.

AR4-Physical: Solomon S., Qin D., Manning M., Chen Z., Marquis M., Averyt K. B., Tignor M., and Miller H. L., eds. (2007) *Climate Change 2007: The Physical Science Basis.* Cambridge Univ., Cambridge.

Bacon D. P., Ahmad N., Boybeyi Z., Dunn T., Hall M. S., Lee P. C. S., Sarma A., Turner M. D., Waight K. T., Young S. H., and Zack J. W. (2000) A dynamically adapting weather and dispersion model: The operational multiscale environment model with grid adaptivity (OMEGA). *Mon. Weather Rev., 128,* 2044–2076.

Bacon D. P., Ahmad N. N., Dunn T. J., Gopalakrishnan S. G., Hall M. S., and Sarma A. (2007) Hurricane track forecasting with OMEGA. *Natl. Hazards, 41,* 457–470.

Benjamin S. G., Grell G. A., Brown J. M., Smirnova T. G., and Bleck R. (2004) Mesoscale weather prediction with the RUC hybrid isentropic-terrain-following coordinate model. *Mon. Weather Rev., 132,* 473–494.

Bradley M. E. and Dowling T. E. (2012) Using 3D finite volume for the pressure-gradient force in atmospheric models. *Q. J. R. Meteor. Soc., 138,* 2126–2135, DOI: 10.1002/qj.1929.

Buras R., Dowling T., and Emde C. (2011) New secondary-scattering correction in DISORT with increased efficiency for forward scattering. *J. Quant. Spectrosc. Radiat. Transfer, 112,* 2028–2034, DOI: 10.1016/j.jqsrt.2011.03.019.

Chen J.-H. and Lin S.-J. (2011) The remarkable predictability of inter-annual variability of Atlantic hurricanes during the past decade. *Geophys. Res. Lett., 38,* L11804, DOI: 10.1029/2011GL047629.

Davies T., Cullen M. J. P., Malcolm A. J., Mawson M. H., Staniforth A., White A. A., and Wood N. (2005) A new dynamical core for the Met Office's global and regional modelling of the atmosphere. *Q. J. R. Meteor. Soc., 131,* 1759–1782, DOI: 10.1256/qj.04.101.

Donner L., Schubert W., and Somerville R., eds. (2011) *The Development of Atmospheric General Circulation Models: Complexity, Synthesis and Computation.* Cambridge Univ., Cambridge.

Dowling T. E., Fischer A. S., Gierasch P. J., Harrington J., LeBeau R. P., and Santori C. M. (1998) The Explicit Planetary Isentropic-Coordinate (EPIC) atmospheric model. *Icarus, 132,* 221–238.

Errico R.M. (1997) What is an adjoint model? *Bull. Am. Meteor. Soc., 78,* 2577–2591.

Evans J. P. (2011) CORDEX — An international climate downscaling initiative. *19th Intl. Congress on Modelling and Simulation*, Perth, Australia, 12–16 December 2011. Available online at *mssanz.org.au/modsim2011.*

Hayashi M. and Itoh J. (2012) The importance of the nontraditional Coriolis terms in large-scale motions in the tropics forced by prescribed cumulus heating. *J. Atmos. Sci.,* DOI: 10.1175/JAS-D-11-0334.1.

Held I. M. (2005) The gap between simulation and understanding in climate modeling. *Bull. Am. Meteor. Soc., 86,* 1609–1614, DOI: 10.1175/BAMS-86-11-1609.

Holton J. R. (2004) *An Introduction to Dynamic Meteorology, 4th edition.* Elsevier, Amsterdam.

Hourdin F., Grandpeix J.-Y., Rio C., Bony S., Jam A., Cheruy F., Rochetin N., Fairhead L., Idelkadi A., Musat I., Dufresne J.-L., Lahellec A., Lefebvre M.-P., and Roehrig R. (2012) LMDZ5: The atmospheric component of the IPSL climate model with revisited parameterizations for clouds and convection. *Climate Dynam.,* DOI: 10.1007/s00382-012-1343-y.

Jeffreys H. (1922) On the dynamics of wind. *Q. J. R. Meteor. Soc., 48,* 29–48, DOI: 10.1002/qj.49704820105

Jolliffe I. T. and Stephenson D. B. (2012) *Forecast Verification: A Practitioner's Guide in Atmospheric Science, 2nd edition.* Wiley and Sons Ltd., Hoboken. 274 pp.

Kalnay E. (2003) *Atmospheric Modeling, Data Assimilation and Predictability.* Cambridge Univ., Cambridge. 341 pp.

Kalnay E., Li H., Miyoshi T., Yang S.-C., and Ballabrera-Poy J. (2007) 4-D-Var or ensemble Kalman filter? *Tellus* A, *59,* 758–773, DOI: 10.1111/j.1600-0870.2007.00261.x.

Kaneda Y., Ishihara T., Iwamoto K., Tamura T., Kawaguchi Y., and Tsukahara T. (2010) Direct numerical simulations of fundamental turbulent flows with the world's largest number of grid-points and application to modeling of engineering turbulent flows. In *Annual Report of the Earth Simulator Center,* pp. 143–148, April 2010–March 2011.

Kinne S., Schulz M., Textor C., Guibert S., Balkanski Y., Bauer S. E., Berntsen T., Berglen T. F., Boucher O., Chin M., Collins W., Dentener F., Diehl T., Easter R., Feichter J., Fillmore D., Ghan S., Ginoux P., Gong S., Grini A., Hendricks J., Herzog M., Horowitz L., Isaksen I., Iversen T., Kirkevåg A., Kloster S., Koch D., Kristjansson J. E., Krol M., Lauer A., Lamarque J. F., Lesins G., Liu X., Lohmann U., Montanaro V., Myhre G., Penner J., Pitari G., Reddy S., Seland O., Stier P., Takemura T., and Tie X. (2006) An AeroCom initial assessment — optical properties in aerosol component modules of global models. *Atmos. Chem. Phys., 6,* 1815–1834.

Kistler R., Kalnay E., Collins W., Saha S., White G., Woollen J., Chelliah M., Ebisuzaki W., Kanamitsu M., Kousky V., van den Dool H., Jenne R., and Fiorino M. (2001) The NCEP-NCAR 50-year reanalysis: Monthly means CD-ROM and documentation. *Bull. Am. Meteor. Soc., 82,* 247–267.

Kravitz B., Robock A., Boucher O., Schmidt H., Taylor K. E., Stenchikov G., and Schulz M. (2011) The Geoengineering Model Intercomparison Project (GeoMIP). *Atmos. Sci. Lett., 12,* 162–167, DOI: 10.1002/asl.316.

Lahoz W., Khattatov B., and Menard R., eds. (2010) *Data Assimilation: Making Sense of Observations*. Springer, New York, DOI: 10.1007/978-3-540-74703-1.

Lewis J. M., Lakshmivarahan S., and Dhall S. K. (2006) *Dynamic Data Assimilation: A Least Squares Approach.* Encycl. Math. App. 104, Cambridge Univ., Cambridge. 654 pp.

Lorenc A. C. and Rawlins F. (2005) Why does 4D-Var beat 3D-Var? *Q. J. R. Meteor. Soc., 131,* 3247–3257.

Marshall J., Adcroft A., Campin J.-M., and Hill C. (2004) Atmosphere-ocean modeling exploiting fluid isomorphisms. *Mon. Weather Rev., 132,* 2882–2894.

Miura H., Tomita H., Nasuno T., Iga S.-I., Satoh M., and Matsuno T. (2005) A climate sensitivity test using a global cloud resolving model under an aquaplanet condition. *Geophys. Res. Lett., 32,* L19717, DOI: 10.1029/2005GL023672.

Lynch P. (2006) *The Emergence of Numerical Weather Prediction: Richardson's Dream.* Cambridge Univ., Cambridge.

Palmer T. and Hagedorn R. (2006) *Predictability of Weather and Climate.* Cambridge Univ., Cambridge.

Pierrehumbert R. T. (2010) *Principles of Planetary Climate.* Cambridge Univ., Cambridge.

Polavarapu S., Ren S., Rochon Y., Sankey D., Ek N., Koshyk J., and Tarasick D. (2005) Data assimilation with the Canadian middle atmosphere model. *Atmos.-Ocean., 43,* 77–100, DOI: 10.3137/ao.430105.

Radebaugh J., Lorenz R. D., Kirk R. L., Lunine J. I., Stofan E. R., Lopes R. M. C., Wall S. D., and the Cassini Radar Team (2007) Mountains on Titan observed by Cassini radar. *Icarus, 192,* 77–91, DOI: 10.1016/j.icarus.2007.06.020.

Randall D. (2011) Chapter 9. The evolution of complexity in general circulation models. In *The Development of Atmospheric General Circulation Models: Complexity, Synthesis and Computation* (L. Donner et al., eds.), Cambridge Univ., Cambridge.

Randall D. A., Wood R. A., Bony S., Colman R., Fichefet T., Fyfe J., Kattsov V., Pitman A., Shukla J., Srinivansan J., Stouffer R. J., Sumi A., and Taylor K. E. (2007) Climate models and their evaluation. In *Climate Change 2007: The Physical Basis. Contribution of Working Group I to the Fourth Assessment Report on the Intergovernmental Panel on Climate Change* (S. Solomon et al., eds.), Chapter 8. Cambridge Univ., Cambridge.

Read P. L., Lewis S. R., Moroz I. M., and Martinez-Alvarado O. (2006) Atmospheric predictability of the martian atmosphere: From low-dimensional dynamics to operational forecasting? In *Second Workshop on Mars Atmosphere Modeling and Observations,* 27 Feb–3 Mar 2006, Granada, Spain.

Richardson M. I., Toigo A. D., and Newman C. E. (2007) Planet WRF: A general purpose, local to global numerical model for planetary atmospheric and climate dynamics. *J. Geophys. Res., 112*, 29, DOI: 10.1029/2006JE002825.

Ritchie H., Temperton C., Simmons A., Hortal M., Davies T., Dent D., and Hamrud M. (1995) Implementation of the semi-Lagrangian method in a high-resolution version of the ECMWF forecast model. *Mon. Weather Rev., 123,* 489–514.

Sandgathe S., Sedlacek D., Iredell M., Black T. L., Henderson T. B., Benjamin S. G., Balaji V., Doyle J. D., Peng M., Stocker R., Campbell T. J., Riishojgaard L. P., Suarez M. J., DeLuca C., Skamarock W., and O'Conner W. P. (2009) *Final Report from the National Unified Operational Prediction Capability (NUOPC) Interim Committee on Common Model Architecture (CMA).* Available online at *gip.noaa.gov/references/CMA_Final_Report_Revised_18Jun.doc*.

Schröder K. P. and Smith R. C. (2008) Distant future of the Sun and Earth revisited. *Mon. Not. R. Astron. Soc., 386*, 155–163, DOI: 10.1111/j.1365-2966.2008.13022.x.

Senior C. A. and Mitchell J. F. B. (1993) Carbon dioxide and climate: The impact of cloud parameterization. *J. Climate, 6*, 393–418.

Shephard M. S., Jansen K. E., Sahni O., and Diachin L. A. (2007) Parallel adaptive simulations on unstructured meshes. *J. Phys. Conf. Ser., 78*, 1–10, DOI: 10.1088/1742-6596/78/1/012053.

Slingo J., Bates K., Nikiforakis N., Piggott M., Roberts M., Shaffrey L., Stevens I., Vidale P. L., and Weller H. (2009) Developing the next-generation climate system models: Challenges and achievements. *Philos. Trans. R. Soc., A367*, 815–831.

Smagorinsky J. (1963) General circulation experiments with the primitive equations. *Mon. Weaher Rev., 91*, 99–164, DOI: 10.1175/1520-0493.

Staniforth A. and Thuburn J. (2012) Horizontal grids for global weather and climate prediction models: A review. *Q. J. R. Meteor. Soc., 138*, 1–26, DOI: 10.1002/qj.958.

Swinbank R., Shutyaev V., and Lahoz W. A. (2003) *Data Assimilation for the Earth System, IV.* Earth and Environmental Sciences Vol. 26, NATO Science Series. IOS, Amsterdam.

Taylor K. E., Stouffer R. J., and Meehl G. A. (2009) A summary of the CMIP5 experiment design. *PCDMI Rept*., 33 pp.

Taylor K. E., Stouffer R. J., and Meehl G. A. (2012) An overview of CMIP5 and the experiment design. *Bull. Am. Meteor. Soc., 93,* 485–498.

Textor C., Schulz M., Guibert S., Kinne S., Balkanski Y., Bauer S., Berntsen T., Berglen T., Boucher O., Chin M., Dentener F., Diehl T., Feichter J., Fillmore D., Ginoux P., Gong S., Grini A., Hendricks J., Horowitz L., Huang P., Isaksen I. S. A., Iversen T., Kloster S., Koch D., Kirkevåg A., Kristjansson J. E., Krol M., Lauer A., Lamarque J. F., Liu X., Montanaro V., Myhre G., Penner J. E., Pitari G., Reddy M. S., Seland Ø., Stier P., Takemura T., and Tie X. (2007) The effect of harmonized emissions on aerosol properties in global models — an AeroCom experiment. *Atmos. Chem. Phys., 7,* 4489–4501.

White A. A., Staniforth A., and Wood N. (2008) Spheroidal coordinate systems for modelling global atmospheres. *Q. J. R. Meteor. Soc., 134,* 261–270, DOI: 10.1002/qj.208.

Wilks D. S. (2005) *Statistical Methods in the Atmospheric Sciences, 2nd edition.* Elsevier, Amsterdam. 627 pp.

Wielicki B. A., Wong T., Loeb N., Minnis P., Priestley K., and Kandel R. (2005) Changes in Earth's albedo measured by satellite. *Science, 308,* 825, DOI: 10.1126/science.1106484.

Wyngaard J. C. (2004) Toward numerical modeling in the "Terra Incognita." *J. Atmos. Sci., 61,* 1816–1826, DOI: 10.1175/1520-0469.

Yokokawa M., Itakura K., Uno A., Ishihara T., and Kaneda Y. (2002) 16.4-Tflops direct numerical simulation of turbulence by a fourier spectral method on the Earth Simulator. *Proc. 2002 ACM/IEEE Conference on Supercomputing,* p. 50. IEEE Conference Publications.

Young R. E., Pfister L., and Houben H. (1984) Baroclinic instability in the Venus atmosphere. *J. Atmos. Sci., 41,* 2310–2333.

Forget F. and Lebonnois S. (2013) Global climate models of the terrestrial planets. In *Comparative Climatology of Terrestrial Planets* (S. J. Mackwell et al., eds.), pp. 213–229. Univ. of Arizona, Tucson, DOI: 10.2458/azu_uapress_9780816530595-ch010.

Global Climate Models of the Terrestrial Planets

François Forget and Sebastien Lebonnois
Laboratoire de Météorologie Dynamique, Institut Pierre Simon Laplace,
Centre National de la Recherche Scientifique

On the basis of the global climate models (GCMs) originally developed for Earth, several teams around the world have been able to develop GCMs for the atmospheres of the other terrestrial bodies in our solar system: Venus, Mars, Titan, Triton, and Pluto. In spite of the apparent complexity of climate systems and meteorology, GCMs are based on a limited number of equations. In practice, relatively complete climate simulators can be developed by combining a few components such as a dynamical core, a radiative transfer solver, a parameterization of turbulence and convection, a thermal ground model, and a volatile phase change code, possibly completed by a few specific schemes. It can be shown that many of these GCM components are "universal" so that we can envisage building realistic climate models for any kind of terrestrial planets and atmospheres that we can imagine. Such a tool is useful for conducting scientific investigations on the possible climates of terrestrial extrasolar planets, or to study past environments in the solar system. The ambition behind the development of GCMs is high: The ultimate goal is to build numerical simulators based only on universal physical or chemical equations, yet able to reproduce or predict all the available observations on a given planet, without any *ad hoc* forcing. In other words, we aim to virtually create in our computers planets that "behave" exactly like the actual planets themselves. In reality, of course, nature is always more complex than expected, but we learn a lot in the process. In this chapter we detail some lessons learned in the solar system: In many cases, GCMs work. They have been able to simulate many aspects of planetary climates without difficulty. In some cases, however, problems have been encountered, sometimes simply because a key process has been forgotten in the model or is not yet correctly parameterized, but also because sometimes the climate regime seems to be result of a subtle balance between processes that remain highly model sensitive, or are the subject of positive feedback and unstability. In any case, building virtual planets with GCMs, in light of the observations obtained by spacecraft or from Earth, is a true scientific endeavor that can teach us a lot about the complex nature of climate systems.

1. INTRODUCTION: BUILDING VIRTUAL PLANETS

As discussed in the previous chapter (Dowling, this volume), many scientific teams around the world have over the past 40 years been developing Earth atmosphere numerical weather prediction models (to predict the weather a few days in advance) and global climate models (GCMs) to simulate the climate system and its long-term evolution. Such models are now used for countless applications, including coupling with the oceans, the biosphere, or the geochemical CO_2 cycles; photochemistry; data assimilation to build data-derived climate databases; etc.

Because these models are built almost entirely on physical equations (rather than empiric parameters), several teams have been able to succesfully adapt them to the other terrestrial planets or satellites that have a solid surface and a sufficiently thick atmosphere. In our solar system, that includes Mars and Venus, but also includes Titan, Triton, and Pluto. For each of these models, and in particular Mars and Titan, the initial GCMs (the goals of which were mostly to predict the thermal structure and the dynamics) are now able to also simulate the climatic cycles of aerosols, clouds, frost, photochemistry, etc., based on planet system models.

The ambition behind the development of such models is high, and goes beyond the study of a limited list of observed phenomena. The ultimate objective is to build numerical simulators only based on universal physical or chemical equations, yet able to reproduce or predict all the available observations on a given planet without any *ad hoc* forcing. We aim to virtually create in our computers planets that "behave" exactly like the actual planets themselves, which is a scientific endeavor by itself. In reality, of course, nature is much more complex than expected, notably because climate systems and atmospheres are controlled by many interacting processes that operate on a large range of spatial and temporal scales and are intrinsically nonlinear. Nevertheless, we can learn a lot in the process.

2. HOW TO BUILD A PLANETARY GLOBAL CLIMATE MODEL

2.1. A Stack of "Bricks"

Global climate models pretend to fully simulate physical systems that exhibit an apparent high level of complexity, resulting from a large number of degrees of freedom, the interaction of various scales, and the fact that the atmosphere tends to propagate many kind of waves. However, the physical and dynamical processes at work can only be described with a limited number of coupled differential equations. In practice, this means that, for a given planetary atmosphere, a GCM is built by stacking together a few independent "bricks," with each brick being designed to resolve the differential equations that control a given process. Fortunately for the modelers, the same equations are often valid in several environments and many of these "bricks" can be reused to build various virtual planets.

Any terrestrial planet GCM must include at least the five following components (see Fig. 1): (1) a three-dimensional hydrodynamical "core" designed to solve the Navier-Stokes fluid dynamical equations, adapted to the case of a rotating spherical envelope; (2) radiative transfer through gas and aerosols; (3) parameterizations of the vertical mixing and transport due to turbulence and convection not resolved by the dynamical core; (4) a thermal model of the surface and near subsurface (storage and conduction of heat); and (5) models to account for the phase change of volatiles on the surface and in the atmosphere (clouds and aerosols). To this list, one could add a catalog of processes that play a secondary role, are only relevant in particular cases, or can be included but are not necessary, e.g., mineral dust lifting, oceanic transport, photochemistry, molecular diffusion, and conduction (at very low pressure), etc.

Below we provide a few key pieces of information and comments on the development of the various components of planetary GCMs (see Table 1 for a summary of key processes and related problems in low-atmosphere terrestrial planet GCMs). The components of Earth GCMs have also been described in the previous chapter (Dowling, this volume). Here we will emphasize the evolution that is needed so that the same tool may be used for the wider range of atmospheric conditions encountered with other planets.

2.2. The Dynamical Core

This central part of the GCM solves the equations describing the explicitly resolved fluid motions. It also includes dissipation for subgrid scales, and provides a transport scheme for tracers. The main problems in the design of the dynamical core have been reviewed in the previous chapter (Dowling, this volume). However, its application to planetary atmospheres in general (not only in the specific case of Earth) introduces some additional constraints that are detailed below.

2.2.1. Equation set and conservation laws. The basic equation set includes three components of the momentum equation, the continuity equation for density, the thermodynamic equation for potential temperature, together with the definition of this variable, which is the temperature of a gas parcel when transported adiabatically to a reference pressure (see, e.g., *Holton,* 2004), the equation of state, and the transport equations for tracers.

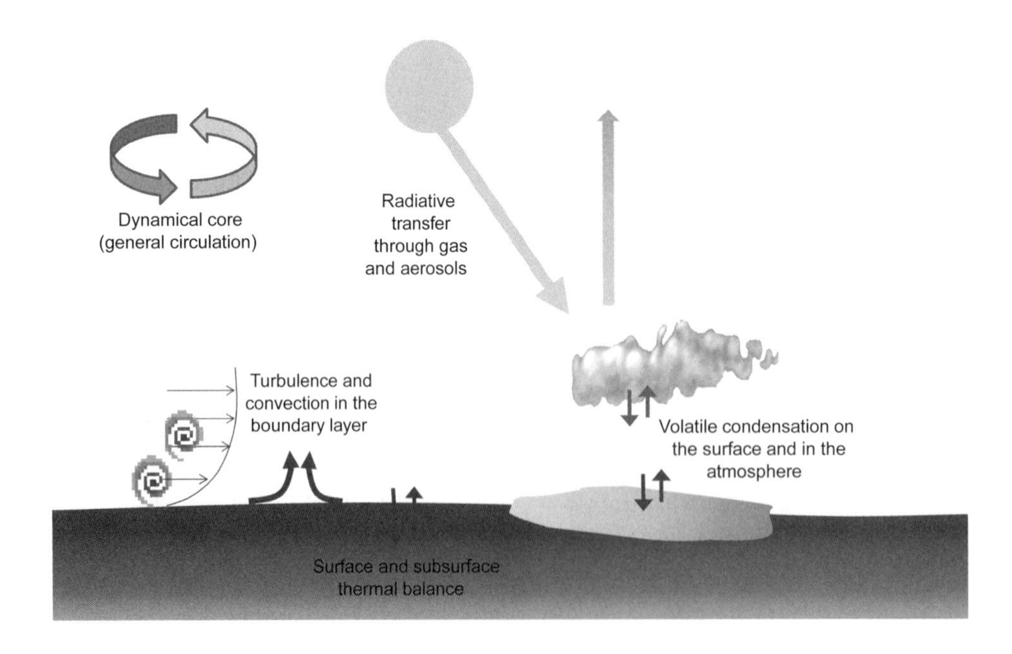

Fig. 1. See Plate 37 for color version. An illustration of the different components that are combined to build a planetary global climate model.

TABLE 1. Summary of key processes and related problems in low-atmosphere terrestrial planet global climate models.

Key physical processes	Earth	Mars	Venus	Titan	Triton/Pluto
Radiative transfer	×	×	Optical thickness*	×	×†
Clouds	H_2O	H_2O ice, CO_2 ice	H_2SO_4	CH_4, C_2H_6	N_2 ice?
Hazes	Aerosols	Mineral dust		Organic haze	Organic haze?
Turbulence and convection	Near-surface	Near-surface	Cloud layers	Near-surface	Near surface
Subsurface heat storage	With oceans	×	×	×	Long-term buffer‡
Dominant gas condensation		CO_2			N_2
Minor species condensation	H_2O	H_2O		CH_4, C_2H_6	CH_4?
Dynamical core					
Deep atmosphere				×§	
Specific heat variations			$C_p(T)$¶		
Composition variations		×§			
Momentum conservation			Critical§	Critical§	

**Eymet et al.* (2009).

†Triton's atmosphere N_2 atmosphere is so pure and teneous that gaseous absorption of radiation can be neglected; molecular conduction then dominates heat transport (*Yelle et al.,* 1991; *Vangvichith et al.,* 2010).

‡See section 2.5.

§See section 2.2.

¶*Lebonnois et al.* (2010a).

These equations are derived from the universal Navier-Stokes equations, simplified for the case of an atmosphere on a rotating planet. To first order the dynamical core developed for the Earth GCMs are therefore usable on other planets. In some cases, however, the simplifications made to simulate the Earth's troposphere are not valid on other planets.

2.2.1.1. Hydrostatic and shallow atmosphere approximations: A first approximation usually made for the equations is to consider hydrostatic equilibrium, eliminating vertically propagating acoustic oscillations. However, this restriction makes it difficult to use the model at a scale below a few kilometers (thus mostly in so-called mesoscale models rather than GCM), or when very strong vertical winds can form, like in the upper atmosphere (thermosphere) of some planets. A second approximation is to consider that the depth of the atmosphere is small compared to the radius of the planet (shallow-atmosphere assumption). With both these approximations, the equations are called "hydrostatic primitive equations." Most of the current GCMs are based on this equation set. The shallow-atmosphere assumption is restrictive in the context of studying planetary atmospheres with a tool as general as possible. Among the four terrestrial atmospheres that are the most studied in the solar system, Titan provides an example for which this approximation is not valid, and where deep-atmosphere equations should be used: Its radius is 2575 km, while the upper limit of the atmosphere taken into account in GCMs reaches 300 km (stratopause) to 500 km (mesopause). Modelers seek to reach even higher.

2.2.1.2. Variations of specific heat C_p: Another additional approximation that may be modified in some planetary applications is the fact that the specific heat C_p is considered as constant with temperature in the definition of potential temperature. An example of this problem is Venus, where C_p varies significantly (around 40%) with temperature from the surface to the atmosphere above the clouds. The effects of this variation on the potential temperature and dynamical core are discussed in *Lebonnois et al.* (2010a).

2.2.1.3. Variations of atmospheric composition and molecular mass: Some processes can create horizontal variations of composition. For instance, this happens when the main constituent of the atmosphere can directly condense on the surface, as on Mars (mostly CO_2), Triton (N_2). or Pluto (N_2). In the martian polar regions, as much as 30% of the atmosphere condenses every year to form CO_2 ice polar caps in both hemispheres. This changes the composition because while carbon dioxide condenses onto the surface, the 5% of noncondensible gases that also form the martian atmosphere (mostly N_2, Ar, and O_2) is left in the atmosphere. Analyzing measurements obtained by the Gamma Ray Spectrometer (GRS) onboard Mars Odyssey, *Sprague et al.* (2004) showed that the mean Ar mixing ratio is enhanced by as much as a factor of 6 during winter and depleted by a factor of 2 to 3 during spring. Because the mean molecular weight of the noncondensible gas is only 32.3 g mol^{-1} compared to 44 g mol^{-1} for CO_2, the enrichment near the surface where most of the CO_2 condenses induces a significant latitudinal and vertical gradient of molecular weight. In practice, the enrichment observed around the time of the winter solstice would have an effect on the circulation as large as a 13 K horizontal temperature gradient, as used in the traditional thermal wind equation, for instance (*Forget and Bertaux,* 2004). Meteorologists have never needed to consider strong gradients in the dry-air component of Earth's atmosphere, and it is still neglected in all existing GCMs. A closer analog would be a gradient of salinity in oceanography.

2.2.1.4. Conservation properties: The equation set defining the dynamical core has conservation properties in its continuous form. The quantities for mass, angular momentum, energy, potential temperature, potential vorticity [for details about the definition and properties of this quantity, see, e.g., *Holton* (2004)], and tracers should all be conserved both in local flux forms and as global invariants when the dynamical core solves the equation set. However, in practice, these equations have to be spatially discretized and solved at given time steps. Therefore, the design of the dynamical core may only guarantee some of the conservation laws. This has to be carefully examined when the equations are tailored for the dynamical core solver, as discussed in *Thuburn* (2008). As an example of questions that arise from the planetary extension of the dynamical core, the conservation of angular momentum is much more critical in superrotating atmospheres (e.g., Venus and Titan) than for Earth.

2.2.2. Discretization. Designing the algorithms for the discretized forms of the equations is constrained by several practical issues. The GCM has to be efficient and run as fast as possible. However, it also needs to respect the conservation laws. These issues are also related to the properties of the advection scheme and to the control of instabilities that tend to grow at the resolution scale. The spatial and timescales involved in a given application of the GCM are then crucial to evaluate the performance of the dynamical core. This is why numerical weather forecast and climate studies have long been distinct applications using two different designs of the dynamical core. There are efforts now made to unify these two applications under the same GCM core, with forecast applications emphasizing very high horizontal resolution, while climate studies need very good conservation performances on very long timescales.

The addition of planetary atmospheres in the covered applications brings further constraints. The conservation issues are crucial for weakly forced atmospheres that evolve on very long timescales (e.g., Venus). The horizontal scales may span from giant planet sizes to very small objects such as Triton, with pressures that range from very dense (Venus) to very thin (Pluto) atmospheres.

Several grid-point methods have been developed and evaluated for the horizontal discretization of the equations. Also widely used is the spectral transform method, which decomposes the fields onto spherical harmonics in order to solve the dynamical equations. Both these approaches have advantages and disadvantages, as reviewed in *Williamson* (2007), for example, and discussed in the previous chapter (Dowling, this volume).

These spatial discretizations are combined with a temporal discretization scheme coupled to an advection scheme. Among the commonly used structures, a dynamical core based on a regular latitude-longitude grid using explicit time steps and a Eulerian advection scheme is confronted with severe problems in the polar regions, due to limits in the time step [the Courant-Fredrich-Levy (CFL) condition] and the consequent use of filters to remove the fastest-moving, computationally unstable waves. Semi-implicit semi-Lagrangian schemes are not sensitive to this pole problem, but are less efficient for conservation. Due to the rise of parallel computing, where thousands of processors are used in a distributed manner, the spectral transform methods encounter increasing difficulties and the most recent efforts have been directed to new grids, such as quasi-uniform grids (cubed sphere, icosahedral grid). Some of the issues associated with currently proposed horizontal grids are reviewed in *Staniforth and Thuburn* (2012).

Although they are very demanding, planetary applications should be kept in mind for these developments, with special attention paid to conservation issues (in particular, angular momentum, which is a critical problem for rapidly rotating atmospheres such as those found on Venus and Titan), the pole problem, and shape-conserving transport schemes and stability over very long timescales.

2.3. Radiative Transfer

Computing the radiative transfer of solar and thermal radiations through the gaseous atmosphere, aerosols, and clouds accurately and quickly enough remains one of the major challenges in the development of a GCM. To first order, the equations are the same as on Earth. The challenge lies in the fact that, as detailed below, the models need to be adapted to specific radiative properties of the atmospheric components, and that a hierarchy of models is necessary to compute the coefficients that are ultimately used in a given planetary atmosphere.

2.3.1. Radiative transfer through gases. 2.3.1.1. Spectroscopic properties: Planetary atmospheres typically have a rich absorption spectrum of hundreds of thousands of lines and continuum features. The first classical challenge, for any model, results from the uncertainties in the exact behavior of the lines, and in particular in the shape of the line profile away from the line center. The sum of the far wings of strong lines tends to create a continuum that is not easy to estimate on a theoretical basis. A continuum can also be created by numerous transitions to nonquantized upper states (e.g., O_2 and O_3 in the visible and ultraviolet region), and by a process known as "collision induced absorption," which occurs at pressure higher than about 1 bar. This results from the fact that during collision, colliding molecules such as CO_2, N_2, O_2, or H_2 can behave differently than a single molecule. On the one hand, collisions induce transitory dipole moments, allowing "forbidden" transitions to take place. On the other hand, dimers with short-lived molecules can briefly form, with more vibrational and rotational models than an isolated molecule, resulting in new spectral absorption bands. Because a very short time is spent during collisions, the lines are very wide, and they overlap to form a featureless continuum. At the opposite end of the spectrum, at pressures lower than about 1 Pa (and in some cases 10 or 100 Pa) the assumption of local thermodynamic equilibrium (LTE) can break down (*Lopez-Puertas and*

Taylor, 2001). Local thermodynamic equilibrium assumes that the internal temperature of molecules as determined by the statistics of the population of vibrational and rotational energy levels is the same than the usual "kinetic temperature." In non-LTE conditions, a variety of complex phenomena occurring at the molecular level (less-efficient transfer between thermal kinetic energy and radiation or, vice versa, fluorescence) must be taken into account.

Overall, atmospheric radiative properties have been well studied in the case of Earth, and numerous spectroscopic databases are available. However, some unknowns remain for observed atmospheres such as those of Mars or Venus, and many more uncertainties affect the modeling of theoretical exotic atmospheres not yet observed (e.g., hot and wet atmospheres with a high partial pressure of water vapor).

2.3.1.2. Radiative transfer algorithms: If the gaseous lines and continuum processes are known, it is possible to accurately solve the monochromatic equation of radiative transfer at a given wavelength by adding the contribution of every line and continuum source, using so-called line-by-line codes. However, these calculations are too slow for GCMs, so the solution is to use band models, in which the radiative flux is integrated with respect to frequency over many lines. In that case, the difficulty lies in statistically representing the behavior of all photons in the band (some are absorbed quickly, other can go through the atmosphere between two lines). These statistics can be stored calculating offline, with a line-by-line model, in each spectral band, the radiative flux between each model layer for a range of pressure and temperature [e.g., *Pollack et al.* (1990); *Hourdin* (1992), for Mars]. An interesting extension of this strategy is the "next exchange formulation" [*Dufresne et al.* (2005) and *Eymet et al.* (2009) for Mars and Venus, respectively], in which the radiative fluxes are no longer considered. Instead, the basic variables are the net exchange rates between each pair of modeled atmospheric layers. This offers a meaningful matrix representation of radiative exchanges, allows qualification of the dominant contributions to the local heating rates (temperature contrast and absorption coefficient are in different terms of the equation), and allows one to compute dominating terms (neighboring layers, cooling to space) with higher frequency than less-important exchanges. Once optimized, this method can be more efficient that other methods. One drawback of this method is that it cannot be used to account for scattering by aerosols and clouds unless the exact cloud properties have been assumed in the offline calculations.

Recently, more and more planetary climate models are using the correlated-k distribution technique. In this method, the individual, monochromatic absorption intensities derived from a line-by-line spectrum at a given reference pressure and temperature are sorted into a smooth function that can be fitted more easily with an analytical equation (e.g., Gaussian quadrature) and therefore described with a limited number of coefficients. Scattering can thus be taken into account within a scattering solver (see below). A detailed description of its use on present-day Mars in the Mars Weather Research and Forecast (WRF) GCM is given by *Mischna et al.* (2012). It is also used in the NASA Ames Mars GCM (see *spacescience.arc.nasa.gov/ mars-climate-modelinggroup/models.html*) and in the generic climate model that we have developed at Laboratoire de Météorologie Dynamique (LMD), notably for exoplanets (*Wordsworth et al.,* 2011) and primitive atmospheres (*Forget et al.,* 2013).

2.3.2. Clouds and aerosols. To compute the interaction between aerosols, cloud particles, and radiation, various methods are available. However, it is interesting to note that a majority of independent groups performing global climate model of the atmosphere now assume a "two-stream approximation" and follow the approach of *Toon et al.* (1989) with the so-called "hemispheric mean approximation" in the thermal range, and usually a delta-Eddington approximation to represent the strong forward scattering at solar wavelengths (e.g., *Hourdin et al.,* 1995a; *Forget et al.,* 1999; *Richardson et al.,* 2007).

An alternative is used on Venus by *Eymet et al.* (2009), who employed a full Monte Carlo model coupled with line-by-line calculations to prepare their "next exchange formulation" coefficient matrix (see above).

2.4. Turbulence and Convection

The dynamical mixing processes that are not resolved by the dynamical core must be parameterized in GCMs. They are especially important in the planetary boundary layer near the surface, where near-surface wind shear and surface heating induce turbulence and convection. Fortunately, there is no reason for the theories and equations developed for Earth not to be valid on other terrestrial planets. For example, turbulent closure schemes such as the well-known *Mellor and Yamada* (1982) model family are now used in many planetary GCMs to compute the turbulent mixing of heat, momentum, and tracers.

In reality, such schemes depend on model parameters that are usually empirically deduced from field experiments on Earth. One should be careful when using them in more extreme conditions, such as nighttime on Mars, when surface radiative cooling can reduce the boundary layer to a thin stably stratified layer with no equivalent on Earth. It is likely that current schemes underestimate the mixing in such conditions (*Petrosyan et al.,* 2011). Conversely, during the martian day the near-surface temperature gradient can be much larger than on Earth, reaching several tens of Kelvins in the first meter.

In fact, when the atmosphere tends to become unstable, it is expected that the vertical motions organize themselves in convection cells that can reach scales of several kilometers. To account for such cells, most GCMs assume that when an unstable (super-adiabatic) temperature lapse rate is produced by the model, mixing is then so efficient that, using a simple enthalpy conserving scheme, the temperature lapse rate is immediately restored to its adiabatic profile with a constant value of the potential temperature (see, e.g., *Holton,* 2004). If the resulting temperature profile is unstable at its upper or lower limit, this mechanism is instantaneously extended in such a way that the final profile is entirely stable or neutral. This so-called "convective adjustment" scheme is not without problems, however. On the one hand, in a real atmosphere the mixing is achieved by parcel exchange through vertical convective motions that remain limited in their efficiency. As a result, convective adjustment tends to overestimate daytime near-surface temperatures on Mars because it forces an adiabatic profile near the surface (*Rafkin and Michaels,* 2003; *Spiga and Forget,* 2009). On Earth, convective adjustment schemes are replaced more and more by "mass flux schemes" aiming at representing the actual convective plumes within the boundary layer. *Colaitis et al.* (2013) has recently adapted such a scheme to Mars. On the other hand, convection can become an even more complex problem when a species present in asignificant amount (e.g., water on Earth, CO_2 on Mars) can condense as a result of adiabatic vertical motions, releasing latent heat in the process. The physical processes involved can become very complex, and this field is among the most active in terrestrial climate science (see Dowling, this volume).

2.5. Surface and Subsurface Thermal Balance

On the solid surface of any planet, temperature evolution is controlled by the balance between incoming fluxes (solar insolation, thermal radiation from the atmosphere and the surface itself, latent heat exchange, sensible heat flux from the atmosphere) and thermal conduction in the soil. This last process is governed by the well-known equation of heat conduction

$$C\frac{\partial T}{\partial t} = \frac{\partial}{\partial z}\left[\lambda\frac{\partial T}{\partial z}\right]$$

with T(z) the ground temperature at depth z, λ the ground heat conductivity (J s^{-1} m^{-1} K^{-1}) and C the ground volumetric specific heat (J m^{-3} K^{-1}). This equation is easy to solve using a finite-volume approach and time integration via an implicit Euler scheme, which allows time steps of several tens of minutes suitable for GCMs. When working on different planets, one must pay attention to the vertical discretization in order to catch both the diurnal thermal wave and the seasonal thermal wave. This can create a problem on a cold planet with a long orbital period such as Triton (165 Earth years) and Pluto (248 Earth years). On these weakly irradiated bodies, the subsurface heat stored during one season can play a major role in the control of the surface temperature during the opposite season (*Hansen and Paige,* 1992, 1996). Therefore, in theory, GCM simulations should be run for hundreds of Earth years to accurately predict surface temperatures. This is computationally too expensive. Alternatively, one can use a simplified model for hundreds of years to prepare the initial subsurface temperature fields to be used by the GCM, which can then be run for only a few years (*Vangvichith et al.,* 2010).

If a liquid ocean is present on the surface (a question that actually arises when dealing with extrasolar planets or early climates on Mars or Venus), the problem becomes more complex. To first order, the heat conduction equation shown above can be used to evaluate the surface temperature assuming a very high conductivity to represent the ocean mixed layer. This creates a so-called "slab ocean," and is a proxy to represent the oceanic high thermal inertia. In reality, however, the heat transported by ocean currents plays a major role that should be represented with a full oceanic transport model. A compromise is to simulate heat transport by adding a horizontal diffusion of temperature and a representation of the wind-driven ("Ekmandriven") heat fluxes (see, e.g., *Codron,* 2012).

2.6. Phase Changes of Volatiles on the Surface and in the Atmosphere

On most planets, a major component of the climate system is the phase change of minor species (water on Earth, Mars, and possibly Venus; methane on Titan and Pluto; ethane and other species on Titan) and sometimes the major atmospheric component (CO_2 on Mars, N_2 on Triton and Pluto). These phase changes affect the energy balance directly through latent heat exchange and indirectly by changing the radiative properties of the surface and atmosphere.

On the surface, for minor species, most models assume that the air just above the surface condensate deposit is in solid-gas equilibrium so that the vapor flux F is controlled by the turbulent transport from the surface to the atmosphere or vice-versa: $F = \rho C_d u(q_1 - q_0)$ with ρ the air density, u the wind velocity in the first model layer, C_d a turbulent drag coefficient that depends on the surface roughness and the height of the first layer, q_0 the saturated mixing ratio at surface temperature, and q_1 the mixing ratio in the first layer (see, e.g., *Montmessin et al.,* 2004).

If the major species can be in condensed form, then the surface budget is controlled by energy balance considerations. The amount of condensate undergoing phase changes depends on the amount of latent heat required to keep the surface in vapor-pressure equilibrium with the atmosphere (see Appendix in *Forget et al.,* 1998).

In the atmosphere, the condensation of ice particles is a complex microphysical phenomenon that involves nucleation, particle growth, coalescence, etc. In practice, however, in most GCMs it has been possible to obtain acceptable results by simply estimating the condensation rate of ice at each timestep by energy balance consideration (i.e., by condensing the amount of vapor that allows the species to be in solid-gas equilibrium at the end of the timestep, taking into account the possible supersaturation effect, the release of latent heat, and possible evaporation or sublimation). Then the predicted mass of ice in a GCM box can be distributed between particles of prescribed size (an assumption made in many Earth GCMs for which observations are available), or on a prescribed number of particle per kilograms of air [an assumption that works well for Mars (e.g., *Montmessin et al.,* 2004)]. The particles are then transported by the general circulation and undergo gravitational sedimentation. Precipitation resulting from coalescence of cloud particles can also be parameterized using a simple cloud condensate content threshold (*Emanuel and Ivkovi-Rothman,* 1999), with precipitation and evaporation also taken into account. Additional complexity can be added (and must be added in the Earth case) by estimating a cloud coverage fraction in each grid box.

2.7. Additional Schemes

In addition to the key processes mentioned above, planetary global climate models are sometimes completed with specific planetary parameterizations.

2.7.1. Photochemistry. To correctly model the atmospheric composition, it is sometimes necessary to couple the GCM with a photochemical model able to compute, at each timestep and in each GCM grid box, the photolysis of key molecules and their chemical reactions. Because the circulation can redistribute the different gases in the atmosphere, coupling a photochemical model within the GCM allows the interpretation of specific patterns that are observed. These patterns result from the balance between photochemical reactions and advection by dynamics. Examples include depletion of methane or nitrous oxide in Earth's winter polar vortex (e.g., *Abrams et al.,* 1996), enrichment in many hydrocarbons and nitriles in Titan's winter polar vortex (e.g., *Coustenis and Bézard,* 1995; *Lebonnois et al.,* 2001), latitudinal variations in CO and OCS abundances below the venusian clouds (e.g., *Tsang et al.,* 2008; *Marcq et al.,* 2008), or seasonal variations in the ozone abundance in the martian atmosphere (e.g., *Perrier et al.,* 2006).

The impact of dynamics on the atmospheric composition may affect the radiative transfer by modifying the distribution of opacity sources, the same as for the haze or clouds. Although it may be of second order compared to haze variability, the composition variability may still affect the circulation in a coupled loop that is worth modeling and understanding. Microphysics and photochemistry are also coupled through the compounds that condense to form the hazes or clouds, such as the water cycle on Mars or the methane cycle on Titan.

Photochemical models included into GCMs have been developed for Mars (*Lefèvre et al.,* 2004, 2008; *Moudden and McConnell,* 2007) and for Titan (*Crespin et al.,* 2008), and are under development for Venus (e.g., for the LMD Venus GCM).

2.7.2. Sources of aerosols. In desert areas on Earth and Mars, winds and convective vortices (i.e., "dust devils") can lift mineral dust particles that can strongly affect the radiative properties of the atmosphere. These processes have been parameterized in several Mars GCM studies (*Murphy et al.,* 1995; *Newman et al.,* 2002; *Basu et al.,* 2004; *Kahre et al.,* 2006; *Mulholland et al.,* 2013). Similarly, the formation and transport of organic aerosols in the upper atmosphere of Titan have been studied by GCMs (e.g., *Rannou et al.,* 2002).

2.7.3. Upper atmosphere processes. To increase the GCM altitude domain, developments are needed to take into account the specific physical processes that are significant at the very low densities that exist in the upper mesosphere and in the thermosphere: (1) non-LTE corrections to radiative balance, as detailed above; (2) extreme UV absorption by atmospheric molecules (which is significant above pressure levels near 10^{-4} Pa); and (3) molecular conduction of temperature and momentum (viscosity), and molecular diffusion of species, which affect the energy balance and the composition of the atmosphere above the homopause also around 10^{-4} Pa. This has been done in Mars GCMs (*Bougher et al.,* 2006; *González-Galindo et al.,* 2009, 2010).

3. LESSONS LEARNED IN THE SOLAR SYSTEM

3.1. Some Successes on Different Planets

What have we learned when applying GCMs to other atmospheres? First, by many measures, GCMs work for other planets. As detailed below, they have been able to predict the behavior of many aspects of several climate systems on the basis of physical equations only. This is a comforting fact that should give confidence in our ability to understand climate changes on Earth, and provides an interesting and teachable refutation of the attacks on climate science (*Hartmann,* 2012).

3.1.1. Mars. Countless studies have been published showing that Mars GCMs can succesfully predict observations. One must admit that, assuming the right amount of dust in the atmosphere, it has been relatively easy to simulate the thermal structure of the atmosphere and the behavior of atmospheric waves such as thermal tides and baroclinic waves (*Pollack et al.,* 1990; *Haberle et al.,* 1993; *Hourdin et al.,* 1993, 1995a; *Collins et al.,* 1996; *Wilson and Hamilton,* 1996; *Wilson,* 1997; *Forget et al.,* 1999; *Lewis et al.,* 1999; *Angelats i Coll et al.,* 2004). In 2000, an unexpected disagreement between models and observations of diurnal variations of atmospheric temperature from the Infrared Thermal Mapping (IRTM) experiment on Viking led modelers *Wilson and Richardson* (2000) to question the observations, and finally demonstrate that one of the IRTM channels was suffering from a leak in the spectral filter.

Similarly, on the basis of simple equations to represent the sublimation, transport, and condensation of water, Mars GCMs have been able to reproduce without difficulty the main seasonal characteristics of the water cycle (*Richardson and Wilson,* 2002; *Montmessin et al.,* 2004). Another example has been the prediction of the variations of ozone (which are controlled by atmospheric dynamics and photochemical processes related to water vapor) before the first observations obtained by the Spectroscopic Investigation of the Characteristics of the Atmosphere of Mars (SPICAM) ultraviolet spectrometer onboard Mars Express (*Lefèvre et al.,* 2004; *Perrier et al.,* 2006; *Lebonnois et al.,* 2006). After the inclusion of unexpected heterogeneous chemical processes, the observations could even be reproduced in impressive detail (*Lefèvre et al.,* 2008). An interesting and spectacular application of such models has also been the simulation of paleoclimates and the prediction of the formation of nonpolar glaciers millions of years ago when Mars obliquity was higher. In such conditions, GCMs were able to form glaciers in specific regions where geologists found the remnants of such glaciers (*Forget et al.,* 2006; *Madeleine et al.,* 2009) as shown in Fig. 2.

3.1.2. Titan. Titan has a very complex climate system, with similarities to both Earth and Venus. Like our hot neighbor, Titan's atmosphere undergoes superrotation and is shrouded by a thick haze layer in altitude. However, it follows an Earth-like seasonal cycle induced by the obliquity of Saturn, and hosts a methane cycle reminiscent of the terrestrial water cycle.

Following the Voyager missions in 1980 and motivated by the preparation for the Cassini-Huygens mission that was launched in 1997, a GCM for Titan was developed as early as 1992 at LMD (*Hourdin et al.,* 1995b). It predicted superrotating wind fields with amplitude and characteristics comparable to observations. In this work, the crucial role of angular momentum conservation was emphasized. However, among the GCMs using realistic radiative forcings and investigating the stratospheric superrotation, the Köln GCM (*Tokano et al.,* 1999, and subsequent studies), as well as a version of the Community Atmosphere Model (CAM) adapted to Titan (*Friedson et al.,* 2009), both had trouble getting high zonal winds in the stratosphere. The Titan WRF GCM (*Newman et al.,* 2011) has succeeded in reproducing Titan's superrotation after investigation and severe reduction of the horizontal dissipation used in this dynamical core.

Titan GCMs have been used for much more than atmospheric dynamics. For example, the LMD GCM of *Hourdin et al.* (1995b) was extended into a climate model including microphysics and photochemistry. This was initially achieved at the price of a reduction of the dynamical core to a two-dimensional version, using a parameterization to take into account the mixing generated by three-dimensional barotropic waves. In spite of this limitation, *Rannou et al.* (2002, 2004) were able to interpret many features of the haze layer, particularly the detached haze observed at 350 km altitude at the time of Voyager 1. It was shown with the model how the large-scale structures in the haze layers are driven by the meridional circulation, and how the haze structures in return affect thermal structure and atmospheric circulation through radiative feedback.

The LMD climate model was further developed to include the effects of the coupling between chemical composition and dynamics (*Lebonnois et al.,* 2003; *Hourdin et al.,* 2004). Results were consistent with and explained the latitudinal profiles of stratospheric composition observed by the Voyager 1 spacecraft (*Coustenis and Bézard,* 1995), and by CIRS onboard the Cassini spacecraft (see *Crespin et al.,* 2008). Overall, the model succesfully reproduced the thermal structure observed in the troposphere and stratosphere, the abundance and vertical profiles of most chemi-

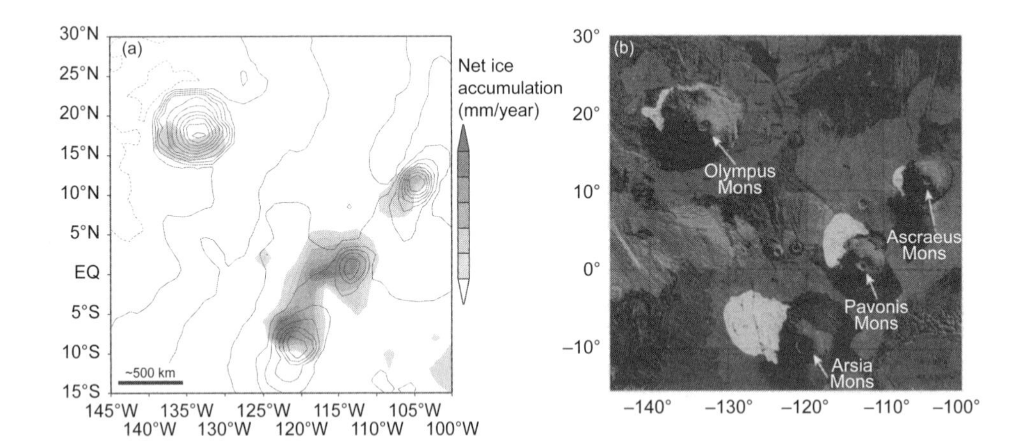

Fig. 2. See Plate 38 for color version. An example of succesful prediction made by a global climate model designed to simulate the details of the present-day Mars water cycle, but assuming a 45° obliquity like on Mars a few millions of years ago. **(a)** Modeled net surface water ice accumulation (in milimeters per martian year) in the Tharsis region, suggesting that glaciers could have formed on the northwest slopes of the Tharsis Montes and Olympus Mons volcanos. **(b)** Geologic map of the same region at the time of the GCM simulations, showing the location of fan-shaped deposits interpreted to be the depositional remains of geologically recent cold-based glaciers. The agreement between observed glacier landform locations and model predictions pointed to an atmospheric origin for the ice and permitted a better understanding of the formation of martian glaciers. From *Forget et al.* (2006).

cal compounds in the stratosphere, and their enrichment in the winter polar region (*Lebonnois et al.,* 2009). Similarly, modeling of tropospheric clouds was also included (*Rannou et al.,* 2006), and comparisons between modeled cloud distributions and clouds observed in Titan's troposphere, with groundbased observations (*Roe et al.,* 2002) and by Cassini instruments (*Griffith et al.,* 2005; *Porco et al.,* 2005), gave insight into the role of dynamics in the tropospheric methane cycle. The LMD model has recently been updated back to a three-dimensional model using the latest version of the LMD GCM dynamical core (*Lebonnois et al.,* 2012). This model spontaneously predicted the detailed thermal structure observed by Huygens in the lowest 5 km, which exhibits strong variations in the lapse rate that could be attributed to diurnal and seasonal boundary layer processes (*Charnay and Lebonnois,* 2012).

3.1.3. Venus. Early developments of GCMs for the atmosphere of Venus started in the 1970s with the first three-dimensional models of *Kalnay de Rivas* (1975) and *Young and Pollack* (1977). A recent review of the history of Venus' atmospheric modeling was recently published by *Lewis et al.* (2013). Many recent GCMs have been developed around the world in the 2000s, with efforts in Japan (*Yamamoto and Takahashi,* 2003, 2004, 2006b; *Takagi and Matsuda,* 2007), the United Kingdom (*Lee et al.,* 2005, 2007), and the United States (*Herrnstein and Dowling,* 2007; *Hollingsworth et al.,* 2007; *Lee and Richardson,* 2010) that were able to produce superrotation, but primarily based on a simplified formulation of temperature forcing (using a relaxation scheme toward a prescribed temperature structure) and adjusting some unrealistic parameters in these radiative schemes (solar heating rates or equator-to-pole temperature contrasts in the deep atmosphere were stronger than observed). Additional studies were done with these models to parameterize the clouds and their interaction with the dynamics (*Yamamoto and Takahashi,* 2006a; *Lee et al.,* 2010).

The development of a "full" GCM for Venus, i.e., at least coupling a three-dimensional dynamical core and the implementation of a more realistic radiative transfer in GCMs, is more recent. This results from the high opacities present in the atmosphere, due to the cloud layer, whose structure is not known with precision, and to the thick atmosphere of CO_2. These opacities and the uncertainties associated with their computation make the radiative transfer (both for solar and infrared wavelengths) a tricky problem. *Ikeda et al.* (2007) presented simulations of Venus atmospheric circulation forced by a fully implemented radiative transfer that showed wind speeds that were interestingly Venus-like. *Lebonnois et al.* (2010a) published a GCM that included a radiative transfer module based on net-exchange rate matrices for the computation of the infrared fluxes (*Eymet et al.,* 2009). Another radiative transfer scheme has also been developed for the Venus version of the planet WRF GCM (*Lee and Richardson,* 2011).

These models successfully reproduced the main feature of the circulation of Venus, the superrotation. However, they also revealed that the superrotation of the Venus atmosphere is the result of a sensitive balance and is thus challenging, as discussed below in section 3.2.

3.1.4. Triton and Pluto. As mentioned above, Triton and Pluto are two distant and cold planetary bodies with a thin atmosphere of N_2 in solid gas equilibrium with nitrogen ice deposits on the surface. Methane and carbon dioxide are also present in the atmospheres of these planetary bodies, although only on Pluto is their concentration high enough to significantly affect the radiative balance of the atmosphere.

The Voyager 2 spacecraft flew by and observed Triton in 1989, during which it obtained one thermal profile and observed wind streaks and plumes in the atmosphere, allowing wind directions to be constrained. Recently, *Vangvichith et al.* (2010) have developed a GCM for Triton, and show that the thermal structure was naturally predicted by the model. Modeling the wind direction, however, required a more detailed analysis of their sensitivity to the distribution of surface ices.

Even fewer observations are available for Pluto. However, the dwarf planet is the objective of the NASA mission New Horizons, which will fly by Pluto on July 14, 2015. Within that context, several Pluto GCM projects have recently been initiated (*Vangvichith and Forget,* 2012; *Zalucha and Gulbis,* 2012). Much will be learned by New Horizons, and it will be very interesting to see how well GCMs can predict the Pluto environment on the basis of the sparse data that we currently have.

3.2. Why and When Global Climate Models Fail

Even more than from the successes, we can learn from the failures of planetary GCMs. On the one hand, the failures help identify areas in which Earth climate models may meet some of the problems. On the other hand, this is key information to take into account when applying planetary climate models to scientific questions for which we have no direct observations, such as for extrasolar planets or the early climates of terrestrial planets (section 4). Some of the sources of errors and challenges are listed below.

3.2.1. Missing physical processes. As can be expected, in many cases GCMs fail to accurately simulate an observed phenomenon simply because a physical process is not included in the GCM. Two striking examples from Mars can be cited.

First, the carbon dioxide martian atmosphere undergoes large-amplitude seasonal variations due to CO_2 condensation and sublimation in the polar regions. This atmospheric mass cycle, present in surface pressure measurements (such as those obtained by the Viking landers), is a major process that must be well represented by GCMs. However, for many years, GCMs were not able to reproduce the cycle without including unrealistically low emissivities to lower the net infrared cooling and reduce the condensation rates by a factor of 2 (see, e.g., *Hourdin et al.,* 1993). The solution for this enigma came from the Mars Odyssey and Phoenix missions, which revealed the presence of a high-thermal-inertia ice-rich layer below a few centimeters of sand in the martian high latitudes. Global climate models had assumed that these regions had a sand-like low thermal inertia, whereas in reality, at seasonal scale the ice layer was able to store a lot of heat during summer and release it in fall and winter, thus significantly reducing the CO_2 condensation rate (*Haberle et al.,* 2008).

Second, for many years, the thin water ice clouds present in the martian atmosphere had been assumed to have a limited impact on the martian climate. Recently, several teams have included the effect of these clouds in GCM simulations (*Madeleine et al.,* 2012; *Wilson,* 2011; *Kahre et al.,* 2012; *Read et al.,* 2011). What they found is that not only do the clouds affect the thermal structure locally, but their radiative effects could solve several long-lasting Mars climate enigmas, such as the pause in baroclinic waves around winter solstice, the intensity of regional dust storms in the northern mid-latitudes, or the strength of a thermal inversion observed above the southern winter pole.

3.2.2. Positive feedbacks and instability. Another challenge for climate modelers presents itself when the system is very sensitive to a parameter because of positive feedbacks. A well-known example is the albedo of snow and sea-ice on Earth. If one tries to model the Earth climate systems from scratch, it is rapidly obvious that this model parameter must be tuned to ensure a realistic climate at high and mid-latitude. An overestimation of the ice albedo results in colder temperature, more ice and snow, etc. When modeling climate systems that are poorly observed, it is necessary to carefully explore the sensitivity of the modeled system to such parameters in order to "bracket" the reality.

3.2.3. Nonlinear behavior and threshold effects. An extreme version of the model sensitivity problem is present when the climate depends on processes that are nonlinear or that depend on poorly known processes. For instance, currently the main source of variability in the Mars climate system is related to the local, regional, and sometimes global dust storms that occur on Mars in seasons and locations that vary from year to year. This dust cycle remains poorly understood, possibly because the lifting of dust occurs above a local given wind threshold stress that may or may not be reached depending on the meteorological conditions. As a result, modeling the dust cycle, and in particular the interannual variability of global dust storms, remains one of the major challenges in planetary climatology — not to mention a hypothetical ability to predict the dust storms (*Newman et al.,* 2002; *Basu et al.,* 2006; *Mulholland et al.,* 2013). Most likely, in addition to the threshold effect, a physical process related to the evolution of the surface dust reservoirs is missing in the models.

3.2.4. Complex subgrid-scale processes. Another variation of the problems mentioned above can be directly attributed to processes that cannot be resolved by the dynamical core, but that play a major role in the planetary climate. Mars dust storms would be once again a good example, but the most striking example is the representation of subgrid-scale clouds in the Earth GCMs. As detailed in the previous chapter (Dowling, this volume), the parameterization of GCM model clouds has been identified as a major source of disagreement between models when predicting the future of our planet (see, e.g., *Dufresne and Bony,* 2008).

3.2.5. Weak forcings, long timescales. While different GCMs can easily agree between themselves and with the observations when modeling a system strongly forced by the variations of, say, insolation, GCM simulations naturally become model sensitive when the evolution of the system primarily depends on a subtle balance between modeled processes. An interesting case is that of Venus general circulation.

As mentioned above, GCM simulations show that the superrotation of the Venus atmosphere is the result of a subtle equilibrium. It involves balance in the exchanges of angular momentum between surface and atmosphere, and balance in the angular momentum transport between the mean meridional circulation and the planetary waves, thermal tides, and gravity waves (although this latest contribution is yet to be studied, as their parameterization in Venus GCMs is still in progress). Modeling this balance is sensitive to the dynamical core details, to the boundary conditions, and possibly also to initial conditions.

The sensitivity to the dynamical core is illustrated by the many differences obtained in the recent modeling of Venus circulation, but has been more formally demonstrated in comparative studies between Venus GCMs under identical physical forcings (*Lee and Richardson,* 2010; *Lebonnois et al.,* 2013). These studies revealed that various dynamical cores that would give very similar results in Earth or Mars conditions can predict very different circulation patterns in Venus-like conditions (Fig. 3).

The wide dispersion of the modeled wind fields in these studies has to be related to the various dynamical core implementations through angular momentum conservation and/or horizontal dissipation processes. Since the planetary waves play a crucial role in angular momentum balance, their representation in the spatial and temporal discretization mechanisms of the dynamical cores may also be part of this sensitivity.

Since they control the exchanges of angular momentum between the atmosphere and the surface, the lower boundary conditions used in the model also have a strong influence on the resulting wind field. The planetary boundary layer scheme has an impact not only on the surface friction, but also on the temperature structure of the deepest atmospheric layer, therefore affecting the circulation close to the surface. The presence of topography introduces the mountain torque in the surface-atmosphere exchanges. This torque results from different surface pressures on the eastern and western (for its component affecting the zonal wind) sides of topographical features. It is a major component of these angular momentum exchanges, as shown for example in *Lebonnois et al.* (2010a).

The sensitivity to initial conditions with such models is also a problem. *Kido and Wakata* (2008) first obtained different strengths for the zonal wind field when starting from a resting point or from preexisting superrotation.

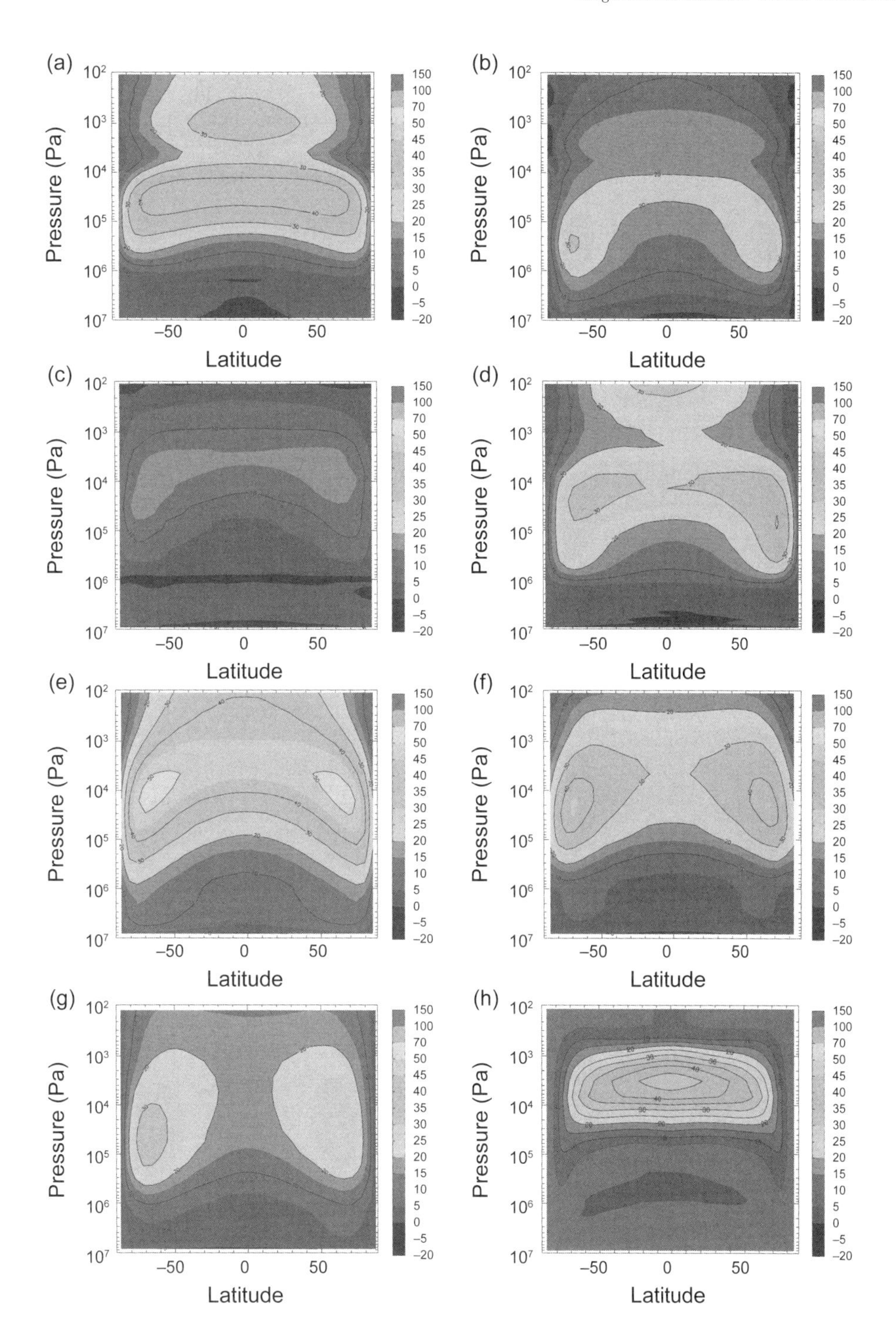

Fig. 3. Zonally and temporally averaged zonal wind fields obtained in Venus-like conditions by various dynamical cores sharing the exact same solar forcing and boundary layer scheme. Unit is m s^{-1}. **(a)** "Spectral" model from CCSR (Japan), **(b)** finite difference model from LMD (France), **(c)** "spectral" model from Open University (UK), **(d)** finite difference model from Oxford (UK), **(e)** "spectral" model from *Lee and Richardson* (2010) (USA), **(f)** finite difference model from *Lee and Richardson* (2010) (USA), **(g)** finite volume model from *Lee and Richardson* (2010) (USA), and **(h)** finite volume model from UCLA (USA). For the UCLA simulation **(h)**, the resolution is much higher than the other baseline runs and the results are averaged over a 10-year period). The results show that even in very similar modeling conditions, the wind speeds obtained with the different dynamical cores (which would give similar results in Earth's case) are widely different. Figure from *Lebonnois et al.* (2013).

Similar results were obtained during the International Space Science Institute (ISSI) intercomparison study (*Lebonnois et al.,* 2013), and also in recent simulations by the LMD Venus GCM (*Lebonnois et al.,* 2010b). More generally, such results motivate the organization of intercomparison between different GCMs, as has been done on Earth within the frame of the successful CMIP program (section 6 in Dowling, this volume).

4. TOWARD A UNIVERSAL CLIMATE MODEL FOR TERRESTRIAL PLANETS

4.1 One Model to Simulate Them All

Our experience in the solar system has shown that the different model components that make a climate model can be applied without major changes to most terrestrial planets. It has also revealed potential weaknesses and inaccuracies of GCMs. On this basis, can we envisage the development of "universal" GCMs able to simulate the various possible climate on planets? Many teams have used GCMs to simulate and explore the circulation regime as a function of the planet's characteristics using models with simplified physics (see Showman et al., this volume; *Read,* 2011; *Heng et al.,* 2011; *Edson et al.,* 2011), including giant planets [notably using the explicit planetary isentropic coordinate EPIC model; *Dowling et al.* (1998, 2006)].

The next step is to design a climate model able to "fully" simulate any type of terrestrial climate, i.e., with any atmospheric cocktail of gases, clouds, and aerosols; for any planetary size; and around any star. At LMD, we have recently developed such a tool, by combining the necessary parameterizations listed in this chapter: (1) the dynamical core; (2) the heating and cooling of the atmosphere and surface by solar and thermal radiation (i.e., the radiative transfer); (3) the storage and diffusion of heat in the subsurface; (4) the mixing of subgrid-scale turbulence and convection; and (5) the formation, transport, and radiative effects of any clouds and aerosols that may be present. As mentioned above, (1), (3), and (4) are almost universal processes. The radiative transfer equations are also universals, but to simulate the three-dimensional climates on a new planet, one challenge has been to develop a radiative transfer code fast enough for three-dimensional simulations and versatile enough to accurately model any atmospheric cocktail or thick atmosphere. For this purpose, we used the the correlated-k distribution technique. We also included a dynamical representation of heat transport and sea-ice formation on a potential ocean from *Codron* (2012).

Why develop such a model? The main purpose is to be able to explore a wide range of possible climates to better understand what may happen on different planets for which we have no direct observations, but are nevertheless the subject of major scientific examination. A first example is the nature of the climate that may have existed in the distant past (several billions years ago) on Earth (*Charnay et al.,* 2012), Mars (*Forget et al.,* 2013; *Wordsworth,* 2012), or Venus (*Leconte et al.,* 2013). Another application is the modeling of extrasolar planets, for example, to prepare for future observations by simulating their spectral signature (*Selsis et al.,* 2011), or to assess planetary habitability (i.e., the range of conditions that allow liquid water, and therefore life).

4.2 Modeling Habitability

Before 2011, nearly all studies of habitability have been performed with simple one-dimensional steady-state radiative convective models that simulate the global mean conditions. Exceptions to this rule have either been parameterized energy-balance models (EBMs) that study the change in surface temperature with latitude only (*Williams and Kasting,* 1997; *Spiegel et al.,* 2008), or three-dimensional simulations with Earth climate models (*Joshi,* 2003). In many cases, one-dimensional models may not be sufficient to estimate the habitability of a planet, and GCMs should be used. First, GCMs allow simulation of local habitability conditions due to the diurnal and seasonal cycles, for example, which lets us investigate the meaning of the habitable zone more precisely than is possible in a globally averaged simulation. They also help us to better understand the distribution and impact of clouds, which are of central importance to both the inner and outer edges of the habitable zone, as discussed earlier. Finally, three-dimensional models allow predictions of the poleward and/or nightside transport of energy by the atmosphere and, in principle, the oceans. This is necessary to assess if the planetary water or a CO_2 atmosphere will collapse on the nightside of a tidally locked planet, or at the poles of a planet with low obliquity.

The model developed at LMD is now applied to better understand the limit of the habitable zone. For instance, *Wordsworth et al.* (2011) applied it to explore the habitability of planet Gliese 581d, discovered in 2007 (*Udry et al.,* 2007). Gliese 581d receives 35% less stellar energy than Mars and is probably locked in tidal resonance, with extremely low insolation at the poles and possibly a permanent nightside. Under such conditions, it was unknown whether any habitable climate on the planet would be able to withstand global glaciation and/or atmospheric collapse, and one-dimensional models were not conclusive. *Wordsworth et al.* (2011) performed three-dimensional climate simulations that demonstrated Gliese 581d would have a stable atmosphere and surface liquid water for a wide range of plausible cases, making it the first confirmed super-Earth (an exoplanet of 2–10 Earth masses) in the habitable zone. Taking into account the formation of CO_2 and water ice clouds, they found that atmospheres with over 10 bar CO_2 and varying amounts of background gas (e.g., N_2) yield global mean temperatures above 0°C for both land- and ocean-covered surfaces (Fig. 4).

Similarly, *Leconte et al.* (2013) have applied the three-dimensional LMD GCM to explore the possible climate on warm tidally locked planets such as Gliese 581c and HD 85512b. With the same stellar flux, a planet like Earth

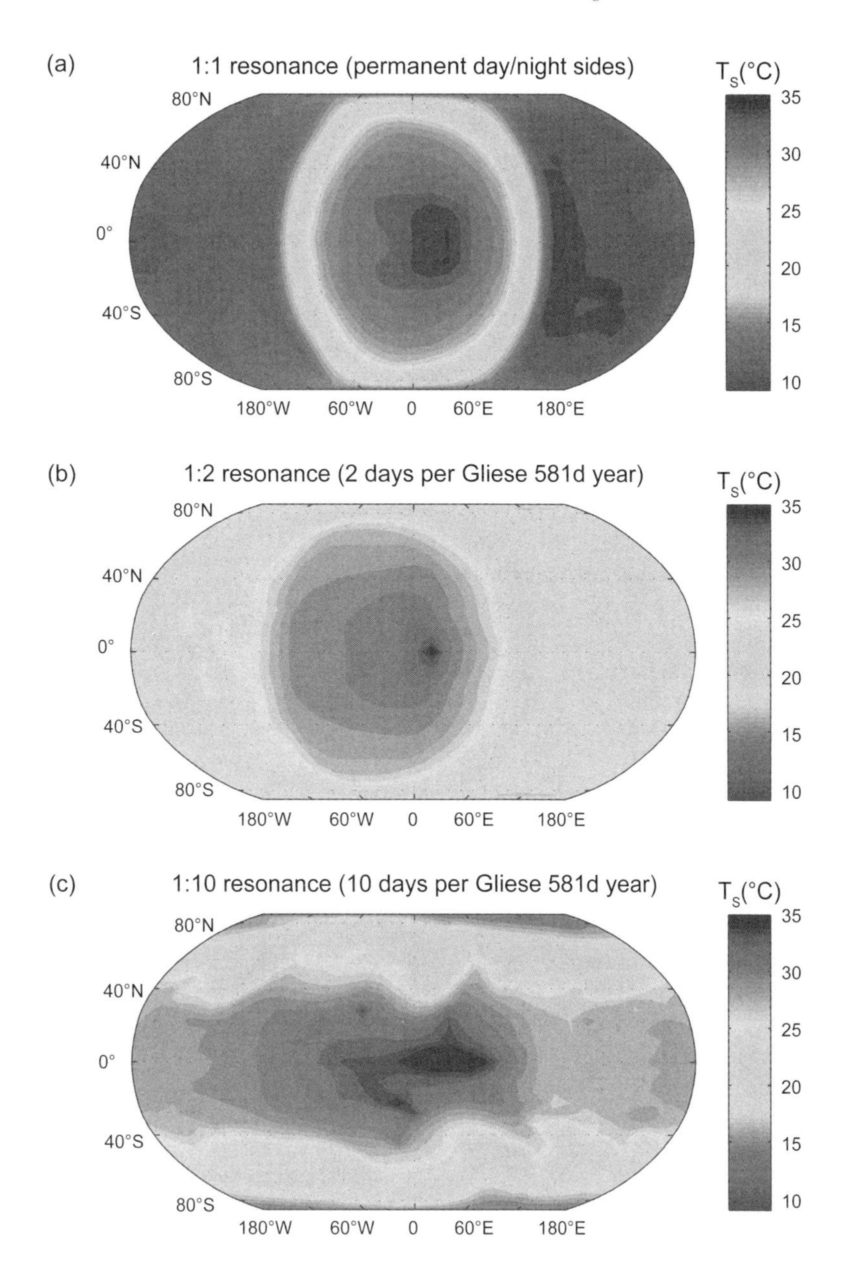

Fig. 4. See Plate 39 for color version. Surface temperature snapshots from three-dimensional global climate model simulations for the extrasolar planet Gliese 581d, assuming a 20-bar CO_2 atmosphere and for three possible rotation rates. Such three-dimensional simulations can help better understand the habitability of exoplanets, in spite of the lack of observations. Figure from *Wordsworth et al.* (2011).

would not be habitable because of the runaway greenhouse instability induced by the positive feedback of water vapor on surface temperature, which leads to the complete evaporation of all liquid water surface reservoirs. On a tidally locked planet, they found that two stable climate regimes could exist. One is the classical runaway state, and the other is a collapsed state where water is captured in permanent cold traps (i.e., on the nightside). If a thick ice cap can accumulate there, gravity-driven ice flows and geothermal flux should come into play to produce long-lived liquid water at the edge and/or bottom of the ice cap, well inside the inner edge of the classical habitable zone.

5. CONCLUSIONS

Realistic GCMs, able to simulate the details of planetary environments on the basis of universal equations, can now be developed by combining a limited number of components such as a dynamical core, a radiative transfer solver, a parameterization of turbulence and convection, a thermal ground model, and a volatile phase change code, possibly completed by a few specific schemes. Our experience in the solar system acquired by simulating the available terrestrial atmosphere (Venus, Earth, Mars, Titan, Triton, Pluto) has, in many cases, confirmed the robustness of the GCM approach originally developed for Earth, notably to study climate change. It can be shown that many of the GCM components are "universal," so that we can envision building realistic climate models for any kind of planet and atmosphere we can imagine. Such a tool is useful for conducting scientific investigations on the possible climates on terrestrial extrasolar planets, or for studying past environments in the solar system.

Nevertheless, the history of planetary GCM modeling has also shown that even assuming the most realistic model input parameters, the simulated climate can be wrong, sometimes simply because a key process has been forgotten in the model or is not yet correctly parameterized, but also sometimes because the climate regime seems to be the result of a subtle balance between processes that remain highly model sensitive or that are the subject of positive feedback. In any case, building virtual planets with GCMs, in light of the observations obtained by spacecraft or from Earth, is a true scientific endeavor that can teach us a great deal about the complex nature of climate systems.

REFERENCES

Abrams M. C., Manney G. L., Gunson M. R., Abbas M. M., Chang A. Y., Goldman A., Irion F. W., Michelsen H. A., Newchurch M. J., Rinsland C. P., Salawitch R. J., Stiller G. P., and Zander R. (1996) Trace gas transport in the Arctic vortex inferred from ATMOS ATLAS-2 observations during April 1993. *Geophys. Res. Lett., 23,* 2345–2348.

Angelats i Coll M., Forget F., López-Valverde M. A., Read P. L., and Lewis S. R. (2004) Upper atmosphere of Mars up to 120 km: Mars Global Surveyor accelerometer data analysis with the LMD general circulation model. *J. Geophys. Res., 109(E1),* E01011.

Basu S., Richardson M. I., and Wilson R. J. (2004) Simulation of the martian dust cycle with the GFDL Mars GCM. *J. Geophys. Res.–Planets, 109(E18),* 11006.

Basu S., Wilson J., Richardson M., and Ingersoll A. (2006) Simulation of spontaneous and variable global dust storms with the GFDL Mars GCM. *J. Geophys. Res.–Planets, 111(E10),* 9004.

Bougher S.W., Bell J. M., Murphy J. R., Lopez-Valverde M. A., and Withers P. G. (2006) Polar warming in the Mars thermosphere: Seasonal variations owing to changing insolation and dust distributions. *Geophys. Res. Lett., 33,* 2203.

Charnay B. and Lebonnois S. (2012) Two boundary layers in Titan's lower troposphere inferred from a climate model. *Nature Geosci., 5,* 106–109.

Charnay B., Forget F., Wordsworth R., Leconte J., Millour E., and Codron F. (2012) Exploring the possible climates of the Archean Earth with a 3D GCM. Abstract P11G-06 presented at 2012 Fall Meeting, AGU, San Francisco, California.

Codron F. (2012) Ekman heat transport for slab oceans. *Climate Dynam.,* 38, 379–389.

Colaitis A., Spiga A., Hourdin F., Rio C., Forget F., and Millour E. (2013) A thermal plume model for the martian convective boundary layer. *J. Geophys. Res.–Planets,* in press.

Collins M., Lewis S. R., Read P. L., and Hourdin F. (1996) Baroclinic wave transitions in the martian atmosphere. *Icarus, 120,* 344–357.

Coustenis A. and Bézard B. (1995) Titan's atmosphere from Voyager infrared observations. IV. Latitudinal variations of temperature and composition. *Icarus, 115,* 126–140.

Crespin A., Lebonnois S., Vinatier S., Bézard B., Coustenis A., Teanby N. A., Achterberg R. K., Rannou P., and Hourdin F. (2008) Diagnostics of Titan's stratospheric dynamics using Cassini/CIRS data and the IPSL General Circulation Model. *Icarus, 197,* 556–571.

Dowling T. E., Fischer A. S., Gierasch P. J., Harrington J., Lebeau R. P., and Santori C. M. (1998) The Explicit Planetary Isentropic-Coordinate (EPIC) atmospheric model. *Icarus, 132,* 221–238.

Dowling T. E., Bradley M. E., Colón E., Kramer J., Lebeau R. P., Lee G. C. H., Mattox T. I., Morales-Juberías R., Palotai C. J., Parimi V. K., and Showman A. P. (2006) The EPIC atmospheric model with an isentropic/terrain-following hybrid vertical coordinate. *Icarus, 182,* 259–273.

Dufresne J.-L., Fournier R., Hourdin C., and Hourdin F. (2005) Net exchange reformulation of radiative transfer in the CO_2 15-μm band on Mars. *J. Atmos. Sci., 62,* 3303–3319.

Dufresne J.-L. and Bony S. (2008) An assessment of the primary sources of spread of global warming estimates from coupled atmosphere ocean models. *J. Climate, 21,* 5135.

Edson A., Lee S., Bannon P., Kasting J. F., and Pollard D. (2011) Atmospheric circulations of terrestrial planets orbiting low-mass stars. *Icarus, 212,* 1–13.

Emanuel K. A. and Ivkovi-Rothman M. (1999) Development and evaluation of a convection scheme for use in climate models. *J. Atmos. Sci., 56,* 1766–1782.

Eymet V., Fournier R., Dufresne J.-L., Lebonnois S., Hourdin F., and Bullock M. A. (2009) Net-exchange parameterization of the thermal infrared radiative transfer in Venus' atmosphere. *J. Geophys. Res., 114,* E11008.

Forget F. and Bertaux J. (2004) *The Martian Atmosphere at the Time of the Beagle 2 Landing: The Density was Not Anomalously Low.* ESA Technical Report 09-2004.

Forget F., Hourdin F., and Talagrand O. (1998) CO_2 snow fall on Mars: Simulation with a general circulation model. *Icarus, 131,* 302–316.

Forget F., Hourdin F., Fournier R., Hourdin C., Talagrand O., Collins M., Lewis S. R., Read P. L., and Huot J.-P. (1999) Improved general circulation models of the martian atmosphere from the surface to above 80 km. *J. Geophys. Res., 104,* 24155–24176.

Forget F., Haberle R. M., Montmessin F., Levrard B., and Head J. W. (2006) Formation of glaciers on Mars by atmospheric precipitation at high obliquity. *Science, 311,* 368–371.

Forget F., Wordsworth R., Millour E., Madeleine J.-B., Kerber L., Leconte J., Marcq E., and Haberle R. M. (2013) 3D modelling of the early martian climate under a denser CO_2 atmosphere: Temperatures and CO_2 ice clouds. *Icarus, 222,* 81–99.

Friedson A. J.,West R. A.,Wilson E. H., Oyafuso F., and Orton G. S. (2009) A global climate model of Titan's atmosphere and surface. *Planet. Space Sci., 57,* 1931–1949.

González-Galindo F., Forget F., López-Valverde M. A., Angelats i Coll M., and Millour E. (2009) A ground-to-exosphere martian general circulation model. 1. Seasonal, diurnal and solar cycle variation of thermospheric temperatures. *J. Geophys. Res.–Planets, 114(E13),* 4001.

González-Galindo F., Bougher S. W., López-Valverde M. A., Forget F., and Murphy J. (2010) Thermal and wind structure of the martian thermosphere as given by two general circulation models. *Planet. Space Sci., 58,* 1832–1849.

Griffith C. A., Penteado P., Baines K., Drossart P., Barnes J., Bellucci G., Bibring J., Brown R., Buratti B., Capaccioni F., Cerroni P., Clark R., Combes M., Coradini A., Cruikshank D., Formisano V., Jaumann R., Langevin Y., Matson D., McCord T., Mennella V., Nelson R., Nicholson P., Sicardy B., Sotin C., Soderblom L. A., and Kursinski R. (2005) The evolution of Titan's mid-latitude clouds. *Science, 310,* 474–477.

Haberle R. M., Pollack J. B., Barnes J. R., Zurek R. W., Leovy C. B., Murphy J. R., Lee H., and Schaeffer J. (1993) Mars atmospheric dynamics as simulated by the NASA/Ames general circulation model, 1, the zonal-mean circulation. *J. Geophys. Res., 98(E2),* 3093–3124.

Haberle R. M., Forget F., Colaprete A., Schaeffer J., Boynton W. V., Kelly N. J., and Chamberlain M. A. (2008) The effect of ground ice on the martian seasonal CO_2 cycle. *Planet. Space Sci., 56,* 251–255.

Hansen C. J. and Paige D. A. (1992) A thermal model for the seasonal nitrogen cycle on Triton. *Icarus, 99,* 273–288.

Hansen C. J. and Paige D. A. (1996) Seasonal nitrogen cycles on Pluto. *Icarus, 120,* 247–265.

Hartmann W. K. (2012) Science of global climate modeling: Confirmation from discoveries on Mars. In *AAS/Division for Planetary Sciences Meeting Abstracts, 44,* 206.12.

Heng K., Frierson D. M. W., and Phillipps P. J. (2011) Atmospheric circulation of tidally locked exoplanets: II. Dual-band radiative transfer and convective adjustment. *Mon. Not. R. Astron. Soc., 418,* 2669–2696.

Herrnstein A. and Dowling T. E. (2007) Effect of topography on the spin-up of a Venus atmospheric model. *J. Geophys. Res., 112,* E04S08.

Hollingsworth J. L., Young R. E., Schubert G., Covey C., and Grossman A. S. (2007) A simple-physics global circulation model for Venus: Sensitivity assessments of atmospheric superrotation. *Geophys. Res. Lett., 34,* L05202.

Holton J. R. (2004) *An Introduction to Dynamic Meteorology, 4th edition.* International Geophysics Series, Elsevier/Academic, Amsterdam.

Hourdin F. (1992) A new representation of the CO_2 15 μm band for a martian general circulation model. *J. Geophys. Res., 97(E11),* 18319–18335.

Hourdin F., Le Van P., Forget F., and Talagrand O. (1993) Meteorological variability and the annual surface pressure cycle on Mars. *J. Atmos. Sci., 50,* 3625–3640.

Hourdin F., Forget F., and Talagrand O. (1995a) The sensitivity of the martian surface pressure to various parameters: A comparison between numerical simulations and Viking observations. *J. Geophys. Res., 100,* 5501–5523.

Hourdin F., Talagrand O., Sadourny R., Régis C., Gautier D., and McKay C. P. (1995b) General circulation of the atmosphere of Titan. *Icarus, 117,* 358–374.

Hourdin F., Lebonnois S., Luz D., and Rannou P. (2004) Titan's stratospheric composition driven by condensation and dynamics. *J. Geophys. Res.–Planets, 109,* 12005.

Ikeda K., Yamamoto M., and Takahashi M. (2007) Superrotation of the Venus atmosphere simulated by an atmospheric general circulation model. In *IUGG/IAMAS Meeting,* July 2–13, Perugia, Italy.

Joshi M. (2003) Climate model studies of synchronously rotating planets. *Astrobiology, 3,* 415–427.

Kahre M. A., Murphy J. R., and Haberle R. M. (2006) Modeling the martian dust cycle and surface dust reservoirs with the NASA Ames general circulation model. *J. Geophys. Res.–Planets, 111(E10),* 6008.

Kahre M. A., Hollingsworth J. L., and Haberle R. M. (2012) Simulating Mars' dust cycle with a Mars general circulation model: Effects of water ice cloud formation on dust lifting strength and seasonality. In *Comparative Climatology of Terrestrial Planets,* Abstract #8062. LPI Contribution 1675, Lunar and Planetary Institute, Houston.

Kalnay de Rivas, E. (1975) Further numerical calculations of the circulation of the atmosphere of Venus. *J. Atmos. Sci., 32,* 1017–1024.

Kido A. and Wakata Y. (2008) Multiple equilibrium states appear in a Venus-like atmospheric general circulation model. *J. Meteorol. Soc. Japan, 86,* 969–979.

Lebonnois S., Toublanc D., Hourdin F., and Rannou P. (2001) Seasonal variations in Titan's atmospheric composition. *Icarus, 152,* 384–406.

Lebonnois S., Hourdin F., Rannou P., Luz D., and Toublanc D. (2003) Impact of the seasonal variations of ethane and acetylene distributions on the temperature field of Titan's stratosphere. *Icarus, 163,* 164–174.

Lebonnois S., Quémerais E., Montmessin F., Lefèvre F., Perrier S., Bertaux J.-L., and Forget F. (2006) Vertical distribution of ozone on Mars as measured by SPICAM/Mars Express using stellar occultations. *J. Geophys. Res.–Planets, 111,* 9.

Lebonnois S., Hourdin F., and Rannou P. (2009) The coupling of winds, aerosols and photochemistry in Titan's atmosphere. *Philos. Trans. R. Soc., A367,* 665–682.

Lebonnois S., Hourdin F., Eymet V., Crespin A., Fournier R., and Forget F. (2010a) Superrotation of Venus' atmosphere analysed with a full general circulation model. *J. Geophys. Res., 115,* E06006.

Lebonnois S., Hourdin F., Forget F., Eymet V., and Fournier R. (2010b) The LMD Venus general circulation model: Improvements and questions. In *Venus Express Science Workshop,* June 20–26, Aussois, France.

Lebonnois S., Burgalat J., Rannou P., and Charnay B. (2012) Titan global climate model: New 3-dimensional version of the IPSL Titan GCM. *Icarus, 218,* 707–722.

Lebonnois S., Lee C., Yamamoto M., Dawson J., Lewis S. R., Mendonca J., Read P. L., Parish H., Schubert G., Bengtsson L., Grinspoon D., Limaye S., Schmidt H., Svedhem H., and Titov D. (2013) Models of Venus atmosphere. In *Towards Understanding the Climate of Venus: Application of Terrestrial Models to Our Sister Planet* (L. Bengtsson et al., eds.), pp. 129–156. ISSI Scientific Report Series 11, Springer, Netherlands.

Leconte J., Forget F., Charnay B., Wordsworth R., Selsis F., and Millour E. (2013) 3D climate modeling of close-in land planets: Circulation patterns, climate moist bistability and habitability. *Astron. Astrophys., 554,* A69.

Lee C. and Richardson M. I. (2010) A general circulation model ensemble study of the atmospheric circulation of Venus. *J. Geophys. Res., 115,* E04002.

Lee C. and Richardson M. I. (2011) Realistic solar and infrared radiative forcing within a Venus GCM. In *AGU Fall Meeting*

Abstracts, A1642.

Lee C., Lewis S. R., and Read P. L. (2005) A numerical model of the atmosphere of Venus. *Adv. Space Res., 36,* 2142–2145.

Lee C., Lewis S. R., and Read P. L. (2007) Superrotation in a Venus general circulation model. *J. Geophys. Res., 112,* E04S11.

Lee C., Lewis S. R., and Read P. L. (2010) A bulk cloud parameterization in a Venus general circulation model. *Icarus, 206,* 662–668.

Lefèvre F., Lebonnois S., Montmessin F., and Forget F. (2004) Three-dimensional modeling of ozone on Mars. *J. Geophys. Res.–Planets, 109,* E07004.

Lefèvre F., Bertaux J.-L., Clancy R. T., Encrenaz T., Fast K., Forget F., Lebonnois S., Montmessin F., and Perrier S. (2008) Heterogeneous chemistry in the atmosphere of Mars. *Nature, 454,* 971–975.

Lewis S. R., Collins M., Read P. L., Forget F., Hourdin F., Fournier R., Hourdin C., Talagrand O., and Huot J.-P. (1999) A climate database for Mars. *J. Geophys. Res., 104,* 24177–24194.

Lewis S. R., Dawson J., Lebonnois S., and Yamamoto M. (2013) Modeling efforts. In *Towards Understanding the Climate of Venus: Application of Terrestrial Models to Our Sister Planet* (L. Bengtsson et al., eds.), pp. 111–127. ISSI Scientific Report Series 11, Springer, Netherlands.

Lopez-Puertas M. and Taylor F. W. (2001) *Non-LTE Radiative Transfer in the Atmosphere.* World Scientific, London.

Madeleine J.-B., Forget F., Head J. W., Levrard B., Montmessin F., and Millour E. (2009) Amazonian northern mid-latitude glaciation on Mars: A proposed climate scenario. *Icarus, 203,* 390–405.

Madeleine J.-B., Forget F., Millour E., Navarro T., and Spiga A. (2012) The influence of radiatively active water ice clouds on the martian climate. *Geophys. Res. Lett., 39,* 23202.

Marcq E., Bézard B., Drossart P., Piccioni G., Reess J. M., and Henry F. (2008) A latitudinal survey of CO, OCS, H_2O, and SO_2 in the lower atmosphere of Venus: Spectroscopic studies using VIRTIS-H. *J. Geophys. Res., 113,* E00B07.

Mellor G. L. and Yamada T. (1982) Development of a turbulence closure model for geophysical fluid problems. *Rev. Geophys., 20(4),* 851–875.

Mischna M. A., Lee C., and Richardson M. (2012) Development of a fast, accurate radiative transfer model for the martian atmosphere, past and present. *J. Geophys. Res.–Planets, 117,* 10009.

Montmessin F., Forget F., Rannou P., Cabane M., and Haberle R. M. (2004) Origin and role of water ice clouds in the martian water cycle as inferred from a general circulation model. *J. Geophys. Res.–Planets, 109(E18),* 10004.

Moudden Y. and McConnell J. C. (2007) Three-dimensional online chemical modeling in a Mars general circulation model. *Icarus, 188,* 18–34.

Mulholland D. P., Read P. L., and Lewis S. R. (2013) Simulating the interannual variability of major dust storms on Mars using variable lifting thresholds. *Icarus, 223,* 344–358.

Murphy J., Pollack J. B., Haberle R. M., Leovy C. B., Toon O. B., and Schaeffer J. (1995) Three-dimensional numerical simulation of martian global dust storms. *J. Geophys. Res., 100,* 26357–26376.

Newman C. E., Lewis S. R., Read P. L., and Forget F. (2002) Modeling the martian dust cycle, 1. Representations of dust transport processes. *J. Geophys. Res.–Planets, 107(E12),* 6-1 to 6-18.

Newman C. E., Lee C., Lian Y., Richardson M. I., and Toigo A. D. (2011) Stratospheric superrotation in the Titan WRF model. *Icarus, 213,* 636–654.

Perrier S., Bertaux J. L., Lefèvre F., Lebonnois S., Korablev O., Fedorova A., and Montmessin F. (2006) Global distribution of total ozone on Mars from SPICAM/MEX UV measurements. *J. Geophys. Res.–Planets, 111(E10),* 9.

Petrosyan A., Galperin B., Larsen S. E., Lewis S. R., Määttänen A., Read P. L., Renno N., Rogberg L. P. H. T., Savijärvi H., Siili T., Spiga A., Toigo A., and Vázquez L. (2011) The martian atmospheric boundary layer. *Rev. Geophys., 49,* 3005.

Pollack J. B., Haberle R. M., Schaeffer J., and Lee H. (1990) Simulations of the general circulation of the martian atmosphere, 1, Polar processes. *J. Geophys. Res., 95,* 1447–1473.

Porco C. C., Baker E., Barbara J., Beurle K., Brahic A., Burns J. A., Charnoz S., Cooper N., Dawson D. D., Del Genio A. D., Denk T., Dones L., Dyudina U., Evans M. W., Giese B., Grazier K., Helfenstein P., Ingersoll A. P., Jacobson R. A., Johnson T. V., McEwen A., Murray C. D., Neukum G., Owen W. M., Perry J., Roatsch T., Spitale J., Squyres S., Thomas P., Tiscareno M., Turtle E., Vasavada A. R., Veverka J., Wagner R., and West R. (2005) Cassini imaging science: Initial results on Saturn's atmosphere. *Science, 307,* 1243–1247.

Rafkin S. C. R. and Michaels T. I. (2003) Meteorological predictions for 2003 Mars Exploration Rover high-priority landing sites. *J. Geophys. Res.–Planets, 108(E12),* 8091.

Rannou P., Hourdin F., and McKay C. P. (2002) A wind origin for Titan's haze structure. *Nature, 418,* 853–856.

Rannou P., Hourdin F., McKay C. P., and Luz D. (2004) A coupled dynamics-microphysics model of Titan's atmosphere. *Icarus, 170,* 443–462.

Rannou P., Montmessin F., Hourdin F., and Lebonnois S. (2006) The latitudinal distribution of clouds on Titan. *Science, 311,* 201–205.

Read P. L. (2011) Dynamics and circulation regimes of terrestrial planets. *Planet. Space Sci., 59,* 900–914.

Read P. L., Montabone L., Mulholland D. P., Lewis S. R., Cantor B., and Wilson R. J. (2011) Midwinter suppression of baroclinic storm activity on Mars: Observations and models. In *Mars Atmosphere: Modelling and Observation* (F. Forget and E. Millour, eds.), pp. 133–135. Available online at *www-mars.lmd.jussieu.fr/paris2011/abstracts/read_paris2011.pdf.*

Richardson M. I. and Wilson R. J. (2002) Investigation of the nature and stability of the martian seasonal water cycle with a general circulation model. *J. Geophys. Res.–Planets, 107(E5),* 7-1.

Richardson M. I., Toigo A. D., and Newman C. E. (2007) Planet WRF: A general purpose, local to global numerical model for planetary atmospheric and climate dynamics. *J. Geophys. Res., 112,* E09001.

Roe H. G., de Pater I., Macintosh B. A., and McKay C. P. (2002) Titan's clouds from Gemini and Keck adaptive optics imaging. *Astrophys. J., 581,* 1399–1406.

Selsis F., Wordsworth R. D., and Forget F. (2011) Thermal phase curves of nontransiting terrestrial exoplanets. I. Characterizing atmospheres. *Astron. Astrophys., 532,* A1.

Spiegel D. S., Menou K., and Scharf C. A. (2008) Habitable climates. *Astrophys. J., 681,* 1609–1623.

Spiga A. and Forget F. (2009) A new model to simulate the martian mesoscale and microscale atmospheric circulation: Validation and first results. *J. Geophys. Res.–Planets, 114,* E02009.

Sprague A. L., Boynton W. V., Kerry K. E., Janes D. M., Hunten D. M., Kim K. J., Reedy R. C., and Metzger A. E. (2004) Mars' south polar Ar enhancement: A tracer for south polar seasonal meridional mixing. *Science, 306,* 1364–1367.

Staniforth A. and Thuburn J. (2012) Horizontal grids for global weather and climate prediction models: A review. *Q. J. R. Meteorol. Soc., 138,* 1–26.

Takagi M. and Matsuda Y. (2007) Effects of thermal tides on the Venus atmospheric superrotation. *J. Geophys. Res., 112,* D09112.

Thuburn J. (2008) Some conservation issues for the dynamical cores of NWP and climate models. *J. Comput. Phys., 227,* 3715–3730.

Tokano T., Neubauer F. M., Laube M., and McKay C. P. (1999) Seasonal variation of Titan's atmospheric structure simulated by a general circulation model. *Planet. Space Sci., 47,* 493–520.

Toon O. B., McKay C. P., Ackerman T. P., and Santhanam K. (1989) Rapid calculation of radiative heating rates and photodissociation rates in inhomogeneous multiple scattering atmospheres. *J. Geophys. Res., 94,* 16287–16301.

Tsang C., Irwin P., Wilson C. F., Taylor F. W., Lee C., de Kok E., Drossart P., Piccioni G., Bézard B., and Calcutt S. (2008) Tropospheric carbon monoxide concentrations and variability on Venus from Venus Express/VIRTIS-M observations. *J. Geophys. Res., 113,* E00B08.

Udry S., Bonfils X., Delfosse X., Forveille T., Mayor M., Perrier C., Bouchy F., Lovis C., Pepe F., Queloz D., and Bertaux J. (2007) The HARPS search for southern extrasolar planets XI. Super-Earths (5 and 8 Earth mass) in a 3-planet system. *Astron. Astrophys., 469,* L43–L47, DOI: 10.1051/0004-6361:20077612.

Vangvichith M. and Forget F. (2012) A Pluto GCM including the CH_4 cycle. In *Pluto/Charon Atmospheres Workshop Abstracts.*

Vangvichith M., Forget F., Wordsworth R., and Millour E. (2010) A 3-D general circulation model of Triton's atmosphere. *Bull. Am. Astron. Soc., 42,* 952.

Williams D. M. and Kasting J. F. (1997) Habitable planets with high obliquities. *Icarus, 129,* 254–267.

Williamson D. L. (2007) The evolution of dynamical cores for global atmospheric models. *J. Meteorol. Soc. Japan, 85B,* 241–269.

Wilson R. J. (1997) A general circulation model of the martian polar warming. *Geophys. Res. Lett., 24,* 123–126.

Wilson R. J. (2011) Water ice clouds and thermal structure in the martian tropics as revealed by Mars Climate Sounder. In *Mars Atmosphere: Modelling and Observation* (F. Forget and E. Millour, eds.), pp. 219–222. Available online at *www-mars.lmd.jussieu.fr/paris2011/abstracts/read_paris2011.pdf.*

Wilson R. W. and Hamilton K. (1996) Comprehensive model simulation of thermal tides in the martian atmosphere. *J. Atmos. Sci., 53,* 1290–1326.

Wilson R. J. and Richardson M. I. (2000) Infrared measurements of atmospheric temperatures revisited. *Icarus, 145,* 555–579.

Wordsworth R. (2012) Transient conditions for biogenesis on low-mass exoplanets with escaping hydrogen atmospheres. *Icarus, 219,* 267–273.

Wordsworth R. D., Forget F., Selsis F., Millour E., Charnay B., and Madeleine J.-B. (2011) Gliese 581d is the first discovered terrestrial-mass exoplanet in the habitable zone. *Astrophys. J. Lett., 733,* L48.

Yamamoto M. and Takahashi M. (2003) The fully developed superrotation simulated by a general circulation model of a Venus-like atmosphere. *J. Atmos. Sci., 60,* 561–574.

Yamamoto M. and Takahashi M. (2004) Dynamics of Venus' superrotation: The eddy momentum transport processes newly found in a GCM. *Geophys. Res. Lett., 31,* L09701.

Yamamoto M. and Takahashi M. (2006a) An aerosol transport model based on a two-moment microphysical parameterization in the Venus middle atmosphere: Model description and preliminary experiments. *J. Geophys. Res., 111(E08),* E08002.

Yamamoto M. and Takahashi M. (2006b) Superrotation maintained by meridional circulation and waves in a Venus-like AGCM. *J. Atmos. Sci., 63(12),* 3296–3314.

Yelle R. V., Lunine J. I., and Hunten D. M. (1991) Energy balance and plume dynamics in Triton's lower atmosphere. *Icarus, 89,* 347–358.

Young R. E. and Pollack J. B. (1977) A three-dimensional model of dynamical processes in the Venus atmosphere. *J. Atmos. Sci., 34,* 1315–1351.

Zalucha A. M. and Gulbis A. A. S. (2012) Comparison of a simple 2-D Pluto general circulation model with stellar occultation light curves and implications for atmospheric circulation. *J. Geophys. Res.–Planets, 117,* 5002.

Krasnopolsky V. A. and Lefèvre F. (2013) Chemistry of the atmospheres of Mars, Venus, and Titan. In *Comparative Climatology of Terrestrial Planets* (S. J. Mackwell et al., eds.), pp. 231–275. Univ. of Arizona, Tucson, DOI: 10.2458/azu_uapress_9780816530595-ch11.

Chemistry of the Atmospheres of Mars, Venus, and Titan

Vladimir A. Krasnopolsky
Catholic University of America

Franck Lefèvre
Laboratoire Atmosphères et Observations Spatiales (LATMOS), Paris

---

Observations and models for atmospheric chemical compositions of Mars, Venus, and Titan are briefly discussed. While the martian CO_2-H_2O photochemistry is comparatively simple, Mars' obliquity, elliptic orbit, and rather thin atmosphere result in strong seasonal and latitudinal variations that are challenging in both observations and modeling. Venus' atmosphere presents a large range of temperature and pressure conditions. The atmospheric chemistry involves species of seven elements, with sulfur and chlorine chemistries dominating up to 100 km. The atmosphere below 60 km became a subject of chemical kinetic modeling only recently. Photochemical modeling is especially impressive for Titan: Using the N_2 and CH_4 densities at the surface and temperature and eddy diffusion profiles, it is possible to calculate vertical profiles of numerous neutrals and ions throughout the atmosphere. The Cassini-Huygens observations have resulted in significant progress in understanding the chemistry of Titan's atmosphere and ionosphere and provide an excellent basis for their modeling.

1. INTRODUCTION

Energies of the solar UV photons and energetic particles may be sufficient to break chemical bonds in atmospheric species and form new molecules, atoms, radicals, ions, and free electrons. These products initiate chemical reactions that further complicate the atmospheric composition, which is also significantly affected by dynamics and transport processes. Photochemical products may be tracers of photochemistry and dynamics in the atmosphere and change its thermal balance. Gas exchange between the atmosphere, space, and solid planet also determines the properties of the atmosphere.

Studies of the atmospheric chemical composition and its variations are therefore essential for all aspects of atmospheric science, and many instruments on spacecraft missions to the planets are designed for this task. Interpretation of the observations requires photochemical models that could adequately simulate photochemical and transport processes.

One-dimensional steady-state global-mean self-consistent models are a basic type of photochemical modeling. These models simulate altitude variations of species in an atmosphere. Vertical transport in the model is described by eddy diffusion coefficient K(z). The solar UV spectrum, absorption cross sections, reaction rate coefficients, eddy and molecular diffusion coefficients K(z) and D(z), and temperature profile T(z) are the input data for the model. The model provides a set of continuity equations for each species in a spherical atmosphere

$$\frac{1}{r^2}\frac{d\left(r^2\Phi\right)}{dr} = P - nL$$

$$\Phi = -(K + D)\frac{dn}{dr} - n\left[K\left(\frac{1}{H_a} + \frac{1}{T}\frac{dT}{dr}\right) + D\left(\frac{1}{H} + \frac{1+\alpha}{T}\frac{dT}{dr}\right)\right]$$

Here r is radius, Φ and n are the flux and number density of species, P and L are the production and loss of species in chemical reactions, α is the thermal diffusion factor, H_a and H are the mean and species scale heights, and T is the temperature. Substitution of the second equation in the first results in an ordinary second-order nonlinear differential equation. Finite-difference analogs for these equations may be solved using methods described by *Allen et al.* (1981) and *Krasnopolsky and Cruikshank* (1999).

Boundary conditions for the equations may be densities, fluxes, and velocities of the species. Fluxes and velocities are equal to zero at a chemically passive surface and at the exobase for molecules that do not escape. Requirements for the boundary conditions are discussed in *Krasnopolsky* (1995). Generally, the number of nonzero conditions should be equal to the number of chemical elements in the system. This type of photochemical modeling is a powerful tool for studying atmospheric chemical composition. For example, in the case of Titan, a model results in vertical profiles of 83 neutral species and 33 ions up to 1600 km using densities of N_2 and CH_4 near the surface (*Krasnopolsky,* 2009a, 2012c).

Photochemical general circulation models (GCMs; see Forget and Lebonnois, this volume) present significant progress in the study and reproduction of atmospheric properties under a great variety of conditions. These GCMs are the best models for studying variations of species with local time, latitude, season, and location.

Some basic properties of the terrestrial planets and their atmospheres are listed in Table 1. They are discussed in detail in other chapters of this book.

Below we will briefly discuss observational data on and photochemical modeling of the atmospheric chemical compositions on Mars, Venus, and Titan. Earth's photochemistry is strongly affected by life and human activity (*Yung and DeMore,* 1999) and is beyond the scope of this chapter. Previous reviews of the photochemistry of Mars, Venus, and Titan may be found in *Levine* (1985), *Krasnopolsky* (1986), and *Yung and DeMore* (1999).

2. MARS

2.1. Chemical Composition of Mars' Atmosphere and Its Variations

Mars' obliquity of 24° and the elliptic orbit with perihelion at 1.381 AU (L_S = 251°) and aphelion at 1.666 AU (L_S = 71°) induce a great variety of conditions in the comparatively thin atmosphere of Mars. A summary of observational data on the chemical composition of the martian atmosphere is given in Table 2. Groundbased detections of the parent species CO_2 (*Kuiper,* 1949) and H_2O (*Spinrad et al.,* 1963) and the main photochemical products CO (*Kaplan et al.,* 1969) and O_2 (*Barker,* 1972; *Carleton and Traub,* 1972) were a basis for the first photochemical models. Observations of O_3 by the Mariner 9 ultraviolet (UV) spectrometer (*Barth et al.,* 1973) revealed strong variations in Mars' photochemistry (Fig. 1). The ozone UV absorption band at 255 nm has a halfwidth of 40 nm, the solar light varies within the band by a factor of ~15, the surface reflection and dust properties may vary as well, and the Mariner team adopted an upper limit of 3 μm-atm for the UV ozone detection (1 μm-atm = 10^{-4} cm × 2.7 × 10^{19} cm^{-3} = 2.7 × 10^{15} cm^{-2}). The observations at low and middle latitudes beyond the polar caps and hoods were below this limit.

The upper atmosphere was studied by the UV spectrometers at Mariner 6, 7, and 9, as well as mass spectrometers and retarding potential analyzers on the Viking landers (*Barth et al.,* 1972; *Nier and McElroy,* 1977; *Hanson et al.,* 1977). Mass spectrometers on the Viking landers and more recently on the Mars Science Laboratory (MSL) measured the compositions of the lower atmosphere, especially N_2 and noble gases Ar, Ne, Kr, and Xe. Helium was later observed using the Extreme Ultraviolet Explorer (*Krasnopolsky and Gladstone,* 2005). The first detailed study of the H_2O global distribution was made by the Viking orbiters (*Jakosky and Farmer,* 1982).

Noxon et al. (1976) detected the $O_2(^1\Delta_g)$ dayglow at 1.27 μm and properly explained its excitation by photolysis of ozone and quenching by CO_2. *Traub et al.* (1979) repeated their observations at three latitude bands. Using groundbased IR heterodyne spectroscopy at 9.6 μm, *Espenak et al.* (1991) detected ozone at low latitudes with the column abundance of ~1 μm-atm at L_S = 208°. These observations have been continued (*Fast et al.,* 2006, 2009). To improve the ozone detection at large airmass, *Clancy et al.* (1999) observed the UV spectra of Mars near the limb using the Hubble Space Telescope (HST). Some data from *Fast et al.* (2006) and *Clancy et al.* (1999) are shown in Fig. 2.

Clancy and Nair (1996) predicted significant seasonal variations of ozone above 20 km at the low and middle latitudes. *Krasnopolsky* (1997) argued that the O_2 dayglow at 1.27 μm is the best tracer for that ozone, because the dayglow is quenched by CO_2 below ~20 km, and started groundbased mapping observations of this dayglow (Figs. 3, 4). Interpretation of the groundbased observations of ozone and the O_2 dayglow at 1.27 μm does not require any assumptions regarding reflectivities of the surface rocks and atmospheric dust.

TABLE 1. Basic properties of Venus, Earth, Mars, and Titan.

Property	Venus	Earth	Mars	Titan
Mean distance from Sun (AU)	0.72	1.0	1.52	9.54
Radius (km)	6052	6378	3393	2575
Mass (Earth = 1)	0.815	1	0.1075	0.0225
Length of year	224.7 d	365.26 d	1.881 yr	29.42 yr
Gravity (cm s^{-2})	887	978	369	135
Mean atmospheric pressure (bar)	92	1.0	0.0061	1.5
Mean temperature near surface (K)	737	288	210	94
Mean exobase altitude (km)	200	400	200	1350
Mean exospheric temperature (K)	260	1000	230	160
Escape velocity (km s^{-1})	10.3	11.2	5.0	2.64
Main gas	CO_2 (0.96)	N_2 (0.78)	CO_2 (0.96)	N_2 (0.95)
Second gas	N_2 (0.034)	O_2 (0.21)	N_2 (0.019)	CH_4 (0.05)
Third gas	SO_2 (1.3 × 10^{-4})	Ar (0.0093)	Ar (0.019)	H_2 (0.001)

Atmospheric composition is given near the surface; mixing ratios are in parentheses.

Hydrogen (H_2) was predicted by photochemistry as a long-living product on Mars, and it had been detected using the Far Ultraviolet Spectroscopic Explorer (FUSE) (*Krasnopolsky and Feldman,* 2001). Continuous monitoring of temperature profiles, H_2O, water ice, and dust atmospheric abundances for 1998–2006 by MGS/TES (*Smith,* 2004) presents a convenient and detailed database for various applications. The observed variations of H_2O are shown in Fig. 5.

Groundbased observations of CO (*Krasnopolsky,* 2003a) revealed variations of long-living species that are induced by condensation and sublimation of CO_2 at the polar caps. This effect and its interpretation were confirmed by the GRS observations of Ar from Mars Odyssey (*Sprague et al.,* 2004, 2012).

Peroxide H_2O_2 was detected using groundbased microwave and infrared instruments, and a summary of the current data is in Fig. 6. Along with ozone and the O_2 dayglow, H_2O_2 is a sensitive tracer of Mars' photochemistry, especially odd hydrogen chemistry.

Tentative detections of CH_4 were made by groundbased instruments (*Krasnopolsky et al.,* 2004, 2012a; *Mumma et al.,* 2009; *Villanueva et al.,* 2013) and the Planetary Fourier Spectrometer (PFS) onboard Mars Express (MEX) (*Formisano et al.,* 2004; *Geminale et al.,* 2011). Traces of the methane absorption were also found in the Mars Global Surveyor (MGS) Thermal Emission Spectrometer (TES) spectra by *Fonti and Marzo* (2010). Methane is so exciting on Mars because of its possible biogenic origin. However, the latest groundbased observations and measurements by

TABLE 2. Chemical composition of the martian atmosphere.

Species	Mixing Ratio	Comments and References
CO_2	0.96	*Kuiper* (1949); global and annually mean pressure 6.1 mbar (*Kliore et al.,* 1973)
N_2	0.019	*Mahaffy et al.* (2013)
Ar	0.019	*Mahaffy et al.* (2013)
Ne	1–2.5 ppm	*Owen et al.* (1977), *Bogard et al.* (2001)
Kr	300–360 ppb	*Owen et al.* (1977), *Bogard et al.* (2001)
Xe	50–80 ppb	*Owen et al.* (1977), *Bogard et al.* (2001)
H	$(3–30) \times 10^4$ cm^{-3}	at 250 km for $T_\infty \approx 300$ and 200 K; *Anderson and Hord* (1971), *Anderson* (1974), *Chaufray et al.* (2008)
O	0.005–0.02	at 125 km; *Strickland et al.* (1972), *Stewart et al.* (1992), *Chaufray et al.* (2009)
O_2	$(1.2–2) \times 10^{-3}$	*Barker* (1972), *Carleton and Traub* (1972), *Owen et al.* (1977), *Trauger and Lunine* (1983), *Hartogh et al.* (2010a), *Mahaffy et al.* (2013)
CO	10^{-3}	*Kaplan et al.* (1969), *Krasnopolsky* (2007a), *Billebaud et al.* (2009), *Smith et al.* (2009), *Hartogh et al.* (2010b), *Sindoni et al.* (2011)
H_2O	0–70 pr. μm	*Spinrad et al.* (1963), *Smith* (2004), *Fedorova et al.* (2006b, 2009), *Fouchet et al.* (2007), *Melchiorri et al.* (2007), *Tschimmel et al.* (2008), *Smith et al.* (2009), *Sindoni et al.* (2011), *Maltagliati et al.* (2013)
O_3	0–60 μm-atm	*Barth and Hord* (1971), *Barth et al.* (1973), *Clancy et al.* (1999), *Fast et al.* (2006, 2009), *Perrier et al.* (2006), *Lebonnois et al.* (2006)
$O_2(^1\Delta_g)$	0.6–35 μm-atm*	*Noxon et al.* (1976), *Krasnopolsky* (2013c), *Fedorova et al.* (2006a), *Altieri et al.* (2009)
He	10 ppm	*Krasnopolsky et al.* (1994), *Krasnopolsky and Gladstone* (2005)
H_2	17 ppm	*Krasnopolsky and Feldman* (2001)
H_2O_2	0–40 ppb	*Clancy et al.* (2004), *Encrenaz et al.* (2004, 2011a)
CH_4	0–40 ppb	*Krasnopolsky et al.* (2004), *Formisano et al.* (2004), *Mumma et al.* (2009), *Fonti and Marzo* (2010), *Geminale et al.* (2011), *Krasnopolsky* (2012a), *Villanueva et al.* (2013), *Webster et al.* (2013)
C_2H_6	<0.2 ppb	*Krasnopolsky* (2012a), *Villanueva et al.* (2013)
SO_2	<0.3 ppb	*Encrenaz et al.* (2011), *Krasnopolsky* (2012a)
NO	<1.7 ppb	*Krasnopolsky* (2006b)
HCl	<0.2 ppb	*Hartogh et al.* (2010a)
H_2CO	<3 ppb	*Krasnopolsky et al.* (1997), *Villanueva et al.* (2013)
NH_3	<5 ppb	*Maguire* (1977)
HCN	<4 ppb	*Villanueva et al.* (2013)

*$O_2(^1\Delta_g)$ dayglow at 1.27 μm is measured in megarayleighs (MR); 1 MR = 1.67 μm-atm of $O_2(^1\Delta_g)$ for this airglow.

First detections along with the latest data are given.

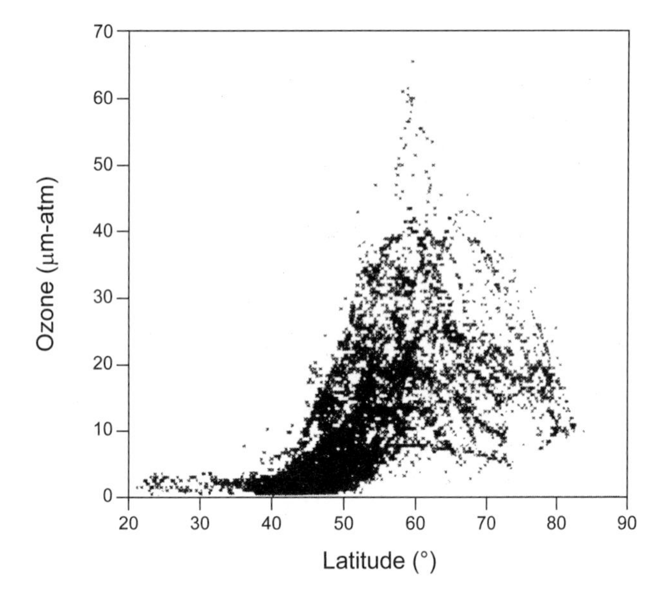

Fig. 1. Mariner 9 observations of ozone at the end of northern winter. From *Traub et al.* (1979).

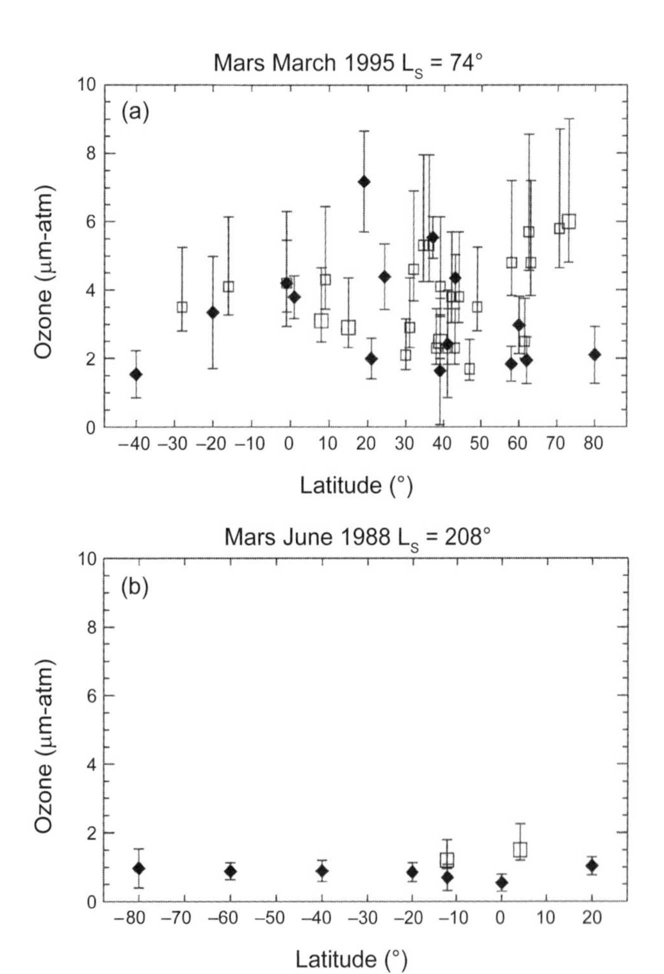

Fig. 2. Groundbased (black symbols) (*Fast et al.,* 2006) and HST (open symbols) (*Clancy et al.,* 1999) observations of ozone near aphelion and perihelion.

MSL's Curiosity rover (*Webster et al.,* 2013) only provide upper limits to methane on Mars.

The latest results on the minor species on Mars have been obtained by the current MEX and Mars Reconnaissance Orbiter (MRO) missions. Ultraviolet spectra at 210–300 nm observed with the MEX ultraviolet and infrared atmospheric spectrometer (SPICAM) for 1.25 martian years were divided by a reference spectrum with no ozone measured at Olympus Mons during perihelion (*Perrier et al.,* 2006). The ratio is then fitted using an ozone column, a linear surface reflection, and a constant dust optical depth between 210 and 300 nm (four parameters). This approach is reasonable, although the retrieved values may have systematic uncertainties ~0.5 μm-atm. The observed seasonal-latitudinal variations of ozone are shown in Fig. 7. SPICAM stellar occultations revealed nighttime vertical profiles of O_3 (*Lebonnois et al.,* 2006) that are highly variable with peak densities up to 10^{10} cm^{-3} (Fig. 8). Variations of the O_2 dayglow at 1.27 μm were observed by the IR channel of SPICAM (Fig. 9). Four MEX and two MRO instruments are measuring H_2O (references to the observations are given in Table 2). Vertical profiles of H_2O, CO_2, and aerosol were measured by solar occultations using SPICAM-IR (*Fedorova et al.,* 2009; *Maltagliati et al.,* 2013). Three instruments — one on MRO and two on MEX — are observing CO (Table 2), and SPICAM is studying the UV dayglow and properties of the upper atmosphere. CO, O_2, and restrictive upper limits to H_2O_2 and HCl were measured by the Herschel Space Observatory (Table 2).

While an attempt to detect Mars' nightglow in the visible range resulted in only upper limits (*Krasnopolsky and Krysko,* 1976), the SPICAM-UV observations revealed emissions of the NO δ- and γ-band system that are excited in the reaction between N and O. Both species

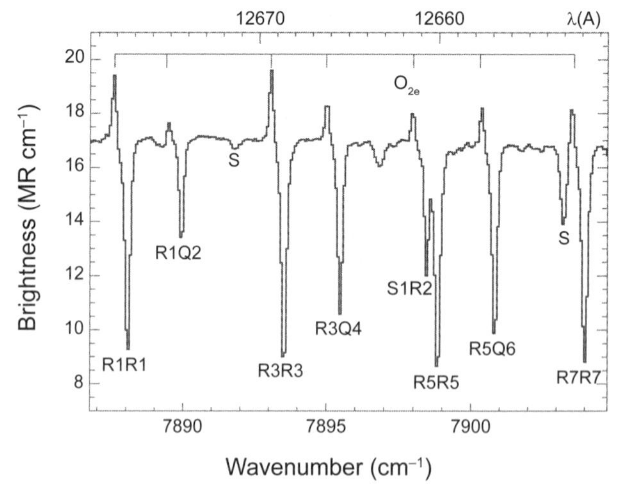

Fig. 3. IRTF/CSHELL spectrum of the O_2 dayglow at 1.27 μm. The dayglow emission lines are Doppler-shifted to the red from the telluric O_2 absorption lines. Solar lines are marked S. From *Krasnopolsky* (2007a).

are transported from the dayside thermosphere; the mean nightglow intensity is 36 R and the peak altitude is ~70 km (*Cox et al.,* 2008). [One Rayleigh (R) is 10^6 photons per (cm^2 s 4π sr).] This nightglow is similar to that observed on Venus (section 3.4.1).

Polar O_2 nightglow at 1.27 μm (Fig. 10) is a new phenomenon observed by the Visible and Infrared Mineralogical Mapping Spectrometer (OMEGA) on MEX (*Bertaux et al.,* 2012), SPICAM-IR (*Fedorova et al.,* 2012), and the Compact Reconnaissance Imaging Spectrometer for Mars (CRISM) on MRO (*Clancy et al.,* 2012b). Mean vertical nightglow intensity is 300 kR, far above that expected for the O_2 nightglow at low latitudes (section 2.3.2). Recently, *Clancy et al.* (2012c) reported their detection of the OH polar nightglow using CRISM.

Thus, the globally and annually mean atmospheric pressure is 6.1 mbar on Mars, and the atmosphere consists of

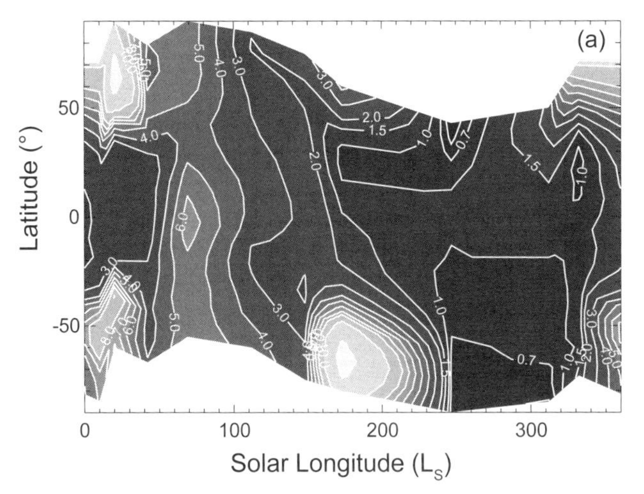

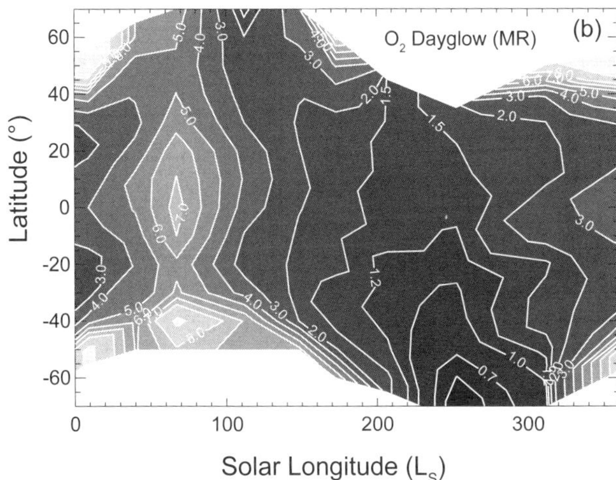

Fig. 4. Seasonal evolution of the $O_2(^1\Delta_g)$ dayglow at 1.27 μm (MR) near local noon. **(a)** Earth-based observations (updated from *Krasnopolsky,* 2007a). **(b)** Calculations by the one-dimensional model of *Krasnopolsky* (2009b). One megarayleigh (MR) is equal to 10^{12} photons cm^{-2} s^{-1} (4π ster)$^{-1}$, i.e., 4.5 × 10^{15} cm^{-2} or 1.7 μm-atm of the $O_2(^1\Delta_g)$ column abundance for the dayglow at 1.27 μm.

CO_2 (96%) and its products [CO, O_2, $O_2(^1\Delta_g)$, O_3, O], H_2O and its observed products (H, H_2, H_2O_2), CH_4, and chemically inactive species N_2, Ar, He, Ne, Kr, and Xe. While the CO_2-H_2O photochemistry is comparatively simple and well studied in the laboratory, it presents some challenging tasks in the martian atmosphere.

2.2. Global-Mean Models

The time required for global-scale mixing in the martian atmosphere is ~0.5 yr (section 2.5). (Hereafter, Earth years are used unless otherwise specified.) Species with lifetimes exceeding this time are well mixed, and their variations are only caused by condensation and sublimation of CO_2 at the polar caps (section 2.1). The long-living photochemical products are O_2, CO, and H_2, and the global-mean one-dimensional models are currently the only means for calculating these species. Short-living photochemical products are variable on Mars and cannot be studied in detail by the global-mean models.

2.2.1. CO_2 stability problem and basic CO_2-H_2O chemistry. The observed low abundances of CO and O_2 of ~10^{-3} were initially puzzling, because the CO_2 dry (without H_2O) chemistry predicts their mole fractions of 0.08 and 0.04 (*Nair et el.,* 1994) in reactions

$$CO_2 + h\nu\ (\lambda < 200\ nm) \rightarrow CO + O$$
$$O + O + M \rightarrow O_2 + M$$
$$O_2 + h\nu \rightarrow O + O$$
$$O + O_2 + M \rightarrow O_3 + M$$
$$O_3 + h\nu \rightarrow O_2 + O$$
$$O + CO + M \rightarrow CO_2 + M$$

Spins of O, CO, and CO_2 are 1, 0, and 0, respectively, and the spins do not conserve in the last reaction; therefore it is very slow, giving rise to CO and O_2. That was a so-called CO_2 stability problem. The problem was solved by *McElroy and Donahue* (1972) and *Parkinson and Hunten* (1972) using a catalytic effect of odd hydrogen (H, OH, HO_2, and H_2O_2 due to its fast photolytic conversion to

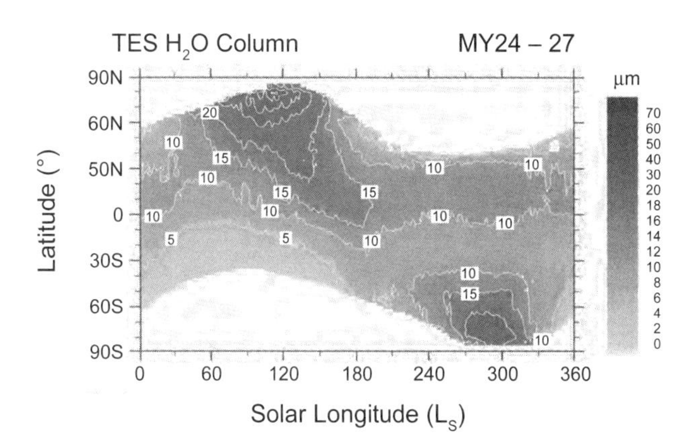

Fig. 5. Seasonal evolution of the H_2O column (per μm) observed by MGS/TES in 1999–2004 (*Smith,* 2004).

OH). Odd hydrogen is formed by photolysis of H_2O and reactions of $O(^1D)$

$$H_2O + h\nu\ (\lambda < 195\ nm) \rightarrow OH + H$$
$$O(^1D) + H_2O \rightarrow OH + OH$$
$$O(^1D) + H_2 \rightarrow OH + H$$

All these reactions and those discussed below, their rate coefficients, and column rates in the mean daytime atmosphere are given in Table 3. $O(^1D)$ is the lowest metastable state of O with energy of 1.97 eV and radiative lifetime of 110 s. It is formed on Mars by photolysis of ozone with minor contributions from photolyses of CO_2 and O_2. $O(^1D)$ is strongly quenched by CO_2.

Odd hydrogen drives the basic photochemical cycle

$$H + O_2 + M \rightarrow HO_2 + M$$
$$HO_2 + O \rightarrow OH + O_2$$
$$CO + OH \rightarrow CO_2 + H$$
$$\text{Net } O + CO \rightarrow CO_2$$

HO_2 is formed by the termolecular reaction; therefore H is the most abundant odd hydrogen species above ~35 km, while HO_2 and H_2O_2 dominate below this altitude. Odd

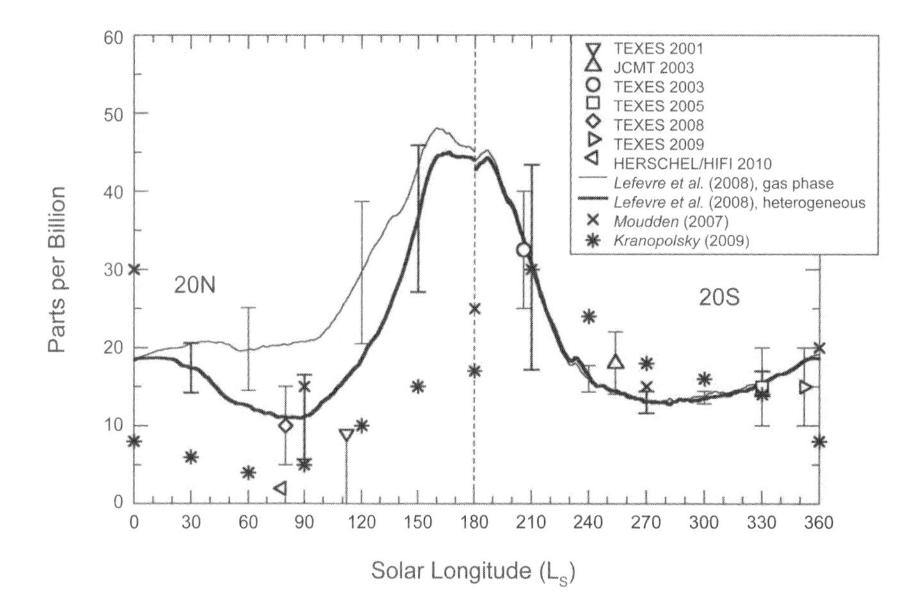

Fig. 6. Observed and calculated H_2O_2 mixing ratios at low latitudes (centered at 20°N before and 20°S after L_S = 180°). Solid and thin lines: updated (reduced water) three-dimensional model by *Lefevre et al.* (2008) with and without heterogeneous chemistry; crosses: three-dimensional gas-phase model by *Moudden* (2007); stars: one-dimensional model by *Krasnopolsky* (2009b). From *Encrenaz et al.* (2012).

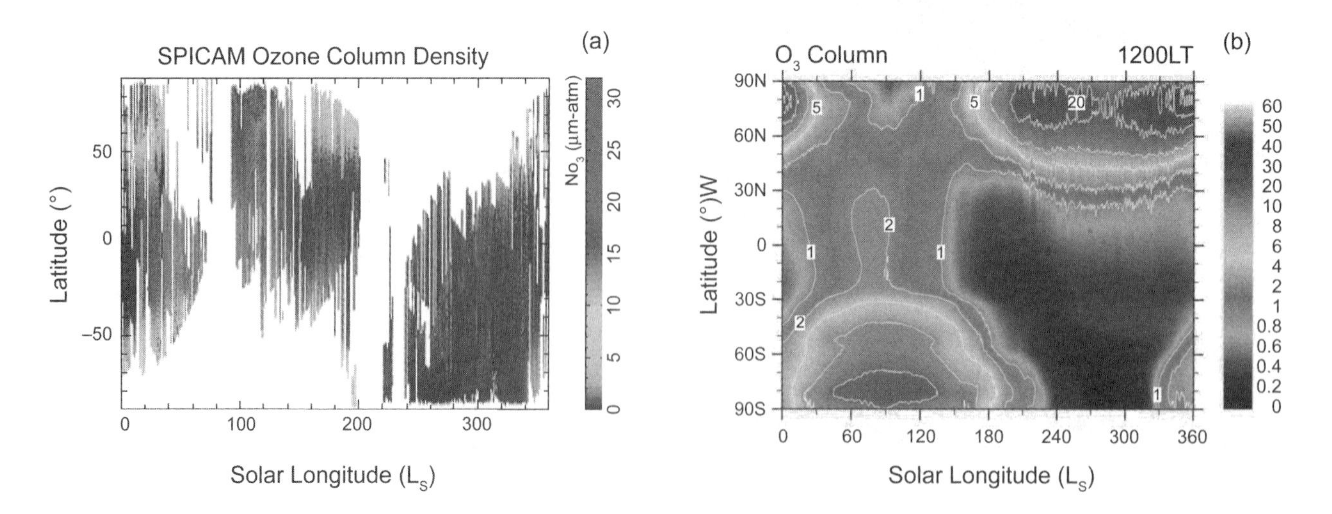

Fig. 7. See Plate 40 for color version. Seasonal evolution of the daytime O_3 column (µm-atm) observed **(a)** by MEX/SPICAM (*Perrier et al.*, 2006) and **(b)** from a three-dimensional model (updated from *Lefevre et al.*, 2004).

hydrogen is not lost in the cycle and therefore acts as a catalyst. OH is also formed by photolysis of H_2O_2 and HO_2 and in reactions of H + O_3 and H + HO_2. Reaction of CO + OH is the major process in formation of the CO=O bonds that balances breaking of these bonds by photolysis of CO_2. Balances of key reactions of formation and breaking bonds of the basic species are convenient for analyses of photochemistry, especially for dense atmospheres like those of Venus and Titan, where the number of cycles may be large.

Another pathway for loss of odd oxygen (O and O_3) is formation of the O=O bonds

$$O + O + M \rightarrow O_2 + M$$
$$O + OH \rightarrow O_2 + H$$

Breaking of the O=O bonds occurs by photolyses of O_2, H_2O_2, HO_2, and NO_2 (section 2.2.3). Odd hydrogen is lost in the reactions

$$OH + HO_2 \rightarrow H_2O + O_2$$
$$H + HO_2 \rightarrow H_2 + O_2$$
$$H_2O + O$$
$$OH + H_2O_2 \rightarrow H_2O + HO_2$$

Production of H_2 in H + HO_2 is balanced by its loss in reactions with OH, O(^{1}D), and flow to the upper atmosphere with subsequent dissociation and escape of H and H_2. This escape is ~2 × 10^8 cm^{-2} s^{-1}(*Anderson,* 1974). Hydrogen escape should be balanced by loss of O from the atmosphere in the proportion 2:1 in the steady-state global-mean models. This proportion is that in H_2O and presumes a surface reservoir of water that replenishes its loss from the atmosphere. If this proportion is broken, accumulation of either H or O would break the steady state. According to *McElroy* (1972), the required loss of O occurs by nonthermal escape of fast oxygen atoms formed by dissociative recombination of O_2^+. Early calculations by *Fox* (1993) did not support that suggestion, and oxidation of the surface rock is an alternative solution. Model results are insensitive to this choice, and typically a flow of oxygen equal to ~10^8 cm^{-2} s^{-1} is adopted as the upper boundary condition. Current models of the oxygen escape induced by dissociative recombination of O_2^+ (*Fox and Hac,* 2010) demonstrate a significant scatter of the results for various

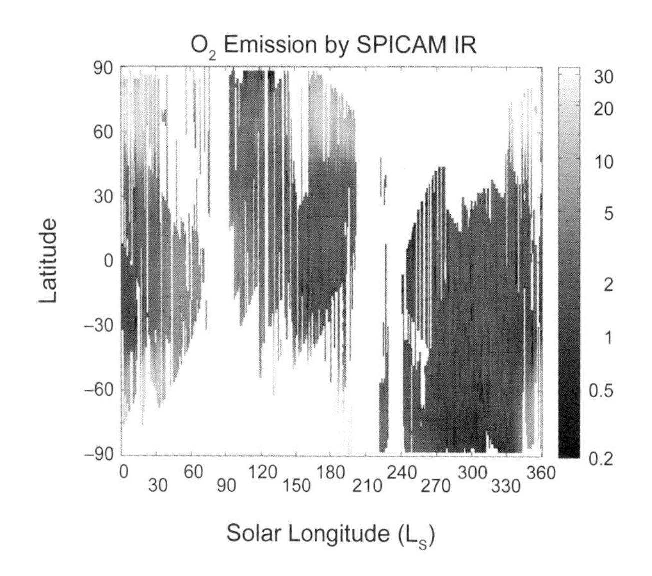

Fig. 9. Seasonal-latitudinal variations of the O_2 dayglow at 1.27 µm (in MR) observed by SPICAM-IR (*Fedorova et al.,* 2006a).

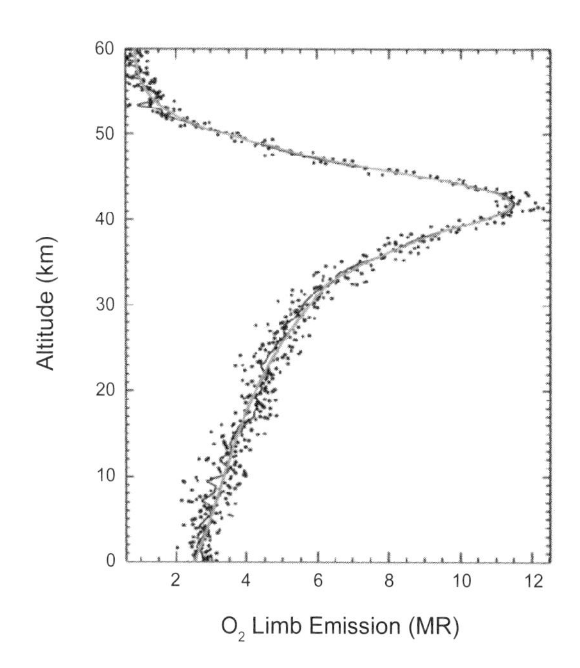

Fig. 10. Limb profile of the polar O_2 nightglow at 1.27 µm observed by MEX/OMEGA at 77°S, L_S = 120° (*Bertaux et al.,* 2012). Airmass is ~40 at the limb, and the nightglow vertical intensity is ~250 kR.

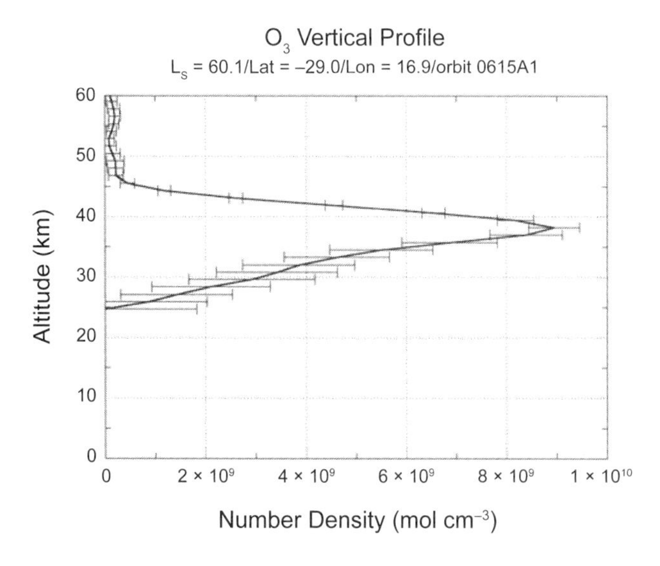

Fig. 8. Ozone nighttime vertical profile observed at 29°S, L_S = 60° (*Lebonnois et al.,* 2006).

TABLE 3. Reactions in the lower and middle atmosphere of Mars, their rate coefficients, and dayside-mean column rates (*Krasnopolsky*, 2010a).

#	Reaction	Rate Coefficient	Column Rate
1	$CO_2 + h\nu \rightarrow CO + O$	—	1.49 + 12
2	$CO_2 + h\nu \rightarrow CO + O(^1D)$	—	2.36 + 10
3	$O_2 + h\nu \rightarrow O + O$	—	1.70 + 11
4	$O_2 + h\nu \rightarrow O + O(^1D)$	—	1.11 + 10
5	$H_2O + h\nu \rightarrow H + OH$	—	1.76 + 10
6	$HO_2 + h\nu \rightarrow OH + O$	2.6×10^{-4}	4.23 + 10
7	$H_2O_2 + h\nu \rightarrow OH + OH$	4.2×10^{-5}	5.68 + 10
8	$O_3 + h\nu \rightarrow O_2(^1\Delta) + O(^1D)$	3.4×10^{-3}	6.38 + 12
9	$O(^1D) + CO_2 \rightarrow O + CO_2$	$7.4 \times 10^{-11} e^{120/T}$	6.42 + 12
10	$O(^1D) + H_2O \rightarrow OH + OH$	2.2×10^{-10}	1.50 + 9
11	$O(^1D) + H_2 \rightarrow OH + H$	1.1×10^{-10}	1.04 + 8
12	$O_2(^1\Delta) + CO_2 \rightarrow O_2 + CO_2$	10^{-20}	4.94 + 12
13	$O_2(^1\Delta) \rightarrow O_2 + h\nu$	2.24×10^{-4}	1.46 + 12
14	$O + CO + CO_2 \rightarrow CO_2 + CO_2$	$2.2 \times 10^{-33} e^{-1780/T}$	5.15 + 7
15	$O + O + CO_2 \rightarrow O_2 + CO_2$	$1.2 \times 10^{-32} (300/T)^2$	2.55 + 10
16	$O + O_2 + CO_2 \rightarrow O_3 + CO_2$	$1.4 \times 10^{-33} (300/T)^{2.4}$	6.45 + 12
17	$H + O_2 + CO_2 \rightarrow HO_2 + CO_2$	$1.7 \times 10^{-31} (300/T)^{1.6}$	1.93 + 12
18	$O + HO_2 \rightarrow OH + O_2$	$3 \times 10^{-11} e^{200/T}$	1.64 + 12
19	$O + OH \rightarrow O_2 + H$	$2.2 \times 10^{-11} e^{120/T}$	3.64 + 11
20	$CO + OH \rightarrow CO_2 + H$	1.5×10^{-13}	1.64 + 12
21	$H + O_3 \rightarrow OH + O_2$	$1.4 \times 10^{-10} e^{-470/T}$	6.17 + 10
22	$H + HO_2 \rightarrow OH + OH$	7.3×10^{-11}	2.36 + 10
23	$H + HO_2 \rightarrow H_2 + O_2$	$1.3 \times 10^{-11} (T/300)^{0.5} e^{-230/T}$	7.91 + 8
24	$H + HO_2 \rightarrow H_2O + O$	1.6×10^{-12}	5.18 + 8
25	$OH + HO_2 \rightarrow H_2O + O_2$	$4.8 \times 10^{-11} e^{250/T}$	1.71 + 10
26	$HO_2 + HO_2 \rightarrow H_2O_2 + O_2$	$2.3 \times 10^{-13} e^{600/T}$	5.75 + 10
27	$OH + H_2O_2 \rightarrow HO_2 + H_2O$	$2.9 \times 10^{-12} e^{-160/T}$	7.08 + 8
28	$OH + H_2 \rightarrow H_2O + H$	$3.3 \times 10^{-13} (T/300)^{2.7} e^{-1150/T}$	4.69 + 8
29	$O + O_3 \rightarrow O_2 + O_2$	$8 \times 10^{-12} e^{-2060/T}$	4.83 + 7
30	$OH + O_3 \rightarrow HO_2 + O_2$	$1.5 \times 10^{-12} e^{-880/T}$	1.30 + 7
31	$HO_2 + O_3 \rightarrow OH + O_2 + O_2$	$10^{-14} e^{-490/T}$	1.77 + 8
32	$H_2O_2 + O \rightarrow OH + HO_2$	$1.4 \times 10^{-12} e^{-2000/T}$	6.76 + 6
33	$NO_2 + h\nu \rightarrow NO + O$	0.0037	7.26 + 10
34	$NO_2 + O \rightarrow NO + O_2$	$5.6 \times 10^{-12} e^{180/T}$	1.78 + 10
35	$NO + HO_2 \rightarrow NO_2 + OH$	$3.5 \times 10^{-12} e^{250/T}$	9.04 + 10
36	Diffusion-limited flux of H_2	$V = 1.14 \times 10^{13}/n_{79}$	2.18 + 8
37	Loss of H_2O_2 at the surface	V = 0.02	3.43 + 7
38	Loss of O_3 at the surface	V = 0.02	6.13 + 7
39	Photolysis of O_2 above 80 km*	V = 0.32	1.20 + 10
40	Photolysis of CO_2 above 80 km*	—	1.20 + 11

*O and CO as the direct and indirect (via ion reactions) photolysis products are returned by their downward fluxes at 80 km.

Photolysis rates and k_{13} are in s^{-1}, velocities V are in cm s^{-1}, second and third order reaction rate coefficients are in $cm^3\ s^{-1}$ and $cm^6\ s^{-1}$, respectively. Photolysis rates for HO_2, H_2O_2, O_3, and NO_2 refer to the lower atmosphere and are calculated for $\lambda > 200$ nm at 1.517 AU. Column reaction rates are in $cm^{-2}\ s^{-1}$ and summed up from 1 to 79 km and multiplied by $(1 + h/R)^2$. $1.49 + 12 = 1.49 \times 10^{12}$.

initial conditions and do not rule out the oxygen flow of $\sim 10^8\ cm^{-2}\ s^{-1}$. The problem remains uncertain and may be solved by the upcoming Mars Atmosphere and Volatile Evolution (MAVEN) mission.

2.2.2. Published global-mean models. Mars' photochemistry critically depends on the catalytic effect of H_2O and its products. Photolysis of H_2O is very sensitive to absorption by the overhead CO_2 that restricts the effective spectral interval to ~10 nm centered at 190 nm. The H_2O photolysis is therefore more efficient at 40 km than near the surface by a factor of 400. The early models (e.g., *Kong and McElroy,* 1977a) were aimed at simulating the observed CO and O_2, and adopted the H_2O layer in the lowest 3–5 km with a steep cutoff above the layer. The odd hydrogen production was low and occurred near the surface, and strong eddy diffusion $K \approx 3 \times 10^8\ cm^2\ s^{-1}$ was required to move up odd hydrogen and reduce CO and O_2.

Krasnopolsky (1993) found that photochemistry weakly

affects water vapor that is mixed up to a saturation level at ~20 km at mean conditions and then follows the saturation density. Eddy diffusion in the lower atmosphere was ~10^6 cm^2 s^{-1} in that and subsequent models based on *Krasnopolsky* (1986, pp. 42–45) and *Korablev et al.* (1993). The calculated CO was significantly smaller in *Krasnopolsky* (1993) than the observed values, and NO_x, SO_x (sections 2.2.3 and 2.2.4), and heterogeneous chemistry were applied to fit the CO observations.

The observed mean CO mixing ratios in the lowest ~20 km are 1000 ppm in *Krasnopolsky* (2007a), 1100 ppm in *Billebaud et al.* (2009), 700 ppm in *Smith et al.* (2009), 940 ppm in *Hartogh et al.* (2010b), and 990 ppm in *Sindoni et al.* (2011). Therefore almost all observations give the mean CO equal to 1000 ppm.

Nair et al. (1994) changed rate coefficients of OH + HO_2 and CO + OH in their model and got the CO mole fraction f_{CO} = 490 ppm. *Krasnopolsky* (1995) achieved the similar value by reduction of the H_2O photolysis cross section near 190 nm to correct for possible impurities in the laboratory observations and a possible temperature effect. However, the suggestions by *Nair et al.* (1994) and *Krasnopolsky* (1995) disagree with the later laboratory studies.

Krasnopolsky and Feldman (2001) detected H_2 in Mars' upper atmosphere; extrapolation to the middle and lower atmosphere resulted in $f_{H_2} \approx 17$ ppm, smaller than the model predictions (*Nair et al.,* 1994; *Krasnopolsky,* 1995) by a factor of 2–3. Detailed modeling of the upper atmosphere (section 2.2.5) confirmed that the measured H_2 abundance agrees with the H distributions and escape observed by Mariner 6, 7, and 9 (*Anderson,* 1974). The measured H_2 became an additional constraint for Mars models.

Krasnopolsky (2006a) calculated models with heterogeneous loss of odd hydrogen on the water ice aerosol. However, he pointed out that the uptake coefficients required to fit the observed CO exceeded the laboratory data. Furthermore, his model for seasonal-latitudinal variations of O_3, H_2O_2, and the O_2 dayglow (section 2.3.1) failed to reproduce the observations with these uptake coefficients.

Zahnle et al. (2008) argued that the diffusion-limiting flow (*Chamberlain and Hunten,* 1987, p. 370) is a good approximation for H_2 as a boundary condition at ~100 km. Molecular diffusion coefficient is $D_i = a_i T^{\alpha_i}/n$, and the diffusion limiting flow is

$$\Phi_i^{DL} = \frac{a_i T^{\alpha_i}}{H} f_i$$

where H is the scale height, α = 0.75 for H_2, T almost cancels out in this relationship, and $\Phi_{H_2}^{DL} = 1.14 \times 10^{13} f_{H_2}$ cm^{-2} s^{-1}. *Zahnle et al.* (2008) suggested that a heterogeneous loss of O_3 and H_2O_2 on the surface with velocity of 0.02 cm s^{-1} may compensate for the escape of H and H_2 (section 2.1.1). Their water was near the standard value of 10 precipitable (pr.) μm in the basic model that gave mixing ratios of O_2, CO, and H_2 at 1300, 470, and 20 ppm, respectively. While the value for CO is still smaller than the observed abundance by a factor of 2, it is greater than those in the gas-phase models with the standard chemistry in *Nair et al.* (1994) and *Krasnopolsky* (2006a) by a factor of 4, and it is not clear how that was achieved. Here we do not consider their model B, which assumed a constant relative humidity of 17% up to 100 km. This assumption disagrees with the expected behavior of H_2O on Mars.

Krasnopolsky (2010a) tested the boundary conditions for H_2, O_3, and H_2O_2 from *Zahnle et al.* (2008) and found that they do not significantly change the model results, with the model CO smaller than the observed value by a factor of 8. *Krasnopolsky* (2010a) concluded that photochemistry is almost frozen out during the nighttime, and a daytime mean model with a mean solar zenith angle of 60° may be more adequate than the global mean model with the solar flux additionally reduced by a factor of 2 to account for the nighttime. However, eddy diffusion occurs in the nighttime, and therefore the model K was doubled to 4 × 10^6 cm^2 s^{-1} in the lower atmosphere.

The photolysis rates are dayside-mean in this model (Table 3). The boundary conditions are the CO_2 and H_2O densities at the surface, which correspond to 6.1 mbar and 9.5 pr. μm, respectively. Other boundary conditions are shown in lines 36–40. The conditions in lines 39–40 are the downward flows of O and CO from dissociation of O_2 and CO_2 above 80 km. Zero fluxes at both boundaries are the conditions for all other species.

The calculated density profiles are shown in Fig. 11. The H_2O densities are restricted by the saturation values. The calculated abundances of the major photochemical products are 1600 ppm for O_2, 20 ppm for H_2, 0.9 μm-atm for O_3, 1.5 MR or 2.5 μm-atm for $O_2(^1\Delta_g)$, and 7.6 ppb for H_2O_2, in good agreement with the observations (Table 2). The CO mixing ratio varies from 120 ppm near the surface, which is still much less than the observed value, to 900 ppm at 80 km.

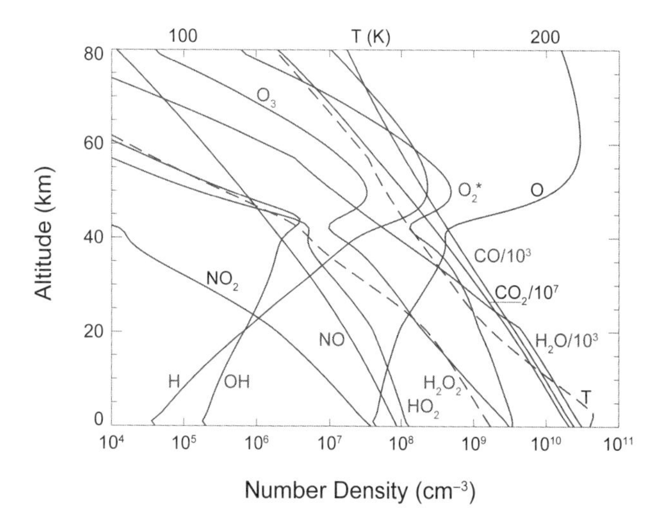

Fig. 11. Vertical profiles of species in the dayside-mean model by *Krasnopolsky* (2010a). The calculated mole fractions of O_2 and H_2 are 1600 and 20 ppm, do not vary with altitude, and are not shown.

The mean lifetime of a species is equal to a ratio of its column abundance (Fig. 11) to its daytime column production/loss rate (Table 3). This ratio should be doubled to account for the nighttime. The lifetimes of O_2 and H_2 are 60 and 370 yr, respectively. The lifetime of the calculated CO is 1 yr, while a ratio of the observed column CO to its production is 8 yr.

Photolysis of CO_2 forms CO and O_2 in the proportion 1:0.5, and O_2 significantly exceeds this proportion for both observed and calculated CO. This means that O_2 is a photochemical product of mostly H_2O, although the H_2O photolysis is weaker than that of CO_2 by 2 orders of magnitude.

2.2.3. Nitrogen chemistry. Nitrogen chemistry may be initiated in the lower and middle atmosphere by either lightning or a flow of NO from the upper atmosphere. There is no evidence of lightning, while the flow of NO is controversial. Atomic nitrogen in the ground state and the lowest metastable state $N(^2D)$ is formed in the upper atmosphere by predissociation of N_2 at 80–100 nm, electron impact dissociation of N_2, and ion reactions. $N(^2D)$ reacts with CO_2 and forms NO or is quenched by O and CO. N and NO are the main product of nitrogen chemistry in the upper atmospheres of Earth, Mars, and Venus. These species recombine in $N + NO \rightarrow N_2 + O$. Some part of the atomic nitrogen is transported to the nightside and excites the UV nightglow of the NO γ and δ bands

$$N + O \rightarrow NO + h\nu$$
$$N + O + M \rightarrow NO + M$$
$$N + NO \rightarrow N_2 + O$$

If a flow of N from the upper atmosphere exceeds that of NO, then nitrogen chemistry below 80 km is actually restricted to these reactions. However, if the downward flux of NO is greater than that of N, then the excess of NO initiates a more complicated chemistry. The situation depends on a fine balance between the production and loss of N and $N(^2D)$. Some data from laboratory experiments favor N > NO and no effective nitrogen chemistry below 80 km. However, NO was detected in the upper atmosphere by the Viking mass spectrometers (*Nier and McElroy,* 1977), and this requires a flow of NO and therefore nitrogen chemistry below 80 km. Detection and observations of the NO nightglow by SPICAM (*Bertaux et al.,* 2005; *Cox et al.,* 2008) show that the transport of N to the nightside is rather effective with a mean downward flow of $\sim 10^8$ cm^{-2} s^{-1} on the nightside.

Nitrogen chemistry in Mars' lower and middle atmosphere was considered in detail by *Nair et al.* (1994) and *Krasnopolsky* (1995). The calculated NO mole fraction was 0.3 ppb, below the observed upper limit of 1.7 ppb (Table 2). Basic reactions of the nitrogen chemistry on Mars are

$$NO + HO_2 \rightarrow NO_2 + OH$$
$$NO_2 + O \rightarrow NO + O_2$$
$$NO_2 + h\nu \rightarrow NO + O$$

The first reaction actually breaks the O=O bond and forms OH that reacts with CO, the second reaction restores the bond, and the third reaction ends the O=O breaking in the first reaction. Therefore it looks like the photolysis of NO_2 breaks the O=O bond, although the NO_2 structure is O=N=O. Nineteen percent of the O=O bonds are removed by the nitrogen chemistry (Table 3), and its role is neither critical nor negligible.

2.2.4. Sulfur chemistry. Except for H_2O and CO_2, SO_2 is the most abundant in terrestrial volcanic outgassing; therefore the stringent upper limit of 0.3 ppb to SO_2 (Table 2) is very important and restricts the outgassing from Mars at a level smaller than that on Earth by a factor of more than 2000 (*Krasnopolsky,* 2012a). Along with the lack of current volcanism and hot spots with endogenic heat sources in the Mars Odyssey Thermal Emission Imaging System (THEMIS) observations, this is an indication of extremely slow geological processes on Mars at the current time.

Our calculations of sulfur chemistry on Mars for the SO_2 abundance of 0.1 ppb are shown in Fig. 12. Sulfur chemistry is discussed in the Venus section; here we add a heterogeneous reaction $SO_2 + H_2O_2 + \text{ice} \rightarrow H_2SO_4 + \text{ice}$ (*Clegg and Abbat,* 2001) that is negligible on Venus. Formation of gaseous H_2SO_4 requires water vapor and occurs below 22 km; it is removed by uptake on the ice haze above 22 km and in the boundary layer near the surface. The sulfur reaction rates in the model are typically smaller than those in the basic model (Table 3) by 2 orders of magnitude and weakly affect the basic atmospheric chemistry on Mars. Therefore the perturbation in the basic chemistry by the sulfur chemistry is small for the SO_2 upper limit of 0.3 ppb, and the results in Fig. 12 may be linearly scaled to other SO_2 abundances.

2.2.5. Upper atmosphere and ionosphere. A boundary between the middle and upper atmosphere at 80 km is convenient for photochemical modeling, because it is possible to neglect ionization processes and molecular diffusion up to this boundary. Transport processes dominate

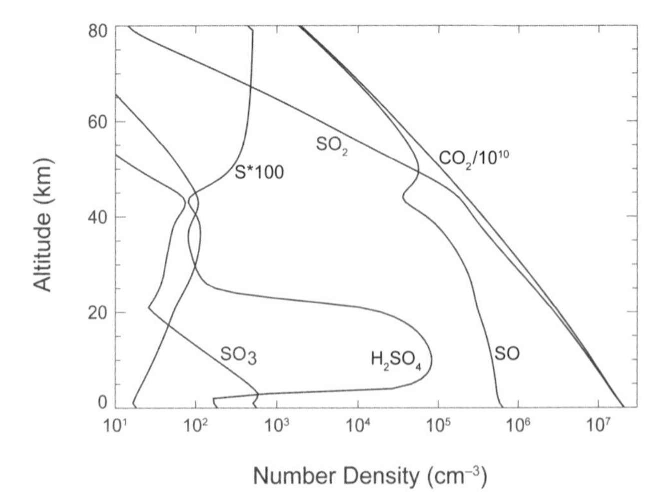

Fig. 12. Sulfur chemistry on Mars calculated for f_{SO_2} = 0.1 ppb.

near 80 km, and photolysis by solar Lyman-α is the only significant chemical reaction in this region. Vacuum is defined where the mean free path $l = (\Sigma\ \sigma_i\ n_i)^{-1}$ is greater than the intrinsic size of the medium. Therefore an upper boundary of an atmosphere is exobase, where $l = 2$ H and the vertical collisional thickness is $\Sigma\ \sigma_i n_i H_i = 0.5$. Here $\sigma_i \approx 3 \times 10^{-15}$ cm^2 is the collisional cross section. The martian exobase varies from 180 to 250 km.

Ionization and dissociative ionization of the atmospheric constituents by the solar extreme ultraviolet (EUV) and photoelectrons, ion-neutral reactions, and dissociative recombination are the major photochemical processes in the upper atmosphere. They form an ionosphere that extends up to an ionopause with a mean dayside altitude of ~300 km on Mars. A perturbed solar wind environment is above the ionopause.

The solar EUV radiation varies with solar activity approximately proportional to its index $F_{10.7\ cm}$ (the solar flux at 10.7 cm) that is typically between 70 and 220 units at 1 AU and 25 to 100 units at Mars' elliptic orbit. Two major solar cycles have periods of 11 yr and 28 d, respectively. Variations of the solar EUV strongly affect density, composition, thermal balance, and dynamics of the upper atmosphere and ionosphere. Variations of the exospheric temperature T_∞ with the solar activity in observations and models are shown in Fig. 13.

The ionization is maximum at $2\sigma N = 1$; here 2 is the mean airmass and $\sigma \approx 2.5 \times 10^{-17}$ cm^2 is the mean EUV absorption cross section of CO_2. This gives the CO_2 column at the ionization peak $N \approx 2 \times 10^{16}$ cm^{-2}, which is at ~130 km. The absorption is weak above ~160 km, and the atmosphere becomes isothermal. The neutral and ion species are subject to separation by molecular and ambipolar diffusion, respectively.

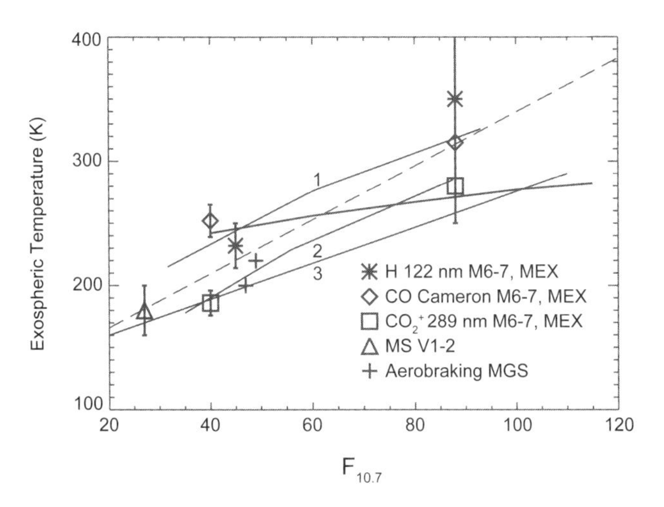

Fig. 13. Exospheric temperature as a function of the solar activity index retrieved from the MGS densities at 390 km (solid line) compared with other observations and models of $T_{infinity}$. Linear fit to the other observations is shown by dashed line. Thin lines are models by *Bougher et al.* (2000, 2008, 2009) (lines 1, 2, and 3, respectively). Model by *Gonzalez-Galindo et al.* (2009) is similar to line 3. All models are shown for the fall equinox. From *Krasnopolsky* (2010a).

Almost all ionization and dissociative ionization events finally result in dissociation of 1 and 2 CO_2 molecules, respectively. Neutral and ion compositions at medium solar activity on Mars are shown in Fig. 14. They were calculated using a code from *Krasnopolsky* (2002) with $F_{10.7\ cm}$ and T_∞ as input parameters. Similar figures were presented in *Krasnopolsky* (2002) for low and high solar activity. The calculated O and CO agree with the Mariner, Viking, and MEX observations (Table 2). Atomic hydrogen profiles are very sensitive to the exospheric temperature and agree with the Mariner and MEX observations of H and the FUSE observation of H_2. HD is chosen at 12 ppb at and below 80 km to fit D observed by HST (*Krasnopolsky et al.,* 1998). Profiles of O_2^+ and CO_2^+ at low solar activity are similar to those measured by the Viking entry probes (*Hanson et al.,* 1977); the observed and calculated profiles of O^+ are significantly different. *Fox et al.* (1996) calculated N and NO chemistry for various solar activity.

2.3. Variations of Mars Photochemistry in One-Dimensional Models

2.3.1. Steady-state models for local conditions. Changes in temperature and dynamics induce variations of the H_2O abundance and its vertical distribution, and H_2O controls Mars' photochemistry. Variations of the martian photochemistry may be studied by one-dimensional models assuming that local productions and losses of species exceed their delivery and removal by winds. These models are designed to reproduce odd oxygen and odd hydrogen, while the long-living products O_2, CO, and H_2 may be adopted at their observed values. Using the data from Fig. 11 and Table 3, lifetimes of odd oxygen and odd hydrogen are ~10^5 and ~2×10^5 s, i.e., on the order of 1 d, and these species may vary on a daily scale. This estimate agrees with the observations.

Liu and Donahue (1976) calculated a model with various abundances of water vapor. They proved anticorrelation between water and ozone that was, however, insufficient to explain the observed abundances of ozone up to 40–60 μm-atm in the subpolar regions at the end of winter. Photolysis of the abundant ozone enhances production of odd hydrogen in the reaction of $O(^1D) + H_2$, preventing a further increase of O_3 even in the lack of H_2O. The study of *Kong and McElroy* (1977b) involved condensation of H_2O and then H_2O_2, reduced O_3 and H_2O_2 photolysis rates at large airmass and high ozone, and influx of atomic oxygen from middle latitudes. They succeeded at explaining the variations in ozone (Fig. 15) observed by Mariner 9.

Clancy and Nair (1996) argued that the enhanced solar heating near perihelion ($L_S = 251°$) stimulates dust storms, and the atmosphere at low and middle latitudes is dust-free and cold near aphelion ($L_S = 71°$), and dusty and therefore warm near perihelion. This changes the condensation level of H_2O from ~10 to ~40 km, respectively, and induces significant seasonal variations of odd hydrogen (by 1 and 2 orders of magnitude at 20 and 40 km, respectively) and the overall photochemistry. They calculated a model of seasonal variations of Mars' photochemistry at 30°N using the adjusted kinetic data from *Nair et al.* (1994, section 2.2.2). The calculated O_3 column varied from 1.0 μm-atm at perihelion to 3.2 μm-atm at aphelion. The growth is mostly due to ozone near 20 km that increases by an order of magnitude. Their model motivated *Krasnopolsky* (1997) to initiate regular groundbased observation of the O_2 dayglow as a photochemical tracer (Fig. 4). The O_2 dayglow at 1.27 μm is excited by photolysis of O_3 and quenched by CO_2 below ~15 km. It is more sensitive to ozone near ~20 km than the O_3 column.

Krasnopolsky (2006a) argued that contrary to the claimed anticorrelation of O_3 with H_2O, both the O_2 dayglow and ozone observations near aphelion ($L_S \approx 40°–130°$) show rather constant values at latitudes from 20°S to 60°N, while the water vapor abundance increases by an order of magnitude in this latitude range. Variations of the condensation level appear to be insufficient to explain this behavior, and heterogeneous loss of H_2O_2 at the water ice particles was suggested and added to the gas-phase chemistry in Table 3 to simulate the observed seasonal and latitudinal variations of Mars photochemistry. The adopted reaction probability was 5×10^{-4} and 3×10^{-4} in two versions of the model (*Krasnopolsky,* 2006a, 2009b), much smaller than the uptake coefficient recommended from laboratory studies (*Crowley et al.,* 2010). However, the measured uptake of H_2O_2 on ice is reversible, and therefore the adopted reaction probabilities (irreversible sinks) appear plausible. The ice aerosol was calculated assuming the particle radius of 2 μm (*Montmessin et al.,* 2004).

Krasnopolsky (2006a, 2009b) combined T(z), H_2O, and aerosol observations from the MGS TES database with variable heliocentric distance, daytime duration, and mean cosine of solar zenith angle to calculate photochemistry at ~100 seasonal-latitudinal points. The latitudinal coverage was ±50° centered at the subsolar latitude, because H_2O becomes smaller than ~1 pr. μm and could not be measured by TES beyond this range. The calculated O_3, $O_2(^1\Delta_g)$, and H_2O_2 columns were interpolated to make seasonal-latitudinal maps.

The first version of the model was calculated before the SPICAM ozone data became available and fitted the groundbased and HST ozone measurements (Fig. 2). These O_3 abundances exceeded those retrieved from the SPICAM observations (Fig. 7) typically by a factor of 1.5–2, and the second version (Figs. 4 and 16) was aimed to fit the SPICAM ozone. The SPICAM-IR observations of the O_2 dayglow (Fig. 9) corrected for the local-time variations agree with the groundbased data (Fig. 4). The agreement between the model and the O_2 dayglow observations is also good except for the observed very bright region at 65°S and $L_S = 173°$.

The adopted heterogenous loss of H_2O_2 on the ice aerosol explains the observed minimum of H_2O_2 near aphelion, when H_2O is very abundant (Figs. 5 and 6). The model is in good agreement with all observed photochemical tracers [O_3, $O_2(^1\Delta_g)$, H_2O_2].

2.3.2. Time-dependent models. These models were calculated by *Krasitsky* (1978), *Shimazaki* (1981), *Garcia Munoz et al.* (2005), *Krasnopolsky* (2006a), and *Zhu and Yee* (2007). Ozone at low latitudes was calculated by *Krasitsky* (1978) at ~10 μm-atm during the nighttime and ~6 μm-atm during the daytime, above the observed values.

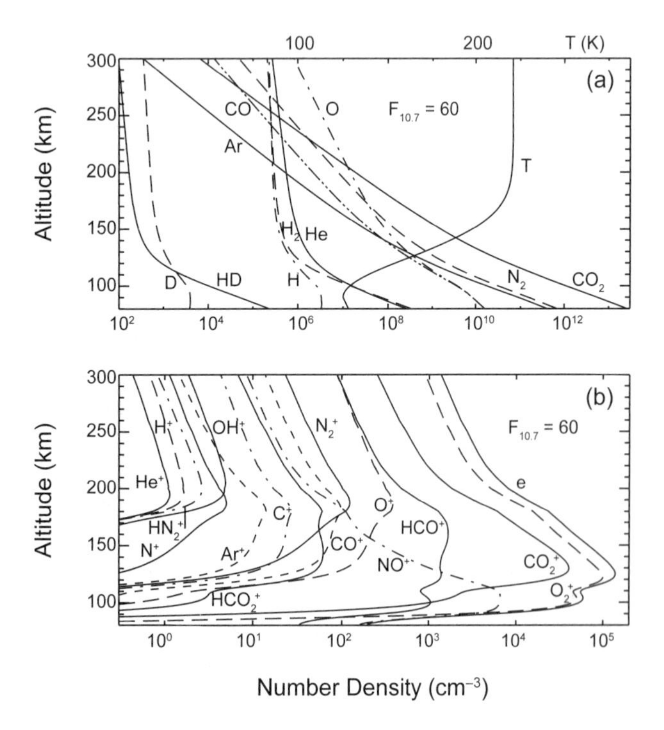

Fig. 14. Model for the composition of Mars' upper atmosphere and ionosphere at medium solar activity. Updated from *Krasnopolsky* (2002, 2010a).

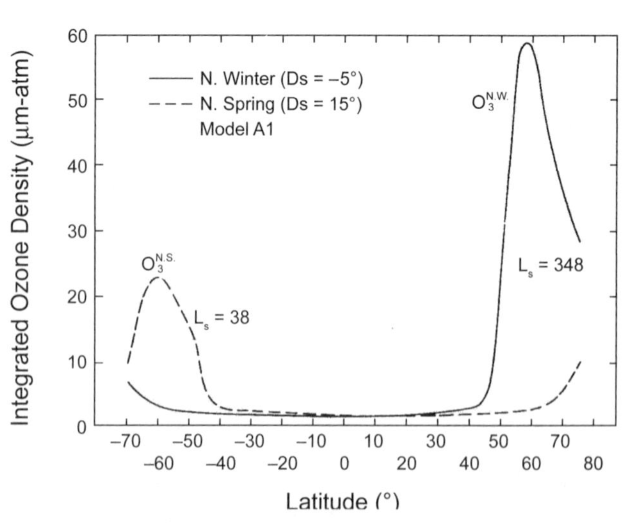

Fig. 15. Model of latitudinal variations of ozone at two seasons (*Kong and McElroy,* 1977b). This model fits the Mariner 9 observations (*Barth et al.,* 1973) in the polar and subpolar regions.

Shimazaki (1981) calculated variations of photochemistry at 65°N for the whole martian year with a time step of 12 min. To gain the computation time and stability of the solution, he assumed that odd hydrogen is not affected by the vertical transport. The calculated ozone increased from 0.2 μm-atm at $L_S \approx 90°$ to 3 μm-atm at 200°, 30 μm-atm at 230°–310°, a peak of 55 μm-atm at 345°, and a steep decrease to ~1 μm-atm at 30°, in reasonable agreement with the current observations and model (Fig. 7).

Garcia Munoz et al. (2005) calculated a model for 20°N at equinox to simulate diurnal variations of the O_2 and OH airglow and ozone. The calculated O_2 dayglow at 1.27 μm peaks near 13 hr, in accordance with the observations (*Krasnopolsky,* 2003b), with an intensity of 0.6 and 0.95 MR for quenching by CO_2 of 2×10^{-20} and 10^{-20} cm^3 s^{-1}, respectively. The O_2 nightglow excited by the O + O + M reaction is constant at 25 kR. The OH nightglow was calculated for excitation by $H + O_3 \rightarrow OH(v) + O_2$ and two extreme cases of quenching by CO_2 to either v = 0 or Δv = 1. The strongest bands for these cases are (4–2) 1.58 μm and (1–0) 2.80 μm with intensities of 0.18 kR and 12 kR, respectively. The calculated ozone peaks at 0.55 μm-atm near 13 hr; the nighttime values are flat at 0.22 μm-atm.

Krasnopolsky (2006a) calculated models for subsolar latitudes near equinox (L_S =173°) and the northern summer maximum (L_S = 112°, July 15 in the terrestrial calendar). Local time variations of the observable species are shown in Fig. 17. Ozone is almost constant at L_S = 173° and demonstrates a nighttime increase by a factor of 3 at L_S = 112°. Time-dependent models should apply a time step that is much shorter than the shortest photolysis lifetime; 25 and 30 s were used in *Garcia Munoz et al.* (2005) and *Krasnopolsky* (2006a), respectively.

Zhu and Yee (2007) adopted photochemical equilibrium as a boundary condition for radicals instead of the closed boundaries in the other papers. This actually presumes unspecified sources and sinks at the surface, and explains why the ozone profile in their model is very different from those in the other papers.

2.4. Three-Dimensional Photochemical General Circulation Models

An attempt to account for the seasonal and latitudinal variations of Mars photochemistry was made in a two-dimensional zonal-mean model by *Moreau et al.* (1991). Recently, the increase in computer power allowed the

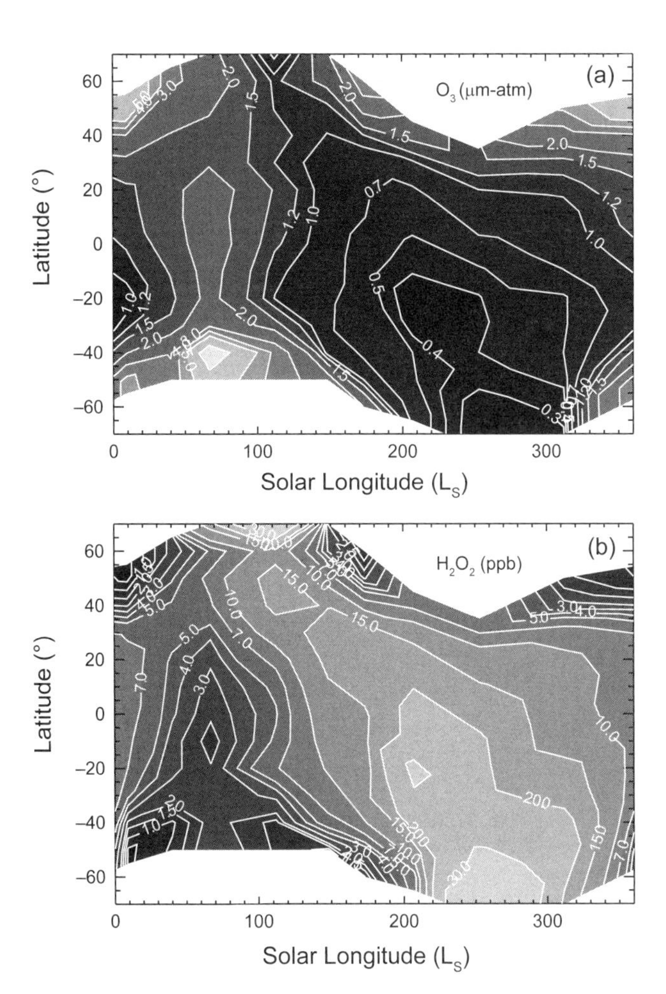

Fig. 16. Calculated seasonal-latitudinal variations of O_3 and H_2O_2 (*Krasnopolsky,* 2009b).

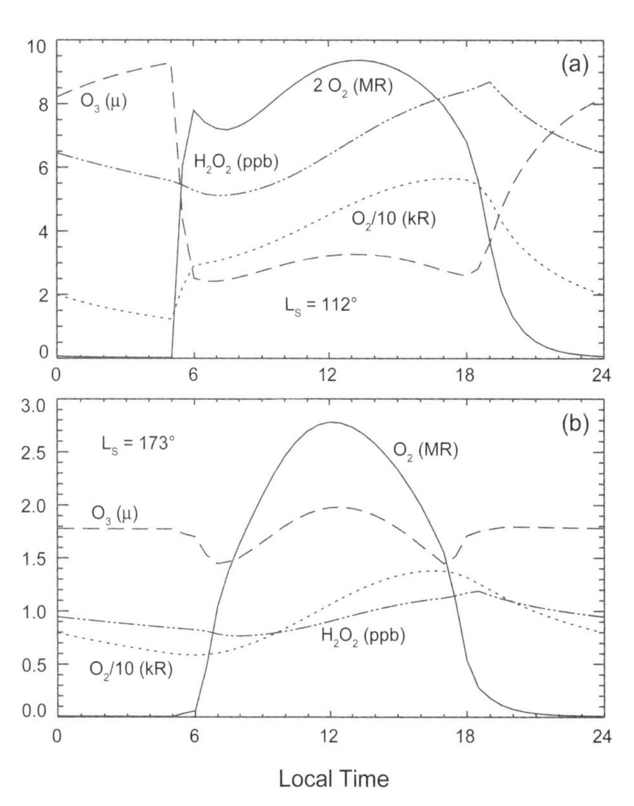

Fig. 17. Variations of the observable photochemical products with local time in two seasons (*Krasnopolsky,* 2006a). The O_2 airglow at 1.27 μm is excited by photolysis of O_3 (solid lines) and by O + O + M (dotted lines); the latter should be scaled by a factor of 1.5 to account for a recent rate coefficient of this reaction.

development of full general circulation models (GCM) of Mars with photochemistry. In addition to an adequate representation of atmospheric transport, martian GCMs are able to provide a realistic description of the three-dimensional field of water vapor and its variations at all scales, which is a crucial advantage for constraining properly the fast chemistry of the lower atmosphere. Martian GCMs are discussed by Forget and Lebonnois (this volume), and here we consider photochemical aspects of those models.

Seasonal variations on Mars are complicated by both the obliquity and the elliptic orbit, and are very prominent because of the comparatively thin atmosphere. Mars' photochemistry is relatively simple and may be involved in GCMs in its full extent. That is why the importance and success of the photochemical GCMs are so evident and impressive for Mars. These martian models are the most advanced tools available for simulating variations of the basic atmospheric properties.

There are currently two martian photochemical GCMs: that of *Lefèvre et al.* (2004, 2008, 2009) (hereafter LMD GCM), and that of *Moudden and McConnell* (2007) and *Moudden* (2007). Because of the long computation time involved, photochemical GCMs have up to now been essentially applied to the analysis of short-lived species. *Lefèvre et al.* (2004) characterized the three-dimensional variations of ozone in the atmosphere of Mars, and later demonstrated the importance of heterogeneous chemistry to reproduce the observations of ozone or H_2O_2 (*Lefèvre et al.,* 2008). The model included the heterogeneous loss of OH and HO_2 on the ice aerosol with uptake coefficients of 0.03 and 0.025 (*Cooper and Abbatt,* 1996), respectively. Recently, the LMD GCM was able to successfully reproduce the polar nightglow of O_2 (section 2.1). Applications of this GCM to the methane problem (section 2.5) are in *Lefèvre and Forget* (2009).

The MGS TES observations of the H_2O column variations are summarized in Fig. 5. Distributions of H_2O calculated by LMD GCM at aphelion and perihelion are shown in Fig. 18. The H_2O saturated vapor pressure varies from 150 to 210 K by 5 orders of magnitude, and the model reflects this behavior, which induces variations of odd hydrogen (Fig. 19). There is a significant increase in the H densities in the winter polar regions by the downwelling circulation. The calculated variations of H_2O_2 at low latitudes are shown and compared with the observations and models by *Krasnopolsky* (2009b) and *Moudden* (2007) in Fig. 6.

Vertical zonally-mean distributions of daytime odd oxygen (O and O_3) computed by LMD GCM are shown in Fig. 20. The calculated seasonal evolution of the daytime O_3 column is compared with the SPICAM observations in Fig. 7.

The chemical lifetime of O exceeds a month above 60 km. O atoms may therefore be transported quasi-passively from low and middle latitudes on the dayside into the polar night, where they are brought toward lower altitudes by the downwelling circulation above the winter pole (Fig. 21) and excite the polar nightglow that was observed by three instruments (section 2.1).

Gagne et al. (2012) computed a map of seasonal-latitudinal variations of the O_2 dayglow based on the LMD GCM. The map reproduces main features of the dayglow variations shown in Fig. 4 and discussed above.

2.5. Methane and Related Problems

Methane was detected on Mars (Fig. 22) by four independent teams using groundbased (*Krasnopolsky et al.,*

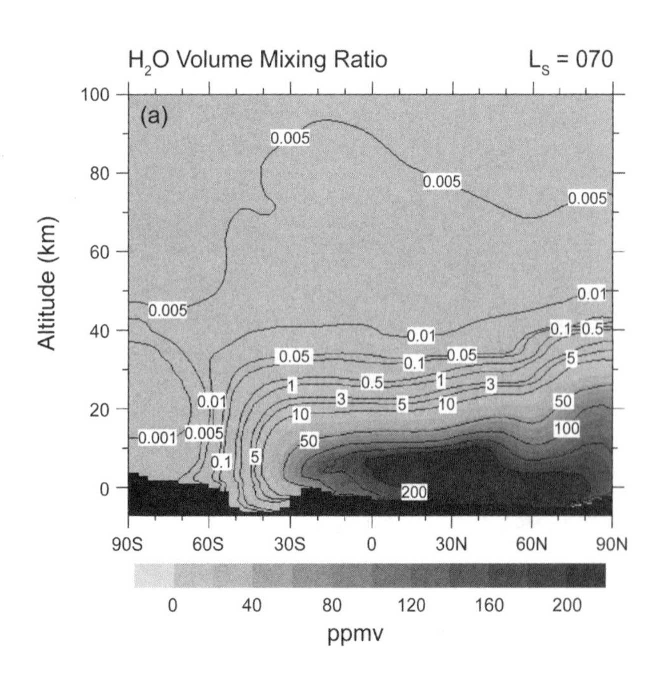

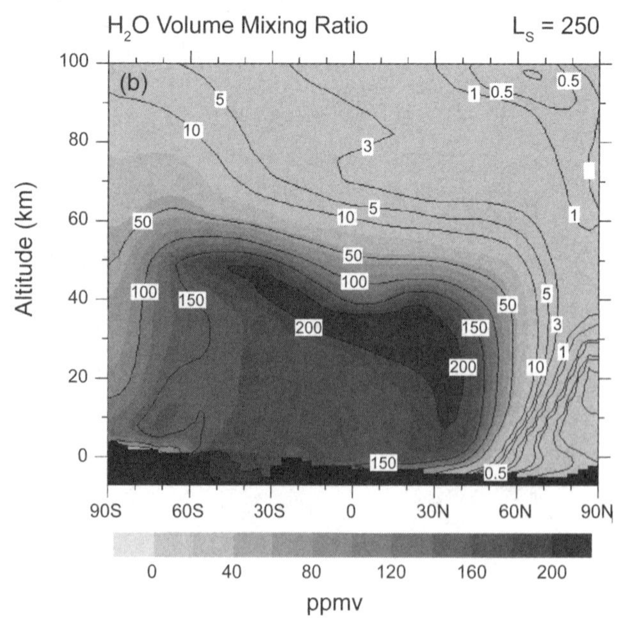

Fig. 18. Variations of H_2O at aphelion and perihelion in the LMD GCM. Updated from *Montmessin et al.* (2004).

2004; *Mumma et al.*, 2009; *Krasnopolsky,* 2011c, 2012a; *Villanueva et al.*, 2013) and spacecraft (*Formisano et al.*, 2004; *Fonti and Marzo,* 2010; *Geminale et al.*, 2011) instruments. Despite the high spectral resolving power and sensitivity in the groundbased observation, the detection of the martian methane against the telluric methane that exceeds it by a factor of ~10^4 is a difficult task. (This is done using a Doppler shift from Mars' geocentric velocity.) The MEX PFS and MGS TES spectrometers that were used to study methane on Mars do not have the required sensitivity and spectral resolution for an unambiguous identification of CH_4 (*Zahnle et al.*, 2011). The detection of CH_4 is made by summing thousands of spectra, but this does not suppress instrumental effects and systematic errors. Overall, the spacecraft observations suggest highly variable but generally stable methane on Mars with a mean abundance of ~15 ppb.

Mean CH_4 abundances from the groundbased observations are shown in Fig. 22. Methane was detected at 10 ppb in January 1999 and 20–30 ppb in January–March 2003. Observations in January 2006 gave an upper limit of 8 ppb (*Villanueva et al.*, 2013), and that in February 2006 showed 10 ppb near Valles Marineris with a mean abundance of ~8 ppb (*Krasnopolsky,* 2012a). Observations by both teams at the end of 2009 and the beginning of 2010 did not detect CH_4 with upper limits of 7–8 ppb. Finally, a tunable laser spectrometer onboard MSL's Curiosity rover did not detect methane with an upper limit of ~3 ppb (*Webster et al.*, 2013). These data favor episodic injection of methane into the atmosphere in 1998 and 2003.

Methane cannot be formed photochemically on Mars, its delivery by comets and interplanetary dust is insignificant (*Krasnopolsky,* 2006c), and geological and biogenic sources are under consideration. *Atreya et al.* (2007) suggested serpentinization, i.e., a set of hydrothermal reactions that involves iron-bearing silicates, H_2O, and CO_2. *Chassefière* (2009) considered the possibility of methane clathrates. Alternatively, *Krasnopolsky* (2006c) argued that the lack

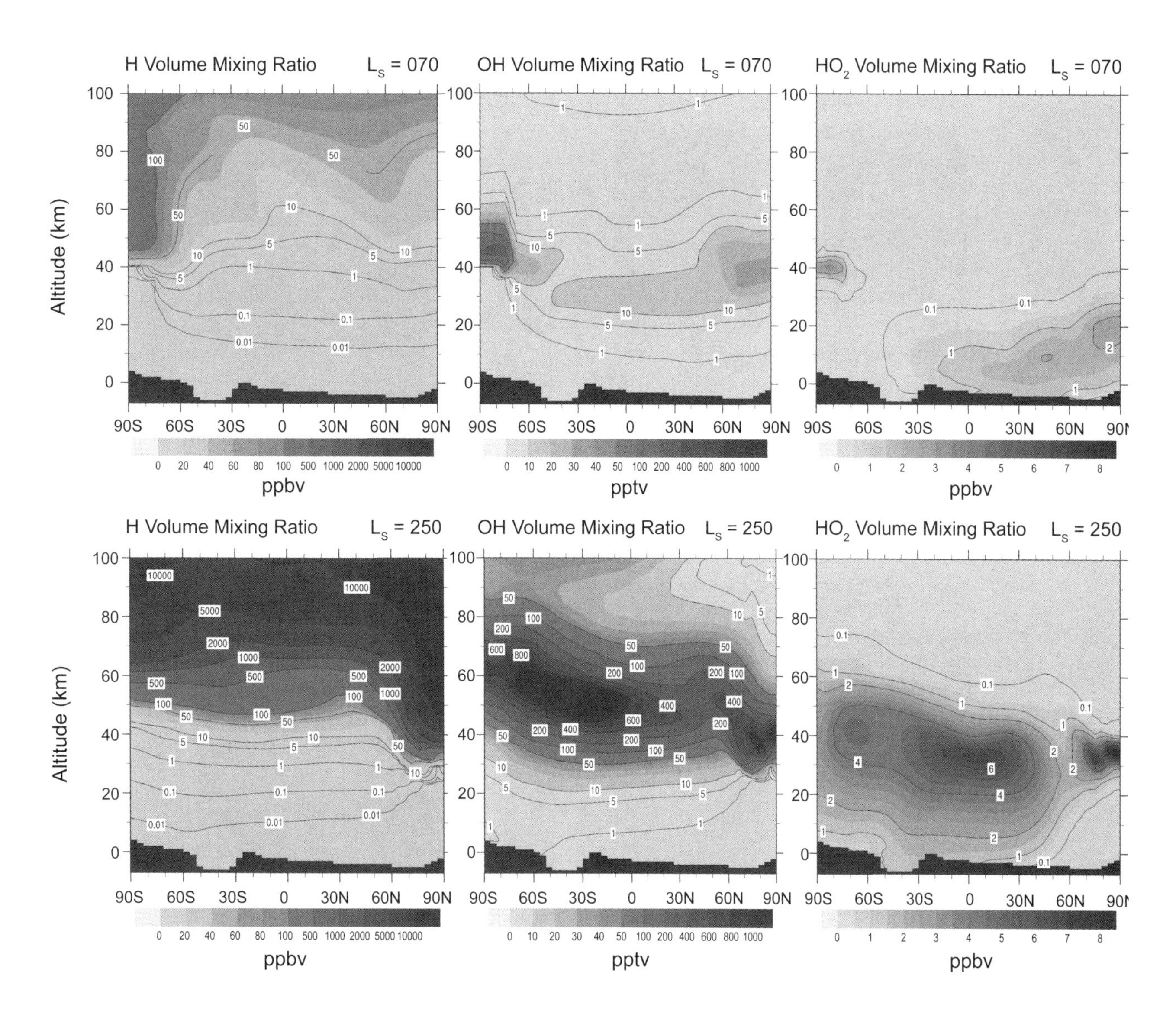

Fig. 19. Vertical distributions of zonally-mean H, OH, and HO_2 at aphelion and perihelion calculated by the LMD GCM.

of volcanism, hot spots with endogenic heat sources, and SO_2 (which is more abundant than CH_4 in terrestrial outgassing) favors biogenic methane. Methanogenic bacteria exist on the Earth, and even prior to the detection of methane methanogenesis in net reactions

$$4\ H_2 + CO_2 \rightarrow CH_4 + 2\ H_2O + 1.71\ eV$$
$$4\ CO + 2\ H_2O \rightarrow CH_4 + 3\ CO_2 + 0.48\ eV$$

was considered as a plausible pathway of metabolism on Mars.

Gas-phase loss of CH_4 on Mars involves its photolysis by solar Lyman-α at ~80 km and reactions with OH and $O(^1D)$ with a total lifetime of ~300 yr. Therefore variations of methane on Mars require an unknown heterogeneous loss with a characteristic time shorter than the global-mixing time of ~0.5 yr (*Krasnopolsky et al.,* 2004; *Lefèvre and Forget,* 2009). *Krasnopolsky* (2006c) concluded that the heterogeneous effect of dust is smaller than that of the surface rocks if they are of the same composition. He considered kinetic data on reactions of CH_4 with metal oxides, superoxide ions, and in the classic Fischer-Tropsch process used to make synthetic gasoline. *Trainer et al.* (2011) established upper limits to trapping of methane by the polar ice analogs. All these reactions are extremely slow at the martian temperatures and cannot explain the observed variations of methane. We do not consider effects of electrochemistry in dust devils that were overestimated by orders of magnitude by applying a steady-state model to those transient events. Search for organics by the Viking instruments resulted in restrictive upper limits (*Biemann et al.,* 1976).

Lefèvre and Forget (2009) and *Mischna et al.* (2011) simulated variations of methane on Mars using their GCMs. The only variations obtained with the LMD GCM and the conventional gas-phase chemistry are those induced by condensation and sublimation of CO_2 at the polar caps, similar to those observed for CO and Ar, with no significant variations at other latitudes. GCMs have tested the behavior of methane under various assumptions on its sources and heterogeneous sinks, their locations, and temporal variations,

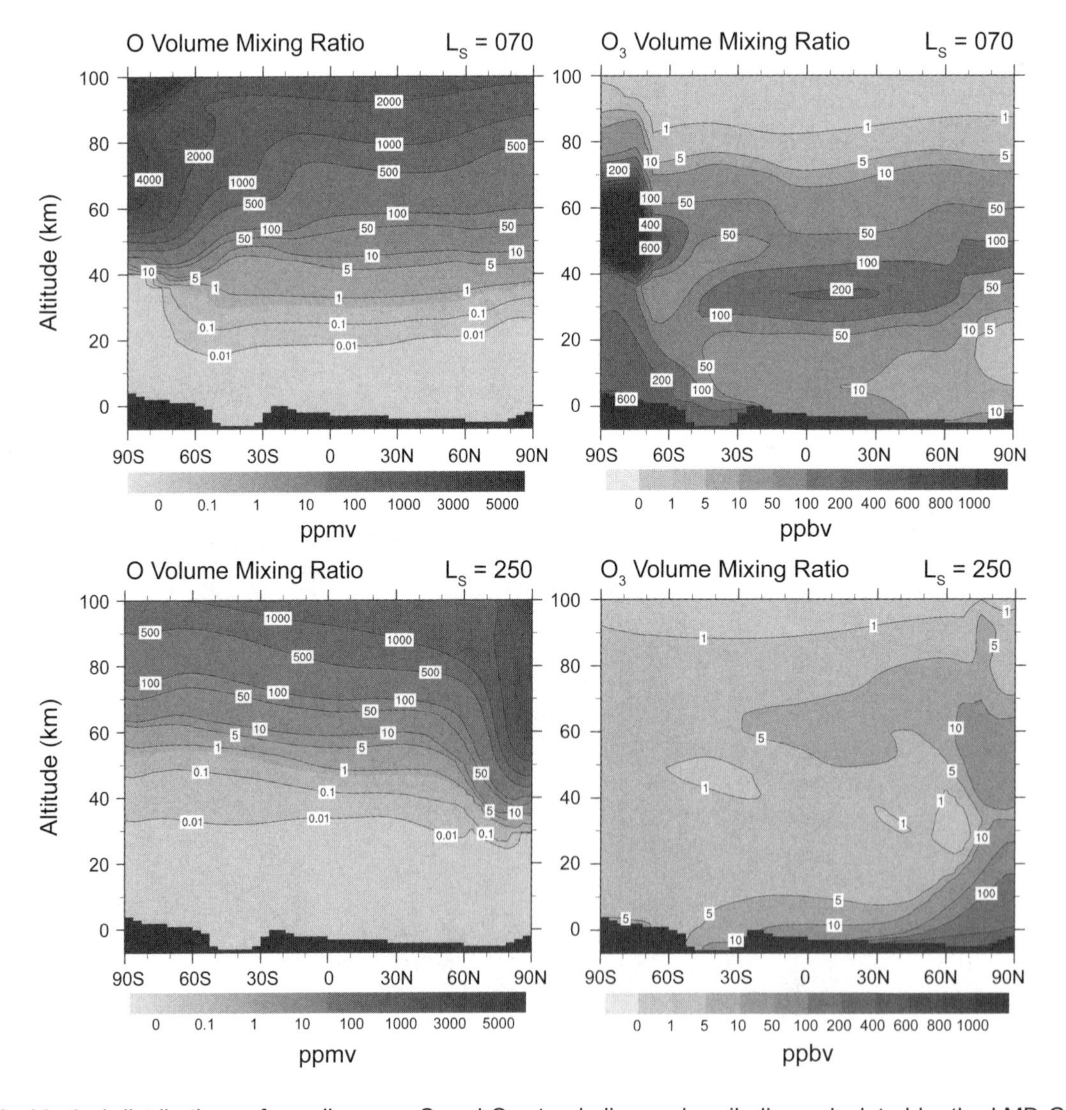

Fig. 20. Vertical distributions of zonally-mean O and O_3 at aphelion and perihelion calculated by the LMD GCM.

without reconciling current observations of methane and the known atmospheric chemistry and dynamics. Overall, the problem of methane on Mars, including its variations, origin, and chemistry, remains an enigma and needs further study.

2.6. Conclusions: Unsolved Problems

Overall, there has been significant progress in the study of Mars' photochemistry and its variations in the last decade. The MGS TES database; observations by the MEX and MRO orbiters; data from the HST, EUVE, FUSE, and Herschel orbiting observatories; and groundbased high-resolution spectroscopy have made valuable contributions to our knowledge of the martian atmosphere. Photochemical GCMs have resulted in a breakthrough in modeling of variations in Mars' photochemistry, but some problems remain.

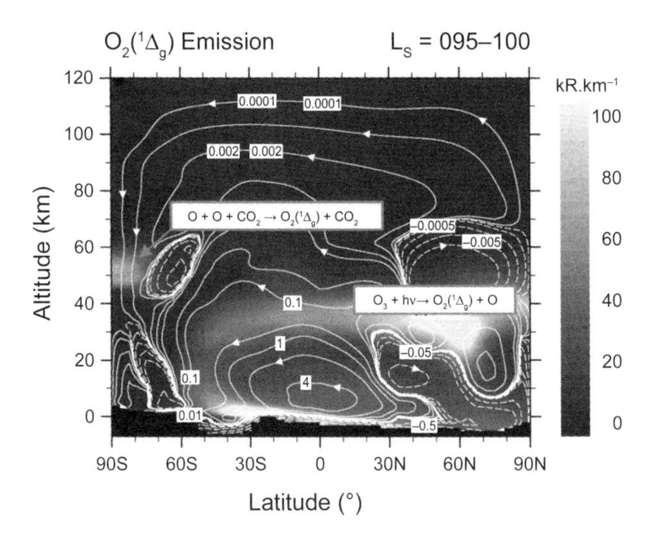

Fig. 21. A schematic of distribution of the zonally-mean $O_2(^1\Delta_g)$ airglow (in kR km^{-1}) at L_S = 95°–100° calculated by the LMD GCM. The airglow is excited by photolysis of O_3 and termolecular association of O_2 below and above ~40 km, respectively. White lines show the meridional stream function of air in 10^9 kg s^{-1}. From *Clancy et al.* (2012b).

The difference of a factor of 8 between the observed and calculated CO abundances is puzzling. The CO abundance is well measured and its production by photolysis of CO_2 is well known. It appears as if the only way to diminish the difference is to reduce odd hydrogen. *Krasnopolsky and Parshev* (1977) (see also *Krasnopolsky,* 1986, pp. 70–72) found that a nighttime cold trap at the surface and in the boundary layer affects the H_2O distribution and reduces its abundance above 8 km by a factor of ~2, while the H_2O daytime column is not changed. This effect may be even stronger using the current data. Photolysis of H_2O is low near the surface, and the odd hydrogen production becomes smaller as well. A similar effect could be for a low quantum yield of the H_2O photolysis at 190 nm. However, a significant reduction in odd hydrogen would change the entire photochemistry to that in the early models.

Advancements in computer power will make it possible in the future to run GCM simulations for a thousand years. This will allow calculations of the evolution of the long-living O_2, CO, and H_2, and will perhaps help to solve the CO problem.

Many aspects of the methane problem are unclear, especially its geographic and temporal variations, which that remain at odds with the known atmospheric chemistry and physics. Solving this enigma will require detailed high-quality spacecraft observations by a specially-designed

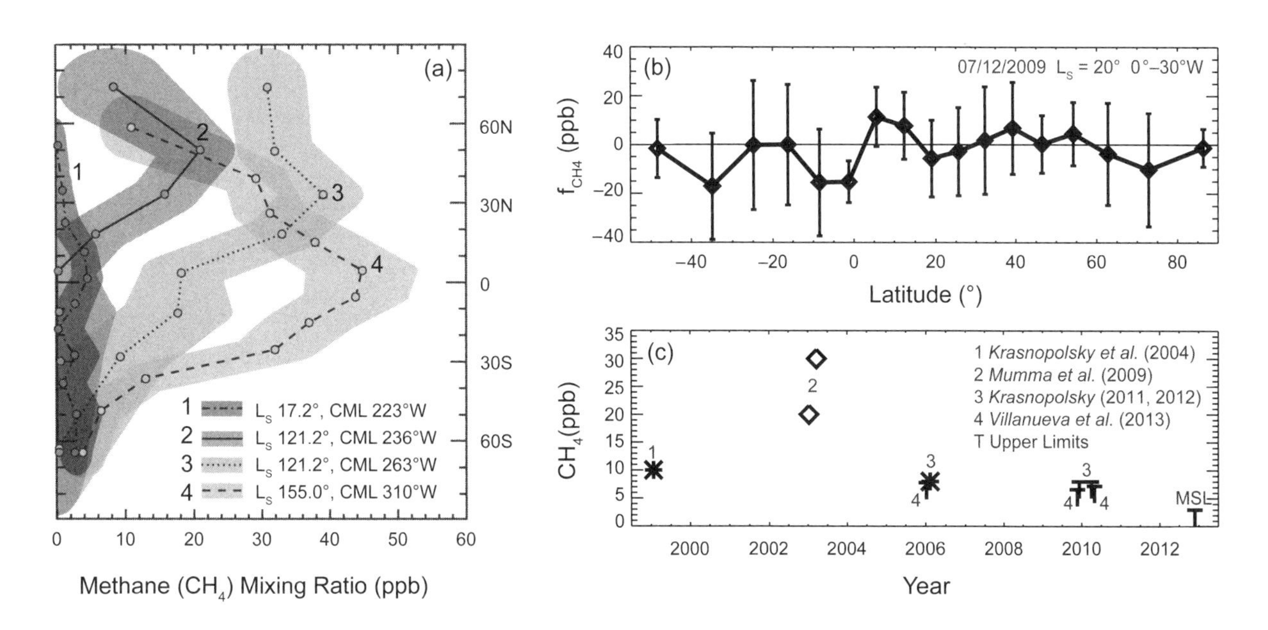

Fig. 22. (a) Variations of methane on Mars observed by *Mumma et al.* (2009) in 2003 and 2006. **(b)** Observations by *Krasnopolsky* (2012a) in December 2009 that resulted in a CH_4 upper limit of 8 ppb. **(c)** Groundbased and MSL (*Webster et al.,* 2013) observations. No methane has been detected since 2006.

instrument and experiments in the laboratory.

The O_2 and OH nightglow at low and middle latitudes has not been yet detected. H_2O profiles were observed in a narrow seasonal and latitudinal range and remain uncertain. The same is true of ozone nighttime profiles; daytime ozone profiles and profiles of the O_2 dayglow at 1.27 μm have not been published. NO has not been detected in the lower atmosphere, and its chemistry is therefore speculative. Some important kinetic data are uncertain: A rate coefficient of termolecular formation of O_3 is poorly known for $M = CO_2$ and is obtained by scaling of that for N_2; quenching of $O_2(^1\Delta_g)$ by CO_2 is uncertain as well. This list of problems may be continued.

3. VENUS

Venus' size (R = 6051 km) and mass are close to those of Earth with gravity $g = 887$ cm s^{-2}, and it is on a circular orbit of 0.72 AU with a small obliquity. The solid Venus exhibits a retrograde (clockwise) rotation with a period of 243 d. Venus' atmosphere is very dense and hot, with surface pressure and temperature of 92 bar and 735 K, respectively. Dynamical processes accelerate rotation of the atmosphere, which has a period of 4 d at the cloud tops (~70 km). There are therefore no seasons on Venus, although latitudinal variations of some atmospheric properties exist. Diurnal variations may be prominent near and above the cloud tops but are not expected in the bulk cloud layer and below the clouds, where the atmosphere is dense and typical lifetimes are long.

We will discuss below observations and photochemical modeling of the Venus chemical composition below 112 km, which was the upper boundary in models by *Yung and DeMore* (1982) and later models. This boundary is similar to that of 80 km on Mars: The atmospheric densities are similar, and it is possible to neglect molecular diffusion and ionization processes below these boundaries. We will also consider the nighttime chemistry and the nightglow excitation on Venus, and the upper boundary will extend to 130 km in that case.

3.1. Observations of the Chemical Composition

A brief summary of the Venus atmospheric composition is given in Table 4. CO_2 was detected by *Adams and Dunham* (1932); much later *Moroz* (1964) observed CO. Fourier transform high-resolution spectroscopy resulted in a breakthrough in the planetary studies and detections of HCl, HF, and a more reliable detection of CO (*Connes et al.,* 1967, 1968). The first Venera landing probes established the abundance of N_2 at a few percent (*Vinogradov and Surkov,* 1970; *Keldysh,* 1977). Mass spectrometers and gas chromatographs on Veneras 11–14 and Pioneer Venus landing probes measured abundances of N_2, CO, SO_2, and the noble gases (Table 4). Variations of SO_2 near the cloud tops were observed from the Pioneer Venus orbiter and in rocket experiments, and variations of SO_2 and H_2O were measured in the upper cloud layer by the Venera 15 orbiter.

Further progress in the studies of the chemical composition was related to a discovery of the strong nighttime emissions at 1.74 and 2.3 μm by *Allen and Crawford* (1984). *Krasnopolsky* (1986, p. 181) argued and *Kamp et al.* (1988) and *Crisp et al.* (1989) proved that these emissions originate from the hot lower atmosphere in the CO_2 transparency windows. Bands of other gases in these windows as well as in those at 1.28, 1.18, and 1.1 μm may be used to measure their abundances. High-resolution spectroscopy of the nighttime emissions revealed absorptions by CO_2, CO, H_2O, HDO, OCS, HCl, and HF; OCS was unambiguously detected for the first time on Venus (*Bezard et al.,* 1990). A composite analysis of the observations by a few teams was made by *Pollack et al.* (1993) and *Taylor et al.* (1997). Radio occultations by the Mariner 10 and Magellan orbiter and microwave observations using the Very Large Array measured vertical profiles of gaseous H_2SO_4 in the lower atmosphere.

The latest achievements in the field are relevant to the Venus Express (VEX) orbiter and groundbased submillimeter and infrared observations. Solar occultations using VEX's Solar Occulation at Infrared (SOIR) and stellar occultations using VEX's Spectroscopy for Investigation of Characteristics of the Atmosphere of Venus (SPICAV) resulted in vertical profiles of H_2O, HDO, CO, HCl, HF, SO_2, and SO at 70–100 km (*Fedorova et al.,* 2008; *Vandaele et al.,* 2008; *Belyaev et al.,* 2012) and detection of the nighttime ozone (*Montmessin et al.* 2011). Variations of SO_2, H_2O, CO, and OCS were measured near the cloud tops and in the lower atmosphere by SPICAV and the Visible and Infrared Thermal Imaging Spectrometer (VIRTIS) (*Marcq et al.,* 2008, 2011; *Belyaev et al.,* 2012).

Vertical profiles of CO at 70–100 km, mesospheric abundances of H_2O, SO_2, SO, and a restrictive upper limit to H_2SO_4 vapor were measured from the groundbased submillimeter observatories (*Clancy et al.,* 2012a; *Sandor and Clancy,* 2005; *Sandor et al.,* 2010, 2012). NO and OCS were detected for the first time at the cloud tops (*Krasnopolsky,* 2006b, 2008). Lightning is the only source of NO in Venus' lower atmosphere, and the detection of NO is an unambiguous indication of lightning on Venus. Variations of CO, HCl, HF, H_2O, OCS, and SO_2 were observed by means of groundbased spatially-resolved high-resolution IR spectroscopy (*Krasnopolsky,* 2010b, 2010c; *Krasnopolsky et al.,* 2013). Observations of the Venus nightglow will be considered in section 3.4.

3.2. Chemistry of the Lower Atmosphere

While the martian chemistry is restricted to mostly the CO_2-H_2O interactions and products, Venus' chemistry involves seven elements (O, C, S, H, Cl, N, F). Their parent species are CO_2, SO_2, H_2O, HCl, NO, and HF. The very dense and hot lower atmosphere with p = 92 bar and T = 735 K significantly complicates its modeling as well.

All published models neglect the fluorine chemistry. It is expected to be rather similar to the chlorine chemistry,

TABLE 4. Chemical composition of Venus' atmosphere below 112 km.

Gas	h (km)	Mixing Ratio	References
CO_2	—	0.965	*Adams and Dunham* (1932), *Connes et al.* (1967), *Moroz* (1968)
N_2	—	0.035	*Gelman et al.* (1979), *Istomin et al.* (1980), *Hoffman et al.* (1980), *Oyama et al.* (1980)
He	—	9 ppm	*Kumar and Broadfoot* (1975), *von Zahn et al.* (1980), *Krasnopolsky and Gladstone* (2005)
Ne	—	7 ppm	*Hoffman et al.* (1980), *Istomin et al.* (1980), *Oyama et al.* (1980)
Ar	—	70 ppm	*Gelman et al.* (1979), *Hoffman et al.* (1980), *Istomin et al.* (1980), *Oyama et al.* (1980)
Kr	—	20 ppb	*Istomin et al.* (1983)
Xe	—	<7 ppb	*Istomin et al.* (1983)
CO	100	10^{-4}–10^{-3}	*Wilson et al.* (1981), *Vandaele et al.* (2008), *Clancy et al.* (2012a)
	75	20–70 ppm	*Wilson et al.* (1981), *Vandaele et al.* (2008), *Clancy et al.* (2012a)
	68	40 ppm	*Moroz* (1964), *Connes et al.* (1968), *Irwin et al.* (2008), *Krasnopolsky* (2010c)
	36–42	30 ppm	*Gelman et al.* (1979), *Oyama et al.* (1980)
	36	27 ppm	*Pollack et al.* (1993), *Taylor et al.* (1997), *Marcq et al.* (2006, 2008), *Tsang et al.* (2009), *Cotton et al.* (2012)
	22	20 ppm	*Oyama et al.* (1980)
	12	17 ppm	*Gelman et al.* (1979)
O_2	65	$<10^{18}$ cm^{-2}	*Trauger and Lunine* (1983)
O_3	95	5×10^7 cm^{-3}	night; *Montmessin et al.* (2011)
H_2O	100	1 ppm	*Fedorova et al.* (2008)
	70–95	1.5 ppm	*Encrenaz et al.* (1995), *Sandor and Clancy* (2005), *Gurwell et al.* (2007), *Fedorova et al.* (2008)
	70	3 ppm	*Fink et al.* (1972), *Bjoraker et al.* (1992), *Fedorova et al.* (2008), *Cottini et al.* (2012), *Krasnopolsky* (2013)
	62	8 ppm	*Ignatiev et al.* (1999)
	0–45	30 ppm	*Pollack et al.* (1993), *Meadows and Crisp* (1996), *Ignatiev et al.* (1997), *Marcq et al.* (2008), *Bezard et al.* (2011)
HCl	70–105	~150 ppb	*Vandaele et al.* (2008)
	94–105	~0 ppb	*Sandor and Clancy* (2012)
	70	400 ppb	*Connes et al.* (1967), *Young* (1972), *Krasnopolsky* (2010c), *Sandor and Clancy* (2012)
	15–30	500 ppb	*Bezard et al.* (1990), *Pollack et al.* (1993), *Iwagami et al.* (2008)
HF	70	4 ppb	*Connes et al.* (1967), *Bjoraker et al.* (1992), *Vandaele et al.* (2008), *Krasnopolsky* (2010c)
	30–40	5 ppb	*Bezard et al.* (1990), *Pollack et al.* (1993)
NO	65	5.5 ppb	*Krasnopolsky* (2006b)
SO_2	90–105	100–700 ppb	*Belyaev et al.* (2012)
	85–100	23 ppb	*Sandor et al.* (2010)
	80	10–70 ppb	*Belyaev et al.* (2012)
	70	20–2000 ppb	*Barker* (1979), *Esposito et al.* (1988, 1997), *Krasnopolsky* (2010d), *Marcq et al.* (2011), *Belyaev et al.* (2012)
	62	400 ppb	*Zasova et al.* (1993)
	40	130 ppm	*Gelman et al.* (1979), *Bezard et al.* (1993a), *Bertaux et al.* (1996), *Marcq et al.* (2008)
	22	180 ppm	*Oyama et al.* (1980)
SO	95–105	50–700 ppb	*Belyaev et al.* (2012)
	85–100	8 ppb	*Sandor et al.* (2010)
OCS	65	1–8 ppb	*Krasnopolsky* (2010d)
	33	3.5 ppm	*Pollack et al.* (1993), *Taylor et al.* (1997), *Marcq et al.* (2008)
H_2SO_4	85–100	<3 ppb	*Sandor et al.* (2012)
	46	2–10 ppm	*Kliore et al.* (1979), *Kolodner and Steffes* (1998), *Butler et al.* (2001), *Jenkins et al.* (2002)
S_3	15	18 ppt	*Krasnopolsky* (2013a)
	6	11 ppt	*Krasnopolsky* (2013a)
S_4	15	6 ppt	*Krasnopolsky* (2013a)
	6	<8 ppt	*Krasnopolsky* (2013a)
H_2S	70	<23 ppb	*Krasnopolsky* (2008)

but HCl is more abundant than HF by a factor of 100. Furthermore, the HCl and HF photolyses begin at 230 and 180 nm, respectively, and the latter is strongly depleted by CO_2 absorption.

3.2.1. H_2O-H_2SO_4 system in Venus' clouds and kinetic problem for OCS and CO. Sulfuric acid is a main product of photochemistry in the middle atmosphere of Venus (see section 3.3). A coupled problem of diffusion and condensation in a two-component aerosol was solved by *Krasnopolsky and Pollack* (1994) for a downward flow of H_2SO_4-H_2O in the Venus clouds at 47–65 km. The model (Fig. 23) properly simulates variations of the H_2O mixing ratio, concentration of sulfuric acid, altitude of the lower cloud boundary, and some features of the cloud structure [the upper, middle, and lower cloud layers (*Knollenberg and Hunten,* 1980)]. It agrees with the observations of H_2O at the cloud tops and H_2SO_4 vapor near the cloud bottom. The model did not involve any assumptions regarding the sulfuric acid concentration.

Krasnopolsky and Pollack (1994) concluded that thermochemical equilibrium calculations may serve as an approximation for the chemical composition near the surface and are generally invalid in the lower atmosphere. Chemical kinetics should therefore be applied to model the chemical composition. They solved a partial chemical kinetic problem for OCS, CO, H_2SO_4, and SO_3 in the lower atmosphere to simulate the observational data from *Pollack et al.* (1993). They suggested that the observed decrease of OCS and increase of CO with altitude may be explained by reactions

$$H_2SO_4 \rightarrow H_2O + SO_3 \quad \text{(R1)}$$
$$SO_3 + OCS \rightarrow CO_2 + (SO)_2 \quad \text{(R2)}$$
$$(SO)_2 + OCS \rightarrow SO_2 + CO + S_2 \quad \text{(R3)}$$
$$\text{Net } H_2SO_4 + 2\ OCS \rightarrow H_2O + CO_2 + SO_2 + CO + S_2 \quad \text{(R4)}$$

A rate coefficient of the basic reaction between SO_3 and OCS was estimated at $10^{-11}\ e^{-10000/T}$ cm^3 s^{-1}, i.e., $\sim 10^{-22}$ cm^3 s^{-1} at 400 K. The calculated abundances agreed with the observations. *Fegley et al.* (1997) later combined thermodynamics with some kinetic considerations to study the chemical composition in the lowest 10 km on Venus.

3.2.2. Self-consistent chemical kinetic model. The next step in the problem was a self-consistent chemical kinetic model for the lower atmosphere (*Krasnopolsky,* 2007b, 2013a). By "chemical kinetic modeling" we mean what is usually referred to as photochemical modeling, but applied to atmospheric regions where photolysis is relatively unimportant. Observations at the Venera 14 landing probe using a filter with a bandpass of 320–400 nm (*Ekonomov et al.,* 1983) showed that the solar light at $\lambda < 400$ nm is depleted at 62 and 60 km by factors of 7 and 45, respectively. Therefore photolyses of S_3 and S_4 are the only important photochemical processes in the model that extends from the surface to the lower cloud boundary at 47 km. Spectra of scattered solar radiation at various altitudes were taken from the Venera 11 observations (*Ekonomov et al.,* 1983).

Apart from the S_3 and S_4 photolyses, there are two other sources that drive chemistry in the lower atmosphere: fluxes of disequilibrium photochemical products from the middle atmosphere and thermochemical reactions in the lowest scale height. Exchange of species at 47 km between the lower and middle atmosphere involves the following net processes

$$CO_2 + SO_2 + H_2O \rightarrow H_2SO_4 + CO \quad 5.7 \times 10^{11}\ \text{cm}^{-2}\ \text{s}^{-1}$$
$$OCS + 2\ CO_2 \rightarrow SO_2 + 3\ CO \quad V[OCS]_{47\ km}$$
$$H_2S + CO_2 \rightarrow H_2O + CO + S \quad V[H_2S]_{47\ km}$$

A flow of chemically neutral CO_2 and SO_2 from the lower atmosphere is returned as a flow of oxidizing sulfuric acid and reduced CO. The column rate of this net process is similar in the photochemical models by *Mills* (1998), *Mills and Allen* (2007), *Zhang et al.* (2012), and *Krasnopolsky* (2012a) and may be used with some confidence as an upper boundary condition in the model for the lower atmosphere. Two other reactions simulate upward flows of OCS and H_2S driven by photolyses of these species in the middle atmosphere. Here V = K/2H, i.e., their rates are half of the limiting diffusion flow. The parent species CO_2, SO_2, H_2O, HCl, and NO are fixed at the lower boundary at the surface by their measured abundances. The boundaries are closed for other species in the model, i.e., their fluxes are zero at the boundaries.

Direct and inverse thermochemical reactions are balanced at thermodynamic equilibrium. However, this balance is broken because of the appearance of the disequilibrium

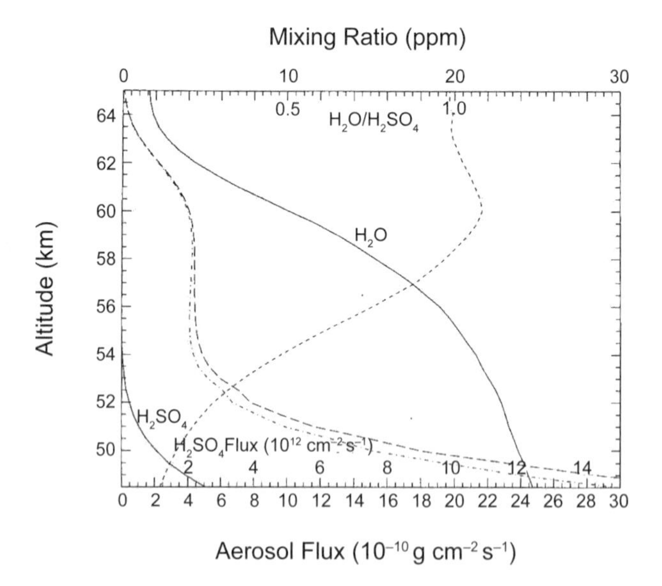

Fig. 23. Calculated mixing ratios of H_2O and H_2SO_4 vapors (solid lines), H_2O/H_2SO_4 ratio in the sulfuric acid aerosol (short dashes), and fluxes of H_2SO_4 in the liquid phase (long dashes, in 10^{12} cm^{-2} s^{-1}) and sulfuric acid aerosol (dash-dotted curve, in 10^{-10} g cm^{-2} s^{-1}). The curves of the aerosol flux remind the structure of Venus' clouds. From *Krasnopolsky and Pollack* (1994).

species and gas exchange with atmospheric layers having different temperatures and compositions.

The model involves 89 reactions of 28 species. Calculated mole fractions of the most abundant species are shown in Fig. 24. The lack of rate coefficients for many reactions was a significant problem that was solved using similar reactions of other species and thermodynamic calculations of inverse processes. R2 and R3 are the main reactions that convert the downward flow of H_2SO_4 and the upward flow of OCS into the flows of CO and S_2. The model includes a S_4 cycle

$$S_2 + S_2 + M \rightarrow S_4 + M$$
$$S_4 + h\nu \rightarrow S_3 + S$$
$$S_3 + h\nu \rightarrow S_2 + S$$
$$2(S + OCS \rightarrow CO + S_2)$$
$$\text{Net } 2\ OCS \rightarrow 2\ CO + S_2$$

that was suggested by *Yung et al.* (2009). Similar to the R1–R4 cycle, it forms CO and S_2 from OCS. This cycle without R2 and R3 cannot fit the observational constraints for OCS, CO, S_3, and S_4. Inclusion of the S_4 cycle in the model improves its agreement with the observations.

Chemistry of the lower atmosphere is driven by sulfur that is formed by

$$(CO + SO_2 \rightarrow CO_2 + SO) \times 2$$
$$SO + SO \rightarrow SO_2 + S$$
$$\text{Net } 2\ CO + SO_2 \rightarrow 2\ CO_2 + S$$

with a rate of 3×10^{10} cm^{-2} s^{-1}. Exchange of S between S_X and OCS is stronger by 2 orders of magnitude and results in significant variations of both species (Fig. 24) while their sum is constant at 20 ppm. S_8 is a dominant sulfur allotrope at 47 km with a mixing ratio of 2.5 ppm. This sulfur should condense near 50 km and form an aerosol layer ~5 km thick.

It was previously thought that aerosol sulfur appears in the cloud layer because of the photochemistry in the middle atmosphere (*Toon et al.,* 1982; *Young,* 1983). The models by *Mills* (1998), *Mills and Allen* (2007), *Zhang et al.* (2012), and *Krasnopolsky* (2012a) do not support this expectation and give a very low production of free sulfur. Now we see that the significant abundances of free sulfur are formed in the lower atmosphere and condense in the middle cloud layer. This sulfur cannot be the second NUV absorber that appears near 60 km (*Ekonomov et al.,* 1983).

A sum of CO and OCS is also almost constant at ~37 ppm. The CO flow from the middle atmosphere is removed by R1–R4 with minor contributions from other reactions. The OCS mixing ratio varies by 2 orders of magnitude within this sum.

The model agrees with the observed values of CO, OCS, H_2SO_4, S_3, and S_4. The predicted abundances and vertical profiles of H_2, H_2S, and SO_2Cl_2 (Fig. 24) may be used for various applications. The values for H_2S and H_2 near the surface are close to those calculated by *Krasnopolsky and Parshev* (1979) and *Fegley et al.* (1997), assuming thermochemical equilibrium. SO_2, H_2O, HCl, and NO are constant in the lower atmosphere at their adopted values at the lower boundary.

3.2.3. Sulfur atmospheric cycles. The nightside spectroscopy results and their simulations in the model are incompatible with the concept of the sulfur atmospheric cycles in *Prinn* (1985) and *von Zahn et al.*(1983). Basic processes of that concept are photolysis of OCS, formation of SO_2, SO_3, and S_x above the clouds, and

$$SO_3 + 4\ CO \rightarrow OCS + 3\ CO_2$$
$$CO + (1/x)\ S_x \rightarrow OCS$$

below the clouds. Here OCS is formed by the species moving down from the middle atmosphere, and its mixing ratio should be rather constant below the clouds, opposite to that observed. Furthermore, the required flux of CO from the middle atmosphere should exceed that of sulfuric acid by more than a factor of 4, while they are almost equal in the current models (section 3.3).

Krasnopolsky (2007b, 2013a) modified the concept of the sulfur atmospheric cycles. Combining this model and that for the middle atmosphere (*Krasnopolsky,* 2012b), the major net effect of photochemistry is

$$CO_2 + SO_2 \rightarrow SO_3 + CO$$

The SO_2 lifetime is on the order of months in the middle atmosphere, and the above reaction may be considered as a fast atmospheric cycle.

Production of reduced sulfur S_X + OCS by

$$SO_2 + 2\ CO \rightarrow 2\ CO_2 + S$$

results in mutual transformations within reduced sulfur, including

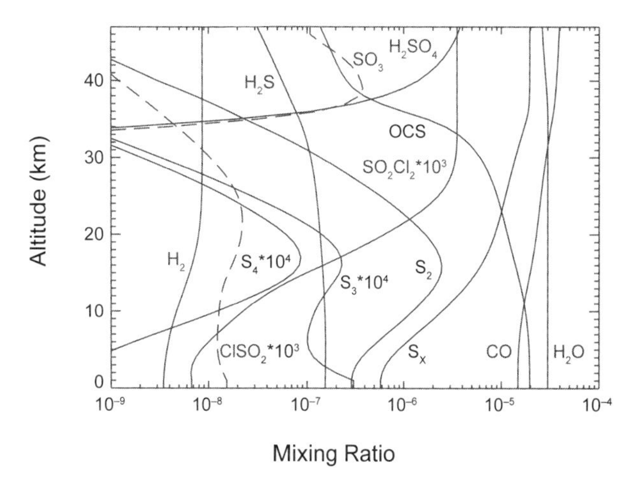

Fig. 24. Profiles of the major species in the model for the lower atmosphere of Venus (*Krasnopolsky,* 2013a). The SO_2, HCl, and NO mixing ratios are constant at 130 ppm, 400 ppb, and 5.5 ppb, respectively, and are not shown.

$SO_3 + 2\ OCS \rightarrow CO_2 + CO + SO_2 + S_2$
$S + OCS \rightarrow CO + S_2$
$S + CO + M \rightarrow OCS + M$

condensation of sulfur aerosol near 50 km, and loss of reduced sulfur in the middle atmosphere by

$$OCS + 2\ CO_2 \rightarrow SO_2 + 3\ CO$$

may be considered as a slow atmospheric cycle, because the lifetime of reduced sulfur is ~3×10^4 yr.

The geological cycle of sulfur remains unchanged (*Prinn,* 1985; *Fegley and Treiman,* 1992) and includes formation of OCS from pyrite

$$FeS_2 + CO_2 + CO \rightarrow FeO + 2\ OCS$$

and return of pyrite via reaction of the abundant SO_2 with carbonate

$SO_2 + CaCO_3 \rightarrow CaSO_4 + CO$
$2\ CaSO_4 + FeO + 7\ CO \rightarrow FeS_2 + 2\ CaCO_3 + 5\ CO_2$

3.2.4. Latitudinal variations of CO and OCS. According to the VEX VIRTIS observations (*Marcq et al.,* 2008), CO near 36 km varies from 24 ppm near the equator to 31 ppm at 60°N and S. Variations of OCS at 33 km are opposite: 4.1 ppm and 2.5 ppm, respectively. *Yung et al.* (2009) applied their two-dimensional transport model and included a conversion of OCS into CO by an unidentified process with a rate of 10^{-8} s^{-1}. The calculated latitudinal variations of OCS agree with the observations while those of CO are smaller than observed. The sum of OCS and CO is constant in the model but varies in the observations.

3.3. Photochemistry of the Middle Atmosphere

3.3.1. History of the problem. Prinn (1971) was the first to recognize the importance of ClO_x chemistry on Venus, before this chemistry was even understood in Earth's atmosphere. *Krasnopolsky and Parshev* (1981, 1983) calculated a photochemical model with a ClCO cycle and related ClCO chemistry that dominate in recombination of CO with O and O_2 on Venus. *Yung and DeMore* (1982) argued that the reaction

$$ClCO + O_2 \rightarrow CO_2 + ClO$$

from the model by *Krasnopolsky and Parshev* (1981, 1983) proceeds in two steps

$ClCO + O_2 + M \rightarrow ClCO_3 + M$
$ClCO_3 + Cl \rightarrow CO_2 + Cl + ClO$
Net $ClCO + O_2 \rightarrow CO_2 + ClO$

with the same net result, and this was later confirmed in the laboratory (*Pernice et al.,* 2004). *Yung and DeMore* (1982) suggested three models: one based on abundant H_2 (model A), one based on NO with a mixing ratio of 30 ppb produced by lightning and significant H_2 in model B, and one based on the ClCO cycle in models B and C. Observations have not confirmed the assumptions of models A and B, and model C was the basic model, further developed by *Mills* (1998) and *Mills and Allen* (2007).

Analysis of the SO_2 and SO profiles at 70–105 km observed by the SPICAV-UV and SOIR solar occultations (*Belyaev et al.,* 2012) revealed a deep minimum in the SO_2 mixing ratio at ~80 km. While the decrease from 70 to 80 km is caused by photolysis of SO_2, the growth to 90–105 km requires a source in this region. *Zhang et al.* (2010, 2012) suggested either photolysis of H_2SO_4 vapor or oxidation of S_x vapor. They adopted $T \approx 250$ K near 100 km on the nightside that is significantly above ~180 K from radio occultations, observations of the O_2 nightglow at 1.27 μm, thermal sounding at 15 μm, and the Pioneer Venus night probe data. Saturated vapor densities of H_2SO_4 and sulfur are higher at this temperature by a few orders of magnitude; *Zhang et al.* (2012) adopted a peak H_2SO_4 mole fraction of 200 ppb at 96 km and the column ratio of 10 ppb above 85 km. *Krasnopolsky* (2011b) calculated a profile of H_2SO_4 vapor by solving the continuity equations that account for condensation and sublimation of H_2O and H_2SO_4 vapors, their photolyses, and vertical transport by eddy diffusion. The calculated H_2SO_4 vapor densities are extremely low and cannot explain the peak of SO_2 near 100 km. *Krasnopolsky* (2011b) proposed that the sulfuric acid aerosol particles of some specific sizes may simulate the observed absorption ascribed to SO_2 in *Belyaev et al.* (2012). Microwave observations established an upper limit of 3 ppb for H_2SO_4 vapor above 85 km (*Sandor et al.,* 2012), which is below the adopted abundance in *Zhang et al.* (2012).

3.3.2. Photochemistry at 47–112 km. Here we will briefly discuss a global-mean model by *Krasnopolsky* (2012b) that is extended down to 47 km to use some data from section 3.2.2 as the lower boundary conditions. Numerical accuracy of the model is improved using a small vertical step; the observations by the Venera 14 landing probe at 320–400 nm (*Ekonomov et al.,* 1983) are applied to simulate the NUV absorption in the clouds; and the standard ClCO kinetic data are used in the model. H_2O densities are calculated (not adopted) and the H_2SO_4 photolysis is negligible in the model. The model involves 153 reactions of 44 species, and the number of reactions was significantly reduced by removal of unimportant processes. Column reaction rates are calculated for all reactions and make it possible to study balances for each species in detail. These are the main features of the model.

Profiles of temperature and eddy diffusion in the model are shown in Fig. 25a. Five versions of the model were calculated with the eddy diffusion breakpoint $h_e = 60 \pm 5$ km and $f_{SO_2} = 9.7 \pm 0.5$ ppm at 47 km. These minor variations result in dramatic changes in photochemistry and abundances of some species in the middle atmosphere.

Photolysis of CO_2 forms CO with a column rate of 4.7 ×

10^{12} cm^{-2} s^{-1}; photolysis of OCS is weaker by 2 orders of magnitude. Recombination of CO_2 proceeds mostly via the ClCO cycle

$$CO + Cl + M \rightarrow ClCO + M$$
$$ClCO + O_2 + M \rightarrow ClCO_3 + M$$
$$\underline{ClCO_3 + X \rightarrow CO_2 + Cl + XO}$$
$$\text{Net } CO + O_2 + X \rightarrow CO_2 + XO \ (3.42 \times 10^{12} \text{ cm}^{-2} \text{ s}^{-1})$$

where X = Cl, O, SO, SO_2, and H. Reaction

$$ClCO + O \rightarrow CO_2 + Cl$$

and photolysis of $ClCO_3$ add 5×10^{11} cm^{-2} s^{-1} to the production of CO_2. The odd hydrogen cycle and termolecular reactions with O and S cover 5% of the CO production, and the remaining CO forms a flux of 6.3×10^{11} cm^{-2} s^{-1} to the lower atmosphere (section 3.2.2). The calculated profile of CO is compared with the observations in Fig. 25b.

Although SO_2 is a minor constituent, its photolysis exceeds that of CO_2 by a factor of 2.5, initiating intense sulfur chemistry. Formation of sulfuric acid aerosol is the main feature of Venus' photochemistry

$$SO_2 + O + M \rightarrow SO_3 + M$$
$$SO_3 + 2\ H_2O \rightarrow H_2SO_4 + H_2O$$

These reactions proceed in a narrow layer near 66 km, where the calculated H_2O mole fraction of 2.5 ppm (Fig. 25c) corresponds to monohydrate $H_2SO_4{\bullet}H_2O$. [The peak altitude is higher than 62 km in *Yung and DeMore* (1982) because the second UV absorber is included in the model by *Krasnopolsky* (2012b).] The formation of sulfuric acid greatly reduces the SO_2 mixing ratio to ~100 ppb near 70 km (Fig. 26). SO

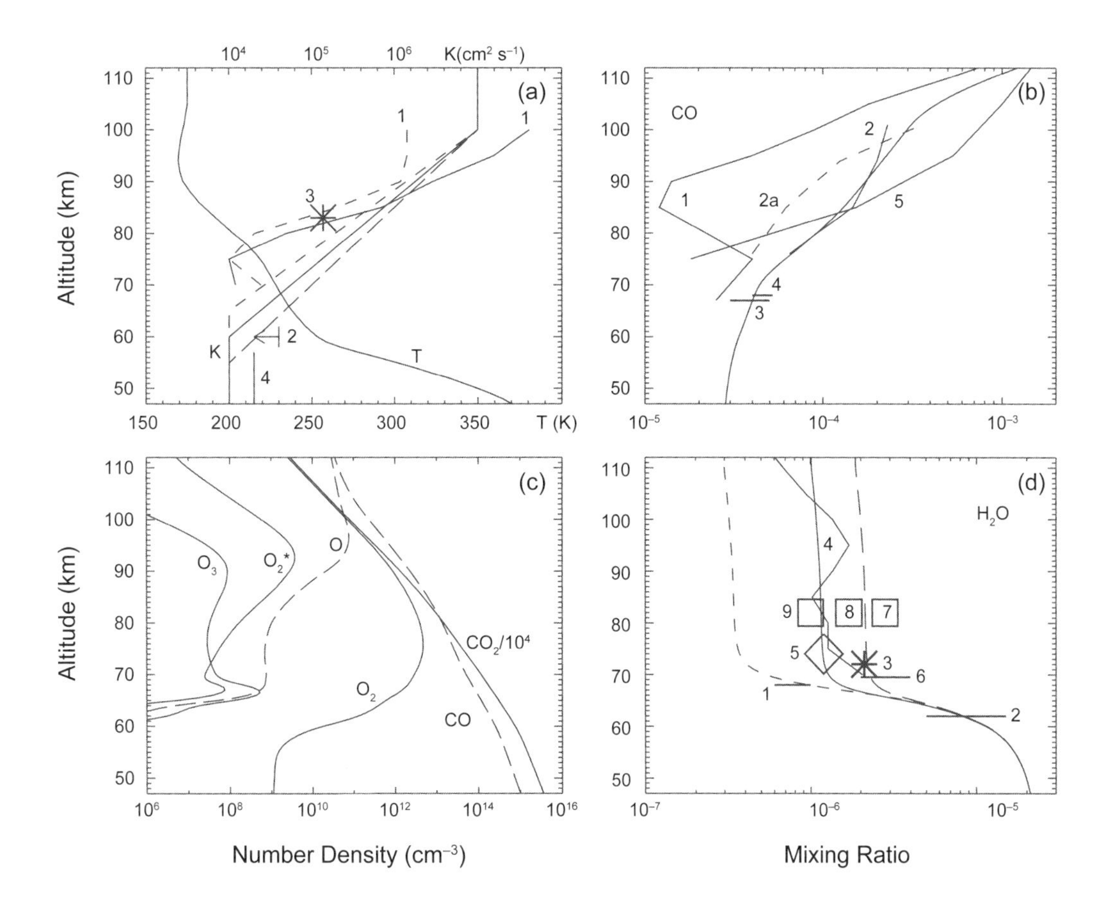

Fig. 25. Some data from photochemical model for Venus atmosphere at 47–112 km (*Krasnopolsky,* 2012b). **(a)** Profiles of temperature and three versions of eddy diffusion in the model. Observations: (1) *Krasnopolsky* (1980, 1983); (2) upper limit from *Woo and Ishimaru* (1981); (3) *Lane and Opstbaum* (1983); (4) *Krasnopolsky* (1985). **(b)** Calculated profile of the CO mixing ratio. Observations: (1) VEX/SOIR (*Vandaele et al.,* 2008); (2) *Clancy et al.* (2012a), dayside observations in 2001–2002; (2a) *Clancy et al.* (2012a), dayside observations in 2007–2009; (3) VEX/VIRTIS (*Irwin et al.,* 2008); (4) *Connes et al.* (1968), *Krasnopolsky* (2010c); (5) *Lellouch et al.* (1994). **(c)** CO_2 and its photochemical products. O_2* is $O_2(^1\Delta_g)$ that emits the airglow at 1.27 µm. **(d)** Profiles of the H_2O mixing ratio: models and observations. Three calculated profiles are for the models with the SO_2 mixing ratio of 10.2 (short dashes), 9.7 (solid), and 9.2 (long dashes) ppm at 47 km. Observations: (1) *Fink et al.* (1972); (2) Venera 15 (*Ignatiev et al.,* 1999); (3) KAO (*Bjoraker et al.,* 1992); (4) VEX/SOIR (*Fedorova et al.,* 2008); (5) *Krasnopolsky* (2010d); (6) VEX/VIRTIS (*Cottini et al.,* 2011). Mean results of the microwave observations: (7) *Encrenaz et al.* (1995); (8) SWAS (*Gurwell et al.,* 2007); (9) *Sandor and Clancy* (2005).

produced by the SO_2 photolysis above ~70 km regenerates SO_2, mostly in the reactions with ClO, NO_2, and SO + O + M. The close balance between the SO_2 photolysis and the regeneration at 75–95 km and the increasing eddy diffusion tends to keep a constant SO_2 mixing ratio of ~30 ppb in the basic model at these altitudes where [SO] < [SO_2]. The regeneration reactions become ineffective above 100 km. SO is the main photochemical product of SO_2 above 75 km. Reaction of SO + NO_2 is mostly responsible for the removal of SO below 75 km. A sum of the SO_2 and SO mixing ratios is nearly constant above 80 km, SO_2/SO ≈ 3 at 80–95 km, and SO dominates above 100 km (Fig. 26).

Photolysis of OCS is strongly affected by absorption by SO_2 and the NUV absorber. [A few candidates were suggested for this absorber, and the solution of ~1% $FeCl_3$ in sulfuric acid is a plausible version (*Zasova et al.,* 1981; *Krasnopolsky,* 1986, 2006d) that is beyond this model.] The calculated OCS abundances agree with the observations.

The aerosol sulfur (Fig. 26) is a minor species in the model with a mixing ratio of 2.5 ppb below 60 km. Photolysis of SO was the main source of free sulfur in the previous models. Reactions

$$SO + SO + M \rightarrow S_2O_2 + M$$
$$S_2O_2 + h\nu \rightarrow SO_2 + S$$

are more effective in this model. Atomic sulfur is mostly lost in the reaction with O_2, which restricts the sulfur aerosol formation.

Effects of minor variations of eddy diffusion on the abundances of the basic sulfur species are shown in Fig. 26. The eddy breakpoint h_e (Fig. 25a) is changed from 60 km in the basic model to 55 and 65 km in the additional versions. Eddy diffusion controls the amount of SO_2 that can be delivered through the bottleneck near 66 km where sulfuric acid forms. SO_2 varies above 70 km within a factor of ~30, and its variations induce variations of SO. Minor variations of the SO_2/H_2O ratio at the lower boundary also result in significant variations of SO_2 and SO above 70 km.

The S atoms from the OCS photolysis react with either CO (in termolecular association of OCS) or O_2. The former dominates below 60 km; that is why the steep decrease in OCS occurs above 60 km (Fig. 26). The delivery of O_2 by eddy diffusion becomes weaker if the eddy breakpoint is at 65 km, and the balance between the two reactions of S moves to 65 km as well with a steep decrease in OCS above 65 km in this case. The abundance of the aerosol sulfur S_a is mostly controlled by the delivery of O_2 that depends on the eddy diffusion breakpoint h_e, and S_a is very sensitive to the variations of h_e (Fig. 26). Overall, there is a good agreement between the model and the observations except

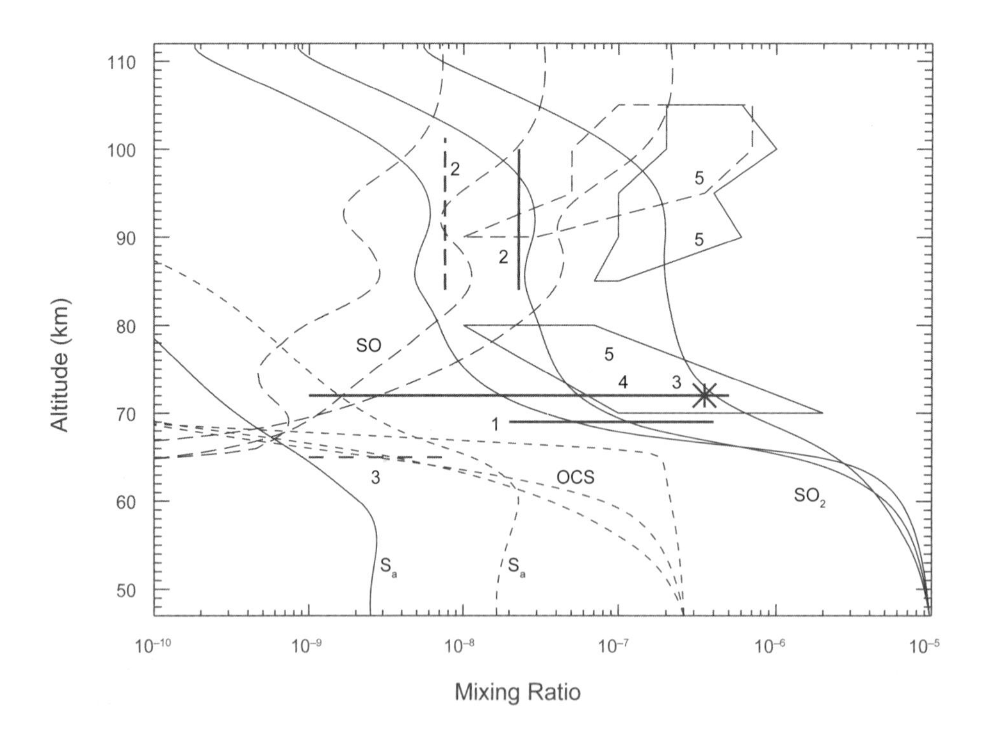

Fig. 26. Basic sulfur species: model results and observations. SO_2 (solid), OCS (short dashes), and SO (long dashes) profiles are shown for the models with the eddy break at h_e = 55, 60, and 65 km. SO_2 and SO are more abundant while OCS and S_a are less abundant for h_e = 55 km. The aerosol sulfur S_a mixing ratio is 4 × 10^{-11} at 47 km for h_e = 55 km and is not shown. Observations: (1) PV, Venera 15, HST, and rocket data (*Esposito et al.,* 1997); (2) mean results of the submillimeter measurements (*Sandor et al.,* 2010); the observed SO_2 varies from 0 to 76 ppb and SO from 0 to 31 ppb; (3) IRTF/CSHELL (*Krasnopolsky,* 2010d); (4) SPICAV-UV, nadir (*Marcq et al.,* 2011); (5) VEX/SOIR and SPICAV-UV occultations (*Belyaev et al.,* 2012).

the problematic SO and SO_2 above 90 km (section 3.3.1).

Water vapor (Fig. 25d) is also greatly affected by the formation of sulfuric acid. The H_2O mixing ratio is very sensitive to small variations of the SO_2/H_2O ratio at the lower boundary. The model H_2O profile is in excellent agreement with the observations.

The calculated vertical profiles of O_2, O, $O_2(^1\Delta_g)$, and O_3 are shown in Fig. 25c. Reactions O + (ClO, NO_2, O + M, and OH) contribute 60, 23, 12, and 5% to the production of O_2, respectively. The O_2 total production is 5.8×10^{12} cm^{-2} s^{-1} and removes 80% of oxygen atoms released by photolyses of CO_2 and SO_2. O_2 is lost in the ClCO cycle as a difference between the reactions

$$ClCO + O_2 + M \rightarrow ClCO_3 + M$$
$$ClCO_3 + O \rightarrow CO_2 + O_2 + Cl$$

This difference accounts for 57% of the O_2 loss, and the remaining 43% is by S + O_2. The calculated O_2 column abundance is 7.6×10^{18} cm^{-2}, exceeding the observed upper limit (Table 3) by an order of magnitude. This is a traditional problem of the O_2 chemistry on Venus that remains unsolved. Attempts to solve it by a significant increase of a ClCO equilibrium constant (*Mills,* 1998, *Mills and Allen,* 2007) are not justified by the laboratory data and result in insufficient improvements for O_2 and significant problems for CO.

Chlorine species in the middle atmosphere of Venus (Fig. 27a) originate from photolysis of HCl that forms equal quantities of H and Cl. Odd hydrogen is also formed by photolysis of H_2O. Its column rate is smaller than that of HCl by a factor of 70. However, the H_2O photolysis exceeds that of HCl above 95 km, and the H densities are greater than Cl above 95 km.

Molecular chlorine is formed in termolecular association Cl + Cl + M and via

$$Cl + X + M \rightarrow ClX + M$$
$$ClX + Cl \rightarrow Cl_2 + X$$
$$\text{Net } Cl + Cl \rightarrow Cl_2$$

Here X = NO, SO_2, O, CO, SO, and O_2, in decreasing order, and Cl_2 immediately dissociates with a column rate of 3.18×10^{13} cm^{-2} s^{-1} that significantly exceeds the rates of the CO_2 and SO_2 photolyses.

A balance between odd hydrogen and chlorine species is established via reactions

$H + HCl \rightarrow H_2 + Cl$	(2.31×10^9)
$Cl + H_2 \rightarrow HCl + H$	(2.40×10^9)
$OH + HCl \rightarrow H_2O + Cl$	(3.16×10^9)

Their column rates in parentheses are in cm^{-2} s^{-1}. The first and second reactions are opposite with almost equal rates. The last reaction converts odd hydrogen into odd chlorine, and this explains why the former is less abundant than the latter below 95 km. The strong depletion of H_2O by the formation of sulfuric acid and the depletion of odd hydrogen in the last reaction diminish odd hydrogen chemistry on Venus.

The first and second reactions forms and removes H_2, respectively. Their altitude distributions are different, and this difference results in a deep minimum of H_2 at 72 km (Fig. 27a). The H_2 mixing ratio is 4.5 ppb below 65 km and 23 ppb above 95 km.

Loss of HCl in reactions

$HCl + h\nu \rightarrow H + Cl$	(9.46×10^{10})
$HCl + OH \rightarrow H_2O + Cl$	(3.16×10^9)
$HCl + O \rightarrow OH + Cl$	(1.60×10^9)
Total	9.94×10^{10} cm^{-2} s^{-1}

is almost balanced by its production

$Cl + HO_2 \rightarrow HCl + O_2$	(6.71×10^{10})
$H + Cl_2 \rightarrow HCl + Cl$	(2.49×10^{10})
$H + ClSO_2 \rightarrow HCl + SO_2$	(3.55×10^9)
Total	9.56×10^{10} cm^{-2} s^{-1}

The difference of 3.8×10^9 cm^{-2} s^{-1} is released as a flow of SO_2Cl_2 into the lower atmosphere (section 3.2.2). This flow exceeds by a factor of 2 the difference between the production of H_2O in R51 HCl + OH and its loss by photolysis. (The factor of 2 is because two odd hydrogen species are formed by photolysis of H_2O.) The calculated SO_2Cl_2 mixing ratio is 30 ppb at 68 km and 8 ppb at 47 km; the Cl_2 mixing ratio is 22 ppb at 70 km.

The total chlorine mixing ratio is constant at 417 ppb in the model, while HCl varies from 400 ppb at 47 km to 340 ppb at 66 km, then to 370 ppb at 90 km and 310 ppb at 110 km. Scaling this profile of HCl to the observed 400 ppb near 70 km (Table 4) results in 490 ppb at 47 km, in excellent agreement with 500 ppb at 15–30 km (Table 4). However, the model does not support HCl $\approx$ 100–150 ppb at 70–105 km in the SOIR observations (*Vandaele et al.,* 2008) and a steep decrease from ~400 ppb at 80 km to 0 at h > 94 km in the submillimeter observations by *Sandor and Clancy* (2012).

Odd nitrogen chemistry was calculated by *Yung and DeMore* (1982) in their model B for 30 ppb of NO and 500 ppb of H_2 with a major cycle

$$H + O_2 + M \rightarrow HO_2 + M$$
$$NO + HO_2 \rightarrow NO_2 + OH$$
$$CO + OH \rightarrow CO_2 + H$$
$$NO_2 + h\nu \rightarrow NO + O$$
$$\text{Net } CO + O_2 \rightarrow CO_2 + O$$

This cycle is weak at 5.5 ppb of NO and 4.5 ppb of H_2 in our model, and more important are the cycles

$$NO + O + M \rightarrow NO_2 + M$$
$$NO_2 + O \rightarrow NO + O_2$$
$$\text{Net } O + O \rightarrow O_2$$

that provides a quarter of the O_2 production

$$NO + Cl + M \rightarrow ClNO + M$$
$$ClNO + Cl \rightarrow NO + Cl_2$$
$$\text{Net } Cl + Cl \rightarrow Cl_2$$

which is responsible for a third of the Cl_2 production, and

$$NO + O + M \rightarrow NO_2 + M$$
$$SO + NO_2 \rightarrow SO_2 + NO$$
$$\text{Net } SO + O \rightarrow SO_2$$

that compensates a quarter of the SO_2 photolysis. The calculated profiles of some odd nitrogen species are shown in Fig. 27b. Overall, NO below 80 km on Venus is important as a convincing proof of lightning (*Krasnopolsky,* 2006b) and an effective catalyst in the above cycles.

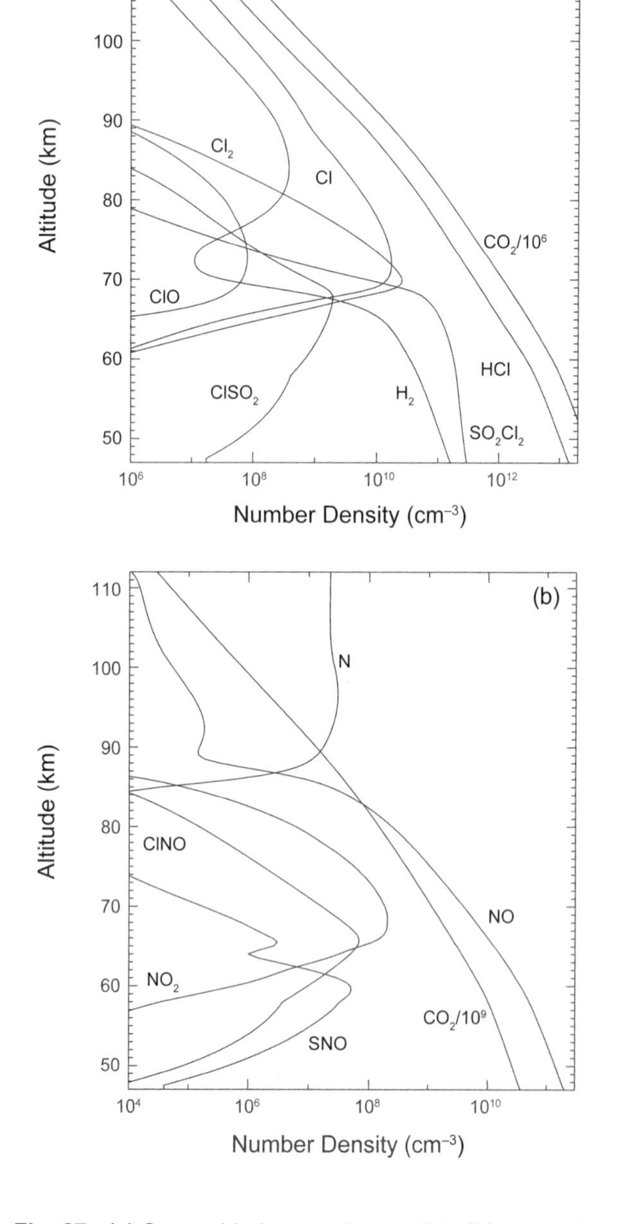

Fig. 27. (a) Some chlorine species and H_2; **(b)** some nitrogen species. From *Krasnopolsky* (2012b).

3.4. Venus Nightglow and Nighttime Photochemistry

3.4.1. Night airglow on Venus. The Venus nightglow was discovered by the Venera 9 and 10 orbiters (*Krasnopolsky et al.,* 1976; *Krasnopolsky,* 1983a) in the visible range and studied in the laboratory (see, e.g., *Slanger and Copeland,* 2003). The visible nightglow consists of the Herzberg II, Herzberg I, and Chamberlain band systems with mean intensities of 2700, 140, and 200 R, respectively. *Connes et al.* (1979) detected a much brighter nightglow of O_2 at 1.27 μm with an intensity of ~1 MR using ground-based high-resolution spectroscopy. Later the O_2 nightglow at 1.27 μm (Fig. 28) was observed by *Crisp et al.* (1996), *Ohtsuki et al.* (2008), *Bailey et al.* (2008), and *Krasnopolsky* (2010b), and by VEX using the VIRTIS spectrograph. The VEX observations (*Piccioni et al.,* 2009) revealed both O_2 (0–0) and (0–1) bands at 1.27 μm and 1.58 μm, respectively.

Morphology of the Venus nightglow may be described by four parameters (Table 5): intensity at a maximum $4\pi I_{max}$ that is near the antisolar point, local time of the maximum LT_{max}, nightside mean intensity $4\pi I$, and mean altitude of the nightglow layer h_{peak}. These parameters for the VEX O_2 nightglow observations at 1.27 μm (*Soret et al.,* 2012) are given in Table 5.

Venus' ultraviolet nightglow was detected by *Feldman et al.* (1979) using the International Ultraviolet Explorer (IUE) and studied by the Pioneer Venus orbiter (*Stewart et al.,* 1980). This nightglow consists of the (0–v″) progressions of the γ (A → X) and δ (C → X) band systems of NO that are excited by the two-body formation of NO (C, v = 0). The δ (0–1) band at 198 nm is the strongest feature, and accounts for 18% of the total emission. The nightglow parameters from the Pioneer Venus observations (*Bougher et*

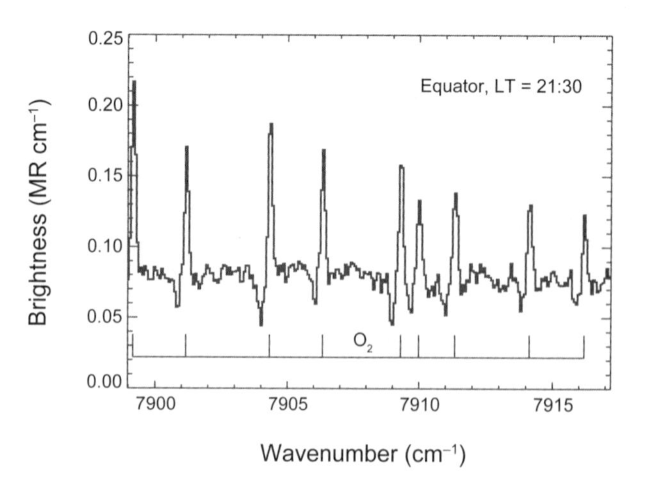

Fig. 28. Spectrum of Venus' O_2 nightglow at 1.27 μm; nine emission lines are seen (*Krasnopolsky,* 2010b).

TABLE 5. Morphology of Venus nightglow.

Nightglow	$4\pi I_{max}$	LT_{max}	$4\pi I_{mean}$	h_{peak} (km)	Comment
$O_2(^1\Delta_g)$ 1.27 μm	1.6 MR	0 hr	0.5 MR	98	VEX, solar min; *Soret et al.* (2012)
NO, γ and δ bands	10.6 kR	2 hr	2.2 kR	115	PV, solar max; *Bougher et al.* (1990)
NO, γ and δ bands	—	—	400 R	113	VEX, solar min; see text
OH 2.9 μm	—	0 hr	7 kR	97	VEX, solar min; *Soret et al.* (2012)

al., 1990) are given in Table 5 and refer to solar maximum.

The NO nightglow is currently observed by VEX using the SPICAV UV spectrograph (*Gerard et al.*, 2008). The published morphology is still insufficient to derive all parameters. The mean peak altitude is 113 km, and the mean peak intensity at the limb is 32 kR. Assuming a layer thickness of 20 km, one gets a mean vertical intensity of 640 R. Most of the observations were made at LT = 0 to 2 hr, and a possible nightside-mean nightglow may be ~400 R (Table 5). The (0–0) transition between the upper states of the δ- and γ-bands was detected at 1.224 μm as well (*Garcia Munoz et al.*, 2009a).

The O 558-nm line and the Herzberg II bands of O_2 were detected and observed by *Slanger et al.* (2001, 2006) using the High Resolution Echelle Spectrometer (HIRES) spectrograph at the Keck Observatory. The Herzberg II bands of O_2 were recently observed by VEX's VIRTIS by summing of the limb spectra (*Garcia Munoz et al.*, 2009b). Similar to the Venera spectra, the O 558-nm line is not seen in the VIRTIS data (*Garcia Munoz et al.*, 2009b) and may probably appear at some rare events of dense nighttime ionosphere.

The VIRTIS observations revealed for the first time the OH rovibrational bands in the Venus nightglow (*Piccioni et al.*, 2008; *Soret et al.*, 2012). The observed nightglow involves the Δv = 1 and 2 band sequences at 2.9 and 1.4 μm, respectively. A mean limb intensity of the Δv = 1 sequence is 350 kR (*Soret et al.*, 2012), i.e., the vertical intensity is ~7 kR. The sequence consists of the (1–0), (2–1), (3–2), and (4–3) bands with relative intensities of 45, 39, 9, and 7%, respectively. The Δv = 2 sequence includes mostly the (2–0) and disagrees with the Δv = 1 band distribution. There is a significant correlation between the OH and O_2 nightglows. *Krasnopolsky* (2010b) found a narrow window for groundbased observations of the OH nightglow on Venus and detected lines of the (1–0) and (2–1) bands that gave the OH nightglow intensities of ~10 kR.

3.4.2. Nightglow excitation problems. The observed O_2 nightglow on Venus is very different from that on Earth, especially in the visible range, where the Herzberg II bands dominate on Venus and Herzberg I and the green line O 558 nm on Earth. The O_2 nightglow on both planets is excited by the termolecular association of O_2, and yields of the O_2 electronic states should not depend on a third body. Therefore, quenching and excitation transfer processes are responsible for the difference. The problem was analyzed by *Krasnopolsky* (2011a), who combined the nightglow observations on both planets with the related laboratory data and some findings in theory. The O_2 electronic states are highly vibrationally excited on Earth, while only v′ = 0 progressions are observed in the Venus nightglow. The latter is caused by rapid vibrational relaxation in CO_2. Quenching reactions depend on vibrational excitation, and this complicates the problem.

Radiative times, yields, and quenching coefficients in the suggested excitation scheme are given in Table 6. Three u-states are excited directly with yields calculated by *Wraight* (1982). Energy transfers from the upper states including $^5\Pi_g$ are essential for the b → X and a → X emissions at 762 nm and 1.27 μm, respectively. These transfers were estimated using spin, sign, and statistical weight considerations. For example, quenching of O_2(c) by CO_2 to the ground state O_2(X) is spin forbidden and to O_2(b) is sign forbidden; therefore O_2(c) is quenched to O_2(a). Quenching of O_2(b) by CO_2 to O_2(X) is spin forbidden; therefore both O_2(c) and O_2(b) populate O_2(a). If a yield of O_2(a) in quenching of $O_2(^5\Pi_g)$ is 0.9, then the effective yield of O_2(a) is 0.7 on Venus and Mars (see details in *Krasnopolsky*, 2011a). The green line $O(^1S)$ 558 nm is excited by

$$O + O_2(A^3\Sigma_u^+, v \geq 6) \rightarrow O_2 + O(^1S)$$
$$k = 1.5 \times 10^{-11}\ cm^3\ s^{-1}$$

Evidently this reaction is insignificant on Venus because of the vibrational relaxation by CO_2.

Another problem is the excitation of the OH nightglow on Venus, because the observed v is ≤4 on Venus and ≤9 on Earth. There are two possible processes for the OH vibrational excitation on the terrestrial planets

$$H + O_3 \rightarrow OH(v \leq 9) + O_2 \quad (1)$$
$$O + HO_2 \rightarrow OH(v \leq 6) + O_2 \quad (2)$$

Kaye (1988) argued that the O-H bond is formed in reaction (1) and large v values are probable. However, OH is the remaining part of HO_2 in reaction (2) with a low vibrational excitation v ≤ 3. *Krasnopolsky* (2013b) adopted the OH transition probabilities, excitation yields in reaction (1), and quenching rate coefficients by CO_2 from *Garcia Munoz et al.* (2005) and proposed collision cascade yields that fit the observed band distribution in the Δv = 1 sequence.

3.4.3. Nighttime chemistry at 80–130 km on Venus. The VEX observations of the O_2 and NO nightglow (*Piccioni et al.*, 2009; *Gerard et al.*, 2008) were initially simulated (*Gerard et al.*, 2008) using a simple set of four reactions that neglected effects of chlorine and hydrogen chemistry.

TABLE 6. Excitation, excitation transfer, and quenching processes in the O_2 nightglow.

State	Bands	τ (s)	α	α_{TE}	α_{TV}	k_O	k_{O_2}	k_{N_2}	k_{CO_2}
$A^3\Sigma_u^+$	HzI	0.14	0.04	0	0	1.3×10^{-11}	4.5×10^{-12}	3×10^{-12}	8×10^{-12}
$A'^3\Delta_u$	Chm	2–4	0.12	0	0	1.3×10^{-11}	3.5×10^{-12}	2.3×10^{-12}	4.5×10^{-13}
$c^1\Sigma_u^-$	HzII	5–7	0.03	0	0	8×10^{-12}	$3 \times 10^{-14}/1.8 \times 10^{-11}$	—	1.2×10^{-16}
$b^1\Sigma_g^+$	762 nm	13	0.02	0.09	0.125	8×10^{-14}	4×10^{-17}	2.5×10^{-15}	3.4×10^{-13}
$a^1\Delta_g$	1.27 μm	4460	0.05	0.35	0.65	—	10^{-18}	$<10^{-20}$	10^{-20}

HzI, HzII, and Chm are the Herzberg I, II, and Chamberlain bands; τ is the radiative lifetime; if two values are given, then they refer to high and low vibrational excitation on the Earth and Venus, respectively; α is the direct excitation yield; α_{TE} and α_{TV} are excitation transfer yields for Earth and Venus from the upper states including $^5\Pi_g$ (see details in *Krasnopolsky,* 2011a); and k_X is quenching rate coefficients in $cm^3\ s^{-1}$. Two values of k_{O_2} are for $c^1\Sigma_g^-$ (v = 0 and 7–11). From *Krasnopolsky* (2011a).

The OH nightglow was initially compared with data from global-mean photochemical models that reflect mostly a daytime chemistry and are generally inapplicable to the nighttime conditions.

To solve these problems, *Krasnopolsky* (2013b) developed a model for nighttime chemistry on Venus. The model involves 86 reactions of 29 species. The lower boundary is at 80 km to avoid the detailed sulfur chemistry. Densities of CO_2, HCl, SO_2, and a flux of NO are the lower boundary conditions, with $V = -K/2H = -0.116$ cm s^{-1} for the remaining species. Fluxes Φ of O, N, H, and Cl at 130 km are the model parameters. These downward fluxes are equal to 3×10^{12}, 1.2×10^9, 10^{10}, and 10^{10} $cm^{-2}\ s^{-1}$, respectively, in the basic version. They are proportional to column abundances of these species at 90 km in the global-mean model (section 3.3.2).

The calculated species altitude profiles are shown in Fig. 29. Major losses of odd hydrogen and odd chlorine are via reactions

$$H + Cl_2 \rightarrow HCl + Cl,\ 6.74 \times 10^9\ cm^{-2}\ s^{-1}$$
$$Cl + HO_2 \rightarrow HCl + O_2,\ 2.72 \times 10^9\ cm^{-2}\ s^{-1}$$

Their sum is almost equal to 10^{10} $cm^{-2}\ s^{-1}$, the adopted fluxes of H and Cl at the upper boundary. Odd hydrogen is partly transformed into odd chlorine via

$$H + HCl \rightarrow H_2 + Cl,\ 5.17 \times 10^8\ cm^{-2}\ s^{-1}$$
$$OH + HCl \rightarrow H_2O + Cl,\ 2.33 \times 10^8\ cm^{-2}\ s^{-1}$$

Therefore odd chlorine is more abundant than odd hydrogen below ≈95 km.

Major odd hydrogen and odd chlorine cycles are

$$X + O_3 \rightarrow XO + O_2$$
$$O + XO \rightarrow X + O_2$$
$$\text{Net } O + O_3 \rightarrow 2\ O_2$$

Here X = H and Cl with the O_2 net production of 2.5×10^{11} and 4.5×10^{11} $cm^{-2}\ s^{-1}$, respectively. These cycles remove almost half the O flux at 130 km, and the remaining half is lost in the termolecular association of O_2. Oxidation of CO is weak above 80 km, and all O atoms form O_2 with a mixing ratio of $\Phi_O/(2n_{80}V) = 85$ ppm at 80 km, increasing to 280 ppm at 100 km (Fig. 29).

The calculated intensity and profile of the O_2 nightglow at 1.27 μm (Fig. 30) agrees with the VEX observations, and the nightglow intensity may be approximated by

$$4\pi I_{O_2} = 127(\Phi_O/10^{12})^{1.22}\ \text{kR}$$

i.e., the O_2 nightglow quantum yield is 0.16 per oxygen atom.

The above H and Cl cycles balance the production of ozone in the termolecular association O + O_2 + M. The calculated O_3 profile (Fig. 29) is in excellent agreement with those observed by stellar occultations from VEX (*Montmessin et al.,* 2011). Variations of the peak O_3 density may be approximated by

$$[O_3]_{max} = \frac{2.2 \times 10^7 \left(\Phi_O/10^{12}\right)^{1.2}}{Y^{0.74+0.3\ln Y}} \text{cm}^{-3};\ Y = \frac{\Phi_{Cl}}{10^{10}}$$

The model profiles of the NO and OH nightglows (Fig. 30) agree with the observations as well, and their vertical intensities are equal to

$$4\pi I_{NO} = 225\left(\Phi_N/10^9\right)\left(\Phi_O/10^{12}\right)^{0.35} \text{R}$$

$$4\pi I_{OH(\Delta v=1)} = 1.2\left(\Phi_O/10^{12}\right)^{1.4} \frac{X^{1.47-0.46\ln X}}{Y^{1.43+0.65\ln Y}} \text{kR};$$

$$X = \frac{\Phi_H}{10^{10}},\ Y = \frac{\Phi_{Cl}}{10^{10}}$$

Reaction O + HO_2 (Fig. 30b) peaks below the observed maximum of the OH nightglow, and its effect in the nightglow excitation is neglected in the basic model. The calculated profile of CO (Fig. 29) agrees with the nighttime microwave observations (*Clancy et al.,* 2012a) as well, and the model fits all observational constraints.

3.4.4. O_2 and NO nightglow in general circulation models. Fluxes of O, N, H, and Cl were actually free parameters in the one-dimensional nightside model by *Krasnopolsky* (2013b). Production of O and N on the dayside and their transport to the nightside with the subsequent

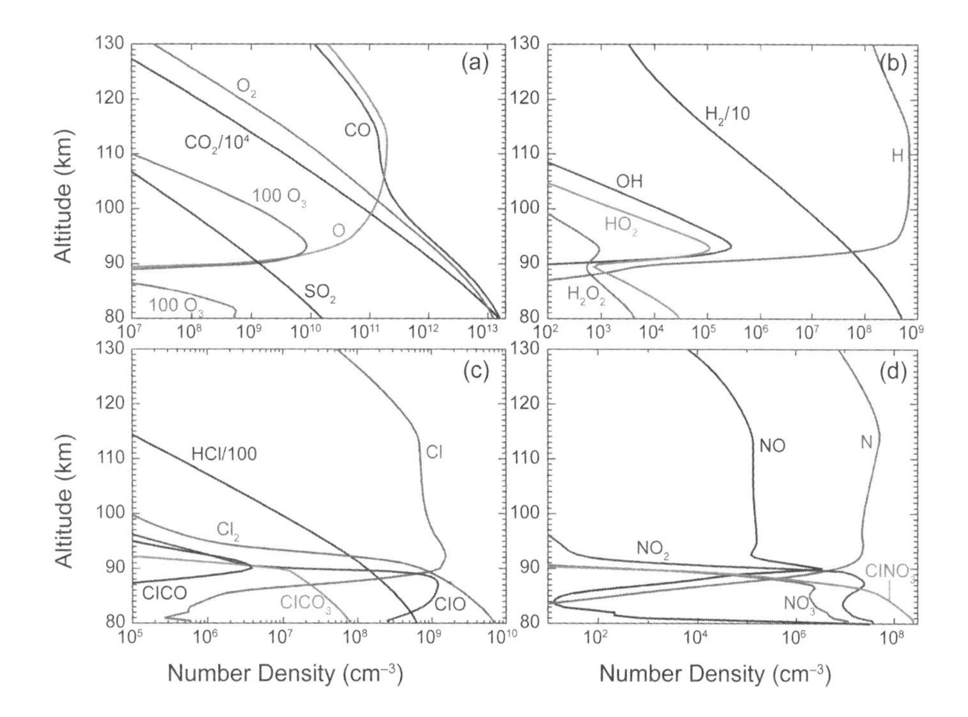

Fig. 29. See Plate 41 for color version. Vertical profiles of species in the model for the nighttime atmosphere of Venus (*Krasnopolsky,* 2013b).

nightglow excitation were simulated by GCMs for the upper atmosphere by *Bougher et al.* (1990) and *Brecht et al.* (2011). Basic ideas of GCMs are discussed by Forget and Lebonnois (this volume).

Dissociation of CO_2 above 80 km proceeds mostly at 130–180 nm and weakly depends on solar activity with the dayside column rate of 9.4×10^{12} cm^{-2} s^{-1} (see section 3.4.3). *Brecht et al.* (2011) calculated the net dayside production of atomic oxygen of 2.3×10^{12} cm^{-2} s^{-1}, i.e., a quarter of the total production. This may be compared with 3×10^{12} cm^{-2} s^{-1} in section 3.4.3, and the mean O_2 nightglow intensity in their model fits the observed value of 0.5 MR (Table 5). The calculated maximum of 1.76 MR and the peak altitude of 100 km also agree with the observations.

Both GCMs involve a detailed chemistry of N, its lowest metastable state N(^{2}D), and NO in the upper atmosphere. According to *Krasnopolsky* (1983b), productions of N and N(^{2}D) are 6×10^9 and 5×10^9 cm^{-2} s^{-1} at solar max and zenith angle z = 60°; N(^{2}D) is either quenched to N by O and CO or to NO by CO_2 with the NO production of 4×10^9 cm^{-2} s^{-1}. NO removes N in the reaction $N + NO \rightarrow N_2 + O$, and the net production of N is 3.2×10^9 cm^{-2} s^{-1} on the dayside at solar maximum. *Bougher et al.* (1990) calculated the total production of N at 8.5×10^9 and 4.2×10^9 and the net production of 3.1×10^9 and 1.05×10^9 cm^{-2} s^{-1} at solar maximum and minimum, respectively, in their preferred models. The value for solar maximum is similar to that in *Krasnopolsky* (1983b) while that at solar minimum is close to 1.2×10^9 cm^{-2} s^{-1} estimated in section 3.4.3 for the VEX observation at solar minimum. Then the expected mean NO nightglow is 1.3 kR at solar maximum, smaller than 2.2 kR in the Pioneer Venus observations (Table 5).

A net production of N is 1.58×10^{10} cm^{-2} s^{-1} at solar minimum in *Brecht et al.* (2011) and exceeds those in *Bougher et al.* (1990) and *Krasnopolsky* (1983b) by an order of magnitude. The mean NO nightglow is 680 R, the maximum is 1.83 kR, and the peak altitude is 106 km (see Table 5 for comparison). The required NO nightglow quantum yield is 0.04, much smaller than 0.35 in section 3.4.3.

3.5. Some Unsolved Problems

The variety of conditions and the complicated chemical composition of the Venus atmosphere suggest many problems that need to be studied. Here we will mention only some critical points.

Mutual conversions between SO_3, OCS, CO, and S_x are the key problem for chemistry in the lower atmosphere. The current model is based on the reactions

$$SO_3 + OCS \rightarrow (SO)_2 + CO_2$$
$$(SO)_2 + OCS \rightarrow CO + SO_2 + S_2$$

from *Krasnopolsky and Pollack* (1994) that have not been studied in the laboratory.

The observed increase in SO_2 above 80 km is highly questionable, and an alternative interpretation of the spectral data should be studied. The lack of detection of O_2 and the very restrictive upper limit by *Trauger and Lunine* (1983) present problems in both observations and modeling. Abundances of Cl_2 and SO_2Cl_2 predicted by the model are significant and may be detected spectroscopically; however,

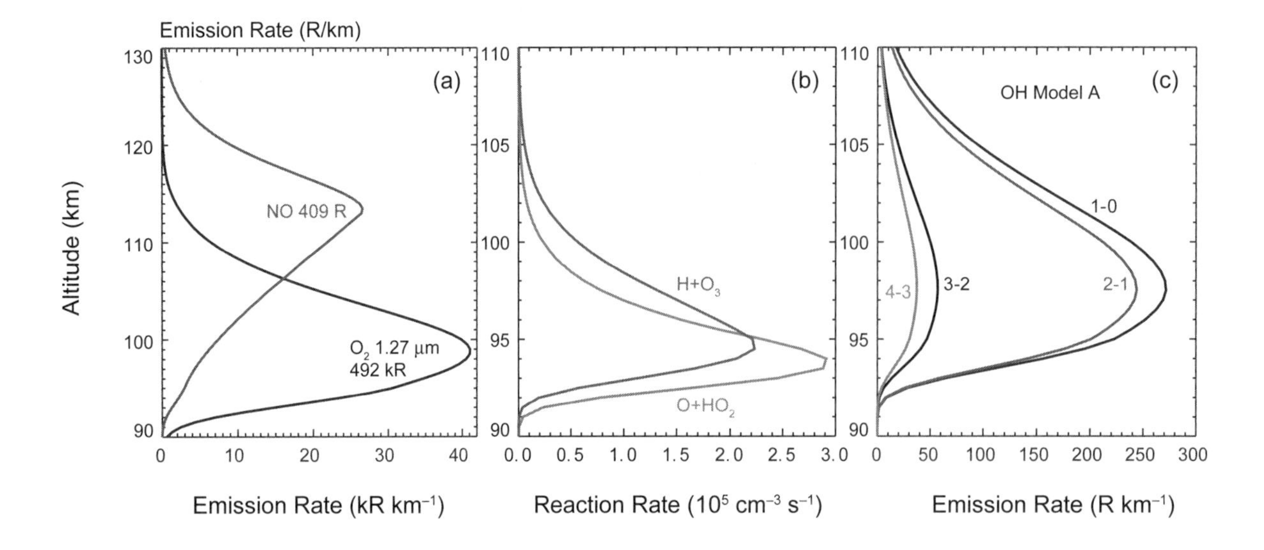

Fig. 30. Calculated vertical profiles of the O_2, NO, and OH nightglow and two reactions of OH excitation in the nighttime model (*Krasnopolsky,* 2013b).

the SO_2Cl_2 chemistry is tentative, and its spectral properties are poorly known. A problem that has not been discussed above is the significant difference between observed HDO and H_2O ratios above the clouds (*Krasnopolsky et al.,* 2013). The submillimeter observations cover large parts of the Venus disk, and the observed significant variations of species abundance are difficult to explain. Many of the problems related to the composition and formation of the Venus clouds are not discussed here.

4. TITAN

There are three bodies with N_2/CH_4 atmospheres in the solar system: Titan, Triton, and Pluto. Their heliocentric distances are 9.5, 30, and 30–50 AU, respectively. Solar heating is weaker on Triton and Pluto by an order of magnitude, and their atmospheres are tenuous, with the bulk N_2 and CH_4 in the solid phase. All nitrogen is in the atmosphere on Titan, and its atmosphere with a surface pressure of 1.5 bar looks very unusual for a satellite. Titan's radius is 2575 km, its surface gravity is 135.4 cm s^{-2}, and its atmospheric mass is 10.8 kg cm^{-2}, an order of magnitude larger than that on Earth. However, Titan's atmosphere would have been similar to those of Triton and Pluto, if Titan could be moved to 30 AU.

Titan orbits Saturn in the equatorial plane with a period of 16 d; its rotational period is equal to the orbit period. Saturn's orbital period is 30 yr and obliquity is 27°, and therefore Titan exhibits an annual cycle with a period of 30 yr as well. Basic data on Titan were obtained by the Voyager 1 flyby in November 1980 and the current Cassini-Huygens mission, which arrived at Saturn in 2004. Both missions refer to summer in the southern hemisphere; Cassini has now been in operation for nine years, and some seasonal changes have been observed. The atmosphere forms one Hadley cell with some indications of superrotation. Here we will not discuss seasonal and latitudinal variations of its chemical composition (see Forget and Lebonnois, this volume). A collection of reviews on Titan may be found in *Brown et al.* (2009).

4.1. Observations of the Chemical Composition

Chemical composition of Titan's atmosphere was studied by Voyager 1 using the Infrared Interferometer Spectrometer and Radiometer (IRIS) (*Coustenis et al.,* 1989), an infrared spectrometer for a range of 180 to 2500 cm^{-1} (4–55 µm) with an apodized spectral resolution of 4.3 cm^{-1}, and the Ultraviolet Spectrometer (UVS) (*Vervack et al.,* 2004), which had a range of 52–170 nm with a resolution of 1.5 nm. Later observations from groundbased (*Marten et al.,* 2002) and Earth-orbiting (*Coustenis et al.,* 2003) observatories in the millimeter and infrared ranges significantly contributed to our understanding of the chemical composition.

The most detailed observations of Titan's chemical composition began in October 2004 and are still being performed by the Cassini mission. The Huygens probe landed on Titan and studied the atmosphere below 150 km. Visits of the Cassini orbiter into the upper atmosphere of Titan down to 880 km gave numerous data from the remote and *in situ* observations. Similar to Voyager 1, both infrared and ultraviolet spectroscopy are used for this purpose. The Composite Infrared Spectrometer (CIRS) covers a range from 10 to 1400 cm^{-1} (7–1000 µm) at an apodized spectral resolution between 0.5 and 15.5 cm^{-1} (*Vinatier et al.,* 2010; *Coustenis et al.,* 2010). The Ultraviolet Imaging Spectrometer (UVIS) operates at 110 to 190 nm, mostly in the stellar and solar occultation mode (*Koskinen et al.,* 2011). Other than the improved capabilities of CIRS and UVIS and the long-term orbiter observations, the major progress of Cassini relative to Voyager 1 is in using the Ion Neutral Mass Spectrometer (INMS) for *in situ* studies of the neutral and ion composition (*Magee et al.,* 2009; *Cui et al.,* 2009; *Vuitton et al.,* 2007; *Cravens et al.,* 2009). The Cassini Plasma Spectrometer (CAPS) measures densities of heavy positive and negative ions (*Coates et al.,* 2012), and the Langmuir probe (LP), Magnetospheric Imaging Instrument (MIMI), and radio occultations (*Kliore et al.,* 2008) study the plasma environment around Titan.

A summary of the Cassini-Huygens observations of Titan's chemical composition is shown in Table 7. The Gas Chromatograph Mass Spectrometer (GCMS) on the Huygens probe measured 5.7% of CH_4 near the surface, decreasing to 1.5% above the tropopause at 45 km (*Niemann et al.,* 2010), in accordance with the condensation conditions. Mixing ratios of H_2 and Ar and isotope ratios of H, N, C, and Ar were measured as well. CIRS observed 11 species in the nadir mode (*Coustenis et al.,* 2010) and 10 species at the limb (*Vinatier et al.,* 2010). The nadir observations refer to 100–200 km, and vertical profiles of species are retrieved from the limb observations. The profiles start at 150 km and extend up to 450 km for some species.

INMS observes the upper atmospheric composition during the Cassini visits down to 880 km. *Yelle et al.* (2008) and *Cui et al.* (2009) compared the observed vertical profiles of N_2, CH_4, H_2, and Ar at 950–1400 km with the GCMS data for these species to retrieve eddy diffusion and escape of H_2 and CH_4. The obtained values are T_∞ = 151 K, K ≈ 3 × 10^7 cm^2 s^{-1}, and escape fluxes Φ_{H_2} = 1.1 × 10^{10} cm^{-2} s^{-1}, Φ_{CH_4} = 2.6 × 10^9 cm^{-2} s^{-1}. Both fluxes are

TABLE 7. GCMS, CIRS, UVIS, and INMS data on Titan's chemical composition (mixing ratios).

h (km)	0*	100–200†	150‡	300‡	400‡	550‡‡	700‡‡	900‡‡	1050§	1077¶	1100**
CH_4	0.057	—	—	—	—	—	—	—	0.022	0.023	—
Ar	3.4–5††	—	—	—	—	—	—	—	1.3–5	1.1–5	—
H_2	1.0–3	—	—	—	—	—	—	—	3.4–3	3.9–3	—
C_2H_2	—	3–6	3–6	3–6	2–6	4–5	4–5	4–4	3.4–4	2.3–4	—
C_2H_4	—	1.3–7	1.5–7	3–8	—	5–6	5–6	4–4	3.9–4	7.5–5	1–3
C_2H_6	—	8–6	1–5	1.4–5	1.7–5	—	—	—	4.6–5	7.3–5	—
C_3H_4	—	5–9	1–8	1–8	—	—	—	—	9.2–6	1.4–4	—
C_3H_6	—	—	—	—	—	—	—	—	2.3–6	—	—
C_3H_8	—	5–7	1.3–6	1–6	5–7	—	—	—	2.9–6	<5–5	—
C_4H_2	—	1.3–9	2–9	6–9	8–9	5–7	1–6	6–5	5.6–6	6.4–5	1–5
C_6H_2	—	—	—	—	—	—	—	—	—	—	8–7
C_6H_6	—	2–10	—	—	—	3–8	6–7	2–5	2.5–6	9–7	3–6
C_7H_4	—	—	—	—	—	—	—	—	—	—	3–7
C_7H_8	—	—	—	—	—	—	—	—	2.5–8	<1.3–7	2–7
NH_3	—	—	—	—	—	—	—	—	—	3–5	7–6
CH_2NH	—	—	—	—	—	—	—	—	—	—	1–5
HCN	—	1–7	2.5–7	5–7	3–7	2–5	4–6	6–4	2.5–4	—	2–4
CH_3CN	—	—	—	—	—	—	—	—	—	3.1–5	3–6
C_2H_3CN	—	—	—	—	—	—	—	—	3.5–7	<1.8–5	1–5
C_2H_5CN	—	—	—	—	—	—	—	—	1.5–7	—	5–7
HC_3N	—	4–10	1–11	1–9	—	1–6	1–6	3–5	1.5–6	3.2–5	4–5
C_4H_3N	—	—	—	—	—	—	—	—	—	—	4–6
HC_5N	—	—	—	—	—	—	—	—	—	—	1–6
C_5H_5N	—	—	—	—	—	—	—	—	—	—	4–7
C_6H_3N	—	—	—	—	—	—	—	—	—	—	3–7
C_6H_7N	—	—	—	—	—	—	—	—	—	—	1–7
C_2N_2	—	—	—	—	—	—	—	—	2.1–6	4.8–5	—
H_2O	—	—	1.9–10	—	—	—	—	—	—	—	<3–7
CO	—	4.5–5	—	—	—	—	—	—	—	—	—
CO_2	—	1.2–8	2–8	1.8–8	3–8	—	—	—	—	—	—

*GCMS (*Niemann et al.,* 2010).
†CIRS nadir (*Coustenis et al.,* 2010).
‡CIRS limb (*Vinatier et al.,* 2010).
§INMS (*Magee et al.,* 2009).
¶INMS (Tables 3 and 4 in *Cui et al.,* 2009).
**Derived from INMS ion spectra (*Vuitton et al.,* 2007).
††3.4–5 = 3.4 × 10^{-5}.
‡‡UVIS stellar occultations (*Koskinen et al.,* 2011).

at their diffusion limits and require hydrodynamic escape. Modeling by *Strobel* (2009) confirmed hydrodynamic escape and the obtained rates. However, direct Monte Carlo simulations, which are equivalent to solving the Boltzmann kinetic equation, result in no hydrodynamic escape from Titan (*Johnson et al.,* 2009). Johnson et al. argued that thermal conduction in the equations for hydrodynamic escape ceases at the exobase, and the equations become invalid. Carbon ions are scarce in the Saturn's magnetospheric plasma, and this does not favor hydrodynamic escape of CH_4 as well.

Analysis of the INMS observations of neutrals is a sophisticated problem that was solved by two independent teams, and some derived results do not coincide (Table 7). The best ion spectra were measured during the T5 flyby with strong precipitation of magnetospheric electrons and simulated by a model with densities of some neutrals as fitting parameters (*Vuitton et al.,* 2007). The retrieved neutral densities are also given in Table 7.

The Voyager UVS solar occultations resulted in vertical profiles of N_2, CH_4, C_2H_2, and C_2H_4 (*Vervack et al.,* 2004). However, the N_2 profiles do not extend to the altitudes of the C_2H_2 and C_2H_4 profiles, and some assumptions are required to obtain mixing ratios for these species. Profiles of the five major hydrocarbons, HCN, and HC_3N were extracted from the Cassini UVIS stellar occultations (*Shemansky et al.,* 2005; *Koskinen et al.,* 2011). However, our model (*Krasnopolsky,* 2009a, 2012c) does not rule out that more species are actually involved in the observed spectra, and the retrieved abundances are therefore rather uncertain.

Groundbased infrared, submillimeter, and millimeter observations were reviewed by *Coustenis et al.* (2009). Among those are the first detections of CO (*Lutz et al.,* 1983) and CH_3CN (*Bezard et al.,* 1993b), as well as vertical profiles of HCN, HC_3N, CH_3CN at 100–450 km (*Marten et al.,* 2002; *Gurwell,* 2004). H_2O and C_6H_6 were detected from the Infrared Space Observatory (*Coustenis et al.,* 1998, 2003), and an H_2O profile at 80–300 km was measured by the Herschel Space Observatory (*Moreno et al.,* 2012).

4.2. Photochemical Modeling of Titan's Atmosphere and Ionosphere

4.2.1. Published models for Titan. Photochemical processes in the dense N_2/CH_4 atmosphere initiate formation of numerous hydrocarbons, nitriles, and their ions. This makes photochemical modeling for Titan a challenging problem. Self-consistent models for Titan's neutral atmosphere were developed by *Yung et al.* (1984), *Toublanc et al.* (1995), *Lara et al.* (1996), *Hebrard et al.* (2007), and *Lavvas et al.* (2008a,b).

Pure ionospheric models adopt a background neutral atmosphere and neglect the effects of ion chemistry on the neutral composition. However, these effects are significant; for example, benzene C_6H_6 is mostly formed by ion reactions. Therefore coupled models of Titan's atmosphere and ionosphere were made by *Banaszkiewicz et al.* (2000), *Wilson and Atreya* (2004), and *Krasnopolsky* (2009a, 2012c).

Here we will briefly discuss the results of *Krasnopolsky* (2009a, 2012c). The model involves ambipolar diffusion and escape of ions and calculates the H_2 and CO densities near the surface that were assigned in some previous models. The number of reactions is significantly reduced relative to that in *Wilson and Atreya* (2004) to remove unimportant processes. Hydrocarbon chemistry is extended to $C_{12}H_{10}$ for neutrals and $C_{10}H_{11}^+$ for ions. The model involves 419 reactions of 83 neutrals and 33 ions, and column rates are given for all reactions. It accounts for effects of magnetospheric electrons, protons, and cosmic rays. Formation of haze by polymerization processes and recombination of heavy ions is considered as well.

There are four versions of the model. Two versions in *Krasnopolsky* (2009a) adopt hydrodynamic escape of light species (m < 20) and either "standard" eddy diffusion or that proposed by *Hörst et al.* (2008). They suggested an increase in $K \sim n^{-2}$ from 75 to 300 km by 5 orders of magnitude to fit better the CIRS limb observations. Our model with this K has both advantages and shortcomings relative to the first version.

It is discussed in section 4.1 that direct Monte Carlo simulations of the escape processes and low abundance of carbon-bearing ions in Saturn's magnetosphere do not support hydrodynamic escape of CH_4. Tests in *Krasnopolsky* (2010e) are not favorable for hydrodynamic escape on Titan as well, and versions 3 and 4 were calculated without hydrodynamic escape. *Yelle et al.* (2006) considered $K = (4 \pm 3) \times 10^9$ cm^2 s^{-1} as an alternative to the hydrodynamic escape, and $K = 10^9$ cm^2 s^{-1} is adopted above 700 km (Fig. 31a) in versions 3 (*Krasnopolsky,* 2010e) and 4 (*Krasnopolsky,* 2012c). Two reactions that stimulate formation of NH_3 (*Yelle et al.,* 2010) were included in these versions. Finally, version 4 is further improved using the Troe approximation for termolecular reactions (see, e.g., *Yung and DeMore,* 1999), some changes in eddy diffusion, and two radiative association reactions proposed by *Vuitton et al.* (2012). Here we will consider the latest version from *Krasnopolsky* (2012c).

4.2.2. Initial data. Basic initial data are the N_2 and CH_4 densities near the surface and profiles of T and K (Fig. 31a). The observed temperature profile is slightly adjusted to bind the GCMS and INMS observations of N_2. The adopted K results in a homopause for CH_4 at 1050 km (Fig. 31a). The model upper boundary is at 1600 km, near the top of the INMS observations. The calculated exobase is at 1360 km.

The UV absorption by the haze significantly affects Titan's photochemistry and was calculated using the Huygens haze data (*Tomasko et al.,* 2008): The haze particles are aggregates of ~3000 monomers with a monomer radius of ~0.05 μm, and the particle number density is 5 cm^{-3} below 80 km decreasing with a scale height of 65 km above 80 km. Scattering properties of the particles were calculated using a code by *Rannou et al.* (1999) and refractive indices from *Khare et al.* (1984). Radiative transfer with multiple scattering in the optically thick haze was calculated using some approximations from *van de Hulst* (1980).

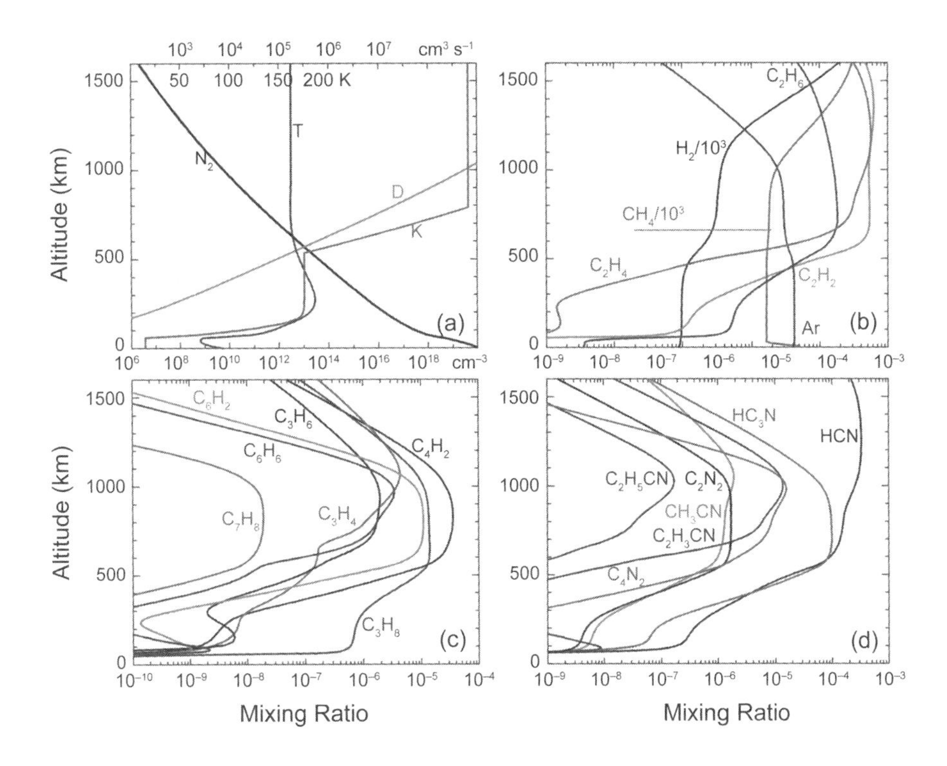

Fig. 31. See Plate 42 for color version. Titan's photochemical model: **(a)** profiles of T, eddy diffusion K, diffusion coefficient of CH_4 in N_2, and N_2 density; **(b)** major hydrocarbons, H_2, and Ar; **(c)** some other hydrocarbons; **(d)** the most abundant nitriles. From *Krasnopolsky* (2012c).

Dissociation and ionization of the atmospheric species by the solar photons (Fig. 32b) were calculated self-consistently using the known cross sections. Precipitations of magntespheric electrons (Fig. 32b) for a typical case and a strong event were taken from Cassini encounters T21 and T5, respectively (*Cravens et al.*, 2009; *Vuitton et al.*, 2007). The ion densities measured by INMS during encounter T5 were especially high and accurate. Productions of N_2^+, N^+, N, and $N(^2D)$ by electron impact were obtained using the data of *Fox and Victor* (1988). Ionization by magnetospheric protons and the cosmic rays (Fig. 32b) were taken from *Cravens et al.* (2008) and *Molina-Cuberos et al.* (1999), respectively. Dissociation and dissociative ionization were calculated combining the data of *Fox and Victor* (1988) and the Born approximation.

Twenty-four species in the model condense near the tropopause. Their saturated vapor pressures were found in the literature and approximated by

$$\ln p_s = a - b/T - c \ln T$$

The model had a vertical step increasing from 2.5 km near the surface to 10 km at the upper boundary.

4.2.3. Hydrocarbons, Ar, and H_2. The calculated CH_4 profile (Fig. 31b) agrees with the observational data (Table 7), while the profile of Ar with the GCMS value near the surface is slightly above the INMS values. Photolysis of CH_4 results in a quarter of the total loss of CH_4. The photolysis products CH_3 and CH_2 radicals are quickly converted to CH, which reacts with CH_4 to form C_2H_4. These reactions double the loss of methane by photolysis.

Photolyses of CH_4, C_2H_2, and C_4H_2 begin at 140, 200, and 260 nm, respectively. Radicals C_2H and C_2 released by photolyses of C_2H_2 and C_4H_2 react with methane. This indirect photolysis results in 30% of the methane loss; the remaining loss is in reactions with other radicals and ions. Chemical production of methane is much smaller than its loss, and the difference is consumed by precipitation of haze, condensation of photochemical products in the troposphere, and escape of H_2 and H. The methane column abundance is 230 g cm^{-2}, and its lifetime is 30 m.y. Surface sources that resupply methane and compensate for its loss are poorly known.

The C:H:N ratio in the total aerosol flux is 1:1.2:0.06 in the numbers of atoms. The ratio in methane is 1:4, which means that 70% of hydrogen from methane escapes to space. The models with and without hydrodynamic escape of H_2 give very similar escape rates that are diffusion-limited in both cases. The H-H bond is stronger than the H bond in some radicals, and many of the H atoms are converted into H_2 by reactions such as

$$H + C_4H_2 + M \rightarrow C_4H_3 + M$$
$$H + C_4H_3 \rightarrow C_4H_2 + H_2$$
$$\text{Net } H + H \rightarrow H_2$$

H_2 is not fixed at the lower boundary in our model; therefore the agreement of the model (Fig. 31b) with the observations

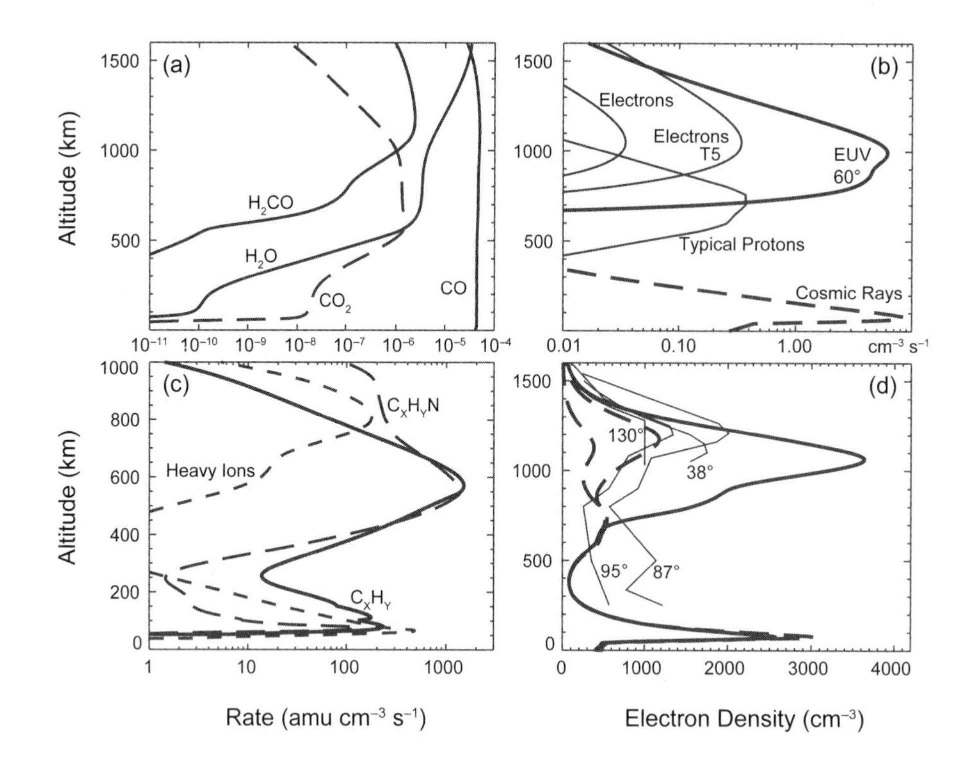

Fig. 32. Titan's photochemical model: **(a)** mixing ratios of oxygen species; **(b)** sources of ionization; ionization by magnetospheric electrons is shown for typical conditions and a strong precipitation event T5; **(c)** production of haze by polymerization of hydrocarbons, nitriles, and recombination of heavy ions; **(d)** calculated (thick lines) and observed (thin lines) electron density profiles. The calculated profiles are for the global-mean, typical nighttime, and nighttime T5 conditions. The observed mean radio occultation dusk (z = 87°) and dawn (z = 95°) profiles (*Kliore et al.,* 2008) and the INMS T5 (z = 130°) and daytime profiles are shown. From *Krasnopolsky* (2009a, 2012c).

(Table 7) may be considered as very good.

Photolysis of methane peaks near 750 km; photolyses of the photochemical products are maximal typically at 300–500 km. Therefore basic reactions that determine species balances typically occur around 500 km. Species mixing ratios are typically constant above 500 km up to the homopause at 1050 km and either decrease or increase or become stable above 1050 km depending on the molecular mass, which may be greater, smaller, or equal to that of N_2, respectively.

Most of the photochemical products condense in the lowest ~100 km, and here their mixing ratios are much smaller than those near ~500 km. Therefore there is an inevitable steep decrease in the product mixing ratios from ~500 to ~100 km. The calculated profiles usually intersect the CIRS nadir values (Table 7) that refer to 100–200 km. However, these profiles are steeper than those from the CIRS limb observation. The chosen steep increase in eddy diffusion at 70–180 km in our model (Fig. 31a) partly compensates for this difference. Even steeper increase in eddy diffusion was suggested by *Hörst et al.* (2008); however, our test shows that the eddy diffusion profile in Fig. 31a is optimal.

We mentioned above that almost all CH_x products of the methane photolysis form CH and then $CH + CH_4 \rightarrow C_2H_4 + H$. This reaction is the major source of ethylene that is mostly lost by photolysis. The calculated densities of C_2H_4 are below the condensation level in the troposphere. C_2H_4 near 300 and 1100 km is within the uncertainties of the CIRS and INMS values, respectively.

Acetylene C_2H_2 is made by photolyses of C_2H_4 and C_4H_{2x} (x = 1, 2, 3), the reactions of C_2H with CH_4 and C_2H_6, and by

$$C_2H_3 + H \rightarrow C_2H_2 + H_2$$
$$C_3H_4 + H \rightarrow C_2H_2 + CH_3$$

It is lost by photolysis (~50%), reactions with radicals, and by condensation (440 g cm^{-2} b.y.$^{-1}$). The calculated mixing ratios near 300 and 1100 km agree with the CIRS and INMS values.

Ethane C_2H_6 is produced by the reaction

$$CH_3 + CH_3 + M \rightarrow C_2H_6 + M$$

and lost by condensation and reactions with radicals (half and half). The calculated densities (Fig. 31b) are smaller and larger than those observed by CIRS and INMS, respectively.

Densities of the other hydrocarbons that are shown in Fig. 31c typically agree with the CIRS and INMS observations within a factor of 3. This is a reasonable agreement, taking into account uncertainties of the observations and the model.

4.2.4. Nitriles. The N≡N triple bond in the N_2 molecule is very strong, with a dissociation energy of 9.76 eV. Ionization of N_2 by the solar photons $\lambda < 80$ nm, photoelectrons, magnetospheric electrons, protons, and cosmic rays (Fig. 32b) form mostly N_2^+ ions that return N_2 after charge exchange with hydrocarbons. Predissociation of N_2 at $\lambda = 80$–100 nm; dissociative ionization at $\lambda < 51$ nm; and electron, proton, and cosmic-ray impact dissociations originate N and $N(^2D)$ atoms and N^+ ions with a total column rate of 7×10^8 cm^{-2} s^{-1}. Most of them (72%) form nitriles C_xH_yCN. The basic reaction of the nitrile formation is $N + CH_3 \rightarrow H_2CN + H$ (63% of the column production of nitriles). The C≡N bonds are strong (7.85 eV in CN and 9.58 eV in HCN); therefore nitriles do not recombine to N_2 in Titan's atmosphere, and condensation and polymerization with precipitation to the surface are the ultimate fate of nitriles on Titan.

H_2CN is converted to HCN in reactions with H, N, NH, and NH_2. Hydrogen cyanide (HCN) is the most abundant nitrile on Titan. Beside H_2CN (25%), reactions of CN with CH_4, C_2H_4, and C_2H_6 (50%), as well as ion reactions (25%), contribute to the production of HCN. Hydrogen cyanide is lost in reactions with CH, C_2H_3, C_3N (62%); ion reactions (17%); condensation (13%); and photolysis (7%). The calculated profile (Fig. 31d) agrees with that observed by *Gurwell* (2004) in the submillimeter range at 100–400 km, CIRS at 100–200 km, UVIS at 550 km, and INMS near 1100 km.

Cyanoacetylene (HC_3N) is another abundant nitrile that is formed by reactions of CN + C_2H_2 (62%) and C_3N + CH_4 (38%) and lost by photolysis (94%). The calculated HC_3N is within uncertainties of the observed INMS and UVIS values at 800–1100 km.

Calculated profiles of some other nitriles are shown in Fig. 31d. The mixing ratios of CH_3CN, C_2H_3CN, C_2H_5CN, and C_2N_2 are in reasonable agreement with the CIRS and INMS observations.

4.2.5. N_xH_y, CH_2NH, and CH_3NH_2. The model involves nine species with N–H bonds. These bonds are ~3.5 eV and comparable with the C–H bonds. Therefore species with the N–H bonds may return N and recombine to N_2, which explains why they are less abundant than nitriles. The model includes reactions of NH and NH_2 with H_2CN (*Yelle et al.,* 2010) that increase the production of ammonia. The calculated profile of CH_2NH agrees with the INMS value, while that of NH_3 is smaller than that observed by INMS.

4.2.6. Oxygen species. CO_2 was the first oxygen species that was discovered on Titan from the Voyager IRIS observations (*Samuelson et al.,* 1983). Subsequently, groundbased observations revealed a significant abundance of CO (*Lutz et al.,* 1983), and CO was later measured using groundbased infrared and microwave instruments and by CIRS on Cassini. A summary of the observations may be found in *Hörst et al.* (2008), and the mean CO is close to the CIRS value of 45 ± 15 ppm (*Flasar et al.,* 2005). CO_2 was detected by IRIS, the Infrared Space Observatory (ISO), and CIRS (Table 7). A search for H_2O was made by the ISO (*Coustenis et al.,* 1998, 2003) using its rotational band at 40 μm, resulting in 8^{+5}_{-4} ppb at 300–500 km. Recently H_2O was observed by the Herschel Space Observatory varying from 2.3×10^{-11} at 80 km to 3.4×10^{-10} at 300 km (*Moreno et al.,* 2012). A detailed study of the CIRS nadir and limb spectra reveals $f_{H_2O} = 1.4 \times 10^{-10}$ at 100 km increasing to 4.5×10^{-10} at 225 km (*Cottini et al.,* 2012).

Oxygen is delivered into Titan's atmosphere as meteoritic water and ions O^+. Using the study by *Pereira et al.* (1997), *Wong et al.* (2002) argued that a basic reaction of the CO production in the prior models, $OH + CH_3 \rightarrow CO + 2H_2$, proceeds with products $H_2O + {}^1CH_2$ and results in recycling of H_2O. They adopted the H_2O influx of 1.5×10^6 cm^{-2} s^{-1} and production of CO at 1.1×10^6 cm^{-2} s^{-1} at the surface to fit the observations.

Hartle et al. (2006) analyzed the Voyager and Cassini CAPS data and found a magnetospheric influx of O^+ on Titan with a rate of ~10^6 cm^{-2} s^{-1}. These ions may originate from Saturn's ring and Enceladus. *Hörst et al.* (2008) calculated a model for the oxygen species on Titan using the influxes of OH and O as fitting parameters. Their results agreed with the observations available at that time, but exceed the recent measurements of H_2O by an order of magnitude.

Our model for oxygen species (Fig. 32a) applied the basic ideas of *Wong et al.* (2002) and *Hörst et al.* (2008) and is calculated for the H_2O and O^+ influxes of 3×10^6 and 1.7×10^6 cm^{-2} s^{-1}. O^+ is neutralized by the charge exchange with CH_4 and then forms CO in reactions with CH_3 and C_2H_4. OH from photolysis of H_2O reacts either with CH_3 and returns H_2O or with CO and forms CO_2. Photolysis of CO_2 returns two CO, because O is converted to CO as well. There is no production of CO from the interior, and CO is not fixed at the surface. Production of formaldehyde H_2CO is a significant branch of the reaction of O + CH_3. Photolysis quickly converts H_2CO to CO. However, the formaldehyde abundance reaches 1 ppm near 1100 km. The model agrees with the observed oxygen species.

4.2.7. Ionosphere. Sources of ionization in Titan's atmosphere are mentioned in section 4.2.2 and shown in Fig. 32b. The profiles of ionization by magnetospheric electrons are approximated for typical conditions and the strong precipitation T5 event using data from *Vuitton et al.* (2007) and *Cravens et al.* (2008, 2009). Ionization by protons is from *Cravens et al.* (2008), and by cosmic rays from *Molina-Cuberos et al.* (1999). Ionization by the solar EUV photons is calculated using the standard cross sections and the solar EUV fluxes reduced by a factor of 2 to account for the nightside and for the solar zenith angle z = 60° (global mean conditions).

Except for 308 neutral reactions, our model involves 111 reactions of 33 ions, much fewer than in the ionospheric models by *Vuitton et al.* (2007) and *Cravens et al.* (2009). Rate coefficients of ion-neutral reactions and dissociative recombinations were taken from *McEwan and Anicich* (2007) and *Woodall et al.* (2007), respectively.

Electron density profiles in Fig. 32d were calculated for the global-mean conditions, typical nighttime conditions, and the T5 event. They are compared with mean radio occultation profiles (*Kliore et al.*, 2008), observed at dusk and dawn, and with the INMS profiles for sums of all ions. Calculations show that our basic profile is in reasonable agreement with the observed dusk profile at z = 87°. The T5 profile also agrees with that observed by INMS. However, the measured daytime profile at z = 38° is smaller than the calculated profile and even that observed at z = 87°. Electron densities at z = 38° measured simultaneously by two other instruments exceed those observed by INMS by a factor of 1.6 as well.

The calculated ionospheric compositions for the global-mean conditions and the T5 event are shown in Fig. 33. Ionospheric models for the T5 event were developed by *Vuitton et al.* (2007) and *Cravens et al.* (2009). *Vuitton et al.* (2007) applied 1250 reactions of 150 ions with densities of 18 neutral species as fitting parameters. *Cravens et al.* (2009) calculated six versions of the model for the adopted various neutral composition. However, a mean difference between the calculated and observed ion densities for the model in Fig. 33 is rather similar to those in *Vuitton et al.* (2007) and *Cravens et al.* (2009).

The recent model for the daytime ion composition by *Westlake et al.* (2012) agrees with the INMS observations at z = 38° better than our model. However, we have discussed above some problems with those observations. All ionospheric models adopt dozens of vertical profiles of neutral species, while our model is based on the two densities of N_2 and CH_4 near the surface. The ionospheric models neglect ambipolar diffusion and escape of ions, which are taken into account in our model.

De la Haye et al. (2008) constructed a model of diurnal variations of the atmosphere and ionosphere above 600 km at 39°N and 74°N. These latitudes correspond to the conditions of the T_A and T5 flybys; z = 62° at noon for 39°N, and polar night conditions at 74°N. The model involved 35 neutral and 44 ion species. Neutrals with lifetimes exceeding 1.5×10^4 s were fixed at 600 km to fit the mean INMS data near 1000 km or taken from *Lebonnois* (2005). All ions were calculated neglecting ambipolar diffusion and escape. The model results will be used in future GCMs.

There are some uncertain data on the significant densities of negative ions near ~1000 km. Negative ions usually appear in dense atmospheric regions where they are formed by three-body collisions. These reactions are ineffective at ~1000 km on Titan, with a density of 7×10^9 cm^{-3}. The problem was studied in detail by *Vuitton et al.* (2009). They involved production of negative ions by radiative ($A + e \rightarrow A^- + h\nu$) and dissociative ($AB + e \rightarrow A^- + B$) electron attachment and ion-pair formation. The loss processes are photo-detachment ($A^- + h\nu \rightarrow A + e$), ion-ion recombination, and ion-neutral associative detachment ($A^- + B \rightarrow AB + e$). Charge exchange reactions were included in the model as well. The observed peaks at m = 22 ± 4 and 44 ± 8 were identified as CN^- and C_3N^-, and the expected total density of negative ions is maximal at ~1 cm^{-3} near

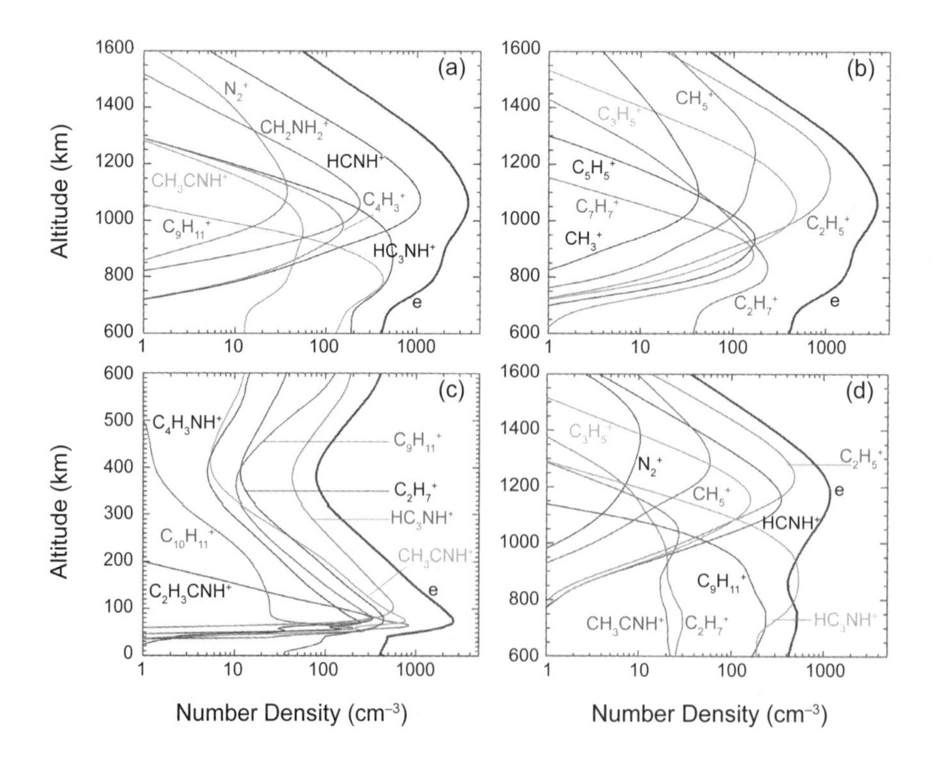

Fig. 33. See Plate 43 for color version. Titan's photochemical model: calculated ionospheric composition for **(a)–(c)** global-mean conditions and **(d)** the strong nighttime electron precipitation T5. From *Krasnopolsky* (2009a, 2012c).

1100 km. The nighttime densities may be significantly higher because of the lack of photo-detachment. They were not calculated by *Vuitton et al.* (2009).

Wahlund et al. (2009) analyzed data of INMS, two CAPS channels, and two Radio and Plasma Wave Science (RPWS) channels during three flybys of Titan with ionospheric peak densities of 900–3000 cm^{-3}. Heavy (>100 amu) positive ions constitute about half of the total ions near 1000 km, and negative ion densities are ~100 cm^{-3}. Heavy ions may dominate below 950 km, and recombination products of heavy ions may contribute to the haze production.

Our model predicts $C_9H_{11}^+$ (119 amu) as a major ion at 500–900 km with densities of ~200 cm^{-3} (Fig. 33). $C_{12}H_{10}$ and $C_{10}H_{11}^+$ with 154 and 131 amu are the heaviest neutral and ion species, respectively, in our model.

4.2.8. Production of haze. There are four basic processes of the aerosol formation in our model (Fig. 32c). Recombination of heavy ions, mostly $C_9H_{11}^+$, peaks at 800 km with a total contribution of 150 g cm^{-2} b.y.$^{-1}$. Polymerization reactions between hydrocarbons (mostly $C_6H + C_4H_2$) and hydrocarbons and nitriles (mostly $C_3N + C_4H_2$) give almost equal haze productions of 1.7 kg cm^{-2} b.y.$^{-1}$ each. Rate coefficients of the reactions of polymerization were taken from *Lavvas et al.* (2008b). Condensation of hydrocarbons and nitriles below ~100 km add 2.8 and 0.3 kg cm^{-2} b.y.$^{-1}$, respectively. C_2H_6 and HCN dominate in those. Some condensation products may sublime near the surface.

The total flow of 6.5×10^{-6} g cm^{-2} yr^{-1} looks negligible compared with precipitation of ~50 g cm^{-2} yr^{-1} of water on Earth. However, the overall sediment is ~300 m for the age of the solar system. We concluded in section 4.2.3 that the lifetime of methane is 30 m.y. on Titan; 30% of the initial hydrogen is retained in the haze, and the remaining 70% escapes to space. Recycling of methane in the interior from the precipitation products therefore requires hydrogen-rich species like H_2O and/or NH_3. The lifetime of N_2 is 40 b.y., much longer than the age of Titan. Here we do not discuss properties of the haze (*Tomasko et al.,* 2008; *Lavvas et al.,* 2011; *Vinatier et al.,* 2012). Various aspects of haze on Titan are discussed in a few chapters in *Brown et al.* (2009).

4.3. Unsolved Problems

Although the Cassini-Huygens mission has been very successful, some unsolved problems related to the chemical composition of Titan's atmosphere and ionosphere remain. One problem is that the atmosphere between 350 and 950 km has not been studied in detail, and this is the region where the basic photochemical processes proceed.

Another problem is that photochemical products on Titan are typically formed near 500 km, and their mixing ratios are rather constant up to ~1000 km due to strong eddy diffusion. The product mixing ratios decrease to ~100 km, where the species condense, and therefore the expected gradients of the mixing ratios are the differences between the INMS observations and the CIRS nadir values divided by ~350 km. However, the CIRS limb observations demonstrate much smaller gradients. This is another general problem.

INMS is the major source of the data on Titan's chemical and ion composition. However, retrieval of species densities from the INMS observations is a difficult process that was performed by two independent teams and has resulted in significant uncertainties. The problem of heavy ions and negative ions also remain uncertain in both observations and modeling, as there are significant differences between the existing observations and models.

Ultimately, one of the difficulties is that the composition of Titan's atmosphere and ionosphere is very complicated, involving many species. There are problems related to some of the species that we have not even discussed in this chapter.

REFERENCES

Adams W. S. and Dunham T. (1932) Absorption bands in the infrared spectrum of Venus. *Publ. Astron. Soc. Pacific, 44,* 243–247.

Allen D. A. and Crawford J. W. (1984) Cloud structure on the dark side of Venus. *Nature, 307,* 222–224.

Allen M., Yung Y. L., and Waters J. W. (1981) Vertical transport and photochemistry in the terrestrial mesosphere and lower thermosphere. *J. Geophys. Res., 86,* 3617–3627.

Altieri F., et al. (2009) O_2 1.27 μm emission maps as derived fromOMEGA/MEX data. *Icarus, 204,* 499–511.

Anderson D. E. (1974) Mariner 6, 7, and 9 ultraviolet spectrometer experiment: Analysis of hydrogen Lyman-alpha data. *J. Geophys. Res., 79,* 1513–1518.

Anderson D. E. and Hord C. W. (1971) Mariner 6 and 7 ultraviolet spectrometer experiment: Analysis of hydrogen Lyman alpha data. *J. Geophys. Res., 76,* 6666–6671.

Atreya S. K., Mahaffy P. R., and Wong A. S. (2007) Methane and related trace species on Mars: Origin, loss, implications for life and habitability. *Planet. Space Sci., 55,* 358–369.

Bailey J., Meadows V. S., Chamberlain S., and Crisp D. (2008) The temperature of the Venus mesosphere from $O_2(a^1\Delta_g)$ airglow observations. *Icarus, 197,* 247–259.

Banaszkiewicz M., Lara L.M., Rodrigo R., Lopez-Moreno J. J., and Molina-Cuberos G. J. (2000) A coupled model of Titan's atmosphere and ionosphere. *Icarus, 147,* 386–404.

Barker E. S. (1972) Detection of molecular oxygen in the martian atmosphere. *Nature, 238,* 447–448.

Barker E. S. (1979) Detection of SO_2 in the UV spectrum of Venus. *Geophys. Res. Lett., 6,* 117–120.

Barth C. A. and Hord C. W. (1971) Mariner 6 and 7 ultraviolet spectrometer experiment: Topography and polar cap. *Science, 173,* 197–201.

Barth C. A., Stewart A. I., Hord C. W., and Lane A. L. (1972) Mariner 9 ultraviolet spectrometer experiment: Mars airglow spectroscopy and variations in Lyman alpha. *Icarus, 17,* 457–468.

Barth C. A., Hord C. W., Stewart A. I., et al. (1973) Mariner 9 ultraviolet spectrometer experiment: Seasonal variation of ozone on Mars. *Science, 179,* 795–796.

Belyaev D., Montmessin F., Bertaux J. L., Mahieux A., Fedorova A., Korablev O., Marcq E., Yung Y. L., and Zhang X. (2012) Vertical profiling of SO_2 and SO above Venus' clouds by SPICAV/SOIR solar occultations. *Icarus, 217,* 740–751.

Bertaux J. L., Widemann T., Hauchecome A., Moroz V. I., and Ekonomov A. P. (1996) Vega 1 and Vega 2 entry probes: An investigation of local UV absorption (220–400 nm) in the atmosphere of Venus (SO_2, aerosols, cloud structure). *J. Geophys Res., 101,* 12709–12745.

Bertaux J. L., Leblanc F., Perrier S., et al. (2005) Nightglow in the upper atmosphere of Mars and implications for atmospheric transport. *Science, 307,* 567–569.

Bertaux J. L., Gondet B., Lefèvre F., et al. (2012) First detection of O_2 1.27 μm nightglow emission at Mars with OMEGA/MEX and comparison with general circulation model prediction. *J. Geophys. Res., 117,* E00J04.

Bezard B., de Bergh C., Crisp D., and Maillard J. P. (1990) The deep atmosphere of Venus revealed by high-resolution nightside spectra. *Nature, 345,* 508–511.

Bezard B., de Bergh C., Fegley B., et al. (1993a) The abundance of sulfur dioxide below the clouds of Venus. *Geophys. Res. Lett., 20,* 1587–1590.

Bezard B., Marten A., and Paubert G. (1993b) Detection of acetonitrile on Titan. *Bull. Am. Astron. Soc., 25,* 1100.

Bezard B., Fedorova A., Bertaux J. L., Rodin A., and Korablev O. (2011) The 1.10 and 1.18-μm nightside windows of Venus observed by SPICAV-IR aboard Venus Express. *Icarus, 216,* 173–183.

Biemann K., Owen T., Rushneck D. R., et al. (1976) Search for organic and volatile inorganic components in two surface samples from the Chryse Planitia region of Mars. *J. Geophys. Res., 82,* 4641–4658.

Billebaud F., Brillet J., Lellouch E., et al. (2009) Observations of CO in the atmosphere of Mars with PFS onboard Mars Express. *Planet. Space Sci., 57,* 1446–1457.

Bjoraker G. L., Larson H. P., Mumma M. J., Timmermann R., and Montani J. L. (1992) Airborne observations of the gas composition of Venus above the cloud tops: Measurements of H_2O, HDO, HF and the D/H and $^{18}O/^{16}O$ isotope ratios. *Bull. Am. Astron. Soc., 24,* 995.

Bogard D. D., Clayton R. N., Marti K., et al. (2001) Martian volatiles: Isotopic composition, origin, and evolution. *Climatol. Evol. Mars, 96,* 425–458.

Bougher S. W., Gerard J. C., Stewart A. I. F., and Fesen C. G. (1990) The Venus nitric oxide airglow: Model calculations based on the Venus thermosphere general circulation model. *J. Geophys. Res., 95,* 6271–6284.

Bougher S. W., Engel S., Roble R. G., and Foster B. (2000) Comparative terrestrial planet thermospheres. 3. Solar cycle variation of global structure and winds at solstices. *J. Geophys. Res., 105,* 17669–17692.

Bougher S. W., Blelly P. L., Combi M., Fox J. L., Mueller-Wodarg I., Ridley A., and Roble R. G. (2008) Neutral upper atmosphere and ionosphere modeling. *Space Sci. Rev., 139,* 107–141.

Bougher S. W., McDunn T. M., Zoldak K. A., and Forbes J. M. (2009) Solar cycle variability of Mars dayside exospheric temperatures: Model evaluation of underlying thermal balance. *Geophys. Res. Lett., 36,* L05201, DOI: 10.1029/2008GL036376.

Brecht A. S., Bougher S. W., Gerard J. C., et al. (2011) Understanding the variability of nightside temperatures, NO UV and O_2 IR nightglow emissions in the Venus upper atmosphere. *J. Geophys. Res., 116,* E08004.

Brown R. H., Lebreton J. P., and Waite J. H., eds. (2009) *Titan from Cassini-Huygens.* Springer, Dordrecht. 535 pp.

Butler B. J., Steffes P. G., Suleiman S. H., Kolodner M. A., and Jenkins J. M. (2001) Accurate and consistent microwave observations of Venus and their implications. *Icarus, 154,* 226–238.

Carleton N. P. and Traub W. A. (1972) Detection of molecular oxygen on Mars. *Science, 177,* 988–992.

Chamberlain J. W. and Hunten D. M. (1987) *Theory of Planetary Atmospheres.* Academic, San Diego. 481 pp.

Chassefière E. (2009) Metastable methane clathrate particles as a source of methane to the martian atmosphere. *Icarus, 204,* 137–144.

Chaufray J. Y., Bertaux J. L., Leblanc F., et al. (2008) Observation of the hydrogen corona with SPICAM on Mars Express. *Icarus, 195,* 598–613.

Chaufray J. Y., Leblanc F., Quémerais E., et al. (2009) Martian oxygen density at the exobase deduced from O I 130.4-nm observations by spectroscopy for the investigation of the characteristics of the atmosphere of Mars on Mars Express. *J. Geophys. Res., 114,* E02006.

Clancy R. T. and Nair H. (1996) Annual (aphelion-perihelion) cycles in the photochemical behavior of the global Mars atmosphere. *J. Geophys. Res., 101,* 12785–12790.

Clancy R. T., Wolff M. J., and James P. B. (1999) Minimal aerosol loading and global increases in atmospheric ozone during the 1996–1997 martian northern spring season. *Icarus, 138,* 49–63.

Clancy R. T., Sandor B. J., and Moriarty-Schieven G. H. (2004) A measurement of the 362 GHz absorption line of Mars atmospheric H_2O_2. *Icarus, 168,* 116–121.

Clancy R.T., Sandor B. J., and Moriarty-Schieven G. (2012a) Thermal structure and CO distribution for the Venus mesosphere/lower thermosphere: 2001–2009 inferior conjunction sub-millimeter CO absorption line observations. *Icarus, 217,* 779–793.

Clancy R. T., et al. (2012b) Extensive MFO CRISM observations of 1.27 μm O_2 airglow in Mars polar night and their comparison to MRO MCS temperature profiles and LMD GCM simulations. *J. Geophys. Res., 117,* E00110.

Clancy R. T., et al. (2012c) OH Meinel band polar nightglow in the Mars atmosphere from CRISM limb observations. Abstract P22A-08 presented at 2012 Fall Meeting, AGU, San Francisco.

Clegg S. M., and Abbatt J. P. D. (2001) Uptake of gas-phase SO_2 and H_2O_2 by ice surfaces: Dependence on partial pressure, temperature, and surface activity. *J. Phys. Chem. A, 105,* 6630–6636.

Coates A. J., et al. (2012) Cassini in Titan's tail: CAPS observations of plasma escape. *J. Geophys. Res., 117,* A5, A05324.

Connes P., Connes J., Benedict W. S., and Kaplan L. D. (1967) Traces of HCl and HF in the atmosphere of Venus. *Astrophys. J., 147,* 1230–1237.

Connes P., Connes J., Kaplan L. D., and Benedict W. S. (1968) Carbon monoxide in the Venus atmosphere. *Astrophys. J., 152,* 731–743.

Connes P., Noxon J. F., Traub W. A., and Carleton N. P. (1979) $O_2(^1\Delta)$ emission in the day and night airglow of Venus. *Astrophys. J. Lett., 233,* L29–L32.

Cooper P. L. and Abbatt J. P. D. (1996) Heterogeneous interactions of OH and HO_2 radicals with surfaces characteristic of atmospheric particulate matter. *J. Phys. Chem., 100,* 2249–2254.

Cottini V., Ignatiev N. I., Piccioni G., Drossart P., Grassi D., and Markiewicz W. J. (2011) Water vapor near the cloud tops of Venus from Venus Express/VIRTIS dayside data. *Icarus, 217(2),* 561–569, DOI: 10.1016/j.icarus.2011.06.018.

Cottini V., et al. (2012) Water vapor on Titan: The stratospheric vertical profile from Cassini/CIRS infrared spectra. In *Workshop on Titan Through Time II,* NASA Goddard Space Flight Center, April 3–5, 2012.

Cotton D. V., Bailey J., Crisp D., and Meadows V. S. (2012) The distribution of carbon monoxide in the lower atmosphere of

Venus. *Icarus, 217,* 570–584.

Coustenis A., Bezard B., and Gautier D. (1989) Titan's atmosphere from Voyager infrared observations. 1. The gas composition of Titan's equatorial region. *Icarus, 80,* 54–76.

Coustenis A., Salama A., Lellouch E., Encrenaz T., Bjoraker G. L., Samuelson R. E., de Grauw T., Feuchtgruber H., and Kessler M. F. (1998) Evidence for water vapor in Titan's atmosphere from ISO/SWS data. *Astron. Astrophys., 336,* L85–L89.

Coustenis A., Salama A., Schulz B., Ott S., Lellouch E., Encrenaz T., Gautier D., and Feuchtgruber H. (2003) Titan's atmosphere from ISO mid-infrared spectroscopy. *Icarus, 161,* 383–403.

Coustenis A., Lellouch E., Sicardy B., and Roe H. (2009) Earth-based perspective and pre-Cassini-Huygens knowledge of Titan. In Titan from Cassini-Huygens (R. H. Brown et al., eds.), pp. 9–34. Springer, Dordrecht.

Coustenis A., et al. (2010) Titan trace gaseous composition from CIRS at the end of the Cassini-Huygens prime mission. *Icarus, 207,* 461–476.

Cox C., Saglam A., Gerard J. C., et al. (2008) Distribution of the ultraviolet nitric oxide martian night airglow: Observations from Mars Express and comparisons with a one-dimensional model. *J. Geophys. Res., 113,* E08012.

Cravens T. E., Robertson I. P., Ledvina S. A., Mitchell D., Krimigis S. M., and Waite J. H. Jr. (2008) Energetic ion precipitation at Titan. *Geophys. Res. Lett., 35,* L03103, DOI: 10.1029/2007GL032451.

Cravens T. E., et al. (2009) Models-data comparisons for Titan's nightside ionosphere. *Icarus, 199,* 174–188.

Crisp D., et al. (1989) The nature of the features on the Venus night side. *Science, 246,* 506–509.

Crisp D., Meadows V. S., Bezard B., de Bergh C., Maillard J. P., and Mills F. P. (1996) Ground-based near-infrared observations of the Venus nightside: 1.27-μm $O_2(a^1\Delta_g)$ airglow from the upper atmosphere. *J. Geophys. Res., 101,* 4577–4594.

Crowley J. N., Ammann M., Cox R. A., et al. (2010) Evaluated kinetic and photochemical data for atmospheric chemistry: Volume V — heterogeneous reactions on solid substrates. *Atmos. Chem. Phys., 10,* 9059–9223.

Cui J., et al. (2009) Analysis of Titan's neutral upper atmosphere from Cassini Ion Neutral Mass Spectrometer measurements. *Icarus, 200,* 581–615.

De La Haye V., Waite J. Jr., Cravens T., Robertson I., and Lebonnois S. (2008) Coupled ion and neutral rotating model of Titan's upper atmosphere. *Icarus, 197(1),* 110–136.

Ekonomov A. P., Moshkin B. E., Moroz V. I., Golovin Yu. M., Gnedykh V. I., and Grigoriev A. V. (1983) UV photometry at the Venera 13 and 14 landing probes. *Cosmic Res., 21,* 194–206.

Encrenaz Th., Lellouch E., Cernicharo J., Paubert G., Gulkis S., and Spilker T. (1995) The thermal profile and water abundance in the Venus mesosphere from H_2O and HDO millimeter observations. *Icarus, 117,* 162–172.

Encrenaz T., Bézard B., Greathouse T. K., et al. (2004) Hydrogen peroxide on Mars: Evidence for spatial and seasonal variations. *Icarus, 170,* 424–429.

Encrenaz T., Greathouse T. K., Richter M. J., et al. (2011) A stringent upper limit to SO_2 in the martian atmosphere. *Astron. Astrophys., 530,* A37.

Encrenaz T., Greathouse T. K., Lefèvre F., and Atreya S. K. (2012) Hydrogen peroxide on Mars: Observations, interpretation, and future plans. *Planet. Space. Sci.,* in press.

Espenak F., Mumma M. J., Kostiuk T., et al. (1991) Ground-based infrared measurements of the global distribution of ozone in the atmosphere of Mars. *Icarus, 92,* 252–262.

Esposito L. W., Copley M., Eckert R., Gates L., Stewart A. I. F., and Worden H. (1988) Sulfur dioxide at the Venus cloud tops, 1978–1986. *J. Geophys. Res., 93,* 5267–5276.

Esposito L. W., Bertaux J.-L., Krasnopolsky V., Moroz V. I., and Zasova L. V. (1997) Chemistry of lower atmosphere and clouds. In *Venus II: Geology, Geophysics, Atmosphere, and Solar Wind Environment* (S. W. Bougher et al., eds.), pp. 415–458. Univ. of Arizona, Tucson.

Fast K., Kostiuk T., Espenak F., et al. (2006) Ozone abundance on Mars from infrared heterodyne spectra. I. Acquisition, retrieval, and anticorrelation with water vapor. *Icarus, 181,* 419–431.

Fast K., Kostiuk T., Lefèvre F., et al. (2009) Comparison of HIPWAC and Mars Express SPICAM observations of ozone on Mars 2006–2008 and variation from 1993 IRHS observations. *Icarus, 203,* 20–27.

Fedorova A., Korablev O., Perrier S., et al. (2006a) Observations of O_2 1.27 μm dayglow by SPICAM IR: Seasonal distribution for the first martian year of Mars Express. *J. Geophys. Res., 111,* E09S07, DOI: 10.1029/2006JE002694.

Fedorova A., Korablev O., Bertaux J. L., et al. (2006b) Mars water vapor abundance from SPICAM IR spectrometer: Seasonal and geographic distributions. *J. Geophys. Res., 111,* E09S08.

Fedorova A., et al. (2008) HDO and H_2O vertical distributions and isotopic ratio in the Venus mesosphere by solar occultation at infrared spectrometer on board Venus Express. *J. Geophys. Res., 113,* E00B22.

Fedorova A. A., Korablev O. I., Bertaux J. L., et al. (2009) Solar infrared occultation observations by SPICAM experiment on Mars-Express: Simultaneous measurements of the vertical distributions of H_2O, CO_2 and aerosol. *Icarus, 200,* 96–117.

Fedorova A. A., Lefèvre F., Guslyakova S., et al. (2012) The O_2 nightglow in the martian atmosphere by SPICAM onboard of Mars-Express. *Icarus, 219,* 596–608.

Fegley B. and Treiman A. H. (1992) Chemistry of atmosphere-surface interactions on Venus and Mars. In Venus and Mars: Atmospheres, Ionospheres, and Solar Wind Interactions (J. Luhmann and R. O. Pepin, eds.), pp. 7–71. AGU Geophys. Monogr. 66, American Geophysical Union, Washington, DC.

Fegley B., Zolotov M. Yu., and Lodders K. (1997) The oxidation state of the lower atmosphere and surface of Venus. *Icarus, 125,* 416–439.

Feldman P. D., Moos H. W., Clarke J. T., and Lane A. L. (1979) Identifications of the UV nightglow from Venus. *Nature, 279,* 221–222.

Fink U., Larson H. P., Kuiper G. P., and Poppe R. F. (1972) Water vapor in the atmosphere of Venus. *Icarus, 17,* 617–631.

Flasar F. M., et al. (2005) Titan's atmospheric temperatures, winds, and composition. *Science, 308,* 975–978.

Fonti S. and Marzo G. A. (2010) Mapping the methane on Mars. *Astron. Astrophys., 512,* A51.

Formisano V., Atreya S. K., Encrenaz T., et al. (2004) Detection of methane in the atmosphere of Mars. *Science, 306,* 1758–1761.

Fouchet T., Lellouch E., Ignatiev N. I., et al. (2007) Martian water vapor: Mars Express PFS/LW observations. *Icarus, 190,* 32–49.

Fox J. L. (1993) The production and escape of nitrogen atoms on Mars. *J. Geophys. Res., 98,* 3297–3310.

Fox J. L. and Hac A. (2010) Isotope fractionation in the photochemical escape of O from Mars. *Icarus, 208,* 176–191.

Fox J. L. and Victor G. A. (1988) Electron energy deposition in N_2 gas. *Planet. Space Sci., 36,* 329–352.

Fox J. L., Zhou P., and Bougher S. W. (1996) The martian thermosphere/ionosphere at high and low solar activities. *Adv. Space. Res., 17(11),* 203–218.

Gagne M. E., Melo S. M. L., Lefèvre F., Gonzalez-Galindo F., and Strong K. (2012) Modeled O_2 airglow distributions in the martian atmosphere. *J. Geophys. Res., 117,* E06005.

García Muñoz A., McConnell J. C., McDade I. C., et al. (2005) Airglow on Mars: Some model expectations for the OH Meinel bands and the O_2 IR atmospheric band. *Icarus, 176,* 75–95.

Garcia Munoz A., Mills F. P., Piccioni G., and Drossart P. (2009a) The near-infrared nitric oxide nightglow in the upper atmosphere of Venus. *Proc. Natl. Acad. Sci., 106,* 985–988.

Garcia Munoz A., Mills F. P., Slanger T. G., Piccioni G., and Drossart P. (2009b) Visible and near-infrared nightglow of molecular oxygen in the atmosphere of Venus. *J. Geophys. Res., 114,* E12002.

Gelman B. G., Zolotukhin V. G., Lamonov N. I., Levchuk B. V., Lipatov A. N., Mukhin L. M., Nenarokov D. F., Okhotnikov B. P., and Rotin V. A. (1979) An analysis of the chemical composition of the atmosphere of Venus on an AMS of the Venera 12 using a gas chromatograph. *Cosmic Res., 17,* 508–518.

Geminale A., Formisano V., and Sindoni G. (2011) Mapping methane in martian atmosphere with PFS-MEX data. *Planet. Space Sci., 59,* DOI: 10.1016/j.pss.201007.011.

Gerard J. C., Cox C., Saglam A., Bertaux J. L., Villard E., and Nehme C. (2008) Limb observations of the ultraviolet nitric oxide nightglow with SPICAV on board Venus Express. *J. Geophys. Res., 113,* E00B03.

Gonzalez-Galindo F., Forget F., Lopez-Valverde M. A., Angelats i Coll M., and Millour E. (2009) A ground-to-exosphere martian general circulation model: 1. Seasonal, diurnal, and solar cycle variation of thermospheric temperatures. *J. Geophys. Res., 114,* E04001.

Gurwell M. (2004) Submillimeter observations of Titan: Global measures of stratospheric temperature, CO, HCN, HC_3N, and the isotopic ratios $^{12}C/^{13}C$ and $^{14}N/^{15}N$. *Astrophys. J. Lett., 616,* L7–L10.

Gurwell M. A., Melnick G. J., Tolls V., Bergin E. A., and Patten B. M. (2007) SWAS observations of water vapor in the Venus mesosphere. *Icarus, 188,* 288–304.

Hanson W. B., Santanini S., and Zuccaro D. R. (1977) The martian ionosphere as observed by Viking retarding potential analyzers. *J. Geophys. Res., 82,* 4351–4363.

Hartle R. E., et al. (2006) Preliminary interpretation of Titan plasma interaction as observed by the Cassini Plasma Spectrometer: Comparisons with Voyager 1. *Geophys. Res. Lett., 33,* 8201, DOI: 10.10129/2005GL024817.

Hartogh P., Jarchow C., Lellouch E., et al. (2010a) Herschel/HIFI observations of HCl, H_2O_2, and O_2 in the martian atmosphere — initial results. *Astron. Astrophys., 521,* DOI: 10.1051/0004-6361/201015160.

Hartogh P., et al. (2010b) First results on martian carbon monoxide from Herschel/HIFI observations. *Astron. Astrophys., 521,* ID L48.

Hebrard E., Dobrijevic M., Benilan Y., and Raulin F. (2007) Photochemical kinetics uncertainties in modeling Titan's atmosphere: First consequences. *Planet. Space Sci., 55,* 1470–1489.

Hoffman J. H., Hodges R. R. Jr., Donahue T. M., and McElroy M. B. (1980) Composition of the Venus lower atmosphere from the Pioneer Venus mass spectrometer. *J. Geophys. Res. 85,* 7882–7890.

Hörst S. M., Vuitton V., and Yelle R. V. (2008) The origin of oxygen species in Titan's atmosphere. *J. Geophys. Res., 113,* E10006, DOI: 10.1029/2008JE003135.

Ignatiev N. I., Moroz V. I., Moshkin B. E., et al. (1997) Water vapor in the lower atmosphere of Venus: A new analysis of optical spectra measured by entry probes. *Planet. Space Sci., 45,* 427–438.

Ignatiev N. I., Moroz V. I., Zasova L. V., and Khatuntsev I. V. (1999) Water vapor in the middle atmosphere of Venus: An improved treatment of the Venera 15 IR spectra. *Planet. Space Sci., 47,* 1061–1075.

Irwin P. G. J., de Kok R., Negrao A., Tsang C. C. C., Wilson C. F., Drossart P., Piccioni G., Grassi D., and Taylor F. W. (2008) Spatial variability of carbon monoxide in Venus' mesosphere from Venus Express/Visible and Infrared Thermal Imaging Spectrometer measurements. *J. Geophys. Res., 113,* E00B01.

Istomin V. G., Grechnev K. V., and Kochnev V. A. (1980) Mass spectrometer measurements of the lower atmosphere of Venus. *Space Res., 20,* 215–218.

Istomin V. G., Grechnev K. V., and Kochnev V. A. (1983) Venera 13 and Venera 14: Mass spectrometry of the atmosphere. *Cosmic Res., 21,* 410–415.

Iwagami N., et al. (2008) Hemispheric distributions of HCl above and below the Venus' clouds by ground-based 1.7 μm spectroscopy. *Planet. Space Sci., 56,* 1424–1434.

Jakosky B. M. and Farmer C. B. (1982) The seasonal and global behavior of water vapor in the Mars atmosphere: omplete global results of the Viking atmospheric water detector experiment. *J. Geophys. Res., 87,* 2999–3019.

Jenkins J. M., Kolodner M. A., Butler B. J., Suleiman S. H., and Steffes P. G. (2002) Microwave remote sensing of the temperature and distribution of sulfur compounds in the lower atmosphere of Venus. *Icarus, 158,* 312–328.

Johnson R. E., Tucker O. J., Michael M., Sittler E. C., Smith H. T., Young D. T., and Waite J. H. (2009) Mass loss processes in Titan's atmosphere. In *Titan from Cassini-Huygens* (R. H. Brown et al., eds.), pp. 373–392. Springer, Dordrecht.

Kamp L. W., Taylor F. W., and Calcutt S. B. (1988) Structure of Venus' atmosphere from modeling of nightside infrared spectra. *Nature, 336,* 360–362.

Kaplan L. D., Connes J., and Connes P. (1969) Carbon monoxide in the martian atmosphere. *Astrophys. J., 157,* 187–192.

Kaye J. A. (1988) On the possible role of the reaction $O + HO_2$ $OH + O_2$ in OH airglow. *J. Geophys. Res., 93,* 285–288.

Keldysh M. V. (1977) Venus exploration with the Venera 9 and Venera 10 spacecraft. *Icarus, 30,* 605–625.

Khare B. N., Sagan C., Arakawa E. T., Suits F., Calcott T. A., and Williams M. W. (1984) Optical constants of organic tholins produced in a simulated titanian atmosphere from soft X-ray to microwave frequencies. *Icarus, 60,* 127–137.

Kliore A. J., Fjeldbo G., Seidel B. L., et al. (1973) S band radio occultation measurements of the atmosphere and topography of Mars with Mariner 9: Extended mission coverage of polar and intermediate latitudes. *J. Geophys. Res., 78,* 4331–4351.

Kliore A. J., Elachi C., Patel I. R., and Cimino J. B. (1979) Liquid content of the lower clouds of Venus as determined from Mariner 10 radio occultation. *Icarus, 37,* 51–72.

Kliore A. J., et al. (2008) First results from the Cassini radio occultations of the Titan ionosphere. *J. Geophys. Res., 113,* A09317, DOI: 10.1029/2007JA012965.

Knollenberg R. G. and Hunten D. M. (1980) Microphysics of the clouds of Venus: Results of the Pioneer Venus particle size spectrometer experiment. *J. Geophys. Res., 85,* 8039–8058.

Kolodner M. A. and Steffes P. G. (1998) The microwave absorption and abundance of sulfuric acid vapor in the Venus atmosphere based on new laboratory measurements. *Icarus, 132,* 151–169.

Kong T. Y. and McElroy M. B. (1977a) Photochemistry of the martian atmosphere. *Icarus, 32,* 168–189.

Kong T. Y. and McElroy M. B. (1977b) The global distribution of O_3 on Mars. *Planet. Space Sci., 25,* 839–857.

Korablev O. I., Krasnopolsky V. A., Rodin A. V., and Chassefiere E. (1993) Vertical structure of martian dust measured by solar infrared occultations from the Phobos spacecraft. *Icarus, 102,* 76–87.

Koskinen T. T., et al. (2011) The mesosphere and thermosphere of Titan revealed by Cassini/UVIS stellar occultations. *Icarus, 216,* 507–534.

Krasitsky O. P. (1978) A model for the diurnal variations of the composition of the martian atmosphere. *Cosmic Res., 16,* 434–442.

Krasnopolsky V. A. (1980) Veneras 9 and 10: Spectroscopy of scattered radiation in overcloud atmosphere. *Cosmic Res., 18,* 899–906.

Krasnopolsky V. A. (1983a) Venus spectroscopy in the 3000–8000 Å region by Veneras 9 and 10. In *Venus* (D. M. Hunten et al., eds.), pp. 459–483. Univ. of Arizona, Tucson.

Krasnopolsky V. A. (1983b) Lightning and nitric oxide on Venus. *Planet. Space Sci., 31,* 1363–1369.

Krasnopolsky V. A. (1985) Chemical composition of Venus clouds. *Planet. Space Sci., 33,* 109–117.

Krasnopolsky V. A. (1986) *Photochemistry of the Atmospheres of Mars and Venus.* Springer-Verlag, Berlin-New York. 334 pp.

Krasnopolsky V. A. (1993) Photochemistry of the martian atmosphere (mean conditions). Icarus, 101, 313–32.

Krasnopolsky V. A. (1995) Uniqueness of a solution of a steady state photochemical problem: Applications to Mars. *J. Geophys. Res., 100,* 3263–3276.

Krasnopolsky V. A. (1997) Photochemical mapping of Mars. *J. Geophys. Res., 102,* 13313–13320.

Krasnopolsky V. A. (2002) Mars' upper atmosphere and ionosphere at low, medium, and high solar activities: Implications for evolution of water. *J. Geophys. Res., 107(E12),* 5128.

Krasnopolsky V. A. (2003a) Spectroscopic mapping of Mars CO mixing ratio: Detection of north-south asymmetry. *J. Geophys. Res., 108,* DOI: 10.1029/2002JE001926.

Krasnopolsky V. A. (2003b) Spectroscopy of Mars O_2 1.27 μm dayglow at four seasonal points. *Icarus, 165,* 315–325.

Krasnopolsky V. A. (2006a) Photochemistry of the martian atmosphere: Seasonal, latitudinal, and diurnal variations. *Icarus, 185,* 153–170.

Krasnopolsky V. A. (2006b) A sensitive search for nitric oxide in the lower atmospheres of Venus and Mars: Detection on Venus and upper limit for Mars. *Icarus, 182,* 80–91.

Krasnopolsky V. A. (2006c) Some problems related to the origin of methane on Mars. *Icarus, 180,* 359–367.

Krasnopolsky V. A. (2006d) Chemical composition of Venus atmosphere and clouds: Some unsolved problems. *Planet. Space Sci., 54,* 1352–1359.

Krasnopolsky V. A. (2007a) Long-term spectroscopic observations of Mars using IRTF/CSHELL: Mapping of O_2 dayglow, CO, and search for CH_4. *Icarus, 190,* 93–102.

Krasnopolsky V. A. (2007b) Chemical kinetic model for the lower atmosphere of Venus. *Icarus, 191,* 25–37.

Krasnopolsky V.A. (2008) High-resolution spectroscopy of Venus: Detection of OCS, upper limit to H_2S, and latitudinal variations of CO and HF in the upper cloud layer. *Icarus, 197,* 377–385.

Krasnopolsky V. A. (2009a) A photochemical model of Titan's atmosphere and ionosphere. *Icarus, 201,* 226–256.

Krasnopolsky V. A. (2009b) Seasonal variations of photochemical tracers at low and middle latitudes on Mars: Observations and models. *Icarus, 201,* 564–569.

Krasnopolsky V. A. (2010a) Solar activity variations of thermospheric temperatures on Mars and a problem of CO in the lower atmosphere. *Icarus, 207,* 638–647.

Krasnopolsky V. A. (2010b) Venus night airglow: Ground-based detection of OH, observations of O_2 emissions, and photochemical model. *Icarus, 207,* 17–27.

Krasnopolsky V. A. (2010c) Spatially-resolved high-resolution spectroscopy of Venus. 1. Variations of CO2, CO, HF, and HCl at the cloud tops. *Icarus, 208,* 539–547.

Krasnopolsky V. A. (2010d) Spatially-resolved high-resolution spectroscopy of Venus. 2. Variations of HDO, OCS, and SO_2 at the cloud tops. *Icarus, 209,* 314–322.

Krasnopolsky V. A. (2010e) The photochemical model of Titan's atmosphere and ionosphere: A version without hydrodynamic escape. *Planet. Space Sci., 58,* 1507–1515.

Krasnopolsky V. A. (2011a) Excitation of the oxygen nightglow on the terrestrial planets. *Planet. Space Sci., 59,* 754–766.

Krasnopolsky V. A. (2011b) Vertical profile of H_2SO_4 vapor at 70–110 km and some related problems. *Icarus, 215,* 197–203.

Krasnopolsky V. A. (2011c) A sensitive search for methane and ethane on Mars. *EPSC Abstracts, 6,* 49.

Krasnopolsky V. A. (2012a) Search for methane and upper limits to ethane and SO_2 on Mars. *Icarus, 217,* 144–152.

Krasnopolsky V. A. (2012b) A photochemical model for the Venus atmosphere at 47–112 km. *Icarus, 218,* 230–246.

Krasnopolsky V. A. (2012c) Titan's photochemical model: Further update, oxygen species, and comparison with Triton and Pluto. *Planet. Space Sci., 73,* 318–326.

Krasnopolsky V. A. (2013a) S_3 and S_4 abundances and improved chemical kinetic model for the lower atmosphere of Venus. *Icarus, 225,* 570–580.

Krasnopolsky V. A. (2013b) Nighttime photochemical model and night airglow on Venus. *Planet. Space Sci.,* in press.

Krasnopolsky V. A. (2013c) Night and day airglow of oxygen at 1.27 μm on Mars. *Planet. Space Sci.,* in press.

Krasnopolsky V. A. and Cruikshank D. P. (1999) Photochemistry of Pluto's atmosphere and ionosphere near perihelion. *J. Geophys. Res., 104,* 21979–21996.

Krasnopolsky V. A. and Feldman P. D. (2001) Detection of molecular hydrogen in the atmosphere of Mars. *Science, 294,* 1914–1917.

Krasnopolsky V. A. and Gladstone G. R. (2005) Helium on Mars and Venus: EUVE observations and modeling. *Icarus, 176,* 395–407.

Krasnopolsky V. A. and Krysko A. A. (1976) On the night airglow in the martian atmosphere. *Space Res., 16,* 1005–1008.

Krasnopolsky V. A. and Parshev V. A. (1977) Altitude profile of water vapor on Mars. *Cosmic Res., 15,* 673–675.

Krasnopolsky V. A. and Parshev V. A. (1979) Chemical composition of Venus' troposphere and cloud layer based on Venera 11, Venera 12, and Pioneer Venus measurements. *Cosmic Res., 17,* 630–637.

Krasnopolsky V. A. and Parshev V. A. (1981) Chemical composition of the atmosphere of Venus. *Nature, 282,* 610–612.

Krasnopolsky V. A. and Parshev V.A. (1983) Photochemistry of the Venus atmosphere. In *Venus* (D. M. Hunten et al., eds.),

pp. 431–458. Univ. of Arizona, Tucson.
Krasnopolsky V. A. and Pollack J. B. (1994) H_2O-H_2SO_4 system in Venus' clouds and OCS, CO, and H_2SO_4 profiles in Venus' troposphere. *Icarus, 109,* 58–78.
Krasnopolsky V. A., Krysko A. A., Rogachev V. N., and Parshev V. A. (1976) Spectroscopy of the Venus night airglow from the Venera 9 and 10 orbiters. *Cosmic Res., 14,* 789–795.
Krasnopolsky V. A., Bowyer S., Chakrabarti S., et al. (1994) First measurement of helium on Mars: Implications for the problem of radiogenic gases on the terrestrial planets. *Icarus, 109,* 337–351.
Krasnopolsky V. A., Bjoraker G. L., Mumma M. J., et al. (1997) High-resolution spectroscopy of Mars at 3.7 and 8 μm: A sensitive search for H_2O_2, H_2CO, HCl, and CH_4, and detection of HDO. *J. Geophys. Res., 102,* 6525–6534.
Krasnopolsky V. A., Mumma M. J., and Gladstone G. R. (1998) Detection of atomic deuterium in the upper atmosphere of Mars. *Science, 280,* 1576–1580.
Krasnopolsky V. A., Maillard J. P., and Owen T. C. (2004) Detection of methane in the martian atmosphere: Evidence for life? *Icarus, 172,* 537–547.
Krasnopolsky V. A., Belyaev D. A., Gordon I. E., Li G., and Rothman L. S. (2013) Observations of D/H ratios in H_2O, HCl, and HF on Venus and new DCl and DF line strengths. *Icarus, 224,* 57–65.
Kuiper G. P. (1949) Survey of planetary atmospheres. In *The Atmospheres of the Earth and Planets* (G. P. Kuiper, ed.), pp. 304–345. Univ. of Chicago, Chicago.
Kumar S. and Broadfoot A. L. (1975) Helium 584 A airglow emission from Venus: Mariner 10 observations. *Geophys. Res. Lett., 2,* 357–360.
Lane W. A. and Optsbaum R. (1983) High altitude Venus haze from Pioneer Venus limb scans. *Icarus, 54,* 48–58.
Lara L. M., Lellouch E., Lopes-Moreno J., and Rodrigo R. (1996) Vertical distribution of Titan's atmospheric neutral constituents. *J. Geophys. Res., 101,* 23261–23283.
Lavvas P. P., Coustenis A., and Vardavas I. M. (2008a) Coupling photochemistry with haze formation in Titan's atmosphere, part I: Model description. *Planet. Space Sci., 56,* 27–66.
Lavvas P. P., Coustenis A., and Vardavas I. M. (2008b) Coupling photochemistry with haze formation in Titan's atmosphere, part II: Results and validation with Cassini/Huygens data. *Planet. Space Sci., 56,* 67–99.
Lavvas P., Sander M., Kraft M., and Imanaka H. (2011) Surface chemistry and particle shape: Processes for the evolution of aerosols in Titan's atmosphere. *Astrophys. J., 728,* Article ID 80.
Lebonnois S. (2005) Benzene and aerosol production in Titan and Jupiter's atmospheres: A sensitivity study. *Planet. Space Sci., 53,* 486–497.
Lebonnois S., Quémerais E., Montmessin F., et al. (2006) Vertical distribution of ozone on Mars as measured by SPICAM/Mars Express using stellar occultations. *J. Geophys. Res., 111,* E09S05, DOI: 10.1029/2005JE002643.
Lefèvre F. and Forget F. (2009) Observed variations of methane on Mars unexplained by known atmospheric chemistry and physics. *Nature, 460,* 720–723.
Lefèvre F., Lebonnois S., Montmessin F., et al. (2004) Three-dimensional modeling of ozone on Mars. *J. Geophys. Res., 109,* E07004, DOI: 10.1029/2004JE002268.
Lefèvre F., Bertaux J. L., Clancy R. T., et al. (2008) Heterogeneous chemistry in the atmosphere of Mars. *Nature, 454,* 971–975.
Lellouch E., Goldstein J. J., Rosenqvist J., Bougher S. W., and Paubert G. (1994) Global circulation, thermal structure, and carbon monoxide distribution in Venus' mesosphere in 1991. *Icarus, 110,* 315–339.
Levine J. S., ed. (1985) *The Photochemistry of Atmospheres: Earth, the Other Planets, and Comets.* Academic, Orlando.
Liu S. C. and Donahue T. M. (1976) The regulation of hydrogen and oxygen escape from Mars. *Icarus, 28,* 231–246.
Lutz B. L., de Bergh C., and Owen T. (1983) Titan: Discovery of carbon monoxide in its atmosphere. *Science, 220,* 1374–1375.
Magee B. A., Waite J. H., Mandt K. E., Westlake J., Bell J., and Gell D. A. (2009) INMS-derived composition of Titan's upper atmosphere: Analysis methods and model comparison. *Planet. Space Sci., 57,* 1895–1916.
Maguire W. C. (1977) Martian isotopic ratios and upper limits for possible minor constituents as derived from Mariner 9 infrared spectrometer data. *Icarus, 32,* 85–97.
Mahaffy P. R., et al. (2013) Abundance and isotopic composition of gases in the martian atmosphere: First results from the Mars Curiosity Rover. *Science,* in press.
Maltagliati L., et al. (2013) Annual survey of water vapor vertical distribution and water-aerosol coupling in the martian atmosphere observed by SPICAM/MEX solar occulations. *Icarus, 223,* 942–962.
Marcq E., Encrenaz T., Bezard B., and Birlan M. (2006) Remote sensing of Venus' lower atmosphere from ground-based IR spectroscopy: Latitudinal and vertical distribution of minor species. *Planet. Space Sci., 54,* 1360–1370.
Marcq E., Bezard B., Drossart P., Piccioni G., Reess M., and Henry F. (2008) A latitudinal survey of CO, OCS, H_2O, and SO_2 in the lower atmosphere of Venus: Spectroscopic studies using VIRTIS-H. *J. Geophys. Res., 113,* E00B07.
Marcq E., Belyaev D., Montmessin F., Fedorova A., Bertaux J. L., Vandaele A. C., and Neefs E. (2011) An investigation of the SO_2 content of the venusian mesosphere using SPICAV-UV in nadir mode. *Icarus, 211,* 58–69.
Marten A., Hidayat T., Biraud Y., and Moreno R. (2002) New millimeter heterodyne observations of Titan: Vertical distributions of nitriles HCN, HC_3N, CH_3CN, and the isotopic ratio $^{15}N/^{14}N$ in its atmosphere. *Icarus, 158,* 532–544.
McElroy M. B. (1972) Mars: An evolving atmosphere. *Science, 175,* 443–445.
McElroy M. B. and Donahue T. M. (1972) Stability of the martian atmosphere. *Science, 177,* 986–988.
McEwan M. J. and Anicich V. G. (2007) Titan's ion chemistry: A laboratory perspective. *Mass Spectrom. Rev., 26,* 281–319.
Meadows V. S. and Crisp D. (1996) Ground-based near-infrared observations of the Venus nightside: The thermal structure and water abundance near the surface. *J. Geophys. Res., 101,* 4595–4622.
Melchiorri R., Encrenaz T., Fouchet T., et al. (2007) Water vapor mapping on Mars using OMEGA/Mars Express. *Planet. Space Sci., 55,* 333–342.
Mills F. P. (1998) I. Observations and photochemical modeling of the Venus middle atmosphere. II. Thermal infrared spectroscopy of Europa and Callisto. Ph.D. thesis, California Institute of Technology.
Mills F. P. and Allen M. (2007) A review of selected issues concerning the chemistry in Venus' middle atmosphere. *Planet. Space Sci., 55,* 1729–1740.
Mischna M. A., Allen M., Richardson M. I., et al. (2011) Atmospheric modeling of Mars methane surface release. *Planet.*

Space Sci., 59, 227–237.

Molina-Cuberos G. J., Lopes-Moreno J. J., Rodrigo R., and Lara L. M. (1999) Chemistry of the galactic cosmic ray induced ionosphere of Titan. *J. Geophys. Res., 104,* 21997–22024.

Montmessin F., Forget F., Rannou P., Cabane M., and Haberle R. M. (2004) Origin and role of water ice clouds in the martian water cycle as inferred from a general circulation model. *J. Geophys. Res., 109,* E10004, DOI: 10.1029/2004JE002284.

Montmessin F., Bertaux J. L., Lefèvre F., et al. (2011) A layer of ozone detected in the nightside upper atmosphere of Venus. *Icarus, 216,* 82–85.

Moreau D., Esposito L. W., and Brasseur G. (1991) The chemical composition of the dust-free martian atmosphere: Preliminary results of a two-dimensional model. *J. Geophys. Res., 96,* 7933–7945.

Moreno R., Lellouch E., Lara L.M., Courtin R., Hartogh P., and Rengel M. (2012) Observations of H_2O in Titan's atmosphere with Herschel. In *Workshop on Titan Through Time II,* NASA Goddard Space Flight Center, April 3–5, 2012.

Moroz V. I. (1964) New observations of Venus infrared spectrum. *Astron. Zh., 41,* 711–714.

Moroz V. I. (1968) The CO_2 bands and some optical properties of the atmosphere of Venus. *Sov. Astron. J., 11,* 653–661.

Moudden Y. (2007) Simulated seasonal variations of hydrogen peroxide in the atmosphere of Mars. *Planet. Space Sci., 55,* 2137–2143.

Moudden Y. and McConnell J. C. (2007) Three-dimensional online modeling in a Mars general circulation model. *Icarus, 188,* 18–34.

Mumma M. J., Villanueva G. L., Novak R. E., et al. (2009) Strong release of methane on Mars in northern summer 2003. *Science, 323,* 1041–1045.

Nair,H., Allen M., Anbar A. D., et al. (1994) A photochemical model of the martian atmosphere. *Icarus, 111,* 124–150.

Niemann H. B., Atreya S. K., Demick J. E., Gautier D., Haberman J. A., Harpold D. N., Kasprzak W. T., Lunine J. I., Owen T. C., and Raulin F. (2010) Composition of Titan's lower atmosphere and simple surface volatiles as measured by the Cassini-Huygens probe gas chromatograph mass spectrometer experiment. *J. Geophys. Res., 115,* E12006.

Nier A. O. and McElroy M. B. (1977) Composition and structure of Mars' upper atmosphere: Results from the neutral mass spectrometers on Viking 1 and 2. *J. Geophys. Res., 82,* 4341–4348.

Noxon J. F., Traub W. A., Carleton N. P., et al. (1976) Detection of O_2 dayglow emission from Mars and the martian ozone abundance. *Astrophys. J., 207,* 1025–1030.

Ohtsuki S., Iwagami N., Sagawa H., Ueno M., Kasaba Y., Imamura T., Yanagisawa K., and Nishihara E. (2008) Distribution of the Venus 1.27-μm O_2 airglow and rotational temperature. *Planet. Space Sci., 56,* 1391–1398.

Owen T., Biemann K., Rushnek D. R., et al. (1977) The composition of the atmosphere at the surface of Mars. *J. Geophys. Res., 82,* 4635–4639.

Oyama V. I., Carle G. C.,Woeller F., Pollack J. B., Reynolds R. T., and Craig R. A. (1980) Pioneer Venus gas chromatography of Venus. *J. Geophys. Res., 85,* 7891–7902.

Parkinson T. D. and Hunten D. M. (1972) Spectroscopy and aeronomy of O_2 on Mars. *J. Atmos. Sci., 29,* 1380–1390.

Pereira R. A., Baulch D. L., Pilling M. J., Robertson S. H., and Zeng G. (1997) Temperature and pressure dependence of the multichannel rate coefficients for the CH_3 + OH system. *J. Phys. Chem. A, 101,* 9681–9693.

Pernice H., Garcia P., Willner H., Francisco J. S., Mills F. P., Allen M., and Yung Y. L. (2004) Laboratory evidence for a key intermediate in the Venus atmosphere: Peroxychloroformyl radical. *Proc. Natl. Acad. Sci., 101,* 14007–14010.

Perrier S., Bertaux J. L., Lefèvre F., et al. (2006) Global distribution of total ozone on Mars from SPICAM/MEX UV measurements. *J. Geophys. Res., 111,* E09S06, DOI: 10.1029/2006JE002681.

Piccioni G., et al. (2008) First detection of hydroxyl in the atmosphere of Venus. *Astron. Astrophys., 483,* L29–L33.

Piccioni G., Zasova L., Migliorini A., Drossart P., Shakun A., Garcia Munoz A., Mills F. P., and Cardesin-Moinelo A. (2009) Near-IR oxygen nightglow observed by VIRTIS in the Venus upper atmosphere. *J. Geophys. Res., 114,* E00B38.

Pollack J. B., et al. (1993) Near-infrared light from Venus' nightside — A spectroscopic analysis. *Icarus, 103,* 1–42.

Prinn R.G. (1971) Photochemistry of HCl and other minor constituents in the atmosphere of Venus. *J. Atmos. Sci., 28,* 1058–1068.

Prinn R. G. (1985) The photochemistry of the atmosphere of Venus. In *The Photochemistry of Atmospheres* (J. S. Levine, ed.), pp. 281–336. Academic, Orlando.

Rannou P., McKay C. P., Botet R., and Cabane M. (1999) Semi-empirical model of absorption and scattering by isotropic fractal aggregates of spheres. *Planet. Space Sci., 47,* 385–396.

Samuelson R. E., Maguire W. C., Hanel R. A., Kunde V. G., Jennings D. E., Yung Y. L., and Aikin A. C. (1983) CO_2 on Titan. *J. Geophys. Res., 88,* 8709–8715.

Sandor B. J. and Clancy R. T. (2005) Water vapor variations in the Venus mesosphere from microwave spectra. *Icarus, 177,* 129–143.

Sandor B. J. and Clancy R. T. (2012) Observations of HCl altitude dependence and temporal variation in the 70–100 km mesosphere of Venus. *Icarus,* in press.

Sandor B. J., Clancy R. T., Moriarty-Schieven G., and Mills F. P. (2010) Sulfur chemistry in the Venus mesosphere from SO_2 and SO microwave spectra. *Icarus, 208,* 49–60.

Sandor B. J., Clancy R. T., and Moriarty-Schieven G. (2012) Upper limits for H_2SO_4 in the mesosphere of Venus. *Icarus,* in press.

Shemansky D. E., Stewart A. I. F., West R. A., Esposito L. W., Hallett J. T., and Liu X. (2005) The Cassini UVIS stellar probe of the Titan atmosphere. *Science, 308,* 978–982.

Shimazaki T. (1981) A model of temporal variations in ozone density in the martian atmosphere. *Planet. Space Sci., 29,* 21–33.

Sindoni G., Formisano V., and Geminale A. (2011) Observations of water vapour and carbon monoxide in the martian atmosphere with the SWC of PFS/MEX. *Planet. Space Sci., 59,* 149–162.

Slanger T.G. and Copeland R. A. (2003) Energetic oxygen in the upper atmosphere and the laboratory. *Chem. Rev., 103,* 4731–4765.

Slanger T. G., Cosby P. C., Huestis D. L., and Bida T. A. (2001) Discovery of the atomic oxygen green line in the Venus night airglow. *Science, 291,* 463–465.

Slanger T. G., Huestis D. L., Cosby P. C., Chanover N. J., and Bida T. A. (2006) The Venus nightglow ground-based observations and chemical mechanisms. *Icarus, 182,* 1–9.

Smith M. D. (2004) Interannual variability in TES atmospheric observations of Mars during 1999–2003. *Icarus, 167,* 148–165.

Smith M. D., Wolff M. J., Clancy R. T., et al. (2009) Compact Reconnaissance imaging spectrometer observations of water vapor and carbon monoxide. *J. Geophys. Res., 114,* E00D03.

Spinrad H., Münch G., and Kaplan L. D. (1963) The detection of water vapor on Mars. *Astrophys. J., 137,* 1319.

Soret L., Gerard J. C., Montmessin F., et al. (2012) Atomic oxygen on the Venus nightside: Global distribution deduced from airglow mapping. *Icarus, 217,* 849–855.

Sprague A. L., Boynton W. V., Kerry K. E., et al. (2004) Mars' south polar Ar enhancement: A tracer for south polar seasonal meridional mixing. *Science, 306,* 1364–1367.

Sprague A. L., Boynton W. V., Forget F., et al. (2012) Interannual similarity and variation in seasonal circulation of Mars' atmospheric Ar as seen by the Gamma Ray Spectrometer on Mars Odyssey. *J. Geophys. Res., 117,* E04005.

Stewart A. I. F., Gerard J. C., Rusch D. W., and Bougher S. W. (1980) Morphology of the Venus ultraviolet night airglow. *J. Geophys. Res., 85,* 7861–7870.

Stewart A. I. F., Alexander M. J., Meier R. R., Paxton L. J., Bougher S. W., and Fesen C. G. (1992) Atomic oxygen in the martian atmosphere. *J. Geophys. Res., 97,* 91–102.

Strickland D. J., Thomas G. E., and Sparks P. R. (1972) Mariner 6 and 7 ultraviolet spectrometer experiment: Analysis of the O I 1304 and 1356 Å emissions. *J. Geophys. Res., 77,* 4052–4058.

Strobel D. F. (2009) Titan's hydrodynamically escaping atmosphere: Escape rates and the structure of the exobase region. *Icarus, 202,* 632–641.

Taylor F. W., Crisp D., and Bezard B. (1997) Near-infrared sounding of the lower atmosphere of Venus. In *Venus II* (S. W. Bougher et al., eds.), pp. 325–352. Univ. of Arizona, Tucson.

Tomasko M. G., Doose L., Engel S., Dafoe L. E., West R., Lemmon M., Karkoschka E., and See C. (2008) A model of Titan's aerosols based on measurements made inside the atmosphere. *Planet. Space Sci., 56,* 669–707.

Toon O. B., Turco R. P., and Pollack J. B. (1982) The ultraviolet absorber on Venus: Amorphous sulfur. *Icarus, 51,* 358–373.

Toublanc D., Parisot J. P., Brillet J., Gautier D., Raulin F., and McKay C. P. (1995) Photochemical modeling of Titan's atmosphere. *Icarus, 113,* 2–26.

Trainer M. G., Tolbert M. A., McKay C. P., et al. (2011) Limits on the trapping of atmospheric CH_4 in martian polar ice analogs. *Icarus, 208,* 192–197.

Traub W. A., Carleton N. P., Connes P., et al. (1979) The latitude variation of O_2 dayglow and O_3 abundance on Mars. *Astrophys. J., 229,* 846–850.

Trauger J. T. and Lunine J. I. (1983) Spectroscopy of molecular oxygen in the atmosphere of Venus and Mars. *Icarus, 55,* 272–281.

Tsang C. C. C., Taylor F. W., Wilson C. F., Liddell S. J., Irwin P. G. J., Piccioni G., Drossart P., and Calcutt S. B. (2009) Variability of CO concentration in the Venus troposphere from Venus Express/VIRTIS using a band ratio technique. *Icarus, 201,* 432–443.

Tschimmel M., Ignatiev N. I., Titov D. V., et al. (2008) Investigation of water vapor on Mars with PFS/SW of Mars Express. *Icarus, 195,* 557–575.

Vandaele A. C., et al. (2008) Composition of the Venus mesosphere measured by solar occultation at infrared on board Venus Express. *J. Geophys Res., 113,* E00B23.

Van de Hulst H. C. (1980) *Multiple Light Scattering: Tables, Formulas, and Applications.* Academic, San Diego.

Vervack R. J. Jr., Sandel B. R., and Strobel D. F. (2004) New perspectives on Titan's upper atmosphere from a reanalysis of the Voyager 1 UVS solar occultations. *Icarus, 170,* 91–112.

Villanueva G. I., et al. (2013) A sensitive search for organics (CH_4, CH_3OH, H_2CO, C_2H_6, C_2H_2, C_2H_4), hydroperoxyl (HO_2), nitrogen compounds (N_2O, NH_3, HCN) and chlorine species (HCl, CH_3Cl) on Mars using ground-based high-resolution infrared spectroscopy. *Icarus, 223,* 11–27.

Vinatier S., et al. (2010) Analysis of Cassini/CIRS limb spectra of Titan acquired during the nominal mission. I. Hydrocarbons, nitriles and CO_2 vertical mixing ratio profiles. *Icarus, 205,* 559–570.

Vinatier S., Rannou P., Anderson C. M., et al. (2012) Optical constants of Titan's stratospheric aerosols in the 70–1500 cm^{-1} spectral range constrained by Cassini/CIRS observations. *Icarus, 219,* 5–12.

Vinogradov A. P. and Surkov Yu. A. (1970) Chemical composition of the atmosphere of Venus. *Cosmic Res., 5,* 8–18.

Von Zahn U., Fricke K. H., Hunten D. M., Krankowsky D., Mauersberger K., and Nier A. O. (1980) The upper atmosphere of Venus during morning conditions. *J. Geophys. Res., 85,* 7829–7840.

Von Zahn U., Kumar S., Niemann H., and Prinn R. (1983) Composition of the Venus atmosphere. In *Venus* (D. M. Hunten et al., eds.), pp. 299–430. Univ. of Arizona, Tucson.

Vuitton V., Yelle R. V., and McEwan M. J. (2007) Ion chemistry and N-containing molecules in Titan's upper atmosphere. *Icarus, 191,* 722–742.

Vuitton V., Lavvas P., Yelle R. V., et al. (2009) Negative ion chemistry in Titan's upper atmosphere. *Planet. Space Sci., 57,* 1558–1572.

Vuitton V., Yelle R.V., Lavvas P., and Kippenstein S. J. (2012) Rapid association reaction at low pressure: Impact on the formation of hydrocarbons on Titan. *Astrophys. J., 744(1),* Article ID 11.

Wahlund J. E., et al. (2009) On the amount of heavy molecular ions in Titan's ionosphere. *Planet. Space Sci., 57,* 1857–1865.

Webster C. R., et al. (2013) Isotope ratios of H, C and O in CO_2 and H_2O of the martian atmosphere. *Science,* in press.

Westlake J. H., Waite J. H. Jr., Mandt K. E., et al. (2012) Titan's ionospheric composition and structure: Photochemical modeling of Cassini INMS data. *J. Geophys. Res., 117,* E01003.

Wilson E. H. and Atreya S. K. (2004) Current state of modeling the photochemistry of Titan's mutually dependent atmosphere and ionosphere. *J. Geophys. Res., 109,* E06002, DOI: 10.1029/2003JE002181.

Wilson W. J., Klein M. J., Kakar R. K., et al. (1981) Venus. I. Carbon monoxide distribution and molecular-line search. *Icarus, 45,* 624–637.

Wong A. S., Morgan C. G., Yung Y. L., and Owen T. C. (2002) Evolution of CO on Titan. *Icarus, 155,* 382–392.

Woo R. and Ishimaru A. (1981) Eddy diffusion coefficient for the atmosphere of Venus from radio scintillation measurements. *Nature, 289,* 383–384.

Woodall J., Agundez M., Markwick-Kemper A. J., and Millar T. J. (2007) The UMIST database for astrochemistry 2006. *Astron. Astrophys., 466,* 1197–1204.

Wraight P. C. (1982) Association of atomic oxygen and airglow excitation mechanisms. *Planet. Space Sci., 30,* 251–259.

Yelle R. V., Borggren N., de la Haye V., Kasprzak W. T., Niemann H. B., Mueller-Wodarg I., and Waite J. H. Jr. (2006) The vertical structure of Titan's upper atmosphere from Cassini ion neutral mass spectrometer measurements. *Icarus, 182,* 567–576.

Yelle R. V., Cui J., and Mueller-Wodarg I. C. F. (2008) Eddy diffusion and methane escape from Titan's atmosphere. *J. Geophys. Res. 113,* E10003, DOI:10.1029/2007JE003031.

Yelle R. V., et al. (2010) Formation of NH_3 and CH_2NH in Titan's upper atmosphere. *Faraday Discussions, 147,* 31–49.

Young A. T. (1983) Venus cloud microphysics. *Icarus, 56,* 568–580.

Young L. D. G. (1972) High resolution spectra of Venus: A review. *Icarus, 17,* 632–658.

Yung Y. L. and DeMore W. B. (1982) Photochemistry of the stratosphere of Venus: Implications for atmospheric evolution. *Icarus, 51,* 199–247.

Yung Y. L. and DeMore W. B. (1999) *Photochemistry of Planetary Atmospheres.* Oxford Univ., Oxford/New York. 456 pp.

Yung Y. L., Allen M., and Pinto J. P. (1984) Photochemistry of the atmosphere of Titan: Comparison between model and observations. *Astrophys. J. Suppl. Ser., 55,* 465–506.

Yung Y. L., Liang M. C., Jiang X., Shia R. L., Lee C., Bezard B., and Marcq E. (2009) Evidence for carbonyl sulfide (OCS) conversion to CO in the lower atmosphere of Venus. *J. Geophys. Res., 114,* E00B34.

Zahnle K., Haberle R. M., Catling D. C., et al. (2008) Photochemical instability of the ancient martian atmosphere. *J. Geophys. Res., 113,* E11004, DOI: 10.1029/2008JE003160.

Zahnle K., Freedman R. S., and Catling D. C. (2011) Is there methane on Mars? *Icarus, 212,* 493–503.

Zasova L. V., Krasnopolsky V. A., and Moroz V. I. (1981) Vertical distribution of SO_2 in the upper cloud layer and origin of the UV absorption. *Adv. Space Res., 1(1),* 13–16.

Zasova L. V., Moroz V. I., Esposito L. W., and Na C. Y. (1993) SO_2 in the middle atmosphere of Venus: IR measurements from Venera 15 and comparison to UV data. *Icarus, 105,* 92–109.

Zhang X., Liang M. C., Montmessin F., Bertaux J. L., Parkinson C., and Yung Y. L. (2010) Photolysis of sulphuric acid as the source of sulphur oxides in the mesosphere of Venus. *Nature Geosci., 3,* 834–837.

Zhang X., Liang M. C., Mills F. P., Belyaev D. A., and Yung Y. L. (2012) Sulfur chemistry in the middle atmosphere of Venus. *Icarus, 217,* 714–739.

Zhu X. and Yee J. H. (2007) Wave-photochemistry coupling and its effect on water vapor, ozone and airglow variations in the atmosphere of Mars. *Icarus, 189,* 136–150.

Showman A. P., Wordsworth R. D., Merlis T. M., and Kaspi Y. (2013) Atmospheric circulation of terrestrial exoplanets. In *Comparative Climatology of Terrestrial Planets* (S. J. Mackwell et al., eds.), pp. 277–326. Univ. of Arizona, Tucson, DOI: 10.2458/azu_uapress_9780816530595-ch12.

Atmospheric Circulation of Terrestrial Exoplanets

Adam P. Showman
University of Arizona

Robin D. Wordsworth
University of Chicago

Timothy M. Merlis
Princeton University

Yohai Kaspi
Weizmann Institute of Science

The investigation of planets around other stars began with the study of gas giants, but is now extending to the discovery and characterization of super-Earths and terrestrial planets. Motivated by this observational tide, we survey the basic dynamical principles governing the atmospheric circulation of terrestrial exoplanets, and discuss the interaction of their circulation with the hydrological cycle and global-scale climate feedbacks. Terrestrial exoplanets occupy a wide range of physical and dynamical conditions, only a small fraction of which have yet been explored in detail. Our approach is to lay out the fundamental dynamical principles governing the atmospheric circulation on terrestrial planets — broadly defined — and show how they can provide a foundation for understanding the atmospheric behavior of these worlds. We first survey basic atmospheric dynamics, including the role of geostrophy, baroclinic instabilities, and jets in the strongly rotating regime (the "extratropics") and the role of the Hadley circulation, wave adjustment of the thermal structure, and the tendency toward equatorial superrotation in the slowly rotating regime (the "tropics"). We then survey key elements of the hydrological cycle, including the factors that control precipitation, humidity, and cloudiness. Next, we summarize key mechanisms by which the circulation affects the global-mean climate, and hence planetary habitability. In particular, we discuss the runaway greenhouse, transitions to snowball states, atmospheric collapse, and the links between atmospheric circulation and CO_2 weathering rates. We finish by summarizing the key questions and challenges for this emerging field in the future.

1. INTRODUCTION

The study of planets around other stars is an exploding field. To date, numerous exoplanets have been discovered, spanning a wide range of masses, incident stellar fluxes, orbital periods, and orbital eccentricities. A variety of observing methods have allowed observational characterization of the atmospheres of these exoplanets, opening a new field in comparative climatology (for introductions, see *Deming and Seager,* 2009; *Seager and Deming,* 2010). Because of their relative ease of observability, this effort to date has emphasized transiting giant planets with orbital semimajor axes of ~0.1 AU or less. The combination of radial velocity and transit data together allow estimates of the planetary mass, radius, density, and surface gravity. Wavelength-dependent observations of the transit and secondary eclipse, when the planet passes in front of and behind its star, respectively, as well as the full-orbit light curves, allow the atmospheric composition, vertical temperature structure, and global temperature maps to be derived — at least over certain ranges of pressure. These observations provide evidence for a vigorous atmospheric circulation on these worlds (e.g., *Knutson et al.,* 2007).

Despite a major emphasis to date on extrasolar giant planets (EGPs), super-Earths and terrestrial planets are increasingly becoming accessible to discovery and characterization. Over 50 super-Earths have been confirmed from groundbased and spacebased planet searches, including planets that are Earth-sized (*Fressin et al.,* 2012) and planets that orbit within the classical habitable zones (HZ) of their stars (*Borucki et al.,* 2012, 2013). Hundreds of additional super-Earth candidates have been found by the NASA Kepler mission (*Borucki et al.,* 2011). For observationally favorable systems, such as super-Earths orbiting M dwarfs (*Charbonneau et al.,* 2009), atmospheric characterization has already begun (e.g., *Bean et al.,* 2010; *Désert et al.,*

2011; *Berta et al.,* 2012), placing constraints on the atmospheric composition of these objects. Methods that are used today for EGPs to obtain dayside infrared (IR) spectra, map the day-night temperature pattern, and constrain cloudiness and albedo will be extended to the characterization of smaller planets in coming years. This observational vanguard will continue over the next decade with attempts to determine the composition, structure, climate, and habitability of these worlds. Prominent next-generation observational platforms include NASA's James Webb Space Telescope (JWST) and Transiting Exoplanet Survey Satellite (TESS), as well as a wide range of upcoming groundbased instruments.

This observational tide provides motivation for understanding the circulation regimes of extrasolar terrestrial planets. Fundamental motivations are threefold. First, we wish to understand current and future spectra, light curves, and other observations, and understand the role of dynamics in affecting these observables. Second, the circulation — and climate generally — play a key role in shaping the habitability of terrestrial exoplanets, a subject of crucial importance in understanding our place in the cosmos. Third, the wide range of conditions experienced by exoplanets allows an opportunity to extend our theoretical understanding of atmospheric circulation and climate to a much wider range of conditions than encountered in the solar system. Many of the dynamical mechanisms controlling the atmospheric circulation of terrestrial exoplanets will bear similarity to those operating on Earth, Mars, Venus, and Titan — but with different details and spanning a much wider continuum of behaviors. Understanding this richness is one of the main benefits to be gained from the study of terrestrial exoplanets.

Here we survey the basic dynamical principles governing the atmospheric circulation of terrestrial exoplanets, discuss the interaction of their circulation with global-scale climate feedbacks, and review the specific circulation models of terrestrial exoplanets that have been published to date. Much of our goal is to provide a resource summarizing basic dynamical processes, as distilled from not only the exoplanet literature but also the solar system and terrestrial literature, that will prove useful for understanding these fascinating atmospheres. Our intended audience includes graduate students and researchers in the fields of astronomy, planetary science, and climate — without backgrounds in atmospheric dynamics — who wish to learn about, and perhaps enter, the field. This review builds on the previous survey of the atmospheric circulation on exoplanets by *Showman et al.* (2010) and complements other review articles in the literature describing the dynamics of solar-system atmospheres (e.g., *Gierasch et al.,* 1997; *Leovy,* 2001; *Read and Lewis,* 2004; *Ingersoll et al.,* 2004; *Vasavada and Showman,* 2005; *Schneider,* 2006; *Flasar et al.,* 2009; *Del Genio et al.,* 2009; *Read,* 2011; as well as the other chapters in this volume).

Section 2 provides an overview of dynamical fundamentals of terrestrial planet atmospheres, including both the rapidly and slowly rotating regimes, with a particular emphasis on aspects relevant to terrestrial exoplanets. Section 3 reviews basic aspects of the hydrological cycle relevant to understanding exoplanets with moist atmospheres. Section 4 addresses the interaction between atmospheric dynamics and global-scale climate, and section 5 summarizes outstanding questions.

2. OVERVIEW OF DYNAMICAL FUNDAMENTALS

Key to understanding atmospheric circulation is understanding the extent to which rotation dominates the dynamics. This can be quantified by the Rossby number, $Ro = U/fL$, defined as the ratio between advection forces and Coriolis forces in the horizontal momentum equation. Here, U is a characteristic horizontal wind speed, $f = 2\Omega \sin \phi$ is the Coriolis parameter, L is a characteristic horizontal scale of the flow ($\sim 10^3$ km or more for the global-scale flows considered here), Ω is the planetary rotation rate (2π over the rotation period), and ϕ is the latitude.

In nearly inviscid atmospheres, the dynamical regime differs greatly depending on whether the Rossby number is small or large. When $Ro \ll 1$, the dynamics are rotationally dominated. Coriolis forces approximately balance pressure-gradient forces in the horizontal momentum equation. This so-called geostrophic balance supports the emergence of large horizontal temperature gradients; as a result, the atmosphere is generally unstable to a type of instability known as baroclinic instability. These instabilities generate eddies that dominate much of the dynamics, controlling the equator-to-pole heat fluxes, temperature contrasts, meridional mixing rates, vertical stratification, and the formation of zonal jets. (The terms zonal and meridional refer to the east-west and north-south directions, respectively; thus, the zonal wind is eastward wind, meridional wind is northward wind, and meridional mixing rates refer to mixing rates in the north-south direction. A zonal average is an average in longitude. Except for regions close to the surface, atmospheres are generally stably stratified, meaning they are stable to dry convection: Air parcels that are displaced upward or downward will return to their original location rather than convecting.)

On the other hand, when $Ro \gtrsim 1$, rotation plays a modest role; the dynamics are inherently ageostrophic, horizontal temperature contrasts tend to be small, and baroclinic instability is less important or negligible. The temperature structure is regulated by a large-scale overturning circulation that transports air latitudinally — i.e., the Hadley circulation — as well as by adjustment of the thermal structure due to atmospheric waves. These two regimes differ sufficiently that they are best treated separately.

It is useful to define terms for these regimes. In Earth's atmosphere, the regime of $Ro \gtrsim 1$ approximately coincides with the tropics, occurring equatorward of ~20°–30° latitude, whereas the regime of $Ro \ll 1$ approximately coincides with the extratropics, occuring poleward of ~30° latitude. Broadening our scope to other planets, we define the "tropics" and "extratropics" as the dynamical regimes —

regardless of temperature — where large-scale circulations exhibit Ro ≳ 1 and Ro ≪ 1, respectively.

Figure 1 illustrates how the extent of the tropics and extratropics depend on rotation rate for a typical terrestrial planet. The sphere on the left depicts an Earth- or Mars-like world where the boundary between the regimes occurs at ~20°–30° latitude. At longer rotation periods, the tropics occupy a greater fraction of the planet, and idealized general circulation model (GCM) experiments show that, for Earth-like planetary radii, gravities, and incident stellar fluxes, planets exhibit Ro ≳ 1 everywhere when the rotation period exceeds ~10 Earth days (e.g., *Del Genio and Suozzo,* 1987) (Fig. 1b). [A GCM solves the global three-dimensional fluid-dynamics equations relevant to a rotating atmosphere, coupled to calculations of the atmospheric radiative-transfer everywhere over the full three-dimensional grid (necessary for determining the radiative heating/cooling rate, which affects the dynamics), and parameterizations of various physical processes including frictional drag against the surface, sub-grid-scale turbulence, and (if relevant) clouds. "Idealized" GCMs refer to GCMs where these components are simplified, e.g., adopting a gray radiative-transfer scheme rather than solving the full nongray radiative transfer.]

Dynamically, such slowly-rotating planets are essentially "all tropics" worlds [this term was first coined by *Mitchell et al.* (2006) in reference to Titan]. Venus and Titan are examples in our own solar system, exhibiting near-global Hadley cells, minimal equator-pole temperature differences, little role for baroclinic instabilities, and a zonal jet structure that differs significantly from those on Earth and Mars. Terrestrial exoplanets characterizable by transit techniques will preferentially be close to their stars and tidally locked, implying slow rotation rates; many of these exoplanets should likewise be "all tropics" worlds.

Figure 2 previews several of the key dynamical processes occurring at large scales on a generic terrestrial exoplanet, which we survey in more detail in the subsections that follow. In the extratropics, the baroclinic eddies that dominate the meridional heat transport (section 2.1.2) generate meridionally propagating Rossby waves (section 2.1.3), which leads to a convergence of momentum into the instability latitude, generating an eddy-driven jet stream (section 2.1.4). Multiple zones of baroclinic instability, and multiple eddy-driven jets, can emerge in each hemisphere if the planet is sufficiently large or the planetary rotation is sufficiently fast. In the tropics, the Hadley circulation (section 2.2.1) dominates the meridional heat transport; in idealized form, it transports air upward near the equator and poleward in the upper troposphere, with a return flow to the equator along the surface. Due to the relative weakness of rotational effects in the tropics, atmospheric waves can propagate unimpeded in longitude, and adjustment of the thermal structure by these waves tends to keep horizontal temperature gradients weak in the tropics (section 2.2.2). Many exoplanets will rotate synchronously and therefore exhibit permanent day- and nightsides; the resulting, spatially locked day-night heating patterns will generate large-scale, standing equatorial Rossby and Kelvin waves, which in many cases will lead to equatorial superrotation, i.e., an eastward flowing jet at the equator (section 2.2.3). Significant communication between the tropics and extratropics can occur, among other mechanisms, via meridionally propagating Rossby waves that propagate from one region to the other.

We review the extratropical and tropical regimes, along with the key processes shown in Fig. 2, in this section.

2.1. Extratropical Regime

2.1.1. Force balances and geostrophy. The extratropical regime corresponds to Ro ≪ 1. For typical terrestrial-planet wind speeds of ~10 m s^{-1} and Earth-like planet sizes, planets will have extratropical zones for rotation periods of a few (Earth) days or shorter. When Ro ≪ 1 and friction is weak, the Coriolis force and pressure-gradient force will

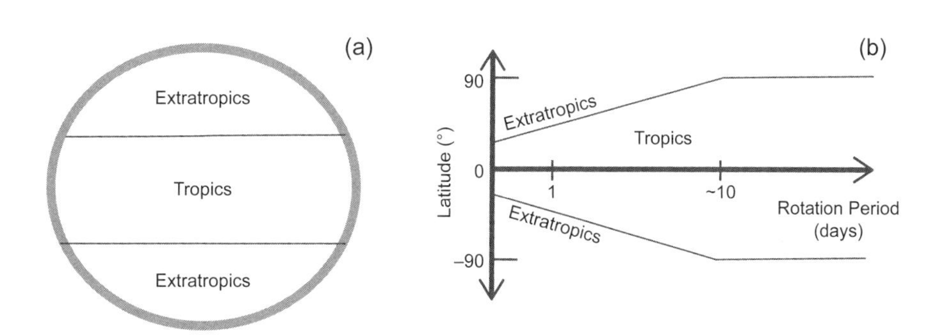

Fig. 1. Schematic illustration of the regimes of extratropics (defined as Ro ≪ 1) and tropics (defined as Ro ≳ 1). For an Earth- or Mars-like planet, the boundary between the regimes occurs at ~20°–30° latitude [**(a)**]; however, the transition occurs at higher latitudes when the rotation period is longer, and terrestrial planets with rotation periods longer than ~10 Earth days may represent "all tropics" worlds [**(b)**].

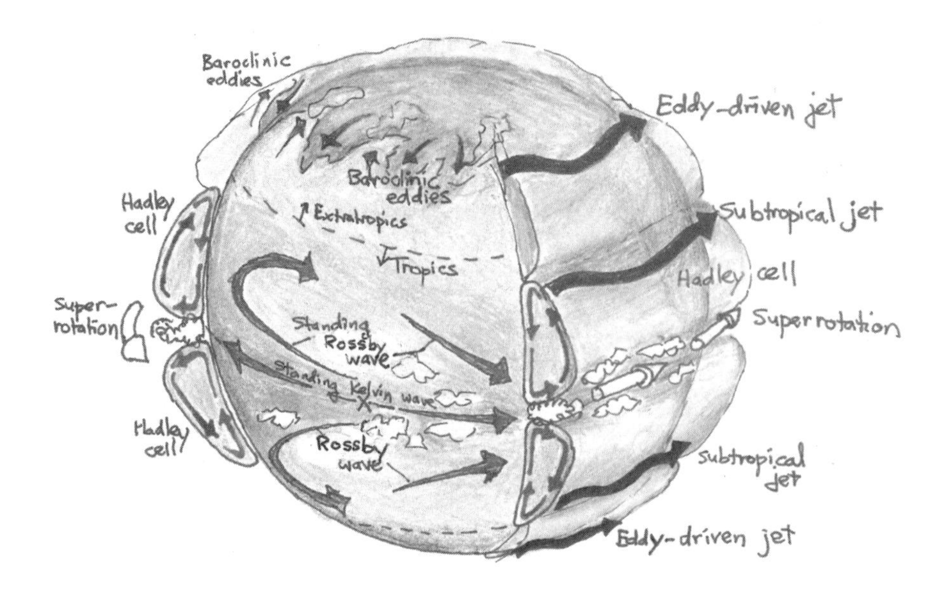

Fig. 2. See Plate 44 for color version. Schematic illustration of dynamical processes occurring on a generic terrestrial exoplanet. These include baroclinic eddies, Rossby waves, and eddy-driven jet streams in the extratropics, and Hadley circulations, large-scale Kelvin and Rossby waves, and (in some cases) equatorial superrotation in the tropics. The "X" at the equator marks the substellar point, which will be fixed in longitude on synchronously rotating planets. Cloud formation, while complex, will likely be preferred in regions of mean ascent, including the rising branch of the Hadley circulation, within baroclinic eddies, and — on synchronously rotating planets — in regions of ascent on the dayside.

approximately balance in the horizontal momentum equation; the resulting balance, called geostrophic balance, is given by (e.g., *Vallis,* 2006, pp. 85–88)

$$fu = -\left(\frac{\partial \Phi}{\partial y}\right)_p \qquad fv = \left(\frac{\partial \Phi}{\partial x}\right)_p \tag{1}$$

where Φ is the gravitational potential, x and y are eastward and northward distance, u and v are zonal and meridional wind speed, f is the Coriolis parameter, and the derivatives are evaluated at constant pressure. The implication is that winds tend to flow along, rather than across, isobars. In our solar system, the atmospheres of Earth, Mars, and all four giant planets exhibit geostrophic balance away from the equator, and the same will be true for a wide range of terrestrial exoplanets [see *Pedlosky* (1987), *Holton* (2004), or *Vallis* (2006) for introductions to the dynamics of rapidly rotating atmospheres in the geostrophic regime]. In a geostrophic flow, there exists a tight link between winds and temperatures. Differentiating the geostrophic equations in pressure (which here acts as a vertical coordinate) and invoking hydrostatic balance and the ideal-gas law, we obtain the thermal-wind equations (*Vallis,* 2006, pp. 89–90)

$$f\frac{\partial u}{\partial \ln p} = \frac{\partial (RT)}{\partial y} \qquad f\frac{\partial v}{\partial \ln p} = -\frac{\partial (RT)}{\partial x} \tag{2}$$

where T is temperature and R is the specific gas constant. The equation implies that, for the geostrophic component of the flow, meridional temperature gradients accompany vertical shear (i.e., vertical variation) of the zonal wind, and zonal temperature gradients accompany vertical shear (i.e., vertical variation) of the meridional wind. Because the surface wind is generally weak compared to that at the tropopause, one can thus obtain an estimate of the wind speed in the upper troposphere — given the equator-pole temperature gradient — by integrating equation (2) vertically.

To order of magnitude, for example, equation (2) implies a zonal wind $\Delta u \sim R\Delta T_{eq-pole}\Delta \ln p/(fa)$, where $\Delta T_{eq-pole}$ is the temperature difference between the equator and pole, $\Delta \ln p$ is the number of scale heights over which this temperature difference extends, and a is the planetary radius. Inserting Earth parameters ($\Delta T_{eq-pole} \approx 20$ K, a $\approx$ 6000 km, R = 287 J kg^{-1} K^{-1}, f $\approx 10^{-4}$ s^{-1}, and $\Delta \ln p = 1$), we obtain $\Delta u \approx 10$ m s^{-1}, which is indeed a characteristic value of the zonal wind in Earth's upper troposphere.

What are the implications of thermal-wind balance for a planet's global circulation pattern? Given the thermal structure expected on typical, low-obliquity, rapidly rotating exoplanets, with a warm equator and cool poles, equation (2) makes several useful statements:

1. It implies that the zonal wind increases (i.e., becomes more eastward) with altitude. Assuming the surface winds are weak, this explains the predominantly eastward nature of the tropospheric winds on Earth and Mars, especially in mid-latitudes, and suggests an analogous pattern on rapidly rotating exoplanets.

2. Because temperature gradients — at least on Earth and Mars — peak in mid-latitudes, thermal-wind balance

helps explain why the zonal wind shear — hence the upper-tropospheric zonal winds themselves — are greater at mid-latitudes than at the equator or poles. This is the latitude of the jet streams, and thermal-wind balance therefore describes how the winds increase with height in the jet streams. Because the tropospheric meridional temperature gradient is greater in the winter hemisphere than the summer hemisphere, thermal-wind balance also implies that the upper-tropospheric, mid-latitude winds should be faster in the winter hemisphere, as in indeed occurs on Earth (*Peixoto and Oort,* 1992) and Mars (*Smith,* 2008).

3. If the mean zonal temperature gradients are small compared to mean meridional temperature gradients, which tends to be true when the rotation is fast (and in particular when the solar day is shorter than the atmospheric radiative time constant), thermal wind implies that the time-mean zonal winds are stronger than the time-mean meridional winds. This provides a partial explanation for the zonal (east-west) banding of the wind structure on Earth and Mars and suggests that a similarly banded wind pattern will occur on rapidly rotating exoplanets. (In the presence of topography or land-ocean contrasts, some local regions may exhibit meridional winds that, even in a time average, are not small compared to zonal winds. In such a case, $u \gg v$ only in the zonal mean.)

The fact that Coriolis forces can balance pressure gradients in a geostrophic flow implies that such flows can sustain larger horizontal pressure and temperature contrasts than might otherwise exist. To an order of magnitude, hydrostatic balance of the dynamical pressure and density fluctuations δp and $\delta\rho$ (see *Holton,* 2004, pp. 41–42) implies that these fluctuations satisfy

$$\delta p \approx \delta\rho g D \tag{3}$$

where D is the vertical scale of the circulation. In the horizontal momentum equation, the pressure-gradient force is approximately $\delta\rho gD/(\rho L)$, where L is a characteristic horizontal length scale of the flow. Noting that $\delta\rho/\rho \sim \delta\theta_h/\theta$, where $\delta\theta_h$ is the characteristic horizontal potential temperature difference and θ is the characteristic potential temperature, geostrophy implies (cf. *Charney,* 1963)

$$\frac{\delta\theta_h}{\theta} \sim \frac{fUL}{gD} \sim \frac{Fr}{Ro}, \qquad Ro \lesssim 1 \tag{4}$$

where $Fr \equiv U^2/gD$ is a dimensionless quantity known as a Froude number. [Potential temperature is defined as the temperature an air parcel would have if transported adiabatically to a reference pressure p_0 (often taken to be 1 bar). When the ratio of gas constant to specific heat R/c_p is constant, as is often approximately true in atmospheres, then it is defined by $\theta = T(p_0/p)^{R/c_p}$. It is conserved following adiabatic reversible processes and is thus a measure of atmospheric entropy.] For a typical, rapidly rotating terrestrial planet where $U \approx 10$ m s^{-1}, $D \approx 10$ km, $g \approx 10$ m s^{-2}, $f \approx 10^{-4}$ s^{-1} (implying rotation periods of an Earth day), the fractional temperature contrasts over distances comparable to the planetary radius approach ~0.1, implying temperature contrasts of ~20K. This is similar to the actual equator-to-pole temperature contrast in Earth's troposphere. In contrast, these values greatly exceed typical horizontal temperature contrasts in the tropical regime of Ro ~ 1 (see equation (16)).

An important horizontal length scale in atmospheric dynamics is the Rossby deformation radius, defined in the extratropics as

$$L_D = \frac{ND}{f} \tag{5}$$

where N is the Brunt-Vaisala frequency, which is a measure of vertical stratification. Because the deformation radius is a natural length scale that emerges from the interaction of gravity (buoyancy) and rotation in stably stratified atmospheres, many phenomena, including geostrophic adjustment, baroclinic instabilities, and the interaction of convection with the environment, produce atmospheric structures with horizontal sizes comparable to the deformation radius. As a result, the circulations in atmospheres (and oceans) often exhibit predominant length scales not too different from the deformation radius.

2.1.2. Baroclinic instabilities and their effect on thermal structure. Planets experience meridional gradients in the net radiative heating rate. At low obliquities, this gradient corresponds to net heating at the equator and net cooling at the poles, leading to meridional temperature contrasts with, generally, a hot equator and cold poles. Even if individual air columns are convectively stable (i.e., if the potential temperature increases with height in the troposphere), potential energy can be extracted from the system if the cold polar air moves downward and equatorward and if the warm equatorial air moves upward and poleward. The question is then whether dynamical mechanisms actually exist to extract this energy and thereby transport heat from the equator to the poles. In the tropics, the dominant meridional heat-transport mechanism is a thermally direct Hadley circulation (see Fig. 2 and section 2.2.1), but such circulations tend to be suppressed by planetary rotation in the extratropics. At $Ro \ll 1$, the meridional temperature gradients are associated with an upward-increasing zonal wind in thermal-wind balance (equation (2)). It is useful to think about the limit where longitudinal gradients of heating and temperature are negligible — in which case the (geostrophic) wind is purely zonal — and where the meridional wind is zero, as occurs when these zonal winds are perfectly balanced. In such a hypothetical solution, the temperature at each latitude would be in local radiative equilibrium, and no meridional heat transport would occur.

It turns out that this hypothetical steady state is not dynamically stable: Small perturbations on this steady solution grow over time, producing eddies that extract potential energy from the horizontal temperature contrast (transporting warm low-latitude air upward and poleward, transporting cool high-latitude air downward and equatorward, thereby lowering the center of mass, flattening isentropes, and reducing the meridional temperature gradient). [Isentropes are surfaces of constant entropy, which are equivalent to surfaces of constant potential temperature. In a stably stratified atmosphere where entropy increases with altitude, isentropes will bow downward (upward) in regions that, as measured on an isobar, are hot (cold).] This is baroclinic instability, so named because it depends on the baroclinicity of the flow — i.e., on the fact that surfaces of constant density incline significantly with respect to surfaces of constant pressure. The instabilities are inherently three-dimensional and manifest locally as tongues of cold and warm air penetrating equatorward and poleward in the extratropics. Figure 2 illustrates such eddies schematically and Fig. 3 provides examples from GCM experiments under Earth-like conditions. The fastest-growing modes have zonal wavelengths comparable to the deformation radius and growth rates proportional to $(f/N)\partial u/\partial z$, where $\partial u/\partial z$ is the vertical shear of the zonal wind that exists in thermal wind balance with the meridional temperature gradient. For Earth-like conditions, these imply length scales of ~4000 km and growth timescales of ~3–5 days for the dominant modes (for reviews, see, e.g., *Pierrehumbert and Swanson,* 1995; *Vallis,* 2006, Chapter 6).

As expected from baroclinic instability theory, the dominant length scale of the baroclinic, heat-transporting eddies in the extratropics scales inversely with the planetary rotation rate. This is illustrated in Fig. 3, which shows instantaneous snapshots of the temperature in idealized Earth-like GCM experiments where the rotation rate is varied between half and four times that of Earth (*Kaspi and Showman,* 2012). The smaller eddies in the rapidly rotating models are less efficient at transporting energy meridionally, leading to a greater equator-to-pole temperature difference in those cases. A similar dependence has been found by other authors as well (e.g., *Schneider and Walker,* 2006; *Kaspi and Schneider,* 2011).

In the extratropics of Earth and Mars, baroclinic instabilities play a key role in controlling the thermal structure, including the equator-to-pole temperature gradient and the vertical stratification (i.e., the Brunt-Vaisala frequency); this is also likely to be true in the extratropics of terrestrial exoplanets. In particular, GCM experiments suggest that the extratropics — when dominated by baroclinic instability — adjust to a dynamic equilibrium in which meridional temperature gradients and tropospheric stratifications scale together (atmospheres with larger tropospheric stratification exhibit larger meridional temperature gradients and vice versa). Figure 4 illustrates this phenomenon from a sequence of idealized GCM experiments from *Schneider and Walker* (2006) for terrestrial planets forced by equator-to-pole heating gradients. The abscissa shows the meridional temperature difference across the baroclinic zone and the ordinate shows the vertical stratification through the troposphere; each symbol represents the equilibrated state of a particular model including the effects of dynamics. Although the radiative-equilibrium thermal structures fill a significant fraction of the parameter space, the baroclinic eddy entropy fluxes adjust the thermal structure to a state where the vertical stratification and meridional temperature difference are comparable, corresponding to a line with a slope of one in Fig. 4. The extent to which this relationship holds in general and parameter regimes where it may break down are under investigation (*Zurita-Gotor and Vallis,* 2009; *Jansen and Ferrari,* 2012).

Most work on this problem has so far investigated planets forced only by equator-pole heating gradients (e.g., *Schneider and Walker,* 2006; *Kaspi and Showman,* 2012), but the zonal (day-night) heating gradients could significantly influence the way that baroclinic eddies regulate the thermal structure, particularly on synchronously rotating planets. *Edson et al.* (2011) compared GCM simulations for a control — Earth-like simulation forced only by an equator-pole heating gradient and an otherwise identical synchronously rotating planet with day-night thermal forcing. They found that the (zonal-mean) meridional isentrope slopes were gentler in the model with day-night forcing than in the control model. This differing behavior presumably results from the fact that the synchronously rotating model exhibits meridional heat fluxes that are primarily confined

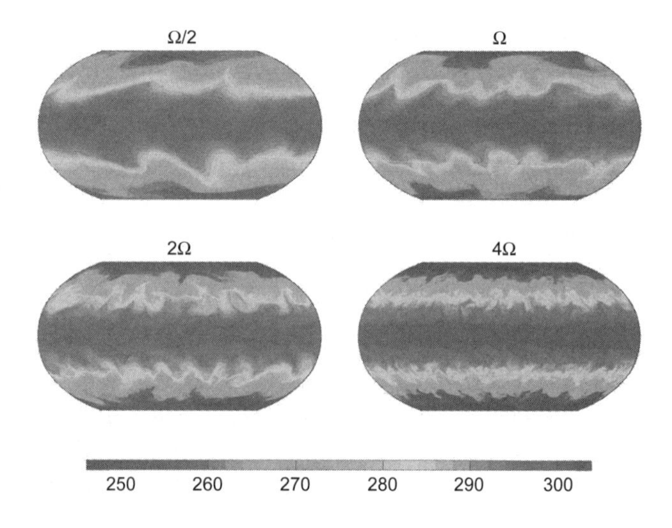

Fig. 3. See Plate 45 for color version. Surface temperature (colorscale, in K) from GCM experiments in *Kaspi and Showman* (2012), illustrating the dependence of temperature and jet structure on rotation rate. Experiments are performed using the Flexible Modeling System (FMS) model analogous to those in *Frierson et al.* (2006) and *Kaspi and Schneider* (2011); radiative transfer is represented by a two-stream, gray scheme with no diurnal cycle (i.e., the incident stellar flux depends on latitude but not longitude). A hydrological cycle is included with a slab ocean. Planetary radius, gravity, atmospheric mass, incident stellar flux, and atmospheric thermodynamic properties are the same as on Earth; models are performed with rotation rates from half (upper left) to four times that of Earth (lower right). Baroclinic instabilities dominate the dynamics in mid- and high-latitudes, leading to baroclinic eddies whose length scales decrease with increasing planetary rotation rate.

to the dayside (rather than occurring at all longitudes as in the control model) and from the fact that the day-night forcing generates planetary-scale standing waves (see section 2.2.3) in the synchronous model that are absent in the control model. Both traits can influence the (zonal-mean) meridional heat transport and therefore the isentrope slopes and zonal-mean temperature differences between the equator and poles. Additional work quantifying the behavior for tidally locked planets over a wider parameter regime would be highly beneficial.

Over the past several decades, many authors have attempted to elucidate theoretically how the meridional heat fluxes due to baroclinic eddies depend on the background meridional temperature gradient, tropospheric stratification, planetary rotation rate, and other parameters. Almost all this work has emphasized planets in an Earth-like regime, forced by equator-pole heating gradients (e.g., *Green,* 1970; *Stone,* 1972; *Larichev and Held,* 1995; *Held and Larichev,* 1996; *Pavan and Held,* 1996; *Haine and Marshall,* 1998; *Barry et al.,* 2002; *Thompson and Young,* 2006; *Thompson and Young,* 2007; *Schneider and Walker,* 2008; *Zurita-Gotor and Vallis,* 2009) (for reviews, see *Held,* 1999a; *Showman et al.,* 2010). If a theory for this dependence could be developed, it would constitute a major step toward a predictive theory for the dependence of the equator-to-pole temperature difference on planetary parameters. However, this is a challenging problem, and no broad consensus has yet emerged.

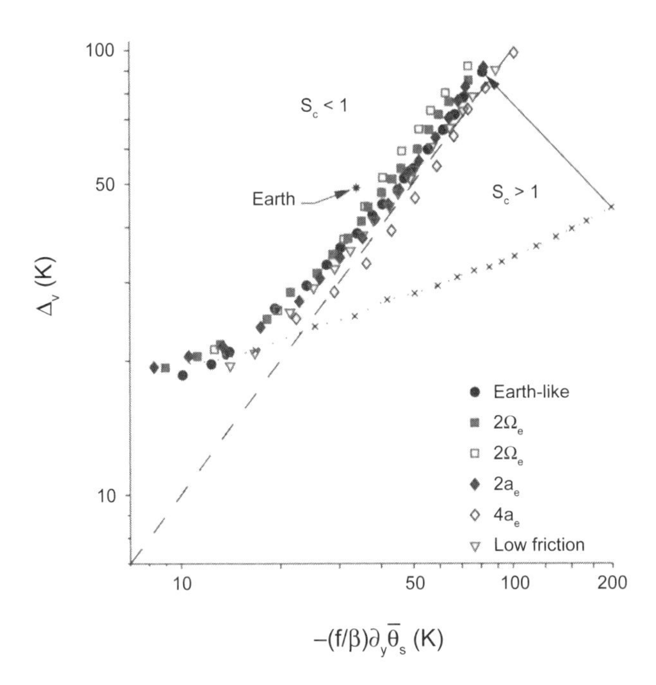

Fig. 4. A measure of the near-surface meridional potential temperature difference across the extratropics (abscissa) and potential temperature difference in the extratropics taken vertically across the troposphere (ordinate) in models of planets forced by equator-to-pole heating gradients from *Schneider and Walker* (2006). When the equator-to-pole forcing is sufficiently great, the extratropical dynamics are dominated by baroclinic eddies, which adjust the thermal structure to a state where horizontal and vertical potential temperature differences in the extratropics are comparable. Different symbols denote models with differing rotation rates, planetary radii, and/or boundary-layer friction, coded in the legend. Crosses depict the radiative-equilibrium states for the Earth-like models. The dashed line denotes a slope of 1.

Nevertheless, recent GCM studies demonstrate how the equator-to-pole temperature differences and other aspects of the dynamics depend on planetary radius, gravity, rotation rate, atmospheric mass, and incident stellar flux (*Kaspi and Showman,* 2012). These authors performed idealized GCM experiments of planets forced by equator-to-pole heating gradients, including a hydrological cycle and representing the radiative transfer using a two-stream, gray approach. *Kaspi and Showman* (2012) found that the equator-to-pole temperature difference decreases with increasing atmospheric mass and increases with increasing rotation rate, planetary radius (at constant interior density), or planetary density (at constant interior mass). Figure 5 shows the zonal-mean surface temperature and meridional eddy energy fluxes vs. latitude for the cases with differing rotation rates (Figs. 5a,c) and atmospheric masses (Figs. 5b,d). Meridional eddy energy fluxes are weaker at faster rotation rates (Fig. 5a), presumably because the smaller eddy length scales (Fig. 3) lead to less efficient energy transport. This helps to explain the fact that larger equator-to-pole temperature differences occur in the faster rotating simulations, as is evident in Fig. 5c and Fig. 3. Likewise, massive atmospheres exhibit a greater thermal storage capacity than less-massive atmospheres, allowing a greater meridional energy flux at a given baroclinic eddy amplitude. Everything else being equal, planets with greater atmospheric mass therefore exhibit smaller equator-to-pole temperature differences than planets with lesser atmospheric mass (Fig. 5d). Additional detailed work is warranted to clarify the physical processes controlling these trends and to seek a predictive theory for the dependences.

2.1.3. Rossby waves. Much of the structure of the large-scale atmospheric circulation can be understood in terms of the interaction of Rossby waves with the background flow; they are the most important wave type for the large-scale circulation. In this section we survey their linear dynamics, and follow in subsequent sections with discussions of how they help to shape the structure of the extratropical circulation via nonlinear interactions.

Rossby waves are best understood through conservation of potential vorticity (PV), which for a shallow fluid layer of thickness h can be written as

$$q = \frac{\zeta + f}{h} \tag{6}$$

where $\zeta = \mathbf{k} \cdot \nabla \times \mathbf{v}$ is the relative vorticity, **k** is the vertical (upward) unit vector, and **v** is the horizontal velocity vector. As written, this is the conserved form of PV for the shallow-water equations (e.g., *Pedlosky,* 1987; *Vallis,* 2006; *Showman et al.,* 2010), but the same form holds in

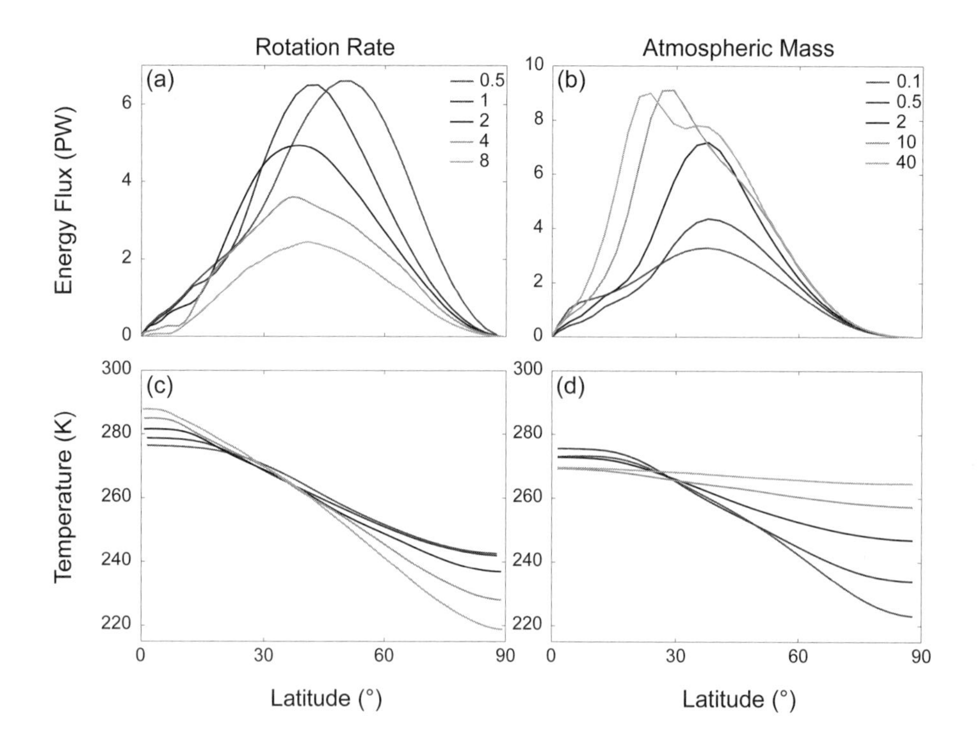

Fig. 5. See Plate 46 for color version. Latitude dependence of the **(a),(b)** vertically and zonally integrated meridional energy flux and **(c),(d)** vertically and zonally averaged temperature in GCM experiments from *Kaspi and Showman* (2012) for models varying rotation rate and atmospheric mass. The energy flux shown in **(a)** and **(b)** is the flux of moist static energy, defined as cpT + gz + Lq, where cp is specific heat at constant pressure, g is gravity, z is height, L is latent heat of vaporization of water, and q is the water vapor abundance. The left column [**(a)** and **(c)**] explores sensitivity to rotation rate; models are shown with rotation rates ranging from half to eight times that of Earth. In these experiments, the atmospheric mass is held constant at the mass of Earth's atmosphere. The right column [**(b)** and **(d)**] explores the sensitivity to atmospheric mass; models are shown with atmospheric masses from 0.1 to 40 times the mass of Earth's atmosphere. In these experiments, the rotation rate is set to that of Earth. The equator-to-pole temperature difference is smaller, and the meridional energy flux is larger, when the planetary rotation rate is slower, and/or when the atmospheric mass is larger. Other model parameters, including incident stellar flux, optical depth of the atmosphere in the visible and infrared, planetary radius, and gravity, are Earth-like and are held fixed in all models.

the three-dimensional primitive equations if the relative vorticity is evaluated at constant potential temperature and h is appropriately defined in terms of the gradient of pressure with respect to potential temperature (*Vallis,* 2006, pp. 187–188). To illuminate the dynamics, we consider the simplest system that supports Rossby waves, namely a one-layer barotropic model governing two-dimensional, nondivergent flow. For this case, h can be considered constant, leading to conservation of absolute vorticity, $\zeta + f$, following the flow. This can be written, adopting Cartesian geometry, as

$$\frac{\partial \zeta}{\partial t} + \mathbf{v} \bullet \nabla \zeta + v\beta = F \tag{7}$$

where v is the meridional velocity, $\beta \equiv df/dy$ is the gradient of the Coriolis parameter f with northward distance y, and F represents any sources or sinks of potential vorticity. For a brief review of equation sets used in atmospheric dynamics, see *Showman et al.* (2010); more detailed treatments can be found in *Vallis* (2006), *McWilliams* (2006), or *Holton* (2004).

To investigate how the β effect allows wave propagation, we consider the version of equation (7) linearized about a state of no zonal-mean flow, which amounts to dropping the term involving advection of ζ. Given the assumption that the flow is horizontally nondivergent, we can define a streamfunction, ψ, such that $u = \frac{-\partial\psi}{\partial y}$ and $v = \frac{\partial\psi}{\partial x}$. This definition implies that $\zeta = \nabla^2\psi$, allowing the linearized equation to be written

$$\frac{\partial \nabla^2 \psi}{\partial t} + \beta \frac{\partial \psi}{\partial x} = 0 \tag{8}$$

This is useful because it is now an equation in only one variable, ψ, which can easily be solved to determine wave behavior. Assuming that β is constant (an approximation known as the "beta plane"), and adopting plane-wave solutions $\psi = \psi_0 \exp[i(kx + ly - \omega t)]$, we obtain a dispersion relation

$$\omega = -\frac{\beta k}{k^2 + l^2} \tag{9}$$

where k and l are the zonal and meridional wavenumbers (just 2π over the longitudinal and latitudinal wavelengths, respectively) and ω is the oscillation frequency. [For introductions to wave equations and dispersion relations, see, e.g., *Vallis* (2006), *Holton* (2004), *Pedlosky* (1987), or *Pedlosky* (2003).] The resulting waves are called Rossby waves. They are large-scale (wavelengths commonly $\sim 10^3$ km or more), low-frequency (periods of order a day or longer), and — away from the equator on a rapidly rotating planet — are in approximate geostrophic balance. As demonstrated by equation (9), they exhibit westward phase speeds.

The wave-induced zonal and meridional velocities are defined, respectively, as $u' = u_0' \exp[i(kx + ly - \omega t)]$ and $v' = v_0' \exp[i(kx + ly - \omega t)]$, where u_0' and v_0' are complex amplitudes. The above solution implies that

$$u_0' = -il\psi_0 \qquad v_0' = -ik\psi_0 \tag{10}$$

The velocities represented by these relations are parallel to lines of constant phase and therefore perpendicular to the direction of phase propagation. Equation (10) implies that the zonal and meridional wave velocities are correlated; this provides a mechanism by which Rossby wave generation, propagation, and dissipation can transport momentum and thereby modify the mean flow, a point we return to in section 2.1.4.

Physically, the restoring force for Rossby waves is the β effect, namely, the variation with latitude of the planetary vorticity. Because the Coriolis parameter varies with latitude, PV conservation requires that any change in latitude of fluid parcels will cause a change in the relative vorticity. The flow associated with these relative vorticity perturbations leads to oscillations in the position of fluid parcels and thereby allows wave propagation. The meridional velocities deform the phase surfaces in a manner that leads to the westward phase velocities captured in equation (9) (see discussion in *Vallis,* 2006; *Holton,* 2004).

2.1.4. Jet formation I: Basic mechanisms. A wide range of numerical experiments and observations show that zonal jets tend to emerge spontaneously on rapidly rotating planets. On many planets, such zonal jets dominate the circulation, and thus understanding them is crucial to understanding the circulation as a whole. The dynamics controlling jets is intimately connected to the dynamics controlling the meridional temperature gradients, vertical stratification, and other aspects of the dynamics. Sufficiently strong jets exhibit sharp gradients of the meridional PV gradient, and as such they can act as barriers to the meridional mixing of heat, moisture, and chemical tracers (e.g., *Beron-Vera et al.,* 2008), significantly affecting the meridional structure of the atmosphere.

While there exist many mechanisms of jet formation, among the most important is the interaction of Rossby waves with the mean flow. This mechanism plays a key role in causing the extratropical jets on Earth, Mars, and perhaps Jupiter and Saturn; the mechanism is similarly expected to play an important role in the atmospheres of terrestrial exoplanets.

As *Thompson* (1971), *Held* (1975), and many subsequent authors have emphasized, a key property of meridionally propagating Rossby waves is that they induce a meridional flux of prograde (eastward) eddy angular momentum into their generation latitude. To illustrate, we again consider the solutions to equation (8). The latitudinal transport of (relative) eastward eddy momentum per mass is $\overline{u'v'}$, where u' and v' are the deviation of the zonal and meridional winds from their zonal mean, and the overbar denotes a zonal average. Using equation (10) shows that

$$\overline{u'v'} = -\frac{1}{2}\hat{\psi}^2 kl \tag{11}$$

Now, the dispersion relation (equation (9)) implies that the meridional group velocity is $\partial\omega/\partial l = 2\beta kl/(k^2 + l^2)^2$. Since the group velocity must point away from the region where the Rossby waves are generated, we must have $kl > 0$ north of this wave source and $kl < 0$ south of the wave source. Combining this information with equation (11) shows that $\overline{u'v'} < 0$ north of the source and $\overline{u'v'} > 0$ south of the source. Thus, the Rossby waves flux eastward momentum into the latitude of the wave source.

This process is illustrated in Fig. 6. Suppose some process generates Rossby wave packets at a specific latitude, which propagate north and south from the latitude of generation. The northward-propagating wave packet exhibits eddy velocities tilting northwest-southeast, while the southward propagating packet exhibits eddy velocities tilting southwest-northeast. The resulting eddy velocities visually resemble an eastward-pointing chevron pattern centered at the latitude of wave generation. The correlation between these velocities leads to nonzero Reynolds stresses (that is, a nonzero $\overline{u'v'}$) and a flux of angular momentum into the wave generation latitude. An example of this schematic chevron pattern in an actual GCM experiment is shown in Fig. 7.

The above reasoning is for free (unforced) waves but can be extended to an atmosphere forced by vorticity sources/sinks and damped by frictional drag (see reviews in *Held,* 2000; *Vallis,* 2006; *Showman and Polvani,* 2011). In the nondivergent, barotropic system, the zonal-mean zonal momentum equation, adopting a Cartesian coordinate system for simplicity, is given by

$$\frac{\partial \bar{u}}{\partial t} = -\frac{\partial\left(\overline{u'v'}\right)}{\partial y} - \frac{\bar{u}}{\tau_{drag}} \tag{12}$$

where we have decomposed the winds into their zonal means (given by overbars) and the deviations therefrom (given by primes), such that $u = \bar{u} + u'$, $v = \bar{v} + v'$, $\zeta = \bar{\zeta} + \zeta'$, etc. Here, y is northward distance and we have parameterized drag by a linear (Rayleigh) friction that relaxes the winds toward zero over a specified drag time constant τ_{drag}. To link the momentum budget to the vorticity sources and sinks, we first note that, for the horizontally nondivergent system, the definition of vorticity implies that $\overline{v'\zeta'} = -\partial(\overline{u'v'})/\partial y$. Second, we multiply the linearized version of equation (7) by ζ' and zonally average, yielding an equation for the so-called pseudomomentum

$$\frac{\partial \mathcal{A}}{\partial t} + \overline{v'\zeta'} = \frac{\overline{\zeta' F'}}{2\left(\beta - \frac{\partial^2 \overline{u}}{\partial y^2}\right)} \tag{13}$$

where $\mathcal{A} = (\beta - \partial^2\overline{u}/\partial y^2)^{-1}\overline{\zeta'^2}/2$ is the pseudomomentum, which characterizes the amplitude of the eddies in a statistical (zonal-mean) sense. Here, F′ is the eddy component of the vorticity source/sink defined in equation (7). We then combine equations (12) and (13). If the zonal-mean flow equilibrates to a statistical steady state (i.e., if $\partial\overline{u}/\partial t$ and $\partial\mathcal{A}/\partial t$ are zero), this yields a relationship between the zonal-mean zonal wind, $\overline{u}$, and the eddy generation of vorticity

$$\frac{\overline{u}}{\tau_{drag}} = \frac{\overline{\zeta' F'}}{2\left(\beta - \frac{\partial^2 \overline{u}}{\partial y^2}\right)} \tag{14}$$

(Note that this equilibration does not require the *eddies themselves* to be steady, but rather simply that their zonally averaged mean amplitude, characterized by $\mathcal{A}$, is steady.)

What are the implications of this equation? Consider a region away from wave sources where the eddies are dissipated. Dissipation acts to reduce the eddy amplitudes, implying that F′ and ζ' exhibit opposite signs (see equation (7). (For example, for the specific case of linear drag, represented in the zonal and meridional momentum equations as $-u/\tau_{drag}$ and $-v/\tau_{drag}$, respectively, then in the absence of forcing $F = -\zeta/\tau_{drag}$.) Therefore, regions of net wave dissipation will exhibit $\overline{\zeta' F'} < 0$ and the zonal-mean zonal wind will be westward. On the other hand, in regions where eddies are generated, the wave sources act to increase the wave amplitudes, implying that F′ and ζ' exhibit the same signs. In such a region, $\overline{\zeta' F'} > 0$ and the zonal-mean zonal wind will be eastward. The two regions are linked because waves propagate (in the sense of their group velocity) from their latitude of generation to their latitude of dissipation; this allows a statistical steady state in both the zonal-mean eddy amplitudes and the zonal-mean zonal wind to be achieved despite the tendency of damping (forcing) to locally decrease (increase) the eddy amplitudes. In summary, we thus recover the result that eastward net flow occurs in regions of net wave generation while westward net flow occurs in regions of net wave damping.

In the extratropics, baroclinic instabilities constitute a primary mechanism for generating Rossby waves in the free troposphere, which propagate from the latitude of instability to surrounding latitudes (see Fig. 7). To the extent that the extratropics of exoplanets are baroclinically unstable, this mechanism thus implies the emergence of the extratropical eastward jets; indeed, the mid-latitude jets in the tropospheres of Earth and Mars result from this process. Such "eddy-driven" jets are illustrated schematically in Fig. 2. [In the terrestrial literature, these are referred to as "eddy-driven" jets to distinguish them from the subtropical jets that occur at the poleward edge of the Hadley cell (see section 2.2).] To be more precise, the eddy-momentum flux convegence should be thought of as driving the surface wind (e.g., *Held,* 2000). The only torque that can balance the vertically integrated eddy-momentum flux convergence — and the associated eastward acceleration — is frictional drag at the surface. Thus, in latitudes where baroclinic instability or other processes leads to strong Rossby wave generation, and the radiation of those waves to other latitudes, eastward zonal-mean winds emerge at the surface. The zonal wind in the upper troposphere is then set by the sum of the surface wind and the vertically integrated wind shear, which is in thermal-wind balance with the meridional temperature gradients.

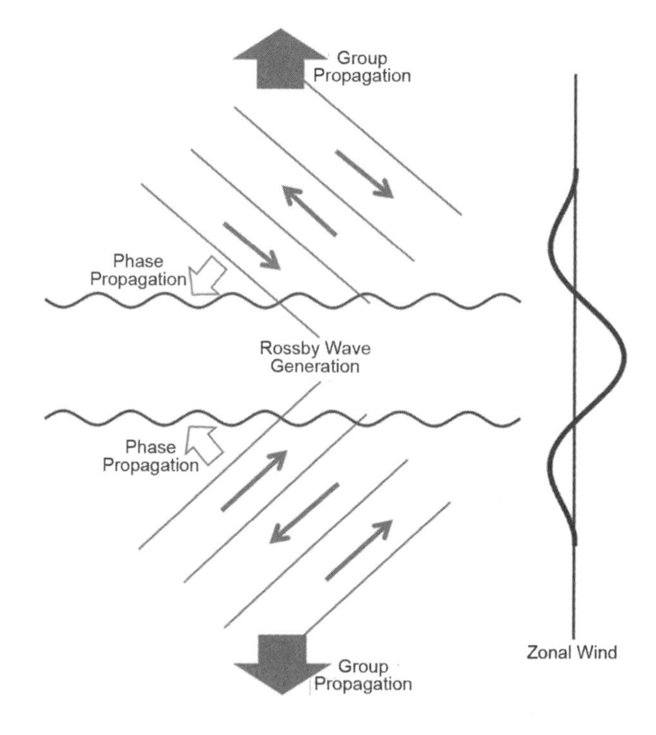

Fig. 6. Schematic illustration of how Rossby wave generation can lead to the formation of zonal jets. Imagine that some process — such as baroclinic instability — generates Rossby waves at a specific latitude, and that the Rossby waves then propagate north and south from their latitude of generation. Rossby waves with northward group propagation exhibit eddy velocities tilting northwest-to-southeast, whereas those with southward group propagation exhibit eddy velocities tilting southwest-to-northeast. These correlations imply that the waves transport eastward eddy momentum into the latitude region where Rossby waves are generated, resulting in formation of an eddy-driven jet.

In the above theory, the only source of explicit damping was a linear drag, but in the real extratropical atmosphere, wave breaking will play a key role and can help explain many aspects of extratropical jets. Indeed, much of jet dynamics can be understood in terms of spatially inhomogeneous mixing of potential vorticity (PV, see equation (6)) caused by this wave breaking (*Dritschel and McIntyre,* 2008). Rossby waves manifest as meridional undulations in PV contours, and Rossby wave breaking implies that the PV contours become so deformed that they curl up and overturn in the longitude-latitude plane (see Fig. 8 for an example). Such overturning generally requires large wave amplitudes; in particular, the local, wave-induced perturbation to the meridional PV gradient must become comparable to the background (zonal-mean) gradient, so that, locally, the meridional PV gradient changes sign. This criterion implies that Rossby wave breaking occurs more easily (i.e., at smaller wave amplitude) in regions of weak PV gradient than in regions of strong PV gradient. Now, the relationship between PV and winds implies that eastward zonal jets comprise regions where the (zonal-mean) meridional PV gradient is large, whereas westward zonal jets comprise regions where the meridional PV gradient is small (Fig. 9). Therefore, vertically and meridionally propagating Rossby waves are more likely to break between (or on the flanks of) eastward jets, where the PV gradient is small, than at the cores of eastward jets, where the PV gradient is large. This wave breaking causes irreversible mixing and homogenization of PV. The mixing is thus spatially inhomogeneous — mixing preferentially homogenizes the PV in the regions where its gradient is already weak, and sharpens the gradients in between. As emphasized by *Dritschel and McIntyre* (2008), this is a positive feedback: By modifying the background PV gradient, such mixing promotes continued mixing at westward jets but inhibits future mixing at eastward jets. Because the PV jumps are associated with eastward jets, this process tends to sharpen eastward jets and leads to broad "surf zones" of westward flow in between. This is precisely the behavior evident in Fig. 8. Because of this positive feedback, one generally expects that when an initially homogeneous system is stirred, robust zonal jets should spontaneously emerge — even if the stirring itself is not spatially organized (e.g., *Dritschel and McIntyre,* 2008; *Dritschel and Scott,* 2011; *Scott and Dritschel,* 2012; *Scott and Tissier,* 2012). Nevertheless, strong radiative forcing and/or frictional damping represent sources and sinks of PV that can prevent a pure PV staircase (and the associated zonal jets) from being achieved. The factors that determine the equilibrium PV distribution in such forced/damped cases is an area of ongoing research.

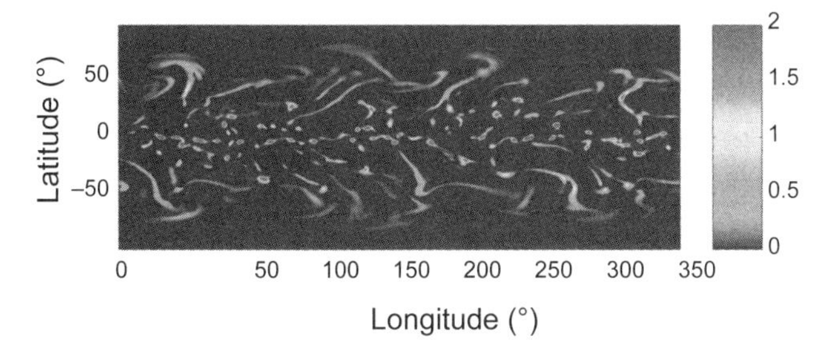

Fig. 7. See Plate 47 for color version. Instantaneous precipitation (units 10^{-3} kg m^{-2} s^{-1}) in an idealized Earth GCM by *Frierson et al.* (2006), illustrating the generation of phase tilts by Rossby waves as depicted schematically in Fig. 6. Baroclinic instabilities in midlatitudes (~40°–50°) generate Rossby waves that propagate meridionally. On the equatorward side of the baroclinically unstable zone (latitudes of ~20°–50°), the waves propagate equatorward, leading to characteristic precipitation patterns tilting southwest-northeast in the northern hemisphere and northwest-southeast in the southern hemisphere. In contrast, the phase tilts are reversed (although less well organized) poleward of ~50°–60° latitude, indicative of poleward Rossby wave propagation. Equatorward of 20° latitude, tropical convection dominates the precipitation pattern.

2.1.5. Jet formation II: Rossby wave interactions with turbulence. Additional insights on the formation of zonal jets emerge from a consideration of atmospheric turbulence. In the Ro ≪ 1 regime, the interaction of large-scale atmospheric turbulence with planetary rotation — and in particular with the β effect — generally leads to the formation of a zonally banded appearance and the existence of zonal jets [for important examples, see *Rhines* (1975), *Williams* (1978), *Maltrud and Vallis* (1993), *Cho and Polvani* (1996), *Huang and Robinson* (1998), and *Sukoriansky et al.* (2007); reviews can be found in *Vasavada and Showman* (2005), *Showman et al.* (2010), and *Vallis* (2006, Chapter 9)]. By allowing Rossby waves, the β effect introduces a fundamental anisotropy between the zonal and meridional direction that often leads to jets.

Consider the magnitudes of the terms in the vorticity equation (7). Being a curl of the wind field, the relative vorticity has characteristic magnitude $\frac{U}{L}$. Therefore, the advection term $\mathbf{v} \cdot \nabla\zeta$ has characteristic magnitude $\frac{U^2}{L^2}$, where L is some horizontal lengthscale of interest. The β term has characteristic magnitude βU. For a given wind amplitude, the advection term dominates at small scales (i.e., as L → 0). At these scales, the β term is unimportant, and the equation therefore describes two-dimensional turbulence. Such turbulence will be isotropic, because (at scales too small for β to be important) the equation contains no terms that would distinguish the east-west from the north-south directions. On the other hand, at large scales, the β term will dominate over the nonlinear advection term, and this implies that Rossby waves dominate the dynamics. The transition between these regimes occurs at the Rhines scale

$$L_R = \left(\frac{U}{\beta}\right)^{1/2} \tag{15}$$

The Rhines scale is traditionally interpreted as giving the transition scale between the regimes of turbulence (at small scales) and Rossby waves (at large scales). Generally, two-dimensional and quasi-two-dimensional fluids exhibit an upscale energy transfer from small scales to large scales (e.g., *Vallis,* 2006, Chapter 8). Therefore, if turbulent energy is injected into the fluid at small scales (e.g., through convection, baroclinic instabilities, or other processes), turbulent interactions can drive the energy toward larger scales where it can be affected by β. At scales close to the Rhines scale, the β effect forces the turbulence to become anisotropic: Since the term $v\beta$ involves only the meridional speed but not the zonal speed, the dynamics becomes different in the east-west and north-south directions.

This anisotropy causes the development of zonal banding in planetary atmospheres. As shown by *Vallis and Maltrud* (1993) and other authors, the transition between turbulence and Rossby waves is anisotropic, occurring at different length scales for different wavevector orientations. (Equation (15) gives the characteristic value ignoring this geometric effect.) Essentially, nonlinear interactions are better able to transfer turbulent energy upscale for structures that are zonally elongated than for structures that are isotropic or meridionally elongated. Preferential development of zonally elongated structures at large scales, often consisting of zonal jets, results. In many cases, the turbulent interactions do not involve the gradual transport of energy to incrementally ever-larger lengthscales (a transport that would be "local" in spectral space) but rather tend to involve spectrally nonlocal interactions wherein small-scale turbulence directly pumps the zonal jets (e.g., *Nozawa and Yoden,* 1997; *Huang and Robinson,* 1998; *Sukoriansky et al.,* 2007; *Read et al.,* 2007; *Wordsworth et al.,* 2008).

These processes are illustrated in Fig. 10, which shows the results of three solutions of the two-dimensional, nondivergent vorticity equation (7) starting from an initial condition containing numerous small-scale, close-packed, isotropic vortices. The model on the left is nonrotating, the model in the middle has intermediate rotation rate, and the model on the right is rapidly rotating. In the nonrotating case (left panel), the two-dimensional turbulence involves vortex mergers that drive turbulent energy from small scales toward larger scales. The flow remains isotropic and no jets form. In the rapidly rotating case, however, the turbulence interacts with Rossby waves at scales comparable to L_R, leading to robust zonal banding (right panel).

A wide variety of studies have shown that zonal jets generated in this way exhibit characteristic meridional widths comparable to the Rhines scale (for reviews, see

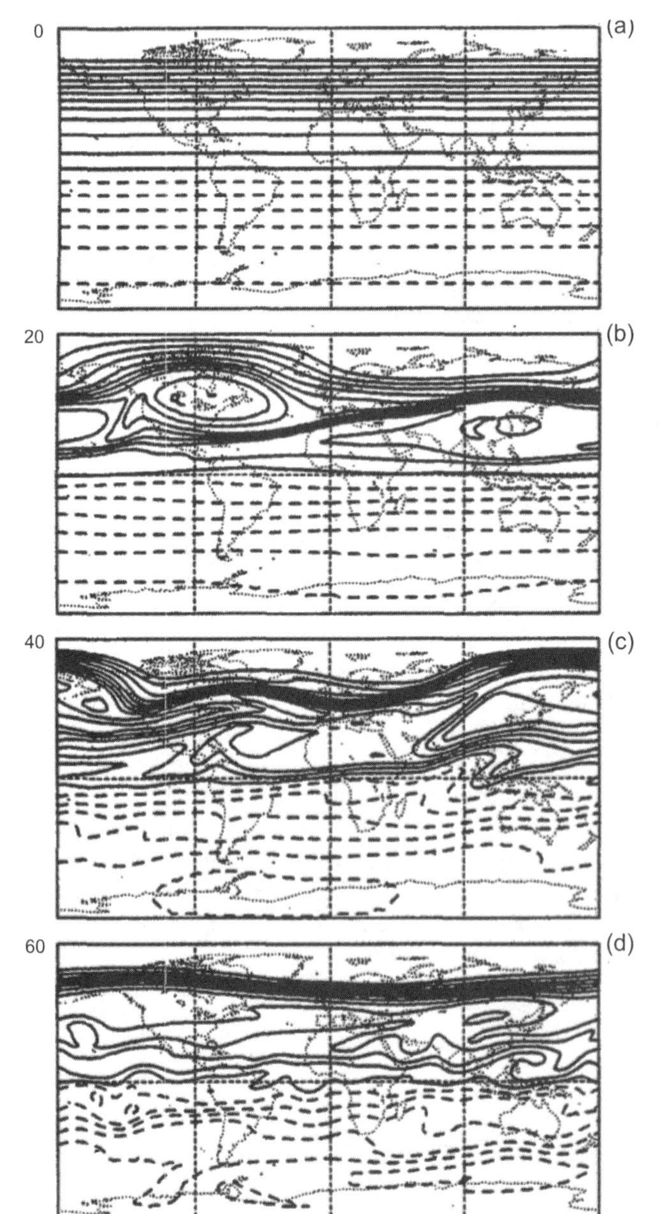

Fig. 8. Demonstration of how breaking Rossby waves can shape jet dynamics. Plots show time evolution of the potential vorticity (in contours, with positive solid, negative dashed, and a contour interval of $0.25 \times 10{-}8$ m^{-1} s^{-1}) in a shallow-water calculation of Earth's stratospheric polar vortex by *Polvani et al.* (1995). The model is initialized from a zonally-symmetric initial condition with broadly distributed PV organized into a polar vortex centered over the north pole [**(a)**]. A topographic perturbation is then introduced to generate a Rossby wave, which manifests as undulations in the PV contours [**(b)**]. The wave amplitude becomes large enough for the undulations to curl up, leading to wave breaking [**(b)** and **(c)**]. The edge of the vortex (corresponding to the region of tightly spaced PV contours) is resistant to such wave breaking, but the regions of weaker meridional PV gradient on either side are more susceptible. This breaking mixes and homogenizes the PV in those regions, thereby lessening the meridional gradient still further [this manifests as a widening in the latitudinal separation of the PV contours, especially from the equator to ~60°N latitude, in **(c)** and **(d)**]. In contrast, stripping of material from the vortex edge by this mixing process causes a sharpening of the PV jump associated with the vortex edge (visible as a tightening of the contour spacing at ~70°–80°N latitude in the fourth panel). This sharp PV jump is associated with a narrow, fast eastward jet — the polar jet — at the edge of the polar vortex.

Vasavada and Showman, 2005; *Vallis,* 2006, Chapter 9; and *Del Genio et al.,* 2009). Nevertheless, there has been significant debate about the relationship of the Rhines scale to other important lengthscales, such as Rossby deformation radius. The characteristic lengthscale for baroclinic instabilities is the deformation radius; therefore, in a flow driven by baroclinic instabilities, turbulent energy is injected at scales close to L_D. In principle, it is possible that L_R greatly exceeds L_D, in which case nonlinear interactions would transfer the energy upscale from the deformation radius to the scale of the jets themselves. However, atmospheric GCMs under Earth-like conditions have generally found

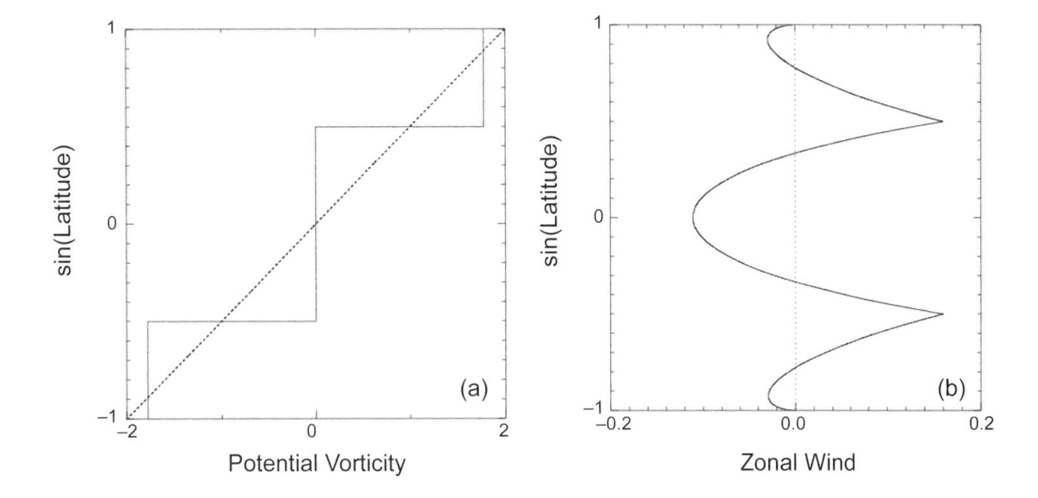

Fig. 9. Relationship between **(a)** PV and **(b)** zonal winds vs. the sine of latitude for a flow consisting of a PV "staircase" with constant-PV strips separated by regions of sharp PV gradients. In a motionless fluid on a spherical planet, the zonal wind is zero, and the PV increases smoothly with sine of latitude (dotted lines). But if PV is homogenized into strips, as shown in **(a)**, then the implied zonal wind structure (demanded by the relationship between PV and winds) is as shown in **(b)** (solid curves). Thus, homogenization of PV into strips on rapidly rotating planets implies the emergence of zonal jets. This is for the specific case of a two-dimensional, horizontally nondivergent flow governed by equation (7), but a similar relationship holds even in more realistic models: Regions of weak PV gradients correspond to regions of broad westward flow, while PV discontinuities correspond to sharp eastward jets. For the specific case of a zonally symmetric flow governed by equation (7), in Cartesian geometry with constant β, it is straightforward to show that the zonal-wind profile consists of parabolas connected end-to-end. For analytic solutions in more general cases, see *Marcus and Lee* (1998), *Dunkerton and Scott* (2008), and *Wood and McIntyre* (2010). From *Scott* (2010).

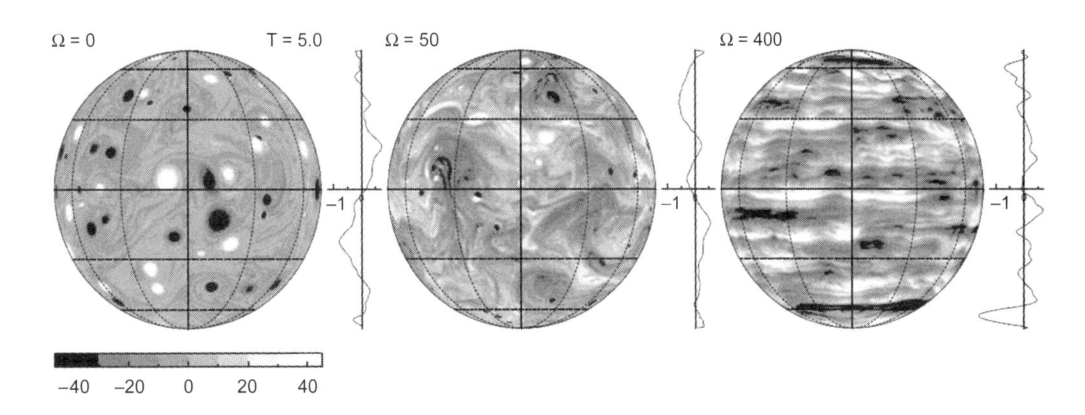

Fig. 10. Solutions of the two-dimensional nondivergent barotropic vorticity equation (7) illustrating the effect of planetary rotation on large-scale atmospheric turbulence. Three simulations are shown, initialized from identical initial conditions containing turbulence at small length scales. Only the rotation rate differs between the three models; from left to right, the rotation rates are zero, intermediate, and rapid. There is no forcing or large-scale damping (save for a numerical viscosity required for numerical stability) so that the total flow energy is nearly constant in all three cases. In the nonrotating case, energy cascades to larger length scales as vortices merge, but the flow remains isotropic. In the rapidly rotating case, the flow is highly anisotropic and zonal banding develops. From *Hayashi et al.* (2000); see also *Yoden et al.* (1999), *Ishioka et al.* (1999), and *Hayashi et al.* (2007).

that the circulation adjusts to a state where the deformation radius is comparable to the Rhines scale (e.g., *Schneider and Walker,* 2006, 2008). In this case, baroclinic instabilties directly inject turbulent energy into the flow at scales comparable to the meridional width of the zonal jets. The injected energy interacts nonlinearly with the mean flow to generate jets, but significant upscale transfer of the energy is not necessarily involved. Still, situations exist (including in Earth's ocean) where the Rhines scale and deformation radius differ substantially and significant upscale energy transfer occurs between the two scales (e.g., *Jansen and Ferrari,* 2012).

These arguments suggest that, for extratropical jets driven by baroclinic instability, the relative sizes of the deformation radius and of the extratropics itself determines the number of jets. When a planet is small, like Earth or Mars, only one strip of baroclinic instabilities — and one eddy-driven jet — can fit into the baroclinic zone (as illustrated schematically in Fig. 2). When the planet is relatively large relative to the eddy size, however, the baroclinic zone breaks up into multiple bands of baroclinic instability, and hence multiple jet streams. Figure 11 shows an example from *Schneider and Walker* (2006) showing an Earth-like case (Fig. 11a) and a case at four times the Earth rotation rate (Fig. 11b). The Earth case exhibits only one, midlatitude eddy-driven jet in each hemisphere. In the rapidly rotating case, the deformation radius is four times smaller, and the baroclinic zone breaks up into three jets in each hemisphere. This process coincides with a steepening of the isentropes (thin gray contours), which is associated with a greater equator-to-pole temperature difference, consistent with those shown in Figs. 3 and 5.

2.2. Tropical Regime

The tropics, defined here as regimes of Ro $\gtrsim$ 1, are inherently ageostrophic, and their dynamics differ significantly from those of the extratropics. Unlike the case of geostrophic flow, horizontal temperature gradients in the Ro $\gtrsim$ 1 regime tend to be modest, which affects the dynamics in myriad ways.

The tendency toward weak temperature gradients can be motivated by considering arguments analogous to those leading up to equation (4). In the absence of a significant Coriolis force, large-scale horizontal pressure-gradient forces tend to be balanced by advection, represented to order of magnitude as U^2/L. The fractional horizontal potential temperature difference is then (*Charney,* 1963)

$$\frac{\delta\theta_h}{\theta} \sim \frac{U^2}{gD} \sim \text{Fr} \qquad \text{Ro} \gtrsim 1 \tag{16}$$

Comparison with equation (4) immediately shows that, in the rapidly rotating regime characterized by Ro $\ll$ 1, the lateral temperature contrasts are a factor of Ro^{-1} bigger than they are in the slowly rotating regime of Ro $\gtrsim$ 1. These trends are evident in the GCM experiments in Figs. 3 and 5c. For a given wind speed, slowly rotating planets tend to be more horizontally isothermal than rapidly rotating planets. In the case of a typical terrestrial planet where U ~ 10 m s^{-1}, D ~ 10 km, and g $\approx$ 10 m s^{-2}, we have Fr ~ 10^{-3}. One might thus expect that, in the tropics of Earth, and globally on Venus, Titan, and slowly rotating exoplanets, the lateral temperature contrasts are ~1 K. Note, however, that because of the quadratic dependence of $\delta\theta_h$ on wind speed, large temperature differences could occur if the winds are sufficiently fast. Of course, additional arguments would be needed to obtain a self-consistent prediction of both $\delta\theta_h$ and U on any given planet.

We here discuss several of the major dynamical mechanisms relevant for the tropical regime.

2.2.1. Hadley circulation and subtropical jets. Planets generally exhibit meridional gradients of the mean incident stellar flux, with, at low obliquity, highly irradiated, warm conditions at low latitudes and poorly irradiated, cooler conditions at high latitudes. The Hadley circulation represents the tropical response to this insolation gradient and is the primary mechanism for meridional heat transport in the tropical atmosphere. Stripped to its essence, the Hadley circulation in an atmosphere forced primarily by an equator-pole heating gradient can be idealized as an es-

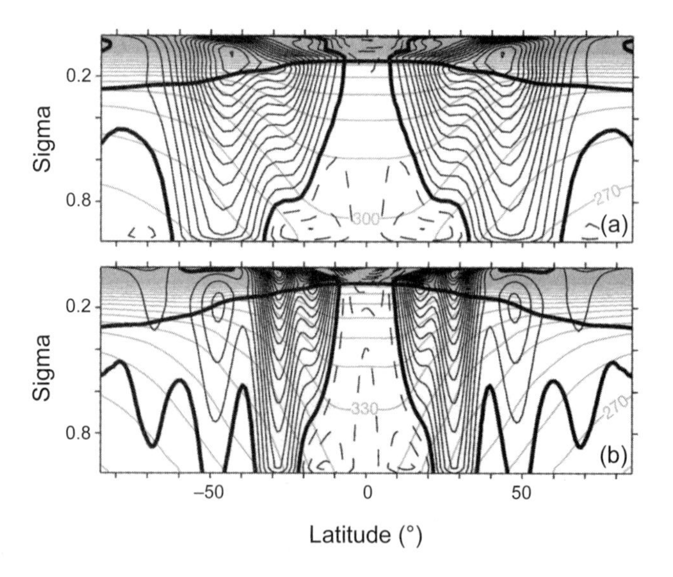

Fig. 11. Idealized GCM simulations of Earth-like planets exhibiting jet formation. These models are forced by equator-pole heating gradients. Black contours show zonal-mean zonal wind (positive solid, negative dashed, zero contour is thick; contour interval is 2.5 m s^{-1}). Gray contours show zonal-mean potential temperature (K, contour interval is 10 K). Heavy line near top denotes the tropopause. Vertical coordinate is ratio of pressure to surface pressure. **(a)** Earth-like case; **(b)** case at four times the Earth rotation rate. The Earth-like case exhibits one eddy-driven jet in each hemisphere. In the rapidly rotating case, this jet is confined much closer to the equator and additional, eddy-driven jets have emerged in each hemisphere. From *Schneider and Walker* (2006).

sentially two-dimensional circulation in the latitude-height plane: Hot air rises near the equator, moves poleward aloft, descends at higher latitudes, and returns to the equator along the surface (see Fig. 2 for a schematic). All the terrestrial planets with thick atmospheres in the solar system — Earth, Venus, Mars, and Titan — exhibit Hadley cells, and Hadley circulations will likewise play an important role on terrestrial exoplanets.

The Hadley circulations exert a significant effect on the mean planetary climate. Because of its meridional energy transport, the meridional temperature gradient across the Hadley circulation tends to be weak. Moreover, on planets with hydrological cycles (section 3), the ascending branch tends to be a location of cloudiness and high rainfall, whereas the descending branches tend to be drier and more cloud-free. In this way, the Hadley circulation exerts control over regional climate. On Earth, most tropical rainforests occur near the equator, at the latitude of the ascending branch, while many of the world's major deserts (the Sahara, the American southwest, Australia, and South Africa) occur near the latitudes of the descending branches. Because the latitudinal and vertical distribution of cloudiness and humidity can significantly influence the planetary albedo and greenhouse effect, the mean planetary surface temperature — as well as the distribution of temperature, cloudiness, and humidity across the planet — depend significantly on the structure of the Hadley circulation. Moreover, by helping to control the equator-pole distribution of temperature, humidity, and clouds, the Hadley circulation will influence the conditions under which terrestrial planets can experience global-scale climate feedbacks, including transitions to globally glaciated "snowball" states, atmospheric collapse, and runaway greenhouses (section 4).

The structure of the Hadley cell is strongly controlled by planetary rotation. Frictional drag acts most strongly in the surface branch of the circulation, and the upper branch, being decoupled from the surface, is generally less affected by frictional drag. A useful limit to consider is one where the upper branch is frictionless, such that individual air parcels ascending at the equator conserve their angular momentum per unit mass about the planetary rotation axis, given by $m = (\Omega a \cos \phi + u) a \cos \phi$, as they move poleward (*Held and Hou,* 1980). If the ascending branch is at the equator and exhibits zero zonal wind, then the zonal wind in the upper (poleward-flowing) branch of such an angular-momentum conserving circulation is

$$u = \Omega a \frac{\sin^2 \phi}{\cos \phi} \tag{17}$$

where a is the planetary radius and ϕ is latitude. Thus, in the upper branch of the Hadley cell, the zonal wind increases rapidly with latitude, the more so the faster the planet rotates or the larger the planetary size. Under modern Earth conditions, this equation implies that the zonal wind is 134 m s^{-1} at 30° latitude, reaches 1000 m s^{-1} at a latitude of 67°, and becomes infinite at the poles. This is of course impossible, and implies that planetary rotation, if sufficiently strong, confines the Hadley circulation to low latitudes. Real Hadley cells do not conserve angular momentum, and exhibit zonal winds increasing more slowly with latitude than expressed by equation (17); nevertheless, rotation generally confines Hadley circulations to low latitudes even in this case.

Over the past 30 years, many GCM studies have been performed to investigate how the Hadley circulation depends on planetary rotation rate and other parameters (*Hunt,* 1979; *Williams and Holloway,* 1982; *Williams,* 1988a,b; *Del Genio and Suozzo,* 1987; *Navarra and Boccaletti,* 2002; *Walker and Schneider,* 2005, 2006; *Kaspi and Showman,* 2012). Figure 12 illustrates an example from *Kaspi and Showman* (2012), showing GCM experiments for planets with no seasonal cycle forced by equator-pole heating gradients. Models were performed for rotation rates ranging from 1/16th to 4 times that of Earth (top to bottom panels in Fig. 12, respectively). Other parameters in these experiments, including solar flux, planetary radius and gravity, and atmospheric mass are all Earth-like. In general, the Hadley circulation consists of a cell in each hemisphere extending from the equator toward higher latitudes (dark red and blue regions in Fig. 12). As the air in the upper branch moves poleward, it accelerates eastward by the Coriolis force, leading to the so-called subtropical jets whose amplitudes peak at the poleward edge of the Hadley cell (such subtropical jets are shown schematically in Fig. 2). At sufficiently low rotation rates (top two panels of Fig. 12), the Hadley circulation is nearly global, extending from equator to pole in both hemispheres, with the subtropical jets peaking at latitudes of ~60°. Such models constitute "all tropics" worlds lacking any high-latitude extratropical baroclinic zone. At faster rotation rates (bottom five panels), the Hadley circulation — and the associated subtropical jets — become confined closer to the equator, and the higher latitudes develop an extratropical zone with baroclinic instabilities. The emergence of mid- and high-latitude eddy-driven jets in the extratropics, with eastward surface wind, can be seen in these rapidly rotating cases (cf. section 2.1.4). As shown in Fig. 12, the tropospheric temperatures are relatively constant with latitude across the Hadley cell and begin decreasing poleward of the subtropical jets near its poleward edge. The resulting equator-to-pole temperature differences are small at slow rotation and increase with increasing rotation rate (Fig. 12).

Seasonality exerts a strong effect on the Hadley circulation. When a planet's obliquity is nonzero, the substellar latitude oscillates between hemispheres, crossing the equator during equinox and reaching a peak excursion from the equator at solstice. The rising branch of the Hadley circulation tends to follow the latitude of maximum sunlight and therefore oscillates between hemispheres as well. (When the atmospheric heat capacity is large, the response will generally be phase lagged with respect to the stellar heating pattern.) Near equinox, the rising branch lies close to the equator, with Hadley cells of approximately equal strength in each hemisphere (cf. Fig. 12). Near solstice, the rising branch lies in the summer hemisphere. As before, two cells exist, with air in one cell (the "winter" cell) flowing across the equator and descending in the winter hemisphere, and air in the other cell (the "summer" cell) flowing poleward toward the summer pole. Generally, for obliquities relevant to Earth, Mars, and Titan, the (cross-equatorial) winter cell is much stronger than the summer cell [for the Earth case, see, e.g., *Peixoto and Oort* (1992, pp. 158–160)]. The zonal wind structure at solstice and equinox also differ

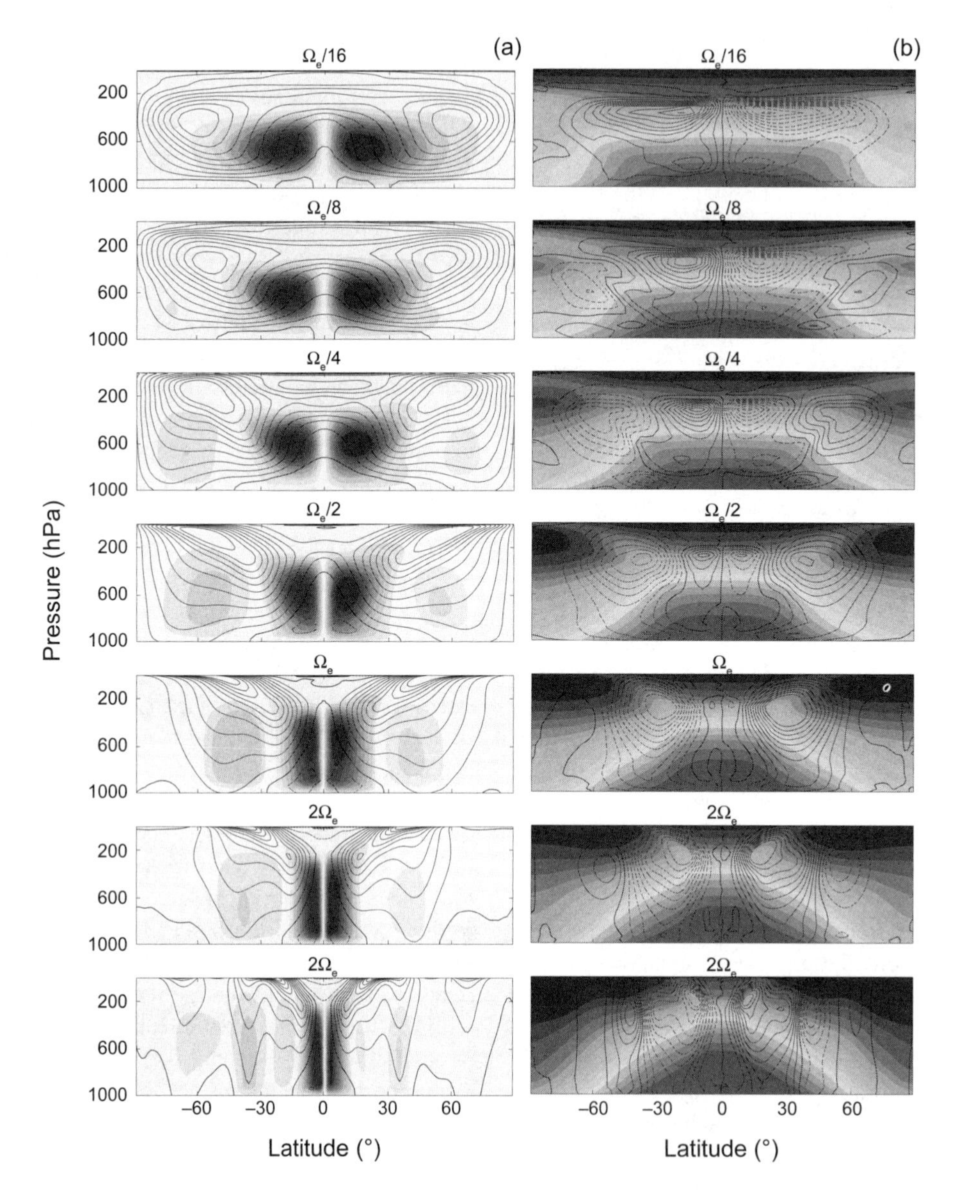

Fig. 12. See Plate 48 for color version. Zonal-mean circulation for a sequence of idealized GCM experiments of terrestrial planets from *Kaspi and Showman* (2012), showing the dependence of the Hadley circulation on planetary rotation rate. The models are driven by an imposed equator-pole insolation pattern with no seasonal cycle. The figure shows seven experiments with differing planetary rotation rates, ranging from 1/16th to four times that of Earth from top to bottom, respectively. **(a)** Thin black contours show zonal-mean zonal wind; the contour interval is 5 m s^{-1}, and the zero-wind contour is shown in a thick black contour. Orange/blue colorscale depicts the mean-meridional streamfunction, with blue denoting clockwise circulation and orange denoting counterclockwise circulation. **(b)** Colorscale shows zonal-mean temperature. Contours show zonal-mean meridional eddy-momentum flux, $\overline{u'v'}\cos\phi$. Solid and dashed curves denote positive and negative values, respectively (implying northward and southward transport of eastward eddy momentum, respectively). At slow rotation rates, the Hadley cells are nearly global, the subtropical jets reside at high latitude, and the equator-pole temperature difference is small. The low-latitude meridional momentum flux is equatorward, leading to equatorial superrotation (eastward winds at the equator) in the upper troposphere. At faster rotation rates, the Hadley cells and subtropical jets contract toward the equator, an extratropical zone, with eddy-driven jets, develops at high latitudes, and the equator-pole temperature difference is large. The low-latitude meridional momentum flux is poleward, resulting from the absorption of equatorward-propagating Rossby waves coming from the extratropics.

significantly. The zonal wind in the ascending branch is generally weak, and, the greater its latitude, the lower its angular momentum per unit mass. If the ascending branch lies at latitude ϕ_0 and exhibits zero zonal wind, and if the upper branch of the Hadley circulation conserves angular momentum, the zonal wind in the upper branch is

$$u = \Omega a \frac{\cos^2 \phi_0 - \cos^2 \phi}{\cos \phi} \tag{18}$$

Notice that this equation implies strong westward (negative) wind speeds near the equator if ϕ_0 is displaced off the equator — a result of the fact that, in the winter cell, air is moving away from the rotation axis as it approaches the equator. Just such a phenomenon of strong westward equatorial wind under solsticial conditions can be seen in both idealized, axisymmetric models of the Hadley cell (e.g., *Lindzen and Hou,* 1988; *Caballero et al.,* 2008; *Mitchell et al.,* 2009) and full GCM simulations performed under conditions of high obliquity (*Williams and Pollard,* 2003). Nevertheless, eddies can exert a considerable effect on the Hadley cell, which will cause deviations from equation (18).

The Hadley circulation can exhibit a variety of possible behaviors depending on the extent to which the circulation in the upper branch is angular-momentum conserving. The different regimes can be illuminated by considering the zonal-mean zonal wind equation, written here for the three-dimensional primitive equations in pressure coordinates on the sphere (cf. *Held,* 2000, section 6)

$$\frac{\partial \bar{u}}{\partial t} = \left(f + \bar{\zeta}\right)\bar{v} - \bar{\omega}\frac{\partial \bar{u}}{\partial p} - \frac{1}{a\cos^2\phi}\frac{\partial\left(\cos^2\phi\,\overline{u'v'}\right)}{\partial \phi} - \frac{\partial\left(\overline{u'\omega'}\right)}{\partial p} \tag{19}$$

where $\omega = dp/dt$ is the vertical velocity in pressure coordinates, d/dt is the advective derivative, and as before, the overbars and primes denote zonal averages and deviations therefrom. In a statistical steady state, the lefthand side of equation (19) is zero. Denoting the eddy terms (i.e., the sum of the last two terms on the righthand side) by –S, we obtain

$$\left(f + \bar{\zeta}\right)\bar{v} = \bar{\omega}\frac{\partial \bar{u}}{\partial p} + S \tag{20}$$

For Earth, it turns out that the vertical advection by the mean flow (first term on righthand side) does not play a crucial role, at least for a discussion of the qualitative behavior, so that the zonal momentum balance for the upper branch of the Hadley circulation can be written (e.g., *Held,* 2000; *Schneider,* 2006; *Walker and Schneider,* 2006)

$$\left(f + \bar{\zeta}\right)\bar{v} = f\left(1 - Ro_H\right)\bar{v} \approx S \tag{21}$$

where the Rossby number associated with the Hadley circulation is defined as $Ro_H = -\bar{\zeta}/f$.

The Hadley circulation exhibits distinct behavior depending on whether the Rossby number, Ro_H, is large or small. Essentially, Ro_H is a nondimensional measure of the effect of eddies on the Hadley cell (e.g., *Held,* 2000; *Schneider,* 2006; *Walker and Schneider,* 2006; *Schneider and Bordoni,* 2008). The meridional width, amplitude, temperature structure, and seasonal cycle of the Hadley circulation depend on the relative roles of thermal forcing (in the form of meridional heating gradients) and mechanical forcing (in the form of S). The Hadley circulation will respond differently to a given change in thermal forcing depending on whether eddy forcing is negligible or important.

Theories currently exist for the $Ro_H \to 1$ and $Ro_H \to 0$ limits, although not yet for the more complex intermediate case that seems to be representative of Earth and perhaps planets generally.

When eddy-induced accelerations are negligible, then S = 0, and for nonzero circulations the absolute vorticity must then be zero within the upper branch, i.e., $f + \bar{\zeta} = 0$, or, in other words, $Ro_H \to 1$. The definitions of relative vorticity and angular momentum imply that $f + \bar{\zeta} = (a^2 \cos \phi)^{-1}\partial\bar{m}/\partial\phi$, from which it follows that a circulation with zero absolute vorticity exhibits angular momentum per mass that is constant with latitude. This is the angular-momentum conserving limit mentioned previously and, for atmospheres experiencing a heating maximum at the equator, would lead to a zonal-wind profile obeying equation (17) in the upper branch. Several authors have explored theories of such circulations, both for annual-mean conditions and including a seasonal cycle (e.g., *Held and Hou,* 1980; *Lindzen and Hou,* 1988; *Fang and Tung,* 1996; *Polvani and Sobel,* 2002; *Caballero et al.,* 2008; *Adam and Paldor,* 2009, 2010). Given the angular-momentum conserving wind in the upper branch (equation (17)), and assuming the near-surface wind is weak — a result of surface friction — the mean vertical shear of the zonal wind between the surface and upper troposphere is therefore known. The tropospheric meridional temperature gradient can then be obtained from the thermal-wind balance (equation (2)), or from generalizations of it that include curvature terms on the sphere (cf. *Held and Hou,* 1980). When the Hadley cell is confined to low latitudes, this leads to a quartic dependence of the temperature on latitude; for example, in the *Held and Hou* (1980) model, the potential temperature at a mid-tropospheric level is

$$\theta = \theta_{equator} - \frac{\Omega^2 \theta_0}{2ga^2 H} y^4 \tag{22}$$

where H is the vertical thickness of the Hadley cell, y is northward distance from the equator, and $\theta_{equator}$ is the potential temperature at the equator. This dependence implies that the temperature varies little with latitude across most of the Hadley cell but plummets rapidly near the poleward edges of the cell. In this $Ro_H \rightarrow 1$ regime, the strength of the Hadley cell follows from the thermodynamic energy equation, and in particular from the radiative heating gradients (with heating near the equator and cooling in the subtropics). The Hadley cell also has finite meridional extent that is determined by the latitude at which the integrated cooling away from the equator balances the integrated heating near the equator. The Hadley circulation in this limit is thermally driven.

On the other hand, eddy-momentum accelerations S are often important and can play a defining role in shaping the Hadley cell properties. In the limit $Ro_H \ll 1$, the strength of the Hadley circulation is determined not by the thermal forcing (at least directly) but rather by the eddy-momentum flux divergences: Because the absolute vorticity at $Ro_H \ll 1$ is approximately f, the meridional velocity in the Hadley cell is given, via equation (21), by $\overline{v} \approx S/f$, at least under conditions where the vertical momentum advection of the mean flow can be neglected. Generally, in the Earth- and Mars-like context, the primary eddy effects on the Hadley cell result from the equatorward propagation of Rossby waves generated in the baroclinincally unstable zone in the mid-latitudes (section 2.1.4). The zonal phase velocities of these waves, while generally westward relative to the peak speeds of the eddy-driven jet, are eastward relative to the ground. These Rossby waves propagate into the subtropics (causing a poleward flux of eddy angular momentum visible in the bottom three panels of Fig. 12), where they reach critical levels on the flanks of the subtropical jets (*Randel and Held,* 1991). The resulting wave absorption generally causes a westward wave-induced acceleration of the zonal-mean zonal flow, which removes angular momentum from the poleward-flowing air in the upper branch of the Hadley circulation. Therefore, the angular momentum decreases poleward with distance away from the equator in the upper branch of the Hadley cell. As a result, although the zonal wind still increases with latitude away from the equator, it does so more gradually than predicted by equation (17); for example, at 20° latitude, the zonal-mean zonal wind speed in Earth's upper troposphere is only ~20 m s^{-1}, significantly weaker than the ~60 m s^{-1} predicted by equation (17). In turn, the weaker vertical shear of the zonal wind implies a weaker meridional temperature gradient through thermal wind (equation (2)), leading to a meridional temperature profile than can remain flatter over a wider range of latitudes than predicted by equation (22) (see *Farrell* (1990) for examples of this phenomenon in the context of a simple, axisymmetric model).

Real planets probably lie between these two extremes. On Earth, observations and models indicate that eddy effects on the Hadley cell are particularly important during the spring and fall equinox and in the summer hemisphere cell, whereas the winter hemisphere (cross-equatorial) Hadley cell is closer to the angular momentum conserving limit (e.g., *Kim and Lee,* 2001; *Walker and Schneider,* 2005, 2006; *Schneider and Bordoni,* 2008; *Bordoni and Schneider,* 2008, 2010). In particular, the Rossby numbers Ro_H vary from $\lesssim$0.3–0.4 in the equinoctal and summer cells to $\gtrsim$0.7 in the winter cell. Physically, the mechanisms for the transition between these regimes involve differences in an eddy-mean-flow feedback between equinoctal and solsticial conditions (*Schneider and Bordoni,* 2008; *Bordoni and Schneider,* 2008). During the equinox, the ascending branch of the Hadley cell lies near the equator, and the upper branch transports air poleward — i.e., toward the rotation axis — into both hemispheres. Angular momentum conservation therefore implies that the upper branch of the Hadley cell exhibits eastward zonal flow (equation (17)), allowing the existence of critical layers and the resultant absorption of the equatorward-propagating Rossby waves from mid-latitudes. Eddies therefore play a crucial role (S is large and Ro_H is small). During solstice, the ascending branch of the Hadley cell is displaced off the equator into the summer hemisphere. Air flows poleward in the upper branch of the summer cell, again leading to eastward zonal flow, the existence of critical layers, and significant eddy influences. On the other hand, air in the winter-hemisphere cell rises in the subtropics of the summer hemisphere and flows across the equator to the winter hemisphere, where it descends in the subtropics. Because this equatorward motion moves the air away from the rotation axis, angular momentum conservation leads to a broad region of westward zonal winds across much of the winter cell (equation (18)) (*Lindzen and Hou,* 1988). Because the equatorward-propagating mid-latitude eddies exhibit eastward zonal phase speeds (relative to the ground), the winter-hemisphere cell therefore lacks critical layers and is largely transparent to these waves. (Indeed, most of them have already reached critical levels and been absorbed before even reaching the latitude of the winter cell.) The result is a much smaller net absorption of eddies, and therefore a winter hemisphere Hadley cell whose upper branch is much closer to the angular-momentum conserving limit.

Other feedbacks between Hadley cells and eddies are possible, particularly on tidally locked exoplanets where the day-night heating pattern dominates. In particular, tidally locked exoplanets exhibit a strong day-night heating pattern that induces standing, planetary-scale tropical waves, which can drive an eastward (superrotating) jet at the equator (section 2.2.3). Such superrotation corresponds to a local maximum at the equator of angular momentum per mass about the planetary rotation axis; by definition, this means that the meridional circulation in a Hadley cell, if any, must cross contours of constant angular momentum. This can only occur in the presence of eddy accelerations that alter the angular momentum of the air in the upper branch as it flows meridionally. Therefore, Hadley cells in the presence of superrotation jets tend to be far from the angular-momentum conserving limit. The term $\overline{\omega}\partial\overline{u}/\partial p$ in equation (20) may be crucial on such planets, unlike the typical case on Earth.

Shell and Held (2004) explored the interaction between superrotation and the Hadley circulation in a zonally symmetric, 1½-layer shallow-water model where the effect of the tropical eddies was parameterized as a specified eastward acceleration at the equator. They found four classes of behavior depending on the strength of the eddy forcing. For sufficiently weak zonal eddy acceleration, the model's Hadley circulation approached the angular-momentum conserving limit. For eastward eddy accelerations exceeding a critical value, however, two stable equilibria emerged for a given eddy acceleration. In one equilibrium, the Hadley cell is strong, and the massive advection from below of air with minimal zonal wind inhibits the ability of the eddy forcing to generate a strongly superrotating jet at the equator. As a result, angular momentum varies only modestly with latitude near the equator, promoting the ability of the Hadley cell to remain strong. In the other equilibrium, the Hadley cell and the associated vertical momentum advection are weak, so the eddy acceleration is therefore able to produce a fast superrotating jet. The corresponding thermal profile is close to radiative equilibrium, maintaining the Hadley cell in its weak configuration. *Shell and Held* (2004) showed that when the eddy forcing exceeds a second critical threshold, only the latter solution exists, corresponding to a weak Hadley cell and a strongly superrotating jet. For eddy forcing exceeding a third critical threshold, the Hadley circulation collapses completely, replaced by an eddy-driven meridional circulation with sinking motion at the equator and ascending motion off the equator. It will be interesting to use more sophisticated models to determine the extent to which these various regimes apply to terrestrial exoplanets, and, if so, the conditions for transitions between them.

2.2.2. Wave adjustment. As discussed earlier in this section, the slowly rotating regime generally exhibits small horizontal temperature contrasts, which results from the existence of dynamical mechanisms that efficiently regulate the thermal structure. The Hadley circulation, discussed previously, is one such mechanism, and is particularly important in the latitudinal direction at large scales. Another key mechanism is adjustment of the thermal structure by gravity waves.

How does such wave adjustment work? When moist convection, radiative heating gradients, or other processes generate horizontal temperature variations, the resulting pressure gradients lead to the radiation of gravity waves. The horizontal convergence/divergence induced by these waves adjusts air columns up or down in such a way as to flatten the isentropes and therefore erase the horizontal temperature constrasts. This process, which is essentially the nonrotating endpoint of geostrophic adjustment, plays a key role in minimizing the horizontal temperature contrasts in the tropics; the resulting dynamical regime is often called the "weak temperature gradient" or WTG regime (*Sobel et al.,* 2001; *Sobel,* 2002).

This adjustment process is most simply visualized in the context of a one-layer fluid. Imagine a shallow, nonrotating layer of water in one spatial dimension, whose initial surface elevation is a step function. This step causes a sharp horizontal pressure gradient force, and this will lead to radiation of gravity waves (manifested here as surface water waves) in either direction. Figure 13 shows an analytical solution of this process from *Kuo and Polvani* (1997). The horizontal convergence/divergence induced by the waves changes the fluid thickness. Assuming the waves can radiate to infinity (or break at some distant location), the final state is a flat layer. When the fractional height variation is small, this adjustment process causes only a small horizontal displacement of fluid parcels; the adjustment is done by waves, not long-distance horizontal advection.

This wave-adjustment mechanism also acts to regulate the thermal structure in three-dimensional, continuously stratified atmospheres (e.g., *Bretherton and Smolarkiewicz,* 1989; *Sobel,* 2002). In a stratified atmosphere, horizontal temperature differences are associated with topographic variations of isentropes, which play a role directly analogous to the topography of the water surface in the example described above. Gravity waves induce horizontal convergence/divergence, which changes the vertical thickness of the fluid columns, thereby pushing the isentropes up or down. Assuming that planetary rotation is sufficiently weak, and that the waves can radiate to infinity, the final state is one with flat isentropes. Note that, if the initial fractional temperature contrast is small, this adjustment causes only a small lateral motion of fluid parcels.

Figure 14 demonstrates this process explicitly in a three-dimensional, continously stratified atmosphere. There we

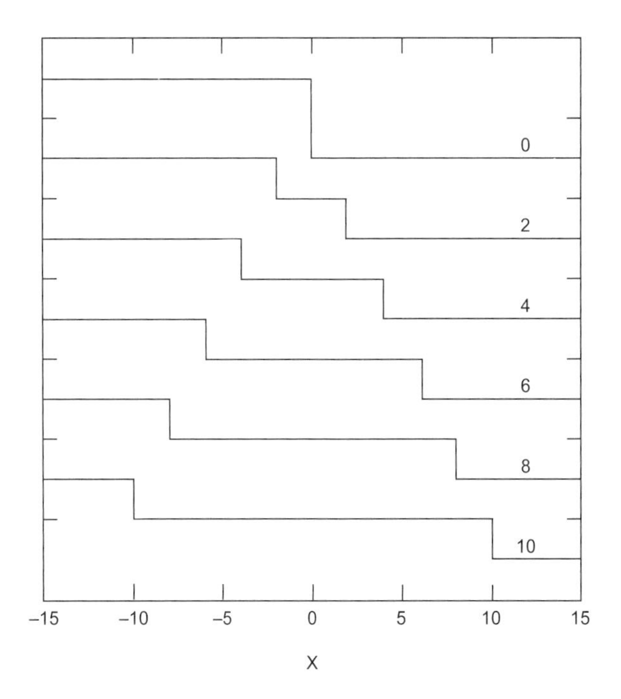

Fig. 13. An analytic solution of the wave adjustment process in a one-dimensional, nonrotating shallow-water fluid. Each curve shows the elevation of the water surface (with an arbitrary vertical offset for clarity) vs. horizontal distance at a particular time. Numbers labeling each curve give time. The initial condition is shown at the top, and subsequent states are shown underneath. Note how all the topography is "carried away" by the waves. From *Kuo and Polvani* (1997).

show a solution of the global, three-dimensional primitive equations for an initial-value problem in which half the planet (the "dayside") is initialized to have a potential temperature that is 20 K hotter than the other half (the "nightside"). The heating/cooling is zero so that the flow is adiabatic and isentropes are material surfaces. The top row shows the initial condition, and subsequent rows show the state at later times during the evolution. At early times (0.5×10^4 s, second row), waves begin to radiate away from the day-night boundary; they propagate across most of the planet by ~2×10^4 s (third row). They also propagate upward where they are damped in the upper atmosphere, leaving the long-term state well-adjusted (fourth row). Note that, in the final state, the isentropes are approximately flat and the day-night temperature differences are greatly reduced over their initial values.

The timescales for this wave-adjustment process can differ significantly from the relevant advection and mixing timescales. Because the wave-adjustment mechanism requires the propagation of gravity waves, the timescale to adjust the temperatures over some distance L is essentially the gravity wave propagation time over distance L, i.e., $\tau_{adjust} \sim L/NH$, where N is the Brunt-Vaisala frequency and H is a scale height. For example, *Bretherton and Smolarkiewicz* (1989) show how gravity waves adjust the thermal structure in the environment surrounding tropical cumulus convection; they demonstrate that the process operates on a timescale much shorter than the lateral mixing timescale. Moreover, if the wave propagation speeds differ significantly from the wind speeds, then τ_{adjust} can differ significantly from the horizontal advection time, $\tau_{adv} \sim L/U$. In the example shown in Fig. 14, for example, the wave speeds are c ~ NH ~ 200 m s^{-1} (using $N \approx 0.02$ s^{-1} and $H \approx 10$ km), but the peak tropospheric wind speeds in this simulation are almost 10 times smaller. This implies that the timescale for wave propagation is nearly 10 times shorter than the timescale for air to advect over a given distance.

Understanding the conditions under which the WTG regime breaks down is critical for understanding the atmospheric stability, climate, and habitability of exoplanets. Synchronously rotating planets exhibit permanent day- and nightsides, and for such atmospheres to remain stable against collapse, they must remain sufficiently warm on the nightside (*Joshi et al.,* 1997). Crudely, one might expect that dynamics fails to erase the day-night temperature contrast when the radiative timescale becomes shorter than the relevant dynamical timescale. Most previous exoplanet literature has assumed that the relevant comparison is between the radiative and advective timescales (e.g., *Showman*

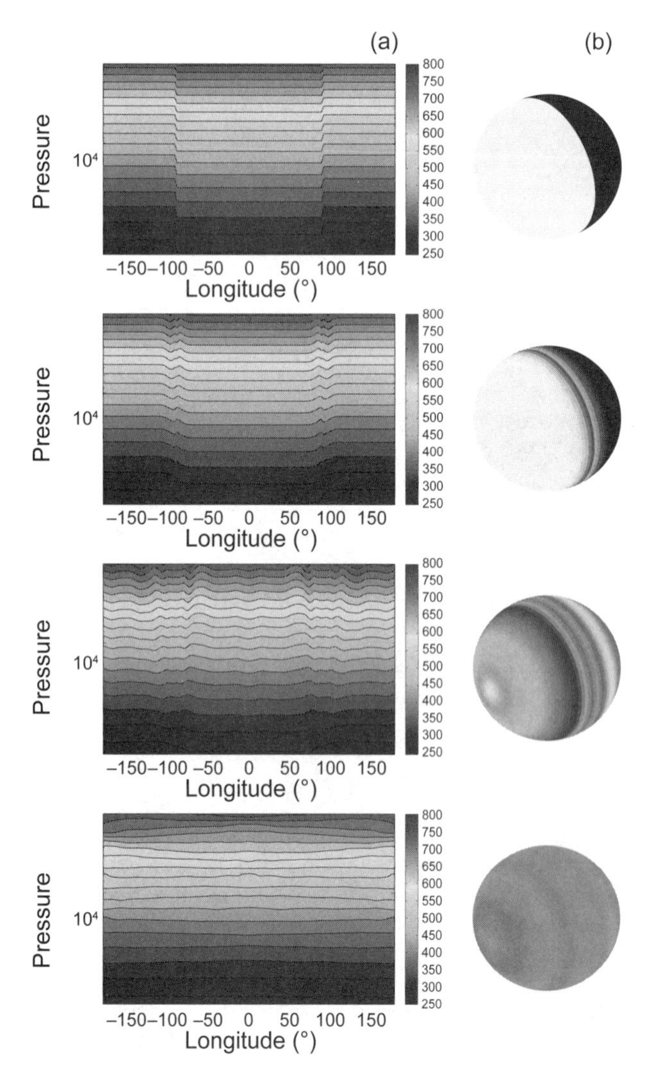

Fig. 14. See Plate 49 for color version. Numerical solution of wave adjustment on a spherical, nonrotating terrestrial planet with the radius and gravity of Earth. We solved the global, three-dimensional primitive equations, in pressure coordinates, using the MITgcm. Half of the planet (the "nightside") was initialized with a constant (isothermal) temperature of T_{night} = 250 K, corresponding to a potential temperature profile $\theta_{night} = T_{night}(p_0 = p)^\kappa$, where $\kappa = R/c_p$ = 2/7 and p_0 = 1 bar is a reference pressure. The other half of the planet (the "dayside") was initialized with a potential temperature profile $\theta_{night}(p) + \Delta\theta$, where $\Delta\theta$ = 20 K is a constant. Domain extends from approximately 1 bar at the bottom to 0.001 bar at the top; equations were solved on a cubed-sphere grid with horizontal resolution of C32 (32 × 32 cells per cube face, corresponding to an approximate resolution of 2.8°) and 40 levels in the vertical, evenly spaced in log-p. The model includes a sponge at pressures less than 0.01 bar to absorb upward-propagating waves. This is an initial value problem; there is no radiative heating/cooling, so the flow is adiabatic. **(a)** Potential temperature (colorscale and contours) at the equator vs. longitude and pressure; **(b)** temperature at a pressure of 0.2 bar over the globe at times of 0 (showing the initial condition), 0.5×10^4 s, 3×10^4 s, and the final long-term state once the waves have propagated into the upper atmosphere. Air parcels move by only a small fraction of a planetary radius during the adjustment process, but the final state nevertheless corresponds to nearly flat isentropes with small horizontal temperature variations on isobars.

et al., 2010; *Cowan and Agol,* 2011). However, when the wave-adjustment time is short, a comparison between the radiative and wave-adjustment timescales may be more appropriate. For typical terrestrial-planet parameters ($N \approx 10^{-2}$ s^{-1}, H = 10 km and taking L to be a typical terrestrial-planet radius of 6000 km) yields $\tau_{adjust} \sim 10^5$ s. This would suggest that on planets with radiative time constants $\lesssim 10^5$ s, the waves are damped, the WTG regime breaks down, and large day-night temperature differences may occur. Earth, Venus, and Titan are safely out of danger, but Mars is transitional, and any exoplanet whose atmosphere is particularly thin and/or hot is also at risk. Nevertheless, subtleties exist. For example, there exists a wide range of wave-adjustment timescales associated with waves of differing wavelengths and phase speeds; moreover, the timescale for waves to propagate vertically out of the troposphere is not generally equivalent to the timescale for them to propagate horizontally across a hemisphere. Although the wave-adjustment process is fundamentally a linear one, nonlinearities may become important at high amplitude as the WTG regime breaks down and the fractional day-night temperature difference becomes large. The horizontal advection time may play a key role in this case. Further work is warranted on the precise conditions for WTG breakdown and the extent to which they can be packaged as timescale comparisons.

For simplicity, we have so far framed the discussion around nonrotating planets; to what extent does the mechanism carry over to rotating planets? Because horizontal Coriolis forces are zero at the equator, the picture has broad relevance for tropical meteorology even on rapidly rotating planets. More specifically, planetary rotation tends to trap tropical wave modes into an equatorial waveguide, whose meridional width is approximately the equatorial Rossby deformation radius, $L_{eq} = (c/\beta)^{1/2}$, where c is a typical gravity wave speed and $\beta = df/dy$ is the derivative of the Coriolis parameter with northward distance; this tends to yield a characteristic waveguide width of order $(NH/\beta)^{1/2}$ in a continuously stratified atmosphere [see *Matsuno* (1966), *Holton* (2004, pp. 394–400, 429–432), or *Andrews et al.* (1987, pp. 200–208) for a discussion of equatorial wave trapping]. Typical values are $L_{eq} \sim 10^3$ km for Earth and Mars. These equatorially trapped modes, including the Kelvin wave, Rossby waves, and mixed Rossby-gravity waves — as well as smaller-scale gravity waves triggered by convection and other processes — can adjust the thermal state in a manner analogous to that described here. Because large-scale, equatorially trapped waves can propagate in longitude and height but not latitude, the adjustment process will occur more efficiently in the zonal than the meridional direction, and it will tend to be confined to within an equatorial deformation radius of the equator. Indeed, the mechanism described here helps to explain why Earth's tropospheric tropical temperatures are nearly zonally uniform, and moreover shows how moist convection can regulate the thermal structure over wide areas of the tropics despite its sporadic occurrence (*Bretherton and Smolarkiewicz,* 1989). On slowly rotating planets — including Venus, Titan, and tidally locked super-Earths, where the deformation radius is comparable to or greater than the planetary radius — this wave-adjustment processes will not be confined to low latitudes but will act globally to mute horizontal temperature contrasts. In large measure, this mechanism is responsible for the small horizontal temperature contrasts observed in the tropospheres of Venus and Titan.

Note that the WTG regime (when it occurs) applies best in the free troposphere, where friction is weak and waves are free to propagate; in the frictional boundary layer near a planet's surface, the scaling will not generally hold and there may exist substantial horizontal temperature gradients. For example, *Joshi et al.* (1997) and *Merlis and Schneider* (2010) present terrestrial exoplanet simulations exhibiting weak temperature gradients in the free troposphere (day-night contrasts $\lesssim 3$ K) but larger day-night temperature contrasts at the surface (reaching ~30–50 K).

2.2.3. Equatorial superrotation. Recent theoretical work suggests that many tidally locked terrestrial exoplanets will exhibit a fast eastward, or superrotating, jet stream at the equator. More specifically, superrotation is defined as atmospheric flow whose angular momentum (per unit mass) about the planet's rotation axis exceeds that of the planetary surface at the equator. (According to this definition, most eastward jets at mid to high latitudes are not superrotating.) In our solar system, the tropical atmospheres of Venus, Titan, Jupiter, and Saturn all superrotate. Even localized layers within Earth's equatorial stratosphere exhibit superrotation, part of the so-called "quasi-biennial oscillation" (*Andrews et al.,* 1987). In contrast, Uranus and Neptune, as well as the tropospheres of Earth and Mars, exhibit mean westward equatorial flow (subrotation). Interestingly, three-dimensional circulation models of synchronously rotating exoplanets, which are subject to a steady, day-night heating pattern, have consistently showed the emergence of such superrotation — both for terrestrial planets (*Joshi et al.,* 1997; *Merlis and Schneider,* 2010; *Heng and Vogt,* 2010; *Edson et al.,* 2011; *Wordsworth et al.,* 2011) and hot Jupiters (*Showman and Guillot,* 2002; *Cooper and Showman,* 2005; *Showman et al.,* 2008, 2009, 2013; *Menou and Rauscher,* 2009; *Rauscher and Menou,* 2010; *Heng et al.,* 2011a,b; *Perna et al.,* 2010, 2012). Figure 15 shows examples from several recent studies.

Equatorial superrotation is interesting for several reasons. First, it influences the atmospheric thermal structure and thus plays an important role in shaping observables. When radiative and advective timescales are similar, the superrotation can cause an eastward displacement of the thermal field that influences IR light curves and spectra (e.g., *Showman and Guillot,* 2002). An eastward displacement of the hottest regions from the substellar point has been observed in light curves of the hot Jupiter HD 189733b (*Knutson et al.,* 2007) and it may also be detectable for synchronously rotating super-Earths with next-generation observatories (*Selsis et al.,* 2011). The thermal structure of the leading and trailing terminators may also differ. Second, superrotation is dynamically interesting; understanding the mechanisms

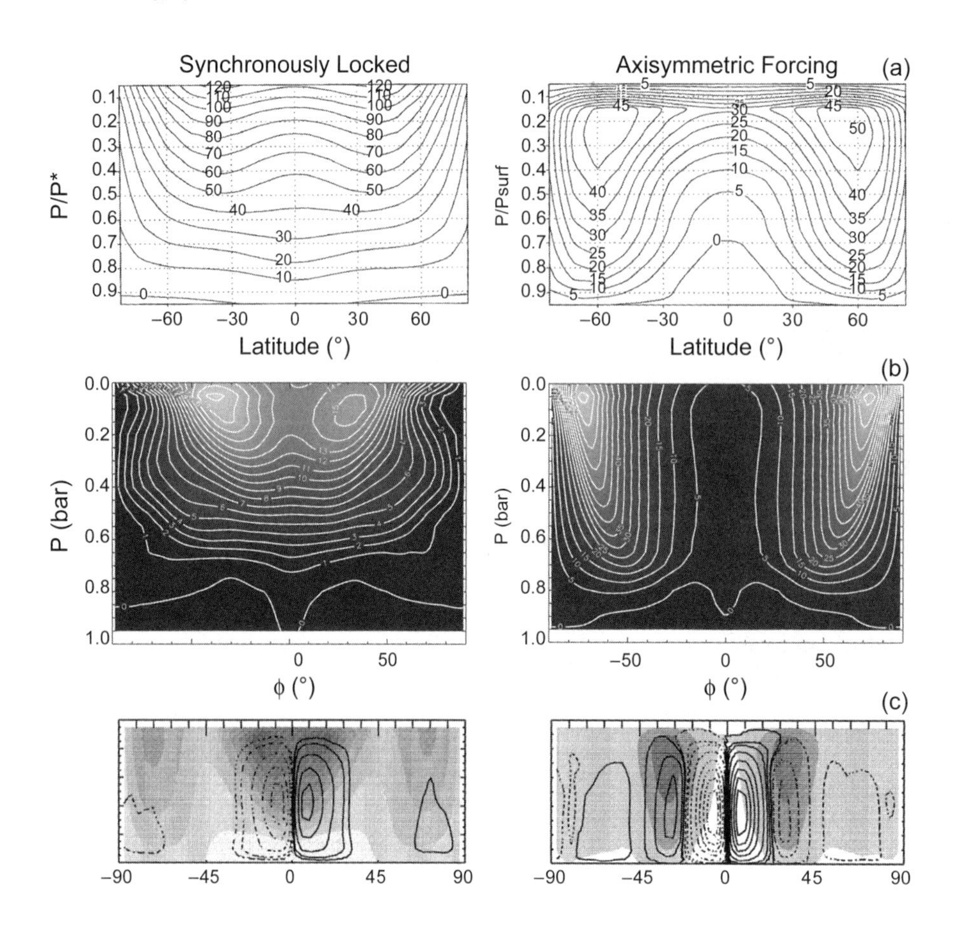

Fig. 15. Zonal-mean zonal winds (contours) vs. latitude (abscissa) and pressure (ordinate) from recent studies illustrating the development of equatorial superrotation in models of synchronously rotating terrestrial exoplanets. In each case, the left column shows a synchronously rotating model with a steady, day-night heating pattern, and the right model shows an otherwise similar control experiment with axisymmetric heating (i.e., no day-night pattern). **(a)** From *Joshi et al.* (1997); rotation period is 16 Earth days. Contours give zonal-mean zonal wind in m s^{-1}. **(b)** *Heng and Vogt* (2010); rotation period is 37 days (left) and 20 days (right). Contours give zonal-mean zonal wind in m s^{-1}. **(c)** *Edson et al.* (2011); rotation period is 1 Earth day. Grayscale gives zonal-mean zonal wind; eastward is shaded, with peak values (in dark shades) reaching ~30 m s^{-1}. Contours give mean-meridional streamfunction. In all the models, the synchronously rotating variants exhibit a strong, broad eastward jet centered at the equator, whereas the axisymmetrically forced variants exhibit weaker eastward or westward flow at the equator, with eastward jets peaking in the mid-to-high latitudes.

that drive superrotation in the exoplanet context may inform our understanding of superrotation within the solar system (and vice versa). The equator is the region farthest from the planet's rotation axis; therefore, a superrotating jet corresponds to a local maximum of angular momentum per mass about the planet's rotation axis. Maintaining a superrotating equatorial jet against friction or other processes therefore requires angular momentum to be transported up-gradient from regions where it is low (outside the jet) to regions where it is high (inside the jet). *Hide* (1969) showed that the necessary up-gradient angular-momentum transport must be accomplished by waves or eddies.

In models of synchronously rotating exoplanets, the defining feature that allows emergence of strong superrotation is the steady day-night heating contrast. Models that include strong dayside heating and nightside cooling — fixed in longitude due to the synchronous rotation — generally exhibit a broad, fast eastward jet centered at the equator. In contrast, otherwise similar models with axisymmetric forcing (i.e., an equator-to-pole heating gradient with no diurnal cycle) exhibit only weak eastward or westward winds at the equator, often accompanied by fast eastward jets in the mid- to high latitudes (Fig. 15).

Although the Earth's troposphere is not superrotating, it does exhibit tropical zonal heating anomalies due to longitudinal variations in the surface type (land vs. ocean), sea-surface temperature, and prevalence of cumulus convection near the equator (*Schumacher et al.,* 2004; *Kraucunas and Hartmann,* 2005; *Norton,* 2006). Qualitatively, these tropical heating/cooling anomalies resemble the daynight heating contrast on a synchronously rotating exoplanet, albeit at higher zonal wavenumber and lower amplitude. Starting in the 1990s, several authors in the terrestrial literature demonstrated using GCMs that, if sufficiently strong, these types of tropical heating anomalies can drive equatorial superrotation (*Suarez and Duffy,* 1992; *Saravanan,* 1993; *Hoskins et al.,* 1999; *Kraucunas and Hartmann,* 2005; *Norton,* 2006). The qualitative similarity in the forcing and the response suggests that the mechanism for superrotation is the same in both the terrestrial and the exoplanet models.

What is the mechanism for the equatorial superrotation occurring in these models? Motivated by arguments analogous to those in section 2.1.4, *Held* (1999b) and *Hoskins et al.* (1999) suggested heuristically that the superrotation results from the poleward propagation of Rossby waves generated at low latitudes by the tropical zonal heating anomalies; subsequent authors have likewise invoked this conceptual framework in qualitative discussions of the topic (e.g., *Tziperman and Farrell,* 2009; *Edson et al.,* 2011; *Arnold et al.,* 2012). At its essence, the hypothesis is attractive, since it is simple, based on the natural relationship of the meridional propagation directions of Rossby waves to the resulting eddy velocity phase tilts, and seemingly links extratropical and tropical dynamics. Nevertheless, the idea remains qualitative, and its relevance to the type of equatorial superrotation seen in exoplanet GCMs has not been demonstrated.

Moreover, challenges exist. First, unlike in the extratropics, large-scale baroclinic equatorial wave modes in the tropics are equatorially trapped, confined to a wave guide whose meridional width is approximately the equatorial Rossby deformation radius. Such waves — including the Kelvin waves, equatorial Rossby waves, and the mixed Rossby-gravity wave — can propagate in longitude and height but not latitude. Unlike the case of the barotropic Rossby waves discussed in section 2.1.3, simple analytic solutions for such waves (see, e.g., *Matsuno,* 1966; *Holton,* 2004; *Andrews et al.,* 1987) exhibit no meridional momentum flux. Second, in the context of slowly rotating exoplanets, these waves exhibit meridional scales typically extending from the equator to the pole (e.g., *Mitchell and Vallis,* 2010), leaving little room for meridional propagation. Given these issues, it is unclear that the paradigm of waves propagating from one latitude to another applies.

It can be shown explicitly that the barotropic theory for the interaction of meridionally propagating Rossby waves with the mean flow — which explains the emergence of eddy-driven jets in the extratropics (section 2.1.4) — fails to explain the equatorial superrotation emerging from exoplanet GCMs (*Showman and Polvani,* 2011). Many GCMs of synchronously rotating exoplanets exhibit circulation patterns — including equatorial superrotation — that, to zeroth order, are mirror symmetric about the equator (e.g., *Showman and Guillot,* 2002; *Cooper and Showman,* 2005; *Showman et al.,* 2008, 2009, 2013; *Heng et al.,* 2011a,b; *Showman and Polvani,* 2010, 2011; *Rauscher and Menou,* 2010, 2012; *Perna et al.,* 2010, 2012). For a flow with such symmetry, the relative vorticity is antisymmetric about, and zero at, the equator — at all longitudes, not just in the zonal mean. Under these conditions, the forcing $\overline{\zeta' F'}$ equals zero in equations (13) and (14). Equation (14) therefore predicts that the zonal-mean zonal wind is zero at the equator — inconsistent with the equatorial superrotation emerging in the GCMs.

Showman and Polvani (2011) offered an alternate theory to overcome these obstacles and show how the equatorial superrotation can emerge from the day-night thermal forcing on synchronously rotating exoplanets. They presented an analytic theory using the 1½-layer shallow-water equations, representing the flow in the upper troposphere; they demonstrated the theory with a range of linear and nonlinear shallow-water calculations and full, three-dimensional GCM experiments. In their theory, the day-night forcing generates standing, planetary-scale Rossby and Kelvin waves, analogous to those described in analytic solutions by *Matsuno* (1966) and *Gill* (1980) (see Fig. 2 for a schematic illustration of these waves in the context of the overall circulation of a synchronously rotating exoplanet). In *Showman and Polvani*'s (2011) model, these baroclinic wave modes are equatorially trapped and, in contrast to the theory described in section 2.1.4, exhibit no meridional propagation. The Kelvin waves, which straddle the equator, exhibit eastward (group) propagation, while the Rossby modes, which lie on the poleward flanks of the Kelvin wave, exhibit westward (group) propagation. Relative to radiative-equilibrium solutions, this latitudinally varying zonal propagation causes an eastward displacement of the thermal and pressure fields at the equator, and a westward displacement of them at higher latitudes (see Fig. 16 for an analytic solution demonstrating this behavior). In turn, these displacements naturally generate an eddy velocity pattern with velocities tilting northwest-southeast in the northern hemisphere and southwest-northeast in the southern hemisphere, which induces an equatorward flux of eddy momentum (Fig. 16) and equatorial superrotation.

It is crucial to note that, in the *Showman and Polvani* (2011) theory, the superrotation results not from wave tilts associated with meridional wave propagation (which cannot occur for these equatorially trapped modes), but rather from the differential zonal propagation of these Kelvin and Rossby modes. Indeed, in full, three-dimensional GCMs exhibiting equatorial superrotation in response to tropical heating anomalies, the tropical eddy response bears striking resemblance to these analytical, "Gill-type" solutions (*Kraucunas and Hartmann,* 2005; *Norton,* 2006; *Caballero and Huber,* 2010; *Showman and Polvani,* 2011; *Arnold et al.,* 2012), providing further evidence that superrotation in these three-dimensional models results from this mechanism. Interestingly, though, *Showman and Polvani* (2011) and *Showman et al.* (2013) showed that, even when thermal damping is sufficiently strong to inhibit significant zonal propagation of these wave modes, the multiway force balance between pressure-gradient forces, Coriolis forces, advection, and (if present) frictional drag can in some cases lead to prograde-equatorward and retrograde-poleward tilts in the eddy velocities, allowing an equatorward momentum convergence and equatorial superrotation. At face value, this latter mechanism does not seem to require appeal to wave

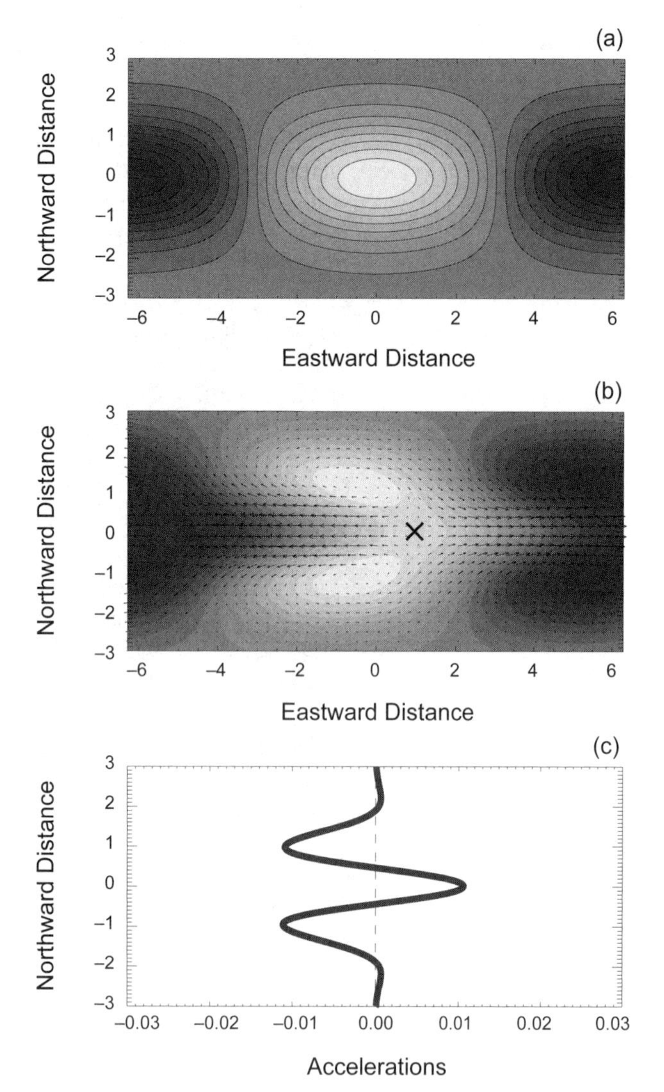

Fig. 16. Analytical solutions of the shallow-water equations from *Showman and Polvani* (2011) showing how day-night thermal forcing on a synchronously rotating exoplanet can induce equatorial superrotation. **(a)** Spatial structure of the radiative-equilibrium height field vs. longitude (abscissa) and latitude (ordinate). **(b)** Height field (shades of gray) and eddy velocities (arrows) for steady, analytic solutions forced by radiative relaxation and linear drag, performed on the equatorial β plane and analogous to the solutions of *Matsuno* (1966) and *Gill* (1980). The equatorial behavior is dominated by a standing, equatorially trapped Kelvin wave and the mid-to-high latitude behavior is dominated by an equatorially trapped Rossby wave. The superposition of these two wave modes leads to eddy velocities tilting northwest-southeast (southwest-northeast) in the northern (southern) hemisphere, as necessary to converge eddy momentum onto the equator and generate equatorial superrotation. **(c)** The zonal accelerations of the zonal-mean wind implied by the linear solution. The acceleration is eastward at the equator, implying that equatorial superrotation will emerge. Note that the equatorward momentum flux that induces superrotation results from differential zonal, rather than meridional, propagation of the wave modes; indeed, these modes are equatorially trapped and exhibit no meridional propagation at all.

propagation at all [see *Showman and Polvani* (2011) and *Showman et al.* (2013) for further discussion].

Nevertheless, real atmospheres contain a wide variety of wave modes, and there remains a need to better test and integrate these various mechanisms. Linear studies show that tropical convection on Earth generates two broad classes of wave mode — baroclinic, equatorially trapped waves that propagate in longitude and height but remained confined near the equator, and barotropic Rossby waves that propagate in longitude and latitude but exhibit no vertical propagation (e.g., *Hoskins and Karoly,* 1981; *Salby and Garcia,* 1987; *Garcia and Salby,* 1987). By focusing on the 1½-layer shallow-water model, *Showman and Polvani*'s (2011) study emphasized the equatorially trapped component and did not include a barotropic (meridionally propagating) mode. It would therefore be worth revisiting this issue with a multilayer model that includes both classes of wave mode; such a model would allow a better test of the *Held* (1999b) hypothesis and allow a determination of the relative importance of the two mechanisms under various scenarios for day-night heating on exoplanets.

Several wave-mean-flow feedbacks exist that may influence the strength and properties of equatorial superrotation on exoplanets. First are the possible feedbacks with the Hadley circulation (*Shell and Held,* 2004) described in section 2.2.1. These feedbacks have yet to be explored in an exoplanet context.

Second, on planets rotating sufficiently rapidly to exhibit baroclinically active extratopical zones, there is a possible feedback involving the effect of mid-latitude eddies on the equatorial flow. In the typical Earth-like (nonsuperrotating) regime, equatorward-propagating Rossby waves generated by mid-latitude baroclinic instabilities can be absorbed on the equatorward flanks of the subtropical jets (section 2.2.1), causing a westward acceleration that helps inhibit a transition to superrotation — even in the presence of tropical forcing like that shown in Fig. 16. On the other hand, if such tropical forcing becomes strong enough to overcome this westward torque, a transition to superrotation nevertheless becomes possible. Once this occurs, the equatorward-propagating Rossby waves no longer encounter critical levels in the tropics, and the tropics become transparent to these waves, eliminating any westward acceleration associated with their absorption. This feedback suggests that, near the transition, modest increases in the tropical wave source can cause sudden and massive changes in the equatorial jet speed. Two-level Earth-like GCMs with imposed tropical eddy forcing confirm this general picture and show that, near the transition, hysteresis can occur (*Suarez and Duffy,* 1992; *Saravanan,* 1993; *Held,* 1999b). Nevertheless, otherwise similar GCM experiments with a more continuous vertical structure suggest that the feedback in practice is not strong (*Kraucunas and Hartmann,* 2005; *Arnold et al.,* 2012), hinting that the two-level models may not properly capture all the relevant dynamics. Additional work is needed to determine the efficacy of this feedback in the exoplanet context.

Third, *Arnold et al.* (2012) pointed out a feedback between the tropical waves (like those in Fig. 16) and the superrotation. In the presence of stationary, zonally asymmetric tropical heating/cooling anomalies, the linear response comprises steady, planetary-scale Rossby waves (cf. Fig. 16), whose thermal extrema are phase shifted to the west of the substellar longitude due to the westward (group) propagation of the waves. In the presence of modest superrotation, the phase shifts are smaller, and the wave amplitudes are larger. A resonance occurs when the (eastward) speed of the superrotation equals the (westward) propagation speed of the Rossby waves, leading to very large Rossby wave amplitude for a given magnitude of thermal forcing. Arnold et al. show that this feedback leads to an increased convergence of eddy momentum onto the equator as the superrotation develops, accelerating the superrotation still further. This positive feedback drives the atmosphere toward the resonance; beyond the resonance, the feedback becomes negative. Arnold et al. demonstrated these dynamics in linear shallow-water calculations and full GCM experiments. Nevertheless, a puzzling aspect of the interpretation is that, for the equatorially trapped wave modes considered by *Showman and Polvani* (2011) and *Arnold et al.* (2012), the ability of the waves to cause equatorward momentum convergence depends not just on the Rossby waves but on both equatorially trapped Rossby and Kelvin waves, whose superposition — given the appropriate zonal phase offset of the two modes — leads to the necessary phase tilts and equatorward momentum fluxes to cause superrotation. The resonance described above would not apply to the Kelvin wave component, since its (group) propagation is eastward. Additional theoretical work may clarify the issue.

Edson et al. (2011) performed GCM simulations of synchronously rotating but otherwise Earth-like exoplanets over a broad range of rotation rates, and they found the existence of a sharp transition in the speed of superrotation as a function of rotation rate. The system exhibited hysteresis, meaning the transition from weak to strong superrotation occurred at a different rotation rate than the transition from strong to weak superrotation. The specific mechanisms remain unclear; nevertheless, the model behavior suggests that positive feedbacks, possibly analogous to those described above, help to control the superrotation in their models. The speed of superrotation likewise differed at differing rotation rates in the GCM simulations of *Merlis and Schneider* (2010), but the transition was a more gradual function of rotation rate, and these authors did not suggest the existence of any hysteresis.

Finally, we note that slowly rotating planets exhibit a natural tendency to develop equatorial superrotation even in the absence of day-night or other imposed eddy forcing. This has been demonstrated in a variety of GCM studies where the imposed thermal forcing does not include a diurnal cycle (i.e., where the radiative equilibrium temperature depends on latitude and pressure but not longitude) (e.g., *Del Genio et al.,* 1993; *Del Genio and Zhou,* 1996; *Yamamoto and Takahashi,* 2003; *Herrnstein and Dowling,* 2007; *Richardson et al.,* 2007; *Lee et al.,* 2007; *Hollingsworth et al.,* 2007; *Mitchell and Vallis,* 2010; *Parish et al.,* 2011; *Lebonnois et al.,* 2012). Superrotation of this type can be seen in the axisymmetrically forced, slowly rotating models from *Kaspi and Showman* (2012) in Fig. 12 (top four panels) and *Joshi et al.* (1997) and *Heng and Vogt* (2010) in Fig. 15. Although the zonal wind at the equator corresponds to a local minimum with respect to latitude, it is eastward, and the equatorial upper troposphere still comprises a local maximum of angular momentum per mass. Thus, even this configuration (unlike a case where the zonal-mean zonal wind at the equator is zero or westward) must be maintained by eddy transport of angular momentum to the equator. Just such eddy-momentum fluxes can be seen in Fig. 12: In the slowly rotating cases exhibiting superrotation, eddy momentum converges onto the equator (opposite to the sign of the low-latitude eddy-momentum fluxes in the faster-rotating cases lacking superrotation). Although the details are subtle, several authors have suggested that the superrotation in this class of model results from equatorward angular momentum transport by a barotropic instability of the subtropical jets (e.g., *Del Genio et al.,* 1993; *Del Genio and Zhou,* 1996; *Mitchell and Vallis,* 2010). This mechanism is relevant to the superrotation on Venus and Titan (although the thermal tides, which represent a day-night forcing that can trigger global-scale wave modes, are relevant to those planets as well).

2.3. Effect of Day-Night Forcing

Many observationally accessible exoplanets will be sufficiently close to their stars to be tidally despun to a synchronous rotation state. For such exoplanets, the slowly rotating, tropical dynamical regime will often go hand-in-hand with a strong day-night thermal forcing arising from the fact that they have permanent daysides and nightsides. A major question is how the atmospheric circulation depends on the strength of this day-night forcing. Will the circulation exhibit an essentially day-night flow pattern? Or, instead, will the circulation respond primarily to the zonal-mean radiative heating/cooling, developing a zonally banded flow in response to the planetary rotation? What is the amplitude of the day-night temperature difference? Is there a danger of the atmosphere freezing out on the nightside?

To pedagogically illustrate the dynamics involved, we present in Fig. 17 four numerical solutions of the shallow-water equations, representing the upper tropospheric flow on a sychronously rotating, terrestrial exoplanet. The shallow-water layer represents the mass above a given isentrope in the mid-troposphere. Since atmospheres are stably stratified, the entropy increases with height, and thus radiative heating (which increases the entropy of air) transports mass upward across isentropes, while radiative cooling (which decreases the entropy of air) transports mass downward across isentropes. Thus, day-night heating in the shallow-water system is parameterized as a mass source/sink that adds mass on the dayside (representing its transport into the upper-tropospheric shallow-water layer from the lower troposphere) and removes it on the nightside (representing its transport from the shallow-water layer into the lower troposphere). Here, we represent this process as a Newtonian relaxation with a characteristic radiative timescale, τ_{rad}, that is a specified free parameter. The models are identical to those in *Showman et al.* (2013) except that the planetary parameters are chosen to be appropriate to a terrestrial exoplanet.

Figure 17 demonstrates that the amplitude of the day-night forcing exerts a major effect on the resulting circulation patterns. The four simulations are identical except for the imposed radiative time constant, which ranges from short in the top panels to long in the bottom panels. When the radiative time constant is extremely short (Fig. 17a), the circulation consists primarily of day-night flow in the upper troposphere, and the thermal structure is close to radiative equilibrium, with a hot dayside and a cold nightside. The radiative damping is so strong that any global-scale waves of the sort shown in Fig. 16 are damped out. This simulation represents a complete breakdown of the WTG regime described in section 2.2.2. Intermediate radiative time constants (Figs 17b,c) lead to large day-night contrasts coexisting with significant dynamical structure, including planetary-scale waves that drive a superrotating equatorial jet. When the radiative time constant is long (Fig. 17d), the circulation exhibits a zonally banded pattern, with little variation of the thermal structure in longitude — despite the day-night nature of the imposed thermal forcing. In this limit, the circulation is firmly in the WTG regime. Interestingly, the day-night forcing is sufficiently weak that superrotation does not occur in this case.

In a planetary atmosphere, the radiative time constant should increase with increasing temperature or decreasing atmospheric mass. Figure 17 therefore suggests that synchronously rotating planets that are particularly hot, or have thin atmospheres, may exhibit day-to-night flow patterns with large day-night temperature differences, while planets that are cooler, or have thicker atmospheres, may exhibit banded flow patterns with smaller day-night temperature differences (e.g., *Showman et al.,* 2013). In the context of hot Jupiters, these arguments suggest that the most strongly irradiated planets should exhibit less efficient day-night heat redistribution (with temperatures closer to radiative equilibrium) than less-irradiated planets, a trend that already

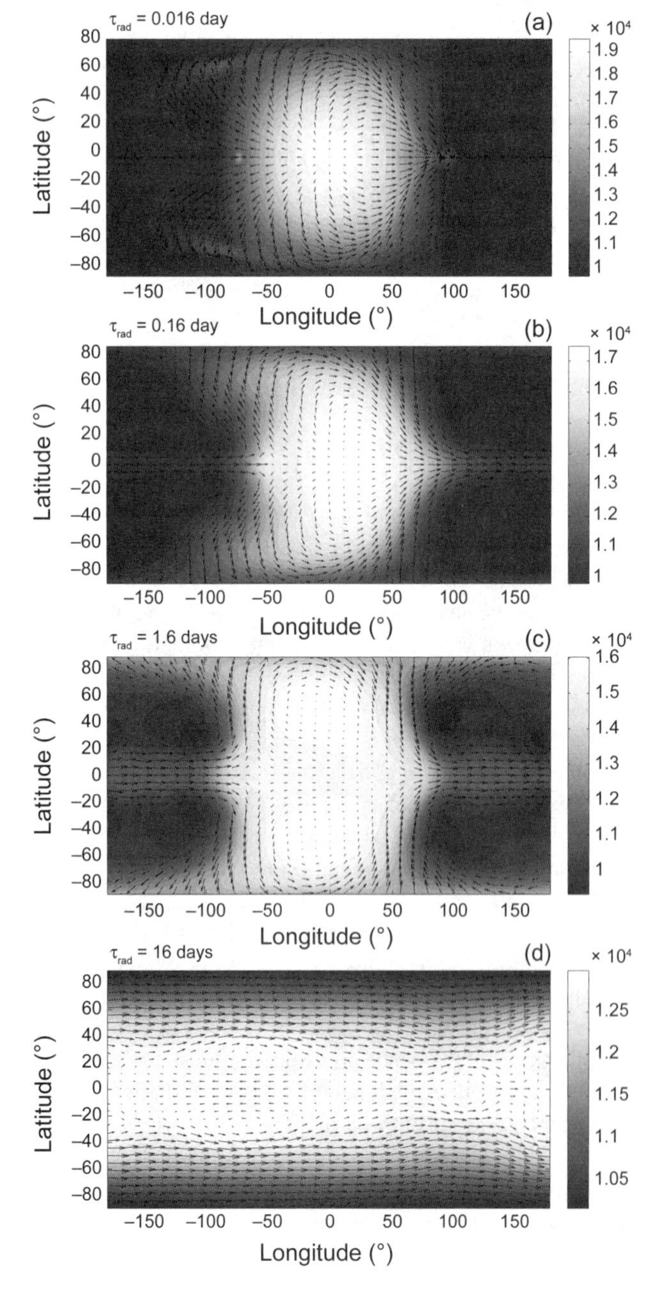

Fig. 17. Layer thickness gh (gray scale, units m^2 s^{-2}) and winds (arrows) for the equilibrated (steady-state) solutions to the shallow-water equations in full spherical geometry for sychronously rotating terrestrial exoplanets driven by a day-night thermal forcing. Specifically, to drive the flow, a mass source/sink term $(h_{eq}-h)/\tau_{rad}$ is included in the continuity equation, where h_{eq} is a specified radiative-equilibrium layer thickness, h is the actual layer thickness (depending on longitude λ, latitude ϕ, and time), and τ_{rad} is a specified radiative time constant. Here h_{eq} equals a constant, H, on the nightside, and equals $H + \cos\lambda\cos\phi$ on the dayside. The substellar point is at longitude, latitude (0°, 0°) (for further details about the model, see *Showman et al.,* 2013). Here, the planet radius is that of Earth, rotation period is 3.5 Earth days, and gH = 10^4 m^2 s^{-2}. These values imply an equatorial deformation radius, $(\sqrt{gH}/\beta)^{1/2}$, which is 60% of the planetary radius. The four models are identical in all ways except the radiative time constant, which is varied from 0.016 to 16 Earth days from top to bottom, respectively. **(a)** When the radiative time constant is short, the flow exhibits a predominantly day-night circulation pattern with a large day-night contrast close to radiative equilibrium. **(b),(c)** Intermediate radiative time constants still exhibit considerable day-night contrasts, yet allow the emergence of significant zonal flows. **(d)** Long radiative time constants lead to a zonally banded flow pattern with little variation of thermal structure in longitude — despite the day-night forcing.

appears to be emerging in secondary-eclipse observations (*Cowan and Agol,* 2011).

In the context of terrestrial planets, the trend in Fig. 17 could likewise describe a transition in circulation regime occurring as a function of incident stellar flux, but it might also describe a transition as a function of atmospheric mass for a given stellar flux. Because of its thin atmosphere, Mars, for example, has a radiative time constant considerably shorter than that on Earth, which helps to explain the much larger day-night temperature differences on Mars compared to Earth. Even more extreme is Jupiter's moon Io, due to its particularly tenuous atmosphere (surface pressure approximately 1 nbar). Sulfur dioxide is liberated by sublimation on the dayside but collapses into surface frost on the nightside, and the resulting day-night pressure differences are thought to drive supersonic flows from day to night (*Ingersoll et al.,* 1985; *Ingersoll,* 1989). (Neither Mars nor Io is synchronously rotating with respect to the Sun, but one can still obtain large day-night temperature differences in such a case if the radiative time constant is not long compared to the solar day.) In the exoplanet context, close-in rocky planets subject to extreme stellar irradiation may lie in a similar regime, with tenuous atmospheres maintained by sublimating silicate rock on the dayside, fast day-night airflows, and condensation of this atmospheric material on the nightside (*Léger et al.,* 2011; *Castan and Menou,* 2011). Prominent examples include CoRoT-7b, Kepler-10b, and 55 Cnc-e. Even terrestrial exoplanets with temperatures similar to those of Earth or Mars could experience sufficiently large day-night temperature differences to cause freeze-out of the atmosphere on the nightside if the atmosphere is particularly thin (*Joshi et al.,* 1997), an issue we return to in section 4.

3. HYDROLOGICAL CYCLE

The hydrological cycle — broadly defined, where the condensate may be water or other chemical species — plays a fundamental role in determining the surface climate of a planet. It directly influences the surface climate by setting the spatial and temporal distributions of precipitation and evaporation and indirectly influences the surface temperature by affecting the horizontal energy transport, vertical convective fluxes, atmospheric lapse rate, and planetary albedo; these are key aspects of planetary habitability. The hydrological cycle is also fundamental in determining the radiative balance of the planet through the humidity and cloud distributions. The planetary radiative balance both affects the surface climate and is what is remotely observed.

Here, we survey the role of the hydrological cycle in affecting the atmospheric circulation and climate of terrestrial planets. We begin with a brief discussion of the role of the ocean (section 3.1), as this is the ultimate source of atmospheric moisture and can influence atmospheric circulation in a variety of ways. We next discuss the thermodynamics of phase change (section 3.2), which has significant implications for how a hydrological cycle affects the circulation. Global precipitation is discussed next (section 3.3), with an emphasis on energetic constraints that are independent of details of the atmospheric circulation. The following section (3.4) highlights the important role of the atmospheric circulation for regional precipitation, in contrast to global precipitation. We then follow with a survey of how the atmospheric circulation helps to control the distribution of humidity (section 3.5) and clouds (section 3.6). Finally, we summarize how the hydrological cycle in turn affects the structure of the atmospheric circulation (section 3.7).

3.1. Oceans

The presence of an ocean affects planetary climate and atmospheric circulations in several key ways. We begin the discussion of the hydrological cycle with the ocean in recognition that it is the source of water vapor in an atmosphere, and the influences of the hydrological cycle on climate (presented in what follows) may be modulated by differences in water availability on other planets. Oceanography is, of course, a vast field of research, and we will only briefly discuss aspects that may be of immediate interest in the exoplanet context. For overviews of physical oceanography and the ocean's effect on climate in the terrestrial context, see, e.g., *Peixoto and Oort* (1992), *Siedler et al.* (2001), *Pedlosky* (2004), *Vallis* (2006, 2011), *Marshall and Plumb* (2007), *Olbers et al.* (2011), and *Williams and Follows* (2011), among others.

Oceans on exoplanets will span a wide range. At one extreme are planets lacking surface condensate reservoirs (Venus) or with liquid surface reservoirs small enough to form only disconnected lakes (Titan). In such a case, the surface condensate should not contribute significantly to the meridional or day-night heat transport, but could still significantly impact the climate via energy exchanges and evaporation into the atmosphere. At the other extreme are super-Earths whose densities are sufficiently low to require significant fractions of volatile materials in their interiors. Prominent examples include GJ 1214b (*Charbonneau et al.,* 2009; *Rogers and Seager,* 2010; *Nettelmann et al.,* 2011) and several planets in the Kepler-11 system (*Lissauer et al.,* 2011; *Lopez et al.,* 2012); many of the hundreds of super-Earths discovered by Kepler also likely fall into this category (e.g., *Rogers et al.,* 2011). Many such planets will contain water-rich fluid envelopes thousands of kilometers thick. If their interior temperature profiles are sufficiently hot (in particular, if, at the pressure of the critical point, the atmospheric temperature exceeds the temperature of the critical point), then these planets will exhibit a continuous transition from a supercritical fluid in the interior to a gas in the atmosphere, with no phase boundary; on the other hand, if their interior temperature profiles are sufficiently cold, then their water-rich envelopes will be capped by an oceanic interface, overlain by an atmosphere (*Léger et al.,* 2004; *Wiktorowicz and Ingersoll,* 2007). [For a pure water system, the critical point pressure and temperature are 221 bar and 647 K respectively, but they depend on composition for a system containing other components such as hydrogen (e.g., *Wiktorowicz and Ingersoll,* 2007).] Earth lies at an intermediate point on this continuum, with an ocean that is a small fraction of the planet's mass and radius, yet is continuously interconnected and thick enough to cover most of the surface area.

Our understanding of ocean dynamics and ocean-atmosphere interactions stem primarily from Earth's ocean, so to provide context we first briefly overview the ocean structure. Earth's oceans have a mean thickness of 3.7 km and a total mass ~260 times that of the atmosphere. At the surface, the ocean is warm at the equator (~27°C in an annual and longitudinal average) and cold at the poles (~0°C). The deep ocean temperature is relatively uniform at a temperature of ~1°–2°C thoughout the world — even in the tropics. The transition from warmer surface water to cooler deep water is called the thermocline and occurs at ~0.5 km depth, depending on latitude. Because low-density, warmer water generally lies atop high-density, cooler water, the ocean is stably stratified and does not convect except at a few localized regions near the poles. However, turbulence caused by wind and waves homogenizes the top ~10–100 m (depending on weather conditions and latitude), leading to profiles of density, salinity, and composition that vary little across this so-called "mixed layer." Because of its efficient communication with the atmosphere, the thickness of the mixed layer exerts a strong effect on climate (e.g., on the extent to which the ocean modulates seasonal cycles). Nevertheless, the deeper ocean also interacts with the atmosphere on a wide range of timescales up to thousands of years.

Earth's oceans affect the climate in numerous ways. For discussion, it is useful to decompose the role of the ocean into time-dependent (e.g., modification of heat capacity) and time-independent (e.g., time-mean energy transports) categories.

Consider the time-dependent energy budget of the ocean mixed layer

$$\alpha \frac{\partial T_{surf}}{\partial t} = R_{sfc} - LE - SH - \nabla \cdot F_o \tag{23}$$

where α is the heat capacity per unit area of the mixed layer (approximately equal to $\rho c_p h$, where ρ is the density, c_p is the specific heat, and h is the thickness of the mixed layer), net surface radiative flux R_{sfc}, latent enthalpy flux LE, sensible enthalpy flux SH, and ocean energy flux divergence $\nabla \cdot F_o$, including vertical advection into the base of the mixed layer. Earth's mixed layer depth has geographic and seasonal structure because it is connected both to the interior ocean's thermal structure and currents from below and is forced by the atmosphere and radiation from above.

From this budget, a clear time-dependent effect of the ocean is that it acts as a thermal surface reservoir, which reduces the seasonal and diurnal cycle amplitude compared to possible land-covered planets [α for the ocean is substantially larger than for land (*Hartmann,* 1994a; *Pierrehumbert,* 2010)]. In addition to the direct modification of the heat capacity, the presence of an ocean leads to weaker temperature fluctuations through evaporation, which is a more sensitive function of temperature than surface longwave radiation or sensible surface fluxes (e.g., *Pierrehumbert,* 2010). Oceans further influence a planet's thermal evolution on long timescales through the storage of heat in the interior ocean (i.e., the vertical component of the ocean energy flux divergence in equation (23)) and through their role as a chemical reservoir (*Sarmiento and Gruber,* 2006); the ocean storage and release of carbon is important for Earth's glacial cycles, for example.

In addition to the effect of the ocean in determining a planet's time-dependent evolution, oceans modify a planet's time-mean climate. Evaporation of water from the ocean is the essential moisture source to the atmosphere and to land regions with net precipitation. Oceans affect the radiation balance of a planet directly through their albedo and indirectly through providing the moisture source for the atmospheric hydrological cycle. Finally, ocean energy transport is important in determining the surface temperature.

Many authors have examined the sensitivity of ocean energy transport to changes in surface wind stress and surface buoyancy gradients, often decomposing the energy transport into components associated with deep meridional overturning circulations and shallow, wind-driven circulations (cf. *Ferrari and Ferreira,* 2011). Here, we note that ocean energy transport can affect surface climate differently than atmospheric energy transport. For example, *Enderton and Marshall* (2009) presented simulations with different simplified ocean basin geometries; the ocean energy transport, surface temperature, and sea ice of the simulated equilibrium climates differed, but the total energy transport was relatively unchanged [i.e., the changes in atmospheric energy transport largely offset those of the ocean; changes in atmospheric and oceanic energy transport need not compensate in general (e.g., *Vallis and Farneti,* 2009)]. An implication of these results is that while observations of a planet's top-of-atmosphere radiation allow the total (ocean and atmosphere) energy transport to be determined, this alone is not sufficient to constrain the surface temperature gradient.

3.2. Thermodynamics of Phase Change

Before describing the ways the hydrological cycle can interact with the atmospheric dynamics (which we do in subsequent subsections), we first summarize the pure thermodynamics of a system exhibiting phase changes. Most atmospheres in the solar system have contituents that can condense: water on Earth, CO_2 on Mars, methane on Titan, N_2 on Triton and Pluto, and several species, including H_2O, NH_3, and H_2S, on Jupiter, Saturn, Uranus, and Neptune. Because atmospheric air parcels change temperature and pressure over time (due both to atmospheric motion and to day-night or seasonal temperature swings), air parcels make large excursions across the phase diagram, and often strike one or more phase boundaries — leading to condensation. This condensation can exert major influence over the circulation.

The starting point for understanding these phase changes is the Clausius-Clapeyron equation, which relates changes in the saturation vapor pressure of a condensable vapor, e_s, to the latent heat of vaporization L, temperature T, density of vapor ρ_{vap}, and density of condensate ρ_{cond}

$$\frac{de_s}{dT} = \frac{1}{T}\frac{L}{\rho_{vap}^{-1} - \rho_{cond}^{-1}} \tag{24}$$

For condensate-vapor interactions far from the critical point, the condensate density greatly exceeds the vapor density and can be ignored in equation (24). Adopting the ideal-gas law, the equation can then be expressed

$$\frac{1}{e_s}\frac{de_s}{dT} = \frac{L}{R_v T^2} \tag{25}$$

where R_v is the gas constant of the condensable vapor.

The saturation vapor pressure is the pressure of vapor if the air were in equilibrium with a saturated surface; it increases with temperature and is a purely thermodynamic quantity. In contrast, the vapor pressure of an atmosphere is not solely a thermodynamic quantity, as atmospheres are generally subsaturated. Therefore, the vapor pressure depends on the relative humidity $\mathcal{H} = e/e_s$, which in turn depends on the atmospheric circulation (discussed in what follows). Condensation occurs when air becomes supersaturated, although there are also microphysical constraints such as the availability of condensation nuclei. A consequence of equation (25) is that atmospheric humidity will increase with temperature, if relative humidity is unchanged.

The vertical structure of water vapor or other condensible species can be quantified by comparing it to that of pressure. The pressure scale height H_p for hydrostatic atmospheres is given by

$$\frac{1}{H_p} = \frac{\partial \ln p}{\partial z} = \frac{g}{RT} \tag{26}$$

For Earth, g ~ 10 m s^{-2}, T ~ 280 K, R ~ 300 J kg^{-1} K^{-1}, and the pressure scale height is $H_p \approx 8$ km. The pressure scale height is a decreasing function of planet mass (through g) and an increasing function of planet temperature.

Because the saturation vapor pressure of condensable species depend strongly on temperature, it is possible for the scale height of water — and other condensables — to differ greatly from the pressure scale height. The scale height of water vapor H_w can be estimated as

$$\frac{1}{H_w} = \frac{\partial \ln e_s}{\partial z} = \frac{\partial \ln e_s}{\partial T}\frac{\partial T}{\partial z} = \frac{L}{R_v T^2}\frac{\partial T}{\partial z} \tag{27}$$

where the chain rule and Clausius-Clapeyron relationship (equation (25)) have been used and vertical variations in relative humidity have been neglected. The scale height of water vapor is the inverse of the product of the lapse rate $\Gamma = -\partial T/\partial z$ and the fractional change of vapor pressure with temperature (i.e., the Clausius-Clapeyron relationship). The scale height of water increases with temperature [through equation (25)] and with decreasing lapse rate (i.e., less rapid vertical temperature decrease).

Using Earth-like values in equations (26) and (27) of g ~ 10 m s^{-2}, T ~ 280 K, R ~ 300 J kg^{-1} K^{-1}, Rv ~ 450 J kg^{-1} K^{-1}, L ~ 2.5 × 10^6 J kg^{-1}, the ratio of the water vapor and pressure scale heights H_w/H_p is ~0.2 for dry adiabatic lapse rate ($\Gamma = g/c_p$). Thus, for Earth-like situations, the scale height of water vapor is significantly less than the pressure scale height. This implies that near-surface water vapor dominates the column water vapor (the mass-weighted vertical integral of water vapor from the surface to the top of atmosphere). Moreover, the saturation vapor pressure at the tropical tropopause is ~10^4 times less than at the surface. As a result, the middle and upper atmosphere (stratosphere and above) are very dry. Still, that the bulk of the atmospheric water vapor resides near the surface does not mean that the upper tropospheric humidity is unimportant: Small concentrations there can be important in affecting the radiation balance as water vapor is a greenhouse gas (e.g., *Held and Soden,* 2000); moreover, upper-tropospheric water controls the formation and distribution of cirrus clouds, which is important in the atmospheric radiation balance.

The condensable gas will be a larger fraction of the pressure scale height if the latent heat is smaller or if the tropospheric stratification is greater. Titan provides an example in our solar system; the methane abundance at the tropopause is only a factor of ~3 smaller than near the surface (*Niemann et al.,* 2005). In such a situation, the condensable vapor abundance in the stratosphere may be considerable, which can have implications for planetary evolution because it will affect the rate at which such species (whether water or methane) can be irreversibly lost via photolytic breakup and escape of hydrogen to space.

3.3. Global Precipitation

A planet's precipitation is a fundamental part of its hydrological cycle and has important implications for global climate feedbacks such as the carbonate-silicate feedback cycle (section 4). At equilibrium, a planet's precipitation equals its evaporation in the global mean. Global precipitation is controlled by energy balance requirements, either of the surface (to constrain evaporation) or of the atmospheric column (to constrain the latent heating from precipitation). More complete discussions and reviews of the control of global precipitation can be found in *Held* (2000), *Pierrehumbert* (2002), *Allen and Ingram* (2002), *Schneider et al.* (2010), and *O'Gorman et al.* (2011).

The dry static energy s = c_pT + gz is the sum of the specific enthalpy and potential energy. The dry static energy is materially conserved in hydrostatic atmospheres in the absence of diabatic processes (e.g., radiation or latent heat release), if the kinetic energy of the atmosphere is neglected (cf. *Betts,* 1974; *Peixoto and Oort,* 1992). The dry static energy budget is

$$\frac{\partial s}{\partial t} + \mathbf{v} \bullet \nabla s = Q_r + Q_c + SH \quad (28)$$

where Q_r is the radiative component of the diabatic tendency, Q_c is the latent-heat release component of the diabatic tendency, and SH is the sensible component of the surface enthalpy flux. Taking the time- and global-mean of the dry static energy budget eliminates the lefthand side of equation (28), so that when it is integrated vertically over the atmospheric column, the global-mean precipitation is a function of the radiation and sensible surface flux

$$L\langle P\rangle = \langle R_{atm}\rangle + \langle SH\rangle = \langle R_{TOA}\rangle - \langle R_{sfc}\rangle + \langle SH\rangle \quad (29)$$

with latent heat of vaporization L, precipitation P, net atmospheric radiation R_{atm}, net top-of-atmosphere (TOA) radiation R_{TOA}, and net surface radiation R_{sfc}. The operator $\langle\bullet\rangle$ denotes the time- and global-mean. The radiation terms are evaluated only at the surface and top-of-atmosphere because of the vertical integration over the atmospheric column, and they can be further decomposed into components associated with the shortwave and longwave parts of the radiation spectrum.

The time- and global-mean of the surface energy budget (equation (23)) is

$$L\langle E\rangle = \langle R_{sfc}\rangle - \langle SH\rangle \quad (30)$$

with evaporation E and other variables defined as in equation (29). In both budgets, the sensible turbulent surface flux (the dry component of the surface enthalpy flux) enters, so the global precipitation is not solely a function of the radiative properties of the atmosphere. The turbulent surface fluxes are represented by bulk aerodynamic formulas in GCMs (*Garratt,* 1994) [e.g., $SH = c_p \rho_a C_D \|\mathbf{v}_{surf}\|(T_{surf} - T_a)$], which depend on the temperature difference between the surface and surface air (T_s–T_a), surface wind speed $\|\mathbf{v}_{surf}\|$, and surface roughness (through c_d).

From these budgets, one expects global precipitation to increase if the solar constant increases (Fig. 18b), the top-of-atmosphere or surface albedo decreases, or sensible surface fluxes decrease. Reductions in longwave radiation at the surface that are greater than those at the top-of-atmosphere will increase the global precipitation (Fig. 18a), as is expected for increased greenhouse gas concentrations on Earth. This illustrates that the vertical structure of radiative changes is important. Another example is that changes in top-of-atmosphere albedo and surface albedo differ in their effect on global mean precipitation because they differ in their effect on shortwave absorption by the atmosphere [a component of R_{atm} in equation (4)]. Thus, it is not possible to infer global-mean precipitation from measurements of the radiances at the top-of-atmosphere (R_{TOA}), because the sensible surface flux SH and net surface radiation R_{sfc} matter.

The surface energy budget suggests the existence of an approximate limit to the magnitude increases in global-mean precipitation due to increased greenhouse gas concentration. Under greenhouse gas warming, the air-sea temperature difference at the surface generally decreases. In the limit of no air-sea temperature difference, the net surface longwave radiation and sensible surface fluxes vanish, which leaves the cooling from latent surface fluxes to balance the warming from net solar radiation at the surface. This limit is approximately attained in idealized GCM simulations (Fig. 18a) (*O'Gorman and Schneider,* 2008). However, sensible surface fluxes may reverse sign (from cooling the surface to warming it), so the warmest simulation in Fig. 18a exceeds the precipitation expected from all the surface solar radiation being being converted into evaporation. The sensitivity of precipitation to greenhouse gas changes has implications for the rate of silicate weathering (see section 4) (*Pierrehumbert,* 2002; *Le Hir et al.,* 2009).

The energetic perspectives on global precipitation illustrate how the rate of change of global precipitation and atmospheric humidity can differ. If the relative humidity does not change, the atmospheric humidity increases with temperature at the rate given by the Clausius-Clapeyron relation (equation (25)). In contrast, the global precipitation depends on radiative fluxes such as the top-of-atmosphere insolation and longwave radiation. The radiative fluxes depend in part on the atmospheric humidity through water vapor's absorption of shortwave and longwave radiation (which gives rise to the water vapor feedback), but this dependence does not constrain the radiative changes to vary with the water vapor concentration at the rate given by equation (25). Precipitation and humidity have different dimension as well: Precipitation and evaporation are fluxes of water, whereas the atmospheric humidity is a concentration of water. The suggestion that precipitation changes in proportion to atmospheric humidity has been termed the "saturation fallacy" (*Pierrehumbert et al.,* 2007).

Another result that can be understood by considering the energy budget is that there is not a unique relationship between the global mean temperature and precipitation. For example, in geoengineering schemes that offset warming due to increased CO_2 by increasing albedo (reducing the solar radiation), two climates with the same global mean temperature may have differing global mean precipitation because the sensitivity of global precipitation per degree temperature change are not equal for different types of radiation changes (*Bala et al.,* 2008; *O'Gorman et al.,* 2011). Figure 18 shows that warming forced by solar constant changes results in a larger increase in global-mean precipitation than warming forced by changes in the longwave optical depth in an idealized GCM (*O'Gorman et al.,* 2011).

While the global mean precipitation is potentially a variable of interest in characterizing a planet's habitability, regional precipitation variations are substantial. Earth has regions of dry deserts and rainy tropical regions, where the annual-mean precipitation differs by a factor of ~20. The regional precipitation is far from the global mean because of water vapor transports by the atmospheric circulation.

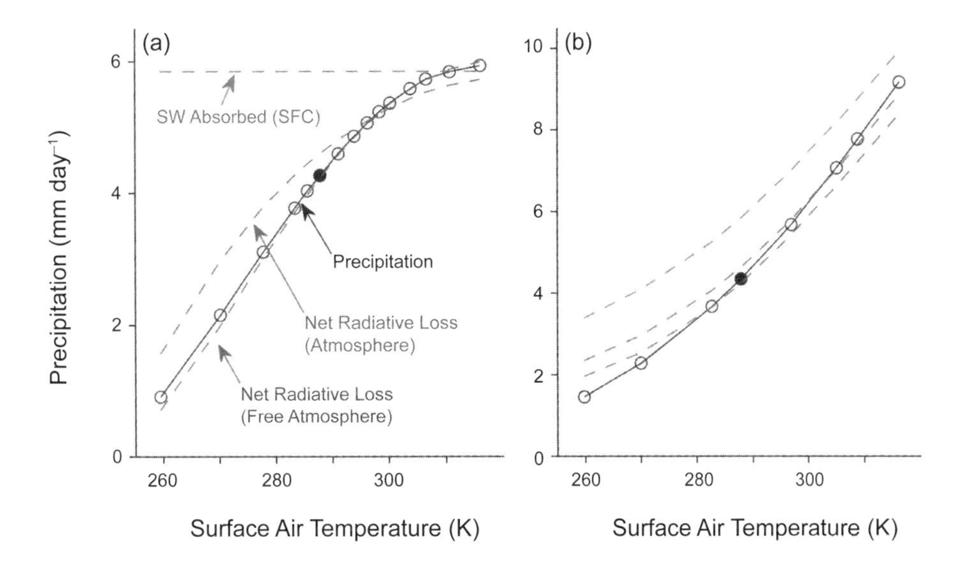

Fig. 18. See Plate 50 for color version. Global-mean precipitation (solid line with circles) vs. global-mean surface air temperature in two series of statistical-equilibrium simulations with an idealized GCM in which **(a)** the optical depth of the longwave absorber is varied and **(b)** the solar constant is varied (from about 800 W m^{-2} to about 2300 W m^{-2}). The filled circles indicate the reference simulation (common to both series) that has the climate most similar to present-day Earth's. The red dashed lines show the net radiative loss of the atmosphere, the blue dashed lines show the net radiative loss of the free atmosphere (above $\sigma = p/p_s = 0.86$), and the green dashed lines show the net absorbed solar radiation at the surface (all in equivalent precipitation units of millimeters per day). From *O'Gorman et al.* (2011).

Furthermore, the sensitivity of regional precipitation to external perturbations can be of opposite sign as the global mean changes [for example, some regions may dry as the global mean precipitation increases (*Chou and Neelin,* 2004; *Held and Soden,* 2006)].

3.4. Regional Precipitation

To understand the variations in regional precipitation, we first illustrate the processes that give rise to condensation (Fig. 19). In tropical latitudes of Earth-like planets, there are time-mean regions of ascent where moist air is continually brought from the surface to the colder upper troposphere, which leads to supersaturation and condensation (near the equator in Fig. 19). There are time-mean regions of subsidence where dry air from the upper troposphere descends toward the surface, so the mean circulation leads to subsaturation and precipitation occurs during transient events, when it does occur (near 30° in Fig. 19, corresponding to the descending branches of the Hadley cells). In the extratropical regions of Earth-like planets, air parcels approximately follow surfaces of constant entropy that slope upward and poleward. As parcels ascend in the warm, poleward-moving branches of extratropical cyclones (i.e., transient baroclinic eddies like those described in section 2.1.2), they become supersaturated and condense in the colder regions of the atmosphere. This is illustrated in Fig. 19, where gray lines indicate entropy surfaces and black arrows indicate the trajectories of air masses. This description is focused on mean meridional circulations in the tropics and transient eddies in the extratropics. Superimposed on these processes affecting the zonal-mean precipitation are stationary eddies (i.e., eddy structures that are fixed rather than traveling in longitude) or zonally asymmetric mean flows that give rise to important longitudinal variations in precipitation on Earth.

While the zonal mean is an excellent starting point for Earth's climate and atmospheric circulations, stationary eddies or zonally asymmetric mean flows are of central importance for tidally locked planets. The mean circulation aspect of the description will have an east-west component: The near-surface branch of the circulation will flow from the nightside to the dayside, ascend and precipitate there, and return to the nightside aloft (*Merlis and Schneider,* 2010). One can rotate the time-mean circulation in Fig. 19 from a north-south orientation (Hadley circulation) to an east-west orientation (Walker circulation) to conceptualize the effect of the nightside to dayside circulations on precipitation; however, the factors controlling the circulations themselves may differ as their angular momentum balances differ (the extent to which the circulation strength is slaved to the momentum balance or responds directly to radiation). In contrast, transient eddies (that is, eddies that travel in longitude) exist because temperature gradients are not aligned with the planet's spin axis, which gives rise to baroclinic instability (section 2.1.2), so the conceptual picture cannot simply be rotated. The affect of zonal asymmetry of background state on baroclinic eddies is not settled; in particular, the extent to which stationary eddies and baroclinic eddies are separable (i.e., the extent to which they are linear) is a research area (*Held et al.,* 2002; *Kaspi and Schneider,* 2011).

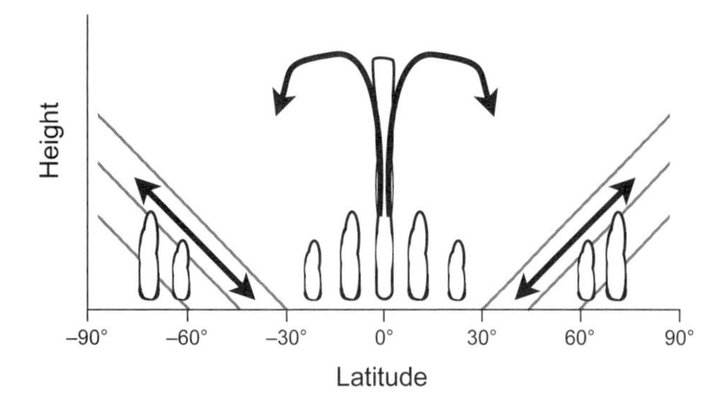

Fig. 19. Processes leading to condensation and affecting atmospheric humidity for Earth-like planets, based on Fig. 8 of *Held and Soden* (2000). Gray lines indicate surfaces of constant entropy that air parcels approximately follow. In the tropics, the ascending branch of the Hadley cell leads to significant condensation and rainfall, whereas the descending branches (at latitudes of ~20°–30° on Earth) exhibit less condensation and fewer clouds. In the extratropics, the rising motion in baroclinic eddies transporting energy poleward and upward (indicated schematically with sloping lines) leads to significant condensation.

Simulations of Earth-like tidally locked exoplanets from *Merlis and Schneider* (2010) illustrate the importance of zonally asymmetric (east-west) mean atmospheric circulations in determining precipitation (Fig. 20). For both slowly and rapidly rotating simulations, the surface zonal wind converges [$\partial_x[u] < 0$, with the [•] indicating a time-mean] on the dayside near the subsolar point, ascends, and diverges ($\partial_x[u] > 0$) in the upper troposphere. The nightside features convergent mean zonal winds in the upper troposphere, subsidence, and divergent mean zonal winds near the surface. These are thermally direct circulations that transport energetic air from the dayside to the nightside. The water vapor is converging near the surface on the dayside where there is substantial precipitation. There is a clear connection between the pattern of convergence in the surface winds and precipitation — the rapidly rotating case, in particular, has a crescent-shaped precipitation field. The shape of the convergence zone is similar to that of the Gill model (*Gill,* 1980; *Merlis and Schneider,* 2010).

To be quantitative about regional precipitation, consideration of the water vapor budget, a conservation equation for the atmospheric humidity, can be useful. For timescales long enough that the humidity tendency is small, the timemean (denoted [•]) net precipitation, [P–E], is balanced by the mass-weighted vertical integral (denoted {•}) of the convergence of the water vapor flux, $-\nabla \cdot [\mathbf{u}q]$

$$[P-E] = -\int_0^{p_s} \nabla \cdot [\mathbf{u}q]\frac{dp}{g} = -\nabla \cdot \{[\mathbf{u}q]\} \qquad (31)$$

with horizontal wind vector **u** and specific humidity q. The atmospheric circulation converges water vapor into regions of net precipitation (P > E) from regions of net evaporation (E > P), where there is a divergence of water vapor. The water vapor budget explicitly shows that the atmospheric circulation is fundamental in determining this basic aspect of climate.

The small scale height of water vapor in Earth-like atmospheres emphasizes the near-surface water vapor concentration, so a conceptually useful approximation to equation (31) is to consider the near-surface region of the atmosphere. For example, the precipitation distribution follows the pattern of the convergence of the surface wind in Fig. 20.

Note that the surface temperature does not directly appear in equation (31). In spite of this, there have been many attempts to relate precipitation to surface temperature for Earth's climate. The surface temperature may vary together with aspects of climate that do directly determine the amount of precipitation (e.g., high near-surface humidity or the largest magnitude near-surface convergence of the horizontal winds may occur in the warmest regions of the tropics). There are conceptual models that relate the surface temperature to pressure gradients that determine regions of convergence through the momentum balance (*Lindzen and Nigam,* 1987; *Back and Bretherton,* 2009a). Likewise, parcel stability considerations (e.g., where is there more convective available potential energy) may be approximately related to regions of anomalously high surface temperature (*Sobel,* 2007; *Back and Bretherton,* 2009b).

3.5. Relative Humidity

Reviews by *Pierrehumbert et al.* (2007) and *Sherwood et al.* (2010b) discuss the processes affecting Earth's relative humidity and its sensitivity to climate changes.

The atmosphere of a planet is generally subsaturated — the relative humidity is less than 100%. The relative humidity of the atmosphere depends on the combined effect of inhomogeneities in temperature and the atmospheric circulation.

For Earth-like planets, the atmospheric circulation produces dry air by advecting air upward and/or poleward to cold temperatures where the air becomes supersaturated and water vapor condenses (Fig. 19). When this air is advected downward or equatorward, the water vapor concentration will be subsaturated with respect to the local temperature.

The nonlocal influence of the temperature field on the atmospheric humidity through the atmospheric circulation can be formalized by the tracer of last saturation paradigm (*Galewsky et al.,* 2005; *Pierrehumbert et al.,* 2007). In this framework, the humidity of a given region in the atmosphere is decomposed into the saturation humidity at the nonlocal (colder) region of last saturation, with weights given by the probability that last saturation occurred there.

In addition to the adiabatic advection, convection affects the relative humidity and can act to moisten or dry the atmosphere (e.g., *Emanuel,* 1994). In subsiding regions of Earth's tropics, convection moistens the free troposphere

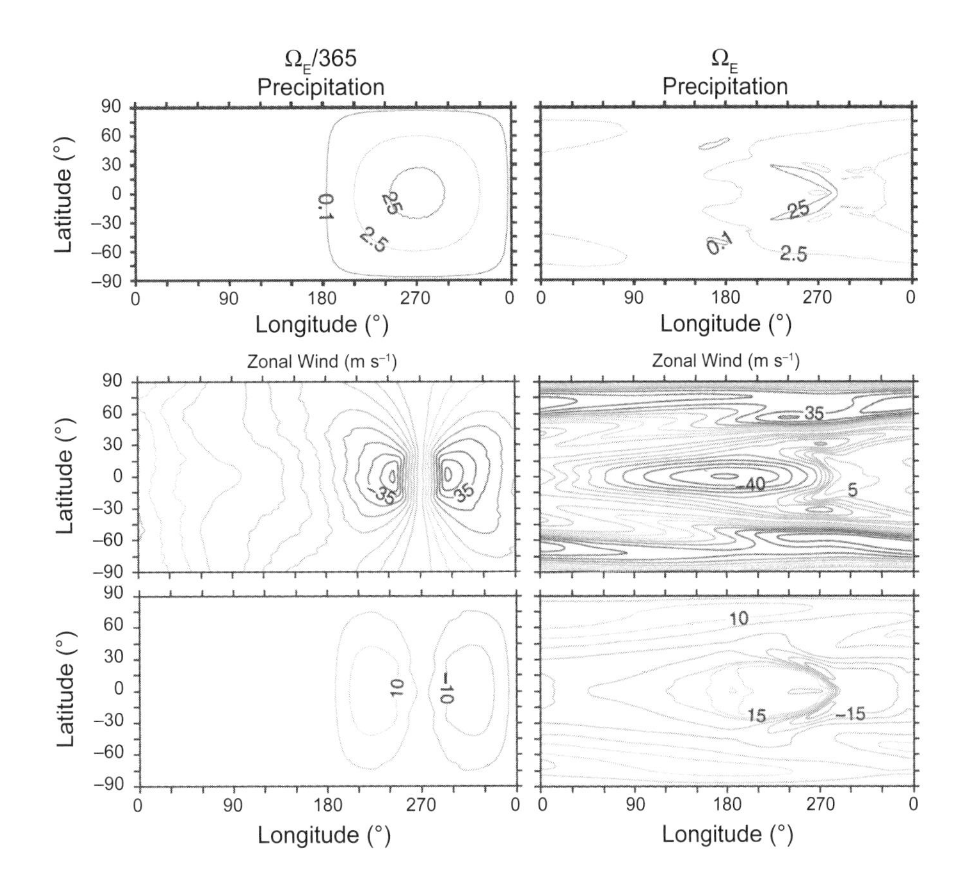

Fig. 20. See Plate 51 for color version. Time-mean circulation in two GCM simulations of Earth-like synchronously rotating exoplanets; models with rotation periods of one Earth year and one Earth day, respectively, are shown on the left and right. Top row shows time-mean precipitation (contours of 0.1, 2.5, and 25.0 mm day^{-1}). Middle and bottom rows show time-mean zonal wind on the $\sigma = p/p_s = 0.28$ model level and at the surface, respectively (contour interval is 5 m s^{-1}; the zero contour is not displayed). The subsolar point is fixed at 270° longitude in both models. Adapted from Figs. 2, 4, and 11 of *Merlis and Schneider* (2010).

to offset the substantial drying associated with subsidence from the cold and dry upper troposphere (e.g., *Couhert et al.,* 2010). Therefore, both the water vapor advection by the large-scale circulations and small-scale convective processes, with the latter parameterized in GCMs, are important in determining the relative humidity.

These relative humidity dynamics are important for various aspects of planetary atmospheres: They influence the radiation balance, atmospheric circulations, and precipitation. The processes controlling the relative humidity are tied directly to three-dimensional atmospheric circulations, which is a difficulty of single-column models. Single-column models do not explicitly simulate the atmospheric eddies or mean circulations that advect water and set the global-mean relative humidity. Furthermore, spatial variations in relative humidity are radiatively important. For example, the same global-mean relative humidity with a different spatial distribution has a different global-mean radiative cooling (*Pierrehumbert,* 1995; *Pierrehumbert et al.,* 2007). Therefore, the magnitude of the dayside to nightside relative humidity contrasts of tidally locked exoplanets may be important in determining the global climate.

For greenhouse gas-forced climate changes on Earth, relative humidity changes (<1% K^{-1}) are typically smaller than specific humidity changes (~7% K^{-1}) (*Held,* 2000; *Sherwood et al.,* 2010a). Therefore, it is a useful state variable for climate feedback analysis that avoids strongly canceling positive water vapor feedback and negative lapse rate feedback found in the conventional feedback analysis, which uses specific humidity as a state variable (*Held and Shell,* 2012). Climate feedback analysis depends on the mean state, which clearly differs between tidally locked exoplanets and Earth-like planets. A climate feedback analysis has not been performed for a tidally locked planet to diagnosing the relative contributions of the water vapor, lapse rate, albedo, and cloud feedbacks to the climate sensitivity.

3.6. Clouds

The formation of clouds from the phase change of condensible species — like water vapor — is often associated with ascending motion. Ascent may be the result of convergent horizontal winds or convective instability, so clouds typically have small space and timescales. Therefore, the spatial distribution of clouds and their sensitivity to external parameters such as the rotation rate or solar constant depend on atmospheric turbulence of a variety of scales: from the scales of the global circulations (~1000 km) down to those of moist convection and boundary layer turbulence (~10 km and smaller).

Clouds can substantially affect the radiation balance. To illustrate, we consider an Earth-like case, although other possibilities exist (e.g.,*Wordsworth et al.,* 2010). First, consider the longwave component of the radiation and assume that clouds are completely opaque in that region of the spectrum. If a cloud is low in the atmosphere (near the surface), the temperature at the top of the cloud, from where the radiation is reemitted after absorption, will be close to the temperature at which the radiation would be emitted in the absence of any clouds (e.g., from the surface or the clear air near the surface). In this case, the cloud will not substantially affect the longwave radiation relative to cloud-free conditions. If a cloud is high in the atmosphere (near the tropopause, for example), the temperature at the top of the cloud will be substantially cooler than the emission temperature in the absence of clouds. So, high clouds enhance the atmospheric greenhouse by absorbing longwave radiation emitted at the higher temperatures of atmosphere and surface below and reemitting longwave radiation at the colder cloud temperature. Simply characterizing the effect of clouds on the shortwave radiation depends more subtly on the optical characteristics. There can be a substantial reflection (e.g., cumulus or stratus) or little reflection (e.g., cirrus) of shortwave radiation. Additionally, the effect of clouds on the planetary albedo depends on the surface albedo; for example, in a snowball Earth state (section 4.2), a planet may be largely ice covered, in which case clouds will not significantly modify the albedo. Considering all these effects, the net radiative effect of clouds (the sum of shortwave and longwave) can be warming, cooling, or close to neutral. In the current Earth climate, these three possibilities are manifest with cirrus, stratocumulus, and convecting cumulus clouds, respectively.

The cloud radiative forcing (the difference in top-of-atmosphere net radiation between cloudy and clear conditions) on Earth is negative: Earth would be warmer without clouds. This does not imply that clouds are expected to damp (negative radiative feedback) rather than amplify (positive radiative feedback) radiative perturbations. In comprehensive general circulation model projections of climate change, the cloud feedback is on average positive (i.e., the change in cloud properties amplifies the radiative perturbation from CO_2 changes) (*Soden and Held,* 2006). However, there is substantial uncertainty in the magnitude, and the physical basis of this projection, if there is one, is still being elucidated (e.g., *Bony et al.,* 2006; *Zelinka and Hartmann,* 2010).

In addition to their important role in determining climate through the radiation balance, clouds likewise affect remote (e.g., spacebased) observations of planets. Interpreting cloud observations may help reveal aspects of the atmospheric circulation, such as the characteristic space and timescales of a planet's "weather" (e.g., cloud tracking on Jupiter), and climate (e.g., clouds on Titan indicate regions of precipitation). Preliminary efforts are being made to explore the effects of clouds on light curves and spectra of terrestrial exoplanets (e.g., *Pallé et al.,* 2008; *Cowan et al.,* 2009; *Fujii et al.,* 2011; *Robinson et al.,* 2011; *Kawahara and Fujii,* 2011; *Fujii and Kawahara,* 2012; *Sanromá and Pallé,* 2012; *Gómez-Leal et al.,* 2012; *Karalidi et al.,* 2012).

3.7. Effect of the Hydrological Cycle on Atmospheric Temperature and Circulations

While much of our discussion has focused on the role of atmospheric circulations in shaping the distributions of precipitation, humidity, and clouds, water vapor is not a passive tracer and can affect the atmospheric temperature and winds (condensible species also affect the atmospheric mass, which is a minor effect on Earth, but is important on Mars and is potentially important on some exoplanets; see section 4.4). Here, we outline some ways in which the latent heat release of condensation affects the atmosphere's thermal structure and water vapor influences the energetics of atmospheric circulations. Additional material may be found in reviews by *Sobel* (2007) and *Schneider et al.* (2010).

3.7.1. Thermal structure. The atmosphere's stratification in low latitudes (or slowly rotating regions more generally) is largely determined by convection. The tropical atmosphere is close to a moist adiabat on Earth (e.g., *Xu and Emanuel,* 1989), although entrainment and ice-phase processes affect the upper troposphere (*Romps and Kuang,* 2010). Simulations of Earth-like tidally locked planets (*Merlis and Schneider,* 2010) also have near-moist adiabatic stratification over large regions of the dayside. The influence of convecting regions is nonlocal: Gravity waves propagate away density variations (see section 2.2.2), whereas gradients can be rotationally balanced in the extratropics (more generally, in regions where the Coriolis force is important in the horizontal momentum balance). The relative importance of moist convection in determining the tropical stratification will be mediated by the geography of condensate reservoirs (e.g., Titan or possible dry exoplanets) and the vertical distribution of radiatively active agents (e.g., Venus and Titan). The convective lapse rate is the starting point for theories of the depth of the troposphere (i.e., the height of the tropopause) in low latitudes (*Held,* 1982; *Thuburn and Craig,* 2000; *Schneider,* 2007), which may have implications for the interpretation of observed radiance.

Latent heat release also affects the stratification of the extratropical atmosphere. The warm and moist sectors of extratropical cyclones condense as they move poleward and upward (Fig. 19). The vertical component of these transports is important in determining the lapse rate in the extratropics. Accounting for this interaction between extratropical cyclones and condensation or convection is an ongoing area of research for Earth-like planets (*Juckes,* 2000; *Frierson et al.,* 2006; *O'Gorman,* 2011).

In addition to the importance of the latent heat of condensation in determining the atmosphere's vertical thermal structure, it helps determine the horizontal temperature gradients. On Earth, about half of the meridional atmospheric energy transport in mid-latitudes is in the form of water vapor (e.g., *Trenberth and Caron,* 2001). This is primarily accomplished by transient baroclinic eddies. The transient eddy moisture flux can be parameterized in terms of mean humidities and characteristic eddy velocities (*Pierrehumbert,* 2002; *O'Gorman and Schneider,* 2008; *Caballero and Hanley,* 2012). In the low latitudes of Earth, the mean meridional circulation transports water vapor equatorward. This thermally indirect latent energy flux is a consequence of the small-scale height of water vapor. The dry component of the energy flux is larger and poleward, so the total energy transport by the Hadley circulation is thermally direct. For Earth-like exoplanets, the energy transports by the dayside to nightside circulation will have a similar decomposition — a thermally indirect latent component and a thermally direct dry component (*Merlis and Schneider,* 2010).

3.7.2. Circulation energetics. Latent heat release affects atmospheric circulations in a number of subtle ways. On the one hand, latent heat release has the potential to directly energize atmospheric circulations. On the other hand, the effect of latent heat release on the thermal structure of the atmosphere may weaken atmospheric circulations by increasing the static stability or decreasing horizontal temperature gradients.

The quantification of the kinetic energy cycle of the atmosphere is typically based on the dry thermodynamic budget, which is manipulated to form the Lorenz energy cycle (*Lorenz,* 1955; *Peixoto and Oort,* 1992). Latent heat release appears as an energy source in these budgets. For Earth's atmosphere, the diabatic generation of eddy kinetic energy from latent heat release is up to ~1 W m^{-2} (*Chang,* 2001; *Chang et al.,* 2002), which is a small fraction of the ≈80 W m^{-2} global-mean latent heat release (e.g., *Trenberth et al.,* 2009). This low efficiency of converting latent heat into kinetic energy is consistent with the overall low efficiency of the kinetic energy cycle (*Peixoto and Oort,* 1992). There are also alternative formulations of the kinetic energy cycle that incorporate moisture into the definition of the available potential energy (*Lorenz,* 1978, 1979; *O'Gorman,* 2010).

The energy transport requirements can play a role in determining the strength of the tropical circulations. For example, in angular momentum-conserving Hadley cell theories, the circulation energy transport is constrained to balance the top-of-atmosphere energy budget (*Held and Hou,* 1980; *Held,* 2000). Such an energy transport requirement does not directly determine the circulation mass transport, however. The ratio of the energy to mass transport is a variable known as the gross moist stability [other related definitions exist (*Raymond et al.,* 2009)]. The gross moist stability expresses that the energy transport of tropical circulations results from the residual between the largely offsetting thermally indirect latent heat transport and the thermally direct dry static energy. Therefore, the gross moist stability is much smaller than the dry static stability for Earth-like atmospheres. The gross moist stability is an important, although uncertain, parameter in theories of tropical circulations (*Neelin and Held,* 1987; *Held,* 2000, 2001; *Sobel,* 2007; *Frierson,* 2007; *Merlis et al.,* 2012).

4. ATMOSPHERIC CIRCULATION AND CLIMATE

On any terrestrial planet, the influence of the atmospheric circulation on the climate is fundamental. On Earth, climate is generally understood as the statistics (long-term averages and variability) of surface pressure, temperature, and wind, along with variations in the atmospheric content of water vapor and precipitation of liquid water or ice (*Hartmann,* 1994b). For terrestrial bodies in general, this definition can readily be extended to include the atmospheric transport of other condensable species, as in the cases of the CO_2 cycle on Mars or the methane cycle on Titan.

The close interaction between atmospheric dynamics and climate is readily apparent if we examine even the most basic features of the inner planets in our solar system. For example, Venus has a mean surface temperature of around 735 K, with almost no variation between the equator and the poles. In contrast, Mars' mean surface temperature is only 210 K, but in summer daytime temperatures can reach 290 K, while in the polar winter conditions are cold enough for atmospheric CO_2 to condense out as ice on the surface. Nonetheless, annual mean temperatures near the top of Olympus Mons are barely different from those in Mars' northern lowland plains. On Earth, surface temperature variations lie between these two extremes — the variation in climate with latitude is significant, but surface temperatures also rapidly decrease with altitude due to thermal coupling with the atmosphere. The differences in insolation, atmospheric pressure, and composition (92 bar CO_2 for Venus vs. around 1 bar N_2/O_2 for Earth and 0.006 bar CO_2 for Mars) go a long way toward explaining these gross climatic differences. In particular, from planetary boundary layer theory it can be shown that the sensible heat exchange between the surface and atmosphere is

$$SH = \rho_a \overline{w'(c_p T')} \approx c_p \rho_a C_D \|\mathbf{v}_{surf}\| (T_{surf} - T_a) \quad (32)$$

where the overbar denotes a time or spatial mean, w′ and T′ are deviations from the mean vertical velocity and temperature, T_{surf} is surface temperature, C_D is a drag coefficient, c_p is the specific atmospheric heat capacity, and ρ_a, $\|\mathbf{v}_{surf}\|$, and T_a are the density, mean wind speed, and temperature of the atmosphere near the surface (*Pierrehumbert,* 2010). Because $F_{sens} \propto \rho_a$, the thermal coupling between the atmosphere and surface will generally increase with the atmospheric pressure. Given the tendency for dense, slowly rotating atmospheres to homogenize horizontal temperature differences rapidly (see section 2.2), it should be intuitively clear why on planets like Venus, variations in annual mean temperature as a function of latitude and longitude tend to be extremely weak.

In detail, the picture can be more complex, because $||\mathbf{v}_{surf}||$ is also a function of the atmospheric composition and insolation pattern, and C_D depends on both the surface roughness and degree of boundary layer stratification. Beyond this, there are many other effects that couple with atmospheric dynamics. As discussed in section 3, a key additional process on Earth is the hydrological cycle, which affects climate through latent heat transport and the radiative properties of water vapor and water ice/liquid clouds. Examples from other solar system planets include the dust and CO_2 cycles on Mars, and the methane/hydrocarbon haze cycle on Titan. These processes, which can be fascinatingly complex, are a large part of the reason why detailed long-term predictions of climate are challenging even for the relatively well-observed planets of the solar system.

Given these complexities, it is unsurprising that for terrestrial exoplanets, study of the coupling between atmospheric circulation and climate in generalized cases is still in its infancy. The present state of knowledge is still heavily based on limited exploration of a few Earth-like or Earth-similar cases. Rather than attempting an overview of the entire subject, therefore, here we review a small selection of the many situations where atmospheric dynamics are expected to have a key influence on planetary climate.

4.1. Influence of Dynamics on the Runaway H_2O Greenhouse

The well-known runaway H_2O greenhouse effect occurs due to the feedback between surface temperature and atmospheric IR opacity on a planet with surface liquid water. Above a given threshold for the incoming stellar flux $F_0 > F_{limit}$ (see Fig. 21), the thermal radiation leaving the planet no longer depends on the surface temperature, and water will continue to evaporate until all surface sources have disappeared (*Kombayashi,* 1967; *Ingersoll,* 1969). Classical runaway greenhouse calculations (e.g., *Kasting et al.,* 1993) were performed in one dimension and assumed homogenous atmospheres with either 100% relative humidity or a fixed vertical profile. In reality, however, the variations in relative humidity with latitude due to dynamical processes should have a fundamental effect on the planet's radiative budget and hence on the critical stellar flux at which the transition to a runaway greenhouse occurs. The global variations of relative humidity in Earth's atmosphere are strongly dependent on the ascending and descending motion of the Hadley cells (section 2.2.1) and subtropical mixing due to synoptic-scale eddies (section 2.1.2). Even on Earth today, without heat transport to higher latitudes by the atmosphere and ocean, the tropics would most likely be in a runaway greenhouse state. (Assuming, that is, that the negative feedback from cloud albedo increases is not so strong as to overwhelm the increase in solar forcing past the conventional runaway greenhouse threshold.) *Pierrehumbert* (1995) proposed that regulation of mean tropical sea surface temperatures is in fact governed by a balance between the radiative heating in saturated upwelling regions and radiative cooling by "radiator fins" in larger regions of net subsidence.

Very few researchers have yet taken on the challenge of modeling the transition of an ocean planet to a runaway greenhouse state in a three-dimensional GCM. *Ishiwatari et al.* (2002) studied the appearance of the runaway greenhouse state in an idealized three-dimensional circulation model with gray-gas radiative transfer. Their results roughly corresponded to those found in one dimension using a global mean relative humidity of 60%. However, the simplicity of their radiative scheme, neglect of clouds, and problems arising from the need for strong vertical damping near the runaway state meant that they were unable to quantitatively constrain the greenhouse transition for cases such as early Venus. Later, *Ishiwatari et al.* (2007) compared results using a GCM with those from a one-dimensional energy balance model (EBM). EBMs are intermediate-complexity models that replace vertically resolved radiative transfer with empirical functions for the fluxes at the top of the atmosphere and represent all latitudinal heat transport processes by a simple one-dimensional diffusion equation. Ishiwatari et al. found that at intermediate values of the solar constant, multiple climate solutions including runaway greenhouse, globally, and partially glaciated states were possible. Clearly, comparison of these results with simulations using a GCM specifically designed to remain physically robust in the runaway limit would be an interesting future exercise. As well as taking into account the nongray radiative transfer of water vapor, such a model would need to account for the locally changing mass of water vapor in the large-scale dynamics and subgrid-scale convection, and include some

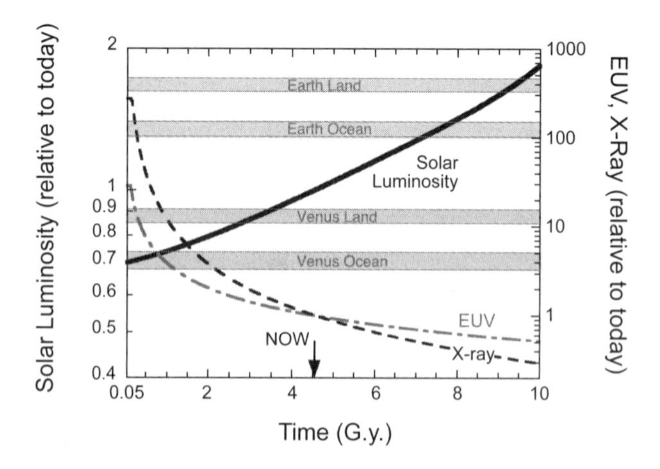

Fig. 21. Solar luminosity as a function of time, alongside runaway H_2O greenhouse thresholds for land and aqua planets at Venus and Earth's orbits (from *Abe et al.,* 2011). With a depleted water inventory, Venus could conceivably have remained a habitable planet until as little as 1 b.y. ago.

representation of the radiative effects of clouds.

The importance of clouds to the runaway greenhouse limit is central, but our understanding of how they behave as the solar constant is increased is still poor. *Kasting* (1988) performed some simulations in one dimension with fixed water cloud layers and concluded that because of their effect on the planetary albedo, Venus could easily have once been able to maintain surface liquid water. However, they did not include dynamical or microphysical cloud effects in their model. Going further, *Rennó* (1997) used a one-dimensional model with a range of empirical parameterizations of cumulus convection to study the runaway transition, although with simplified cloud radiative transfer. Perhaps unsurprisingly, they found that surface temperature depends sensitively on cloud microphysical assumptions when solar forcing is increased. They also found that in general, mass flux schemes (i.e., convective parameterizations that capture the effect of environmental subsidence associated with cumulus convection) caused a more rapid runaway transition than adjustment schemes (i.e., convective parameterizations that simply represent the effect of convection as a modification of the local vertical temperature profile). Future three-dimensional GCM studies of the runaway greenhouse will be forced to confront these uncertainties in subgridscale physical processes.

Cases where the initial planetary content of water is limited are also of interest for the runaway greenhouse. *Abe et al.* (2011) studied inner habitable zone limits for "land planets," which they defined as planets with limited total H_2O inventories (less than 1 m global average liquid equivalent in their model) and no oceans. They showed that in these cases, the tendency of the limited water inventory to become trapped at the poles causes extreme drying of equatorial regions. This allows the global mean outgoing longwave radiation to continue increasing with temperature, delaying the transition to a runaway greenhouse state. The critical average solar flux values they calculated in their model were 415 W m^{-2} and 330 W m^{-2} for land and aqua (global ocean) planets, respectively. Their results suggest planets with limited water inventories could remain habitable much closer to their host stars than the classical inner edge of the habitable zone suggests. Because the Sun's luminosity has increased with time, their results also suggest that a mainly dry early Venus could have stayed cool enough to maintain regions of surface liquid water until as recently as 1 b.y. ago (see Fig. 21). However, scenarios where liquid water was never present on the surface of Venus are also possible, and have been argued to be more consistent with Ne/Ar isotopic ratios and the low oxygen content in the present-day atmosphere (*Gillmann et al.*, 2009; *Chassefière et al.*, 2012).

The insights of *Abe et al.* (2011) on the importance of relative humidity to the runaway greenhouse have recently been extended to the case of tidally locked exoplanets around M stars by *Leconte et al.* (2013). On a tidally locked terrestrial planet, the darkside may act as a cold trap for volatiles, which can lead to total collapse of the atmosphere in extreme cases, as we discuss later. For planets close to or inside the inner edge of the habitable zone, however, the effects of this process on H_2O lead to an interesting bistability in the climate. One state consists of a classical runaway greenhouse where all the water is in the atmosphere. In the other state, the vast majority of the H_2O is present as ice on the planet's darkside, while the planet's dayside is relatively hot and extremely dry, allowing it to effectively radiate the incoming solar radiation back to space (see Fig. 22). In this scenario, depending on the thickness of the ice sheet and the planet's thermal history, liquid water could be present in some amount near the ice sheet's edges, allowing marginal conditions for habitability well inside the classical inner edge of the habitable zone. Further work may be required to assess whether such a collapsed state would remain stable under the influence of transient melting events due to, e.g., meteorite impacts on the planet's darkside.

4.2. Snowball Earth Dynamics and Climate

Runaway greenhouse states occur because of the effectiveness of gaseous H_2O as an IR absorber. However, the

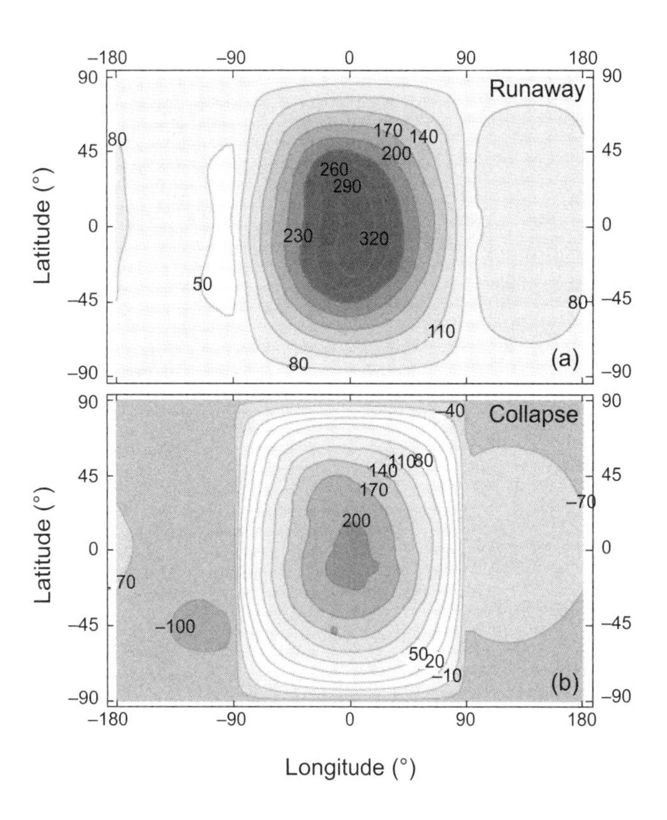

Fig. 22. Surface temperature maps in degrees Centigrade from GCM simulations of a tidally locked, inner-edge exoplanet with a 200-mbar equivalent incondensible atmospheric component and fixed total water inventory. **(a)** The planet was initialized with a mean water vapor column of 250 kg m^{-2} and remains in a runaway greenhouse state. **(b)** The smaller water vapor starting inventory (150 kg m^{-2}) has led to a transition to a collapsed state, with lower surface temperatures and ice on the planet's dark side.

ability of water to cause fundamental changes in climate systems also extends to colder conditions. Solid H_2O on the surface of a planet can have an equally drastic effect on circulation and climate, in this case due to its properties in the visible part of the spectrum.

Since the 1960s, it has been known that Earth's climate has an alternative equilibrium state to present-day conditions. If the ice sheets that are currently confined to the poles instead extended all the way to the equator, the elevation in surface albedo would cause so much sunlight to be reflected back to space that mean global temperatures would drop to around 230–250 K. As a result, surface and ocean ice would be stable even in the tropics, and Earth could remain in a stable "snowball" state unless the greenhouse effect of the atmosphere increased dramatically. Snowball events are believed to have occurred at least twice previously in Earth's history: once around 640 Ma and again at 710 Ma, in the so-called Marinoan and Sturtian ice ages. Geological evidence (specifically, carbon-isotope ratio excursions and observations of cap carbonates above glacial deposits) suggests that these events were ended by the buildup of atmospheric CO_2 to extremely high levels due to volcanic outgassing (*Hoffman and Schrag,* 2002; *Pierrehumbert et al.,* 2011). This scenario is particularly plausible because the land weathering of silicates and associated drawdown of CO_2 into carbonates should be greatly reduced in a snowball climate (see section 4.3).

Snowball Earth transitions are interesting in an exoplanetary context first because they represent a clear challenge to the concept of the habitability zone. If a habitable planet falls victim to snowball glaciation and cannot escape it, it may be irrelevant if it continues to receive the same flux from its host star. Identifying how these events can occur and how they end is hence of key importance in understanding the uniqueness of Earth's current climate. Snowball glaciation is also interesting from a purely physical viewpoint, as it is another example of a situation where the interplay between atmospheric dynamics and climate is fundamental to the problem. As we have seen in section 2, heat transport between a planet's regions of high and low mean insolation is strongly dependent on the properties of the atmosphere and planetary rotation rate. As a result, the physics of the snowball transition should vary considerably between different exoplanet cases.

Recently, it has been hypothesized that changes in surface albedo due to atmospheric transport of dust may be critical to terminating snowball Earth episodes (*Abbot and Pierrehumbert,* 2010). The reduced surface temperature and stellar energy absorption at the equator and absence of ocean heat transport on a globally frozen planet mean that typically, the tropics become a region of net ablation (i.e., mean evaporation exceeds precipitation). As a result, dust is slowly transported to the surface at the equator by a combination of horizontal glacial flow and vertical ice advection (see Fig. 23). After a long time period, the buildup of dust on the surface lowers the albedo sufficiently to melt the surrounding ice and exit the planet from a snowball state. The fingerprints of the "mudball" state that presumably would follow this may have been imprinted on Earth's geological record, in the form of variations in thickness with (paleo)-latitude of clay drape deposits.

Other possibilities for snowball deglaciation come from water ice clouds. *Pierrehumbert* (2002) demonstrated using a one-dimensional radiative model that clouds have a net warming effect on a snowball planet. Because the surface albedo is high, clouds do not significantly increase the planetary albedo by reflecting incoming radiation to space. Nonetheless, they still effectively absorb outgoing IR radiation and hence cause greenhouse warming. These conclusions were broadly confirmed by a later three-dimensional GCM intercomparison (*Abbot et al.,* 2012), although differences in cloud radiative forcing between models was found to be significant. Just as for the runaway greenhouse, future detailed studies of the physics of cloud formation and radiative transfer under snowball conditions, possibly using cloud-resolving models, will be important for elucidating their role.

Further interesting couplings between atmospheric dynamics and climate arises if a planet near the snowball transition has a limited H_2O inventory. *Abe et al.* (2011) studied idealized planets where the total amount of water (atmospheric and surface) only amounted to a 20–60-cm globally averaged layer. They showed that for these so-called "land" planets, the drying of tropical regions by a dynamical process similar to that responsible for dust accumulation would prevent the formation of stable surface ice there for a solar flux as low as 80% of that received by Earth today. As a result, the planetary albedo could remain low even if global mean surface temperatures dropped below the freezing point of H_2O, and transient regions of liquid water could persist longer than if the planet possessed as much water as Earth.

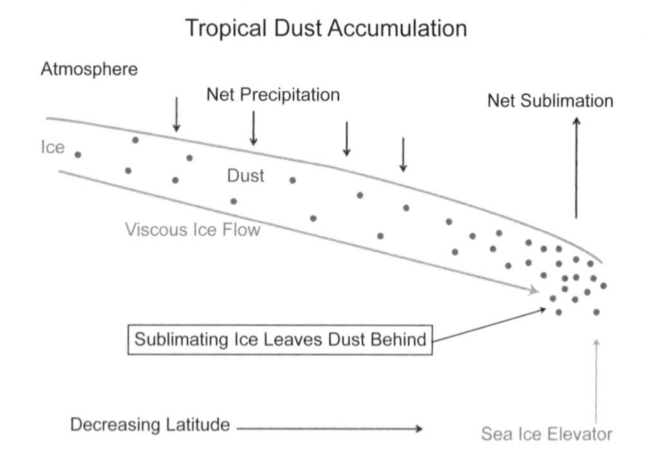

Fig. 23. Schematic of the process by which tropical dust accumulates on a globally glaciated planet. The net sublimation at the tropics and precipitation in subtropical regions occurs primarily due to the reduced surface heat capacity of the surface when a thick ice layer is present (from *Abbot and Pierrehumbert,* 2010).

Finally, if the planet orbits a star of a different spectral type than the Sun, further complications emerge. Recent work (*Joshi and Haberle,* 2012) has demonstrated that around red dwarf M stars, the mean albedos of snow and ice will be significantly reduced due to the red-shifted incident spectrum. As a result, the ice-albedo feedback is expected to be weaker, and snowball events may not be as difficult to exit as on Earth. *Joshi and Haberle* (2012) suggested that this effect may also help extend the outer edge of the habitable zone around M stars. However, it is likely to be of lesser importance for planets with dense CO_2 atmospheres, because the increased visible optical depth of the atmosphere and presence of CO_2 clouds means the dependence of the planetary albedo on surface properties is already weak in these cases (*Wordsworth et al.,* 2011).

4.3. Carbonate-Silicate Weathering and Atmospheric Dynamics

So far, we have focused on the coupling between circulation and climate in the context of H_2O. However, the role of other gases in climate may be equally significant. Carbon dioxide is of particular interest, given that it is the majority constituent of the atmospheres of both Venus and Mars. It is also the key greenhouse gas in our own planet's atmosphere, because of its special role in maintaining habitable surface conditions.

On Earth, the carbonate-silicate cycle is fundamental to the long-term evolution of CO_2 in the atmosphere. In particular, it is believed that the dependence of land silicate weathering on temperature leads to a negative feedback between atmospheric CO_2 and mean surface temperature, rendering the climate stable to relatively large variations in solar luminosity (*Walker et al.,* 1981). Indeed, the assumption of an efficient carbonate-silicate cycle lies behind the classical definition of the habitable zone (*Kasting et al.,* 1993).

The idea that climate could naturally self-regulate on rocky planets due to an abiotic process is a highly attractive one, and probably a major part of the explanation for why Earth has maintained a (mostly) clement climate throughout its lifetime, despite significant increases in the solar flux (around 30% since 4.4 Ga). Nonetheless, there are still major uncertainties in the nature of Earth's carbon cycle, including the importance of seafloor weathering (*Sleep and Zahnle,* 2001; *Le Hir et al.,* 2008), the exact role of plants in land weathering, and the dependence of the cycle on the details of plate tectonics.

Even given these uncertainties, it is already clear that substantial differences from the standard carbon cycle arise when a planet becomes synchronously locked. For example, *Edson et al.* (2012) investigated the relationship between weathering, atmospheric CO_2 concentrations, and the distribution of continents for tidally locked planets around M-class stars. They coupled Earth GCM simulations with a simple parameterization for the global CO_2 weathering rate, which following *Walker et al.* (1981) was written as

$$\frac{dCO_2|_{tot}}{dt} = \frac{1}{2}\int W_0\left(\frac{f_{CO_2}}{355\text{ ppmv}}\right)^{0.3} \times \left(\frac{\mathcal{R}}{0.665\text{ mm d}^{-1}}\right)\exp\frac{T-T_0}{T_U}dA \tag{33}$$

where f_{CO_2} is the atmospheric CO_2 volume mixing ratio, $W_0 = 8.4543 \times 10^{-10}$ C s^{-1} m^{-2} is the estimated present-day terrestrial weathering rate, $\mathcal{R}$ is the runoff rate, T is temperature, $T_0 = 288$ K, $T_U = 17.7$ K, and the integral is over the planet's weatherable (i.e., land) surface. By iterating between the simulations and equation (33), *Edson et al.* (2012) derived self-consistent f_{CO_2} values as a function of the (constant) substellar longitude, given Earth's present-day geography. They found that f_{CO_2} could vary by a factor of $\sim 1 \times 10^4$ (7 ppmv vs. 60331 ppmv), depending on whether the substellar point was located over the Atlantic or Pacific oceans. As a result, the mean surface air temperatures in their simulations varied between 247 and 282 K, with the climate in both cases similar to the "eyeball" state discussed in *Pierrehumbert* (2011) (see next section).

The differences in temperature and equilibrium atmospheric CO_2 abundance between the two cases were caused by the differences in land area in the illuminated hemisphere. The illuminated hemisphere of a tidally locked planet is warmer, with higher precipitation rates (see Fig. 20), so a greater land area there implies enhanced global weathering rates and hence more rapid drawdown of CO_2. Assuming that global CO_2 levels are controlled by an expression like equation (33) is clearly an oversimplification given the complexity of Earth's real carbon cycle. In particular, basalt carbonization on the seafloor, which has a much weaker dependence on surface temperatures, is still poorly understood, but could have played a vital role in CO_2 weathering throughout Earth's history. Nonetheless, the study highlights the dramatic differences that can be caused by three-dimensional effects, and shows that a more detailed understanding of dynamical couplings with the carbonate-silicate cycle will be vital in the future.

Kite et al. (2011) also explored the possible nature of climate-weathering feedbacks on tidally locked planets using an idealized energy balance climate model coupled to a simplified parameterization of silicate weathering. They noted that because heat transport efficiency depends on the atmospheric mass, in some cases reducing atmospheric pressure can increase the weathering rate and/or the liquid water volume on the dayside of the planet. This can cause positive feedbacks that lead to a decrease in atmospheric mass. The mechanism described by *Kite et al.* (2011) is most likely to be important for planets like Mars, with thin, CO_2-dominated atmospheres.

4.4. Collapse of Condensable Atmospheres

In section 4.1, we briefly discussed situations where water vapor can become trapped as ice on the darksides of tidally locked planets. Here, we discuss the extension of this process to the majority constituent of a planet's atmosphere. As we will see, atmospheric collapse is a problem of major importance to planetary habitability that sensitively depends on the details of the coupling between circulation and climate.

The interest in the habitability of tidally locked planets stems from the fact that cool, faint red dwarf stars (M and K spectral class) are significantly more common in the galaxy than stars like the Sun, and they host some of the lowest-mass and best-characterized exoplanets discovered so far (e.g., *Udry et al.,* 2007; *Mayor et al.,* 2009; *Charbonneau et al.,* 2009; *Bean et al.,* 2010). In the earliest detailed study of exoplanet habitability (*Kasting et al.,* 1993), it was shown that planets around M-dwarf stars could sustain surface liquid water if they were in sufficiently close orbits. As discussed previously, the tidal interaction with their host stars means that such planets will in most cases have resonant or synchronous rotation rates and low obliquities, and hence permanent regions where little or no starlight reaches the surface. Initially, this was thought to be a potentially insurmountable obstacle to habitability, because the regions receiving no light could become so cold that any gas (including nitrogen) would freeze out on the surface there, depleting the atmosphere until the planet eventually became completely airless.

The problem of atmospheric collapse was first investigated quantitatively in a series of pioneering papers by M. Joshi and R. Haberle (*Joshi et al.,* 1997; *Joshi,* 2003). Using a combination of basic scale analysis and three-dimensional atmospheric modeling, they showed that the collapse of the atmosphere was only inevitable if it was inefficient at transporting heat. As we described earlier, the efficiency of atmospheric heat transport depends primarily on the composition (via the specific heat capacity and the IR opacity), the average wind speed, and (most critically) the total surface pressure. Assuming a surface emissivity of unity, the surface heat budget for a planet with condensible atmospheric species can be written

$$\alpha \frac{\partial T_{surf}}{\partial t} = F_{sw}^{dn} + \left(F_{lw}^{dn} - \sigma T_{surf}^4\right) - LE - SH \tag{34}$$

where σ is the Stefan-Boltzmann constant, α is the surface heat capacity, F_{sw}^{dn} and F_{lw}^{dn} are the net downwelling fluxes of short- and longwave radiation from above, LE is the latent heat flux, and SH is the sensible heat flux, as described in equation (32). Equation (34) is essentially the unaveraged version of equation (29), except that for a single-component atmosphere, LE only becomes important when the atmosphere begins to condense on the surface (as on present-day Mars, for example). On the darkside of a tidally locked planet $F_{sw}^{dn} = 0$, so radiative cooling of the surface until the atmosphere begins to condense can be prevented in two ways: tight thermal coupling with the atmosphere via convection (the SH term), or radiative heating of the surface by an optically thick atmosphere (F_{lw}^{dn}). Both effects can be important in principle, although the temperature inversion caused by contact of warmer air from the dayside with the nightside surface may shut down convection, reducing the magnitude of the sensible heat flux or even reversing its sign. The strength of F_{lw}^{dn} will depend on both the gas opacity of the atmosphere and the effects of clouds and aerosols, if present, and hence can be expected to vary with composition as well as atmospheric pressure. In any case, the calculations of *Joshi et al.* (1997) showed that for CO_2 atmospheres with surface pressures of around 100 mbar or more on Earth-like planets, both atmospheric transport and surface coupling become efficient enough to avoid the possibility of collapse and the climate can remain stable.

While Joshi and Haberle's simulations showed that synchronous rotation does not rule out the possibility of habitable climates around M-class stars, their work was focused on planets of Earth's mass and net stellar flux, and hence could not be applied to more general cases. Interest in this issue was reignited with the discovery of potentially habitable planets in the Gliese 581 system in 2007 (*Udry et al.,* 2007; *Mayor et al.,* 2009). Two planets, GJ581c and d, orbit near the outer and inner edges of their system's habitable zone, and have estimated masses that suggest terrestrial rather than gas giant compositions. Simple one-dimensional models indicated that with a CO_2-rich atmosphere, the "d" planet in particular could support surface liquid water (*Wordsworth et al.,* 2010; *von Paris et al.,* 2010; *Hu and Ding,* 2011; *Kaltenegger et al.,* 2011). Because M stars have red-shifted spectra compared to the Sun, CO_2 clouds typically cause less warming for planets around them, because the increase in planetary albedo they cause becomes nearly as significant as the greenhouse effect caused by their IR scattering properties. However, in CO_2-rich atmospheres this is more than compensated for by increased near-IR absorption in the middle and lower atmosphere, which lowers the planetary albedo and hence increases warming.

The results from one-dimensional models for GJ581d were suggestive, but the increase in condensation temperature with pressure means that atmospheric collapse becomes more of a threat as the amount of CO_2 in the atmosphere increases. For example, reference to vapor pressure data for CO_2 shows p_{sat} = 10 bar at ~235 K, which is comparable to annual mean surface temperatures at Earth's south pole. It was therefore initially unclear if high-CO_2 scenarios were plausible for GJ581d.

Wordsworth et al. (2011) investigated the stability of CO_2-rich atmospheres on GJ581d using a GCM with band-resolved radiative transfer, simplified cloud physics, and evaporation/condensation cycles for CO_2 and H_2O included. They found that for this planet, which receives only around 30% of Earth's insolation, the atmosphere could indeed be unstable to collapse even for pressures as high as 2–5 bar. Nonetheless, at higher pressures the simulated atmospheres stabilized due to a combination of the increased greenhouse warming and homogenization of the surface temperature by atmospheric heat transport. For atmospheric CO_2 pressures of 10–30 bar (which are plausible given Venus and Earth's total CO_2 inventories), Wordsworth et al. modeled surface temperatures for GJ581d in the 270–320 K range, with horizontal temperature variations on the order of 10–50 K (see Fig. 24). Sensitivity studies indicated that while the exact transition pressure from unstable to stable atmospheres was dependent on the assumed cloud microphysical properties, the general conclusion of stability at high CO_2 pressure was not. The authors concluded that despite the low stellar flux it receives, GJ581d could therefore potentially support surface liquid water. As such, it is one of the most interesting currently known targets for follow-up characterization.

Any atmospheric observations of planets around Gliese 581 will be challenging, because the system does not appear to be aligned correctly for transit spectroscopy, and the planet-star contrast ratio constraints required for direct observations are severe. Distinguishing between habitable scenarios for GJ581d and other possibilities, such as a thin/collapsed atmosphere or a hydrogen-helium envelope, may be best accomplished via a search for absorption bands of CO_2/H_2O in the planet's emitted IR radiation that do not significantly vary over the course of one orbit (see Fig. 25). As described in *Selsis et al.* (2011), estimates of a non-transiting planet's atmospheric density are also possible, in principle, via analysis of the IR "variation spectrum" derived from phase curves. Despite the challenges, at a distance of ~20 light-years from the Sun, GJ581 is still a relatively close neighbor, and a future dedicated mission such as NASA's proposed Terrestrial Planet Finder (TPF) or ESA's Darwin would have the necessary capability to perform such observations.

Other recent studies have noted that even if global habitability is ruled out by the freezing of surface water on a planet's nightside, it is still possible to maintain local regions of liquid water. *Pierrehumbert* (2011) investigated a series of possible climates for a 3.1 $M_\oplus$ tidally locked exoplanet around an M star receiving 63% of Earth's incident solar flux. He hypothesized that if the planet's composition was dominated by H_2O, it would most likely have an icy surface and hence a high albedo. However, if the atmosphere were extremely thin and hence inefficient at transporting heat, surface temperatures near the substellar point could allow some H_2O to melt, even though the global mean temperature would be around 192 K if heat transport were efficient. Given a denser mixed N_2-CO_2-H_2O atmosphere, an "eyeball" climate could also be maintained where permanent ice is present on the nightside, but temperatures are high enough on the dayside to allow an ocean to form (see Fig. 26). Although such a scenario is admittedly hypothetical, it raises many interesting questions on the general nature of exoplanet climate and may help with the interpretation of observations in the future.

Clearly, the likelihood of atmospheric collapse also depends strongly on the condensation temperature of the

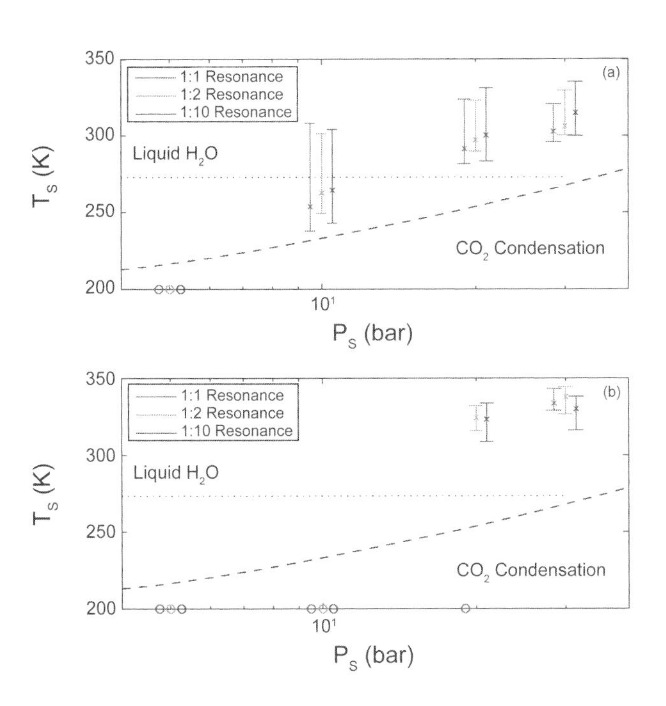

Fig. 24. See Plate 52 for color version. Simulated annual mean surface temperature (maximum, minimum, and global average) as a function of atmospheric pressure and rotation rate for GJ581d assuming **(a)** a pure CO_2 atmosphere and **(b)** a mixed CO_2-H_2O atmosphere with infinite water source at the surface (from *Wordsworth et al.*, 2011). Data plotted with circles indicate where the atmosphere had begun to collapse on the surface in the simulations, and hence no steady-state temperature could be recorded. In the legend, 1:1 resonance refers to a synchronous rotation state, and 1:2 and 1:10 resonances refer to despun but asynchronous spin-orbit configurations.

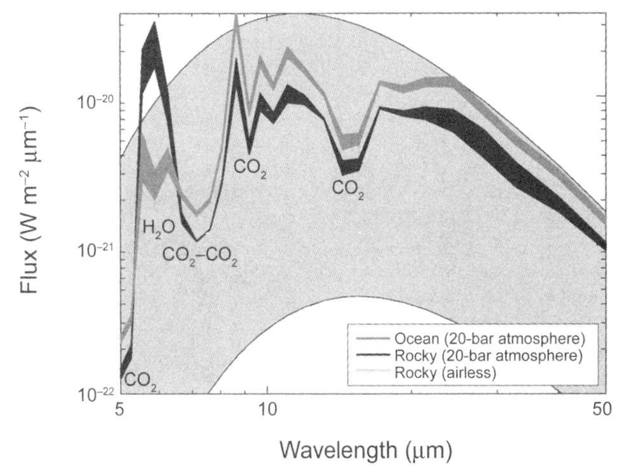

Fig. 25. Simulated emission spectra for GJ581d given various atmospheric scenarios (from *Wordsworth et al.*, 2011). In all cases the thicknesses of the lines correspond to the range of variation predicted over the course of one planetary orbit. The relatively low variation in the cases shown by the darker gray lines is a characteristic signature of a planet with a stable, dense atmosphere with efficient horizontal heat redistribution.

majority gas in the atmosphere. Despite this progress in modeling exotic exoplanet atmospheres in specific cases, the general problem of atmospheric condensation in a planet's cold trap(s) for arbitrary conditions remains unsolved. Nonetheless, scale analysis has recently been used by *Heng and Kopparla* (2012) to place some basic general constraints on atmospheric stability. These authors assumed the criterion $\tau_{rad} > \tau_{adv}$ as a condition for stability, where τ_{rad} and τ_{adv} are representative timescales for radiative cooling and advection, respectively. The radiative timescale can be approximately defined as $\tau_{rad} = c_p p/g\sigma T^3_{rad}$, with p the total pressure, g gravity, and T_{rad} the characteristic emission temperature, while $\tau_{adv} \sim L/U$, with L and U a characteristic horizontal length and wind speed, respectively. Evaluating τ_{rad} is straightforward for a known atmospheric composition, but U and hence τ_{adv} are difficult to assess *a priori* without a general theory for the response of the atmospheric circulation to different forcing scenarios. Heng and Kopparla neatly circumvented this problem by simply taking U to be the speed of sound, allowing a potentially strict criterion for atmospheric collapse based entirely on known properties of the atmosphere.

Despite its elegance, their approach was limited by its neglect of the role of thermodynamics in the problem. In particular, a moving gas parcel with a majority constituent that condenses at low temperatures may continue to radiate until the local value of τ_{rad} becomes large, which tends to make an atmosphere much more resistant to collapse. There is still considerable scope for theoretical development of the atmospheric collapse problem, and given its fundamental importance to the understanding of both atmospheric circulation and habitability around M stars, it is likely that it will continue to receive attention for some time.

The dependence of collapse on saturation vapor temperature is particularly interesting when considered in the context of more exotic scenarios for terrestrial planet atmospheric composition. In particular, the saturation vapor temperatures of hydrogen and helium at 1 bar are under 25 K, rendering them essentially incondensible on planets receiving even very little energy from their host stars.

Many rocky planets are believed to form with thin primordial envelopes of these gases, which have large optical depths in the IR for pressures above a fraction of a bar. For planets of Earth's mass or lower, these envelopes generally escape rapidly under the elevated stellar wind and extreme ultraviolet (XUV) fluxes from young stars. However, higher-mass or more-distant planets may experience a range of situations where hydrogen and helium can remain in their atmospheres for longer periods. *Stevenson* (1999) suggested that some planetary embryos ejected from their system during formation could sustain liquid oceans if they kept their primordial hydrogen envelopes. *Pierrehumbert and Gaidos* (2011) extended this idea to planets in distant orbits from their stars. These planets could be detectable in theory due to gravitational microlensing, although given their low equilibrium temperatures, atmospheric characterization would be extremely challenging even with the most ambitious planned missions. *Wordsworth* (2012) noted that in general, very few planets are left with just the right amount of hydrogen in their atmospheres to maintain surface liquid water after the initial period of intense XUV fluxes and rapid hydrogen escape. Nonetheless, they showed that transient periods of surface habitability (between ~10,000 yr and ~100 m.y.) still occur for essentially any planets receiving less flux than Earth that form with a hydrogen envelope but later lose it. In addition, XUV photolysis in hydrogen-rich atmospheres can lead to the formation of a range of prebiotic compounds (*Miller,* 1953; *DeWitt et al.,* 2009). Conditions on young terrestrial planets that at least allow life to form could hence be much more common than previously assumed.

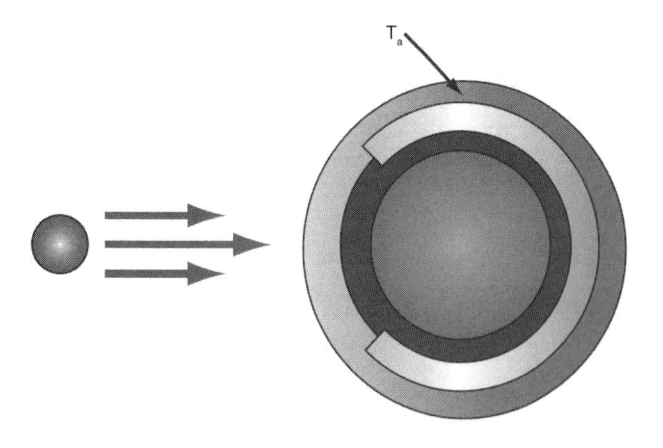

Fig. 26. Schematic of the "eyeball" climate state for a synchronously rotating terrestrial planet (*Pierrehumbert,* 2011). Sea ice is present across the permanent nightside of the planet, but stops at a stellar zenith angle primarily determined by the balance between the local stellar flux and horizontal atmospheric and oceanic heat transport.

5. CONCLUSIONS

Given the diversity emerging among more massive exoplanets, it is almost certain that terrestrial exoplanets will occupy an incredible range of orbital and physical parameters, including orbital semimajor axis, orbital eccentricity, incident stellar flux, incident stellar spectrum, planetary rotation rate, obliquity, atmospheric mass, composition, volatile inventory (including existence and mass of oceans), and evolutionary history. Since all these parameters affect the atmospheric dynamics, it is likewise probable that terrestrial exoplanets will exhibit incredible diversity in the specific details of their atmospheric circulation patterns and climate. Only a small fraction of this diversity has yet been explored in GCMs and similar models. As we have emphasized in this review, existing theory suggests that a number of unifying themes will emerge governing the atmospheric circulation on this broad class of bodies. This theory — summarized here — will provide a foundation for understanding, and provide a broad context for, the results of GCM investigations of particular objects.

Exoplanets that rotate sufficiently rapidly will exhibit extratropics at high latitudes, where the dynamics are approximately geostrophic, and where eddies resulting from baroclinic instabilities control the meridional heat transport, equator-to-pole temperature differences, meridional mixing rates, existence of jet streams, and thermal stratification in the troposphere. Regions where rotation is less dominant — near the equator, and globally on slowly rotating exoplanets — will exhibit tropical regimes where wave adjustment and Hadley circulations typically act to minimize horizontal temperature differences. Significant interactions between the tropics and extratropics can occur by a variety of mechanisms, perhaps the most important of which is the propagation of Rossby waves between the two regions. The hydrological cycle on many terrestrial exoplanets will exert significant effects both on the mean climate — through the greenhouse effect and clouds — and on the circulation patterns. Existing studies demonstrate that the circulation can influence global-scale climate feedbacks, including the runaway greenhouse, transitions to snowball Earth states, and atmospheric collapse, leading to a partial control of atmospheric circulation on the mean climate and therefore planetary habitability.

Exoplanets can exhibit greatly different regimes of thermal forcing than occur on solar system planets, a topic that has yet to be fully explored. The day-night thermal forcing on sychronously rotating and despun exoplanets will play a more important role than on Earth, leading in many cases to equatorial superrotation and other dynamical effects. When the heating rates are sufficiently great (in the limit of thin and/or hot atmospheres), the heating and cooling gradients can overwhelm the ability of the atmosphere to transport heat horizontally, leading to freezeout of the more refractory components (such as water or carbon dioxide) on the nightside or poles. This regime remains poorly understood, although it is important for understanding the structure, and even existence, of atmospheres on hot terrestrial planets and super-Earths.

Observations will be needed to move the field forward. These in many cases will represent extensions to terrestrial planets of techniques currently being applied to hot Jupiters. Short-term gains are most likely for transiting planets, particularly those around small M stars: A terrestrial planet orbiting such a star exhibits a planet/star radius ratio similar to that of a gas giant orbiting a Sun-like star, thereby making characterization (relatively) easier. Transit spectroscopy of such systems will yield constraints on atmospheric gas composition and existence of hazes (e.g., *Barman,* 2007; *Sing et al.,* 2008; *Pont et al.,* 2012; *Bean et al.,* 2010; *Désert et al.,* 2011; *Berta et al.,* 2012). In the longer-term future, such spectroscopy could also provide direct measurements of atmospheric wind speeds at the terminator (*Snellen et al.,* 2010; *Hedelt et al.,* 2011). Secondary eclipse detections and, eventually, full-orbit light curves will provide information on the emission spectrum as a function of longitude, allowing constraints on the vertical temperature profile and day-night temperature difference to be inferred. This will first be possible for hot systems (*Demory et al.,* 2012), but with significant investment of resources from the James Webb Space Telescope will also be possible for planets in the habitable zones (*Seager et al.,* 2008). Eclipse mapping may also eventually be possible, as is now being performed for hot Jupiters (*Majeau et al.,* 2012; *de Wit et al.,* 2012). Using direct imaging to obtain spectroscopy of planets is another observational avenue — not limited to transiting systems — that could be performed from a platform like the TPF. All these observations will be a challenge, but the payoff will be significant: the first characterization of terrestrial worlds around stars in the solar neighborhood.

In the meantime, theory and models can help to address open questions concerning the behavior of atmospheric circulation and climate across a wider range of parameter space than encountered in the solar system. Major questions emphasized in this review include the following: What is the dependence of three-dimensional temperature structure, humidity, and horizontal heat flux on planetary rotation rate, incident stellar flux, atmospheric mass, atmospheric composition, planetary radius and gravity, and other conditions? What are the regimes of atmospheric wind? What is the influence of an ocean in modulating the atmospheric climate? Under what conditions can clouds form, and what is their three-dimensional distribution for planets of various types? How do seasonal cycles (due to nonzero obliquity and/or orbital eccentricity) generally affect the atmospheric circulation and climate on planets? What is the role of the atmospheric circulation — through its control of temperature, humidity, cloudiness, and precipitation — in affecting longer-term climatic processes including runaway greenhouses, ice-age cycles, transitions to snowball Earth states, the carbonate-silicate feedback, and collapse of condensable atmospheric constituents onto the poles or nightside? And what is the continuum of atmospheric behaviors on exoplanets ranging from sub-Earth-sized terrestrial planets to super-Earths and mini-Neptunes? While a detailed understanding of particular exoplanets must await observations, significant insights into the fundamental physical mechanisms controlling atmospheric circulation and climate — of planets in general — can be made now with current theoretical and modeling tools. This will not only lay the groundwork for understanding future observations but will place the atmospheric dynamics and climate of solar-system worlds, including Earth, into its proper planetary context.

Acknowledgments. We thank J. Mitchell and an anonymous referee for thorough reviews of the manuscript. This paper was supported by NASA grants NNX10AB91G and NNX12AI79G to A.P.S. T.M.M. was supported by a Princeton Center for Theoretical Science fellowship.

REFERENCES

Abbot D. S. and Pierrehumbert R. T. (2010) Mudball: Surface dust and snowball Earth deglaciation. *J. Geophys. Res.–Atmos., 115(D14),* 3104.

Abbot D. S., Voigt A., Branson M., Pierrehumbert R. T., Pollard D., Le Hir G., and Koll D. D. B. (2012) Clouds and snowball Earth deglaciation. *Geophys. Res. Lett., 39,* 20711.

Abe Y., Abe-Ouchi A., Sleep N. H., and Zahnle K. J. (2011) Habitable zone limits for dry planets. *Astrobiology, 11,* 443–460.

Adam O. and Paldor N. (2009) Global circulation in an axially symmetric shallow water model forced by equinoctial differential heating. *J. Atmos. Sci., 66,* 1418.

Adam O. and Paldor N. (2010) Global circulation in an axially symmetric shallow-water model, forced by off-equatorial differential heating. *J. Atmos. Sci., 67,* 1275–1286.

Allen M. R. and Ingram W. J. (2002) Constraints on future changes in climate and the hydrologic cycle. *Nature, 419,* 224–232.

Andrews D. G., Holton J. R., and Leovy C. B. (1987) *Middle Atmosphere Dynamics.* Academic, San Diego.

Arnold N. P., Tziperman E., and Farrell B. (2012) Abrupt transition to strong superrotation driven by equatorial wave resonance in an idealized GCM. *J. Atmos. Sci., 69,* 626–640.

Back L. E. and C. S. Bretherton (2009a) On the relationship between SST gradients, boundary layer winds, and convergence over the tropical oceans. J. Climate, 22, 4182–4196.

Back L. E. and Bretherton C. S. (2009b) A simple model of climatological rainfall and vertical motion patterns over the tropical oceans. *J. Climate, 22,* 6477–6497.

Bala G., Duffy P. B., and Taylor K. E. (2008) Impact of geoengineering schemes on the global hydrological cycle. *Proc. Natl. Acad. Sci., 105,* 7664–7669.

Barman T. (2007) Identification of absorption features in an extrasolar planet atmosphere. *Astrophys. J. Lett., 661,* L191–L194.

Barry L., Craig G. C., and Thuburn J. (2002) Poleward heat transport by the atmospheric heat engine. *Nature, 415,* 774–777.

Bean J. L., Kempton E., and Homeier D. (2010) A ground-based transmission spectrum of the super-Earth exoplanet GJ 1214b. *Nature, 468,* 669–672.

Beron-Vera F. J., Brown M. G., Olascoaga M. J., Rypina I. I., Koçak H., and Udovydchenkov I. A. (2008) Zonal jets as transport barriers in planetary atmospheres. *J. Atmos. Sci., 65,* 3316.

Berta Z. K., et al. (2012) The flat transmission spectrum of the super-Earth GJ1214b fromWide Field Camera 3 on the Hubble Space Telescope. *Astrophys. J., 747,* 35.

Betts A. K. (1974) Further comments on comparison of the equivalent potential temperature and the static energy. *J. Atmos. Sci., 31,* 1713–1715.

Bony S., et al. (2006) How well do we understand and evaluate climate change feedback processes? *J. Climate, 19,* 3445–3482.

Bordoni S. and T. Schneider (2008) Monsoons as eddy-mediated regime transitions of the tropical overturning circulation. *Nature Geosci., 1,* 515–519.

Bordoni S. and Schneider T. (2010) Regime transitions of steady and time-dependent Hadley circulations: Comparison of axisymmetric and eddy-permitting simulations. *J. Atmos. Sci., 67,* 1643–1654.

Borucki W. J., et al. (2011) Characteristics of planetary candidates observed by Kepler. II. Analysis of the first four months of data. *Astrophys. J., 736,* 19.

BoruckiW. J., et al. (2012) Kepler-22b: A 2.4 Earth-radius planet in the habitable zone of a Sun-like star. *Astrophys. J., 745,* 120.

Borucki W. J., et al. (2013) Kepler-62: A five-planet system with planets of 1.4 and 1.6 Earth radii in the habitable zone. *Science, 340(6132),* 587–590, DOI: 10.1126/science.1234702.

Bretherton C. S. and Smolarkiewicz P. K. (1989) Gravity waves, compensating subsidence and detrainment around cumulus clouds. *J. Atmos. Sci., 46,* 740–759.

Caballero R. and Hanley J. (2012) Midlatitude eddies, stormtrack diffusivity and poleward moisture transport in warm climates. *J. Atmos. Sci.,* in press.

Caballero R. and Huber M. H. (2010) Spontaneous transition to superrotation in warm climates simulated by CAM3. *Geophys. Res. Lett.,* in press.

Caballero R., Pierrehumbert R. T., and Mitchell J. L. (2008) Axisymmetric, nearly inviscid circulations in non-condensing radiative-convective atmospheres. *Q. J. R. Meteor. Soc., 134,* 1269–1285.

Castan T. and Menou K. (2011) Atmospheres of hot super- Earths. *Astrophys. J. Lett., 743,* L36.

Chang E. K. M. (2001) GCM and observational diagnoses of the seasonal and interannual variations of the Pacific storm track during the cool season. *J. Atmos. Sci., 58,* 1784–1800.

Chang E. K. M., Lee S., and Swanson K. L. (2002) Storm track dynamics. *J. Climate, 15,* 2163–2183.

Charbonneau D., et al. (2009) A super-Earth transiting a nearby low-mass star. *Nature, 462,* 891–894.

Charney J. G. (1963) A note on large-scale motions in the tropics. *J. Atmos. Sci., 20,* 607–608.

Chassefière E., Wieler R., Marty B., and Leblanc F. (2012) The evolution of Venus: Present state of knowledge and future exploration. *Planet. Space Sci., 63,* 15–23.

Cho J. Y.-K. and Polvani L. M. (1996) The morphogenesis of bands and zonal winds in the atmospheres on the giant outer planets. *Science, 8(1),* 1–12.

Chou C. and Neelin J. D. (2004) Mechanisms of global warming impacts on regional tropical precipitation. *J. Climate, 17,* 2688–2701.

Cooper C. S. and Showman A. P. (2005) Dynamic meteorology at the photosphere of HD 209458b. *Astrophys. J. Lett., 629,* L45–L48.

Couhert A., Schneider T., Li J., Waliser D. E., and Tompkins A. M. (2010) The maintenance of the relative humidity of the subtropical free troposphere. *J. Climate, 23,* 390–403.

Cowan N. B. and Agol E. (2011) The statistics of albedo and heat recirculation on hot exoplanets. *Astrophys. J., 729,* 54.

Cowan N. B., et al. (2009) Alien maps of an ocean-bearing world. *Astrophys. J., 700,* 915–923.

de Wit J., Gillon M., Demory B.-O., and Seager S. (2012) Towards consistent mapping of distant worlds: Secondary-eclipse scanning of the exoplanet HD 189733b. *Astron. Astrophys., 548,* A128.

Del Genio A. D. and Suozzo R. J. (1987) A comparative study of rapidly and slowly rotating dynamical regimes in a terrestrial general circulation model. *J. Atmos. Sci., 44,* 973–986.

Del Genio A. D. and Zhou W. (1996) Simulations of superrotation on slowly rotating planets: Sensitivity to rotation and initial condition. *Icarus, 120,* 332–343.

Del Genio A. D., Zhou W., and Eichler T. P. (1993) Equatorial superrotation in a slowly rotating GCM — Implications for Titan and Venus. *Icarus, 101,* 1–17.

Del Genio A. D., Achterberg R. K., Baines K. H., Flasar F. M., Read P. L., Sánchez-Lavega A., and Showman A. P. (2009) Saturn atmospheric structure and dynamics. In *Saturn from Cassini-Huygens* (M. K. Dougherty et al., eds.), p. 113. Springer, Dordrecht.

Deming D. and Seager S. (2009) Light and shadow from distant worlds. *Nature, 462,* 301–306.

Demory B.-O., Gillon M., Seager S., Benneke B., Deming D.,

and Jackson B. (2012) Detection of thermal emission from a super-Earth. *Astrophys. J. Lett., 751,* L28.

Désert J.-M., Bean J., Miller-Ricci Kempton E., Berta Z. K., Charbonneau D., Irwin J., Fortney J., Burke C. J., and Nutzman P. (2011) Observational evidence for a metal-rich atmosphere on the super-Earth GJ1214b. *Astrophys. J. Lett., 731,* L40.

DeWitt H. L., Trainer M. G., Pavlov A. A., Hasenkopf C. A., Aiken A. C., Jimenez J. L., McKay C. P., Toon O. B., and Tolbert M. A. (2009) Reduction in haze formation rate on prebiotic Earth in the presence of hydrogen. *Astrobiology, 9,* 447–453.

Dritschel D. G. and McIntyre M. E. (2008) Multiple jets as PV staircases: The Phillips effect and the resilience of eddy-transport barriers. *J. Atmos. Sci., 65,* 855.

Dritschel D. G. and Scott R. K. (2011) Jet sharpening by turbulent mixing. *Philos. Trans. R. Soc. A, 369,* 754–770.

Dunkerton T. J. and Scott R. K. (2008) A barotropic model of the angular momentum conserving potential vorticity staircase in spherical geometry. *J. Atmos. Sci., 65,* 1105.

Edson A., Lee S., Bannon P., Kasting J. F., and Pollard D. (2011) Atmospheric circulations of terrestrial planets orbiting low-mass stars. *Icarus, 212,* 1–13.

Edson A. R., Kasting J. F., Pollard D., Lee S., and Bannon P. R. (2012) The carbonate-silicate cycle and CO_2/climate feedbacks on tidally locked terrestrial planets. *Astrobiology, 12,* 562–571.

Emanue K. A. (1994) *Atmospheric Convection.* Oxford Univ., New York.

Enderton D. and Marshall J. (2009) Explorations of atmosphere ocean-ice climates on an aquaplanet and their meridional energy transports. *J. Atmos. Sci., 66,* 1593–1611.

Fang M. and Tung K. K. (1996) A simple study of nonlinear Hadley circulation with an ITCZ: Analytic and numerical solutions. *J. Atmos. Sci., 53,* 1241–1261.

Farrell B. F. (1990) Equable climate dynamics. *J. Atmos. Sci., 47,* 2986–2995.

Ferrari R. and Ferreira D. (2011) What processes drive the ocean heat transport? *Ocean Modelling, 38,* 171–186.

Flasar F. M., Baines K. H., Bird M. K., Tokano T., and West R. A. (2009) Atmospheric dynamics and meteorology. In *Titan from Cassini-Huygens* (R. H. Brown et al., eds.), pp. 323–352. Springer, Dordrecht.

Fressin F., et al. (2012) Two Earth-sized planets orbiting Kepler-20. *Nature, 482,* 195–198.

Frierson D. M. W. (2007) The dynamics of idealized convection schemes and their effect on the zonally averaged tropical circulation. *J. Atmos. Sci., 64,* 1959–1976.

Frierson D. M. W., Held I. M., and Zurita-Gotor P. (2006) A gray-radiation aquaplanet moist GCM. Part I: Static stability and eddy scale. *J. Atmos. Sci., 63,* 2548–2566.

Fujii Y. and Kawahara H. (2012) Mapping Earth analogs from photometric variability: Spin-orbit tomography for planets in inclined orbits. *Astrophys. J., 755,* 101.

Fujii Y., Kawahara H., Suto Y., Fukuda S., Nakajima T., Livengood T. A., and Turner E. L. (2011) Colors of a second Earth. II. Effects of clouds on photometric characterization of Earthlike exoplanets. *Astrophys. J., 738,* 184.

Galewsky J., Sobel A., and Held I. (2005) Diagnosis of subtropical humidity dynamics using tracers of last saturation. *J. Atmos. Sci., 62,* 3353–3367.

Garcia R. R. and Salby M. L. (1987) Transient response to localized episodic heating in the tropics. Part II: Far-field behavior. *J. Atmos. Sci., 44,* 499–532.

Garratt J. R. (1994) *The Atmospheric Boundary Layer.* Cambridge Univ., New York.

Gierasch P. J., et al. (1997) The general circulation of the Venus atmosphere: An assessment. In *Venus II: Geology, Geophysics, Atmosphere, and Solar Wind Environment* (S. W. Bougher et al., eds.), pp. 459–500. Univ. of Arizona, Tucson.

Gill A. E. (1980) Some simple solutions for heat-induced tropical circulation. *Q. J. R. Meteor. Soc., 106,* 447–462.

Gillmann C., Chassefière E., and Lognonné P. (2009) A consistent picture of early hydrodynamic escape of Venus atmosphere explaining present Ne and Ar isotopic ratios and low oxygen atmospheric content. *Earth Planet. Sci. Lett., 286,* 503–513.

Gómez-Leal I., Pallé E., and Selsis F. (2012) Photometric variability of the disk-integrated thermal emission of the Earth. *Astrophys. J., 752,* 28.

Green J. S. A. (1970) Transfer properties of the large-scale eddies and the general circulation of the atmosphere. *Q. J. R. Meteor. Soc., 96,* 157–185.

Haine T. W. N. and Marshall J. (1998) Gravitational, symmetric, and baroclinic instability of the ocean mixed layer. *J. Phys. Oceanogr., 28,* 634–658.

Hartmann D. L. (1994a) A PV view of zonal mean flow vacillation. J. Atmos. Sci., submitted.

Hartmann D. L. (1994b) *Global Physical Climatology, Vol. 56.* Academic, San Diego.

Hayashi Y.-Y., Ishioka K., Yamada M., and Yoden S. (2000) Emergence of circumpolar vortex in two dimensional turbulence on a rotating sphere. In *Fluid Mechanics and Its Applications, Vol. 58: Proceedings of the IUTAM Symposium on Developments in Geophysical Turbulence* (R. M. Kerr and Y. Kimura, eds.), pp. 179–192. Kluwer, Dordrecht.

Hayashi Y.-Y., Nishizawa S., Takehiro S.-I., Yamada M., Ishioka K., and Yoden S. (2007) Rossby waves and jets in two-dimensional decaying turbulence on a rotating sphere. *J. Atmos. Sci., 64,* 4246–4269.

Hedelt P., Alonso R., Brown T., Collados Vera M., Rauer H., Schleicher H., Schmidt W., Schreier F., and Titz R. (2011) Venus transit 2004: Illustrating the capability of exoplanet transmission spectroscopy. *Astron. Astrophys., 533,* A136.

Held I. M. (1975) Momentum transport by quasi-geostrophic eddies. *J. Atmos. Sci., 32,* 1494–1497.

Held I. M. (1982) On the height of the tropopause and the static stability of the troposphere. *J. Atmos. Sci., 39,* 412–417.

Held I. M. (1999a) The macroturbulence of the troposphere. *Tellus, 51A-B,* 59–70.

Held I. M. (1999b) Equatorial superrotation in Earthlike atmospheric models. Bernhard Haurwitz Memorial Lecture, American Meteorological Society. Available online at *www.gfdl.noaa.gov/cms-filesystem-action/user_files/ih/lectures/super.pdf.*

Held I. M. (2000) The general circulation of the atmosphere. Paper presented at 2000 Woods Hole Oceanographic Institute Geophysical Fluid Dynamics Program, Woods Hole Oceanographic Institute, Woods Hole, Massachusetts. Available online at *www.whoi.edu/page.do?pid=13076.*

Held I. M. (2001) The partitioning of the poleward energy transport between the tropical ocean and atmosphere. *J. Atmos. Sci., 58,* 943–948.

Held I. M. and Hou A. Y. (1980) Nonlinear axially symmetric circulations in a nearly inviscid atmosphere. *J. Atmos. Sci., 37,* 515–533.

Held I. M. and Larichev V. D. (1996) A scaling theory for horizontally homogeneous, baroclinically unstable flow on a beta plane. *J. Atmos. Sci., 53,* 946–952.

Held I. M. and Shell K. M. (2012) Using relative humidity as a state variable in climate feedback analysis. *J. Climate, 25,* 2578–2582.

Held I. M. and Soden B. J. (2000) Water vapor feedback anad global warming. *Annu. Rev. Energy Environ., 25,* 441–475.

Held I. M. and Soden B. J. (2006) Robust responses of the hydrological cycle to global warming. *J. Climate, 19,* 5686.

Held I. M., Ting M., and Wang H. (2002) Northern winter stationary waves: Theory and modeling. *J. Climate, 15,* 2125–2144.

Heng K. and Kopparla P. (2012) On the stability of super-Earth atmospheres. *Astrophys. J., 754,* 60.

Heng K. and Vogt S. S. (2010) Gliese 581g as a scaled-up version of Earth: Atmospheric circulation simulations. ArXiv e-prints, arXiv:1010.4719.

Heng K., Frierson D. M. W., and Phillipps P. J. (2011a) Atmospheric circulation of tidally locked exoplanets: II. Dual-band radiative transfer and convective adjustment. *Mon. Not. R. Astron. Soc., 418,* 2669–2696.

Heng K., Menou K., and Phillipps P. J. (2011b) Atmospheric circulation of tidally locked exoplanets: A suite of benchmark tests for dynamical solvers. *Mon. Not. R. Astron. Soc., 413,* 2380–2402.

Herrnstein A. and Dowling T. E. (2007) Effects of topography on the spin-up of a Venus atmospheric model. *J. Geophys. Res.–Planets, 112(E11),* DOI: 10.1029/2006JE002804.

Hide R. (1969) Dynamics of the atmospheres of the major planets with an appendix on the viscous boundary layer at the rigid bounding surface of an electrically-conducting rotating fluid in the presence of a magnetic field. *J. Atmos. Sci., 26,* 841–853.

Hoffman P. F. and Schrag D. (2002) Review article: The snowball Earth hypothesis: Testing the limits of global change. *Terra Nova, 14,* 129–155.

Hollingsworth J. L., Young R. E., Schubert G., Covey C., and Grossman A. S. (2007) A simple-physics global circulation model for Venus: Sensitivity assessments of atmospheric superrotation. *Geophys. Res. Lett., 34,* 5202.

Holton J. R. (2004) *An Introduction to Dynamic Meteorology, 4th edition.* Academic, San Diego.

Hoskins B. J. and Karoly D. J. (1981) The steady linear response of a spherical atmosphere to thermal and orographic forcing. *J. Atmos. Sci., 38,* 1179–1196.

Hoskins B., Neale R., Rodwell M., and Yang G. (1999) Aspects of the large-scale tropical atmospheric circulation. *Tellus B–Chem. Phys. Meteor., 51,* 33–44.

Hu Y. and Ding F. (2011) Radiative constraints on the habitability of exoplanets Gliese 581c and Gliese 581d. *Astron. Astrophys., 526,* A135.

Huang H.-P. and Robinson W. A. (1998) Two-dimensional turbulence and persistent zonal jets in a global barotropic model. *J. Atmos. Sci., 55,* 611–632.

Hunt B. G. (1979) The influence of the Earth's rotation rate on the general circulation of the atmosphere. *J. Atmos. Sci., 36,* 1392–1408.

Ingersoll A. P. (1969) The runaway greenhouse: A history of water on Venus. *J. Atmos. Sci., 26,* 1191–1198.

Ingersoll A. P. (1989) Io meteorology — How atmospheric pressure is controlled locally by volcanos and surface frosts. *Icarus, 81,* 298–313.

Ingersoll A. P., Summers M. E., and Schlipf S. G. (1985) Supersonic meteorology of Io — Sublimation-driven flow of SO_2. *Icarus, 64,* 375–390.

Ingersoll A. P., Dowling T. E., Gierasch P. J., Orton G. S., Read P. L., Sánchez-Lavega A., Showman A. P., Simon- Miller A. A., and Vasavada A. R. (2004) Dynamics of Jupiter's atmosphere. In *Jupiter: The Planet, Satellites and Magnetosphere* (F. Bagenal et al., eds.), pp. 105–128. Cambridge Univ., New York.

Ishioka K., Yamada M., Hayashi Y.-Y., and Yoden S. (1999) Pattern formation from two-dimensional decaying turbulence on a rotating sphere. Nagare Multimedia, The Japan Society of Fluid Mechanics. Available online at *www.nagare.or.jp/mm/99/ishioka/.*

Ishiwatari M., Takehiro S.-I., Nakajima K., and Hayashi Y.-Y. (2002) A numerical study on appearance of the runaway greenhouse state of a three-dimensional gray atmosphere. *J. Atmos. Sci., 59,* 3223–3238.

Ishiwatari M., Nakajima K., Takehiro S., and Hayashi Y.-Y. (2007) Dependence of climate states of gray atmosphere on solar constant: From the runaway greenhouse to the snowball states. *J. Geophys. Res.–Atmos., 112(D11),* 13120.

Jansen M. and Ferrari R. (2012) Macroturbulent equilibration in a thermally forced primitive equation system. *J. Atmos. Sci., 69,* 695–713.

Joshi M. (2003) Climate model studies of synchronously rotating planets. *Astrobiology, 3,* 415–427.

Joshi M. M. and Haberle R. M. (2012) Suppression of the water ice and snow albedo feedback on planets orbiting red dwarf stars and the subsequent widening of the habitable zone. *Astrobiology, 12,* 3–8.

Joshi M. M., Haberle R. M., and Reynolds R. T. (1997) Simulations of the atmospheres of synchronously rotating terrestrial planets orbiting M dwarfs: Conditions for atmospheric collapse and the implications for habitability. *Icarus, 129,* 450–465.

Juckes M. N. (2000) The static stability of the midlatitude troposphere: The relevance of moisture. *J. Atmos. Sci., 57,* 3050–3057.

Kaltenegger L., Segura A., and Mohanty S. (2011) Model spectra of the first potentially habitable super-Earth — Gl581d. *Astrophys. J., 733,* 35.

Karalidi T., Stam D. M., and Hovenier J. W. (2012) Looking for the rainbow on exoplanets covered by liquid and icy water clouds. *Astron. Astrophys., 548,* A90.

Kaspi Y. and Schneider T. (2011) Downstream self-destruction of storm tracks. *J. Atmos. Sci., 68,* 2459–2464.

Kaspi Y. and Showman A. P. (2012) Three-dimensional atmospheric circulation and climate of terrestrial exoplanets and super Earths. *AAS/Division for Planetary Sciences Meeting Abstracts, Vol. 44,* #208.04.

Kasting J. F. (1988) Runaway and moist greenhouse atmospheres and the evolution of earth and Venus. *Icarus, 74,* 472–494.

Kasting J. F., Whitmire D. P., and Reynolds R. T. (1993) Habitable zones around main sequence stars. *Icarus, 101,* 108–128.

Kawahara H. and Fujii Y. (2011) Mapping clouds and terrain of Earth-like planets from photometric variability: Demonstration with planets in face-on orbits. *Astrophys. J. Lett., 739,* L62.

Kim H.-K. and Lee S. (2001) Hadley cell dynamics in a primitive equation model. Part II: Nonaxisymmetric flow. *J. Atmos. Sci., 58,* 2859–2871.

Kite E. S., Gaidos E., and Manga M. (2011) Climate instability on tidally locked exoplanets. *Astrophys. J., 743,* 41.

Knutson H. A., Charbonneau D., Allen L. E., Fortney J. J., Agol E., Cowan N. B., Showman A. P., Cooper C. S., and Megeath S. T. (2007) A map of the day-night contrast of the extrasolar planet HD 189733b. *Nature, 447,* 183–186.

Kombayashi M. (1967) Discrete equilibrium temperatures of a

hypothetical planet with the atmosphere and the hydrosphere of one component-two phase system under constant solar radiation. *J. Meteor. Soc. Japan, 45,* 137–138.

Kraucunas I. and Hartmann D. L. (2005) Equatorial superrotation and the factors controlling the zonal-mean zonal winds in the tropical upper troposphere. *J. Atmos. Sci., 62,* 371–389.

Kuo A. C. and Polvani L. M. (1997) Time-dependent fully nonlinear geostrophic adjustment. *J. Phys. Oceanogr., 27,* 1614–1634.

Larichev V. D. and Held I. M. (1995) Eddy amplitudes and fluxes in a homogenous model of fully developed baroclinic instability. *J. Phys. Oceanogr., 25,* 2285–2297.

Le Hir G., Ramstein G., Donnadieu Y., and Goddéris Y. (2008) Scenario for the evolution of atmospheric pCO_2 during a snowball Earth. *Geology, 36(1),* 47–50.

Le Hir G., Donnadieu Y., Goddéris Y., Pierrehumbert R. T., Halverson G. P., Macouin M., Nédélec A., and Ramstein G. (2009) The snowball Earth aftermath: Exploring the limits of continental weathering processes. *Earth Planet. Sci. Lett., 277(3),* 453–463.

Lebonnois S., Covey C., Grossman A., Parish H., Schubert G., Walterscheid R., Lauritzen P., and Jablonowski C. (2012) Angular momentum budget in general circulation models of superrotating atmospheres: A critical diagnostic. *J. Geophys. Res.–Planets, 117(E16),* 12004.

Leconte J., Forget F., Charnay B., Wordsworth R., Selsis F., Millour E., and Spiga A. (2013) 3D climate modeling of closein land planets: Circulation patterns, climate moist bistability, and habitability. *Astron. Astrophys.,* in press.

Lee C., Lewis S. R., and Read P. L. (2007) Superrotation in a Venus general circulation model. *J. Geophys. Res.–Planets, 112(E4),* DOI: 10.1029/2006JE002874.

Léger A., et al. (2004) A new family of planets? "Ocean-Planets." *Icarus, 169,* 499–504.

Léger A., et al. (2011) The extreme physical properties of the CoRoT-7b super-Earth. *Icarus, 213,* 1–11.

Leovy C. (2001) Weather and climate on Mars. *Nature, 412,* 245–249.

Lindzen R. S. and Hou A. V. (1988) Hadley circulations for zonally averaged heating centered off the equator. *J. Atmos. Sci., 45,* 2416–2427.

Lindzen R. S. and Nigam S. (1987) On the role of sea surface temperature gradients in forcing low-level winds and convergence in the tropics. *J. Atmos. Sci., 44,* 2418–2436.

Lissauer J. J., et al. (2011) A closely packed system of low-mass, low-density planets transiting Kepler-11. *Nature, 470,* 53–58.

Lopez E. D., Fortney J. J., and Miller N. (2012) How thermal evolution and mass-loss sculpt populations of super-Earths and sub-Neptunes: Application to the Kepler-11 system and beyond. *Astrophys. J., 761,* 59.

Lorenz E. N. (1955) Available potential energy and the maintenance of the general circulation. *Tellus, 7,* 157–167.

Lorenz E. N. (1978) Available energy and the maintenance of a moist circulation. *Tellus, 30,* 15–31.

Lorenz E. N. (1979) Numerical evaluation of moist available energy. *Tellus, 31,* 230–235.

Majeau C., Agol E., and Cowan N. B. (2012) A two-dimensional infrared map of the extrasolar planet HD 189733b. *Astrophys. J. Lett., 747,* L20.

Maltrud M. E. and Vallis G. K. (1993) Energy and enstrophy transfer in numerical simulations of two-dimensional turbulence. *Phys. Fluids, 5,* 1760–1775.

Marcus P. S. and Lee C. (1998) A mode! for eastward and westward jets in laboratory experiments and planetary atmospheres. *Phys. Fluids, 10,* 1474–1489.

Marshall J. and Plumb R. A. (2007) *Atmosphere, Ocean, and Climate Dynamics.* Academic, San Diego.

Matsuno T. (1966) Quasi-geostrophic motions in the equatorial area. *J. Meteorol. Soc. Japan, 44,* 25–43.

Mayor M., et al. (2009) The HARPS search for southern extrasolar planets. XVIII. An Earth-mass planet in the GJ 581 planetary system. *Astron. Astrophys., 507,* 487–494.

McWilliams J. C. (2006) *Fundamentals of Geophysical Fluid Dynamics.* Cambridge Univ., Cambridge.

Menou K. and Rauscher E. (2009) Atmospheric circulation of hot Jupiters: A shallow three-dimensional model. *Astrophys. J., 700,* 887–897.

Merlis T. M. and Schneider T. (2010) Atmospheric dynamics of Earth-like tidally locked aquaplanets. *J. Adv. Model. Earth Syst., 2,* 13, DOI: 10.3894/JAMES.2010.2.13.

Merlis T. M., Schneider T., Bordoni S., and Eisenman I. (2012) Hadley circulation response to orbital precession. Part I: Aquaplanets. *J. Climate, 26,* 740–753.

Miller S. L. (1953) A production of amino acids under possible primitive Earth conditions. *Science, 117,* 528–529.

Mitchell J. L. and Vallis G. K. (2010) The transition to superrotation in terrestrial atmospheres. *J. Geophys. Res.–Planets, 115(E14),* 12008.

Mitchell J. L., Pierrehumbert R. T., Frierson D. M.W., and Caballero R. (2006) The dynamics behind Titan's methane clouds. *Proc. Natl. Acad. Sci., 103,* 18421–18426.

Mitchell J. L., Pierrehumbert R. T., Frierson D. M.W., and Caballero R. (2009) The impact of methane thermodynamics on seasonal convection and circulation in a model Titan atmosphere. *Icarus, 203,* 250–264.

Navarra A. and Boccaletti G. (2002) Numerical general circulation experiments of sensitivity to Earth rotation rate. *Climate Dyn., 19,* 467–483.

Neelin J. D. and Held I. M. (1987) Modeling tropical convergence based on the moist static energy budget. *Mon. Weather Rev., 115,* 3–12.

Nettelmann N., Fortney J. J., Kramm U., and Redmer R. (2011) Thermal evolution and structure models of the transiting super-Earth GJ 1214b. *Astrophys. J., 733,* 2.

Niemann H. B., et al. (2005) The abundances of constituents of Titan's atmosphere from the GCMS instrument on the Huygens probe. *Nature, 438,* 779–784.

Norton W. A. (2006) Tropical wave driving of the annual cycle in tropical tropopause temperatures. Part II: Model results. *J. Atmos. Sci., 63,* 1420–1431.

Nozawa T. and Yoden S. (1997) Formation of zonal band structure in forced two-dimensional turbulence on a rotating sphere. *Phys. Fluids, 9,* 2081–2093.

O'Gorman P. A. (2010) Understanding the varied response of the extratropical storm tracks to climate change. *Proc. Natl. Acad. Sci., 107,* 19176–19180.

O'Gorman P. A. (2011) The effective static stability experienced by eddies in a moist atmosphere. *J. Atmos. Sci., 68,* 75–90.

O'Gorman P. A. and Schneider T. (2008) The hydrological cycle over a wide range of climates simulated with an idealized GCM. *J. Climate, 21,* 3815–3832.

O'Gorman P. A., Allan R. P., Byrne M. P., and Previdi M. (2011) Energetic constraints on precipitation under climate change. *Surv. Geophys., 33,* 585–608, DOI: 10.1007/s10712-011-9159-6.

Olbers R. G., Willebrand J., and Eden C. (2011) *Ocean Dynamics.* Springer-Verlag, New York.

Pallé E., Ford E. B., Seager S., Montañés-Rodríguez P., and Vazquez M. (2008) Identifying the rotation rate and the presence of dynamicweather on extrasolar Earth-like planets from photometric observations. *Astrophys. J., 676,* 1319–1329.

Parish H. F., Schubert G., Covey C., Walterscheid R. L., Grossman A., and Lebonnois S. (2011) Decadal variations in a Venus general circulation model. *Icarus, 212,* 42–65.

Pavan V. and Held I. M. (1996) The diffusive approximation for eddy fluxes in baroclinically unstable flows. *J. Atmos. Sci., 53,* 1262–1272.

Pedlosky J. (1987) *Geophysical Fluid Dynamics, 2nd edition.* Springer-Verlag, New York.

Pedlosky J. (2003) *Waves in the Ocean and Atmosphere.* Springer-Verlag, New York.

Pedlosky J. (2004) *Ocean Circulation Theory.* Springer-Verlag, New York.

Peixoto J. P. and Oort A. H. (1992) *Physics of Climate.* American Institute of Physics, New York.

Perna R., Menou K., and Rauscher E. (2010) Magnetic drag on hot Jupiter atmospheric winds. *Astrophys. J., 719,* 1421–1426.

Perna R., Heng K., and Pont F. (2012) The effects of irradiation on hot jovian atmospheres: Heat redistribution and energy dissipation. *Astrophys. J., 751,* 59.

Pierrehumbert R. and Gaidos E. (2011) Hydrogen greenhouse planets beyond the habitable zone. *Astrophys. J. Lett., 734,* L13.

Pierrehumbert R. T. (1995) Thermostats, radiator fins, and the local runaway greenhouse. *J. Atmos. Sci., 52,* 1784–1806.

Pierrehumbert R. T. (2002) The hydrologic cycle in deep-time climate problems. *Nature, 419,* 191–198.

Pierrehumbert R. T. (2010) *Principles of Planetary Climate.* Cambridge Univ., Cambridge.

Pierrehumbert R. T. (2011) A palette of climates for Gliese 581g. *Astrophys. J. Lett., 726,* L8.

Pierrehumbert R. T. and Swanson K. L. (1995) Baroclinic instability. *Annu. Rev. Fluid Mech., 27,* 419–467.

Pierrehumbert R. T., Brogniez H., and Roca R. (2007) On the relative humidity of the atmosphere. In *The Global Circulation of the Atmosphere* (T. Schneider and A. H. Sobel, eds.), pp. 143–185. Princeton Univ., Princeton.

Pierrehumbert R. T., Abbot D. S., Voigt A., and Koll D. (2011) Climate of the Neoproterozoic. *Annu. Rev. Earth Planet. Sci., 39,* 417–460.

Polvani L. M. and Sobel A. H. (2002) The Hadley circulation and the weak temperature gradient approximation. *J. Atmos. Sci., 59,* 1744–1752.

Polvani L. M., Waugh D. W., and Plumb R. A. (1995) On the subtropical edge of the stratospheric surf zone. *J. Atmos. Sci., 52,* 1288–1309.

Pont F., Sing D. K., Gibson N. P., Aigrain S., Henry G., and Husnoo N. (2012) The prevalence of dust on the exoplanet HD 189733b from Hubble and Spitzer observations. ArXiv e-prints, arXiv:1210.4163.

Randel W. J. and Held I. M. (1991) Phase speed spectra of transient eddy fluxes and critical layer absorption. *J. Atmos. Sci., 48,* 688–697.

Rauscher E. and Menou K. (2010) Three-dimensional modeling of hot Jupiter atmospheric flows. *Astrophys. J., 714,* 1334–1342.

Rauscher E. and Menou K. (2012) The role of drag in the energetics of strongly forced exoplanet atmospheres. *Astrophys. J., 745,* 78.

Raymond D. J., Sessions S., Sobel A. H., and Fuchs Z. (2009) The mechanics of gross moist stability. *J. Adv. Model. Earth Syst., 1,* Article #9, 20 pp.

Read P. L. (2011) Dynamics and circulation regimes of terrestrial planets. *Planet. Space Sci., 59,* 900–914.

Read P. L. and Lewis S. R. (2004) *The Martian Climate Revisited.* Springer/Praxis, New York.

Read P. L., Yamazaki Y. H., Lewis S. R., Williams P. D., Wordsworth R., Miki-Yamazaki K., Sommeria J., and Didelle H. (2007) Dynamics of convectively driven banded jets in the laboratory. *J. Atmos. Sci., 64,* 4031.

Rennó N. O. (1997) Multiple equilibria in radiative-convective atmospheres. *Tellus Series A, 49,* 423.

Rhines P. B. (1975) Waves and turbulence on a beta-plane. *J. Fluid Mech., 69,* 417–443.

Richardson L. J., Deming D., Horning K., Seager S., and Harrington J. (2007) A spectrum of an extrasolar planet. *Nature, 445,* 892–895.

Robinson T. D., et al. (2011) Earth as an extrasolar planet: Earth model validation using EPOXI Earth observations. *Astrobiology, 11,* 393–408.

Rogers L. A. and Seager S. (2010) Three possible origins for the gas layer on GJ 1214b. *Astrophys. J., 716,* 1208–1216.

Rogers L. A., Bodenheimer P., Lissauer J. J., and Seager S. (2011) Formation and structure of low-density exo-Neptunes. *Astrophys. J., 738,* 59.

Romps D. M. and Kuang Z. (2010) Do undiluted convective plumes exist in the upper tropical troposphere? *J. Atmos. Sci., 67,* 468–484.

Salby M. L. and Garcia R. R. (1987) Transient response to localized episodic heating in the tropics. Part I: Excitation and short-time near-field behavior. *J. Atmos. Sci., 44,* 458–498.

Sanromá E. and Pallé E. (2012) Reconstructing the photometric light curves of Earth as a planet along its history. *Astrophys. J., 744,* 188.

Saravanan R. (1993) Equatorial superrotation and maintenance of the general circulation in two-level models. *J. Atmos. Sci., 50,* 1211–1227.

Sarmiento J. L. and Gruber N. (2006) *Ocean Biogeochemical Dynamics.* Princeton Univ., Princeton.

Schneider T. (2006) The general circulation of the atmosphere. *Annu. Rev. Earth Planet. Sci., 34,* 655–688.

Schneider T. (2007) The thermal stratification of the extratropical troposphere. In *The Global Circulation of the Atmosphere* (T. Schneider and A. H. Sobel, eds.), pp. 47–77. Princeton Univ., Princeton.

Schneider T. and Bordoni S. (2008) Eddy-mediated regime transitions in the seasonal cycle of a Hadley circulation and implications for monsoon dynamics. *J. Atmos. Sci., 65,* 915.

Schneider T. and Walker C. C. (2006) Self-organization of atmospheric macroturbulence into critical states of weak nonlinear eddy eddy interactions. *J. Atmos. Sci., 63,* 1569–1586.

Schneider T. and Walker C. C. (2008) Scaling laws and regime transitions of macroturbulence in dry atmospheres. *J. Atmos. Sci., 65,* 2153–2173.

Schneider T., O'Gorman P. A., and Levine X. J. (2010) Water vapor and the dynamics of climate changes. *Rev. Geophys., 48,* 3001.

Schumacher C., Houze R. A. Jr., and Kraucunas I. (2004) The tropical dynamical response to latent heating estimates derived from the TRMM precipitation radar. *J. Atmos. Sci., 61,* 1341–1358.

Scott R. K. (2010) The structure of zonal jets in shallow water turbulence on the sphere. In *IUTAM Symposium on Turbulence in the Atmosphere and Oceans* (D. G. Dritschel, ed.), pp. 243–252. Springer, Dordrecht.

Scott R. K. and Dritschel D. G. (2012) The structure of zonal jets in geostrophic turbulence. *J. Fluid Mech., 711,* 576–598.

Scott R. K. and Tissier A.-S. (2012) The generation of zonal jets by large-scale mixing. *Phys. Fluids, 24(12),* 126601.

Seager S. and Deming D. (2010) Exoplanet atmospheres. *Annu. Rev. Astron. Astrophys., 48,* 631–672.

Seager S., Deming D., and Valenti J. A. (2008) Transiting exoplanets with JWST. ArXiv e-prints, arXiv:0808.1913.

Selsis F., Wordsworth R. D., and Forget F. (2011) Thermal phase curves of nontransiting terrestrial exoplanets. I. Characterizing atmospheres. *Astron. Astrophys., 532,* A1.

Shell K. M. and Held I. M. (2004) Abrupt transition to strong superrotation in an axisymmetric model of the upper troposphere. *J. Atmos. Sci., 61,* 2928–2935.

Sherwood S. C., Ingram W., Tsushima Y., Satoh M., Roberts M., Vidale P. L., and O'Gorman P. A. (2010a) Relative humidity changes in a warmer climate. *J. Geophys. Res., 115,* 09104.

Sherwood S. C., Roca R., Weckwerth T. M., and Andronova N. G. (2010b) Tropospheric water vapor, convection, and climate. *Rev. Geophys., 48,* RG2001.

Showman A. P. and Guillot T. (2002) Atmospheric circulation and tides of "51 Pegasus b-like" planets. *Astron. Astrophys., 385,* 166–180.

Showman A. P. and Polvani L. M. (2010) The Matsuno-Gill model and equatorial superrotation. *Geophys. Res. Lett., 37,* 18811.

Showman A. P. and Polvani L. M. (2011) Equatorial superrotation on tidally locked exoplanets. *Astrophys. J., 738,* 71.

Showman A. P., Cooper C. S., Fortney J. J., and Marley M. S. (2008) Atmospheric circulation of hot Jupiters: Three-dimensional circulation models of HD 209458b and HD 189733b with simplified forcing. *Astrophys. J., 682,* 559–576.

Showman A. P., Fortney J. J., Lian Y., Marley M. S., Freedman R. S., Knutson H. A., and Charbonneau D. (2009) Atmospheric circulation of hot Jupiters: Coupled radiative-dynamical general circulation model simulations of HD 189733b and HD 209458b. *Astrophys. J., 699,* 564–584.

Showman A. P., Cho J. Y.-K., and Menou K. (2010) Atmospheric circulation of exoplanets. In *Exoplanets* (S. Seager, ed.), pp. 471–516. Univ. of Arizona, Tucson.

Showman A. P., Fortney J. J., Lewis N. K., and Shabram M. (2013) Doppler signatures of the atmospheric circulation on hot Jupiters. *Astrophys. J., 762,* 24.

Siedler G., Church J., and Gould J. (2001) *Ocean Circulation and Climate: Observing and Modeling the Global Ocean.* International Geophysics Series, Vol. 77, Academic, New York.

Sing D. K., Vidal-Madjar A., Lecavelier des Etangs A., Desert J. M., Ballester G., and Ehrenreich D. (2008) Determining atmospheric conditions at the terminator of the hot-Jupiter HD209458b. ArXiv e-prints, arXiv:0803.1054, DOI: 10.1086/590076.

Sleep N. H. and Zahnle K. (2001) Carbon dioxide cycling and implications for climate on ancient Earth. *J. Geophys. Res.–Planets, 106(E1),* 1373–1399.

Smith M. D. (2008) Spacecraft observations of the martian atmosphere. *Annu. Rev. Earth Planet. Sci., 36,* 191–219.

Snellen I. A. G., de Kok R. J., de Mooij E. J. W., and Albrecht S. (2010) The orbital motion, absolute mass and high-altitude winds of exoplanet HD209458b. *Nature, 465,* 1049–1051.

Sobel A. H. (2002) Water vapor as an active scalar in tropical atmospheric dynamics. *Chaos, 12,* 451–459.

Sobel A. H. (2007) Simple models of ensemble-averaged tropical precipitation and surface wind, given the sea surface temperature. In *The Global Circulation of the Atmosphere* (T. Schneider and A. H. Sobel, eds.), pp. 219–251. Princeton Univ., Princeton.

Sobel A. H., Nilsson J., and Polvani L. M. (2001) The weak temperature gradient approximation and balanced tropical moisture waves. *J. Atmos. Sci., 58,* 3650–3665.

Soden B. J. and Held I. M. (2006) An assessment of climate feedbacks in coupled ocean-atmosphere models. *J. Climate, 19,* 3354–3360.

Stevenson D. J. (1999) Life-sustaining planets in interstellar space? *Nature, 400,* 32.

Stone P. H. (1972) A simplified radiative-dynamical model for the static stability of rotating atmospheres. *J. Atmos. Sci., 29,* 406–418.

Suarez M. J. and Duffy D. G. (1992) Terrestrial superrotation: A bifurcation of the general circulation. *J. Atmos. Sci., 49,* 1541–1556.

Sukoriansky S., Dikovskaya N., and Galperin B. (2007) On the "arrest" of inverse energy cascade and the Rhines scale. *J. Atmos. Sci., 64,* 3312–3327.

Thompson A. F. and Young W. R. (2006) Scaling baroclinic eddy fluxes: Vortices and energy balance. *J. Phys. Oceanogr., 36,* 720.

Thompson A. F. and Young W. R. (2007) Two-layer baroclinic eddy heat fluxes: Zonal flows and energy balance. *J. Atmos. Sci., 64,* 3214–3231.

Thompson R. O. R. Y. (1971) Why there is an intense eastward current in the North Atlantic but not in the South Atlantic? *J. Phys. Oceanogr., 1,* 235–238.

Thuburn J. and Craig G. C. (2000) Stratospheric influence on tropopause height: The radiative constraint. *J. Atmos. Sci., 57,* 17–28.

Trenberth K. and Caron J. (2001) Estimates of meridional atmosphere and ocean heat transports. *J. Climate, 14,* 3433–3443.

Trenberth K. E., Fasullo J. T., and Kiehl J. (2009) Earth's global energy budget. *Bull. Am. Meteor. Soc., 90,* 311–323.

Tziperman E. and Farrell B. (2009) Pliocene equatorial temperature: Lessons from atmospheric superrotation. *Paleoceanography, 24,* PA1101, DOI: 10.1029/2008PA001652.

Udry S., et al. (2007) The HARPS search for southern extra-solar planets. XI. Super-Earths (5 and 8 $M_{\oplus}$) in a 3-planet system. *Astron. Astrophys., 469,* L43–L47.

Vallis G. K. (2006) *Atmospheric and Oceanic Fluid Dynamics: Fundamentals and Large-Scale Circulation.* Cambridge Univ., Cambridge.

Vallis G. K. (2011) *Climate and the Oceans.* Princeton Primers in Climate Series, Princeton Univ., Princeton.

Vallis G. K. and Farneti R. (2009) Meridional energy transport in the coupled atmosphere-ocean system: Scaling and numerical experiments. *Q. J. R. Meteor. Soc., 135,* 1643–1660.

Vallis G. K. and Maltrud M. E. (1993) Generation of mean flows and jets on a beta plane and over topography. *J. Phys. Oceanogr., 23,* 1346–1362.

Vasavada A. R. and Showman A. P. (2005) Jovian atmospheric dynamics: An update after Galileo and Cassini. *Rept. Progr. Phys., 68,* 1935–1996.

von Paris P., Gebauer S., Godolt M., Grenfell J. L., Hedelt P., Kitzmann D., Patzer A. B. C., Rauer H., and Stracke B. (2010) The extrasolar planet Gliese 581d: A potentially habitable planet? *Astron. Astrophys., 522,* A23.

Walker C. C. and Schneider T. (2005) Response of idealized Hadley circulations to seasonally varying heating. *Geophys. Res. Lett., 32,* L06813.

Walker C. C. and Schneider T. (2006) Eddy influences on Hadley circulations: Simulations with an idealized GCM. *J. Atmos. Sci., 63,* 3333–3350.

Walker J. C. G., Hays P. B., and Kasting J. F. (1981) A negative feedback mechanism for the long-term stabilization of the Earth's surface temperature. *J. Geophys. Res., 86,* 9776–9782.

Wiktorowicz S. J. and Ingersoll A. P. (2007) Liquid water oceans in ice giants. *Icarus, 186,* 436–447.

Williams D. M. and Pollard D. (2003) Extraordinary climates of Earth-like planets: Three-dimensional climate simulations at extreme obliquity. *Intl. J. Astrobiology, 2,* 1–19.

Williams G. P. (1978) Planetary circulations. I — Barotropic representation of jovian and terrestrial turbulence. *J. Atmos. Sci., 35,* 1399–1426.

Williams G. P. (1988a) The dynamical range of global circulations — I. *Climate Dyn., 2,* 205–260.

Williams G. P. (1988b) The dynamical range of global circulations — II. *Climate Dyn., 3,* 45–84.

Williams G. P. and Holloway J. L. (1982) The range and unity of planetary circulations. *Nature, 297,* 295–299.

Williams R. G. and Follows M. J. (2011) *Ocean Dynamics and the Carbon Cycle.* Cambridge Univ., Cambridge.

Wood R. B. and McIntyre M. E. (2010) A general theorem on angular-momentum changes due to potential vorticity mixing and on potential-energy changes due to buoyancy mixing. *J. Atmos. Sci., 67,* 1261–1274.

Wordsworth R. (2012) Transient conditions for biogenesis on low-mass exoplanets with escaping hydrogen atmospheres. *Icarus, 219,* 267–273.

Wordsworth R. D., Read P. L., and Yamazaki Y. H. (2008) Turbulence, waves, and jets in a differentially heated rotating annulus experiment. *Phys. Fluids, 20(12),* 126602.

Wordsworth R. D., Forget F., Selsis F., Madeleine J.-B., Millour E., and Eymet V. (2010) Is Gliese 581d habitable? Some constraints from radiative-convective climate modeling. *Astron. Astrophys., 522,* A22.

Wordsworth R. D., Forget F., Selsis F., Millour E., Charnay B., and Madeleine J.-B. (2011) Gliese 581d is the first discovered terrestrial-mass exoplanet in the habitable zone. *Astrophys. J. Lett., 733,* L48.

Xu K.-M. and Emanuel K. A. (1989) Is the tropical atmosphere conditionally unstable? *Mon. Weather Rev., 117,* 1471–1479.

Yamamoto M. and Takahashi M. (2003) The Fully developed superrotation simulated by a general circulation model of a Venus-like atmosphere. *J. Atmos. Sci., 60,* 561–574.

Yoden S., Ishioka K., Hayashi Y.-Y., and Yamada M. (1999) A further experiment on two-dimensional decaying turbulence on a rotating sphere. *Nuovo Cimento C Geophys. Space Phys. C, 22,* 803–812.

Zelinka M. D. and Hartmann D. L. (2010) Why is longwave cloud feedback positive? *J. Geophys. Res., 115,* D16117.

Zurita-Gotor P. and Vallis G. K. (2009) Equilibration of baroclinic turbulence in primitive equations and quasigeostrophic models. *J. Atmos. Sci., 66,* 837–863.

Part III:
Clouds and Hazes

Esposito L. W., Colaprete A., English J., Haberle R. M., and Kahre M. A. (2013) Clouds and aerosols on the terrestrial planets. In *Comparative Climatology of Terrestrial Planets* (S. J. Mackwell et al., eds.), pp. 329–353. Univ. of Arizona, Tucson, DOI: 10.2458/azu_uapress_9780816530595-ch13.

Clouds and Aerosols on the Terrestrial Planets

L. W. Esposito
University of Colorado

A. Colaprete
NASA Ames Research Center

J. English
National Center for Atmospheric Research

R. M. Haberle and M. A. Kahre
NASA Ames Research Center

Clouds and aerosols are common on the terrestrial planets, highly variable on Earth and Mars, and completely covering Venus. Clouds form by condensation and photochemical processes. Nucleation of cloud droplets by certain aerosols provides an indirect linkage. Earth clouds cover over half of the planet, are composed of mainly liquid water or ice, and are a significant component of Earth's surface and top of atmosphere energy balance. On Venus, H_2SO_4 is the dominant cloud constituent, produced by chemical cycles operating on SO_2, likely produced from geologic activity. Martian water ice clouds generally have smaller particles than on Earth, although they form by the same processes. Mars clouds affect the deposition of radiation, drive photochemical reactions, and couple to the dust cycle. In the past, Mars clouds may have produced a significant greenhouse effect at times of high obliquity and early in its history. Mars atmospheric dust has both a seasonal cycle and great dust storms. Dust significantly influences the thermal and dynamical structure of the martian atmosphere. Mars CO_2 clouds provide both latent heat and radiative effects on the atmosphere, possibly more important on the early, wet, and warmer Mars climate.

1. INTRODUCTION

Clouds and aerosols play a key role in climate models, and uncertainties in the effects of clouds provide the largest error source for predicting global warming. Earth aerosols affect climate through direct radiative changes, aerosol-cloud interactions (indirect effects), atmospheric chemistry, snow albedo, and ocean biogeochemistry (*Mahowald et al.,* 2011). Clouds on the other planets expose us to alternate scenarios and examples of climate histories that are different from Earth (for a comparison of mesosphere clouds on Mars and Earth, see the chapter by Määttänen et al. in this volume). An early comparison of terrestrial clouds to those of Jupiter was made by *Rossow* (1978). Although Earth clouds are the best studied and known, we can still learn from the other planets. *Kahn* (1989) provides examples of how comparative planetology can illuminate our understanding of Earth by providing more extreme cases of more subtle effects on Earth, by providing better data that records early planet history now erased on the Earth, and providing inspiration leading to new insights.

2. EARTH CLOUDS AND AEROSOLS

2.1. Measurement Techniques

As can be expected, we have a deeper understanding of Earth's clouds and aerosols than other planets due to a larger array of measurements and more sophisticated climate models. Measurements of clouds and aerosols on Earth can be broadly categorized into *in situ* (instruments that locally measure properties) and remote sensing (instruments that remotely measure properties, typically using the electromagnetic spectrum). Common *in situ* techniques involve balloons, aircraft, and surface stations. Common remote sensing techniques involve active lidars and radars and passive radiometers and photometers onboard satellites, aircraft, and surface stations. Since the late 1970s, we have gained a near-global dataset of many cloud and aerosol properties from satellites. Additionally, the increase in computing power in recent years has enabled development of sophisticated climate models that include treatment of clouds and aerosols, although many of their processes are still parameterized as their spatial and temporal scales are smaller than the grid resolution allowed by current supercomputers.

2.2. Clouds

Clouds are ubiquitous and regularly cover over half of Earth's surface. They are almost entirely composed of liquid water droplets or ice crystals. Clouds form when air becomes sufficiently supersaturated with respect to liquid water or ice to overcome the energy barrier of changing the curvature of the cloud drop, which almost exclusively occurs in the troposphere (between Earth's surface and 18 km above the surface). Supersaturation is commonly produced through the ascent of air parcels, which is usually caused by surface heating. Therefore clouds most often occur over water and between the surface and 2 km (Fig. 1). Clouds are also prevalent in the tropical upper troposphere from strong convection and polar middle troposphere from advection and radiative cooling in the winter. Zonally, cloud fraction is associated with large-scale atmospheric circulation patterns, with cloud fraction higher in the ascending branches of the Hadley and polar cells (around the equator and 50° latitude, respectively), and lower at the descending branches (20° latitude and at the poles) (Figs. 2, 3). The Arctic sees more clouds than expected from large-scale circulation patterns due to radiative cooling in the winter (*Morrison et al.,* 2011). The clustering of deserts near 20° latitude and the presence of rain forests near the equator illustrate the significant impacts of global circulation patterns and clouds on climate, as clouds are associated with precipitation.

Nucleation of cloud drops or ice crystals from vapor often requires significant supersaturation to occur, due to the surface tension energy barrier involved with changing the curvature of a drop. Indeed, homogeneous nucleation of pure water usually requires temperatures below –40°C and is therefore uncommon. Heterogeneous nucleation on cloud condensation nuclei (CCN) or ice nuclei (IN) can occur at warmer temperatures. Only certain aerosols act as CCN. The likelihood of aerosols acting as CCN increases with aerosol size and solubility, with typical CCN being 50–100 nm (*McMurry,* 2000). Once CCN are "activated" (vapor begins

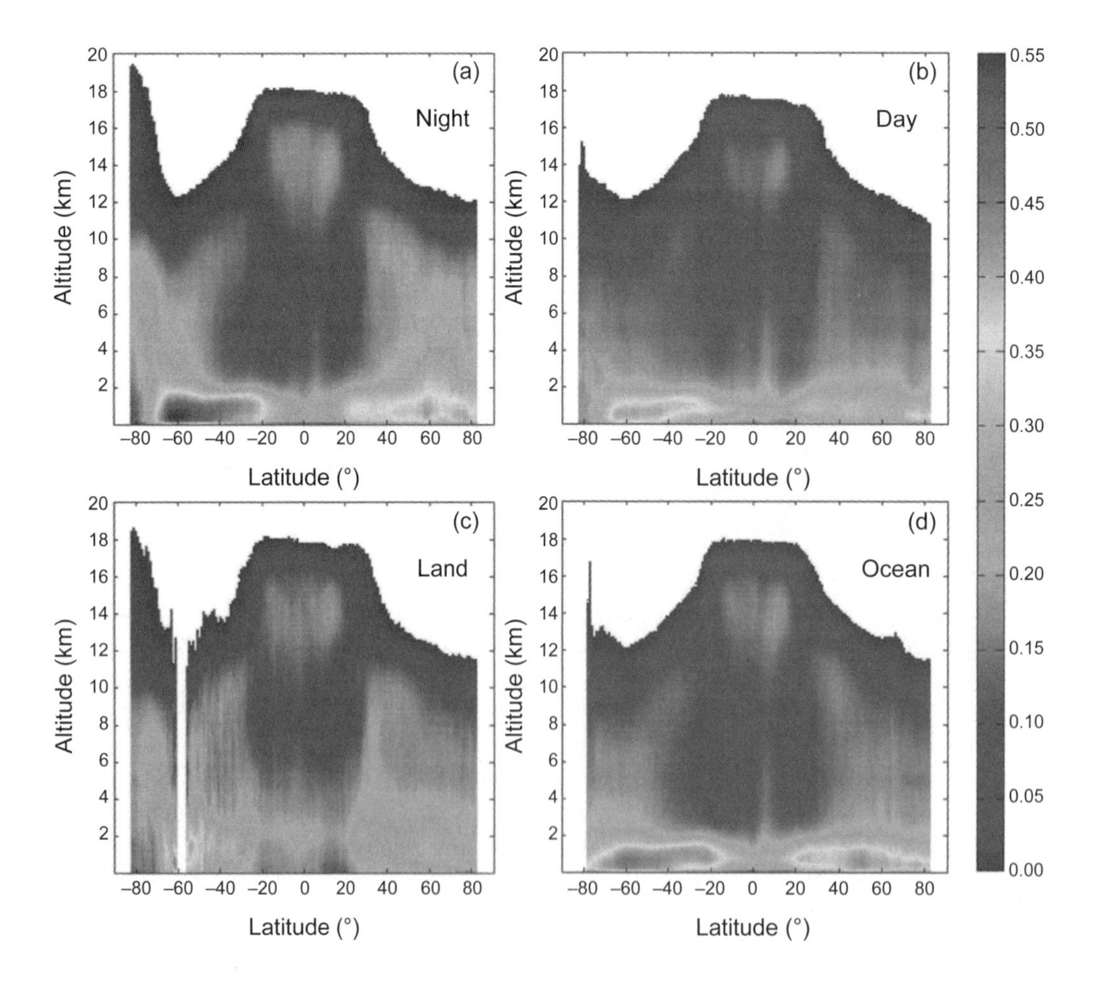

Fig. 1. See Plate 53 for color version. Latitude zonally-averaged vertical distribution of the cloud-occurrence frequencies for the CALIPSO observations from June 2006 to May 2007. A value of 0.5 means a cloud is present one half the time. **(a)** Nighttime measurements; **(b)** daytime measurements; **(c)** measurements over land; and **(d)** measurements over ocean. From *Wu et al.* (2011).

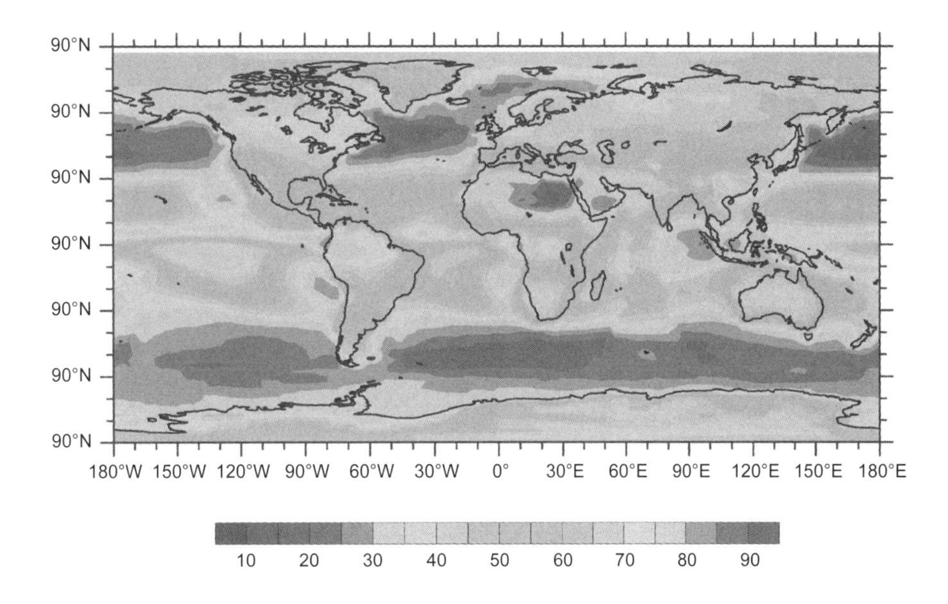

Fig. 2. See Plate 54 for color version. Annual average total cloud amount (percent) from July 1983 to December 2009 from the International Satellite Cloud Climatology Project (ISCCP). From *Rossow et al.* (1996).

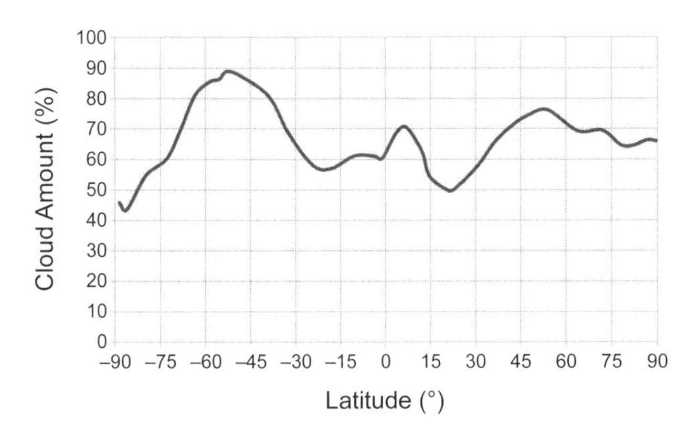

Fig. 3. Zonal average cloud amount from ISCCP. From *Rossow et al.* (1996).

to condense), cloud drops grow rapidly to about 10 μm. At this point a cloud is born. Cloud drops will grow further by coagulation with other droplets or by continued condensation of water vapor. Once a cloud drop's radius reaches about 100 μm, it begins to fall relatively rapidly and may fall as a raindrop, the size of which is typically about 1 mm. Clouds can then dissipate through precipitation, or through evaporation as the cloud moves to drier regions or dry air is entrained. The lifetime of a cloud is typically hours to days, but can range from minutes to months.

Clouds on Earth are classified into different types based on their phase, formation mechanism, altitude, and size. Broadly speaking, there are three main types of clouds: cumulus, stratus, and cirrus. Cumulus clouds are typically formed by convection; stratus clouds are formed by advection, radiative cooling, or large-scale weak uplift; and cirrus clouds are formed by freezing of supercooled droplets. While cumulus clouds and stratus clouds can contain liquid water or ice crystals, cirrus clouds are composed exclusively of ice crystals. Liquid-containing clouds are possible above temperatures of –40°C, and ice-containing clouds are possible below temperatures of 0°C. At temperatures between 0° and –40°C clouds are commonly, but not always, mixed-phase. Mixed-phase stratus clouds can persist for days or weeks in places such as the Arctic. This is surprising because ice has a lower equilibrium vapor pressure than liquid, meaning ice crystals will grow at the expense of liquid drops in subfreezing temperatures (*Wegener,* 1911; *Bergeron,* 1935; *Findeisen,* 1938). The persistence of mixed-phase clouds appears to be due to a complex interaction between turbulence, radiative cooling, microphysics, entrainment, and surface fluxes of heat and moisture (*Morrison et al.,* 2011). Cumulus clouds are common in the tropics and in summer months.

2.3. Aerosols

Aerosols are suspensions of solid and/or liquid particles in the atmosphere (*Mahowald et al.,* 2011; *Chin et al.,* 2009). As with clouds, the vast majority of Earth's aerosols are located in the troposphere. Aerosols can be composed of many different species, including dust, sea salt, sulfate, black carbon, organic carbon, and nitrates (*Seinfeld and Pandis,* 1998) (Table 1). Individual aerosol particles can be composed of a mixture of different species. Aerosols are produced from both natural and anthropogenic processes, with some dominated by natural processes in most locations (e.g., sea salt) and others dominated by anthropogenic processes in most locations (e.g., black carbon). Aerosols can be emitted directly from Earth's surface as primary particles

TABLE 1. Earth aerosols.

Species	Production (Tg yr^{-1})*	Lifetime (days)*	Primary Pathway(s)	Relative Size	Phase(s)	Radiative Impact	Key Source(s)
Dust	2000	4	Direct	Coarse	Solid	Cooling	Deserts
Sea salt	8000	0.5	Direct	Coarse	Both	Cooling	Oceans
Sulfate	200	5	*In situ*	Fine	Both	Cooling	Fossil fuel combustion, marine plankton, volcanos
Organic matter	70	7	Both	Fine	Liquid	Cooling	Biomass burning, fossil fuel combustion
Black carbon	10	7	Direct	Fine	Solid	Warming	Biomass burning, fossil fuel combustion
Nitrates	20	5	*In situ*	Fine	Both	Cooling	Agriculture, biomass burning, fossil fuel combustion

*Values are an approximate average of many observational and modeling studies; uncertainty is about 50%.

(e.g., dust, sea salt, and black carbon) or *in situ* by nucleation from the gas phase (e.g., sulfates, nitrates). Aerosols formed *in situ* are typically smaller in size and responsible for the vast majority of the number concentration present in Earth's atmosphere. Aerosols can also be produced from evaporated clouds. Aerosol number concentration varies from as low as 10 cm^{-3} in clean polar regions to 10^5 cm^{-3} in urban continental regions (*McMurry,* 2000). The stratosphere contains some aerosols at a much lower concentration than the troposphere, including micrometeoritic smoke (*Hervig et al.,* 2009) and volcanic sulfate aerosols. Concentrations of volcanic sulfate aerosol vary greatly depending on episodic volcanic eruptions. The lifetime of stratospheric aerosol is a year or two, which is about two orders of magnitude longer than the lifetime of tropospheric aerosol.

Aerosols are typically present in various log-normal size modes with diameters that can vary by four orders of magnitude, from 1 nm to 10 μm or so. The smallest particles are in the nucleation mode, which can be smaller than several nanometers radius. By definition, only aerosols formed *in situ* contain this mode. The accumulation mode has the longest lifetime in Earth's atmosphere, consisting of 100-nm to 1-μm particles that are too small to sediment rapidly and too large to be lost to larger particles. The coarse mode consists of particles greater than 1 μm that sediment relatively rapidly.

Aerosol size distributions are influenced by microphysical processes including nucleation, condensation, evaporation, coagulation, sedimentation, and deposition. Nucleation is only relevant to particles formed *in situ*. Sulfate aerosols produced from the binary homogeneous nucleation of sulfuric acid and water are responsible for the majority of aerosol number concentration. Depending on the availability of sulfuric acid vapor, particle size will increase by condensation of vapor and decrease by evaporation. Coagulation is a loss process of particle number and increases the size mode peak. Aerosol sinks include wet and dry deposition, with wet deposition typically dominating the loss process, therefore aerosol lifetime is shorter in places with relatively high precipitation. Aerosol lifetime ranges from less than one day to more than a week, depending on its composition, size, and location.

2.4. Climate Impacts

Both clouds and aerosols play a significant role in Earth's climate, and they interact with one another. Aerosols are the building blocks of most clouds, and aerosol number concentration can influence cloud drop size, optical properties, and lifetime. Conversely, clouds influence precipitation, which is a major loss process for aerosols, and evaporating clouds can leave aerosols behind (*Koren and Feingold,* 2011). Additionally, perturbations to radiative forcing from greenhouse gases have a complex influence on cloud properties (*Andrews and Gregory,* 2012).

2.4.1. Impact of clouds on climate. While clouds are known to strongly influence climate, quantifying their contributions to Earth's radiative budget remains difficult (*Stevens and Schwartz,* 2012). Clouds generally reduce both incoming shortwave and outgoing longwave radiation, with the net effect a function of cloud drop number concentration, phase, size distribution, thickness, altitude, latitude, and surface albedo. Globally averaged estimates are that clouds reduce incoming shortwave radiation by about 49 W m^{-2} and reduce outgoing longwave radiation by about 29 W m^{-2}, resulting in a globally averaged net cooling of about 20 W m^{-2} (*Loeb et al.,* 2009). The radiative impact of individual clouds can vary significantly. The cloud impact on incoming shortwave radiation is strongly affected by the albedo of the underlying surface. Over snow-covered surfaces, cloud shortwave forcing is small, while it is large over dark surfaces such as oceans and forests. The cloud impact on outgoing longwave radiation is strongly affected by cloud thickness and altitude, as temperature generally decreases with altitude. Clouds with smaller water droplets/crystals are more efficient at interacting with both shortwave and longwave radiation. Liquid-containing clouds usually have a stronger radiative effect, as liquid drops are usually smaller than ice crystals. There are some exceptions; for example, liquid droplets in tropical water clouds are sometimes larger than ice crystals in polar clouds. In the tropics over water or forest, optically thick clouds usually have a strong cooling impact. In the Arctic winter, when shortwave radiation is near zero and the surface undergoes radiative cooling, the presence of mixed-phase clouds limits surface cooling more than ice clouds. Very thin clouds, such as high cirrus clouds, can allow most shortwave radiation through while absorbing some longwave radiation, causing surface warming. Water vapor, while needed to form clouds, induces a warming effect on climate by absorbing in the longwave but not in the shortwave. Additionally, clouds control other aspects of Earth's climate including diurnal temperature variations and precipitation.

2.4.2. Impact of aerosols on climate. Once thought to play a radiatively insignificant role in Earth's energy balance compared to greenhouse gases, in recent decades the radiative forcing of aerosols has become more appreciated (*Charlson et al.,* 1992; *IPCC,* 2007). Aerosols can themselves interact with shortwave and longwave radiation (the "direct" effect), as well as modify cloud properties (the "indirect" effect; discussed in section 2.4.3). The direct effect is determined by three primary aerosol optical properties: single-scattering albedo (the preference for reflecting rather than absorbing), specific extinction coefficient (the efficiency of interacting with radiation), and scattering phase function (the preferred direction of altering photons). These optical properties vary by aerosol composition, size, and spatial distribution, as well as radiation wavelength and relative humidity (*Ramaswamy et al.,* 2001; *Ramanathan et al.,* 2001; *Kaufman et al.,* 2002), hence aerosols can cool or warm the climate. Generally most aerosols such as sulfates, nitrates, some organics, and mineral dust reflect and scatter incoming shortwave radiation, inducing a cooling effect on climate. This is especially true with mineral dust and sulfates, which have a single-scattering albedo of nearly 1 and higher atmospheric burdens. Black carbon, smoke, and some organics can be net absorbers, causing warming. As with clouds, the radiative impact of aerosols depends not only on the single-scattering albedo, but also the albedo of the underlying surface. Also as with clouds, the radiative impacts of aerosols are strongly dependent upon particle size. Mass extinction efficiency, the efficiency of an aerosol per unit mass to scatter and absorb radiation, is a way to translate between aerosol mass and its resulting direct effect on radiative forcing. Mass extinction efficiency can vary by an order of magnitude or more across the size range known to exist in the atmosphere. For example, sulfate aerosols of radius 0.5 μm are most efficient at absorbing shortwave radiation, while particles with a radius of 0.8 μm are most efficient at scattering longwave radiation. Aerosols are generally more efficient at interacting with shortwave radiation than longwave radiation. The overall radiative effect is therefore a function of mass scattering efficiency and column number concentration. Various assumptions regarding aerosol size are made by instruments to more accurately measure the aerosol direct radiative effect, and this is an area of active research (*Kahn,* 2012).

2.4.3. Interactions between clouds and aerosols. The interactions between clouds and aerosols are very complex and an area of active research. Aerosols indirectly affect climate by modifying the microphysical properties of clouds, which can alter cloud radiative properties, amount, and lifetime. The "first indirect effect," also called the "cloud albedo effect" or "Twomey effect," is the microphysical effect of increasing cloud droplet number concentration (and decreasing droplet size) when liquid water content is held constant (*Twomey,* 1977). This increases mass scattering efficiency. The "second indirect effect," also called the "cloud lifetime effect" or "Albrecht effect," is the effect of the change in cloud droplet number concentration on liquid water content, cloud height, and lifetime of clouds (*Albrecht,* 1989). The second indirect effect can be attributed to microphysical, dynamical, convective, or radiative changes. For instance, the Twomey effect can increase the number and decrease the size of cloud droplets, slowing down precipitation processes and increasing cloud lifetime.

2.5. Future Investigations

Clouds remain one of the largest uncertainties in future projections of climate change by global climate models, owing to the physical complexity of cloud processes and the small scale of individual clouds relative to the size of the model computational grid. There remain many unanswered questions regarding cloud-aerosol interactions, which are very complex.

There remains uncertainty regarding aerosol source emissions, composition of aerosol mixtures, and the resulting optical properties. Better quantification of the mass, size, and composition of aerosol mixtures can improve climate model predictions of aerosol radiative effects. The properties that determine which aerosols become CCN or IN are not well understood. Finally, there remain unknowns regarding the formation of new aerosols, as their 1-nm size is below the approximate 3-nm detection limit of instruments.

While the past few decades have brought near-global satellite coverage of aerosol optical depth and cloud fraction, there remain knowledge gaps regarding some key properties such as cloud phase (liquid vs. ice), aerosol/cloud particle size distributions, and vertical profiles. Cloud detection is also difficult in places where albedo or thermal contrasts are small, such as in the Arctic. Progress continues toward the ability to remotely measure some of these microphysical details.

3. VENUS CLOUDS AND HAZES

3.1. Measurement Techniques and Description

Venus' atmosphere can be divided into regions based on its composition, chemistry, and clouds. The upper atmosphere, above about 110 km, has low densities and overlaps with the ionosphere so that photodissociation, ion-neutral, and ion-ion reactions are increasingly dominant as one goes to higher altitudes. The middle atmosphere, in the altitude region at about 60–110 km, receives sufficiently intense ultraviolet (UV) radiation from the Sun that its chemistry is dominated by photon-driven processes, termed "photochemistry." The lower atmosphere, below about 60 km, receives little UV radiation from the Sun, and due to the high atmospheric temperatures the chemistry is controlled increasingly by thermal processes, termed thermodynamic equilibrium chemistry or "thermochemistry."

Cloud and haze layers extend from ~30 to 90 km with the main cloud deck lying at ~45 to 70 km. The clouds define a transition region that reflects the competition between the middle atmosphere, dominated by photochemistry, and the lower atmosphere, dominated by thermochemistry (*Esposito et al.,* 1997). The clouds and hazes of Venus cover the entire planet (Fig. 4) and have enormous vertical extent, depth of ~60 km, with an average visibility in the Venus clouds better than several kilometers (see Table 2). The main cloud deck extends from ~70 km (the level of unit optical depth in the UV) down to altitudes between 45 and 50 km.

Spacecraft *in situ* measurements allow us to divide the cloud system into upper, middle, and lower clouds (see review by *Esposito et al.,* 1983) (Fig. 5). Based on the Pioneer Venus and Venera nephelometer results and the Pioneer Venus Cloud Particle Size Spectrometer (LCPS) measurements, it appears that the middle and upper cloud structure are planetwide features. In all cases the opacity is higher in the middle than upper cloud, typically by a factor of 2. The lower cloud is well defined and highly variable from location to location. Sharp layers are evident at the Pioneer Venus Large and Small (Night) Probe sites. The clouds within the main deck would all be thin stratoform in terrestrial classification. Instabilities are slight and latent convection potential is negligible (see *Knollenberg et al.,* 1980). Only the middle cloud region appears to have any potential for convective overturning. This is in contrast to Earth, where convection dominates in the troposphere.

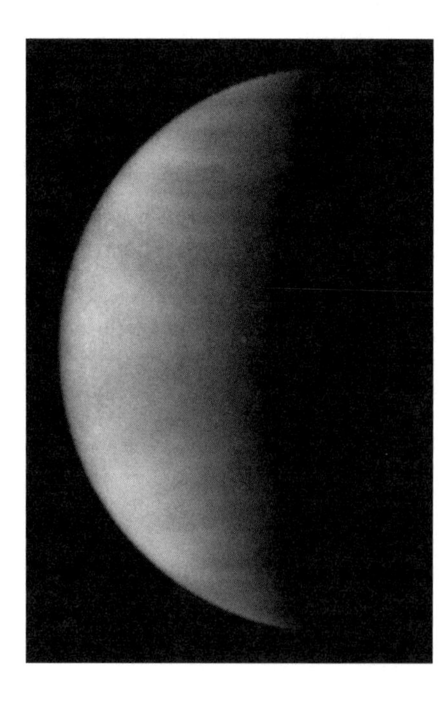

Fig. 4. Venus image from HST at 218 nm. Credit: L. W. Esposito, University of Colorado, Boulder/NASA.

3.2. Key Chemical and Physical Processes

Three dominant chemical processes have been identified in the venusian atmosphere (*Mills et al.,* 2007): the CO_2 cycle, the sulfur oxidation cycle, and the polysulfur cycle. The CO_2 cycle includes the photodissociation of CO_2 on the dayside, transport of a significant fraction of the CO and O to the nightside, production of O_2, emission of highly variable oxygen airglow on both the dayside and nightside, and conversion of CO and O_2 into CO_2 via catalytic processes. The sulfur oxidation cycle involves the upward transport of SO_2, oxidation of a significant fraction of the SO_2 to form H_2SO_4, condensation of H_2SO_4 and H_2O to form a majority of the mass comprising the cloud and haze layers, downward transport of sulfuric acid in the form of cloud droplets, evaporation of the cloud droplets, and decomposition of H_2SO_4 to produce SO_2. There is solid observational evidence for both the CO_2 and the sulfur oxidation cycles. The polysulfur cycle is more speculative but plausible based on existing laboratory data and limited observations. It involves the upward transport of sulfur as either SO_2 or OCS, photodissociation to produce S, formation of polysulfur (S_x) via a series of association reactions, downward transport of S_x, thermal decomposition of S_x, and reactions with oxygen and CO to produce SO_2 and OCS, respectively.

The three chemical cycles couple at a number of key points (*Esposito et al.,* 1997). First, the middle and lower atmosphere portions of the two sulfur cycles are believed to balance each other, on the assumption that the atmosphere is near a long-term equilibrium point. However, little modeling has been done across the boundary between photochemistry and thermochemistry and there appear to be inconsistencies between the best-fit solutions derived for each altitude region.

TABLE 2. Venus cloud properties.

Altitude (km)	Name	Constituents*	Composition
Above 70	Upper hazes	Mode 1	H_2SO_4
57–70	Upper cloud	Modes 1, 2	H_2SO_4
50–57	Middle cloud	Modes 2′, 3	H_2SO_4 + ?
47–50	Lower cloud	Modes 2′, 3	H_2SO_4 + ?
45–48	Pre-cloud layer	?	H_2SO_4
30–48	Lower thin hazes	?	?

*Size modes: Mode 1, mean diameter <0.6 μm; Mode 2, mean diameter 2 μm; Mode 2′, mean diameter 3 μm; Mode 3, mean diameter 7 μm.

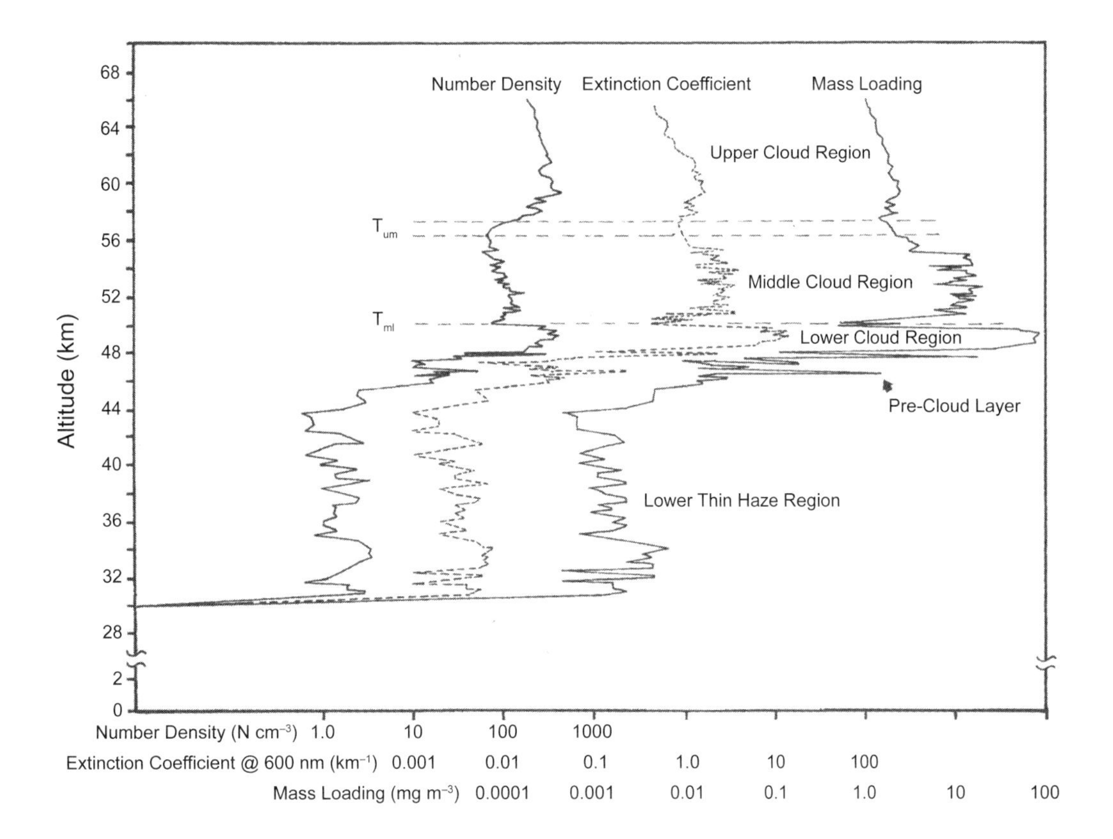

Fig. 5. Cloud property vertical profiles. T_{m1} is the middle-lower cloud transition; T_{um} is the upper-middle cloud transition. From *Knollenberg and Hunten* (1980).

Sulfur dioxide was first detected in the atmosphere of Venus by *Barker* (1979) from the ground, and it was subsequently confirmed by *Stewart et al.* (1979) and *Conway et al.* (1979). These observations indicated that the abundances of SO_2 in the 1978–1979 period were larger than the previously established upper limits (*Owen and Sagan,* 1972) by orders of magnitude. Continuous observations by Pioneer Venus from 1978 to 1986 show a steady decline in the cloud top SO_2 abundance toward values consistent with previous upper limits (*Esposito et al.,* 1988). This decline has been confirmed by International Ultraviolet Explorer (IUE) observations (*Na et al.,* 1990) and by Hubble observations (*Na and Esposito,* 1995) (see Figs. 4, 6). Analysis of UV spectra from the Hubble Space Telescope (HST) Goddard High Resolution Spectrometer (GHRS) give an SO_2 abundance of less than 25 ppb at the cloud tops (*Na and Esposito,* 1995) (see Fig. 6 for the time history of SO_2 cloud top measurements). Explanations that have been advanced for the likely rapid increase and observed slow decline of SO_2 include active volcanism (*Esposito,* 1984), changes in the effective eddy diffusion within the cloud layers (*Krasnopolsky,* 1986, p. 147), and changes in atmospheric dynamics (*Clancy and Muhleman,* 1991). Venus Express observations (*Marcq et al.,* 2013) have also measured time variations in SO_2 since 2006, showing another increase and decline similar to the large increase and decline seen in the 1970s and 1980s (*Esposito et al.,* 1997).

The latest Venus Express images (*Titov et al.,* 2012) characterize the cloud morphology: mottled clouds near the subsolar point, streaky clouds at 50° latitude, and a featureless bright polar hood with occasional spiral structures (Fig. 7).

3.3. Comparisons Among Venus, Earth, and Mars

Catalytic chemistry plays fundamental roles in the atmospheres of all three planets (see, e.g., *Yung and DeMore,* 1999). This means trace abundances of highly reactive radicals govern the primary chemical cycles on each planet. On Venus, a small amount of Cl has a major impact on the production of CO_2 from CO and O_2. On Earth, hydrogen, nitrogen, and halogen radicals play an analogous role. The most prominent example is the regulation of the stratospheric ozone layer. On Mars, catalytic cycles even control the abundance of its main atmospheric constituent, CO_2 (*Stock et al.,* 2012); CO_2 production is dominated by chemical pathways including HO_x, O_x, and NO_x.

The greatest difference between Venus and Earth concerns the amount of water on these two planets (see, e.g., *Yung and DeMore,* 1999). The mixing ratio of water vapor in the lower atmosphere of Venus is variable, with a maximum value of 1.5×10^{-4}. This is equivalent to a layer of 2–10 cm of water, uniformly spread over the surface of the planet. For comparison, Earth contains an average layer of

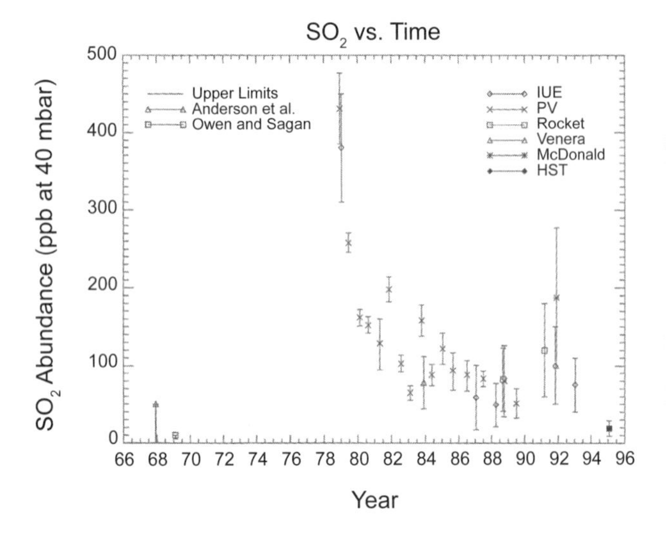

Fig. 6. SO_2 abundance near the Venus cloud tops. From *Esposito et al.* (1997).

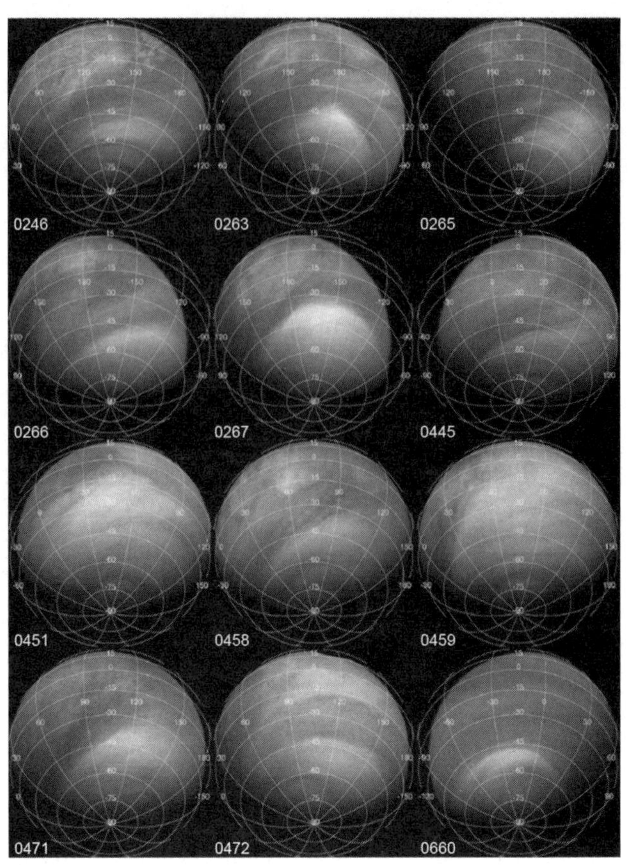

Fig. 7. Global views of Venus clouds from Venus Express. South pole is at the bottom. Orbit number to lower left. From *Titov et al.* (2012).

2.7 km of water, residing mostly in the oceans. The lack of an ocean on Venus has at least three dramatic consequences for the atmosphere. First, most of the planet's CO_2 remains in the atmosphere, in contrast to Earth, where most of the 50 bar of CO_2 are sequestered as carbonate rock in the sediments. Second, the atmosphere of Venus contains large quantities of SO_2. On Earth, most of the volatile sulfur resides in the ocean as sulfate ions. The presence of this large amount of SO_2 in the atmosphere is responsible for the production of a dense H_2SO_4 cloud on Venus, which has significant climatic effects. Third, the atmosphere of Venus contains large amounts of HCl. On Earth, the bulk of chlorine is in the form of salt (NaCl) in the oceans.

It is a remarkable fact that Earth has remained habitable throughout its planetary history even though the solar luminosity has increased substantially. The major energy source for the solar system is the Sun. The Sun's luminosity has gradually increased by about 40% since the origin of the solar system. The reduced solar constant during the nascent period of Earth would imply that the planet was completely frozen, a result in conflict with known geological evidence (e.g., sedimentary rocks). Although the abiotic carbonate/silicate weathering feedback can provide a "thermostat" for global temperature, it cannot overcome the steady increase in the solar constant. A resolution of this paradox may be to postulate that the CO_2 content of the atmosphere has been changing with time to compensate for the changing solar constant. According to the Gaia hypothesis, the biosphere may indeed have evolved since its origin to counteract the problem of the increasing luminosity of the Sun. However, there is also a finite limit to the power of Gaia. The abundance of CO_2 is now very low, while the luminosity of the Sun continues to increase. Further decrease of CO_2 by biological activity may be difficult because photosynthesis itself stops when the CO_2 mixing ratio falls below a threshold level of about 150 ppmv. Thus, there is a point beyond which the Gaian control of the global environment would fail and the Earth would become Venus-like in 30–300 m.y. (*Lovelock and Whitfield,* 1982). Venus may be the ultimate graveyard of all terrestrial planets above a certain critical size.

3.4. Future Investigations and Comparisons

The Pioneer Venus LCPS measurement of larger, so called "Mode 3" particles has provided a controversy that is still unresolved (see *Esposito et al.,* 1983). Do these large particles really exist as a separate size mode, and if so, what is their composition? The starting point for the Mode 3 controversy comes from direct evidence for asymmetric (possible crystalline) particles provided by *Knollenberg and Hunten* (1980). *Knollenberg et al.* (1980) further state that only such crystals of high aspect ratio could satisfy the Pioneer Venus LCPS, Solar Flux Radiometer (LSFR), and Nephelometer (LN) results simultaneously. *Toon et al.* (1982) argue that this mode may be an artifact of a calibration error. In this case, all the Venus cloud aerosols could be sulfuric acid, with no need for a larger, solid cloud aerosol compound. Since the largest amount of mass [~80% according to *Knollenberg and Hunten* (1980)] is within the

Mode 3 particles, it is extremely important to verify their existence and determine their composition.

Currently, the only positively identified species in the visible atmosphere that absorb in the near UV are SO_2 and SO. These gases can account for the absorption observed at wavelengths short of 320 nm at all altitudes (*Esposito,* 1980; *Pollack et al.,* 1980) and for the absorption observed at 320–390 nm below 55 km (*Ekonomov et al.,* 1983). However, they cannot account for the absorption observed in the upper cloud at wavelengths longward of 320 nm, nor can they account for the absorption observed at 400–500 nm (*Esposito,* 1980; *Pollack et al.,* 1980; *Ekonomov et al.,* 1983). In addition, these other absorbers must explain the phase angle dependence of the UV dark markings (*Barker et al.,* 1975) as well as their short lifetime above the clouds [from hours to days; see *Esposito et al.* (1983)]. They must also be consistent with the visible-wavelength solar flux observations of *Tomasko et al.* (1980), which show absorption at 58–62 km, and little absorption below. Near-UV solar flux absorption results from Venera 14 (*Ekonomov et al.,* 1983, 1984) provide an additional constraint. *Esposito and Travis* (1982) noted the correlation between dark markings seen longward of 320 nm and SO_2 enhancements seen at 207 nm. This means that in addition to the absorption spectrum, a good candidate must also match constraints on its vertical distribution, lifetime, and correlation with SO_2 enhancement. This last correlation could be either chemical or dynamical because the SO_2 observable in the middle UV is likely the result of local upwelling (*Esposito and Travis,* 1982). More recent studies of the unknown UV absorber using Venus Express results (*Molaverdikhani et al.,* 2012) find similar properties. The identification and distribution of this absorber are important because it is the most important absorber of shortwave radiation.

Better observations of cloud particle properties and their variation are also required. The size distribution, shape, and composition of the majority of the aerosol mass are still unclear, despite our assurance that "Mode 2" (the aerosols visible at the cloud tops) are spherical droplets of concentrated sulfuric acid.

Determination of the mineralogy on the surface of Venus is needed for significant progress on understanding the chemical interaction between the atmosphere and surface and how this interaction may have changed over the course of Venus' evolution. Potentially, the surface activity and composition may interact with Venus' climate (*Bullock and Grinspoon,* 2001; *Taylor and Grinspoon,* 2009) (see Fig. 8). Like the climate of Earth, the Venus climate system involves numerous complicated feedbacks. Hints of a few of these complications can be seen in Fig. 8.

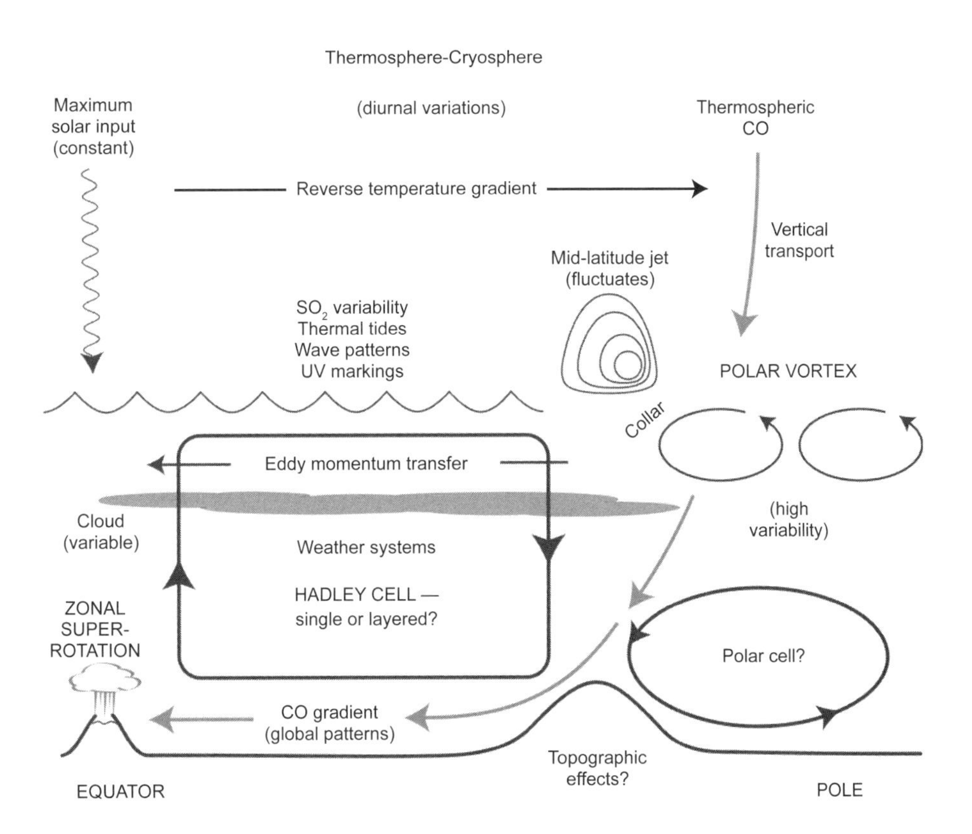

Fig. 8. Some of the global-scale meteorological features observed on Venus that may be coupled to the general circulation and affect the climate. From *Taylor and Grinspoon* (2009).

4. MARTIAN WATER ICE CLOUDS

4.1. Observations

4.1.1. Measurement techniques. Mars water ice clouds have been observed from Earth-based telescopes, the Hubble Space Telescope, spacecraft in orbit around Mars, and spacecraft that have landed on the surface. Figure 9 shows an image of martian clouds captured by the Mars Color Imager (MARCI) on the Mars Reconnaissance Orbiter (MRO), which is still operating at Mars.

The data acquired from all these platforms are based on remote sensing instruments that sample in the UV, visible, near-infrared, and infrared portions of the electromagnetic spectrum. Depending on the technique used, these data reveal information on the temporal and spatial variation of martian water ice clouds, their column optical depth, and their particle sizes and shapes.

Five orbital missions are responsible for most of our understanding of the global and seasonal behavior of martian water ice clouds: Viking, Mars Global Surveyor (MGS), Mars Odyssey (MO), Mars Express (MEx), and MRO. The Phobos mission also acquired limited but useful data (e.g., *Rodin et al.,* 1997). The imaging cameras on the Viking orbiters provided the first glimpse into global and seasonal cloud variability on Mars (*Kahn,* 1984). Wide-angle images from the Mars Orbiter Camera (MOC) on MGS returned the first daily global maps of clouds (*Cantor et al.,* 2002), while water ice column optical depths were retrieved from the Thermal Emission Spectrometer (TES) radiance spectra (*Pearl et al.,* 2001). Similar data were retrieved from the Thermal Emission Imaging System (THEMIS) on MO (*Smith et al.,* 2003). Ultraviolet solar and stellar occultation observations by the Spectroscopy for Investigation of the Characteristics of the Atmosphere of Mars (SPICAM) instrument on MEx provided high-resolution point profiles of cloud properties (*Federova et al.,* 2009).

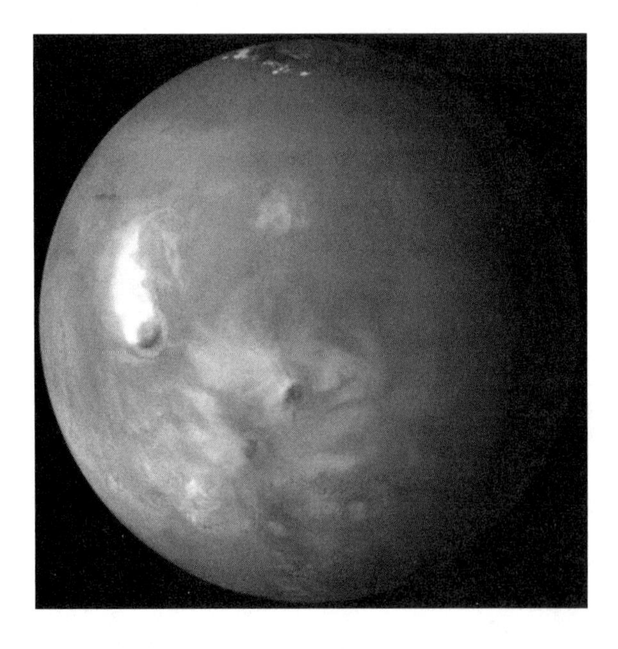

Fig. 9. Spherical projection of MARCI images from July 15 to 17, 2010 (L_s = 118°–119°), centered over the Tharsis region. Lighting conditions are those for 1500 Mars local time at the center of the disk. Figure created by Bruce Cantor of Malin Space Science Systems (MSSS). Credit: MSSS/NASA/JPL.

More recently, the Mars Climate Sounder (MCS) on MRO provided dedicated limb-sounding capability at Mars and has revealed the vertical structure of clouds up to about 80 km altitude with 5–10-km vertical resolution (*McCleese et al.,* 2010), and MARCI has extended the MOC global mapping data by acquiring daily global maps in seven different wavelengths including two in the UV (*Malin et al.,* 2008). It is worth noting that MGS, MO, and MRO were in Sun synchronous orbits at fixed local times ranging from 2:00 p.m. to 5:30 p.m., while Viking and MEx were in highly elliptical orbits that precessed in local time.

Finally, from the surface, the two Viking Landers, Pathfinder, the two Mars Exploration Rovers Spirit and Opportunity, and the Phoenix Lander acquired sky images of martian clouds. Phoenix also carried a Light Detection and Ranging (LIDAR) instrument to probe the near-surface vertical structure of clouds (*Whiteway et al.,* 2009).

4.1.2. Global distribution. Martian water ice clouds fall into three general categories: mid-altitude tropical clouds (10–40 km), low-altitude polar hoods (<10 km), and high-altitude hazes (>40 km). We focus here on clouds below 40 km. Mid-altitude tropical clouds are most pronounced during northern summer when Mars is farthest from the Sun. Hence, these clouds are sometimes referred to as the aphelion cloud belt (ACB). The morphology of clouds in the ACB exhibit a variety of forms ranging from diffuse hazes to localized optically thick assemblages with convective features. These later clouds are often associated with prominent volcanos (see Fig. 9). Modeling studies indicate that the ACB forms when water subliming for the north polar residual cap (NPRC) is swept up in the rising branch of the Hadley cell (*Montmessin et al.,* 2004). MCS data indicate large diurnal variability in cloud opacities of the ACB (*Heavens et al.,* 2010).

The polar hoods are present at high latitudes of both hemispheres and have been observed from Earth for many years (*Martin et al.,* 1992). In general, they form during fall, peak in thickness during winter, retreat in spring, and disappear during summer. These clouds closely resemble fogs, although lee waves and frontal clouds and have also been observed (*Briggs and Leovy,* 1974; *Wang and Fisher,* 2009). The north polar hood is thicker (τ ~ 0.2 at 12.1 μm) and more extensive than the south polar hood (~0.1). MCS data have revealed additional significant hemispheric asymmetries in their vertical structure and latitudinal extent (*Benson et al.,* 2010, 2011).

Phoenix LIDAR observations have revealed some insights into the behavior of clouds in the planetary boundary layer at a high northern latitude during summer (*Whiteway et al.,* 2009; *Daerden et al.,* 2010). Clouds form near the top of the daytime boundary layer (about 4 km altitude) in the early morning and grow sufficiently thick to fall nearly all the way to the surface before subliming (Fig. 10). This downward transfer of water balances the upward flux during the day and helps maintain high water vapor concentrations near the surface.

The global distribution of 12.1-μm column cloud opacities for a single Mars year is shown in Fig. 11. The tropical ACB is readily apparent. Peak opacities reach ~0.15 during early northern summer. The edge of the polar hood is also evident in each hemisphere as it closely tracks the edge of the seasonal CO_2 ice cap. In these retrievals northern summer (L_s ~ 90°–120°) is the cloudiest time of year as the planet is furthest from the Sun, temperatures are cooler, and water is rapidly subliming from the NPRC. Mars is least cloudy (but very dusty) at the opposite season (L_s ~ 270°–300°) when the planet is closest to the Sun, temperatures are warmer, and the supply of water is limited.

4.1.3. Interannual variability. Observations of column cloud opacities constructed from a combination of TES (*Smith,* 2004) and THEMIS (*Smith,* 2009) data spanning a period of over 7 Mars years are shown in Fig. 12. As is clearly evident, zonal-mean global cloud patterns are remarkably repeatable from year to year.

The thin martian atmosphere is tightly coupled to the low thermal inertia surface, which rapidly responds to the steady repeatable seasonal variations in solar insolation. However, there are year-to-year variations in the spatial structure of clouds, particularly over the Tharsis volcanos and north polar hoods (*Benson et al.,* 2006, 2011). The main source of interannual variability is variations in atmospheric dust content, which are most pronounced during the perihelion season when global dust events sometimes occur (e.g., *Cantor et al.,* 2008). Such variations lead to dramatic changes in atmospheric temperatures and wind systems (e.g., *Haberle et al.,* 1993), which can strongly affect clouds.

4.2. Properties

4.2.1. Physical. The size and shape of martian water ice clouds are not well constrained due to the lack of *in situ* measurements and the heavy reliance on retrievals of remote sensing observations not primarily intended to characterize clouds. However, cloud particles are generally much smaller than those in Earth's atmosphere. Retrieved effective particle radii (r_{eff}) are typically in the 1–4-μm range and cluster into two distinct populations between 1 and 2 μm and between 2 and 4 μm, with the former found mostly in southern hemisphere and the latter dominating the ACB (*Clancy et al.,* 2003; *Wolff and Clancy,* 2003) (see Table 3). These small sizes are a direct consequence of the low vapor pressures prevailing in the martian environment. Larger particles have been inferred from near0infrared spectra [R_{eff} ~ 5 μm (*Madeliene et al.,* 2012a)], Phoenix LIDAR observations [r_{eff} ~ 35 μm (*Whiteway et al.,* 2009)], and model simulations over volcanos [r_{eff} ~ 8 μm (*Michaels*

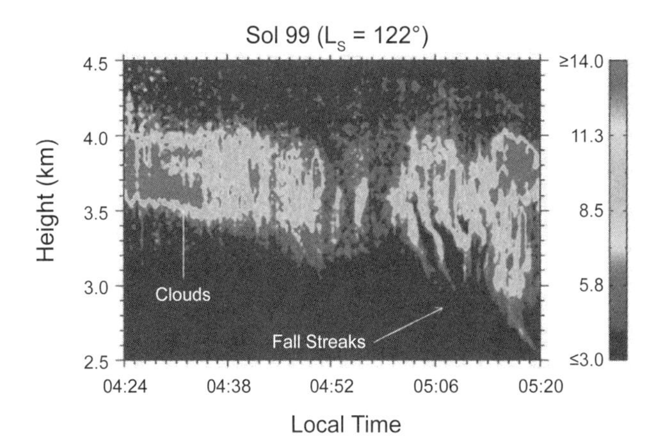

Fig. 10. See Plate 55 for color version. Time/height contour plot of the backscatter coefficient (× 10^{-6} m^{-1} sr^{-1}) derived from the Phoenix LIDAR backscatter signal at 532 nm for the seasonal date L_s = 122°. From *Whiteway et al.* (2009).

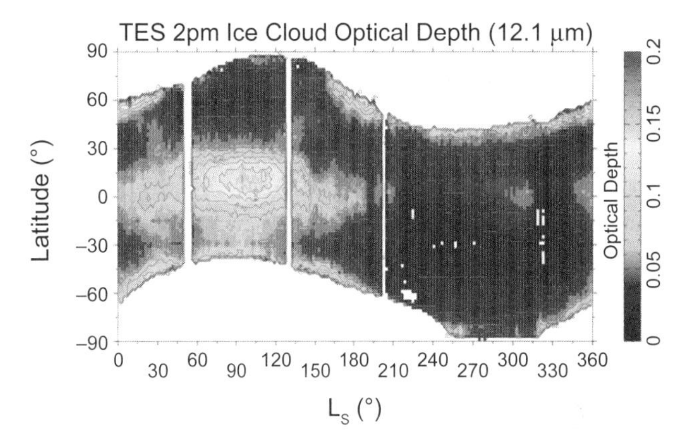

Fig. 11. See Plate 56 for color version. Seasonal and latitudinal variation of the 12.1-μm zonal-mean column water ice cloud opacity for a combination of Mars Year (MY) 26 and 27 (L_s = 0°–31° from MY27; L_s = 31°–360° from MY26). Data courtesy of Michael Smith.

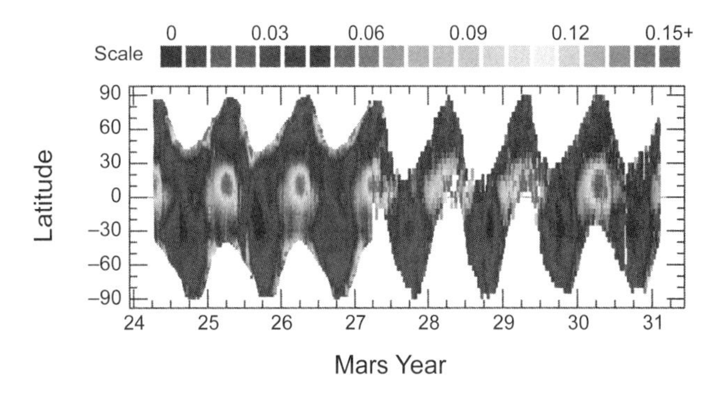

Fig. 12. See Plate 57 for color version. Seven plus Mars years of cloud observations. Figure constructed from TES data up to MY27 and THEMIS data thereafter. Courtesy of Michael Smith.

TABLE 3. Mars water ice clouds.

Cloud Type	Altitude (km)	Mean Particle Radius r_{eff} (μm)	Composition
Low-altitude polar hoods	0–10	4–6	H_2O
Mid-altitude tropical clouds (ACB)	10–40	1–2	H_2O
High-altitude hazes	>40	0.1–1	H_2O
Volcanic clouds	10–40	8	H_2O
Top of PBL clouds at Phoenix Lander site	4–6	35	H_2O

et al., 2006)], but these are localized and not representative of broad regions. Smaller particles have also been retrieved at high altitudes [r_{eff} < 0.1 (*Montmessin et al.*, 2006b); r_{eff} ~ 0.2–1.0 μm (*Clancy et al.*, 2009)]. Thus martian water ice cloud particles are generally small, although there have been detections of submicrometer particles and particles with sizes similar to terrestrial cirrus clouds.

In most of the retrievals small effective variances are needed to fit the data. Typical values are v_{eff} = 0.1, which suggests that there is very little size dispersion in martian clouds. The shape of martian water ice cloud particles is even less well constrained. However, typical terrestrial crystal habits do not fit well the observed scattering properties of martian clouds, and are instead better fit with multifaceted equidimensional shapes such as droxtals (*Wolff et al.*, 2011).

4.2.2. Radiative. The well-known optical properties of water ice show very little absorption at solar wavelengths such that the particles are nearly perfect scatterers and single-scattering albedos are ϖ_0 ~ 1. At these wavelengths, the size and shape of the ice crystals strongly favors scattering in the forward direction with asymmetry parameters g > 0.8. While there is some absorption in the near-infrared at 1.5, 2.0, 2.5, and 3.1 μm, it is of little consequence to atmospheric heating and instead plays a more prominent role in the identification of water ice in surface deposits (*Langevin et al.*, 2007). In the infrared, however, water ice has broad absorption features between 11 and 16 μm and between 35 and 45 μm, and these have been used to retrieve opacities from nadir and limb TES spectra (e.g., *Pearl et al.*, 2001). The infrared absorption properties of water ice particles have a very important influence on atmospheric heating rates and hence dynamical motions.

4.3. Microphysics

Martian water ice clouds form by the same processes as those on Earth. A comprehensive treatment can be found in *Michaelangeli et al.* (1993), *Colaprete et al.* (1999), *Montmessin et al.* (2002), and *Määttänen et al.* (2005). Before ice crystals grow they first nucleate. Given the abundance of dust particles in the martian atmosphere, heterogeneous nucleation must certainly dominate. Nucleation is nearly instantaneous and begins when the supersaturation exceeds a critical value, which is typically 20–100%. This can be achieved, for example, by cooling a saturated air parcel at 200 K by several degrees. However, laboratory experiments with martian dust analogs suggest that higher supersaturations may be required at lower temperatures (*Iraci et al.*, 2010). There is some evidence of very high supersaturations at certain seasons and locations (*Maltagliati et al.*, 2011).

Once nucleated, ice crystals grow by molecular and thermal diffusion of water vapor through a background gas of CO_2. Growth is also a very fast process, but less so than nucleation. Growth by coagulation is negligible on Mars because of the low concentration of ice particles and narrow size distributions. Growth rates increase as the background pressure increases. Since the timescales of nucleation and growth (seconds to minutes) are much shorter than those associated with temperature changes due to radiation, convection, and dynamical transport (hours to days), ice crystals will grow (or sublimate) at a rate fast enough to maintain vapor pressures close to the saturation vapor pressure. However, above 40 km the mean free path exceeds the particle size and growth times increase. This may also play a role in the observed high supersaturations.

As the ice particles grow they gravitationally settle at speeds that increase with altitude. Since the timescales for growth and settling have opposite dependencies on altitude, ice crystals will have more time to grow at low altitudes where settling rates are low and growth rates are high, and less time to grow at higher altitudes where the opposite holds true. Thus, ice crystal particle sizes should roughly increase with decreasing altitude. Observations and modeling studies are consistent with this general trend.

4.4. Climatic Effects

4.4.1. Present. Although precipitation and latent heating are negligible, clouds on Mars nevertheless play an important role in its present climate system. They control the exchange of water between hemispheres, alter the deposition of solar and infrared energy in the atmosphere and on the surface, affect photochemical reactions, and are coupled to the dust cycle, a major component of the planet's climate system.

The interplay between Mars' eccentric orbit, its general circulation, and clouds has a strong effect on the meridional transport of atmospheric water vapor (see *Clancy et al.*, 1996). As mentioned earlier, water subliming from the NPRC during summer is swept up in the rising branch of the Hadley cell and condenses as it crosses the equator, forming the ACB. This cloud belt restricts the flow of water into the southern hemisphere. During southern summer the water returns to the northern hemisphere by the same process. However, because the planet is much closer to the Sun at this season, and because the atmosphere is dustier, atmospheric temperatures are warmer and the altitude to saturation is higher. Consequently cloudiness is greatly reduced. Figure 13 illustrates this so-called "Clancy effect."

The seasonal variation in the cloud condensation level favors a net annual mean transport toward the northern hemisphere and helps explain the present stability of ice at the north pole.

The radiative effects of clouds are also important. They have a direct effect on surface temperatures by reflecting sunlight back to space during the day, and blocking the loss of infrared radiation from the surface at night. Consequently, they can reduce the amplitude of the diurnal cycle (*Wilson et al.,* 2007). Cloud radiative effects also affect time and zonal mean temperatures by absorbing thermal infrared radiation coming up from the ground in the ACB (thus warming the atmosphere) and emitting radiation in the polar hoods (thus cooling the atmosphere). Several modeling studies show improved agreement with TES observations when the radiative effects of clouds are included (*Wilson et al.,* 2008; *Madeleine et al.,* 2011; *Haberle et al.,* 2011). The subsequent tightening of the equator-to-pole temperature gradient can also significantly intensify baroclinic systems and dust lifting (*Kahre et al.,* 2011), suggesting a potentially important link between the dust and water cycles.

4.4.2. Past climates. A unique aspect of Mars' past is the large predicted variation in its orbit parameters. For example, its obliquity, presently at 25.2°, has varied between 15° and 45° during the past 20 m.y. (*Laskar et al.,* 2004). General circulation model (GCM) studies suggest that at times of high obliquity (>35°) the martian climate is much wetter and much cloudier, and that its general circulation can mobilize and redistribute surface ice, possibly forming glaciers at nonpolar latitudes (*Haberle et al.,* 2000; *Mischna et al.,* 2003; *Forget et al.,* 2006; *Madeleine et al.,* 2012b).

However, these studies have ignored the radiative effects of clouds. More recent work with radiatively active clouds in high obliquity conditions indicates they can produce a significant greenhouse effect where the negative forcing caused by reflected sunlight is more than compensated for by a positive forcing in the infrared (*Haberle et al.,* 2012). This is made possible on Mars because the simulated circulation forms clouds at high altitudes where temperatures are cold (Fig. 14). Cold high clouds are very effective at reducing the outgoing longwave radiation, which requires higher surface temperatures to maintain energy balance. On Earth clouds have a net overall cooling effect because of the prominence of low clouds associated with a global ocean.

A cloud greenhouse effect may also have helped provide warm and wet conditions on early Mars ca. 3.5 Ga when a variety of lines of evidence suggest flowing water on the surface (*Segura et al.,* 2002; *Colaprete et al.,* 2003a; *Urata and Toon,* 2012). The simulations in these studies indicate that optically thick high level water ice clouds can form in a thick atmosphere and sustain surface temperatures very near the melting point of water in spite of the reduced solar luminosity at that time, provided there is a significant source at the surface (such as an ocean, low albedo caps, or a large impactor) and the cloud particle sizes are large enough to block the outgoing infrared.

4.5. Future Investigations

Clouds will clearly be a major focus of the next generation of Mars GCMs. In the near-term, the goal will be to reproduce the present water cycle and temperature fields with cloud microphysics packages that include the processes of nucleation, growth, and sedimentation, as well as radiative algorithms that account for their predicted sizes and compositions. In the long term, developing models that fully couple the dust, water, and CO_2 cycles to each other is the ultimate goal. When the models can successfully

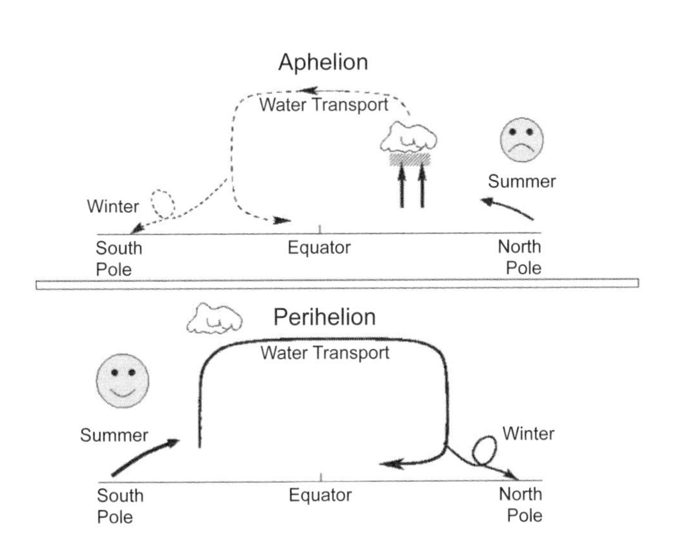

Fig. 13. Cartoon illustrating the "Clancy effect" (see text for details). Courtesy of Franck Montmessin.

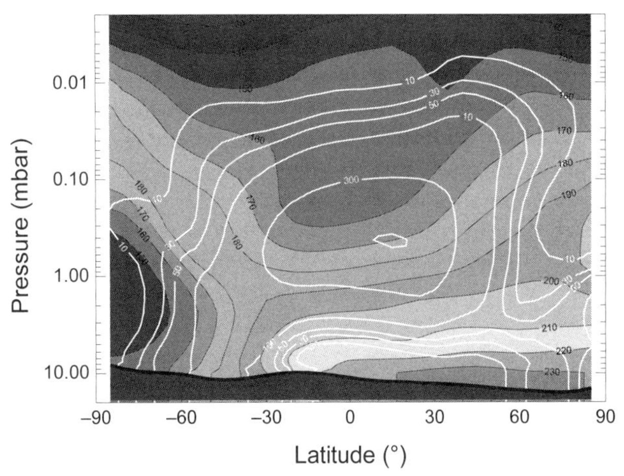

Fig. 14. See Plate 58 for color version. Time and zonally averaged temperatures (color coded with black contours) and water ice cloud volume-mixing ratios in ppmv (white contours) as simulated by the NASA/Ames Mars General Circulation Model for orbital conditions at 632 Ka (see *Haberle et al.,* 2012).

reproduce the present climate system, they can be more confidently applied to the study of past climates. Much work clearly remains.

From an observational perspective, clouds are not the main thrust of future Mars missions. At the time of this writing the only existing approved future mission with a payload capable of observing and characterizing clouds is NASA's Mars Science Laboratory (MSL), which landed on Mars on August 5, 2012. MSL has camera systems and a meteorology package that can observe clouds from the surface for at least one Mars year and possibly longer. The European Space Agency (ESA) and Russia are planning to launch an orbiter in 2016 and lander in 2018 that may have elements in their payloads relevant to clouds. Later in this decade, Japan is considering an equatorial orbiter with an imaging capability that would provide new perspectives on cloud activity.

5. MARS ATMOSPHERIC DUST

5.1. Observations

5.1.1. Measurement techniques. The presence of dust in the martian atmosphere has been inferred from observations dating back to the eighteenth century, when early astronomers noted the obscuration of surface features (*McKim,* 1999). Since then, the spatial and temporal behavior of dust in Mars' atmosphere and its affect on climate and weather has been monitored with observations made by Earth-based telescopes, Mars orbiting platforms, and Mars surface-based landers and rovers. Data obtained with instruments on these platforms utilize several portions of the electromagnetic spectrum including the infrared, near-infrared, visible, and UV. Taken together, these data yield information on the spatial and temporal patterns of atmospheric dust content, the physical characteristics of atmospheric dust grains, and the impact that atmospheric dust has on the climate of Mars.

Observations of the seasonal cycle of atmospheric dust from Mars orbit have been collected from instruments on Mariner 9, Viking, MGS, MO, MEx, and MRO. Extensive spatial and temporal mapping of the atmospheric column dust opacity has been derived from observations of the 9-μm silicate band absorption feature from infrared spectrometers on Mariner 9 [Infrared Interferometric Spectrometer (IRIS)], Viking [Infrared Thermal Mapper (IRTM)], and MGS [Thermal Emission Spectrometer (TES)]. Data acquired with the broadband IR Thermal Emission Imaging System (THEMIS) on MO overlap in time with those obtained with the MGS TES, allowing for a continuous record of derived atmospheric dust optical depth from the beginning of the MGS mission to the present day (Fig. 15) (*Smith,* 2008, 2009).

Observations made with the Compact Reconnaissance Imaging Spectrometer for Mars (CRISM) on MRO are used to derive column dust optical depths and to constrain the radiative properties of airborne dust. Visible imaging observations from cameras on Viking, MGS [Mars Orbiter Camera (MOC)], and MRO [Mars Color Imager (MARCI)] yield a variety of information regarding dust storm frequency, duration, size, evolution, and morphology. Ultraviolet solar and stellar occultation observations by the SPICAM instrument on MEx provide vertical profiles of dust at specific locations and seasons (*Montmessin et al.,* 2006a). The MCS onboard MRO continues to observe the limb of Mars, providing unprecedented information on the vertical distribution of atmospheric dust (Fig. 16) (*Heavens et al.,* 2010).

Observations of dust from the surface of Mars have been acquired from the Viking landers; Mars Pathfinder; Mars Exploration Rovers (MERs), Spirit and Opportunity; and Mars Phoenix. Observations of the Sun with visible imagers on all landers and rovers have yielded quantitative measurements of the visible line-of-sight dust optical depth at the location of the landers/rovers (*Colburn et al.,* 1989; *Smith and Lemmon,* 1999; *Lemmon et al.,* 2004). Additional quantitative measurements of dust optical depth have been derived from data obtained with the mini-TES instruments on the MERs (*Smith et al.,* 2006) and the LIDAR instrument on Mars Phoenix (*Dickinson et al.,* 2011).

5.1.2. Global distribution. Dust is present in the martian atmosphere throughout the year but the atmospheric dust load varies with season (Fig. 15). The total atmospheric dust loading is generally characterized by low-level atmos-

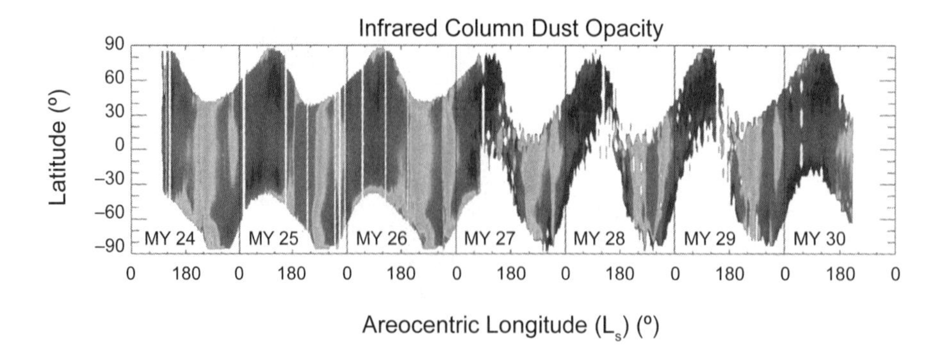

Fig. 15. See Plate 59 for color version. Zonally averaged 9-μm column dust opacity for approximately six of the most recent martian years, as observed by MGS/TES (MY24–26) and Mars Odyssey (MY27–30). Data courtesy of Michael Smith.

pheric dust content during northern spring and summer, and increased levels of atmospheric dust loading during northern autumn and winter (*Smith,* 2004; *Smith,* 2008). Although year-to-year variability does exist, many features of the observed behavior of atmospheric dust repeat from one year to the next.

The peak of the annual atmospheric dust optical depth occurs during southern spring and summer when Mars is near perihelion. During these seasons, dust storms of different sizes and duration produce local, regional, and planet-encircling dust clouds (*Cantor et al.,* 2001). The most dramatic and thermodynamically significant dust events are hemispheric- or planet-encircling dust storms, which occur when local and regional storms coalesce (*Martin and Zurek,* 1993; *Cantor et al.,* 2006). During these events, large dust opacities over extended regions are accompanied by atmospheric temperatures that are as much as 20 K warmer than temperatures representative of years in which no significant dust storms are observed (*Smith et al.,* 2002; *Smith,* 2004, 2008). In each of the recorded years in which a planet-encircling dust storm did not occur, two peaks in dust loading are observed — one before (at approximately L_s = 230°) and one after (at approximately L_s = 340°) southern summer solstice. Increased dust opacities during these time periods are likely related to dust storms associated with northern hemisphere traveling weather systems [i.e., frontal storms (*Wang et al.,* 2005; *Wang and Fisher,* 2009)].

An optically thin dust haze is maintained at low and middle latitudes during northern spring and summer (L_s = 0°–180°; Fig. 15). Short-lived localized dust storms produce distinct dust clouds along the growing and receding edges of the seasonal CO_2 caps in both the north and the south. While these cap-edge local storms are abundant, they are less efficient at increasing the overall atmospheric dust load during these seasons compared to northern autumn and winter (*Cantor et al.,* 2001). Based on spacecraft lander and orbiter observations and GCM results, it has been suggested that dust devils are responsible for maintaining the background dust haze during these seasons (*Fisher et al.,* 2005; *Cantor et al.,* 2006; *Kahre et al.,* 2006). Martian dust devils were first identified in Viking lander meteorology data (*Ryan and Lucich,* 1983) and in Viking Orbiter images (*Thomas and Gierasch,* 1985). Dust devils are observed to be frequent dust-raising events, with peak activity occurring during local spring and summer (*Cantor et al.,* 2006; *Fisher et al.,* 2005; *Balme et al.,* 2003).

Information about the vertical distribution of dust in the martian atmosphere has been gathered from measurements made with several different instruments. Lander-observed line-of-sight dust abundances from the surface upward at different airmasses yield insight into the vertical distribution of dust within the bottom 1–3 scale heights. Observations suggest that dust is likely well mixed in the lowest atmospheric scale height at low to middle latitudes (*Pollack et al.,* 1977; *Smith et al.,* 1997; *Lemmon et al.,* 2004; *Smith,* 2008), but may not be at higher latitudes (*Dickinson et al.,* 2011). Information on the vertical distribution of dust above the lowest scale height has been derived from orbiter observations of the limb made with the cameras on Mariner 9 and the Viking orbiters, MRO's MCS, MGS's TES, and MEx's SPICAM (*Anderson and Leovy,* 1978; *Jaquin et al.,* 1986; *Heavens et al.,* 2011; *Clancy et al.,* 2010; *Montmessin et*

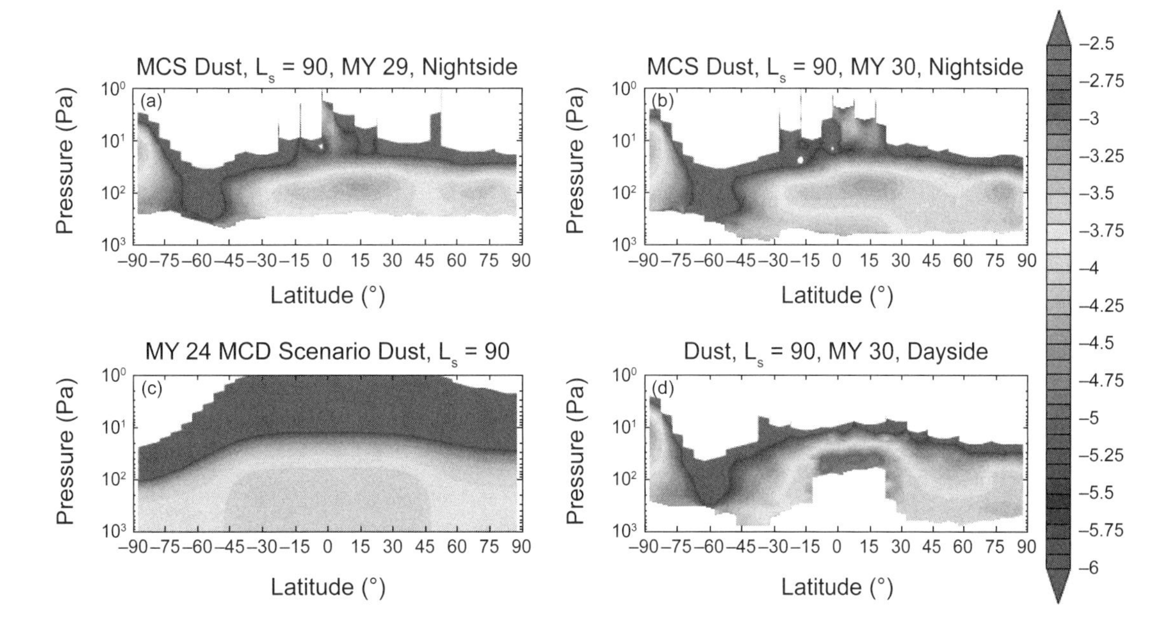

Fig. 16. See Plate 60 for color version. Zonal average density-scaled dust opacity (m^2 kg^{-1}) at L_s = 90° as observed by MRO/MCS during the night from two different Mars years [**(a)**,**(b)**] and during the day from one Mars year [**(d)**]. Also shown is the dust distribution used in the Mars Climate Database at L_s = 90° that does not include an enhanced dust layer aloft [**(c)**]. From *Heavens et al.* (2011).

al., 2006a). Observations indicate that the vertical extent of dust varies with both season and latitude. Dust tends to be more deeply mixed at seasons and locations of higher dust loading. This results in dust extending to the highest altitudes at equatorial and subtropical latitudes during northern autumn and winter. During the 2001 global dust storm, TES observed dust up to approximately 60 km (*Clancy et al.*, 2010). Compelling evidence has recently come from MCS that enhanced layers of dust that are disconnected from the surface exist over prolonged periods of time and longitude at northern tropical latitudes during northern spring and summer (Fig. 16).

5.1.3. Interannual variability. The dust cycle exhibits a substantial degree of interannual variability during some seasons. Observations indicate that there is significant year-to-year variability on the atmospheric dust content between L_s 180° and 360°, but there is little year-to-year variability in the dust cycle between L_s 0° and 180°. The most notable example of this year-to-year variability during northern autumn and winter is the existence of global dust events during some years but not in others (Fig. 15). Global-scale dust events occur approximately one out of every three Mars years, with season of initiation ranging from L_s 184°–320° (*Martin and Zurek,* 1993; *Smith et al.,* 2002; *Cantor,* 2007). Variations in the atmospheric dust loading lead to substantial changes in atmospheric temperatures and the general circulation (*Smith,* 2004; *Haberle et al.,* 1993). Through radiative-dynamical feedbacks and interactions with the water and CO_2 cycles (i.e., the seasonal exchange of CO_2 between the atmosphere and the surface), variations in the dust cycle are likely responsible for the interannual variability observed in atmospheric temperatures and the water and CO_2 cycles.

5.2. Properties

5.2.1. Size, shape, and composition. Information regarding the physical properties of dust in the atmosphere of Mars comes from retrievals from remote sensing observations from orbit and the surface. Observations at different wavelengths (e.g., visible and infrared) of scattered and absorbed light by dust are used to place constraints on particle size and shape. Retrieved effective particle sizes (r_{eff}) during low and moderate dust loading conditions typically range from 1.4 to 1.7 μm. Larger dust particle sizes (r_{eff} = 1.8–2.5 μm) have been observed during high dust loading conditions (e.g., the 2001 global dust event) (see Table 4). These particle size retrievals are based on column-integrated properties. Only recently has information on the vertical distribution of particle sizes been retrieved. These retrievals suggest that during times of large-scale, highly elevated dust optical depths, particle sizes are very similar in the middle and lower atmospheres. By contrast, during times of low dust loading, a trend in particle size is seen as altitude increases, whereby r_{eff} = 1.4–1.8 μm in the lowest 10 km, r_{eff} = 1.0 μm near 20 km, r_{eff} = 0.5 μm in the 30–40-km range, and r_{eff} < 0.3 μm above 40 km (*Montmessin et al.,* 2006a; *Rannou et al.,* 2006; *Federova et al.,* 2009; *Clancy et al.,* 2010).

It is challenging to derive information on the shapes of dust particles in the martian atmosphere. However, near-infrared and visible wavelength observations that measure the scattering phase function from an orbiting platform or surveys of sky brightness from a surface platform can be used to place constraints on particle shapes. *Clancy et al.* (2003) found that TES emission phase functions were best fit with disk-shaped particles with axial ratios ranging from 0.5 to 2.0 instead of simple spheres.

Information regarding the composition of martian airborne dust has been gleaned from matching the shape of the observed infrared spectral features against those expected from the spectra of known minerals. Early work suggested that the minerals montmorillonite and palagonite were good Mars dust analogs (*Toon et al.,* 1977; *Clancy et al.,* 1995). More recent studies suggest that three framework silicates likely dominate the primary composition of the dust: plagioclase, feldspar, and zeolite (*Bandfield and Smith,* 2003; *Hamilton et al.,* 2005, *Smith,* 2008).

5.2.2. Radiative. Dust grains in the martian atmosphere absorb and scatter solar energy. The broadband visible wavelength single-scattering albedo of airborne dust is approximately 0.92 (*Wolff et al.,* 2009). Forward scattering is favored by dust particles of the size and shape observed in the martian atmosphere, with asymmetry parameters ≥0.6 over the visible region of the spectrum. In the infrared, silicate dust has broad absorption features at 9 and 22 μm that have been used to retrieve dust content in the martian atmosphere (IRTM, TES, THEMIS, MCS). Because dust efficiently absorbs in the visible and radiates in the infrared, the presence of atmospheric dust greatly affects atmospheric heating rates and the thermal and dynamical structure of the atmosphere and affects the temperature of the surface. The most dramatic example of the effect of atmospheric dust on atmospheric and surface temperatures occurs during global dust storms. Atmospheric temperatures have been observed to be more than 20 K warmer during these events relative to years without global-scale events (*Smith et al.,* 2002, 2004; *Smith,* 2008). During times of large atmospheric dust loading, daytime surface temperatures are suppressed and nighttime surface temperatures are increased, which results in a reduced diurnal cycle in surface temperature (*Ryan and Henry,* 1979). Thus, airborne dust is a critical component of the radiative balance of the martian atmosphere and surface.

TABLE 4. Mars dust aerosols.

Atmospheric Conditions	Mean Particle Radius (μm)	Vertical Profile
Low dust loading	r_{eff} = 1.4–1.7	Size decreasing with altitude
High dust loading	r_{eff} = 1.8–2.5	Same at all altitudes

Shapes: Ellipsoidal; composition: framework silicates, like plagioclase, feldspar, and zeolite.

5.3. Physics of Dust Lifting, Transport, and Removal

The physical processes of dust lifting, transport, and removal ultimately regulate the quantity and distribution of dust in Mars' atmosphere. Dust lifting occurs through the exchange of momentum and heat between the atmosphere and the surface. Two major processes are thought to be responsible for dust lifting from the surface of Mars: dust lifting due to near-surface wind stress and dust lifting due to convective vortices (i.e., dust devils).

Surface winds on Mars have not been observed (*Zurek et al.,* 1992) nor predicted by GCMs (*Pollack et al.,* 1990) to be strong enough to lift dust-sized particles from the surface directly by near-surface wind. Thus, dust particles that are lifted by near-surface wind stress are likely injected into the atmosphere through the process of saltation. During saltation, sand-sized particles are lifted from the surface and travel only a short distance before falling back to the surface under the influence of gravity. Upon returning to the surface, these ~100-μm particles kick up dust-sized particles that can then enter into suspension.

Convective vortices are ubiquitous on Mars, and it is clear that they are capable of lifting dust from the surface. The process by which dust devils lift dust on Mars is not yet well understood. One possibility is that the tangential wind around the vortex core exceeds the threshold necessary for wind stress dust lifting by saltation. A second possibility is that the pressure drop that occurs in the horizontal direction across the vortex produces a "suction" effect that lifted dust from surface even though the near-surface wind stress is not high enough for saltation. Laboratory studies of vortex generation under Mars conditions using NASA's Martian Surface Wind Tunnel (MARSWIT) suggest that dust lifting is dependent only on the magnitude of the pressure drop across the vortex instead of their physical size (*Neakrase and Greeley,* 2010). These results suggest that the suction effect dominates.

Dust lifted from the surface travels some distance before returning to the ground. The distance that a dust particle travels depends on the balance between that grain's gravitational fall velocity and the strength of vertical and horizontal atmospheric transport. Dust is removed from the atmosphere through the process of gravitational sedimentation. The gravitational sedimentation velocity of a particle is reached when the downward force of gravity and upward drag and buoyancy forces are in balance. At the temperatures and air densities in the martian atmosphere, the mean-free path of the gas is long compared to a typical dust particle's radius. Thus, the Cunningham slip-flow correction is required to account for a particle's increase in fall velocity due to relatively infrequent collisions with gas molecules.

5.4. Climatic Effects

Airborne dust is critical for the current climate of Mars. The presence of atmospheric dust has a significant effect on the distribution of visible and infrared radiant energy in the atmosphere and on the surface, and thus on the thermal and dynamical state of the atmosphere. Additionally, couplings with the water and CO_2 cycles significantly influence those climate cycles.

The thermal and dynamical state of Mars' low-mass, short-radiative time-constant atmosphere is very sensitive to the quantity and distribution of airborne dust. Airborne dust efficiently absorbs and scatters energy at visible wavelengths, which results in the heating of the dust grains themselves and the surrounding atmospheric gas (*Gierasch and Goody,* 1972). At infrared wavelengths, airborne dust efficiently absorbs in the 9-μm silicate band, increases the emissivity of the CO_2 gas at 15 μm, and locally cools or warms the atmosphere depending upon the environmental conditions (*Toon et al.,* 1977; *Pollack et al.,* 1979, 1990).

Interactions between the dust, CO_2, and water cycles are important for the current climate of Mars. The dust, water, and CO_2 cycles are coupled through cloud condensation processes. Suspended dust particles are thought to provide seed nuclei for heterogeneously nucleated water ice and CO_2 ice clouds (*Montmessin et al.,* 2002; *Michaelangeli et al.,* 1993; *Määttänen et al.,* 2005). Ice-covered dust particles fall at speeds different than the dust particle alone, and depend on the ratio of dust to ice mass and the density of each material (*Rossow,* 1978). Thus, cloud formation can change the vertical distribution (and by extension the horizontal distribution) of dust, water, and CO_2 in the atmosphere. Additional coupling between the dust and CO_2 cycles include the radiative effects of polar airborne dust, the influence of dust upon atmospheric heat transport into the polar regions, and the modification of seasonal CO_2 ice cap properties.

5.5. Future Investigations

More modeling work is needed to understand the martian dust cycle and how atmospheric and surface dust affects the climate of Mars. Questions regarding how, where, and when dust enters the atmosphere remain largely unanswered. To date, Mars GCM simulations that include a fully interactive dust cycle are only able to reproduce broad features of the observed annual cycle of atmospheric dust loading (*Newman et al.,* 2002; *Basu et al.,* 2004; *Kahre et al.,* 2006). Further work is needed to understand what controls the observed level of interannual variability in the dust cycle. This work will likely include more sophisticated dust-lifting parameterizations and the use of finite dust reservoirs. Additional modeling work is also needed to understand how the dust cycle couples to and affects the water and CO_2 cycles.

In the near future, observations of dust from the martian surface retrieved from instruments on MSL should add to our understanding of dust in Mars' atmosphere. Imaging observations of dust-lifting events acquired with the camera systems coupled with observations of near-surface winds with the meteorology package may help us better understand dust-lifting processes. The main focus of NASA's Mars program after MSL does not include characterizing dust in the lower and middle atmosphere of Mars. Missions by ESA and Russia later in this decade will likely include instruments relevant for the study of atmospheric dust.

6. MARS CARBON DIOXIDE CLOUDS

As discussed in the chapter by Rafkin et al. in this volume, each martian year nearly 30% of the atmospheric mass condenses to form the polar caps. During the polar night, the radiative cooling of the surface and atmosphere is balanced by the release of latent heat from condensing CO_2. The formation of CO_2 clouds within the polar hood alters the thermodynamic state of the atmosphere through the distribution of latent heat release and infrared radiative effects. MGS radio occultation, TES, and MCS measurements in the polar regions show temperature profiles that frequently follow the saturation curve for CO_2 in the lower portion (below ~20 km) of the atmosphere (*Hinson et al.,* 1999, *Colaprete et al.,* 2005). These observations suggest that the temperature in this region is maintained by atmospheric condensation of CO_2. However, in many places temperatures do fall below the saturation temperature of CO_2 and result in an environment in which convective instability may result. The percentage of polar cap ice that forms from the direct condensation of CO_2 on the surface, relative to that which forms through the consolidation of precipitating CO_2 snow, is an important aspect of the energy balance of the cap itself. Furthermore, the process of ice deposition via cloud precipitation may lend clues to the nature and distribution of polar frosts, including, e.g., the position of the south pole residual cap (SPRC).

6.1. Observations of Carbon Dioxide Clouds

Carbon dioxide clouds most commonly form during the polar nights, thus making direct visual observations of them difficult. However, several spectral observations in the infrared have confirmed their presence. Mariner 6 and 7 made the first observations of CO_2 clouds with their infrared spectrometers. Limb spectra from these spacecraft revealed sharp reflectance spikes at 4.26 μm that were attributed to CO_2 ice crystals.

Other observations in the polar nights are less direct. Observations by Mariner 9 and Viking Orbiter have shown evidence for surface brightness temperatures below the expected CO_2 saturation temperature in the polar regions. Similar observations were made by the TES onboard MGS (*Kieffer et al.,* 2000) and the MCS flying onboard MRO (*Hayne et al.,* 2011).

The Mars Orbiter Laser Altimeter (MOLA), while making laser ranging measurements of the martian surface, detected laser returns (reflections) from altitudes extending from the surface to as high as 40 km (*Ivanov et al.,* 2001). These reflections are believed to be from CO_2 clouds (based on the limited availability of water vapor in the polar night) and provide a unique dataset as to the vertical structure of CO_2 clouds and their particle properties. An intriguing aspect of the MOLA observations is that there were distinct differences between the number and type of reflections seen. MOLA had the ability to capture reflections across four time gates, with each gate corresponding to the width in the returned laser pulse spread over time. In the south many more channel 1 echoes, or reflections from a highly extinctive surface, were observed relative to the north. It has been argued that these channel 1 reflections could be the result of either large flat CO_2 crystals (*Neumann et al.,* 2003) or higher concentrations of much smaller CO_2 ice grains (*Colaprete et al.,* 2005).

The TES and MCS instruments have both made measurements of brightness temperatures below the expected saturation temperature. MCS made these observations looking both nadir and at the martian limb. In the south, a large ~500-km-diameter cloud with visible optical depth ~0.1–1.0 persists over the SPRC for much of the southern winter (Fig. 17). In the latitude range 70°–80°S clouds and the associated low emission regions are characterized by short lifetimes and smaller sizes (~10 km) and appear to correspond to the channel 1 clouds observed by MOLA. These observations extend earlier similar measurements by IRTM and TES, and support the idea that these low brightness temperatures are associated with solid CO_2 ice grains, both in the atmosphere and on the surface (*Hayne et al.,* 2011).

Carbon dioxide clouds have also been observed at high (mesospheric) altitudes at mid and low latitudes. The first indication of the potential for high-altitude CO_2 clouds came during atmospheric entry of Mars Pathfinder. During entry, Mars Pathfinder measured temperatures below the vapor pressure of CO_2 ice at altitudes around 80 km (*Schofield*

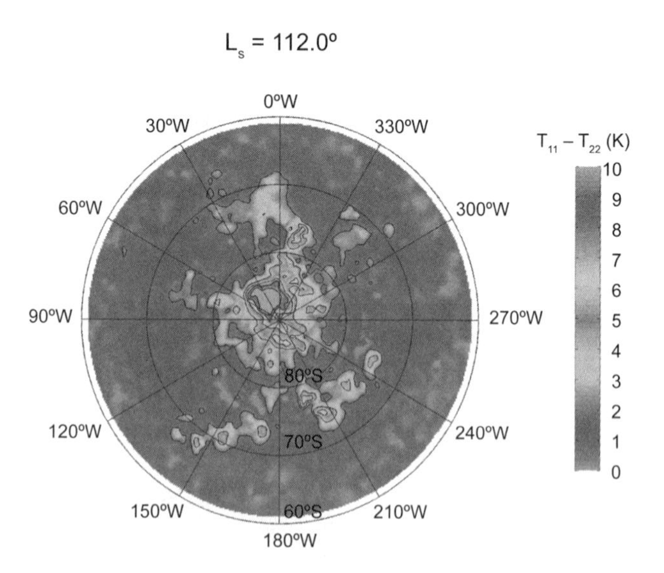

Fig. 17. See Plate 61 for color version. Anomalously low brightness temperatures as measured by MRO MCS. These low brightness temperatures are believed to be the result of infrared scattering by CO_2 clouds (*Hayne et al.,* 2011).

et al., 1997). Once on the surface, the Mars Pathfinder lander observed on occasion pre-dawn, blue-colored clouds. Together with the temperature observations, *Clancy and Sandor* (1998) suggested that these pre-dawn clouds were high-altitude CO_2 clouds. Similar temperature excursions have been observed by the SPICAM instrument on MEx during stellar occultation measurements (*Montmessin et al.,* 2006a) and in MEx OMEGA and HRSC observations (*Montmessin et al.,* 2007). In the case of the SPICAM observations, cloud layers were also observed in proximity to the regions where temperatures fell below the saturation temperature of CO_2, suggesting that the composition of the clouds was CO_2.

Additional observations of high-altitude clouds have come from TES and the MOC on MGS. The TES instrument made solar (single broad channel) and thermal infrared observations across the limb that revealed detached layers at altitudes of approximately 40 and 80 km. The composition of the clouds is not definitive due to the lack of thermal emission observations at these higher altitudes; however, the optical depth ($\tau \sim 0.01$ in the nadir) and vertical position of the higher-altitude layer is consistent with the clouds being composed of CO_2. In several MOC images of the limb these detached hazes are clearly evident (Fig. 18). Likewise, the THEMIS instrument on MO has also made observations of high-altitude clouds at both equatorial and mid-latitude regions with optical depths between 0.05 and 0.5 (*McConnochie et al.,* 2010). Radiative transfer analysis of the observations by TES is consistent with CO_2 particles with radii of r < 1.5 μm. Similar analysis of the OMEGA observations results in somewhat different cloud properties, with optical depths ($\tau < 0.5$) and particle radii (r < 3 μm). Analysis of SPICAM data suggest optical depths of $\tau < 0.01$ and smaller radii, with r < 0.1 μm. The discrepancies may be due to difference in observational and analysis techniques, or associated with different local times of day when the observations were made. Table 5 summarizes the locations and characteristics of martian CO_2 clouds.

Fig. 18. Mesosphere CO_2 clouds as imaged by MGS MOC (*Clancy et al.,* 2007).

6.2. Carbon Dioxide Cloud Formation Microphysics

The formation (nucleation) and growth of CO_2 clouds in the martian atmosphere represent a special case of cloud formation, one in which the primary atmosphere constituent (CO_2) is condensing to form clouds. In the case in which the minor constituent is condensing, e.g., water vapor, diffusion of the minor gas through the primary gas can be an important factor in cloud nucleation and growth. On Mars, CO_2 cloud growth is limited only by the release of latent heat and self-diffusion of the primary gas to the growing crystal (*Colaprete et al.,* 2003b). The formation of CO_2 clouds is generally believed to occur via heterogeneous nucleation — the formation of a CO_2 ice embryo on the surface of a substrate such as a dust grain (*Colaprete et al.,* 2003b; *Määttänen et al.,* 2005). Even in the case of heterogeneous nucleation some saturation of the CO_2 gas is required before ice embryo energy barriers are overcome and nucleation occurs. *Glandorf et al.* (2002) measured in the lab the saturation at which CO_2 cloud nucleation occurs — the critical supersaturation — to be about 35%. This level of saturation is consistent with observations of saturation in the winter polar regions (*Hinson and Wilson,* 2002; *Colaprete et al.,* 2008). The critical supersaturation can depend strongly on the nuclei size (*Colaprete et al.,* 2003b; *Määttänen et al.,* 2005), and therefore a CO_2 cloud's properties, including the number and size of particles in the cloud, will depend on the local population of nuclei, e.g., dust. It is possible that in certain circumstances the formation and growth of CO_2 clouds within a supersaturated region can provide sufficient local latent heating to

TABLE 5. Martian CO_2 cloud properties.

Cloud Location/Season	Typical Vertical Extent	Visible Optical Depth	Mean Particle Radius (μm)	Source
North polar night troposphere	0–40 km	0.1–1+	~5–60	MOLA, modeling
South polar night troposphere	0–40 km	0.1–1+, most highest extinction clouds	~1–30	MOLA, modeling, MCS
Equatorial and mid-latitude mesosphere	60–100 km	0.01–1	0.1–2	TES, SPICAM, OMEGA, CRISM, THEMIS, MOC, modeling

produce buoyant convection (*Colaprete et al.,* 2005). The cumuliform morphology of some mesospheric CO_2 clouds suggests convection may be involved. While in the mesosphere the total amount of potential convective energy may be small, in the polar night this convection may play a significant role in the redistribution of energy and affect the overall temperature structure of the polar night atmosphere (*Colaprete et al.,* 2008).

6.3. Implications for Current and Early Mars Climates

Carbon dioxide cloud formation is important for maintaining the martian polar night temperatures near the saturation temperature and is thus a critical source of atmospheric energy. As mentioned, the release of latent heat may result in buoyancy and convection, resulting in a source of vertical mixing in an otherwise stable atmosphere. It is apparent from thermal observations (e.g., TES and MCS) that these clouds are having an impact on the total observed thermal emission; however, the radiative effects these cloud particles have is still not entirely understood and has not been studied as thoroughly as martian water ice clouds (see section 4). In any case, the radiative importance of CO_2 clouds could be important as cloud optical depths are expected at times to be significant ($\tau \sim 1$).

Carbon dioxide cloud formation may also have played a significant role in the early martian climate. One explanation for the fluvial features seen in the ancient cratered terrain of Mars is that during their formation, Mars was warm enough for liquid water to be stable at the surface. Climate models show that the greenhouse effects of a thick, 2–5-bar CO_2 atmosphere would have been sufficient to raise the mean surface temperature to above 273 K. The formation of CO_2 clouds within these thick atmospheres would almost be a certainty, and their effects may cause the surface temperature to be lowered (*Kasting,* 1991) or raised (*Forget and Pierrehumbert,* 1997), depending on the characteristics of the clouds themselves (*Colaprete et al.,* 2003b).

6.4. Future Work and Observations

As described, there are a variety of important aspects of CO_2 clouds that require further investigation. Continued theoretical work on CO_2 cloud formation, including microphysics and radiative effects, is important. With respect to radiative effects, studies of the optical coefficients for CO_2 ice grains with irregular shapes and composition should be made, as only studies using spherical grains have been made to date (*Colaprete et al.,* 2003b). This study could also benefit from laboratory measurements of how CO_2 ice grains nucleate and grow.

As a critical part of the martian climate system, the continued inclusion of CO_2 clouds in martian climate models is important. These models have begun to approach the sophistication of their water ice cloud cousins, and with the most recent slew of observations, there now exists a reasonable set of constraints on the model predicts. Future modeling work should include the interactive microphysics and cloud radiative effects. To date most observations of CO_2 clouds are limited by the fact that the bulk of CO_2 cloud formation occurs during the polar nights on Mars. MOLA observation provided a tantalizing look into the darkness, but a dedicated LIDAR instrument, similar to that used to observe terrestrial clouds, would provide the necessary data on cloud distribution and properties (e.g., particle size) that is needed to truly understand their role in the martian climate system.

7. COMPARISONS AND CONCLUSIONS

Clouds and aerosols have important climate effects — both direct and indirect — on all terrestrial planets. The complex processes present on the different planets allow multiple connections between the processes and significant feedbacks. Past climates, e.g., on Mars, may also have been significantly affected by clouds.

Venus is constantly covered in clouds, Mars has fewer clouds due to low concentrations of H_2O, and Earth falls somewhere in between. Earth clouds provide a negative feedback for global warming, and the same is true for Venus. On Mars, the feedback is unknown but likely near neutral (*Harberle et al.,* 2011). On Earth and Mars, most clouds nucleate on aerosols. On Venus, there is a much larger role for homogenous nucleation, particularly at the lower cloud boundary. On both Mars and Earth, aerosols influence the clouds strongly by providing cloud condensation nuclei. Comparisons are shown in Table 6.

Some useful comparisons are possible. As an example, CCN on Earth are formed *in situ* at many locations and are very tiny (10–100 nm). This is very different than the situation for Mars, where the CCN are lifted from the surface and are in the micrometer size range. The Venus climate shows multiple feedback cycles. Is the feedback on Earth climate enhancing, or mitigating climate change? The direct evidence from the more extreme Venus system may provide some perspective.

Because current Earth climate models are generally too specialized to be applied directly to the other terrestrial planets, climate models for these planets should be built up from basic physical principles, and then developed to explore thresholds and nonlinear effects. This is especially true for treating clouds and aerosols in climate models. Feedback effects are hard to test on Earth, and may be more evident on the other planets. These findings could provide insights for improving Earth models. On Venus, volcanic activity, cloud formation, and the greenhouse effect provide a rich opportunity to search for possible feedbacks.

NASA is undertaking a major effort to find terrestrial planets around other stars. The first observations will not resolve an extrasolar planet; its global brightness, variability, and spectrum will be the first measurements. The mass, radius, and albedo can also be inferred from the observations. It is important to remember that the observed temperature

TABLE 6. Comparison of clouds across terrestrial planets.

	Earth	Mars	Venus
Cloud composition	Water (liquid, ice)	Water and CO_2 ice	Sulfuric acid and unknown absorbers
Typical cloud droplet size	20 μm (liquid), 200 μm (ice crystal)	1–5 μm (water ice) 10–50 μm (CO_2 ice)	0.1–10 μm
Global average cloud amount	60%	30%	100%
Common cloud locations	Most common at mid/high latitudes from 0–2 km; also common in tropics up to 18 km	Polar regions during fall/winter (<10 km); tropics during northern summer (20–40 km)	48–70 km with hazes above and below
Overall cloud climate effect on TOA energy budget	Cooling	Near neutral	Cooling
Overall cloud climate effect on surface energy budget	Cooling	Near neutral	Cooling

may differ from the effective temperature and also from the surface temperature. Venus provides an excellent example in our own solar system.

Venus-like planets could be clearly distinguished from potentially habitable Earth-like planets. The models we have developed for Venus' clouds and chemistry (along with the coupled radiation and dynamics) provide a starting point for understanding those planets.

REFERENCES

Albrecht B. A. (1989) Aerosols, cloud microphysics, and fractional cloudiness. *Science, 245(4923),* 1227–1230.

Anderson E. and Leovy C. (1978) Mariner 9 television limb observations of dust and ice hazes on Mars. *J. Atmos. Sci., 35,* 723–734.

Andrews T. and Gregory J. M. (2012) Cloud adjustment and its role in CO_2 radiative forcing and climate sensitive: A review. *Surv. Geophys., 33,* 619–634.

Balme M. R., Whelley P. L., and Greeley R. (2003) Mars: Dust devil track survey in Argyre Planitia and Hellas Basin. *J. Geophys. Res., 108(E8),* DOI: 10.1029/2003JE002096.

Bandfield J. L. and Smith M. D. (2003) Multiple emission angle surface-atmosphere separations of thermal emission spectrometer data. *Icarus, 161,* 47–65.

Barker E. S. (1979) Detection of SO_2 in the UV spectrum of Venus. *Geophys. Res. Lett., 6,* 117–120.

Barker E. S., et al. (1975) Relative spectrophotometry of Venus from 3067 to 5960 Å. *J. Atmos. Science, 32,* 1205.

Basu S., Richardson M. I., and Wilson R. J. (2004) Simulation of the martian dust cycle with the GFDL Mars GCM. *J. Geophys. Res, 109(E11),* E11006, DOI: 10.1029/2004JE002,243.

Benson J. L., James P. B., Cantor B. A., and Remigio R. (2006) Interannual variability of water ice clouds over major martian volcanoes observed by MOC. *Icarus, 184,* 365–371.

Benson J. L., Kass D. M., Keinböhl A., McCleese D. J., Schofield J. T., and Taylor F. W. (2010) Mars' south polar hood as observed by the Mars Climate Sounder. *J. Geophys. Res., 115,* E12015.

Benson J. L., Kass D. M., and Keinböhl A. (2011) Mars' north polar hood as observed by the Mars Climate Sounder. *J. Geophys. Res., 116,* E03008.

Bergeron T. (1935) *Proces Verbaux de l'Association de Meteorologie* (P. Duport, ed.), pp. 156–178. International Union of Geodesy and Geophysics, Karlsruhe, Germany.

Briggs G. A. and Leovy C. B. (1974) Mariner 9 observations of the Mars north polar hood. *Bull. Am. Meteor. Soc., 55,* 278–296.

Bullock M. and Grinspoon D. (2001) The recent evolution of climate on Venus. *Icarus, 150,* 19–37.

Cantor B. (2007) MOC observations of the 2001 Mars planet-encircling dust storm. *Icarus, 186,* 60–96, DOI: 10.1016/j.icarus.2006.08.019.

Cantor B. A., James P. B., Caplinger M., and Wolff M. J. (2001) Martian dust storms: 1999 Mars orbiter camera observations. *J. Geophys. Res., 106,* 23653–23687.

Cantor B., Malin M., and Edgett K. S. (2002) Multiyear Mars Orbiter Camera (MOC) observations of repeated martian weather phenomena during the northern summer season. *J. Geophys. Res., 107(E3),* 5014, DOI: 10.1029/2001JE001588.107.

Cantor B. A., Kanak K. M., and Edgett K. S. (2006) Mars Orbiter Camera observations of Martian dust devils and their tracks (September 1997 to January 2006) and evaluation of theoretical vortex models. *J. Geophys. Res., 111,* E12002, DOI: 10.1029/2006JE002700.111.

Cantor B. A., Malin M. C., Wolff M. J., et al. (2008) Observations of the martian atmosphere by MRO-MARCI, an overview of 1 Mars year (abstract). In *Third International Workshop on the Mars Atmosphere: Modeling and Observations,* Abstract #9075, LPI Contrib. No. 1447, Lunar and Planetary Institute, Houston.

Charlson R. J., Schwartz S. E., Hales J. M., Cess R. D., Coakley J. A. Jr., Hansen J. E., and Hoffman D. J. (1992) Climate forcing by anthropogenic aerosols. *Science, 255,* 423–430.

Chin M., Kahn R. A., Remer L. A., Yu H., Rind R., Feingold G., Quinn P. K., Schwartz S. E., Streets D. G., and DeCola P. (2009) Atmospheric aerosol properties and climate impacts. In *A Report by the U.S. Climate Change Science Program and the Subcommittee on Global Change Research* (M. Chin et al., eds.). NASA, Washington, DC.

Clancy R. T. and Muhleman D. O. (1991) Long-term (1979–1990) changes in the thermal, dynamical, and compositional structure of the Venus mesosphere as inferred from microwave spectral observations of ^{12}CO, ^{13}CO and $C^{18}O$. *Icarus, 89,* 129–146.

Clancy R. T. and Sandor B. J. (1998) CO_2 ice clouds in the upper atmosphere. *Geophys. Res. Lett., 25,* 492.

Clancy R. T., Lee S. W., Gladstone G. R., McMillan W. W., and Rousch T. (1995) A new model for Mars atmospheric dust based upon analysis of ultraviolet through infrared observations from Mariner 9, Viking, and Phobos. *J. Geophys. Res, 100,* 5251–5263.

Clancy R. T., Grossman A. W., Wolff M. J., James P. B., Billawala Y. N., Sandor B. J., Lee S. W., and Rudy D. J. (1996) Water vapor saturation at low altitudes around Mars aphelion: A key to Mars climate? *Icarus, 122,* 36–62.

Clancy R. T., Wolff M. J., and Christensen P. R. (2003) Mars aerosol studies with the MGS TES emission phase function observations: Optical depths, particle sizes, and ice cloud types versus latitude and solar longitude. *J. Geophys. Res., 108(E9),* DOI: 10.1029/2003JE002058.

Clancy R. T., Wolff M. J., Whitney B. A., Cantor B. A., and Smith M. D. (2007) Mars equatorial mesospheric clouds: Global occurrence and physical properties from Mars Global Surveyor TES and MOC limb observations. *J. Geophys. Res., 112,* E04004.

Clancy R. T., Wolff M. J., Malin M. C., Cantor B. A., and Michaels T. I. (2009) Valles Marineris cloud trails. *J. Geophys. Res., 114(E11),* E11002, DOI: 10.1029/2008JE003323.

Clancy R. T., Wolff M. J., Whitney B. A., Cantor B. A., Smith M. D., and McConnochie T. H. (2010) Extension of atmospheric dust loading to high altitudes during the 2001 Mars dust storm: MGS TES limb observations. *Icarus, 207(1),* 98–109.

Colaprete A., Toon O. B., and Magalhaes J. A. (1999) Cloud formation under Mars Pathfinder conditions. *J. Geophys. Res., 104,* 9043–9053.

Colaprete A., Haberle R. M., Segura T. L., Toon O. B., and Zahnle K. (2003a) Post impact Mars climate simulations using a GCM (abstract). In *Sixth International Conference on Mars,* Abstract #3281. LPI Contrib. No. 1164, Lunar and Planetary Institute, Houston.

Colaprete A., Haberle R. M., and Toon O. B. (2003b) Formation of convective carbon dioxide clouds near the south pole of Mars. *J. Geophys. Res., 108(E7),* 5081.

Colaprete A., Barnes J. R., Haberle R. M., Hollingsworth J. L., Kieffer H. H., and Titus T. N. (2005) Albedo of the south pole of Mars determined by topographical forcing of atmosphere dynamics. *Nature, 435,* 184–188.

Colaprete A., Barnes J. R., Haberle R. M., and Montmessin F. (2008) CO_2 clouds, CAPE and convection on Mars: Observations and general circulation modeling. *Planet. Space Sci., 56,* 150–180.

Colburn D., Pollack J. B., and Haberle R. M. (1989) Diurnal variations in optical depth at Mars. *Icarus, 79,* 159–189.

Conway R. R., McCoy R. P., Barth C. A., and Lane A. L. (1979) IUE detection of sulfur dioxide in the atmosphere of Venus. *Geophys. Res. Lett., 6,* 629–631.

Daerden F., Whiteway A., Davy R., Verhoeven C., Komguem L., Dickinson C., Taylor P. A., and Larsen N. (2010) Simulating observed boundary layer clouds on Mars. *Geophys. Res. Lett., 37,* L04203.

Dickinson C., Komguem L., Whiteway J. A., Illnicki M., Popovici V., Junkermann W., Connolly P., and Hacker J. (2011) Lidar atmospheric measurements on Mars and Earth. *Planet. Space Sci., 59(10),* 942–951.

Ekonomov A. P., et al. (1983) UV photometry at the Venera 13 and 14 landing probes. *Cosmic Res., 21,* 194–206.

Ekonomov A. P., Moroz V. I., Moshkin B. E., Gnedykh V. I., Golovin Y. M., and Crigoryev A. V. (1984) Scattered UV solar radiation within the clouds of Venus. *Nature, 307,* 345–347.

Esposito L. W. (1980) Ultraviolet contrasts and the absorbers near the Venus cloud tops. *J. Geophys. Res., 85,* 8151–8157.

Esposito L. W. (1984) Sulfur dioxide: Episodic injection shows evidence for active Venus volcanism. *Science, 223,* 1072–1074.

Esposito L. W. and Travis L. D. (1982) Polarization studies of the Venus UV contrasts: Cloud height and haze variability. *Icarus, 51,* 374–390.

Esposito L. W., Knollenberg R. G., Marov M. Y., Toon O. B., and Turco R. P. (1983) The clouds and hazes of Venus. In *Venus* (D. M. Hunten et al., eds.), pp. 484–564. Univ. of Arizona, Tucson.

Esposito L. W., Copley M., Eckert R., Gates L., Stewart A. I. F., and Worden H. (1988) Sulfur dioxide at the Venus cloud tops, 1978–1986. *J. Geophys. Res., 93,* 5267–5276.

Esposito L. W., Bertaux J. L., Krasnopolsky V., Moroz V. I., and Zasova L. V. (1997) Chemistry of lower atmosphere and clouds. In *Venus II* (S. W. Bougher et al., eds.), pp. 415–458. Univ. of Arizona, Tucson.

Federova A. A., Korablev O. I., Bertaux J. L., Rodin A. V., Montmessin F., Belyaev D., and Reberac A. (2009) Solar infrared occultation observations by SPICAM experiment on Mars-Express: Simultaneous measurements of the vertical distributions of H_2O, CO_2, and aerosol. *Icarus, 200,* 96–117.

Findeisen W. (1938) *Kolloid-Meteorologische, 2nd edition.* American Meteorological Society, Boston.

Fisher J. A., Richardson M. I., Newman C. E., Szwast M. A., Graf C., Basu S., Ewald S. P., Toigo A. D., and Wilson R. J. (2005) A survey of martian dust devil activity using Mars Global Surveyor Mars Orbiter Camera images. *J. Geophys. Res., 110(E3),* E03004, DOI: 10.1029/2003JE002165.

Forget F. and Pierrehumbert R. T. (1997) Warming early Mars with carbon dioxide clouds that scatter infrared radiation. *Science, 278,* 1273–1276.

Forget F., Haberle R. M., Montmessin F., and Levrard B. (2006) Formation of glaciers on Mars by atmospheric precipitation at high obliquity. *Science, 311,* 368–370.

Gierasch P. J. and Goody R. M. (1972) The effect of dust on the temperature of the martian atmosphere. *J. Atmos. Sci., 29,* 400–402.

Glandorf D., Colaprete A., Toon O. B., and Tolbert M. (2002) CO_2 snow on Mars and early Earth. *Icarus, 160,* 66–72.

Haberle R. M., Pollack J. B., Barnes J. R., Zurek R. W., Leovy C. B., Murphy J. R., Lee H., and Schaeffer J. (1993) Mars atmospheric dynamics as simulated by the NASA Ames General Circulation Model. I — The zonal-mean circulation. *J. Geophys. Res., 98,* 3093–3123.

Haberle R. M., McKay C. P., Schaeffer J., Joshi M., Cabrol N. A., and Grin E. A. (2000) Meteorological control on the formation of martian paleolakes (abstract). In *Lunar Planet. Sci. XXXI,* Abstract #1509. Lunar and Planetary Institute, Houston.

Haberle R. M., Montmessin F., Kahre M. A., Hollingsworth J. L., Schaefer J., Wolff M., and Wilson R. J. (2011) Radiative effects of water ice clouds on the martian seasonal water cycle (abstract). In *Fourth International Workshop on the Mars Atmosphere: Modeling and Observation,* Paris, France. Available online at *www-mars.lmd.jussieu.fr/paris2011/abstracts/haberle_paris2011.pdf.*

Haberle R. M., Kahre M. A., Hollingsworth J. L., Schaeffer J., Montmessin F., and Phillips R. J. (2012) A cloud greenhouse effect on Mars: Significant climate change in the recent past? (abstract). In *Lunar Planet. Sci. XLIII,* Abstract #1665. LPI

Contrib. No 1659, Lunar and Planetary Institute, Houston.
Hamilton V. E., McSween H. Y., and Hapke B. (2005) Mineralogy of martian atmospheric dust inferred from thermal infrared spectra of aerosols. *J.Geophys. Res., 110.*
Hayne P. O., Aharonson O, and Paige D. A. (2011) The role of carbon dioxide snowfall in the formation and energy balance of the seasonal ice caps of Mars (abstract). In *Fifth International Conference on Mars Polar Science and Exploration,* Abstract #6069. LPI Contrib. No. 1323, Lunar and Planetary Institute, Houston.
Heavens N. G., Benson J. L., Kass D. M., Kleinböhl A., Abdou W. A., McCleese D. J., Richardson M. I., Schofield J. T., Shirley J. H., and Wolkenberg P. M. (2010) Water ice clouds over the martian tropics during northern summer. *Geophys. Res. Lett., 37,* L18202.
Heavens N. G., Richardson M. I., Kleinböhl A., Kass D. M., McCleese D. J., Abdou W., Benson J. L., Schofield J. T., Shirley J. H., and Wolkenberg P. M. (2011) The vertical distribution of dust in the martian atmosphere during northern spring and summer: Observations by the Mars Climate Sounder and analysis of zonal average vertical dust profiles. *J. Geophys. Res., 116,* E04003, DOI: 10.1029/2010JE003691.
Hervig M. E., Gordley L. L., Deaver L. E., Siskind D. E., Stevens M. H., Russell J. M. III, Bailey S. M., Megner L., and Bardeen C. G. (2009) First satellite observations of meteoric smoke in the middle atmosphere. *Geophys. Res. Lett., 36,* L18805.
Hinson D. P. and Wilson R. J. (2002) Transient eddies in the southern hemisphere of Mars. *Geophys. Res. Lett., 29,* 58-1, DOI: 10.1029/ 2001GL014103.
Hinson D. P., Flasar F. M., Simpson R. A., Twicken J. D., and Tyler G. L. (1999) Initial results from radio occultation measurements with Mars Global Surveyor. *J. Geophys. Res., 104,* 26997–27012.
IPCC (2007) *Climate Change 2007: The Physical Science Basis, Contribution of Working Group I to the Fourth Assessment Report of the Intergovernmental Panel on Climate Change* (S. Solomon et al., eds.). Cambridge Univ., Cambridge.
Iraci L. T., Phebus B. D., Stone B. M., and Colaprete A. (2010) Water ice cloud formation on Mars is more difficult than presumed: Laboratory studies of ice nucleation on surrogate materials. *Icarus, 201,* 985–991.
Ivanov A. B. and Muhleman D. O. (2001) Cloud reflection observations: Results from the Mars Orbiter Laser Altimeter (MOLA). *Icarus, 154,* 190–206.
Jaquin F., Gierasch P., and Kahn R. (1986) The vertical structure of limb hazes in the martian atmosphere. *Icarus, 68,* 442–461.
Kahn R. (1984) The spatial and seasonal distribution of martian clouds and some meteorological implications. *J. Geophys. Res., 89,* 6671–6688.
Kahn R. A. (1989) *Comparative Planetology and the Atmosphere of Earth: A Report to the Solar System Exploration Division.* NASA, Washington, DC.
Kahn R. A. (2012) Reducing the uncertainties in direct aerosol radiative forcing. *Surv. Geophys., 33,* 701–721, DOI: 10.1007/ s10712-011-9153-z.
Kahre M. A., Murphy J. R., and Haberle R. M. (2006) Modeling the martian dust cycle and surface dust reservoirs with the NASA Ames general circulation model. *J. Geophys. Res., 111,* E06008, DOI: 10.1029/2005JE002588.
Kahre M. A., Hollingsworth J. L., Haberle R. M., and Montmessin F. (2011) Coupling Mars' dust and water cycles: Effects on dust lifting vigor, spatial extent and seasonality (abstract). In *Fourth International Workshop on the Mars Atmosphere: Modeling and Observation,* Paris, France. Available online at *www-mars.lmd.jussieu.fr/paris2011/abstracts/kahre_paris2011.pdf.*
Kasting J. F. (1991) CO_2 condensation and the climate of early Mars. *Icarus, 94,* 1–13.
Kaufman Y. J., Tanre D., and Boucher O. (2002) A satellite view of aerosols in the climate system. *Nature, 419,* 215–223.
Kieffer H. H., Titus T. N., Mullins K. F., and Christensen P. R. (2000) Mars south polar spring and summer behavior observed by TES: Seasonal cap evolution controlled by frost grain size. *J. Geophys. Res., 105,* 9653–9699.
Knollenberg R. G. and Hunten D. M. (1980) The microphysics of the clouds of Venus: Results of the Pioneer Venus particles and size spectrometer experiment. *J. Geophys. Res, 85,* 8039–8058.
Knollenberg R. G., Travis L., Tomasko M., Smith P., Ragent B., Esposito L. W., McCleese D., Martonchik J., and Beer R. (1980) The clouds of Venus: A synthesis report. *J. Geophys. Res, 85,* 8059–8011.
Koren I. and Feingold G. (2011) Aerosol-cloud-precipitation system as a predator-prey problem. *Proc. Natl. Acad. Sci., 108,* 12227–12232.
Krasnopolsky V. A. (1986) *Photochemistry of the Atmosphere of Mars and Venus.* Springer-Verlag, Berlin.
Langevin Y., Bibring J.-P., Montmessin F., Forget F., Vincendon M., Douté S., Poulet F., and Gondet B. (2007) Observations of the south seasonal cap of Mars during recession in 2004–2006 by OMEGA visible/near-infrared imaging spectrometer on board Mars Express. *J. Geophys. Res., 112,* E08S12.
Laskar J., Correia A. C. M., Gastineau M., Joutel F., Levrard B., and Robutel P. (2004) Long term evolution and chaotic diffusion of the insolation quantities of Mars. *Icarus, 170,* 343–364.
Lemmon M. T., Wolff M. J., Smith M. D., et al. (2004) Atmospheric imaging results from the Mars Exploration Rovers: Spirit and Opportunity. *Science, 306(5702),* 1753–1756.
Loeb N. G., Wielicki B. A., Doelling D. R., Smith G. L., Keyes D. F., Kato S., Manalo-Smith N., and Wong T. (2009) Toward optimal closure of the Earth's top-of-atmosphere radiation budget. *J. Climate, 22(3),* 748–766.
Lovelock J. E. and Whitfield M. (1982) Life-span of the biosphere. *Nature, 296,* 561–563.
Määttänen A., Vehkamäki H., Lauri S., Merikallio S., Kauhanen J., Savijärvi H., and Kulmala M. (2005) Nucleation studies in the martian atmosphere. *J. Geophys. Res., 110,* E02002.
Madeleine J.-B., Forget F., and Millour E. (2011) Modeling radiatively active water-ice clouds: Impact on the thermal structure and water cycle (abstract). In *Fourth International Workshop of the Mars Atmosphere: Modeling and Observations,* Paris, France. Available online at *www-mars.lmd.jussieu.fr/paris2011/ abstracts/madeleine2_paris2011.pdf.*
Madeleine J.-B., Forget F., Spiga A., Wolff M. J., Montmessin F., Vincendon M., Jouglet D., Gondet B., Bibring J. P., Langevin Y., and Schmitt B. (2012a) Aphelion water-ice cloud mapping and property retrieval using the OMEGA imaging spectrometer onboard Mars Express. *J. Geophys. Res., 117,* E00J07, DOI: 10.1029/2011JE003940.
Madeleine J.-B., Forget F., Head J. W., Navarro T., Millour E., Spiga A., Colaitis A., Montmession F., and Määtänen A. (2012b) Amazonian glacial cycles on Mars: Response of the new LMD global climate model to orbital variations (abstract). In *Lunar Planet. Sci. XLIII,* Abstract #1661. LPI Contrib. No. 1659, Lunar and Planetary Institute, Houston.
Mahowald N., Ward D., Kloster S., Flanner M. G., Heald C. L.,

Heavens N. G., Hess P. G., Lamarque J.-F., and Chuang P. Y. (2011) Aerosol impacts on climate and biogeochemistry. *Annu. Rev. Environ. Resources, 36,* 45–74.

Malin M. C., Calvin W. M., Cantor B. A., Clancy R. T., Haberle R. M., James P. B., Thomas P. C., Wolff M. J., Bell J. F. III, and Lee S. W. (2008) Climate, weather, and north polar observations from the Mars Reconnaissance Orbiter Mars Color Imager. *Icarus, 194,* 501–512.

Maltagliati L., Montmessin F., Fedorova A., Forget F., Bertaux J. -L., and Korablev O. (2011) Evidence of water vapor in excess of saturation in the atmosphere of Mars. *Science, 333,* 1868–1870.

Marcq E., Bertaux J. L., Montmessin F., and Belyaev D. (2013) Long term variations of sulphur dioxide above Venus' clouds: A case for active volcanism? *Nature Geosci., 6,* 25–28, DOI: 10.1038/ngeo1650.

Martin L. J., James P. B., Dollfus A., Iwasaki K., and Beish J. D. (1992) Telescopic observations — Visual, photographic, polarimetric. In *Mars* (H. H. Kieffer et al., eds.), pp. 34–70. Univ. of Arizona, Tucson.

Martin T. Z. and Zurek R. W. (1993) An analysis of the history of dust activity on Mars. *J. Geophys. Res., 98,* 3221–3246.

McCleese D. J., Heavens N. G., Schofield J. T., et al. (2010) Structure and dynamics of the martian lower and middle atmosphere as observed by the Mars Climate Sounder: Seasonal variations in zonal mean temperature, dust, and water ice aerosols. *J. Geophys. Res., 115,* E12016.

McConnochie T. H., Bell J. F., Savransky D., Wolff M. J., Toigo A. D., Wang H., Richardson M. I., and Christensen P. R. (2010) THEMIS-VIS observations of clouds in the martian mesosphere: Altitudes, wind speeds, and decameter-scale morphology. *Icarus, 210(2),* 545–565.

McKim R. (1999) *Telescopic Martian Dust Storms: A Narrative and Catalogue.* Mem. British Astron. Soc., Vol. 44, British Astronomical Association, London.

McMurry P. H. (2000) A review of atmospheric aerosol measurements. *Atmos. Environ., 34,* 1959–1999.

Michaelangeli D. V., Toon O. B., Haberle R. M., and Pollack J. B. (1993) Numerical simulations of the formation and evolution of water ice clouds in the martian atmosphere. *Icarus, 100,* 261–285.

Michaels T. I., Colaprete A., and Rafkin S. C. R. (2006) Significant vertical water transport by mountain-induced circulations on Mars. *Geophys. Res. Lett., 33,* L16201.

Mills F. P., Esposito L. W., and Yung Y. L. (2007) Atmospheric composition, chemistry and clouds. In *Exploring Venus as a Terrestrial Planet* (L. W. Esposito et al., eds.), pp. 73–101. AGU Geophys. Monogr. 176, American Geophysical Union, Washington, DC.

Mischna M. A., Richardson M. I., Wilson R. J., and McCleese D. J. (2003) On the orbital forcing of martian water and CO_2 cycles: A general circulation model study with simplified volatile schemes. *J. Geophys., Res., 108(E6),* 5062, DOI: 10.1029/2003JE002051.

Molaverdikhani K., McGouldrick K., and Esposito L. W. (2012) The abundance and distribution of the unknown ultraviolet absorber in the venusian atmosphere. *Icarus, 217(2),* 648–660.

Montmessin F., Rannou P., and Cabane M. (2002) New insights into martian dust distribution and water-ice clouds. *J. Geophys. Res., 107(E6),* 4-1 to 4-14, DOI: 10.1029/2001JE001520.

Montmessin F., Forget F., Rannou P., Cabane M., and Haberle R. M. (2004) Origin and role of water ice clouds in the martian water cycle as inferred from a general circulation model. *J. Geophys. Res., 105,* 4109–4121.

Montmessin F., Quemérais E., Bertaux J. -L., Korablev O., Rannou P., and Lebonnois S. (2006a) Stellar occultations at UV wavelengths by the SPICAM instrument: Retrievals and analysis of martian haze profiles. *J. Geophys. Res., 111,* E09S09.

Montmessin F., Bertaux J. -P., Quémerais E., Korablev O., Rannou P., Forget F., Perrier S., Fussen D., Lebonnois S., Rébérac A, and Dimarellis E. (2006b) Subvisible CO_2 clouds detected in the mesosphere of Mars. *Icarus, 183,* 403–410.

Montmessin F., Gondet B., Bibring J. -P., Langevin Y., Drossart P., Forget F., and Fouchet F. (2007) Hyperspectral imaging of convective CO_2 ice clouds in the equatorial mesosphere of Mars. *J. Geophys. Res., 112,* E11S90, DOI: 10.1029/2007JE002944.

Morrison H., Boer G. D., Feingold G., Harrington J., Shupe M. D., and Sulia K. (2011) Resilience of persistent Arctic mixed-phase clouds. *Nature, 5,* 11–17.

Na C. Y. and Esposito L. W. (1995) UV observations of Venus with HST (abstract). *Bull. Am. Astron. Soc., 27,* 1071.

Na C. Y., Esposito L. W., and Skinner T. E. (1990) International Ultraviolet Explorer observation of Venus SO_2 and SO. *J. Geophys. Res., 95,* 7485–7491.

Neakrase L. D. V. and Greeley R. (2010) Dust devil sediment flux on Earth and Mars: Laboratory simulations. *Icarus, 206(1),* 306–318.

Neumann G. A., Smith D. E., and Zuber M. T. (2003) Two Mars years of cloud detected by the Mars Orbiter Laser Altimeter. *J. Geophys. Res., 108(E4),* 5023, DOI: 10.1029/2002JE001849.

Newman C. E., Lewis S. R., Read P. L., and Forget F. (2002) Modeling the martian dust cycle 2. Multiannual radiatively active dust transport simulations. *J. Geophys. Res., 107(E12),* 5124, DOI: 10.1029/2002JE001920.

Owen T. and Sagan C. (1972) Minor constituents in planetary atmospheres: Ultraviolet spectroscopy from the Orbiting Astronomical Observatory. *Icarus, 16,* 557–568.

Pearl J. C., Smith M. D., Conrath B. J., Bandfield J. S., and Christensen P. R. (2001) Observations of martian ice clouds by the Mars Global Surveyor Thermal Emission Spectrometer: The first Mars year. *J. Geophys. Res., 106,* 12325–12338.

Pollack J. B., Colburn D., Kahn R., Hunter J., van Camp W., Carlston C. E., and Wolf M. R. (1977) Properties of aerosols in the martian atmosphere, as inferred from Viking Lander imaging data. *J. Geophys. Res., 82,* 4479–4496.

Pollack J. B., Colburn D. S., Flasar F. M., Kahn R., Carlston C. E., and Pidek D. C. (1979) Properties and effects of dust suspended in the martian atmosphere. *J. Geophys. Res., 84,* 2929–2945.

Pollack J. B., et al. (1980) Distribution and source of the UV absorption in Venus atmosphere. *J. Geophys. Res., 85,* 8141–8150.

Pollack J. B., Haberle R. M., Schaeffer J., and Lee H. (1990) Simulations of the general circulation of the martian atmosphere. 1: Polar processes. *J. Geophys. Res., 95,* 1447–1473.

Ramanathan V., Crutzen P. J., Kiehl J. T., and Rosenfeld D. (2001) Aerosols, climate, and the hydrological cycle. *Science, 294,* 2119–2124.

Ramaswamy V., et al. (2001) Radiative forcing of climate change. In *Climate Change 2001: The Scientific Basis. Contribution of Working Group I to the Third Assessment Report of the Intergovernmental Panel on Climate Change* (J. T. Houghton et al., eds.), pp. 349–416. Cambridge Univ., Cambridge.

Rannou P., Perrier S., Bertaux J.-L., Montmessin F., Korablev O., and Reberac A. (2006) Dust and cloud detection at the Mars limb with UV scattered sunlight with SPICAM. *J. Geophys. Res., 111,* E9.

Rodin A. V., Korblev O. I., and Moroz V. I. (1997) Vertical distribution of water in the near-equatorial troposphere of Mars: Water vapor and clouds. *Icarus, 125,* 212–229.

Rossow W. B. (1978) Cloud microphysics — Analysis of the clouds of Earth, Venus, Mars, and Jupiter. *Icarus, 36,* 1–50.

Rossow W. B., Walker A. W., Beuschel D. E., and Roiter M. D. (1996) *International Satellite Cloud Climatology Project (ISCCP) Documentation of New Cloud Datasets.* WMO/TD-No. 737, World Meteorological Organization, Geneva, Switzerland. 115 pp.

Ryan J. A. and Henry R. M. (1979) Mars atmospheric phenomena during major dust storms, as measured at the surface. *J. Geophys. Res., 84,* 2821–2829.

Ryan J. A. and Lucich R. D. (1983) Possible dust devils vortices on Mars. *J. Geophys. Res., 88(C15),* 11005–11011, DOI: 10.1029/JC088iC15p11005.

Schofield J. T., Barnes J. R., Crisp D., Haberle R. M., Larsen S., Magalhaes J. A., Murphy J. R., Seiff A., and Wilson G. (1997) The Mars Pathfinder atmospheric structure investigation/meteorology. *Science, 278,* 1752.

Segura T. L., Toon O. B., Colaprete A., and Zahnle K. (2002) Environmental effects of large impacts on Mars. *Science, 298,* 1977–1980.

Seinfeld J. H. and Pandis S. N. (1998) *Atmospheric Chemistry and Physics: From Air Pollution to Climate Change, 1st edition.* Wiley, New York.

Smith M. D. (2004) Interannual variability in TES atmospheric observations of Mars during 1999–2003. *Icarus, 167,* 148–165.

Smith M. D. (2008) Spacecraft observations of the martian atmosphere. *Annu. Rev. Earth Planet. Sci., 36,* 191–219.

Smith M. D. (2009) THEMIS observations of Mars aerosol optical depth from 2002–2008. *Icarus, 202,* 444–452.

Smith M. D., Conrath B. J., Pearl J. C., and Christensen P. R. (2002) Thermal Emission Spectrometer observations of martian planet-encircling dust storm 2001A. *Icarus, 157(1),* 259–263.

Smith M. D., Bandfield J. L, Christensen P. R., and Richardson M. I. (2003) Thermal Emission Imaging System (THEMIS) infrared observations of atmosphere dust and water ice cloud optical depth. *J. Geophys. Res., 108(E11),* 5115.

Smith M. D., Wolff M. J, Spanovich N., et al. (2006) One martian year of atmospheric observations using MER Mini-TES. *J. Geophys. Res., 111(12),* E12S13.

Smith P. H. and Lemmon M. (1999) Opacity of the martian atmosphere measured by the Imager for Mars Pathfinder. *J. Geophys. Res., 104(E4),* 8975–8986.

Smith P. H., Bell J. F. III, Bridges N. T., et al. (1997) Results from the Mars Pathfinder Camera. *Science, 278,* 5344.

Stevens B. and Schwartz S. E. (2012) Observing and modeling Earth's energy flows. *Surv. Geophys, 33,* 779–816.

Stewart A. I., Anderson D. E., Esposito L. W., and Barth C. A. (1979) Ultraviolet spectroscopy of Venus: Initial results from the Pioneer Venus orbiter. *Science, 203,* 777–779.

Stock J. W., Boxe C. S., Lehmann R., Grenfell J. L., Patzer A. B. C., Rauer H., and Yung Y. L. (2012) Chemical pathway analysis of the martian atmosphere: CO_2 formation pathways. *Icarus, 219(1),* 13–24.

Taylor F. and Grinspoon D. (2009) Climate evolution of Venus. *J. Geophys. Res., 114,* E00B40.

Thomas P. and Gierasch P. J. (1985) Dust devils on Mars. *Science, 230,* 175–177.

Titov D. V., Markiewicz W. J., Ignatiev N. I., et al. (2012) Morphology of the cloud tops as observed by the Venus Express Monitoring Camera. *Icarus, 217,* 682–701.

Tomasko M. G., Doose L. R., Smith P. H., and Odell A. P. (1980) Measurements of the flux of sunlight in the atmosphere of Venus. *J. Geophys. Res., 85,* 8167–8186.

Toon O. B., Pollack J. B., and Sagan C. (1977) Physical properties of the particles comprising the martian dust storm of 1971–1972. *Icarus, 30,* 663–696.

Toon O. B., Turco R. P., and Pollack J. B. (1982) The ultraviolet absorber on Venus: Amorphous sulfur. *Icarus, 51,* 358.

Twomey S. A. (1977) The influence of pollution on the shortwave albedo of clouds. *J. Atmos. Science, 34,* 1149–1152.

Urata R. A. and Toon O. B. (2012) Simulations of the martian water cycle with a GCM and implications for the early climate (abstract). In *Third Conference on Early Mars: Geologic, Hydrologic, and Climatic Evolution and the Implications for Life,* Abstract #7041. LPI Contrib. No. 1680, Lunar and Planetary Institute, Houston.

Wang H. and Fisher J. A. (2009) North polar frontal clouds and dust storms on Mars during spring and summer. *Icarus, 204,* 103–113.

Wang H., Zurek R. W., and Richardson M. I. (2005) Relationship between frontal dust storms and transient eddy activity in the northern hemisphere of Mars as observed by Mars Global Surveyor. *J. Geophys. Res., 110,* E07005, DOI: 10.1029/2005JE002423110.

Wegener A. (1911) *Thermodynamik der Atmosphare.* Verlag von Johann Ambrosius Barth, Leipzig. 331 pp.

Whiteway J. A., Komguem L., Dickinson C., et al. (2009) Mars water-ice clouds and precipitation. *Science, 325,* 68–70.

Wilson R. J., Neumann G. A., and Smith M. D. (2007) Diurnal variation and radiative influence of martian water ice clouds. *Geophys. Res. Lett., 34,* L02710.

Wilson R. J., Lewis S. R., Montabone L., and Smith M. D. (2008) Influence of water ice clouds on martian tropical atmospheric temperatures. *Geophys. Res. Lett., 35,* L07202, DOI: 10.1029/2007GL032405.

Wolff M. J. and Clancy R. T. (2003) Constraints on the size of martian aerosols from Thermal Emission Spectrometer observations. *J. Geophys. Res., 108,* 1–22.

Wolff M. J., Smith M. D., Clancy R. T., Arvidson R., Kahre M., Seelos F., Murchie S., and Savijärvi H. (2009) Wavelength dependence of dust aerosol single scattering albedo as observed by the Compact Reconnaissance Imaging Spectrometer. *J. Geophys. Res., 114,* E00D04.

Wolff M. J., Clancy R. T., Cantor B., and Madeleine J.-B. (2011) Mapping water ice clouds (and ozone) with MRO/MARCI. In *4th International Workshop of the Mars Atmosphere: Modelling and Observations,* Paris, France. Available online at *www-mars.lmd.jussieu.fr/paris2011/abstracts/wolff_paris2011.pdf.*

Wu D., Hu Y. X., McCormick M. P., and Yan F. Q. (2011) Global cloud-layer distribution statistics from 1 year CALIPSO lidar observations. *Intl. J. Remote Sensing, 32(5),* 1269–1288.

Yung Y. L. and DeMore W. B., eds. (1999) *Photochemistry of Planetary Atmospheres.* Oxford Univ., New York. 145 pp.

Zurek R. W., Barnes F. R., Haberle R. M., Pollack J. B., Tillman J. E., and Leovy C. B. (1992) Dynamics of the atmosphere of Mars. In *Mars* (H. H. Kieffer et al., eds.), pp. 835–933. Univ. of Arizona, Tucson.

Renno N. O., Halleaux D., Elliott H., and Kok J. F. (2013) The lifting of aerosols and their effects on atmospheric dynamics.
In *Comparative Climatology of Terrestrial Planets* (S. J. Mackwell et al., eds.), pp. 355–365. Univ. of Arizona, Tucson,
DOI: 10.2458/azu_uapress_9780816530595-ch14.

The Lifting of Aerosols and Their Effects on Atmospheric Dynamics

Nilton O. Renno, Douglas Halleaux, Harvey Elliott
University of Michigan

Jasper F. Kok
University of California at Los Angeles

This chapter reviews theoretical and observational studies of the effects of aerosols on atmospheric dynamics. It starts with a brief review of studies of saltation, the physical process that ejects mineral dust aerosols from the surface of the Earth and other planets into their atmospheres. Saltation on Earth, Mars, Venus, and Titan is briefly discussed, focusing on the main differences and similarities. The weather systems that lift dust on Earth and Mars are discussed, with a focus on their role in dust lifting. Then, dust electrification is reviewed with a focus on the dynamical effects of electric discharges. Finally, the effects of aerosols on clouds and their dynamics are reviewed. This chapter ends with a summary of our current understanding of the effects of aerosols on clouds and climate.

1. INTRODUCTION

Aerosols produce a direct effect on climate by scattering and absorbing solar and infrared radiation, and an indirect effect by altering cloud processes via increases in cloud droplet number and ice particle concentration. The indirect effect increases the cloud albedo (*Twomey,* 1974) and decreases the precipitation efficiency of convective clouds (*Albrecht,* 1989), affecting cloud dynamics. On Earth, dust aerosols lifted in Africa inhibit the formation of hurricanes in the Atlantic Ocean by cooling the sea surface (*Lau and Kim,* 2007). On Mars, dust storms can grow rapidly and become global in extent (*Cantor,* 2007), producing large changes in the amplitude of thermal tides (*Wilson and Hamilton,* 1996). These observations suggest that dust aerosols play a significant role in large-scale atmospheric circulations and dynamics.

On Earth, the concentration of atmospheric aerosols has increased significantly with human activity (e.g., *Ginoux et al.,* 2012; *Penner et al.,* 2001), making aerosol-climate interactions an important component of anthropogenic climate change. Despite considerable progress, aerosol-climate interactions remain one of the most uncertain processes in our current understanding of the Earth's climate and its variability (*IPCC,* 2007). Therefore comparative studies of the processes that lift and transport dust aerosols on Earth and other planets have the potential to improve our understanding of one of the most uncertain climate processes.

Atmospheric dust-lifting processes include dusty plumes and vortices, dust storms (gust fronts), and large-scale winds (windblown dust). The sources of mineral dust and their contribution to the terrestrial aerosol budget have been studied extensively (e.g., *Gill and Gillette,* 1991; *Guelle et al.,* 2000; *Ginoux et al.,* 2001; *Prospero et al.,* 2002; *Koch and Renno,* 2005; *Engelstaedter et al.,* 2006; *Washington et al.,* 2006a,b). *Koch and Renno* (2005) showed evidence that small-scale processes such as dusty plumes and vortices contribute to about one-third of the terrestrial global mineral dust budget, while *Engelstaedter and Washington* (2007) showed that variations in the annual dust cycle of West African dust "hot spots" correlate with variations in small-scale wind gusts caused by boundary layer convection. Moreover, Engelstaedter and Washington showed that the annual dust cycle is in phase with changes in near-surface convergence and therefore the intensity of convective circulations. These results are consistent with the idea that on Earth, convective plumes and vortices play an important role in the dust cycle (*Koch and Renno,* 2005).

Past and current missions have shown that aeolian processes in the form of wind erosion features, dust devils, and dust storms have been actively modifying the surface of Mars (*Cantor et al.,* 1999; *Edgett and Malin,* 2000; *Zimbelman,* 2000; *Wilson and Zimbelman,* 2004; *Greeley et al.,* 2008; *Sullivan et al.,* 2008; *Bridges et al.,* 2012). The Mars Global Surveyor detected orbit-to-orbit variations in atmospheric density by factors of at least 2 at an altitude of about 120 km, probably caused by variations in atmospheric dust content and temperature (*Keating et al.,* 1998). Thus, a better characterization of dust lifting processes is important for improving our understanding of the most important geological process actively modifying the surface of Mars (*Greeley and Iversen,* 1985), and producing short-term atmospheric variability that affects aerobraking; aerocapture; and entry, descent, and landing (*Braun and Manning,* 2007).

There is evidence that, besides dust storms, dust devils

play an important role in the martian dust cycle. This idea is consistent with the fact that the atmospheric dust opacity increased throughout the Mars Pathfinder (MPF) mission in spite of low wind conditions and the absence of dust storms on the planet (*Smith and Lemmon,* 1999). Indeed, *Ferri et al.* (2003) showed that the dust flux injected into the atmosphere by dust devils on an active martian day is an order of magnitude larger than the mission-mean dust deposition rate observed at the MPF landing site. This result confirms that dust devils contribute significantly to the maintenance of the dust content of the atmosphere of Mars, and perhaps are even the primary suppliers of dust to the atmosphere.

2. SALTATION AND DUST LIFTING

2.1. Saltation

A good understanding of dust and sand transport by wind is essential for a wide range of planetary processes. As wind speed increases, particles with diameters of ~100–200 μm, referred to as sand, start to move. As these particles move, they leap ("saltus" in Latin) across the surface in a process known as saltation (e.g., *Greeley and Iversen,* 1985; *Kok and Renno,* 2009). The impact of saltating particles ejects the smaller, harder to lift, dust particles into the air (*Bagnold,* 1941; *Greeley and Iversen,* 1985; *Shao et al.,* 1993) and cause particles with diameters larger than about 500 μm to skip or roll along the surface, in processes referred to as "reptation" and "creep," respectively (*Bagnold,* 1941). Dust particles are not usually lifted directly by the wind because the cohesion forces between them and other soil particles are large compared to the aerodynamic forces caused by the wind (e.g., *Shao and Lu,* 2000). Furthermore, dust particles are too small to stick out of the viscous boundary layer, which extends only ~0.5 mm from the surface on Earth and ~2 mm on Mars, making them less likely to be lifted directly by the wind (*Iversen and White,* 1982). After being ejected into the air by the impact of saltating particles on the surface, dust particles are transported upward by turbulent eddies and convective flows (e.g., *Renno et al.,* 2004).

Saltation begins when the wind shear stress at the surface (τ) exceeds a critical value known as the "fluid threshold" τ_t, which for loose sand on Earth is ~0.05 N/m^2 (*Bagnold,* 1941; *Greeley and Iversen,* 1985). The saltating sand particles follow ballistic trajectories while being accelerated by the wind. After a few leaps on the surface, these saltating particles accelerate sufficiently to eject other sand particles when "splashing" on the surface (*Bagnold,* 1973; *Ungar and Haff,* 1987). These newly ejected particles are then accelerated by the wind and eject more particles when impacting on the surface. This process causes an exponential increase in the number of airborne particles while saltation develops (*Anderson and Haff,* 1988, 1991; *Shao and Raupach,* 1992; *McEwan and Willetts,* 1993; *Kok and Renno,* 2009; *Andreotti et al.,* 2010). Saltation reaches a statistically steady state when the momentum available from the wind is transferred to the saltating particles, which, in turn, transfer it to the surface (*Kok and Renno,* 2009). Indeed, saltation modifies the near-surface wind profile substantially (e.g., *Owen,* 1964).

When saltation reaches a statistically steady state, the wind rarely directly forces surface particles to move. Indeed, in steady state, surface particles are forced to move predominantly by the impacts of saltating particles on them. That is, momentum is transferred to the surface mostly by the impacts of saltating particles, and to a lesser extent by wind drag or shear stress (*Bagnold,* 1937, 1973; *Ungar and Haff,* 1987; *Anderson and Haff,* 1988, 1991; *Shao and Raupach,* 1992; *McEwan and Willetts,* 1991, 1993; *Kok and Renno,* 2009; *Kok et al.,* 2012). As a result, once saltation starts, it can be maintained at shear stress values below the "fluid threshold velocity." The minimum shear velocity at which saltation can be maintained is referred to as the "impact threshold" (*Bagnold,* 1941). On Earth, the impact threshold velocity is approximately 80–85% of the fluid threshold velocity (*Bagnold,* 1937; *Kok and Renno,* 2009). As explained below, on Mars this difference is much larger (e.g., *Claudin and Andreotti,* 2006; *Kok,* 2010a).

Numerical models of saltation developed during the past few decades have provided great insights into this important geological process (*White and Schulz,* 1977; *Anderson and Hallet,* 1986; *Ungar and Haff,* 1987; *Werner,* 1990; *McEwan and Willetts,* 1991, 1993; *Shao and Li,* 1999; *Almeida et al.,* 2006; *Zheng et al.,* 2006; *Almeida et al.,* 2008; *Kok and Renno,* 2009; *Carneiro et al.,* 2011; *Duran et al.,* 2012). Most of these models were restricted to the simulation of the saltation of monodisperse particles, and perhaps partially because of this they were not able to accurately reproduce the wide range of processes occurring in natural saltation. In order to mitigate this problem, *Kok and Renno* (2009) developed COMSALT, a physically based numerical model of saltation that includes the splashing of particles of various sizes from the surface and a detailed treatment of the influence of turbulence on particle trajectories. More recently, *Carneiro et al.* (2011) and *Duran et al.* (2012) developed direct element method models of saltation, which require a reduced number of assumptions and parameterizations than previous models.

Saltating sand transported by the wind creates sand dunes and ripples, and causes erosion (*Bagnold,* 1941; *Greeley and Iversen,* 1985; *Claudin and Andreotti,* 2006; *Sullivan et al.,* 2008; *Kok et al.,* 2012). On Earth, mineral dust ejected into the air by saltation affects clouds and climate (*Sokolik et al.,* 2001; *Renno et al.,* 2013) and provides nutrients to ecosystems (e.g., *Mahowald et al.,* 2009). On Mars, dust plays a major role in climate (*Fenton et al.,* 2007) and affects dynamical processes such as thermal tides (*Wilson and Hamilton,* 1996). On Titan, Venus, and Mars, saltation and dust transport by the wind are important geological processes (*Greeley and Iversen,* 1985; *Weitz et al.,* 1994; *Claudin and Andreotti,* 2006; *Lorenz et al.,* 2006; *Greeley,* 2008; *Sullivan et al.,* 2008; *Bourke,* 2010; *Bridges et al.,* 2012). *Kok et al.* (2012) provides a detailed review of saltation on these planetary bodies. As on Earth, dust aerosols on Venus might have a significant impact on cloud processes (*Esposito et al.,* 2007).

Large differences between the density of the air at the surface and the gravitational acceleration produce large differences in saltation between Earth, Mars, Venus, and Titan. Larger wind speeds are required to start saltation on Mars than on Earth because of the lower density of the air at the surface of Mars, but once saltation starts on Mars, the lower gravity and the lower drag of the thinner air allows particles to saltate higher than on Earth (*Almeida et al.,* 2008; *Kok,* 2010a). On Venus and Titan the density of the air at the surface is much higher than on Earth. This reduces the speed of the impact of saltating particles on the surface and limits the depth of the saltation layer (*White,* 1981; *Williams and Greeley,* 1994). *Kok et al.* (2012) show that these large differences in environmental conditions cause large differences in saltation and consequently in some of the properties of sand dunes between these planetary bodies, such as the sand dune length scales.

Recent studies (*Claudin and Andreotti,* 2006; *Almeida et al.,* 2008; *Kok,* 2010a,b; *Pähtz et al.,* 2012) have used numerical and analytical models to demonstrate that on Mars, saltation can likely be maintained at wind speeds an order of magnitude smaller than that required for initiating it. This hysteresis effect indicates that on Mars saltation is possible at much lower wind speeds than previously thought. This might help resolve the interesting puzzle that numerical models and measurements suggest that the threshold friction velocity necessary to initiate saltation on Mars is rarely exceeded, in spite of accumulating evidence that saltation is relatively common on Mars (*Sullivan et al.,* 2000; *Fenton et al.,* 2005; *Bourke et al.,* 2008; *Kok,* 2010a; *Bridges et al.,* 2012, 2013).

As discussed above, saltation is sensitive to environmental parameters such as gravity, air density, viscosity, and sand properties. Consequently, the properties of saltation vary substantially among the planetary bodies on which it occurs (*Kok et al.,* 2012). An overview of these differences is given in Fig. 1, which presents results of simulations of saltation on Earth, Mars, Venus, and Titan with COMSALT (*Kok and Renno,* 2009). The simulations are performed using monodisperse particles of diameters equal to 100, 250, and 500 µm, with the values of the thermodynamic and sand parameters given by *Kok et al.* (2012). The friction velocity is set to values equal to twice the minimum value for which saltation can be sustained (i.e., twice the impact threshold u_{*it} on Earth and Mars, and twice the fluid threshold u_{*ft} on Venus and Titan as indicated in Table 1) on each planetary body studied. Figure 1 indicates that sand particles saltate an order of magnitude higher on Mars than on Earth, as found in previous investigations (*Almeida et al.,* 2008; *Kok,* 2010a). Furthermore, the large air drag produced by Venus' dense air causes sand particles to saltate an order of magnitude lower there than on Earth (*White,* 1981; *Williams and Greeley,* 1994), whereas on Titan the lower gravity and higher air density combine to produce saltation layers with depths similar to those observed on Earth.

Our simulations also indicate that on Earth, Mars, Venus, and Titan the number density of saltating particles increases approximately exponentially toward the surface, with an even larger increase in the concentration of saltating particle near the surface because of the presence of low-energy reptators. These results are in agreement with measurements (*Namikas,* 2003; *Creyssels et al.,* 2009). This strong maximum in the number density at the surface causes the horizontal flux of saltator kinetic energy to peak at the surface, despite the gradual increase of particle speeds with height (e.g., *Rasmussen and Sorensen,* 2008; *Creyssels et al.,* 2009). These results seem counter to observations that the maximum abrasion of fence posts and rocks occurs above the surface (e.g., *Sharp,* 1980; *Anderson,* 1986). This discrepancy could be due to several factors. First, the flow diverges around obstacles and therefore only particles that cannot follow the streamlines impact the obstacles. Consequently, particles with small speeds, such as reptating particles close to the surface, are unlikely to impact an obstacle, whereas faster particles traveling higher in the saltation layer could have sufficient momentum to impact the obstacle despite the divergence of streamlines around it. This effect would create a maximum abrasion rate some distance from the surface. A second explanation is that some observations of abrasion by saltators in nature are made in *supply limited* areas, where the flux of saltators is limited by the availability of loose sand-sized particles. In these areas, particles achieve much greater speeds, and the increase of particle concentration toward the surface is much weaker (*Ho et al.,* 2011), such that the horizontal flux of saltator kinetic energy likely peaks above the surface. Finally, abrasion might depend nonlinearly on the maximum value of the kinetic energy of individual particles that peaks above the surface.

Figure 1 indicates that the results of the simulations for Titan are more complex than those for Earth, Mars, and Venus. This happens because the particles that saltate most easily (the particle size at which the threshold friction velocity is minimum) on Earth, Mars, and Venus are those with diameters of ~100 µm (*Iversen and White,* 1982), while the small gravitational acceleration on Titan causes the minimum in the fluid threshold to occur at ~250 µm (*Kok et al.,* 2012). Since the results presented in Fig. 1 are all for simulations done with friction velocity equal to twice the threshold friction velocity for each particle size, this causes the results for the simulation of Titan to look different. The fluid and impact thresholds for saltation on Earth, Mars, Venus, and Titan for particles of diameters equal to 100, 250, and 500 µm are listed in Table 1. It also lists the gravitational constants and the values of the physical parameters of the various planetary bodies that largely determine the fluid and impact thresholds. The fluid threshold is calculated following *Iversen and White* (1982), whereas the impact threshold is obtained using the numerical saltation model COMSALT and the procedure described by *Kok and Renno* (2009).

2.2. Dust Lifting

Convective vortices play an important role in the vertical transport of heat, momentum, and tracer species (*Renno et al.,* 2004; *Renno,* 2008). Measurements show that the amount of dust lifted by strong terrestrial dust devils is

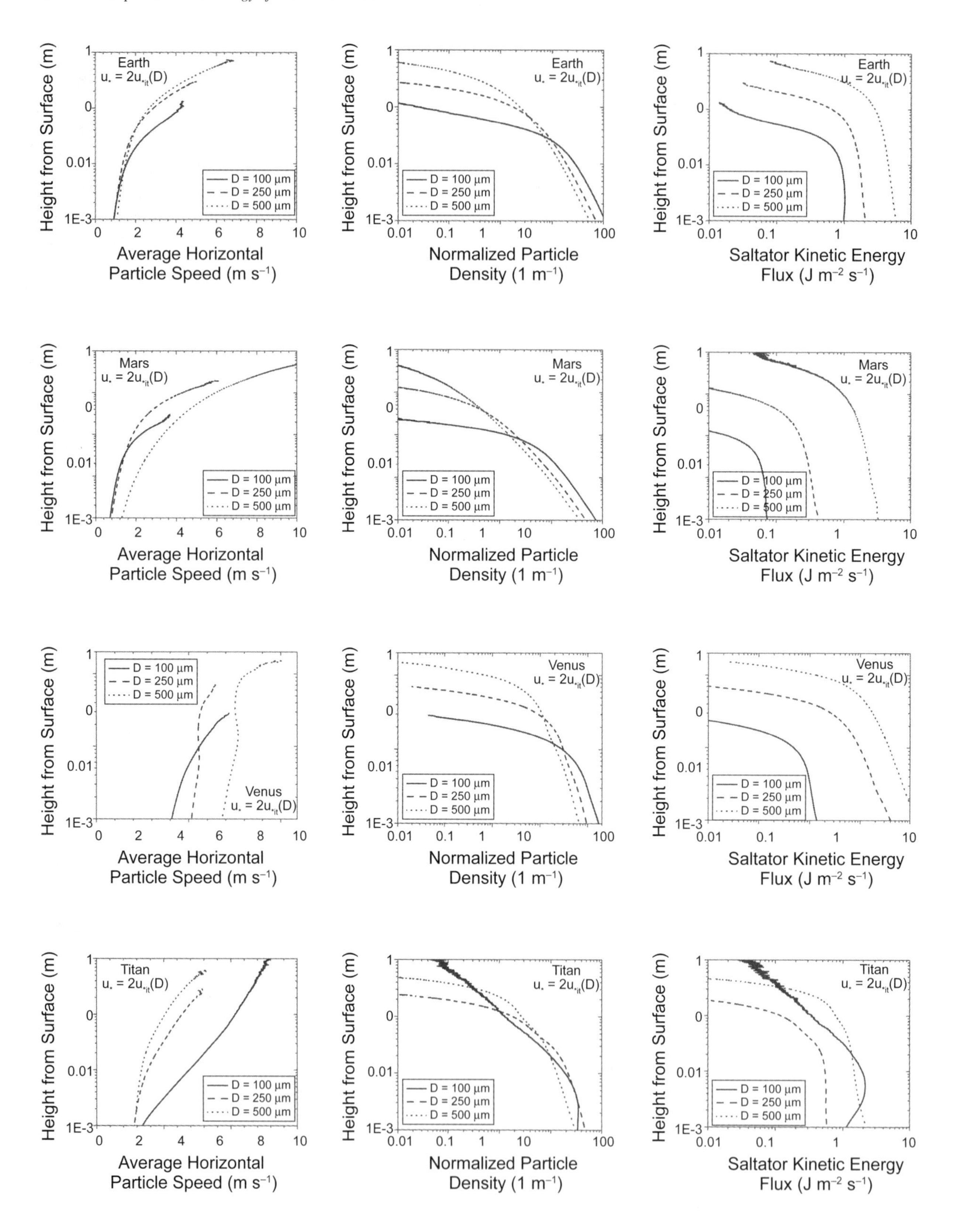

Fig. 1. Results of numerical simulations of saltation of monodisperse particles of diameters (D) equal to 100, 250, and 500 µm, friction velocity equal to twice the minimum value for saltation to occur [$u_* = 2\min(u_{*ft}, u_{*it})$] for particles with each of these three diameters, and thermodynamic and sand properties as specified in *Kok et al.* (2012). The vertical profiles of the average particle velocity (left), the normalized particle density (center), and the kinetic energy flux (right) for saltation on Earth, Mars, Venus, and Titan (from top to bottom) are shown. The simulations were performed with the numerical saltation model COMSALT (*Kok and Renno,* 2009).

TABLE 1. The fluid and impact thresholds for saltation on Earth, Mars, Venus, and Titan (for particles of diameters 100, 250, and 500 μm) and the values of the physical parameters that largely determine the fluid and impact thresholds on these planetary bodies.

Planetary body	Gravitational constant	Air density (kg m^{-3})	Dynamic viscosity (kg m^{-1} s^{-1})	Particle density (kg m^{-3})	Fluid threshold for D = 100 μm (m s^{-1})	Impact threshold for D = 100 μm (m s^{-1})	Fluid threshold for D = 250 μm (m s^{-1})	Impact threshold for D = 250 μm (m s^{-1})	Fluid threshold for D = 500 μm (m s^{-1})	Impact threshold for D = 500 μm (m s^{-1})
Earth	1	1.2	1.8×10^{-5}	2650	0.21	0.14	0.28	0.20	0.37	0.29
Mars	0.378	0.02	1.2×10^{-5}	3000	1.6	0.12	1.8	0.22	2.3	0.45
Venus	0.904	66	3.2×10^{-5}	3000	0.025	0.17	0.037	0.12	0.054	0.15
Titan	0.138	5.1	6.3×10^{-5}	~1000	0.045	0.15	0.035	0.094	0.043	0.082

about 1 g m^{-2} s^{-1}), and that the amount of dust lifted by convective plumes in arid places is an order of magnitude smaller (*Koch and Renno,* 2005). It follows from these measurements that convective plumes and vortices contribute to about one-third of the global budget of mineral dust. The remaining two-thirds is contributed by dust storms and windblown dust (*Koch and Renno,* 2005). On Mars, dust devils might play an even more important role on dust lifting than on Earth (*Ferri et al.,* 2003).

Renno (2008) proposed an expression for the pressure drop along streamlines that is a generalization of Bernoulli's equation to convective circulations. This expression explains the basic dynamical features of convective vortices such as the formation of dust-lifting bands. The theory predicts that the static pressure decreases with increases in kinetic energy. It shows that the maximum pressure gradients occur in the regions of maximum wind, where large amounts of dust are lifted by the resulting updrafts (*Renno,* 2008). This dynamical effect might also play an important role in the formation of dust devils by increasing the surface heat flux.

3. DUST ELECTRIFICATION

Saltating dust particles collide with the surface and with each other, get charged, and produce bulk electrical fields ranging from ~1 to ~100 kV m^{-1} (*Schmidt et al.,* 1998; *Farrell et al.,* 2004; *Renno et al.,* 2004; *Renno and Kok,* 2008). In terrestrial dust devils and dust storms, the electric fields that have been measured in the saltation layer (*Schmidt et al.,* 1998) are larger than those that have been measured above it and are illustrated in Figs. 2 and 3 (*Renno et al.,* 2004; *Kok and Renno,* 2008). This is probably mainly because the dust particle concentrations are much larger near the surface than aloft. Martian dust devils are larger and stronger than terrestrial dust devils (*Thomas and Gierasch,* 1985; *Renno et al.,* 2000; *Cantor et al.,* 2002) and therefore could also be capable of producing substantial electric fields. However, large increases in the conductivity of the thin martian atmosphere with the electric field limits the electric fields in the saltation layer to less than a few 10 kV m^{-1} (*Kok and Renno,* 2009).

In martian dust storms, electric discharges might also limit the electric fields to values on the order of a few 10 kV m^{-1}. Indeed, there is evidence that martian dust storms of rapid vertical growth produce powerful electric discharges that force electrodynamical processes of planetary scales (*Ruf et al.,* 2009; *Renno and Ruf,* 2012).

Measurements in terrestrial dust devils and dust storms indicate that negative charges are found aloft (*Renno et al.,* 2004), and this is consistent with the idea that during collisions of sand with dust particles, negative charges are transferred to the smaller dust particles (*Melnik and Parrot,* 1998; *Kok and Renno,* 2008; *Kok and Lacks,* 2009). Since the larger sand particles stay in the saltation layer while the smaller dust particles are lifted by convective updrafts, charges are separated and produce large bulk electric fields. *Renno et al.* (2003) estimated the maximum charge in individual dust particles by assuming that, after energetic collisions between dust and sand particles, the particles' charge is limited by field emission (the emission of electrons from the surface of a conductor by a large electric field). Then, *Renno et al.* (2004) assumed that microdischarges occur while the particles move away from each other and while being left with a residual charge on the order of that necessary to produce electric discharges (*Renno et al.,* 2003). *Renno et al.* (2004) showed that this residual charge is large enough to produce bulk electric fields of a few 10 kV m^{-1} in dust devils and convective dust storms. Then, microdischarges produce nonthermal radiation (*Renno et al.,* 2003; *Ruf et al.,* 2009; *Renno and Ruf,* 2012). This allows dust electrification to be studied remotely via measurements of the emission of broadband nonthermal radiation by colliding dust particles.

Ruf et al. (2009) used the Deep Space Network (DSN) to search for the emission of nonthermal radiation by martian dust storms. They found evidence for it, but were surprised to discover spectral peaks suggesting modulation at various frequencies and their harmonics as illustrated in Fig. 4. They hypothesized that the emission of nonthermal radiation that they observed was caused by electric discharges in a deep convective dust storm, modulated by a planetary-scale phenomena known as Schumann Resonances (*Ruf et al.,* 2009; *Renno and Ruf,* 2012). In contrast to *Ruf et al.* (2009), *Gurnett et al.* (2010) and *Anderson et al.* (2012) searched for signs of electric activity on Mars but did not find evidence for it.

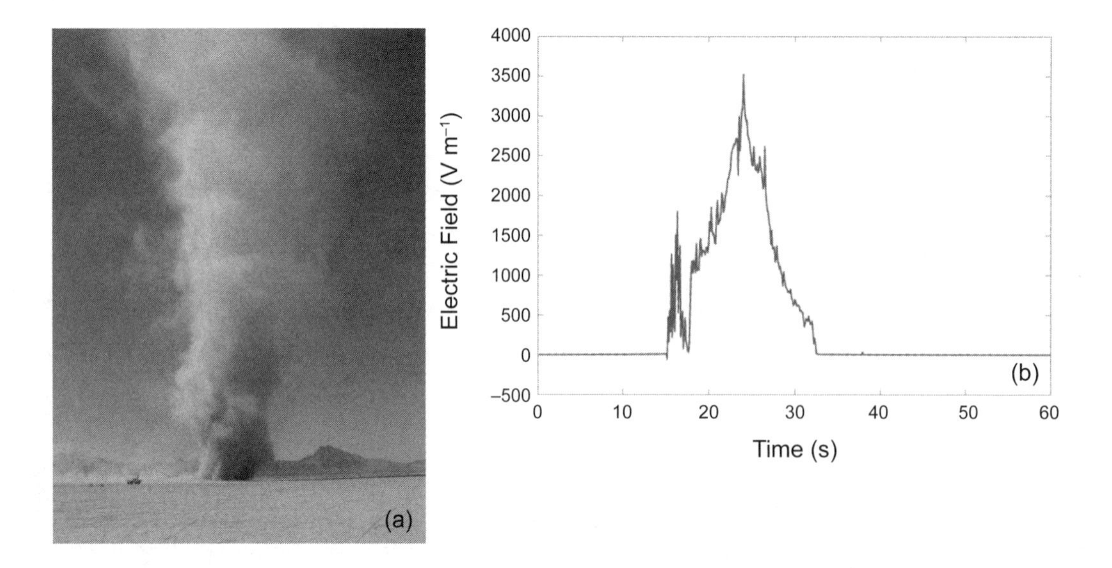

Fig. 2. (a) See Plate 62 for color version. Dust devil photographed during a field campaign in Nevada. **(b)** Electric field (~10 m above the surface) measured (*Renno et al.,* 2008) during the passage of a similar dust devil over the instruments. The abrupt increase in the electric field indicates the passage of a dusty spiral band over the sensor.

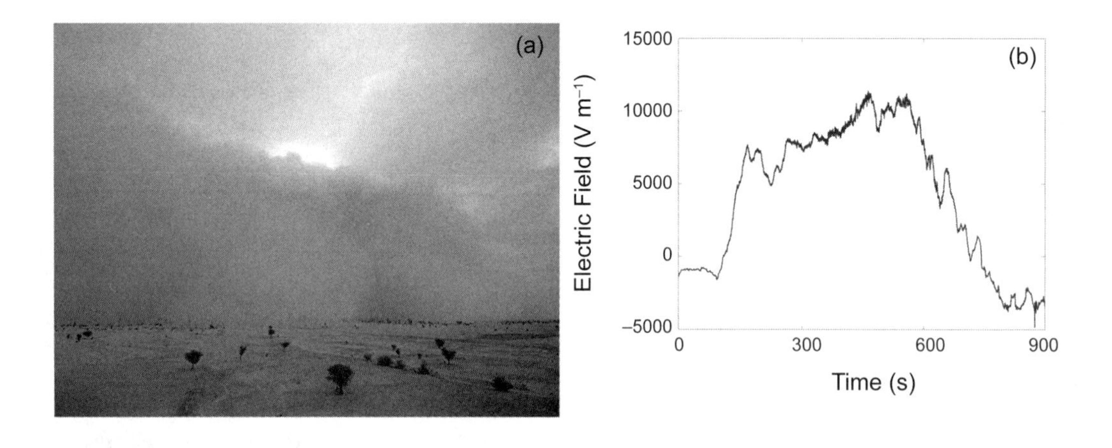

Fig. 3. (a) See Plate 63 for color version. Gust front photographed during a field campaign in Niger. **(b)** Electric field (~10 m above the surface) measured (*Renno et al.,* 2008) in a similar gust front.

The forcing of Schumann Resonances (SR) by electric discharges is proportional to the change in charge moment that they produce (*Renno and Ruf,* 2012). The maximum charge moment of a dust storm is $M_{Max} = zA\varepsilon E_{Max}$, where z is the depth of the dust storm, A is the area covered by the dust plume, $\varepsilon \approx 8.85 \times 10^{-12}$ F m^{-1} is the electric permittivity of the martian air (~free space), and E_{Max} is the maximum electric field in the dust storm, taken as the nearly critical (close to the breakdown value) electric field amplitude $E_{max} \sim 20$ kV m^{-1}. *Ruf et al.* (2009) estimated the charge moment to be $M_{Max} \approx 10^9$ C m in the convective martian dust storm they observed on June 8, 2006. They calculated the SR forcing by postulating that the dust storm would be completely discharged during the minute long bursts seen in the kurtosis. In this case, the averaged rate of charge transfer squared would be 10^{16} (C m)^2 s^{-1}.

Renno and Ruf (2012) showed that the forcing of SR by dust devils is many order of magnitudes smaller than that of the June 2006 dust storm. They argued that typical windblown dust storms are shallow, no more than a few kilometers deep, and much less likely to produce charge separation than convective storms with rapid vertical development such as the June 8, 2006, dust storm; therefore they produce a charge moment at least an order of magnitude smaller than the June 8, 2006, dust storm. For this reason, the forcing of SR by shallow windblown dust storms of similar area is at least ~10^2 times smaller than that of the June 8, 2006, dust storm. Shallower dust storms of larger

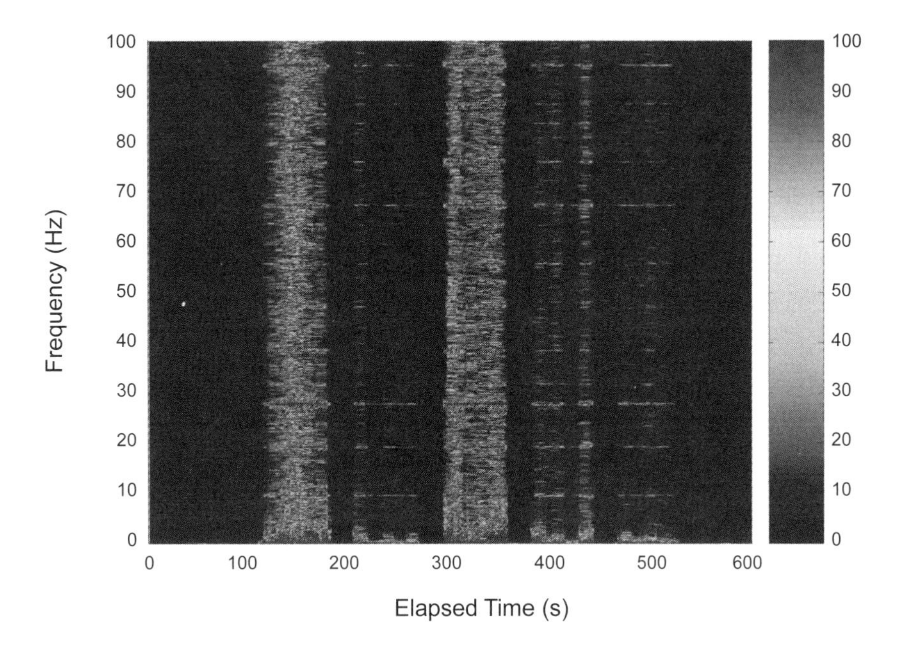

Fig. 4. See Plate 64 for color version. Time series of the power spectrum of the nonthermal radiation collected over 10 minutes on June 8, 2006, beginning at 21:58 UTC indicating modulation by SRs. Color bar is linearly proportional to power spectral density. After *Ruf et al.* (2009).

area are not uncommon, but it is unlikely that the entire dust storm would be electrically active. Therefore, convective dust storms of large vertical growth and capable of being strongly electrified — analogous to terrestrial thunderstorms — might be a relatively rare event on Mars.

4. EFFECTS OF AEROSOLS ON CLOUDS

Aerosols play an important role in clouds by providing nuclei for the condensation of water vapor in cloud droplets. The nucleation of these droplets depends both on the composition and size of the aerosol particles that seed them as well as on the water vapor pressure on them (*Seinfeld and Pandis,* 2006; *Andreae and Rosenfeld,* 2008). Therefore, an increase in the number of these aerosol particles, known as cloud condensation nuclei (CCN), leads to an increase in the number of cloud droplets and a decrease in their size. The increase in cloud droplet number, in turn, increases cloud reflectance (*Twomey,* 1977).

Albrecht (1989) suggested that because the coalescence efficiency of small cloud droplets is lower than that of larger droplets, aerosols decrease precipitation and increase cloud lifetime and area coverage. This idea is supported by recent aircraft measurements on Earth indicating higher aerosol concentrations in areas of broken marine stratocumulus clouds than in nearby overcast areas (*Wood et al.,* 2011). The cloud albedo effect is well accepted, whereas cloud lifetime and area coverage are recognized but have not received sufficient attention. This is unfortunate since remote-sensing measurements suggest that this effect is larger than the cloud albedo effect (*Sekiguchi et al.,* 2003; *Kaufman et al.,* 2005; *Rosenfeld et al.,* 2006).

It has been hypothesized that convective clouds forming in clean maritime environments with CCN ~200 cm^{-3} distribute their liquid water into large droplets that can rain out (warm rain) before freezing, inhibiting the development of thunderstorms. In contrast, it has been hypothesized that convective clouds forming in continental environments with CCN ~2000 cm^{-3} contain a large number of small droplets that make the clouds brighter. According to the hypothesis, coalescence — and therefore the formation of warm rain — is inhibited in these clouds. In addition, the latent heat released when small droplets freeze aloft produces intense thunderstorms (*Williams et al.,* 2002). This is illustrated in Fig. 5.

The effects of aerosols and thermodynamics on clouds can independently explain a surprisingly large number of weather phenomena (*Williams et al.,* 2002, 2005; *Williams and Renno,* 1993; *Williams and Stanfill,* 2002; *Rosenfeld et al.,* 2008). Examples include the dominance of lightning activity over continents, the dependence of lightning on temperature on several timescales, the dominance of clouds producing "warm rain" over oceans, the dominance of large hail over continents, the dominance of upper tropospheric ice processes in continental convective clouds, and the occasional onset of explosive lightning in hurricanes.

Why are thermodynamics and aerosol effects on moist convection so tightly interrelated and, at the same time, so poorly distinguished? The physical differences between land and ocean surfaces play a major role in explaining this situation and may explain, via both thermodynamic and aerosol effects, why moist convection over land is distinctly

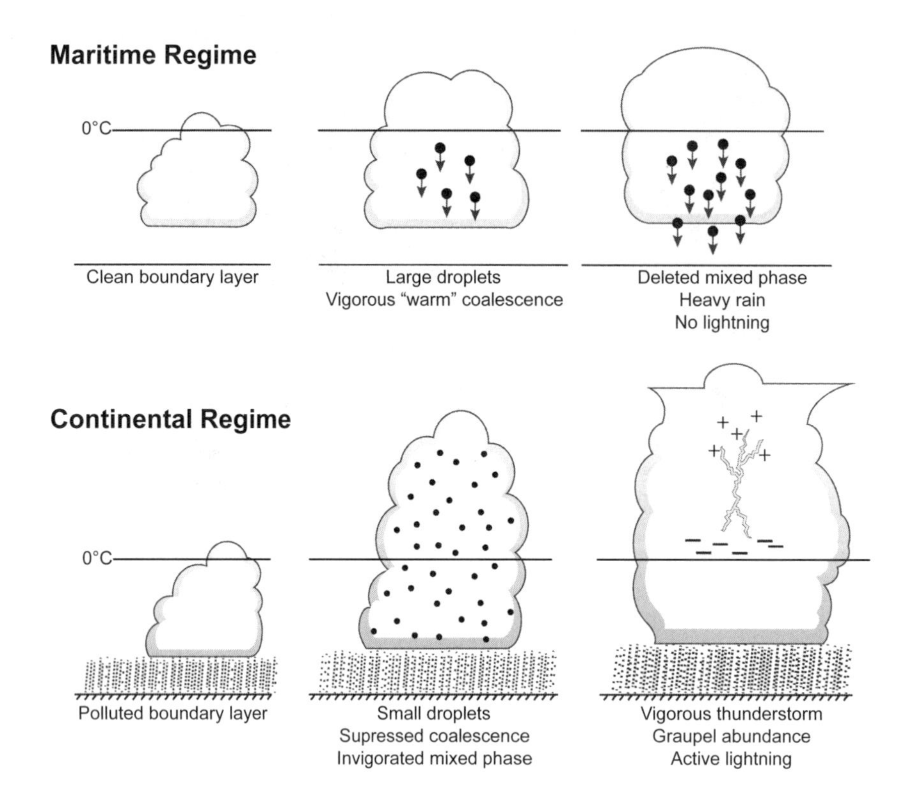

Fig. 5. Aerosols increase the number of cloud droplets, inhibit precipitation, and invigorate clouds and thunderstorms (graupel is soft hail formed by the freezing of supercooled water droplets on snow). After *Williams et al.* (2002).

different from that over the oceans. Land surfaces possess lower heat capacity material than the mobile liquid ocean water, and therefore become hotter than the oceans for the same solar forcing. However, the land surface is also the primary source for aerosols caused by fire and smoke, mineral dust storms, and volcanic eruptions, resulting in aerosol concentrations in the continental boundary layer that are an order of magnitude greater than those over the oceans (*Andreae and Rosenfeld,* 2008; *Schaefer and Day,* 1998; *Renno et al.,* 2013). The concept for a space mission capable of measuring all quantities necessary to produce global maps of activated CCN and the properties of the clouds associated with them have been recently proposed (*Rosenfeld et al.,* 2012; *Renno et al.,* 2013).

5. CONCLUSIONS

As wind speed increases, sand particles start to move and leap across the surface in a process known as saltation (*Bagnold,* 1941; *Greeley and Iversen,* 1985; *Kok and Renno,* 2009). The impact of these leaping particles on the surface ejects the smaller, harder to lift, dust particles into the air (*Greeley and Iversen,* 1985). Dust particles are generally not lifted directly by the wind because the cohesion forces between them and the surface are usually large compared to the drag and shear stress forces caused by the wind (e.g., *Kok and Renno,* 2009). After being ejected into the air by the impact of saltating particles on the surface, dust particles are transported upward by turbulent eddies and convective flows (e.g., *Renno et al.,* 2004).

Convective plumes and vortices play an important role in the vertical transport of dust particles away from the surface (*Renno et al.,* 2004; *Renno,* 2008). There is evidence that they contribute to about one-third of the global budget of mineral dust. The remaining two-thirds is contributed by large-scale dust storms and windblown dust (*Koch and Renno,* 2005). On Mars, dust devils might play an even more important role on dust lifting than on Earth (*Ferri et al.,* 2003).

Saltation and dust lifting produces bulk electrical fields ranging from ~1 to ~100 kV m^{-1} (e.g., *Schmidt et al.,* 1998; *Renno and Kok,* 2008). There is evidence that these electric fields affect dust lifting (*Kok and Renno,* 2006, 2008). There is also evidence that the collision of sand and dust particles with each other causes the emission of nonthermal radiation (*Renno et al.,* 2003), and that electric discharges in convective martian dust storms produce a planetary-scale phenomena known as Schumann Resonances (*Ruf et al.,* 2009; *Renno and Ruf,* 2012).

Aerosols affect climate by scattering and absorbing solar and infrared radiation, and by altering cloud processes via increases in cloud droplet number and ice particle concentration. These effects increase the cloud albedo and decrease the precipitation efficiency of convective clouds (*Twomey,* 1974; *Albrecht,* 1989), affecting cloud dynamics. On Earth, dust aerosols inhibit the formation of hurricanes in the Atlantic Ocean, while on Mars dust storms can grow rapidly and become global in extent, producing large changes in the amplitude of thermal tides (*Wilson and Hamilton,* 1996; *Cantor,* 2007; *Lau and Kim,* 2007). This suggests that dust aerosols play a significant role in large-scale atmospheric circulations and dynamics.

Acknowledgments. This work was supported by the National Science Foundation under Awards AGS 1118467 and AGS 1137716.

REFERENCES

Albrecht B. A. (1989) Aerosols, cloud microphysics, and fractional cloudiness. *Science, 245,* 1227–1230, DOI: 10.1126/science.245.4923.1227.

Almeida M. P., Andrade J. S., and Herrmann H. J. (2006) Aeolian transport layer. *Phys. Rev. Lett., 96,* DOI: 10.1103/PhysRevLett.96.018001.

Almeida M. P., Parteli E. J. R., Andrade J. S., and Herrmann H. J. (2008) Giant saltation on Mars. *Proc. Natl. Acad. Sci., 105(17),* 6222–6226.

Anderson M. M., Siemion A. P. V., Barott W. C., Bower G. C., Delory G. T., de Pater I., and Werthimer D. (2012) The Allen Telescope Array search for electrostatic discharges on Mars. *Astrophys. J., 744,* DOI: 10.1088/0004-637X/744/1/15.

Anderson R. S. (1986) Erosion profiles due to particles entrained by wind: Application of an eolian sediment-transport model. *Geol. Soc. Am. Bull., 97,* 1270-1278.

Anderson R. S. and Haff P. K. (1988) Simulation of eolian saltation. *Science, 241,* 820–823.

Anderson R. S. and Haff P. K. (1991) Wind modification and bed response during saltation of sand in air. *Acta Mech. Suppl., 1,* 21–51.

Anderson R. S. and Hallet B. (1986) Sediment transport by wind — toward a general model. *Geol. Soc. Am. Bull., 97,* 523–535.

Andreae M. O. and Rosenfeld D. (2008) Aerosol-cloud-precipitation interactions. Part 1. The nature and sources of cloud-active aerosols. *Earth-Sci. Rev., 89,* 13–41.

Andreotti B., Claudin P., and Pouliquen O. (2010) Measurements of the aeolian sand transport saturation length. *Geomorphology, 123,* 343–348.

Bagnold R. A. (1937) The transport of sand by wind. *Geographical J., 89,* 409–438.

Bagnold R. A. (1941) *The Physics of Blown Sand and Desert Dunes.* Methuen, New York.

Bagnold R. A. (1973) Nature of saltation and bed-load transport in water. *Proc. R. Soc. Lond., A332,* 473–504.

Bourke M. C. (2010) Barchan dune asymmetry: Observations from Mars and Earth. *Icarus, 205,* 1983–1997.

Bourke M. C., Edgett K. S., and Cantor B. A. (2008) Recent aeolian dune change on Mars. *Geomorphology, 94,* 247–255.

Braun R. D. and Manning R. M. (2007) Mars entry, descent and landing challenges. *J. Spacecraft Rockets, 44(2),* 310–323.

Bridges N. T., Ayoub F., Avouac J-P., Leprince S., Lucas A., and Mattson S. (2012) Earth-like sand fluxes on Mars. *Nature, 485,* 339–342, DOI: 10.1038/nature11022.

Bridges N., Geissler P., Silvestro S., and Banks M. (2013) Bedform migration on Mars: Current results and future plans. *Aeolian Res., 9,* 133–151, DOI: 10.1016/j.aeolia.2013.02.004.

Cantor B. A. (2007) MOC observations of the 2001 Mars planet-encircling dust storm. *Icarus, 186,* 60–96.

Cantor B. A., James P. B., Caplinger M., and Wolff M. J. (1999) Martian dust storms: 1999 Mars Orbiter Camera observations. *J. Geophys. Res., 106,* 2156–2202, DOI: 10.1029/2000JE001310.

Cantor B., Malin M., and Edgett K. S. (2002) Multiyear Mars Orbiter Camera (MOC) observations of repeated martian weather phenomena during the northern summer season. *J. Geophys. Res., 107,* DOI: 10.1029/2001JE001588.

Carneiro M. V., Pahtz T., and Herrmann H. J. (2011) Jump at the onset of saltation. *Phys. Rev. Lett., 107,* 098001.

Claudin P. and Andreotti B. (2006) A scaling law for aeolian dunes on Mars, Venus, Earth, and for subaqueous ripples. *Earth Planet. Sci. Lett., 252,* 30–44.

Creyssels M., Dupont P., El Moctar A. O., Valance A., Cantat I., Jenkins J. T., Pasini J. M., and Rasmussen K. R. (2009) Saltating particles in a turbulent boundary layer: Experiment and theory. *J. Fluid Mech., 625,* 47–74, DOI: 10.1017/S0022112008005491.

Duran O., Andreotti B., and Claudin P. (2012) Numerical simulation of turbulent sediment transport, from bed load to saltation. *Phys. Fluids, 24,* 103306.

Edgett K. S. and Malin M. C. (2000) New views of Mars eolian activity, materials, and surface properties: Three vignettes from the Mars Global Surveyor Mars Orbiter Camera. *J. Geophys. Res., 105(E1),* 1623–1650, DOI: 10.1029/1999JE001152.

Engelstaedter S. and Washington R. (2007) Atmospheric controls on the annual cycle of North African dust. *J. Geophys. Res., 112,* DOI: 10.1029/2006JD007195.

Engelstaedter S., Washington R., and Tegen I. (2006) North African dust emissions and transport. *Earth-Sci. Rev., 79,* 73–100.

Esposito L. W., Stofan E. R., and Cravens T. E., eds. (2007) *Exploring Venus as a Terrestrial Planet.* AGU Geophys. Monogr. 176, American Geophysical Union, Washington, DC. 225 pp.

Farrell W. M., Smith P. H., Delory G. T., Hillard G. B., Marshall J. R., Catling D., Hecht M., Tratt D. M., Rennó N., Desch M. D., Cummer S. A., Houser J. G., and Johnson B. (2004) Electric and magnetic signatures of dust devils from the 2000–2001 MATADOR desert tests. *J. Geophys. Res., 109,* DOI: 10.1029/2003JE002088.

Fenton L. K., Toigo A. D., and Richardson M. I. (2005) Aeolian processes in Proctor crater on Mars: Mesoscale modeling of dune-forming winds. *J. Geophys. Res., 110,* DOI: 10.1029/2004JE002309.

Fenton L. K., Geissler P. E., and Haberle R. M. (2007) Global warming and climate forcing by recent albedo changes on Mars. *Nature, 446,* 646–649.

Ferri F., Smith P. H., Lemmon M., and Renno N. O. (2003) Dust devils as observed by Mars Pathfinder. *J. Geophys. Res., 108,* 5133, DOI: 10.1029/2000JE001421.

Gill T. E. and Gillette D. A. (1991) Owens Lake: A natural laboratory for aridification, playa desiccation and desert dust. *GSA Abstr. Prog., 23,* 462.

Ginoux P., Chin M., Tegen I., Prospero J. M., Holben B., Dubovik O., and Lin S.-J. (2001) Sources and distributions of dust aerosols simulated with GOCART model. *J. Geophys. Res.,*

106, 20255–20273.

Ginoux P., Prospero J. M., Gill T. E., Hsu N. C., and Zhao M. (2012) Global-scale attribution of anthropogenic and natural dust sources and their emission rates based on MODIS Deep Blue aerosol products. *Rev. Geophys., 50,* RG3005.

Greeley R. and Iversen J. D. (1985) *Wind as a Geological Process on Earth, Mars, Venus, and Titan.* Cambridge Univ., New York.

Greeley R., Whelley P. L., Neakrase L. D. V., et al. (2008) Columbia Hills, Mars: Aeolian features seen from the ground and orbit. *J. Geophys. Res., 113,* E06S06, DOI: 10.1029/2007JE002971.

Guelle W., Balkanski Y. J., Schulz M., et al. (2000) Modeling the atmospheric distribution of mineral aerosol: Comparison with ground measurements and satellite observations for yearly and synoptic timescales over the North Atlantic. *J. Geophys. Res., 105,* 1997–2012.

Gurnett D. A., Morgan D. D., Granoth L. J., et al. (2010) Non-detection of impulsive radio signals from lightning in martian dust storms using the radar receiver on the Mars Express spacecraft. *Geophys. Res. Lett., 37,* L17802.

Ho T. D., Valance A., Dupont P., and El Moctar A. O. (2011) Scaling laws in aeolian sand transport. *Phys. Rev. Lett., 106,* 094501.

IPCC (2007) *IPCC (The Intergovernmental Panel on Climate Change) Fourth Assessment Report on Scientific Aspects of Climate Change for Researchers, Students, and Policymakers.* Cambridge Univ., New York.

Iversen J. D. and White B. R. (1982) Saltation threshold on Earth, Mars and Venus. *Sedimentology, 29,* 111–119.

Kaufman Y. J., Koren I., Remer L. A., Rosenfeld D., and Rudich Y. (2005) The effect of smoke, dust, and pollution aerosol on shallow cloud development over the Atlantic Ocean. *Proc. Natl. Acad. Sci., 102,* 11207–11212.

Keating G. M., Bougher S. W., Zurek R. W., et al. (1998) The structure of the upper atmosphere of Mars: In situ accelerometer measurements from Mars Global Surveyor. *Science, 279,* 1672–1676, DOI: 10.1126/science.279.5357.1672.

Koch J. and Renno N. O. (2005) The role of convective plumes and vortices on the global aerosol budget. *Geophys. Res. Lett., 32,* DOI: 10.1029/2005GL023420.

Kok J. F. (2010a) Difference in the wind speeds required for initiation versus continuation of sand transport on Mars: Implications for dunes and dust storms. *Phys. Rev. Lett., 104,* 074502, DOI: 10.1103/PhysRevLett.104.074502.

Kok J. F. (2010b) An improved parameterization of wind-blown sand flux on Mars that includes the effect of hysteresis. *Geophys. Res. Lett., 37,* L12202.

Kok J. F. and Lacks D. J. (2009) Electrification of granular systems of identical insulators. *Phys. Rev. E, 79,* DOI: 10.1103/PhysRevE.79.051304.

Kok J. F. and Renno N. O. (2006) The effects of electric fields on dust lifting. *Geophys. Res. Lett., 33,* L19S10, DOI: 10.1029/2006GL026284.

Kok J. F. and Renno N. O. (2008) Electrostatics in wind-blown sand. *Phys. Rev. Lett., 100,* 014501.

Kok J. F. and Renno N. O. (2009) A comprehensive numerical model of steady-state saltation (COMSALT). *J. Geophys. Res., 114,* DOI: 10.1029/2009JD011702.

Kok J. F., Parteli E. J. R., Michaels T. I., and Karam D. B. (2012) The physics of wind-blown sand and dust. *Rept. Prog. Phys., 75,* 106901, DOI: 10.1088/0034-4885/75/10/106901.

Lau K.-M. and Kim K.-M. (2007) How nature foiled the 2006 hurricane forecasts. *Eos Trans. AGU, 88(26),* 271, DOI: 10.1029/2007EO260010.

Lorenz R. D., Wall S., Radebaugh J., et al. (2006) The sand seas of Titan: Cassini RADAR observations of longitudinal dunes. *Science, 312,* 724–727.

Mahowald N. M., Engelstaedter S., Luo C., et al. (2009) Atmospheric iron deposition: Global distribution, variability, and human perturbations. *Annu. Rev. Marine Sci., 1,* 245–278.

Melnik O. and Parrot M. (1998) Electrostatic discharge in martian dust storms. *J. Geophys. Res., 103,* 29107–29118.

McEwan I. K. and Willetts B. B. (1991) Numerical model of the saltation cloud. *Acta Mech. Suppl., 1,* 53–66.

McEwan I.K. and Willetts B. B. (1993) Adaptation of the near-surface wind to the development of sand transport. *J. Fluid Mech., 252,* 99–115.

Namikas S. L. (2003) Field measurement and numerical modelling of aeolian mass flux distributions on a sandy beach. *Sedimentology, 50,* 303–326.

Owen P. R. (1964) Saltation of uniform grains in air. *J. Fluid Mech., 20,* 225–242.

Pähtz T., Kok J. F., and Herrmann H. J. (2012) The apparent surface roughness of a sand surface blown by wind from an analytical model of saltation. *New J. Phys., 14,* 043035.

Penner J. E., Andreae M., Annegarn H., Barrie L., Feichter J., Hegg D., Jayaraman A., Leaitch R., Murphy D., Nganga J., and Pitari G. (2001) Aerosols, their direct and indirect effects. In *Intergovernmental Panel on Climate Change, Report to IPCC from the Scientific Assessment Working Group (WGI),* pp. 289–348. Cambridge Univ., New York.

Prospero J. M., Ginoux P., Torres O., Nicholson S. E., and Gill T. E. (2002) Environmental characterization of global sources of atmospheric soil dust identified with the Nimbus 7 Total Ozone Mapping Spectrometer (TOMS) absorbing aerosol product. *Rev. Geophys., 40,* 1002, DOI: 10.1029/2000RG000095.

Rasmussen K. R. and Sorensen M. (2008) Vertical variation of particle speed and flux density in aeolian saltation: Measurement and modeling. *J. Geophys. Res., 113,* F02S12.

Renno N. O. (2008) A general theory for convective plumes and vortices. *Tellus, 60A,* 688–699.

Renno N. O. and Kok J. F. (2008) Electric activity and dust lifting on earth and beyond. *Space Sci. Rev., 137,* 419–434.

Renno N. O. and Ruf C. (2012) Comments on the search for electrostatics on Mars. *Astrophys. J., 761(2),* 88, DOI: 10.1088/0004-637X/761/2/88.

Renno N. O., Nash A. A., Lunine J., and Murphy J. (2000) Martian and terrestrial dust devils: Test of a scaling theory using Pathfinder data. *J. Geophys. Res.–Planets, 105,* 1859–1865.

Renno N. O., Wong A. S., Atreya S. K., de Pater I., and Roos-Serote M. (2003) Electrical discharges and broadband radio emission by martian dust devils and dust storms. *Geophys. Res. Lett., 30,* DOI: 10.1029/2003GL017879.

Renno N. O., Abreu V., Koch J., Smith P. H., Hartogenisis O., de Bruin H. A. R., Burose D., Delory G. T., Farrell W. M., Parker M., Watts C. J., and Carswell A. (2004) MATADOR 2002: A field experiment on convective plumes and dust devils. *J. Geophys. Res., 109,* E07001, DOI: 10.1029/2003JE002219.

Renno N. O., Kok J. F., Kirkham H., and Rogacki S. (2008) A miniature sensor for electrical field measurements in dusty planetary atmospheres. *J. Physics Conf. Ser., 142,* 012075.

Renno N. O., Williams E., Rosenfeld D., Fischer D. G., Fischer J., Kremic T., Agrawal A., Andreae M. O., Bierbaum R., Blakeslee R., Boerner A., Bowles N., Christian H., Dunion

J., Horvath A., Huang X., Khain A., Kinne S., Lemos M. C., Penner J. E., Pöschl U., Quaas J., Seran E., Stevens B., Walati T., and Wagner T. (2013) CHASER: An innovative satellite mission concept to measure the effects of aerosols on clouds and climate. *Bull. Am. Meteor. Soc., 94,* in press.

Rosenfeld D., Kaufman Y., and Koren I. (2006) Switching cloud cover and dynamical regimes from open to closed Benard cells in response to aerosols suppressing precipitation. *Atmos. Chem. Phys., 6,* 2503–2511.

Rosenfeld D., Lohmann U., Raga G. B., O'Dowd C. D., Kulmata M., Reissell A., and Andreae M. O. (2008) Flood or drought: How do aerosols affect precipitation? *Science, 321,* 1309–1313.

Rosenfeld D., Williams E., Andreae M. O., Freud E., Puschl U., and Renno N. O. (2012) The scientific basis for a satellite mission to retrieve CCN and their impacts on convective clouds. *Atmos. Meas. Tech., 5,* 2039–2055, DOI: 10.5194/amt-5-2039-2012.

Ruf C., Renno N. O., Kok J. F., Bandelier E., Sander M. J., Gross S., Skjerve L., and Cantor B. (2009) The emission of non-thermal microwave radiation by a martian dust storm. *Geophys. Res. Lett., 36,* L13202, DOI: 10.1029/2009GL038715.

Schaefer V. J. and Day J. A. (1998) *A Field Guide to the Atmosphere.* Houghton Mifflin Harcourt, Boston.

Schmidt D. S., Schmidt R. A., and Dent J. D. (1998) Electrostatic force on saltating sand. *J. Geophys. Res., 103,* 8997–9001.

Seinfeld J. H. and Pandis S. N. (2006) *Atmospheric Chemistry and Physics: From Air Pollution to Climate Change, 2nd edition.* Wiley, New York. 1203 pp.

Sekiguchi M., Nakajima T., Suzuki K., et al. (2003) A study of the direct and indirect effects of aerosols using global satellite data sets of aerosol and cloud parameters. *J. Geophys. Res., 108(D22),* 4699.

Shao Y. P. and Li A. (1999) Numerical modelling of saltation in the atmospheric surface layer. *Boundary-Layer Meteorol., 91,* 199–225.

Shao Y. P. and Lu H. (2000) A simple expression for wind erosion threshold friction velocity. *J. Geophys. Res., 105,* 22437–22443.

Shao Y. and Raupach M. R. (1992) The overshoot and equilibration of saltation. *J. Geophys. Res., 97,* 20559–20564.

Shao Y., Raupach M. R., and Findlater P. A. (1993) Effect of saltation bombardment on the entrainment of dust by wind. *J. Geophys. Res., 98,* 12719–12726.

Sharp R. P. (1980) Wind-driven sand in Coachella Valley, California: Further data. *Geol. Soc. Am. Bull., 91,* 724–730.

Smith P. H. and Lemmon M. (1999) Opacity of the martian atmosphere measured by the Imager for Mars Pathfinder. *J. Geophys. Res., 104(E4),* 8975–8985, DOI: 10.1029/1998JE900017.

Sokolik I. N., Winker D. M., Bergametti G., Gillette D. A., Carmichael G., Kaufman Y. J., Gomes L., Schuetz L., and Penner J. E. (2001) Introduction to special section: Outstanding problems in quantifying the radiative impacts of mineral dust. *J. Geophys. Res., 106,* 18015–18027, DOI: 10.1029/2000JD900498.

Sullivan R., Greeley R., Kraft M., Wilson G., Golombek M., Herkenhoff K., Murphy J., and Smith P. (2000) Results of the imager for Mars Pathfinder windsock experiment. *J. Geophys. Res., 105,* 24547–24562.

Sullivan R., Arvidson R., Bell J. F. III, et al. (2008) Wind-driven particle mobility on Mars: Insights from Mars Exploration Rover observations at "El Dorado" and surroundings at Gusev Crater. *J. Geophys. Res., 113,* E06S07, DOI: 10.1029/2008JE003101.

Thomas P. and Gierasch P. J. (1985) Dust devils on Mars. *Science, 230,* 175–177.

Twomey S. (1974) Pollution and the planetary albedo. *Atmos. Environ., 8,* 1251–1256, DOI: 10.1016/0004-6981(74)90004-3.

Twomey S. (1977) The influence of pollution on the short wave albedo of clouds. *J. Atmos. Sci., 34,* 1149–1152.

Ungar J. E. and Haff P. K. (1987) Steady-state saltation in air. *Sedimentology, 34,* 289–299.

Washington R. W., Todd M. C., Engelstaedter S., M'bainayel S., and Mitchell F. (2006a) Dust and the low level circulation over the Bodele Depression, Chad: Observations from BoDEX 2005. *J. Geophys. Res., 111,* DOI: 10.1029/2005JD006502.

Washington R., Todd M. C., Lizcano G., Tegen I., Flamant C., Koren I., Ginoux P., Engelstaedter S., Bristow C. S., Zender C. S., Goudie A., Warren A., and Prospero J. (2006b) Links between topography, wind, deflation, lakes and dust: The case of the Bodélé Depression, Chad. *Geophys. Res. Lett., 33,* DOI: 10.1029/2006GL025827.

Weitz C. M., Plaut J. J., Greeley R., and Saunders R. S. (1994) Dunes and microdunes on Venus: Why were so few found in the Magellan data? *Icarus, 112,* 282–295.

Werner B. T. (1990) A steady-state model of wind-blown sand transport. *J. Geol., 98,* 1–17.

White B. R. (1981) Venusian saltation. *Icarus, 46,* 226–232.

White B. R. and Schulz J. C. (1977) Magnus effect in saltation. *J. Fluid Mech., 81,* 497–512.

Williams E. R. and Renno N. O. (1993) An analysis of the conditional instability of the tropical atmosphere. *Mon. Weather Rev., 121,* 21–36.

Williams E. and Stanfill S. (2002) The physical origin of the land-ocean contrast in lightning activity. *Comptes Rendus–Phys., 3,* 1277–1292.

Williams E. R., Rosenfeld D., Madden M., et al. (2002) Contrasting convective regimes over the Amazon: Implications for cloud electrification. *J. Geophys. Res., Spec. Issue, 107(D20),* 8082, DOI: 10.1029/2001JD000380.

Williams E. R., Mushtak V. C., Rosenfeld D., Goodman S. J., and Boccippio D. J. (2005) Thermodynamic conditions favorable to superlative thunderstorm updraft, mixed phase microphysics and lightning flash rate. *Atmos. Res., 76,* 288–306.

Williams S. H. and Greeley R. (1994) Windblown sand on Venus — the effect of high atmospheric density. *Geophys. Res. Lett., 21,* 2825–2828.

Wilson R. J. and Hamilton K. (1996) Comprehensive model simulation of thermal tides in the martian atmosphere. *J. Atmos. Res., 53,* 1290–1326.

Wilson S. A. and Zimbelman J. R. (2004) Latitude-dependent nature and physical characteristics of transverse aeolian ridges on Mars. *J. Geophys. Res., 109,* E10003, DOI: 10.1029/2004JE002247.

Wood R., Mechoso C. R., Bretherton C. S., et al. (2011) The VAMOS ocean-cloud-atmosphere-land study regional experiment (VOCALS-REx): Goals, platforms, and field operations. *Atmos. Chem. Phys. Discuss., 11,* 627–654.

Zheng X. J., Huang N., and Zhou Y. (2006) The effect of electrostatic force on the evolution of sand saltation cloud. *Eur. Phys. J., E19,* 129–138.

Zimbelman J. R. (2000) Non-active dunes in the Acheron Fossae region of Mars between the Viking and Mars Global Surveyor eras. *Geophys. Res. Lett., 27(7),* 1069–1072, DOI: 10.1029/1999GL008399.

Marley M. S., Ackerman A. S., Cuzzi J. N., and Kitzmann D. (2013) Clouds and hazes in exoplanet atmospheres. In *Comparative Climatology of Terrestrial Planets* (S. J. Mackwell et al., eds.), pp. 367–391. Univ. of Arizona, Tucson, DOI: 10.2458/azu_uapress_9780816530595-ch15.

Clouds and Hazes in Exoplanet Atmospheres

Mark S. Marley
NASA Ames Research Center

Andrew S. Ackerman
NASA Goddard Institute for Space Studies

Jeffrey N. Cuzzi
NASA Ames Research Center

Daniel Kitzmann
Zentrum für Astronomie und Astrophysik, Technische Universität Berlin

---

Clouds and hazes are commonplace in the atmospheres of solar system planets and are likely ubiquitous in the atmospheres of extrasolar planets as well. Clouds affect every aspect of a planetary atmosphere, from the transport of radiation, to atmospheric chemistry, to dynamics, and they influence — if not control — aspects such as surface temperature and habitability. In this review we aim to provide an introduction to the role and properties of clouds in exoplanetary atmospheres. We consider the role clouds play in influencing the spectra of planets as well as their habitability and detectability. We briefly summarize how clouds are treated in terrestrial climate models and consider the far simpler approaches that have been taken so far to model exoplanet clouds, the evidence for which we also review. Since clouds play a major role in the atmospheres of certain classes of brown dwarfs, we briefly discuss brown dwarf cloud modeling as well. We also review how the scattering and extinction efficiencies of cloud particles may be approximated in certain limiting cases of small and large particles in order to facilitate physical understanding. Since clouds play such important roles in planetary atmospheres, cloud modeling may well prove to be the limiting factor in our ability to interpret future observations of extrasolar planets.

1. INTRODUCTION

Clouds and hazes are found in every substantial solar system atmosphere and are likely ubiquitous in extrasolar planetary atmospheres as well. They provide sinks for volatile compounds and influence both the deposition of incident flux and the propagation of emitted thermal radiation. Consequently they affect the atmospheric thermal profile, the global climate, the spectra of scattered and emitted radiation, and the detectability by direct imaging of a planet. As other chapters in this book attest, clouds and hazes are a complex and deep subject. In this chapter we will broadly discuss the roles of clouds and hazes as they relate to the study of exoplanet atmospheres. In particular, we will focus on the challenge of exoplanet cloud modeling and discuss the impact of condensates on planetary climates.

The terms "clouds" and "hazes" are sometimes used interchangeably. Here we use the term "cloud" to refer to condensates that grow from an atmospheric constituent when the partial pressure of the vapor exceeds its saturation vapor pressure. Such supersaturation is typically produced by atmospheric cooling, and cloud particles will generally evaporate or sublimate in unsaturated conditions. A general framework for such clouds in planetary atmospheres is provided by *Sánchez-Lavega et al.* (2004). By "haze" we refer to condensates of vapor produced by photochemistry or other nonequilibrium chemical processes. This usage is quite different from that of the terrestrial water cloud microphysics literature where the distinction depends on water droplet size and atmospheric conditions.

Because exoplanetary atmospheres can plausibly span such a wide range of compositions as well as temperature and pressure conditions, a large number of species may form clouds. Depending on conditions, clouds in a solar composition atmosphere can include exotic refractory species such as Al_2O_3, $CaTiO_3$, Mg_2SiO_4, and Fe at high temperature, and Na_2S, MnS, and of course H_2O at lower temperatures. Many other species condense as well, including CO_2 in cold,

Mars-like atmospheres and NH_3 in the atmospheres of cool giants, such as Jupiter and Saturn. In Earth-like atmospheres water clouds are likely important, although Venus-like conditions and sulfuric-acid or other clouds are possibilities as well. Depending on atmospheric temperature, pressure, and composition the range of possibilities is very large. Furthermore, not all clouds condense directly from the gas to a solid or liquid phase as the same species. For example, in the atmosphere of a gas giant exoplanet solid MnS cloud particles are expected to form around 1400 K from the net reaction $H_2S + Mn \rightarrow MnS(s) + H_2$ (*Visscher et al.,* 2006).

Clouds strongly interact with incident and emitted radiative fluxes. The clouds of Earth and Venus increase the planetary Bond albedo (the fraction of all incident flux that is scattered back to space) and consequently decrease the equilibrium temperature. Clouds can also "trap" infrared radiation and heat the atmosphere. Hazes, in contrast, because of their usually smaller particle sizes, can scatter incident light away from a planet but not strongly affect emergent thermal radiation, and thus predominantly result in a net cooling. The hazes of Titan play such an "anti-greenhouse effect" role in the energy balance of the atmosphere. For these reasons global atmospheric models of exoplanets, including those aiming to define the habitable zone given various assumptions, must consider the effects of clouds. However, clouds are just one ingredient in such planetary atmosphere models. Bulk atmospheric composition, incident flux, gravity, chemistry, molecular and atomic opacities, and more must all be integrated along with the effect of clouds in order to construct realistic models. Introductory reviews by *Burrows and Orton* (2010) and *Seager and Deming* (2010) cover the important fundamentals of atmospheric modeling and place cloud models in their broader context.

In the remainder of this chapter we discuss the importance of clouds to exoplanet atmospheres, particularly considering their impact on habitability, discuss cloud modeling in general and the types of models developed for exoplanet studies, and finally briefly review observations of exoplanet clouds. We include an overview of how the Mie opacity of particles behaves in various limits and consider the case of fluffy particles. Because clouds have played such a large role in efforts to understand the atmospheres of brown dwarfs, we also briefly review the findings in this field.

2. IMPORTANCE OF CLOUDS AND HAZES IN EXOPLANET ATMOSPHERES

2.1. Albedo, Detectability, and Characterization

Before discussing the role of clouds on the spectra of extrasolar planets, it is worthwhile to review the various albedos that enter the discussion. The Bond or bolometric albedo is the fraction of all incident light, integrated over the entire stellar spectrum, that is scattered back to space by a planet. This albedo is a single number and enters into the computation of a planet's equilibrium temperature (the temperature an airless planet would have if its thermal emission were equal to the incident radiation that it absorbs). It is also useful to know the monochromatic ratio of all scattered to incident light as a function of wavelength, which is the spherical albedo. For historical reasons it is more common to discuss the geometric albedo, which is specifically the wavelength-dependent ratio of the light scattered by the entire planet in the direction directly back toward its star compared to that which would be so scattered by a perfectly reflecting Lambert disk of the same radius as the planet. Care must be taken to distinguish all these albedos, as they can differ markedly from one another even for the same planet, and the literature is rife with confusion between them.

Perhaps the single most important effect of clouds is to brighten the reflected light spectra of exoplanets, particularly at optical wavelengths ($0.38 < \lambda < 1$ μm). For planets with appreciable atmospheres, Rayleigh scattering is most efficient in the blue. At longer wavelengths, however, absorption by either the planetary solid surface or oceans (for terrestrial planets) or by atmospheric gaseous absorbers becomes important in the red. This is because for common molecules at planetary temperatures, such as H_2O, CO_2, or CH_4, vibrational-rotational transitions become important at wavenumbers below about 15,000 cm^{-1} or wavelength $\lambda > 0.6$ μm. Except for diatomic species (notably O_2), planetary atmospheres are generally not warm enough to exhibit strong electronic absorption features in the optical. Thus the reflected light or geometric albedo spectrum of a generic cloudless planet with an atmosphere would be bright at blue wavelengths from Rayleigh scattering and dark in the red and at longer wavelengths from gaseous molecular or surface absorption.

Clouds, however, tend to be bright with a fairly gray opacity through the optical. Thus a thick, scattering cloud can brighten a planet in the far red by scattering more light back to space than a cloudless planet. As a result, two similar planets, one with and one without cloud cover, will have very different geometric albedos in the red, and consequently differing brightness contrasts with their parent stars. The mean Earth water clouds increase the contrast between the reflection spectrum of an Earth-like planet and its host star by one order of magnitude (*Kitzmann et al.,* 2011b). This effect, first noted for giant planets by *Marley et al.* (1999) and *Sudarsky et al.* (2000) and further explored in *Cahoy et al.* (2010), is illustrated in Fig. 1, which plots geometric albedo as a function of λ for giant planets with and without clouds. A warm, cloudless atmosphere is dark in reflected light while a cooler atmosphere, sporting water clouds, is much brighter. In this case clouds affect the model spectra far more than a factor of three difference in the atmospheric abundance of heavy elements, which is also shown.

In the case of searches for planets by direct coronagraphic imaging in reflected light, clouds may even control whether or not a planet is discovered. Depending on the spectral bandpass used for planet discovery at a given distance from its primary star a cloudy planet may be brighter and more detectable than a cloudless one. Discussions of

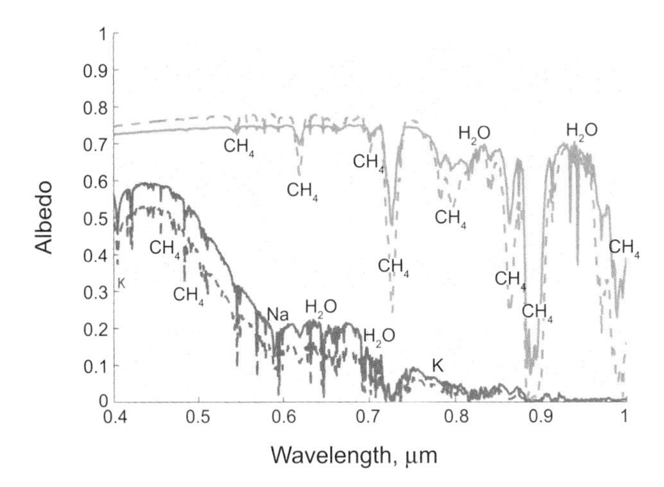

Fig. 1. Model geometric albedo spectra for Jupiter-like planets at 2 AU (upper curves) and 0.8 AU (lower curves) from their parent star. Solid and dashed lines lines show models with a solar abundance and a 3× enhanced abundance of heavy elements respectively. Prominent absorption features are labeled. The 2-AU-model planets possess a thick water cloud, while the 0.8-AU models are warmer and cloudless and consequently darker in reflected light. Figure modified from *Cahoy et al.* (2010).

the influence of clouds and albedo on detectability include those by *Tinetti et al.* (2006a) and *Kitzmann et al.* (2011a).

Once a planet is detected, spectra are needed to characterize atmospheric abundances of important molecules. For terrestrial extrasolar planets this is best done by the analysis of the thermal emission spectrum (*Selsis,* 2004; *Tinetti et al.,* 2012). Clouds, however, may conceal the thermal emission from the surface and dampen spectral features of molecules (e.g., the bio-indicators N_2O or O_3). Indeed, clouds on Earth have a larger impact on the emitted infrared flux than the differences between night and day (*Hearty et al.,* 2009; *Tinetti et al.,* 2006a). Thermal emission spectra are therefore very sensitive to the types and fractional coverages of clouds present in the atmosphere. At high spectral resolution the most important terrestrial molecular spectral features in the mid-infrared, such as O_3, remain detectable even for cloud-covered conditions in many cases. Figure 2 shows thermal emission spectra affected by low-level water droplet and high-level water ice clouds of an Earth-like planet orbiting different main-sequence dwarf stars (adopted from *Vasquez et al.,* 2013). With increasing cloud cover of either cloud type, the important 9.6-μm absorption band of ozone is strongly dampened, along with an overall decrease in the thermal radiation flux. For some cases presented in Fig. 2 (e.g., for the F-type star and 100% high-level cloud cover), the ozone band seems to be completely absent or even appears in emission rather than absorption (F-type star, 100% low-level clouds). At lower spectral resolution, such as could be obtained for terrestrial exoplanets in the near future, clouds render the molecular features even less detectable; thus clouds will strongly affect the determination of their atmospheric composition. For example, as shown by *Kitzmann et al.* (2011a), a substantial amount of water clouds in an Earth-like atmosphere can completely hide the spectral signature of the bio-indicator ozone in low-resolution thermal emission spectra.

Clouds can also obscure spectral features originating from the surface of a planet. In principle, signals of surface vegetation ("vegetation red edge") are present in the reflected light spectra of Earth-like planets. However, as often pointed out (e.g., *Arnold et al.,* 2002; *Hamdani et al.,* 2006), this spectral feature can easily be concealed by clouds. For the detectability of possible vegetation signatures of terrestrial extrasolar planets, *Montañés-Rodríguez et al.* (2006) and *Tinetti et al.* (2006b) concluded that clouds play a crucial role for these signatures in the reflection spectra. Thus, apart from the scattering characteristics of different planetary surface types, the presence of clouds has been found to be one of the most important factors determining reflection spectra.

Transmission spectra of transiting planets can be used to obtain many atmospheric properties, such as atmospheric composition and temperature profiles. Transmission spectroscopy of extrasolar giant planets has already proven to be a successful method for the characterization of giant exoplanets. As shown by *Pallé et al.* (2009), the major atmospheric constituents of a terrestrial planet remain detectable in transmission spectra even at very low signal-to-noise ratios. Thus, transmission spectra can provide more information about the atmospheres of exoplanets than reflection spectra. Theoretical transmission spectra of Earth-sized transiting planets have been studied by *Ehrenreich et al.* (2006), including the effects of optically thick cloud layers. Their results show that the transmission spectra only contain information about the atmosphere above the cloud layer and that clouds can effectively increase the apparent radius of the planet. The impact of clouds on the transmission spectra therefore depends strongly on their atmospheric height. If cloud layers are only located in the lower atmosphere, which is already opaque due to absorption and scattering by gas species, their overall effect on the spectrum will be small (*Kaltenegger and Traub,* 2009).

2.2. Habitability of Extrasolar Terrestrial Planets

From the terrestrial bodies in the solar system that have an atmosphere, we know that clouds are a common phenomenon, and they should also be expected to occur in atmospheres of terrestrial extrasolar planets. Apart from the usual well-known greenhouse gases, clouds have the most important climatic impact in the atmospheres of terrestrial planets by affecting the energy budget in several ways. First, clouds can scatter incident stellar radiation back to space, resulting in atmospheric cooling (the albedo effect). On the other hand, clouds can trap thermal radiation within the lower atmosphere by either absorption and reemission

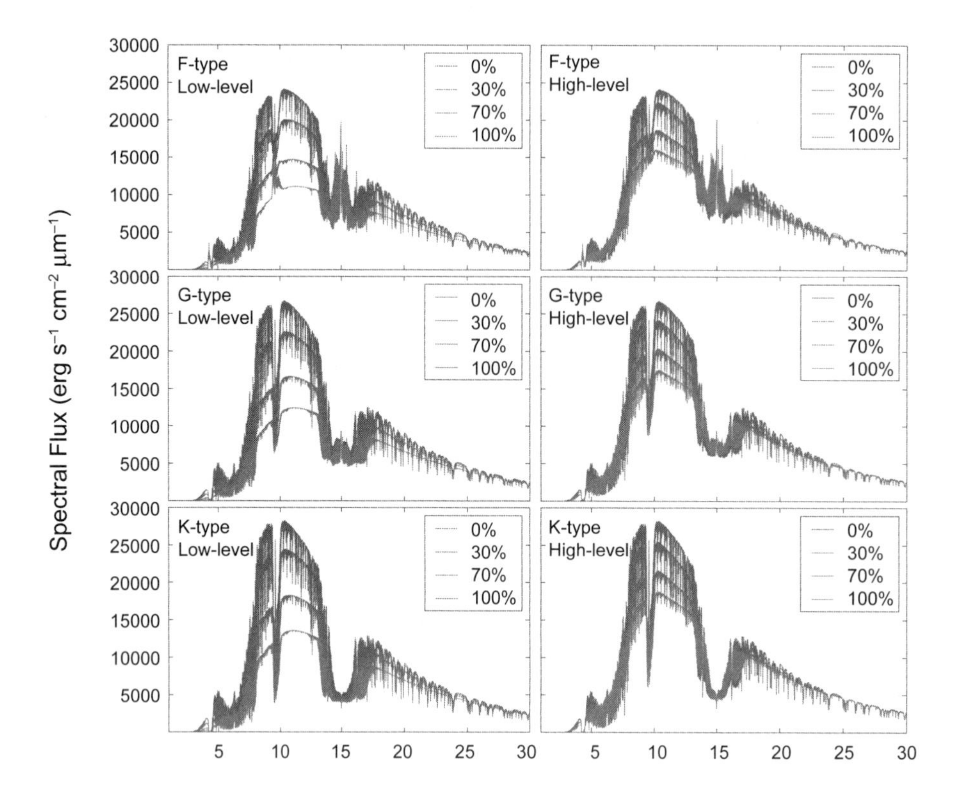

Fig. 2. See Plate 65 for color version. Planetary thermal emission spectra influenced by low-level water droplet (left panel) and high-level ice clouds (right panel) for an Earth-like planet orbiting different kinds of main-sequence host stars (adopted from *Vasquez et al.,* 2013). For each central star the spectra are shown for different cloud coverages. Note especially the strong impact of the cloud layers on the 9.6-μm absorption band of ozone.

at their local temperature (the classical greenhouse effect) or by scattering thermal radiation back toward the surface (the scattering greenhouse effect), which heats the lower atmosphere and planetary surface. All these effects are determined by the wavelength-dependent optical properties of the cloud particles (absorption and scattering cross-sections, single-scattering albedo, asymmetry parameter, scattering phase function). These properties can differ considerably for different cloud-forming species (owing to their refractive indices) and atmospheric conditions (composition and temperature structure). Note that the single-scattering albedo, yet a fourth type of albedo (see section 2.1 for a description of the others), measures the fraction of all incident light scattered by a single cloud particle.

Life as we know it requires the presence of liquid water to form and survive. In the context of terrestrial exoplanets we are primarily concerned with habitable conditions on the planetary surface. Life forms may of course also exist in other environments, such as deep below the planetary surface or within a subsurface ocean (*Lammer et al.,* 2009). Since there is no possibility of detecting the presence of such habitats by remote observations, our current definition of a habitable terrestrial planet assumes a reservoir of liquid water somewhere on its surface. Liquid surface water implies that these planets would also have abundant water vapor in their atmospheres. We can therefore safely assume that water (and water ice) clouds will naturally occur in the atmospheres of these exoplanets throughout the classical habitable zone.

As such, H_2O and CO_2 clouds are the prime focus for habitable exoplanets. Other possible condensing species also found in our solar system include C_2H_6 and CH_4 (e.g., in the atmosphere of Titan), N_2 (Triton), or H_2SO_4 (Venus, Earth). Any chemical species in liquid form on a planetary surface can in principle be considered as a potential cloud-forming species in the atmosphere if its atmospheric partial pressure is high enough and the temperature low enough.

Given these considerations, it is clear that clouds are important for the determination of the boundaries of the classical habitable zone around different kinds of stars. In the next two subsections we discuss some of the ways that clouds influence the habitable zone boundaries.

2.2.1. Inner edge of the habitable zone. The closer a terrestrial planet orbits its host star, the greater the stellar insolation and consequently surface temperature. This leads to enhanced evaporation of surface water and increases the amount of water vapor in the atmosphere. The strong greenhouse effect by this additional water vapor further increases the surface temperature and therefore the evaporation of surface water, resulting in a positive feedback cycle (see Covey et al., this volume).

The inner boundary of the classical habitable zone is usually defined either as the distance from a host star at which an Earth-like planet completely loses its liquid surface water by evaporation (runaway greenhouse limit), or as the distance from the host star when water vapor can reach the upper atmosphere (water loss limit). The latter is the definition favored by *Kasting et al.* (1993); the former definition is used by *Hart* (1979). In the present Earth atmosphere water escape is limited by the cold trap at the tropopause, which loses its efficiency in a sufficiently warm, moist atmosphere (*Kasting et al.,* 1993).

Using a one-dimensional atmospheric model, *Kasting et al.* (1993) estimated that the inner boundary of the habitable zone for the runaway greenhouse scenario for an Earth around the Sun would be located at 0.84 AU. This distance, however, was calculated with the clouds treated as a surface albedo effect and neglecting any feedback effects. For example, once formed, clouds increase the planetary albedo, which in turn partially reduces the temperature increase due to the higher stellar insolation and the cloud greenhouse effect. *Kasting* (1988) investigated some feedback effects and concluded that with a single layer of thick water clouds, Earth could be moved as close as 0.5 AU (for a cloud coverage of 100%) or 0.67 AU (50% cloud coverage) from the Sun before all liquid surface water would be lost. *Goldblatt and Zahnle* (2011) explored the various feedback issues in detail and concluded that more sophisticated modeling approaches are necessary to explore habitability.

Water clouds also contribute to the greenhouse effect. Depending on the temperature at which the cloud emits thermal radiation, its greenhouse effect can match or exceed the albedo effect. Cold clouds, such as cirrus, are net greenhouse warmers. Low clouds with temperatures near the effective radiating temperature affect climate almost entirely by their albedo, which depends on such factors as patchiness, cloud thickness, and the size distributions and composition of cloud particles. The exact climatic impact of clouds therefore crucially depends on the balance between their greenhouse and albedo effects. Whether the greenhouse or albedo effect dominates for clouds forming under runaway greenhouse conditions cannot be easily determined *a priori*. A better understanding of the cloud microphysics and convection processes in moist atmospheres during a runaway greenhouse process is needed to determine the cloud properties and their fractional coverage near the inner habitable zone (HZ) boundary. The effect of clouds under runaway greenhouse conditions represents one of the most important unresolved issues in planetary climates.

Note that this discussion is relevant to wet planets with moist atmospheres fed by extensive seas. Dry — land or desert — planets will have unsaturated atmospheres in the tropics and thus radiate at a higher temperature and cool more efficiently than planets with a water-saturated atmosphere. As a result, the habitable zone for such planets may be larger (*Abe et al.,* 2011).

2.2.2. *Outer edge of the habitable zone.* The outer edge of the habitable zone is set by the point at which there is no longer liquid water available at the surface as it is locked in ice. With falling insolation, planets found progressively farther away from their central star have lower atmospheric and surface temperatures. With lower surface temperatures, the removal of CO_2 from the atmosphere due to the carbonate-silicate cycle, which controls the amount of CO_2 in Earth's atmosphere on timescales on the order of 1 m.y., becomes less efficient (see Covey et al., this volume). Thus, if the terrestrial planet is still geologically active, CO_2 can accumulate in the atmosphere by volcanic outgassing. With decreasing atmospheric temperatures, CO_2 will condense at some point to form clouds composed of CO_2 ice crystals. Because condensation nuclei can be expected to be available in atmospheres of terrestrial planets, the most dominant nucleation process for the formation of CO_2 clouds is usually assumed to be heterogeneous nucleation (*Glandorf et al.,* 2002), in which cloud particles form on existing seed particles.

Just like water clouds, the presence of CO_2 clouds will result in an increase of the planetary albedo by scattering incident stellar radiation back to space. However, in contrast to water, CO_2 ice is almost transparent in the infrared (*Hansen,* 1997, 2005) except within some strong absorption bands. Thus CO_2 clouds are unlikely to trigger a substantial classical greenhouse effect by absorption and reemission of thermal radiation.

On the other hand, CO_2 ice particles can efficiently scatter thermal radiation. This allows for a scattering greenhouse effect in which a fraction of the outgoing thermal radiation is scattered back toward the planetary surface (*Forget and Pierrehumbert,* 1997). Depending on the cloud properties, this scattering greenhouse effect can outweigh the albedo effect and can in principle increase the surface temperature above the freezing point of water. However, the scattering greenhouse effect is a complex function of the cloud optical depth and particle size (*Colaprete and Toon,* 2003). Furthermore, because the greenhouse effect of CO_2 clouds depends on the scattering properties of the ice particles, the particle shape (which cannot be expected to be spherical) or particle surface roughness may play an important role. Such effects cannot be easily quantified because neither the particle shapes nor their surface properties are known. These and other uncertainties in the cloud microphysics of CO_2 ice in cool CO_2-dominated atmospheres makes the calculation of the position of the outer HZ boundary complicated. More details regarding the formation of CO_2 clouds in the present and early martian atmosphere can be found in *Glandorf et al.* (2002), *Colaprete and Toon* (2003), and *Määttänen et al.* (2005), as well as in Esposito et al. (this volume). Inferences about CO_2 cloud particle sizes in the current martian atmosphere as constrained by a variety of datasets are presented by *Hu et al.* (2012).

For a fully cloud-covered early Mars with a thick CO_2-dominated atmosphere and CO_2 clouds composed of spherical ice particles, *Forget and Pierrehumbert* (1997) estimated that the outer boundary of the HZ is located at 2.4 AU, in contrast to the cloud-free boundary of 1.67 AU estimated by *Kasting et al.* (1993). This greater value has been further used by *Selsis et al.* (2007) to extrapolate the effects of CO_2 clouds on the outer HZ boundary toward other main-sequence central stars (see also *Kaltenegger and Sasselov,* 2011).

For terrestrial super-Earths it has been suggested by *Pierrehumbert and Gaidos* (2011) that these planets could have retained much of their primordial H_2 atmosphere because of their greater mass. According to their model study, the classical habitable zone might be far larger than expected for a CO_2-dominated atmosphere, although this study did not explore the impact of clouds.

3. CLOUDS AND RADIATION

Clouds are important for planetary atmospheres because they interact with both incident short-wavelength radiation from the parent star and emergent thermal radiation emitted by the planetary surface, if present, and the atmosphere itself. Perhaps the simplest example of such interaction occurs for spherical cloud particles composed of a single constituent. In this case the wavelength-dependent optical properties can be computed from Mie calculations that depend only on particle size and the wavelength-dependent complex refractive index of the bulk condensate. In this section we briefly review the basics of the interaction of cloud particles with radiation and summarize the use of Mie theory for modeling this interaction as well as point out some useful simplifications that can be applied in certain limiting cases.

3.1. Basic Radiative Properties of Cloud Particles

Cloud opacity ultimately depends upon the radiative properties of the constituent particles. A particle with radius r has a cross section to scatter losslessly or to absorb incident radiation at some wavelength λ, given by C_s or C_a respectively. These cross sections are defined as $C_s = Q_s\pi r^2$ and $C_a = Q_a\pi r^2$, and their sum is the extinction cross section C_e. The scattering and absorption efficiencies Q_s and Q_a, and, from them, the extinction efficiency $Q_e = Q_s + Q_a$, are thus defined. The particle single scattering albedo is $\omega = Q_s/Q_e$, and the scattering phase function is $P(\theta)$. These quantities can be computed by a Mie code, which computes the electromagnetic wave propagation through spherical particles, given the optical properties of the cloud material. Good physically-based introductions to radiative properties of particles can be found in *van de Hulst* (1957), *Hansen and Travis* (1974), and *Liou* (2002).

Modeling cloud particle radiative properties can seem forbidding; workers often think a Mie scattering code, with all its exotic predictions, is needed. However, in many applications, much simpler approaches not only can provide very good quantitative fidelity — and greater physical insight — but also can be applied directly to more complex kinds of particles than the spheres for which Mie theory is derived.

The direct beam of radiant flux vertically traversing a cloud *layer* of particles composed of some material q is reduced by a factor $\exp(-\tau_q)$, where the layer optical depth is $\tau_q = nQ_e\pi r^2 H_q = nC_eH_q$, and n is the particle number density, r the particle radius, and H_q the vertical thickness of the layer (assuming a uniform vertical distribution for simplicity). Q_e and thus τ_q are functions of the wavelength λ, through the λ-dependent real and imaginary refractive indices of the material in question [n_r , n_i; see below, and *Draine and Lee* (1984), *Pollack et al.* (1994), or *www.astro.uni-jena.de/Laboratory/Database/databases.html* for typical values]. The particle opacity κ (in units of length-squared per mass) is the effective particle cross section per unit mass of either solids or gas.

If the phase function is strongly forward scattering, as can be the case for particles with $r/\lambda > 10$ or so (*Hansen and Travis,* 1974), much of the "scattering" can be approximated as unattenuated radiation by reducing the layer optical depth and renormalizing the particle albedo and phase function. This is done by applying simple corrections to both ω and τ_q known as similarity relations. *Irvine* (1975), *van de Hulst* (1980), and *Liou* (2002) review these, and present a number of ways to solve the radiative transfer problem in general.

3.2. Refractive Indices and Opacity

All calculations of particle radiative properties [Q_e, Q_s, Q_a, ω and $P(\theta)$] begin with the refractive indices of the material and the particle size. While Mie codes are available for download (e.g., *ftp://climate1.gsfc.nasa.gov/wiscombe/Single_Scatt/*), analytical approximations that are valid in all the relevant limits can often be of great use. Such approximations rely on the fact that realistic clouds have size and shape distributions that average away the exotic fluctuations shown by Mie theory for particles of specific sizes (see *Hansen and Travis,* 1974, for examples). With current computational capabilities, Mie calculations are not onerous, but if many wavelengths and/or grids of numerous models are of interest, the burden is compounded, so one needs to understand whether such detailed predictions are actually needed.

Even simpler approaches are feasible. In many cases of interest, only globally averaged emergent fluxes or reflectivities, or perhaps heating calculations, are needed; here, the details of the phase function are of less interest than the optical depth and particle albedo, which are based only on the efficiencies. For these cases one can do fairly well using asymptotic expressions for efficiencies in the limiting regimes $r/\lambda \ll 1$ and $r/\lambda \gg 1$, connecting them with a suitable bridging function. These expressions depend on the refractive indices of the material in question.

The simplest limit is when $r/\lambda \gg 1$ (geometrical optics); in fact, many of the cloud models discussed in this chapter fall into this regime at 1–10 μm wavelengths. In this range it is convenient to neglect diffraction, which is strongly concentrated in the forward direction and can be lumped with the direct beam as noted above. Then, assuming the particle has density ρ and is itself opaque, $Q_e \leq 1$ and C_e reduces to the physical cross section. Thus the solids-based opacity is simply $\kappa = 3\pi r^2/4\rho\pi r^3 = 3/4r\rho$. In this regime, particle growth drastically reduces the opacity because the surface-to-mass ratio decreases linearly with radius (*Miyake and Nakagawa,* 1993; *Pollack et al.,* 1994).

At the opposite extreme, when $r/\lambda \ll 1$ (the Rayleigh limit), simple analytical expressions also apply (*Draine and Lee,* 1984; *Bohren and Huffman,* 1983). For simplicity below we give the expressions for $n_i \ll n_r \sim 1$ (appropriate for silicates, oxides, water, but not iron), but the general expressions are not much more complicated (*Draine and Lee,* 1984). Thus $Q_a = 24xn_rn_i/(n_r^2 + 2)^2$ and $Q_s = 8x^4(n_r^2-1)^2/3(n_r^2 + 2)^2$, where $x = 2\pi r/\lambda$ (see also *van de Hulst,* 1957, p. 70). Because Q_s decreases much faster than Q_a with decreasing r/λ in this regime, small particles become not only less effective at interacting with radiation, but increasingly become absorbers/emitters rather than scatterers. In this limit, their cross section $C_a = Q_a\pi r^2$ per unit particle mass becomes constant since Q_a is proportional to r.

3.3. Shape and Porosity

In the case of terrestrial cloud droplets, in which condensation dominates growth, and raindrops, in which coagulation dominates growth, spherical particles of constant density are assumed, provided they fall slowly (although drag-induced deformation is taken into account when computing terminal fall-speeds for large raindrops). However, if particles condense from their vapor phase as solids, tiny initial monomers may instead coagulate by sticking into porous aggregates, perhaps having some fractal properties where the density may depend on the size. Nonspherical particle shape adds complexity to the computation of optical properties, which is commonly the case for solid particles. For instance, the single-scattering properties of ice particles in the terrestrial atmosphere depend on not only the geometric shape of the crystals (which depend on the temperature), but also the microscopic surface roughness. As noted by *Fu et al.* (1998), the extinction and absorption cross sections depend mainly on projected areas and particle volumes (note that random orientation is typically assumed), while the scattering phase function is largely determined by the aspect ratio of the crystal components and their small-scale roughness (*Fu,* 2007).

One simple approximation to scattering by equidimensional, but still irregular, particles was developed by *Pollack and Cuzzi* (1980). This approach uses Mie theory for particles with small-to-moderate r/λ, where shape irregularities are indiscernible to the waves involved and Mie calculations converge rapidly. For larger r/λ a combination of diffraction, external reflection, and internal transmission is used as adjusted for shape and parameterized by laboratory experiments. The approach was validated with experimental data and is easy to apply in cases where the angular distribution of scattered radiation is important, as long as a particle size distribution smears away the significant oscillations in scattering properties that Mie theory predicts for monodisperse spheres (*Hansen and Travis,* 1974). This method saves on computational effort for particles with large r/λ but does involve a Mie code for the smaller particles.

In the general exoplanet case for tiny particles in gas of typical densities, plausible interparticle collision velocities are very small and lead to sticking but only minor compaction (see, e.g., *Cuzzi and Hogan,* 2003; *Dominik and Tielens,* 1997), until particles become large enough (10–100 μm) where sedimentation velocity can exceed several meters per second and compaction or bouncing arise (section 6.3) (see *Dominik et al.,* 2007; *Zsom et al.,* 2010). The thermodynamic properties of condensates and the temperature-pressure (T-P) structure of the atmosphere in question determine whether the condensate appears as a liquid or a solid. Figure 3 shows condensation curves of a number of important cloud-forming compounds, along with the T-P profiles of a range of exoplanetary and substellar objects.

Models of growing aggregates have been fairly well developed in the literature of protoplanetary disks (e.g., *Weidenschilling,* 1988; *Dominik and Tielens,* 1997; *Beckwith et al.,* 2000; *Dominik et al.,* 2007; *Blum,* 2010). The properties of clouds made of this rather different kind of particle will differ in their opacity and vertical distribution from predictions of the simplest models (see section 6.1); these properties in turn will impact, and can in principle be constrained by, remote observations of reflected and emitted radiation.

3.4. Porous Particles and Effective Medium Theory

The most straightforward way of modeling porous aggregates is to model their effective refractive indices based on their constituent materials and porosity. If the monomers from which the particles are made are smaller than the wavelength in question, they act as independent dipoles immersed in an enveloping medium (the medium can be another material; here we assume it is vacuum). The aggregate as a whole can then be modeled as having *effective* refractive indices that depend only on the porosity of the aggregate and the refractive indices (but not the size) of the monomers. This is the so-called effective medium theory (EMT); several variants are discussed by *Bohren and Huffman* (1983), *Ossenkopf* (1991), *Stognienko et al.* (1995), and *Voschinnikov et al.* (2006).

For most combinations of materials, EMTs can be even further approximated by simple volume averages such that the refractive indices of the particle as a whole can be written (for a simple two-component particle with component 2 being vacuum) as $n_i = fn_{i1} + (1-f)n_{i2}$ and $(n_r-1) = f(n_{r1}-1) + (1-f)(n_{r2}-1)$, where n_r and n_i are the effective real and imaginary indices of the aggregate as a whole, and component 1 has volume fraction $f = 1-\phi$, where ϕ is the porosity. If a material that has large refractive indices, such as iron, is involved, the full expressions are needed

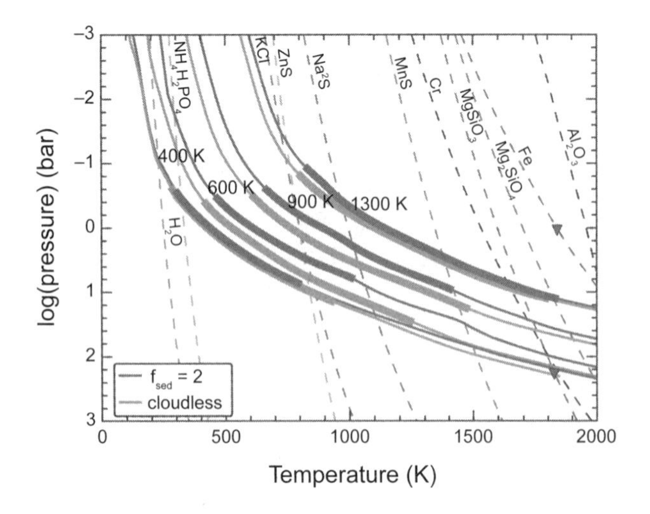

Fig. 3. See Plate 66 for color version. Model brown dwarf atmospheric temperature-pressure profiles for several effective temperatures (solid) are plotted along with condensation curves (dashed) for a variety of compounds. Diamonds along the iron and enstatite curves denote the melting point of each compound with the liquid phase lying to the right. Thicker portions of temperature profiles denote the model photospheres, the regions of the atmospheres from which most of the emitted radiation emerges. For each atmosphere model two curves are shown, corresponding to a cloudless calculation (blue) and a cloudy model with sedimentation efficiency, f_{sed} = 2 (red). Figure modified from *Morley et al.* (2012).

(see *Wright,* 1987; *Bohren and Huffman,* 1983; or other basic references). We will assume the porous aggregates are roughly equidimensional, not needle-like structures, but even that added complication can often be tractable in semi-analytical ways for most materials (*Wright,* 1987; *Bohren and Huffman,* 1983). Different mixing rules can introduce uncertainties on the order of 1% and 10% respectively in the real and imaginary refractive indices for black carbon inclusions in liquid water droplets (*Lesins et al.,* 2002).

In the $r/\lambda \ll 1$ regime, porosity plays little role, both because the opacity is simply proportional to the total mass regardless of how it is distributed (see section 3.2), and because ever-smaller particles are unlikely to be aggregates of ever-smaller monomer constituents, but are more likely to be monomers themselves. *Ferguson et al.* (2007) present numerical calculations that show that for the tiny particles they modeled (0.35 nm–0.17 μm), this expectation is indeed fulfilled to first order, also showing that mixing of materials within these tiny aggregates has little effect (see also *Allard et al.,* 2001; *Helling et al.,* 2008; *Witte et al.,* 2009; and references therein). Particles this small are likely to be well-mixed at all levels of typical exoplanet atmospheres (see below). Some targets of interest may have high haze layers where this regime is of interest.

However, in the $r/\lambda \gg 1$ regime, which covers typical planetary clouds observed in the optical or near-infrared, porosity does affect opacity. Specifically, for a particle of mass M and porosity ϕ, the opacity $\kappa = C_e/M = 3/4\rho r = 3/4\rho_s(1-\phi)r$ where $r = (3M/4\rho_s(1-\phi))^{1/3}$ is the actual radius. Comparing the opacity of this particle to that of a "solid" particle with the same mass having radius r_s gives $\kappa/\kappa_s = (1-\phi)^{-2/3}$. The porous particle thus has a larger opacity for the same mass, and indeed reaches the geometrical optics (λ independent) limit at lower mass than a nonporous particle. In this regime, effective medium theory should be a convenient and valid way of modeling porous and/or aggregate grain properties (section 3.2) (see also *Helling et al.,* 2008, and references therein).

4. CLOUD MODELS

The cloud properties required to compute radiative fluxes are rather extensive. Thermal emission requires the emissivity and temperature of the cloud particles. Vertical fluxes can be computed from knowledge of the vertical distribution of particle cross-section (for each species of particle) together with the wavelength-dependent scattering phase function and the single-scattering albedo. For a basic two-stream approach these quantities can be boiled down to a vertical profile of extinction coefficient, asymmetry parameter, and single-scattering albedo.

Complexities can arise if particle temperature differs from the local atmospheric temperature by virtue of latent heat exchange and radiative heating of the particles, but this should be rare in exoplanet applications (*Woitke and Helling,* 2003), and the particles are typically assumed to be at ambient atmospheric air temperature. Clouds that are not horizontally uniform are another possible complication that can be considered.

The required scattering and absorption total cross sections and overall asymmetry parameter needed for input to a radiative transfer model are found by integration of the single-scattering properties and emissivities over the particle size distributions. The size distributions in turn can be computed by a so-called bin model, which tracks the number of particles in multiple different bin sizes as the particles interact with the atmosphere. Such an approach is computationally expensive, and a more efficient treatment is to assume a particular functional form with a small number of free parameters.

For terrestrial applications, log-normal distributions, with three parameters (total number, geometric mean radius, and geometric standard deviation), and gamma distributions, also with three parameters (total number, slope parameter, and shape parameter), are commonly used. Separate size distributions are used for particles of different phases and bulk densities (e.g., raindrops and snowflakes) and for each mode of a multimodal size distribution. For example, the parameterized size distribution of cloud droplets, which grow principally through condensation in the terrestrial atmosphere, are treated distinctly from raindrops, which grow principally through collisions. Although a poor approximation for cloud droplets, it is often assumed that the shape parameter is zero for other cloud species, which allows the gamma distribution to collapse to an exponential distribution, thereby reducing the free parameters from three to two.

For the computation of heating rates from the divergence of radiative fluxes, the vertical distribution of any absorbing and emitting species is obviously critical. Also, any vertical redistribution by scatterers in bands with emission or absorption requires that their vertical distribution is accurate. If horizontal photon transport is unimportant, it is feasible to represent the global atmosphere — or the atmosphere within any model column of finite horizontal extent, for that matter — with two columns, one clear and one cloudy [for example, see the discussion in the context of brown dwarfs in *Marley et al.* (2010)]. Assumptions about the vertical overlap of clouds are critical and can induce large errors in top-of-atmosphere fluxes as well as heating rates (e.g., *Barker et al.,* 1999).

In the context of extrasolar planet atmosphere modeling we must connect a simple, usually one-dimensional, model of the atmosphere with what is potentially a complex brew of cloud properties that we would ideally need to know. However, it is clear that the number of variables can quickly grow to an unmanageable extent. In this chapter we discuss various approaches that have been taken to address this problem.

4.1. Conceptual Framework for Cloud Modeling

Given a profile of atmospheric temperature as a function of elevation or pressure we can ask where clouds form and what are their radiative properties. Here we give a simplified discussion of the problem following *Sánchez-Lavega et al.* (2004).

The abundance of a given atmospheric species a in the vapor can be given by its mass mixing ratio $m_a = \rho_a/\rho_g = \varepsilon P_a/P$ where ρ_g and P are the density and pressure of the "background" atmosphere (i.e., not including species a) and ρ_a and P_a are the density and partial pressure of vapor species a. Here $\varepsilon = \mu_a/\mu$ or the ratio of the molecular weight of species a, μ_a, to the background gas mean molecular weight μ. A not-unreasonable assumption is that any vapor in excess of the saturation vapor pressure of a, $P_{v,a}(T)$, condenses out. We can define the saturation ratio of a as $f_a = P_a/P_{v,a}(T)$.

Deep in the atmosphere, below cloud base, $P_a \ll P_{v,a}$ so the species is found in the gas phase. In a Lagrangian framework one can imagine a rising parcel of gas that cools adiabatically as it expands, and at some point it may reach $P_a = P_{v,a}$ and a saturation ratio $f_a = 1$. In a Eulerian framework this condition requires that the thermal profile intersects with the vapor pressure curve for a given constituent, for instance, as the 400-K model intersects the H_2O vapor pressure curve in Fig. 3. However, the thermal profile for the 1300-K model does not intersect the H_2O vapor pressure curve. Thus H_2O would be expected to condense in the cooler, but not the warmer, atmosphere. To this point the problem is relatively straightforward for those species that condense directly from the gas phase, such as H_2O. For other species, however, condensation instead proceeds by a net chemical reaction when the condensed phase has a lower Gibb's free energy than the vapor phase. One such example is $H_2S + 2Na \rightleftharpoons Na_2S(s) + H_2$.

Above the condensation level a number of issues arise. First, the assumption that all vapor in excess of saturation condenses may be faulty. Condensation may require a supersaturation ($f_a > 1$) before it proceeds, owing to the thermodynamic energy barrier of forming new particles. If so, what degree of supersaturation is required? Above cloud base, which is the lowest level at which condensation occurs, abundance of the condensed phase will depend on a balance between the convective mixing of particles from below and the downward sedimentation of the condensate particles. If the sedimentation rate of some portion of the size distribution of condensate is greater than the vertical mixing velocity scale, the scale height for the condensed phase, H_a, will be smaller than the atmospheric scale height H evaluated at cloud base. In the solar system typical values of H_a/H range from 0.05 to 0.20 (*Sánchez-Lavega et al.,* 2004).

Nucleation is the starting process for the formation of a condensed phase (either liquid or solid) from a gaseous state, or the formation of solid from a liquid state. It creates an initial distribution of nuclei (embryos) that, if large enough, are stable with respect to the higher-entropy phase and tend to grow into larger particles by condensation or freezing. Such a phase transition can only occur spontaneously under thermodynamically favorable conditions (for the following discussion we first focus on the process of condensate nucleation from the vapor phase). Such conditions require the saturation ratio f_a to exceed unity ($f_a = 1$ characterizes phase equilibrium).

The simplest nucleation mechanism is homogeneous nucleation, where the initial nuclei are formed by random spontaneous collisions of monomers in the vapor phase (e.g., single H_2O molecules) into a molecular cluster. This process is, however, connected to an energy barrier that can prevent the formation of a stable new phase. A supersaturated gas phase possesses a high Gibb's free energy, whereas at the same time molecules in the condensed phase would be at a lower potential energy. Thus, removing molecules from the vapor phase and adding them into a condensed phase would in principle lower the total free energy of the entire thermodynamical system. The corresponding net change in free energy depends mainly on the volume of an embryo (as a function of the particle radius r) and the supersaturation of the gas.

However, the creation and growth of nuclei also implies that a new surface is generated that additionally contributes to the free energy owing to surface tension. It is only thermodynamically favorable to form a condensed phase when the net reduction of free energy in the system ($\sim r^3$) is greater than (or equal to) the free energy required from the surface tension ($\sim r^2$). Therefore there exists a critical radius for which these two contributions balance each other. Embryos smaller than the critical size are unstable and will tend to evaporate again quickly, whereas those with sizes larger than the critical radius will tend to grow freely. The critical radius is, in particular, a function of the saturation ratio. Low supersaturations yield very large critical radii while, on the other hand, for increasing values of f_a the size of the critical embryos decreases.

Homogeneous nucleation usually requires a very high supersaturation and is therefore unlikely to occur in atmospheres of terrestrial planets on a large scale (observed supersaturations for water vapor in the atmosphere of Earth are normally in the range of a few percent, for example), because heterogeneous nucleation occurs at much lower supersaturations and thus quenches the supersaturation well before homogeneous nucleation occurs.

The predominant nucleation process from the vapor for terrestrial planetary atmospheres is thought to be heterogeneous nucleation. Here, the initial distribution of nuclei is formed on preexisting surfaces. The presence of such surfaces substantially lowers the supersaturation required to form the critical clusters. Depending on the properties of these surfaces (e.g., whether they are soluble with respect to the condensed phase), nucleation is already possible for saturation ratios close to unity. Such surfaces can be provided by dust, sea salt, pollen, or even bacteria. In case of terrestrial planets these particles (condensation nuclei) are usually largely available because of mechanical process associated with the planetary surface — such as wave breaking, bubble bursting, and dust saltation — and from the formation of haze particles (such as sulfuric acid or sulfate droplets) that result from photochemical processes. Therefore heterogeneous nucleation can reasonably be expected as the dominant nucleation mechanism for terrestrial extrasolar planets. This assumption, however, complicates the treatment of cloud formation, because details on such condensation nuclei (composition, number density, and size distribution) are not available in the case of exoplanets.

Additionally, the formation of a solid phase (e.g., ice crystals) can in principle occur by different pathways. It can form directly from the gas phase by homogeneous or heterogeneous nucleation. On the other hand, it is also possible to form the liquid phase as an intermediate step, followed by freezing of the supercooled liquid into the solid phase afterward. Whether this indirect or direct pathway occurs depends on the properties of the condensing species and on the local atmospheric conditions. Water ice cloud crystals in Earth's atmosphere form by homogeneous and heterogeneous freezing of liquid in mixed-phase clouds (such as cumulonimbus and Arctic stratiform clouds) and heterogeneous nucleation in cirrus clouds. In the martian atmosphere, CO_2 clouds form directly by heterogeneous nucleation from the gas phase.

Given a composition of the cloud, the task for any cloud model then becomes one of computing cloud particle sizes and their number distribution through the atmosphere above the cloud base. A very simple solution would be to simply assume a mean particle size and a cloud scale height, and this is effectively the approach many investigations have taken to explore the effect of exoplanet clouds. For the remainder of this section we consider efforts to more rigorously derive expected particle sizes and the vertical distribution of cloud particles. We begin by considering the most thoroughly modeled clouds, terrestrial clouds of liquid water and water ice. We then move on to cloud models that have been constructed for terrestrial and giant extrasolar planets and conclude by considering the lessons learned from efforts to model clouds expected in brown dwarf atmospheres.

4.2. Perspective from Earth Science

Cloud modeling of terrestrial clouds comes in a wide assortment of classes. For detailed cloud studies of limited spatial extent, dynamical frameworks range from zero-dimensional parcel models, to one-dimensional column models, to two-dimensional eddy-resolving models, to three-dimensional large-eddy simulations and cloud-resolving models. The difference between such models is the number of spatial dimensions represented in the governing equations. Another varying aspect of cloud models is the degree of detail in describing cloud microphysics. The simplest models simply assume that all vapor in excess of saturation condenses and assume a fixed size for the cloud particles. Others assume a functional shape for the cloud particle size distributions and parameterize the microphysical processes, prognosing one, two, or three moments of the size distribution for each condensate species (such as cloud water, rain, cloud ice, snow, graupel, and hail). The most detailed approach is to resolve the cloud particle size distributions without making any assumptions about the functional shape of the size distributions.

Global climate model frameworks similarly range from one-dimensional radiative convective models (e.g., *Manabe and Wetherald,* 1967), to two-dimensional, zonally averaged models (e.g., *Schneider,* 1972), to modern three-dimensional general circulation models (GCMs). [See *Schneider and Dickinson* (1974) for an early, comprehensive review of approaches to global climate modeling.] With respect to clouds in global models, one-dimensional radiative convective models suffer from a major deficiency, namely, predicting plausible estimates of horizontal cloud coverage (see *Ramanathan and Coakley,* 1978). A two-dimensional framework is an intermediate step, although modern climate studies rely principally on three-dimensional GCMs. The rest of this section will focus on clouds in modern GCMs.

The representation of clouds in GCMs remains a major challenge, as cloud feedbacks constitute a leading source of uncertainty in current model-derived estimates of overall climate sensitivity, which are typically cast in terms of the sensitivity of globally averaged surface temperature to changes in radiative forcings (see *Hansen et al.,* 1984). The response of tropical cirrus clouds to increasing sea surface temperatures has been a topic of great debate in the last two decades. At one extreme is the thermostat hypothesis of *Ramanathan and Collins* (1991), which suggests that increased water vapor leads to more extensive, thicker anvils that will on net cool the planet through increased albedo, while at the other extreme is the argument of *Lindzen et al.* (2001) that greater condensate loading in a warmer climate precipitates more efficiently and leads to a dryer upper atmosphere that traps less infrared energy, driving a negative water vapor feedback. Both hypotheses produce a negative climate feedback for tropical clouds, one employing increased solar reflection, the other relying on increased infrared emission. While these are provocative ideas that have spawned countless evaluations of the hypothesis (with neither withstanding scrutiny), the predominant concept currently is that the tropical climate feedback for modern GCMs is determined primarily by the response of clouds in the marine boundary layer (*Bony and Dufresne,* 2005), of which the transition from overcast stratocumulus to broken cloud fields of trade cumulus is a leading primary candidate responsible for that response. With respect to cloud feedbacks in general, attention on the climate feedback of cirrus clouds formed from the detrainment of deep tropical convection has been supplanted to a large degree by more recent focus on the climate feedback of shallow clouds.

A fundamental problem in representing clouds in the GCMs is that the native GCM grid cells are very coarse, on the order of 100 km horizontally and 1 km vertically. Although model resolution steadily improves as computing power advances, the problem of convection and clouds being unresolved persists in models designed to span the globe and simulate climate over timescales of decades to millennia. The basic assumption in a conventional treatment of convection and clouds in a GCM is that cloud properties and precipitation rates in a model grid cell, which is much larger horizontally than any cloud, can be computed based on the mean properties of the grid cell. The purpose of a cloud parameterization is to compute cloud properties and precipitation rates from those mean properties.

A somewhat recent version of the NASA Goddard Institute for Space Studies (GISS) GCM (*Schmidt et al.,* 2006) includes a conventional treatment of clouds and convection. The atmospheric model grid spacing is roughly 2 × 2.5° with 20 or 23 vertical layers, and thus all convection and cloud physics are necessarily highly parameterized. Deep convection is parameterized based on the convective instability of model columns using idealized updrafts and downdrafts, which detrain air into stratiform cloud layers. The stratiform cloud cover is computed as a diagnostic function of grid-scale relative humidity, and the relative humidity thresholds used to compute cloud cover in that diagnostic function serve as the principle tuning knobs for the GCM, with the dual-tuning targets of top-of-atmosphere radiative balance and an overall albedo reasonably close to satellite-based estimates (note that such tuning can easily result in exaggerated cloudiness in some regions that make up for insufficient cloudiness in others). In the GISS GCM the only prognostic cloud variable is the mass mixing ratio of cloud condensate, which is a fundamental component of the cloud parameterization. Any precipitation is assumed to evaporate or fall out in one time step (30 min) and the phase of the condensate is probabilistically determined from temperature to allow for a modest amount of supercooling (liquid colder than the melting point) on average. The standard version of the GCM assumes different cloud droplet concentrations over ocean and land and also assumes a fixed number concentration for ice particles, and a fixed effective variance of the condensate size distributions is assumed for computing cloud optical thickness.

A more complex approach for stratiform cloud microphysics is used in the most recent version of the National Center for Atmospheric Research (NCAR) GCM. That scheme uses two moments (mass and number) for two prognostic (cloud water and cloud ice) and two diagnostic (rain and snow) hydrometeor species (*Morrison and Gettelman,* 2009). The rapidly sedimenting species are treated diagnostically with a tridiagonal solver to allow for long time steps (20 min with 2 microphysics substeps) in a manner that avoids numerical instability associated with falling through more than one layer during a time step. A novel aspect of this microphysics scheme is that by assuming a particular subgrid-scale distribution of cloud water, microphysical processes that involve cloud water take into account the problem of grid averaging over nonlinear process rates (*Pincus and Klein,* 2000). Taking into account joint subgrid-scale distributions of just two moments for two species complicates the math considerably (*Larson and Griffin,* 2012).

An alternative to parameterizing convection within grid scales O(100 km) is the multiscale modeling framework (*Randall et al.,* 2003), in which two-dimensional cloud-resolving models are embedded within each GCM column. While some convection-related aspects of the global circulation are treated well by such an approach, it shares some deficiencies with the more traditional approach to GCM cloud parameterization, namely the poor representation of the pervasive and climatologically important regime of shallow marine convection (e.g., *Marchand and Ackerman,* 2010). [Explicit resolution of such clouds requires horizontal resolution O(100 m) and vertical resolution O(10 m).] Avoiding issues related to embedding two-dimensional slices within GCM columns (problems including how to orient the slices and shortcomings including the use of periodic lateral boundary conditions for each embedded two-dimensional submodel), the most expensive, yet perhaps straightforward, approach to climate modeling is the Earth Simulator, a global cloud-resolving model with simulations run on a 3.5-km horizontal grid (*Satoh et al.,* 2008). The computational demands of such an approach are vast, with on the order of 10^{19} grid cells in such a model. Even such a brute-force approach falls far short of the grid resolution required to explicitly simulate shallow marine convection, however. It is safe to say that even on the most powerful computing platforms that global simulations of Earth will be saddled with cloud and convective parameterizations for the foreseeable future.

4.3. Exoplanet Clouds

In extreme contrast to the situation for Earth outlined above, there are currently no observational constraints for atmospheres of terrestrial exoplanets that would provide information about what kind of clouds may have to be considered for a particular exoplanet. Available observables are confined to basic planetary parameters, such as radius and orbital inclination (if a transit event can be observed), planetary mass (from the radial velocity method), orbital eccentricity and distance, and additionally the type of the central star. Consequently, we are faced with the difficult problem of modeling clouds in an environment without actually knowing any further details.

Cloud formation can only be treated theoretically in compliance with the chemical composition of the atmosphere, because the atmosphere determines the condensing species forming a cloud. Considering how diverse atmospheres of terrestrial exoplanets can be expected to be, the self-consistent modeling of cloud formation in such atmospheres without observational constraints or theoretical predictions is somewhat ambiguous. The composition of a terrestrial planet cannot be easily deduced from simple theoretical arguments. In the case of a planet that has lost its primordial atmosphere, the atmospheric composition is determined by the combination of the outgassed chemical species from the planet's interior and the volatiles delivered by impacts of asteroids and comets and will therefore depend on the planet's mantle composition, the physical processes in the planetary interior, and the composition and sizes of impactors. Another factor with a huge impact on the atmospheric composition is also the possible existence of a biosphere that interacts chemically with the atmosphere.

The long-term evolution of the atmospheric composition, such as that resulting from the carbonate-silicate cycle as on Earth, depends also on the occurrence of plate tectonics. To date, the only known planet with plate tectonics is Earth. It is currently not quite well understood under which conditions a planetary crust will start plate tectonics and how this process is maintained over an extended period of time. This is especially true for more massive terrestrial planets like super-Earths, where there is much controversy regarding plate tectonics (see, e.g., *Valencia et al.,* 2007; *O'Neill et al.,* 2007).

Other important processes determining the atmospheric composition are the escape mechanisms of atmospheric gas to space. While thermal escape is a function of the atmospheric temperature, planetary mass, and molecular weight, the nonthermal escape processes (erosion of the atmosphere by a stellar wind, for example) are much more complicated (e.g., *Lammer et al.,* 2008; Tian et al., this volume). They not only depend on the activity of the central star — which itself is a function of the stellar type and its particular stellar evolution — but also on the possible existence of a planetary magnetic field (linked to the rotation rate and interior dynamics of the planet), which can protect the planetary atmosphere against loss processes.

In contrast to a low-mass planet like Earth, a more massive terrestrial super-Earth might retain a part of its primordial hydrogen-dominated atmosphere, partly enriched by volcanic outgassing or additional external delivery. Such atmospheres may be vastly different from those known within our solar system. A discussion of possible atmospheric compositions can be found in *Seager and Deming* (2010).

The secondary atmosphere of an Earth-like planet (Earth-like with respect to the chemical composition of the planetary mantle) would most likely be rich in H_2O and CO_2 (*Schaefer et al.,* 2012). Therefore, clouds composed of these species are of prime interest for habitable Earth-like planets. Cooler atmospheres can also contain significant amounts of CH_4 and NH_3, or SO_2 in the case of high temperatures (*Schaefer et al.,* 2012).

In this environment effective cloud models must match the problem at hand. For example, highly sophisticated terrestrial cloud microphysics and dynamics models are not required in order to ascertain the range of plausible albedos for a hypothetical Earth-like terrestrial planet. However, more sophisticated approaches than simple *ad hoc* models may be needed to interpret the colors of a directly imaged planet.

4.4. Cloud Models for Terrestrial Exoplanets

In principle the basic mathematical description of cloud microphysics in atmospheres of terrestrial exoplanets do not deviate from their solar system counterparts. The temporal and spatial evolution of the cloud particle size distribution can be described by means of a master equation ("general dynamic equation") incorporating all relevant gain and loss processes. This includes nucleation, evaporation, sedimentation, coagulation/coalescence, diffusion, or hydrodynamical transport (see *Pruppacher and Klett,* 1997; *Lamb and Verlinde,* 2011). While the numerical solution of the master equation is still quite challenging, it can be efficiently performed by the methods summarized in *Williams and Loyalka* (1991) or by applying more advanced methods [e.g., continuous and discontinuous Galerkin methods by *Sandu and Borden* (2003)].

Self-consistent modeling of the formation and temporal and spatial evolution of clouds is an unsolved problem. An ideal treatment would require a thorough knowledge of the state of the atmosphere, including its composition, the spatial distribution of chemical species, atmospheric temperature and dynamics, and the size distribution and density of potential cloud condensation nuclei and heterogeneous freezing nuclei. However, even in the terrestrial atmosphere the formation of ice at temperatures warmer than the homogeneous freezing temperature for water (about 233 K for typical drop sizes) is not well understood (e.g., *Fridlind et al.,* 2007, 2012); far less, if anything, is known about heterogeneous freezing and condensation nuclei in other atmospheres and would hardly be detectable by remote observations of terrestrial extrasolar planets. Lack of laboratory data to derive the necessary microphysical rates under atmospheric conditions more "exotic" than found in the solar system further complicates the problem of describing cloud formation in atmospheres of exoplanets.

Additionally, many atmospheric models for exoplanets are restricted to one spatial dimension and are often considered to be stationary, which makes a detailed description of cloud microphysics very difficult. While the climatic effects of clouds can be approximately treated in a one-dimensional model atmosphere, a consistent modelling of cloud formation would, in principle, require a three-dimensional dynamical atmospheric model as described in section 4.2. In comparison to one-dimensional models, however, three-dimensional GCMs contain many more free parameters and are very computationally intensive. Properties such as the surface orography, the distribution of surface types (fractions of oceans and land mass), and the local distribution of chemical species (affected by volcanism, for example) play a major role for the dynamics and chemistry of the atmosphere (*Joshi,* 2003) and directly affect the formation and evolution of clouds. Since none of these detailed properties are known for a terrestrial exoplanet, many additional assumptions about the planet and its atmosphere have to enter the calculations. On the other hand, three-dimensional models are the only opportunity to obtain information about the possible temporal variation and fractional coverage of clouds and their distribution throughout the atmosphere. Such results might be required to analyze transmission spectra of terrestrial exoplanetary atmospheres containing patchy clouds.

Given the aforementioned challenges due to the lack of observational constraints, clouds in atmospheres of terrestrial exoplanets are usually treated in a simplified way. The simplest method to account for the effects of clouds in an atmospheric model is a modified surface albedo. This approach has been widely used in the past (e.g., *Kasting et al.,* 1993; *Segura et al.,* 2003, 2005; *Grenfell et al.,* 2007). The surface albedo of these kind of models is modified to yield a specified surface temperature for a given reference scenario. For example, a common reference scenario is an Earth-like planet around the Sun at a known orbital distance of 1 AU. The surface albedo is then adjusted to obtain the mean surface temperature of Earth (288 K), thereby mimicking the climatic effects of clouds. This adjusted surface albedo value is then used in all subsequent model calculations, assuming that the net effect of the clouds is invariant from changes of the atmospheric conditions, or type of central star. This approach makes no assumptions about the physical nature of the clouds, or their composition, size, or optical properties, but instead estimates their effects based on the original tuning to surface temperature. While a modified surface albedo can crudely describe the climatic effect of clouds, it cannot simulate their impact on the planetary spectra.

A more detailed approach is to consider model scenarios where the properties of clouds can be assumed to be approximately known. For a completely Earth-like planet, one could expect that the properties of clouds in such an atmosphere would closely resemble those found on Earth. This approach was used by *Kitzmann et al.* (2010) to study the impact of mean Earth water clouds in the atmospheres of Earth-like exoplanets. In contrast to a modified surface albedo, the wavelength-dependent optical properties of the cloud particles are explicitly taken into account to study their effects for different incident stellar spectra and other parameters. Thus, in addition to the influence on the atmospheric and surface temperatures, their impact on the planetary spectra can also be studied in detail using this modeling approach (e.g., *Kaltenegger et al.,* 2007). If the properties of the clouds are not known (e.g., for a CO_2 cloud in a thick CO_2 dominated atmosphere of a super-Earth), parameter studies can be performed, varying the cloud properties over a wide range to estimate the possible effect of clouds. This approach has been used by *Forget and Pierrehumbert* (1997) for CO_2 ice clouds in an early martian atmosphere, for example.

More detailed treatments of cloud formation in exoplanetary atmospheres include simplified air parcel and vertically resolved one-dimensional cloud models based on those originally developed for Earth's atmosphere. Such models can be used to determine mean cloud properties under different atmospheric conditions [see *Neubauer et al.* (2011, 2012) for several cloud species (e.g., H_2O and H_2SO_4) or *Zsom et al.* (2012) for water (droplet and ice) clouds, for example]. However, these more detailed descriptions of cloud formation already need to include many additional assumptions, such as the distribution of cloud condensation nuclei, which strongly influence the resulting cloud properties.

While cloud models for terrestrial exoplanets so far lack many of the sophisticated and detailed cloud microphysics needed to reproduce the complicated cloud structures known from Earth observations, they nonetheless offer an important first-order estimate of cloud effects in exoplanetary atmospheres. One of the largest uncertainties for one-dimensional models is the treatment of fractional clouds. Unless the atmosphere is globally supersaturated, thus resulting in a completely cloud-covered planet, the fraction of the atmosphere where clouds are present has to be introduced as a free parameter. Thus, one-dimensional models are incapable of organically describing planets with fractional cloud cover.

Although one-dimensional models are commonly employed for many terrestrial exoplanet applications, three-dimensional models have also been used. For the terrestrial super-Earth Gliese 581d, *Wordsworth et al.* (2011) adapted a Mars global circulation model that included a simplified treatment of CO_2 ice cloud formation. This microphysical model assumes a certain size and number density of cloud condensation nuclei and equally distributes the condensable material among them, accounting also for the sedimentation and hydrodynamical transport of the cloud particles within the atmosphere. The corresponding formation of CO_2 clouds lead to an increase of the surface temperature of Gliese 581d in their model calculations. The same approach was also used by *Wordsworth et al.* (2010) in a one-dimensional atmospheric model for the same planet.

4.5. Giant Planet Cloud Models

Unlike the case for Earth-like planets where water clouds are the greatest concern, a wide variety of species may condense in the hydrogen-helium-dominated atmospheres of giant planets. *Sánchez-Lavega et al.* (2004) review the standard framework for cloud formation in giant planets. Homogeneous condensation occurs when the partial pressure of a species in the gas phase exceeds its saturation vapor pressure at a given temperature in the atmosphere. Sánchez-Lavega et al. tabulate the vapor pressures for many relevant species. Other expressions for additional species can be found in *Ackerman and Marley* (2001) and *Morley et al.* (2012). Curves tracing the set of pressure and temperature conditions at which a given species condenses assuming equilibrium chemistry ("condensation equilibrium curves") are shown for many species in Fig. 3 and a schematic illustration of the resulting cloud decks is shown in Fig. 4.

Once a cloud layer forms, the condensate is removed from the overlying atmosphere and thus is no longer available to react at lower temperatures higher in the atmosphere. Thus the calculation of the chemical equilibrium state for the atmosphere must account for the presence of the cloud. Such a "condensation chemistry" is distinct from equilibrium chemistry calculations in which the condensate remains in communication with the gas phase and is available for reaction at lower temperatures. Condensation chemistry is discussed in detail by *Fegley and Lodders* (1996), *Lodders and Fegley* (2002), and *Visscher et al.* (2006) as well as by *Burrows and Sharp* (1999) in the context of brown dwarf models and *Sudarsky et al.* (2003) for extrasolar giant planets. The schematic shown in Fig. 4 accounts for condensation chemistry. If Fe were not sequestered into a deep cloud layer in Jupiter's atmosphere, the Fe would react with gaseous H_2S to form FeS, thus removing H_2S from the observable atmosphere, in contradiction to observations (*Fegley and Lodders,* 1996).

Once the cloud base pressure is found, the challenge is to describe the variation in cloud particle sizes and number densities above this level. Early attempts to develop cloud models for use in giant solar system atmospheres included the work of *Lewis* (1969), *Rossow* (1978), and *Carlson et al.* (1988). These and other early works are reviewed by *Ackerman and Marley* (2001). In this subsection we focus on more recent modeling approaches that are in use today, in particular the cloud models of Ackerman and Marley and of Helling and collaborators.

4.5.1. Ackerman and Marley. Iron and silicates condense in the atmospheres of warm brown dwarfs (Fig. 3), and these clouds must be accounted for in models of brown dwarf emergent spectra. Early modeling attempts for such clouds simply computed the mass of dust that would be found in chemical equilibrium for a given assumed initial abundance of gas (as if the gas were isolated at a given pressure and temperature from the rest of the atmosphere). The lower the atmospheric temperature, the more dust that would be present. Atmospheric models following such a prescription [such as the "DUSTY" models of *Allard et al.* (2001)] adequately reproduced the near-infrared colors of the warmest brown dwarfs but predicted far too great of a dust load in cooler objects. Thus it was apparent that an accounting for sedimentation of grain particles was required. One approach used in the literature was to set a variable "critical" temperature for a given cloud such that cloud particles would only be found between cloud base and the specified temperature (*Tsuji,* 2002). Another approach was to limit the cloud to be confined within an arbitrary distance, usually one scale-height, of the cloud base. Both approaches required the choice of an arbitrary particle size for the grains. The advantage of such approaches is that they are computationally very tractable for modeling and thus allow the exploration of a large parameter space. One disadvantage is that it is difficult to consider particle size effects and other complexities.

In order to allow for vertically varying particle number densities and sizes, a second approach was suggested by *Ackerman and Marley* (2001). In their formulation downward transport of particles by sedimentation is balanced by upward mixing of vapor and condensate (either solid grains or liquid drops)

$$K_{zz}\frac{\partial q_t}{\partial z} + f_{sed}w^*q_c = 0 \qquad (1)$$

where K_{zz} is the vertical eddy diffusion coefficient, q_t is the mixing ratio of condensate plus vapor, q_c is the mixing ratio of condensate, w^* is the convective velocity scale, and f_{sed} is a dimensionless parameter that describes the efficiency of sedimentation. Together with a constraint on maximum supersaturation as well as assumptions regarding particle size distributions constrained by f_{sed}, w^*, and q_c (as described by Ackerman and Marley) the solution of this equation allows computation of a self-consistent variation in condensate number density and particle size with altitude above an arbitrary cloud base.

In their model the cloud base is found by determining at which point in the atmosphere the local gas abundance exceeds the local condensate saturation vapor pressure $P_{v,a}$ at which point the atmosphere becomes saturated. In cases where the formation of condensates does not proceed by homogeneous condensation, an equivalent vapor pressure curve is computed, as described by *Morley et al.* (2012).

The Ackerman and Marley cloud model has the advantage of not requiring knowledge of microphysical processes to compute particle sizes. For a given sedimentation efficiency, clouds are simply assumed to have grown large enough to provide the required downward mass flux that balances equation (1). Since the solution is numerically rapid, a large number of models can be computed and compared with data within a tractable amount of time. Sample model temperature-pressure profiles along with equilibrium condensation curves are illustrated in Fig. 3.

Considering the simplicity of this approach, the model has fared fairly well in comparisons with data. *Stephens et al.* (2009), for example, compared model spectra for L and T dwarfs to a large database of near- to mid-infrared spectra. They found that cloudy L dwarfs can generally be well fit by clouds computed with f_{sed} = 1 to 2, while early

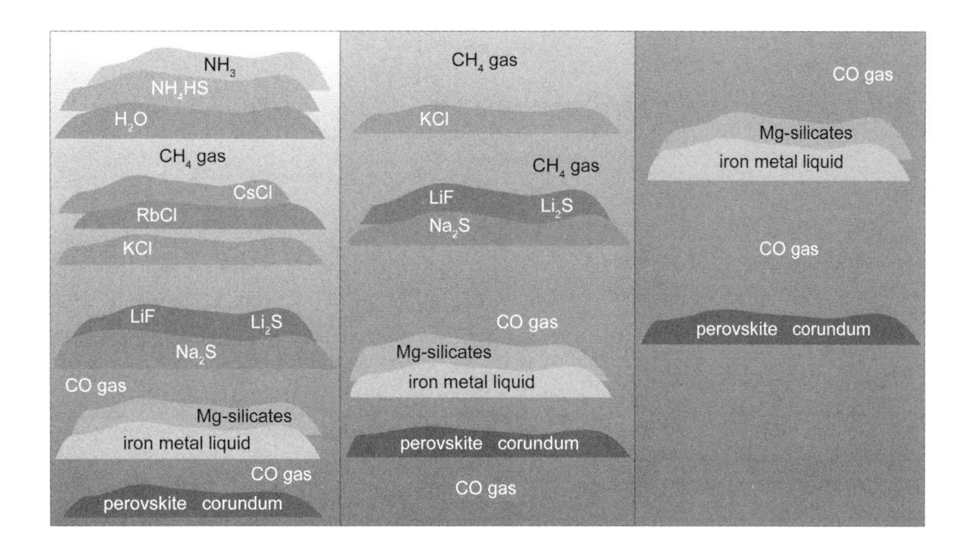

Fig. 4. Schematic illustration of cloud layers expected in extrasolar planet atmospheres based on consideration of equilibrium chemistry in the presence of precipitation. The three panels correspond roughly to effective temperatures T_{eff} of approximately 120 K (Jupiter-like, left), to 600 K (middle), to 1300 K (right). Note that with falling atmospheric temperature the more refractory clouds form at progressively greater depth in the atmosphere and new clouds composed of more volatile species form near the top of the atmosphere. Figure courtesy of K. Lodders.

T dwarfs, which exhibit thinner clouds, are better fit with f_{sed} = 3 to 4. The model thus provides a framework for describing mean global clouds in a one-dimensional sense, but the model lacks the ability to explain why the sedimentation efficiency might change at effective temperatures around 1200 K, where the near-infrared spectra of L dwarfs evolves over a small temperature range.

Marley et al. (2010) considered the effect of partial cloudiness on L dwarf spectra computed with the *Ackerman and Marley* (2001) cloud model. Their method assumed that clear and cloudy columns of atmosphere had the same temperature profile and together emitted the flux corresponding to a specified effective temperature. They found that partially cloudy L dwarfs would have emergent spectra comparable to standard models with homogeneous cloud cover but with larger values of f_{sed}. Thus a dwarf with 50% clear skies and 50% cloudy skies with f_{sed} = 2 ends up with a model spectrum comparable to that of a homogeneous cloud cover with f_{sed} = 4.

4.5.2. Helling and collaborators. The most extensive body of work on cloud formation in giant exoplanet and brown dwarf atmospheres has been undertaken by Helling and her collaborators (*Helling and Woitke,* 2006; *Helling et al.,* 2008; *Witte et al.,* 2009, 2011; *de Kok et al.,* 2011), who follow the trajectory of seed particles from the top of their model atmospheres as they sink downward. The seeds grow and accrete condensate material as they fall, resulting in "dirty" or compositionally layered grains. This work extends the dust-moment method of *Gail et al.* (1984) and *Gail and Sedlmayr* (1988). It accounts for the microphysics of grain growth given these conditions and available relevant laboratory data. Particle nucleation is explicitly computed, taking into account barriers to grain formation.

Because condensation is envisioned in this framework to proceed downward from the top of the atmosphere rather than upward from the deep atmosphere, and because grains are allowed to interact with the surrounding gas, the cloud composition predicted by the Helling approach differs substantially from that employed in the Ackerman and Marley framework. An example is shown in Fig. 5 for a model brown dwarf (log g = 5 in CGS units, T_{eff} = 1600 K). Here *Helling et al.* (2008) predict that in addition to the usual Fe, $MgSiO_3$,and Mg_2SiO_4 cloud layers, additional condensates including SiO_2 and MgO will form. These latter species are not predicted by equilibrium condensation for a cooling gas mixed upward from higher temperature and pressure conditions The TiO_2 cloud seeds arising at the model at the top of the atmosphere in this approach is also evident.

In fact, the presence of these initial TiO_2 seed particles at the top of the model atmosphere in the Helling framework deserves some discussion. In the equilibrium chemistry condensation framework, Ti-bearing condensates (e.g., $CaTiO_3$) would form much deeper in the atmosphere as a gas parcel rises vertically and cools. Precipitation of such particles would remove the condensate from the gas phase and notable amounts of refractory TiO_2 seed particles would not be expected to arrive at the top of the atmosphere. Conversely, in the Helling conceptual framework, the formation of $CaTiO_3$ at the expected equilibrium cloud base is thought to be kinetically inhibited (since multiple collisions of molecules would be required), resulting in the refractory TiO_2 clusters being mixed further upward to ultimately seed condensa-

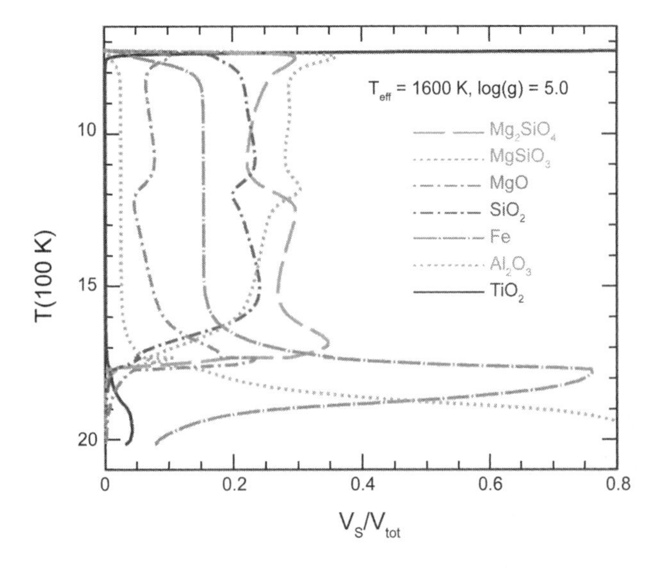

Fig. 5. See Plate 67 for color version. Composition of atmospheric cloud layers for a T_{eff} = 1600 K, log(g) = 5 brown dwarf as computed by the dust model of Helling and collaborators (*Helling et al.,* 2008). The vertical axis is atmospheric temperature with the top of the atmosphere to the top of the figure. The horizontal axis gives the relative volumes V_s of each dust species indicated by line labels as a ratio to the total dust volume V_{tot}. Unlike the Ackerman and Marley condensation equilibrium approach, this model predicts that MgO and SiO_2 are important condensates along with the TiO_2 seed particles that are formed at the top of the model.

tion in the cooler upper reaches of the atmosphere. Thus the Helling approach fundamentally assumes that vertical mixing timescales, at least in localized columns, are faster than condensation timescales. As the seeds eventually fall from the top of the atmosphere, they then encounter other condensable molecules that are likewise assumed to not have been cold trapped below and the seeds then accrete these species and grow. The model iterates to find a self-consistent solution. This top-down approach thus differs from most of the other cloud modeling approaches discussed in the literature, which generally conceive of a condensation sequence operating from the depths of the atmosphere upward with species sequentially condensing, as conceptually shown in Fig. 4. Atmospheric mixing by breaking gravity waves (*Freytag et al.,* 2010) might provide a mechanism to stir the atmosphere sufficiently to deliver the seed particles.

Because of the complexity of the computational approach required to compute cloud properties in this framework, there have been fewer comparisons between model spectra computed with the Helling clouds and data than has been the case with the Ackerman and Marley cloud model. Some direct comparisons between different cloud models are shown by *Helling et al.* (2008). Ultimately, only a thorough comparison of the predictions of all cloud modeling schools and data will be required to establish which conceptual framework is a better approximation over which ranges of conditions. An application of the Helling framework to the clouds of Jupiter would be enlightening.

4.5.3. Other approaches. More simplified approaches have also been taken for cloud models, such as specifying particle sizes and cloud heights. *Sudarsky et al.* (2000, 2003) computed model exoplanet albedo spectra given various particle size and cloud height assumptions. In particular, they utilized the *Deirmendjian* (1964) size distribution and explored the effects of changing mean cloud particle sizes and widths of the size distribution. *Sudarsky et al.* (2000) also considered the effects of various plausible photochemical hazes on giant planet albedo spectra.

Cooper et al. (2003) employed the timescale comparison framework pioneered by *Rossow* (1978) to compute cloud models for brown dwarfs and extrasolar giant planets. In this approach various timescales for particle nucleation, growth, and sedimentation are compared to derive expected condensate particle sizes. As discussed by *Ackerman and Marley* (2001), a difficulty with the Rossow method is that some of the critical timescales depend upon unknown factors, particularly the assumed supersaturation. Nevertheless, the *Cooper et al.* (2003) model provides a useful survey of likely particle sizes for species of interest expected under various combinations of gravity and effective temperature. For example, in agreement with most of the other cloud models, Cooper et al. predict typical silicate grain sizes in the range of 10–200 μm.

Tsuji and collaborators (*Tsuji,* 2002, 2005; *Tsuji et al.,* 2004) have computed brown dwarf models by specifying cloud-top and cloud-base temperatures. For the directly imaged planets, *Currie et al.* (2011), *Madhusudhan et al.* (2011), and *Bowler et al.* (2010) employ a variety of approaches to specify cloud properties and explore parameter space. Approaches such as these offer the advantage of quickly exploring the phase space of possible models and establishing the effect of plausible cloud models on spectra. The lack of physical complexity in such models is offset by their useful ability to offer qualitative understanding of the effect of various condensate properties.

4.6. Lessons Learned from Cloudy Brown Dwarfs

Brown dwarfs — hydrogen-helium-rich objects with masses between about 12× and 80× that of Jupiter (M_J) — have been a proving ground for understanding the role of clouds in exotic atmospheres. This is because the class of brown dwarfs known as the L dwarfs have atmospheric temperatures in the regime in which iron and silicate grains condense from the gas phase. It is apparent from the available data that these refractory condensates do not form a pall of particles mixed through the entire atmosphere, but rather form discrete cloud layers. As such, L dwarfs were the first objects outside the solar system for which a detailed description of clouds was required in order to interpret their emitted spectra.

There have been several comparisons of cloudy brown dwarf atmosphere models to observations. The most extensive to date are presented by *Cushing et al.* (2008) and *Stephens et al.* (2009). These authors compared atmosphere models of *Saumon and Marley* (2008) computed using the *Ackerman and Marley* (2001) cloud model to a variety of L- and T-type brown dwarf spectra from 0.8 to 15 μm. For the L dwarfs and early T dwarfs the cloudy models clearly did a better job reproducing the data than cloudless models. The tunable f_{sed} parameter, with typical values between 1 and 3, allowed sufficient dynamic range to generally reproduce most of the observed spectra. Meanwhile, *Witte et al.* (2011) compared model spectra computed with the *Helling et al.* (2008) cloud formulation to spectra of L-type brown dwarfs. Their cloud model likewise shows much better agreement with data than either cloudless or very cloudy models with no dust sedimentation. In all these studies the matches between models and data are very good in many cases, but nevertheless there remain notable spectral mismatches and it is clear that a more sophisticated cloud model would be required to fit all objects.

The most important lesson learned from the campaign to model brown dwarfs may be that large grids of atmosphere models — including a variety of cloud descriptions — are required. Models should not be so complex that the creation of such grids are a challenge. In the next section we review the available exoplanet data relevant to clouds. As we will see, the exoplanet data do not yet require large systematic model grids, but such models will undoubtedly be required as more data become available. Ideally, more models will be available than has been the case for brown dwarfs, thereby permitting more systematic comparisons of various cloud modeling frameworks to large exoplanet datasets.

5. OBSERVATIONS OF EXOPLANET CLOUDS

5.1. Transiting Planets

5.1.1. Transmission spectra. Perhaps the most convincing evidence of high-altitude clouds or hazes on any extrasolar planet to date is found in the case of the transiting planet HD 189733b. This 1.1-M_J planet orbits a bright nearby K star and is thus an excellent target for detailed studies of its atmosphere. This planet is notable because its transit radius — the apparent size of the planet as a function of wavelength — follows a smooth power law as first measured by *Pont et al.* (2008). Signatures of molecular or atomic absorption expected from a clear, solar composition atmosphere are almost absent, although Na is detected at high spectral resolution (*Huitson et al.,* 2012). Figure 6 illustrates the smooth variation in atmospheric transmission as measured from 0.3 through 1 μm (*Pont et al.,* 2008; *Sing et al.,* 2009; *Gibson et al.,* 2011). The red curve presents a pure Rayleigh scattering model while a gas-opacity-only model is also shown. With the possible exception of a spectral feature at 1.5 μm, the smooth variation of planet radius with wavelength extends to at least 2.5 μm (*Gibson et al.,* 2011).

The most natural explanation for the HD 189733b transmission spectrum is that a population of small, high-albedo particles is present in the atmosphere at low pressures (*Lecavelier des Etangs et al.,* 2008). In the terminology of Mie scattering this requires that the scattering efficiency is large compared to the absorption efficiency, or $Q_{abs} \ll Q_{scat}$. Lecavelier des Etangs et al. demonstrate that this in turn requires a material with an imaginary index of refraction that is small compared to the real index, and suggest $MgSiO_3$ as a possible candidate. In the condensation chemistry framework (section 3.4), however, silicates are expected to condense much deeper in the atmosphere, and the mechanism by which small grains could be transported to the upper atmosphere has yet to be fully explored, although vigorous mixing is a likely explanation.

Another transiting planet for which particulate opacity may be important is GJ 1214b. This 6.5-$M_\oplus$ (Earth-mass) planet orbits an M star. Planetary structure models that fit the observed mass and radius of the planet can be found with either a thick hydrogen atmosphere comprising less than 3% of the mass of the planet, or a water-rich planet surrounded by a steam atmosphere. Transmission spectra of a large-scale-height, hydrogen-dominated atmosphere (Fig. 7) are predicted to exhibit well-defined absorption bands, while a water-dominated atmosphere would have a much smaller scale height and consequently a very smooth transmission spectrum (*Miller-Ricci and Fortney,* 2010).

The observed flat transmission spectrum from the optical to perhaps 5 μm (*Bean et al.,* 2010, 2011; *Berta et al.,* 2012; *Désert et al.,* 2011) is consistent with a small scale-height atmosphere, thus apparently favoring the water-rich alternative. However, high-altitude clouds or hazes, as with HD 189733b, could also be concealing atmospheric absorption bands if they lie at altitudes above 200 mbar (*Bean et al.,* 2010) (Fig. 7). Since the nominal model pressure-temperature profile does not cross the condensation curve of any major species at solar abundance (although ZnS and KCl do condense around 1 bar), Bean et al. proposed photochemical products as the likely source of the haze rather than clouds. However, as shown in Fig. 7, *Morley et al.* (2013) find that very extended (f_{sed} = 0.1) equilibrium sulfide clouds, including Na_2S, in an atmosphere enriched in heavy elements can also attenuate the flux sufficiently to reproduce the data. Such small values of f_{sed} are seen in some terrestrial regimes (*Ackerman and Marley,* 2001), but whether or not this would be plausible in the atmosphere of GJ 1214b remains to be investigated.

Miller-Ricci Kempton et al. (2012) explored possible photochemical pathways in the atmosphere of this planet to explore possible mechanisms for haze production and found that second-order hydrocarbons, including C_2H_2, C_2H_4, and C_2H_6, are efficiently produced by UV photolysis of methane in the atmosphere. While their model did not explore the chemistry to hydrocarbons on the order of higher than C_2H_6, they argue that polymerization of the initial photochemical products are likely and that complex hydrocarbons including tholins and soots are likely to form. *Morley et al.* (2013) demonstrate that such a hydrocarbon haze could well explain the flat transmission spectrum observed for this planet

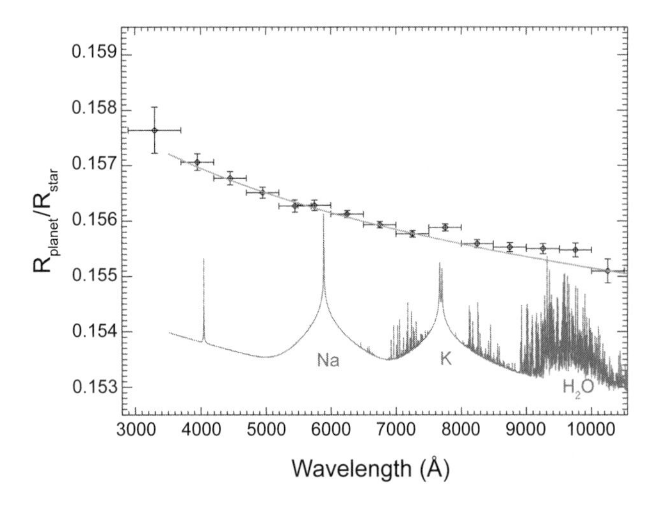

Fig. 6. See Plate 68 for color version. Observed transmission spectrum (points) for transiting hot Jupiter planet HD 189733b (*Sing et al.*, 2011). The width of the wavelength bin for each measurement is indicated by the x-axis error bars. The y-axis error bars denote the 1-σ error. The smooth curve denotes the prediction of a simple haze Rayleigh scattering-only model while the lower curve is a model for gaseous absorption only from *Fortney et al.* (2010). Further details in *Sing et al.* (2011). Figure courtesy of J. Fortney.

if the atmosphere is indeed hydrogen rich with efficient methane photolysis (with results very similar to the cloudy case shown in Fig. 7). In an unpublished manuscript, *Zahnle et al.* (2009) argue that for solar composition atmospheres in general there is a range of atmospheric temperatures near 1000 K that would be expected to favor methane photolysis and the formation of higher-order hydrocarbon soots or hazes. A definitive exploration of this point would require new generations of computer codes that can follow the fate of hydrocarbon species produced by photochemistry.

While it is still too soon to definitely characterize the atmospheres of either HD 189733b or GJ 1214b, it is apparent that cloud or haze opacity is likely important in at least some transiting planet atmospheres. The James Webb Space Telescope will obtain transit spectra of dozens of planets and characterize atmospheric absorbers as a function of planet mass and composition and the degree of stellar insolation. Such comprehensive studies will help to map out the conditions under which clouds and hazes are found.

5.1.2. Transiting planets in reflected light. In addition to the light transmitted through the atmosphere of transiting planets, measurements have also been made of the apparent brightness of the dayside of transiting planets. By comparing the brightness of a transiting system immediately before and after a planet is occulted by its primary star, the combination of light reflected and emitted by the planet can be measured. In the limit in which the planet does not emit but rather shines only by reflected light within the passband, such a measurement provides the mean geometric albedo of the planet. However, since such a measurement is most tractable for large planets orbiting close to their stars, i.e., the hot Jupiter class of planets, thermal emission cannot generally be disregarded.

Upper limits on the geometric albedo have been placed on many planets [see the summary in *Demory et al.* (2011)], generally finding the passband-averaged geometric albedo to be less than about 30% and in some cases much less (*Kipping and Spiegel,* 2011). Such a finding is not surprising for hot giant planet atmospheres (*Marley et al.,* 1999; *Sudarsky et al.,* 2000, 2003), since in the absence of a cloud layer most of the incident flux is absorbed rather than scattered. Despite these predictions, as constraints became available from studies of transiting planets, the preponderance of low albedos was often treated as surprising when compared to the bright albedo of Jupiter.

Persuasive evidence has been found, however, indicating that at least one hot Jupiter has a large geometric albedo that is likely attributable to a haze or cloud layer. Kepler-7b (*Latham et al.*, 2010) has a geometric albedo averaged over the Kepler bandpass [423 to 897 nm (*Koch et al.,* 2010)] of 0.32 ± 0.03 (*Demory et al.,* 2011). Such a high albedo is more typical of the cloudy solar system giants (e.g., *Karkoschka,* 1994) than a deep scattering atmosphere. *Demory et al.* (2011) thus argue that the most likely explanation is that this particular planet has a bright photochemical haze or cloud layer, perhaps similar to that seen in the transit spectra of HD 189733b. More detailed modeling to constrain the properties of the scattering layer has not yet been done, and to date transit spectra have not been obtained for Kepler 7b.

As albedos are measured for more planets with appropriate corrections for thermal emission, it may become apparent which particular combinations of planet mass, composition, and stellar incident flux (particularly including UV flux) conspire to produce high-albedo planets.

5.2. Directly Imaged Planets

5.2.1. Young Jupiters. Giant planets start their lives in a warm, extended state with luminosities much greater than at later times after they have cooled. For this reason giant planets are easier to detect at young ages of a few hundred million years or less, and several have already been detected, including 2M 1207b and the planets orbiting the A star HR 8799 (*Marois et al.,* 2008, 2010). The near-infrared thermal emission of all these objects (which have effective temperatures near 1000 K) points to the presence of substantial refractory cloud decks, most likely comprised of iron and silicate grains (e.g., *Marois et al.,* 2008; *Barman et al.,* 2011). Much of the literature on these objects has focused on constraining the properties of these clouds because in field brown dwarfs, clouds have largely dissipated by 1000 K. *Currie et al.* (2011), *Madhusudhan et al.* (2011), and *Bowler et al.* (2010) also all explored models for the directly imaged planets and agreed that clouds played a critical role in shaping their thermal emission at a lower effective temperature than is typical for more-massive field brown dwarfs.

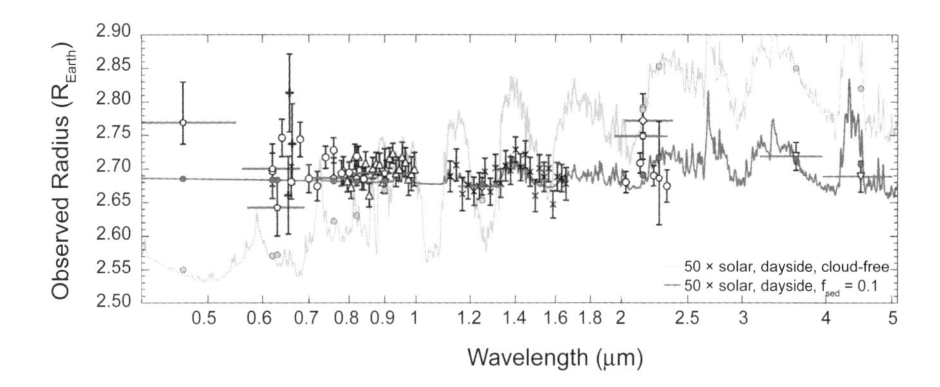

Fig. 7. See Plate 69 for color version. Observed (points) and model (lines) transit radius of GJ 1214b. Datapoints are from multiple observations as described in *Morley et al.* (2013). Models are for a cloudless H_2-He-rich atmosphere with 50× the solar abundance of heavy elements (gray) and the same atmosphere with the equilibrium abundance of clouds computed with f_{sed} = 0.1 (red). Figure modified from *Morley et al.* (2013).

Marley et al. (2012) also constructed model atmospheres for these planets and concurred that clouds are present in the atmospheres of HR 8799b, c, and d. They applied the cloud model of *Ackerman and Marley* (2001) and found that they could reproduce much of the available data by using cloud parameters typically seen in warmer L dwarfs. They argue from mass balance considerations that mean cloud particle size likely varies inversely with $\sqrt{g}$ where g is the gravitational acceleration, and that all else being equal the column optical depth of a cloud varies proportionately to $\sqrt{g}$. The net result of these scaling relationships is that clouds in lower-gravity objects tend to be similar to clouds found at warmer effective temperatures in higher gravity, more-massive brown dwarfs. They find that as objects cool their spectra can be modeled by increasing the sedimentation efficiency, f_{sed} in equation (1). The effective temperature at which f_{sed} begins to increase varies with gravity such that lower-gravity objects begin to lose their refractory clouds at warmer effective temperatures than low-mass objects.

At cooler effective temperatures near 600 K, another set of condensates become important in field dwarfs. *Morley et al.* (2012) have demonstrated that Na_2S and other clouds (Fig. 3) moderately redden the near-infrared colors of late T-type brown dwarfs. Once young giant planets in this effective temperature range are discovered, it will be possible to test if the scaling relationships seen at higher effective temperatures persist.

Ultimately understanding the nature of clouds in the warm, young giant planets hinges on understanding why the cloud-clearing effective temperature varies as it does. In the next few years many more young giant planets are expected to be discovered by the upcoming Gemini Planet Imager (GPI) and Spectro-Polarimetric High-contrast Exoplanet REsearch (SPHERE) exoplanet surveys.

5.2.2. Giants in reflected light. No giant planet has yet been imaged in reflected light, although such an observation is less technically difficult than imaging a terrestrial planet in reflected starlight and a number of space telescope missions have been proposed. As with terrestrial planets, clouds play a large role in affecting the reflection spectra of a giant planet (*Marley et al.,* 1999; *Sudarsky et al.,* 2000; *Cahoy et al.,* 2010).

One way of visualizing the effect of clouds in a giant planet atmosphere is to imagine a Jupiter twin lying at progressively closer distances to the Sun (Fig. 4). The optical geometric albedo spectrum of Jupiter generally consists of a bright continuum punctuated by methane absorption bands. The bright continuum is formed by scattering from gas and stratospheric hazes in the blue and the bright ammonia clouds at longer wavelengths. The methane bands are formed from the column of gas overlying the clouds. Below the ammonia cloud-tops lie cloud decks of NH_4SH and H_2O.

A Jupiter twin (with the same internal heat flux) lying closer to its star at 2 AU would have a warmer atmosphere in which the ammonia and ammonium hydrosulfide clouds would not condense. Instead, the planet would be covered by a bright global layer of water clouds that would give a very high albedo (Fig. 1). At even closer distances to its star, the atmosphere would be too warm for water clouds to form. Gas absorption thus overwhelms gaseous Rayleigh scattering and the planet becomes much darker (Fig. 1) than the cloudless case.

6. ADDITIONAL TOPICS

6.1. Clouds of Low-Porosity Aggregates

Most of the discussion in this chapter has focused on cloud or haze particles as fully dense spheres. However, fluffy or porous aggregate particles may very well form, and such particles behave differently as they interact with both the atmosphere and radiation. In this section we simplify the

Ackerman and Marley (2001) approach to illustrate how low-porosity aggregates would behave. In this simplified version, we neglect the fraction of the condensible species present as vapor above the cloud base, so $q_t = q_c$. We stress that this treatment is not a replacement for the complete model.

When the particles are too small to settle (see below), the condensate has the same scale height as the atmosphere. Thus the condensate *mixing ratio* $q = q_c = c/g$ is constant and so has an infinite scale height H_q. However, when the cloud is significantly settled such that the condensate mixing ratio scale height $H_q \ll H$, and if it is assumed that the vapor mixing ratio is negligible compared to the condensate ratio, then equation (1) governing the vertical cloud distribution can be approximated as $K_{zz}dq/dz + v_f q = 0$, where $K_{zz} = Lw^*$ is the vertical eddy diffusion coefficient, a property of macroscopic turbulence with typical lengthscale L and large (energy-containing) eddy velocity w^*; L is usually taken to be the atmospheric scale height H, although this might not be valid if the associated large eddy timescale L/w^* is much longer than the rotation period of the object (*Schubert and Zhang,* 2000) or the temperature profile is stable (Ackerman and Marley use an *ad hoc* stability correction). Then, for a settled cloud with uniform mass mixing ratio q and effective thickness H_q, the above equation is approximated by $K_{zz}q/H_q + v_f q = 0$, thus the cloud thickness is roughly $H_q = Hw^*/v_f$. The scale height in the simplified exponential solution for condensate mass mixing ratio of *Ackerman and Marley* (2001) is essentially the same. That is, when the convective eddy velocity is much larger than the settling speed, the cloud particles are not able to settle ($f_{sed} = 0$).

Particle settling velocities depend upon the dynamical regime. The settling speed v_f is the product of the local gravity g and the gas drag stopping time of the particle, t_s. Equation B2 in *Ackerman and Marley* (2001) contains a bridging expression β that covers both the so-called Stokes and Epstein drag regimes, in which *r* is respectively greater than, or less than, the gas molecule mean free path $l_m = m_{H_2}/\sigma_{H_2}\rho_g$, where m_{H_2} and σ_{H_2} are the mass and cross section of a hydrogen molecule and ρ_g is the gas density. In equations B1 and B3 of *Ackerman and Marley* (2001), the density contrast is essentially the particle density because $\rho \gg \rho_g$ for all applications of interest. So, v_f is proportional to ρ, and for particles with porosity ϕ, $\rho = \rho_s(1-\phi)$. In the Epstein regime ($r/l_m < 1$) the expression for t_s is very simple: $t_s = r\rho/c\rho_g$ where c is the sound speed. *Cuzzi and Weidenschilling* (2006) show how the stopping time in the Stokes regime ($r/l_m > 1$) is essentially larger by a factor of r/l_m.

We can thus show that for porous particles under local gravity g, having the same mass as a particle with no porosity, $v_f = gr\rho/c\rho_g = gr_s\rho_s(1-\phi)^{2/3}/c\rho_g$, and thus $H_q = Hw^*/v_f = H_{q(solid)}(1-\phi)^{-2/3}$. Porous particles are lofted to greater heights than nonporous particles of the same mass, because of their slower settling speed. In this sense they behave like smaller particles; however, their radiative behavior is not that of smaller particles, as noted in section 3.4; porous particles in the regime $r/\lambda \gg 1$ indeed have large and wavelength-independent opacity.

6.2. Polarization

Polarization may provide an additional avenue for characterizing cloudy planets. Unlike the stellar radiation emitted by the central star, the light scattered by clouds will in general be polarized. Thus, investigating the polarized radiation scattered by clouds in contrast to the nonpolarized stellar radiation may be an opportunity for characterizing cloudy exoplanetary atmospheres at short wavelengths in the future (see *Stam,* 2008).

Giant planets can also be polarized in the thermal infrared. In this case the required asymmetry in radiation emitted across the apparent planetary disk must be provided either by rotational flattening or irregularities in the global cloud cover. *Marley and Sengupta* (2011) investigated the former mechanism and *de Kok et al.* (2011) the latter. Both sets of authors found that in the most favorable circumstances polarization fractions of a few percent were plausible, and that in such cases polarization confirms the presence of a scattering condensate layer.

While polarization undoubtedly provides additional constraints on cloud particle sizes (e.g., *Bréon and Colzy,* 2000), condensate phase (e.g., *van Diedenhoven et al.*, 2012a), and even asymmetry parameter (*van Diedenhoven et al.,* 2012b) and atmospheric structure, it may in practice be of limited value for studies of exoplanets. Within the solar system the most well known discovery attributable to a polarization measurement is the particle size of the clouds of Venus (*Hansen and Hovenier,* 1974). An imaging photopolarimeter carried on the Pioneer 10 and 11 spacecraft also constrained the vertical structure of Jupiter's clouds (*Gehrels et al.,* 1974). However, in general, other techniques have proven more valuable. Especially given the low signal to noise and difficulty inherent in any measurement of an extrasolar planet, the value of further dividing the light into polarization channels must be weighed against other potential measurements (e.g., obtaining higher-resolution spectroscopy).

7. CONCLUSIONS

As with the planetary atmospheres of solar system planets, clouds are expected to play major roles in the vertical structure, chemistry, and reflected and emitted spectra of extrasolar planets. That said, the most extensive experience to date with clouds outside the solar system has been with the brown dwarfs. The presence of refractory clouds in L dwarfs and sulfide and salt clouds in late T dwarfs has been well established, and a number of methods have been developed to model these clouds.

Compared to cloud modeling approaches within the solar system or particularly on Earth, exoplanet cloud models are still in their infancy. In many cases arbitrary clouds are employed that specify a range of plausible cloud properties that are usually sufficient to explore parameter space. More sophisticated attempts to derive particle sizes, composition, and number density as a function of height through the atmosphere from a given set of assumptions will be needed once higher-resolution, broad-wavelength spectral data become available.

As of the time of this chapter's writing, the best evidence for clouds in extrasolar planet atmospheres lies in the spectra of the planets orbiting the nearby A star HR 8799. Planets b, c, and d each have red near-infrared colors that are best explained by global refractory cloud decks that have persisted to lower effective temperatures than in higher-mass field brown dwarfs. This persistence of clouds to lower effective temperatures for lower-gravity objects continues a trend that has already been recognized among brown dwarfs.

Among the transiting planets there is convincing evidence for high-altitude clouds — or perhaps a photochemical haze — in the atmosphere of HD 189733b. The transit spectra of this planet lacks the deep absorption bands expected for a clear atmosphere of absorbing gas, but rather exhibits a smoothly varying absorption profile likely caused by small grain scattering. A second planet, GJ 1214b, also has a transit spectrum lacking absorption features. In this case the spectrum may be attributable either to a small atmospheric scale height resulting from a high mean molecular weight composition or from clouds.

These early detections are like distant clouds seen at sunset presaging a coming storm. The GPI, SPHERE, and many other direct-imaging planet searches are expected to discover dozens of young, self-luminous extrasolar giant planets over the coming decade (*Oppenheimer and Hinkley,* 2009; *Traub and Oppenheimer,* 2010). Many of these planets will exist in the effective temperature range in which clouds shape their emergent spectra. Meanwhile, continued studies of transiting planets, particularly by the upcoming James Webb Space Telescope, will probe the atmospheres of the transiting planets and test the hypothesis that hazes of photochemical origin are important in some regimes. Finally, future space-based coronagraphic telescopes may eventually image extrasolar giant and terrestrial planets in reflected light. All such efforts to discover and characterize extrasolar planets will hinge upon an understanding of the role clouds play in shaping the climate and atmospheres of planets.

REFERENCES

Abe Y., Abe-Ouchi A., Sleep N. H., and Zahnle K. J. (2011) Habitable zone limits for dry planets. *Astrobiology, 11(5),* 443–460.

Ackerman A. S. and Marley M. S. (2001) Precipitating condensation clouds in substellar atmospheres. *Astrophys. J., 556,* 872–884.

Allard F., Hauschildt P. H., Alexander D. R., Tamani A., and Schweitzer A. (2001) The limiting effects of dust in brown dwarf model atmospheres. *Astrophys. J., 556,* 357–372.

Arnold L., Gillet S., Lardière O., Riaud P., and Schneider J. (2002) A test for the search for life on extrasolar planets. Looking for the terrestrial vegetation signature in the Earthshine spectrum. *Astron. Astrophys., 392,* 231–237.

Barker H. W., Stephens G. L., and Fu Q. (1999) The sensitivity of domain-averaged solar fluxes to assumptions about cloud geometry. *Q. J. R. Meteor. Soc., 125,* 2127–2152.

Barman T. S., Macintosh B., Konopacky Q. M., and Marois C. (2011) Clouds and chemistry in the atmosphere of extrasolar planet HR 8799b. *Astrophys. J., 733,* 65.

Bean J. L., Miller-Ricci Kempton E., and Homeier D. (2010) A ground-based transmission spectrum of the super-Earth exoplanet GJ 1214b. *Nature, 468,* 669–672.

Bean J.L., Désert J.-M., Kabath P., et al. (2011) The optical and near-infrared transmission spectrum of the super-Earth GJ 1214b: Further evidence for a metal-rich atmosphere. *Astrophys. J., 743,* 92.

Beckwith S. V. W., Henning T., and Nakagawa Y. (2000) Dust properties and assembly of large particles in protoplanetary disks. In *Protostars and Planets IV* (V. Mannings et al., eds.), p. 533. Univ. of Arizona, Tucson.

Berta Z. K., Charbonneau D., Désert J.-M., et al. (2012) The flat transmission spectrum of the super-Earth GJ1214b from Wide Field Camera 3 on the Hubble Space Telescope. *Astrophys. J., 747,* 35.

Blum J. (2010) Dust growth in protoplanetary disks — a comprehensive experimental/theoretical approach. *Res. Astron. Astrophys., 10,* 1199–1214.

Bohren C. and Huffman D. R. (1983) *Absorption and Scattering of Light by Small Particles.* Wiley, New York.

Bony S. and Dufresne J.-L. (2005) Marine boundary layer clouds at the heart of tropical cloud feedback uncertainties in climate models. *Geophys. Res. Lett., 32,* 20806.

Bowler B. P., Liu M. C., Dupuy T. J., and Cushing M. C. (2010) Near-infrared spectroscopy of the extrasolar planet HR 8799b. *Astrophys. J., 723,* 850.

Bréon F. and Colzy S. (2000) Global distribution of cloud droplet effective radius from POLDER polarization measurements. *Geophys. Res. Lett., 27(24),* 4065–4068.

Burrows A. and Orton G. (2010). Giant planet atmospheres and spectra. In *Exoplanets* (S. Seager, ed.), p. 419. Univ. of Arizona, Tucson.

Burrows A. and Sharp C. M. (1999) Chemical equilibrium abundances in brown dwarf and extrasolar giant planet atmospheres. *Astrophys. J., 512,* 843.

Cahoy K., Marley M., and Fortney J. (2010) Exoplanet albedo spectra and colors as a function of planet phase, separation, and metallicity. *Astrophys. J., 724,* 189–214.

Carlson B. E., Rossow W. B., and Orton G. S. (1988) Cloud microphysics of the giant planets. *J. Atmos. Sci., 45,* 2066.

Colaprete A. and Toon O. B. (2003) Carbon dioxide clouds in an early dense martian atmosphere. *J. Geophys. Res.–Planets, 108,* 5025.

Cooper C., Sudarsky D., Milsom J., Lunine J., and Burrows A. (2003) Modeling the formation of clouds in brown dwarf atmospheres. *Astrophys. J., 586,* 1320.

Currie T., Burrows A., Itoh Y., et al. (2011) A combined Subaru/VLT/MMT 1–5 μm study of planets orbiting HR 8799: Implications for atmospheric properties, masses, and formation. *Astrophys. J., 729,* 128.

Cushing M. C., Marley M. S., Saumon D., Kelly B., Vacca W., Rayner J., Freedman R., Lodders K., and Roellig, T. (2008) Atmospheric parameters of field L and T dwarfs. *Astrophys. J., 678,* 1372.

Cuzzi J. N. and Hogan R. C. (2003) Blowing in the wind: I. Velocities of chondrule-sized particles in a turbulent protoplanetary nebula. *Icarus, 164,* 127–138.

Cuzzi J. N. and Weidenschilling S. (2006) Particle-gas dynamics and primary accretion. In *Meteorites and the Early Solar*

System II (D. S. Lauretta and H. Y. McSween Jr., eds.), pp. 353–381. Univ. of Arizona, Tucson.

Deirmendjian D. (1964) Scattering and polarization of water clouds and hazes in the visible and infrared. *Appl. Optics IP, 3,* 187.

de Kok R. J., Helling C., Stam D. M., Woitke P., and Witte S. (2011) The influence of non-isotropic scattering of thermal radiation on spectra of brown dwarfs and hot exoplanets. *Astron. Astrophys., 531,* A67.

Demory B.-O., Seager S., Madhusudhan N., et al. (2011) The high albedo of the hot Jupiter Kepler-7b. *Astrophys. J. Lett., 735,* L12.

Désert J.-M., Bean J., Miller-Ricci Kempton E., et al. (2011) Observational evidence for a metal-rich atmosphere on the super-Earth GJ1214b. *Astrophys. J. Lett., 731,* L40.

Dominik C. and Tielens A. G. G. M. (1997) The physics of dust coagulation and the structure of dust aggregates in space. *Astrophys. J., 480,* 647.

Dominik C., Blum J., Cuzzi J. N., and Wurm G. (2007) Growth of dust as the initial step toward planet formation. In *Protostars and Planets V* (B. Reipurth et al., eds.), pp. 783–800. Univ. of Arizona, Tucson.

Draine B. T. and Lee H. M. (1984) Optical properties of interstellar graphite and silicate grains. *Astrophys. J., 285,* 89–108.

Ehrenreich D., Tinetti G., Lecavelier des Etangs A., et al. (2006) The transmission spectrum of Earth-size transiting planets. *Astron. Astrophys., 448,* 379–393.

Fegley B. and Lodders K. (1996) Atmospheric chemistry of the brown dwarf Gliese 229B: Thermochemical equilibrium predictions. *Astrophys. J., 472,* L37.

Ferguson J. W., Heffner-Wong A., Penley J. L., Barman T. S., and Alexander D. R. (2007) Grain physics and Rosseland mean opacities. *Astrophys. J., 666,* 261–266.

Freytag B., Allard F., Ludwig H.-G., Homeier D., and Steffen M. (2010) The role of convection, overshoot, and gravity waves for the transport of dust in M dwarf and brown dwarf atmospheres. *Astron. Astrophys., 513,* A19.

Fu Q. (2007) A new parameterization of an asymmetry factor of cirrus clouds for climate models. *J. Atmos. Sci., 64,* 4140–4150.

Fu Q., Yang P., and Sun W. B. (1998) An accurate parameterization of the infrared radiative properties of cirrus clouds for climate models. *J. Climate, 11,* 2223–2237.

Forget F. and Pierrehumbert R. T. (1997) Warming early Mars with carbon dioxide clouds that scatter infrared radiation. *Science, 278,* 1273.

Fortney J., Shabram M., Showman A., Lian Y., Freedman R., Marley M., and Lewis N. (2010) Transmission spectra of three-dimensional hot Jupiter model atmospheres. *Astrophys. J., 709,* 1396–1406.

Fridlind A. M., Ackerman A. S., McFarquhar G., Zhang G., Poellot M. R., DeMott P. J., Prenni A. J., and Heymsfield A. J. (2007) Ice properties of single-layer stratocumulus during the Mixed-Phase Arctic Cloud Experiment (M-PACE): Part II, model results. *J. Geophys. Res., 112,* D24202.

Fridlind A. M., van Diedenhoven B., Ackerman A. S., Avramov A., Mrowiec A., Morrison H., Zuidema P., and Shupe M. D. (2012) A FIRE-ACE/SHEBA case study of mixed-phase Arctic boundary-layer clouds: Entrainment rate limitations on rapid primary ice nucleation processes. *J. Atmos. Sci., 69,* 365–389.

Gail H.-P. and Sedlmayr E. (1988) Dust formation in stellar winds. IV — Heteromolecular carbon grain formation and growth. *Astron. Astrophys., 206,* 153–168.

Gail H.-P., Keller R., and Sedlmayr E. (1984) Dust formation in stellar winds. I — A rapid computational method and application to graphite condensation. *Astron. Astrophys., 133,* 320–332.

Gehrels T., Coffeen D., Tomasko M., et al. (1974) The imaging photopolarimeter experiment on Pioneer 10. *Science, 183,* 318–320.

Gibson N. P., Pont F., and Aigrain S. (2011) A new look at NICMOS transmission spectroscopy of HD 189733, GJ-436 and XO-1: No conclusive evidence for molecular features. *Mon. Not. R. Astron. Soc., 411,* 2199.

Glandorf D. L., Colaprete A., Tolbert M. A., and Toon O. B. (2002) CO_2 snow on Mars and early Earth: Experimental constraints. *Icarus, 160,* 66–72.

Goldblatt C. and Zahnle K. (2011) Clouds and the faint young sun paradox. *Clim. Past, 7,* 203–220.

Grenfell J. L., Stracke B., von Paris P., Patzer B., Titz R., Segura A., and Rauer H. (2007). The response of atmospheric chemistry on earthlike planets around F, G and K stars to small variations in orbital distance. *Planet. Space Sci., 55,* 661–671.

Hamdani S., Arnold L., Foellmi C., et al. (2006) Biomarkers in disk-averaged near-UV to near-IR Earth spectra using Earthshine observations. *Astron. Astrophys., 460,* 617–624.

Hansen G. B. (1997) The infrared absorption spectrum of carbon dioxide ice from 1.8 to 333 μm. *J. Geophys. Res., 102,* 21569–21588.

Hansen G. B. (2005) Ultraviolet to near-infrared absorption spectrum of carbon dioxide ice from 0.174 to 1.8 μm. *J. Geophys. Res.–Planets, 110,* E11003.

Hansen J. E. and Hovenier J. W. (1974) Interpretation of the polarization of Venus. *J. Atmos. Sci., 31,* 1137–1160.

Hansen J. E. and Travis L. (1974) Light scattering in planetary atmospheres. *Space Sci. Rev., 16,* 527–610.

Hansen J., Lacis A., Rind D., et al. (1984) Climate sensitivity: Analysis of feedback mechanisms. In *Climate Processes and Climate Sensitivity* (J. E. Hansen and T. Takahashi, eds.), pp. 130–163. AGU Geophys. Monogr. 29, Maurice Ewing Vol. 5. American Geophysical Union, Washington DC.

Hart M. H. (1979) Habitable zones about main sequence stars. *Icarus, 37,* 351–357.

Hearty T., Song I., Kim S., and Tinetti G. (2009) Mid-infrared properties of disk averaged observations of Earth with AIRS. *Astrophys. J., 693,* 1763–1774.

Helling Ch. and Woitke P. (2006) Dust in brown dwarfs. V. Growth and evaporation of dirty dust grains. *Astron. Astrophys., 455,* 325–338.

Helling Ch., Ackerman A., Allard F., et al. (2008) A comparison of chemistry and dust cloud formation in ultracool dwarf model atmospheres. *Mon. Not. R. Astron. Soc., 391,* 1854–1873.

Hu R., Cahoy K., and Zuber M. (2012) Mars atmospheric CO_2 condensation above the north and south poles as revealed by radio occultation, climate sounder, and laser ranging observations. *J. Geophys. Res., 117,* 7002.

Huitson C. M., Sing D. K., Vidal-Madjar A., et al. (2012) Temperature-pressure profile of the hot Jupiter HD 189733b from HST sodium observations: Detection of upper atmospheric heating. *Mon. Not. R. Astron. Soc., 422,* 2477.

Irvine W. M. (1975) Multiple scattering in planetary atmospheres. *Icarus, 25,* 175–204.

Joshi M. (2003) Climate model studies of synchronously rotating planets. *Astrobiology, 3,* 415–427.

Kaltenegger L. and Sasselov D. (2011) Exploring the habitable zone for Kepler planetary candidates. *Astrophys. J., 736,* L25.

Kaltenegger L. and Traub W. A. (2009) Transits of Earth-like planets. *Astrophys. J., 698,* 519–527.

Kaltenegger L., Traub W. A., and Jucks K. W. (2007) Spectral evolution of an Earth-like planet. *Astrophys. J., 658,* 598–616.

Karkoschka E. (1994) Spectrophotometry of the jovian planets and Titan at 300- to 1000-nm wavelength: The methane spectrum. *Icarus, 111,* 174–192.

Kasting J. F. (1988) Runaway and moist greenhouse atmospheres and the evolution of Earth and Venus. *Icarus, 74,* 472–494.

Kasting J. F., Whitmire D. P., and Reynolds R. T. (1993) Habitable zones around main sequence stars. *Icarus, 101,* 108–128.

Kipping D. M. and Spiegel D. S. (2011) Detection of visible light from the darkest world. *Mon. Not. R. Astron. Soc., 417,* L88.

Kitzmann D., Patzer A. B. C., von Paris P., et al. (2010) Clouds in the atmospheres of extrasolar planets. I. Climatic effects of multi-layered clouds for Earth-like planets and implications for habitable zones. *Astron. Astrophys., 511,* A66.

Kitzmann D., Patzer A. B. C., von Paris P., Godolt M., and Rauer H. (2011a) Clouds in the atmospheres of extrasolar planets. II. Thermal emission spectra of Earth-like planets influenced by low and high-level clouds. *Astron. Astrophys., 531,* A62.

Kitzmann D., Patzer A. B. C., von Paris P., Godolt M., and Rauer H. (2011b) Clouds in the atmospheres of extrasolar planets. III. Impact of low and high-level clouds on the reflection spectra of Earth-like planets. *Astron. Astrophys., 534,* A63.

Koch D., Borucki W. J., Basri G., et al. (2010) Kepler mission design, realized photometric performance, and early science. *Astrophys. J., 713,* L79–L86.

Lamb D. and Verlinde J. (2011) *Physics and Chemistry of Clouds.* Cambridge Univ., Cambridge.

Lammer H., Kasting J. F., Chassefière E., Johnson R. E., Kulikov Y. N., and Tian F. (2008) Atmospheric escape and evolution of terrestrial planets and satellites. *Space Sci. Rev., 139,* 399–436.

Lammer H., Bredehöft J. H., Coustenis A., et al. (2009) What makes a planet habitable? *Astron. Astrophys. Rev., 17,* 181–249.

Larson V. E. and Griffin B. M. (2012) Analytic upscaling of a local microphysics scheme. Part I: Derivation. *Q. J. R. Met. Soc.,* DOI: 10.1002/qj.1967.

Latham D., Borucki W. J., Koch D. G., et al. (2010) Kepler-7b: A transiting planet with unusually low density. *Astrophys. J., 713,* L140–L144.

Lecavelier des Etangs A., Pont F., Vidal-Madjar A., and Sing D. (2008) Rayleigh scattering in the transit spectrum of HD 189733b. *Astron. Astrophys., 481,* L83–L86.

Lesins G., Chylek P., and Lohmann U. (2002) A study of internal and external mixing scenarios and its effect on aerosol optical properties and direct radiative forcing. *J. Geophys. Res., 107(D10),* 4094.

Lewis J. S. (1969) The clouds of Jupiter and the NH_3-H_2O and NH_3-H_2S systems. *Icarus, 10,* 365–378.

Lindzen R. S., Chou M.-D., and Hou A. Y. (2001) Does the Earth have an adaptive infrared iris? *Bull. Am. Meteor. Soc., 82,* 417–432.

Liou K.-N. (2002) *An Introduction to Atmospheric Radiation.* Elsevier, Amsterdam.

Lodders K. and Fegley B. (2002) Atmospheric chemistry in giant planets, brown dwarfs, and low-mass dwarf stars. I. Carbon, nitrogen, and oxygen. *Icarus, 155,* 393.

Manabe S. and Wetherald R. T. (1967) Thermal equilibrium of the atmosphere with a given distribution of relative humidity. *J. Atmos. Sci., 24,* 241–259.

Määttänen A., Vehkamäki H., Lauri A., et al. (2005) Nucleation studies in the martian atmosphere. *J. Geophys. Res.–Planets, 110,* 2002.

Madhusudhan N., Burrows A., and Currie T. (2011) Model atmospheres for massive gas giants with thick clouds: Application to the HR 8799 planets and predictions for future detections. *Astrophys. J., 737,* 34.

Marchand R. and Ackerman T. (2010) An analysis of cloud cover in multiscale modeling framework global climate model simulations using 4 and 1 km horizontal grids. *J. Geophys. Res., 115,* D16207.

Marley M. S. and Sengupta S. (2011) Probing the physical properties of directly imaged gas giant exoplanets through polarization. *Mon. Not. R. Astron. Soc., 417,* 2874–2881.

Marley M., Gelino C., Stephens D., Lunine J., and Freedman R. (1999) Reflected spectra and albedos of extrasolar giant planets. I. Clear and cloudy atmospheres. *Astrophys. J., 513,* 879–893.

Marley M., Saumon D., and Goldblatt C. (2010) A patchy cloud model for the L to T dwarf transition. *Astrophys J., 723,* L117–L121.

Marley M. S., Saumon D., Cushing M., et al. (2012) Masses, radii, and cloud properties of the HR 8799 planets. *Astrophys J., 754,* 135.

Marois C., Macintosh B., Barman T., et al. (2008) Direct imaging of multiple planets orbiting the star HR 8799. *Science, 322,* 1348.

Marois C., Zuckerman B., Konopacky Q. M., Macintosh B., and Barman T. (2010) Images of a fourth planet orbiting HR 8799. *Nature, 468,* 1080.

Miller-Ricci E. and Fortney J. J. (2010) The nature of the atmosphere of the transiting super-Earth GJ 1214b. *Astrophys J. Lett., 716,* L74–L79.

Miller-Ricci Kempton E., Zahnle K., and Fortney J. J. (2012) The atmospheric chemistry of GJ 1214b: Photochemistry and clouds. *Astrophys J., 745,* 3.

Miyake K. and Nakagawa Y. (1993) Effects of particle size distribution on opacity curves of protoplanetary disks around T Tauri stars. *Icarus, 106,* 20.

Montañés-Rodríguez P., Pallé E., Goode P. R., and Martín-Torres F. J. (2006) Vegetation signature in the observed globally integrated spectrum of Earth considering simultaneous cloud data: Applications for extrasolar planets. *Astrophys. J., 651,* 544–552.

Morley C., Fortney J., Marley M., Visscher C., Saumon D., and Leggett S. (2012) Neglected clouds in T and Y dwarf atmospheres. *Astrophys. J., 756,* 172.

Morley C., Fortney J. J., Kempton E. M.-R., Marley M., and Visscher C. (2013) Quantitatively assessing the role of clouds in the transmission spectrum of GJ 1214B. *Astrophys. J.,* in press.

Morrison H. and Gettelman A. (2009) A new two-moment bulk stratiform cloud microphysics scheme in the Community Atmosphere Model, version 3 (CAM3). Part I: Description and numerical tests. *J. Climate, 21,* 3642–3659.

Neubauer D., Vrtala A., Leitner J. J., Firneis M. G., and Hitzenberger R. (2011) Development of a model to compute the extension of life supporting zones for Earth-like exoplanets. *Origins Life Evol. Biosph., 41,* 545–552.

Neubauer D., Vrtala A., Leitner J. J., Firneis M. G., and Hitzenberger R. (2012) The life supporting zone of Kepler-22b and the Kepler planetary candidates: KOI268.01, KOI701.03, KOI854.01 and KOI1026.01. *Planet. Space Sci., 73,* 397–406.

O'Neill C., Jellinek A. M., and Lenardic A. (2007) Conditions for the onset of plate tectonics on terrestrial planets and moons. *Earth Planet. Sci. Lett., 261,* 20–32.

Oppenheimer B. and Hinkley S. (2009) High-contrast observations in optical and infrared astronomy. *Annu. Rev. Astron. Astrophys., 47,* 253–289.

Ossenkopf V. (1991) Effective-medium theories for cosmic dust grains. *Astron. Astrophys., 251,* 210–219.

Pallé E., Zapatero Osorio M. R., Barrena R., et al. (2009) Earth's transmission spectrum from lunar eclipse observations. *Nature, 459,* 814–816.

Pierrehumbert R. and Gaidos E. (2011) Hydrogen greenhouse planets beyond the habitable zone. *Astrophys. J. Lett., 734,* L13.

Pincus R. and Klein S. A. (2000) Unresolved spatial variability and microphysical process rates in large-scale models. *J. Geophys. Res., 105,* 27059–27065.

Pollack J. B. and Cuzzi J. N. (1980) Scattering by nonspherical particles of size comparable to the wavelength — A new semi-empirical theory and its application to tropospheric aerosols. *J. Atmos. Sci., 37,* 868–881.

Pollack J. B., Hollenbach D., Beckwith S., et al. (1994) Composition and radiative properties of grains in molecular clouds and accretion disks. *Astrophys. J., 421,* 615–639.

Pont F., Knutson H., Gilliland R. L., Moutou C., and Charbonneau D. (2008) Detection of atmospheric haze on an extrasolar planet: The 0.55–1.05 μm transmission spectrum of HD 189733b with the Hubble Space Telescope. *Mon. Not. R. Astron. Soc., 385,* 109.

Pruppacher H. and Klett J. (1997) *Microphysics of Clouds and Precipitation, 2nd edition.* Kluwer, Dordrecht.

Ramanathan V. and Coakley J. A. Jr. (1978) Climate modeling through radiative-convective models. *Rev. Geophys. Space Phys., 14,* 465–489.

Ramanathan V. and Collins W. (1991) Thermodynamic regulation of ocean warming by cirrus clouds deduced from observations of the 1987 El Nino. *Nature, 351,* 27–32.

Randall D., Khairoutdinov M., Arakawa A., and Grabowski W. (2003) Breaking the cloud parameterizations deadlock. *Bull. Am. Meteorol. Soc., 84,* 1547–1564.

Rossow W.B. (1978) Cloud microphysics — Analysis of the clouds of Earth, Venus, Mars, and Jupiter. *Icarus, 36,* 1–50.

Sánchez-Lavega A., Pérez-Hoyos S., and Hueso R. (2004) Clouds in planetary atmospheres: A useful application of the Clausius-Clapeyron equation. *Am. J. Phys., 72,* 767–774.

Sandu A. and Borden C. (2003) A framework for the numerical treatment of aerosol dynamics. *Appl. Numer. Math., 45(4),* 475–497.

Satoh M., Matsuno T., Tomita H., et al. (2008) Nonhydrostatic icosahedral atmospheric model (NICAM) for global cloud resolving simulations. *J. Comp. Phys., 227,* 3486–3514.

Saumon D. and Marley M. (2008) The evolution of L and T dwarfs in color-magnitude diagrams. *Astrophys. J., 689,* 1327–1344.

Schaefer L., Lodders K., and Fegley B. (2012) Vaporization of the Earth: Application to exoplanet atmospheres. *Astrophys. J., 755,* 41.

Schmidt G. A., Ruedy R., Hansen J. E., et al. (2006) Present day atmospheric simulations using GISS ModelE: Comparison to in-situ, satellite and reanalysis data. *J. Climate, 19,* 153–192.

Schneider S. H. (1972) Cloudiness as a global climatic feedback mechanism: The effects on the radiation balance and surface temperature of variations in cloudiness. *J. Atmos. Sci., 29,* 1413–1422.

Schneider S. H. and Dickinson R. E. (1974) Climate modeling. *Rev. Geophys. Space Phys., 12,* 447–493.

Schubert G. and Zhang K. (2000) Dynamics of giant planet interiors. In *From Giant Planets to Cool Stars* (C. Griffith and M. Marley, eds.), pp. 210–222. ASP Conf. Series 212, Astronomical Society of the Pacific, San Francisco.

Seager S. and Deming D. (2010) Exoplanet atmospheres. *Annu. Rev. Astron. Astrophys., 48,* 631.

Segura A., Krelove K., Kasting J. F., et al. (2003) Ozone concentrations and ultraviolet fluxes on Earth-like planets around other stars. *Astrobiology, 3,* 689–708.

Segura A., Kasting J. F., Meadows V., et al. (2005) Biosignatures from Earth-like planets around M dwarfs. *Astrobiology, 5,* 706–725.

Selsis F. (2004) The atmosphere of terrestrial exoplanets: Detection and characterization. In *Extrasolar Planets: Today and Tomorrow* (J. Beaulieu et al., eds.), p. 170. ASP Conf. Series 321, Astronomical Society of the Pacific, San Francisco.

Selsis F., Kasting J. F., Levrard B., et al. (2007) Habitable planets around the star Gliese 581? *Astron. Astrophys., 476,* 1373–1387.

Sing D. K., Désert J.-M., Lecavelier des Etangs A., et al. (2009) Transit spectrophotometry of the exoplanet HD 189733b. I. Searching for water but finding haze with HST NICMOS. *Astron. Astrophys., 505,* 891–899.

Sing D. K., Pont F., Aigrain S., et al. (2011) Hubble Space Telescope transmission spectroscopy of the exoplanet HD 189733b: High-altitude atmospheric haze in the optical and near-ultraviolet with STIS. *Mon. Not. R. Astron. Soc., 416,* 1443.

Stam D. M. (2008) Spectropolarimetric signatures of Earth-like extrasolar planets. *Astron. Astrophys., 482,* 989–1007.

Stephens D., Leggett S., Cushing M., Marley M., Saumon D., Geballe T., Golimowski D., Fan X., and Noll K. (2009) The 0.8–14.5 μm spectra of mid-L to mid-T dwarfs: Diagnostics of effective temperature, grain sedimentation, gas transport, and surface gravity. *Astrophys. J., 702,* 154.

Stognienko R., Henning Th., and Ossenkopf V. (1995) Optical properties of coagulated particles. *Astron. Astrophys., 296,* 797–809.

Sudarksy D., Burrows A., and Pinto P. (2000) Albedo and reflection spectra of extrasolar giant planets. *Astrophys. J., 538,* 885–903.

Sudarsky D., Burrows A., and Hubeny I. (2003) Theoretical spectra and atmospheres of extrasolar giant planets. *Astrophys. J., 588,* 1121–1148.

Tinetti G., Meadows V. S., Crisp D., et al. (2006a) Detectability of planetary characteristics in disk-averaged spectra. I: he Earth model. *Astrobiology, 6,* 34–47.

Tinetti G., Rashby S., and Yung Y. L. (2006b) Detectability of red-edge-shifted vegetation on terrestrial planets orbiting M stars. *Astrophys. J. Lett., 644,* L129–L132.

Tinetti G., Beaulieu J. P., Henning T., et al. (2012) EChO — Exoplanet Characterisation Observatory. *Exp. Astron., 34,* 311–353.

Traub W. and Oppenheimer B. (2010) Direct imaging of exoplanets. In *Exoplanets* (S Seager., ed.), pp. 111–156. Univ. of Arizona, Tucson.

Tsuji T. (2002) Dust in the photospheric environment: Unified cloudy models of M, L, and T dwarfs. *Astrophys. J., 621,* 1033–1048.

Tsuji T. (2005) Dust in the photospheric environment. III. A fundamental element in the characterization of ultracool dwarfs. *Astrophys. J., 621,* 1033–1048.

Tsuji T., Nakajima T., and Yanagisawa K. (2004) Dust in the photospheric environment. II. Effect on the near-infrared spectra of L and T dwarfs. *Astrophys. J., 607,* 511–529.

Valencia D., O'Connell R. J., and Sasselov D. D. (2007) Inevitability of plate tectonics on super-Earths. *Astrophys. J. Lett., 670,* L45–L48.

van de Hulst H. C. (1957) *Light Scattering by Small Particles.* Wiley and Sons, New York.

van de Hulst H. C. (1980) *Multiple Light Scattering, Vols. 1 and 2.* Academic, New York.

van Diedenhoven B., Fridlind A. M., Ackerman A. S., and Cairns B. (2012a) Evaluation of hydrometeor phase and ice properties in cloud-resolving model simulations of tropical deep convection using radiance and polarization measurements. *J. Atmos. Sci., 69,* 3290–3314.

van Diedenhoven B., Cairns B., Geogdzhayev I. V., Fridlind A. M., Ackerman A. S., Yang P., and Baum B. A. (2012b) Remote sensing of ice crystal asymmetry parameter using multi-directional polarization measurements. Part I: Methodology and evaluation with simulated measurements. *Atmos. Meas. Tech., 5,* 2361–2374.

Vasquez M., Schreier F., Gimeno García S., Kitzmann D., Patzer A. B. C., Rauer H., and Trautmann T. (2013) Infrared radiative transfer in atmospheres of Earth-like planets around F, G, K, and M-type stars — II. Thermal emission spectra influenced by clouds. *Astron. Astrophys.,* in press.

Visscher C., Lodders K., and Fegley B. (2006) Atmospheric chemistry in giant planets, brown dwarfs, and low-mass dwarf stars. II. Sulfur and phosphorus. *Astrophys. J., 648,* 1181.

Voshchinnikov N. V., Il'in V. B., Henning Th., and Dubkova D. N. (2006) Dust extinction and absorption: The challenge of porous grains. *Astron. Astrophys., 445,* 167–177.

Weidenschilling S. (1988) Formation processes and time scales for meteorite parent bodies. In *Meteorites and the Early Solar System* (J. F Kerridge and M. S. Matthews, eds.), pp. 348–371. Univ. of Arizona, Tucson.

Williams M. and Loyalka S. (1991) *Aerosol Science: Theory and Practice: With special Applications to the Nuclear Industry.* Pergamon, New York.

Witte S., Helling Ch., and Hauschildt P. (2009) Dust in brown dwarfs and extra-solar planets. II. Cloud formation for cosmologically evolving abundances. *Astron. Astrophys., 506,* 1367–1380.

Witte S., Helling Ch., Barman T., Heidrich N., and Hauschildt P. (2011) Dust in brown dwarfs and extra-solar planets. III. Testing synthetic spectra on observations. *Astron. Astrophys., 529,* A44.

Woitke P. and Helling C. (2003) Dust in brown dwarfs. II. The coupled problem of dust formation and sedimentation. *Astron. Astrophys., 399,* 297–313.

Wordsworth R. D., Forget F., Selsis F., et al. (2010) Is Gliese 581d habitable? Some constraints from radiative-convective climate modeling. *Astron. Astrophys., 522,* A22.

Wordsworth R. D., Forget F., Selsis F., et al. (2011) Gliese 581d is the first discovered terrestrial-mass exoplanet in the habitable zone. *Astrophys. J. Lett., 733,* L48.

Wright E. L. (1987) Long-wavelength absorption by fractal dust grains. *Astrophys. J., 320,* 818–824.

Zahnle K., Marley M. S., and Fortney J. J. (2009) Thermometric soots on warm Jupiters? Available online at *arxiv.org/abs/0911.0728.*

Zsom A., Ormel C. W., Guettler C., Blum J., and Dullemond C. P. (2010) The outcome of protoplanetary dust growth: Pebbles, boulders, or planetesimals? II. Introducing the bouncing barrier. *Astron. Astrophys., 513,* A57.

Zsom A., Kaltenegger L., and Goldblatt C. (2012) A 1D microphysical cloud model for Earth, and Earth-like exoplanets: Liquid water and water ice clouds in the convective troposphere. *Icarus, 221,* 603–616.

Määttänen A., Pérot K., Montmessin F., and Hauchecorne A. (2013) Mesospheric clouds on Mars and on Earth. In *Comparative Climatology of Terrestrial Planets* (S. J. Mackwell et al., eds.), pp. 393–413. Univ. of Arizona, Tucson, DOI: 10.2458/azu_uapress_9780816530595-ch16.

Mesospheric Clouds on Mars and on Earth

Anni Määttänen
Centre National de la Recherche Scientifique/Université Versailles St Quentin

Kristell Pérot
Université Pierre et Marie Curie/Centre National de la Recherche Scientifique
(now at Chalmers University of Technology)

Franck Montmessin and Alain Hauchecorne
Centre National de la Recherche Scientifique/Université Versailles St Quentin

Mesospheric clouds form high in the atmospheres of Mars and Earth. The terrestrial mesospheric clouds form out of water vapor, whereas the martian clouds are often composed of CO_2, the major component of the atmosphere. Particle size analyses show that the ice crystals in the martian mesosphere are larger than cloud crystals in Earth's mesosphere. The formation seasons of these clouds are similar on the two planets, but the three-dimensional spatial distributions are different. The plausible formation mechanisms of the mesospheric clouds reveal interesting similarities. The observational history of this phenomenon is much longer for Earth than for Mars, but a reasonable comparison of cloud properties and formation processes can be conducted. However, the shortness of the martian dataset inhibits the analysis of long-term trends, whereas on Earth possible links of cloud properties with the solar cycle and other periodic and long-term forcings can be evaluated. We will present an overview of mesospheric clouds both on Earth and on Mars, revealed by a growing body of observational evidence, as well as a comparison revealing interesting similarities and differences.

1. INTRODUCTION

A multitude of clouds and hazes are observed in the atmospheres of terrestrial planets; these clouds and hazes are formed by the particular vapors that are available in the atmosphere in question. Cloud droplets in Earth, Venus, and Titan are composed of a variety of species, and form mainly in the lower part of the atmosphere. Ice crystals form as well, at least on Earth, Titan, and Mars, through freezing of droplets or direct deposition of vapor into ice. Hazes observed high above the clouds on Venus and Titan are formed though photochemical processes, resulting in haze particles of sulfur species and complex organic compounds. For a summary of clouds and hazes in planetary climates, see Esposito et al. (this volume), and for a comparative review of sulfuric acid aerosols in terrestrial planet atmospheres, see *McGouldrick et al.* (2011).

The highest clouds on Earth and Mars are observed in their mesospheres, high above the lower cloud systems considered climatologically important. The pressure and density in the mesosphere are only fractions of those in the near-surface environment. Trace gases that have their sources in the lower atmosphere are nearly absent, unless the atmospheric dynamics drive strong advection or if there is a local source. The formation of mesospheric clouds on Earth and on Mars reveals that this region of the atmosphere is dynamically and microphysically active, and that the activity is coupled to the processes taking place in the layers below.

Terrestrial polar mesospheric clouds (PMCs) are the highest clouds in Earth's atmosphere. Also called noctilucent clouds (NLCs), they can be observed visually only at evening or morning twilight, when the lower atmosphere is already in the dark but these high-altitude clouds are still sunlit. Extreme conditions of temperature and humidity are needed to make their formation possible. These clouds have been observed with satellites, lidars, and rockets for several decades.

The highest clouds in the martian atmosphere, the martian mesospheric clouds (MMCs), have been identified and observed with imagery and spectroscopic methods only in the last decade. The increasing amount of satellite data has enabled detailed studies of their properties. A unique characteristic of these clouds is that in most cases their composition has been confirmed as CO_2, the major component of the martian atmosphere.

Even though the histories of mesospheric cloud observations on the two planets are fairly different, similar questions have been raised. How can these clouds form at such high altitudes? What are the atmospheric dynamics involved? What are the clouds composed of, and what is the source of this condensible? How do the cloud particles nucleate? What information can their observation reveal about their environment? Mesospheric clouds can help us to obtain information about this elevated region of the atmosphere, and their observations have pushed forward mesospheric studies on both planets. Unanswered questions on Mars may be solved through similar problems or phenomena on Earth, and mutual comparison can lead to advances in both fields.

The motivation for the review and comparison focused on these specific clouds and presented in this chapter is twofold. First, mesospheric clouds form in a part of the atmosphere where very few parameters influence their formation. Similar formation conditions on the two planets enable a very robust and interesting comparison. It will be seen that the plausible formation mechanisms of the mesospheric clouds reveal interesting dynamical and microphysical similarities. Second, and more generally, studying a similar phenomenon, or the same region of the atmosphere, on two different planets can result in more insight and feedback between the scientific communities.

A summary of the mesospheric cloud properties on the two planets can be found in Table 1. First, we provide an overview of the mesospheres and the mesospheric cloud observations for both planets (sections 2 and 3). Section 4 contains the comparison, where we discuss the similarities and differences of the clouds observed on the two planets, discuss their mesospheres, and raise some microphysical considerations, such as the origin of the condensation nuclei (CN) and the formation mechanisms of the clouds.

2. OVERVIEW OF THE TERRESTRIAL MESOSPHERIC CLOUDS AND THE TERRESTRIAL MESOSPHERE

Terrestrial mesospheric clouds, usually called polar mesospheric clouds (PMCs) or noctilucent clouds (NLCs), were identified for the first time more than 120 years ago. Since their discovery, much effort has been put into understanding the existence and the properties of these clouds, but many issues still remain unresolved.

2.1. Observations of Terrestrial Mesospheric Clouds

In 1883, the small volcanic island of Krakatoa, located in Indonesia, was shaken by an extremely violent eruption, considered to be the largest explosion in recorded human history. The volcanic plume reached altitudes of tens of kilometers. The large amounts of volcanic materials (dust and volatiles) that were injected into the stratosphere and the mesosphere had significant effects on the Earth system. Average global temperatures decreased significantly for many years following the eruption (by as much as 1°C at northern high latitudes) (*Rampino and Self,* 1982). This cooling was due to the scattering of a part of the incident solar radiation back to space by sulfate aerosols, formed from volcanic sulfur gases (*Robock,* 2000). In addition, the interaction between solar radiation and these volcanic aerosols produced various global optical phenomena all around the world. For example, spectacular sunsets were observed for years. This period of time was particularly favorable for twilight sky observations, and in 1885, R. C. Leslie mentioned the NLCs in literature for the first time (*Leslie,* 1885).

These particular clouds are so high that they can only be seen when the Sun is below the horizon (see Fig. 1), which is why they were called noctilucent (or "night-shining") clouds by groundbased observers. The first photographs of NLCs were taken by O. Jesse in 1887 (*Schröder,* 2001). Shortly thereafter, the first photographic station was built in Germany, and research on NLCs began. The average altitude of the clouds was estimated at 82 km, making them the highest meteorological phenomenon on Earth.

There were many reports of sightings in the years following the eruption of Krakatoa, but soon thereafter, observations of NLCs became much less frequent. *Vestine* (1934) published an inventory of all the observations recorded from 1885 to 1933. *Witt* (1957) and *Ludlam* (1957) summarized the current knowledge on this subject in the 1960s, based purely on photographs. For a long time, visual observation was the prevailing observation method, and large networks of observers were established in order to monitor the evolution of NLCs.

In 1962, the first sounding rocket was launched into an NLC. These *in situ* measurements provided important

TABLE 1. Summary of mesospheric cloud properties on Mars and Earth.

Property	Mars	Earth
Latitudes	–25°–25°N, 50°S, 40°–60°N	50°–90°N, 55°–90°S
Longitudes	–130°–30°E, 120°–170°E	all
Altitudes	40–100 km	80–86 km
Season	$L_S \approx$ 330°–80°/90°–150°/200°–300°	Jun–Aug (NH), mid-Nov to mid-Feb (SH)
Morphology	Cirriform, wavy structure/Cumuliform?	Cirriform, wavy structure
Composition	CO_2/H_2O ice	H_2O ice
Visible optical depth	0.01–0.6	$<10^{-4}$
Radius	80–3000 nm	5–90 nm

Fig. 1. See Plate 70 for color version. Noctilucent cloud (or terrestrial polar mesospheric cloud) observed on July 14, 2009, at Djurö Island, Stockholm Archipelago, Sweden. Photo by Kristell Pérot (*Pérot et al.,* 2010).

information about the actual physical environment in which the cloud formation took place, thus enabling a better understanding of their microphysical properties and formation process (*Hemenway et al.,* 1964). Since then, several rocket programs have been conducted, providing local "snapshot" measurements made in the polar summer mesopause region. Together with other observation methods and theoretical studies, they led (and continue to lead) to a more complete picture of the microphysics of NLCs and their environment [see *Gumbel et al.* (2001) and *Rapp et al.* (2010) for concrete examples of studies based on sounding rocket campaigns].

Presently, PMCs are mainly observed by satellite instruments. Contrary to groundbased or *in situ* measurements, satellite observations have the advantage of providing global and continuous measurements, ensuring reliable statistics. Their contribution is essential in establishing the long-term climatology of the cloud properties. An important step was reached in 1972 when one of the Orbiting Geophysical Observatories, OGO-6, detected for the very first time a scattering layer in the polar summer mesopause (*Donahue et al.,*1972). Since 1978, the Solar Backscatter Ultraviolet Radiometer (SBUV) and SBUV/2 instruments have continued to observe the polar mesospheric clouds. This continuous 34-year record has provided the unique opportunity to study changes in PMC properties over an extended period. This series of instruments is characterized by nadir-viewing geometry, which has provided accurate geolocation, but no information on the PMC altitude.

Most of the other instruments observing the PMCs look at Earth's limb. These instruments can be subdivided into two groups: those using the solar occultation method [e.g., Stratospheric Aerosol and Gas Experiment (SAGE) or Polar Ozone and Aerosol Measurement (POAM)], and those measuring scattered solar radiation [e.g., Student Nitric Oxide Explorer (SNOE), Optical Spectrograph and Infra-Red Imager System (OSIRIS), or SCanning Imaging Absorption spectroMeter for Atmospheric CHartographY (SCIAMACHY)]. We refer the reader to *Deland et al.* (2006) for an overview of the satellite datasets through the year 2006.

More recently, Global Ozone Monitoring by Occultation of Stars (GOMOS) has also been used to study NLCs. The particular viewing geometry associated with the stellar occultation technique ensures highly accurate altitude determination and a large amount of data with good temporal and spatial coverage. The GOMOS dataset will be predominantly used to illustrate the properties of terrestrial mesospheric clouds in this chapter. The detection and retrieval algorithms, as well as the first climatology obtained, are described in *Pérot et al.* (2010).

In 2007, NASA launched the Aeronomy of Ice in the Mesosphere (AIM), the first satellite mission entirely dedicated to the study of the PMCs and related processes. The spacecraft carries three instruments: the Solar Occultation for Ice Experiment (SOFIE), which measures ice water content, temperature, and a number of other atmospheric constituents in the mesosphere using solar occultation; the Cloud Imaging and Particle Size (CIPS), a nadir-viewing camera that provides panoramic images of the PMCs; and finally the Cosmic Dust Experiment (CDE), which measures the dust particle influx in the upper atmosphere [see *Russell et al.* (2009) for a general description of the AIM mission].

These clouds can also be observed by lidar in both hemispheres (see, e.g., *Fiedler et al.,* 2011; *Baumgarten et al.,* 2010). This technique enables a very accurate determination of the cloud properties. It only provides local observations, but the obtained vertical profiles are characterized by good resolution and diurnal coverage. This allows monitoring of the local time dependence of PMC properties and the role of tidal effects on the mesospheric environment.

2.2. Spatial and Seasonal Distribution of Clouds

Table 1 summarizes the spatial and seasonal distribution of the clouds, illustrated by the GOMOS/Environmental Satellite (Envisat) observations shown in Fig. 2. This plot shows the seasonal evolution of the PMC occurrence frequency averaged over the full eight-year GOMOS PMC dataset. The frequency of occurrence is a good indicator of PMC activity over the season. Noctilucent clouds form in both hemispheres, at high latitudes during summer. The formation of NLCs is strongly dependent on latitude, and only slightly dependent on longitude. Thus zonal symmetry is usually assumed when studying the properties of PMCs. Noctilucent cloud formation is clearly a high-latitude phenomenon: They appear around the pole at the beginning of the cloud season, then extend to 50° latitude approximately 20 days after the solstice, then once again concentrating near the pole in late summer. Cloud coverage typically reaches 100% during the core of the cloud season at latitudes above

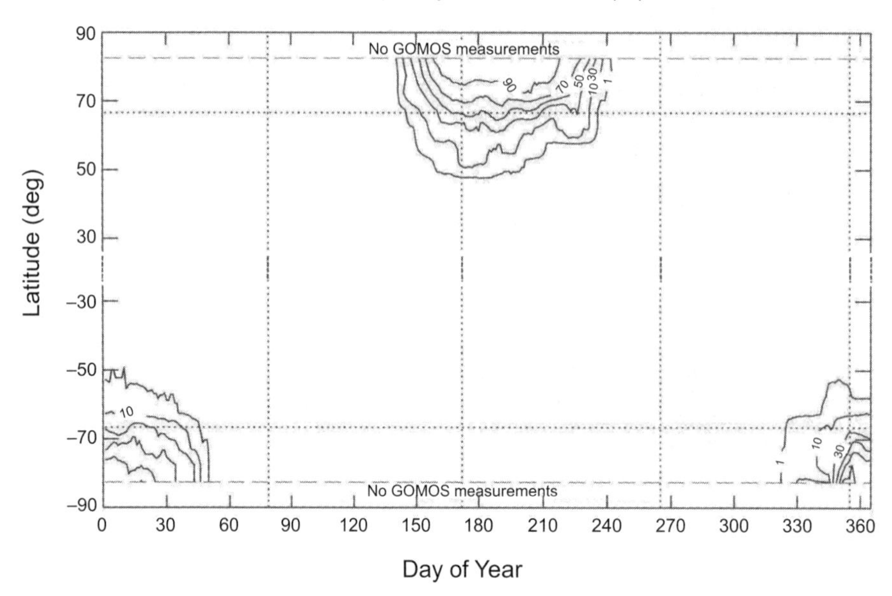

Fig. 2. Daily PMC occurrence frequency, calculated for latitudinal bins of 5°. This quantity is simply obtained by dividing the number of clouds detected by the total number of measurements. Included are all measurements made by GOMOS/ENVISAT between 2002 and 2010 (300,000 observations performed and about 21,000 PMCs detected during this period). The horizontal dotted lines represent the polar circles. The vertical dotted lines indicate the equinoxes and solstices. The horizontal dashed lines near the poles mark the maximum latitudinal extent of the GOMOS observational coverage.

80°. A more detailed look reveals that the clouds are highly localized both in time and in space. They appear over a period of approximately three months during the local summer, from the end of May to the end of August in the northern hemisphere, and from mid-November to mid-February in the southern hemisphere. Their evolution is very fast, appearing and increasing in number in only a few days and disappearing rapidly at the end of the season. The peak frequency is reached 10–20 days after the summer solstice.

The altitude of PMCs can be accurately determined from lidar measurements or from limb observations by spacebased instruments. Altitude retrieval from GOMOS data is highly accurate because of the particular viewing geometry of the stellar occultation technique and the very good vertical resolution of GOMOS photometers.

Most of the terrestrial mesospheric clouds form slightly below the mesopause, between 80 and 86 km, with an average altitude of about 83 km. The seasonal variation of PMC altitude for both hemispheres is shown in Fig. 3. In the northern hemisphere, the minimum altitude is reached in the middle of the season. In the southern hemisphere, the minimum value is reached later and greater variability is observed early in the season. Figure 4 illustrates the latitudinal variation of PMC altitude at both hemispheres. The daily means show a clear, linear increase in altitude toward the pole, with a slope that is slightly steeper in the north.

Several differences between the two hemispheres are noteworthy. First, PMCs form more frequently in the north than in the south. Their formation zone extends further as well: The minimum latitude where PMCs are observed is about 50° in the north and 55° in the south. The formation altitude is, on average, slightly higher in the south. Moreover, more variability is observed in the south, whatever the parameter under consideration.

2.3. Mesospheric Structures Favoring Cloud Formation

The chemical, radiative, and dynamical mesospheric features are essential to understand why cloud formation is possible at such high altitudes. Our description here is based on *Brasseur and Solomon* (2005), who present a review of the middle atmosphere.

In the mesosphere the temperature decreases with altitude, which is a consequence of the relatively low O_3 concentration. Energy losses, due to radiative cooling by CO_2, overcome the energy absorption by molecular oxygen or ozone. The mesopause region, situated between 85 and 90 km (the altitude is variable according to the season and latitude), therefore corresponds to a temperature minimum.

One could expect the summer mesopause to be warmer than its winter counterpart since it receives solar radiation while the winter pole is in the dark. Interestingly, the polar summertime mesopause is actually the coldest place on Earth, with temperatures that can drop below 120 K (*Rapp et al.,* 2002). This can be explained through dynamical pro-

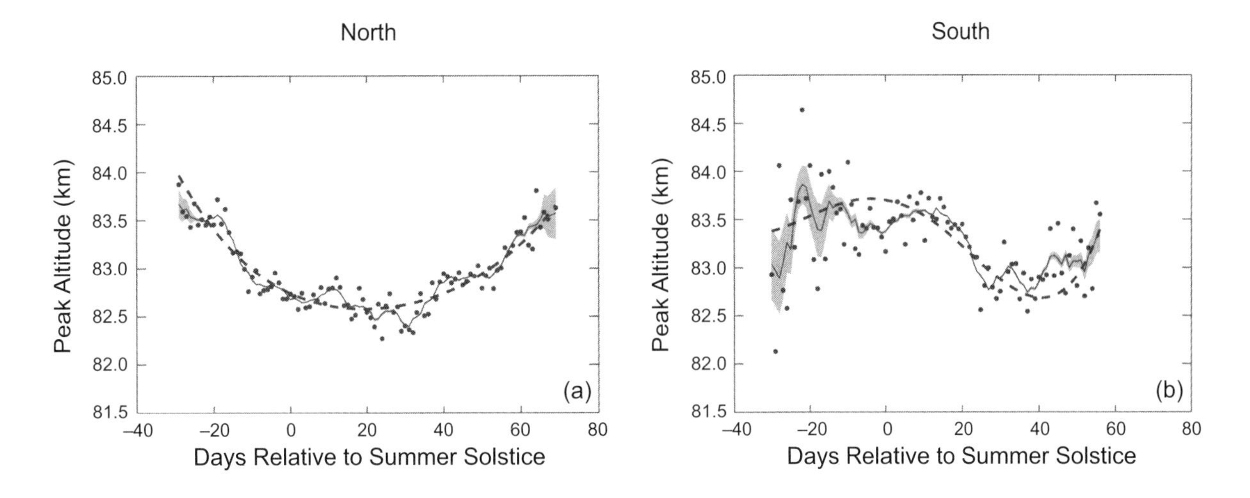

Fig. 3. Seasonal variation of daily mean PMC altitudes for both hemispheres retrieved from GOMOS data (2002–2010). Time is expressed in number of days relative to summer solstice. The error of the mean is displayed as the gray area around the curves, which have been smoothed with a five-day running mean (error values are very low due to the large number of measurements). The dashed curves correspond to a degree 4 polynomial fit.

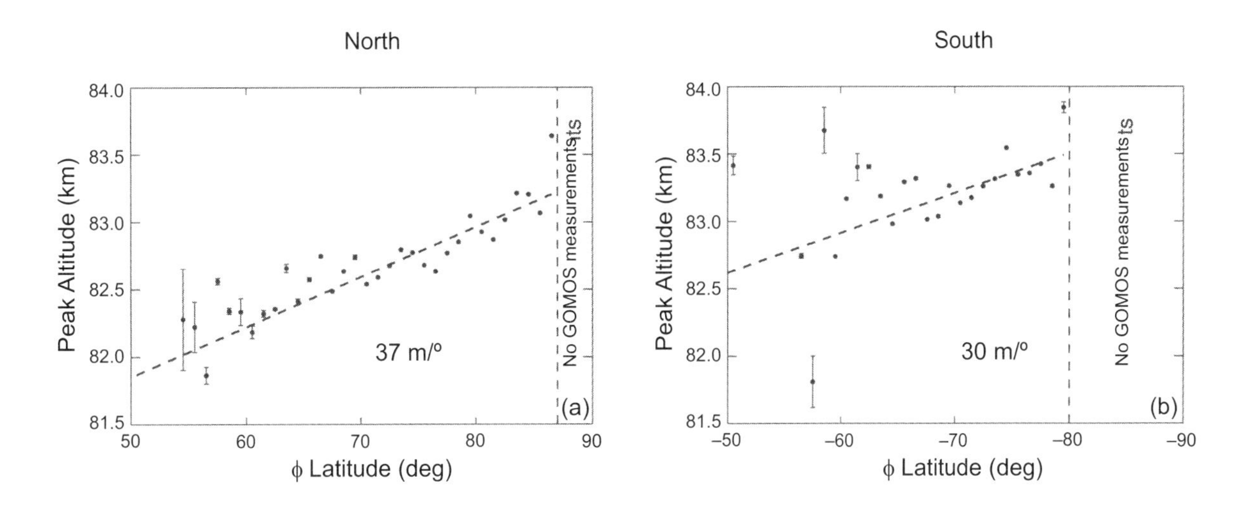

Fig. 4. PMC peak altitudes (derived from GOMOS observations) as a function of latitude for both hemispheres. Each symbol corresponds to a mean value calculated for latitudal bins of 1°. The vertical bars represent the error of the mean, and the dashed line is a weighted linear fit of the data. The vertical dashed lines mark the maximum latitudinal extent of GOMOS observational coverage.

cesses driving the summer mesosphere far from radiative equilibrium.

The mean zonal flow in the middle atmosphere is westward in the summer hemisphere and eastward in the winter hemisphere, a consequence of the geostrophic balance between the Coriolis force and the pressure gradient. However, this balanced situation is strongly disturbed by internal gravity waves (GWs). Internal GWs are oscillations arising in a stably stratified atmosphere when an air parcel is displaced vertically. They are generated in the lower atmosphere, by orography or weather disturbances, and can propagate vertically with eastward or westward phase speed. The background zonal wind acts as a filter for their vertical propagation. Only waves with a phase speed opposite to (or in the same direction but larger than) the wind speed can propagate higher in the atmosphere. Consequently, mostly GWs with eastward phase speed will be able to reach the highest altitudes above the summer pole, since the zonal wind is westward. As they reach high altitudes, where the atmospheric density becomes very low, their amplitude grows and the waves become unstable. They break and deposit momentum at a certain altitude, resulting in a wave drag, which acts against the mean zonal flow. This situation induces a meridional component of the wind from the summer pole to the equator. At the winter pole, the reverse process occurs and results in a meridional wind from the equator toward the winter pole. Finally, the interaction between GWs and the background flow leads to a global meridional circulation in the mesosphere, from the summer to the winter pole. By continuity, an upwelling takes place in the summer hemisphere and a downwelling in the winter hemisphere. These processes cause a strong adiabatic cooling in summer and warming in winter, which explains the observed mesospheric structure.

Moreover, these dynamical phenomena are asymmetric between hemispheres. The wave activity differs between the hemispheres, mostly due to the asymmetric distribution of continents and oceans. Thus the filtering effect of GWs is different, and consequently interhemispheric differences in temperature appear (*Siskind et al.,* 2003). The final result of the asymmetry is a colder polar summer mesopause in the north than in the south.

The terrestrial mesosphere also exhibits global-scale periodic oscillations in temperature, called migrating thermal tides. Although these tides are not necessary for PMC formation, since the clouds can form at all local times, they have an effect on the variability of their occurrence frequency and other properties over the diurnal cycle. This effect can be modeled (*Stevens et al.,* 2010) and observed by groundbased lidar (*Fiedler et al.,* 2005).

The general circulation of the mesosphere affects not only the temperature but also the composition of the mesosphere: The high-latitude summertime mesopause is particularly cold and wet. Constituents, which are more abundant in the lower layers, can be advected upward during the summer, leading to an increase in mesospheric concentration. This is particularly the case for water vapor and for methane, which also contributes to the increase in H_2O concentration. Oxidation of methane occurs above 30 km, and once it has been advected to these altitudes, the process results in a local production of water vapor. Methane vertical transport and oxidation provides roughly half the water content in the polar middle atmosphere (*Thomas et al.,* 1989).

The extreme conditions of temperature and humidity enable ice crystal formation in this region. The mesosphere is predominantly supersaturated (with respect to water ice) throughout the entire summer season, during a period of about 3 months, showing a peak around 10 or 20 days after the solstice. Supersaturation is reached in both hemispheres at all longitudes between latitudes 55° and the pole. Saturation ratios are generally lower in the southern hemisphere, but in both hemispheres the saturation ratios increase with latitude, and with altitude, from about 81–83 km up to the mesopause.

2.4. Morphology of the Mesospheric Clouds

R. Leslie, who first identified these clouds, described them as *"a sea of luminous silvery white clouds [. . .] wavelike in form, and evidently at a great elevation"* (*Leslie,* 1885). Polar mesospheric clouds display bright pearly white or electric-blue radiance (Plate 70). At first glance, this thin layer of ice crystals may look like cirrostratus clouds, but they exhibit a particular wavy structure.

Observers classified these clouds into different categories, depending on their morphology. Four types of NLCs exist: veils, bands, billows, and whirls (*Gadsden and Parviainen,* 2006). The two most common categories are bands and billows. Bands are characterized by long streaks, often occurring in groups arranged roughly parallel to each other or interwoven. Billows correspond to arrangements of closely spaced, roughly parallel short streaks, often exhibiting a wave-like structure with undulations.

Gravity waves have a strong dynamical effect in the polar mesosphere and make PMC formation possible (see section 2.3). They also create local thermal disturbances that can directly affect the cloud structure, since ice particles grow sufficiently to be observed only in the cold phase of the waves. Thus, in most cases, GW signature clearly appears when looking at the clouds. Images of PMCs can therefore be an exceptionally rich resource for quantifying GW activity and their interaction with the environment in high-latitude regions.

Taylor et al. (2011) showed how polar mesospheric cloud observations can be used for GW studies. They used high-resolution images from the AIM CIPS instrument, a panoramic UV imager designed to measure radiance and morphology of the PMCs (see section 2.1). Information on wave properties and variability can be derived from the very distinctive and regular wave patterns that often occur in PMCs. It is indeed possible to investigate the statistical properties and the seasonal characteristics of the waves. *Taylor et al.* (2011) found a broad spectrum of GWs, with horizontal wavelengths ranging from 20 to 400 km in the PMC field. The small-scale wave events (<50 km) appeared to be more frequent than the largest-scale events. The directions of motion of the waves could also be determined, as well as their velocities (see also *Chandran et al.,* 2009).

The results presented by *Baumgarten et al.* (2009) are another example showing how the NLC structures can be used as a tracer of wave phenomena. Their investigation of polar mesospheric cloud morphology was based on simultaneous observations from lidar, satellites, and cameras on different scales. Local-scale structures provide information about GW properties. Global-scale structures, observed from space, can provide information about planetary waves that affect the mesospheric thermal properties and the PMC large-scale morphology (see also *Merkel et al.,* 2008). Polar mesospheric cloud observations can therefore contribute to a better understanding of wave phenomena in the terrestrial middle atmosphere in polar regions.

2.5. Cloud Properties: Composition, Optical Depth, and Crystal Size

Hervig et al. (2001) showed that the main constituent of polar mesospheric cloud particles is H_2O, but the measurements are generally inconsistent with pure ice. Although the mesospheric dynamics lead to particularly high water vapor saturation ratios in the high-latitude summertime mesopause region, it is highly unlikely that homogeneous nucleation (direct formation of water cluster from the vapor phase) will occur. Instead, ice particle formation and growth requires condensation nuclei (CN). The presence of preexisting cores significantly lowers the nucleation barrier, therefore making heterogeneous nucleation much more probable. A number of CN candidates have been suggested, such as proton hydrate ion clusters, soot, sulfuric acid particles, sodium bicarbonate, or sodium hydroxide (*Rapp and Thomas,* 2006). The most popular candidates are currently meteoric smoke particles. These particles, which are found in the stratosphere and mesosphere, are a product of the condensation of vapors resulting from meteoroid ablation. A recent work by *Hervig et al.* (2012) provides the first observational evidence, based on multiwavelength observations from SOFIE (cf. section 2.1), that PMC particles consist of a mixture of water ice and meteoric smoke. The results indicate that crystals contain small amounts of meteoric smoke (0.01–3 vol.%), composed of carbon and metal compounds (*Hervig et al.,* 2012).

Saturation ratios are the highest at the mesopause level, between 86 and 90 km (*Lübken,* 1999). The ice particles form in this layer, and then grow by direct deposition of water vapor on their surface (*Rapp and Thomas,* 2006). In fact, ice particles are persistently present above the summer pole, from approximately 81 to 90 km. Yet, PMC observations above 86 km are extremely rare. It is now known that NLCs are actually only the visible part of this thicker layer containing smaller particles, which can produce very strong radar echoes called polar mesosphere summer echoes (PMSEs) [for a review of PMSEs, see *Rapp and Lübken* (2004)]. These nanometer-sized PMSE particles continue to grow while falling down through the atmosphere under the action of gravity. Once they reach sizes of some tens of nanometers, they are able to scatter light efficiently, and finally become visible to optical instruments and ground observers. Below this level the crystals eventually sublimate when reaching subsaturated regions.

Noctilucent clouds are very tenuous objects, characterized by an extremely small vertical optical depth in the visible wavelengths of only 10^{-6} to 10^{-4} (Table 1), which is why they cannot be visually observed during daytime [for a review of NLC optical properties, see *Kokhanovsky* (2005)]. The UV optical depth is somewhat higher (the values can reach 4×10^{-3}).

Despite the significant increase in the number and quality of the measurements of PMCs, their formation mechanism and crystal size distribution are still poorly known. Because of the diversity of observation methods, a comparison between published crystal size distribution results is difficult and must be done on a statistical basis. The size distribution is difficult to measure because its complete description is associated with multiple parameters (shape and width of the distribution, shape and composition of the crystals, etc.). Results can therefore be very sensitive to assumptions made for these parameters. In general, it is assumed that PMC crystals are composed of water ice, and the small proportion of meteoric material that is also present is generally not taken into account.

Model studies, including all the currently known microphysical processes, lead to size distributions that are well represented by Gaussian distributions (*Berger and Von Zahn,* 2002; *Rapp and Thomas,* 2006). *Rapp et al.* (2007) also showed that log-normal size distribution, which is the most commonly used distribution in the case of usual tropospheric clouds, fails to explain the PMC observations. Moreover, a recent study by *Hervig and Gordley* (2010) showed strong evidence that ice particles are nonspherical, with axial ratios (defined as the axial axis divided by the equatorial axis) most likely being 2 or 0.5.

Most of the published results about PMC particle sizes suggest effective radii from approximately 20 to 100 nm, but the values vary according to the method used and the assumptions chosen. The most recent dataset gives values between 5 and 90 nm (Table 1), with an average of 35.4 nm (*Hervig et al.,* 2009b). This result was derived from SOFIE observations in the northern hemisphere in 2007 that are characterized by a particularly high sensitivity. These effective radii are in agreement with those obtained from Arctic Lidar Observatory for Middle Atmosphere Research (ALOMAR) lidar measurements, ranging from 14 to 80 nm with a mean value of 37.9 nm for the same season (*Hervig et al.,* 2009b). The differences between datasets can be generally explained by differences in sensitivity.

2.6. Variability and Long-Term Trends

The PMCs are important tracers for the various processes taking place in the atmosphere at higher altitudes. Their study is therefore of great interest in many respects.

Polar mesospheric cloud properties are characterized by significant seasonal, interannual, and interhemispheric variability through various coupling processes that play a decisive role at all levels. *Karlsson et al.* (2007) showed that a dynamical coupling between the summer mesosphere and the winter stratosphere, referred to as interhemispheric coupling, significantly affects the interannual variability of PMC properties. This is explained by the fact that the planetary wave activity in the winter stratosphere has a major influence on the summer mesosphere temperature. This coupling mechanism is based on the effect of the lower atmospheric circulation on the filtering of GWs and, hence, on the forcing of the global meridional circulation in the mesosphere. *Karlsson et al.* (2009) showed that this interhemispheric coupling is also efficient at modulating seasonal variations of PMC properties. Moreover, the greater variability observed in the cloud characteristics in the southern summer mesopause is also consistent with the fact that wave activity is more important in the northern winter stratosphere. Recent work by *Karlsson et al.* (2011) and *Gumbel and Karlsson* (2011) showed that intrahemispheric coupling processes could also have a leading role in variability of the cloud properties. Indeed, the onset of the PMC season could be controlled by the persistence of the stratospheric polar vortex in the same hemisphere. This result revealed that the state of the mesosphere is strongly connected to lower altitude dynamics. Both inter- and intrahemispheric coupling processes prove to have a major influence on the state of the mesosphere.

The interannual variability of PMC properties is also known to be modulated by the 11-year solar cycle. The cyclic variation of solar activity strongly influences water vapor abundance and temperature in the upper mesosphere. Lyman-α radiation, which is responsible for the photodissociation of H_2O molecules, varies by almost a factor of 2 over a cycle (*Woods and Rottman,* 1997). In addition, in solar maximum conditions, the increased intensity of solar radiation creates diabatic heating of the middle atmosphere, due to energy absorption by molecular oxygen and ozone. This results in a warmer and drier mesosphere. Inversely, this region is colder and more humid when solar activity is at a minimum. In these conditions, PMC formation is therefore favored. A clear anticorrelation between their occurrence frequency/albedo and Ly-α flux has been established by several ground- and spacebased instruments (e.g., *Deland et al.,* 2003; *Hervig and Siskind,* 2006). This result is also well reproduced by models (*Khosravi et al.,* 2002; *Sonnemann and Grygalashvyly,* 2005). However, time lags between solar intensity and PMC response on the order of 0.5 to 2 years have been reported, but are not yet understood.

Besides this periodic component, satellite observations indicate that the PMCs have increased in frequency and brightness over recent decades. This is particularly the case for the Solar Backscattered UltraViolet (SBUV) series of instruments, which provide a long-term dataset (*Deland et al.,* 2003, 2007). Moreover, PMCs seem to form at lower latitudes than ever before, which could reveal an extension of the geographical area in which they form (*Taylor et al.,* 2002). Such behavior suggests the PMCs could be considered as a possible indicator of long-term global change in the mesosphere. This idea was formulated for the first time more than 20 years ago (*Thomas et al.,* 1989), and was the source of intense yet controversial discussion (e.g., *Thomas,* 1996, 2003; *Thomas and Olivero,* 2001; *Von Zahn,* 2003; *Thomas et al.,* 2003, 2010).

This potential climate change in the mesosphere could be connected to anthropogenic emissions of greenhouse gases. Although the increase of such gases causes an increase in temperature in the lowest parts of the atmosphere, it leads to a global cooling at higher altitudes (*Akmaev and Fomichev,* 2000). In addition, the atmospheric methane concentration has more than doubled since the preindustrial era, and methane oxidation is an important source of water vapor in the middle atmosphere. Changes in the atmospheric chemical composition by human industrial activities would therefore tend to cool and humidify the mesosphere, which would favor the formation of PMCs (*Thomas et al.,* 1989; *Olivero and Thomas,* 2001).

Are NLCs really an indicator of climate change in the middle atmosphere? A better understanding of the significant sources of natural variability is essential to answer this question.

3. OVERVIEW OF THE MARTIAN MESOSPHERIC CLOUDS AND THE MARTIAN MESOSPHERE

The story of the martian mesospheric clouds (MMCs) started decades ago with speculations on the possibility of the atmospheric CO_2 to condense not only directly at the surface at the poles, but also as clouds in the near-surface polar atmosphere and in the equatorial mesosphere (above 40 km). Carbon dioxide (and not H_2O) was the most likely candidate to condense as ice at mesospheric altitudes, since temperatures below CO_2 condensation temperature were observed. In addition, CO_2 will not need a particular source, since it is the main atmospheric constituent, whereas water vapor should be mostly confined in the lower atmosphere and therefore water ice clouds should mainly form at lower altitudes. This scenario seems to prevail, but exceptions exist. Our discussion here will focus on the mesospheric CO_2 ice clouds, but we will mention in our discussion unconfirmed composition or confirmed CO_2 or H_2O ice clouds. We will first detail the history of mesospheric cloud observations and then describe their properties and related open questions.

The martian seasons are described with the help of the solar longitude, L_S, given in degrees: $L_S = 0°$ corresponds to the northern hemisphere (NH) spring equinox, $L_S = 90°$ to the NH summer solstice, $L_S = 180°$ to the NH autumn equinox, and $L_S = 270°$ to the NH winter solstice.

3.1. Observations of Martian Mesospheric Clouds

Even as early as the dawn of satellite observations of Mars, *Herr and Pimentel* (1970) announced a plausible detection of CO_2 clouds in the atmosphere of Mars through Mariner 6 and 7 observations in the infrared. The spectra displayed a scattering peak in the 4.3-μm region, recognized from laboratory spectra as a signature of solid CO_2. However, the Mariner CO_2 ice signature appeared in the lower atmosphere (around 20–30 km) at low latitudes, where CO_2 is unlikely to be condensing. Nearly two decades later, *Schofield et al.* (1997) reported Mars Pathfinder lander descent profiles revealing subcondensation temperatures at high altitudes. Soon thereafter, *Clancy and Sandor* (1998) argued that sufficient evidence had been accumulated to verify the existence of CO_2 ice clouds on Mars. They based their claim on Mariner observations, temperatures below CO_2 condensation observed by Pathfinder and groundbased submillimeter observations, and blue predawn clouds observed by Pathfinder from the surface of Mars. They noted that uncertainties in the altitude information of Mariner observations had encouraged discarding the evident CO_2 ice observation. Both the Mariner and Pathfinder observations were explained by *Clancy and Sandor* (1998) through MMCs composed of small CO_2 ice crystals (radii of some hundreds of nanometers). They also concluded that these clouds should form in tidal and gravity wave temperature minima and be quite a common phenomenon at low martian latitudes. In fact, later the Mariner observations were associated with the CO_2 non-local-thermal-equilibrium (non-LTE) emission at 4.3 μm (*Lopez-Valverde et al.,* 2005).

No further evidence on the existence of MMCs came until they were indisputably observed by new satellite missions [Mars Global Surveyor (MGS), Mars Express (MEx), Mars Odyssey (MOd)] in the equatorial mesosphere in the northern hemisphere summer (*Clancy et al.,* 2003, 2004, 2007; *Formisano et al.,* 2006; *Montmessin et al.,* 2006a,b, 2007; *Inada et al.,* 2007; *McConnochie et al.,* 2010; *Määttänen et al.,* 2010; *Scholten et al.,* 2010; *Vincendon et al.,* 2011). They were also discovered at northern and southern midlatitudes in the local late autumn (*Inada et al.,* 2007; *Montmessin et al.,* 2007; *McConnochie et al.,* 2010; *Määttänen et al.,* 2010; *Scholten et al.,* 2010). These studies suggested that the composition of MMCs was mainly (but not exclusively) CO_2 ice.

The series of mesospheric cloud discoveries listed above started with the imaging and thermal emission observations of the martian limb by *Clancy et al.* (2003, 2004). The stellar occultation observations of *Montmessin et al.* (2006a) brought up further evidence on MMCs, and gave strong indications about the possibility of CO_2 cloud formation through the simultaneous observations of CO_2 supersaturation. The supersaturated pockets did not coincide exactly with the observed aerosol layers but were found mostly above them, sometimes adjacent to, but mainly detached from, the aerosol layer.

Montmessin et al. (2007) showed that mesospheric CO_2 ice clouds indeed exist, since they spectroscopically confirmed clouds to consist of CO_2 ice crystals (through observations of a scattering peak at 4.24 μm caused by CO_2 ice) and published the first study of their spectral signature and properties using data from the MEx Visible and Infrared Mineralogical Mapping Spectrometer (OMEGA). The spectroimager OMEGA dataset of *Montmessin et al.* (2007) confirmed the previous mesospheric cloud observations by the MGS Thermal Emission Spectrometer (TES) and Mars Orbiting Camera (MOC) (*Clancy et al.,* 2003, 2004) and was supported by new limb observations by TES and MOC (*Clancy et al.,* 2007) of high-altitude aerosol layers with very similar spatial and temporal distribution as the clouds observed by OMEGA. *Inada et al.* (2007) also reported the detection of MMCs (of unverified composition) in MOd Thermal Emission Imaging System Visible Imaging Subsystem (THEMIS-VIS) data.

After the confirmation of the existence of MMCs by these first publications, including the identification of CO_2 ice, additional climatologies were published with datasets from different instruments. *Määttänen et al.* (2010) and *Scholten et al.* (2010) investigated these clouds more thoroughly, again using spectroimaging data from the OMEGA instrument, along with complementary MEx High Resolution Stereo Camera (HRSC) observations. *Määttänen et al.* (2010) conducted a companion study on OMEGA and HRSC data presenting three martian years of OMEGA observations of CO_2 clouds complemented by a selection of HRSC high-altitude cloud observations. *Määttänen et al.* (2010) analyzed the OMEGA observations to map the spatial and seasonal distribution of the clouds and to study their internal structures, publishing for the first time mapping of variations of optical depth and effective radius inside a cloud. *Scholten et al.* (2010) presented the HRSC dataset of high-altitude (≥70 km) clouds, which were discovered from a distinct parallax of high-altitude atmospheric features in HRSC's stereo images. By comparing these with simultaneous observations and spectroscopic identification by OMEGA, *Scholten et al.* (2010) were unambiguously able to determine that certain HRSC-imaged clouds were formed of CO_2 ice, thus resulting in publication of the first verified high-resolution images of this phenomenon (some examples of HRSC clouds can be seen in Fig. 5). The HRSC stereo observations enabled analysis of cloud altitudes (through the apparent cloud parallax) and estimates of mesospheric wind speeds (through cloud displacement in subsequent images). A large THEMIS-VIS dataset was published as well (*McConnochie et al.,* 2010), including about a dozen equatorial mesospheric cloud observations and a larger set of midlatitude observations, with estimates of cloud altitude and prevailing wind speed. *McConnochie et al.* (2010) analyzed some of their equatorial cloud observations with a radiative transfer model and concluded that CO_2 was the most likely candidate for the cloud composition based on their estimates of ice mass (through analyzed particle radii and optical depths) and realistic quantities of condensable vapors (CO_2 or H_2O) in the mesosphere. Unfortunately, they could not apply their simple, plane-parallel radiative transfer model to the midlatitude observations because of the demanding geometry (high incidence angles). Compact Reconnaissance Imaging Spectrometer for Mars (CRISM) data from the Mars Reconnaissance Orbiter (MRO) were analyzed by *Vincendon et al.* (2011) to map mesospheric CO_2 clouds, fitting well the spatial and seasonal distribution of previous datasets. They also published OMEGA nadir data covering the Mars year 30 cloud season. In addition, based on one OMEGA limb observation of a high-altitude H_2O ice cloud, *Vincendon et al.* (2011) argued that the clouds observed by the MEx SPICAM instrument (*Montmessin et al.,* 2006a) might instead be composed of H_2O ice. The climatology that can be compiled from all the aforementioned datasets presently covers a total of seven martian years (more than 13 Earth years) of data on the MMCs.

Mesospheric cloud observations are important indicators of the mesospheric state, since they reveal very cold mesospheric temperatures (particularly cold temperatures are needed for CO_2 to condense). A completely new, unique dataset of mesospheric wind speeds has emerged from the MMC observations. Parallaxes of the clouds in subsequent images (available from HRSC through its stereo capabilities, and from THEMIS-VIS, whose imaging sequences allow similar analysis) provide information about their altitude,

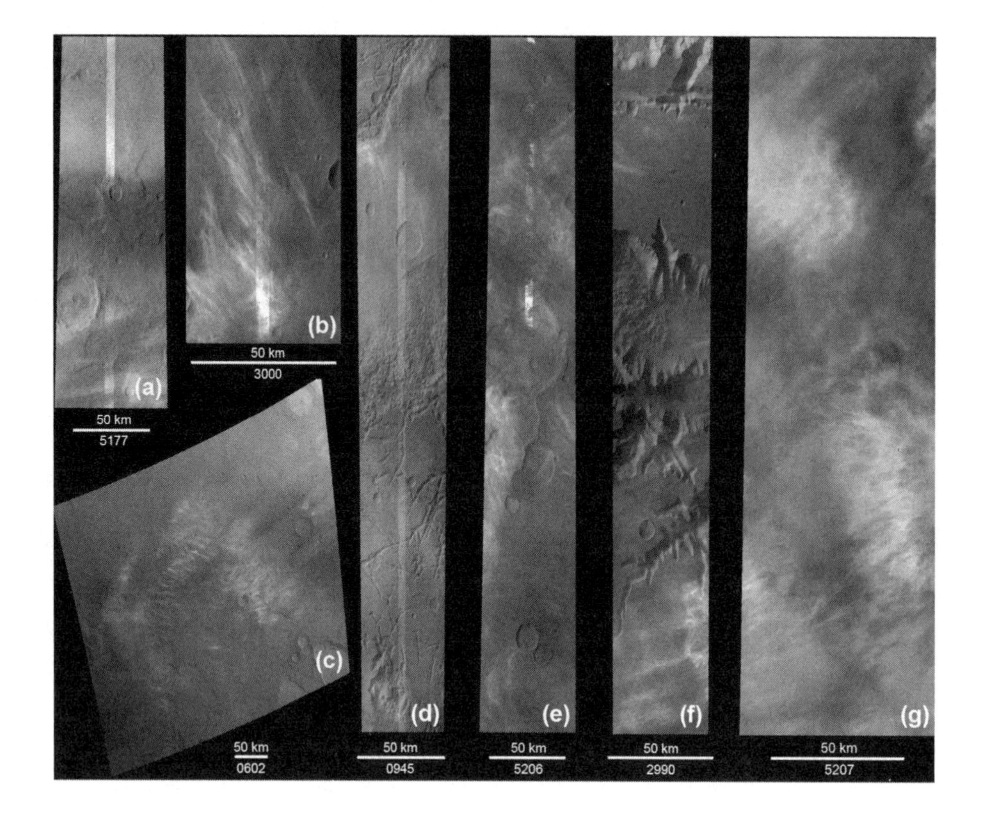

Fig. 5. HRSC image examples of mesospheric ice clouds, with four simultaneous observation frames [(a), (b), (d), and (e)] by OMEGA (white, narrow streaks) and HRSC. Reprinted from *Määttänen et al.* (2010), with permission from Elsevier.

but the apparent parallax of the clouds in the direction perpendicular to the orbit reveals information on the cloud speeds. The cloud speeds were analyzed in 20 cases of HRSC imaging by *Scholten et al.* (2010). *McConnochie et al.* (2010) published an extensive THEMIS-VIS dataset of northern midlatitude clouds with altitude and speed determinations.

The Sun-synchronous satellite orbits and nadir observations constrained only to near pericenter for most instruments limit the local time coverage. Limb observations and occultations do not suffer as much from this limitation since they are off-nadir observations. The Mars Climate Sounder (MCS) on MRO is a radiometer that primarily observes the martian limb in the thermal infrared, providing atmospheric profiles with good spatial and temporal coverage. High-altitude clouds are visible in the MCS data, but their high-altitude cloud observations were only being analyzed during the writing of this chapter (*Kass et al.*, 2011; *Sefton-Nash et al.*, 2012). The very first MCS results on high-altitude detached layers have just been published (*Sefton-Nash et al.*, 2013).

3.2. Spatial and Seasonal Distribution of Clouds

The near-equatorial MMCs are typically observed during the first half of the martian year, the first clouds appearing right at the spring equinox ($L_S = 0°$) or immediately after, and the last cloud disappearing around $L_S = 150°$, well before the autumn equinox (Table 1). Figures 6a and 6b show the spatial and seasonal distribution of MMCs from all published datasets, the details of which will be discussed in the following text.

The first extensive MMC dataset was the one published by *Clancy et al.* (2003, 2004, 2007). It was based on observations of TES and MOC on MGS that observed equatorial MMCs (which they referred to as Mars equatorial mesospheric, or MEM, clouds) during martian years (MY) 24–26 (April 1999–May 2004). The seasonal distribution that emerged revealed equatorial MMCs forming onward from the spring equinox ($L_S = 0°$) until $L_S = 55°$, with a pause around aphelion ($L_S = 71°$), and new detections between $L_S = 105°$ and $L_S = 180°$. Peak frequencies of occurrence were observed at $L_S = 20°–35°$ and a secondary peak at $L_S = 145°–160°$. *Clancy et al.* (2007) did not mention anything about interannual variability in the dataset. Their equatorial MMCs were observed in a latitude range from 20°N to 15°S and they were confined to certain longitude ranges (60°–120°W, 320°–350°W, and 30°W–10°E). Limb observations allowed for cloud altitude determinations as well, and a radiative transfer model was used to estimate composition and particle sizes. This first multiannual dataset revealed the basic seasonal behavior of the equatorial MMCs, which was confirmed and complemented by the later datasets.

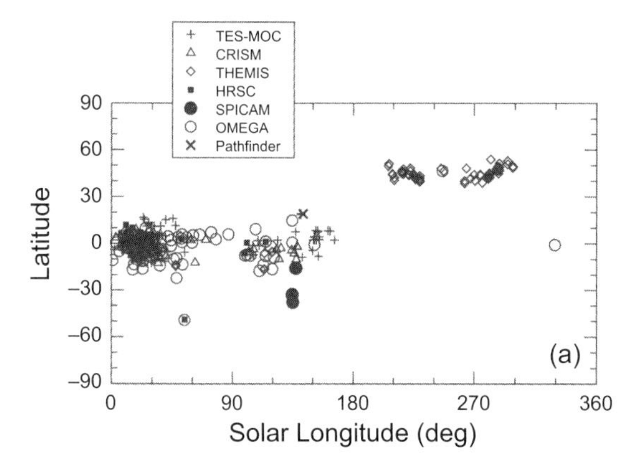

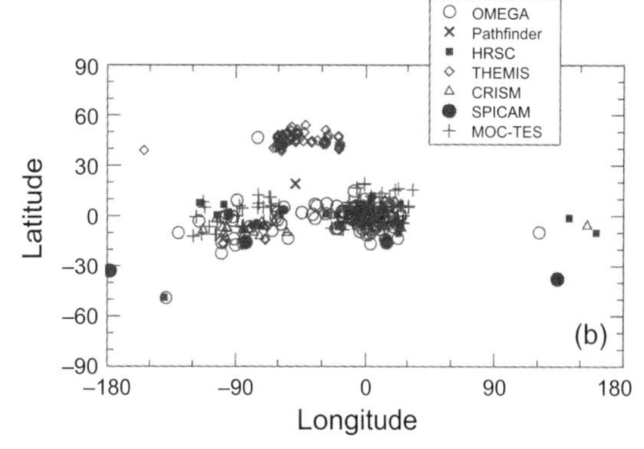

Fig. 6. (a) Latitude-solar longitude map of datasets on mesospheric clouds, as noted in the legend. **(b)** Latitude-longitude map of datasets on mesospheric clouds, as noted in the legend. The Pathfinder rover observation of bluish clouds is also marked.

The next long dataset that emerged was the OMEGA dataset (*Montmessin et al.,* 2007; *Määttänen et al.,* 2010; *Vincendon et al.,* 2011), in which the MMCs were mostly observed near the equator (±20°), but also at midlatitudes. The start of the cloud season was very regular (at the spring equinox, L_S = 0°) during the first three years of observations (MY27–29), but in the end of the third year (MY29) an exceptionally early season start was observed when a spring cloud formed at L_S = 330°, one martian month before the equinox. Generally, the clouds prevailed until L_S = 60°, after which a pause of about 30° of L_S (equivalent to one martian month) took place (MY27–29). The clouds reappeared at L_S = 90° or slightly before, and continued to form until L_S = 140°–150°. The clouds were mostly found in a limited longitudinal range that spans from –120°E through the prime meridian to 25°E, with some clouds observed at around 150°–170°E.

In MY30 the cloud formation continued until L_S = 65° (*Vincendon et al.,* 2011) and clouds were observed also at L_S = 74°–78°. These are the closest observations to aphelion made so far, and thus the first part of this cloud season was exceptionally long, starting at L_S = 330° (*Määttänen et al.,* 2010) and continuing up to L_S = 78°. A short aphelion break was observed between L_S = 78°–115°, after which clouds were again observed until L_S = 150°. No peak in the frequency of occurrence after L_S = 150° has been detected in OMEGA data, contrary to the TES and MOC data of *Clancy et al.* (2007).

THEMIS-VIS observed a handful of equatorial clouds that fit well the distributions derived from the other datasets (*McConnochie et al.,* 2010). Most of their clouds were found around Valles Marineris, in one of the longitude ranges where the other instruments have found clouds as well.

CRISM observed equatorial clouds of both H_2O and CO_2 composition, but *Vincendon et al.* (2011) focused on the CO_2 ice identification. The spatial and seasonal distribution agreed well with the previous datasets, with second-order differences in the timing of the aphelion pause and the maxima of occurrence frequencies. Cloud after L_S = 140° were not observed, differing (like OMEGA) from the MOC TES MEM cloud observation peak at this season.

The MMCs are not only observed near the equator, but also at midlatitudes, although at lower altitudes. OMEGA observed two midlatitude clouds: one in the southern hemisphere [MY27, reported by *Montmessin et al.* (2007)], observed simultaneously by HRSC (*Scholten et al.,* 2010), and another in the northern hemisphere (*Määttänen et al.,* 2010). Both clouds formed around 45°–50° latitude during the local autumn (L_S = 54° at 50°S, L_S = 250° at 45°N). Local times (LT) of these cloud observations vary, the southern one being observed at 08 LT and the northern one at 14 LT. THEMIS-VIS observed the northern midlatitudes at twilight (large solar zenith angles) and discovered high-altitude clouds at latitudes 39°–51°N, resembling those observed by OMEGA and HRSC.

The clouds observed in nadir are mostly daytime clouds, with the exception of a dozen cases observed by OMEGA in the morning (8–11 LT) and the midlatitude clouds observed by THEMIS-VIS at twilight (around 17 LT), but it should be kept in mind that the observational coverage in local time is quite limited.

Altitude determination of the MMCs can be challenging. Published altitude datasets are plotted in Fig. 7 (see also Table 1). To analyze the cloud altitude from OMEGA nadir observations, a full spectral inversion of several unknowns with radiative transfer modeling is required (*Montmessin et al.,* 2007) and has not yet been done. What has been shown is that the clouds need to reside above 40 km to be seen at 4.24 μm on OMEGA nadir data (*Montmessin et al.,* 2007), giving a lower limit for the observed altitudes. However, two observations of cloud shadows enabled *Montmessin et al.* (2007) to calculate the cloud altitudes from the observation geometry; these clouds resided around 80 km. The limb-pointing observation modes provide direct estimates of the cloud altitudes through the observation geometry determination. SPICAM observed high-altitude hazes at 90–100 km,

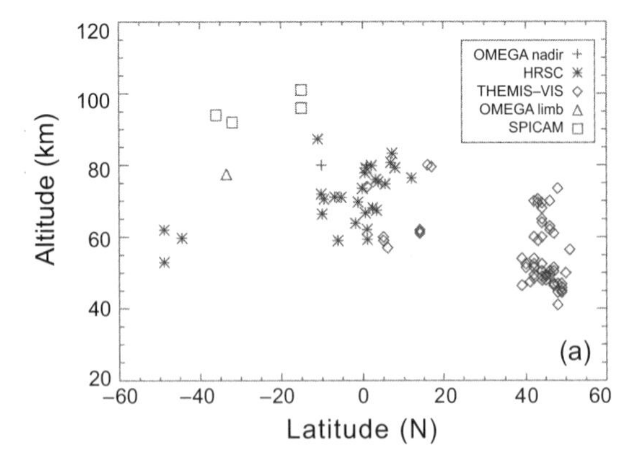

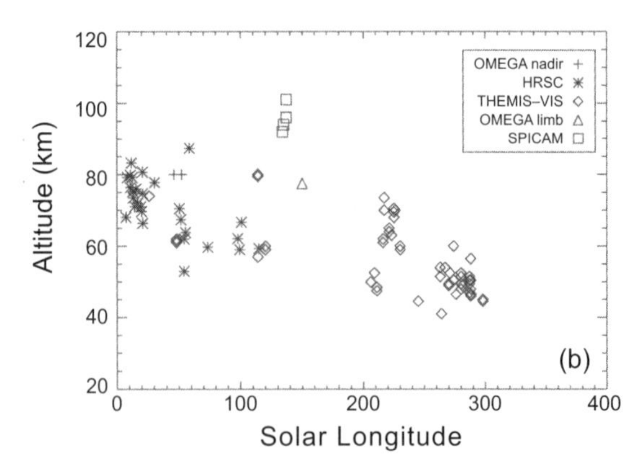

Fig. 7. (a) Altitude-latitude plot of datasets on mesospheric clouds, as noted in the legend. **(b)** Altitude-solar longitude map of datasets on mesospheric clouds, as noted in the legend. Note that possible trends may be caused by observational coverage on different years (see discussion in section 3.2).

immediately after local midnight. MOC and TES observations in the afternoon (13–15 LT) displayed mesospheric haze altitudes of 60–80 km. *Vincendon et al.* (2011) reported an OMEGA observation of an H_2O cloud at 80 km altitude.

The stereo imagers (HRSC) and other multifilter imagers (THEMIS-VIS) can provide precise altitude estimates from nadir observations without radiative transfer modeling (*Scholten et al.,* 2010; *McConnochie et al.,* 2010). The altitudes of the clouds were analyzed in 28 HRSC observations: The MMCs formed mainly at 59–87-km altitudes. These observations (Fig. 7) showed a slight trend for higher altitudes near the equator, and the midlatitude clouds were observed at the lower end of the range, at 53–62 km. A similar trend of higher formation altitudes in early spring than later in spring and summer appeared (Fig. 7). However, the dataset was compiled with observations from different years, with all the early spring observations being from MY29, so the trend could not be distinguished from an interannual effect. Two HRSC-observed clouds in the southern hemisphere have associated altitude determinations (53–62 km). The altitudes observed by MOC and TES and by HRSC agree, even if the observed range of HRSC altitudes is larger. THEMIS altitudes for the near-equatorial clouds are in the same range as those observed by HRSC, and the northern midlatitude cloud altitudes are similar to those observed in the southern midlatitudes by HRSC (Fig. 7).

3.3. Mesospheric Structures Favoring Cloud Formation

Atmospheric models were not able to predict these clouds even though their existence was speculated based on very cold temperatures attainable in the martian atmosphere (*Schofield et al.,* 1997; *Montmessin et al.,* 2006b; *Forget et al.,* 2009). Even presently, modeling efforts have not been able to conclusively explain the atmospheric circulations involved in the cloud formation, or the particle sizes observed (*Colaprete et al.,* 2008; *González-Galindo et al.,* 2011; *Spiga et al.,* 2012). The spatial and seasonal distribution of the MMCs present distinct patterns. Our knowledge of the structure of the martian mesosphere is largely based on modeling, since observations of this part of the atmosphere are scarce. The models have been validated with observations to an extent, using mostly lower-atmosphere observations. The predicted mesospheric temperature structures and winds were subjected to scrutiny though mesospheric cloud observations that imply low temperatures at certain altitudes, and provided rare mesospheric wind measurements. In the following we will try to understand the origin of the mesospheric cloud patterns by drawing from our current understanding of atmospheric processes, and state-of-the-art atmospheric modeling.

Colaprete et al. (2008) discussed CO_2 cloud formation through an extensive treatment of the convective available potential energy (CAPE), which they derived from both observations and modeling. They conducted a comparison, which led them to conclude that the CAPE magnitudes predicted by the model could not explain the observed ones if the so-called moist convection processes were omitted. In moist convection the latent heat released in condensation fuels the vertical motions (through buoyancy) and maintains the convective cell. They focused on polar atmosphere studies, but since they performed global modeling, MMCs appeared in the results as well. At that time, only the observations of *Clancy et al.* (2007) and *Montmessin et al.* (2006a) were available. *Colaprete et al.* (2008) modeled mesospheric tropical clouds at altitudes of 60–70 km displaying very large particle sizes (1–10 μm) and significant opacities ($\tau > 0.1$). Nevertheless, the spatial and temporal distributions differed from what was seen by *Clancy et al.* (2007). *Colaprete et al.* (2008) noted also that the modeled particle sizes and those observed by SPICAM differed greatly, and explained this through the assumptions made on the CN in the model. However, the top of the atmospheric column in the model of *Colaprete et al.* (2008) was at 80 km, lower than any of the possible CO_2 clouds observed by SPICAM (*Montmessin et al.,* 2006a), and thus the model of *Colaprete et al.* (2008) could not have modeled the SPICAM high-altitude clouds. This low model top might be the main reason for the discrepancies between cloud predictions and observations: The tides and waves that propagate to the upper atmosphere will not be properly modeled with such a low top (*González-Galindo et al.,* 2011). Nevertheless, the work of *Colaprete et al.* (2008) was the first global modeling attempt to include CO_2 ice microphysics to describe equatorial mesospheric CO_2 clouds.

Määttänen et al. (2010) performed CAPE calculations based on condensed ice mass in a cloud estimated from observed cloud optical depths. Their conclusions were that the CAPE values in the mesosphere derived from the cloud observations were too small to induce convective scale updrafts, but were in a realistic range at the poles.

Clancy et al. (2007) suggested that the mesospheric cloud formation around the equator might be driven by the well-known thermal tide temporal and spatial variations that are particularly strong at this region (but reach higher latitudes as well). The martian atmosphere is very strongly heated during daytime by absorption of solar radiation by mineral dust and gases (CO_2). The day-night variations in the atmosphere are strong and they drive a thermal tide appearing as a very typical node-antinode structure propagating vertically through the atmosphere. *González-Galindo et al.* (2011) showed through climate modeling with the Laboratoire de Météorologie Dynamique (LMD-MGCM) that the thermal tide is the dominant effect in the thermal structure of the martian low-latitude mesosphere, especially in the mesospheric cloud formation altitudes. In particular, they showed that the diurnal variation of formation altitudes (daytime clouds: 60–85 km, nighttime clouds: 90–110 km) followed the predicted altitude of the temperature minimum (antinode) of the thermal tide. The longitudinal confinement of the equatorial MMCs was also explained well by the model through the superposition of the diurnal migrating thermal tide and nonmigrating tides. However, *González-Galindo et al.* (2011) noted that these effects were not sufficient to decrease the modeled temperatures below the CO_2 condensation level, but other perturbations, such as GWs, were required.

The study of *González-Galindo et al.* (2011) had less success in trying to explain formation of CO_2 clouds in the northern midlatitudes, where the temperature minima predicted by the model were found in the morning at high altitudes, whereas the cloud observations of THEMIS-VIS focused on late afternoon and below 60 km. Possible explanations were identified, one of which simply implied that the composition of the THEMIS-VIS clouds might be H_2O [the composition analysis of *McConnochie et al.* (2010) could not be applied to their midlatitude observations]. Another possibility is that the LMD-MGCM overestimates mesospheric temperatures.

The additional perturbations required for subcondensation temperatures mentioned by *González-Galindo et al.* (2011) were theoretically studied by *Spiga et al.* (2012), who showed that gravity wave propagation to the mesosphere is favored in some regions and completely suppressed in others, depending on the structure of the lower atmosphere zonal winds and atmospheric stability. They showed that most of the equatorial MMCs have been observed in regions of favored gravity wave propagation (where GWs are not filtered out by the thermal structure or the large-scale winds). This study supports the hypothesis of the MMCs forming through an interplay between atmospheric general circulation and GWs.

Model comparisons with the rare dataset of mesospheric wind speeds were performed by *Määttänen et al.* (2010), *McConnochie et al.* (2010), and *González-Galindo et al.* (2011). The wind directions and speeds deduced from observations were mainly in the range expected by the LMD-MGCM predictions (*Määttänen et al.,* 2010; *González-Galindo et al.,* 2011). *McConnochie et al.* (2010) found some discrepancies between their data and the Weather Research and Forecasting (WRF) global climate model related to the westerly winds at midlatitudes above 65 km. These model-observation comparisons confirmed that the equatorial mesosphere is dominated by easterly winds with fairly large wind speeds, whereas in the northern midlatitudes the major wind component is westerly. In some cases the longitudinal variations of the wind speeds were quite well reproduced by the model, even though some very strong easterlies could not be explained even through daily variability.

3.4. Morphology of the Mesospheric Clouds

Most of the clouds seen by the high-(spatial)-resolution instruments (HRSC, THEMIS-VIS, CRISM) have a filamentary, cirrus-like character, with elongated, thin threads sometimes exhibiting wavelike structures forming the cloud bulk. Even if these cirriform clouds show significant variation in shapes and internal complexity, the conclusion about the cirrus-like nature of the clouds remains the same. Even on Earth, cirrus clouds, also formed of ice crystals, exhibit a variety of morphologies and subtypes.

McConnochie et al. (2010) classified their midlatitude clouds in different subtypes: "clumpy," "linear," and "linear periodic" clouds. In their classification, the clumpy clouds showed small-scale structures but no large-scale organization. The linear clouds were organized into linear features in the southwest-to-northeast direction. The linear periodic clouds had multiple organized linear features with regular spacing. Their equatorial clouds were described (following *Clancy and Sandor,* 1998) to be similar to the Pathfinder "discrete linear clouds" composed of "filaments" and "lineations."

The HRSC equatorial clouds showed mostly "ripple-like elongated habit oriented roughly in east-west, sometimes northwest-southeast directions" (*Scholten et al.,* 2010). Scholten et al. also noted that the cloud morphologies were not limited to this case, but a multitude of forms were found in the images.

CRISM CO_2 ice clouds were found to present linear stripes (with predominantly east-west orientation). The compression in the orbital direction of the image by their observation technique (where during an imaging sequence the instrument looks at one point at the surface from different observation angles) was deemed to partly cause this persistent feature (*Vincendon et al.,* 2011).

Montmessin et al. (2007) published OMEGA images with clouds of clearly roundish, cumuliform appearance seen in the first set of OMEGA data published (MY27), and explained these as a manifestation of moist convection fueled by the latent heat release in condensation [a process that had been suggested for polar CO_2 clouds by *Colaprete et al.* (2008)]. In addition, moist convection would also be able to create strong updrafts that are needed to keep the observed micrometer-sized crystals aloft (section 3.4), which seems to defy sedimentation. Such clouds were found in the rest of the dataset as well [MY27–29 (*Määttänen et al.,* 2010)], but mostly in conditions of wide imaging angles when the instrument was observing far from the pericenter (and from the planet surface). No simultaneous wide-angle images from OMEGA and HRSC were acquired before 2010, since HRSC had stronger constraints on observations far from periapsis. *Vincendon et al.* (2011) published a simultaneous observation from OMEGA and HRSC, and stated that the previous observations claimed to be cumuliform clouds had been biased by the OMEGA spatial resolution, which is not high enough to distinguish the filamentary inner structure of the clouds seen in the HRSC image.

The imaging instrument studies (*McConnochie et al.,* 2010; *Scholten et al.,* 2010) have noted that the MMCs often (but not exclusively) exhibit aligned wave structures. In the midlatitude clouds the "linear" class of *McConnochie et al.* (2010) presented southwest-to-northeast oriented linear structures. HRSC often observed equatorial clouds that were oriented in the west-east or northwest-to-southeast direction (*Scholten et al.,* 2010). CRISM clouds were predominantly east-west oriented, but *Vincendon et al.* (2011) also showed that the east-west orientation was somewhat overestimated due to the compression effect of the observation projection. In other words, any southeast-northwest-oriented cloud will become predominantly east-west oriented due to the CRISM observation geometry and image spatial compression. Nevertheless, a combination of these imaging datasets to study the waveforms found in the clouds might lend clues to the mesospheric dynamics and winds, the formation mechanisms of the clouds, and possibly provide a tool for gravity-wave studies.

3.5. Cloud Properties: Composition, Optical Depth, and Crystal Size

The observation methods and wavelengths largely define whether the cloud composition can or cannot be determined. In the case of OMEGA observations (*Montmessin et al.,* 2007; *Määttänen et al.,* 2010; *Vincendon et al.,* 2011), the spectral recognition of CO_2 ice at the wavelength of 4.24 μm is unambiguous. Other instruments have not conducted observations at the same wavelength, but radiative transfer modeling enabled *Vincendon et al.* (2011) to discriminate between H_2O and CO_2 ice in CRISM spectra.

The TES and MOC observations of *Clancy et al.* (2007) were more difficult to interpret, and they concluded through radiative transfer modeling that a water ice composition of $r_{eff} < 1$-μm crystals or CO_2 ice composition of $r_{eff} < 1.5$-μm crystals could be consistent with modeling of radiances from TES solar and IR bands.

McConnochie et al. (2010) performed radiative transfer modeling on their THEMIS-VIS observations, but in only some equatorial cases, since their plane-parallel model was too simplified for conditions at twilight. Their preferred composition was heterogeneous CO_2 ice crystals (crystals containing a dust core). The condensed ice masses were so large based on their estimates on optical depth and particle sizes that water was ruled out because of its small abundance in the mesosphere.

The visible optical depths (Table 1) of the MMCs may be in the range 0.3–0.6 as determined in some nadir observations (*Montmessin et al.,* 2007; *McConnochie et al.,* 2010; *Määttänen et al.,* 2010; *Vincendon et al.,* 2011). Subvisible clouds were observed by *Montmessin et al.* (2006a) with optical depths in the range 0.006–0.05 (in the UV, 200 nm), and by *Clancy et al.* (2007), whose TES limb radiance analysis gave 0.005 for the visible (0.7 μm) optical depth of the MEM clouds.

The estimation of particle sizes in the clouds (Table 1) can be acquired in many ways, with the most used but not the simplest being full radiative transfer modeling and spectral inversion of the data. A simpler way to achieve such estimates is through observation of cloud shadows, giving a good estimate of the attenuation of insolation by the cloud. The wavelength dependence of this attenuation can give an estimate of the particle size.

OMEGA observed twice the cloud shadows, which enabled, as shown by *Montmessin et al.* (2007), the estimation of the cloud altitude from geometry, as well as the opacity and the effective radii of the particles from the observed attenuation of sunlight in the shadow. *Montmessin et al.* (2007) performed the calculations for the cloud on average, and *Määttänen et al.* (2010) refined the study by calculating the properties (opacity and effective radius) for each cloud pixel (mapping these properties and, for the first time, their variation inside the cloud). This is, to our knowledge, the first ever within-cloud mapping of the horizontal variation of cloud properties. These clouds were relatively opaque, since they cast rarely observed shadows on the planet's surface. Indeed, the cloud opacities were found to go up to 0.5–0.6 with mean values of 0.14 and 0.25 for the two observations. The particle sizes were modeled with Mie scattering using a gamma size distribution. The retrieved effective radii were mainly between 1 and 3 μm.

Three equatorial cloud observations from THEMIS-VIS (*McConnochie et al.,* 2010) were analyzed with a full radiative transfer model using several candidate compositions (dust, H_2O ice, CO_2 ice, or dust core covered with one of the ices). The CO_2 ice results (with or without a dust core) gave in three cases an effective radius close to the ones observed by OMEGA (0.7–1.5 μm vs. 1–3 μm). Five cases were found in a lower radius range range of 50–100 nm and one observation could not be analyzed at all with pure CO_2 ice composition. *McConnochie et al.* (2010) mentioned estimates on the upper limit of the dust core size in one case (0.5 μm). Had the dust cores been larger, they would have also been seen in the infrared filter.

The THEMIS-VIS observations of small radii agree with the SPICAM stellar occultation particle sizes [80–130 nm (*Montmessin et al.,* 2006a)], but THEMIS-VIS observations were made during daytime, and SPICAM observations 01 LT in the southern subtropics at 90–100 km altitude. One daytime OMEGA observation of *Vincendon et al.* (2011) showed that small-grain water ice clouds can be observed at high altitudes in the southern low latitudes, suggesting that the SPICAM observations might be related to these high-altitude water ice hazes. CRISM CO_2 ice cloud observations (*Vincendon et al.,* 2011) were analyzed with radiative transfer modeling, giving particle size estimates at the lower end of OMEGA retrievals, 0.5–2.0 μm, and their visible optical depths (at 0.5 μm) were typically around 0.2. These daytime values agree well with those of OMEGA.

3.6. Variability and Long-Term Trends

All the published datasets sum up to seven martian years (MY24–30) of mesospheric cloud observations. However, the different datasets have been acquired with varying instruments and methods. The spatial and temporal coverage of the data is far from ideal, and none of these datasets are obtained from an instrument designed to monitor the martian atmosphere with good temporal and spatial coverage. Such an instrument exists, the MCS on MRO, and their preliminary results (*Kass et al.,* 2011; *Sefton-Nash et al.,* 2012) were recently reinforced by a paper on the first results of high-altitude layer observations (*Sefton-Nash et al.,* 2013). However, this chapter only discusses the other datasets, and in the following we will list briefly the possible daily and interannual variability detected in the other datasets.

Determination of daily variability suffers from the limited local time distribution of the observations and, in particular, the rarity of nighttime observations. According to the available observations, the daytime MMCs form at lower altitudes (60–85 km) and they can be characterized by very large particle sizes (up to 3 μm). The nighttime clouds form higher in the atmosphere (90–100 km, following the propagation of the thermal tide antinode), and the nighttime cloud particle sizes are smaller (around 100 nm).

Interannual variations should be easier to detect, but the problem is the multitude of the observation methods, along with the insufficient spatial and temporal coverage of the instruments that have provided datasets so far. The occurrence frequencies of equatorial MMCs were reported to have two peaks in the MY24–26 dataset of *Clancy et al.* (2007), but OMEGA and CRISM (MY26–30) do not see the second peak in mesospheric CO_2 cloud observations in late summer (*Määttänen et al.,* 2010; *Vincendon et al.,* 2011). *Vincendon et al.* suggested that the late summer MMCs of *Clancy et al.* (2007) might be an example of high-altitude H_2O clouds similar to the one *Vincendon et al.* (2011) observed with OMEGA at the martian limb. OMEGA observations show apparent interannual variations of the start and length of the cloud season. The first clouds have been observed on varying dates within one month around the spring equinox (see section 3.2 for details). The length of the aphelion pause seems to vary as well, with solitary clouds observed sometimes even at summer solstice (L_S = 86.7°, MY29), and cloud formation resuming immediately afterward (L_S = 99.3°, MY27); alternatively, as in the case of MY30, there could be a clear pause starting a bit before the solstice and lasting until almost L_S = 115° (*Määttänen et al.,* 2010; *Vincendon et al.,* 2011). The midlatitude clouds have been extensively mapped only by *McConnochie et al.* (2010), who did not evoke interannual variations in their dataset.

It is not possible to extract long-term trends because of the fairly short time series and the different observational methods and coverages.

4. COMPARISON OF THE MESOSPHERIC CLOUDS, THEIR PROPERTIES, AND FORMATION MECHANISMS ON THE TWO PLANETS

In this section we will summarize and compare the properties of the mesospheric clouds on the two planets and discuss their differences and similarities. The discussion will be based on the reviews of the two cloud types presented in the preceding sections, which are critical for following this comparison.

The properties of mesospheric clouds on both planets are grouped in Table 1, and constitute the core of this chapter. At first glance it might seem that these clouds are very different on the two planets: On Mars they are equatorial or midlatitude clouds, and on Earth they are found at high latitudes. On Earth they are also observed to be homogeneous (on average) at all longitudes, whereas on Mars the longitudinal distribution is very distinct. On Earth, there is little variation in the PMC formation altitude, whereas on Mars the MMC altitude range is much wider. Interestingly enough, both the PMCs and the equatorial MMCs form during the local summer. In addition, autumn MMCs form at the midlatitudes. On both planets the cloud morphology seems similar, mostly cirriform with wavy structures, but some martian observations suggest cumuliform-type morphologies (unconfirmed). A big difference between the planets is the cloud composition, since on Mars the atmospheric CO_2 can condense as ice crystals, but on Earth the PMCs are composed of water ice. However, on Mars, certain observations suggest water ice composition for the MMCs as well. The optical depths are generally (both on Earth and Mars) in the subvisible range, but in rare martian observations the clouds have cast observable shadows on the planet's surface, revealing surprisingly high optical depths. The crystals sizes are clearly different: The largest terrestrial PMC crystals are approximately the same size as the smallest MMC particles, and the martian size range spans several orders of magnitude and measuring up to few micrometers (in effective radius).

Some interesting points arise throughout this comparison, and in what follows we will discuss these points based on the present-day knowledge (which may be largely based on unconfirmed hypotheses or theoretical work, particularly for Mars). We will discuss the formation mechanism related to mesospheric dynamics and to microphysics. We will also discuss the role of mesospheric clouds as possible indicators of climate change.

4.1. Mesospheric Structures Enabling Cloud Formation

Since the martian atmosphere has been observed only for some decades and there are much less data than for Earth, it is clear that our understanding of the Mars mesospheric processes is not as advanced as for the terrestrial mesosphere. Only in recent years have important advances on mesospheric modeling on Mars attempted to explain the mesospheric cloud seasonality and spatial distribution. On Earth, the structure and dynamics of the mesosphere are much better known, thanks to a larger observational database and more advanced models.

On both planets, GWs propagating up to mesospheric altitudes are crucial for the mesospheric cloud formation. Their vertical propagation is limited by the thermal and wind structure of the lower atmosphere, since large-scale circulation can effectively filter them out. On both planets the seasonal variation of this filtering effect defines the timing of the mesospheric cloud season. However, the effect of the GWs on the mesospheric circulation is not the same on Mars as it is on Earth. As previously mentioned, on Mars, general circulation (mainly from the thermal tide and nonmigrating tides) defines the coldest areas in the mesosphere where cloud formation can occur. However, GWs are required to locally perturb the mesospheric temperature structure so that a supersaturated state can be attained and cloud formation can be initiated. On Earth, the GWs are responsible for the large-scale mesospheric circulation that makes cloud formation possible.

The properties of GWs have been studied on Earth based on observations of wave patterns of the PMCs/NLCs. The wave patterns observed (alignment of cloud features; see section 3.4) in MMCs have not yet been exploited: Similar methods might be usable on Mars as well and might help shed light on GW properties in the martian mesosphere. It should be noted that on Earth the NLC features used for GW studies are a result of the combination of the phase speed of the GWs and the local wind. The effect of GW phase speed has not been taken into account in the interpretation of MMC speeds (relating their speeds directly to local wind speeds).

Another aspect related to the structure of the mesosphere is the local vapor supply. On Mars, the vapor source for CO_2 clouds could be described as "infinite," since CO_2 is the major component of the atmosphere and thus will not be depleted rapidly during cloud formation. However, for H_2O on Mars the question of the source is an important one. It was thought that water vapor would be confined to the lowest atmosphere, since the vapor condensation level (and cloud formation level) would function as an effective cold trap for the vapor. However, water vapor supersaturation on Mars (*Maltagliati et al.,* 2011) and high-altitude water ice clouds (*Vincendon et al.,* 2011) have recently been observed. The general circulation is most probably responsible for the vertical advection of water vapor, and the local ineffectiveness of the cold trap in lack of CN (inhibiting cloud formation at lower altitudes) is the factor that allows the increased vertical extent of water vapor. In the case of Earth, the vapor source is also an important factor. On the one hand, water vapor is advected from lower altitudes by general circulation through strong upward motions that occur in the summer polar region, but on the other hand, the same upward motions can advect methane, which is photodissociated above 30 km, leading to local formation of water vapor. These two mesospheric water sources are presently thought to be of equal importance.

4.2. Cloud Microphysics

Even if the mesospheric clouds form in the coldest regions of the atmosphere, the observed temperatures and supersaturations are not sufficient for homogeneous nucleation to occur, and heterogeneous nucleation is considered as the main formation pathway. Thus, CN are needed, but their source is unknown. The martian atmosphere is extremely dusty, but the thinness of the atmosphere promotes the sedimentation of particles, no matter what altitude they are lofted to. In other words, even if during a dust storm strong vertical motions could lift dust particles up to high altitudes, they would sediment to lower levels relatively fast. Even if the smallest dust particles sediment more slowly, their number is limited, thus reducing their efficiency as the main source of activated ice nuclei. Thus dust particles, despite their ubiquitousness in the lower atmosphere, are an improbable mesospheric CN candidate on Mars. However, other ideas might come from the terrestrial community, where meteoric smoke particles are currently the most popular candidate as CN.

The meteor fluxes and subsequent meteoric smoke formation rates, along with the importance of meteoric smoke on both planets, could be estimated. Discussions have been presented for Earth (with few comments on the significance of meteoric smoke on other planets) by *Saunders and Plane* (2011), and *Plane* (2011) and *Saunders et al.* (2010) discussed and experimentally evaluated the role of meteoric smoke as ice nuclei. The meteoric smoke particles have been detected, for the first time, from an instrument orbiting Earth (*Hervig et al.,* 2009a), but on Mars such detections have not yet been made, and definite confirmation on the ice nuclei identity will need to wait for dedicated observations, laboratory experiments on ice nuclei candidates, and modeling of nucleation microphysics. The terrestrial community has made more progress toward this end, and even if definitive answers have not yet been acquired, they may perhaps help to shed light on the mystery of martian mesospheric ice nuclei.

Another microphysical parallel between the two mesospheres may be represented by the observed offset in altitude of the maximum supersaturation and the visible clouds. On Earth, water ice particles, with radii ranging from a few nanometers to about 100 nm, are persistently present in the high-latitude summertime mesopause region. The smallest particles produce strong radar echoes called PMSEs (see section 2.3), which are always observed in the areas where the highest supersaturations are found, right below the mesopause. However, the clouds themselves are found on average some kilometers below. On Mars, SPICAM stellar occultation observations (*Montmessin et al.,* 2006a) revealed detached cloud layers below supersaturated pockets, sometimes adjacent to these pockets, sometimes kilometers below them. On Earth, the reason for this altitude difference is well understood. Polar mesosphere summer echoes appear in the region where nucleation (ice particle formation) begins, but the crystal sizes are too small to be observed with optical instruments. The particles sediment, growing simultaneously and reaching observable sizes at lower altitudes. This scenario may also be the explanation for the offset of clouds and supersaturated pockets on Mars. However, on Earth the PMSE altitude is directly related to the mesopause altitude, which is fairly stable. On Mars, the probable explanation for the supersaturated pocket formation is propagating GWs, which render the supersaturated pocket altitudes much more variable. In any case, the microphysical explanation of the offset (sedimentation and simultaneous growth) may also play a role on Mars.

4.3. Mesospheric Clouds: Climate Change Indicators?

As discussed in section 2.6, there are some indications of long-term changes in the PMC observations, revealed by a particularly long observational time. The PMCs seem to have increased in frequency and brightness over a timescale of decades, and they have been observed at unprecedentedly low latitudes. Whether these changes are due to long-term global changes in the atmosphere is still an open and debated question.

It is known that greenhouse gas emissions, which have a warming effect on the lower atmosphere, lead to an overall cooling of the upper layers (strato- and mesosphere). This in itself should have a direct effect on increased PMC formation. An indirect effect of anthropogenic emissions is the increase of water vapor in the upper atmosphere through increased global methane concentration (and the consequential photodissociation of water vapor in the middle atmosphere). In principle, more water and colder temperatures mean more clouds. Thus it seems plausible that changes in the thermal structure and chemical composition of the atmosphere should lead to changes in the PMCs as well.

However, despite the decades-long time series of observations of PMCs, their natural variability is not yet fully understood. Thus a final answer to whether the PMCs are the harbingers of global climate change or not requires better insight into the full PMC variability and its sources. In a nonlinear system as the atmosphere, this remains a difficult task, but dedicated missions and observations through scientific analysis and modeling will undoubtedly cast light on the unresolved issues.

However, on Mars, this deduction is not straightforward. First, the MMC datasets are much smaller than in the case of PMCs, and even solar cycle effects might still be difficult to see, despite a sufficiently long-term dataset. In addition, present knowledge indicates that no rapid global change is happening on Mars at the moment. Moreover, at present the radiative effects of the MMC are negligible. In a past climate the CO_2 clouds (which did not necessarily form in the mesosphere) may have had an important radiative role through their scattering greenhouse effect (see, for example, *Forget and Pierrehumbert,* 1997). Carbon dioxide ice strongly scatters infrared radiation and thus may act as a warming climate component, but as recent climate modeling results show, this effect may only have been moderate (*Forget et al.,* 2013).

5. CONCLUSIONS

In this chapter, we have summarized and compared the present knowledge regarding mesospheric clouds on Mars and on Earth. Any such comparison will reveal both similarities and differences, since no atmospheric phenomenon is perfectly reproduced on another planet. The main goals of this chapter were to present reviews of the martian and terrestrial mesospheric clouds, compare the common factors, and briefly discuss the ways in which research on the mesospheric clouds on one planet can contribute to research on the other.

The datasets on the martian clouds published so far sum up to seven martian years of MMC observations. We are still far from complete coverage that would allow drawing definitive conclusions on short- or long-term variations, but a picture of the main features of the MMCs has emerged. The PMCs on Earth have been monitored for decades by a multitude of instruments. They have recently become a subject of great interest because they have proved to be very useful tracers of the complex dynamical processes that control the state of the terrestrial mesosphere, and this is also true of the MMCs. The evolution of the PMCs may also be related to global climate change at upper altitudes.

The comparison of mesospheric clouds on the two planets has revealed certain similarities regarding the formation mechanisms of these clouds. On both planets, the interplay of general circulation and GWs is considered a critical ingredient in the development of an atmospheric state that enables the formation of PMCs, even if the method that leads to this end result is not the same on the two planets. On both planets the exact cloud formation mechanism remains an open question: If the cloud crystals form by heterogeneous nucleation, what are the CN composed of, and where do they come from?

Significant differences are revealed as well. The terrestrial mesospheric clouds form at a remarkably narrow altitude range dictated by the position of the mesopause, whereas on Mars the clouds have been observed in an altitude range of tens of kilometers, modulated at least in part by the vertical distribution of the nodes and antinodes of the diurnal thermal tide. The particle sizes on Earth are clearly smaller than on Mars; one of the reasons for this might be vapor concentrations, which are much smaller on Earth (water vapor is a trace gas) than on Mars (CO_2 condensation, since CO_2 is a major species).

Terrestrial PMC studies are at a more advanced state thanks to longer observational datasets. As an outlook, we note some possibilities where PMC studies can help to clarify the mysteries of the martian mesosphere and clouds. Research on the possible CN for both planets can advance at the same time, since a good candidate, meteoric smoke, probably shares similar properties on both planets. The PMCs have been used to extract information on GW properties in the mesosphere, and similar methods could be applied to Mars as well.

Acknowledgments. The authors thank the reviewers for careful reading of the manuscript and their constructive comments. AM gratefully acknowledges mesospheric cloud project funding from the French Space Agency (CNES). We also thank the European Space Agency (ESA) and the company ACRI-ST, respectively, for funding the GOMOS mission and for the data processing, and we thank the CNES for funding the LATMOS participation in GOMOS.

REFERENCES

Akmaev R. and Fomichev V. (2000) A model estimate of cooling in the mesosphere and lower thermosphere due to the CO_2 increase over the last 3–4 decades. *Geophys. Res. Lett., 27(14),* 2113–2116, DOI: 10.1029/1999GL011333.

Baumgarten G., Fiedler J., Fricke K., Gerding M., Hervig M., Hoffmann P., Müller N., Pautet P.-D., Rapp M., Robert C., Rusch D., Von Savigny C., and Singer W. (2009) The noctilucent cloud (NLC) display during the ECOMA/MASS sounding rocket flights on 3 August 2007: Morphology on global to local scales. *Ann. Geophys., 27(3),* 953–965, DOI: 10.5194/angeo-27-953-2009.

Baumgarten G., Fiedler J., and Rapp M. (2010) On microphysical processes of noctilucent clouds (NLC): Observations and modeling of mean and width of the particle size distribution. *Atmos. Chem. Phys., 10,* 6661–6668, DOI: 10.5194/acp-10-6661-2010.

Berger U. and Von Zahn U. (2002) Icy particles in the summer mesopause region: Three-dimensional modeling of their environment and two-dimensional modeling of their transport. *J. Geophys. Res., 107(A11),* 1366, DOI: 10.1029/2001JA000316.

Brasseur G. P. and Solomon S. (2005) *Aeronomy of the Middle Atmosphere: Chemistry and Physics of the Stratosphere and Mesosphere.* Springer, Berlin. 644 pp.

Chandran A., Rusch D., Palo S., Thomas G., and Taylor M. (2009) Gravity wave observations in the summertime polar mesosphere from the Cloud Imaging and Particle Size (CIPS) experiment on the AIM spacecraft. *J. Atmos. Solar-Terrestrial Phys., 71(3–4),* 392–400, DOI: 10.1016/j.jastp.2008.09.041.

Clancy R. T. and Sandor B. J. (1998) CO_2 ice clouds in the upper atmosphere of Mars. *Geophys. Res. Lett., 25,* 489–492.

Clancy R. T., Wolff M. J., Whitney B. A., and Cantor B. A. (2003) Vertical distributions of dust optical depth during the 2001 planet encircling storm from a spherical radiative transfer analysis of MOC limb images. In *Sixth International Conference on Mars* (A. L. Albee and H. H. Kieffer, eds.), p. 3205.

Clancy R. T., Wolff M., Whitney B., and Cantor B. (2004) The distribution of high altitude (70 km) ice clouds in the Mars atmosphere from MGS TES and MOC limb observations. *AAS/Division for Planetary Sciences Meeting Abstracts #36, Bull. AAS, 36,* 1128.

Clancy R. T., Wolff M. J., Whitney B. A., Cantor B. A., and Smith M. D. (2007) Mars equatorial mesospheric clouds: Global occurrence and physical properties from Mars Global Surveyor Thermal Emission Spectrometer and Mars Orbiter Camera limb observations. *J. Geophys. Res., 112,* E04004, DOI: 10.1029/2006JE002805.

Colaprete A., Barnes J. R., Haberle R. M., and Montmessin F. (2008) CO_2 clouds, CAPE and convection on Mars: Observations and general circulation modeling. *Planet. Space Sci., 56,* 150–180, DOI: 10.1016/j.pss.2007.08.010.

Deland M., Shettle E., Thomas G., and Olivero J. (2003) Solar backscattered ultraviolet (SBUV) observations of polar mesospheric clouds (PMCs) over two solar cycles. *J. Geophys. Res., 108(D8),* 8445, DOI: 10.1029/2002JD002398.

Deland M., Shettle E., Thomas G., and Olivero J. (2006) A quarter-century of satellite polar mesospheric cloud observations. *J. Atmos. Solar-Terr. Phys., 68(1),* 9–29, DOI: 10.1016/j.jastp.2005.08.003.

Deland M., Shettle E., Thomas G., and Olivero J. (2007) Latitude-dependent long-term variations in polar mesospheric clouds from SBUV version 3 PMC data. *J. Geophys. Res., 112,* D10315, DOI: 10.1029/2006JD007857.

Donahue T., Guenter B., and Blamont J. (1972) Noctilucent clouds in daytime: Circumpolar particulate layers near the summer mesopause. *J. Atmos. Sci., 29(6),* 1205–1209.

Fiedler J., Baumgarten G., and Von Cossart G. (2005) Mean diurnal variations of noctilucent clouds during 7 years of lidar observations at ALOMAR. *Ann. Geophys., 23,* 1175–1181.

Fiedler J., Baumgarten G., Berger U., Hoffmann P., Kaifler N., and Lübken F.-J. (2011) NLC and the background atmosphere above ALOMAR. *Atmos. Chem. Phys., 11,* 5701–5717, DOI: 10.5194/acp-11-5701-2011.

Forget F. and Pierrehumbert R. T. (1997) Warming early Mars with carbon dioxide clouds that scatter infrared radiation. *Science, 278,* 1273, DOI: 10.1126/science.278.5341.1273.

Forget F., Hourdin F., Fournier R., Hourdin C., Talagrand O., Collins M., Lewis S. R., Read P. L., and Huot J.-P. (1999) Improved general circulation models of the martian atmosphere from the surface to above 80 km. *J. Geophys. Res., 104,* 24155–24176.

Forget F., Montmessin F., Bertaux J.-L., González-Galindo F., Lebonnois S., Quémerais E., Reberac A., Dimarellis E., and López-Valverde M. A. (2009) Density and temperatures of the upper martian atmosphere measured by stellar occultations with Mars Express SPICAM. *J. Geophys. Res., 114(E13),* E01004, DOI: 10.1029/2008JE003086.

Forget F., Wordsworth R., Millour E., Madeleine J.-B., Kerber L., Leconte J., Marcq E., and Haberle R. M. (2013) 3D modelling of the early martian climate under a denser CO_2 atmosphere: Temperatures and CO_2 ice clouds. *Icarus, 222,* 81–99, DOI: 10.1016/j.icarus.2012.10.019.

Formisano V., Maturilli A., Giuranna M., D'Aversa E., and López-Valverde M. A. (2006) Observations of non-LTE emission at 4–5 microns with the Planetary Fourier Spectrometer aboard the Mars Express mission. *Icarus, 182,* 51–67.

Gadsden M. and Parviainen P. (2006) *Observing Noctilucent Clouds.* The International Association of Geomagnetism and Aeronomy. 39 pp.

González-Galindo F., Määttänen A., Forget F., and Spiga A. (2011) The martian mesosphere as revealed by CO_2 cloud observations and general circulation modeling. *Icarus, 216,* 10–22.

Gumbel J. and Karlsson B. (2011) Intra- and inter-hemispheric coupling effects on the polar summer mesosphere. *Geophys. Res. Lett., 38,* L14804, DOI: 10.1029/2011GL047968.

Gumbel J., Stegman J., Murtagh D., and Witt G. (2001) Scattering phase functions and particle sizes in noctilucent clouds. *Geophys. Res. Lett., 28(8),* 1415–1418, DOI: 10.1029/2000GL012414.

Hemenway C., Soberman R., and Witt G. (1964) Sampling of noctilucent cloud particles. *Tellus, 16(1),* 84–88, DOI: 10.1111/j.2153-3490.1964.tb00146.x.

Herr K. C. and Pimentel G. C. (1970) Evidence for solid carbon dioxide in the upper atmosphere of Mars. *Science, 167,* 47–49.

Hervig M. and Gordley L. (2010) Temperature, shape, and phase of mesospheric ice from Solar Occultation For Ice Experiment observations. *J. Geophys. Res., 115,* D15208, DOI: 10.1029/2010JD013918.

Hervig M. and Siskind D. (2006) Decadal and inter-hemispheric variability in polar mesospheric clouds, water vapor, and temperature. *J. Atmos. Solar-Terrestrial Phys., 68(1),* 30–41, DOI: 10.1016/j.jastp.2005.08.010.

Hervig M., Thompson R., McHugh M., Gordley L., Russell J. III, and Summers M. (2001) First confirmation that water ice is the primary component of polar mesospheric clouds. *Geophys. Res. Lett., 28(6),* 971–974.

Hervig M. E., Gordley L. L., Deaver L. E., Siskind D. E., Stevens M. H., Russell J. M. III, Bailey S. M., Megner L., and Bardeen C. G. (2009a) First satellite observations of meteoric smoke in the middle atmosphere. *Geophys. Res. Lett., 36,* L18805, DOI: 10.1029/2009GL039737.

Hervig M., Gordley L., Stevens M., Russell J. III, Bailey S., and Baumgarten G. (2009b) Interpretation of SOFIE PMC measurements: Cloud identification and derivation of mass density, particle shape, and particle size. *J. Atmos. Solar-Terrestrial Phys., 71(3–4),* 316–330, DOI: 10.1016/j.jastp.2008.07.009.

Hervig M., Deaver L., Bardeen C., Russell J. III, Bailey S., and Gordley L. (2012) The content and composition of meteoric smoke in mesospheric ice particles from SOFIE observations. *J. Atmos. Solar-Terrestrial Phys., 84–85,* 1–6, DOI: 10.1016/j.jastp.2012.04.005.

Inada A., Richardson M. I., McConnochie T. H., Strausberg M. J., Wang H., and Bell J. F. III (2007) High-resolution atmospheric observations by the Mars Odyssey Thermal Emission Imaging System. *Icarus, 192,* 378–395.

Karlsson B., Körnich H., and Gumbel J. (2007) Evidence for interhemispheric stratosphere-mesosphere coupling derived from noctilucent cloud properties. *Geophysical Res. Lett., 34,* L16806, DOI: 10.1029/2007GL030282.

Karlsson B., McLandress, C., and Shepherd T. (2009) Interhemispheric mesospheric coupling in a comprehensive middle atmosphere model. *J. Atmos. Solar-Terrestrial Phys., 71(3–4),* 518–530, DOI: 10.1016/j.jastp.2008.08.006.

Karlsson B., Randall C. E., Shepherd T. G., Harvey V. L., Lumpe J., Nielsen K., Bailey S. M., Hervig M., and Russell J. M. III (2011) On the seasonal onset of polar mesospheric clouds and the breakdown of the stratospheric polar vortex in the southern hemisphere. *J. Geophys. Res., 116,* D18107, DOI: 10.1029/2011JD015989.

Kass D. M., Abdou W. A., McCleese D. J., Schofield J. T., and Määttänen A. (2011) MCS climatology of detached, localized haze layers. In *Mars Atmosphere: Modelling and Observations Workshop Proceedings,* pp. 400–403. Grenada, Spain, CNES/ESA.

Khosravi R., Brasseur G., Smith A., Rusch D., Walters S., Chabrillat S., and Kockarts G. (2002) Response of the mesosphere to human-induced perturbations and solar variability calculated by a 2-D model. *J. Geophys. Res., 107(D18),* 4358, DOI: 10.1029/2001JD001235.

Kokhanovsky A. (2005) Microphysical and optical properties of noctilucent clouds. *Earth-Sci. Rev., 71,* 127–146, DOI: 10.1016/j.earscirev.2005.02.005.

Leslie R. C. (1885) Sky glows. *Nature, 32,* 245, DOI: 10.1038/032245a0.

López-Valverde M. A., López-Puertas M., López-Moreno J. J., Formisano V., Grassi D., Maturilli A., Lellouch E., and Drossart P. (2005) Analysis of CO_2 non-LTE emissions at 4.3 μm in the martian atmosphere as observed by PFS/Mars Express and SWS/ISO. *Planet. Space Sci., 53,* 1079–1087.

Lübken F.-J. (1999) Thermal structure of the Arctic summer mesosphere. *J. Geophys. Res., 104(D8),* 9135–9149.

Ludlam F. (1957) Noctilucent clouds. *Tellus, 9(3),* 341–364, DOI: 10.1111/j.2153-3490.1957.tb01890.x.

Määttänen A., Montmessin F., Gondet B., Scholten F., Hoffmann H., González-Galindo F., Spiga A., Forget F., Hauber E., Neukum G., Bibring J.-P., and Bertaux J.-L. (2010) Mapping the mesospheric CO_2 clouds on Mars: MEx/OMEGA and MEx/HRSC observations and challenges for atmospheric models. *Icarus, 209,* 452–469.

Maltagliati L., Montmessin F., Fedorova A., Korablev O., Forget F., and Bertaux J.-L. (2011) Evidence of water vapor in excess of saturation in the atmosphere of Mars. *Science, 333,* 1868–1871.

McConnochie T. H., Bell J. F., Savransky D., Wolff M. J., Toigo A. D., Wang H., Richardson M. I., and Christensen P. R. (2010) THEMIS-VIS observations of clouds in the martian mesosphere: Altitudes, wind speeds, and decameter-scale morphology. *Icarus, 210,* 545–565.

McGouldrick K., Toon O. B., and Grinspoon D. H. (2011) Sulfuric acid aerosols in the atmospheres of the terrestrial planets. *Planet. Space Sci., 59,* 934–941, DOI: 10.1016/j.pss.2010.05.020.

Merkel A., Garcia R., Bailey S., and Russell J. III (2008) Observational studies of planetary waves in PMCs and mesospheric temperature measured by SNOE and SABER. *J. Geophys. Res., 113,* D14202, DOI: 10.1029/2007JD009396.

Montmessin F., Bertaux J.-L., Quémerais E., Korablev O., Rannou P., Forget F., Perrier S., Fussen D., Lebonnois S., Reberac A., and Dimarellis E. (2006a) Subvisible CO_2 clouds detected in the mesosphere of Mars. *Icarus, 183,* 403–410.

Montmessin F., Quémerais E., Bertaux J.-L., Korablev O., Rannou P., and Lebonnois S. (2006b) Stellar occultations at UV wavelengths by the SPICAM instrument: Retrieval and analysis of martian haze profiles. *J. Geophys. Res., 111,* E09S09, DOI: 10.1029/2005JE002662.

Montmessin F., Gondet B., Bibring J.-P., Langevin Y., Drossart P., Forget F., and Fouchet T. (2007) Hyper-spectral imaging of convective CO_2 ice clouds in the equatorial mesosphere of Mars. *J. Geophys. Res., 112,* DOI: 10.1029/2007JE002944.

Olivero J. and Thomas G. (2001) Evidence for changes in greenhouse gases in the mesosphere. *Adv. Space Res., 28(7),* 931–936, DOI: 10.1016/S0273-1177(01)00457-4.

Pérot K., Hauchecorne A., Montmessin F., Bertaux J.-L., Blanot L., Dalaudier F., Fussen D., and Kyrölä E. (2010) First climatology of polar mesospheric clouds from GOMOS/ENVISAT stellar occultation instrument. *Atmos. Chem. Phys., 10,* 2723–2735, DOI: 10.5194/acp-10-2723-2010.

Plane J. M. C. (2011) On the role of metal silicate molecules as ice nuclei. *J. Atmos. Solar-Terr. Phys., 73,* 2192–2200.

Rampino M. and Self S. (1982) Historic eruptions of Tambora (1815), Krakatau (1883) and Agung (1963), their stratospheric aerosols and climatic impact. *Quaternary Res., 18,* 127–143.

Rapp M. and Lübken F.-J. (2004) Polar mesosphere summer echoes (PMSE): Review of observations and current understanding. *Atmos. Chem. Phys., 4,* 2601–2633, DOI: 10.5194/acp-4-2601-2004.

Rapp M. and Thomas G. (2006) Modeling the microphysics of mesospheric ice particles: Assessment of current capabilities and basic sensitivities. *J. Atmos.Solar-Terrestrial Phys., 68,* 715–744, DOI: 10.1016/j.jastp.2005.10.015.

Rapp M., Lübken F.-J., and Müllemann A. (2002) Small-scale temperature variations in the vicinity of NLC: Experimental and model results. *J. Geophys. Res., 107(D19),* 4392, DOI: 10.1029/2001JD001241.

Rapp M., Thomas G., and Baumgarten G. (2007) Spectral properties of mesospheric ice clouds: Evidence for nonspherical particles. *J. Geophys. Res., 112,* D03211, DOI: 10.1029/2006JD007322.

Rapp M., Strelnikova I., Strelnikov B., Hoffmann P., Friedrich M., Gumbel J., Megner L., Hoppe U.-P., Robertson S., Knappmiller S., Wolff M., and Marsh D. (2010) Rocketborne *in situ* measurements of meteor smoke: Charging properties and implications for seasonal variation. *J. Geophys. Res., 115,* D00I16, DOI: 10.1029/2009JD012725.

Robock A. (2000) Volcanic eruptions and climate. *Rev. Geophys., 38(2),* 191–219, DOI: 10.1029/1998RG000054.

Russell J. III, Bailey S., Gordley L., Rush D., Horányi M., Hervig M., Thomas G., Randall C., Siskind D., Stevens M., Summers M., Taylor M., Englert C., Espy P., McClintock W., and Merkel A. (2009) The Aeronomy of Ice in the Mesosphere (AIM) mission: Overview and early science results. *J. Atmos. Solar-Terr. Phys., 71(3–4),* 289–299, DOI: 10.1016/j.jastp.2008.08.011.

Saunders R. W. and Plane J. M. C. (2011) A photo-chemical method for the production of olivine nanoparticles as cosmic dust analogues. *Icarus, 212,* 373–382.

Saunders R. W., Möhler O., Schnaiter M., Benz S., Wagner R., Saathoff H., Connolly P. J., Burgess R., Murray B. J., Gallagher M., Wills R., and Plane J. M. C. (2010) An aerosol chamber investigation of the heterogeneous ice nucleating potential of refractory nanoparticles. *Atmos. Chem. Phys., 10,* 1227–1247.

Schofield J. T., Barnes J. R., Crisp D., Haberle R. M., Larsen S., Magalhaes J. A., Murphy J. R., Seiff A., and Wilson G. (1997) The Mars Pathfinder Atmospheric Structure Investigation/Meteorology (ASI/MET) experiment. *Science, 278,* 1752–1757.

Scholten F., Hoffmann H., Määttänen A., Montmessin F., Gondet B., and Hauber E. (2010) Concatenation of HRSC colour and OMEGA data for the determination and 3D parameterization of high-altitude CO_2 clouds in the martian atmosphere. *Planet. Space Sci., 58,* 1207–1214.

Schröder W. (2001) Otto Jesse and the investigation of noctilucent clouds 115 years ago. *Bull. Am. Meteor. Soc., 82(11),* 2457–2468.

Sefton-Nash E., Teanby N. A., Calcutt S. B., Hurley J., and Irwin P. G. J. (2012) Detection and mapping of ice clouds in Mars' mesosphere. In *Lunar Planet Sci. Conf. 43,* Abstract #1817, Lunar and Planetary Institute, Houston.

Sefton-Nash E., Teanby N. A., Montabone L., Irwin P. G. J., Hurley J., and Calcutt S. B. (2013) Climatology and first-order composition estimates of mesospheric clouds from Mars Climate Sounder limb spectra. *Icarus, 222,* 342–356, DOI: 10.1016/j.icarus.2012.11.012.

Siskind D., Eckermann S., and McCormack J. (2003) Hemispheric differences in the temperature of the summertime stratosphere and mesosphere. *J. Geophys. Res., 108(D2),* 4051, DOI: 10.1029/2002JD002095.

Sonnemann G., and Grygalashvyly M. (2005) Solar influence on mesospheric water vapor with impact on NLCs. *J. Atmos. Solar-Terrestrial Phys., 67,* 177–190, DOI: 10.1016/j.jastp.2004.07.026.

Spiga A., González-Galindo F., López-Valverde M.-A., and Forget F. (2012) Gravity waves, cold pockets and CO_2 clouds in the martian mesosphere. *Geophys. Res. Lett., 39,* L02201, DOI: 10.1029/2011GL050343.

Stevens M., Siskind D., Eckermann S., Coy L., McCormack J., Englert C., Hoppel K., Nielsen K., Kochenash A., Hervig M., Randall C., Lumpe J., Bailey S., Rapp M., and Hoffmann P. (2010) Tidally induced variations of polar mesospheric cloud altitudes and ice water content using a data assimilation system. *J. Geophys. Res., 115,* D18209, DOI: 10.1029/2009JD013225.

Taylor M., Gadsden M., Lowe R., Zalcik M., and Brausch J. (2002) Mesospheric cloud observations at unusually low latitudes. *J. Atmos. Solar-Terrestrial Phys., 64(8–11),* 991–999, DOI: 10.1016/S1364-6826(02)00053-6.

Taylor M., Pautet P.-D., Zhao Y., Randall C., Lumpe J., Bailey S., Carstens J., Nielsen K., Russell J. III, and Stegman J. (2011) High-latitude gravity wave measurements in noctilucent clouds and polar mesospheric clouds. In *Aeronomy of the Earth's Atmosphere and Ionosphere* (M. A. Abdu *et al.,* eds.), pp. 93–105. IAGA Special Sopron Book Series, Vol. 2, Springer Science, DOI: 10.1007/978-94-007-0326-1_7.

Thomas G. (1996) Is the polar mesosphere the miner's canary of global change? *Adv. Space Res., 18(3),* 149–158.

Thomas G. (2003) Are noctilucent clouds harbingers of global change in the middle atmosphere? *Adv. Space Res., 32(9),* 1737–1746, DOI: 10.1016/S0273-1177(03)00674-4.

Thomas G. and Olivero J. (2001) Noctilucent clouds as possible indicators of global change in the mesosphere. *Adv. Space Res., 28(7),* 937–946, DOI: 10.1016/S0273-1177(01)00456-2.

Thomas G., Olivero J., Jensen E., Schroeder W., and Toon O. (1989) Relation between increasing methane and the presence of ice clouds at the mesopause. *Nature, 338,* 490–492, DOI: 10.1038/338490a0.

Thomas G., Olivero J., Deland M., and Shettle E. (2003) Comment on "Are noctilucent clouds truly a 'miner canary' for global change?". *EOS Trans. AGU, 84(36),* 352–353.

Thomas G., Marsh D., and Lübken F.-J. (2010) Mesospheric ice clouds as indicators of upper atmosphere climate change: Workshop on Modeling Polar Mesospheric Cloud Trends. *EOS Trans. AGU, 91(20),* 183.

Vestine E. (1934) Noctilucent clouds. *J. R. Astron. Soc. Canada, 28,* 249.

Vincendon M., Pilorget C., Gondet B., Murchie S., and Bibring J.-P. (2011) New near-IR observations of mesospheric CO_2 and H_2O clouds on Mars. *J. Geophys. Res., 116,* E00J02.

Von Zahn U. (2003) Are noctilucent clouds truly a 'miner's canary' for global change? *EOS Trans. AGU, 84(28)*, 261–264.

Witt G. (1957) Noctilucent cloud observations. *Tellus 9(3)*, 365–371, DOI: 10.1111/j.2153-3490.1957.tb01891.x.

Woods T. and Rottman G. (1997) Solar Lyman α irradiance measurements during two solar cycles. *J. Geophys. Res. 102,* 8769–8779.

Part IV:
Surface and Interior

Segura T. L., Zahnle K., Toon O. B., and McKay C. P. (2013) The effects of impacts on the climates of terrestrial planets. In *Comparative Climatology of Terrestrial Planets* (S. J. Mackwell et al., eds.), pp. 417–437. Univ. of Arizona, Tucson, DOI: 10.2458/azu_uapress_9780816530595-ch17.

The Effects of Impacts on the Climates of Terrestrial Planets

Teresa L. Segura
Space Systems/Loral

Kevin Zahnle
NASA Ames Research Center

O. Brian Toon
University of Colorado

Christopher P. McKay
NASA Ames Research Center

The impacts of large asteroids and comets have contributed to the climate histories of terrestrial planets. Large impact events deliver volatiles and kinetic energy to a planet, and launch debris and volatiles into its atmosphere through formation of craters. Impacts are responsible for global climate effects such as warm temperatures, release of subsurface ice, precipitation, mass extinctions, and possible runaway greenhouse atmospheres, which may last for at least centuries.

1. INTRODUCTION

The surfaces of Venus, Mars, Titan, and Earth collectively show thousands of craters and impact basins. These impacts of asteroids and comets have affected planetary climates since the formation of the solar system. Indeed, the bodies of the planets originated from planetesimal collisions. Impacts after planetary formation may have brought in most of the water and other atmospheric constituents on terrestrial planets.

Impacts of objects larger than a few hundred kilometers, particularly during the late heavy bombardment period around 4 Ga, may have heated Earth's oceans to the boiling point (*Sleep et al.,* 1989), thereby creating a challenge for any life on the planet, and may also have generated many of the river valleys that we observe on Mars (e.g., *Segura et al.,* 2008). Impacts by objects of about 10 km in size on Earth may have triggered extinction events several times in the geologic record. Firm evidence shows that the global mass extinction event at ~65 m.y. ago, one of six known mass extinctions in Earth history, was caused by an impact in the Yucatán Peninsula near the village of Chicxulub (*Schulte et al.,* 2010). Such impacts are expected every few hundred million years (*Michel and Morbidelli,* 2007). An impact of an object around 50 m in diameter, which released about 10 Mt of energy, equivalent to a very large nuclear weapon, occurred in Tunguska, Siberia, at the beginning of the twentieth century. (1 Mt is equivalent to 4×10^{15} J). Such an impact is expected every few thousand years (*Michel and Morbidelli,* 2007). Smaller impacts occur frequently; there is an energy release equivalent to the Hiroshima nuclear explosion about once every three years on Earth. Fortunately, these explosions mainly occur at high altitude, so they do not generate shock waves that reach the ground. Due to its lower atmospheric pressure, Mars is not so lucky. The surface of Mars is bombarded by small objects, which created at least five craters with diameters from 4 to 12 m over a four-year period (*Byrne et al.,* 2009). Impacts are expected to cause effects on the climates and atmospheres of terrestrial planets that vary qualitatively and quantitatively with the size of the impactor (e.g., *Toon et al.,* 1997).

The surfaces of the terrestrial planets are highly cratered, representing the surviving remnants of the countless impacts of comets and asteroids or their secondaries. Previous research (e.g., *Toon et al.,* 1997) has shown that impacts can cause global climate phenomena such as cooling, caused by decreased sunlight reaching the surface due to suspended ejecta and smoke from impact-induced wildfires, as well as atmospheric chemistry changes such as ozone destruction and the injection of gases such as water vapor, CO_2, and SO_2. The effects of some of these changes could last for months to centuries; the effects of extinctions, of course, last forever.

When a large asteroid or comet strikes a target planet, it transfers its tremendous kinetic energy to the target. This delivery of energy leads to the destruction of the impactor itself, creation of a crater, and production of ejecta. This process is illustrated below in Fig. 1.

Previous research (e.g., *Sleep and Zahnle,* 1998; *Toon et al.,* 1997; *Segura et al.,* 2002, 2008) has shown that large impacts produce globally extensive, hot debris layers on the

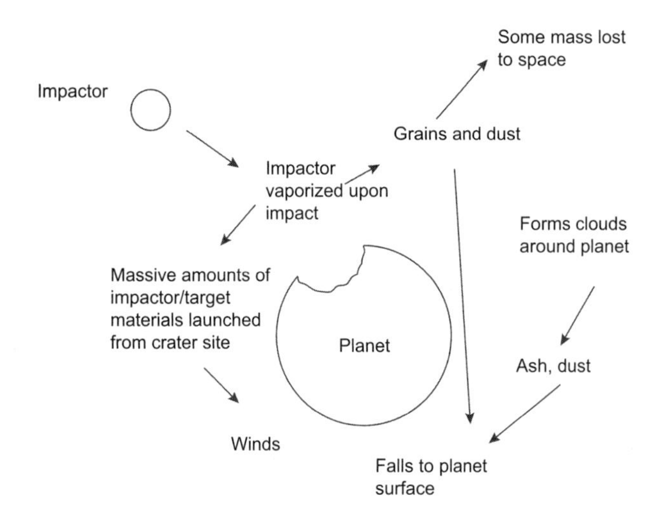

Fig. 1. Illustration showing the origin and fate of ejecta following impact on a rocky planet.

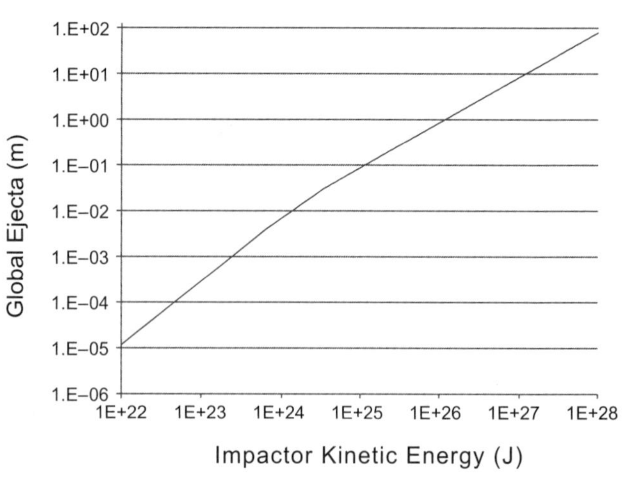

Fig. 2. The equivalent global thickness of the total ejecta as a function of impactor kinetic energy. For large impact events, the global debris layer may be tens or hundreds of meters in thickness.

surface of a rocky planet. These hot layers are distinct from the ejecta blanket, which is composed of pulverized target material, thickens toward the crater, and usually maintains some contact with it. Instead, the hot layers result from the expansion of the fireball containing melted and vaporized rock, and are more or less globally distributed. Figure 2 shows the equivalent global thickness of total (solid, vapor, melt) globally dispersed ejecta computed using the methods described in *Zahnle* (1990), *Sleep and Zahnle* (1998), and *Segura et al.* (2008), for a given impactor kinetic energy. The thickness of the layer is not a strong function of the particular terrestrial planet. This global layer of debris may be centimeters to hundreds of meters thick for the largest impactors. For the Chicxulub impactor, an Earth example we can use for comparison, the Fig. 2 shows several millimeters of globally distributed ejecta, assuming the impact had an energy near 4×10^{23} J (about 10^8 Mt). This is consistent with the few-millimeter-thick Cretaceous-Paleogene (K-Pg) distal layer existing today (*Smit,* 1999).

Since the fireball ejecta are melted and/or vaporized, any water or other volatiles contained in the ejecta, from impactor or target materials, would also be vaporized (*Toon et al.,* 1997). The injected water vapor and hot debris layer can provide an extensive hydrological cycle on the planet for a large impact. Heat from the debris layer will conduct downward, and may melt subsurface ice; the injected water will rain out, causing erosion of the surface; and the planet as a whole can take many years to finally radiate to space the massive amount of energy delivered by a large impact (*Sleep et al.,* 1989; *Segura et al.,* 2008). Thus impacts provide heat, small particles, and new volatiles capable of temporarily altering the climate on the impacted terrestrial planet. The relative contributions of injected volatiles, global debris layers, atmospheric chemistry, and dust are planet-dependent and will be discussed in the following sections.

2. THE EFFECTS OF IMPACTS ON EARTH

The environmental consequences of cosmic impacts on Earth were a subject of informed speculation from *Whiston* (1696) to *Urey* (1973). Nonetheless, impacts did not become mainstream Earth science until *Alvarez et al.* (1980) showed that the age of the dinosaurs ended when Earth was struck by an ~10-km-diameter asteroid. Alvarez et al.'s discovery revealed cosmic impacts as genuine agents of catastrophe on Earth. This realization had consequences on several scales, from the appreciation that there are impacts of all sizes affecting Earth, to the recognition that large impacts can lead to mass extinctions, to a fresh assessment of what the craters of the Moon might mean for Earth.

Impacts come in all sizes, with larger impacts being less common, as discussed by *Stuart and Binzel* (2004) and *Michel and Morbidelli* (2007). For example, there is a continuous rain of micrometeoroids on Earth. Micrometer-sized meteoroids do not ablate, and have long been detected in the stratosphere as relatively rare micrometeorites. The ablation of particles 100-μm and larger produces a molecular vapor trail, or meteor, that recondenses to form a faint haze of nanometer-sized particles. Such particles were predicted by *Rosinski and Snow* (1961) and *Hunten et al.* (1980), shown to be present in the lower stratosphere by *Murphy et al.* (1998), detected in the upper atmosphere by *Hervig et al.* (2009), and most recently been found to have been affecting satellite and groundbased observations of particles above 30 km by *Neely et al.* (2011).

Sand-sized grains and boulders produce shooting stars, which contribute to a layer of metals in the upper atmosphere, but are otherwise mainly curiosities. Larger objects, such as the approximately 50-m-diameter iron meteoroid that created Meteor Crater in Arizona, reach the surface if they are sufficiently strong. Comets and stony objects in this size range are more likely than irons to disintegrate in the atmosphere, and leave little behind on the surface beyond pulverized rock and dust, such as was the case at Tunguska (*Chyba et al.,* 1993). While most objects disintegrate so far above the surface that they do not produce shock waves that reach the surface, the Tunguska impact released an energy equivalent to about a 2–20-Mt nuclear blast. A forest was destroyed by shock waves, but the effects were isolated to a small area of Earth, except for glowing skies and possibly some ozone loss perhaps caused by water vapor lofted to the upper atmosphere by the energy release. Such a blast, which might be expected every few thousand years (*Michel and Morbidelli,* 2007), would obliterate an area of about 2000 km^2 and would be devastating in the unlikely event that the explosion occurred over a city. The most dramatic impact for which we have evidence of environmental changes is that at Chicxulub, but Chicxulub is not the largest impact to have occurred on Earth.

2.1. The Cretaceous-Paleogene Impact (Chicxulub)

The impact 65 m.y. ago, at the Cretaceous-Paleogene (K-Pg) geologic boundary, into a shallow sea straddling the current shoreline near the village of Chicxulub in the Yucatán Peninsula of Mexico, demonstrates the deadly effects of impacts (*Alvarez et al.,* 1980). The diameter of the impactor has usually been estimated to have been around 10 km, with a transient crater diameter of 100 km (*Artemieva and Morgan,* 2009). This estimate is based on the amount of Ir in the global boundary layer, described below. The impact released 10^8 to 10^9 Mt of energy, 10,000 to 100,000 times the energy in the world's nuclear arsenals, and the equivalent of 1 to 10 1-Mt weapons per 25 km^2 of Earth's surface area, assuming 20% of the impact energy was released globally. The impact was first recognized from the ~1-cm-thick global layer of debris from the impact, which is often referred to as the K-Pg layer. This layer, which directly separates Cretaceous and Paleogene sediments, contains several materials (e.g., Ir, Os, and shocked quartz) that tell us clearly that the extinction event was caused by an impact and not a volcano (*Schulte et al.,* 2010). Moreover, it contains three materials that provide direct clues to the mechanisms of the extinction. First, the K-Pg layer contains a large number, around 10^4 cm^{-2}, of melted submillimeter-sized spherules (*Smit,* 1999). Second, it contains vaporized impactor material, mixed with pulverized target material. Most of this material is small, ranging in size from nanometers to tens of micrometers. Finally, the global debris layer contains a large amount of smoke. The role that each of these materials played in the global mass extinction is controversial, subject to heated debate and ongoing research.

The submillimeter spherules are thought to arise from the high-pressure fireball of vaporized impactor and target material, which blows away from the crater. As discussed by *Johnson and Melosh* (2012), as the vapor from the impactor and target cools, about half of it condenses to form 200-μm-sized spherules, a process first suggested by *O'Keefe and Ahrens* (1982). The remainder may remain as nanometer-sized particles, which are also recognized in the fireball layer (*Wdowiak et al.,* 2001). After the fireball expands into space and returns to Earth under gravity, the spherules reenter the atmosphere and deposit their kinetic energy. The deposited energy heats the air, which then radiates energy toward the surface. Numerous studies have been conducted of this radiated energy, all of which conclude it is sufficient to ignite flammable materials such as leaf litter on the surface (*Melosh et al.,* 1990; *Goldin and Melosh,* 2009). More work is warranted on this problem, which is very complex. In particular, we show below that if the Chicxulub event were at the high end of energy estimates, the atmosphere even at the ground would have been heated by tens of degrees, which clearly would contribute to the propagation of wildfires. Further discussion on the implications of igniting the material is below. The light impulse lasted only a few hours, and the spherules themselves fell out within a few days due to their large size, so they are not important contributors to climate change. However, the radiated energy and the fires it started are likely to have broiled large land creatures alive, serving as a major extinction mechanism (e.g., *Robertson et al.,* 2004).

While the submillimeter spherules are easy to detect in the K-Pg boundary layer sediments, it is much more difficult to identify the vaporized impactor and the pulverized debris from the target. Some of the pulverized material is recognizable as shocked quartz, but most of it is difficult to distinguish from ambient sediments or debris from the disintegration of the spherules. The submicrometer fraction is of greatest interest because it might remain in the atmosphere for years and perturb the climate. *Toon et al.* (1997) computed the amount of submicrometer-pulverized debris by extrapolating over many orders of magnitude in energy from nuclear weapons tests, combined with impact crater studies. They concluded that a significant amount of dust was placed into the atmosphere; in fact, enough to reduce sunlight below the levels needed for photosynthesis in the oceans and cause major climate changes on the surface. *Pope* (2002) estimated the submicrometer dust from the mass of the shocked quartz, and concluded that there was about 10^5 times less than estimated by *Toon et al.* (1997). However, *Pope*'s (2002) estimate of clastic material is about three orders of magnitude below that estimated from impact cratering calculations (*Artemieva and Morgan,* 2009). *Pope* (2002) used data on shocked quartz grains but assumed the quartz is 50% of all the pulverized debris. However, sandstones and shale, which might contain some quartz, composes only 3–4% of the first 3 km of sediments at Chicxulub (*Pope et al.,* 1997). The underlying basement rock is granite, which might contain about 23% quartz, but it probably contributes little material to the distal layer (*Artemieva and Morgan,* 2009). The assumption that quartz dominates the clastic material is a likely source of the factor of 1000 needed to make *Pope*'s (2002) estimate of the total clastic debris agree with *Artemieva and Morgan*'s (2009) estimate. *Toon et al.*'s (1997) estimate of pulverized rock did not account for the fact that 90% of it may not have had sufficiently high velocity to leave the crater; however, they likely underestimated the fraction of the debris that is submicrometer-sized by an order of magnitude. Further work is needed to determine the amount of submicrometer debris in the K-Pg distal layer.

Pope's (2002) work on shocked quartz is not relevant to the vaporized impactor, which *Toon et al.* (1997) showed was also an important source of submicrometer particles. This vaporized debris is clearly observed in the distal layer (e.g., *Wdowiak et al.,* 2001; *Verma et al.,* 2002). The particles in this layer are Fe-rich, contain Ir, and have typical particle sizes near 10–20 nm. They clearly formed in a different process from the spherules, probably by condensation from the vapor containing the impactor and target (*Johnson and Melosh,* 2012). This material was predicted by *Toon et al.* (1997) to have a significant thickness, and it is observable by visible inspection of the K-Pg layer. This material is likely to have been significant to the global climate and to light transmission.

The third interesting component of the K-Pg layer is carbon spherules, soot, with sizes in the range of tens to hundreds of nanometers (*Wolbach et al.,* 1985, 1988, 1990, 2003). The abundance of soot in the K-Pg layer indicates that the majority of Earth's biosphere burned. *Belcher et al.* (2003, 2004, 2005, 2009), *Belcher* (2009), and *Harvey et al.* (2008) challenged the idea that there were global wildfires on the basis of the absence of charcoal, the presence of noncharred organic matter, and the lack of post-impact erosion at some sites. Instead of global fires producing the soot, they suggested that the soot in the debris layer originated from the impact site itself because they claimed the morphology of the soot, the chain length of polycyclic aromatic hydrocarbons (PAHs), and the presence of carbon cenospheres were inconsistent with burning the biosphere. *Robertson et al.* (2012), among others, argue these assertions are either incorrect or have alternate explanations that are consistent with global firestorms initiated by an infrared heat pulse. In particular, Robertson et al. show that the observed charcoal depletion in the K-Pg layer has been misinterpreted due to a failure to correct properly for sediment deposition rate. For instance, the PAH data were corrected, which converts a PAH depletion to an excess, but the same corrections were not applied to the charcoal data. Robertson et al. also show that the mass of soot potentially released from the impact site is one or two orders of magnitude too low to supply the observed soot. Finally, soot morphology, PAHs, and cenospheres are in fact consistent with forest fires. Irrespective of the origin of the soot, its abundance is enough to create massive climate changes and to stop photosynthesis in the oceans.

There are additional possible ejecta that may be important, but have not been detected clearly. Several authors (*Kring et al.,* 1996; *Pope et al.,* 1997; *Toon et al.,* 1997) independently suggested that there would be large S injections either from S in the body of the impactor, or from S in the target rocks in the Yucatán. Since sulfate — a possible final form of the injected S — is water soluble, common in the unperturbed environment, and geochemically mobile, it would be difficult to detect an impact-related enhancement in the K-Pg layer. While dust and soot might have relatively short residence times in the atmosphere, it is possible that sulfate might take longer to form because its formation requires water and sunlight. The slow formation of sulfate might make the perturbation to the climate last longer than just from soot and dust (*Pope et al.,* 1997).

Despite the many ongoing controversies it is clear that there is an abundance of particulate matter that was injected, or might have been injected. Submicrometer soot particles observed in the K-Pg layer would create an optical depth, a measure of the radiative importance of the material, on the order of 100. Submicrometer and smaller rock particles, either from the pulverized target or vaporized impactor and target, could also have produced optical depths well above 100. Finally, sulfate particles could have contributed modest optical depths. Optical depths of 100 would prevent virtually all light from reaching the surface, eliminating photosynthesis and cooling the ground.

Surprisingly, there are no numerical studies of the global climate effects of impacts with modern climate models. *Toon et al.* (1982) and *Pollack et al.* (1983) used one-dimensional microphysical, climate models to show that the injections of dust particles would lead to a period of perhaps six months in which light levels were too low to support photosynthesis in the oceans, and that a rapid and dramatic temperature drop would occur. *Covey et al.* (1994) conducted the only three-dimensional simulations to date. *Pierazzo et al.* (2003) and her coworkers conducted several one-dimensional studies of the climate effects due to S injections. Perhaps the most interesting result of these simulations was that light levels remained below those needed for photosynthesis for about two years, assuming that dust was present in the sulfate to provide some absorption. Possibly the most relevant study is that of *Robock et al.* (2007), who simulated an injection of soot of about 0.1% of that expected for the K-Pg extinction event. In this study temperatures plunge to ice age conditions within weeks globally, precipitation declines by about 50% globally, and the soot self-lofts so that the event is prolonged over a period of about a decade. We now have models that are capable of simulating the chemistry; interactions of dust, smoke, and sulfate; as well as the implications for an ozone-loss climate. It would be useful to apply them.

2.2. Thermal Climatic Effects of Very Large Impacts on Earth

The major climatic effects of impacts smaller than Chicxulub are indirect, in the sense that the impact modifies how the planet interacts with sunlight rather than heating the planet directly. As discussed above, the most important of these changes result from the dust, soot, and sulfates made, raised, and suspended in the upper atmosphere by the impact. By blocking sunlight these particles cause impact winters. Photosynthesis can be shut off and surface temperatures can plummet by tens of degrees over months or years.

Impacts larger than Chicxulub inject enough energy into the environment that direct heating is significant. For great impacts the heating sterilizes the surface and evaporates significant fractions of the ocean. Direct thermal heating is the subject of Figs. 3–7. Chicxulub itself, if at the high end of the range of plausible energies, would have been large enough to produce important direct thermal effects.

To illustrate the consequences of very large impacts on Earth (Figs. 3–7) we use a simple but general model to track the global flow of energy from the impact into the environment. The model is based on discussions presented elsewhere by *Zahnle* (1990), *Toon et al.* (1997), *Zahnle and Sleep* (1997, 2006), and *Sleep and Zahnle* (1998). The model starts from the presumption that a significant fraction (here, 50%) of the impact energy goes into impact ejecta that interact with the atmosphere. Much of this energy is initially in rock vapor and melt that explode to spread condensates worldwide. Added to the vapor and melt are high-velocity ejecta that were never melted and water in the ejecta if the impact took place in an ocean. For modeling we make the following assumptions:

1. The upper atmosphere is heated by impact ejecta when the ejecta reenter the atmosphere (*Zahnle,* 1990; *Melosh et al.,* 1990). The resulting mix of hot air, hot silicates, and hot rock vapor cools by thermal radiation, outward to space and inward to the lower atmosphere. The rock vapor condenses and rains out as the upper atmosphere cools. The temperature ("upper atm" on Figs. 3 and 4) can be high for relatively modest impacts because the thermal inertia of the upper atmosphere is small.

2. The lower atmosphere is treated as silicate-free. Its energy balance is determined by radiative exchange with the upper atmosphere and with the surface, and by evaporation of seawater. Its temperature is given as "lower atm" in Figs. 3 and 4 and simply "atm" on Figs. 5–7. The lower atmosphere fills with water vapor as the oceans evaporate. When it cools, the pace of cooling is governed by condensation of water vapor. Radiative cooling takes into account the physics of water vapor atmospheres as discussed by *Nakajima et al.* (1992).

3. The energy balance of the oceans includes radiative exchange with the lower atmosphere, evaporation of water from the surface, heating by rock raindrops, and the growth of a mixed surface layer of hot saline water residual to evaporation. The sea surface temperature (labeled "ocean" in Figs. 3–7) determines the partial pressure of atmospheric water vapor. During evaporation the air temperature is much higher than the sea surface temperature. Eventually the Earth cools enough that water vapor condenses and the resultant rains replenish the oceans with hot and relatively fresh water (how fresh would depend on continental weathering, which might be rather rapid under the circumstances). Where the oceans evaporate, a thermal wave propagates conductively into the seafloor. The depth to which heating exceeds 400 K is denoted "sterilized ocean crust" in Fig. 7.

4. The surface temperature of dry land (emergent continents) is presumed to be the same as the lower atmosphere. A thermal wave propagates conductively into the continents assuming conductivity and heat capacity appropriate to dry rock. The temperature in the dry ground and the depth of heating are denoted "cont" and "heated continental crust" in Figs. 3–6; in Fig. 7 sterilized continental crust refers specifically to the depth to which temperatures exceed 400 K, a proxy for sterilization.

The illustrative models shown here (Figs. 3–7) assume that continents cover 30% of Earth's surface and that uniform-depth oceans cover the rest. The atmosphere is 1 bar and the oceans have the same volume as the modern Earth oceans. Detailed hypsometry is neglected.

Figure 3 illustrates a Chicxulub impact, into a shallow sea, whose energy is derived by scaling from the 180-km-diameter of the crater. It represents the large end of the range of possibilities. It assumes a 5×10^8 Mt (2×10^{24} J) impact and puts 50% of this energy into the ejecta, atmosphere, and ocean. For comparison, before discovery of the Chicxulub crater, *Melosh et al.* (1990) estimated that the

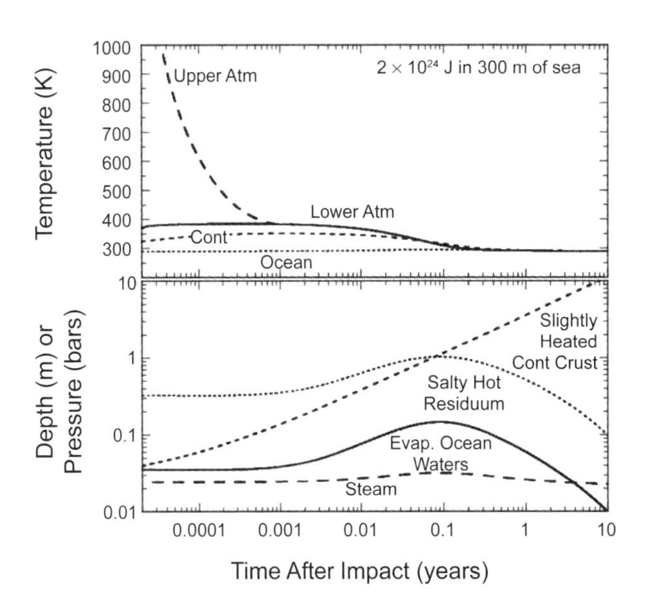

Fig. 3. Short-term heating by a high-energy Chicxulub impact. The modeled impact is of a 20-km asteroid at 15 km/s striking a 300-m-deep shallow sea. Energy released is 2×10^{24} J. Temperatures are shown for the top of the atmosphere at heights where ballistically launched impact ejecta came to rest ("upper atm"); in the lower atmosphere ("lower atm"); of the sea surface ("ocean"); and in dry ground near the surface of dry land ("cont"). The lower panel plots the depth of ocean water evaporated; the depth of residual warm salty water left behind by evaporation ("residuum"); the depth to which dry land is heated to the continental temperature; and the partial pressure of water vapor ("steam") in the atmosphere. Impact winter, which dominates the impact's climatic effects on timescales of months to years, has not been included here.

ejecta would contain 6–25 × 10^{22} J, factors of 4–16 times less energy. *Goldin and Melosh* (2009) are not specific about the impactor energy they assume, but it appears to be at the low end of the *Melosh et al.* (1990) range.

Figure 3 shows that the energy impulse from Chicxulub could have been big enough to raise the surface temperature of Earth's 1-bar atmosphere by many tens of degrees (limited by the atmosphere's heat capacity) for several days, and to evaporate ~10 cm of water from the oceans. The upper atmosphere grew very hot from the energy of infalling ejecta, so that the sky would have glowed like fresh lava for a few hours, and radiative heating of surfaces of low thermal inertia (e.g., dry leaves) would have been intense. The impact would have made the air both very hot and very dry, and probably very windy. These effects would have kindled global wildfires, and the general atmospheric heating would likely have contributed to their propagation (*Melosh et al.,* 1990). On longer timescales (weeks to years), Fig. 3 is misleading because it ignores the blocking of sunlight by the ejecta, by sulfates from the anhydrite that the impactor struck, and by soot from the wildfires that the impact set, all of which would cool Earth.

Earth retains recognizable remnants of two craters bigger than Chicxulub. The Vredefort (~300 km, 2.02 Ga) and Sudbury (~250 km, 1.85 Ga) craters record impacts that would have released on the order of 1 × 10^{25} J. The thermal effects of an impact on this scale on Earth are illustrated by Fig. 4. Figure 4, which is for impacts on land, is specific to Vredefort and Sudbury. The thermal effects are markedly more severe than for Chicxulub. In particular, the atmosphere and dry surfaces briefly become very hot, and the topmost 5 m of the seas heat to 50°C. The consequences for life at the surface would be horrific, but much less so only tens of meters below the sea surface.

There is an indirect record of bigger craters on Earth that extends deep into the Archean. Beds of impact-generated spherules record several impacts (as many as nine) as big as or bigger than the K-Pg event that took place on Earth between ~3.5 Ga and ~2.5 Ga (*Lowe and Byerly,* 1986; *Sleep et al.,* 1989; *Byerly et al.,* 2002; *Bottke et al.,* 2012). The spherule beds give a lower limit to the number of big impacts on Earth at this time. We can use the lunar record to set an upper bound. The Moon samples the same population of impactors but with only about 4–5% of the effective cross-section of Earth (the effective cross section takes into account that Earth's geometric cross-section is enhanced by gravitational focusing), but there do not appear to be any lunar craters bigger than Chicxulub that date to Earth's Archean. Thus we can state with some confidence that Earth experienced on the order of 10 impacts greater than the K-Pg event in the billion years between 2.5 Ga and 3.5 Ga, and further state that the biggest of the 10 was likely some 30–100 times greater (by energy release) than the K-Pg event, comparable to the energy released by the Orientale impact on the Moon (1 × 10^{26} J). Earth experienced perhaps two or three such impacts in the mid to late Archean. These matters are discussed by *Sleep et al.* (1989) and *Zahnle and Sleep* (1997, 2006), and more comprehensively by *Bottke et al.* (2012).

Figure 5 illustrates the direct thermal consequences of an Orientale-scale 1 × 10^{26} J impact on Earth. This particular illustration is for an impact in a deep ocean but the environmental consequences are much the same for a land impact. Events occur on two basic timescales: a day or so during which rock vapor is present in the atmosphere and dry land melts to the depth of a meter or so, and a longer period of a decade during which the atmosphere contains abundant hot steam. The melt would drain into the ground or flow downhill and pool; these consequences are neglected in Fig. 5. Within a few days the melt would freeze and the lands begin to cool, but it would take a decade for Earth to fully cool and the thermal effects would be imprinted several meters into the ground. Tens of meters of water would have evaporated off the oceans, leaving about 10 m of very hot (~400 K) saline waters at the surface. Surface waters would thus be rendered very nearly uninhabitable. Conditions in deep water would remain relatively benign — there is no particular reason to expect life's demise. Even so, photosynthetic communities may have fared ill. As *Bottke et al.* (2012) point out, an event that very nearly wipes out life at the surface is likely to be followed by a new and different surface flora than had existed previously. The intuition that impacts might provide the impetus for biological revolu-

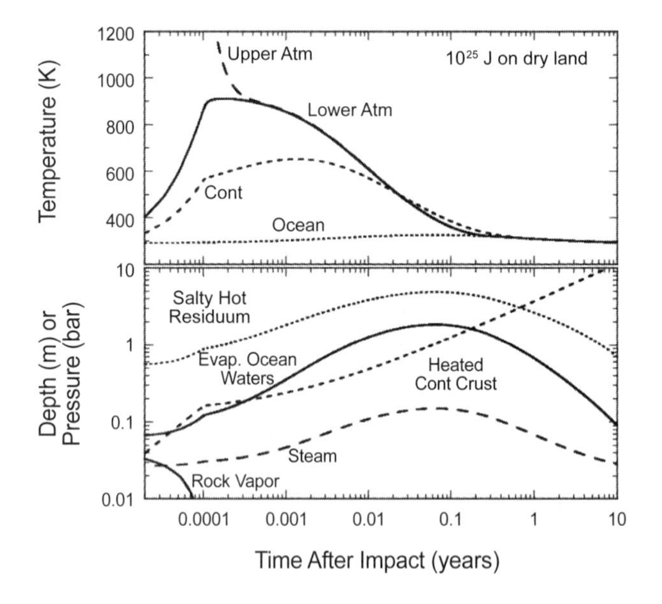

Fig. 4. Heating after a Vredefort- or Sudbury-scale impact on Earth. This model assumes an impact on dry land. Air temperatures ("lower atm") and ground temperatures ("cont") are briefly hot, and the sea surface temperature ("ocean") rises to 50°C to a depth on the order of 5 m. Deeper waters remain temperate. A meter of water evaporates from the oceans, leaving a residuum of warm salty water a few meters deep. "Heated cont crust" refers to the depth to which dry ground is heated to the continental temperature shown in the top panel.

tions may not be universal but it is certainly widespread, in principle testable, and attempts are currently being made to link particular spherule beds to particular paleontological and geochemical horizons.

There are even bigger craters on the Moon. The largest ones predate Earth's geological record, but we can be confident that Earth, because it is bigger than the Moon, would have been hit 20 times for each time the Moon was struck. Moreover, of the 20 biggest bodies to hit the Earth-Moon system, it is likelier than not that the dozen biggest all hit Earth. We can assert with a high degree of confidence that Earth was hit by several projectiles that were much bigger than any that ever hit the Moon (*Sleep et al.,* 1989; *Chyba,* 1991; *Zahnle and Sleep,* 2006; *Bottke et al.,* 2010). The biggest of these would have sufficed to evaporate the oceans, thus posing a serious hazard to life on Earth (*Maher and Stevenson,* 1988; *Sleep et al.,* 1989; *Oberbeck and Fogleman,* 1990; *Abramov and Mojzsis,* 2009). Only extreme thermophiles would have been likely to survive, in deep niches underground that the heat of the impact does not reach (*Zahnle and Sleep,* 1997; *Abramov and Mojzsis,* 2009). The general topic has been reviewed several times since with somewhat different emphasis and increasing sophistication, especially regarding the astronomical and geological data and their interpretation (*Zahnle and Sleep,* 1997, 2006; *Sleep and Zahnle,* 1998; *Abramov and Mojzsis,* 2009; *Bottke et al.,* 2010).

The impact modeled in Fig. 6 is comparable to the largest impacts recorded in the extant rocky crusts of the solar system: those that excavated the South Pole-Aitken basin on the Moon (2500 km wide, 13 km deep) or the Hellas Planitia (2100 km wide, 8 km deep) basin on Mars. The energy released from these impacts was on the order of 10^{27} J. The figure plots ocean depth, sea surface temperature, atmospheric temperature, and the amount of steam and rock vapor in the atmosphere as a function of time.

As the rock vapor cools it condenses into rock raindrops that fall out and quench in the oceans. The energy absorbed by the oceans — the thermal radiation and the cooled molten rain — evaporates water. The water vapor builds up in the atmosphere until the energy of the rock vapor is exhausted. Thereafter the steam atmosphere cools radiatively until the excess water vapor has fully rained out. How long this takes is governed by the thermal energy of the atmosphere, the latent heat of condensation of the water vapor, and the radiative physics of the runaway greenhouse effect. More details describing these sorts of models are given by *Zahnle and Sleep* (1997, 2006) and *Sleep and Zahnle* (1998).

The response occurs on two basic timescales: a period of a few days during which rock vapor is present in the atmosphere, and a longer period of more than 100 years

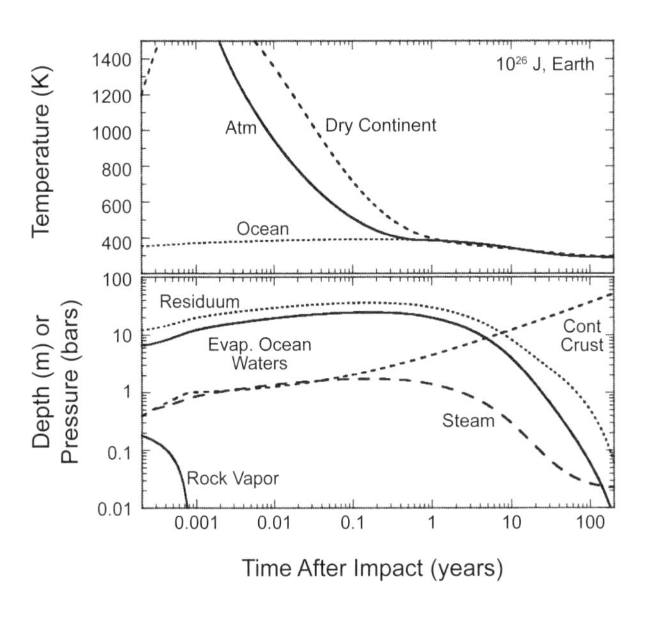

Fig. 5. Aftermath of an Orientale-scale impact on Earth (1-bar atmosphere). Air temperatures ("atm") get very high. The sea surface ("ocean") reaches 380 K to a depth of 30 m ("residuum"), but the deep ocean (not shown) remains cool. The ground temperature on dry land (dry continent) also gets very hot — briefly hot enough to melt or fuse the surface — but the depth of heating ("cont crust," lower panel) is only on the order of a few meters. Partial pressures of water and rock vapors are shown on the lower panel. This impact took place in a deep ocean; the difference between land and ocean impact is not very great for an impact on this scale.

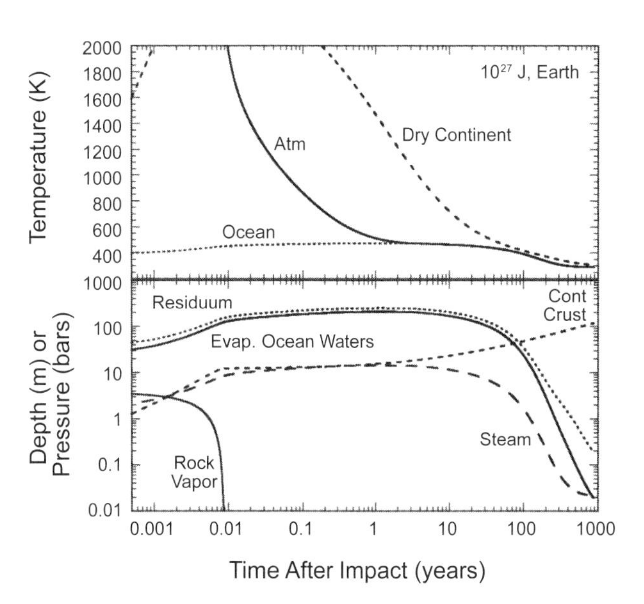

Fig. 6. Aftermath of a South Pole-Aitken- or Hellas-scale impact on Earth. The upper panel shows atmospheric temperature, sea surface temperature ("ocean"), and the near-surface ground temperature ("dry continent"). The lower panel shows the depth of ocean water evaporated, the depth of water evaporated and the depth of the hot saline residue of evaporation, the depth to which dry ground is heated to the continental temperature shown on the top panel, and the amounts of steam and rock vapor in the atmosphere.

during which the atmosphere contains abundant hot steam. Energy in the rock vapor evaporates a few hundred meters of seawater. The surface waters of the ocean heat to nearly 500 K, but the deep waters can remain cool. Hot fresh rainwater pools at the top of the ocean, so that the ocean is stably stratified both by temperature and by composition.

An impact on this scale would have been disastrous for photosynthetic life: In short, the photic zone would be eliminated. However, deep waters remain habitable and hydrothermally hosted autotrophs as well as a wide variety of heterotrophs ought to have survived. Sterilization of the upper tens of meters of land is probable.

Figure 7 considers the full range of impact energies most likely to be pertinent to the survival of life on Earth. For impacts of 10^{28} J — comparable to what would be released if Vesta or Ceres were to hit Earth — there is energy enough to evaporate kilometers of seawater. The surface waters of the ocean heat to 500–600 K. Hot boiled brine fouls most of what was not evaporated. Whether habitable niches survive in the deep oceans may depend on the presence of trenches and thus the details of mantle convection. The deep waters are at first cold and dense and therefore stable against convection. Underground niches are more survivable (*Sleep et al.,* 1989; *Zahnle and Sleep*, 1997), but these too become too hot for life as we know it unless protected by several hundreds of meters of cover. Ecological consequences would be severe and only thermophilic ecosystems are likely to survive.

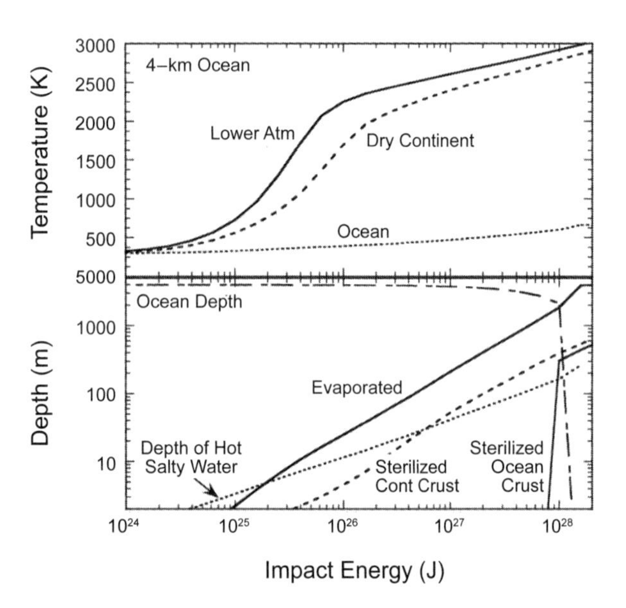

Fig. 7. Effects of impacts on Earth as a function of impact energy. The low end shown here corresponds to Chicxulub; the high end corresponds to an impact by Ceres, the largest known asteroid. On the order of one impact on this scale is likely on Earth at the time when the lunar Imbrium and Orientale impact basins formed. The top panel shows the highest temperatures reached in the troposphere ("lower atm"), at the ocean's surface ("ocean"), and in dry ground ("dry continent"). The bottom panel shows the minimum depth of the ocean, the maximum depth of ocean water evaporated, the depth to which the remnant ocean is sterilized by hot salty water, the depth to which dry land is sterilized ("cont. crust"), and the depth to which the exposed oceanic crust (after the ocean has evaporated) is sterilized. Sterilization is defined as temperature exceeding 400 K. A more nuanced model would show that adverse environmental effects are patchy and thus the planet may be more habitable after enormous impacts than evident here.

3. THE EFFECTS OF IMPACTS ON MARS

The climate effects of large impacts on Mars have been studied by previous researchers including *Sleep and Zahnle* (1998), *Segura et al.* (2002, 2008, 2012), *Segura and Colaprete* (2009), and *Toon et al.* (2010). The primary sources of climatic effects for Mars are the global debris layers and vaporized water resulting from the impact event as described in the introduction and section 2.2. Mars, like the other terrestrial planets, probably acquired surface and subsurface water over time from comets and carbonaceous asteroid impacts during the period of heavy bombardment (*Chyba,* 1990). Indeed, observations have determined that ice may be ubiquitous in the martian soil (*Boynton et al.,* 2002) and that water ice resides in both the northern (*Langevin et al.,* 2005) and southern (*Bibring et al.,* 2004) polar caps. Thus climate effects should result from the vaporization and redistribution of this subsurface and polar water caused by the kinetic energy of the impact event, as well as water in the impactor itself.

Direct heating by impacts of two different scales, 10^{24} J (Lyot) and 10^{26} J (Argyre) is addressed in Fig. 8. The model is essentially the same as that used for Earth impacts (Figs. 3–7). The model considers ice-rich (wet) ground, dry ground, and CO_2-ice-covered ground, although the heat sink of evaporating CO_2 proved relatively unimportant for impacts on the scales considered here and is not further addressed here or in Fig. 8. The models shown here assume that 80% of the surface was covered by ground that was 50% ice by mass — this is not an unreasonable assumption for early Mars although it would not be a good assumption today. Figure 8 shows that thermal inertia of the atmosphere is important for a Lyot-scale impact: Lyot in a 10-mbar CO_2 atmosphere would raise the surface temperature to the melting point of silicate dust and thus would produce fusion crusts on dry land, but a 100-mbar CO_2 atmosphere would not get hot enough to do that although the effects last slightly longer. Figure 8 shows that the bigger, Argyre-scale impact raises the temperature of ground waters to 100°C, thus providing conditions suitable to formation of specular hematite near the surface (*Catling and Moore,* 2003).

Segura et al. (2002, 2008) numerically modeled the climate effects of martian impacts in one dimension, limiting their computations to those resulting from large impactors that can produce global effects. The initial conditions for the simulation include the average global hot debris layer after it has condensed and rained out, with an injected average column amount of water still suspended in the atmosphere (see section 2.2). There are three primary sources of water that make up the total injected water. Following *Segura et al.* (2002), these include the water in the impactor itself, the water/ice in the target material that makes up the crater, and any surface water/ice that evaporates while the hot rock vapor is suspended in the atmosphere. It is also possible that secondary craters may release a substantial amount of water by excavating near-surface ground ice that can then evaporate either in the heat of a big impact or in sunlight for a more modest impact — this redirection of solar energy amplifies the climatic effects of the latter. The injected water remains as vapor in the hot atmosphere because as the rock vapor condenses, it gives off its latent heat to the atmosphere and the atmosphere remains hot.

Using a time-dependent one-dimensional radiative-convective atmospheric model coupled to a subsurface thermal model, *Segura et al.* (2002) computed the climate effects of impactors 100–250 km in diameter, corresponding to craters 500–1500 km in diameter. Implementing the initial conditions described above, and letting the model run out

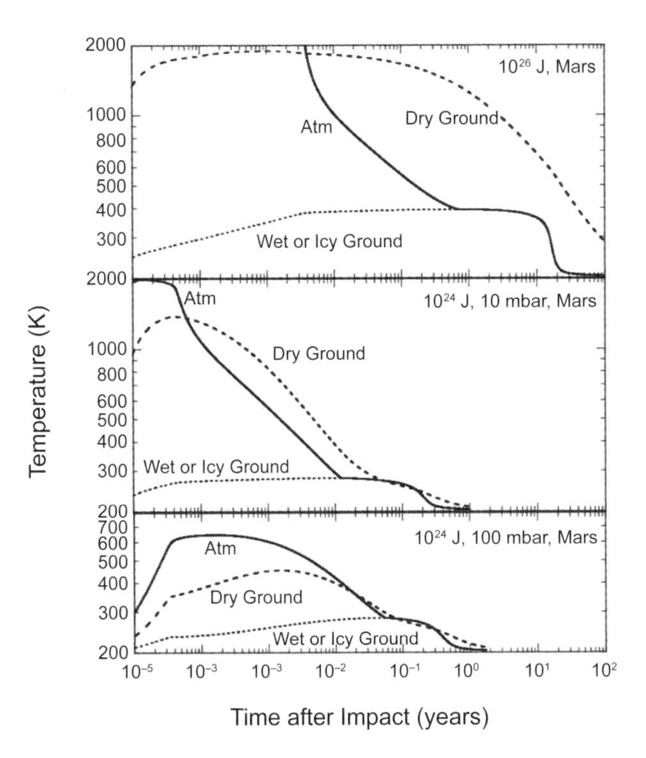

Fig. 8. A comparison of the short-term thermal consequences (global average) of three impacts on Mars: two Lyot-scale impacts, one in a 10-mbar CO_2 atmosphere (middle panel, 10^{24} J) and the other in a 100-mbar CO_2 (bottom panel, 10^{24} J) atmosphere; and an Argyre-scale impact (top panel, 10^{26} J) in a 10-mbar CO_2 atmosphere. Temperatures are shown in wet/icy ground, dry ground, and air. These models address only the direct thermal effects of the impacts; they do not include the indirect effects that stem from sunlight and the resulting greenhouse warming.

to equilibrium, Segura et al. found that the surface of Mars may be kept above the freezing point of water for decades to centuries, with precipitation totals of meters to tens of meters. In a follow-on paper, *Segura et al.* (2008) computed the effects of impactors 30–100 km in diameter, corresponding to craters 200–600 km in diameter, representing the lower limit on impactor sizes that can be accurately simulated in one dimension. In this updated numerical model, latent heating due to cloud condensation and evaporation and a hydrological cycle, to account for the precipitation and evaporation of water at the surface, were both added. They found that the surface of Mars would be kept above the freezing point for a period of time from three months to several years, depending on the assumed background CO_2 pressure (which varied from 150 mbar to 2 bar), with global average martian precipitation rates of 2 m/yr, about twice the current global average for Earth.

In separate simulations, *Segura et al.* (2008) also considered the radiative effects of water clouds in the modeled atmosphere. Water clouds scatter in the visible but absorb in the infrared, creating a greenhouse effect. The results of these simulations showed that water cloud greenhouse warming doubled the time Mars remained above the freezing point as compared to simulations that did not include the radiative effects of clouds. Furthermore, their results showed that the planet did not return to the same equilibrium temperature of 212 K as it did in previous results. Instead, the model settled into a stable climate with temperature near 250 K, and did not return to the 212 K equilibrium temperature even after *centuries* of simulation time, suggesting that the impact-induced greenhouse climate would last for many years hence. The limitation here is the long running time of simulation, not the duration of the greenhouse, as after centuries of simulation time the model had yet to reach equilibrium. Similar results have since been reported from some GCM models in which the injection of meteorologically active water is caused by changing Mars' obliquity (e.g., *Haberle et al.*, 2012). In *Segura et al.* (2008) creation of a sustained impact-induced greenhouse depended on the size of the impact, the removal rate of water at the poles, and details of the water clouds such as altitude and particle size. Although the resulting martian global mean temperature of 250 K is below freezing, it is likely that there were regions above freezing seasonally. *Urata and Toon* (2012) showed that surface temperatures are highly sensitive to the assumptions made for cloud properties. They find large areas of the tropics that are above freezing on an annual average in a climate that appears to be both internally consistent and stable. They also find that significant precipitation rates can occur if there are tropical sources of water, such as snowfields, or frozen seas. In this model warm global temperatures (near 260 K) vanish once CO_2 levels decline below about 200 mbar, mainly because there is not enough poleward heat transport to keep the polar ice relatively warm. Thus the duration of the greenhouse depends on several atmospheric variables, and ultimately is yet to be determined.

Building on the results by *Segura et al.* (2002, 2008) using a one-dimensional model, *Segura and Colaprete* (2009) studied the effects of martian impacts in three dimensions using the Ames Mars general circulation model (MGCM). They incorporated a hydrological cycle into the Ames MGCM to include the formation of clouds, precipitation, and surface and regolith reservoirs. In these computations, impacts by objects smaller than 10 km diameter can be simulated as the size of the modeled impactor is not limited to those with global average effects, as was the case for one-dimensional simulations. For each simulation, an impact debris layer and atmospheric thermal plume, defined as a function of impact diameter (*Sleep and Zahnle,* 1998), were emplaced at any location in the model after a period of model spin-up time. Soil water abundance and distribution were specified for each simulation and water was allowed to diffuse to the surface at a fixed rate if regolith temperatures rose above freezing; however, infiltration of the subsurface by surface water was not modeled. Both water clouds (wet and cold microphysics) and CO_2 clouds were modeled. The radiative effects of water clouds (warming and cooling effect), CO_2 clouds (warming and cooling effect), and dust (predominantly cooling effect) were included, as well as water cloud coalescence, which induces the formation of large-scale convective cumulus clouds, a strong regional cooling forcing. Latent heat effects for both CO_2 and water cloud condensation and surface frosts were also included. An early Mars atmosphere containing 300 mbar of CO_2 and a solar flux that is 75% current levels was assumed.

Several key results stemmed from these simulations. The first is the confirmation of the results found by *Segura et al.* (2008) that a metastable warm climate is sustained for *at least* decades (Fig. 9), maintained by the greenhouse effects of water vapor, CO_2, and the radiative effects of clouds of both types. Thus, although the dust and clouds can induce cooling, in these simulations the overall effect is to warm. The three-dimensional model simulates seasonal effects and sustains the water in the atmosphere: If the water precipitates out and then freezes in the winter hemisphere, for example, the seasonal return of warmth and humidity allows the cycle to continue. Interhemispherical atmospheric transport appears important to maintaining sufficient water vapor concentrations for greenhouse warming. This elevated state of humidity (around 40–60% on average globally), along with the existing 300 mbar of CO_2, results in a stable but temporary "warm and wet" climate. In the MGCM, after 20 years there was no indication of the climate returning to its previous "cold and dry" state. The system will finally collapse and return to pre-impact conditions when sufficient water is removed from the system (by surface infiltration, for example), but computing limitations have so far prohibited running of the three-dimensional model to this state, so this duration is also yet to be determined. The second result is that tens of meters of rainfall, produced in several locations across the martian globe in the years following impact, were able to carve river valleys, transport sediment, and form mineralogical signatures (Fig. 10).

Impacts may not have contributed to significant global climate effects on Mars since the late Noachian. First, the impact rate has declined over time. Today a 5-km object may hit Earth roughly once every 10 m.y. The current impact rate for a 5-km diameter object on Mars, with a higher average crater-production rate 1.3 times that of Earth (*Hartmann,* 1999) but only ~25% of Earth's surface area, is roughly once every 30 m.y. A 5-km-diameter impactor (60-km crater) will raise less than 10 cm of water, unless it impacts a polar cap. The partial pressure associated with 10 cm of water is near the triple point for water, so given these relatively small effects, such events have probably not been recognized in the martian geologic record.

Second, the long-lasting warming effects of impact events (as opposed to the immediate direct heating described in Fig. 8) were reduced as the martian surface pressure declined due to impact erosion and escape to space. This is shown in Fig. 11, where both the maximum global surface temperature and the average global surface temperature observed in simulations decline as the surface pressure declines. This is due to three factors: (1) a higher surface pressure permits greater interhemispherical mass transport of water vapor, from the summer hemisphere to the winter hemisphere, thereby maintaining atmospheric water vapor concentrations; (2) the higher heat capacity of a more massive atmosphere slows down the radiative cooling of the atmosphere, so the climate effects last much longer in a more massive atmosphere; and (3) decreased CO_2 pressure decreases absorption in the CO_2 bands, resulting in cooler temperatures and lower humidity, reducing the water vapor greenhouse effect as well. To simulate an impact in the current martian environment, *Segura et al.* (2008) modeled a 30-km-diameter impactor (with cloud radiative effects excluded) in one dimension with a 6.1-mbar background CO_2 atmosphere. They found that the global surface temperature fell below freezing after just one day, as all energy radiated to space rather quickly in the thin atmosphere. Thus the effects of impacts are not as dramatic today as they were in the martian past when the surface pressure was probably higher.

The heating and rainfall following impacts should result in erosion on Mars, such as the martian river valley networks, especially as the climate effects should last for *at least* centuries. It would be useful for geologists to determine the characteristics of the terrain in which the martian rivers are cut. After a large impact, a global debris layer that can be tens of hundreds of meters thick will be created, and rain will quickly fall on it. Massive water releases onto what is essentially pulverized rock should lead to rapid erosion. In contrast, rivers cut into basalt or other bedrock would take much longer to erode. Distinguishing which of these scenarios occurred on Mars would help clarify the processes forming the river valleys.

A number of geological studies have suggested that very long periods of time were required to create the martian river valleys (e.g., *Barnhart et al.,* 2009; *Hoke et al.,* 2011). However, these are opposed by other studies. *Fassett and*

Head (2005) show that breached craters may have been caused by flows of <20 years duration, and *Stepinski et al.* (2004) conclude that the martian terrain experienced enough limited rainfall to begin the carving of channels, but not enough to form a complex slope distribution. Even *Hoke et al.* (2011) state they are unable to rule out massive floods. Furthermore, there are data for Earth, such as the formation of the New River in California, where 43 miles of channels were formed, with an average width of 1000 feet and depth of 50 feet, in less than 2 years by the accidental diversion of the Colorado River (*Austin,* 1984). In order to determine the time to form rivers, we need to know if they are in bedrock, or regolith from the impact; how erosion occurs with hot water; and what the duration of the flow would be after a large impact. Regarding this last point, several models show long-lasting warm climates persisting for long periods of time, and with substantial recycling of water and substantial rainfall rates (*Segura et al.,* 2008; *Urata and Toon,* 2012).

4. THE EFFECTS OF IMPACTS ON VENUS

Venus currently has a very thick atmosphere with an enormous thermal inertia that makes it difficult for cosmic impacts to affect it directly. The indirect effect of adding important greenhouse gases, water in particular, is more likely to matter to the climate (*Bullock and Grinspoon,* 2001).

The biggest crater on Venus is Mead, a well-preserved ~280-km-diameter two-ring basin. Conventional crater scaling suggests that Mead was created by a 40-km-diameter, 1×10^{17}-kg asteroid striking at 25 km/s and releasing 2×10^{25} J on impact. The energy released is about 10 times that of Chicxulub. The Sudbury and Vredefort impact structures on Earth, both ca. 1.9 Ga, are comparable.

The heat capacity of Venus' CO_2 atmosphere is 5.7×10^{23} J/K (or 1.2×10^9 J/m²/K using Cp appropriate to CO_2 at 750 K). If 50% of Mead's impact energy eventually reached the atmosphere, the atmosphere as a whole could have heated up by 18 K. But in practice relatively little of this energy reaches the lower atmosphere. Most of the impact energy that reaches the atmosphere is distributed globally by impact ejecta that reenter the atmosphere at high altitudes. What results is that the atmosphere above 0.1 bar approaches the 1500–2000 K temperature where rock vapors condense. Outwardly directed thermal radiation from the hot silicates is lost to Venus, while downwardly directed thermal radiation evaporates sulfuric acid clouds and heats the thick air. Some 10–20% of the thermal radiation reaches the surface through opacity gaps, especially at ~1 μm, which is near the Wien peak. With these effects

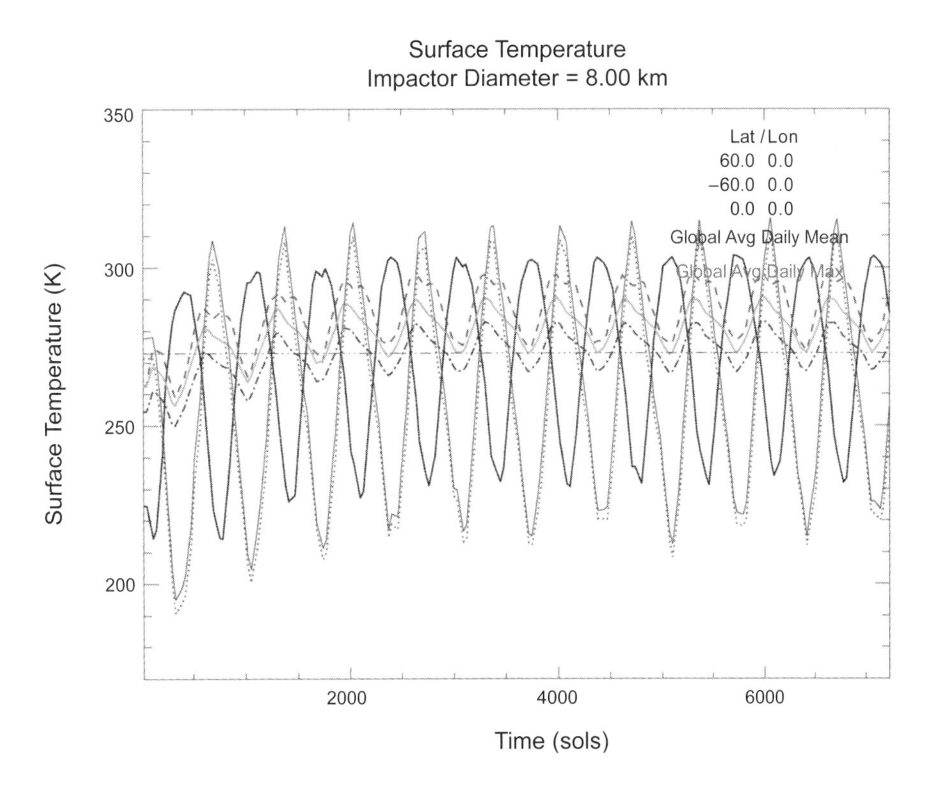

Fig. 9. See Plate 71 for color version. Surface temperature at three locations (black, blue, red) on Mars plus global average daily mean temperature (purple) and global average daily max temperature (aqua) for several years following impact with an 8-km-diameter object. Note that at the end of simulation time the model has yet to return to equilibrium.

taken into account, we estimate that Mead ejecta radiatively heated the lower atmosphere by ~2 K. Given a characteristic hot-silicate radiating temperature of 2000 K, the heating would take place over a few hours.

After the silicates cool, the dust provides an anti-greenhouse effect that cools the lower atmosphere. This cooling is very slow. At present the lower atmosphere is in energy balance with heating by ~10% of the incident sunlight, or ~60 W m^{-2}. At this rate it takes about 15 months for the lower atmosphere to cool by 2 K.

Long-term perturbations to the albedo and the greenhouse effect would likely be more important. The atmosphere at the nominal 30 ppmv H_2O contains 6 × 10^{15} kg of H_2O. A volatile-rich Mead impactor that is 6% water would therefore double the atmosphere's water. This increases the greenhouse effect. The perturbation on the albedo is likely to be more important, but the sign of the albedo effect is uncertain. Venus' albedo from H_2SO_4 clouds is very high, at ~78%, so one expectation is that any perturbation is likelier to decrease this high albedo than to increase it. As noted, the impact would evaporate the H_2SO_4 clouds and cover the old atmosphere under a shroud of dust. When the dust clears, it is not obvious what sort of atmosphere would remain, or when bright sulfuric acid clouds would reform. *Bullock and Grinspoon* (2001) have discussed some of the complexities consequent to volcanic eruptions of SO_2 and H_2O and pointed out that similar consequences follow large asteroid or comet impacts. Their results, although not directly applicable to the present problem, imply that Mead's indirect effects are likely to result in long-term surface temperature changes on the order of 10–30 K, which could be of either sign. This would be enough to force the volatile, high-dielectric material [probably lead sulfide (*Schaefer and Fegley,* 2004)] responsible for the "snowline" that is so prominent on mountains in the Magellan RADAR images to move up and down mountains, for example, but may not have other major consequences.

To the extent that anything is revealed by available observations, the object that created Mead in fact had no discernible effect on Venus, apart from the crater itself. Evidently it was too small, too dry, or both. Bigger impacts would be more potent, but Venus does not preserve any sign of impacts as big as those of the late bombardment, or even as big as the impacts inferred from the thickest Archean spherule beds on Earth before 2.5 Ga. An impact 10 times bigger than Mead would likely have more or less the same consequences for Venus' climate that *Bullock and Grinspoon* (2001) and *Solomon et al.* (1999) ascribe to volcanic outgassing of a global ocean of basaltic magma ~10 km deep formed by catastrophic global resurfacing. Whether this means that a great impact did play a role in sparking Venus' resurfacing is an open question.

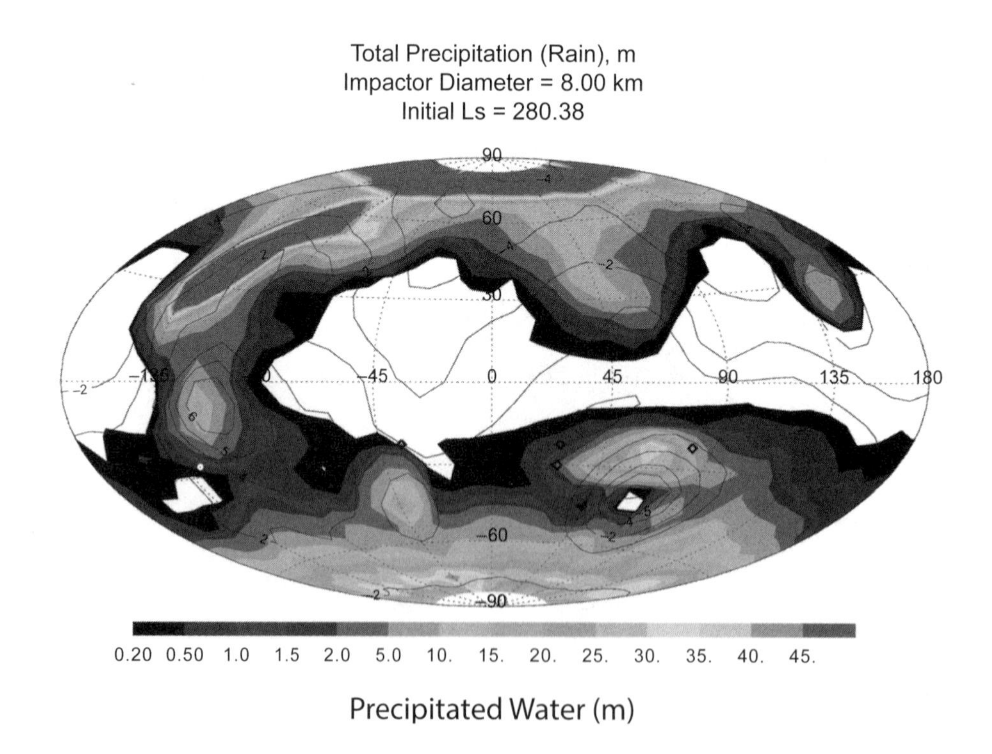

Fig. 10. See Plate 72 for color version. Total precipitated rain (meters) across the martian globe for an 8-km-diameter impactor near 0,0° (*Segura and Colaprete,* 2009). Longitude is marked globally and contours represent topography in kilometers referenced to mean Mars elevation.

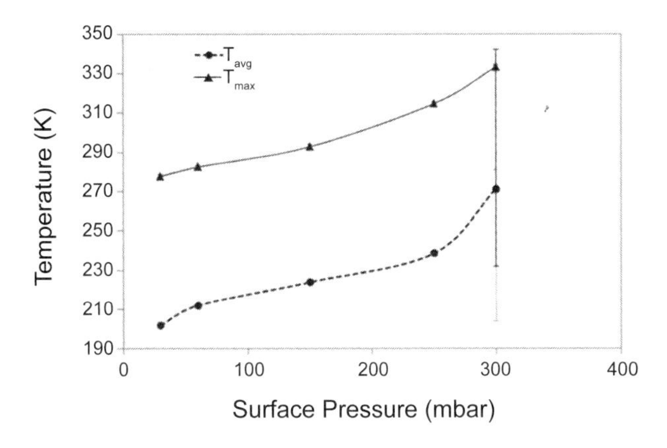

Fig. 11. The global daily average (circles) and daily maximum (triangles) surface temperatures as a function of total CO_2 surface pressure following an 8-km impact.

5. THE EFFECTS OF IMPACTS ON TITAN

Titan is the only moon in the solar system with an appreciable atmosphere. The surface pressure is 1.5 atm with a surface temperature of 90–94 K (*Flaser et al.,* 2005). The atmosphere is composed primarily of N_2 with ~5% CH_4, 0.1% H_2, and trace organic compounds (*Flaser et al.,* 2005). Titan has an active hydrological cycle of liquid CH_4 including clouds, storms, rain, and surface indications of fluvial features. Large lakes composed of liquid CH_4 and C_2H_6 are present in the polar regions (e.g., *Hayes et al.,* 2008). The climate system of Titan and the surface features of lakes, dunes, and fluvial channels present many similarities to the corresponding features on Earth. In this way, Titan can be considered as a terrestrial planet. It is therefore not surprising that impacts have played an important role in Titan's history, just as they have for Earth and Mars.

5.1. Impact Origin of Volatiles

Titan's volatile inventory may be the result of impacts (*Zahnle et al.,* 1992). *Griffith and Zahnle* (1995) explored this in detail and concluded that the N_2 present in Titan's atmosphere could readily have come directly from cometary organic N sources. *Griffith and Zahnle* (1995) predict CO in abundance on Titan as well, based on the assumption that this is the stable form of C in gas that has experienced high-temperature shocks. More detailed impact simulations (e.g., *Pierazzo et al.,* 1997), and estimates of the cometary impact flux onto the outer planet satellites (e.g., *Zahnle et al.,* 2003; *Gomes et al.,* 2005) support the suggestion that Titan's atmosphere could have been produced by cometary impacts. In the first 700 m.y. after its formation, ~5 × 10^{23} g of cometary material impacted Titan (*Gomes et al.,* 2005) at an average velocity of ~10 km/s (*Zahnle et al.,* 2003). Assuming a cometary Ni abundance in the range of 0.001–0.04 by mass (*Zahnle et al.,* 1992; *Griffith and Zahnle,* 1995), this is more than enough N_2 to supply Titan's present inventory. We note in passing that N_2 from impact shocks implies that CO is also forged (from organics and H_2O) by impact shocks. The latter implies after some photochemistry that CO_2 must be an important constituent ice of Titan.

Nitrogen and H isotopes provide additional constraints on the origin of Titan's volatiles. Titan's N_2 is heavy, with a $^{14}N/^{15}N$ ratio of 167.7 ± 0.6 (*Neimann et al.,* 2010). Heavy N can be interpreted as evidence for massive fractionating N escape beginning from an Earth-like (272) or Sun-like (~430) ratio. However, theory suggests that it is very difficult for Titan to have lost enough N slowly enough to fractionate the remnant (*Mandt et al.,* 2009). An alternative story begins from the observation that Titan's N_2 is very similar to N on Mars (~170) and to N in HCN and CN in Jupiter-family comets [155 ± 25 in 103P/Hartley 2 (*Meech et al.,* 2011) and 160 ± 40 in 17P/Holmes (*Bockelee-Morvan et al.,* 2008)]. This similarity can be interpreted as evidence in favor of a cometary source.

It has been argued that because N_2 and Ar have similar volatility, the low $^{36}Ar/N$ ratio on Titan is evidence against a cometary source (cf. *Lunine et al.,* 2010). However, N_2 has not been observed in comets (*Biver et al.,* 2012). Rather, N in comets is present either as NH_3, a nitrile (e.g., HCN), or in more complex organic molecules. The $^{36}Ar/N$ constraint does not apply.

The D/H ratio in Titan's CH_4 (1.3e–4) and in H_2 (1.35 ± 0.3e–4) measured by the Huygens probe (*Niemann et al.,* 2010) is strikingly similar to what it is on Earth, in most meteorites, and in the Jupiter-family comet 103P/Hartley 2 [1.6 ± 0.2e–4 (*Hartogh et al.,* 2011)]. This also is consistent with a Jupiter-family cometary source delivered by impacts. Traditional models that endow Titan with volatiles from a putative saturnian subnebula equilibrate these volatiles with saturnian H_2, which is expected to have a low, subsolar D/H ratio on the order of 2e–5. This does not preclude a subnebular origin of Titan's volatiles: It may be that H is fractionated by escape and the coincidence with Earth and meteorites is an accident, or it may be that the same physical conditions and physical process that set the D/H ratio in ice in the solar nebula acted in the saturnian subnebula.

An alternative to the view that the N_2 in Titan's atmosphere came directly from cometary impact is the suggestion that the N_2 formed from the dissociation of a NH_3-rich primordial atmosphere. *Atreya et al.* (1978) suggested photodissociation as the process that could form N_2. *Jones and Lewis* (1987) and *McKay et al.* (1988) proposed impact shock processing of atmospheric NH_3 to form N_2. As an impactor enters the atmosphere, it creates a high-temperature shock wave that results in both UV irradiation and compressional heating of the surrounding gas (e.g., *McKay and Borucki,* 1997). At high temperatures (>10,000 K) in the shocked gases, reactions are thought to be so rapid that the mixture is in chemical equilibrium. As the gas cools, a temperature is reached at which the gas phase reactions become too slow to maintain equilibrium, and the chemical composition of the mixture is quenched or frozen at the equilibrium value determined by the freeze-out temperature. The composition of cometary gases as a function of freeze-out temperature is shown in Fig. 12 from *McKay and Borucki* (1997). As shown in this figure, gas-phase reactions could account for N_2 in Titan's atmosphere but not CH_4.

While the gas-phase chemistry may be slow at low temperatures and therefore not produce CH_4, *Kress and McKay* (2004) suggested that the dust that condenses out in the wake of a large comet impact is likely to have very effective catalytic properties, opening up reaction pathways to convert CO and H_2 to CH_4 and CO_2, at temperatures of a few hundred Kelvin. *Kress and McKay* (2004) conclude that CH_4 could have formed in Titan's atmosphere via catalysis in the aftermath of large impacts. When catalyzed reactions are taken into consideration in impact chemistry, production of both CH_4 and N_2 results (as well as CO_2, which would condense on Titan).

Sekine et al. (2011) suggested an alternative scenario for the impact production of the N_2 in Titan's atmosphere. They considered cometary impacts during the late heavy bombardment after the formation of Titan. In their scenario, impact processing of an ammonium hydrate (NH_3-H_2O) ice surface is the source of atmospheric N_2. Sekine et al. conducted shock simulation experiments to show that ammonia ice converts to N_2 very efficiently during impacts. Numerical calculations based on their experimental results indicate that Titan would acquire sufficient N_2 to produce the current atmosphere and that most of the atmosphere present before the late heavy bombardment would have been replaced by impact-induced N_2. An advantage of their scenario is that it is capable of generating a N_2-rich atmosphere without production of primordial Ar — consistent with observations from the Huygens probe.

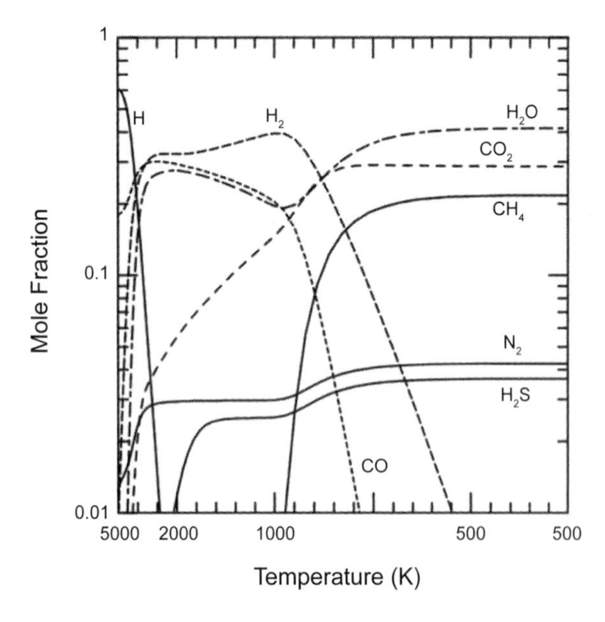

Fig. 12. Impact shock chemistry. Thermodynamic equilibrium mixture of cometary volatiles as a function of temperature for gas phase reactions for a total pressure of 1 bar from *McKay and Borucki* (1997). At typical freeze out of 2000 to 3000 K, N is in the form of N_2 and C is in the form of CO_2 and CO.

5.2. Climate Effects of Impacts

Titan's thick atmosphere and volatile surface cause it to respond to large impacts in a somewhat Earth-like manner. It has long been appreciated that the impact of a comet big enough to completely penetrate Titan's atmosphere would generate impact melts of liquid water that could persist for considerable periods of time (*Thompson and Sagan,* 1992; *Artemieva and Lunine,* 2003, 2005). These studies emphasize the chemistry that would take place when the organic debris known to be widespread and abundant on Titan's surface (*Lorenz et al.,* 2008a) are dissolved in liquid water. Here we focus on the thermal consequences of two big craters on Titan.

The model for impacts on Titan somewhat resembles the model for impacts on Earth. The model and its consequences are more fully described elsewhere (K. Zahnle, in preparation). Methane replaces water as the active volatile, and a widely accepted but hypothetical water ice crust replaces Earth's rocky crust. One quantitative difference is that Titan's atmosphere is much thicker than Earth's and has much more thermal inertia, which means that only the greatest impacts can do much to raise its temperature. Another difference is that an ice crust is relatively easy to melt, and if it melts, the melt is denser than the ice underneath, which means that icebergs rise up from the deep and any open water would soon be choked with ice or turned to slush or sink into the interior.

A key qualitative difference is that Titan's known methane is mostly in the atmosphere, whereas water on Earth is mostly condensed. Lakes dot the polar regions that between them have a volume on the order of 10^5 km^3 (*Lorenz et al.,* 2008b), or the equivalent of 0.01 bar if evaporated. This volume is equivalent to lakes with an average depth of 40 m covering 3% of Titan's surface, which is the approximation we make here. Methane is also present in the crust near the surface at the Huygens landing site. Heat from the Huygens probe evaporated methane (*Niemann et al.,* 2005) and other volatiles [including C_2H_6 and CO_2 (*Lorenz et al.,* 2006)] from the surface. We will arbitrarily assume that the crust contains 5% methane (g/g), which corresponds to a porosity of 10% (liquid methane's density is about half that of the crust), and that the porosity and methane extend to 1.5 km beneath the surface.

Most of the impact's energy is initially invested in target materials close to where the comet strikes, and in the materials of the comet itself. These materials are heated, melted, evaporated, and they also acquire significant kinetic energy, such that the bulk of the most strongly heated materials are ejected from the crater. A significant fraction of this energy is quickly shared with the atmosphere, through drag during ejection and then later, globally, when the more far-flung ejecta reenter the atmosphere.

Rather than fully model these events, we assume that 50% of the impact energy goes into the atmosphere and the near-surface environment. This energy will include almost all the kinetic energy of impact ejecta that escape from Titan, as these are mostly swept up again by Titan, where they will deposit their energy in the atmosphere or the near-surface environment.

Some energy remains localized at the crater, much of this in a crater lake. Using scaling rules developed for hypervelocity cratering in ice by *Kraus et al.* (2011) and *Senft and Stewart* (2011), we find that the crater lake takes up about 10% of the impact's energy. Detailed numerical modeling (*Senft and Stewart,* 2011) using a new, more fully featured equation of state for the water substance shows that the lake is deeply buried, with the liquid water pooling at the bottom of the transient crater rather than near the surface as had been assumed by *Artemieva and Lunine* (2003, 2005). As a consequence the energy in the lake is inaccessible on short timescales. The balance of the impact energy is also presumed inaccessible, either very deeply buried or lost through radiative cooling from the top of the atmosphere when the bulk of the ejecta are first falling back to Titan and the upper atmosphere is both very hot and relatively opaque.

For simplicity we presume that the climatically available energy raises the ejecta (mostly water) and the atmosphere to a uniform temperature T. The ejecta contain methane that is vaporized. Methane lakes are partially vaporized by infalling warm ejecta. If there is enough energy in the impact, water can melt and evaporate. For the present purpose we restrict our consideration to N_2, H_2O, and CH_4. It is not unlikely that C_2H_6 and CO_2 are abundant, and if abundant, important. In particular, the presence of other ices more volatile than H_2O will tend to limit surface temperatures and thus prevent water ice from melting.

We track the energy as it flows into or out of six reservoirs until Titan has returned to the state it was in before the impact came. The six reservoirs are the excess thermal energy in the atmosphere, the excess thermal energy in the crust, the latent heat associated with melting crustal water ice, the latent heat of evaporating water, the latent heat associated with evaporating crustal methane, and the latent heat of methane from lakes.

Menrva, at 444 km nominal diameter, is Titan's biggest indubitable crater (*Wood et al.,* 2010). Menrva is comparable to the Gilgamesh basin on Ganymede, and like Gilgamesh it is not ancient, but otherwise hard to date geologically. Conventional crater scaling relations suggest that it was formed by an impact releasing on the order of 1.6×10^{24} J, closely comparable to the Chicxulub event on Earth. This energy estimate is unlikely to be good to better than a factor of 2. Thermal effects of a Menrva-sized impact are illustrated in Fig. 13.

There is evidence of at least one older, bigger crater on Titan. Hotei Regio is a 700-km-diameter quasicircular IR albedo feature that appears to lie within a larger basin that appears, at least on the side mapped by radar, to be an arc of a circle (Hotei Arcus). The albedo feature has been a leading candidate for a cryovolcanic flow (*Soderblom et al.,* 2009). But Soderblom et al. suggest that the basin itself is an ancient impact feature, perhaps on the order of 800–1200 km diameter, that has been severely degraded. Crater scaling suggests an impact energy on the order of $1–5 \times 10^{25}$ J. Here we will consider an impact at the bottom of this range. Thermal effects of a 1.2×10^{25} J impact are illustrated in Fig. 14.

In these models we have ignored clathrates. The absence of detectable Kr and Xe in the atmosphere, along with the high Ne/Ar ratio (Ne does not enter clathrates but Ar does), provide evidence that clathrates play a role in Titan (*Niemann et al.,* 2010). Clathrate formation is severely kinetically inhibited at low temperatures (*Pietrass et al.,* 1995). Methane clathrate is stable at Titan's surface for temperatures below 200 K, but experiments at 230 K and 60 bar that imply formation times on the order of 10^6 s, and the measured activation energy of 46 kJ/mol (*Staykova et al.,* 2002), together suggest that it could take decades to make clathrates from methane and water ice at 200 K. This is comparable to the timescale during which Titan

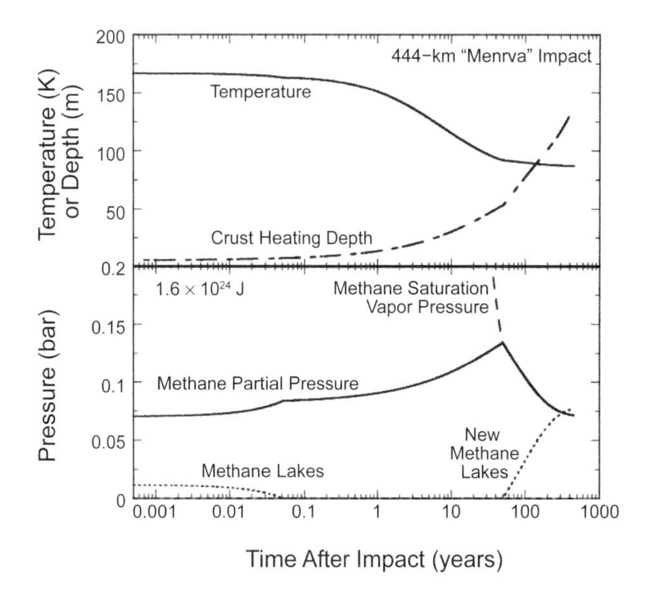

Fig. 13. Some global thermal consequences of the Menrva impact if it were to occur on Titan today. Plotted are the atmospheric (surface) temperature, the depth of crustal heating (which here corresponds to the depth from which cryptic crustal methane is evaporated), methane's saturation vapor pressure and the actual partial pressure of methane in the atmosphere, and the effective volume of the methane lakes (here plotted as bar-equivalents, i.e., the partial pressure the methane would contribute to the atmosphere if evaporated). Methane is released from pores by the propagation of the thermal wave into the crust. "Crust heating depth" shows the progress of the thermal wave. Methane begins to drizzle out 50 years after the impact. The new methane lakes are bigger because methane has been moved from the crust to the lakes. The evolution of methane shown here ignores the slight possibility of clathrate formation.

would remain this warm from a Hotei-scale impact. Other clathrates form more easily — ethane, CO_2, and Xe in particular — and it is plausible that large impacts could make them.

Thompson and Sagan (1992) suggested that impacts on Titan could create crater lakes in the ice-rich crust. These would become ice-covered, but the subice liquid habitat would persist for a long period of time depending on the size of the impact crater. *Artemieva and Lunine* (2003, 2005) and *O'Brien et al.* (2005) have further considered this. *O'Brien et al.* (2005) conclude that 15-km-diameter craters can sustain liquid water or water-ammonia environments for ~100–1000 years and 150-km craters can sustain them for ~1000–10,000 years. *Artemieva and Lunine* (2005) conclude that water ice melting and exposure of organics to liquid water has been widespread because of impacts. Organics processed in water and water-ammonia mixtures can alter and incorporate O and form important prebiological molecules such as amino acids (*Neish et al.*, 2008, 2009).

However, detailed numerical modeling using a more realistic equation of state for the water substance by *Senft and Stewart* (2011) appear to show that impact melt sinks to the bottom of the transient crater more-or-less simultaneously with the creation of the crater. This should not be unexpected — the crater lake is, after all, denser than ice, and the ice in which the lake is nested is very warm (and thus of low viscosity) because it too was heated by the impact. For all but Titan's smallest craters, Rayleigh-Taylor instabilities at the interface between the liquid water and the warm ice grow quickly and overturn. This continues until the water reaches the cold ice of the transient crater's floor. Although not discussed in their paper, this behavior is clearly seen in the movies that make up the online supplemental material to *Senft and Stewart* (2011). Whether the deeply buried lakes are long-lived is an open question — the falling waters could not avoid mixing with a great deal of ice, and any water that reaches the bottom of the bowl would still be subject to draining into any cracks the impact made in the cold ice. It is possible that crater lakes, even the biggest, freeze very quickly.

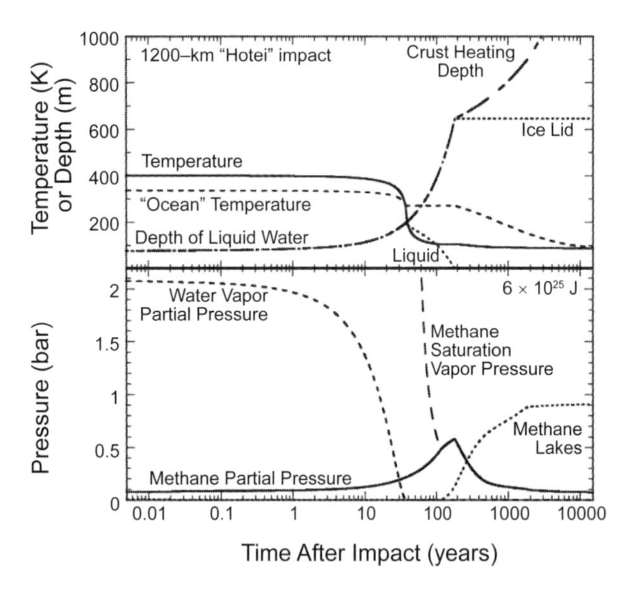

Fig. 14. Global thermal consequences of an impact comparable to the one that formed the putative Hotei impact basin on Titan. The temperature reaches 400 K and the partial pressure of water vapor in the atmosphere exceeds 2 bar. The hypothetical ice crust melts to a depth of 600 m on global average. Liquid water is never more than 200 m thick but the ice sheet that freezes on top of the ocean eventually reaches 600 m thick. Water rain and water snow occur. The liquid waters last about 200 years, although for most of that time the ocean is under thickening ice and probably rather slushy. The volume of methane lakes is given in terms of bar-equivalents rather than as depth. Methane begins to drizzle after 100 years. The results shown here ignore the strong probability of clathrate formation as the liquid water ocean freezes.

Liquid water generated globally by a Hotei-scale impact is quite different. First, both the surface and the atmosphere are raised above water's melting point. Thus the surface is less hostile to liquid water. Second, the ice under the meltwater is cold. Rayleigh-Taylor instabilities are suppressed by the high viscosity of the cold crust. The Rayleigh-Taylor instabilities that do grow grow very slowly because they are limited to growing in the thin layer of conductively heated ice. Thus overturn is frustrated. Rather, the meltwater would be subject to topographic control, flowing downhill and ponding, and of course the waters will be subject to choking and crusting with loose country materials that float. The result might resemble lava tubes more than it resembles rivers.

Impacts on the surface and the resulting liquid water melt could have other effects that are important to surface processes. These could include providing a mechanism for producing the sand-sized particles from organic solids. These sand-sized particles are clearly present in the large dunes (*Radebaugh et al.*, 2008). Impacts could be an effective source of subsurface volatiles such as CH_4 and NH_3, thereby resupplying the atmosphere. This possible source of volatiles is particularly interesting in that there is no clear evidence of cryovolcanism. Perhaps impact-induced volatile sources are important.

6. IMPACT-TRIGGERED RUNAWAY GREENHOUSE ATMOSPHERES

Recently, *Segura et al.* (2012) showed that impacts may push a terrestrial planet into a temporary "runaway" greenhouse state. They explored the limits of flux-temperature space with a one-dimensional numerical model and found that for a sample planet with 6.1 mbar CO_2 and 500 mbar water, there is a range of incoming solar flux such that for one given flux, two stable temperatures may exist, one warm and one cool. This is shown by the "s-curve" in Fig. 15. Curves A and C in Fig. 15 form the runaway curve defined by the climatological equations described in *Nakajima et al.* (1992) and replicate Fig. 2 in that paper. The turnover point near 280 W m^{-2} (denoted RA) shows the traditional runaway point found by them and computed by many others (e.g., *Kasting,* 1988); beyond this point the planet's inventory of water is in the atmosphere and the surface temperature is very warm. Note that the RA runaway point in this curve is lower than the limit found by *Nakajima et al.* (1992) for the gray case; this is expected for two reasons. First, as shown by section 3b in Nakajima et al., a radiative-convective model gives a lower value of the outgoing infrared flux for a saturated atmosphere. Second, the ambient greenhouse produced by the CO_2 pressure of 6.1 mbar moves the runaway point to the left. Curve C is unstable, as described by Nakajima et al. Curve B is the runaway branch computed for a given planetary water inventory, which in this sample planet is 500 mbar. The triangles and squares are found by running a one-dimensional radiative-convective model out to equilibrium for each flux point, yielding an equilibrium temperature from both a warm starting temperature (squares) and a cold starting temperature (triangles). To the left of point L, the numerical solutions for both warm and cold temperatures starts converge to the same point in flux-temperature space. However, in the range between point L and the RA point, for one equilibrium flux and a cold and warm starting temperature, two *different* temperature solutions exist. This range defines the region where a planet could run away at

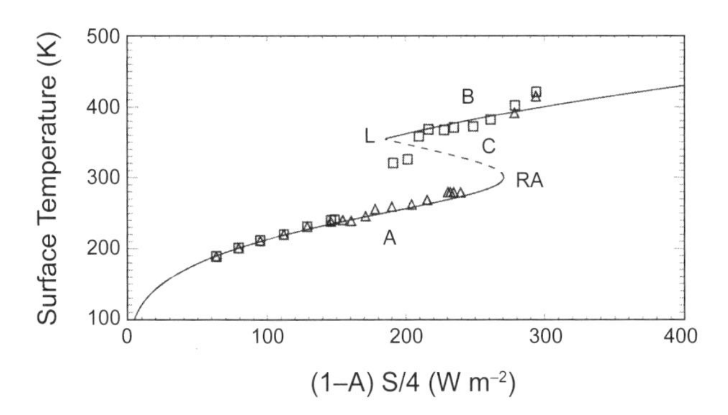

Fig. 15. The analytic equation solutions (curves) following *Nakajima et al.* (1992) compared to detailed numerical computations (triangles and squares) for a sample planet with a 500-mbar water inventory. The triangles show the equilibrium temperature for a given flux computed for a "cool start," at a starting temperature less than Curve A, while the squares show the equilibrium temperature for a given flux computed for a "warm start," at a temperature warmer than Curve B. Point L is set by the water inventory, which is the flux-temperature point at which the planet will move to the runaway branch (Curve B). The numerical solutions are irregular near Point L because this represents the boundary of the unstable region and therefore may not lie on the stable Curve B as predicted.

an incoming solar flux less than the traditional runaway point. One solution, Curve A, is the "cold" solution, and represents a planet's "normal" stable state where most of the water is in the liquid/ice phase at the surface. The second solution, Curve B, is the "warm" temperature solution, and represents the runaway state where the entire inventory of water is in the atmosphere. The water inventory and relative effects of other greenhouse gases, such as CO_2, determine point L in Fig. 15. This point may have a lower limit; this is an extrapolation that remains to be tested with further simulations. This point at lower flux where a runaway could occur is not a runaway point in the conventional sense; it is the point where a runaway would continue once large-scale evaporation is initiated, e.g., by a large temperature perturbation. It is not the point in flux where a runaway can be initiated by increasing solar flux, S.

Segura et al. (2012) concluded that the kinetic energy delivered by a large impact could push a planet to this warmer temperature solution where it may be maintained until water loss (e.g., hydrodynamic escape to space) reduces the water inventory such that the upper solution is no longer viable. The system would then collapse to the lower stable solution, albeit with much less water than it had at the start. The necessary energy pulse is defined by the temperature difference between dual solutions; the perturbation must be large enough to increase the temperature past Curve C. This flux range where dual solutions exist and where a planet is at risk of running away at a flux much lower than the traditional runaway point depends on the planet's water inventory; the energy pulse must be capable of perturbing the planet enough to vaporize all its water in order to run away.

In a more specific example, *McKay et al.* (1999) showed that these multiple stable climate states can exist on Titan. Similar to the case described above, the low-temperature case corresponds to the majority of the CH_4 on the surface, while for the high-temperature case all the CH_4 is in the atmosphere. McKay et al. considered this existence of dual climate solutions and the ability to move between them in the context of geothermal heating, but the energy source could also be associated with discrete large-impact events. Based on the results of McKay et al., we can estimate that an impact-induced runaway would occur on Titan if the impact energy was sufficient to warm the surface by ~20 K. Detailed models have not yet been constructed to test this suggestion.

The research in *Segura et al.* (2012) shows that there may be a set of bistable climate solutions for terrestrial planets and that a planet's movement between the two states could be triggered by large impacts. Their use of a radiative-convective model to explore the flux-temperature space, using a sample planet, is a first step in understanding the possible impact-triggered runaway greenhouse. These results raise many interesting questions that should be explored with more sophisticated, multidimensional models, and the feedbacks due to cloud formation and atmospheric chemistry should be considered. S-curves may be defined for any terrestrial planet given the water inventory. Further research on this topic could reveal that terrestrial planets in other solar systems may have experienced a runaway greenhouse due to impacts, and it may be that some planets are still locked in a runaway due to planet-forming impacts. Others may have their water abundances adjusted by escape to the amount required to no longer be in the runaway state. Likewise, the consideration of late-occurring impacts as triggers for runaway greenhouse atmospheres decreases the likelihood of finding worlds with life on them.

7. CONCLUSIONS AND OPPORTUNITIES FOR FUTURE RESEARCH

We have shown here that impacts can cause interesting climate effects on terrestrial planets, including prolonged warm temperatures, release of subsurface ice, precipitation, mass extinctions, and possible runaway greenhouses. These effects may last for centuries or longer and may be responsible for many geological features we see on the terrestrial planets such as river valleys on Mars and the K-Pg layer on Earth. Determination of the exact duration of such climate effects remains to be studied.

For the Earth, uncertainties remain about the column abundances, residence times, and temperature effects of soot, S, and other materials that may have been injected into the atmosphere following impact. For example, the amount of injected water that remains in the upper atmosphere and the effects of falling ice particles on dust removal have yet to be quantified. Aspects of atmospheric chemistry are highly uncertain, including the rate of conversion of S to sulfate, the formation of acid rain, and the abundances of toxins. The K-Pg layer has been utilized extensively as a tool to study the climatic effects of large impacts on Earth, possibly capable of mass extinctions, yet proper modeling (preferably with a GCM) of the effects of these materials on Earth's climate remains to be performed.

There have been many studies on the effects of impacts on Mars, predominantly in the context of whether the warm temperature/rainfall impulses would be strong enough to form fluvial features contemporaneous with the impact events themselves. Research remains to be completed on the duration of the temporary warm climate caused by impacts, and the relative roles of cloud type/height and surface recycling of water. Geologic landforms such as alluvial fans and stepped deltas (*Kraal et al.,* 2008) have been studied within the context of brief climate episodes such as those following impact events, but questions still remain.

Venus' thick CO_2 atmosphere and its large heat capacity severely limit the climate effects of impacts on Venus. The impact that formed the largest remaining impact crater on Venus, Mead, likely had little effect on Venus even though the energy released was about 10 times that of the Chicxulub impact. Longer-term effects may have been more important, but even still may amount to less than 50 K of temperature increase or decrease. Since Venus does not reveal any craters larger than Mead, prediction of their climate effects remain moot. Additional insight into the history of Venus and its resurfacing and other geological events must be attained.

Titan's thick atmosphere and rocky surface imply that climate effects from impacts will be similar to those on the Earth, although methane rather than water is the precipitating volatile. Menrva, Titan's largest well-preserved impact basin observed to date, was not big enough to melt the putative water ice crust, but the bigger (but speculative) Hotei impact would have been big enough to melt water globally. There has been little published on the modeling of the effects of impacts on Titan and the subject would benefit greatly from future research.

Study of the climate effects of large asteroid and comet impacts remains a captivating topic in planetary science research. Many of the climate effects that have been discussed have been investigated by only a single researcher or group, and often in only a preliminary study. Further work will undoubtedly narrow many of the uncertainties.

REFERENCES

Abramov O. and Mojzsis S. J. (2009) Microbial habitability of the Hadean Earth during the late heavy bombardment. In *Lunar Planet. Sci. Conf. 40,* Abstract #2379. Lunar and Planetary Institute, Houston.

Alvarez L., Alvarez W., Asaro F., and Michel H. V. (1980) Extraterrestrial cause for the Cretaceous-Tertiary extinction. *Science, 208,* 1095–1108.

Artemieva N. and Lunine J. I. (2003) Cratering on Titan: Impact melt, ejecta, and the fate of surface organics. *Icarus, 164,* 471–480.

Artemieva N. and Lunine J. I. (2005) Impact cratering on Titan II. Global melt, escaping ejecta, and aqueous alteration of surface organics. *Icarus, 175,* 522–533.

Artemieva N. and Morgan J. (2009) Modeling the formation of the K-Pg boundary layer. *Icarus, 201,* 768–780.

Atreya S. K., Donahue T. M., and Kuhn W. R. (1978) Evolution of a nitrogen atmosphere on Titan. *Science, 201,* 611–613.

Austin S. A. (1984) Rapid erosion at Mount St. Helens. *Origins, 11(2),* 90–98.

Barnhart C. J., Howard A. D., and Moore J. M. (2009) Long-term precipitation and late-stage valley network formation: Landform simulations of Parana Basin, Mars. *J. Geophys. Res., 114,* DOI: 10.1029/2008JE003122.

Belcher C. M. (2009) Reigniting the Cretaceous-Palaeogene firestorm 400 debate. *Geology, 37(12),* 1147–1148, DOI: 10.1130/focus122009.1.

Belcher C. M., Collinson M. E., Sweet A. R., Hildebrand A. R., and Scott A. C. (2003) Fireball passes and nothing burns — The role of thermal radiation in the Cretaceous-Tertiary event: Evidence from the charcoal record of North America. *Geology, 31,* 1061–1064, DOI: 10.1130/G19989.1.

Belcher C. M., Collinson M. E., Sweet A. R., Hildebrand A. R., and Scott A. C. (2004) Fireball passes and nothing burns. The role of thermal radiation in the K/T event: Evidence from the charcoal record of North America: Comment and Reply. *Geology, 32(1),* e50–e51, DOI: 10.1130/0091-7613-32.1.e51.

Belcher C. M., Collinson M. E., and Scott A. C. (2005) Constraints on the thermal energy released from the Chicxulub impactor:

New evidence from multimethod charcoal analysis. *J. Geol. Soc. London, 162,* 591–602, DOI: 10.1144/0016764904-104.

Belcher C. M., Finch P., Collinson M .E., Scott A. C., and Grassineau N. V. (2009) Geochemical evidence for combustion of hydrocarbons during the K-T impact event. *Proc. Natl. Acad. Sci., 106,* 4112–4117, DOI: 10.1073/pnas.0813117106.

Bibring J.-P., Langevin Y., Poulet F., et al. (2004) Perennial water ice identified in the south polar cap of Mars. *Nature, 428,* 627–630.

Biver N., Crovisier J., Bockelee-Morvan D., et al. (2012) Ammonia and other parent molecules in Comet 10P/Tempel 2 from Herschel/HIFI and ground-based radio observations. *Astron. Astrophys., 539,* A68, arXiv:1201.4318v1.

Bockelee-Morvan D., Biver N., Jehin E., et al. (2008) Large excess of heavy nitrogen in both hydrogen cyanide and cyanogen from comet 17P/Holmes. *Astrophys. J. Lett., 679,* L49–L52.

Bottke W. F., Walker R. J., Day, J. M. D., Nesvorny D., and and Elkins-Tanton L. (2010) Stochastic late accretion to Earth, the Moon, and Mars. *Science, 330,* 1527–1530.

Bottke W. F., Vokrouhlický D., Minton D., and Nesvorný D. (2012) An Archaean heavy bombardment from a destabilized extension of the asteroid belt. *Nature, 485,* 78–81.

Boynton W. V., Feldman W. C., Squyres S. W., et al. (2002) Distribution of hydrogen in the near surface of Mars: Evidence for subsurface ice deposits. *Science, 297,* 81–85.

Bullock M. A. and Grinspoon D. H. (2001) The recent evolution of climate on Venus. *Icarus, 150,* 19–37.

Byerly G. R., Lowe D. R., Wooden J. L., and Xie X. (2002) An Archean impact layer from the Pilbara and Kaapvaal cratons. *Science, 297,* 1325–1327.

Byrne S., Dundas C. M., Kennedy M. R., et al. (2009) Distribution of mid-latitude ground ice on Mars from new impact craters. *Science, 325,* 1674–1676, DOI: 10.1126/science.1175307.

Catling D. C. and Moore J. M. (2003) The nature of coarse-grained crystalline hematite and its implications for the early environment of Mars. *Icarus, 165,* 277–300.

Chyba C. F. (1990) Impact delivery and erosion of planetary oceans in the early inner solar system. *Nature, 343,* 129–133.

Chyba C. (1991) Terrestrial mantle siderophiles and the lunar impact record. *Icarus, 92,* 217–233.

Chyba C. F., Thomas P. J., and Zahnle K. J. (1993) The 1908 Tunguska explosion: Atmospheric disruption of a stony asteroid. *Nature, 361,* 40–44.

Covey C., Thompson S. L., Weissman P. R., and MacCracken M. C. (1994) Global climate effects of atmospheric dust from an asteroid or comet impact on Earth. *Global Planet. Change, 9,* 263–273.

Fassett C. I. and Head J. W. (2005) Fluvial sedimentary deposits on Mars: Ancient deltas in a crater lake in the Nili Fossae region. *Geophys. Res. Lett., 32,* L14201, DOI: 10.1029/2005GL023456.

Flaser F. M., Achterberg R. K., Conrath B. J., et al. (2005) Titan's atmospheric temperatures, winds, and composition. *Science, 308,* 975–978, DOI: 10.1126/science.1111150.

Goldin T. J. and Melosh H. J. (2009) Self-shielding of thermal radiation by Chicxulub impact ejecta: Firestorm or fizzle? *Geology, 37,* 1135–1138, DOI: 10.1130/G30433A.1.

Gomes R., Levison H. F., Tsiganis K., and Morbidelli A. (2005) Origin of the cataclysmic late heavy bombardment period of the terrestrial planets. *Nature, 435,* 466–469.

Griffith C. A. and Zahnle K. (1995) Influx of cometary volatiles to planetary moons: The atmospheres of 1000 possible Titans. *J. Geophys. Res., 100,* 16907–16922.

Haberle R. M., Kahre M. A., and Hollingsworth J. L. (2012) A cloud greenhouse effect on Mars: Significant climate change in the recent past? (abstract). Presented at the Mars Recent Climate Change Workshop, May 15–17, 2012, Moffett Field, California. Available online at *spacescience.arc.nasa.gov/mars-climate-workshop-2012/documents/extendedabstracts/Haberle_RM_ExAbstRev.pdf.*

Hartmann W. K. (1999) *Moons and Planets.* Wadsworth, Stamford, Connecticut.

Hartogh P., Lis D. C., Bockelée-Morvan D., et al. (2011) Ocean-like water in the Jupiter-family comet 103P/Hartley 2. *Nature, 478,* 218–220.

Harvey M. C., Brassell S. C., Belcher C. M., and Montanari A. (2008) Combustion of fossil organic matter at the Cretaceous-Paleogene (K-P) boundary. *Geology, 36,* 355–358, DOI: 10.1130/G24646A.1.

Hayes A., Aharonson O., Callahan P., et al. (2008) Hydrocarbon lakes on Titan: Distribution and interaction with a porous regolith. *Geophys. Res. Lett., 35,* L09204.

Hervig M. E., Gordley L. L., Deaver L. E., Siskind D. E., Stevens M. H., Russell J. M. III, Bailey S. M., Megner L., and Bardeen C. G. (2009) First satellite observations of meteoritic smoke in the middle atmosphere. *Geophys. Res. Lett., 36,* L18805, DOI: 10.1029/2009GL039737.

Hoke M. R., Hynek B. M., and Tucker G. E. (2011) Formation timescales of large martian valley networks. *Earth Planet. Sci. Lett., 312,* 1–12.

Hunten D. M., Turco R. P., and Toon O. B. (1980) Smoke and dust particles of meteoric origin in the mesosphere and stratosphere. *J. Atmos. Sci., 37,* 1342–1357.

Johnson B. C. and Melosh H. J. (2012) Formation of spherules in impact produced vapor plumes. *Icarus, 217,* 416–430.

Jones T. D. and Lewis J. S. (1987) Estimated impact shock production of N_2 and organic compounds on early Titan. *Icarus, 72,* 381–393.

Kasting J. F. (1988) Runaway and moist greenhouse atmospheres and the evolution of Earth and Venus. *Icarus, 74(3),* 472–494.

Kraal E.R., van Dijk M., Postma G., and Kleinhans M. G. (2008) Martian stepped-delta formation by rapid water release. *Nature, 451(7181),* 973–976.

Kraus R. G., Senft L. E., and Stewart S. T. (2011) Impacts onto H_2O ice: Scaling laws for melting, vaporization, excavation, and final crater size. *Icarus, 214,* 724–738.

Kress M. E. and McKay C. P. (2004) Formation of methane in comet impacts: Implications for Earth, Mars, and Titan. *Icarus, 168,* 475–483.

Kring D. A., Melosh H. J., and Hunten D. M. (1996) Impact-induced perturbations of atmospheric sulfur. *Earth Planet. Sci. Lett., 140,* 201–212.

Langevin Y., Poulet F., Bibring J.-P., et al. (2005) Summer evolution of the north polar cap of Mars as observed by OMEGA/Mars Express. *Science, 307(5715),* 1581–1584.

Lorenz R. D., Niemann H. B, Harpold D. N., Way S. H., and Zarnecki J. C. (2006) Titan's damp ground: Constraints on Titan surface thermal properties from the temperature evolution of the Huygens GCMS inlet. *Meteoritics & Planet. Sci., 41,* 1705–1714.

Lorenz R. D., Mitchell K. L., Kirk R. L., et al. (2008a) Titan's inventory of organic surface materials. *Geophys. Res. Lett., 35,* L02206, DOI: 10.1029/2007GL032118.

Lorenz R.D., Lopes R. M., Paganelli F., et al. (2008b) Fluvial channels on Titan: Initial Cassini RADAR observations. *Planet. Space Sci., 56,* 1132–1144.

Lowe D. R. and Byerly G .R. (1986) Early Archean silicate spherules of probable impact origin South Africa and Western Australia. *Geology, 11,* 668–671.

Lunine J. I., Choukroun M., and Stevenson D. J. (2010) The origin and evolution of Titan. In *Titan from Cassini-Huygens* (R. H. Brown et al., eds.), pp. 35–85. Springer, New York.

Maher K. A. and Stevenson D. J. (1988) Impact frustration of the origin of life. *Nature, 331,* 612–614.

Mandt K. E., Waite J. H., Lewis W., et al. (2009) Isotopic evolution of the major constituents of Titan's atmosphere based on Cassini data. *Planet. Space Sci., 57,* 1917–1930.

McKay C. P. and Borucki W. R. (1997) Organic synthesis in experimental impact shocks. *Science, 276,* 390–392.

McKay C. P., Scattergood T. W., Pollack J. B., Borucki W. J., and Van Ghyseghem H. T. (1988) High temperature shock formation of N_2 and organics on primordial Titan. *Nature, 332,* 520–522.

McKay C. P., Lorenz R. D., and Lunine J. I. (1999) Analytic solutions for the antigreenhouse effect: Titan and the early Earth. *Icarus, 137,* 56–61.

Meech K. J. and 196 co-authors (2011) EPOXI: Comet 103P/Hartley 2 observations from a worldwide campaign. *Astrophys. J. Lett., 734(1),* L1.

Melosh H. J., Schneider N. M., Zahnle K. J., and Latham D. (1990) Ignition of global wildfires at the Cretaceous Tertiary boundary. *Nature, 343,* 251–254.

Michel P. and Morbidelli A. (2007) Review of the population of impactors and the impact cratering rate in the inner solar system. *Meteoritics & Planet. Sci., 42,* 1861–1869.

Murphy D. M., Thomson D. S., and Mahoney T. M. J. (1998) In situ measurements of organics, meteoritic material, mercury, and other elements in aerosols at 5 to 19 kilometers. *Science, 282,* 1664–1669.

Nakajima S., Hayashi Y.-Y., and Abe Y. (1992) A study on the 'runaway greenhouse effect' with a one dimensional radiative-convective equilibrium model. *J. Atmos. Sci., 49,* 2256–2266.

Neely R. R. III, English J. M., Toon O. B., Solomon S., Mills M., and Thayer J. (2011) Implications of extinction due to meteoritic smoke in the upper stratosphere. *Geophys. Res. Lett., 38,* L24808, DOI: 10.1029/2011GL049865.

Neish C. D., Somogyi Á., Imanaka H., Lunine J. I., and Smith M. A. (2008) Rate measurements of the hydrolysis of complex organic macromolecules in cold aqueous solutions: Implications for prebiotic chemistry on the early Earth and Titan. *Astrobiology, 8(2),* 273–287.

Neish C. D., Somogyi Á., Lunine J. I., and Smith M. A. (2009) Low temperature hydrolysis of laboratory tholins in ammonia-water solutions: Implications for prebiotic chemistry on Titan. *Icarus, 201,* 412–421.

Niemann H. B., Atreya S. K., Bauer S. J., et al. (2005) The abundances of constituents of Titan's atmosphere from the GCMS instrument on the Huygens probe. *Nature, 438,* 779–784.

Niemann H. B., Atreya S. K., Demick J. E., et al. (2010) Composition of Titan's lower atmosphere and simple surface volatiles as measured by the Cassini-Huygens probe gas chromatograph mass spectrometer experiment. *J. Geophys. Res., 115,* E12006.

Oberbeck V. and Fogleman G. (1990) Estimates of the maximum time required for the origin of life. *Origins Life Evol. Biosph., 19,* 549–560.

O'Brien D. P., Lorenz R. D., and Lunine J. I. (2005) Numerical calculations of the longevity of impact oases on Titan. *Icarus, 173,* 243–253.

O'Keefe J. D. and Ahrens T. J. (1982) The interaction of the Cretaceous/Tertiary extinction bolide with the atmosphere, ocean, and solid Earth. In *Geological Implications of Impacts of Large Asteroids and Comets on the Earth* (L. T. Silver and P. H. Schultz, eds.), pp. 103–120. GSA Special Paper 190, Geological Society of America, Boulder, Colorado.

Pierazzo E., Vickery A. M., and Melosh H. J. (1997) A re-evaluation of impact melt production. *Icarus, 127,* 408–423.

Pierazzo E., Hahmann A. N., and Sloan L. (2003) Chicxulub and climate: Radiative perturbations of impact-produced S-bearing gases. *Astrobiology, 3,* 99–118.

Pietrass T., Gaede H. C., Bifone A., Pines A., and Ripmeester J. A. (1995) Monitoring xenon clathrate formation on ice surfaces with optically enhanced ^{129}Xe, NMR. *J. Am. Inst. Chem. Soc., 117,* 7520–7525.

Pollack J. B., Toon O. B., Ackerman T. P., McKay C., and Turco R. P. (1983) Environmental effects of an impact generated dust cloud: Implications for the Cretaceous Tertiary extinctions. *Science, 219,* 287–289.

Pope K. O. (2002) Impact dust not the cause of the Cretaceous-Tertiary mass extinction. *Geology, 30,* 99–102.

Pope K. O., Baines K. H., Ocampo A. C., and Ivanov B. A. (1997) Energy, volatile production, and climate effects of the Chicxulub Cretaceous/Tertiary impact. *J. Geophys. Res., 102,* 21645–21664.

Radebaugh J., Lorenz R. D., Lunine J. I., et al. (2008) Dunes on Titan observed by Cassini RADAR. *Icarus, 194,* 690–703.

Robertson D. S., McKenna M. C., Toon O. B., et al. (2004) Survival in the first hours of the Cenozoic. *Geol. Soc. Am. Bull., 116,* 760–768.

Robertson D. S., Lewis W. M., Sheehan P. M., and Toon O. B. (2012) K/Pg extinction: Re-evaluation of the heat/fire hypothesis. *J. Geophys. Res.,* in press, DOI: 10.1002/jgrg.20018.

Robock A., Oman L., and Stenchikov G. L. (2007) Nuclear winter revisited with a modern climate model and current nuclear arsenals: Still catastrophic consequences. *J. Geophys. Res., 112,* D13107, DOI: 10.1029/2006jd008235.

Rosinski J. and Snow R. H. (1961) Secondary particulate matter from meteor vapors. *J. Meteor., 18,* 736–745.

Schaefer L. and Fegley B. (2004) Heavy metal frost on Venus. *Icarus, 168,* 215–219.

Schulte P., Alegret L., Arenillas I., et al. (2010) The Chicxulub asteroid impact and mass extinction at the Cretaceous-Paleogene boundary. *Science, 327,* 1214–1218, DOI: 10.1126/science.1177265.

Segura T. L. and Colaprete A. (2009) Global modeling of impact-induced greenhouse warming on early Mars (abstract). In *Lunar Planet. Sci. Conf. 40,* Abstract #1056. Lunar and Planetary Institute, Houston.

Segura T. L., Toon O. B., Colaprete A., and Zahnle K. (2002) Environmental effect of large impacts on Mars. *Science, 290,* 1976–1980.

Segura T.L., Toon O.B., Colaprete A. (2008) Modeling the environmental effects of moderate-sized impacts on Mars. J. Geophys. Res. 113:E11007.

Segura T. L., McKay C. P., and Toon O. B. (2012) An impact-induced, stable, runaway climate on Mars. *Icarus, 220(1),* 144–148.

Sekine Y., Genda H., Sugita S., Kadono T., and Matsui T. (2011) Replacement and late formation of atmospheric N_2 on undifferentiated Titan by impacts. *Nature Geosci., 4(6),* 359–362.

Senft L. E. and Stewart S. T. (2011) Modeling the morphological diversity of impact craters on icy satellites. *Icarus, 214,* 67–81.

Sleep N. H. and Zahnle K. (1998) Refugia from asteroid impacts on early Mars and the early Earth. *J. Geophys. Res., 103,* 28529–28544.

Sleep N. H., Zahnle K. J., Kasting J. F., and Morowitz H. J. (1989) Annihilation of ecosystems by large asteroid impacts on the early Earth. *Nature, 342,* 139–142.

Smit J. (1999) The global stratigraphy of the Cretaceous-Tertiary boundary layer impact ejecta. *Annu. Rev. Earth Planet. Sci., 27,* 75–113.

Soderblom L. A., Brown R. H., Soderblom J. M., et al. (2009) The geology of Hotei Regio, Titan: Correlation of Cassini VIMS and RADAR. *Icarus, 204(2),* 610–618.

Solomon S. C., Bullock M. A., and Grinspoon D. H. (1999) Climate change as a regulator of tectonics on Venus. *Science, 286,* 87–89.

Staykova D. K., Hansen T., Salamatin A. N., and Kuhs W. F. (2002) Kinetic diffraction experiments and the formation of porous gas hydrates. *Proc. Fourth Intl. Conf. on Gas Hydrates,* pp. 537–642.

Stepinski T. F., Collier M. L., McGovern P. J., and Clifford S. M. (2004) Martian geomorphology from fractal analysis of drainage networks *J. Geophys. Res., 109(E2),* E02005, DOI: 10.1029/2003JE002098.

Stuart J. S. and Binzel R. P. (2004) Bias-corrected population, size distribution and impact hazard for near-Earth objects. *Icarus, 170,* 295–311.

Thompson W. R. and Sagan C. (1992) Organic chemistry on Titan — surface interactions. In Proceedings of the Symposium on Titan (B. Kaldeich, ed.), pp. 167–176. ESA Special Paper 338, Noordwijk, The Netherlands.

Toon O. B., Pollack J. P., Ackerman T. P., Turco R. P., McKay C. P., and Liu M. S. (1982) Evolution of an impact generated dust cloud and its effects on the atmosphere. In *Geological Implications of Impacts of Large Asteroids and Comets on the Earth* (L. T. Silver and P. H. Schultz, eds.), pp. 187–200. GSA Special Paper 190, Geological Society of America, Boulder, Colorado.

Toon O. B., Zahnle K., Morrison D., Turco R. P., and Covey C. (1997) Environmental perturbations caused by the impacts of asteroids and comets. *Rev. Geophys., 35,* 41–78.

Toon O. B., Segura T. L., and Zahnle K. (2010) The formation of martian river valleys by impacts. *Annu. Rev. Earth Planet. Sci., 38,* 303–322.

Urata R. A. and Toon O. B. (2012) Simulations of the martian water cycle with a GCM and implications for the early climate (abstract). In *Third Conference on Early Mars: Geologic, Hydrologic, and Climatic Evolution and the Implications for Life,* Abstract #7041. Lunar and Planetary Instgitute, Houston.

Urey H. C. (1973) Cometary collisions and geological periods. *Nature, 242,* 32–33, DOI: 10.1038/242032a0.

Verma H. C., Upadhyay C., Tripathi A., et al. (2002) Thermal decomposition pattern and particle size estimation of iron minerals associated with the Cretaceous-Tertiary boundary at Gubbio. *Meteoritics & Planet. Sci., 37,* 901.

Wdowiak T. J., Armendarez L. P., Agresti D. G., et al. (2001) Presence of an iron-rich nanophase material in the upper layer of the Cretaceous-Tertiary boundary clay. *Meteoritics & Planet. Sci., 36,* 123–133.

Whiston W. (1696) *A New Theory of the Earth, From its Original, to the Consummation of All Things, Where the Creation of the World in Six Days, the Universal Deluge, And the General Conflagration, As laid down in the Holy Scriptures, Are Shewn to be perfectly agreeable to Reason and Philosophy.* Benjamin Tooke, London.

Wolbach W. S., Lewis R. S., and Anders E. (1985) Cretaceous extinctions: Evidence for wildfires and search for meteoritic material. *Science, 230,* 167–170.

Wolbach W. S., Gilmour I., Anders E., Orth C. J., and Brooks R. R. (1988) Global fire at the Cretaceous-Tertiary boundary. *Nature, 334,* 665–669.

Wolbach W. S., Gilmour I., and Anders E. (1990) Major wildfires at the Cretaceous/Tertiary boundary. In *Global Catastrophes in Earth History* (V. Sharpton and P. Ward, eds.), pp. 391–400. GSA Special Paper 247, Geological Society of America, Boulder, Colorado.

Wolbach W. S., Widicus S., and Kyte F. T. (2003) A search for soot from global wildfires in Central Pacific Cretaceous-Tertiary boundary and other extinction and impact horizon sediments. *Astrobiology, 3,* 91–97.

Wood C. A., Lorenz R., Kirk R. L., et al. (2010) Impact craters on Titan. *Icarus, 206,* 334–344.

Zahnle K. J. (1990) Atmospheric chemistry by large impacts. In *Global Catastrophes in Earth History* (V. Sharpton and P. Ward, eds.), pp. 271–287. GSA Special Paper 247, Geological Society of America, Boulder, Colorado.

Zahnle K. J and Sleep N. H. (1997) Impacts and the early evolution of life. In *Comets and the Origin and Evolution of Life, 1st edition* (P. J. Thomas et al., eds.), pp. 175–208. Springer, New York.

Zahnle K. J. and Sleep N. H. (2006) Impacts and the early evolution of life. In *Comets and the Origin and Evolution of Life, 2nd edition* (P. J. Thomas et al., eds.), pp. 207–252. Springer, Berlin-Heidelberg.

Zahnle K., Pollack J. B., Grinspoon D., and Dones L. (1992) Impact-generated atmospheres over Titan, Ganymede, and Callisto. *Icarus, 95,* 1–23.

Zahnle K., Schenk P., Dones L., and Levison H. F. (2003) Cratering rates in the outer solar system. *Icarus, 163,* 263–289.

Grotzinger J. P., Hayes A. G., Lamb M. P., and McLennan S. M. (2013) Sedimentary processes on Earth, Mars, Titan, and Venus. In *Comparative Climatology of Terrestrial Planets* (S. J. Mackwell et al., eds.), pp. 439–472. Univ. of Arizona, Tucson, DOI: 10.2458/azu_uapress_9780816530595-ch18.

Sedimentary Processes on Earth, Mars, Titan, and Venus

J. P. Grotzinger
California Institute of Technology

A. G. Hayes
Cornell University

M. P. Lamb
California Institute of Technology

S. M. McLennan
State University of New York at Stony Brook

The production, transport and deposition of sediment occur to varying degrees on Earth, Mars, Venus, and Titan. These sedimentary processes are significantly influenced by climate that affects production of sediment in source regions (weathering), and the mode by which that sediment is transported (wind vs. water). Other, more geological, factors determine where sediments are deposited (topography and tectonics). Fluvial and marine processes dominate Earth both today and in its geologic past, aeolian processes dominate modern Mars although in its past fluvial processes also were important, Venus knows only aeolian processes, and Titan shows evidence of both fluvial and aeolian processes. Earth and Mars also feature vast deposits of sedimentary rocks, spanning billions of years of planetary history. These ancient rocks preserve the long-term record of the evolution of surface environments, including variations in climate state. On Mars, sedimentary rocks record the transition from wetter, neutral-pH weathering, to brine-dominated low-pH weathering, to its dry current state.

1. INTRODUCTION

The atmospheres of solid planets exert a fundamental control on their surfaces. Interactions between atmospheric and geologic processes influence the morphology and composition of surfaces, and over the course of geologic time determine the historical evolution of the planet's surface environments including climate. In the case of Earth, the origin of life likely occurred within surface environments that were tuned to favor prebiotic chemical reactions. Microbial evolution and the advent of oxygenic photosynthesis created Earth's unique atmospheric fingerprint that would be visible from other solar systems.

Earth, Mars, Venus, and Titan are examples of solid planets that have relatively dense atmospheres (Table 1). Yet each planet has shown divergent evolution over the course of geologic history, and their atmospheres are a testimony to this evolution. A second manifestation of this evolution is preserved in each planet's sedimentary record. Processes that operate at planetary surfaces have the potential to record a history of their evolution in the form of sedimentary rocks. Both Earth and Mars have a well-defined atmosphere, hydrosphere, cryosphere, and lithosphere. Interactions among these elements give rise to the flow of currents of air, liquid water, and ice, that in turn transport sediments from sites of weathering and erosion where they are formed, to sites of deposition where they are stored, to create a stratigraphic record of layered sediments and sedimentary rocks. This gives rise to the "source-to-sink" concept by which sedimentary systems on Earth have been characterized (Fig. 1). This unifying concept allows efficient comparison between planets. Our study of Mars has matured to the point where source-to-sink has found remarkable analogous application, and even Titan shows strong evidence for source-to-sink systems, albeit formed under very different conditions than either Earth or Mars. Venus, while recently resurfaced by volcanism and lacking a hydrosphere, also shows evidence for sediment transport in the form of surficial aeolian deposits. This provides clear evidence of the general validity of the approach (Fig. 2).

Time and history also are critically important cornerstones of the sedimentary record. Our understanding of the evolution of Earth's very ancient climate derives from detailed examination of the mineralogic, textural, and geochemical signatures preserved in the sedimentary rock record. Furthermore, Earth's sedimentary record attests to the formation and evolution of continents, plate tectonics, surface temperature, the composition of the atmosphere and oceans, the evolution of life and the biogeochemical cycle of carbon, and the occurrence of extraordinary events in the history of climate — to name just a few examples (Fig. 3a). On Mars, we are beginning to uncover a similarly impressive list of evolutionary milestones, including billion-year-scale changes in aqueously formed mineral assemblages, the role of large impacts on surface evolution, river systems that once extended at almost hemispheric scales, and recognition of a rock cycle featuring deposition and erosion of vast sequences of strata (Fig. 3b).

The goals of this paper are to provide an overview of the fundamental aspects of source-to-sink systems on Earth, Mars, Venus, and Titan. Each planet can be discussed in terms of weathering and erosion, transport, and deposition — all key parts of source-to-sink systems. Although evidence of deposition and a stratigraphic record for Venus and Titan are very limited, both planets can be evaluated with reference to weathering, erosion, and sediment transport. We then close with a discussion of the history of deposition on these planets, and what information is embedded within these ancient records.

2. PLANETARY SOURCE-TO-SINK SYSTEMS

Weathering and erosion sculpt planetary landscapes and generate sediment, which is redistributed to alluvial plains, deltas, lakes, and ocean bottoms, to form diverse eolian landforms (Fig. 1). This transfer of sediment and solute mass from source to sink plays a key role in the cycling of elements, water, fractionation of biogeochemically important isotopes, and the formation of soils, groundwater reservoirs, masses of sediment, and even the composition of the atmosphere. The objective of source-to-sink studies is to identify and attempt to quantify the mass fluxes of sediments and solutes across planetary surfaces (e.g., *Allen,* 2008).

This framework creates a convenient starting point in which to understand the broader significance of weathering, erosion, transport, and deposition of sediments. Whether or not sediment is ever deposited ultimately depends on two factors: weathering and erosion to produce a *supply* of sediments, and the capacity of the flow to maintain the *transport* of those sediments (see below). The simplest approximation for sedimentary timescales (e.g., *Paola and Voller,* 2005) assumes the conservation of mass per unit area of the land surface, and the balance of sediment being transported vs. that being deposited is illustrated by the expression

$$\frac{d\eta}{dt} = -\frac{dq_s}{dx} - \sigma + \Omega \tag{1}$$

where η is the elevation of the bed of sediment, q_s is the flux of sediment (both solid and dissolved loads), σ the

TABLE 1. Sedimentologically useful parameters.

	Venus	Earth	Mars	Titan
Gravity	8.9 m s^{-2}	9.8 m s^{-2}	3.7 m s^{-2}	1.35 m s^{-2}
Mean surface temperature	735 K	288 K	214 K	93.6 K
Density of hydrosphere	N/A	1030 kg m^{-3}	N/A	445–662 kg m^{-3}
Submerged specific density	N/A	1.65	1.65 1.04 (viscous brines)	0.5–1.2 (water ice) up to 2 (organics)
Viscosity of hydrosphere (kinematic)	N/A	10^{-6} m^2 s^{-1}	10^{-6} m^2 s^{-1} (water flows) 4 × 10^{-5} m^2 s^{-1} (viscous brines)	10^{-7}–10^{-6} m^2 s^{-1}
Viscosity of atmosphere	33.4 μPa s	18.1 μPa s	10.8 μPa s	6.2 μPa s
Atmospheric density	71.92 kg m^{-3}	1.27 kg m^{-3}	0.027 kg m^{-3}	5.3 kg m^{-3}
Atmospheric composition				
Nitrogen (N_2)	3.5%	78.08%	2.7%	95%
Oxygen (O_2)	almost zero	20.95%	almost zero	zero
Carbon Dioxide (CO_2)	96.5%	0.035%	95.3%	zero
Water vapor (H_2O)	0.003%	~1%	0.03%	zero
Other gases	almost zero	almost zero	2% (mostly Ar)	5% (methane)
Composition of crust				
Exposed crust (SiO_2)	47%	66%	49%	~0%
Exposed crust dominant material	Plagioclase, pyroxene, olivine	Quartz, plagioclase, K-feldspar	Plagioclase, pyroxene, olivine	Water ice and clathrate hydrate overlain by organic veneer
Dune material	Basalt	Quartz	Basalt	Organics (lower density)

Data sources: *Nesbitt and Young* (1984), *Surkov et al.* (1984), *Yung et al.* (1984), *Tobie et al.* (2006), *McLennnan and Grotzinger* (2008), *Taylor and McLennan* (2009), *Lorenz et al.* (2010), *Tosca et al.* (2011), *Choukroun and Sotin* (2012).

local rate of surface subsidence or uplift (negative) due to tectonics, and Ω is the net accumulation rate of sediment from the atmosphere (e.g., settling of photolysis particles on Titan). In this one-dimensional formulation x is taken as the down-current direction and sediment-bed porosity is neglected. In the absence of tectonic uplift or subsidence, as is likely the case on Mars, Titan, and Venus, and in the absence of atmospheric deposition (precipitation of solids) as is likely the case on Earth, Mars, and Venus (excluding regions where ice deposits form), equation (1) simplifies to

$$\frac{d\eta}{dt} = -\frac{dq_s}{dx} \quad (2)$$

The relation states that bed elevation increases proportionally to the amount of sediment that drops out of transport, and conversely decreases proportionally to the amount of sediment that becomes entrained by the flow. In sediment source areas, sediment flux is typically limited by the conversion of rock to sediment (i.e., a sediment supply limited landscape); these landscapes tend to be net erosional and q_s is set by the erosional processes at work including weathering, river incision into bedrock, and abrasion by wind, for example. In sediment bypass and depositional areas, sediment supply is typically abundant, and therefore q_s is set by the ability of a flow to transport sediment [i.e., a transport limited landscape (*Howard et al.,* 1994)]. For these cases, sediment flux and therefore deposition rate (equation (2)) is governed by the flow dynamics alone; in effect, wherever the current slows — usually due to a divergence in the flux of fluid — net sedimentation will result. On the other hand, when the current velocity increases, net erosion will result.

In one other simplification, in many cases it can be assumed that flows of liquid across planetary surfaces will move downhill, in response to gravity, and flows of air will move in response to regional gradients in atmospheric pressure. The influence of tides and geostrophic flows associated with Earth's global ocean are a very important process that we will ignore in this paper. The simplification results in the formulation of a straightforward constitutive law for sediment transport

$$q_s = -K\frac{d\eta}{dx} \quad (3)$$

where K is a transport coefficient that takes into account all the detailed mechanics of sediment transport (see below), and may not always be independent of x as assumed here. This transport law states that sediment transport is proportional to slope and is based on theoretical, experimental, and empirical analyses of downslope transport by creep, overland flow, and channel flow processes.

Substituting equation (3) into equation (2) yields

$$\frac{\partial\eta}{\partial t} = K\frac{\partial^2\eta}{\partial x^2} \quad (4)$$

which states that topographic evolution through sediment transport follows a linear diffusive process. This simple model of erosion and sedimentation can effectively account for a vast range of source-to-sink systems on Earth (e.g., *Kenyon and Turcotte,* 1985; *Flemings and Grotzinger,* 1996; *Paola,* 2000), and likely on other planets as well where gravity results in net downslope transport of liquid and

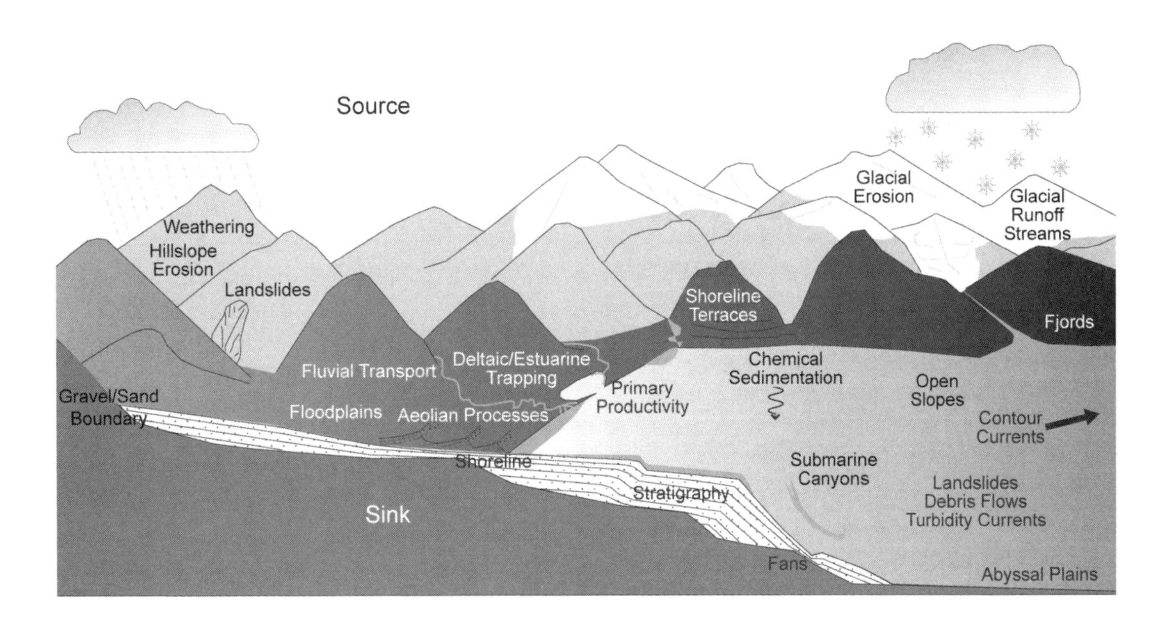

Fig. 1. Surface processes on Earth give rise to sediments in source areas that are transported to sites of deposition, or sinks. This "source-to-sink" concept is applicable, with modifications, to Mars and Titan.

sediment. Figure 4a shows a simple model of an alluvial fan based on a two-dimensional diffusion model of sediment transport (STRATA) (see *Flemings and Grotzinger,* 1996). In this example the source (flux) of sediment is provided through the lefthand boundary and redistribution of that flux is regulated by the topography of the fan as it builds through time. Over time the profile adjusts so that the time lines of the model output converge to indicate the flux into the leftmost cell is mostly balanced by the flow out of the rightmost cell. The morphology of the modeled fan surface is close to equilibrium. One must envisage the righthand boundary passing, in reality, from the fan into an adjacent geomorphic feature such as a playa lake.

Such a simple system as shown in Fig. 4a illustrates the source-to-sink concept. We can implement a simple perturbation, taken to simulate climate change — say an increase in water flux through enhanced precipitation — that results in an increase in the flux of sediment to the fan (Fig. 4b). The system response is apparent and this forward model of the history of deposition can be used to illustrate how the inverse problem — recovery of climate history — could be addressed using the record of sedimentation.

We now turn to more detailed descriptions of the components of planetary source-to-sink systems. Sediment production by weathering and erosion is first described, and then the discussion moves on to sediment transport and deposition, and we end with a discussion of the planetary stratigraphic record (sedimentary history), with emphasis on events that represent past changes in climate.

3. WEATHERING

The weathering process initiates changes, separation, and redistribution of rock fragments, minerals, and chemical elements within the surficial environment of a planet and accordingly is the starting place for the source-to-sink sedimentary paradigm. Although classical weathering takes place mainly within soil profiles, the transport of sediment from the initial source to the final depositional basin may involve many intervening short-term stages of deposition and reerosion during which both physical and chemical weathering processes act further on the sedimentary material (*Johnsson and Meade,* 1990). For longer timescales ($>\sim 10^5$–10^6 yr), it is worth remembering that on Earth approximately 70% of sediment on average is derived from weathering and erosion of preexisting sedimentary rocks, and accordingly the influences of weathering on mineralogy and geochemistry can accumulate over several sedimentary cycles (*Veizer and Mackenzie,* 2003). A key

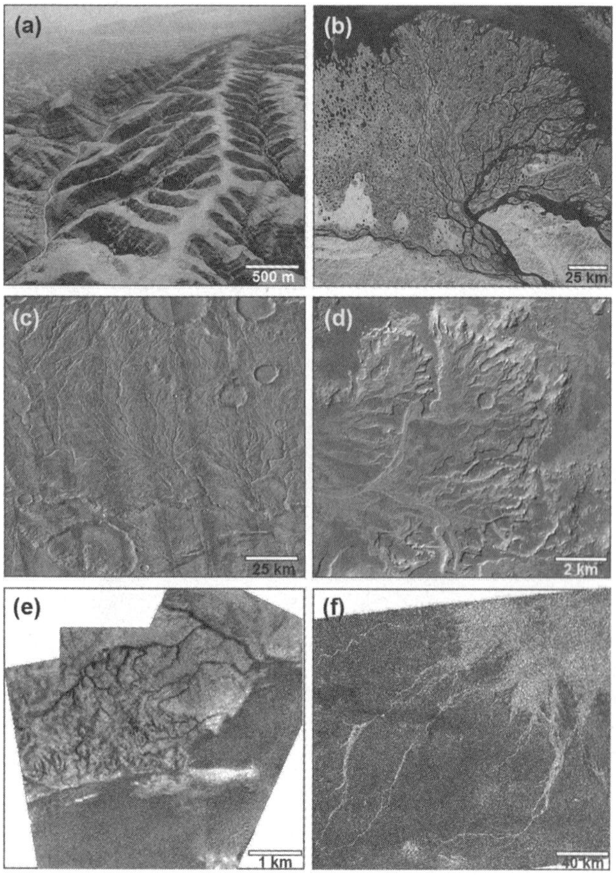

Fig. 2. Source-to-sink systems on Earth, Mars, and Titan. **(a)** Source region on Earth, where surface runoff results in bedrock channel incision, and formation of stream networks, Mirbat escarpment, southern Oman. Oblique aerial photograph, Petroleum Development Oman. **(b)** River delta represents a sediment sink where divergence in the flux at the edge of the continent results in sediment deposition. Lena river delta, northern Siberia. Landsat 7 image, NASA. **(c)** Source region on Mars at Solis Planum (–41.7 N, 266.8 E) with well-developed branched stream network developed in bedrock. It has been inferred that such networks are consistent with ancient atmospheric precipitation of water. CTX images, from east to west: P21_009263_1379_XN_42S091W, B05_011531_1378_XI_42S091W, P07_003633_1378_XI_42S092W, B04_011386_1388_XI_41S092W, P04_002710_1375_XI_42S092W, P07_003699_1373_XI_42S093W, P22_009540_1376_XN_42S093W, B02_010318_1381_XI_41S094W, P08_004121_1379_XI_42S094W, B19_016898_1377_XN_42S094W, P10_005110_1383_XI_41S095W. **(d)** Eberswalde delta represents a local sediment sink on Mars within an impact crater. Flow from a broad region surrounding the crater gathered water and sediment that entered the impact basin as a single channel stream; divergence of this flux at the entry point, in what was inferred to have been a lake, resulted in creation of the delta. CTX image B18_016777_1580_XN_22S034W. **(e)** Source region at Huygens probe landing area on Titan (10°S, 192°W). The dendritic network drains into an alluvial plain that is strewn with decimeter-scale water-ice cobbles. Near-infrared (660–1000 nm) mosaic of frames from the Huygens Descent Imager and Spectral Radiometer (DISR). **(f)** Potential alluvial fan at Elvigar Flumina (19.3°N,78.5°E) on Titan, which may represent a local sediment sink. Ku-band Synthetic Aperture Radar (SAR) image acquired by the Cassini spacecraft on February 15, 2005.

issue in interpreting the history of weathering in terms of source-to-sink processes is distinguishing the *intensity* of weathering (controlled largely by climate) and the *duration* of weathering [controlled largely by physical characteristics such as slope, relief, mass wasting rates (*Johnsson,* 1993)]. This issue becomes especially relevant on Mars and other planetary surfaces where age and duration of most processes are very poorly constrained.

On Earth, the intensity of weathering provides a direct link to climate through the carbon biogeochemical cycle, for example, by consuming CO_2 from the atmosphere during weathering reactions and transporting dissolved carbon (mainly as bicarbonate) to the oceans where it may be precipitated and stored in carbonate rocks. There is some relationship between weathering rates and surface temperatures, but the details are complex (e.g., *West et al.,* 2005). On Mars and Venus, any link between weathering and climate may also be strongly influenced by an analogous sulfur cycle (see further discussion below). In addition, the mineralogical and geochemical (including isotopic) composition of sediments

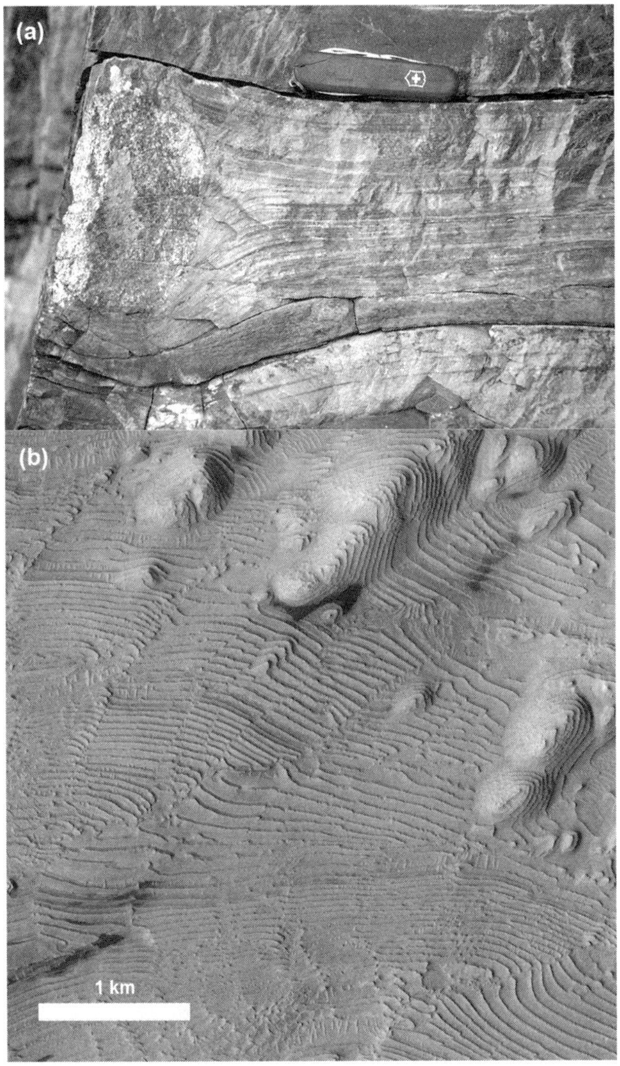

Fig. 3. Inferred records of climate preserved in the very ancient stratigraphic records of Earth and Mars. **(a)** Large stone (on left) dropped from melting glacier, and termed a "dropstone," pierces and then is overlain by fresh sediment in ~635-m.y.-old marine deposits, Sultanate of Oman. Penknife is 7 cm long. **(b)** Highly rhythmic strata are interpreted to record obliquity changes on Mars over 2 b.y. ago. Danielson crater, Arabia Terra region, HiRISE image PSP_002733_1800.

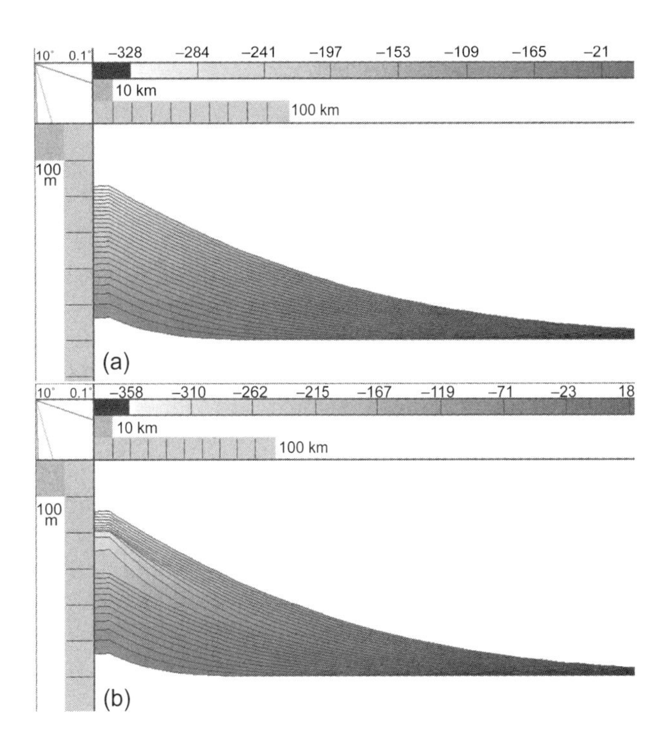

Fig. 4. STRATA is a forward model of sedimentation based on solving simple sediment transport (diffusion) equation for different initial conditions. Output displays cross-sections through synthetic alluvial fan deposits with black lines marking the geomorphic evolution of the fan at constant time values. Scale bars shown, vertical exaggeration indicated by slopes in upper lefthand corners of plots, and gray scale is water depth — with negative values corresponding to elevation of alluvial above datum. **(a)** Sediment is supplied from the lefthand boundary with a flux of 50 m^2/y. With time the fan acquires a steady-state profile at which point the flux of sediment in from the left boundary (into the fan) is equal to the flux out of the right boundary (out of the fan). **(b)** All parameters as in **(a);** however, the flux is abruptly changed from 50 to 100 m^2/y about halfway through the run. This simulates a change in climate (e.g., from relatively dry to wetter), which in turns stimulates increased sediment production. The profile of the fan changes (steepens) in response but then relaxes as the system once again approaches steady state. This is important because it means that, on the basis of geometry alone, ancient strata can be used to infer the very ancient climate of past planetary history. Such an approach has worked well on Earth (*Dorn,* 2009) and should be applicable to Mars and perhaps Titan.

being deposited at any given time provides a proxy for the penecontemporaneous climate and so a record of climate change is preserved within the stratigraphic record.

3.1. A Planetary Perspective on Weathering

Weathering conventionally is subdivided into chemical, physical (or mechanical), and biological processes (*Brantley et al.,* 2007; *Brantley and Lebedeva,* 2011), but in reality these boundaries are largely artificial and in detail these processes act in a highly synergistic manner. For example, salts may be chemically deposited within rock fractures that in turn undergo repeated thermal expansion during diurnal cycles, thus promoting physical disintegration. Impact processes result in the generation of large amounts of mechanically generated particles, thus providing great surface area upon which chemical reactions may take place. In a uniquely terrestrial example, root systems of plants serve to physically disrupt invaded rocks, thus promoting physical weathering, but also influence chemical gradients within soils that can greatly facilitate chemical weathering reactions.

The sedimentary cycles of atmosphere-bearing rocky planetary bodies, whether the familiar terrestrial cycle — intimately related to plate tectonics and biological activity — or a less-familiar surface such as Venus, weathering mostly accomplishes two fundamental roles. Chemical weathering begins the process of reequilibrating surface rocks and minerals, formed at relatively high pressures and temperatures and under differing volatile conditions (e.g., f_{O_2}, f_{CO_2}, f_{SO_2}, a_{H_2O}), with planetary surface conditions that typically are at lower temperature and pressure and with different but highly variable aqueous and volatile conditions. As part of this process, when a dynamic hydrological cycle also exists, constituents from the primary rocks and minerals may be liberated and dissolved into aqueous solutions (or held as colloids) to be chemically deposited later by various processes (evaporation, biological activity, diagenesis, etc.). In more exotic cases rock-atmosphere-hydrosphere interaction may be impeded by icy crusts that in turn may interact with any available near-surface fluids, which in the case of Titan is methane and ethane, and/or undergo "sublimation weathering" in cases where temperatures are too low for liquid water (*Melosh,* 2011). Weathering also promotes mechanical breakdown of rocks and minerals (and in the case of Titan, ices and solid organics), typically facilitated by chemical (and in the case of Earth, biological) processes, into sedimentary particles of widely varying size and composition that can then be physically transported from their sources and deposited into sedimentary basins and also become the agents for further physical weathering and erosion.

We are concerned here with sedimentation on Earth, Mars, Titan, and Venus where substantial atmospheres exist and sedimentary records have been established to varying degrees. However, weathering in some form takes place on all planetary surfaces, even on airless bodies (*Melosh,* 2011). For example, asteroidal parent bodies from which water-rich carbonaceous chondrites (CI, CM, CR, CV, and CH types) were derived are characterized by varying degrees of low-temperature, largely isochemical aqueous alteration resulting in a wide variety of secondary phases (e.g., phyllosilicates, carbonates, sulfates). On airless bodies such as the Moon, Mercury, and asteroids, ionizing particle radiation from solar winds and cosmic rays and meteorite, micrometeorite, and cosmic dust bombardment result in a set of diverse processes collectively termed space weathering. The effects of space weathering include formation of glasses and agglutinates in planetary regolith, implantation of gases, damage to crystal structures, reduction of ferrous iron (e.g., in olivine) to nanophase Fe^0, and general darkening/reddening of planetary surfaces (*Gaffey,* 2010). Although space weathering affects only the outermost few nanometers of exposed surfaces, impact gardening redistributes and effectively sequesters affected material into the deeper regolith.

Terrestrial chemical weathering is controlled by the carbon cycle in which weak carbon-based acids [i.e., carbonic ($pK_a \sim +6.5$) and organic acids (pK_a mostly >+2)] dominate chemical weathering processes under mildly acidic to circumneutral pH (~5–8) conditions (*Berner,* 1995). Although sulfur forms important volcanic gases, the sulfur cycle plays a lesser role in global weathering because sulfur is readily transferred through the ocean into the sedimentary rock record (facilitated by microorganisms, for example, through the process of bacterial sulfate reduction) and further recycled through the mantle by plate tectonics (*Canfield,* 2004). Earth's atmosphere contains only sub-parts per billion levels of total sulfur as various gaseous compounds. However, from a planetary perspective, sulfur cycles rather than carbon cycles may be more influential in controlling weathering processes. In the absence of plate tectonics and large water masses, sulfur-bearing minerals accumulate on the surface of planets (or as sulfur-bearing atmospheric compounds), as witnessed on the surfaces of Mars (*King and McLennan,* 2010) and Mercury (*Nittler et al.,* 2011) and in the atmosphere of Venus where the concentration of SO_2 in the 9.7 MPa atmosphere is about 150 ppmv (*Johnson and Fegley,* 2002). Since sulfur forms strong acids (e.g., sulfuric acid; $pK_a \sim -3$), oxidative weathering processes dominated by a sulfur cycle are more likely to take place under much lower pH conditions (*Johnson and Fegley,* 2002; *Halevy et al.,* 2007; *McLennan,* 2012).

In the sections below, we will discuss some major influences on physical and chemical weathering on Earth, Mars, Venus, and Titan. From a broader planetary perspective, another key factor that must also be kept in mind is the role of impacts in producing sedimentary particles. Many planetary surfaces, including large regions of Mars, Mercury, and the Moon, are very ancient (≥3.8 Ga) and accordingly are strongly influenced by impact processes on many scales, which serve an important role in physically fracturing and disrupting rocks to expose large surface areas. Impacts also serve to vertically mix surface regolith through the process of impact gardening, thus influencing the overall weathering process (*Melosh,* 2011). On airless bodies, impact-induced regolith remains locally distributed, whereas on planetary bodies with atmospheres and hydrospheres, such as those considered here, impact-produced debris may be redistributed over large regions by aeolian and aqueous processes (*Grotzinger et al.,* 2011). Weathering acts to further facilitate the production of particles that form sediment. These sediment particles can cause erosion by acting as tools for abrasion.

3.2. Terrestrial Weathering

On Earth, weathering takes place within the so-called "critical zone," the near-surface region of the planet, between the deepest groundwater and the outer extent of vegetation, where interacting geological, chemical, physical, and biological activity supports life.

3.2.1. Chemical weathering. Chemical weathering includes various processes involving the interaction of aqueous solutions, mostly shallow groundwater, with rocks at and near the Earth's surface (Fig. 5). The overall result is (1) transformation of primary igneous-metamorphic rocks and minerals into a suite of secondary minerals that are more stable under ambient surficial conditions (e.g., clays and insoluble oxides); (2) transfer into aqueous solution of soluble components (e.g., Ca^{2+}, Mg^{2+}, K^+, Na^+, HCO_3^-, SO_4^{2-}, Cl^-) that may be redistributed within the weathering profile or carried out of the soil system to be deposited elsewhere; and (3) weakening and disruption of the critical zone, contributing to episodes of mass wasting and exposure of fresh rock to weathering solutions. Numerous processes are involved, including hydrolysis, dissolution, oxidation, hydration-dehydration, carbonation, ion exchange, and chelation. Of these reactions, hydrolysis is the most important volumetrically for generating the terrestrial sedimentary rock record. Hydrolysis involves the movement of free H^+ (or H_3O^+) in aqueous solutions into rocks and minerals, first along fractures and mineral/grain boundaries and then by diffusion into crystal structures. Protons release and/or replace other ions in mineral structures to either dissolve the primary mineral, e.g., olivine dissolution

$$Mg_2SiO_{4(\text{forsterite})} + 4H^+ \Leftrightarrow 2Mg^{2+} + H_4SiO_{4(\text{orthosilicic acid})} \quad (5)$$

or alter the primary mineral to a more thermodynamically stable mineral, e.g., anorthite ⇒ kaolinite (Fig. 5)

$$CaAl_2Si_2O_{8(\text{anorthite})} + 2H^+ + H_2O \Leftrightarrow Al_2Si_2O_5(OH)_{4(\text{kaolinite})} + Ca^{2+} \quad (6)$$

An important development in quantitatively understanding the chemical and mineralogical changes associated with weathering processes is the development of the chemical index of alteration (CIA) concepts (e.g., *Nesbitt and Young,* 1984; *Nesbitt and Wilson,* 1992; *Nesbitt and Markovics,* 1997; *Nesbitt,* 2003). This can be seen on the "feldspar ternary" that plots mole proportions of Al_2O_3–(CaO^* + Na_2O)–K_2O, or A-CN-K (Fig. 6a), where CaO^* represents the calcium in silcate minerals and is thus corrected for carbonates and phosphates. Such a diagram captures most of the major-element (and thus mineralogical) changes observed in the weathering of granitic (sensu lato) rocks. The CIA scale [CIA = 100 × Al_2O_3/(Al_2O_3 + CaO^* + Na_2O + K_2O)] is shown on the left of Fig. 6a. Typical weathering trends for different igneous rock compositions, constrained from thermodynamic-kinetic studies of feldspar and glass alteration (e.g., *Nesbitt and Young,* 1984) and field studies of weathering profiles, are also shown and illustrate the dissolution of mafic minerals and reaction of plagioclase to clay followed by the reaction of K-feldspar to clay, leading in the most extreme examples to highly aluminous clay-rich soils.

The susceptibility of igneous minerals to weathering processes is well known to approximate the reverse of the Bowen's igneous reaction series (*Goldich,* 1938). However, a more quantitative expression of this relationship can be seen from the Gibbs free energies of various weathering reactions expressed on a per atom basis (*Curtis,* 1976) (Table 2). Thus mafic minerals such as olivine, pyroxene, and Ca-plagioclase are more readily dissolved and altered than are minerals from more evolved igneous settings, such

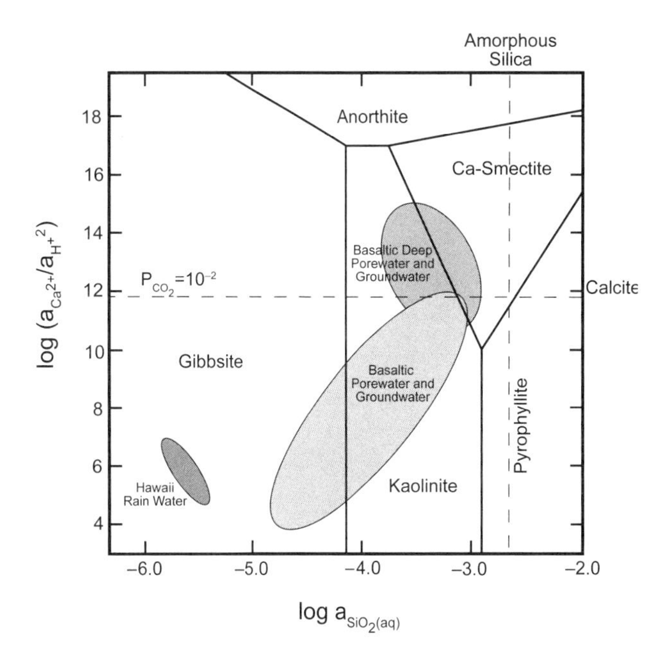

Fig. 5. Mineral stability diagram plotting $a_{Ca^{2+}}/a_{H^+}{}^2$ vs. $a_{SiO_2(aq)}$ (where a is activity) showing the stability relationships among pure anorthite and various clay minerals at standard temperature and pressure (25°C; 1 bar). Dashed lines are solubility limits for calcite (P_{CO_2} = 10^{-2} atm) and amorphous silica. Also shown are typical compositions for Hawaiian rainwater and groundwaters taken from different depths in selected basaltic terrains. Adapted from *Nesbitt and Wilson* (1992). This diagram illustrates the fundamental instability of a common basaltic igneous mineral in the presence of near-surface waters.

as K-feldspar. In detail many other factors, such as pH and other ion activities, can also affect mineral stabilities.

Much work has been carried out to evaluate the kinetics of weathering reactions and timescales for the formation of weathering profiles (see numerous reviews in *White and Brantley,* 1995; *Oelkers and Schott,* 2009). A long-standing issue is the lack of agreement between chemical weathering rates determined in the field compared to those determined in the laboratory, the former typically being several orders of magnitude slower. The major sources of this discrepancy have to do with the changing character of mineral surfaces (e.g., lowering of reactive surface areas as heterogeneities diminish; development of physical barriers due to leaching and/or deposition of precipitates) and changes in evolving aqueous fluids (e.g., changing saturation states) as natural weathering proceeds. It is important to keep the differences between natural and laboratory rates in mind when considering weathering from a planetary perspective, since on most planets, detailed field studies are either limited (e.g., Mars) or unavailable (e.g., Venus, Titan) and experiments are heavily relied on to gain insight.

3.2.2. The importance of basaltic weathering. From a planetary perspective, many studies of chemical weathering on Earth are not particularly informative. Terrestrial weathering takes place mostly on igneous/metamorphic rocks of the upper continental crust or on sedimentary rocks ultimately derived from the upper continental crust (*Taylor and McLennan,* 1985). The exposed upper continental crust of Earth approximates to a granodioritic composition on average and the "tertiary" continental crust is unique within the solar system (*Taylor and McLennan,* 2009). Instead, planetary surfaces are predominantly characterized by "primary" and "secondary" crusts composed of mafic and ultramafic rocks. This can be seen from Table 3, which tabulates estimates of the chemical composition (and mineralogy) of the terrestrial upper continental crust and martian exposed crust. There is no reliable estimate for the average exposed venusian crust, but remote sensing and available surface analyses (Table 3) also indicate a basaltic composition (*Taylor and McLennan,* 2009).

Accordingly, terrestrial studies that focus on the weathering of basaltic rocks (Fig. 6b) and minerals are far more relevant (e.g., *Nesbitt and Wilson,* 1992; *Brantley and Chen,* 1995; *Louvat and Allegre,* 1997; *Stefannsson and Gislason,* 2001; *Navarre-Sitchler and Brantley,* 2007). Primary mineralogy in such weathering systems is dominated by olivine, pyroxenes, plagioclase, and Fe-Ti-oxides (*McLennan and Grotzinger,* 2008; *Taylor et al.,* 2008). Accordingly, the "mafics ternary" plotting Al_2O_3–(CaO^* + Na_2O + K_2O)–(FeO_T + MgO), or A-CNK-FM, better captures the geochemical variation. During basaltic weathering the relative importance of alteration to clays, compared to primary mineral dissolution, is considerably less than for weathering of granitic compositions (Table 2).

3.2.3. Physical weathering. Most physical weathering processes involve differential volume changes whereby

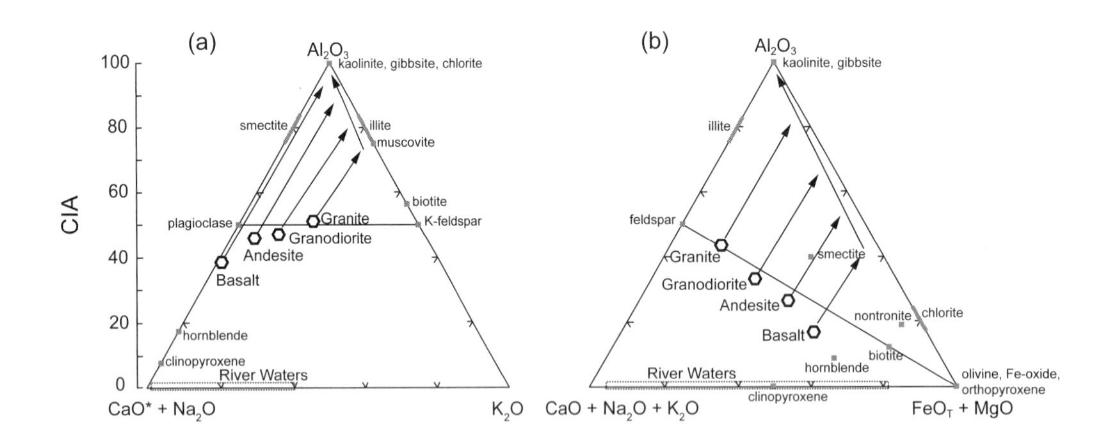

Fig. 6. Ternary diagrams plotting molar proportions **(a)** Al_2O_3–(CaO^* + Na_2O)–K_2O, or A-CN-K, and **(b)** Al_2O_3–(CaO^* + Na_2O + K_2O)–(FeO_T + MgO), or A-CNK-FM, illustrating weathering trends in terrestrial settings (adapted from *Nesbitt and Young,* 1984; *Nesbitt and Wilson,* 1992). CaO^* refers to calcium in the silicate components only (i.e., corrected for carbonates and phosphates). The CIA scale is plotted on the left side of the A-CN-K diagram. Plotted are selected igneous and sedimentary minerals, average compositions of major igneous rock types and the range for most natural waters. The thin horizontal line on the A-CN-K diagram (connecting plagioclase and K-feldspar) and the thin diagonal line on the A-CNK-FM diagram (connecting feldspar to FM apex) effectively separate the lower parts of the diagrams dominated by primary igneous minerals (unweathered) from the upper parts dominated by clays (weathered). The arrows schematically indicate general trends for increasing degrees of weathering for various rock types, based on field studies of weathering profiles and thermodynamic/kinetic modeling.

TABLE 2. Some common simplified weathering reactions with their Gibb's free energy of reaction at standard temperature (25°C) and pressure (1 bar) (after *Curtis,* 1976).

		ΔGr^0 (Kcal/mole)	ΔGr^0 (Kcal/gram atom)
Sulfide			
Pyrite oxidation	$2FeS_{2(s)} + 4H_2O_{(l)} + 7.5O_{2(g)} \Leftrightarrow Fe_2O_{3(s)} + 4SO_4{}^{2-}{}_{(aq)} + 8H^+{}_{(aq)}$	–583.5	–17.68
Olivine			
Fayalite ⇒ hematite	$Fe_2SiO_{4(s)} + 0.5O_{2(g)} \Leftrightarrow Fe_2O_{3(s)} + SiO_{2(s)}$	–52.7	–6.58
Forsterite dissolution	$Mg_2SiO_{4(s)} + 4H^+{}_{(aq)} \Leftrightarrow 2Mg^{2+}{}_{(aq)} + 2H_2O_{(l)} + SiO_{2(s)}$	–44.0	–4.00
Pyroxene			
Enstatite dissolution	$MgSiO_{3(s)} + 2H^+{}_{(aq)} \Leftrightarrow Mg^{2+}{}_{(aq)} + H_2O_{(l)} + SiO_{2(s)}$	–20.9	–2.98
Diopside dissolution	$CaMg(SiO_3)_{2(s)} + 4H^+{}_{(aq)} \Leftrightarrow Ca^{2+}{}_{(aq)} + Mg^{2+}{}_{(aq)} + 2H_2O_{(l)} + 2SiO_{2(s)}$	–38.1	–2.72
Amphibole			
Anthopyllite dissolution	$Mg_7Si_8O_{22}(OH)_{2(s)} + 14H^+{}_{(aq)} \Leftrightarrow 7Mg^{2+}{}_{(aq)} + 8H_2O_{(l)} + 8SiO_{2(s)}$	–137.2	–2.49
Tremolite dissolution	$Ca_2Mg_5Si_8O_{22}(OH)_{2(s)} + 14H^+{}_{(aq)} \Leftrightarrow 2Ca^{2+}{}_{(aq)} + 5Mg^{2+}{}_{(aq)} + 8H_2O_{(l)} + 8SiO_{2(s)}$	–123.2	–2.24
Feldspar			
Anorthite ⇒ kaolinite	$CaAl_2Si_2O_{8(s)} + 2H^+{}_{(aq)} + H_2O_{(l)} \Leftrightarrow Al_2Si_2O_5(OH)_{4(s)} + Ca^{2+}{}_{(aq)}$	–23.9	–1.32
Albite ⇒ kaolinite	$2NaAlSi_3O_{8(s)} + 2H^+{}_{(aq)} + H_2O_{(l)} \Leftrightarrow Al_2Si_2O_5(OH)_{4(s)} + 2Na^+{}_{(aq)} + 4SiO_{2(s)}$	–23.1	–0.75
Microcline ⇒ kaolinite	$2KAlSi_3O_{8(s)} + 2H^+{}_{(aq)} + H_2O_{(l)} \Leftrightarrow Al_2Si_2O_5(OH)_{4(s)} + 2K^+{}_{(aq)} + 4SiO_{2(s)}$	–15.9	–0.51

TABLE 3. Chemical and mineralogical comparison of the terrestrial upper continental crust, martian upper crust, and available analyses from the Venus surface.

	Earth Upper Crust	Mars Upper Crust	Venus		
			Vega 2	Venera 13	Venera 14
SiO_2	65.9	49.3	45.6	45.1	48.7
TiO_2	0.65	0.98	0.2	1.59	1.25
Al_2O_3	15.2	10.5	16.0	15.8	17.9
$FeO_T{}^*$	4.52	18.2	7.7	9.3	8.8
MnO	0.08	0.36	0.14	0.2	0.16
MgO	2.21	9.06	11.5	11.4	8.1
CaO	4.20	6.92	7.5	7.1	10.3
Na_2O	3.90	2.97	(2.0)	(2.0)	(2.4)
K_2O	3.36	0.45	0.48	4.0	0.2
P_2O_5	0.16	0.90	—	—	—
Other	—	—	5.0	1.9	1.3
Sum	100.2	99.6	96.1	98.4	99.1
Olivine	—	17	19	29	10
Pyroxene	1	22	22	11	27
Fe-Ti oxides	1	10	<1	3	2
Glass	13	21	—	—	—
Plagioclase	35	29	55	24	59
K-feldspar	11	—	3	25	1
Sheet silicates	14	—	—	—	—
Quartz	20	—	—	—	—
Nepheline	—	—	—	9	
Other	4	—	—	—	—

*Total iron as FeO_T.

Chemical compositions in weight percent. Data sources: Earth upper crust chemistry and mineralogy from *Nesbitt and Young* (1984) and *Taylor and McLennan* (2009); martian upper crust chemistry and mineralogy from *McLennan and Grotzinger* (2008) and *Taylor and McLennan* (2009); Venus chemical analyses from *Surkov et al.* (1984, 1986, 1987) (note that Na_2O was not analyzed and values are estimates from Surkov et al. and that "Other" includes SO_3 and Cl), and Venus mineralogy are CIPW norms based on volatile-free compositions assuming all Fe as FeO.

shear stresses, on a variety of scales ranging from crystals/grains to large outcrops, cause expansion/contraction that in turn destabilizes some part of a rock mass. *Melosh* (2011) recently reviewed physical weathering processes on planetary surfaces, including Earth [his terminology and organization differ somewhat from terrestrially oriented geomorphology texts (e.g., *Easterbrook,* 1999)]. The processes identified include (*Melosh,* 2011)

1. Stress corrosion cracking involving the propagation of fractures assisted by chemically reactive fluids (providing an environmental and kinetic framework) and influencing formation of fractures on a broad range of scales from microscopic cracks on mineral surfaces through to larger scale vertical joints.

2. Sheeting and exfoliation joints, which are near-surface, subhorizontal features formed as a result of vertical stress release giving rise to upward rock expansion and fracture growth due to compressive stresses parallel to the planetary surface.

3. Spheroidal weathering, which results from chemical weathering where fluids percolate along preexisting joint-bounded fractures leading to corestones within soils (e.g., Fig. 7) and to free-standing monoliths, called tors, that are preserved on some mass-wasted weathered surfaces.

4. Frost shattering resulting from growth of ice crystals in fractures fed by thin films of fluid (water in the case of Earth) that have significantly depressed freezing temperatures. On Earth such processes can exert pressures of ~10 MPa. Frost shattering currently is probably of negligible importance on Mars (atmospheric pressure below triple point of water), Venus (always above melting temperature), or Titan (where the liquid is a hydrocarbon; see section 3.5).

5. Salt weathering resulting from crystallization of dissolved salts (typically halite on Earth) within rock fractures and is thus a similar process to frost shattering. Crystallization pressures of common evaporite minerals are in the range of ~10–55 MPa.

6. Rock disintegration in response to volume changes that typically result from various processes (e.g., mineral reactions, clay swelling/shrinking). On Earth this process generates quartz sand grains, but its role in producing sand-sized particles on planets with basaltic surfaces is less well understood. For example, on Mars olivine sands appear fairly common and may form by rock disintegration leaving behind the relatively hard olivine. Plagioclase is also a common basaltic phenocryst mineral, is more resistant to chemical weathering than olivine, but has not yet been identified as an especially abundant sand grain. On the other hand, plagioclase cleavage probably makes it more susceptible to physical breakdown than olivine.

7. Insolation weathering resulting from thermal stresses generated from diurnal temperature changes. The lack of evidence for this process on the Moon, where temperature changes are extreme, suggests that the process also requires aqueous activity.

Additional erosional processes are also important in understanding physical weathering. As weathering profiles develop on slopes from chemical-physical-biological activity they inevitably become gravitationally unstable and are transported downslope by soil creep and mass wasting. Transport of particles, by either aeolian or fluvial processes, further lead to saltation abrasion of substrate surfaces by bouncing particles, and highly energetic transport (e.g., rock falls) results in rock shattering. During glacial transport large amounts of fine-grained sediment are produced by physical grinding, and rock plucking is also a common process influencing the weathering surface over which glaciers travel.

What is the relative role of chemical vs. physical weathering processes in the formation of sediment? By comparing the particulate and dissolved loads within rivers, *Millot et al.* (2002) derived an empirical relationship between chemical (R_{Chem}) and physical (R_{Phys}) weathering rates (in tons/km^2/yr)

Fig. 7. (a) Photograph showing corestone weathering features in a soil horizon formed on the Kohala volcano on the north side of Hawai'i Island. **(b)** Close up of corestones showing the exfoliating surface. The circular corestone on the right is approximately 1 m in diameter.

$$R_{Chem} = 0.39 \times \left(R_{phys}\right)^{0.66} \quad (7)$$

that appears to apply to sediments derived from both granitic and basaltic terranes. This relationship is broadly consistent with that determined from an evaluation of chemical/physical denudation rates in soils (*Anderson et al.,* 2007).

3.2.4. Biological processes. Biological activity is well known to strongly influence physical weathering (e.g., root growth, bioturbation of soils), but a more recent development is recognition of the importance of biological activity in also promoting chemical weathering related to chemical gradients and microenvironments (e.g., in pore space) set up both by macroscopic organisms (e.g., root systems) and microbial activity (*Nesbitt,* 1997; *Berner et al.,* 2004; *Chorover et al.,* 2007; *Amundson et al.,* 2007). On the other hand, from a planetary perspective, it is not yet possible to evaluate the significance of such processes beyond Earth.

3.3. Weathering on Mars

3.3.1. Chemical processes. The red color of Mars has long sent the message that some form of secondary (oxidative) weathering affected the surface and this was confirmed by early remote sensing that indicated the presence of ferric iron (*Burns,* 1993). Evidence for the occurrence of aqueous alteration first came from *in situ* chemical measurements of soils that have high sulfur and chlorine abundances and can be modeled to represent mixtures of silicate minerals and sulfate/chloride salts using mass balance calculations (e.g., *Clark et al.,* 1976; *Clark, 1993; Foley et al.,* 2008; *Haskin et al.,* 2005; *Ming et al.,* 2006; *Hurowitz and McLennan,* 2007; *McLennan and Grotzinger,* 2008; *Hecht et al.,* 2009; *McLennan,* 2012). Recent acquisition of Mössbauer spectrometry, thermal infrared spectroscopy, and microscopic imaging of textures allow for considerable refinement of such models. For example, *McSween et al.* (2010) (see also *McGlynn et al.,* 2012) modeled Meridiani and Gusev soils as mixtures of ~75 ± 5% unaltered igneous components (plagioclase, pyroxene, olivine, oxides, phosphates) and ~25 ± 5% of a secondary alteration assemblage (sulfates, silica, clays, secondary oxides, chlorides). These values provide possible insight into the overall scale of chemical alteration that has affected the martian surface.

Many lines of evidence indicate that Mars has experienced a long-lived chemically dynamic sedimentary rock cycle (see reviews in *McLennan and Grotzinger,* 2008; *Grotzinger et al.,* 2011; *McLennan,* 2012). Orbital remote sensing across a wide wavelength scale (VNIR–TIR) indicates numerous occurrences of sedimentary minerals in ancient layered sequences including a variety of clays (e.g., nontronite, kaolinite) and chemically deposited constituents of likely sedimentary origin (Mg-, Ca-, Fe-sulfates, amorphous silica, iron oxides, chlorides, perchlorates) (*Bibring et al.,* 2006, 2007; *Murchie et al.,* 2009). The late Noachian–early Hesperian Burns formation at Meridiani Planum consists of Mg-Ca-Fe-sulfate cemented aeolian sandstones also containing amorphous silica and altered basalt (*Squyres et al.,* 2004; *Clark et al.,* 2005; *McLennan et al.,* 2005; *Glotch et al.,* 2006) and this secondary mineralogy is broadly consistent with that predicted from experimental and theoretical modeling studies of basalt alteration under low pH conditions (*Tosca et al.,* 2004, 2005; *Tosca and McLennan,* 2006).

Accordingly, two aspects of martian weathering appear to differ fundamentally from our terrestrial experience. As discussed above, the basaltic martian upper crust is profoundly different from the granodioritic terrestrial upper continental crust, both chemically and mineralogically, and this has direct implications for the composition of weathering profiles, clastic sediments, aqueous weathering fluids, and chemical sediments throughout the geological history of Mars. In post-Noachian terrains there is evidence that low-pH aqueous conditions typically were prevalent, resulting in a number of very distinct weathering reactions that typically are not observed during terrestrial weathering.

The presence of clay-bearing strata in Noachian terrains suggests that some form of weathering has proceeded throughout martian geological time. Noachian clays are dominated by Fe-Mg smectites (e.g., nontronite) with lesser more aluminous smectites and highly aluminous kaolinite. Their origin, and thus implications for understanding source-to-sink processes, is not well constrained and could be due to a variety of processes, including pedogenic weathering, *in situ* alteration (analogous to terrestrial bentonites), diagenesis, and/or hydrothermal alteration associated with volcanism and/or impacts. In addition, comparisons with terrestrial conditions are also complicated by a variety of other differences between the two planets that could influence aqueous alteration styles and rates, such as tectonics, vegetation, and atmospheric composition.

Assuming that at least some martian clays were derived from surficial weathering, the Fe-Mg smectites could be consistent with relatively low water/rock ratios (i.e., arid conditions) since in well-drained basaltic soils with high throughput of weathering fluids, more aluminous clays, notably kaolinite, would be expected (e.g., *White,* 1995; *van der Weijden and Pacheco,* 2003; *Milliken and Bish,* 2010; *Ehlmann et al.,* 2011). Another possibility is that clay minerals, formed in low water/rock ratio aqueous environments, could be then later transported and deposited in water-dominated fluvial and lacustrine environments (*Milliken and Bish,* 2010).

Toward the end of Noachian time the character of secondary mineralogy changed dramatically from being clay-dominated to being sulfate- and iron oxide-dominated (*Bibring et al.,* 2006) and this may have had an important effect on chemical weathering and other sedimentary processes. The change has been attributed to the acidification, oxidation, and desiccation of the martian surface and may be preserved in the stratigraphic record of Mars, e.g., at Gale Crater's Mount Sharp (*Milliken et al.,* 2010; *Grotzinger and Milliken,* 2012). One of the major objectives of the Mars Science Laboratory mission to Gale Crater is to investigate this transition (*Grotzinger,* 2009; *Grotzinger et al.,* 2012). The nature of chemical weathering thought to dominate from the late Noachian is illustrated on A-CNK-FM ternary diagrams and compared to weathering of basalts under terrestrial conditions in Fig. 8. Under the circumneutral pH and oxidizing terrestrial conditions, weathering residues are concentrated in insoluble Al and Fe^{3+} and natural waters in turn are Al- and Fe-depleted (Figs. 8a,b). Accordingly, terrestrial weathering profiles and siliciclastic sediments derived from basaltic weathering regimes have compositions that scatter above the feldspar–(FeO_T + MgO) join on the A-CNK-FM diagram.

On the other hand, most altered rock and sediment analyzed on Mars (soils, altered rock surfaces, and sedimentary rocks) scatter parallel to and below the feldspar–(FeO_T + MgO) join (Fig. 8c). This trend is consistent with acidic alteration experiments (pH < 5) where mafic minerals (olivine, Fe-Ti oxides, pyroxene, glass) are more readily dissolved rather than altered to clays (Fig. 8d) (*Tosca et al.,* 2004; *Golden et al.,* 2005; *Hurowitz et al.,* 2005, 2006). Lack of evidence for Al-mobility, e.g., absence of Al-sulfates that might be expected in acidic environments that produced Fe^{3+}-sulfates (e.g., jarosite), led *Hurowitz and McLennan* (2007) to suggest that low water/rock ratios were involved so that only relatively soluble minerals (e.g, olivine but not plagioclase) were mainly affected.

3.3.2. Physical processes. Physical weathering on Mars differs as much in style as in degree compared to Earth. In addition to the increased role of impact processes in generating sedimentary particles, discussed above, the common occurrence of aeolian features and presence of polar ice caps also point to the importance of aeolian saltation abrasion and various glacial and permafrost processes. These were also likely important in the geological past as suggested, e.g., by the aeolian origin for the late Noachian–early Hesperian Burns formation (*Grotzinger et al.,* 2005). Widespread evidence for ancient channels on the martian surface also point to fluvial erosion of bedrock driven by saltation abrasion during aqueous transport. Section 4 will discuss these aspects in more detail.

It may be possible to provide some qualitative constraints on the relative roles of physical vs. chemical weathering on Mars if we assume that present-day soils reflect some time-integrated record of sedimentary processes. *McGlynn et al.* (2012) recently modeled the mineralogy of martian soils at Gusev Crater and Meridiani Planum and concluded that the chemical constituents (sulfates, chlorides, amorphous silica) represent ~15% of the soil. This is perhaps a surprising result since the terrestrial Phanerozoic sedimentary rock record (*Garrels and Mackenzie,* 1971) similarly contains ~18% chemical constituents (carbonate, evaporite, silica). As discussed above, impact processes were likely an efficient mechanism for producing sediment on Mars, suggesting that chemical weathering processes may be of less importance. However, balanced against that is the fact that basaltic minerals such as olivine, pyroxene, and Fe-Ti-oxides tend to dissolve more readily during chemical weathering (Table 2) and form a combination of dissolved constituents (Mg, Fe, Ca) and amorphous silica. In comparison, the bulk of the minerals in terrestrial continental crust (quartz, plagioclase, K-feldspar) are either resistant to chemical weathering (quartz) or form a lesser proportion of dissolved constituents and a larger proportion of resistant clays (plagioclase, K-feldspar) (*McLennan,* 2003).

3.4. Weathering on Venus

On Venus, where surface water is essentially absent, experiments indicate that the hot (~735 K), dense (72 kg/m^3), and dry (~30 ppm H_2O) CO_2-N_2-SO_2 atmosphere (Table 1) will interact directly with surface basalts to form a host of potential secondary minerals (e.g., anhydrite, pyrite, magnesite, magnetite), although details have yet to be confirmed by *in situ* measurements (*Johnson and Fegley,* 2002). The large ranges in surface temperature (627–753 K), pressure (4.5–11 MPa), and air density (36–74 kg/m^3) associated with different altitudes (–2 to +11 km) may allow for differences in the rates and types of chemical weathering across the venusian surface (*Basilevsky and Head,* 1993). Images from the Venera landers indicate substantial physical disruption of the basaltic surface (*Florensky et al.,* 1977; *Basilevsky et al.,* 1985), perhaps facilitated by chemical reactions, and formation of sedimentary particles that in turn may be involved in aeolian processes (*Craddock,* 2011), although yardangs and dune deposits appear uncommon (*Greeley et al.,* 1995) (see section 4. 2 below for further discussion).

It is likely that the role of impacts in facilitating physical weathering on Venus is limited, although the two dune fields identified on the surface are found in close association with large impacts, suggesting their ejecta supplied the dune-forming sediment (*Greeley et al.,* 1995). Regardless, Venus was entirely resurfaced by basaltic volcanism at about 1.0–0.85 Ga and there are only about 940 craters on the surface, approximately equal to the terrestrial record over the same period of time (*Taylor and McLennan,* 2009). In addition, the filtering effects of the very thick atmosphere results in no basins <5 km in diameter and few <30 km being formed. The thick atmosphere also limits the transport of ejecta blankets and for many of the craters that are present, parts of the ejecta blankets are entirely missing due to complex atmospheric interactions. Venusian craters are mostly highly pristine with little evidence for significant erosion and degradation.

3.5. Weathering on Titan

Titan's cold (~95 K) surface (Table 1) is composed of an unknown mixture of crystalline water ice, clathrate hydrate, and solid organics, with perhaps a minor component of ammonia-rich ices (*Lorenz and Lunine,* 2005). There are few impact craters on the surface, indicating that it is relatively young (*Wood et al.,* 2010; *Neish et al.,* 2013). The atmosphere is dense (0.15 MPa, 5 kg/m^3) and nitrogen-based, with a volumetric concentration of ~5% methane at the surface (*Lunine and Atreya,* 2008) (Table 1). A low adiabatic lapse rate (~1 K/km) and 2–3 K diurnal and seasonal temperature variations (*Jennings et al.,* 2009, 2011) suggest that physical weathering mechanisms involving freeze/thaw cycles and/or variations in thermal stress are thermodynamically improbable. The insolubility of water ice in liquid methane and ethane ($<10^{-11}$ mole fraction) argues against dissolution-based chemical weathering of water-ice bedrock (*Lorenz and Lunine,* 1996; *Perron et al.,* 2006), despite the observation of karst-like morphologies (e.g., *Hayes et al.,* 2008; *Mitchell et al.,* 2007) and an abundance of topographically closed depressions in the polar regions (*Hayes et al.,* 2013). Regardless, the abundance of aeolian dunes in Titan's equatorial regions indicates that sedimentary particles are being produced (see section 5.2.3 and Fig. 14d below).

A unique sediment production mechanism on Titan

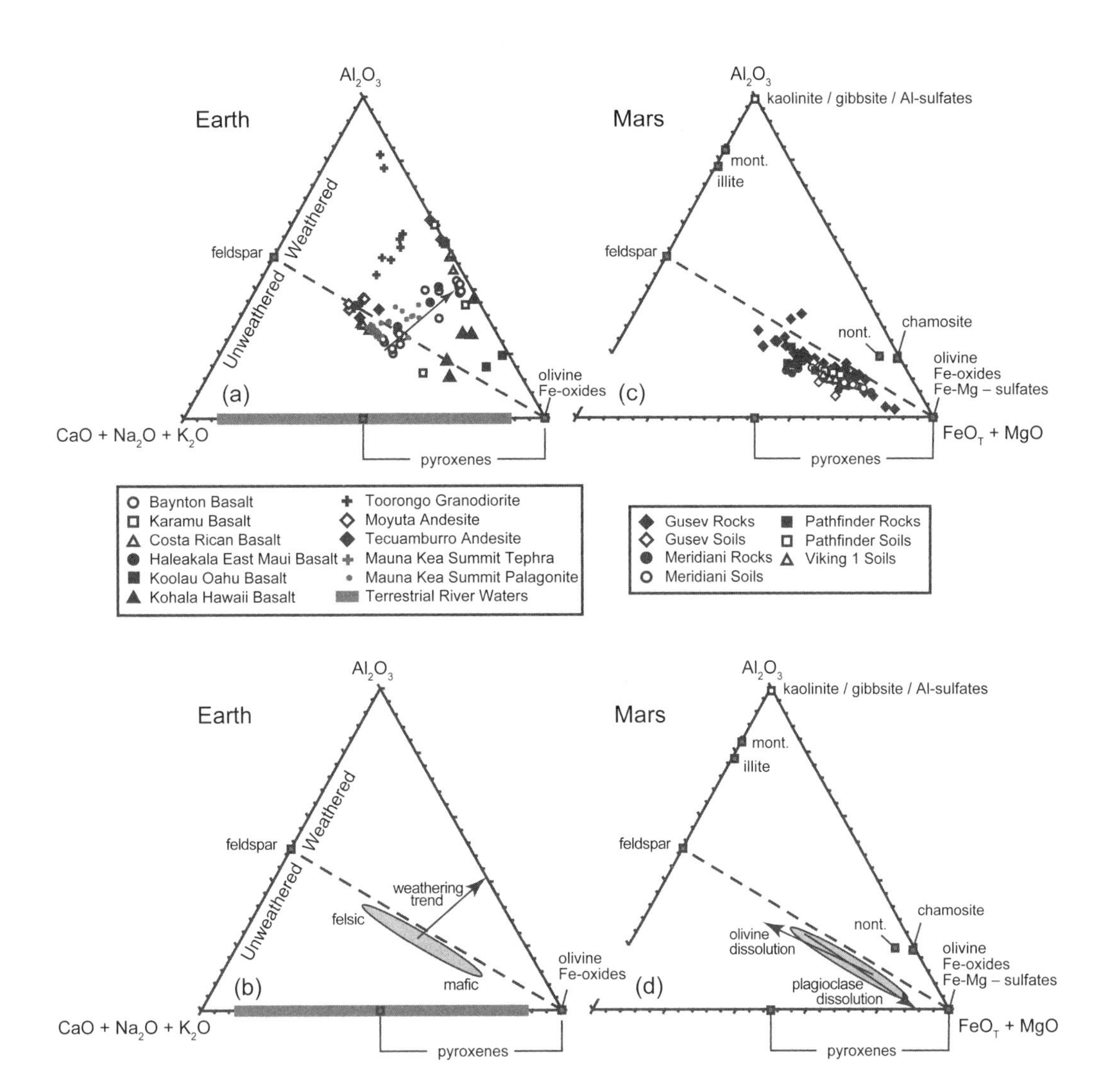

Fig. 8. Ternary diagrams plotting molar proportions Al_2O_3–($CaO + Na_2O + K_2O$)–($FeO_T + MgO$), or A-CNK-FM, comparing basaltic weathering trends on Earth and Mars. In this case, no corrections are made for nonsilicate components. Selected igneous and sedimentary minerals are plotted and the dashed diagonal lines connect feldspars to the FM apex, effectively separating the lower part of the diagrams dominated by primary igneous minerals (unweathered) from the upper parts dominated by clays (weathered). **(a)** Weathering and alteration profiles in terrestrial basaltic-andesitic settings illustrating that increased alteration results in enrichment insoluble Al and Fe and concentration in residual clays and oxides. **(b)** Schematic diagram showing compositional variation in primary terrestrial igneous rocks with an arrow indicating a typical terrestrial weathering trend. **(c)** Martian soils, igneous rocks, altered rock surfaces, and sedimentary rocks. Almost all data fall below and parallel to the diagonal line in spite of widespread evidence for alteration. **(d)** Schematic diagram showing compositional variation in martian mafic to ultramafic igneous rocks with arrows showing trends for olivine and plagioclase dissolution from typical basalt. Differing trends for terrestrial and martian alteration patterns can be explained by low-pH, low-water/rock ratio conditions on Mars whereby mineral dissolution dominates over alteration to clays. Adapted from *Hurowitz and McLennan* (2007) and *McLennan* (2012).

involves organic particulates that are generated through methane photolysis in the upper atmosphere (*Yung et al.,* 1984). While the particles settling out from the atmosphere are likely too small to saltate, they may form aggregates or crusts that can be eroded into saltatable-sized particles of ~200 μm (*Lorenz et al.,* 2006). Another Titan-specific process involves the interaction between ice and liquid alkane. Recent experiments have shown water-ice is wettable (lyophilic) to methane and ethane under Titan-relevant conditions, allowing liquid alkanes to be absorbed into the ice and potentially cause changes to both its optical and physical properties (*Sotin et al.,* 2009). If there is an associated volume change, repeated wetting and drying events may weaken water ice in a physical weathering mechanism analogous to terrestrial rock disintegration. While water ice is insoluble in liquid methane and ethane, constituents of the complex organic sediments generated through photolysis (e.g., acetylene) may have sufficient solubility to generate dissolution features given sufficient abundance and time (*Mitchell and Malaska,* 2011). The recent observation of potential evaporite deposits found in the floors of empty lake basins (*Barnes et al.,* 2011) further support the plausibility of dissolution chemistry on Titan.

In addition to the soluble components of the complex hydrocarbons and nitriles generated in the upper atmosphere, the solubility of pure ammonia ice may approach that of calcium carbonate on Earth (10^{-5}–10^{-6} mole fraction), although measurement uncertainties of ammonia ice solubility at 94 K cover three orders of magnitude (*Perron et al.,* 2006). While thermodynamic equilibrium would require ammonia ice to be in the form of ammonia-water hydrates on Titan (*Perron et al.,* 2006), disequilibrium mixtures containing pure ammonia ice can be formed by flash freezing, as might be expected during an ammonia-water cryovolcanic event (*Yarger et al.,* 1993). Finally, methanol has been suggested as a minor component of Titan's primordial ocean (*Deschamps et al.,* 2010). If methanol can be brought to the surface, it would also be soluble in liquid alkane. If Titan's surface contains a significant contribution of any of these soluble materials, chemical weathering may be important.

4. SEDIMENT PRODUCTION

Sediment production is the process of breakdown of rock or immobile regolith into mobile particles. There are a host of potential processes involved in sediment production. In addition to chemical and physical weathering (see above), scour by flowing fluids (e.g., wind, rivers, or glaciers) and the sediment they transport, hillslope processes (e.g., landslides, debris flows, and soil creep), and impact cratering can be important for producing sediment. Although many processes may be important, Earth, Mars, and Titan all show convincing evidence for sediment production and large-scale transport by rivers, and including Venus, all four bodies show evidence for wind erosion and transport. Here we focus on sediment production and erosion by rivers and wind.

4.1. Fluvial Bedrock Erosion

Fluvial bedrock erosion is one of the main drivers of landscape evolution on Earth; in addition, similar processes appear active on Titan, and were likely important in the martian past. For example, patterns of ancient river networks on Mars have been used to infer volatile sources and as evidence for precipitation. Dendritic drainage patterns suggest distributed fluid source areas such as rainfall or snowmelt in some places (e.g., *Mangold et al.,* 2004), whereas in other places large anabranching bedrock valleys with inner gorges, cataracts, and streamlined islands suggest catastrophic outburst flooding (*Baker,* 1982). Longitudinal valley networks with stubby amphitheater-headed tributary channels often are inferred to represent erosion from groundwater seepage (*Sharp and Malin,* 1975; *Carr and Clow,* 1981). However, similar valley shapes more often occur on Earth due to headwall undermining or collapse from waterfall processes (*Lamb et al.,* 2006, 2008a). Like Mars and Earth, the surface of Titan shows evidence for dendritic drainage networks and valley networks with stubby tributary heads (e.g., *Burr et al.,* 2013), but it lacks apparent outburst-flooding features (*Burr,* 2010).

Fluvial processes have been considered a possibility on the surface of Venus. Although there are more than 200 channels preserved on the plains of Venus, many >500 km in length and a few with deltas, they typically are smooth at a centimeter scale with no boulders, as characterized by radar. These channels are best interpreted to represent erosion by low-viscosity lava flows (*Baker et al.,* 1992).

Rivers can be divided into those that are supply-limited or transport-limited with respect to sediment supply (section 2.1). Supply-limited rivers typically have a flux of sediment q_s (per unit channel width) that is less than the transport capacity (i.e., $q_s < q_{sc}$), and partial exposure of bedrock on their beds. These rivers tend to be net erosional and this is where sediment is produced. River erosion into bedrock typically occurs by plucking of jointed fragments of rock and abrasion from impacting sediment (*Whipple et al.,* 2000). Although dissolution may play a role in some cases (e.g., carbonates on Earth, sulfates on Mars, and ice bedrock on Titan may dissolve in certain hydrocarbon mixtures), it is generally thought to be a minor component of fluvial bedrock erosion on all three bodies (e.g., *Whipple,* 2004; *Perron et al.,* 2006). A host of incision mechanisms can occur at migrating knickpoints and waterfalls, including plunge-pool scour, undercutting, seepage, and block toppling (e.g., *Whipple,* 2004; *Williams and Malin,* 2004; *Lamb et al.,* 2006; *Lamb and Dietrich,* 2009; *Warner et al.,* 2010).

The mechanics of river bed incision by plucking and abrasion were recently reviewed by *Burr et al.* (2013) for Earth and Titan, and are summarized here. Plucking is a process of removal of blocks of rock from a river bed and is typically modeled as a function of bed shear stress (τ_b) in excess of the threshold stress (τ_c) required for motion (*Hancock,* 1998; *Whipple et al.,* 2000). This makes erosion by plucking extremely sensitive to joint patterns in rock at meter or submeter scale (e.g., *Whipple et al.,* 2000; *Lamb and Fonstad,* 2010), and therefore predictions are difficult on Titan and Mars where fine-scale joint patterns cannot readily be measured. Models for river-bed abrasion are based on particle impacts of the bed by saltating and suspended sediment (*Sklar and Dietrich,* 2004; *Lamb et al.,* 2008b). Abrasion ceases when the sediment supply exceeds the transport capacity (i.e., $q_s > q_{sc}$) because deposition protects the underlying bedrock from erosion. In abrasion models, the erosion rate is a function of rock strength, bed shear stress, sediment supply, and the capacity of the river to transport sediment (*Sklar and Dietrich,* 2001). Abrasion models can be scaled for Mars and Titan (*Collins,* 2005; *Burr et al.,* 2013), and estimates for Titan, for example, show that incision rates predicted by a saltation-abrasion model, for fixed channel geometry and sediment conditions at Titan temperatures, are about 40 times less than on Earth for terrestrial rock of similar strength (*Sklar,* 2012). The strength of ice, and thus its resistance to erosion, also depends on crystal grain size and the concentration of solid impurities, both of which are poorly constrained on Titan (*Litwin et al.,* 2012).

4.2. Wind Erosion

Aeolian erosion occurs primarily through deflation, the removal and transport of loosened material, and abrasion, the mechanical wear of coherent material (*Anderson,* 1986). Morphologies that result from aeolian erosion include wind streaks, ventifacts, yardangs, deflation basins and pans, and inverted relief (*Laity,* 2011). Ventifacts, or wind-abraded rocks, have been observed on Earth and Mars (*Laity and Bridges,* 2009), while wind streaks, yardangs, and deflation basins have been reported on Venus, Earth, Mars, and Titan (*Greeley and Iversen,* 1987; *Greeley et al.,* 1995; *Bourke et al.,* 2010). Sediment piles that collected on the Venera 13 spacecraft during landing on Venus were observed to deflate due to near-surface winds during the hour following touchdown (*Selivanov et al.,* 1982), also suggesting active aeolian processes. Inverted relief resulting from differential erosion between fluvial and/or volcanic channel deposits and their surrounding terrain are observed on Earth and Mars (*Pain et al.,* 2007). Aeolian erosion is also one of the primary mechanisms of dust formation. Dust concentrations have far-reaching importance on a planetary surface, influencing geomorphologic processes, atmosphere conditions, and the planetary stratigraphic record (*Grotzinger and Milliken,* 2012; *Farmer,* 1993).

Physical weathering rates on Mars are difficult to determine with any precision and likely changed dramatically over geological time with an episodic reduction in rates at about the end of Noachian time. Similarly, relatively little work has been done to evaluate the full range of processes that may have been responsible for erosion and denudation of the planet (*Malin,* 1974; *Chan et al.,* 2008; *Viles et al.,* 2010). Diurnal thermal cycling, involving low temperatures and temperature differences of >75°C or more, undoubtedly plays a significant role (*Leask and Wilson,* 2003; *Viles et al.,* 2010), comparable to the most extreme deserts on Earth.

The rate of aeolian abrasion can be estimated by the multiplication of materials susceptibility to erosion (S_a), the particle flux of entrained aeolian material (q), and the wind frequency above the saltation threshold (f). *Greeley et al.* (1982) predicted the abrasion rate at the Viking 1 landing site on Mars to be up to ~210 μm per year using measured meteorological data, significantly greater than expected considering the age of the surface determined by crater counting and interpreted as evidence for a lack of sand-sized material. *In situ* estimates of current erosion rates in Gusev Crater and on Meridiani Planum, using a variety of approaches (e.g., lags of resistant hematitic concretions, deflation under rocks), indicate <10 nm/yr and mostly <1 nm/yr or >10^2 times (and up to 10^5 times) slower than any denudation rate measured on Earth (*Golombek et al.,* 2006). *Carr and Head* (2010) reviewed the literature on martian denudation rates over geological time and concluded that the highest average rates in the Noachian (which includes both fluvial and aeolian erosion), at ~5 m/10^6yr (~5000 nm/yr), were comparable to or less than denudation rates on highly stabilized terrestrial cratons. The end of Noachian time marked a fundamental shift in physical weathering processes with a drop of some 4–5 orders of magnitude, leading to average Hesperian/Amazonian rates of ~2–3 × 10^{-5} m/10^6yr (~0.02–0.03 nm/yr). Most recently, *Bridges et al.* (2012) used measurements of dune migration in the Nili Patera dune field, Mars, to model abrasion and derived modest rates of 1–10 μm/yr for flat ground and of 10–50 μm/yr for vertical rock faces, similar to basalt abrasion rates in Antarctica's Victoria Valley on Earth of 30–50 μm/yr (*Malin,* 1986).

5. SEDIMENT TRANSPORT

Earth, Mars, Titan, and Venus show evidence for large-scale transport of sediment from source areas to depositional sinks. Fluvial and aeolian processes are likely the most significant mechanisms for large-scale sediment transport and also represent important components of planetary climate. Detailed study of both modern and ancient fluvial and aeolian deposits can reveal information regarding both past and present atmospheric conditions.

5.1. Fluvial Sediment Transport

River valleys on Titan and Mars, and observations of rounded grains in a putative river plain at the Huygens landing site on Titan (*Tomasko et al.,* 2005), suggest that sediment has moved within rivers on both planetary bodies. The best-developed quantitative theories for sediment transport are for rivers with their bed and banks composed of unconsolidated sediment (i.e., alluvial rivers), rather than bedrock. Unlike net-erosional, supply-limited rivers discussed in section 4.1, alluvial rivers tend to be net depositional or bypass sediment over geologic timescales. In this case, the sediment flux is determined by the capacity of the flow to transport the material (i.e., $q_s = q_{sc}$), rather than erosion from bedrock, and sediment flux can be predicted relatively accurately from flow hydraulics and sediment size. It is instructive to consider the dynamics of channelized fluid flow before discussing sediment transport.

5.1.1. Flow dynamics. Flow of a Newtonian fluid in a channel can be modeled through conservation equations for fluid mass and momentum and a parameterization for bed friction (e.g., a turbulence drag law) (e.g., see *Burr et al.,* 2013, for a recent review). For depth-averaged, Reynolds-averaged conditions in a wide channel, these equations are $Q = Uhw$, $\tau_{bd} = \rho ghS$, and $\tau_b = \rho C_f U^2$, where Q is the volumetric fluid flux, U is the depth-averaged flow velocity, h is the flow depth, w is the channel width, τ_{bd} is the driving stress acting on the river bed due to gravitational acceleration of the fluid, τ_b is the shear stress acting on the bed due to friction, ρ is the fluid density, g is gravitational acceleration, S is the dimensionless bed slope gradient, and C_f is a dimensionless friction coefficient (e.g., *Chow,* 1959). For many river systems, the flow is approximately steady and uniform flow, i.e., $\tau_b = \tau_{bd}$ (e.g., *Parker et al.,* 2007), and therefore the conservation equations can be combined as $Q = wh(ghS/C_f)^{1/2}$. This analysis shows that, at bankfull conditions, the fluid discharge can be calculated from measurements of channel depth, width and slope, and the bed friction coefficient. The bed friction coefficient is a function of the fluid viscosity only for laminar or hydraulically smooth flows (*Nikuradse,* 1933; *Schlichting,* 1979; *Garcia,* 2007), conditions that are unlikely for most rivers on Earth, Mars, and Titan where w > 1 m and D > 1 mm (or the bed contains bedforms; section 5.1.2) (e.g., *Perron et al.,* 2006; *Burr et al.,* 2013). For turbulent, hydraulically rough flow, C_f depends on the ratio of the flow depth to the bed-roughness length scale (e.g., *Parker,* 1991). Together, these equations can be used to calculate possible river discharges on Earth, Mars, and Titan (*Burr et al.,* 2002; *Irwin et al.,* 2004; *Perron et al.,* 2006; *Lorenz et al.,* 2008). Hydraulic equations developed for water on Earth should be directly applicable to Mars and Titan when differences in gravity and fluid density are taken into account. One of the biggest limitations in applying these hydraulic equations is identification of river channel width (w) rather than valley width, the latter of which can be orders of magnitude larger than the former.

5.1.2. Sediment transport. The motion of fluid in a river imparts a shear stress on the river bed. Once a critical shear stress is surpassed, sediment begins to move as bed load by rolling and bouncing along the riverbed. For noncohesive sediment, the criterion for sediment motion can be calculated from the nondimensional bed-shear stress, or Shields stress, $\tau_* = \tau_b/[(\rho_s - \rho)gD]$. For small particle Reynolds numbers [$Re_p = [(rgD)^{1/2}D/\nu]$, where $r = (\rho_s - \rho)/\rho$ is the submerged specific density of the sediment, and ρ_s and ρ are the densities of sediment and fluid, respectively], the critical value of the Shields stress, τ_{*c}, for incipient motion, varies with the particle Reynolds number, and therefore fluid viscosity (*Shields,* 1936). For large particle Reynolds numbers ($Re_p > 10^2$) and for shallow channel slopes ($S < 0.05$), $\tau_{*c} \approx 0.045$ (*Miller et al.,* 1977; *Wilcock,* 1993), and sediment motion is independent of fluid viscosity. This corresponds to sediment motion for D > ~1 mm (i.e., $Re_p > 10^2$) for conditions on Earth, Mars, and Titan, assuming values for siliceous grains in water on Earth and Mars [r = 1.65 (e.g., quartz on Earth; plagioclase on Mars), $\nu = 10^{-6}$ m²/s, and g = 9.81 m/s² and g = 3.73 m/s² for Earth and Mars, respectively], and r = 1.18, g = 1.35 m/s², and $\nu = 4 \times 10^{-7}$ m²/s for transport of water-ice sediment by flowing methane on Titan (e.g., *Burr et al.,* 2006, 2013; *Perron et al.,* 2006) (see also Table 1). On steep slopes (S > 5%), τ_{*c} can increase by as much as an order of magnitude due to particle emergence and changes to the structure of the flow (e.g., *Lamb et al.,* 2008c). In river beds with mixtures of sediment sizes, which may exist on Titan (*Tomasko et al.,* 2005), the statistics of the mixture (e.g., median particle size) and particle-size bimodality can influence incipient motion conditions (e.g., *Parker,* 1990; *Wilcock and Crowe,* 2003). In addition, cohesion may influence incipient sediment motion for very fine particles (e.g., D > 0.05 mm), and for organic sediment particles on Titan (*Rubin and Hesp,* 2009).

Sediment is suspended within the flow at large flow velocities and for small particles, when the settling velocity of a particles is less than the magnitude of turbulent velocity fluctuations within the fluid, or $u_* \equiv \sqrt{\tau_b/\rho} >\sim 0.8w_s$, where u_* is the bed-shear velocity and w_s is the particle-settling velocity (*Bagnold,* 1966). Empirical settling-velocity equations (e.g., *Dietrich,* 1982) explicitly account for particle and fluid densities, fluid viscosity, and gravity, making them straightforward to apply to different fluids and planetary bodies.

Stability fields for each of the three sediment transport conditions (no motion, bed load, and suspended load) are functions of the particle diameter and bed shear stress for a given transport system (Fig. 9). Stability fields are shifted in different systems as a function of gravity, particle density, fluid density, and viscosity (e.g., *Komar,* 1980; *Burr et al.,* 2006). For Earth and Mars, we make the calculation for silicious grains in water (r = 1.65, $\nu = 10^{-6}$ m²/s), where the shift in transport conditions is due only to differences in gravity (g = 9.81 m/s² and g = 3.73 m/s² for Earth and Mars, respectively) (Table 1). This results in a reduction of the necessary bed shear stress to transport particles and initiate suspension on Mars relative to Earth. Following *Lamb et al.* (2012), a second case is shown for Mars for the case of viscous brines, which have been inferred to be important in transporting sediment in some environments on Mars due to the observation of sediment mineral phases that are highly soluble in fresh water (*McLennan et al.,* 2005; *Tosca et al.,* 2011). Given one of the most viscous brines considered by *Tosca et al.* (2011) (r = 1.04 and $\nu = 4 \times 10^{-5}$ m²/s; Table 1), this shifts the transport thresholds to substantially coarser particle sizes. Thus, grains as large as a few millimeters transported in turbulent brines may be suspended once moved, owing to the reduced settling velocity in the high-viscosity fluid. Finally, following *Burr et al.* (2006), we show a case for water-ice particles transported by liquid methane on Titan (r = 1.18, g = 1.35 m/s^2, and $\nu = 4 \times 10^{-7}$ m^2/s; Table 1). Here the transport boundaries are shifted to much lower bed shear stresses primarily due to lower gravity on Titan as compared to Earth and Mars. The density of particles is uncertain on Titan, however, as Titan's river networks may contain organic tholin compounds (*Schröder and Keller,* 2008; *Janssen et al.,* 2009) with different transport properties depending on their porosity (e.g., *Burr et al.,* 2013).

5.1.3. Fluvial bedforms. In low-gradient (S < 1%), sand-bedded channels on Earth, river beds are not often flat, but instead contain bedforms such as ripples and dunes (e.g., *Southard,* 1991). These bedforms are important to consider for fluvial processes on Mars and Titan because they influence fluid-flow and sediment transport rates, and they leave distinct signatures in the sedimentary record (see section 6). Note that only conceptual models and empirical correlations are available for bedforms in steep (S > 1%) channels with coarse beds (e.g., *Montgomery and Buffington,* 1997), and their application to other planetary bodies is uncertain.

Lamb et al. (2012) compiled results from five sediment transport and bedform studies (*Chabert and Chauvin,* 1963; *White,* 1970; *Brownlie,* 1983; *Southard and Boguchwal,* 1990a; *van den Berg and van Gelder,* 1993) to build a comprehensive bedform stability diagram for subcritical flows (Fr < 1, where $Fr = U/\sqrt{gh}$ is the Froude number) that spans a large range in particle Reynolds number, Re_p (Fig. 10a). The compilation shows that bedform phase space, like incipient sediment motion, can be cast as a function of the Shields stress, τ_*, and Re_p. At low Shields stresses, sediment is immobile (Figs. 9, 10a). At higher Shields stresses sediment is mobile and the bed may take on a number of states: Ripples form at low Re_p and moderate values of τ_*, whereas at higher Re_p lower plane bed and dunes are stable (Fig. 10a). For all Re_p, the bed is planar (upper plane bed) for large values of the Shields stress.

Like the modes of sediment transport (Fig. 9), bedform state depends on fluid and particle densities, fluid viscosity, and gravity, which are different for fluvial processes on Mars, Titan, and Earth (e.g., *Southard and Bogouchwal,* 1990b). Following application of the bedform state diagram (Fig. 10a) to Mars by *Lamb et al.* (2012) and Titan by *Burr et al.* (2013), we illustrate these differences for the cases of silicious grains in water on Earth (r = 1.65, $\nu = 10^{-6}$ m^2/s, and g = 9.81 m/s^2), silicious grains in viscous brines on Mars (r = 1.04, $\nu = 4 \times 10^{-5}$ m^2/s, and g = 3.73 m/s^2), and water-ice grains in liquid methane on Titan (r = 1.18, g = 1.35 m/s^2, and $\nu = 4 \times 10^{-7}$ m^2/s) (Table 1) (Figs. 10b,c). Bedform stability under flowing martian brines is shifted to coarser sediment sizes as compared to freshwater flows on Earth (Fig. 10b). This is primarily due to the 40-fold increase in fluid viscosity, which affects the particle Reynolds number. For example, ripples on Earth only occur for relatively fine particles (D < 1 mm), whereas martian brines may produce ripples in sediment sizes up to 20 mm (Fig. 10b) (*Lamb et al.,* 2012). The bedform stability is shifted to much lower bed stresses on Titan as compared to Earth (Fig. 10c). This results from the 10-fold reduction in the submerged specific weight of sediment (i.e., rg) that affects the Shields stress. For example, at the bed stresses necessary for initial motion of 0.1-mm diameter sediment on Earth, the bed state on Titan is predicted to be upper-plane bed (Fig. 10c) (*Burr et al.,* 2013).

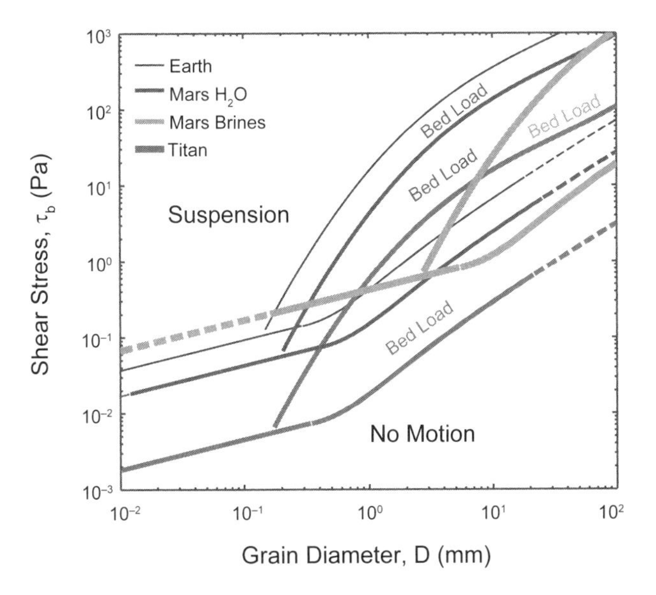

Fig. 9. Sediment transport diagram showing the bed shear stress required for initial motion and suspension for siliceous grains (e.g., quartz, feldspar) transported by water on Earth, siliceous grains (e.g., plagioclase) transported by water and viscous brines on Mars, and water-ice grains transported in methane on Titan. Labels "No Motion" and "Suspension" apply to all four cases. Following *Komar* (1980) and *Burr et al.* (2006).

5.2. Aeolian Sediment Transport

Aeolian landforms, such as dunes, wind streaks, and yardangs, have been observed on the surfaces of Earth, Mars, Titan, and Venus (e.g., *Bourke et al.,* 2010). The ubiquitous presence of these features suggests that the entrainment, transport, and deposition of sediment by wind are important geomorphic processes that operate in arid regions across the solar system. Unlike the fluvial processes described above, which are currently active on Earth and Titan, it is likely that aeolian processes are active on all four bodies. Aeolian transport involves interactions between the wind and ground surface, resulting in a strong dependence on textural and surficial conditions (e.g., surface crusts, surface moisture, and vegetation cover). Grain size is also important, leading to fundamental differences in the processes responsible for the entrainment, transport, and deposition of large (sand-sized or greater) vs. small (silt, clay, and dust) particles. Similar to flow in rivers, aeolian systems operate in both transport and supply limited regimes; however, most aeolian environments on Earth tend to be supply limited (*Nickling and Neuman,* 2009). In supply-limited systems flux is determined by sediment availability, or the rate at which the surface can supply grains to the air stream. Many natural surfaces exhibit spatial and/or temporal variations in surface and/or atmospheric conditions that can limit sediment availability (*Kocurek and Lancaster,* 1999).

5.2.1. Flow dynamics. Like the river processes described in section 5.1, the wind also exerts a bed shear stress on a planet's surface. The bed shear stress, τ_b, or bed shear velocity $\left(u_* \equiv \sqrt{\tau_b/\rho}\right)$, can be related to the wind velocity, U, at a given height above the bed, z, using a drag law, e.g., the commonly used log law $(U/u_*) = [(1/\kappa)\ln(z/z_0)]$, where z_0 is a function of the bed roughness for hydraulically rough flow (see section 5.1.2). Once the threshold for grain entrainment is surpassed, sediment moves in both bed load (saltation and creep) and suspended load, although saltation accounts for ~95% of the total mass transport in a typical aeolian system (*Nickling and Neuman,* 2009). This sediment is ultimately deposited to form bedforms such as loess, sand sheets, ripples, and dunes. During transport, momentum exchange between a developing saltation cloud and the wind results in a reduction of wind speed as the transport rate approaches a dynamic equilibrium known as steady-state saltation (*Anderson and Haff,* 1988). This equilibrium is typically reached within a few seconds of initial grain movement (*Nickling and Neuman,* 2009).

5.2.2. Sediment entrainment. Bagnold (1941) classified wind-driven particle motion into three modes with increasing wind speed: creep, saltation, and suspension (Fig. 11). There are no sharp distinctions between these transport modes; rather, they represent end members of a continuum. In aeolian systems, the specific mode of transport is often determined by the particle setting velocity (w_s) relative to the bed shear velocity (u_*). For low-particle Reynolds numbers ($Re_p < \sim 10$), the settling velocity is given by $w_s = \{[(\rho_p/\rho)-1]gd^2/(18\nu)\}$ (i.e., Stoke's Law). For high Reynolds numbers ($Re_p > \sim 100$), u_*/w_s can be recast in terms of $\tau_* = u_*^2/[(\rho_s-\rho)/\rho]gD$, which is typical in river sediment transport theories, since for this regime the particle settling velocity is proportional to $[(\rho_s-\rho)/\rho]gD$ and independent of viscosity (e.g., *Dietrich,* 1982).

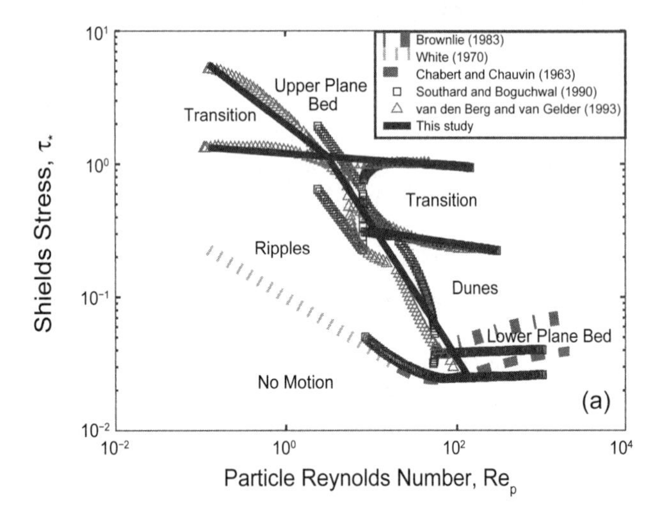

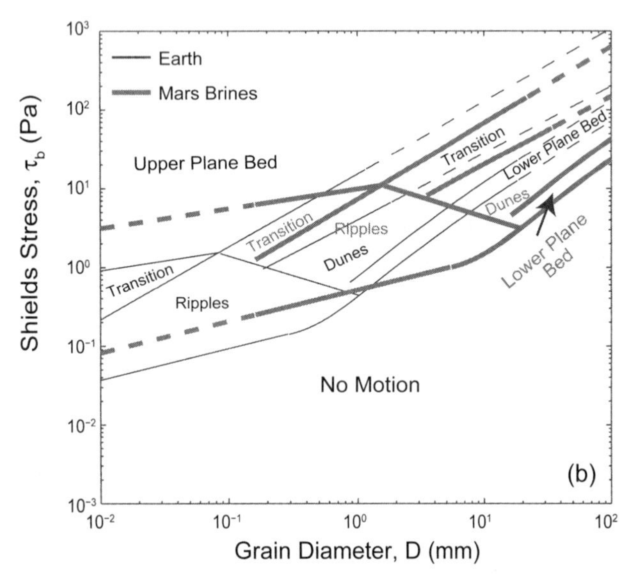

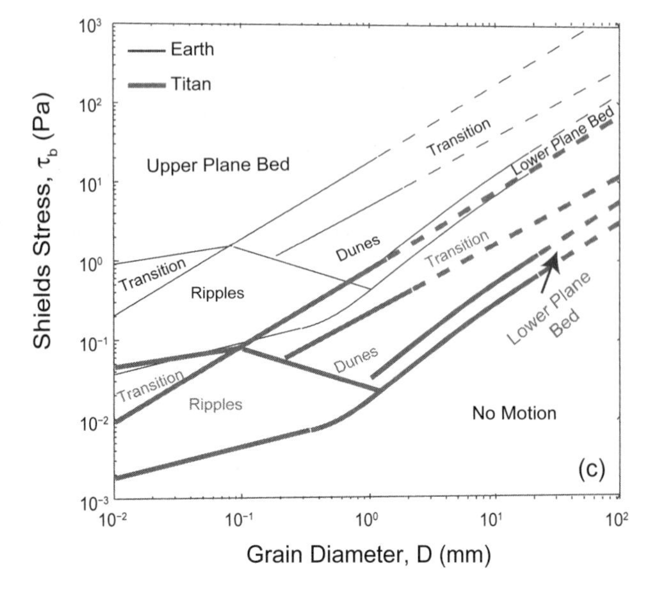

Fig. 10. Stable bedform states. **(a)** Bed states as a function of the Shields stress and particle Reynolds number (modified after *Lamb et al.,* 2012; *Burr et al.,* 2013). Solid lines are best-fit boundaries between bed states that summarize findings from five previous studies (*Chabert and Chauvin,* 1963; *White,* 1970; *Brownlie,* 1983; *Southard and Boguchwal,* 1990a; *van den Berg and van Gelder,* 1993). Stable bed states as a function of bed-shear stress and particle diameter for the cases of **(b)** viscous brines transporting siliceous grains on Mars and freshwater flows transporting siliceous grains on Earth (modified after *Lamb et al.,* 2012), and **(c)** water-ice particles on Titan and freshwater flows transporting siliceous grains on Earth (modified after *Burr et al.,* 2013). Dashed lines are power-law extrapolations of stability boundaries from **(a)**. Labels "Upper Plane Bed" and "No Motion" apply to all three cases.

Initial particle motion by saltation occurs when aerodynamic drag, lift, and moment surpass the restoring forces of gravity and interparticle forces (e.g., Van der Waals forces, soil moisture, surface crusts, etc.). For example, *Iversen and White* (1982) and *Iversen et al.* (1987) used wind tunnel experiments to empirically derive the critical Shields stress for grain entrainment, τ_{*c}, over a range of environments and found that it was primarily a function of the particle-friction Reynolds number, particle-to-fluid density ratio, and interparticle forces. *Shao and Lu* (2000) use the data from *Iversen and White* (1982) to derive an expression for τ_{*c} that is relevant for grain sizes that include sand: $\tau_{*c} \approx 0.012\,[1 + 3 \times 10^{-4}\,(\mathrm{kg/s^2})/(D^2\rho_s g)]$. Figure 12 shows this expression for a range of silicate grain sizes on Venus, Earth, and Mars, and for hydrocarbon grains for Titan. In all cases interparticle forces were assumed to be proportional to particle diameter following *Jordan* (1954), which was calibrated using silicate grains as described in *Iversen et al.* (1987). It should be noted that there is little known in regard to the cohesive behavior of water ice and/or hydrocarbon grains under Titan conditions and that *in situ* observation suggests that additional electrostatic forces may be important on Mars (*Sullivan et al.,* 2008). For large particles (D > ~200 μm), interparticle forces are negligible compared to gravity, whereas for small particles (D < ~70 μm), interparticle forces dominate (*Iversen et al.,* 1987). Venus and Titan have significantly lower thresholds for grain entrainment than Earth and Mars (Fig. 12). This difference results mainly from the high atmospheric density and viscosity on Venus and the high atmospheric density and low gravity on Titan, both of which reduce the effective settling velocity of particles. Titan's low atmospheric viscosity actually increases the settling velocity for small particles, although its effects are marginalized by the high density and low gravity.

Large particles ($w_s/u_* > 10$) (*Nino et al.,* 2003) are too big to be lifted by the wind. These particles are moved short distances by pushing and/or rolling in response to the impact of smaller grains in a process known as surface creep. At greater wind speeds (or smaller particle sizes) ($2.5 < w_s/u_* < 10$) (*Nishimura and Hunt,* 2000), grains are lifted into the wind and bounce across the surface following parabolic trajectories characteristic of aeolian saltation. Assuming the initial take off speed is on the order of the critical shear velocity (u_{*c}), *Almeida et al.* (2008) found that the characteristic saltation height, H_s, and length, L_s, are approximately $[H_s,L_s] \approx [81,1092](\nu^2/g)^{1/3}(u_*-u_{*c})/\sqrt{gD}$. On Earth and Titan, saltation heights are up to ~15 cm and lengths are ~2 m, while on Mars particles can follow trajectories with heights of up to ~5 m and lengths of ~120 m (*Almedia et al.,* 2008). Venus, as expected from its extremely low atmospheric viscosity (Table 1), has predicted saltation heights of ~1 cm and lengths of ~30 cm. Saltation often accounts for greater than 95% of the total mass transport in an aeolian system (*Nickling and Neuman,* 2009).

Initial saltating and creeping grains generate impact forces that set off a cascade effect that exponentially increases the number of grains entrained in the flow (*Nickling and Neuman,* 2009). Particle motion initiated by impact from saltating particles is known as splash, and the distribution of ejected particles is known as the splash function (*Ungar and Haff,* 1987). Saltating particles can also instigate suspension by ejecting finer-grained particles, which may be too cohesive to saltate on their own, into the atmosphere upon impact (*Chepil and Woodruff,* 1963).

Like particles in water, significant sediment suspension can occur when $w_s/u_* < \sim 1$ (*Nino et al.,* 2003). In wind, particles can be kept aloft for several days to weeks and

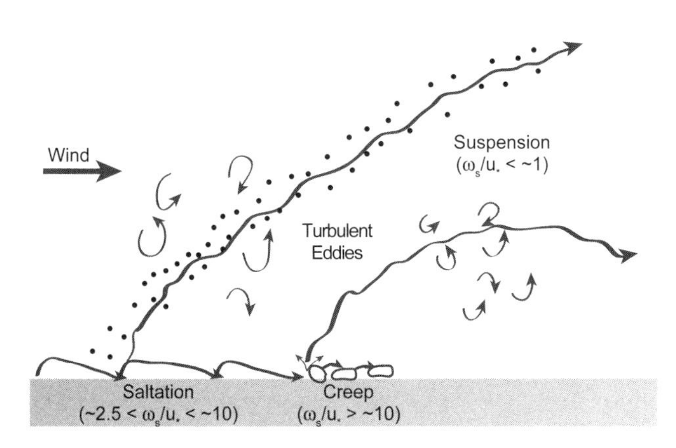

Fig. 11. Eolian sediment transport modes by wind (modified after *Pye,* 1987).

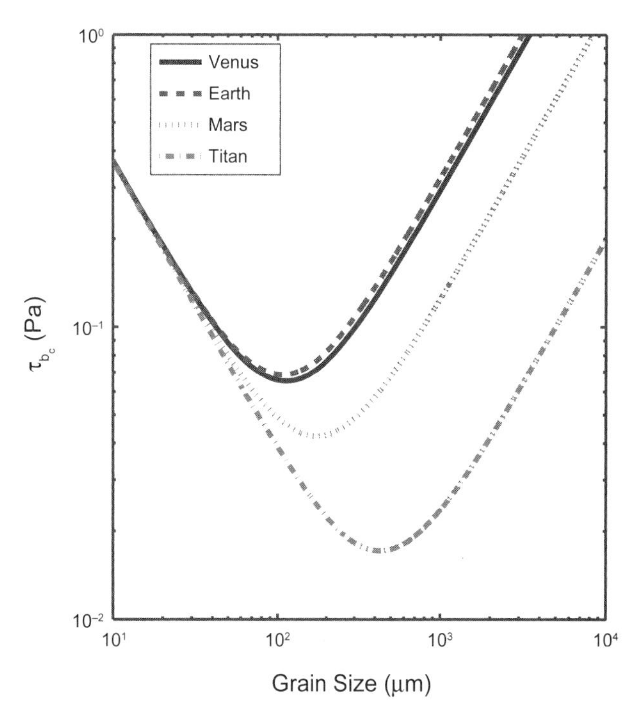

Fig. 12. Critical shear stress for eolian grain entrainment by saltation using the data from *Iversen and White* (1982) and equations presented in *Shao and Lu* (2000).

travel thousands of kilometers in a process called long-term suspension (*Tsoar and Pye,* 1987). On Mars, suspended particles of this type generate local, regional, and global dust storms that drape the surface in fine-grained material (e.g., *Greeley and Iversen,* 1987). Dust storms have significant environmental impacts on Earth, which include contaminating air quality, modifying soil composition, and contributing to the desertification of arid and semi-arid landscapes (e.g., the Dust Bowl drought in the North American Great Plains). Interparticle forces make lifting small particles difficult and, as a result, suspended particles are typically entrained through secondary processes such as collisions and dust devils (*Bagnold,* 1941; *Nishimura and Hunt,* 2000).

5.2.3. Aeolian deposits and bedforms. Aeolian deposits include fine-grained material such as loess, desert pavement, sand sheets, dune fields, and sand seas or ergs. Loess, which are deposits of wind-blown silt-sized particles, cover up to ~10% of Earth's surface and are the closest analog to the aeolian dust deposits found at all elevations on Mars (*O'Hara-Dhand et al.,* 2010). Deposition of silt and clay-sized material plays a major role in the inflation of desert pavement surfaces. Desert pavement, which is common in Earth's arid regions and similar to surfaces observed at Gusev Crater on Mars, consist of a lag deposit of gravel or larger clasts that armor underlying fine-grained material against deflation. Fine-grained material may be incorporated into the underlying mantle through wetting-drying cycles, freeze-thaw cycles, and/or bioturbation (*Pelletier et al.,* 2007). Sand seas, or ergs (the names are used interchangeably), are areas of wind-lain sand deposits that cover a minimum of 125 km^2 (*Fryberger and Alhbrandt,* 1979). Smaller regions of sand cover are termed dune fields or, if they are relatively dune free, sand sheets.

Aeolian bedforms are known to occur on Earth, Mars, Venus, and Titan (Fig. 13) and emerge from grain interactions and self-organization regardless of differences in the fundamental variables affecting sediment transport such as gravity, fluid density, and particle composition (*Anderson,* 1990; *Werner and Kocurek,* 1999; *Kocurek et al.,* 2010). Remarkably, individual dune patterns characteristics (e.g., length and spacing) appear to be independent of these environmental parameters and are self-similar over a wide range of scales (Fig. 14) (*Hayes et al.,* 2012). Dune fields that are not in equilibrium with respect to wind regime, sediment supply, or sediment availability have been shown to have pattern variables that are distinct from those that are in equilibrium (e.g., *Ewing et al.,* 2010). For a given crest spacing, dune fields that have significantly shorter average crest lengths are either still evolving toward equilibrium or have been modified by a change in climatic or geomorphic conditions, resulting in crestlines that are either truncated or broken up by a new generation of dunes (*Beveridge et al.,* 2006; *Derickson et al.,* 2007; *Ewing et al.,* 2010).

Similar to fluvial bedforms, aeolian dunes occur at different scales as part of a hierarchy of bedforms that comprise, in order of increasing size, wind ripples (spacing 0.1–1 m), simple dunes (either isolated or superimposed on top of larger bedforms), compound or complex dunes, and draa (spacing greater than ~500 m). Dune morphology responds to both wind regime (magnitude and orientation) and sediment supply. The main morphologic dune classifications are crescentic (barchans and transverse), crescentic ridges (laterally linked crescentic dunes), linear, star, and parabolic (*McKee,* 1979). Barchan dunes (Fig. 13a) typically occur in areas of limited sediment supply, align their ridges perpendicular to their transport direction, and coalesce laterally to form crescentic ridges as sand supply increases. Linear dunes (Figs. 13b,d) are characterized by their long sinuous crestlines, parallel orientation, and regular spacing. They form in areas of a bimodal wind regime, although linear dunes can also form in a unimodal wind regime if the sediment is highly cohesive (*Rubin and Hesp,* 2009). Vast sand seas of linear dunes, presumably composed of dark organic sediment, are found within 30° of Titan's equator (*Lorenz et al.,* 2006). Star dunes have multiple crestlines (Fig. 13b) and form in multidirectional or complex wind regimes. The largest dunes observed on Earth are examples of star dunes and, in some cases, linear dunes are observed to transition into star dunes when sediment availability is reduced. Crescentic, linear, and star dunes have been observed on Earth, Mars, and Titan. Dunes orient their crestlines such that they are as perpendicular as possible to all dune-forming winds [i.e., the concept of maximum gross bedform normal transport described by *Rubin and Hunter* (1987)] and, in the case of large linear dunes, can take hundreds of thousands of years to react to changes in the wind regime. On the other hand, smaller bedforms, such as wind ripples and isolated simple dunes, can reorient within much short time periods (hours to hundreds of years). As a result of these characteristics, dunes, especially multiscale dune fields, represent unique and robust records of both past and present climatic conditions on any planetary body (e.g., *Ewing et al.,* 2010).

6. SEDIMENT DEPOSITION AND STRATIGRAPHIC RECORD

6.1. Earth

On Earth, source-to-sink systems are part of the rock cycle driven by plate tectonics, and regional subsidence in large sedimentary basins is what creates and preserves the stratigraphic record. Uplift and denudation of mountains generate large volumes of sediment that are transported to oceans almost entirely by fluvial processes. Deposition typically occurs at numerous intermediate points in sediment routing pathways, in addition to significant accumulation in major basins that act as terminal sinks (Fig. 1). In each case, deposition is due to a divergence in the flux of water due to local conditions including, for example, overbank flooding of streams, flow expansion where channelized flows enter bodies of standing water, or decrease in channel slope.

Examples of intermediate sediment repositories include soils of mountainous regions, where weathering occurs and sediment particles are generated. Once captured by streams,

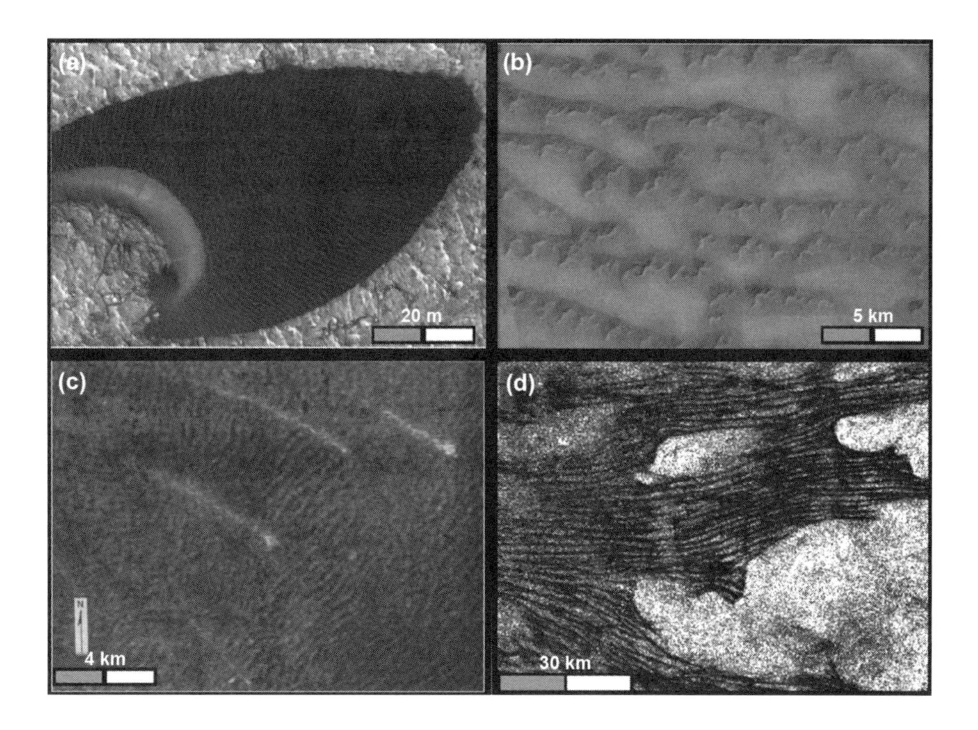

Fig. 13. Aeolian bedforms on Mars, Earth, Venus, and Titan. **(a)** Barchan dune with superimposed wind ripples in Nili Patera, Mars (*Ewing et al.,* 2010). **(b)** Linear and star dunes in Rub' al Khali, Oman, Earth. **(c)** Fortuna-Meshkenet dune field on Venus. Note that the wind streaks suggest a northwest wind direction, while the dunes are oriented between east-west and northeast-southwest (*Greeley et al.,* 1992). **(d)** Linear dunes diverted around topography in the Belet dune field on Titan.

temporary deposition may occur in abandoned channels or floodplains. In arid climates sediments may be transported from dry river beds to form sheets of windblown sand and silt. However, most sediments that are produced on Earth end up being deposited in ocean basins — the terminal sink (Fig. 1). This might include deltas, continental margins, or submarine fans at the base of the continental rise. There they will reside for tens to hundreds of millions of years before those sediments are recycled back into the sedimentary system after uplift due to plate convergence (e.g., continental collision) or into the mantle during subduction of lithospheric plates. Accordingly, uplift of previously deposited sediments and sedimentary rocks will result in their erosion and conversion back into sedimentary particles. In this regard, Earth is different from all other planets because sediment transport pathways are part of a longer-term tectonic cycle. On Venus, Mars, and Titan sediment transport is driven, at least in large part, by topographic gradients generated through other processes such as impacts and volcanism. Earth also differs in that the overwhelming majority of sediments were deposited in oceans and, to a lesser degree, streams. Eolian and ice-related deposits are relatively rare.

Earth's stratigraphic record preserves evidence of these processes at earlier times in its history. Sedimentary basins are sites of preferentially thick accumulation of sediments, where they become lithified through cementation and filling of initial pore space to form sedimentary rocks. Sedimentary basins on Earth are created by tectonic subsidence driven by two geodynamic processes: thinning due to stretching of the lithosphere, or dynamic depression of the lithosphere due to tectonic loading during mountain building. In either case broad depressions are created into which sediments are transported, creating accumulations that can range up to 20 km in thickness, but generally are on the order of 5–10 km. Significantly, thickness is proportional to the amount of tectonic subsidence — this is not something that necessarily holds true for either Mars or Titan.

Sedimentary strata that formed in these intermediate and terminal sinks can preserve signals of past climates, which can be revealed once the influence of tectonics and intrinsic dynamics in the transport system (e.g., *Jerolmack and Paola,* 2010) have been removed. In some cases climate variability is expressed as high-frequency variations in stratigraphic thickness, which may map back to changes in sea level and ice volume (*Goldhammer et al.,* 1987). Changes in sedimentary rock composition may also relate to climate change (*Warren,* 2010); in some rare cases dramatic shifts may be inferred (*Hoffman et al.,* 1998). But in many cases ancient changes in climate are recovered from variations in other proxies that are embedded within the stratigraphic record, which serves as a sort of "carrier signal." Sediments that accumulate as a result of (bio)chemical precipitation show this best. For example, carbonate strata may show variability in silicate or oxide mineral content, organic carbon concentration, or oxygen isotope ratios, to name just a few. While over the past million years of Earth history the best climate records may be found in ice cores, or cave deposits such as stalagmites, only sediments and sedimentary rocks can preserve evidence of Earth's earlier climate history — extending back ~3.5 b.y. (Fig. 3).

6.2. Mars

In contrast to Earth, Mars has never had plate tectonics that affected the cycling of materials, nor has it developed large sedimentary basins that actively subsided during sediment accumulation (*McLennan and Grotzinger,* 2008). In the absence of plate tectonics, it appears that the flux of sediment has declined over time (*Grotzinger et al.,* 2011; *Grotzinger and Milliken,* 2012). Early on, primary sediments may have consisted mostly of impact- and volcanic-generated particles that could have been transported by fluvial and aeolian processes. Chemical weathering of fragmented bedrock in the presence of near-neutral pH fluids would have generated clays and rare carbonates; weathering under more acidic conditions generated dissolved salts that precipitated as sulfates, halides, and rare carbonates (*McLennan et al.,* 2005; *Ehlmann et al.,* 2011; *Morris et al.,* 2010; *Milliken and Bish,* 2010). Over time, Mars is regarded to have undergone an evolution from more neutral pH weathering, to more acidic weathering, to eventual desiccation (*Bibring et al.,* 2006). As the flux of impactors and volcanism declined, and the planet lost its hydrologic cycle, the formation of sedimentary rocks also declined. Today Mars appears to be in a net state of erosion and outcrops of sedimentary rocks are exposed as a result of wind-driven denudation.

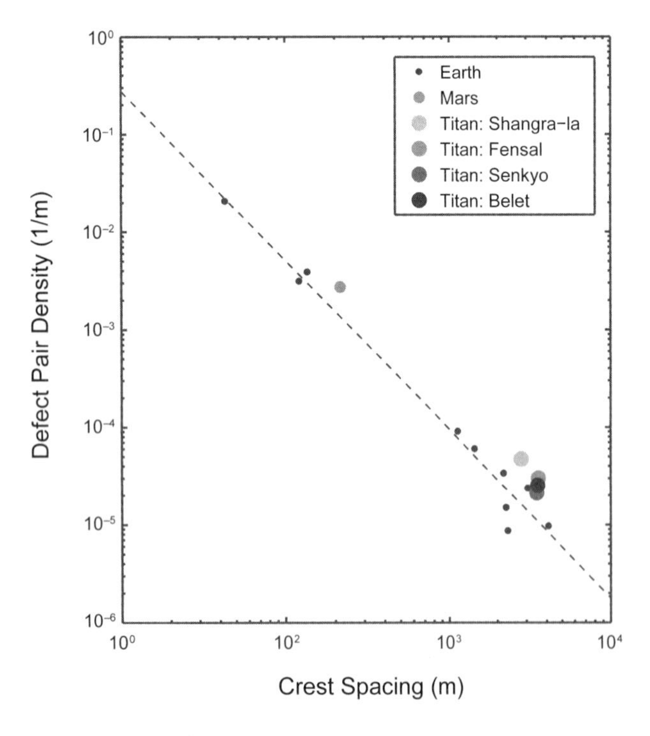

Fig. 14. Logarithmic plot of dune crest spacing vs. defect density on Earth, Mars, and Titan. Defect density is the inverse of the average dune crest length. Note that the data points follow a similar power law relationship, irrespective of environmental conditions. Dune fields that do not follow this relationship are observed to be out of equilibrium with either the local wind regime or sediment availability.

Nevertheless, some sedimentary deposits were formed and deposited locally, whereas other, more substantial, deposits accumulated as vast sheets that can be correlated for at least hundreds of kilometers (*Grotzinger and Milliken,* 2012). Local deposits were formed in alluvial fan, deltaic, sublacustrine fan, and lacustrine environments, in addition to canyon/valley-filling fluvial deposits of likely catastrophic origin. These former deposits suggest a more gradualistic view of sedimentation on Mars, perhaps even involving meteoric precipitation (e.g., *Mangold et al.,* 2004; *Hynek et al.,* 2010), which complements the Viking-era view of Mars that highlighted great catastrophic floods of more regional extent that were formed during sudden outbursts of groundwater. Regionally extensive sedimentary deposits have less obvious origins, but significant quantities of hydrated sulfate minerals suggest that some of these may have formed as lacustrine evaporites, particularly in the Valles Marineris canyon/chasma network of basins. Others may have involved aeolian reworking of previously deposited sulfates, or perhaps aqueous (groundwater) alteration of previously deposited basaltic sediments (*Grotzinger et al.,* 2005; *McLennan et al.,* 2005). The Burns formation at Meridiani Planum is the type example of this sort of deposit having been studied in detail by the Opportunity rover at multiple locations. At every locality visited it features largely or entirely aeolian traction deposits representing ancient sand dunes and draas, but punctuated by brief episodes of fluvial transport (*Grotzinger et al.,* 2005, 2006; *Metz et al.,* 2009; *Hayes et al.,* 2011b; *Edgar et al.,* 2012). Only recently has evidence been presented for a possible lacustrine mudstone, significantly different in grain size and diagenesis from other facies of the Burns formation (*Edgar,* 2013). The Burns formation has been alternatively interpreted to represent an impact or volcanic base surge deposit (*Knauth et al.,* 2005; *McCollom and Hynek,* 2005); however, this model is inconsistent with available data that support aeolian/fluvial interpretations (*McLennan and Grotzinger,* 2008; *Fralick et al.,* 2012).

Another major type of regionally extensive sedimentary deposit occurs as meter-scale stratification with highly rhythmic organization (Fig. 3b) and is well developed over much of Terra Arabia. The significant lateral continuity of relatively thin beds, their distribution over broadly defined highs as well as lows, and the lack of strong spectral absorption features suggests these rocks may be "duststones," formed by settling of fine particulates from the ancient martian atmosphere (*Bridges and Muhs,* 2012; *Grotzinger and Milliken,* 2012). Sedimentation may have been driven by climate cycles in the Milankovitch bands (*Lewis et al.,* 2008). The most ancient sedimentary deposits on Mars may be dominated by stacked impact-generated debris sheets, similar to the Moon, and possibly including impact melt sheets.

In the past decade it has become clear that Mars has sediment sources and sinks distributed at both local and regional scale (see review in *Grotzinger and Milliken,* 2012). One of the longest largest source-to-sink systems on Mars has its source in the Argyre region and extends to the northern lowlands, inferred to be the terminal sink. The total length of the outflow system, from where Uzboi Vallis begins at the rim of Argyre impact basin to where the outflow system discharges onto the northern plains at Chryse Planitia, is over 4000 km. It is hypothesized that fine-grained sediments transported through this system accumulated along the fringe of the northern lowlands. A series of valleys and basins, known as the Uzboi-Ladon-Morova network, link Argyre with the northern lowlands and trapped sediments in alluvial and lacustrine environments along the way to the northern lowlands (Fig. 15). The basins along this network are interpreted to once have been lakes, created when large impacts blocked drainage networks (*Grant and Wilson,* 2011). These impact basins are regarded as significant local sinks for sediments, including clay minerals (*Milliken and Bish,* 2010). The Ladon basin is the largest potential lake basin (Fig. 15b), although others are also developed, including the craters Holden and Eberswalde (Fig. 15c). Clay-bearing strata of possible lacustrine origin are preserved in the Ladon basin (Fig. 15d) and the Eberswalde-Holden Craters (Figs. 15a,b).

An enduring and controversial hypothesis asserts that the northern lowlands of Mars may represent a terminal sediment sink of a scale comparable to the ocean basins on Earth; the flattest surface in the solar system may reflect sedimentation at the bottom of a vanished ocean basin on Mars (*Aharonson et al.,* 1998). On the other hand, this very flat surface could owe its origin to accumulation of atmospheric dust over billions of years, in a manner perhaps similar to what has been suggested for the duststones of Terra Arabia (*Bridges and Muhs,* 2012; *Grotzinger and Milliken,* 2012). Other suggestions include lacustrine and lava flow deposits on the northern plains. In any case, the low-lying northern plains would still be a terminal sediment sink involving settling of suspended sediment from a fluid, either liquid water or air. Furthermore, in these scenarios, it is likely that the terminal sink represented by the northern lowlands is in part filled — at least at its margins — with sediments derived from the Argyre-to-Chryse fluvial system.

6.3. Titan

The evidence for extensive atmospheric precipitation, standing bodies of liquid, and channelized flow makes Titan the only extraterrestrial body in the solar system with an active hydrologic cycle. Similar to Mars, Titan has never been subject to plate tectonics and does not, at least currently, have a surface ocean to act as terminal sediment sink. Unlike Mars and Venus, Titan is subject to active pluvial, fluvial, and lacustrine processes. Titan's methane-based hydrologic cycle drives a source-to-sink sediment transport system that generates morphologic features, including dunes, rivers, and lakes, that are strikingly similar to morphologies found on Earth and Mars (Figs. 2e,f). Unlike Earth and Mars, however, there is no obvious recycling method for these deposits. Unfortunately, the low resolution at which we have observed Titan's surface (~300 m) makes it difficult to study any stratigraphic record deposited by this system.

Sediment on Titan can consist of both water-ice and organic particulates. Sediments can be generated through fluvial and aeolian erosion of bedrock, or by impact and possible cryovolcanic processes that generate sediment that is then relocated through fluvial and aeolian transport (section 3). The anomalously high radar backscatter of a subset of the fluvial valleys observed by the Cassini RADAR (predominantly those found at equatorial latitudes) has been interpreted as resulting from the deposition of water-ice cobbles that act as semitransparent Mie scatterers (*Le Gall et al.,* 2010), and the low backscatter and infrared signature of Titan's dunes suggests transport and deposition of sand-sized organic sediment; it is possible that these deposits form stratigraphic units, although their extent and thickness are not currently known. Organic sediment consists of higher-order hydrocarbons (C_xH_y) and nitriles ($C_xH_yN_z$) that are generated through methane photolysis in Titan's upper atmosphere (*Lorenz and Lunine,* 1996). This material settles through the atmosphere to coat the surface in a veneer of organic material. The thickness of this organic layer is dependent on both the specific photolysis pathways active in the atmosphere, which determine the flux of sediment, and the persistence of methane, which determines the timescales for production. At its present rate, photolysis would deplete the atmospheric methane content in ~26 m.y. (*Yung et al.,* 1984). Without methane, which is Titan's primary greenhouse gas, the nitrogen-based atmosphere would partially collapse and likely shut down active sediment transport and deposition (*Lorenz et al.,* 1997). The rate and nature (episodic vs. continuous) of methane outgassing from Titan's interior (e.g., *Tobie et al.,* 2006; *Nelson et al.,* 2009; *Moore and Pappalardo,* 2011; *Choukroun and Sotin,* 2012) will control source terms in the sediment transport system.

The largest sedimentary repositories on Titan are vast equatorial dune fields found within 30° of the equator. Dunes encompass ~12.5% of Titan's surface (*Le Gall et al.,* 2012). If the dunes are composed of organics, they represent the largest reservoir of hydrocarbon on Titan, estimated at 250,000 km^3, assuming dunes heights of 100 m (*Le Gall et al.,* 2012). However, this volume is still smaller than most estimates of photolysis production, assuming a continuous supply of methane, which suggests the presence of an additional unseen hydrocarbon reservoir, overestimation of photolysis production, or discontinuous (i.e., episodic) production (*Lorenz et al.,* 2008). Unseen hydrocarbon reservoirs could include a subsurface ocean stored in the porous upper crust, large crustal clathrate reservoirs, or the incorporation of thick sand sheet deposits in the amorphous plains that dominate the midlatitudes (*Lopes et al.,* 2010). The detection of radiogenic ^{40}Ar suggests geologically recent interaction between Titan's interior and atmosphere (*Niemann et al.,* 2005) and has motivated the development of episodic methane outgassing models that predict methane-free periods in Titan's climatic history (e.g., *Tobie et al.,* 2006).

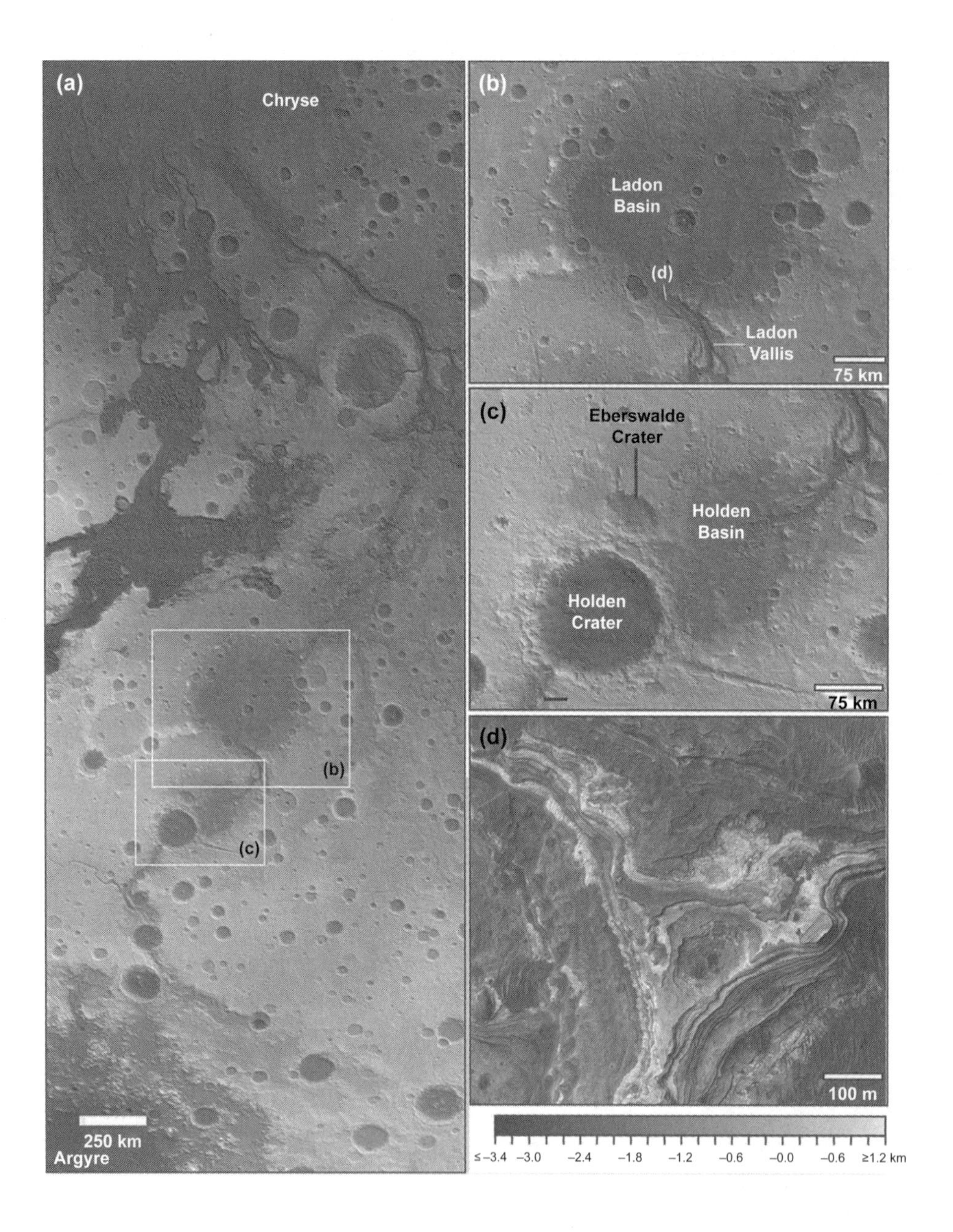

Fig. 15. The Uzboi-Ladon-Morava (ULM) source-to-sink system. **(a)** MOLA elevation shows the low elevation of the ULM system relative to the surrounding ancient highlands. The ULM system was once throughgoing from the northern rim of the Argyre impact basin to Chryse Planitia at the edge of the northern lowlands, but it was blocked at one point by the formation of Holden Crater [see **(c)**]. Much of the channel system exhibits evidence for sedimentary fill and multiple episodes of incision. **(b)** Close-up of the ancient Ladon Basin, which has been filled with sediment of an unknown thickness; numerous fluvial features are found along the heavily degraded rim and wall. **(c)** Close-up of Eberswalde Crater (see also Fig. 2d), Holden Crater, and the more ancient Holden Basin; breach of Uzboi Vallis into Holden Crater is visible in the southwest portion of the image; note that the formation of Holden Crater blocked water flow through the larger ULM system. **(d)** Close-up of finely stratified sedimentary outcrops exposed in a terrace in Ladon Vallis where it enters Ladon Basin along the southern rim; location marked by "**(d)**" in **(b)**; HiRISE image PSP_006637_1590. After *Grotzinger and Milliken* (2012).

In the polar regions, lakebeds are the primary sediment sinks. The presence of these features, which include both small (lakes) and large (seas or mare) standing bodies of liquid as well as empty depressions (empty lakes), imply mechanisms that can generate topographic lows at multiple scales (*Hayes et al.,* 2008). These processes can include, but are certainly not limited to, tectonic subsidence and sediment loading at large scales, and dissolution or solution processes at all scales. Impact and volcanic processes are also possible, but are generally not considered the primary mechanism for basin formation (see the recent review by *Aharonson et al.,* 2013).

Lacustrine features are observed both connected to and independent of regional and local drainage networks. The largest lakes, or seas, in the north, consisting of Ligeia, Kraken, and Punga Mare, have shorelines that include shallow bays with drowned river valleys (Figs. 16a,b). Such drowned valleys indicate that once well-drained upland landscapes became swamped by rising fluid levels where sedimentation did not keep pace with the rising relative base level. The absence of sedimentary deposits at the terminus of these valleys provides further evidence for relatively rapid increases in liquid level. At Ligeia Mare, nearby fluvial systems include a drainage network that runs parallel to the sea shoreline (Fig. 16a). As the modern slope is perpendicular to the shoreline, such an orientation may suggest a large-scale, and apparently nonuniform, down-dropping of topography associated with sea formation. As base level rose, the shores of Ligeia Mare encroached on this network, and in some cases shoreline retreat appears to be erasing topography as evidenced by the abrupt termination of large river valleys at the shoreline.

While the areal abundance of lakes in the south polar region is ~25 times smaller than in the north, albedo-dark regions observed in the lowest elevations of the south may represent preserved paleoshorelines of past seas that could have once encompassed areas comparable to their northern counterparts (*Hayes et al.,* 2011a). The asymmetric distribution of filled lakes between Titan's north and south polar regions has been attributed to the 25% increase in peak solar insolation between northern and southern summer, arising from Titan's current orbital configuration (*Schneider et al.,* 2012). Analogous to Croll-Milankovich cycles on Earth, the orbital parameters that determine Titan's insolation pattern cycle on timescales of tens of thousands of years (*Aharonson et al.,* 2009), potentially driving variations in polar sediment transport and deposition.

In addition to the dunes and mare, smaller-scale depositional morphologies have also been observed on Titan. RADAR-bright alluvial fans (e.g., Elvigar Flumina) (*Lorenz et al.,* 2008), lobate structures interpreted as deltas at Ontario Lacus (*Wall et al.,* 2010), isolated lacustrine environments (*Hayes et al.,* 2008), and other fluvial deposits (*Burr et al.,* 2013) have been documented. These deposits appear to be associated with local deposition and may or may not represent terminal sediment sinks. Empty lake basins are a few hundred meters deep and have significantly different microwave scattering characteristics than surrounding terrain (*Hayes et al.,* 2008). These empty lakebeds are also

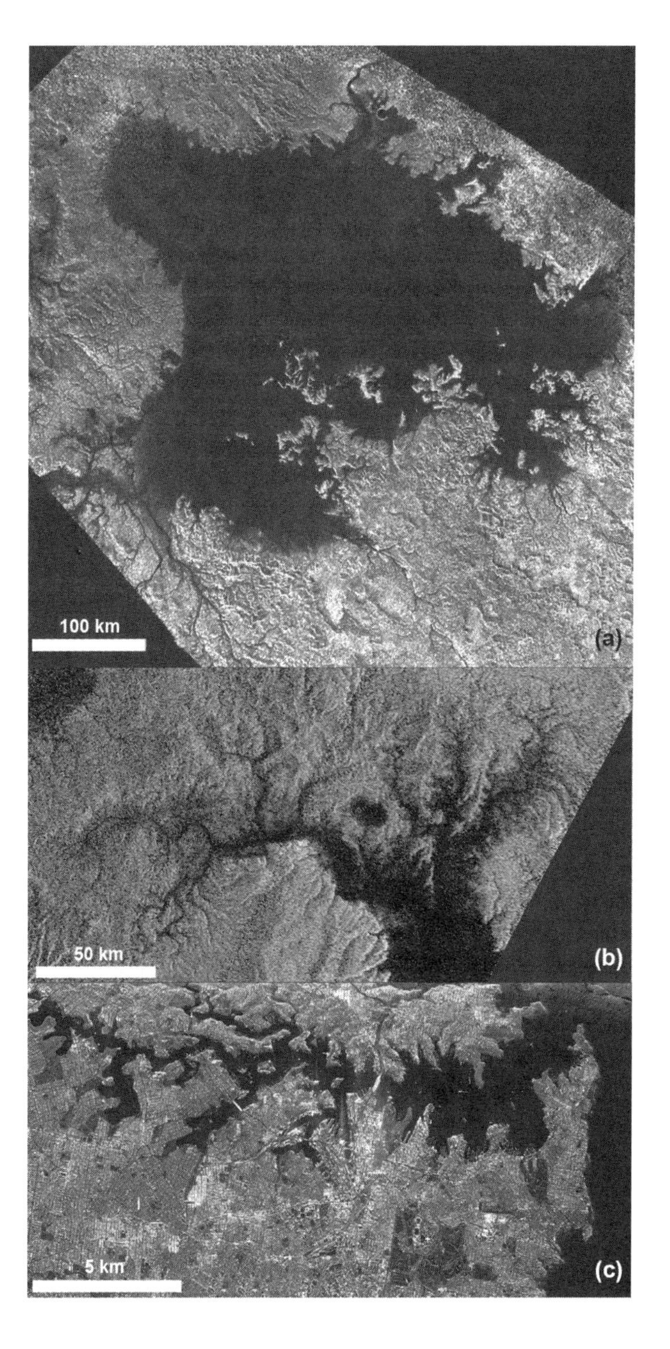

Fig. 16. (a) Polar stereographic mosaic of Titan's Ligeia Mare using Synthetic Aperture Radar (SAR) data acquired between February 2006 and April 2007. The crenulated shoreline morphology and shallow bays are characteristic of drowned topography. North is approximately up. **(b),(c)** Examples of drowned river valleys on Titan [**(b)**] and Earth [**(c)**]; **(b)** Ku-Band SAR image of a drowned-river valley on Titan acquired in April 2007, centered at (76°N, 80°E) near the westernmost shores of Kraken Mare; **(c)** C-Band AIRSAR image of Georges River, a drowned river valley located in Sydney, Australia. AIRSAR data was obtained from the Alaska Satellite Facility (*www.asf.alaska.edu*).

observed to be bright at 5 μm, which *Barnes et al.* (2011) interpreted as evidence of organic evaporite deposits. The equatorial basins Hotei and Tui Regio, which are similarly bright at 5 μm, have also been identified as candidate paleolakes partially filled in by evaporites (*Moore and Howard,* 2010). Lobate flow-like morphologies observed in both regions, however, have been used to suggest a cryovolcanic origin for these deposits (e.g., *Lopes et al.,* 2013). Regardless, it is unclear how these precipitants could be transported out of the closed basins in which they are observed, suggesting that they may act as terminal sediment sinks. Spectrally distinct rings observed around Ontario Lacus further suggest that evaporites are deposited on Titan (*Barnes et al.,* 2011).

Titan also may have intermediate sediment repositories analogous to mountainous soils found on Earth. *Turtle et al.* (2011) observed a darkening of ~500,000 km^2 of terrain following a large equatorial cloudburst in September 2010. This was interpreted as evidence of surface wetting and/or ponding from a precipitation event. In the year following the storm event, the region returned to its original albedo, with the exception of a specific morphologic feature — VIMS Equatorial Bright Terrain (*Soderblom et al.,* 2007) — which appeared to brighten prior to returning to their original spectrum (*Barnes et al.,* 2013). This brightening was interpreted as evidence for volatile precipitation (*Barnes et al.,* 2013).

6.4. Venus

The surface of Venus was entirely resurfaced by volcanism and related processes at about 750 Ma in a process that probably took on the order of <100 m.y. (see *Taylor and McLennan,* 2009, for a recent review). Accordingly, any evidence for whether or not Venus ever had an ancient stratigraphic record of sedimentary deposits is permanently obscured. There is no evidence to suggest a more recent (<750 Ma) stratigraphic record apart from the limited occurrence of surficial aeolian deposits.

7. CONCLUSIONS

1. Interactions between atmospheric and geologic processes influence the morphology and composition of planetary surfaces, and over the course of geologic time determine the historical evolution of the planet's surface environments including climate. The flow of currents of air, liquid water, and ice in turn transport sediments from sites of weathering and erosion where they are formed, to sites of deposition where they are stored, to create a stratigraphic record of layered sediments and sedimentary rocks.

2. This transfer of sediment and solute mass from source to sink plays a key role in the cycling of elements, water, and fractionation of biogeochemically important isotopes, as well as the formation of soils, groundwater reservoirs, masses of sediment, and the composition of the atmosphere. Models based on conservation of mass and simple transport laws show how the ancient rock record may preserve evidence of surface processes, including the influence of climate.

3. The form, intensity, and duration of weathering vary greatly and are influenced by surface mineralogy (e.g., basalt vs. granite vs. ice) and the presence, extent, and composition of atmosphere (e.g., S-cycle vs. C-cycle) and hydrosphere, which in turn are controlled by climate, the nature of tectonics (e.g., influencing topography), and impact history.

4. Chemical and physical weathering, which initiates reequilibration of minerals formed at high P-T to those stable under surface conditions, generates solute-bearing solutions (including brines) and promotes formation of sedimentary particles. Weathering influences all planetary surfaces including those on Earth, Mars, Venus, and Titan, occurring even on planetary bodies that lack atmospheres (albeit at very low rates).

5. Weathering on Earth is a result of chemical, physical, and biological processes operating on silica-rich continental crust. Mars differs in the apparent absence of biological influences, as well as the preponderance of basaltic crust. After Noachian time, weathering on Mars was likely dominated by low-pH and possibly briny fluids. Venus likely experiences limited but very heterogeneous weathering due to lack of surface water but extremes in atmospheric temperature, pressure, and density. There is no evidence for fluvial processes and the entire planet was resurfaced ~1 b.y. ago. Titan shows only trivial temperature variation at its surface, which limits physical weathering. Chemical weathering of water-ice substrates are likely caused by interactions with alkane-rich fluids. Sediments also are contributed by the novel mechanism of methane photolysis in the atmosphere, and aggregate to form particles at the surface.

6. For low-gradient river systems and aeolian transport we have robust semi-empirical equations for the inception of sediment motion, sediment fluxes, and equilibrium bed states. These formulae have been thoroughly tested such that their application to planetary bodies with different gravitational acceleration, particle densities, fluid densities, and fluid viscosities is well founded. For many cases (e.g., channelized fluid flow and transport of coarse particles) fluid deformation is governed by turbulence and molecular viscosity ceases to play a role in transport dynamics. These equations can be used in combination with observations of active sediment transport and bed states (e.g., existence of dunes), or records of past processes preserved within the sedimentary strata, to reconstruct wind and river flow rates, which in turn can constrain atmospheric processes.

7. Such robust, quantitative theory is lacking for a number of first-order processes in sediment production and transport including rock abrasion by rivers and wind, plucking of jointed rock, waterfall and knickpoint retreat, groundwater seepage erosion, and sediment transport and equilibrium bed states in steep mountain channels. For these processes experiments and field observations on Earth are needed to build a quantitative understanding that can be applied to other planets.

8. Aeolian sediment transport and deposition is the only sedimentologically important process that is shared in common by Earth, Mars, Venus, and Titan. The threshold for grain entrainment by the wind is significantly lower for Venus and Titan as compared to Earth and Mars. Earth and Titan are the wettest planets and so have correspondingly small volumes of aeolian sediment as compared to Mars, which has vast deposits of both modern and ancient wind-blown deposits. Venus shows very limited aeolian deposits, perhaps because of the low production rate of saltatable sediment particles, which in turn limits supply and availability.

9. Processes that operate at planetary surfaces have the potential to record a history of their evolution in the form of sedimentary rocks. Earth's record is strongly influenced by plate tectonics and biological processes, and deposition occurs predominantly in marine settings, and with extensive recycling.

10. Mars preserves a perhaps simpler but surprisingly vast record, likely dominated by aeolian and, earlier on, impact processes; almost all sediments represent primary deposits with very limited recycling. Interaction with water was more extensive prior to ~ 3 b.y. ago, and even then it is possible that transport and alteration occurred dominantly in the presence of brines, with limited water/rock interaction. The sedimentary rock record bears evidence for global transitions from clay, to sulfate, to anhydrous ion oxide-bearing sediments.

11. Titan does not have an observed stratigraphic record, yet patterns of sedimentation involving water-ice and organic particles suggest one may exist. Broad alluvial valleys are covered with bimodally sized sediments suggesting that deposits underlie these surfaces. Organic sediment generated through methane photolysis in the upper atmosphere may coat the surface of Titan, and its accumulation rate depends on methane degassing from the planet's interior. This organic sediment is reworked to form aeolian bedforms. Lakebeds in polar regions may represent significant sinks for Titan's sediments. The current undersupply of sediment relative to rising liquid base levels in the north polar region results in flooded drainage networks; however, the south polar basins may preserve paleo-shorelines and sedimentary deposits.

Acknowledgments. Funding was provided by the NASA Astrobiology Institute and Mars Science Laboratory Project to J.P.G., and by the Miller Institute for Basic Research in Science to A.G.H. S.M.M. was supported by NASA Mars Data Analysis and Mars Fundamental Research grants. J. Griffes and K. Stack helped with construction of Figs. 1, 2, and 4. Thanks to D. Burr, L. Sklar, and two anonymous reviewers for comments on an earlier version of this chapter.

REFERENCES

Aharonson O., Zuber M. T., Neumann G. A., and Head J. W. (1998) Mars: Northern hemisphere slopes and slope distributions. *Geophys. Res. Lett., 25,* 4413–4416.

Aharonson O., Hayes A. G., Lunine J., Lorenz R., Allison M., and Elachi C. (2009) An asymmetric distribution of lakes on Titan as a possible consequence of orbital forcing. *Nature Geosci., 2,* 851–854.

Aharonson O., Hayes A. G., Lopes R. M. C., Lucas A., Hayne P., and Perron T. (2013) Titan's surface geology. In *Titan: Surface, Atmosphere, and Magnetosphere* (I. Mueller-Wodarg, ed.), Cambridge Univ., Cambridge. ISBN 13:978052119926.

Allen P. A. (2008) From landscapes into geological history. *Nature, 105,* 274–276.

Almeida M. P., Parteli E. J. R., Andrade J. S., and Herrmann H. J. (2008) Giant saltation on Mars. *Proc. Natl. Acad. Sci., 105,* 6222–6226.

Amundson R., Richter D. D., Humphyreys G. S., Jobbagy E. C., and Gaillardet J. (2007) Coupling between biota and Earth materials in the critical zone. *Elements, 3,* 327–332.

Anderson R. S. (1986) Erosion profiles due to particles entrained by wind: Application of an eolian sediment-transport model. *Geol. Soc. Am. Bull., 97,* 1270–1278.

Anderson R. S. (1990) Eolian ripples as examples of self-organization in geomorphological systems. *Earth Sci. Rev., 29,* 77–96.

Anderson R. S. and Haff P. K. (1988) Simulation of eolian saltation. *Science, 241,* 820–823.

Anderson S. P., von Blanckenburg F., and White A. F. (2007) Physical and chemical controls on the critical zone. *Elements, 3,* 315–319.

Bagnold R. A. (1941) *The Physics of Blown Sand and Desert Dunes.* Methuen, London. 256 pp.

Bagnold R. A. (1966) *An Approach to the Sediment Transport Problem from General Physics.* U.S. Geological Survey, Washington, DC. 37 pp.

Baker V. R. (1982) *The Channels of Mars.* Univ. of Texas, Austin. 198 pp.

Baker V. R., Komatsu G., Parker T. J., Gulick V. C., Kargel J. S., and Lewis J. S. (1992) Channels and valleys on Venus: Preliminary analysis of Magellan data. *J. Geophys. Res., 97,* 13421–13444.

Barnes J. W., Bow J., Schwartz J., Brown R. H., Soderblom J., Hayes A. G., Le Mouelic S., Rodriguez S., Sotin C., Jaumann R., Stephan K., Soderblom L. A., Clark R. N., Buratti B. J., Baines K. H., and Nicholson P. D. (2011) Organic sedimentary deposits in Titan's dry lake beds. *Icarus, 216(1),* 136–140.

Barnes J., Burrati B., Turtle E., Bow J., Dalba P., Perry J., Brown R., Rodriguez S., Le Mouelic S., Baines K., Sotin C., Lorenz R., Malaska M., McCord T., Clark R., Jaumann R., Hayne P., Nicholson P., Soderblom J., and Soderblom L. (2013) Precipitation-induced surface brightenings seen on Titan by Cassini VIMS and ISS. *Planetary Sci., 2:1.*

Basilevsky A. T. and Head J. W. (1993) The surface of Venus. *Rep. Prog. Phys., 96,* 1699–1734.

Basilevsky A. T., Kuzmin R. O., Nikolaeva O. V., Pronin A. A., Ronca L. B., Avduevsky V. S., Uspensky G. R., Choremukhina Z. P., Semenchenko V. V., and Ladygin V. M. (1985) The surface of Venus as revealed by the Venera landings: Part II. *Geol. Soc. Am. Bull., 96,* 137–144.

Berner E. K., Berner R. A., and Moulton K. L. (2004) Plants and mineral weathering: Past and present. In *Treatise on Geochemistry, Vol. 5: Surface and Ground Water, Weathering , and Soils* (J. I. Drever, ed.), pp. 169–188. Elsevier-Pergamon, Oxford.

Berner R. A. (1995) Chemical weathering and its effect on atmospheric CO_2 and climate. *Rev. Mineral., 31,* 565–583.

Beveridge C., Kocurek R., Ewing R. C., Lancaster N., Morthekai

P., Singhvi A. K., and Mahan A. (2006) Development of spatially diverse and complex dune-field patterns: Gran Desierto Dune Field, Sonora, Mexico. *Sedimentology, 53,* 1391–1409.

Bibring J.-P., Langevin Y., Mustard J. F., Poulet F., Arvidson R., Gendrin A., Gondet B., Mangold N., Pinet P., Forget F., and the OMEGA Team (2006) Global mineralogical and aqueous Mars history derived from OMEGA/Mars Express data. *Science, 312,* 400–404.

Bibring J.-P., Arvidson R. E., Gendrin A., Gondet B., Langevin Y., Le Mouelic S., Mangold N., Morris R. V., Mustard J. F., Poulet F., Quantin C., and Sotin C. (2007) Coupled ferric oxides and sulfates on the martian surface. *Science, 317,* 1206–1210.

Bourke M. C., Lancaster N., Fenton L. K., Parteli E. J. R., Zimbelman J. R., and Radebaugh J. (2010) Extraterrestrial dunes: An introduction to the special issue on planetary dune systems. *Geomorphology, 121,* 1–14.

Brantley S. L. and Chen Y. (1995) Chemical weathering rates of pyroxenes and amphiboles. *Rev. Mineral., 31,* 119–172.

Brantley S. L. and Lebedeva M. (2011) Learning to read the chemistry of regolith to understand the critical zone. *Annu. Rev. Earth Planet. Sci., 39,* 387–416.

Brantley S. L., Goldhaber M. B., and Ragnarsdottir K. V. (2007) Crossing disciplines and scales to understand the critical cone. *Elements, 3,* 307–314.

Bridges N. and Muhs D. (2012) Duststones on Mars: Source, transport, deposition and erosion. In *Sedimentary Geology of Mars* (J. Grotzinger and R. Milliken, eds.), pp. 169–182. SEPM Spec. Publ. 102, Society for Sedimentary Geology, Tulsa.

Bridges N. T., Ayoub F., Avouac J.-P., Leprince S., Lucas A., and Mattson S. (2012) Earth-like sand fluxes on Mars. *Nature, 485,* 339–342.

Brownlie W. R. (1983) Flow depth in sand-bed channels. *J. Hydraul. Eng.-ASCE, 109,* 959–990.

Burr D. M. (2010) Palaeoflood-generating mechanisms on Earth, Mars, and Titan. *Global Planet. Change, 70,* 5–13.

Burr D. M., McEwen A. S., and Sakimoto S. E. H. (2002) Recent aqueous floods from the Cerberus Fossae, Mars. *Geophys. Res. Lett., 29,* 1–4.

Burr D. M., Emery J. P., Lorenz R. D., Collins G. C., and Carling P. A. (2006) Sediment transport by liquid surficial flow: Application to Titan. *Icarus, 181,* 235–242.

Burr D. M., Perron J. T., Lamb M. P., Irwin R. P., Collins G., Howard A. D., Sklar L. S., Moore J. M., Ádámkovics M., Baker V., Drummond S. A., and Black B. A. (2013) Fluvial feature on Titan: Insights from morphology and modeling. *Geol. Soc. Am. Bull., B30612.1,* 1–23, DOI: 10.1130/B30612.1.

Burns R. G. (1993) Rates and mechanisms of chemical weathering of ferromagnesian silicate minerals on Mars. *Geochim. Cosmochim. Acta, 57,* 4555–4574.

Canfield D. E. (2004) The evolution of the Earth surface sulfur reservoir. *Am. J. Sci., 304,* 839–861.

Carr M. H. and Clow G. D. (1981) Martian channels and valleys — their characteristics, distribution, and age. *Icarus, 48,* 91–117.

Carr M. H. and Head J. W. (2010) Geologic history of Mars. *Earth Planet. Sci. Lett., 294,* 185–203.

Chabert J. and Chauvin J. L. (1963) Formation de Dunes et de Rides Dans Les Modeles Fluviaux. *Bull. Cen. Rech. Ess. Chatau, 4,* 31–51.

Chan M., Yonkee W. A., Netoff D. I., Seiler W. M., and Ford R. L. (2008) Polygonal cracks in bedrock on Earth and Mars: Implications for weathering. *Icarus, 194,* 65–71.

Chepil W. S. and Woodruff N. P. (1963) The physics of wind erosion and its control. *Adv. Agronomy, 15,* 211–302.

Chorover J., Kretzschmar R., Garcia-Pichel F., and Sparks D. L. (2007) Soil biogeochemical processes within the critical zone. *Elements, 3,* 321–326.

Choukroun M. and Sotin C. (2012) Is Titan's shape caused by its meteorology and carbon cycle? *Geophy. Res. Lett., 39,* L04201, DOI: 10.1029/2011GL050747,2012.

Chow V. T. (1959) *Open Channel Hydraulics.* McGraw Hill, New York. 680 pp.

Clark B. C. (1993) Geochemical components in martian soil. *Geochim. Cosmochim. Acta, 57,* 4575–4581.

Clark B. C., Baird A. K., Rose H. J., Toulmin P., Keil K., Castro A. J., Kelliher W. C., Rowe C. D., and Evans P. H. (1976) Inorganic analyses of martian surface samples at Viking landing sites. *Science, 194,* 1283–1288.

Clark B. C., Morris R. V., McLennan S. M., Gellert R., Jolliff B., Knoll A., Lowenstein T. K., and 19 colleagues (2005) Chemistry and mineralogy of outcrops at Meridiani Planum. *Earth Planet. Sci. Lett., 240,* 73–94.

Collins G. C. (2005) Relative rates of fluvial bedrock incision on Titan and Earth. *Geophys. Res. Lett., 32,* L22202, DOI: 10.1029/2005GL024551.

Craddock R. A. (2011) Aeolian processes on the terrestrial planets: Recent observations and future focus. *Progr. Phys. Geog., 36,* 110–124.

Curtis C. D. (1976) Stability of minerals in surface weathering reactions: A general thermochemical approach. *Earth Surf. Proc., 1,* 63–70.

Derickson D., Kocurek G., Ewing R. C., and Bristow C. (2007) Origin of a complex and spatially diverse dune-field pattern, Algodones, southeastern California. *Geomorphology, 99,* 186–204.

Deschamps F., Mousis O., Sanchez-Valle C., and Lunine J. I. (2010) The role of methanol in the crystallization of Titan's primordial ocean. *Astrophys. J., 724,* 887–894.

Dietrich W. E. (1982) Settling velocity of natural particles. *Water Res. Res., 18,* 1615–1626.

Dorn R. I. (2009) The role of climatic change in alluvial fan development. In *Geomorphology of Desert Environments* (A. D. Parsons and A. D. Abrahams, eds), pp. 723–742. Springer, London.

Easterbrook D. J. (1999) *Surface Processes and Landforms, 2nd edition.* Prentice-Hall, New Jersey. 546 pp.

Edgar L. A., Grotzinger J. P., Hayes A. G., Rubin D. M., Squyres S. W., Bell J. F., and Herkenhoff K. E. (2012) Stratigraphic architecture of bedrock reference section, Victoria Crater, Meridiani Planum, Mars. In *Sedimentary Geology of Mars* (J. Grotzinger and R. Milliken, eds.), pp. 195–210. SEPM Spec. Publ. 102, Society for Sedimentary Geology, Tulsa.

Edgar L. A. (2013) Identifying and interpreting stratification in sedimentary rocks on Mars: Insight from rover and orbital observations and terrestrial field analogs. Ph.D. thesis, California Institute of Technology, Pasadena.

Ehlmann B. L., Mustard J. F., Murchie S. L., Bibring J.-P., Meunier A., Fraeman A. A., and Langevin Y. (2011) Subsurface water and clay mineral formation during the early history of Mars. *Nature, 479,* 53–60.

Ewing R. C., Peyret A.-P. B., Kocurek G., and Bourke M. (2010) Dune field pattern formation and recent transporting winds in the Olympia Undae Dune Field, north polar region of Mars. *J. Geophys. Res.–Planets, 115,* 8005, DOI:

10.1029/2009JE003526.

Farmer A. M. (1993) The effects of dust on vegetation: A review. *Environ. Poll., 79,* 63–75.

Flemings P. B. and Grotzinger J. P. (1996) STRATA: Freeware for analyzing classic stratigraphic problems. *GSA Today, 6,* 1–7.

Florensky C. P., Ronca L. B., Basilevsky A. T., Burba G. A., Nikolaeva O. V., Pronin A. A., Trackhtman A. M., Volkov V. P., and Zasetsky V. V. (1977) The surface of Venus as revealed by Soviet Venera 9 and 10. *Geol. Soc. Am. Bull., 88,* 1537–1545.

Foley C. N., Economou T. E., Clayton R. N., Brückner J., Dreibus G., Rieder R., and Wänke H. (2008) Martian surface chemistry: APXS results from the Pathfinder landing site. In *The Martian Surface: Composition, Mineralogy, and Physical Properties* (J. F. Bell III, ed.), pp. 35–57. Cambridge Univ., Cambridge.

Fralick P., Grotzinger J., and Edgar L. (2012) Potential recognition of accretionary lapilli in distal impact deposits on Mars: A facies analog provided by the 1.85 Ga Sudbury impact deposit. In *Sedimentary Geology of Mars* (J. Grotzinger and R. Milliken, eds.), pp. 211–228. SEPM Spec. Publ. 102, Society for Sedimentary Geology, Tulsa.

Fryberger S. G. and Alhbrandt T. S. (1979) Mechanisms for the formation of eolian sand seas. *Z. Geomorphol., 23,* 440–460.

Gaffey M. J. (2010) Space weathering and the interpretation of asteroid reflectance spectra. *Icarus, 209,* 564–574.

Garcia M. H. (2007) *Sedimentation Engineering: Process, Measurement, Modeling, and Practice*. American Society of Civil Engineers, Reston, Virginia. 1132 pp.

Garrels R. M. and Mackenzie F. T. (1971) *Evolution of Sedimentary Rocks.* Norton, New York. 397 pp.

Glotch T. D., Bandfield J. L., Christensen P. R., Calvin W. M., McLennan S. M., Clark B. C., Rogers A. D., and Squyres S. W. (2006) Mineralogy of the light-toned outcrop rock at Meridiani Planum as seen by the Miniature Thermal Emission Spectrometer and implications for its formation. *J. Geophys. Res., 111,* E12S03, DOI: 10.1029/2005JE002672.

Golden D. C., Ming D. W., Morris R. V., and Mertzmann S. M. (2005) Laboratory-simulated acid-sulfate weathering of basaltic materials: Implications for formation of sulfates at Meridiani Planum and Gusev crater, Mars. *J. Geophys. Res., 110,* E12S07, DOI: 10.1029/2005JE002451.

Goldhammer R. K., Dunn D. A., and Hardie L. A. (1987) High frequency glacio-eustatic sealevel oscillations with Milankovitch characteristics recorded in Middle Triassic platform carbonates in northern Italy. *Am. J. Sci., 287,* 853–892.

Goldich S. S. (1938) A study in rock weathering. *J. Geol., 46,* 17–58.

Golombek M. P., Grant J. A., Crumpler L. S., Greeley R., Arvidson R. E., Bell J. F. III, Weitz C. M., Sullivan R., Christensen P. R., Soderblom L. A., and Squyres S. W. (2006) Erosion rates at the Mars Exploration Rover landing sites and long-term climate change on Mars. *J. Geophys. Res., 111,* E12S10, DOI: 10.1029/2006JE002754.

Grant J. A. and Wilson S. A. (2011) Late alluvial fan formation in southern Margaritifer Terra, Mars. *Geophys. Res. Lett.*, 38, L08201, DOI: 10.1029/2011GL046844.

Greeley R. and Iverson J. D. (1987) *Wind as a Geological Process on Earth, Mars, Venus, and Titan.* Cambridge Univ., Cambridge. 348 pp.

Greeley R., Bender K., Thomas P. E., Schubert G., Limonadi D., and Weitz C. M. (1995) Wind-related features and processes on Venus — Summary of Magellan results. *Icarus, 115,* 399–420.

Greeley R., Leach R. N., Williams S. H., Krinsley D. H., Marshall J. R., White B. R., and Pollack J. B. (1982) Rate of wind abrasion on Mars. *J. Geophys. Res., 87,* 10009–10024.

Grotzinger J. (2009) Beyond water on Mars. *Nature Geosci., 2,* 231–233.

Grotzinger J. P. and Milliken R. E. (2012) Sedimentary rock record of Mars: Distributions, origins, and global stratigraphy. In *Sedimentary Geology of Mars* (J. Grotzinger and R. Milliken, eds.), pp. 1–48. SEPM Spec. Publ. 102, Society for Sedimentary Geology, Tulsa.

Grotzinger J. P., Bell J. F. III, Calvin W., Clark B. C., Fike D., Golombek M., Greeley R., Herkenhoff K. E., Jolliff B., Knoll A. H., Malin M., McLennan S. M., and 6 colleagues (2005) Stratigraphy and sedimentology of a dry to wet eolian depositional system, Burns formation, Meridiani Planum, Mars. *Earth Planet. Sci. Lett., 240,* 11–72.

Grotzinger J., Bell J., Herkenhoff K., Johnson J., Knoll A., McCartney E., McLennan S., Metz J., Moore J., Squyres S., Sullivan R., Aharonson O., Arvidson R., Jolliff B., Golombek M., Lewis K., Parker T., and Soderblom J. (2006) Sedimentary textures formed by aqueous processes, Erebus crater, Meridiani Planum, Mars. *Geology, 34,* 1085–1088.

Grotzinger J., Beaty D., Dromart G., Gupta S., Harris M., Hurowitz J., Kocurek G., McLennan S., Milliken R., Ori G. G., and Sumner D. (2011) Mars sedimentary geology: Key concepts and outstanding questions. *Astrobiology, 11,* 77–87.

Grotzinger J. P., Crisp J., Vasavada A. R., Anderson R. C., Baker C. J., Barry R., Blake D. F., Conrad P., Edgett K. S., Ferdowski B., Gellert R., Golombek M., Gomez-Elvira J., Hassler D. M., Jandura L., Litvak M., Mahaffy P., Maki J., Meyer M., Malin M. C., Mitrofanov. I., Simmonds J. J., Vaniman D., Welch R. V., and Wiens R. (2012) Mars Science Laboratory Mission and science investigation. *Space Sci. Rev., 170,* 5–56.

Halevy I., Zuber M. T., and Schrag D. P. (2007) A sulfur dioxide climate feedback on early Mars. *Science, 318,* 1903–1907.

Hancock G. S., Anderson R. S., and Whipple K. X. (1998) Beyond power: Bedrock river incision process and form. In *Rivers Over Rock: Fluvial Processes in Bedrock Channels* (J. Tinkler and E. Wohl, eds.), pp. 35–60. AGU Geophys. Mono. 107, American Geophysical Union, Washington, DC.

Haskin L. A., Wang A., Jolliff B. L., McSween H. Y., Clark B. C., Des Marais D. J., McLennan S. M., and 23 colleagues (2005) Water alteration of rocks and soils on Mars at the Spirit rover site in Gusev crater. *Nature, 436,* 66–69.

Hayes A. G., Aharonson O., Callahan P., Elachi C., Gim Y., Kirk R., Lewis K., Lopes R., Lorenz R., Lunine J., Mitchell K., Mitri G., Stofan E., and Wall S. (2008) Hydrocarbon lakes on Titan: Distribution and interaction with a porous regolith. *Geophys. Res. Lett., 35,* L09204, DOI: 10.1029.2008GL033409.

Hayes A. G., Aharonson O., Lunine J. I., Kirk R. L., Zebker H. A., Wye L. C., Lorenz R. D., Turtle E. P., Paillou P., Mitri G., Wall S. D., Stofan E. R., Mitchell K. L., Elachi C., and the Cassini RADAR Team (2011a) Transient surface liquid in Titan's polar regions from Cassini. *Icarus, 211,* 655–671.

Hayes A. G., Grotzinger J. P., Edgar L. A., Squyres S. W., Watters W. A., and Sohl-Dickstein J. (2011b) Reconstruction of eolian bed forms and paleocurrents from cross-bedded strata at Victoria Crater, Meridiani Planum, Mars. *J. Geophys. Res., 16,* E00F21, DOI: 10.1029/2010JE003688.

Hayes A. G., Ewing R. C., Lucas A., McCormick C., Troy S., and Balard C. (2012) Determining the timescales of the dune forming winds on Titan. In *Third International Planetary Dunes Workshop: Remote Sensing and Data Analysis of Planetary*

Dunes, pp. 46–47. LPI Contribution No. 1673, Lunar and Planetary Institute, Houston.

Hayes A. G., Dietrich W. E. , Kirk R. L., Turtle E. P., Barnes J. W., Lucas A., and Mitchell K. L. (2013) Morphologic analysis of polar landscape evolution on Titan. In *Lunar Planet. Sci. XLIV,* Abstract #2000. Lunar and Planetary Institute, Houston.

Hecht M. H., Kounaves S. P., Quinn R. C., West S. J., Young S. M. M., Ming D. W., Catling D. C., Clark B. C., Boynton W. V., Hoffman J., DeFlores L. P., Gospodinova K., Kapit J., and Smith P. H. (2009) Detection of perchlorate and the soluble chemistry of martian soil at the Phoenix site. *Science, 325,* 64–67.

Hoffman P. F., Kaufman A. J., Halverson G. P., and Schrag D. P. (1998) A Neoproterozoic snowball Earth. *Science, 281,* 1342–1346.

Howard A. D., Dietrich W. E., and Seidl M. A. (1994) Modelling fluvial erosion on regional and continental scales. *J. Geophys. Res.–Solid Earth, 99,* 13971–13986.

Hurowitz J. A. and McLennan S. M. (2007) A ~3.5 Ga record of water-limited, acidic conditions on Mars. *Earth Planet. Sci. Lett., 260,* 432–443.

Hurowitz J. A., McLennan S. M., Lindsley D. H., and Schoonen M. A. A. (2005) Experimental epithermal alteration of synthetic Los Angeles meteorite: Implications for the origin of martian soils and the identification of hydrothermal sites on Mars. *J. Geophys. Res.–Planets, 110,* E07002, DOI: 10.1029/2004JE002391.

Hurowitz J. A., McLennan S. M., Tosca N. J., Arvidson R. E., Michalski J. R., Ming D. W., Schöder C., and Squyres S. W. (2006) In-situ and experimental evidence for acidic weathering on Mars. *J. Geophys. Res.–Planets, 111,* E02S19, DOI: 10.1029/2005JE002515.

Hynek B. M., Beach M., and Hoke M. R. T. (2010) Updated global map of martian valley networks and implications for climate anc hydrologic processes. *J. Geophys. Res.–Planets, 115,* E09008, DOI: 10.1029/2009JE003548.

Irwin R. P., Howard A. D., and Maxwell T. A. (2004) Geomorphology of Ma'adim Vallis, Mars, and associated paleolake basins. *J. Geophys. Res.–Planets, 109,* E12009, DOI: 10.1029/2004JE002287.

Iversen J. D. and White B. R. (1982) Saltation threshold on Earth, Mars and Venus. *Sedimentology, 29,* 111–119.

Iversen J. D., Greeley R., Marshall J. R., and Pollack J. B. (1987) Aeolian saltation threshold: The effect of density ratio. *Sedimentology, 34,* 699–706.

Janssen M. A., Lorenz R. D., West R., Paganelli F., Lopes R. M., Kirk R. L., Elachi C., Wall S. D., Johnson W. T. K., and Anderson Y. (2009) Titan's surface at 2.2-cm wavelength imaged by the Cassini RADAR radiometer: Calibration and first results. *Icarus, 200,* 222–239.

Jennings D. E., Flasar F. M., Kunde V. G., Samuelson R. E., Pearl J. C., Nixon C. A., Carlson R. C., Mamoutkine A. A., Brasunas J. C., Guandique E., Achterberg R. K., Bjoraker G. L., Romani P. N., Segura M. E., Albright S. A., Elliott M. H., Tingley J. S., Calcutt S., Coustenis A., and Courtin R. (2009) Titan's surface brightness temperatures. *Astrophys. J. Lett., 691,* L103.

Jennings D. E., Cottini V., Nixon C. A., Flasar F. M., Kunde V. G., Samuelson R. E., Romani P. N., Hesman B. E., Carlson R. C., Gorius N. J. P., Coustenis A., and Tokano T. (2011) Seasonal changes in Titan's surface temperatures. *Astrophys. J. Lett., 737,* L15, DOI: 10.1088/2041-8205/737/1/L15.

Jerolmack D. J. and Paola C. (2010) Shredding of environmental signals by sediment transport. *Geophys. Res. Lett., 37,* L19401, DOI: 10.1029/2010GL044638.

Johnson N. M. and Fegley B. Jr. (2002) Experimental studies of atmosphere-surface interactions on Venus. *Adv. Space Res., 29,* 233–241.

Johnsson M. J. (1993) The system controlling the composition of clastic sediments. In *Processes Controlling the Composition of Clastic Sediments* (M. J. Johnsson and A. Basu, eds.), pp. 1–19. Geol. Soc. Am. Spec. Paper 284, Geological Society of America, Boulder.

Johnsson M. J. and Meade R. H. (1990) Chemical weathering of fluvial sediments during alluvial storage: The Macuapanim Island pont bar, Solimoes River, Brazil. *J. Sed. Petrol., 60,* 827–842.

Jordan D. W. (1954) The adhesion of dust particles. *British J. Appl. Phys., 5,*194–198.

Kenyon P. M. and Turcotte D. L. (1985) Morphology of a delta prograding by bulk sediment transport. *Geol. Soc. Am. Bull., 96,* 1457–1465.

King P. L. and McLennan S. M. (2010) Sulfur on Mars. *Elements, 6,* 107–112.

Knauth L. P., Burt D. M., and Wohletz K. H. (2005) Impact origin of sediments at the Opportunity landing site on Mars. *Nature, 438,* 1123–1128.

Kocurek G. and Lancaster N. (1999) Aeolian system sediment state: Theory and Mojave desert Kelso dune field example. *Sedimentology, 46,* 505–515.

Kocurek G., Ewing R. C., and Mohrig D. (2010) How do bedform patterns arise? New views on the role of bedform interactions within a set of boundary conditions. *Earth Surf. Proc. Landforms, 35,* 51–63.

Komar P.D. (1980) Modes of sediment transport in channelized water flows with ramifications to the erosion of the martian outflow channels. *Icarus, 32,* 317–329.

Laity J. E. (2011) *Wind Erosion in Drylands,* pp. 539–568. John Wiley and Sons, Ltd.

Laity J. E. and Bridges N. T. (2009) Ventifacts on Earth and Mars: Analytical field, and laboratory studies supporting sand abrasion and windward feature development. *Geomorphology, 105,* 202–217.

Lamb M. P. and Dietrich W. E. (2009) The persistence of waterfalls in fractured rock. *Geol. Soc. Am. Bull., 121,* 1123–1134.

Lamb M. P. and Fonstad M. A. (2010) Rapid formation of a modern bedrock canyon by a single flood event. *Nature Geosci., 3,* 477–481.

Lamb M. P., Howard A. D., Johnson J., Whipple K. X., Dietrich W. E., and Perron J. T. (2006) Can springs cut canyons into rock? *J. Geophys. Res.–Planets, 111,* E07002, DOI: 10.1029/2005JE002663.

Lamb M. P., Dietrich W. E., Aciego S. M., DePaolo D. J., and Manga M. (2008a) Formation of Box Canyon, Idaho, by megaflood: Implications for seepage erosion on Earth and Mars. *Science, 320,* 1067–1070.

Lamb M. P., Dietrich W. E., and Sklar L. S. (2008b) A model for fluvial bedrock incision by impacting suspended and bed load sediment. *J. Geophys. Res.–Earth Surface, 113,* F03025, DOI: 10.1029/2007JF000915.

Lamb M. P., Dietrich W. E., and Venditti J. G. (2008c) Is the critical Shields stress for incipient sediment motion dependent on channel-bed slope? *J. Geophys. Res.–Earth Surface, 113,* F02008, DOI: 10.1029/2007JF000831.

Lamb M. P., Grotzinger J. P., Southard J. B., and Tosca N. J.

(2012) Were aqueous ripples on Mars formed by flowing brines? In *Sedimentary Geology of Mars* (J. Grotzinger and R. Milliken, eds.), pp. 139–150. SEPM Spec. Publ. 102, Society for Sedimentary Geology, Tulsa.

Leask H. J. and Wilson L. (2003) Heating and cooling of rocks on Mars: Consequences for weathering. In *Lunar Planet. Sci. XXXIV,* Abstract #1804, Lunar and Planetary Institute, Houston.

Le Gall A., Janssen M. A., Paillou P., Lorenz R. D., and Wall S. D. (2010) Radar-bright channels on Titan. *Icarus, 207,* 948–958.

Le Gall A., Hayes A. G., Ewing R., Janssen M. A., Radebaugh J., Savage C., Encrenaz P., and the Cassini RADAR Team (2012) Latitudinal and altitudinal controls of Titan's dune field morphometry. *Icarus, 217,* 231–242.

Lewis K. W., Aharonson O., Grotzinger J. P., Kirk R. L., McKewan A. S., and Suer T.-A. (2008) Quasi-periodic bedding in the sedimentary rock record of Mars. *Science, 322,* 1532–1535.

Litwin K. L., Zygielbaum B. R., Polito P. J., Sklar L. S., and Collins G. C. (2012) Influence of temperature, composition, and grain size on the tensile failure of water ice: Implications for erosion on Titan. *J. Geophys. Res.–Planets, 117,* E08013, DOI: 10.1029/2012JE004101.

Lopes R. M. C., Stofan E. R., Peckyno R., Radebaugh J., Mitchell K. L., Mitri G., Wood C. A., Kirk R. L., Wall S. D., Lunine J. I., Hayes A., Lorenz R., Farr T., Wye L., Craig J., Ollerenshaw R. J., Janssen M., Legall A., Paganelli F., West R., Stiles B., Callahan P., Anderson Y., Valora P., Soderblom L., and the Cassini RADAR Team (2010) Distribution and interplay of geologic processes on Titan from Cassini radar data. *Icarus, 205,* 540–558.

Lopes R. M. C., Kirk R. L., Michell K. L., LeGall A., Barnes J. W., Hayes A. G., Kargel J., Wye L., Radebaugh J., Stofan E. R., Janssen M. A., Neish C. D., Wall S. D., Wood C. A., Lunine J. I., and Malaska M. (2013) Cryovolcanism on Titan: New results from the Cassini RADAR and VIMS. *J. Geophys. Res.–Planets,* in press.

Lorenz R. D. and Lunine J. I. (1996) Erosion on Titan: Past and present. *Icarus, 122,* 79–91.

Lorenz R. D. and Lunine J. I. (2005) Titan's surface before Cassini. *Planet. Space Sci., 53,* 557–576.

Lorenz R. D., McKay C. P., and Lunine J. I. (1997) Photochemically induced collapse of Titan's atmosphere. *Science, 275,* 642–644.

Lorenz R. D., Wall S., Radebaugh J., Boubin G., Reffet E., Janssen M., Stofan E., Lopes R., Kirk R., Elachi C., Lunine J., Mitchell K., Paganelli F., Soderblom L., Wood C., Wye L., Zebker H., Anderson Y., Ostro S., Allison M., Boehmer R., Callahan P., Encrenaz P., Ori G. G., Francescetti G., Gim Y., Hamilton G., Hensley S., Johnson W., Kelleher K., Muhleman D., Picardi G., Posa F., Roth L., Seu R., Shaffer S., Stiles B., Vetrella S., Flamini E., and West R. (2006) The sand seas of Titan: Cassini RADAR observations of longitudinal dunes. *Science, 312,* 724–727.

Lorenz R. D., Lopes R. M., Paganelli F., Lunine J. I., Kirk R. L., Mitchell K. L., Soderblom L. A., Stofan E. R., Ori G., Myers M., Miyamoto H., Radebaugh J., Stiles B., Wall S. D., and Wood C. A. (2008) Fluvial channels on Titan: Initial Cassini RADAR observations. *Planet. Space Sci., 56,* 1132–1144.

Lorenz R. D., Newman J. I., and Lunine J. I. (2010) Threshold of wave generation on Titan's lakes and seas: Effects of viscosity and implications for Cassini observations. *Icarus, 207,* 932–937.

Lunine J. I. and Atreya S. (2008) The methane cycle on Titan. *Nature Geosci., 1,* 159–164.

Malin M. C. (1974) Salt weathering on Mars. *J. Geophys. Res., 79,* 3888–3894.

Malin M. C. (1986) Rates of geomorphic modification in ice-free areas, southern Victoria Land, Antarctica. *Antarctic J. U.S., 20,* 18–21.

Mangold N., Quantin C., Ansan V., Delacourt C., and Allemand P. (2004) Evidence for precipitation on Mars from dendritic valleys in the Valles Marineris Area. *Science, 305,* 78–81.

McCollom T. M. and B. M. Hynek (2005) A volcanic environment for bedrock diagenesis at Meridiani Planum on Mars. *Nature, 438,* 1129–1131.

McGlynn I. O., Fedo C. M., and McSween H. Y. (2012) Soil mineralogy at the Mars Exploration Rover landing sites: An assessment of the competing roles of physical sorting and chemical weathering. *J. Geophys. Res., 117,* E01006, DOI: 10.1029/2010JE003712.

McKee E. D. (1979) *A Study of Global Sand Seas.* U.S. Geol. Surv. Prof. Paper 1052, U.S. Govt. Printing Office, Washington, DC. 429 pp.

McLennan S. M. (2003) Sedimentary silica on Mars. *Geology, 31,* 315–318.

McLennan S. M. (2012) Geochemistry of sedimentary processes on Mars. In *Sedimentary Geology of Mars* (J. P. Grotzinger and R. E. Milliken, eds.) pp. 119–138. SEPM Spec. Publ. 102, Society for Sedimentary Geology, Tulsa.

McLennan S. M. and Grotzinger J. P. (2008) The sedimentary rock cycle of Mars. In *The Martian Surface: Composition, Mineralogy, and Physical Properties* (J. F. Bell III, ed.), pp. 541–577. Cambridge Univ., Cambridge.

McLennan S. M., Bell J. F. III, Calvin W., and 29 colleagues (2005) Provenance and diagenesis of the evaporite-bearing Burns formation, Meridiani Planum, Mars. *Earth Planet. Sci. Lett., 240,* 95–121.

McSween H. Y., McGlynn I. O., and Rogers A. D. (2010) Determining the modal mineralogy of martian soils. *J. Geophys. Res., 115,* E00F12, DOI: 10.1029/2010JE003582.

Melosh H. J. (2011) *Planetary Surface Processes.* Cambridge Univ., New York. 500 pp.

Metz J. M., Grotzinger J., Rubin D. M., Lewis K. W., Squyres S. W., and Bell J. F. III (2009) Sulfate-rich eolian and wet interdune deposits, Erebus crater, Meridiani Planum, Mars. *J. Sed. Res., 79,* 247–264.

Miller M. C., McCave I. N., and Komar P. D. (1977) Threshold of sediment motion under unidirectional currents. *Sedimentology, 41,* 883–903.

Milliken R.E. and Bish D.L. (2010) Sources and sinks of clay minerals on Mars. *Philos. Mag., 90,* 2293–2308.

Milliken R. E., Grotzinger J. P., and Thompson B. J. (2010) Paleoclimate of Mars as captured by the stratigraphic record in Gale Crater. *Geophys. Res. Lett., 37,* L04201, DOI: 10.1029/2009GL041870.

Millot R., Gaillardet J., Dupré B., and Allègre C. J. (2002) The global control of silicate weathering rates and the coupling with physical erosion: New insights from rivers of the Canadian Shield. *Earth Planet. Sci. Lett., 196,* 83–98.

Ming D. W., Mittlefehldt D. W., Morris R. V., Golden D. C., Gellert R., Yen A., Clark B. C., Squyres S. W., Farrand W. H., Ruff S. W., Arvidson R. E., Klingelhöfer G., McSween H. Y., Rodionov D. S., Schröder C., de Souza P. A., and Wang A. (2006) Geochemical and mineralogical indicators for aqueous processes in the Columbia Hills of Gusev crater, Mars. *J.*

Geophys. Res., 111, E02S12, DOI: 10.1029/2005JE002560.

Mitchell K. L. and Malaska M. (2011) Karst on Titan. In *First International Planetary Caves Workshop: Implications for Astrobiology, Climate, Detection, and Exploration,* p. 15. LPI Contribution No. 1640, Lunar and Planetary Institute, Houston.

Mitchell K. L., Kargel J. S., Wood C. A., Radebaugh J., Lopes R. M. C., Lunine J. I., Stofan E. R., Kirk R. L., and the Cassini RADAR Team (2007) Titan's crater lakes: Caldera vs. karst. In *Lunar Planet. Sci. XXXVIII,* Abstract #2064. Lunar and Planetary Institute, Houston.

Montgomery D. R. and Buffington J. M. (1997) Channel-reach morphology in mountain drainage basins. *Geol. Soc. Am. Bull., 109,* 596–611.

Moore J. M and Howard A. D. (2010) Are the basins of Titan's Hotei Regio and Tui Regio sites of former low latitude seas? *Geophys. Res. Lett., 37,* L22205, DOI: 10.1029/2010GL046753.

Moore J. M. and Pappalardo R. T. (2011) Titan: An exogenic world? *Icarus, 212,* 790–806.

Morris R. V., Ruff S. W., Gellert R., Ming D. W., Arvidson R. E., Clark B. C., Golden D. C., Siebach K., Klingelhöfer G., Schroder C., Fleischer I., Yen A. S., and Squyres S. W. (2010) Identification of carbonate-rich outcrops on Mars by the Spirit rover. *Science, 329,* 421–424.

Murchie S. L., Mustard J. F., Ehlmann B. L., Milliken R. E., Bishop J. L., McKeown N. K., Noe Dobrea E. Z., Seelos F. P., Buczkowski D. L., Wiseman S. M., Arvidson R. E., Wray J. J., Swayze G., Clark R. N., Des Marais D. J., McEwen A. S., and Bibring J.-P. (2009) A synthesis of martian aqueous mineralogy after 1 Mars year of observations from the Mars Reconnaissance Orbiter. *J. Geophys. Res.–Planets, 114,* E00D06, DOI: 10.1029/2009JE003342.

Navarre-Sitchler A. and Brantley S. (2007) Basalt weathering across scales. *Earth Planet. Sci. Lett., 261,* 321–334.

Neish C. D., Kirk R. L., Lorenz R. D., Bray V., Schenk P., Stiles B., Turtle E. P., Mitchell K., and Hayes A. G. (2013) Crater topography on Titan: Implications for landscape evolution. *Icarus, 223(1),* 82–90, DOI: 10.1016/j.icarus.2012.11.030.

Nelson R. M., Kamp L. W., Matson D. L., Irwin P. G. J., Baines K. H., Boryta M. D., Leader F. E., Jaumann R., Smythe W. D., Sotin C., Clark R. N., Cruikshank D. P., Drossart P., Pearl J. C., Hapke B. W., Lunine J., Combes M., Bellucci G., Bibring J.-P., Capaccioni F., Cerroni P., Coradini A., Formisano V., Filacchione G., Langevin R. Y., McCord T. B., Mennella V., Nicholson P. D., and Sicardy B. (2009) Saturn's Titan: Surface change, ammonia, and implications for atmospheric and tectonic activity. *Icarus, 199,* 429–441.

Nesbitt H. W. (1997) Bacterial and inorganic weathering processes and weathering of crystalline rocks. In *Biological-Mineralogical Interactions* (J. M. McIntosh and L. A. Grout, eds.), pp. 113–142. Mineralogical Association of Canada Short Course Series 25.

Nesbitt H. W. (2003) Petrogenesis of siliciclastic sediments and sedimentary rocks. In *Geochemistry of Sediments and Sedimentary Rocks: Evolutionary Considerations to Mineral Deposit-Forming Environments* (D. R. Lentz, ed.), pp. 39–51. Geol. Assoc. Canada GeoText, Vol. 4.

Nesbitt H. W. and Markovics G. (1997) Weathering of granodioritic crust, long-term storage of elements in weathering profiles, and petrogenesis of siliciclastic sediments. *Geochim. Cosmochim. Acta, 61,* 1653–1670.

Nesbitt H. W. and Wilson R. E. (1992) Recent chemical weathering of basalts. *Am. J. Sci., 292,* 740–777.

Nesbitt H. W. and Young G. M. (1984) Prediction of some weathering trends of plutonic and volcanic rocks based on thermodynamic and kinetic considerations. *Geochim. Cosmochim. Acta, 48,* 1523–1534.

Nickling W. G. and Neuman C. M. (2009) *Aeolian Sediment Transport,* pp. 539–568. Springer, Berlin.

Niemann H. B., Atreya S. K., Bauer S. J., Carignan G. R., Demick J. E., Frost R. L., Gautier D., Haberman J. A., Harpold D. N., Hunten D. M., Israel G., Lunine J. I., Kasprzak W. T., Owen T. C., Paulkovich M., Raulin F., Raaen E., and Way S. H. (2005) The abundances of constituents of Titan's atmosphere from the GCMS instrument on the Huygens probe. *Nature, 438,* 779–784.

Nikuradse J. (1933) Stromungsgesetze in rauhen Rohren. *Forschung auf dem Gebiete des Ingenieurwesens, Forschungsheft 361.* VDI Verlag, Berlin, Germany (in German). (Translated in *Laws of Flow in Rough Pipes,* NACA TM 1292, National Advisory Committee for Aeronautics, Washington, DC, 1950).

Nino Y., Lopez F., and Garcia M. (2003) Threshold for particle entrainment into suspension. *Sedimentology, 50,* 247–263.

Nishimura K. and Hunt J. C. R. (2000) Saltation and incipient suspension above a at particle bed below a turbulent boundary layer. *J. Fluid Mech., 417,* 77–102.

Nittler L. R., Starr R. D., Weider S. Z., McCoy T. J., Boynton W. V., Ebel D. S., Ernst C. M., Evans L. G., Goldstein J. O., Hamara D. K., Lawrence D. J., McNutt R. L., Schlemm C. E., Solomon S. C., and Sprague A. L. (2011) The major-element composition of Mercury's surface from MESSENGER X-ray spectrometry. *Science, 333,* 1847–1850.

Oelkers E. H. and Schott J., eds. (2009) *Thermodynamics and Kinetics of Water-Rock Interaction.* Rev. Mineral. Geochem., Vol. 70. 569 pp.

O'Hara-Dhand K., Taylor R. L. S., Smalley I. J., Krinsley D. H., and Vita- Finzi C. (2010) Loess and dust on Earth and Mars: Particle generation by impact mechanisms. *Central European J. Geosci., 2,* 45–51.

Pain C. F., Clarke J. D. A., and Thomas M. (2007) Inversion of relief on Mars. *Icarus, 190,* 478–491.

Paola C. (2000) Quantitative models of sedimentary basin filling. *Sedimentology, 47, Suppl. 1,* 121–178.

Paola C. and Voller V. R. (2005) A generalized Exner equation for sediment mass balance. *J. Geophys. Res.–Earth Surface, 110,* F04014, DOI: 10.1029/2004JF000274.

Parker G. (1990) Surface-based bedload transport relation for gravel rivers. *J. Hydr. Res., 28,* 417–436.

Parker G. (1991) Selective sorting and abrasion of river gravel. II: Applications. *J. Hydr. Engin., 117,* 150–171.

Parker G., Wilcock P. R., Paola C., Dietrich W. E., and Pitlick J. (2007) Physical basis for quasi-universal relations describing bankfull hydraulic geometry of single-thread gravel bed rivers. *J. Geophys. Res.–Earth Surface, 112,* F04005, DOI: 10.1029/2006JF000549.

Pelletier J. D., Cline M., and Delong S. B. (2007) Desert pavement dynamics: Numerical modeling and field-based calibration. *Earth Surf. Proc. Landforms, 32,* 1913–1927.

Perron J. T., Lamb M. P., Koven C. D., Fung I. Y., Yager E., and Adamkovics M. (2006) Valley formation and methane precipitation rates on Titan. *J. Geophys. Res.–Planets, 111,* E11001, DOI: 10.1029/2005JE002602.

Pye K. (1987) *Aeolian Dust and Dust Deposits.* Academic, London.

Rubin D. M. and Hesp P. A. (2009) Multiple origins of linear dunes on Earth and Titan. *Nature Geosci., 2,* 653–658.

Rubin D. M. and Hunter R. E. (1987) Bedform alignment in directionally varying flows. *Science, 237,* 276–278.

Schlichting H. (1979) *Boundary-Layer Theory.* McGraw-Hill, New York. 817 pp.

Schneider T., Graves S. D. B., Schaller E. L., and Brown M. E. (2012) Polar methane accumulation and rainstorms on Titan from simulations of the methane cycle. *Nature, 481,* 58–61.

Schröder S. and Keller H. (2008) The reflectance spectrum of Titan's surface at the Huygens landing site determined by the descent imager/spectral radiometer. *Planet. Space Sci., 56,* 753–769.

Selivanov A. S., Gektin Yu. M., Naraeva M. K., Panfilov A. S., and Fokin A. B. (1982) Evolution of the Venera 13 imagery. *Sov. Astron. Lett., 8,* 235–237.

Shao Y. and Lu H. (2000) A simple expression for wind erosion threshold friction velocity. *J. Geophys. Res., 105,* 22437–22444.

Sharp R. P. and Malin M. C. (1975) Channels on Mars. *Geol. Soc. Am. Bull., 86,* 593–609.

Shields A. (1936) *Application of Similarity Principles and Turbulence Research to Bed-Load Movement.* U.S. Dept. of Agriculture Soil Conservation Service Cooperative Laboratory, California Institute of Technology, Pasadena.

Sklar L.S. (2012) Erodibility of Titan ice bedrock constrained by laboratory measurements of ice strength and erosion by sediment impacts. *AGU Fall Meeting,* Abstract #EP44A-06.

Sklar L. S. and Dietrich W. E. (2001) Sediment and rock strength controls on river incision into bedrock. *Geology, 29,* 1087–1090.

Sklar L. S. and Dietrich W. E. (2004) A mechanistic model for river incision into bedrock by saltating bed load. *Water Resour. Res., 40,* W06301, DOI: 10.1029/2012WR012267.

Soderblom L. A., Kirk R. L., Lunine J. I., Anderson J. A., Baines K. H., Barnes J. W., Barrett J. M., Brown R. H., Buratti B. J., Clark R. N., Cruikshank D. P., Elachi C., Janssen M. A., Jaumann R., Karkoschka E., Mouélic S. L., Lopes R. M., Lorenz R. D., McCord T. B., Nicholson P. D., Radebaugh J., Rizk B., Sotin C., Stofan E. R., Sucharski T. L., Tomasko M. G., and Wall S. D. (2007) Correlations between Cassini VIMS spectra and ADAR SAR images: Implications for Titan's surface composition and the character of the Huygens probe landing site. *Planet. Space Sci., 55,* 2025–2036.

Sotin C., Mielke R., Choukroun M., Neish C., Barmatz M., Castillo J., Lunine J., and Mitchell K. (2009) Ice-hydrocarbon interactions under Titan-like conditions: Implications for the carbon cycle on Titan. In *Lunar Planet. Sci. XL,* Abstract #2088. Lunar and Planetary Institute, Houston.

Southard J. B. (1991) Experimental determination of bed-form stability. *Annu. Rev. Earth Planet. Sci., 19,* 423–455.

Southard J. B. and Boguchwal L. A. (1990a) Bed configurations in steady unidirectional water flows. 2. Synthesis of flume data. *J. Sed. Petrol., 60,* 658–679.

Southard J. B. and Boguchwal L. A. (1990b) Bed configurations in steady unidirectional water flows. 3. Effects of temperature and gravity. *J. Sed. Petrol., 60,* 680–686.

Squyres S. W., Grotzinger J. P., Arvidson R. E., Bell J. F. III, Christensen P. R., Clark B. C., Crisp J. A., Farrand W. H., Herkenhoff K. E., Johnson J. R., Klingelhöfer G., Knoll A. H., McLennan S. M., and 5 colleagues (2004) In-situ evidence for an ancient aqueous environment on Mars. *Science, 306,* 1709–1714.

Stefannsson A. and Gislason S. (2001) Chemical weathering of basalts, southwest Iceland: Effect of rock crystallinity and secondary minerals on chemical fluxes to the ocean. *Am. J. Sci., 301,* 513–556.

Sullivan R., Arvidson R., Bell J. F., Gellert R., Golombek M., Greeley R., Herkenhoff K., Johnson J., Thompson S., Whelley P., and Wray J. (2008) Wind-driven particle mobility on Mars: Insights from Mars Exploration Rover observations at El Dorado and surroundings at Gusev Crater. *J. Geophys. Res., 113,* E06S07, DOI: 10.1029/2008JE003101.

Surkov Yu. A., Barsukov V. L., Moskalyeva L. P., Kharyukova V. P., and Kemurdzhian A. L. (1984) New data on the composition, structure, and properties of Venus rock obtained by Venera 13 and Venera 14. *Proc. Lunar Planet. Sci. Conf. 14th, Part 2, J. Geophys. Res., 89,* B393–B402.

Surkov Yu. A., Moskalyeva L. P., Kharyukova V. P., Dudin A. D., Smirnov G. G., and Zaitseva S. Ye. (1986) Venus rock composition at the Vega 2 landing site. *Proc. Lunar Planet. Sci. Conf. 17th, Part 1, J. Geophys. Res., 91,* E215–E218.

Surkov Yu. A., Kirnozov F. F., Glazov V. N., Dunchenko A. G., Tatsy L. P., and Sobornov O. P. (1987) Uranium, thorium, and potassium in the venusian rocks at the landing sites of Vega 1 and 2. *Proc. Lunar Planet. Sci. Conf. 17th, Part 2, J. Geophys. Res., 92,* E537–E540.

Taylor G. J., McLennan S. M., McSween H. Y., Wyatt M. B., and Lentz R. C. F. (2008) Implications of observed primary lithologies. In *The Martian Surface: Composition, Mineralogy, and Physical Properties* (J. F. Bell III, ed.), pp. 501–518. Cambridge Univ., Cambridge.

Taylor S. R. and McLennan S. M. (1985) *The Continental Crust: Its Composition and Evolution.* Blackwell, Oxford. 312 pp.

Taylor S. R. and McLennan S. M. (2009) *Planetary Crusts: Their Composition, Origin, and Evolution.* Cambridge Univ., Cambridge. 378 pp.

Tobie G., Lunine J., and Sotin C. (2006) Episodic outgassing as the origin of atmospheric methane on Titan. *Nature, 440,* 61–64.

Tomasko M. G., Archinal B., Becker T., Bézard B., Bushroe M., Combes M., Cook D., Coustenis A., de Bergh C., Dafoe L. E., Doose L., Douté S., Eibl A., Engel S., Gliem F., Grieger B., Holso K., Howington-Kraus A., Karkoschka E., Keller H. U., Kirk R., Kramm R., Kuppers M., Lellouch E., Lemmon M., Lunine J., McFarlane E., Moores J., Prout M., Rizk B., Rosiek M., Ruffer P., Schroeder S., Schmitt B., See C., Smith P., Soderblom L., Thomas N., and West R. (2005) Rain, wind, and haze during the Huygens probe's descent to Titan's surface. *Nature, 438,* 765–778.

Tosca N. J. and McLennan S. M. (2006) Chemical divides and evaporite assemblages on Mars. *Earth Planet. Sci. Lett., 241,* 21–31.

Tosca N. J., McLennan S. M., Lindsley D. H., and Schoonen M. A. A. (2004) Acid-sulfate weathering of synthetic martian basalt: The acid fog model revisited. *J. Geophys. Res., 109,* E05003, DOI: 10.1029/2003JE002218.

Tosca N. J., McLennan S. M., and 6 colleagues (2005) Geochemical modeling of evaporation processes on Mars: Insight from the sedimentary record at Meridiani Planum. *Earth Planet. Sci. Lett., 240,* 122–148.

Tosca N. J., McLennan S. M., Lamb M. P., and Grotzinger J. P. (2011) Physicochemical properties of concentrated martian surface waters. *J. Geophys. Res.–Planets, 116,* E05004, DOI: 10.1029/2010JE003700.

Tsoar H. and Pye K. (1987) Dust transport and the question of desert loess formation. *Sedimentology, 34,* 139–153.

Turtle E. P., Perry J. E., Hayes A. G., Lorenz R. D., Barnes J. W., McEwen A. S., West R. A., Ray T. L., Del Genio A. D., Barbara J. M., and Schaller E. L. (2011) Extensive and rapid surface changes near Titan's equator: Evidence for April showers? *Science, 18,* 331.

Ungar J. E. and Haff P. K. (1987) Steady state saltation in air. *Sedimentology, 34,* 289–299.

van den Berg J. H. and van Gelder A. (1993) A new bedform stability diagram, with emphasis on the transition of ripples to plane bed in flows over fine sand and silt. *Spec. Publ. Intern. Assoc. Sediment., 17,* 11–21.

van der Weijden C. H., and Pacheco F. A. L. (2003) Hydrochemistry, weathering and weathering rates on Madeira Island. *J. Hydrol., 283,* 122–145.

Veizer J. and Mackenzie F. T. (2003) Evolution of sedimentary rocks. In *Treatise on Geochemistry, Vol. 7: Sediments, Diagenesis, and Sedimentary Rocks* (F. T. Mackenzie, ed.), pp. 369–407. Elsevier-Pergamon, Oxford.

Viles H., Ehlmann B., Wilson C. F., Cebula T., Page M., and Bourke M. (2010) Simulating weathering of basalt on Mars and Earth by thermal cycling. *Geophys. Res. Lett., 37,* L18201, DOI: 10.1029/2010GL043522.

Wall S., Hayes A., Bristow C., Lorenz R., Stofan E., Lunine J., Le Gall A., Janssen M., Lopes R., Wye L., Soderblom L., Paillou P., Aharonson O., Zebker H., Farr T., Mitri G., Kirk R., Mitchell K., Notarnicola C., Casarano D., and Ventura B. (2010) Active shoreline of Ontario Lacus, Titan: A morphological study of the lake and its surroundings. *Geophys. Res. Lett., 37,* L05252, DOI: 10.1029/2009GL041821.

Warner N. H., Gupta S., Kim J. R., Lin S. Y., and Muller J. P. (2010) Retreat of a giant cataract in a long-lived (3.7–2.6 Ga) martian outflow channel. *Geology, 38,* 791–794.

Warren J. K. (2010) Evaporites through time: Tectonic, climatic, and eustatic controls in marine and nonmarine deposits. *Earth-Sci. Rev., 98,* 217–268.

Werner B. T. and Kocurek G. (1999) Bedform spacing from defect dynamics. *Geology, 27,* 727–730.

West A. J., Galy A., and Bickle M. (2005) Tectonic and climatic controls on silicate weathering. *Earth Planet. Sci. Lett., 235,* 211–228.

Whipple K. X. (2004) Bedrock rivers and the geomorphology of active orogens. *Annu. Rev. Earth Planet. Sci., 32,* 151–185.

Whipple K. X., Hancock G. S., and Anderson R. S. (2000) River incision into bedrock: Mechanics and relative efficacy of plucking, abrasion, and cavitation. *Geol. Soc. Am. Bull., 112,* 490–503.

White A. F. (1995) Chemical weathering rates of silicate minerals in soils. *Rev. Mineral., 31,* 407–461.

White A. F. and Brantley S. L., eds. (1995) *Chemical Weathering Rates of Silicate Minerals.* Rev. Mineral., Vol. 31. 583 pp.

White S. J. (1970) Plane bed thresholds of fine grained sediments. *Nature, 228,* 152–153.

Wilcock P. R. (1993) Critical shear-stress of natural sediments. *J. Hydrol. Eng., 119,* 491–505.

Wilcock P. R. and Crowe J. C. (2003) Surface-based transport model for mixed-size sediment. *J. Hydrol. Eng., 129,* 120–131.

Williams R. M. E. and Malin M. C. (2004) Evidence for late stage fluvial activity in Kasei Valles, Mars. *J. Geophys. Res., 109,* E06001, DOI: 10.1029/2003JE002178.

Wood C. A., Lorenz R., Kirk R., Lopes R., Mitchell K., and Stofan E. (2010) Impact craters on Titan. *Icarus, 206,* 334–344.

Yarger J., Lunine J. I., and Burke M. (1993) Calorimetric studies of the ammonia-water system with application to the outer solar system. *J. Geophys. Res., 98,* 13109–13117.

Yung Y. L., Allen M., and Pinto J. P. (1984) Photochemistry of the atmosphere of Titan — Comparison between model and observations. *Astrophys. J., 55,* 465–506.

O'Neill C., Lenardic A., Höink T., and Coltice N. (2013) Mantle convection and outgassing on terrestrial planets. In *Comparative Climatology of Terrestrial Planets* (S. J. Mackwell et al., eds.), pp. 473–486. Univ. of Arizona, Tucson, DOI: 10.2458/azu_uapress_9780816530595-ch19.

Mantle Convection and Outgassing on Terrestrial Planets

Craig O'Neill
Macquarie University

Adrian Lenardic and Tobias Höink
Rice University

Nicolas Coltice
École Normale Supérieure de Lyon

Argon-40 degassing from the mantle during volcanism places an important constraint on the tectonic evolution of a planet through time. While Earth has degassed approximately 50% of the ^{40}Ar produced over its history, Venus has only degassed ~24%. Here we explore the effect of tectonic style on ^{40}Ar degassing using numerical models of mantle convection. We find a strong sensitivity of mantle degassing rates to tectonic regime, but also to Rayleigh number and internal heating, such that stagnant-lid models may either degas far more or less than mobile-lid simulations, depending on the relative internal temperatures and lid thicknesses. Evolutionary models of degassing rates through time imply finite degassing efficiencies of terrestrial planets in any tectonic regime. Earth's canonical 50% ^{40}Ar degassing efficiency is in line with model predictions of a tectonically active planet over its lifetime, and Venus' low ^{40}Ar atmospheric concentration is consistent with planetary modes with long periods of tectonic quiescence.

1. INTRODUCTION

A fundamental question in planetary science is why two astronomically similar planets, like Venus and Earth, with similar mass, composition, and orbits (*Kaula,* 1999), might diverge down very different tectonic and atmospheric evolutionary paths. Geological evidence can give us insights into the tectonic regime of a planet through time, but is limited by the finite crustal record in Earth's deep past and the lack of geochronology, lithological constraints, or ancient exposed crust on Venus.

Noble gases, such as helium, argon, and xenon, can provide important constraints on the degassing history and atmospheric evolution of terrestrial planets. Due to its relative lightness, helium has a short residence in planetary atmospheres [around 0.5 m.y. for Earth (*Schubert et al.,* 2001)]. Helium-3/helium-4 ratios have been used to constrain contemporary degassing fluxes, although atmospheric loss of helium is poorly constrained (e.g., *Schubert et al.,* 2001).

Primordial, nonradiogenic noble gas isotopes (e.g., ^{36}Ar, ^{38}Ar, ^{132}Xe) are likely to be degassed during the earliest history of the planet, and for Earth-sized planets, the heavier isotopes may survive over planetary lifetimes. In contrast, radiogenic noble gases track the degassing history of a planet as it evolves. The two radiogenic noble gases of particular interest for planetary degassing evolution are ^{40}Ar and ^{129}Xe. Krypton-40 decays to both ^{40}Ar and ^{40}Ca with a half-life of 1.2 G.y. (*Pollack and Black,* 1982). Xenon-129 is produced from ^{129}I, with a half-life of 15.7 m.y., and thus records the earliest degassing record. Xenon-129 is retained in the atmosphere of Earth-sized planets, although not in the atmosphere of Mars (*Musselwhite et al.,* 1991).

1.1. Noble Gases and Primordial Atmospheres

The formation and loss of a primordial atmosphere is one of most critical factors in subsequent planetary evolution (*Lenardic et al.,* 2008). Early degassing, from the magma ocean and hot early mantle, probably expelled most nonsoluble noble gases, although *Shcheka and Keppler* (2012) make a case for argon and krypton retention, relative to xenon, in perovskite during magma ocean crystallization. During this degassing, high rates of hydrodynamic escape, intense meteorite impacts, and ultraviolet (UV) radiation from the early Sun probably resulted in the loss of a large fraction of the primordial atmosphere (*Pepin,* 1991), although the relative effects of these on Earth, Venus, and Mars are poorly constrained.

Impact erosion and hydrodynamic escape were probably both more significant loss mechanisms on early Mars than on either Earth or Venus, due to Mars' lower gravity (*Musselwhite et al.,* 1991). The high ^{129}Xe/^{132}Xe ratios in the atmosphere of Mars (2.4) compared to Earth (0.985) suggests large losses of primordial ^{132}Xe, as well as early produced ^{129}Xe (*Musselwhite et al.,* 1991). Musselwhite et al.

suggest that expulsion of the parent ^{129}I into a near-surface aqueous reservoir might effectively replenish atmospheric ^{129}Xe as bombardment losses waned. Mars' ^{40}Ar/^{36}Ar ratio is ~3000 ± 500 (*Turcotte and Schubert,* 1988), which is consistent with the loss of most of its primordial ^{36}Ar during early bombardment, and replenishment of ^{40}Ar since then.

Xenon isotope systematics indicate significant Xe loss during atmospheric degassing and early bombardment on Earth and Mars. Xenon is depleted relative to argon and krypton on Earth, which may be due to the higher solubility of argon and krypton in perovskite (*Shchenka and Keppler,* 2012). This would have allowed argon and krypton to survive early atmospheric loss, and their subsequent degassing would have concentrated them relative to xenon, which was largely lost. On Earth, while only about 2% of the ^{129}Xe produced by ^{129}I decay (half-life 15.7 m.y.) remains in the atmosphere-mantle system (*Coltice et al.,* 2009), approximately 30% of the $^{131-136}$Xe produced by ^{244}Pu (half-life 82 m.y.) survived — indicating a decreased degassing rate, but also a decreasing atmospheric loss rate over the first few hundred million years of Earth's history (*Coltice et al.,* 2009).

Pepin (1991) note that despite large uncertainties in the measurements (e.g., *von Zahn et al.,* 1983), Venus has nearly solar-like Ar:Kr:Xe ratios. Venus has ~80 times more primordial $^{36+38}$Ar than Earth (*Kaula,* 1999), although only around 3–6 times more krypton and xenon (*Pepin,* 1991). These high concentrations of noble gases imply Venus has retained a large fraction of its primordial atmosphere (*Pepin,* 1991), with only moderate hydrodynamic loss. *Cameron* (1983) suggests that Earth (and Mars) may have lost most of their primordial ^{36}Ar due to impact collisions — in particular, the Moon-forming impact (*Turcotte and Schubert,* 1988), and Turcotte and Schubert suggest the characteristics of Venus' noble gas profiles suggest it has avoided this scale of impact-induced atmospheric loss.

1.2. Argon-40 and Planetary Evolution

Argon-40 can play an important part in tracking the degassing history of a planet through time (*Kaula,* 1999). Argon-40 is produced by the decay of ^{40}K in the mantle or crust. As a noble gas, it behaves incompatibly, and is lost from the mantle during melting events (*Coltice et al.,* 2000). It diffuses from the crustal reservoir on a timescale on the order of ~1 G.y. (*Bender et al.,* 2008). As it is reasonably heavy and inert, it should be stable in the atmospheres of Earth and Venus over the lifetimes of these planets (*Kaula,* 1999), and therefore provides an important constraint on the melting and degassing history, and thus tectonic evolution, of a planet through time.

Assuming a chrondritic ^{40}K concentration for Earth, Earth's atmosphere possesses approximately half of the ^{40}Ar that should have been generated over its ~4.56-G.y. history [~6.6 × 10^{16} kg, vs. 13.4 × 10^{16} kg generated by ^{40}K decay (*Allegre et al.,* 1996)]. Previous modeling has suggested that plate tectonics is extremely efficient at degassing the mantle (*Davies,* 1999; *van Keken and Ballantine,* 1999), and so the question becomes, where is the missing ^{40}Ar? The crustal reservoir degasses on too short a timescale to harbor the required degree of ^{40}Ar, and has a reasonably well constrained ^{40}K abundance (*Bender et al.,* 2008). Historically, the lower mantle has been taken to represent a poorly degassed reservoir, although the required isolation has increasingly been considered unlikely (*Xie and Tackley,* 2004; *van Keken and Ballantine,* 1999; *Gonnermann and Mukhopadhyay,* 2009). Other suggestions include unprocessed reservoirs within the lower mantle, potassium in the core, or a nonchrondritic potassium concentration for Earth (e.g., *Davies,* 1999). *Sleep* (1979) explored the problem for an evolving Earth, and suggested a high concentration of potassium in the core to explain the argon deficit. More recently, *Watson et al.* (2007) suggested that argon may be more compatible than previously assumed; however, this result has provoked controversy (e.g., *Ballentine and Holland,* 2008; *Bouhifd et al.,* 2008). Alternatively, *Phipps Morgan* (1988) and *Xie and Tackley* (2004) have shown that there is an "ingrowth" period of ^{40}Ar, early in a planet's history, before it can be degassed. However, the most efficient degassing epoch is early in a planet's evolution, and so a temporal lag can be expected between ^{40}Ar production and its degassing into the atmosphere — which has quantitatively been shown as a possible explanation for the "hidden" reservoir (e.g., *Xie and Tackley,* 2004).

Venus' atmospheric argon was measured by three mass spectrometers and two gas chromatographs during the Pioneer mission, as discussed by *von Zahn et al.* (1983). *Kaula* (1999) presents a comprehensive summary of these observations and potassium abundance from the Venera mission samples, and concludes that the atmospheric abundance of ^{40}Ar of 1.4 ± 0.46 × 10^{16} kg indicates that Venus has lost ~24% of its radiogenic ^{40}Ar. The difference between Earth's and Venus' atmospheric ^{40}Ar cannot be attributed to differences in atmospheric loss, as non-radiogenic (i.e., primordial) ^{36}Ar is ~80 times more abundant in Venus' atmosphere than in Earth's (*Kaula,* 1999). Instead, it has been suggested that the differences in ^{40}Ar abundances between Earth and Venus may in fact reflect differences in either the degassing/volcanic history of these planets (*Kaula,* 1999) or fundamental differences in mantle viscosity (*Xie and Tackley,* 2004). Even if argon does not behave incompatibly in the mantle, this fundamental difference between Earth and Venus still merits explanation.

Mars has an absolute abundance of ^{40}Ar that is a factor of 16 ± 3 times less than that of Earth [i.e., 4.5 ± 1 × 10^{14} kg (*Pollack and Black,* 1982; *Turcotte and Schubert,* 1988)]. Mars has lost primordial ^{36}Ar — the Earth/Mars absolute abundance ratio for ^{36}Ar is 165 ± 45 (*Pollack and Black,* 1982), and Mars possesses ~20% more potassium relative to Earth, exacerbating the ^{40}Ar deficit (*Pollack and Black,* 1982; *Dreibus and Wanke,* 1985; *Kiefer and Li,* 2009). *Pollack and Black* (1982) and *Turcotte and Schubert* (1988) both suggest the limited degassing efficiency of Mars' current stagnant-lid tectonic regime may explain the ^{40}Ar atmospheric deficit.

The aim of this paper is to quantify the difference that

tectonic regime has on ^{40}Ar degassing efficiency in fully dynamic mantle convection simulations. We will then compare ^{40}Ar degassing predictions to proposed tectonic evolutionary histories of Earth and Venus, and postulate that the deficits and differences in atmospheric ^{40}Ar between these two planets may in fact be due to their individual tectonic evolution.

2. TECTONIC HISTORIES

The discovery of ancient Hadean zircons from Jack Hills in Western Australia (e.g., *Wilde et al.,* 2001), with evidence of water and differentiated continental-like crust, has led to the suggestion that plate tectonics has operated continuously since Hadean times (*Harrison et al.,* 2005). The alternative view is that little direct evidence for plate tectonics extends beyond ~1 Ga, and therefore is a recent phenomenon (*Stern,* 2008). Alternatively, *O'Neill et al.* (2007b) suggest, based on modeling results and paleomagnetic analysis, that Earth may have been in episodic overturn regime for much of the Precambrian, and only more recently (i.e., late Archaean to early Proterozoic) progressed into a modern plate tectonic regime — somewhat counterintuitive to traditional wisdom (cf. *Condie et al.,* 2009; *Silver and Behn,* 2008). This model has found support in ^{142}Nd analyses of Archaean rocks, suggesting on the basis of mantle mixing times that the Hadean/early Archaean was largely stagnant (*Debaille et al.,* 2009). The tectonic evolution of the Earth is a subject with a long history (e.g., *Condie and Pease,* 2008) that cannot be fully covered here. However, each of these scenarios has distinct implications for the ^{40}Ar degassing history of Earth.

The cratering record on Venus suggests a rapid burst of resurfacing around ~750 Ma (*Phillips and Hansen,* 1998) and very little surface activity since then, and it has been postulated that Venus may be in an episodic regime due to a lack of water and therefore higher fault strength (*Moresi and Solomatov,* 1998). Alternatively, a missing sublithospheric low-viscosity channel, akin to Earth's asthenosphere, can place an otherwise mobile-lid planet into the episodic regime (*Höink et al,* 2012). A nonequilibrium current state is supported by lithospheric thickness estimates based on surface deformation (*Brown and Grimm,* 1999), which suggest a gradual thickening through time — consistent with a past resurfacing event.

In contrast, an "equilibrium" model for Venus involving random volcanic resurfacing has also been posited to explain the surface age distribution (*Bjonnes et al.,* 2012; *Hauck et al.,* 1998; *Phillips et al.,* 1992). While recent volcanic resurfacing has been suggested from emissivity data (*Smrekar et al.,* 2010), the rates are rather low (~1 km^3 yr^{-1}) — consistent with endmember "equilibrium" resurfacing models, but not with outgassing required to match current SO_2 rates [0.2–11 km^3 yr^{-1} (*Fegley and Prinn,* 1989); 4.6–9.2 km^3 yr^{-1} (*Bullock and Grinspoon,* 2001)].

The effect of these differing evolutionary scenarios is that they have an impact on the atmospheric composition and temperature, which, in turn, may result in tectonic effects due to the coupling between the atmosphere and plate/mantle systems. An example was provided by *Solomon et al.* (1999), who showed that a large deviation in surface temperature may result from a near-global volcanic event, as inferred for Venus. The increase in H_2O and SO_2 could potentially increase surface temperatures by almost ~100 K — enough to impart significant (~100 MPa) thermal stress on the crust, which *Solomon et al.* (1999) argue may have resulted in the formation of widespread wrinkle ridges on Venus. Similarly, *Anderson and Smrekar* (1999) demonstrate that step changes in surface temperature may account for the evolution of tectonic features such as polygonal and gridded terrains to wrinkle ridges, as a function of propagating crustal temperatures. *Lenardic et al.* (2008) showed that the temperature swings due to climate evolution may propagate into the mantle. As atmospheric temperature rises, high temperature fronts propagate into the mantle. The result is decreased plate-mantle coupling due to lower viscosities, which may even cause a switch in tectonics from an active-lid to a stagnant-lid regime. Evidence for this coupled atmosphere/plate system may be found in the noble gas systematics of the atmosphere.

Despite the comparative paucity of geological context on Venus, a number of scenarios may be explored for its tectonic evolution. The first scenario is that plate tectonics once operated on Venus, and shut down at 750 Ma to 1 Ga, leaving the planet in a stagnant-lid mode since that time (e.g., *Phillips and Hansen,* 1998). An alternative scenario is that Venus may have indefinitely been in some sort of episodic overturn regime (*Moresi and Solomatov,* 1998). The work of *O'Neill et al.* (2007b) implies that Venus may have once been in a true stagnant-lid regime, and transferred into an episodic regime as early radioactive heating waned. While a classic stagnant-lid mode is ruled out by restrictively low heat flows (*Reese et al.,* 1999), a stagnant-lid "heat-pipe" model — where heat is transported through the lithosphere by volcanic pipes (e.g., *Spohn,* 1991; *van Thienen et al.,* 2005) — is another possibility. This mode of behavior is implied in "equilibrium" resurfacing models (*Phillips et al.,* 1992; *Smrekar et al.,* 2010). Again, each of these scenarios has implications for Venus' accumulated ^{40}Ar.

While Mars may have had short-lived plate tectonics (*Sleep,* 1994; *Connerney et al.,* 1999), or perhaps mantle overturn events (*Debaille et al.,* 2009) very early in its evolution that probably contributed to its early atmosphere (*Jakosky and Phillips,* 2001), this activity occurred largely before the ingrowth of ^{40}Ar in its mantle. From the perspective of ^{40}Ar degassing, Mars has, effectively, behaved largely as a classic stagnant-lid planet for most of its evolution.

3. GEODYNAMIC MODELS

Models developed to simulate plate tectonics on Earth require a strongly temperature-dependent mantle viscosity, and thus highly viscous rigid plates, but also some near-surface brittle failure mechanism that allowed these rigid plates to fault and be recycled. Three regimes of mantle convection have been identified in such models (*Moresi and Solomatov,* 1998) (Fig. 1). Stagnant-lid convection occurs when the stresses generated by the convecting system are insufficient to overcome the brittle resistive strength of the plate, and convection occurs under a thick, immobile lid (Fig. 1). Such a regime has been postulated to be applicable to Mars, the Moon, and Mercury (*O'Neill et al.,* 2007a).

The example in Fig. 1 demonstrates some of the characteristics of classic stagnant-lid convection. Due to its immobility, the lid has thickened and cooled, and is on average much thicker than an equivalent mobile-lid system. The lid acts as an impediment to further heat loss, and mantle temperatures rise. Hotter internal temperatures tend to modulate further thickening of the lid, and for very hot systems (with high internal heating) relative lid thicknesses may be thinned substantially, allowing voluminous melting. In the case in Fig. 1, despite a hot interior, the thick lid prevents the warm mantle from ascending and melting adiabatically, retarding melt production as seen in Fig. 1b. The lower-temperature contrast within the actively convecting mantle also affects plume morphology, which does not have the buoyancy contrast seen in cooler active-lid models (*Jellinek et al.,* 2002). The result is that convection assumes a very steady planform, dominated by long-lived and rather immobile columnar upwelling and downwelling plumes.

Mobile-lid convection, of which plate tectonics is an example, involves the mobility and recycling of the lid (Figs. 1c–f). It efficiently removes heat from the system, keeping mantle temperatures cooler than in comparable stagnant-lid regimes. Mobile-lid convection requires that the convective stresses generated by the system exceed the resistive strength of the plate, allowing brittle deformation and faulting (shown in orange in Fig. 1c), and thus plate mobility. The mobility of the surface results in a far more time-dependent flow, and mixing of generated anomalies such as depletion are more efficient (Fig. 1f). As the lid is

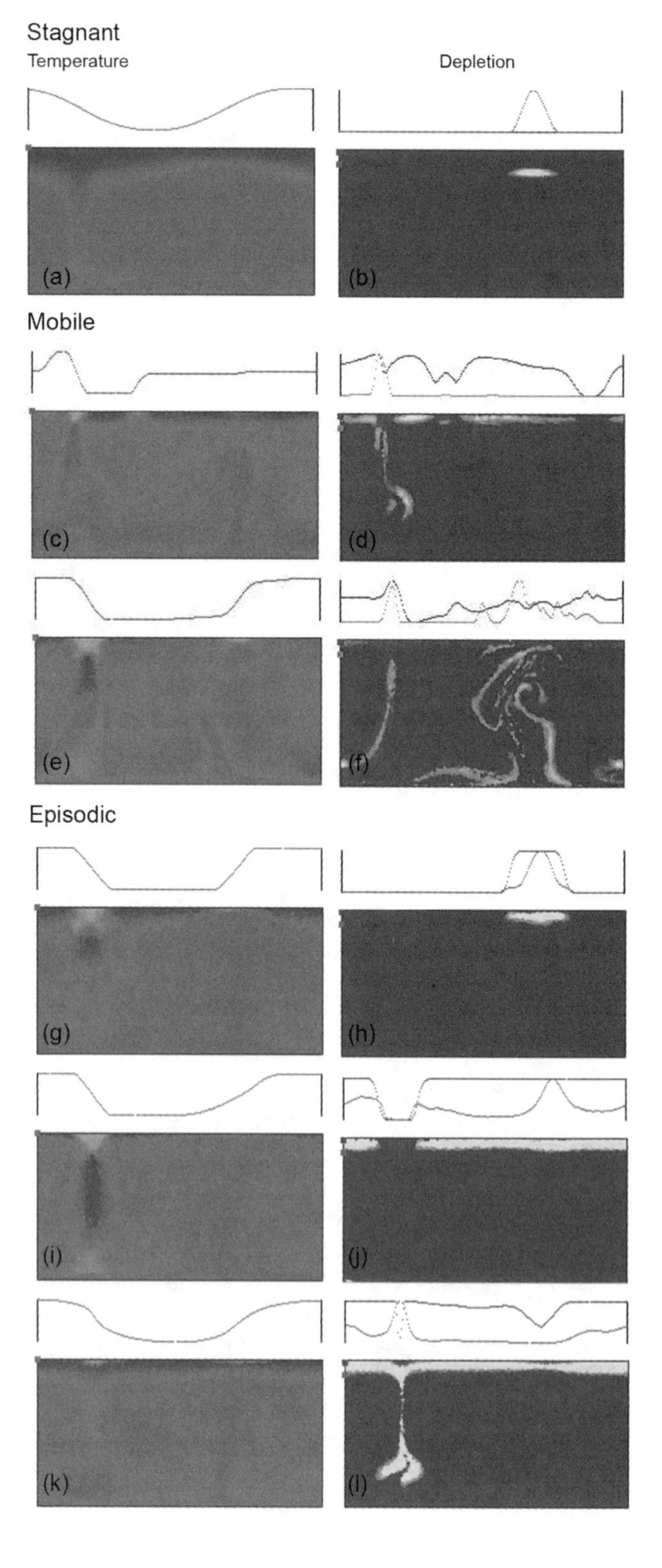

Fig. 1. See Plate 73 for color version. Snapshots of temperature field and depletion for the time evolution of three tectonic regimes. [For upper mantle convection (670 km depth); $Ra_{basal} = 3 \times 10^7$, 64 × 32 grid, with refinement in the boundary layers, periodic side boundaries, and free-slip, constant temperature top and bottom boundaries. Nondimensional internal heating is 0.5, temperature-dependent viscosity and brittle failure included as per text and Table 1.]. Hot temperatures are red, cold temperatures are blue, and brittle failure is shown as orange. In the depletion plots, blue represents material than has not had melt extracted; yellow is material that has undergone some degree of partial melting (plastic deformation is purple). **(a)**,**(b)** Temperature and depletion snapshots of a stagnant lid regime at the same time interval. Profiles above these sections are normalized surface velocity (red, above the temperature plot, normalized by maximum amplitude at each time step) and horizontal depletion profiles [mustard and blue lines above depletion sections, at depth indicated by markers on the lefthand side of the depletion plot; see arrows in **(b)**]. **(c)**–**(f)** Snapshots at model times of 10 m.y. [**(c)** and **(d)**] and 100 m.y. [**(e)** and **(f)**] of evolution in the mobile lid regime. **(g)**–**(l)** Time evolution of temperature and depletion fields during an episodic overturn, at model times of 12 m.y. [**(g)** and **(h)**], 20 m.y. [**(i)** and **(j)**], and 50 m.y. [**(k)** and **(l)**].

actively being generated at spreading centers, the average thickness is less than in the stagnant example. Additionally, most melt in this model is generated at spreading centers, where mantle is able to rise to the near surface and adiabatically decompress efficiently — resulting in high melt generation rates despite lower internal temperatures than the stagnant-lid example. Since the lid in mobile downwellings takes the form of entire coherent slabs of lithosphere, while upwellings are in the form of erratic, time-dependent plumes, there is a large temperature/buoyancy contrast compared to the surrounding mantle.

A third tectonic regime identified by *Moresi and Solomatov* (1998) lays at the transition between lid stagnation and mobility and is known as episodic overturn. This regime is effectively an oscillation between the mobile- and stagnant-lid modes, where convective stresses are only periodically able to exceed lid strength, enabling a short-lived, rapid burst of subduction before convection settles into a stagnant-lid mode again. Prior to an overturn event, mantle temperatures are high due to inefficient heat loss during the stagnant phase. Together with extreme rapid spreading velocities during the overturn (an order of magnitude above the mobile-lid average), this results in immense volcanism (e.g., Fig. 1j), reaching very high melt fractions. Subsequent to this, the subducted slabs cool the mantle and convection reestablishes a steady planform, resulting in subdued melt rates primarily related to plumes impinging on the thickening lithosphere. The lithospheric oscillates in this case from very thin during and immediately postoverturn, to exceptionally thick during the stagnant epoch.

To model the efficiency of ^{40}Ar degassing in stagnant, mobile, and episodic tectonic regimes, we have used the finite-element codes Ellipsis (*Moresi et al.,* 2003) (two-dimensional, history-dependent depletion), Underworld (two- and three-dimensional Cartesian), and CitcomS (spherical geometry). We adopt the model setup described in Table 1 and shown in Fig. 1 for convection in a 2 × 1 aspect ratio box for an Earth-scale system. We include a default temperature-dependent viscosity using the Frank-Kamenetski approximation that varies over five orders of magnitude, and a Byerlee-style friction criterion. The numerical techniques and implementations are more fully described in previous work (*Moresi and Solomatov,* 1998; *Moresi et al.,* 2003; *O'Neill et al.,* 2005; *Zhong et al.,* 2000; *Tan et al.,* 2006). We consider only one reservoir (i.e., no separate geochemical components), and calculate mantle melting using the formulism of *O'Neill et al.* (2005), where the melt fraction is calculated from the supersolidus temperature using a parameterized melt function (*McKenzie and Bickle,* 1988). Depletion of mantle residuum in considered in some models (Fig. 1), and in these cases it is assumed ^{40}Ar is partitioned into the earliest melt fractions. Parameters used in calculating melting are listed in Table 1. We do not consider melt migration explicitly here, assuming it is effectively stripped from the mantle and erupts. We calculate melting and gas extraction on the particle swarm, the smallest subdivision of our model. Previous work (*O'Neill et al.,* 2007c) suggests that melt extraction, and therefore degassing rates, are fairly insensitive to the actual particle size and/or problem resolution. We assume that once a portion of the mantle is melted, all local ^{40}Ar is partitioned into the melt and then erupted into the atmosphere. Clearly this is simplistic, but it allows us to compare tectonic effects without introducing additional unconstrained complexities, such as partitioning coefficient uncertainties, magma degassing efficiency, and separate crustal ^{40}Ar reservoirs, which we will return to in the discussion (section 5). For the purposes of comparing different tectonic regimes, we have simply increased the yield strength in the different models in Fig. 1 to simulate stagnant, episodic, and mobile-lid convection, leaving all other system parameters the same. The two-dimensional models were started from a previous steady-state simulation, and run for ~250 m.y. During this interval we simply used a present-day ^{40}Ar mantle concentration for the mantle of 1.33×10^{-4} kg m^{-3} (*Turcotte and Schubert,* 1982), and assumed this value was constant for the duration of the length of simulation (i.e., ^{40}Ar was not produced, only extracted during melting). Argon-40 is considered in the evolutionary models later.

TABLE 1. Physical properties for numerical models.

Symbol	Property	Value
ρ_m	Mantle density	3400 kg m^{-3}
g	gravitation acceleration	9.81 m s^{-2}
α	Thermal expansivity	3×10^{-5} K^{-1}
ΔT	Nonadiabatic T contrast	2555 K
d_m	Depth of convecting mantle	2895 km
κ	Thermal diffusivity	10^{-6} m^2 s^{-1}
k	Thermal conductivity	3.5 W m^{-1} K^{-1}
C_p	Specific heat	1200 J K^{-1} kg^{-1}
H	Mantle heat production	5×10^{-12} W kg^{-1} (varies)
$\eta(T)$	T-dependent viscosity	10^{19}–10^{24} Pa s^{-1} (FK approx*)
—	Lower mantle viscosity increase	$5*\eta(T)$
dP/dT	Clayperon slope at 670 km	–2.5 MPa K^{-1}
$\Delta\rho/\rho$	Density change across transition	10%
C	Cohesion	2×10^5, varies†
μ	Friction coefficient	0.15–0.6
$T_{sol}(0)$	Solidus at 0 GPa	1150°C
M_{sol}	Linear solidus slope	0.9°C MPa^{-1}
$\Delta T_{liq-sol}$	Solidus/liquidus ΔT	500°C
L	Latent heat of melting	200 kJ kg^{-1}

*Frank-Kamenetski viscosity profile varies (in nondimensional case) from 1 when T = 1, to 10^5 when T = 0 (i.e., where $\eta = \eta_0 \times 10^{-nT}$, $\eta_0 = 10^5$, n = 11.514; e.g., *Lenardic et al.* (2003). Three-dimensional simulations use a viscosity relationship of $\eta = \eta_0 \times 10^{[E/(T+1)-E/2]}$, with a viscosity contrast of 10^4, where E is the nondimensional activation energy.

†The Byerlee frictional law defines a yield stress above which brittle yielding commences as $\tau_{yield} = C + \mu P$, where C is the cohesion (2×10^5 for mobile-lid behavior in Fig. 1), μ is the dimensionless coefficient of friction, and P is pressure.

For references see *Schubert et al.* (2001) and references therein.

4. RESULTS

4.1. Parameter Exploration: Mantle Melting and Degassing

This first section explores numerical models of mantle convection in different convective regimes, which calculate mantle melting but do not track the depleted residuum (i.e., melting at each step melts a fertile mantle). An important consideration is the sensitivity of the results to Rayleigh number (Ra), internal heating, and the solidus used. In Fig. 2, the effect of internal heating and Rayleigh number is demonstrated for internally heated stagnant-lid models. These models do not include depleted mantle residuum, and so overestimate mantle melting, but are useful as a first comparison over a wide parameter range. Using a Frank-Kamenetski approximation, the viscosity contrast for these models is 3×10^4. It should be noted that as we are exploring steady-state behavior of the system, and comparing regimes, we are not considering melt depletion, and for simplicity in this suite of calculations we are not considering latent heat (we include it in later examples).

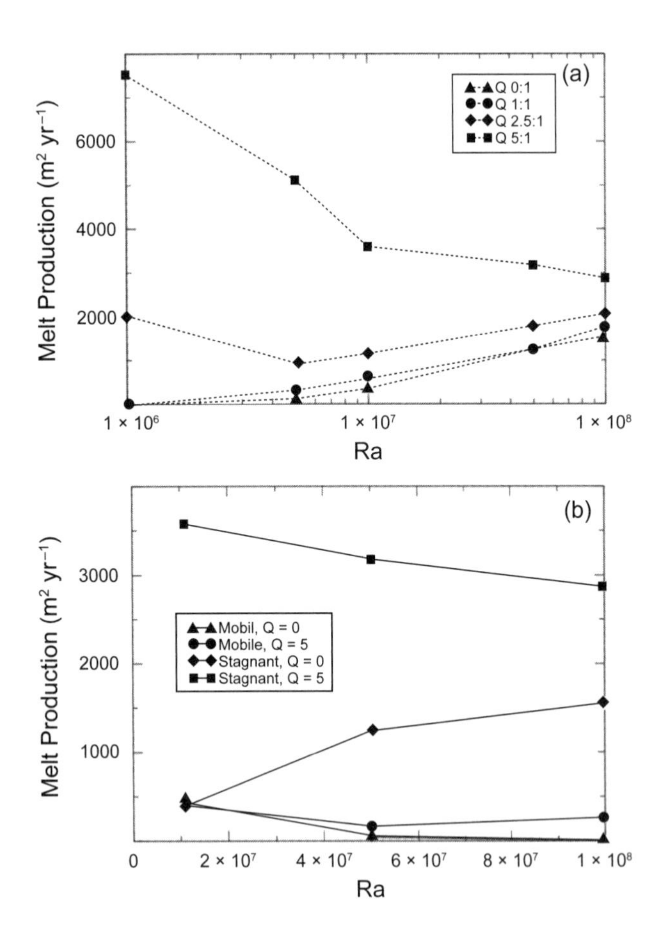

Fig. 2. (a) Variation in instantaneous two-dimensional melt production (m^2 yr^{-1}) with basal Rayleigh number (Ra) and internal heating ratio Q (0–5), for two-dimensional 1 × 1 stagnant lid simulations, with a viscosity contrast of 3 × 10^4. Simulations performed using *Underworld* (*Moresi et al.*, 2007), melt calculations *a posteri* in these examples, and depletion not considered. **(b)** Comparison in melt production between mobile and stagnant lid convection, for varying Ra and internal heating, otherwise same as for **(a)**.

A large difference in melting behavior is seen for the varying ratio of internal to bottom heating. For low internal heating ratios ($H_{internal}/H_{bottom} < \sim 2–3$), the internal temperatures are fundamentally set by the system and do not vary significantly with Ra. The net effect is that for low Ra, the large thickness of the lid precludes upwelling mantle from passing the solidus. For higher Ra, and thus smaller lid thicknesses, upwelling material may pass the solidus, and thus melting increases with Ra. For larger heating ratios ($H_{internal}/H_{bottom} > \sim 3$), internal temperatures are governed largely by heating rate. For low (basal) Ra, the inherently thick lid is thinned by the higher internal temperatures. However, extremely high internal temperatures are required to sustain a heat flux across the lid capable of losing the heat that is internally generated, and as a result significant internal melting occurs. For higher basal Ra, thinner lids and higher internal velocities effectively distribute the heat, the net result being lower internal temperatures and therefore lower melting rates (although still elevated compared to the lower heating ratio cases).

Mobile-lid cases differ from stagnant-lid cases in that the constraints of lid thickness are not as relevant as in stagnant-lid convection, and melt rates are mostly dependent on plate velocities and internal temperatures. In Fig. 2b, for low internal heating, melting rates decrease for increasing Ra — primarily in response to the lower internal temperatures in the more well-mixed system, and despite thinner lid thicknesses. For higher heating ratios, internal temperatures are higher, and thus melting rates somewhat elevated. There is little systematic variation with basal Ra, however. For higher Ra, mobile-lid convection in these cases results in less melt than stagnant-lid convection. This is primarily a result of the high internal temperatures in the stagnant-lid models, which, coupled with thin lithospheres at high Ra, permit voluminous melting. The inclusion of melt depletion would result in the development of a thick chemical boundary layer, which would impede the ascent of hot mantle and thus melting in the stagnant-lid case, but not for mobile-lid convection where melting occurs at spreading ridges. This would reduce or shut off melting in a hot stagnant regime — a point we return to in section 4.3.

The appropriate heating ratio range for whole mantle convection is on the order of ~10–15 (Table 1). However, *O'Farrell and Lowman* (2010) compared the equivalence of heating rates between spherical and planar geometries and found that whole-mantle Earth-like spherical systems are, in fact, best represented by low (<3) heating ratios in planar, Cartesian simulations, to give consistent internal temperatures. The Ra that gives us lithospheric thicknesses most appropriate to Earth (~100 km) is around 5×10^6–1×10^7. For this range, and lower heating rates, mobile-lid melt production exceeds that of stagnant-lid. However, this is contingent on the important caveats that changes in heating ratio, Rayleigh number, and lithosphere thickness will affect the melting ratios significantly — as will the development of a thick depleted boundary layer in models that include melt depletion, shown in section 4.3. Additionally, the effect on whole-mantle convection models is best addressed in three-dimensional spherical models, which are necessary to properly constrain the thermal balance.

4.2. Three-Dimensional Models

A large assumption in geometry is involved in extrapolating two-dimensional melt production calculations and degassing rates to three-dimensional volumes, and it is important to assess the robustness of our results in a three-dimensional spherical geometry. Styles of upwelling (tubular vs. assumed sheet-like in two dimensions), melt localization, and global melt distribution are also, critically, three-dimensional effects.

Figure 3 illustrates the effect of three-dimensional spherical geometry on melting. In these preliminary examples we have used CitcomS with a mantle-depth equivalent Ra of 1×10^5 and a nondimensional internal heating of 60, and an asthenospheric low-viscosity channel for the mobile-lid regime. The three-dimensional parameters are chosen to give a comparable temperature profile in the mantle to the two-dimensional case [see *O'Farrell and Lowman* (2010) for a discussion of heat production equivalence between three-dimensional spherical models and two-dimensional Cartesian models; our value of 60 is the planar-geometry equivalent of ~3–4 nondimensional heat production rate, according to their analysis]. While a continual, steady melting rate is exhibited by the mobile-lid model, the episodic model alternates between no melt production in the stagnant phase, to almost global melting during the active subduction phase (Fig. 4). In contrast, the thick lid in the stagnant model precludes any melting, and the melt production is zero in this example. Despite the erratic melt production in the episodic regime, the extreme melting during the active phase means the time-averaged melt production over one whole episodic cycle exceeds, in this example, the time-averaged mobile-lid melt production over a similar period by a factor of almost 9. This is strongly dependent on the period of activity, as well as heating ratio, and would also be affected by the inclusion of history-dependent depletion. The latter would decrease the melt production in the episodic regime, as a lesser amount of fertile mantle material is processed through the melting zone in episodic convection, which instead exhibits large melt fractions from limited source volumes. This similarly impacts argon

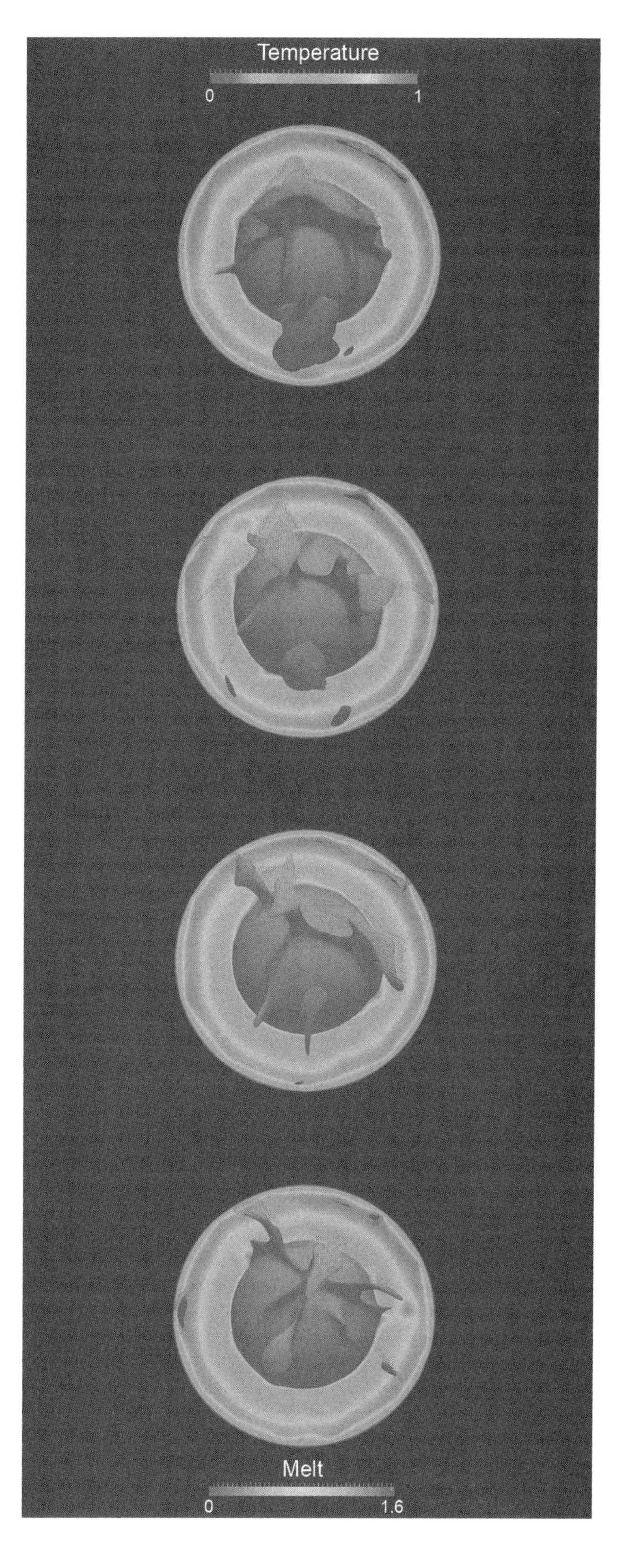

Fig. 3. See Plate 74 for color version. Temperature isosurface (T = 0.9, orange), temperature slice, and melt production (colored glyphs) in a three-dimensional spherical mobile lid convection model using CitcomS, divided into 96 subdomains with a grid spacing of 65 × 65 × 65 each (3,449,952 nodes total, average vertical spacing 22 km). Rayleigh number is 1×10^5 (assuming an upper mantle depth scale; higher in the CitcomS input, which assumes planetary radii), viscosity contrast is 1×10^4, heating ratio is 60 (see text for discussion on spherical and Cartesian internal heating), and the model contains a low-viscosity asthenosphere. Other parameters as in Table 1. Model time for the snapshots are at 0, 114, 214, and 382 m.y., respectively.

degassing from these models, which is dependent on the melt history of the mantle.

As the melt calculations were performed *a posteri* in three dimensions, the main thing lacking is a calculation of the progressive depletion of the mantle source zone as melt is extracted. This is why episodic melting exceeds mobile-lid melting in this example; extremely large volumes are extracted from comparatively smaller volumes of mantle than in the mobile-lid case, which continually processes mantle through spreading zones. We next address the problem of the depletion of the mantle residuum.

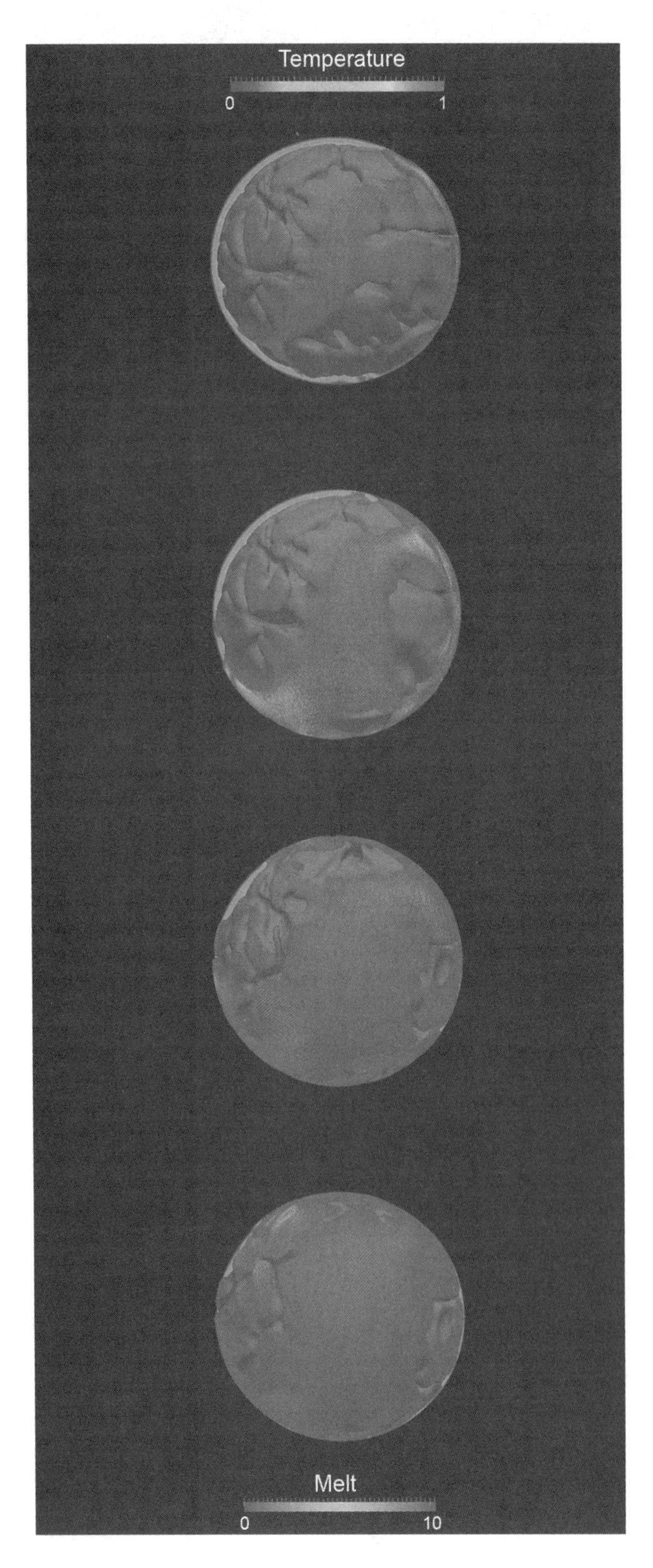

Fig. 4. See Plate 75 for color version. Temperature isosurface (T = 0.9, orange), temperature slice, and melt production (glyphs) in a three-dimensional spherical episodic convection model using CitcomS. Other parameters as per Fig. 3 and Table 1. The time slices track the evolution of melt over one overturn, and begin in a stagnant lid mode, with high internal temperatures, and thus the temperature isosurface is near the top. After yielding and plastic failure, the melt is generated over a wide area, correlated with active spreading (discrete glyphs). Downwellings are evident by depressions in the isocontour surface and low melt generation. Modeled times for the snapshots are 83, 178, 246, and 303 m.y., respectively.

4.3. Mantle Depletion

Figure 1 shows the time evolution of temperature and melt-depletion fields for simulations of stagnant-lid, episodic overturn, and mobile-lid convection, using Ellipsis. Melting in the mobile-lid model takes place largely at spreading centers. Although the mantle is comparatively cool in mobile-lid models, a large degree of melt is produced as upwelling mantle at divergent margins can adiabatically rise to the near surface and generate large amounts of decompression melt. This melting is more or less continuous, and thus mobile-lid convection efficiently processes the mantle over long periods.

In contrast, the absence of subduction results in a hot interior in the stagnant-lid model. In this particular case, little melt is generated because of the thick immobile lid. Upwelling mantle cannot rise to sufficiently low pressures to generate significant melt, and very little mantle is processed through melt zones in this regime. This contrasts with the previous hot stagnant-lid example (Fig. 2b), and also highlights an important difference in stagnant-lid models with depleted residuum compared to those without — the depleted residuum can act as a thick "boundary layer" to impede flow to shallow depths; it can be significantly thicker than the thermal boundary layer in "hot" stagnant models, and can shut off melt production in these cases. As such, it is critically important to calculate mantle depletion in mantle convection models incorporating melting. Figure 5b,c show a time series during an episodic overturn event. During overturn, the hot deep mantle (with elevated temperatures due to a lack of subduction in the stagnant phase) can ascend quickly due to rapid subduction and very fast seafloor spreading. As a result, large amounts of melt are generated by hot mantle rising adiabatically to low pressures in the spreading zones. The time-averaged melt generation rate in episodic regime over a few overturns is comparable to mobile-lid models, and during each overturn very high degrees of partial melting are reached. However, this melting is localized in space and time. While generating a large volume of melt from a localized portion of mantle during an overturn, the rest of the mantle is not efficiently processed in the interim. Additionally, since argon in these models is partitioned into the very first phase of melt, generating more melt from already depleted material will not add to the argon budget, and the degassing efficiency of episodic convection is less than that of mobile-lid convection, as it is not processing the same amount of mantle.

The time evolution of melting and ^{40}Ar degassing from these models are shown in Figs. 5a,b. We have scaled the two-dimensional results up to Earth's range by scaling the average melt rate of the mobile-lid model (0.1 km^3 yr^{-1} assuming unit thickness, as shown in Fig. 5a) by a multiplier to get the current global melt rate [~19 km^3 yr^{-1} (*Schubert et al.,* 2001)]. We then use this same multiplier to convert the model ^{40}Ar produced in an Earth-scale global ^{40}Ar. These scaled values are reflected in the axes of Figs. 5a,b. Quantitatively, the amount of ^{40}Ar is still less than that of Earth as we are not considering crustal reservoirs, which make up most of Earth's present ^{40}Ar flux.

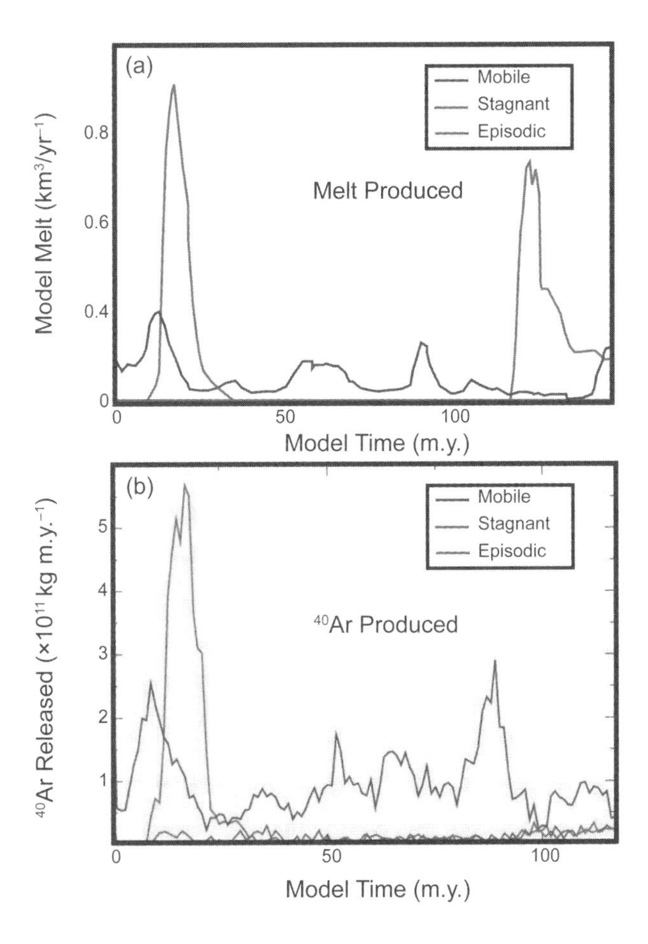

Fig. 5. See Plate 76 for color version. **(a)** Melt produced by the models shown in Fig. 1 vs. model time. **(b)** ^{40}Ar degassed from the models in Fig. 1, against model time. Note the axis was clipped to compare models over an appropriate time interval (i.e., one episodic overturn).

To obtain the average rate of ^{40}Ar degassing in these models, we took the time average of all regimes from Fig. 5b over the period it took for one episodic overturn. For these cases, mobile-lid convection is on average 1.8 times more efficient than episodic convection in degassing ^{40}Ar, and 20 times more efficient than stagnant-lid convection. These differences will be very significant over the lifetime of a planet. The ratio of atmospheric ^{40}Ar degassed by these models is 1:11:20 for stagnant:episodic:mobile regimes respectively. In the following section, we integrate these values over the lifetime of an evolving planet.

4.4. Evolutionary Models

Argon degassing rates have likely changed significantly with time as (1) volcanism was greater in the past due to higher mantle temperatures, and has declined through time, and (2) the ^{40}Ar concentration of the mantle has not been constant; a lot more ^{40}Ar was generated in the past than is currently being generated. With this in mind we use the ratio of degassing efficiencies between tectonic regimes to explore evolutionary models of atmospheric ^{40}Ar.

The present-day degassing rate of ^{40}Ar has been constrained to be 4.4×10^{12} kg m.y.$^{-1}$ for the combined crust-mantle reservoir (*Bender et al.*, 2008). Since the residence time of ^{40}Ar in the crust [~1 G.y. (*Bender et al.,* 2008)] is much less than the age of Earth (4.56 G.y.), we consider only one combined crust-mantle reservoir. Since we are not calculating melt generation here, we have fit our current ^{40}Ar degassing rate (from section 4.3) to this value (e.g., Fig. 6b). The evolution of volcanic degassing rate through time is not well studied. *O'Neill et al.* (2007c) calculated that it should decline by a factor of ~12 for Mars — we have assumed a similar decline here. We assume ^{40}Ar degassing follows this, and apply an exponentially decreasing function that declines by this factor. To calculate our degassing rate curves, we then apply the degassing efficiency ratios for different tectonic models from the numerical simulations. The results are shown in Fig. 6. The mantle ^{40}Ar production curve (dashed line) is based on the assumed mantle ^{40}K concentration of *Turcotte and Schubert* (1982). Using the present as t = 0, the amount of ^{40}K present in the crust mantle reservoir is

$$^{40}K_{mass}(t) = e^{\lambda t} C_{^{40}K} M_e \quad (1)$$

where λ is the decay constant of ^{40}K, M_e the mass of the crust + mantle, and $C_{^{40}K}$ the present concentration of ^{40}K (in kg kg^{-1}). The mass of ^{40}Ar produced is

$$d^{40}Ar_{mass\_produced}/dt = 0.105\, d^{40}K/dt \quad (2)$$

The 0.105 factor comes from the amount of ^{40}K that decays to ^{40}Ar (*Xie and Tackley,* 2004). This ^{40}Ar production rate can then be coupled with the loss rate to calculate the evolution of the ^{40}Ar reservoir

$$^{40}Ar_t = {}^{40}Ar_{t-1} + d^{40}Ar_{mass\_produced}/dt - R_{t-1}C_{t-1}{}^{40}Ar_{t-1} \quad (3)$$

Here the subscript t refers to current timestep, t–1 the previous timestep, $^{40}Ar_{t-1}$ the previous mass of ^{40}Ar in the reservoir, $d^{40}Ar_{mass\_produced}/dt$ the amount of ^{40}Ar produced by decay of ^{40}K over the previous timestep, R the degassing rate (in kilograms per timestep), and C the concentration.

Figure 6b illustrates the evolution of the ^{40}Ar degassing rate, assuming the ^{40}Ar production curve and volcanic degassing rate of Fig. 6a. Earth's current ^{40}Ar degassing rate is used to peg the mobile-lid curve at t = 0 Ga; the other curves assume the relative efficiencies deduced from the numerical simulations. While volcanic degassing is presumed to be very high on the early Earth, ^{40}Ar required a significant period of time to build up to substantial levels within the mantle. There is a peak in ^{40}Ar degassing where substantial amounts of ^{40}Ar have been generated in the mantle and volcanic degassing rates are still high, at around 3–4 Ga. In more recent times, the decline in ^{40}Ar production, previous degassing of mantle ^{40}Ar, and declining volcanic degassing rates conspire to give a gradual tapering in the degassing rate of ^{40}Ar since ~3.5 Ga.

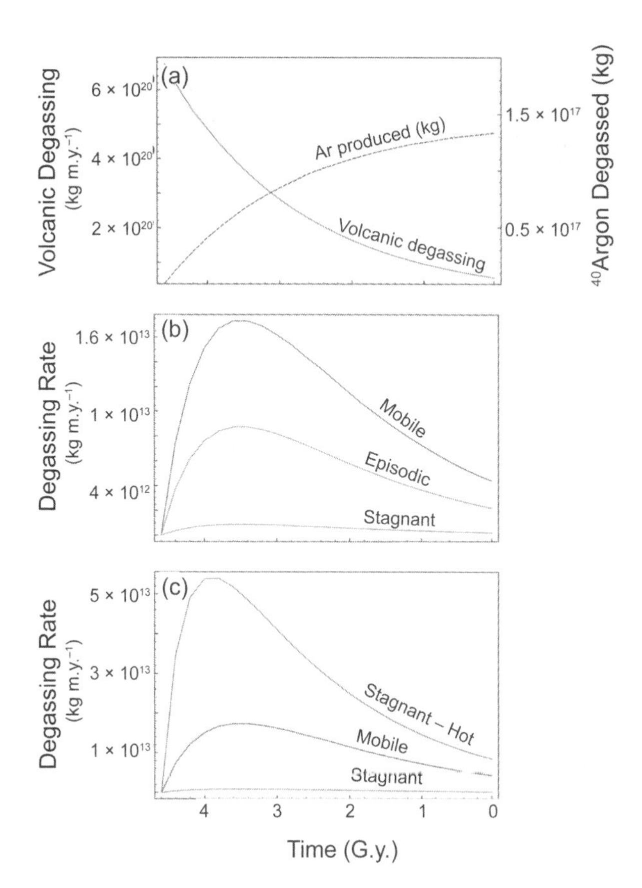

Fig. 6. (a) ^{40}Ar production and volcanic degassing rates for the mantle (note separate axes). ^{40}Ar production in the mantle is calculated from the ^{40}K decay over time. Modeled using the parameters listed in *Turcotte and Schubert* (1982). Also shown is the calculated volcanic degassing rate (dashed line). **(b)** Degassing rate of ^{40}Ar, calculated from the mantle concentration and volcanic degassing rate for each time, for mobile, episodic, and stagnant lid regimes, based on the relative efficiencies observed in section 3.1. **(c)** Same as for **(b)**, but with a hot stagnant lid case representing an endmember explored in Fig. 3. Depending on the relative degassing efficiency of mobile/stagnant lid convection — which itself depends strongly on the internal temperatures and lid thickness — the rate of volcanism and degassing in stagnant lid models can either greatly exceed, or fall short of, the mobile lid case, highlighting the sensitivity to system parameters.

In this example, the mobile-lid degassing efficiency is high compared to other regimes (Fig. 6), but it does not approach ^{40}Ar production. The implication of this is that while plate tectonics might efficiently degas the mantle, it is far from 100% efficient and, coupled with the time-dependent ^{40}Ar production curve, the effective efficiency could be on the order of 50%. Figure 6c is a similar plot, but includes an inferred degassing rate for a "hot" stagnant case, explored in section 4.1, which potentially degasses the mantle more effectively than even mobile lid (although a depleted residuum is again not included in this model).

The implication of these degassing rates to current atmospheric ^{40}Ar abundances is shown in Fig. 7. Note these results are scaled to an Earth-sized planet with an Earth-like composition, and therefore a 24% degassing efficiency on this plot (3.2×10^{16} kg) is not numerically the same as a 24% efficiency on Venus (which actually possesses $1.4 \pm 0.46 \times 10^{16}$ kg of atmospheric ^{40}Ar), but is dimensionally equivalent. These results can be directly scaled to Venus' size, but it is easier to compare them for a constant-sized planet.

In these particular scenarios, the relative degassing efficiency of episodic convection compared to mobile-lid convection is around half, meaning that the final degassing efficiency is consistent with Venus' final atmospheric ^{40}Ar concentration. This provides some resolution to Venus' ^{40}Ar conundrum: A sporadically active planet can combine efficient heat loss and periodically active tectonics with a low degassing efficiency of ^{40}Ar (on the order of ~24%).

Of course these results are strongly dependent on the relative degassing ratios, which we have shown are sensitive to Ra and internal heating. To provide a comparison, Fig. 7 illustrates the dependence of relative degassing efficiency for a thick stagnant lid (as per Figs. 1 and 6b) and a thin stagnant lid. The primary physical difference between these systems is lithospheric thickness; in the hot stagnant-lid case, the upwelling convecting mantle can rise far beyond the solidus due to the thinness of the lithosphere (provided a depleted mantle residuum is absent). The combination of thin lithosphere/accessible melting zone and hot internal temperatures results in voluminous melting. The melting is likely to be exaggerated, however, as the latter model does not consider depletion of the mantle residuum — rather simply the instantaneous melting of a fertile mantle. Regardless, the sensitivity is marked, and highlights the first-order effect of melting style on ^{40}Ar degassing. The amount of ^{40}Ar integrated over the planet's lifetime is implausible for Venus (almost 100% degassed). In addition, the lithospheric

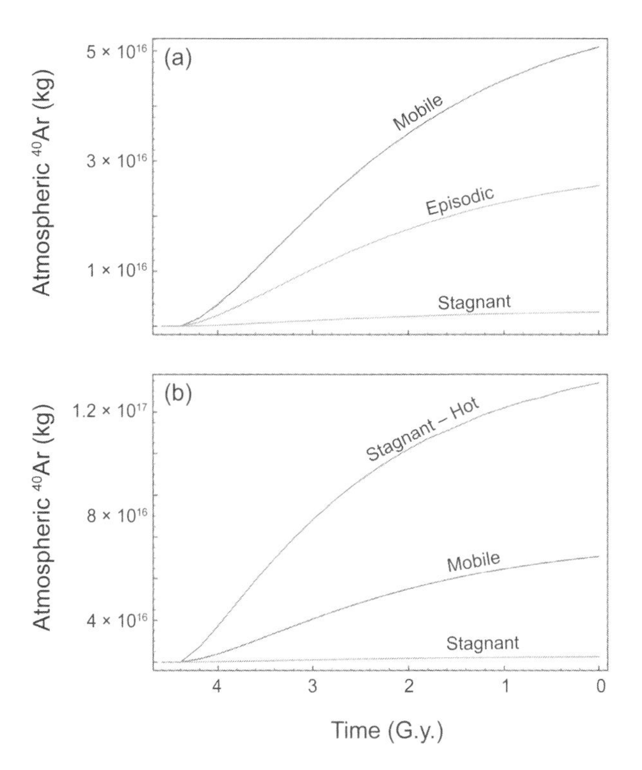

Fig. 7. (a) Accumulated atmospheric ^{40}Ar over time, for the integrated degassing rates shown in **(b)**. For the relative efficiencies of the models shown in Fig. 1, mobile lid convection accumulates almost present-day levels of ^{40}Ar (~6.6 × 10^{16} kg), while episodic convection is around half as efficient. **(b)** Similar to **(a)**, but including an endmember "hot" stagnant lid example from Fig. 3. For hot, thin lithosphere, stagnant lid convection efficiently degasses the mantle, in contrast to cooler stagnant models where the thick lid inhibits melting and degassing. Given Venus' ^{40}Ar deficit and large elastic lithosphere thickness estimates, this endmember is probably not appropriate to Venus.

thickness (<100 km) required to generate this melting implies lithospheric thicknesses much less than those estimated [200–400 km (*Moore and Schubert,* 1997); 125–225 km (*Smrekar,* 1994)], and elastic lithospheric thicknesses of around <10 km, far less than that inferred currently for Venus [>20–30 km in places (*Moore and Schubert,* 1997; *Smrekar,* 1994)]. Melting is also contingent on uncertainties in volatile components; local lithospheric thickness variations and heat transfer by volcanic piping would almost certainly modulate these simplistic modeled lithospheric thicknesses. Figures 1 and 6b show examples of stagnant-lid convection with venusian-scale lithospheric thicknesses (~300 km), and exhibit low rates of volcanism and mantle degassing — too low to explain Venus' current atmospheric ^{40}Ar.

5. DISCUSSION AND CONCLUSIONS

The main purpose of this paper is to explore the tectonic controls on ^{40}Ar degassing through time. One of our primary results is that the tectonic evolution of a planet is crucial in understanding its atmospheric ^{40}Ar record. The corollary is also true, that atmospheric ^{40}Ar can be an important constraint on the evolution of a planet. However, a number of caveats deserve explicit mention. The first of these is the uncertainty in the partitioning coefficients of argon. Traditionally it has been assumed to be incompatible, as modeled here, and this is consistent with the bulk of laboratory results (e.g., *Heber et al.,* 2007). However, some recent work has suggested argon is far more compatible than has been assumed (*Watson et al.,* 2007). This work has been fairly controversial (*Ballantine and Holland,* 2008; *Bouhifd et al.,* 2008). If true, then the ability of Earth to degas ^{40}Ar at all becomes extremely problematic, and accumulating Earth's current ^{40}Ar atmospheric abundance becomes very difficult. The problem is compounded for Venus. However, the basic result, i.e., the *relative* effect of tectonic regime on degassing efficiency, is unaffected.

A large simplification here was the consideration of only one combined crust-mantle reservoir. On Venus, crustal temperatures are so high that the crust is open to argon loss, and so this is not an issue. On Earth, extrusive melts degas at nearly 100% efficiency with respect to argon, and intrusive melts degas at around ~92% efficiency (*Namiki and Solomon,* 1998), so there is a reasonable, but finite, efficiency of argon degassing from mantle melts from the outset. More important is ^{40}Ar trapped within the continental crust. *Bender et al.* (2008) calculate that ^{40}Ar degassing from the crust currently exceeds that of the mantle by a factor of 4. This factor is due to a combination of ^{40}K enrichment in the crust, the current areal extent of continental crust, and spreading centers currently processing a fairly depleted mantle reservoir. The ratio was probably smaller in the past due to (1) continental growth and (2) processing of less-depleted mantle. Additionally, crustal recycling results in the return of significant amounts of ^{40}K to the mantle. The current areal extent of Archaean cratons means that most of Earth's surface (even if it was not continental material) was recycled for most of Earth's history; long-lived crustal reservoirs of ^{40}Ar are volumetrically small (even if they are enriched in ^{40}K). What's more (and probably most) important is that *Bender et al.* (2008) calculate the crustal residence time of ^{40}Ar to be ~1 G.y. So even if we have significant sequestration of ^{40}K, and therefore trapped ^{40}Ar, into the continental crust over the timescales considered here (4.56 G.y.), these reservoirs behave as if open to the atmosphere. Given these arguments and the complexities in modeling crustal growth and recycling, the choice of one combined reservoir was both less *ad hoc* and more useful in unraveling the tectonic controls on ^{40}Ar crustal degassing.

Within the simplifications made in this analysis, it is apparent that the tectonic regime of a planet exerts a fundamental control on its degassing history. The dynamics of degassing tend to result in an extremely degassed upper mantle and comparatively enriched lower mantle — in line with observations (*Coltice et al.,* 2000) (see also Fig. 1). However, tectonically inefficient degassing in effect acts as a "temporal reservoir" in that the total current atmospheric

^{40}Ar budget is less due to the lower integrated degassing efficiency through time. The ^{40}Ar atmospheric budget can also place important constraints on the tectonic history of a planet.

Mantle melting and volcanic degassing are weakly dependent on Ra and internal heating rates in mobile-lid or episodic overturn convection, as the internal temperatures are largely modulated by convective-lid overturn and, since melting occurs primarily at spreading centers, it is weakly influenced by lithospheric thickness. In contrast, the stagnant-lid models we present show a strong dependence on both Ra and internal heating. While the internal temperatures are generally higher in all stagnant-lid models, the melting is largely sensitive to lid thickness, ranging from no melt for thick lids, to massive melting for very thin stagnant lids. *Smrekar and Parmentier* (1996) suggested that a thick, depleted thermochemical lithospheric layer on Venus — essentially a thick stagnant-lid regime — might impede dynamic melting, as observed here. Elastic lithospheric thickness estimates (e.g., *Moore and Schubert,* 1997; *Smrekar,* 1994) suggest that either a thick stagnant lid or an episodic regime is appropriate for Venus, and an episodic regime can self-consistently explain Venus' outgassing and current lithospheric state.

These alternative histories for Venus may have very different climatic implications. On the one hand, a stagnant-lid planet with a thick lithosphere may generate little volcanism. A thin, hot stagnant lithosphere will initially produce more melt, but if mantle depletion is considered, this will rapidly result in the development of a thick, depleted boundary layer, hindering upwelling and further melting (*Smrekar and Parmentier,* 1996). *Reese et al.* (1999) suggested a model whereby temperatures within a stagnant lid might escalate, resulting in widespread melting and a global volcanic event, although this also would be shut off by the development of a depleted layer. The injection of voluminous volcanic gases (e.g., SO_2, H_2O) into the atmosphere of Venus is most easily accomplished in a tectonically driven episodic scenario, and could potentially (i.e., postoverturn) account for the sequence in deformation styles from polygonal terrains to wrinkle ridges described in *Solomon et al.* (1999) and *Anderson and Smrekar* (1999). The predicted temperature fluctuation could alter not only the surface tectonic style, but also the style of global tectonics. *Lenardic et al.* (2008) demonstrated that surface temperature increases on the order of ~100 K could cause the transition of a mobile-lid system into a stagnant mode. Thus is it conceivable that the volcano-tectonic event on Venus ca. 750 Ma precipitated its own end.

Mars' evolution has been characterized by a low-Ra stagnant-lid regime, and Fig. 6b suggests a very low volcanic and degassing rate through time — consistent with martian constraints (e.g., *O'Neill et al.,* 2007c). From Fig. 7a, this would result in an argon deficit of approximately 25 compared to a mobile-lid planet — of a similar magnitude to Mars' observed absolute abundance deficit of ~16 (*Pollack and Black,* 1982).

Given the model sensitivity and geometrical extrapolations made in these models, full resolution of the planetary volcanism/outgassing problem in the future will minimally require three things: (1) modeling the melt-depleted mantle residuum to constrain ^{40}Ar sourcing in the modeled mantle; (2) three-dimensional spherical geometry to self-consistently handle geometry of melting and global heat-flux and energetics constraints; and (3) long-term evolutionary history, including appropriate initial conditions, declining heat production, and ^{40}Ar production. These are fairly intense computational requirements.

The results presented here suggest that Venus' tectonic and degassing history and current lithospheric state are consistent with an episodic overturn regime throughout most of its history. While there is still a range of behavior in such models (e.g., hot, thin stagnant-lid models can degas more than even equivalent mobile-lid models), our preferred models, incorporating mantle depletion effects and also consistent with most venusian geophysical constraints, suggests an episodic mode for much of Venus' history. This scenario is supported by nonradiogenic argon abundances, which suggest Venus retained much of its primordial atmosphere (*Kaula,* 1999). This contrasts with suggestions that Venus may have had plate tectonics throughout much of its history (e.g., *Arkani-Hamed et al.,* 1993), although a plate tectonic scenario is possible if, for instance, internal viscosities on Venus are much higher due to dehydration (e.g., *Xie and Tackley,* 2004). The implication of this result is that Venus and Earth may have diverged tectonically very early in their evolution. The finite degassing efficiency of episodic convection, coupled with time-dependent ^{40}Ar production, means that total ^{40}Ar degassing of Venus (24% of that produced) is consistent with this style of convection over most of Venus' history.

Acknowledgments. C.O. acknowledges ARC support (DP0880801, DP110104 145, and FT100100717). This is GEMOC publication 896 and CCFS Publication 331.

REFERENCES

Allegre C. J., Hofmann A., and O'Nions K. (1996) The argon constraints on mantle structure. *Geophys. Res. Lett., 23,* 3555–3557.

Anderson F. S. and Smrekar S. E. (1999) Tectonic effects of climate change on Venus. *J. Geophys. Res., 104,* 30743–30756.

Arkani-Hamed J., Schaber G. G., and Strom R. G. (1993) Constraints on the thermal evolution of Venus inferred from Magellan data. *J. Geophys. Res., 98,* 5309–5315.

Ballentine C. J. and Holland G. (2008) What CO_2 well gases tell us about the origin of noble gases in the mantle and their relationship to the atmosphere. *Philos. Trans. R. Soc. A, 366,* 4183–4203.

Bender M. L., Barnett B., Dreyfus G., Jouzel J., and Porcelli D. (2008) The contemporary degasing rate of ^{40}Ar from the solid Earth. *Proc. Natl. Acad. Sci., 105,* 8232–8237.

Bjonnes E. E., Hansen V. L., James B., and Swenson J. B. (2012) Equilibrium resurfacing of Venus: Results from new Monte Carlo modeling and implications for Venus surface histories.

Icarus, 217(2), 451–461, DOI: 10.1016/j. icarus.2011.03.033.
Bouhifd M. A., Jephcoat A. P., and Kelley S. P. (2008) Argon solubility drop in silicate melts at high pressures: A review of recent experiments. *Chem. Geol., 256,* 252–258.
Brown C. D. and Grimm R. E. (1999) Recent tectonic and lithospheric thermal evolution of Venus. *Icarus, 139,* 40–48, DOI: 10.1006/icar.1999.6083.
Bullock M. A. and Grinspoon D. H. (2001) The recent evolution of climate on Venus. *Icarus, 150(1),* 19–37, DOI: 10.1006/icar.2000.6570.
Cameron A. G. W. (1983) Origin of the atmospheres of the terrestrial planets. *Icarus, 56,* 195–201.
Coltice N., Albarede F., and Gillet Ph. (2000) ^{40}K-^{40}Ar constraints on recycling continental crust in the mantle. *Science, 288,* 845–847.
Coltice N., Marty B., and Yokochi R. (2009) Xenon isotope constraints on the thermal evolution of the early Earth. *Chem. Geol., 266,* 4–9.
Condie K. C. and Pease V., eds. (2008) *When Did Plate Tectonics Begin on Planet Earth?* GSA Special Paper 440, Geological Society of America, Boulder.
Condie K. C., O'Neill C., and Aster R. (2009) Evidence and implications of a widespread magmatic shut down for 250 Myr on Earth. *Earth Planet. Sci. Lett., 282,* 294–298.
Connerney J. E. P., Acuna M. H., Wasilewski P. J., Ness N. F., Reme H., Mazelle C., Vignes D., Lin R. P., Mitchell D. L., and Cloutier P. A. (1999) Magnetic lineations in the ancient crust of Mars. *Science, 284,* 794–798.
Davies G. F. (1999) Geophysically constrained mantle mass flows and the ^{40}Ar budget: A degasssed lower mantle? *Earth Planet. Sci. Lett., 166,* 149–162.
Debaille V., Brandon A. D., O'Neill C., Yin Q.-Z., and Jacobsen B. (2009) Isotopic evidence for mantle overturn in early Mars and its geodynamic consequences. *Nature Geosci., 2,* 548–551.
Dreibus G. and Wanke H. (1985) Mars, a volatile-rich planet. *Meteoritics, 20,* 367–381.
Fegley B. and Prinn R. G. (1989) Estimation of the rate of volcanism on Venus from reaction rate measurements. *Nature, 337(6202),* 55–58, DOI: 10.1038/337055a0.
Gonnermann H. M. and Mukhopadhyay S. (2009) Preserving noble gases in a convecting mantle. *Nature, 459,* 560–563.
Harrison T. M., Blichert-Toft J., Muller W., Albarede F., Holden P., and Mojzsis S. J. (2005) Heterogenous Hadean hafnium: Evidence of continental crust at 4.4 to 4.5 Ga. *Science, 310,* 1947–1950.
Hauck S. A. II, Phillips R. J., and Price M. H. (1998) Venus: Crater distribution and plains resurfacing models. *J. Geophys. Res., 103(E6),* 13635–13642, DOI: 10.1029/98JE00400.
Heber V. S., Brooker R. A., Kelley S. P., and Wood B. J. (2007) Crystal-melt partitioning of noble gases (helium, neon, argon, krypton, and xenon) for olivine and clinopyroxene. *Geochim. Cosmochim. Acta, 71,* 1041–1061.
Höink T., Lenardic A., and Richards M. A. (2012) Depth-dependent viscosity and mantle stress amplification: Implications for the role of the asthenosphere in maintaining plate tectonics. *Geophys. J. Intl., 191,* 30–41.
Jakosky B. M. and Phillips R. J. (2001) Mars' volatile and climate history. *Nature, 412,* 237–244.
Jellinek A. M., Lenardic A., and Manga M. (2002) The influence of interior mantle temperature on the structure of plumes: Heads for Venus, tails for the Earth. *Geophys. Res. Lett., 29(11),* 27-1 to 27-4, DOI: 10.1029/2001GL014624.
Kaula W. M. (1999) Constraints on Venus evolution from radiogenic argon. *Icarus, 139,* 32–39.
Kiefer W. S. and Li Q. (2009) Mantle convection controls the observed lateral variations in lithospheric thickness on present-day Mars. *Geophys. Res. Lett., 36(18),* DOI: 10.1029/2009GL039827.
Lenardic A., Jellinek A. M., and Moresi L. N. (2008) A climate induced transition in the tectonic style of a terrestrial planet. *Earth Planet. Sci. Lett., 271,* 34–42.
McKenzie D. and Bickle M. J. (1988) The volume and composition of melt generated by extension of the lithosphere. *J. Petrol., 29,* 625–679.
Moore W. B. and Schubert G. (1997) Venusian crustal and lithospheric properties from nonlinear regressions of highland geoid and topography. *Icarus, 128,* 415–428.
Moresi L. and Solomatov V. S. (1998) Mantle convection with a brittle lithosphere: Thoughts on the global tectonic styles of the Earth and Venus. *Geophys. J. Intl., 133,* 669–682.
Moresi L., Dufour F., and Mulhaus H. B. (2003) A Lagrangian integration point finite element method for large deformation modeling of viscoelastic geomaterials. *J. Comp. Phys., 184,* 476–497.
Moresi L., Quenette S., Lemiale V., Meriaux C., Appelbe B., and Mühlhaus H. B. (2007) Computational approaches to studying non-linear dynamics of the crust and mantle. *Phys. Earth Planet. Inter., 163,* 69–82, DOI: 10.1016/j.pepi.2007.06.009.
Musselwhite D. S., Drake M. J., and Swindle T. D. (1991) Early outgassing of Mars supported by differential water solubility of iodine and xenon. *Nature, 352,* 697–699.
Namiki N. and Solomon S. C. (1998) Volcanic degassing of argon and helium and the history of crustal production on Venus. *J. Geophys. Res., 103,* 3655–3677.
O'Farrell K. A. and Lowman J. P. (2010) Emulating the thermal structure of spherical shell convection in plane-layer geometry mantle convection models. *Phys. Earth Planet. Inter., 182,* 73–84, DOI: 10.1016/j.pepi.2010.06.010.
O'Neill C., Moresi L., and Lenardic A. (2005) Insulation and depletion due to thickened crust: Effects on melt production on Mars and Earth. *Geophs. Res. Lett., 32,* DOI: 10.1029/2005GL022855.
O'Neill C., Jellinek A. M., and Lenardic A. (2007a) Conditions for the onset of plate tectonics on terrestrial planets and moons. *Earth Planet. Sci. Lett., 261,* 20–32.
O'Neill C., Lenardic A., Moresi L., Torsvik T. H., and Lee C.-T. A. (2007b) Episodic Precambrian subduction. *Earth Planet. Sci. Lett., 262,* 552–562.
O'Neill C., Lenardic A., Jellinek A. M., and Kiefer W. S. (2007c) Melt propagation and volcanism in mantle convection simulations, with applications for martian volcanic and atmospheric evolution. *J. Geophys. Res., 112(E7),* DOI: 10.1029/2006JE002799.
Pepin R. O. (1991) On the origin and early evolution of terrestrial planet atmospheres and meteoritic volatiles. *Icarus, 92,* 2–79.
Phillips R. J. and Hansen V. L. (1998) Geological evolution of Venus: Rises, plains, plumes and plateaus. *Science, 279,* 1492–1497.
Phillips R. J., Raubertas R. F., Arvidson R. E., Sarkar I. C., Herrick R. R., Izenberg N., and Grimm R. E. (1992) Impact craters and Venus resurfacing history. *J. Geophys. Res., 97(E10),* 15923–15948, DOI: 10.1029/ 92JE01696.
Phipps Morgan J. (1988) Thermal and rare gas evolution of the mantle. *Chem. Geol., 145,* 431–445.
Pollack J. B. and Black D. C. (1982) Noble gases in planetary atmospheres: Implications for the origin and evolution of atmospheres. *Icarus, 51,* 169–198.
Reese C. C., Solomatov V. S., and Moresi L. N. (1999) Non-Newtonian stagnant lid convection and magmatic resurfacing on Venus. *Icarus, 139,* 67–80.

Schubert G., Turcotte D. L., and Olson P. (2001) *Mantle Convection in the Earth and Planets.* Cambridge Univ., Cambridge. 940 pp.
Shcheka S. S. and Keppler H. (2012) The origin of the terrestrial noble-gas signature. *Nature, 490,* 531–534.
Silver P. G. and Behn M. D. (2008) Intermittent plate tectonics? *Science, 319,* 85–88.
Sleep N. H. (1979) Thermal history and degassing of the Earth: Some simple calculations. *J. Geol., 87,* 671–686.
Sleep N. H.(1994) Martian plate tectonics. *J. Geophys. Res., 99(E3),* 5639–5655.
Smrekar S. E. (1994) Evidence for active hotspots on Venus from analysis of Magellan gravity data. *Icarus, 112,* 2–26.
Smrekar S. E. and Parmentier E. M. (1996) The interaction of mantle plumes with surface thermal and chemical boundary layers: Applications to hotspots on Venus. *J. Geophys. Res.– Solid Earth, 101,* 5397–5410.
Smrekar S. E., Stofan E. R., Mueller N., Treiman A., Elkins-Tanton L., Helbert J., Piccioni G., and Drossart P. (2010) Recent hotspot volcanism on Venus from VIRTIS emissivity data. *Science, 328(5978),* 605–608, DOI: 10.1126/science.1186785.
Solomon S. C., Bullock M. A., and Grinspoon D. H. (1999) Climate change as a regulator of tectonics on Venus. *Science, 286,* 87–90.
Spohn T. (1991), Mantle differentiation and thermal evolution of Mars, Mercury, and Venus. *Icarus, 90(2),* 222–236, DOI: 10.1016/0019-1035(91) 90103-Z.
Stern R. J. (2008) Modern-style plate tectonics began in Neoproterozoic time: An alternative interpretation of Earth's tectonic history. In *When Did Plate Tectonics Begin?* (K. Condie and V. Pease, eds.), pp. 265–280, GSA Special Paper 440, Geological Society of America, Boulder.
Tan E., Choi E., Thoutireddy P., Gurnis M., and Aivazis M. (2006) Geoframework: Coupling multiple models of mantle convection within a computational framework. *Geochem. Geophys. Geosyst., 7(6),* DOI: 10.1029/2005GC001155.
Turcotte D. L. and Schubert G. (1982) *Geodynamics: Applications of Continuum Physics to Geological Problems.* Wiley, New York. 450 pp.
Turcotte D. L. and Schubert G. (1988) Tectonic implications of radiogenic noble gases in planetary atmospheres. *Icarus, 74,* 36–46.
Van Keken P.E. and Ballentine C.J. (1999) Dynamical models of mantle volatile evolution and the role of phase transitions and temperature-dependent rheology. *J. Geophys. Res., 104,* 7137–7151.
Van Thienen P., Vlaar N. J., and van den Berg A. P. (2005) Assessment of the cooling capacity of plate tectonics and flood volcanism in the evolution of Earth, Mars and Venus. *Phys. Earth Planet. Inter., 150,* 287–315.
von Zahn U., Kumar S., Niemann H., and Prinn R. (1983) Composition of the Venus atmosphere. In *Venus* (D. M. Hunten et al., eds.), pp. 299–430. Univ. of Arizona, Tucson.
Watson E. B., Thomas J. B., and Cherniak D. J. (2007) ^{40}Ar retention in the terrestrial planets. *Nature, 449,* 299–304.
Wilde S. A., Valley J. W., Peck W. H., and Graham C. M. (2001) Evidence from detrital zircons for the existence of continental crust and oceans on the Earth 4.4 Gyr ago. *Nature, 409,* 175–178.
Xie S. and Tackley P. (2004) Evolution of helium and argon isotopes in a convecting mantle. *Phys. Earth Planet. Inter., 146,* 417–439.
Zhong S., Zuber M. T., Moresi L., and Gurnis M. (2000) Role of temperature-dependent viscosity and surface plates in spherical shell models of mantle convection. *J. Geophys. Res., 105,* 11063–11082.

Brain D. A., Leblanc F., Luhmann J. G., Moore T. E., and Tian F. (2013) Planetary magnetic fields and climate evolution. In *Comparative Climatology of Terrestrial Planets* (S. J. Mackwell et al., eds.), pp. 487–501. Univ. of Arizona, Tucson, DOI: 10.2458/azu_uapress_9780816530595-ch20.

Planetary Magnetic Fields and Climate Evolution

D. A. Brain
University of Colorado

F. Leblanc
Centre National de la Recherche Scientifique/Institut Pierre Simon Laplace

J. G. Luhmann
University of California, Los Angeles

T. E. Moore
NASA Goddard Space Flight Center

F. Tian
Tsinghua University

We explore the possible connections between magnetic fields and climate at the terrestrial bodies Venus, Earth, Mars, and Titan. Magnetic fields are thought to have negligible effects on the processes that change a planet's climate, except for processes that alter the abundance of atmospheric gases. Particles can be added or removed at the top of an atmosphere, where collisions are infrequent and a more substantial fraction of particles are ionized (and therefore subject to magnetic forces) than at lower altitudes. The absence of a global magnetic field at Mars for much of its history may have contributed to the removal of a substantial fraction of its atmosphere to space. The persistence of a global magnetic field should have decreased both ionization and removal of atmospheric ions by several processes, and may have indirectly decreased the loss rate of neutral particles as well. While it is convenient to think of magnetic fields as shields for planetary atmospheres from impinging plasma (such as the solar wind), observations of ions escaping from Earth's polar cusp regions suggest that magnetic shielding effects may not be as effective as previously thought. One explanation that requires further testing is that magnetic fields transfer momentum and energy from incident plasma to localized regions of the atmosphere, resulting in similar (or possibly greater) escape rates than if the momentum and energy were imparted more globally to the atmosphere in the absence of a magnetic field. Trace gases can be important for climate despite their low relative abundance in planetary atmospheres. At Venus, removal of O^+ over the history of the planet has likely contributed to the loss of water from the atmosphere, leading to a runaway greenhouse situation and having implications for the chemistry of atmosphere-surface interactions. Conversely, Titan's robust atmospheric chemistry may result from the addition of trace amounts of oxygen from Saturn's magnetosphere, which then participate in chemical reactions that produce carbon monoxide (CO) and carbon dioxide (CO_2). Models of the entire atmosphere system (including planetary plasma interactions) should continue to shed light on the connections between magnetic fields and climate, as well as models that consider a single planetary body in both magnetized and unmagnetized states. Future measurements, such as those that will be made by the Mars Atmosphere and Volatile Evolution (MAVEN) spacecraft to Mars, will provide better constraints on the importance of magnetic fields in the evolution of atmospheres.

1. INTRODUCTION

Planets and their atmospheres are influenced by magnetic fields. Aurora are perhaps the most visually striking example of this connection, and have been observed at many planetary bodies including Earth, Mars, the jovian planets, and three jovian moons (see review by *Mauk and Bagenal,* 2012). Magnetic fields may also influence the evolution of planetary surfaces in the absence of a substantial atmosphere, as demonstrated by Jupiter's moon Ganymede, which is noticeably brighter in polar regions than at low and mid-latitudes. This color variation results from differences in the weathering (bombardment) of the surface ice by energetic charged particles, which are deflected by Ganymede's global magnetic field at low latitudes but access the surface along field lines in polar regions that connect to Jupiter's ionosphere (*Khurana et al.,* 2007).

A more difficult question is whether magnetic fields influence planetary climate. On one hand, the demonstrated connections between magnetic fields and processes like those described above that occur at or above planetary surfaces suggest that the effects should extend to climate. On the other hand, magnetic fields are generated in the deep interiors of planets and their influence on their surroundings for planets with robust atmospheres is most noticeable at many atmospheric scale heights above the surface, where particle densities are a small fraction of those at the surface. Further, climate is determined by the properties of a planet's atmosphere, which is primarily composed of neutral particles whose motion is unaffected by ambient magnetic fields.

It is difficult to directly compare different planets in order to infer whether or how planetary magnetic fields influence climate. While the solar system has both magnetized and unmagnetized bodies, no two bodies differ from each other *only* in terms of their magnetic field. Objects also differ in size, distance from the Sun, atmospheric composition, and the plasma environment in which they are embedded — and all these factors may influence climate. For example, Earth (magnetized) and Venus (unmagnetized) are similar in size, but differ in distance from the Sun and atmospheric composition. Earth and Titan (unmagnetized) both have atmospheres with nitrogen as the most abundant species, but have different size and are embedded in very different plasma environments. Mars has likely had periods where it possessed a global magnetic field, while today it does not — but it is difficult at present to isolate the influence that the magnetic field had on climate since many other factors that influence climate have changed as well.

To investigate the influence that magnetic fields have on climate, we are then forced to investigate the effects of magnetic fields on the individual processes that change climate. We can consider that there are four basic ways to alter a planet's climate, and examine how the presence or absence of global scale magnetic fields might alter them. Changes in the amount of sunlight incident at the top of the atmosphere (i.e., long-term evolution of solar luminosity) can alter climate, but photons are impervious to magnetic fields. Periodic and permanent changes in a planet's orbital parameters (e.g., ellipticity, obliquity, rotation rate) can alter climate, but orbital parameters are determined through gravitational interactions, which are not influenced by magnetic fields. Climate can be altered by variations in the amount of sunlight reflected or absorbed by the atmosphere and surface (i.e., albedo). Albedo is determined by a combination of surface features and cloud cover. There have been some proposals that some cloud formation processes can be influenced by global magnetic fields, but these mechanisms are still vigorously debated and the connections are not entirely clear (*Courtillot et al.,* 2007; *Bard and Delaygue,* 2008). Finally, climate can be altered by changes in the amount of particles in the atmosphere. We proceed by considering the links between magnetic fields and atmospheric abundance variations.

Atmospheric abundance at any given moment is determined by the combination of source and loss processes that have acted on an atmosphere over time. Source mechanisms include outgassing, impacts, and surface interactions such as evaporation, sublimation, or chemistry. Both surface interactions and outgassing can be considered to be independent of planetary magnetic fields, although the amount of internal heat in an object should be correlated both with its outgassing rate and its ability to sustain the convection in its outer core required to generate a magnetic field. Impacts of macroscopic objects such as asteroids and comets are also not influenced by planetary magnetic fields. However, the motion (and hence impact) of microscopic objects, when charged, can be influenced by magnetic fields.

Loss mechanisms for atmospheres include surface interactions, impacts, and escape to space (Tian et al., this volume). We established above that both surface interactions and impact of macroscopic objects are unaffected by magnetic fields, while microscopic impacts may be influenced. The term "escape to space" refers to a suite of thermal and nonthermal processes that accelerate both neutral particles and ions away from a planet's atmosphere. These processes include Jeans (thermal) escape, photochemical escape, and sputtering (also a form of impact) for neutral particles, and pickup, ion outflow, and bulk escape for ions. As described in section 2 below, all the ion removal processes should be influenced by the presence or absence of a planetary magnetic field, and photochemical escape and sputtering should be indirectly affected by magnetic fields as well.

From the above discussion, it is apparent that the strongest link between planetary magnetic fields and climate evolution occurs at the top of an atmosphere, where significant quantities of atmospheric particles can be electrically charged, and where atmospheric densities (and hence collision frequencies) are low. This is also a boundary region for the atmosphere, through which both particles and energy can be added or removed from the planet. To this point our discussion has been framed in general terms, without reference to specific observations, models, or planetary objects. In the remainder of this chapter we narrow our focus to the "terrestrial" atmospheres at Earth, Venus, Mars, and Titan.

All four planetary bodies identified above have magnetic fields generated at the planet that prevent some fraction of the incident plasma from directly entering the atmosphere. The region of space dominated by magnetic fields of planetary origin is termed a "magnetosphere" (Fig. 1). Intrinsic magnetospheres, like those of Earth and the giant planets, result from magnetic fields generated in the interior of the planet, via convective motion of electrically conducting fluid. The strong magnetic moment associated with this motion generates a global-scale magnetic field (typically dominated by the dipole term) that in many cases is capable

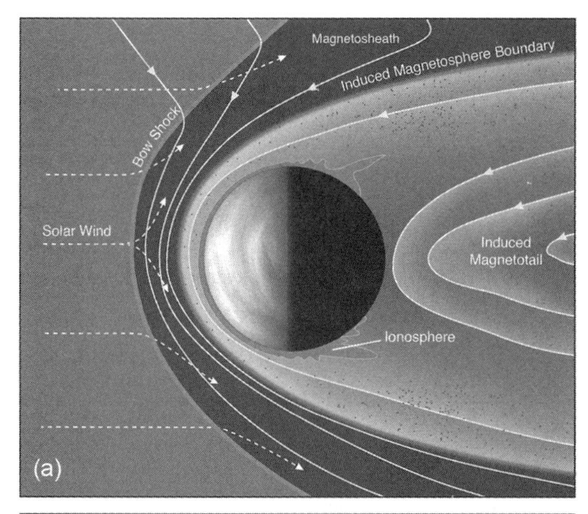

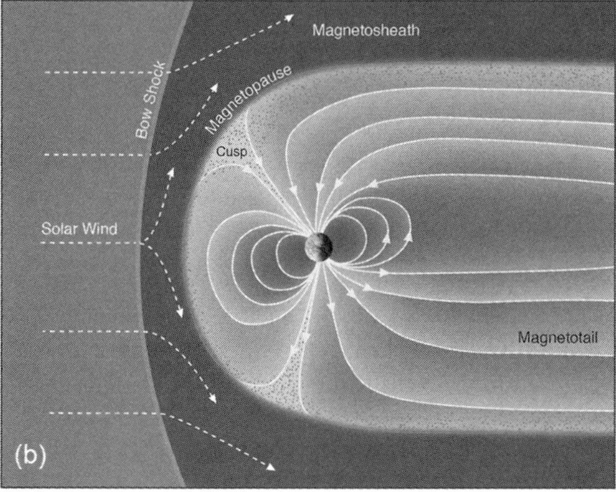

Fig. 1. See Plate 77 for color version. Cartoon diagram showing the structure and main features of **(a)** induced and **(b)** intrinsic magnetospheres. Note the difference in the size of the two interaction regions relative to the size of the planet. Venus, Titan, and Mars have primarily induced magnetospheres, while Earth has an intrinsic magnetosphere. Diagram courtesy of S. Bartlett.

of diverting incident plasma at distances of many times the planetary radius. Induced magnetospheres, by contrast, result from magnetic fields generated by currents induced at the planet by the incident plasma itself. In the case of planets with significant atmospheres (such as Venus, Mars, and Titan), the currents are induced in the ionosphere, and incident plasma is diverted around the planet at much smaller distances than for a typical intrinsic magnetosphere, if at all.

In the following sections we present four examples from the terrestrial planets that highlight the ways in which magnetic fields and climate may be related. We present the specific examples in the context of broader questions, the answers to which might be applied to other planetary objects in our solar system and beyond. In section 2 we ask whether the lack of a magnetic field can lead to important changes in bulk atmospheric abundance by examining the results of observations and models at Mars. In section 3 we ask whether global magnetic fields reduce atmospheric escape rates by examining measurements of ion loss at Earth in comparison with loss rates at other planets. In section 4 we ask whether the lack of a global magnetic field can contribute to the removal of climatically important atmospheric trace gases by exploring the dehydration of the Venus atmosphere. In section 5 we ask whether the lack of a global magnetic field can contribute to the addition of climatically important atmospheric trace gases by presenting observations of Titan's upper atmosphere and bulk atmospheric chemistry. Finally, in section 6, we discuss the implications of these examples for climate evolution at all planetary bodies, and identify promising areas of research for the near future.

2. CAN LACK OF A PLANETARY MAGNETIC FIELD LEAD TO IMPORTANT CHANGES IN ATMOSPHERIC ABUNDANCE?

The planet Mars contains geomorphologic and geochemical features that suggest that liquid water was readily available at the surface at some time in the past (*Jakosky and Phillips,* 2001; *Craddock and Howard,* 2002). Geomorphological features suggesting long-lived liquid surface water include extended dendritic valley networks most likely formed via precipitation, river deltas, and regions of layered rock reminiscent of terrestrial sedimentary layers (*Carr and Clow,* 1981; *Malin and Edgett,* 2001, 2003). Geochemical features include minerals at the surface that require liquid water to form (*Christensen et al.,* 2000; *Poulet et al.,* 2005; *Ehlmann et al.,* 2008), although some of these minerals could have formed in subsurface water. Liquid water is not stable at the surface for long periods of time today, suggesting that some change has occurred at Mars over the last few billion years. An enticing inference is that the martian atmosphere has evolved over time from a more substantial greenhouse atmosphere capable of keeping surface temperatures above freezing for sufficient time to create the many features that remain today (*Pollack et al.,* 1987).

The removal of a climatically significant portion of the martian atmosphere may have been made possible by the absence of a shielding global magnetic field for much of its history. Mars lacks a significant global dynamo magnetic field today, but localized regions of strong crustal magnetization provide compelling evidence that Mars possessed a global dynamo at one time (*Acuña et al.,* 1999). The timing of both the turn-on and shut-off of the dynamo is debated. Correlation of the strength of magnetized regions with surface crater densities (including buried craters) suggests that most regions of crustal magnetization were emplaced prior to ~4.1 b.y. ago (*Lillis et al.,* 2008). However, other studies that model individual anomalies or that consider the lack of magnetization in ancient large impact basins have concluded that a martian dynamo could have persisted until

3.6 b.y. ago or longer (*Schubert et al.,* 2000; *Milbury et al.,* 2012). Regardless, it seems likely that Mars had a global magnetic field that shut off billions of years ago, leaving only crustal magnetic fields that influenced portions of the upper atmosphere. If atmospheric loss processes were active on Mars, what role has its magnetic field played?

An abundance of observational evidence suggests that loss processes are active for the martian atmosphere, and more specifically, that escape to space occurs today. The Phobos 2 spacecraft measured O^+ in the shadow of Mars (*Lundin et al.,* 1989), demonstrating that ionized atmospheric particles are escaping from the atmosphere into the solar wind. Today, the Mars Express spacecraft measures planetary heavy ions escaping from the planet as well (*Barabash et al.,* 2007b), and has begun to probe how ion escape rates vary with external conditions (Fig. 2) (*Courtillot et al.,* 2007; *Bard and Delaygue,* 2008; *Lundin et al.,* 2008a; *Edberg et al.,* 2010; *Nilsson et al.,* 2010, 2011). Based on the above references, escape rates via ion loss processes range from highs of nearly 10^{26} particles per second, down to more meager loss rates of 10^{24} particles per second (an early estimate from Mars Express of a few times 10^{23} particles per second did not account for ions escaping with energies of tens of electron volts or less). In addition, spacecraft observations, theoretical calculations, and computer simulations suggest that both ion and neutral loss processes are active at Mars today, with net escape rates from the atmosphere of 10^{24}–10^{26} particles per second (*Jakosky and Phillips,* 2001; *Craddock and Howard,* 2002; *Lammer et al.,* 2008).

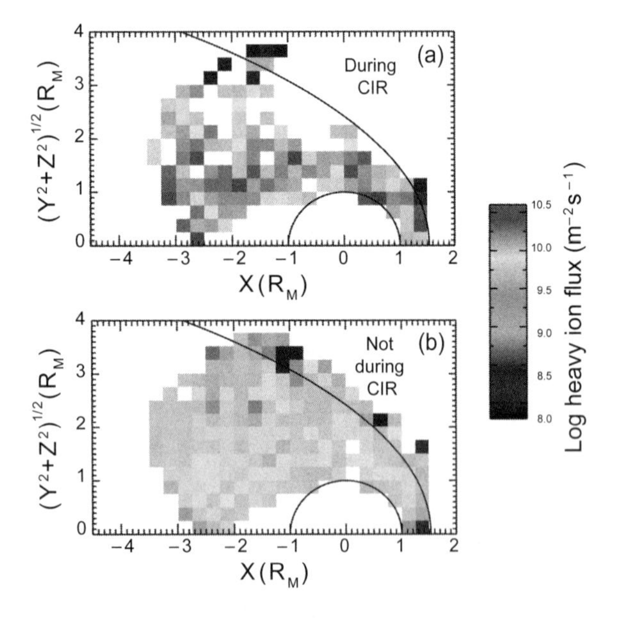

Fig. 2. See Plate 78 for color version. Heavy ion fluxes measured by the Mars Express ASPERA IMA sensor as a function of location with respect to Mars in cylindrical MSO coordinates during co-rotating interaction region (CIR) events associated with **(a)** high solar wind pressure and **(b)** all other times. As solar wind pressure increases, atmospheric escape rates increase. From *Edberg et al.* (2010).

While it is clear that escape to space occurs today at Mars, it is less clear whether this escape, integrated over martian history back to the time of formation of the evidence for liquid surface water, was significant. In other words, did atmospheric escape to space play a prominent role in the climate change inferred at Mars? One piece of observational evidence that suggests that the loss has been significant comes from atmospheric isotope ratios. Above an atmosphere's homopause the atmospheric species are diffusively separated, so the uppermost layers of an atmosphere are enriched in lighter atoms and molecules. Therefore, lighter species are preferentially removed to space via loss processes, leaving the atmosphere enriched in heavier isotopes. Ratios of D/H, $^{38}Ar/^{36}Ar$, $^{13}C/^{12}C$, $^{15}N/^{14}N$, and $^{18}O/^{16}O$ can be used to infer that 50–90% of the martian atmosphere has been lost to space over time (*Jakosky and Phillips,* 2001). Impacts may have contributed to the loss of an additional 90% of the martian atmosphere (*Pollack et al.,* 1987; *Brain and Jakosky,* 1998).

Are the present-day loss rates consistent with the conclusion that loss to space has been significant? The answer depends upon which estimate of loss is accurate. According to *Lundin et al.* (1989, 1990), an O^+ loss rate of 10^{25} is sufficient to strip the entire martian atmosphere of its O (including O incorporated in atmospheric CO_2) in ~100 m.y. Thus, assuming constant loss rates over time and considering only the ion loss processes above, the ancient martian atmosphere could have been about 40 times thicker than it is today — on the order of 250–300 mbar of CO_2. More modest loss rates indicate a correspondingly smaller amount of atmosphere has been stripped away to space. However, there are strong indications that loss rates have declined over time. Present-day observations (Fig. 3) indicate that ion loss rates increase by an order of magnitude or more during moderate solar storm events (*Futaana et al.,* 2008). Solar storms are characterized by periods of disturbed and elevated solar input at Mars, and are characteristic of conditions that may have been prevalent at the planet earlier in solar system history (see discussion in *Luhmann et al.,* 2007). Studies of Sun-type stars suggest that young stars have higher extreme ultraviolet (EUV) fluxes, more flare and storm events, and stronger solar wind (*Ayres,* 1997; *Ribas et al.,* 2005; *Wood et al.,* 2005). Therefore, use of loss rates today to extrapolate backward in time likely underestimates the important of escape to space over martian history. However, the amount of loss is still uncertain. We need stronger constraints on both the present-day loss rates and the processes that control them. The upcoming Mars Atmosphere and Volatile Evolution (MAVEN) spacecraft mission should provide these constraints by measuring atmospheric loss simultaneously with the drivers that lead to loss.

We are still left with the question of whether the absence of a global magnetic field at Mars has influenced escape rates over martian history. This question is not yet

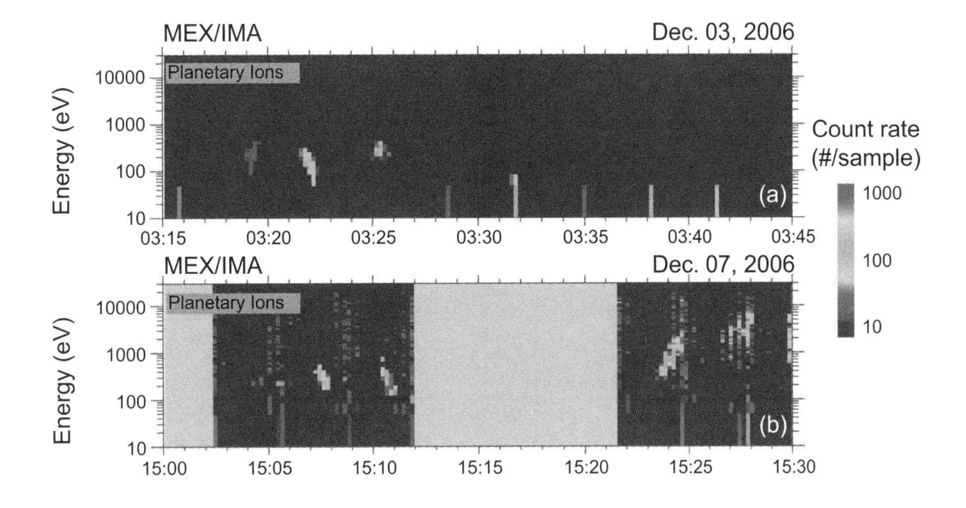

Fig. 3. See Plate 79 for color version. Comparison of fluxes of escaping ions at Mars **(a)** during quiet times and **(b)** during a solar storm. Ion escape rates increased by approximately an order of magnitude during this event. From *Futaana et al.* (2008).

answered for Mars. However, we are able to speculate on how the loss rates from different processes (Fig. 4) might have been different if a global dynamo magnetic field at Mars had persisted to the present day (e.g., *Hutchins et al.,* 1997). The ion loss processes of pickup, outflow, and bulk plasma escape are likely to be different if Mars possessed a magnetic field. Loss rates from the pickup process could be smaller for two reasons. First, exospheric neutrals would be ionized at a smaller rate in regions inside the magnetosphere, where charge exchange and electron impact would be inhibited due to the fact that the solar wind did not have access to a large portion of high-altitude neutrals. Second, exospheric neutrals ionized (by sunlight) on closed magnetospheric field lines could not be carried away from Mars by the convective electric field of the solar wind. If Mars possessed a global magnetic field, outflow would necessarily occur in the two polar cap regions formed in the global magnetosphere. Outflow occurs at Mars today in locations where solar wind magnetic field lines thread the ionosphere, and in cusp regions of the martian remnant crustal magnetic fields. Loss rates of O^+ from outflow range from lows of $\sim 10^{23}$ s^{-1} (from cusp regions only) to highs of 10^{25} s^{-1} (globally) (*Lammer et al.,* 2003; *Ergun et al.,* 2006; *Lundin et al.,* 2008b; *Dubinin et al.,* 2009; *Andersson et al.,* 2010). A global magnetic field would reduce the total area of the atmosphere that is magnetically connected to the solar wind, thereby reducing the size of the atmospheric region from which outflow can occur. However, as argued in section 3 below, the ion outflow rates from Mars may correspondingly increase as pickup ion loss rates decrease, despite the smaller source region. Finally, bulk escape rates are likely to be reduced in the presence of a global field. Bulk loss processes, such as stripping of the ionosphere via the Kelvin-Helmholtz instability, occur when there is shear (magnetic and/or velocity) between the ionosphere and the overlying plasma environment. In the presence of a global field, the plasma flowing past Mars would be deflected at much higher altitudes, well above the ionosphere. Hence the shear would be reduced and bulk escape processes would be inhibited.

Perhaps surprisingly, neutral loss rates may also be changed in the presence of a martian global magnetic field. Neutral loss occurs vis three main processes: Jeans (thermal) escape, photochemical escape, and sputtering. Jeans escape could largely be unaffected, since the process involves the escape of neutral particles. However, the extent to which atmospheric neutrals receive their energy from the solar wind may have an indirect influence on Jeans escape rates. Neutrals can receive energy from the solar wind via impact with incident plasma, or via joule heating processes. Changes in these energy sources caused by the presence or absence of a global magnetic field are likely to be negligible compared to the energy input from incident solar photons. Photochemical escape rates may be slightly reduced as well. For example, photochemical escape of O at Mars is dominated by dissociative recombination of O_2^+. In the presence of a global magnetic field, the production of O_2^+ is likely to be reduced (especially at high altitude where charge exchange and electron impact are presently important at Mars), leading to lower escape rates for the energetic atomic oxygen dissociation products. Additionally, the rate of dissociation of O_2^+ depends upon the ambient electron temperature. Electron temperature, in turn, is influenced by the configuration of magnetic fields in the ionosphere. At Mars, for example, higher electron temperatures are inferred in regions of localized magnetic field (*Krymskii et al.,* 2003; *Nielsen et al.,* 2007). High electron temperatures correspond to low recombination rates, reducing photochemical escape. Thus, the influence of both global and localized magnetic fields at Mars should influence the photochemical escape

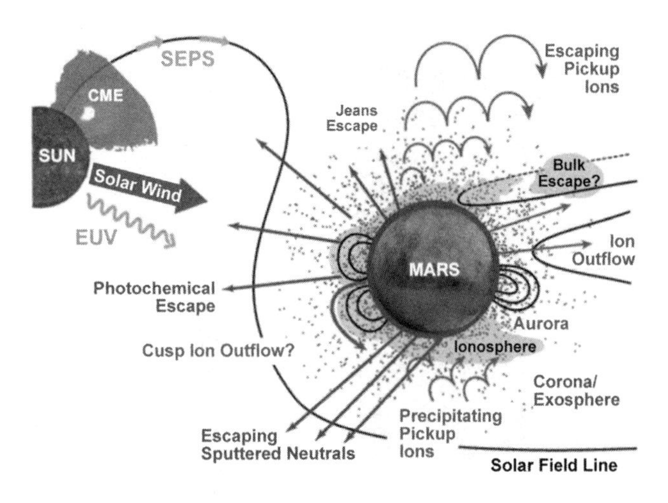

Fig. 4. See Plate 80 for color version. Diagram showing the interaction of the martian upper atmosphere with its surroundings. Neutral atmospheric reservoirs and processes are shown in blue, and ionized reservoirs and processes are shown in red. Solar inputs include EUV photons, solar wind plasma, and solar energetic particles (SEPs). The image is available at *lasp.colorado.edu/home/maven/*.

of O. Finally, atmospheric sputtering rates would likely decrease if Mars had a magnetic field. Sputtering occurs when energetic neutrals or ions impact the collisional exobase region from above, transferring energy via coulomb collisions to atmospheric particles that subsequently escape. Few energetic ions (other than a small fraction of highly energetic solar storm particles, and galactic cosmic rays) would have access to the martian atmosphere if it were protected by a global magnetic field; thus sputtering loss rates would be lower. However, present sputtering rates are thought to be much lower than loss rates from Jeans or photochemical processes (*Lammer et al.,* 2003), although the relative importance of the two processes should vary by species and with time. In fact, sputtering may have been the dominant escape process at the moment the dynamo ceased (*Chassefière et al.,* 2007).

In summary, it is not clear whether the absence of a magnetic field at Mars for much of its history has significantly influenced the atmospheric abundance. We have evidence that escape to space is occurring today, and may have been significant over martian history. And we can speculate on the ways in which a global magnetic field would qualitatively alter present-day loss rates, but don't have good quantitative constraints on the magnitude of these effects. Detailed modeling of the martian plasma interaction and neutral escape in the presence of a magnetic field, including all the appropriate physics, would provide useful constraints on the answer to this question. A final intriguing possibility is that by comparing today's loss rates from regions with crustal fields to loss rates from regions lacking crustal fields we might learn about the importance of magnetic fields for atmospheric escape. Initial studies suggest that crustal fields reduce escape rates by a small but measurable amount (*Lundin et al.,* 2011; *Nilsson et al.,* 2011), although the escape rates were measured regionally and globally (as opposed to the more difficult measurement of escape rates above individual crustal fields), and thus crustal field influences were likely diluted.

3. DOES AN INTRINSIC MAGNETOSPHERE REDUCE ATMOSPHERIC ESCAPE RATES AT ALL?

That the presence of an intrinsic magnetosphere endows a planet such as Earth with a certain shielding effect seems an entirely plausible proposition. Clearly, incident charged particles are deflected by the magnetic field creating the magnetosphere (via the Lorentz force, $\mathbf{v} \times \mathbf{B}$, where $\mathbf{v}$ is velocity and $\mathbf{B}$ is magnetic field), and the planet is thereby directly shielded from the impact of such particles on its atmosphere. In practice the shielding is most effective in the equatorial regions and least effective in the polar regions due to the relative orientation of the planetary magnetic field and the incident plasma flow. For a given particle energy there will be an impact cutoff below a certain magnetic latitude that is higher for lower-energy particles, with the lowest-energy particles unable to reach the planet at any latitude. Particles of sufficiently extreme energy, on the other hand, are incident all over the planet.

However, atmospheric escape at Earth is not dominantly driven by energetic particle impact on the atmosphere, and the single particle picture above is only part of the story. Most of the plasma energy flux received at Earth from the Sun is in the form of solar wind kinetic energy flux (ρv^2, where ρ and v are the mass density and velocity of the solar wind), with a much smaller contribution from electromagnetic (Poynting) energy flux ($\mathbf{E} \times \mathbf{B}/\mu_0$, where μ_0 is the permeability of free space) associated with the interplanetary magnetic ($\mathbf{B}$) and electric ($\mathbf{E}$) fields. However, when the interplanetary field (re)connects with the terrestrial magnetic field at the dayside magnetopause, a magnetohydrodynamic dynamo is created that converts flow energy into electromagnetic energy that is dissipated in the ionosphere, where it drags the plasma through the neutral atmospheric gas (Fig. 5). A substantial fraction of the solar wind kinetic energy flux is converted to Poynting flux and transmitted along magnetic flux tubes linked to the ionosphere to accomplish this dissipation, with corresponding deceleration applied to the solar wind plasma driver flow.

Strangeway et al. (2005) have observed that in the dayside cusp region, enhanced ionospheric outflow flux (over and above polar wind H^+ fluxes) scales directly and strongly with the incident Poynting flux and the incident precipitating soft electron flux (Fig. 6). Simple theoretical calculations, based on a generalized form of Jeans escape (*Moore and Khazanov,* 2010) suggest that the observed outflow scalings can be understood as a combination of enhanced ambipolar diffusion and transverse heating of O^+,

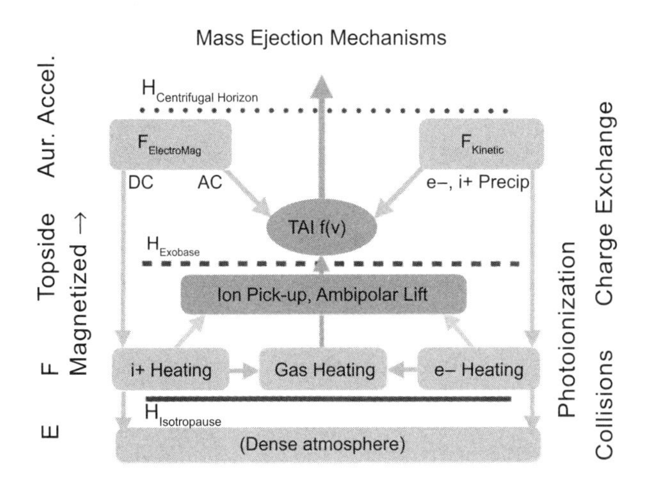

Fig. 5. Schematic diagram of ionospheric outflow mechanisms illustrating the role of ion pickup, ambipolar electric field, and centrifugal forcing in the outflow of ionospheric plasma. Important altitudes (isotropause, exobase, centrifugal horizon) are shown as horizontal lines in different styles. Different ionospheric regions are identified at far left, and important processes for particles are identified at far right. "F" refers to a force, "H" to a height, "AC" and "DC" to alternating and direct currents, and "TAI" to transversely accelerated ions. After *Moore and Khazanov* (2010) and *Strangeway* (2005).

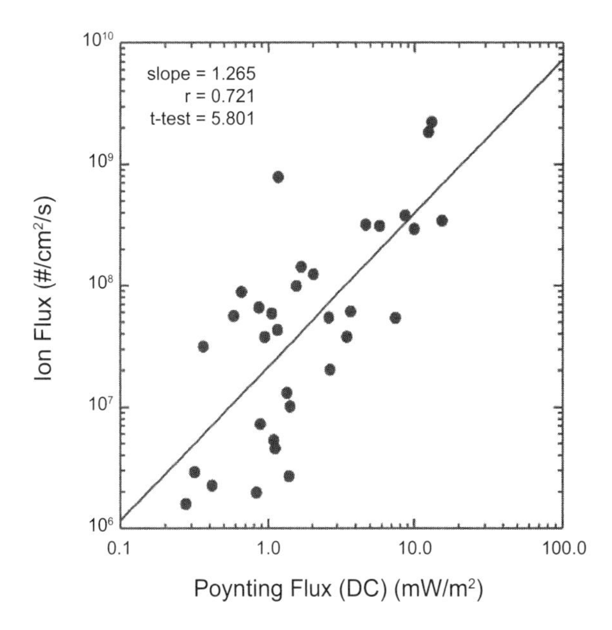

Fig. 6. Escape flux of atomic O ions from Earth vs. Poynting flux at 4000-km altitudes. From *Strangeway et al.* (2005).

with the latter doing most of the work to lift plasma out of Earth's gravity. The enhanced ambipolar diffusion results from increased electron temperatures in the F region of the ionosphere caused by soft electron precipitation. If we attribute atmospheric outflow to ionization and heating by solar wind kinetic energy flux, then we must ask how the flux of such energy into the atmosphere is influenced by the presence or absence of a magnetosphere.

The relevant comparison appears then to be the effect of a magnetosphere on the energy flux delivered to the upper atmosphere. With no magnetosphere, the solar wind plasma is incident almost directly upon the atmosphere, or rather with its partially ionized ionospheric layer, similar to the case at Venus, Mars, or Titan. For this case, we may then consider the energy flux of the solar wind to be rather evenly distributed over the entire dayside of the Earth, although with a varying angle of incidence analogous to that of sunlight, so that the greatest incident fluxes are delivered near the subsolar point, falling smoothly to zero as the solar zenith angle falls to 90° at the terminators. The peak solar wind energy flux depends on solar wind intensity of course, but on average it is typically less than 0.2 mW/m^2 (*Phillips et al.,* 1995).

In contrast, when a magnetosphere is present, a magnetopause is formed at the boundary with the solar wind and interplanetary magnetic field. The magnetic field lines that thread that layer connect to the auroral zones as rings around each pole. Reconnection of the fields occurs continuously, although the coupling to the atmosphere depends strongly on the orientation of the interplanetary field (*Palmroth et al.,* 2003). For maximal (re)connection of the two fields, and maximal energy transfer, the net effect is to concentrate the solar wind energy incident upon the magnetopause into the much smaller auroral zones, and the energy flux incident upon the atmosphere in those ring-like haloes around the Earth's poles will typically range from a few milliwatts per meters squared to 100 mW/m^2 (*Strangeway et al.,* 2005).

Thus, we can summarize the net effect of the presence of a magnetosphere as the focusing or concentration of the energy flux of the solar wind into a small fraction of the total atmosphere, typically reaching values 10–100 times larger than in the case with no magnetosphere. The presence of a magnetosphere therefore causes the concentration of solar wind energy flux into much smaller regions having much larger energy flux. This means that the effect is to greatly enhance the outflow flux of gravitationally bound heavier species (O^+), with minimal effect on the outflow of lighter, unbound species (H^+). On a planet like Mars, where O^+ is much less tightly bound than at Earth (the escape velocity from the exobase is ~25% that of Earth), the presence or absence of a magnetosphere should have made less difference to the long-term escape of the atmosphere than at Earth (here we have assumed comparable thermal ion velocities at both planets), where the magnetosphere substantially increases the amount of O^+ that escapes, compared with an unmagnetized Earth.

All ionized outflows from planetary atmospheres must have a source in the planetary neutral gas atmosphere. We note that the presence or absence of magnetospheres should have little effect on escaping neutral particles, except where neutral escape is caused by ion acceleration, followed by sputtering or charge exchange. For planetary objects with neutral escape rates that are comparable to or exceed ion escape rates, then, whether a planet has a magnetosphere should provide only a perturbation to the overall escape flux. Estimates of ion and neutral escape rates at Mars, for example, suggest that this is the case (*Lammer et al.,* 2008). Observations at Earth, however, suggest that enhanced neutral escape rates accompany and are comparable with enhanced ion escape rates caused by solar wind energy dissipation (*Gardner and Schunk,* 2004, 2005). Regardless, the mixing ratios of escaping particles may be considerably affected even when neutral escape is considered, thereby influencing the chemistry that occurs in the remaining atmosphere.

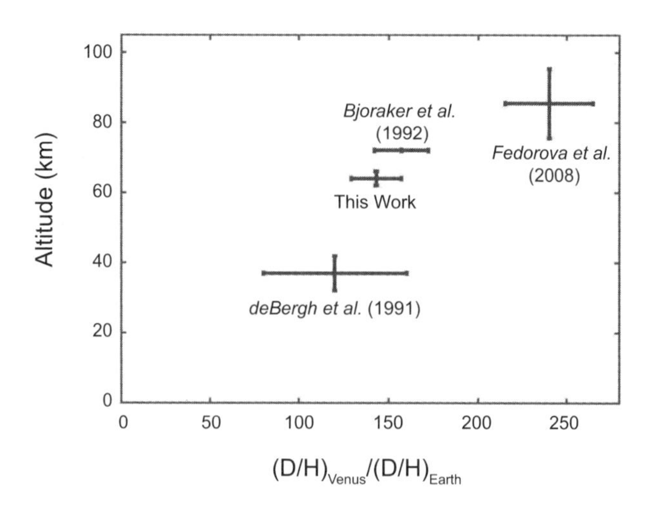

Fig. 7. Measurements of the Venus D/H ratio at various altitudes. From *Matsui et al.* (2012).

4. CAN LACK OF A PLANETARY MAGNETIC FIELD CONTRIBUTE TO THE REMOVAL OF CLIMATICALLY IMPORTANT TRACE GASES?

The argument was made above that the size of a planet's effective cross section for collecting solar wind energy available to power atmosphere escape processes can be greatly increased by an intrinsic planetary magnetosphere, thus diminishing its shielding role. While this seems straightforward, it is in fact greatly complicated by what happens to the collected energy. For example, at Earth its fate involves a number of different dissipative processes, including particle energization and resistive heating of the thermosphere, which is a large sink. In addition, not all the ions that are observed flowing out of the polar atmosphere in response to the entering energy escape into space. Many become part of the magnetospheric particle population, including the plasma sheet and the ring current associated with geomagnetic storms, while others precipitate back into the atmosphere. What fraction of the observed ion outflow escapes is still debated (*Seki et al.,* 2011). In contrast, largely unmagnetized planets such as Venus have a relatively simple and direct solar wind interaction, where there is unavoidable interpenetration of the upper atmosphere and solar wind and a dynamic boundary layer between the solar wind and ionospheric plasmas. In principle, any atmospheric ionized atom or molecule that gains enough energy to escape Venus' gravitational field and makes it to the collisionless region above the exobase region is free to leave.

The arrival of spacecraft, including atmospheric probes, at Venus cemented the case for a "twin Earth" whose crushing CO_2 atmosphere and caustic near-surface environment render it extremely inhospitable to life as we know it. Among the discoveries made by the Pioneer Venus (PV) probe, which made *in situ* isotopic measurements of the atmosphere, was a very high ratio of deuterium to hydrogen (D/H) compared to Earth's atmosphere (factor of ~100 or more; see Fig. 7) (*Donahue et al.,* 1982). This immediately suggested that Venus had experienced much greater loss of the lighter atmospheric constituent H than Earth. Along with this measurement, water was found to be practically absent (*Hoffman et al.,* 1980). Solar system formation models favor scenarios where the early terrestrial planets shared a comparable inventory of water. Yet the Earth's oceans contain kilometers global equivalent depth, whereas Venus has only a few precipitable micrometers. Although Mars can hide substantial water ice in its polar caps and regolith, obvious surface temperature arguments prevent Venus from having that alternative.

While the D/H discovery was made by the PV probe, the Pioneer Venus orbiter made observations of what appeared to be energized atmospheric ions escaping in the solar wind wake in comet-tail-like fashion. Such ions had been seen before, in the 1970s, on the Venera missions (e.g., *Vaisberg et al.,* 1977). The identification of their composition was only indirectly possible in these early measurements, but the evidence pointed to a large contribution from O^+ ions that was later confirmed (Fig. 8) on Venus Express (VEX) (*Barabash et al.,* 2007a). Although Venus receives more solar radiant energy than Earth at its 0.73 AU heliocentric distance, and experiences photochemical processes in its CO_2-dominated atmosphere that, as at Mars, produce a suprathermal "corona" of atomic oxygen around it, the fraction of hot O that is at and above the escape speed of ~11 km/s is in this case insignificant. Apparently the solar wind interaction accelerates initially gravitationally bound heavy ions, including O^+, to sufficient speeds to be lost to the planet. Together with the D/H ratio, these observations hinted that perhaps the relative dearth of water at Venus could be due to the combination of H escape, accomplished in large part in neutral form by the Jeans or thermal escape process that was likely enhanced early in Venus' life, together with solar-wind-interaction-enabled, O^+ escape. A challenge to researchers has been to establish whether currently observed processes can account for the loss of

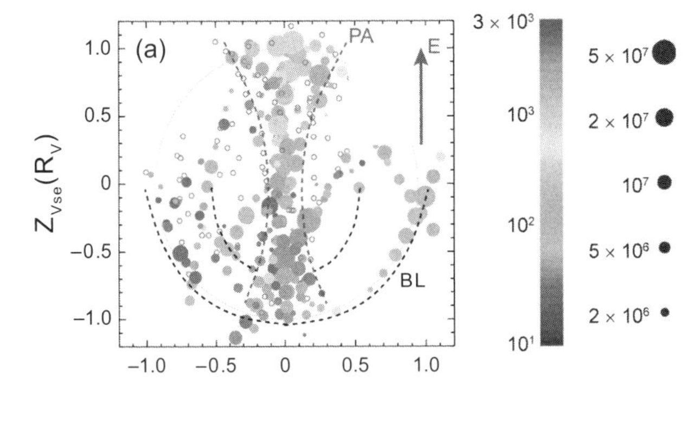

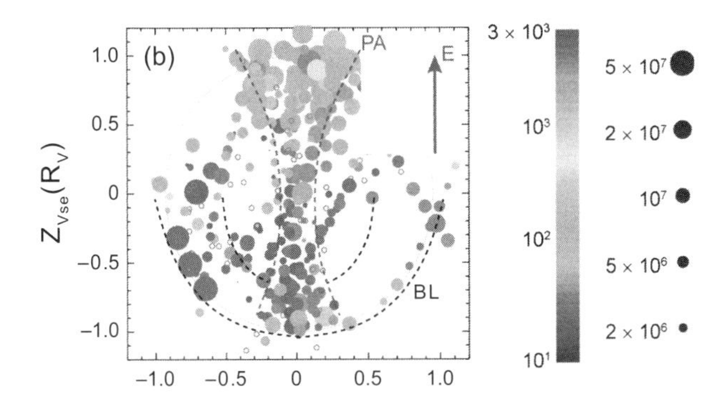

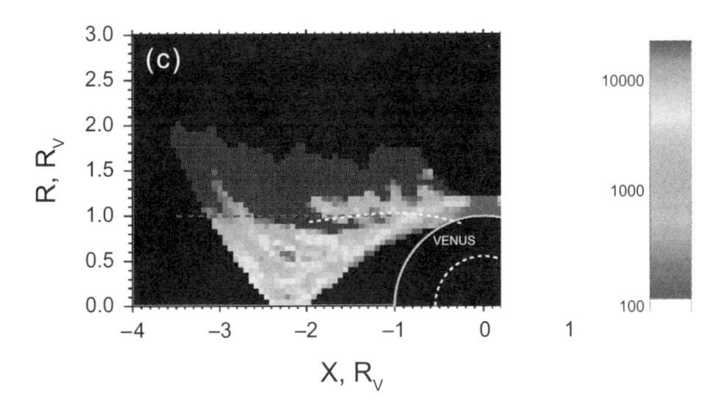

Fig. 8. See Plate 81 for color version. Measurements of ion escape by Venus Express. Fluxes (size) and energies (color) of **(a)** O^+ and **(b)** H^+ ions escaping from the Venus atmosphere are shown for a cross-section of the Venus magnetotail downstream from the planet (*Barabash et al.,* 2007a). The coordinate system used has the electric field in the solar wind directed along the y-axis. Fluxes of escaping ions of planetary origin (with mass to charge ratio >14) are shown in cylindrical coordinates in **(c)**. From *Fedorov et al.* (2008).

an Earth-like inventory of water, or whether the O left behind from water molecule dissociation and H escape was somehow sequestered by crustal material oxidation — an option difficult to realize without an unreasonable degree of overturning and resurfacing episodes over time (*Fegley et al.,* 1997).

The typically measured O^+ escape rates reported in the literature, based on spacecraft observations, are in the neighborhood of 10^{23}–10^{25} O^+/s (e.g., *Hartle and Grebowsky,* 1995; *Barabash et al.,* 2007a; *Fedorov et al.,* 2011). These rates fall short by several orders of magnitude compared to those needed to lose an ocean's worth of oxygen over ~3.5–4.0 b.y. — assuming that is when the early water inventory started its downward trend. An episode of hydrodynamic outflow of H driven by early high solar EUV fluxes (e.g., *Sekiya et al.,* 1980) could have managed the H loss, but the O remains problematic. Was it collisionally dragged out by the early H outflow, or is the answer in what we are observing today? Several theories have been proposed for enhancing current O^+ loss rates within the framework of the modern-day processes. In particular, enhanced solar EUV fluxes, solar wind, and solar activity in combination can be argued to increase loss rates significantly (*Luhmann et al.,* 2007, 2008). One proposal invokes a process not particularly significant today, atmospheric sputtering, that may have been dominant under early solar system conditions (*Russell et al.,* 2007). The question is whether any of these hypotheses can be proven by the analysis of ongoing observations.

But let us return to the issue of the currently observed comparable losses of O^+ ions at Earth described above, where water has clearly been retained. Earth's O^+ escape rates have also been shown to respond to disturbed solar wind conditions (e.g., see review by *Moore and Horwitz,* 2007, and references therein; see also *Cully et al.,* 2003). Are the details of each planet's history and how the losses occurred the answer to the ultimate impact? Early in Earth's history the CO_2 in its atmosphere was transformed into carbonate rock by the presence of liquid water — a process that did not happen to anything approaching the same extent on Venus — likely because of its slightly closer distance to the Sun and resulting "runaway greenhouse" conditions (*Pollack,* 1971; *Kasting,* 1988). The resulting CO_2 composition of Venus' atmosphere has provided the photochemistry that keeps O in the upper atmosphere and a major constituent of the ionosphere-necessary ingredients for solar wind erosion to be effective. Could this chemistry path, *in combination* with the weak planetary field of Venus that enables direct solar wind energy deposition and energization of O^+, be the answer? Could the O^+ loss processes at Venus have been enhanced by orders of magnitude over time, while equivalent enhancements did not occur on Earth under similar solar conditions? Perhaps the more pertinent question to ask in the context of this chapter is how loss of H_2O and O from Venus over time would have altered its climate evolution in the first place. Further work is clearly needed to better understand how two initially quite similar terrestrial planetary bodies took greatly divergent paths because of the differences of their intertwined thermal, interior, atmospheric, and space environment histories. The significance of the weak planetary field of Venus in the overall picture of the planet's fate, including its climate, remains to be figured out.

5. CAN LACK OF A PLANETARY MAGNETIC FIELD CONTRIBUTE TO THE ADDITION OF CLIMATICALLY IMPORTANT ATMOSPHERIC GASES?

Planetary magnetic fields shield a planet from incident charged particles, thereby preventing them from being absorbed by the atmosphere. However, even if all the incident plasma on an object were absorbed (in the absence of a global magnetic field and a conducting ionosphere), it is not clear whether it would be important for determining the bulk atmospheric composition. For example, the total mass of charged particles from the solar wind encountering an unmagnetized Earth *over its entire history* would be ~2% of Earth's present-day atmospheric mass. Furthermore, unmagnetized planets with atmospheres form induced obstacles to incident flowing plasma in their ionosphere, so that most incident plasma is deflected around the planet even in the absence of a global field. However, there is some suggestion that the abundance of certain atmospheric trace gases can be influenced by the presence or absence of a planetary magnetic field. For example, the helium abundance in the martian atmosphere has proven difficult to explain based on internal processes (*Krasnopolsky et al.,* 1994; *Barabash et al.,* 1995), with one suggestion being that the atmosphere captures helium ions from the solar wind that subsequently recombine to form neutral helium (*Krasnopolsky and Gladstone,* 1996). Both observations and plasma simulations support this idea (*Chanteur et al.,* 2009; *Stenberg et al.,* 2011).

Saturn's moon Titan is another place where trace gas abundances may be influenced by the absence of a global magnetic field. Titan's chief atmospheric constituents are N_2 and CH_4, with trace amounts of hydrogen and a wide variety of hydrocarbons. The Cassini plasma spectrometer (CAPS) observed O^+ precipitating into Titan's atmosphere from Saturn's magnetosphere (*Hartle et al.,* 2006). Once in the atmosphere, O^+ is likely to recombine, forming neutral O that can participate in a wide variety of chemical reactions with other atmospheric constituents. Some of the most important reactions lead to the production of atmospheric CO, whose abundance has proved difficult to explain at Titan (*Hörst et al.,* 2008). If H_2O or OH also precipitate into Titan's atmosphere (which is likely considering Enceladus' prolific production of H_2O products in Saturn's magnetosphere), then the observed abundance of CO_2 at Titan can also be explained through the reaction $CO + OH \rightarrow CO_2 + H$ (Fig. 9). *Hörst et al.* therefore proposed that the CO in Titan's atmosphere need not be primordial or subsequently outgassed. It is important to note, based on the above description, that if Titan possessed a global magnetic field then its atmosphere would lack relevant amounts of both CO and CO_2.

The O^+ ions absorbed by Titan's atmosphere are deposited in the ionosphere. Titan's ionosphere has the most complex known composition in the solar system (*Cravens et al.,* 2009). With almost 50 species (essentially hydrocarbons and nitriles) below 100 amu/q detected by the Cassini Ion

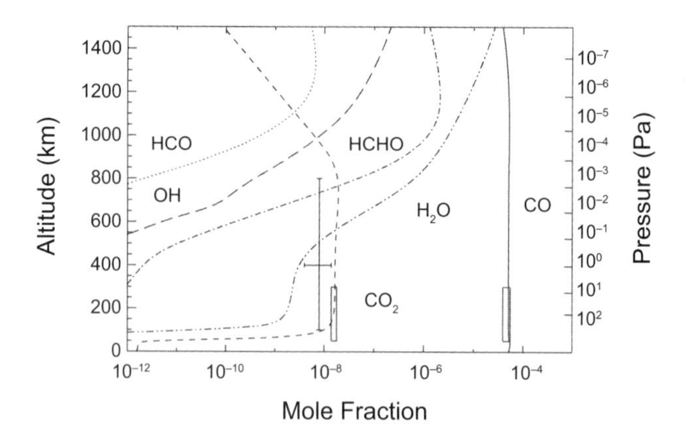

Fig. 9. Modeled mole fractions of various oxygen-bearing species in Titan's atmosphere, assuming an external source of O^+ from Saturn's magnetosphere, and subsequent atmospheric chemistry. Measured values (and uncertainties) of CO_2 and CO are shown as boxes. From *Hörst et al.* (2008).

and Neutral Mass Spectrometer (INMS) (Cravens et al.), Titan's ionosphere is also composed of ion species with mass up to several hundred amu/q [consistent with polycyclic aromatic hydrocarbons (*Crary et al.,* 2009)]. Negative ions with masses up to a few thousands of amu/q were also discovered in the ionosphere by the Cassini Plasma Spectrometer (CAPS) (*Coates et al.,* 2007). These heavy negative ions could be transported deeper in the atmosphere, serving as precursors for Titan's haze, which lies a few hundred kilometers below the ionosphere (*Waite et al.,* 2007).

What causes the complex chemistry in the ionosphere? Titan's ionospheric density peaks between 950 and 1250 km altitude, nearly half a planetary radius above the surface (*Agren et al.,* 2009). The ionosphere is asymmetric, with densities larger by ~50% on the dusk side compared to dawn (*Kliore et al.,* 2008). Titan is nearly always embedded in Saturn's magnetosphere, so that Titan's ionosphere is produced by a combination of solar EUV radiation incident on the dayside, with a significant and possibly dominant contribution on the nightside from particles originating in Saturn's magnetosphere (*Galand et al.,* 2013; *Kliore et al.,* 2011). The lack of a global magnetic field impacts the ionospheric chemistry in three ways. First, the lack of a magnetic field enables the nightside ionospheric source of O (and perhaps H_2O) described above, and thus plays a role in the availability of important species. Next, magnetic fields influence the diffusion times of ions. Depending upon the difference between timescales for chemical reactions and timescales for diffusion, magnetic fields may play a role in controlling the availability of reactants at given altitudes, and thus the overall chemistry of the ionosphere. Finally, electron transport is influenced by the orientation and strength of magnetic fields. In an intrinsic magnetosphere, vertical electron transport is efficient in auroral regions, where local magnetic fields are nearly vertical, and vertical transport is inhibited at lower latitudes where local fields are horizontal. Since electrons contribute to ionospheric chemistry, magnetic fields may additionally influence chemistry by influencing electron motion.

6. SUMMARY AND FUTURE NEEDS

Above, we have established that a number of possible links may exist between magnetic fields and climate. This is because many of the processes that can alter climates on the terrestrial planets act at the upper boundary of a planet's atmosphere. Significant quantities of atmospheric particles are ionized in these regions, and their motion is therefore governed in part by ambient magnetic and electric fields. Thus, it is possible that the presence of a planetary magnetic field can affect the addition or removal of atmospheric particles at a planet, and therefore the abundance and composition of both the bulk atmosphere and trace gases important for climate.

The atmosphere of Mars is thought to have changed from a state capable of supporting liquid surface water to today's cold, thin climate. One means of making this change is via the escape of atmospheric particles to space, through interaction with the Sun and solar wind. The escape of ionized atmospheric particles may be appreciably reduced in the presence of a global martian magnetic field because magnetic fields deflect moving charged particles — both the incident solar wind and atmospheric ions that might have escaped the planet in the absence of a global field to trap them. The escape rate of neutrals may be indirectly reduced as well in the presence of a global field because the solar wind may provide a small fraction of the energy driving neutral escape.

Observations from Earth suggest that the flux of atmospheric ions leaving the polar regions is directly proportional to the solar wind energy encountering the planet's magnetosphere. This, in turn, suggests that the presence of an intrinsic magnetosphere may not prevent atmospheric erosion, but rather transfers energy and momentum from the solar wind to localized regions of the atmosphere, leaving global escape rates largely unchanged. However, a number of issues are not clear at present, including whether all the observed upflowing ions escape the planet's magnetosphere and whether all the energy collected by the magnetosphere goes into driving escape. There are also other proposed indirect climate impacts under investigation, such as the role of magnetic fields in cloud formation processes (and consequently albedo alteration).

At Venus, the lack of a global magnetic field may have contributed to the dehydration of the planet's atmosphere. Dissociation of atmospheric water into hydrogen and oxygen likely resulted in the hydrogen being removed to space via Jeans escape. However, the lack of a magnetic field may have also led to the stripping of oxygen (and additional hydrogen) from the atmosphere, an ongoing process that is still observed by spacecraft today. This removal of a chemically reactive greenhouse gas-contributing species from the atmosphere, perhaps made possible by the lack of a global magnetic field, could have profoundly affected the climate of the planet.

At Titan, the lack of a global magnetic field allows O^+ from Saturn's magnetosphere (likely originating at Enceladus) to impact and be absorbed by the atmosphere. The O^+ makes chemical reactions possible in the atmosphere that lead to the formation of CO and CO_2, which are important trace gases in the lower atmosphere. The lack of a magnetic field also makes complex chemistry possible in the ionosphere and upper atmosphere, possibly leading to the high-altitude haze on Titan that affects energy balance in the atmosphere.

The examples above demonstrate that the relationship between planetary magnetic fields and climate evolution is likely to be different at each object. Each planet has a different magnetic history, and the timing and relative importance of different atmospheric source and loss processes has been different at each body. It is therefore possible or even likely that the presence or absence of a magnetic field at one body has different consequences than the presence or absence at another.

What is needed to better constrain the importance of the connections between magnetic fields and climate? Insight is likely to come from the continually improving models for the atmospheres of these planets. While state-of-the-art detailed models exist for lower atmospheres, upper atmospheres, and plasma environments of these planets, there is considerable challenge in coupling these models from the ground to space and applying them to time periods long enough to be important for global climate. Several groups, however, have begun this task. Even without "whole atmosphere" models, significant progress could be made by using existing models in novel ways. For example, *Kallio et al.* (2008) implemented a global model for the martian interaction with the solar wind run both with and without a martian dynamo magnetic field of varying strengths and compared the results (Fig. 10). While their work focused primarily on the structure of the magnetosphere, model exercises such as this could provide insight into the rate of change of atmospheric abundance on a planet due to its interaction with its surrounding plasma environment. Similarly, models of planetary plasma interactions at earlier epochs in solar system history, constrained by observations of Sun-like stars, may reveal the role that early magnetospheres played in the evolution of paleomagnetospheres (*Sterenborg et al.,* 2011). One must be cautious, however, that a chosen model is appropriate for examining the processes that influence atmospheres and climate. For this reason, activities that compare the results of different models using the same input and boundary conditions, such as those undertaken by *Brain et al.* (2010), are also a useful tool.

Observations are also an important way to make progress. Many relevant observations of Venus, Earth, Mars, and Titan have been made over the past decades, and have led to important insight into the physical processes described in this chapter. But in order to evaluate how these pro-

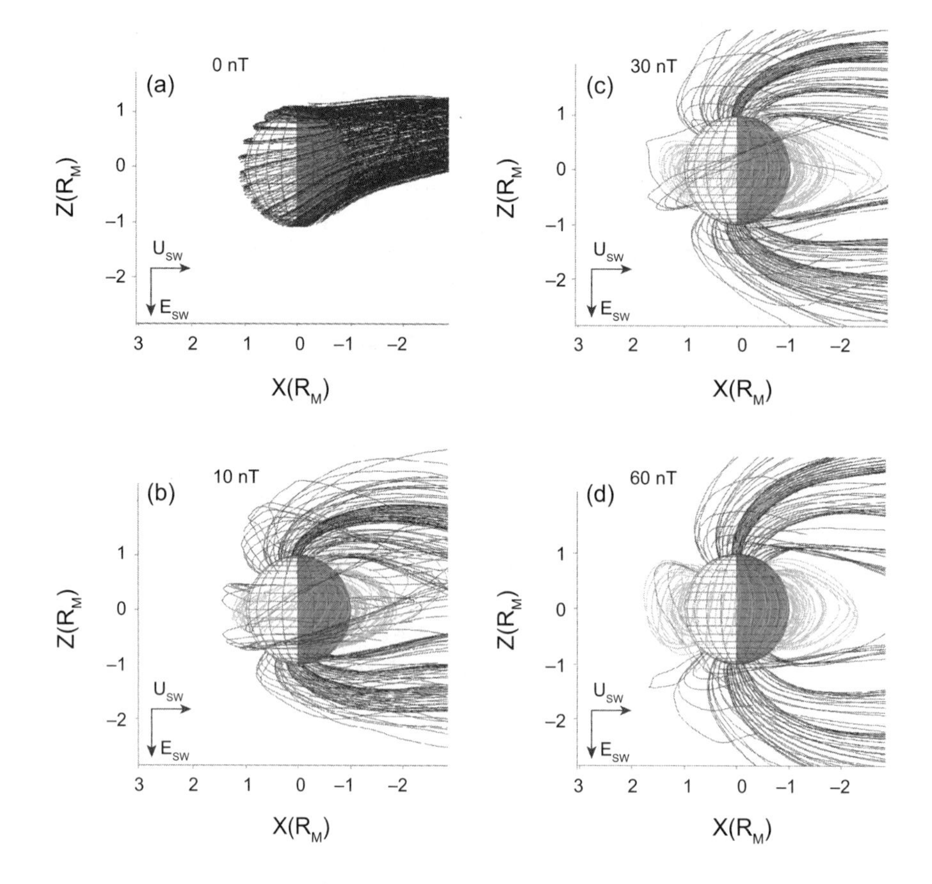

Fig. 10. Modeled magnetic field configuration for Mars, assuming varying strengths of a global dynamo magnetic field. The fraction of field lines that are closed (connect at both ends to the planet) increases with the strength of the planetary dynamo field. From *Kallio et al.* (2008).

cesses are influenced by the presence or absence of global magnetic fields, two kinds of observing strategies would be most useful. First, a comprehensive set of observations made at a single planet is required to evaluate atmospheric escape processes and how they vary with external drivers. The relevant measurements to date for a given planet have come from many spacecraft, separated temporally and/or physically. Spacecraft missions focused comprehensively on atmospheric escape, such as the upcoming MAVEN mission to Mars, should provide a means of evaluating both the absolute and relative importance of all the relevant escape processes for Mars. Second, comparisons of the net atmospheric escape at both magnetized and unmagnetized planets are critical to evaluating the importance of magnetic fields. Some progress is being made in this area already, but it can be difficult to compare the measurements made in Earth's magnetosphere, for example, to those made downstream from Venus.

A third possibility is the comparison of the exchange of energy and mass between the solar wind and the martian atmosphere, comparing regions near localized crustal magnetic fields with unmagnetized regions. By examining the differences in processes in these two regions we may be able to evaluate the role that magnetic fields play in controlling atmospheric abundance when all other controlling variables are identical. One caveat, however, is the fact that atmospheres are primarily composed of neutral particles that are free to move, unconstrained by magnetic fields. So differences in atmospheric escape near and far from crustal fields may be at least partially influenced by the flow of neutral particles between these two regions.

Finally, observations of exoplanet atmospheres can both inform and be informed by our knowledge of the solar system bodies discussed here. The number and diversity of known exoplanet atmospheres is rapidly increasing, along with our capability to study them in detail. It may be that the information required to determine the link between planetary and magnetic fields and climate comes from outside our own solar system.

REFERENCES

Acuña M. H., Connerney J. E. P., Ness N. F., et al. (1999) Global distribution of crustal magnetization discovered by the Mars Global Surveyor MAG/ER experiment. *Science, 284,* 790, DOI: 10.1126/science.284.5415.790.

Agren K., Wahlund J. E., Garnier P., Modolo R., Cui J., Galand M., and Muller-Wodarg I. (2009) On the ionospheric structure of Titan. *Planet. Space Sci., 57(1),* 1821–1827, DOI: 10.1016/j.pss.2009.04.012.

Andersson L., Ergun R. E., and Stewart A. I. F. (2010) The combined atmospheric photochemistry and ion tracing code: Reproducing the Viking Lander results and initial outflow results. *Icarus, 206(1),* 120–129, DOI: 10.1016/j.icarus.2009.07.009.

Ayres T. R. (1997) Evolution of the solar ionizing flux. *J. Geophys. Res., 102(E),* 1641–1652, DOI: 10.1029/96JE03306.

Barabash S., Kallio E., Lundin R., and Koskinen H. (1995) Measurements of the nonthermal helium escape from Mars. *J. Geophys. Res., 100(A),* 21307–21316, DOI: 10.1029/95JA01914.

Barabash S., Fedorov A., Sauvaud J. J., et al. (2007a) The loss of ions from Venus through the plasma wake. *Nature, 450(7170),* 650–653, DOI: 10.1038/nature06434.

Barabash S., Fedorov A., Lundin R., and Sauvaud J.-A. (2007b) Martian atmospheric erosion rates. *Science, 315(5811),* 501–503, DOI: 10.1126/science.1134358.

Bard E. and Delaygue G. (2008) Comment on "Are there connections between the Earth's magnetic field and climate?" by V. Courtillot, Y. Gallet, J.-L. Le Mouël, F. Fluteau, and A. Genevey, EPSL 253, 328, 2007. *Earth Planet. Sci. Lett, 265(1),* 302–307, DOI: 10.1016/j.epsl.2007.09.046.

Brain D. A. and Jakosky B. M. (1998) Atmospheric loss since the onset of the martian geologic record: Combined role of impact erosion and sputtering. *J. Geophys. Res., 103(E),* 22689–22694, DOI: 10.1029/98JE02074.

Brain D., Barabash S., Boesswetter A., et al. (2010) A comparison of global models for the solar wind interaction with Mars. *Icarus, 206(1),* 139–151, DOI: 10.1016/j.icarus.2009.06.030.

Carr M. H. and Clow G. D. (1981) Martian channels and valleys — Their characteristics, distribution, and age. *Icarus, 48,* 91–117, DOI: 10.1016/0019-1035(81)90156-1.

Chanteur G. M., Dubinin E., Modolo R., and Fraenz M. (2009) Capture of solar wind alpha-particles by the martian atmosphere. *Geophys. Res. Lett., 36,* L23105, DOI: 10.1029/2009GL040235.

Chassefière E., Leblanc F., and Langlais B. (2007) The combined effects of escape and magnetic field histories at Mars. *Planet. Space Sci., 55(3),* 343–357, DOI: 10.1016/j.pss.2006.02.003.

Coates A. J., Crary F. J., Lewis G. R., Young D. T., Waite J. H., and Sittler E. C. (2007) Discovery of heavy negative ions in Titan's ionosphere. *Geophys. Res. Lett., 34(2),* 22103, DOI: 10.1029/2007GL030978.

Courtillot V., Gallet Y., Le Mouel J.-L., Fluteau F., and Genevey A. (2007) Are there connections between the Earth's magnetic field and climate? *Earth Planet. Sci. Lett., 253(3),* 328–339, DOI: 10.1016/j.epsl.2006.10.032.

Craddock R. A. and Howard A. D. (2002) The case for rainfall on a warm, wet early Mars. *J. Geophys. Res.-Planet, 107(E),* 5111, DOI: 10.1029/2001JE001505.

Crary F. J., Magee B. A., Mandt K., Waite J. H., Westlake J., and Young D. T. (2009) Heavy ions, temperatures and winds in Titan's ionosphere: Combined Cassini CAPS and INMS observations. *Planet. Space Sci., 57(1),* 1847–1856, DOI: 10.1016/j.pss.2009.09.006.

Cravens T. E., Yelle R. V., Wahlund J. E., Shemansky D. E., and Nagy A. F. (2009) Composition and structure of the ionosphere and thermosphere. In *Titan from Cassini Huygens* (R. H. Brown et al., eds.), pp. 259–295. Springer, New York, DOI: 10.1007/978-1-4020-9215-2_11.

Cully C. M., Donovan E. F., Yau A. W., and Arkos G. G. (2003) Akebono/Suprathermal Mass Spectrometer observations of low-energy ion outflow: Dependence on magnetic activity and solar wind conditions. *J. Geophys. Res., 108,* A1093, DOI: 10.1029/2001JA009200.

Donahue T. M., Hoffman J. H., Hodges R. R., and Watson A. J. (1982) Venus was wet — A measurement of the ratio of deuterium to hydrogen. *Science, 216,* 630–633, DOI: 10.1126/science.216.4546.630.

Dubinin E., Fraenz M., Woch J., Barabash S., and Lundin R. (2009) Long-lived auroral structures and atmospheric losses through auroral flux tubes on Mars. *Geophys. Res. Lett., 36(8),* 08108, DOI: 10.1029/2009GL038209.

Edberg N. J. T., Nilsson H., Williams A. O., Lester M., Milan S. E., Cowley S. W. H., Fränz M., Barabash S., and Futaana Y. (2010) Pumping out the atmosphere of Mars through solar wind pressure pulses. *Geophys. Res. Lett., 37(3),* 03107, DOI: 10.1029/2009GL041814.

Ergun R. E., Andersson L., Peterson W. K., Brain D., Delory G. T., Mitchell D. L., Lin R. P., and Yau A. W. (2006) Role of plasma waves in Mars' atmospheric loss. *Geophys. Res. Lett., 33(1),* 14103, DOI: 10.1029/2006GL025785.

Fedorov A., Ferrier C., Sauvaud J. A., et al. (2008) Comparative analysis of Venus and Mars magnetotails. *Planet. Space Sci., 56(6),* 812–817, DOI: 10.1016/j.pss.2007.12.012.

Fedorov A., Barabash S., Sauvaud J. A., Futaana Y., Zhang T. L., Lundin R., and Ferrier C. (2011) Measurements of the ion escape rates from Venus for solar minimum. *J. Geophys. Res., 116(A),* 07220, DOI: 10.1029/2011JA016427.

Fegley B., Zolotov M. Y., and Lodders K. (1997) The oxidation state of the lower atmosphere and surface of Venus. *Icarus, 125(2),* 416–439, DOI: 10.1006/icar.1996.5628.

Futaana Y., Barabash S., Yamauchi M., McKenna Lawlor S., Lundin R., Luhmann J., Brain D., Carlsson E., Sauvaud J., and Winningham J. (2008) Mars Express and Venus Express multi-point observations of geoeffective solar flare events in December 2006. *Planet. Space Sci., 56(6),* 873–880, DOI: 10.1016/j.pss.2007.10.014.

Galand M., Coates A., Cravens T., and Wahlund J.-E. (2013) Titan's ionosphere. In *Titan: Interior, Surface, Atmosphere, and Space Environment* (I. Mueller-Wodarg et al., eds.), Chapter 11, in press. Cambridge Univ., Cambridge.

Gardner L. C. and Schunk R. W. (2004) Neutral polar wind. *J. Geophys. Res., 109(A),* 05301, DOI: 10.1029/2003JA010291.

Gardner L. C. and Schunk R. W. (2005) Global neutral polar wind model., *J. Geophys. Res., 110(A),* 10302, DOI: 10.1029/2005JA011029.

Hartle R. E. and Grebowsky J. M. (1995) Planetary loss from light ion escape on Venus. *Adv. Space Res., 15,* 117, DOI: 10.1016/0273-1177(94)00073-A.

Hartle R. E., Sittler E. C., Neubauer F. M., et al. (2006) Initial interpretation of Titan plasma interaction as observed by the Cassini plasma spectrometer: Comparisons with Voyager 1. *Planet. Space Sci., 54(1),* 1211–1224, DOI: 10.1016/j.pss.2006.05.029.

Hoffman J. H., Hodges R. R., Donahue T. M., and McElroy M. M. (1980) Composition of the Venus lower atmosphere from the Pioneer Venus mass spectrometer. *J. Geophys. Res., 85,* 7882–7890, DOI: 10.1029/JA085iA13p07882.

Hörst S. M., Vuitton V., and Yelle R. V. (2008) Origin of oxygen species in Titan's atmosphere. *J. Geophys. Res., 113(E),* 10006, DOI: 10.1029/2008JE003135.

Hutchins K. S., Jakosky B. M., and Luhmann J. G. (1997) Impact of a paleomagnetic field on sputtering loss of martian atmospheric argon and neon. *J. Geophys. Res., 102(E),* 9183–9190, DOI: 10.1029/96JE03838.

Jakosky B. M. and Phillips R. J. (2001) Mars' volatile and climate history. *Nature, 412(6),* 237–244.

Kallio E., Barabash S., Janhunen P., and Jarvinen R. (2008) Magnetized Mars: Transformation of Earth-like magnetosphere to Venus-like induced magnetosphere. *Planet. Space Sci., 56(6),* 823–827, DOI: 10.1016/j.pss.2007.12.005.

Kasting J. F. (1988) Runaway and moist greenhouse atmospheres and the evolution of Earth and Venus. *Icarus, 74,* 472–494, DOI: 10.1016/0019-1035(88)90116-9.

Khurana K. K., Pappalardo R. T., Murphy N., and Denk T. (2007) The origin of Ganymede's polar caps. *Icarus, 191(1),* 193–202, DOI: 10.1016/j.icarus.2007.04.022.

Kliore A. J., Nagy A. F., Marouf E. A., et al. (2008) First results from the Cassini radio occultations of the Titan ionosphere. *J. Geophys. Res., 113,* A09317, DOI: 10.1029/2007JA012965.

Kliore A. J., Nagy A. F., Cravens T. E., Richard M. S., and Rymer A. M. (2011) Unusual electron density profiles observed by Cassini radio occultations in Titan's ionosphere: Effects of enhanced magnetospheric electron precipitation? *J. Geophys. Res., 116(A),* 11318, DOI: 10.1029/2011JA016694.

Krasnopolsky V. A. and Gladstone G. R. (1996) Helium on Mars: EUVE and PHOBOS data and implications for Mars' evolution. *J. Geophys. Res., 101(A),* 15765–15772, DOI: 10.1029/96JA01080.

Krasnopolsky V. A., Bowyer S., Chakrabarti S., Gladstone G. R., and McDonald J. S. (1994) First measurement of helium on Mars: Implications for the problem of radiogenic gases on the terrestrial planets. *Icarus, 109,* 337–351, DOI: 10.1006/icar.1994.1098.

Krymskii A. M., Breus T. K., Ness N. F., Hinson D. P., and Bojkov D. I. (2003) Effect of crustal magnetic fields on the near terminator ionosphere at Mars: Comparison of in situ magnetic field measurements with the data of radio science experiments on board Mars Global Surveyor. *J. Geophys. Res., 108(A),* 1431, DOI: 10.1029/2002JA009662.

Lammer H., Lichtenegger H. I. M., Kolb C., Ribas I., Guinan E. F., Abart R., and Bauer S. J. (2003) Loss of water from Mars: Implications for the oxidation of the soil. *Icarus, 165(1),* 9–25, DOI: 10.1016/S0019-1035(03)00170-2.

Lammer H., Kasting J. F., Chassefière E., Johnson R. E., Kulikov Y. N., and Tian F. (2008) Atmospheric escape and evolution of terrestrial planets and satellites. *Space Sci. Rev., 139(1),* 399–436, DOI: 10.1007/s11214-008-9413-5.

Lillis R. J., Frey H. V., and Manga M. (2008) Rapid decrease in martian crustal magnetization in the Noachian era: Implications for the dynamo and climate of early Mars. *Geophys. Res. Lett., 35(1),* 14203, DOI: 10.1029/2008GL034338.

Luhmann J. G., Kasprzak W. T., and Russell C. T. (2007) Space weather at Venus and its potential consequences for atmosphere evolution. *J. Geophys. Res., 112(E4),* DOI: 10.1029/2006JE002820.

Luhmann J. G., Fedorov A., Barabash S., Carlsson E., Futaana Y., Zhang T. L., Russell C. T., Lyon J. G., Ledvina S. A., and Brain D. A. (2008) Venus Express observations of atmospheric oxygen escape during the passage of several coronal mass ejections. *J. Geophys. Res., 113,* DOI: 10.1029/2008JE003092.

Lundin R., Borg H., Hultqvist B., Zakharov A., and Pellinen R. (1989) First measurements of the ionospheric plasma escape from Mars. *Nature, 341,* 609–612, DOI: 10.1038/341609a0.

Lundin R., Zakharov A., Pellinen R., Barabasj S. W., Borg H., Dubinin E. M., Hultqvist B., Koskinen H., Liede I., and Pissarenko N. (1990) ASPERA/Phobos measurements of the ion outflow from the martian ionosphere. *Geophys. Res. Lett., 17,* 873–876, DOI: 10.1029/GL017i006p00873.

Lundin R., Barabash S., Fedorov A., Holmström M., Nilsson H., Sauvaud J. A., and Yamauchi M. (2008a) Solar forcing and planetary ion escape from Mars. *Geophys. Res. Lett., 35(9),* 09203, DOI: 10.1029/2007GL032884.

Lundin R., Barabash S., Holmström M., Nilsson H., Yamauchi M., Fraenz M., and Dubinin E. M. (2008b) A comet-like escape of ionospheric plasma from Mars. *Geophys. Res. Lett., 35(1),* 18203, DOI: 10.1029/2008GL034811.

Lundin R., Barabash S., Yamauchi M., Nilsson H., and Brain D. (2011) On the relation between plasma escape and the martian crustal magnetic field. *Geophys. Res. Lett., 38(2),* 02102, DOI: 10.1029/2010GL046019.

Malin M. C. and Edgett K. S. (2000) Sedimentary rocks of early Mars. *Science, 290(5),* 1927–1937, DOI: 10.1126/science.290.5498.1927.

Malin M. C. and Edgett K. S. (2003) Evidence for persistent flow and aqueous sedimentation on early Mars. *Science, 302(5),* 1931–1934, DOI: 10.1126/science.1090544.

Matsui H., Iwagami N., Hosouchi M., Ohtsuki S., and Hashimoto G. L. (2012) Latitudinal distribution of HDO abundance above Venus' clouds by ground-based 2.3 μm spectroscopy. *Icarus, 217(2),* 610–614, DOI: 10.1016/j.icarus.2011.07.026.

Mauk B. and Bagenal F. (2012). Comparative auroral physics: Earth and other planets. In *Auroral Phenomenology* (A. Keiling et al., eds.), pp. 3–26. AGU Geophysical Monograph 197, AGU, Washington, DC.

Milbury C., Schubert G., Raymond C. A., Smrekar S. E., and Langlais B. (2012) The history of Mars' dynamo as revealed by modeling magnetic anomalies near Tyrrhenus Mons and Syrtis Major. *J. Geophys. Res., 117(E),* 10007, DOI: 10.1029/2012JE004099.

Moore T. E. and Horwitz J. L. (2007) Stellar ablation of planetary atmospheres. *Rev. Geophys., 45(3),* 3002, DOI: 10.1029/2005RG000194.

Moore T. E. and Khazanov G. V. (2010) Mechanisms of ionospheric mass escape. *J. Geophys. Res., 115,* DOI: 10.1029/2009JA014905.

Nielsen E., Fraenz M., Zou H., Wang J., Gurnett D., Kirchner D., Morgan D., Huff R., Safaeinili A., and Plaut J. (2007) Local plasma processes and enhanced electron densities in the lower ionosphere in magnetic cusp regions on Mars. *Planet. Space Sci., 55(14),* 2164–2172, DOI: 10.1016/j.pss.2007.07.003.

Nilsson H., Carlsson E., Brain D. A., Yamauchi M., Holmström M., Barabash S., Lundin R., and Futaana Y. (2010) Ion escape from Mars as a function of solar wind conditions: A statistical study. *Icarus, 206(1),* 40–49, DOI: 10.1016/j.icarus.2009.03.006.

Nilsson H., Edberg N., Stenberg G., and Barabash S. (2011) Heavy ion escape from Mars, influence from solar wind conditions and crustal magnetic fields. *Icarus, 215,* 474–484, DOI: 10.1016/j.icarus.2011.08.003.

Palmroth M., Pulkkinen T. I., Janhunen P., and Wu C. C. (2003) Stormtime energy transfer in global MHD simulation. *J. Geophys. Res.-Space, 108(A),* 1048, DOI: 10.1029/2002JA009446.

Phillips J. L., Bame S. J., Barnes A., et al. (1995) Ulysses solar wind plasma observations from pole to pole. *Geophys. Res.*

Lett., 22(2), 3301–3304, DOI: 10.1029/95GL03094.

Pollack J. B. (1971) A nongrey calculation of the runaway greenhouse: Implications for Venus' past and present. *Icarus, 14,* 295, DOI: 10.1016/0019-1035(71)90001-7.

Pollack J. B., Kasting J. F., Richardson S. M., and Poliakoff K. (1987) The case for a wet, warm climate on early Mars. *Icarus, 71(2),* 203–224, DOI: 10.1016/0019-1035(87)90147-3.

Ribas I., Guinan E. F., Güdel M., and Audard M. (2005) Evolution of the solar activity over time and effects on planetary atmospheres. I. High-energy irradiances (1–1700 Å). *Astrophys. J., 622(1),* 680–694, DOI: 10.1086/427977.

Russell C. T., Luhmann J. G., Cravens T. E., Nagy A. F., and Strangeway R. J. (2007) Venus upper atmosphere and plasma environment: Critical issues for future exploration. In *Exploring Venus as a Terrestrial Planet* (L. W. Esposito et al., eds), pp. 139–156. AGU Geophysical Monograph 176, AGU, Washington, DC, DOI: 10.1029/176GM09.

Schubert G., Russell C. T., and Moore W. B. (2000) Geophysics: Timing of the martian dynamo. *Nature, 408(6),* 666–667.

Seki K., Elphic R. C., Hirahara M., Terasawa T., and Mukai T. (2001) On atmospheric loss of oxygen ions from Earth through magnetospheric processes. *Science, 291(5),* 1939–1941, DOI: 10.1126/science.1058913.

Sekiya M., Nakazawa K., and Hayashi C. (1980) Dissipation of the rare gases contained in the primordial Earth's atmosphere. *Earth Planet. Sci. Lett., 50,* 197–201, DOI: 10.1016/0012-821X(80)90130-2.

Stenberg G., Nilsson H., Futaana Y., Barabash S., Fedorov A., and Brain D. (2011) Observational evidence of alpha-particle capture at Mars. *Geophys. Res. Lett., 38(9),* 09101, DOI: 10.1029/2011GL047155.

Sterenborg M. G., Cohen O., Drake J. J., and Gombosi T. I. (2011) Modeling the young Sun's solar wind and its interaction with Earth's paleomagnetosphere. *J. Geophys. Res., 116,* A01217, DOI: 10.1029/2010JA016036.

Strangeway R. J., Ergun R. E., Su Y.-J., Carlson C. W., and Elphic R. C. (2005) Factors controlling ionospheric outflows as observed at intermediate altitudes. *J. Geophys. Res., 110(A),* 03221, DOI: 10.1029/2004JA010829.

Vaisberg O. L., Romanov S. A., Smirnow V. N., Karpinskii I. P., Khazanov B. I., Polenov B. V., Bogdanov A. V., and Antonov N. M. (1977) Structure of the region of interaction of solar wind with Venus inferred from measurements of ion-flux characteristics on Venera 9 and Venera 10. *Cosmic Res., 14(6),* 709–718. (Published in Russian in *Kosmicheskie Issledovaniia, 14,* Nov.–Dec. 1976, 827–838.)

Waite J. H., Young D. T., Cravens T. E., Coates A. J., Crary F. J., Magee B., and Westlake J. (2007) The process of tholin formation in Titan's upper atmosphere. *Science, 316(5),* 870, DOI: 10.1126/science.1139727.

Wood B. E., Müller H. R., Zank G. P., Linsky J. L., and Redfield S. (2005) New mass-loss measurements from astrospheric Lyα absorption. *Astrophys. J., 628(2),* L143–L146, DOI: 10.1086/432716.

Part V:

Solar Influences

Zent A. P. (2013) Orbital drivers of climate change on Earth and Mars. In *Comparative Climatology of Terrestrial Planets* (S. J. Mackwell et al., eds.), pp. 505–538. Univ. of Arizona, Tucson, DOI: 10.2458/azu_uapress_9780816530595-ch21.

Orbital Drivers of Climate Change on Earth and Mars

Aaron P. Zent
NASA Ames Research Center

Oscillations of orbital elements and spin axis orientation affect the climate of both Earth and Mars by redistributing solar power both latitudinally and seasonally, often resulting in secondary changes in reflected and emitted radiation (radiative forcing). Multiple feedback loops between different climatic elements operate on both planets, with the result that climate response is generally nonlinear with simple changes in solar energy. Both insolation history and geochemical climate proxies can be treated as time series data, and analyzed in terms of component frequencies. The correspondence between frequencies measured in climate proxies and orbital oscillations is the key to relating orbital cause and climatic effect. Discussions of both Earth and Mars focus on the last 5–10 m.y., because this is the period in which the orbital history and geologic record are best understood. The terrestrial climate is an extraordinarily complex system, and a vast amount of data is available for analysis. While the geologic record strongly supports the role of Milankovitch cycles as the underlying cause of glacial cycles, orbitally driven insolation changes alone cannot explain the observations in detail. Early Pleistocene glacial cycles responded linearly to the 41-k.y. oscillations in obliquity. However, over the last 1 m.y., glacial/interglacial oscillations have become more extreme as the climate has cooled. Long cooling intervals marked by an oscillating buildup of ice sheets are now followed by brief, intense periods of warming. At the same time, glacial/interglacial cycles have shifted from 41 k.y. to ~100 k.y. No such changes occurred in the solar forcing due to orbital oscillations. While orbital oscillations still appear to pace glacial cycles, their subtle interplay with ice-sheet dynamics and shifts in ocean circulation have come to dominate the late Pleistocene climate system. In contrast to Earth, the martian climate is ostensibly a much simpler system about which we have almost no quantitative data. Lacking climate proxies and chronological data, we are forced to rely on climate modeling and whatever constraints can be extracted from the predominantly remote sensing data available. Obliquity oscillations account for most of the power in historical insolation. Unfortunately, the last 5 m.y. are an anomalous period in Mars' climate history due to a secular decrease in Mars' obliquity. Subsequent to that, however, models and observations are consistent with the hypothesis that during periods of higher obliquity, enhanced polar summer insolation increases atmospheric water vapor and dust content, and ice stability shifted toward the equator. Polar caps become thermodynamically unstable, and much of the surface H_2O inventory migrates from high latitudes to the tropics. As obliquity decreases, ice returns to the poles, leaving unstable ice-rich deposits in the mid latitudes that are mantled by dust. Low-obliquity periods entail — at least on occasion — collapse of the atmosphere onto the poles and high-latitude CO_2 glaciers. During protracted nodes in obliquity, mid-latitude ice undergoes slow but sustained sublimation and redistribution to the poles. Because of the tremendous breadth of the subject matter, this chapter necessarily presents a high-level overview, and the reader will be compelled to investigate the copious references for a more rigorous explanation of most topics.

1. INTRODUCTION

Terrestrial climate science, the precursor of all planetary climatology, began in the Swiss Alps. Nineteenth-century naturalists who were studying alpine glaciers and their effects on the landscape could not help but notice, when they came down the mountain, familiar features in places far removed from any glacier. It took decades to convince the geological community that the European climate had indeed changed substantially. Likewise for Mars, climate change was first inferred from the existence of features that are incompatible with its current climate; the attribution of that climate change to astronomical causes proceeded largely by analogy to what we know of Earth.

A planets' climate is ultimately determined by its radiation balance, which can change in three fundamental ways: (1) the incident solar radiation may change or be redistributed (e.g., by orbital and axial oscillations), (2) the reflected radiation (i.e., the planetary albedo) may change (e.g., by changes in ice cover or atmospheric opacity), or

(3) the emitted radiation may change (e.g., by changes in greenhouse gas concentrations). Any external factor that changes the radiation balance results in a *climate forcing*, which is measured in W m^{-2}, and quantifies the change in net radiation through the top of the atmosphere that results (*Hansen et al.,* 2005). We will see that orbital oscillations result in direct orbital climate forcing, often resulting in secondary changes in reflected and emitted radiation (radiative forcing). Feedback loops, both positive and negative, have been identified between many forcing agents, which amplify or curtail initial climate forcings. Thus the problem of assigning a unique forcing value (and attributing its climatic consequence) to a finite change in a specific orbital parameter is both physically complex and inherently subjective. Obviously, then, we cannot rigorously limit our discussion to those climate changes directly attributable to orbital control.

Orbital oscillations are the result of gravitational interactions between planets, the Sun, and, in Earth's case, the Moon. A rapidly rotating body is distorted from a spherical shape through development of an equatorial bulge that — all other things being equal — is proportional to the square of its rotational angular velocity. Rapidly rotating planets (e.g., Earth and Mars) are therefore better characterized as oblate spheroids, while slowly rotating planets (e.g., Mercury and Venus) are for all practical purposes spheres (*Seidelmann et al.,* 2007). It is the deviation from sphericity that allows a planet to experience gravitational torques exerted by the other bodies of the solar system. By inference then, any rapidly rotating planet in any multibody solar system is likely to experience orbital oscillations throughout its existence.

Understanding how such oscillations might have affected climate requires knowledge of a planet's orbital history and access to interpretable paleoclimate data, generally found in the geologic record. On Earth at least, that record becomes increasingly sparse with age. Furthermore, over periods longer than 10–20 m.y. for Mars and 50 m.y. for Earth, the orbits of the inner solar system become impossible to calculate deterministically (*Laskar et al.,* 2011). However, "floating" astronomical timescales with 0.10- to 0.40-m.y. resolution are defined for entire epochs and stages in the Paleogene and all three Mesozoic periods (*Hinnov and Ogg,* 2007). Nonetheless, this chapter will be weighted toward events of the last few million years, a period for which both the orbital histories and geologic records of Earth and Mars are clear and most of the analysis to date has focused.

Section 2 briefly enumerates the various lines of evidence for climate change and defines the orbital variations that alter planetary radiation balance. Sections 3 and 4 provide individual discussions of Earth and Mars, respectively.

2. BACKGROUND

2.1. Evidence for Climate Change

2.1.1. In the field. The case for widespread continental glaciation, and by implication climate change, was finally made by Louis Agassiz, based in part on his own fieldwork, and in part on the ideas of many earlier geologists (*Agassiz,* 1840a,b). Throughout the nineteenth century, these geologists mapped recognizable moraines, U-shaped valleys, and striations in exposed bedrock across much of Europe (Fig. 1). They identified large erratic boulders and coarse, poorly sorted gravels that could not be traced to local sources. Agassiz was the first to recognize glacial tills, the poorly sorted and extremely heterogeneous sediments deposited directly by glaciers, and over the course of a few decades he eventually brought a skeptical geologic community to accept glaciation as the most likely explanation for this puzzling suite of features (*McPhee,* 1998).

In the course of their mapping, geologists found plant fragments and fossil teeth between till layers, demonstrating that there had been multiple ice advances separated by sustained warm intervals. Four cycles of glacial advance and retreat were eventually identified; in North America they were referred to as the Nebraskan, Kansan, Illinoian, and Wisconsin. However, it was not possible to construct a comprehensive glacial history, in part because each glacial advance destroyed evidence of previous ice sheets, and also because it was difficult to determine precise and reliable ages. This state of affairs persisted until the 1960s, when new dating techniques were developed and the locus of climate science moved from the field to the laboratory.

2.1.2. Geochemical climatology. Climate forcing occurs over a variety of timescales driven by many different mechanisms, and each component of the climate responds to forcing at different rates. To understand these relationships, we need records of the climate (i.e., climate proxies) that can be easily measured, reliably dated, confidently interpreted, and record many cycles of climate change. Fortunately, for thermodynamic or kinetic reasons, many naturally occurring processes leave chemical signatures that can be transformed into estimates of climate parameters (*Urey,* 1947). Examples

Fig. 1. Moraine ridge of the Nithsdale Valley glacier, near Thornhill, Dumfrieshire, Scotland: ". . . probably ranks as the first glacial locality to be fully recognized and understood in Britain." From *Boylan* (1981).

of climate proxies include stable O- and H-isotope ratios in $CaCO_3$, ice, or trapped air (paleothermometry and continental ice volume); Mg/Ca in foraminiferal calcite and Sr/Ca in corals (paleothermometry) (*Burton and Walter,* 1991; *Beck et al.,* 1992); and N-Ar isotopes in trapped air (glacial firn processes) (*Craig et al.,* 1988; *Schwander et al.,* 1988).

With the development of marine isotopic stratigraphy (*Emiliani,* 1966; *Shackleton,* 1967; *Shackleton and Opdyke,* 1973), it became possible to analyze deep-ocean sediments that recorded several reversals of Earth's magnetic field, making those reversals available as chronological markers. The remarkable correspondence between the sedimentary record and astronomical cycles established the importance of orbital oscillations in climate history (*Hays et al.,* 1976).

Marine sediments have been the most useful samples in establishing global climate history, and the key climate proxy has been the stable O-isotope chemistry of biogenic $CaCO_3$. The ratio of $^{18}O/^{16}O$ in ocean water increases with the extent of continental ice because $H_2{}^{16}O$ evaporates preferentially, and $H_2{}^{18}O$ precipitates preferentially during atmospheric transport. Thus, during the rapid growth of ice sheets, ^{18}O is enriched in ocean waters and ^{16}O is enriched in the polar ice. The ^{18}O enrichment is mirrored by organisms that fractionate ocean water in forming their shells. Hence, continental ice mass is positively correlated with $\delta^{18}O$ in marine sediments, where

$$\delta^{18}O = \left[\frac{\left({}^{18}O/{}^{16}O \right)_{smp}}{\left({}^{18}O/{}^{16}O \right)_{smow}} - 1 \right] \times 1000‰ \tag{1}$$

A complicating factor is that $\delta^{18}O$ is also a function of deep-water temperature (e.g., *Maslin et al.,* 1998). Here, the distinction between planktic (column-dwelling) and benthic (bottom-dwelling) foraminifera is of use, as are a variety of independent deep-sea temperature proxies such as Mg/Ca in benthic foraminiferal calcite or atmospheric samples trapped in continental ice cores (*Shackleton,* 2000).

So key has this technique been that it is now standard procedure to refer to glacial periods by their Marine Isotopic Stage (MIS) number. Working backward from the present (MIS 1), stages with even numbers have high $\delta^{18}O$ and represent cold glacial periods, while odd-numbered stages (low $\delta^{18}O$) represent warm interglacials (Fig. 2).

As alluded to, cores from continental ice sheets are a second invaluable source of data. Proxies measured in ice cores include the deuterium abundance in ice [δD_{ice}, a temperature proxy for the H_2O source region (*Lewis et al.,* 2013)], $\delta^{18}O_{ice}$ (a proxy for condensation temperature), dust content (desert aerosols), and Na abundance (marine aerosols). Stable-isotope ratios in polar ice are interpreted oppositely from the marine sediment case, as would be expected in a source/sink relationship; models predict that δD should be linear with temperature in high-latitude ice sheets.

Ice cores also yield a unique opportunity to directly sample air trapped in firn as it was laid down and compressed, allowing direct measurement of ancient atmospheric isotopes as well as the abundance of greenhouse gases (e.g., CO_2 and CH_4) (*Petit et al.,* 1999; *Jansen et al.,* 2007). Thus it is now possible to lay multiple climate proxies along a common timeline (Fig. 2), allowing synergistic evaluation of many independent datasets.

2.1.3. Remote sensing. Roughly simultaneously with the geochemical revolution in paleoclimatology, spacecraft began to characterize the martian surface and environment. As with Earth, it quickly became clear that the visible geology and current climate could not be reconciled. An ever-increasing variety of features are now recognized as relicts of earlier climate states (Fig. 3).

Enormous polar layered deposits (PLD) were discovered at both poles in Mariner 9 imagery (Fig. 3a) (*Murray et al.,* 1972; *Soderblom et al.,* 1973). An absence of impact craters showed that the deposits were geologically young and their uniform, rhythmic layering suggested a cyclic process (*Murray et al.,* 1973; *Ward,* 1973), and so the conceptual link between the PLD and quasi-periodic orbital variations was quickly made.

The Viking orbiters made global measurements of surface temperatures, regolith thermophysical properties, and atmospheric H_2O (*Kieffer et al.,* 1977; *Farmer et al.,* 1977), while the Viking landers recorded annual atmospheric pressure cycles (*Hess et al.,* 1980). These data gave modelers the first quantitative measures of the CO_2, H_2O, and dust cycles that define the current climate.

With the arrival of Mars Global Surveyor (MGS) and subsequent orbiters, observational evidence for recent climate change mushroomed, challenging the existing models. Very young gullies formed by fluid-mobilized flow (*Malin and Edgett,* 2000) and circular pits in the south residual cap (*Thomas et al.,* 2000) that are eroding at up to 3–5 m per martian year (m yr^{-1}♂) were unanticipated (*Malin et al.,* 2001a). [Note: In the literature, rates are often given relative to either terrestrial *or* martian years, without specifying which; we annotate them here by using the appropriate symbol (⊕ or ♂) when necessary.]

At high latitudes, deposits that mantled the underlying topography were known from Viking, but they were now seen to be layered, many meters thick, and in some instances ice-rich, notably in locations where ground ice is not stable (Fig. 3b) (*Mustard et al.,* 2001). Deposits at the base of the Olympus Mons escarpment and the Tharsis volcanos were seen to be glacial in origin, and showed evidence of multiple phases of activity, each requiring that hundreds of meters of ice repeatedly accumulate at elevations ≥7 km (Fig. 3c) (*Neukum et al.,* 2004).

Essentially pure H_2O ice was excavated by the Phoenix lander, and small impacts that have occurred in the course of the Mars Reconnaissance Orbiter mission have likewise excavated clean ice, in some instances in locations where it is not currently stable (Fig. 3d) (*Byrne et al.,* 2009). The discovery (*Balme et al.,* 2009) of very young sorted stone circles that are diagnostic of periglacial freeze-thaw (*Kessler and Werner,* 2003) indicate processes that were not predicted *a priori* by any models, and the annual appearance of dark,

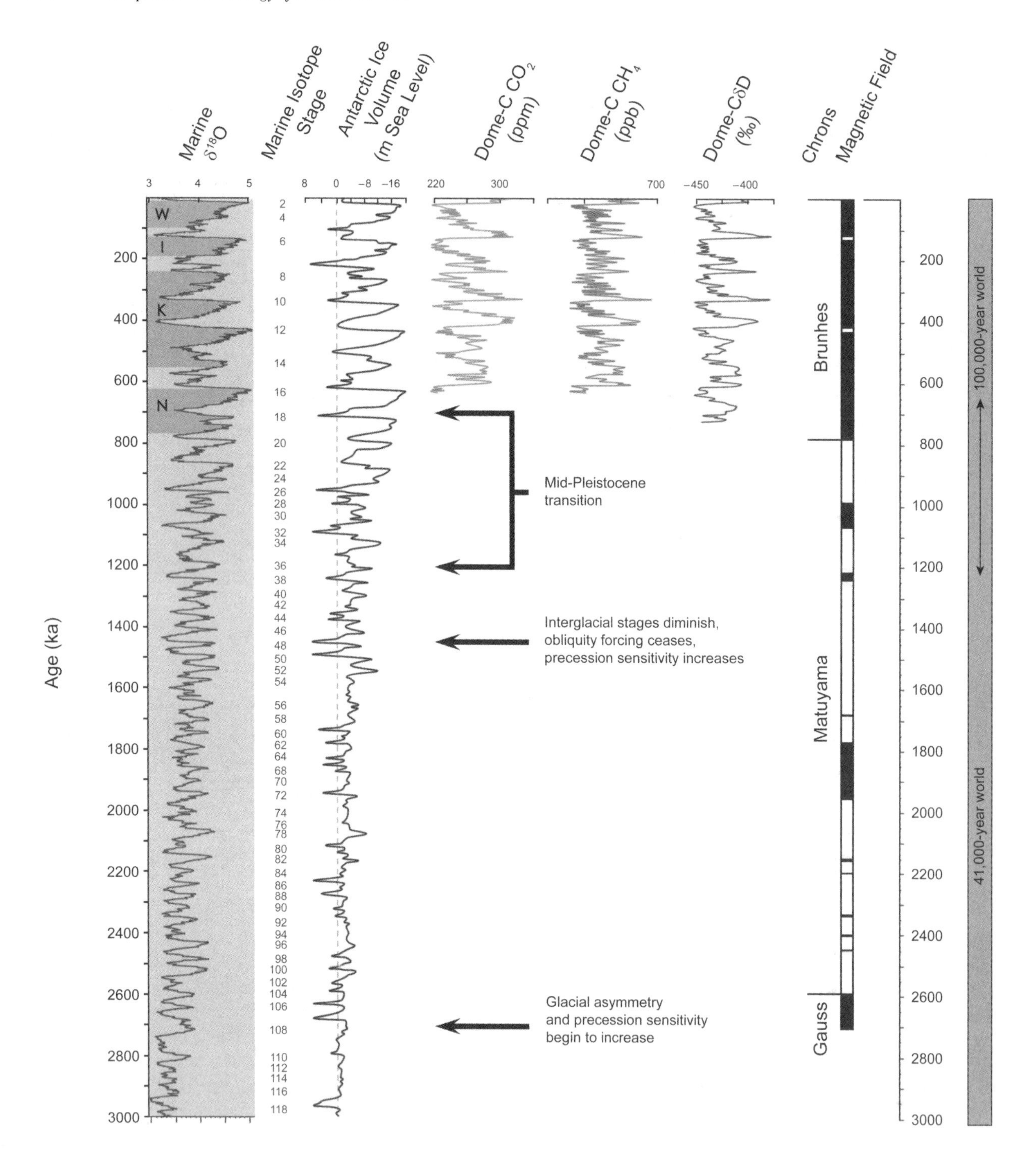

Fig. 2. *Left to right:* Marine sediment O-isotope record (*left*) for the past 3 m.y. The four most recent glacial advances are shaded. Even-numbered (glacial) Marine Isotope Stages are shown for reference. Ice core data from the Dome C site (*Augustin et al.,* 2004) record 800 k.y. of local temperatures (δD), as well as the co-varying CO_2 and CH_4 abundance. The magnetic chrons are on the right. At around 1 Ma, the principle periodicity in the $\delta^{18}O$ record transitions from 41 ka (the period of the obliquity) to ~100 ka. From *Gibbard and van Kolfschoten* (2005).

slowly moving linear features on warm slopes (*McEwen et al.,* 2011) apparently forces us to accommodate the contemporary flow of low-eutectic brines just beneath the surface. Earlier theories of Mars' climate history must be substantially revised to come to grips with these discoveries.

2.2. Astronomical Cycles

2.2.1. Orbital and axial oscillations. Changes in the annual distribution of insolation across a planet may result from variations in the solar flux or from changes in

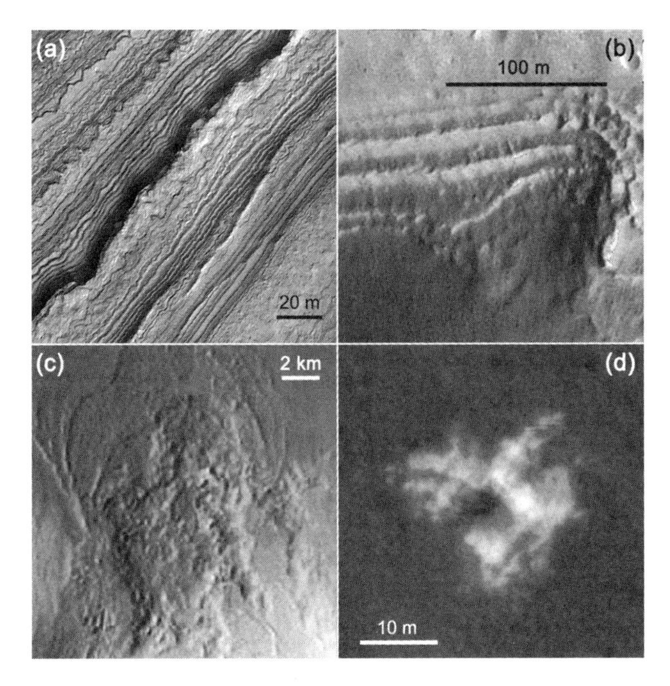

Fig. 3. Array of landforms that are indicative of recent changes in the martian climate. **(a)** Polar layered terrain (HiRISE image PSP_005011_0885); **(b)** ice-rich, mid-latitude mantle deposits (HiRISE image PSP_001507_1400); **(c)** evidence of low-latitude Alpine glaciers (HRSC image from *Neukum et al.,* 2004); **(d)** excess ice exposed by small impacts at mid latitudes (HiRISE Image ESP_011548_2360).

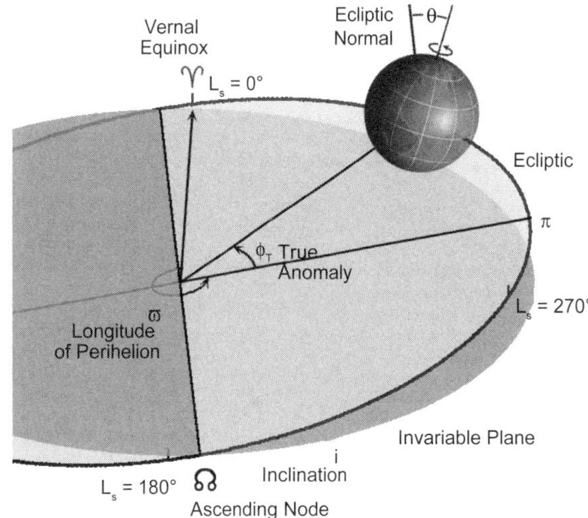

Fig. 4. Diagram of relevant parameters and obliquity, θ. The vernal equinox is the prime reference direction, and the longitude of perihelion is measured counterclockwise from it. The letter π denotes perihelion, from which the true anomaly is measured.

the planet's orbit. The average solar radiation at 1 AU is 1368 W m^{-2}, and 588 W m^{-2} at the mean martian orbital distance (1.524 AU). These values vary over the course of an 11-year sunspot cycle by approximately 0.1% (*Kopp and Lean,* 2011). The record is unclear over 10^6-yr timescales, and absent evidence to the contrary we assume solar output was effectively constant.

Our goal is to calculate the intensity and distribution of solar radiation over a planet as a function of time. To do this, we require knowledge of only three parameters (*Milanković,* 1938). Obliquity (θ) is the angle between a planet's rotational axis and the normal to its orbital plane (Fig. 4). For Earth, the orbital plane is known as the ecliptic. Because the ecliptic is inclined ~1.57° to the invariable plane (the center of mass for the entire solar system), a torque arises, primarily from the outer planets, that can move the ecliptic in inertial space. Gravitational torques on the equatorial bulge can also reorient the rotational axis (*Lissauer et al.,* 2012). Obliquity changes result in latitudinal redistribution of annual average insolation, $\overline{I_Y}(\lambda)$, without affecting the global mean insolation. As obliquity increases, $\overline{I_Y}(\lambda)$ increases at the poles and decreases in the tropics. Conversely, lower obliquity increases the annual mean meridional insolation gradient.

Orbital eccentricity ($\varepsilon = \sqrt{a^2 - b^2}/a$, where a and b are the orbit's semimajor and semiminor axes) describes an orbit's departure from circularity (Fig. 5); ε ranges from 0 for a circular orbit to 1 for a parabolic (i.e., open) orbit. As ε increases, the variation in global mean insolation throughout the year also increases. Where θ determines the *magnitude* of seasonal variations, ε determines the seasonal *asymmetry* between hemispheres.

The final variable, ϖ, is longitude of perihelion, which denotes the season at which perihelion occurs (Fig. 6). It is the angle between the vernal equinox and the point of perihelion, measured counterclockwise in the orbit plane (see also Fig. 4). The point of perihelion slowly migrates around the orbit due to two precessional cycles. *Axial precession* is the slow and continuous reorientation of a planet's rotational axis, tracing out a cone over a full cycle. The result is the movement of the vernal equinox relative to the fixed stars. Axial precession occurs because solar gravity pulls a planet's equatorial bulge toward the ecliptic, resulting in a torque that is perpendicular to the rotational axis. *Orbital precession* is the precession of the orbits' major axis around the Sun. It is the result of the radial accelerations on a planet due to relatively weak interplanetary gravitational forces.

A composite parameter, the *precession index* (ε sin ϖ), is a measure of the combined effect of eccentricity and precession, and correlates with the summertime diurnally averaged insolation ($\overline{I_d}$) at high northern latitudes.

2.2.2. Insolation. Calculating the top-of-the-atmosphere insolation from the orbital parameters proceeds by finding the planet in its orbit at time t after perihelion. From this, the distance to the Sun can be found, and spherical geometry identities allow the calculation of insolation at any latitude.

To find the position of the planet at time t, the mean anomaly (ϕ_M) is first found; it is the time since perihelion

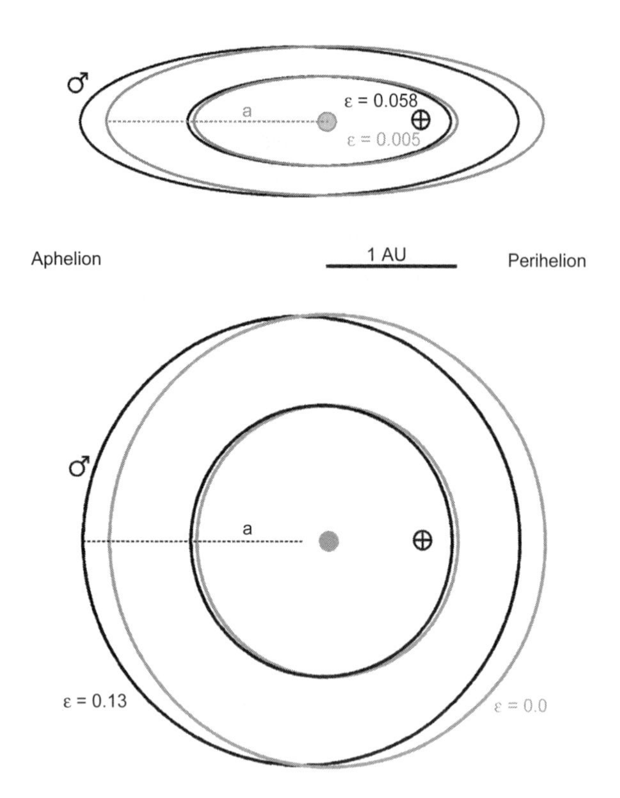

Fig. 5. The orbits of Earth and Mars at maximum (black) and minimum (gray) eccentricity, drawn to scale. The Sun is at one focus of each ellipse, with aphelion to the left and aphelion to the right. The upper diagram is an oblique view of the two orbits, and shows the semimajor axis of Mars' orbit, a, at minimum eccentricity. Because Mars minimum eccentricity is zero, the Sun is at the center of the orbit. The lower diagram shows a at the maximum eccentricity. Here the foci of the orbital ellipse are clearly separated. The length of a is identical in the two figures.

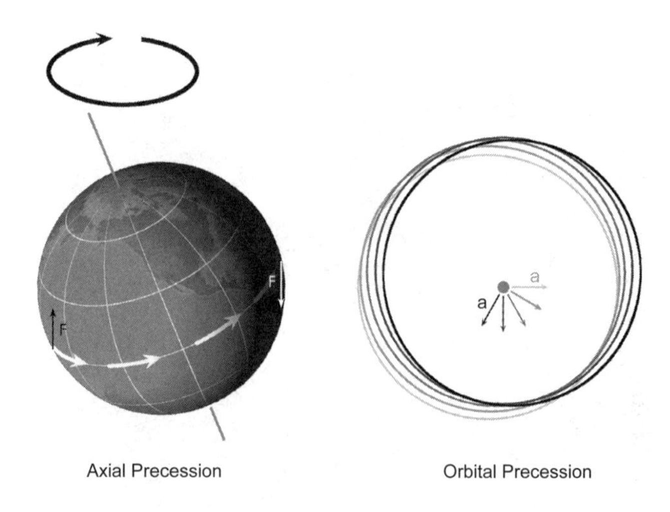

Fig. 6. Axial precession and orbital precession combine to account for the precession of the equinoxes. The vertical arrows indicate the gravitational forces acting on the equatorial bulge to produce the torque responsible for axial precession. On the right, the arrows indicate the semimajor axis of the orbit at five different times, separated by 30°, with reference to the Sun.

multiplied by the mean motion ($n_M = P/2\pi$), where P is the orbital period

$$\phi_M = n_M t = t\sqrt{\frac{G(M_\odot + M_P)}{a^3}} \tag{2}$$

where G is the gravitational constant, $M_\odot$ and M_P are the mass of the Sun and planet, respectively, and a is the semimajor axis of the orbit.

The eccentric anomaly, ϕ_E, can then be found by solving Kepler's equation

$$\phi_M = \phi_E - \sin\phi_E \tag{3}$$

This equation does not have a closed-form solution; it is usually solved via a numerical root-finding method (e.g., Newton-Raphson). The radial distance between the planet and the Sun is then

$$r = a(1 - \varepsilon\cos\phi_E) \tag{4}$$

We also calculate the true anomaly ϕ_T, which is the physical geometric angle in the counterclockwise direction as seen from the Sun, measured in the plane of the orbit, between the direction of perihelion and the position of the planet (Fig. 4)

$$\cos\phi_T = \frac{\cos\phi_E - \varepsilon}{1 - \varepsilon\cos\phi E} \tag{5}$$

If we let the solar longitude L_s define the angle along the orbit, where $L_s = 0°$ at the vernal equinox, then $\phi_T = L_s - \varpi$ and the solar declination δ is

$$\delta = \theta\sin(\phi_T + \varpi) \tag{6}$$

Then for a point at latitude λ and hour angle η, the solar zenith angle (z_a) can be found from

$$\cos(z_a) = \sin\lambda\sin\delta + \cos\lambda\cos\delta\cos\eta \tag{7}$$

which finally allows the instantaneous insolation at latitude λ, season ϕ_T, and hour η to be determined from

$$I(\lambda, \phi_T, \eta) = S\left(\frac{r_m}{r}\right)^2 \cos(z_a) \tag{8}$$

where r_m is the mean annual heliocentric distance, S is the solar constant at r_m, and I = 0 if $\cos(z_a) < 0$.

Two integral versions of this expression are commonly used in paleoclimate analyses: the annually averaged and diurnally averaged insolation. The diurnal average insolation is just the integral of the instantaneous insolation over η. Since the average solar distance can be approximated (for present purposes) as a, the diurnal average insolation can be expressed as

$$\overline{I_d}(\lambda,\phi_T)=\frac{S}{\pi}\frac{a^2}{r^2}\left[\eta_0\sin\lambda\sin\delta+\cos\lambda\cos\delta\sin(\eta_0)\right] \quad (9)$$

where η_0 is the hour angle of sunrise, $\cos(\eta_0) = -\tan\lambda\tan\delta$. In the case that $\tan\lambda\tan\delta > 1$, the Sun never sets, and $\eta_0 = \pi$. Where $\tan\lambda\tan\delta < 1$, the Sun does not rise, and $\eta_0 = 0$.

The annual average insolation at latitude λ is equivalent to taking the double integral of equation (8) over η and ϕ_T. *Ward* (1974) gives the expression

$$\overline{I_Y}(\lambda)=\frac{S}{2\pi^2\sqrt{1-\varepsilon^2}}\times \int_0^{2\pi}\left[1-(\sin\lambda\cos\theta-\cos\lambda\sin\theta\sin\eta)^2\right]^{\frac{1}{2}}d\eta \quad (10)$$

In Fig. 7 we plot the meridional profile of $\overline{I_Y}(\lambda)$ for the highest and lowest θ achieved during the last 5 m.y. for both Earth and Mars. In the terrestrial case, increasing θ from 22.041° to 24.465° slightly decreases $\overline{I_Y}$ at the equator by 3.5 W m^{-2} but causes $\overline{I_Y}$ to increase at the poles by 17 W m^{-2}.

In all cases, the values of I calculated here are valid only at the top of the atmosphere. Radiative transfer to the surface, and the planet's radiative balance, requires a case-by-case consideration of each climate system.

For geometrical reasons, there are some generalizations that can be made about how orbital changes will affect insolation, and these can be useful to bear in mind. First, we can note that when θ is varying, it causes the annual average insolation at the two poles to vary in phase. However, as precession causes ϖ to shift, one pole will experience summer perihelion and the other summer aphelion, so insolation changes at the two poles will be out of phase.

Changing ε affects the diurnally averaged insolation, $\overline{I_d}(\lambda,\phi_T)$; however, averaged over a full year, $\overline{I_Y}(\lambda)$ (or its integral over a year, the total solar energy) is much less sensitive to ε because heliocentric distance is inversely proportional to angular velocity: When summer insolation is high, summers are short.

2.3. Age Models

Every attempt to correlate a sedimentary sequence with an historical record such as insolation must inevitably confront the murky problem of relating a time coordinate (e.g., million years before present, or Ma) to a spatial one (e.g., depth below a surface). The concept is straightforward; those strata that can be dated accurately are used to establish chronological markers via an appropriate technique, and then a suitable means of interpolating between the markers is chosen. The result, through which a specific age can be ascribed to a sample from anywhere in the sequence, is known as an *age model.* Although a few types of samples naturally incorporate both time and climate signals (e.g., tree rings, corals, varved sediments, some ice cores), laboratory measurements are generally required to assign ages.

Uranium-series dating is commonly used to date $CaCO_3$-bearing materials, and is based upon how close their $^{234}U/^{230}Th$ ratio is to its secular equilibrium value. It has been applied to establish chronologies of sea-level variations by dating corals in raised beaches (e.g., *Stirling et al.,* 2009). Uranium-series dating has an upper age limit of just over 5×10^5 yr, fixed by the half-life of ^{230}Th, and in some environments there are open-system artifacts that present severe challenges (*Andersen et al.,* 2009). In ocean sediments, palaeomagnetic dating is frequently used, relying upon the presence of magnetic grains in the sample and temporal variations in the direction and intensity of Earth's magnetic field. Reference data from sites with established age control have been used to determine the temporal pattern of field reversals for the last 10^7 yr, as well as amplitude variations on timescales $\leq 10^3$ yr (*Hambach et al.,* 2008). Volcanic ash layers, where present, can also be dated via K/Ar or $^{40}Ar/^{39}Ar$ isotopes (e.g., *Thornalley et al.,* 2011).

Age interpolation between chronologic control points in ocean sediments is frequently accomplished via an approach known as orbital tuning. In early studies where age control was good, it was observed that the $\delta^{18}O$ signal seemed to maintain a fixed-phase relationship with orbital parameters. Orbital tuning makes the explicit assumption that the same is true where age control is poor, and uses an astronomical forcing function to adjust the age model (e.g., *Shackleton and Imbrie,* 1990). An example we will refer to below is the "LR04" benthic $\delta^{18}O$ stack (*Lisiecki and Raymo,* 2005) (see Fig. 2), which is a tuned, composite stack that provides continuous and global $\delta^{18}O$ values from prior to the development of northern hemisphere ice sheets through the present.

While tuning enforces coherence between orbital drivers and climate signals, it obviously precludes quantitative evaluation of the orbit-climate relationship. Such tests require sample ages that are derived independently. *Huybers and Wunsch* (2004) and *Huybers* (2007) provided a planktic/benthic O-isotope stack that spans 0–2.58 m.y. where interpolation is based on the assumption of constant sedimentation rates between geomagnetic age controls.

3. CLIMATE CHANGE ON EARTH

Croll (1864) seems to have been the first to suggest that glaciation was associated with changes in Earth's orbit. He proposed that glaciation could occur in either hemisphere when (1) eccentricity is large and (2) winters coincide with aphelion; this combination would result in long, cold winters. His theories did not impress his contemporaries, and little was heard of his work subsequently. Instead, the theory has come to be associated with Milutin Milanković, who argued (*Milanković,* 1938) that northern hemisphere glaciers advance when arctic summer insolation is weak enough to allow snow and ice to persist into the following winter. These conditions are more common when θ is low and aphelion coincides with northern summer.

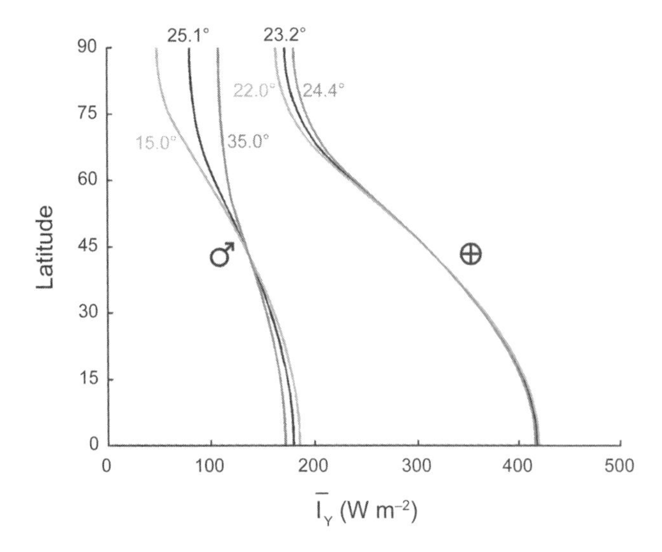

Fig. 7. The annual average insolation ($\bar{I}_Y$) at the top of the atmosphere as a function of latitude and obliquity for Earth and Mars. The planet's current obliquity (black) is bracketed by the maximum (dark gray) and minimum (light gray) θ achieved over the last 3 m.y.

3.1. Earth's Orbit

Earth's obliquity undergoes variations of ~±1.212° around its mean value (23.253°) with a period of ~4.1 × 10^4 yr (Fig. 8). The current value is θ = 23.439° and decreasing (*Laskar et al.,* 2011). The amplitude modulations that yield obliquity nodes every 1.3 m.y. indicate that the θ cycle is actually a doublet, with components at 41 k.y. and 39 k.y. Minor components occur also near 54 k.y. and 28 k.y. (*Hinnov,* 2000).

The axial precession of Earth completes a 360° cycle over 2.6 × 10^4 yr. The orbital precession has a period of 1.13 × 10^5 yr; the cumulative effect of the two precessions leads to a complex suite of frequencies, including a doublet at 23.8 and 22.3 k.y., and a singlet at 19 k.y. Solar forcing thus oscillates at these frequencies, even though they do not correspond to the oscillations of any orbital elements. Currently, perihelion occurs around January 3 ($\varpi \approx$ 282.89°) and aphelion around July 4 ($\varpi \approx$ 102.89°).

Eccentricity varies from 0.005 to 0.0578 with two primary periodicities: 10^5 and 4.05 × 10^5 yr. The 10^5 frequency component is forced by the orbital perihelia of Venus and Jupiter and is itself a doublet, with frequencies of 127 k.y. and 97 k.y.; each of those frequencies are also doublets. Currently, ε = 0.016704 and is decreasing. The precession index, ε sin ϖ, which we have already noted is proportional to summertime high-latitude diurnal insolation, is dominated by these precession frequencies.

Orbital solutions for the past 3 m.y. (*Laskar et al.,* 2011) are shown in Fig. 8. In addition, the diurnally averaged insolation at 65°N on the day of the summer solstice ($\overline{I_{65}}$) is plotted. $\overline{I_{65}}$ is widely used in interpreting palaeoclimate records because it measures insolation at a key place and time; northern hemisphere glaciation appears to initiate at approximately that latitude (*Andrews and Barry,* 1978), and survival of surface snow through the summer is the prerequisite to further ice expansion. There is also considerable land mass at 65°N, and land surfaces have a much lower heat capacity than the oceans, and therefore are more sensitive to insolation changes (*Petit et al.,* 1999). The proportionality between ε sin ϖ and $\overline{I_{65}}$ is evident in Fig. 8.

3.2. Cenozoic Cooling

Before taking up the discussion of Pliocene and Pleistocene orbital oscillations and their climatic consequences, it is important to be aware of the secular climate drift known as the Cenozoic cooling. Pleistocene glacial cycles are superimposed on this ongoing cooling trend, and cannot be understood in isolation.

Orbital climate forcing typically varies on 10^4–10^5-yr timescales; the Cenozoic cooling has been underway for several 10^7 yr, and clearly is driven by nonorbital factors. Attempts to explain the Cenozoic cooling can be crudely divided into two camps. The first posits changes in continental and ocean basin geography, driven largely by plate tectonics; geographic changes force changes in global scale ocean circulation, which in turn has a major effect on climate.

The oceans transport about half as much heat as the atmosphere from equator to pole, but some heat is also transferred to the deep oceans (Fig. 9). Wind-driven currents carry warm waters from low latitude toward the poles. In the North Atlantic and Southern Ocean, radiative cooling is rapid, and sea ice formation excludes salts, increasing the density of the cold surface waters. The dense water sinks, becoming the North Atlantic Deep Water (NADW) and Antarctic Bottom Water (AABW). The NADW flows toward Antarctica at depth; the AABW flows north and east. In the Indian and Pacific Oceans, upwelling returns deep waters to the surface. This global-scale process is referred to as the "thermohaline circulation"; it transports 10^{15} W into the North Atlantic, warming northern Atlantic regional air temperatures by up to 10°C (*Rahmstorf,* 2002). In addition, the ocean exchanges gases with the atmosphere, thus influencing its greenhouse gas content.

Supporters of tectonic hypotheses point to a number of key events that could have resulted in substantial changes in oceanic circulation and heat transport, as well as ocean/atmosphere exchange. For example, the undocking of Australia from Antarctica 41 m.y. ago and the opening of Drake's Passage permitted the development of the Antarctic Circumpolar Current (ACC). Essentially lacking a meridional component, the ACC keeps cold water circulating around Antarctica, strengthening the formation of the AABW and leading to decreases in sea surface and continental temperatures (*Toggweiler and Bjornsson,* 2000; *Scher and Martin,* 2006).

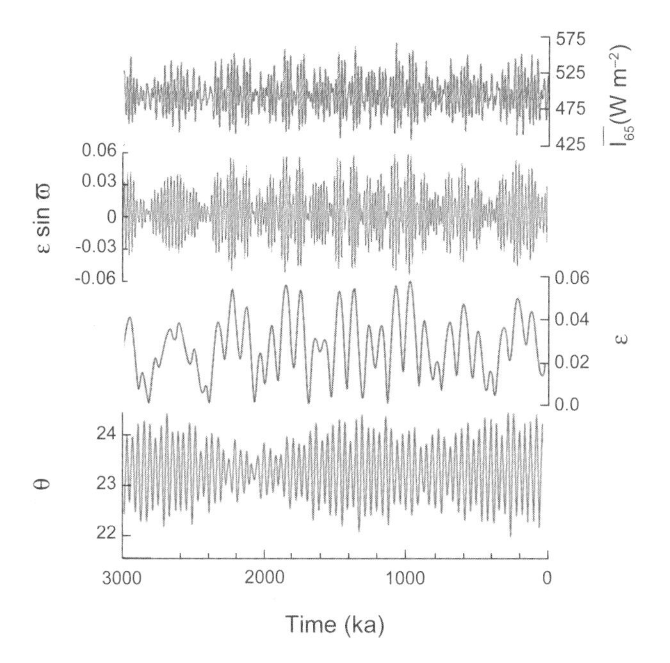

Fig. 8. Earth's orbital parameters for the last 3 m.y., and the average summer solstice insolation at 65°N (top), a season and latitude that are key for predicting ice sheet behavior (*Berger and Loutre,* 1991).

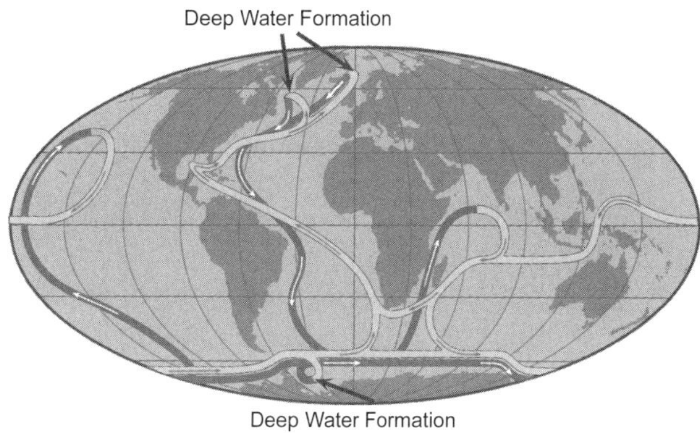

Fig. 9. Schematic of the ocean's thermohaline circulation. Near-surface currents are shown in light gray, while deep currents are shown in dark gray. Sites of North Atlantic Deep Water (NADW) formation in the North Atlantic are denoted for both the interglacial circulation (higher-latitude site) and the circulation during glacials (south of Greenland).

Benthic foraminiferal Mg/Ca and $\delta^{18}O$ records show that deep ocean waters have cooled by ~12°C since the Eocene in four main cooling events: ~50 Ma, 33.5 Ma, 13.9 Ma, and 2.6 Ma (Fig. 10) (*Zachos et al.,* 1993; *Lear et al.,* 2000). However, discrete tectonically driven circulation changes cannot be uniquely correlated with these cooling events.

An alternative hypothesis points to substantial changes in the atmospheric partial pressure of CO_2 (P_{CO_2}); atmospheric CO_2 has decreased since the Eocene from >2000 ppm to preindustrial levels of <250 ppm (*Beerling and Royer,* 2011; *Pearson and Palmer,* 2002; *DeConto and Pollard,* 2003). Abrupt decreases in atmospheric CO_2 concentrations correlate with the early Eocene (50 Ma) and Eocene-Oligocene (33.5 Ma) cooling events, the latter of which was the most extreme oceanographic and climatic change of the past 50 m.y. and contemporaneous with the appearance of the East Antarctic Ice Sheet (*Pearson et al.,* 2009; *Zachos et al.,* 2008). *DeConto and Pollard* (2003) have argued that CO_2 decreases preceded ocean cooling and Antarctic glaciation (*Pagani et al.,* 2011), and that the opening of Drake's Passage could only have had a role in triggering glaciation if atmospheric CO_2 was already near some critical threshold.

Benthic $\delta^{18}O$ time-series dating from the early Oligocene through the Miocene have shown evidence that the climate varied in a quasi-periodic fashion during all periods characterized by glaciation. Much of the spectral power since the early Oligocene is concentrated in the obliquity band (*Zachos et al.,* 2001). Interestingly, only in the early Miocene is a clear record of spectral power in both the 100- and 400-k.y. bands detectable, and confidently ascribed to eccentricity. As we shall see, there is considerable uncertainty regarding the role of eccentricity, if any, in subsequent Pliocene-Pleistocene climate forcing.

3.3. Pliocene-Pleistocene Glaciation

The Pliocene-Pleistocene glaciation is the most recent of the four Cenozoic cooling events, and notable for the significant expansion of northern hemisphere glaciation (NHG). The cause of NHG expansion remains uncertain. Global temperatures at the time of onset, ~3 Ma, were ~3°C higher than today, and P_{CO_2} was ~ 30% higher but falling (*Ravelo et al.,* 2004). Sea levels were 10–20 m higher, reflecting the absence of northern hemisphere ice sheets, and thermohaline circulation was more vigorous. One factor was almost certainly a drop in radiative forcing due to continuing decrease of P_{CO_2}. Another commonly invoked factor is the closure of the Isthmus of Panama. As the waterway between the two Americas narrowed, warm surface Atlantic currents were probably deflected northward (*Keigwin,* 1982). The timing of the final stages of closure between about 3.6 and 2.7 Ma (*Duque-Caro,* 1990; *Haug and Tiedemann,* 1998) is sufficiently close to the onset of NHG to make an attractive hypothesis — that the tectonically driven change in the Atlantic circulation leads to increased moisture transport to high latitudes and to the development of the northern ice sheets.

3.3.1. Orbital forcing. The fundamental approach used to relate orbital drivers to climate change is time-series analysis (*Chatfield,* 2003). The orbital history and sedimentary stacks are time-series data, and thus their discrete Fourier transforms are frequencies. Implicit in the concept of time series is the notion that data are separated by uniform time intervals, which is the reason that age models are essential.

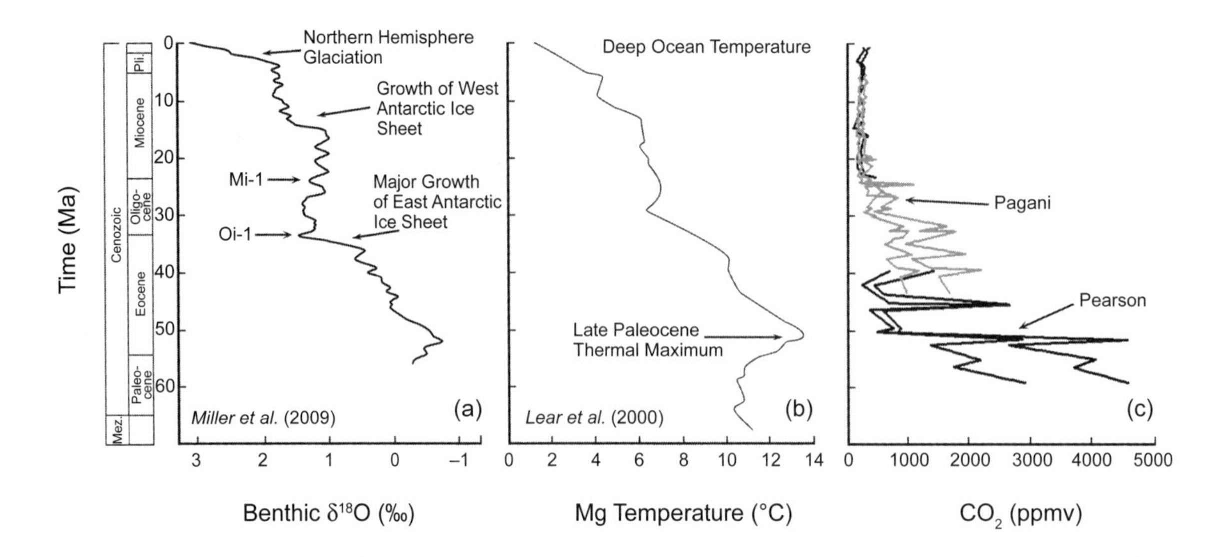

Fig. 10. Cenozoic cooling. **(a)** Benthic $\delta^{18}O$ over the past ~70 m.y. Major ice sheet expansions are indicated. Oi-1 is a positive $\delta^{18}O$ excursion associated with the onset of continental ice accumulation on Antarctica. Mi-1 is a similar major transient global glaciation event at the beginning of the Miocene (*Miller et al.,* 2009). **(b)** Deep sea temperatures relative to the present determined by Mg/Ca thermometry (*Lear et al.,* 2000). **(c)** P_{CO_2} for the period 0 to 65 Ma. Data are a compilation of marine and lacustrine proxy records from *Pearson and Palmer* (2002) (black) and *Pagani et al.* (2005) (gray).

The correspondence, or lack thereof, between frequencies measured (and modeled) in climate proxies and orbital drivers is the key to relating cause and effect in climate models (see, e.g., *Storch and Zwiers,* 2001).

Probably the most commonly used climate proxy is marine $\delta^{18}O$, and for illustrative purposes we will focus on that, and in particular, on a 5.3-m.y.-old stack of benthic $\delta^{18}O$ records from 57 globally distributed sites, the so-called LR04 stack (*Lisecki and Raymo,* 2005).

The LR04 stack for the past 3 m.y. is shown in Fig. 11a. As we have seen, benthic $\delta^{18}O$ is indicative of average global ice volume and deep-water temperature (which in turn reflects high-latitude sea surface temperatures). Several important conclusions are immediately obvious. Over the last 5 m.y., the average benthic $\delta^{18}O$ has increased nearly linearly, reflecting increasing ice volume through the Pleistocene, mostly in the northern hemisphere. Indeed, from 4.2 to 1.7 Ma, the climate was actually dominated by interglacials; from ~1.4 Ma onward, interglacial conditions have prevailed only ~20% of the time, usually lasting just 10–30 k.y. The variance of $\delta^{18}O$ about its running mean has also increased considerably, indicating that as the climate has cooled, glacial/interglacial oscillations have become more extreme.

The shape of the $\delta^{18}O$ oscillations has also evolved. With higher amplitude oscillations, $|d\delta^{18}O/dt|$ increased, but particularly so during deglaciations, eventually creating the pronounced sawtooth shape of the late Pleistocene. Long cooling intervals, marked by an oscillating buildup of ice sheets to maximum volume, are followed by brief, intense periods of warming. Simultaneously, the dominant $\delta^{18}O$ periodicity increased to ~10^5 yr. This puzzling climate shift is known as the mid-Pleistocene transition (MPT), and it is problematic to assign it a definitive beginning and end; the mean, variance, skewness, and timescale associated with the glacial cycles all exhibit an approximately linear trend over the last 2 m.y. (*Huybers,* 2007).

The last 3 m.y. of $\overline{I_{65}}$, which is generally assumed to be a metric of the main forcing controlling glacial/interglacial variations, is shown in Fig. 11b for comparison. It is immediately apparent that that there is no secular drift, and that neither variance, nor asymmetry, nor periodicity increase in the manner of $\delta^{18}O$ (see also Fig. 8). The inference to be drawn is that insolation must be considered more globally, and that additional forcing, such as feedbacks between climate components, must have become increasingly important since the MPT.

In Figures 11c,d, the power spectral densities of $\delta^{18}O$ and $\overline{I_{65}}$ show how the power in each time series is distributed over frequency. (Here power, by analogy to signal processing applications, is simply the squared value of the signal, and is not a measure of energy per unit time.) Two intervals are considered in Fig. 11c to elucidate the nature of the MPT frequency shift: the entire 5-m.y. interval, and the most recent 1 m.y.

Looking at the full stack, the 41-k.y. θ component dominates $\delta^{18}O$ throughout the late Pliocene and most of the Pleistocene. A smaller complex of bands at ~10^5 yr is evident, but there is very little spectral power in $\delta^{18}O$ at frequencies corresponding to precession.

These same frequencies are present in the DFT of insolation, but with very different relative spectral power (Fig. 11d). The peak power in $\overline{I_{65}}$ occurs in the 19–23-k.y.

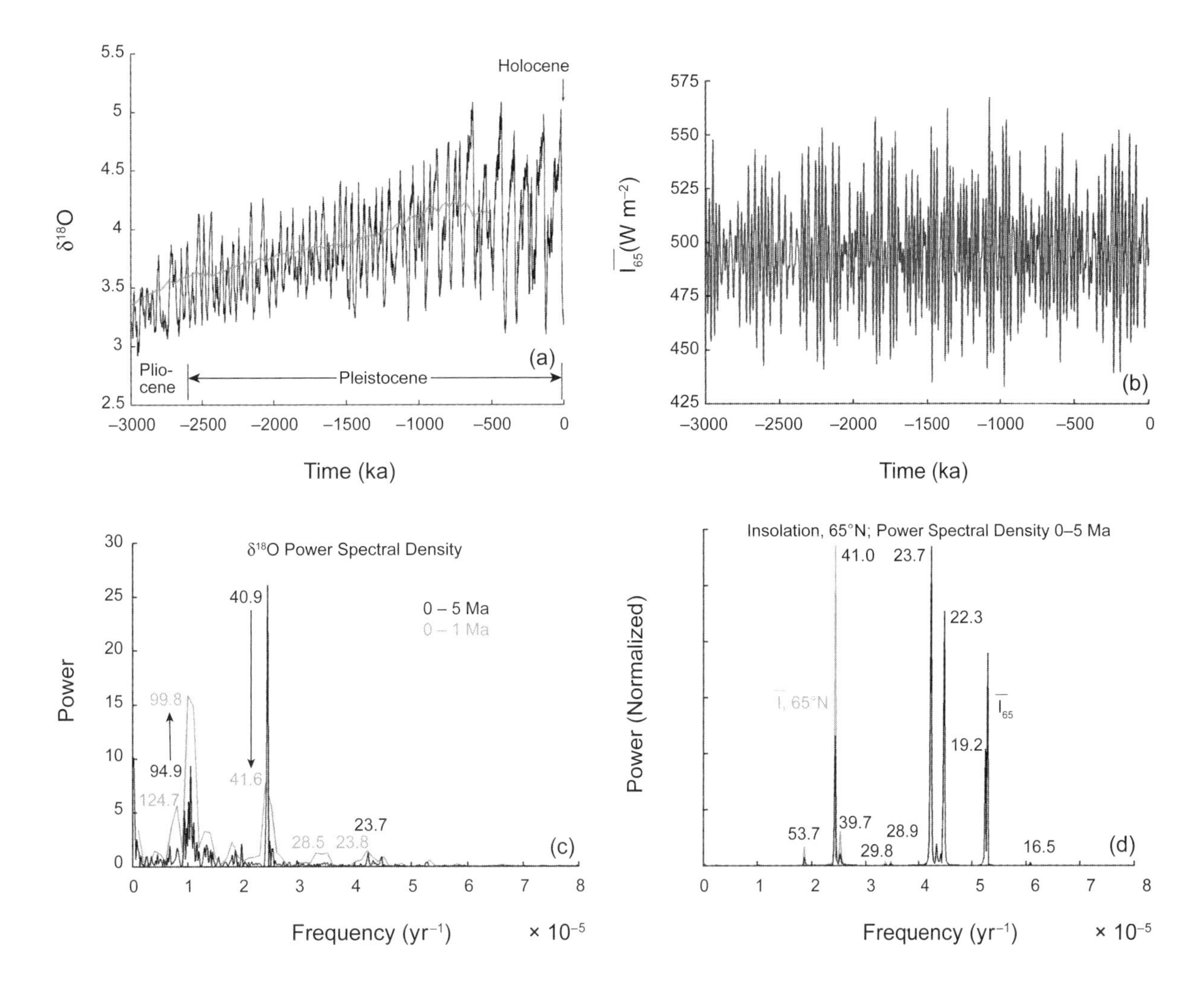

Fig. 11. (a) The LR04 $\delta^{18}O$ Pliocene-Pleistocene stack and its 500-k.y. moving average (gray line) over the last 3 m.y. **(b)** The diurnally averaged insolation at latitude 65°N at the summer solstice (L_s = 90°) for the last 3 m.y. **(c)** Power spectral density of the LR04 $\delta^{18}O$ data over the full 5-m.y. extent of the stack (black line), and for the last 1 m.y. only (gray line). The 1-m.y. time series was not padded to the same length as the 5-m.y. time series, so the frequency resolution is lower. The arrows indicate the decrease in power in the 41-k.y. θ bands and increase in the 100-k.y. bands. Numbers indicate the period equivalent to the frequency, in k.y. **(d)** The power spectral density of insolation at 65°N over the last 5 m.y. The summer solstice daily average insolation, $\overline{I}_{65}$ (black line), is dominated by ϖ, and to a lesser extent θ. The annually averaged insolation $\overline{I}_Y$ at the same latitude (gray line) depends only on θ. Numbers indicate the period equivalent to the frequency, in k.y. Spectra normalized to the same peak power.

ϖ bands, with somewhat less in the 41-k.y. θ bands; the weaker 54- and 28-k.y. θ bands appear as well. In fact, daily insolation at all latitudes between the limits of polar night is dominated by ϖ. Eccentricity variations contribute <0.2% of the total power in the $\overline{I}_{65}$ spectrum. However, if we return to the LR04 $\delta^{18}O$ stack, and examine the DFT over the most recent 1 m.y. (gray line in Fig. 11c), it is clear that the $\delta^{18}O$ signal has shifted toward a dominant 10^5-yr cycle. The $\overline{I}_{65}$ spectrum over the same restricted period is unchanged from Fig. 11d, and we must conclude that the MPT occurred without a meaningful change in orbital forcing (*Meyers and Hinnov,* 2010).

Even this cursory overview of astronomical forcing and climate response raises several questions: (1) Why have recent glacial/interglacial cycles become strongly asymmetric while simultaneously growing more extreme? (2) What accounts for the appearance and recent dominance of climate variability in the 100-k.y. bands when no corresponding power exists in the radiative forcing? (3) Why do global ice sheet volumes respond so weakly in the 19–23-k.y. bands that dominate the variability in $\overline{I}_{65}$?

To address these questions, the scope of most investigations has been expanded from simple time-series analysis. To a greater or lesser degree, most now incorporate (1) astronomically controlled solar forcing, consistent with Milanković's hypothesis, and (2) nonlinear feedbacks within the climate system, which amplify or dampen the effects of solar forcing. These feedbacks change the relative power in the astronomical frequency bands, but astronomical forcing remains the pacemaker, and no new frequencies are introduced. Hypotheses of this sort can only be evaluated through consideration of the entire climate system, rather than mere frequency analysis. While no theory comprehensively answers these questions, it is not for want of candidates.

While the increase in mean $\delta^{18}O$ has been nearly linear throughout most of the Pleistocene, the rapid increase in variance is best fit as an exponential over the past 5 m.y. The trend toward increasing variance is seen even in $d\delta^{18}O/dt$. Taken together, these results indicate that the shift to lower frequencies *per se* does not account for the increasing amplitude of $\delta^{18}O$ oscillations. The implication is that the climatic sensitivity to external forcing and/or internal variability has increased through time. For example, *Lisiecki and Raymo* (2007) point out that the variation in $\delta^{18}O$ was proportional to forcing in the θ bands until 1.4 Ma; the amplitudes of obliquity forcing and $\delta^{18}O$ response are uncorrelated thereafter. This lack of correlation indicates that obliquity is no longer driving glacial cycles. However, it may continue to pace glacial cycles, a possibility that we will discuss further below.

The increasingly asymmetric shape of late Pleistocene $\delta^{18}O$ oscillations is most likely due to a change in the internal dynamics of the climate system because asymmetry is not found in any orbital or insolation curves. Numerous explanations have been put forward for the asymmetry between rates of glaciation and deglaciation. Several lines of evidence support the hypothesis that sawtooth asymmetry results from different response times for the growth and decay of northern hemisphere ice sheets (e.g., *Imbrie and Imbrie,* 1980; *Paillard,* 1998): (1) The appearance of glacial asymmetry is approximately coeval with the onset of major northern hemisphere glaciation; (2) glacial asymmetry is observed before 1.4 Ma, at a time when cycle amplitudes responded proportionally to obliquity; (3) there are many mechanisms that allow for rapid ice sheet decay (e.g., *Clark et al.,* 1999; *Tarasov and Peltier,* 2004), including basal sliding, iceberg calving, a nonlinear sensitivity to temperature, and a positive feedback with sea level. Conversely, long-term ice sheet growth is subject to negative feedback between ice sheet height and accumulation rate (*Lisiecki and Raymo,* 2007). Finally, it is worth bearing in mind that the shape of glacial cycles in the $\delta^{18}O$ record could be distorted if sedimentation rates varied systematically and globally with changes in the global ice volume.

Turning to the related question of why post-MPT glacial cycles display a dominant 100-k.y. period, we begin by noting that there is only a very weak 100-k.y. signal in $\overline{I_{65}}$, but climate dynamics are very complex, involving much more than insolation and certainly much more than insolation distilled to a single metric such as $\overline{I_{65}}$.

Berger and Loutre (2010) point out that dominant frequencies in ε are actually related to the four dominant ϖ frequencies

$$1.053\times10^{-5}\left(yr^{-1}\right)=\frac{1}{9.494\times10^{4}}=\frac{1}{1.897\times10^{4}}-\frac{1}{2.372\times10^{4}}$$

$$8.110\times10^{-6}\left(yr^{-1}\right)=\frac{1}{1.233\times10^{5}}=\frac{1}{1.897\times10^{4}}-\frac{1}{2.243\times10^{4}}$$

$$1.004\times10^{-5}\left(yr^{-1}\right)=\frac{1}{9.959\times10^{4}}=\frac{1}{1.915\times10^{4}}-\frac{1}{2.372\times10^{4}}$$

Thus, they argue for a nonlinear response to ϖ forcing, characterized by frequencies that are exactly those of the eccentricity.

Many workers have invoked the idea of a threshold in the climate system, which must be crossed to permit the expression of the 10^5-yr periodicity through some nonlinear amplification mechanism. A linearly decreasing CO_2 concentration from 320 ppm at 3 Ma to 200 ppm at the Last Glacial Maximum (LGM) was prescribed to drive models by *Berger et al.* (1999) that qualitatively reproduced the observed frequency shift to 100 k.y. as a result of a beat between the main precessional frequencies. However, the Antarctic CO_2 data do not appear to be compatible with a linear decrease, at least from 800 k.y. onward (*Jouzel et al.,* 2007).

Huybers and Wunsch (2005) examined the statistical validity of the apparent 100-k.y. periodicity, testing the null hypothesis that glacial terminations are independent of obliquity over the past 700 k.y. They found that they could reject that hypothesis at the 5% significance level. Whatever else may be happening, glacial terminations continue to be paced by θ; over the last 2 m.y., 33 out of 36 terminations have occurred when θ was anomalously large. The simplest inference is that during the early Pleistocene, deglaciations occurred nearly every obliquity cycle, while late-Pleistocene deglaciations more often skip one or two obliquity beats. The authors postulate the existence of a threshold for deglaciation that slowly increases through the Pleistocene, causing increasingly frequent skips. They cite evidence that the intervals between successive deglaciations cluster into 80-k.y. or 120-k.y. periods, indicative of two or three obliquity cycles, and leading to the appearance of a 100-k.y. period.

Other workers have invoked shifts in the controls on glaciation to activate new sources of low-frequency variability. These have included the inherent properties of ice sheet dynamics (*Imbrie and Imbrie,* 1980), crustal isostacy under ice loads, the carbon cycle (*Shackleton,* 2000) meridional heat transport, and/or deep ocean circulation (*Imbrie et al.,* 1993).

Finally, it is worth noting that the 10^5-yr periodicity is a relatively recent phenomenon, and by almost any measure, there have been fewer than 10 strongly asymmetric 100-k.y. glacial cycles. Thus, the length of the climate record may be insufficient to establish statistically significant relationships between climate and orbital forcing.

The apparent weakness or absence of power at ϖ frequencies in the climate record is in part a consequence of examining such long time series. In the pre-MPT Pleistocene, ϖ bands are essentially absent; however, precession is clearly observed in post-MPT ice-volume and sea-level records, and has increased progressively over the last 1 m.y. (*Berger and Loutre,* 2010). Indeed, the correlation of power in forcing and response suggest that post-MPT precession responses are directly forced, either linearly or with nonlinear positive feedback.

Explanations for the pre-MPT weakness in the ϖ bands tend to hinge on the question of whether $\overline{I_{65}}$ is really the correct measure of forcing. The annual mean insolation at any latitude depends only on θ, and hence meridional gradients in annual average insolation also vary only with θ. Likewise, the integrated insolation between any two points along a planets orbit depend primarily on obliquity (Fig. 11d). So perhaps variations in the summer insolation gradient between high and low latitudes, driven by obliquity, exert the dominant control on the high-latitude climate. *Huybers and Denton* (2008) proposed that ice sheet ablation is better determined by integrating the total insolation exceeding a melting threshold. Certainly the equator-to-pole insolation gradient is strongly correlated to glacial-interglacial ice-volume variations from 3.0 to 0.8 Ma (*Raymo and Nisancioglu,* 2003).

Another explanation for the lack of a precession signal in records of ice volume was proposed by *Raymo et al.* (2006). They put forward a model in which northern hemisphere ice sheets wax and wane at precession periods, driven by the strongly nonlinear response of ice ablation to summer insolation intensity. In this model, however, the precession component of changes in ice volume is missing from marine records of $\delta^{18}O$ because it is "canceled out" by changes in southern hemisphere ice volume that are of opposite phase.

And a final caution here as well: The ϖ bands, having the highest frequency, are most susceptible to noise and small errors in age models. This might contribute to the apparent decrease in ϖ band power with age.

3.3.2. Climate response. We conclude this section by shifting from a focus on astronomical forcing to the glacial/deglacial response and the feedback mechanisms that may amplify (or dampen) climate changes. The discussion will make frequent reference to Fig. 12, which shows the most recent 500 k.y. of orbital parameters, integrated and averaged insolation history, and a suite of climate proxy data (*Lisiecki and Raymo,* 2005; *Muhs et al.,* 2012; *Jouzel et al.,* 2007; *Lüthi et al.,* 2008). Five glacial maxima and five glacial terminations (T-I through T-V) are captured in Fig. 12.

3.3.2.1. Glaciation: In its broadest outlines, the Milanković hypothesis that glacial conditions arise as a result of diminished insolation at high northern latitudes (low θ and aphelion in NH summer) is supported by the data.

Both $\overline{I_Y}$ and its equator-to-pole gradient are greatest when θ is minimized (cf. Fig. 7), leading to strengthening of the atmospheric and ocean circulations that transport heat and moisture poleward (*Mantsis et al.,* 2011). Consequently, precipitation increases in the high latitudes for two reasons: (1) An increase in high-latitude winter insolation is enhanced through low-cloud feedback; the extent of

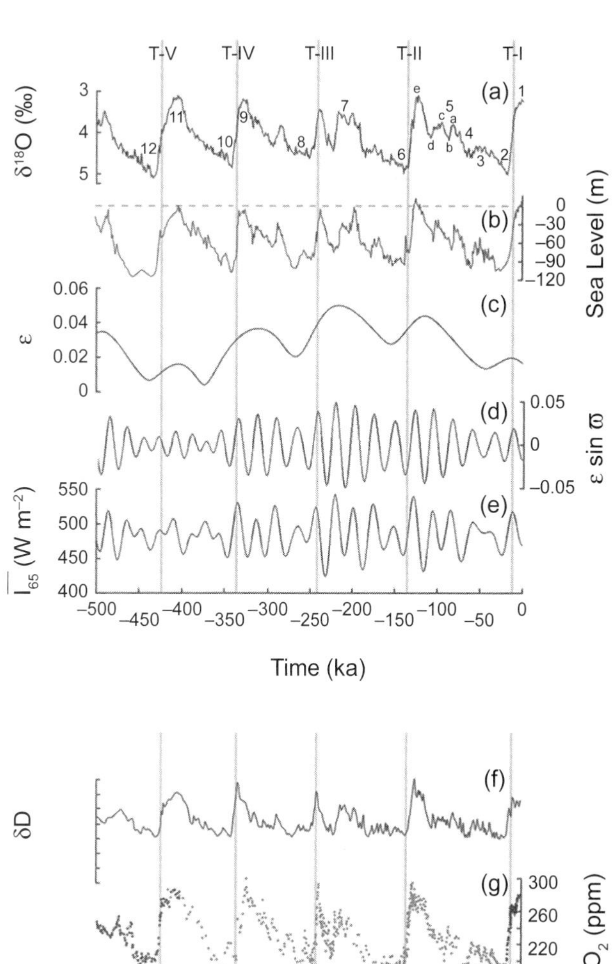

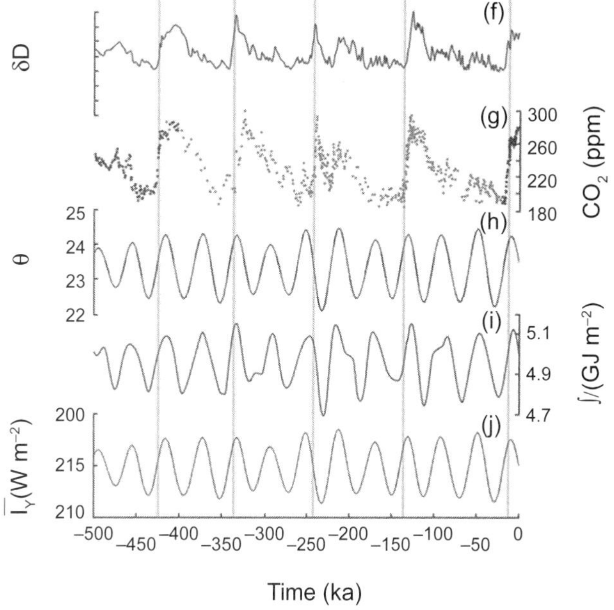

Fig. 12. Climate proxy data, orbital drivers, and insolation for the last 500 k.y., including the last five glacial terminations (vertical gray lines). **(a)** $\delta^{18}O$ composite record from the LR04 stack (*Lisiecki and Raymo,* 2005); numbers refer to Marine Isotope Stage. **(b)** Red Sea record of sea-level fluctuations (*Muhs et al.,* 2012); current sea level denoted by dashed line. **(c)** Orbital eccentricity. **(d)** Precession index. (e) $\overline{I_{65}}$, summer solstice diurnally averaged insolation at 65°N. **(f)** δD from EPICA Dome C (*Jouzel et al.,* 2007). **(g)** Compilation of CO_2 records, EPICA Dome C (*Lüthi et al.,* 2008). **(h)** Obliquity. **(i)** Integrated summer insolation at 65°N (*Huybers and Denton,* 2008). **(j)** $\overline{I_Y}$, annually averaged insolation at 65°N.

low clouds strongly influences the local radiation budget, generally decreasing the downward shortwave flux and increasing the upward longwave flux. This results in colder air temperatures and increased winter snowfall. (2) An increase in the summer meridional insolation gradient enhances summer eddy activity, and increases the intensity summer snowfall (*Loutre et al.,* 2004; *Lee et al.,* 2008). Due to increases in polar ice and global cloud coverage, mean global temperatures also decrease, even though the global mean insolation is unchanged.

The last glacial cycle is divided into marine oxygen isotope stages (MIS) 1–5, including the last interglacial, substage MIS-5e (Fig. 12a). Sea-level reconstructions show that climate moved rapidly from this state into the last glacial stage, reaching almost half the glacial-maximum ice volume within a few thousand years (*Rahmstorf,* 2002).

Bonelli et al. (2009) used a two-and-a-half-dimensional climate model coupled with a three-dimensional ice sheet model to simulate the initiation of major northern hemisphere ice sheets at the end of MIS-5e, 118–116 ka (Fig. 12). The model was driven by solar forcing and a P_{CO_2} history prescribed from the Vostok ice core (*Barnola et al.,* 1987). While decreasing NH summer insolation alone was sufficient to establish the Laurentide ice sheet in North America, they found that both low insolation and low P_{CO_2} were necessary to trigger glaciation over Eurasia. The difference arises from the heat transported to the North Atlantic by the Atlantic Meridional Overturning Circulation (AMOC), the Atlantic component of the global thermohaline circulation (Fig. 9), and its moderating effect on European climate. Indeed, when NH summer insolation is low, NADW formation is significantly intensified due to greater radiative cooling and reduced freshwater runoff. This is balanced by increased flow of warm equatorial surface water (i.e., a corresponding intensification of the AMOC), which increases high-latitude evaporation. This allows more snow to accumulate during winter (*Lambeck et al.,* 2002; *Wang and Mysak,* 2002; *Bintanja and van de Wal,* 2008), but also increases ablation, requiring P_{CO2} to fall further before perennial ice can become established in Europe. In contrast, the Laurentide ice sheet originated at mid-continent where only atmospheric heat transport is effective and surface temperatures respond more strongly to insolation.

Carbon dioxide — To first order, the correlation between ice volume ($\delta^{18}O$) and atmospheric CO_2 (Figs. 12a,g) is striking, in both the timing and asymmetry of changes. Agreement between these records suggests a cause and effect relationship, but establishing their phasing (i.e., which leads and which lags) is problematic. The fundamental reason is that Antarctic and Greenland snow eventually compacts, trapping air, but the process can take several thousand years, during which time pores can remain diffusively connected to the atmosphere (*Brook,* 2013), and therefore there is uncertainty in the age of the trapped air.

Polar ice also records past local temperatures in both δD and $\delta^{18}O$ incorporated into the H_2O molecule. From this it is possible to establish that Antarctic temperatures, along with CO_2, CH_4, and NO_2, have varied in phase with both $\delta^{18}O_{atm}$ and $\overline{I_{65}}$ (i.e., the *northern* hemisphere insolation) over the past 800 k.y. (*Petit et al.,* 1999; *Lüthi et al.,* 2008). But why does insolation at high northern latitudes apparently control Antarctic temperatures, especially since the precession signal, the strongest in NH insolation, is out of phase between north and south?

The ocean contains about 50 times more C than the atmosphere, and hypotheses that link northern insolation to global CO_2 and global temperatures invariably invoke changes in the oceanic C sink, either through physical or biological mechanisms (*Rahmstorf,* 2002). For example, warming the sea surface releases CO_2, which contributes to further warming. Changes to the thermohaline circulation alter the amount of CO_2 sequestered in the deep ocean, in turn influencing surface temperature. The relationships are sufficiently complex that many models simply take CO_2 along with insolation as independent variables when investigating paleoclimates.

Ice sheets — As they become established and begin to grow, the high albedo and increasing surface elevation of ice sheets begin to feed back into the climate over broad spatial scales (*Abe-Ouchi et al.,* 2007). A simple example is the ice-albedo effect. The higher albedo of ice reflects more solar energy to space, leading to cooling, which causes more ice to precipitate and survive. This is a straightforward example of a positive feedback mechanism.

Negative feedbacks progressively develop as well. As the thickness of ice sheets increase, their surfaces grow colder due to the adiabatic lapse rate of the atmosphere. Initially, this slows ablation, leading to greater accumulation. But because colder air cannot hold as much moisture, precipitation eventually begins to decline over a growing ice sheet. As older snow and ice surfaces are darker than fresh snow, they begin to warm when precipitation declines, increasing ablation. These processes progressively slow the buildup of the ice sheets up to the time they start to retreat.

In North America, ice sheets 3 km thick force a high-pressure anticyclonic circulation centered over the western portion of the ice sheet. This disrupts the prevailing westerlies and brings up warm southerly air on the western flank. Over most of the ice sheet, however, northerlies decrease temperatures to well below the zonal mean climate. There is an east-west temperature difference across North America of about 10°C, which results in an east-west asymmetry in the ice sheet; the Laurentide ice sheets commonly extended up to 1000 km further south along the eastern margin of the continent. The thick ice also forces a large-amplitude atmospheric stationary wave that likely affects the climate over northwestern Europe and the Fennoscandian ice sheet (*Roe and Lindzen,* 2001).

Ocean circulation — Growth of large ice sheets requires warm temperatures during winters to enhance moisture transport to high latitudes, and cool temperatures during summers to prevent melting of snow. Surface water transports heat efficiently and a northward penetration of warm surface currents can encourage ice-sheet development, while suppression of these currents discourages ice growth. However, as ice sheets reach their mature size, and develop marine margins, they play a role in changing ocean circulation, whose response to external forcing contains both short (~1 k.y.) and longer (3–5 k.y.) time constants. Sea-level analyses for Svalbard, Greenland, and East Antarctica lead to similar conclusions that, at the height of glaciation, the ice sheets extended offshore, in most cases as far as the shelf margin, where they are susceptible to rapid reductions in volume (*Lambeck et al.,* 2002).

Coupled atmosphere-ocean models generally show that the location of deep water formation in the northern hemisphere shifts from the Labrador and Nordic seas to the south of Iceland and Greenland. Consensus regarding the rate of circulation and the role of the AMOC is more problematic. Models predict various changes to the thermohaline circulation, and in fact do not even agree on whether it becomes stronger or weaker (*Otto-Bliesner et al.,* 2007). $^{231}Pa/^{230}Th$ measurements (a kinematic proxy for the meridional overturning circulation) in North Atlantic cores suggest a 30–40% decrease in AMOC circulation at the last glacial maximum (*McManus et al.,* 2004).

3.3.2.2. Deglaciation: Glacial maxima do not occur as an established state. Ice sheets do not reach a "mature" size, stabilize, and subsequently ablate; they continue to grow until their sudden collapse. Terminations are autocatalytic processes that quickly move to completion, and it is problematic to identify the climate's unique entry point for each of the most recent terminations.

Milanković's hypothesis that increasing insolation at high northern latitudes should lead to deglaciation is supported, in a statistical sense, by analysis of paleoclimate proxies from globally distributed marine sediments (*Lisiecki and Raymo,* 2005; *Huybers and Wunsch,* 2005), cap ice in both hemispheres (*Kawamura et al.,* 2007), and speleothems (cave deposits) that are precisely dated but reflect only local conditions (*Cheng et al.,* 2009). Clearly the most recent termination, T-I, was initiated by an increase in insolation that began at ~24 ka; a 130-m rise in sea level between 19 and 7 ka denotes the subsequent disappearance of the glacial ice. Yet whether calculated as $\overline{I_{65}}$, $\overline{I_Y}$, or $\int I_S$ (the integrated summertime insolation), glacial conditions persisted through equivalent insolation maxima 108, 85, and 51 ka (depending on which measurement one uses). Thus it appears that one or more additional conditions are required for termination.

While T-I demonstrates that $\overline{I_{65}}$ maxima alone are not sufficient to bring about a glacial termination, the penultimate deglaciation (T-II: MIS-5e → MIS-5d) demonstrates that they are not necessary either. Disentangling the mechanisms of T-II is more complex, because dating is less precise relative to the rapid sequence of events and the geologic record is more sparse. Yet data from multiple independent sources confirm that T-II was well-advanced by 135 ka, predating the $\overline{I_{65}}$ maximum by at least 9 k.y. (*Spötl et al.,* 2002; *Gallup et al.,* 2002). *Drysdale et al.* (2009) presented a speleothem-based North Atlantic marine chronology that dates T-II to 141 ka, too early to be explained by $\overline{I_{65}}$, but potentially consistent with $\overline{I_Y}$ or $\int I_S$. *Huybers and Wunsch* (2004) made the argument that obliquity may play a role in fixing Late Pleistocene terminations because obliquity controls the annual integrated total energy per unit area reaching the poles. If the relatively early dates of *Drysdale et al.* (2009) for T-II are accepted, then it could be argued that T-I and -II are separated by three obliquity cycles and that they started at near-identical obliquity phases.

Likewise, $\overline{I_{65}}$ appears an unlikely candidate for T-V, which began at ~427 ka, and initiated the most recent of the so-called "super-interglacials." High latitudes were ice-free for at least twice as long as other interglacials, and the absence of ice-rafted debris for ~23 k.y. suggests interglacial conditions endured in the NH through a full precession cycle (*Loutre,* 2003). Again, $\overline{I_Y}$ and $\int I_S$ were nearer appreciable maxima, and seem to be better candidates.

None of these terminations appear to be linear with orbital forcing; by the time ice sheets reach the continental shelves and begin to become unstable, the subtle interplay between Milankovitch cycles, ice-sheet dynamics, and shifts in ocean circulation seems to drive the climate system.

Carbon dioxide — T-I is the most recent climate shift, with by far the best record, so *Parrenin et al.* (2013) set out to address the chicken and egg problem of CO_2 and temperature during T-I by using a unique method to establish the gas age–ice age difference and by creating a composite record of Antarctic temperature from several ice cores from the Antarctic interior. They use the ratio of $^{15}N/^{14}N$ in nitrogen (N_2), which is enriched in firn due to gravitational settling. The enrichment depends on the depth of the firn column. Once this depth is determined from the nitrogen data, a simple model can predict the offset in depth between gas and coeval ice and the amount of time this represents (*Brook,* 2013). Their analysis indicates that CO_2 concentrations and Antarctic temperature were tightly coupled throughout T-I, within an uncertainty of <200 yr. Support for this conclusion comes from recent independent work of *Pedro et al.* (2012), who concluded that CO_2 lagged temperature by <400 yr over T-I, and they could not exclude the possibility of a slight lead. Thus, we remain unable to determine how the climate entered even the most recent temperature-CO_2 feedback loop, but it is clear that the climate responds extremely quickly.

Ice sheets — One potential prerequisite for complete glacial termination is the existence of a massive, isostatically compensated ice sheet (*Denton et al.,* 2010). Models suggest that the largest ice sheets can throttle NADW formation because they supply fresh water to the North Atlantic and simultaneously reduce evaporation through cooling downstream of the Laurentide ice sheet stationary wave. When ice sheets reach the open ocean and establish marine margins they begin to cool the oceans by calving icebergs, which extract much of the latent heat of melting directly from ocean waters.

In addition, marine-based ice sheets are inherently more unstable than those with terrestrial margins, because of their susceptibility to changes in sea level. During T-I and

T-II, the southernmost margins of land-based NH ice sheets responded nearly instantaneously to orbitally driven solar forcing. Land-based ice sheets subsequently retreated at a rate proportional to global warming. Ice sheets with marine margins did not respond immediately, but then experienced rapid collapse (*Carlson and Winsor,* 2012). This probably has to do with the greater thermal mass of the oceans, which respond to insolation changes far more slowly than land, and the consequences of ice shelf collapse, which allow the rapid flow of the land-based ice.

The albedo feedback associated with ice sheets operates during terminations as well. When insolation starts to increase, enhanced ablation lowers the zonal surface albedo. It also induces replacement of tundra by taiga, which reduces the albedo of continental surfaces covered by snow (*Berger,* 2001).

Ocean circulation — The role of the ocean circulation is that of a highly nonlinear amplifier of climatic changes (*Rahmstorf,* 2002). Collapse of NADW formation and thermohaline circulation forced by the largest ice sheets has multiple consequences (*Clark et al.,* 1999). It decreases meridional heat transport to the North Atlantic. The colder North Atlantic surface causes an expansion of winter sea ice and a highly seasonal climate in the North Atlantic. These profound disruptions of oceanic heat transport likely play a role in the millennial-scale cold stadials that comprise the low-temperature compliment to warm interstadials (*Cheng et al.,* 2009).

4. CLIMATE CHANGE ON MARS

The study of orbitally driven climate variations on Mars has unfolded through a sequence very different from the terrestrial case. Solutions for martian orbital parameters are available to at least 10 Ma (*Laskar et al.,* 2004), so solar climate forcing is known. And it is thought that a sedimentary record of climate change, analogous to terrestrial marine sediments, exists in the PLD, so the problem appears tractable at some level.

There are two difficulties in modeling astronomically driven climate change on Earth that Mars does not share. First, insolation changes caused by Mars' orbital variations are enormous relative to Earth's, in principle resulting in more profound climate change. Second, Earth's climate is extremely complex due to sensible heat transport by the oceans and atmosphere, large latent heat budgets, and an active biosphere. Unfortunately, there are two even more critical and pragmatic factors standing in the way of a detailed understanding of martian climate history: our inability to access and measure martian climate proxies, and of course our inability to access and date martian samples, leaving us without meaningful age models. Consideration of the prime importance these capabilities have in terrestrial climatology should convey the profoundly primitive nature of our understanding of Mars.

Instead, Mars climatology has proceeded by efforts to (1) establish the nature of the PLD, their sedimentary sequence, and age, and any other detectable indicators of climate change; to date, this has been accomplished almost entirely by remote sensing; and (2) independently model variations in climate that can be predicted from the known insolation history and volatile reservoirs, testing model predictions against what can be seen of the geologic record. While a quantitative and robust understanding of Mars' climate history must await a future era of martian geochemical climatology, some early efforts are underway to try to establish crude age models by correlating stratigraphy with insolation history and attempting to identify meaningful climate proxies in the available data.

4.1. Orbit and Forcing

Mars' current orbital parameters are $\theta = 25.19°$, $\varepsilon = 0.093$, and $\varpi = 251.1°$, but their oscillations are extraordinary (Fig. 13); θ excursions are 10×, and total ε variations 2.5× greater than Earth's.

Mars' spin axis and orbital plane precess at nearly the same rate (cf. Fig. 6), producing a secular spin-orbit resonance that causes θ oscillations $\geq \pm 10°$. The obliquity oscillates with a period of $\sim 1.25 \times 10^5$ yr⊕ (6.64×10^4 yr♂), with an amplitude modulation of $\sim 1.3 \times 10^6$ yr⊕ (6.9×10^5 yr♂). A secular decrease in θ at ~5 Ma⊕ has resulted in lower polar temperatures than were characteristic of the preceding 1 G.y.⊕; the current climate is thus anomalous in historical terms. Precise orbital solutions are not available earlier than 20 Ma⊕, beyond which only statistical treatment of orbital probability functions is possible (*Laskar et al.,* 2004). Such studies indicate (*Armstrong et al.,* 2004) that θ has historically been even higher, averaging ~35° over the past 1 G.y.⊕, and regularly reaching 60°.

Eccentricity ranges from 0.0 to 0.13 with 0.05 amplitude oscillation every $\sim 9.6 \times 10^4$ yr⊕ (5.1×10^4 yr♂) and 0.1 amplitude oscillation of $\sim 2 \times 10^6$ yr⊕ (1.06×10^6 yr♂). Currently, annual heliocentric distances range from 1.38 AU in late northern autumn to 1.66 AU in late northern spring, resulting in a ~40% insolation over a year; the southern summer is hotter and shorter insolation variation over the northern summer. As on Earth, the highest frequency variations derive from precession, $\sim 5.1 \times 10^4$ yr⊕ (2.7×10^4 yr ♂), reflected in the precession index.

The uppermost plot in Fig. 13, by loose analogy to the terrestrial metric for insolation, is the diurnally averaged summer solstice insolation at 90°N, $\overline{I_{90}}$. The choice of 65°N for Earth was motivated principally by the increasing fraction of the surface covered by water at higher latitudes. Lacking a similar constraint for Mars, we will here use both 90°N and 80°N, which effectively span the latitude range of the current permanent polar cap. The excellent correlation between $\varepsilon \sin \varpi$ and $\overline{I_{65}}$ that we noted for the terrestrial case does not hold for Mars, principally because of the extraordinarily large range of the θ oscillation.

In Fig. 14, a simple examination of the orbit and high latitude solar forcing is shown. The normalized power of each parameter is given in Fig. 14a, where the complex fre-

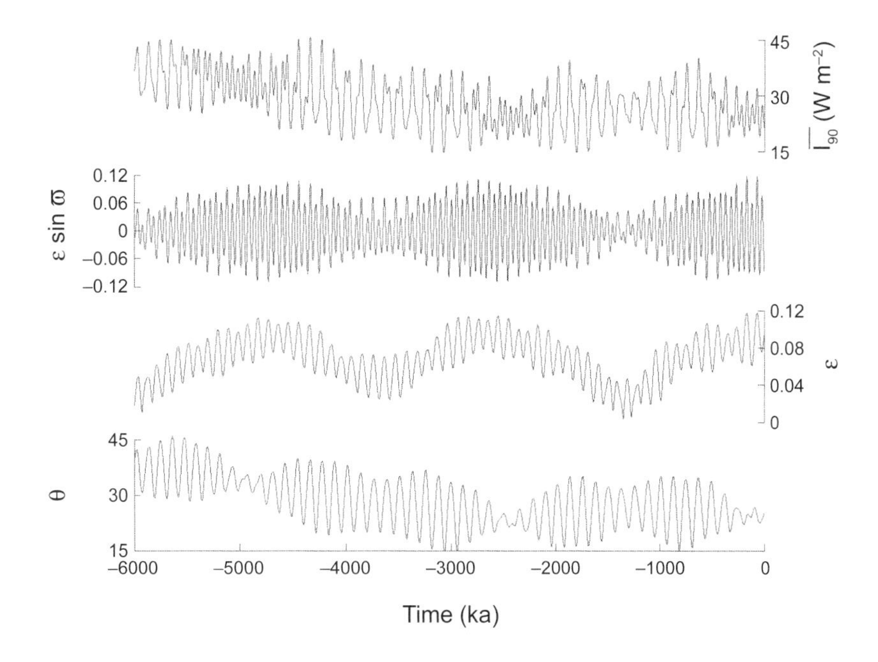

Fig. 13. Mars' orbital elements over 6 m.y.⊕ and the diurnally averaged summer solstice insolation at the north pole.

quency structure of each of the orbital parameters is evident. The θ power is contributed by four bands between 106.4 and 123.4 ka. Precession likewise has four bands between 54.9 and 51.5 ka, with weak bands in the 112–119-ka range. Eccentricity separates into three frequencies between 95 and 103 ka. The peak of the lowest frequency modulation is not well resolved at ~2.4 Ma, because only 10 m.y. of data were analyzed.

The relative variations in solar forcing, whether averaged diurnally at the peak of summer or over the entire year, dwarf those of Earth (Fig. 14b). The polar diurnal summer solstice insolation (here given for 80°N) varies by a factor of three over the 6 m.y.⊕ shown here, and by a factor of 2.5 for $\overline{I_Y}$ when annually averaged. The comparable ranges for Earth are 14% and 1%, respectively.

Summer polar insolation reflects significant contributions from all three orbital parameters (Fig. 14c). During nodes in θ (including the past 400 k.y.) that occur every ~2.5 m.y.⊕, $\overline{I_{90}}$ is dominated by precession; at other times (*Laskar et al.,* 2002), the large oscillations in θ dominate polar insolation, with low-frequency components from ε, which is merged into the 2.5-m.y. θ line. As with Earth, for reasons deriving from Kepler's second law, the annually averaged insolation at high latitude is solely a function of θ (Fig. 14d).

4.2. The Geologic Record

The astonishingly large variations in solar forcing, and the fact that the relevant frequencies are much lower than Earth's lead to the expectation (or at least the hope) that the geological record of their history will be proportionately stark. As it happens, nature may have cooperated in this instance; the PLD show cyclic layering that can be characterized even from orbit (Fig. 15).

There is also a suite of geomorphic features that, taken as a whole, suggest much of the surface inventory of H_2O has repeatedly migrated between high- and low-latitude reservoirs. Furthermore, although the current global distribution of H_2O ground ice is largely thermodynamically stable, significant outliers of unstable ice suggest recent changes in its equilibrium distribution. Finally, the relative abundance of H_2O in some mid- to high-latitude ground ice is inconsistent with formation in the current climate.

4.2.1. Polar layered deposits. The north PLD (NPLD) and the south PLD (SPLD), tremendous sedimentary stacks that cover polar latitudes in both hemispheres, are similar in gross morphology and dimension (*Smith et al.,* 2001). Both are roughly domical, ~1000 km across, with ~3.5 km relief relative to the surroundings (*Phillips et al.,* 2008). Both are dissected by spiral troughs, and both are composed largely of H_2O ice (*Picardi et al.,* 2005; *Plaut et al.,* 2007). Layering as fine as 10 cm is observed (*Malin et al.,* 2001; *Limaye et al.,* 2012), and is likely due to varying dust/ice ratios or degrees of cementation (*Murray,* 1972; *Cutts,* 1973). Unconformities pervade both the NPLD and SPLD (e.g., *Howard et al.,* 1982; *Tanaka,* 2005), but faulting is uncommon (*Byrne,* 2009). These parallels suggest that the PLD share a similar origin. They differ however, in stratigraphic context, elevation, and age, and thus record and preserve Mars' climate history differently.

4.2.1.1. North polar layered deposits: At the top of the north polar sedimentary stack is the northern residual ice cap (NRIC), a meters-thick H_2O deposit that unconformably overlies the NPLD and is seasonally draped by

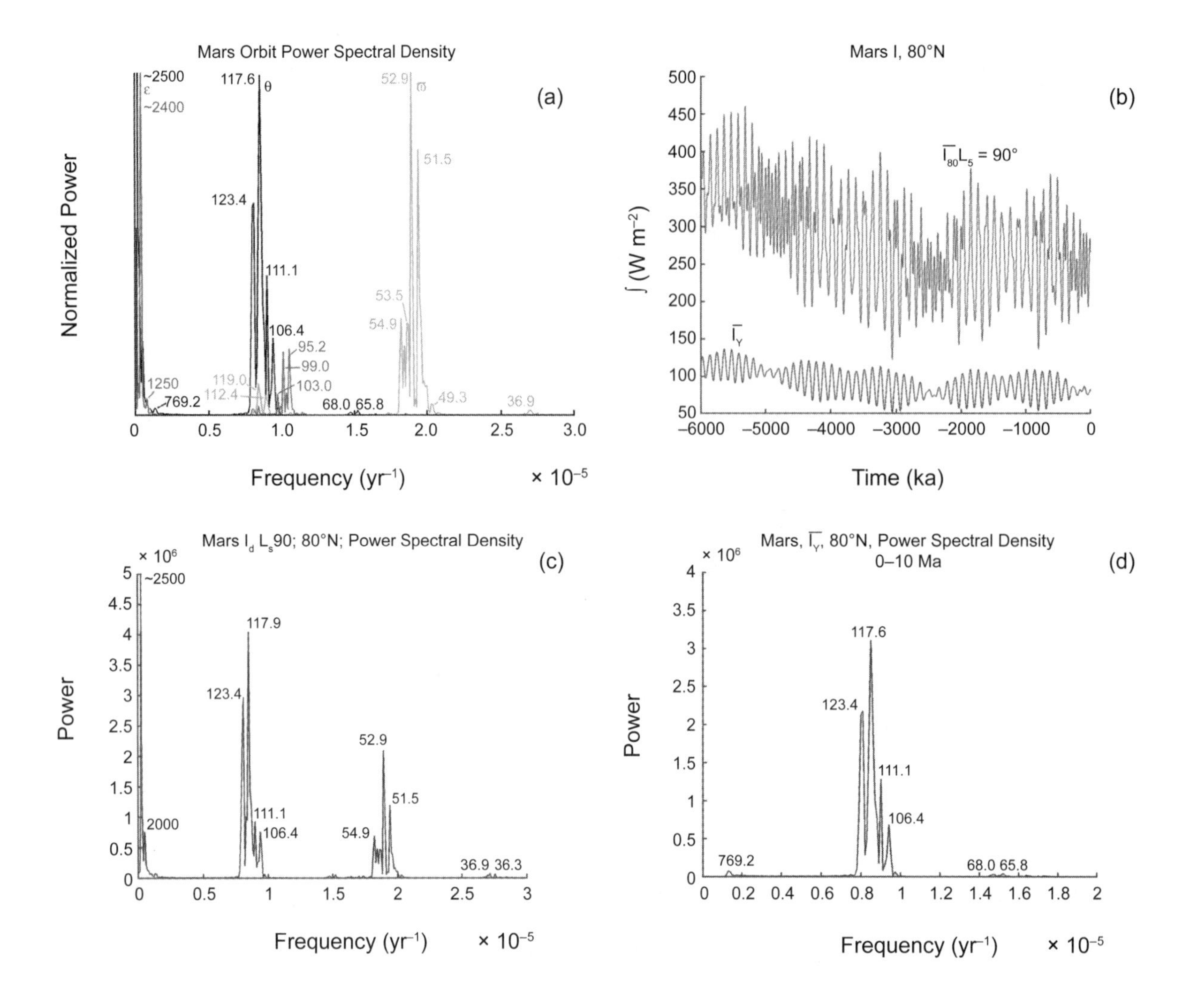

Fig. 14. (a) Fast Fourier transform (FFT) of Mars' orbital parameters. θ (black); ε (dark gray), ϖ (light gray). All powers normalized to facilitate comparison of frequencies. The eccentricity variations are not well determined because too few cycles are captured in 10 m.y.⊕. Numbers indicate period of oscillation in k.y. **(b)** Insolation at 80°N. The diurnally averaged insolation at summer solstice (top) and the annually averaged insolation (bottom). **(c)** Power spectral density of diurnally averaged summer solstice insolation at 80°N for the past 10 m.y.⊕. **(d)** Power spectral density of annually averaged insolation at 80°N for the past 10 m.y.⊕. As with Earth, $\overline{I_Y}$ varies only with θ.

CO_2 and water frosts (*Kieffer et al.,* 1977; *Farmer et al.,* 1977). Large, dust-free (i.e., old) ice grains are exposed in summer, implying that the NRIC is currently losing mass (*Byrne,* 2009), although the NPLD surface is geologically young; ≤10^5 yr⊕ (*Herkenhoff and Plaut,* 2000; *Tanaka,* 2005). Relative ice accumulation rates have been spatially variable (*Fishbaugh and Hvidberg,* 2006) and crater size frequency curves indicate continuous resurfacing (*Galla et al.,* 2008), all of which preclude a single age estimate for the entire NPLD.

Beneath the NRIC, NPLD layers are contiguous over large areas. Distinctively dark and coherent marker beds have been correlated over hundreds of kilometers (*Malin and Edgett,* 2001; *Kolb and Tanaka,* 2001). The surface expression of layers in both PLD are partially obscured by dust and ice mantles (*Herkenhoff et al.,* 2007). Hence, layer brightness in an image is not determined uniquely by composition, but is also affected by the mantling deposits, and the shape, slope, and azimuthal orientation of the outcrop (*Fishbaugh and Hvidberg,* 2006). For this reason, comparing spectral analyses of PLD images with insolation time series is not straightforward (*Fishbaugh et al.,* 2010; *Hvidberg et al.,* 2012). Nonetheless, dark marker beds appear to have a ~30-m periodicity (*Laskar et al.,* 2002; *Milkovich and Head,* 2005). *Perron and Huybers* (2009) saw, but did not confirm, the 30-m signal, citing concerns about image noise, but did identify another spectral peak at ~1.6 m. In HiRISE imagery, 1–2-m sets of very thin beds and dark, 3–4-m topographically prominent marker beds are separated by undifferentiated materials (Fig. 15) (*Fishbaugh et al.,*

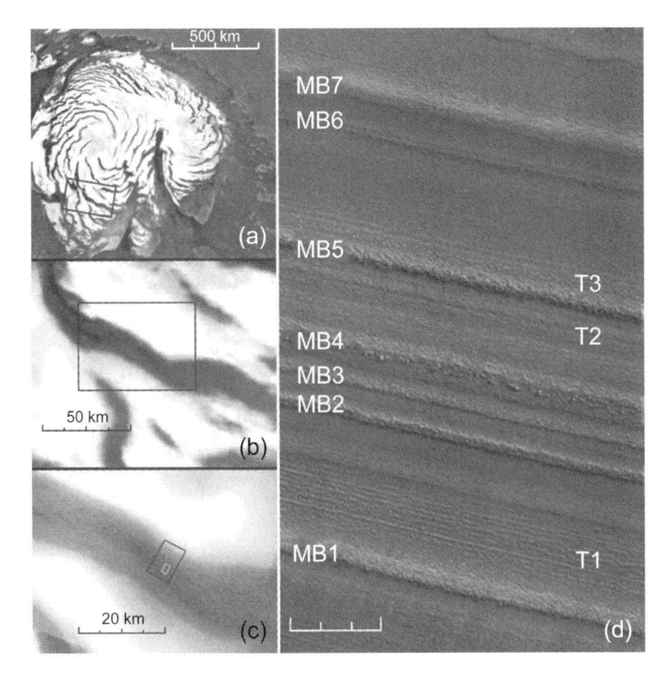

Fig. 15. Layering in the north polar layered deposits at 84°N–101°W. Marker beds are labeled MB, and thinly-bedded units are labeled T. Upslope and illumination direction are both toward upper right. HiRISE Image PSP_001636_2760.

2010; *Limaye et al.,* 2012). Imagery and spectral analysis thus seem to tell stories that are at least consistent.

Two ground penetrating radars have also acquired PLD data: the 1.8–5-MHz Mars Advanced Radar for Subsurface and Ionospheric Sounding (MARSIS) on Mars Express, and the 20-MHz Shallow Radar (SHARAD) on Mars Reconnaissance Orbiter (Fig. 16).

NPLD subsurface reflectors mapped by SHARAD are laterally continuous for up to 1000 km, suggesting the mechanisms that created them were homogeneous over that scale. They are closely spaced down to ~500 m and more widely spaced at greater depth. Reflectors cluster into four distinct packets, separated by three interpacket layers of few reflections (*Phillips et al.,* 2008).

Radar reflections result from abrupt changes of dielectric constant, which depends on the proportions of ice and dust. The dielectric transparency of the NPLD implies that the dust content is quite small; the dustiest layers likely contain ≤10% refractories, and most are considerably cleaner. The three intrapacket strata are likely essentially pure ice.

MARSIS detects the interface between the NPLD and the underlying terrain; it rests unconformably on a dark basal unit (BU) (*Malin et al.,* 2001b) that has thicker layers. The BU is dome-shaped, reaching a thickness of ~1 km, whereas the NPLD has a roughly constant thickness, except near its boundaries. Thus, elevation variations across the pole are largely attributable to thickness variations in the BU; depositional and erosional processes have been largely uniform across the cap throughout the accumulation of the NPLD (*Phillips et al.,* 2008).

4.2.1.2. South polar layered deposits: The SPLD differ substantially from the NPLD; the base of the deposit is ~6 km higher, and it is offset from the pole by ~2°, forming a broad, asymmetric plateau (*Kolb and Tanaka,* 2001). Craters counts imply the SPLD surface age is 10–100 m.y.⊕, orders of magnitude older than the NPLD (*Herkenhoff and Plaut,* 2000; *Koutnik et al.,* 2002). The SPLD is covered in places by a residual CO_2 cap a few meters thick, and seasonal deposits of CO_2 and trace water (*Byrne,* 2009).

Unconformities in the SPLD occur at elevations from 1500 to 3200 m and imply several erosional episodes were followed by further deposition (*Kolb and Tanaka,* 2006). *Milkovich and Plaut* (2008) identified three distinct sequences in the SPLD, the lowest of which is thicker in the center and is draped asymmetrically by the upper units. This is also very different from the NPLD, which is composed of nearly flat and continuous units. The uppermost layer is heavily pitted in exposure and forms distinctive benches along the SPLD scarps.

The layers within the SPLD differ in appearance from those of the NPLD; individual SPLD layers are more uniform in albedo, and are distinguished by surface texture or erosional style. Layer thickness varies widely, with a larger mean and variance than the NPLD (*Byrne and Ivanov,* 2004; *Milkovitch and Plaut,* 2008; *Limaye et al.,* 2012).

MARSIS radar signals also detect the interface between the SPLD and its substrate. The reflected power again suggests a composition of nearly pure water ice (*Seu et al.,* 2007).

Phillips et al. (2011) discovered four deposits at depth within the SPLD that return virtually no signal, and identified at least one as CO_2 ice based on its permittivity and transparency. The volume of this previously unknown CO_2 reservoir is ~30× greater than the residual CO_2 cap. When and how this CO_2 was sequestered beneath more refractory SPLD material is unknown. However, it implies a previous state of Mars' climate that had higher atmospheric pressure, likely in excess of the triple point of H_2O.

4.2.2. Ground ice. Mars has a substantial inventory of H_2O ground ice (*Carr and Schaber,* 1977; *Squyres and Carr,* 1986). Leakage neutron and gamma-ray emission data from the Mars Odyssey Gamma Ray Spectrometer (GRS) are remarkably consistent with the geographic distribution and burial depth predicted for diffusive equilibration with an atmosphere containing ~20 pr μm H_2O [the thickness of H_2O that would result from condensing the entire atmospheric column; 1 pr μm of H_2O ≈ 1 g m^{-2} (*Leighton and Murray,* 1966; *Farmer and Doms,* 1979; *Clifford and Hillel,* 1983; *Mellon and Jakosky,* 1993; *Mellon et al.,* 2004)]. The ice table lies within a few centimeters of the surface at high latitudes, but drops to ~1 m at ±45°, falling off rapidly equatorward. However, several lines of evidence indicate that some ice occurs well equatorward of where it is stable, and must have been emplaced under different conditions or by some process other than vapor diffusion.

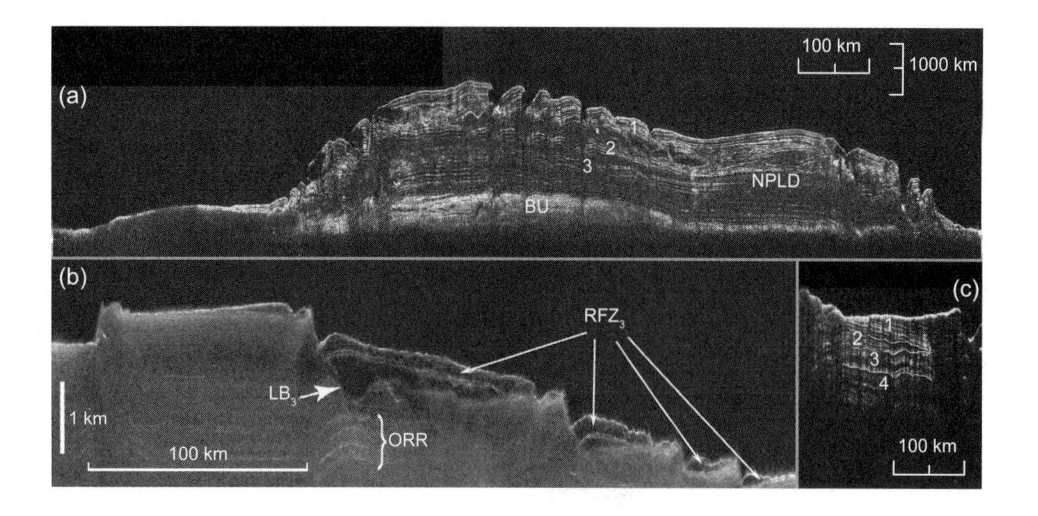

Fig. 16. (a) Radargram from SHARAD orbit 5192. Range time delay, the usual ordinate in a radargram, has been converted to depth by assigning real permittivities of 1 and 3 above and below the detected ground surface, respectively. The north polar layered deposits and basal units are labeled. Packet regions are numbered (*Phillips et al.,* 2008). **(b)** One of the reflection-free zones (RFZ3) in the south polar layered deposits (SPLD) that appears to be composed of CO_2 ice. The lower boundary of RFZ3, LB3, is an irregular buried erosional surface that truncates subhorizontal reflectors. Extending several hundred meters beneath RFZ3 is a zone of unorganized, weak radar reflectors that in turn is underlain by a coherent sequence of organized (layered) radar reflectors (ORR) (*Phillips et al.,* 2011). **(c)** SHARAD radargram of Promethei Lingula in the SPLD from orbit 2202. Four main reflectors, which separate different dielectric sequences, are indicated. Sequence 2 pinches out toward the edge of the cap, and may represent an unconformity (*Seu et al.,* 2007).

Vapor phase transport is likewise unable to account for localized but surprisingly high ice concentrations. GRS data indicate that ice occupies >90% of regolith volume across large regions poleward of ±50° (*Boynton et al.,* 2002). Essentially pure subsurface ice was excavated at the Phoenix site (*Smith et al,* 2009) and at lower latitudes by small, recent impact craters (Fig. 3d) (*Byrne et al.,* 2009). Although distributed heterogeneously at both local and regional scales, excess ground ice (i.e., ice that exceeds the undisturbed pore volume of its host soil) is significant because it appears to be incompatible with the current climate; it can only result from precipitation or *in situ* segregation (*Zent,* 2008).

One clue as to a possible emplacement mechanism is the presence of a young, ice-rich mantle deposit that is concentrated in two bands: 30°–60°N and 25°–65°S (*Mustard et al.,* 2001). It exhibits a low thermal inertia and smooth surface even at 25–50 cm/pixel (*Raack et al.,* 2012), and consists of a succession of meter-scale strata that drape underlying topography. The margins are often highly degraded and display periglacial features such as polygons, sublimation pits, gullies, and lobate flow features (e.g., *Milliken et al.,* 2003). These characteristics are consistent with material that has been deposited from the atmosphere, cemented, and then partially disaggregated (*Mustard et al.,* 2001). Their low crater retention ages and distribution into temperate latitudes led *Head et al.* (2003) to identify them as deposits relict of martian "ice ages."

Although some, and perhaps most, excess ice may derive from snowpacks now covered by a dusty sublimation lag, the snowpack hypothesis has difficulty accounting for cases where large rocks, and in some instances boulder fields, lie stratigraphically above the excess ice. *In situ* segregation of ice via thin films of unfrozen H_2O has been suggested as an explanation for the nearly pure ice excavated by Phoenix (*Zent,* 2008; *Gallagher et al.,* 2011; *Zent et al.,* 2011). However, such a process would likely operate only over small depth scales.

4.3. Climate Models

By the end of the Viking era, a coherent climate model had emerged, based on straightforward physics and consistent with observational constraints available at that time (*Toon et al.,* 1980; *Pollack and Toon,* 1982; *Carr,* 1982; *Kieffer and Zent,* 1992). Then, as now, the working hypothesis was that the PLD is a mixture of ice and dust, the latter arriving in atmospheric suspension, although there was considerable uncertainty regarding their relative proportions (*Toon et al.,* 1980; *Pollack and Toon,* 1982).

Laboratory studies had shown that CO_2, which makes up 95% of Mars' atmosphere, is readily adsorbed onto fine-grained minerals at martian regolith conditions, and models suggested the regolith may contain more CO_2 than the atmosphere (*Fanale and Cannon,* 1971, 1979). Thus a fundamental assumption was that CO_2 cap ice, regolith adsorbate, and the atmosphere are the three reservoirs for exchangeable CO_2. What follows is a consequence of the fact that vapor pressure over ice is an intensive (mass-independent) property, while vapor pressure over adsorbate is an extensive (mass-dependent) property. CO_2 will partition itself between the atmosphere and regolith by maintaining adsorptive equilibrium at every point in the regolith that is in diffusive contact with the atmosphere. (The CO_2 cap at the south pole holds only a small fraction of the atmospheric mass and cannot buffer atmospheric pressure.) As solar forcing changes with the orbit, surface and regolith temperatures respond, and CO_2 exchanges between the atmosphere and regolith to maintain adsorptive equilibrium. If, as sometimes happens, this mechanism drives atmospheric P_{CO_2} to exceed its saturation vapor pressure at the poles, then a permanent CO_2 cap was assumed to exist. Control of P_{CO_2} then shifts from adsorptive equilibrium to the cap radiative temperature. The regolith temperature and cap-controlled pressure then fix the adsorbed regolith inventory, and any remaining CO_2 would migrate to the cap.

Low θ favors this scenario; during obliquity minima, the cold cap would maintain $P_{CO_2} \ll 100$ Pa. Such low pressures preclude dust-lifting, and atmospheric dust opacity, which plays a key role in the current climate, would approach zero below some θ/P_{CO_2} threshold. The albedo and emissivity of CO_2 ice are extremely sensitive to small amounts of dust, which can only reach the cap surface in atmospheric suspension. The result would be a cleaner, brighter (hence colder) cap, completing a feedback loop that drives P_{CO_2} lower. To maintain adsorptive equilibrium, CO_2 would desorb from the regolith and eventually accumulate as polar ice. Polar CO_2 is an effective cold trap, and atmospheric P_{H_2O} must also decrease by orders of magnitude. Low θ also results in greater insolation, hence higher temperatures, between $\pm 43°$ latitude, forcing the retreat of ground ice to higher latitudes and greater depth (*Toon et al.,* 1980; *Fanale et al.,* 1982; *Pollack and Toon,* 1982).

Conversely, high θ increases polar insolation; no perennial CO_2 caps are expected, but the seasonal CO_2 caps would be substantially larger than at present due to the expansion of polar night. Under the reasonable assumption that the cold polar regolith should adsorb proportionately more CO_2 than the low-latitude regolith, it was thought that polar warming at high θ would cause desorption and force P_{CO_2} to rise. Quantitatively, ΔP_{CO_2} over a θ cycle would depend on the composition, distribution, and total regolith surface area accessible to the atmosphere.

The current atmosphere is just capable of dust lifting, so any pressure increase should lead to a dustier atmosphere, enhanced dust deposition at the poles, and therefore to darker perennial H_2O caps. Along with increasing polar insolation, this would yield much higher atmospheric P_{H_2O}. Net cooling between $\pm$ 43° and higher P_{H_2O} emplaces ground ice at lower latitudes and shallower depths (*Jakosky and Carr,* 1985).

This model has required modification in the last decade, although the gross outlines remain unchanged. In describing the recent evolution of the climate models outlined above, we provide more detailed background on the current martian climate and geologic record as needed, but attempt no stand-alone comprehensive review of either.

4.3.1. Low obliquity. The salient characteristic of the low θ climate was very low P_{atm} resulting from the stabilization of perennial CO_2 caps. It was recognized at the time that this scenario was critically dependent on the albedo (A_{CO_2}) and emissivity (ε_{CO_2}) of the seasonal caps.

While modelers initially assumed static values that resulted in a stable CO_2 cap below some θ threshold, it has become evident that the radiative properties of the current seasonal CO_2 caps are neither uniform nor static, and that their behavior is exceedingly complex. A_{CO_2} and ε_{CO_2} are dependent on CO_2 grain size, dust size, and mixing ratio, and the abundance and grain size of H_2O. Pure CO_2 ice is very bright, $A_{CO_2} > 0.8$, at all sizes <50 μm, but darkens significantly in intimate mixtures of dust. Likewise, the emissivity decreases substantially. CO_2 is precipitated to the cap as very bright snow in some areas (*Forget et al.,* 1998), yet occurs elsewhere as partially transparent slab ice with very long photon path lengths (*Kieffer et al.,* 2000; *Eluszkiewicz and Moncet,* 2003).

On emerging from polar night, the Lambertian albedo of the spherical harmonic (SH) cap ranges from ~0.2 to ~0.4, and then evolves nonuniformly (*Paige and Ingersoll,* 1985; *Kieffer et al.,* 2000). After an initial increase in albedo that is proportional to insolation, different regions of the cap reach plateaus ranging from ~0.2 to ~0.6. Bright regions then decrease in albedo with mild warming until $A_{CO_2} \approx 0.35$, at which point sublimation occurs. However, one section of the southern seasonal cap, the so-called Cryptic Region, remains cold with low albedo until it sublimates (*Kieffer et al.,* 2000). Adding to the difficulty of deterministically predicting A_{CO_2} and ε_{CO_2} under even current conditions, the northern seasonal cap, unlike the southern, incorporates both H_2O and CO_2 frosts.

The initial brightening of the cap with insolation is likely the result of surface dust removal. Mechanisms that might clean the surface include preferential heating of surface dust grains, which then volatilize a thin envelope of CO_2 and sink into the ice, as well as lofting off the surface by the net vertical wind associated with sublimation of CO_2 (*Kieffer,* 1990).

It is beyond the scope of this chapter to review the complex physics of the seasonal caps in more detail, but the consequences for orbitally driven climate are summarized in Fig. 17. The contours show the obliquity below which a permanent CO_2 cap is stable as a function of A_{CO_2} and ε_{CO_2}, along with the values assumed for those parameters by different modelers over the years.

Whether or not low θ is characterized by atmospheric collapse, low dust opacity and clean ice appears to be contingent on processes that control the albedo of the CO_2

caps. Some recent climate simulations, which all use lower A_{CO_2} and ε_{CO_2} than did early models (although none attempt to treat the seasonal evolution of the cap properties), have failed to predict perennial low-θ CO_2 caps for any orbital configurations in the last 10 m.y. (e.g., *Haberle et al.,* 2003; *Armstrong et al.,* 2004; *Zent,* 2008).

There is evidence, however, that large-scale atmospheric collapse must occur on occasion: The exceptional cleanliness of the recently discovered CO_2 deposit within the SPLD suggests that it precipitated in the absence of dust, which indicates low atmospheric pressure. Additionally, *Kreslavsky and Head* (2011) report instances of a distinctive high-latitude glacial morphology that they argue derived from CO_2 glaciers (Fig. 18). The deposits are small ridges forming narrow overlapping loops that extend distally for up to 15 km. These looped ridges are very similar to terrestrial drop moraines that form when cold-based glaciers dynamically stabilize (i.e., the flow due to internal deformation is balanced by frontal ice ablation) (*Marchant et al.,* 1993).

The identification of CO_2 as the glacial material is based on constraining the flow thickness using the relationship of flow features to surrounding topography, allowing the basal shear stress to be bracketed between 10^4 and 10^5 Pa (*Milliken et al.,* 2003). This is far below the shear stress required for flow of H_2O ice at Mars polar temperatures, a conclusion supported by the fact that the nearby NPLD does not

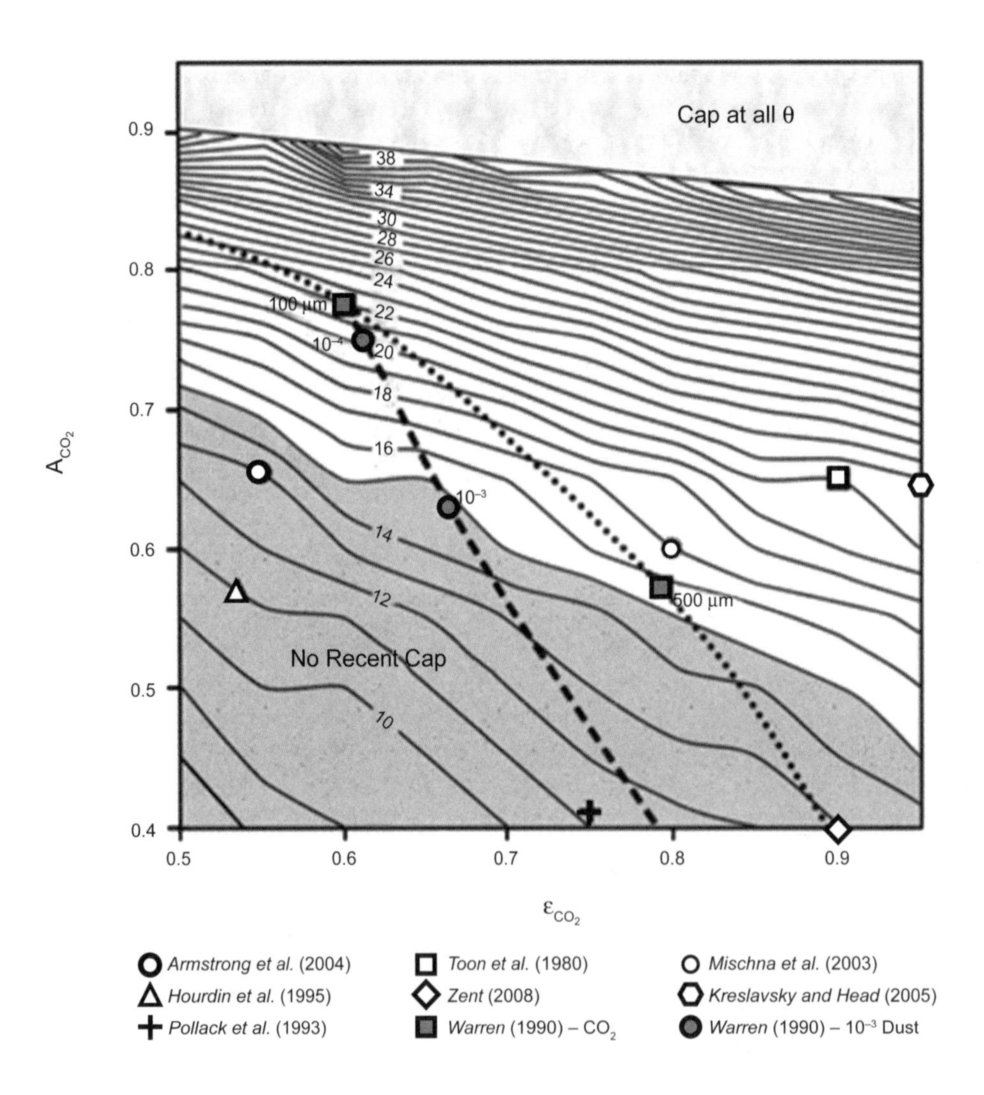

Fig. 17. Perennial cap formation as a function of CO_2 emissivity and albedo. The labeled contours indicate the critical obliquity below which permanent caps will form based on a model by *Armstrong et al.* (2004). Other models give very similar results. The lowest obliquity attained in the last 10 m.y. is 15°, so no permanent caps form if A_{CO_2} and ε_{CO_2} lie with the dark gray shaded area. Data points indicate A_{CO_2} and ε_{CO_2} values assumed by several recent models. Values calculated by *Warren* (1990) for pure CO_2 ice lie along the dotted line, with grain size indicated at the points. The dashed line represents 100-µm dusty ice with the weight fraction of dust indicated at the points.

flow. *Kreslavsky and Head* (2011) suggest CO_2 ice, which could be expected to flow ~4 m y^{-1}⊕ on the measured slope. Such a rate is of the correct order of magnitude to produce the observed flows in the course of a low-θ excursion. CO_2 ice of the indicated thickness as far south as 75°N could only come about as a consequence of atmospheric collapse.

Two additional factors that will come into play should the atmosphere collapse at low obliquity have yet to be assessed quantitatively in terms of quasi-periodic climate variations. First, precipitating a substantial fraction of atmospheric CO_2 will inevitably increase the relative abundance of N_2 and Ar in the atmosphere. If all atmospheric CO_2 were to condense, the residual Ar-N_2 atmosphere would have a pressure of ~25 Pa. *Zahnle et al.* (2008) showed that if H_2O abundance in the atmosphere is significantly lower than the present (which is inevitable in the presence of perennial CO_2 caps), then CO_2 is photolyzed more efficiently, and partly replaced by O_2 and CO in the atmosphere. Thus the composition of the martian atmosphere may also be variable due to orbital oscillations, with Ar, N_2, O_2, and CO dominating the lowest-θ atmosphere.

Second, the near-surface regolith, in which thermal conduction is dominated by pore gases, will become much more insulating. For silt-sized grains, the thermal conductivity would likely decrease by an order of magnitude or more, leading to higher subsurface temperatures as the planetary heat flow is trapped below a more insulating regolith. Depending on the magnitude of the pressure drop and strength of the heat flow, it is possible that subsurface ice could melt and redistribute at low θ (*Wood and Griffiths,* 2009).

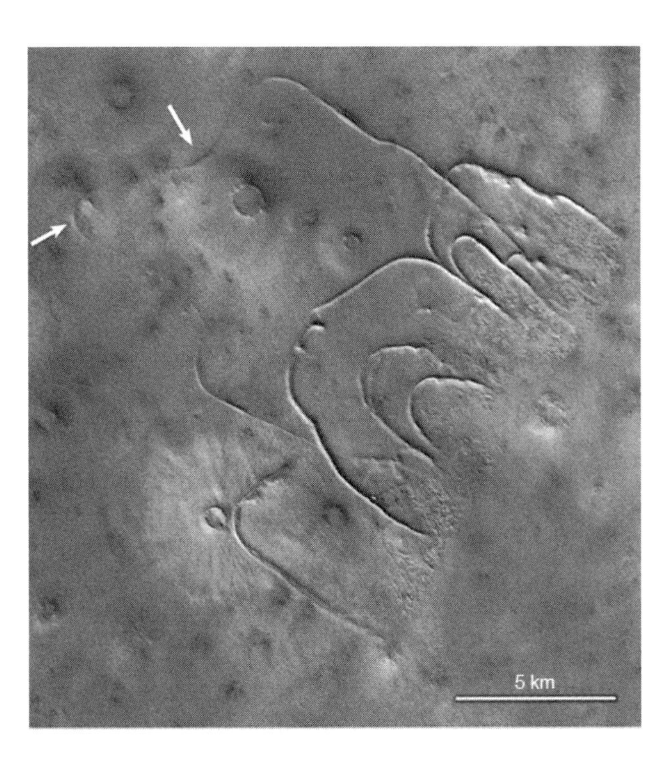

Fig. 18. Apparent drop moraines in the Olympia Mensae region, 74°N, 267°W. Based on the flow distance and low slope, *Kreslavsky and Head* (2011) identified these as the likely products of CO_2 glaciers. Mars Reconnaissance Orbiter Context Image P16_007357_2541_XN_74N266W.

4.3.2. High obliquity. The high-obliquity climate has received the bulk of recent attention because it is more conspicuous in the geologic record. It is during high θ that a variety of non-polar-ice-related deposits, impossible to produce in today's climate, are thought to form. These include the ice-rich mantle deposits, mid-latitude glacial features, concentric crater fill, pedestal craters, lobate debris aprons, lineated valley fill, gullies, and tropical mountain glacier deposits (Fig. 19). As we shall see, it useful to distinguish between two different climatic regimes that may occur in the course of a high-obliquity excursion.

4.3.2.1. Temperature and general circulation: With no oceans and little sensible heat transport by the atmosphere, large oscillations in orbital geometry have a profound effect on Mars' surface temperatures (Fig. 20). The condensation of the seasonal cap places a lower limit on surface temperatures that is unlikely to be substantially different from the 150 K limit of today (see below). As obliquity increases, the latitudinal extent of the cap also increases as more atmosphere collapses onto the winter pole. While latitudes <±43° cool slightly (Fig. 7), the summertime diurnally averaged temperature for latitudes ±75° increases from ~220 K at present to >280 K for obliquities as high as 45°, which is the maximum θ for the last 10 m.y.

The residual H_2O caps receive $\overline{I_Y}$ some 40% greater at high obliquity than currently, potentially increasing the atmospheric H_2O column up to 5000 pr μm at θ = 45°, compared to ≤100 pr μm today (*Richardson and Wilson,* 2002; *Mischna et al.,* 2003; *Levrard et al.,* 2004; *Forget et al.,* 2006).

Increased θ also has a first-order effect on atmospheric circulation, since it affects the latitudinal temperature gradient that in turn drives much of the circulation. Specifically, Mars global circulation models (GCMs) predict strong intensification of the meridional overturning Hadley cell, which results in higher wind velocities, including at the planets' surface (*Fenton and Richardson,* 2001, *Haberle et al.,* 2003).

4.3.2.2. Radiative effects of clouds and dust: As the obliquity increases, the water cycle intensifies and the atmosphere becomes wetter and cloudier. The radiative effects of both H_2O vapor and clouds are usually slight in the current climate, but models suggest their importance with increasing θ.

H_2O clouds reflect some solar energy, but they also absorb and reradiate upwelling IR from the surface. Thus, daytime clouds tend to warm the atmosphere and cool the surface. At night, atmospheric temperatures are lowered by IR emission to space, and the surface is warmed by downwelling IR emission from the clouds. Fully radiatively active H_2O clouds are a recent feature in Mars GCM studies, but they indicate a strong positive feedback between cloud formation and surface ice sublimation. Thin polar clouds warm the cap through IR emission, which increases

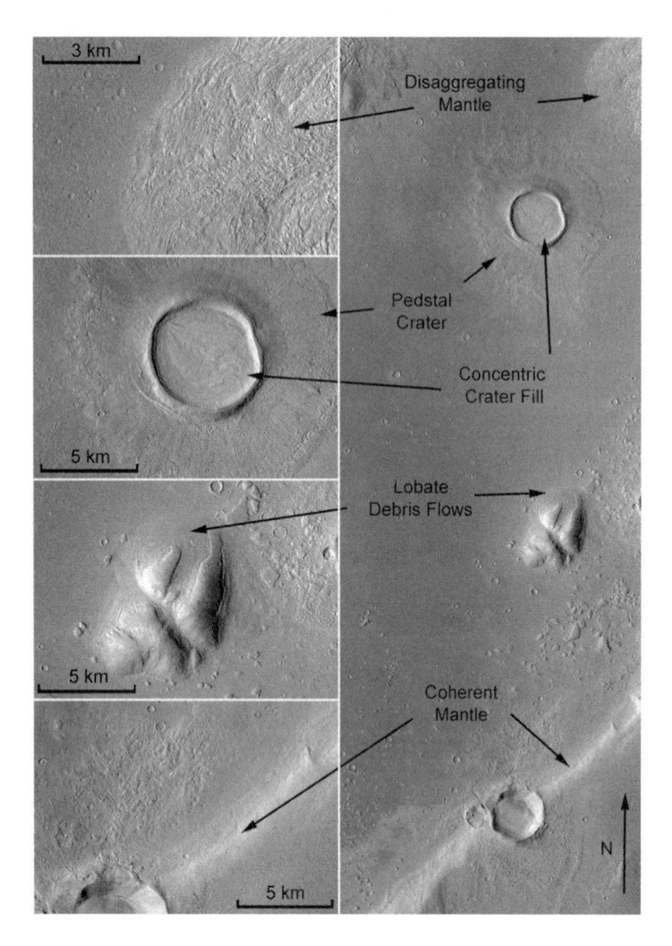

Fig. 19. Mid-latitude ice-related features in the southern Phlegra Montes, 34°N, 197°W. Ice is currently unstable at this latitude; however, conspicuous flow patterns and radar data indicate substantial quantities remain in the subsurface. Mars Reconnaissance Context Image B02_010506_2142_XN_34N196W.

sublimation and cloud cover. The result is warming of the high latitudes by up to 30 K relative to a cloud-free climate (*Haberle et al.,* 2012; *Madeleine et al.,* 2012).

Airborne dust is a key component of the martian climate because it is highly variable, strongly alters the atmospheric radiation balance, and is a potent thermal driver of atmospheric circulation (*Zurek et al.,* 1992). Dust injection from the surface occurs via near-surface wind stresses, as well as convective vortices (i.e., dust devils). Airborne dust alters the radiative balance of the current atmosphere by absorbing and scattering solar radiation, reducing the solar flux reaching the surface. The effect, which is very sensitive to dust opacity, is to cool the surface and warm the atmosphere (*Savijarvi et al.,* 2005).

A feedback is established when increasing θ changes the peak forcing latitude of the meridional circulation, strongly amplifying near-surface winds. Thus, the dust lifting potential increases sharply with θ; the addition of radiatively active dust dramatically intensifies this process due to positive feedbacks between circulation strength and dust lifting. There is broad concensus that high atmospheric opacity is likely at both solstices for high θ (*Haberle et al.,* 2003; *Newman et al.,* 2005), and that the PLD will accumulate the vast majority of its dust complement at high θ (see below). Ultimately, the dust opacity may be limited by the supply of surface dust at sites of high lifting stress.

Until very recently, climate models treated either dust or H_2O clouds, but not both; the interaction of radiatively active clouds and dust is the current cutting edge of Mars climate modeling. It appears however, that H_2O column abundances on the order of 10^3 pr μm or greater are possible at high obliquity (*Madeleine et al.,* 2009; *Haberle et al.,* 2012; *Forget,* 2012).

4.3.2.3. Carbon dioxide: *Atmospheric pressure* — Crater-based estimates that the martian regolith may be 100–1000 m thick informed early calculations of the CO_2 inventory (*Fanale and Cannon,* 1979); that estimate has subsequently decreased by a factor of 5–10. Volcanic and sedimentary layering can isolate underlying material; H_2O displaces CO_2 from adsorption sites (*Zent and Quinn,* 1995), and the highest-capacity reservoir, the high-latitude regolith, is now known to be widely saturated with H_2O ice. Recalculating the exchangeable regolith adsorbate inventory subject to these assumptions yields estimates of less than half the mass of the atmosphere. If this is the case, average P_{atm} decreases at higher θ because the adsorbed inventory is too small to substantially increase pressure, but the seasonal pressure minima decrease substantially as a greater fraction of the winter hemisphere experiences polar night and more CO_2 condenses (see below).

As with the low-θ case, we know that the canonical high-P_{atm}, high-θ correlation must occasionally hold; the recently discovered CO_2 reservoir within the SPLD implies an earlier climate with higher atmospheric pressure. How so much CO_2 came to be sequestered beneath the SRIC is a new and puzzling question, as is its role in recent climate oscillations. If released into the atmosphere, it would increase P_{atm} by 400–500 Pa. While increased dust lifting and atmospheric opacity would likely result, the additional radiative forcing would be inadequate to increase surface temperature (*Phillips et al.,* 2008).

Seasonal cycle — The seasonal cycle of CO_2 is found to become more extreme at high obliquity (Fig. 21). Currently (θ = 25°), atmospheric mass varies annually by ~25%. At θ = 35°, the mass varies by 36%; at 45°, it is 47%; and at 60°, $\Delta P_{atm} \approx 58\%$. Because so much more CO_2 precipitates onto the winter cap, the annually averaged atmospheric pressure drops with θ.

4.3.2.4. Water: In contrast to glacial periods on Earth, which form due to *lower* polar insolation, martian glaciation is most extensive at times of high polar insolation (i.e., high θ) (*Jakosky and Carr,* 1985; *Head et al.,* 2003). Above a θ threshold that ranges from ~28° to 35° (depending on the other orbital elements), polar insolation becomes large enough that ice begins to accumulate on the surface outside the polar regions (Fig. 22).

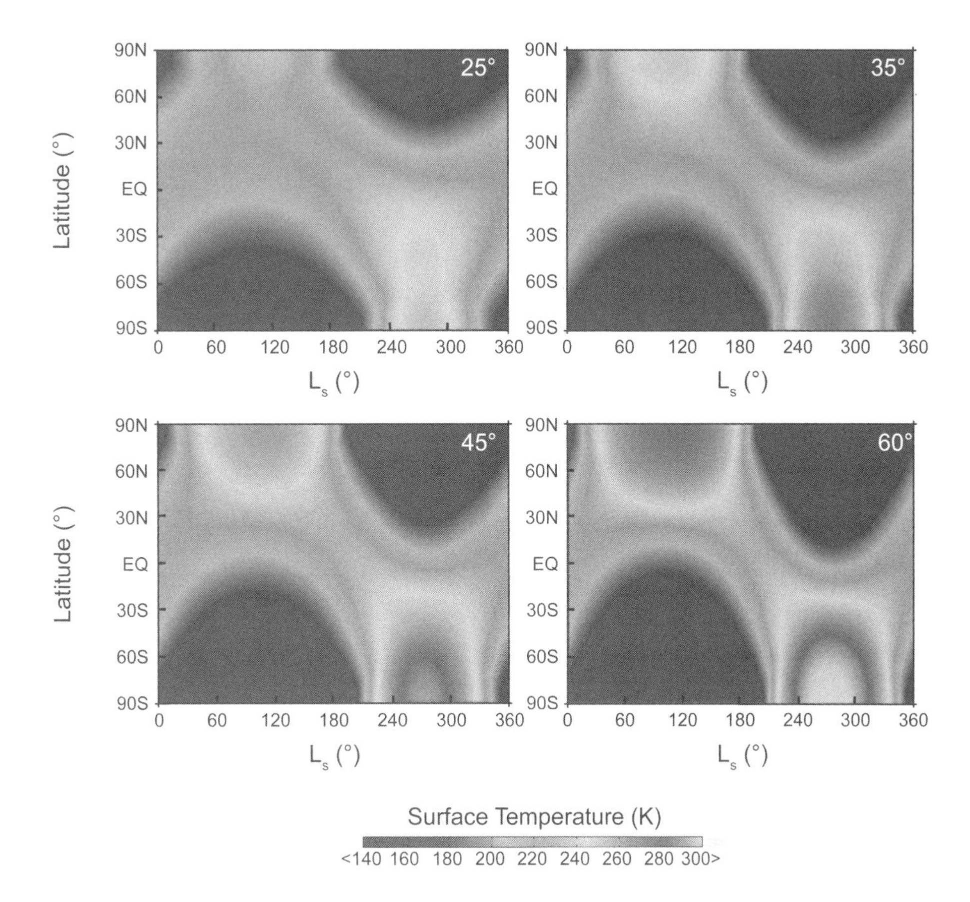

Fig. 20. See Plate 82 for color version. Diurnally and zonally averaged surface temperature over a martian year calculated for four obliquities (θ ≥ 25°). For all models, ε = 0.093 and ϖ = 251°. Surface emissivity assumed = 1, and surface albedo is assumed to be 0.21 everywhere (including the current permanent caps). Although 45° is the highest obliquity attained in the last several million years, q regularly reached 60° prior to 10 Ma.

Mischna et al. (2003), assuming sublimation from the north cap only and constant cloud particle diameters of 2 μm, found that ice accumulated in regions of high thermal inertia or high topography, and that these deposits were highly mobile, migrating to the tropics at θ = 45°. *Forget et al.* (2006) allowed cloud particles to grow and precipitate, and predicted up to 70 mm yr^{-1}♂ of ice accumulation in four localized areas on the western flanks of the Tharsis Montes, Olympus Mons, and Elysium Mons (and nowhere else). Prevailing westerlies cool adiabatically as they move upslope, resulting in orographic precipitation. If the south cap is instead assumed to supply all H_2O, some ice still accumulates in southern Tharsis, but the highest rates are found east of Hellas.

Comparing these predictions to the geologically mapped sites of the glacier-related deposits yields remarkable agreement. The largest glacial deposits are observed on the western flanks of Arsia and Pavonis Mons, and at lower elevations on Ascraeus and Olympus Mons (*Shean,* 2005, 2007). In the southern hemisphere, the heavily dissected terrain east of Hellas is likewise the site of a unique concentration of ice-related landforms (*Crown et al.,* 1992).

4.3.2.5. Transient and equilibrium high-θ climate: As long as the poles retains an inventory of surface ice, the atmosphere will serve as a conduit for transferring that H_2O to low-latitude reservoirs. However, if sustained long enough, the small amounts of dust contained in the ice might accumulate to yield a lag deposit that would insulate the underlying ice, and place a diffusive barrier between it and the atmosphere (*Jakosky et al.,* 1993, 1995; *Mischna and Richardson,* 2005). If this indeed were to happen, the cap would switch off as a vapor source, and the only source of atmospheric water would be the low-latitude ice deposits. Even if no mantling occurs, polar ice would be exhausted after ~10^4 yr with the same result. This could be considered the equilibrium high θ-climate. Annual average water vapor abundances under those circumstances would likely be only ~20–80 pr μm (*Mischna and Richardson,* 2005).

Between the time the polar caps become unstable and their subsequent depletion, a transient high-θ climate would reflect the existence of an immense and highly unstable H_2O reservoir at the caps. In order to obtain a diffusively insulated sublimation lag and establish the equilibrium climate state, the thickness of the dust needs to be on the order of 1 m (*Hofstadter and Murray,* 1990). If we were to assume 1% dust concentration (*Phillips et al.,* 2008), 100 m of polar ice would need to be removed to establish such a lag. For the past 4 m.y., high θ is reached only fleetingly (thousands of years), and models that incorporate radiatively active clouds as well as dust have yet to address the lifetime of this transient high-θ climate. Nonetheless, it seems inevitable that high-θ excursions likely yield the most vigorous water cycles of the past few million years, with ice stable in the tropics, unstable but present at the poles, and exceptionally high atmospheric H_2O abundances, at least for several 10^3 yr.

As θ decreases from peak values, low-latitude ice again becomes unstable relative to the poles. Initially, GCM models show that broad-scale glaciation in the northern mid-latitudes occurs, especially in the Deuteronilus-Protonilus Mensae region (0°–80°E, 30°–50°N), where large concentrations of lobate debris aprons and lineated valley fills are observed (Fig. 19). Because the increase in low-latitude temperatures is nowhere near as great as the increase in polar temperatures with increasing θ, the reestablishment of polar H_2O is unlikely to involve the same climatic extremes. Once equatorial reservoirs are exhausted, these mid-latitude deposits are also returned to the pole (*Levrard et al.,* 2004; *Madeleine et al.,* 2009). If a sufficient thickness of dust is deposited on this ice, or a lag deposit forms, the ice may be preserved long after it is no longer stable. This is the canonical working hypothesis for the origin and persistence of the ice-rich mantle deposits discovered by MGS.

4.3.3. Eccentricity and precession. The climatic effects of variations in ε and ϖ are less dramatic, in part because

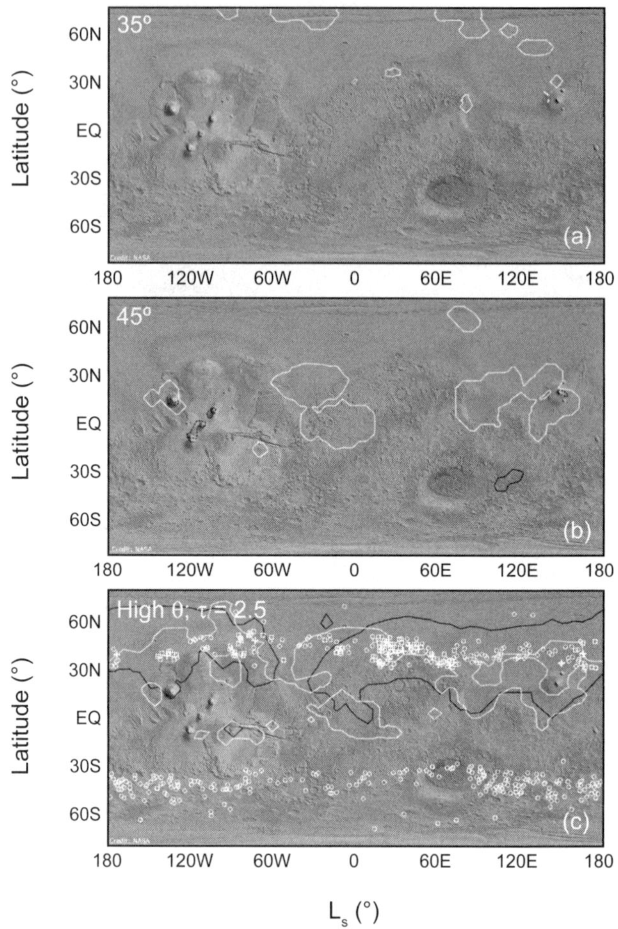

Fig. 22. Predicted location of high-obliquity ice sheets from several models, and mid-latitude ice-related features. **(a)** θ = 35°. Perimeter of ice sheets assuming a northern cap source (*Mischna et al.,* 2003). **(b)** θ = 45°. White line: Perimeter of ice sheets, assuming a northern cap source (*Mischna et al.,* 2003). Black lines: Same, from *Forget et al.* (2006). The dashed line is for a sublimating northern cap; the solid line is for an H_2O source at the south cap. **(c)** τ = 2.5. Effects of high dust opacity. White line: θ = 35°; perimeter of ice sheets, assuming a northern cap source (*Madeleine et al.,* 2009). Black line: θ = 45°. Same from *Forget* (2012). Equivalent predictions for a southern cap source are not yet available. The circles (○) represent locations where ice-rich mantling deposits have been mapped. The squares (□) are instances of lineated valley fill, crosses (+) denote regions showing evidence of glaciation, and lobate debris aprons are indicated by diamonds (◇). Data from *Bleamaster and Crown* (2010), *Head and Marchant* (2006), *Squyres* (1979), *Mest and Crown* (2003), *Hubbard et al.* (2011), *Plaut et al.* (2009), and *Li et al.* (2005).

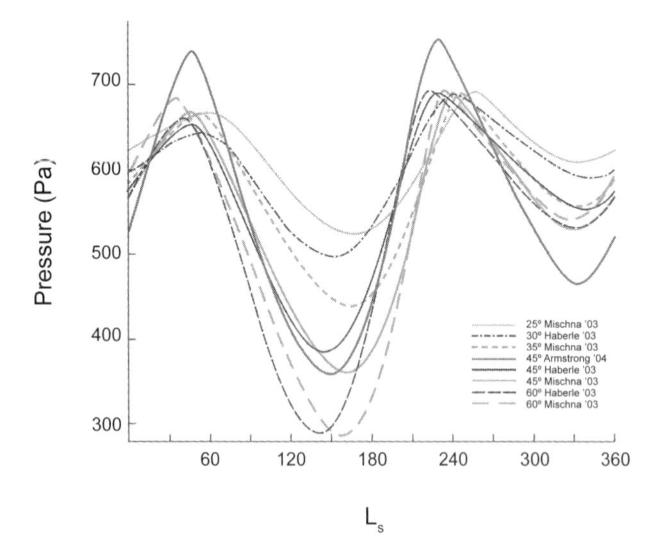

Fig. 21. Seasonal variation of Mars' global mean surface pressure at five different obliquities ($\theta \geq 25°$) from three different models. For all models, ε = 0.093 and ϖ = 251°. Pressure minima are due to condensation of the southern (~L_s 150°) and northern (~L_s 340°) seasonal caps; they are offset from the solstices due to sublimation of the summer hemisphere cap.

of the magnitude of θ oscillations, and in part because they vary at comparable or higher frequencies. If we hold ϖ constant (251°) and increase ε, northern hemisphere winters grow shorter and warmer, and the northern hemisphere summers become longer and cooler. The effect of high ε is thus to decrease the polar cap heating rate, and hence to reduce northern polar cap surface temperatures. Due to the nonlinearity of the saturation vapor pressure with temperature, the total amount of water sublimed from the northern polar cap is reduced compared to that lost at present eccentricity. The shift of the zonally averaged surface ice deposits to the north at high eccentricity is also readily attributed to the modified polar cap heating rates.

If instead we move ϖ by 180° to northern spring/summer, which occurred as recently as 2.1×10^4 and 7.5×10^4 yr⊕ ago, sublimation from the north cap is enhanced. *Montmessin et al.* (2007) found that under these conditions, the south (aphelion summer) hemisphere became the more stable reservoir, and accumulated ice as rapidly as 1 mm yr^{-1}♂ (assuming $\varepsilon = 0.1$). The southern accumulation of water ice at reversed ϖ is likely limited to a few meters per cycle since Mars' topography asymmetry favors the accumulation ice in the north (*Forget,* 2012).

4.4. Reading the Martian Climate Record

The regularity of fine bed thickness in the NPLD is consistent with a quasi-periodic process, albeit with unknown periods; the SPLD thickness measurements show no such regularity (*Limaye et al.,* 2012). Thus, attempts to tie PLD stratigraphy to orbital drivers have focused on the NPLD. The stratigraphic column reveals that a simple rhythmic or bundled layer sequence is not immediately apparent, implying that, in spite of Mars' comparatively simple climate system, nonlinear responses to orbital forcing must play a role (*Fishbaugh et al.,* 2010).

Considering the instability of polar ice at high θ in concert with the secular decrease in obliquity ~5 Ma, most workers believe that the NPLD postdates that transition. Figure 23 shows three models of the growth of the NPLD with time (*Levrard et al.,* 2007; *Zent,* 2008; *Greve et al.,* 2010). Although significant differences exist between models, they all predict that no permanent and substantial NPLD existed until ~4.2 Ma, that permanent polar ice was firmly established by 3 Ma (coincidently the same time that Earth's northern polar ice became established), and that the last significant nonpolar surface ice occurred at ~490 ka.

The polar ice deposits grow relatively rapidly to their present-day thicknesses, interrupted only by brief θ maxima at ~3.2, 1.9, and 0.7 Ma. If the NPLD is ~4 m.y.⊕ old, its minimum rate of growth would be ~1.5 mm yr^{-1}♂; since the cap has been periodically unstable for some fraction of that time, instantaneous accumulation rates must be somewhat in excess of that.

The task of correlating the radar stratigraphy with the orbital history is only beginning, but a few models have been published. The multiple reflectors within packets and

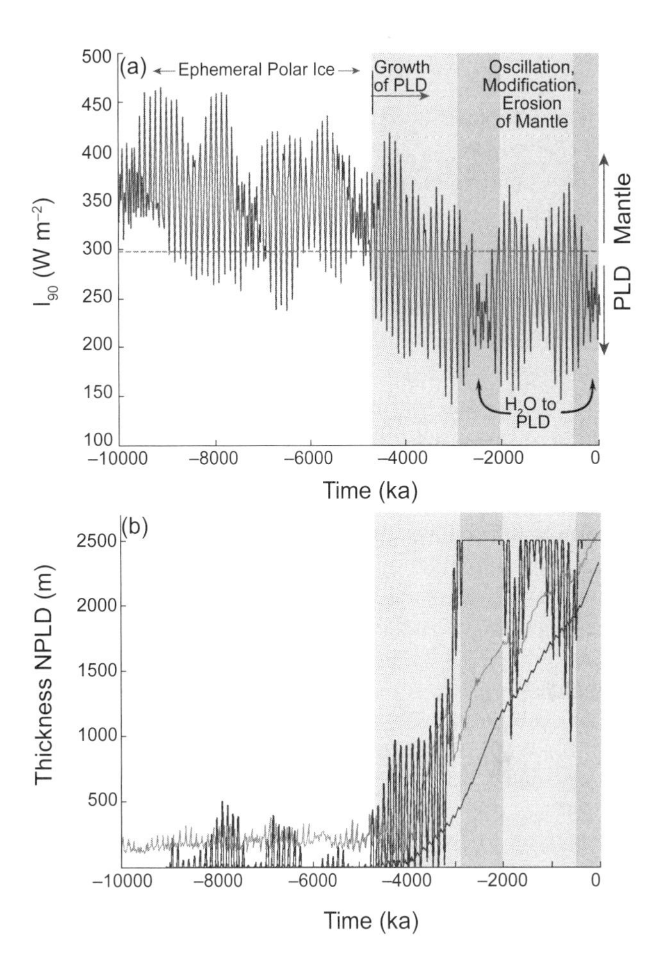

Fig. 23. (a) Mars' annual average polar insolation over the past 10 m.y. Major phases in the development of the north polar layered deposits (NPLD) are indicated by the gray zones. The dashed line represents the boundary between stable polar ice and stable low-latitude ice deposits that eventually form mantling deposits. **(b)** Modeled thickness of the NPLD over the last 1 m.y. from *Zent* (2008), *Greve et al.* (2010), and *Levrard et al.* (2007).

the packet-interpacket sequences represent two periodicities, which have been related to orbital history in several ways.

Levrard et al. (2007) argued in favor of a variable dust supply over a background of overall ice growth. Citing the expected correlation between obliquity and dust opacity, they argued that packets could form during times of maximum θ amplitude, with the packet layers resulting from sublimation lags formed during θ maxima. Interpacket zones would then correspond to recent nodes in the θ oscillation, when the atmosphere was relatively dust-free and polar ice was consistently the thermodynamically preferred reservoir over periods of 300–500 k.y. This hypothesis would seem to predict an interpacket zone at the top of the current NPLD, as θ has been in a node for the past 400 k.y., but SHARAD data indicate a packet with strong reflectors is at the top of the NPLD column (*Putzig et al.,* 2009).

Phillips et al. (2008) suggested interpackets correspond to brief excursions to θ minima at ~0.8, ~2.0, and ~3.2 Ma when ice may have been accumulating more rapidly at the poles (*Levrard et al.,* 2007). This model must explain why the corresponding excursions to maximum θ during interpacket deposition do not produce a detectable signature in the SHARAD data.

Putzig et al. (2009) proposed a very similar model, but referenced theirs against the polar insolation history rather than θ. They also explicitly identify packets as originating during nodes in the insolation curve. Packet layering would then reflect variations in dust content during ice accumulation rather than by the production of lags during sublimation. It is also difficult in this case to explain why periods of large-amplitude insolation variations would produce uniform, reflection-free beds, while high-frequency packets correspond to extended periods of climatic stability. Considerable work remains to read the problematic record of martian climate and assemble it into a coherent story.

Acknowledgments. Partial support for the time spent writing this chapter was provided by the Mars Fundamental Research Program (203959.02.02.20.01). We are grateful to the Space Science and Astrobiology Division at NASA Ames Research Center and the Planetary Exploration Division at NASA Headquarters for their support. Thanks to R. Haberle, K. Zahnle, M. Bullock, A. Simon-Miller, S. Mackwell, and P. Read for essential discussions. Two anonymous reviewers improved the chapter immensely. Finally, thanks to J. Cotton Thomas for help with references, proofing, hot food, and indispensible moral and psychological support.

REFERENCES

Abe-Ouchi A., Segawa T., and Saito F. (2007) Climatic conditions for modeling the northern hemisphere ice sheets throughout the ice age cycle. *Climate Past, 3,* 423–438.

Agassiz L. (1840a) *Etudes sur les glaciers. Ouvrage accompagné d'un atlas de 32 planches.* Jent et Gassmann, Neuchâtel.

Agassiz L. (1840b) Glaciers and the evidence of their having once existed in Scotland, Ireland and England. *Proc. Geol. Soc. Lond., 3,* 327–332.

Andersen M. B., Gallup C. D., Scholz D., Stirling C. H., and Thompson W. G. (2009) U-series dating of fossil coral reefs: Consensus and controversy. *PAGES (Past Global Changes) News, 17,* 55–57.

Andrews J. T. and Barry R. G. (1978) Glacial inception and disintegration during last glaciation. *Ann. Rev. Earth. Planet. Sci., 6,* 205–228.

Armstrong J. C., Leovy C. B., and Quinn T. (2004) A 1 Gyr climate model for Mars: New orbital statistics and the importance of seasonally resolved polar processes. *Icarus, 171,* 255–271.

Augustin L., Barbante C., Barnes P. R. F., et al. (2004) Eight glacial cycles from an Antarctic ice core. *Nature, 429,* 623–629.

Balme M. R., Gallagher C. J., Page D. P., Murray J. B., and Muller J. P. (2009) Sorted stone circles in Elysium Planitia, Mars: Implications for recent martian climate. *Icarus, 200,* 30–38.

Barnola J. M., Raynaud D., Korotkevitch V. S., and Lorius C. (1987) Vostok ice core: A 160,000 year record of atmospheric CO_2. *Nature, 329,* 408–414.

Beck J. W., Edwards R. L., Ito E., Taylor F. W., Recy J., Rougerie F., Joannot P., and Henin C. (1992) Sea-surface temperature from coral skeletal strontium/calcium ratios. *Science, 257,* 644–647.

Beerling D. B. and Royer D. L. (2011) Convergent Cenozoic CO_2 history. *Nature Geosci., 4,* 417–419.

Berger A. (2001) The role of CO_2, sea-level and vegetation during the Milankovitch forced glacial-interglacial cycles. In *Geosphere-Biosphere Interactions and Climate* (L. Bengtsson and Cl. U. Hammer, eds.), pp. 119–146. Cambridge Univ., New York.

Berger A. and Loutre M. F. (1993) Insolation values for the climate of the last 10 million years. *Quaternary Sci. Rev., 10,* 297–317.

Berger A. and Loutre M. F. (2010) Modeling the 100-kyr glacial–interglacial cycles. *Global Planet. Change, 72,* 275–281.

Berger A., Li X. S., and Loutre M. F. (1999) Modeling the northern hemisphere ice volume over the last 3 Ma. *Quaternary Sci. Rev., 18,* 1–11.

Bintanja R. and van de Wal R. S. W. (2008) North American ice-sheet dynamics and the onset of 100,000-year glacial cycles. *Nature, 454,* 869–872.

Bleamaster L. F. III and Crown D. A. (2010) Geologic map of MTM -40277, -45277, -40272, and -45272 quadrangles, eastern Hellas Planitia region of Mars. *U.S. Geol. Surv. Sci. Inv. Map 3096.* Available online at *pubs.usgs.gov/sim/3096.*

Bonelli S., Charbit S., Kageyama M., Woillez M.-N., Ramstein G., Dumas C., and Quiquet A. (2009) Investigating the evolution of major northern hemisphere ice sheets during the last glacial-interglacial cycle. *Climate Past, 5,* 329–345.

Boylan P. J. (1981) The role of William Buckland (1784–1856) in the recognition of glaciations in great Britain. In *The Quaternary in Great Britain* (J. Neale and J. Flenley eds.), pp. 1–8. Pergamon, Oxford.

Boynton W. V., Feldman W. C., Squyres S. W. Prettyman T. H., Brückner J., Evans L. G., Reedy R. C., Starr R., Arnold J. R., Drake D. M., Englert P. A. J., Metzger A. E., Mitrofanov I., Trombka J. I., d'Uston C., Wänke H., Gasnault O., Hamara D. K., Janes D. M., Marcialis R. L., Maurice S., Mikheeva I., Taylor G. J., Tokar R., and Shinohara C. (2002) Distribution of hydrogen in the near surface of Mars: Evidence for subsurface ice deposits. *Science, 297,* 81–85.

Brook E. J. (2013) Leads and lags at the end of the last ice age. *Science, 339,* 1042–1044.

Burton E. A. and Walter L. M. (1991) The effects of P_{CO_2} and temperature on magnesium incorporation in calcite in seawater and $MgCl_2$-$CaCl_2$ solutions. *Geochim Cosmochim. Acta, 55,* 777–785.

Byrne S. (2009) The polar deposits of Mars. *Annu. Rev. Earth Planet. Sci., 37,* 535–560.

Byrne S. and Ivanov A. B. (2004) Internal structure of the martian south polar layered deposits. *J. Geophys. Res., 109,* E11001, DOI: 10.1029/2004JE002267.

Byrne S., Dundas C. M., Kennedy M. R., Mellon M. T., McEwen A. S., Cull S. C., Daubar I. J., Shean D. E., Seelos K. D., Murchie S. L., Cantor B. A., Arvidson R. E., Edgett K. S., Reufer A., Thomas N., Harrison T. N., Posiolova L. V., and Seelos F. P. (2009) Distribution of mid-latitude ground ice on Mars from new impact craters. *Science, 325,* 1674–1676.

Carlson A. E. and Winsor K. (2012) Northern hemisphere ice-sheet responses to past climate warming. *Nature Geosci., 5,* 607–613.

Carr M. H. (1982) Periodic climate change on Mars: Review of evidence and effects on distribution of volatiles. *Icarus, 50,* 129–139.

Carr M. H. and Schaber G. G. (1977) Martian permafrost features. *J. Geophys. Res., 82,* 4039–4054.

Chatfield C. (2003) *The Analysis of Time Series, 6th edition.* CRC, Boca Raton, Florida. 352 pp.

Cheng H., Edwards R. L., Broecker W. S., Denton G. H., Kong X., Wang Y., Zhang R., and Wang X. (2009) Ice age terminations. *Science, 326,* 248–252.

Clark P., Alley R., and Pollard D. (1999) Northern hemisphere ice-sheet influences on global climate change. *Science, 286,* 1104–1111.

Clifford S. M. and Hillel D. (1983) The stability of ground ice in the equatorial region of Mars. *J. Geophys. Res., 88,* 2456–2474.

Craig H., Horibe Y., and Sowers T. (1988) Gravitational separation of gases and isotopes in polar ice caps. *Science, 242,* 1675–1678.

Croll J. (1864) On the physical cause of the change of climate during geological epochs. *Philos. Mag., 28,* 121–137.

Crown D. A., Price K. H., andGreeley R. (1992) Geologic evolution of the east rim of the Hellas basin Mars. *Icarus, 100,* 1–25.

Cutts J. A. (1973) Nature and origin of layered deposits of the martian polar regions. *J. Geophys. Res., 78,* 4231–4291.

DeConto R. M. and Pollard D. (2003) Rapid Cenozoic glaciation of Antarctica induced by declining atmospheric CO_2. *Nature, 421,* 245–250.

Denton G. H., Anderson R. F., Toggweiler J. R., Edwards R. L., Schaefer J. M., and Putnam A. E. (2010) The last glacial termination. *Science, 328,* 1652–1657.

Drysdale R. N., Hellstrom J. C., Zanchetta G., Fallick A. E., Sánchez Goñi M. F., Couchoud I., McDonald J., Maas R., Lohmann G., and Isola I. (2009) Evidence for obliquity forcing of glacial termination II. *Science, 325,* 1527–1531.

Duque-Caro H. (1990) Neogene stratigraphy, paleoceanography and paleobiogeography in northwest South America and the evolution of the Panama Seaway. *Palaeogeogr., Palaeoclimatol., Palaeoecol., 77,* 203–234.

Eluszkiewicz J. and Moncet J.-L. (2003) A coupled microphysical/radiative transfer model of albedo and emissivity of planetary surfaces covered by volatile ices. *Icarus, 166,* 375–384.

Emiliani C. (1966) Paleo-temperature analysis of Caribbean cores P6304-8 and P6304-9 and a generalized temperature curve for the past 425,000 years. *J. Geol., 74,* 109–124.

Fanale F. P. and Cannon W. A. (1971) Adsorption on the martian regolith. *Nature, 230,* 502–504.

Fanale F. P. and Cannon W. A. (1979) Mars: CO_2 adsorption and capillary condensation on clays — Significance for volatile storage and atmosphere history. *J. Geophys. Res., 84,* 8404–8414.

Fanale F. P., Salvail J. R., Banerdt W. B., and Saunders R. S. (1982) Mars: The regolith-atmosphere-cap system and climate change. *Icarus, 50,* 381–407.

Farmer C. B. and Doms P. E. (1979) Global seasonal variation of water vapor on Mars and implications for permafrost. *J. Geophys. Res., 84,* 2881–2888.

Farmer C. B., Davies D. W., Holland A. L., LaPorte D. D., and Doms P. E. (1977) Mars: Water vapor observations from the Viking orbiters. *J. Geophys. Res., 82,* 4225–4248.

Fenton L. and Richardson M. I. (2001) Martian surface winds: Insensitivity to orbital changes and implications for aeolian processes. *J. Geophys. Res., 106,* 32885–32909.

Fishbaugh K. and Hvidberg C. (2006) Martian north polar layered deposits stratigraphy: Implications for accumulation rates and flow. *J. Geophys. Res., 111,* E06012, DOI: 10.1029/2005JE002571.

Fishbaugh K. E., Hvidberg C. S., Byrne S., Russell P. S., Herkenhoff K. E., Winstrup M., and Kirk R. (2010) First high-resolution stratigraphic column of the martian north polar layered deposits. *Geophys. Res. Lett., 37,* L07201.

Forget F. (2012) Mars climate changes as simulated by traditional global climate models. In *Mars Recent Climate Change Workshop*, available online at *spacescience.arc.nasa.gov/mars-climate-workshop-2012/documents/abstracts/Forget_F_InAbst.doc*.

Forget F., Hourdin F., and Talagrand O. (1998) CO_2 snow fall on Mars: Simulation with a general circulation model. *Icarus, 131,* 302–316.

Forget F., Haberle R. M., Montmessin F., Levrard B., and Head J. W. (2006) Formation of glaciers on Mars by atmospheric precipitation at high obliquity. *Science, 311,* 368–371.

Galla K. G., Byrne S., Murray B., McEwen A., and the HiRISE Team (2008) Craters and resurfacing of the martian north polar cap (abstract). *Eos Trans. AGU, 89(53),* Fall Meet. Suppl., Abstract P41B-1360.

Gallagher C., Balme M. R., Conway S.J., and Grindrod P.M. (2011) Sorted clastic stripes, lobes and associated gullies in high-latitude craters on Mars: Landforms indicative of very recent, polycyclic ground-ice thaw and liquid flows. *Icarus, 211,* 458–471.

Gallup C. D., Cheng H., Taylor F. W., and Edwards R. L. (2002) Direct determination of the timing of sea level change during termination II. *Science, 295,* 310–314.

Gibbard P. and van Kolfschoten T. (2005) The pleistocene and holocene epochs. In *A Geologic Time Scale 2004* (F. M. Gradstein et al., eds.), p. 86. Cambridge Univ., Cambridge.

Greve R., Grieger B., and Stenzel O. J. (2010) MAIC-2, a latitudinal model for the martian surface temperature, atmospheric water transport and surface glaciation. *Planet. Space Sci., 58,* 931–940.

Haberle R. M., Murphy J. R., and Schaeffer J. (2003) Orbital change experiments with a Mars general circulation model, *Icarus, 161,* 66–89.

Haberle R. M., Kahre M. A., Hollingsworth J. L., Schaeffer J., Montmessin F., and Phillips R. J. (2012) A cloud greenhouse effect on Mars: Significant climate change in the recent past? (abstract). In *Lunar Planet. Sci. Conf. 43rd,* Abstract #1665. Lunar and Planetary Institute, Houston.

Hambach U., Rolf C., and Schnepp E. (2008) Magnetic dating of Quaternary sediments, volcanites and archaeological materials: An overview. *J. Quaternary Sci., 57,* 25–51.

Hansen J., Sato M., Ruedy R., Nazarenko L., Lacis A., Schmidt G. A., Russell G., Aleinov I., Bauer M., Bauer S., Bell N., Cairns B., Canuto V., Chandler M., Cheng Y., Del Genio A., Faluvegi G., Fleming E., Friend A., Hall T., Jackman C., Kelley M., Kiang N., Koch D., Lean J., Lerner J., Lo K., Menon S., Miller R., Minnis P., Novakov T., Oinas V., Perlwitz Ja., Perlwitz Ju., Rind D., Romanou A., Shindell D., Stone P., Sun S., Tausnev N., Thresher D., Wielicki B., Wong T., Yao M., and Zhang S. (2005) Efficacy of climate forcings. *J. Geophys. Res., 110,* DOI: 10.1029/2005JD005776.

Haug G. H. and Tiedemann R. (1998) Effect of the formation of the Isthmus of Panama on Atlantic Ocean thermohaline circulation. *Nature, 393,* 673–676.

Hays J. D., Imbrie J., and Shackleton N. J. (1976) Variations in the Earth's orbit: Pacemaker of the ice ages. *Science, 194,* 1121–1132.

Head J. W. and Marchant D. R. (2006) Evidence for global-scale northern mid-latitude glaciation in the Amazonian period of Mars: Debris-covered glacier and valley glacier deposits in the 30–50 latitude band (abstract). In *Lunar Planet. Sci. XXXVII,* Abstract #1127. LPI Contrib. No. 1303, Lunar and Planetary Institute, Houston.

Head J. W., Mustard J. F., Kreslavsky M. A., Milliken R. E., and Marchant D. R. (2003) Recent ice ages on Mars. *Nature, 426,* 797–802.

Herkenhoff K. E. and Plaut J. J. (2000) Surface ages and resurfacing rates of the polar layered deposits on Mars. *Icarus, 144,* 243–253.

Herkenhoff K. E., Byrne S., Russell P. S., Fishbaugh K. E., and McEwen A. S. (2007) Meter-scale morphology of the north polar region of Mars. *Science, 317,* 1711–1716.

Hess S. L., Ryan J. A., Tillman J. E., Henry R. M., and Leovy C. B. (1980) The annual cycle of pressure on Mars measured by Viking Landers 1 and 2. *Geophys. Res. Lett., 7,* 197–200.

Hinnov L. A. (2000) New perspectives on orbitally forced stratigraphy. *Annu. Rev. Earth Planet. Sci., 28,* 419–475.

Hinnov L. A. and Ogg J. G. (2007) Cyclostratigraphy and the astronomical time scale. *Stratigraphy, 4,* 239–251.

Hofstadter M. D. and Murray B. C. (1990) Ice sublimation and rheology: Implications for the martian polar layered deposits. *Icarus, 84,* 352–361.

Hourdin F., Forget F., and Talagrand O. (1995) The sensitivity of the martian surface pressure and atmospheric mass budget to various parameters: A comparison between numerical simulations and Viking observations. *J. Geophys. Res., 100,* 5501–5523.

Howard A. D., Cutts J. A., and Blasius K. R. (1982) Stratigraphic relationships within martian polar cap deposits. *Icarus, 50,* 161–215.

Hubbard B., Milliken R. E., Kargel J. S., Limaye A., and Souness C. (2011) Geomorphological characterisation and interpretation of a mid-latitude glacier-like form: Hellas Planitia, Mars. *Icarus, 211,* 330–346.

Huybers P. (2007) Glacial variability over the last two million years: An extended depth-derived age model, continuous obliquity pacing, and the Pleistocene progression, *Quaternary Sci. Rev., 26,* 37–55.

Huybers P. and Denton G. (2008) Antarctic temperature at orbital timescales controlled by local summer duration. *Nature Geosci., 1,* 787–792.

Huybers P. and Wunsch C. (2004) A depth-derived Pleistocene age model: Uncertainty estimates, sedimentation variability, and nonlinear climate change. *Paleoceanogr., 19,* 1028.

Huybers P. and Wunsch C. (2005) Obliquity pacing of the late Pleistocene glacial terminations. *Nature, 434,* 491–494.

Hvidberg C. S., Fishbaugh K. E., Winstrup M., Svensson A., Byrne S., and Herkenhoff K. E. (2012) Reading the climate record of the martian polar layered deposits. *Icarus, 221,* 405–419.

Imbrie J. and Imbrie J. Z. (1980) Modeling the climatic response to orbital variations. *Science, 207,* 943–953.

Imbrie J., Mix A. C., and Martinson D. G. (1993) Milankovitch theory viewed from Devils Hole. *Nature, 363,* 531–533.

Jakosky B. M. and Carr M. H. (1985) Possible precipitation of ice at low latitudes of Mars during periods of high obliquity. *Nature, 315,* 559–561.

Jakosky B. M., Henderson B. G., and Mellon M. T. (1993) The Mars water cycle at other epochs: Recent history of the polar caps and layered terrain. *Icarus, 102,* 286–297.

Jakosky B. M., Henderson B. G., and Mellon M. T. (1995) Chaotic obliquity and the nature of the martian climate. *J. Geophys. Res., 100,* 1579–1584.

Jansen E., Overpeck J., Briffa K. R., Duplessy J.-C., Joos F., Masson-Delmotte V., Olago D., Otto-Bliesner B., Peltier W. R., Rahmstorf S., Ramesh R., Raynaud D., Rind D., Solomina O., Villalba R., and Zhang D. (2007) Palaeoclimate. In *Climate Change 2007: The Physical Science Basis. Contribution of Working Group I to the Fourth Assessment Report of the Intergovernmental Panel on Climate Change* (S. D. Solomon et al., eds). Cambridge Univ., Cambridge.

Jouzel J., Masson-Delmotte V., Cattani O., Dreyfus G., Falourd S., Hoffmann G., Minster B., Nouet J., Barnola J. M., Chappellaz J., Fischer H., Gallet J. C., Johnsen S., Leuenberger M., Loulergue L., Luethi D., Oerter H., Parrenin F., Raisbeck G., Raynaud D., Schilt A., Schwander J., Selmo E., Souchez R., Spahni R., Stauffer B., Steffensen J. P., Stenni B., Stocker T. F., Tison J. L., Werner M., and Wolff E. W. (2007) Orbital and millennial antarctic climate variability over the past 800,000 years. *Science, 317,* 793–796.

Kawamura K., Parrenin F., Lisiecki L., Uemura R., Vimeux F., Severinghaus J. P., Hutterli M. A., Nakazawa T., Aoki S., Jouzel J., Raymo M. E., Matsumoto K., Nakata H., Motoyama H., Fujita S., Goto-Azuma K., Fujii Y., and Watanabe O. (2007) Northern hemisphere forcing of climatic cycles in Antarctica over the past 360,000 years. *Nature, 448,* 912–916.

Keigwin L. D. (1982) Isotope paleoceanography of the Caribbean and east Pacific: Role of Panama uplift in late Neogene time. *Science, 217,* 350–353.

Kessler M. A. and Werner B. T. (2003) Self-organization of sorted patterned ground. *Science, 299,* 380–383.

Kieffer H. H. (1990) H_2O grain size and the amount of dust in Mars' residual north polar cap. *J. Geophys. Res., 95,* 1481–1493.

Kieffer H. H. and Zent A. P. (1992) Quasi-periodic climate change on Mars. In *Mars* (H. H. Kieffer et al., eds.), pp. 1180–1218. Univ. of Arizona, Tucson.

Kieffer H. H., Martin T. Z., Peterfreund A. R., Jakosky B. M., Miner E. D., and Palluconi F. D. (1977) Thermal and albedo mapping of Mars during the Viking primary mission. *J. Geophys. Res., 82,* 4249–4291.

Kieffer H. H., Titus T. N., Mullins K. F., and Christensen P. (2000) Mars south polar spring and summer behavior observed by TES: Seasonal cap evolution controlled by frost grain size. *J. Geophys. Res., 105,* 9653–9699.

Kolb E. J. and Tanaka K. L. (2001) Geologic history of the polar regions of Mars based on Mars Global Surveyor data: II. Amazonian Period. *Icarus,* 154, 22–39.

Kolb E. J. and Tanaka K L. (2006) Accumulation and erosion of south polar layered deposits in the Promethei Lingula region, Planum Australe, Mars. *Mars, Intl. J. Mars Sci. Explor., 2,* 1–9.

Kopp G. and Lean J. L. (2011) A new, lower value of total solar irradiance: Evidence and climate significance. *Geophys. Res. Lett., 38,* L01706, DOI: 10.1029/2010GL045777.

Koutnik M., Byrne S., and Murray B. (2002) South polar layered deposits of Mars: The cratering record. *J. Geophys. Res., 107,* 5100.

Kreslavsky M. A. and Head J. W. (2005) Mars at very low obliquity: Atmospheric collapse and the fate of volatiles. *Geophys. Res. Lett., 32,* L12202, DOI: 10.1029/2005GL022645.

Kreslavsky M. A. and Head J. W. (2011) Carbon dioxide glaciers on Mars: Products of recent low obliquity epochs (?). *Icarus,*

216, 111–115.

Lambeck L., Yokoyama Y., and Purcell T. (2002) Into and out of the last glacial maximum: Sea-level change during oxygen isotope stages 3 and 2. *Quaternary Sci. Rev., 21,* 343–360.

Laskar J., Levrard B., and Mustard J. F. (2002) Orbital forcing of the martian polar layered deposits. *Nature, 419,* 375–377.

Laskar J., Correia A. C. M., Gastineau M., Joutel F., Levrard B., and Robutela P. (2004) Long term evolution and chaotic diffusion of the insolation quantities of Mars. *Icarus, 170,* 343–264.

Laskar J., Fienga A., Gastineau M., and Manche H. (2011) La2010: A new orbital solution for the long-term motion of the Earth. *Astron. Astrophys., 532,* DOI: 10.1051/0004-6361/201116836.

Lear C. H., Elderfield H., and Wilson P. A. (2000) Cenozoic deep-sea temperatures and global ice volumes from Mg/Ca in benthic foraminiferal calcite. *Science, 287,* 269–272.

Lee S.-Y. and Poulsen J. (2008) Amplification of obliquity forcing through mean-annual and seasonal atmospheric feedbacks. *Climate Past Discussions,4,* 515–534.

Leighton R. B. and Murray B. M. (1966) Behavior of carbon dioxide and other volatiles on Mars. *Science, 153,* 136–144.

Levrard B., Forget F., Montmessin F., and Laskar J. (2004) Recent ice-rich deposits formed at high latitudes on Mars by sublimation of unstable equatorial ice during low obliquity. *Nature, 431,* 1072–1075.

Levrard B., Forget F., Montmessin F., and Laskar J. (2007) Recent formation and evolution of northern martian polar layered deposits as inferred from a global climate model. *J. Geophys. Res., 112,* E06012, DOI: 10.1029/2006JE002772.

Lewis S. C., LeGrande A. N., Kelley M., and Schmidt G. A. (2013) Modeling insights into deuterium excess as an indicator of water vapor source conditions. *J. Geophys. Res., 118,* 243–262, DOI: 10.1029/2012JD017804.

Li H., Robinson M. S., and Jurdy D. M. (2005) Origin of martian northern hemisphere mid-latitude lobate debris aprons. *Icarus, 176,* 382–394.

Limaye A. B. S., Aharonson O., and Perron J. T. (2012) Detailed stratigraphy and bed thickness of the Mars north and south polar layered deposits. *J. Geophys. Res., 117,* E06009, DOI: 10.1029/2011JE003961.

Lisiecki L. E. and Raymo M. E. (2005) A Pliocene-Pleistocene stack of 57 globally distributed benthic $\delta^{18}O$ records. *Paleoceanogr., 20,* PA1003.

Lisiecki L. E. and Raymo M. E. (2007) Plio-Pleistocene climate evolution: Trends and transitions in glacial cycle dynamics. *Quaternary Sci. Rev., 26,* 56–69.

Lissauer J. J., Barnes J. W., and Chambers J. E. (2012) Obliquity variations of a moonless Earth. *Icarus, 217,* 77–87.

Loutre M. F. (2003) Clues from MIS 11 to predict the future climate: A modeling point of view. *Earth Planet. Sci. Lett., 212,* 213–224.

Loutre M.-F., Paillard D., Vimeux F., and Cortijo E. (2004) Does mean annual insolation have the potential to change the climate? *Earth Planet. Sci. Lett., 221,* 1–14.

Lüthi D., Le Floch M., Bereiter B., Blunier T., Barnola J.-M., Siegenthaler U., Raynaud D., Jouzel J., Fischer H., Kawamura K., and Stocker T. F. (2008) High-resolution carbon dioxide concentration record 650,000–800,000 years before present. *Nature, 453,* 378–382.

Madeleine J.-B., Forget F., Head J. W., Levrard B., Montmessin F., and Millour E. (2009) Amazonian northern mid-latitude glaciation on Mars: A proposed climate scenario. *Icarus 203,* 390–405.

Madeleine J.-B., Forget F., Head J. W., Navarro T., Millour E., Spiga A., Colaitis A., Montmessin F., and Määttänen A. (2012) LMD Mars GCM simulations at high obliquity using an improved cloud model. In *Workshop on Recent Climate Change on Mars,* available online at *spacescience.arc.nasa.gov/mars-climate-workshop-2012/documents/abstracts/Madeleine_JB_InAbs.doc.*

Malin M. C. and Edgett K. S. (2000) Evidence for recent groundwater seepage and surface runoff on Mars. *Science, 288,* 2330–2335.

Malin M. C. and Edgett K.S. (2001) Mars Global Surveyor Mars Orbiter Camera: Interplanetary cruise through primary mission. *J. Geophys. Res., 106,* 23429–23570, DOI: 10.1029/2000JE001455.

Malin M. C., Caplinger M. A., and Davis S. D. (2001) Observational evidence for an active surface reservoir of solid carbon dioxide on Mars. *Science, 294,* 2146–2148.

Mantsis D. F., Clement A. C., Broccoli A. J., and Eeb M. P. (2011) Climate feedbacks in response to changes in obliquity. *J. Climate, 24,* 2830–2845.

Marchant D. R., Denton G. H., and Swisher C. C. III (1993) Miocene-Pliocene-Pleistocene glacial history of Arena Valley, Quartermain Mountains, Antarctica. *Geograf. Ann. Ser. A, Phys. Geogr., 4,* 269–302.

Maslin M., Li A., X. S., Loutre M.-F., and Berger A. (1998) The contribution of orbital forcing to the progressive intensification of northern hemisphere glaciation. *Quaternary Sci. Rev., 17,* 411–426.

McEwen A. S., Ojha L., Dundas C. M., Mattson S. S., Byrne S., Wray J. J., Cull S. C., Murchie S. L., Thomas N., and Gulick V. C. (2011) Seasonal flows on warm martian slopes. *Science, 333,* 740–744.

McManus J. F., Francois R., Gherardi J.-M., Keigwin L. D., and Brown-Leger S. (2004) Collapse and rapid resumption of Atlantic meridional circulation linked to deglacial climate changes. *Nature, 428,* 834–837.

McPhee J. (1998) *Annals of the Former World.* Farrar, Straus and Giroux, New York. 696 pp.

Mellon M. T. and Jakosky B. M. (1993) Geographic variations in the thermal and diffusive stability of ground-ice on Mars. *J. Geophys. Res, 98,* 3345–3364.

Mellon M. T., Feldman W. C., and Prettyman T. H. (2004) The presence and stability of ground-ice in the southern hemisphere of Mars. *Icarus, 169,* 324–340.

Mest S. C. and Crown D. A. (2003) Geologic map of MTM-40252 and-40257 quadrangles, Reull Vallis region of Mars. *U.S. Geol. Surv. Inv. Ser. I-2730,* available online at *pubs.usgs.gov/imap/i2730.*

Meyers S. R. and Hinnov L. A. (2010) Northern hemisphere glaciation and the evolution of Plio-Pleistocene climate noise. *Paleoceanogr., 25,* PA3207.

Milanković M. (1938) Astronomische Mittel zur Erforschung der erdgeschichtlichen Klimate. In *Handbuch der Geophysik, Vol. 9* (B. Gutenberg, ed.), Gebrüder Borntraeger, Berlin.

Milkovich S. M. and Head J. W. III (2005) North polar cap of Mars: Polar layered deposit characterization and identification of a fundamental climate signal. *J Geophys. Res., 110,* E01005, DOI: 10.1029/2004JE002349.

Milkovich S. M. and Plaut J. J. (2008) Martian south polar layered deposit stratigraphy and implications for accumulation history. *J. Geophys. Res., 113,* E06007, DOI: 10.1029/2007JE002987.

Miller K. G., Wright J. D., Katz M. E., Wade B. S., Browning J. V., Cramer B. S., and Rosenthal Y. (2009) Climate threshold at the Eocene-Oligocene transition: Antarctic ice sheet influence on ocean circulation. In *The Late Eocene Earth — Hothouse, Icehouse, and Impacts* (C. Koeberl et al., eds.), pp. 1–10. Geol. Soc. Am. Spec. Paper 452, Geological Society of America, Denver, DOI: 10.1130/2009.2452.

Milliken R. E., Mustard J. F., and Goldsby D. L. (2003) Viscous flow features on the surface of Mars: Observations from high resolution Mars Orbiter Camera (MOC) images. *J. Geophys. Res., 108,* 5057, DOI: 10.1029/2002JE002005.

Mischna M. A. and Richardson M. I. (2005) A reanalysis of water abundances in the martian atmosphere at high obliquity. *Geophys. Res. Lett., 32,* L03201.

Mischna M. A., Richardson M. I., Wilson R. J., and McCleese D. J. (2003) On the orbital forcing of martian water and CO_2 cycles: A general circulation model study with simplified volatile schemes. *J. Geophys. Res, 108,* DOI: 10.1029/2003JE002051.

Montmessin F., Haberle R. M., Forget F., Langevin Y., Clancy R. T., and Bibring J.-P. (2007) On the origin of perennial water ice at the south pole of Mars: A precession-controlled mechanism? *J. Geophys. Res., 112,* E8, DOI: 10.1029/2007JE002902.

Muhs D. R., Pandolfi J. M., Simmons K. R., and Schumann R. R. (2012) Sea-level history of past interglacial periods: New evidence from uranium-series dating of corals from Curaçao, Leeward Antilles islands. *Quaternary Res., 78,* 157–169.

Murray B. C., Soderblom L. A., Cutts J. A., Sharp R. P., Milton D. J., and Leighton R. B. (1972) Geological framework of the south polar region of Mars. *Icarus, 17,* 328–345.

Murray B. C., Ward W. R., and Yeung S. C. (1973) Periodic insolation variations on Mars. *Science, 180,* 638–640.

Mustard J. F., Cooper C. D., and Rifkin M. K. (2001) Evidence for recent climate change on Mars from the identification of youthful near-surface ground ice. *Nature, 412,* 411–414.

Neukum G., Jaumann R., Hoffmann H., Hauber E., Head J. W., Basilevsky A. T., Ivanov B. A., Werner S. C., van Gasselt S., Murray J. B., McCord T. and the HRSC Co-Investigator Team (2004) Recent and episodic volcanic and glacial activity on Mars revealed. *Nature, 432,* 971–979.

Newman C. E., Lewis S. R., and Read P. L. (2005) The atmospheric circulation and dust activity in different orbital epochs on Mars. *Icarus, 174,* 135–160.

Otto-Bliesner B. L., Hewitt C. D., Marchitto T. M., Brady E., Abe-Ouchi A., Crucifix M., Murakami S., and Weber S. L. (2007) Last glacial maximum ocean thermohaline circulation: PMIP2 model intercomparisons and data constraints. *Geophys. Res. Lett., 34,* L12706.

Pagani M., Zachos J. C., Freeman K. H., Tipple B., and Bohaty S. (2005) Marked decline in atmospheric carbon dioxide concentrations during the Paleogene. *Science, 309,* 600–603.

Pagani M., Huber M., Liu Z., Bohaty S. M., Henderiks J., Sijp W., Krishnan S., and DeConto R. M. (2011) The role of carbon dioxide during the onset of Antarctic glaciation. *Science, 334,* 1261–1265.

Paige D. A. and Ingersoll A. P. (1985) The annual heat balance of the martian polar caps from Viking observations. Ph.D. thesis, California Institute of Technology, Pasadena.

Paillard D. (1998) The timing of Pleistocene glaciations from a simple multiple-state climate model. *Nature, 391,* 378–391.

Parrenin F., Masson-Delmotte V., Köhler P., Raynaud D., Paillard D., Schwander J., Barbante C., Landais A., Wegner A., and Jouzel J. (2013) Synchronous change of atmospheric CO_2 and Antarctic temperature during the last deglacial warming. *Science, 339,* 1060–1063.

Pearson P. N. and Palmer M. R. (2002) Atmospheric carbon dioxide concentrations over the past 60 million years. *Nature, 406,* 695–699.

Pearson P. N., Foster G. L., and Wade B. S. (2009) Atmospheric carbon dioxide through the Eocene-Oligocene climate transition. *Nature, 461,* 1110–1113.

Pedro J. B., Rasmussen, S. O., and van Ommen T. D. (2012) Tightened constraints on the time-lag between Antarctic temperature and CO_2 during the last deglaciation. *Climate Past, 8,* 1213–1221, DOI: 10.5194/cp-8-1213-2012.

Perron J. T. and Huybers P. J. (2009) Is there an orbital signal in the polar layered deposits on Mars? *Geology, 37,* 155–158.

Petit J. R., Jouzel J., Raynaud D., Barkov N. I., Barnola J.-M., Basile I., Bender M., Chappellaz J., Davis M., Delaygue G., Delmotte M., Kotlyakov V. M., Legrand M., Lipenkov V. Y., Lorius C., Pépin L., Ritz C., Saltzman E., and Stievenard M. (1999) Climate and atmospheric history of the past 420,000 years from the Vostok ice core, Antarctica. *Nature, 399,* 429–436.

Phillips R. J., Zuber M. T., Smrekar S. E., Mellon M. T., Head J. W., Tanaka K. L., Putzig N. E., Milkovich S. M., Campbell B. A., Plaut J. J., Safaeinili A., Seu R., Biccari D., Carter L. M., Picardi G., Orosei R., Mohit P. S., Heggy E., Zurek R. W., Egan A. F., Giacomoni E., Russo F., Cutigni M., Pettinelli E., Holt J. W., Leuschen C. J., and Marinangeli L. (2008) Mars north polar deposits: Stratigraphy, age, and geodynamical response. *Science, 320,* 1182–1186.

Phillips R. J., et al. (2011) Massive CO_2 ice deposits sequestered in the south polar layered deposits of Mars. *Science, 332,* 838–841.

Picardi G., Plaut J. J., Biccari D., Bombaci O., Calabrese D., Cartacci M., Cicchetti A., Clifford S. M., Edenhofer P., Farrell W. M., Federico C., Frigeri A., Gurnett D. A., Hagfors T., Heggy E., Herique A., Huff R. L., Ivanov A. B., Johnson W. T. K., Jordan R. L., Kirchner D. L., Kofman W., Leuschen C. J., Nielsen E., Orosei R., Pettinelli E., Phillips R. J., Plettemeier D., Safaeinili A., Seu R., Stofan E. R., Vannaroni G., Watters T. R., and Zampolini E. (2005) Radar soundings of the subsurface of Mars. *Science, 310,* 1925–1928.

Plaut J. J., Picardi G., Safaeinili A., Ivanov A. B., Milkovich S. M., Cicchetti A., Kofman W., Mouginot J., Farrell W. M., Phillips R. J., Clifford S. M., Frigeri A., Orosei R., Federico C., Williams I. P., Gurnett D. A., Nielsen E., Hagfors T., Heggy E., Stofan E. R., Plettemeier D., Watters T. R., Leuschen C. J., Edenhofer P. (2007) Subsurface radar sounding of the south polar layered deposits of Mars. *Science, 316,* 93–97.

Plaut J. J., Safaeinili A., Holt J. W., Phillips R. J., Head J. W. III, Seu R., Putzig N. E., and Frigeri A. (2009) Radar evidence for ice in lobate debris aprons in the mid-northern latitudes of Mars. *Geophys. Res. Lett., 36,* L02203.

Pollack J. B. and Toon O. B (1982) Quasi-periodic climate change on Mars: A review. *Icarus, 50,* 259–287.

Pollack J. B., Haberle R. M., Murphy J. R., Schaeffer J., and Lee H. (1993) Simulations of the general circulation of the martian atmosphere 2. Seasonal pressure variations. *J. Geophys. Res., 98,* 3149–3181.

Putzig N. E., Phillips R. J., Campbell B. A., Holt J. W., Plaut J. J., Carter L. M., Egan A. F., Bernardini F., Safaeinili A., and Seu R. (2009) Subsurface structure of Planum Boreum from Mars Reconnaissance Orbiter shallow radar soundings.

Icarus, 204, 443–457.
Raack J., Reiss D., and Hiesinger H. (2012) Gullies and their relationships to the dust-ice mantle in the northwestern Argyre Basin, Mars. *Icarus, 219,*129–141.
Rahmstorf S. (2002) Ocean circulation and climate during the past 120,000 years. *Nature, 419,* 207–214.
Ravelo A., Andreasen D., Lyle M., Lyle A., and Wara M. (2004) Regional climate shifts caused by gradual global cooling in the Pliocene epoch. *Nature, 429,* 263–267.
Raymo M. E. and Nisancioglu K. (2003) The 41 kyr world: Milankovitch's other unsolved mystery. *Paleoceanogr., 18(1),* DOI: 10.1029/2002PA000791.
Raymo M. E., Lisiecki L. K., and Nisancioglu K. (2006) Plio-Pleistocene ice volume, Antarctic climate, and the global $\delta^{18}O$ record. *Science, 313,* 492–495.
Richardson M. I. and Wilson R. J. (2002) Investigation of the nature and stability of the martian seasonal water cycle with a general circulation model. *J. Geophys. Res., 107,* 5031, DOI: 10.1029/ 2001JE001536.
Roe G. H. and Lindzen R. S. (2001) The mutual interaction between continental-scale ice sheets and atmospheric stationary waves. *J. Climate, 14,* 1450–1465.
Savijarvi H., Crisp D., and Harri A.-M. (2005) Effects of CO_2 and dust on present-day solar radiation and climate on Mars. *Q. J. R. Meteor. Soc., 131,* 2907–2922.
Scher H. D. and Martin E. E. (2006) Timing and climatic consequences of the opening of Drake Passage. *Science, 312,* 428–431.
Schwander J., Stauffer B., and Sigg A. (1988) Air mixing in firn and the age of the air at pore close-off. *Ann. Glaciol., 10,* 141–145.
Seidelmann P. K., Archinal B. A., A'Hearn M. F., Conrad A., Consolmagno G. J., Hestroffer D., Hilton J. L., Krasinsky G. A., Neumann G., Oberst J., Stooke P., Tedesco E. F., Tholen D. J., Thomas P. C., and Williams I. P. (2007) Report of the IAU/IAG Working Group on cartographic coordinates and rotational elements: 2006. *Cel. Mech. Dyn. Astron., 98,* 155–180.
Seu R., Phillips R. J., Alberti G., Biccari D., Bonaventura F., Bortone M., Calabrese D., Campbell B. A., Cartacci M., Carter L. M., Catallo C., Croce A., Croci R., Cutigni M., Di Placido A., Dinardo S., Federico C., Flamini E., Fois F., Frigeri A., Fuga O., Giacomoni E., Gim Y., Guelfi M., Holt J. W., Kofman W., Leuschen C. J., Marinangeli L., Marras P., Masdea A., Mattei S., Mecozzi R., Milkovich S. M., Morlupi A., Mouginot J., Orosei R., Papa C., Paternò T., Persi del Marmo P., Pettinelli E., Pica G., Picardi G., Plaut J. J., Provenziani M., Putzig N. E., Russo F., Safaeinili A., Salzillo G., Santovito M. R., Smrekar S. E., Tattarletti B., and Vicari D. (2007) Accumulation and erosion of Mars' south polar layered deposits. *Science, 317,* 1716–1720.
Shackleton N. J. (1967) Oxygen isotope analyses and Pleistocene temperature re-assessed. *Nature, 215,* 15–17.
Shackleton N. J. (2000) The 100,000-year ice age cycle identified and found to lag temperature, carbon dioxide, and orbital eccentricity. *Science, 289,* 1897–1902.
Shackleton N. J. and Imbrie J. (1990) The $\delta^{18}O$ spectrum of oceanic deep water over a five-decade band. *Climatic Change, 16,* 217–230.
Shackleton N. J. and Opdyke N. D. (1973) Oxygen isotope and palaeomagnetic stratigraphy of Equatorial Pacific core V28-238: Oxygen isotope temperatures and ice volumes on a 10^5 year and 10^6 year scale. *Quatenary Res., 3,* 39–55.
Shean D. E., Head J. W., and Marchant D. R. (2005) Origin and evolution of a cold-based tropical mountain glacier on Mars: The Pavonis Mons fan-shaped deposit. *J. Geophys. Res., 110,* E05001.
Shean D. E., Head J. W. III, Fastook J. L., and Marchant D. R. (2007) Recent glaciation at high elevations on Arsia Mons, Mars: Implications for the formation and evolution of large tropical mountain glaciers. *J. Geophys. Res., 112,* E03004.
Smith M. D., Zuber M. T., and Neumann G. A. (2001) Seasonal variations of snow depth on Mars. *Science, 294,* 2141–2146.
Smith P. H., Tamppari L. K., Arvidson R. E., Bass D., Blaney D., Boynton W. V., Carswell A., Catling D. C., Clark B., Duck T., DeJong E., Fisher D., Goetz W., Gunnlaugsson P., Hecht M., Hipkin V., Hoffman J., Hviid S., Keller H., Kounaves S., Lange C. F., Lemmon M., Madsen M., Markiewicz W., Marshall J., McKay C., Mellon M. T., Ming D., Morris R., Renno N., Pike W. T., Staufer U., Stoker C., Taylor P., Whiteway J., and Zent A. P. (2009) Water at the Phoenix landing site. *Science, 325,* 58–62.
Soderblom L. A., Malin M. C., Cutts J. A., and Murray B. C. (1973) Mariner 9 observations of the surface of Mars in the north polar region. *J. Geophys. Res., 78,* 4197–4210.
Spötl C., Mangini A., Frank N., Eichstädter R., and Burns S. (2002) Start of the last interglacial period at 135 ka: Evidence from a high Alpine speleothem. *Geology, 30,* 815–818.
Squyres S. W. (1979) The distribution of lobate debris aprons and similar flows on Mars. *J. Geophys. Res., 84,* 8087–8096.
Squyres S. W. and Carr M. H. (1986) Geomorphic evidence for the distribution of ground ice on Mars. *Science, 231,* 249–252.
Stirling C. H. and Andersen M. B. (2009) Uranium-series dating of fossil coral reefs: Extending the sea-level record beyond the last glacial cycle. *Earth Planet. Sci. Lett., 284,* 269–283.
Tanaka K. L. (2005) Geology and insolation-driven climatic history of Amazonian north polar materials on Mars. *Nature, 437,* 991–994.
Tarasov L. and Peltier W. R. (2004) A geophysically constrained large ensemble analysis of the deglacial history of the North American ice-sheet complex. *Quaternary Sci. Rev., 23,* 359–388.
Thomas P. C., et al. (2000) North-south geological differences between the residual polar caps on Mars. *Nature, 404,* 161–164.
Thornalley D. Jr., Barker S., Broecker W. S., Elderfield H., and McCave I. N. (2011) The deglacial evolution of North Atlantic deep convection. *Science, 331,* 202–205.
Toggweiler J. R. and Bjornsson H. (2000) Drake Passage and paleoclimate. *J. Quaternary Sci., 15,* 319–328.
Toon O. B., Pollack J. B., Ward W., Burns J. A., and Bilski K. (1980) The astronomical theory of climatic change on Mars. *Icarus, 44,* 552–607.
Urey H. C. (1947) The thermodynamic properties of isotopic substances. *J. Chem. Soc., 69,* 562–581.
Von Storch H. and Zwiers F.W. (2001) *Statistical Analysis in Climate Research.* Cambridge Univ., Cambridge.
Wang Z. and Mysak L. A. (2002) Simulation of the last glacial inception and rapid ice sheet growth in the McGill paleoclimate model. *Geophys. Res. Lett., 29(23),* 2102.
Ward W. R. (1973) Large-scale variations in the obliquity of Mars. *Science, 181,* 260–262.
Ward W. R. (1974) Climatic variations on Mars: 1. Astronomical theory of insolation. *J. Geophys. Res., 79,* 3375–3386.
Warren S. G., Wiscombe W. J., and Firestone J. F. (1990) Spectral albedo and emissivity of CO_2 in martian polar caps: Model results. *J. Geophys. Res., 95,* 14717–14741.

Wood S. E. and Griffiths S. D. (2009) Epochal seasonal thermal modeling of Mars' polar surface energy balance: Perennial CO_2 ice and atmospheric collapse (abstract). In *Third International Workshop on Mars Polar Energy Balance and the CO_2 Cycle*, Abstract #7032. LPI Contrib. No. 1494, Lunar and Planetary Institute, Houston.

Zachos J. C., Lohmann K. C., Walker J. C. G., and Wise S. W. (1993) Abrupt climate change and transient climates during the Paleogene: A marine perspective. *J. Geol., 101,* 191–213.

Zachos J., Pagani M., Sloan L., Thomas E., and Billups K. (2001) Trends, rhythms, and aberrations in global climate 65 Ma to present. *Science, 292,* 686–693.

Zachos J. C., Dickens G. R., and Zeebe R. E. (2008) An early Cenozoic perspective on greenhouse warming and carbon-cycle dynamics. *Nature, 451,* 279–283.

Zahnle K., Haberle R. M., Catling D. C., and Kasting J. F. (2008) Photochemical instability of the ancient martian atmosphere. *J. Geophys. Res., 113,* E11004.

Zent A. P. (2008) An historical search for thin H_2O films at the Phoenix landing site. *Icarus, 196,* 385–408.

Zent A. P. and Quinn R. C. (1995) Simultaneous adsorption of CO_2 and H_2O under Mars-like conditions and application to the evolution of the martian climate. *J. Geophys. Res., 100,* 5341–5349.

Zent A. P., Sizemore H. G., and Rempel A. W. (2011) Ice lens formation and frost heave at the Phoenix landing site (abstract). In *Lunar Planet. Sci. Conf. 42nd,* Abstract #2543. Lunar and Planetary Institute, Houston.

Zurek R. W., Barnes J. R., Haberle R. M., Pollack J. B., Tillman J. E., and Leovy C. B. (1992) Dynamics of the atmosphere of Mars. In *Mars* (H. H. Kieffer et al., eds.), pp. 835–933. Univ. of Arizona, Tucson.

Harder J. W. and Woods T. N. (2013) Solar irradiance variability and its impacts on the Earth climate system. In *Comparative Climatology of Terrestrial Planets* (S. J. Mackwell et al., eds.), pp. 539–565. Univ. of Arizona, Tucson, DOI: 10.2458/azu_uapress_9780816530595-ch22.

Solar Irradiance Variability and Its Impacts on the Earth Climate System

J. W. Harder and T. N. Woods
University of Colorado

The Sun plays a vital role in the evolution of the climates of terrestrial planets. Observations of the solar spectrum are now routinely made that span the wavelength range from the X-ray portion of the spectrum (5 nm) into the infrared to about 2400 nm. Over this very broad wavelength range, accounting for about 97% of the total solar irradiance, the intensity varies by more than 6 orders of magnitude, requiring a suite of very different and innovative instruments to determine both the spectral irradiance and its variability. The origins of solar variability are strongly linked to surface magnetic field changes, and analysis of solar images and magnetograms show that the intensity of emitted radiation from solar surface features in active regions has a very strong wavelength and magnetic field strength dependence. These magnetic fields produce observable solar surface features such as sunspots, faculae, and network structures that contribute in different ways to the radiated output. Semi-empirical models of solar spectral irradiance are able to capture much of the Sun's output, but this topic remains an active area of research. Studies of solar structures in both high spectral and spatial resolution are refining this understanding. Advances in Earth observation systems and high-quality three-dimensional chemical climate models provide a sound methodology to study the mechanisms of the interaction between Earth's atmosphere and the incoming solar radiation. Energetic photons have a profound effect on the chemistry and dynamics of the thermosphere and ionosphere, and these processes are now well represented in upper atmospheric models. In the middle and lower atmosphere the effects of solar variability enter the climate system through two nonexclusive pathways referred to as the top-down and bottom-up mechanisms. The top-down mechanism proceeds through the alteration of the photochemical rates that establish the middle atmospheric temperature structure and circulation patterns. In the bottom-up mechanism, the increased solar cycle forcing at Earth's surface increases the latent heat flux and evaporation processes, thereby altering the tropical wind patterns.

1. INTRODUCTION

1.1. Description of the Sun-Earth Connection Problem

In the 1965 edition of the classic book *Physical Climatology*, W. D. Sellers (*Sellers*, 1965) presents an estimate of the energy from the Sun intercepted by Earth and compares it with other large-scale energy sources that act continuously or quasi-continuously in the atmosphere and at its boundaries. He found that these combined sources are only about 0.025% of the input from the total solar irradiance (TSI). The typical peak-to-peak amplitude change in the TSI over the course of the 11-year solar irradiance cycle is about 1.1 W m^{-2} out of 1361 W m^{-2}, or ~800 parts per million (ppm). This exceeds all the non-TSI natural contributions by a factor of 3. Solar irradiance variability is put into the context of anthropogenic and natural climate forcing in Chapter 2 of the *Fourth Assessment Report of the Intergovernmental Panel on Climate Change* (IPCC) (*Forster et al.*, 2007). In this report, the authors estimate that the combined radiative forcing due to anthropogenic contributions [+1.6 (–1.0,+0.8) W m^{-2}] is likely to be at least five times larger than the secular change in TSI since the dawn of the industrial age in the year 1750. However, the report ranks the scientific understanding of the solar contribution as "low" and notes that additional climate forcing through the Sun's ultraviolet (UV) contributions and other solar-related mechanisms cannot be ruled out. The recent review of solar forcing in Earth's atmosphere by *Gray et al.* (2010) delineates a number of potential mechanisms that might contribute significantly to the Earth climate observations that were beyond the scope of the IPCC report. Both the *Gray et al.* (2010) review and the IPCC assessment emphasize what can be learned from proxy-based studies of solar variability; this is a very natural tendency since long-term records are needed to discern the lowest-frequency modes of the solar forcing component (*Douglass and Clader,* 2002). For this chapter, however, we will emphasize the role of solar spectral irradiance (SSI) variability, which was given lower priority in these two studies. Despite the difficulty of the spectral measurements with their inherently larger uncertainties, greater insight into the mechanisms that produce both variability in the solar output and the altitude-dependent response of Earth's atmosphere to this variability can be gained through studies of SSI variability.

1.2. Structure of Earth and Solar Atmospheres

An understanding of the emitted spectrum of the Sun and Earth's responses to this solar input begins with an understanding of atmospheric radiation. The temperature structures of these two atmospheres are displayed in Fig. 1. We will first discuss this in terms of the solar atmosphere, and then return to the topic in section 3.1 when we discuss the effects of solar variability on Earth's atmosphere. The study of radiative transfer explored in the next section is highly simplified and detailed discussion is beyond the scope of the material in this chapter. There are a number of very high quality publications on this topic, notably the graduate text on stellar atmospheres by *Rutten* (2003) and the work by *Goody and Yung* (1989) on atmospheric radiative transfer.

An understanding of atmospheric temperature and structure proceeds through the analysis of a black-body radiator that follows local thermodynamic equilibrium (LTE); the radiant flux (or radiancy) is the energy emitted from a light source per unit area of the emitter into an angle of 2π steradians. In the frequency band between ν and $\nu + d\nu$ and at the temperature T this is given by (*Penner,* 1959)

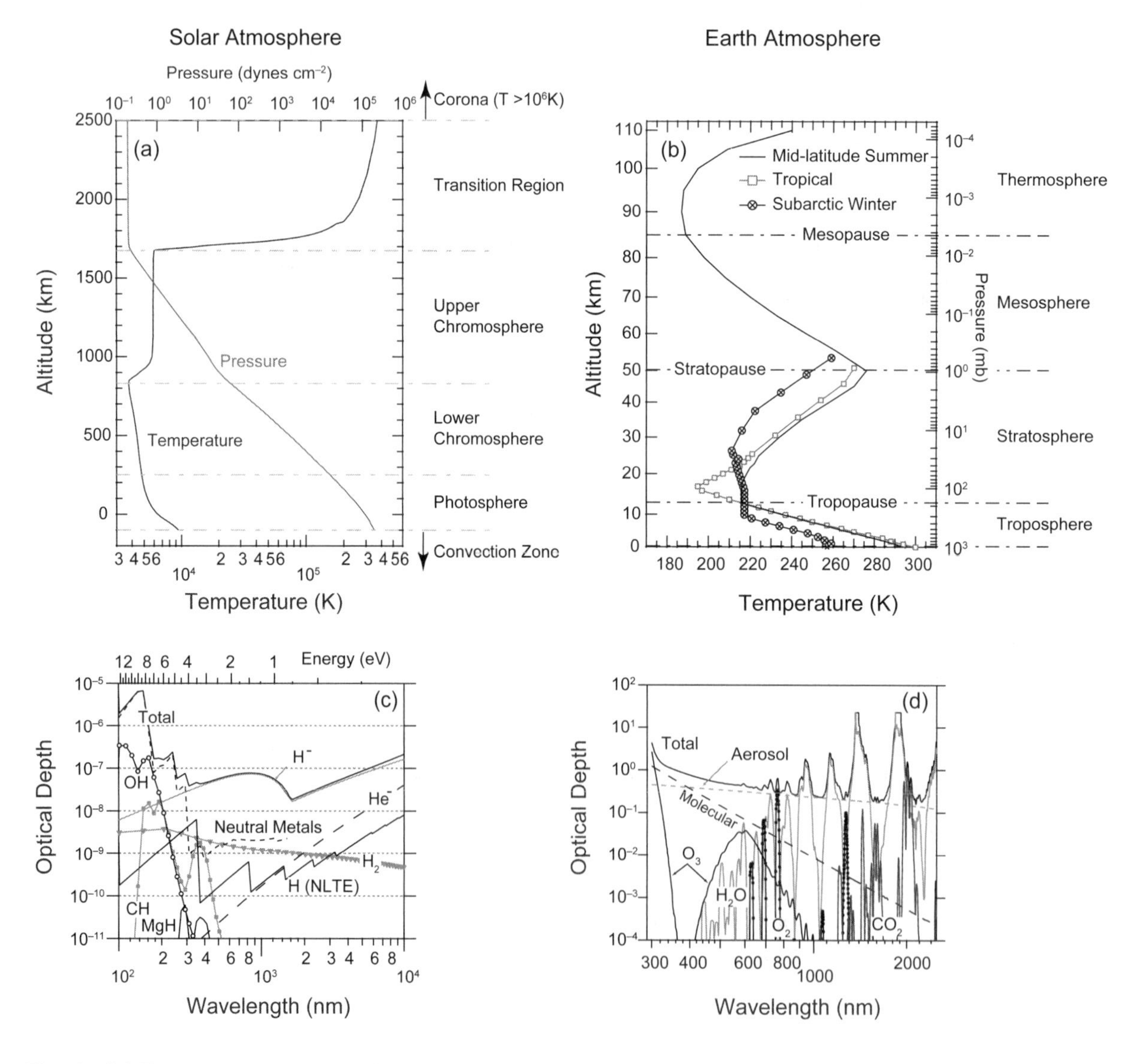

Fig. 1. (a) Temperature structure of the solar atmosphere with distinguishable regions of the solar atmosphere identified. **(b)** The same for Earth's atmosphere. **(c),(d)** Major sources of opacity in the atmospheres of the Sun and Earth that attenuate light as it passes through the atmosphere.

$$B_\nu d\nu = \frac{2\pi h\nu^3}{c^2} \frac{d\nu}{\left[\exp\left(\frac{h\nu}{kT}\right)\right] - 1} \tag{1}$$

where h and k are the Planck and Boltzmann constants, c the speed of light, and ν the radiation frequency (s^{-1}).

As seen in Fig. 2, we must account for both the absorption and emission of radiation from atomic and molecular material along an oblique ray path as light passes through the solar atmosphere (*Foukal,* 1990). The change in the beam intensity, $I_\lambda(\theta,z)$, along a slant path of angle θ through a slab of atmosphere of thickness, dz, and monochromatic extinction and emission coefficients $\kappa_\lambda(z)$ and $\varepsilon_\lambda(z)$ can be written

$$I_\lambda(\theta, z + dz) - I_\lambda(\theta, z) = \left[\varepsilon_\lambda(z) - \kappa_\lambda(z) I_\lambda(\theta, z)\right] \sec\theta \, dz \tag{2}$$

or

$$\mu \frac{dI_\lambda(\theta,z)}{dz} = \varepsilon_\lambda(z) - \kappa_\lambda(z) I_\lambda(\theta,z);$$

$\mu = \cos\theta$, the heliocentric angle for the Sun

The optical depth can be defined as $d\tau_\lambda = -\kappa_\lambda dz$ as a measure of the opacity of the medium; κ_λ is the absorption coefficient per unit length (cm^{-1}) which is proportional to the gas density of the material, ρ. The composition and physical properties of the material give the mass absorption coefficient, k_λ, in units of gm^{-1}, so that $\kappa_\lambda = k_\lambda \rho$. In the absence of emission sources the differential equation can be readily solved to give the familiar Lambert absorption law in terms of the emergent $[I(\lambda)]$ and incident intensities $[I_0(\lambda)]$

$$\frac{I(\lambda)}{I_0(\lambda)} = \exp(-\tau_\lambda) \tag{3}$$

A layer of optical depth $\tau_\lambda = 1$ reduces the intensity of a transmitted light beam to $e^{-1} = 0.368$; media with optical depths less that unity are referred to as "optically thin" whereas those that exceed unity are "optically thick." In the case where the medium also emits radiation, this equation can be written in terms of a source function with $S_\lambda = \varepsilon_\lambda/\kappa_\lambda$

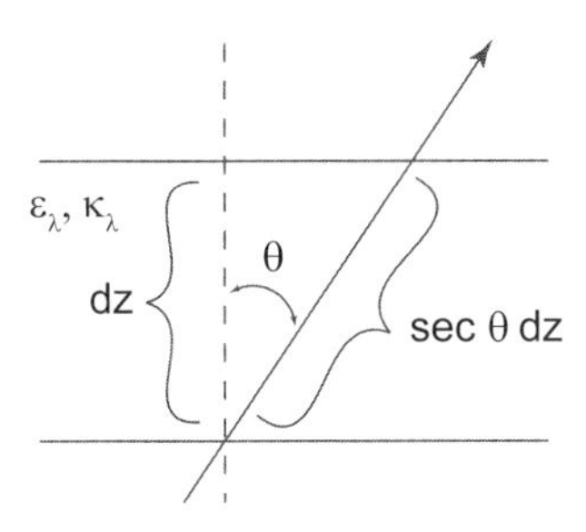

Fig. 2. Geometry of a light ray path (drawn as an arrow) through a slab of atmosphere.

$$\mu \frac{dI_\lambda}{d\tau_\lambda} = I_\lambda - S_\lambda \tag{4}$$

These equations are set up for radiative transfer studies of the Sun where the path through the atmosphere is indexed to the heliospheric angle, but this can be replaced by an integral over a path (l) through the atmosphere by accounting for the altitude-dependent density of the material and the mass absorption coefficient

$$\frac{I(\lambda)}{I_0(\lambda)} = \exp\left[-\int_0^L k_\gamma \rho(l) dl\right] \tag{5}$$

The majority of the radiation emitted from the Sun (Fig. 1a) comes from the photosphere, the lowest layer of detectable radiation that can escape the solar atmosphere; below this level the convection zone is characterized as being optically thick and responsible for the fluid-like character of the Sun giving rise to the granulation structure evident in high-resolution images. Superficially, the photosphere resembles a black-body radiator with an emission temperature of 5770 K, but deviations from this temperature give evidence of the structure of the atmosphere and the sources of opacity that arise from the material composition of the star. Figure 1c shows the most important sources of opacity in the Sun. In the visible and IR regions the opacity is dominated by the negative hydrogen ion (H^-) that reaches local minimum near 1600 nm (*Stallcop,* 1974, and references therein). In the near-UV both neutral metals and the molecular fragments CH, OH, NH, and other metal hydrides play a dominant role. The photospheric layers are defined as the region of the solar atmosphere where the visible continuum is formed. It is customary to define the arbitrary zero height of the atmosphere as the radius at which $\tau_{500} = 1$, where τ_{500} is the continuum optical depth at 500 nm for a radial line of sight. Thus, the photospheric layers extend from about 80 km below to about 300 km above this zero level. Further information about the temperature structure of the solar atmosphere is gleaned from the quantum mechanical properties of the observed atoms and molecules in both absorption and emission with a calibrated spectral radiometer providing in-depth information through the equations of radiative transfer.

The average measured solar disk intensity, I_a, multiplied by the area of the solar disk, A_s, gives the total amount of radiation that the Sun emits toward a unit area collecting surface of a detector, A_c. I_a has units of power per unit source area per steradian per unit spectral bandwidth, such as W m^{-2} $ster^{-1}$ nm^{-1} in MKS units. The radiative flux R (radiated energy per unit area and unit wavelength) that crosses the collecting surface is

$$R = \left[I_a A_s \left(\frac{A_c}{D^2}\right)\right] \frac{1}{A_c} = \Omega I_a \tag{6}$$

where D is the Sun-Earth distance at one astronomical unit (1 AU) and (A_c/D^2) is the solid angle subtended by the collecting area. Since the collecting area appears in both the denominator and numerator, it cancels out to give the simple relationship between irradiance and the average solar disk intensity through the solid angle subtended from the Sun to Earth (Ω). With the measured irradiance of the Sun, the brightness temperature (T_b) can be found from inverting the Planck radiation law and solving for the temperature of the black body (with λ giving the wavelength)

$$T_b = \frac{h\nu}{k}\ln^{-1}\left(\frac{2h\nu^4}{c^2\lambda}\frac{1}{1_a}+1\right) \quad (7)$$

The existence of the chromosphere and the corona (see Fig. 1a) has been known and studied for a long time owing to the remarkable coincidence that the Sun and the Moon have almost exactly the same angular diameters as seen from Earth's surface. The relatively dim emissions from these layers of the Sun were observed when the overwhelming contribution from the photosphere was removed during total eclipses. The low chromospheric layers are defined here as extending from the top of the photosphere (at about 200 km height) up to the temperature minimum at about 800 km height. The lower chromospheric layers are mostly transparent in IR and visible continuum spectra, but they have significant continuum opacity in the near-UV. In particular, several bound-free absorption edges in the 110–400-nm spectral region, particularly for Mg and Al, raises the height at which the continuum radiation forms to near the temperature minimum. The upper chromosphere is the region where the temperature rises from the minimum and reaches an extended plateau with temperatures of 6000–8000 K. The upper chromosphere extends up to heights of 1500–2200 km, and is far more variable than the lower chromosphere, and this variability is strongly linked to magnetic field strength. Solar images in narrowband H_α and extreme UV wavelengths show complicated elongated structures resembling short loops and patterns of dark "rosette" and "spicules" in the magnetic network. In plage structures long "fibrils" connecting regions of opposite magnetic field polarity are evident. In the near-UV, the upper-chromosphere layers produce a complicated mix of absorption and emission lines that are hard to discern in the region from 170 to 200 nm. Space observations show this upper-chromospheric structure in Lyman α, the Mg II h and k emission cores, and the free-bound Lyman, C I, and He I continua (*Varnazza et al.,* 1981). The temperature structure derived from these lines and continua form in low-density layers with large departures, thereby requiring non-LTE calculations to model their radiative response. The transition region of the atmosphere extends from a temperature of about 100,000 K up to coronal values of about 1.5 × 10^6 K; in this region the logarithmic gradient temperature decreases monotonically from the steep gradient to the nearly uniform coronal temperature. This region produces a number of emission lines in the UV and extreme UV (EUV) from elements in a highly ionized state; it also has significant effects on the H and He ionization in the upper chromosphere. The contribution from transition region lines to the irradiance is small, but their large variability has observable effects for Earth's upper atmosphere.

As seen in Fig. 1b, Earth's atmosphere is characterized by regions where there is a sign change in the gradient of temperature ($\Delta T/\Delta z$) with respect to altitude (see *Brasseur and Solomon,* 2005); regions where the sign of the gradient is the same are called the "spheres," as in troposphere, stratosphere, and mesosphere. The regions where the signs reverse are referred to as a "pause," as in tropopause, stratopause, and mesopause. Sunlight penetrating Earth's atmosphere is absorbed at different altitudes, and Fig. 1d shows the optical depth in the visible and near-infrared (NIR) regions as viewed from a spectral radiometer looking up through the atmosphere from the surface. Absorption at short wavelengths is dominated by the very strong absorption due to ozone protecting the biosphere from UV radiation; at longer wavelengths water vapor is the dominant source and shows a more banded structure as a function of wavelength. In the "window" regions where the water absorption is weak, additional absorption occurs due to carbon dioxide that is well mixed throughout the entire atmosphere. The importance of these contributions will be discussed more in section 3 in the context of atmospheric heating rates and the influence of solar variability on the atmosphere.

In the troposphere, air parcels are buoyant and undergo adiabatic expansion with net cooling, and there is a constant decrease in temperature until the tropopause is reached. The highest tropopause height is achieved in the tropics for dynamical reasons, and here the tropopause temperature reaches its lowest values of about 195 K. Above the tropopause the stratosphere warms due to the exothermic reactions that produce ozone. This continues up to the stratopause, where the rate of ozone production is no longer able to overcome the temperature lapse rate of atmospheric cooling. The coldest temperatures in the atmosphere are achieved at the mesopause. In the thermosphere, temperatures increase again with height and can reach values of 500–2000 K depending on the level of solar activity. The composition of the atmosphere finally changes at these altitudes due to significant production of O and N whose concentration becomes comparable to O_2 and N_2 at altitudes of about 130 km.

1.3. Observations of the Solar Spectrum

1.3.1. Groundbased measurements of the solar spectrum. Concerted efforts to construct a radiometrically calibrated solar spectrum began in the mid 1960s with an effort by *Arvensen et al.* (1969, see references therein for the early history of spectral irradiance measurements) covering the range of 300–2400 nm with a variable resolution ranging from 0.1 to 5 nm in the IR, and the calibration of the instrument was tied to the National Bureau of Standards measurements at the time. The measurements were performed from an aircraft platform (Convair CV-990), and solar measurements were performed at varying solar zenith angles and extrapolated to zero air mass via the Langley method. While this resolution was considered acceptable for Earth-atmospheric transmission and radiative transfer studies, advances in solar physics demanded higher spectral resolution to understand the radiative properties of the Sun, so a higher-resolution visible spectrum was constructed using the Fourier Transform spectrometer at the Kitt Peak National Observatory in Arizona. *Neckel and Labs* (1984) constructed a calibration curve to apply to the high-resolution spectrum valid in the 330–660-nm region to about a 1.5% absolute accuracy over that spectral range. The calibration is valid except in the neighborhood of telluric lines, predominantly due to water vapor. Based on this effort and including concurrent spacebased observations in the UV, *Kurucz* (1992a,b,c) constructed an extra-atmospheric spectrum that has been used in many studies of atmospheric transmission and as a basis for estimating solar irradiance variability. Additional high-quality information about the UV portion of the spectrum was obtained by *Anderson and Hall* (1989) and *Hall and Anderson* (1991) from a balloon-borne spectrometer with a resolution of 0.01 nm operable in the 200–310-nm spectral range. These measurements were corrected for the strong atmospheric absorption of ozone in the Hartley-Huggins bands, important throughout the full measurement range, and oxygen in the Schumann-Runge bands in the 193–210-nm range. In spite of the difficulties of determining the absolute radiant output of the Sun from ground, aircraft, and balloon platforms, these measurements remain important even in the time period of high-quality spacebased measurements mostly because of the quality of the wavelength calibration, and, particularly in the case of the Kitt Peak atlas, the high spectral resolution over broad wavelength ranges remains a standard against which spacebased instrumentation and spectral synthesis models can be compared.

1.3.2. Spacebased measurements of the solar spectrum. Measurements of the solar spectrum from spacebased platforms date from the earliest years of the space program, starting with sounding rocket measurements of the solar UV radiation (*Baum et al.,* 1946). Measurements from the Orbiting Solar Observatory (OSO) emphasized the important X-ray and EUV portions of the spectrum over the course of eight flights dating from 1962 to 1975. These experiments were continued and extended into the Skylab era (*Reeves et al.,* 1977). By the late 1970s–early 1980s, dedicated missions began monitoring long-term changes in the solar irradiance, particularly in the 110–400-nm spectral range because of its importance to stratospheric ozone chemistry. The NOAA SBUV series of experiments (*Cebula et al.,* 1991) were started in 1978 to monitor global total ozone, but had a mode to measure the daily solar spectrum from backscatter off a diffuse scattering screen at the entrance aperture of the spectrometer. Changes in the solar spectrum were determined from wavelength-dependent scaling factors derived from the Mg II core-to-wing (*Heath and Schlesinger,* 1986). Beginning on October 13, 1981 (solar cycle 21) a two-channel spectrometer onboard the Solar Mesosphere Explorer (SME) performed daily measurements of full disk solar irradiance. These observations cover the spectral interval 120 to 305 nm with spectral resolution of about 0.75 nm and a relative accuracy of ~1%. In the early 1990s (solar cycles 22 and 23), the launch of the NASA Upper Atmospheric Research Satellite (UARS) employed two independent instruments to measure the UV component of solar spectral irradiance. The Solar Ultraviolet Spectral Irradiance Monitor (SUSIM) (*Brueckner et al.,* 1993) and the Solar Stellar Irradiance Comparison Experiment (SOLSTICE) (*Rottman et al.,* 1993) were spectrometers designed to directly address issues related to instrument degradation observed in the SBUV and SME measurements. Degradation corrections for SUSIM were done by using a series of interchangeable calibration lamps, optical elements, and detectors to construct reference spectra that could be used to correct the working daily channel (*Brueckner et al.,* 1993). For SOLSTICE an ensemble of highly stable bright UV stars were used as standard light sources, interchanging entrance apertures and using photon-counting photomultipliers as detectors to bridge an 9-order-of-magnitude difference in signal strength to allow the comparison of the Sun to these UV stars (*Rottman et al.,* 1993). In the time period from 2003 to present (solar cycles 23 and 24) the NASA Solar Radiation and Climate Experiment (SORCE) extended the UARS SOLSTICE experiment (*McClintock et al.,* 2005), and the Spectral Irradiance Monitor (SIM) was developed to replace and extend the UARS SOLSTICE N-channel spectrometer that was used to study solar variability in the 280–420-nm region. The SIM instrument covers a much wider spectral range (200–2400 nm) and uses an electrical substitution radiometer (ESR) as its primary detector in the 265–2400-nm range (*Harder et al.,* 2005a,b). The SIM instrument consists of two identical mirror image spectrometers of the Fèry prism design. A Fèry prism has a concave front surface and a convex back surface that allows focus and dispersion with only one refractive optical element; this is an advantage over most spectral radiometers, which are based on grating designs requiring multiple optical elements. The two spectrometer channels are exposed to solar radiation at different rates throughout the mission; the different exposure rate permits the determination of the time and wavelength dependencies of the degradation function [see the auxiliary material of *Harder et al.* (2009) for a discussion of this analysis]. Each SIM spectrometer employs four detectors in its focal plane: n–p silicon photodiodes are used to record the UV and visible wavelengths (200–308 and 310–950 nm, respectively), an InGaAs photodiode covers the 950–1620-nm range, and the ESR covers the 258–2423-nm range. The ESR is a thermal detector with inherently long integration times, so the fast-response photodiodes are used twice daily to acquire the spectrum and the radiation hard ESR measurements are used to recalibrate the radiometric sensitivity of the photodiodes over their operating ranges and to provide a daily 1620–2423-nm data product. Overlapping with the

SORCE experiment is a measurement of solar variability from the European Space Agency (ESA) SCanning Imaging Absorption spectroMeter for Atmospheric CHartographY (SCIAMACHY) (*Pagaran et al.,* 2011), which operated from 2002 until 2012 but lacked a method to internally measure instrument degradation.

Through a series of well-documented calibration papers, Thuillier and co-workers (*Thuillier et al.,* 2004, and references therein) have presented the calibration and flight observations of the SOLar SPECtrum (SOLSPEC) instrument for its Spacelab (1982), ATmospheric Laboratory for Applications and Science (ATLAS) missions 1–2–3 flights (1992–1994), and the SOlar SPectrum (SOSP) instrument on the EURECA mission (1993–1994). For these ATLAS experiments, comparisons of the space-shuttle-based instruments SOLSPEC, SSBUV, and SUSIM were performed relative to the operating satellite instruments SBUV, UARS SUSIM, and UARS SOLSTICE (*Woods et al.,* 1996). This intercomparison activity established the SOLSPEC spectrum as the *de facto* standard for solar spectral irradiance in the 400–2400-nm range and serves as the solar cycle 22 solar minimum reference spectrum. The differences between ATLAS 1 and 3 are confined to wavelengths less than 410 nm and the visible and IR portions of the two composites are identical. Differences for the UV portion of the spectrum for the two composites are mostly due to solar variability between the high solar activity ATLAS 1 time period (March 1992) and the quiet Sun ATLAS 3 period (November 1994). A similar comparison between the ATLAS 3 reference spectrum and the SORCE SIM and SOLSTICE experiment was reported by *Harder et al.* (2010). *Woods et al.* (2009) produced the Solar Irradiance Reference Spectrum (SIRS), a solar cycle 23 solar minimum spectrum analogous to the ATLAS 3 spectrum through the activities of the Whole Heliosphere Interval.

Routine monitoring of the EUV portion (0.1–120 nm) of the spectrum began with the NASA Thermosphere Ionosphere Mesosphere Energetics Dynamics (TIMED) Solar EUV Explorer (SEE) instrument (*Woods et al.,* 2005). Prior to this time, much of the understanding of the EUV portion of the spectrum was based on observations from the Atmosphere Explorer E (AE-E) satellite experiment (*Hinteregger et al.,* 1981) and rocket underflights to maintain the absolute irradiance scale of the instrument. With the termination of the AE-E measurements in 1981, routine daily measurements did not resume until the launch of TIMED SEE in 2002. This long gap in the data has been referred to in the literature as the "EUV Hole" and the routine measurements from SEE have ameliorated this difficulty. The TIMED spacecraft with both Earth-viewing and Sun-viewing requirements had only a 3% measurement duty cycle, thereby underestimating the role of important solar flare events, so the key data product from SEE is the solar cycle variability of the descending phase of solar cycle 23. Multiple flare events were captured by SEE, but a full account of the irradiance in this portion of the spectrum with information about flare events could not begin until the launch of the Solar Dynamics Observatory (SDO) Extreme Ultraviolet Variability Experiment (EVE) (*Woods et al.,* 2012). The EVE instrument covers the 5–105-nm range with three spectrometer channels with a spectral resolution of 0.1 nm and a cadence of 10 s from a geostationary orbit. Complete coverage of flare events from this instrument supersedes the Solar and Heliospheric Observatory (SoHO) Solar EUV Monitor (SEM) (*Judge et al.,* 1998) and the GOES XRS (*Garcia,* 2000), which had high cadence for flare detection but with the limitations of broadband filters. The measurements of the EUV spectral region have very practical use for space weather applications and understanding the thermospheric and ionospheric processes of Earth (*Woods et al.,* 2005). Understanding the variability in this portion of the spectrum is also of great importance to the aeronomy of Mars and Venus since it provides the energy needed to initiate photochemical reactions (*Bauer and Taylor,* 1981; *Bauer and Hantsch,* 1989).

Figure 3a is the composite solar minimum 23 spectra acquired during the Whole Heliosphere Interval (*Woods et al.,* 2009), and attempts to give an idea of the highly structured nature of the solar spectrum spanning a range of about 7 orders of magnitude in intensity and about 3 orders of magnitude in wavelength. Figure 3b is from the far UV channel of SORCE SOLSTICE at a resolution of 0.1 nm with a detail of the Lyman α in Fig. 3c. Similarly, Fig. 2d is the SORCE SOLSTICE middle-UV channel, showing a detail of the important Mg II feature with the h and k chromospheric emission lines identified (Fig. 3e). Figure 3e also shows the high-resolution spectrum convolved with a 1.1-nm full-width half-maximum (FWHM) triangular function to simulate the response function of the NOAA SBUV instruments (dashed line). Also identified are the wavelengths used to produce the Mg II index (*Snow et al.,* 2005). Figure 3f is a high-resolution spectrum taken from the Kitt Peak Fourier Transform interferometer indicating the deep and highly structured Fraunhofer absorption lines of the solar photosphere. Figure 3g details the Ca II K absorption feature. Variable structure in the bottom of this solar line produces bright emissions that are discernible in groundbased images of the Sun, which is discussed in the next section.

Considering all the instruments discussed in this section, the absolute irradiance scales over this broad wavelength range are valid to only a few percent. The instrumentation used to acquire solar data represents a compromise between resolution, absolute irradiance, and acquisition time in all spectral regions. Most observations of the Sun do not completely resolve its spectral structure, and issues related to wavelength registration, scattered light rejection, and incomplete specification of instrument profile and response are not uniformly reported throughout the literature. As discussed in a later section, the ability to correct for inflight response degradation for spacebased instrumentation is also a critical matter. These shortcomings have necessitated highly detailed intercomparison studies, and the results have been widely published. In spite of these shortcomings, the efforts to measure the solar spectrum have provided a sound basis for understanding the nature of not only of our own star but also the physical understanding of other stars and the development of modern astrophysics.

2. ORIGINS OF SOLAR ACTIVITY AND IRRADIANCE VARIABILITY

In section 2.1 we will describe the nature and time evolution of magnetic structures that are observable in a wide range of solar images made from both groundbased and satellite platforms. Aided by the discussion of these solar surface features, the discussion of section 2.2 will then turn to measurements of solar variability that provide the longest record of the solar output and give the most reliable indicator of how solar variability enters into the Earth climate system. Section 2.3 is a discussion of how radiant properties of these solar surface features can provide a basis for calculating the solar irradiance over the course of the solar cycle, which can then be compared with observations.

2.1. Role of Surface Magnetism and Identification of Solar Surface Features

From the perspective of solar spectral irradiance, the starting point for understanding the origins of variability begin with the series of highly cited papers by J. Vernazza, E. Avrett, and R. Loeser (VAL) (*Vernazza et al.,* 1973, 1976, 1981). In a historical sense, the VAL models were refinements, based on improved observations from the Harvard-Smithsonian Reference Atmosphere (*Gingerich et al.,* 1971) and the earlier Bilderberg Continuum Atmosphere (*Gingerich and de Jager,* 1968). For the VAL models, a one-dimensional model of the solar atmosphere that included the photosphere, chromosphere, and transition region was produced for six identifiable magnetic features observable in solar images. In each case, the model atmosphere was determined by using non-local thermodynamic equilibrium (non-LTE) radiative transfer, statistical equilibrium, and hydrostatic equilibrium equations to determine the

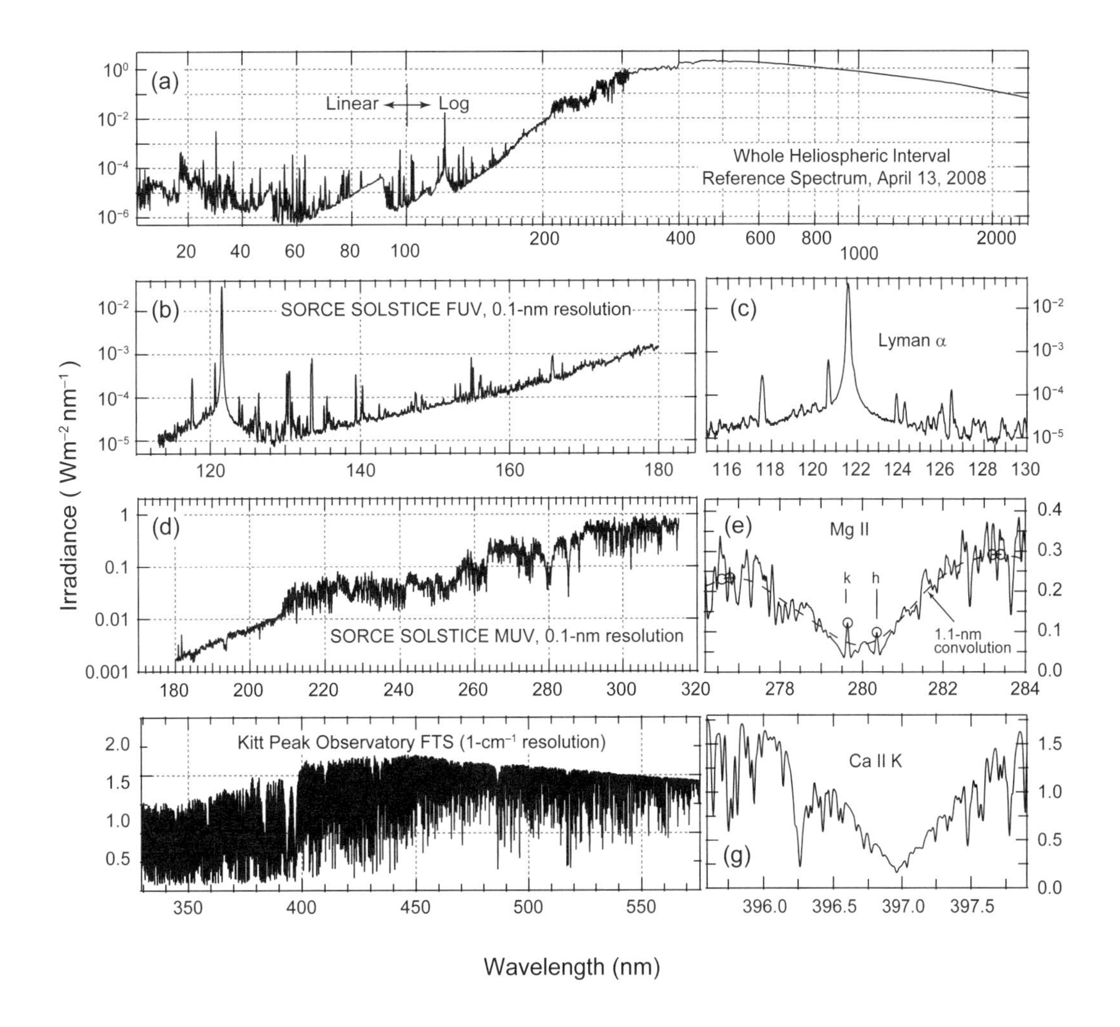

Fig. 3. Representative spectra in the ultraviolet and visible portions of the spectrum; **(e)** shows the wavelength region of the Mg II. The dashed line represents a 1.1-nm convolution of the 0.1-nm resolution SORCE SOLSTICE spectrum, and the circled data points are the ones used to calculate the Mg II core-to-wing ratio proxy.

temperature-density stratification that produces a calculated spectrum that matched the Skylab observations of the quiet Sun in the EUV wavelength range of 40–140 nm (*Reeves et al.,* 1977). Subsequent work by J. M. Fontenla, Avrett, and Loeser (FAL) (*Fontenla et al.,* 1990, 1991, 1993) improved the atmospheric models of the solar transition region by explaining the origin of the observed UV line emission of the Lyman α spectrum (121.6 nm) by including ambipolar diffusion in the energy balance of the lower transition region.

As indicated in the VAL papers and subsequent publications, solar structures identifiable in images have different atmospheric properties — i.e., vertical profiles of temperature, neutral density, electron and proton densities, etc. — and therefore a different radiative output. These solar features have the classification listed in Table 1.

Analysis of these solar structures has proceeded from groundbased studies of images at selected wavelengths with relatively narrow bandpass filters near the Ca II emission feature at 393.4 nm along with other images in H_α (656.46 nm) and other continuum wavelengths. An example of the content of a solar active region is shown in Fig. 4 for June 7, 2006, a time of moderate activity during the descending phase of solar cycle 23. The data source for these images is the Precision Solar Photometric Telescope (PSPT) (*Rast et al.,* 2008) and it shows the same active region as seen through three different filters. Below the full-disk images are magnifications of the active region using three different filters: a filter at 607.1 nm (Fig. 4a) measuring clean solar continuum to identify sunspot umbra and penumbra (models S and R), a broadband Ca II K filter (Fig. 4b) to identify bright active regions (F, H, P), and a narrow-band Ca II K filter (Fig. 4c) that has sufficiently high contrast to distinctly show differences between the quiet Sun (B) and quiet network (E).

The interpretation of the photometric information of active regions is reinforced by concurrent magnetograms (Fig. 4d), in this case from the SoHO Michelson Doppler Imager (MDI) instrument (*Scherrer et al.,* 1995). The following paragraphs describe the solar features identified in Table 1 and shown in Fig. 4, aided by the exceptional descriptive narratives of the evolution of solar features presented in *Zwaan* (1985, 1987). The existence of all the magnetic surface structures described here result from the emergence of twisted magnetic flux elements into the photosphere from subsurface motions that generate toroidal magnetic flux ropes. After these flux ropes emerge to form active regions, they ultimately produce filaments and coronal helmets in the upper part of the solar atmosphere (*Rust,* 2001). Figure 5a shows a greatly simplified cartoon of the process, but an examination of recent movies of the emergence of magnetic flux elements from the SDO Atmospheric Imaging Assembly (AIA) instrument (*Lemen et al.,* 2012) indicates the rich and complex nature of this process as depicted in Fig. 5b. Video clips of this process are updated frequently throughout the mission and can be found at *sdo.gsfc.nasa.gov.*

TABLE 1. Solar surface features observable in solar images.

Model Name	Description
C (or B)	Average supergranule cell interior (quiet Sun)
E (or D)	Average network
F	Bright network/faint plage
H	Average plage
P	Bright plage
S	Sunspot umbra
R	Sumspot penumbra

Plage (P) is specifically the brightest facular region associated with a magnetic bipolar active region. Active regions also contain one or more sunspots. Active region sizes range from 20,000 km to 100 megameters (Mm). The number and extent of these active regions varies with the solar activity cycle; they frequently tend to be isolated regions on the solar disk and with a synodic rotation period of 27.2753 days these regions are identifiable in irradiance measurements (both total and spectral) because they modulate the Sun's radiated output. The red continuum image in Fig. 4a and magnetogram in Fig. 4d show the presence of two isolated sunspots (S) surrounded by penumbral (R) regions within this active region. The continuum images can be used to define the specific area of sunspots because the contrasts between facular regions and the quiet Sun is very low at these wavelengths while contrast with sunspots is large. The brightest portions of the active region surrounding the sunspots is identified as plage (P) in the chromospheric Ca images (Figs. 4b,c), and appears as saturated black-and-white areas of bipolar structure in the MDI images. The MDI images are produced from the observer's line-of-sight component of circularly polarized light either pointing into the Sun (coded in black) and out of the Sun (areas in white); the gray portion corresponds to regions of weak magnetic field strength of less than the 20 Gauss (G) noise limit. MDI uses a refracting telescope to feed sunlight through a series of filters to produce images of the Sun with a 94 mÅ bandpass centered on the Ni I 6768-Å solar absorption line. A magnetogram is constructed by measuring the Doppler shift separately in right and left circularly polarized light. The difference between these two is a measure of the Zeeman splitting and can be calibrated to give the magnetic flux density (*Scherrer et al.,* 1995) over about 3 orders of magnitude of field strength. A contour of magnetic flux levels of 50 G defines a suitable outline for an active region; however, within that contour the flux density varies widely but the mean flux density is about 100 ± 20 G regardless of the size or age of the active region. At the resolution of these images plage regions appear as unstructured bright areas of size similar to supergranular dimensions. Supergranules that are apparent in both intensity and magnetic field strength (a clearly apparent example is labeled SG in Fig. 4b) are attributed to large-scale convection cells (*Leighton et al.,* 1962), and the separation between the plage regions and supergranules indicates that the interaction between convection and magnetic field is reciprocal, so field and convection tend to expel each other. When the mean flux density is in excess of about 100 G, supergranular cells cannot develop. Another manifestation of this same phenomena occurs in the so-called "moat cell," which may surround part of a mature sunspot (*Harvey and Harvey,* 1973). The interaction between the active region and surface and subsurface processes is responsible for much of the distribution of photospheric magnetic flux (*Schrijver,* 1989; *Wang and Sheeley,* 1991). For example, after an active region emerges onto the surface, differential rotation, meridional flow, supergranular convection cell formation, and chance encounters between other magnetic flux elements act to fragment and disperse an active region's magnetic flux. Fragmentation of active regions results in the formation of other solar features as discussed next.

Enhanced network or facula (H) arises from the cancellation of magnetic flux along a magnetic neutral line within a bipolar active region and appears to separate the active region into two separate regions (*Schrijver,* 1989). Each subsequent region is primarily of one or the other magnetic polarity. Sunspots have mostly dissipated, and supergranular convection is no longer completely suppressed within the region. However, the resulting structure is still large (with sizes about half that of active regions), compact, and bright in chromospheric images (*Worden et al.,* 1998). These particular structures are defined as enhanced network (*Zwaan,* 1987; *Wang et al.,* 1991).

Active network (F) is produced when supergranular convection cells form randomly on the surface; they have diameters between 15,000 and 24,000 km (*Hagenaar et al.,* 1997; *Berrilli et al.,* 1998), and deform or dissipate approximately every 24 to 48 hr (*Wang et al.,* 1991). The flows associated with supergranular convection cells merge and fragment magnetic concentrations (e.g., *Leighton,* 1964; *Schrijver et al.,* 1998). Supergranular convection cells appear to be suppressed in active regions and mildly suppressed in enhanced network *(Zwaan,* 1987; *Schrijver,* 1989; *Wang et al.,* 1991). Despite the suppression of supergranular convection within large-scale magnetic regions, supergranular flows can still remove small-scale magnetic flux concentrations from the edges of the larger, compact regions (*Leighton,* 1964; *Schrijver,* 1989). These small-scale (approximate diameters of 15,000 km) bright features are called active network by *Lean et al.* (1982) and *Skumanich et al.* (1984). Like the enhanced network, they are primarily of one magnetic polarity or the other. Transition region emissions, such as from the He II 30.4 nm, are sometimes used to identify the active network because the contrast is stronger for active network in transition region emissions. Chromospheric and coronal emissions can also be used to identify the active network, but it is challenging to identify active network in photospheric images. The migration of the active network to the poles (via meridional flow and the random walk diffusion associated with the supergranular cells) is thought to be responsible for reversing the magnetic polarity of the solar polar magnetic fields (*Leighton,* 1964) every solar cycle.

Sunspot umbra and penumbra (S and P) characteristically show reduced temperature and intense magnetic fields. Following the descriptive commentaries of *Parker* (1974, 1979), it is observed that sunspots are assembled over a

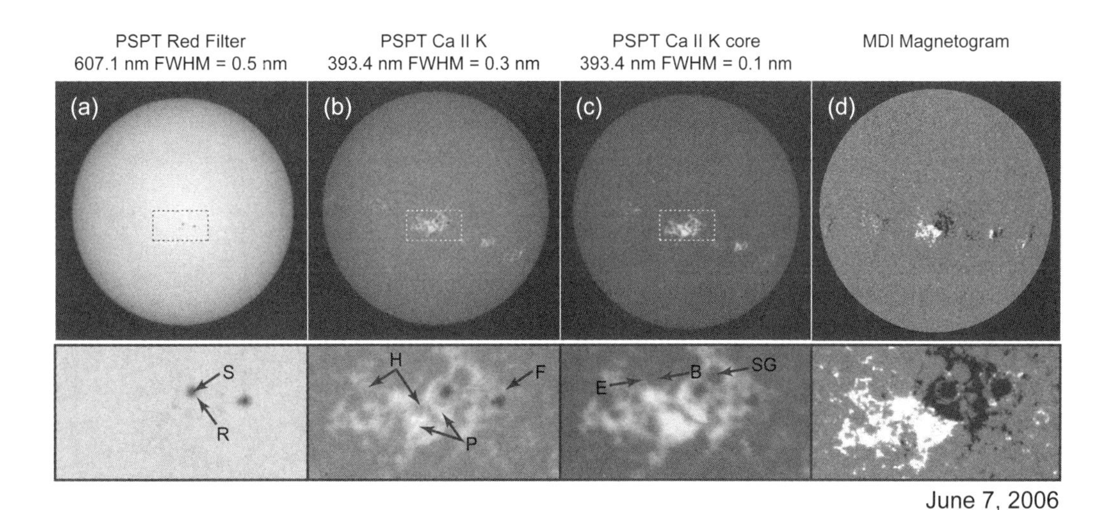

Fig. 4. A comparative study of the content of an active region during the descending phase of solar cycle 23. Below the full-disk image is a detail of the active region outlined in the full-disk image. The active region contains two small sunspots with surrounding facular and plage contributions that are observable in three different PSPT filter bandpasses [**(a)**–**(c)**] with a SoHO MDI magnetogram in **(d)** to show the extent of the magnetic field that corresponds to what is seen in the images. The PSPT images are available at *lasp.colorado.edu/pspt_access/* and MDI images available at *soi.stanford.edu/data/.*

period of hours to days through the progressive gathering of multiple small flux tubes prior to the coalescence into a single large tube. These small flux tubes are observed as the magnetic knots in the solar photosphere with individual fluxes of about 10^{19} Maxwells (1500 G across a diameter of 1000 km). The coalescence of three or four magnetic knots is sufficient to produce a visible cool area, referred to as a pore, on the surface of the Sun with a diameter of 1500–2000 km and a field intensity of 1500 G. Pores are short-lived and decompose within a few hours back into individual magnetic knots. In a small fraction of cases the magnetic knots continue to move up to the pore, merging with it and increasing its size. The pore grows into a sunspot with the accumulation of about 30 magnetic knots (~4×10^{20} Maxwells), forming a single tube of 3×10^3 G across a diameter of 4000 km. The temperature of the visible sunken surface of the sunspot is then only 4×10^3 K, so that the radiative energy flux from the sunspot is about one-fifth the intensity of the normal photospheric quiet Sun (6×10^{10} ergs cm^{-2}). At this point, the spot begins to show the surrounding filamentary penumbral structure.

Further accumulation of the magnetic knots may occur; the spot can then grow to a diameter of 6×10^4 km or more in extreme cases. The field intensity and temperature at the visible surface remain near 3×10^3 G and 4×10^3 K. Growth of the sunspot may continue so long as new flux (in the form of the small individual tubes) continues to emerge through the surface of the Sun. When new flux ceases to arrive at the surface, the growth ceases and the spot begins to erode away at its edges into the small flux tubes again. In a small fraction of the cases, however, the decomposition is enormously delayed, the spot taking on a circular form surrounded by a field-free ring, or "moat." The field-free moat is produced by a radial outflow of gas from the spot at the visible surface of the Sun. The spot may survive this form for one or more rotations of the Sun before slowly decomposing back into individual magnetic knots. Photospheric images are used to identify the dark sunspots because the upper solar atmosphere layers are normally bright above the sunspots.

The convective transport of heat up to the surface of the Sun is strongly inhibited by the nearly vertical magnetic field within the sunspot, thereby providing a direct explanation for the reduced temperature at the visible surface. Apparently, a field of 3×10^3 G would reduce the convective heat flux to very low levels, raising the question as to why the observed energy flux is as large as one-fifth the normal photospheric value. It was recognized that the emerging energy flux would be channeled around the sunspot, and appearing at the surface of the Sun as a bright ring around the sunspot. A conclusive observation of this phenomenon did not occur until 1999; *Rast et al.* (1999, 2001) demonstrated that these rings are about 0.5–1.0% brighter in red and blue continuum (10 K warmer) than the surrounding photosphere and extend about one sunspot radius outward from the outer penumbral boundary. Most of the excess radiation is not directly associated with the strongest regions of Ca II K emission surrounding the spots or with measurable vertical magnetic fields.

A comprehensive theory explaining the formation of sunspots has not emerged until very recent times. *Rempel et al.* (2009) presented this model and explained that our physical understanding of sunspot structure has been hampered for decades by (1) insufficient observational resolution of the fine structure, (2) lack of information about how the atmospheric layers below the visible surface contribute to

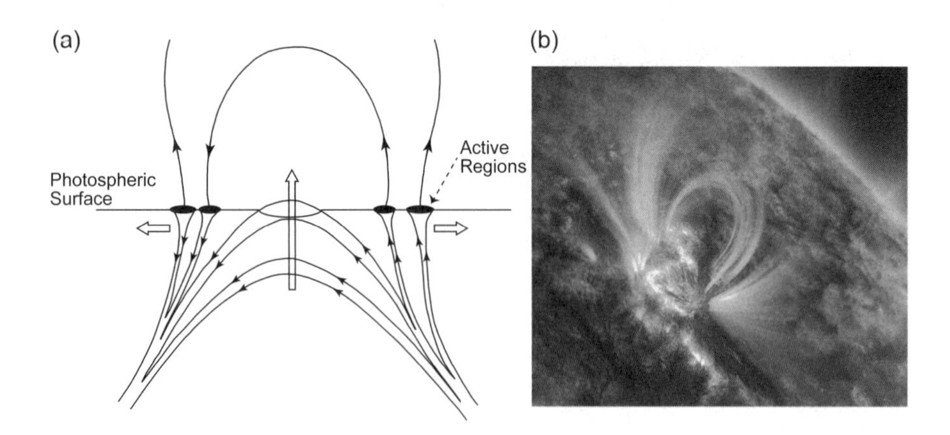

Fig. 5. (a) Cartoon showing the emergence of magnetic flux, the separation of polarities, and coalescence of sunspots. Unfilled arrows indicate local displacements of flux tubes. Real emerging flux regions consist of many more separate flux loops, as indicated in **(b)**, an image from the SDO AIA imager at 171 nm that is indicative of the solar coronal region rather than photosphere and therefore does not show the underlying dark sunspot regions that are best observed in the visible spectrum, as seen in Fig. 4. **(a)** Adapted from *Zwaan* (1985); **(b)** from *sdo.gsfc.nasa.gov*.

their formation, and (3) insufficient computational power to perform *ab initio* three-dimensional MHD simulation of a full sunspot that includes the surrounding granulation. Progress on all three of these fronts has been made in recent years because of (1) improvements in adaptive optics, image selection, and reconstruction for groundbased telescopes, and the advent of spectropolarimetry in the visible from space as observed by the Hinode satellite (*Ichimoto et al.,* 2008); (2) advances in local helioseismology to probe the subsurface structure of sunspots (*Cameron et al.,* 2008); and (3) the ever-increasing computational power of parallel computers, which has finally made *ab initio* simulations of full sunspots a viable activity.

2.2. Observations of Total Solar Irradiance and Solar Spectral Irradiance Variability Through Proxies of Solar Variability

Our understanding of the Sun-Earth connection relies on records of a sufficient length that secular trends in the Sun can be compared to changes in Earth's climate. On climate-length records, direct observations of solar irradiance are limited to a record commensurate with the length of the space age. Certainly a number of valuable proxies have been provided and these are discussed here. By far the longest record of direct solar observations is the sunspot record. J. A. Eddy made one of the strongest cases that the variability of the Sun enters into Earth's climate in his landmark 1976 paper describing the line of evidence that the mid-17th century Little Ice Age coincided with a time period of very low sunspot number (*Eddy,* 1976). Eddy's paper also recounts the very rich and interesting history of the sunspot record. This sunspot number represents one of the most critical climate data records and the longest direct record of the role of the Sun in Earth's climatic response. This sunspot data can be found at *spidr.ngdc.noaa.gov/spidr/dataset.do*. The sunspot record dating back to the year 1610 is shown in Fig. 6a.

The occurrence of sunspots in the visible provides an indication of photospheric activity on the Sun, but important multicycle indicators of chromospheric and coronal origin must also be considered. Foremost of these is the F10.7-cm radio flux (*Tapping,* 1987, and references therein) that provides an indicator of solar activity back to 1947. The solar centimetric emissions, like the F10.7-cm flux, are not uniformly distributed over the solar disk; there is a component associated with active region structures, and a background that may also vary over the disk, thereby showing rotational modulation. Underlying these nonuniform contributions is a steady level that is associated with a quiet Sun component. Free-free (bremsstrahlung) emission and gyroresonant absorption are the two physical mechanisms invoked to describe the solar microwave emission, and both of these processes are based upon characteristics of the thermal plasma. The F10.7 data is available via FTP from *ftp://ftp.ngdc.noaa.gov/STP/SOLAR_DATA/SOLAR_RADIO/FLUX/Penticton_Observed/daily/DAILYPLT.OBS*.

Additional monitors of solar cycle response come from satellite-based observation of the Lyman α (Fig. 2c) (*Woods et al.,* 2000) characteristic of the upper chromosphere and transition region and the Mg II index (Fig. 2e) (*Viereck et al.,* 2004; *Snow et al.,* 2005) for the lower chromosphere. These monitors can be constructed from observations back to 1978, and *Woods et al.* (2000) explain how the Lyman α measurement can be extended back to 1947 through a proxy-model based on the coronal F10.7 radio flux measurement. Composite Lyman α and Mg II core-to-wing data is available at *lasp.colorado.edu/lisird/mgii*. These F10.7 and Lyman α measurements and the Mg II proxy are shown in Figs. 6b–d.

Concurrent with the observations of bright UV observations of the chromosphere and transition region are the measurements of the total solar irradiance (TSI) based on the principle of electrical substitution radiometry (ESR). The physical basis of the ESR is rooted in the notion of equivalence between the radiant power emitted from a light source (such as the Sun) and power generated from electrical resistors (*Hengstberger,* 1989). Historically, the first radiometers date back to the concurrent efforts of F. Kurlbaum and K. Ångstrom in the late 19th century to measure the Stephan-Boltzmann constant in the laboratory and the total power emitted from the Sun respectively. Adaptation of this technology for spaceflight measurements of the TSI began in the late 1970s by *Willson* (1979) using an Active Cavity Radiometer (ACR) flown on the Solar Maximum Mission (SMM) (1980–1989) and by *Hickey et al.* (1988, and references therein) using a thermopile-based radiometer on the NIMBUS-7 satellite; these experiments initiated the first solar cycle length record of the TSI (*Willson and Hudson,* 1991). This record is continued to the present time with more advanced radiometers based on the ACRIM design (*Willson,* 1979), the VIRGO DIARAD and the PMOV6 instruments on SoHO (*Fröhlich,* 2003; *Mekaoui et al.,* 2004), and the SORCE Total Irradiance Monitor (TIM) instrument. Since 2003, the SORCE TIM instrument (*Kopp and Lawrence,* 2005) set a new standard for TSI measurements because it employed Ni phosphorous black as the light-absorbing material in the radiometer cone along with phase sensitive detection to eliminate thermal drifts in the observations.

Figure 7 shows the accumulated record of TSI measurements since the NIMBUS-7 measurement started in 1978 with their reported native calibration scales; note that the spread in the TSI values shown on this plot exceeds a 1000-ppm change — about the expected level of change over a solar cycle. Until the launch of the SORCE TIM instrument the accepted value of the solar cycle minimum TSI value was about 1366 W m^{-2}; *Kopp and Lean* (2011) report on the laboratory investigations used to demonstrate that this lower TSI value is a more plausible result. The dominant source of discrepancy between these different instruments is related to the contribution from light scattered from the view-limiting baffles entering the radiometer. An ACRIM spare radiometer and a PREMOS preflight instrument were also compared at the TSI Radiometer Facility (*Kopp et al.,* 2012) and now report lower TSI values commensurate with the SORCE TIM. The accumulated set of missions that constitute the TSI record require detailed analysis of instrument performance to construct a composite TSI record. From the set of measurements depicted in Fig. 7, three composite time series have been constructed, referred to as the PMOD (*Fröhlich,* 2009), ACRIM (*Willson and Mordvinov,* 2003), and IRMB (*Dewitte et al.,* 2004) composites. Differences between the three composites stems from the manner in which the gap between the end of ACRIM I and the beginning of ACRIM II observations are bridged and the methodologies used to construct the composites are described in detail in these documents. In the two most commonly cited studies, the ACRIM composite indicates that the TSI trend between minima during solar cycles 21–24 approaches the next minima at a rate of +0.008% per decade, whereas the PMOD composite TSI trend between minima during solar cycles 21–24 approaches next minima at –0.011% per decade. With the improved accuracy and agreement between TSI instrumentation leveraged during the SORCE era, the ability to establish a solar cycle 24 minimum value should facilitate a decision regarding which of these three composites produces the most plausible long-term trend in the TSI. The PMOD and the ACRIM composites are shown in Fig. 6e along with the other records of variability for comparison.

2.3. Measurements of the Solar Spectral Irradiance Time Series in the Ultraviolet/Visible/Infrared

Measurements of the spectral irradiance variability using instruments with the internal capabilities to correct for inflight instrument degradation induced from the harsh space environment have been conducted over the last 30 years, with the majority of the effort centered on the UV portions

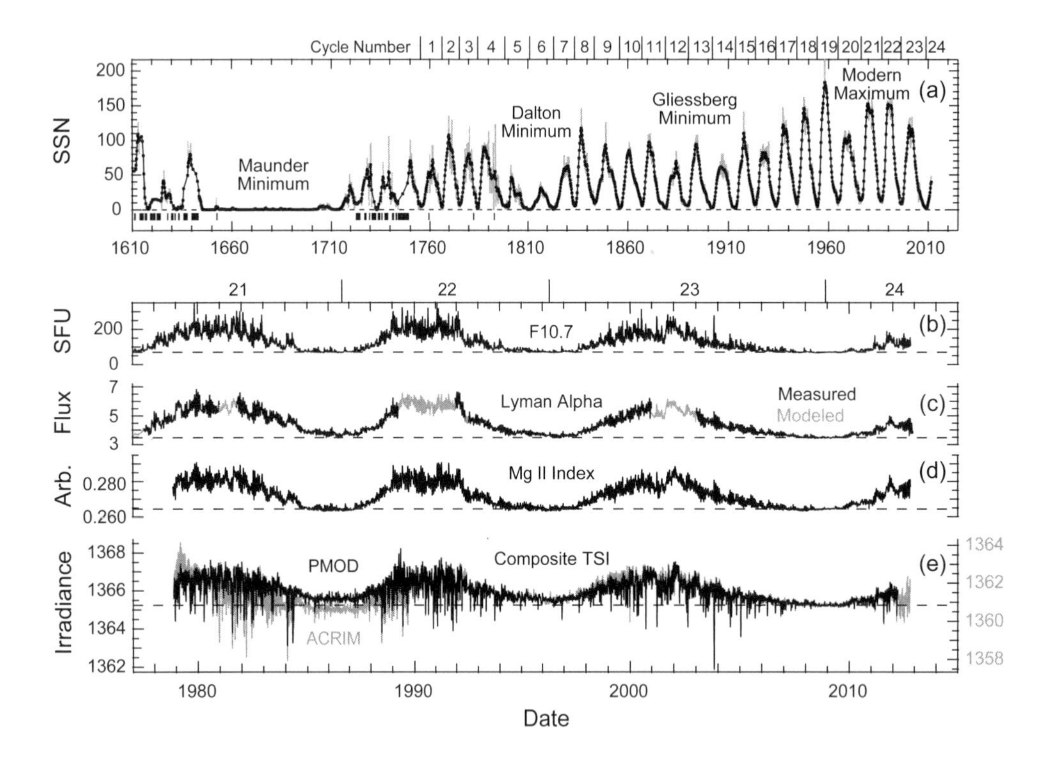

Fig. 6. Established monitors of solar variability. **(a)** Historical record of sunspot group number; also identifies the cycle number and named maximum and minimum time periods. Hash marks below the zero line indicate 81-day time periods with no data and the gray trace shows the 81-day average; the black trace is a smoothed version of the 81-day average. Over the modern space era (1978–present, solar cycles 21–24), **(b)**,**(c)**,**(d)** show the coronal and chromospheric monitors for F10.7-cm radio flux (units of solar flux units, SFU, 10^{-22} W m^{-2} Hz^{-1}), measured Lyman α flux (photons s^{-1} cm^{-2} nm^{-1}), and Mg II core-to-wing ratio (arbitrary units). In **(c)**, missing Lyman α data are filled in with a proxy based on Mg II and/or F10.7. **(e)** Most commonly reported TSI composites, PMOD (lefthand axis) and ACRIM (righthand axis, gray trace). The graph scale for each axis spans ±3.5 W m^{-2}, but note the different reported absolute values; the two time series are aligned to coincide at the solar cycle 23 minimum to enhance the differences in the two time series in cycles 21 and 22.

of the spectrum owing to its importance to understanding solar cycle effects on the photochemistry in Earth's atmosphere, such as ozone change. More recently for Earth climate change studies, the SORCE SSI instruments have been providing daily SSI measurements since 2003 in the UV, visible, and near-IR (NIR). Like the long-standing TSI record spanning several solar cycles, attempts to construct a full spectral irradiance composite time series at any given wavelength has been hampered by varying absolute calibration and inconsistent degradation corrections; this situation is particularly true for the mid-UV spectral range, and for the visible and IR spectral ranges. The SORCE SIM is the only instrument reporting full-wavelength-range spectral variability based on instrument-only corrections. Furthermore, consider that spacecraft subsystems (such as pointing capabilities, data recorders, battery operations, etc.) have a finite lifetime and these failures have contributed to termination of the data records. Figure 8 gives an example of the situation in the critical 230–250-nm range that has significant contributions to the interpretation of solar-cycle-related trends in ozone. This figure shows the time series as measured by SME, NOAA SBUV, UARS SOLSTICE and SUSIM, and SORCE SOLSTICE and SIM. It shows the differences in absolute calibration in this range on the order of about 8% with significant gaps in the record that are difficult to fill. It is interesting to note that the UARS SUSIM is the only instrument to measure contiguously across a solar maximum time period (solar cycle 23) and most of these records began during the descending phase of a solar cycle. When each time series is viewed independently in a solar maximum to solar minimum context, SME, UARS, and SORCE SOLSTICE and SIM all produce a descending phase irradiance change of about 0.06 W m^{-2} in this band with SUSIM indicating a 0.035 W m^{-2} change. During the rising phases of solar cycles 21, 22, and 23 over the limited range of comparability, SME, SUSIM, UARS SOLSTICE, and SIM produce similar trends with SORCE SOLSTICE giving a much weaker increase in irradiance, but between cycles 22 and 23 the SUSIM and the SORCE data cannot be reconciled. Currently none of these trends can be accurately combined into a composite time series as effectively as the TSI or Lyman α.

The situation in the visible and IR is even more unexpected. *Harder et al.* (2009) report that the integrated 1.1 W m^{-2} increase in the UV irradiance between 200 and 400 nm reported by SIM during the descending phase of cycle 23 is compensated by decrease in the visible and IR that sums to give the TSI. This effect observed in SIM is explained in terms of solar brightness temperature defined in equation (6) and shown in Fig. 9a. In addition to irradiance modulation from the active region passage, the SSI values for wavelengths with a brightness temperature greater than 5770 K show a brightening with decreasing solar activity, whereas those with lower brightness temperatures show a

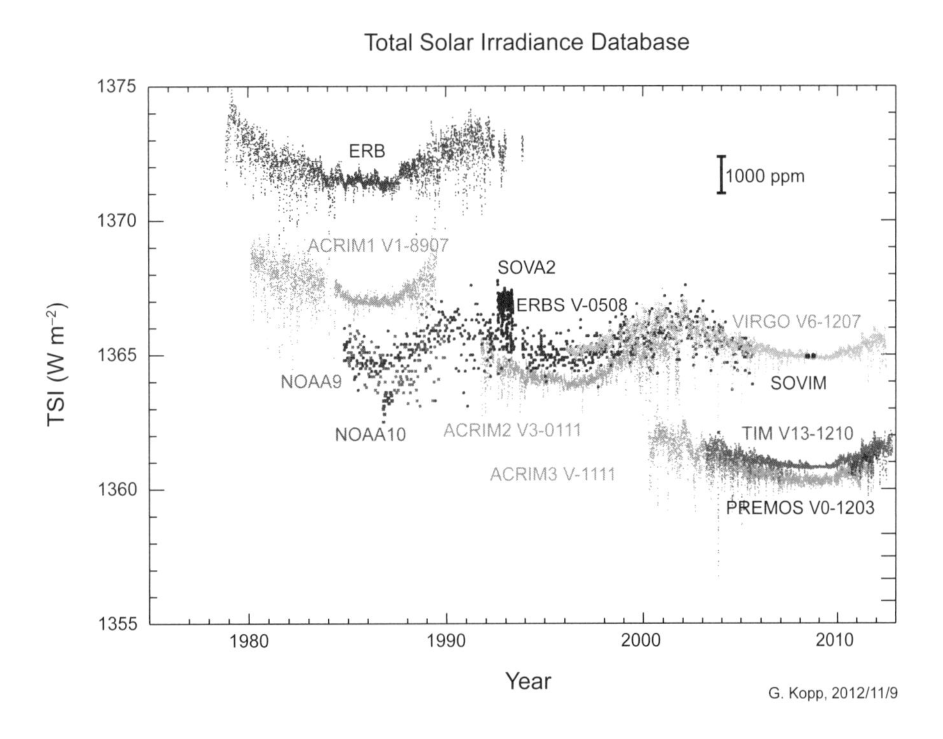

Fig. 7. Time record of measurements of the total solar irradiance (TSI). The different time series show a span of about 10 W m^{-2}, or approximately 0.7%, indicating the level of calibration accuracy. Within any individual time series the scatter of the data generally shows the passage of sunspots, and is not representative of the noise level of the instrument. Figure courtesy of Greg Kopp, University of Colorado.

dimming. Note that the 5770-K brightness temperature is derived from solving the Stephan-Boltzmann equation with an irradiance of 1361 W m^{-2} and is representative of the opacity-weighted mean output of the Sun — i.e., the TSI. Figure 9b shows the trends in the discrete integrated bands identified according to whether they are greater or lesser than the 5770-K temperature divide line. These results demonstrate that different parts of the solar atmosphere contribute differently to the TSI with the behavior in the deep photospheric layers giving an opposing and nearly compensating trend to that in the upper photospheric and lower chromospheric layers.

This observation is consistent with a slightly higher temperature at the top of the photosphere that increases irradiance at some wavelengths during solar maximum and a slightly lower temperature at the bottom of the photosphere that induces dimming at other wavelengths. This effect might be explained by a slight reduction of the Rosseland mean opacity (*Mihalas,* 1978) in the photosphere during solar maximum, or from a slightly increased mechanical energy transport through the photosphere, for instance, by magnetic free-energy flow (*Trujillo Bueno et al.,* 2004) or by convective motions (*Henoux and Somov,* 1991); see the discussion in section 2.4 for more information about the SRPM models that can partially explain the SIM observations.

Woods et al. (2000) noted that constructing the Lyman α time series (see Fig. 6c), with about a factor of 2 solar cycle variability, from instruments with different degradation effects spanning the AE-E, SME, and UARS missions contributed to the uncertainties associated with the construction of the composite. The measurement uncertainties and degradation corrections associated with longer wavelengths important for lower chromospheric and photospheric processes where solar variability is on the order of 0.01–10% have a much larger impact and the uncertainties become commensurate with the magnitude

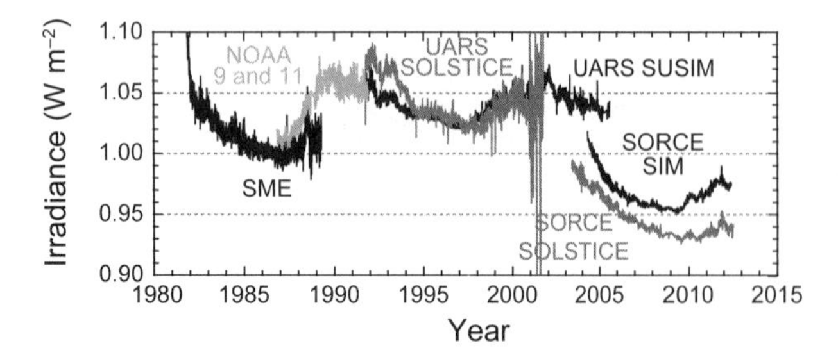

Fig. 8. Time series of measured spectral irradiance integrated from 230 to 250 nm. This plot shows the record in solar cycles 21, 22, and 23. The peak-to-peak absolute calibrations spread for these values is about 9%; over the mid-UV (200–300 nm) wavelength range it is more like 5%. The solar cycle irradiance over this wavelength band is on the order of 0.035–0.06 W m^{-2}.

of the solar signal itself. Degradation corrections rely on the ability to intercompare different optical elements and detectors, such as the methodology employed by SORCE SIM and UARS SUSIM instruments, and/or the usage of standard light sources such as the usage of standard stars for SOLSTICE or internal calibration lamps (UARS SUSIM) must continue for future efforts. However, the ability to track changes of critical components and the transfer of lamp (or star) calibrations to the solar observation geometry will remain uncertainty-limiting effects. The development of ultraclean optical cavities to minimize hydrocarbon contamination of optical surfaces, such as that done for the SDO EVE instrument, and the usage of degradation impervious detectors such as the SIM ESR will pave the way for an improved understanding of the long-term trends in solar spectral irradiance.

2.4. Modeling of Irradiance Variability Based on Image and Magnetogram Analysis

In the last two decades, a concerted effort has gone into producing models of solar variability. Based on information from datasets related to measured X-ray and UV variability, the TSI, and measures of solar activity based on magnetograms and photometric images, these models are able to synthesize spectral variability through a variety (and a combination) of proxy-based and physics-based models. The degree to which models agree among themselves and observations is dependent on the target wavelength and solar atmospheric regions. In the X-ray and vacuum UV spectral regions most representative of the solar corona and important for the Earth ionospheric and mesospheric processes, realistic proxy-based models such as the Flare Irradiance Spectral Model (FISM) (*Chamberlin et al.,* 2007) can be constructed to apply on solar cycle and rotational timescales based on readily available space weather products like Lyman α, Mg II, 30.5 nm, 36.5 nm, 0–4 nm, and F10.7 with a high degree of reliability. FISM now includes important flare components that account for solar variability on timescales from hours to minutes to improve important space weather predictive skills (*Chamberlin et al.,* 2008). In a similar manner and more germane to the transition and upper chromospheric regions, *Woods et al.* (2000) and *Woods and Rottman* (2001) demonstrate that most chromospheric emissions show a simple linear relation to the chromospheric Mg index, but Lyman α irradiance, as well as other transition region emissions, behave differently owing to about a factor of 2 increased contribution from active network to Lyman α variability.

Lean et al. (1997) and *Lean* (2000) developed an approach for calculating the solar spectral irradiance variability that parameterizes the observed irradiance time series at any given wavelength in terms of proxy indicators (hereafter referred to as NRL SSI). For wavelengths longer than 300 nm the two main sources of variability are assumed to be dark sunspots and the bright faculae, which are represented by, respectively, a sunspot blocking function derived in *Lean et al.* (1997) and the Mg index compiled by NOAA (*Viereck et al.,* 2004). Likewise, for wavelengths from about 30 to 300 nm, the variations are represented by the Mg II core-to-wing ratio. For the parameterizations of the shortest EUV and X-ray emissions (mainly at wavelengths below about 27 nm) the 10.7-cm radio flux is used in addition to the Mg index. Multiple regression determines the association of the de-trended time series of the proxies, implying that the associations are determined from a subset of the range of possible variations. The reconstruction of the irradiance from the as-reported proxy indicators (not de-trended) therefore assumes that the proxies and the irradiance behave in similar ways over both rotational and solar cycle timescales.

A different approach was adopted by N. Krivova and coworkers (*Krivova et al.,* 2003, 2009, and references therein) in the development of the Spectral And Total Irradiance Reconstruction (SATIRE) model assuming that all irradiance variations on timescales of days and longer are solely due to the evolution of the solar surface magnetic features. A four-component model of the solar photosphere was constructed from models of the quiet Sun (set by a threshold of magnetic field strength), sunspot umbrae and penumbrae, and bright magnetic features representative of faculae and the network. Each of the four components are described by the time-independent spectrum calculated in the LTE approximation from the corresponding model atmospheres described in *Solanki and Unruh* (1998). A filling factor defining the fraction of the solar surface covered by each component is determined from daily magnetograms and continuum images from the SoHO MDI (*Scherrer et al.,* 1995) for solar cycle 23 and extended back to solar cycles 21 and 22 using NSO Kitt Peak images (*Wenzler et al.,* 2005). The model has a single free parameter, B_{sat}, representing the field strength below which the facular contrast is proportional to the magnetogram signal, while it

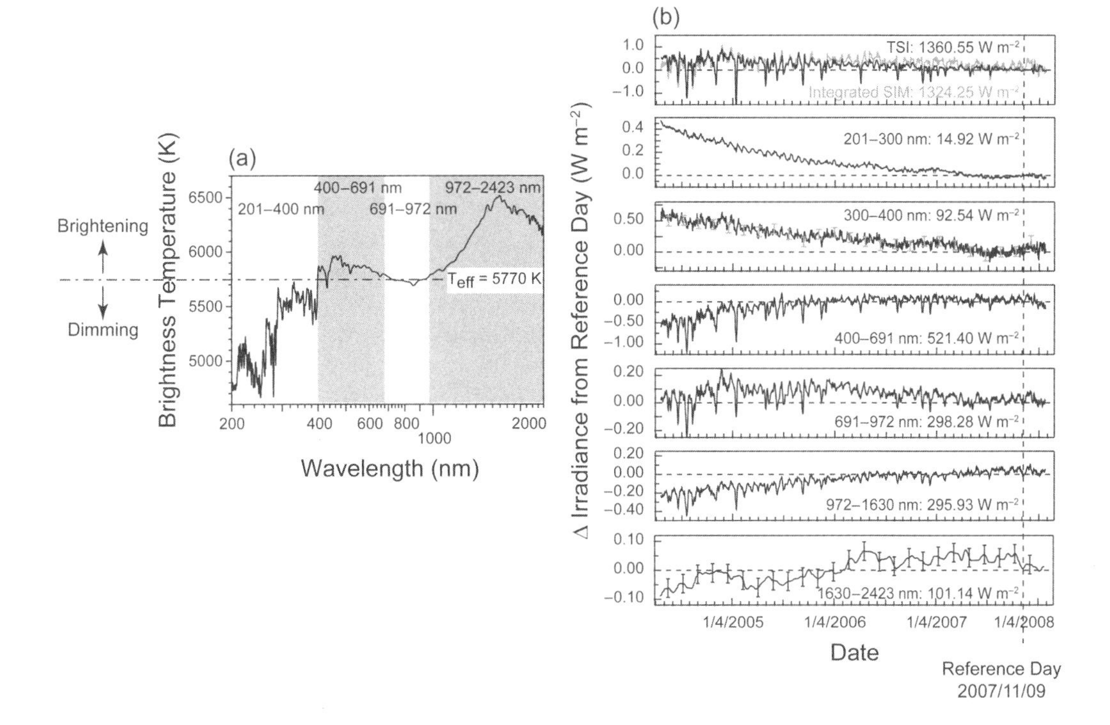

Fig. 9. **(a)** Brightness temperature measured by the SORCE SIM. **(b)** Time series in integrated bands reported in *Harder et al.* (2009). Based on both these panels, wavelengths with a brightness temperature less than T_{eff} = 5770 K will show variations in phase with the TSI and is consistent with observations in the UV such as seen in Fig. 8. In **(b)**, the time series is integrated in bands that correspond to the different regions shaded in the brightness temperature plot. The top graph in **(b)** corresponds to the integrated SIM data compared to the TSI measured by SORCE TIM after accounting for the ~36 W m^{-2} irradiance deficit not measured for wavelengths longer than 2423 nm. The SIM and TIM agree to approximately 200 ppm after accounting for this difference. A solar cycle 23 solar minimum reference day was selected in November 2007; the value listed for each graph in **(b)** corresponds to the integrated irradiance at the time of the minimum. The solar minimum occurred in February 2009, but for this long solar mimimum time period, the irradiance levels in 2007 are still representative of minimum conditions.

is independent (saturated) above that. *Krivova et al.* (2003) obtained a value of 280 G for B_{sat} from a fit to the VIRGO TSI time series. The inadequacies of the LTE approximation affecting the UV portion of the spectrum were addressed by *Krivova et al.* (2006) by using the observed SUSIM UV spectra to extrapolate available models to shorter wavelengths. The SUSIM time series was empirically corrected to account for the exposure-dependent degradation of the sensitivity through the proportionality to the Mg II core-to-wing ratio. Using this model *Krivova et al.* (2006) estimate that up to 60% of the total irradiance variations over the solar cycle might be produced at wavelengths less than 400 nm.

As an outgrowth of the FAL series of solar models and the National Science Foundation (NSF) Radiative Inputs from Sun to Earth (RISE) program (*Fontenla et al.*, 1999), J. Fontenla and co-workers developed the Solar Radiation Physical Modeling (SRPM) program to perform spectral synthesis and irradiance variability estimates over a very broad wavelength range. The SRPM is based on the seven components listed in Table 1 by developing solar atmospheric models to characterize radiative properties of these features. Daily solar irradiance spectra were constructed for most of solar cycle 23 based on a set of physical models of the solar features and non-LTE calculations of their emitted spectra as a function of viewing angle with solar images specifying the distribution of the features on the solar disk. SRPM emphasizes the effects on the SSI of the upper chromosphere and full non-LTE radiative transfer calculation of atomic level populations and ionizations that are essential for physically consistent results at UV wavelengths and for deep lines in the visible and IR. The radiance for ten observation angles is computed for seven components that correspond to the main features observed in ~2-arcsec-resolution images of the solar disk. The SSI is computed by adding the contributions of these components and using their distribution over the solar disk determined from PSPT images (*Ermolli et al.*, 2007; *Rast et al.*, 2008). The PSPT telescopes achieve ~0.1% pixel-to-pixel relative photometric precision and acquire full-disk solar images through narrow-band interference filters, including two filters centered on the red continuum (607.09 nm, ~0.46 nm FWHM) and the Ca II K line (393.41 nm, FWHM ~0.27 nm). The images at the two wavelengths are obtained within minutes to allow good feature registration with sunspot umbra and penumbra (models S and P) determined from the red filter and with the other features (B, D, F, H, and P) determined from the Ca II K filter. The solar features seen in each observation pair were identified using the decomposition method described by *Fontenla and Harder* (2005). The method utilizes a contrast threshold scheme derived from partitioning intensity histograms constructed from the images as a function of heliocentric angle.

Figure 10a shows the one-dimensional solar atmosphere models for each of the seven solar surface features identified in PSPT solar images used in the SRPM spectral synthesis up to the onset of the transition region. The sunspot umbra (S) and penumbra (R) models are based on *Maltby et al.* (1986) and *Kjeldseth-Moe and Maltby* (1969, 1974a,b) respectively, and the quiet Sun and active region models (B, D, F, H, and P) are derived in a series of papers by Fontenla and co-workers (*Fontenla et al.*, 2006, 2007, 2009). Figure 10b shows a detail of photospheric temperature gradient for model atmospheres B (quiet Sun) and H (facula). Note that these two models cross one another so that the quiet Sun component indicates a slightly steeper temperature gradient in the photosphere relative to the active regions with higher magnetic fields. It is important to note that these differences in the photospheric temperature gradient are not captured in any of the other models discussed in this section.

The temperature structure in the SRPM atmospheric models is commensurate with the observations of *Topka et al.* (1997). They show that at some continuum wavelengths in the visible regions facula show a negative contrast relative to quiet Sun at disk center and only show positive contrast as the active regions approach the solar limb (see discussion related to Fig. 4 for more details). Additional supporting evidence in the visible comes from *Preminger et al.* (2011) showing similar results from image analysis at San Fernando Observatory. They measured the photometric sum, which gives the relative contribution of solar features to the disk-integrated intensity of the image. The photometric sums at red and blue continuum wavelengths are negatively correlated with solar activity with strong short-term variability and weak solar-cycle variability. Their Ca II K-line photometric sum measurement shows the expected positive correlation with solar activity with strong variations on solar-cycle timescales. *Sánchez Cuberes et al.* (2002) reported observations of anti-solar-cycle trends in the IR, particularly in the neighborhood of the minimum in the H^- opacity at 1623 nm.

Figure 11a is a study of solar variability as calculated by SRPM and measured by SORCE SIM (figure adapted from Figs. 7 and 9 of *Fontenla et al.*, 2011) to show different contributions of solar features over the course of the descending phase of solar cycle 23 ranging from late 2000 through the peak of the solar cycle in 2002, and then in the descending phase out to near solar minimum in September 2008. The six mask images on the left show the feature composition of the Sun as produced from SRPM masks using Rome Observatory PSPT data. These features are identified in Table 1, and their respective solar atmospheric temperature profiles are shown in Fig. 10a. Times of high solar activity (mask images labeled high 2, high 1, and peak) are characterized by increased areas of facula and plage (models H and P) as well as larger contributions from the active network (model F). Figure 11b shows the irradiance differences for each image with respect to the solar minimum time period of late 2008 labeled as low in Fig. 11a. Note that the spectral synthesis is enhanced in the UV with significant variability for wavelengths out to the Ca II opacity edge near 400 nm. The predicted negative solar cycle differences are found in the visible portion of the spectrum, thereby reproducing, for the most part, the trends shown in the brightness temperature depicted in Fig. 9. This is to be compared with the SIM observations (Fig. 11c) during the descending phase of the solar cycle that correspond to the last three SRPM masks labeled as mid 2, mid 1, and low. The SIM spectra shown as a red trace in Fig. 11c and the mid 2 SRPM mask represent a case where the solar disk is dominated by contributions from facula and plage and the mid 1 case (and black trace in Fig. 11c) is where a large, complex, active region with significant sunspot umbral and penumbral contributions are apparent on the disk. Notice that the mid 2 and mid 1 cases are separated by only 11 days. The SIM observations of Fig. 11c generally show larger spectral irradiance differences in the UV than the synthesis — more in line with the SRPM high 1 and high 2 cases. The structure in the irradiance curves for the two cases in the visible show a high degree of agreement. The greater suppression of irradiance in the mid 1 case (black curve in Fig. 11c) is due to the contribution from the large sunspot area near the disk center. Another critical observation seen in all the SRPM masks is the slow decay of the average and active network components (models D and F represented as white and green colors, respectively); these components are distributed quasi-uniformly over the solar disk and therefore do not produce rotational modulation as dramatically as isolated active regions as seen in the sunspot and bright facular and plage components. The slow decay of these quieter components will contribute to solar cycle length irradiance trends but not as significantly to variability seen on rotational timescales.

Finally, in terms of the predicted SSI response, *Fontenla et al.* (2011, section 6) argued that the integrated SRPM spectrum (referred to as the bolometric flux) has a significantly smaller solar cycle modulation during solar cycle 23 so some component of the solar cycle modulation occurs that is not directly related to the rotational modulation and is not currently captured by the present SRPM calculations, even when these calculations capture most details of the solar rotation induced by the passage of active regions. One factor that affects the solar cycle trend of the bolometric flux but not its rotational modulation is the difficult to detect "quiet Sun" network change apparent in PSPT images (Fig. 11a) and correspond to small spatial scales and magnetic field magnitudes that have been often overlooked in SSI and TSI studies. The present SRPM method for constructing the SSI is based on a feature discrimination scheme that relies on contrast levels with respect to the median brightness as a function of heliocentric angle. This also applies to the so-called "quiet Sun" features that have a small contrast in the PSPT Ca II image, thus it is entirely

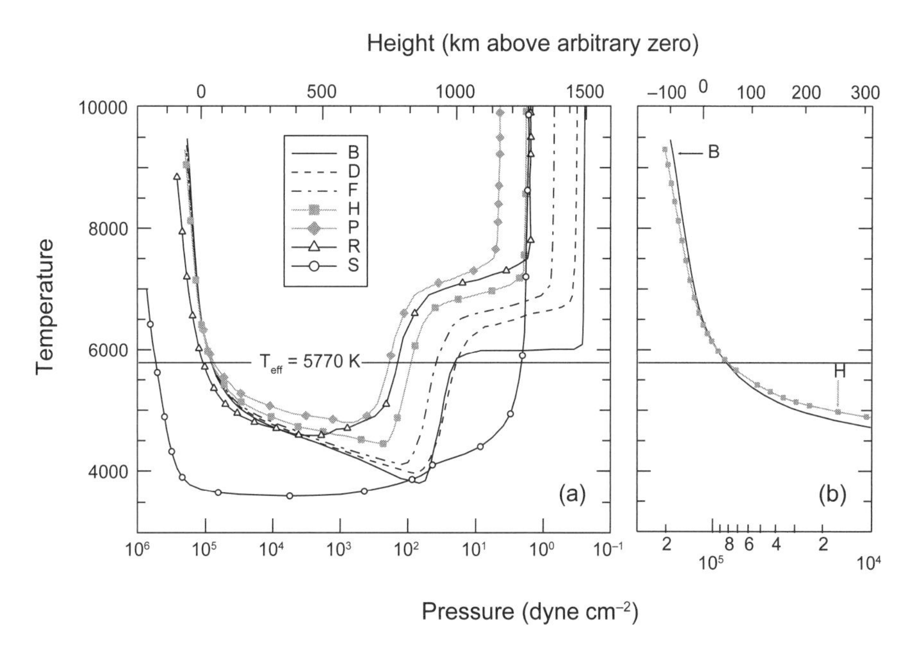

Fig. 10. Solar atmospheric models represented in SRPM (*Fontenla et al.,* 2007). These models show two distinct differences from those represented in the VAL series of papers (*Vernazza et al.,* 1981): (1) The SRPM models show a lower-temperature chromospheric temperature minimum rising faster to a flatter upper chromospheric plateau before rising into the transition region [shown in **(a)**]; and (2) as shown in **(b)**, the SRPM models for facula (H) and quiet Sun (B) have a crossover in the photosphere close to T_{eff} that accounts for the center-to-limb contrast differences reported in *Topka et al.* (1997) for the visible spectral regions and in the infrared shown in *Sánchez Cuberes et al.* (2002).

plausible that the median could change as the temperature structure and magnetic field evolve.

Figure 12 compares the solar cycle variability of the NRL and the SRPM model. Two cases of the SRPM are shown; the blue curve corresponds to the variability reported in *Fontenla et al.* (2011) and the red curve is from a study where the median is shifted to correct the integrated SRPM spectrum to match the TSI. In both cases, the SRPM spectra produce larger variations in the MUV spectral range and anti-solar-cycle trends in the visible and IR (notice that SRPM reverts to an in-phase solar cycle trend near 3500 nm — outside the SIM observation range). The NRL model produces in-phase solar cycle trends everywhere in the spectrum except in the neighborhood of the 1600-nm H^- opacity region.

3. MODELING AND OBSERVATIONS OF EARTH'S ATMOSPHERIC RESPONSE TO SOLAR VARIABILITY

Our discussion now turns to the atmospheric response of planetary atmospheres to forcing from the solar spectrum and its intrinsic variability. Before beginning this, we must first discuss the distance to the Sun; the solar intensity at a planet varies inversely as the distance squared. This is generally referred to as the $1/R^2$ effect and studies of the intrinsic variability are normalized to mean Earth-Sun distance of 1 AU. The resulting change in intensity due to changes in the orbital eccentricity is generally called insolation, as it is not an intrinsic variation of the solar radiation. Earth's orbital eccentricity (e = 0.017) gives rise to a 7% insolation variation with a period of one year. As an example, the solar intensity relative to the irradiance at Earth is a factor of 2 at Venus, 1% at Saturn, and 0.1% at Neptune.

Following the arguments presented in *Reid* (1999) the radiation balance of Earth (or any other planet) can be expressed as terms of

$$(1-\alpha)\frac{S}{4} = \varepsilon\sigma T^4 \tag{8}$$

In this equation S is the Sun's total irradiance and S/4 is the average irradiance at the top of the atmosphere (the ratio of the planet's surface area to its cross section), α is the planetary albedo, and ε ($\approx$1) and σ are the effective emissivity and the Stephan-Boltzman constant respectively;

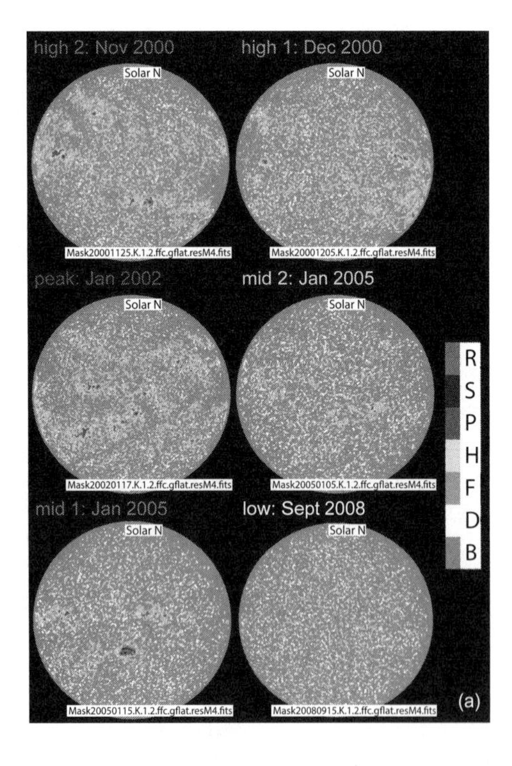

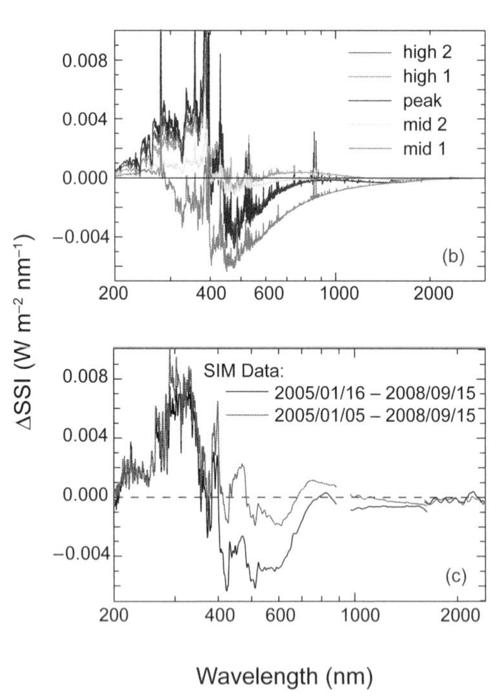

Fig. 11. See Plate 83 for color version. **(a)** Six time periods over solar cycle 23. The masks are constructed from Rome Observatory PSPT images and the surface feature contributions are color coded and identified in Table 1. **(b)** Change in irradiance as predicted by SRPM from the 2008 reference spectrum. **(c)** SIM difference spectra from 2005 relative to 2008. The SRPM masks in **(a)** labeled as "mid 2" corresponds to the red trace in **(c)**, and "mid 1" is represented as the black curve in **(c)**. The offsetting solar cycle trends reported in *Harder et al.* (2009) are generally reproduced by the model but with lower amplitude.

T is the effective radiation temperature of the planet, and for Earth this is about 255 K (–18°C) corresponding to a tropospheric height of about 6 km above the surface. The atmosphere is optically thick at this temperature for altitudes below this height and will not radiate out to space. Differentiating and rearranging this equation yields

$$\frac{\Delta T}{T} = \frac{1}{4}\frac{\Delta S}{S};\ \Delta T = 0.6\ \text{K for}\ \frac{\Delta S}{S} = 1\% \tag{9}$$

In terms of the observations of TSI (see section 2.2), the canonical 0.08% change in the TSI only produces a 0.05-K change in the planetary temperature, but this highly simplified calculation belies the complexity of the problem. An assessment of this highly complex radiation balance problem in Earth's atmosphere was recently presented by *Wild et al.* (2012) and employed both spacebased and surface observations in an extensive model intercomparison study for the IPCC CMIP5 assessment. From this study, the authors conclude that estimates for the global mean downward solar and thermal radiation coincide within 2 W m^{-2} with most satellite-derived estimates. But uncertainties remain for the major surface energy balance components with a 2σ uncertainty on the order of 10 W m^{-2}, roughly double the corresponding range of the top of the atmosphere net solar and thermal flux uncertainties.

The spectral dependence plays a crucial role in this process as shown in Fig. 13, which shows the heating rate

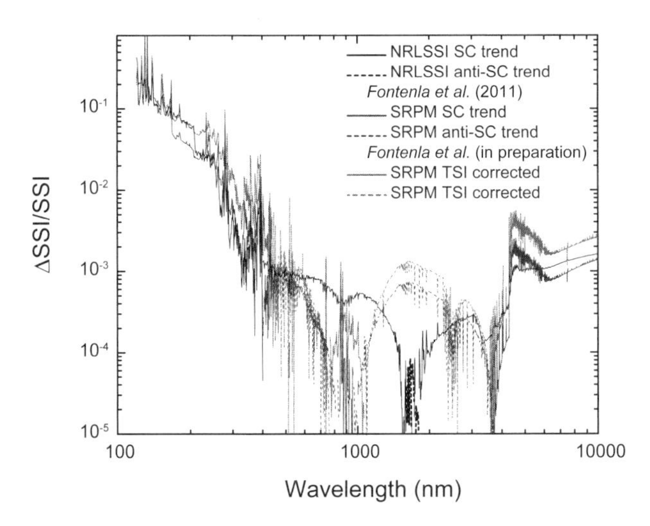

Fig. 12. See Plate 84 for color version. Estimates of solar maximum to solar minimum SSI variability from the NRL SSI and two versions of SRPM. These two models currently represent the two extremes of solar spectral variability. The SRPM calculation represented in *Fontenla et al.* (2011) is shown in blue, and an estimated change in irradiance that matches the TSI based on shifts in the median of the brightness distribution in PSPT images is shown in red.

of the atmosphere as a function of altitude and wavelength in Earth's atmosphere generated from the MODTRAN radiative transfer model (*Berk et al.,* 2006). This plot is appropriate for the middle atmosphere from the top of the stratosphere (~1 mb) to the surface and is representative of cloud-free equatorial conditions. Cooler colors (shades of blue) represent cooling (measured in K/day/cm^{-1}), with the warmer colors (shades of red and yellow) showing atmospheric heating, or equivalently, negative cooling. The cooling (heating) rate is defined as the energy lost (gained) per unit volume, per unit time, into a 2π steradian hemisphere along Earth's radial direction (*Harries et al.,* 2008). Essentially, this is a measure of the rate of the loss of energy by that volume to space. As is the typical practice in atmospheric science, the heating rate per wavenumber (1 wavenumber = 1 cm^{-1} = $10^7/\lambda$, with λ = wavelength in nanometers) has units of J m^{-3} s^{-1} $(cm^{-1})^{-1}$, and corresponds to the vertical gradient in the net radiative flux: net radiative flux = upwelling flux–downwelling flux, for a cooling rate plot such as that shown in Fig. 13. Recall that this radiative flux is derived from the Planck radiation law discussed in the context of equation (1). For an atmosphere in hydrostatic equilibrium, the cooling rate as a function of wavenumber and altitude, $h_v(z)$, can be written (*Andrews,* 2010)

$$h_v(z) = \frac{dT_v(z)}{dt} = \frac{1}{\rho(z)c_p}\frac{dF_{v,N(z)}}{dz} \tag{10}$$

Here $\rho(z)$ is the atmospheric density, c_p is the heat capacity at constant pressure, and $F_{v,N(z)}$ is the flux associated with the concentration of the atmospheric absorbers. Total heating rates can be derived by integrating over wavenumber and/or altitude, as shown in Fig. 13.

A number of atmospheric absorbers critical to determining the heating rate appear in this plot (and also appear in Fig. 1d); the Hartley-Huggins Bands of ozone (0.2–0.37 μm) warm the upper layers of the atmosphere and below these altitude levels (appear as black regions in the plot) the radiation is completely absorbed and cannot contribute to the heating. The weaker Chappius bands of ozone (0.43–0.74 μm) absorb radiation throughout the stratosphere and troposphere, and the water vapor overtone bands from ~0.7 to 5 μm induce positive heating rates (i.e., negative cooling rates), while emissions in the IR from carbon dioxide and water vapor produce a net cooling. In IR spectral regions of low absorption, i.e., the window regions discussed in the context of Fig. 1d, a fraction of thermal radiation (defined in equation (1)) is emitted upward by the surface and has a high probability of escaping to space. In wavelength regions with higher absorption cross sections (see equation (5)), emitted radiation has a lower probability of escaping to space and is reabsorbed in a higher and colder layer of the atmosphere. Some of the absorbed energy will then be reemitted at the lower local thermodynamic temperature. This process continues until the radiation eventually escapes to space, which forms the physical basis of the greenhouse effect (for more discussion see *Harries et al.,* 2008).

Because of this strong wavelength-dependent spectral response in the atmosphere, a number of climatically important mechanisms have been proposed to contribute to the evolution of climate variables observed over the course of an 11-year solar cycle through enhanced atmospheric response at wavelengths where solar variability is fractionally larger than the 0.08% change in the TSI. These mechanisms proceed through two different, but perhaps not mutually exclusive, mechanisms that will be discussed in the following paragraphs. The first of these potential mechanisms is a *top-down* mechanism, where UV radiation is absorbed in the upper atmosphere predominantly by ozone and this exothermal process then generates both vertical and horizontal atmospheric temperature gradients that can drive solar-cycle-dependent circulation patterns. A second mechanism is a *bottom-up* mechanism that accounts for an ocean response in the sea surface temperature to solar variations that can be another factor providing an amplifying link for the solar influence on the tropospheric circulation.

The study of the top-down mechanism begins with the observations first published by K. Labitzke (*Labitzke,* 1987; *Geller,* 2006) noting the association between the polar stratospheric temperature in the northern winter and solar cycle (as determined through sunspot number) in the winters when the 50-mb equatorial winds are in the westerly phase of the quasibiennial oscillation (QBO). This finding was initially questioned because the observation was based on only three solar cycles of data, but this finding has been verified in subsequent solar cycles since the first report (*Labitzke,* 2006). *Haigh* (1996) and *Shindell et al.* (1999) both explain this observation in terms of the combined changes in the spectral composition of the solar variations and the photochemical production of stratospheric ozone leading to additional positive feedback with ozone amplifying the direct solar UV influences on the zonal mean state of the stratosphere; circulation changes initially induced in the stratosphere then subsequently penetrate into the troposphere. At solar maximum, a warming of the summer stratosphere was found to strengthen easterly winds that penetrated into the equatorial upper troposphere, causing poleward shifts in the positions of the subtropical westerly jets that broadened the tropical Hadley circulation and shifted the storm tracks in a poleward direction. *Kodera* (2006) discusses two processes that have consequences for the mean flow interaction in the upper stratosphere. The first process involves changes in the vertical propagation of planetary waves and the resultant tropospheric circulation change in the extratropics of the winter hemisphere. The

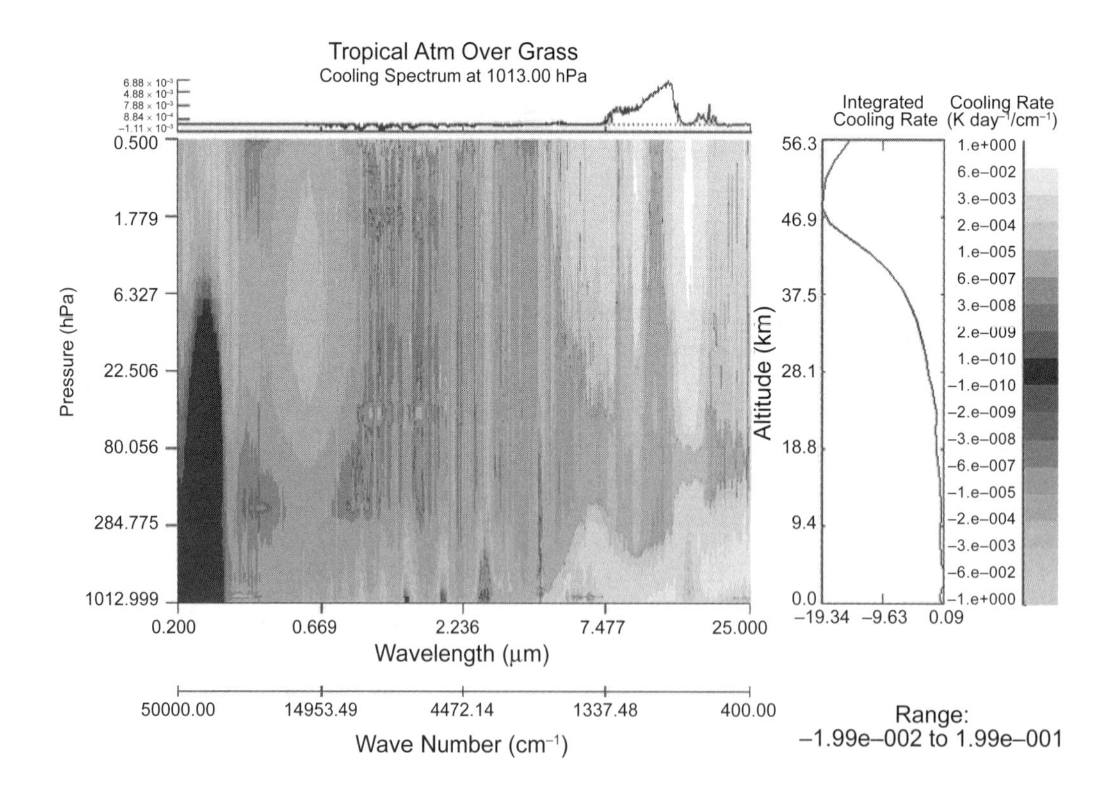

Fig. 13. See Plate 85 for color version. A MODTRAN®5.2 radiative transfer model calculation showing the cooling rate of Earth's lower atmosphere in response to solar (mostly shortward of 5 μm) and thermal (mostly longward of 5 μm) energy, based on a tropical atmosphere over a "grasslands" albedo. Figure courtesy of G. P. Anderson (Air Force Research Laboratory, Kirtland AFB, New Mexico) and A. Berk (Spectral Sciences, Inc.).

second involves changes in the global meridional circulation in the stratosphere and associated convective activity change in the tropics. These processes are operable on the 11-year solar cycle, but could also have effects on centennial-scale solar influences. *Haigh and Blackburn* (2006) also suggest from model studies that heating of the lower stratosphere (in this case from solar cycle irradiance change) tends to weaken the subtropical jets and the tropospheric mean meridional circulations. The positions of the jets, and the extent of the Hadley cells, respond to the distribution of the stratospheric heating, with low-latitude heating displacing them poleward, and uniform heating displacing them equatorward.

Much of the information that forms the basis of this mechanism is based on measurements of stratospheric ozone spanning the 1980–2005 time period; measurements from NOAA SBUV, the Stratospheric Aerosol and Gas Experiment II (SAGE II), and the UARS Halogen Occultation Experiment (HALOE) provide a contiguous dataset that can constrain these proposed mechanisms. *Soukharev and Hood* (2006) show that during the 1992–2003 period, with no major volcanic contamination, the vertical structure of the tropical ozone solar cycle response can be characterized by statistically significant response in-phase with the solar cycle in the upper (1–3 hPa) and lower (30–50 hPa) stratospheric regions and by statistically insignificant responses in the middle stratosphere (5–10 hPa). Analyses of the SBUV and SAGE records with lengths as long as 16 years indicated that the mean low-latitude response in the upper stratosphere (1–3 hPa) is in the range of 2–4% but decreases to zero or slightly negative values in the middle stratosphere (5–10 hPa) before increasing again in the lower stratosphere. Reanalysis of SAGE II and HALOE data by *Remsberg and Lingenfelser* (2010) used the 14-year time series and applied multiple linear regression analysis that include seasonal, 28- and 21-month, 11-year sinusoidal components, as well as linear trend terms. Their analysis did not apply proxies for the 28-month QBO-like component, 11-year solar UV flux, or reactive chlorine terms in recognition of the regression study of *Lee and Smith* (2003), showing that the effect of decadal-scale periodicities in the stratosphere at low latitudes can mimic the solar cycle forcing due to dynamical interactions with the QBO and volcanic forcings. In the place of these proxies, the Remsberg analysis focused on the periodic 11-year terms to see whether they are in-phase with that of a direct, UV flux forcing or are dominated by some other decadal-scale influence. Like the *Soukharev and Hood* (2006) study, this analysis showed in-phase behavior over most of the latitude/altitude domain and that they have maximum minus minimum variations between 25 S and 25 N that peak near 4% between 30 and 40 km, giving further credence to the observation of the UV influence.

Based on analysis by *Hood et al.* (2010) the observed tropical El Niño-Southern Oscillation (ENSO) response explains this dual peaked vertical structure of the tropical solar cycle ozone response. The ENSO ozone response is negative in the lower stratosphere due to increased upwelling but changes sign, becoming positive in the middle stratosphere (5–10 hPa) due mainly to advective decreases of temperature and odd nitrogen chemistry, which photochemically increases ozone. A similar mechanism may explain the observed lower stratospheric solar cycle ozone and temperature response and the absence of a significant response in the tropical middle stratosphere.

The bottom-up mechanism (*Meehl et al.,* 2008, and references therein) involves the increased solar forcing at the surface through increased latent heat flux and evaporation. The resulting moisture is carried to the convergence zones by the trade winds, strengthening the intertropical convergence zone (ITCZ) and the South Pacific convergence zone (SPCZ). Once these precipitation regimes begin to intensify, an amplifying set of coupled feedbacks similar to that in La Niña-like cold events occurs. There is a strengthening of the trades and greater upwelling of colder water that extends the equatorial cold tongue farther west and reduces precipitation across the equatorial Pacific, while further increasing precipitation in the ITCZ and SPCZ. One of the important consequences of this bottom-up mechanism is that changes in solar irradiance over the course of the 11-year cycle influence the strength and extent of the Walker circulation (east-west tropical circulation pattern), which then couples to the north-south tropical Hadley circulation. Because of this, both the top-down and bottom-up mechanisms enhance the amplification of the solar influence signal (*Meehl et al.,* 2009).

Understanding the consequence of solar variability in Earth's atmosphere, particularly the 11-year solar cycle, requires in-depth modeling to determine the atmospheric response in key variables such as the ozone concentration, temperature, and shifts in the wind field. In-depth comparisons of climate models with observations is required to achieve closure in the role of solar variability in climate. Currently multiple models have the capability to address the important issues discussed in this section. A very thorough examination of climate models to capture potential solar-related phenomena, conducted by *Austin et al.* (2008), expanded the scope of the model intercomparison conducted by *Eyring et al.* (2006). The coupled chemistry-climate models (CCM) used in this study were sponsored through the Chemistry-Climate Model Validation Activity (CCMVal) (*Eyring et al.,* 2005) for the World Climate Research Program (WCRP) Stratospheric Processes and their Role in Climate (SPARC) program. The models were all transient simulations covering the period 1950–2004 with almost identical forcings (sea surface temperatures, greenhouse gases, and halocarbons). In the case of the *Austin et al.* (2008) study, the models with differing inputs to the solar cycle, inclusion/forcing of the QBO, and length of the model runs produced disparate results. However, the fully interactive chemistry employed in these models (see, e.g., *Marsh et al.,* 2007; *Schmidt and Brasseur,* 2006) improves the atmospheric response to radiative heating and ozone photolysis caused by irradiance variations. The models now produce improved vertical structure of the annual mean ozone signal in the tropics, including the lower stratospheric maximum discussed in *Soukharev and Hood* (2006). For models in this study that covered the full temporal range 1960–2005, the ozone solar signal below 50 hPa changes substantially from the first two solar cycles to the last two solar cycles, suggesting that the difference is due to an aliasing between the sea surface temperatures and the solar cycle during the first part of the period, thereby emphasizing the importance of how sea surface temperatures are handled in the models.

Meehl et al. (2009) provides further insights into the importance of whole atmosphere models through a comparative study that used three different simulations to emphasize different aspects of the top-down and bottom-up mechanisms. The first used the National Center for Atmospheric Research (NCAR) CCSM3 with coupled components of atmosphere, ocean, land, and sea ice, but without a resolved stratosphere and interactive ozone chemistry. The second was a version of WACCM with no dynamical coupled air-sea interaction but a resolved stratosphere and interactive chemistry. The third model combines these two different modules to give a full atmospheric response. The findings from these combined models indicate that each mechanism acting alone can produce a weak signature of the observed enhancement of the tropical precipitation maxima, but when both act in concert the two mechanisms work together to produce climate anomalies closer to the observed values but still generally underestimate the observations.

In Earth's upper atmosphere the aeronomic physical and chemical processes that affect the energy balance of the thermosphere and mesosphere are currently well-represented in a number of models, including the ones discussed in previous paragraphs to describe climatic effects in the lower and middle atmosphere. These processes have been assimilated into globally averaged models representative of the coupled mesosphere, thermosphere, and ionosphere to quantitatively calculate a thermal and compositional structure consistent with observations. *Roble* (1995) described these processes for the NCAR Thermospheric-Ionospheric-Electrodynamics general circulation model (TIE-GCM) that can then be unified with models of the middle and lower atmosphere (*Marsh et al.*, 2007; *Garcia et al.*, 2007). Similar high-quality modeling efforts came from the HAMMONIA chemistry climate model (*Schmidt and Brasseur,* 2006). Because of the quality of the chemical and dynamical schemes in these models, they have been adapted for usage in other planetary bodies besides Earth; see *https://wiki.ucar.edu/display/etcam/Extraterrestrial+CAM* for additional details and publications on these models.

4. FUTURE DIRECTIONS FOR SOLAR INFLUENCES RESEARCH

In the concluding remarks of *Harder et al.* (2009), the authors noted that the impacts of the larger solar variability observed in the SORCE SIM and SOLSTICE instruments could have a testable impact on Earth climate studies. The first of these studies, conducted by *Haigh et al.* (2010), looked at the impact on ozone and temperature in Earth's atmosphere. Using SORCE spectral changes led to a significant decrease from 2004 to 2007 in stratospheric ozone below an altitude of 45 km, with an increase above this altitude. Their simulation with a radiative-photochemical model was consistent with contemporaneous measurements of ozone from the Aura-MLS satellite, with the caveat of a short data record making precise attribution to solar effects difficult (*Garcia,* 2010). *Merkel et al.* (2011) extended and confirmed this study by (1) utilizing a three-dimensional GCM (WACCM) with chemistry rather than a static two-dimensional stratospheric model, (2) using nine years of TIMED SABER ozone data that covered the full descending phase of solar cycle 23 in place of the four-year MLS record, (3) detecting the out-of-phase ozone response in an independent instrument observation with similar linear trends, and (4) segregating the data by day and night to isolate photochemical effects. The effect in the lower mesosphere described in these two papers is primarily the loss of odd oxygen due to HO_x reactions. More UV under solar active conditions leads to an enhancement of O_3 photolysis. This leads to more $O(^1D)$ and therefore more OH (hydroxyl radical) through the reaction of $O(^1D)$ with H_2O. Ozone is depleted through several catalytic processes with OH and H (*Grenfell et al.,* 2006). More UV is also transmitted to the stratosphere, leading to greater O_2 photolysis and therefore more stratospheric O_3. A further suggestion of the plausibility of this mechanism was found by *Wang et al.* (2013) analyzing the hydroxyl radical (OH). Additional modeling sensitivity studies using different chemistry climate models of the upper stratosphere to changes in UV were conducted by *Swartz et al.* (2012) and *Shapiro et al.* (2013). Dynamical impacts of larger solar cycle changes in the UV from SORCE were invoked in a study by *Ineson et al.* (2011) showing that the British Met Office Hadley Centre model (HadGEM3) responded to the solar minimum levels with patterns in surface pressure and temperature that resembled the negative phase of the North Atlantic or Arctic Oscillation with similar magnitude to observations. If these measurements of solar UV irradiance are correct, the low solar activity during the solar cycle 23 minimum time period drives cold winters in northern Europe and the United States, and mild winters over southern Europe and Canada with little direct change in globally averaged temperature. *Matthes* (2011) cautioned that these findings await confirmation of the large amplitude of variability in solar UV radiation with SIM measurements taken over a longer period, and an incorporation of the full spectral variability in climate model simulations that consider the stratosphere, ozone chemistry, and full ocean dynamics.

The standard observations of fundamental data records must be continued without interruption owing to the need for long-term records required to discern atmospheric trends; these include such basic things as the UV and visible spectra of the Sun and TSI, ozone, atmospheric trace gases, and basic parameters of dynamics and temperature. Without these continuing efforts, we will remain in a regime where we do not have sufficiently long records to determine the solar cycle effects. In addition, the absolute calibration of the observations is also of great interest. This is particularly true for the absolute levels of the solar spectrum that can have as large as, or even greater, influence on the outcome of simulations as the relative change over the solar cycle (*Zhong et al.,* 2008). With improvements in radiometric calibration capabilities (e.g., *Brown et al.,* 2004) advanced by national and international standards laboratories, compliance to absolute radiometric scales will aid in the standardization of modeling efforts.

The advances in observational and theoretical solar physics will continue to play a vital role in understanding the influence of the Sun in Earth sciences. Both groundbased and spacebased observations of the Sun are needed. The Solar Dynamics Observatory (see special issue of *Solar Physics, 275(1–2),* 2012) will provide crucial very broad contributions to our understanding of solar physics particularly for coronal and transition region processes (AIA and EVE) and the critical understanding of the origins of solar magnetic fields from the HMI instrument. Additional resources to understand the Sun on finer spatial scales are planned and in development. The upcoming IRIS mission (see discussion at *www.nasa.gov/mission_pages/iris/index.html*) will image chromospheric and transition region areas on the Sun in high spatial and temporal timescales. Development of the National Science Foundation Advanced Technology Solar Telescope will provide versatile and high-quality solar images from a large-aperture telescope over a very broad wavelength range applying adaptive optics to overcome atmospheric turbulence problems from groundbased systems (*atst.nso.edu/files/press/ATST_book.pdf*). These improved observations along with advances in one-dimensional modeling of the solar atmosphere, such as those in SRPM (*Fontenla et al.,* 2011), and three-dimensional modeling (*Uitenbroek and Criscuoli,* 2011) will continue to improve the prediction and interpretation skills of irradiance variability.

Acknowledgments. The authors gratefully acknowledge contributions to this manuscript from several of our colleagues. G. Anderson (AFGL) provided Fig. 13, G. Kopp (Univ. Colorado) provided Fig. 7, and J. Fontenla (Northwest Research Associates) provided Fig. 12. We benefited from advice and discussions from M. Rast (Univ. Colorado), J. Fontenla, and R. Willson (ACRIM). We would like to thank V. George (Univ. Colorado) for editing and document preparation. We also appreciate helpful manuscript improvements from our reviewers. This research was supported by NASA contract NAS5-97045. Correspondence and requests for materials should be addressed to J.W.H. (*jerald.harder@lasp.colorado.edu*).

REFERENCES

Anderson G. P. and Hall L. A. (1989) Solar irradiance between 2000 and 3100 angstroms with spectral band pass of 1.0 angstroms. *J. Geophys. Res., 94,* 6435–6441.

Andrews D. G. (2010) *An Introduction to Atmospheric Physics, 2nd edition.* Cambridge Univ., Cambridge.

Arvensen J. C., Griffin R. N., and Pearson B. D. (1969) Determination of extraterrestrial solar spectral irradiance from a research aircraft. *Appl. Opt., 8,* 2215–2232.

Austin J., Tourpali K., Rozanov E., et al. (2008) Coupled chemistry climate model simulations of the solar cycle in ozone and temperature. *J. Geophys. Res., 113,* D11306, DOI: 10.1029/2007JD009391.

Bauer S. J. and Hantsch M. H. (1989) Solar cycle variations of the upper atmosphere temperature of Mars. *Geophys. Res. Lett., 16,* 373–376.

Bauer S. J. and Taylor H. A. (1981) Modulation of Venus ion densities with solar variations. *Geophys. Res. Lett., 8,* 840–842.

Baum W. A., Johnson F. S., Oberly J. J., Rockwood E. E., Strain C. V., and Tousey R. (1946) Solar ultraviolet spectrum to 88 kilometers. *Phys. Rev., 70,* 781–782.

Berk A., Anderson G. P., Acharya P. K., et al. (2006) *MODTRAN®5: 2006 Update, Proceedings of Algorithms and Technologies for Multispectral, Hyperspectral, and Ultraspectral Imagery XI* (S. S. Shen and P. E. Lewis, eds.), p. 62331F. SPIE Conf. Series 6233, DOI: 10.1117/12.665077.

Berrilli F., Ermolli I., Florio A., and Peitropaolo E. (1998) Geometrical properties of the chromospheric network cells from OAR/PSPT images. *Mem. Soc. Astron. Ital., 69,* 635–638.

Brasseur G. and Solomon S. (2005) *Aeronomy of the Middle Atmosphere, 3rd edition.* Springer, Berlin, ISBN-10: 1402032846.

Brown S. W., Eppeldauer G., Rice J. P., Zhang J., and Lykke K. (2004) Spectral Irradiance and Radiance Responsivity Calibrations using Uniform Sources (SIRCUS) facility at NIST. In *Earth Observing Systems* (W. L. Barnes and J. J. Butler, eds.), p. 363. SPIE Conf. Series 5542.

Brueckner G. E., Edlow K. L., Floyd L. E., Lean J. L., and VanHoosier M. E. (1993) The Solar Ultraviolet Spectral Irradiance Monitor (SUSIM) experiment on board the Upper Atmosphere Research Satellite (UARS). *J. Geophys. Res., 98,* 10695–10711.

Cameron R., Gizon L., and Duvall T. L. (2008) Helioseismology of sunspots: Confronting observations with three-dimensional HMD simulations of wave propagation. *Solar Phys., 251,* 291–308.

Cebula R. P., DeLand M. T., Hilsenrath E., Schlesinger B. M., Hudson R. D., and Heath D. F. (1991) Intercomparison of the solar irradiance measurements from the Nimbus-7 SBUV, the NOAA-9 and NOAA-11 SBUV/2, and the STS-34 SSBUV instruments: A preliminary study. *J. Atmos. Terr. Phys., 53,* 993–997.

Chamberlin P. C., Woods T. N., and Eparvier F. C. (2007) Flare Irradiance Spectral Model (FISM): Daily component algorithms and results. *Space Weather, 5,* S07005, DOI: 10.1029/2007SW000316.

Chamberlin P. C., Woods T. N., and Eparvier F. C. (2008) Flare Irradiance Spectral Model (FISM): Flare component algorithms and results. *Space Weather, 6,* S05001, DOI: 10.1029/2007SW000372.

Dewitte S., Crommelynck D., Mekaoui S., and Joukoff A. (2004) Measurement and uncertainty of the long-term total solar irradiance trend. *Solar Phys., 224,* 209–216.

Douglass D. H. and Clader B. D. (2002) Climate sensitivity of the Earth to solar irradiance. *Geophys. Res. Lett., 29,* DOI: 10.1029/2002GL015345.

Eddy J. A. (1976) The Maunder minimum. *Science, 192,* 1189–1202.

Ermolli I., Criscuoli S., Centrone M., Giorgi F., and Penza V. (2007) Photometric properties of facular features over the activity cycle. *Astron. Astrophys., 465,* 305–314, DOI: 10.1051/0004-6361:20065995.

Eyring V., Harris N. R. P., Rex M., et al. (2005) A strategy for process-oriented validation of coupled chemistry-climate models. *Bull. Am. Meteorol. Soc., 86,* 1117–1133.

Eyring V., Butchart N., Waugh D. W., et al. (2006) Assessment of temperature, trace species, and ozone in chemistry-climate model simulations of the recent past. *J. Geophys. Res., 111,* D22308, DOI: 10.1029/2006JD007327.

Fontenla J. M., Avrett E. H., and Loeser R. (1990) Energy balance in the solar transition region. I. Hydrostatic thermal models with ambipolar diffusion. *Astrophys. J., 355,* 700–718.

Fontenla J. M., Avrett E. H., and Loeser R. (1991) Energy balance in the solar transition region. II. Effects of pressure and energy input on hydrostatic models. *Astrophys. J., 377,* 712–725.

Fontenla J. M., Avrett E. H., and Loeser R. (1993) Energy balance in the solar transition region. III. Helium emission in hydrostatic, constant-abundance models with diffusion. *Astrophys. J., 406,* 319–345.

Fontenla J., White O. R., Fox P. A., Avrett E. H., and Kurucz R. L. (1999) Calculation of solar irradiances. I. Synthesis of the solar spectrum. *Astrophys. J., 518,* 480–499.

Fontenla J. and Harder J. (2005) Physical modeling of spectral irradiance variations. *Mem. Soc. Astron. Ital., 76,* 826–833.

Fontenla J. M., Avrett E., Thuillier G., and Harder J. (2006) Semiempirical models of the solar atmosphere. I. The quiet- and active Sun photosphere at moderate resolution. *Astrophys. J., 639,* 441–458, DOI: 10.1086/499345.

Fontenla J. M., Balasubramaniam K. S., and Harder J. (2007) Semiempirical models of the solar atmosphere. II. The quiet-Sun low chromosphere at moderate resolution. *Astrophys. J., 667,* 1243–1257, DOI: 10.1086/520319.

Fontenla J. M., Curdt W., Haberreiter M., Harder J., and Tian H. (2009) Semi-empirical models of the solar atmosphere. III. Set of non-LTE models for far-ultraviolet/extreme-ultraviolet irradiance computation. *Astrophys. J., 707,* 482–502, DOI: 10.1088/0004-637X/707/1/482.

Fontenla J. M., Harder J., Livingston W., Snow M., and Woods T. (2011) High-resolution solar spectral irradiance from extreme ultraviolet to far infrared. *J. Geophys. Res., 116,* D20108, DOI: 10.1029/2011JD016032.

Forster P., Ramaswamy V., Artaxo P., et al. (2007) Changes in atmospheric constituents and in radiative forcing. *In Climate Change 2007: The Physical Science Basis. Contribution of Working Group I to the Fourth Assessment Report of the Intergovernmental Panel on Climate Change* (S. D. Solomon et al., eds.). Cambridge Univ., Cambridge.

Foukal P. (1990) *Solar Astrophysics,* pp. 42–44. Wiley-Interscience, New York.

Fröhlich C. (2003) Long-term behavior of space radiometers. *Metrologia, 40,* S60–S65.

Fröhlich C. (2009) Evidence of a long-term trend in total solar irradiance. *Astron. Astrophys., 501,* L27–L30, DOI: 10.1051/0004-6361/200912318.

Garcia H. (2000) Thermal-spatial analysis of medium and large solar flares, 1976 to 1996. *Astrophys. J. Suppl., 127,* 189.

Garcia R. R. (2010) Solar surprise? *Nature, 467,* 668–669, DOI: 10.1038/467668a.

Garcia R. R., Marsh D. R., Kinnison D. E., Boville B. A., and Sassi F. (2007) Simulation of secular trends in the middle atmosphere, 1950–2003. *J. Geophys. Res., 112,* D09301, DOI: 10.1029/2006JD007485.

Geller M. A. (2006) Discussion of the solar UV/planetary wave mechanism. *Space Sci. Rev., 125,* 237–246.

Gingerich O. and de Jager C. (1968) The Bilderberg model of the photosphere and low chromosphere. *Solar Phys., 3,* 5–25.

Gingerich O., Noyes R. W., Kalkofen W., and Cuny Y. (1971) The Harvard-Smithsonian reference atmosphere. *Solar Phys., 18,* 347–365.

Goody R. M. and Yung Y. L. (1989) *Atmospheric Radiation, 2nd edition.* Oxford Univ., New York.

Gray L. J., Beer J., Geller M., et al. (2010) Solar influences on climate. *Rev. Geophys., 48,* RG4001, DOI: 10.1029/2009RG000282.

Grenfell J. L., Lehmann R., Mieth P., Langematz U., and Steil B. (2006) Chemical reaction pathway affecting stratospheric and mesospheric ozone. *J. Geophys. Res., 111,* D17311, DOI: 10.1029/2004JD005713.

Hagenaar H. J., Schrijver C. J., and Title A. M. (1997) The distribution of cell sizes of the solar chromomspheric network. *Astrophys. J., 481,* 988–995.

Haigh J. D. (1996) The impact of solar variability on climate. *Science, 272,* 981–984.

Haigh J. D. and Blackburn M. (2006) Solar influences on dynamical coupling between the stratosphere and the troposphere. *Space Sci. Rev., 125,* 331–344.

Haigh J. D., Winning A. R., Toumi R., and Harder J. W. (2010) An influence of solar spectral variations on radiative forcing of climate. *Nature, 467,* 696–699, DOI: 10.1038/nature09426.

Hall L. A. and Anderson G. P. (1991) High-resolution solar spectrum between 2000 and 3100 Å. *J. Geophys. Res., 96,* 12927–12931.

Harder J. W., Lawrence G., Fontenla J., Rottman G. J., and Woods T. N. (2005a) The Spectral Irradiance Monitor: Scientific requirements, instrument design, and operation modes. *Solar Phys., 230,* 141–167.

Harder J. W., Fontenla J., Lawrence G., Woods T., and Rottman G. (2005b) The Spectral Irradiance Monitor: Measurement equations and calibration. *Solar Phys., 230,* 169–204.

Harder J. W., Fontenla J. M., Pilewskie P., Richard E. C., and Woods T. N. (2009) Trends in solar spectral irradiance variability in the visible and infrared. *Geophys. Res. Lett., 36,* L07801, DOI: 10.1029/2008GL036797.

Harder J. W., Thuillier G., Richard E. C., Brown S. W., Lykke K. R., Snow M., McClintock W. E., Fontenla J. M., Woods T. N., and Pilewskie P. (2010) The SORCE SIM solar spectrum: Comparison with recent observations. *Solar Phys., 263,* 3–24, DOI: 10.1007/s11207-010-9555-y.

Harries J., Carli B., Rizzi R., Serio C., Mlynczak M., Palchetti L., Maestri T., Brindley H., and Masiello G. (2008) The far-infrared Earth. *Rev. Geophys., 46,* RG4004, DOI: 10.1029/2007RG000233.

Harvey K. and Harvey J. (1973) Observations of moving magnetic features near sunspots. *Solar Phys., 28,* 61–71.

Heath D. F. and Schlesinger B. M. (1986) The Mg 280-nm doublet as a monitor of changes in solar ultraviolet irradiance. *J. Geophys. Res., 91,* 8672–8682.

Hengstberger F., ed. (1989) *Absolute Radiometry: Electrically Calibrated Thermal Detectors of Optical Radiation.* Academic, San Diego, ISBN 978-0123408105.

Henoux J. C. and Somov B. V. (1991) The photospheric dynamo. *Astron. Astrophys., 241,* 613–617.

Hickey J. R., Alton B. M., Kyle H. L., and Hoyt D. (1988) Total solar irradiance measured by ERB/NIMBUS-7. *Space Sci. Rev., 48,* 321–342.

Hinteregger H. E., Fukui K., and Gilson B. R. (1981) Observational, reference and model data on solar EUV form measurmentnts on AE-E. *Geophys. Res. Lett., 8,* 1147–1150.

Hood L. L., Soukharev B. E., and McCormack J. P. (2010) Decadal variability of the tropical stratosphere: Secondary influence of the El Niño–Southern Oscillation. *J. Geophys. Res., 115,* D11113, DOI: 10.1029/2009JD012291.

Ichimoto K., Tsuneta S., Suematsu Y., et al. (2008) Net circular polarization of sunspots in high spatial resolution. *Astron. Astrophys., 481,* L9–L12.

Ineson S., Scaife A. A., Knight J. R., Manners J. C., Dunstone N. J., Gray L. J., and Haigh J. D. (2011) Solar forcing of winter climate variability in the northern hemisphere. *Nature Geosci., 4,* 753–757, DOI: 10.1038/ngeo1282.

Judge D. L., McMullin D. R., Ogawa H. S., et al. (1998) First solar EUV irradiances obtained from SOHO by the CELIAS/SEM. *Solar Phys., 177,* 161.

Kjeldseth-Moe O. and Maltby P. (1969) A model for the penumbra of sunspots. *Solar Phys., 8,* 275–283.

Kjeldseth-Moe O. and Maltby P. (1974a) The temperature of penumbral filaments. *Solar Phys., 36,* 101–108.

Kjeldseth-Moe O. and Maltby P. (1974b) Models for different sunspot umbrae. *Solar Phys., 36,* 109–114.

Kodera K. (2006) The role of dynamics in solar forcing. *Space Sci. Rev., 125,* 319–330.

Kopp G. and Lawrence G. (2005) The Total Irradiance Monitor (TIM): Instrument design. *Solar Phys., 230,* 91–109.

Kopp G. and Lean J. L. (2011) A new, lower value of total solar irradiance: Evidence and climate significance. *Geophys. Res. Lett., 38,* L01706, DOI: 10.1029/2010GL045777.

Kopp G., Fehlmann A., Finsterle W., Harber D., Heuerman K., and Willson R. (2012) Total solar irradiance data record accuracy and consistency improvements. *Metrologia, 49,* S29–S33, DOI: 10.1088/0026-1394/49/2/S29.

Krivova N. A., Solanki S. K., Fligge M., and Unruh Y. C. (2003) Reconstruction of solar total and spectral irradiance variations in cycle 23: Is solar surface magnetism the cause? *Astron. Astrophys., 399,* L1–L4.

Krivova N. A., Solanki S. K., and Floyd L. (2006) Reconstruction of solar UV irradiance in cycle 23. *Astron. Astrophys., 452,* 631–639.

Krivova N. A., Solanki S. K., Wenzler T., and Podlipnik B. (2009) Reconstruction of solar UV irradiance since 1974. *J. Geophys. Res., 114,* D00I04, DOI: 10.1029/2009JD012375.

Kurucz R. L. (1992a) Atomic and molecular data for opacity calculations. *Rev. Mex. Astron. Astrofis., 23,* 45–48.

Kurucz R. L. (1992b) "Finding" the "missing" solar ultraviolet opacity. *Rev. Mex. Astron. Astrofis., 23,* 181–186.

Kurucz R. L. (1992c) Remaining line opacity problems for the solar spectrum. *Rev. Mex. Astron. Astrofis., 23,* 187–194.

Labitzke K. (1987) Sunspots, the QBO, and the stratospheric temperature in the north polar region. *Geophys. Res. Lett., 14,* 535–537.

Labitzke K. (2006) Solar variation and stratospheric response. *Space Sci. Rev., 125,* 247–260.

Lean J. (2000) Evolution of the Sun's spectral irradiance since the Maunder minimum. *Geophys. Res. Lett., 27,* 2425–2428.

Lean J. L., White O. R., Livingston W. C., Heath D. F., Donnelly R. F., and Skumanich A. (1982) A three-component model for the variability of the solar ultraviolet flux: 145–200 nm. *J. Geophys. Res., 87,* 10307–10317.

Lean J. L., Rottman G. J., Kyle H. L., Woods T. N., Hickey J. R., and Puga L. C. (1997) Detection and parameterization of variations in solar mid and near-ultraviolet radiation (200–400 nm). *J. Geophys. Res., 102,* 29939–29956.

Lee H. and Smith A. K. (2003) Simulation of the combined effects of solar cycle, quasi-biennial oscillation, and volcanic forcing on stratospheric ozone changes in recent decades. *J. Geophys. Res., 108,* 4049, DOI: 10.1029/2001JD001503.

Leighton R. (1964) Transport of magnetic fields on the Sun. *Astrophys. J., 140,* 1547–1562.

Leighton R. B., Noyes R. W., and Simon G. W. (1962) Velocity fields in the solar atmosphere. I. Preliminary report. *Astrophys. J., 135,* 474–499.

Lemen J. R., Title A. M., Akin D. J., et al. (2012) The Atmospheric Imaging Assembly (AIA) on the Solar Dynamics Observatory (SDO). *Solar Phys., 275,* 17–40.

Maltby P., Avrett E. H., Carlsson M., Kjeldseth-Moe O., Kurucz R. L., and Loeser R. (1986) A new sunspot umbral model and its variation with the solar cycle. *Astrophys. J., 306,* 284–303.

Marsh D. R., Garcia R. R., Kinnison D. E., Boville B. A., Sassi F., Solomon S. C., and Matthes K. (2007) Modeling the whole atmosphere response to solar cycle changes in radiative and geomagnetic forcing. *J. Geophys. Res., 112,* D23306, DOI: 10.1029/2006JD008306.

Matthes K. (2011) Solar cycle and climate predictions. *Nature Geosci., 4,* 735–736.

McClintock W. E., Rottman G. J., and Woods T. N. (2005) Solar-Stellar Irradiance Comparison Experiment II (SOLSTICE II): Instrument concept and design. *Solar Phys., 230,* 225–258.

Meehl G. A., Arblaster J. M., Branstator G., and van Loon H. (2008) A coupled air-sea response mechanism to solar forcing in the Pacific region. *J. Climate, 21,* 2883–2897.

Meehl G. A., Arblaster J. M., Matthes K., Sassi F., and van Loon H. (2009) Amplifying the Pacific climate system response to a small 11-year solar cycle forcing. *Science, 325,* 1114–1118.

Mekaoui S., Dewitte S., Crommelynck D., Chavalier A., Conscience C., and Joukoff A. (2004) Absolute accuracy and repeatability of the RMIB radiometers for TSI measurements. *Solar Phys., 224,* 237–246.

Merkel A. W., Harder J. W., Marsh D. R., Smith A. K., Fontenla J. M., and Woods T. N. (2011) The impact of solar spectral irradiance variability on middle atmospheric ozone. *Geophys. Res. Lett., 38,* L13802, DOI: 10.1029/2011GL047561.

Mihalas D. (1978) *Stellar Atmospheres.* W. H. Freeman, San Francisco. 76 pp.

Neckel H. and Labs D. (1984) The solar radiation between 3300 and 12500 Å. *Solar Phys., 90,* 205–258.

Pagaran J., Harder J. W., Webber M., Floyd L. E., and Burrows J. P. (2011) Intercomparison of SCIAMACHY and SIM vis-IR irradiance over several solar rotational timescales. *Astron. Astrophys., 528,* A67.

Parker E. N. (1974) The nature of the sunspot phenomenon. I: Solutions of the heat transport equation. *Solar Phys., 36,* 249–274.

Parker E. N. (1979) Sunspots and the physics of magnetic flux tubes. I. The general nature of the sunspot. *Astrophys. J., 230,* 905–913.

Penner S. S. (1959) *Quantitative Molecular Spectroscopy and Gas Emissivities,* pp. 1–15. Addison-Wesley, Reading, Massachusetts.

Preminger D. G., Chapman G. A., and Cookson A. M. (2011) Activity-brightness correllations for the Sun and Sun-like stars. *Astrophys. J. Lett., 739,* L45, DOI: 10.1088/2041-8205/739/2/L45.

Rast M. P., Fox P. A., Lin H., Lites B. W., Meisner R. W., and White O. R. (1999) Bright rings around sunspots. *Nature, 401,* 678–679.

Rast M. P., Meisner R. W., Lites B. W., Fox P. A., and White O. R. (2001) Sunspot bright rings: Evidence from case studies. *Astrophys. J., 557,* 864–879.

Rast M. P., Ortiz A., and Meisner R. W. (2008) Latitudinal variation of the solar photospheric intensity. *Astrophys. J., 673,* 1209–1217.

Reeves E. M., Humber M. C. E., and Timothy J. G. (1977) Extreme UV spectroheliometer on the Apollo Telescope Mount. *Appl. Optics, 16,* 837–848.

Reid G. C. (1999) Solar variability and its implications for the human environment. *J. Atmos. Solar Terr. Phys., 61,* 3–14.

Rempel M., Schüssler M., and Knölker M. (2009) Radiative magnetohydrodynamic simulation of sunspot structure. *Astrophys. J., 691,* 640–649.

Remsberg E. and Lingenfelser G. (2010) Analysis of SAGE II ozone of the middle and upper stratosphere for its response to a decadal-scale forcing. *Atmos. Chem. Phys., 10,* 11779–11790, DOI: 10.5194/acp-10-11779-2010.

Roble R. G. (1995) Energetics of the mesosphere and thermosphere. In *The Upper Mesosphere and Lower Thermosphere: A Review of Experiment and Theory* (R. M. Johnson and T. L. Killeen, eds.), pp. 1–21. AGU Geophys. Monogr. 87, American Geophysical Union, Washington, DC.

Rottman G. J., Woods T. N., and Sparn T. P. (1993) Solar-Stellar Irradiance Comparison Experiment I: 1. Instrument design and operation. *J. Geophys. Res., 98,* 10667–10677.

Rust D. M. (2001) A new paradigm for solar filament eruptions. *J. Geophys. Res., 106,* 25075–25088.

Rutten R. J. (2003) *Radiative Transfer in Stellar Atmospheres.* Available online at *www.staff.science.uu.nl/~rutte101/Course_notes.html.*

Sánchez Cuberes M., Vaquez M., Bonet J. A., and Sobotka M. (2002) Infrared photometry of solar photospheric structures. II. Center-to-limb variation of active regions. *Astrophys. J., 570,* 886–899.

Scherrer P. H., Bogart R. S., Bush R. I., et al. (1995) The solar oscillations investigation — Michelson Doppler Imager. *Solar Phys., 162,* 129–188.

Schmidt H. and Brasseur G. P. (2006) The response of the middle atmosphere to solar cycle forcing in the Hamburg model of the neutral and ionized atmosphere. *Space Sci. Rev., 125,* 345–356, DOI: 10.1007/s11214-006-9068-z.

Schrijver C. J. (1989) The effect of an interaction of magnetic flux and supergranulation on the decay of magnetic plages. *Solar Phys., 122,* 193–208.

Schrijver C. J., Title A.M., Harvey K. L., et al. (1998) Large-scale coronal heating by the small-scale magnetic field of the Sun. *Nature, 394,* 152–154.

Sellers W. D. (1965) *Physical Climatology.* Univ. of Chicago, Chicago.

Shapiro A. V., Rozanov E. V., Shapiro A. I., et al. (2013) The role of the solar irradiance variability in the evolution of the middle atmosphere during 2004–2009. *J. Geophys. Res.–Atmos., 118,* DOI: 10.1002/jgrd.50208.

Shindell D., Rind D., Balachandran N., Lean J., and Lonergan P. (1999) Solar cycle variability, ozone, and climate. *Science, 284,* 305–308.

Skumanich A., Lean J. L., White O. R., and Livingston W. C. (1984) The Sun as a star: Three-component analysis of chromospheric variability in the calcium K line. *Astrophys. J., 282,* 776–783.

Snow M., McClintock W. E., Woods T. N., White O. R., Harder J. W., and Rottman G. (2005) The Mg II index from SORCE. *Solar Phys., 230,* 325–344.

Solanki S. K. and Unruh Y. C. (1998) A model of the wavelength dependence of solar irradiance variations. *Astron. Astrophys., 329,* 747–753.

Soukharev B. E. and Hood L. L. (2006) Solar cycle variation of stratospheric ozone: Multiple regression analysis of longterm satellite data sets and comparisons with models. *J. Geophys. Res., 111,* D20314, DOI: 10.1029/2006JD007107.

Stallcop J. R. (1974) Absorption of infrared radiation by electrons in the field of a neutral hydrogen atom. *Astrophys. J., 187,* 179–183.

Swartz W. H., Stolarski R. S., Oman L. D., Fleming E. L., and Jackman C. H. (2012) Middle atmosphere response to different descriptions of the 11-yr solar cycle in spectral irradiance in a chemistry-climate model. *Atmos. Chem. Phys., 12,* 5937–5948, DOI: 10.5194/acp-12-5937-2012.

Tapping K. F. (1987) Recent solar radio astronomy at centimeter wavelengths: The temporal variability of the 10.7-cm flux. *J. Geophys. Res., 92,* 829–838.

Thuillier G., Floyd L., Woods T. N., Cebula R., Hilsenrath E., Hersé M., and Labs D. (2004) Solar irradiance reference spectra for two solar active levels. *Adv. Space Res., 34,* 256–261.

Topka K. P., Tarbell T. D., and Title A. M. (1997) Properties of the smallest solar magnetic elements. II. Observations versus hot wall models of faculae. *Astrophys. J., 484,* 479, DOI: 10.1086/304295.

Trujillo Bueno J., Shchukina N., and Asensio Ramos A. (2004) A substantial amount of hidden magnetic energy in the quiet Sun. *Nature, 430,* 326–329.

Uitenbroek H. and Criscuoli S. (2011) Why one-dimensional models fail in the diagnosis of average spectra from inhomogeneous stellar atmospheres. *Astrophys. J., 736,* 69, DOI: 10.1088/0004-637X/736/1/69.

Vernazza J. E., Avrett E. H., and Loeser R. (1973) Structure of the solar chromosphere. I. Basic computations and summary of the results. *Astrophys. J., 184,* 605–631.

Vernazza J. E., Avrett E. H., and Loeser R. (1976) Structure of the solar chromosphere. II. The underlying photosphere and temperature-minimum region. *Astrophys. J. Suppl. Ser., 30,* 1–60.

Vernazza J. E., Avrett E. H., and Loeser R. (1981) Structure of the solar chromosphere. III. Models of the EUV brightness components of the quiet Sun. *Astrophys. J. Suppl. Ser., 45,* 635–725.

Viereck R. A., Floyd L. E., Crane P. C., et al. (2004) A composite Mg II index spanning from 1978 to 2003. *Space Weather, 2,* S10005, DOI: 10.1029/2004SW000084.

Wang H., Zirin H., and Ai G. (1991) Magnetic flux transport of decaying active regions and enhanced magnetic network. *Solar Phys., 131,* 53–68.

Wang S., Li K.-F., Pongetti T. J., et al. (2013) Midlatitude atmospheric OH response to the most recent 11-y solar cycle. *Proc. Natl. Acad. Sci., 110(6),* 2023–2028, DOI: 10.1073/pnas.1117790110.

Wang Y.-M. and Sheeley N. R. (1991) Magnetic flux transport and the Sun's dipole moment: New twists to the Babcock-Leighton model. *Astrophys. J., 375,* 761–770.

Wenzler T., Solanki S. K., and Krivova N. A. (2005) Can surface magnetic fields reproduce solar irradiance variations in cycles 22 and 23? *Astron. Astrophys., 432,* 1057–1061.

Wild M., Folini D., Schär C., Loeb N., Dutton E., and König-Langlo G. (2012) The global energy balance from a surface perspective. *Climate Dynam., 323,* DOI: 10.1007/s00382-012-1569-8.

Willson R. C. (1979) Active cavity radiometer type IV. *Appl. Opt., 18,* 179–188.

Willson R. C. and Hudson H. S. (1991) The Sun's luminosity over a complete solar cycle. *Nature, 351,* 42.

Willson R. C. and Mordvinov A. V. (2003) Secular total solar irradiance trend during solar cycles 21–23. *Geophys. Res. Lett., 30,* 1199, DOI: 10.1029/2002GL016038.

Woods T. N. and Rottman (2001) Solar ultraviolet variability over time periods of aeronomic interest. In *Comparative Aeronomy in the Solar System* (M. Mendillo et al., eds.), AGU Geophys. Monogr. 130, American Geophysical Union, Washington, DC.

Woods T. N., Prinz D. K., Rottman G. J., et al. (1996) Validation of the UARS solar ultraviolet irradiances: Comparison with the ATLAS 1 and 2 measurements. *J. Geophys. Res., 101,* 9541–9569.

Woods T. N. , Tobiska W. K., Rottman G. J., and Worden J. R. (2000) Improved solar Lyman α irradiance modeling from 1947 through 1999 based on UARS observations. *J. Geophys. Res., 105,* 27195–27215.

Woods T. N., Eparvier F. G., Bailey S. M., Chamberlin P. C., Lean J., Rottman G. J., Solomon S. C., Tobiska W. K., and Woodraska D. L. (2005) Solar EUV Experiment (SEE): Mission overview and first results. *J. Geophys. Res., 110,* A01312, DOI: 10.1029/2004JA010765.

Woods T. N., Chamberlin P. C., Harder J. W., Hock R. A., Snow M., Eparvier F. G., Fontenla J., McClintock W. E., and Richard E. C. (2009) Solar irradiance reference spectra (SIRS) for the 2008 Whole Heliosphere Interval (WHI). *Geophys. Res. Lett., 36,* L01101, DOI: 10.1029/2008GL036373.

Woods T. N., Eparvier F. G., Hock R., et al. (2012) Extreme Ultraviolet Variability Experiment (EVE) on the Solar Dynamics Observatory (SDO): Overview of science objectives, instrument design, data products, and model developments. *Solar Phys., 275,* 115–143, DOI: 10.1007/s11207-009-9487-6.

Worden J. R., White O. R., and Woods T. N. (1998) Evolution of chromospheric structures derived from Ca II K spectroheliograms: Implications for solar ultraviolet irradiance variability. *Astrophys. J., 496,* 998–1014.

Zhong W., Osprey S. M., Gray L. J., and Haigh J. D. (2008) Influence of the prescribed solar spectrum on calculations of atmospheric temperature. *Geophys. Res. Lett., 35,* L22813, DOI: 10.1029/2008GL035993.

Zwaan C. (1985) The emergence of magnetic flux. *Solar Phys., 100,* 397–414.

Zwaan C. (1987) Elements and patterns in the solar magnetic field. *Annu. Rev. Astron. Astrophys., 25,* 83–111.

Tian F., Chassefière E., Leblanc F., and Brain D. A. (2013) Atmosphere escape and climate evolution of terrestrial planets. In *Comparative Climatology of Terrestrial Planets* (S. J. Mackwell et al., eds.), pp. 567–581. Univ. of Arizona, Tucson, DOI: 10.2458/azu_uapress_9780816530595-ch23.

Atmosphere Escape and Climate Evolution of Terrestrial Planets

Feng Tian
National Astronomical Observatories/Tsinghua University

Eric Chassefière
Universite Paris

Francois Leblanc
Centre National de la Recherche Scientifique/Institut Pierre Simon Laplace

David A. Brain
University of Colorado

The climate of a planet is primarily determined by its orbital distance from its star, the luminosity of the star, the existence of oceans, the pressure of its atmosphere, and the composition of its atmosphere. The last two components are what could be impacted by atmosphere escape. The Sun, as the dominant energy source driving the climate of terrestrial planets, was not always as bright as it is today. Stellar evolution theory predicts that the luminosity of the young Sun was 75% of its present luminosity, at approximately 4 b.y. ago (4 Ga) (*Gough,* 1981). Although the Sun could have lost some of its mass, thus making the very young Sun somewhat more massive than it is now and therefore could have emitted more energy, most of this mass loss was completed prior to 4 Ga (*Wood et al.,* 2005). Thus the Sun has provided increasingly more energy to solar system planets during the past 4 b.y. Contrary to the evolutionary trend of the total luminosity increasing with time, the young Sun should have emitted much stronger EUV, soft X-ray, and far-UV photons than at present. These photons are from the upper atmosphere of the Sun and are linked to solar magnetic activity. Generally speaking, a young star rotates much faster and thus has stronger magnetic activity. Observations of solar-type stars with different ages show that the EUV energy flux from a 0.5-b.y.-old solar-type star could be as much as 20 times that of the present Sun (*Ribas et al.,* 2005). Accompanying this much-enhanced solar extreme ultraviolet (XUV) radiation is a much stronger solar wind, with mass flux up to 1000 times more intense than the present solar wind flux (*Wood et al.,* 2005). It can be expected that many more energetic-particle events were caused by the young Sun. The fate of the atmospheres of terrestrial planets in such an environment and the consequences for their climates are the focus of this chapter.

1. ATMOSPHERE ESCAPE PROCESSES

From a kinetic point of view, any particle at the exobase with outgoing velocity exceeding the escape velocity of a planet can escape. The exobase is an altitude in the atmosphere beyond which few collisions between particles occur. High particle velocity can be achieved through a number of processes. Those processes in which the high velocity is linked to the exobase temperature are called thermal escape. An extreme limit of thermal escape is Jeans escape, in which the escaping particles do not influence the velocity distribution near the exobase or the temperature in the upper thermosphere (for an illustration of Jeans escape, see Fig. 1). The theory of planetary exospheres and Jeans escape was elaborated upon in the 1960s (*Chamberlain,* 1963; *Öpik and Singer,* 1961). The other extreme of thermal escape is hydrodynamic escape (discussed in more detail later in this chapter), which is typically driven by strong energy deposition in the thermosphere, leading to large escape rates.

Those processes in which the high velocity is not linked to the exobase temperature are called nonthermal escape (some of these processes are illustrated in Fig. 1). In most cases these processes are associated with the presence of ion species and their behavior in the electric and magnetic field. The effect of magnetic field on atmosphere escape and the climate change of terrestrial planets are also discussed in Brain et al. (this volume). Neutral atoms and molecules in the atmosphere of the planet may be ionized through photoionization, impact ionization, or charge exchange with solar ions. Once formed around a nonmagnetized planet (or a weakly magnetized one), the ions can be dragged along by solar magnetic field lines wrapping the planet and could partially escape the planet (under the control of an interplanetary magnetic field some of these ions will recollide with the planetary atmosphere and thus will not be lost). This ion pickup process is less efficient on a planet with a strong intrinsic magnetic field, such as Earth, because the solar or interplanetary electric and magnetic fields are kept at distances far from the planet. On planets with strong intrinsic magnetic fields, ions could escape the planet by flowing outward to space through open magnetic field lines in the polar region (the polar wind). Both ionospheric outflow and pick-up ion escape result in the direct loss of ions to space, with the ion outflow process starting to operate at altitudes much lower than the exobase. Those energetic ions trapped by the magnetic field of the planet can also escape by exchanging charges with atoms (charge exchange). On the other hand, sputtering is a process in which loss of neutral atoms occurs as a result of collisions between exospheric neutral and energetic ions formed in the exosphere or magnetosphere. Once formed, these ions can be accelerated by either the solar wind electric field or other processes in the planetary magnetosphere, and some of them can impact the upper atmosphere. Usually a great number of neutral atoms can be produced per incident energetic ion (see, e.g., *Luhmann and Kozyra,* 1991). Note that although the solar wind is a strong source of energetic ions, such as H^+ and He^{++}, only a few percent of incoming solar wind ions could reach the exobase of a planetary atmosphere. Thus the sputtering process is mainly driven by energetic ions produced locally.

Photochemical escape is another type of nonthermal escape process, the best example of which is the production of O atoms from dissociative recombination of O_2^+ ions in the Mars atmosphere with kinetic energy higher than their potential energy in the Mars gravity field. Because the collisional cross sections of high-energy O atoms are small,

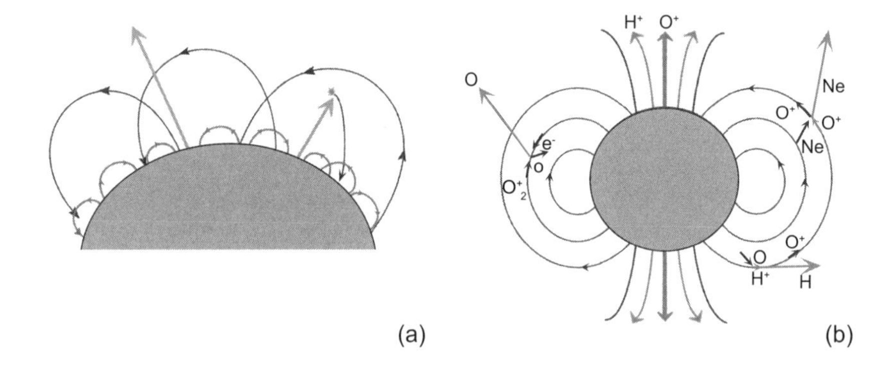

Fig. 1. Cartoons of some atmospheric escape processes. **(a)** Thermal escape. The gray arrows represent particles with sufficient kinetic energy to overcome the gravity field of the planet (here the curvature of the particle trajectory is ignored). The black curves represent particles with insufficient energy to escape the planet (long darker ones have more kinetic energy than the short dashed ones). These particles are at the same temperature, but probability distribution determines that only a fraction of the total population can obtain high kinetic energy. Some particles with high kinetic energy may experience collisions on their way toward space. Collisions will take kinetic energy away from these particles and cause them to stay in the planetary atmosphere. The collisional frequency is proportional to atmospheric density, so for pressure altitudes greater than the exobase height, as depicted in **(a)**, collisions are rare. **(b)** Some nonthermal escape processes. The dark black curves represent the magnetic field of a planet. The long straight gray arrows pointing upward and downward near the polar regions represent the polar wind. Ions with high kinetic energy can escape a planet through the open magnetic field lines. The magnetic field lines are closed at mid-low-latitude regions. There the ions cannot escape directly unless they are highly energetic. The lower right of the cartoon shows the charge exchange reaction, in which a proton with high kinetic energy collides with a slowly moving oxygen atom. They exchange charge so that the proton becomes a hydrogen atom with high kinetic energy and can thus escape the planet and its magnetic field. The oxygen atom becomes an oxygen ion and moves along the magnetic field line. The right side shows the sputtering process, in which an energetic oxygen ion, accelerated by either ionospheric or magnetospheric processes, moves along the magnetic field line and collides with a slowly moving Ne atom. The result of the collision is the acceleration of the Ne atom, which can then escape from the planet. The oxygen ion loses part of its kinetic energy and becomes a slowly moving particle along the magnetic field line. In reality a highly energetic oxygen ion may experience many collisions and thus cause many particles to escape before its energy is indistinguishable from the ions in the environment. The left side shows an example of photochemical escape. In this case a molecular oxygen ion is moving along the magnetic field line and collides with an electron. The dissociative recombination reaction leads to the formation of two oxygen atoms with one having more energy than the other. On a small mass planet such as Mars, the kinetic energy of one of the oxygen atoms is high enough for the particle to escape. On a more massive planet this process is less efficient. Cartoons courtesy of the Lunar and Planetary Institute.

the escaping O atoms can be produced somewhat deep in the thermosphere of Mars. For more massive planets, the kinetic energy of atoms from dissociative recombination processes may not be sufficient for photochemical escape to be important.

On Mars and Earth, the main escape process for H during solar maximum is Jeans escape. The altitude of the exobase is 250 km on Mars and 500 km on Earth (*Chassefière and Leblanc,* 2004). During solar minimum, H escapes from Earth and Mars through nonthermal escape processes because of reduced exobase temperatures. On Earth, the sum of H escape rates through both thermal and nonthermal escape processes remain constant under different solar activity levels. Such a phenomenon can be explained by the diffusion-limited escape theory (*Hunten,* 1973), i.e., the total escape from the exobase is controlled by the diffusion of the escaping species from the background gases at the homopause, an altitude below which different species are well mixed by turbulence. As we will see later, the situations on other terrestrial planets are more complicated.

On Venus, thermal escape is negligible even at solar maximum because the atmosphere of Venus is dominated by CO_2, which is an efficient infrared cooling agent, and the venusian exobase is always much colder than that of Earth. The exobase of Mars is also cold. But the weak gravitational field of Mars allows substantial thermal escape. A thermal escape parameter λ_{ex} can be defined as the ratio of the gravitational energy of a particle (GMm/r_{ex}, where M, m, and r_{ex} are the mass of the planet, the mass of the particle, and the radius of the planet respectively) to its thermal energy (kT_{ex}, where k is the Boltzmann constant and T_{ex} the exobase temperature). For H atoms, $\lambda_{ex} \approx 22$ for Venus, but only ≈5 and 7 for Mars and Earth, respectively, during solar maximum (*Chassefière and Leblanc,* 2004). Because the thermal escape flux is proportional to $\exp(-\lambda_{ex})$, the present thermal escape flux at Venus is ≈6 orders of magnitude smaller than on Mars and Earth.

Hydrodynamic escape occurs for an escape parameter λ_{ex} on the order of unity. Low values of λ_{ex} may be reached for H or H_2-rich thermospheres heated by a strong XUV solar flux, e.g., in the following cases: (1) primordial H_2/He atmospheres; or (2) outgassed H_2O-rich atmosphere during an episode of runaway and/or the wet greenhouse effect. Hydrodynamic escape may have been quite efficient in removing large amounts of water from the primitive Venus atmosphere (*Kasting and Pollack,* 1983), with a possible significant contribution of the solar wind as an additional source of energy (*Chassefière,* 1996, 1997). The concept of hydrodynamic escape was first proposed in *Parker* (1964) and numerical studies were first carried out by *Watson et al.* (1981). Because the escape rate of certain atmospheric species depends exponentially on its mass, isotopic fractionation of H naturally occurs in Jeans escape and nonthermal escape processes. In addition, D/H escape rates are strongly dependent on gravity fractionation in the thermosphere (*Krasnopolsky and Feldmann,* 2001). On the other hand, hydrodynamic escape does not result in any significant isotopic fractionation of H. But isotopic fractionation of heavier elements like noble gases during hydrodynamic escape could have occurred (*Hunten,* 1973; *Zahnle and Kasting,* 1986; *Hunten et al.,* 1987).

It was proposed that at intermediate escape parameter $\lambda_{ex} \sim 10$, a slow hydrodynamic escape could occur, in which the outflow could reach the sonic point at a distance further than the exobase level. This condition is satisfied by atmosphere escape from Pluto and Titan (*Krasnopolsky,* 1999; *Strobel,* 2008). A recent molecular model (*Volkov et al.,* 2011) using the direct simulation Monte Carlo (DSMC) method shows that the transition between hydrodynamic escape and Jeans escape occurs in a rather narrow range of escape parameter (between 2 and 3) and the escape rate obtained by the DSMC model is much smaller than those obtained by the slow hydrodynamic escape model (*Tucker and Johnson,* 2009; *Tucker et al.,* 2013).

When the escape parameter is in the intermediate range, the energy budget in the upper thermosphere could be strongly influenced by the escaping particles. *Tian et al.* (2008) showed that if the energy associated with the outward advection of the upper thermosphere, which is a result of enhanced Jeans escape at the exobase, is neglected, the temperature and exobase level can both grow rapidly with increasing XUV until traditional hydrodynamic escape occurs. However, when the advection cooling is included in the energy budget, the occurrence of the traditional hydrodynamic escape is delayed until much higher XUV radiation levels. Based on the importance of the escape on the energy budget, the upper thermosphere could be considered in a hydrodynamic regime instead of a hydrostatic regime, in which case it is necessary to couple an upper atmosphere model with a kinetic model near the exobase level in order to appropriately evaluate the escape rate. Nonthermal escape processes would help the transition of planetary upper atmospheres from a hydrostatic regime to a hydrodynamic regime.

In the following we will describe different studies on atmosphere escape rates through different physical processes on Venus, Mars, and Earth. In section 2 we will describe the current knowledge regarding atmosphere escape histories of these terrestrial planets. In section 3 we will summarize our understanding of how climate changed on the solar system terrestrial planets based on their atmosphere escape history. But before we dive in, it is useful to mention that atmosphere escape is difficult to measure. Ideally, one would like to build a "net" covering the whole sky and monitor all different species going through the net all at one time. In reality, one or more satellites are launched into certain orbits and measure the flux (number density + velocity) of certain species moving in certain directions. Such measurements, when integrated over time and included into three-dimensional models, assuming that different atmosphere escape processes are working in a time-independent fashion, can hopefully provide a global picture. Certainly atmosphere escape processes change over time and may actually interact with each other. For example, the sputtering process depends on the formation of energetic ions, which themselves could be escaping through the ion pickup process from Mars. Thus the atmosphere escape problem is poorly constrained by observations, and every analysis the readers will find is likely to come to a different conclusion without further constraining measurements. Extrapolating the knowledge on atmosphere escape from terrestrial planets at the current time to their histories in order to infer their climate histories normally depends on computer simulations. And *all* models/calculations must by their nature make many simplifying assumptions and must fill in missing information about the basic parameters and/or the physics. A good example is that even with all our *in situ*, detailed observation capabilities on Earth, we still have a poor understanding of Earth's atmosphere and climate evolution.

1.1. Venus

At present it is thought that atmospheric escape from Venus is dominated by the loss of ions, with a smaller contribution from the escape of neutral species. The loss rate of neutral O atoms by sputtering from Venus' atmosphere is estimated to be on the order of 25% of the loss rate of O^+ ions (*Lammer et al.,* 2006). Neutral H atoms are also expected to escape through charge exchange and photochemical reactions at a rate of 50% of the H^+ loss (*Lammer et al.,* 2006). These values are highly uncertain. Indeed, neutral escape (in the form of a few electron volts or tens of electron volts of H and O atoms) has never been directly measured from orbiters, neither on Mars nor on Venus, and sputtering models present uncertainties by typically one or two orders of magnitude.

From recent measurements by the Analyzer of Space Plasmas and Energetic Atoms (ASPERA) instrument package on Venus Express, the H^+ escape flux from Venus' upper atmosphere by nonthermal escape processes at minimum solar activity is $\approx 0.3–1 \times 10^{25}$ s^{-1} (*Barabash et al.,* 2007; *Lundin,* 2011). This lower limit corresponds to an average surface flux of $\approx 0.6–2 \times 10^6$ cm^{-2} s^{-1} and a solar cycle averaged value of $2–6 \times 10^6$ cm^{-2} s^{-1}. The plasma analyzer and ion-neutral spectrometer of the Pioneer Venus mission obtained a value from 7×10^6 to 4×10^7 cm^{-2} s^{-1} in the solar cycle average (*Donahue,* 1999). According to *Donahue et al.* (1997), H^+ ion densities at solar maximum are about six times their values at solar minimum, explaining why Venus Express values are smaller than Pioneer Venus values.

Interestingly, the O^+ escape flux measured by ASPERA (*Barabash et al.,* 2007) is compatible with a stoichiometric proportion of the loss rates of the two species (2 H for 1 O), as expected from theoretical photochemical considerations for planetary atmospheres where water vapor is the main source of H and O (*McElroy and Donahue,* 1972). Despite large uncertainties about the extrapolation to the whole planet and averaged solar cycle conditions of ASPERA measurements, and about the relative fractions of neutral escape with respect to ion escape, this result suggests that the present estimates of H and O global escape rates are self-consistent and that there is no need for O left behind by H escape to react with the surface. Whether H and O escape from early Venus' history at the stoichiometric proportion is unclear. From Venus Express and Pioneer Venus measurements, these escape rates may be translated in an escape rate of H_2O in the solar cycle average in the range from 10^6 to 10^7 cm^{-2} s^{-1}.

The escape rate of He^+ has been also measured by ASPERA and represents a little less than 10% of the O^+ escape rate. Note that the He escape rate could be partially balanced by the implantation of He^{++} from the solar wind (*Chanteur et al.,* 2009). The D fractionation factor is on the order of 0.1 (D escapes 10 times less than H relative to their global content), which results in H fractionation and a progressive increase of the D/H ratio in the atmosphere.

1.2. Earth

Hydrogen escape from Earth today is limited by diffusion at the homopause level at a rate of 3×10^8 cm^{-2} s^{-1}. Unlike Venus, where neutral loss is always less efficient than ion loss, loss of H from present Earth is always dominated by neutrals during solar maximum by Jeans escape. During solar minimum it is dominated by charge exchange, followed by polar wind as the second most important escape channel (*Chamberlain and Hunten,* 1987). Ion pickup, explained in an earlier section, is unimportant on Earth because the solar wind is deflected by the planet's intrinsic magnetic field at more than 10 Earth radii distance and does not interact with the planet's upper atmosphere directly. Photochemical escape is unimportant on Earth or Venus because the kinetic energy of neutral particles resulting from photochemical processes is usually far less than their gravitational energy.

On the other hand, the escape of O from Earth is dominated by ion loss through magnetospheric processes. Current estimate of the ionospheric outflow rate of O^+ is $\sim 7 \times 10^{25}$ s^{-1}. However, the combined O loss rate from Earth's magnetosphere is 10 times smaller (*Seki et al.,* 2001). One explanation for this difference is that there is a strong returning flux of low-energy O ions from the magnetosphere back to the low-latitude ionosphere. In this case, the escape of O ions from Earth is much less efficient than for Venus or Mars — Earth's magnetic field is protecting Earth's O reservoir. An alternative explanation is that there are unknown magnetospheric processes that can cause O ions to escape.

1.3. Mars

Phobos 2 ion measurements suggested a rate of escaping ions between 5×10^{24} ions/s (*Verigin et al.,* 1991) and 2.5×10^{25} ions/s (*Lundin et al.,* 1989) for solar maximum conditions. At solar minimum, ASPERA-3/Mars Express observed within the 10 eV–30 keV range an escape flux of 10^{24} O^+/s, 2×10^{23} CO_2^+/s, and 7×10^{23} ($O_2^+ + CO^+$)/s (*Nilsson et al.,* 2011). It is estimated that ion escape rates increased by a factor of 1.6 when solar wind flux increases by more than a factor of 5 (*Nilsson et al.,* 2011). When the co-rotating interaction regions (CIR) encountered Mars, the ion loss rate increased by a factor of 2.5 (*Edberg et al.,* 2010; *Nilsson et al.,* 2011). Co-rotating interaction regions are typically formed between slow and fast solar wind streams, and observations show that magnetic storms occur when CIRs interact with the magnetosphere of Earth. It is shown that heavy ion precipitating flux increased during CIR passages of Mars, which could potentially explain the enhancement of ion loss rate during CIR events (*Hara et al.,* 2011). These observations suggest that ion loss should depend strongly on solar wind conditions. However, because of the limited temporal and spatial coverages of observations during the missions, a clear conclusion regarding the dependency of ion escape with respect to solar activity cannot be reached at this moment (*Nilsson et al.,* 2010).

Lundin et al. (2008) found a strong positive correlation between EUV/UV flux and ion escape using ASPERA-3/Mars Express measurements. However, a more recent analysis based on a larger set of data did not confirm this result, probably because of a lack of significant EUV/UV variation during the period covered by these works (*Nilsson et al.,* 2011). An increase of the ion escape with increasing EUV/UV flux was predicted by models (e.g., *Ma and Nagy,* 2007; *Chaufray et al.,* 2007). *Ma and Nagy* (2007) suggested an increase by a factor 3 to 4 from solar minimum to maximum conditions. These authors also calculated an ion escape rate increase by up to one order of magnitude for a solar dynamical pressure increase by a factor of 30. It is interesting to note that similar observations at Venus with an increase by a factor of 1.9 of the ion escape during high dynamical pressure periods were also reported (*Edberg et al.,* 2011).

The observed O^+ escape rate from Mars [~0.2×10^{25} O^+/s (*Nilsson et al.,* 2011)] is not much smaller than the O^+ escape rate from Venus [0.1–0.5×10^{25} O^+/s (*Lundin,* 2011)]. Such a small difference in ion escape rate needs to be reconciled with the facts that (1) both planets have similar upper atmospheres (in terms of composition and temperature) but very different atmospheric surface pressure (90 bar at Venus and only 7 mbar at Mars); (2) the planets are at different distances from the Sun (implying a solar EUV/UV flux approximately four times more intense at Venus than at Mars); (3) the planets have different masses and therefore gravity (roughly three times less intense at Mars than at Venus); and (4) Mars has a crustal magnetic field that might impact the ion escape (e.g., *Lundin et al.,* 2011).

An atmosphere can also be lost when neutral atmospheric particles are accelerated above the energy escape and the exobase. Such neutral particles are usually particularly difficult to detect because of their low density and because they are mixed with a nonescaping neutral population. The H martian exosphere was observed by Mariner 6 and 7 (*Barth et al.,* 1971; *Anderson and Hord,* 1971) during solar maximum conditions. *Anderson and Hord* (1971) derived an exospheric temperature of 350 ± 100 K, with density of H at the exobase of $3.0 \pm 1 \times 10^4$ H cm^{-3}. The corresponding Jeans escape was calculated to be 1.8×10^8 H/cm²/s. At solar minimum, using the Copernicus Orbiting Astronomical Observatory to observe from Earth the Lyman α emission from the martian exosphere, *Levine et al.* (1978) derived an exobase density equal to 1.5×10^6 H cm^{-3} implying a Jeans escape of 1.6×10^8 H cm^{-2} s^{-1}. More recently, SPICAM UV spectrometer onboard Mars Express captured the first vertical profiles of the martian Lyman α emission for solar minimum conditions (*Chaufray et al.,* 2008). These authors concluded that the H exosphere was produced from the exobase with a temperature between 200 and 250 K and a density between 1 and 4×10^5 H cm^{-3}, corresponding to a Jeans escape of $1.4 \pm 0.6 \times 10^8$ H cm^{-2} s^{-1}. They also discussed the possibility of a second exospheric population, nonthermally produced in the upper martian atmosphere (with an equivalent temperature larger than 350 K). The lack of solar activity dependency of the H escape is interpreted as the evidence of H escape controlled by its diffusion rate at the homopause (*Hunten,* 1973). The diffusion rate at the homopause is equal to 3.7×10^8 H cm^{-2} s^{-1} at 110 km [including H and H_2 (*Zahnle et al.,* 2008)] and is essentially dependent on the atmospheric conditions at the homopause, which are not strongly variable with respect to local time, season, and solar cycles (*González-Galindo et al.,* 2009). If correct, the maximum expected escape rate, 3.7×10^8 H cm^{-2} s^{-1} based on diffusion-limited assumption, is therefore inconsistent with the present estimate of the escape rate based on analysis of the exospheric profile, $<2 \times 10^8$ H cm^{-2} s^{-1}, when supposing that the H exosphere is controlled by thermal processes.

Up until very recently, the exospheric O neutral component was known only up to 700 km in altitude (*Barth et al.,* 1971; *Chaufray et al.,* 2008). The UV spectrometer ALICE onboard the Rosetta mission extended for the first time this profile up to 1200 km in altitude. This measurement underlined the existence of two O populations in the martian exosphere, a thermal component at the exobase temperature and a nonthermal component with a transition around 700 km altitude (*Feldman et al.,* 2011). This is the first unambiguous observation of nonthermal processes in Mars' atmosphere. It also confirmed the validity of the analogy with Venus where this nonthermal component of the O exosphere had been observed three decades ago (*Bertaux et al.,* 1978; *Nagy et al.,* 1981). *Yagi et al.* (2012) built a three-dimensional model of the O exosphere by describing both the thermal component of this exosphere and by taking into account the dissociative recombination of the O_2^+ molecules in Mars' upper atmosphere to describe the nonthermal exospheric component (*McElroy and Donahue,* 1972). *Yagi et al.* (2012) showed that a reasonable agreement between modeled and observed profiles could be obtained. This suggests that the Yagi et al. estimate of the escape rate due to dissociative recombination is probably of the right order of magnitude. As a matter of fact, the rate calculated by Yagi et al. was also in agreement with the most complete study of Mars' neutral O escape performed by *Valeile et al.* (2010). These authors concluded that from perihelion to aphelion a few times 10^{25} O/s at solar minimum are escaping and above 10^{26} O/s at solar maximum, i.e., significantly more than the equivalent O ion escape.

The first observations of the thermospheric emission of the neutral C martian atmospheric species were reported by *Barth et al.* (1971), highlighting the CO_2, CO, and C profiles up to 200 km at solar maximum. Thirty years later, *Leblanc et al.* (2006) reported similar profiles for solar minimum conditions. There are so far no direct measurements of any C species above the exobase. Several processes of formation of C exospheric molecules and atoms were described, among which photodissociation of CO, dissociative recombination of CO^+, and electron impact dissociation of CO are thought to be the most important (*Fox,* 2004). Fox made the most complete calculation of all possible photochemical channels for the production of C escaping particles and found that between 7.5×10^{23} C/s and 4.4×10^{24} C/s may escape from solar minimum to maximum conditions. *Cipriani et al.* (2007) calculated that 9×10^{24} C/s should escape by sputtering at solar maximum and 1.4×10^{23} C/s at solar minimum conditions. Sputtering of Mars' atmosphere by exospheric ions picked up by the solar wind may also induce the loss of atmospheric species (*Luhmann et al.,* 1992; *Leblanc and Johnson,* 2002) but is now thought to be less efficient than dissociative recombination (*Chaufray et al.,* 2007). As Mars' atmosphere is essentially composed of CO_2 molecules, it is unclear how Mars' atmospheric composition could have been maintained throughout Mars' history with an O escape rate that was one order of magnitude larger than the C escape rate. The escape rates described above are summarized in Table 1.

2. ATMOSPHERE ESCAPE HISTORIES

As described in the previous section, most atmospheric escape mechanisms operate near the exobase region or beyond, which is above the homopause of the terrestrial planets. In the region between the homopause and the exobase, collisions are rare enough that various species are separated by the gravity of the planet so that the vertical density structure of each species depends on the molecular/atomic weight of that species. Heavier species (such as CO_2) have small scale heights, while lighter species (such as H) have large scale heights. Thus, as altitude increases, an atmosphere is increasingly enriched in light species. Since escape proceeds from the top of the atmosphere, lighter species should be preferentially removed to space, leaving the atmosphere enriched in heavier species. Such a process is called isotopic fractionation for isotopes of the same elements. Examination of the ratios of isotopes, such as D/H, $^{15}N/^{14}N$, $^{18}O/^{16}O$, $^{13}C/^{12}C$, and isotopes of the noble gases Ne, Kr, Ar, and Xe can provide an indication of the importance of atmospheric escape at a given planet over its history.

While the mechanism for enriching atmospheres in heavier isotopes is straightforward to understand, interpretation of isotope measurements requires great care. First, some processes in the atmosphere can change isotope ratios. Most known processes enrich an atmosphere in lighter isotopes, so that inferences drawn from isotope ratios provide in the worst case a minimum constraint on atmospheric escape. However, some processes, such as isotopic self-shielding of N_2 photodissociation and subsequent removal of compounds with heavier nitrogen from Titan's atmosphere (*Liang et al.,* 2007) or UV excitation of N_2 (*Muskatel et al.,* 2011), might leave an atmosphere enriched in lighter isotopes. Note that the fractionation effect in the N_2 photoionization process is much less prominent than that in N_2 photodissociation (*Croteau et al.,* 2011) and the detailed modeling of the isotopic effect of UV excitation of N_2 has not been worked out (*Muskatel et al.,* 2011). Thus the isotopic fractionation effect of photochemistry in planetary atmospheres is not fully understood. Second, atmospheric escape (and therefore isotopic fractionation) acts only on the gases in an atmosphere at any given time, while an atmosphere may have its isotope ratios buffered or even entirely reset by significant exchanges with volatile reservoirs at the surface (e.g., oceans, polar caps), at the subsurface (e.g., volcanism), or from space (e.g., impacts). Third, since even an atmosphere lacking any source or sink mechanisms should have a greater abundance of one isotope relative to another (e.g., H is more common than D), isotope ratios are measured relative to some reference. For D/H the reference is often terrestrial standard mean ocean water (SMOW). This reference can provide a constraint on the relative importance of atmospheric escape at, e.g., Venus and Earth. But an absolute assessment of the importance of atmospheric escape relies on the assumption that the reference is indicative of the "starting point" for both atmospheres. Despite these potential issues, the study of isotope ratios in solar system objects and their atmospheres is relatively robust, and it is quite possible to draw meaningful inferences about the evolution of the atmospheres of the terrestrial planets.

2.1. Venus

One of the major observational clues to the significant role of atmospheric escape in shaping the present Venus atmosphere is its high H-isotopic fractionation factor. The D/H ratio in Venus' atmosphere has been measured repeatedly by means of *in situ* and Earth-based measurement of the HDO/H_2O ratio and by inference from *in situ* Pioneer Venus Orbiter mass spectrometer measurements of ion and neutral species (H^+, D^+, O, O^+) in the high atmosphere (e.g., *Donahue et al.,* 1997). All these measurements show a high value of the D/H ratio of 150 ± 30 times the SMOW value, and the measured ratio increases with altitude (*Matsui et al.,* 2012). This high value is unique among the atmospheres of the solar system planets and shows that at some point H efficiently escaped from the atmosphere. Despite careful work to include the effect of H escape and the photoinduced isotopic fractionation effect (PHIFE) of H_2O and HCl, the measured altitude distribution of HDO/H_2O ratio cannot be reproduced (*Liang and Yung,* 2009).

At the present level of H and O escape (a solar cycle average escape rate in the range from 10^6–10^7 cm^{-2} s^{-1}), the

TABLE 1. Present-day atmospheric escape processes on terrestrial planets.

	Venus	Earth	Mars
Jeans escape	Small	H: 10^8 (solmax) (*Chamberlain and Hunten,* 1987)	H: 10^8 (*Anderson and Hord,* 1971; *Levine et al.,* 1978; *Chaufray et al.,* 2008; *Zahnle et al.,* 2008; *González-Galindo et al.,* 2009)
Charge exchange	H: 10^6–10^7 (*Lammer et al.,* 2006)	H: 10^8 (solmin) (*Chamberlain and Hunten,* 1987)	10^4–10^5 (*Chaufray et al.,* 2007)
Ion pickup	H^+: 10^6–10^7 (*Barabash et al.,* 2007; *Lundin,* 2011; *Donahue,* 1999; *Donahue et al.,* 1997) O^+: 10^6–10^7 (*Barabash et al.,* 2007); He^+: 10^5–10^6 (*Chanteur et al.,* 2009)	Small	O^+: 10^6~10^7 (*Verigin et al.,* 1991; *Lundin et al.,* 1989; *Nilsson et al.,* 2011) C^+: 10^5 (*Nilsson et al.,* 2011)
Sputtering	O: 10^5–10^6 (*Lammer et al.,* 2006)	Small	C: 10^5 (solmin); 10^7 (solmax) (*Cipriani et al.,* 2007)
Photochemical escape	Small	Small	O: 10^7–10^8 (*Yagi et al.,* 2012; *Valeile et al.,* 2010) C: 10^6 (*Fox,* 2004)
Polar wind	N/A	H: 10^7 (*Chamberlain and Hunten,* 1987)	N/A
Ionospheric outflow	?	O^+: 10^6 (*Seki et al.,* 2001)	?

All fluxes are in unit of cm^{-2} s^{-1}.

amount of water lost per billion years is a global equivalent layer (GEL) of a few centimeters depth. Assuming a global average mixing ratio of 30 ppmv of water in the atmosphere (*Marcq et al.,* 2008), the amount of water in the venusian atmosphere is equivalent to a GEL of 1.3 cm depth. Since D in the venusian atmosphere is enriched by a factor of 150 with respect to terrestrial ocean water, the amount of D present today in the atmosphere corresponds to an initial GEL of water of $\approx 1.3 \times 150 = 200$ cm = 2 m depth (assuming no D escape for simplicity). At the present rate, the time required for such a quantity to be removed from the atmosphere is on the order of tens to hundreds of billions of years, longer than the age of the solar system. It implies that escape rates of H and O must have been larger in the past, which agrees with the fact that the solar XUV flux and solar wind were both stronger at earlier epochs. A 2-m-deep GEL of water present at the surface of Venus 4 b.y. ago may have been lost by nonthermal escape, at rates comparable with model predictions (*Kumar et al.,* 1983), and the generated H fractionation factor is on the order of the value observed today (*Gillmann et al.,* 2009). For example, by taking into account the decrease of the solar EUV flux since solar system formation, it is found that the loss of a GEL of 2 m depth is compatible with the O ion escape process being continuously operative during the last 4 b.y. (*Kulikov et al.,* 2006).

On the other hand, the high D/H ratio in Venus' atmosphere today does not necessarily prove that early Venus had an ocean. It is pointed out (*Grinspoon,* 1987) that the timescale for Venus to lose H through nonthermal escape processes is on the order of 0.1 b.y. and continuous supply of water to Venus through outgassing or cometary delivery, followed by subsequent nonthermal escape, is thus a possible solution to the high D/H measured in Venus atmosphere without the need for a global ocean.

That being said, much more water could have been present at the beginning of Venus' evolution (*Shimazu and Urabe,* 1968; *Rasool and DeBergh,* 1970; *Donahue et al.,* 1982, 1997), 4.6 b.y. ago. It has been shown (*Kasting and Pollack,* 1983) that photolysis of water molecules followed by hydrodynamic escape of H is able to remove the total H content of one terrestrial ocean in less than a few hundred million years. Such a rapid loss of water after the formation of Venus is also consistent with the present Ne- and Ar-isotopic ratios in the venusian atmosphere, as measured by Pioneer Venus and various Venera missions (e.g., *Donahue and Pollack,* 1983, and references therein).

To finally understand how much water Venus had in its early days, measurements of noble and stable isotopes in Venus' atmosphere will be necessary. Unfortunately, the concentrations of heavy noble gases (Kr, Xe) and their isotopes in Venus' atmosphere are mostly unknown, and our knowledge of light noble gases and stable isotopes is still incomplete and inaccurate. A crucial link of the future space exploration of Venus, allowing us to improve our understanding of the early escape and more generally the early evolution of Venus, is a *in situ* mission, such as a balloon probe deployed at cloud altitude in the atmosphere of Venus, that is able to provide accurate measurements of all noble gases and their isotopes, as well as stable isotopes (*Chassefière et al.,* 2012, and references therein).

2.2. Earth

There is no clear evidence that Earth experienced early Venus-like massive loss of water. The D/H ratio of the bulk Earth and that of the chondrules are six times higher than that in the proto-Sun or Jupiter (*Robert,* 2006). The CO_2 reservoir on Earth is mainly in the form of carbonate rocks. If released into the atmosphere, Earth would have a CO_2 atmosphere with a pressure similar to that of Venus today, therefore Earth's CO_2 probably did not escape the planet.

On the other hand, Earth probably lost H to space rapidly during its early history. Earth is the only terrestrial planet we know of so far with an atmosphere containing a significant pressure of O. It is generally believed that this O-rich atmosphere is the result of the evolution of the biosphere, in particular the oxygenic photosynthesis. However, the production of O during the oxygenic photosynthesis is accompanied by the formation of organic materials, most of which will decay by reacting with O eventually. Thus oxygenic photosynthesis by itself cannot oxidize the planet. One theory involves the burial of organic materials into the interior of Earth, which is a much larger reservoir in comparison with the atmosphere, the ocean, and the crust. The other way to oxidize the surface layers of Earth is H escape, perhaps assisted by the evolving biosphere (*Catling et al.,* 2001).

Hydrogen escape from the atmosphere of the early Earth could have been limited by diffusion if the energy used to drive the escape is negligible or can be readily supplied by the absorption of solar XUV radiation and other energy injection mechanisms in the thermosphere. However, because the diffusion-limited H escape is proportional to the total concentration of H-bearing species in the stratosphere, and the available energy to drive escape is not unlimited, H escape will change from diffusion limited to energy limited at certain total H concentration. *Tian et al.* (2005a,b) showed that this transition probably is at a H concentration level of 1% or higher in the hydrodynamic escape scenario. Where this transition should be when considering other nonthermal escape processes remains unknown. No matter how efficient atmosphere escape is, the escape is ultimately controlled by how much H there is in the atmosphere, which means the outgassing of H from the interior of Earth needs to be balanced. The current estimate for the H outgassing rate during the Archean Earth is between a few times 10^{10} and 10^{11} cm^{-2} s^{-1} (*Tian et al.,* 2005b).

Oxygen probably did not escape rapidly from early Earth prior to the appearance of an oxic atmosphere on the planet around 2.4–2.2 Ga because the atmosphere would have been depleted of O. After the atmosphere became oxic, between 1 and 10% of present Earth's atmospheric O could have been lost through ionospheric outflow and magnetospheric processes (*Seki et al.,* 2001). Note that this estimate is based on measurements of present ionospheric outflow, and the loss rate could have been much higher under stronger solar XUV radiation and stronger solar wind conditions. In particular, atmosphere escape could have been much enhanced during magnetic inverse periods, which occurred repeatedly through Earth's history, although to what degree the lack of a strong dipole magnetic field could have affected Earth's atmosphere and climate evolution has not been studied.

Nitrogen could have escaped from early Earth as well. Interestingly, the evidence could be found on lunar soil, in which the N-isotopic ratio is different from that in the solar wind (*Ozima et al.,* 2005). Thus some early "Earth wind" must have implanted atmospheric material on the lunar surface. Because it is difficult for N to escape Earth, the current explanation is that Earth might not have had an intrinsic magnetic field during its early history, which allowed a more direct interaction between the early Earth's atmosphere with the solar wind, and ion pickup process dragged N ions from Earth and deposited them on the Moon (*Ozima et al.,* 2005). Unfortunately, the lunar soil N-isotopic ratio measurement can only place a lower bound for the Earth wind. Whether Earth's magnetic field was absent, how strong the early Earth wind was, and how much N was lost by Earth remain interesting scientific questions.

2.3. Mars

The D/H ratio in the martian atmosphere is 5.2 times higher than the SMOW value (*Owen et al.,* 1988; *Yung et al.,* 1988; *Bjoraker et al.,* 1989; *Krasnopolsky et al.,* 1997). Fractionation of H by thermal escape might explain this ratio (*Owen et al.,* 1988), as the thermal escape of D is much smaller than the H escape rate (*Krasnopolsky,* 2002). Models predict that a GEL of 4–50 m of water would have been lost from Mars since 3.5 Ga (*Yung et al.,* 1988; *Lammer et al.,* 1996; *Kass and Yung,* 1999; *Lammer et al.,* 2003). Trapping of O by serpentinization in the crust (*Chassefière and Leblanc,* 2011b) could help to reconcile a strong H escape with a lower O escape rate suggested by models.

The Xe-isotopic ratio in Mars' atmosphere suggests fractionation during an early hydrodynamic escape phase (the first few hundred millions years after Mars' differentiation). In the meanwhile, the Kr-isotopic ratio is almost unfractionated with respect to Earth (*Pepin,* 1994), so that elements lighter than Xe should not have been fractionated during this early period of hydrodynamic escape (*Jakosky and Jones,* 1997). As a consequence, the present isotopic ratios of Ne, Ar, and N in Mars atmosphere are taken as signatures of atmospheric evolution during the last 4 b.y.

Because light elements like Ar and Ne are thought to be lost to space rather easily, *Jakosky et al.* (1994) concluded that the fact that the martian $^{20}Ne/^{22}Ne$ is so close to the Earth ratio cannot have been solely driven by loss to space and must have also been partially driven by outgassing episodes. Because ^{40}Ar is produced by the decay of long-lived ^{40}K, the present atmospheric ^{40}Ar abundance and $^{40}Ar/^{36}Ar$ ratio should also depend on the degassing history from the crust (*Hamano and Ozima,* 1978; *Leblanc et al.,* 2012). Similarly, the $^{12}C/^{13}C$ and $^{16}O/^{18}O$ ratios at Mars are close to those on Earth (*Owen,* 1992), and thus exchanges between the atmosphere and nonfractionated surface reservoirs of C and O are necessary. The $^{15}N/^{14}N$ ratio observed in Mars' atmosphere is ~1.6 times that of Earth's value and can be easily produced by atmosphere escape processes such as sputtering and dissociative recombination. It was believed that Mars atmospheric isotopes represented a loss of a massive atmosphere during the same period of time (*Jakosky and Phillips,* 2001). More recent work suggests that Mars atmospheric isotopes are more consistent with a weak atmospheric escape during the past 4.1 b.y. (e.g., *Chassefière and Leblanc,* 2011a,b).

The timing of massive atmosphere loss from Mars is both important and highly uncertain. Although most earlier model calculations typically assume that escape to space dominated loss processes after the end of the late heavy bombardment period and the time at which Mars lost its magnetic field (*Hutchins et al.,* 1997; *Brain and Jakosky,* 1998), studies of apatite in martian meteorites suggest that the majority of water escaped early in martian history (*Greenwood et al.,* 2008). It is interesting to note that these two views are not necessarily in conflict — the bulk of atmospheric escape could have occurred early on, and removal to space could have dominated a weak overall loss rate for the past few billion years as well. The climate consequences of different escape scenarios will be addressed in the next section.

On the modeling side, how neutral species escape Mars atmosphere today is still poorly constrained by observations. In particular, no observations are presently available to constrain the dependency of the neutral escape on solar activity (EUV/UV variation along a solar cycle as an example) or on solar wind forcing (during solar energetic particle events or with respect to the solar dynamical pressure). For the time being, we are left with modeling to guess what these dependencies could be. But this still remains a delicate exercise with large uncertainties. As an example, in *Chassefière and Leblanc* (2004) and *Chassefière et al.* (2007), the escape rate of O atoms due to sputtering was estimated to be strongly nonlinear in function of the EUV solar flux following a logarithmic slope of ~7, implying a very large escape rate at an early stage of the solar system when the EUV/UV flux was much larger (*Luhmann et al.,* 1992). *Chaufray et al.* (2007) used a hybrid magnetospheric approach coupled to a three-dimensional exospheric model to describe Mars' interaction with the solar wind for solar minimum and maximum EUV/UV fluxes and derived a variation of the sputtered flux following a logarithmic slope of only 1.8. Chaufray et al. also concluded that sputtering should be less efficient than the dissociative recombination of O_2^+ during most of Mars' history. *Valeille et al.* (2010) calculated the evolution of the O escape rate due to dissociative recombination and obtained a rate varying from 9.0×10^{25} O s^{-1} average at present along one solar cycle to 5.9×10^{26} O s^{-1} at 3.6 Ga (when the solar flux EUV flux was six times larger than the present solar minimum flux). The dependency of the escape rate of O due to dissociative recombination follows therefore a logarithmic slope of only 1.4 with respect to EUV intensity.

With the most optimistic estimate regarding how atmospheric escape processes depend on XUV and other solar forcing, and taking into account the evolution of solar XUV and solar wind, *Chassefière and Leblanc* (2011a,b) concluded that only ~10 mbar of CO_2 and ~5 m of water (depth of an ocean covering the whole surface of Mars) could have escaped since 4.1 Ga, which corresponds to a period during which it is thought that Mars had an atmosphere at least a few hundred millibars thick (*Bibring et al.,* 2006; *Bouley et al.,* 2009) and an intrinsic magnetic field (*Acuña et al.,* 1999; *Lillis et al.,* 2008). An intrinsic magnetic field should strongly impact the solar wind interaction with Mars. However, how an intrinsic magnetic field dynamo might change the global erosion of Mars' atmosphere is unclear when comparing the atmospheric thickness and erosion rates of Mars and Venus as well as the role of Mars' crustal field (*Lundin et al.,* 2011; *Edberg et al.,* 2011). *Chassefière and Leblanc*'s (2011a,b) estimate of Mars' accumulated atmospheric escape is somewhat in contradiction with the numerous geological and mineralogical clues suggesting that the atmosphere was a few hundred millibars thick during the Noachian (*Carr and Head,* 2003). Other processes of escape could have been more efficient at an early stage of Mars. For example, impact erosion (*Melosh and Vickery,* 1989) could have occurred during the late heavy bombardment period around 3.9 Ga (*Gomes et al.,* 2005) and helped to remove a dense early Mars atmosphere. However, recent modeling of impact erosion process (*Pham et al.,* 2009) suggests that this process is not as efficient as suggested by *Melosh and Vickery* (1989). However, a strong atmospheric escape from Mars during the past 4 b.y. might be possible if the unobserved neutral escape of Mars' atmosphere would be a few orders larger and would follow a dependency with solar activity much deeper than presently modeled.

On the other hand, solar XUV flux should have been strong enough to heat the upper atmosphere of Mars before 4.1 Ga, leading to a strong atmosphere escape (*Tian et al.,* 2009). In fact, the calculated escape rate was so great prior to 4.1 Ga that the timescale for early Mars to lose 1 bar of CO_2 was as short as 10 m.y. and 1 m.y. at 4.1 and 4.5 Ga respectively. Note that this calculation is for C loss from a CO_2-dominant atmosphere, and loss of H is not included. If early Mars were in a Venus-like situation, water could have entered the middle atmosphere of early Mars and H, a photolysis product of water, would have been abundant in the upper atmosphere. In this situation it could have been H that was lost instead of C. However, there are two problems with this scenario. First, there is no evidence for early Mars to have been anywhere close to any type of greenhouse state. Second, the timescale for water to be lost from early Mars would have been so short that C would have escaped after the epoch of rapid H loss. Tian et al. concluded that a warm and wet atmosphere could not have been maintained for a geologically long time prior to 4.1 Ga and a stable, dense martian atmosphere could only have been possible after the solar XUV radiation level decreased to values incapable of sustaining a highly expanded early martian upper atmosphere. This theory is in sharp contrast to the traditional idea that early Noachian Mars had a dense atmosphere until the planet lost its intrinsic magnetic field and much of the atmosphere then escaped due to interaction with the solar wind.

3. CLIMATE IMPACT OF ATMOSPHERE ESCAPE AND DISCUSSIONS

Among the three terrestrial planets in the solar system, Earth is the only planet with a long-lasting stable climate suitable for maintaining the existence of liquid water on its surface and supporting life. That being said, it does not mean that Earth's climate has always been like that of today over its entire history. There were epochs of global glaciations — Snowball Earth (*Kirschvink,* 1992; *Hoffman et al.,* 1998) and some episodes with warmer climate (*Knauth and Lowe,* 2003; *Hren et al.,* 2009; *Blake et al.,* 2010). But these climate changes are mild compared with the climate change on Venus or Mars. Early Venus likely started with large amount of water. Water vapor is a greenhouse gas and thus early Venus was probably even hotter than it is today. There are ongoing debates about whether early Mars was warm and wet or cold and dry. In the first school of thought, Mars started with a dense atmosphere that supported a warm climate. This atmosphere disappeared because of atmosphere escape and caused Mars to become progressively colder over time. In the second school of thought, early Mars was about as cold as it is today and there were considerable climate changes over time due to variations in atmospheric pressure, composition, solar luminosity evolution, and Mars' orbital parameters.

The history of climate at the terrestrial planets can be constrained using a variety of methods and techniques, each with its own associated uncertainties and limits. On Earth, observations of trapped gases in ice cores and rocks can help to establish atmospheric conditions at times relatively close to now, complemented by indirect means of assessing climate conditions such as analysis of tree rings, fossils preserved in sediments, layering in ice sheets, and minerals in rocks that require certain temperatures to form. However, the ice record does not exist before ~1 m.y. ago on Antarctica or Greenland. Trees did not exist on Earth prior to 400 m.y. ago. Few fossils are found prior to the Cambrian age, ~0.5 Ga. Therefore geological evidence such as mineralogy is crucial for determining the climate during these periods of time. The same idea applies to Mars, where the presence or lack of certain minerals at the surface provides some insights into the martian climate during different epochs. Venus, with a cloud-covered atmosphere that is difficult to see through, and a surface that has completely reformed over the past 0.5 b.y. or less, provides little in the way of physical observational evidence for climate conditions throughout its history. For all planets, sophisticated models of the atmospheres and drivers of climate can help place boundaries on climate conditions at different times, but such models are limited by the reliability of their input conditions.

Mars is probably the best example of how severe climate change occurs as a result of atmosphere escape. One of the explanations for the existence of the valley networks on the surface of Mars today is a stable and long-lasting Earth-like hydrological cycle — a warm and wet early Mars. The alternative is a normally cold and dry early Mars with transient warm periods and flash floods. Climate modeling of early Mars has shown that at least 3 bar of CO_2, along with the greenhouse effect from other gases or clouds, would have been needed in order to raise the surface temperature of early Mars to a maximum temperature, which is still below the freezing point of water (*Kasting,* 1991; *Tian et al.,* 2010). Water with high salinity could flow under much lower temperatures, but is much more viscous than pure water. Thus whether the valley networks observed on Mars today could have been formed by viscous salty water at low surface temperatures is unknown.

Let's suppose that Mars had a dense atmosphere at some time during its early evolutionary history. The dense atmosphere could have been either lost to space or reabsorbed by progressive oxidation and hydradation of the crust. As we described earlier, the former explanation still lacks support from theoretical modeling for the period after 4.1 Ga. The upcoming Mars Atmosphere and Volatile Evolution (MAVEN) mission might help by discovering more efficient escape processes. The latter explanation needs to be reconciled with the lack of identification of important reservoirs of carbonate on or below Mars' surface.

On the other hand, atmosphere escape from Mars prior to 4.1 Ga could have been very rapid because of the stronger XUV radiation from the young Sun. As a result of rapid atmosphere escape, repeated warm episodes during the early Noachian could only have lasted for no longer than 1–10 m.y. after massive volcanic eruption events, which would have supplied CO_2 and other greenhouse gases into the atmosphere. The increase of atmospheric greenhouse gases would have warmed up the surface, sending more greenhouse gases from the surface and subsurface cryosphere to the atmosphere, which could then warm the surface even more. Surface fluvial features could have been formed during these short warm episodes. The warm periods would have ended with the loss of the CO_2 atmosphere, returning the planet to its cold state, which was likely the normal climate state between volcanic eruptions. Interestingly, volcanic eruptions on Earth today primarily cause surface cooling through the formation of sulfate aerosols in the stratosphere. This is because volcanic eruptions on the present Earth do not change the atmosphere pressure, while massive volcanic eruptions on early Mars did, especially during the rapid atmosphere loss epoch. In support of these ideas, analyses and formation models for the ancient martian meteorite ALH 84001 suggest that the martian atmosphere should have been at most 400 mbar by ~4.16 Ga (*Cassata et al.,* 2012). Additional support for the rapid loss of martian atmosphere during the early Noachian can be found in the ages of valley networks, which rarely exceed the late Noachian to early Hesperian time (*Hoke and Hynek,* 2009; *Fassett and Head,* 2009; *Hoke et al.,* 2011). One possible explanation is that all older valley networks were erased during the late heavy bombardment period and/or by heavy weathering during the later wet period of time. The alternative is that the distribution of valley networks was much more limited during the early Noachian due to the lack of a thick atmosphere and cold-dry climate.

If early Mars was a cold and dry place, how could it have lost most of its water early as proposed by *Greenwood et al.* (2008)? Episodic warming periods after giant impacts could have mobilized surface and subsurface water during the transient period of time. And if C loss from early Noachian Mars could have been efficient, loss of water in the form of H during such short yet repeated warm periods of time should be possible, although detailed models have not yet been worked out.

If Venus had oceans after its formation, they probably did not last for longer than a few hundred million years. Because of its orbital distance from the Sun, Venus receives 1.4 times as much solar radiation as Earth. This higher solar constant would have raised the surface temperature relative to Earth (all other things being equal), which would have increased evaporation of water into the atmosphere, which would have increased the surface temperature even further. This positive feedback between temperature and water vapor content in the atmosphere would have resulted in a moist greenhouse state on early Venus and a large amount of water vapor in its stratosphere (*Kasting,* 1988). The subsequent loss of H from early Venus would have caused the planet to lose its oceans. During this process, the surface pressure would have been reduced gradually and the surface temperature of early Venus would have changed from a wet and hot climate to a dry and hot climate, such as it has today. Such a cooling related to atmosphere escape would have occurred during the first several hundred million years after the formation of Venus.

Note that in both the early Venus and early Mars cases, atmosphere escape could have influenced the major component of the atmosphere and thus altered the pressure. On Mars it was CO_2 and on Venus it was water vapor for which the pressure could have reached a couple of hundred bars. It is unclear whether atmosphere escape could have caused early Earth's atmosphere pressure to change this much, except perhaps for the transition from a H-dominated atmosphere to a more neutral atmosphere (H content less than 30%) right after the planet's formation. During the magma ocean scenario, the interaction between water and hot rocks would have been an abundant source of H, and thus Earth could have had a massive H envelope. Recently it has been proposed that the collision-induced absorption of H_2 molecules could cause significant greenhouse warming on exoplanets (*Pierrehumbert and Gaidos,* 2011; *Wordsworth,* 2012). Thus atmosphere escape of H could have accelerated the collapse of the super greenhouse on very early Earth.

Hydrogen escape could have impacted the climate of Archean Earth as well. Despite the much-reduced solar luminosity, geological records of glaciations are scare during the Archean (*Haqq-misra et al.,* 2008). In fact, there is some debatable evidence supporting a somewhat warmer than present Archean Earth (*Knauth and Lowe,* 2003; *Gaucher et al.,* 2008; *Hren et al.,* 2009; *Blake et al.,* 2010). The typical proposed solution to this so called "faint young Sun" (FYS) problem is a much-enhanced greenhouse effect on Archean Earth. Current geological and geochemical evidence supports a limited level of atmospheric CO_2 during the Archean (*Rosing et al.,* 2010; *Sheldon,* 2006). Methane has been proposed as an important greenhouse gas, but the formation of organic haze from methane photolysis in a neutral atmosphere would have impaired a methane solution to the FYS problem. However, laboratory experiments show that increasing atmospheric H concentration delays the formation of organic haze (*Trainer et al.,* 2006). A more recent proposal that the collision-induced absorption by N_2 and H_2 under high atmospheric pressure and H abundance could have helped early Earth to maintain a warm climate highlights the important role that H could have played in early Earth's climate (*Wordsworth and Pirrehumbert,* 2013). Thus a less-efficient H escape could have helped solve the FYS problem, and H loss could have played a role in determining Archean Earth's climate.

Up to the present, Earth is the only body in the solar system where life is found. In this chapter we have summarized our current understanding of the relationship between atmosphere escape and the climate evolution of solar system terrestrial planets. Therefore there is possibly a link between atmosphere escape and the habitability of planets. The nature of this link is currently unclear, but should be associated with the stability of planet climate. The detection of more terrestrial-like exoplanets and the knowledge of their atmospheres will help to better address this problem.

Acknowledgments. The authors are grateful to an anonymous reviewer and the editors for providing helpful and constructive comments on this chapter. We also thank the Lunar and Planetary Institute for providing the cartoon drawings used in Fig. 1.

REFERENCES

Acuña M. H., Connerney J. E. P., Ness N. F., Lin R. P., Mitchell D., Carlson C. W., McFadden J., Anderson K. A., Reme H., Mazelle C., Vignes D., Wasilewski P., and Cloutier P. (1999) Global distribution of crustal magnetization discovered by the Mars Global Surveyor MAG/ER Experiment. *Science, 284,* 790.

Anderson D. E. and Hord C. W. (1971) Mariner 6 and 7 ultraviolet spectrometer experiment: Analysis of hydrogen Lyman-alpha data. *J. Geophys. Res., 76,* 6666–6673.

Barabash S., Fedorov A., Sauvaud J. J.,. et al. (2007) The loss of ions from Venus through the plasma wake. *Nature, 450,* 650–653.

Barth C. A., Hord C. W., Pearce J. B., Kelly K. K., Anderson G. P., and Stewart A. I. (1971) Mariner 6 and 7 ultraviolet spectrometer experiment: Upper atmosphere data. *J. Geophys. Res., 76,* 2213–2227.

Bertaux J.-L., Blamont J. E., Marcelin M., Kurt V. G., Romanova N. N., and Smirnov A. S. (1978) Lyman-alpha observations of Venera 9 and 10. The non-thermal hydrogen population in the exosphere of Venus. *Planet. Space Sci., 26,* 817–832.

Bibring J. P., Langevin Y., Mustard J. F., et al. (2006) Global mineralogical and aqueous Mars history derived from OMEGA/Mars Express data. *Science, 312,* 400–404.

Bjoraker G. L., Mumma M. J., and Larson H. P. (1989) Isotopic abundance ratios for hydrogen and oxygen in the martian atmosphere. *Bull. Am. Astron. Soc., 21,* 991.

Blake R. E., Chang S. J., and Lepland A. (2010) Phosphate oxygen isotopic evidence for a temperate and biologically active Archaean ocean. *Nature, 464,* 1029–1032.

Brain D. A. and Jakosky B. M. (1998) Atmospheric loss since the onset of the martian geologic record: Combined role of impact erosion and sputtering. *J. Geophys. Res., 103(E),* 22689–22694, DOI: 10.1029/98JE02074.

Bouley S., Ansan V., Mangold N., Masson Ph., and G. Neukum (2009) Fluvial morphology of Naktong Vallis: A late activity with multiple processes. *Planet. Space Sci., 57,* 982–999.

Carr M. H. and Head J. W. (2003) Oceans on Mars: An assessment of the observational evidence and possible fate. *J. Geophys. Res., 108,* DOI: 10.1029/2002JE001963.

Cassata W. S., Shuster D. L., Renne P. R., and Weiss B. P. (2012) Trapped Ar isotopes in meteorite ALH 84001 indicate Mars did not have a thick ancient atmosphere. *Icarus, 1–5,* DOI: 10.1016/j.icarus.2012.05.005.

Catling D. C., Zahnle K. J., and McKay C. P. (2001) Biogenic methane, hydrogen escape, and the irreversible oxidation of early Earth. *Science, 293,* 839–843.

Chamberlain J. W. (1963) Planetary coronae and atmospheric evaporation. *Planet. Space Sci., 11,* 901–960.

Chamberlain J. W. and Hunten D. M. (1987) *Theory of Planetary Atmospheres.* Academic, New York.

Chanteur G. M., Dubinin E., Modolo R., and Fraenz M. (2009) Capture of solar wind alpha particles by the martian atmosphere. *Geophys. Res. Lett., 36,* L23105, DOI: 10.1029/2009GL040235.

Chassefière E. (1996) Hydrodynamic escape of hydrogen from a hot water-rich atmosphere: The case of Venus. *J. Geophys. Res., 101,* 26039–26056.

Chassefière E. (1997) Loss of water on the young Venus: The effect of a strong primitive solar wind. *Icarus, 126,* 229–232.

Chassefière E. and Leblanc F. (2004) Mars atmospheric escape and evolution; interaction with the solar wind. *Planet. Space Sci., 52,* 1039–1058.

Chassefière E., Leblanc F., and Langlais B. (2007) The combined effects of escape and magnetic field histories at Mars. *Planet. Space Sci., 55(3),* 343–357.

Chassefière E. and Leblanc F. (2011a) Methane release and the carbon cycle on Mars. *Planet. Space Sci., 59,* 207–217.

Chassefière E. and Leblanc F. (2011b) Constraining methane release due to serpentinization by the observed D/H ratio on Mars. *Earth Planet. Sci. Lett., 310,* 262–271.

Chassefière E., Wieler R., Marty B., and Leblanc F. (2012) The evolution of Venus: Present state of knowledge and future exploration. *Planet. Space Sci., 63–64,* 15–23.

Chaufray J. Y., Modolo R., Leblanc F., Chanteur G., Johnson R. E., and Luhmann J. G. (2007) Mars solar wind interaction: Formation of the martian corona and atmospheric loss to space. *J. Geophys. Res., 112(E9),* DOI: 10.1029/2007JE002915.

Chaufray J. Y., Bertaux J. L., Leblanc F., and Quémerais E. (2008) Observation of the hydrogen corona with SPICAM on Mars Express. *Icarus, 195(2),* 598–613.

Cipriani F., Leblanc F., and Berthelier J. J. (2007) Martian corona: Nonthermal sources of hot heavy species. *J. Geophys. Res., 112,* E07001, DOI: 10.1029/2006JE002818.

Croteau P., Randazzo J. B., Kostko O., Ahmed M., Liang M.-C., Yung Y. L., and Boering K. A. (2011) Measurements of isotope effects in the photoionization of N_2 and implications for Titan's atmosphere. *Astrophys. J. Lett., 728(2),* L32, DOI: 10.1088/2041-8205/728/2/L32.

Donahue T. M. and Pollack J. B. (1983) Origin and evolution of the atmosphere of Venus. In *Venus* (D. M. Hunten et al., eds.), pp. 1003–1036. Univ. of Arizona, Tucson.

Donahue T. M. (1999) New analysis of hydrogen and deuterium escape from Venus. *Icarus, 141,* 226–235.

Donahue T. M., Grinspoon D. H., Hartle R. E., and Hodges R. R. Jr. (1997) Ion/neutral escape of hydrogen and deuterium: Evolution of water. In *Venus II* (S. W. Bougher et al., eds.), pp. 385–414. Univ. of Arizona, Tucson.

Donahue T. M., Hoffman J. H., Hodges R. R., and Watson A. J. (1982) Venus was wet — A measurement of the ratio of deuterium to hydrogen. *Science, 216,* 630–633.

Edberg N. J. T., Nilsson H., Williams A. O., Lester M., Milan S. E., Cowley S. W. H., Fränz M., Barabash S., and Futaana Y. (2010) Pumping out the atmosphere of Mars through solar wind pressure pulses. *Geophys. Res. Lett., 37,* L03107, DOI: 10.1029/2009GL041814.

Edberg N. J. T., Nilsson H., Futaana Y., et al. (2011) Atmospheric erosion of Venus during stormy space weather. *J. Geophys. Res., 116,* A09308, DOI: 10.1029/2011JA016749.

Fassett C. I. and Head J. W. (2009) The timing of martian valley network activity: Constraints from buffered crater counting. *Icarus, 195,* 61.

Feldman P. D., Steffl A. J., Parker J. W., et al. (2011) Rosetta-Alice observations of exospheric hydrogen and oxygen on Mars. *Icarus, 214,* 394–399.

Fox J. L. (2004) CO_2^+ dissociative recombination: A source of thermal and nonthermal C on Mars. *J. Geophys. Res., 109,* A08306, DOI: 10.1029/2004JA010514.

Gaucher E. A., Govindarajan S., and Ganesh O. K. (2008) Palaeotemperature trend for Precambrian life inferred from resurrected proteins. *Nature, 451,* 704–707.

Gillmann C., Chassefière E., and Lognonné Ph. (2009) A consistent picture of early hydrodynamic escape of Venus atmosphere explaining present Ne and Ar isotopic ratios and low oxygen atmospheric content. *Earth Planet. Sci. Lett., 3–4,* 503–513.

Gomes R., Levison H., Tsiganis K., and Morbidelli A. (2005) Origin of the cataclysmic late heavy bombardment of the terrestrial planets. *Nature, 435,* 466–469, DOI: 10.1038/nature03676.

González-Galindo F., Forget F., López-Valverde M. A., Angelats i

Coll M., and Millour E. (2009) A ground-to-exosphere martian general circulation model: 1. Seasonal, diurnal, and solar cycle variation of thermospheric temperatures. *J. Geophys. Res., 114,* E04001, DOI: 10.1029/2008JE003246.

Gough D. O. (1981) Solar interior structure and luminosity variations. *Solar Phys., 74,* 21–34.

Greenwood J. P., Itoh S., Sakamoto N., Vicenzi E. P., and Yurimoto H. (2008) Hydrogen isotope evidence for loss of water from Mars through time. *Geophys. Res. Lett., 35(5),* 05203, DOI: 10.1029/2007GL032721.

Grinspoon D. H. (1987) Was Venus wet? Deuterium reconsidered. *Science, 238,* 1702–1704.

Hamano Y. M. and Ozima M. (1978) Earth-atmosphere evolution model based on Ar isotopic data. In *Terrestrial Rare Gases* (E. C. Alexander Jr. and L. Ozima, eds.), pp. 155–177. Japan Scientific Societies, Tokyo.

Haqq-Misra J. D., Domagal-Goldman S. D., Kasting P. J., and Kasting J. F. (2008) A revised, hazy methane greenhouse for the Archean Earth. *Astrobiology, 8,* 1127.

Hara T., Seki K., Futaana Y., Yamauchi M., Yagi M., Matsumoto Y., Tokumaru M., Fedorov A., and Barabash S. (2011) Heavy-ion flux enhancement in the vicinity of the martian ionosphere during CIR passage: Mars Express ASPERA-3 observations. *J. Geophys. Res., 116,* 2309.

Hoffman P. F., Kaufman A. J., Halverson G. P., and Schrag D. P. (1998) A Neoproterozoic Snowball Earth. *Science, 5381,* 1342–1346.

Hoke M. R. T. and Hynek B. M. (2009) Roaming zones of precipitation on ancient Mars as recorded in valley networks. *J. Geophys. Res., 114,* E08002, DOI: 10.1029/2008JE003247.

Hoke M. R. T., Hynek B. M., and Tucker G. E. (2011) Formation timescales of large martian valley networks. *Earth Planet Sci. Lett., 312,* 1.

Hren M. T., Tice M. M., and Chamberlain C. P. (2009) Oxygen and hydrogen isotope evidence for a temperate climate 3.42 billion years ago. *Nature, 462,* 205–208.

Hunten D. M. (1973) The escape of light gases from planetary atmospheres. *J. Atmos. Sci., 30,* 1481–1494.

Hunten D. M., Pepin R. O., and Walker J. C. G. (1987) Mass fractionation in hydrodynamic escape. *Icarus, 69,* 532–549.

Hutchins K. S., Jakosky B. M., and Luhmann J. G. (1997) Impact of a paleomagnetic field on sputtering loss of martian atmospheric argon and neon. *J. Geophys. Res., 102(E),* 9183–9190, DOI: 10.1029/96JE03838.

Jakosky B. M., Pepin R. O., Johnson R. E., and Fox J. L. (1994) Mars atmospheric loss and isotopic fractionation by solar-wind-induced sputtering and photochemical escape. *Icarus, 111,* 271–288.

Jakosky B. M. and Jones J. H. (1997) The history of martian volatiles. *Rev. Geophys., 35,* 1–16.

Jakosky B. M. and Phillips R. J. (2001) Mars' volatile and climate history. *Nature, 412(6),* 237–244.

Kass D. M. and Yung Y. L. (1999) Water on Mars: Isotopic constraints on exchange between the atmosphere and surface. *Geophys. Res. Lett., 26,* 3653–3656.

Kasting J. F. (1988) Runaway and moist greenhouse atmospheres and the evolution of Earth and Venus. *Icarus, 74,* 472–494.

Kasting J. F. (1991) CO_2 condensation and the climate of early Mars. *Icarus, 94,* 1–13.

Kasting J. F. and Pollack J. B. (1983) Loss of water from Venus. I. Hydrodynamic escape of hydrogen. *Icarus, 53,* 479–508.

Kirschvink J. L. (1992) Late Proterozoic low-latitude global glaciation: The Snowball Earth. In *The Proterozoic Biosphere* (J. W. Schopf and and C. Klein, eds.), pp. 51–52. Cambridge Univ., Cambridge.

Knauth L. P. and Lowe D. R. (2003) High Archean climate temperature inferred from oxygen isotope geochemistry of cherts in the 3.5 Ga Swaziland Supergroup, South Africa. *Geol. Soc. Am. Bull., 115,* 566–580.

Krasnopolsky V. A. (1999) Hydrodynamic flow of N_2 from Pluto. *J. Geophys. Res., 104,* 5955–5962.

Krasnopolsky V. A. (2002) Mars' upper atmosphere and ionosphere at low, medium and high solar activities: Implications for evolution of water. *J. Geophys. Res., 107,* 5128.

Krasnopolsky V. A. and Feldman P. D. (2001) Detection of molecular hydrogen in the atmosphere of Mars. *Science, 294,* 1914.

Krasnopolsky V. A., Bjoraker G. L., Mumma M. J., and Jennings D. E. (1997) High-resolution spectroscopy of Mars at 3.7 and 8 µm: A sensitive search of H_2O_2, H_2CO, HCl, and CH_4, and detection of HDO. *J. Geophys. Res., 102,* 6525.

Kulikov Yu. N., Lammer H., Lichtenegger H. I. M., Terada N., Ribas I., Kolb C., Langmayr D., Lundin R., Guinan E. F., Barabash S., and Biernat H. K. (2006) Atmospheric and water loss from early Venus. *Planet. Space Sci., 54(13–14),* 1425–1444.

Kumar S., Hunten D. M., and Pollack J. B. (1983) Nonthermal escape of hydrogen and deuterium from Venus and implications for loss of water. *Icarus, 55,* 369–389.

Lammer H., Stumptner W., and Bauer S. J. (1996) Loss of H and O from Mars: Implications for the planetary water inventory. *Geophys. Res. Lett., 23,* 3353–3356.

Lammer H., Lichtenegger H. I. M., Kolb C., Ribas I., Guinan E. F., Abart R., and Bauer S. J. (2003) Loss of water from Mars: Implications for the oxidation of the soil. *Icarus, 165,* 9–25.

Lammer H., Kulikov Y. N., and Lichtenegger H. I. M. (2006) Thermospheric X-ray and EUV heating by the young Sun on early Venus and Mars. *Space Sci. Rev., 122,* 189–196.

Leblanc F. and Johnson R. E. (2002) Role of molecules in pick-up ion sputtering of the martian atmosphere. *J. Geophys. Res.–Planets,* DOI: 10209/2000JE001473.

Leblanc F., Chaufray J. Y., Witasse O., Lilensten J., and Bertaux J.-L. (2006) The martian dayglow as seen by SPICAM UV spectrometer on Mars Express. *J. Geophys. Res., 111(E9),* E09S11, DOI: 10.1029/2005JE002664.

Leblanc F., Chassefière E., Gillmann C., and D. Breuer (2012) Mars' atmospheric ^{40}Ar: A tracer for past crustal erosion. *Icarus, 218,* 561.

Levine J. S., McDougal D. S., Anderson D. E., and Barker E. S. (1978) Atomic hydrogen on Mars: Measurements at solar minimum. *Science, 200,* 1048–1051.

Liang M.-C. and Yung Y. L. (2009) Modeling the distribution of H_2O and HDO in the upper atmosphere of Venus. *J. Geophys. Res., 114(2),* DOI: 10.1029/2008JE003095.

Liang M. C., Heays A. N., Lewis B. R., Gibson S. T., and Yung Y. L. (2007) Source of nitrogen isotope anomaly in HCN in the atmosphere of Titan. *Astrophys. J. Lett., 664,* L115.

Lillis R. J., Frey H. V., and Manga M. (2008) Rapid decrease in martian crustal magnetization in the Noachian era: Implications for the dynamo and climate of early Mars. *Geophys. Res. Lett., 35,* L14203, DOI: 10.1029/2008GL034338.

Luhmann J. G. and Kozyra J. U. (1991) Dayside pickup oxygen ion precipitation at Venus and Mars: Spatial distributions, energy deposition and consequences. *J. Geophys. Res., 96,* 5457–5468.

Luhmann J. G., Johnson R. E., and Zhang M. H. G. (1992) Evolutionary impact of sputtering of the martian atmosphere by

O^+ pick-up ions. *Geophys. Res. Lett., 19,* 2151–2154.

Lundin R. (2011) Ion acceleration and outflow from Mars and Venus: An overview. *Space Sci. Rev., 162,* 309–334.

Lundin R., Zakharov A., Pellinen R., Borg H., Hultqvist B., Pissarenko N., Dubinin E. M., Barabash S., Liede I., and Koskinen H. (1989) First measurement of the ionospheric plasma escape from Mars. *Nature, 341,* 609.

Lundin R., Barabash S., Fedorov A., Holmström M., Nilsson H., Sauvaud J.-A., and Yamauchi M. (2008) Solar forcing and planetary ion escape from Mars. *Geophys. Res. Lett., 35,* L09203, DOI: 10.1029/2007GL032884.

Lundin R., Barabash S., Yamauchi M., Nilsson H., and Brain D. (2011) On the relation between plasma escape and the martian crustal magnetic field. *Geophys. Res. Lett., 38,* L02102, DOI: 10.1029/2010GL046019.

Ma Y.-J. and Nagy A. F. (2007) Ion escape fluxes from Mars. *Geophys. Res. Lett., 34,* L08201, DOI: 10.1029/2006GL029208.

Marcq E., Bézard B., Drossart P., Piccioni G., Reess J. M., and Henry F. (2008) A latitudinal survey of CO, OCS, H_2O, and SO_2 in the lower atmosphere of Venus: Spectroscopic studies using VIRTIS-H. *J. Geophys. Res., 113,* E00B07, DOI: 10.1029/2008JE003074.

Matsui H., Iwagami N., Hosouchi M., Ohtsuki S., and Hashimoto G. L. (2012) Latitudinal distribution of HDO abundance above Venus' clouds by ground-based 2.3 μm spectroscopy. *Icarus, 217(2),* 610–614, DOI: 10.1016/j.icarus.2011.07.026.

McElroy M. B. and Donahue T. M. (1972) Stability of the martian atmosphere. *Science, 177,* 986–988.

Melosh H. J. and Vickery A. M. (1989) Impact erosion of the primordial atmosphere of Mars. *Nature, 338,* 487–489.

Muskatel B. H., Remacle F., Thiemens M. H., and Levine R. D. (2011) On the strong and selective isotope effect in the UV excitation of N_2 with implications toward the nebula and martian atmosphere. *Proc. Natl. Acad. Sci. U.S.A., 108,* 6020–6025.

Nagy A. F., Cravens R. E., Yee J.-H., and Stewart A. I. F. (1981) Hot oxygen atoms in the upper atmosphere of Venus. *Geophys. Res. Lett., 8(6),* 629–632.

Nilsson H., Carlsson E., Brain D., Yamauchi M., Holmström M., Barabash S., Lundin R., and Futaana Y. (2010) Ion escape from Mars as a function of solar wind conditions: A statistical study. *Icarus, 206,* 40–49, DOI: 10.1016/j.icarus.2009.03.006.

Nilsson H., Edberg N. J. T., Stenberg G., Barabash S., Holmstrom M., Futaana Y., Lundin R., and Fedorov A. (2011) Heavy ion escape from Mars, influence from solar wind conditions and crustal magnetic fields. *Icarus, 215,* 475–484.

Öpik E. J. and Singer S. F. (1961) Distribution of density in a planetary exosphere. II. *Phys. Fluids, 4,* 221–233.

Owen T., Maillard J. P., DeBergh C., and Lutz B. L. (1988) Deuterium on Mars: The abundance of HDO and the value of D/H. *Science, 240,* 1767.

Owen T. (1992) The composition and early history of the atmosphere of Mars. In *Mars* (H. H. Kieffer et al., eds.), pp. 818–833. Univ. of Arizona, Tucson.

Ozima M., Seki K., Terada N., Miura Y. N., Podosek F. A., and Shinagawa H. (2005) Terrestrial nitrogen and noble gases in lunar soils. *Nature, 436,* 655.

Parker E. N. (1964) Dynamical properties of stellar coronas and stellar winds. I. Integration of the momentum equation. *Astrophys. J., 139,* 72.

Pepin R. O. (1994) Evolution of the martian atmosphere. *Icarus, 111,* 289–304.

Pham L. B. S., Karatekin O., and Dehant V. (2009) Effects of meteorite impacts on the atmospheric evolution of Mars. *Astrobiology, 9,* 45–54.

Pierrehumbert R. and Gaidos E. (2011) Hydrogen greenhouse planets beyond the habitable zone. *Astrophys. J. Lett., 734,* L13.

Rasool S. I. and de Bergh C. (1970) The runaway greenhouse and accumulation of CO_2 in the Venus atmosphere. *Nature, 226,* 1037–1039.

Ribas I., Guinan E. F., Güdel M., and Audard M. (2005) Evolution of the solar activity over time and effects on planetary atmospheres, I. High energy irradiances (1–1700 A). *Astrophys. J., 622,* 680–694.

Robert F. (2006) Solar system deuterium/hydrogen ratio. In *Meteorites and the Early Solar System II* (D. S. Lauretta and H. Y. McSween Jr., eds.), pp. 341–351. Univ. of Arizona, Tucson.

Rosing M. T., Bird D. K., Sleep N. H., and Bjerrum C. J. (2010) No climate paradox under the faint early Sun. *Nature, 464,* 744.

Seki K., Elphic R. C., Hirahara M., Terasawa T., and Mukai T. (2001) On atmospheric loss of oxygen ions from Earth through magnetospheric processes. *Science, 291,* 1939–1941.

Sheldon N. D. (2006) Precambrian paleosols and atmospheric CO_2 levels. *Precambrian Res., 147,* 148–155.

Shimazu Y. and Urabe T. (1968) An energetic study of the evolution of the terrestrial and cytherean atmospheres. *Icarus, 9,* 498–506.

Strobel D. F. (2008) Titan's hydrodynamically escaping atmosphere. *Icarus, 193,* 588.

Tian F., Toon O. B., Pavlov A. A., and DeSterck H. (2005a) Transonic hydrodynamic escape of hydrogen from extrasolar planetary atmospheres. *Astrophys. J., 621,* 1049.

Tian F., Toon O. B., Pavlov A. A., and DeSterck H. (2005b) A hydrogen rich early Earth atmosphere. *Science, 308,* 1014.

Tian F., Kasting J. F., Liu H., and Roble R. G. (2008) Hydrodynamic planetary thermosphere model: 1. Response of the Earth's thermosphere to extreme solar EUV conditions and the significance of adiabatic cooling. *J. Geophys. Res., 113,* E05008, DOI: 10.1029/2007JE002946.

Tian F., Kasting J. F., and Solomon S. C. (2009) Thermal escape of carbon from the early martian atmosphere. *Geophys. Res. Lett., 36,* L02205, DOI: 10.1029/2008GL036513.

Tian F., Claire M.W., Haqq-Misra J. D., et al. (2010) Photochemical and climate consequences of sulfur outgassing on early Mars. *Earth Planet. Sci. Lett., 295,* 412.

Trainer M. G., Pavlov A. A., Dewitt H. L., Jimenez J. L., McKay C. P., Toon O. B., and Tolbert M. A. (2006) Organic haze on Titan and the early Earth. *Proc. Natl. Acad. Sci., 103,* 18035.

Tucker O. J. and Johnson R. E. (2009) Thermally driven atmospheric escape: Monte Carlo simulations for Titan's atmosphere. *Planet. Space Sci., 57,* 1889–1894.

Tucker O. J., Johnson R. E., Deighan J. I., and Volkov A. N. (2013) Diffusion and thermal escape of H_2 from Titan's atmosphere: Monte Carlo simulations. *Icarus, 222,* 149.

Valeille A., Bougher S. W., Tenishev V., Combi M. R., and Nagy A. F. (2010) Water loss and evolution of the upper atmosphere and exosphere over martian history. *Icarus, 206,* 28–39.

Verigin M. I., Shutte N. M., Galeev A. A., et al. (1991) Ions of planetary origin in the martian magnetosphere (Phobos 2/TAUS experiment). *Planet. Space Sci., 39,* 131.

Volkov A. N., Johnson R. E., Tucker O. J., and Erwin J. T. (2011) Thermally-driven atmospheric escape: Transition from hydrodynamic to Jeans escape. *Astrophys. J. Lett., 729,* L24, DOI: 10.1088/2041-8205/729/2/L24.

Watson A. J., Donahue T. M., and Walker J. C. G. (1981) The dynamics of a rapidly escaping atmosphere — Applications to

the evolution of Earth and Venus. *Icarus, 48,* 150.
Wood B. E., Müller H.-R., Zank G. P., Linsky J. L., and Redfield S. (2005) New mass-loss measurements from astrospheric Lyα absorption. *Astrophys. J. Lett., 628,* L143.
Wordsworth R. (2012) Transient conditions for biogenesis on low-mass exoplanets with escaping hydrogen atmospheres. *Icarus, 219,* 267.
Wordsworth R. and Pierrehumbert R. (2013) Hydrogen-nitrogen greenhouse warming in Earth's early atmosphere. *Science, 339,* 64, DOI: 10.1126/science.1225759.
Yagi M., Leblanc F., Chaufray J. Y., Gonzalez-Galindo F., and Modolo R. (2012) Mars exospheric thermal and non-thermal components: Seasonal and local variations. *Icarus, 221(2),* 682–693.
Yung Y. L., Chen J.-S., Pinto J. P., Allen M., and Paulsen S. (1988) HDO in the martian atmosphere — Implications for the abundance of crustal water. *Icarus, 76,* 146–159.
Zahnle K. J. and Kasting J. F. (1986) Mass fractionation during transonic escape and implications for loss of water from Mars and Venus. *Icarus, 68,* 462–480.
Zahnle K., Haberle R. M., Catling D. C., and Kasting J. F. (2008) Photochemical instability of the ancient martian atmosphere. *J. Geophys. Res., 113(E11),* E11004.1–E11004.16.

Des Marais D. J. (2013) Planetary climate and the search for life. In *Comparative Climatology of Terrestrial Planets* (S. J. Mackwell et al., eds.), pp. 583–601. Univ. of Arizona, Tucson, DOI: 10.2458/azu_uapress_9780816530595-ch24.

Planetary Climate and the Search for Life

David J. Des Marais
NASA Ames Research Center

The search for evidence of life in the context of planetary environments beyond Earth is a key aspect of the emerging field of astrobiology. This chapter reviews life's basic attributes, habitable environments, and biosignatures. It then considers potential strategies to search for life under some example scenarios that represent a variety of target planets, planetary climates, and missions. Life might be envisioned as a self-sustaining system, capable of Darwinian evolution, that utilizes free energy to sustain and propagate an automaton (a self-replicator), a metabolic network, and functionally related larger structures. These functions specify a level of molecular complexity that in turn defines requirements for chemical ingredients, energy, and environmental conditions that are essential for life. The chemical versatility and pervasive cosmic distribution of organic compounds indicate that carbon plays a prominent role in many examples of extraterrestrial life, even if alternative chemistries also exist. Liquid water is an ideal solvent for life because it enhances noncovalent molecular interactions, it facilitates catalysis, and it imparts stability over remarkably wide ranges of temperature and pressure. Planetary climates play key roles in maintaining habitable environments by determining the availability of liquid water, biochemical raw materials, energy, and favorable conditions. Accordingly, strategies to explore for evidence of life must first consider how planetary climates determined the distribution of any habitable environments and the conditions that favored the preservation of any evidence of life in accessible deposits.

1. INTRODUCTION

An ongoing key goal in space exploration is to determine whether life has ever existed anywhere beyond Earth. Finding life on another world would have an enormous impact both scientifically and socially. There is a broad societal interest especially in areas such as achieving a deeper understanding of life, searching for extraterrestrial biospheres, assessing the societal implications of discovering other examples of life, and extending the human presence in space to other worlds. Finding a second example of life would trigger follow-up investigations to understand how that life functioned; how its structure, biochemistry, and physiology might resemble that of terrestrial life; how those attributes might differ and how they function; and whether any record of how that life began might be accessed. Even an apparent negative result would contribute significantly to our understanding of life as an emergent property of planetary systems. If a careful search finds no evidence of life in environments that could have sustained life and preserved evidence of it, we would need to understand whether this absence reflects differences between that planet and our own in the nature, extent, and duration of conditions that could have sustained a biosphere. Perhaps chemical processes that are precursors to life ("prebiotic chemistry") can be characterized and therefore provide insights into our potential origins. Clearly a consideration of planetary climates and their changes through time is a key aspect of any such investigation.

2. ASTROBIOLOGY

NASA's Astrobiology Roadmap (*Des Marais et al.,* 2008) articulates the following three fundamental questions that capture the breadth of the field and reflect the public interest: How does life begin and evolve? Does life exist elsewhere in the universe? What is the future of life on Earth and beyond? These questions address the search for habitable and inhabited planets beyond our solar system; the exploration of our own solar system; laboratory and field investigations related to the origins and early evolution of our biosphere; and studies of the potential of life to adapt to future challenges, both on Earth and beyond. The broad interdisciplinary character of astrobiology compels investigators to participate in research teams that pursue the most comprehensive and insightful understanding possible of biological, planetary, and cosmic phenomena.

The search for life in the context of planetary environments beyond Earth is a key aspect of the emerging field of astrobiology. This chapter addresses astrobiology by reviewing life's basic attributes, habitable environments, and biosignatures, and then by considering potential strategies to search for life under some example scenarios that represent the diversity of target planets, planetary climates, and missions.

The significant roles played by climate and other environmental parameters are apparent in virtually all of the seven Astrobiology Roadmap Goals (*Des Marais et al.,* 2008). The first two goals seek to determine the nature and distribution of habitable environments in our solar system and beyond. In order to comprehend the cosmic distribution of life we must characterize other planetary climates and assess them in the context of the resources and conditions that are universally required for life. The third goal is to understand how life emerges from cosmic and planetary environments and processes. For example, understanding Earth's earliest surface environments is essential for constraining the prebiotic processes that led to the origins of our biosphere. The fourth through sixth goals focus upon biological processes and evolution, but always in the context of environmental constraints and changes. Clearly important are the nature of planetary climates, how they change through time, their interactions with the surface and subsurface, and the geological and geochemical environment of the surface and subsurface. And we must better comprehend climate change in order to comprehend the future of our own biosphere. The seventh goal seeks to identify signatures of life, termed "biosignatures" (indicators of life), recognizing that any such features must be resolved from nonbiological features that might arise in the environment.

Life-related investigations have become a unifying theme for the planetary sciences. Habitability and the potential origins and survival of life are intimately linked to evolving planetary environments. An essential aspect of the search for evidence of life on other planets is understanding the interplay of processes that range from geophysical phenomena to climates. Regarding that search, this review focuses on Mars, Titan, and exoplanets as astrobiology targets for several reasons. First, these particular objects are widely viewed as promising hosts of past or present habitable environments. Second, each has an atmosphere that might have facilitated a habitable climate sometime in the past. Third, these objects as a group represent a contrasting range of scenarios under which habitable conditions might have been maintained.

3. ATTRIBUTES AND REQUIREMENTS OF LIFE

The recent expansion of efforts to characterize planetary environments has accelerated the search for life beyond Earth, but this search also requires a working concept of life's fundamental attributes. This concept helps to identify the "services" that an environment must provide in order to sustain life, and it helps to identify and interpret any biosignatures. This in turn helps to identify past or present planetary climates as promising candidates for exploration.

Any working concept of life must admit the possibility that life on Earth and elsewhere might differ in fundamental ways. Without a second known example of life, it is probably not possible to determine which characteristics are unique to terrestrial life and which are common or required for life in general. Accordingly, some have proposed that this might even preclude our enunciating a unique definition of life (*Cleland and Chyba,* 2007). But to travel any path toward a better understanding is to start taking steps, however uncertain, along that path. It is therefore worthwhile to hypothesize which aspects of life as we know it might be universal and which might represent local solutions to survival on Earth.

3.1. Universal Concepts and Attributes?

Recent studies (e.g., *Baross,* 2007) have identified the following potentially universal attributes of life:

- Life must exploit thermodynamic disequilibrium in the environment in order to perpetuate its own disequilibrium state.
- Life most probably consists of interacting sets of covalently bonded molecules that include a diversity of heteroatoms (e.g., N, O, P, S, etc., as in Earth-based life) that promote chemical reactivity.
- Life requires a liquid solvent that supports these molecular interactions.
- Life employs a molecular system capable of Darwinian evolution.

This list implicates the following basic universal functions (Fig. 1):

- Life harvests energy from its environment and converts it to forms of chemical energy that directly sustain its other functions.
- Life sustains "metabolism," namely a network of chemical reactions that synthesize all the key

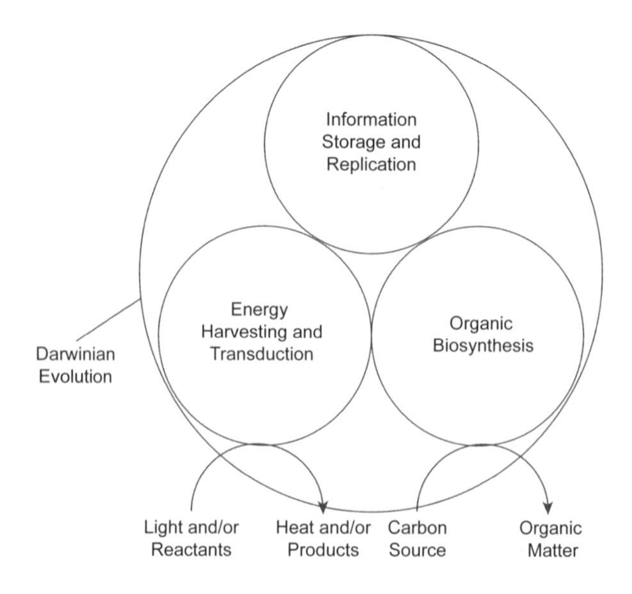

Fig. 1. Life's basic functions.

chemical compounds required for maintenance, growth, and self-replication.

- Life sustains an "automaton," a multicomponent system that is essential for self-replication and self-perpetuation (*Von Neumann,* 1966).

Perhaps the most characteristic feature of life is its capacity for self-replication to create populations upon which natural selection can act to maintain Darwinian evolution. When Von Neumann first presented his theory of the automaton, he predicted the fundamental components involved in biological self-replication (Fig. 2) that must exist, well before molecular biologists discovered the essential components of the DNA-RNA-protein system during the mid-twentieth century. The complexity implied by the multiple components engaged in these synergistic processes of self-replication requires that several relatively complex molecules can remain functional in environments where they can self-replicate more rapidly than they are degraded. This perspective has implications regarding the unique virtues of carbon-based chemistry (section 3.2), as well as the conditions that are required for an environment to remain habitable (section 3.5).

Accordingly, life might be envisioned as a self-sustaining system, capable of Darwinian evolution, that utilizes free energy to sustain and propagate an automaton, a metabolic network, and functionally related larger structures. These functions specify a level of molecular complexity that in turn defines requirements for chemical ingredients, energy, and environmental conditions that are essential for life.

Because astronomers and flight missions must necessarily conduct specific chemical observations to search for life, our working concept of life must provide more than simply a short list of generic universal attributes. However, as we cite our own biosphere to specify additional details, we risk overlooking extraterrestrial life that might differ

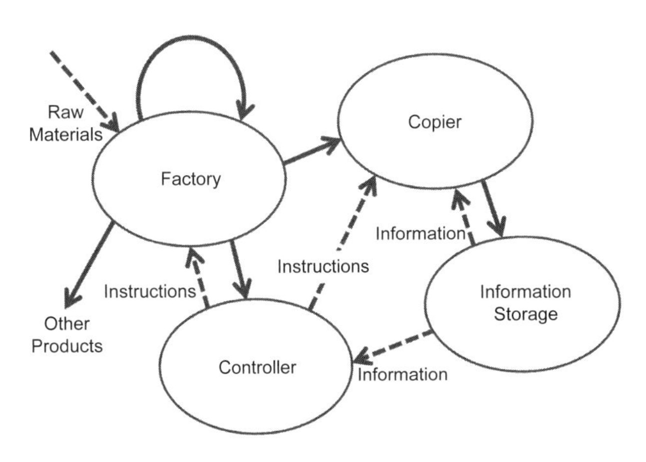

Fig. 2. An automaton, a system capable of self-replication (e.g., *Von Neumann,* 1966), has several complex components. For example, in biological systems, DNA stores and provides information (dashed arrows), and ribosome "factories" replicate other components as well as proteins involved in other cellular processes.

substantially from our own. For example, because exotic life might not employ familiar molecules to store genetic information and perform self-replication, we should not simply equate the search for life with a search for specific Earthly biochemicals such as DNA and RNA. Clearly we must identify a path that leads from our incomplete Earth-centric concept of life ultimately to a concept that is more universal. But we have little choice but to begin by identifying additional attributes of life that are universal among living systems as we know them.

3.2. Carbon

The hypothesis that carbon commonly plays prominent roles in the chemistry of life in the universe seems quite valid. This is true even if alternative life forms that are based on other elements also exist in some local environments. A key requirement of life is that compounds involved in self-replication and catalytic enzymes must be at least moderately complex. The unique ability of carbon to form highly stable chains, rings, heteroatomic molecules, etc., and thereby create stable complex molecules is thus a crucially important attribute.

Silicon has been considered the most viable alternative to carbon because it is the lightest element having a half-filled outer shell with four unpaired valence electrons, analogous to the outer shell of the carbon atom. Thus, for example, both carbon and silica form CH_4 and SiH_4, respectively, and both of these are gases under ambient conditions. Also, chains consisting of alternating carbon and oxygen atoms form chain-like compounds called polyacetals, whereas silicon and oxygen can form silicone chains. But silicon's strong affinity for oxygen favors the formation of SiO_2, the simplest oxide of silicon, which readily leads to a wide variety of relatively insoluble solids. In contrast, because CO_2 is a gas and is quite soluble in water, it is readily available to supply, participate in, and be excreted by biological processes.

Living systems must be able to store energy from their environment, and carbohydrates are centrally important storage media. Indeed, these "polyformaldehyde" compounds function as "biochemical batteries" whose polymer lengths represent their "state of charge."

Silicon is incapable of forming stable compounds that are analogous to carbohydrates. Longer chains of silicon atoms, especially those having six atoms or more, become increasingly more susceptible to aqueous hydrolysis. In general, silicon-based molecules having the size and complexity required for biological functions would be highly unstable in the presence of water or other polar solvents. Thus, for example, silicon cannot form compounds analogous to enzymes, whose chirality and complexity enables them to possess the attributes of chemical selectivity and regulation that are essential for sustaining metabolism and genetic functions.

Also, carbon is the fourth most abundant element in the universe because it is produced along with oxygen (third most abundant element) and nitrogen (fifth most abundant by atomic fraction) during nuclear reactions in aging stars. Carbon compounds are observed in the envelopes of these stars and in cooler, diffuse dark clouds. Because complex carbon chemistry is indeed pervasive in the universe, complex organic compounds were probably incorporated into protoplanetary disks and planets.

In contrast, only SiO_2 and more complex silicates have been detected in the outer layers of cool stars. Silicones, silanes, or other reduced silicon compounds might be important precursors of silicon-based life, but so far none of these compounds have been detected.

Although it is indeed impossible to prove that *all* life in the universe is necessarily based on carbon compounds, the chemical versatility and pervasive cosmic distribution of organic compounds indicate that carbon plays a prominent role in many examples of extraterrestrial life. Whether Earth-like conditions exist elsewhere in the universe remains to be demonstrated. But if life ever arose in environments similar to that on Earth, its biochemistry is probably based upon carbon.

Therefore, seeking carbon-based life is a logical, high-priority search strategy because, if life exists elsewhere, many if not most of its representatives will very likely be carbon-based. This by no means precludes future astrobiologists from pursuing truly exotic life forms, but such searches should be accompanied by laboratory and theoretical studies that explore the feasibility of such exotic life and also determine the specific features to be sought.

3.3. Water

Life's key functions require not only compounds that are relatively complex and stable. Such compounds also must assume certain molecular orientations in order to become functional and interact with other molecules in specific ways. It has long been recognized that a solvent plays critical roles in promoting biological organization (e.g., *Tanford,* 1978). Water is the solvent for life on Earth, but how can we assess how prominently it might assume this role elsewhere?

Noncovalent interactions between molecules strongly modulate the structures and functions of living cells (*Pohorille and Pratt,* 2012). For example, protein folding, interactions between enzymes and ligands, the regulation of gene expression, the self-assembly of boundary structures, and ion transport across membranes are all governed by noncovalent interactions. Such interactions must be in the right energy range in order to be effective. Noncovalent attractions between molecules must be sufficiently strong that inherent thermal noise cannot dismantle functional configurations. The system should also exhibit sufficient stability that it could function properly over the range of environmental temperatures. But if these interactions are too strong, then a potentially prohibitive expenditure of energy might be required to achieve the regulation of local thermodynamic equilibria that are essential to maintain a functional metabolism. Critical biomolecular interactions would become essentially irreversible.

Accordingly, the electrostatic interactions between biomolecules cannot be too strong. This requires that the solvent should have a high dielectric constant. Also, interactions between nonpolar molecules or groups should be sufficiently strong to favor their organization. Water meets all these requirements and therefore is an excellent solvent for life. Water has a high dielectric constant and it facilitates hydrophobic interactions between nonpolar species. The attributes of water that are most critical for supporting life remain remarkably constant over a wide range of temperatures and pressures. Biomolecular structures, interactions, and regulation exploit the characteristic similarities between the energies associated with hydrophobic and electrostatic interactions that can be achieved in liquid water.

Water does not necessarily enjoy a virtual monopoly on the attributes that make a solvent ideal for life (e.g., *Baines,* 2004). Alternatives such as formamide and dialcohols also have high dielectric constants. Vesicles can form in some of these alternative solvents due to solvophobic effects (*McDaniel et al.,* 1983). But very few solvents have all these favorable properties. One candidate liquid that has not yet been sufficiently explored is formamide. However, it seems inconceivable that any alternative solvent could approach the sheer abundance and broad distribution of water in the known universe. Therefore, even if alternative solvents support exotic examples of life in some cosmic niches, one can also surmise that water is the solvent of choice for the vast majority of occurrences of extraterrestrial life.

3.4. Basic Attributes of Life Based on Carbon and Water

In his book *Beginnings of Cellular Life,* H. Morowitz articulated a set of attributes, listed below, that are shared by all life on Earth (*Morowitz,* 1992):

1. *"The chemistry of life is carried out in aqueous solutions or at water interfaces. Cells can survive the removal and restoration of cellular water, but water is essential to cellular function."* Liquid water is therefore essential to maintain life's structures, functions, and long-term survival.

2. *"The major atomic components in the covalently bonded portions of all functioning biological systems are C, H, N, O, P and S."* Those particular chemical compounds of these elements that can engage in genetic, metabolic, and energy-harvesting functions therefore deserve particular attention.

3. *"A cell is the most elementary unit that can sustain life."* Single-celled organisms are by far the most diverse and most ancient forms of life known, and they can tolerate greater environmental extremes than can plants and animals. But we remain uncertain of whether the earliest life was necessarily compartmentalized in cells.

4. *"There is a universal set of small organic molecules that constitutes the total mass of all cellular systems."* The set of key organic biomolecules on Earth represents an in-

finitesimally small subset of all possible organic molecules. A very small array of molecules participates in intermediary metabolism and provides the building blocks of cellular structures and macromolecules. That this array is small perhaps reflects an evolutionary selection for speed and efficiency of biochemical processes. Living systems must efficiently maintain high information contents at molecular and structural levels.

5. *"Those reactions that proceed at appreciable rates in living cells are catalyzed by enzymes."* Catalysis enables cells to maintain their metabolism and to repair damage caused by environmental challenges such as radiation. Enzymes help to acquire essential energy by harvesting sunlight and by accelerating reactions between oxidized and reduced chemical species in the environment.

6. *"Sustained life is a property of an ecological system rather than a single organism or species."* For example, biofilms and microbial mats are highly successful ecological systems (e.g., *Des Marais,* 2010). Fossilized biofilms have been identified as stromatolites in rocks more than 3.4 b.y. old (*Walter,* 1976).

7. *"All populations of replicating biological systems give rise to altered phenotypes that are the result of mutated genotypes."* All known living systems evolve by Darwinian-like natural selection and survival. "Phenotype" is a cell's molecular and structural "machinery"; genotype is the information preserved in genomic molecules (DNA and RNA are examples in our biosphere). This is the automaton that is capable of Darwinian evolution.

In summary, the case is made that a substantial and pervasive fraction of any life in the universe is likely based upon organic chemical reactions occurring in water. This argument does not necessarily preclude the possibility that radical alternatives might exist. But the fundamental characteristics of carbon- and water-based life impose certain basic requirements upon a planetary environment that must be met in order for that environment to be habitable.

3.5. Biosignatures

3.5.1. Definition. The search for life beyond Earth rests upon the premise that signatures of life (termed "biosignatures") will be recognizable in the context of their planetary environments. A biosignature is an object, substance, and/or pattern whose origin specifically requires a biological agent (*Des Marais et al.,* 2008). Examples of biosignatures might be complex organic molecules and/or atmospheric constituents whose formation is virtually unachievable in the absence of life. Thus the usefulness of a biosignature is determined not only by the probability of life creating it, but also by the *improbability* of nonbiological processes producing it.

3.5.2. Categories of biosignatures. Basic attributes of life such as those enumerated above have allowed biochemists, paleontologists, and astrobiologists to identify and interpret categories of biosignatures in ancient geologic deposits on Earth. Categories of biosignatures that might also be applicable to the search for life beyond Earth include the following.

3.5.2.1. Organic compounds with characteristic molecular structures: Anomalously high relative abundances of specific organic molecules in a rock *might* constitute evidence of a biological origin. Molecules such as porphyrins, fatty acids, amino acids, and polysaccharides help to sustain key biochemical functions and therefore can be relatively abundant. Some of these molecules are relatively robust and can survive burial and storage in sedimentary rocks. For example, 2.7-b.y.-old ancient sediments contain higher molecular weight hydrocarbons having molecular structures that reveal their ancestry as hopanoids, isoprenoids, and steroids (*Brocks et al.,* 2003). These compounds are key constituents of cellular membranes and they can even be diagnostic for particular groups of microorganisms.

3.5.2.2. Remnants of cells or other microscopic biological structures: Microorganisms have been preserved as micrometer-scale spheres, rods, and filaments in fine-grained mineral and sedimentary deposits as old as 3.5 b.y. (*Schopf et al.,* 1992). Their biological origin can be demonstrated by the presence of cellular structures or particular chemical compositions. Their size distributions might match those of modern microbial populations.

3.5.2.3. Patterns of stable isotopic compositions: Both the biosynthesis of organic matter and biologically catalyzed oxidation-reduction reactions can discriminate against the heavier isotopes of carbon or sulfur to create patterns that indicate biological origins. For example, the bimodal distribution of $^{13}C/^{12}C$ abundances in ancient sedimentary reservoirs of carbonates and reduced carbon indicate that our biosphere existed since at least 3.5 b.y. ago (*Des Marais,* 2001). The geologic record of sedimentary sulfides reveals a comparable antiquity for sulfate-reducing microorganisms (*Canfield,* 2001).

3.5.2.4. Rock fabrics and structures: Biofilms can permeate and stabilize accreting sediment surfaces and create distinctive fabrics. For example, ancient rocks can preserve curved and/or contorted sheet-like structures indicating the former presence of flexible organic biofilms (*Tice and Lowe,* 2006). Stacks of laminations can accumulate in aqueous sedimentary environments to form stromatolites, which are microbial reef-like structures that represent the oldest, most widespread fossil evidence of early microbial communities (*Walter,* 1976).

3.5.2.5. Minerals: Microorganisms can precipitate minerals such as magnetite, sulfides, or phosphates during their metabolic or energy-harvesting activities. The crystal morphology and chemical composition of such minerals can indicate their biological origins (e.g., *Bazylinski and Moskowitz,* 1997).

3.5.2.6. Products of energy harvesting and storage: Life processes energy by promoting reactions between reduced and oxidized compounds. Products of these reactions can become biosignatures. For example, atmospheric O_2 is a biosignature on Earth because its high abundance requires both a photosynthetic source and the burial in sediments of a substantial quantity of photosynthetically produced organic matter (*Des Marais,* 2001). This sedimentary organic matter and its oxidized equivalents in the atmosphere and elsewhere are biosignatures. They are stored energy that we exploit when we burn fossil fuels. Microorganisms can harvest chemical energy either by reducing sulfate or by oxidizing Fe^{2+}, creating sulfide (*Canfield,* 2001) and iron oxide mineral products (e.g., *Bazylinski and Moskowitz,* 1997) that can become biosignatures in sedimentary rocks.

3.6. Multiple Environmental Requirements of Life

We can categorize the range of environmental conditions in which life on Earth occurs and discuss those aspects of the environment that provide support for the metabolic pathways actually utilized by life and therefore allow life to persist. According to the concepts and attributes of life articulated above, the principal determinants of habitability are (*Des Marais et al.,* 2008):

- The presence, chemical activity, and persistence of liquid water;
- The availability of thermodynamic disequilibria (i.e., usable energy sources);
- Physicochemical environmental factors (e.g., temperature, pH, salinity, radiation) that bear on the stability of covalent and hydrogen bonds in biomolecules as well as the energy required for a living system to maintain biological molecules and structures in a functional state;
- The availability of bioessential elements, principally C, H, N, O, P, S, and a variety of metals.

All measurements of these parameters must necessarily be placed within geological and environmental context to the greatest extent possible. These requirements, in turn, have direct implications for planetary climates. Accordingly, the parameters indicated in Fig. 3 will be explored further in the context of Mars exploration (section 4, below).

Recognizing that we have only one example of a biosphere that might indeed be unique in fundamental ways, we wish to extrapolate from our one example to understand the conditions suitable for life anywhere. This is a more challenging task that must confront and resolve many speculations and uncertainties, particularly at the outset of our voyages of exploration.

4. MARS

During its history, the climate of Mars has been more similar to that of Earth than has the climate of any other planet in our solar system (see also Rafkin et al., this volume). This similarity and the proximity of Mars have made Mars the principal focus of our search for a second example of life in our solar system. Accordingly, NASA's Viking lander missions to Mars during the 1970s searched for evidence of life in samples of the surface regolith (e.g., *Klein and De Vincenzi,* 1995). Although the near-unanimous consensus of the scientific community was that the Viking experiments did *not* detect definitive evidence of life, Viking's observations made valuable contributions to our understanding of the martian surface and set the stage for its subsequent exploration. Observations by the Viking landers, the Mars Exploration Rovers, the Phoenix Lander, and the Mars Science Laboratory Curiosity rover have characterized the environmental conditions and the availability of resources as a basis for understanding whether habitable conditions exist there today. Laboratory analyses of martian meteorites have provided additional insights. Mars therefore provides a valuable example of how planetary climates are essential to the potential presence of life beyond Earth, and therefore how our exploration strategies must reflect the linkages between climates and life.

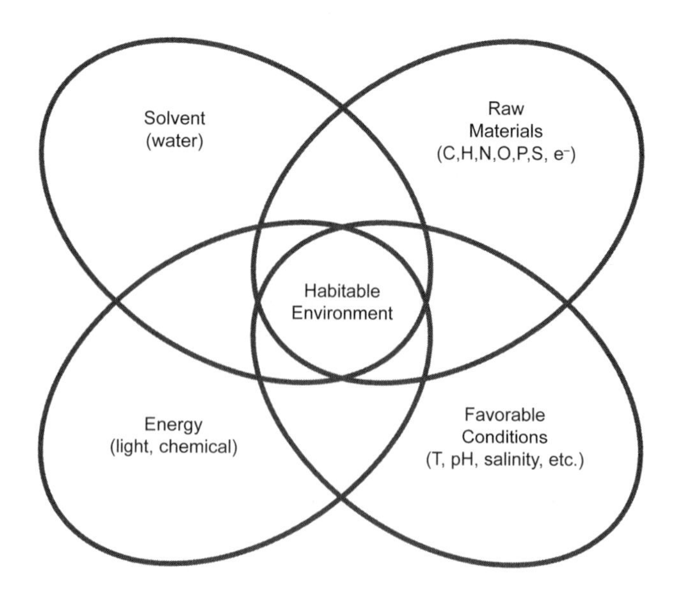

Fig. 3. Resources and conditions that a habitable environment must provide simultaneously. Adapted from graphic by T. Hoehler.

4.1. Present-Day Climate and Habitability

4.1.1. Liquid water. Ambient temperatures and pressures must be within the limits that allow liquid water to exist. Water can occur in its liquid state at temperatures below 0°C either as thin films at grain contacts or when solutes depress its freezing point (*Jakosky et al.,* 2003). Microbial growth has been observed below –10°C and metabolic activity typically associated with growth has been observed at –20°C (*Bakermans et al.,* 2003). Although life might persist at lower temperatures, the scarcity of liquid water and the much slower rate of biochemical reactions at those temperatures would probably restrict the occurrences of any ecosystems in such environments. Of course, organisms can remain viable for very long periods at lower temperatures in a suspended state, as occurs when microorganisms survive as spores or are stored at liquid nitrogen temperatures.

Organisms can also persist in seasonally or episodically dry environments by developing strategies to survive periods of prolonged dryness. However, all organisms require that a minimum level of water activity (*Grant,* 2004) be maintained at least intermittently so they can become active metabolically and repair damage caused by radiation, chemical degradation, or physical disruption that might have occurred during their dormancy. In order to mitigate detrimental effects of certain dissolved salts, microorganisms can maintain organic "compatible solutes" and/or compatible salts inside their cells in order to exclude disruptive solutes and to solve the challenges of osmoregulation (*Galinski and Trueper,* 1994).

In order to achieve self-repair and other functions that are critical for their survival, microorganisms require access to water molecules to be available above a critical threshold. The availability of liquid water to biological systems can quantified in terms of its chemical activity, which in turn can be measured as its vapor pressure above an aqueous solution. When water is equilibrated with the relative humidity, rh, of a gas in an overlying headspace, the water activity, a_w, is

$$a_w = rh/100$$

For pure water, $a_w = 1.0$. Water activity decreases as solute concentrations increase and as water becomes increasingly absorbed into surfaces, as occurs during desiccation in a porous medium such as unconsolidated sediment.

The Mars Exploration Program Analysis Group (MEPAG) is a science-community-based forum that provides input to NASA on science planning and prioritization for Mars exploration. The MEPAG Special Regions Science Analysis Group (MEPAG SR-SAG) reviewed our understanding of conditions at the martian surface as part of a study to support planetary protection requirements for exploring that planet. Figure 4 illustrates the equilibrium water activity of martian soil as a function of temperature, assuming an absolute humidity of 0.8 μbar, in equilibrium with the atmosphere (*MEPAG,* 2006). The box in the upper right corner of Fig. 4 delineates the conditions under which terrestrial life could propagate, i.e., where it can metabolize, grow, repair, and replicate itself. The SR-SAG concluded that, wherever the surface and shallow subsurface are at or close to being equilibrated thermodynamically with the atmosphere, any combination of temperature and water activity in the martian shallow subsurface is considerably below the threshold conditions required for terrestrial life to propagate. In warm soil, the chemical activity of water, a_w, is several orders of magnitude too low to support life — there simply is insufficient water to dampen the soil by the required amount. Water activity can attain a value of unity at the frost point, but then the temperature is far too low to support propagation.

Calculations of the water activity by SR-SAG have compensated for the effects of thin films and solute freezing point depression. Liquid water cannot persist at the surface under equilibrium conditions.

Gully-like features, namely recurring slope lineae (RSL),

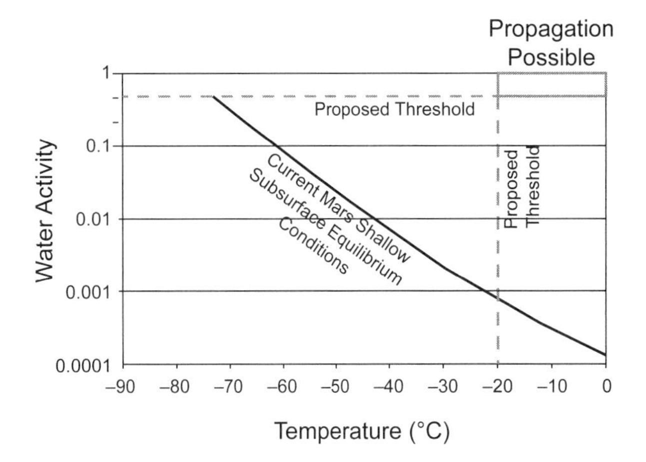

Fig. 4. Water activity vs. temperature in equilibrium with the atmosphere, at the surface of Mars today (*MEPAG,* 2006). The conditions in which life can grow and repair and replicate itself are represented by the box at the upper right corner. Note that the water activity of pure ice is less than unity over the entire range of temperatures that occur on present-day Mars.

indicate that aqueous fluids might emerge and flow even under present-day conditions (e.g., *Dundas et al.,* 2010; *McEwen et al.,* 2011). These relatively dark, 0.5- to 5-m-wide features extend down steep (25° to 45°) slopes from bedrock outcrops. They elongate during warm seasons and fade during cold seasons. Their formation is consistent with the presence of near-surface fluids. Highly saline brines can remain fluid at temperatures that are achieved seasonally at the martian surface. But microbial life as we know it cannot be metabolically active in highly saline brines [for further discussion of life and water activity, see *Tosca et al.* (2008)].

Microorganisms can thrive under highly acidic conditions such as those inferred for the deposits at the Meridiani site (*Knoll et al.,* 2005). Acid-mine drainage occurs where iron oxides, jarosite, and other sulfate minerals have precipitated abundantly. Some acid-tolerant eukarya, bacteria, and archea can maintain their cell interiors at near-neutral pH values by utilizing membrane-bound pumps to export hydrogen ions (e.g., *Booth,* 1985). Acidic solutions impose additional challenges by mobilizing toxic metals (*Rothschild and Mancinelli,* 2001) and by limiting certain sources of essential nutrients. For example, biological nitrogen fixation is not known to occur within acid mine drainages having low pH (*Baker and Bandfield,* 2003). However, commonly occurring phosphates such as apatite are more soluble in acidic waters. Phosphate is apparently relatively abundant in martian olivine basalts (e.g., *Fisk and Giovannoni,* 1999).

But proteins and nucleic acids are rapidly degraded in strongly acidic solutions. Furthermore, acidic conditions tend to inhibit HCN condensation reactions to form purines as well as the Strecker synthesis of amino acids. These two processes are considered to be crucial in certain scenarios for prebiotic chemistry and the origins of life (*Miller and*

Orgel, 1974). However, the acidic highly oxidizing environment inferred from observations at Meridiani Planum might represent the waning stages of an earlier period when conditions might have been more favorable for life to begin (*Knoll et al.,* 2005; *Bibring et al.,* 2006).

4.1.2. Raw materials. Mars certainly has abundant H, C, and O in atmospheric CO_2 and H_2O, and in carbonates (*Ehlmann et al.,* 2008) and bound H (*Boynton et al.,* 2002). Non-atmospheric carbon also occurs as carbonates in dust (*Bandfield et al.,* 2000) and in martian meteorites. The abundance of carbonate might be a few percent in ALH 84001 (*Mittlefehldt,* 1994). However, the total carbon inventory sequestered in the crust is unknown and might be sufficiently small to make its availability for life problematic. The global nitrogen inventory is also not known. The abundance of atmospheric N_2 is low, and nitrogen-bearing compounds have not been detected on the surface. Thus the global inventory of nitrogen might be sufficiently low that its availability for life is uncertain. Life on Earth requires several additional elements, including P, S, Fe, Mn, Ca, Mg, etc. These elements are generally considered to be available in silicate igneous rocks and related materials. Landed spacecraft and laboratory analyses of martian meteorites confirm that these elements occur in martian crustal rocks (e.g., *McSween,* 1994) and therefore would presumably be available to any martian biota.

4.1.3. Energy. Most of the energy that fuels Earth's biosphere is harvested from sunlight via photosynthesis. Mars orbits the Sun at typically 1.4 times the Earth's distance, and therefore Mars' surface still receives abundant sunlight. However, sunlight as a source of energy might be problematic today because any organisms would have difficulty surviving in the current surface environment. Due to the low atmospheric density and the absence of a magnetic field, the radiation environment at the martian surface is problematic for either origin of life or its survival.

4.1.4. Radiation. The radiation environment at the martian surface is often mentioned as being problematic for either origin of life or its continued evolution. High-energy galactic and solar cosmic rays and solar ultraviolet radiation have sufficient energy to break molecular bonds in cellular components, and can damage or kill living organisms (*Kminek and Bada,* 2006). These processes have the potential to affect life at the martian surface, although subsurface life (shielded beneath only millimeters of regolith for ultraviolet light or decimeters to meters for cosmic rays) would be protected. The presence of an early intrinsic magnetic field (e.g., *Solomon et al.,* 2005) created localized crustal magnetism that generated "mini-magnetospheres" (*Brain et al.,* 2003) that might provide substantial protection in some regions. Also, some organisms on Earth have developed the capability to survive in high-radiation environments (likely as a side benefit of living in very dry, desiccating environments), and might be able to do so on Mars as well (*Ghosal et al.,* 2005). The key is whether the constructive processes of metabolism in organisms occurred more quickly than the destructive processes involving radiation; it is unknown which would win out on Mars.

4.1.5. Subsurface environments. Due to the harsh conditions on the surface of Mars today, any modern habitable environments are probably restricted to the deep subsurface where liquid water is present and cosmic radiation is greatly attenuated.

A plausible source of energy for potential martian organisms is chemical reactions at and beneath the surface (*Shock,* 1997). Some terrestrial organisms make use of such chemical reactions. The reactions take advantage of the chemical disequilibrium that exists in many environments. For example, rocks emplaced by geological processes often find themselves out of chemical equilibrium with water. Useful reactions can include those between water and rock (equivalent to geochemical weathering reactions) or involving oxidation or reduction of Fe, Mn, S, or C, and can occur in aqueous systems that are at relatively low-temperature conditions (as long as liquid water is available) or that have been heated by nearby volcanic or impact activity (*Nealson,* 1997). An inventory of the energy available via such reactions indicates that there is sufficient energy to support a martian biosphere (*Jakosky and Shock,* 1998; *Varnes et al.,* 2003; *Link et al.,* 2005).

Chemotrophic microorganisms require redox gradients where oxidized and reduced constituents can react to provide energy. For example, the weathering of olivine basalt by acidic fluids readily releases Fe^{2+} and makes it available for oxidation by microorganisms. Aqueous alteration of olivine-rich rocks also can release H_2 (*Sleep et al.,* 2004), which is utilized by virtually all microorganisms as a substrate to produce energy and to synthesize organic matter. These and other sources of reduced compounds will diminish as weathering reactions exhaust the supply of unaltered rocks. Chemotrophs require that reduced materials be provided at the minimum rate needed to maintain their viability. All the S and Fe in the Meridiani bedrock apparently resides in oxidized forms as sulfates and ferric minerals, therefore once these bedrock materials achieved their current state of alteration, they probably no longer provided the redox gradient necessary to support chemotrophic microorganisms. However, both S and the olivine in basalts clearly could have been available in the source regions of these deposits. Examples of such regions include volcanic vents, hydrothermal systems, and basalts subjected to acidic weathering.

4.2. Past Climates and Habitability

Orbital observations during the Mariner and Viking missions mapped numerous features indicating that water once flowed abundantly across the surface. These features as well as evidence of more vigorous geologic activity in the past indicate that early climates were wetter and perhaps also somewhat warmer. A denser atmosphere was required for liquid water to be stable at the surface and also would have provided substantial protection from radiation, as it does on Earth today. Redox energy from volcanism, hydrothermal activity, and weathering of crustal materials would have been more readily available. The likelihood of any life at the surface would have been greatest during early wetter epochs. Life also might have persisted in long-lived hydrothermal systems, particularly those that had been associated with intrusive magmatism or volcanos (*Gulick,* 1998). The long-term stability of the martian crust has exceeded that of Earth and has allowed Mars to preserve an excellent record of its early history.

4.2.1. Diverse examples of sedimentation in ancient aqueous environments. Cameras and spectrometers onboard orbiters from several space agencies continue to map the surface of Mars for evidence of liquid water and geologic processes that might have sustained habitable environments. These campaigns have already been highly successful and have identified diverse sedimentary deposits from aqueous environments that might have been habitable in the distant past (*Bibring et al.,* 2006; *Murchie et al.,* 2009). The following categories of deposits represent thousands of occurrences throughout broad expanses of the southern highlands. These categories are listed below in the general order of oldest to youngest inferred ages.

4.2.1.1. Layered phyllosilicates: These deposits are prominent in the vicinity of Nili Fossae and Mawrth Vallis and consist of multiple layers that exhibit stratified compositions and are commonly polygonally fractured. They typically exhibit a lower layer of Fe/Mg-rich clay, a middle layer of Al-rich smectite, possibly montmorillonite, with hydrated silica, and a thin upper layer with Al-rich phyllosilicates of the kaolinite group. These deposits are apparently late Noachian in age (roughly 4 to 3.5 Ga).

4.2.1.2. Phyllosilicates in intracrater fans: These fans occur where large channels enter some craters, e.g., Holden crater, that might have hosted open basin lakes. The fans exhibit strong spectral signatures of Fe/Mg phyllosilicates that are overlain by deposits having no distinctive spectral features. The fan morphologies indicate deposition in quiescent lacustrine environments and the phyllosilicates might have derived from geologic units elsewhere in the watershed.

4.2.1.3. Plains sediments: These flat-lying sedimentary deposits occupy topographic lows within highlands closed basins and can exhibit phyllosilicates that are sometimes overlain by deposits that have been interpreted as chlorides. These deposits might represent playas and lacustrine environments that hosted the deposition of evaporites.

4.2.1.4. Deep phyllosilicates: Outcrops of phyllosilicates are observed in the central peaks, walls, rims, and ejecta of highland craters as well as in the walls of Valles Marineris and Nili Fossae. Orbital observations have confirmed that these deposits are widely distributed in the martian highlands (*Ehlmann et al.,* 2011). Their occurrences are consistent with their origins in materials exhumed from depth. Chlorite and saponite are common, although Al smectite, kaolinite, and other hydrated phases are also observed. Multiple modes of origin seem required to produce the diverse compositions of phyllosilicates and their geologic settings. Diagenetic processes can alter deposits buried by volcanism, sedimentation, or impacts. In other cases, hydrothermal activity associated with volcanism or impacts can alter igneous rocks to produce such assemblages (*Murchie et al.,* 2009).

These deposits might be very abundant in Noachian crust. Whether the phyllosilicates initially formed at the surface or at depth remains unresolved.

4.2.1.5. Carbonate-bearing deposits: These approximately 20-m-thick deposits have been identified at Nili Fossae and elsewhere in craters and escarpments around the perimeter of Isidis basin (*Ehlmann et al.,* 2008). Their spectral features are consistent with an Mg carbonate composition and they tend to overlie a regionally extensive olivine-rich deposit. Ehlmann et al. proposed that either the carbonates formed during the alteration of olivine, or the carbonates were deposited as aqueous sediments from surface waters.

Using observations by the Mars Exploration Rover Spirit, *Morris et al.* (2010) identified Mg-Fe carbonates in outcrops in the Columbia Hills, Gusev Crater. They postulate that these carbonates were deposited during the Noachian era by carbonate-bearing solutions associated with volcanic hydrothermal activity.

4.2.1.6. Intracrater clay-sulfate deposits: These have been observed as indurated beds in highland craters in Terra Sirenum (*Wray et al.,* 2009). Their spectral features indicate kaolinite and Fe and Mg sulfates, consistent with acid weathering of basaltic materials and chemical sedimentation in closed basin acid-saline lakes.

4.2.1.7. Meridiani-type layered deposits: These deposits have been studied extensively, both from orbit and by the Mars Exploration Rover Opportunity, and have been interpreted as late Noachian to early Hesperian in age (ca. 3.5 to 3.0 Ga) (*Squyres et al.,* 2006). They exhibit the composition of highly altered basaltic material with highly soluble sulfates (including jarosite) and ferric oxides (most prominently as hematitic concretions).

The bedrock experienced acidic conditions during its formation and/or later diagenesis. Conceivably four or more episodes of cementation occurred (*McLennan et al.,* 2005). Some of the primary minerals, possibly sulfates, were dissolved, producing millimeter-scale vugs and crystal-shaped voids (Fig. 5). Rounded millimeter-sized hematitic concretions precipitated within the sediment pore spaces (Fig. 5). These concretions displayed internal laminations that were aligned with laminations in the surrounding matrix. Ferric iron is more soluble in more acidic waters. The mineral jarosite is stable in the pH range 1 to 5. The bedrock at the Meridiani site had been modified over an extended period of time by a slowly fluctuating, chemically evolving groundwater system having fluids that persistently maintained relatively high ionic strengths (*McLennan et al.,* 2005).

Rising groundwater fed evaporitic playa lakes that, upon drying, provided sulfate-rich detritus that was entrained in dune deposits that are evident across an area of some 300,00 km^2 (*Squyres et al.,* 2006). These deposits indicate that ancient martian climates allowed bodies of water to persist at the surface.

Fig. 5. This Microscopic Imager mosaic image of the 3-cm-diameter "London" RAT hole was obtained on Sol 149 during the Karatepe ingress into Endurance crater, Meridiani Planum. This hole shows sedimentary laminations (that locally exhibit ripples), the granular nature of the sediment, blade-shaped vugs created by the dissolution of a soluble mineral, and a cross section of a hematite-rich concretion (upper left center). These features are consistent with "dirty evaporite" sediment that experienced a standing body of water at the surface and then was altered at depth by fluids that dissolved some minerals and precipitated others.

4.2.1.8. Valles-type layered deposits: These sulfate- and iron oxide/hydroxide-bearing layered deposits occur within canyon walls and as mounds within the chasmata of the Valles Marineris region. The layered mound deposits frequently exhibit a monohydrated sulfate (probably kieserite) that transitions upward into polyhydrated sulfates (e.g., *Mangold et al.,* 2008). These deposits might represent evaporite deposits from a combination of aeolian debris and groundwater and/or hydrothermal activity within these topographic depressions.

4.2.1.9. Hydrated silica-bearing deposits: These <1-m- to ~10-m-thick deposits were observed on the Hesperian plains surrounding Valles Marineris and the chasmata. They might represent products of acid weathering of volcanic ash, hydrothermal deposits, or chemical and detrital precipitation in a fluvial or lacustrine environment.

4.2.1.10. Gypsum plains: Gypsum is associated with dune deposits that surround the northern polar region in the vicinity of the polar ice cap. They have been interpreted as a wind-modified component of the north polar basal unit. Several origins have been proposed, including interactions between pyroxene with acidic snow, or precipitation from sulfate-rich groundwater created from basal melting of the polar cap (*Fishbaugh et al.,* 2007; *Langevin et al.,* 2005). The origins of these deposits remain an enigma, however.

4.2.2. Transitions in ancient climates and potentially habitable environments. Bibring et al. (2006) proposed that the most ancient sediments observed were deposited under relatively wet, neutral pH conditions that were conducive to the formation of the extensive phyllosilicates, whereas the sulfates were deposited somewhat later under global climates that were drier and hosted depositional environments that were more acidic. This marked deterioration of the earlier wetter and perhaps somewhat warmer climate might have been accentuated by the shutdown of the martian interior dynamo and the weakening of the magnetic field. This event allowed the solar wind to ablate the early atmosphere, thus cooling and drying the climate. Volcanic emanations provided sulfur species that, upon oxidation, created acidic environments and sulfate-bearing deposits (*Solomon et al.,* 2005). The alternation of phyllosilicate- and sulfate-bearing deposits is occasionally observed and is consistent with interpretation that, during this major transition in climate, the environmental conditions alternated repeatedly between favoring phyllosilicate deposition vs. favoring sulfates. The occasional deposition of acid-labile carbonates indicates that acidic conditions also waxed and waned during this climate transition.

4.2.3. Chaotic obliquity changes and climate. Mars' orbital elements change on timescales longer than ~10^4 years in response to gravitational forcing by Jupiter and the other planets (*Ward,* 1974; *Toon et al.,* 1980). These changes redistribute incoming sunlight across the latitudes and seasons and thereby alter the climate and the processes that distribute volatile species. In some scenarios, transient or steady-state liquid water might arise at or near the surface at low and high latitudes. Therefore liquid water might have occurred episodically relatively recently.

The obliquity of Mars varies on a timescale of about 10^5 years, with the amplitude of the oscillations changing on a timescale of about 10^6 years. On timescales longer than about 10^7 years, the system is chaotic (*Touma and Wisdom,* 1993). Combined with the multiple frequencies of forcing, this means that the obliquity cannot be predicted and that it can only be determined on a statistical basis (*Laskar et al.,* 2004). On the long timescales, the obliquity is thought to have varied from as low as nearly 0° to as high as 70° or more. The most likely value is near 40°, as compared with the present-day value of 25.2° (*Laskar et al.,* 2004).

The cooling or heating of the poles would also allow them to act as either sinks or sources for water vapor. At low obliquity, more water from the atmosphere would freeze onto the polar cap, and water would also be drawn from low- and mid-latitude sources of ground ice. At high obliquity, the polar caps would warm and release additional water vapor into the atmosphere. This water could exchange between the polar caps seasonally, it could diffuse into the regolith and condense out as additional ground ice, or it could even saturate the atmosphere at low latitudes and allow ice to condense onto the surface. In an extreme scenario, all the water ice in the polar deposits could move to low latitudes, covering the surface globally with up to tens of meters of ice (*Levrard et al.,* 2004).

Surface ice, if deposited in layers 1 km thick at high obliquity, could flow as a glacier, and we might expect to see glacial morphologies. Such features are observed and at the same locations where ice is predicted to deposit preferentially during the periods of high obliquity (*Forget et al.,* 2006).

Although obliquity forcing can result in deposition and removal of ice, can it produce the liquid water that might be relevant for potential organisms? Polar surface temperatures at moderate and high obliquity can rise to the melting point of ice and give rise to liquid water directly. At slightly lower obliquity, temperatures beneath a thin covering of ice can reach the melting point, or liquid water could occur where the proximity to grain boundaries depresses the melting temperature (*Jakosky et al.,* 2003). Geological evidence for liquid water has been seen associated with the polar deposits, suggesting that such a melting process actually might have occurred (*Fishbaugh and Head,* 2002).

At lower latitudes, the occurrence of deposits of ice thick enough to allow glacial movement allows the possibility of melting to form liquid water. Temperatures at a depth of 1 cm to several tens of centimeters can be tens of degrees warmer than at the surface, allowing ice to melt (*Clow,* 2003). Where surface ice is absent but subsurface ice is present, peak daily and annual temperatures could allow small quantities of ice to melt.

4.2.4. Early cold, "somewhat moist" early martian climates? Recent models of early climates illustrate the uncertainties that remain in our understanding of the temperatures and water cycling that accompanied early martian climates. Recent models (*Forget et al.,* 2013; *Wordsworth et al.,* 2013) indicate that conditions were too cold for liquid water to remain stable at the surface for extended intervals of time. For a wide range of orbital obliquities, water ice can still accumulate in the highlands. Episodic meteorite impacts and volcanism could cause episodic extensive melting and provide a robust mechanism for recharging highland water sources. Accordingly, they proposed a "globally subzero, icy highlands" scenario in order to explain most of the observed ancient fluvial and lacustrine features.

Consistent with these concepts, *Ehlmann et al.* (2011) proposed that, during the early Noachian and later periods, groundwater circulation was sustained by hydrothermal and other processes and formed the extensive subsurface phyllosilicate deposits. Perhaps subsurface environments maintained the most persistent habitable conditions on early Mars.

4.3. Search for Life

4.3.1. Search strategy constrained by climates. Whereas climates in the distant past might have favored habitable environments at least at some localities, clearly much of the martian surface for most of its history has been markedly less favorable for life. The combination of dry conditions, oxidizing surface environments, and typically low rates of sedimentation are not conducive to the preservation of information about ancient environments and any biota. Demonstrating fossil evidence in ancient rocks is quite difficult even on Earth, and confirming any martian biosignatures might be made even more challenging if any life on Mars was truly different. Accordingly, the following sections summarize a sequential strategy whereby the most promising sites are identified and characterized as a prelude to life detection efforts.

4.3.2. Habitability. Orbital observations provide the basis for the selection of landing sites that appear the most promising for having witnessed habitable conditions sometime in the past. The selection of Meridiani Planum, Gusev Crater, and Gale Crater are excellent examples of such efforts. Observations by landers/rovers would continue this evaluation as traverses and sample selections are determined. *MEPAG* (2010) identified the following key objectives to characterize the past habitability of a site: "(1) Establish overall geological context. (2) Constrain prior water availability with respect to duration, extent, and chemical activity. (3) Constrain prior energy availability with respect to type (e.g., light, specific redox couples), chemical potential (e.g., Gibbs energy yield), and flux. (4) Constrain prior physicochemical environment, emphasizing temperature, pH, water activity, and chemical composition; and (5) constrain the abundance and characterize potential sources of biologically essential elements. For ancient surface environments, these observations basically attempt to reconstruct the ancient climate and its associated processes.

4.3.3. Preservation. To be useful, biosignatures must be sequestered and preserved in geologic deposits. For example, preservation is favored by rapid burial in fine-grained, clay-rich sediments (*Farmer and Des Marais,* 1999; *Summons et al.,* 2011). Preservation in chemical precipitates is favored by rapid entombment in fine-grained, stable minerals. Biosignatures are preserved better in deposits that are relatively impermeable to gases and fluids. Rocks composed of silica (e.g., cherts) and phosphates (e.g., phosphorites) afford exceptional preservation. Carbonates and shales are less resistant but they are far more abundant than cherts and phosphates and so they have preserved most of the information about Earth's early biosphere.

Climates influence the nature of sedimentary deposits in ways that can determine their capacity to preserve biosignatures. For example, aqueous weathering releases ingredients that can precipitate to form cements that lower the permeability of sedimentary rocks. Phyllosilicates formed in aqueous environments effectively bind and sequester organic matter. Relatively high rates of aqueous sedimentation rapidly remove biosignatures from degradative processes in the environment.

MEPAG identified the following key objectives to assess the extent to which a site might preserved evidence of ancient environments and any life (*MEPAG, 2010*): "(1) Determine the major processes that degrade or preserve complex organic compounds, focusing particularly on characterizing oxidative effects in surface and near-surface environments (including determination of the 'burial depth' in regolith or rocks that may shield from such effects), the prevalence, extent, and type of metamorphism, and potential mechanisms and rates for obscuring isotopic or stereochemical information. (2) Identify the processes and environments that preserve or degrade physical structures on micron to meter scales. (3) Characterize processes that preserve or degrade environmental imprints of metabolism, including blurring of chemical or mineralogical gradients and loss of stable isotopic and/or stereochemical information."

4.3.4. Biosignatures. Finally, MEPAG advocates that evidence of ancient life should be sought in those environments that have been determined to exhibit a high combined potential for prior habitability and preservation of biosignatures. Potential biosignatures could be indicated by the following efforts (*MEPAG,* 2010): "(1) Characterize organic chemistry, including (where possible) stable isotopic composition and stereochemical information. Characterize co-occurring concentrations of possible bioessential elements. (2) Seek evidence of possibly biogenic physical structures, from microscopic (micron-scale) to macroscopic (meter-scale), combining morphological, mineralogical, and chemical information where possible. (3) Seek evidence of the past conduct of metabolism, including stable isotopic composition of prospective metabolites; mineral or other indicators of prior chemical gradients; localized concentrations or depletions of potential metabolites (especially biominerals); and evidence of catalysis in chemically sluggish systems."

5. TITAN

5.1. Environment

Titan, the largest moon of Saturn, is uniquely intriguing in the context of planetary climates, prebiotic chemistry, and life. Its dense atmosphere consists principally of nitrogen and methane and shrouds the planet in an opaque organic haze. The chapters by Griffith et al. and Krasnopolsky and Lefèvre in this volume provide further details about Titan and its atmosphere. Titan's average density of 1.87 g cm^{-3} is consistent with its interior being composed of approximately equal masses of silicate rocks and water ice; ammonia is also probably present (*Lunine and Rizk,* 2007). Titan is relatively large, compared to other ice-rich worlds in our solar system, and its size indicates that it might sustain tectonic activity in its interior. The temperature estimated to prevail at the surface is about 100 K; water ice exhibits the physical characteristics of solid rock at this temperature.

Titan hosts several geological and atmospheric processes that are analogous to those on Earth, albeit at much lower temperatures. The landscape exhibits aeolian features consistent with the action of winds (e.g., *Lorenz et al.,* 2006). Stream systems (Fig. 6) and lakes indicate that volatile constituents such as hydrocarbons evaporate and precipitate in cycles that are analogous to Earth's hydrologic cycle (*Schneider et al.,* 2012). The landscape has been resurfaced at rates sufficient to obliterate evidence for virtually all impact craters (*Lorenz et al.,* 2007). Processes that have been proposed as resurfacing agents include (1) windborne particles consisting of ice and condensed hydrocarbons; (2) particles ejected by cryovolcanism, in which the fluids involved are dominated by water as well as ammonia and other low-melting point constituents; (3) flowing liquid methane; and (4) perpetual deposition of solid (polymer-like) particles produced by photochemical reactions of hydrocarbons and nitrogen-bearing species in the atmosphere (*Lunine and Rizk,* 2007). Accordingly, the composition of the landscape is dominated by water ice, solid polymer like substances, and fluids consisting principally of light hydrocarbons. These climate-driven processes apparently have buried silicate rocks that might reside in Titan's crust, causing them to become relatively scarce near the surface.

The probable mix of complex hydrocarbons and nitrogen-bearing species in the atmosphere and surface make Titan an intriguing environment in the context of prebiotic organic chemistry. Cryovolcanism and other processes that vent the interior have replenished the atmosphere with methane and ammonia, which in turn have reacted to produce solid particles and nitrogen-containing organic compounds, including nitriles, amines, and amino acids (*Lunine and Rizk,* 2007; *Dougherty,* 2001). The solar UV promotes radical reactions involving methane and ammonia, leading to higher molecular weight species as well as highly unsaturated hydrocarbons such as acetylene. The molecular hydrogen produced during such photochemical reactions can escape to space. Thus although the atmosphere is in a

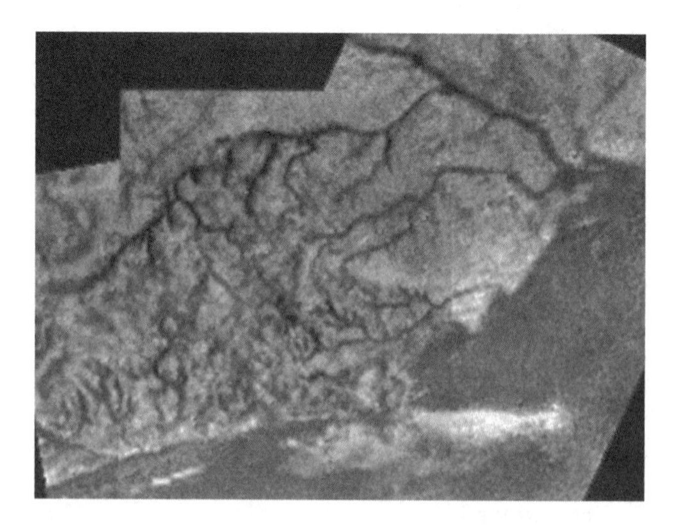

Fig. 6. Image showing tributaries feeding a major river channel in the high ridge area on Titan. Credit: NASA/JPL/ESA/University of Arizona.

chemically reduced state, the escape of molecular hydrogen prevents the atmosphere from becoming highly reducing. Highly unsaturated compounds such as acetylene conceivably could function as oxidants in redox reactions that are relevant to prebiotic chemistry.

5.2. Prebiotic Chemistry and Life

Unfortunately, the surface environment does not appear to be conducive to life as we know it. First, the very low temperatures keep water frozen as solid as rock and therefore unavailable as a solvent for life. As discussed earlier in this chapter in section 3.3, nonpolar solvents such as hydrocarbons lack several of the key attributes deemed essential for a biologically useful solvent. Molecules having complexity sufficient to conduct essential processes of self-replication and metabolism would be solids at these very low temperatures. Finally, the scarcity of silicate rocks and other metal-bearing crustal constituents would deprive any life of key nutrients.

Liquid water might conceivably form under certain circumstances, leading to the synthesis of important prebiotic compounds. Larger impacts could melt Titan's water-ice-rich crust, forming a zone of liquid water at the base of the impact crater (*Artemieva and Lunine,* 2003) that might persist for hundreds of years or more and promote the formation of carboxylic acids and amino acids from the ammonia-water mixtures. Alternatively, liquids bearing ammonia and water might exist beneath Titan's icy crust, interact with silicate minerals, and thereby promote prebiotic reactions involving hydrocarbons and other species (*Lunine and Rizk,* 2007). If some kind of transport mechanism delivered highly unsaturated hydrocarbons (produced in Titan's atmosphere) to such subsurface environments, these hydrocarbons could serve as oxidants in energy-yielding redox reactions that, in turn, could promote prebiotic chemical reactions. It is interesting to speculate that such an environment might foster a Titan-unique example of life that differs fundamentally from life as we know it.

6. EXOPLANETS

Exoplanets are planets that orbit stars other than our own. We are on the verge of a new era in which exoplanets, even Earth-like habitable worlds, will be observed directly and thus will enable a far more comprehensive search for other examples of life. Our enhanced observational capabilities will enable us to address the question of origins in a truly comprehensive fashion, from the origins of galaxies, stars, planetary systems, and planets, to the origins of habitable planetary environments and biospheres. While planetary systems similar to our own have yet to be found, we expect planetary systems to be common in our celestial neighborhood. The following discussion provides only a few examples of the many factors that can affect the habitability of exoplanets and the variety of strategies to search for evidence of life.

6.1. Habitable Climates

6.1.1. Circumstellar habitable zones and continuously habitable zones. The availability of liquid water at or below the surface of a planet is one of the fundamental requirements for life (section 3.3). In order to evaluate the presence of liquid water by astronomical methods, the zone where it occurs must affect the composition of a planet's surface environment and atmosphere. Therefore, unlike certain planets in our own solar system where we can send robotic landers to potentially explore beneath their surface, only evidence of climates and visible chemical constituents are accessible. This boundary condition frames the operational definition of the circumstellar habitable zone of other planetary systems (*Kasting et al.,* 1993).

A circumstellar habitable zone (HZ) is the locus of orbital radii of planets within which liquid water is stable at their surfaces. Beyond the outer radius of the HZ, any water near the surface of a planet would freeze. This occurs because water, CO_2, and other key greenhouse gases cannot compensate for the lower insolation. The inner edge of a HZ is determined by the loss of water due to photolysis followed by the loss of hydrogen to space. Earth has retained its water because its stratosphere is cold and water is trapped at the tropopause. The stratospheric water mixing ratio is very low, around 2 to 4 ppmv, and therefore little water participates in photolysis and hydrogen escape.

The luminosity of stars that evolve along the main sequence of stellar evolution will exhibit an increase over time, and therefore the HZ surrounding such a star will slowly migrate outward. For example, our own Sun began with a brightness of about 70% of its current value, increasing steadily thereafter (*Gough,* 1981). Accordingly, a continuously habitable zone (CHZ) is the locus of orbital radii of planets within which liquid water has been stable at their surfaces throughout the entire history of that planetary system. For example, the orbital radius of Earth has remained within the CHZ boundaries of our solar system, whereas that of Venus has not.

Although orbital radius and stellar luminosity clearly influence strongly the location of HZ and CHZ, the characteristics of a particular planet also influence its climate and therefore the HZ and CHZ.

6.1.2. Stellar mass. For stars having greater mass and luminosity, the associated HZ and CHZ will have greater radii as required to achieve levels of insolation that are optimal for habitable planets. However, there are limits to the range of stellar masses (*Kasting et al.,* 1993). Stars larger than 1.5 times our Sun's mass will have very large ultraviolet fluxes and relatively short lifetimes. Stars having less than 0.5 times our Sun's mass have very long lifetimes and tend to brighten more slowly, and therefore they have longer-lived HZs and CHZs.

Because any potentially habitable planets must orbit close to these smaller stars to receive adequate insolation, they are susceptible to becoming tidally locked, meaning that one side of the planet always faces the star as it orbits. This potentially creates the risk that water and other volatiles necessary to maintain an equable climate might freeze on the dark, cold side, rendering the planet uninhabitable. However, with some atmospheric compositions, global-scale circulation could deliver heat to the cold side and mitigate the hazards of condensation of water and other greenhouse gases. Even relatively tenuous atmospheres of noncondensable gases might prevent atmospheric collapse on the nightside of the planet (e.g., *Joshi,* 2003).

6.1.3. Planet size and tectonic activity. Our newfound ability to detect and characterize extrasolar planets allows us to explore relationships between their size, climates, and potential for habitability. The size and composition of a planet can modulate the intensity and style of its tectonic activity, which is driven principally by the flow of radiogenic heat from the interior (*Des Marais et al.,* 2002). Examples of tectonic processes are volcanism, mountain building, basin formation, and lateral movements such as plate tectonics.

The quantity of heat that must escape the interior, per unit area of the planet's surface, scales linearly with the planet's radius. Because the hot interiors of larger planets will cool much more slowly during their lifetimes, their tectonic processes will also be more vigorous and persist for much longer. To the extent that the interior is in a more reduced state than the crust, the more active tectonic processes on larger planets will favor more reduced surface environments that persist longer. The stronger gravitational field on a larger planet will lower the rate of hydrogen escape to space, thus generally favoring a relatively more reduced atmosphere.

Tectonic activity enhances chemical disequilibria among rocks and fluids and thereby provides fresh materials for weathering and erosion and sources of chemical energy for life (*Lunine,* 1999). Tectonics affects climates by continually modifying planetary landscapes and drives the geochemical cycles that modulate greenhouse gas inventories that, in turn, are critical to ensure climate stability over geologic timescales (*Walker et al.,* 1981; *Kasting et al.,* 1984). Tectonics also regulates the oxidation state of the surface environment by controlling rates of delivery and burial of oxidized and reduced chemical species between the mantle, crust, and surface environment. During a planet's lifetime, as internal sources of accretionary and radiogenic heat decline, heat flow and tectonic activity will also decrease inexorably.

Mars provides an illustrative example of the consequences associated with smaller size. Tectonic processes have declined more rapidly on Mars than on Earth. Mars' thicker, deeply fractured solid crust has assimilated water, CO_2, and other volatiles. Due to its lower gravity, the substantial loss of its volatiles to space has created an oxidized surface environment.

6.1.4. Volatile inventories. Volatile species such as water, CO_2, and other greenhouse gases tend to modulate climate and enhance habitable environments (*Kasting et al.,* 1993). Greenhouse gases warm a planet's surface to temperatures that can greatly exceed that of an airless body. Earth's oceans, hydrologic cycle, and feedback controls involving weathering, volcanism, and greenhouse gases have maintained remarkably stable climates for billions of years. Recent models indicate that the HZ for a planet having a larger fraction of its surface covered by water will extend to greater orbital radii (*Von Bloh et al.,* 2009). Weathering of continents is an effective sink for CO_2, a critical greenhouse gas, and therefore a planet having a relatively greater fraction of land area will lose its atmospheric CO_2 inventory more rapidly at orbital radii near the outer edge of the HZ. Models of "pseudo-Earths" having different land-ocean fractions have been investigated with respect to their habitable environments (*Spiegel et al.,* 2008; *Abbot et al.,* 2012).

However, a water-rich habitable planet is relatively more susceptible to falling into a "snowball" (ice-covered) state due to the high albedo of snow and ice and the inherent climatic stability associated with a global ice cover. The habitability of a snowball planet is ultimately rescued by volcanic CO_2 emanations that warm the atmosphere, triggering a global-scale melting event and additional greenhouse warming due to increasing levels of water vapor. Exoplanets having more eccentric orbits would be able to exit a snowball state sooner than those in more circular orbits (*Spiegel et al.,* 2010).

6.2. Observations of Exoplanets

6.2.1. Abundance and size distribution of candidate planets. Current models for the formation of planetary systems cannot yet predict the cosmic distribution of planetary systems having habitable planets, thus an observational program is required to obtain a statistically robust assessment of the frequency and size of planets accompanying stars that are approximately the size of our Sun. The Kepler mission (*www.kepler.nasa.gov*) has validated this approach by revealing an abundance and diversity of planetary systems that far exceed even the most optimistic predictions made before the mission began.

Kepler soon identified a remarkably diverse set of candidate planets in systems that differed substantially from our own solar system (*Borucki et al.,* 2010, 2011). Although larger planets orbiting close to their stars (e.g., the "hot Jupiters") were soon readily identified because of their robust and frequent transit signals, the climates of such planets are most certainly uninhabitable (e.g., *Desert et al.,* 2011).

Kepler's discoveries of more than 1100 candidate planets have confirmed that planetary systems are remarkably diverse and that most stars similar to our own probably harbor planetary systems (*Batalha et al.,* 2013). About 20% of stars that exhibit transits actually host multiple candidates. The rate of discovery of progressively smaller planets at longer orbital periods indicates that Earth-sized planets in circumstellar habitable zones indeed exist and conceivably might be quite abundant. Planets whose estimated masses range from those of "super-Earths" (>1–$10\ M_\oplus$) to "Neptunes" (14–$17\ M_\oplus$) are particularly abundant. This rapidly expanding field promises many additional significant and exciting discoveries.

6.2.2. Climates and habitability. Efforts to detect evidence of life must always be conducted in the context of characterizing the environment being explored. As discussed above, the size of a planet is important for assessing its habitability. The relationships between mass, radius, and thermal environment for planets in our solar system can help in estimating the likely mass of an exoplanet, and thereby inferring whether this planet is likely to exhibit geological (tectonic) activity conducive to sustaining a favorable environment for life (*Des Marais et al.,* 2002).

But characterizing the climates of newly discovered candidate planets remains far more challenging than their discovery. The currently most promising hosts of habitable climates are super-Earths located adjacent to or within their circumstellar habitable zones (e.g., *Tuomi et al.,* 2013; *Anglada-Escudé et al.,* 2013).

Environments that are habitable for life as we know it must provide sources of water and raw materials such as carbon. Furthermore, water vapor and gas carbon species such as CO_2 and CH_4 are important greenhouse gases, and therefore detecting their presence in an atmosphere is generally consistent with the presence of conditions favorable for life. Spectral features can, in principle, quantify the presence and abundance of water vapor. For a planetary atmosphere resembling that of Earth, observation of the 8–12-μm infrared continuum would allow the detection of water vapor and also provide an estimate of a planet's surface temperature (*Des Marais et al.,* 2002). Unfortunately, atmospheres significantly denser and warmer than Earth's will obscure their surfaces and preclude such estimations. In contrast, CO_2 can be readily detected in the infrared over a broader range of atmospheric conditions.

6.2.3. Biosignatures. Because astronomical observations of exoplanets will consist of measurements averaged across the entire visible disk of a planet, life must be global in extent and on the surface. Furthermore, the biosphere must have persisted sufficiently long to have substantially affected environmental conditions. Remotely sensed biosignatures (indicators of life) on Earth fall into the following three categories: (1) atmospheric species produced by life that are in disequilibrium with the surrounding environment; (2) surface features attributable to life (e.g., spectral features created by photosynthetic organisms); and (3) features that vary over time, e.g., biologically driven seasonal variations in albedo, atmospheric species, or features on the surface.

The most robust atmospheric species indicating life is O_2 or its photolytic byproduct, O_3. Molecular oxygen is a major byproduct of photosynthesis on Earth, and it might be expected in other comparably advanced biospheres. Oxygen and O_3 are readily detected in the visible–near-infrared and infrared, respectively (*Des Marais et al.,* 2002). The strongest O_3 features in Earth's spectrum are the 200–360-nm Huggins and Hartley bands; however, their interpretation is complicated by the fact that many gases absorb in the ultraviolet. Care must be taken to recognize potential occurrences of O_2 formed by nonbiological processes. For example, O_2 could be produced by H_2O photolysis, followed by escape of H_2 to space. This O_2 could accumulate on a cold planet with low levels of tectonic activity because O_2 removal by weathering and reactions with reduced volcanic gases would be minimal. Accordingly, the detection of H_2O vapor and CO_2 would indicate that these processes are sufficiently robust to overwhelm abiotic sources of O_2. Therefore a detection of O_2 under these circumstances might constitute compelling evidence of life.

Methane is produced by methanogenic archea but it also has nonbiological sources, e.g., in hydrothermal systems. Nitrous oxide (N_2O) is promising because it is produced during microbial oxidation-reduction reactions, it apparently has negligible nonbiological sources, and it is relatively stable photochemically. However, because CH_4 and H_2O features can readily obscure its key spectral feature near 8 μm, N_2O might be detectible only at relatively high abundances (*Des Marais et al.,* 2002). Reduced sulfur gases can also be biogenic (*Domagal-Goldman et al.,* 2011) but typically suffer from relatively short photochemical lifetimes. The most promising of these is methanethiol (CH_3SH), which exhibits a strong spectral feature and, on Earth, typically originates from a biological source and enjoys a photochemical lifetime that is long relative to that of other reduced sulfur species (*Pilcher,* 2003).

Characterizing the climates of exoplanets is an essential part of the search for life on these bodies. In addition to indicating whether a planet is habitable, environmental conditions are also important for actually observing evidence of habitation. The technical challenges associated with detecting and interpreting key spectral features of exoplanets indicate that understanding climates will be critical for confirming the presence of biosignatures among the spectral noise that arises within habitable planetary climates.

7. FUTURE DIRECTIONS

7.1. Mars

Recent discoveries have revealed the former presence of diverse environments that might have been habitable sometime in the past. The following questions pose key remaining challenges to our efforts to demonstrate whether life actually existed on Mars.

Does Mars meet the environmental requirements that would allow either an origin or the continued existence of life on Mars? Current evidence hints that present-day subsurface environments might have the capacity to support life, but this remains undemonstrated. Mars could have sustained life at least for some period of time at some time in the past. The apparently highly saline and acidic ancient waters inferred at the Meridiani Planum landing site seem unfavorable for the origins of life (*Knoll et al.,* 2005), although any established life as we know it might have been able to persist there (*Tosca et al.,* 2008). Environmental conditions at other places and times might have been more favorable. For example, Curiosity Rover found gravels several kilometers from the rim of Gale Crater that could have been transported by a rapidly flowing stream (*Williams et al.,* 2013). Even if life could not have originated on Mars, it might conceivably have survived if it had been transferred from Earth in a meteorite.

How long did any habitable conditions persist, and would these conditions have allowed any life to begin or persist? Was early Mars significantly warmer and wetter or was it actually cold and only episodically "moist?" The Curiosity rover will explore this question in Gale Crater, as will any future rovers, including those tasked to collect samples for eventual return to Earth-based laboratories.

Many additional questions address the evolution of Mars as a planet that, in essence, are critical in regard to its habitability: *What is the history of geological processes at the surface (e.g., the history of volcanism and its relation to liquid water)? What are the thermal history and interior composition and structure that drive surface and crustal processes such as volcanism? What was the initial volatile inventory of Mars and the subsequent evolution of that inventory?*

7.2. Titan

The evidence for hydrocarbon cycles analogous to Earth's hydrologic cycle indicates that organic chemical reactions might create compounds that might be key for life. But water-mediated chemical reactions would proceed very slowly, if at all, on the frigid surface. Therefore warmer deep-subsurface conditions hold the greatest potential to sustain liquid water and host habitable environments. Remote observations and landed missions to characterize the cycling of volatile species, organic chemical processes, and the deep subsurface promise the greatest return for astrobiology.

7.3. Exoplanets

This area of exploration is rapidly expanding in many directions, all of which are highly promising (e.g., see Domagal-Goldman and Segura, this volume). For example, we need to devise ways to infer the climates associated with the numerous, diverse candidate planets found by Kepler and other observers. How might such climates affect both the habitability of those candidates and our strategies for characterizing them remotely? Using Earth as an analog, relationships can be explored between the remotely observable features of its atmosphere and surface and the planetary processes that can affect those features (e.g., *Robinson et al.,* 2011). We can utilize our understanding of planetary processes to develop models for other planetary systems where planets, orbits, and stars differ from our own. Ultimately, of course, additional observations *must* be made of other planetary systems, both by utilizing current capabilities and by developing more highly capable observational platforms; e.g., the James Webb Space Telescope could observe potentially habitable exoplanets orbiting nearby M-dwarf stars. We should develop new spacecraft to perform more detailed spectroscopic observations of potentially habitable exoplanets.

Acknowledgments. The author acknowledges support from the NASA Astrobiology Institute and the excellent comments of two anonymous reviewers.

REFERENCES

Abbot D. S., Cowan N. B., and Ciesla F. J. (2012) Indication of insensitivity of planetary weathering behaviour and habitable zone to surface land fraction. *Astrophys. J., 756,* 178–190, DOI: 10.1088/0004-637X/756/2/178.

Anglada-Escudé G., Tuomi M., Gerlach E., Barnes R., Heller R., Jenkins J. S., Wende S., Vogt S. S., Butler R. P., Reiners A., and Jones H. R. A. (2013) **A** dynamically-packed planetary system around GJ 667C with three super-Earths in its habitable zone. *Astron. Astrophys.,* in press.

Artemieva N. and Lunine J. I. (2003) Cratering on Titan: Impact melt and the fate of surface organics. *Icarus, 164,* 471–480.

Bains W. (2004) Many chemistries could be used to build living systems. *Astrobiology, 4,* 137–167.

Baker B. J. and Banfield J. F. (2003) Microbial communities in acid mine drainage. *FEMS Microb. Ecol., 44,* 139–152.

Bakermans C., Tsapin A. I., Souza-Egipsy V., Gilichinisky D. A., and Nealson K. H. (2003) Reproduction and metabolism at –10 degrees C of bacteria isolated from Siberian permafrost. *Environ. Microbiol., 5,* 321–326.

Bandfield J. L., Glotch T. D., and Christensen P. R. (2000) Spectroscopic identification of carbonate minerals in the martian dust. *Science, 301,* 1084–1087.

Baross J. A. (2007) *The Limits of Organic Life in Planetary Systems.* National Academies, Washington, DC. 100 pp.

Batalha N. M., Rowe J. F., Bryson S. T., Barclay T., Burke C. J., Caldwell D. A., Christiansen J. L., Mullally F., Thompson S. E., Brown T. M., Dupree A. K., Fabrycky D. C., Ford E. B., Fortney J. J., Gilliland R. L., Isaacson H., Latham D. W., Marcy G. W., Quinn S., Ragozzine D., Shporer A., Borucki W. J., Ciardi D. R., Gautier T. N. III, Haas M. R., Jenkins J. M., Koch D. G., Lissauer J. J., Rapin W., Basri G. S., Boss A. P., Buchhave L. A., Charbonneau D., Christensen-Dalsgaard J., Clarke B. D., Cochran W. D., Demory B.-O., Devore E., Esquerdo G. A., Everett M., Fressin F., Geary J. C., Girouard F. R., Gould A., Hall J. R., Holman M. J., Howard A. W., Howell S. B., Ibrahim K. A., Kinemuchi K., Kjeldsen H., Klaus T. C., Li J., Lucas P. W., Morris R. L., Prsa A., Quintana E., Sanderfer D. T., Sasselov D., Seader S. E., Smith J. C., Steffen J. H., Still M., Stumpe M. C., Tarter J. C., Tenenbaum P., Torres G., Twicken J. D., Uddin K., Van Cleve J., Walkowicz L., and Welsh W. F. (2013) *Astrophys. J. Suppl., 204,* 24–44, DOI: 10.1088/0067-0049/204/2/24.

Bazylinski D. A. and Moskowitz B. M. (1997) Microbial biomineralization of magnetic iron minerals: Microbiology, magnetism and environmental significance. In *Geomicrobiology: Interactions Between Microbes and Minerals* (J. F. Banfield and K. H. Nealson, eds.), pp. 181–223. Mineralogical Society of America, Washington, DC.

Bibring J.-P., Langevin Y., Mustard J. F., Poulet F., Arvidson R., Gendrin A., Gondet B., Mangold N., Pinet P., Forget F., and the OMEGA Team (2006) Global mineralogical and aqueous

Mars history derived from OMEGA/Mars Express data. *Science, 312,* 400–404.

Booth I. R. (1985) Regulation of cytoplasmic pH in bacteria. *Microbiol. Rev., 49,* 359–378.

Borucki W. J., Koch D., Basri G., Batalha N., Brown T., Caldwell D., Caldwell J., Christensen-Dalsgaard J., Cochran W. D., DeVore E., Dunham E. W., Dupree A. K., Gautier T. N., Geary J. C., Gilliland R., Gould A., Howell S. B., Jenkins J. M., Kondo Y., Latham D. W., Marcy G. W., Meibom S., Kjeldsen H., Lissauer J. J., Monet D. G., Morrison D., Sasselov D., Tarter J., Boss A., Brownlee D., Owen T., Buzasi D., Charbonneau D., Doyle L., Fortney J., Ford E. B., Holman M. J., Seager S., Steffen J. H., Welsh W. F., Rowe J., Anderson H., Buchhave L., Ciardi D., Walkowicz L., Sherry W., Horch E., Isaacson H., Everett M. E., Fischer D., Torres G., Johnson J. A., Endl M., MacQueen P., Bryson S. T., Dotson J., Haas M., Kolodziejczak J., Van Cleve J., Chandrasekaran H., Twicken J. D., Quintana E. V., Clarke B. D., Allen C., Li J., Wu H., Tenenbaum P., Verner E., Bruhweiler F., Barnes J., Prsa A. (2010) Kepler planet-detection mission: Introduction and first results. *Science, 327,* 977–980, DOI: 10.1126/science.1185402.

Borucki W. J., Koch D. G., Basri G., Batalha N., Brown T. M., Bryson S. T., Caldwell D., Christensen-Dalsgaard J., Cochran W. D., DeVore E., Dunham E. W., Dupree A. K., Gautier Y. N. III, Geary J. C., Gilliland R., Gould A., Howell S. B., Jenkins J. M., Latham D. W., Lissauer J. J., Marcy G. W., Rowe J., Sasselov D., Boss A., Charbonneau D., Ciardi D., Doyle L., Dupree A. K., Ford E. B., Fortney J., Holman M. J., Seager S., Steffen J. H., Tarter J., Welsh W. F., Allen C., Buchhave L. A., Christiansen J. L., Clarke B. D., Das S., Dasert J-M., Endl M., Fabrycky D., Fressin F., Haas M., Horch E., Howard A., Isaacson H., Kjeldsen H., Kolodziejczak J., Kulesa C., Li J., Lucas P. W., Machalek P., McCarthy D., MacQueen P., Meibom S., Miquel T., Prsa A., Quinn S. N., Quintana E. V., Ragozzine D., Sherry W., Shporer A., Tenenbaum P., Torres G., Twicken J. D., Van Cleve J., Walkowicz L., Witteborn F. C., and Still M. (2011) *Astrophys. J., 736,* 19–40, DOI: 10.1088/0004-637X/736/1/19.

Boynton W., Feldman W. C., Squyres S. W., et al. (2002) Distribution of hydrogen in the near surface of Mars: Evidence for subsurface ice deposits. *Science, 297(5578),* 81–85.

Brain D. A., Bagenal F., Acuna M. H., and Connerney J. E. P. (2003), Martian magnetic morphology: Contributions from the solar wind and crust. *J. Geophys. Res., 108,* DOI: 10.1029/2002JA009482.

Brocks J. J., Buick R., Summons R. E., and Logan G. A. (2003) A reconstruction of Archean biological diversity based on molecular fossils from the 2.78–2.45 billion year old Mount Bruce Supergroup, Hamersley Basin, Western Australia. *Geochim. Cosmochim. Acta, 67,* 4321–4335.

Canfield D. E. (2001) Biogeochemistry of sulfur isotopes. In *Stable Isotope Geochemistry* (J. W. Valley and D. R. Cole, eds.), pp. 607–636. Reviews in Mineralogy, Vol. 43, Mineralogical Society of America, Washington, DC.

Cleland C. E. and Chyba C. F. (2007) Does 'life' have a definition? In *Astrobiology and Life* (W. T. Sullivan and J. A. Baross, eds.), pp. 119–131. Cambridge Univ., Cambridge.

Clow G. D. (2003) Generation of liquid water on Mars through melting of a dusty snowpack. *Icarus, 72,* 95–127.

Desert J.-M., Charbonneau D., Fortney J. J., Madhusudhan N., Knutson H. A., Fressin F., Deming D., Borucki W. J., Brown T. M., Caldwell D., Ford E. B., Gilliland R. L., Latham D. W., Marcy G. W., Seager S., and the Kepler Science Team (2011) The atmospheres of the hot Jupiters Kepler-5b and Kepler 6-b observed during occultations with Warm-Spitzer and Kepler. *Astrophys. J. Suppl., 197,* 11–21.

Des Marais D. J. (2001) Isotopic evolution of the biogeochemical cycle during the Precambrian. In *Stable Isotope Geochemistry* (J. W. Valley and D. R. Cole, eds.), pp. 555–578. Reviews in Mineralogy, Vol. 43, Mineralogical Society of America, Washington, DC.

Des Marais D. J. (2010) Marine hypersaline *Microcoleus*-dominated cyanobacterial mats in the saltern at Guerrero Negro, Baja California Sur, Mexico. In *Microbial Mats* (J. Seckbach, ed.), pp. 401–420. Cellular Origins, Life in Extreme Habitats and Astrobiology (COLE) Series, Springer, Berlin.

Des Marais D. J., Harwit M. O., Jucks K. W., Kasting J. F., Lin D. N. C., Lunine J. I., Schneider J., Seager S., Traub W. A., and Woolf N. J. (2002) Remote sensing of planetary properties and biosignatures on extrasolar terrestrial planets. *Astrobiology, 2,* 153–181.

Des Marais D. J., Nuth J. A. III, Allamandola L. J., Boss A. P., Farmer J. D., Hoehler T. M., Jakosky B. M., Meadows V. S., Pohorille A., and Spormann A. M. (2008) The NASA Astrobiology Roadmap. *Astrobiology, 8,* 715–730.

Domagal-Goldman S. D., Meadows V. S., Claire M. W., and Kasting J. F. (2011) Using biogenic sulfur gases as remotely detectable biosignatures on anoxic planets. *Astrobiology, 11,* 419–441.

Dougherty D. (2001) Polymer stereochemistry: An opportunity and challenge on Titan. *Enantiomer, 6,* 101–106.

Dundas C. M., McEwen A. S., Diniega S., Byrne S. and Martinez-Alonso S. (2010) New and recent gully activity on Mars as seen by HiRISE. *Geophys. Res. Lett., 37,* DOI: 10.1029/2009GL041351.

Ehlmann B. L., Mustard J. F., Murchie S. L., Poulet F., Bishop J. L., Brown A. J., Calvin W. M., Clark R. N., Des Marais D. J., Milliken R. E., Roach L. H., Roush T. L., Swayze G. A., and Wray J. J. (2008) Orbital identification of carbonate-bearing rocks on Mars. *Science, 322,* 1828–1832.

Ehlmann B. L., Mustard J. F., Murchie S. L., Bibring J.-P., Meunier A., Fraeman A. A., and Langevin Y. (2011) Subsurface water and clay mineral formation during the early history of Mars. *Nature, 479,* 53–60.

Farmer J. D. and Des Marais D. J. (1999) Exploring for a record of ancient martian life. *J. Geophys. Res., 104,* 26977–26995.

Fishbaugh K. E. and Head J. W. III (2002) Topographic characterization from Mars Orbiter Laser Altimeter data and implications for mechanisms of formation. *J. Geophys. Res., 107,* 5013, DOI: 5010.1029/2000 JE001351.

Fishbaugh K. E., Poulet F., Chevrier V., Langevin Y., and Bibring J.-P. (2007) On the origin of gypsum in the Mars north polar region. *J. Geophys. Res., 112,* E07002, DOI: 10.1029/2006JE002862.

Fisk M. R. and Giovannoni S. J. (1999) Sources of nutrients and energy for a deep biosphere on Mars. *J. Geophys. Res., 104,* 11805–11815.

Forget F., Haberle R. M., Montmessin F., Levrard B., and Head J. W. (2006) Formation of glaciers on Mars by atmospheric precipitation at high obliquity. *Science, 311,* 368–371.

Forget F., Wordsworth R., Millour E., Madeleine J.-B., Kerber L., Leconte J., Marcq E., and Haberle R. M. (2013) 3D modeling of the early martian climate under a denser CO_2 atmosphere: Temperatures and CO_2 ice clouds. *Icarus, 222,* 81–99.

Galinski E. A. and Trueper H. G. (1994) Microbial behaviour in salt-stressed ecosystems. *FEMS Microbiol. Rev., 15,* 95–108.

Ghosal D., Omelchenko M. V., Gaidamakova E. K., Matrosova V. Y., Vasilenko A., Venkateswaran A., Kostandarithes H. M., Brim H., Makarova K. S., Wackett L. P., Fredrickson J. K., and Daly M. J. (2005) How radiation kills cells: Survival of *Deinococcus radiodurans* and *Shewanella oneidensis* under oxidative stress. *FEMS Microbiol. Rev., 29,* 361–375.

Gough D. O. (1981) Solar interior structure and luminosity variations. *Solar Phys., 74,* 21.

Grant J. A. (2004) Life at low water activity. *Philos. Trans. R. Soc. Lond. Series B–Biological Sciences, 359,* 1249–1266.

Gulick V. C. (1998) Magmatic intrusions and a hydrothermal origin for fluvial valleys on Mars. *J. Geophys. Res., 103,* 19365–19388.

Jakosky B. M. and Shock E. L. (1998) The biological potential of Mars, the early Earth, and Europa. *J. Geophys. Res, 103,* 19359–19364.

Jakosky B. M., Nealson K. H., Bakermans C., Ley R. E., and Mellon M. T. (2003) Sub-freezing activity of microorganisms and the potential habitability of Mars' polar regions. *Astrobiology, 3,* 343–350.

Joshi M. (2003) Climate model studies of synchronously rotating planets. *Astrobiology, 3,* 415–427.

Kasting J. F., Pollack J. B., and Ackerman T. P. (1984) Response of Earth's atmosphere to increases in solar flux and implications for loss of water from Venus. *Icarus, 57,* 335–355.

Kasting J. F., Whitmire D. P., and Reynolds R. T. (1993) Habitable zones around main sequence stars. *Icarus, 101,* 472–494.

Kminek D. and Bada J. L. (2006) The effect of ionizing radiation on the preservation of amino acids on Mars. *Earth Planet. Sci. Lett., 245,* 1–5.

Klein H. P. and De Vincenzi D. L. (1995) Exobiological exploration of Mars. *Adv. Space Res., 15,* (3)151–(3)156.

Knoll A. H., Carr M., Clark B., Des Marais D. J., Farmer J. D., Fischer W. W., Grotzinger J. P., Hayes A., McLennan S. M., Malin M., Schröder C., Squyres S., Tosca N. J., and Wdowiak T. (2005) An astrobiological perspective on Meridiani Planum. *Earth Planet. Sci. Lett., 240,* 179–189.

Langevin Y., Poulet F., Bibring J.-P., and Gondet B. (2005) Sulfates in the north polar region of Mars detected by OMEGA/Mars Express. *Science, 307,* 1584–1587, DOI: 10.1126/science.1109091.

Laskar J., Correia A. C. M., Gastineau M., Joutel F., Levrard B., and Robutel P. (2004) Long term evolution and chaotic diffusion of the insolation quantities of Mars. *Icarus, 170,* 343–364.

Levrard B., Forget F., Montmessin F., and Laskar J. (2004) Recent ice-rich deposits formed at high latitudes on Mars by sublimation of unstable equatorial ice during low obliquity. *Nature, 431,* 1072–1075.

Link L. S., Jakosky B. M., and Thyne G. D. (2005) Biological potential of low-temperature aqueous environments on Mars. *Intl. J. Astrobiology, 4,* 155–164.

Lorenz R. D., Wall S., Radebaugh J., Boubin G., Reffet E., Janssen M., Stofan E., Lopes R., Kirk R., Elachi C., Lunine J., Mitchell K., Paganelli F., Soderblom L., Wood C., Wye L., Zebker H., Anderson Y., Ostro S., Allison M., Boehmer R., Callahan P., Encrenaz P., Ori G. G., Francescetti G., Gim Y., Hamilton G., Hensley S., Johnson W., Kelleher K., Muhleman D., Picardi G., Posa F., Roth L., Seu R., Shaffer S., Stiles B., Vetrella S., Flamini E., and West R. (2006) The sand seas of Titan: Cassini RADAR observations of longitudinal dunes. *Science, 312,* 724–727.

Lorenz R. D., Wood C. A., Lunine J. I., Wall S. D., Lopes R. M., Mitchell K. L., Paganelli F., Anderson Y. Z., Wye L., Tsai C., Zebker H., and Stofan E. R. (2007) Titan's young surface: Initial impact crater survey by Cassini RADAR and model comparison. *Geophys. Res. Lett., 34,* L07204, DOI: 10.1029/2006 GL028971.

Lunine J. I. (1999) *Earth: Evolution of a Habitable World.* Cambridge Univ., Cambridge.

Lunine J. I. and Rizk B. (2007) Titan. In *Planets and Life* (W. T. Sullivan III and J. A. Baross, eds.), pp. 425–443. Cambridge Univ., Cambridge.

Mangold N., Gendrin A., Gondet B., LeMouelic S., Quantin C., Ansan V., Bibring J.-P., Langevin Y., Masson P., and Neukum G. (2008) Spectral and geological study of the sulfate-rich region of West Candor Chasma, Mars. *Icarus, 194,* 519–543, DOI: 10.1016/j.icarus.2007.10.021.

McDaniel R. V., McIntosh T. J., and Simon S. A. (1983) Glycerol substitutes for water in lecithin bilayers. *Biophys. J., 41,* A116.

McEwen A. S., Ojha L., Dundas C. M., Mattson S. S., Byrne S., Wray J. J., Cull S. C., Murchie S. L., Thomas N., and Gulick V. C. (2011) Seasonal flows on warm martian slopes. *Science, 5,* 740–743.

McLennan S. M., Bell J. F., Calvin W. M., Christensen P. R., Clark B. C., de Souza P. A., Farmer J., Farrand W. H., Fike D. A., Gellert R., Ghosh A., Glotch T. D., Grotzinger J. P., Hahn B., Herkenhoff K. E., Hurowitz J. A., Johnson J. R., Johnson S. S., Jolliff B., Klingerhofer G., Knoll A. H., Learner Z., Malin M. C., McSween H. Y. Jr., Pocock J., Ruff S. W., Soderblom L. A., Squyres S. W., Tosca N. J., Watters W. A., Wyatt M. B., and Yen A. (2005) Provenance and diagenesis of the evaporite-bearing Burns formation, Meridiani Planum, Mars. *Earth Planet. Sci. Lett., 240,* 95–121.

McSween H. Y. Jr. (1994) What have we learned about Mars from C/SNC meteorites. *Meteoritics, 29,* 757–779.

MEPAG (2006) Findings of the Mars Special Regions Science Analysis Group. *Astrobiology, 6(5),* 677–732. Available online at *mepag.jpl.nasa.gov/reports/ast_2006_6_677.pdf.*

MEPAG (2010) *Mars Science Goals, Objectives, Investigations, and Priorities: 2010.* Available online at *mepag.jpl.nasa.gov/reports/MEPAG_Goals_Document_2010_v17.pdf.*

Miller S. L. and Orgel L. E. (1974) *The Origins of Life on the Earth.* Prentice-Hall, Englewood Cliffs.

Mittlefehldt D. W. (1994) ALH84001, a cumulate orthopyroxenite member of the martian meteorite clan. *Meteoritics, 29,* 214–221.

Morowitz H. (1992) *Beginnings of Cellular Life.* Yale Univ., New Haven.

Morris R. V., Ruff S. W., Ming D. W., Arvidson R. E., Benton B. C., Golden D. C., Siebach K., Klingelhöfer G., Schröder C., Fleischer I., Yen A. S., and Squyres S. W. (2010) Identification of carbonate-rich outcrops on Mars by the Spirit rover. *Science, 329,* 421–424.

Murchie S. L., Mustard J. F., Ehlmann B. L., Milliken R. E., Bishop J. L., McKeown N. K., Noe Dobrea E. Z., Seelos F. P., Buczkowski D. L., Wiseman S. M., Arvidson R. E., Wray J. J., Swayze G., Clark R. N., Des Marais D. J., McEwen A. S., and Bibring J.-P. (2009) A synthesis of martian aqueous mineralogy after 1 Mars year of observations from the Mars Reconnaissance Orbiter. *J. Geophys. Res., 114,* E00D06, DOI: 10.1029/2009JE003342,2009.

Nealson K. H. (1997) The limits of life on Earth and searching for life on Mars. *J. Geophys. Res., 102,* 23675–23686.

Pilcher C. B. (2003) Biosignatures of early Earths. *Astrobiology, 3,* 471–486.

Pohorille A. and Pratt L. R. (2012) Is water the universal solvent for life? *Origins Life Evol. Biosph., 42(5),* 405–409, DOI: 10.1007/s11084-012-9301-6.

Robinson T. D., Meadows V. S., Crisp D., Deming D., A'Hearn M. F., Charbonneau D., Livengood T. A., Seager S., Barry R. K., Hearty T., Hewagama T., Lisse C. M., McFadden L. A., and Wellnitz D. D. (2011) Earth as an extrasolar planet: Earth model validation using EPOXI Earth observations. *Astrobiology, 11,* 393–408.

Rothschild L. J. and Mancinelli R. L. (2001) Life in extreme environments. *Nature, 409,* 1092–1101.

Schneider T., Graves S. D. B., Schaller E. L., and Brown M. E. (2012) Polar methane accumulation and rainstorms on Titan from simulations of the methane cycle. *Nature, 481,* 58–61.

Schopf J. W., Chang S., Ernst W. G., Holland H. D., Kasting J. F., and Lowe D. R. (1992) Geology and paleobiology of the Archean Earth. In *The Proterozoic Biosphere* (J. W. Schopf and C. Klein, eds.), pp. 5–42. Cambridge Univ., Cambridge.

Shock E. L. (1997) High-temperature life without photosynthesis as a model for Mars. *J. Geophys. Res., 102,* 23687–23694.

Sleep N. H., Meibom A., Fridriksson Th., Coleman R. G., and Bird D. K. (2004) H_2-rich fluids from serpentinization: Geochemical and biotic implications. *Proc. Natl. Acad. Sci., 101,* 12818–12823.

Solomon S. C., Aharonson O., Aurnou J. M., Banerdt W. B., Carr M. H., Dombard A. J., Frey H. V., Golombek M. P., Hauck S. A. II, Head J. W. III, Jakosky B. M., Johnson C. L., McGovern P. J., Neumann G. A., Phillips R. J., Smith D. E., and Zuber M. T. (2005) New perspectives on ancient Mars. *Science, 307,* 1214–1220.

Spiegel D. S., Menou K., and Scharf C. A. (2008) Habitable climates. *Astrophys. J., 681,* 1609–1623.

Spiegel D. S., Raymond S. N., Dressing C. D., Scharf C. A., and Mitchell J. L. (2010) Generalized Milankovitch cycles and long-term climatic habitability. *Astrophys. J., 721,* 1308–1318.

Squyres S. W., Arvidson R. E., Bollen D., Bell J. F. III, Brückner J., Cabrol N. A., Calvin W. M., Carr M. H., Christensen P. R., Clark B. C., Crumpler L., Des Marais D. J., d'Uston C., Economou T., Farmer J., Farrand W. H., Folkner W., Gellert R., Glotch T. D., Golombek M., Gorevan S., Grant J. A., Greeley R., Grotzinger J., Herkenhoff K. E., Hviid S., Johnson J. R., Klingelhöfer G., Knoll A. H., Landis G., Lemmon M., Li R., Madsen M. B., Malin M. C., McLennan S. M., McSween H. Y., Ming D. W., Moersch J., Morris R. V., Parker T., Rice J. W. Jr., Richter L., Rieder R., Schröder C., Sims M., Smith M., Smith P., Soderblom L. A., Sullivan R., Tosca N. J., Wänke H., Wdowiak T., Wolff M., and Yen A. (2006) Overview of the Opportunity Mars Exploration Rover mission to Meridiani Planum: Eagle Crater to Purgatory Ripple. *J. Geophys. Res., 111,* E12S12, DOI: 10.1029/2006JE002771.

Summons R. E., Amend J. P., Bish D., Buick R., Cody G. D., Des Marais D. J., Dromart G., Eigenbrode J. L., Knoll A. H., and Sumner D. Y. (2011) Preservation of martian organic and environmental records. *Astrobiology, 11,* 157–181.

Tanford C. (1978) The hydrophobic effect and the organization of living matter. *Science, 200,* 1012–1018.

Tice M. M. and Lowe D. R. (2006) The origin of carbonaceous matter in pre-3.0 Ga greenstone terrains: A review and new evidence from the 3.42 Ga Buck Reef Chert. *Earth Sci. Rev., 76,* 259–300.

Toon O. B., Pollack J. B., Ward W., Burns J. A., and Bilski K. (1980) The astronomical theory of climatic change on Mars. *Icarus, 44,* 552–607.

Tosca N. J., Knoll A. H., and McLennan S. M. (2008) Water activity and the challenge for life on early Mars. *Science, 320,* 1204–1207.

Touma J. and Wisdom J. (1993) The chaotic obliquity of Mars. *Science, 259,* 1294–1297.

Tuomi M., Anglada-Escude G., Gerlach E., Jones H. R. A., Reiners A., Rivera E. J., Vogt S. S., and Butler P. (2013) Habitable-zone super-Earth candidate in a six-planet system around the K2.5V star HD 40307. *Astron. Astrophys., 549,* A48, DOI: 10.1051/0004-6361/201220268.

Varnes E. S., Jakosky B. M., and McCollom T. M. (2003) Biological potential of martian hydrothermal systems. *Astrobiology, 3,* 407–414.

Von Bloh W., Cuntz M., Schroeder K.-P., Bounama C., and Franck S. (2009) Habitability of super-Earth planets around other Suns: Models including red giant branch evolution. *Astrobiology, 9,* 593–602.

Von Neumann J. (1966) *Theory of Self-Reproducing Automata* (A. Burks, ed.). Univ. of Illinois, Urbana.

Walker J. C. G., Hays P. B., and Kasting J. F. (1981) A negative feedback mechanism for the long-term stabilization of Earth's surface temperature. *J. Geophys. Res., 86,* 9776–9782.

Walter M. R. (1976) *Stromatolites.* Elsevier, Amsterdam.

Ward W. R. (1974) Astronomical theory of insolation. *J. Geophys. Res., 79,* 3375–3386.

Williams R. M. E., Grotzinger J. P., Dietrich W. E., Gupta S., Sumner D. Y., Wiens R. C., Mangold N., Malin M. C., Edgett K. S., Maurice S., Forni O., Gasnault O., Ollila A., Newsom H. E., Dromart G., Palucis M. C., Yingst R. A., Anderson R. B., Herkenhoff K. E., Le Mouélic S., Goetz W., Madsen M. B., Koefoed A., Jensen J. K., Bridges J. C., Schwenzer S. P., Lewis K. W., Stack K. M., Rubin D., Kah L. C., Bell J. F. III, Farmer J. D., Sullivan R., Van Beek T., Blaney D. L., Pariser O., Deen R. G., and the MSL Science Team (2013) Martian fluvial conglomerates at Gale Crater. *Science, 340,* 1068–1072.

Wordsworth R., Forget F., Millour E., Head J. W., Madeleine J.-B., and Charnay B. (2013) Global modeling of the early climate under a denser CO_2 atmosphere: Water cycle and ice evolution. *Icarus, 222,* 1–19.

Wray J. J., Murchie S. L., Squyres S. W., Seelos F. P., and Tornabene L. L. (2009) Diverse aqueous environments on ancient Mars revealed in the southern highlands. *Geology, 37,* 1043–1046.

Color Section

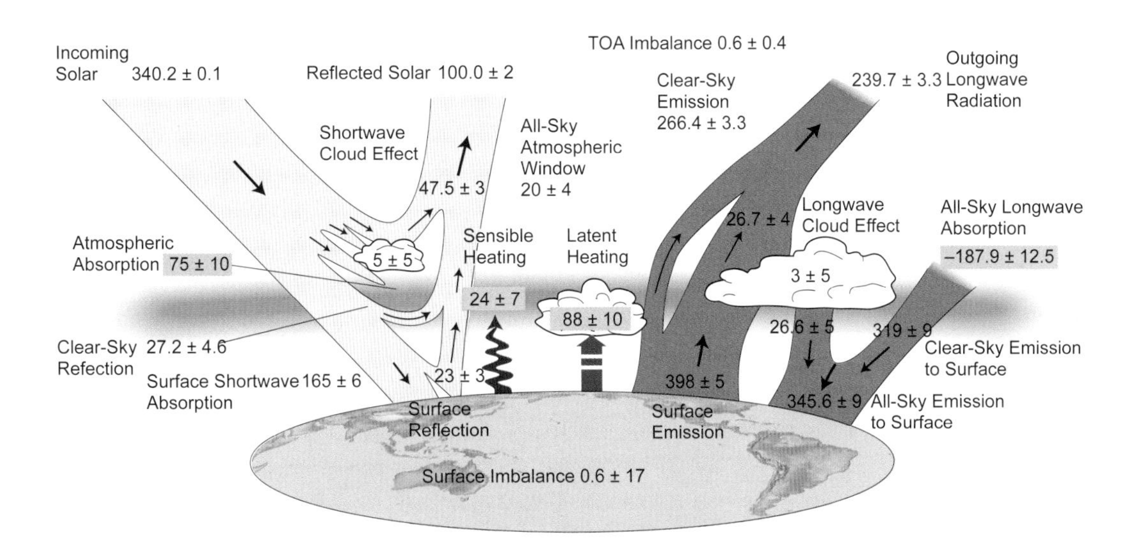

Plate 1. Estimate of the components of the current annual mean energy balance of Earth at TOA (upper row), within the atmosphere (middle row), and at the surface (bottom row). SW fluxes are in yellow, LW fluxes in magenta, and surface turbulent fluxes in red and violet. From *Stephens et al.* (2012). ©Copyright Nature Publishing Group; reprinted with permission.

Accompanies chapter by Del Genio (pp. 3–18).

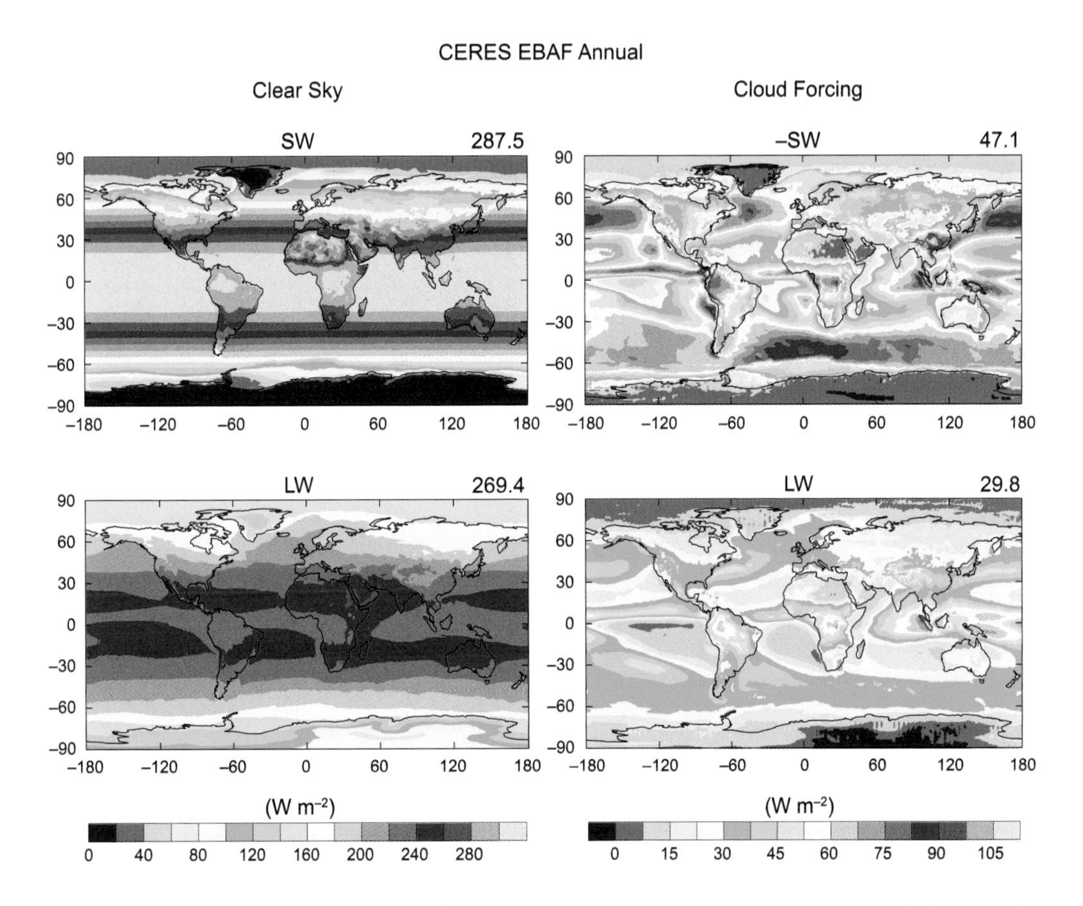

Plate 2. TOA Earth SW (upper panels) and LW (lower panels) annual mean clear-sky fluxes (left panels) and cloud forcing (right panels) derived from the NASA Clouds and the Earth's Radiant Energy System Energy Balanced and Filled (CERES EBAF) data product. The global mean values are given in the upper right corner of each panel. The SW cloud forcing has been multiplied by –1 for display purposes. Figure courtesy of J. Jonas.

Accompanies chapter by Del Genio (pp. 3–18).

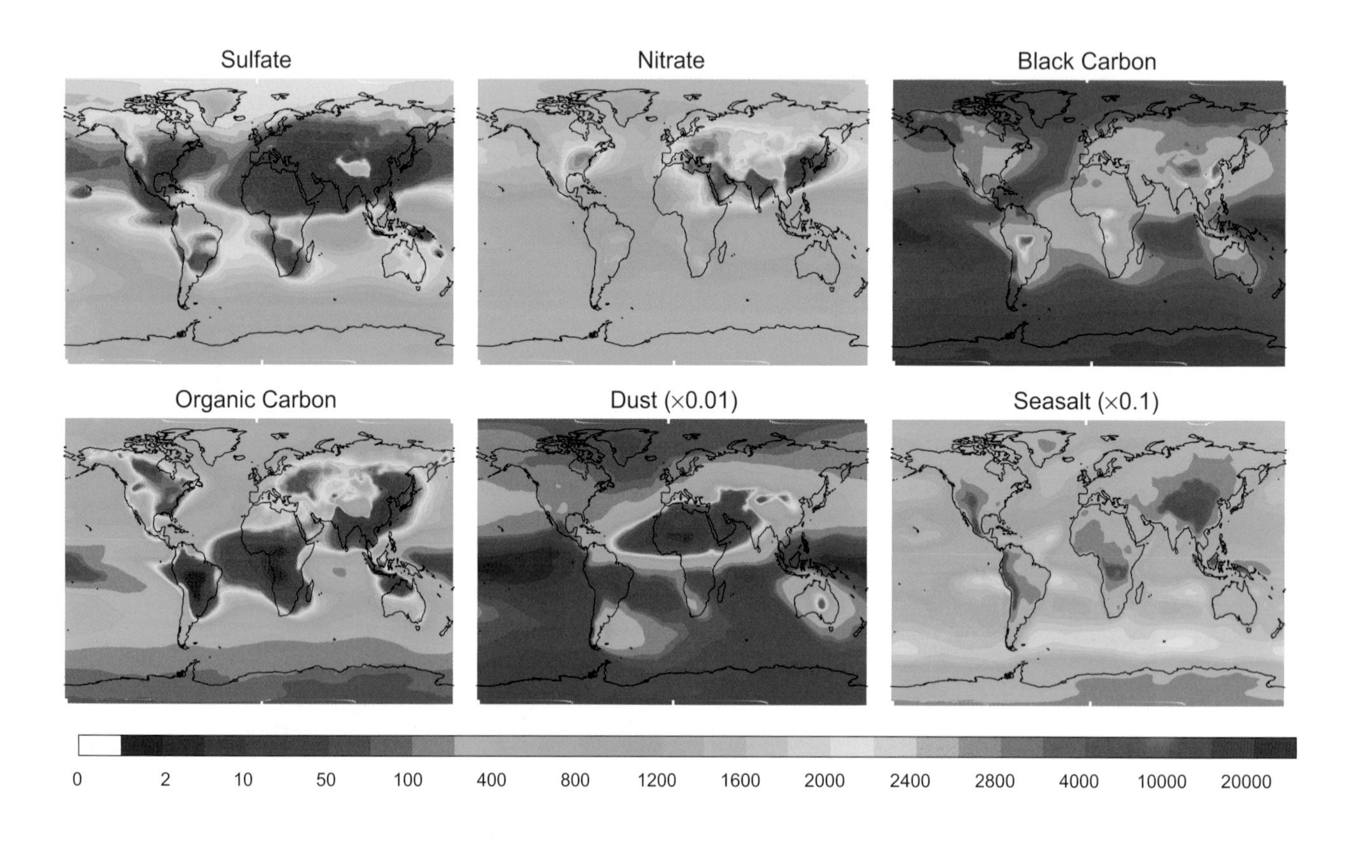

Plate 3. Annual mean column mass concentrations (μg/m²) of the six major Earth aerosol types, calculated from the MATRIX model in the Goddard Institute for Space Studies GCM (*Bauer et al.,* 2008). The dust and sea salt concentrations have been multiplied by factors of 0.01 and 0.1, respectively. Figure courtesy of S. Bauer.

Accompanies chapter by Del Genio (pp. 3–18).

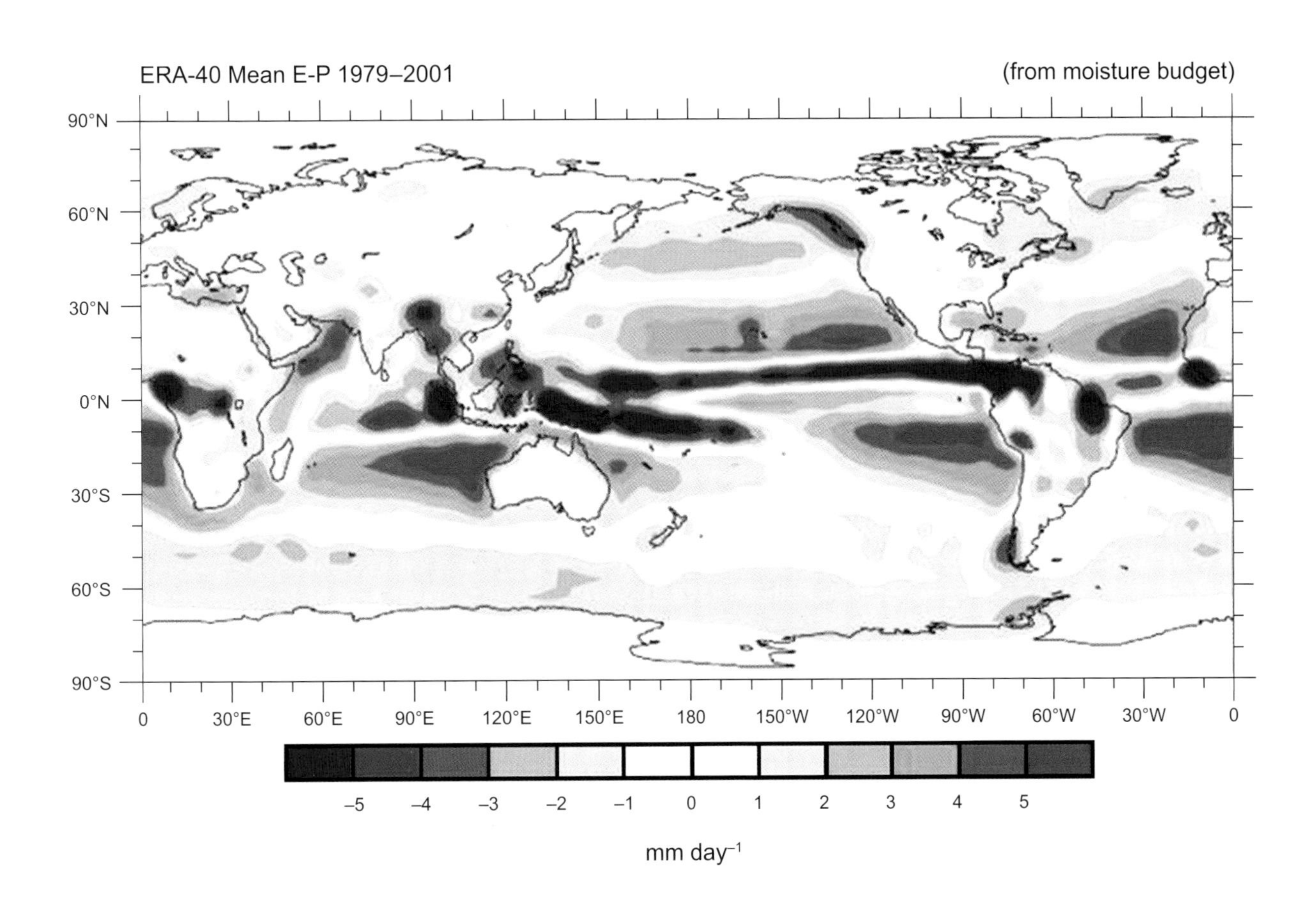

Plate 4. The annual mean Earth surface water budget E-P calculated from monthly means of the vertically integrated moisture budget of the atmosphere in the European Centre for Medium Range Weather Forecasts 40-year reanalysis (ERA-40). From *Trenberth et al.* (2007). ©Copyright 2007 American Meteorological Society; reprinted with permission.

Accompanies chapter by Del Genio (pp. 3–18).

Hydrological Cycle

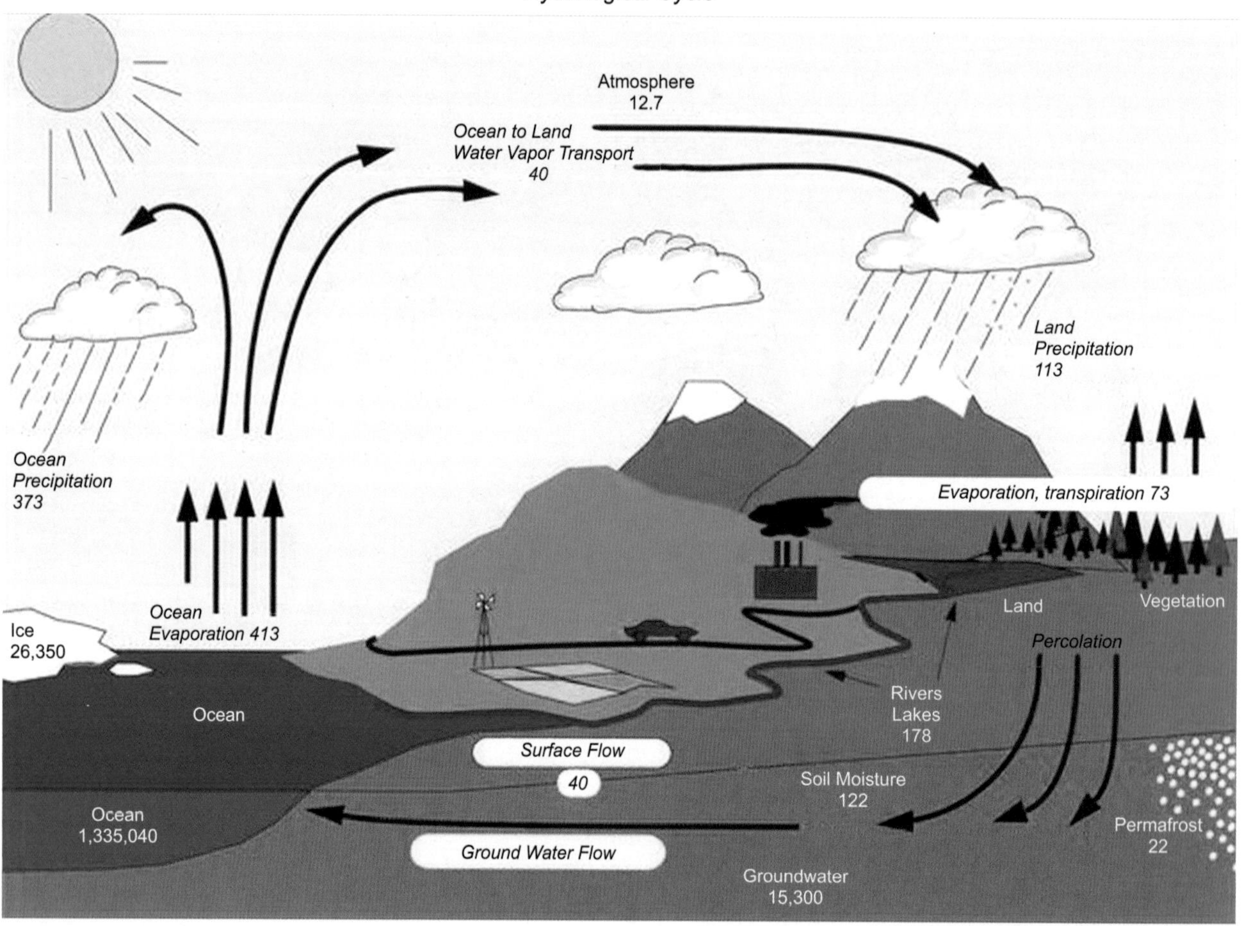

Plate 5. Global annual mean components of Earth's hydrological cycle. Storage amounts in the reservoirs are given in roman font, and flows/exchanges between reservoirs in italics (units are 1000 km^3). From *Trenberth et al.* (2007).

Accompanies chapter by Del Genio (pp. 3–18).

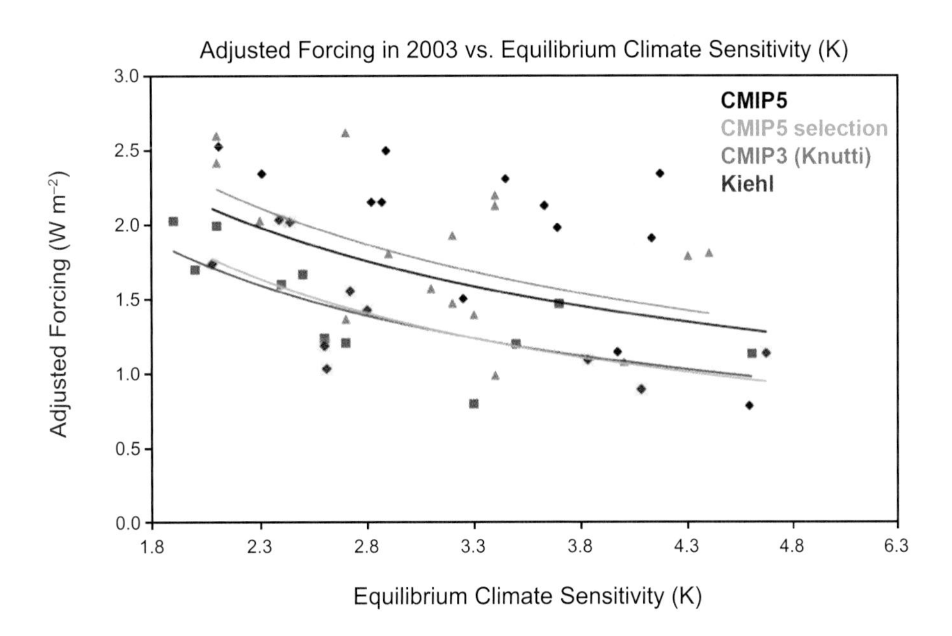

Plate 6. The relationship between 2003 climate forcing and equilibrium climate sensitivity for the turn-of-century GCMs analyzed by *Kiehl* (2007) (red), the CMIP3 GCMs analyzed by *Knutti* (2008) (blue), the CMIP5 GCMs (black), and a subset of CMIP5 GCMs that are within the 90% uncertainty range of the observed 100-year linear temperature trend (green). From *Forster et al.* (2013). ©Copyright American Geophysical Union; reprinted with permission.

Accompanies chapter by Del Genio (pp. 3–18).

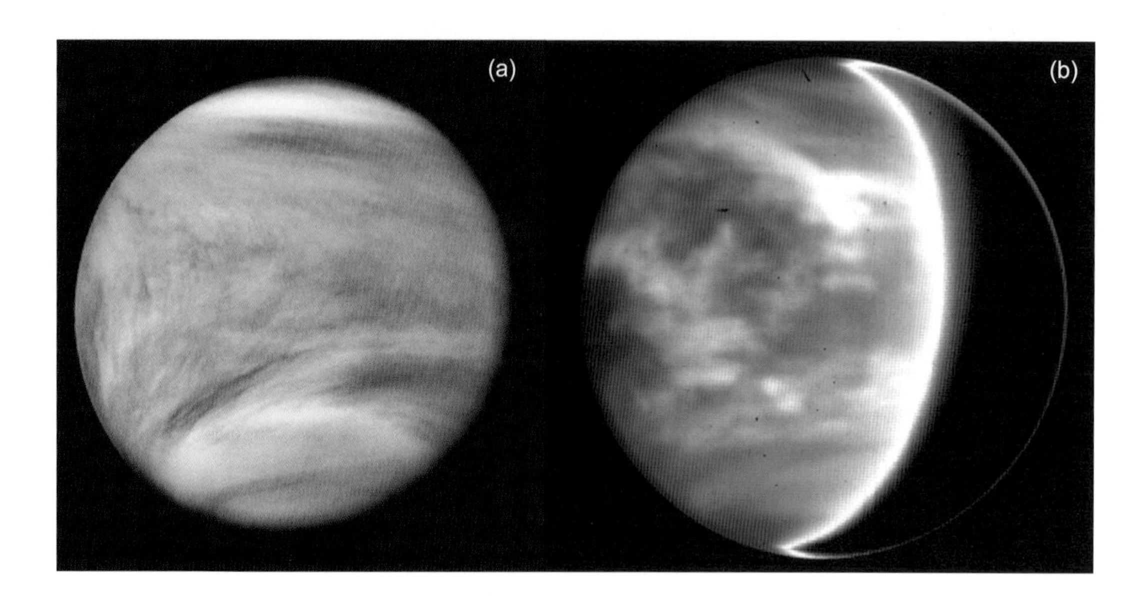

Plate 7. (a) Venus in reflected UV sunlight. Acquired on February 26, 1979, by the Pioneer Venus Orbiter. The distinctive tilted bands at the cloudtops (68 km) are produced by low wavenumber planetary waves that circle the planet in about four days. **(b)** The nightside of Venus glowing at 2.3 μm (*Young et al.,* 2007, 2010). This image was taken with the SpeX imager/spectrometer at NASA's 3-m Infrared Telescope Facility on Mauna Kea, Hawaii (*Rayner et al.,* 2003) on May 6, 2004. Images and spectra are taken simultaneously; the dark vertical line is the spectrometer slit. The plate scale is 0.115″/pixel. Seeing that morning was about 0.5″, with about 2 pr. μm of H_2O vapor above the observatory — superb conditions for infrared observations. Cloud features as small as 80 km were resolved. The lower clouds are silhouetted by upwelling heat radiation from below. Bright regions are holes in the lower cloud, while dark regions are where the lower cloud is thickest. The haze that makes up the upper cloud and the atmosphere are transparent to radiation at this wavelength. Venus' clouds are highly variable from night to night at this level of the atmosphere (about 52 km).

Accompanies chapter by Bullock and Grinspoon (pp. 19–54).

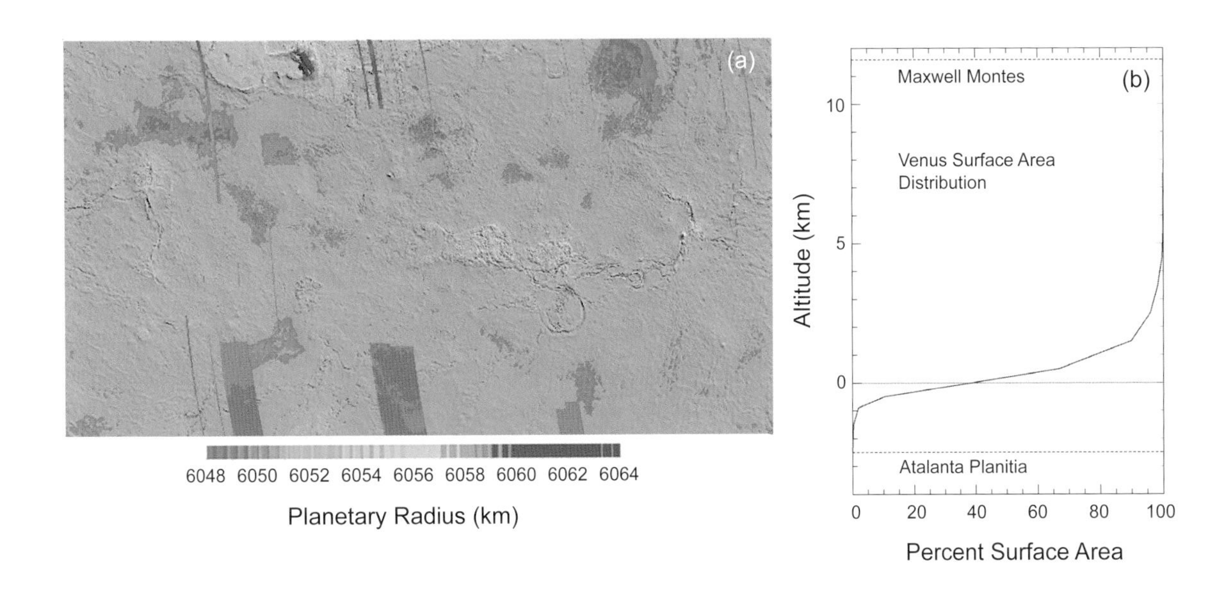

Plate 8. (a) Altimetry of the surface of Venus obtained by the Magellan radar altimeter (*Ford and Pettengill,* 1992). "Sea level" datum (shown in green) is 6051 km from the center of the planet. Blue areas are lowland volcanic plains, and yellow regions are the two major continent-like highlands Aphrodite (center, equator) and Ishtar Terra (upper left). The highest region on Venus is Maxwell Montes on top of Ishtar, shown in orange. Large volcanic shields that form at the intersections of major rift systems are also yellow. (**b)** Cumulative hypsometric curve for Venus' topography. It shows the percentage of the surface area above a given altitude. Venus' topography is unimodal, with deep chasmata 2 km below the datum and Maxwell Montes 12 km above the datum. This is unlike Earth, which has a bimodal topography consisting of distinct oceanic and continental crust. Data from *Ford and Pettengill* (1992).

Accompanies chapter by Bullock and Grinspoon (pp. 19–54).

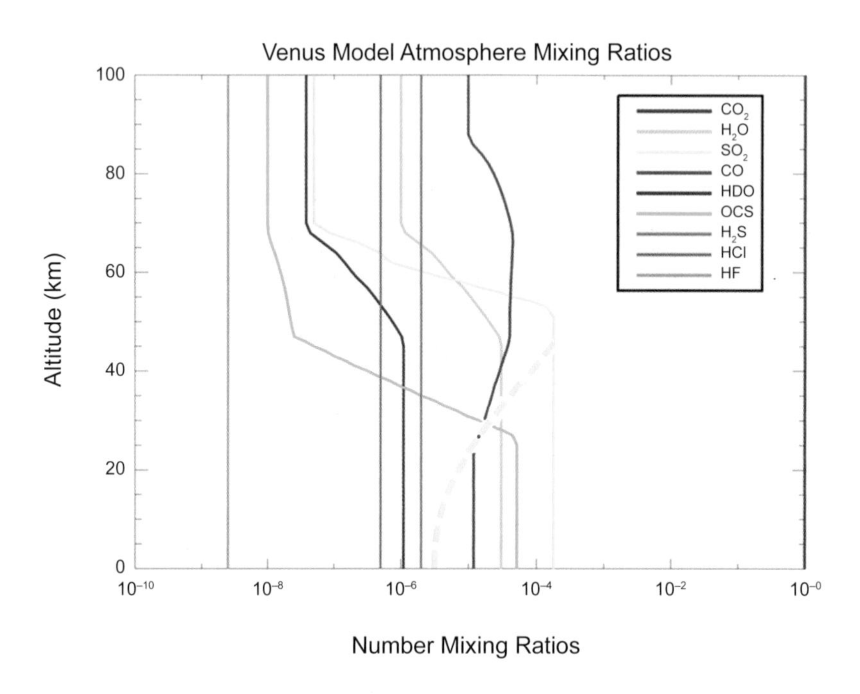

Plate 9. Molar mixing ratios of nine radiatively active gases in Venus' atmosphere. Data from Table 2, with N_2 removed and deuterated water (HDO) added. The plotted mixing ratios are from the Pioneer Venus Gas Chromatograph (*Oyama et al.,* 1980) and several groundbased spectroscopic datasets. The data are from *Kliore et al.* (1986), updated to include the best available retrievals since then. Clouds occupy the region from 48 to 68 km. Alternate Vega SO_2 abundance (*Bertaux et al.,* 1996) is shown by the dashed line, as in *Taylor and Grinspoon* (2009). Note that H_2O and SO_2 are both depleted in the clouds as they are removed by condensation to form H_2SO_4/H_2O cloud aerosols. Note also that the CO and OCS mixing ratios are complementary, with CO increasing with altitude from well below the clouds to well above them, and OCS decreasing over the same range.

Accompanies chapter by Bullock and Grinspoon (pp. 19–54).

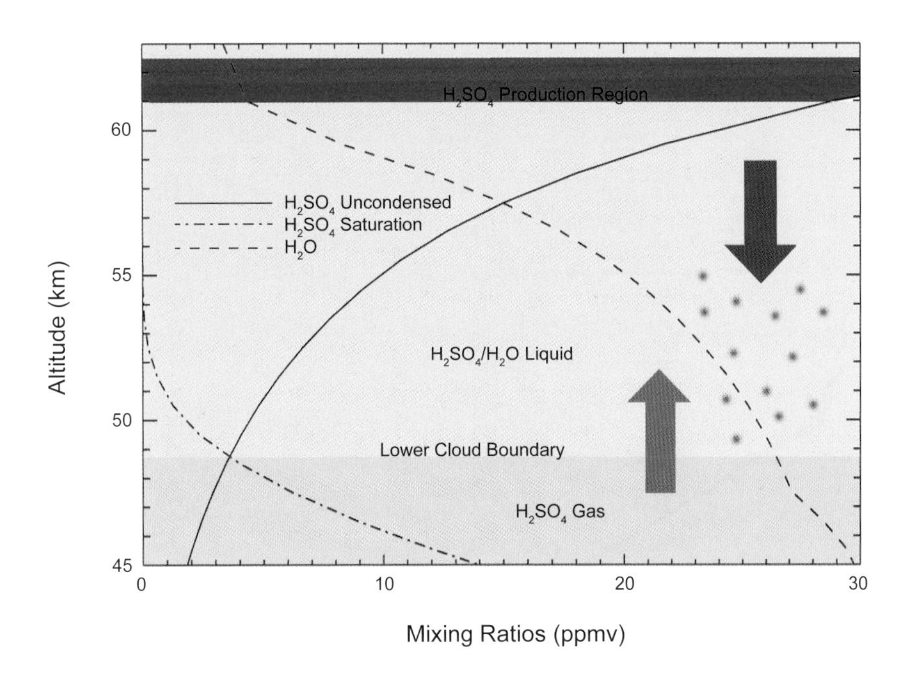

Plate 10. Schematic of the physical and chemical processes that create the Venus clouds. H_2SO_4 is produced photochemically between 62 and 64 km (red band). The saturation vapor pressure curves for H_2SO_4 (dot dashed line, lower left) and H_2O (dashed line, center) determine how cloud aerosols grow as they fall through the three cloud layers. The lower cloud boundary is formed where the vapor pressure of H_2SO_4 equals its saturation vapor pressure, at about 48 km (boundary between green and yellow regions). There is an ~200-m break in the clouds at about 52 km (not shown) between the lower and middle cloud layers. The flux of liquid H_2SO_4 is balanced by an upward flux of H_2SO_4 gas, resupplying vapor from which cloud particles can grow. The rate of vapor mixing and hence the resupply of H_2SO_4 is highly dependent on the eddy diffusion coefficient in this one-dimensional model. In reality, three-dimensional advection by all processes controls the distribution of vapor constituents.

Accompanies chapter by Bullock and Grinspoon (pp. 19–54).

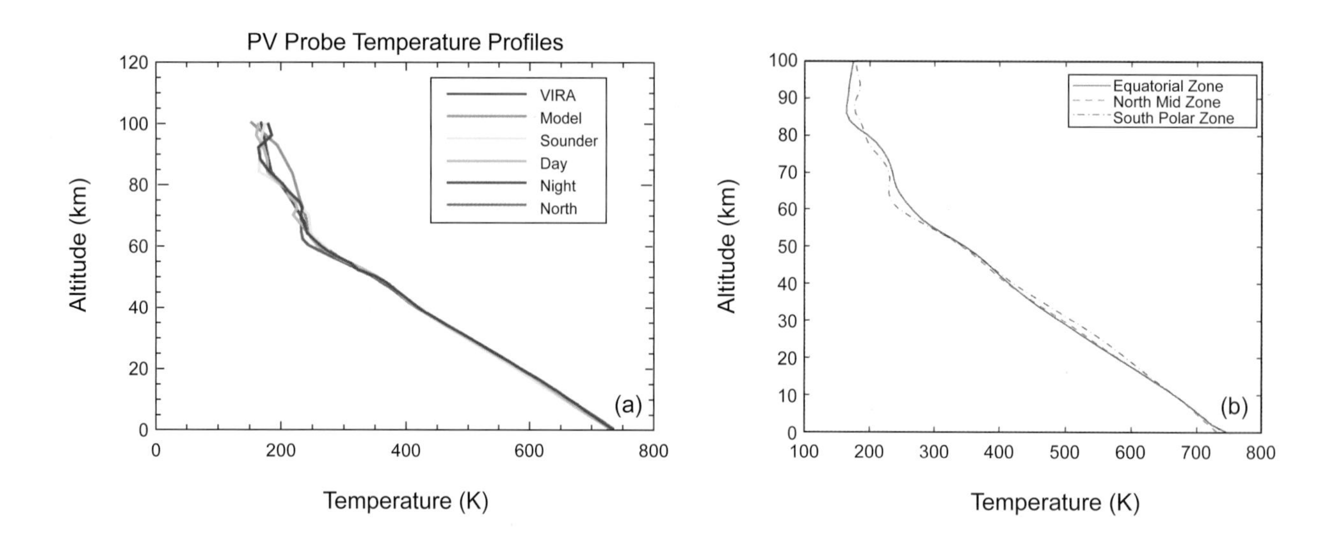

Plate 11. (a) The globally averaged equatorial atmospheric temperature profile of Venus (red line) (*Kliore et al.,* 1986) compared with the radiative transfer calculations of *Bullock and Grinspoon* (2001) (orange line). *In situ* temperatures measured by the four Pioneer Venus atmospheric entry probes are shown in other colors. Yellow: sounder probe, green: day probe, purple: night probe, and magenta: north probe. **(b)** Temperature retrievals from Magellan occultations at three latitude regions (*Jenkins et al.,* 2002).

Accompanies chapter by Bullock and Grinspoon (pp. 19–54).

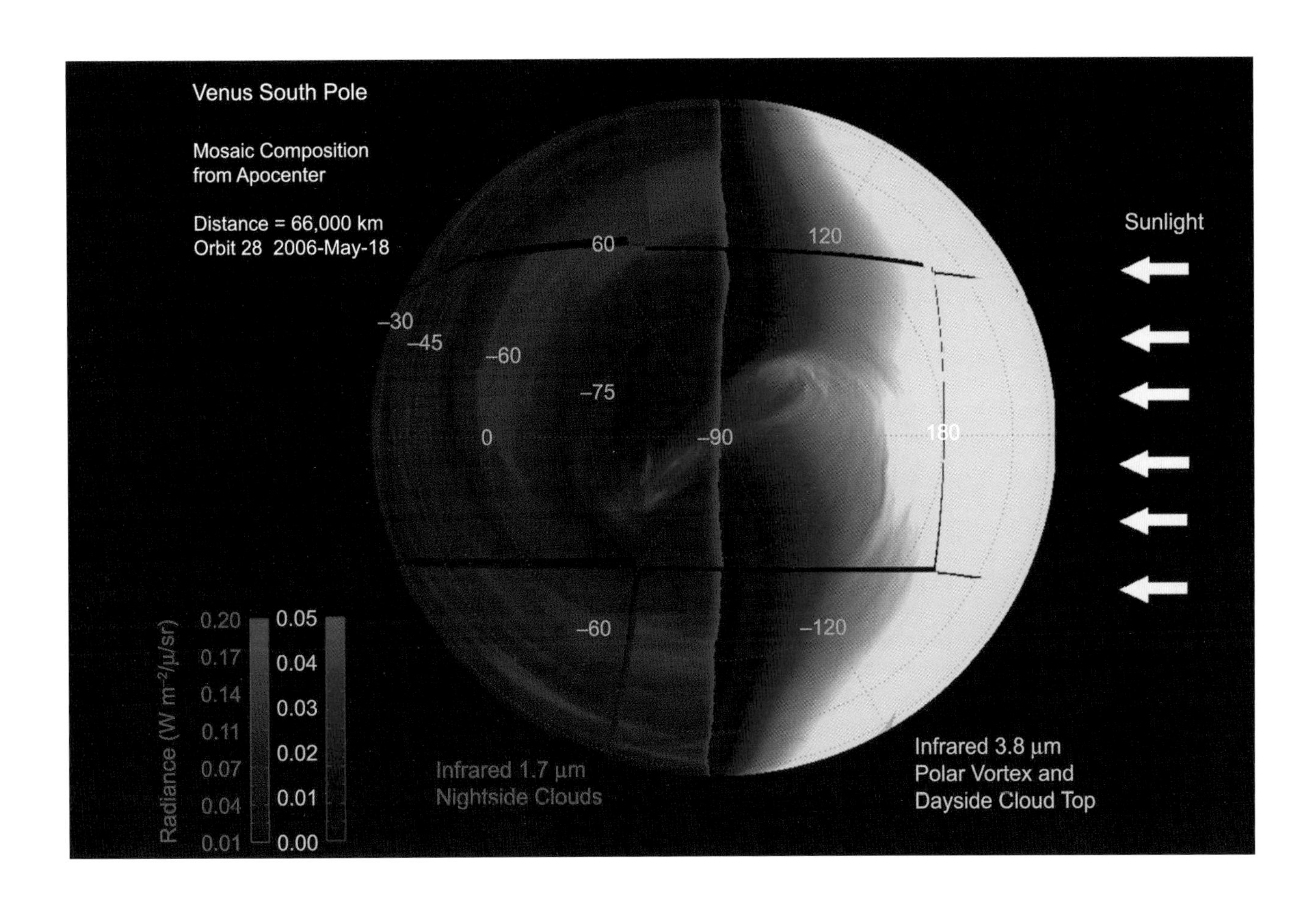

Plate 12. The south pole of Venus as imaged by VIRTIS at two different wavelengths. The vortex on May 18, 2006, was bipolar, rotating about once every four days. The sunlit (blue) image of the cloudtops was acquired at 3.8 μm, and the nightside (red) image at 1.7 μm is of the lower clouds silhouetted by infrared radiation from below. The nightside measurement probes the same level of the atmosphere as the groundbased infrared image in Fig. 1b. The red image shows the clouds at a level that is 10–15 km below that shown in the blue image. Press release image ESA/VIRTIS-VenusX IASF-INAF, Observatoire de Paris (A. Cardesín Moinelo, IASF-INAF).

Accompanies chapter by Bullock and Grinspoon (pp. 19–54).

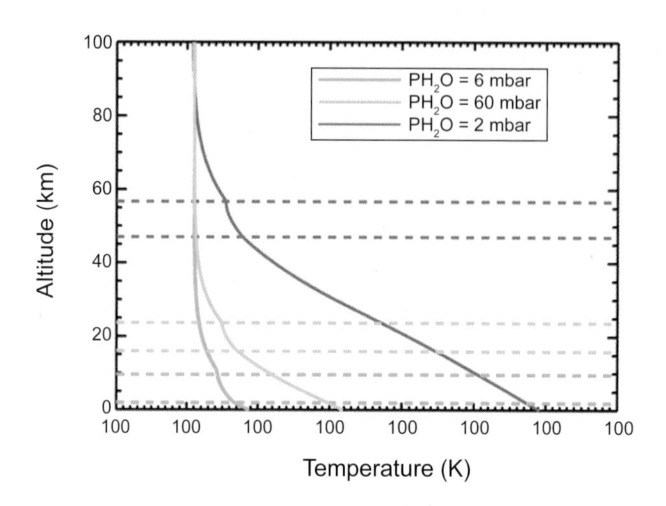

Plate 13. Radiative-convective equilibrium calculations for a 1-bar N_2 atmosphere with 350 ppmv of CO_2, available surface water, and cloud formation. The model is a one-dimensional two-stream radiative calculation with a two-component gray absorption. Calculations were performed for 6 (green), 60 (blue), and 2000 mbar (orange) of H_2O vapor in the atmosphere under present-day solar luminosity at 1 AU. Clouds form where the vapor pressure of water equals its saturation vapor pressure, and grow until they are no more than one scale height thick (horizontal dashed lines). A wet adiabatic lapse rate prevails in the cloud and the Eddington approximation is used to calculate the reflection and absorption of sunlight in the clouds. The 6-mbar case predicts a surface temperature of 285 K with cloud formation in the bottom 10 km of the atmosphere, much like Earth's present-day atmosphere. 2 bar of H_2O in the atmosphere results in a Venus-like surface temperature and clouds from 45 to 55 km.

Accompanies chapter by Bullock and Grinspoon (pp. 19–54).

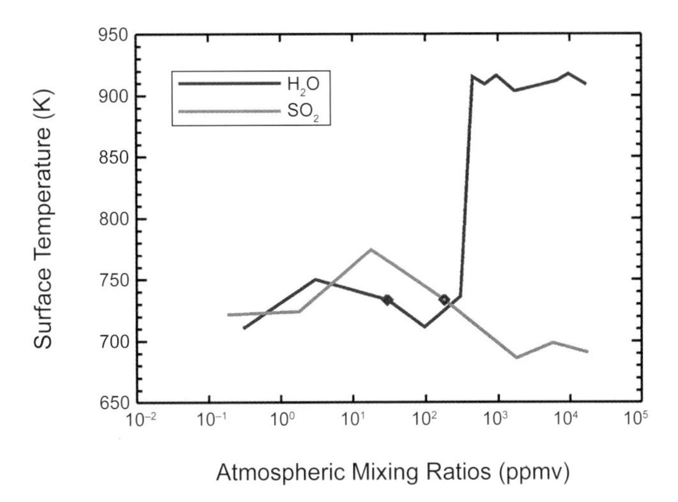

Plate 14. A suite of several hundred runs of the Venus climate model of *Bullock and Grinspoon* (2001). Atmospheric H_2O is varied from 1/100th the present abundance to 1000 times the present abundance (blue curve), and atmospheric SO_2 is varied from 1/1000th the present abundance to 100 times the present abundance (orange curve). The model calculates that the clouds will disappear if atmospheric water drops below about 2 ppmv, or if SO_2 drops below about 5 ppmv (not shown). The red dots indicate current conditions. These data are used in section 4 to assess the sensitivity of the climate model to perturbations in H_2O and SO_2. Around present-day conditions the slope is negative for both species, indicating that albedo feedback is more important than the greenhouse effect to Venus' energy balance. However, if atmospheric water abundance is increased by more than about 10 times its present value, the enhanced greenhouse initiates a cloud collapse runaway. Warmer temperatures raise the cloud base, thinning the clouds, which lowers the albedo and increases radiative forcing. This in turn warms the atmosphere, further thinning the clouds. The clouds rapidly erode from below until all that is left is a thin, high water cloud with a stable surface temperature that is 200 K hotter than Venus today.

Accompanies chapter by Bullock and Grinspoon (pp. 19–54).

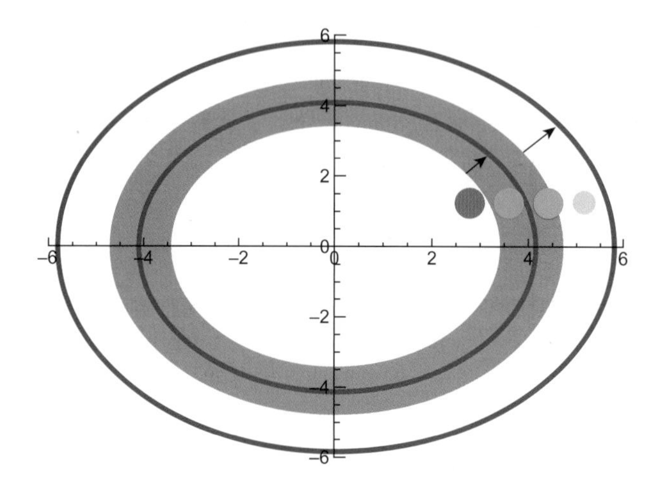

Plate 15. Evolution of the habitable zone around an F star. The orange ring represents the region of the stellar nebula where liquid water was stable on a forming planet. As the star brightens, the habitable zone moves outward (black arrows) to the zone outlined by the red ellipses. The intersection of these two regions is the continuously habitable zone. The blue circle represents an Earth-like planet that remains in the continuously habitable zone for billions of years. The dark brown circle depicts a planet that would undergo a runaway greenhouse early in its history, while the lighter brown circle is more representative of a Venus-like planet that undergoes a moist greenhouse. The gray circle is a planet that is frozen for much of its existence, like Mars, but where liquid water may one day become stable at the surface as the luminosity of the star increases over time. Units on the axes are in AU.

Accompanies chapter by Bullock and Grinspoon (pp. 19–54).

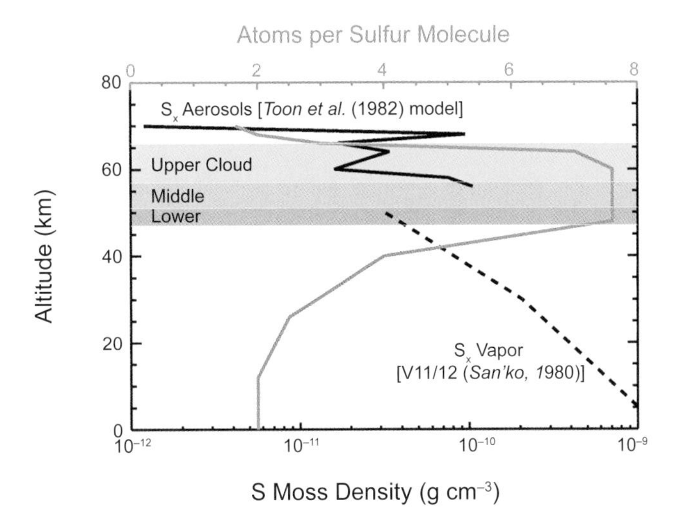

Plate 16. Elemental S in the Venus atmosphere. The only measurements of elemental S on Venus were made with the scanning spectrophotometers on Veneras 11 and 12 (*San'ko,* 1980) (black dashed line). These detected only the total amount of S_x vapor. The thermodynamics of S allotropes require that S is mostly in the form of S_2 near the surface and above the clouds (orange line). Within the clouds, however, S_8 is stable and other allotropes also appear. *Toon et al.* (1982) used a cloud chemical/microphysical model to estimate the abundance of S allotropes in the upper Venus cloud. Their data showed that conversion between S_4 and S_3 in the upper clouds is plausible (where the near-UV absorber must reside), and could explain the absorption of almost half the sunlight that enters the clouds. *Schulze-Makuch et al.* (2004) hypothesized that the mild and persistent conditions in the Venus clouds make them a possible abode for life that uses the abundant S cleverly. From *Schulze-Makuch et al.* (2004).

Accompanies chapter by Bullock and Grinspoon (pp. 19–54).

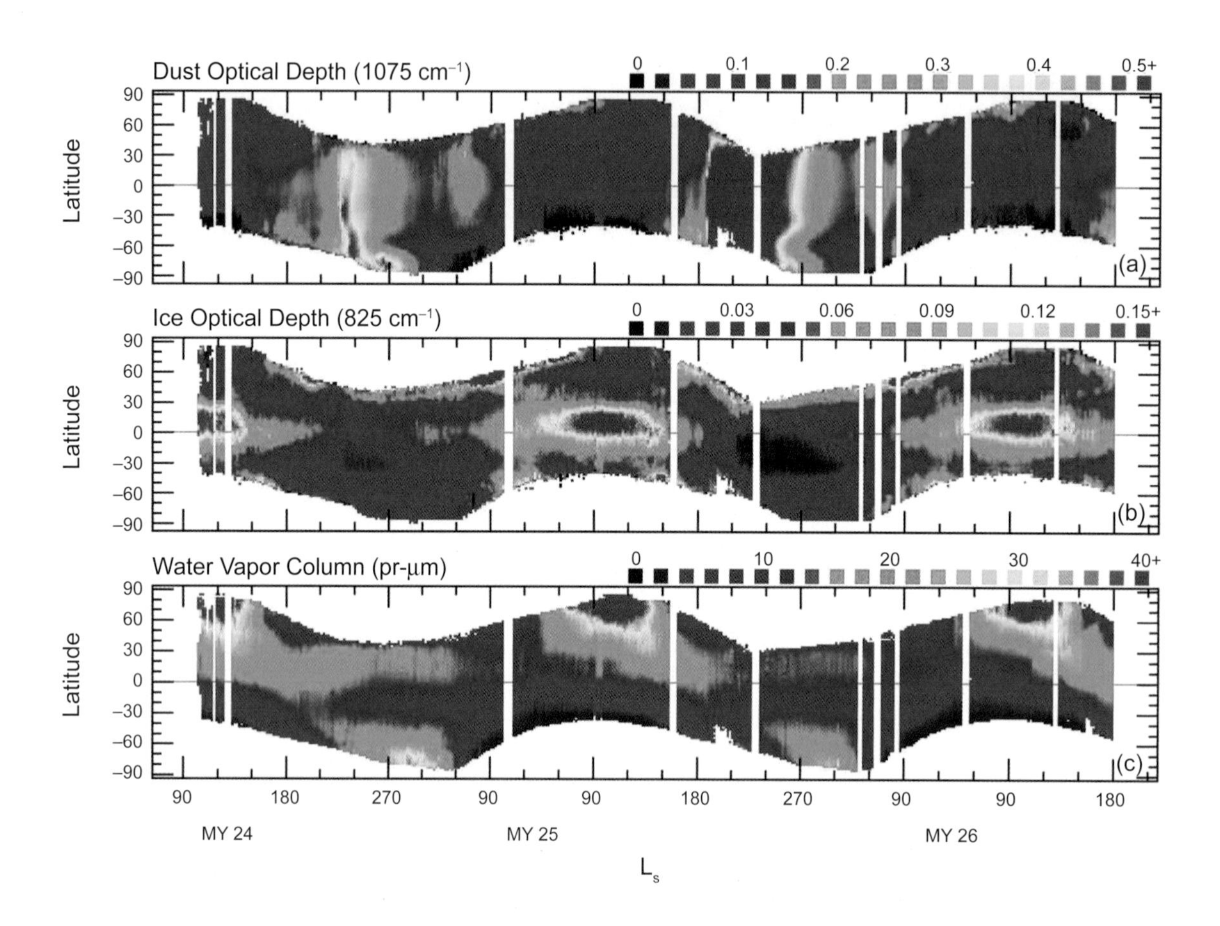

Plate 17. Zonally averaged dust optical depth at 1075 cm^{-1} scaled to an equivalent 6.1-mbar pressure surface (top), water ice optical depth at 825 cm^{-1} (middle), and water vapor column abundance in precipitable micrometers (bottom), as a function of season (L_s) and latitude.

Accompanies chapter by Rafkin et al. (pp. 55–89).

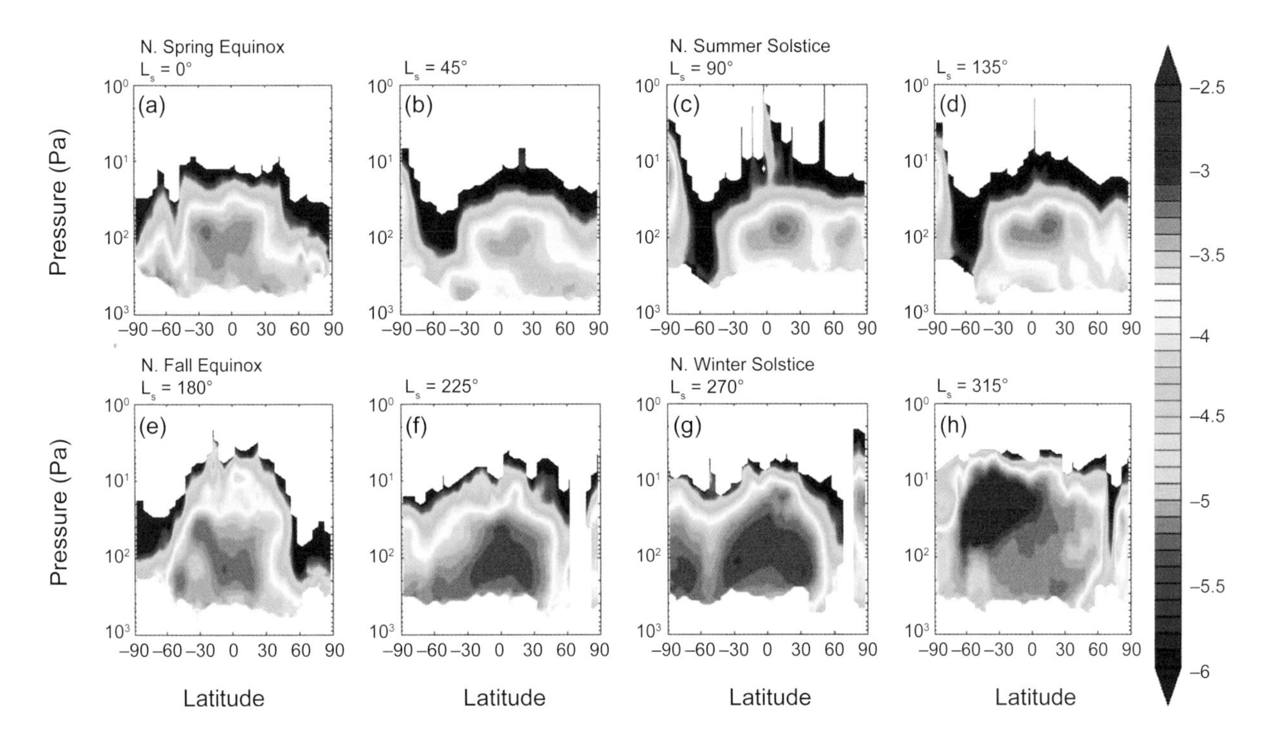

Plate 18. Detached or elevated dust layers have been identified from MCS limb retrievals. These layers persist throughout the year and are in contrast to the monotonically decreasing profile described by the Conrath-υ profile. Shaded values are $\log_{10}$ of the density scale opacity. From *McCleese et al.* (2010).

Accompanies chapter by Rafkin et al. (pp. 55–89).

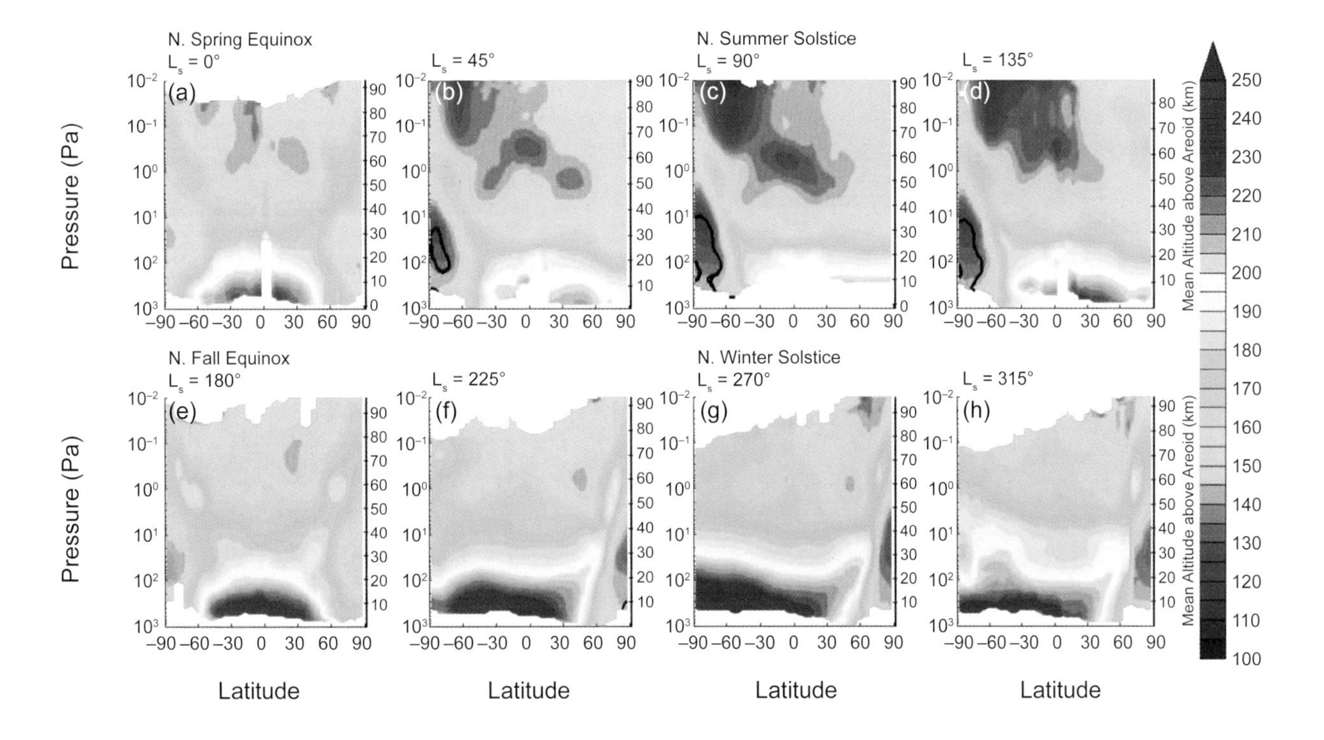

Plate 19. Zonally averaged temperature profiles obtained from MCS limb retrievals.

Accompanies chapter by Rafkin et al. (pp. 55–89).

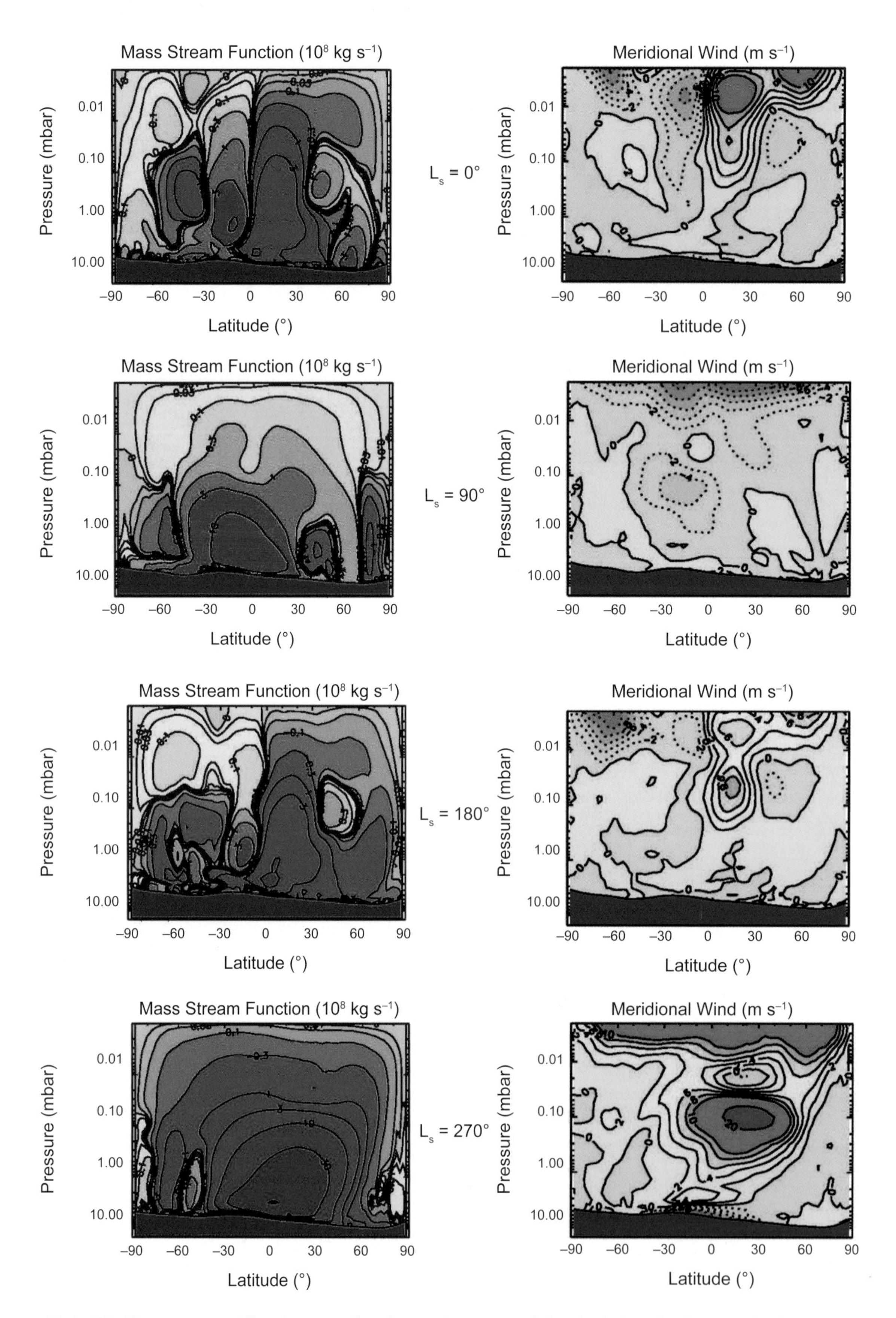

Plate 20. The mean meridional stream function and mean meridional wind at the four cardinal seasons as determined from the NASA Ames GCM (*Hollingsworth et al.,* 2011).

Accompanies chapter by Rafkin et al. (pp. 55–89).

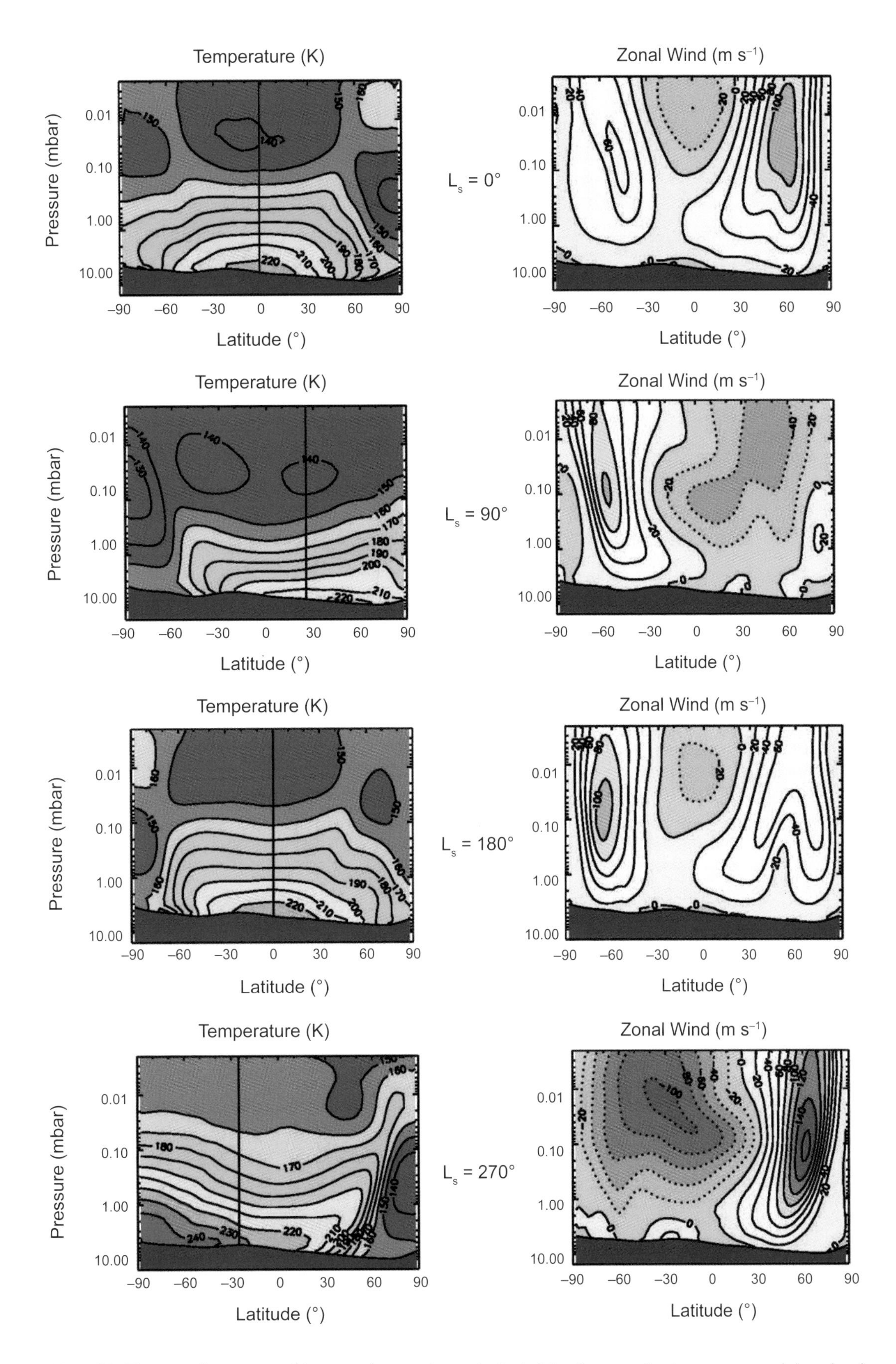

Plate 21. The zonally averaged temperature and zonal wind at the four cardinal seasons as determined from the NASA Ames GCM (*Hollingsworth et al.,* 2011).

Accompanies chapter by Rafkin et al. (pp. 55–89).

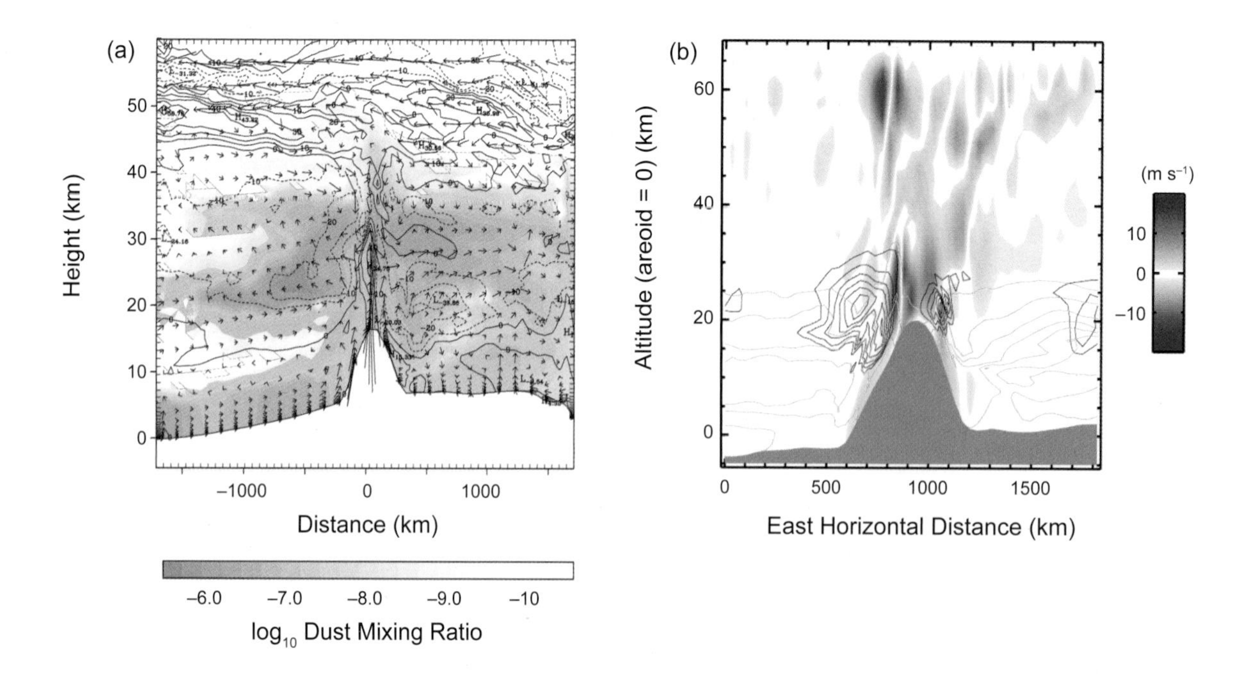

Plate 22. (a) Dust and **(b)** water transport at Olympus Mons, Mars, as predicted from a mesoscale model (*Rafkin et al.,* 2002; *Michaels et al.,* 2006). Dust (brown shading) is lifted into the boundary layer and advected vertically along the slopes of the mountain where it is ejected in upper level outflow. Water vapor (green contours) follows a similar trajectory, except that the rising moist air (red shades are positive vertical velocity, blue is negative) can also produce clouds (black contours). The thermal circulation from the mountain produces elevated or detached layers of dust and water vapor.

Accompanies chapter by Rafkin et al. (pp. 55–89).

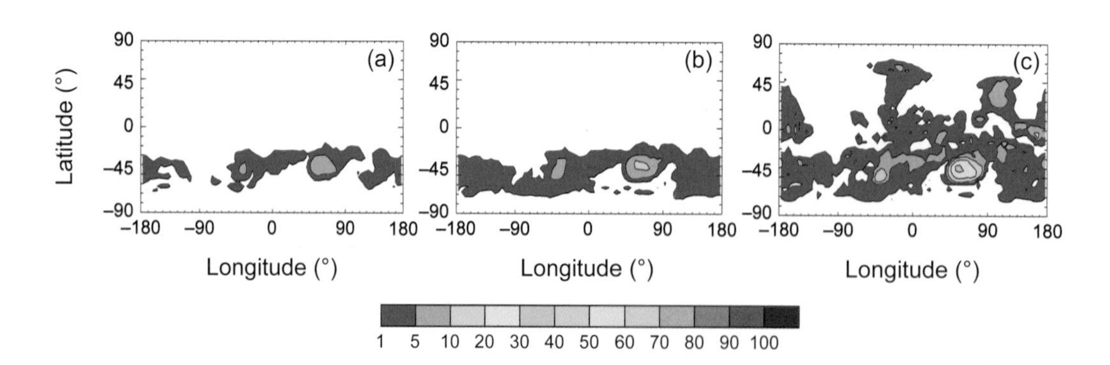

Plate 23. Map view of fraction of the year for which surface temperatures exceed 273 K on Mars for various atmospheric compositions. **(a)** 500-mbar CO_2 atmosphere; **(b)** 500-mb CO_2 atmosphere with water vapor from nominal martian water cycle; **(c)** 500-mbar CO_2 atmosphere with water vapor from nominal martian water cycle plus 245 ppm SO_2. All runs use present-day orbital configuration. Results from *Mischna et al.* (2013).

Accompanies chapter by Rafkin et al. (pp. 55–89).

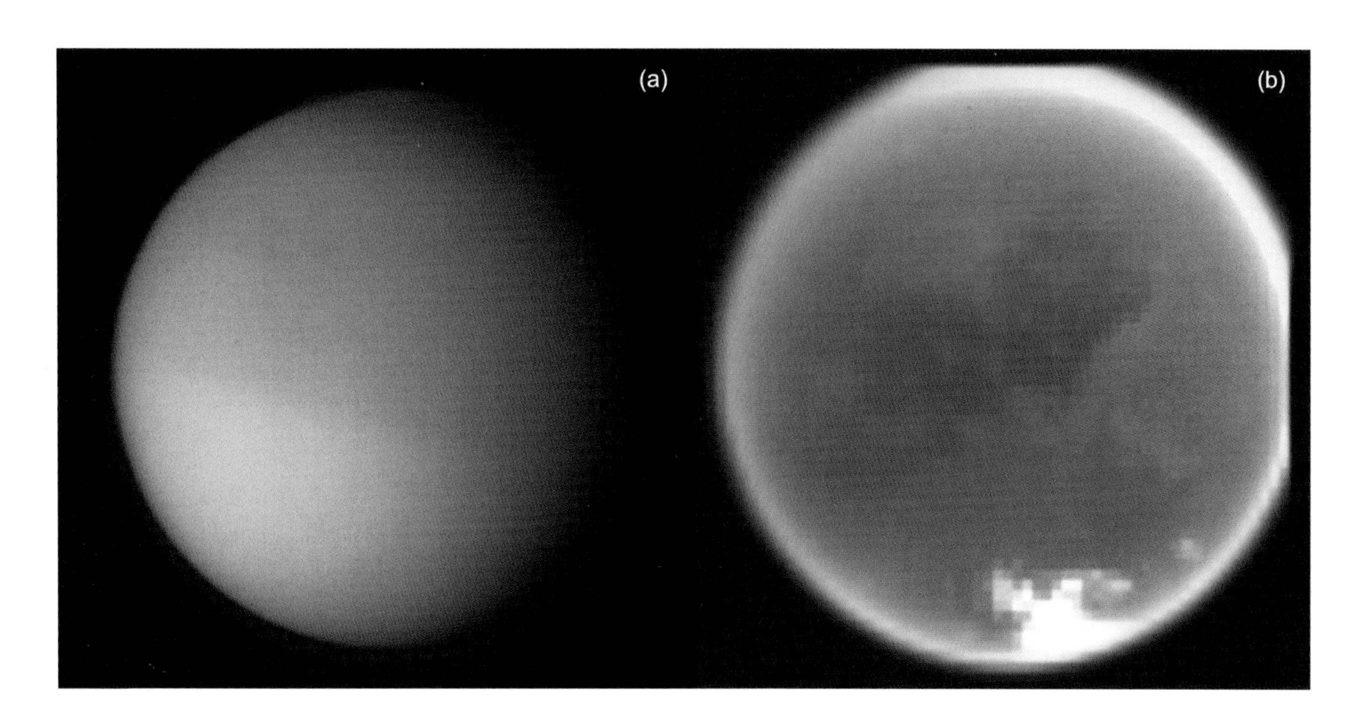

Plate 24. (a) At optical wavelengths, Titan's atmosphere below 100 km is shrouded by a complex organic haze. **(b)** At near-IR wavelengths between strong methane bands, Titan's surface and south polar tropospheric clouds (shown in white) can be seen through the haze. Credit: Voyager, Cassini, University of Arizona.

Accompanies chapter by Griffith et al. (pp. 91–119).

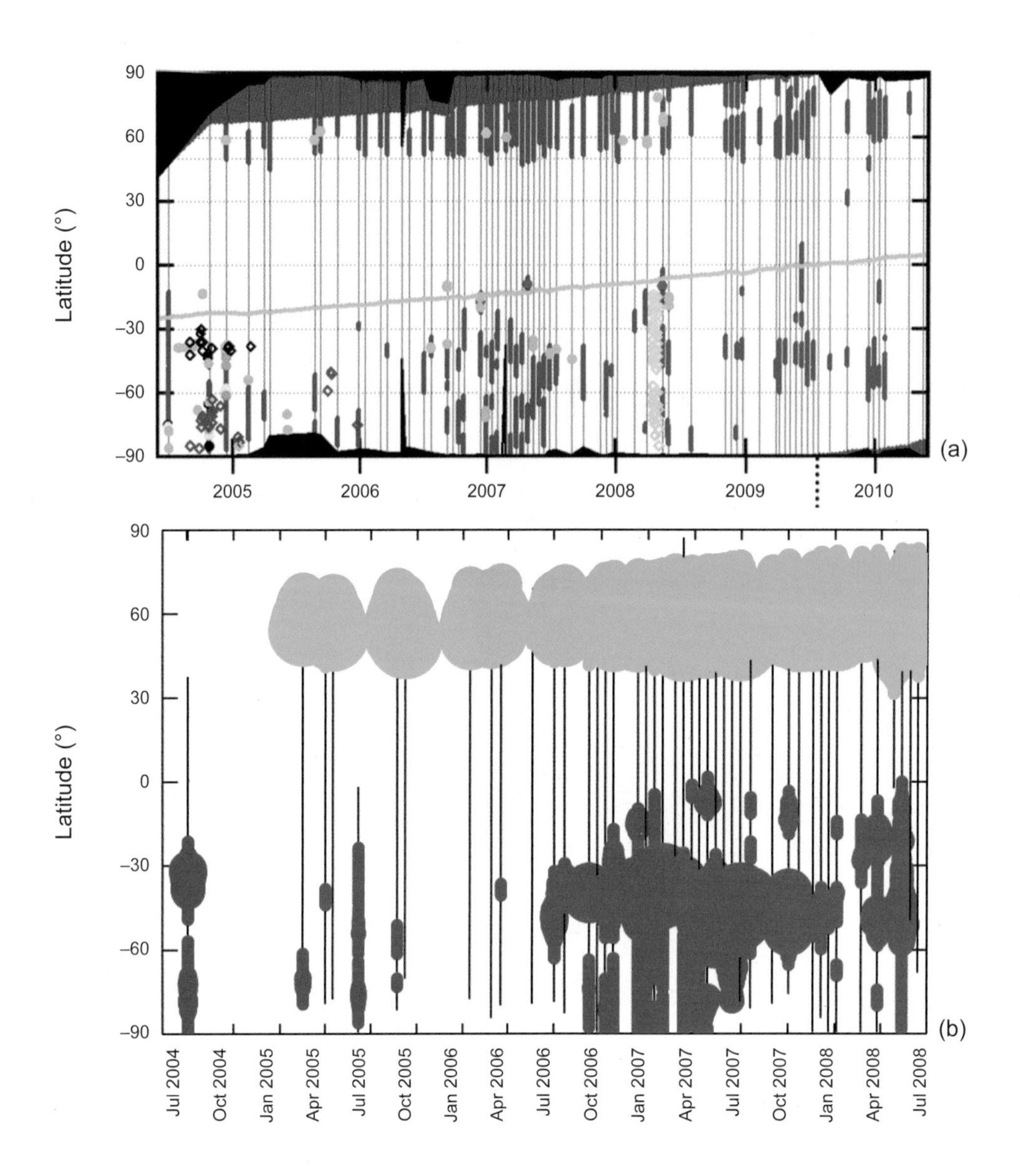

Plate 25. Time evolution of cloud latitudes from July 2004 to April 2010. Vertical thin lines mark the time and latitude coverage of VIMS observations. **(a)** Thicker blue lines show the latitude extension of detected clouds, summed over all longitudes. Dots and diamonds are Cassini and Earth-based detections (*Porco et al.,* 2005; *Baines et al.,* 2005; *Griffith et al.,* 2005a, 2006, 2009; *LeMouelic et al.,* 2012; *Turtle et al.,* 2009; *Roe et al.,* 2005; *Schaller et al.,* 2006a, 2009; *de Pater et al.,* 2006; *Hirtzig et al.,* 2006). The green line indicates the latitude of the maximum of insolation, with north spring equinox in August 2009. Gray areas are night; black areas no observations. From *Rodriguez et al.* (2011). **(b)** Circles show latitudes of tropospheric methane clouds (red) and tropopause ethane clouds (green) recorded by VIMS; the size is proportional to areal coverage. From *Brown et al.* (2010).

Accompanies chapter by Griffith et al. (pp. 91–119).

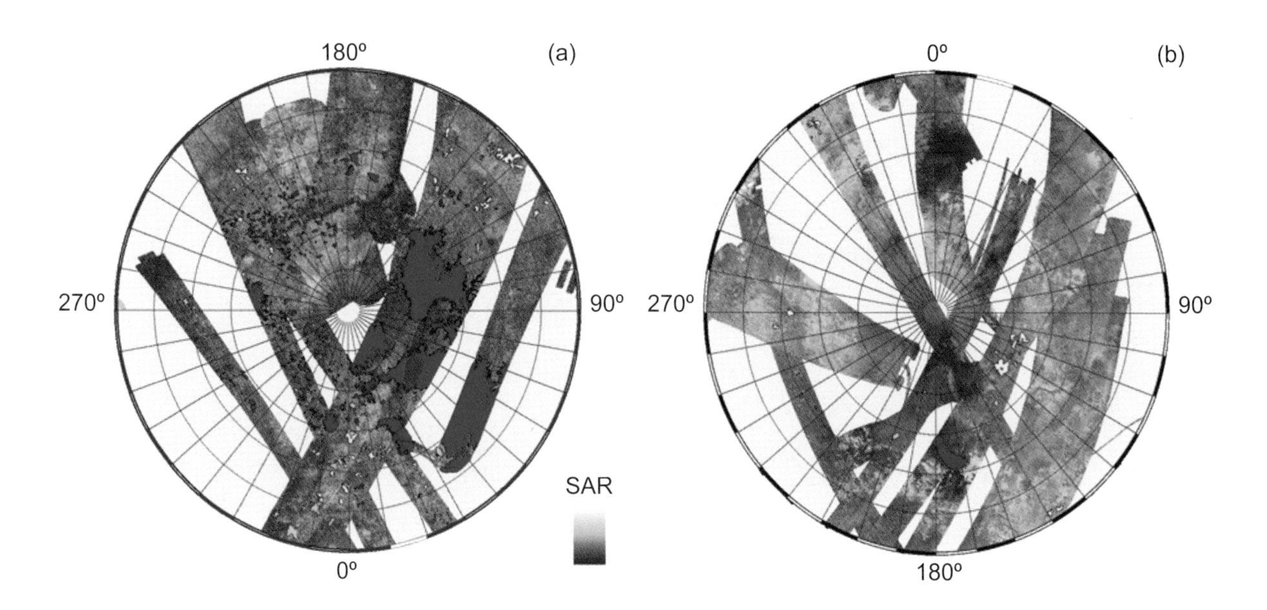

Plate 26. Distribution of Titan's lakes for the **(a)** northern and **(b)** southern hemispheres. The colors indicate whether the lake is filled (blue), intermediate (cyan), or empty (pale blue). The backscatter brightness of Cassini's RADAR is shown in shades of brown. The diameter of the image is ~2400 km. From *Aharonson et al.* (2009).

Accompanies chapter by Griffith et al. (pp. 91–119).

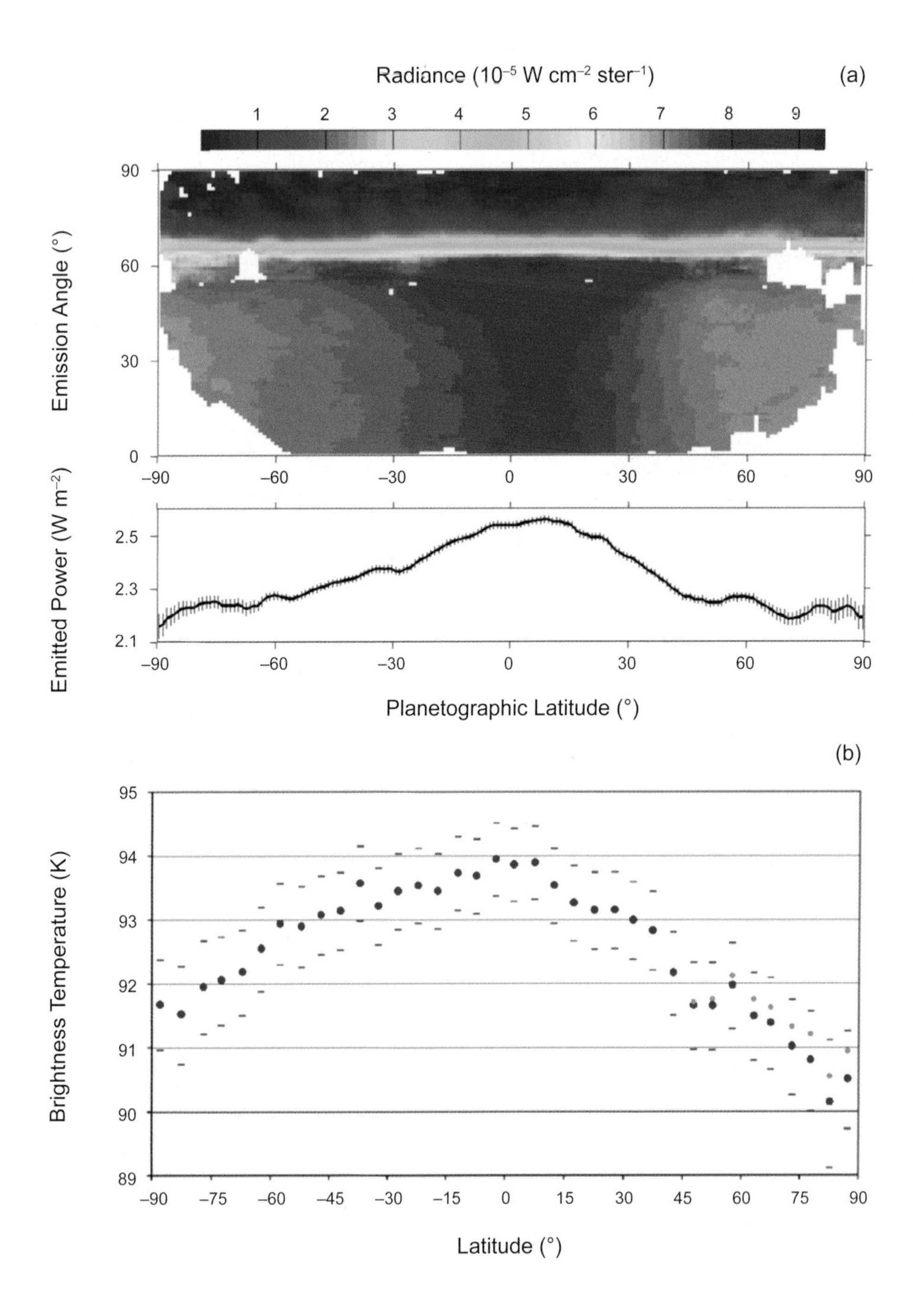

Plate 27. (a) Titan's outgoing longwave radiation as measured by Cassini CIRS (*Li et al.*, 2011). **(b)** The brightness temperature of Titan's surface (*Jennings et al.*, 2009).

Accompanies chapter by Griffith et al. (pp. 91–119).

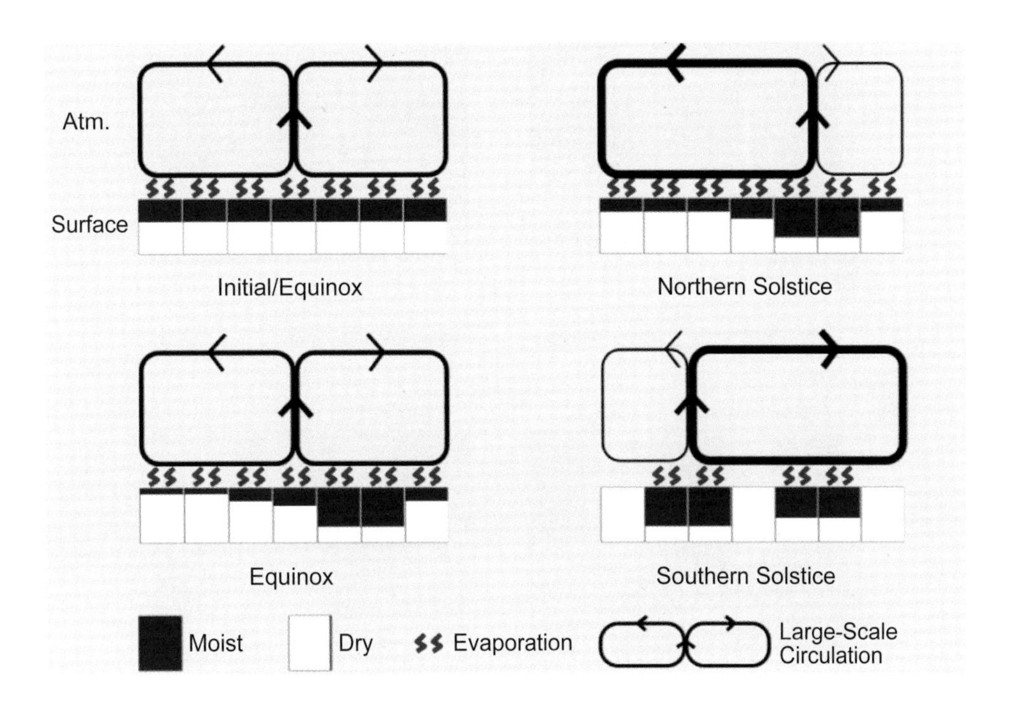

Plate 28. Titan's global overturning circulation transports methane to the convergence zone, where lifting deposits it locally as precipitation. A strong seasonal cycle leads climatologically dry conditions at the equator.

Accompanies chapter by Griffith et al. (pp. 91–119).

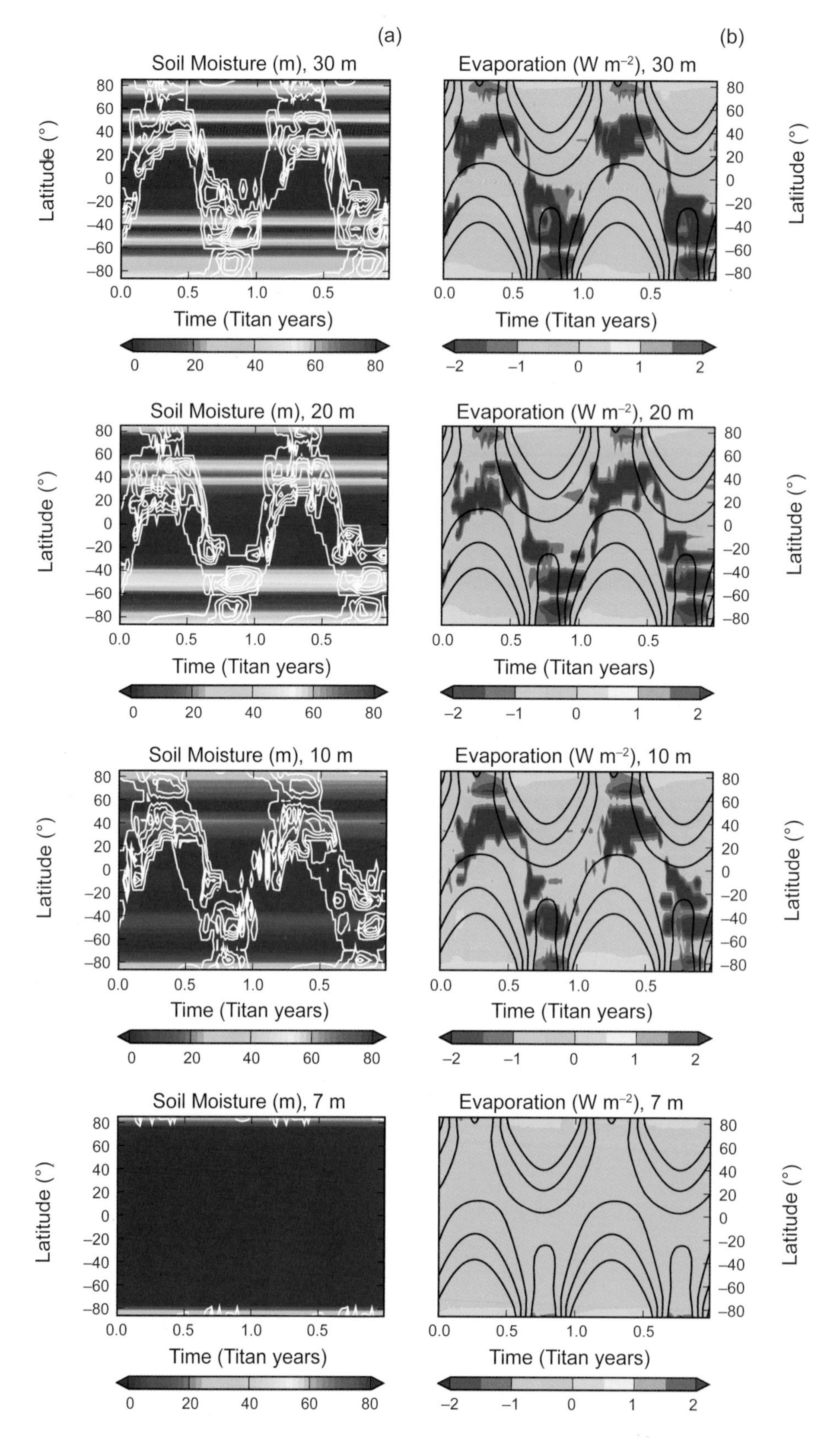

Plate 29. Climate simulations show the influence of a limited methane reservoir, with inventories ranging between 30 and 7 m global liquid equivalent (*Mitchell,* 2008). **(a)** Seasonal pattern of precipitation (white) and surface reservoir depth (colorbar; m) over the final two years of a 45-Titan-year simulation. **(b)** Seasonal patterns of insolation (black) and surface evaporation (colorbar; W m^{-2}) over the same two Titan years.

Accompanies chapter by Griffith et al. (pp. 91–119).

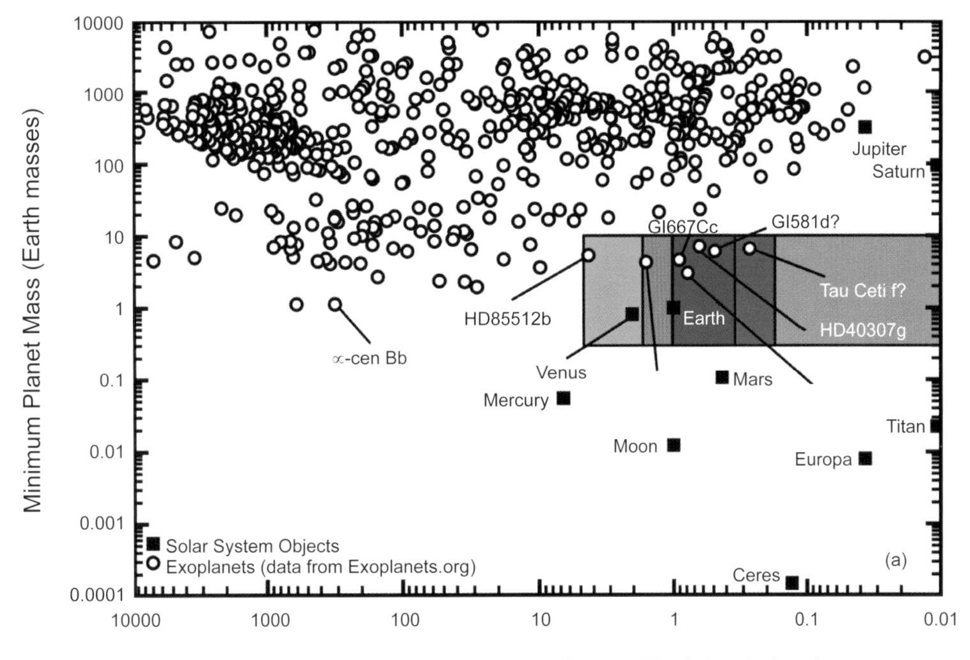

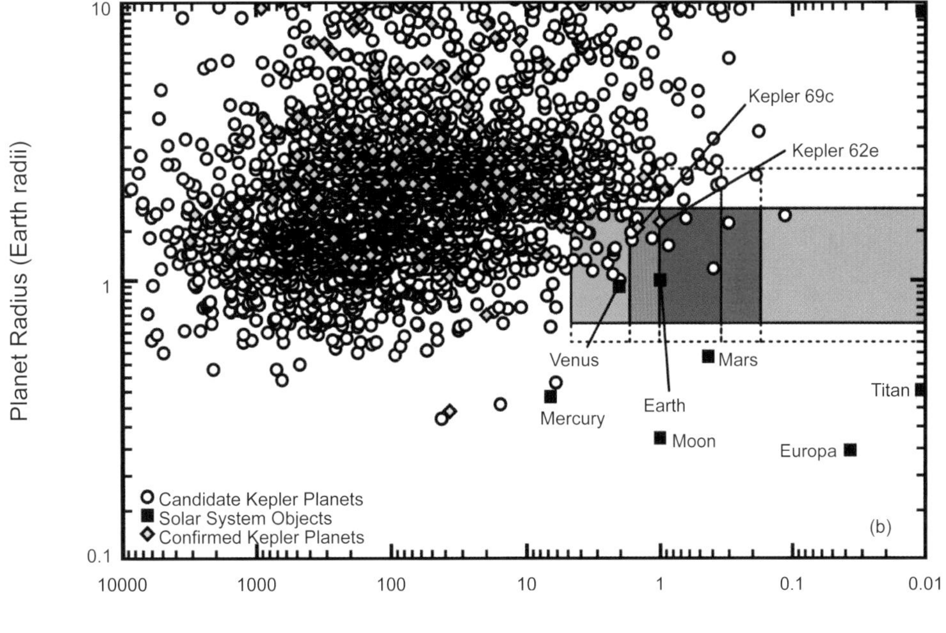

Plate 30. Solar system objects (squares), exoplanets and candidate Kepler planets (circles), and confirmed Kepler planets (diamonds) plotted as a function of energy incident upon the top of the planetary atmosphere and either **(a)** planet mass or **(b)** planet radius. All units are scaled to Earth. In both panels, the colored boxes are representative of different habitable zones defined in the literature. Blue areas represent the "conservative habitable zone," according to updated calculations (*Kopparapu et al.,* 2013). Brown areas show an extension of the inner edge of the habitable zone for dry planets (*Abe et al.,* 2011). Orange areas represent further extension inward based on the cloud effects, and represent the "100% cloud inner limit" (*Selsis et al.,* 2007). Similarly, red boxes represent an extension of the outer part of the habitable zone by clouds according to the "100% cloud outer limit" (*Selsis et al.,* 2007). Gray boxes show an extension of the outer edge of the habitable zone for H_2-He greenhouse planets (*Pierrehumbert and Gaidos,* 2011). In **(b),** vertical extensions of these boxes are also included with dashed lines, showing planetary radii for which objects could be potentially habitable, but only if planets in those areas have compositions that are "just right" (see text).

Accompanies chapter by Domagal-Goldman and Segura (pp. 121–135).

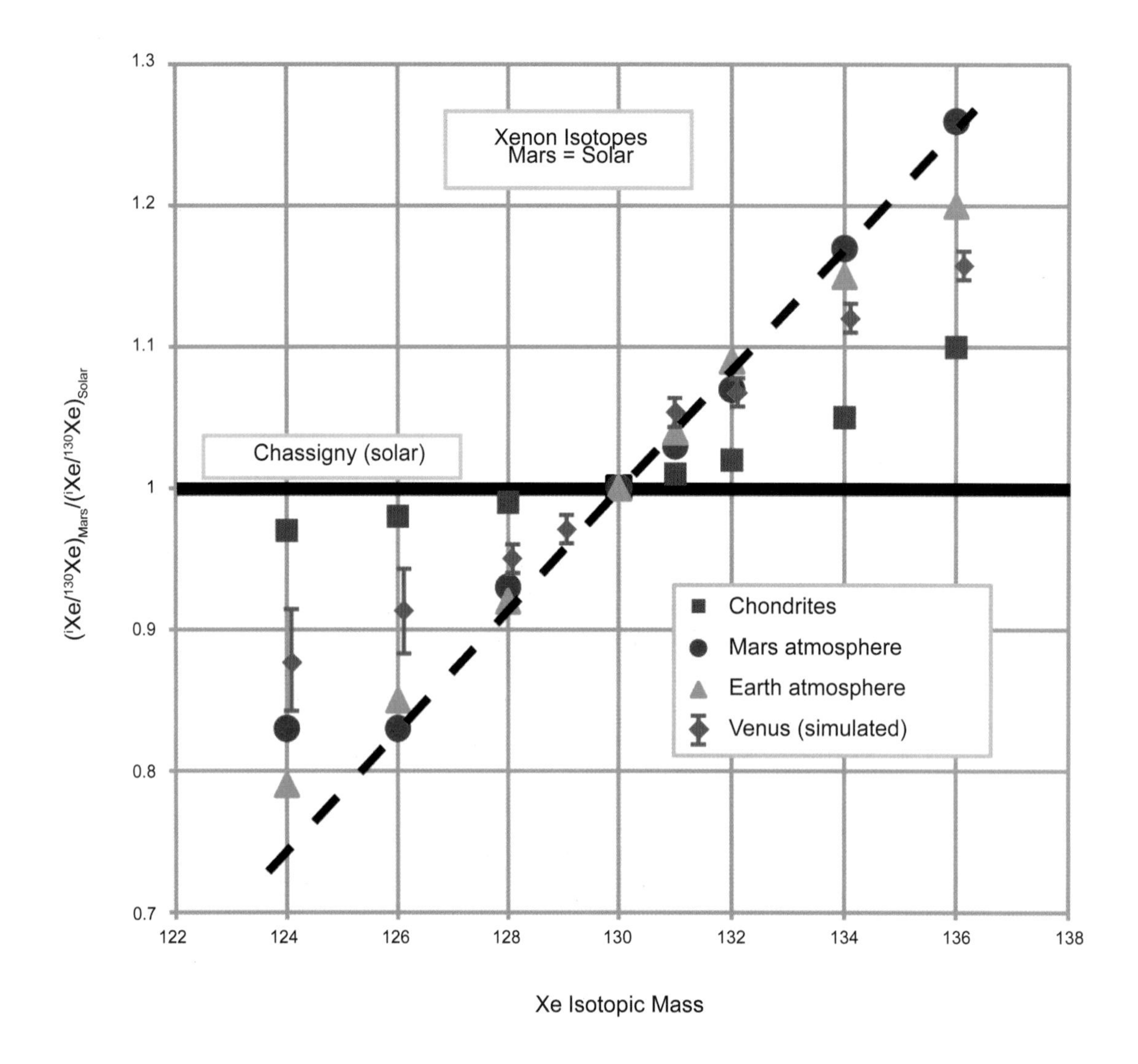

Plate 31. Fractionation of Xe isotopes. Possible Venus Xe fractionation pattern observed by a future *in situ* mission is depicted (green) compared to the patterns of Earth, Mars, chondrites, and the Sun (after *Bogard et al.,* 2001). A common Venus/Earth/Mars pattern would bolster the hypothesis that a common source of comets or large planetesimals delivered volatiles throughout the inner solar system. A different pattern for Venus would strengthen the solar EUV blowoff theory.

Accompanies chapter by Baines et al. (pp. 137–160).

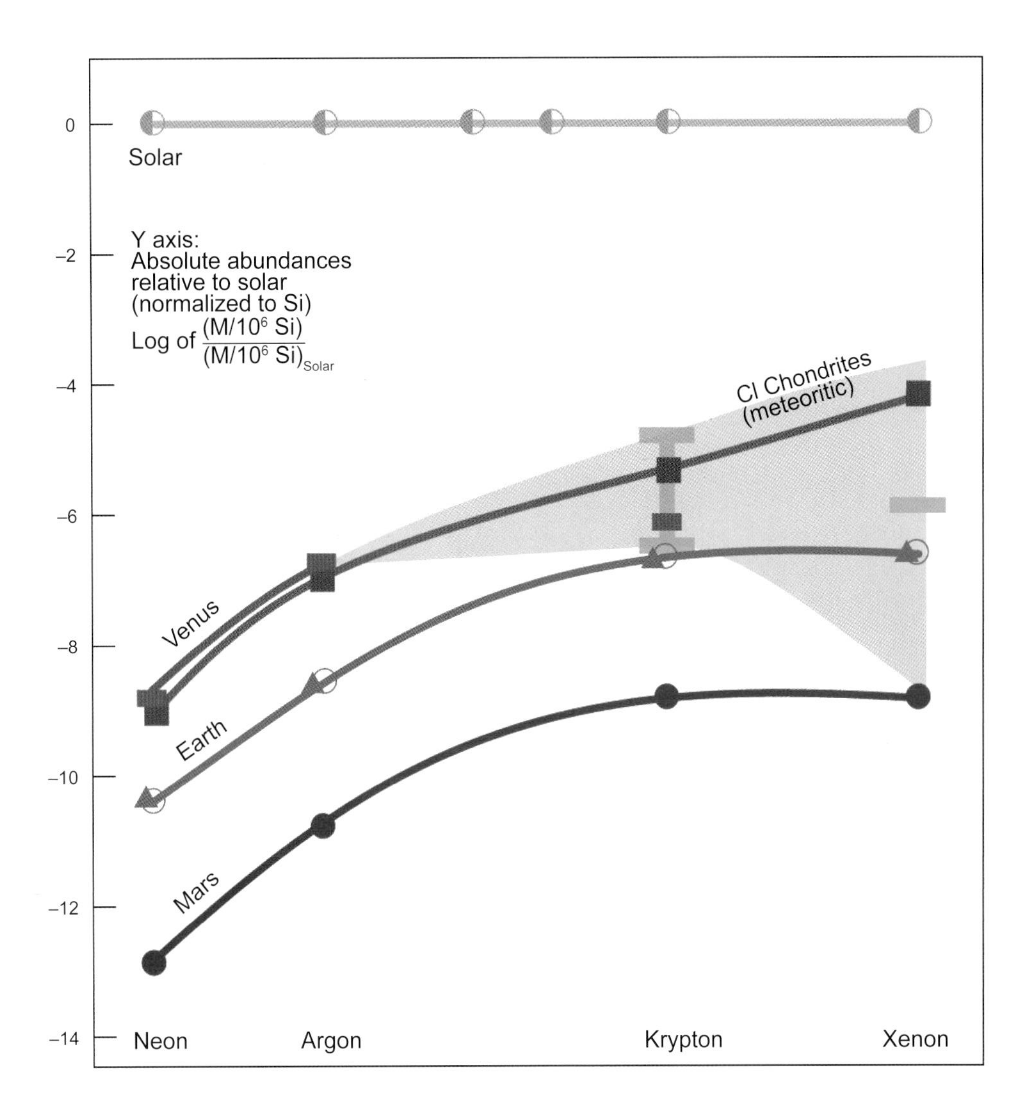

Plate 32. Noble gas abundances for Earth, Mars, Venus, chondrites, and the Sun (after *Pepin,* 1991). Missing Xe and poorly constrained Kr data for Venus are critical for understanding the history of its early atmosphere. For Ne and Ar, and likely Kr, Venus is more enhanced in noble gases vs. Earth and Mars, likely indicating (1) a large blowoff of the original atmospheres of Mars and Earth, and/or (2) enhanced delivery of noble elements to Venus from comets and/or solar-wind-implanted planetesimals. For Venus, the unknown/poorly constrained Xe/Kr and Kr/Ar ratios — denoted by the range of slopes of the aqua area — are consistent with all solar system objects shown, including (1) a solar-type (Jupiter and cometary) composition (orange line), (2) a chondritic composition (brown line), and (3) the composition of Earth and Mars (blue and red lines).

Accompanies chapter by Baines et al. (pp. 137–160).

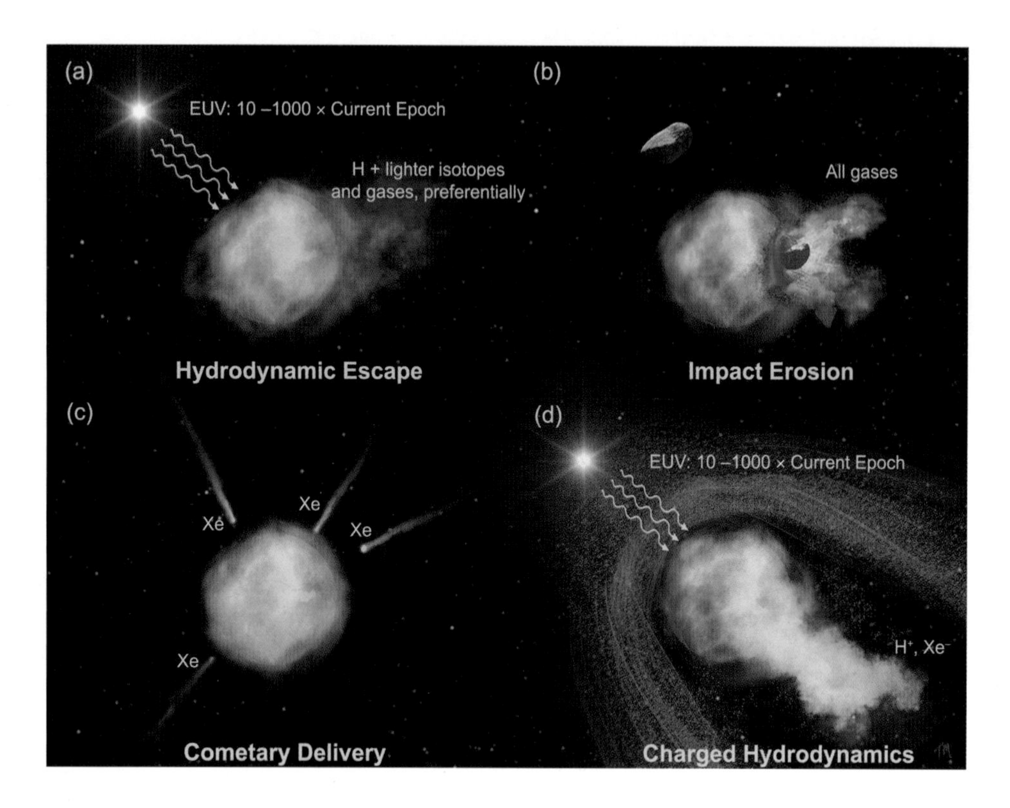

Plate 33. Schematics showing major features of the four main hypotheses of early planetary evolution that led to atmospheric Xe-isotopic abundances on the terrestrial planets. **(a)** Hydrodynamic escape. Early solar EUV was 100–1000 times more intense than today, heating the atmosphere to the point where H began to flow out of the exosphere. It dragged lighter molecules with it, fractionating the atmosphere so that it was isotopically heavier. **(b)** Impact erosion. Small impacts both added material and eroded the atmosphere. Large impacts blew off most of the atmosphere above the plane of impact, indiscriminate of mass, so no fractionation occurred. **(c)** Cometary delivery. Cold comets from the outer solar system with trapped Xe incorporating the isotopic ratios of the cold outer solar system impacted the surface. **(d)** Charged hydrodynamics. Hydrogen and Xe are both easily ionized, unlike other molecules. Thus ionization by the UV or solar wind preferentially allowed H and Xe to escape, while also allowing somewhat smaller portions of other ionized atoms and molecules to go along for the ride. The process is fractionating, dependent upon both mass and ionization energy.

Accompanies chapter by Baines et al. (pp. 137–160).

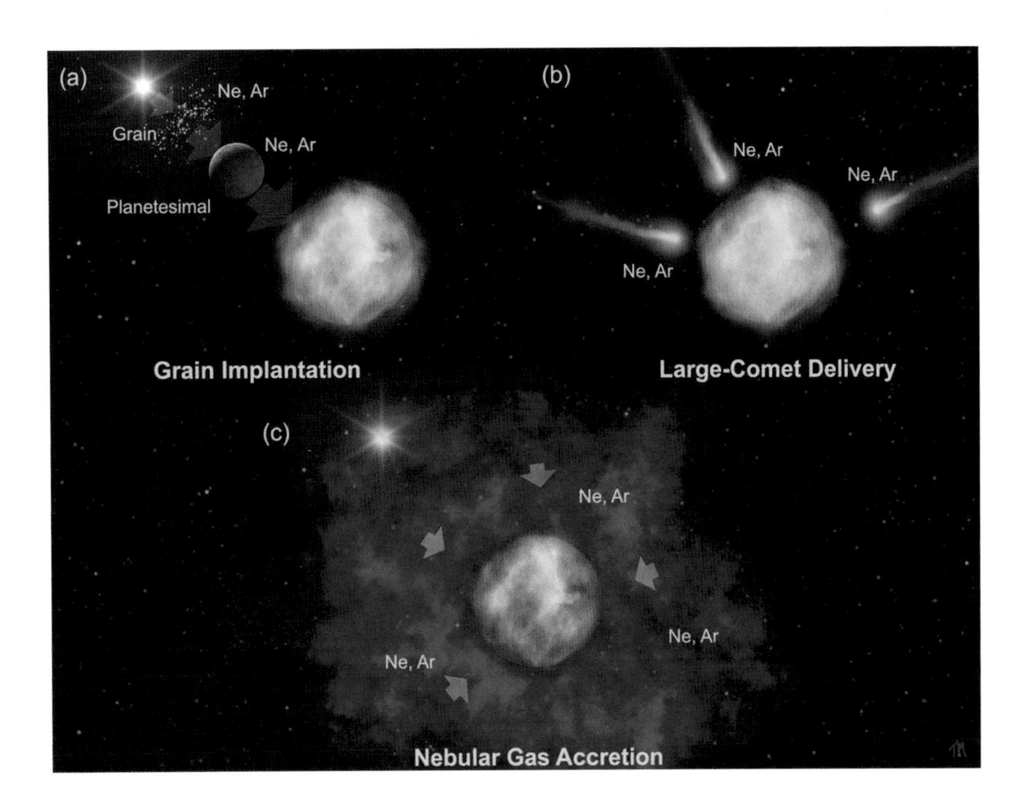

Plate 34. Schematics of the three main hypotheses for the origin of large Ne and Ar abundances in Venus's atmosphere. **(a)** Grain implantation. Early solar wind fluxes were many orders of magnitude higher than today. Solar-wind Ar and Ne implanted into grains around the Sun, which coelesced into Ar,Ne-rich planetesimals that impacted Venus. **(b)** Large-comet delivery. Argon and Ne frozen and trapped in cometary ice latices was delivered by one or more large comets (≥200 km diameter) from the cold reaches of the outer silar system, leaving Venus with an Ar,Ne-rich atmosphere. **(c)** Nebular gas. Venus' present atmosphere came directly from the nebula in which it was formed. The atmosphere survived relatively intact against subsequent impactors that blew off the original atmospheres of Earth and Mars, resulting in a relatively Ar,Ne-rich atmosphere on Venus.

Accompanies chapter by Baines et al. (pp. 137–160).

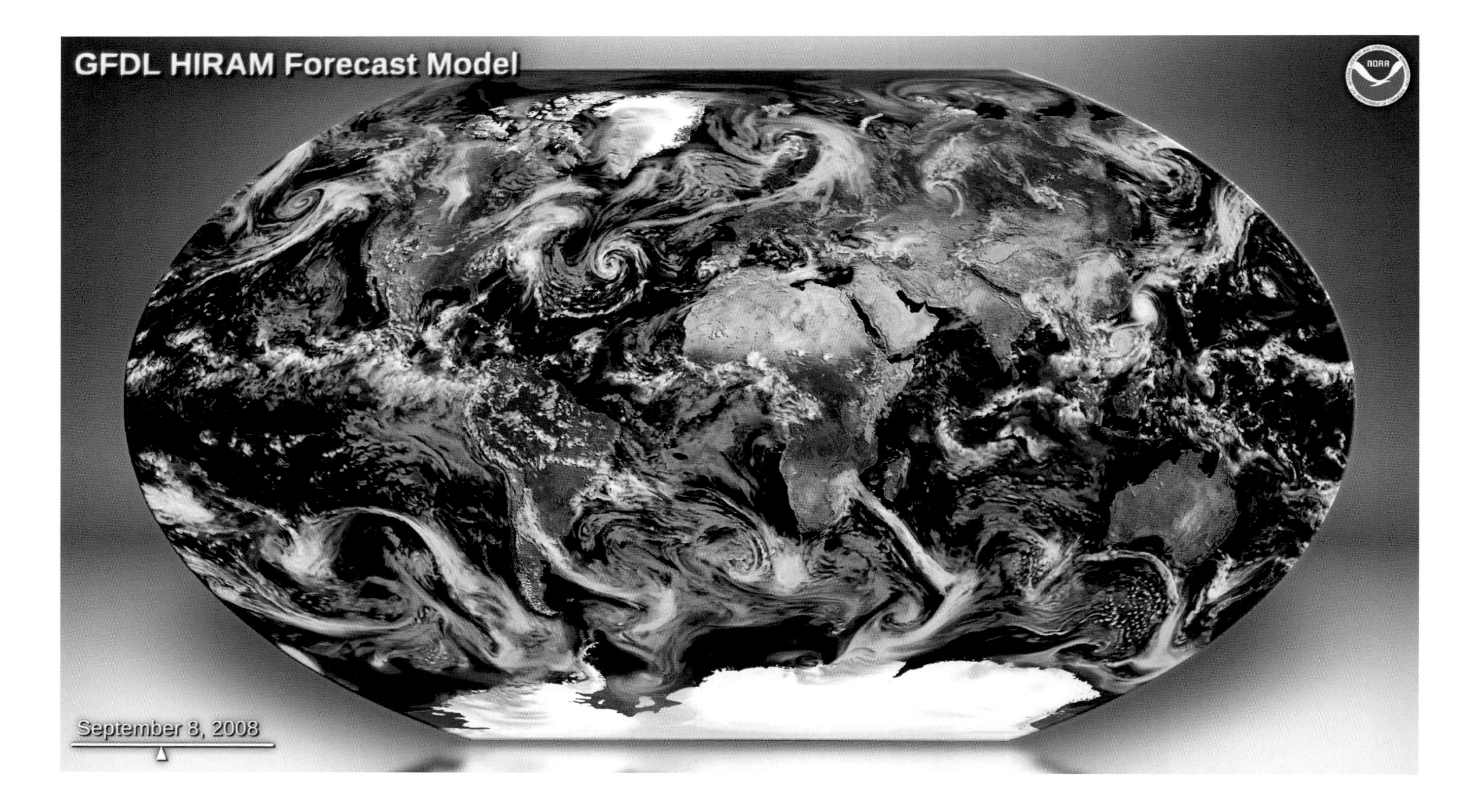

Plate 35. Cloud-system resolving Earth GCM. Shown is a snapshot (simulated time: September 8, 2008) of the Geophysical Fluid Dynamics Laboratory (GFDL) High Resolution Atmosphere Model (HiRAM), run in 2011 at 12.5-km (0.1°) resolution to simulate the peak of the 2008 hurricane-typhoon season (the full movie is available at www.gfdl.noaa.gov/visualizations-hurricanes). Both large-scale and vortex-scale features are nudged toward the analysis (see section 4) as the simulation proceeds. These results indicate that the Atlantic hurricane season has significant predictability, as long as ocean temperatures can be forecast precisely (*Chen and Lin,* 2011). Image supplied by Shian-Jiann Lin, GFDL.

Accompanies chapter by Dowling (pp. 193–211).

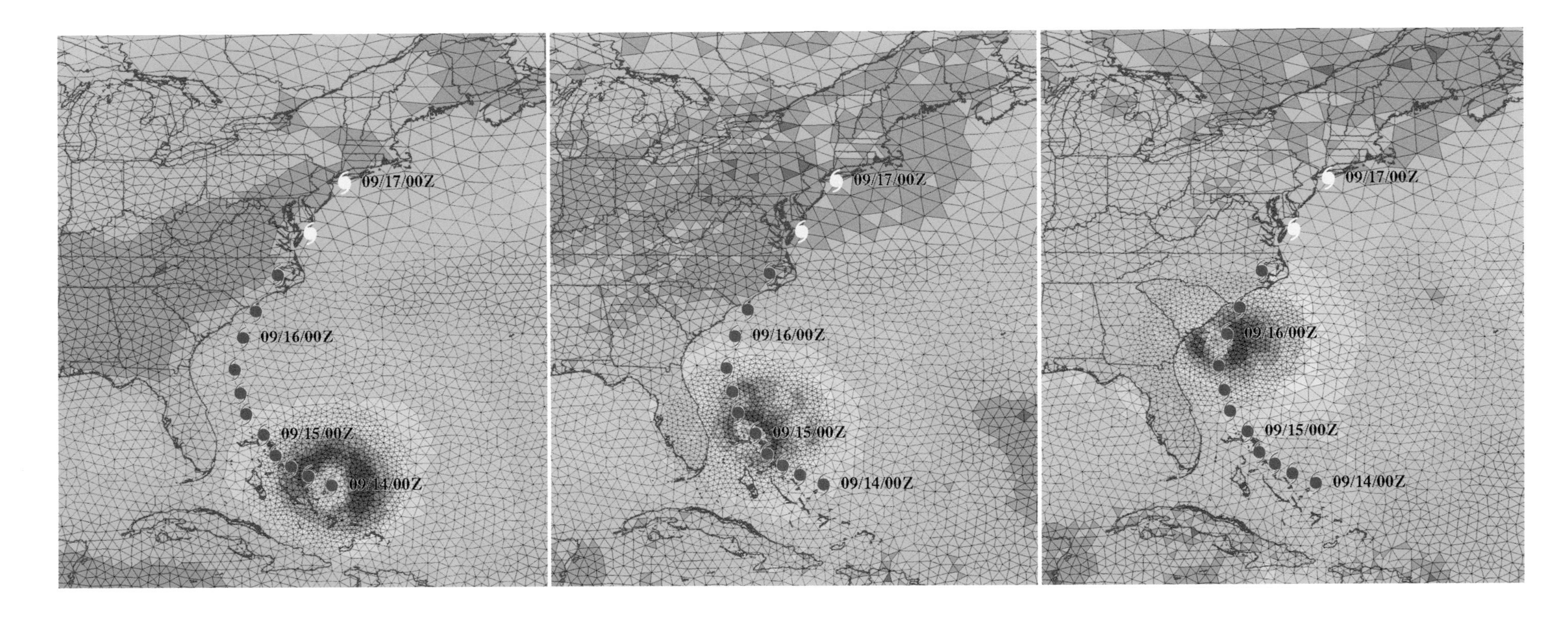

Plate 36. The OMEGA atmospheric model's unstructured, triangular horizontal mesh with dynamic adaptation optimizes grid efficiency based on local conditions in space and time. Shown is an OMEGA simulation of Category 4 Hurricane Floyd from 1999, with wind speed indicated by color. The cyclone symbols mark the actual storm track (time is indicated in MM/DD/HH format at zero hour UTC).

Accompanies chapter by Dowling (pp. 193–211).

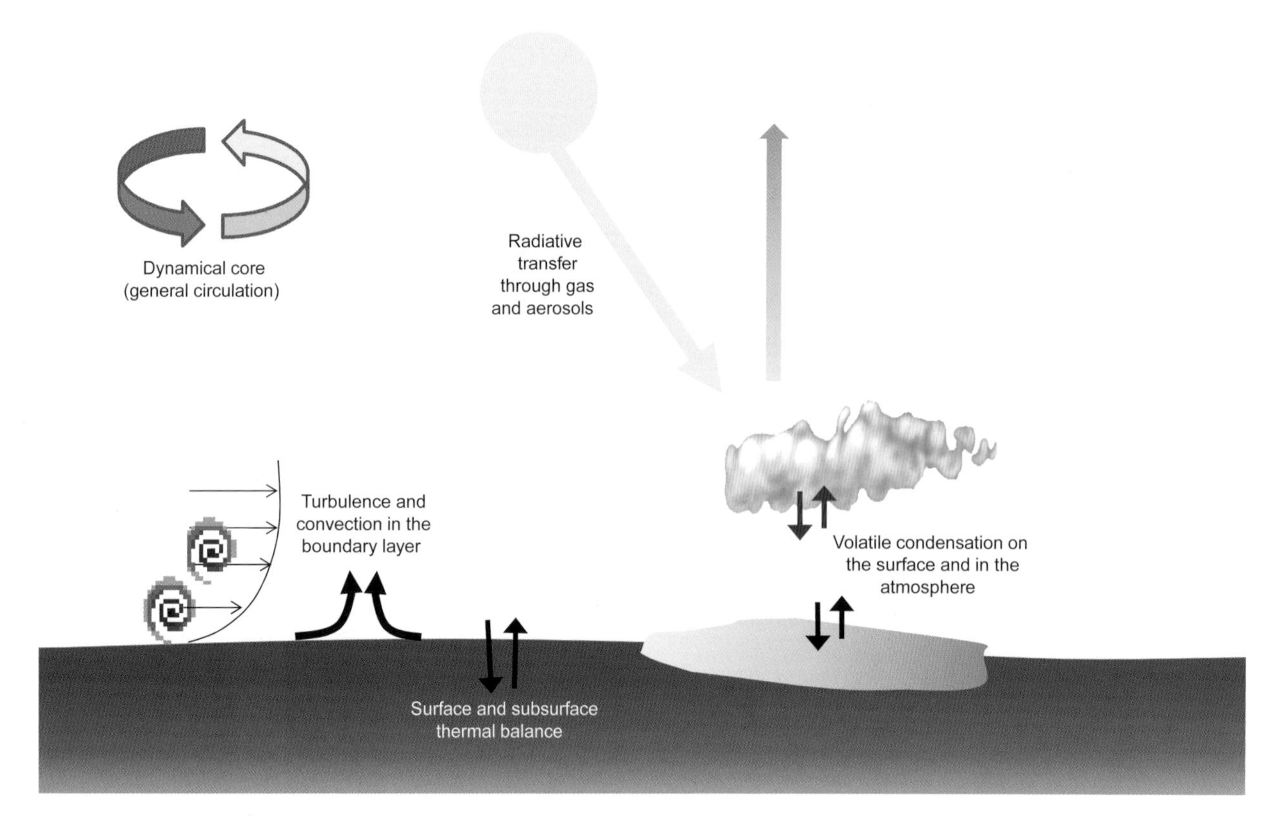

Plate 37. An illustration of the different components that are combined to build a planetary global climate model.

Accompanies chapter by Forget and Lebonnois (pp. 213–229).

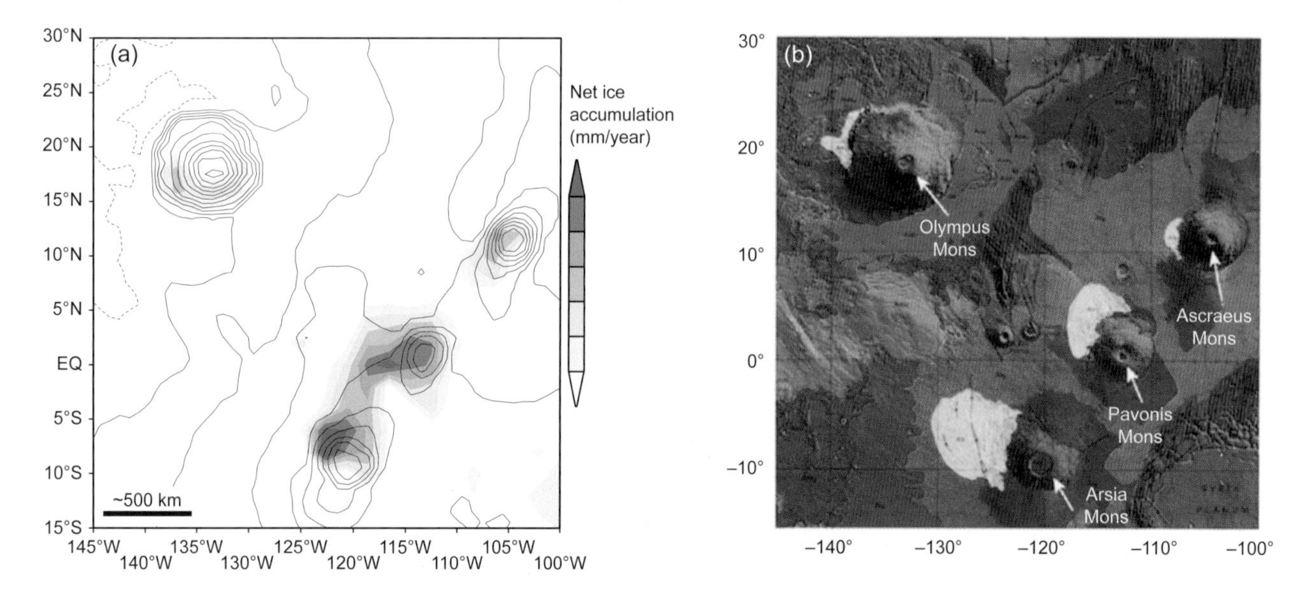

Plate 38. An example of succesful prediction made by a global climate model designed to simulate the details of the present-day Mars water cycle, but assuming a 45° obliquity like on Mars a few millions of years ago. **(a)** Modeled net surface water ice accumulation (in milimeters per martian year) in the Tharsis region, suggesting that glaciers could have formed on the northwest slopes of the Tharsis Montes and Olympus Mons volcanos. **(b)** Geologic map of the same region at the time of the GCM simulations, showing the location of fan-shaped deposits interpreted to be the depositional remains of geologically recent cold-based glaciers. The agreement between observed glacier landform locations and model predictions pointed to an atmospheric origin for the ice and permitted a better understanding of the formation of martian glaciers. From *Forget et al.* (2006).

Accompanies chapter by Forget and Lebonnois (pp. 213–229).

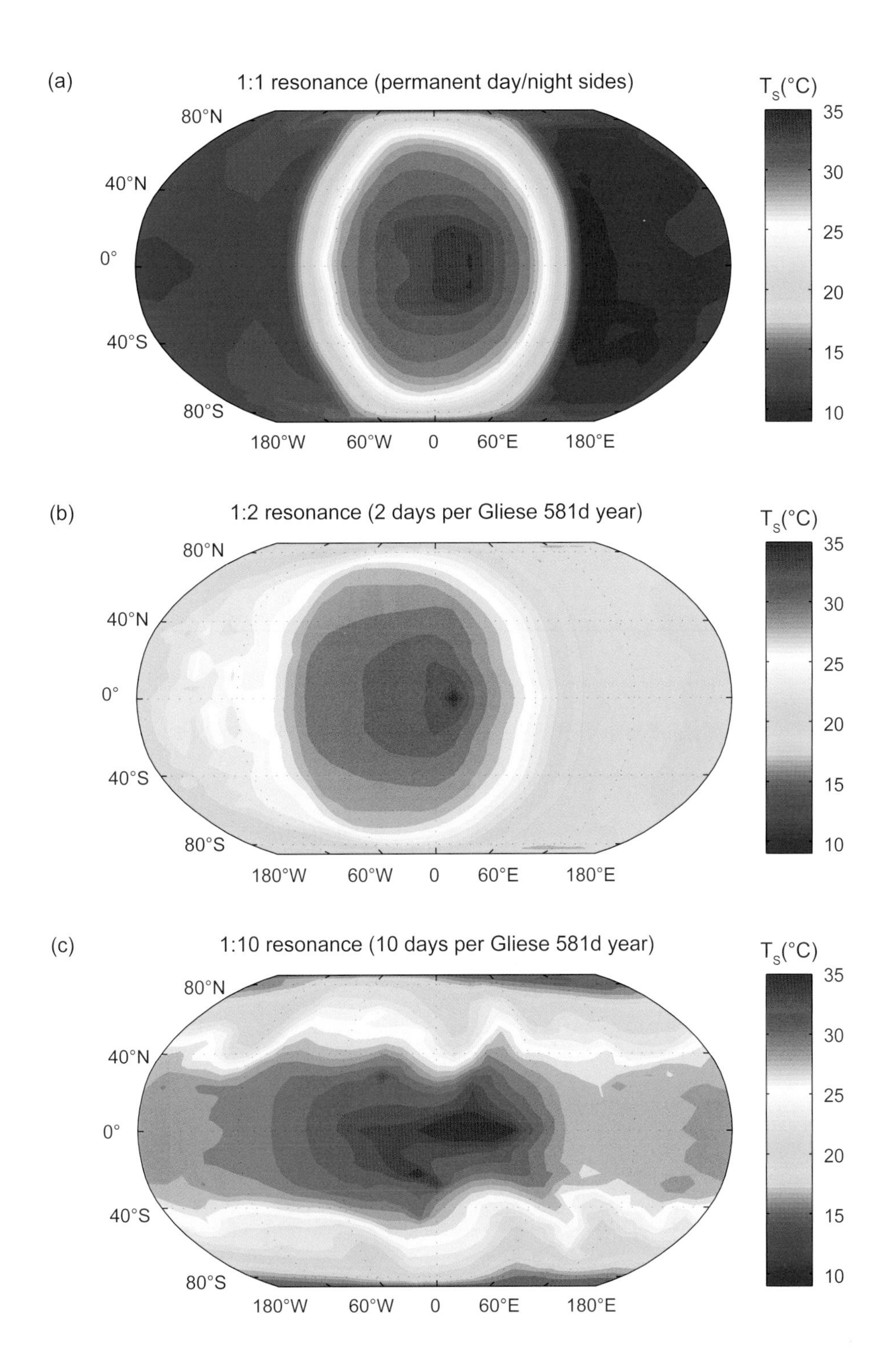

Plate 39. Surface temperature snapshots from three-dimensional global climate model simulations for the extrasolar planet Gliese 581d, assuming a 20-bar CO_2 atmosphere and for three possible rotation rates. Such three-dimensional simulations can help better understand the habitability of exoplanets, in spite of the lack of observations. Figure from *Wordsworth et al.* (2011).

Accompanies chapter by Forget and Lebonnois (pp. 213–229).

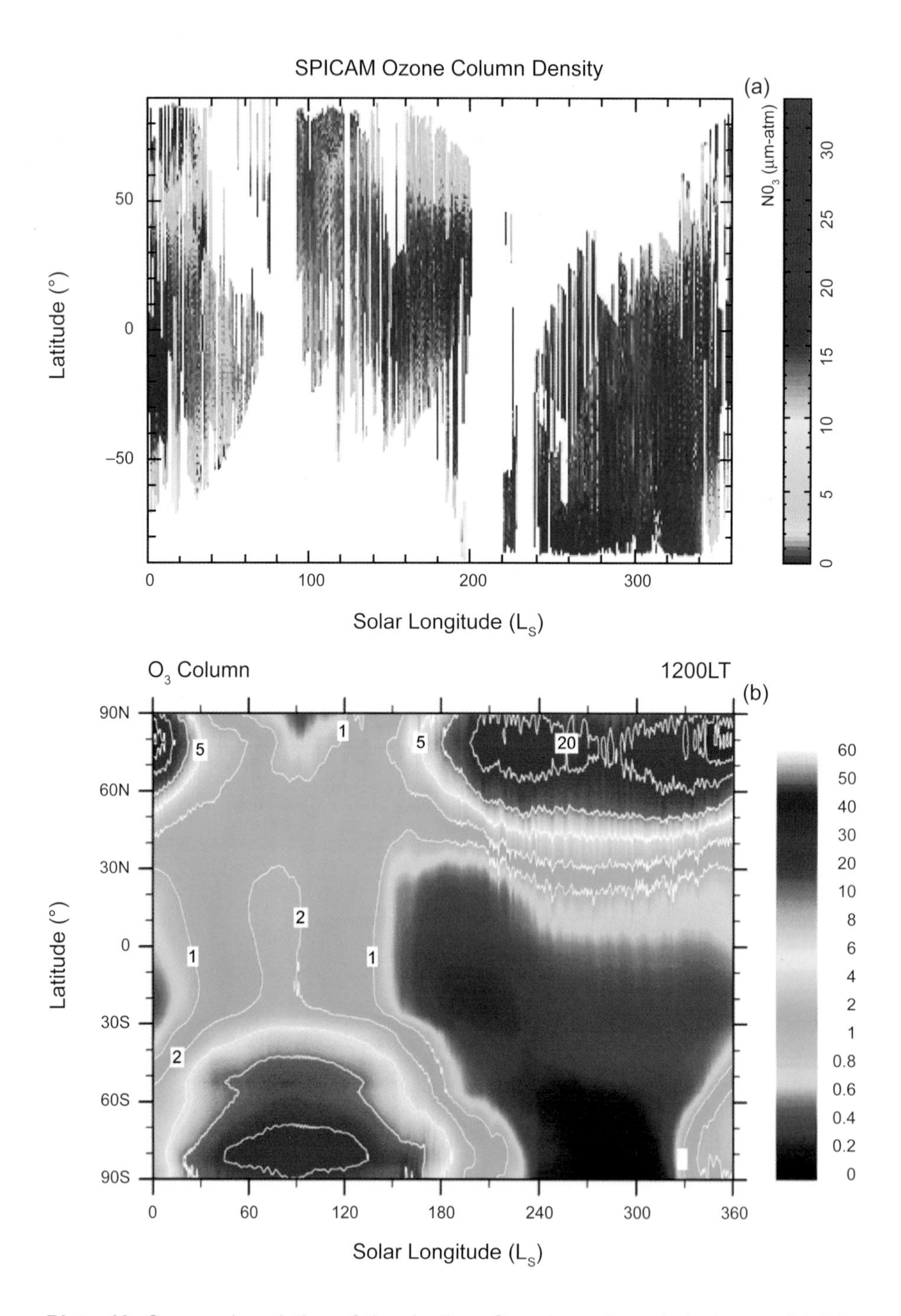

Plate 40. Seasonal evolution of the daytime O_3 column (µm-atm) observed **(a)** by MEX/SPICAM (*Perrier et al.,* 2006) and **(b)** from a three-dimensional model (updated from *Lefevre et al.,* 2004).

Accompanies chapter by Krasnopolsky and Lefèvre (pp. 231–275).

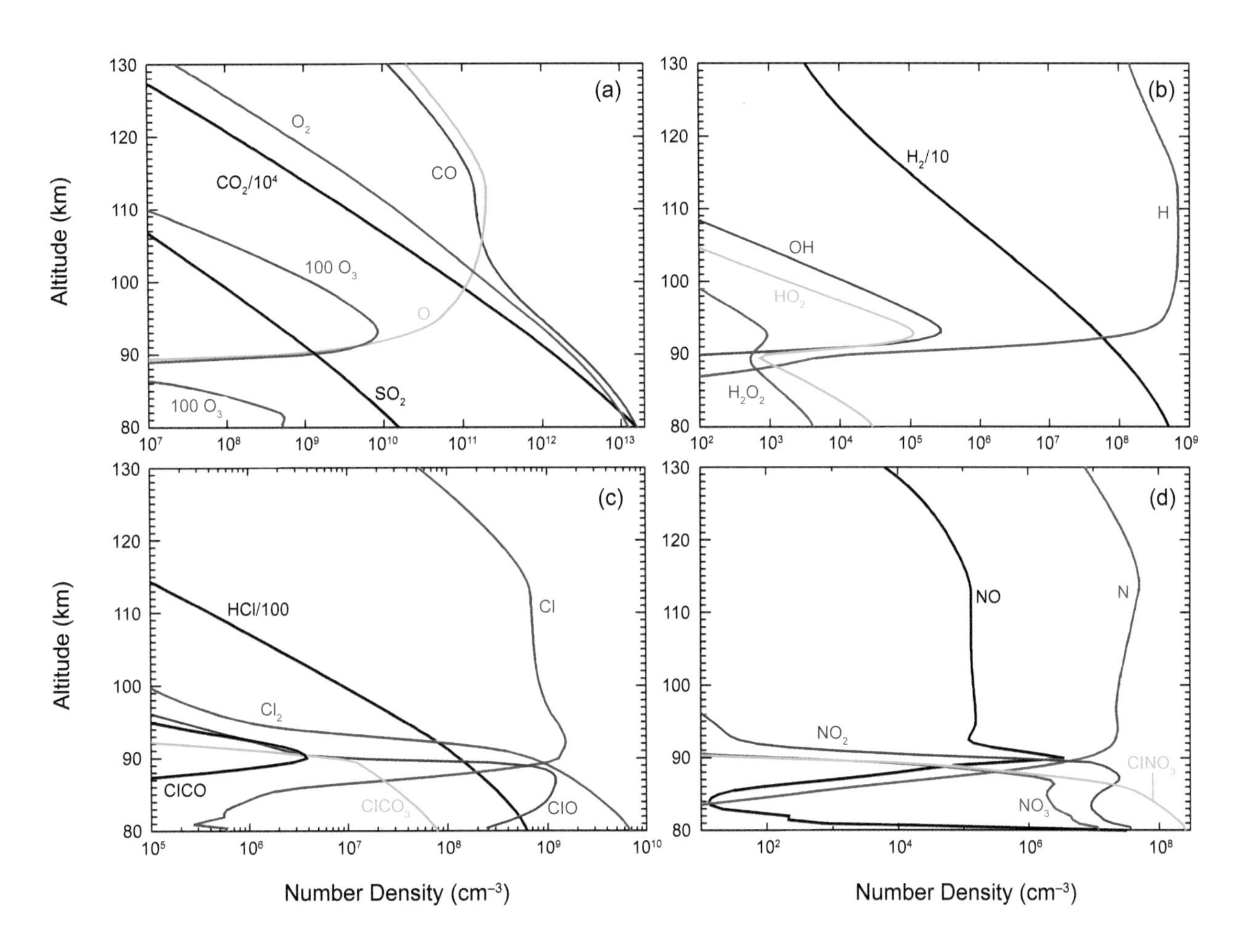

Plate 41. Vertical profiles of species in the model for the nighttime atmosphere of Venus (*Krasnopolsky,* 2013b).

Accompanies chapter by Krasnopolsky and Lefèvre (pp. 231–275).

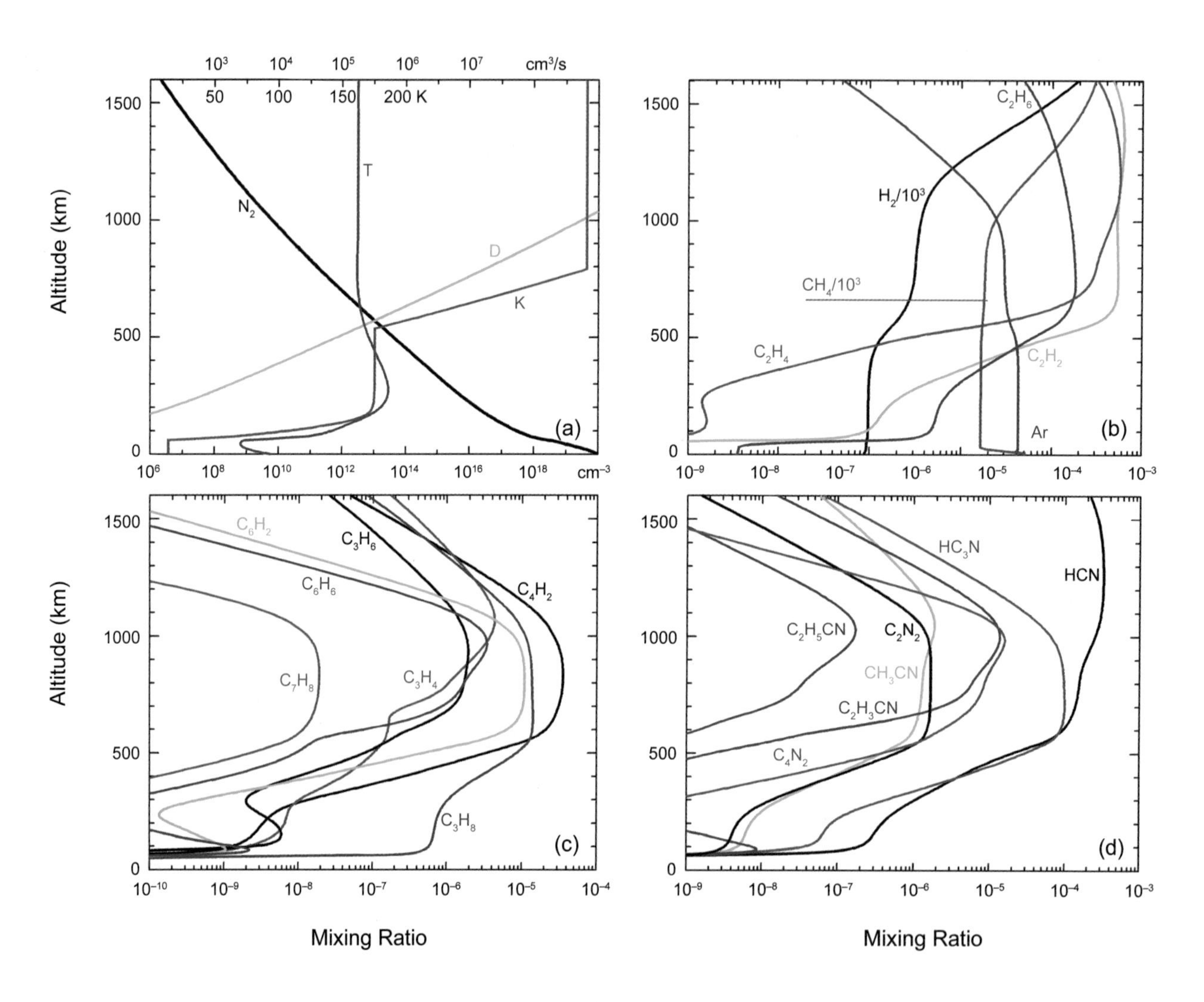

Plate 42. Titan's photochemical model: **(a)** profiles of T, eddy diffusion K, diffusion coefficient of CH_4 in N_2, and N_2 density; **(b)** major hydrocarbons, H_2, and Ar; **(c)** some other hydrocarbons; **(d)** the most abundant nitriles. From *Krasnopolsky* (2012c).

Accompanies chapter by Krasnopolsky and Lefèvre (pp. 231–275).

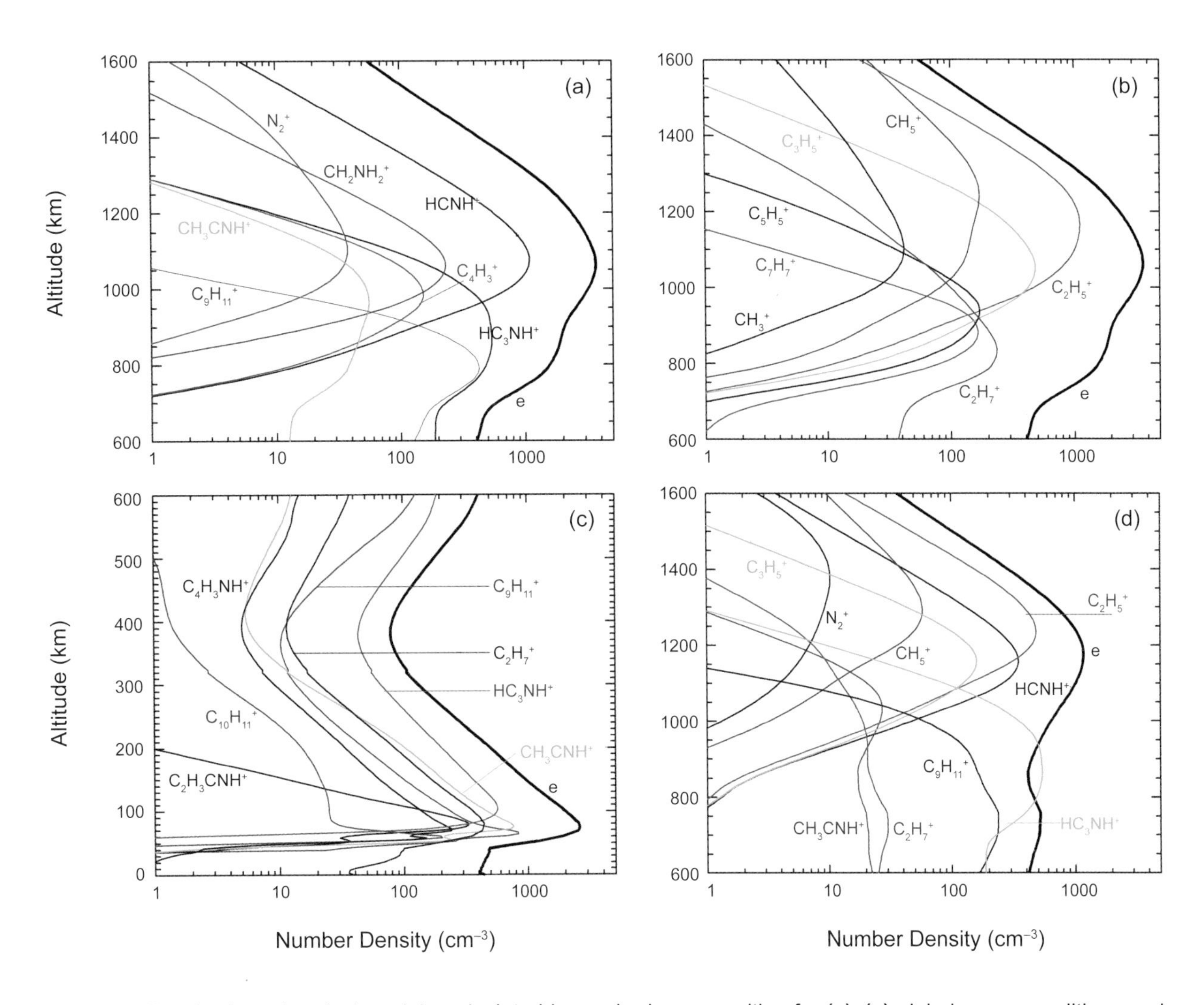

Plate 43. Titan's photochemical model: calculated ionospheric composition for **(a)**–**(c)** global-mean conditions and **(d)** the strong nighttime electron precipitation T5. From *Krasnopolsky* (2009a, 2012c).

Accompanies chapter by Krasnopolsky and Lefèvre (pp. 231–275).

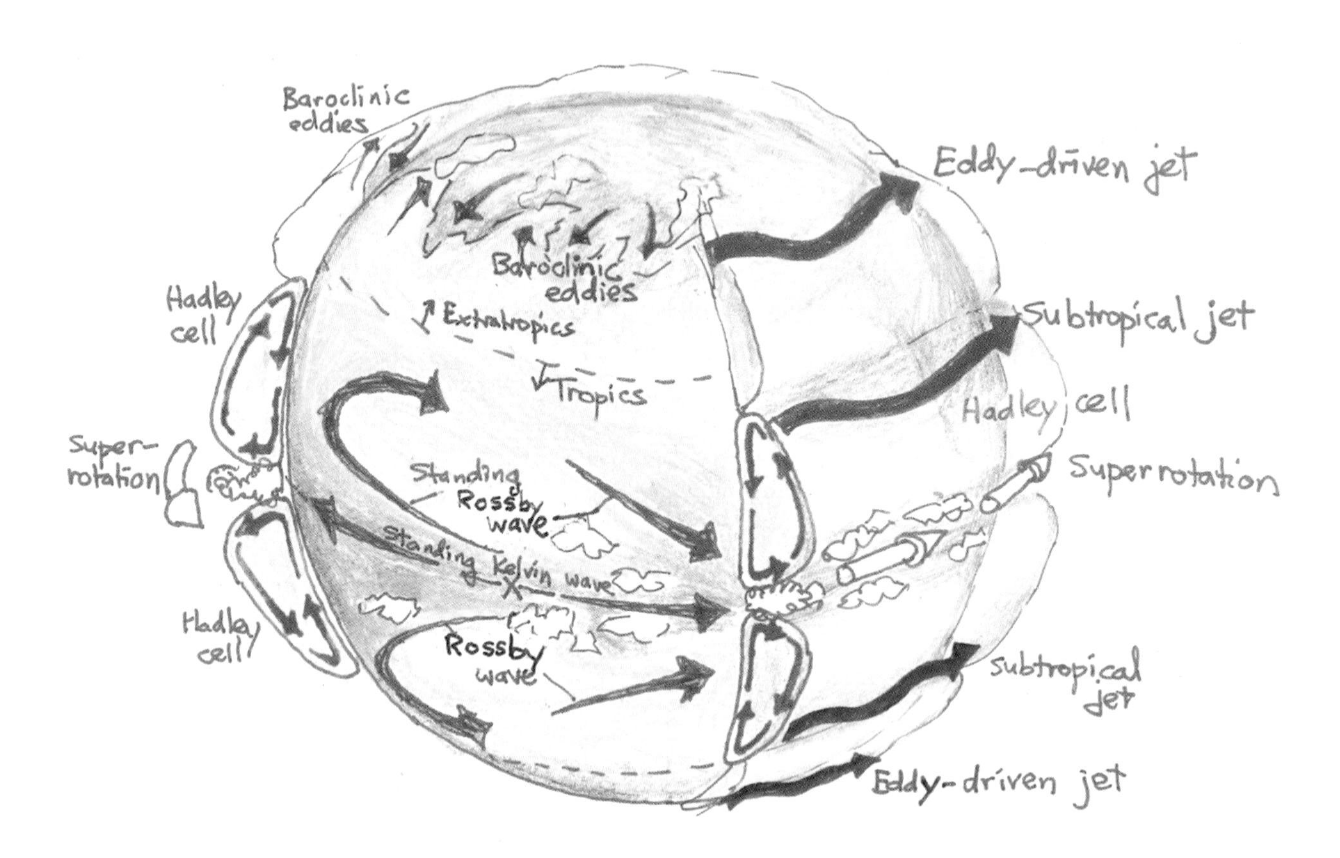

Plate 44. Schematic illustration of dynamical processes occurring on a generic terrestrial exoplanet. These include baroclinic eddies, Rossby waves, and eddy-driven jet streams in the extratropics, and Hadley circulations, large-scale Kelvin and Rossby waves, and (in some cases) equatorial superrotation in the tropics. The "X" at the equator marks the substellar point, which will be fixed in longitude on synchronously rotating planets. Cloud formation, while complex, will likely be preferred in regions of mean ascent, including the rising branch of the Hadley circulation, within baroclinic eddies, and — on synchronously rotating planets — in regions of ascent on the dayside.

Accompanies chapter by Showman et al. (pp. 277–326).

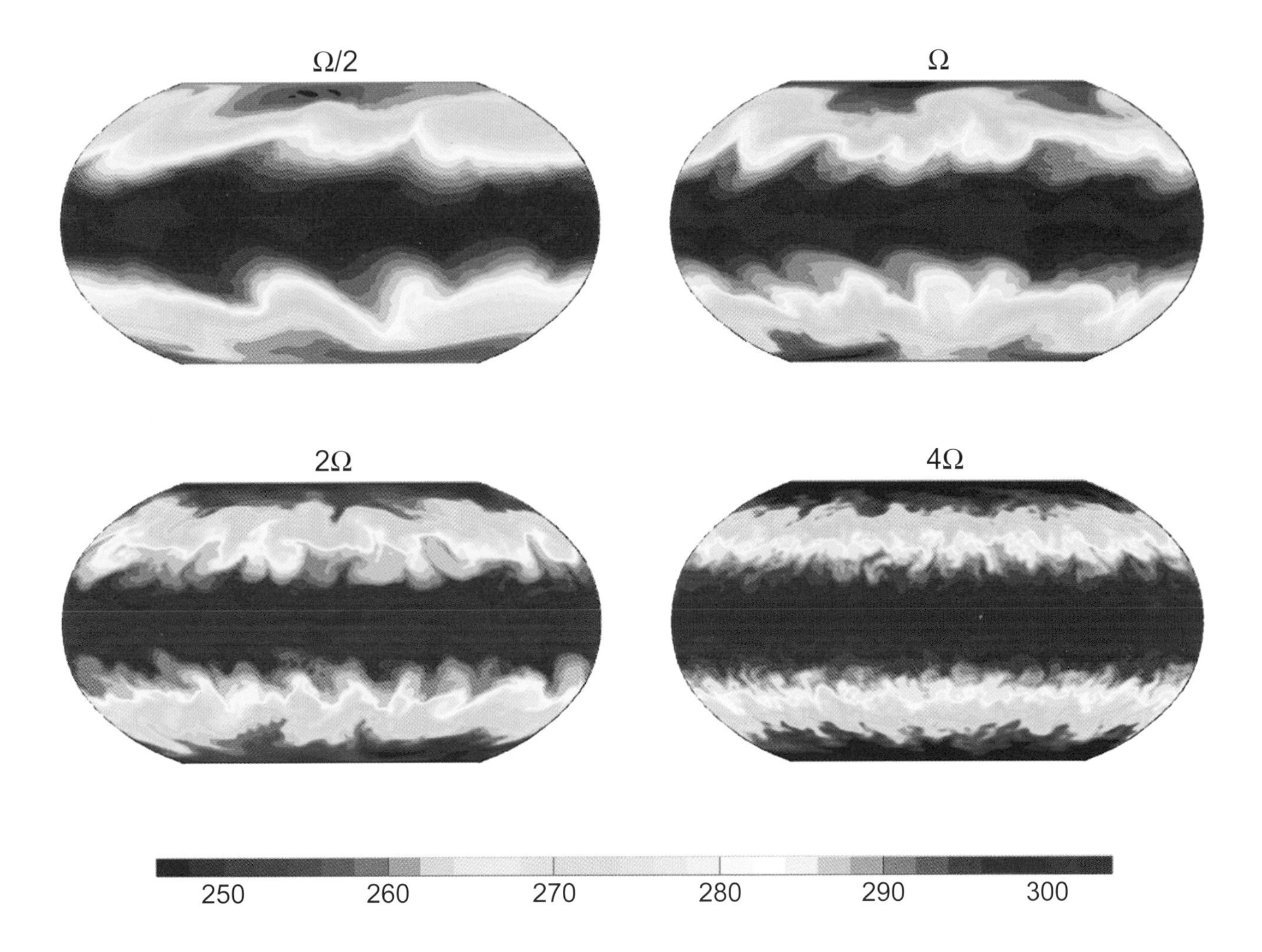

Plate 45. Surface temperature (colorscale, in K) from GCM experiments in *Kaspi and Showman* (2012), illustrating the dependence of temperature and jet structure on rotation rate. Experiments are performed using the Flexible Modeling System (FMS) model analogous to those in *Frierson et al.* (2006) and *Kaspi and Schneider* (2011); radiative transfer is represented by a two-stream, gray scheme with no diurnal cycle (i.e., the incident stellar flux depends on latitude but not longitude). A hydrological cycle is included with a slab ocean. Planetary radius, gravity, atmospheric mass, incident stellar flux, and atmospheric thermodynamic properties are the same as on Earth; models are performed with rotation rates from half (upper left) to four times that of Earth (lower right). Baroclinic instabilities dominate the dynamics in mid- and high-latitudes, leading to baroclinic eddies whose length scales decrease with increasing planetary rotation rate.

Accompanies chapter by Showman et al. (pp. 277–326).

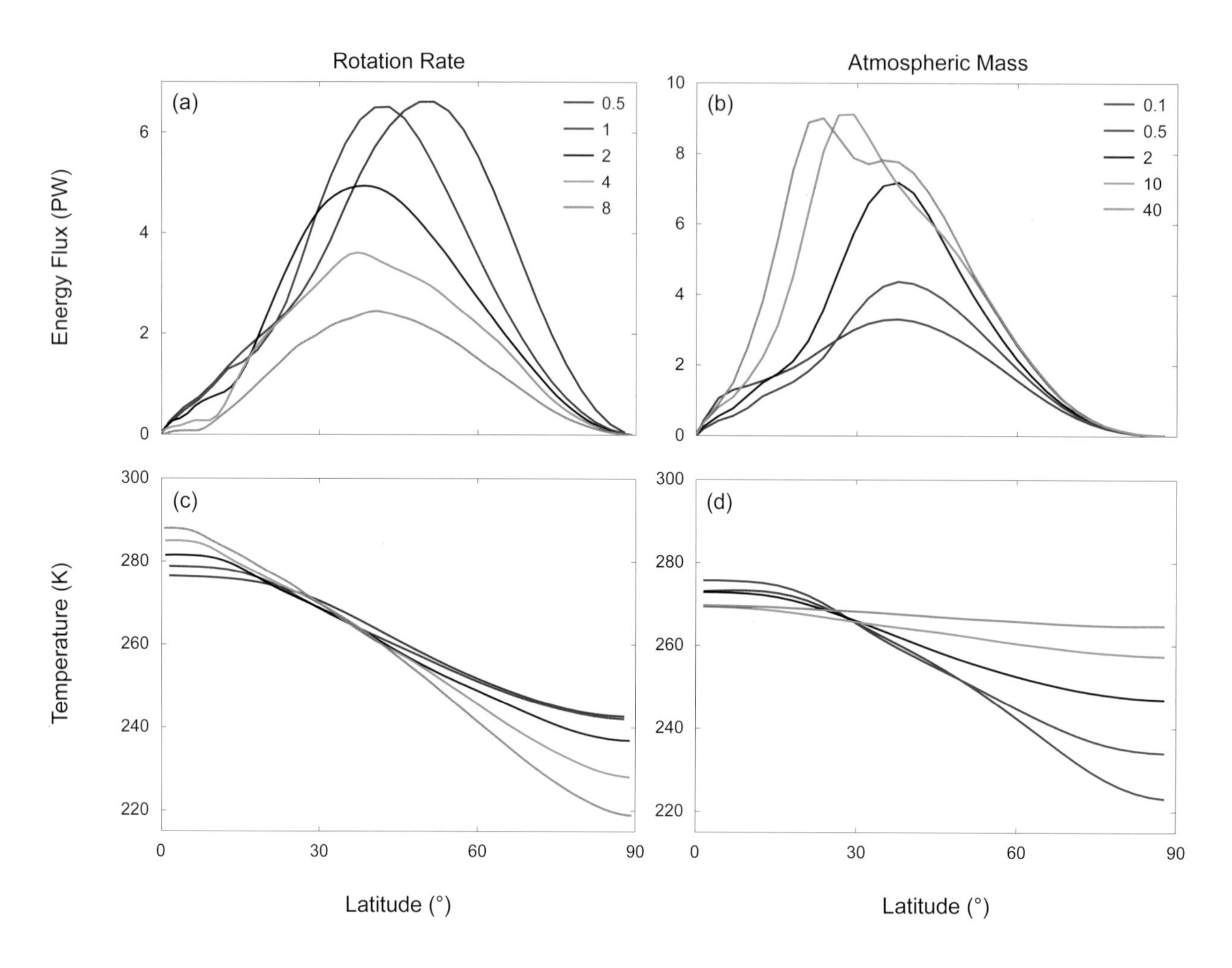

Plate 46. Latitude dependence of the **(a),(b)** vertically and zonally integrated meridional energy flux and **(c),(d)** vertically and zonally averaged temperature in GCM experiments from *Kaspi and Showman* (2012) for models varying rotation rate and atmospheric mass. The energy flux shown in **(a)** and **(b)** is the flux of moist static energy, defined as cpT + gz + Lq, where cp is specific heat at constant pressure, g is gravity, z is height, L is latent heat of vaporization of water, and q is the water vapor abundance. The left column [**(a)** and **(c)**] explores sensitivity to rotation rate; models are shown with rotation rates ranging from half to eight times that of Earth. In these experiments, the atmospheric mass is held constant at the mass of Earth's atmosphere. The right column [**(b)** and **(d)**] explores the sensitivity to atmospheric mass; models are shown with atmospheric masses from 0.1 to 40 times the mass of Earth's atmosphere. In these experiments, the rotation rate is set to that of Earth. The equator-to-pole temperature difference is smaller, and the meridional energy flux is larger, when the planetary rotation rate is slower, and/or when the atmospheric mass is larger. Other model parameters, including incident stellar flux, optical depth of the atmosphere in the visible and infrared, planetary radius, and gravity, are Earth-like and are held fixed in all models.

Accompanies chapter by Showman et al. (pp. 277–326).

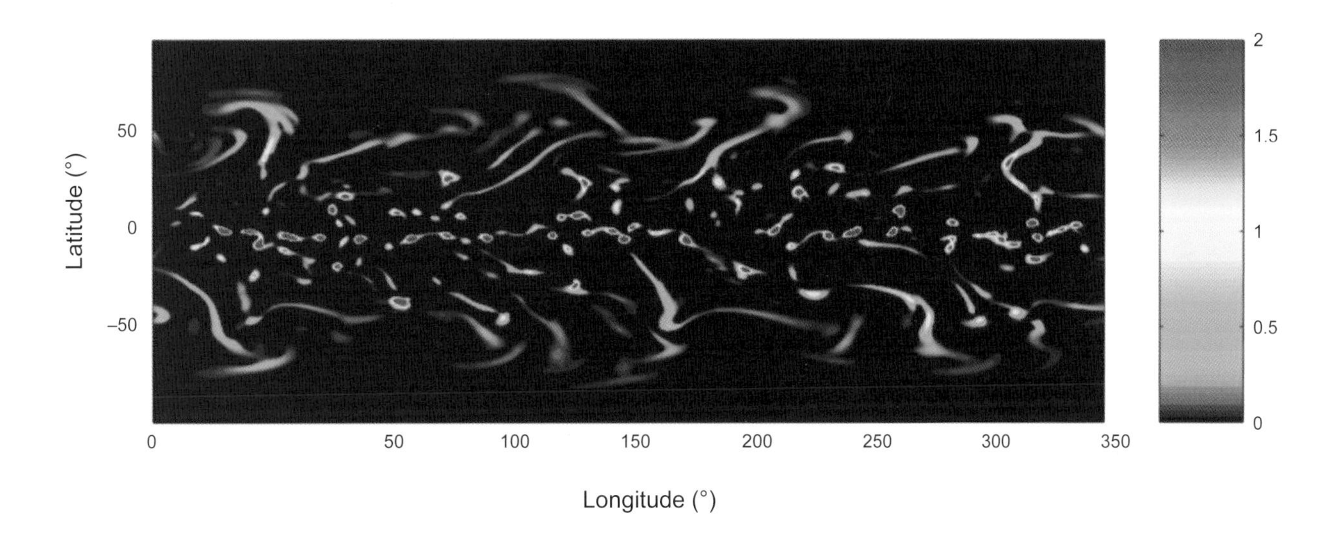

Plate 47. Instantaneous precipitation (units 10^{-3} kg m^{-2} s^{-1}) in an idealized Earth GCM by *Frierson et al.* (2006), illustrating the generation of phase tilts by Rossby waves as depicted schematically in Fig. 6. Baroclinic instabilities in midlatitudes (~40°–50°) generate Rossby waves that propagate meridionally. On the equatorward side of the baroclinically unstable zone (latitudes of ~20°–50°), the waves propagate equatorward, leading to characteristic precipitation patterns tilting southwest-northeast in the northern hemisphere and northwest-southeast in the southern hemisphere. In contrast, the phase tilts are reversed (although less well organized) poleward of ~50°–60° latitude, indicative of poleward Rossby wave propagation. Equatorward of 20° latitude, tropical convection dominates the precipitation pattern.

Accompanies chapter by Showman et al. (pp. 277–326).

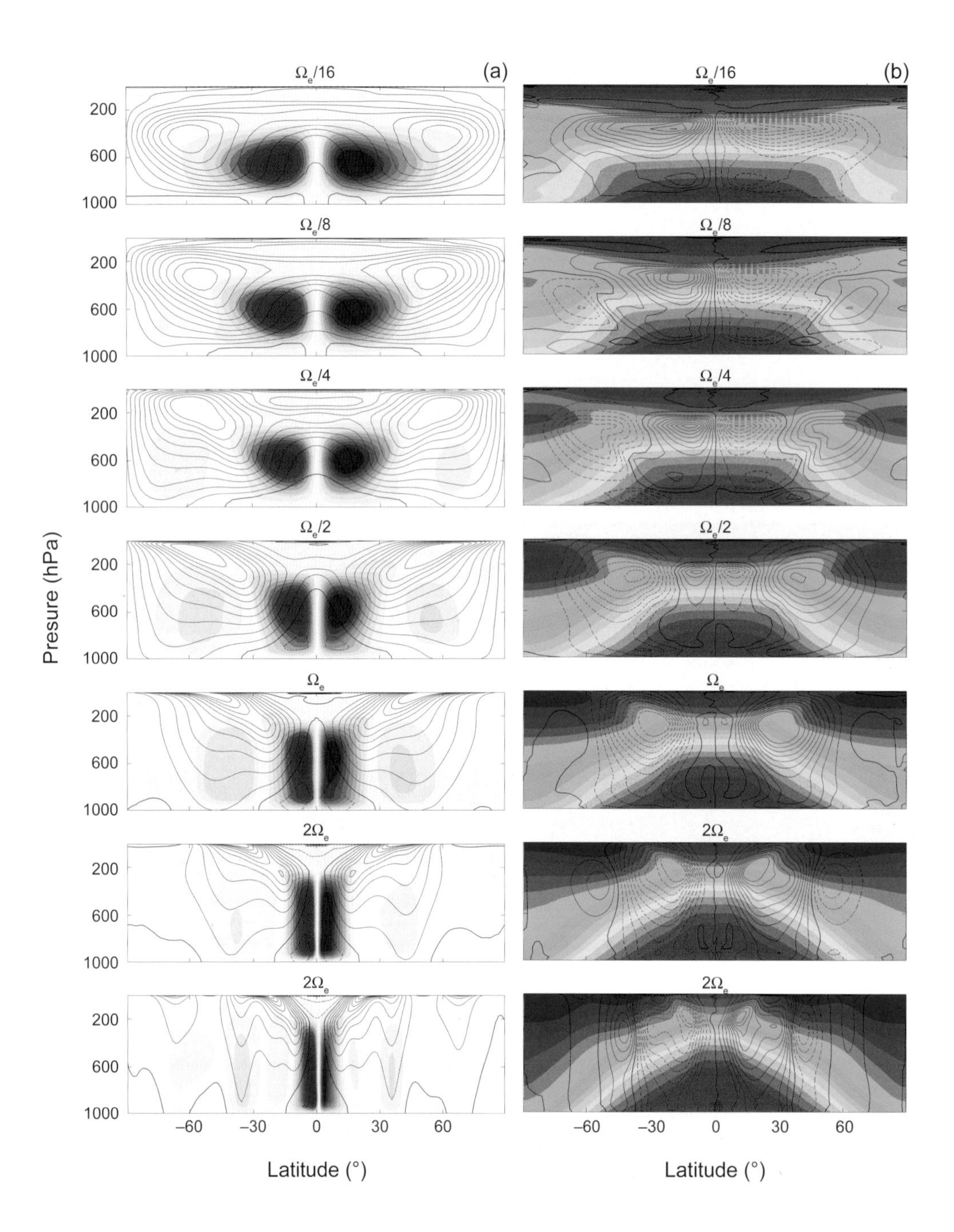

Plate 48. Zonal-mean circulation for a sequence of idealized GCM experiments of terrestrial planets from *Kaspi and Showman* (2012), showing the dependence of the Hadley circulation on planetary rotation rate. The models are driven by an imposed equator-pole insolation pattern with no seasonal cycle. The figure shows seven experiments with differing planetary rotation rates, ranging from 1/16th to four times that of Earth from top to bottom, respectively. **(a)** Thin black contours show zonal-mean zonal wind; the contour interval is 5 m s^{-1}, and the zero-wind contour is shown in a thick black contour. Orange/blue colorscale depicts the mean-meridional streamfunction, with blue denoting clockwise circulation and orange denoting counterclockwise circulation. **(b)** Colorscale shows zonal-mean temperature. Contours show zonal-mean meridional eddy-momentum flux, $\overline{u'v'}\cos\phi$. Solid and dashed curves denote positive and negative values, respectively (implying northward and southward transport of eastward eddy momentum, respectively). At slow rotation rates, the Hadley cells are nearly global, the subtropical jets reside at high latitude, and the equator-pole temperature difference is small. The low-latitude meridional momentum flux is equatorward, leading to equatorial superrotation (eastward winds at the equator) in the upper troposphere. At faster rotation rates, the Hadley cells and subtropical jets contract toward the equator, an extratropical zone, with eddy-driven jets, develops at high latitudes, and the equator-pole temperature difference is large. The low-latitude meridional momentum flux is poleward, resulting from the absorption of equatorward-propagating Rossby waves coming from the extratropics.

Accompanies chapter by Showman et al. (pp. 277–326).

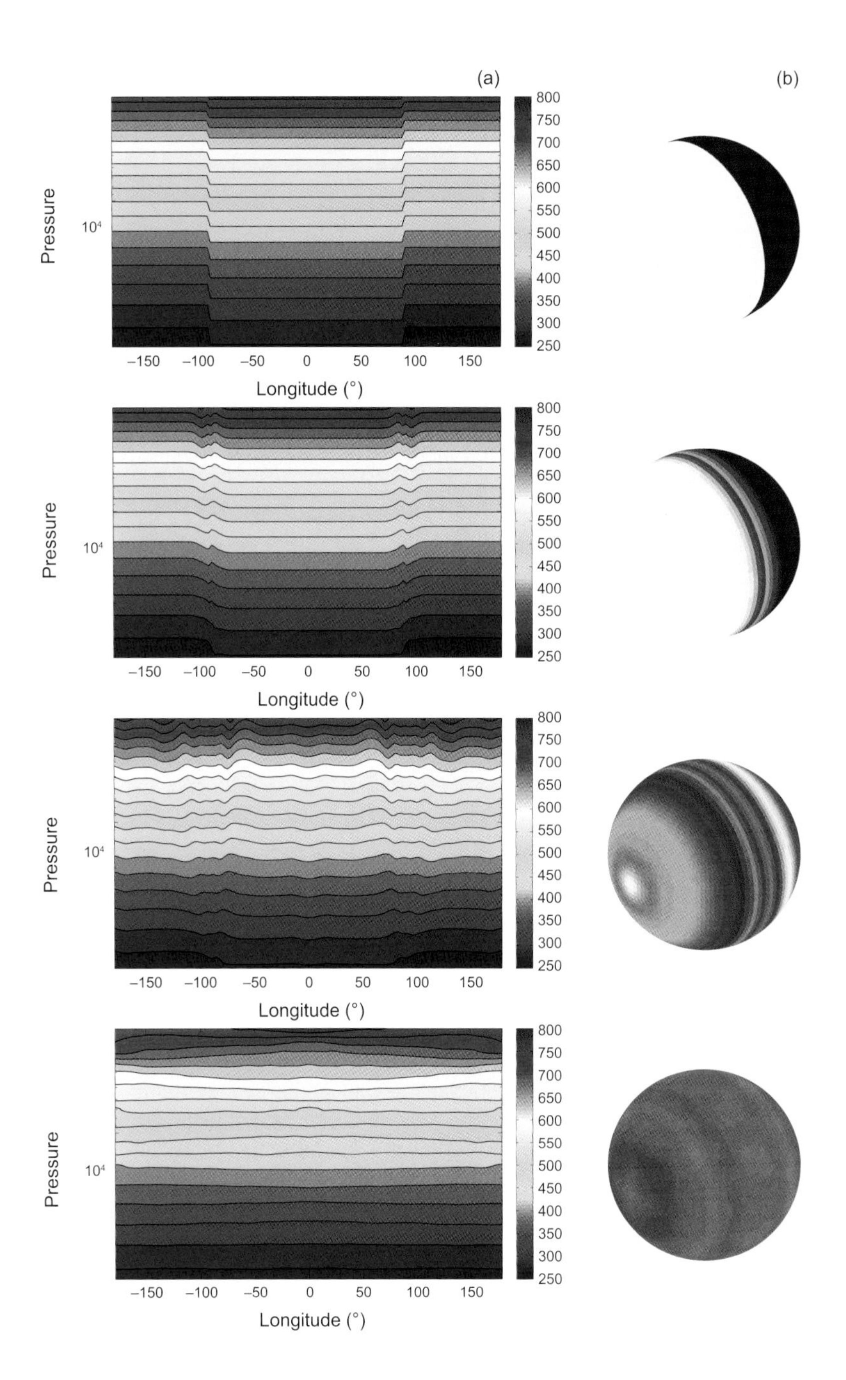

Plate 49. Numerical solution of wave adjustment on a spherical, nonrotating terrestrial planet with the radius and gravity of Earth. We solved the global, three-dimensional primitive equations, in pressure coordinates, using the MITgcm. Half of the planet (the "nightside") was initialized with a constant (isothermal) temperature of T_{night} = 250 K, corresponding to a potential temperature profile $\theta_{night} = T_{night}(p_0 = p)^{\kappa}$, where $\kappa = R/c_p = 2/7$ and p_0 = 1 bar is a reference pressure. The other half of the planet (the "dayside") was initialized with a potential temperature profile $\theta_{night}(p) + \Delta\theta$, where $\Delta\theta$ = 20 K is a constant. Domain extends from approximately 1 bar at the bottom to 0.001 bar at the top; equations were solved on a cubed-sphere grid with horizontal resolution of C32 (32 × 32 cells per cube face, corresponding to an approximate resolution of 2.8°) and 40 levels in the vertical, evenly spaced in log-p. The model includes a sponge at pressures less than 0.01 bar to absorb upward-propagating waves. This is an initial value problem; there is no radiative heating/cooling, so the flow is adiabatic. **(a)** Potential temperature (colorscale and contours) at the equator vs. longitude and pressure; **(b)** temperature at a pressure of 0.2 bar over the globe at times of 0 (showing the initial condition), 0.5 × 10^4 s, 3 × 10^4 s, and the final long-term state once the waves have propagated into the upper atmosphere. Air parcels move by only a small fraction of a planetary radius during the adjustment process, but the final state nevertheless corresponds to nearly flat isentropes with small horizontal temperature variations on isobars.

Accompanies chapter by Showman et al. (pp. 277–326).

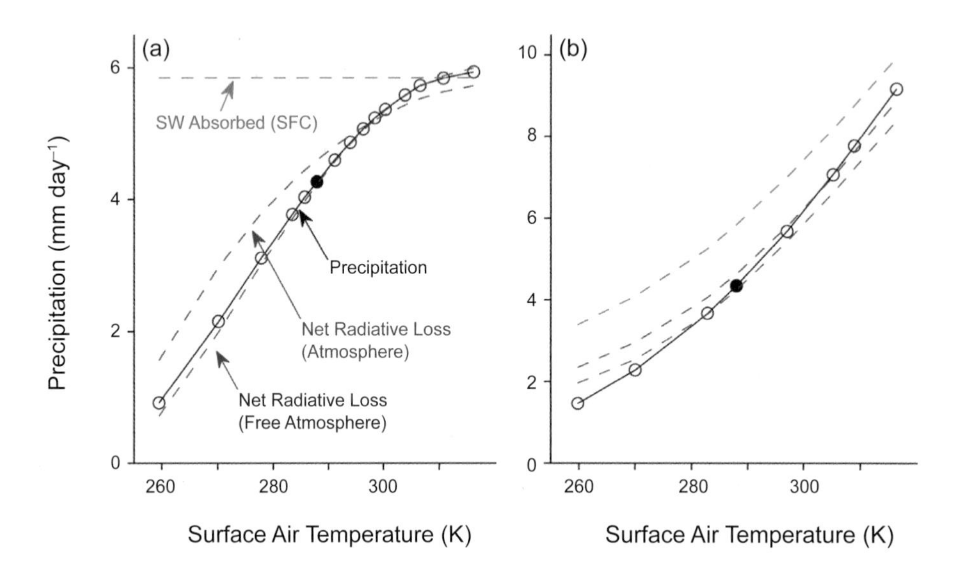

Plate 50. Global-mean precipitation (solid line with circles) vs. global-mean surface air temperature in two series of statistical-equilibrium simulations with an idealized GCM in which **(a)** the optical depth of the longwave absorber is varied and **(b)** the solar constant is varied (from about 800 W m^{-2} to about 2300 W m^{-2}). The filled circles indicate the reference simulation (common to both series) that has the climate most similar to present-day Earth's. The red dashed lines show the net radiative loss of the atmosphere, the blue dashed lines show the net radiative loss of the free atmosphere (above $\sigma = p/p_s = 0.86$), and the green dashed lines show the net absorbed solar radiation at the surface (all in equivalent precipitation units of millimeters per day). From *O'Gorman et al.* (2011).

Accompanies chapter by Showman et al. (pp. 277–326).

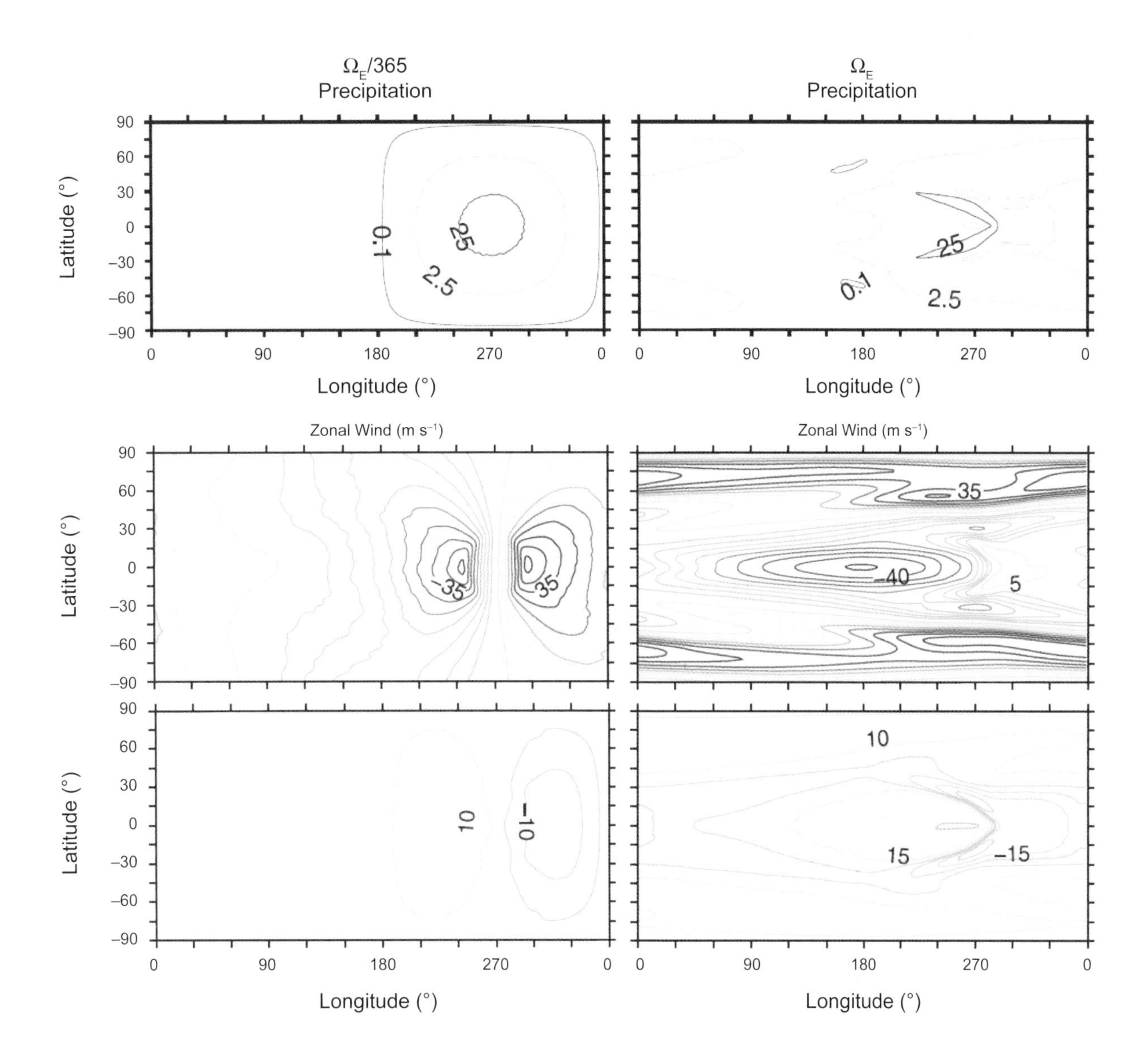

Plate 51. Time-mean circulation in two GCM simulations of Earth-like synchronously rotating exoplanets; models with rotation periods of one Earth year and one Earth day, respectively, are shown on the left and right. Top row shows time-mean precipitation (contours of 0.1, 2.5, and 25.0 mm day^{-1}). Middle and bottom rows show time-mean zonal wind on the $\sigma = p/p_s = 0.28$ model level and at the surface, respectively (contour interval is 5 m s^{-1}; the zero contour is not displayed). The subsolar point is fixed at 270° longitude in both models. Adapted from Figs. 2, 4, and 11 of *Merlis and Schneider* (2010).

Accompanies chapter by Showman et al. (pp. 277–326).

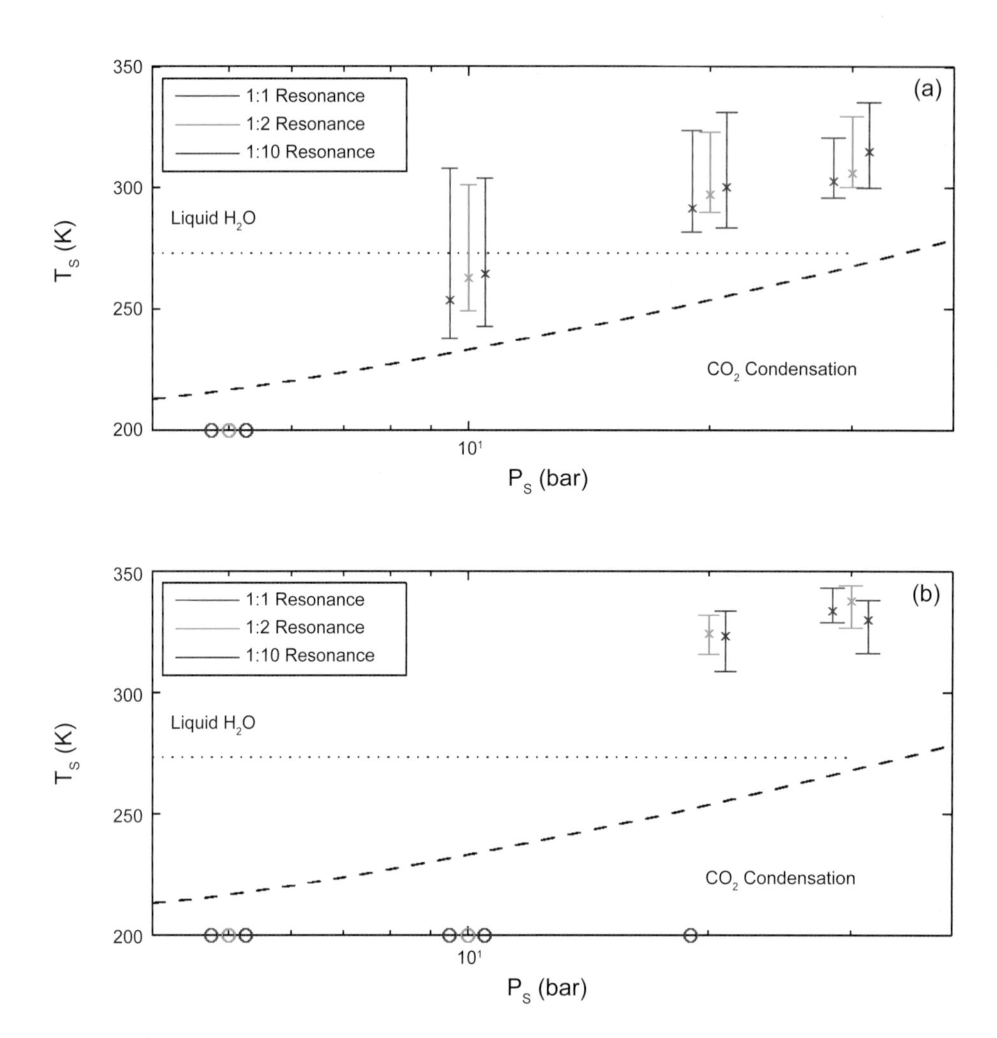

Plate 52. Simulated annual mean surface temperature (maximum, minimum, and global average) as a function of atmospheric pressure and rotation rate for GJ581d assuming **(a)** a pure CO_2 atmosphere and **(b)** a mixed CO_2-H_2O atmosphere with infinite water source at the surface (from *Wordsworth et al.,* 2011). Data plotted with circles indicate where the atmosphere had begun to collapse on the surface in the simulations, and hence no steady-state temperature could be recorded. In the legend, 1:1 resonance refers to a synchronous rotation state, and 1:2 and 1:10 resonances refer to despun but asynchronous spin-orbit configurations.

Accompanies chapter by Showman et al. (pp. 277–326).

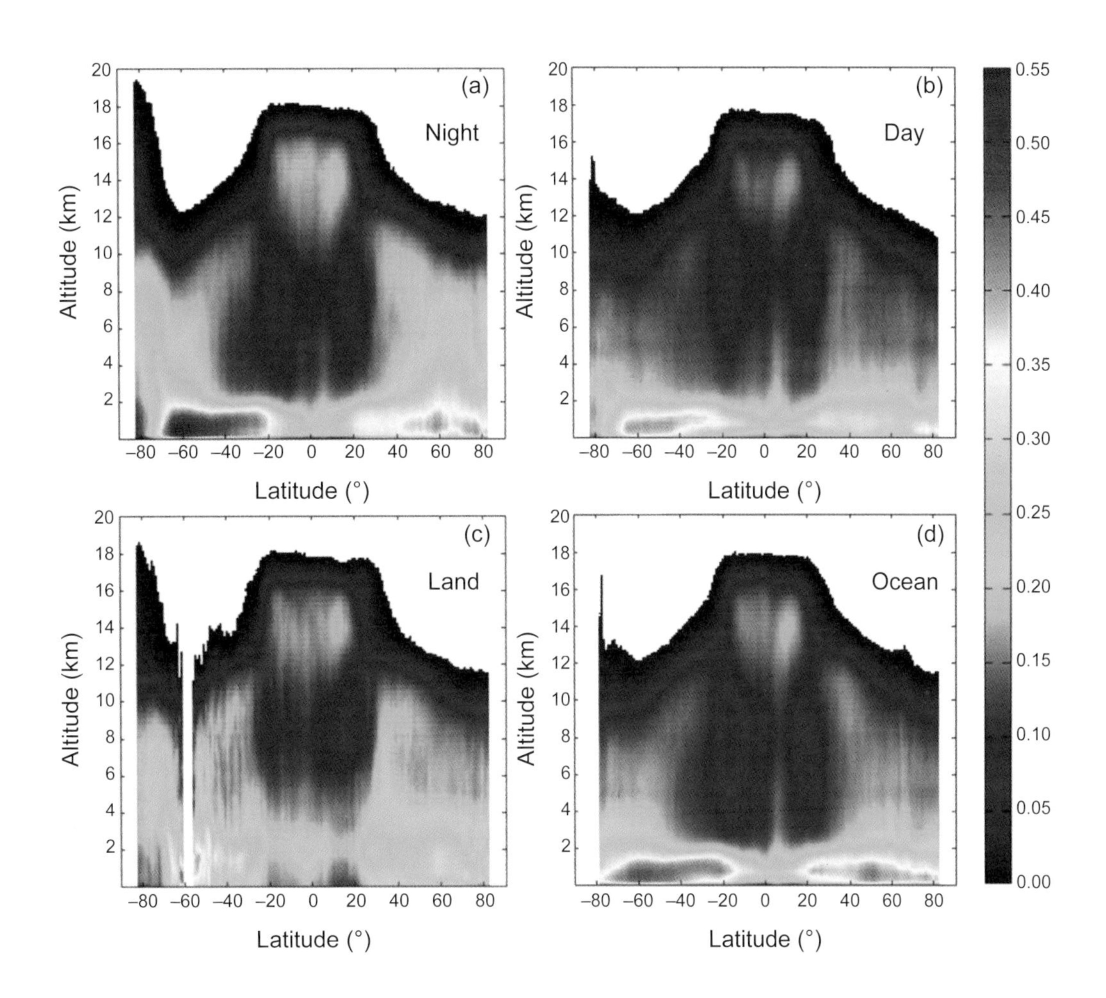

Plate 53. Latitude zonally-averaged vertical distribution of the cloud-occurrence frequencies for the CALIPSO observations from June 2006 to May 2007. A value of 0.5 means a cloud is present one half the time. **(a)** Nighttime measurements; **(b)** daytime measurements; **(c)** measurements over land; and **(d)** measurements over ocean. From *Wu et al.* (2011).

Accompanies chapter by Esposito et al. (pp. 329–353).

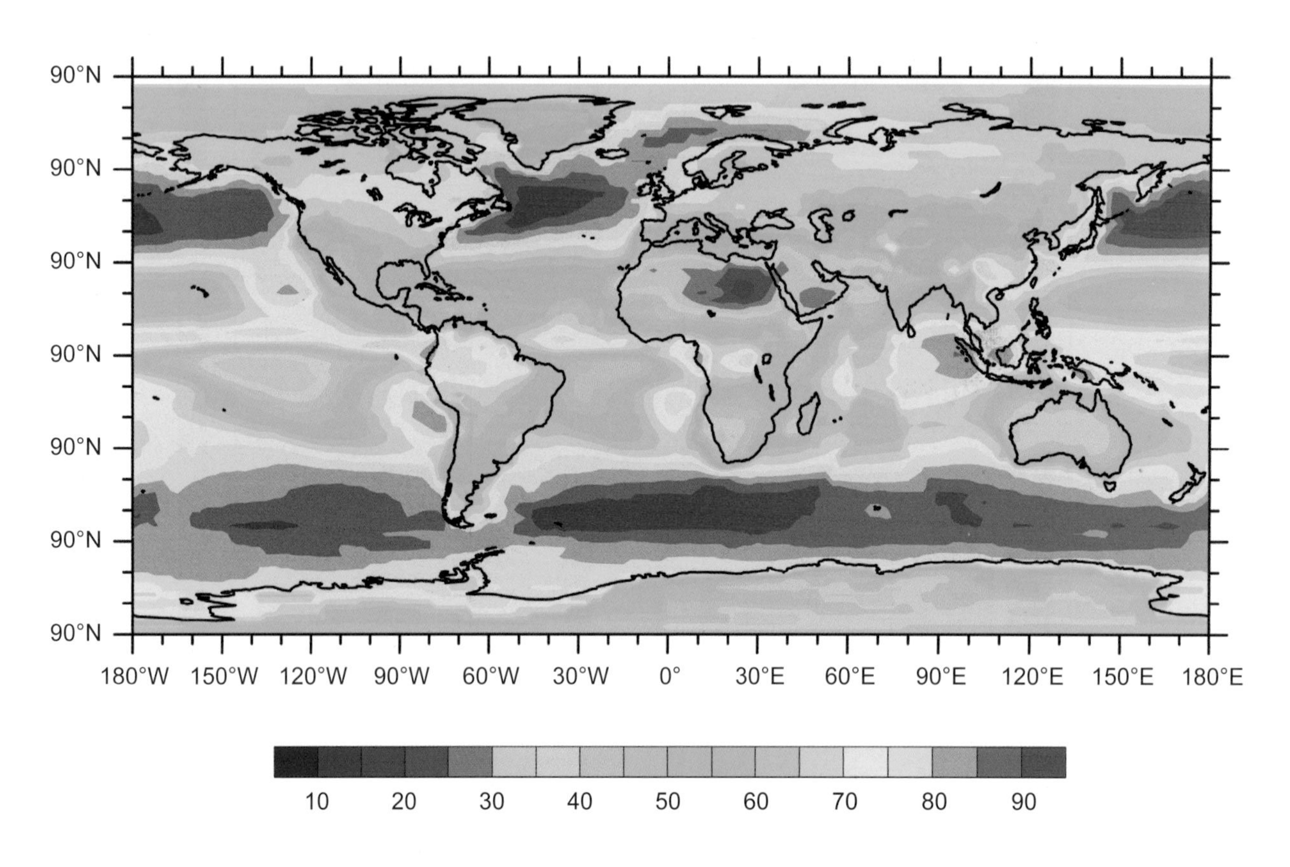

Plate 54. Annual average total cloud amount (percent) from July 1983 to December 2009 from the International Satellite Cloud Climatology Project (ISCCP). From *Rossow et al.* (1996).

Accompanies chapter by Esposito et al. (pp. 329–353).

Plate 55. Time/height contour plot of the backscatter coefficient (× 10^{-6} m^{-1} sr^{-1}) derived from the Phoenix LIDAR backscatter signal at 532 nm for the seasonal date L_s = 122°. From *Whiteway et al.* (2009).

Accompanies chapter by Esposito et al. (pp. 329–353).

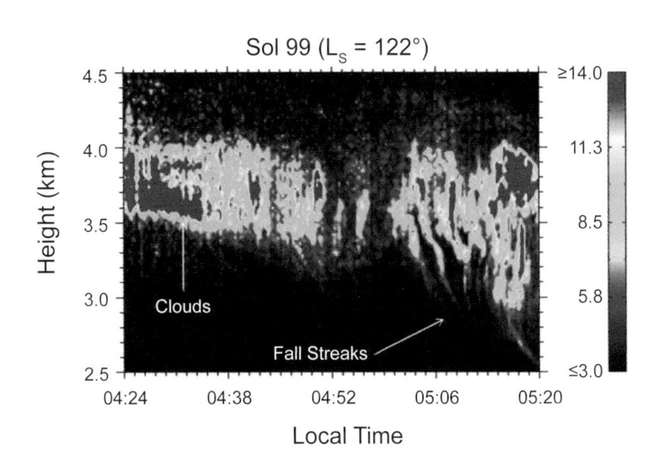

Plate 56. Seasonal and latitudinal variation of the 12.1-μm zonal-mean column water ice cloud opacity for a combination of Mars Year (MY) 26 and 27 (L_s = 0°–31° from MY27; L_s = 31°–360° from MY26). Data courtesy of Michael Smith.

Accompanies chapter by Esposito et al. (pp. 329–353).

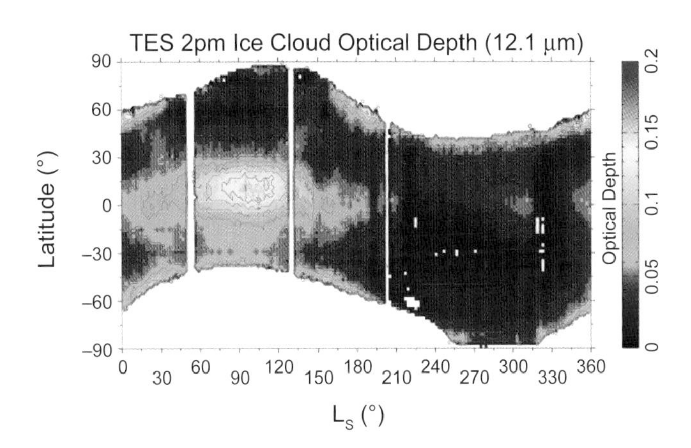

Plate 57. Seven plus Mars years of cloud observations. Figure constructed from TES data up to MY27 and THEMIS data thereafter. Courtesy of Michael Smith.

Accompanies chapter by Esposito et al. (pp. 329–353).

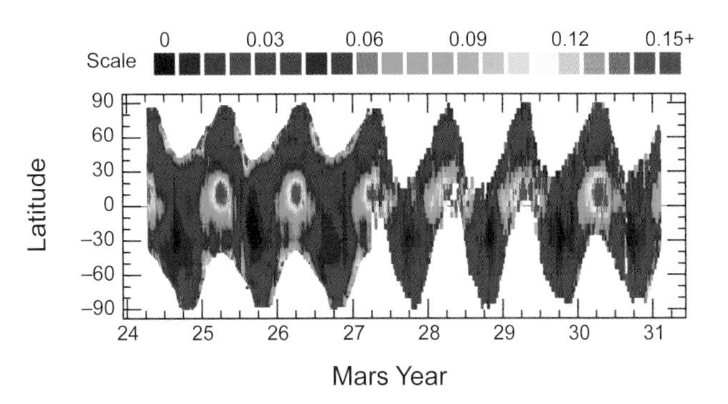

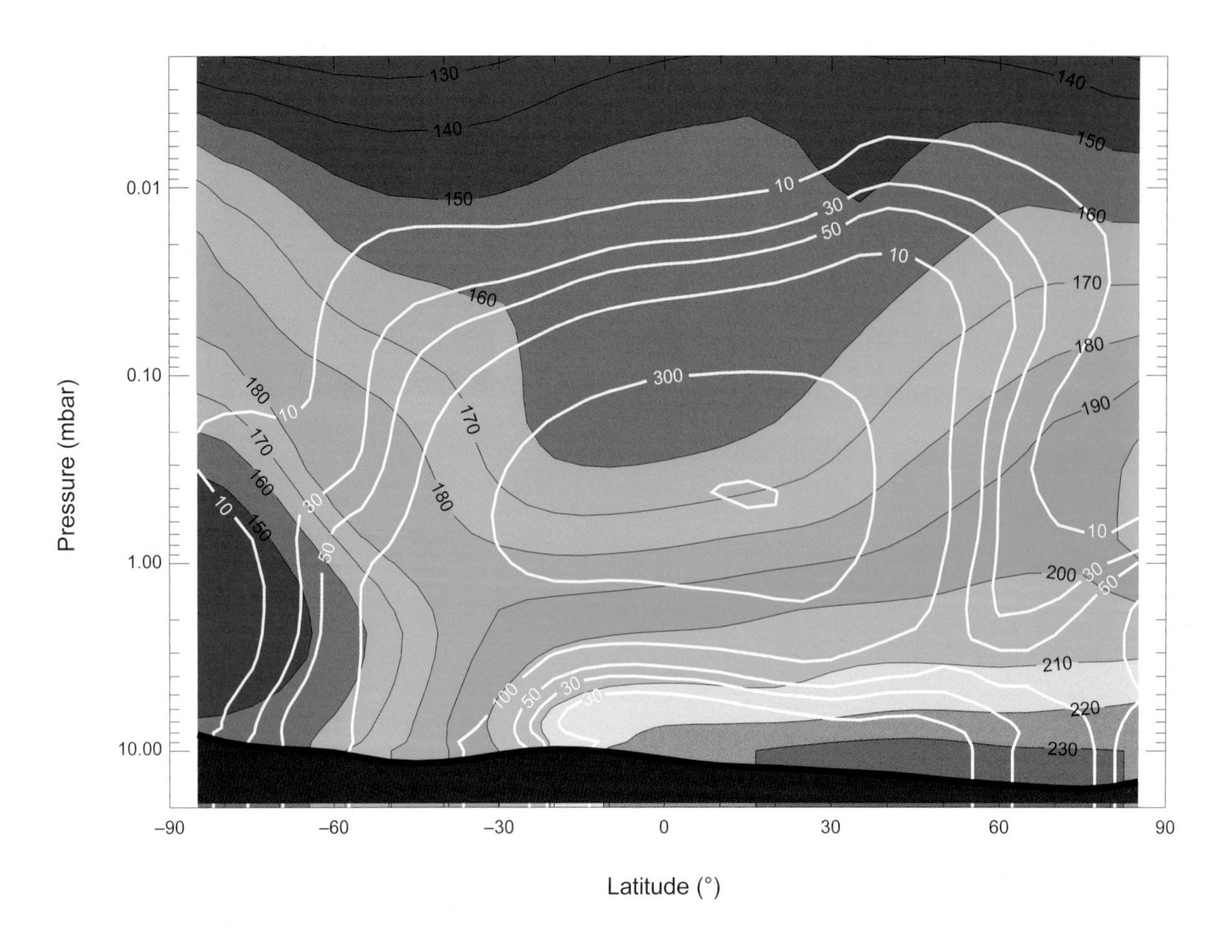

Plate 58. Time and zonally averaged temperatures (color coded with black contours) and water ice cloud volume-mixing ratios in ppmv (white contours) as simulated by the NASA/Ames Mars General Circulation Model for orbital conditions at 632 Ka (see *Haberle et al.,* 2012).

Accompanies chapter by Esposito et al. (pp. 329–353).

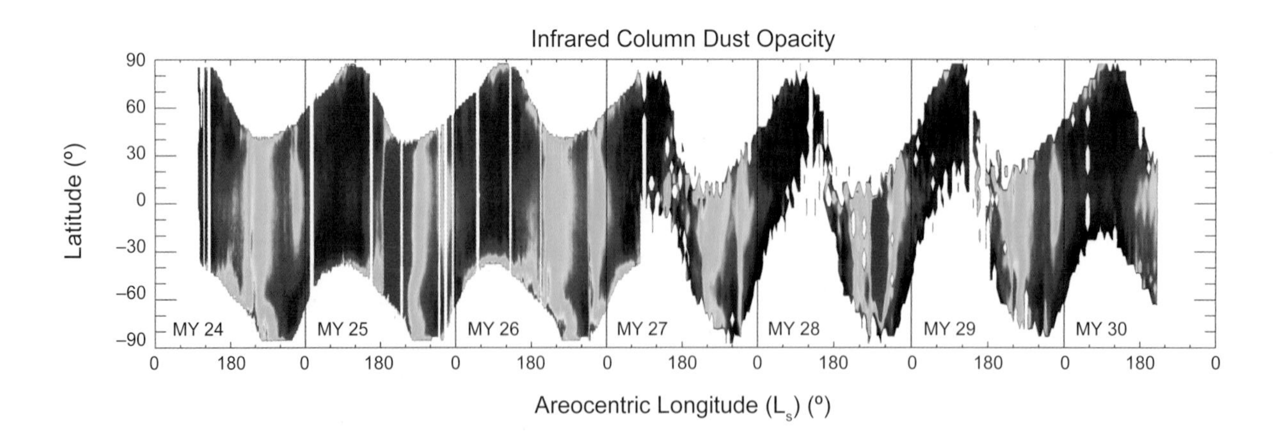

Plate 59. Zonally averaged 9-μm column dust opacity for approximately six of the most recent martian years, as observed by MGS/TES (MY24–26) and Mars Odyssey (MY27–30). Data courtesy of Michael Smith.

Accompanies chapter by Esposito et al. (pp. 329–353).

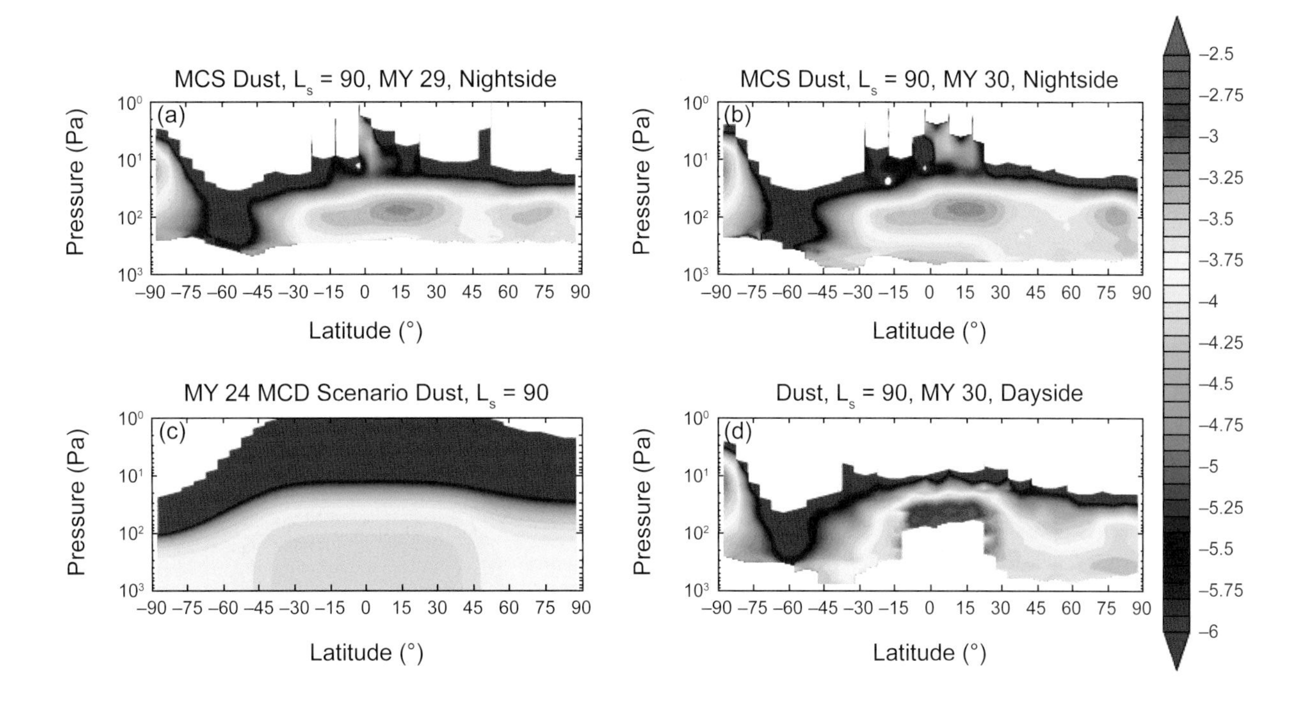

Plate 60. Zonal average density-scaled dust opacity (m^2 kg^{-1}) at L_s = 90° as observed by MRO/MCS during the night from two different Mars years [**(a)**,**(b)**] and during the day from one Mars year [**(d)**]. Also shown is the dust distribution used in the Mars Climate Database at L_s = 90° that does not include an enhanced dust layer aloft [**(c)**]. From *Heavens et al.* (2011).

Accompanies chapter by Esposito et al. (pp. 329–353).

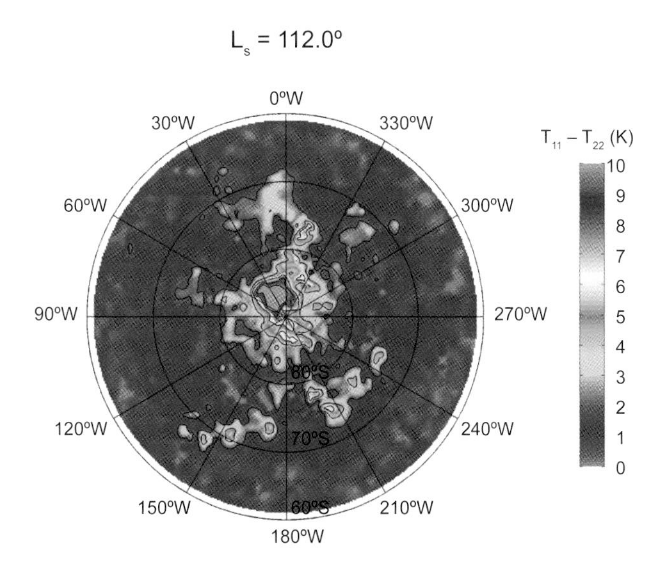

Plate 61. Anomalously low brightness temperatures as measured by MRO MCS. These low brightness temperatures are believed to be the result of infrared scattering by CO_2 clouds (*Hayne et al.,* 2011).

Accompanies chapter by Esposito et al. (pp. 329–353).

Plate 62. Dust devil photographed during a field campaign in Nevada.

Accompanies chapter by Renno et al. (pp. 355–365).

Plate 63. Gust front photographed during a field campaign in Niger.

Accompanies chapter by Renno et al. (pp. 355–365).

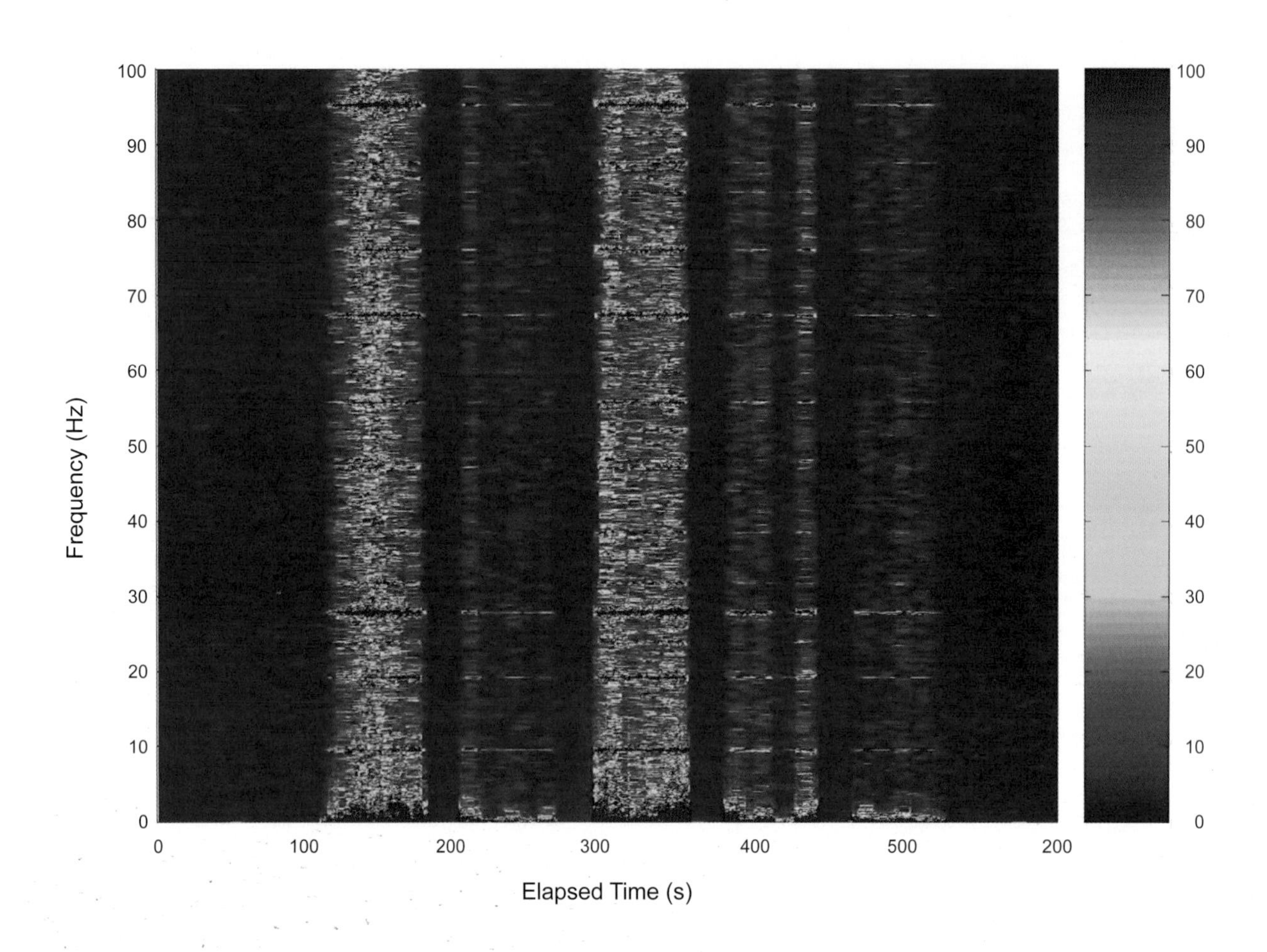

Plate 64. Time series of the power spectrum of the nonthermal radiation collected over 10 minutes on June 8, 2006, beginning at 21:58 UTC indicating modulation by SRs. Color bar is linearly proportional to power spectral density. After *Ruf et al.* (2009).

Accompanies chapter by Renno et al. (pp. 355–365).

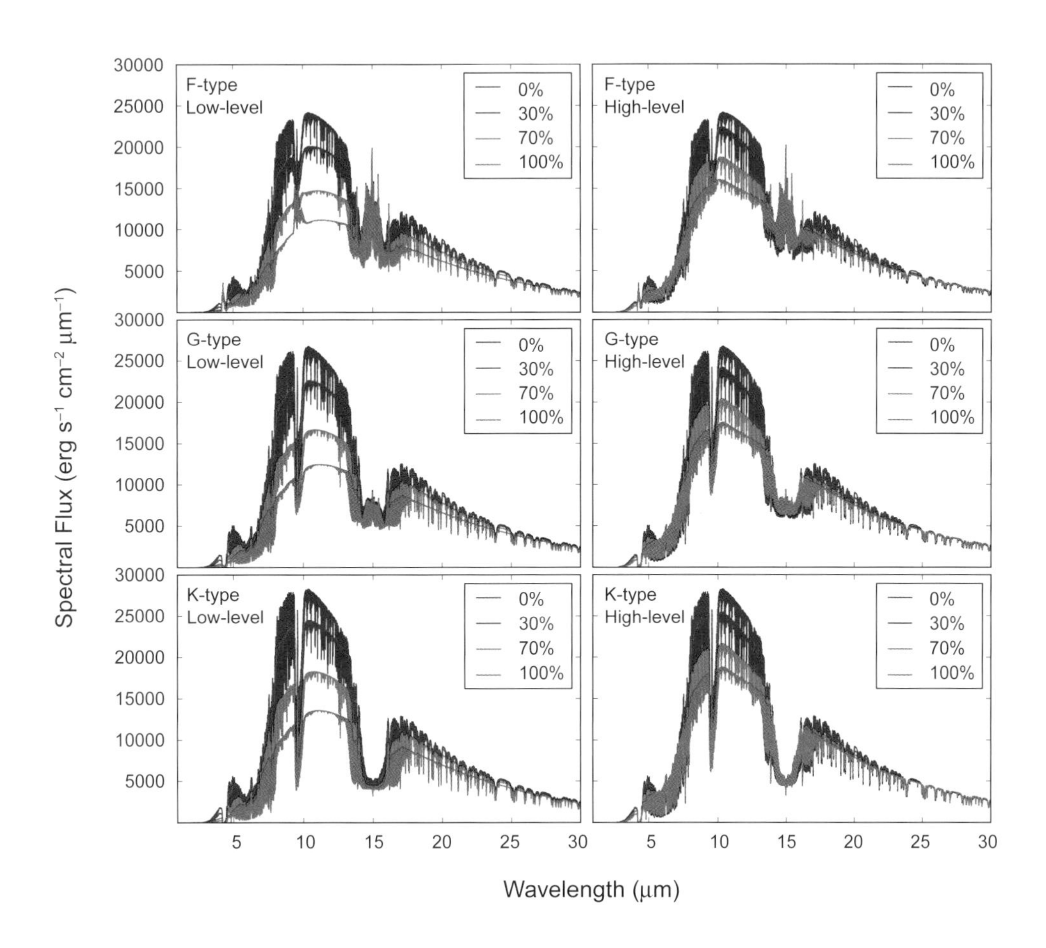

Plate 65. Planetary thermal emission spectra influenced by low-level water droplet (left panel) and high-level ice clouds (right panel) for an Earth-like planet orbiting different kinds of main-sequence host stars (adopted from *Vasquez et al.*, 2013). For each central star the spectra are shown for different cloud coverages. Note especially the strong impact of the cloud layers on the 9.6-μm absorption band of ozone.

Accompanies chapter by Marley et al. (pp. 367–391).

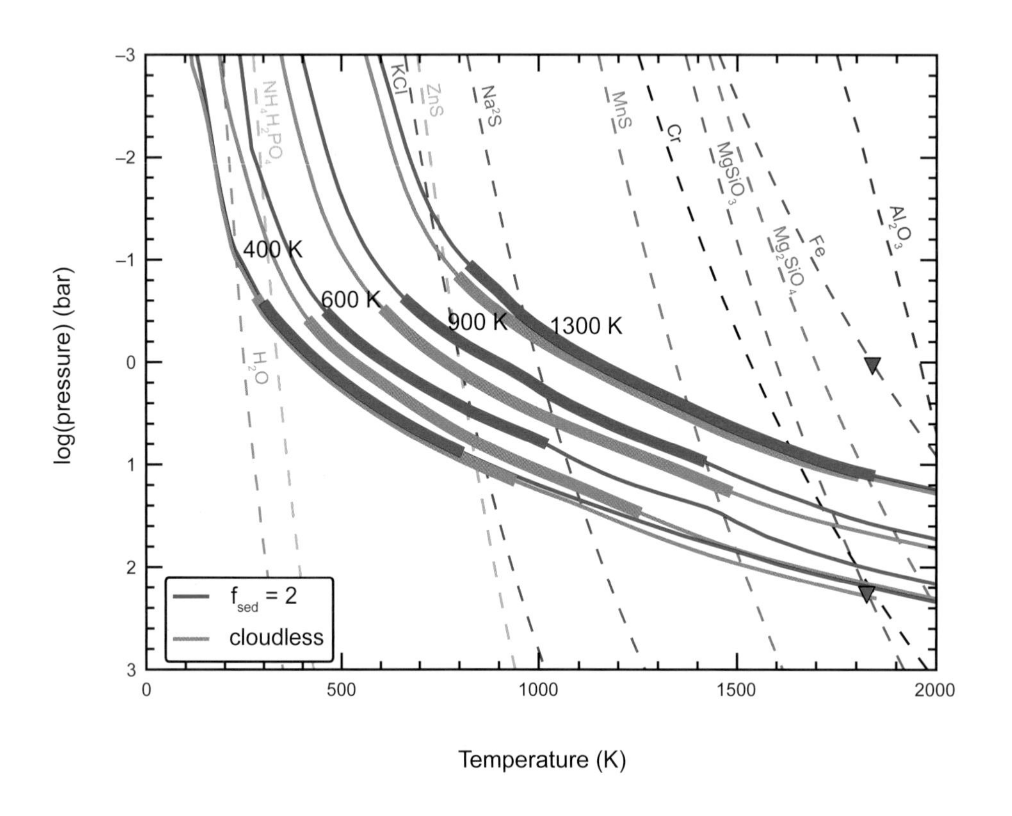

Plate 66. Model brown dwarf atmospheric temperature-pressure profiles for several effective temperatures (solid) are plotted along with condensation curves (dashed) for a variety of compounds. Diamonds along the iron and enstatite curves denote the melting point of each compound with the liquid phase lying to the right. Thicker portions of temperature profiles denote the model photospheres, the regions of the atmospheres from which most of the emitted radiation emerges. For each atmosphere model two curves are shown, corresponding to a cloudless calculation (blue) and a cloudy model with sedimentation efficiency, f_{sed} = 2 (red). Figure modified from *Morley et al.* (2012).

Accompanies chapter by Marley et al. (pp. 367–391).

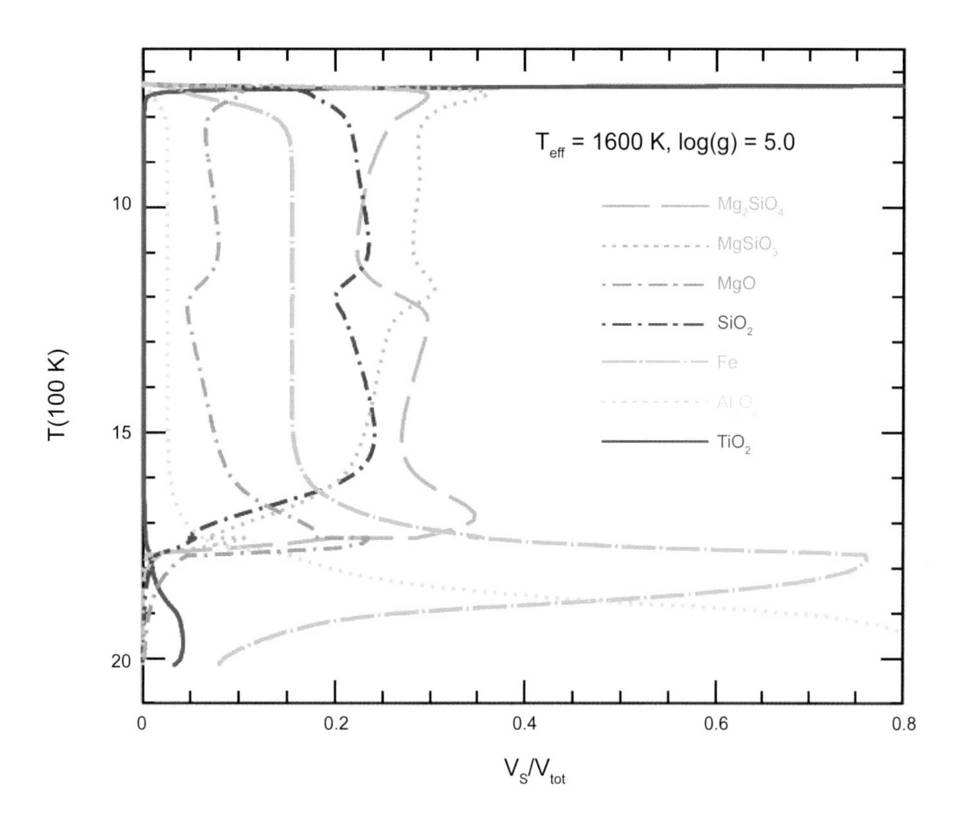

Plate 67. Composition of atmospheric cloud layers for a T_{eff} = 1600 K, log(g) = 5 brown dwarf as computed by the dust model of Helling and collaborators (*Helling et al.,* 2008). The vertical axis is atmospheric temperature with the top of the atmosphere to the top of the figure. The horizontal axis gives the relative volumes V_s of each dust species indicated by line labels as a ratio to the total dust volume V_{tot}. Unlike the Ackerman and Marley condensation equilibrium approach, this model predicts that MgO and SiO_2 are important condensates along with the TiO_2 seed particles that are formed at the top of the model.

Accompanies chapter by Marley et al. (pp. 367–391).

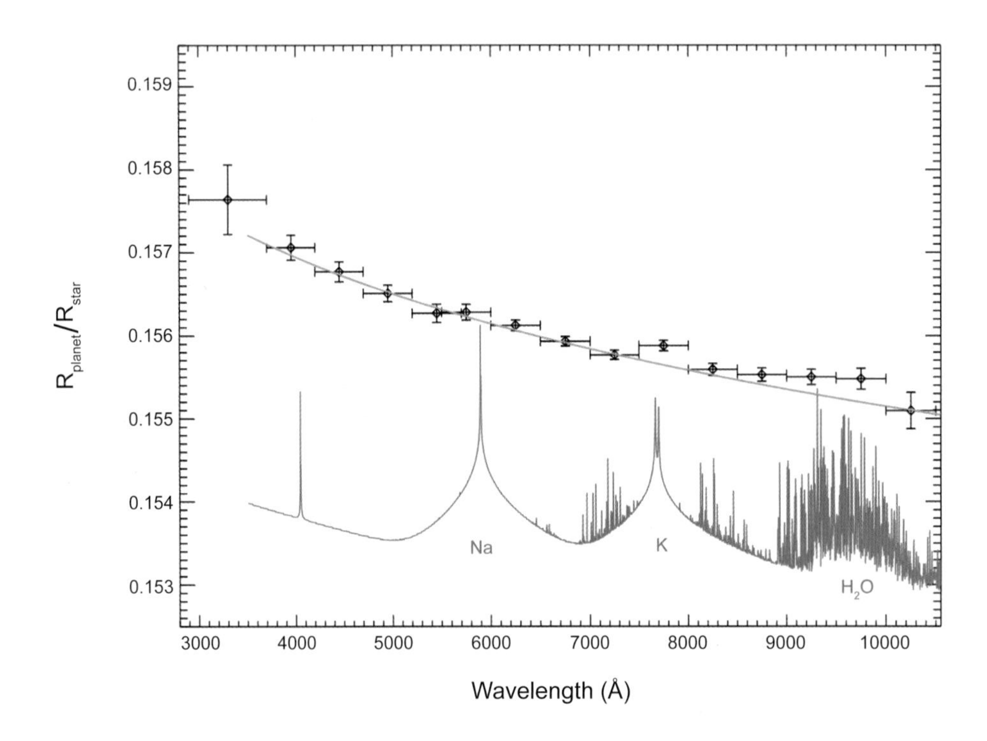

Plate 68. Observed transmission spectrum (points) for transiting hot Jupiter planet HD 189733b (*Sing et al.,* 2011). The width of the wavelength bin for each measurement is indicated by the x-axis error bars. The y-axis error bars denote the 1-σ error. The smooth curve denotes the prediction of a simple haze Rayleigh scattering-only model while the lower curve is a model for gaseous absorption only from *Fortney et al.* (2010). Further details in *Sing et al.* (2011). Figure courtesy of J. Fortney.

Accompanies chapter by Marley et al. (pp. 367–391).

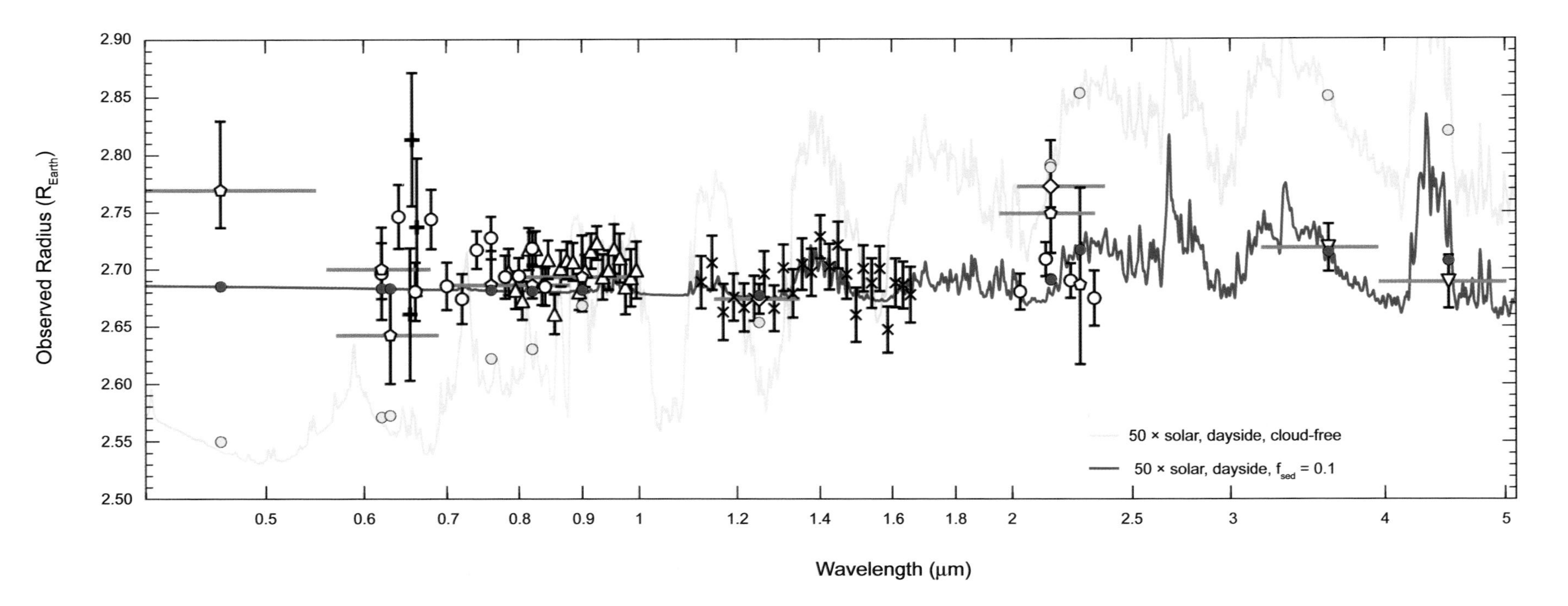

Plate 69. Observed (points) and model (lines) transit radius of GJ 1214b. Datapoints are from multiple observations as described in *Morley et al.* (2013). Models are for a cloudless H_2-He-rich atmosphere with 50× the solar abundance of heavy elements (gray) and the same atmosphere with the equilibrium abundance of clouds computed with f_{sed} = 0.1 (red). Figure modified from *Morley et al.* (2013)

Accompanies chapter by Marley et al. (pp. 367–391).

Plate 70. Noctilucent cloud (or terrestrial polar mesospheric cloud) observed on July 14, 2009, at Djurö Island, Stockholm Archipelago, Sweden. Photo by Kristell Pérot (*Pérot et al.*, 2010).

Accompanies chapter by Määttänen et al. (pp. 393–413).

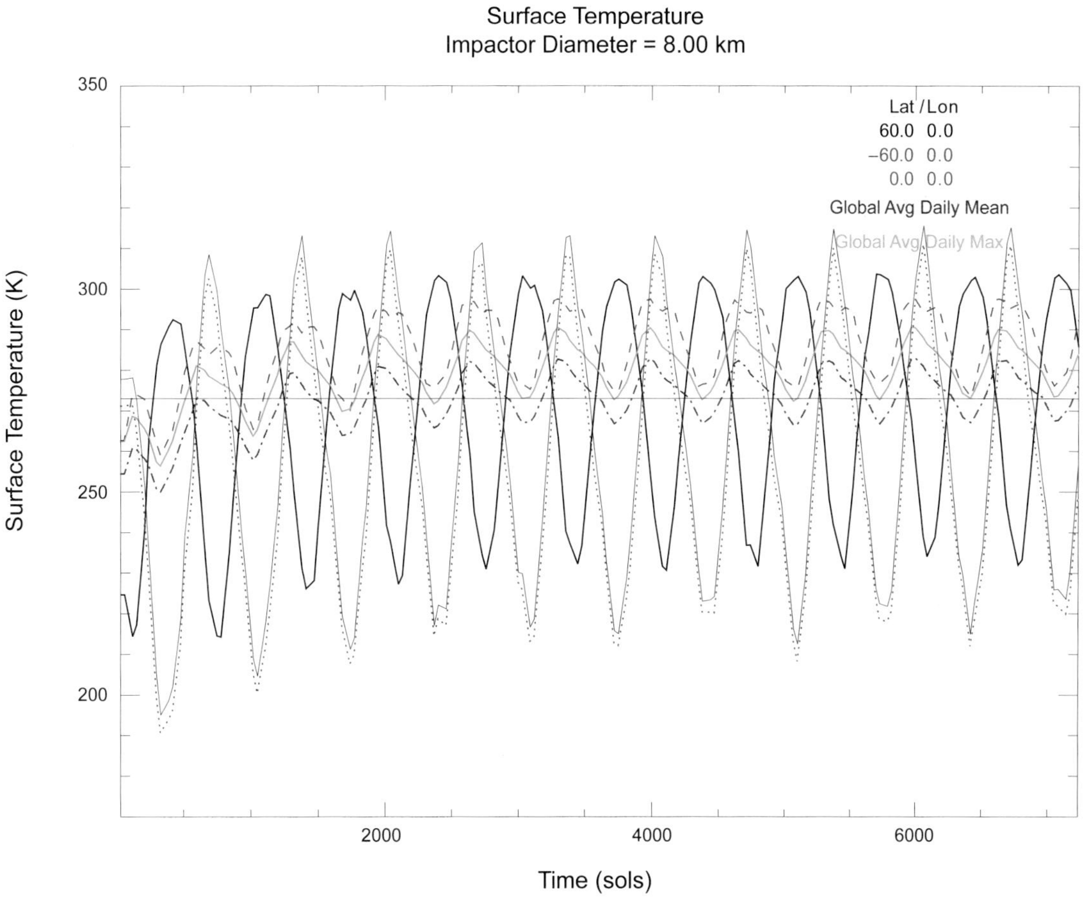

Surface Temperature
Impactor Diameter = 8.00 km

Lat / Lon
60.0 0.0
−60.0 0.0
0.0 0.0
Global Avg Daily Mean
Global Avg Daily Max

Surface Temperature (K)
350
300
250
200

2000
4000
6000

Time (sols)

Plate 71. Surface temperature at three locations (black, blue, red) on Mars plus global average daily mean temperature (purple) and global average daily max temperature (aqua) for several years following impact with an 8-km-diameter object. Note that at the end of simulation time the model has yet to return to equilibrium.

Accompanies chapter by Segura et al. (pp. 417–437).

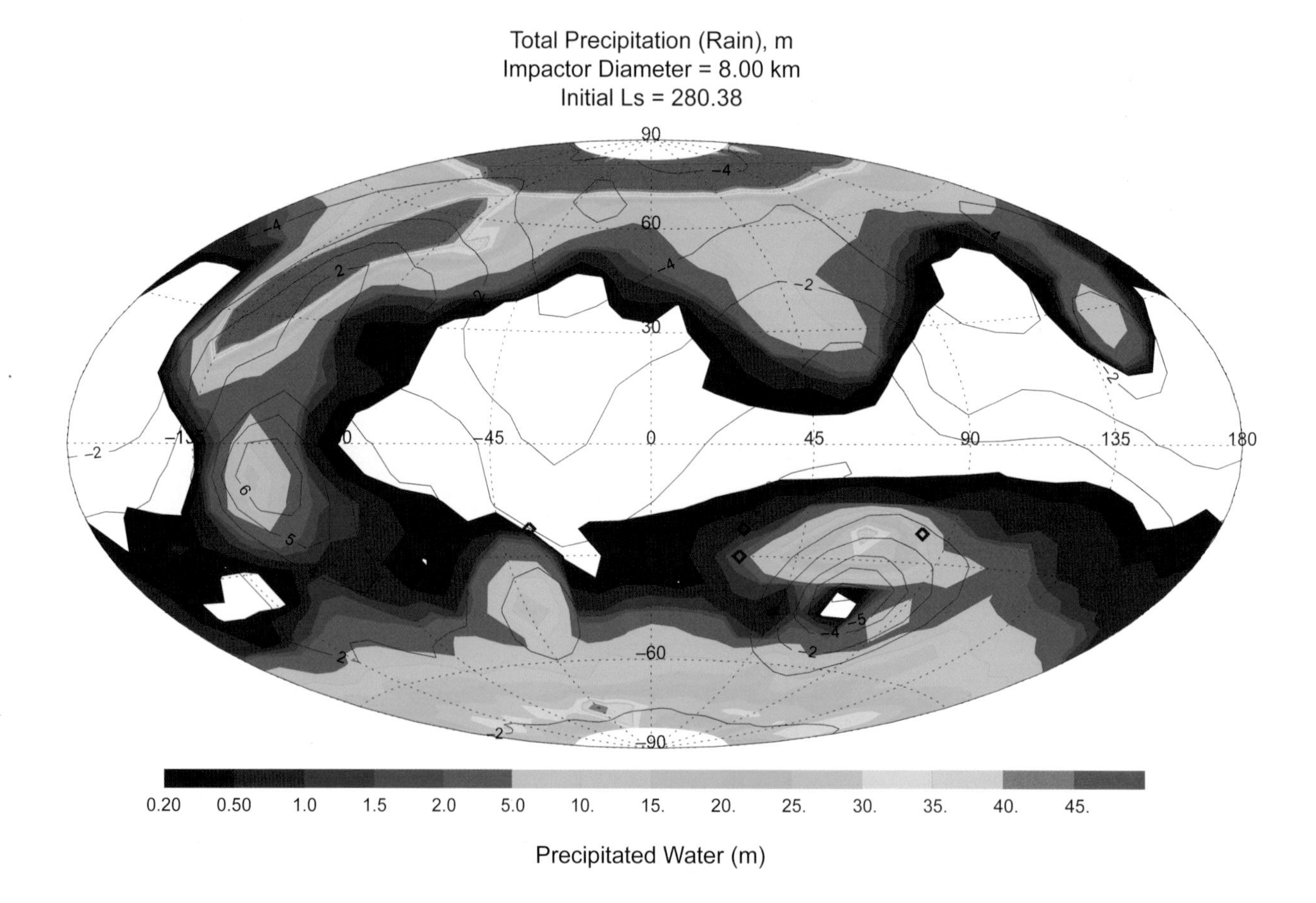

Plate 72. Total precipitated rain (meters) across the martian globe for an 8-km-diameter impactor near 0,0° (*Segura and Colaprete,* 2009). Longitude is marked globally and contours represent topography in kilometers referenced to mean Mars elevation.

Accompanies chapter by Segura et al. (pp. 417–437).

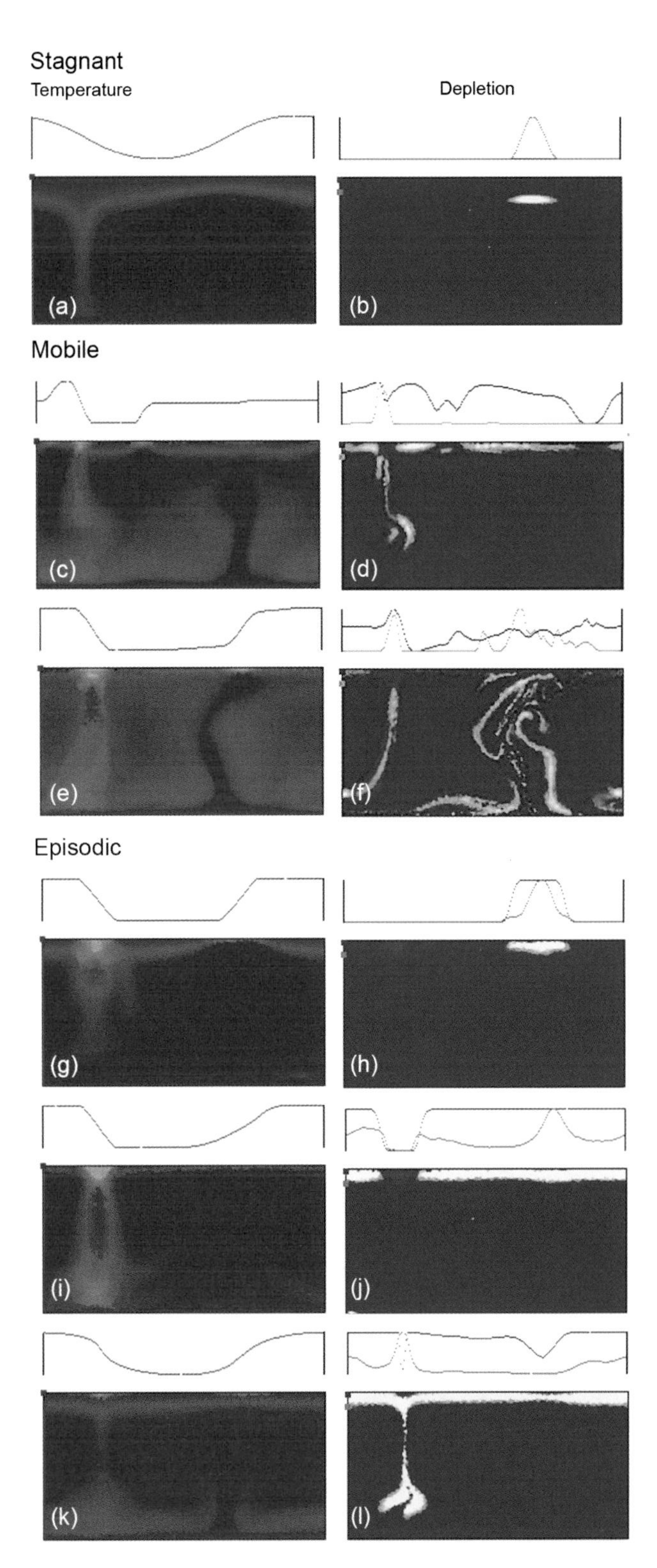

Plate 73. Snapshots of temperature field and depletion for the time evolution of three tectonic regimes. [For upper mantle convection (670 km depth); $Ra_{basal} = 3 \times 10^7$, 64×32 grid, with refinement in the boundary layers, periodic side boundaries, and free-slip, constant temperature top and bottom boundaries. Nondimensional internal heating is 0.5, temperature-dependent viscosity and brittle failure included as per text and Table 1.]. Hot temperatures are red, cold temperatures are blue, and brittle failure is shown as orange. In the depletion plots, blue represents material than has not had melt extracted; yellow is material that has undergone some degree of partial melting (plastic deformation is purple). **(a)**,**(b)** Temperature and depletion snapshots of a stagnant lid regime at the same time interval. Profiles above these sections are normalized surface velocity (red, above the temperature plot, normalized by maximum amplitude at each time step) and horizontal depletion profiles [mustard and blue lines above depletion sections, at depth indicated by markers on the lefthand side of the depletion plot; see arrows in **(b)**]. **(c)–(f)** Snapshots at model times of 10 m.y. [**(c)** and **(d)**] and 100 m.y. [**(e)** and **(f)**] of evolution in the mobile lid regime. **(g)–(l)** Time evolution of temperature and depletion fields during an episodic overturn, at model times of 12 m.y. [**(g)** and **(h)**], 20 m.y. [**(i)** and **(j)**], and 50 m.y. [**(k)** and **(l)**].

Accompanies chapter by O'Neill et al. (pp. 473–486).

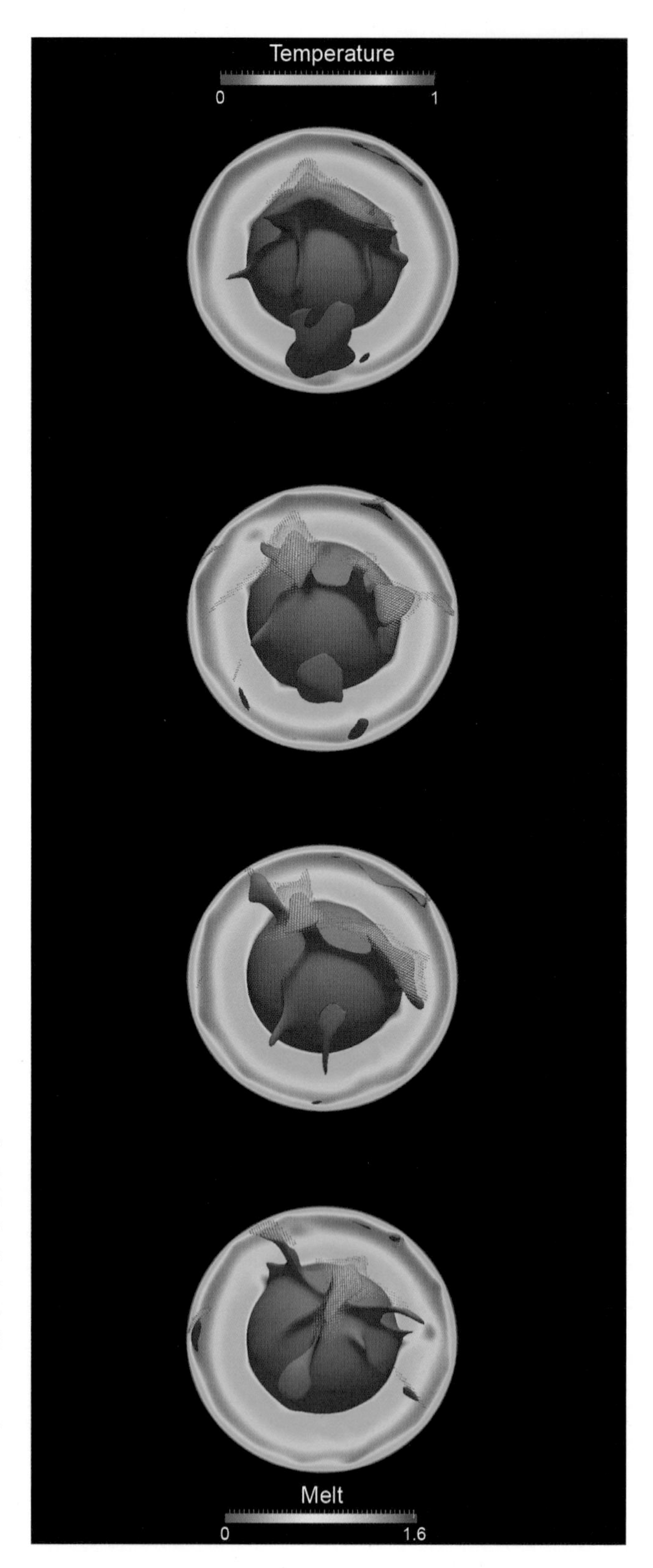

Plate 74. Temperature isosurface (T = 0.9, orange), temperature slice, and melt production (colored glyphs) in a three-dimensional spherical mobile lid convection model using CitcomS, divided into 96 subdomains with a grid spacing of 65 × 65 × 65 each (3,449,952 nodes total, average vertical spacing 22 km). Rayleigh number is 1×10^5 (assuming an upper mantle depth scale; higher in the CitcomS input, which assumes planetary radii), viscosity contrast is 1×10^4, heating ratio is 60 (see text for discussion on spherical and Cartesian internal heating), and the model contains a low-viscosity asthenosphere. Other parameters as in Table 1. Model time for the snapshots are at 0, 114, 214, and 382 m.y., respectively.

Accompanies chapter by O'Neill et al. (pp. 473–486).

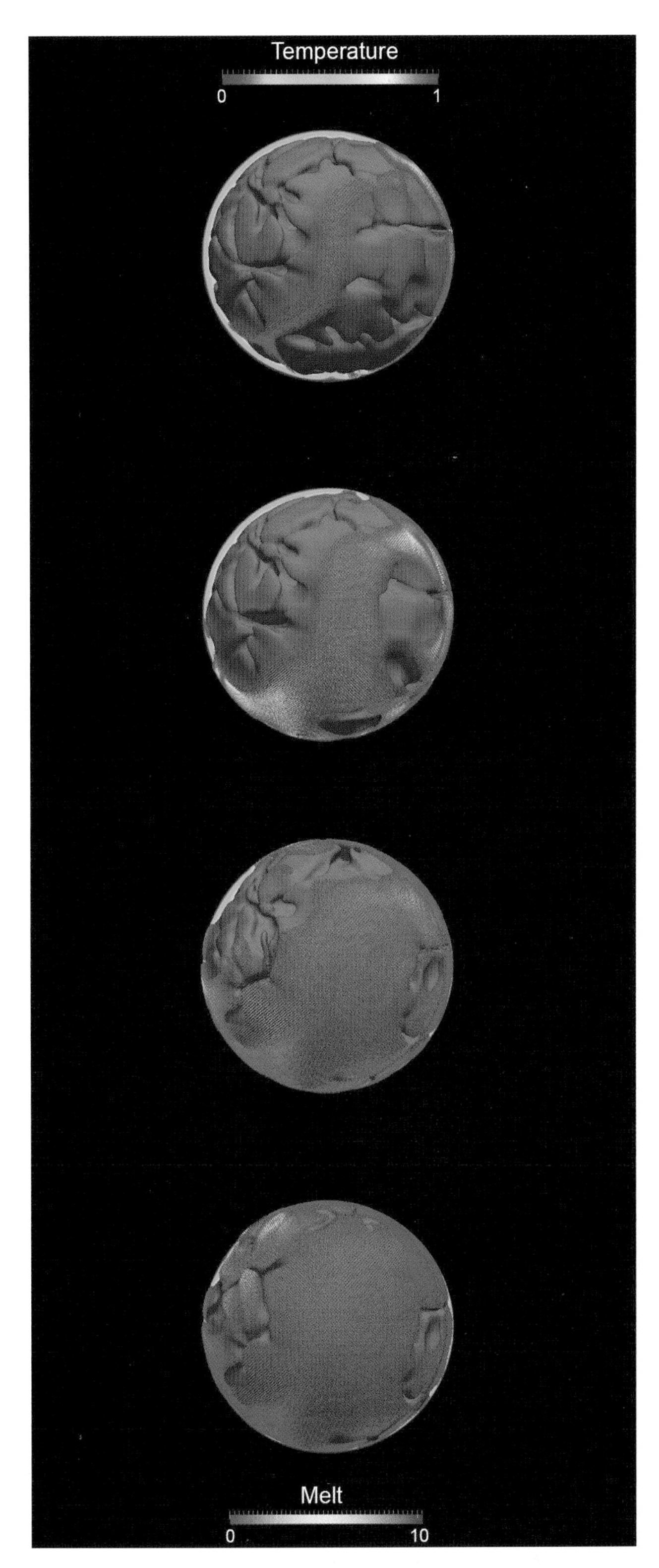

Plate 75. Temperature isosurface (T = 0.9, orange), temperature slice, and melt production (glyphs) in a three-dimensional spherical episodic convection model using CitcomS. Other parameters as per Fig. 3 and Table 1. The time slices track the evolution of melt over one overturn, and begin in a stagnant lid mode, with high internal temperatures, and thus the temperature isosurface is near the top. After yielding and plastic failure, the melt is generated over a wide area, correlated with active spreading (discrete glyphs). Downwellings are evident by depressions in the isocontour surface and low melt generation. Modeled times for the snapshots are 83, 178, 246, and 303 m.y., respectively.

Accompanies chapter by O'Neill et al. (pp. 473–486).

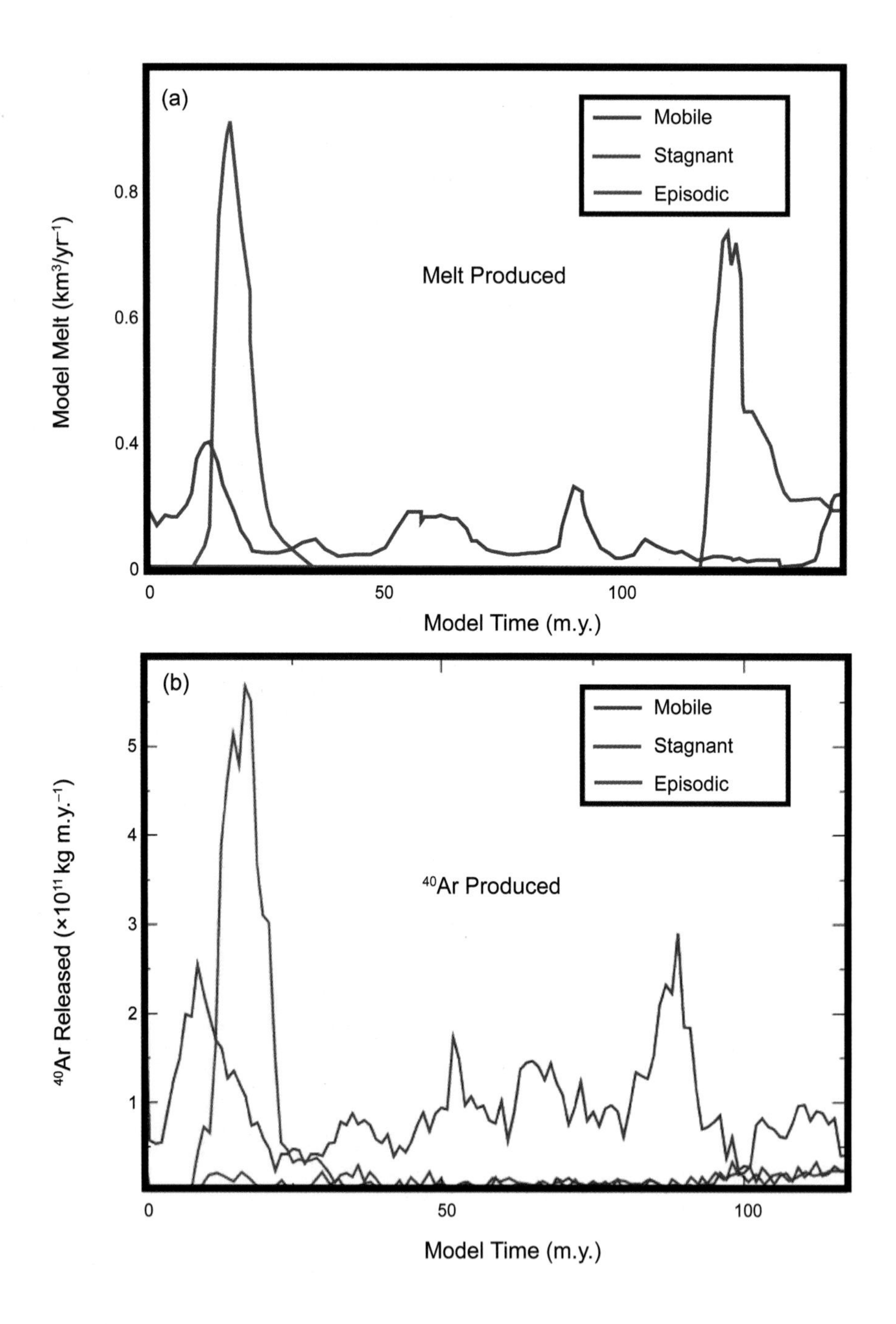

Plate 76. (a) Melt produced by the models shown in Fig. 1 vs. model time. **(b)** ^{40}Ar degassed from the models in Fig. 1, against model time. Note the axis was clipped to compare models over an appropriate time interval (i.e., one episodic overturn).

Accompanies chapter by O'Neill et al. (pp. 473–486).

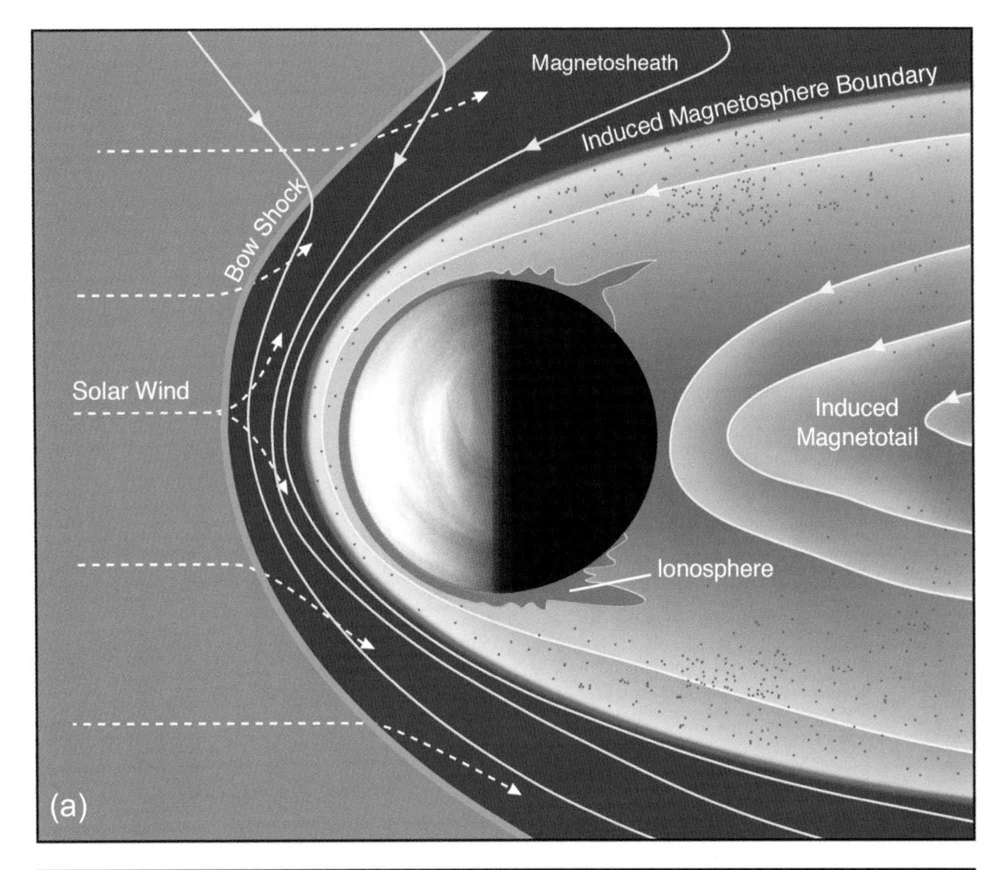

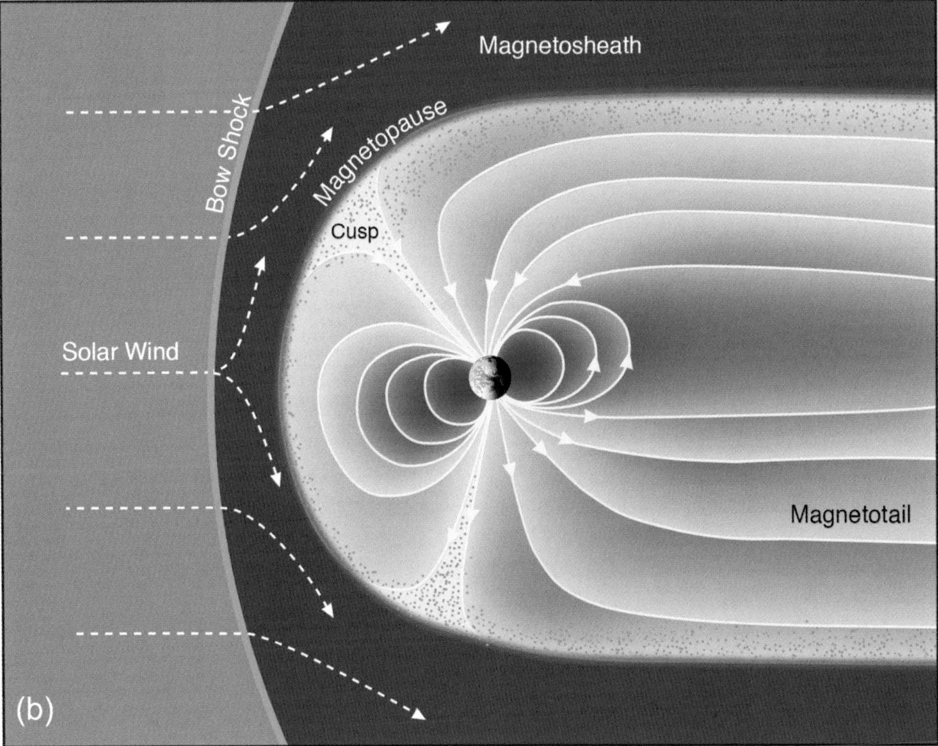

Plate 77. Cartoon diagram showing the structure and main features of **(a)** induced and **(b)** intrinsic magnetospheres. Note the difference in the size of the two interaction regions relative to the size of the planet. Venus, Titan, and Mars have primarily induced magnetospheres, while Earth has an intrinsic magnetosphere. Diagram courtesy of S. Bartlett.

Accompanies chapter by Brain et al. (pp. 487–501).

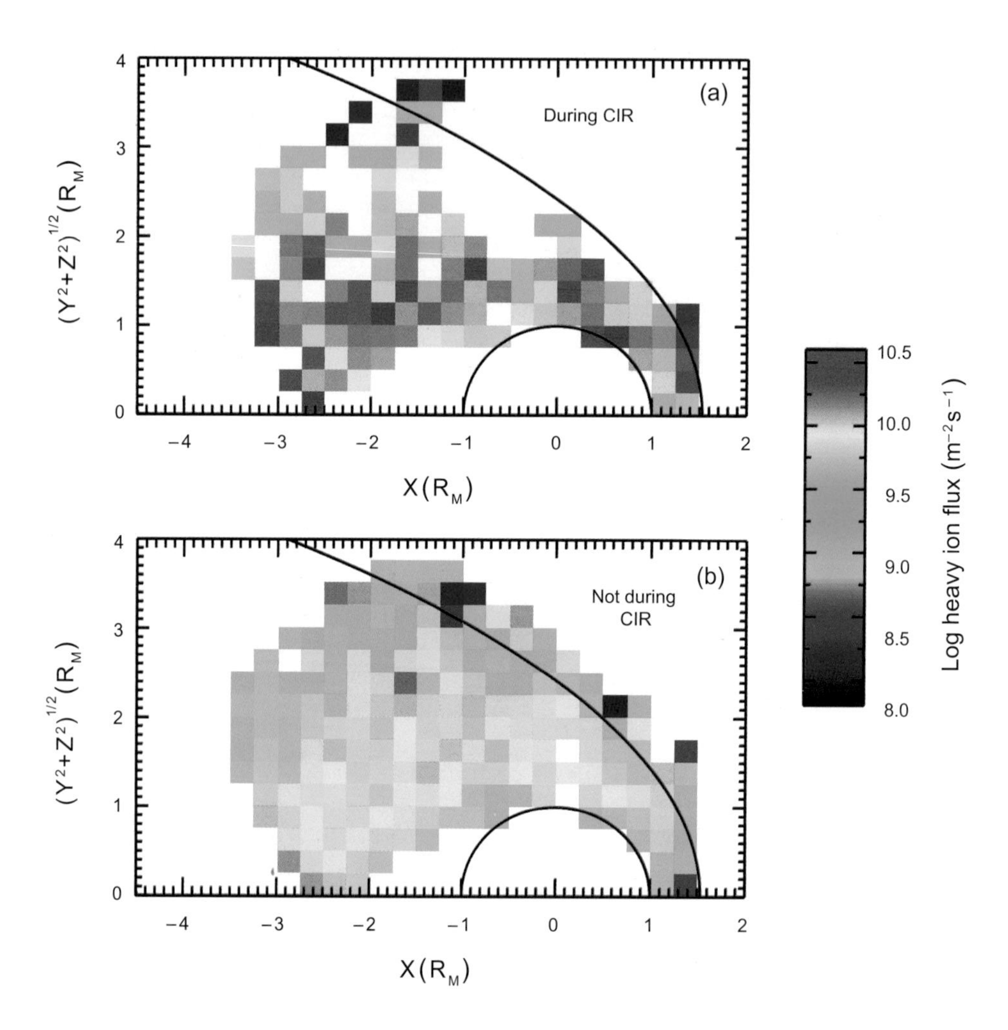

Plate 78. Heavy ion fluxes measured by the Mars Express ASPERA IMA sensor as a function of location with respect to Mars in cylindrical MSO coordinates during co-rotating interaction region (CIR) events associated with **(a)** high solar wind pressure and **(b)** all other times. As solar wind pressure increases, atmospheric escape rates increase. From *Edberg et al.* (2010).

Accompanies chapter by Brain et al. (pp. 487–501).

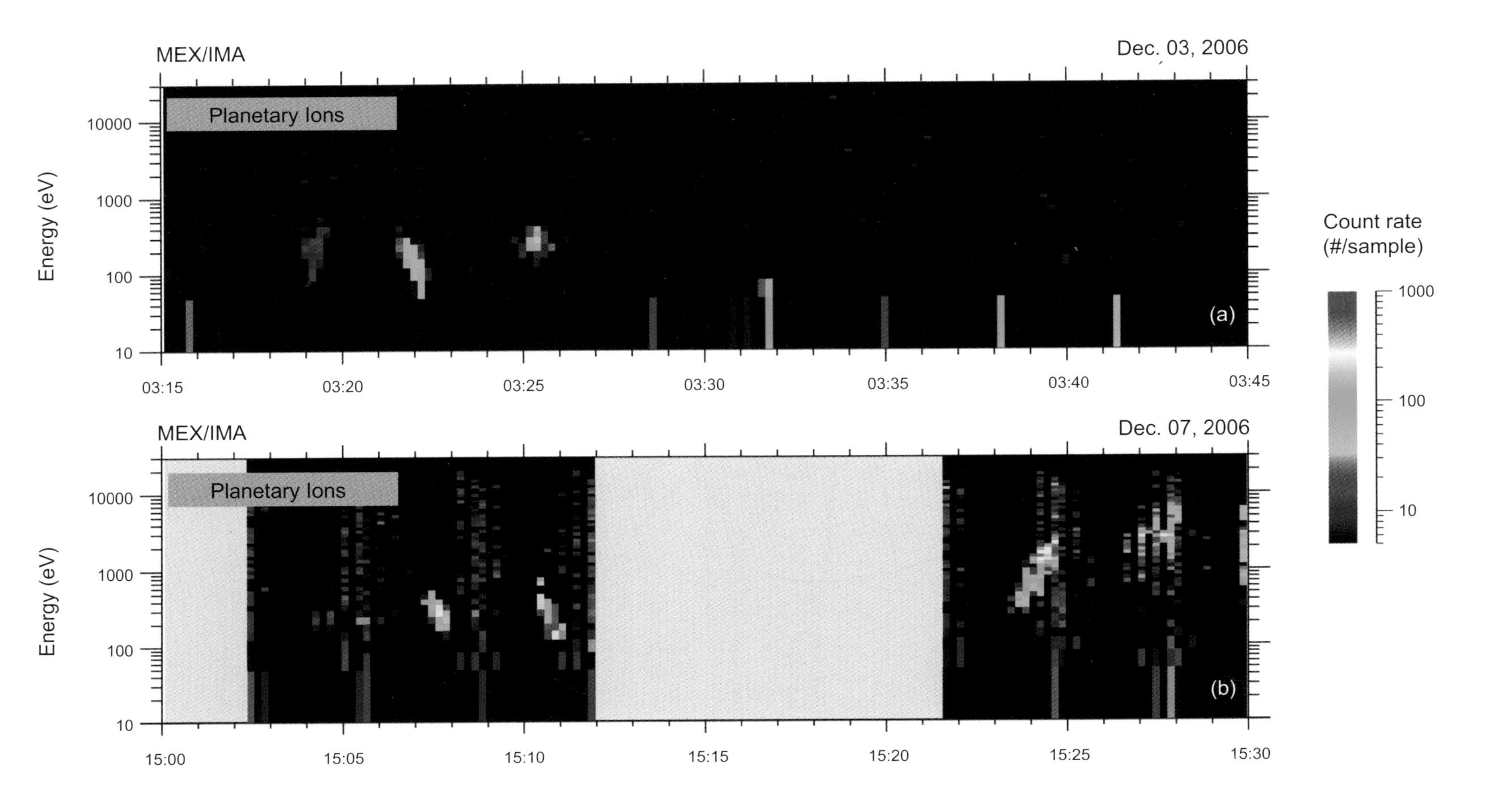

Plate 79. Comparison of fluxes of escaping ions at Mars **(a)** during quiet times and **(b)** during a solar storm. Ion escape rates increased by approximately an order of magnitude during this event. From *Futaana et al.* (2008).

Accompanies chapter by Brain et al. (pp. 487–501).

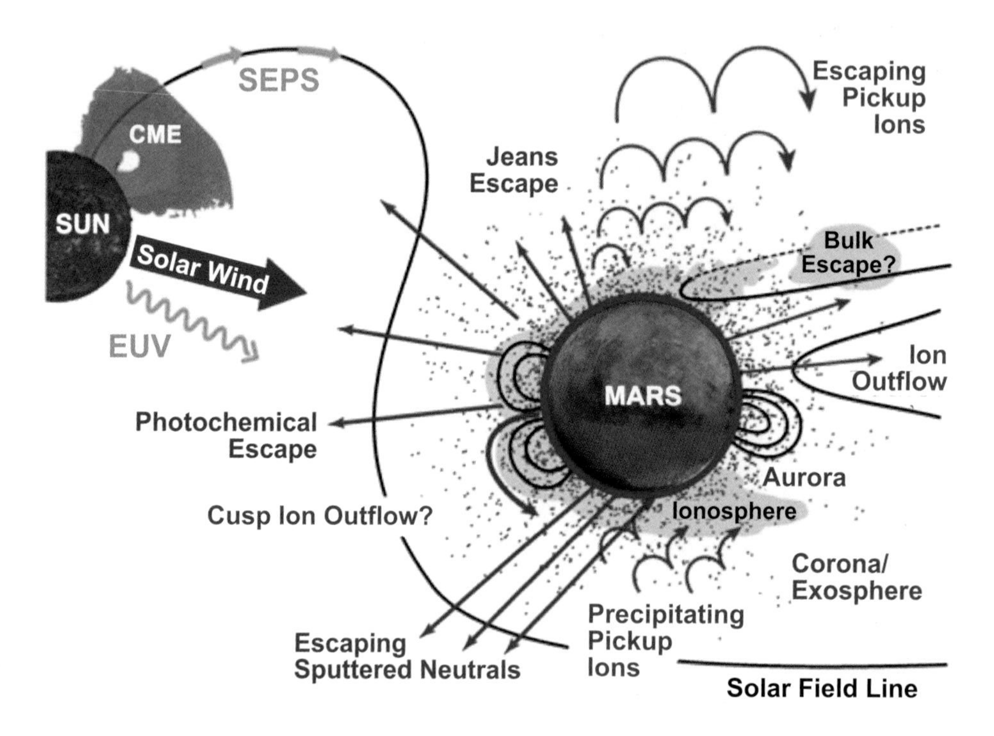

Plate 80. Diagram showing the interaction of the martian upper atmosphere with its surroundings. Neutral atmospheric reservoirs and processes are shown in blue, and ionized reservoirs and processes are shown in red. Solar inputs include EUV photons, solar wind plasma, and solar energetic particles (SEPs). The image is available at *lasp.colorado.edu/home/maven/*.

Accompanies chapter by Brain et al. (pp. 487–501).

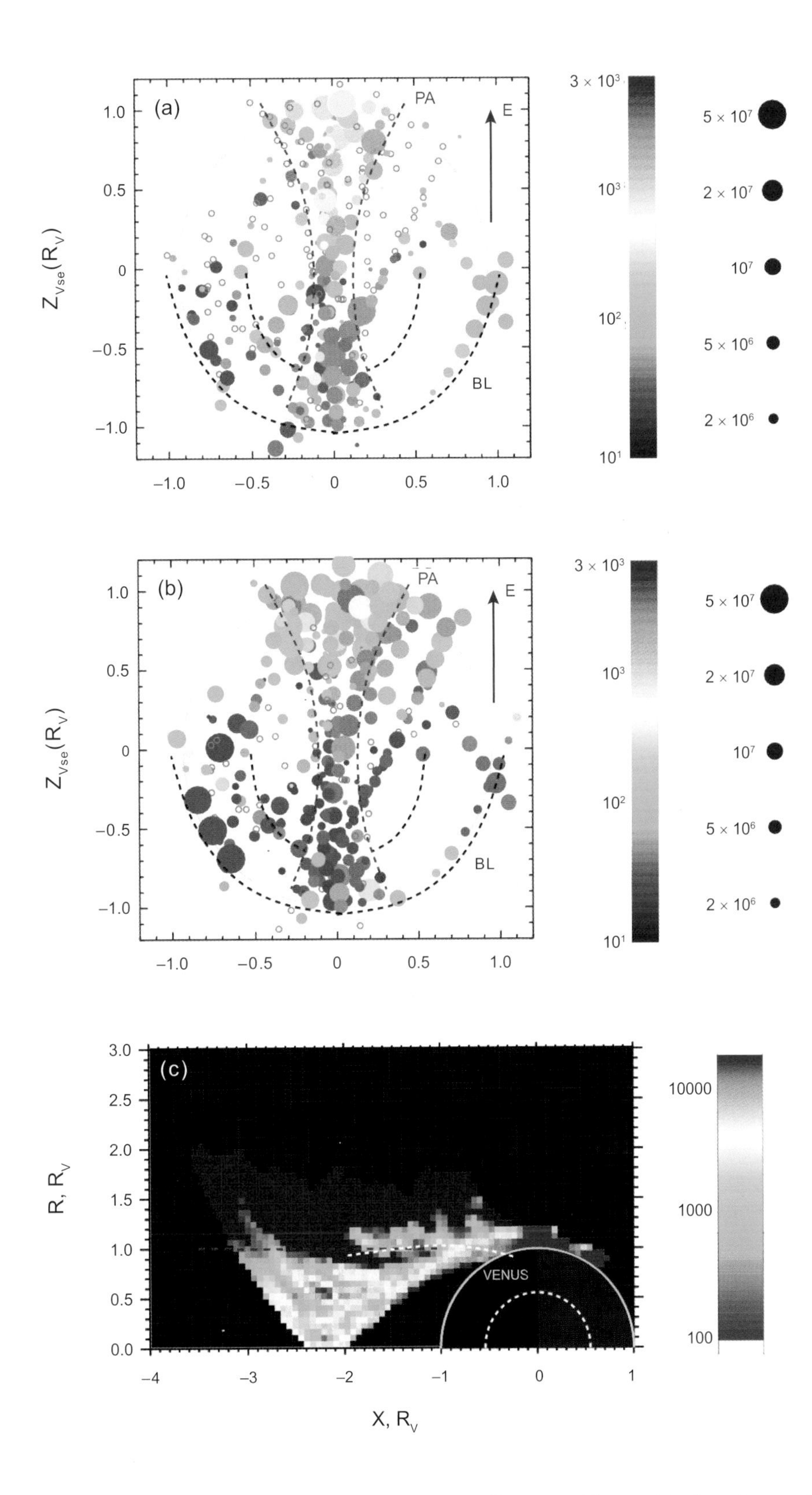

Plate 81. Measurements of ion escape by Venus Express. Fluxes (size) and energies (color) of **(a)** O+ and **(b)** H+ ions escaping from the Venus atmosphere are shown for a cross-section of the Venus magnetotail downstream from the planet (*Barabash et al.*, 2007a). The coordinate system used has the electric field in the solar wind directed along the y-axis. Fluxes of escaping ions of planetary origin (with mass to charge ratio >14) are shown in cylindrical coordinates in **(c)**. From *Fedorov et al.* (2008).

Accompanies chapter by Brain et al. (pp. 487–501).

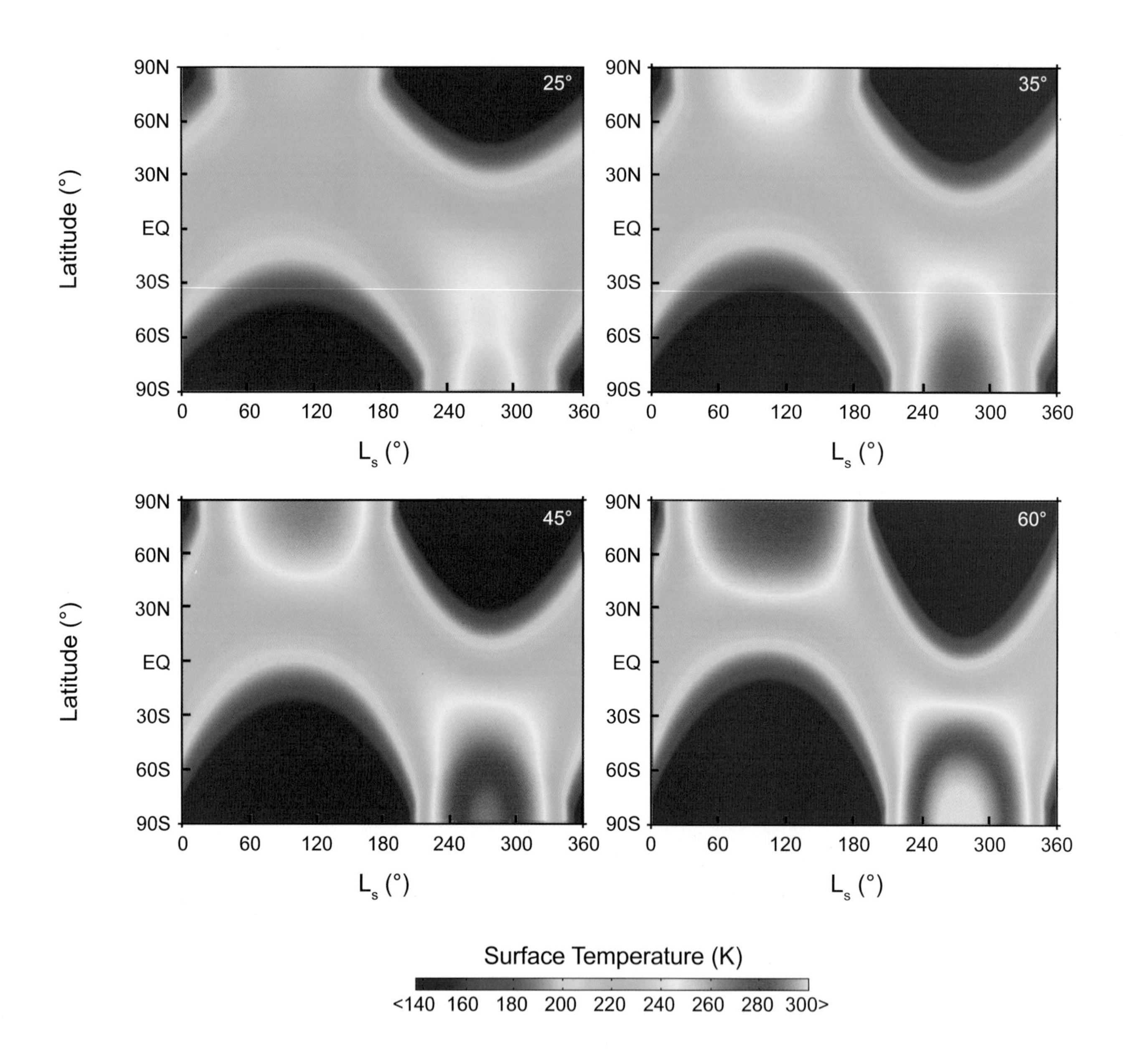

Plate 82. Diurnally and zonally averaged surface temperature over a martian year calculated for four obliquities ($\theta \geq 25°$). For all models, $\varepsilon = 0.093$ and $\varpi = 251°$. Surface emissivity assumed = 1, and surface albedo is assumed to be 0.21 everywhere (including the current permanent caps). Although 45° is the highest obliquity attained in the last several million years, q regularly reached 60° prior to 10 Ma.

Accompanies chapter by Zent (pp. 505–538).

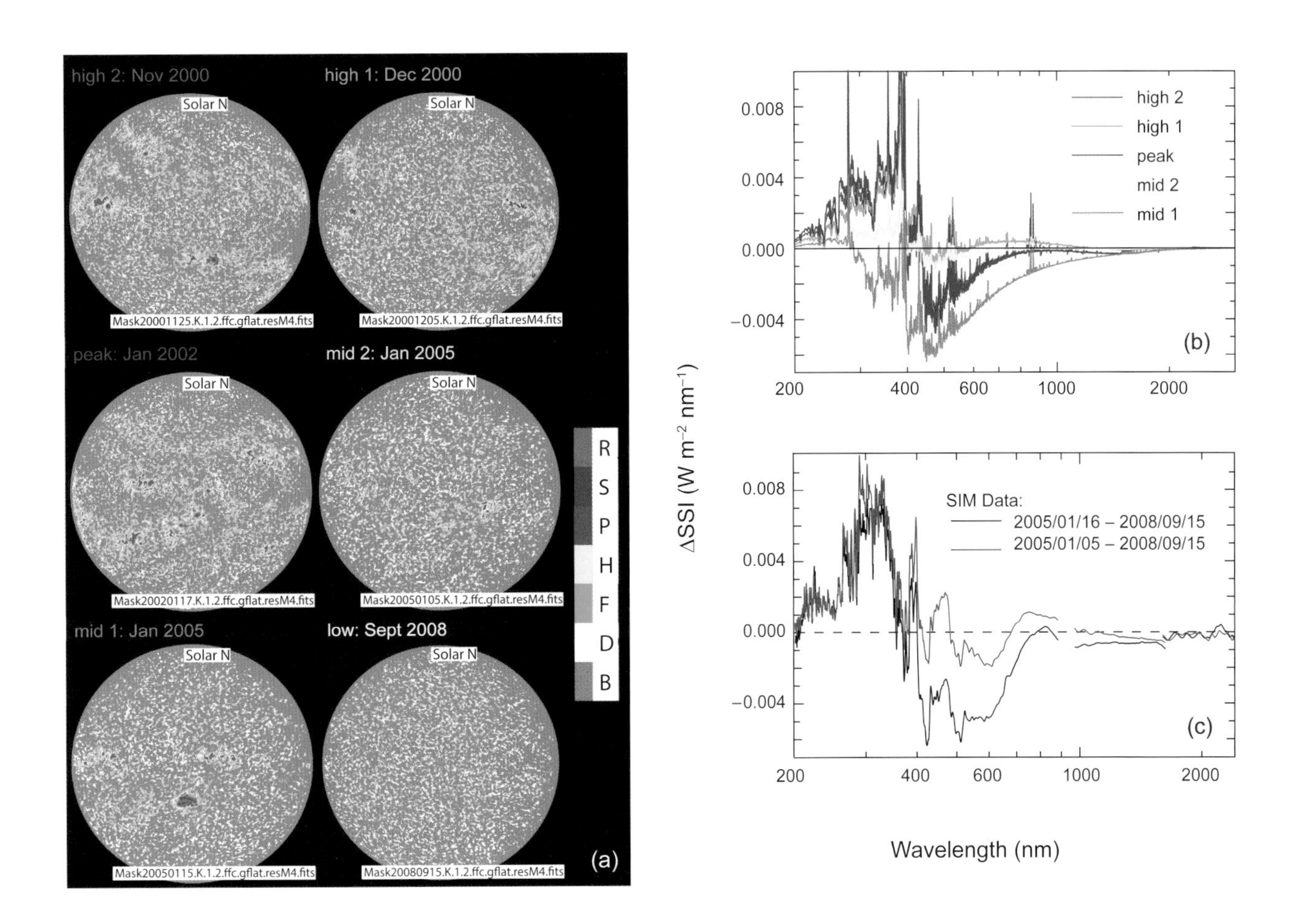

Plate 83. (a) Six time periods over solar cycle 23. The masks are constructed from Rome Observatory PSPT images and the surface feature contributions are color coded and identified in Table 1. **(b)** Change in irradiance as predicted by SRPM from the 2008 reference spectrum. **(c)** SIM difference spectra from 2005 relative to 2008. The SRPM masks in **(a)** labeled as "mid 2" corresponds to the red trace in **(c)**, and "mid 1" is represented as the black curve in **(c)**. The offsetting solar cycle trends reported in *Harder et al.* (2009) are generally reproduced by the model but with lower amplitude.

Accompanies chapter by Harder and Woods (pp. 539–565).

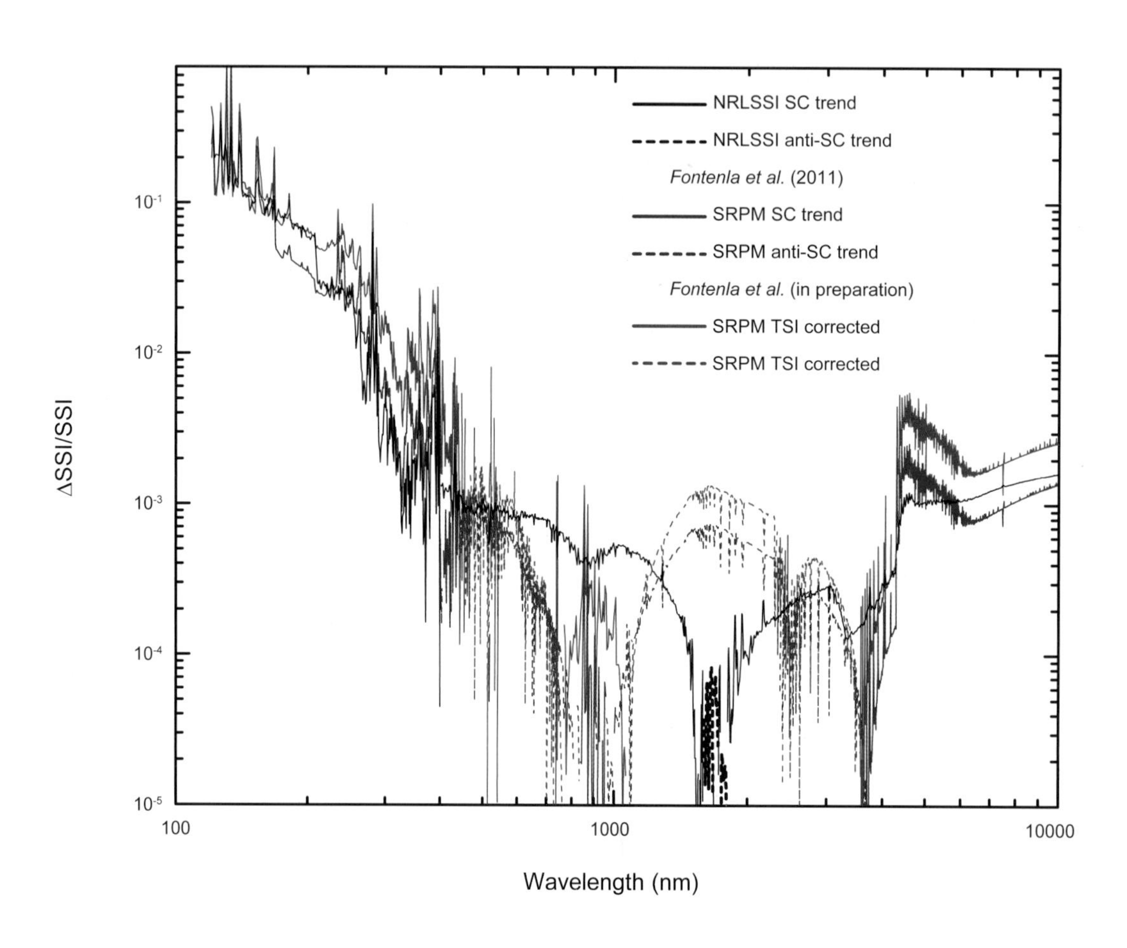

Plate 84. Estimates of solar maximum to solar minimum SSI variability from the NRL SSI and two versions of SRPM. These two models currently represent the two extremes of solar spectral variability. The SRPM calculation represented in *Fontenla et al.* (2011) is shown in blue, and an estimated change in irradiance that matches the TSI based on shifts in the median of the brightness distribution in PSPT images is shown in red.

Accompanies chapter by Harder and Woods (pp. 539–565).

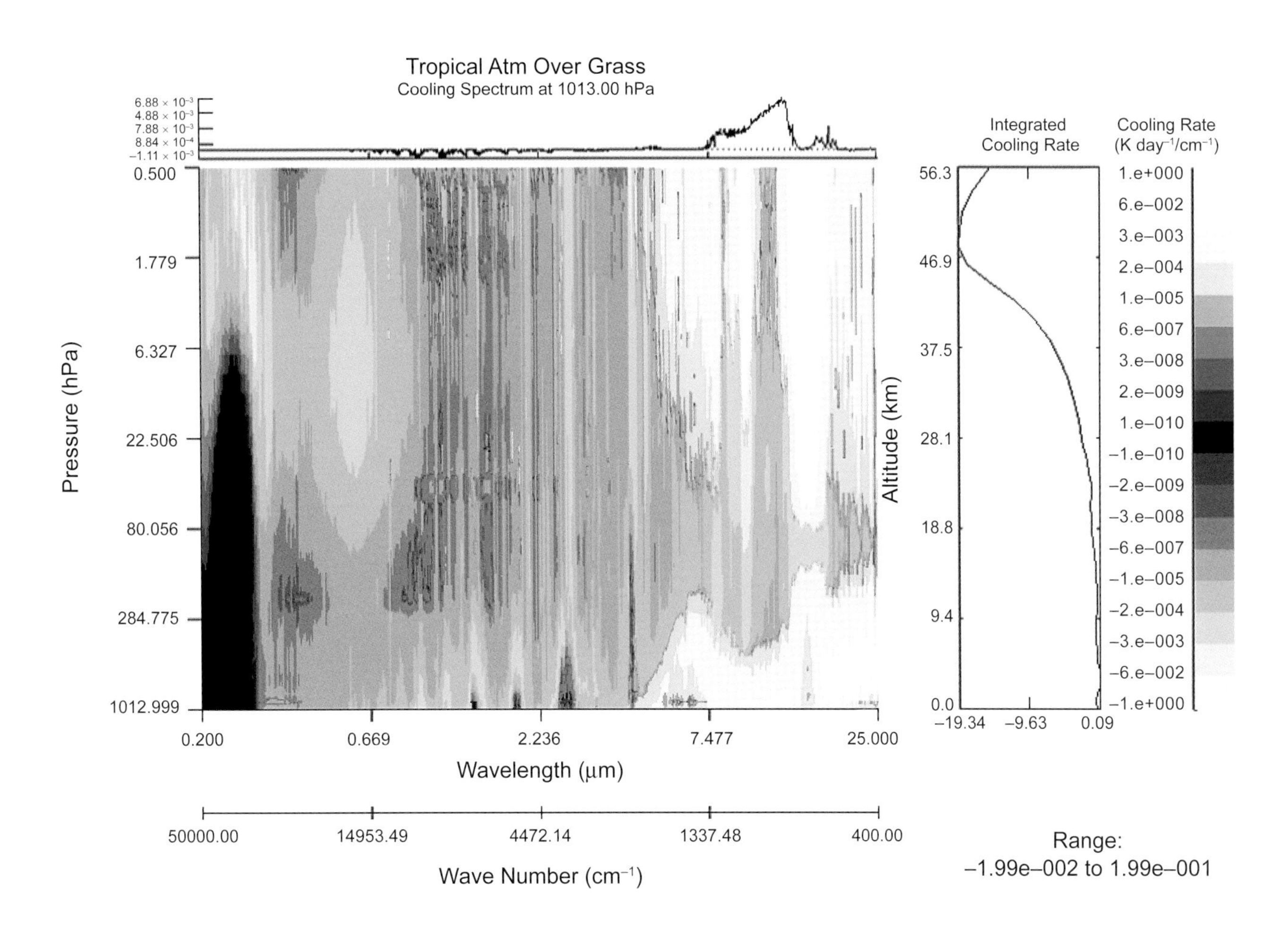

Plate 85. A MODTRAN®5.2 radiative transfer model calculation showing the cooling rate of Earth's lower atmosphere in response to solar (mostly shortward of 5 μm) and thermal (mostly longward of 5 μm) energy, based on a tropical atmosphere over a "grasslands" albedo. Figure courtesy of G. P. Anderson (Air Force Research Laboratory, Kirtland AFB, New Mexico) and A. Berk (Spectral Sciences, Inc.).

Accompanies chapter by Harder and Woods (pp. 539–565).

Index

Page numbers refer to specific pages on which an index term or concept is discussed, or the first use of the word or term in a section or chapter where it is used frequently.